D1288443

MAMMAL SPECIES OF THE WORLD

MAMMAL SPECIES OF THE WORLD

A TAXONOMIC AND GEOGRAPHIC REFERENCE

SECOND EDITION

Edited by Don E. Wilson
and DeeAnn M. Reeder

Smithsonian Institution Press • Washington and London
in association with the American Society of Mammalogists

This volume was produced from a database created under the supervision of the editors, who assume full responsibility for the contents and form

Book designer: Kathleen M. Sims
Typesetter: Ralph S. Walker, Office of Information Resource Management, Smithsonian Institution

Library of Congress Cataloging-in-Publication Data

Mammal species of the world: a taxonomic and geographic reference / edited by Don E. Wilson and DeeAnn M. Reeder. — 2nd ed.
p. cm.
Includes bibliographical references and index.
ISBN 1-56098-217-9
1. Mammals—Classification. 2. Mammals—Geographical distribution. I. Wilson, Don E. II. Reeder, DeeAnn M.
QL708.M35 1992
599'.0012—dc20 92-22703
CIP

British Library Cataloguing-in-Publication Data available

99 98 97 5 4

The paper used in this publication meets the minimum requirements of the American National Standard for Permanence of Paper for Printed Library Materials Z39.48–1984

Contents

Preface xiii

Acknowledgments xv

List of Contributors xvii

Introduction 1

Checklist of Mammal Species of the World
- Class Mammalia 13
  - Order Monotremata by **Colin P. Groves** 13
    - Family Tachyglossidae 13
    - Family Ornithorhynchidae 13
  - Order Didelphimorphia by **Alfred L. Gardner** 15
    - Family Didelphidae 15
      - Subfamily Caluromyinae 15
      - Subfamily Didelphinae 16
  - Order Paucituberculata by **Alfred L. Gardner** 25
    - Family Caenolestidae 25
  - Order Microbiotheria by **Alfred L. Gardner** 27
    - Family Microbiotheriidae 27
  - Order Dasyuromorphia by **Colin P. Groves** 29
    - Family Thylacinidae 29
    - Family Myrmecobiidae 29
    - Family Dasyuridae 29
  - Order Peramelemorphia by **Colin P. Groves** 39
    - Family Peramelidae 39
    - Family Peroryctidae 40
  - Order Notoryctemorphia by **Colin P. Groves** 43
    - Family Notoryctidae 43
  - Order Diprotodontia by **Colin P. Groves** 45
    - Family Phascolarctidae 45
    - Family Vombatidae 45
    - Family Phalangeridae 46
    - Family Potoroidae 48
    - Family Macropodidae 50
    - Family Burramyidae 57
    - Family Pseudocheiridae 58
    - Family Petauridae 60
    - Family Tarsipedidae 62
    - Family Acrobatidae 62
  - Order Xenarthra by **Alfred L. Gardner** 63
    - Family Bradypodidae 63
    - Family Megalonychidae 63

Subfamily Choloepinae 63
Family Dasypodidae 64
Subfamily Chlamyphorinae 64
Subfamily Dasypodinae 64
Family Myrmecophagidae 67
Order Insectivora by **Rainer Hutterer** 69
Family Solenodontidae 69
Family Nesophontidae 69
Family Tenrecidae 70
Subfamily Geogalinae 70
Subfamily Oryzoryctinae 70
Subfamily Potamogalinae 72
Subfamily Tenrecinae 73
Family Chrysochloridae 74
Family Erinaceidae 76
Subfamily Erinaceinae 76
Subfamily Hylomyinae 79
Family Soricidae 80
Subfamily Crocidurinae 81
Subfamily Soricinae 105
Family Talpidae 124
Subfamily Desmaninae 124
Subfamily Talpinae 124
Subfamily Uropsilinae 130
Order Scandentia by **Don E. Wilson** 131
Family Tupaiidae 131
Subfamily Tupaiinae 131
Subfamily Ptilocercinae 133
Order Dermoptera by **Don E. Wilson** 135
Family Cynocephalidae 135
Order Chiroptera by **Karl F. Koopman** 137
Family Pteropodidae 137
Subfamily Pteropodinae 137
Subfamily Macroglossinae 153
Family Rhinopomatidae 155
Family Craseonycteridae 155
Family Emballonuridae 156
Family Nycteridae 161
Family Megadermatidae 162
Family Rhinolophidae 163
Subfamily Rhinolophinae 163
Subfamily Hipposiderinae 169
Family Noctilionidae 176
Family Mormoopidae 176
Family Phyllostomidae 177
Subfamily Phyllostominae 177

Subfamily Lonchophyllinae 181
Subfamily Brachyphyllinae 182
Subfamily Phyllonycterinae 182
Subfamily Glossophaginae 183
Subfamily Carolliinae 186
Subfamily Stenodermatinae 187
Subfamily Desmodontinae 194
Family Natalidae 194
Family Furipteridae 195
Family Thyropteridae 195
Family Myzopodidae 196
Family Vespertilionidae 196
Subfamily Kerivoulinae 196
Subfamily Vespertilioninae 198
Subfamily Murininae 228
Subfamily Miniopterinae 230
Subfamily Tomopeatinae 231
Family Mystacinidae 231
Family Molossidae 232
Order Primates by **Colin P. Groves** 243
Family Cheirogaleidae 243
Subfamily Chaeirogaleinae 243
Subfamily Phanerinae 244
Family Lemuridae 244
Family Megaladapidae 245
Family Indridae 246
Family Daubentoniidae 247
Family Loridae 247
Family Galagonidae 248
Family Tarsiidae 250
Family Callitrichidae 251
Family Cebidae 254
Subfamily Alouattinae 254
Subfamily Aotinae 255
Subfamily Atelinae 256
Subfamily Callicebinae 258
Subfamily Cebinae 259
Subfamily Pitheciinae 260
Family Cercopithecidae 262
Subfamily Cercopithecinae 262
Subfamily Colobinae 269
Family Hylobatidae 274
Family Hominidae 276
Order Carnivora by **W. Christopher Wozencraft** 279
Family Canidae 279
Family Felidae 288

Subfamily Acinonychinae 288
Subfamily Felinae 288
Subfamily Pantherinae 297
Family Herpestidae 299
Subfamily Galidiinae 299
Subfamily Herpestinae 300
Family Hyaenidae 308
Subfamily Hyaeninae 308
Subfamily Protelinae 309
Family Mustelidae 309
Subfamily Lutrinae 309
Subfamily Melinae 313
Subfamily Mellivorinae 315
Subfamily Mephitinae 315
Subfamily Mustelinae 317
Subfamily Taxidiinae 325
Family Odobenidae 325
Family Otariidae 326
Family Phocidae 329
Family Procyonidae 332
Subfamily Potosinae 333
Subfamily Procyoninae 334
Family Ursidae 336
Subfamily Ailurinae 336
Subfamily Ursinae 337
Family Viverridae 340
Subfamily Cryptoproctinae 340
Subfamily Euplerinae 340
Subfamily Hemigalinae 341
Subfamily Nandiniinae 342
Subfamily Paradoxurinae 342
Subfamily Viverrinae 344
Order Cetacea by **James G. Mead** and **Robert L. Brownell, Jr.** 349
Family Balaenidae 349
Family Balaenopteridae 349
Family Eschrichtiidae 350
Family Neobalaenidae 351
Family Delphinidae 351
Family Monodontidae 357
Family Phocoenidae 358
Family Physeteridae 359
Family Platanistidae 360
Family Ziphiidae 361
Order Sirenia by **Don E. Wilson** 365
Family Dugongidae 365
Family Trichechidae 365

Order Proboscidea by **Don E. Wilson** 367
Family Elephantidae 367
Order Perissodactyla by **Peter Grubb** 369
Family Equidae 369
Family Tapiridae 370
Family Rhinocerotidae 371
Order Hyracoidea by **Duane A. Schlitter** 373
Family Procaviidae 373
Order Tubulidentata by **Duane A. Schlitter** 375
Family Orycteropodidae 375
Order Artiodactyla by **Peter Grubb** 377
Family Suidae 377
Subfamily Babyrousinae 377
Subfamily Phacochoerinae 377
Subfamily Suinae 377
Family Tayassuidae 379
Family Hippopotamidae 380
Family Camelidae 381
Family Tragulidae 382
Family Giraffidae 383
Family Moschidae 383
Family Cervidae 384
Subfamily Cervinae 384
Subfamily Hydropotinae 388
Subfamily Muntiacinae 388
Subfamily Capreolinae 389
Family Antilocapridae 392
Family Bovidae 393
Subfamily Aepycerotinae 393
Subfamily Alcelaphinae 393
Subfamily Antilopinae 395
Subfamily Bovinae 400
Subfamily Caprinae 404
Subfamily Cephalophinae 410
Subfamily Hippotraginae 412
Subfamily Peleinae 413
Subfamily Reduncinae 413
Order Pholidota by **Duane A. Schlitter** 415
Family Manidae 415
Order Rodentia 417
Suborder Sciurognathi 417
Family Aplodontidae by **Don E. Wilson** 417
Family Sciuridae by **Robert S. Hoffmann, Charles G. Anderson, Richard W. Thorington, Jr.**, and **Lawrence R. Heaney** 419
Subfamily Sciurinae 419

Subfamily Pteromyinae 459
Family Castoridae by **Don E. Wilson** 467
Family Geomyidae by **James L. Patton** 469
Family Heteromyidae by **James L. Patton** 477
Subfamily Dipodomyinae 477
Subfamily Heteromyinae 481
Subfamily Perognathinae 482
Family Dipodidae by **Mary Ellen Holden** 487
Subfamily Allactaginae 487
Subfamily Cardiocraniinae 491
Subfamily Dipodinae 492
Subfamily Euchoreutinae 495
Subfamily Paradipodinae 495
Subfamily Sicistinae 495
Subfamily Zapodinae 498
Family Muridae by **Guy G. Musser** and **Michael D. Carleton** 501
Subfamily Arvicolinae 501
Subfamily Calomyscinae 535
Subfamily Cricetinae 536
Subfamily Cricetomyinae 540
Subfamily Dendromurinae 541
Subfamily Gerbillinae 546
Subfamily Lophiomyinae 563
Subfamily Murinae 564
Subfamily Myospalacinae 675
Subfamily Mystromyinae 676
Subfamily Nesomyinae 677
Subfamily Otomyinae 679
Subfamily Petromyscinae 683
Subfamily Platacanthomyinae 684
Subfamily Rhizomyinae 685
Subfamily Sigmodontinae 687
Subfamily Spalacinae 753
Family Anomaluridae by **Fritz Dieterlen** 757
Subfamily Anomalurinae 757
Subfamily Zenkerellinae 757
Family Pedetidae by **Fritz Dieterlen** 759
Family Ctenodactylidae by **Fritz Dieterlen** 761
Family Myoxidae by **Mary Ellen Holden** 763
Subfamily Graphiurinae 763
Subfamily Leithiinae 765
Subfamily Myoxinae 769
Suborder Hystricognathi by **Charles A. Woods** 771
Family Bathyergidae 771
Family Hystricidae 773

Family Petromuridae 774
Family Thryonomyidae 775
Family Erethizontidae 775
Family Chinchillidae 777
Family Dinomyidae 778
Family Caviidae 778
Subfamily Caviinae 778
Subfamily Dolichotinae 780
Family Hydrochaeridae 781
Family Dasyproctidae 781
Family Agoutidae 783
Family Ctenomyidae 783
Family Octodontidae 787
Family Abrocomidae 789
Family Echimyidae 789
Subfamily Chaetomyinae 789
Subfamily Dactylomyinae 790
Subfamily Echimyinae 791
Subfamily Eumysopinae 793
Subfamily Heteropsomyinae 799
Family Capromyidae 800
Subfamily Capromyinae 800
Subfamily Hexolobodontinae 803
Subfamily Isolobodontinae 803
Subfamily Plagiodontinae 804
Family Heptaxodontidae 804
Subfamily Clidomyinae 805
Subfamily Heptaxodontinae 805
Family Myocastoridae 805
Order Lagomorpha by **Robert S. Hoffmann** 807
Family Ochotonidae 807
Family Leporidae 813
Order Macroscelidea by **Duane A. Schlitter** 829
Family Macroscelididae 829

Appendix I. Remarks on Bibliography and Publication Dates 831

Appendix II. List of *Mammalian Species* Accounts 835

Literature Cited 843

Index 1001

Preface

"A checklist of species is an invaluable tool for both researchers and the interested public." Thus began the first edition of this work, published by the Association of Systematics Collections (ASC) and Allen Press in 1982. That first edition was prepared by 189 professional mammalogists from 23 countries. It was coordinated by a special Checklist Committee of the American Society of Mammalogists. During the ensuing decade, it became the industry standard for mammalian taxonomy, providing an authoritative reference for nonspecialists and establishing an overall taxonomic hypothesis for testing by systematic mammalogists.

Students of mammalian taxonomy have made significant advances in recent years. One hundred seventy one new species have been described since the 1982 edition, and numerous taxonomic changes have been suggested. A revised edition of the checklist is clearly warranted.

The American Society of Mammalogists anticipated the need for revision and established a Standing Checklist Committee under the chairmanship of Karl F. Koopman in June 1982, concurrent with the publication of the first edition. Duane A. Schlitter joined Koopman as co-chair in 1985, and they coordinated the committee's efforts until 1990. At that time, Don E. Wilson assumed the chairmanship of the committee, with a mandate to expand the committee and produce a second edition of the checklist. The authors of this edition constitute that expanded committee. With support from the Office of Biodiversity Programs at the Smithsonian Institution's National Museum of Natural History, DeeAnn M. Reeder joined the project as a full-time research assistant in August 1991.

The new edition differs in several significant ways from the original. Perhaps most important of these is the abandonment of the egalitarian approach of the first edition that allowed any mammalogist the opportunity to have dissenting opinions published along with the consensus taxonomic arrangement. Each section of the new edition carries authorship, and taxonomic decisions are solely the responsibility of the appropriate author(s). Authors were chosen on the basis of expertise with their respective groups and were allowed to adopt the taxonomic arrangement they felt best reflected the current state of knowledge. This allows for more cohesive arrangements of taxa, and forces conflicting opinions to be expressed in the scientific literature where they can be substantiated by data.

As with the original, this edition should not be viewed as final. Mammalian taxonomy is constantly being refined as phylogenetic hypotheses are tested, and we anticipate continued change in the arrangement presented here. The checklist will be maintained in electronic form to simplify future updates. We welcome your suggestions, comments, and additions, and would particularly appreciate receiving copies of pertinent literature for preparation of future editions.

Don E. Wilson
DeeAnn M. Reeder
National Museum of Natural History
Smithsonian Institution

Acknowledgments

When a work as complex as *Mammal Species of the World* is finished, it is almost impossible to identify all of the many contributions deserving mention that have made the feat possible. We are ultimately most grateful to the hundreds of professional mammalogists who devoted their careers to the systematic study of Mammalia. Each has made a particular contribution to this synthesis.

The project owes much to Stephen R. Edwards and Robert S. Hoffmann, who initiated the effort that led to the first edition. That first edition was capably edited by James H. Honacki, Kenneth E. Kinman, and James W. Koeppl. The original Checklist Committee, chaired by Alfred L. Gardner and including Robert L. Brownell, Jr., Robert S. Hoffmann, Karl F. Koopman, Guy G. Musser, and Duane A. Schlitter, worked intensively to complete the first edition.

Elaine Hoagland, who succeeded Steve Edwards as executive director of the Association of Systematics Collections, facilitated transfer of the copyright to the American Society of Mammalogists. Duane Schlitter and Karl Koopman served as co-chairmen of the Checklist Committee for many years, keeping the project functioning during the decade between editions.

Special thanks are extended to the staff of the Smithsonian Institution's Office of Information Resource Management. Joe Russo generously made staff time available to assist us in many ways. Barbara Weitbrecht provided exceptional service in developing the database program for the project, and Mignon Erixon-Stanford helped with the fine-tuning. Ralph Walker of that office prepared the camera-ready copy used in the final publication. IBM Corporation, University Relations and Academic Information Systems, through the BioCIS project, generously provided the hardware and software. Peter Cannell of the Smithsonian Institution Press worked with us at all stages to facilitate final production. W. Christopher Wozencraft with the assistance of Craig Ludwig developed the database program used for the literature cited. W. Christopher Wozencraft devoted countless hours to converting, editing, and managing the incoming citations for the database. Robert S. Hoffmann helped us in ways too numerous to list; his constant encouragement and enthusiasm, as well as his willingness to prepare substantial portions of the text and to review others, was greatly appreciated. Portions of the manuscript were reviewed by R. Angermann, L. Barnes, A. Berta, G. Bronner, R.L. Brownell, Jr., G. B. Corbet, F. Dieterlen, L.H. Emmons, R.D. Fisher, T. Flannery, R. García-Perea, H. Griffiths, C.P. Groves, C.O. Handley, Jr., J.E. Heyning, R.S. Hoffmann, M.E. Holden, T. Holmes, C. Jones, M. Lawrence, Lin Yong-lie, C.A. Long, R.D.E. MacPhee, J.T. Marshall, G.G. Musser, P. Myers, F. Palacios, W.F. Perrin, G. Peters, R.H. Pine, D. Rice, O. Rossolimo, D.T. Rowe-Rowe, G.I. Shenbrot, I. Stirling, G. Storch, M.E. Taylor, R.H. Tedford, R.W. Thorington, Jr., H. van Rompaey, C. van Zyll de Jong, J. Wahlert, J. Watson, C. Watts, R. Wayne, L. Werdelin, A. Wyss, and P. Youngman.

The Division of Mammals at the National Museum of Natural History, Smithsonian Institution, provided an intellectual home for the editors, and many of the staff assisted in numerous ways. The collections and library facilities were invaluable to the completion of our efforts. In addition, curators and staff at other institutions aided individual authors in their efforts through the loan of specimens or by providing information. These include S. Anderson,

R. Voss, and the library staff at the American Museum of Natural History (New York); R. Baker at Texas Tech University (Lubbock); C. Carmichael at Michigan State University (East Lansing); S. Chakraborty at the Zoological Survey of India (Calcutta); M. Hafner at Louisiana State University Museum (Baton Rouge); P. Myers at the University of Michigan (Ann Arbor); P. Jenkins at the Natural History Museum (London); M. Rutzmoser at the Museum of Comparative Zoology (Cambridge); S. Goodman, J. Kerbis, and B. Patterson at the Field Museum of Natural History (Chicago); J. Patton at the Museum of Vertebrate Zoology (Berkeley); and D. Schmidly at Texas A&M University (College Station).

Finally, we would like to express our gratitude to the National Museum of Natural History and the Seidell Fund of the Smithsonian Institution for making funds available for the completion of this project.

List of Contributors

CHARLES G. ANDERSON
Department of Zoology
University of Tennessee
Knoxville, TN 37916

ROBERT L. BROWNELL, JR.
Southwest Fisheries Science Center
8604 La Jolla Shores Rd
La Jolla, CA 92038

MICHAEL D. CARLETON
Division of Mammals
National Museum of Natural History
Washington, DC 20560

FRITZ DIETERLEN
Staatliches Museum für Naturkunde
Schloss Rosenstein
7000 Stuttgart 1
Germany

ALFRED L. GARDNER
National Biological Survey
National Museum of Natural History
Washington, DC 20560

COLIN P. GROVES
Department of Archaeology and Anthropology
The Australian National University
GPO Box 4
Canberra ACT 0200
Australia

PETER GRUBB
35 Downhills Park Road
London N17 6PE
England

VIRGINIA HAYSSEN
Department of Biological Sciences
Smith College
Northampton, MA 01063

LAWRENCE R. HEANEY
Field Museum of Natural History
Roosevelt Road at Lake Shore Drive
Chicago, IL 60605

ROBERT S. HOFFMANN
Assistant Secretary for the Sciences
National Museum of Natural History
Washington, DC 20560

MARY ELLEN HOLDEN
Museum of Vertebrate Zoology
1120 Life Sciences Building
University of California
Berkeley, CA 94720

RAINER HUTTERER
Zoologisches Forschunginstitut und Museum Alexander Koenig
Adenauerallee 150-164
5300 Bonn 1
Germany

KARL F. KOOPMAN
Department of Mammalogy
American Museum of Natural History
New York, NY 10024

JAMES G. MEAD
Division of Mammals
National Museum of Natural History
Washington, DC 20560

GUY G. MUSSER
Department of Mammalogy
American Museum of Natural History
New York, NY 10024

JAMES L. PATTON
Museum of Vertebrate Zoology
University of California
Berkeley, CA 94720

DUANE A. SCHLITTER
Section of Mammals
Carnegie Museum of Natural History
5800 Baum Boulevard.
Pittsburgh, PA 15206

RICHARD W. THORINGTON, JR.
Division of Mammals
National Museum of Natural History
Washington, DC 20560

DON E. WILSON
Office of Biodiversity Programs
National Museum of Natural History
Washington, DC 20560

CHARLES A. WOODS
Florida Museum of Natural History
University of Florida
Gainesville, FL 32611-2035

W. CHRISTOPHER WOZENCRAFT
Division of Natural Sciences
Lewis-Clark State College
Lewiston, ID 83601

Introduction

The dynamic, rapidly changing state of mammalian taxonomy, documented by an enormous literature, long hampered the compilation of a detailed, complete world checklist. During the past century, the first complete appraisal of all mammals of the world was produced by Trouessart (1898-99, 1904-05). A more modern compilation was provided when E.P. Walker and colleagues brought out, in 1964, the first edition of *Mammals of the World*. This three volume compendium, now in its fifth edition (Nowak, 1991), was taxonomically arranged (with considerable supplementary natural history data) to the generic level, and later editions enumerated the species in each genus in addition to furnishing an illustration of at least one member of each genus. V.E. Sokolov based his *Systematics of Mammals*, published in Russian (1973-79), on Walker's *Mammals of the World*. He provided a list of species with a brief summary of geographic distribution in each generic account. Finally, Corbet and Hill (1980) listed the species of the world, abbreviated distributions, common names, literature citations to major regional distributional works, and some additional revisionary works where appropriate. Their latest revision (1991) differs in some respects with this volume, reflecting both differing interpretation and, in some cases, additional literature appearing since their work.

This volume will undoubtedly be used by many readers who are not systematic mammalogists. Do not be alarmed or disheartened by the debate over definition of species limits within many groups of mammals. Authors have noted differences of opinion in order to emphasize areas needing additional taxonomic study. Mammals are no worse off in this regard than other groups of animals, and in fact are probably better known than most, with the possible exception of birds.

THE PROCESS OF COMPILATION

Knowledge of the systematics of mammals is distributed over an extensive assortment of works. The process of compiling and editing the information contained in this edition differed in several respects from that of the first edition.

First Edition

The first edition evolved from three sources: a manuscript written by Kenneth E. Kinman, a preliminary list developed by the Gainesville Field Station of the U.S. Fish and Wildlife Service, and the International Species Inventory (ISIS) List (Seal and Makey, 1974). The editors (J.H. Honacki, K.E. Kinman, and J.W. Koeppl) developed a draft checklist that was then made available to the professional mammalogy community. Members of the American Society of Mammalogists were invited to contribute to the list at whatever level they wished. As a result, appropriate portions of the draft checklist were sent to 255 individuals. These reviewers were asked to provide the following information: author of the scientific name of the species, and citation; type locality; distribution; citations of revisions or reviews; important synonyms; and explanatory comments if necessary. One hundred fifty mammalogists provided reviews, covering 85 percent of the species in the list. Additional reviewers were eventually found for the missing species. Compiling the various drafts included the necessity of comparing the various contributions and including information on which there was agreement. In cases where there was not agreement, errors may have been retained. To determine valid names of taxa in cases of disagreement, the most recent reviewer was followed. Some of these names have stood the test of time, and others have not.

The penultimate draft was forwarded to the Checklist Committee of the American Society of Mammalogists, consisting of Alfred L. Gardner (chair), Robert L. Brownell, Jr., Robert S. Hoffmann, Karl F. Koopman, Guy G. Musser, and Duane A. Schlitter. This group furnished the final review of the text and provided much useful additional information. The committee detected many problems resulting from the egalitarian effort, not all of which were satisfactorily resolved in the first edition.

Two categories of information were added after the review process. The protected status of affected taxa was listed, based on the Federal Register. The ISIS numbers for each species were listed, based on Seal and Makey (1974). In this second edition the protected status has been updated and retained, but the ISIS numbers have not, due to their limited usefulness to most users of the first edition.

Second Edition

The second edition was compiled and edited in a distinctly different fashion from the first. Although the original decision to consult many professional mammalogists was theoretically sensible, it introduced a practical information management dilemma for the first edition. The wealth of material and the challenge of uniting numerous different, frequently conflicting opinions caused weaknesses in the original volume. Nevertheless, the taxonomic hypotheses outlined in that volume provided the basis for the refinements apparent in this edition. This edition is envisioned as the second step in organizing a continuing taxonomic database for mammals.

Because comprehensive survey of the literature is now prohibitive for a single mammalogist (or perhaps because they don't make us like they use to), the labor of composing this edition was partitioned among 20 authors, specialists on their respective taxa. Although some authors had been maintaining records for their groups of interest for some time, preparation of the second edition began in earnest in the fall of 1990, when electronic and paper copies of the appropriate portions of the original text were distributed to the members of the committee along with a mandate to produce an up-to-date revision. Most of 1991 was spent in the production of that updated text. The authors have examined the literature on mammals, from the initial works to the most recent, and have provided references published in many languages. As each portion was received from the authors, it was converted to an electronic database, edited, outputted in word processing format, and returned to the authors for revisions. After a second round of editing, reviews were sought from other members of the community of systematic mammalogists. Although both editors and reviewers made suggestions to the authors, each section is the product of an individual author's scholarship, and represents that author's best hypothesis of the relationships of a particular group as currently understood. Obvious errors of omission or commission should be pointed out to the editors; differences of opinion should be argued directly with the authors.

The information leading to this edition was compiled with computer technology that permitted rapid manuscript revision and organization. This allowed automatic indexing of the many names used and standardization of literature citations, as well as considerably more consistency in citations, punctuation, format, and of spelling of place names, although infelicities undoubtedly remain. Additionally, the electronic database affords the possibility of future production of revised editions with minimal investment of time and resources.

Details on the 4,629 species of mammals covered by this edition (up from 4,170 in the first edition) are the product of myriad former scholars united by a common curiosity about the diversity of mammals (Figure 1, Table 1). A major feature of this edition, like the original, is that it identifies gaps in our knowledge in need of further study, and will serve as a starting point for the third edition.

Second Edition, With Corrections

Corrections were made to the text of the second edition at its third printing. These emendations did not alter the scientific content, nor did they alter taxonomic arrangements. Errors corrected included, typographical mistakes, misspellings, incorrect dates, and overlooked synonyms. Literature citations were further standardized. Although pagination of species accounts did not change, corrections to the index increased the total page number. We thank all of the authors for their continued dedication to improving the quality of the text. Particular thanks are due to Alan Peterson, Bruce Patterson, Duane Schlitter, and Gordon Corbet for assistance in finding and correcting errors in the original text of the second edition.

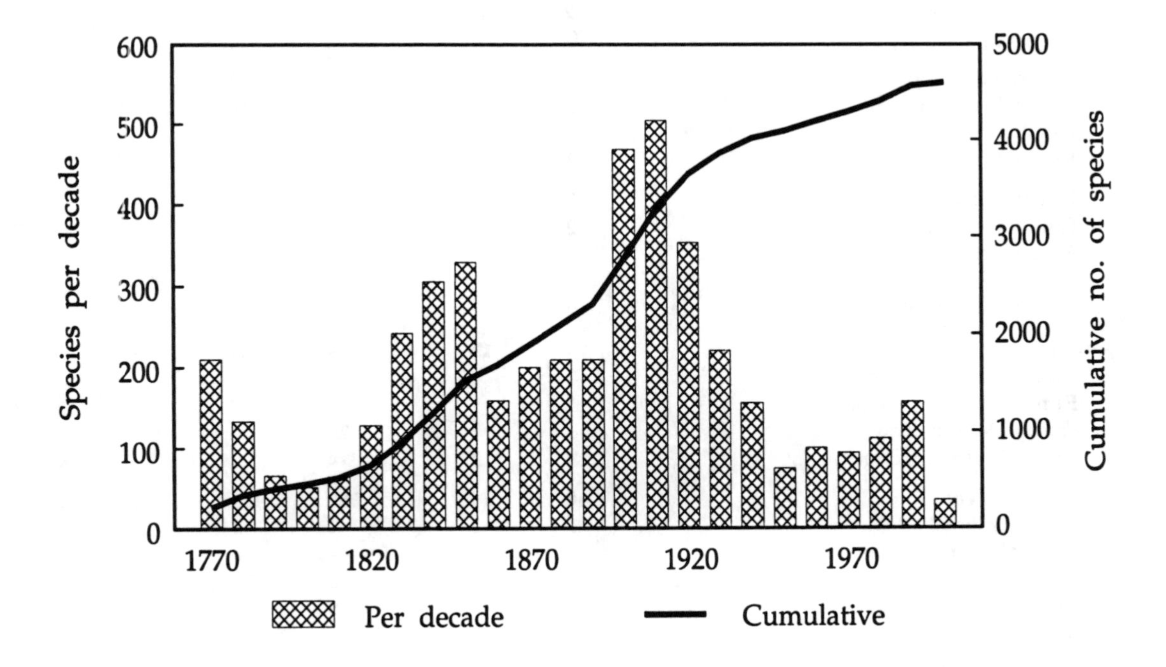

FIGURE 1. The description of the world's mammal fauna by decade (bars) and cumulatively (solid line).

Table 1. Comparisons of Genera and Species between the First and Second Editions.

	Number of Genera		Number of Species[1]		
	First Edition	Second Edition	First Edition	Second Edition	Described since the first edition
Class Mammalia	**1033**	**1135**	**4170**	**4629**	**172**
Order Monotremata	**3**	**3**	**3**	**3**	—
Family Tachyglossidae	2	2	2	2	—
Family Ornithorhynchidae	1	1	1	1	—
Order Didelphimorphia[2]	**11**	**15**	**76**	**63**	—
Family Didelphidae	11	15	76	63	—
Order Paucituberculata[2]	**3**	**3**	**7**	**5**	—
Family Caenolestidae	3	3	7	5	—
Order Microbiotheria[2]	**1**	**1**	**1**	**1**	—
Family Microbiotheriidae	1	1	1	1	—
Order Dasyuromorphia[2]	**15**	**17**	**52**	**63**	**10**
Family Thylacinidae	1	1	1	1	—
Family Myrmecobiidae	1	1	1	1	—
Family Dasyuridae	13	15	50	61	10
Order Peramelemorphia[2]	**8**	**8**	**19**	**21**	**2**
Family Peramelidae[3]	4	4	10	10	—
Family Peroryctidae[4]	4	4	9	11	2
Order Notoryctemorphia[2]	**1**	**1**	**1**	**2**	—
Family Notoryctidae	1	1	1	2	—
Order Diprotodontia[2]	**32**	**38**	**107**	**117**	**3**
Family Phascolarctidae	1	1	1	1	—
Family Vombatidae	2	2	3	3	—
Family Phalangeridae	3	6	15	18	1
Family Potoroidae[5]	5	5	9	9	—
Family Macropodidae	11	11	48	54	2
Family Burramyidae	2	2	5	5	—
Family Pseudocheiridae[6]	2	5	14	14	—
Family Petauridae	3	3	9	10	—
Family Tarsipedidae	1	1	1	1	—
Family Acrobatidae[7]	2	2	2	2	—
Order Xenarthra	**13**	**13**	**29**	**29**	—
Family Bradypodidae	1	1	3	3	—
Family Megalonychidae[8]	1	1	2	2	—
Family Dasypodidae	8	8	20	20	—
Family Myrmecophagidae	3	3	4	4	—
Order Insectivora	**64**	**66**	**389**	**428**	**23**
Family Solenodontidae	1	1	2	3	—
Family Nesophontidae[9]	—	1	—	8	—
Family Tenrecidae	12	10	33	24	1
Family Chrysochloridae	7	7	18	18	—
Family Erinaceidae	8	7	17	21	1
Family Soricidae	21	23	288	312	19
Family Talpidae	15	17	31	42	2
Order Scandentia	**5**	**5**	**16**	**19**	—
Family Tupaiidae	5	5	16	19	—

Table 1.—cont.

	Number of Genera		Number of Species[1]		
	First Edition	Second Edition	First Edition	Second Edition	Described since the first edition
Order Dermoptera	**1**	**1**	**2**	**2**	—
Family Cynocephalidae	1	1	2	2	—
Order Chiroptera	**175**	**177**	**918**	**925**	**28**
Family Pteropodidae	42	42	161	166	9
Family Rhinopomatidae	1	1	3	3	—
Family Craseonycteridae	1	1	1	1	—
Family Emballonuridae	12	13	48	47	—
Family Nycteridae	1	1	14	12	—
Family Megadermatidae	4	4	5	5	—
Family Rhinolophidae	10	10	127	130	3
Family Noctilionidae	1	1	2	2	—
Family Mormoopidae	2	2	8	8	—
Family Phyllostomidae	47	49	138	143	5
Subfamily Phyllostominae	11	11	30	33	1
Subfamily Lonchophyllinae	3	3	8	9	1
Subfamily Brachyphyllinae	1	1	2	2	—
Subfamily Phyllonycterinae	2	2	3	3	—
Subfamily Glossophaginae	10	10	23	22	1
Subfamily Carolliinae	2	2	7	7	—
Subfamily Stenodermatinae	16	17	62	62	2
Subfamily Desmodontinae	2	3	3	3	—
Family Natalidae	1	1	4	5	—
Family Furipteridae	2	2	2	2	—
Family Thyropteridae	1	1	2	2	—
Family Myzopodidae	1	1	1	1	—
Family Vespertilionidae	36	35	315	318	11
Family Mystacinidae	1	1	1	2	—
Family Molossidae	12	12	86	80	—
Order Primates	**51**	**60**	**181**	**233**	**11**
Family Cheirogaleidae	4	4	7	7	—
Family Lemuridae	3	4	9	10	1
Family Megaladapidae[10]	1	1	7	7	—
Family Indridae[11]	3	3	4	5	1
Family Daubentoniidae	1	1	1	1	—
Family Loridae[12]	4	4	5	6	—
Family Galagonidae[13]	2	4	8	11	—
Family Tarsiidae	1	1	3	5	1
Family Callitrichidae[14]	5	4	16	26	1
Family Cebidae	11	11	31	58	6
Family Cercopithecidae	11	18	76	81	1
Family Hylobatidae	1	1	9	11	—
Family Hominidae[15]	4	4	5	5	—
Order Carnivora	**108**	**129**	**270**	**271**	**1**
Family Canidae	11	14	35	34	—
Family Felidae	5	18	37	36	—

Table 1.—cont.

	Number of Genera		Number of Species[1]		
	First Edition	Second Edition	First Edition	Second Edition	Described since the first edition
Family Herpestidae	17	18	36	37	1
Family Hyaenidae[16]	3	4	4	4	—
Family Mustelidae	23	25	63	65	—
Family Odobenidae	1	1	1	1	—
Family Otariidae	7	7	14	14	—
Family Phocidae	10	10	19	19	—
Family Procyonidae[17]	6	6	18	18	—
Family Ursidae[17]	6	6	9	9	—
Family Viverridae	19	20	34	34	—
Order Cetacea	**39**	**41**	**77**	**78**	1
Family Balaenidae	1	2	2	3	—
Family Balaenopteridae	2	2	6	6	—
Family Eschrichtiidae	1	1	1	1	—
Family Neobalaenidae[18]	1	1	1	1	—
Family Delphinidae	17	17	33	32	—
Family Monodontidae	2	2	2	2	—
Family Phocoenidae	3	4	6	6	—
Family Physeteridae	2	2	3	3	—
Family Platanistidae	4	4	5	5	—
Family Ziphiidae	6	6	18	19	1
Order Sirenia	**3**	**3**	**5**	**5**	—
Family Dugongidae	2	2	2	2	—
Family Trichechidae	1	1	3	3	—
Order Proboscidea	**2**	**2**	**2**	**2**	—
Family Elephantidae	2	2	2	2	—
Order Perissodactyla	**6**	**6**	**18**	**18**	—
Family Equidae	1	1	9	9	—
Family Tapiridae	1	1	4	4	—
Family Rhinocerotidae	4	4	5	5	—
Order Hyracoidea	**3**	**3**	**7**	**6**	—
Family Procaviidae	3	3	7	6	—
Order Tubulidentata	**1**	**1**	**1**	**1**	—
Family Orycteropodidae	1	1	1	1	—
Order Artiodactyla	**77**	**81**	**187**	**220**	**5**
Family Suidae	5	5	8	16	1
Family Tayassuidae	2	3	3	3	—
Family Hippopotamidae	2	2	2	4	—
Family Camelidae	3	3	6	6	—
Family Tragulidae	2	3	4	4	—
Family Giraffidae	1	2	2	2	—
Family Moschidae[19]	1	1	4	4	1
Family Cervidae	14	16	34	43	2
Family Antilocapridae[20]	1	1	1	1	—
Family Bovidae	45	45	123	137	1
Order Pholidota	**1**	**1**	**7**	**7**	—

Table 1.—cont.

	Number of Genera		Number of Species[1]		
	First Edition	Second Edition	First Edition	Second Edition	Described since the first edition
Family Manidae	1	1	7	7	—
Order Rodentia	**394**	**443**	**1719**	**2015**	**86**
Suborder Sciurognathi	**338**	**375**	**1535**	**1786**	**75**
Family Aplodontidae	1	1	1	1	—
Family Sciuridae	49	50	261	273	2
Family Castoridae	1	1	2	2	—
Family Geomyidae	5	5	35	35	1
Family Heteromyidae	5	6	63	59	—
Family Dipodidae[21]	15	15	44	51	4
Family Muridae	246	281	1100	1326	67
Subfamily Arvicolinae[22]	19	26	128	143	3
Subfamily Calomyscinae	1	1	5	6	1
Subfamily Cricetinae[22]	4	7	19	18	1
Subfamily Cricetomyinae	3	3	5	6	—
Subfamily Dendromurinae	8	8	20	23	—
Subfamily Gerbillinae	15	14	81	110	1
Subfamily Lophiomyinae	1	1	1	1	—
Subfamily Murinae	107	122	433	529	47
Subfamily Myospalacinae	1	1	5	7	—
Subfamily Mystromyinae	1	1	1	1	—
Subfamily Nesomyinae	7	7	10	14	—
Subfamily Otomyinae	2	2	12	14	1
Subfamily Petromyscinae	2	2	3	5	—
Subfamily Platacanthomyinae	2	2	2	3	—
Subfamily Rhizomyinae[22]	3	3	6	15	—
Subfamily Sigmodontinae	69	79	366	423	13
Subfamily Spalacinae[22]	1	2	3	8	—
Family Anomaluridae	3	3	7	7	—
Family Pedetidae	1	1	1	1	—
Family Ctenodactylidae	4	4	5	5	—
Family Myoxidae[23]	8	8	16	26	1
Suborder Hystricognathi	**56**	**68**	**184**	**229**	**11**
Family Bathyergidae	5	5	9	12	—
Family Hystricidae	4	3	11	11	—
Family Petromuridae	1	1	1	1	—
Family Thryonomyidae	1	1	2	2	—
Family Erethizontidae	5	4	12	12	1
Family Chinchillidae	3	3	6	6	—
Family Dinomyidae	1	1	1	1	—
Family Caviidae	5	5	14	14	1
Family Hydrochaeridae	1	1	1	1	—
Family Dasyproctidae	2	2	13	13	—
Family Agoutidae	1	1	2	2	—
Family Ctenomyidae	1	1	33	38	4
Family Octodontidae	5	6	8	9	1

Table 1.—cont.

	Number of Genera		Number of Species[1]		
	First Edition	Second Edition	First Edition	Second Edition	Described since the first edition
Family Abrocomidae	1	1	2	3	1
Family Echimyidae	15	20	55	78	1
Family Capromyidae	4	8	13	20	2
Family Heptaxodontidae[9]	—	4	—	5	—
Family Myocastoridae	1	1	1	1	—
Order Lagomorpha	**12**	**13**	**62**	**80**	**2**
Family Ochotonidae	2	2	19	26	2
Family Leporidae	10	11	43	54	—
Order Macroscelidea	**4**	**4**	**15**	**15**	**—**
Family Macroscelididae	4	4	15	15	—

[1] All numbers for the First edition are entered in terms of the taxonomy of the Second edtion.
[2] Included in Marsupialia in the First Edition.
[3] Includes Thylacomyidae, listed as distinct in the First Edition.
[4] Included in Peramelidae in the First Edition.
[5] Included in Macropodidae in the First Edition.
[6] Included in Petauridae in the First Edition.
[7] Included in Burramyidae in the First Edition.
[8] Includes Choloepidae, listed as distinct in the First Edition.
[9] Genera and species not listed in the First Edition.
[10] Included in Lemuridae in the First Edition.
[11] Spelled Indriidae in the First Edition.
[12] Spelled Lorisidae in the First Edition.
[13] Spelled Galagidae in the First Edition.
[14] Includes Callimiconidae, listed as distinct in the First Edition.
[15] Includes Pongidae, listed as distinct in the First Edition.
[16] Includes Protelidae, listed as distinct in the First Edition.
[17] Ursidae now includes *Ailurus* and *Ailuropoda; Ailurus* was included in Procyonidae in the First Edition.
[18] Included in Balaenidae in the First Edition.
[19] Included in Cervidae in the First Edition.
[20] Included in Bovidae in the First Edition.
[21] Includes Zapodidae, listed as distinct in the First Edition.
[22] Considered distict families in the First Edition.
[23] Called Gliridae in the First Edition; includes Selviniidae, listed as distinct in the First Edition.

ORGANIZATION OF THE BOOK

Species recognized herein are limited to existing or recently extinct species (possibly alive during the preceding 500 years); in instances where the persistence of a species is doubtful, the comment section so indicates.

Taxonomic Arrangement

Various workers have reviewed the higher categories of mammals since Simpson (1945) produced his definitive classification, including Anderson and Jones (1967, 1984) and McKenna (1975). These treatments were modified by Corbet and Hill (1980, 1986, 1991), and their arrangement of the orders has been generally followed here. Modifications to this arrangement have been duly noted and documented in the text, and include, among others, placement of Pinnipedia (Families Otariidae and Phocidae) in Carnivora, and reorganization of Rodentia primarily following Carleton (1984). Perhaps the most significant departure is the recognition of seven orders that were formerly contained in Marsupialia (Table 1). Families and subfamilies (when used) are presented alphabetically, although in a few cases the author has presented them in phylogenetic sequence. Generic names are ordered alphabetically within subfamilies, and species names are ordered alphabetically within genera, without exception. If subgenera are thought to be useful, they are listed in the comments section of each species.

Scientific names applied to domesticated mammals, and to the natural ancestors of domesticated forms, are the earliest valid names as called for by the Code of the International Commission on Zoological Nomenclature (ICZN, 1985).

Scientific Name and Authority

Each currently used scientific name is followed by the name of the author(s) and the year in which it was described, for example, *Vulpes vulpes* (Linnaeus, 1758). A Species originally named by its author in a genus other than the one in which it is now placed has parentheses surrounding the author and date. In this example Linnaeus, when he first named the red fox *vulpes*, placed it in the genus *Canis* instead of *Vulpes*. Should it be returned to *Canis*, the parentheses would be removed, that is, *Canis vulpes* Linnaeus, 1758. Following the species name, authority, and date is the citation for the work in which the original description appeared.

In some instances, especially in older literature, the date printed on a publication was not the actual date of publication. For example, many of the "parts" of the *Proceedings of the Zoological Society of London* were published in the year following that on the text. The actual date of publication is that used as the date of authority, and is found in brackets after the citation. Appendix I discusses publication dates and other bibliographic details for a variety of pertinent literature.

Usually, only the first page on which the species name appears follows the title citation. In some cases, for a variety of reasons, more than one page may be cited, as well as references to figures or plates. We have attempted to cite the first page on which the name appears, but if it is not listed unequivocally, the reader is advised to consult the original source.

Family names follow the principle of coordination outlined in Article 36 of the International Code of Zoological Nomenclature. Generic accounts include the type species for which the generic name was proposed. In general, the name appears as it was originally described, including authority. When the type species is no longer a recognized species, the species with which it is currently synonymized is listed in parentheses in its original form with its authority.

Type Locality

The type locality is the geographical site where the type material of a species was obtained. Type localities quoted exactly from the original description are enclosed in quotation marks. Information not surrounded by quotation marks has been arranged where possible with the current country name followed by state, province, or district, and specific locality. Elevation above sea level has been included when available, as have global coordinates in some cases. When appropriate, restrictions of the type locality made by revisers have been included as well. We have followed a variety of sources for current country names, given the rapidly changing conditions in many parts of the world.

Distribution

The geographical range of each species is summarized using contemporary political units or, in some cases, geographical names. However, geographical names are usually used only when the entire area is included in the range of the species. We have attempted to standardize usage and spelling, but some inconsistencies may remain, particularly in transliterations from other alphabets. Country names have changed drastically during the time the text was being prepared; we have attempted to give the most current name whenever possible. In cases where it was not possible to determine the current country name, we have used "former" (e.g., former USSR, if the currently recognized republic is indeterminate). Distribution records resulting from human introduction are noted. Maps of the distributions of many species are provided in the cited literature. If a species is known only from the type locality, that is noted.

Status

Mammal species covered by the regulations for the U.S. Endangered Species Act (U.S. ESA) as of July 15, 1991; those listed in the 1990 International Union for Conservation of Nature and Natural Resources (IUCN) Red List of Threatened Animals; and those listed in the appendices of the Convention on International Trade in Endangered Species of Wild Fauna and Flora (CITES) as of June 11, 1992, are noted in the text in this category. U.S. ESA endangered and threatened categories and IUCN extinct, endangered, vulnerable, rare, indeterminate (known to be endangered, vulnerable, or rare, but information is not complete enough to assign a category), insufficiently known (suspected to belong to the above categories), threatened, and commercially threatened categories are provided. In addition, Appendices I, II, and III of CITES are listed, where Appendix I includes species threatened with extinction that are or may be affected by trade, Appendix II includes species that although not necessarily threatened with extinction may become so unless trade in them is strictly controlled, as well as nonthreatened species that must be subject to regulation in order to control threatened species, and Appendix III includes species that any Party identifies as being subject to regulation within its jurisdiction for purposes of preventing or restricting exploitation, and for which it needs the cooperation of other Parties in controlling trade. The Federal Register (for U.S. ESA) and CITES appendices should be consulted for updates on current status.

Synonyms

Authors have attempted to provide taxonomic lists consistent with recent literature, tempered by their own individual judgments. Considerable effort has been expended to compile a complete list of synonyms that have been used in the scientific literature for each taxon. These are usually either names of later origin than that used (junior synonyms) or names that are systematically invalid for various reasons. Also included here are subspecies names, including any that might be currently recognized. Not included are subsequent emendations, misspellings, incorrect allocations, or partial synonyms. Theoretically, any scientific name used for a mammal should be found in this volume, either as a currently recognized species or as a synonym. The editors would appreciate notice of any omissions.

Comments

Taxonomic and nomenclatorial alternatives are accompanied by appropriate documentation, including opinions of the International Commission on Zoological Nomenclature (ICZN); revisions and additional literature sources are also cited. In the interest of brevity, secondary reference sources are sometimes cited to document taxonomic evidence, and reference to the primary sources can be found therein. Personal opinions of the author(s) and unpublished information are sometimes included here as well. When appropriate, other data such as references discussing type locality, occurrence of hybridization, and species known only from a single or few specimens are also included in the comments section. Absence of a comments section may indicate either that the species is taxonomically noncontroversial or that it is too poorly known to require comment.

Appendices

Appendix I discusses publication dates, particularly for older and problematic literature, and other bibliographic details. Appendix II lists the cumulative index (numbers 1-402) for the *Mammalian Species* accounts as compiled by Virginia Hayssen. This list has been ammended to reflect the higher taxonomic changes included in this edition. In some instances, the status of the species listed have changed, and these changes are depicted in the main text. The index for the entire work should be consulted to locate changes in specific or generic allocation.

Bibliographic Treatment

The literature cited section contains the works consulted in the compilation of this text. Works appearing after July 1, 1992, are not given. References cited as authorities for original descriptions are not included in the literature cited but are given after each currently recognized name in the text in sufficient detail to allow the reader to locate them. Author and date citations in the synonym section are not included in the literature cited as they are provided only as authorities for the names.

Index

The index contains all taxonomic names contained in this volume, including synonyms. Page references to all currently recognized generic and species names employed in this volume are in boldface type. Species names are individually listed alphabetically and are also associated with the genus in the alphabetical listing.

CHECKLIST OF MAMMAL SPECIES OF THE WORLD

ORDER MONOTREMATA

by Colin P. Groves

ORDER MONOTREMATA

COMMENTS: Reviewed by Griffiths (1978).

Family Tachyglossidae Gill, 1872. Smithson. Misc. Coll., 11:27.

Tachyglossus Illiger, 1811. Prodr. Syst. Mammal. Avium., p. 114.
TYPE SPECIES: *Echidna novaehollandiae* Lacépède, 1799.
SYNONYMS: *Acanthonotus* Goldfuss, 1809; *Echidna* G. Cuvier, 1797; *Echinopus* G. Fischer, 1814; *Syphonia* Rafinesque, 1815.

Tachyglossus aculeatus (Shaw, 1792). Nat. Misc., 3, pl. 109.
TYPE LOCALITY: Australia, New South Wales, New Holland (= Sydney).
DISTRIBUTION: S and E New Guinea; Australia, including Kangaroo Isl (off South Australia) and Tasmania.
STATUS: Abundant throughout its range.
SYNONYMS: *acanthion, acanthrous, australiensis, breviaculeata, corealis, hobartensis, hystrix, ineptus, lawesi, longiaculeata, longirostris, multiaculeata, myrmecophragris, novaehollandiae, setosa, sydneiensis, typicus.*
COMMENTS: Includes *lawesi, typicus,* and *setosus*; see Ride (1970:231). Name commonly attributed to Shaw and Nodder, but Nodder was the publisher, not an author.

Zaglossus Gill, 1877. Ann. Rec. Sci. Indus., May:171.
TYPE SPECIES: *Tachyglossus bruijni* Peters and Doria, 1876.
SYNONYMS: *Acanthoglossus, Bruynia, Prozaglossus.*

Zaglossus bruijni (Peters and Doria, 1876). Ann. Mus. Civ. Stor. Nat. Genova, 9:183.
TYPE LOCALITY: Indonesia, Irian Jaya, Vogelkop, Manokwari Div., Arfak Mtns.
DISTRIBUTION: Interior New Guinea; Salawati Isl (Indonesia).
STATUS: CITES - Appendix II; IUCN - Vulnerable.
SYNONYMS: *bartoni, bubuensis, dunius, galaris, goodfellowi, nigroaculeatus, pallidus, tridactyla, villosissima.*
COMMENTS: Includes *bartoni* and *bubuensis*; see Van Deusen and George (1969:22).

Family Ornithorhynchidae Gray, 1825. Ann. Philos., n.s., 10:343.

Ornithorhynchus Blumenbach, 1800. Gotting. Gelehrt. Anz., 1:609-610.
TYPE SPECIES: *Ornithorhynchus paradoxus* Blumenbach, 1800 (= *Platypus anatinus* Shaw, 1799).
SYNONYMS: *Dermipus, Platypus.*
COMMENTS: *Platypus* Shaw, 1799 was preoccupied by *Platypus* Herbst, 1793, a genus of Coleoptera. *Ornithorynchus* is the next available name.

Ornithorhynchus anatinus (Shaw, 1799). Nat. Misc., 10, pl. 385-386.
TYPE LOCALITY: Australia, New South Wales, New Holland (= Sydney).
DISTRIBUTION: Queensland, New South Wales, SE South Australia, Victoria, and Tasmania (Australia).
STATUS: Common but vulnerable to local extinction.
SYNONYMS: *brevirostris, crispus, fuscus, laevis, novaehollandiae, paradoxus, phoxinus, rufus, triton.*
COMMENTS: Name is commonly attributed to Shaw and Nodder, but Nodder was the publisher, not an author.

ORDER DIDELPHIMORPHIA

by Alfred L. Gardner

ORDER DIDELPHIMORPHIA

SYNONYMS: Marsupialia, Ameridelphia.

COMMENTS: Includes Ameridelphia (see Aplin and Archer, 1987; Marshall et al., 1990; and Szalay, 1982); but not Microbiotheriidae (Marshall et al., 1990; *contra* Reig et al., 1987).

Family Didelphidae Gray, 1821. London Med. Repos., 15:308.

COMMENTS: Placed in the order Polyprotodontia by Kirsch (1977); also see Aplin and Archer (1987). Does not include *Dromiciops*; see Kirsch and Calaby (1977).

Subfamily Caluromyinae Kirsch, 1977. Aust. J. Zool., Suppl. ser., 52:111.

Caluromys J. A. Allen, 1900. Bull. Am. Mus. Nat. Hist., 13:189.

TYPE SPECIES: *Didelphis philander* Linnaeus, 1758, by original designation.

SYNONYMS: *Mallodelphys* Thomas, 1920 (type species *Didelphis laniger* Desmarest, 1820, by original designation; valid as a subgenus to include *C. derbianus* and *C. lanatus*); *Philander* Burmeister, 1856 (preoccupied by *Philander* Tiedemann, 1808); *Sarigua* Muirhead, 1819 (part).

COMMENTS: Comparatively uncommon to rare in collections, perhaps due to nocturnal and arboreal habits; but probably common in suitable habitat. Vulnerable to loss of tropical forest habitat.

Caluromys derbianus (Waterhouse, 1841). Jardine's Natur. Libr., 11:97.

TYPE LOCALITY: None given; restricted to Colombia, Cauca, Cauca Valley (Cabrera, 1958:2).

DISTRIBUTION: México, Central America, W Colombia, and W Ecuador.

SYNONYMS: *aztecus* Thomas, 1913; *canus* Matschie, 1917; *centralis* Hollister, 1914; *fervidus* Thomas, 1913; *guayanus* Thomas, 1899; *nauticus* Thomas, 1913; *pallidus* Thomas, 1899; *pictus* Thomas, 1913; *pulcher* Matschie, 1917; *pyrrhus* Thomas, 1901; *senex* Thomas, 1913.

COMMENTS: Reviewed by Bucher and Hoffmann (1980, Mammalian Species, 140).

Caluromys lanatus (Olfers, 1818). *In* W. L. Eschwege, J. Brasilien, Neue Bibliothek. Reisen., 15:206.

TYPE LOCALITY: "Paraguay;" restricted to Caazapá, Caazapá (Cabrera, 1916).

DISTRIBUTION: N and C Colombia, NW and S Venezuela, E Ecuador, E Perú, E Bolivia, E and S Paraguay, N Argentina (Provincia Misiones), and W and S Brazil.

SYNONYMS: *antioquiae* Matschie, 1917; *bartletti* Matschie, 1917; *cahyensis* Matschie, 1917; *cicur* Bangs, 1898; *jivaro* Thomas, 1913; *juninensis* Matschie, 1917; *lanigera* Desmarest, 1820; *meridensis* Matschie, 1917; *modesta* Miranda-Ribeiro, 1936; *nattereri* Matschie, 1917; *ochropus* Wagner, 1842; *ornata* Tschudi, 1845; *vitalina* Miranda-Ribeiro, 1936.

Caluromys philander (Linnaeus, 1758). Syst. Nat., 10th ed., 1:54.

TYPE LOCALITY: "America;" restricted to Surinam (Thomas, 1911*a*).

DISTRIBUTION: Venezuela (including Margarita Isl), Trinidad and Tobago, Guyana, Surinam, French Guiana, and Brazil.

SYNONYMS: *affinis* Wagner, 1842; *cajopolin* Müller, 1776; *cayopollin* Schreber, 1777; *dichura* Wagner, 1842; *leucurus* Thomas, 1904; *trinitatis* Thomas, 1894; *venezuelae* Thomas, 1903.

Caluromysiops Sanborn, 1951. Fieldiana Zool., 31:474.

TYPE SPECIES: *Caluromysiops irrupta* Sanborn, 1951.

COMMENTS: Monotypic.

Caluromysiops irrupta Sanborn, 1951. Fieldiana Zool., 31:474.

TYPE LOCALITY: Perú, Cuzco, "Quincemil, Province of Quispicanchis."

DISTRIBUTION: SE Perú and W Brazil.

COMMENTS: Uncommon to rare; reports from Leticia, Colombia and Iquitos, Perú (see Izor and Pine, 1987) probably are based on captives acquired in live animal trade.

Glironia Thomas, 1912. Ann. Mag. Nat. Hist., ser. 8, 9:239.
TYPE SPECIES: *Glironia venusta* Thomas, 1912, by original designation.

Glironia venusta Thomas, 1912. Ann. Mag. Nat. Hist., ser. 8, 9:240.
TYPE LOCALITY: Perú, Huánuco, "Pozuzo."
DISTRIBUTION: Amazonian Brazil, Ecuador, Perú, and Bolivia.
STATUS: Rare.
SYNONYMS: *aequatorialis* Anthony, 1926; *criniger* Anthony, 1926.
COMMENTS: Reviewed by Marshall (1978*c*, Mammalian Species, 107).

Subfamily Didelphinae Gray, 1821. London Med. Repos., 15:308.

Chironectes Illiger, 1811. Prodr. Syst. Mammal. Avium., p. 76.
TYPE SPECIES: *Lutra minima* Zimmermann, 1780, by monotypy.
SYNONYMS: *Gamba* Liais, 1872; *Memina* Fischer, 1814; *Sarigua* Muirhead, 1819 (part).

Chironectes minimus (Zimmermann, 1780). Geogr. Gesch. Mensch. Vierf. Thiere, 2:317.
TYPE LOCALITY: "Gujana;" restricted to Cayenne, French Guiana (Cabrera, 1958:44).
DISTRIBUTION: Oaxaca and Tabasco, México, south through Central America to Colombia, Ecuador, Brazil, Perú, Venezuela, the Guianas, Paraguay, and NE Argentina.
SYNONYMS: *argyrodytes* Dickey, 1928; *bresslaui* Pohle, 1927; *cayennensis* Turton, 1802; *guianensis* Kerr, 1792; *gujanensis* Link, 1795; *langsdorffi* Boitard, 1845; *palmata* Daudin, 1799; *panamensis* Goldman, 1914; *paraguensis* Kerr, 1792; *sarcovienna* Shaw, 1800; *variegatus* Olfers, 1818; *yapock* Desmarest, 1820.
COMMENTS: Reviewed by Marshall (1978*d*, Mammalian Species, 109).

Didelphis Linnaeus, 1758. Syst. Nat., 10th ed., 1:54.
TYPE SPECIES: *Didelphis marsupialis* Linnaeus, 1758, by subsequent selection (Thomas, 1911*a*).
SYNONYMS: *Dimerodon* Ameghino, 1889; *Leucodidelphis* Ihering, 1914; *Opossum* Schmid, 1818 (part); *Sarigua* Muirhead, 1819 (part).

Didelphis albiventris Lund, 1840. K. Dansk. Vid. Selsk. Afhandl., p. 20 [preprint of Lund, 1841].
TYPE LOCALITY: Brazil, Minas Gerais, "Rio das Velhas," Lagoa Santa.
DISTRIBUTION: A Guyana Highland isolate in S Venezuela, SW Suriname, and N Brazil; and the major distribution in Colombia, Ecuador, Perú, Brazil, Bolivia, Paraguay, Uruguay, and the northern half of Argentina.
SYNONYMS: *andina* J. A. Allen, 1902; *antigua* Ameghino, 1889; *bonariensis* Marelli, 1930; *brasiliensis* Liais, 1872 (part); *dennleri* Marelli, 1830; *imperfecta* Mondolfi and Pérez-Hernández, 1984; *lechei* Ihering, 1892; *leucotis* Wagner, 1847; *meridensis* J. A. Allen, 1902; *paraguayensis* J. A. Allen, 1902 (not available from Oken, 1816); *pernigra* J. A. Allen, 1900; *poecilotis* Wagner, 1842.
COMMENTS: Formerly known as *D. azarae*; see Hershkovitz (1969).

Didelphis aurita Wied-Neuwied, 1826. Beitr. Naturgesch. Brasil., 2:395.
TYPE LOCALITY: Brazil, Bahia, "Villa Viçosa am Flusse Paruhype."
DISTRIBUTION: E Brazil, SE Paraguay, and NE Argentina.
SYNONYMS: *azarae* Temminck, 1824; *brasiliensis* Liais, 1872 (part); *koseritzi* Ihering, 1892; *longipilis* Miranda-Ribeiro, 1935; *melanoidis* Miranda-Ribeiro, 1935.
COMMENTS: Previously considered a disjunct population of *D. marsupialis* (see Cerqueira, 1985). The senior synonym is *D. azarae* Temminck, 1824 (see Hershkovitz, 1969); however, the name had been misapplied to *D. albiventris* for over 160 years.

Didelphis marsupialis Linnaeus, 1758. Syst. Nat., 10th ed., 1:54.
TYPE LOCALITY: "America;" restricted to Surinam (Thomas, 1911*a*).
DISTRIBUTION: Tamaulipas, México, south throughout Central and South America to Perú, Bolivia, and Brazil.

SYNONYMS: *austroamericana* J. A. Allen, 1902 (not available from Oken, 1816); *battyi* J. A. Allen, 1902; *cancrivora* Gmelin, 1788; *caucae* J. A. Allen, 1900; *colombica* J. A. Allen, 1900; *etensis* J. A. Allen, 1902; *insularis* J. A. Allen, 1902; *karkinophaga* Zimmermann, 1780; *mesamericana* J. A. Allen, 1902 (part; not available from Oken, 1816); *particeps* Goldman, 1917; *richmondi* J. A. Allen, 1901; *tabascensis* J. A. Allen, 1901; *typica* Thomas, 1888 (part).
COMMENTS: Revised by Gardner (1973).

Didelphis virginiana Kerr, 1792. *In* Linnaeus, Anim. Kingdom, 1:193.
TYPE LOCALITY: "Virginia, Louisiana, Mexico, Brasil, and Peru;" restricted to Virginia (J. A. Allen, 1901*c*:160).
DISTRIBUTION: S Canada; E and C United States (with introduced populations in Pacific states); México; and in Central America south into N Costa Rica.
SYNONYMS: *boreoamericana* J. A. Allen, 1902 (not available from Oken, 1816); *breviceps* Bennett, 1833; *californica* Bennett, 1833; *cozumelae* Merriam, 1901; *illinensium* Link, 1795; *mesamericana* J. A. Allen, 1902 (part; not available from Oken, 1816); *pigra* Bangs, 1898; *pilosissima* Link, 1795; *pruinosa* Wagner, 1843; *texensis* J. A. Allen, 1901; *typica* Thomas, 1888 (part); *woapink* Barton, 1806; *yucatanensis* J. A. Allen, 1901.
COMMENTS: Revised by Gardner (1973, 1982); reviewed by McManus (1974, Mammalian Species, 40).

Gracilinanus Gardner and Creighton, 1989. Proc. Biol. Soc. Wash., 102:4.
TYPE SPECIES: *Didelphys microtarsus* Wagner, 1842, by original designation.
COMMENTS: Previously in *Marmosa* (*sensu lato*; see Gardner and Creighton (1989).

Gracilinanus aceramarcae (Tate, 1931). Am. Mus. Novit., 493:12.
TYPE LOCALITY: Bolivia, La Paz, "Rio Aceramarca, tributary of Rio Unduavi, Yungas."
DISTRIBUTION: Bolivia (type locality).

Gracilinanus agilis (Burmeister, 1854). Syst. Uebers. Thiere Bras., 1:139.
TYPE LOCALITY: Brazil, Minas Gerais, "Lagoa Santa."
DISTRIBUTION: Brazil, E Perú, E Bolivia, Paraguay, Uruguay, and adjacent Argentina.
SYNONYMS: *beatrix* Thomas, 1910; *blaseri* Miranda-Ribeiro, 1936; *buenavistae* Tate, 1931; *chacoensis* Tate, 1931; *muscula* Shamel, 1930 (preoccupied; replaced by *formosa* Shamel, 1930); *peruana* Tate, 1931; *rondoni* Miranda-Ribeiro, 1936; *unduaviensis* Tate, 1931.
COMMENTS: The forms *agilis* and *microtarsus* may prove to be conspecific.

Gracilinanus dryas (Thomas, 1898). Ann. Mag. Nat. Hist., ser. 7, 1:456.
TYPE LOCALITY: Venezuela, Mérida, "Culata."
DISTRIBUTION: Andes of W Venezuela.

Gracilinanus emiliae (Thomas, 1909). Ann. Mag. Nat. Hist., ser. 8, 3:379.
TYPE LOCALITY: Brazil, "Para."
DISTRIBUTION: NE Brazil.
SYNONYMS: *agricolai* Moojen, 1943.

Gracilinanus marica (Thomas, 1898) Ann. Mag. Nat. Hist., ser. 7, 1:455.
TYPE LOCALITY: Venezuela, Mérida, "R. Albarregas."
DISTRIBUTION: N Colombia and Venezuela.

Gracilinanus microtarsus (Wagner, 1842). Arch. Naturgesch., 8(1):359.
TYPE LOCALITY: Brazil, São Paulo, "Ypanema."
DISTRIBUTION: SE Brazil.
SYNONYMS: *guahybae* Tate, 1931; *herhardti* Miranda-Ribeiro, 1936.

Lestodelphys Tate, 1934. J. Mammal., 15:154.
TYPE SPECIES: *Notodelphys halli* Thomas, 1921, by original designation.
SYNONYMS: *Notodelphys* Thomas, 1921 (preoccupied).
COMMENTS: Monotypic.

Lestodelphys halli (Thomas, 1921). Ann. Mag. Nat. Hist., ser. 9, 8:137.
TYPE LOCALITY: Argentina, Santa Cruz, "Cabo Tres Puntas;" subsequently emended to "Estancia Madujada, not far from Puerto Deseado" (Thomas, 1929:45).

DISTRIBUTION: Provincia Mendoza south to Provincia de Santa Cruz, Argentina.
COMMENTS: Reviewed by L. G. Marshall (1977, Mammalian Species, 81).

Lutreolina Thomas, 1910. Ann. Mag. Nat. Hist., ser. 8, 5:247.
TYPE SPECIES: *Didelphis crassicaudata* Desmarest, 1804, by monotypy.
SYNONYMS: *Sarigua* Muirhead, 1819 (part).

Lutreolina crassicaudata (Desmarest, 1804). Tabl. Méth. Hist. Nat., *in* Nouv. Dict. Hist. Nat., 24:19.
TYPE LOCALITY: Paraguay, Asunción, by subsequent restriction (Cabrera, 1958:39).
DISTRIBUTION: South America in two populations: E Colombia, Venezuela, and W Guyana; E Bolivia, SE Brazil, Paraguay, Uruguay, and N Argentina.
SYNONYMS: *bonaria* Thomas, 1923; *crassicaudis* Olfers, 1818; *ferruginea* Larrañaga, 1923; *lutrilla* Thomas, 1923; *paranalis* Thomas, 1923; *travassosi* Miranda-Ribeiro, 1936; *turneri* Günther, 1879.
COMMENTS: Reviewed by Marshall (1978*a*, Mammalian Species, 91).

Marmosa Gray, 1821. London Med. Repos., 15:308.
TYPE SPECIES: *Didelphis marina* Gray, 1821, by monotypy (incorrect subsequent spelling of *Didelphis murina* Linnaeus, 1758).
SYNONYMS: *Asagis* Gloger, 1841; *Cuica* Liais, 1872; *Grymaeomys* Burmeister, 1854; *Notagogus* Gloger, 1841; *Opossum* Schmid, 1818 (part); *Sarigua* Muirhead, 1819 (part); *Stegomarmosa* Pine, 1972.

Marmosa andersoni Pine, 1972. J. Mammal., 53:279.
TYPE LOCALITY: Perú, Cuzco, "Hda. Villa Carmen, Cosnipata."
DISTRIBUTION: Known only from the holotype at the type locality.
COMMENTS: Type species of *Stegomarmosa*.

Marmosa canescens (J. A. Allen, 1893). Bull. Amer. Mus. Nat. Hist., 5:235.
TYPE LOCALITY: México, Oaxaca, "Santo Domingo de Guzman, Isthmus of Tehuantepec."
DISTRIBUTION: México from S Sonora to Oaxaca, Yucatán, and Tres Marías Isls.
SYNONYMS: *gaumeri* Osgood, 1913; *insularis* Merriam, 1908; *oaxacae* Merriam, 1897; *sinaloae* J. A. Allen, 1898.

Marmosa lepida (Thomas, 1888). Ann. Mag. Nat. Hist., ser. 6, 1:158.
TYPE LOCALITY: "Peruvian Amazons;" identified as Perú, Loreto, Santa Cruz, Huallaga River, by Thomas (1888*a*:348).
DISTRIBUTION: Surinam and E Colombia, Ecuador, Perú, Bolivia, and probably Brazil.
SYNONYMS: *grandis* Tate, 1931.

Marmosa mexicana Merriam, 1897. Proc. Biol. Soc. Wash., 11:44.
TYPE LOCALITY: México, Oaxaca, "Juquila."
DISTRIBUTION: Tamaulipas, México to W Panamá.
SYNONYMS: *mayensis* Osgood, 1913; *savannarum* Goldman, 1917; *zeledoni* Goldman, 1917.

Marmosa murina (Linnaeus, 1758). Syst. Nat., 10th ed., 1:55.
TYPE LOCALITY: "Asia, America;" restricted to Surinam by Thomas (1911*a*).
DISTRIBUTION: Colombia, Venezuela, Trinidad and Tobago, Guyana, Surinam, French Guiana, Brazil, E Ecuador, E Perú, and E Bolivia.
SYNONYMS: *bombascarae* Anthony, 1922; *chloe* Thomas, 1907; *dorsigera* Linnaeus, 1758; *duidae* Tate, 1931; *guianensis* Kerr, 1792; *klagesi* J. A. Allen, 1900; *macrotarsus* Wagner, 1842; *madeirensis* Cabrera, 1913; *maranii* Thomas, 1924; *meridionalis* Miranda-Ribeiro, 1936; *moreirae* Miranda-Ribeiro, 1936; *muscula* Cabanis, 1848; *musicola* Osgood, 1913; *parata* Thomas, 1911; *quichua* Thomas, 1899; *roraimae* Tate, 1931; *tobagi* Thomas, 1911; *waterhousei* Tomes, 1860.

Marmosa robinsoni Bangs, 1898. Proc. Biol. Soc. Wash., 12:95.
TYPE LOCALITY: Venezuela, Nueva Esparta, "Margarita Island."
DISTRIBUTION: Belize, Honduras (Isla Ruatán), Panamá, Colombia, W Ecuador, NW Perú, N Venezuela, Trinidad and Tobago, and Grenada (Lesser Antilles).
SYNONYMS: *casta* Thomas, 1911; *chapmani* J. A. Allen, 1900; *fulviventer* Bangs, 1901; *grenadae* Thomas, 1911; *isthmica* Goldman, 1912; *luridavolta* Goodwin, 1961; *mimetra* Thomas,

1921; *mitis* Bangs, 1898; *nesaea* Thomas, 1911; *pallidiventris* Osgood, 1912; *ruatanica* Goldman, 1911; *simonsi* Thomas, 1899.
COMMENTS: Reviewed by O'Connell (1983, Mammalian Species, 203).

Marmosa rubra Tate, 1931. Am. Mus. Novit., 493:6.
TYPE LOCALITY: Perú, Loreto, "mouth of Rio Curaray."
DISTRIBUTION: E Ecuador and Perú.

Marmosa tyleriana Tate, 1931. Am. Mus. Novit., 493:6.
TYPE LOCALITY: Venezuela, Amazonas, "Central Camp, Mt. Duida Plateau, Upper Rio Orinoco."
DISTRIBUTION: Guayanan Highland tepuis of Venezuela.
SYNONYMS: *phelpsi* Tate, 1939.

Marmosa xerophila Handley and Gordon, 1979. *In* J. F. Eisenberg (ed.), Vertebrate ecology in the northern Neotropics, Smithson. Inst. Press., p. 68.
TYPE LOCALITY: Colombia, Guajira, "La Isla, 15 m, near Cajoro, 37 km NNE Paraguaipoa."
DISTRIBUTION: NE Colombia and NW Venezuela.

Marmosops Matschie, 1916. Sitzb. Ges. Naturf. Fr., Berlin, 1916(1):267.
TYPE SPECIES: *Didelphis incana* Lund, 1840, by original designation.
COMMENTS: Previously in *Marmosa* (*sensu lato*; see Gardner and Creighton, 1989).

Marmosops cracens Handley and Gordon, 1979. *In* J. F. Eisenberg (ed.), Vertebrate ecology in the northern Neotropics, Smithson. Inst. Press., p. 66.
TYPE LOCALITY: Venezuela, Falcón, "near La Pastora (11°12'N, 68°37'W), 150 m, 14 km ENE Mirimire."
DISTRIBUTION: Known only from vicinity of the type locality.

Marmosops dorothea (Thomas, 1911). Ann. Mag. Nat. Hist., ser. 8, 7:516.
TYPE LOCALITY: Bolivia, La Paz, "Rio Solocame, 67°W., 16°S."
DISTRIBUTION: Yungas, Beni, and W Chaco regions of Bolivia.
SYNONYMS: *ocellata* Tate, 1931; *yungasensis* Tate, 1931.

Marmosops fuscatus (Thomas, 1896). Ann. Mag. Nat. Hist., ser. 6, 18:313.
TYPE LOCALITY: Venezuela, Mérida, "Rio Abbarregas [= Río Alvarregas]."
DISTRIBUTION: E Andes of Colombia, N Venezuela, and Trinidad and Tobago.
SYNONYMS: *carri* J. A. Allen and Chapman, 1897; *perfuscus* Thomas, 1924.

Marmosops handleyi (Pine, 1981). Mammalia, 45:67.
TYPE LOCALITY: Colombia, Antioquia, "9 km S Valdivia."
DISTRIBUTION: Known only from the vicinity of the type locality.
COMMENTS: Known from only two specimens.

Marmosops impavidus (Tschudi, 1845). Fauna Peruana, 1:149.
TYPE LOCALITY: "Der mittleren und tiefen Waldregion;" interpreted by Cabrera (1958:16) as Perú, Junín, "Montaña de Vitoc, cerca de Chanchamayo".
DISTRIBUTION: Mountains of W Panamá to Venezuela and Colombia, Ecuador, and Perú, with an isolated population in S Venezuela.
SYNONYMS: *albiventris* Tate, 1931; *caucae* Thomas, 1900; *celicae* Anthony, 1922; *madescens* Osgood, 1913; *neblina* Gardner, 1990; *oroensis* Anthony, 1922; *sobrina* Thomas, 1913; *ucayaliensis* Tate, 1931.

Marmosops incanus (Lund, 1840). K. Dansk. Vid. Selsk. Afhandl., p. 21 [preprint of Lund, 1841].
TYPE LOCALITY: Brazil, Minas Gerais, "Rio das Velhas," Lagoa Santa.
DISTRIBUTION: E Brazil from the states of Bahia south to Paraná.
SYNONYMS: *bahiensis* Tate, 1931; *paulensis* Tate, 1931; *scapulatus* Burmeister, 1856.

Marmosops invictus (Goldman, 1912). Smithson. Misc. Coll., 60(2):3.
TYPE LOCALITY: Panamá, Darien, "Cana."
DISTRIBUTION: Panamá.
COMMENTS: Reviewed by Pine (1981).

Marmosops noctivagus (Tschudi, 1845). Fauna Peruana, 1:148.
TYPE LOCALITY: "Der mittleren und tiefen Waldregion;" restricted by Tate (1933) to Perú, Junín, Montaña de Vitoc, near Chanchamayo, Río Perené drainage.
DISTRIBUTION: Amazonian Brazil, Ecuador, Perú, and Bolivia.
SYNONYMS: *collega* Thomas, 1920; *keaysi* J. A. Allen, 1900; *leucastra* Thomas, 1927; *lugenda* Thomas, 1927; *neglecta* Osgood, 1915; *polita* Cabrera, 1913; *purui* Miller, 1913; *stollei* Miranda-Ribeiro, 1936.

Marmosops parvidens Tate, 1931. Am. Mus. Novit., 493:13.
TYPE LOCALITY: Guyana, East Demerara-West Coast Berbice, "Hyde Park, 30 miles up the Demerara River."
DISTRIBUTION: Colombia, Perú, Venezuela, Guyana, Surinam, and scattered populations in Brazil.
SYNONYMS: *bishopi* Pine, 1981; *juninensis* Tate, 1931; *pinheiroi* Pine, 1981; *woodalli* Pine, 1981.
COMMENTS: Reviewed by Pine (1981) who recognized five subspecies.

Metachirus Burmeister, 1854. Syst. Uebers. Thiere Bras., 1:135.
TYPE SPECIES: *Didelphis myosuros* Temminck, 1824.
COMMENTS: Hall (1981) followed Pine (1973) in using the name *Philander*; see Hershkovitz (1976, 1981) who reaffirmed the use of *Philander* for the gray four-eyed opossums.

Metachirus nudicaudatus (Desmarest, 1817). Nouv. Dict. Hist. Nat., Nouv. ed., 9:424.
TYPE LOCALITY: French Guiana, "Cayenne."
DISTRIBUTION: Nicaragua to Paraguay and N Argentina.
SYNONYMS: *antioquiae* J. A. Allen, 1916; *bolivianus* J. A. Allen, 1901; *colombianus* J. A. Allen, 1900; *dentaneus* Goldman, 1912; *inbutus* Thomas, 1923; *infuscus* Thomas, 1923; *modestus* Thomas, 1923; *personatus* Miranda-Ribeiro, 1936; *phaeurus* Thomas, 1901; *tschudii* J. A. Allen, 1900.
COMMENTS: *M. nudicaudatus* is usually cited from É. Geoffroy St.-Hilaire's (1803) catalog of the mammals in the Paris Museum, but that work was not actually published; see Appendix I.

Micoureus Lesson, 1842. Nouv. Tabl. Regn. Anim. Mammifères, p. 186.
TYPE SPECIES: *Didelphis cinerea* Temminck, 1824, by subsequent designation (Thomas, 1888*a*).
COMMENTS: Previously in *Marmosa* (*sensu lato*; see Gardner and Creighton, 1989).

Micoureus alstoni (J. A. Allen, 1900). Bull. Am. Mus. Nat. Hist., 13:189.
TYPE LOCALITY: Costa Rica, Cartago, "Tres Rios."
DISTRIBUTION: E Central America from Belize to Panamá and adjacent Caribbean islands.

Micoureus constantiae (Thomas, 1904). Proc. Zool. Soc. Lond., 1903(2):243 [1904].
TYPE LOCALITY: Brazil, Mato Grosso, "Chapada."
DISTRIBUTION: E Bolivia and adjacent Brazil south into N Argentina.
SYNONYMS: *budini* Thomas, 1920.

Micoureus demerarae (Thomas, 1905). Ann. Mag. Nat. Hist., ser. 7, 16:313.
TYPE LOCALITY: Guyana, East Demerara-West Coast Berbice, "Comaccka, 80 miles up Demerara River."
DISTRIBUTION: Colombia, Venezuela, French Guiana, Guyana, Surinam, Brazil, and E Paraguay.
SYNONYMS: *areniticola* Tate, 1931; *cinerea* Temminck, 1824; *domina* Thomas, 1920; *esmeraldae* Tate, 1931; *limae* Thomas, 1920; *meridae* Tate, 1931; *paraguayana* Tate, 1931; *pfrimeri* Miranda-Ribeiro, 1936; *travassosi* Miranda-Ribeiro, 1936.
COMMENTS: Previously known as *Marmosa cinerea*; *cinerea* preoccupied by *D. cinerea* Goldfuss, 1812.

Micoureus regina (Thomas, 1898). Ann. Mag. Nat. Hist., ser. 7, 2:274.
TYPE LOCALITY: Colombia, Cundinamarca, "West Cundinamarca (Bogotá Region)."
DISTRIBUTION: Colombia, Ecuador, Perú, and Bolivia.
SYNONYMS: *germana* Thomas, 1904; *mapiriensis* Tate, 1931; *parda* Tate, 1931; *perplexa* Anthony, 1922; *phaea* Thomas, 1899; *rapposa* Thomas, 1899; *rutteri* Thomas, 1924.

Monodelphis Burnett, 1830. Quart. J. Sci. Lit. Art., 1829:351 [1830].
TYPE SPECIES: *Monodelphis brachyura* Burnett, 1830 (= *Didelphis brevicaudata* Erxleben, 1777) by subsequent designation (Matschie, 1916).
SYNONYMS: *Hemiurus* Gervais, 1855; *Microdelphys* Burmeister, 1856; *Minuania* Cabrera, 1919; *Monodelphiops* Matschie, 1916; *Peramys* Lesson, 1842.

Monodelphis adusta (Thomas, 1897). Ann. Mag. Nat. Hist., ser. 6, 20:219.
TYPE LOCALITY: Colombia, "W. Cundinamarca, in the low-lying hot regions."
DISTRIBUTION: E Panamá, Colombia, Ecuador, and Perú.
SYNONYMS: *melanops* Goldman, 1912; *peruvianus* Osgood, 1913.

Monodelphis americana (Müller, 1776). Linné's Vollständ. Natursyst., Suppl., p. 36.
TYPE LOCALITY: "Brasilien," restricted to Brazil, Pernambuco, Pernambuco, by Cabrera (1958:7).
DISTRIBUTION: E Brazil from the states of Pará south to Santa Catarina.
SYNONYMS: *brasiliensis* Erxleben, 1777; *brasiliensis* Daudin, *in* Lacépède, 1799; *trilineata* Lund, 1840; *tristriata* Illiger, 1815.

Monodelphis brevicaudata (Erxleben, 1777). Syst. Regni Anim., 1:80.
TYPE LOCALITY: "In Americae australis silvis;" restricted to Surinam by Matschie (1916).
DISTRIBUTION: Venezuela, Surinam, French Guiana, and Amazon Basin of Brazil and Bolivia.
SYNONYMS: *brachyuros* Schreber, 1777; *dorsalis* J. A. Allen, 1904; *glirina* Wagner, 1842; *hunteri* Waterhouse, 1841; *orinoci* Thomas, 1899; *palliolatus* Osgood, 1914; *sebae* Gray, 1827; *surinamensis* Zimmermann, 1780; *touan* Daudin, *in* Lacépède, 1799; *touan* Bechstein, 1800; *touan* Shaw, 1800; *tricolor* Desmarest, 1820.

Monodelphis dimidiata (Wagner, 1847). Abh. Akad. Wiss., München, 5(1):151, footnote.
TYPE LOCALITY: Uruguay, Maldonado, "Maldonado am la Plata."
DISTRIBUTION: Uruguay, SE Brazil, and NE Argentina.
SYNONYMS: *fosteri* Thomas, 1924.
COMMENTS: See Pine et al. (1985).

Monodelphis domestica (Wagner, 1842). Arch. Naturgesh., 8:359.
TYPE LOCALITY: Brazil, Mato Grosso, "Cuyaba."
DISTRIBUTION: Brazil, Bolivia, and Paraguay.
SYNONYMS: *concolor* Gervais, 1856.

Monodelphis emiliae (Thomas, 1912). Ann. Mag. Nat. Hist., ser. 8, 9:89.
TYPE LOCALITY: Brazil, Pará, "Boim, R. Tapajoz."
DISTRIBUTION: Amazon Basin of Perú and Brazil.
COMMENTS: Previously treated as a subspecies of *M. touan*. Reviewed by Pine and Handley (1984).

Monodelphis iheringi (Thomas, 1888). Ann. Mag. Nat. Hist., ser. 6, 1:159.
TYPE LOCALITY: "Rio Grande do Sul;" identified as Brazil, Rio Grande do Sul, Taquara, by Thomas (1888*a*).
DISTRIBUTION: SE Brazil (Espírito Santo, São Paulo, Santa Catarina, and Rio Grande do Sul).
COMMENTS: Previously considered a subspecies of *M. americana*. Reviewed by Pine (1977).

Monodelphis kunsi Pine, 1975. Mammalia, 39:321.
TYPE LOCALITY: Bolivia, Beni, "La Granja, W bank of Río Itonamas, 4 k N Magdalena."
DISTRIBUTION: Known only from four localities, two in Bolivia and two in Brazil.
COMMENTS: See Anderson (1982, Mammalian Species, 190).

Monodelphis maraxina Thomas, 1923. Ann. Mag. Nat. Hist., ser. 9, 12:157.
TYPE LOCALITY: Brazil, Pará, "Caldeirão."
DISTRIBUTION: Brazil, Pará, Marajó Isl.
COMMENTS: Eight specimens known. See Pine (1980*a*).

Monodelphis osgoodi Doutt, 1938. J. Mammal., 19:100.
TYPE LOCALITY: Bolivia, Cochabamba, "Incachaca."
DISTRIBUTION: SE Perú and C Bolivia.
COMMENTS: Previously included in *M. adusta*.

Monodelphis rubida (Thomas, 1899). Ann. Mag. Nat. Hist., ser. 7, 4:155.
TYPE LOCALITY: Brazil, "Bahia."
DISTRIBUTION: E Brazil from Goiás south to São Paulo.
SYNONYMS: *goyana* Miranda-Ribeiro, 1936; *umbristriata* Miranda-Ribeiro, 1936.
COMMENTS: Formerly under *M. umbristriata*.

Monodelphis scalops (Thomas, 1888). Ann. Mag. Nat. Hist., ser. 6, 1:158.
TYPE LOCALITY: "Brazil;" restricted to Rio de Janeiro, Therezôpolis, by Vieira (1949).
DISTRIBUTION: SE Brazil from Espírito Santo south to Santa Catarina.
COMMENTS: Reviewed by Pine and Abravaya (1978).

Monodelphis sorex (Hensel, 1872). Abh. König. Akad. Wiss. Berlin, 1872:122.
TYPE LOCALITY: "Provinz Rio Grande do Sul;" restricted to Brazil, Rio Grande do Sul, Taquara, by Cabrera (1958).
DISTRIBUTION: SE Brazil, S Paraguay, and NE Argentina.
SYNONYMS: *henseli* Thomas, 1888; *itatiayae* Miranda-Ribeiro, 1936; *lundi* Matschie, 1916; *paulensis* Vieira, 1950.
COMMENTS: See Pine et al. (1985).

Monodelphis theresa Thomas, 1921. Ann. Mag. Nat. Hist., ser. 9, 8:441.
TYPE LOCALITY: Brazil, Rio de Janeiro, "Theresopolis, Organ Mts."
DISTRIBUTION: E Brazil and Peruvian Andes.

Monodelphis unistriata (Wagner, 1842). Arch. Naturgesch., 8(1):360.
TYPE LOCALITY: Brazil, São Paulo, "Ytarare" (= Itararé).
DISTRIBUTION: State of São Paulo, Brazil.
COMMENTS: Apparently only known from the holotype (Pine, 1992, in litt.).

Philander Tiedemann, 1808. Zool., v. 1, Allgemeine Zool., Mensch Säugthiere, Landshut, p. 426.
TYPE SPECIES: *Philander virginianus* Tiedemann, 1808, by subsequent designation (Hershkovitz, 1949*a*).
SYNONYMS: *Hylothylax* Cabrera, 1919; *Metachirops* Matschie, 1916; *Philander* Gray, 1843 (based on *Philander* Brisson, 1762); *Sarigua* Muirhead, 1919 (part).
COMMENTS: Pine (1973) used *Metachirops* for this genus, as did Hall (1981), Husson (1978), and Corbet and Hill (1980; but not 1991 when they used *Philander*). *Philander* Tiedemann is used here as the correct generic name because, as Hershkovitz (1949*a*; 1981) pointed out, *Philander virginianus* Tiedemann was a replacement name for *Didelphis opossum* Linnaeus. Hershkovitz (1976) designated the female illustrated by Seba (1734) as lectotype of *D. opossum* and, thereby, also making Seba's specimen the lectotype of *Philander virginianus*. Therefore, Husson's (1978) attempt to synonymize *Philander* Tiedemann under *Didelphis* Linnaeus by synonymizing *P. virginianus* Tiedemann with *Didelphis virginiana* Kerr was too late and is invalid. If Husson's (1978) action had preceded Hershkovitz's (1976), *Metachirops* would be the correct name for this genus.

Philander andersoni (Osgood, 1913). Field Mus. Nat. Hist. Publ., Zool Ser., 10:95.
TYPE LOCALITY: Perú, Loreto, "Yurimaguas."
DISTRIBUTION: S Venezuela, W Brazil, E Colombia, Ecuador, and Perú.
SYNONYMS: *mcilhennyi* Gardner and Patton, 1972; *nigratus* Thomas, 1923.
COMMENTS: See account by Emmons and Feer (1990).

Philander opossum (Linnaeus, 1758). Syst. Nat., 10th ed., 1:55.
TYPE LOCALITY: "America;" restricted to Surinam by J. A. Allen (1900) and further restricted to Paramaribo, Surinam, by Matschie (1916).
DISTRIBUTION: Tamaulipas, México, through Central and South America to Paraguay and NE Argentina.
SYNONYMS: *azaricus* Thomas, 1923; *canus* Osgood, 1913; *crucialis* Thomas, 1923; *frenata* Olfers, 1818; *fuscogriseus* J. A. Allen, 1900; *grisescens* J. A. Allen, 1901; *melantho* Thomas, 1923; *melanurus* Thomas, 1899; *pallidus* J. A. Allen, 1901; *quica* Temminck, 1824; *superciliaris* Olfers, 1818; *virginianus* Tiedemann, 1808.
COMMENTS: Corbet and Hill (1980), Hall (1981), Husson (1978), and Pine (1973) used *Metachirops opossum* for this species.

Thylamys Gray, 1843. List Specimens Mamm. Coll. Brit. Mus., p. 101.
TYPE SPECIES: *Didelphis elegans* Waterhouse, 1939, by monotypy.
SYNONYMS: *Sarigua* Muirhead, 1819 (part).
COMMENTS: Previously a subgenus under *Marmosa*.

Thylamys elegans (Waterhouse, 1839). Zool. H.M.S. "Beagle," Mammalia, p. 95.
TYPE LOCALITY: Chile, Coquimbo, "Valparaiso."
DISTRIBUTION: C Perú south to Región Valdivia, Chile, on Pacific side of Andes and from Cochabamba, Bolivia, to Neuquén, Argentina, on the east.
SYNONYMS: *cinderella* Thomas, 1902; *coquimbensis* Tate, 1931; *janetta* Thomas, 1926; *soricina* Philippi, 1894; *sponsoria* Thomas, 1921; *tatei* Handley, 1957; *venusta* Thomas, 1902.

Thylamys macrura (Olfers, 1818). *In* W. L. Eschwege, J. Brasilien, Neue Bibliothek. Reisen. 15:205.
TYPE LOCALITY: "Sudamerica;" restricted to Paraguay, Presidente Hayes, "Tapoua" (= Tapua); see comments.
DISTRIBUTION: Paraguay and S Brazil.
SYNONYMS: *grisea* Desmarest, 1827.
COMMENTS: Based on Azara's (1801:290) "Micouré à queue longue;" therefore, the type locality is Tapua. Previously called *T. griseus*.

Thylamys pallidior (Thomas, 1902). Ann. Mag. Nat. Hist., ser. 7, 10:161.
TYPE LOCALITY: Bolivia, Oruro, "Challapata."
DISTRIBUTION: Argentina and E and S Bolivia.
SYNONYMS: *bruchi* Thomas, 1921; *fenestrae* Marelli, 1932; *pulchella* Cabrera, 1934.

Thylamys pusilla (Desmarest, 1804). Tabl. Méth. Hist. Nat., *in* Nouv. Dict. Hist. Nat., 24:19.
TYPE LOCALITY: Not stated; restricted to Paraguay, Misiones, "San Ignacio," by Tate (1933).
DISTRIBUTION: C and S Brazil, Paraguay, SE Bolivia, and N Argentina.
SYNONYMS: *citella* Thomas, 1912; *karimii* Petter, 1968; *marmota* Thomas, 1902; *nana* Olfers, 1818; *verax* Thomas, 1921.

Thylamys velutinus (Wagner, 1842). Archiv Naturgesch., 8(1):360.
TYPE LOCALITY: Brazil, São Paulo, "Ypanema."
DISTRIBUTION: SE Brazil.
SYNONYMS: *pimelura* Reinhardt, 1851.

ORDER PAUCITUBERCULATA

by Alfred L. Gardner

ORDER PAUCITUBERCULATA

SYNONYMS: Marsupialia.

Family Caenolestidae Trouessart, 1898. Cat. Mamm. Viv. Foss., 2(5):1205.
COMMENTS: Reviewed by Marshall (1980) and Bublitz (1987).

Caenolestes Thomas, 1895. Ann. Mag. Nat. Hist., ser. 6, 16:367.
TYPE SPECIES: *Hyracodon fuliginosus* Tomes, 1863, by monotypy.
SYNONYMS: *Hyracodon* Tomes, 1863 (preoccupied).

Caenolestes caniventer Anthony, 1921. Am. Mus. Novit., 20:6.
TYPE LOCALITY: Ecuador, El Oro, "El Chiral."
DISTRIBUTION: SW Ecuador and NW Perú.

Caenolestes convelatus Anthony, 1924. Am. Mus. Novit., 120:1.
TYPE LOCALITY: Ecuador, Pichincha, "Las Maquinas, Western Andes 7000 feet altitude, on trail from Aloag to Santo Domingo de los Colorados."
DISTRIBUTION: W Colombia and NW Ecuador.
SYNONYMS: *barbarensis* Bublitz, 1987.

Caenolestes fuliginosus (Tomes, 1863). Proc. Zool. Soc. Lond., 1863:51.
TYPE LOCALITY: "Ecuador."
DISTRIBUTION: Colombia, Ecuador, and NW Venezuela.
SYNONYMS: *centralis* Bublitz, 1987; *obscurus* Thomas, 1895; *tatei* Anthony, 1923.

Lestoros Oehser, 1934. J. Mammal., 15:240.
TYPE SPECIES: *Orolestes inca* Thomas, 1917, by original designation.
SYNONYMS: *Cryptolestes* Tate, 1934 (preoccupied); *Orolestes* Thomas, 1917 (preoccupied).

Lestoros inca (Thomas, 1917). Smithson. Misc. Coll., 68(4):3.
TYPE LOCALITY: Perú, Cuzco, "Torontoy."
DISTRIBUTION: S Andean Perú.
SYNONYMS: *gracilis* Bublitz, 1987.

Rhyncholestes Osgood, 1924. Field Mus. Nat. Hist. Publ., Zool. Ser., 14:170.
TYPE SPECIES: *Rhyncholestes raphanurus* Osgood, 1924, by original designation.

Rhyncholestes raphanurus Osgood, 1924. Field Mus. Nat. Hist. Publ., Zool. Ser., 14:170.
TYPE LOCALITY: Chile, Biobio, "mouth of Rio Inio, south end of Chiloé Island."
DISTRIBUTION: SC Chile including Chiloé Isl.
SYNONYMS: *continentalis* Bublitz, 1987.
COMMENTS: See Patterson and Gallardo (1987, Mammalian Species, 286).

ORDER MICROBIOTHERIA
by Alfred L. Gardner

ORDER MICROBIOTHERIA

SYNONYMS: Marsupialia.
COMMENTS: Included in order Polyprotodonta by Reig et al. (1987); considered a separate order by Aplin and Archer (1987) and Marshall et al. (1990).

Family Microbiotheriidae Ameghino, 1887. Bol. Mus. la Plata, 1:6.
COMMENTS: Usually considered a subfamily of the Didelphidae.

Dromiciops Thomas, 1894. Ann. Mag. Nat. Hist., ser. 6, 14:186.
TYPE SPECIES: *Dromiciops gliroides* Thomas, 1894, by monotypy.

Dromiciops gliroides Thomas, 1894. Ann. Mag. Nat. Hist., ser. 6, 14:187.
TYPE LOCALITY: Chile, Biobio, "Huite, N.E. Chiloe Island."
DISTRIBUTION: Chile and adjacent Argentina from about 36°S to near 43°S.
SYNONYMS: *australis* F. Philippi, 1893 (preoccupied by *Didelphys australis* Goldfuss, 1812).
COMMENTS: Reviewed by Marshall (1978*b*, Mammalian Species, 99, as *D. australis*).

ORDER DASYUROMORPHIA

by Colin P. Groves

ORDER DASYUROMORPHIA

SYNONYMS: Dasyuroidea, Dasyuriformes.

COMMENTS: Recognized as an order by Aplin and Archer (1987) who proposed a new syncretic classification of the marsupials. Includes the Australian component of Marsupicarnivora (see Ride, 1964*b*).

Family Thylacinidae Bonaparte, 1838. Nuovi Ann. Sci. Nat., 2(1):112.

COMMENTS: Some authors include this family in the Dasyuridae, see Vaughan (1978:39), but also see Ride (1964*b*), Archer and Kirsch (1977), and Kirsch and Calaby (1977:15) who retained this family.

Thylacinus Temminck, 1824. Monogr. Mamm., 1:23.

TYPE SPECIES: *Didelphis cynocephala* Harris, 1808.

SYNONYMS: *Lycaon* (preoccupied by *Lycaon* Brooks, 1827, a canid), *Paracyon, Peralopex*.

Thylacinus cynocephalus (Harris, 1808). Trans. Linn. Soc. Lond., 9:174.

TYPE LOCALITY: Australia, Tasmania.

DISTRIBUTION: Tasmania.

STATUS: CITES - Appendix I pe [possibly extinct]; U.S. ESA - Endangered; IUCN - Extinct.

SYNONYMS: *breviceps, communis, harrisii, lucocephalus, striatus*.

COMMENTS: Probably extinct; but tracks and sightings continue to be reported; see Ride (1970:201).

Family Myrmecobiidae Waterhouse, 1841. Nat. Hist. Marsup. or Pouched Animals (Naturalist's Libr., 10):60.

COMMENTS: This citation is usually listed as "Cat. Mamm. Mus. Zool Soc., 1838" but Myrmecobiidae is not used in this catalogue; see Palmer (1904). Some authors include this family in the Dasyuridae; see Vaughan (1978:39); but also see Ride (1964*b*), Archer and Kirsch (1977), and Kirsch and Calaby (1977:15), who retained this family.

Myrmecobius Waterhouse, 1836. Proc. Zool. Soc. Lond., 1836:69.

TYPE SPECIES: *Myrmecobius fasciatus* Waterhouse, 1836.

Myrmecobius fasciatus Waterhouse, 1836. Proc. Zool. Soc. Lond., 1836:69.

TYPE LOCALITY: Australia, Western Australia, Mt. Kokeby, S of Beverley.

DISTRIBUTION: SW Western Australia; formerly in NW South Australia and SW New South Wales.

STATUS: U.S. ESA and IUCN - Endangered.

SYNONYMS: *rufus*.

Family Dasyuridae Goldfuss, 1820. Handb. Zool., II:447.

COMMENTS: Some authors include Thylacinidae and Myrmecobiidae in this family; see Vaughan (1978:39); but also see Ride (1964*b*), Archer and Kirsch (1977), Kirsch and Calaby (1977:15), and Archer (1982) who retained these families. Revised by Tate (1947).

Antechinus Macleay, 1841. Ann. Mag. Nat. Hist., [ser. 1], 8:242.

TYPE SPECIES: *Antechinus stuartii* Macleay, 1841.

COMMENTS: For the exclusion of *Parantechinus* and *Pseudantechinus* see Haltenorth (1958:18) and Ride (1964*a*); and of *Dasykaluta* see Archer (1982:434).

Antechinus bellus (Thomas, 1904). Nov. Zool., 11:229.

TYPE LOCALITY: Australia, Northern Territory, South Alligator River.

DISTRIBUTION: N Northern Territory (Australia).

STATUS: Unknown.

Antechinus flavipes (Waterhouse, 1838). Proc. Zool. Soc. Lond., 1837:75 [1838].
TYPE LOCALITY: Australia, New South Wales, north of Hunter river.
DISTRIBUTION: Cape York Peninsula (Queensland) to Victoria and SE South Australia, SW Western Australia.
STATUS: Common.
SYNONYMS: *leucogaster, rubeculus, rufogaster.*

Antechinus godmani (Thomas, 1923). Ann. Mag. Nat. Hist., ser. 9, 11:174.
TYPE LOCALITY: Australia, Queensland, Ravenshoe, Dinner Creek, 2900 ft. (884 m.), 17°40'S, 145°30'E.
DISTRIBUTION: NE Queensland.
STATUS: Common.
COMMENTS: Included in *flavipes* by Haltenorth (1958:18); but see Kirsch and Calaby (1977:15).

Antechinus leo Van Dyck, 1980. Aust. Mamm., 3:1.
TYPE LOCALITY: Australia, Queensland, Cape York Penninsula, Nesbit River, Buthen Buthen (13°21'S, 143°28'E).
DISTRIBUTION: Cape York Peninsula from the Iron Range to the southern limit of the McIlwraith Range.
STATUS: Unknown.

Antechinus melanurus (Thomas, 1899). Ann. Mus. Civ. Stor. Nat. Genova, 20:191.
TYPE LOCALITY: Papua New Guinea, Central Prov., Astrolabe Range, Moroka, 1300 m.
DISTRIBUTION: New Guinea.
STATUS: Unknown.
SYNONYMS: *modesta.*
COMMENTS: Generic allocation uncertain.

Antechinus minimus (É. Geoffroy, 1803). Bull. Sci. Soc. Philom. Paris, 81:159.
TYPE LOCALITY: Australia, Tasmania; probably Waterhouse Isl, Bass Strait (see Wakefield and Warneke, 1963:209-210).
DISTRIBUTION: Coastal SE South Australia to Tasmania.
STATUS: Probably common.
SYNONYMS: *affinis, concinnus, maritima, rolandensis.*

Antechinus naso (Jentink, 1911). Notes Leyden Mus., 33:236.
TYPE LOCALITY: Indonesia, Irian Jaya, Djajawidjaja Div., Helwig Mtns, south of Mt. Wilhelmina, about 2000 m.
DISTRIBUTION: Interior New Guinea, 5,000-9,000 ft.
STATUS: Common.
SYNONYMS: *centralis, habbema, mayeri, misim, tafa.*
COMMENTS: Generic allocation uncertain.

Antechinus stuartii Macleay, 1841. Ann. Mag. Nat. Hist., [ser. 1], 8:242.
TYPE LOCALITY: Australia, New South Wales, Manly (Spring Cove, Sydney Harbour); neotype from Waterfall, Royal National Park.
DISTRIBUTION: E Queensland, E New South Wales, Victoria (Australia).
STATUS: Common.
SYNONYMS: *adusta, burrelli, unicolor.*
COMMENTS: Van Dyck (1982) stated that *Antechinus stuartii adustus*, from the vine forest region between Paluma (19°00'S, 146°12'E) and Mt. Spurgeon (16°25' S, 145°12'E), is probably a valid species. Dickman et al. (1988) showed that *A. stuartii* in E New South Wales is actually divided into two quite distinct species: *A. stuartii* north of about 35°S, and an undescribed species mainly south of this latitude (but recurring farther north and inland at Mt. Canobolas, 33°10'S, 149°00'E; and the two have been found together at Kioloa, 35°32'S, 150°23'E).

Antechinus swainsonii (Waterhouse, 1840). Mag. Nat. Hist. [Charlesworth's], 4:299.
TYPE LOCALITY: Australia, Tasmania.
DISTRIBUTION: SE Queensland, E New South Wales, E and SE Victoria, coastal SE Australia, and Tasmania.
STATUS: Common.

SYNONYMS: *assimilis, mimetes, moorei, niger.*

Antechinus wilhelmina Tate, 1947. Bull. Am. Mus. Nat. Hist., 88:130.
TYPE LOCALITY: Indonesia, Irian Jaya, Djajawidjaja Div., 9 km. N. Lake Habbema, north of Mt. Wilhelmina, 2800 m.
DISTRIBUTION: C New Guinea, 7000-9000 ft.
STATUS: Unknown.
SYNONYMS: *hageni.*
COMMENTS: Generic allocation uncertain.

Dasycercus Peters, 1875. Sitzb. Ges. Naturf. Fr. Berlin, 1875:73.
TYPE SPECIES: *Chaetocercus cristicauda* Krefft, 1867.
SYNONYMS: *Amperta, Chaetocercus, Dasyuroides*
COMMENTS: Original name *Chaetocercus* Kreft, 1867 is preoccupied. Includes *Dasyuroides;* see Mack (1961) and Mahoney and Ride (*in* Walton, 1988:18).

Dasycercus byrnei (Spencer, 1896). Rept. Horn Sci. Exped. Cent. Aust., Zool., 2:36.
TYPE LOCALITY: Australia, Northern Territory, Charlotte Waters.
DISTRIBUTION: Junction of Northern Territory, South Australia, and Queensland (C Australia).
STATUS: Rare.
SYNONYMS: *pallidius.*
COMMENTS: Included in *Dasycercus* by Mack (1961); accepted as a member of *Dasycercus* by Mahoney and Ride (*in* Walton, 1988).

Dasycercus cristicauda (Krefft, 1867). Proc. Zool. Soc. Lond., 1866:435 [1867].
TYPE LOCALITY: Australia, South Australia, probably Lake Alexandrina.
DISTRIBUTION: Arid Australia from NW Western Australia to SW Queensland, N South Australia.
STATUS: Rare or indeterminate.
SYNONYMS: *blighi, blythi, hillieri.*

Dasykaluta Archer, 1982. *In* M. Archer, Carnivorous Marsupials, 2:434.
TYPE SPECIES: *Antechinus rosamondae* Ride, 1964.

Dasykaluta rosamondae (Ride, 1964). W. Aust. Nat., 9:58.
TYPE LOCALITY: Australia, Western Australia, Woodstock Station (via Marble Bar), 21°35'S, 119°E.
DISTRIBUTION: NW Western Australia.
STATUS: Rare.

Dasyurus É. Geoffroy, 1796. Mag. Encyclop., ser. 2, 3:469.
TYPE SPECIES: *Didelphis maculata* Anon., 1791.
SYNONYMS: *Dasyurinus, Dasyurops, Nasira, Notoctonus, Satanellus, Stictophonus.*
COMMENTS: See Haltenorth (1958:20).

Dasyurus albopunctatus Schlegel, 1880. Notes Leyden Mus., 2:51.
TYPE LOCALITY: Indonesia, Irian Jaya, Vogelkop, Manokwari Div., Arfak Mtns, Sapoea.
DISTRIBUTION: New Guinea.
STATUS: Uncommon.
SYNONYMS: *daemonellus, fuscus.*

Dasyurus geoffroii Gould, 1841. Proc. Zool. Soc. Lond., 1840:151 [1841].
TYPE LOCALITY: Australia, New South Wales, Liverpool Plains.
DISTRIBUTION: Western Australia; formerly in South Australia, Northern Territory, S Queensland, W New South Wales, and NW Victoria (Archer, *in* Tyler, 1979; Waithman, 1979).
STATUS: Rare.
SYNONYMS: *fortis.*
COMMENTS: Formerly included in *Dasyurinus.* The New Guinea records of this species actually refer to *D. spartacus.*

Dasyurus hallucatus Gould, 1842. Proc. Zool. Soc. Lond., 1842:41.
TYPE LOCALITY: Australia, Northern Territory, Port Essington.
DISTRIBUTION: Australia: N Northern Territory, N and NE Queensland, and N Western Australia.
STATUS: Common.
SYNONYMS: *exilis, nesaeus, predator.*
COMMENTS: Sometimes assigned to *Satanellus*. The original name *Mustela quoll* Zimmermann, 1783, was suppressed under Article 80 of the International Code of Zoological Nomenclature (International Commission on Zoological Nomenclature, 1985), now correctly *Dasyurus hallucatus.*

Dasyurus maculatus (Kerr, 1792). *In* Linnaeus, Anim. Kingdom, 1:170.
TYPE LOCALITY: Australia, New South Wales, Port Jackson.
DISTRIBUTION: Australia: E Queensland, E New South Wales, E and S Victoria, SE South Australia, Tasmania. Formerly occurred in South Australia.
STATUS: Widespread but rare; locally common in Tasmania.
SYNONYMS: *gracilis, macrourus, novaehollandiae, ursinus.*
COMMENTS: Formerly included in *Dasyurops;* see Haltenorth (1958).

Dasyurus spartacus Van Dyck, 1987. Aust. Mamm., 11:145.
TYPE LOCALITY: Papua New Guinea, Trans-fly Plains, Marehead, 8°41′S, 141°39′E.
DISTRIBUTION: Fly Plains, Papua New Guinea.
STATUS: Unknown.
COMMENTS: Formerly included in *D. geoffroii.*

Dasyurus viverrinus (Shaw, 1800). Gen. Zool. Syst. Nat. Hist., 1(2), Mammalia, p. 491.
TYPE LOCALITY: Australia, New South Wales, Sydney.
DISTRIBUTION: Probably survives only in Tasmania; formerly South Australia, New South Wales, and Victoria (Archer, *in* Tyler, 1979).
STATUS: U.S. ESA - Endangered.
SYNONYMS: *alboguttata, guttatus, maugei.*
COMMENTS: *Dasyurus quoll* Zimmermann, 1777 (not *Mustela quoll* Zimmermann, 1783), is invalid: this work was rejected by Opinion 257 of the International Commission on Zoological Nomenclature (1954*a*). Also the name *Dasyurus maculata* Anon., 1791, was suppressed under Article 80 of the International Code of Zoological Nomenclature (International Commission on Zoological Nomenclature, 1985). "Original" name is now *Didelphis viverrina* Shaw, 1800.

Murexia Tate and Archbold, 1937. Bull. Am. Mus. Nat. Hist., 73:335 (footnote), 339.
TYPE SPECIES: *Phascogale murex* Thomas, 1913 (= *Phascogale longicaudata* Schlegel, 1866).

Murexia longicaudata (Schlegel, 1866). Ned. Tijdschr. Dierk., 3:356.
TYPE LOCALITY: Indonesia, Aru Islands, Wonoumbai.
DISTRIBUTION: New Guinea, Aru Islands.
STATUS: Uncommon.
SYNONYMS: *aspera, maxima, murex, parva.*

Murexia rothschildi (Tate, 1938). Nov. Zool., 41:58.
TYPE LOCALITY: Papua New Guinea, Central Prov., head of Aroa River, about 1220 m.
DISTRIBUTION: SE New Guinea.
STATUS: Uncommon.

Myoictis Gray, 1858. Proc. Zool. Soc. Lond., 1858:112.
TYPE SPECIES: *Myoictis wallacei* Gray, 1858 (= *Phascogale melas* Müller, 1840).

Myoictis melas (Müller, 1840). *In* Temminck, Verh. Nat. Ges. Ned. Overz. Bezitt., Land-en Volkenkunde, p. 20[1840], see comments.
TYPE LOCALITY: "Nieuw-Guinea, in de triton's baai (op 3°39′Z. breedte)" = Indonesia, Irian Jaya, Fakfak Div., Lobo Dist., near Triton Bay, Mt. Lamantsjieri.
DISTRIBUTION: New Guinea; Salawati Isl and Aru Isls (Indonesia).
STATUS: Rare.
SYNONYMS: *bruijni, pilicauda, senex, thorbeckiana, wallacei, wavicus.*

COMMENTS: This species was further described by Müller and Schlegel, *in* Temminck, Verh. Nat. Gesch. Nederland. Overz. Bezitt., Zool., Mammalia, p. 149[1845], pl. 25[1843].

Neophascogale Stein, 1933. Z. Säugetierk., 8:87.
TYPE SPECIES: *Phascogale venusta* Thomas, 1921.

Neophascogale lorentzi (Jentink, 1911). Notes Leyden Mus., 33:234.
TYPE LOCALITY: Indonesia, Irian Jaya, Djajawidjaja (= Jayawijaya) Div., Helwig Mtns, south of Mt. Wilhelmina, 2600 m.
DISTRIBUTION: C New Guinea (highlands).
STATUS: Uncommon.
SYNONYMS: *nouhuysii, rubrata, venusta.*

Ningaui Archer, 1975. Mem. Queensl. Mus., 17(2):239.
TYPE SPECIES: *Ningaui timealeyi* Archer, 1975.
COMMENTS: An undescribed species of *Ningaui* occurs in Northern Territory (Australia); see Johnson and Roff (1980).

Ningaui ridei Archer, 1975. Mem. Queensl. Mus., 17(2):246.
TYPE LOCALITY: Australia, Western Australia, 38.6 km ENE Laverton (28°30'S, 122°47'E).
DISTRIBUTION: Northern Territory, South Australia, and Western Australia (deserts).
STATUS: Common.

Ningaui timealeyi Archer, 1975. Mem. Queensl. Mus., 17(2):244.
TYPE LOCALITY: Australia, Western Australia, 32.2 km SE Mt. Robinson.
DISTRIBUTION: NW Western Australia.
STATUS: Common.

Ningaui yvonnae Kitchener, Stoddart, and Henry, 1983. Aus. J. Zool., 31:366.
TYPE LOCALITY: "Mt. Manning Area, Western Australia Goldfields, 29°58'S.,119°32'E".
DISTRIBUTION: Australia: Western Australia to New South Wales, Victoria.
STATUS: Common.

Parantechinus Tate, 1947. Bull. Am. Mus. Nat. Hist., 88:137.
TYPE SPECIES: *Phascogale apicalis* Gray, 1842.
COMMENTS: Kitchener and Caputi (1988) restricted this genus to *P. apicalis.*

Parantechinus apicalis (Gray, 1842). Ann. Mag. Nat. Hist., [ser. 1], 9:518.
TYPE LOCALITY: Australia, SW Western Australia.
DISTRIBUTION: Inland periphery of SW Western Australia.
STATUS: U.S. ESA - Endangered; IUCN - Indeterminate.

Parantechinus bilarni (Johnson, 1954). Proc. Biol. Soc. Wash., 67:77.
TYPE LOCALITY: Australia, Northern Territory, Oenpelli (12°20'S, 133°3'E).
DISTRIBUTION: Northern Territory (Australia).
STATUS: Rare.
COMMENTS: Included in *Antechinus* (= *Pseudantechinus*) *macdonnellensis* by Ride (1970:116), but see Kirsch and Calaby (1977:15). According to Kitchener and Caputi (1988), this species should be transferred to *Pseudantechinus.*

Phascogale Temminck, 1824. Monogr. Mamm., 1:23, 56.
TYPE SPECIES: *Didelphis penicillata* Shaw, 1800 (= *Vivera tapoatafa* Meyer, 1793).
SYNONYMS: *Ascogale, Phascologale, Phascoloictis, Tapoa.*

Phascogale calura Gould, 1844. Proc. Zool. Soc. Lond., 1844:104.
TYPE LOCALITY: Australia, Western Australia, Williams River, Military Station.
DISTRIBUTION: Inland SW Western Australia, formerly in Northern Territory, South Australia, NW Victoria, SW New South Wales, but probably extinct in all places except the Western Australia wheat belt.
STATUS: IUCN - Indeterminate; limited distribution, vulnerable.

Phascogale tapoatafa (F. Meyer, 1793). Zool. Entdeck., p. 28.
TYPE LOCALITY: Australia, New South Wales, Sydney.

DISTRIBUTION: SW Western Australia, SE South Australia, S Victoria, E New South Wales, SE and N Queensland, Northern Territory.
SYNONYMS: *penicillata, pirata, tafa.*

Phascolosorex Matschie, 1916. Mitt. Zool. Mus. Berlin, 8:263.
TYPE SPECIES: *Phascogale dorsalis* Peters and Doria, 1876.

Phascolosorex doriae (Thomas, 1886). Ann. Mus. Civ. Stor. Nat. Genova, 4:208.
TYPE LOCALITY: Indonesia, Irian Jaya, Vogelkop, Manokwari Div., Arfak Mtns, Mori.
DISTRIBUTION: W interior New Guinea.
STATUS: Uncommon.
SYNONYMS: *pan, umbrosa.*

Phascolosorex dorsalis (Peters and Doria, 1876). Ann. Mus. Civ. Stor. Nat. Genova, 8:335.
TYPE LOCALITY: Indonesia, Irian Jaya, Vogelkop, Manokwari Div., Arfak Mtns, Hatam.
DISTRIBUTION: W and E interior New Guinea (not known from central region).
STATUS: Common.
SYNONYMS: *brevicaudata, whartoni.*

Planigale Troughton, 1928. Rec. Aust. Mus., 16:282.
TYPE SPECIES: *Planigale brunneus* Troughton, 1928 (= *Phascogale ingrami* Thomas, 1906).
COMMENTS: Revised by Archer (1976).

Planigale gilesi Aitken, 1972. Rec. S. Aust. Mus., 16(10):1.
TYPE LOCALITY: Australia, South Australia, Ann Creek Station (No. 3 bore) (28°18'S, 136°29'40"E).
DISTRIBUTION: NE South Australia, NW New South Wales, and SW Queensland (Australia).
STATUS: Unknown.

Planigale ingrami (Thomas, 1906). Abstr. Proc. Zool. Soc. Lond., 1906(32):6.
TYPE LOCALITY: Australia, Northern Territory, Alexandria.
DISTRIBUTION: Australia: N and E Queensland, NE Northern Territory, NE Western Australia.
STATUS: U.S. ESA - Endangered as *P. i. subtilissima.*
SYNONYMS: *brunnea, subtilissima.*
COMMENTS: See Archer (1976:351).

Planigale maculata (Gould, 1851). Mamm. Aust., 1, pl. 44.
TYPE LOCALITY: Australia, New South Wales, Clarence River.
DISTRIBUTION: E Queensland, NE New South Wales, and N Northern Territory (Australia).
STATUS: Unknown.
SYNONYMS: *minutissimus, sinualis.*
COMMENTS: Transferred to *Planigale* from *Antechinus* by Archer (1976:346).

Planigale novaeguineae Tate and Archbold, 1941. Am. Mus. Novit., 1101:7.
TYPE LOCALITY: Papua New Guinea, Central Prov., Rona Falls, Laloki River (vicinity of Port Moresby), 250 m.
DISTRIBUTION: S New Guinea.
STATUS: Uncommon.

Planigale tenuirostris Troughton, 1928. Rec. Aust. Mus., 16:285.
TYPE LOCALITY: Australia, "Collected at Bourke or Wilcannia, New South Wales".
DISTRIBUTION: NW New South Wales, and SC Queensland (Australia).
STATUS: U.S. ESA - Endangered.

Pseudantechinus Tate, 1947. Bull. Am. Mus. Nat. Hist., 88:139.
TYPE SPECIES: *Phascogale macdonnellensis* Spencer, 1896.
COMMENTS: Separated from *Antechinus* by Archer (1982:434).

Pseudantechinus macdonnellensis (Spencer, 1896). Rept. Horn Sci. Exped. Cent. Aust., Zool., 2:27.
TYPE LOCALITY: Australia, Northern Territory, south of Alice Springs.
DISTRIBUTION: N Western Australia, Northern Territory, central deserts (Australia).
STATUS: Unknown.

SYNONYMS: *mimulus*.

Pseudantechinus ningbing Kitchener, 1988. Rec. W. Aust. Mus., 14:62.
TYPE LOCALITY: Australia, Western Australia, Kimberley region, Mitchell Plateau, *ca*. 220m, 14°53'40"S, 125°45'20"E.
DISTRIBUTION: Kimberley region, Western Australia.

Pseudantechinus woolleyae Kitchener and Caputi, 1988. Rec. W. Aust. Mus., 14:39.
TYPE LOCALITY: Australia, Western Australia, near Newligunn bore, 10km.117° from Errabiddy Homestead, 25°33'00"S, 117°08'00"E.
DISTRIBUTION: Western Australia: Pilbara region and further south - between *ca*. 21° and 28°S, 115° and 122°E.

Sarcophilus F. Cuvier, 1837. *In* E. Geoffroy and F. Cuvier, Hist. Nat. Mammifères, pt. 4, 7(70):1-6, "Sarcophile oursin".
TYPE SPECIES: *Didelphis ursina* Harris, 1808 (= *Dasyurus laniarius* Owen, 1838).
SYNONYMS: *Diabolus, Ursinus*.

Sarcophilus laniarius (Owen, 1838). *In* T. L. Mitchell, Three Exped. into the Interior of E. Australia. p. 363.
TYPE LOCALITY: Australia, Wellington Caves (Pleistocene).
DISTRIBUTION: Australia: Tasmania, perhaps S Victoria, where it is known as a subfossil.
STATUS: Common.
SYNONYMS: *dixonae, harrisii, satanicus, ursina*.
COMMENTS: Usually called *S. harrisii*, but see Werdelin (1987:9).

Sminthopsis Thomas, 1887. Ann. Mus. Civ. Stor. Nat. Genova, ser. 2, 4:503.
TYPE SPECIES: *Phascogale crassicaudata* Gould, 1844.
SYNONYMS: *Antechinomys, Podabrus*.
COMMENTS: Includes *Antechinomys* as a subgenus; formerly considered a valid genus by Archer (1977), but considered a subgenus by Archer (1979:329); see also Kirsch and Calaby (1977:15). On the basis of other anatomic and isozymic data, Lidicker (1983:1317) considered *Antechinomys* a distinct genus, despite similarity in dental morphology with *Sminthopsis*. An additional undescribed species has been reported from S New Guinea and N Australia; see Archer (*in* Tyler, 1979) and Waithman (1979). Revised by Archer (1981). Original name *Podabrus*, Gould, 1845, is preoccupied.

Sminthopsis aitkeni Kitchener, Stoddart and Henry, 1984. Rec. W. Aust. Mus., 11:204.
TYPE LOCALITY: Australia, South Australia, Kangaroo Isl, Section 46, Cassini.
DISTRIBUTION: Kangaroo Isl (South Australia).
STATUS: Unknown.
COMMENTS: Separated from *S. murina* by Kitchener et al. (1984*b*:204).

Sminthopsis archeri Van Dyck, 1986. Aust. Mamm., 9:112.
TYPE LOCALITY: Papua New Guinea, Trans-fly Plains, Morehead (8°04'S, 141°39'E).
DISTRIBUTION: S Papua New Guinea; Northern Gulf, Queensland (Australia).
STATUS: Common?

Sminthopsis butleri Archer, 1979. Aust. Zool., 20(2):329.
TYPE LOCALITY: Australia, Western Australia, Kalumburu (14°15'S, 126°40'E).
DISTRIBUTION: In Australia known only from the type locality; also in Papua New Guinea.
COMMENTS: Name first mentioned by Kirsch (1977:47), but first made available by Archer (1979).

Sminthopsis crassicaudata (Gould, 1844). Proc. Zool. Soc. Lond., 1844:105.
TYPE LOCALITY: Australia, Western Australia, Williams River.
DISTRIBUTION: South Australia, SW Queensland, SE Northern Territory, S Western Australia, W New South Wales, W Victoria.
STATUS: Common?
SYNONYMS: *centralis, ferruginea*.
COMMENTS: See Archer (1979:329, 1981:176).

Sminthopsis dolichura Kitchener, Stoddart and Henry, 1984. Rec. W. Aust. Mus., 11:204.
TYPE LOCALITY: Western Australia, 6 km SSE of Buningonia Spring, 32°28'S, 123°36'E.
DISTRIBUTION: Western Australia, South Australia.
STATUS: Unknown.
COMMENTS: Separated from *S. murina* by Kitchener et al. (1984*b*).

Sminthopsis douglasi Archer, 1979. Aust. Zool., 20(2):337.
TYPE LOCALITY: Australia, Queensland, Cloncurry River Watershed, Julia Creek (20°40'S, 141°40'E).
DISTRIBUTION: Known only from type locality and Richmond in Cloncurry River Watershed, Queensland, and possibly Mitchell Plateau, Western Australia.
STATUS: IUCN - Indeterminate.

Sminthopsis fuliginosus (Gould, 1852). Mamm. Aust., 1, pl. 41.
TYPE LOCALITY: Western Australia, King George Sound.
DISTRIBUTION: SW Western Australia.
COMMENTS: Separated from *S. murina* by Kitchener et al. (1984*b*).

Sminthopsis gilberti Kitchener, Stoddart, and Henry, 1984. Rec. W. Aust. Mus., 11:204.
TYPE LOCALITY: Western Australia, Mt. Saddleback, 32°58'S, 116°20'E.
DISTRIBUTION: SW Western Australia.
STATUS: Unknown.

Sminthopsis granulipes Troughton, 1932. Rec. Aust. Mus., 18:350.
TYPE LOCALITY: Australia, Western Australia, King George Sound (Albany).
DISTRIBUTION: SW Western Australia.
STATUS: Unknown.

Sminthopsis griseoventer Kitchener, Stoddart, and Henry, 1984. Rec. W. Aust. Mus., 11:204.
TYPE LOCALITY: Western Australia, Bindoon, 31°18'S, 116°01'E.
DISTRIBUTION: SW Western Australia.
STATUS: Unknown.
SYNONYMS: *caniventer*.

Sminthopsis hirtipes Thomas, 1898. Nov. Zool., 5:3.
TYPE LOCALITY: Australia, Northern Territory, Charlotte Waters.
DISTRIBUTION: Central deserts in Northern Territory and Western Australia; also coastal scrub 500 km N of Perth.
STATUS: Common.

Sminthopsis laniger (Gould, 1856). Mamm. Aust., 1, pl. 33.
TYPE LOCALITY: Australia, interior New South Wales.
DISTRIBUTION: Western Australia, S Northern Territory, N Victoria, W New South Wales, SW Queensland, N South Australia.
STATUS: U.S. ESA - Endangered.
SYNONYMS: *spenceri*.
COMMENTS: Subgenus *Antechinomys*; see Archer (1979:329, 1981:187). However, see Lidicker (1983:1317), who considered *Antechinomys* a distinct genus. Includes *spenceri*; see Archer (1977:19).

Sminthopsis leucopus (Gray, 1842). Ann. Mag. Nat. Hist., [ser. 1], 10:261.
TYPE LOCALITY: Australia, Tasmania.
DISTRIBUTION: S and SE Victoria, Tasmania, New South Wales, and Queensland (Australia).
STATUS: Common?
SYNONYMS: *ferruginifrons, leucogenys, mitchelli*.
COMMENTS: See Archer (1979:329, 1981:102).

Sminthopsis longicaudata Spencer, 1909. Proc. R. Soc. Victoria, 21 (n.s.):449.
TYPE LOCALITY: Australia, Western Australia.
DISTRIBUTION: Western Australia.
STATUS: CITES - Appendix I; U.S. ESA - Endangered; IUCN - Insufficiently known.
COMMENTS: Known from only four specimens; see Ride (1970:201).

Sminthopsis macroura (Gould, 1845). Proc. Zool. Soc. Lond., 1845:79.
TYPE LOCALITY: Australia, Queensland, Darling Downs.

DISTRIBUTION: Australia: NW New South Wales, W Queensland, S Northern Territory, N South Australia, N Western Australia.
STATUS: Common.
SYNONYMS: *froggatti, larapinta, monticola, stalkeri.*
COMMENTS: See Archer (1979:329, 1981:148).

Sminthopsis murina (Waterhouse, 1838). Proc. Zool. Soc. Lond., 1837:76 [1838].
TYPE LOCALITY: Australia, New South Wales, N of Hunter River.
DISTRIBUTION: SW Western Australia, SE South Australia, Victoria, New South Wales, E Queensland.
STATUS: Common.
SYNONYMS: *albipes, tatei.*
COMMENTS: See Archer (1979:329, 1981:94-99).

Sminthopsis ooldea Troughton, 1965. Proc. Linn. Soc. N.S.W., 1964, 89:316 [1965].
TYPE LOCALITY: Australia, South Australia, Ooldea.
DISTRIBUTION: Edge of Nullarbor Plain (South Australia), Western Australia, S Northern Territory.
STATUS: Unknown.
COMMENTS: Originally described as a subspecies of *murina*, but considered a distinct species by Archer (1975:243) and Kirsch and Calaby (1977:15).

Sminthopsis psammophila Spencer, 1895. Proc. R. Soc. Victoria, 7 (n.s.):223.
TYPE LOCALITY: Australia, Northern Territory, Lake Amadeus.
DISTRIBUTION: Australia: SW Northern Territory (vicinity of Ayer's Rock) and Eyre Peninsula (South Australia).
STATUS: CITES - Appendix I; U.S. ESA - Endangered; IUCN - Insufficiently known.
COMMENTS: Known only from five specimens; see Archer (1981:215).

Sminthopsis virginiae (de Tarragon, 1847). Rev. Zool. Paris, p. 177.
TYPE LOCALITY: None given; Archer (1981:132) designated Australia, Queensland, Herbert Vale.
DISTRIBUTION: N Queensland, N Northern Territory (Australia); Aru Isls (Indonesia); S New Guinea.
STATUS: Unknown.
SYNONYMS: *lumholtzi, nitela, rona, rufigenis.*
COMMENTS: See Archer (1979:329, 1981:132) and Kirsch and Calaby (1977:15). De Tarragon's 1847 description of *S. virginiae* did not specify a type locality and his type specimen, now lost, had no known locality. Collett (1886[1887]:548) named *S. nitela* from Herbert Vale and it was subsequently renamed *S. lumholtzi*, both of which are referable to *virginiae*; see Archer (1981:136).

Sminthopsis youngsoni McKenzie and Archer, 1982. Aust. Mamm., 5:267.
TYPE LOCALITY: Western Australia, Edgar Ranges, 18°50'S, 123°05'E.
DISTRIBUTION: Western Australia, Northern Territory.
STATUS: Unknown.

ORDER PERAMELEMORPHIA
by Colin P. Groves

ORDER PERAMELEMORPHIA

SYNONYMS: Perameliformes, Perameloidea.

COMMENTS: Recognized as an order by Aplin and Archer (1987) who proposed a new syncretic classification of the marsupials.

Family Peramelidae Gray, 1825. Ann. Philos., n.s., 10:336.

SYNONYMS: Thylocomyidae.

COMMENTS: Includes the family Thylacomyidae; see Vaughan (1978:39) and Groves and Flannery (1990); but also see Archer and Kirsch (1977). Revised by Tate (1948*b*).

Chaeropus Ogilby, 1838. Proc. Zool. Soc. Lond., 1838:26.

TYPE SPECIES: *Perameles ecaudatus* Ogilby, 1838.

SYNONYMS: *Choeropus.*

Chaeropus ecaudatus (Ogilby, 1838). Proc. Zool. Soc. Lond., 1838:25.

TYPE LOCALITY: Australia, New South Wales, banks of Murray River, south of the junction with Murrumbridge River.

DISTRIBUTION: Australia: SW New South Wales, Victoria, S Northern Territory, N South Australia, Western Australia.

STATUS: CITES - Appendix I pe [possibly extinct]; U.S. ESA - Endangered; IUCN - Extinct.

SYNONYMS: *castanotis, occidentalis.*

COMMENTS: Probably extinct, last taken in 1907; see Ride (1970:200).

Isoodon Desmarest, 1817. Nouv. Dict. Hist. Nat., Nouv. ed., 16:409.

TYPE SPECIES: *Didelphis obesula* Shaw, 1797.

SYNONYMS: *Thylacis.*

COMMENTS: Includes *Thylacis* of Haltenorth, 1958, which was an incorrect usage; see Van Deusen and Jones (1967:74) and Lidicker and Follett (1968). Revised by Lyne and Mort (1981).

Isoodon auratus (Ramsay, 1887). Proc. Linn. Soc. N.S.W., ser. 2, 2:551.

TYPE LOCALITY: Australia, Western Australia, Derby.

DISTRIBUTION: Australia: Formerly Northern Territory and N Western Australia, survives in NW of Western Australia and on Barrow Isl.

STATUS: Rare.

SYNONYMS: *arnhemensis, barrowensis.*

COMMENTS: See Ride (1970:96) and Lyne and Mort (1981).

Isoodon macrourus (Gould, 1842). Proc. Zool. Soc. Lond., 1842:41.

TYPE LOCALITY: Australia, Northern Territory, Port Essington.

DISTRIBUTION: NE Western Australia, N Northern Territory, E Queensland, and NE New South Wales (Australia); S and E New Guinea.

STATUS: Common.

SYNONYMS: *macrura, moresbyensis, torosa.*

Isoodon obesulus (Shaw, 1797). Nat. Misc., 8:298.

TYPE LOCALITY: Australia, New South Wales, Sydney, Ku-ring-gai Chase Natl. Park, 33°36′S, 151°16′E, see Dixon (1981).

DISTRIBUTION: SE New South Wales, S Victoria, SE South Australia, N Queensland, SW Western Australia, Nuyts Arch. (Great Australian Bight, S Australian coast), and Tasmania.

STATUS: Locally common.

SYNONYMS: *affinis, fusciventer, nauticus, peninsulae.*

Macrotis Reid, 1837. Proc. Zool. Soc. Lond., 1836:131 [1837].

TYPE SPECIES: *Perameles lagotis* Reid, 1837.

SYNONYMS: *Paragalia, Phalacomys, Thylacomys.*

COMMENTS: Not preoccupied by *Macrotis* Dejean, 1833, a *nomen nudum* (Troughton, 1932*b*). Archer and Kirsch (1977) placed *Macrotis* (including its junior synonym *Thylacomys*) in a separate family (the name available being Thylacomyidae), rather than in Peramelidae. Groves and Flannery (1990) placed *Macrotis* back in Peramelidae.

Macrotis lagotis (Reid, 1837). Proc. Zool. Soc. Lond., 1836:129 [1837].
TYPE LOCALITY: Australia, Western Australia, Swan River.
DISTRIBUTION: Formerly in Western Australia, South Australia, Northern Territory, W New South Wales, SW Queensland. Survives only in SW Queensland, Northern Territory/ Western Australia border region and Kimberleys.
STATUS: CITES - Appendix I; U.S. ESA and IUCN - Endangered.
SYNONYMS: *cambrica, grandis, interjecta, nigripes, sagitta.*

Macrotis leucura (Thomas, 1887). Ann. Mag. Nat. Hist., ser. 5, 19:397.
TYPE LOCALITY: Undesignated; as the specimen was sent by the South Australian Museum's taxidermist to London, Thomas (1887*a*) thought it might have originated near Adelaide or in the northern part of South Australia, whence others in the same collection had come.
DISTRIBUTION: C Australia.
STATUS: CITES - Appendix I; U.S. ESA - Endangered; IUCN - Extinct.
SYNONYMS: *minor, miselius.*
COMMENTS: Possibly extinct; see Ride (1970:200).

Perameles É. Geoffroy, 1804. Ann. Mus. Hist. Nat. Paris, 4:56.
TYPE SPECIES: *Perameles nasuta* E. Geoffroy, 1804.
SYNONYMS: *Thylacis.*
COMMENTS: *Perameles* was also used by Geoffroy, 1804, Bull. Sci. Soc. Philom. Paris, 3(80):249.

Perameles bougainville Quoy and Gaimard, 1824. *In* de Freycinet, Voy. autour du monde...l'Uranie et al Physicienne, Zool., p. 56.
TYPE LOCALITY: Australia, Western Australia, Shark Bay, Peron Peninsula.
DISTRIBUTION: Formerly in S South Australia, NW Victoria, W New South Wales, S Western Australia, Bernier and Dorre Isls, survives only on Bernier and Dorre Isls (off Western Australia).
STATUS: CITES - Appendix I; U.S. ESA - Endangered; IUCN - Rare.
SYNONYMS: *arenaria, fasciata, myosuros, notina.*
COMMENTS: See Ride (1970:100).

Perameles eremiana Spencer, 1897. Proc. R. Soc. Victoria, 9 (n.s.):9.
TYPE LOCALITY: Australia, Northern Territory, Burt Plain (N of Alice Springs).
DISTRIBUTION: N South Australia, S Northern Territory, Great Victoria Desert (Western Australia).
STATUS: U.S. ESA - Endangered; IUCN - Extinct.
COMMENTS: Possibly extinct; see Ride (1970:200).

Perameles gunnii Gray, 1838. Ann. Nat. Hist., 1:107.
TYPE LOCALITY: Australia, Tasmania.
DISTRIBUTION: Australia: S Victoria, where restricted to Hamilton, and Tasmania.
STATUS: Common in Tasmania, Endangered in Victoria.

Perameles nasuta É. Geoffroy, 1804. Ann. Mus. Hist. Nat. Paris, 4:62.
TYPE LOCALITY: Australia, New South Wales, Sydney.
DISTRIBUTION: Australia: E Queensland, E New South Wales, E Victoria.
STATUS: Common.
SYNONYMS: *lawsoni, major, musei, pallescens.*

Family Peroryctidae Groves and Flannery, 1990. *In* Seebeck et al. (eds.), Bandicoots and Bilbies, p. 2.
COMMENTS: See Groves and Flannery (1990).

Echymipera Lesson, 1842. Nouv. Tabl. Regn. Anim. Mammifères, p. 192.
TYPE SPECIES: *Perameles kalubu* J. Fischer, 1829.
COMMENTS: See Groves and Flannery (1990).

Echymipera clara Stein, 1932. Z. Säugetierk., 7:256.
TYPE LOCALITY: Indonesia, Irian Jaya, Tjenderawasih Div., Japan Isl.
DISTRIBUTION: NC New Guinea.
STATUS: IUCN - Rare.
COMMENTS: See Flannery (1990*a*) for an assessment of its affinities.

Echymipera davidi Flannery, 1990. *In* Seebeck et al. (eds.), Bandicoots and Bilbies, p. 29.
TYPE LOCALITY: Papua New Guinea, Trobriand Isls, Kiriwina Isl (08°30'S, 151°00'E).
DISTRIBUTION: Kiriwina Isl.
STATUS: Unknown.
COMMENTS: Not closely related to any other species of *Echymipera* (Flannery, 1990*a*).

Echymipera echinista Menzies, 1990. Science in New Guinea, 16:92.
TYPE LOCALITY: Papua New Guinea, Western (Fly River) Province, Wipim, near Iamega (08°51'S, 142°58'E).
DISTRIBUTION: Known only from Western (Fly River) Province, Papua New Guinea.
STATUS: Rare.

Echymipera kalubu (J. Fischer, 1829). Synopsis Mamm., p. 274.
TYPE LOCALITY: Indonesia, Irian Jaya, Sorong Div., Waigeo Isl.
DISTRIBUTION: New Guinea and adjacent small islands including Bismarck Arch., Mysol and Salawati Isls.
STATUS: Common.
SYNONYMS: *albiceps, breviceps, cockerelli, doreyanus, garagassi, hispida, myoides, oriomo, philipi, rufiventris.*
COMMENTS: The name *kalubu* has been attributed to Lesson, 1828, Dict. Class. Hist. Nat., 13:200; but see Husson (1955:290).

Echymipera rufescens (Peters and Doria, 1875). Ann. Mus. Civ. Stor. Nat. Genova, 7:541.
TYPE LOCALITY: Indonesia, Kei Isls.
DISTRIBUTION: Cape York Peninsula (Queensland, Australia); New Guinea and certain small islands off SE coast; Kei and Aru Isls (Indonesia).
STATUS: Uncommon.
SYNONYMS: *aruensis, australis, gargàntua, keiensis, welsianus.*

Microperoryctes Stein, 1932. Z. Säugetierk., 7:256.
TYPE SPECIES: *Microperoryctes murina* Stein, 1932.
SYNONYMS: *Ornoryctes.*
COMMENTS: For synonymy of *Ornoryctes* with *Microperoryctes* instead of with *Peroryctes*, see Groves and Flannery (1990).

Microperoryctes longicauda (Peters and Doria, 1876). Ann. Mus. Civ. Stor. Nat. Genova, 8:335.
TYPE LOCALITY: Indonesia, Irian Jaya, Vogelkop, Manokwari Div., Arfak Mtns, Hatam, 1520 m.
DISTRIBUTION: Interior New Guinea.
STATUS: Common.
SYNONYMS: *dorsalis, magnus, ornatus.*
COMMENTS: Formerly included in *Peroryctes*, but see Groves and Flannery (1990).

Microperoryctes murina Stein, 1932. Z. Säugetierk., 7:257.
TYPE LOCALITY: Indonesia, Irian Jaya, Paniai Div., Weyland Mtns, Sumuri Mtn, 2500 m.
DISTRIBUTION: W interior New Guinea.
STATUS: Rare.

Microperoryctes papuensis (Laurie, 1952). Bull. Brit. Mus. (Nat. Hist.), Zool., 1:291.
TYPE LOCALITY: Papua New Guinea, Milne Bay Prov., Mt. Mura, (30 mi NW Mt. Simpson), Boneno, 1220-1525 m.
DISTRIBUTION: SE interior New Guinea.
STATUS: Rare.
COMMENTS: Formerly included in *Peroryctes*, but see Groves and Flannery (1990).

Peroryctes Thomas, 1906. Proc. Zool. Soc. Lond., 1906:476.
TYPE SPECIES: *Perameles raffrayana* Milne-Edwards, 1878.

Peroryctes broadbenti (Ramsay, 1879). Proc. Linn. Soc. N.S.W., 3:402, pl. 27.
TYPE LOCALITY: Papua New Guinea, Central Prov., banks of Goldie River (a tributary of the Laloki River) inland from Port Moresby.
DISTRIBUTION: SE New Guinea.
STATUS: Rare.
COMMENTS: Included in *raffrayana* by Laurie and Hill (1954:10), but considered a distinct species by Van Deusen and Jones (1967:74).

Peroryctes raffrayana (Milne-Edwards, 1878). Ann. Sci. Nat. Zool. (Paris), ser. 6, 7(art. 11):1.
TYPE LOCALITY: Indonesia, Irian Jaya, Vogelkop, Manokwari Div., Amberbaki.
DISTRIBUTION: New Guinea.
STATUS: Uncommon.
SYNONYMS: *mainois, rothschildi.*
COMMENTS: Laurie and Hill (1954:10) included *broadbenti* in this species; but see Van Deusen and Jones (1967:74).

Rhynchomeles Thomas, 1920. Ann. Mag. Nat. Hist., ser. 9, 6:429-430.
TYPE SPECIES: *Rhynchomeles prattorum* Thomas, 1920.

Rhynchomeles prattorum Thomas, 1920. Ann. Mag. Nat. Hist., ser. 9, 6:429-430.
TYPE LOCALITY: Indonesia, Ceram Isl, Mt. Manusela, 1800 m.
DISTRIBUTION: Ceram Isl (Indonesia).
STATUS: Rare.
COMMENTS: See Groves and Flannery (1990).

ORDER NOTORYCTEMORPHIA

by Colin P. Groves

ORDER NOTORYCTEMORPHIA

SYNONYMS: Syndactyliformes.

COMMENTS: Recognized as an order by Aplin and Archer (1987) who proposed a new syncretic classification of the marsupials.

Family Notoryctidae Ogilby, 1892. Cat. Aust. Mammalia, p. 5.

COMMENTS: Relationships unknown.

Notoryctes Stirling, 1891. Trans. R. Soc. S. Aust., 14:154.

TYPE SPECIES: *Psammoryctes typhlops* Stirling, 1889.

SYNONYMS: *Neoryctes, Psammoryctes*.

COMMENTS: *Psammoryctes* is preoccupied (see Iredale and Troughton, 1934).

Notoryctes caurinus Thomas, 1920. Ann. Mag. Nat. Hist., ser. 9, 6:111.

TYPE LOCALITY: Western Australia, Wollal (= Wallal), Ninety Mile Beach.

DISTRIBUTION: NW Western Australia.

STATUS: Unknown.

COMMENTS: Separated from *N. typhlops* by Walton (1988:47).

Notoryctes typhlops (Stirling, 1889). Trans. R. Soc. S. Aust., 12:158.

TYPE LOCALITY: Australia, Northern Territory, Indracowrie, 100 mi. (161 km) from Charlotte Waters.

DISTRIBUTION: Western deserts from Ooldea (South Australia) to Charlotte Waters and NW Western Australia, Northern Territory.

STATUS: Thought to be rare, but no real data exist.

ORDER DIPROTODONTIA

by Colin P. Groves

ORDER DIPROTODONTIA

SYNONYMS: Phalangeriformes.

COMMENTS: Recognized as an order by Aplin and Archer (1987) who proposed a new syncretic classification of the marsupials.

Family Phascolarctidae Owen, 1839. Proc. Zool. Soc. Lond., 1839:19.

COMMENTS: Formerly included in the Phalangeridae; see Ride (1970:225).

Phascolarctos Blainville, 1816. Nouv. Bull. Sci. Soc. Philom. (Paris), p. 108.

TYPE SPECIES: *Lipurus cinereus* Goldfuss, 1817 (seen by Blainville in ms., published 1817).

SYNONYMS: *Draximenus, Koala, Lipurus, Liscurus, Morodactylus.*

Phascolarctos cinereus (Goldfuss, 1817). Die Säugethiere, pt. 65, pl. 155, Aa, Ac.

TYPE LOCALITY: Australia, New South Wales.

DISTRIBUTION: Australia: SE Queensland, E New South Wales, SE South Australia, and Victoria. Introduced on Kangaroo Isl, South Australia and at Yanchep, Western Australia.

STATUS: Vulnerable.

SYNONYMS: *adustus, flindersii, fuscus, koala, subiens, victor.*

Family Vombatidae Burnett, 1830 (1820). Quart. J. Sci. Arts, 1829:351 [1830].

SYNONYMS: Phascolomyidae.

COMMENTS: Phascolomyidae Goldfuss, 1820, is based on *Phascolomis*, a junior synonym (Haltenorth, 1958:32). Because Phascolomyidae was replaced with Vombatidae before 1961, and because Vombatidae has won general acceptance, it is to be maintained (Art. 40b of the Code of Nomenclature, International Commission on Zoological Nomenclature, 1985).

Lasiorhinus Gray, 1863. Ann. Mag. Nat. Hist., ser. 3, 11:458.

TYPE SPECIES: *Lasiorhinus mcoyi* Gray, 1863 (= *Phascolomys latifrons* Owen, 1845).

SYNONYMS: *Wombatula.*

COMMENTS: This genus needs revision.

Lasiorhinus krefftii (Owen, 1873). Philos. Trans. R. Soc. London, 162:178, pl. 17, 20.

TYPE LOCALITY: Australia, New South Wales, Wellington Caves, Breccia Cavern.

DISTRIBUTION: Australia: SE and E Queensland, Deniliquin (New South Wales).

STATUS: CITES - Appendix I; U.S. ESA and IUCN - Endangered.

SYNONYMS: *barnardi, gillespiei.*

COMMENTS: Includes *gillespiei* and *barnardi* according to Kirsch and Calaby (1977:23), who stated that only a single remnant population of *krefftii* remained at the type locality of *barnardi*. However, populations historically known as *barnardi* may not be referable to *krefftii*.

Lasiorhinus latifrons (Owen, 1845). Proc. Zool. Soc. Lond., 1845:82.

TYPE LOCALITY: Australia, South Australia.

DISTRIBUTION: S South Australia, SE Western Australia.

STATUS: Locally common.

SYNONYMS: *lasiorhinus, mcoyi.*

Vombatus É. Geoffroy, 1803. Bull. Sci. Soc. Philom. Paris, 72:185.

TYPE SPECIES: *Didelphis ursina* Shaw, 1800.

SYNONYMS: *Amblotis, Opposum, Phascolomis, Wombatus.*

COMMENTS: *Phascolomis* É. Geoffroy, 1803 is a junior synonym (Haltenorth, 1958:32).

Vombatus ursinus (Shaw, 1800). Gen. Zool. Syst. Nat. Hist., 1(2), Mammalia, p. 504.

TYPE LOCALITY: Australia, Tasmania, Bass Strait, Cape Barren Island.

DISTRIBUTION: E New South Wales, S Victoria, SE South Australia, Tasmania, islands in the Bass Strait, and extreme SE Queensland (Australia).
STATUS: Common.
SYNONYMS: *angasii, assimilis, bassii, fossor, fuscus, hirsutus, mitchelli, niger, platyrhinus, setosus, tasmaniensis, vombatus, wombat.*

Family Phalangeridae Thomas, 1888. Cat. Marsup. Monotr. Brit. Mus., p. 126.
COMMENTS: Distinct from Phascolarctidae (see Ride, 1970:22) and does not include Pseudocheiridae, Petauridae, Burramyidae, or Acrobatidae (see Aplin and Archer, 1987). A provisional classification was given by Flannery et al. (1987).

Ailurops Wagler, 1830. Naturliches Syst. Amphibien, p. 26.
TYPE SPECIES: *Phalangista ursina* Temminck, 1824.
SYNONYMS: *Ceonix, Eucuscus.*
COMMENTS: Flannery et al. (1987) placed this genus in its own subfamily, Ailuropinae.

Ailurops ursinus (Temminck, 1824). Monogr. Mamm., 1:10.
TYPE LOCALITY: Indonesia, Sulawesi, Sulawesi Utara, Minahasa, Manado.
DISTRIBUTION: Sulawesi, Peleng Isl, Talaud Isls, Togian Isl, Muna Isl, Butung Isl, and Lembeh Isl (Indonesia).
STATUS: Common.
SYNONYMS: *flavissimus, furvus, melanotis, togianus.*
COMMENTS: Formerly included in *Phalanger.*

Phalanger Storr, 1780. Prodr. Meth. Mammal., p. 38.
TYPE SPECIES: *Didelphis orientalis* Pallas, 1766.
SYNONYMS: *Balantia, Cuscus, Phalangista, Sipalus.*
COMMENTS: Does not include *Spilocuscus* (see Ride, 1970:248). Revised by Tate (1945), Feiler (1978*a-c*), and G. G. George (1979).

Phalanger carmelitae Thomas, 1898. Ann. Mus. Civ. Stor. Nat. Genova, 19:5.
TYPE LOCALITY: Papua New Guinea, Central Prov., upper Vanapa River.
DISTRIBUTION: Interior New Guinea.
STATUS: Common.
SYNONYMS: *coccygis.*
COMMENTS: Formerly included in *vestitus* (see G. G. George, 1979:94).

Phalanger lullulae Thomas, 1896. Novit. Zool., 3:528.
TYPE LOCALITY: Papua New Guinea, Milne Bay Prov., Woodlark Isl.
DISTRIBUTION: Woodlark Isl (Papua New Guinea).
STATUS: IUCN - Endangered; rare.
COMMENTS: Formerly included in *orientalis* (see G. G. George, 1979:97).

Phalanger matanim Flannery, 1987. Rec. Aust. Mus., 39:183.
TYPE LOCALITY: Papua New Guinea, West Sepik Province, Telefomin area, Upper Sol River, 5°06'S, 141°42'E, 2600 m.
DISTRIBUTION: Telefomin area, W Papua New Guinea.
STATUS: Rare.

Phalanger orientalis (Pallas, 1766). Misc. Zool., p. 61.
TYPE LOCALITY: Indonesia, Amboina (= Ambon) Isl, Maluku.
DISTRIBUTION: Timor and Ceram Isls (Indonesia) to New Guinea, and adjacent small islands; Bismarck Arch.; Solomon Isls; and E Cape York (Queensland, Australia).
STATUS: CITES - Appendix II.
SYNONYMS: *alba, albidus, ambonensis, breviceps, brevinasus, ducatoris, fusca, intercastellanus, kiriwinae, matsika, meeki, microdon, mimicus, minor, molucca, moluccensis, peninsulae, rufa, vulpecula.*
COMMENTS: Formerly included *interpositus* and *lullulae* (see G. G. George, 1979). Flannery et al. (1987) suggested that the southern races may prove to be specifically, even generically distinct, as *Strigocuscus mimicus*, but Flannery (1990) did not adopt this course.

Phalanger ornatus (Gray, 1860). Proc. Zool. Soc. Lond., 1860:374.
TYPE LOCALITY: Indonesia: Bachian (= Bacan or Batjan) Isl.
DISTRIBUTION: Halmahera, Ternate, Tidore, Bacan, Morotai Isls (Indonesia).
STATUS: Apparently common.
COMMENTS: For status see Groves (1987). Provisionally allotted to *Strigocuscus* by Flannery et al. (1987) although they did not examine specimens.

Phalanger pelengensis Tate, 1945. Am. Mus. Novit., 1283:3.
TYPE LOCALITY: Indonesia: Peleng Isl.
DISTRIBUTION: Peleng and Sulu Isls (Indonesia).
STATUS: Unknown.
SYNONYMS: *mendeni*.
COMMENTS: For status see Groves (1987). Flannery et al. (1987) doubted that *pelengensis* really belongs to *Phalanger*.

Phalanger rothschildi Thomas, 1898. Novit. Zool., 5:433.
TYPE LOCALITY: Moluccas, Pulau Obi Besar, Loiwij.
DISTRIBUTION: Pulau (Isl) Obi Besar.
STATUS: Common.
COMMENTS: Commonly included in *Strigocuscus celebensis*, but see Groves (1987).

Phalanger sericeus Thomas, 1907. Ann. Mag. Nat. Hist., ser. 7, 20:74.
TYPE LOCALITY: Papua New Guinea, Angabunga Range, Owgarra, 6,000 ft (= 1829 m).
DISTRIBUTION: C and E New Guinea, higher elevations.
STATUS: Common.
SYNONYMS: *occidentalis*.
COMMENTS: Called *P. vestitus* by most authors, but see Menzies and Pernetta (1986:594).

Phalanger vestitus (Milne-Edwards, 1877). C. R. Acad. Sci. Paris, 85:1080.
TYPE LOCALITY: Indonesia, Irian Jaya, Vogelkop, Sorong Div., Tamrau Range, Karons Mtns.
DISTRIBUTION: Interior New Guinea.
STATUS: IUCN - Rare.
SYNONYMS: *interpositus, permixtio*.
COMMENTS: Formerly known as *P. interpositus* (see Flannery, 1990).

Spilocuscus Gray, 1862. Proc. Zool. Soc. Lond., 1861:316 [1862].
TYPE SPECIES: *Phalangista maculata* Desmarest, 1818.
COMMENTS: Separated from *Phalanger* by G. G. George (1979).

Spilocuscus maculatus (Desmarest, 1818). Nouv. Dict. Hist. Nat., Nouv. ed., 25:472.
TYPE LOCALITY: Indonesia, Irian Jaya, Vogelkop, Manokwari Div., Manokwari.
DISTRIBUTION: New Guinea and adjacent small islands; Aru and Kei Isls, Ceram, Amboina and Selayar Isls (Indonesia); Cape York Peninsula (Queensland, Australia).
STATUS: CITES - Appendix II; but "common" in New Guinea (Flannery, 1990).
SYNONYMS: *brevicaudatus, chrysorrhous, goldiei, kraemeri?, nudicaudatus, ochropus, quoy, variegata*.
COMMENTS: Commonly cited from "E. Geoffroy, 1803. Cat. Mamm. Mus. Hist. Nat., Paris., p. 149"; but see comments in Appendix I. Feiler (1978*a*) included *atrimaculatus* in this species, but G. G. George (1979:98) placed it in *rufoniger*.

Spilocuscus rufoniger (Zimara, 1937). Anz. Akad. Wiss. Wien, 74:35.
TYPE LOCALITY: Papua New Guinea, Morobe Prov., Sattelberg.
DISTRIBUTION: N New Guinea.
STATUS: IUCN - Rare.
SYNONYMS: *atrimaculatus*.
COMMENTS: Includes *atrimaculatus* (see G. G. George, 1979:98), but also see Feiler (1978*a*), who placed it in *maculatus*.

Strigocuscus Gray, 1862. Proc. Zool. Soc. Lond., 1861:319 [1862].
TYPE SPECIES: *Cuscus celebensis* Gray, 1858.
COMMENTS: Flannery et al. (1987) resurrected this genus for *S. celebensis*, and provisionally for *S. gymnotis*.

Strigocuscus celebensis (Gray, 1858). Proc. Zool. Soc. Lond., 1858:105.
TYPE LOCALITY: Indonesia, Sulawesi, Sulawesi Selatan, Ujung Pandang (= Macassar).
DISTRIBUTION: Sulawesi, Sanghir Isls, and Taliabu (= Sula) Isls (Indonesia).
STATUS: Common.
SYNONYMS: *feileri, sangirensis*.

Strigocuscus gymnotis (Peters and Doria, 1875). Ann. Mus. Civ. Stor. Nat. Genova, 7:543.
TYPE LOCALITY: Indonesia, Aru Isls, Gialnhegen Isl (restricted by Van der Feen, 1962:40).
DISTRIBUTION: New Guinea; Aru Isls, Wetar Isl, Timor Isl, and other small Indonesian islands.
STATUS: Common.
SYNONYMS: *leucippus*.
COMMENTS: Distribution poorly known. Included in *Strigocuscus* by Flannery et al. (1987).

Trichosurus Lesson, 1828. *In* Bory de Saint-Vincent (ed.), Dict. Class. Hist. Nat. Paris, 13:333.
TYPE SPECIES: *Didelphis vulpecula* Kerr, 1792.
SYNONYMS: *Cercaertus, Psilogrammurus, Tapoa, Trichurus*.

Trichosurus arnhemensis Collett, 1897. Proc. Zool. Soc. Lond., 1897:328.
TYPE LOCALITY: Australia, Northern Territory, Daly River.
DISTRIBUTION: N Northern Territory, NE Western Australia, Barrow Isl (Australia).
STATUS: Common.

Trichosurus caninus (Ogilby, 1836). Proc. Zool. Soc. Lond., 1835:191 [1836].
TYPE LOCALITY: Australia, New South Wales, Hunter River.
DISTRIBUTION: Australia: SE Queensland, E New South Wales, E Victoria.
STATUS: Common.
SYNONYMS: *nigrans*.

Trichosurus vulpecula (Kerr, 1792). *In* Linnaeus, Anim. Kingdom, 1:198.
TYPE LOCALITY: Australia, New South Wales, Sydney.
DISTRIBUTION: Australia: E Queensland, E New South Wales, Victoria, Tasmania, SE and N South Australia, SW Western Australia; introduced to New Zealand (Wodzicki, 1950).
STATUS: Common, in many cities, lives commensally.
SYNONYMS: *bougainvillei, cookii, cuvieri, eburacensis, felina, fuliginosa, grisea, hypoleucus, johnstonii, lemurina, melanura, mesurus, novaehollandiae, raui, ruficollis, selma, tapouaru, vulpina, xanthopus*.
COMMENTS: *Phalangista johnstonii* Ramsay, 1988, may be a distinct species.

Wyulda Alexander, 1918. J. R. Soc. West. Aust. (1917-1918), 4:31.
TYPE SPECIES: *Wyulda squamicaudata* Alexander, 1918.

Wyulda squamicaudata Alexander, 1918. J. R. Soc. West. Aust. (1917-1918), 4:31.
TYPE LOCALITY: Australia, Western Australia, Wyndham.
DISTRIBUTION: NE Western Australia, Kimberleys.
STATUS: U.S. ESA - Endangered.
COMMENTS: Provisionally placed in *Trichosurus* by Flannery et al. (1987).

Family Potoroidae Gray, 1821. London Med. Repos., 15:308.
SYNONYMS: Hypsiprymnodontinae.
COMMENTS: Separated from Macropodidae by Archer and Bartholamai (1978).

Aepyprymnus Garrod, 1875. Proc. Zool. Soc. Lond., 1875:59.
TYPE SPECIES: *Bettongia rufescens* Gray, 1837.

Aepyprymnus rufescens (Gray, 1837). Mag. Nat. Hist. [Charlesworth's], 1:584.
TYPE LOCALITY: Australia, New South Wales.
DISTRIBUTION: Australia: NE Victoria, E New South Wales, E Queensland.
STATUS: Rare, localized.
SYNONYMS: *melanotis*.

Bettongia Gray, 1837. Mag. Nat. Hist. [Charlesworth's], 1:584.
TYPE SPECIES: *Bettongia setosa* Gray, 1837 (= *Kangurus gaimardi* Desmarest, 1822).
SYNONYMS: *Bettongiops*.
COMMENTS: Formerly included in Macropodidae.

Bettongia gaimardi (Desmarest, 1822). Mammalogie. *In* Encycl. Méth., 2(Suppl.):542.
TYPE LOCALITY: Australia, New South Wales, Port Jackson.
DISTRIBUTION: Formerly coastal SE Queensland and N New South Wales, south to SW Victoria; now extinct on mainland Australia; survives in Tasmania.
STATUS: CITES - Appendix I; U.S. ESA - Endangered.
SYNONYMS: *cuniculus, formosus, hunteri, lepturus, minimus, phillippi, setosa, whitei*.
COMMENTS: See Corbet and Hill (1980:16).

Bettongia lesueur (Quoy and Gaimard, 1824). *In* de Freycinet, Voy. autour du monde... l'Uranie et al Physicienne, Zool., p. 64.
TYPE LOCALITY: Australia, Western Australia, Dirk Hartog Island (Shark Bay).
DISTRIBUTION: Formerly in Dampier Land (Western Australia), South Australia, Dirk Hartog Isl, Barrow Isl, Bernier and Dorre Isls, Northern Territory, and SW New South Wales (Australia); now extinct except on W Australian Isls.
STATUS: CITES - Appendix I; U.S. ESA - Endangered; IUCN - Rare.
SYNONYMS: *anhydra, graii, harveyi*.
COMMENTS: Commonly misspelt "*lesueuri*", but the original spelling is *lesueur*, with no indication that it is an error.

Bettongia penicillata Gray, 1837. Mag. Nat. Hist. [Charlesworth's], 1:584.
TYPE LOCALITY: Australia, New South Wales.
DISTRIBUTION: SW Western Australia, S South Australia including St. Francis Isl, NW Victoria, C New South Wales, E Queensland.
STATUS: CITES - Appendix I; U.S. ESA and IUCN - Endangered; *B. p. tropica* may be extinct (see Ride, 1970:199).
SYNONYMS: *francisca, gouldii, ogilbyi, tropica*.
COMMENTS: See Sharman et al. (1980).

Caloprymnus Thomas, 1888. Cat. Marsup. Monotr. Brit. Mus., p. 114.
TYPE SPECIES: *Bettongia campestris* Gould, 1843.

Caloprymnus campestris (Gould, 1843). Proc. Zool. Soc. Lond., 1843:81.
TYPE LOCALITY: Australia, South Australia.
DISTRIBUTION: South Australia/Queensland border country.
STATUS: CITES - Appendix I pe [possibly extinct]; U.S. ESA - Endangered; IUCN - Indeterminate; possibly extinct, see Ride (1970:198).

Hypsiprymnodon Ramsay, 1876. Proc. Linn. Soc. N.S.W., 1:33.
TYPE SPECIES: *Hypsiprymnodon moschatus* Ramsay, 1876.
SYNONYMS: *Pleopus*.
COMMENTS: Formerly included in Macropodidae.

Hypsiprymnodon moschatus Ramsay, 1876. Proc. Linn. Soc. N.S.W., 1:34.
TYPE LOCALITY: Australia, Queensland, Rockingham Bay.
DISTRIBUTION: NE Queensland (Australia).
STATUS: Rare, localized.
SYNONYMS: *nudicaudatus*.

Potorous Desmarest, 1804. Tabl. Méth Hist. Nat., *in* Nouv. Dict. Hist. Nat., 24:20.
TYPE SPECIES: *Didelphis murina* Cuvier, 1798 (= *Didelphis tridactyla* Kerr, 1792).
SYNONYMS: *Hypsiprymnus, Potoroiis, Potoroo, Potoroops*.
COMMENTS: Formerly included in Macropodidae.

Potorous longipes Seebeck and Johnston, 1980. Aust. J. Zool., 28:121.
TYPE LOCALITY: Australia, Victoria, Bellbird Creek, 32 km E. Orbost.
DISTRIBUTION: NE Victoria (Australia).
STATUS: IUCN - Indeterminate; endangered.
COMMENTS: Known from very few specimens; first collected in 1968.

Potorous platyops (Gould, 1844). Proc. Zool. Soc. Lond., 1844:103.
TYPE LOCALITY: Australia, Western Australia, Swan River, L. Walyormouring.
DISTRIBUTION: Formerly SW Western Australia and Kangaroo Isl, South Australia.
STATUS: IUCN - Extinct.
SYNONYMS: *morgani*.
COMMENTS: Possibly extinct (Ride, 1970:199).

Potorous tridactylus (Kerr, 1792). *In* Linnaeus, Anim. Kingdom, 1:198.
TYPE LOCALITY: Australia, New South Wales, Sydney.
DISTRIBUTION: SE Queensland, coastal New South Wales, NE Victoria, SE South Australia, SW Western Australia, Tasmania, and King Island (Australia).
STATUS: Endangered on mainland Australia, common in Tasmania.
SYNONYMS: *apicalis, apicalis, benormi, gilbertii, micropus, minor, murina, muscola, myosurus, peronii, rufus, setosus, trisulcatus, tuckeri*.
COMMENTS: See Kirsch and Calaby (1977:21).

Family Macropodidae Gray, 1821. London Med. Repos., 15:308.
COMMENTS: Revised by Tate (1948*a*).

Dendrolagus Müller, 1840. *In* Temminck, Verh. Nat. Gesch. Nederland Overz. Bezitt., Land-en Volkenkunde, p. 20, footnote[1840].
TYPE SPECIES: *Dendrolagus ursinus* Müller, 1840 (*recte Hypsiprymnus ursinus* Temminck, 1836; designated by Thomas, 1888).

Dendrolagus bennettianus De Vis, 1887. Proc. R. Soc. Queensl., 3(1886):11 [1887].
TYPE LOCALITY: Australia, Queensland, Daintree River.
DISTRIBUTION: NE Queensland (Australia).
STATUS: CITES - Appendix II.
COMMENTS: Considered a subspecies of *dorianus* by Haltenorth (1958); but see Ride (1970:223) and Kirsch and Calaby (1977:17).

Dendrolagus dorianus Ramsay, 1883. Proc. Linn. Soc. N.S.W., 8:17.
TYPE LOCALITY: Papua New Guinea, "ranges behind Mt. Astrolabe."
DISTRIBUTION: Interior New Guinea: Wondiwoi Peninsula, Irian Jaya; Papua New Guinea/ Indonesian border to extreme SE of mainland Papua New Guinea.
STATUS: IUCN - Vulnerable as *D. d. notatus*; uncommon.
SYNONYMS: *aureus, mayri, notatus, palliceps, profugus, stellarum*.
COMMENTS: Does not include *bennettianus* (see Ride, 1970:223, Kirsch and Calaby, 1977:17).

Dendrolagus goodfellowi Thomas, 1908. Ann. Mag. Nat. Hist., ser. 8, 2:452.
TYPE LOCALITY: Papua New Guinea, Owen Stanley Range, vic. Mt. Obree, 8000 ft. (2438 m).
DISTRIBUTION: E New Guinea.
STATUS: IUCN - Vulnerable as *D. g. shawmayeri*; uncommon.
SYNONYMS: *buergersi, pulcherrimus, shawmayeri*.
COMMENTS: Does not include *spadix*. Groves (1982*d*) regarded this species as a subspecies of the earlier named *matschiei*; but see Flannery (1990:100-104), also Ganslosser (1980).

Dendrolagus inustus Müller, 1840. *In* Temminck, Verh. Nat. Gesch. Nederland Overz. Bezitt., Land-en Volkenkunde, p. 20, footnote[1840], see comments.
TYPE LOCALITY: Indonesia, Irian Jaya, Fakfak Div., Lobo Dist., near Triton Bay, Mt. Lamantsjieri.
DISTRIBUTION: N and extreme W New Guinea; Yapen Isl.
STATUS: CITES - Appendix II.
SYNONYMS: *finschi, keiensis, maximus, schoedei, sorongensis*.
COMMENTS: This species was further described by Schlegel and Müller, *in* Temminck, Verh. Nat. Gesch. Nederland. Overz. Bezitt., Zool., Mammalia, p. 131, 143[1845], pl. 20, 22, 23[1841]. Considered a subspecies of *ursinus* by Haltenorth (1958); but see Kirsch and Calaby (1977:17) and Groves (1982*d*).

Dendrolagus lumholtzi Collett, 1884. Proc. Zool. Soc. Lond., 1884:387.
TYPE LOCALITY: Australia, Queensland, Herbert Vale.

DISTRIBUTION: NE Queensland (Australia).
STATUS: CITES - Appendix II.
SYNONYMS: *fulvus*.

Dendrolagus matschiei Forster and Rothschild, 1907. Nov. Zool., 14:506.
TYPE LOCALITY: Papua New Guinea, Morobe Prov., Rawlinson Mtns.
DISTRIBUTION: Extreme NE interior New Guinea (Huon Peninsula); Umboi Isl (Introduced).
STATUS: Uncommon.
SYNONYMS: *deltae, flavidior, xanthotis*.
COMMENTS: See Kirsch and Calaby (1977:21) and Lidicker and Ziegler (1968). See also comments under *goodfellowi*.

Dendrolagus scottae Flannery and Seri, 1990. Rec. Aust. Mus., 42:237.
TYPE LOCALITY: Papua New Guinea, West Sepik Prov., Torricelli Mtns, Sweipini, 1400 m. (3°23'S, 142°06'E).
DISTRIBUTION: Torricelli Mtns only (Papua New Guinea).
STATUS: Endangered.
COMMENTS: May be a subspecies of *D. dorianus*.

Dendrolagus spadix Troughton and Le Souef, 1936. Aust. Zool., 8:194.
TYPE LOCALITY: Papua, Western Division, between Bamu, upper Awarra and Strickland Rivers.
DISTRIBUTION: S New Guinea.
STATUS: Rare.
COMMENTS: A subspecies of *D. matschiei* according to Groves (1982*d*), but made a full species by Flannery (1990).

Dendrolagus ursinus (Temminck, 1836). Discours preliminaire destine a servir d'introduction al faune du Japon, p. 6 (footnote 2).
TYPE LOCALITY: Indonesia, Irian Jaya, Fakfak Div., Lobo Dist., near Triton Bay, Mt. Lamantsjieri.
DISTRIBUTION: Extreme NW New Guinea.
STATUS: CITES - Appendix II.
SYNONYMS: *leucogenys*.
COMMENTS: Does not include *inustus* (see Kirsch and Calaby, 1977:17). Correct original citation presented by Husson (1955).

Dorcopsis Schlegel and Müller, 1845. *In* Temminck, Verh. Nat. Gesch. Nederland Overz. Bezitt., Zool., p. 130[1845].
TYPE SPECIES: *Didelphis brunii* Quoy and Gaimard, 1830 (= *Macropus muelleri* Lesson, 1827).
COMMENTS: Does not include *Dorcopsulus* (see Flannery, 1990:89-92). The name *Conoyces* Lesson, 1842 was used for this genus by Troughton (1937), but Tate (1948*a*) showed that the type species of *Conoyces* is *Didelphis brunii* Gmelin, 1788 [= *Thylogale brunii* (Schreber, 1778)]; see under *Thylogale*.

Dorcopsis atrata Van Deusen, 1957. Am. Mus. Novit., 1826:5.
TYPE LOCALITY: Papua New Guinea, Milne Bay Prov., Goodenough Island, eastern slopes, near "Top Camp", about 1600 m.
DISTRIBUTION: Goodenough Isl (Papua New Guinea).
STATUS: IUCN - Rare; very rare, localized, probably endangered.

Dorcopsis hageni Heller, 1897. Abh. Zool. Anthrop.-Ethnology. Mus. Dresden, 6(8):7.
TYPE LOCALITY: Papua New Guinea, Madang Prov., near Astrolabe Bay, Stefansort.
DISTRIBUTION: NC New Guinea.
STATUS: Common.
SYNONYMS: *caurina, eitape*.

Dorcopsis luctuosa (D'Albertis, 1874). Proc. Zool. Soc. Lond., 1874:110.
TYPE LOCALITY: "Southeast of New Guinea".
DISTRIBUTION: S New Guinea.
STATUS: Common.
SYNONYMS: *beccarii, chalmersi, phyllis*.

COMMENTS: Usually included in *D. veterum* (= *D. muelleri*), but see Groves and Flannery (1989).

Dorcopsis muelleri (Lesson, 1827). *In* Duperry (Lesson and Garnot, eds.), Voy. autour du Monde...la Coquille, Zool., 1:164.

TYPE LOCALITY: Indonesia, Irian Jaya, Vogelkop, Manokwari Div., Dorei (= Manokwari), Lobo Bay.

DISTRIBUTION: W New Guinea; Misool and Salawati Isls, Aru Isls, and Yapen Isl (Indonesia).

STATUS: Common.

SYNONYMS: *brunii, lorentzii, mysoliae, rufolateralis, veterum, yapeni.*

COMMENTS: *D. muelleri* was regarded as a junior synonym of *D. veterum* by Kirsch and Calaby (1977:21) and Husson (1955:299). George and Schuerer (1978) rejected *veterum* as based on a *Dendrolagus* (probably *inustus*), and employed *muelleri;* Groves and Flannery (1989) agreed. The original name *Didelphis brunii* Quoy and Gaimard, was preoccupied and is now *Macropus muelleri.*

Dorcopsulus Matschie, 1916. Sitzb. Ges. Naturf. Fr., Berlin, 57.

TYPE SPECIES: *Dorcopsis macleayi* Miklouho-Maclay, 1885.

COMMENTS: Formerly included in *Dorcopsis* but revived as a full genus by Flannery (1990).

Dorcopsulus macleayi (Miklouho-Maclay, 1885). Proc. Linn. Soc. N.S.W., 10:145, 149.

TYPE LOCALITY: Papua New Guinea, Central Prov., "inland from Port Moresby".

DISTRIBUTION: Extreme SE New Guinea.

STATUS: IUCN - Rare.

Dorcopsulus vanheurni (Thomas, 1922). Ann. Mag. Nat. Hist., ser. 9, 9:264.

TYPE LOCALITY: Indonesia, Irian Jaya, Djajawidjaja Div., Doormanpad-bivak (3°30'S, 138°30'E), 1410 m.

DISTRIBUTION: Interior New Guinea.

STATUS: Common.

SYNONYMS: *rothschildi.*

COMMENTS: Regarded as conspecific with *macleayi* by Kirsch and Calaby (1977:21).

Lagorchestes Gould, 1841. Monogr. Macropodidae, pt. 1, pl. 12 (text).

TYPE SPECIES: *Macropus leporides* Gould, 1841.

SYNONYMS: *Lagocheles* (*nomen nudum*).

COMMENTS: This genus is probably polyphyletic.

Lagorchestes asomatus Finlayson, 1943. Trans. R. Soc. South Aust., 67:319.

TYPE LOCALITY: Australia, Northern Territory, between Mt. Farewell and Lake Mackay.

DISTRIBUTION: Known only from the type locality.

STATUS: IUCN - Extinct; probably extinct.

COMMENTS: Known from a single unsexed skull (Kirsch and Calaby, 1977:22).

Lagorchestes conspicillatus Gould, 1842. Proc. Zool. Soc. Lond., 1841:82 [1842].

TYPE LOCALITY: Australia, Western Australia, Barrow Island.

DISTRIBUTION: N Western Australia and adjacent islands, N Northern Territory, N and W Queensland (Australia).

STATUS: Common locally.

SYNONYMS: *pallidior, leichardti.*

COMMENTS: May consist of two or three distinct species.

Lagorchestes hirsutus Gould, 1844. Proc. Zool. Soc. Lond., 1844:32.

TYPE LOCALITY: Australia, Western Australia, York district.

DISTRIBUTION: C Western Australia, C Australia, Dorre Isl and Bernier Isl (Western Australia). Survives only on Bernier and Dorre Isls, and a tiny area NW of Alice Springs (Northern Territory, Australia).

STATUS: CITES - Appendix I; U.S. ESA - Endangered; IUCN - Rare.

SYNONYMS: *bernieri, dorreae.*

Lagorchestes leporides (Gould, 1841). Proc. Zool. Soc. Lond., 1840:93 [1841].

TYPE LOCALITY: Australia, interior New South Wales.

DISTRIBUTION: Formerly W New South Wales, E South Australia, NW Victoria.

STATUS: IUCN - Extinct.
COMMENTS: Almost certainly extinct; not recorded for more than a century (Kirsch and Calaby, 1977:22).

Lagostrophus Thomas, 1887. Proc. Zool. Soc. Lond., 1886:544 [1887].
TYPE SPECIES: *Kangurus fasciatus* Péron and lesueur, 1807.
COMMENTS: Placed in subfamily Sthenurinae by Flannery (1983). The other Sthenurinae are giant fossil kangaroos.

Lagostrophus fasciatus (Péron and Lesueur, 1807). *In* Péron, Voy. Decouv. Terres. Austral., Atlas, pl. 27, 1:114.
TYPE LOCALITY: Australia, Western Australia, Bernier Island (Shark Bay).
DISTRIBUTION: Survives only on Bernier and Dorre Isls (Western Australia); formerly in SW Western Australia, perhaps South Australia.
STATUS: CITES - Appendix I; U.S. ESA - Endangered; IUCN - Rare.
SYNONYMS: *albipilis, elegans, striatus.*

Macropus Shaw, 1790. Nat. Misc., 1, pl. 23 (text).
TYPE SPECIES: *Macropus giganteus* Shaw, 1790.
SYNONYMS: *Boriogale, Dendrodorcopsis, Gerboides, Gigantomys, Halmatopus, Halmaturus, Kangurus, Megaleia, Notamacropus, Osphranter, Phascolagus, Prionotemmus.*
COMMENTS: Includes *Megaleia* and *Protemnodon* (*sensu* Haltenorth, 1958); see Kirsch and Calaby (1977:17). Rationale for present usage of *Macropus* given by Calaby (1966). Ride (1962) discussed generic nomenclature for all Macropodinae. Van Gelder (1977*b*) included *Thylogale* and *Wallabia* in this genus, but see Kirsch and Calaby (1977:17) and Corbet and Hill (1980:17-18). Dawson and Flannery (1985) divided this genus into three subgenera: *Macropus, Notamacropus* and *Osphranter.*

Macropus agilis (Gould, 1842). Proc. Zool. Soc. Lond., 1841:81 [1842].
TYPE LOCALITY: Australia, Northern Territory, Port Essington.
DISTRIBUTION: NE Western Australia, Northern Territory, Queensland; S New Guinea; Kiriwina Isls and other islands off the SE coast of New Guinea.
STATUS: Common.
SYNONYMS: *aurantiacus, aurescens, binoe, crassipes, jardinii, nigrescens, papuanus, siva.*
COMMENTS: Subgenus *Notamacropus.*

Macropus antilopinus (Gould, 1842). Proc. Zool. Soc. Lond., 1841:80 [1842].
TYPE LOCALITY: Australia, Northern Territory, Port Essington.
DISTRIBUTION: N Queensland, Northern Territory, NE Western Australia.
STATUS: Common.
COMMENTS: Subgenus *Osphranter.*

Macropus bernardus Rothschild, 1904. Nov. Zool., 10:543.
TYPE LOCALITY: Australia, Northern Territory, head of South Alligator River.
DISTRIBUTION: Interior of N Northern Territory.
STATUS: Rare.
SYNONYMS: *woodwardi.*
COMMENTS: Subgenus *Osphranter.* The original name *Dendrodorcopsis woodwardi* was preoccupied.

Macropus dorsalis (Gray, 1837). Mag. Nat. Hist. [Charlesworth's], 1:583.
TYPE LOCALITY: Australia, New South Wales, probably interior (Namoi Hills), according to Iredale and Troughton (1934).
DISTRIBUTION: Australia: E Queensland, E New South Wales.
STATUS: Common.
COMMENTS: Subgenus *Notamacropus.*

Macropus eugenii (Desmarest, 1817). Nouv. Dict. Hist. Nat., Nouv. ed., 17:38.
TYPE LOCALITY: Australia, South Australia, Nuyt's Arch., St. Peter's Isl.
DISTRIBUTION: SW Western Australia, South Australia, Kangaroo Isl, Wallaby Isl and other islands.
SYNONYMS: *bedfordi, dama, decres, derbyanus, emiliae, gracilis, houtmanni, obscurior, flindersi.*
COMMENTS: Subgenus *Notamacropus.*

Macropus fuliginosus (Desmarest, 1817). Nouv. Dict. Hist. Nat., Nouv. ed., 17:35.
TYPE LOCALITY: Australia, South Australia, Kangaroo Island.
DISTRIBUTION: SW New South Wales, NW Victoria, South Australia, SW Western Australia, Tasmania, King Isl, and Kangaroo Isl (Australia).
STATUS: Abundant, yet U.S. ESA - Threatened.
SYNONYMS: *melanops, ocydromus.*
COMMENTS: Subgenus *Macropus;* see Kirsch and Poole (1972) for discussion of specific limits and subspecies included in this taxon and in *giganteus.*

Macropus giganteus Shaw, 1790. Nat. Misc., 1, pl. 33 (text).
TYPE LOCALITY: Australia, Queensland, Cooktown (= "New Holland"), King's Plains.
DISTRIBUTION: E and C Queensland, Victoria, New South Wales, SE South Australia, and Tasmania (Australia).
STATUS: U.S. ESA - Endangered as *M. g. tasmaniensis;* otherwise Threatened; yet abundant throughout eastern Australia.
SYNONYMS: *griseofuscus, labiatus, major, tasmaniensis, tridactylus.*
COMMENTS: Subgenus *Macropus.* Opinion 760 of the International Commission on Zoological Nomenclature (1966) placed this name on the Official List of Specific Names in Zoology, see Calaby et al. (1963) for discussion. Revised by Kirsch and Poole (1972) who discussed specific limits and the subspecies included in this taxon. See Poole (1982, Mammalian Species, 187).

Macropus greyi Waterhouse, 1846. Nat. Hist. Mamm., 1:122.
TYPE LOCALITY: Australia, South Australia, Coorong.
DISTRIBUTION: Formerly SE South Australia and adjacent Victoria.
STATUS: IUCN - Extinct.
COMMENTS: Subgenus *Notamacropus.* Almost certainly extinct (Kirsch and Calaby, 1977:22; Ride, 1970:47).

Macropus irma (Jourdan, 1837). C. R. Acad. Sci. Paris, 5:523.
TYPE LOCALITY: Australia, Western Australia, Swan River.
DISTRIBUTION: SW Western Australia.
STATUS: Rare.
SYNONYMS: *manicatus, melanopus.*
COMMENTS: Subgenus *Notamacropus.*

Macropus parma Waterhouse, 1846. Nat. Hist. Mamm., 1:149.
TYPE LOCALITY: Australia, New South Wales.
DISTRIBUTION: E New South Wales; introduced to Kawau Isl (New Zealand), see Wodzicki and Flux (1967).
STATUS: U.S. ESA - Endangered.
COMMENTS: Subgenus *Notamacropus* (Dawson and Flannery, 1985).

Macropus parryi Bennett, 1835. Proc. Zool. Soc. Lond., 1834:151 [1835].
TYPE LOCALITY: Australia, New South Wales, Stroud (near Port Stephens).
DISTRIBUTION: Australia: E Queensland, NE New South Wales.
STATUS: Common.
SYNONYMS: *pallida.*
COMMENTS: Subgenus *Notamacropus* (Dawson and Flannery, 1985). Formerly included in *Protemnodon,* see Haltenorth (1958:39); but also see Kirsch and Calaby (1977).

Macropus robustus Gould, 1841. Proc. Zool. Soc. Lond., 1840:92 [1841].
TYPE LOCALITY: Australia, New South Wales, interior (summit of mountains).
DISTRIBUTION: Australia: Western Australia, South Australia, S Northern Territory, Queensland, New South Wales, Barrow Isl.
STATUS: Abundant.
SYNONYMS: *alligatoris, alexandriae, argentatus, bracteator, cervinus, erubescens, hagenbecki, isabellinus, magnus, reginae, rubens, woodwardi.*
COMMENTS: Subgenus *Osphranter.* See Richardson and Sharman (1976). McAllan and Bruce (1989) would date *robustus* from: The Athenaeum, 670:685 [29 August 1840].

Macropus rufogriseus (Desmarest, 1817). Nouv. Dict. Hist. Nat., Nouv. ed., 17:36.
TYPE LOCALITY: Australia, Tasmania, King Island.

DISTRIBUTION: SE South Australia, Victoria, SE Queensland, E New South Wales, Tasmania, King Isl and adjacent islands (Australia); introduced in England (Corbet and Hill, 1980:18).
STATUS: Common.
SYNONYMS: *banksianus, bennetti, fruticus, griseus, leptonyx, ruficollis, rutilus, vinosus.*
COMMENTS: Subgenus *Notamacropus.*

Macropus rufus (Desmarest, 1822). Mammalogis. *In* Encycl. Méth., 2(Suppl.):541.
TYPE LOCALITY: Australia, New South Wales, Blue Mtns.
DISTRIBUTION: Mainland, mid-latitude Australia.
STATUS: U.S. ESA - Threatened; yet very abundant.
SYNONYMS: *dissimulatus, griseolanosus, lanigerus, occidentalis, pallidus, pictus, ruber.*
COMMENTS: Subgenus *Osphranter.*

Onychogalea Gray, 1841. Appendix C *In* J. Two Exped. Aust., 2:402.
TYPE SPECIES: *Macropus unguifer* Gould, 1841.

Onychogalea fraenata (Gould, 1841). Proc. Zool. Soc. Lond., 1840:92 [1841].
TYPE LOCALITY: Australia, New South Wales, interior.
DISTRIBUTION: Formerly in S Queensland, interior New South Wales; survives only near Taunton, Queensland (Australia).
STATUS: CITES - Appendix I; U.S. ESA and IUCN - Endangered.
COMMENTS: McAllan and Bruce (1989) argued that the original publication of this name was in The Athenaeum, 670:685 [29 August 1840], as [*Macropus*] *frenatus.*

Onychogalea lunata (Gould, 1841). Proc. Zool. Soc. Lond., 1840:93 [1841].
TYPE LOCALITY: Australia, Western Australia, coast.
DISTRIBUTION: SC and SW Western Australia, S Northern Territory.
STATUS: CITES - Appendix I; U.S. ESA - Endangered; IUCN - Extinct; probably extinct.
COMMENTS: Extinct throughout most or all of its former range. McAllan and Bruce (1989) argued that the original publication of this name was in The Athenaeum, 670:685 [29 August 1840].

Onychogalea unguifera (Gould, 1841). Proc. Zool. Soc. Lond., 1840:93 [1841].
TYPE LOCALITY: Australia, Western Australia, Derby (King Sound).
DISTRIBUTION: N Australia: Western Australia, Northern Territory, Queensland.
STATUS: Secure.
SYNONYMS: *annulicauda.*
COMMENTS: McAllan and Bruce (1989) argued that the original description of this name was in The Athenaeum, 670:685 [29 August 1840].

Petrogale Gray, 1837. Mag. Nat. Hist. [Charlesworth's], 1:583.
TYPE SPECIES: *Kangurus penicillatus* Gray, 1827.
SYNONYMS: *Heteropus, Peradorcas.*
COMMENTS: Revision of this genus is underway by Sharman, Eldridge et al.; a preliminary account of their arrangement was provided by Poole (1979) and Briscoe et al. (1982). Kitchener and Sanson (1978) considered this genus as probably congeneric with *Peradorcas.*

Petrogale assimilis Ramsay, 1877. Proc. Linn. Soc. N.S.W., 1:360.
TYPE LOCALITY: Australia, Queensland, Palm Isl.
DISTRIBUTION: Queensland.
STATUS: Locally common.
SYNONYMS: *puella.*

Petrogale brachyotis (Gould, 1841). Proc. Zool. Soc. Lond., 1840:128 [1841].
TYPE LOCALITY: Australia, Western Australia, Hanover Bay.
DISTRIBUTION: Coast of NW Australia, N Northern Territory.
STATUS: Common.
SYNONYMS: *longmani, signata, venustula, wilkinsi.*

Petrogale burbidgei Kitchener and Sanson, 1978. Rec. W. Aust. Mus., 6:269-285.
TYPE LOCALITY: Australia, Western Australia, Mitchell Plateau, Crystal Creek (14°30'S, 125°47'20"E).
DISTRIBUTION: Kimberleys (Western Australia), Bonaparte Arch., and adjacent islands.
STATUS: Rare.

Petrogale concinna Gould, 1842. Proc. Zool. Soc. Lond., 1842:57.
TYPE LOCALITY: Australia, Western Australia, Wyndham.
DISTRIBUTION: Australia: NE and NW Northern Territory, NE Western Australia.
SYNONYMS: *canescens, monastria.*
COMMENTS: Formerly included in a separate genus *Peradorcas*, see Kitchener and Samson (1978).

Petrogale godmani Thomas, 1923. Abstr. Proc. Zool. Soc. Lond., 1923(235):13.
TYPE LOCALITY: Australia, Queensland, Cooktown (Black Mtn).
DISTRIBUTION: Cape York Peninsula, N Queensland.
STATUS: Threatened by genetic introgression from *P. assimilis.*
COMMENTS: Included in *penicillata* in a preliminary account by Poole (1979:21). According to Eldridge et al. (1989) this species is composite: *P. godmani* from the region between Cooktown and Mareeba, and an undescribed species from the central Cape York Peninsula.

Petrogale inornata Gould, 1842. Monogr. Macropodidae, pt. 2, pl. 25.
TYPE LOCALITY: Australia, Queensland, Cape Upstart.
DISTRIBUTION: N Queensland (Australia).
STATUS: Localized.

Petrogale lateralis Gould, 1842. Monogr. Macropodidae, pt. 2, pl. 24.
TYPE LOCALITY: Australia, Western Australia, Swan River.
DISTRIBUTION: Australia: Western Australia, South Australia, Northern Territory, W Queensland.
STATUS: Common.
SYNONYMS: *hacketti, pearsoni, purpureicollis.*
COMMENTS: McAllan and Bruce (1989) argued that the original publication of this name is in The Athenaeum, 670:685 [29 August 1840].

Petrogale penicillata (Gray, 1827). *In* Griffith et al., Anim. Kingdom, Mamm., 3, plate only.
TYPE LOCALITY: Australia, New South Wales, Sydney.
DISTRIBUTION: E Australia.
STATUS: Common.
SYNONYMS: *albogularis, herberti, longicauda.*

Petrogale persephone Maynes, 1982. Aust. Mamm., 5:47.
TYPE LOCALITY: Australia, Queensland, 9.6 km N of Proserpine, base of Mt. Dryander, 20°19'S, 148°33'E.
DISTRIBUTION: Restricted to district around Proserpine.
STATUS: IUCN - Rare; vulnerable.

Petrogale rothschildi Thomas, 1904. Nov. Zool., 11:166.
TYPE LOCALITY: Australia, Western Australia, Cossack.
DISTRIBUTION: NW Western Australia.
STATUS: Vulnerable.

Petrogale xanthopus Gray, 1855. Proc. Zool. Soc. Lond., 1854:259 [1855].
TYPE LOCALITY: Australia, South Australia, Flinders Range.
DISTRIBUTION: Australia: SW Queensland, South Australia, NW New South Wales.
STATUS: U.S. ESA - Endangered.
SYNONYMS: *celeris, xanthopygus.*

Setonix Lesson, 1842. Nouv. Tabl. Regn. Anim. Mammifères, p. 194.
TYPE SPECIES: *Kangurus brachyurus* Quoy and Gaimard, 1830.

Setonix brachyurus (Quoy and Gaimard, 1830). *In* Dumont d'Urville, Voy...de Astrolabe, Zool., 1(L'Homme, Mamm. Oiseaux):114.
TYPE LOCALITY: Australia, Western Australia, King George Sound (Albany).
DISTRIBUTION: SW Western Australia, Rottnest Isl, and Bald Isl (Australia).
STATUS: U.S. ESA - Endangered.
SYNONYMS: *brevicaudatus.*

Thylogale Gray, 1837. Mag. Nat. Hist. [Charlesworth's], 1:583.
TYPE SPECIES: *Halmaturus (Thylogale) eugenii* Gray, 1837 (= *Halmaturus thetis* Lesson, 1828).
SYNONYMS: *Conoyces.*
COMMENTS: Included in *Macropus* by Van Gelder (1977*b*), but see Kirsch and Calaby (1977:17). See comments under *Dorcopsis.*

Thylogale billardierii (Desmarest, 1822). Mammalogie. *In* Encycl. Méth., 2(Suppl.):542.
TYPE LOCALITY: Australia, Tasmania.
DISTRIBUTION: Australia: SE South Australia, Victoria, Tasmania, islands in Bass Strait; probably survives only in Tasmania.
SYNONYMS: *brachytarsus, rufiventer, tasmanei.*

Thylogale brunii (Schreber, 1778). Die Säugethiere, 3:551.
TYPE LOCALITY: Indonesia, Aru Isls.
DISTRIBUTION: C and E New Guinea and adjacent small islands; Bismarck Arch. (Papua New Guinea); Aru Isls.
STATUS: Uncommon.
SYNONYMS: *browni, gracilis, jukesii, keysseri, lauterbachi, lugens, lanatus, tibol.*
COMMENTS: *T. bruijni* is a later spelling (Haltenorth, 1958:38).

Thylogale stigmatica (Gould, 1860). Mamm. Aust., 2, pt. 12, pl. 33-34.
TYPE LOCALITY: Australia, Queensland, Point Cooper (N of Rockingham Bay).
DISTRIBUTION: E Queensland, E New South Wales (Australia); SC New Guinea.
STATUS: Uncommon.
SYNONYMS: *coxenii, gazella, oriomos, temporalis, wilcoxi.*
COMMENTS: Citation for original description given as Proc. Zool. Soc. Lond., 1860:375, by some authors, but this is dated Nov. 13, while Mammal. Aust., Part 12 was published Nov. 1.

Thylogale thetis (Lesson, 1828). Monogr. Mamm., p. 229.
TYPE LOCALITY: Australia, New South Wales, Sydney.
DISTRIBUTION: E Queensland, E New South Wales (Australia).
STATUS: Uncommon.
SYNONYMS: *eugenii, nuchalis.*

Wallabia Trouessart, 1905. Cat. Mamm. Viv. Foss., Suppl. fasc., 4:834.
TYPE SPECIES: *Kangurus ualabatus* Lesson and Garnot, 1826 (= *Kangurus bicolor* Desmarest, 1804).
COMMENTS: Included in *Macropus* by Van Gelder (1977*b*), but see Kirsch and Calaby (1977:17).

Wallabia bicolor (Desmarest, 1804). Tabl. Méth Hist. Nat., *in* Nouv. Dict. Hist. Nat., 24:357.
TYPE LOCALITY: Unknown.
DISTRIBUTION: Australia: E Queensland, E New South Wales, Victoria, SE South Australia, Stradbroke Isl, Fraser Isl.
STATUS: Common.
SYNONYMS: *apicalis, ingrami, lessonii, mastersii, ualabatus, welsbyi.*

Family Burramyidae Broom, 1898. Proc. Linn. Soc. N.S.W., 10:564.

Burramys Broom, 1896. Proc. Linn. Soc. N.S.W., 10:564.
TYPE SPECIES: *Burramys parvus* Broom, 1896.

Burramys parvus Broom, 1896. Proc. Linn. Soc. N.S.W., 10:564 [fig. in pl. 25, p. 273].
TYPE LOCALITY: Australia, New South Wales, Taralga (fossil).

DISTRIBUTION: Mountains of NE Victoria and S New South Wales (Australia).
STATUS: CITES - Appendix II; U.S. ESA - Endangered.

Cercartetus Gloger, 1841. Gemein Hand.-Hilfsbuch. Nat., 1:85.
TYPE SPECIES: *Phalangista nana* Desmarest, 1818.
SYNONYMS: *Dromicia, Dromiciella, Dromiciola, Eudromicia.*
COMMENTS: Includes *Eudromicia* (see Kirsch and Calaby, 1977:16).

Cercartetus caudatus (Milne-Edwards, 1877). C. R. Acad. Sci. Paris, 85:1079.
TYPE LOCALITY: Indonesia, Irian Jaya, Vogelkop, Manokwari Div., Arfak Mtns.
DISTRIBUTION: Interior New Guinea; Fergusson Isl (Papua New Guinea); NE Queensland (Australia).
STATUS: Common.
SYNONYMS: *macrura.*
COMMENTS: Includes *macrura* (see Ride, 1970:224). Formerly included in *Eudromicia*; see comment under genus.

Cercartetus concinnus (Gould, 1845). Proc. Zool. Soc. Lond., 1845:2.
TYPE LOCALITY: Australia, Western Australia, Swan River.
DISTRIBUTION: SW Western Australia, S and SE South Australia, W Victoria, SW New South Wales.
STATUS: Common.
SYNONYMS: *minor, neillii.*

Cercartetus lepidus (Thomas, 1888). Cat. Marsup. Monotr. Brit. Mus., p. 142.
TYPE LOCALITY: Australia, Tasmania.
DISTRIBUTION: Australia: Tasmania, NW Victoria/South Australia border, and Kangaroo Island (South Australia).
STATUS: Common.
COMMENTS: Formerly included in *Eudromicia*; see comment under genus.

Cercartetus nanus (Desmarest, 1818). Nouv. Dict. Hist. Nat., Nouv. ed., 25:477.
TYPE LOCALITY: Australia, Tasmania, Ile Maria.
DISTRIBUTION: Australia: SE South Australia, E New South Wales, Victoria, and Tasmania.
STATUS: Common.
SYNONYMS: *britta, gliriformis, unicolor.*

Family Pseudocheiridae Winge, 1893. Med. Udsigt over Pungdyrenes Slaegtskab. E. Mus. Lundii, 11. pt. 2:89.
COMMENTS: Separated from Petauridae by Archer (1984).

Hemibelideus Collett, 1884. Proc. Zool. Soc. London, 1884:385.
TYPE SPECIES: *Phalangista (Hemibelideus) lemuroides* Collett, 1884.
COMMENTS: Formerly included in *Pseudocheirus.*

Hemibelideus lemuroides (Collett, 1884). Proc. Zool. Soc. Lond., 1884:385.
TYPE LOCALITY: Australia, North Queensland.
DISTRIBUTION: NE Queensland (Australia).
STATUS: Localized but not endangered where it occurs.
SYNONYMS: *cervinus.*

Petauroides Thomas, 1888. Cat. Marsup. Monotr. Brit. Mus., p. 163.
TYPE SPECIES: *Didelphis volans* Kerr, 1792.
SYNONYMS: *Petaurista, Schoinobates, Volucella.*
COMMENTS: Formerly *Schoinobates* Lesson, 1842. This name was used by Lesson only for *Petaurista leucogenys* Temminck, 1823, a giant flying squirrel (McKay, 1982). The names *Volucella* and *Petaurista* were both preoccupied.

Petauroides volans (Kerr, 1792) *In* Linnaeus, Anim. Kingdom, 1:199.
TYPE LOCALITY: Australia, New South Wales, Sydney.
DISTRIBUTION: E Australia, from Dandenong Ranges (Victoria) to Rockhampton (Queensland).

STATUS: Common.
SYNONYMS: *armillatus, cinereus, didelphoides, incanus, macroura, maximus, minor, peronii, taguanoides, volucella.*

Petropseudes Thomas, 1923. Ann. Mag. Nat. Hist., ser. 9, 11:250.
TYPE SPECIES: *Pseudochirus dahli* Collett, 1895.
COMMENTS: Separated from *Pseudocheirus* by McKay (1988:94).

Petropseudes dahli (Collett, 1895). Zool. Anz., 18(490):464.
TYPE LOCALITY: Australia, Northern Territory, Mary River.
DISTRIBUTION: N Northern Territory, NW Western Australia.
STATUS: Localized.

Pseudocheirus Ogilby, 1837. Mag. Nat. Hist. (Charlesworth), 1:457.
TYPE SPECIES: *Phalangista cookii* Desmarest, 1818 (= *Didelphis peregrinus* Boddaert, 1785), see Thomas (1888).
SYNONYMS: *Hepoona, Pseudochirulus, Pseudochirus, Ptenos.*
COMMENTS: *Pseudochirus* Ogilby, 1836. Proc. Zool. Soc. Lond., 1836:26 is a *nomen nudum*, see Palmer (1904).

Pseudocheirus canescens (Waterhouse, 1846). Nat. Hist. Mamm., 1:306.
TYPE LOCALITY: Indonesia, Irian Jaya, Fakfak Div., Triton Bay.
DISTRIBUTION: New Guinea and Salawati Isl.
STATUS: Uncommon.
SYNONYMS: *avarus, bernsteini, dammermani, grisescenti, gyrator.*

Pseudocheirus caroli (Thomas, 1921). Ann. Mag. Nat. Hist., ser. 9, 8:357.
TYPE LOCALITY: Indonesia, Irian Jaya, Paniai Div., Weyland Range, Menoo Valley, Mt. Kunupi, 1830 m.
DISTRIBUTION: WC New Guinea.
STATUS: Rare.
SYNONYMS: *versteagi.*

Pseudocheirus forbesi (Thomas, 1887). Ann. Mag. Nat. Hist., ser. 5, 19:146.
TYPE LOCALITY: Papua New Guinea, Central Prov., Astrolabe Range, near Port Moresby, Sogeri, 458 m.
DISTRIBUTION: Interior E New Guinea.
STATUS: Common.
SYNONYMS: *barbatus, capistratus, larvatus, lewisi, longipilis.*

Pseudocheirus herbertensis (Collett, 1884). Proc. Zool. Soc. Lond., 1884:383.
TYPE LOCALITY: Australia, Queensland, Herbert Vale.
DISTRIBUTION: NE Queensland (Australia).
STATUS: Localized but not uncommon.
SYNONYMS: *cinereus, colletti, mongan.*
COMMENTS: *P. h. cinereus* may be a distinct species.

Pseudocheirus mayeri (Rothschild and Dollman, 1932). Abstr. Proc. Zool. Soc. Lond., 1932(353):15.
TYPE LOCALITY: Indonesia, Irian Jaya, Paniai Div., Weyland Range, Gebroeders Mtns, 1830 m.
DISTRIBUTION: C interior New Guinea.
STATUS: Common.
SYNONYMS: *pygmaeus.*
COMMENTS: See Laurie and Hill (1954:21).

Pseudocheirus peregrinus (Boddaert, 1785). Elench. Anim., p. 78.
TYPE LOCALITY: Australia, Queensland, Endeavour River.
DISTRIBUTION: Australia: Cape York Peninsula (Queensland) to SE South Australia and SW Western Australia, Tasmania, islands of the Bass Straits.
STATUS: Common.

SYNONYMS: *banksii, bassianus, caudivolvula, convolutor, cookii, incana, incanens, laniginosa, modestus, notialis, novaehollanidiae, occidentalis, oralis, pulcher, rubidus, victoriae, viverrina.*

COMMENTS: Includes *laniginosus, cookii* (= *convolutor*), *victoriae, rubidus,* and *occidentalis;* see Ride, (1970:246).

Pseudocheirus schlegeli (Jentink, 1884). Notes Leyden Mus., 6:110.

TYPE LOCALITY: Indonesia, Irian Jaya, Vogelkop, Manokwari Div., Arfak Mtns.

DISTRIBUTION: Extreme NW New Guinea.

STATUS: Rare.

Pseudochirops Matschie, 1915. Sitzb. Ges. Naturf. Fr. Berlin, 4:86.

TYPE SPECIES: *Phalangista (Pseudochirus) albertisii* Peters, 1874.

COMMENTS: Separated from *Pseudocheirus* by McKay (1988).

Pseudochirops albertisii (Peters, 1874). Ann. Mus. Civ. Stor. Nat. Genova, 6:303.

TYPE LOCALITY: Indonesia, Irian Jaya, Vogelkop, Manokwari Div., Arfak Mtns, Hatam, 1520 m.

DISTRIBUTION: N and W New Guinea, including Yapen Isl (Indonesia).

STATUS: Uncommon.

SYNONYMS: *coronatus, insularis, paradoxus, schultzei.*

Pseudochirops archeri (Collett, 1884). Proc. Zool. Soc. Lond., 1884:381.

TYPE LOCALITY: Australia, N Queensland, Herbert River District.

DISTRIBUTION: NE Queensland (Australia).

STATUS: Locally common, but restricted.

Pseudochirops corinnae (Thomas, 1897). Ann. Mus. Civ. Stor. Nat. Genova, 18:142.

TYPE LOCALITY: Papua New Guinea, Central Prov., upper Vanapa River.

DISTRIBUTION: Interior New Guinea.

STATUS: Common.

SYNONYMS: *argenteus, buergersi, caecias, fuscus.*

Pseudochirops cupreus (Thomas, 1897). Ann. Mus. Civ. Stor. Nat. Genova, 18:145.

TYPE LOCALITY: Papua New Guinea, Owen Stanley Range.

DISTRIBUTION: Interior New Guinea.

STATUS: Common.

SYNONYMS: *beauforti, obscurior.*

Family Petauridae Bonaparte, 1838. Nuovi Ann. Sci. Nat., 2(1):112.

Dactylopsila Gray, 1858. Proc. Zool. Soc. Lond., 1858:109.

TYPE SPECIES: *Dactylopsila trivirgata* Gray, 1858.

SYNONYMS: *Dactylonax.*

COMMENTS: See Haltenorth (1958:28).

Dactylopsila megalura Rothschild and Dollman, 1932. Abstr. Proc. Zool. Soc. Lond., 1932(353):14.

TYPE LOCALITY: Indonesia, Irian Jaya, Paniai Div., Weyland Range, Gebroeders Mtns.

DISTRIBUTION: Interior New Guinea.

STATUS: Rare.

COMMENTS: Considered a subspecies of *trivirgata* by Ziegler (*in* Stonehouse and Gilmore, 1977:131).

Dactylopsila palpator Milne-Edwards, 1888. Mem. Cent. Soc. Philom. Paris, p. 174.

TYPE LOCALITY: "South coast of New Guinea".

DISTRIBUTION: Interior New Guinea.

STATUS: Common.

SYNONYMS: *ernstmayri, palpator.*

COMMENTS: Formerly included in *Dactylonax* (see Haltenorth, 1958).

Dactylopsila tatei Laurie, 1952. Bull. Br. Mus. (Nat. Hist.), Zool., 1:278.
TYPE LOCALITY: Papua New Guinea, Milne Bay Prov., Fergusson Isl, Faralulu Dist., mountains above Taibutu Village, 610-915 m.
DISTRIBUTION: Fergusson Isl; Papua New Guinea.
STATUS: Rare.
COMMENTS: Considered a subspecies of *trivirgata* by Ziegler (*in* Stonehouse and Gilmore, 1977:131); considered a distinct species by G. G. George (1979:94).

Dactylopsila trivirgata Gray, 1858. Proc. Zool. Soc. Lond., 1858:111.
TYPE LOCALITY: Indonesia, Aru Isls.
DISTRIBUTION: New Guinea and adjacent small islands; Aru Isls; NE Queensland (Australia).
STATUS: Common.
SYNONYMS: *albertisii, angustivittis, arfakensis, biedermanni, hindenburgi, infumata, kataui, malampus, occidentalis, picata.*

Gymnobelideus McCoy, 1867. Ann. Mag. Nat. Hist., ser. 3, 20:287.
TYPE SPECIES: *Gymnobelideus leadbeateri* McCoy, 1867.

Gymnobelideus leadbeateri McCoy, 1867. Ann. Mag. Nat. Hist., ser. 3, 20:287.
TYPE LOCALITY: Australia, Victoria, Bass River.
DISTRIBUTION: NE Victoria.
STATUS: U.S. ESA - Endangered; IUCN - Vulnerable.

Petaurus Shaw, 1791. Nat. Misc., 2, pl. 60.
TYPE SPECIES: *Petaurus australis* Shaw, 1791.
SYNONYMS: *Belideus, Petaurella, Petaurula, Ptilotus, Xenochirus.*

Petaurus abidi Ziegler, 1981. Aust. Mamm., 4:81.
TYPE LOCALITY: Papua New Guinea, West Sepik Prov., Mt. Somoro, 3°25'S, 142°05'E.
DISTRIBUTION: NC New Guinea.
STATUS: Rare.

Petaurus australis Shaw, 1791. Nat. Misc., 2, pl. 6.
TYPE LOCALITY: Australia, New South Wales, Sydney.
DISTRIBUTION: Coastal Queensland, New South Wales, and Victoria (Australia).
STATUS: Common locally.
SYNONYMS: *cunninghami, flaviventer, petaurus, reginae.*

Petaurus breviceps Waterhouse, 1839. Proc. Zool. Soc. Lond., 1838:152 [1839].
TYPE LOCALITY: Australia, New South Wales.
DISTRIBUTION: SE South Australia to Cape York Peninsula (Queensland), Tasmania (introduction), N Northern Territory, NE Western Australia; New Guinea and adjacent small islands, including Bismarck Arch.; Aru Isls and N Moluccas (Indonesia).
STATUS: Common.
SYNONYMS: *ariel, biacensis, flavidus, kohlsi, longicaudatus, notatus, papuanus, tafa.*
COMMENTS: See Smith (1973, Mammalian Species, 30). McAllan and Bruce (1989) argued that the original publication of this name was in The Athenaeum, 580:880 [8 Dec 1838].

Petaurus gracilis (de Vis, 1883). Abstr. Proc. Linn. Soc. N.S.W., 20 Dec. 1882, ii.
TYPE LOCALITY: Australia, Queensland, Cardwell region.
DISTRIBUTION: Known only from the region of Barrett's Lagoon, near Tully, Queensland, Australia.
STATUS: Probably endangered.
COMMENTS: History of description given by Van Dyck (1990). Species resurrected from synonomy with *P. norfolcensis* by Van Dyck (1991).

Petaurus norfolcensis (Kerr, 1792). *In* Linnaeus, Anim. Kingdom, 1:270.
TYPE LOCALITY: Australia, New South Wales, Sydney.
DISTRIBUTION: Australia: E Queensland, E New South Wales, E Victoria.
STATUS: Common.
SYNONYMS: *sciurea.*

Family Tarsipedidae Gervais and Verreaux, 1842. Proc. Zool. Soc. Lond., 1842:1.

Tarsipes Gervais and Verreaux, 1842. L'Institut, l'ere Section, Sci., Math, Phys., Nat., 427:75.
TYPE SPECIES: *Tarsipes rostratus* Gervais and Verreaux, 1842.
COMMENTS: For correct authorship see Mahoney (1981).

Tarsipes rostratus Gervais and Verreaux, 1842. L'Institut, l'ere Section, Sci., Math, Phys., Nat., 427:75.
TYPE LOCALITY: Australia, Western Australia, King George Sound (Albany), see Gray (1842).
DISTRIBUTION: SW Western Australia.
STATUS: Rare.
SYNONYMS: *spencerae, spenserae.*
COMMENTS: The name *T. spenserae* is considered a misspelling because it was presented as a patronym for Spencer (Gray, 1842:40). Ride (1970) emended the name to *spencerae;* see Mahoney (1981) for details. Mahoney (1981) presented evidence that *Tarsipes rostratus* Gervais and Verreaux, 1842 predates *T. spenserae* Gray, 1842.

Family Acrobatidae Aplin, 1987. *In* M. Archer (ed.), Possums and Opossums, xxii.
COMMENTS: Separated from Burramyidae by Aplin (*in* Aplin and Archer, 1987).

Acrobates Desmarest, 1818. Nouv. Dict. Hist. Nat., Nouv. ed., 25:405.
TYPE SPECIES: *Didelphis pygmaea.*
SYNONYMS: *Ascobates, Cercoptenus.*

Acrobates pygmaeus (Shaw, 1793). Zool. New Holland, 1:5.
TYPE LOCALITY: Australia, New South Wales, Sydney.
DISTRIBUTION: Australia: E Queensland to SE South Australia, inland to Deniliquin (New South Wales).
STATUS: Common.
SYNONYMS: *frontalis, pulchellus.*
COMMENTS: Tate (1938:60) believed the single specimen (of *A. pulchellus* which is considered a synonym of *pygmaeus*) obtained in NW New Guinea was probably an introduction as a pet.

Distoechurus Peters, 1874. Ann. Mus. Civ. Stor. Nat. Genova, 6:303.
TYPE SPECIES: *Phalangista* (*Distoechurus*) *pennata* Peters, 1874.

Distoechurus pennatus (Peters, 1874). Ann. Mus. Civ. Stor. Nat. Genova, 6:303.
TYPE LOCALITY: Indonesia, Irian Jaya, Vogelkop, Manokwari Div., "Andai" (Probably = Arfak Mtns, Hatam, 1520 m). See Van der Feen (1962:52).
DISTRIBUTION: New Guinea.
STATUS: Abundant.
SYNONYMS: *amoenus, dryas, neuhassi.*

ORDER XENARTHRA
by Alfred L. Gardner

ORDER XENARTHRA

SYNONYMS: Edentata Vicq-d'Azyr, 1792.
COMMENTS: Reviewed by Wetzel (1985*a*).

Family Bradypodidae Gray, 1821. London Med. Repos., 15:304.

Bradypus Linnaeus, 1758. Syst. Nat., 10th ed., 1:34.
TYPE SPECIES: *Bradypus tridactylus* Linnaeus, 1758, by subsequent designation (Miller and Rehn, 1901).
SYNONYMS: *Acheus* F. Cuvier, 1825; *Arctopithecus* Gray, 1850 (preoccupied); *Eubradypus* Lönnberg, 1942; *Hemibradypus* Anthony, 1906; *Ignavus* Blumenbach, 1779; *Scaeopus* Peters, 1864.
COMMENTS: Avila-Pires (*in* Wetzel and Avila-Pires, 1980) considered *Scaeopus* a separate genus.

Bradypus torquatus Illiger, 1811. Prodr. Syst. Mamm. Avium, p. 109.
TYPE LOCALITY: "Brasilia;" restricted to the Atlantic drainage of the Brazilian states of Bahia, Espírito Santo, and Rio de Janeiro, by Wetzel and Avila-Pires (1980).
DISTRIBUTION: Coastal forests of SE Brazil.
STATUS: U.S. ESA and IUCN - Endangered.
SYNONYMS: *crinitus* Gray, 1850; *cristatus* Hamilton-Smith, 1827; *mareyi* Anthony, 1907; *melanotis* Swainson, 1835.
COMMENTS: Some authors believe erroneously that *B. torquatus* Illiger, 1811, is a *nomen nudum* and attribute the name to Desmarest (1816*a*).

Bradypus tridactylus Linnaeus, 1758. Syst. Nat., 10th ed., 1:34.
TYPE LOCALITY: "Americæ meridionalis arboribus;" restricted to Surinam by Thomas (1911*a*).
DISTRIBUTION: Guyana, Surinam, French Guiana, Venezuela south of the Orinoco, and N Brazil (south to the southern bank of the Amazonas/Solimões).
SYNONYMS: *ai* Lesson, 1827; *blainvillii* Gray, 1850; *cummunis* Lesson, 1841; *flaccidus* Gray, 1850; *cuculliger* Wagler, 1831; *gularis* Rüppell, 1842.

Bradypus variegatus Schinz, 1825. Das Thierreich, 4:510.
TYPE LOCALITY: "Sudamerika;" restricted to Brazil by Mertens (1925) who suggested that the type may have come from Bahia.
DISTRIBUTION: Honduras to Colombia, Ecuador, Brazil, W Venezuela, E Perú and Bolivia, Paraguay, and N Argentina.
STATUS: CITES - Appendix II.
SYNONYMS: *ai* Wagler, 1831 (preoccupied); *beniensis* Lönnberg, 1942; *boliviensis* Gray, 1871; *brachydactylus* Wagner, 1855; *brasiliensis* Blainville, 1840; *castaneiceps* Gray, 1871; *codajazensis* Lönnberg, 1942; *dorsalis* Fitzinger, 1871; *ecuadorianus* Spillmann, 1927; *ephippiger* Philippi, 1870; *gorgon* Thomas, 1926; *griseus* Gray, 1871; *infuscatus* Wagler, 1831; *ignavus* Goldman, 1913; *macrodon* Thomas, 1917; *marmoratus* Gray, 1850; *miritibae* Lönnberg, 1942; *nefandus* Spillmann, 1927; *pallidus* Wagner, 1844; *problematicus* Gray, 1850; *subjuruanus* Lönnberg, 1942; *tocantinus* Lönnberg, 1942; *trivittatus* Cornalia, 1849; *unicolor* Fitzinger, 1871; *ustus* Lesson, 1840; *violeta* Thomas, 1917.

Family Megalonychidae Ameghino, 1889. Actas Acad. Nac. Cienc. Cordoba, Buenos Aires, 6:690.
COMMENTS: Includes *Choloepus* and approximately 12 genera of extinct sloths.

Subfamily Choloepinae Gray, 1871. Proc. Zool. Soc. Lond., 1871:430.
COMMENTS: Formerly included in Bradypodidae (see Hoffstetter, 1969; Patterson and Pascual, 1968*a*) or Choloepidae (see Honacki et al., 1982). Placed in Megalonychidae by Webb (1985) and Wetzel (1985*a*).

Choloepus Illiger, 1811. Prodr. Syst. Mamm. Avium., p. 108.
TYPE SPECIES: *Bradypus didactylus* Linnaeus, 1758, by subsequent designation (Gray, 1827).
SYNONYMS: *Unaues* Rafinesque, 1815; *Unaus* Gray, 1821.
COMMENTS: Reviewed by Wetzel (1985*a*).

Choloepus didactylus (Linnaeus, 1758). Syst. Nat., 10th ed., 1:35.
TYPE LOCALITY: "Zeylona;" corrected to Surinam by Thomas (1911*a*). Not British Guiana as stated by Tate (1939).
DISTRIBUTION: Guianas and Venezuela (delta and south of Río Orinoco) south into Brazil (Maranhão west along Rio Amazonas/Solimões) and west into upper Amazon Basin of Ecuador and Perú.
SYNONYMS: *brasiliensis* Fitzinger, 1871; *curi* Link, 1795; *florenciae* J. A. Allen, 1913; *kouri* Daudin, 1799; *napensis* Lönnberg, 1922; *unau* Link, 1795.
COMMENTS: Reviewed by Wetzel and Avila-Pires (1980).

Choloepus hoffmanni Peters, 1858. Monatsb. K. Preuss. Akad. Wiss. Berlin, 1858:128.
TYPE LOCALITY: "Costa Rica;" restricted to San José, Escazú, by Goodwin (1946); corrected to Heredia, Volcán Barbara, by Wetzel and Avila-Pires (1980).
DISTRIBUTION: Central America (Nicaragua) into South America east to W Venezuela and south to Brazil (Mato Grosso) and E Bolivia.
STATUS: CITES - Appendix III (Costa Rica).
SYNONYMS: *andinus* J. A. Allen, 1913; *augustinus* J. A. Allen, 1913; *capitalis* J. A. Allen, 1913; *florenciae* J. A. Allen, 1913; *juruanus* Lönnberg, 1942; *pallescens* Lönnberg, 1928; *peruvianus* Menegaux, 1906.

Family Dasypodidae Gray, 1821. London Med. Repos., 15:305.

Subfamily Chlamyphorinae Bonaparte, 1850. Conspectus Syst., p. 4.

Chlamyphorus Harlan, 1825. Ann. Lyc. Nat. Hist. N.Y., 1:235.
TYPE SPECIES: *Chlamyphorus truncatus* Harlan, 1825, by monotypy.
SYNONYMS: *Burmeisteria* Gray, 1865 (preoccupied); *Calyptophractus* Fitzinger, 1871.

Chlamyphorus retusus Burmeister, 1863. Abhandl. Gesd. Naturf. Halle, 7:167.
TYPE LOCALITY: Bolivia, Santa Cruz, "Sta. Cruz de la Sierra."
DISTRIBUTION: Gran Chaco of N Argentina, W Paraguay, and SE Bolivia.
STATUS: IUCN - Insufficiently known.
SYNONYMS: *clorindae* Yepes, 1939.
COMMENTS: Commonly listed under *Burmeisteria retusa*; reviewed by Wetzel (1985*b*).

Chlamyphorus truncatus Harlan, 1825. Ann. Lyc. Nat. Hist. N.Y., 1:235.
TYPE LOCALITY: "Mendoza . . . interior of Chili, on the east of the Cordilleras, in lat. 33°25' and long. 69°47', in the province of Cuyo;" restricted to Río Tunuyán, 33°25'S, 69°45'W, Mendoza, Argentina, by Cabrera (1958).
DISTRIBUTION: Argentina.
STATUS: U.S. ESA - Endangered; IUCN - Insufficiently known.
SYNONYMS: *minor* Lahille, 1895; *ornatus* Lahille, 1895; *patquiensis* Yepes, 1931; *typicus* Lahille, 1895.

Subfamily Dasypodinae Gray, 1821. London Med. Repos., 15:305.
COMMENTS: Wetzel (1985*b*) divided the subfamily into the four tribes: Dasypodini, Euphractini, Priodontini, and Tolypeutini.

Cabassous McMurtrie, 1831. Anim. Kingdom, 1:164.
TYPE SPECIES: *Dasypus unicinctus* Linnaeus, 1758, by monotypy.
SYNONYMS: *Arizostus* Gloger, 1841; *Lysiurus* Ameghino, 1891; *Tatusia* Lesson, 1827 (part); *Tatoua* Gray, 1865; *Xenurus* Wagler, 1830 (preoccupied); *Ziphila* Gray, 1873.
COMMENTS: Revised by Wetzel (1980).

Cabassous centralis (Miller, 1899). Proc. Biol. Soc. Washington, 13:4.
TYPE LOCALITY: Honduras, Cortés, "Chamelecon."
DISTRIBUTION: México (Chiapas) to N Colombia.

STATUS: CITES - Appendix III (Costa Rica).

Cabassous chacoensis Wetzel, 1980. Ann. Carnegie Mus., 49(2):335.
TYPE LOCALITY: Paraguay, Presidente Hayes, "5-7 km W Estancia Juan de Zalazar."
DISTRIBUTION: Gran Chaco of W Paraguay and NW Argentina. Known from Mato Grosso, Brazil, on the basis of one zoological park specimen (Wetzel, 1980).

Cabassous tatouay (Desmarest, 1804). Tabl. Méth. Hist. Nat., *in* Nouv. Dict. Hist. Nat., 24:28.
TYPE LOCALITY: Paraguay, restricted to "a 27° de lat. sur" by Cabrera (1958).
DISTRIBUTION: Uruguay, S Brazil, SE Paraguay, and NE Argentina.
STATUS: CITES - Appendix III (Uruguay).
SYNONYMS: *dasycercus* G. Fischer, 1814; *gymnurus* Olfers, 1818; *nudicaudus* Lund, 1839.

Cabassous unicinctus (Linnaeus, 1758). Syst. Nat., 10th ed., 1:50.
TYPE LOCALITY: "Africa;" restricted to "l'Amérique" by Buffon (1763), and to Surinam by Thomas (1911*a*).
DISTRIBUTION: South America east of the Andes from Colombia to Mato Grosso do Sul, Brazil.
SYNONYMS: *duodecimcinctus* Schreber, 1774; *hispidus* Burmeister, 1854; *latirostris* Gray, 1873; *loricatus* Wagner, 1855; *multicinctus* Thunberg, 1818; *octodecimcinctus* Erxleben, 1777; *squamicaudis* Lund, 1845; *verrucosus* Wagner, 1844.

Chaetophractus Fitzinger, 1871. Sitzb. Kaiserl. Akad. Wiss., Wein, 64(1):268.
TYPE SPECIES: *Dasypus villosus* (Desmarest, 1804) by subsequent designation (Yepes, 1928).
SYNONYMS: *Dasyphractus* Fitzinger, 1871; *Loricatus* Desmarest, 1804 (part); *Tatus* Olfers, 1818 (part); *Tatusia* Lesson, 1827 (part).
COMMENTS: Formerly included in *Euphractus* (see Wetzel, 1985*b*)

Chaetophractus nationi (Thomas, 1894). Ann. Mag. Nat. Hist., ser. 6, 13:70.
TYPE LOCALITY: Bolivia, Oruro, "Orujo."
DISTRIBUTION: Cochabamba, Oruro, and La Paz (Bolivia).
COMMENTS: Distribution and status uncertain, may be a subspecies of *vellerosus* (see Wetzel, 1985*b*).

Chaetophractus vellerosus (Gray, 1865). Proc. Zool. Soc. Lond., 1865:376.
TYPE LOCALITY: Bolivia, Santa Cruz, "Santa Cruz de la Sierra."
DISTRIBUTION: Chaco Boreal of Bolivia and Paraguay south to C Argentina and west to the Puna de Tarapacá of Chile.
SYNONYMS: *boliviensis* Grandidier and Neveu-Lemaire, 1908; *brevirostris* Fitzinger, 1871; *desertorum* Krumbiegel, 1940; *pannosus* Thomas, 1902.

Chaetophractus villosus (Desmarest, 1804). Tabl. Méth. Hist. Nat., *in* Nouv. Dict. Hist. Nat., 24:28.
TYPE LOCALITY: Argentina, Buenos Aires, "Les Pampas" south of Río de la Plata between 35° and 36° south (Azara, 1801:164).
DISTRIBUTION: Gran Chaco of Bolivia, Paraguay, and Argentina south to Santa Cruz, Argentina, and Magallanes, Chile.
SYNONYMS: *octocinctus* Molina, 1782 (preoccupied).

Dasypus Linnaeus, 1758. Syst. Nat., 10th ed., 1:50.
TYPE SPECIES: *Dasypus novemcinctus* Linnaeus, 1758, by Linnaean tautonomy.
SYNONYMS: *Cachicamus* McMurtrie, 1831; *Cataphractus* Storr, 1780; *Hyperoambon* Peters, 1864; *Loricatus* Desmarest, 1804; *Muletia* Gray, 1874; *Praopus* Burmeister, 1854; *Tatu* Blumenbach, 1779; *Tatusia* Lesson, 1827; *Zonoplites* Gloger, 1841.
COMMENTS: Reviewed by Wetzel (1985*b*) and Wetzel and Mondolfi (1979).

Dasypus hybridus (Desmarest, 1804). Tabl. Méth. Hist. Nat., *in* Nouv. Dict. Hist. Nat., 24:28.
TYPE LOCALITY: Paraguay, Misiones, San Ignacio (as restricted by Cabrera, 1958).
DISTRIBUTION: Argentina, Paraguay, and S Brazil south to Río Negro, Argentina.
SYNONYMS: *auritus* Olfers, 1818.
COMMENTS: Includes a paratype of *mazzai* Yepes, 1933 (Wetzel and Mondolfi, 1979).

Dasypus kappleri Krauss, 1862. Archiv Naturgesch., 28(1):20.
TYPE LOCALITY: "Den Urwäldern des Marowiniflusse in Surinam;" restricted to the neighborhood of Albina near the mouth of the Marowijne River by Husson (1978).
DISTRIBUTION: Colombia (east of the Andes), Venezuela (south of the Orinoco), Guyana, Surinam, and south through the Amazon Basin of Brazil, Ecuador and Perú.
SYNONYMS: *beniensis* Lönnberg, 1942; *pastasae* Thomas, 1901; *pentadactylus* Peters, 1864.

Dasypus novemcinctus Linnaeus, 1758. Syst. Nat., 10th ed., 1:51.
TYPE LOCALITY: "America Meridionali;" restricted to Pernambuco, Brazil, by Cabrera (1958).
DISTRIBUTION: S USA, México, Central and South America to N Argentina, the Lesser Antilles (Grenada), and Trinidad and Tobago.
SYNONYMS: *aequatorialis* Lönnberg, 1913; *boliviensis* Gray, 1873; *brevirostris* Gray, 1873; *davisi* Russell, 1953; *fenestratus* Peters, 1864; *granadiana* Gray, 1873; *hoplites* G. M. Allen, 1911; *leptocephala* Gray, 1873; *leptorhynchus* Gray, 1873; *longicaudatus* Kerr, 1792; *longicaudatus* Daudin, *in* Lacépède, 1799; *longicaudus* Wied, 1826; *lundi* Fitzinger, 1871; *mazzai* Yepes, 1933; *mexianae* Hagmann, 1908; *mexicanus* Peters, 1864; *niger* Desmarest, 1804; *niger* Olfers, 1818 (preoccupied); *niger* Lichtenstein, 1818 (preoccupied); *octocintus* Schreber, 1774; *peba* Desmarest, 1822; *serratus* G. Fischer, 1814; *texanum* Bailey, 1905; *uroceras* Lund, 1839.
COMMENTS: Reviewed by McBee and Baker (1982, Mammalian Species, 162).

Dasypus pilosus (Fitzinger, 1856). Versamml. Deutsch. Nat. Arzte, Wien, Tageblatt, 32:123.
TYPE LOCALITY: "Peru;" restricted to montane Perú by Wetzel and Mondolfi (1979).
DISTRIBUTION: Known only from the Peruvian Andes in San Martín, La Libertad, Huánuco, and Junín.
SYNONYMS: *hirsutus* Burmeister, 1862.

Dasypus sabanicola Mondolfi, 1968. Mem. Soc. Cienc. Nat. La Salle, 27:151.
TYPE LOCALITY: Venezuela, Apure, "Hato Macanillal."
DISTRIBUTION: Llanos of Venezuela and Colombia.

Dasypus septemcinctus Linnaeus, 1758. Syst. Nat., 10th ed., 1:51.
TYPE LOCALITY: "Indiis;" restricted to Pernambuco, Brazil, by Hamlett (1939).
DISTRIBUTION: Lower Amazon Basin of Brazil to the Gran Chaco of Bolivia, Paraguay, and N Argentina.
SYNONYMS: *megalolepis* Cope, 1889; *propalatum* Rhoads, 1894.

Euphractus Wagler, 1830. Naturliches Syst. Amphibien, p. 36.
TYPE SPECIES: *Dasypus sexcinctus* Linnaeus, 1758, by subsequent designation (Palmer, 1904).
SYNONYMS: *Encoubertus* McMurtrie, 1831 (part); *Loricatus* Desmarest, 1804 (part); *Pseudotroctes* Gloger, 1841; *Scleropleura* Milne-Edwards, 1871; *Tatus* Olfers, 1818 (part).
COMMENTS: Moeller (1968) included *Chaetophractus* and *Zaedyus* in this genus, *contra* Wetzel (1985*b*) whose usage is followed here.

Euphractus sexcinctus (Linnaeus, 1758). Syst. Nat., 10th ed., 1:51.
TYPE LOCALITY: "America meridionale;" restricted to Pará, Brazil, by Thomas (1907*b*).
DISTRIBUTION: S Surinam and adjacent Brazil as a northern isolated segment; E Brazil to Bolivia, Paraguay, Uruguay, and N Argentina as the main population.
SYNONYMS: *boliviae* Thomas, 1907; *bruneti* Milne-Edwards, 1871; *encoubert* Desmarest, 1822; *flavimanus* Desmarest, 1804; *flavipes* G. Fischer, 1814; *gilvipes* Lichtenstein, 1818; *mustelinus* Fitzinger, 1871; *poyu* Larrañaga, 1923; *setosus* Wied, 1826; *tucumanus* Thomas, 1907.
COMMENTS: Reviewed by Wetzel (1985*b*) and Redford and Wetzel (1985, Mammalian Species, 252).

Priodontes F. Cuvier, 1825. Dentes des Mamm., p. 257.
TYPE SPECIES: *Dasypus gigas* G. Cuvier, 1817 (= *Dasypus maximus* Kerr, 1792), by monotypy.
SYNONYMS: *Cheloniscus* Wagler, 1830; *Loricatus* Desmarest, 1804 (part); *Polygomphius* Gloger, 1841; *Priodon* McMurtrie, 1831; *Prionodos* Gray, 1865.
COMMENTS: Reviewed by Wetzel (1985*b*).

Priodontes maximus (Kerr, 1792). *In* Linnaeus, Anim. Kingdom, 1:112.
TYPE LOCALITY: French Guiana, "Cayenne."
DISTRIBUTION: South America east of the Andes from N Venezuela and the Guianas south to Paraguay and N Argentina.
STATUS: CITES - Appendix I; U.S. ESA - Endangered; IUCN - Vulnerable as *P. giganteus*.
SYNONYMS: *giganteus* G. Fischer, 1814; *gigas* G. Cuvier, 1817; *grandis* Olfers, 1818.

Tolypeutes Illiger, 1811. Prodr. Syst. Mamm. Avium., p. 111.
TYPE SPECIES: *Dasypus tricinctus* Linnaeus, 1758, by subsequent designation (Yepes, 1928).
SYNONYMS: *Apara* McMurtrie, 1831; *Cheloniscus* Gray, 1873 (preoccupied); *Sphaerocormus* Fitzinger, 1871; *Tatusia* Lesson, 1827; *Tolypoides* Grandidier and Neveu-Lemaire, 1905.
COMMENTS: Reviewed by Wetzel (1985*b*).

Tolypeutes matacus (Desmarest, 1804). Tabl. Méth. Hist. Nat., *in* Nouv. Dict. Hist. Nat., 24:28.
TYPE LOCALITY: No locality mentioned; restricted to Argentina, Tucumán, Tucumán, by Sanborn (1930).
DISTRIBUTION: E Bolivia and SW Brazil south through the Gran Chaco of Paraguay to Argentina (Buenos Aires).
SYNONYMS: *apar* Desmarest, 1822; *bicinctus* Grandidier and Neveu-Lemaire, 1905; *brachyurus* G. Fischer, 1814; *conurus* I. Geoffroy, 1847; *muriei* Garrod, 1878.

Tolypeutes tricinctus (Linnaeus, 1758). Syst. Nat., 10th ed., 1:56.
TYPE LOCALITY: "In India orientali;" redefined as Pernambuco, Brazil, by Sanborn (1930).
DISTRIBUTION: Brazilian states of Bahia, Ceará, and Pernambuco.
STATUS: IUCN - Indeterminate; considered to be extinct.
SYNONYMS: *globulus* Olfers, 1818; *quadricinctus* Linnaeus, 1758; *quadricinctus* Olfers, 1818 (preoccupied).

Zaedyus Ameghino, 1889. Actas Acad. Nac. Cienc. Cordoba, Buenos Aires, 6:867.
TYPE SPECIES: *Dasypus minutus* Desmarest, 1822 (= *Loricatus pichiy* Desmarest, 1804), by original designation.
SYNONYMS: *Loricatus* Desmarest, 1804 (part); *Tatusia* Lesson, 1827 (part).
COMMENTS: Sometimes considered a subgenus of *Euphractus*; reviewed by Wetzel (1985*b*).

Zaedyus pichiy (Desmarest, 1804). Tabl. Méth. Hist. Nat., *in* Nouv. Dict. Hist. Nat., 24:28.
TYPE LOCALITY: Argentina, Buenos Aires, Bahia Blanca, as restricted by Cabrera (1958).
DISTRIBUTION: Mendoza, San Luis, and Buenos Aires, Argentina, south through Argentina and E Chile to the Straits of Magellan.
SYNONYMS: *caurinus* Thomas, 1928; *ciliatus* G. Fischer, 1814; *fimbriatus* Olfers, 1818; *marginatus* Wagler, 1830; *minutus* Desmarest, 1822; *patagonicus* Desmarest, 1819; *quadricinctus* Molina, 1782 (preoccupied).

Family Myrmecophagidae Gray, 1825. Ann. Philos., n.s., 10:343.

Cyclopes Gray, 1821. London Med. Repos., 15:305.
TYPE SPECIES: *Myrmecophaga didactyla* Linnaeus, 1758.
SYNONYMS: *Cyclothurus* Lesson, 1842; *Didactyla* Liais, 1872; *Didactyles* F. Cuvier, 1829; *Eurypterna* Gloger, 1841; *Myrmydon* Wagler, 1830.

Cyclopes didactylus (Linnaeus, 1758). Syst. Nat., 10th ed., 1:35.
TYPE LOCALITY: "America australi;" restricted to Surinam by Thomas (1911*a*).
DISTRIBUTION: México (Veracruz and Oaxaca) to Colombia and west of Andes to S Ecuador, east of Andes to Venezuela, Trinidad, Guyana, Surinam, French Guiana, and S Colombia and Venezuela, south to Bolivia (Santa Cruz) and Brazil (Acre east to Alagoas).
SYNONYMS: *catellus* Thomas, 1928; *codajazensis* Lönnberg, 1942; *dorsalis* Gray, 1865; *eva* Thomas, 1902; *ida* Thomas, 1900; *juruanus* Lönnberg, 1942; *melini* Lönnberg, 1928; *mexicanus* Hollister, 1914; *monodactyla* Kerr, 1792; *unicolor* Desmarest, 1822.

Myrmecophaga Linnaeus, 1758. Syst. Nat., 10th ed., 1:35.
TYPE SPECIES: *Myrmecophaga tridactyla* Linnaeus, 1758, by subsequent selection (Thomas, 1901*a*).
SYNONYMS: *Falcifer* Rehn, 1900.

Myrmecophaga tridactyla Linnaeus, 1758. Syst. Nat., 10th ed., 1:35.
TYPE LOCALITY: "America *meridionali;*" restricted to Brazil, Pernambuco, Pernambuco [= Recife], by Thomas (1911*a*).
DISTRIBUTION: Belize and Guatemala through South America to Uruguay and the Gran Chaco of Bolivia, Paraguay, Argentina.
STATUS: CITES - Appendix II; IUCN - Vulnerable.
SYNONYMS: *artata* Osgood, 1912; *centralis* Lyon, 1906; *jubata* Linnaeus, 1758.

Tamandua Gray, 1825. Ann. Philos., n.s., 10:343.
TYPE SPECIES: *Myrmecophaga tamandua* G. Cuvier, 1798 (= *Tamandua tetradactyla* Linnaeus, 1758), by monotypy.
SYNONYMS: *Dryoryx* Gloger, 1841; *Uroleptes* Wagler, 1830.

Tamandua mexicana (Saussure, 1860). Rev. Mag. Zool. Paris, ser. 2, 12:9.
TYPE LOCALITY: México, "Tabasco."
DISTRIBUTION: E México (Tamaulipas), Central America, South America to NW Perú and NW Venezuela.
SYNONYMS: *chiriquensis* J. A. Allen, 1904; *hesperia* Davis, 1955; *instabilis* J. A. Allen, 1904; *opistholeuca* Gray, 1873; *punensis* J. A. Allen, 1916; *sellata* Cope, 1889; *tambensis* Lönnberg, 1937; *tenuirostris* J. A. Allen, 1904.
COMMENTS: Revised by Wetzel (1975).

Tamandua tetradactyla (Linnaeus, 1758). Syst. Nat., 10th ed., 1:35.
TYPE LOCALITY: "America meridionali;" restricted to Brazil, Pernambuco, Pernambuco [= Recife], by Thomas (1911*a*).
DISTRIBUTION: South America east of the Andes from Colombia, Venezuela, Trinidad, and the Guianas, south to Uruguay and N Argentina.
STATUS: CITES - Appendix III (Guatemala).
SYNONYMS: *bivittata* Desmarest, 1817; *chapadensis* J. A. Allen, 1904; *crispa* Rüppell, 1842; *kriegi* Krumbiegel, 1940; *longicaudata* Wagner, 1844; *longicaudata* Turner, 1853 (preoccupied); *myosura* Pallas, 1766; *nigra* Desmarest, 1804; *opisthomelas* Gray, 1873; *quichua* Thomas, 1927; *straminea* Cope, 1889; *tamandua* G. Cuvier, 1798.

ORDER INSECTIVORA

by Rainer Hutterer

ORDER INSECTIVORA

COMMENTS: Formerly included elephant shrews and tree shrews which, since Butler (1972) are placed in two separate orders, Macroscelidea and Scandentia. Reviewed by Cabrera (1925). Phylogeny of living and fossil insectivores treated by Van Valen (1967). For basic data on brain structure and evolution see Stephan et al. (1991). For a synopsis of karyotype data see Reumer and Meylan (1986).

Family Solenodontidae Gill, 1872. Smithson. Misc. Coll., 11(1):19.

COMMENTS: Dobson (1882:82) was the first to raise Gill's subfamily to family level.

Solenodon Brandt, 1833. Mem. Acad. Imp. Sci., St. Petersbourg, ser. 6, 2:459.

TYPE SPECIES: *Solenodon paradoxus* Brandt, 1833.

SYNONYMS: *Antillogale, Atopogale.*

COMMENTS: Includes *Antillogale* and *Atopogale;* see Patterson (1962:2) and Varona (1974:6). Besides the two extant species, two presumably extinct species have been described from Cuba ("Giant Solenodon") and Hispaniola (*Antillogale marcanoi* Patterson, 1962); see Morgan and Woods (1986). Remains of *Solenodon marcanoi* have been found in a horizon of "Late Pleistocene or Recent" age (Patterson, 1962).

Solenodon cubanus Peters, 1861. Monatsb. K. Preuss. Akad. Wiss. Berlin, 1861:169.

TYPE LOCALITY: Cuba, Oriente Prov., Bayamo.

DISTRIBUTION: Oriente Prov. (Cuba).

STATUS: U.S. ESA and IUCN - Endangered.

SYNONYMS: *poeyanus.*

COMMENTS: Sometimes placed in a distinct genus or subgenus, *Atopogale,* see Hall and Kelson (1959:22) and Hall (1981:22), but see Poduschka and Poduschka (1983:225-238) who regarded *Atopogale* as a synonym of *Solenodon.* For biological information see Varona (1983*b*).

Solenodon marcanoi (Patterson, 1962). Breviora, 165:2.

TYPE LOCALITY: Dominican Republic, San Rafael Prov., Hondo Valle Mun.; unnamed cave 2 km SW of Rancho La Guardia.

DISTRIBUTION: Known only from the type locality.

STATUS: Extinct.

COMMENTS: See comment under *Solenodon.*

Solenodon paradoxus Brandt, 1833. Mem. Acad. Imp. Sci., St. Petersbourg, ser. 6, 2:459.

TYPE LOCALITY: "Hispaniola", Dominican Republic.

DISTRIBUTION: Haiti, Dominican Republic (Hispaniola).

STATUS: U.S. ESA and IUCN - Endangered.

Family Nesophontidae Anthony, 1916. Bull. Am. Mus. Nat. Hist., 35:725.

COMMENTS: Known only from sub-Recent fossils from the Greater Antilles. One genus with eight taxa have been named, of which Hall (1981) listed six as valid species. Morgan and Woods (1986) recognized eight species. Recent efforts to locate surviving populations have been unsuccessful (Woods et al., 1985). See also comments under *Nesophontes.*

Nesophontes Anthony, 1916. Bull. Am. Mus. Nat. Hist., 35:725.

TYPE SPECIES: *Nesophontes edithae* Anthony, 1916.

COMMENTS: All species of *Nesophontes* appear to have survived the late Pleistocene extinction, at least five species are known to have existed into post Columbian times, and several species apparently did not go extinct until the early part of this century (Morgan and Woods, 1986). Includes an undescribed species from the Cayman Isls which was found in post-Columbian deposits (Morgan and Woods, 1986; Varona, 1974).

Nesophontes edithae Anthony, 1916. Bull. Am. Mus. Nat. Hist., 35:725.
TYPE LOCALITY: Puerto Rico, Cueva Cathedral, near Morovis.
DISTRIBUTION: Puerto Rico.
STATUS: Extinct.

Nesophontes hypomicrus Miller, 1929. Smithson. Misc. Coll., 81:4.
TYPE LOCALITY: Haití, 4 mi east of St. Michel, cave near the Atalaya plantation.
DISTRIBUTION: Haití and Gonave Isl.
STATUS: Extinct.

Nesophontes longirostris Anthony, 1919. Bull. Am. Mus. Nat. Hist., 41:633.
TYPE LOCALITY: Cuba, Oriente, cave near the beach at Daiquirí.
DISTRIBUTION: Cuba.
STATUS: Extinct.

Nesophontes major Arredondo, 1970. Memoria, Soc. Cienc. Nat. La Salle, 30(86):126.
TYPE LOCALITY: Cuba, Habana, Bacuranao, Cueva de la Santa.
DISTRIBUTION: Cuba.
STATUS: Extinct.

Nesophontes micrus G. M. Allen, 1917. Bull. Mus. Comp. Zool., 61:5.
TYPE LOCALITY: Cuba, Matanzas, Sierra de Hato Neuvo.
DISTRIBUTION: Cuba, Haití, and Pinos Isl.
STATUS: Extinct.

Nesophontes paramicrus Miller, 1929. Smithson. Misc. Coll., 81(9):3.
TYPE LOCALITY: Haití, cave approximately 4 mi E St. Michel.
DISTRIBUTION: Haití.
STATUS: Extinct.

Nesophontes submicrus Arredondo, 1970. Memoria, Soc. Cienc. Nat. La Salle, 30(86):137.
TYPE LOCALITY: Cuba, Habana, Bacuranao, Cueva de la Santa.
DISTRIBUTION: Cuba.
STATUS: Extinct.

Nesophontes zamicrus Miller, 1929. Smithson. Misc. Coll., 81:7.
TYPE LOCALITY: Haití, 4 mi east of St. Michel, cave near Atalaya plantation.
DISTRIBUTION: Haití.
STATUS: Extinct.

Family Tenrecidae Gray, 1821. London Med. Repos., 15:301.
COMMENTS: Includes Potamogalinae (see Corbet, 1974).

Subfamily Geogalinae Trouessart, 1879. Rev. Mag. Zool., Paris, ser. 3, 7:275.

Geogale Milne-Edwards and A. Grandidier, 1872. Ann. Sci. Nat. Zool., ser. 5, 15 (art. 19):1.
TYPE SPECIES: *Geogale aurita* Milne-Edwards and A. Grandidier, 1872.
SYNONYMS: *Cryptogale* (see Genest and Petter, 1975).

Geogale aurita Milne-Edwards and A. Grandidier, 1872. Ann. Sci. Nat. Zool., ser. 5, 15 (art. 19):1.
TYPE LOCALITY: Madagascar, Morondava.
DISTRIBUTION: NE and SW Madagascar, in Lamboharana, Tulear [Toliary], and Fenerive.
STATUS: IUCN - *G. a. aurita* Insufficiently known; *G. a. orientalis* Indeterminate.
SYNONYMS: *australis, orientalis*.

Subfamily Oryzorictinae Dobson, 1882. Monogr. Insectivora, 1:71.

Limnogale Major, 1896. Ann. Mag. Nat. Hist., ser. 6, 18:318.
TYPE SPECIES: *Limnogale mergulus* Major, 1896.

Limnogale mergulus Major, 1896. Ann. Mag. Nat. Hist., ser. 6, 18:318.
TYPE LOCALITY: Madagascar, NE Betsileo, Imasindrary.
DISTRIBUTION: E Madagascar, freshwater streams; see map in Nicoll and Rathbun (1990:9).
STATUS: IUCN - Indeterminate.

Microgale Thomas, 1882. J. Linn. Soc., Zool., 16:319.
TYPE SPECIES: *Microgale longicaudata* Thomas, 1882.
SYNONYMS: *Leptogale, Nesogale; Oryzorictes* Major, 1896 (not Grandidier, 1870), *Paramicrogale.*
COMMENTS: See MacPhee (1987*a*:4), who revised the entire genus and whose conclusions are followed here.

Microgale brevicaudata G. Grandidier, 1899. Bull. Mus. Hist. Nat. Paris, 5:349.
TYPE LOCALITY: "environs of Mahanara, NE coast of Madagascar", 78 km S of Iharana [Vohimarina], Antsiranana, Antalaha, Madagascar.
DISTRIBUTION: Madagascar, forest.
STATUS: IUCN - Insufficiently known as *M. brevicaudata;* Indeterminate as *M. occidentalis.*
SYNONYMS: *breviceps, occidentalis.*

Microgale cowani Thomas, 1882. J. Linn. Soc., Zool., 16:319.
TYPE LOCALITY: "Ankafina forest, eastern Betsileo", hill 10 km S of Ambohimahasoa, 3 km W of Tsarafidy town, Fianarantsoa, Fianarantsoa, E Madagascar.
DISTRIBUTION: N, E, and EC Madagascar.
STATUS: IUCN - Insufficiently known as *M. crassipes, M. drouhardi, M. longirostris, M. melanorrhachis,* and *M. taiva.*
SYNONYMS: *crassipes, drouhardi, longirostris, melanorrhachis, nigrescens, taiva.*

Microgale dobsoni Thomas, 1884. Ann. Mag. Nat. Hist., ser. 5, 14:337.
TYPE LOCALITY: "Nandesen forest, central Betsileo", uncertain location, perhaps a patch of forest E of Nandihizana village (1340 m), S of Ambositra, Fianarantsoa, Fianarantsoa, Madagascar.
DISTRIBUTION: Forests of E and EC Madagascar.
COMMENTS: Formerly included in *Nesogale,* see Thomas (1918*a*:302).

Microgale dryas Jenkins, 1992. Bull. Brit. Mus. (Nat. Hist.) Zool., 58:53.
TYPE LOCALITY: NE Madagascar, primary forest in Ambatovaky Special Reserve (16°51'S, 49°08'E), 600-750 m.
DISTRIBUTION: Known only from the type locality.
COMMENTS: This species occurs sympatrically with *M. cowani, M. principula,* and *M. talazaci* (Jenkins, 1992).

Microgale gracilis (Major, 1896). Ann. Mag. Nat. Hist., ser. 6, 18:318.
TYPE LOCALITY: "Ambohimitombo forest", 43 km by road SE of Ambositra, Fianarantsoa, Fianarantsoa, Madagascar.
DISTRIBUTION: E forest of Madagascar.
STATUS: IUCN - Insufficiently known.
COMMENTS: Formerly included in *Leptogale,* see Thomas (1918*a*).

Microgale longicaudata Thomas, 1882. J. Linn. Soc., Zool., 16:319.
TYPE LOCALITY: "Ankafina forest, eastern Betsileo", hill 10 km S of Ambohimahasoa, 3 km W of Tsarafidy town, Fianarantsoa, Fianarantsoa, E Madagascar.
DISTRIBUTION: E and N Madagascar.
STATUS: IUCN - Insufficiently known as *M. longicaudata* and *M. majori;* Indeterminate as *M. prolixacaudata.*
SYNONYMS: *majori, prolixacaudata.*

Microgale parvula G. Grandidier, 1934. Bull. Mus. Hist. Nat. Paris, 6:476.
TYPE LOCALITY: "Environs of Diego Suarez", Antsiranana, Antsiranana, Madagascar.
DISTRIBUTION: N Madagascar.
STATUS: IUCN - Insufficiently known.
COMMENTS: Known only from the holotype.

Microgale principula Thomas, 1926. Ann. Mag. Nat. Hist., ser. 9, 17:250.
TYPE LOCALITY: "Midongy du Sud, SE Madagascar", Midongy Atsimo, Fianarantsoa, Farafangana, Madagascar.
DISTRIBUTION: E and extreme E part of S Madagascar.
STATUS: IUCN - Insufficiently known as *M. principula* and *M. sorella.*
SYNONYMS: *decaryi, sorella.*

Microgale pulla Jenkins, 1988. Am. Mus. Novit., 2910:2.
TYPE LOCALITY: "Foret d'Andrivola, ca. 10 km southwest of Maintimbato Village, ca. 40 km southwest of Maroantsetra", NE Madagascar.
DISTRIBUTION: NE Madagascar.
STATUS: IUCN - Insufficiently known.
COMMENTS: Known only from the holotype; may represent an adult specimen of *M. parvula*, a species of which only the juvenile holotype is known.

Microgale pusilla Major, 1896. Ann. Mag. Nat. Hist., ser. 6, 18:461.
TYPE LOCALITY: "Neighbourhood of Vinanitelo", 50 km SE of Fianarantsoa town and 10 km SSE of Vohitrafeno town, W margin of E forest, Fianarantsoa, Fianarantsoa, Madagascar.
DISTRIBUTION: E, EC, S, and SW Madagascar.

Microgale talazaci Major, 1896. Ann. Mag. Nat. Hist., ser. 6, 18:318.
TYPE LOCALITY: "Neighbourhood of Vinanitelo", 50 km SE of Fianarantsoa town and 10 km SSE of Vohitrafeno town, W margin of E forest, Fianarantsoa, Fianarantsoa, Madagascar.
DISTRIBUTION: N, E, and EC Madagascar.
COMMENTS: Formerly included in *Nesogale*, see Thomas (1918*a*).

Microgale thomasi Major, 1896. Ann. Mag. Nat. Hist., ser. 6, 18:318.
TYPE LOCALITY: "Ampitambe forest (N.E. Betsileo)", Madagascar; uncertain locality, see MacPhee (1987*a*:5-6).
DISTRIBUTION: E Madagascar.
STATUS: IUCN - Insufficiently known.

Oryzorictes A. Grandidier, 1870. Rev. Mag. Zool. Paris, ser. 2, 22:49.
TYPE SPECIES: *Oryzorictes hova* A. Grandidier, 1870.
SYNONYMS: *Nesoryctes*.
COMMENTS: *Nesorycte*s is considered a subgenus of *Oryzorictes*, see Heim de Balsac (1972).

Oryzorictes hova A. Grandidier, 1870. Rev. Mag. Zool. Paris, ser. 2, 22:49.
TYPE LOCALITY: Madagascar, near rice fields of Ankay and Antsihanaka.
DISTRIBUTION: C Madagascar.

Oryzorictes talpoides G. Grandidier and Petit, 1930. Bull. Mus. Hist. Nat. Paris, ser. 2, 2(5):498.
TYPE LOCALITY: Madagascar, Majunga Prov., coastal plain of Marovoay.
DISTRIBUTION: NW Madagascar.

Oryzorictes tetradactylus Milne-Edwards and A. Grandidier, 1882. Le Naturaliste, 4:55.
TYPE LOCALITY: Madagascar, Ampitambe, Sirabe, Imerina.
DISTRIBUTION: C Madagascar.
SYNONYMS: *niger*.
COMMENTS: *O. niger* Major, 1896, is considered a melanistic form of *tetradactylus*, see Thomas (1918*a*:302). Formerly included in *Nesoryctes*, see Heim de Balsac (1972).

Subfamily Potamogalinae Allman, 1865. Proc. Zool. Soc. Lond., 1865:467.

Micropotamogale Heim de Balsac, 1954. C.R. Acad. Sci. Paris, 239:102.
TYPE SPECIES: *Micropotamogale lamottei* Heim de Balsac, 1954.
SYNONYMS: *Mesopotamogale* (see Corbet, 1974).

Micropotamogale lamottei Heim de Balsac, 1954. C.R. Acad. Sci. Paris, 239:103.
TYPE LOCALITY: Guinea, Mt. Nimba, Ziela.
DISTRIBUTION: Environs of Mt. Nimba in Guinea, Liberia, and Ivory Coast.
STATUS: IUCN - Endangered.
COMMENTS: For a survey of the distribution and ecology, see Vogel (1983).

Micropotamogale ruwenzorii (de Witte and Frechkop, 1955). Bull. Inst. Roy. Sci. Nat. Belg., 31(84):1.
TYPE LOCALITY: Zaire, W slopes of Mt. Ruwenzori.

DISTRIBUTION: Ruwenzori region (Uganda, Zaire), and W of Lake Edward and Lake Kivu (Zaire).
STATUS: IUCN - Indeterminate.
COMMENTS: Heim de Balsac (1956) proposed for this species a new genus, *Mesopotamogale*, which is currently regarded as a subgenus; see Corbet (1974).

Potamogale du Chaillu, 1860. Proc. Boston Soc. Nat. Hist., 7:363.
TYPE SPECIES: *Cynogale velox* du Chaillu, 1860.
SYNONYMS: *Bayonia, Mythomys.*

Potamogale velox (du Chaillu, 1860). Proc. Boston Soc. Nat. Hist., 7:363.
TYPE LOCALITY: Gabon, Ogowe River.
DISTRIBUTION: Tropical Africa; from Nigeria to Angola and east to the Rift valley.
SYNONYMS: *allmani, argens.*

Subfamily Tenrecinae Gray, 1821. London Med. Repos., 15:301.

Echinops Martin, 1838. Proc. Zool. Soc. Lond., 1838:17.
TYPE SPECIES: *Echinops telfairi* Martin, 1838.
SYNONYMS: *Echinogale.*

Echinops telfairi Martin, 1838. Proc. Zool. Soc. Lond., 1838:17.
TYPE LOCALITY: Madagascar.
DISTRIBUTION: S Madagascar.
SYNONYMS: *miwarti, nigrescens, pallens.*

Hemicentetes Mivart, 1871. Proc. Zool. Soc. Lond., 1871:72.
TYPE SPECIES: *Erinaceus madagascariensis* Shaw, 1800 (= *Ericulus semispinosus* G. Cuvier, 1798).

Hemicentetes semispinosus (G. Cuvier, 1798). Tabl. Elem. Hist. Nat. Anim., 1798:108.
TYPE LOCALITY: Madagascar.
DISTRIBUTION: Madagascar, in E forests.
SYNONYMS: *buffoni; madagascariensis* Shaw, 1800 (not Zimmermann), *nigriceps, variegatus.*
COMMENTS: Includes *nigriceps*, see Genest and Petter (1975). Eisenberg and Gould (1970:78) believed *nigriceps* is distinct from *semispinosus.*

Setifer Froriep, 1806. *In* Dumeril, Analit. Zool. mit Zusätzen, p. 15.
TYPE SPECIES: *Erinaceus setosus* Schreber, 1777.
SYNONYMS: *Dasogale, Ericulus.*
COMMENTS: Includes *Ericulus*, see Eisenberg and Gould (1970:49); and *Dasogale*, see Poduschka and Poduschka (1982:253).

Setifer setosus (Schreber, 1777). Die Säugethiere, 3:583, pl. 164.
TYPE LOCALITY: Madagascar.
DISTRIBUTION: Madagascar, C plateau.
SYNONYMS: *acanthurus, fontoynonti, nigrescens, spinosus.*
COMMENTS: *Dasogale fontoynonti* was based on a juvenile *Setifer setosus*, see Poduschka and Poduschka (1982:253) and MacPhee (1987*b*:133).

Tenrec Lacépède, 1799. Tabl. Mamm., p.7.
TYPE SPECIES: *Erinaceus ecaudatus* Schreber, 1777.
SYNONYMS: *Centetes* (see Cabrera, 1925:193).

Tenrec ecaudatus (Schreber, 1777). Die Säugethiere, 3:584.
TYPE LOCALITY: Madagascar.
DISTRIBUTION: Madagascar, Comoro Isls, introduced on Reunion, Mauritius, and the Seychelle Isls.
SYNONYMS: *armatus, tanrec.*

Family Chrysochloridae Gray, 1825. Ann. Philos., n.s., 10:335.

COMMENTS: For widely divergent treatments see Simonetta (1968), Meester (1974), and Petter (1981*a*). The generic treatment follows Meester et al. (1986:15-24).

Amblysomus Pomel, 1848. Arch. Sci. Phys. Nat. Geneve, 9:247.

TYPE SPECIES: *Chrysochloris hottentotus* A. Smith, 1829.

SYNONYMS: *Neamblysomus* (see Ellerman et al., 1953).

Amblysomus gunningi (Broom, 1908). Ann. Transvaal Mus., 1:14.

TYPE LOCALITY: South Africa, Transvaal, Woodbush.

DISTRIBUTION: Woodbush Forest and New Agatha Forest Reserve, E Transvaal, South Africa.

STATUS: IUCN - Indeterminate.

COMMENTS: Formerly in the monotypic genus *Neamblysomus* Roberts, 1924.

Amblysomus hottentotus (A. Smith, 1829). Zool. J., 4:436.

TYPE LOCALITY: "Interior parts of South Africa", Grahamstown, E Cape Province, South Africa.

DISTRIBUTION: Natal, Lesotho, Swaziland and Transvaal to S Cape Prov. (South Africa); also NE Orange Free State.

SYNONYMS: *albifrons, devilliersi, drakensbergensis, garneri, longiceps, marleyi, natalensis, orangiensis, pondoliae.*

COMMENTS: Includes *devilliersi* and *marleyi* as subspecies, see Meester et al. (1986:23).

Amblysomus iris Thomas and Schwann, 1905. Abstr. Proc. Zool. Soc. Lond., 1905(18):23.

TYPE LOCALITY: South Africa, Zululand, Umfolozi Station.

DISTRIBUTION: S Cape Prov. to Transkei, Natal, including Zululand, and SE Transvaal (South Africa).

STATUS: IUCN - Indeterminate.

SYNONYMS: *corriae, littoralis, septentrionalis.*

COMMENTS: Includes *corriae* and *septentrionalis* as subspecies, see Meester et al. (1986:23).

Amblysomus julianae Meester, 1972. Ann. Transvaal Mus., 28(4):35.

TYPE LOCALITY: South Africa, Transvaal, Pretoria, The Willows.

DISTRIBUTION: Pretoria, Nylstroom/Nylsvley, and Kruger Nat. Park (Transvaal, South Africa).

STATUS: IUCN - Indeterminate.

Calcochloris Mivart, 1867. J. Anat. Physiol., London, 2:133.

TYPE SPECIES: *Chrysochloris obtusirostris* Peters, 1851.

SYNONYMS: *Chrysotricha.*

COMMENTS: Includes *Chrysotricha*, see Meester et al. (1986:23). Ellerman et al. (1953) included *Calcochloris* in *Amblysomus.*

Calcochloris obtusirostris (Peters, 1851). Bericht. Verhandl. K. Preuss. Akad. Wiss. Berlin, 16:467.

TYPE LOCALITY: Coastal Mozambique, Inhambane, 24°S.

DISTRIBUTION: Zululand and E Transvaal (South Africa), S Zimbabwe, and S Mozambique.

STATUS: IUCN - Rare.

SYNONYMS: *chrysillus, limpopoensis.*

COMMENTS: Includes *chrysillus* and *limpopoensis* as subspecies, see Roberts (1951:114-115).

Chlorotalpa Roberts, 1924. Ann. Transvaal Mus., 10:64.

TYPE SPECIES: *Chrysochloris duthieae* Broom, 1907.

SYNONYMS: *Amblysomus, Carpitalpa.*

COMMENTS: Included in *Amblysomus* by Ellerman et al. (1953) and by Petter (1981*a*). Lundholm (1955*a*:285) described *Carpitalpa* and *Kilimitalpa* (here included in *Chrysochloris*) as subgenera; *Carpitalpa* was regarded by Simonetta (1968) as a valid genus. Both included in *Amblysomus* by Meester (1974).

Chlorotalpa arendsi Lundholm, 1955. Ann. Transvaal Mus., 22:285.

TYPE LOCALITY: E escarpment of Zimbabwe, Inyanga, Pungwe Falls.

DISTRIBUTION: E Zimbabwe and adjacent Mozambique.

COMMENTS: Formerly included in *Carpitalpa* by Simonetta (1968).

Chlorotalpa duthieae (Broom, 1907). Trans. S. Afr. Philos. Soc., 18:292.
TYPE LOCALITY: South Africa, S Cape Prov., Knysna.
DISTRIBUTION: S Cape Prov., South Africa.
STATUS: IUCN - Rare.

Chlorotalpa leucorhina (Huet, 1885). Nouv. Arch. Mus. Hist. Nat. Paris, Bull., 8:8.
TYPE LOCALITY: "Gulf of Guinea Coast, Congo."
DISTRIBUTION: N Angola, Zaire, Cameroon, Central African Republic.
SYNONYMS: *cahni, congicus, luluanus.*
COMMENTS: Includes *cahni* as a subspecies, see Meester (1974). Included in *Chrysochloris* by Allen (1939); included in *Amblysomus* by Simonetta (1968) and Petter (1981*a*).

Chlorotalpa sclateri (Broom, 1907). Ann. Mag. Nat. Hist., ser. 7, 19:263.
TYPE LOCALITY: South Africa, Cape Prov., Beaufort West.
DISTRIBUTION: Cape Prov., E Orange Free State, and S Transvaal (South Africa); Lesotho.
STATUS: IUCN - Indeterminate.
SYNONYMS: *guillarmodi, montana, shortridgei.*
COMMENTS: Meester et al. (1986:21) listed *guillarmodi, shortridgei,* and *montana* as subspecies. Included in *Amblysomus* by Petter (1981*a*).

Chlorotalpa tytonis (Simonetta, 1968). Monitore Zool. Ital., n.s., 2(suppl.):31.
TYPE LOCALITY: Somalia, Giohar (= Villaggio Duca degli Abruzzi).
DISTRIBUTION: Known only from the type locality.
STATUS: IUCN - Indeterminate.
COMMENTS: Assigned to *Amblysomus* by Simonetta (1968:31) and Petter (1981*a*); Meester (1974) placed this species in *Chlorotalpa*.

Chrysochloris Lacépède, 1799. Tabl. Mamm., p. 7.
TYPE SPECIES: *Chrysochloris capensis* Lacépède, 1799 (= *Talpa asiatica* Linnaeus, 1758).
SYNONYMS: *Kalimitalpa.*

Chrysochloris asiatica (Linnaeus, 1758). Syst. Nat., 10th ed., 1:53.
TYPE LOCALITY: "In Sibiria"; usually taken as Cape of Good Hope, South Africa. See Ellerman et al. (1953).
DISTRIBUTION: W Cape Prov. and Robben Isl. (South Africa); perhaps Damaraland, Namibia.
SYNONYMS: *auratus, aurea, capensis, bayoni, calviniae, concolor, damarensis, dixoni, elegans, inaurata, minor, namaquensis, rubra, shortridgei, taylori, tenuis, visserae* (see Meester et al., 1986).

Chrysochloris stuhlmanni Matschie, 1894. Sitzb. Ges. Naturf. Fr. Berlin, p. 123.
TYPE LOCALITY: Uganda, Ruwenzori region, "Ukondjo und Kinyawanga".
DISTRIBUTION: Cameroon, N Zaire, Uganda, Kenya, Tanzania.
SYNONYMS: *balsaci, fosteri, tropicalis, vermiculus.*
COMMENTS: See Meester (1974) who placed *stuhlmanni* in *Chrysochloris*. Lundholm (1955*a*) proposed the name *Chlorotalpa* (*Kilimitalpa*) for this species. Simonetta (1968:31) regarded it as a synonym of *Carpitalpa*. He placed *arendsi, stuhlmanni* and *fosteri* in *Carpitalpa* and *tropicalis* in *Chlorotalpa*. Lamotte and Petter (1981) described *balsaci* from Mt. Oku, Cameroon, a form which may deserve full specific status. Also the isolated *tropicalis* should be re- studied.

Chrysochloris visagiei Broom, 1950. Ann. Transvaal Mus., 21:238.
TYPE LOCALITY: South Africa, Cape Prov., Gouna (54 mi. [87 km] E Calvinia).
DISTRIBUTION: Known only from the type.
STATUS: IUCN - Indeterminate.
COMMENTS: Possibly an aberrant *asiatica*; see Meester (1974). Simonetta (1968:31) included it in *asiatica* as a subspecies.

Chrysospalax Gill, 1883. Standard Nat. Hist., 5 (Mamm.):137.
TYPE SPECIES: *Chrysochloris trevelyani* Günther, 1875.
SYNONYMS: *Bematiscus* (see Ellerman et al., 1953).

Chrysospalax trevelyani (Günther, 1875). Proc. Zool. Soc. Lond., 1875:311.
TYPE LOCALITY: South Africa, Cape Prov., Pirie Forest, near King William's Town.
DISTRIBUTION: Cape Prov. (South Africa).
STATUS: IUCN - Rare.

Chrysospalax villosus (A. Smith, 1833). S. Afr. Quart. J., 2:81.
TYPE LOCALITY: "Towards Natal", near Durban, South Africa; see Roberts (1951:121).
DISTRIBUTION: Transvaal and Natal (South Africa).
STATUS: IUCN - Vulnerable.
SYNONYMS: *dobsoni, leschae, pratensis, rufopallidus, rufus, transvaalensis.*
COMMENTS: Meester et al. (1986:16-17) listed *dobsoni, leschae, rufopallidus, rufus,* and *transvaalensis* as subspecies.

Cryptochloris Shortridge and Carter, 1938. Ann. S. Afr. Mus., 32:284.
TYPE SPECIES: *Cryptochloris zyli* Shortridge and Carter, 1938.
COMMENTS: Simonetta (1968:31) regarded *Cryptochloris* as a synonym of *Chrysochloris.*

Cryptochloris wintoni (Broom, 1907). Ann. Mag. Nat. Hist., ser. 7, 19:264.
TYPE LOCALITY: South Africa, Cape Prov., Little Namaqualand, Port Nolloth.
DISTRIBUTION: Little Namaqualand, Cape Prov., South Africa.
STATUS: IUCN - Indeterminate.

Cryptochloris zyli Shortridge and Carter, 1938. Ann. S. Afr. Mus., 32:284.
TYPE LOCALITY: South Africa, NW Cape Prov., Compagnies Drift, 16 km inland from Lamberts Bay.
DISTRIBUTION: Known only from the type locality.
STATUS: IUCN - Indeterminate.
COMMENTS: Considered a subspecies of *wintoni* by Ellerman et al. (1953); however, Meester et al. (1986:18) argued for specific status.

Eremitalpa Roberts, 1924. Ann. Transvaal Mus., 10:63.
TYPE SPECIES: *Chrysochloris granti* Broom, 1907.

Eremitalpa granti (Broom, 1907). Ann. Mag. Nat. Hist., ser. 7, 19:265.
TYPE LOCALITY: South Africa; Garies, south of Kamiesberg, Little Namaqualand, Cape Prov.
DISTRIBUTION: Coastal dunes from Cape Prov., South Africa, to Namib Desert, Namibia.
STATUS: IUCN - Rare.
SYNONYMS: *cana, namibensis.*
COMMENTS: Includes *namibensis* as a subspecies; see Meester et al. (1986:19).

Family Erinaceidae G. Fischer, 1817. Mem. Soc. Imp. Nat., Moscow, 5:372.
COMMENTS: Name often accredited to Bonaparte, 1838. Reviewed by Corbet (1988) and Frost et al. (1991).

Subfamily Erinaceinae G. Fischer, 1817. Mem. Soc. Imp. Nat., Moscow, 5:372.
COMMENTS: Reviewed by Robbins and Setzer (1985) and Corbet (1988).

Atelerix Pomel, 1848. Arch. Sci. Phys. Nat. Geneve, 9:251.
TYPE SPECIES: *Erinaceus albiventris* Wagner, 1841.
SYNONYMS: *Aethechinus, Peroechinus.*
COMMENTS: Formerly in *Erinaceus,* but see Robbins and Setzer (1985) and Corbet (1988:149).

Atelerix albiventris (Wagner, 1841). *In* Schreber, Die Säugethiere, Suppl. 2:22.
TYPE LOCALITY: Probably Senegal or Gambia; see Allen (1939:20).
DISTRIBUTION: Savanna and steppe zones from Senegal to Ethiopia and south to the Zambezi River.
SYNONYMS: *adansoni, atratus, diadematus, faradjius, heterodactylus, hindei, kilimanus, langi, lowei, oweni, pruneri, sotikae, spiculus, spinifex* (see Corbet, 1988:149 and Ansell, 1974*b*).

Atelerix algirus (Lereboullet, 1842). Mem. Soc. Hist. Nat. Strasbourg, 3(2), art. QQ:4.
TYPE LOCALITY: Algeria, "provient de Oran".

DISTRIBUTION: Coastal Western Sahara to Algeria, Tunisia, and N Libya; introduced into Canary Isls, Balearic Isls, Malta, and Mediterranean France and Spain; one historical record from Puerto Rico.

SYNONYMS: *caniculus, diadematus, fallax, girbaensis, krugi, lavaudeni, vagans.*

COMMENTS: Authorship is often credited to Duvernoy and Lereboullet, 1842, but Saint-Girons (1972) showed that Lereboullet was the only author. Includes *vagans* and *girbaensis* as subspecies, see Hutterer (1983c).

Atelerix frontalis (A. Smith, 1831). S. Afr. Quart. J., 2:10,29.

TYPE LOCALITY: "Cape Colony"; restricted to northern parts of the Graaff Reinet district, Cape Prov., South Africa, by Ellerman et al. (1953).

DISTRIBUTION: Cape Province to E Botswana and W Zimbabwe; and Namibia to SW Angola.

STATUS: IUCN - Rare.

SYNONYMS: *angolae; angolensis, capensis, diadematus* Dobson, 1882, *fractilis.*

COMMENTS: Includes *angolae* as a subspecies, see Meester et al. (1986:15).

Atelerix sclateri Anderson, 1895. Proc. Zool. Soc. Lond., 1895:415.

TYPE LOCALITY: "Taf in Central Somaliland." [Somalia].

DISTRIBUTION: N Somalia.

COMMENTS: Closely related to *albiventris* and might be only a subspecies, see Corbet (1988:152).

Erinaceus Linnaeus, 1758. Syst. Nat., 10th ed., 1:52.

TYPE SPECIES: *Erinaceus europaeus* Linnaeus, 1758.

COMMENTS: Formerly included *Atelerix* and *Aethechinus;* see Corbet (1988) and comments under *Atelerix*. Does not include *Mesechinus*, see comments therein.

Erinaceus amurensis Schrenk, 1859. Reisen im Amur-Lande, 1, pl. 4, fig. 2:100.

TYPE LOCALITY: "In der Nähe der Stadt Aigun, im mandschurischen Dorfe Gulssoja am Amur", E Siberia.

DISTRIBUTION: Russia; Amur River and tributaries, from Zeya eastward, then south through E China to Hunan Prov.; Korea.

SYNONYMS: *chinensis, dealbatus, hanensis, koreanus, koreensis, kreyenbergi, orientalis, tschifuensis, ussuriensis.*

COMMENTS: Formerly included in *europaeus* (see Corbet, 1978c, and Gromov and Baranova, 1981); but considered distinct by Corbet (1984). Range and subspecific boundaries uncertain, partly due to confusion with *Hemiechinus*, see Corbet (1988:144).

Erinaceus concolor Martin, 1838. Proc. Zool. Soc. Lond., 1837:103 [1838].

TYPE LOCALITY: Near Trabzon, Turkey.

DISTRIBUTION: E Europe; S Russia and W Siberia to River Ob; Asia Minor to Israel and Iran; Greek and Adriatic islands including Crete, Corfu, and Rhodes.

SYNONYMS: *abasgicus, bolkayi, danubicus, drozdovskii, kievensis, nesiotes, pallidus, ponticus, rhodius, roumanicus, sacer, transcaucasicus.*

COMMENTS: Formerly included in *europaeus;* but see Kratochvíl (1975), Král (1967), and Orlov (1969:6). Geographic variation is pronounced and some names (see above) possibly represent valid subspecies; see Giagia and Ondrias (1980) and Corbet (1988:142).

Erinaceus europaeus Linnaeus, 1758. Syst. Nat., 10th ed., 1:52.

TYPE LOCALITY: "Europa", Sweden.

DISTRIBUTION: W Europe; Spain to Italy and Istrian Peninsula; north to Poland, Scandinavia and NW European Russia. Islands of Ireland, Britain, Corsica, Sardinia, Sicily, and many smaller islands. European range mapped by Holz and Niethammer (1990:37). Introduced to New Zealand, see King (1990).

SYNONYMS: *caniceps, caninus, consolei, centralrossicus, dissimilis, echinus, erinaceus, hispanicus, italicus, meridionalis, occidentalis, pallidus, suillus, typicus,.*

COMMENTS: Formerly included *amurensis* and *concolor*, see comments therein. Subspecific boundaries are uncertain, see Corbet (1988:137).

Hemiechinus Fitzinger, 1866. Sitzb. Akad. Wiss. Wien, 54, 1:565.

TYPE SPECIES: *Erinaceus platyotis* Sundevall, 1842 (= *Erinaceus auritus* Gmelin, 1770).

SYNONYMS: *Ericius, Erinaceolus, Macroechinus, Paraechinus.*
COMMENTS: Regarded as a subgenus of *Erinaceus* by Gureev (1979:168) and Gromov and Baranova (1981:9). Corbet (1978c:15) considered *Hemiechinus* a distinct genus, later reviewed by Corbet (1988), who included *Mesechinus*, see comments therein. Pavlinov and Rossolimo (1987:12-13) included *Paraechinus* in *Hemiechinus* as a valid subgenus, as did Frost et al. (1991:27), while Corbet (1988) argued for a generic separation of *Paraechinus*. The taxonomy of this group is unsettled, and different opinions about generic and specific boundaries still exist.

Hemiechinus aethiopicus (Ehrenberg, 1832). Symb. Phys. Mamm., 2, sig. k, footnote.
TYPE LOCALITY: Sudan, "In desertis dongolanis habitat".
DISTRIBUTION: Sahara from Mauritania to Egypt and Awash, Ethiopia; Arabia deserts; insular populations on Djerba (Tunisia), Bahrain and Tanb (Persian Gulf).
SYNONYMS: *albatus, albior, blancalis, brachydactylus, deserti, dorsalis, ludlowi, oniscus, pallidus, pectoralis, senaariensis, wassifi.*
COMMENTS: Subgenus *Paraechinus*. Species and subspecies arrangement unclear; Corbet (1988:153-154) retained Arabian *dorsalis* (= *pectoralis*) as a subspecies, while Osborn and Helmy (1980) regarded *aethiopicus, deserti* and *dorsalis* as distinct species.

Hemiechinus auritus (Gmelin, 1770). Nova Comm. Acad. Sci. Petropoli, 14:519.
TYPE LOCALITY: S Russia, "in regione Astrachanensi", (= Astrakhan, 46°21'N, 48°03'E).
DISTRIBUTION: Steppe zone from E Ukraine to Mongolia in the north and from Libya to W Pakistan in the south.
SYNONYMS: *albulus, aegyptius, alaschanicus, brachyotis, calligoni, caspicus, chorassanicus, dorotheae, frontalis* Dobson, 1882, *holdereri, homalacanthus, insularis, libycus, major, megalotis, metwallyi, microtus, minor, persicus, platyotis, russowi, syriacus, turanicus, turfanicus, turkestanicus* (see Corbet, 1988; Frost et al., 1991).
COMMENTS: Subgenus *Hemiechinus*. Corbet (1988:159) accepted *albulus, auritus*, and *megalotis* as valid subspecies; *megalotis* was formerly regarded as a distinct species but intergrades with *auritus* in Afghanistan; see Niethammer (1973). Osborn and Helmy (1980:57-64) recognized two subspecies within Egypt, *aegyptius* and *libycus*. The form of Cyprus (*dorotheae*) may be also distinct, according to Boye (1991:115).

Hemiechinus collaris (Gray, 1830). *In* Hardwicke, Illust. Indian Zool., 1, pl.8.
TYPE LOCALITY: "Doab", restricted by Wroughton (1910:81) to "between Jumna and Ganges Rivers", India; see discussion in Wroughton (1910:81).
DISTRIBUTION: Pakistan and NW India.
SYNONYMS: *blanfordi, grayi, indicus, spatangus.*
COMMENTS: Subgenus *Hemiechinus*. Formerly included in *auritus*, but Roberts (1977) indicated that there is discontinuity in distribution and morphology between *collaris* (which he called *auritus collaris*) and *auritus* (which he called *megalotis*). Includes *blanfordi*, see Corbet and Hill (1992).

Hemiechinus hypomelas (Brandt, 1836). Bull. Sci. Acad. Imp. Sci. St. Petersbourg, 1:32.
TYPE LOCALITY: "Pays de Turcomans", somewhere in S Kazakhstan. See Ognev (1927) for discussion.
DISTRIBUTION: Arid steppe and desert zones from Iran and Turkmenistan east almost to Tashkent (Uzbekistan), to the Indus R. and N Pakistan; isolates in Oman, near Aden and on the islands of Tanb and Kharg in the Persian Gulf.
SYNONYMS: *amir, eversmanni, jerdoni, macracanthus, niger, sabaeus, seniculus.*
COMMENTS: Subgenus *Paraechinus*. Includes *eversmanni, sabaeus* and *seniculus* as possible subspecies; see Corbet (1988:155). Does not include *blanfordi*, see Corbet and Hill (1992)

Hemiechinus micropus (Blyth, 1846). J. Asiat. Soc. Bengal, 15:170.
TYPE LOCALITY: "Bhawulpore" = Bahawalpur, Punjab, Pakistan.
DISTRIBUTION: The arid zones of Pakistan and NW India.
SYNONYMS: *intermedius, kutchicus, mentalis, pictus.*
COMMENTS: Type species of subgenus *Paraechinus*. Biswas and Ghose (1970) regarded *intermedius* as a species but Corbet (1988:156-157) included it in *Paraechinus micropus* as a synonym.

Hemiechinus nudiventris (Horsfield, 1851). Cat. Mamm. Mus. E. India Co., p. 136.
TYPE LOCALITY: "Madras" = Madras city or Tamil Nadu province, India.

DISTRIBUTION: Few records from the S Indian provinces Madras (= Tamil Nadu) and Travancore (= Kerala).
COMMENTS: Subgenus *Paraechinus*. Biswas and Ghose (1970) gave *nudiventris* specific rank while Corbet (1988:156-157) regarded it as a distinct subspecies of *micropus*. Provisonally listed as a species, following Frost et al. (1991:29).

Mesechinus Ognev, 1951. Byull. Moskow. Ova. Ispyt. Prir. Otd. Biol., 56:8.
TYPE SPECIES: *Erinaceus dauuricus* Sundevall, 1842.
COMMENTS: Pavlinov and Rossolimo (1987:11) proposed to place *Mesechinus* as subgenus in *Erinaceus* while Corbet (1988:163) included it in *Hemiechinus*. Most recently, Frost et al. (1991:30) concluded that *Mesechinus* deserves full generic status.

Mesechinus dauuricus (Sundevall, 1842). K. Svenska Vet.-Akad. Handl. Stockholm, 1841:237 [1842].
TYPE LOCALITY: "Dauuria" = Dauryia, Transbaikalia, Russia (49°57'N, 116°55'E).
DISTRIBUTION: NE Mongolia east to upper Amur Basin in Russia and adjacent parts of Inner Mongolia and W Manchuria, China.
SYNONYMS: *manchuricus, przewalskii, sibiricus.*
COMMENTS: Includes *sibiricus;* see Corbet (1978*c*:15, *a*). A considerable confusion of names has occurred in the literature; see Corbet (1988:163). Possibly includes *miodon,* see comments under *M. hughi.*

Mesechinus hughi (Thomas, 1908). Abstr. Proc. Zool. Soc. Lond., 1908(63):44.
TYPE LOCALITY: "Paochi, Shen-si" = Baoji, Shaanxi Prov., China.
DISTRIBUTION: Known from around two localities in Shaanxi and Shanxi Prov., C China.
SYNONYMS: *miodon, sylvaticus.*
COMMENTS: Formerly included in *Erinaceus europaeus* by Ellerman and Morrison-Scott (1951:21), and in *Hemiechinus dauuricus* (here called *Mesechinus dauuricus*) by Corbet (1978*c*:15). Includes *H. sylvaticus* described by Ma (1964:35). The form *miodon,* known from an isolated population in the Ordos desert, Shaanxi, has been alternatively assigned to *M. dauuricus* or to *M. hughi,* see discussion in Frost et al. (1991) for tenative placement here.

Subfamily Hylomyinae Anderson, 1879. Anat. Zool. Res., Yunnan, 1:138.
SYNONYMS: Echinosoricinae, Galericinae, Gymnurinae.
COMMENTS: Better known as Echinosoricinae or Galericinae; but see Frost et al. (1991:23), whose taxonomic proposal is adopted here.

Echinosorex Blainville, 1838. C.R. Acad. Sci. Paris, 6:742.
TYPE SPECIES: *Viverra gymnura* Raffles, 1822.
SYNONYMS: *Gymnura.*
COMMENTS: *Gymnura* Lesson, 1827, is preoccupied by *Gymnura* Kuhl, 1824 (a fish); see Ellerman and Morrison-Scott (1951:17) and Medway (1977:15).

Echinosorex gymnura (Raffles, 1822). Trans. Linn. Soc. Lond., 13:272.
TYPE LOCALITY: Not given; "Sumatra" implied.
DISTRIBUTION: Malayan Peninsula, Borneo and Sumatra, Labuan island.
SYNONYMS: *albus, birmanica, borneotica, candida, minor, rafflesii.*
COMMENTS: Two subspecies, *gymnura* (Sumatra and Malay Peninsula) and *albus* (Borneo) are recognized; see Corbet (1988:128). The common spelling of the specific epithet as *gymnurus* is incorrect; see Frost et al. (1991:24).

Hylomys Müller, 1840. *In* Temminck, Verh. Nat. Gesch. Nederland Overz. Bezitt., Zool., Zoogd. Indisch. Archipel, p. 50[1840].
TYPE SPECIES: *Hylomys suillus* Müller, 1840.
SYNONYMS: *Neohylomys, Neotetracus* (according to Frost et al., 1991:23).
COMMENTS: For date of publication see Appendix I.

Hylomys hainanensis (Shaw and Wong, 1959). Acta Zool. Sinica, 11:422.
TYPE LOCALITY: China, "Pai-sa Hsian, Hainan Island" [= Baisha Xian, an administrative unit at 19°13'N, 109°26'E].

DISTRIBUTION: Island of Hainan, China.

Hylomys sinensis (Trouessart, 1909). Ann. Mag. Nat. Hist., ser. 8, 4:390.
TYPE LOCALITY: "Ta-tsien-lou, province of Se-tchouen (China Occidental) at an altitude of 2454 meters" [= Kangding, Sichuan Sheng, 30°07'N, 102°02'E].
DISTRIBUTION: S China in Sichuan and Yunnan, and adjacent parts of Burma and N Vietnam.
SYNONYMS: *cuttingi, fulvescens.*
COMMENTS: Two subspecies, *fulvescens* and *cuttingi,* have been described; see Corbet (1988:127).

Hylomys suillus Müller, 1840. *In* Temminck, Verh. Nat. Gesch. Nederland Overz. Bezitt., Zool., Zoogd. Indisch. Archipel., p. 50[1840].
TYPE LOCALITY: "Java en het andere van Sumatra" Indonesia.
DISTRIBUTION: Peninsular Malaysia to Indochina and the Yunnan/Burmese border; islands of Borneo, Java, Sumatra and Tioman.
SYNONYMS: *dorsalis, maxi, microtinus, parvus, peguensis, siamensis, tionis.*
COMMENTS: Includes *dorsalis* and *tionis* as discrete subspecies; see Corbet (1988:122). For date of publication see Appendix I.

Podogymnura Mearns, 1905. Proc. U.S. Natl. Mus., 28:436.
TYPE SPECIES: *Podogymnura truei* Mearns, 1905.
COMMENTS: Reviewed by Heaney and Morgan (1982), Corbet (1988), and Frost et al. (1991).

Podogymnura aureospinula Heaney and Morgan, 1982. Proc. Biol. Soc. Washington, 95:14.
TYPE LOCALITY: "Plaridel, Albor Municipality, Dinagat Island, Surigao del Norte Province," Philippines.
DISTRIBUTION: Dinagat Island, Philippines.
COMMENTS: Heaney and Morgan (1982) suggested "golden-spined gymnure" as an English name, but Poduschka and Poduschka (1985) argued that the stiff dorsal hairs are not always spiny and golden, and Corbet and Hill (1991:27) suggested "spiny moonrat" as a common name. Heaney and Morgan (1982) considered that generic rank might be justified for this species but decided to include it in *Podogymnura* in order to emphasize the close relationship between the two species of Philippine gymnures; see also Corbet (1988:130-131).

Podogymnura truei Mearns, 1905. Proc. U.S. Natl. Mus., 28:437.
TYPE LOCALITY: Philippines, Mindanao, Mount Apo, Davao.
DISTRIBUTION: Mindanao Isl, Philippines.
STATUS: IUCN - Vulnerable.
SYNONYMS: *minima.*
COMMENTS: Includes *minima* Sanborn, 1953; see data of Heaney and Morgan (1982).

Family Soricidae G. Fischer, 1817. Mem. Soc. Imp. Nat. Moscow, 5:372.
SYNONYMS: Heterosoricidae.
COMMENTS: Revised by Repenning (1967) and Gureev (1971, 1979). For conflicting views of phylogeny, see Jammot (1983), George (1986), and Reumer (1987). Geological age currently believed to be Miocene, but recently the genera *Cretasorex* (Nesov and Gureev, 1981) from the Upper Cretaceous of Uzbekistan and *Ernosorex* (Wang and Li, 1990) from the Eocene of China have been assigned to this family; while the former clearly represents a shrew the latter does not and may be a member of Plesiosoricidae. Currently accepted limits of subfamilies and tribes are very tentative. Most authors do not follow Reumer (1987) and include the extinct Heterosoricinae as a subfamily (Engesser, 1975; Storch and Qiu, 1991), a view also accepted here. As a consequence, all living shrews as the sister group of Heterosoricidae should be classified in a single subfamily Soricinae, with Soricini and Crocidurini as tribes. However, as a convincing new phylogenetic system of fossil and living shrews has not yet been elaborated and as the application of two extant subfamilies proposed by Repenning (1967) is very much in use, it is provisionally applied here until new evidence has been presented.

Subfamily Crocidurinae Milne-Edwards, 1872. Rech. Hist. Nat. Mammifères, p. 256.
SYNONYMS: Crocidosoricinae, Myosoricinae.
COMMENTS: Perhaps not a monophyletic group. Reumer (1987) postulated a new subfamily, Crocidosoricinae, for a number of fossil genera previously included in Crocidurinae. By application of the characters given for the new clade, a number of recent genera such as *Congosorex*, *Myosorex* and *Surdisorex* would go with it. The entire group is in need of revision.

Congosorex Heim de Balsac and Lamotte, 1956. Mammalia, 20:167.
TYPE SPECIES: *Myosorex polli* Heim de Balsac and Lamotte, 1956.
COMMENTS: Described as a subgenus of *Myosorex* by Heim de Balsac and Lamotte (1956), but differs in its tooth formula, long tail and large ears and was therefore treated as a full genus by Heim de Balsac (1967), a view followed here.

Congosorex polli (Heim de Balsac and Lamotte, 1956). Mammalia, 20:155.
TYPE LOCALITY: Zaire, "Lubondai (Kasai)".
DISTRIBUTION: Known only from type locality in S Zaire.
STATUS: IUCN - Insufficiently known.
COMMENTS: Since the discovery in 1955 this distinct species has not been collected again.

Crocidura Wagler, 1832. Isis, p. 275.
TYPE SPECIES: *Sorex leucodon* Herman, 1780.
SYNONYMS: *Afrosorex*, *Heliosorex*, *Leucodon*, *Paurodus*, *Praesorex*, *Rhinomus* (see Allen, 1939; Heim de Balsac and Meester, 1977; and Hutterer, 1986*a*).
COMMENTS: Eurasian species revised by Jenkins (1976). Phylogenetic relationships of African and Palearctic species studied by Maddalena (1990). Gureev (1979) listed *Praesorex* as a distinct genus.

Crocidura aleksandrisi Vesmanis, 1977. Bonn. Zool. Beitr., 28:3.
TYPE LOCALITY: Libya, Cyrenaica, 5 km W. Tocra.
DISTRIBUTION: Restricted to Cyrenaica, Libya.
COMMENTS: Sometimes included in *C. suaveolens* but currently regarded as a valid species (Hutterer, 1991).

Crocidura allex Osgood, 1910. Field Mus. Nat. Hist. Publ., Zool. Ser., 10(3):20.
TYPE LOCALITY: Kenya, "Naivasha, British East Africa".
DISTRIBUTION: Higlands of SW Kenya; Mt. Kilimanjaro, Meru and Ngorogoro, N Tanzania.
SYNONYMS: *alpina*, *zinki* (see Heim de Balsac and Meester, 1977).
COMMENTS: Gureev (1979) listed *alpina* as a distinct species without comment.

Crocidura andamanensis Miller, 1902. Proc. U.S. Natl. Mus., 24:777.
TYPE LOCALITY: India, Andaman Isls, South Andaman Isl.
DISTRIBUTION: Andaman Isls, Bay of Bengal.
COMMENTS: Erroneously attributed to genus *Suncus* by Krumbiegel (1978:71).

Crocidura ansellorum Hutterer and Dippenaar, 1987. Bonn. Zool. Beitr., 38:1, 269.
TYPE LOCALITY: Zambia, Mwinilunga Distr., Kasombu stream (= Isombu River), 4100 ft.
DISTRIBUTION: N Zambia.
STATUS: IUCN - Insufficiently known.
COMMENTS: Known from only two specimens.

Crocidura arabica Hutterer and Harrison, 1988. Bonn. Zool. Beitr., 39:64.
TYPE LOCALITY: Oman, Dhofar, Khadrafi [16°42'N, 53°09'E].
DISTRIBUTION: Coastal plains of S Arabian Peninsula (Yemen, Oman).
COMMENTS: Previous to the recognition of *arabica*, specimens have been assigned to *russula* or *suaveolens*; see Harrison and Bates (1991).

Crocidura armenica Gureev, 1963. *In* Mammal Fauna of the U.S.S.R., 1:118.
TYPE LOCALITY: Armenia, 14 km down river from Garni.
DISTRIBUTION: Armenia, Caucasus.
COMMENTS: Revised by Gureev (1979), who considered *armenica* as distinct from *pergrisea*; but see Dolgov and Yudin (1975), who considered it a subspecies; Gromov and Baranova (1981) listed it as a distinct species.

Crocidura attenuata Milne-Edwards, 1872. Rech. Hist. Nat. Mamm., p. 263.
TYPE LOCALITY: China, Szechwan Prov., Moupin.
DISTRIBUTION: Assam, India; Nepal; Bhutan; Burma; Thailand; Vietnam; Hainan, China; Taiwan; Peninsular Malaysia; Sumatra; Java; Christmas Isl (Indian Ocean); Batan Isl, Philippines.
SYNONYMS: *aequicaudata, grisea, kingiana, rubricosa, tanakae, trichura.*
COMMENTS: Reviewed by Heaney and Timm (1983*b*) and Jenkins (1976). Jenkins (1982) included the long-tailed *aequicaudata*, which may not be justified.

Crocidura attila Dollman, 1915. Ann. Mag. Nat. Hist., ser. 8, 15:141.
TYPE LOCALITY: Cameroon, Bitye.
DISTRIBUTION: Cameroon Mtns to E Zaire.
COMMENTS: Formerly included in *buettikoferi*, but separated by Hutterer and Joger (1982).

Crocidura baileyi Osgood, 1936. Field Mus. Nat. Hist. Publ., Zool. Ser., 20:225.
TYPE LOCALITY: Ethiopia, Simien Mtns, Ras Dashan (= Mt. Geech).
DISTRIBUTION: Ethiopian highlands west of the Rift Valley.
STATUS: IUCN - Insufficiently known.
COMMENTS: Revised by Dippenaar (1980).

Crocidura batesi Dollman, 1915. Ann. Mag. Nat. Hist., ser. 8, 15:143.
TYPE LOCALITY: "Como River, Gabon."
DISTRIBUTION: Lowland forest in S Cameroon and Gabon.
COMMENTS: Often included in *poensis*; specimens from Cameroon and Gabon have been reported as *wimmeri*; but see Brosset (1988).

Crocidura beatus Miller, 1910. Proc. U.S. Natl. Mus., 38:392.
TYPE LOCALITY: Philippines, Mindanao, Summit of Mt. Bliss, 1,461 m.
DISTRIBUTION: Philippines: Mindanao, Leyte, Maripipi.
SYNONYMS: *parvacauda.*
COMMENTS: Includes *parvacauda*; see Heaney et al. (1987:36). Distribution reviewed by Heaney (1986).

Crocidura beccarii Dobson, 1886. Ann. Mus. Civ. Stor. Nat. Genova, ser. 2, 4:556.
TYPE LOCALITY: Indonesia, Sumatra, Mt. Singalang.
DISTRIBUTION: Sumatra.
COMMENTS: Species identity unresolved; see Jenkins (1982:277).

Crocidura bottegi Thomas, 1898. Ann. Mus. Civ. Stor. Nat. Genova, ser. 2, 18:677.
TYPE LOCALITY: Ethiopia, north-east of Lake Turkana, "between Badditu and Dime".
DISTRIBUTION: Scattered records from Guinea to Ethiopia and N Kenya.
SYNONYMS: *eburnea.*
COMMENTS: Includes *eburnea*; see Heim de Balsac and Meester (1977). Previously also included *obscurior* which is now treated as a distinct species.

Crocidura bottegoides Hutterer and Yalden, 1990. *In* Peters and Hutterer (eds.), Vertebrates in the Tropics, Bonn, p. 67.
TYPE LOCALITY: Ethiopia, Bale Mtns, Harenna Forest, Katcha Camp, 2400 m.
DISTRIBUTION: Bale Mtns and Mt. Albasso, Ethiopia.

Crocidura buettikoferi Jentink, 1888. Notes Leyden Mus., 10:47.
TYPE LOCALITY: Robertsport, Liberia.
DISTRIBUTION: West African high forest; Guinea-Bissau to Liberia; Nigeria.
COMMENTS: Formerly included *attila*; see Hutterer and Joger (1982).

Crocidura caliginea Hollister, 1916. Bull. Am. Mus. Nat. Hist., 35:664.
TYPE LOCALITY: Zaire, Medje.
DISTRIBUTION: Known only from two localities in NE Zaire.
COMMENTS: The species was recently rediscovered by Hutterer and Dudu (1990).

Crocidura canariensis Hutterer, Lopez-Jurado and Vogel, 1987. J. Nat. Hist., 21:1354.
TYPE LOCALITY: Spain, Canary Isls, Fuerteventura, Tiscamanita.
DISTRIBUTION: E Canary Islands.
STATUS: Protected by Spanish law.
COMMENTS: Related to *sicula*; see Maddalena and Vogel (1990).

Crocidura cinderella Thomas, 1911. Ann. Mag. Nat. Hist., ser. 8, 8:119.
TYPE LOCALITY: "Gemenjulla, French Gambia."
DISTRIBUTION: Senegal and Gambia, Mali and Niger.
COMMENTS: May be related to *tarfayensis* of Morocco and Mauritania; see Hutterer (1987).

Crocidura congobelgica Hollister, 1916. Bull. Am. Mus. Nat. Hist., 35:670.
TYPE LOCALITY: Zaire, "Lubila, near Bafwasende".
DISTRIBUTION: NE Zaire.
STATUS: IUCN - Insufficiently known.
COMMENTS: For a discussion of relationships, see Heim de Balsac (1968*a*).

Crocidura cossyrensis Contoli, 1989. Hystrix, N.S., 1:121, footnote.
TYPE LOCALITY: Italy, Pantelleria Isl.
DISTRIBUTION: Only Pantelleria Isl (Italy).
COMMENTS: The species was first reported as *russula*; see Contoli and Amori (1986); then validly named in a footnote (Contoli et al., 1989) and later redescribed by Contoli (1990). Closely related to *russula* and may be part of it (Sará et al., 1990).

Crocidura crenata Brosset, Dubost, and Heim de Balsac, 1965. Mammalia, 29:268.
TYPE LOCALITY: Gabon, Belinga.
DISTRIBUTION: High forest in S Cameroon, N Gabon, and E Zaire.
STATUS: IUCN - Insufficiently known.
COMMENTS: The specific epithet obviously was choosen because the species has extremily long feet and tail; Brosset (1988) showed that they aid in jumping rather than climbing.

Crocidura crossei Thomas, 1895. Ann. Mag. Nat. Hist., ser. 6, 16:53.
TYPE LOCALITY: Nigeria, "Asaba, 150 mi. [241 km] up the Niger River".
DISTRIBUTION: Lowland forest from Sierra Leone to W Cameroon.
SYNONYMS: *ebriensis, ingoldbyi, jouvenetae*.
COMMENTS: Includes *ebriensis, ingoldbyi* and *jouvenetae*; see Heim de Balsac and Meester (1977). May be composite of two species, *crossei* and *jouvenetae*.

Crocidura cyanea (Duvernoy, 1838). Mem. Soc. Hist. Nat. Strasbourg, 2:2.
TYPE LOCALITY: "La riviere des Elephants, au sud de l'Afrique" = Citrusdal, South Africa *fide* Shortridge (1942:27).
DISTRIBUTION: South Africa, Namibia, Angola, Botswana, Mozambique; records further north uncertain.
SYNONYMS: *argentatus, electa, infumatus, martensi, pondoensis, vryburgensis*.
COMMENTS: The species concept applied by Heim de Balsac and Meester (1977), and Meester et al. (1986) included a number of names which evidently do not belong to *cyanea* but to species such as *parvipes* and *smithii*; see Hutterer (1986*a*) and Hutterer and Joger (1982). The limits of distribution of *cyanea* have not yet been established; the taxon *erica* which has been included in *cyanea* may be related to *hirta*.

Crocidura denti Dollman, 1915. Ann. Mag. Nat. Hist., ser. 8, 16:377.
TYPE LOCALITY: Between Mawambi and Avakubi, Ituri Forest, Zaire.
DISTRIBUTION: NE Zaire, Gabon, Cameroon.
COMMENTS: Considered a distinct species by Heim de Balsac (1959:216).

Crocidura desperata Hutterer, Jenkins and Verheyen, 1991. Oryx, 25:165.
TYPE LOCALITY: S Tanzania, Rungwe Mtns, mountain bamboo zone above 2000 m.
DISTRIBUTION: Relict forest patches at Rungwe and Uzungwe Mtns, S Tanzania.
STATUS: Extremely localized; endangered by forest clearing.

Crocidura dhofarensis Hutterer and Harrison, 1988. Bonn. Zool. Beitr., 39:68.
TYPE LOCALITY: Oman, Dhofar, Khadrafi, 620 m.
DISTRIBUTION: Known only from the type locality.
COMMENTS: Originally described as a subspecies of C. *somalica*, but Hutterer et al. (1992) provided arguments for full specific status.

Crocidura dolichura Peters, 1876. Monatsb. K. Preuss. Akad. Wiss. Berlin, 1876:475.
TYPE LOCALITY: Cameroon, "Bonjongo".

DISTRIBUTION: High forest in Nigeria, S Cameroon, Bioko, Gabon, Central African Republic, Congo Republic, Zaire, and adjacent Uganda and Burundi.
COMMENTS: Does not include *latona*, *ludia*, *muricauda*, and *polia*; see under these species.

Crocidura douceti Heim de Balsac, 1958. Mém. Inst. Fr. Afr. Noire, 53:329.
TYPE LOCALITY: Ivory Coast, Adiopodoume.
DISTRIBUTION: Forest-savanna border of Guinea, Ivory Coast, and Nigeria.
COMMENTS: Reviewed by Hutterer and Happold (1983).

Crocidura dsinezumi (Temminck, 1843). *In* Siebold, Fauna Japonica, 2(Mamm.):26.
TYPE LOCALITY: Japan, Kyushu.
DISTRIBUTION: Japan; Quelpart Isl (Korea); possibly Taiwan.
SYNONYMS: *chisai, hosletti, intermedia, okinoshimae, quelpartis, umbrina.*
COMMENTS: The spelling of the name was clarified by Corbet (1978*b*); *dsinezumi* was placed on the Official List of Specific Names; see the International Commission on Zoological Nomenclature (1983). Includes *chisai* and *quelpartis*, but not *orii*; see Corbet (1978*c*). Formerly included in *russula*; see Jameson and Jones (1977). The subspecies *hosletti* described by these authors from Taiwan is tentatively included in *dsinezumi* rather than *russula* on the basis of published measurements and descriptions.

Crocidura eisentrauti Heim de Balsac, 1957. Zool. Jahrb. Abt. Syst. Oekol. Geogr. Tiere, 85:616.
TYPE LOCALITY: Cameroon, Mt. Cameroon, "Johann-Albrecht-Hütte, 2900 m".
DISTRIBUTION: Higher elevations of Mt. Cameroon (Cameroon).
STATUS: IUCN - Insufficiently known.
COMMENTS: Only known from Mount Cameroon. Not conspecific with *C. vulcani*; see under *C. hildegardeae*.

Crocidura elgonius Osgood, 1910. Ann. Mag. Nat. Hist., ser. 8, 5:369.
TYPE LOCALITY: Kenya, Mt. Elgon, Kirui.
DISTRIBUTION: Mt. Elgon (W Kenya); NE Tanzania.
COMMENTS: Regarded as a distinct species by Heim de Balsac and Meester (1977) and Hutterer (1983*b*).

Crocidura elongata Miller and Hollister, 1921. Proc. Biol. Soc. Washington, 34:101.
TYPE LOCALITY: Indonesia, Sulawesi, Temboan (SW from Tondano Lake).
DISTRIBUTION: N and C Sulawesi.
COMMENTS: See Musser (1987) for ecological notes and a photograph.

Crocidura erica Dollman, 1915. Ann. Mag. Nat. Hist., ser. 8, 15:145.
TYPE LOCALITY: Angola, Pungo Andongo.
DISTRIBUTION: W Angola.
COMMENTS: Resembles *hirta* in cranial dimensions; see Heim de Balsac and Meester (1977). Related to *nigricans*, according to Crawford-Cabral (1987).

Crocidura fischeri Pagenstecher, 1885. Jb. Hamburger Wiss. Anst., 2:34.
TYPE LOCALITY: "Nguruman"; northwest of Lake Natron, close to Mt. Sambo, Kenya (near border to Tanzania); see discussion by Moreau et al. (1946) and Aggundey and Schlitter (1986).
DISTRIBUTION: Nguruman (Kenya), and Himo (Tanzania).
COMMENTS: Type species of subgenus *Afrosorex*. Revised by Hutterer (1986*a*).

Crocidura flavescens (I. Geoffroy, 1827). Dict. Class. Hist. Nat., 11:324.
TYPE LOCALITY: "La Cafrerie et le pays des Hottentots" = King William's Town, South Africa.
DISTRIBUTION: South Africa.
SYNONYMS: *capensis, cinnamomeus, knysnae, rutilus.*
COMMENTS: For correct original citation see Ellerman et al. (1953). Does not include *olivieri*; see Maddalena et al. (1987) and comments under that species. Reviewed by Meester (1963).

Crocidura floweri Dollman, 1915. Ann. Mag. Nat. Hist., ser. 8, 15:515.
TYPE LOCALITY: "Giza, Egypt."

DISTRIBUTION: Environs of Upper Nile valley and Wadi el Natrun, Egypt.
COMMENTS: Mummified shrews from Ancient Egypt have been identified as C. *floweri*; see Heim de Balsac and Mein (1971). Possibly related to *crossei* and *arabica*; see Hutterer and Harrison (1988).

Crocidura foxi Dollman, 1915. Ann. Mag. Nat. Hist., ser. 8, 15:514.
TYPE LOCALITY: Nigeria, Panyam.
DISTRIBUTION: Known only from Jos Plateau, Nigeria, but possibly has a wider distribution in the Sudan savanna zone of West Africa from Senegal to S Sudan.
SYNONYMS: *tephra*.
COMMENTS: A member of the *poensis* group, *foxi* may be conspecific with *theresae*, which it antedates; see Hutterer and Happold (1983). A series from Owerri, S Nigeria, referred to *foxi* by these authors, was later, upon re-examination, identified as a dark form of *lamottei*. The holotype of *tephra* Setzer, 1956 has been recently examined and is regarded as representing *foxi* in S Sudan; a previous allocation to *viaria* (Hutterer, 1984) was based upon examination of a paratype skin; however, the holotype represents a different species.

Crocidura fuliginosa (Blyth, 1856). J. Asiat. Soc. Bengal, ser. 2, 24:362.
TYPE LOCALITY: Burma, Schwegyin, near Pegu.
DISTRIBUTION: N India, Burma, adjacent China, Malaysian Peninsula and adjacent isls; perhaps also Borneo, Sumatra and Java; exact distribution unknown.
SYNONYMS: *baluensis, brevicauda, brunnea, doriae, dracula, foetida, grisescens, kelabit, lawuana, lepidura, macklotii, mansumensis, melanorhyncha, orientalis, praedax, pudjonica, tenuis, villosa, vosmaeri*.
COMMENTS: The taxonomy of this common S Asian shrew is in urgent need of revision. Ruedi et al. (1990) recently demonstrated unrecognized sympatry of two entirely cryptic but chromosomally distinct forms, one of which has been provisionally labeled C. cf. *malayana*. For taxa currently included in *fuliginosa* see above and Jenkins (1976, 1982). Medway (1977) and Heaney and Timm (1983*b*) also included *dracula*, which Lekagul and McNeely (1977) considered a distinct species. The list of synonyms is very provisional; see also under *malayana*.

Crocidura fulvastra (Sundevall, 1843). K. Svenska Vet.-Akad. Handl. Stockholm, 1842:172 [1843].
TYPE LOCALITY: Sudan, "Bahr el Abiad".
DISTRIBUTION: Sudan savanna from Kenya to Mali.
SYNONYMS: *arethusa, beta, diana, macrodon, marrensis, sericea, strauchii* (see Hutterer, 1984; Hutterer and Kock, 1983; and Hutterer and Happold, 1983).
COMMENTS: Gureev (1979) listed *beta* as a distinct species without comment.

Crocidura fumosa Thomas, 1904. Ann. Mag. Nat. Hist., ser. 7, 14:238.
TYPE LOCALITY: Kenya, "Camp 18, western slope of Mt. Kenya, 2,600 m".
DISTRIBUTION: Mt. Kenya and Aberdare Range (Kenya).
SYNONYMS: *alchemillae*.
COMMENTS: Includes *alchemillae*; see Dippenaar and Meester (1989) who revised the species.

Crocidura fuscomurina (Heuglin, 1865). Leopoldina, 5, *in* Nouv. Acta Acad. Caes. Leop.-Carol., 32:36.
TYPE LOCALITY: Sudan, Bahr-el-Ghazal, Meshra-el-Req.
DISTRIBUTION: Sudan and Guinea savanna from Senegal to Ethiopia, and south to South Africa.
SYNONYMS: *bicolor, cuninghamei, hendersoni, marita, sansibarica, tephragaster, woosnami*.
COMMENTS: Revised by Hutterer (1983*b*). C. *planiceps* may belong here but relationships are yet unsolved. See comments under C. *pasha*.

Crocidura glassi Heim de Balsac, 1966. Mammalia, 30:448.
TYPE LOCALITY: Ethiopia, "Camp in Gara Mulata Mts, Harar".
DISTRIBUTION: Ethiopian highlands east of Rift Valley.
STATUS: IUCN - Insufficiently known.
COMMENTS: Often confused with *fumosa* or *thalia*; see Dippenaar (1980).

Crocidura goliath Thomas, 1906. Ann. Mag. Nat. Hist., ser. 7, 17:177.
TYPE LOCALITY: "Efulen, Cameroons."
DISTRIBUTION: High forest of S Cameroon, Gabon, and Zaire.
STATUS: Listed by IUCN as Extinct, but this is incorrect.
COMMENTS: Type species of subgenus *Praesorex* Thomas, 1913. Often included in *flavescens* or *olivieri*, but apparently represents a distinct species which lives in sympatry with *C. olivieri* in the Central African forest; see Hutterer (*in* Colyn, 1986:22).

Crocidura gracilipes Peters, 1870. Monatsb. K. Preuss. Akad. Wiss. Berlin, 1870:584.
TYPE LOCALITY: "Auf der Reise nach dem Kilimandscharo"; unidentifiable but usually taken as "Kilimanjaro, Tanzania"; see Moreau et al. (1946:395).
DISTRIBUTION: Known only from the type specimen with unknown origin.
COMMENTS: Does not include *hildegardeae*; see Demeter and Hutterer (1986:201). A recent examination of the type specimen indicated a conspecificy with *C. cyanea*.

Crocidura grandiceps Hutterer, 1983. Rev. Suisse Zool., 90:699.
TYPE LOCALITY: Ghana, Sefwi-Wiawso, Krokosua Hills, N of Asempanaya (Asampaniye).
DISTRIBUTION: High forest regions of Guinea, Ivory Coast, Ghana, Nigeria, and possibly Cameroon.

Crocidura grandis Miller, 1911. Proc. U.S. Natl. Mus., 38:393.
TYPE LOCALITY: Philippines, Mindanao, Grand Malindang Mt., 6100 ft.
DISTRIBUTION: Known only from Mt. Malindang, Mindanao, Philippines.
COMMENTS: Status unknown; probably confined to primary forest (Heaney et al., 1987:38).

Crocidura grassei Brosset, Dubost, and Heim de Balsac, 1965. Biologia Gabonica, 1:165.
TYPE LOCALITY: Gabon, Belinga.
DISTRIBUTION: Recorded from high forest regions at Belinga (Gabon), Boukoko (Central African Republic), and Yaounde (Cameroon; see Heim de Balsac, 1968*c*).
STATUS: IUCN - Insufficiently known.

Crocidura grayi Dobson, 1890. Ann. Mag. Nat. Hist., ser. 6, 6:494.
TYPE LOCALITY: Philippines, Luzon.
DISTRIBUTION: Luzon and Mindoro, Philippines, in primary forest.
SYNONYMS: *halconus*.
COMMENTS: Heaney et al. (1987) included *halconus* as a synonym.

Crocidura greenwoodi Heim de Balsac, 1966. Monitore Zool. Ital., 74(suppl.):215.
TYPE LOCALITY: Somalia, "Gelib".
DISTRIBUTION: S Somalia.
COMMENTS: Species confined to the Horn of Africa; apparently related to *fulvastra* and *hirta*.

Crocidura gueldenstaedtii (Pallas, 1811). Zoogr. Rosso-Asiat., 1:132.
TYPE LOCALITY: Georgia, near Dushet (N of Tbilisi).
DISTRIBUTION: Transcaucasia.
SYNONYMS: *aralychensis*, *bogdanowii*, *fumigatus*, *longicaudata*.
COMMENTS: This name has produced much confusion. Ellerman and Morrison-Scott (1966) listed it as a subspecies of *russula* and were followed by Corbet (1978*c*), among others. Richter (1970) applied *gueldenstaedtii* even to Mediterranean populations of *suaveolens* and was followed in that action by Kahmann and Vesmanis (1976). Hutterer (1981*d*) suggested that all these populations represent *suaveolens*; this was supported by karyological and biochemical data (Catzeflis et al., 1985). Despite convincing evidence some Russian authors (e.g., Grafodatsky et al., 1988) still claim the existence of *gueldenstaedtii* as a species in the Caucasus. This species is almost certainly conspecific with *suaveolens*.

Crocidura harenna Hutterer and Yalden, 1990. *In* Peters and Hutterer (eds.), Vertebrates in the Tropics, Bonn, p. 64.
TYPE LOCALITY: Ethiopia, Bale Mtns, Harenna Forest.
DISTRIBUTION: Known only from the type locality.
COMMENTS: Related to *C. phaeura*.

Crocidura hildegardeae Thomas, 1904. Ann. Mag. Nat. Hist., ser. 7, 14:240.
TYPE LOCALITY: "Fort Hall, Kenya Colony."

DISTRIBUTION: Localized in West Africa (Nigeria, Cameroon), common in C and E Africa; forest.

SYNONYMS: *altae, ibeana, lutreola, maanjae, phaios, procera, rubecula, virgata* (see Heim de Balsac and Meester, 1977).

COMMENTS: Does not include *gracilipes*; see Dieterlen and Heim de Balsac (1979) and Demeter and Hutterer (1986). Gureev (1979) listed *ibeana, lutreola,* and *maanjae* as distinct species without comment. Species group in need of revision.

Crocidura hirta Peters, 1852. Reise nach Mossambique, Säugethiere, p. 78.

TYPE LOCALITY: Mozambique, Tette, 17° S.

DISTRIBUTION: Angola, Zaire, Uganda, Kenya, Somalia, Tanzania, Malawi, Zimbabwe, Zambia, Mozambique, Botswana, Namibia, South Africa.

SYNONYMS: *annellata, beirae, bloyeti, canescens, deserti, langi, luimbalensis, velutina* (see Heim de Balsac and Meester, 1977).

COMMENTS: Gureev (1979) listed *beirae* and *deserti* as distinct species; the latter may well be separable. *C. bloyeti*, formerly listed as a species, is included here because it was based on a juvenile *hirta*. The Angolan *erica* may also belong here.

Crocidura hispida Thomas, 1913. Ann. Mag. Nat. Hist., ser. 8, 11:468.

TYPE LOCALITY: India, Andaman Isls, Middle Andaman Isl (northern end).

DISTRIBUTION: Middle Andaman Isl (Andaman Isls, India).

Crocidura horsfieldii (Tomes, 1856). Ann. Mag. Nat. Hist., ser. 2, 17:23.

TYPE LOCALITY: "Ceylon."

DISTRIBUTION: Sri Lanka; N Thailand to Vietnam; Nepal; Mysore and Ladak (India); Yunnan, Fukien, and Hainan Isl (China); Taiwan; Ryukyu Isls (Japan).

SYNONYMS: *indochinensis, kurodai, myoides, retusa, tadae, watasei, wuchihensis.*

COMMENTS: Subspecies or synonyms discussed by Jenkins (1976) and Jameson and Jones (1977). Usually spelled *horsfieldi* but Corbet and Hill (1991) correctly used *horsfieldii.*

Crocidura jacksoni Thomas, 1904. Ann. Mag. Nat. Hist., ser. 7, 14:238.

TYPE LOCALITY: Kenya, "Ravine Station".

DISTRIBUTION: E Zaire, Uganda, Kenya, N Tanzania.

SYNONYMS: *amalae.*

COMMENTS: Includes *amalae*; see Heim de Balsac and Meester (1977:17).

Crocidura jenkinsi Chakraborty, 1978. Bull. Zool. Surv. India, 1:303.

TYPE LOCALITY: "Wright Myo, South Andaman Isl., India."

DISTRIBUTION: Known only from the type locality.

COMMENTS: Included in *nicobarica* by Corbet and Hill (1991), without comment.

Crocidura kivuana Heim de Balsac, 1968. Biologia Gabonica, 4:319.

TYPE LOCALITY: Zaire, Kivu, Tschibati.

DISTRIBUTION: Kahuzi-Biega National Park (Zaire).

STATUS: IUCN - Insufficiently known.

COMMENTS: Very localized species occurring in montane swamps; see Dieterlen and Heim de Balsac (1979).

Crocidura lamottei Heim de Balsac, 1968. Mammalia, 32:386.

TYPE LOCALITY: Ivory Coast, "Lamto (savane)".

DISTRIBUTION: Sudan and Guinea savanna from Senegal to W Cameroon.

SYNONYMS: *elegans.*

COMMENTS: Includes *elegans* as a subspecies; see Hutterer (1986*a*).

Crocidura lanosa Heim de Balsac, 1968. Biologia Gabonica, 4:309.

TYPE LOCALITY: Zaire, Kivu, "Lemera".

DISTRIBUTION: Uinka (Rwanda); Kivu, Lemera and Irangi (Zaire).

STATUS: IUCN - Insufficiently known.

COMMENTS: Present knowledge summarized by Dieterlen and Heim de Balsac (1979).

Crocidura lasiura Dobson, 1890. Ann. Mag. Nat. Hist., ser. 6, 5:31.

TYPE LOCALITY: NE China, (Manchuria), Ussuri River.

DISTRIBUTION: Ussuri Region (Russia) and NE China to Korea; Kiangsu (China).

SYNONYMS: *campuslincolnensis, lizenkani, sodyi, thomasi, yamashinai* (see Corbet, 1978*c*:29).

Crocidura latona Hollister, 1916. Bull. Am. Mus. Nat. Hist., 35:667.
TYPE LOCALITY: Zaire, Medje.
DISTRIBUTION: Lowland rainforest of NE Zaire.
STATUS: IUCN - Insufficiently known.

Crocidura lea Miller and Hollister, 1921. Proc. Biol. Soc. Washington, 34:102.
TYPE LOCALITY: Indonesia, Sulawesi, Temboan.
DISTRIBUTION: N and C Sulawesi, tropical rain forest (Musser, 1987).

Crocidura leucodon (Hermann, 1780). *In* Zimmermann, Geogr. Gesch. Mensch. Vierf. Thiere, 2:382.
TYPE LOCALITY: France, Bas Rhin, vicinity of Strasbourg.
DISTRIBUTION: France to the Volga and Caucasus; Elburz Mtns; Asia Minor; Israel; Lebanon; Lesbos Isl (Aegean Sea).
SYNONYMS: *albipes, caspica, judaica, lasia, leucodus, microurus, narentae, persica*.
COMMENTS: Reviewed by Richter (1970) and Gureev (1979). Includes *persica*; see Dolgov (1979). Gureev (1979) and Gromov and Baranova (1981) listed *persica* as a distinct species without comment. Includes *lasia*; see Catzeflis et al. (1985), Gureev (1979), and Jenkins (1976); but also Corbet (1978*c*). Includes *caspica* from Iran and *judaica* from Palestine. European range reviewed by Krapp (1990), Arabian range by Harrison and Bates (1991).

Crocidura levicula Miller and Hollister, 1921. Proc. Biol. Soc. Washington, 34:103.
TYPE LOCALITY: Indonesia, Sulawesi, Pinedapa.
DISTRIBUTION: Tropical rain forest of C and SE Sulawesi (Musser, 1987).

Crocidura littoralis Heller, 1910. Smithson. Misc. Coll., 56(15):5.
TYPE LOCALITY: Uganda, Butiaba, east shore of Lake Albert.
DISTRIBUTION: Rain forest of Zaire, Uganda and Kenya.
SYNONYMS: *oritis*.
COMMENTS: This species was included in *monax*, but is now regarded as distinct (Dieterlen and Heim de Balsac, 1979).

Crocidura longipes Hutterer and Happold, 1983. Bonn. Zool. Monogr., 18:53.
TYPE LOCALITY: Nigeria, "Dada, 11°34'N, 04°29'E".
DISTRIBUTION: Known from two swamps in Guinea savanna in W Nigeria.
STATUS: IUCN - Insufficiently known.
COMMENTS: May be related to *foxi*.

Crocidura lucina Dippenaar, 1980. Ann. Transvaal Mus., 32:134-138.
TYPE LOCALITY: Ethiopia, "Web River, near Dinshu".
DISTRIBUTION: Montane moorlands of E Ethiopia.
STATUS: IUCN - Insufficiently known.
COMMENTS: Species confined to the Afro-Alpine moorland (Hutterer and Yalden, 1990).

Crocidura ludia Hollister, 1916. Bull. Am. Mus. Nat. Hist., 35:668.
TYPE LOCALITY: Zaire, Medje.
DISTRIBUTION: Medje and Tandala (N Zaire).
STATUS: IUCN - Insufficiently known.
COMMENTS: Included in *dolichura* by Heim de Balsac and Meester (1977), but regarded as a full species by Hutterer and Dippenaar (1987).

Crocidura luna Dollman, 1910. Ann. Mag. Nat. Hist., ser. 8, 5:175.
TYPE LOCALITY: Zaire, "Bunkeya River, Shaba Province".
DISTRIBUTION: Mozambique, Zambia, Zimbabwe, E Angola, Zaire, Malawi, Tanzania, Kenya, Uganda, Rwanda.
SYNONYMS: *electa, garambae, inyangai, johnstoni, schistacea, umbrosa*.
COMMENTS: Revised by Dippenaar and Meester (1989). Does not include *macmillani, raineyi*, and *selina*. In a biochemical comparison, specimens from Rwanda grouped outside all other African *Crocidura* studied (Maddalena, 1990).

Crocidura lusitania Dollman, 1915. Ann. Mag. Nat. Hist., ser. 8, 15:516.
TYPE LOCALITY: Mauritania, "Trarza country".

DISTRIBUTION: Sahelian zone from S Morocco to Senegal, Nigeria, Sudan and Ethiopia; a Saharan record from Mali.

COMMENTS: For a summary of distributional records, see Hutterer (1986*a*) and Sidiyene (1989).

Crocidura macarthuri St. Leger, 1934. Ann. Mag. Nat. Hist., ser. 10, 13:559.

TYPE LOCALITY: Kenya, Tana River, Merifano (32 km from mouth of Tana River).

DISTRIBUTION: Savanna plains of Kenya and Somalia.

COMMENTS: The species has been recorded from Somalia as *smithii* (e.g., Heim de Balsac, 1966*a*); see Hutterer (1986*a*).

Crocidura macmillani Dollman, 1915. Ann. Mag. Nat. Hist., ser. 8, 16:361.

TYPE LOCALITY: Ethiopia, "Kotelee, Walamo".

DISTRIBUTION: Known only from the type locality.

COMMENTS: Formerly included in *fumosa* (Yalden et al., 1976) or *luna* (Heim de Balsac and Meester, 1977; Hutterer, 1981*b*), but Dippenaar (1980) has shown that in Ethiopia two endemic species, *macmillani* and *thalia*, were covered under these names.

Crocidura macowi Dollman, 1915. Ann. Mag. Nat. Hist., ser. 8, 16:378.

TYPE LOCALITY: Kenya, "Mt. Nyiro, S. of Lake Rudolf [Lake Turkana]".

DISTRIBUTION: Known only from the type locality.

COMMENTS: Regarded as a synonym of *hildegardeae* by Osgood (1936), but retained as a species by Heim de Balsac and Meester (1977), who noticed similarities to *niobe*.

Crocidura malayana Robinson and Kloss, 1911. J. Fed. Malay St. Mus., 4:241-247.

TYPE LOCALITY: Malaysia, Perak, Maxwell's Hill.

DISTRIBUTION: Peninsular Malaysia and offshore islands; exact distribution unknown.

SYNONYMS: *aagaardi, aoris, gravida, klossi, negligens, maporensis, tionis, weberi.*

COMMENTS: This species was included in *fuliginosa* by Jenkins (1976, 1982), but Ruedi et al. (1990) reported two different karyotypes from sympatric populations in Peninsular Malaysia. They provisionally used *malayana* for the sibling species. Jenkins (1976, 1982), who revised this group, listed the synonyms given here under subspecies *fuliginosa malayana*. The proper allocation of all described names has still to be elaborated; other synonyms are listed under *fuliginosa*.

Crocidura manengubae Hutterer, 1982. Bonn. Zool. Beitr., 32:242.

TYPE LOCALITY: Cameroon, "Lager III, 1800m, Manenguba-See, Bamenda-Hochland".

DISTRIBUTION: Bamenda, Adamaoua, and Yaounde highlands, Cameroon.

STATUS: IUCN - Insufficiently known.

Crocidura maquassiensis Roberts, 1946. Ann. Transvaal Mus., 20:312.

TYPE LOCALITY: South Africa, W Transvaal, Maquassi, Klipkuil.

DISTRIBUTION: Transvaal (South Africa); Nyamaziwa Falls, and Matopo Hills (Zimbabwe).

STATUS: IUCN - Insufficiently known.

SYNONYMS: *malani.*

COMMENTS: Includes *malani*; and may be related to *pitmani*; see Meester (1963) and Meester et al. (1986).

Crocidura mariquensis (A. Smith, 1844). Illustr. Zool. S. Afr. Mamm., pl. 44, fig. 1.

TYPE LOCALITY: South Africa, "A wooded ravine near the tropic of Capricorn" = Marico River, near its junction with Limpopo.

DISTRIBUTION: Swamps and forest from South Africa to Mozambique, W Zimbabwe, and Zambia; NW Botswana and NE Namibia to SC Angola; perhaps SE Zaire.

SYNONYMS: *neavei, pilosa, shortridgei, sylvia.*

COMMENTS: Includes *pilosa* and *sylvia* as synonyms and *shortridgei* and *neavei* as subspecies; see Dippenaar (1977, 1979), who reviewed the species and selected a lectotype. May also include *nigricans*, which Crawford-Cabral (1987) considered distinct.

Crocidura maurisca Thomas, 1904. Ann. Mag. Nat. Hist., ser. 7, 14:239.

TYPE LOCALITY: Uganda, Entebbe.

DISTRIBUTION: Entebbe, Echuya Swamp (Uganda); Kaimosi (Kenya); in swamps and primary forest.

Crocidura maxi Sody, 1936. Natuurk. Tijdschr. Ned.-Ind., 96:53.
TYPE LOCALITY: Indonesia, Java, East Java.
DISTRIBUTION: Java, Lesser Sunda Isls, and Amboina (Moluccas, Indonesia).
COMMENTS: Occurs sympatrically with *monticola* in Java; see Jenkins (1982).

Crocidura mindorus Miller, 1910. Proc. U.S. Natl. Mus., 38:392.
TYPE LOCALITY: Philippines, Mindoro, Mt. Halcon, 1,938 m.
DISTRIBUTION: Known only from Mt. Halcon, Mindoro, Philippines.

Crocidura minuta Otten, 1917. Med. Burgerl. Geneesk. Dienst. Ned. Ind., 6:103.
TYPE LOCALITY: Indonesia, Java, East Java.
DISTRIBUTION: Java.
STATUS: Uncertain.
COMMENTS: May be conspecific with *monticola*, or an earlier name for *maxi*; see Jenkins (1982). However, *minuta* Otten, 1917, is preoccupied by *minuta* Lyddeker, 1902, and thus not available. This problem needs to be resolved.

Crocidura miya Phillips, 1929. Spolia Zeylan., 15:113.
TYPE LOCALITY: Sri Lanka, Kandyan Hills, Nilambe Dist., Moolgama, 3,000 ft. [914 m].
DISTRIBUTION: Highlands of C Sri Lanka.
COMMENTS: A very distinctive species, resembling *C. elongata* of Sulawesi, or *C. dolichura* of Africa. Known by a handful of specimens; see Phillips (1980) for further information.

Crocidura monax Thomas, 1910. Ann. Mag. Nat. Hist., ser. 8, 6:310.
TYPE LOCALITY: Tanzania, Mt. Kilimanjaro, Rombo, 6,000 ft. (1,828 m).
DISTRIBUTION: Montane forests in W Kenya and N Tanzania.
STATUS: IUCN - Insufficiently known.
COMMENTS: Part of the the *littoralis* group; see Dieterlen and Heim de Balsac (1979). Does not includes *oritis* (part of *littoralis*) and *ultima* (treated as full species here) as suggested by Heim de Balsac and Meester (1977).

Crocidura monticola Peters, 1870. Monatsb. K. Preuss. Akad. Wiss. Berlin, 1870:584.
TYPE LOCALITY: Indonesia, Java, Mount Lawu, near Surakarta.
DISTRIBUTION: Borneo, Java, Peninsular Malaysia.
SYNONYMS: *bartelsii*.
COMMENTS: Revised by Jenkins (1982).

Crocidura montis Thomas, 1906. Ann. Mag. Nat. Hist., ser. 7, 18:138.
TYPE LOCALITY: Uganda, "Ruwenzori East, 12 500'" = Bujongolo, Mubuku Valley, eastern slope of Mt. Ruwenzori.
DISTRIBUTION: Montane forest in C and E Africa; Mt. Ruwenzori (Uganda); Mt. Meru (Tanzania), Imatong Mtns (Sudan); presumably also in Kenya.
COMMENTS: Formerly a subspecies of *fumosa* but see Demeter and Hutterer (1986) and Dippenaar and Meester (1989), who revised the species.

Crocidura muricauda (Miller, 1900). Proc. Washington Acad. Sci., 2:645.
TYPE LOCALITY: "Mount Coffee, Liberia".
DISTRIBUTION: West African high forest from Guinea to Ghana.
COMMENTS: Usually included in *dolichura* as a subspecies but constantly differs in its hairy tail while *dolichura* never shows any pilosity of the tail.

Crocidura mutesae Heller, 1910. Smithson. Misc. Coll., 56(15):3.
TYPE LOCALITY: Uganda, Kampala.
DISTRIBUTION: Uganda; perhaps more widely distributed.
STATUS: Taxonomic status unsolved.
COMMENTS: A large species, alternatively assigned to *hirta* (Allen, 1939) or *suahelae* (Heim de Balsac and Meester, 1977).

Crocidura nana Dobson, 1890. Ann. Mag. Nat. Hist., ser. 6, 5:225.
TYPE LOCALITY: Somalia, Dollo.
DISTRIBUTION: Somalia, Ethiopia.
COMMENTS: The name *nana* has been applied to various small shrews of Somalia, Ethiopia, and Egypt, leading to the proposal (Setzer, 1957) that *nana* is conspecific with *religiosa* (which it does not antedate); a conclusion followed by Heim de Balsac and

Mein (1971) and Osborn and Helmy (1980). Personal examination of the holotype of *nana* revealed that it represents a juvenile (skull inside the skin) of a species larger that *religiosa;* this conclusion was supported by better preserved topotypical specimens from Somalia in the British Museum (Natural History), which were also compared with the neotype of *religiosa* (Corbet, 1978*c*:27). The proposed conspecificy is therefore not accepted, and *religiosa* remains an endemic of the Nile valley in Egypt. The relation of *nana* with other small species has yet to be studied.

Crocidura nanilla Thomas, 1909. Ann. Mag. Nat. Hist., ser. 8, 4:99.
TYPE LOCALITY: Uganda, "probably Entebbe".
DISTRIBUTION: Dry and moist savanna from West Africa (Mauritania) to Kenya and Uganda; perhaps further south.
SYNONYMS: *rudolfi.*
COMMENTS: Includes *rudolfi;* see Heim de Balsac and Meester (1977). Often confused with other small species such as *fuscomurina* and *pasha.* For a discussion of "small *Crocidura*", see Heim de Balsac (1968*d*).

Crocidura neglecta Jentink, 1888. Notes Leyden Mus., 10:165.
TYPE LOCALITY: Indonesia, Sumatra.
DISTRIBUTION: Sumatra.
COMMENTS: May be conspecific with, and in that case a prior name for *maxi;* see Jenkins (1982).

Crocidura negrina Rabor, 1952. Chicago Acad. Sci. Nat. Hist. Misc., 96:6.
TYPE LOCALITY: Philippines, Negros Isl, Cuernos de Negros Mtn, Dayongan, 1,300 m.
DISTRIBUTION: Primary forest at 500 to 1450m on S Negros Isl (Philippines).
STATUS: Threatened by habitat destruction, according to Heaney et al. (1987).

Crocidura nicobarica Miller, 1902. Proc. U.S. Natl. Mus., 24:776.
TYPE LOCALITY: India, Nicobar Isls, Great Nicobar Isl.
DISTRIBUTION: Great Nicobar Isl. (Nicobar Isls, India).
COMMENTS: Not a species of *Suncus,* as suggested by Krumbiegel (1978:71). Corbet and Hill (1991) included *jenkinsi* which is retained as distinct until more evidence is presented.

Crocidura nigeriae Dollman, 1915. Ann. Mag. Nat. Hist., ser. 8, 15:511.
TYPE LOCALITY: Nigeria, "Asaba, 150 miles up the Niger".
DISTRIBUTION: Rainforest in Nigeria, Cameroon, and Bioko; exact distribution unknown.
COMMENTS: Formerly included in *poensis;* but see Heim de Balsac (1957), Meylan and Vogel (1982), and Hutterer and Happold (1983).

Crocidura nigricans Bocage, 1889. J. Sci. Math. Phys. Nat. Lisboa, ser. 2, 1:28.
TYPE LOCALITY: Angola, Benguela Dist., Quindumbo.
DISTRIBUTION: Angola.
COMMENTS: Regarded unidentifiable by Heim de Balsac and Meester (1977), but specific status upheld by Crawford-Cabral (1987).

Crocidura nigripes Miller and Hollister, 1921. Proc. Biol. Soc. Washington, 34:101.
TYPE LOCALITY: Indonesia, Sulawesi, Temboan, SW from Tondano Lake.
DISTRIBUTION: N and C Sulawesi, in tropical rain forest (Musser, 1987).
SYNONYMS: *lipara.*

Crocidura nigrofusca Matschie, 1895. Säugethiere Deutsch-Ost-Afrikas, p. 33.
TYPE LOCALITY: Zaire, Semliki Valley, "Wukalala, Kinyawanga im Westen des Semliki".
DISTRIBUTION: S Ethiopia and Sudan through E Africa to Zambia and Angola, Zaire, perhaps Cameroon.
SYNONYMS: *ansorgei, cabrerai, kempi, lakiundae, luluae, nilotica, nyikae, provocax, soricoides* (?), *zaodon, zena* (see Heim de Balsac and Meester, 1977).
COMMENTS: Includes *luluae* Matschie, 1926 (Luluabourg, Zaire) and *zaodon* Osgood, 1910 (Nairobi, Kenya) which were listed as separate species by Heim de Balsac and Meester (1977) and Dippenaar and Meester (1989); see Hutterer et al. (1987*b*). The holotypes of *nigrofusca, luluae* and *zaodon* have been studied. Gureev (1979) listed *ansorgei, nilotica,* and *zena* as distinct species without comment.

Crocidura nimbae Heim de Balsac, 1956. Mammalia, 20:131.
TYPE LOCALITY: Guinea, Mt. Nimba, "baraque de Zouguépo".
DISTRIBUTION: Mt. Nimba (Guinea, Liberia); Sierra Leone (specimen in the National Museum of Natural History).
STATUS: IUCN - Insufficiently known.
COMMENTS: A very distinct species; not conspecific with *wimmeri* as previously suggested (see Hutterer, 1983*a*).

Crocidura niobe Thomas, 1906. Ann. Mag. Nat. Hist., ser. 7, 18:138.
TYPE LOCALITY: Uganda, "Ruwenzori East, 6000 ft." [= Mubukee Valley, 1828 m].
DISTRIBUTION: Montane forests of EC Africa (Uganda, Zaire); perhaps Ethiopia.
COMMENTS: Ethiopian records (Corbet and Yalden, 1972; Yalden et al., 1976) uncertain; see Hutterer and Yalden (1990).

Crocidura obscurior Heim de Balsac, 1958. Mém. Inst. Fr. Afr. Noire, 53:328.
TYPE LOCALITY: Guinea, Mt. Nimba, montane prairie.
DISTRIBUTION: Sierra Leone to Ivory Coast; possibly Nigeria.
COMMENTS: Described with some doubt as a subspecies of *bottegi*, but its almost sympatric distribution, a longer skull (Hutterer and Happold, 1983), and a different karyotype (Maddalena, pers. comm.) clearly distinguish it.

Crocidura olivieri (Lesson, 1827). Manuel de Mammalogie, p. 121.
TYPE LOCALITY: Egypt, Sakkara; the neotype designated by Corbet (1978*c*:30) was collected "near Giza".
DISTRIBUTION: Egypt; Senegal to Ethiopia, and southwards to N South Africa.
SYNONYMS: *anchietae, atlantis, bueae, cara, daphnia, darfurea, deltae, ferruginea, fuscosa, giffardi, guineensis, doriana, hansruppi, hedenborgiana, hera, herero, kijabae, kivu, luluana, manni, martiensseni, nyansae, occidentalis, odorata, spurelli, sururae, tatiana, toritensis, zuleika*.
COMMENTS: *Crocidura olivieri* is the valid name for the large African shrews previously known as *flavescens* (which is now the valid name for a species restricted to South Africa); see Maddalena et al. (1987). This group of giant shrews was reviewed by Heim de Balsac and Barloy (1966). Well known subspecies names are *anchietae, doriana, ferruginea, fuscosa, giffardi, guineensis, hansruppi, hedenborgiana, kivu, manni, martiensseni, nyansae, occidentalis, odorata, spurelli*, and *sururae*. Some of these were considered allospecies of a *flavescens* superspecies by Hutterer and Happold (1983). Many authors also distinguished pale (*occidentalis, manni, spurelli*) and black (*giffardi, hedenborgiana, martiensseni, odorata*) color morphs as different species but biochemical evidence showed that they are merely color morphs of a single and highly variable species (Maddalena, 1990). *Crocidura olivieri* may also include *zaphiri*; see Yalden et al. (1976).

Crocidura orii Kuroda, 1924. [New Mammals from the Ryukyu Islands], p. 3.
TYPE LOCALITY: Japan, Ryukyu Islands, Amami-Oshima, Komi.
DISTRIBUTION: Ryukyu Isls, Japan.
COMMENTS: Provisionally included in *dsinezumi* (Corbet, 1978*c*); but regarded as a separate species by Imaizumi (1961, 1970*b*), Abe (1967), and Jenkins (1976). The species was first described by Kuroda (1924) in a publication which, although privately published, has been regarded as available by all subsequent authors.

Crocidura osorio Molina and Hutterer, 1989. Bonn. Zool. Beitr., 40:86.
TYPE LOCALITY: Spain, Canary Isls, Gran Canaria, Finca de Osorio.
DISTRIBUTION: N cloud zone of Gran Canaria Isl, Canary Islands, Spain.
STATUS: Protected by Spanish law.

Crocidura palawanensis Taylor, 1934. Monogr. Bur. Sci. Manila, 30:88.
TYPE LOCALITY: Philippines, Palawan, Sir J. Brooke Point.
DISTRIBUTION: Palawan Isl, Philippines.
COMMENTS: May belong to *fuliginosa* (Heaney et al., 1987).

Crocidura paradoxura Dobson, 1886. Ann. Mus. Civ. Stor. Nat. Genova, 4:566.
TYPE LOCALITY: Indonesia, Sumatra, Mt. Singalang, 2,000 m.
DISTRIBUTION: Sumatra.

STATUS: Unknown.
COMMENTS: A large species with a long tail, possibly related to *fuliginosa;* specific identity unresolved (Jenkins, 1982:277).

Crocidura parvipes Osgood, 1910. Field Mus. Nat. Hist. Publ., Zool. Ser., 10:19.
TYPE LOCALITY: Kenya, "Voi, British East Africa".
DISTRIBUTION: Africa; Guinea and Sudan savanna from Cameroon to S Sudan, Ethiopia (Hutterer and Yalden, 1990), Kenya, Tanzania, S Zaire, Zambia to Angola.
SYNONYMS: *boydi, chitauensis, cuanzensis, katharina, lutrella, nisa.*
COMMENTS: Revised and included in subgenus *Afrosorex* by Hutterer (1986*a*).

Crocidura pasha Dollman, 1915. Ann. Mag. Nat. Hist., ser. 8, 15:517.
TYPE LOCALITY: "Atbara River, Sudan."
DISTRIBUTION: Sudan and Sahelian savanna of Sudan; a single record from Ethiopia (Demeter, 1982).
COMMENTS: Often confused with *nanilla* and *lusitania;* does not include *glebula* which is a synonym of *fuscomurina* or *planiceps;* see Hutterer and Kock (1983) and Hutterer and Happold (1983).

Crocidura pergrisea Miller, 1913. Proc. Biol. Soc. Washington, 26:113.
TYPE LOCALITY: Kashmir, Baltistan, Shigar, Skoro Loomba, 9,500 ft. (2900 m).
DISTRIBUTION: Mountains of W Himalaya (Kashmir).
COMMENTS: Some authors have included *armenica, serezkyensis,* and *zarudnyi* (see Spitzenberger, 1971, and Corbet, 1978*c*, for a review of literature); but all are now considered separate species. A considerable diversity of opinions exists in the literature on the allocation of the different forms. Following Jenkins (1976), the name *pergrisea* is applied only to the largest species, as represented by the type series from Baltistan.

Crocidura phaeura Osgood, 1936. Field Mus. Nat. Hist. Publ., Zool. Ser., 20:228.
TYPE LOCALITY: Ethiopia, Sidamo, west base of Mt. Guramba, NE of Allata.
DISTRIBUTION: Known only from the type locality.
STATUS: IUCN - Insufficiently known. Has not been collected since its discovery.
COMMENTS: Considered a full species by Dippenaar and Meester (1989). Related to *harenna;* see discussion in Hutterer and Yalden (1990).

Crocidura picea Sanderson, 1940. Trans. Zool. Soc. Lond., 24:682.
TYPE LOCALITY: Cameroon, Mamfe Div., Assumbo, Tinta.
DISTRIBUTION: Known only from the type locality.
COMMENTS: Status uncertain; holotype figured by Heim de Balsac and Hutterer (1982:142, fig. 3).

Crocidura pitmani Barclay, 1932. Ann. Mag. Nat. Hist., ser. 10, 10:440.
TYPE LOCALITY: Zambia, "Maluwe-Serenje Distr., 3800 ft."
DISTRIBUTION: C and N Zambia.

Crocidura planiceps Heller, 1910. Smithson. Misc. Coll., 56(15):5.
TYPE LOCALITY: Uganda, Lado Enclave, Rhino Camp.
DISTRIBUTION: Ethiopia, Uganda, Sudan, Zaire, Nigeria.
COMMENTS: Closely related to *fuscomurina,* if not conspecific; see Heim de Balsac (1968*d*) and Hutterer (1983*b*). See comments under *C. pasha.*

Crocidura poensis (Fraser, 1843). Proc. Zool. Soc. Lond., 1842:200 [1843].
TYPE LOCALITY: Equatorial Guinea, Bioko (Fernando Po), Clarence.
DISTRIBUTION: Bioko, Principe Isl, Cameroon to Liberia.
SYNONYMS: *calabarensis, pamela, schweitzeri, soricoides, stampflii* (see Heim de Balsac and Meester, 1977, and Hutterer and Happold, 1983).

Crocidura polia Hollister, 1916. Bull. Am. Mus. Nat. Hist., 35:669.
TYPE LOCALITY: Zaire, Medje.
DISTRIBUTION: Known only from the type locality.
STATUS: IUCN - Insufficiently known. Known only from a single specimen.
COMMENTS: Included in *dolichura* by Heim de Balsac and Meester (1977) but represents a distinct species.

Crocidura pullata Miller, 1911. Proc. Biol. Soc. Washington, 24:241.
TYPE LOCALITY: Kashmir, Kotihar, 7,000 ft.
DISTRIBUTION: Kashmir, India, Afghanistan, Pakistan, Yunnan (China), Thailand, full range unknown.
SYNONYMS: *rapax, vorax*.
COMMENTS: The name *pullata* is provisonally used as a label to include the Asian populations of what has been called *russula* by Jenkins (1976) and many other authors. It can be seen from the measurements provided by Jameson and Jones (1977) that the forms *pullata, rapax* and *vorax* differ from the European *russula* by a longer tail; all have been assigned to the West European species; see Lekagul and McNeely (1977), among others.

Crocidura raineyi Heller, 1912. Smithson. Misc. Coll. 60(12):7-8.
TYPE LOCALITY: Kenya, "North Creek, Mt. Garguez".
DISTRIBUTION: Known only from the type locality.
STATUS: IUCN - Insufficiently known.
COMMENTS: Since its description *C. raineyi* has been considered a valid species, but was synonymized in 1977 with *C. luna*, an error recently corrected by Dippenaar and Meester (1989).

Crocidura religiosa (I. Geoffroy, 1827). Mem. Mus. Hist. Nat. Paris, 15:128.
TYPE LOCALITY: Egypt, Giza.
DISTRIBUTION: Nile Valley (Egypt).
COMMENTS: Described from embalmed specimens from tombs at Thebes; holotype not preserved. Corbet (1978*c*:27) selected a neotype from Giza.

Crocidura rhoditis Miller and Hollister, 1921. Proc. Biol. Soc. Washington, 34:102.
TYPE LOCALITY: Indonesia, Sulawesi, Temboan.
DISTRIBUTION: Tropical rainforest of N, C, and SW Sulawesi (Musser, 1987).

Crocidura roosevelti (Heller, 1910). Smithson. Misc. Coll., 56(15):6.
TYPE LOCALITY: Uganda, Lado Enclave, Rhino Camp.
DISTRIBUTION: Forest-savanna margin of the Central African forest block; records from Angola, Cameroon, Central African Republic, Zaire, Uganda, Rwanda, and Tanzania (Hutterer, 1981*a*).
COMMENTS: Type species of subgenus *Heliosorex* Heller, 1910.

Crocidura russula (Hermann, 1780). *In* Zimmermann, Geogr. Gesch. Mensch. Vierf. Thiere, 2:382.
TYPE LOCALITY: France, Bas Rhin, near Strasbourg.
DISTRIBUTION: S and W Europe including some Atlantic island off France; Mediterranean islands (Ibiza, Sardinia, Pantelleria?); N Africa (Morocco, Algeria, Tunisia).
SYNONYMS: *albiventris, agilis, anthonyi, candidus, chaouianensis, chrysothorax, cinereus, cintrae, constrictus, fimbriatus, foucauldi, heljanensis, hydruntina, ibicensis, ichnusae, inodorus, leucurus, major, moschata, musaraneus, peta, poliogastra, pulchra, rufa, safii, thoracicus, unicolor, yebalensis*.
COMMENTS: Reviewed by Genoud and Hutterer (1990). The species is confined to W Europe and N Africa. Many populations from Asia and even Africa have been erroneously assigned to *russula* (see Ellermann and Morrison-Scott, 1951). Allozyme and karyotype analyses by Catzeflis et al. (1985) have shown that animals from E Europe, Asia Minor and Israel formerly identified as *russula* instead belong to *suaveolens*. This may also be true for other populations further east. Does not include *hosletti, rapax*, or *vorax* (see Ellerman and Morrison-Scott, 1966:81; Jameson and Jones, 1977:465), which are here included in *dsinezumi* and *pullata*, respectively. May include *cossyrensis*; see under that species.

Crocidura selina Dollman, 1915. Ann. Mag. Nat. Hist., ser. 8, 16:371-372.
TYPE LOCALITY: "Mabira Forest, Chagwe, Uganda."
DISTRIBUTION: Known only from three lowland forests in Uganda.
STATUS: IUCN - Insufficiently known.
COMMENTS: Previously included in *fumosa* or *luna*, but considered a distinct species by Dippenaar and Meester (1989).

Crocidura serezkyensis Laptev, 1929. Opred. Mlekopitay. Sredney Asyy, Tashkent, 1:16.
TYPE LOCALITY: Tadshikistan, Pamir Mtns, Lake Sarezskoye.
DISTRIBUTION: Asia Minor, Azerbaijan, Turkmenistan, Tadshikistan and Kazakhstan.
SYNONYMS: *arispa*.
COMMENTS: Previously included in *pergrisea* (Spitzenberger, 1971; Jenkins, 1976), but considered a distinct species by Stogov and Bondar (1966) and Stogov (1985). Populations in Asia Minor (*arispa* Spitzenberger, 1971) are linked with the typical ones in Kazakhstan and Tadshikistan by records from Azerbaijan (Grafodatsky et al., 1988) and Turkmenistan (Stogov and Bondar, 1966). Grafodatsky et al. (1988) reported on the karyotype of a specimen from Dzhulfa, SW Azerbaijan (under the name *pergrisea*); with 2n=22 *serezkyensis* has the lowest chromosome number ever recorded for a shrew.

Crocidura sibirica Dukelsky, 1930. Zool. Anz., 88:75.
TYPE LOCALITY: Russia, Siberia, S Krasnoyarsky Krai, upper Yenisei River, 96 km S of Minusinsk, Oznatchenoie.
DISTRIBUTION: C Asia from Lake Issyk Kul to Upper Ob River; Lake Baikal; perhaps also Sinkiang (China) and Mongolia (see Sokolov and Orlov, 1980:50).
SYNONYMS: *ognevi*.
COMMENTS: Includes *ognevi*; see Yudin (1989).

Crocidura sicula Miller, 1900. Proc. Biol. Soc. Washington, 14:41.
TYPE LOCALITY: Italy, Sicily, Palermo.
DISTRIBUTION: Sicily, Egadi Isls (Italy) and Gozo (Malta).
SYNONYMS: *aegatensis, calypso, caudata, esuae*.
COMMENTS: Revised by Hutterer (1991), who recognized one extinct and three extant subspecies. Formerly included in *leucodon, russula*, or *suaveolens*; but the species has a distinct karyotype (Vogel, 1988) and morphology (Vogel et al., 1989).

Crocidura silacea Thomas, 1895. Ann. Mag. Nat. Hist., ser. 6, 16:53.
TYPE LOCALITY: South Africa, E Transvaal, Barberton dist., De Kaap, Figtree Creek.
DISTRIBUTION: Occurs in most of South Africa, and parts of Botswana, Mozambique, and Zimbabwe; possibly has a wider distribution.
SYNONYMS: *holobrunneus*.
COMMENTS: This species was formerly assigned to *gracilipes* or *hildegardeae*, but is not conspecific with either of these; see Meester et al. (1986) for a discussion.

Crocidura smithii Thomas, 1895. Ann. Mag. Nat. Hist., ser. 6, 16:51.
TYPE LOCALITY: Ethiopia, Webi Shebeli, near Finik.
DISTRIBUTION: Arid regions of Senegal, Ethiopia, and probably Somalia.
SYNONYMS: *debalsaci*.
COMMENTS: Revised by Hutterer (1986*a*). Specimens reported from Somalia by Heim de Balsac (1966*a*) represent *macarthuri*; see under that species. Includes *debalsaci* as a subspecies; see Hutterer (1981*b*).

Crocidura somalica Thomas, 1895. Ann. Mag. Nat. Hist., ser. 6, 16:52.
TYPE LOCALITY: Ethiopia, Middle Webi Shebeli (about 5°30'N, 44°E) near Geledi (Galadi).
DISTRIBUTION: Dry savannas and semi-desert areas of Ethiopia, Sudan, and probably Somalia; Mali.
COMMENTS: Revised by Hutterer and Jenkins (1983). Recently recorded from the Sahara (Mali) by Hutterer et al. (1992), who regarded the subspecies *dhofarensis* from Oman as specifically distinct; see under *dhofarensis*.

Crocidura stenocephala Heim de Balsac, 1979. Säugetierkdl. Mitt., 27:258.
TYPE LOCALITY: E Zaire, "Kahuzi-Biega N.P."
DISTRIBUTION: Montane *Cyperus* swamps at Mt. Kahuzi, E Zaire.
STATUS: IUCN - Insufficiently known.
COMMENTS: Described as a subspecies of *littoralis* but regarded as a full species by Hutterer (1982*a*) and Dippenaar (pers. comm.).

Crocidura suaveolens (Pallas, 1811). Zoogr. Rosso-Asiat., 1:133.
TYPE LOCALITY: Russia, Crimea, Khersones, near Sevastopol.

DISTRIBUTION: Entire Palearctic from Spain to Korea; Atlantic islands (Scilly, Jersey, Sark, Ushant, Yeu); many Mediterranean islands including Corsica, Crete, Cyprus, and Menorca; Tsushima and Ullong Do between Korea and Japan.

SYNONYMS: *antipae, ariadne, astrabadensis, avicennai, balcanica, balearica, bruecheri, cantanbra, caneae, cassiteridum, coreae, corsicana, cypria, cyrnensis, debeauxi, dinniki, enezsizunensis, heptapotamica, hyrcania, iculisma, ilensis, italica, lar, lignicolor, longicauda, mimula, mimuloides, minuta, monacha, mordeni, oayensis, orientis, pamirensis, phaeopus, portali, praecypria, sarda, shantungensis, tristami, utsuryoensis, uxantisi.*

COMMENTS: A widespread and variable species which has often been confused with *russula*; the taxonomic status of many E Asian forms is still unsolved; see also under *gueldenstaedtii*. The European and Arabian range was reviewed by Vlasák and Niethammer (1990) and Harrison and Bates (1991), respectively; a discussion of *suaveolens* in Korea and Taiwan was given by Jones and Johnson (1960) and Jameson and Jones (1977).

Crocidura susiana Redding and Lay, 1978. Z. Säugetierk., 43:307.

TYPE LOCALITY: Iran, Khuzistan Province, 8 km SSW of Dezful (32°19′N, 48°21′E).

DISTRIBUTION: Known only from the vicinity of Dezful (SW Iran), but may have a wider distribution.

Crocidura tansaniana Hutterer, 1986. Bonn. Zool. Beitr., 37:27.

TYPE LOCALITY: Tanzania, Tanga Region, E Usambara Mtns, Amani.

DISTRIBUTION: Usambara Mtns (Tanzania).

STATUS: IUCN - Insufficiently known.

COMMENTS: Previously known only by the holotype, but recently more specimens have been identified.

Crocidura tarella Dollman, 1915. Ann. Mag. Nat. Hist., ser. 8, 17:135.

TYPE LOCALITY: "Chaya, near Ruchuru, Congo Belge."

DISTRIBUTION: Uganda.

COMMENTS: Formerly a subspecies of *turba* but Dippenaar (1980) regarded it a distinct species.

Crocidura tarfayensis Vesmanis and Vesmanis, 1980. Zool. Abh. Mus. Tierk. Dresden, 36:47.

TYPE LOCALITY: Morocco, Agadir Prov., 8 km south Tarfaya, 27°50′N, 12°30′W.

DISTRIBUTION: Atlantic coast of Sahara; south of Agadir (Morocco) through Western Sahara into Mauritania.

SYNONYMS: *agadiri, gouliminensis, tiznitensis* (see Hutterer, 1987).

COMMENTS: Recorded as *whitakeri* from Western Sahara by Heim de Balsac (1968*e*).

Crocidura telfordi Hutterer, 1986. Bonn. Zool. Beitr., 37:28.

TYPE LOCALITY: Tanzania, Uluguru Mtns, Morningside, 1150 m.

DISTRIBUTION: Known only from the type locality, in relict montane forest.

STATUS: IUCN - Insufficiently known.

Crocidura tenuis (Müller, 1840). *In* Temminck, Verh. Nat. Gesch. Nederland Overz. Bezitt., Zool., Zoogd. Indisch. Archipel, p. 26, 50[1840].

TYPE LOCALITY: Indonesia, "Timor."

DISTRIBUTION: Timor (Indonesia).

COMMENTS: Jenkins (1982:273) considered conspecificy of *tenuis* with *fuliginosa* but stated that present evidence is not sufficient; in case of conspecificy, *tenuis* would be the earliest name for the group. See Appendix I for date of publication.

Crocidura thalia Dippenaar, 1980. Ann. Transvaal Mus., 32:138-147.

TYPE LOCALITY: Ethiopia, NW Bale Province, Gedeb Mtns, SE Dodola, 2,600 m (06°55′N, 39°10′E).

DISTRIBUTION: Forest and moorland of the Ethiopian highlands on both sides of the Rift Valley.

STATUS: IUCN - Insufficiently known.

COMMENTS: Previous to its description, *thalia* was known as *C. luna macmillani* (e.g., Hutterer, 1981*c*) or *C. fumosa*; see Yalden (1988), who studied the altitudinal distribution.

Crocidura theresae Heim de Balsac, 1968. Mammalia, 32:398.
TYPE LOCALITY: Guinea, Nzerekore.
DISTRIBUTION: Guinea savanna from Ghana to Guinea.
COMMENTS: May be a subspecies of *foxi*, but *theresae* from Ivory Coast are distinctly smaller and grayer.

Crocidura thomensis (Bocage, 1887). J. Sci. Math. Phys. Nat. Lisboa, 11:212.
TYPE LOCALITY: Sao Tome and Princepe, São Tomé Isl.
DISTRIBUTION: Endemic to São Tomé.
STATUS: IUCN - Insufficiently known.
COMMENTS: For description of the species and designation of a neotype, see Heim de Balsac and Hutterer (1982).

Crocidura turba Dollman, 1910. Ann. Mag. Nat. Hist., ser. 8, 5:176.
TYPE LOCALITY: "Chilui Island, Lake Bangweolo", = Chilubi Isl, Zambia.
DISTRIBUTION: Angola, Zambia, Zaire, Malawi, Tanzania, Kenya, Uganda, Cameroon.
SYNONYMS: *angolae*.
COMMENTS: Includes *angolae*; see Heim de Balsac and Meester (1977:24-25). Range not exactly known, due to confusion with *zaodon* (= *nigrofusca*).

Crocidura ultima Dollman, 1915. Ann. Mag. Nat. Hist., ser. 8, 15:517.
TYPE LOCALITY: "Jombeni Range, Nyeri District", Kenya.
DISTRIBUTION: Known only from the type locality.
COMMENTS: Dippenaar (1980:126), following Allen (1939:46), recognized *ultima* as a full species within the *littoralis-monax* group.

Crocidura usambarae Dippenaar, 1980. Ann. Transvaal Mus., 32:128.
TYPE LOCALITY: Tanzania, Western Usambara Mtns, Shume, 16 mi N. Lushoto.
DISTRIBUTION: Magamba, Shume (Usambara Mtns), perhaps also Ngozi Crater, SW Tanzania.
STATUS: IUCN - Insufficiently known.

Crocidura viaria (I. Geoffroy, 1834). *In* Zool. Voy. de Belanger Indes-Orient., p. 127.
TYPE LOCALITY: "Senegal", restricted to region between Dakar and St. Louis by Hutterer (1984).
DISTRIBUTION: Sahelien and Sudan savanna from S Morocco to Senegal and east to Sudan, Ethiopia and Kenya; perhaps further south.
SYNONYMS: *bolivari, hindei, suahelae* (?), *tamrinensis*.
COMMENTS: Revised by Hutterer (1984); Possibly includes *suahelae*, which may alternatively belong to *zaphiri*. A member of the *flavescens* species group (Maddalena, 1990).

Crocidura voi Osgood, 1910. Field Mus. Nat. Hist. Publ., Zool. Ser., 10:18.
TYPE LOCALITY: Kenya, "Voi, British East Africa".
DISTRIBUTION: Sudan savanna from Kenya and Somalia to Ethiopia and Sudan; single records from Nigeria and Mali.
SYNONYMS: *aridula, butleri, percivali* (see Hutterer, 1986*a*).

Crocidura whitakeri De Winton, 1898. Proc. Zool. Soc. Lond., 1897:954 [1898].
TYPE LOCALITY: Morocco, between Morocco City and Mogador, Sierzet.
DISTRIBUTION: Atlantic and Mediterranean parts of Morocco, Algeria and Tunisia; one record from coastal Egypt.
SYNONYMS: *essaouiranensis, mesatanensis, matruhensis, zaianensis* (see Hutterer, 1987, 1991).
COMMENTS: Range in Morocco mapped by Aulagnier and Thévenot (1987); in Algeria by Rzebik-Kowalska (1988).

Crocidura wimmeri Heim de Balsac and Aellen, 1958. Rev. Suisse Zool., 65:952.
TYPE LOCALITY: Ivory Coast, Adiopodoume.
DISTRIBUTION: S Ivory Coast.
STATUS: IUCN - Insufficiently known. Rare and very localized in distribution.
COMMENTS: Has been assigned to *nimbae*; but see Hutterer (1983*a*). Records outside Ivory Coast are based on misidentifcations; specimen recorded from Cameroon and Gabon refer to *batesi*; see Brosset (1988).

Crocidura xantippe Osgood, 1910. Field Mus. Nat. Hist. Publ., Zool. Ser., 10:19.
TYPE LOCALITY: Kenya, "Voi, British East Africa".
DISTRIBUTION: Nyiru, Voi, Tsavo (SE Kenya); Usambara Mtns (Tanzania).
COMMENTS: Status uncertain; probably related to *hirta*. Not to be confused with *Crocidura xanthippe* Bate, 1937, a Pleistocene shrew from Palestine.

Crocidura yankariensis Hutterer and Jenkins, 1980. Bull. Brit. Mus. (Nat. Hist.) Zool., 39:305.
TYPE LOCALITY: Nigeria, Bauchi State, 16 km E of Yankari Game Reserve boundary, Futuk [9°50'N, 10°55'E].
DISTRIBUTION: Sudan savanna zone in Cameroon, Nigeria, Sudan, Ethiopia, Kenya, and Somalia.
COMMENTS: Previously confused with *somalica;* see Hutterer and Jenkins (1983).

Crocidura zaphiri Dollman, 1915. Ann. Mag. Nat. Hist., ser. 8, 15:509.
TYPE LOCALITY: Ethiopia, "Charada Forest, Kaffa".
DISTRIBUTION: Kaffa Prov. (S Ethiopia); Kaimosi, Kisumu (Kenya).
SYNONYMS: *simiolus*.
COMMENTS: Includes *simiolus*; see Osgood (1936:224). May also include *mutesae* and *suahelae* (here questionably listed in *viaria*), in which case it would be a widely distributed species; see Hutterer and Yalden (1990:70).

Crocidura zarudnyi Ognev, 1928. [Mammals of Eastern Europe and Northern Asia], 1:341.
TYPE LOCALITY: Iran, Baluchistan (border).
DISTRIBUTION: SE Iran, SE Afghanistan, SW Pakistan (Spitzenberger, 1971).
SYNONYMS: *streetorum, tatianae*.
COMMENTS: The species was first named *tatianae* by Ognev (1921), but later (1828) replaced by *zarudnyi;* Ognev argued that *tatianae* was preoccupied by *tatiana* Dollman, 1915 (now a synonym of the African *olivieri*). Strictly following the International Code of Zoological Nomenclature (1985, art. 58), this is not the case, and *zarudnyi* would be an unjustified replacement name. However, since the species has always been called *zarudnyi* it would be justified to present the case to the commission in favor of stability. The definition of *zarudnyi* follows Spitzenberger (1971) and Hassinger (1970), but not Jenkins (1976) who included *arispa* which is now regarded as part of *serezkyensis;* see under that species. As Spitzenberger (1971) pointed out, *zarudnyi* has a shorter rostrum and a heavier mandible than both *pergrisea* and *serezkyensis*. The status of *streetorum* is not clear although it is included here as suggested by Hassinger (1970). The distribution and morphology of *pergrisea, serezkyensis*, and *zarudnyi* should be carefully studied in the Hindukush, Karakoram and Pamir where their ranges may overlap.

Crocidura zimmeri Osgood, 1936. Field Mus. Nat. Hist. Publ., Zool. Ser., 20:223.
TYPE LOCALITY: Zaire, Katanga Prov., near Bukama, "Lualaba River, Katobwe".
DISTRIBUTION: Environs of Upemba National Park, Zaire.
COMMENTS: A large and striking species which is known only by the type series.

Crocidura zimmermanni Wettstein, 1953. Z. Säugetierk., 17:12.
TYPE LOCALITY: Greece, Crete, Ida Mtns, Nida plateau.
DISTRIBUTION: Highlands of the island of Crete.
STATUS: IUCN - Rare.
COMMENTS: Formerly regarded as a subspecies of *russula* but differs in morphology and karyotype; see Vesmanis and Kahmann (1978), Vogel (1986), and Pieper (1990).

Diplomesodon Brandt, 1852. Beitr. Kenntn. Russ. Reiches, 17:299.
TYPE SPECIES: *Sorex pulchellus* Lichtenstein, 1823.
COMMENTS: Subfamily Crocidurinae; see Repenning (1967:15).

Diplomesodon pulchellum (Lichtenstein, 1823). *In* Eversmann, Reise von Orenburg nach Bokhara, Berlin, p. 124.
TYPE LOCALITY: Kazakhstan, eastern bank of Ural River, sands "Bolshie Barsuki".
DISTRIBUTION: W and S Kazakhstan, Uzbekistan, Turkmenistan.
SYNONYMS: *pallidus*.
COMMENTS: Biology and distribution reviewed by Heptner (1939), who also specified the type locality.

Feroculus Kelaart, 1852. Prodr. Faun. Zeylanica, p. 31.
TYPE SPECIES: *Sorex macropus* Blyth, 1851 (= *Sorex feroculus* Kelaart, 1850).
COMMENTS: Repenning (1967:15) placed *Feroculus* in the subfamily Crocidurinae.

Feroculus feroculus (Kelaart, 1850). J. Ceylon Branch Asiat. Soc., 2(5):211.
TYPE LOCALITY: Sri Lanka, central mountains at 6,000 ft., Nuwara Eliya.
DISTRIBUTION: Primary swamps and forests in the central highlands of Sri Lanka.
SYNONYMS: *macropus, newera, newera-ellia.*
COMMENTS: A rare and little-known species; available information summarized by Phillips (1980).

Myosorex Gray, 1838. Proc. Zool. Soc. Lond., 1837:124 [1838].
TYPE SPECIES: *Sorex varius* Smuts, 1832.
COMMENTS: Subfamily status uncertain. Repenning (1967) grouped *Myosorex* in the Crocidurinae; Reumer's (1987) Crocidosoricinae would fit as well. Kretzoi (1965) based the tribe Myosoricini on this genus; the name is available for any taxonomic unit above the genus level. Generic status sometimes questioned; but see Meester (1954). *Surdisorex* and *Congosorex* are often included as subgenera but are treated here as full genera, following Thomas (1906*b*), Hollister (1918), Meester (1953), Heim de Balsac (1966*b*), and my own studies. Partial reviews of *Myosorex* were provided by Heim de Balsac (1967, 1968*b*), Heim de Balsac and Lamotte (1956), and Meester and Dippenaar (1978). The formerly listed *Myosorex preussi* (Matschie, 1893), described from "Mount Cameroun", is not listed here, because a recent examination of the type series has shown that the type series was based on mismatched parts of three different genera (*Crocidura, Sorex, Sylvisorex*), and that *preussi* does not represent a biological species. Species of conservation concern are listed in Nicoll and Rathbun (1990:21).

Myosorex babaulti Heim de Balsac and Lamotte, 1956. Mammalia, 20:150.
TYPE LOCALITY: Zaire, "Kivu".
DISTRIBUTION: Mountains west and east of Lake Kivu, including Idjwi Isl (Zaire, Rwanda, Burundi).
COMMENTS: Formerly included in *blarina*; but see Dieterlen and Heim de Balsac (1979).

Myosorex blarina Thomas, 1906. Ann. Mag. Nat. Hist., ser. 7, 18:139.
TYPE LOCALITY: Uganda, Ruwenzori East, Mubuku Valley, 10,000 ft.
DISTRIBUTION: Montane forest at Mt. Ruwenzori (Uganda, Zaire).

Myosorex cafer (Sundevall, 1846). Ofv. K. Svenska Vet.-Akad. Forhandl. Stockholm, 3:119.
TYPE LOCALITY: South Africa, "E Caffraria interiore et Port-Natal".
DISTRIBUTION: South Africa, eastern escarpment and north to the Transvaal; extreme W Mozambique and E Zimbabwe, in higher elevations above 1,000 m.
SYNONYMS: *swinnyi.*
COMMENTS: Meester (1958) described the geographic variation of the species. Heim de Balsac and Meester (1977) included *affinis, sclateri, swinnyi, talpinus* and *tenuis* in *cafer*, while Wolhuter (in Smithers, 1983:3) and Dippenaar et al. (1983) regarded *sclateri* and *tenuis* as distinct, partly based on new karyotype information. Although no additonal data have yet been published, this view is provisionally accepted here as it better reflects existing variation within the southern African representatives of the genus.

Myosorex eisentrauti Heim de Balsac, 1968. Bonn. Zool. Beitr., 19:20.
TYPE LOCALITY: Equatorial Guinea, Bioko, Pic Santa Isabel, 2400 m.
DISTRIBUTION: Montane forest of Bioko (Fernando Po).
STATUS: IUCN - Insufficiently known.
COMMENTS: The forms *okuensis* and *rumpii* were included in *eisentrauti* by Heim de Balsac and Meester (1977); both are regarded as distinct species in this account.

Myosorex geata (Allen and Loveridge, 1927). Proc. Boston Soc. Nat. Hist., 38:417.
TYPE LOCALITY: Tanzania, Uluguru Mtns, Nyingwa.
DISTRIBUTION: Forests of the Tanzania mountain arc.
STATUS: IUCN - Insufficiently known.

COMMENTS: Formerly in *Crocidura*; see Heim de Balsac (1967:610).

Myosorex longicaudatus Meester and Dippenaar, 1978. Ann. Transvaal Mus., 31:30.
TYPE LOCALITY: South Africa, Cape Province, 14 km NNE Knysna, Diepwalle State Forest Station, 33°57'S, 23°10'E.
DISTRIBUTION: Escarpment forests of the SE Cape Province, South Africa.
STATUS: IUCN - Insufficiently known.

Myosorex okuensis Heim de Balsac, 1968. Bonn. Zool. Beitr., 19:20.
TYPE LOCALITY: Cameroon, Bamenda Highlands, "Oku-See, 2100 m".
DISTRIBUTION: Forested mountains of the Bamenda plateau, Cameroon (Lake Manenguba, Lake Oku, Mt. Lefo).
COMMENTS: Formerly included in *eisentrauti* (see Heim de Balsac and Meester, 1977), but cranially very distinct.

Myosorex rumpii Heim de Balsac, 1968. Bonn. Zool. Beitr., 19:20.
TYPE LOCALITY: Cameroon, "Rumpi-Hills, 1100 mètres".
DISTRIBUTION: Known only from the type locality.
COMMENTS: The holotype and only known specimen is so unique (Heim de Balsac, 1968*b*, fig. 4) that it is considered to represent a valid species. Heim de Balsac (1968*b*) himself was uncertain about the status of this taxon; while he formally named it *M. eisentrauti rumpii*, he labeled all figures and the map with "*Myosorex rumpii*".

Myosorex schalleri Heim de Balsac, 1966. C.R. Acad. Sci. Paris, 263:889.
TYPE LOCALITY: E Zaire, Itombwe Mtns, "Nzombe (Mwenga)".
DISTRIBUTION: Known only from the type locality.
STATUS: IUCN - Insufficiently known.
COMMENTS: Provisionally named by Heim de Balsac (1966*b*); full description by Heim de Balsac (1967). The type locality was later erroneously shifted to the "Albert N. P." (Heim de Balsac and Meester, 1977); Nzombe is located in the Itombwe Mountains (Hutterer, 1986*c*).

Myosorex sclateri Thomas and Schwann, 1905. Abstr. Proc. Zool. Soc. Lond., 1905(15):10.
TYPE LOCALITY: South Africa, Natal, Zululand, Ngoye hills, 250 m.
DISTRIBUTION: Wet habitats in Kwazulu (South Africa).
SYNONYMS: *affinis, talpinus*.
COMMENTS: Provisionally regarded as a distinct species by Wolhuter (in Smithers, 1983:3); occurs in sympatry with *cafer* and has a different karyotype. Meester et al. (1986) included *sclateri* in *cafer*.

Myosorex tenuis Thomas and Schwann, 1905. Proc. Zool. Soc. Lond., 1905:131-132.
TYPE LOCALITY: South Africa, Transvaal, near Wakkerstroom, Zuurbron.
DISTRIBUTION: Transvaal (South Africa) and possibly W Mozambique.
COMMENTS: Provisionally regarded as a distinct species by Wolhuter (in Smithers, 1983:3) because of sympatry with *cafer* and a different karyotype. Meester et al. (1986) included *tenuis* in *cafer*.

Myosorex varius (Smuts, 1832). Enumer. Mamm. Capensium, p. 108.
TYPE LOCALITY: South Africa, Cape of Good Hope, Algoa Bay (Port Elizabeth).
DISTRIBUTION: South Africa, from NW Cape Province to E Transvaal; Lesotho and Orange Free State.
SYNONYMS: *capensis, herpestes, pondoensis, transvaalensis* (see Heim de Balsac and Meester, 1977).
COMMENTS: Revised by Meester (1958).

Paracrocidura Heim de Balsac, 1956. Rev. Zool. Bot. Afr., 54:137.
TYPE SPECIES: *Paracrocidura schoutedeni* Heim de Balsac, 1956.
COMMENTS: Subfamily status uncertain. Revised by Hutterer (1986*c*).

Paracrocidura graueri Hutterer, 1986. Bonn. Zool. Beitr., 37:81.
TYPE LOCALITY: "Urwald hinter den Randbergen des Nord-Westufers des Tanganjika" = Sibatwa, 2,000 m, Itombwe Mtns, Zaire.
DISTRIBUTION: Known only from the type locality.

STATUS: IUCN - Insufficiently known. *P. graueri* is of conservation concern (Nicoll and Rathbun, 1990).
COMMENTS: Known only from the holotype which was collected in 1908.

Paracrocidura maxima Heim de Balsac, 1959. Rev. Zool. Bot. Afr., 59:26.
TYPE LOCALITY: Zaire, Tshibati.
DISTRIBUTION: Zaire, Rwanda, Uganda.
STATUS: IUCN - Insufficiently known. *P. maxima* is of conservation concern (Nicoll and Rathbun, 1990).
COMMENTS: Regarded as a full species by Hutterer (1986*c*:79).

Paracrocidura schoutedeni Heim de Balsac, 1956. Rev. Zool. Bot. Afr., 54:137.
TYPE LOCALITY: Zaire, Kasai, Lubondaie (75 km south of Luluabourg), Tshimbulu (Dibaya).
DISTRIBUTION: Lowland primary forest in S Cameroon, Gabon, Congo Republic, Zaire, and Central African Republic.
SYNONYMS: *camerunensis*.
COMMENTS: A subspecies *camerunensis* was named by Heim de Balsac (1968*b*), based on a specimen from Mt. Cameroon.

Ruwenzorisorex Hutterer, 1986. Z. Säugetierk., 51:260.
TYPE SPECIES: *Sylvisorex suncoides* Osgood, 1936.
COMMENTS: Subfamily uncertain. New data on the brain structure support generic separation; see Stephan et al. (1991).

Ruwenzorisorex suncoides (Osgood, 1936). Field Mus. Nat. Hist. Publ., Zool. Ser., 20:217.
TYPE LOCALITY: Zaire, western slope of Ruwenzori Mountains, Kalongi.
DISTRIBUTION: Montane forest in W Zaire, Uganda, Rwanda, and Burundi.
STATUS: IUCN - Indeterminate. Very localized; listed in Nicoll and Rathbun (1990:21).
COMMENTS: The species has also been found in Burundi (Kerbis, pers. comm.).

Scutisorex Thomas, 1913. Ann. Mag. Nat. Hist., ser. 8, 11:321.
TYPE SPECIES: *Sylvisorex somereni* Thomas, 1910.
COMMENTS: Subfamily Crocidurinae; see Repenning (1967:15).

Scutisorex somereni (Thomas, 1910). Ann. Mag. Nat. Hist., ser. 8, 6:113.
TYPE LOCALITY: Uganda, near Kampala, Kyetume.
DISTRIBUTION: Tropical rainforest of the Zaire Basin and adjacent mountains in Uganda, Rwanda, and Burundi.
SYNONYMS: *congicus*.
COMMENTS: Includes *congicus*; see Heim de Balsac and Meester (1977:7).

Solisorex Thomas, 1924. Spolia Zeylan., 13:94.
TYPE SPECIES: *Solisorex pearsoni* Thomas, 1924.
COMMENTS: Subfamily Crocidurinae; see Repenning (1967:15).

Solisorex pearsoni Thomas, 1924. Spolia Zeylan., 13:94.
TYPE LOCALITY: Sri Lanka, Central Province, near Nuwara Eliya, Hakgala.
DISTRIBUTION: Central highlands of Sri Lanka.
STATUS: Species rare and little-known.
COMMENTS: Inhabits "virgin forest" in the mountains of C Sri Lanka (Phillips, 1980).

Suncus Ehrenberg, 1832. *In* Hemprich and Ehrenberg, Symb. Phys. Mamm., 2:k.
TYPE SPECIES: *Suncus sacer* Ehrenberg, 1832 (= *Sorex murinus* Linnaeus, 1766).
SYNONYMS: *Pachyura, Paradoxodon, Plerodus, Podihik, Sunkus*.
COMMENTS: Subfamily Crocidurinae; see Repenning (1967:15). Occasionally regarded as part of *Crocidura* (e.g. Lekagul and McNeely, 1977:35), but accepted as a full genus by most authors. Includes *Pachyura, Paradoxodon*, and *Plerodus*; see Meester and Lambrechts (1971), who revised the southern African species.

Suncus ater Medway, 1965. J. Malay. Branch R. Asiat. Soc., 36:38.
TYPE LOCALITY: Malaysia, Sabah, Gunong (= Mt.) Kinabalu, Lumu-Lumu, 5,500 ft. (1,676 m).
DISTRIBUTION: Known only from the type locality.
COMMENTS: Reviewed by Medway (1977:16-17).

Suncus dayi (Dobson, 1888). Ann. Mag. Nat. Hist., ser. 6, 1:428.
TYPE LOCALITY: India, Cochin, Trichur.
DISTRIBUTION: S India.
STATUS: Rare.
COMMENTS: A very distinct species resembling *Sylvisorex morio*.

Suncus etruscus (Savi, 1822). Nuovo Giorn. de Letterati, Pisa, 1:60.
TYPE LOCALITY: Italy, Pisa.
DISTRIBUTION: S Europe and N Africa (Morocco, Algeria, Tunisia, Egypt); Arabian Peninsula and Asia Minor to Iraq, Turkmenistan, Afghanistan, Pakistan, Nepal, Bhutan, Burma, Thailand and Yunnan (China); also India and Sri Lanka. West and East African records (Guinea, Nigeria, Ethiopia) are doubtful and need confirmation.
SYNONYMS: *assamensis, atratus, bactrianus, hodgsoni, kura, macrotis, melanodon, micronyx, nanula, nilgirica, nitidofulva, nudipes, pachyurus, perrotteti, pygmaeoides, pygmaeus, suaveolens, travancorensis*.
COMMENTS: European and Asian range reviewed by Spitzenberger (1970, 1990*c*); N African distribution mapped by Vesmanis (1987). Heim de Balsac and Meester (1977) discussed the African records south of the Sahara. Probably includes *Podihik kura*; see Nowak and Paradiso (1983:141). The records east of Afghanistan, particularly from S India (*macrotis, nilgirica*) are only tentatively included; Corbet (1978*c*:31) expressed doubt on the conspecificy of the Indian forms. Many authors included *fellowesgordoni, hosei, madagascariensis*, and *malayanus* in *etruscus*, however, in the present list they are all treated as valid species.

Suncus fellowesgordoni Phillips, 1932. Spolia Zeylan., 17:124.
TYPE LOCALITY: Sri Lanka, Central Province, Ohiya, West Haputale Estate (6,000 ft.).
DISTRIBUTION: Central highlands of Sri Lanka.
COMMENTS: Although usually included in *S. etruscus*, this taxon represents a species endemic to Sri Lanka. *Podihik kura* Deraniyagala, 1958, which was included in this species by Phillips (1980), does not represent *fellowesgordoni*, but is more similar to *etruscus*.

Suncus hosei (Thomas, 1893). Ann. Mag. Nat. Hist., ser. 6, 11:343.
TYPE LOCALITY: Sarawak, Bakong River.
DISTRIBUTION: Lowland forest of Borneo and Sarawak.
COMMENTS: Often included in *etruscus* (e.g. Medway, 1977) but represents a distinct forest species.

Suncus infinitesimus (Heller, 1912). Smithson. Misc. Coll., 60(12):5.
TYPE LOCALITY: Kenya, Laikipia Plateau, Rumruti, 7,000 ft. (2,134 m).
DISTRIBUTION: South Africa to Kenya; Central African Republic; Cameroon.
SYNONYMS: *chriseos, ubanguiensis*.
COMMENTS: Includes *chriseos* and *ubanguiensis*; see Heim de Balsac and Meester (1977). Gureev (1979:383) listed *chriseos* as a distinct species without comment.

Suncus lixus (Thomas, 1898). Proc. Zool. Soc. Lond., 1897:930 [1898].
TYPE LOCALITY: Malawi, Nyika Plateau (between 10 and 11° S and 33°40' to 34°10'E).
DISTRIBUTION: Savanna zones of Kenya, Tanzania, Malawi, Zaire, Zambia, Angola, Botswana, and Transvaal (South Africa).
SYNONYMS: *aequatoria, gratula*.
COMMENTS: Includes *aequatoria* and *gratula*; see Heim de Balsac and Meester (1977). Gureev (1979:383) listed *gratulus* as a distinct species without comment.

Suncus madagascariensis (Coquerel, 1848). Ann. Sci. Nat., Zool. (Paris), ser. 3, 9:194, pl. 11, fig. 1.
TYPE LOCALITY: Madagascar, Nossi-Bé.
DISTRIBUTION: Madagascar and Comores Isls.
STATUS: Unresolved.
SYNONYMS: *coquerelii*.
COMMENTS: This species is often included in *etruscus* but treated as a full species in most reports on the fauna of Madagascar (e.g., Eisenberg and Gould, 1984).

Suncus malayanus (Kloss, 1917). J. Nat. Hist. Soc. Siam, 2:282.
TYPE LOCALITY: Thailand, "Bang Nara, Patani, Peninsular Siam".

DISTRIBUTION: Malaysian peninsula.
COMMENTS: Commonly included in *etruscus* but inhabits tropical forest and does not fit morphologically with the diagnosis of that species; *malayanus* is therefore regarded as a species, as was done by Corbet and Hill (1991:36).

Suncus mertensi Kock, 1974. Senckenbergiana Biol., 55:198.
TYPE LOCALITY: Indonesia, "Rana Mese, Flores".
DISTRIBUTION: Flores Isl, Indonesia.
COMMENTS: A distinct, long-tailed forest shrew.

Suncus montanus (Kelaart, 1850). J. Ceylon Branch Asiat. Soc., 2:211.
TYPE LOCALITY: Sri Lanka, "Nuwara Eliya, Pidurutalagala".
DISTRIBUTION: Forested highlands in Sri Lanka and S India.
SYNONYMS: *ferrugineus, kelaarti, niger.*
COMMENTS: Commonly included in *murinus* (Ellerman and Morrison-Scott, 1966:66), but represents a much smaller and always blackish species of primary forest habitats. Listed as a species by Corbet and Hill (1991:36). The Indian populations may represent a valid subspecies (*niger*).

Suncus murinus (Linnaeus, 1766). Syst. Nat., 12th ed., 1:74.
TYPE LOCALITY: Indonesia, Java.
DISTRIBUTION: Afghanistan, Pakistan, India, Sri Lanka, Nepal, Bhutan, Burma, China, Taiwan, Japan, continental and peninsular Indomalayan Region; introduced into Guam, the Maldive Islands, and probably many other islands; introduced in historical times into coastal Africa (Egypt to Tanzania), Madagascar, the Comores, Mauritius, and Réunion, and into coastal Arabia (Iraq, Bahrain, Oman, Yemen, Saudi Arabia).
SYNONYMS: *albicauda, albinus, andersoni, auriculata, beddomei, blanfordii, blythii, caerulaeus, caerulescens, caeruleus, celebensis, ceylanica, crassicaudus, duvernoyi, edwardsiana, fulvocinerea, fuscipes, geoffroyi, giganteus, griffithii, heterodon, indicus, kandianus, kroonii, kuekenthali, leucura, luzoniensis, malabaricus, mauritiana, media, melanodon, microtis, mulleri, muschata, myosurus, nemorivagus, nitidofulva, occultidens, palawanensis, pealana, pilorides, riukiuana, rubicunda, sacer, saturatior, semmelincki, semmeliki, serpentarius, sindensis, soccatus, sonneratii, swinhoei, temminckii, tytleri, unicolor, viridescens, waldemarii.*
COMMENTS: A very variable species with a number of genetically distinct populations which almost behave like semispecies (Hasler et al., 1977; Yamagata et al., 1987; Yoshida, 1985). A number of laboratory strains have been established (Oda et al., 1985). Much of the present distribution is the result of human agency (Hutterer and Tranier, 1990). Includes *albicauda, auriculata, crassicaudus, duvernoyi, leucura, mauritiana, sacer,* and *geoffroyi*; see Heim de Balsac and Meester (1977). Includes *edwardsiana* (formerly in *Crocidura*), *luzoniensis, occultidens,* and *palawanensis*; see Heaney et al. (1987).

Suncus remyi Brosset, Dubost and Heim de Balsac, 1965. Biologia Gabonica, 1:170.
TYPE LOCALITY: Gabon, Makokou.
DISTRIBUTION: Two localities in rainforest of NE Gabon, Belinga and Makokou.
STATUS: IUCN - Insufficiently known. Very localized, listed in Nicoll and Rathbun (1990:21).
COMMENTS: Ecology described by Brosset (1988). One of the smallest shrews; species not recorded again since its description.

Suncus stoliczkanus (Anderson, 1877). J. Asiat. Soc. Bengal, 46:270.
TYPE LOCALITY: India, Bombay.
DISTRIBUTION: Deserts and arid country in Pakistan, Nepal, India, and Bangladesh.
SYNONYMS: *bidiana, leucogenys, subfulva.*

Suncus varilla (Thomas, 1895). Ann. Mag. Nat. Hist., ser. 6, 16:54.
TYPE LOCALITY: South Africa, Cape Prov., East London.
DISTRIBUTION: Savannahs from the Cape (South Africa) to Zimbabwe, Zambia, Tanzania, E Zaire, Malawi; an isolated record from Nigeria.
SYNONYMS: *meesteri, minor, natalensis, orangiae, tulbaghensis, warreni* (see Heim de Balsac and Meester, 1977:6).

COMMENTS: Closely associated with termite mounds (Lynch, 1986). Gureev (1979:383) listed *orangiae* and *warreni* as distinct species without comment. Common in the Pleistocene of Kenya (Butler and Greenwood, 1979).

Suncus zeylanicus Phillips, 1928. Spolia Zeylan., 14:313.
TYPE LOCALITY: Sri Lanka, "Gonagama Estate, Kitulgala, 900 ft."
DISTRIBUTION: Higlands of Sri Lanka.
COMMENTS: Phillips (1980) stressed that *zeylanicus* differs distinctly from *murinus* in the flesh, particularly by its long and almost naked tail, and that it lives in primary forest. However, its relation to *montanus* has still to be studied.

Surdisorex Thomas, 1906. Ann. Mag. Nat. Hist., ser. 7, 18:223.
TYPE SPECIES: *Surdisorex norae* Thomas, 1906.
COMMENTS: This genus is commonly included in *Myosorex* but was retained as a full genus by Hollister (1918), Meester (1953), and Heim de Balsac (1966*b*). Subfamily uncertain; see under *Myosorex*.

Surdisorex norae Thomas, 1906. Ann. Mag. Nat. Hist., ser. 7, 18:223.
TYPE LOCALITY: Kenya, east side of Aberdare Range, near Nyeri.
DISTRIBUTION: Aberdare Range (Kenya).
COMMENTS: Formerly in *Myosorex*; see Heim de Balsac and Meester (1977). Ecology and distribution described by Duncan and Wrangham (1971).

Surdisorex polulus Hollister, 1916. Smithson. Misc. Coll., 66(1):1.
TYPE LOCALITY: Kenya, west side of Mt. Kenya, 10,700 ft. (3,261 m).
DISTRIBUTION: Mount Kenya (Kenya).
COMMENTS: Included in genus *Myosorex* and regarded as a subspecies of *norae* by Heim de Balsac and Meester (1977); however, both species form a quite distinct clade. For ecology and distribution see Duncan and Wrangham (1971).

Sylvisorex Thomas, 1904. Abstr. Proc. Zool. Soc. Lond., 1904(10):12.
TYPE SPECIES: *Crocidura morio* Gray, 1862.
COMMENTS: Subfamily Crocidurinae; see Repenning (1967:15). The genus was regarded as part of *Suncus* by Smithers and Tello (1976), but was retained by Ansell (1978); it may be polyphyletic and its relation to *Suncus* requires further study. Jenkins (1984) figured and discussed most of the species listed.

Sylvisorex granti Thomas, 1907. Ann. Mag. Nat. Hist., ser. 7, 19:118.
TYPE LOCALITY: Uganda, Ruwenzori East, Mubuku Valley, 10,000 ft. (3,048 m).
DISTRIBUTION: Mountain forests of C (Zaire, Uganda, Rwanda) and E Africa (Kenya, Tanzania); an isolated population in Cameroon.
SYNONYMS: *camerunensis, mundus*.
COMMENTS: The westernmost population may represent a distinct species, *camerunensis*; see Hutterer et al. (1987*b*).

Sylvisorex howelli Jenkins, 1984. Bull. Brit. Mus. (Nat. Hist). Zool., 47:65.
TYPE LOCALITY: Tanzania, Uluguru Mtns, Morningside.
DISTRIBUTION: Usambara and Uluguru Mtns (Tanzania).
STATUS: IUCN - Insufficiently known.
SYNONYMS: *usambarensis*.
COMMENTS: Includes *usambarensis*, which may represent a distinct species; see Hutterer (1986*b*).

Sylvisorex isabellae Heim de Balsac, 1968. Bonn. Zool. Beitr., 19:31.
TYPE LOCALITY: Equatorial Guinea, Bioko (Fernando Po), "Pic Santa Isabel, Refugium, 2000 m."
DISTRIBUTION: Bioko; a similar form occurs in the Bamenda Highlands, Cameroon.
COMMENTS: Included in *morio* by Heim de Balsac and Meester (1977), but represents a distinctly smaller species.

Sylvisorex johnstoni (Dobson, 1888). Proc. Zool. Soc. Lond., 1887:577 [1888].
TYPE LOCALITY: Cameroon, Rio del Rey.

DISTRIBUTION: Lowland forest of the Zaire Basin, SW Cameroon, Gabon, Bioko, Congo Republic, Zaire, Uganda, Tanzania, Burundi.
SYNONYMS: *dieterleni*.
COMMENTS: Species reviewed by Hutterer (1986*b*); recently found in the Congo Republic (Dowsett and Granjon, 1991) and Burundi (Kerbis, pers. comm.).

Sylvisorex lunaris Thomas, 1906. Ann. Mag. Nat. Hist., ser. 7, 18:139.
TYPE LOCALITY: Uganda, "Mubuku Valley, Ruwenzori East, 12,000 ft." (3,810 m).
DISTRIBUTION: The high mountain zone of C Africa up to 4,500 m; Ruwenzori (Uganda, Zaire), Virunga Volcanoes (Rwanda), and on both sides of Lake Kivu (Zaire, Burundi).
SYNONYMS: *ruandae*.
COMMENTS: Includes *ruandae* but not *oriundus*; both were listed as distinct species by Gureev (1979:380-381).

Sylvisorex megalura (Jentink, 1888). Notes Leyden Mus., 10:48.
TYPE LOCALITY: Liberia, Junk River, Schieffelinsville.
DISTRIBUTION: Tropical forest zone of Africa from Upper Guinea to Ethiopia and south to Mozambique and Zimbabwe.
SYNONYMS: *angolensis*, *gemmeus*, *infuscus*, *irene*, *phaeopus*, *sheppardi*, *sorella*, *sorelloides* (see Heim de Balsac and Meester, 1977:7-8).
COMMENTS: *S. megalua* is the most common species of the genus, and enters forested savannas; range mapped by Hutterer et al. (1987*b*). Gureev (1979:381) listed *sorella* as a distinct species without comment. Some geographic variation exists, the Central African forest populations being smallest and darkest.

Sylvisorex morio (Gray, 1862). Proc. Zool. Soc. Lond., 1862:180.
TYPE LOCALITY: "Cameroon Mountains".
DISTRIBUTION: Confined to Mount Cameroon (Cameroon).
COMMENTS: Does not include *isabellae*; see under that species.

Sylvisorex ollula Thomas, 1913. Ann. Mag. Nat. Hist., ser. 8, 11:321.
TYPE LOCALITY: "Cameroons, Bitye, Ja River, 2,000 feet" (610 m).
DISTRIBUTION: S Cameroon and adjacent Nigeria; Gabon; S Zaire.
STATUS: IUCN - Insufficiently known. *S. ollula* is of conservation concern (Nicoll and Rathbun, 1990:21).
COMMENTS: The largest species of the genus; discussed in some detail by Dieterlen and Heim de Balsac (1979).

Sylvisorex oriundus Hollister, 1916. Bull. Am. Mus. Nat. Hist., 35:672.
TYPE LOCALITY: Zaire, Medje.
DISTRIBUTION: NE Zaire.
COMMENTS: Often included in *ollula* but as regarded distinct by Dieterlen and Heim de Balsac (1979), a view supported by personal examination of the holotype.

Sylvisorex vulcanorum Hutterer and Verheyen, 1985. Z. Säugetierk., 50:266.
TYPE LOCALITY: Rwanda, "Karisoke (0°28'S., 29°29'E., 3100 m), Parc National des Volcans".
DISTRIBUTION: High altitude rainforest of E Zaire, Uganda, Rwanda, and Burundi.
STATUS: *S. vulcanorum* is of conservation concern (Nicoll and Rathbun, 1990:21).
COMMENTS: One of the smallest species in the genus; rather similar to *S. granti*.

Subfamily Soricinae G. Fischer, 1817. Mem. Soc. Imp. Nat. Moscow, 5:372.
COMMENTS: The recognition of two subfamilies within the Soricidae is mainly based on Repenning (1967), and has been widely accepted; see George (1986), but see also comments under family. Reumer (1984) modified the tribal subdivision. Work in progress, however, raises doubt on the validity of current concepts.

Anourosorex Milne-Edwards, 1872. Rech. Hist. Nat. Mamm., p. 264.
TYPE SPECIES: *Anourosorex squamipes* Milne-Edwards, 1872.
SYNONYMS: *Pygmura*.
COMMENTS: Tribe Neomyini (Repenning, 1967:61) or Anourosoricini. Reumer (1984:17) placed the genus in the tribe Amblycoptini Kormos, 1926, but this is antedated by

Anourosoricini Anderson, 1879. Specific taxonomy is in need of revision. For the fossil history, see Zheng (1985) and Storch and Qiu (1991).

Anourosorex squamipes Milne-Edwards, 1872. Rech. Hist. Nat. Mamm., p. 264.
TYPE LOCALITY: China, Sichuan Prov., probably Moupin (= Baoxing).
DISTRIBUTION: Shaanxi and Hubei, south to Yunnan (China); Taiwan; N and W Burma; Assam (India) and Bhutan; North Vietnam; Thailand.
SYNONYMS: *assamensis, capito, capnias, schmidi, yamashinai.*
COMMENTS: Includes *schmidi* and *yamashinai* as subspecies; see Petter (1963*b*) and Jameson and Jones (1977).

Blarina Gray, 1838. Proc. Zool. Soc. Lond., 1837:124 [1838].
TYPE SPECIES: *Corsira (Blarina) talpoides* Gray, 1838 (= *Sorex talpoides* Gapper, 1830 = *Sorex brevicaudus* Say, 1823).
SYNONYMS: *Anotus, Blaria, Brachysorex, Corsia; Talposorex* Pomel (not Lesson).
COMMENTS: Tribe Blarinini (Repenning, 1967:37). Reviewed by George et al. (1982, 1986).

Blarina brevicauda (Say, 1823). *In* Long, Account Exped. Pittsburgh to Rocky Mtns, 1:164.
TYPE LOCALITY: USA, Engineer cantonment, west bank of the Missouri R.; restricted to Nebraska, Washington Co., approximately 2 miles east Ft. Calhoun by Jones (1964:68).
DISTRIBUTION: S Canada west to C Saskatchewan and east to SE Canada, south to Nebraska and N Virginia (USA).
SYNONYMS: *aloga, angusta, angusticeps, churchi, compacta, costaricensis, dekayi, fossilis, hooperi, kirtlandi, manitobensis, micrurus, ozarkensis, pallida, simplicidens, talpoides, telmalestes.*
COMMENTS: Includes *telmalestes* (see review by George et al., 1986, Mammalian Species, 261), which Hall (1981:57) listed as a distinct species.

Blarina carolinensis (Bachman, 1837). J. Acad. Nat. Sci. Philadelphia, 7:366.
TYPE LOCALITY: USA, "in the upper and maritime districts of South Carolina".
DISTRIBUTION: S Illinois east to N Virginia, and south through E Texas and N Florida (USA).
SYNONYMS: *peninsulae, shermani.*
COMMENTS: For specific status see Genoways and Choate (1972) and Tate et al. (1980). Hall (1981:54) listed *carolinensis* as a subspecies of *brevicauda*. The Florida population (*peninsulae*) may represent a valid species (George et al., 1982).

Blarina hylophaga Elliot, 1899. Field Columb. Mus. Publ., Zool. Ser., 1:287.
TYPE LOCALITY: USA, Oklahoma, Murray Co., Dougherty.
DISTRIBUTION: USA: S Nebraska and SW Iowa south to S Texas; east to Missouri and NW Arkansas; Oklahoma; extending into Louisiana.
SYNONYMS: *mimina, plumbea.*
COMMENTS: Original spelling *hulophaga* Elliot, 1899, corrected to *hylophaga* by Elliot (1905). Formerly included in *carolinensis*, but separated as a distinct species by George et al. (1981).

Blarinella Thomas, 1911. Proc. Zool. Soc. Lond., 1911:166.
TYPE SPECIES: *Sorex quadraticauda* Milne-Edwards, 1872.
COMMENTS: Tribe Soricini; see Repenning (1967:61). The genus is known from the Late Miocene of China (Storch and Qiu, 1991), and was also recorded from the Pleistocene of Europe (Reumer, 1984).

Blarinella quadraticauda (Milne-Edwards, 1872). Rech. Hist. Nat. Mamm., p. 261.
TYPE LOCALITY: China, Sichuan, "Moupin, Thibet oriental".
DISTRIBUTION: Montane taiga forest of Gansu, Shaanxi, Sichuan, and Yunnan (China).
SYNONYMS: *griselda.*
COMMENTS: Includes *griselda*; see Ellerman and Morrison-Scott (1951) and Hoffmann (1987).

Blarinella wardi Thomas, 1915. Ann. Mag. Nat. Hist., ser. 8, 15:336.
TYPE LOCALITY: "Hpimaw, Upper Burma, about 26°N., 98°35'E. Alt. 8000'."
DISTRIBUTION: Upper Burma and Yunnan (China).

COMMENTS: Included in *quadraticauda* by Ellerman and Morrison-Scott (1951) and subsequent authors, but the species has a much smaller and narrower skull (see measurements in Hoffmann, 1987:134) and is therefore regarded as distinct. Differences were also recognized by Corbet (1978*c*:26).

Chimarrogale Anderson, 1877. J. Asiat. Soc. Bengal, 46:262.

TYPE SPECIES: *Crossopus himalayicus* Gray, 1842.

SYNONYMS: *Chimmarogale, Crossogale.*

COMMENTS: Tribe Neomyini; see Repenning (1967:61). Because of the presence of white teeth the genus was occasionally included in subfamily Crocidurinae, but since Repenning (1967), overwhelming evidence has been accumulated showing that *Chimarrogale* is a soricine shrew (Vogel and Besancon, 1979; Mori et al., 1991). Gureev (1971:226) included *Chimarrogale* in his subtribe Nectogalina within the Blarinini, while Reumer (1984:14) included it in the tribe Soriculini; see comments under genus *Neomys*. Includes *Crossogale*; see Harrison (1958), who also revised the genus. His arrangement was found to be more realistic than the present practice of lumping all forms together in one or two species.

Chimarrogale hantu Harrison, 1958. Ann. Mag. Nat. Hist., ser. 13, 1:282.

TYPE LOCALITY: "banks of a stream at low altitude (under 1,000 ft.) in the Ulu Langat Forest Reserve, Selangor, Malaya, about 20 km. east of Kuala Lumpur."

DISTRIBUTION: Tropical forest of the Malaysian peninsula.

COMMENTS: Included in *himalayica* by Medway (1977) and other authors but retained by Jones and Mumford (1971). The species differs considerably in its morphology and ecology from the species which inhabit the Himalayan region. The photograph of a live animal in Nowak (1991:156) depicts this species.

Chimarrogale himalayica (Gray, 1842). Ann. Mag. Nat. Hist., [ser. 1], 10:261.

TYPE LOCALITY: "India", Punjab, Chamba.

DISTRIBUTION: Kashmir through SE Asia to Indochina; C and S China; Taiwan.

SYNONYMS: *leander, varennei.*

COMMENTS: Corbet (1978*c*) included *leander, platycephala, varennei,* and probably *hantu* in *himalayica*. Gureev (1979) listed *leander, hantu, platycephala,* and *varennei* as distinct species without comment; both views are only partially accepted here. Species reviewed by Jones and Mumford (1971) and Hoffmann (1987).

Chimarrogale phaeura Thomas, 1898. Ann. Mag. Nat. Hist., ser. 7, 2:246.

TYPE LOCALITY: Malaysia, Sabah, "Saiap, Mount Kina Balu".

DISTRIBUTION: Streams in tropical forest of Borneo island.

COMMENTS: Medway (1977) considered *phaeura* as a subspecies of *himalayica* but Corbet (1978*c*) and Jones and Mumford (1971) maintained *styani* and *phaeura* as separate species. Ellerman and Morrison-Scott (1966:87) included *sumatrana* in this species, but Gureev (1979:458) listed it as a distinct species, a view followed here.

Chimarrogale platycephala (Temminck, 1842). Fauna Japon., 1(Mamm.), p. 23, pl. V, fig. 1.

TYPE LOCALITY: Japan, Kyushu, near Nagasaki and Bungo.

DISTRIBUTION: Most of the Japanese Islands.

COMMENTS: Included in *himalayica* since Ellerman and Morrison-Scott (1951), but retained as a separate species by Harrison (1958), Hutterer and Hürter (1981), Hoffmann (1987), and Corbet and Hill (1991). Arai et al. (1985) reported on clinal size variation in Japan. For date of publication see Holthuis and Sakai (1970).

Chimarrogale styani De Winton, 1899. Proc. Zool. Soc. Lond., 1899:574.

TYPE LOCALITY: China, "Yangl-iu-pa, N.W. Sechuen [= Sichuan]."

DISTRIBUTION: Shensi and Sichuan (China), and N Burma.

COMMENTS: Certainly a distinct species, and regarded as such by Jones and Mumford (1971), Corbet (1978*c*), and Hoffmann (1987). Occurs nearly sympatrically with *himalayica* in N Burma.

Chimarrogale sumatrana (Thomas, 1921). Ann. Mag. Nat. Hist., ser. 9, 7:244.

TYPE LOCALITY: Indonesia, Sumatra, "Pager Alam, Padang Highlands".

DISTRIBUTION: Streams in tropical forest of Sumatra.

COMMENTS: Regarded as a race of *phaeura* by Ellerman and Morrison-Scott (1966:87), but considered distinct by Harrison (1958) and Gureev (1979).

Cryptotis Pomel, 1848. Arch. Sci. Phys. Nat. Geneve, 9:249.
TYPE SPECIES: *Sorex cinereus* Bachman, 1837 (= *Sorex parvus* Say, 1823).
SYNONYMS: *Brachysorex, Soriciscus, Xenosorex.*
COMMENTS: Tribe Blarinini; see Repenning (1967:37). Revised in part by Choate (1970) and Choate and Fleharty (1974); the South American species still call for a thorough study. Gureev (1979:433-437) listed many species which Choate (1970) considered synonyms. Formerly included *C. surinamensis* which was transferred to *Sorex araneus* by Husson (1963).

Cryptotis avia G. M. Allen, 1923. Proc. New England Zool. Club, 8:37.
TYPE LOCALITY: "El Verjón, in the Andes east of Bogotá, Colombia."
DISTRIBUTION: E Cordillera of Colombia.
COMMENTS: Accepted as a species by Choate and Fleharty (1974).

Cryptotis endersi Setzer, 1950. J. Washington Acad. Sci., 40:300.
TYPE LOCALITY: "Cylindro, above 4000 ft., Bocas del Toro, Panamá."
DISTRIBUTION: Known only from the type locality; status discussed by Choate (1970:285).
COMMENTS: Considered a relict species by Choate (1970).

Cryptotis goldmani (Merriam, 1895). N. Am. Fauna, 10:25.
TYPE LOCALITY: "mountains near Chilpancingo, Guerrero, Mexico" (altitude 10,000 ft).
DISTRIBUTION: Highlands of Estado de México, Jalisco and Oaxaca to Chiapas (Mexico), and WC Guatemala.
SYNONYMS: *alticola, euryrhynchis, fossor, frontalis, griseoventris, guerrerensis, machetes.*
COMMENTS: Choate (1970) recognized two distinct subspecies, *alticola* and *goldmani.*

Cryptotis goodwini Jackson, 1933. Proc. Biol. Soc. Washington, 46:81.
TYPE LOCALITY: "Calel, altitude 10200 feet, Guatemala."
DISTRIBUTION: S Guatemala, W El Salvador, and S Mexico.
COMMENTS: Reviewed by Choate and Fleharty (1974, Mammalian Species, 44) who included the species in the *mexicana* group (*mexicana, goldmani, goodwini*); recorded from Mexico by Hutterer (1980).

Cryptotis gracilis Miller, 1911. Proc. Biol. Soc. Washington, 24:221.
TYPE LOCALITY: "head of Larí River, Talamanca [= Limon], Costa Rica", near base of Pico Blanco.
DISTRIBUTION: SE Costa Rica and W Panama.
SYNONYMS: *jacksoni.*
COMMENTS: Includes *jacksoni*; considered a relict species by Choate (1970). Specimens from Honduras previously included in *gracilis* were described as a new species, *C. hondurensis* by Woodman and Timm (1992).

Cryptotis hondurensis Woodman and Timm, 1992. Proc. Biol. Soc. Washington, 105:2.
TYPE LOCALITY: "Honduras: Francisco Morazán Department; 12 km WNW of El Zamorano, W slope of Cerro Uyuca [= Cerro Oyuca; ca. 14°05'N, 87°06'W], 1680 m."
DISTRIBUTION: Pine, mixed pine, and oak forests on highlands east of Tegucicalpa, Honduras; possibly also in adjacent regions of Guatemala, El Salvador, and Nicaragua.
COMMENTS: Formerly included in *gracilis*, see comments therein.

Cryptotis magna (Merriam, 1895). N. Am. Fauna, 10:28.
TYPE LOCALITY: "Totontepec, Oaxaca" (altitude 6800 ft), Mexico.
DISTRIBUTION: NC Oaxaca (Mexico).
COMMENTS: Reviewed by Robertson and Rickart (1975, Mammalian Species, 61). A relict species, according to Choate (1970).

Cryptotis meridensis Thomas, 1898. Ann. Mag. Nat. Hist., ser. 7, 1:457.
TYPE LOCALITY: Venezuela, "Merida, alt. 2165 m."
DISTRIBUTION: Cordillera de Merida, and mountains near Caracas, Venezuela, see Tello (1979).

COMMENTS: This species was commonly included in *thomasi* (Handley, 1976; Eisenberg, 1989) but is much larger and has a more robust dentition. Choate (pers. comm., 1983) and Hutterer (1986*d*) therefore considered *meridensis* a valid species.

Cryptotis mexicana (Coues, 1877). Bull. U.S. Geol. Geogr. Surv. Terr., 3:652.
TYPE LOCALITY: "Xalapa, Mexico" [= Jalapa, ca. 1520 m, Veracruz].
DISTRIBUTION: Humid upper tropical zone from Tamaulipas to Chiapas (Mexico); altitudinal range 520 to 3200 m.
SYNONYMS: *madrea, nelsoni, obscura, peregrina, phillipsii.*
COMMENTS: *Notiosorex (Xenosorex) phillipsii* is a synonym of *Cryptotis mexicana*; see Choate (1969). Reviewed by Choate (1973, Mammalian Species, 28), who recognized four subspecies, *mexicana, nelsoni, obscura,* and *peregrina.*

Cryptotis montivaga (Anthony, 1921). Am. Mus. Novit., 20:5.
TYPE LOCALITY: "Bestion, Prov. del Azuay, Ecuador; altitude 10,000 ft." [3,000 m].
DISTRIBUTION: Andean zone of S Ecuador.

Cryptotis nigrescens (J. A. Allen, 1895). Bull. Am. Mus. Nat. Hist., 7:339.
TYPE LOCALITY: "San Isidro (San José), Costa Rica".
DISTRIBUTION: Tropical lowland of Yucatan Peninsula and highlands of Guerrero, Chiapas and Las Margaritas (Mexico), also highlands of Guatemala, El Salvador, Honduras, Costa Rica, and Panama.
SYNONYMS: *mayensis, merriami, merus, micrura, tersus, zeteki.*
COMMENTS: There are three distinct subspecies (*nigrescens, mayensis,* and *merriami*; see Choate, 1970); N. Woodman and R. Timm (pers. comm.) believe the latter two are valid species.

Cryptotis parva (Say, 1823). *In* Long, Account Exped. Pittsburgh to Rocky Mtns, 1:163.
TYPE LOCALITY: "Engineer Cantonment," west bank of Missouri River; restricted by Jones (1964:68) to USA, Nebraska, Washington Co., approximately 2 mi. east Ft. Calhoun.
DISTRIBUTION: Extreme SE Canada through EC and SW USA, Mexico and Central America south to Panama.
SYNONYMS: *berlandieri, celatus, cinereus, elasson, exilipes, eximius, floridana, harlani, macer, nayaritensis, micrurus, olivaceus, orophila, pergracilis, pueblensis, soricina, tropicalis.*
COMMENTS: Reviewed by Whitaker (1974, Mammalian Species, 43), who recognized 9 subspecies, 5 of which occur in Middle America (Choate, 1970). Handley (pers. comm., 1989) suggested that *floridana* may be a distinct taxon; if this proves correct, then the other southern subspecies should be restudied.

Cryptotis squamipes (J. A. Allen, 1912). Bull. Am. Mus. Nat. Hist., 31:93.
TYPE LOCALITY: "crest of Western Andes (alt. 10,340 ft.), 40 miles west of Popayan, Cauca, Colombia."
DISTRIBUTION: S Cordillera Occidental of Colombia and Ecuador.

Cryptotis thomasi (Merriam, 1897). Proc. Biol. Soc. Washington, 11:227.
TYPE LOCALITY: "Plains of Bogota, Colombia (on G. O. Child's estate, near city of Bogota, alt. about 9000 ft)."
DISTRIBUTION: Cordillera Oriental of Colombia, Ecuador, and N Peru.
SYNONYMS: *equatoris, medellinius, osgoodi.*

Megasorex Hibbard, 1950. Contrib. Mus. Paleontol. Univ. Michigan, 8:129.
TYPE SPECIES: *Notiosorex gigas* Merriam, 1897.
COMMENTS: Tribe Neomyini; see Repenning (1967) and George (1986).

Megasorex gigas (Merriam, 1897). Proc. Biol. Soc. Washington, 11:227.
TYPE LOCALITY: Mexico, Jalisco, near San Sebastián, mountains at Milpillas.
DISTRIBUTION: Nayarit to Oaxaca (Mexico).
COMMENTS: Formerly included in *Notiosorex* and still done so by Hall (1981:65); but Repenning (1967:56) and Armstrong and Jones (1972*a*, Mammalian Species, 16) considered *Megasorex* a distinct genus; a view supported by George (1986) on the basis of allozyme data.

Nectogale Milne-Edwards, 1870. C.R. Acad. Sci. Paris, 70:341.
TYPE SPECIES: *Nectogale elegans* Milne-Edwards, 1870.
COMMENTS: Subfamily Soricinae; see Vogel and Besancon (1979); tribe Neomyini; see Repenning (1967:45). Gureev (1971:226) placed *Nectogale* in a new subtribe Nectogalina within the Blarinini, a view not followed by other authors.

Nectogale elegans Milne-Edwards, 1870. C.R. Acad. Sci. Paris, 70:341.
TYPE LOCALITY: China, Sichuan, "Moupin" (= Baoxing).
DISTRIBUTION: Cold mountain streams across the Himalayas and in W and C China; Tibet (Xizang Aut. Region), Nepal, Sikkim (India), Bhutan, N Burma, and Yunnan, Sichuan and Shaanxi (China).
SYNONYMS: *sikhimensis*.
COMMENTS: Includes *sikhimensis*, see Ellerman and Morrison-Scott (1951) and Hoffmann (1987). Species highly adapted for a semi-aquatic life (Hutterer, 1985).

Neomys Kaup, 1829. Skizz. Entwickel.-Gesch. Nat. Syst. Europ. Thierwelt, 1:117.
TYPE SPECIES: *Sorex daubentonii* Erxleben, 1777 (= *Sorex fodiens* Pennant, 1771).
SYNONYMS: *Crossopus, Hydrogale, Leucorrhynchus, Pinalea*.
COMMENTS: Type genus of tribe Neomyini Repenning, 1967, for which Reumer (1984:14) used Soriculini Kretzoi, 1965. However, both are antedated by Neomyini Matschie, 1909.

Neomys anomalus Cabrera, 1907. Ann. Mag. Nat. Hist., ser. 7, 20:214.
TYPE LOCALITY: Spain, "San Martin de la Vega, Jarama River, Madrid Prov."
DISTRIBUTION: Temperate woodlands of Europe, from Portugal to Poland and east to Voronesh, Russia. Records from N Asia Minor and Iran uncertain.
SYNONYMS: *amphibius, josti, milleri, mokrzeckii, rhenanus, soricoides* (see Spitzenberger, 1990*b*).
COMMENTS: *Sorex amphibius* Brehm, 1826 is probably an earlier name for the species (von Knorre, pers. comm.), although it has to be treated as a *nomen oblitum*.

Neomys fodiens (Pennant, 1771). Synopsis Quadrupeds, p. 308.
TYPE LOCALITY: Germany, Berlin.
DISTRIBUTION: Most of Europe including the British Isls and eastwards to Lake Baikal, Yenise River (Russia), Tien Shan (China), and NW Mongolia; disjunct in Sakhalin Isl and adjacent Siberia, Jilin (China), and N Korea.
SYNONYMS: *albus, aquaticus, argenteus, bicolor, brachyotus, canicularius, carinatus, ciliatus, collaris, constrictus, dagestanicus, daubentonii, eremita, fimbriatus, fluviatilis, griseogularis, hermanni, hydrophilus, ignotus, intermedius, limchjnhunii, lineatus, linneana, liricaudatus, longobarda, macrourus, minor, musculus, naias, natans, newtoni, niethammeri, nigripes, orientalis, orientis, pennantii, psilurus, remifer, rivalis, sowerbyi, stagnatilis, stresemanni, teres, watasei*.
COMMENTS: Includes *teres, orientis*, and *watasei* as possible subspecies (Ognev, 1928; Hoffmann, 1987; Yudin, 1989). Many of the listed synonyms have never been properly studied and identified; recently, Lehmann (1983) referred *constrictus* to *Crocidura russula*. The form *niethammeri* from NE Spain may represent a valid species (López-Fuster et al., 1990).

Neomys schelkovnikovi Satunin, 1913. Trud. Obshch. Izuch. Chernomorsk. Poberezh., 3:24.
TYPE LOCALITY: "Svanetiya," = Georgia, Mestiiskii r-n., Ushkul (see Pavlinov and Rossolimo, 1987:29).
DISTRIBUTION: Caucasus (Armenia, Azerbaijan, Georgia); and perhaps adjacent Turkey and Iran.
SYNONYMS: *balkaricus, leptodactylus*.
COMMENTS: Left *incertae sedis* by Ellerman and Morrison-Scott (1951), but given specific rank by most recent Russian authors. Reviewed by Sokolov and Tembotov (1989).

Notiosorex Coues, 1877. Bull. U.S. Geol. Geogr. Surv. Terr., 3:646.
TYPE SPECIES: *Sorex (Notiosorex) crawfordi* Coues, 1877.
COMMENTS: Tribe Neomyini; see Repenning (1967:45). Reumer (1984:14) created a new tribe Notiosoricini to include *Notiosorex*, but George (1986:160) could find no evidence to support this separation. Hall (1981:65) included also *Megasorex gigas*, but Repenning

(1967:56), Armstrong and Jones (1972), and George (1986) considered *Megasorex* a distinct genus. *Notiosorex (Xenosorex) phillipsii* is a synonym of *Cryptotis mexicana*; see Choate (1969). Lindsay and Jacobs (1985) described an extinct species from Pliocene sediments of Chihuahua, Mexico.

Notiosorex crawfordi (Coues, 1877). Bull. U.S. Geol. Geogr. Surv. Terr., 3:631.

TYPE LOCALITY: USA, Texas, El Paso Co., 2 mi. above El Paso, "near Fort Bliss, New Mexico (Practically El Paso Texas)." (Merriam, 1895*b*:32).

DISTRIBUTION: SW and SC USA to Baja California and N and C Mexico.

SYNONYMS: *evotis*.

COMMENTS: Includes *evotis*; see Armstrong and Jones (1971*a*). Reviewed by Armstrong and Jones (1972*b*, Mammalian Species, 17).

Sorex Linnaeus, 1758. Syst. Nat., 10th ed., 1:53.

TYPE SPECIES: *Sorex araneus* Linnaeus, 1758.

SYNONYMS: *Amphisorex, Atophyrax, Corsira, Eurosorex, Homalurus, Microsorex, Musaraneus, Neosorex, Ognevia, Otisorex, Oxyrhin, Soricidus, Stroganovia*.

COMMENTS: Type genus of Soricidae. The systematic relationships of a large number of Holarctic species were studied by George (1988); her proposals for subgeneric allocation are mainly followed here. Keys and/or reviews are available for the species of various geographical areas: Canada (van Zyll de Jong, 1983*a*); North and Middle America (Junge and Hoffmann, 1981; Carraway, 1990); China (Hoffmann, 1987); Siberia (Yudin, 1989); and Europe (Niethammer and Krapp, 1990). *Microsorex* was formerly regarded as a full genus, then reduced to a subgenus of *Sorex* by Diersing (1980*b*), and is now regarded as a synonym of subgenus *Otisorex* (see George, 1988). The subgenus *Amphisorex* (type species *Sorex hermanni* Duvernoy, 1834) was alternatively listed under *Sorex* and *Neomys* by Miller (1912*a*), Ellermann and Morrison-Scott (1951), and Corbet (1978*c*). Miller (1912*a*) stated that the type of *Sorex hermanni* consisted of a skin of *Sorex araneus* and a skull of *Neomys fodiens*. To avoid further confusion, I herewith designate the skin of *Sorex hermanni* Duvernoy, 1834 as the lectotype, thus making *hermanni* a synonym of *araneus*, and *Amphisorex* a synonym of *Sorex*. Besides subgenera a number of species groups have been distinguished such as the *araneus-arcticus* group (Meylan and Hausser, 1973; Hausser et al., 1985), the *cinereus* group (van Zyll de Jong, 1991*b*), and the *vagrans* group (Carraway, 1990), the boundaries and contents of which are still highly controversial. Old World species of *Sorex* were reviewed by Dannelid (1991*b*) who provided a phylogenetic hypothesis of relationships.

Sorex alaskanus Merriam, 1900. Proc. Washington Acad. Sci., 2:18.

TYPE LOCALITY: USA, "Point Gustavus, Glacier Bay, Alaska".

DISTRIBUTION: Known only from the type locality.

COMMENTS: Subgenus *Otisorex*. The species was tentatively included in *palustris* by Junge and Hoffmann (1981), but retained as a species by Hall (1981), Jones et al. (1982), and George (1988); a view supported by the skull figures and measurements given by Jackson (1928). Apparently the species has not been collected again since 1899.

Sorex alpinus Schinz, 1837. Neue Denkschr. Allgem. Schweiz. Gesell. Naturwiss. Neuchatel, 1:13.

TYPE LOCALITY: Switzerland, Canton Uri, St. Gotthard Pass.

DISTRIBUTION: Montane forests of C Europe; including Pyrenees, Carpathians, Tatra, Sudeten, Harz, and Jura Mtns.

SYNONYMS: *hercynicus, intermedius, longobarda, tatricus*.

COMMENTS: Subgenus *Sorex* or *Homalurus*; see Hutterer (1982*b*). Reviewed by Spitzenberger (1990*a*).

Sorex araneus Linnaeus, 1758. Syst. Nat., 10th ed., 1:53.

TYPE LOCALITY: "*in* Europe *cryptis*"; restricted to Uppsala, Sweden by Thomas (1911*a*:143).

DISTRIBUTION: C, E, and N Europe including the British Isls (with some isolated populations in France, Italy and Spain), east to Siberia.

SYNONYMS: *alticola, antinorii, bergensis, bohemicus, bolkayi, carpathicus, castaneus, concinnus, crassicaudatus, csikii, daubentonii, eleonorae, grantii, hermanni, huelleri, ignotus, iochanseni, labiosus, macrotrichus, marchicus, melanodon, mollis, monsvairani, nigra, nuda, ryphaeus, pallidus, personatus, petrovi, peucinius, preussi, pulcher, pyrenaicus, pyrrhonota, quadricaudatus, rhinolophus, silanus, surinamensis, tetragonurus, uralensis, vulgaris, wettsteini.*

COMMENTS: Type species of subgenus *Sorex*. *S. araneus* is the preferred Palearctic species for studies in ecology and evolution; see Hausser et al. (1990) and Hausser (1991) for reviews. The species is well known for its Robertsonian chromosome polymorphism (Meylan, 1964) and for the tendency to establish local karyotype races (Hausser et al., 1985; Searle, 1984; Zima and Král, 1984*b*); in Switzerland, two karyotype races occur which behave like parapatric species (Hausser et al., 1986). Includes *Blarina pyrrhonota* Jentink, 1910, a name assigned to *Cryptotis surinamensis* by Cabrera (1958); however, Husson (1963) showed that the locality information was incorrect and that it was based on a *Sorex araneus*. The holotype skin (skull lost) of *Sylvisorex preussi* Matschie, formerly thought to represent an endemic *Myosorex* of Mt. Cameroon (Heim de Balsac, 1968*b*), is a *Sorex araneus* and is therefore included as a synonym. *Sorex isodon marchicus*, recently described from E Germany (Passarge, 1984), is also tentatively included in *araneus* as no clear characters are known to distinguish it from the latter.

Sorex arcticus Kerr, 1792. Animal Kingdom, p. 206.

TYPE LOCALITY: Canada, Ontario, settlement on Severn River (now Fort Severn), Hudson Bay.

DISTRIBUTION: Yukon and Northwest Territory to Quebec, Nova Scotia, and New Brunswick (Canada); North Dakota, South Dakota, Minnesota, and Wisconsin (USA).

SYNONYMS: *laricorum, maritimensis, pachyurus, richardsonii, spagnicola.*

COMMENTS: Subgenus *Sorex*. Palearctic species currently referred to *arcticus* (Gromov and Baranova, 1981:18) represent *tundrensis* (Junge et al., 1983; Ivanitskaya et al., 1986); see also Sokolov and Orlov (1980) and Hoffmann (1985*a*). Van Zyll de Jong (1983*b*) and Volobouev and van Zyll de Jong (1988) suggested that *maritimensis* may be an independent species.

Sorex arizonae Diersing and Hoffmeister, 1977. J. Mammal., 58:329.

TYPE LOCALITY: USA, "upper end of Miller Canyon, 15 mi S [= 10 mi S, 4¾mi E] Fort Huachuca [near spring at lower edge of Douglas fir zone, Huachuca Mts.] Cochise County, Arizona".

DISTRIBUTION: Mountains of SE Arizona and SW New Mexico (USA; see Conway and Schmitt, 1978 and Hoffmeister, 1986); Chihuahua (Mexico; see Caire et al., 1978).

COMMENTS: Refered to unnamed subgenus by George (1988). Close to *emarginatus* (see Diersing and Hoffmeister, 1977).

Sorex asper Thomas, 1914. Ann. Mag. Nat. Hist., ser. 8, 13:565.

TYPE LOCALITY: "Thian-shan [Tien-shan], Tekes Valley". Note on type specimen tag says "Jigalong" (= Dzhergalan?, see Hoffmann, 1987:119); Narynko'skii r-n., Alma-Ata Obl., Kazakhstan.

DISTRIBUTION: Tien Shan Mountains (Kazakhstan and Sinkiang, China).

COMMENTS: Subgenus *Sorex*. Type locality discussed by Hoffmann (1987) and Pavlinov and Rossolimo (1987). Does not include *excelsus* as suggested by Corbet (1978*c*); see under that species. Reviewed by Hoffmann (1987), who discussed the relationship between *asper* and *tundrensis*.

Sorex bairdii Merriam, 1895. N. Am. Fauna, 10:77.

TYPE LOCALITY: USA, "Astoria, [Clatsop Co.], Oregon".

DISTRIBUTION: NW Oregon (USA).

SYNONYMS: *permiliensis.*

COMMENTS: Subgenus *Otisorex*. This taxon has been alternatively referred to *obscurus*, *vagrans*, and *monticolus*, but was recently given specific rank by Carraway (1990). Includes *permiliensis* as a valid subspecies.

Sorex bedfordiae Thomas, 1911. Abstr. Proc. Zool. Soc. Lond., 1911(90):3.

TYPE LOCALITY: "Omi-san, Sze-chwan" [= China, Sichuan, Emei Shan].

DISTRIBUTION: Montane forests of S Gansu and W Shensi to Yunnan (China); adjacent Burma and Nepal.

SYNONYMS: *gomphus, fumeolus, nepalensis, wardii.*

COMMENTS: Subgenus *Sorex*. Formerly a subspecies of *cylindricauda* but recognized as a full species by Corbet (1978*c*) and Hoffmann (1987).

Sorex bendirii (Merriam, 1884). Trans. Linnean Soc. New York, 2:217.
TYPE LOCALITY: USA, "Klamath Basin, Oregon" = Oregon, Klamath Co., l mile (1.6 km) from Williamson River, l8 miles (29 km) SE of Fort Klamath.
DISTRIBUTION: A narrow coastal area from NW California to Washington (USA); a few records from SE British Columbia (Canada).
SYNONYMS: *albiventer, palmeri*.
COMMENTS: Originally described in the monotypic genus *Atophyrax* Merriam; now in subgenus *Otisorex*. Reviewed by Pattie (1973, Mammalian Species, 27).

Sorex buchariensis Ognev, 1921. Ann. Mus. Zool. Acad. Sci. St. Petersbourg, 22:320.
TYPE LOCALITY: Tadzhikistan, Pamir Mountains, Davan-su River Valley, "Gornaya Bukhara, drevyaya morena lednika Oshanina, dol. p. Davan-Su (khrebet' Petra Velikavo)" [Montane Bukhara, ancient moraine of Oshanin glacier, Peter the Great range].
DISTRIBUTION: Pamir Mtns (Tadzhikistan).
COMMENTS: Referred to subgenus *Eurosorex* by Yudin (1989). Considered a subspecies of *thibetanus* by Dolgov and Hoffmann (1977) and Hoffmann (1987), but retained as a distinct species by Ivanitskaya et al. (1977), Hutterer (1979), Zaitsev (1988), and Yudin (1989). The karyotype of two specimen from Tadshikistan was similar to that of *volnuchini* (Ivanitskaya et al., 1977).

Sorex caecutiens Laxmann, 1788. Nova Acta Acad. Sci. Petropoli, 1785, 3:285 [1788].
TYPE LOCALITY: Russia, Buryatskaya ASSR, SW shore of Lake Baikal (Pavlinov and Rossolimo, 1987:17).
DISTRIBUTION: Taiga and tundra zones from E Europe to E Siberia, south to C Ukraine, N Kazakhstan, Altai Mtns, Mongolia, Gansu and NE China, to Korea and Sakhalin.
SYNONYMS: *altaicus, annexus, araneoides, buxtoni, centralis, koreni, lapponicus, macropygmaeus, karpinskii, pleskei, rozanovi, tasicus, tungussensis*.
COMMENTS: Subgenus *Sorex*. This species still offers many unsolved problems, along with the species of the *tundrensis* and *arcticus* groups. Names like *annexus, cansulus, granarius*, and *shinto* have been included in *caecutiens* in the past but are presently included in other species or treated as separate species; see Hoffmann (1987) for a discussion of problems. The European range was reviewed by Sulkava (1990).

Sorex camtschatica Yudin, 1972. Teriologiya, 1:48.
TYPE LOCALITY: Russia, "Kamchatka, Kambal'naya Bay".
DISTRIBUTION: Russia, S Kamchatka Peninsula.
COMMENTS: Subgenus *Otisorex*. Formerly included in *cinereus* (van Zyll de Jong, 1982) but now recognized as a full species (Ivanitskaya and Kozlovskii, 1983; van Zyll de Jong, 1991*b*).

Sorex cansulus Thomas, 1912. Ann. Mag. Nat. Hist., ser. 8, 10:398.
TYPE LOCALITY: China, Gansu, "46 miles south-east of SE Taochou" (= Lintan).
DISTRIBUTION: Known only from the type locality.
COMMENTS: Subgenus *Sorex*, related to *tundrensis*. The species was recognized by Hoffmann (1987); no specimens other than the type series are known.

Sorex cinereus Kerr, 1792. Animal Kingdom, p. 206.
TYPE LOCALITY: Canada, Ontario, Fort Severn.
DISTRIBUTION: North America throughout Alaska and Canada and southward along the Rocky and Appalachian Mtns to 45°.
SYNONYMS: *acadicus, cooperi, fimbripes, fontinalis, forsteri, hollisteri, idahoensis, frankstounensis, lesueurii, miscix, nigriculus, ohionensis, personatus, platyrhinus, streatori*.
COMMENTS: Type species of subgenus *Otisorex*. Does not occur in Siberia as previously suggested; the taxa *haydeni, jacksoni, ugyunak, portenkoi, leucogaster, beringianus* and *camtschatica* have been included previously but are now considered as separate species; see comments under these taxa and Junge and Hoffmann (1981, and references cited therein), van Zyll de Jong (1982, 1991*b*), van Zyll de Jong and Kirkland (1989), and Pavlinov and Rossolimo (1987). *S. fontinalis* was separated from *cinereus* by Kirkland (1977), Junge and Hoffmann (1981), and Jones et al. (1992), but

is considered, together with *lesueurii*, as a subspecies (van Zyll de Jong and Kirkland, 1989). However, George's (1988) data indicate it is a sister taxon to both *cinereus* and *haydeni*.

Sorex coronatus Millet, 1828. Faune de Maine-et-Loire, I, p. 18.
TYPE LOCALITY: France, Main-et-Loire, Blou.
DISTRIBUTION: W Europe from The Netherlands and NW Germany to France and Switzerland, south to N Spain; also in Jersey (Channel Isls), Liechtenstein and westernmost tip of Austria.
SYNONYMS: *euronotus, fretalis, gemellus, personatus, santonus*.
COMMENTS: Subgenus *Sorex*. A sibling species of *araneus* (Meylan and Hausser, 1978), characterized mainly by the karyotype. Its distribution broadly overlaps with that of *araneus* in Germany. Revised by Hausser (1990).

Sorex cylindricauda Milne-Edwards, 1872. *In* David, Nouv. Arch. Mus. Hist. Nat. Paris, Bull. for 1871, 7(4):92 [1872].
TYPE LOCALITY: China, Sichuan, Moupin (= Baoxing).
DISTRIBUTION: Montane forests of N Sichuan.
COMMENTS: Subgenus *Sorex*. Revised by Hoffmann (1987).

Sorex daphaenodon Thomas, 1907. Proc. Zool. Soc. Lond., 1907:407.
TYPE LOCALITY: Russia, Sakhalin Isl, "Dariné, 25 miles [40 km] N.W. of Korsakoff, Saghalien".
DISTRIBUTION: Ural Mountains to the Kolyma River (Siberia); Sakhalin Isl; Kamchatka Peninsula; Paramushir Isl (N Kuriles); Jilin and Nei Mongol Aut. Region (China).
SYNONYMS: *orii, sanguinidens, scaloni*.
COMMENTS: Type species of subgenus *Stroganovia*, see Yudin (1989), who recognized three subspecies, *daphaenodon, sanguinidens*, and *scaloni*.

Sorex dispar Batchelder, 1911. Proc. Biol. Soc. Washington, 24:97.
TYPE LOCALITY: USA, "Beede's (sometimes called Keene Heights), in the township of Keene, Essex county, New York". Redescribed by Martin (1966:131) as 0.6 mi S, 0.5 mi E St. Huberts, Essex Co., New York, lat. 44°09', long. 73°46'.
DISTRIBUTION: Appalachian Mtns from North Carolina to Maine; S New Brunswick, Nova Scotia (Canada).
SYNONYMS: *blitchi; macrurus* (Batchelder, not of Lehmann).
COMMENTS: Subgenus *Otisorex*. For comparison with *gaspensis* see Kirkland and Van Deusen (1979). Reviewed by Kirkland (1981, Mammalian Species, 155).

Sorex emarginatus Jackson, 1925. Proc. Biol. Soc. Washington, 38:129.
TYPE LOCALITY: "Sierra Madre, near Bolanos, altitude 7,600 feet, State of Jalisco, Mexico".
DISTRIBUTION: Durango, Zacatecas, and Jalisco (Mexico).
COMMENTS: Referred to unnamed subgenus by George (1988:456). Findley (1955*b*) considered this a subspecies of *oreopolus*; however, *oreopolus* belongs to subgenus *Otisorex* (Diersing and Hoffmeister, 1977). For biological and distributional information, see Alvarez and Polaco (1984) and Matson and Baker (1986).

Sorex excelsus G. M. Allen, 1923. Am. Mus. Novit., 100:4.
TYPE LOCALITY: "summit of Ho-shan (=Xue Shan), Pae-tai, 30 miles (48 km) south of Chung-tien (=Zhongdian), Yunnan, China, altitude 13000 feet."
DISTRIBUTION: Yunnan and Sichuan (China), and possibly Nepal.
COMMENTS: Subgenus *Sorex*. Considered as a possible subspecies of *asper* (Corbet, 1978*c*) but retained as a full species related to *tundrensis* by Hoffmann (1987) who also suggested that a specimen from Nepal recorded by Agrawal and Chakraborty (1971) may represent *excelsus*.

Sorex fumeus G. M. Miller, 1895. N. Am. Fauna, 10:50.
TYPE LOCALITY: USA, "Peterboro [Madison Co.], New York."
DISTRIBUTION: S Ontario, S Quebec, New Brunswick, and Nova Scotia (Canada); all of New England and Appalachian Mtns and adjacent areas to NE Georgia (USA).
SYNONYMS: *umbrosus*.

COMMENTS: Subgenus *Otisorex*. Reviewed by Owen (1984, Mammalian Species, 215). Overlaps in distribution and may be easily confused with *arcticus* in part of its range (Junge and Hoffmann, 1981).

Sorex gaspensis Anthony and Goodwin, 1924. Am. Mus. Novit., 109:1.
TYPE LOCALITY: Canada, "Mt. Albert, Gaspé Peninsula, Quebec, 2000 feet elevation".
DISTRIBUTION: Gaspe Peninsula, N New Brunswick, Nova Scotia, and Cape Breton Isl (Canada).
COMMENTS: Subgenus *Otisorex*. For comparison with *dispar*, see Kirkland and Van Deusen (1979). Reviewed by Kirkland (1981, Mammalian Species, 155).

Sorex gracillimus Thomas, 1907. Proc. Zool. Soc. Lond., 1907:408.
TYPE LOCALITY: Russia, Sakhalin Isl, "Dariné, 25 miles [40 km] N.W. of Korsakoff, Saghalien".
DISTRIBUTION: SE Siberia from S shore of the Sea of Okhotsk to N Korea and probably Manchuria; Sakhalin Isl; Hokkaido (Japan).
SYNONYMS: *hyojironis*.
COMMENTS: Subgenus *Sorex*. This species has long been included in *minutus* but its specific status is now widely accepted on the basis of penial (Dolgov and Lukanova, 1966) and cranial (Hutterer, 1979) morphology, karyotype (Orlov and Bulatova, 1983), and allozyme data (George, 1988). The inclusion of *hyojironis* follows Corbet (1978*c*) and is tentative.

Sorex granarius Miller, 1910. Ann. Mag. Nat. Hist., ser. 8, 6:458.
TYPE LOCALITY: "La Granja, Segovia, Spain".
DISTRIBUTION: NW Iberian Peninsula (Portugal and Spain).
COMMENTS: Subgenus *Sorex*, group *araneus*. Afforded specific rank by Hausser et al. (1975); reviewed by Hausser (1990).

Sorex haydeni Baird, 1857. Mammalia, *in* Repts. U.S. Expl. Surv., 8(1):29.
TYPE LOCALITY: USA, "Fort Union, Nebraska" (later Fort Buford, now Mondak, Montana, near Buford, Williams Co., North Dakota).
DISTRIBUTION: SE Alberta, S Saskatchewan, SW Manitoba (Canada); NW Montana southeast to Kansas, east to W and S Minnesota (USA).
COMMENTS: Subgenus *Otisorex*. Formerly included in but now separated from *cinereus* by van Zyll de Jong (1980) and Junge and Hoffmann (1981); both species are closely related (George, 1988). *S. haydeni* occurs in grassy habitats while *S. cinereus* prefers forest and woodland (van Zyll de Jong, 1980).

Sorex hosonoi Imaizumi, 1954. Bull. Natl. Sci. Mus. Tokyo, 35:94.
TYPE LOCALITY: "Tokiwa Mura (Maneki, about 900 m altitude, foot of Mt. Gaki, Japan Alps), Kita-Azumi Gun, Nagano Pref., Central Honsyû [= Honshu], Japan".
DISTRIBUTION: Montane forests of C Honshu (Japan).
SYNONYMS: *shiroumanus*.
COMMENTS: Subgenus *Sorex*. Imaizumi (1970*b*) reported that *hosonoi* occurs sympatrically with *shinto* and therefore should be considered as separate species (Corbet, 1978*c*).

Sorex hoyi Baird, 1857. Mammalia, *in* Repts. U.S. Expl. Surv., 8(1):32.
TYPE LOCALITY: USA, "Racine, Wisconsis."
DISTRIBUTION: N taiga zone of Alaska, Canada and the USA, with S outliers in the montane forests of the Appalachian and Rocky Mtns.
SYNONYMS: *alnorum, eximius, intervectus, montanus, thompsoni, washingtoni, winnemana*.
COMMENTS: Formerly in *Microsorex*, which is a synonym of subgenus *Otisorex*, according to George (1988). Includes *thompsoni* (Diersing, 1980*b*). Reviewed by Long (1974, Mammalian Species, 33) and Junge and Hoffmann (1981).

Sorex hydrodromus Dobson, 1889. Ann. Mag. Nat. Hist., ser. 6, 4:373.
TYPE LOCALITY: USA, Alaska, "Unalaska Islands, Aleutian Islands" (probably in error, presumably from St. Paul, Pribilof Isls).
DISTRIBUTION: Known only from St. Paul in the Pribilof Isls, Bering Sea.
SYNONYMS: *pribilofensis*.
COMMENTS: Subgenus *Otisorex*. There is some discrepancy in the literature on the correct name for this species. Dobson's *hydrodromus* has priority, but because of an

apparently incorrect type locality information and further inconsistencies in the original description, Hoffmann and Peterson (1967) proposed to suppress *hydrodromus* in favour of *pribilofensis* Merriam, 1895, a suggestion followed by van Zyll de Jong (1991*b*). However, Yudin (1969), Baranova et al. (1981), Hall (1981), Junge and Hoffmann (1981), and Honacki et al. (1982) retained *hydrodromus*, while Gureev (1979) listed both *hydrodromus* and *pribilofensis* as species. As the holotype of *hydrodromus* still exists in the St. Petersburg Museum, there seems to be no reason for not following the rule of priority.

Sorex isodon Turov, 1924. C.R. Acad. Sci. Paris, p. 111.
TYPE LOCALITY: Russia, Siberia, NE of Lake Baikal, Barguzinsk taiga, River Sosovka.
DISTRIBUTION: SE Norway and Finland through Siberia to the Pacific coast; Kamchatka; Sakhalin Isl; Kurile Isls; probably also NE China and Korea.
SYNONYMS: *gravesi, princeps, ruthenus*.
COMMENTS: Subgenus *Sorex*. Probably not conspecific with *sinalis* as suggested by Corbet (1978*c*) and Dolgov (1985); see Siivonen (1965) and Hoffmann (1987). Because the well established name *isodon* is antedated by *gravesi*, Hoffmann (1987) suggested that *isodon* be declared the valid name; the case needs to be submitted to the International Commission on Zoological Nomenclature. The species was reviewed by Sulkava (1990). The recently described *isodon marchicus* (Passarge, 1984) is provisionally included in *araneus*; see comments under that species.

Sorex jacksoni Hall and Gilmore, 1932. Univ. California Publ. Zool., 38:392.
TYPE LOCALITY: USA, "Sevoonga, 2 miles east of North Cape, St. Lawrence Island, Bering Sea, Alaska."
DISTRIBUTION: Known only from St. Lawrence Isl (Bering Sea).
COMMENTS: Subgenus *Otisorex*. Placed in the *arcticus* species group by Hall and Gilmore (1932) and in the *cinereus* species group by Hoffmann and Peterson (1967). Separated from *cinereus* by Junge and Hoffmann (1981). Van Zyll de Jong (1982) included *leucogaster* (= *beringianus*), *portenkoi*, and *ugyunak* in this species, but van Zyll de Jong (1991*b*) retained all three as distinct.

Sorex kozlovi Stroganov, 1952. Byull. Moscow Ova. Ispyt. Prir. Otd. Biol., 57:21.
TYPE LOCALITY: "Tibet" (= Qinghai), Dze-Chyu (Zi Qu) River, tributary of Mekong River (= Lancang Jiang).
DISTRIBUTION: Known only from the type locality.
COMMENTS: Type species of subgenus *Eurosorex* Stroganov, 1952. Known from a single specimen, until near-topotype (National Museum of Natural History - 449080) obtained in 1987. Regarded as a subspecies of *thibetanus* by some authors (Dolgov and Hoffmann, 1977; Hoffmann, 1987) or included in *buchariensis* by others (Corbet, 1978*c*; Gureev, 1979). Hutterer (1979) recognized inconsistencies in the various published figures and descriptions of the same holotype specimen and regarded *kozlovi* as a doubtful taxon; see also under *buchariensis* and *thibetanus*.

Sorex leucogaster Kuroda, 1933. Bull. Biogeogr. Soc. Japan, 3,3:155.
TYPE LOCALITY: Russia, Paramushir Isl; given by Ellerman and Morrison-Scott (1951:48) as "Nasauki, Amamu-shiru, 200 ft., North Kurile Islands".
DISTRIBUTION: Probably confined to Paramushir Isl, south of Kamchatka Peninsula.
SYNONYMS: *beringianus*.
COMMENTS: Subgenus *Otisorex*. Formerly included in *cinereus* or *gracillimus* (Corbet, 1978*c*); includes *beringianus* Yudin, 1967. On the status, authorship and valid date of publication see Pavlinov and Rossolimo (1987). Related to *jacksoni* and *ugyunak* (van Zyll de Jong, 1982, 1991*b*).

Sorex longirostris Bachman, 1837. J. Acad. Nat. Sci. Philadelphia, 7:370.
TYPE LOCALITY: USA, "in the swamps of Santee [River], South Carolina"; restricted to Hume Plantation (Cat Island in the mouth of Santee River) by Jackson (1928:85).
DISTRIBUTION: SE USA (except S Florida) west to Louisiana, Arkansas, Missouri, Illinois, and Indiana.
STATUS: U.S. ESA - Threatened as *Sorex longirostris fisheri*.
SYNONYMS: *bachmani, eionis, fisheri, wagneri*.

COMMENTS: Subgenus *Otisorex*. As pointed out by Junge and Hoffmann (1981), this species is inappropriately named because it has one of the shortest rostra of North American *Sorex*. Junge and Hoffmann (1981) also suggested that shrews of the Great Dismal Swamp described as *fisheri* and traditionally included in *longirostris* as a subspecies are much larger and may represent a valid species. Reviewed by French (1980, Mammalian Species, 143). Part of range mapped in detail by Pagels and Handley (1989) and Pagels et al. (1982).

Sorex lyelli Merriam, 1902. Proc. Biol. Soc. Washington, 15:75.

TYPE LOCALITY: USA, "Mt. Lyell, Tuolumne Co., California".

DISTRIBUTION: Altitudes above 2000 m in the Sierra Nevada, California (USA).

COMMENTS: Subgenus *Otisorex*; member of the *cinereus* species group. Related to *milleri*, according to van Zyll de Jong (1991*b*).

Sorex macrodon Merriam, 1895. N. Am. Fauna, 10:82.

TYPE LOCALITY: "Orizaba, Veracruz, Mexico (altitude 4,200 feet)."

DISTRIBUTION: Veracruz, in mountains from 4000-9500 ft (1676-2896 m) and Puebla (Mexico). See Heaney and Birney (1977).

COMMENTS: Subgenus *Otisorex*. Similar to, and possibly conspecific with, *veraepacis* (see Junge and Hoffmann, 1981).

Sorex merriami Dobson, 1890. Monogr. Insectivora, pt. 3 (Soricidae), fasc. l, pl. 23.

TYPE LOCALITY: USA, "Fort Custer, Montana" = Bighorn Co., Little Bighorn River, ca. l mile above Fort Custer (= Hardin).

DISTRIBUTION: Xeric habitats in EC Washington to N and E California, Arizona, northeastward to Nebraska, Wyoming and Montana (USA).

SYNONYMS: *leucogenys*.

COMMENTS: Referred to unnamed subgenus by George (1988:456). Reviewed by Armstrong and Jones (1971*b*, Mammalian Species, 2).

Sorex milleri Jackson, 1947. Proc. Biol. Soc. Washington, 60:131.

TYPE LOCALITY: "Madera Camp, altitude 8,000 feet, Carmen Mountains, Coahuila, Mexico".

DISTRIBUTION: Restricted to the Sierra Madre Oriental of Coahuila and Nuevo Leon, Mexico.

COMMENTS: Subgenus *Otisorex*. Controversial opinions on the systematic status of *milleri* exist; Findley (1955*a*) regarded it as a morphologically distinct relict population allied to *cinereus* and accepted its specific status, as did Hall (1981) and Junge and Hoffmann (1981); while van Zyll de Jong and Kirkland (1989) suggested that *milleri* may not merit full specific status.

Sorex minutissimus Zimmermann, 1780. Geogr. Gesch. Mensch. Vierf. Thiere, 2:385.

TYPE LOCALITY: "Yenisei"; given by Stroganov (1957:176) as "iz raiona sela Kiiskow chto na r. Kie (nyne g. Mariinsk Kemerovskoi oblasti)" [= Russia, Kemerovsk. Obl., Mariinsk (= Kiiskoe), bank of Kiia River (near Yenesei River)]; Restricted by Pavlinov and Rossolimo (1987:23) to "Krasnoyarskii kr., Krasnoyarsk."

DISTRIBUTION: Taiga zone from Norway, Sweden and Estonia to E Siberia; Sakhalin; Hokkaido, and perhaps Honshu (Japan); Mongolia; China; South Korea.

SYNONYMS: *abnormis, barabensis, burneyi, czekanovskii, caudata, exilis, hawkeri, ishikawai, karelicus, minimus, neglectus, perminutus, stroganovi, tscherskii, tschuktschorum, ussuriensis*.

COMMENTS: Subgenus *Sorex* or *Eurosorex*. Yoshiyuki (1988*a*) recognized nine subspecies.

Sorex minutus Linnaeus, 1766. Syst. Nat., 12th ed., l:73.

TYPE LOCALITY: "Yenisei"; restricted by Pavlinov and Rossolimo (1987:15) to "Krasnoyarskii kr., Krasnoyarsk." According to Ellerman and Morrison-Scott (1951:47), Linnaeus' name is based on Laxmann's ms. of *Sibir. Briefe*, and the type locality is Barnaul, Russia.

DISTRIBUTION: Europe to Yenesei River and Lake Baikal, south to Altai and Tien Shan Mtns; populations of Nepal and China have been alternatively identified as *minutus* or *thibetanus*; populations of Turkey and the Caucasus as *minutus* or *volnuchini*; populations of Kashmir and N Pakistan as *minutus, planiceps*, or *thibetanus*.

SYNONYMS: *abnormis, barabensis, becki, canaliculatus, carpetanus, exiguus, exilis, gmelini, gymnurus, heptapotamicus, hibernicus, insulaebellae, kastchenkoi, lucanius, melanderi, minimus, pumilio, pumilus, pygmaeus, rusticus, stroganovi, tschuktschorum*.

COMMENTS: Subgenus *Sorex*. Formerly included *gracillimus*, which is now accepted as specifically distinct; see comments therein. Corbet (1978*c*) included also *planiceps* and *thibetanus*; but see Dolgov and Hoffmann (1977) and Hutterer (1979). May include *volnuchini*, see comments therein. The European populations of *minutus* were revised by Hutterer (1990).

Sorex mirabilis Ognev, 1937. Byull. Moscow Ova. Ispyt. Prir. Otd. Biol., 46(5):268.

TYPE LOCALITY: Russia, Primorskii Krai, Ussuriiskii r-n., Kamenka River (specified by Pavlinov and Rossolimo, 1987).

DISTRIBUTION: N Korea, NE China, and Ussuri region (Russia).

SYNONYMS: *kutscheruki*.

COMMENTS: Placed in monotypic subgenus *Ognevia* by Heptner and Dolgov (1967), who demonstrated that *mirabilis* is not conspecific with *pacificus*, as had been suggested earlier (Bobrinskii et al., 1965). Hutterer (1982*b*) suggested a closer relationship with *Sorex* (*Homalurus*) *alpinus* because of shared derived features of genital morphology.

Sorex monticolus Merriam, 1890. N. Am. Fauna, 3:43.

TYPE LOCALITY: USA, "San Francisco Mountain, Coconino Co., Arizona...altitude 3,500 meters (11,500 feet)".

DISTRIBUTION: Montane boreal and coastal coniferous forest and alpine areas from Alaska to California and New Mexico, east to Montana, Wyoming, and Colorado (USA) and to W Manitoba (Canada); Chihuahua, Durango (Mexico).

SYNONYMS: *alascensis, calvertensis, dobsoni, durangae, elassodon, glacialis, insularis, isolatus, longicauda, longiquus, malitiosus, melanogenys, mixtus, neomexicanus, obscurus, obscuroides, prevostensis, shumaginensis, setosus; similis* (Merriam, not of Hensel), *soperi*.

COMMENTS: Subgenus *Otisorex*. Includes *obscurus* and *durangae*, which were previously included in *vagrans* and *saussurei* respectively; see Hennings and Hoffmann (1977) and map in Junge and Hoffmann (1981); other synonyms follow van Zyll de Jong (1983*a*) and George and Smith (1991). Related to *pacificus* (see George, 1988).

Sorex nanus Merriam, 1895. N. Am. Fauna, 10:81.

TYPE LOCALITY: USA, "Estes Park [Larimer Co.], Colorado".

DISTRIBUTION: Rocky Mountains from Montana to New Mexico; South Dakota; Arizona (USA).

COMMENTS: Subgenus *Otisorex*. Very similar to, and perhaps conspecific with, *tenellus*, see review by Hoffmann and Owen (1980, Mammalian Species, 131); but George (1988) retained both as distinct species.

Sorex oreopolus Merriam, 1892. Proc. Biol. Soc. Washington, 7:173.

TYPE LOCALITY: "Sierra de Colima, Jalisco, Mexico (altitude 10,000 feet) [3,048 m]".

DISTRIBUTION: Jalisco (Mexico); perhaps east to Puebla and Veracruz (Mexico).

SYNONYMS: *orizabae*.

COMMENTS: Subgenus *Otisorex*. Contrary to Findley (1955*b*), this species does not include *emarginatus* or *ventralis*, which Diersing and Hoffmeister (1977) placed in the subgenus *Sorex*. *S. orizabae* was included in *vagrans* by Hennings and Hoffmann (1977:8) but later included in *oreopolus* by Junge and Hoffmann (1981:43).

Sorex ornatus Merriam, 1895. N. Am. Fauna, 10:79.

TYPE LOCALITY: USA, "San Emigdio Canyon, Mt. Piños [Kern Co.], California".

DISTRIBUTION: California coastal ranges from N of San Francisco Bay to N part and S tip of Baja California; Santa Catalina Isl.

SYNONYMS: *californicus, juncensis, lagunae, oreinus, relictus, salarius, salicornicus, sinuosus, willetti*.

COMMENTS: Subgenus *Otisorex*. Reviewed by Owen and Hoffmann (1983, Mammalian Species, 212), who recognized 9 subspecies. For further distributional information see Williams (1979) and Junge and Hoffmann (1981); for karyotype and allozyme data see Brown and Rudd (1981) and George (1988).

Sorex pacificus Coues, 1877. Bull. U.S. Geol. Geogr. Surv. Terr., 3(3):650.

TYPE LOCALITY: USA, "Fort Umpqua [mouth Umpqua River, Douglas County], Oregon."

DISTRIBUTION: Forests of coastal Oregon (USA).

SYNONYMS: *cascadensis, yaquinae*.

COMMENTS: Subgenus *Otisorex*. Not conspecific with *mirabilis*; see Yudin (1969) and Hoffmann (1971). Related to *monticolus*; see Findley (1955*b*), Junge and Hoffmann (1981), and George (1988). Reviewed by Carraway (1985, Mammalian Species, 231), who later (1990) removed *sonomae* from synonymy; see comments under that species.

Sorex palustris Richardson, 1828. Zool. J., 3:517.

TYPE LOCALITY: Canada, "marshy places, from Hudson's Bay to the Rocky Mountains."; not specified.

DISTRIBUTION: Montane and boreal areas of North America below the tree line from Alaska to the Sierra Nevada, Rocky and Appalachian Mtns.

SYNONYMS: *acadicus* (Allen, not of Gilpin), *albibarbis, brooksi, gloveralleni, hydrobadistes, labradorensis, navigator, punctulatus, turneri.*

COMMENTS: Formerly placed in genus *Neosorex* Baird; now in *Sorex* (*Otisorex*). Reviewed by Beneski and Stinson (1987, Mammalian Species, 296), who recognized 9 subspecies. They did not include *alaskanus* as suggested by Junge and Hoffmann (1981:28) and Hall (1981:43); George (1988) also treated *alaskanus* as distinct.

Sorex planiceps Miller, 1911. Proc. Biol. Soc. Washington, 24:242.

TYPE LOCALITY: India, "Dachin, Khistwar, Kashmir (altitude, 9000 feet)".

DISTRIBUTION: Kashmir (India) and N Pakistan.

COMMENTS: Considered a subspecies of *thibetanus* by Dolgov and Hoffmann (1977) and Hoffmann (1987), but retained by Hutterer (1979) because of larger skull measurements. The problem still remains unresolved.

Sorex portenkoi Stroganov, 1956. Proc. Inst. Biol. W. Siberian Branch Acad. Sci. USSR, Zool., 1:11-14.

TYPE LOCALITY: Russia, Koryaksk. Auv. Okr. "bliz pos Anadyr', poberejhe Anadyrsk limana [near Anadyr' settlement, shore of Anadyr' estuary]."

DISTRIBUTION: NE Siberia.

COMMENTS: Subgenus *Otisorex*. Originally described as a subspecies of *cinereus* and treated as such by Yudin (1972) and Okhotina (1977), then included in *ugyunak* (Ivanitskaya and Kozlovskii, 1985), but recently recognized as a distinct species by Zaitsev (1988), and van Zyll de Jong (1991*b*), who, however, pointed out its close relationship to *jacksoni* and *ugyunak*.

Sorex preblei Jackson, 1922. J. Washington Acad. Sci., 12:263.

TYPE LOCALITY: USA, "Jordan Valley, altitude 4,200 feet, Malheur County, Oregon."

DISTRIBUTION: Columbia Plateau of Washington, Oregon and Nevada to W Great Plains of Montana, Utah, Wyoming, and Colorado (specimen 74262 in the National Museum of Natural History) (USA). For reviews of distributional records, see Tomasi and Hoffmann (1984) and Long and Hoffmann (1992).

COMMENTS: Subgenus *Otisorex*.

Sorex raddei Satunin, 1895. Arch. Naturgesch., l:109.

TYPE LOCALITY: Georgia, near Kutais.

DISTRIBUTION: Transcaucasia and N Turkey.

SYNONYMS: *batis, caucasicus.*

COMMENTS: Subgenus *Sorex*; externally similar to *alpinus*. Includes *batis* (Corbet, 1978*c*) and *caucasicus* which in turn now must be called *satunini* (see comments therein and Pavlinov and Rossolimo, 1987).

Sorex roboratus Hollister, 1913. Smithson. Misc. Coll., 60(24):2.

TYPE LOCALITY: Russia, Gorno-Altaisk A.O., "5 mi S Dapuchu [Altai Mtns, Tapucha]".

DISTRIBUTION: Russia east of River Ob to Ussuri River, south to Altai Mtns, N Mongolia, and Primorsk Krai.

SYNONYMS: *aranoides, dukelskiae, jacutensis, platycranius, tomensis, turuchanensis, thomasi, vir.*

COMMENTS: Subgenus *Sorex*. Formerly known as *vir* but *roboratus* has priority (Hoffmann, 1985*a*; Zaitsev, 1988). Taxonomy and distribution revised by Hoffmann (1985*a*).

Sorex sadonis Yoshiyuki and Imaizumi, 1986. Bull. Natl. Sci. Mus. Tokyo, ser. A (Zool.), 12:185.

TYPE LOCALITY: Japan, "Ikari, Sawada-machi, Sado-gun, Sado Island, alt. 20m".

DISTRIBUTION: Sado Isl (Japan).
COMMENTS: Subgenus presumably *Sorex*; Yoshiyuki and Imaizumi (1986) assigned the species to "the *caecutiens-arcticus* section of the *minutus* group." It thus resembles the taxa *annexus, cansulus,* and *shinto* of E Asia; see comments under *caecutiens*.

Sorex samniticus Altobello, 1926. Bol. Inst. Zool. Univ. Roma, 3:102.
TYPE LOCALITY: Italy, Campobasso Prov., Molise, 700 m.
DISTRIBUTION: Italy.
SYNONYMS: *garganicus*.
COMMENTS: Subgenus *Sorex*. Formerly included in *araneus*; considered a distinct species by Graf et al. (1979). Reviewed by Hausser (1990).

Sorex satunini Ognev, 1922. Ann. Zool. Mus. Russ. Acad. Sci., 22:311.
TYPE LOCALITY: Turkey, Kars, Goele, "Gel'skaya kotlovina [depression], Mvuzaret".
DISTRIBUTION: N Turkey and Caucasus.
COMMENTS: Subgenus *Sorex*. Formerly referred to as *caucasicus* Satunin, which is now synonymized with *raddei* Satunin; see Pavlinov and Rossolimo (1987) and Zaitsev (1988). Sokolov and Tembotov (1989), who reviewed the distribution in Caucasus, used *caucasicus* for this species. Considered a distinct species by Graf et al. (1979).

Sorex saussurei Merriam, 1892. Proc. Biol. Soc. Washington, 7:173.
TYPE LOCALITY: "Sierra de Colima, Jalisco, Mexico, (altitude 8000 feet)".
DISTRIBUTION: Coahuila and Durango to Chiapas (Mexico); Guatemala.
SYNONYMS: *cristobalensis, godmani, oaxacae, salvini, veraecrucis*.
COMMENTS: Referred to unnamed subgenus by George (1988:456). Populations from Guatemala provisionally included by Junge and Hoffmann (1981) may be distinct and should be carefully studied.

Sorex sclateri Merriam, 1897. Proc. Biol. Soc. Washington, 11:228.
TYPE LOCALITY: "Tumbala, Chiapas, Mexico (alt. 5000 ft.)"
DISTRIBUTION: Known only from the type locality.
COMMENTS: Referred to unnamed subgenus by George (1988:456).

Sorex shinto Thomas, 1905. Abstr. Proc. Zool. Soc. Lond., 1905(23):19.
TYPE LOCALITY: "Makado, near Nohechi, N Hondo [Honshu, Japan]".
DISTRIBUTION: Honshu, Shikoku, and Hokkaido (Japan).
SYNONYMS: *chouei, saevus, shikokensis*.
COMMENTS: Subgenus *Sorex*. Included in *caecutiens* by Abe (1967) and Corbet (1978*c*), but Imaizumi (1970*b*) treated *shinto* as a separate species, a view supported by Pavlinov and Rossolimo (1987), and by the allozyme data of George (1988).

Sorex sinalis Thomas, 1912. Ann. Mag. Nat. Hist., ser. 8, 10:398.
TYPE LOCALITY: China, Shaanxi, "45 miles S.E. of Feng-siang-fu [Feng Xian], Shen-si, 10,500'".
DISTRIBUTION: C and W China.
COMMENTS: Subgenus *Sorex*. Formerly regarded as conspecific with *isodon*; see Corbet (1978*c*) and Hoffmann (1987) for discussion and specific boundaries.

Sorex sonomae Jackson, 1921. J. Mammal., 2:162.
TYPE LOCALITY: USA, "Sonoma Country side of Gualala River, Gualala, California".
DISTRIBUTION: Pacific coast from Oregon to N California (USA).
SYNONYMS: *tenelliodus*.
COMMENTS: Subgenus *Otisorex*. Includes *tenelliodus* as a distinct subspecies; revised by Carraway (1990).

Sorex stizodon Merriam, 1895. N. Am. Fauna, 10:98.
TYPE LOCALITY: "San Cristobal, Chiapas, Mexico, [9,000 ft.= 2,743 m]".
DISTRIBUTION: Known only from the type locality.
COMMENTS: Referred to unnamed subgenus by George (1988:456). Similar to *ventralis* (Junge and Hoffmann, 1981).

Sorex tenellus Merriam, 1895. N. Am. Fauna, 10:81.
TYPE LOCALITY: USA, "summit of Alabama Hills near Lone Pine, Owens Valley [Inyo Co.], Calif[ornia, about 45000 ft]."

DISTRIBUTION: Mountains of WC Nevada and EC California (USA).
SYNONYMS: *myops*.
COMMENTS: Subgenus *Otisorex*. Similar to *nanus* with which it may form an allospecies (Hoffmann and Owen, 1980, Mammalian Species, 131), but George (1988) retained both as separate species on the basis of allozyme frequencies.

Sorex thibetanus Kastschenko, 1905. Izv. Tomsk. Univ., 27:93.
TYPE LOCALITY: "Tsaidam" [NE Tibet].
DISTRIBUTION: Himalyas and NE Tibet.
COMMENTS: The pygmy shrews of the Himalayas are the subject of controversy. The original description of *thibetanus* (as a subspecies of *minutus*) is not very helpful and the holotype in the Tomsk Academy was considered to be lost (Yudin, pers. comm. 1977 to the author); Hutterer (1979) therefore regarded *thibetanus* as a *nomen dubium*. Dolgov and Hoffmann (1977) and later Hoffmann (1987) used *thibetanus* to define a Himalayan species in which they included *buchariensis*, *kozlovi*, *planiceps*, and specimens from Nepal and China reported as *minutus* by various authors. Hutterer (1979) instead recognized three species, *buchariensis*, *planiceps*, and *minutus* as occurring in the Himalayas and regarded *kozlovi* and *thibetanus* as indeterminable. Zaitsev (1988) pointed out differences between *buchariensis* and *thibetanus*. Surprisingly, the holotype of *thibetanus* turned up in the Zoological Museum of Moscow (Baranova et al., 1981) and Hoffmann (1987) reported on its measurements. These did not solve the problem but instead matched well with *minutus*. The controversy still remains unresolved and can be solved presumably only with new material and a more complete data set. Provisionally all the mentioned forms are listed separately.

Sorex trowbridgii Baird, 1857. Mammalia, *in* Repts. U.S. Expl. Surv., 8(1):13.
TYPE LOCALITY: USA, "Astoria [mouth of the Columbia River, Clatsop Co.], Oregon".
DISTRIBUTION: Coastal ranges from Washington (including Destruction Isl) to California (USA); SW British Columbia (Canada).
SYNONYMS: *destructioni*, *humboldtensis*, *mariposae*, *montereyensis*.
COMMENTS: Referred to unnamed subgenus by George (1988:456). Reviewed by George (1989, Mammalian Species, 337).

Sorex tundrensis Merriam, 1900. Proc. Washington Acad. Sci., 2:16.
TYPE LOCALITY: USA, "St. Michaels, Alaska."
DISTRIBUTION: Sakhalin Isl; Siberia, from the Pechora River to Chukotka, south to the Altai Mtns; Mongolia and NE China; Alaska (USA); Yukon, Northwest Territories (Canada).
SYNONYMS: *amasari*, *baikalensis*, *borealis*, *centralis*, *irkutensis*, *jenissejensis*, *margarita*, *middendorfi*, *parvicaudatus*, *petschorae*, *schnitnikovi*, *sibiriensis*, *transrypheus*, *ultimus*.
COMMENTS: Subgenus *Sorex*. Youngman (1975) provided evidence that *tundrensis* is specifically distinct from *arcticus*. Palearctic populations formerly referred to *arcticus* were included in *tundrensis* by Junge et al. (1983). Hoffmann (1987) and van Zyll de Jong (1991*b*) discussed additional aspects of its taxonomy and distribution. Kozlovskii (1976) found *irkutensis* and *sibiriensis* to be karyotypically distinct; possibly two sibling species occur throughout the Palearctic range. Meylan and Hausser (1991) described a karyotype from Canada that is "identical" to some in Siberia.

Sorex ugyunak Anderson and Rand, 1945. Canadian Field Nat., 59:62.
TYPE LOCALITY: "Tuktuk (Tuktuyaktok), norhteast side of Mackenzie River delta, south of Toker Point, Mackenzie District, Northwest Territories, Canada."
DISTRIBUTION: Mainland tundra west of Hudson Bay (Canada), and N Alaska (USA).
COMMENTS: Subgenus *Otisorex*. Formerly included in *cinereus*, but van Zyll de Jong (1976, 1991*b*) provided arguments for a specific destinction of *ugyunak*; the taxon appears related to *jacksoni* and *portenkoi*. See Junge and Hoffmann (1981) and van Zyll de Jong (1983*a*) for further information.

Sorex unguiculatus Dobson, 1890. Ann. Mag. Nat. Hist., ser. 6, 5:155.
TYPE LOCALITY: Russia, "Saghalien [Sakhalin] Island; Nikolajewsk, at the mouth of the Amur River." Ognev (1928:204) and Ellerman and Morrison-Scott (1951:52) both restricted the type locality to Sakhalin Isl.

DISTRIBUTION: Pacific coast of Siberia from Vladivostok to the Amur, and the islands of Sakhalin (Russia) and Hokkaido (Japan); from Corbet (1978*c*).
SYNONYMS: *yesoensis*.
COMMENTS: Subgenus *Sorex*. The inclusion of *yesoensis* follows Abe (1967). Skaren (1964) suggested a relationship with *obscurus* (= *monticolus*) but this was rejected by Siivonen (1965) and Hoffmann (1971).

Sorex vagrans Baird, 1857. Mammalia, *in* Repts. U.S. Expl. Surv., 8(1):15.
TYPE LOCALITY: USA, "Shoalwater Bay, W.T. [= Willapa Bay, Pacific Co., Washington]."
DISTRIBUTION: Riparian and montane areas of the N Great Basin and Columbia Plateau, north to S British Columbia and Vancouver Island (Canada); east to W Montana, W Wyoming, and Wasatch Mtns (Utah); C Nevada to Sierra Nevada (California).
SYNONYMS: *amoenus, halicoetes, nevadensis, nigriculus, paludivagus, parvidens, shastensis, sukleyi, trigonirostris, vancouverensis*.
COMMENTS: Subgenus *Otisorex*. Findley's (1955*b*) wide concept of the *vagrans* group was substantially modified by Hennings and Hoffmann (1977) and Junge and Hoffmann (1981).

Sorex ventralis Merriam, 1895. N. Am. Fauna, 10:75.
TYPE LOCALITY: "Cerro San Felipe, Oaxaca, Mexico (altitude 1000 feet)."
DISTRIBUTION: NW Puebla to Oaxaca (Mexico).
COMMENTS: Referred to unnamed subgenus by George (1988:456). Similar to *saussurei* but smaller; see Junge and Hoffmann (1981), who allocated the species to subgenus *Sorex*. Hall (1981) included *ventralis* in *oreopolus*.

Sorex veraepacis Alston, 1877. Proc. Zool. Soc. Lond., 1877:445.
TYPE LOCALITY: "Coban (Vera Paz) [Alta Verapaz], Guatemala."
DISTRIBUTION: Montane forests of C Guerrero, Puebla, and Veracruz, south through the highlands of Oaxaca and Chiapas (Mexico), to SW Guatemala.
SYNONYMS: *chiapensis; caudatus* (Merriam, not of Horsfield), *mutabilis, teculyas*.
COMMENTS: Subgenus *Otisorex*. May be conspecific with *macrodon* (see Junge and Hoffmann, 1981:43).

Sorex volnuchini Ognev, 1922. Ann. Mus. Zool. Akad. Sci. St. Petersbourg, 22:322.
TYPE LOCALITY: Ukraine, Krasnodarskii kr., Adygeiskaya A.O. [middle course], r. Kisha (see Pavlinov and Rossolimo, 1987).
DISTRIBUTION: S Ukraine and Caucasus; possibly Turkey and N Iran.
COMMENTS: Subgenus *Sorex*. Formerly included in *minutus* but specimens from Caucasus have a slighly different karyotype (2n=40, NF=60) which led Kozlovskii (1973) and Sokolov and Tembotov (1989) to regard *volnuchini* as a full species. The karyotype of *S. buchariensis* is very similar (Ivanitskaya et al., 1977). Morphologically, *volnuchini* is not distinguishable from *minutus;* also, the karyotype of topotypical specimens needs to be studied to clarify the status of *volnuchini*.

Soriculus Blyth, 1854. J. Asiatic Soc. Bengal, 23:733.
TYPE SPECIES: *Corsira nigrescens* Gray, 1842.
SYNONYMS: *Chodsigoa, Episoriculus*.
COMMENTS: Subfamily Soricinae, tribe Neomyini; see Repenning (1967:45). Reumer (1984:14) claimed that Soriculini Kretzoi, 1965, has priority over Neomyini Repenning, 1965, however, Neomyini Matschie, 1909, has priority over both. Includes *Chodsigoa* and *Episoriculus* as subgenera; see Ellerman and Morrison-Scott (1951) and Corbet (1978*c*:24); but also see Repenning (1967:52) and Jameson and Jones (1977:474-475), who considered *Episoriculus* a distinct genus. Gureev (1979:450-452) erroneously listed *Chodsigoa* as a subgenus of *Notiosorex*. Genus reviewed by Hoffmann (1985*b*).

Soriculus caudatus (Horsfield, 1851). Cat. Mamm. Mus. E. India Co., p. 135.
TYPE LOCALITY: "Sikkim", no exact locality.
DISTRIBUTION: Kashmir to N Burma and SW China.
SYNONYMS: *gracilicauda, sacratus, soluensis, umbrinus*.
COMMENTS: Subgenus *Episoriculus*. Includes *sacratus* and *umbrinus* as subspecies; see Gruber (1969) and Hoffmann (1985*b*).

Soriculus fumidus Thomas, 1913. Ann. Mag. Nat. Hist., ser. 8, 11:216.
TYPE LOCALITY: Taiwan, Chiai Hsien, "Mt. Arisan (= Alishan); Central Formosa. Alt. 8,000'".
DISTRIBUTION: Montane forests of Taiwan.
SYNONYMS: *sodalis*.
COMMENTS: Subgenus *Episoriculus*. Formerly included in *caudatus* by Ellerman and Morrison-Scott (1951:59), but see Jameson and Jones (1977:474) and Hoffmann (1985*b*), who included *sodalis* in *fumidus*. Additional specimens now suggest that *sodalis* may prove distinct (Hoffmann, in litt.).

Soriculus hypsibius de Winton, 1899. Proc. Zool. Soc. Lond., 1899:574.
TYPE LOCALITY: China, Sichuan, "Yang-liu-pa".
DISTRIBUTION: SW and C China, Yunnan, Sichuan and Shaanxi; apparently disjunct population (*larvarum*) in Hebei.
SYNONYMS: *berezowski, larvarum*.
COMMENTS: Type species (as *berezowski*) of subgenus *Chodsigoa*. Does not include *parva* and *lamula*; see Hoffmann (1985*b*).

Soriculus lamula (Thomas, 1912). Ann. Mag. Nat. Hist., ser. 8, 10:399.
TYPE LOCALITY: China, Gansu, "40 miles S.E. of Tao-chou [Lintan]. Alt. 9500'".
DISTRIBUTION: C China, from Yunnan, Sichuan, and Gansu to Fujian.
SYNONYMS: *parva*.
COMMENTS: Subgenus *Chodsigoa*. Includes *parva* as a subspecies. Formerly in *hypsibius* but occurs sympatrically with that species; see Hoffmann (1985*b*).

Soriculus leucops (Horsfield, 1855). Ann. Mag. Nat. Hist., ser. 2, 16:111.
TYPE LOCALITY: "Nepal."
DISTRIBUTION: C Nepal, Sikkim and Assam to S China, N Burma and N Vietnam.
SYNONYMS: *baileyi, gruberi*.
COMMENTS: Subgenus *Episoriculus*. Includes *baileyi* as a subspecies; see Hoffmann (1985*b*).

Soriculus macrurus Blanford, 1888. Fauna Brit. India, 1:231.
TYPE LOCALITY: "Darjeeling, India."
DISTRIBUTION: C Nepal to W and S China and to N Burma and Vietnam.
SYNONYMS: *irene*.
COMMENTS: Subgenus *Episoriculus*. Formerly confused with *leucops*, but shown to be a distinct species by Hoffmann (1985*b*).

Soriculus nigrescens (Gray, 1842). Ann. Mag. Nat. Hist., [ser. 1], 10:261.
TYPE LOCALITY: "India", West Bengal, Darjeeling.
DISTRIBUTION: Middle altitudes of the Himalaya from Tibet and Nepal to Assam and SW China.
SYNONYMS: *caurinus, centralis, pahari, radulus* (see Ellerman and Morrison-Scott, 1951).
COMMENTS: Subgenus *Soriculus*. The form *radulus* is distinctly smaller (Hoffmann, 1985*b*).

Soriculus parca (G. M. Allen, 1923). Am. Mus. Novit., 100:6.
TYPE LOCALITY: "Ho-mu-shu Pass, Western Yunnan, China, 8000 feet."
DISTRIBUTION: SW China, N Burma and Thailand and N Vietnam.
SYNONYMS: *furva, lowei*.
COMMENTS: Formerly included in *smithii*, but retained as a separate species by Hoffmann (1985*b*), with *lowei* and *furva* as tentative subspecies.

Soriculus salenskii Kastschenko, 1907. Ann. Mus. Zool. Acad. Sci. St. Petersbourg, 10:253.
TYPE LOCALITY: China, Sichuan, "Lun-ngan'-fu" (= Liangfu).
DISTRIBUTION: Known only from the type locality in N Sichuan.
COMMENTS: Subgenus *Chodsigoa*. Related to *smithii*.

Soriculus smithii (Thomas, 1911). Abstr. Proc. Zool. Soc. Lond., 1911(90):4.
TYPE LOCALITY: China, Sichuan, "Ta-tsien-lu".
DISTRIBUTION: C Sichuan to W Shaanxi (China).
COMMENTS: Subgenus *Chodsigoa*. Regarded as a subspecies of *salenskii* by Ellerman and Morrison-Scott (1951:60), but as a separate species by Corbet (1978*c*:24). Formerly included *parca* and *furva*; but see Hoffmann (1985*b*).

Family Talpidae G. Fischer, 1817. Mem. Soc. Imp. Nat., Moscow, 5:372.

SYNONYMS: Desmanidae.

COMMENTS: Subfamily systematics very provisional. Cabrera (1925) proposed a division into five subfamilies, three of which are applied here, following Yates (1984), among others. Scalopinae and Condylurinae are tentatively treated as tribes, following Hutchison (1968). *Desmana* and *Galemys* are sometimes placed in a separate family, Desmanidae; see Bobrinskii et al. (1965). Family reviewed by Gureev (1979); see also Gorman and Stone (1990). Relationships of recent moles discussed by Ziegler (1971). Systematics of North American forms reviewed by Yates and Greenbaum (1982); of Palearctic forms by Corbet (1978*c*); of Siberian forms by Yudin (1989); and of Japanese forms by Abe (1988). Phylogeny of Urotrichini discussed by Storch and Qiu (1983).

Subfamily Desmaninae Thomas, 1912. Ann. Mag. Nat. Hist., ser. 8, 9:397.

COMMENTS: Sometimes regarded as a separate family; see Barabasch-Nikiforow (1975). Hutchinson (1974) concluded that Desmaninae and Talpidae were separated since the Eocene. Reviewed by Rümke (1985).

Desmana Güldenstaedt, 1777. Beschaft. Berliner Ges. Naturforsch. Fr., 3:108.

TYPE SPECIES: *Castor moschatus* Linnaeus, 1758.

Desmana moschata (Linnaeus, 1758). Syst. Nat., 10th ed., 1:59.

TYPE LOCALITY: "Habitat in Russiae aquosis."

DISTRIBUTION: Republics of the former USSR; Don, Volga, and S. Ural rivers and their tributaries; introduced into Tachan and Tartas rivers (Ob basin) and Dnepr River.

STATUS: IUCN - Vulnerable.

Galemys Kaup, 1829. Skizz. Entwickel.-Gesch. Nat. Syst. Europ. Thierwelt, 1:119.

TYPE SPECIES: *Mygale pyrenaica* E. Geoffroy, 1811.

Galemys pyrenaicus (E. Geoffroy, 1811). Ann. Mus. Hist. Nat. Paris, 17:193.

TYPE LOCALITY: France, "Les montagnes pres de Tarbes (Hautes-Pyrenees)".

DISTRIBUTION: Streams of the Pyrenees and the northern mountains of the Iberian Peninsula (France, Spain and Portugal).

STATUS: IUCN - Vulnerable.

SYNONYMS: *rufulus*.

COMMENTS: Includes *rufulus* as a possible subspecies; reviewed by Palmeirim and Hoffmann (1983, Mammalian Species, 207) and Juckwer (1990).

Subfamily Talpinae G. Fischer, 1817. Mem. Soc. Imp. Nat., Moscow, 5:372.

SYNONYMS: Condylurinae, Scalopinae, Urotrichinae.

COMMENTS: Includes Condylurini, Scalopini, Scaptonychini, Urotrichini and Talpini as tribes; see Hutchison (1968) and Storch and Qiu (1983). Some of these are sometimes regarded as subfamilies; see comments under family. Generic and specific limits are highly controversial, particularly in the *Talpa* group. No data-based consensus has yet been reached on the taxonomy of many Palearctic and East Asiatic moles.

Condylura Illiger, 1811. Prodr. Syst. Mamm. Avium, p. 125.

TYPE SPECIES: *Sorex cristatus* Linnaeus, 1758.

COMMENTS: Tribe Condylurini. Reviewed by Peterson and Yates (1980). Recorded from the Pliocene of Europe; see Skoczen (1976).

Condylura cristata (Linnaeus, 1758). Syst. Nat., 10th ed., 1:53.

TYPE LOCALITY: USA, Pennsylvania.

DISTRIBUTION: Georgia and NW South Carolina (USA) to Nova Scotia and Labrador (Canada); Great Lakes region to SE Manitoba.

SYNONYMS: *longicaudata, macroura, nigra, parva, prasinatus, radiata*.

COMMENTS: Includes *parva* as a subspecies; see review by Peterson and Yates (1980, Mammalian Species, 129).

Euroscaptor Miller, 1940. J. Mammal., 21:443.
TYPE SPECIES: *Talpa klossi* Thomas, 1929.
COMMENTS: Tribe Talpini. Corbet (1978*c*:32), and subsequent work, included *Euroscaptor* in *Talpa* while Russian and Japanese authors retained it as a genus; most recently Abe et al. (1991). Species allocations and limits are tentative.

Euroscaptor grandis Miller, 1940. J. Mammal., 21:444.
TYPE LOCALITY: China, Sichuan, "Mount Omei, alt. 5000 feet", = Omei-Shan.
DISTRIBUTION: N and S Bakbo and Cha-pa (Vietnam); S China.
COMMENTS: Often included in *Talpa;* but see Gureev (1979:272). Regarded as a synonym of [*E*]. *micrura longirostris* by Ellerman and Morrison-Scott (1966:40).

Euroscaptor klossi (Thomas, 1929). Ann. Mag. Nat. Hist., ser. 10, 3:206.
TYPE LOCALITY: Thailand.
DISTRIBUTION: Highlands of Thailand, Laos and Peninsular Malaysia.
COMMENTS: Corbet (1978*c*:33) and Corbet and Hill (1991:38) included *klossi* in *micrura;* but see Yoshiyuki (1988*b*). May include *malayana* which Harrison (1974:57) included in *micrura*.

Euroscaptor longirostris (Milne-Edwards, 1870). C.R. Acad. Sci. Paris, 70:341.
TYPE LOCALITY: China, Sichuan, Moupin.
DISTRIBUTION: S China.
COMMENTS: Formerly included in *micrura* by Ellerman and Morrison-Scott (1966:40) and Corbet (1978*c*:35). In the *Euroscaptor* group of *Talpa;* see Gureev (1979:272).

Euroscaptor micrura (Hodgson, 1841). Calcutta J. Nat. Hist., 2:221.
TYPE LOCALITY: Nepal, C and N hills.
DISTRIBUTION: E Himalaya; doubtfully in Peninsular Malaysia.
COMMENTS: Does not include *klossi;* see Yoshiyuki (1988*b*). Does not include *malayana,* which Harrison (1974:57) included in *micrura*.

Euroscaptor mizura (Gunther, 1880). Proc. Zool. Soc. Lond., 1880:441.
TYPE LOCALITY: Japan, Honshu, "In the neighbourhood of Yokohama".
DISTRIBUTION: Mountains of Honshu (Japan).
SYNONYMS: *hiwaensis, othai.*
COMMENTS: Imaizumi (1970*b*) and Abe et al. (1991) included this species in the genus *Euroscaptor,* while Corbet (1978*c*) placed it in *Talpa*. Three populations have been named, of which *othai* represents "probably a distinct species", according to Imaizumi (1970*b*); a view supported by Yoshiyuki (1988*b*).

Euroscaptor parvidens Miller, 1940. J. Mammal., 21:203.
TYPE LOCALITY: Vietnam, Di Linh, Blao Forest Station.
DISTRIBUTION: Known from type locality and Rakho on the Chinese border.
COMMENTS: Ellerman and Morrison-Scott (1966:40) included this species in [*E*]. *micrura leucura*. Corbet (1978*c*:33) mentioned *leucura* as a species, but Gureev (1979:274) also listed *parvidens* in the *Euroscaptor* group of *Talpa,* where Miller (1940*b*:444) put his species soon after description. Corbet and Hill (1991:38) did not list *parvidens* and one may assume that they included it in *micrura*.

Mogera Pomel, 1848. Arch. Sci. Phys. Nat. Geneve, 9:246.
TYPE SPECIES: *Talpa wogura* Temminck, 1842.
COMMENTS: Tribe Talpini. Formerly included in *Talpa* by Corbet (1978*c*); but see Imaizumi (1970*b*), Gureev (1979), Yudin (1989), and Abe et al. (1991).

Mogera etigo Yoshiyuki and Imaizumi, 1991. Bull. Natl. Sci. Mus. Tokyo, Ser. A (Zool.), 17:101.
TYPE LOCALITY: "Inugaeshi-shinden, Shirone-shi, Echigo Plain, Niigata Prefecture, Chubu District, Honshu, Japan."
DISTRIBUTION: Echigo Plain, Honshu, C Japan.
COMMENTS: Previously included in *tokudae;* but this was restricted to Sado Isl by Yoshiyuki and Imaizumi (1991).

Mogera insularis Swinhoe, 1863. Proc. Zool. Soc. Lond., 1862:356 [1863].
TYPE LOCALITY: "Formosa (China)" = Taiwan.

DISTRIBUTION: Taiwan, Hainan, SE China.
SYNONYMS: *latouchei*.
COMMENTS: Includes *latouchei*; see Corbet and Hill (1991*c*:38). Included in *Talpa* [*Euroscaptor*] *micrura* by Ellerman and Morrison-Scott (1951:40); but see Corbet (1978*c*:33).

Mogera kobeae Thomas, 1905. Ann. Mag. Nat. Hist., ser. 7, 15:487.
TYPE LOCALITY: Japan, Hondo, Kobe.
DISTRIBUTION: Kyushu, Shikoku and southern part of Honshu, Japan.
COMMENTS: Included in *Talpa robusta* by Ellerman and Morrison-Scott (1966) and Corbet (1978*c*); but retained as a separate species by Abe (1970), Yoshiyuki (1986), and Yoshiyuki and Imaizumi (1991).

Mogera minor Kuroda, 1936. [Botany and Zoology], Tokyo, 4(1), p. 74.
TYPE LOCALITY: Japan, Honshu, Tochigi Pref., Shiobara.
DISTRIBUTION: Honshu, Japan.
SYNONYMS: *imaizumii*.
COMMENTS: Included in *Talpa* [*Euroscaptor*] *micrura* by Ellerman and Morrison-Scott (1966); but retained as a separate species by Yoshiyuki (1986). Renamed *Talpa wogura imaizumii* by Kuroda (1957) for presumed homonymy with *Talpa europaea* var. *minor*.

Mogera robusta Nehring, 1891. Sitzb. Ges. Naturf. Fr. Berlin, 6:95.
TYPE LOCALITY: Russia, Vladivostok.
DISTRIBUTION: Korea to NE China and adjacent Siberia.
SYNONYMS: *coreana*.
COMMENTS: Includes *coreana*; see Corbet (1978*c*). European authors often include *kobeae* and *tokudae*; however, Japanese authors (Imaizumi 1970*b*; Yoshiyuki 1988*b*) treat these as separate species. Formerly included in *Talpa*; but see Imaizumi (1970*b*), Gureev (1979), and Gromov and Baranova (1981).

Mogera tokudae Kuroda, 1940. [A monograph of Japanese mammals ...], Tokyo and Osaka, p. 196.
TYPE LOCALITY: Japan, Sado Isl.
DISTRIBUTION: Restricted to Sado Isl, Japan according to Yoshiyuki and Imaizumi (1991).
COMMENTS: Overlooked by Ellerman and Morrison-Scott (1966); included in *Talpa robusta* by Corbet (1978*c*); but retained as a separate species by Yoshiyuki (1986) and Abe et al. (1991).

Mogera wogura (Temminck, 1842). *In* Siebold, Fauna Japonica, 1(Mamm.), 1:19.
TYPE LOCALITY: Japan, Yokohama, Honshu; restricted by Thomas (1905*b*).
DISTRIBUTION: Japan (Honshu, Kyushu, Tane, Amakusa, Tsushima and other Isls.).
COMMENTS: For a taxonomic discussion see Corbet (1978*c*:35).

Nesoscaptor Abe, Shiraishi and Arai, 1991. J. Mammal. Soc. Japan, 15:48.
TYPE SPECIES: *Nesoscaptor uchidai* Abe, Shiraishi and Arai, 1991.
COMMENTS: Tribe Talpini.

Nesoscaptor uchidai Abe, Shiraishi and Arai, 1991. J. Mammal. Soc. Japan, 15:53.
TYPE LOCALITY: Japan, Ryukyu Isls, Senkaku Isls, west coast of Uotsuri-jima.
DISTRIBUTION: Known only from the type locality.

Neurotrichus Günther, 1880. Proc. Zool. Soc. Lond., 1880:441.
TYPE SPECIES: *Urotrichus gibbsii* Baird, 1858.
COMMENTS: Tribe Urotrichini; see Storch and Qiu (1983:100).

Neurotrichus gibbsii (Baird, 1857). Mammalia, *in* Repts. U.S. Expl. Surv., 8(1):76.
TYPE LOCALITY: USA, Washington, Pierce Co., "Naches Pass, 4,500 ft." (1,372 m).
DISTRIBUTION: SW British Columbia (Canada) to WC California (USA).
SYNONYMS: *hyacinthinus, major, minor*.
COMMENTS: Hall (1981:67) listed *hyacinthinus* and *minor* as subspecies. Reviewed by Carraway and Verts (1991*b*, Mammalian Species, 387).

Parascalops True, 1894. Diagnoses New N. Am. Mamm., p. 2. (preprint of Proc. U.S. Natl. Mus., 17:242).
TYPE SPECIES: *Scalops breweri* Bachman, 1842.
COMMENTS: Tribe Scalopini. Reviewed by Hallett (1978).

Parascalops breweri (Bachman, 1842). Boston J. Nat. Hist., 4:32.
TYPE LOCALITY: "Martha's Vineyard."; restricted to "E. North America" by Hall and Kelson (1959).
DISTRIBUTION: NE United States and SE Canada.
COMMENTS: Reviewed by Hallett (1978, Mammalian Species, 98).

Parascaptor Gill, 1875. Bull. U.S. Geol. Geogr. Surv. Terr., I, 2:110.
TYPE SPECIES: *Talpa leucura* Blyth, 1850.
COMMENTS: Tribe Talpini. Included in *Talpa* by Corbet and Hill (1991:38); but retained as a genus by Abe et al. (1991).

Parascaptor leucura (Blyth, 1850). J. Asiat. Soc. Bengal, 19:215, pl. 4.
TYPE LOCALITY: India, Assam, Khasi Hills, Cherrapunji.
DISTRIBUTION: Burma, Assam (India), and Yunnan (China).
COMMENTS: Formerly included in *T. micrura;* see Ellerman and Morrison-Scott (1951:40); but also see Corbet (1978*c*:33).

Scalopus Desmarest, 1804. Tabl. Méth. Hist. Nat., *in* Nouv. Dict. Hist. Nat., 24:14.
TYPE SPECIES: *Sorex aquaticus* Linnaeus, 1758.
SYNONYMS: *Talpasorex* Lesson (not Schinz).
COMMENTS: Tribe Scalopini. Reviewed by Yates and Schmidly (1978). Commonly cited from "É. Geoffroy, 1803. Cat. Mamm. Mus. d'Hist. Nat., p. 77", but this work was never published (see Appendix I).

Scalopus aquaticus (Linnaeus, 1758). Syst. Nat., 10th ed., 1:53.
TYPE LOCALITY: E USA; fixed by Jackson (1915:33) to Pennsylvania, Philadelphia.
DISTRIBUTION: N Tamaulipas and N Coahuila (Mexico) through E USA to Massachusetts and Minnesota.
SYNONYMS: *aereus, alleni, anastasae, argentatus, australis, bassi, caryi, cryptus, cupreata, howelli, inflatus, intermedius, machrinoides, machrinus, montanus, nanus, parvus, pennsylvanica, porteri, pulcher, sericea, texanus, virginianus.*
COMMENTS: Includes *inflatus* and *montanus;* see Yates and Schmidly (1977). Yates and Schmidly (1978, Mammalian Species, 105) listed sixteen subspecies. Gureev (1979:254) listed *aereus* and *inflatus* as distinct species without comment. Hall (1981:72) included *aereus* and *inflatus* in *aquaticus.*

Scapanulus Thomas, 1912. Ann. Mag. Nat. Hist., ser. 8, 10:396.
TYPE SPECIES: *Scapanulus oweni* Thomas, 1912.
COMMENTS: Tribe Scalopini; see Storch and Qiu (1983:118).

Scapanulus oweni Thomas, 1912. Ann. Mag. Nat. Hist., ser. 8, 10:397.
TYPE LOCALITY: China, Kansu, "23 miles (37 km) S.E. of Tao-chou, 9000" (2,743 m).
DISTRIBUTION: Montane forest in C China: Kansu, Shensi and Sichuan.

Scapanus Pomel, 1848. Arch. Sci. Phys. Nat. Geneve, 9:247.
TYPE SPECIES: *Scalops townsendii* Bachman, 1839.
COMMENTS: Tribe Scalopini. Revised by Jackson (1915:54-76) and Hutchison (1987).

Scapanus latimanus (Bachman, 1842). Boston J. Nat. Hist., 4:34.
TYPE LOCALITY: Probably Santa Clara, Santa Clara Co., California, USA; *fide* Osgood (1907:52).
DISTRIBUTION: SC Oregon (USA) to N Baja California (Mexico).
SYNONYMS: *alpinus, anthonyi, californicus, campi, caurinus, dilatus, grinnelli, insularis, minusculus, monoensis, occultus, parvus, sericatus, truei.*
COMMENTS: Hall (1981:69-70) listed 12 subspecies.

Scapanus orarius True, 1896. Proc. U.S. Natl. Mus., 19:52.
TYPE LOCALITY: USA, Washington, Pacific Co., Shoalwater Bay (= Willapa Bay).

DISTRIBUTION: SW British Columbia (Canada) to NW California, WC Idaho, N Oregon, C and SE Washington (USA).
SYNONYMS: *schefferi, yakimensis.*
COMMENTS: Includes *schefferi* as a subspecies; see Hartman and Yates (1985, Mammalian Species, 253).

Scapanus townsendii (Bachman, 1839). J. Acad. Nat. Sci. Philadelphia, 8:58.
TYPE LOCALITY: USA, Washington, Clark Co., vicinity of Vancouver.
DISTRIBUTION: SW British Columbia (Canada) to NW California (USA).

Scaptochirus Milne-Edwards, 1867. Ann. Sci. Nat. Zool. (Paris), ser. 5, 7:375.
TYPE SPECIES: *Scaptochirus moschatus* Milne-Edwards, 1867.
COMMENTS: Tribe Talpini. Included in *Talpa* by Corbet (1978*c*:36) and Corbet and Hill (1991:38). Retained a genus by Abe et al. (1991) and Gureev (1979:282).

Scaptochirus moschatus Milne-Edwards, 1867. Ann. Sci. Nat. Zool. (Paris), ser. 5, 7:375.
TYPE LOCALITY: "En Mongolie"; Swanhwafu, 100 mi. (161 km) NW of Peking, China.
DISTRIBUTION: NE China: Hopei, Shantung, Shansi, Shensi.
SYNONYMS: *davidianus* (?).
COMMENTS: Included in *micrura* by Ellerman and Morrison-Scott (1966:40); but see Corbet (1978*c*:36). Grulich (1982) pointed out that *Scaptochirus davidianus* Milne-Edwards, 1884, so far regarded as a synonym of *moschatus*, may be an earlier name for *Talpa streeti* Lay, 1965.

Scaptonyx Milne-Edwards, 1872. *In* David, Nouv. Arch. Mus. Hist. Nat. Paris, Bull. for 1871, 7(4):92 [1872].
TYPE SPECIES: *Scaptonyx fusicauda* Milne-Edwards, 1872.
COMMENTS: Tribe Scaptonychini; see Van Valen (1967).

Scaptonyx fusicaudus Milne-Edwards, 1872. *In* David, Nouv. Arch. Mus. Hist. Nat. Paris, Bull. for 1871, 7(4):92 [1872].
TYPE LOCALITY: "Frontière du Kokonoor", vicinity of Kukunor (Lake), China.
DISTRIBUTION: N Burma; S China, Tsinghai, Shensi, Sichuan and Yunnan.
SYNONYMS: *affinis, fusicaudatus* (*sic*).
COMMENTS: Includes *affinis*; see Ellerman and Morrison-Scott (1966:35).

Talpa Linnaeus, 1758. Syst. Nat., 10th ed., 1:52.
TYPE SPECIES: *Talpa europaea* Linnaeus, 1758.
SYNONYMS: *Asioscalops.*
COMMENTS: Includes *Asioscalops* which was retained as a full genus by Yudin (1989:60). Schwarz (1948), Corbet (1978*c*:36), and Corbet and Hill (1991:38) included *Euroscaptor, Parascaptor, Mogera,* and *Scaptochirus*; but these are retained here as full genera, see Abe et al. (1991) and Gureev (1979:256-285). For phylogenetic considerations based on morphology, see Stein (1960) and Grulich (1971); based on allozyme variation, see Filippucci et al. (1987). For a review of European species, see Niethammer and Krapp (1990).

Talpa altaica Nikolsky, 1883. Trans. Soc. Nat. St. Petersburg, 14:165.
TYPE LOCALITY: Siberia, Altai Mtns, Valley of Tourak.
DISTRIBUTION: Taiga zone of Siberia between Ob and Lena rivers; south to N Mongolia.
COMMENTS: Placed by Yudin (1989:52) in genus *Asioscalops*; but see Corbet (1978*c*:33). Kratochvíl and Král (1972) provided karyological evidence for a separation of *altaica* from the remaining *Talpa* species.

Talpa caeca Savi, 1822. Nuovo Giorn. de Letterati, Pisa, 1:265.
TYPE LOCALITY: Italy, Pisa.
DISTRIBUTION: S Europe and (doubtfully) Asia Minor; Alps, Apennines, Balkan, Thrazia.
SYNONYMS: *augustana, beaucournui, dobyi, hercegovinensis, olympica, steini* (see Niethammer, *in* Niethammer and Krapp, 1990).
COMMENTS: Relationships to the Caucasus moles (*caucasica, levantis*) unsolved.

Talpa caucasica Satunin, 1908. Mitt. Kaukas. Mus., 4:5.
TYPE LOCALITY: Russia, Stavropol Krai, Stavropol.
DISTRIBUTION: NW Caucasus.

COMMENTS: Included in *europaea* by Ellerman and Morrison-Scott (1951), but considered a distinct species by Gromov et al. (1963). Reviewed by Sokolov and Tembotov (1989).

Talpa europaea Linnaeus, 1758. Syst. Nat., 10th ed., 1:52.
TYPE LOCALITY: Sweden, Kristianstad, Engelholm.
DISTRIBUTION: Temperate Europe including Britain to the Ob and Irtysh rivers (Russia) in the east.
SYNONYMS: *cinerea, brauneri, ehiki, frisius, kratochvili, pancici, uralensis, velessiensis.*
COMMENTS: Does not include *altaica, caucasica, romana,* and *stankovici;* see comments under these species. Does include *cinerea* as a subspecies, and the names above as synonyms; see Niethammer (*in* Niethammer and Krapp, 1990).

Talpa levantis Thomas, 1906. Ann. Mag. Nat. Hist., ser. 7, 17:416.
TYPE LOCALITY: Turkey, Trabzon, Scalita (= Altyn Dereh).
DISTRIBUTION: Bulgaria, Thracia and N Anatolia (Turkey), and adjacent Caucasus.
SYNONYMS: *minima, orientalis, talyschensis, transcaucasica.*
COMMENTS: On specific status, see Grulich (1972) and Felten et al. (1973). Reviewed by Sokolov and Tembotov (1989). European records by Vohralík (1991).

Talpa occidentalis Cabrera, 1907. Ann. Mag. Nat. Hist., ser. 7, 20:212.
TYPE LOCALITY: C Spain, Guadarrama Mtns, 1200-1300 m, "La Granja, Segovia".
DISTRIBUTION: W and C Iberian Peninsula (Portugal, Spain).
COMMENTS: Formerly regarded as a subspecies of *caeca,* but see Ramalhinho (1985) and Filippucci et al. (1987).

Talpa romana Thomas, 1902. Ann. Mag. Nat. Hist., ser. 7, 10:516.
TYPE LOCALITY: Italy, Ostia near Rome.
DISTRIBUTION: Apennines, Italy, and extreme SE France; a historical record from Sicily.
SYNONYMS: *adamoi, aenigmatica, brachycrania, montana, wittei* (see Niethammer, *in* Niethammer and Krapp, 1990).
COMMENTS: Does not include *stankovici,* see comments therein.

Talpa stankovici V. Martino and E. Martino, 1931. J. Mammal., 12:53.
TYPE LOCALITY: Yugoslavia, Pelister Mtns, "Magarevo Mts., Perister, S. Serbia (Macedonia). Alt. 1000 m."
DISTRIBUTION: European Balkan, Greece including Corfu Isl, S Yugoslavia, probably Albania.
COMMENTS: Formerly included in *romana;* specific status supported by Filippucci et al. (1987). Reviewed by Niethammer (*in* Niethammer and Krapp, 1990).

Talpa streeti Lay, 1965. Fieldiana Zool., 44:227.
TYPE LOCALITY: Iran, Kurdistan, Hezar Darreh.
DISTRIBUTION: N Iran.
SYNONYMS: *streetorum.*
COMMENTS: Grulich (1982) claimed that *Scaptochirus davidianus* Milne-Edwards, 1884, is the earliest valid name for this species.

Urotrichus Temminck, 1841. Het. Instit. K. Ned. Inst., p. 212.
TYPE SPECIES: *Urotrichus talpoides* Temminck, 1841.
SYNONYMS: *Dymecodon.*
COMMENTS: Tribe Urotrichini; see Storch and Qiu (1983:100). Includes *Dymecodon;* see Ellerman and Morrison-Scott (1951:33-34). See also Imaizumi (1970*b*:123) who considered *Dymecodon* a distinct genus.

Urotrichus pilirostris (True, 1886). Proc. U.S. Natl. Mus., 9:97.
TYPE LOCALITY: Japan, Honshu, Enoshima (Yenosima), at mouth of Bay of Yeddo.
DISTRIBUTION: Montane forests of Honshu, Shikoku, Kyushu (Japan).
COMMENTS: Formerly included in *Dymecodon;* see Corbet (1978*c*:37).

Urotrichus talpoides Temminck, 1841. Het. Instit. K. Ned. Inst., p. 215.
TYPE LOCALITY: Japan, Kyushu, Nagasaki.
DISTRIBUTION: Grassland and forest of Honshu, Shikoku, Kyushu (Japan); Dogo Isl, N Tsushima Isl (Japan).
SYNONYMS: *adversus, centralis, hondoensis, minutus.*

COMMENTS: Includes *adversus, centralis, hondonis* and *minutus* as subspecies; see Imaizumi (1970*b*:128).

Subfamily Uropsilinae Dobson, 1883. Monogr. Insectivora, 2:126-172.

Uropsilus Milne-Edwards, 1872. *In* David, Nouv. Arch. Mus. Hist. Nat. Paris, Bull. for 1871, 7(4): 92 [1872].
TYPE SPECIES: *Uropsilus soricipes* Milne-Edwards, 1872.
SYNONYMS: *Nasillus, Rhynchonax.*
COMMENTS: Includes *Nasillus* and *Rhynchonax*; see Ellerman and Morrison-Scott (1966:31) and Corbet and Hill (1980:33); but also see Gureev (1979:201-204), who listed *Rhynchonax* and *Nasillus* as distinct genera. Reviewed by Hoffmann (1984).

Uropsilus andersoni (Thomas, 1911). Abstr. Proc. Zool. Soc. Lond., 1911(100):49.
TYPE LOCALITY: China, Sichuan, "Omi-san" = Emei-Shan.
DISTRIBUTION: Central Sichuan (China).
COMMENTS: Formerly included in *soricipes*, but see Hoffmann (1984).

Uropsilus gracilis (Thomas, 1911). Abstr. Proc. Zool. Soc. Lond., 1911(100):49.
TYPE LOCALITY: China, Sichuan, near Nan-chwan (Nanchuan), Mt. Chin-fu-san (Jingfu Shan).
DISTRIBUTION: Sichuan and Yunnan (China) and N Burma.
SYNONYMS: *atronates, nivatus.*
COMMENTS: Formerly included in *soricipes*; but see Hoffmann (1984).

Uropsilus investigator (Thomas, 1922). Ann. Mag. Nat. Hist., ser. 9, 10:393.
TYPE LOCALITY: China, Yunnan, Kui-chiang-Salween divide at 28°N, 11,000 ft.
DISTRIBUTION: Yunnan (China).
COMMENTS: Hoffmann (1984) included *investigator* in *gracilis* but on morphological and distributional grounds Wang and Yang (1989) and Storch (pers. comm.) concluded that both are sympatric in Yunnan and must therefore be regarded as distinct species.

Uropsilus soricipes Milne-Edwards, 1872. *In* David, Nouv. Arch. Mus. Hist. Nat. Paris, Bull. for 1871, 7(4):92 [1872].
TYPE LOCALITY: China, Sichuan, Moupin.
DISTRIBUTION: C Sichuan (China).
COMMENTS: Formerly included *andersoni, gracilis,* and *investigator,* according to Ellerman and Morrision-Scott (1966), but see Hoffmann (1984). Gureev (1979) listed these as distinct species without comment.

ORDER SCANDENTIA

by Don E. Wilson

ORDER SCANDENTIA

SYNONYMS: Tupaioidea.

COMMENTS: Sometimes included in Insectivora; but see McKenna (1975:41).

Family Tupaiidae Gray, 1825. Ann. Philos., n.s., 10:339.

SYNONYMS: Cladobatae, Glisoricina.

COMMENTS: The classification of this family is controversial, but most evidence suggests a coherent natural group; see Campbell (1966, 1974), Dene et al. (1978), Elliott (1971), and Luckett (1980).

Subfamily Tupaiinae Gray, 1825. Ann. Philos., n.s., 10:339.

Anathana Lyon, 1913. Proc. U.S. Natl. Mus., 45:120.

TYPE SPECIES: *Tupaia ellioti* Waterhouse, 1850.

Anathana ellioti (Waterhouse, 1850). Proc. Zool. Soc. Lond., 1849:107 [1850].

TYPE LOCALITY: India, Andhra Pradesh, "hills between Cuddapah and Nellox," (= Velikanda Range).

DISTRIBUTION: India, south of Ganges River.

SYNONYMS: *pallida, wroughtoni.*

COMMENTS: Ellerman and Morrison-Scott (1966) lumped the three species included by Lyon (1913).

Dendrogale Gray, 1848. Proc. Zool. Soc. Lond., 1848:23.

TYPE SPECIES: *Hylogalea murina* Schlegel and Müller, 1843.

Dendrogale melanura (Thomas, 1892). Ann. Mag. Nat. Hist., ser. 6, 9:252.

TYPE LOCALITY: Malaysia, Sarawak, Mt. Dulit, 5,000 ft. (1,524 m).

DISTRIBUTION: Mountains of NE Sarawak and Kinabalu and Trus Madi, Sabah, nowhere below 3,000 ft. (914 m).

SYNONYMS: *baluensis.*

Dendrogale murina (Schlegel and Müller, 1843). *In* Temminck, Verh. Nat. Gesch. Nederland. Overz. Bezitt., Zool., p. 167[1845], pls. 26, 27[1843].

TYPE LOCALITY: Indonesia, Kalimantan Barat Prov., "Pontianak" (Probably erroneous, see Lyon, 1913).

DISTRIBUTION: From E Thailand, Chatraburi and Trat Provinces, through Cambodia to Vietnam.

SYNONYMS: *frenata.*

Tupaia Raffles, 1821. Trans. Linn. Soc. Lond., 13:256.

TYPE SPECIES: *Tupaia ferruginea* Raffles, 1821 (= *Sorex glis* Diard, 1820).

SYNONYMS: *Lyonogale, Tana.*

COMMENTS: This group was last revised by Lyon (1913), and is badly in need of review. The arrangement presented here represents an hypothesis based on Chasen (1940), Dene et al. (1978), Lekagul and McNeely (1977), Luckett (1980), Lyon (1913), Medway (1961), and Napier and Napier (1967).

Tupaia belangeri (Wagner, 1841). Schreber's Die Säugthiere, Suppl., 2:42.

TYPE LOCALITY: Burma, Pegu, near Rangoon, Siriam.

DISTRIBUTION: Malaysia N of 10° N latitude, Thailand, Burma, India, China, Cambodia, Laos, Vietnam and associated coastal islands.

SYNONYMS: *annamensis, assamensis, brunetta, chinensis, clarissa, cambodiana, cochinchinensis, concolor, dissimilis, gaoligongensis, gonshanensis, kohtauensis, laotum, lepcha, modesta, olivacea, peguanus, pingi, siccata, sinus, tenaster, tonquinia, versurae, yaoshanensis, yunalis.*

COMMENTS: This arrangement places all named forms north of the Isthmus of Kra in *belangeri*, the oldest named form from the region. Some may prove to be distinct species, and some names attributed to *glis* may actually prove to be *belangeri*. Immunological (Dene et al., 1978) evidence supports this arrangement.

Tupaia chrysogaster Miller, 1903. Smithson. Misc. Coll., 45:58.

TYPE LOCALITY: North Pagi Island, off southwest coast of Sumatra.

DISTRIBUTION: N and S Pagi, and Sipora of the Mentawei Islands, off the southwest coast of Sumatra.

COMMENTS: May prove to be only a subspecies of *glis*. May also include *siberu* and possibly *tephrura*, both currently in the synonymy of *glis*.

Tupaia dorsalis Schlegel, 1857. Handl. Beoef. Dierk., 1:59, 447, pl. 3.

TYPE LOCALITY: Borneo.

DISTRIBUTION: The mainland of Borneo at low to moderate elevations (except SE).

COMMENTS: United with *tana* in the genus *Tana* by Lyon (1913), but separation into two genera is not supported by immunological evidence; see Dene et al. (1978).

Tupaia glis (Diard, 1820). Asiat. J. Mon. Reg., 10:478.

TYPE LOCALITY: Malaysia, Penang Isl.

DISTRIBUTION: SE Asia from below 10° N on the Isthmus of Kra (Thailand) through mainland Malaysia and Sumatra (Malaysia) to Java (Indonesia) and various surrounding islands.

SYNONYMS: *anambae, batamana, castanea, chrysomalla, cognata, demissa, discolor, ferruginea, hypochrysa, jacki, lacernata, longicauda, obscura, operosa, pemangilis, penangensis, phaeniura, phaeura, pulonis, raviana, redacta, riabus, siaca, siberu, sordida, tephrura, ultima, umbratilis, wilkinsoni.*

COMMENTS: See Chasen (1940), Dene et al. (1978), Lekagul and McNeely (1977), Lyon (1913), Medway (1961), Napier and Napier (1967), and comments under *belangeri*.

Tupaia gracilis Thomas, 1893. Ann. Mag. Nat. Hist., ser. 6, 12:53.

TYPE LOCALITY: Malaysia, Sarawak, Baram Dist., Apoh River at base of Mt. Batu Song.

DISTRIBUTION: Borneo (except SE), west to Karimata Isl, Belitung Isl, and Banka Isl; north to Banggi Isl.

SYNONYMS: *edarata, inflata.*

Tupaia javanica Horsfield, 1822. Zool. Res. Java, part 3:*Tupaia Javanica*, pls.(col. and figs. g,h,o,p,q) and 4 unno. pp.

TYPE LOCALITY: Indonesia, Java, Jawa Timur Prov., perhaps near Banjuwangi. See Lyon (1913).

DISTRIBUTION: Indonesian Isls of Nias, Sumatra, Java, and Bali.

SYNONYMS: *balina, bogoriensis, occidentalis, tjibruniensis.*

Tupaia longipes (Thomas, 1893). Ann. Mag. Nat. Hist., ser. 6, 11:343.

TYPE LOCALITY: Sarawak.

DISTRIBUTION: Borneo.

SYNONYMS: *salatana.*

COMMENTS: Frequently considered a subspecies of *glis*, but see Dene et al. (1978).

Tupaia minor Günther, 1876. Proc. Zool. Soc. Lond., 1876:426.

TYPE LOCALITY: Malaysia, Sabah, mainland "opposite the island of Labuan."

DISTRIBUTION: S peninsular Thailand, peninsular Malaysia, Sumatra and Lingga Archipelago (Indonesia), Borneo and offshore islands of Laut (Indonesia), Banggi and Balambangan (Malaysia).

SYNONYMS: *caedis, humeralis, malaccana, sincipis.*

Tupaia montana Thomas, 1892. Ann. Mag. Nat. Hist., ser. 6, 9:252.

TYPE LOCALITY: Malaysia, Sarawak, Mt. Dulit, 5,000 ft. (1,524 m).

DISTRIBUTION: Mountains of Sarawak and W Sabah; recorded from 1,200 to 10,400 ft. (366-3,170 m) on Mt. Kinabalu.

SYNONYMS: *baluensis.*

Tupaia nicobarica (Zelebor, 1869). Reise Oesterr. Fregatte Novara Zool. 1(Wirbelthiere), 1(Säugeth.):17, pl. 1.

TYPE LOCALITY: India, Nicobar Isls, Great Nicobar Isl.

DISTRIBUTION: Great and Little Nicobar Isls.
SYNONYMS: *surda.*

Tupaia palawanensis Thomas, 1894. Ann. Mag. Nat. Hist., ser. 6, 9:251.
TYPE LOCALITY: Philippines, Palawan.
DISTRIBUTION: Palawan, Busuanga, Cuyo, and Culion (Philippines).
SYNONYMS: *busuangae, cuyonis, moellendorffi.*

Tupaia picta Thomas, 1892. Ann. Mag. Nat. Hist., ser. 6, 9:251.
TYPE LOCALITY: Malaysia, Sarawak, Baram Dist., Apoh. See Lyon (1913).
DISTRIBUTION: N Sarawak and East Kalimantan (Borneo).
SYNONYMS: *fuscior.*

Tupaia splendidula Gray, 1865. Proc. Zool. Soc. Lond., 1865:322, pl. 12.
TYPE LOCALITY: Borneo.
DISTRIBUTION: S Borneo; Bunguran and Laut (N Natuna Isls) and Karimata (Indonesia).
SYNONYMS: *carimatae, lucida, muelleri, natunae, ruficaudata.*
COMMENTS: See Medway (1961).

Tupaia tana Raffles, 1821. Trans. Linn. Soc. Lond., 13:257.
TYPE LOCALITY: Indonesia, Sumatra, Bencoolen (= Bengkulu).
DISTRIBUTION: Indonesia: Sumatra, Tuanku, Batu group, Lingga group, Banga, Belitung, Tambelon, and Serasan groups, Banggi; Borneo.
SYNONYMS: *banguei, besara, bunoae, cervicalis, chrysura, griswoldi, kelabit, kretami, lingae, masae, nainggolani, nitida, paitana, speciosus, sirhassenensis, tuancus, utara.*
COMMENTS: See Dene et al. (1978) for comment on generic status of *Tana* Lyon (= *Lyonogale* Conisbee).

Urogale Mearns, 1905. Proc. U.S. Natl. Mus., 28:435.
TYPE SPECIES: *Urogale cylindrura* Mearns, 1905 (= *Tupaia everetti* Thomas, 1892).

Urogale everetti (Thomas, 1892). Ann. Mag. Nat. Hist., ser. 6, 9:250.
TYPE LOCALITY: Philippines, Mindanao, Zamboanga.
DISTRIBUTION: Mindanao, Dinigat, and Siargao (Philippines).
SYNONYMS: *cylindrura.*

Subfamily Ptilocercinae Lyon, 1913. Proc. U. S. Natl. Mus., 45:4.
COMMENTS: See Campbell (1974).

Ptilocercus Gray, 1848. Proc. Zool. Soc. Lond., 1848:24 [publ. 1 Aug. 1848].
TYPE SPECIES: *Ptilocercus lowii* Gray, 1848.
COMMENTS: McAllan and Bruce (1989) argued that the original publication should be: The Literary Gazette, 1624:167 [publ. 4 March 1848].

Ptilocercus lowii Gray, 1848. Proc. Zool. Soc. Lond., 1848:24.
TYPE LOCALITY: Malaysia, Sarawak, "caught in the Rajah's house", *i.e.* Kuching.
DISTRIBUTION: S peninsular Thailand; peninsular Malaysia; Sumatra, Riau Isls, Batu Isls, Banka, and Serasan Isl (Indonesia); Borneo; Labuan Isl (Malaysia).
SYNONYMS: *continentis.*

ORDER DERMOPTERA

by Don E. Wilson

ORDER DERMOPTERA

Family Cynocephalidae Simpson, 1945. Bull. Am. Mus. Nat. Hist., 85:54.
SYNONYMS: Colugidae, Galeopithecidae, Galeopteridae.

Cynocephalus Boddaert, 1768. Dierk. Meng., 2:8.
TYPE SPECIES: *Lemur volans* Linnaeus, 1758.
SYNONYMS: *Colugo, Dermopterus, Galeolemur, Galeopithecus, Galeopterus, Galeopus, Pleuropterus.*
COMMENTS: See Ellerman and Morrison-Scott (1966).

Cynocephalus variegatus (Audebert, 1799). Hist. Nat. Singes Makis, sig. Rr. Java.
TYPE LOCALITY: Indonesia, Java.
DISTRIBUTION: Indochina to Java (Indonesia) and Borneo.
SYNONYMS: *abbotti, aoris, borneanus, chombolis, gracilis, hantu, lautensis, lechei, natunae, peninsulae, pumilis, saturatus, taylori, tellonis, ternatensis, terutaus, tuancus, varius.*

Cynocephalus volans (Linnaeus, 1758). Syst. Nat., 10th ed., 1:30.
TYPE LOCALITY: Philippine Isls, S Luzon, Pampanga Prov.
DISTRIBUTION: Philippine Isls: Dinagat, Mindanao, Basilan, Samar, Siargao, Leyte, and Bohol.
SYNONYMS: *marmoratus, philippinensis, rufus, temminckii, undatus.*

ORDER CHIROPTERA
by Karl F. Koopman

ORDER CHIROPTERA

COMMENTS: Includes as suborders Megachiroptera, with the single family Pteropodidae, and Microchiroptera, with the remainder.

Family Pteropodidae Gray, 1821. London Med. Repos., 15:299.

Subfamily Pteropodinae Gray, 1821. London Med. Repos., 15:299.
SYNONYMS: Harpyionycterinae, Nyctimeninae.
COMMENTS: Corbet and Hill (1980) and Hill and Smith (1984) recognized the subfamilies Harpyionycterinae and Nyctimeninae, which are included here.

Acerodon Jourdan, 1837. L'Echo du Monde Savant, 4, No. 275, p. 156.
TYPE SPECIES: *Pteropus jubatus* Eschscholtz, 1831.
COMMENTS: Very closely related to and probably congeneric with *Pteropus* (see Musser et al., 1982*a*).

Acerodon celebensis (Peters, 1867). Monatsb. K. Preuss. Akad. Wiss. Berlin, 1867:333.
TYPE LOCALITY: Indonesia, Sulawesi.
DISTRIBUTION: Sulawesi, Saleyer Isl, and Sula Mangoli (Indonesia).
STATUS: CITES - Appendix II.
SYNONYMS: *arquatus*.
COMMENTS: Includes *arquatus* and Sulawesi specimens formerly in *Pteropus argentatus* (see Musser et al., 1982*a*); also see *P. argentatus*.

Acerodon humilis K. Andersen, 1909. Ann. Mag. Nat. Hist., ser. 7, 3:24-25.
TYPE LOCALITY: Indonesia, Talaud Isls, Lirong.
DISTRIBUTION: Talaud Isls (Indonesia).
STATUS: CITES - Appendix II.

Acerodon jubatus (Eschscholtz, 1831). Zool. Atlas, Part 4:1.
TYPE LOCALITY: Philippines, Luzon, Manila.
DISTRIBUTION: Philippines.
STATUS: CITES - Appendix II.
SYNONYMS: *aurinuchalis, mindanensis, pyrrhocephalus*.

Acerodon leucotis (Sanborn, 1950). Proc. Biol. Soc. Washington, 63:189.
TYPE LOCALITY: Philippines, Calamian Isls, Busuanga Isl, Singay.
DISTRIBUTION: Balabac, Palawan, Busuanga Isl (Philippines).
STATUS: CITES - Appendix II.
SYNONYMS: *obscurus*.
COMMENTS: Formerly included in *Pteropus* (see Musser et al., 1982*a*).

Acerodon lucifer Elliot, 1896. Field Columb. Mus. Publ., Zool. Ser., 1:78.
TYPE LOCALITY: Philippines, Panay, Concepcion.
DISTRIBUTION: Philippines (known only from the type locality).
STATUS: CITES - Appendix II; IUCN - Extinct; probably extinct.

Acerodon mackloti (Temminck, 1837). Monogr. Mamm., 2:69.
TYPE LOCALITY: Indonesia, Timor.
DISTRIBUTION: Lombok, Sumbawa, Flores, Alor Isl, Sumba, and Timor (Indonesia).
STATUS: CITES - Appendix II.
SYNONYMS: *alorensis, floresianus, floresii, gilvus, ochraphaeus, prajae*.

Aethalops Thomas, 1923. Proc. Zool. Soc. Lond., 1923:178.
TYPE SPECIES: *Aethalodes alecto* Thomas, 1923.
SYNONYMS: *Aethalodes*.
COMMENTS: Replacement name for *Aethalodes* (preoccupied).

Aethalops alecto (Thomas, 1923). Ann. Mag. Nat. Hist., ser. 9, 11:251.
TYPE LOCALITY: Indonesia, Sumatra, Indrapura Peak, 7,300 ft. (2,225 m).
DISTRIBUTION: W Malaysia, Borneo, Sumatra, Java, and Lombok.
SYNONYMS: *aequalis, ocypete.*
COMMENTS: Includes *aequalis* (see Hill, 1961*a*).

Alionycteris Kock, 1969. Senckenberg. Biol., 50:319.
TYPE SPECIES: *Alionycteris paucidentata* Kock, 1969.

Alionycteris paucidentata Kock, 1969. Senckenberg. Biol., 50:322.
TYPE LOCALITY: Philippines, Mindanao, Bukidion Prov., Mt. Katanglad.
DISTRIBUTION: Mindanao (Philippines).
STATUS: IUCN - Insufficiently known.

Aproteles Menzies, 1977. Aust. J. Zool., 25:330.
TYPE SPECIES: *Aproteles bulmerae* Menzies, 1977.

Aproteles bulmerae Menzies, 1977. Aust. J. Zool., 25:331.
TYPE LOCALITY: Papua New Guinea, Chimbu Prov., 2 km SE Chuave Govt. Sta., 1,530 m.
DISTRIBUTION: New Guinea.
STATUS: U.S. ESA - Endangered; IUCN - Insufficiently known.
COMMENTS: Originally described from fossil material, but since found living (Hyndman and Menzies, 1980).

Balionycteris Matschie, 1899. Flederm. Berliner Mus. Naturk., p. 80.
TYPE SPECIES: *Cynopterus maculatus* Thomas, 1893.

Balionycteris maculata (Thomas, 1893). Ann. Mag. Nat. Hist., ser. 6, 11:341.
TYPE LOCALITY: Malaysia, Sarawak.
DISTRIBUTION: Thailand; W Malaysia; Borneo; Durian and Galang Isls (Riau Arch., Indonesia).
SYNONYMS: *seimundi.*

Boneia Jentink, 1879. Notes Leyden Mus., 1:117.
TYPE SPECIES: *Boneia bidens* Jentinck, 1879.
COMMENTS: Recognition as a separate genus follows Andersen (1912:55-60), but see Bergmans and Rozendaal (1988), who included it in *Rousettus*.

Boneia bidens Jentink, 1879. Notes Leyden Mus., 1:117.
TYPE LOCALITY: Indonesia, N Sulawesi, Bone (near Gerontalo).
DISTRIBUTION: N Sulawesi (Indonesia).
SYNONYMS: *menadensis.*

Casinycteris Thomas, 1910. Ann. Mag. Nat. Hist., ser. 8, 6:111.
TYPE SPECIES: *Casinycteris argynnis* Thomas, 1910.
COMMENTS: Revised by Bergmans (1990).

Casinycteris argynnis Thomas, 1910. Ann. Mag. Nat. Hist., ser. 8, 6:111.
TYPE LOCALITY: Cameroon, Ja River, Bitye.
DISTRIBUTION: Cameroon to E Zaire.

Chironax K. Andersen, 1912. Cat. Chiroptera Brit. Mus., 2nd ed., p. 658.
TYPE SPECIES: *Cynopterus melanocephalus* Temminck, 1825.

Chironax melanocephalus (Temminck, 1825). Monogr. Mamm., 1:190.
TYPE LOCALITY: Indonesia, W Java, Bantam.
DISTRIBUTION: Thailand, W Malaysia, Borneo, Sumatra, Java, Nias Isl, and Sulawesi.
SYNONYMS: *tumulus.*

Cynopterus F. Cuvier, 1824. Dentes des Mammiferes, p. 248.
TYPE SPECIES: *Pteropus marginatus* E. Geoffroy, 1810 (= *Vespertilio sphinx* Vahl, 1797).

Cynopterus brachyotis (Müller, 1838). Tijdschr. Nat. Gesch. Physiol., 5:146.
TYPE LOCALITY: Borneo, Dewei River.
DISTRIBUTION: Sri Lanka, India, SE Asia, Malaysia, Philippines, Nicobar and Andaman Isls, Borneo, Sumatra, Sulawesi, and Talaud Isls and adjacent small islands.
SYNONYMS: *altitudinis, andamanensis, archipelagus, brachysoma, ceylonensis, concolor, hoffeti, insularum, javanicus, luzoniensis; minor* Revilliod; *minutus.*
COMMENTS: Does not include *angulatus*, which was transferred to *sphinx* (see Hill and Thonglongya, 1972). Includes *minor*, which was listed as distinct by Corbet and Hill (1980:41); see also Hill (1983). Includes *archipelagus* (see Heaney et al., 1987). Kitchener and Moharadatunkamsi (1991) treated *luzoniensis* and *minutus* as separate species, but I follow Hill (1983).

Cynopterus horsfieldi Gray, 1843. List Specimens Mamm. Coll. Brit. Mus., p. 38.
TYPE LOCALITY: Indonesia, Java.
DISTRIBUTION: Thailand, W Malaysia, Borneo, Java, Sumatra, Lesser Sunda Isls, and adjacent small islands.
SYNONYMS: *harpax, lyoni; minor* Lyon; *persimilis, princeps.*
COMMENTS: Includes *harpax* (see Hill, 1961*a*).

Cynopterus nusatenggara Kitchener and Maharadatunkamsi, 1991. Rec. W. Aust. Mus., 15:312.
TYPE LOCALITY: Indonesia, Lesser Sunda Isls, Sumbawa, Jerewah, Desa Belo.
DISTRIBUTION: Lombok, Sumbawi, Flores, Sumba, and adjacent small islands.

Cynopterus sphinx (Vahl, 1797). Skr. Nat. Selsk. Copenhagen, 4(1):123.
TYPE LOCALITY: India, Madras, Tranquebar.
DISTRIBUTION: Sri Lanka, India, S China, SE Asia, W Malaysia, Sumatra, adjacent small islands; perhaps Borneo.
SYNONYMS: *angulatus, babi, gangeticus, pagensis, scherzeri, serasani.*
COMMENTS: Includes *angulatus* (see Hill and Thonglongya, 1972). Does not include *titthaecheilus* (see Hill, 1983).

Cynopterus titthaecheilus (Temminck, 1825). Monogr. Mamm. 1:198.
TYPE LOCALITY: Indonesia, Java, Bogor.
DISTRIBUTION: Sumatra, Java, Lombok, Timor, and adjacent small islands.
SYNONYMS: *major, terminus.*
COMMENTS: Formerly included in *C. sphinx*, but see Hill (1983).

Dobsonia Palmer, 1898. Proc. Biol. Soc. Washington, 12:114.
TYPE SPECIES: *Cephalotes peroni* E. Geoffroy, 1810.
COMMENTS: Reviewed by Jong and Bergmans (1981).

Dobsonia beauforti Bergmans, 1975. Beaufortia, 23(295):3.
TYPE LOCALITY: Indonesia, Irian Jaya, So Rong Div., Waigeo Isl, Njanjef.
DISTRIBUTION: Waigeo Isl (off west end of New Guinea).
COMMENTS: Closely related to *viridis* (see Bergmans, 1975).

Dobsonia chapmani Rabor, 1952. Nat. Hist. Misc., Chicago Acad. Sci., 96:2.
TYPE LOCALITY: Philippines, Negros, Bais, Pagabonin.
DISTRIBUTION: Philippines.
STATUS: IUCN - Extinct?; probably extinct.
COMMENTS: See Bergmans (1978) for taxonomic status.

Dobsonia emersa Bergmans and Sarbini, 1985. Beaufortia, 34:185.
TYPE LOCALITY: New Guinea, Irian Jaya, Biak, Sorido.
DISTRIBUTION: Biak and Owii Isls (in Geelvink Bay).

Dobsonia exoleta K. Andersen, 1909. Ann. Mag. Nat. Hist., ser. 8, 4:533.
TYPE LOCALITY: Indonesia, Sulawesi, Minahassa, Tomohon.
DISTRIBUTION: Sulawesi and adjacent small islands.

Dobsonia inermis K. Andersen, 1909. Ann. Mag. Nat. Hist., ser. 8, 4:532.
TYPE LOCALITY: Solomon Isls, San Cristobal Isl.
DISTRIBUTION: Solomon Isls.

SYNONYMS: *minimus, nesea.*

Dobsonia minor (Dobson, 1879). Proc. Zool. Soc. Lond., 1878:875 [1879].
TYPE LOCALITY: Indonesia, Irian Jaya, Manokwari Div., Amberbaki.
DISTRIBUTION: C and W New Guinea and adjacent small islands; Sulawesi.
STATUS: IUCN - Insufficiently known.

Dobsonia moluccensis (Quoy and Gaimard, 1830). *In* d'Urville, Voy...de Astrolabe, Zool., 1(L'Homme, Mamm. Oiseaux):86.
TYPE LOCALITY: Indonesia, Molucca Isls, Amboina Isl.
DISTRIBUTION: Bismarck Arch.; New Guinea; Aru Isls, Batanta, and Mysol (off W New Guinea) and Molucca Isls (including Waigeo Isl); N Queensland (Australia).
SYNONYMS: *anderseni, magna.*
COMMENTS: Does not include *pannietensis* (see Bergmans, 1979). Includes *anderseni* and *magna*; see Koopman (1979:6), but see also Bergmans and Sarbini (1985).

Dobsonia pannietensis (De Vis, 1905). Ann. Queensl. Mus., 6:36.
TYPE LOCALITY: Papua New Guinea, Louisiade Arch., Panniet Isl.
DISTRIBUTION: Louisiade Arch., D'Entrecasteaux Isls, and Trobriand Isls (Papua New Guinea).
SYNONYMS: *remota.*
COMMENTS: Considered a subspecies of *moluccensis* by Laurie and Hill (1954:42), but as a separate species by Bergmans (1979). Includes *remota* (see Koopman, 1982:6). A record of *remota* from Bougainville Isl is based on a misidentified *inermis* (see Bergmans, 1979).

Dobsonia peroni (E. Geoffroy, 1810). Ann. Mus. Hist. Nat. Paris, 15:104.
TYPE LOCALITY: Indonesia, Lesser Sunda Isls, Timor.
DISTRIBUTION: Sumba, Timor, Flores, Sumbawa, Nusa Penida (near Bali), Komodo, Alor, Wetar, Babar (Indonesia).
SYNONYMS: *grandis, sumbanus.*
COMMENTS: Reviewed by Bergmans (1978).

Dobsonia praedatrix K. Andersen, 1909. Ann. Mag. Nat. Hist., ser. 8, 4:532.
TYPE LOCALITY: Papua New Guinea, Bismarck Arch., "Duke of York group".
DISTRIBUTION: Bismarck Arch.

Dobsonia viridis (Heude, 1896). Mem. Hist. Nat. Emp. Chin., 3:176.
TYPE LOCALITY: Indonesia, Molucca Isls, Kei Isls.
DISTRIBUTION: Sulawesi, Moluccas, and nearby islands.
SYNONYMS: *crenulata, umbrosa.*
COMMENTS: Does not include *chapmani* (see Bergmans, 1978:12). Includes *crenulata* (see Hill, 1991:171); but see Bergmans and Rozendaal (1988:33).

Dyacopterus K. Andersen, 1912. Cat. Chiroptera Brit. Mus., 1:651.
TYPE SPECIES: *Cynopterus spadiceus* Thomas, 1890.

Dyacopterus spadiceus (Thomas, 1890). Ann. Mag. Nat. Hist., ser. 6, 5:235.
TYPE LOCALITY: Malaysia, Sarawak, Baram.
DISTRIBUTION: Sumatra; Borneo; W Malaysia; Luzon and Mindanao (Philippines).
STATUS: IUCN - Insufficiently known.
SYNONYMS: *brooksi.*
COMMENTS: Includes *brooksi*; see Hill (1961*a*) and Kock (1969*b*); but also see Peterson (1969).

Eidolon Rafinesque, 1815. Analyse de la Nature, p. 54.
TYPE SPECIES: *Vespertilio vampyrus helvus* Kerr, 1792.
COMMENTS: Revised by Bergmans (1990).

Eidolon dupreanum (Schlegel, 1867). Proc. Zool. Soc. Lond., 1866:419 [1877].
TYPE LOCALITY: Madagascar, Nossi Bé.
DISTRIBUTION: Madagascar.
COMMENTS: See comments under *E. helvum.*

Eidolon helvum (Kerr, 1792). *In* Linnaeus, Anim. Kingdom, 1(1):xvii, 91.
TYPE LOCALITY: Senegal.
DISTRIBUTION: Senegal to Ethiopia to South Africa; SW Arabia; islands in Guinea Gulf and off E Africa.
SYNONYMS: *buettikoferi, leucomelas, mollipilosus, paleaceus, palmarum, sabaeum, stramineus.*
COMMENTS: Includes *sabaeum*; see Hayman and Hill (1971), who also included *dupreanum*; but also see Bergmans (1990) who retained *dupreanum* as a separate species. See DeFrees and Wilson (1988, Mammalian Species, 312), who also included *dupreanum*.

Epomophorus Bennett, 1836. Proc. Zool. Soc. Lond., 1835:149 [1836].
TYPE SPECIES: *Pteropus gambianus* Ogilby, 1835.
COMMENTS: Revised by Bergmans (1988), who transferred *Micropteropus grandis* to this genus. A key to this genus is presented in Boulay and Robbins (1989).

Epomophorus angolensis Gray, 1870. Cat. Monkeys, Lemurs, Fruit-eating Bats Brit. Mus., p. 125.
TYPE LOCALITY: Angola, Benguella.
DISTRIBUTION: W Angola, NW Namibia.

Epomophorus gambianus (Ogilby, 1835). Proc. Zool. Soc. Lond., 1835:100.
TYPE LOCALITY: Gambia.
DISTRIBUTION: Senegal to W Ethiopia; S Tanzania to Angola and South Africa.
SYNONYMS: *crypturus, epomophorus, guineensis, macrocephalus, megacephalus, parvus, pousarguesi, reii, whitei, zechi.*
COMMENTS: See Boulay and Robbins (1989, Mammalian Species, 344).

Epomophorus grandis (Sanborn, 1950). Publ. Cult. Comp. Diamantes Angola, 10:55.
TYPE LOCALITY: Angola, Lunda, Dundo.
DISTRIBUTION: N Angola, S Congo Republic.
COMMENTS: Transferred from *Micropteropus* to *Epomophorus* by Bergmans (1988).

Epomophorus labiatus (Temminck, 1837). Monogr. Mamm., 2:83.
TYPE LOCALITY: Sudan, Blue Nile Prov., Sennar.
DISTRIBUTION: Nigeria to Ethiopia and south to Congo Republic and Malawi. Senegal records are probably erroneous (see Bergmans, 1988).
SYNONYMS: *anurus, doriae, minor, schoensis.*
COMMENTS: Includes *anurus* and *minor*; see Claessen and de Vree (1991), but see also Bergmans (1988).

Epomophorus minimus Claessen and de Vree, 1991. Senckenberg. Biol., 71:216.
TYPE LOCALITY: Ethiopia, Shewa, Bahadv.
DISTRIBUTION: Ethiopia to Uganda and Tanzania.
COMMENTS: Included in *E. minor* by Bergmans (1988).

Epomophorus wahlbergi (Sundevall, 1846). Ofv. K. Svenska Vet.-Akad. Forhandl. Stockholm, 3(4):118.
TYPE LOCALITY: South Africa, Natal, near Durban.
DISTRIBUTION: Cameroon to Somalia, south to Angola and South Africa (Liberian record probably erroneous); Pemba and Zanzibar Isls.
SYNONYMS: *haldemani, neumanni, stuhlmanni, zenkeri.*
COMMENTS: Reviewed by Acharya (1992, Mammalian Species, 394).

Epomops Gray, 1870. Cat. Monkeys, Lemurs, Fruit-eating Bats Brit. Mus., p. 126.
TYPE SPECIES: *Epomophorus franqueti* Tomes, 1860.
COMMENTS: Reviewed by Bergmans (1989).

Epomops buettikoferi (Matschie, 1899). Megachiroptera Berlin Mus., p. 45.
TYPE LOCALITY: Liberia, Junk River, Schlieffelinsville.
DISTRIBUTION: Guinea to Nigeria.

Epomops dobsoni (Bocage, 1889). J. Sci. Math. Phys. Nat. Lisboa, ser. 2, 1:1.
TYPE LOCALITY: Angola, Benguela, Quindumbo.
DISTRIBUTION: Angola to Rwanda, Tanzania, Malawi, and N Botswana.

Epomops franqueti (Tomes, 1860). Proc. Zool. Soc. Lond., 1860:54.
TYPE LOCALITY: Gabon.
DISTRIBUTION: Ivory Coast to Sudan, Uganda, NW Tanzania, N Zambia, and Angola.
SYNONYMS: *comptus, strepitans.*

Haplonycteris Lawrence, 1939. Bull. Mus. Comp. Zool., 86:31.
TYPE SPECIES: *Haplonycteris fischeri* Lawrence, 1939.

Haplonycteris fischeri Lawrence, 1939. Bull. Mus. Comp. Zool., 86:33
TYPE LOCALITY: Philippines, Mindoro, Mt. Halcyon.
DISTRIBUTION: Philippines.

Harpyionycteris Thomas, 1896. Ann. Mag. Nat. Hist., ser. 6, 18:243.
TYPE SPECIES: *Harpionycteris whiteheadi* Thomas, 1896.

Harpyionycteris celebensis Miller and Hollister, 1921. Proc. Biol. Soc. Washington, 34:99.
TYPE LOCALITY: Indonesia, Sulawesi, middle Sulawesi, Gimpoe.
DISTRIBUTION: Sulawesi.
COMMENTS: Considered a subspecies of *whiteheadi* by Laurie and Hill (1954), but as a separate species by Peterson and Fenton (1970).

Harpyionycteris whiteheadi Thomas, 1896. Ann. Mag. Nat. Hist., ser. 6, 18:244.
TYPE LOCALITY: Philippines, Mindoro Isl, 5000 ft. (1524 m).
DISTRIBUTION: Philippines.
SYNONYMS: *negrosensis.*

Hypsignathus H. Allen, 1861. Proc. Acad. Nat. Sci. Philadelphia, p. 156.
TYPE SPECIES: *Hypsignathus monstrosus* H. Allen, 1861.
COMMENTS: Revised by Bergmans (1989).

Hypsignathus monstrosus H. Allen, 1861. Proc. Acad. Nat. Sci. Philadelphia, p. 157.
TYPE LOCALITY: Gabon.
DISTRIBUTION: Sierra Leone to W Kenya, south to Zambia and Angola; Bioko. Records from Gambia and Ethiopia are doubtful.
SYNONYMS: *labrosus.*
COMMENTS: See Langevin and Barclay (1990, Mammalian Species, 357).

Latidens Thonglongya, 1972. J. Bombay Nat. Hist. Soc., 69:151.
TYPE SPECIES: *Latidens salimalii* Thonglongya, 1972.

Latidens salimalii Thonglongya, 1972. J. Bombay Nat. Hist. Soc., 69:153.
TYPE LOCALITY: India, Tamil Nadu, Madurai Dist., High Wavy Mtns, 2,500 ft. (762 m).
DISTRIBUTION: S India.
STATUS: IUCN - Insufficiently known.

Megaerops Peters, 1865. Monatsb. K. Preuss. Akad. Wiss. Berlin, 1865:256.
TYPE SPECIES: *Pachysoma ecaudata* Temminck, 1837.

Megaerops ecaudatus (Temminck, 1837). Monogr. Mamm., 2:94.
TYPE LOCALITY: Indonesia, Sumatra, Padang.
DISTRIBUTION: Borneo, Sumatra, W Malaysia, Thailand, perhaps Vietnam.

Megaerops kusnotoi Hill and Boedi, 1978. Mammalia, 42:427.
TYPE LOCALITY: Indonesia, Java, Sukabumi, Lengkong, Hanjuang Ciletuh, 700 m.
DISTRIBUTION: Java.
STATUS: IUCN - Insufficiently known.

Megaerops niphanae Yenbutra and Felten, 1983. Senckenberg. Biol., 64:2.
TYPE LOCALITY: Thailand, Nakhon Ratchasima, Amphoe Pak Thong Chai.
DISTRIBUTION: Thailand, Vietnam, NE India.

Megaerops wetmorei Taylor, 1934. Monogr. Bur. Sci. Manila, p. 191.
TYPE LOCALITY: Philippines, Mindanao, Cotabato, Tatayan.
DISTRIBUTION: Philippines, Borneo, W Malaysia.

SYNONYMS: *albicollis*.

Micropteropus Matschie, 1899. Megachiroptera Berlin Mus., p. 36, 57.
TYPE SPECIES: *Epomophorus pusillus* Peters, 1867.
COMMENTS: Revised by Bergmans (1989), who transferred *grandis* from this genus to *Epomophorus*.

Micropteropus intermedius Hayman, 1963. Publ. Cult. Comp. Diamantes Angola, 66:100.
TYPE LOCALITY: Angola, Lunda, Dundo.
DISTRIBUTION: N Angola, SE Zaire.

Micropteropus pusillus (Peters, 1867). Monatsb. K. Preuss. Akad. Wiss. Berlin, 1867:870.
TYPE LOCALITY: Nigeria, Yoruba (see Bergmans, 1989).
DISTRIBUTION: Gambia to Ethiopia, south to Angola, Zambia, Burundi, and Tanzania.

Myonycteris Matschie, 1899. Megachiroptera Berlin Mus., p. 61, 63.
TYPE SPECIES: *Cynonycteris torquata* Dobson, 1878.
COMMENTS: Revised by Bergmans (1976).

Myonycteris brachycephala (Bocage, 1889). J. Sci. Math. Phys. Nat. Lisboa, ser. 2, 1:198.
TYPE LOCALITY: São Tomé and Principe, São Tomé Isl.
DISTRIBUTION: São Tomé Isl (Gulf of Guinea).
COMMENTS: Subgenus *Phrygetis*.

Myonycteris relicta Bergmans, 1980. Zool. Meded. Rijks. Mus. Nat. Hist. Leiden, 14:126.
TYPE LOCALITY: Kenya, Coast Prov., Shimba Hills, Lukore area, Mukanda River.
DISTRIBUTION: Shimba Hills, Kenya; Nguru and Usambara Mtns, Tanzania.
COMMENTS: Subgenus *Myonycteris*.

Myonycteris torquata (Dobson, 1878). Cat. Chiroptera Brit. Mus., p. 71, 76.
TYPE LOCALITY: N Angola.
DISTRIBUTION: Sierra Leone to Uganda, south to Angola and Zambia; Bioko.
SYNONYMS: *leptodon*, *wroughtoni*.
COMMENTS: Subgenus *Myonycteris*. Includes *leptodon* and *wroughtoni*; see Hayman and Hill (1971).

Nanonycteris Matschie, 1899. Megachiroptera Berlin Mus., p. 36, 58.
TYPE SPECIES: *Epomophorus veldkampi* Jentinck, 1888.
COMMENTS: Revised by Bergmans (1989).

Nanonycteris veldkampi (Jentink, 1888). Notes Leyden Mus., 10:51.
TYPE LOCALITY: Liberia, Fisherman Lake, Buluma.
DISTRIBUTION: Guinea to Central African Republic.

Neopteryx Hayman, 1946. Ann. Mag. Nat. Hist., ser. 11, 12:569.
TYPE SPECIES: *Neopteryx frosti* Heyman, 1946.

Neopteryx frosti Hayman, 1946. Ann. Mag. Nat. Hist., ser. 11, 12:571.
TYPE LOCALITY: Indonesia, W Sulawesi, Tamalanti, 3,300 ft. (1,006 m).
DISTRIBUTION: W and N Sulawesi.

Nyctimene Borkhausen, 1797. Deutsche Fauna, 1:86.
TYPE SPECIES: *Vespertilio cephalotes* Pallas, 1767.

Nyctimene aello (Thomas, 1900). Ann. Mag. Nat. Hist., ser. 7, 5:216.
TYPE LOCALITY: Papua New Guinea, Milne Bay Prov., Milne Bay.
DISTRIBUTION: New Guinea.
COMMENTS: Does not include *celaeno*; see Hill (1983).

Nyctimene albiventer (Gray, 1863). Proc. Zool. Soc. Lond., 1862:262 [1863].
TYPE LOCALITY: Indonesia, Molucca Isls, Morotai Isl.
DISTRIBUTION: New Guinea, Molucca and Kei Isls, Solomon Isls, N Queensland (Australia), Bismarck Arch.
SYNONYMS: *papuanus*.

COMMENTS: Formerly included *draconilla*; see Hill (1983:132). Does not include *bougainville*, which Smith and Hood (1983) placed in *vizcaccia*. Includes *papuanus*, but see Peterson (1991).

Nyctimene celaeno Thomas, 1922. Ann. Mag. Nat. Hist., ser. 9, 9:283.
TYPE LOCALITY: New Guinea, Irian Jaya, Geelvink Bay, Legarer River.
DISTRIBUTION: Known only from the type locality; a Halmahera record is incorrect.
COMMENTS: Formerly included in *N. aello* but see Hill (1983).

Nyctimene cephalotes (Pallas, 1767). Spicil. Zool., 3:10.
TYPE LOCALITY: Indonesia, Molucca Isls, Amboina; see Andersen (1912:707) for discussion.
DISTRIBUTION: Sulawesi, Timor, Molucca Isls, and Numfoor Isl (off N coast New Guinea); extreme S New Guinea.
SYNONYMS: *melinus, pallasi*.
COMMENTS: Does not include *vizcaccia*; see Smith and Hood (1983).

Nyctimene certans K. Andersen, 1912. Ann. Mag. Nat. Hist., ser. 9, 8:95.
TYPE LOCALITY: New Guinea, Irain Jaya, Mount Goliath.
DISTRIBUTION: New Guinea.
COMMENTS: Formerly included in *N. cyclotis* but see Peterson (1991).

Nyctimene cyclotis K. Andersen, 1910. Ann. Mag. Nat. Hist., ser. 7, 6:623.
TYPE LOCALITY: Indonesia, Irian Jaya, Manokwari Div., Arfak Mtns.
DISTRIBUTION: New Guinea; New Britain (Bismarck Arch.).
COMMENTS: Formerly included *certans*, but see Peterson (1991).

Nyctimene draconilla Thomas, 1922. Nova Guinea, 13:725.
TYPE LOCALITY: Indonesia, Irian Jaya, Southern Div., Lorentz River, Bivak Isl.
DISTRIBUTION: New Guinea.
COMMENTS: Considered a subspecies of *albiventer* by Laurie and Hill (1954:46), but also see Hill (1983) and Koopman (1982).

Nyctimene major (Dobson, 1877). Proc. Zool. Soc. Lond., 1877:117.
TYPE LOCALITY: Papua New Guinea, Bismarck Arch., Duke of York Isl.
DISTRIBUTION: D'Entrecasteaux Isls, Trobriand Isls, Bismarck and Louisiade Archs. (Papua New Guinea), Solomon Isls, and small islands off the north coast of New Guinea. A New Guinea mainland record is almost certainly erroneous; see Koopman (1979).
SYNONYMS: *geminus, lullulae, scitulus*.

Nyctimene malaitensis Phillips, 1968. Univ. Kansas Publ. Mus. Nat. Hist., 16:822.
TYPE LOCALITY: Solomon Isls, Malaita Isl.
DISTRIBUTION: Malaita Isl (Solomon Isls).

Nyctimene masalai Smith and Hood, 1983. Occas. Pap. Mus. Texas Tech Univ., 81:1.
TYPE LOCALITY: Bismarcks, New Ireland, Ralum.
DISTRIBUTION: New Ireland (Bismarck Arch.).

Nyctimene minutus K. Andersen, 1910. Ann. Mag. Nat. Hist., ser. 7, 6:622.
TYPE LOCALITY: Indonesia, Sulawesi, Minahassa, Tondano.
DISTRIBUTION: Sulawesi, C Moluccas.
SYNONYMS: *varius*.

Nyctimene rabori Heaney and Peterson, 1984. Occas. Pap. Mus. Zool. Univ. Michigan, 7083.
TYPE LOCALITY: Philippines, Negros, Sibulan, Lake Balinsasayo.
DISTRIBUTION: Known only from Negros Isl.
STATUS: IUCN - Endangered.

Nyctimene robinsoni Thomas, 1904. Ann. Mag. Nat. Hist., ser. 7, 14:196.
TYPE LOCALITY: Australia, Queensland, Cooktown.
DISTRIBUTION: E Queensland (Australia).
SYNONYMS: *tryoni*.

Nyctimene sanctacrucis Troughton, 1931. Proc. Linn. Soc. N.S.W., 56:206.
TYPE LOCALITY: Solomon Isls, Santa Cruz Isls.
DISTRIBUTION: Santa Cruz Isls.

Nyctimene vizcaccia Thomas, 1914. Ann. Mag. Nat. Hist., ser. 8, 13:436.
TYPE LOCALITY: New Guinea, Umboi Isl (off NE coast)
DISTRIBUTION: Umboi Isl, Bismarck Arch., Solomon Isls.
SYNONYMS: *bougainville, minor.*
COMMENTS: Includes *bougainville;* see Smith and Hood (1983). Includes *minor,* but see Peterson (1991). Formerly included in *N. cephalotes,* but see Smith and Hood (1983).

Otopteropus Kock, 1969. Senckenberg. Biol., 50:329.
TYPE SPECIES: *Otopteropus cartilagonodus* Kock, 1969.

Otopteropus cartilagonodus Kock, 1969. Senckenberg. Biol., 50:333.
TYPE LOCALITY: Philippines, Luzon, Mountain Prov., Sitio Pactil.
DISTRIBUTION: Luzon (Philippines).
STATUS: IUCN - Insufficiently known.

Paranyctimene Tate, 1942. Am. Mus. Novit., 1204:1.
TYPE SPECIES: *Paranyctimene raptor* Tate, 1924.

Paranyctimene raptor Tate, 1942. Am. Mus. Novit., 1204:1.
TYPE LOCALITY: Papua New Guinea, Western Prov., Fly River, Oroville Camp. (ca. 4 mi. (6 km) below Elavala River mouth).
DISTRIBUTION: New Guinea.

Penthetor K. Andersen, 1912. Cat. Chiroptera Brit. Mus., p. 665.
TYPE SPECIES: *Cynopterus (Ptenochirus) lucasi* Dobson, 1880.

Penthetor lucasi (Dobson, 1880). Ann. Mag. Nat. Hist., ser. 5, 6:163.
TYPE LOCALITY: Malaysia, Sarawak.
DISTRIBUTION: W Malaysia, Borneo, Riau Arch. (Indonesia).

Plerotes K. Andersen, 1910. Ann. Mag. Nat. Hist., ser. 8, 5:97.
TYPE SPECIES: *Epomophorus anchietae* Seabra, 1900.
COMMENTS: Revised by Bergmans (1989).

Plerotes anchietai (Seabra, 1900). J. Sci. Math. Phys. Nat. Lisboa, ser. 2, 6:116.
TYPE LOCALITY: Angola, Benguela, Galanga.
DISTRIBUTION: Angola, Zambia, S Zaire.

Ptenochirus Peters, 1861. Monatsb. K. Preuss. Akad. Wiss. Berlin, 1861:707.
TYPE SPECIES: *Pachysoma (Ptenochirus) jagori* Peters, 1861.

Ptenochirus jagori (Peters, 1861). Monatsb. K. Preuss. Akad. Wiss. Berlin, 1861:707.
TYPE LOCALITY: Philippines, Luzon, Albay, Daraga.
DISTRIBUTION: Philippines.

Ptenochirus minor Yoshiyuki, 1979. Bull. Natl. Sci. Mus. Tokyo, Ser. A (Zool.), 5:75.
TYPE LOCALITY: Philippines, Mindanao, Davao, Mt. Talomo, Baracatan.
DISTRIBUTION: Philippines.

Pteralopex Thomas, 1888. Ann. Mag. Nat. Hist., ser. 6, 1:155.
TYPE SPECIES: *Pteralopex atrata* Thomas, 1888.
COMMENTS: Reviewed by Hill and Beckon (1978), with a key to the species.

Pteralopex acrodonta Hill and Beckon, 1978. Bull. Brit. Mus. (Nat. Hist.) Zool., 34:68.
TYPE LOCALITY: Fiji Isls, Taveuni Isl, Des Voeux Peak, ca. 3,840 ft. (1,170 m).
DISTRIBUTION: Fiji Isls.

Pteralopex anceps K. Andersen, 1909. Ann. Mag. Nat. Hist., ser. 7, 3:266.
TYPE LOCALITY: Papua New Guinea, Bougainville Isl.
DISTRIBUTION: Buka, Bougainville and Choiseul Isls (Solomon Isls).

Pteralopex atrata Thomas, 1888. Ann. Mag. Nat. Hist., ser. 6, 1:155.
TYPE LOCALITY: Solomon Isls, Guadalcanal, Aola.
DISTRIBUTION: Ysabel and Guadalcanal (Solomon Isls).

COMMENTS: Does not include *anceps*; see Hill and Beckon (1978:67, 68).

Pteralopex pulchra Flannery, 1991. Rec. Aust. Mus., 43:125.
TYPE LOCALITY: Solomon Isls, Guadacanal, Mount Makarakomburu, 1230 m.
DISTRIBUTION: Known only by the holotype.

Pteropus Erxleben, 1777. Syst. RegniAnim., p. 130.
TYPE SPECIES: *Vespertilio vampirus niger* Kerr, 1792.
COMMENTS: Originally named *Pteropus* by Brisson, 1762, *in* Regnum Animale, which is not available (Hopwood, 1947). Formerly included *arquatus* and *leucotis* which were transferred to *Acerodon* by Musser et al. (1982*a*).

Pteropus admiralitatum Thomas, 1894. Ann. Mag. Nat. Hist., ser. 6, 13:293.
TYPE LOCALITY: Papua New Guinea, Bismarck Arch., Admiralty Isls.
DISTRIBUTION: Solomon Isls; Admiralty Isls, New Britain, and Tabar Isls (Bismarck Arch.).
STATUS: CITES - Appendix II.
SYNONYMS: *colonus, goweri, solomonis* (see Laurie and Hill, 1954:33).

Pteropus aldabrensis True, 1893. Proc. U.S. Natl. Mus., 16:533.
TYPE LOCALITY: Seychelles, Aldabra Isl.
DISTRIBUTION: Known only from the type locality.
STATUS: CITES - Appendix II; IUCN - Vulnerable as *P. s. aldabrensis*.
COMMENTS: Included in *P. seychellensis* by Hill (1971*b*), but see Bergmans (1990).

Pteropus alecto Temminck, 1837. Monogr. Mamm., 2:75.
TYPE LOCALITY: Indonesia, Sulawesi, Menado.
DISTRIBUTION: Sulawesi, Saleyer Isl, Lombok, Bawean Isl, Kangean Isls, Sumba Isl, and Savu Isl (Indonesia); N and E Australia; S New Guinea.
STATUS: CITES - Appendix II.
SYNONYMS: *aterrimus, baveanus, gouldi, morio*.
COMMENTS: Includes *gouldi*; see Tate (1942*b*:336, 337).

Pteropus anetianus Gray, 1870. Cat. Monkeys, Lemurs, Fruit-eating Bats Brit. Mus., p. 101.
TYPE LOCALITY: Vanuatu (= New Hebrides), Aneiteum (= Aneityum).
DISTRIBUTION: Vanuatu (= New Hebrides) including Banks Isls.
STATUS: CITES - Appendix II.
SYNONYMS: *aorensis, bakeri, banksiana, eotinus, motalavae, pastoris*.
COMMENTS: Includes *eotinus, bakeri*, and *banksiana*; see Felten and Kock (1972:182-185).

Pteropus argentatus Gray, 1844. Mammalia, *in* Voy. "Sulphur", Zool., 1:30.
TYPE LOCALITY: Indonesia, Moluccas Isls, Amboina Isl (uncertain).
DISTRIBUTION: Perhaps Amboina Isl.
STATUS: CITES - Appendix II.
COMMENTS: Uncertain status; known only by the type. Sulawesi specimens were allocated to *Acerodon celebensis* by Musser et al. (1982*a*).

Pteropus brunneus Dobson, 1878. Cat. Chiroptera Brit. Mus., p. 37.
TYPE LOCALITY: Australia, Queensland, Percy Island.
DISTRIBUTION: Known from the type locality only.
STATUS: CITES - Appendix II.
COMMENTS: See Koopman (1984*c*:2-3) for status.

Pteropus caniceps Gray, 1870. Cat. Monkeys, Lemurs, Fruit-eating Bats Brit. Mus., p. 107.
TYPE LOCALITY: Indonesia, Molucca Isls, Halmahera Isls, Batjan.
DISTRIBUTION: Halmahera Isls, Sulawesi, and Sula Isls (Indonesia). The Sulawesi record is dubius. A Sangihe Isl record is erroneus.
STATUS: CITES - Appendix II.
SYNONYMS: *affinis, batchiana, dobsoni*.
COMMENTS: Includes *dobsoni*; see Laurie and Hill (1954:34).

Pteropus chrysoproctus Temminck, 1837. Monogr. Mamm., 2:67.
TYPE LOCALITY: Indonesia, Molucca Isls, Amboina.
DISTRIBUTION: Amboina, Buru, Seram, and small islands east of Seram (Indonesia). A Sangihe Isl record is erroneous; see Bergmans and Rozendaal (1988:65).

STATUS: CITES - Appendix II.

Pteropus conspicillatus Gould, 1850. Proc. Zool. Soc. Lond., 1849:109 [1850].
TYPE LOCALITY: Australia, Queensland, Fitzroy Isl.
DISTRIBUTION: Halmahera Isls; New Guinea and adjacent islands; NE Queensland (Australia).
STATUS: CITES - Appendix II.
SYNONYMS: *chrysauchen, mysolensis.*

Pteropus dasymallus Temminck, 1825. Monogr. Mamm., 1:180.
TYPE LOCALITY: Japan, Ryukyu Isls, Kuchinoerabu Isl.
DISTRIBUTION: Taiwan; Ryukyu Isls, Daito Isls and extreme S Kyushu (Japan).
STATUS: CITES - Appendix II.
SYNONYMS: *daitoensis, formosus, inopinatus, yamagatai, yayeyamae.*
COMMENTS: Includes *daitoensis*; see Kuroda (1933:314).

Pteropus faunulus Miller, 1902. Proc. U.S. Natl. Mus., 24:785.
TYPE LOCALITY: India, Nicobar Isls, Car Nicobar Isl.
DISTRIBUTION: Nicobar Isls (Bay of Bengal, India).
STATUS: CITES - Appendix II.

Pteropus fundatus Felten and Kock, 1972. Senckenberg. Biol., 53:186.
TYPE LOCALITY: Vanuatu (= New Hebrides), Banks Isls, Mota Isl.
DISTRIBUTION: Banks Isls (N Vanuatu).
STATUS: CITES - Appendix II.

Pteropus giganteus (Brünnich, 1782). Dyrenes Historie, 1:45.
TYPE LOCALITY: India, Bengal.
DISTRIBUTION: Maldive Isls, India (incl. Andaman Isls), Sri Lanka, Pakistan, Burma, Tsinghai (China). The Tsinghai record requires confirmation.
STATUS: CITES - Appendix II.
SYNONYMS: *ariel, assamensis, kelaarti, leucocephalus, medius.*
COMMENTS: Includes *ariel*; see Hill (1958:5, 6).

Pteropus gilliardi Van Deusen, 1969. Am. Mus. Novit., 2371:5.
TYPE LOCALITY: Papua New Guinea, Bismarck Arch., New Britain, Whiteman Mtns, Wild Dog Ridge, ca. 1,600 m.
DISTRIBUTION: New Britain Isl (Bismarck Arch.).
STATUS: CITES - Appendix II.

Pteropus griseus (E. Geoffroy, 1810). Ann. Mus. Hist. Nat. Paris, 15:94.
TYPE LOCALITY: Indonesia, Lesser Sunda Isls, Timor.
DISTRIBUTION: Timor, Samao Isl, Dyampea Isl, Bonerato Isl, Saleyer Isl, Sulawesi, and Banda Isls (Indonesia); perhaps S Luzon (Philippines).
STATUS: CITES - Appendix II.
SYNONYMS: *mimus, pallidus.*
COMMENTS: Includes *mimus*; see Laurie and Hill (1954:33).

Pteropus howensis Troughton, 1931. Proc. Linn. Soc. N.S.W., 56:204.
TYPE LOCALITY: Solomon Isls, Ontong Java Isl.
DISTRIBUTION: Ontong Java Isl (Solomon Isls).
STATUS: CITES - Appendix II.

Pteropus hypomelanus Temminck, 1853. Esquisses Zool. sur la Côte de Guine, p. 61.
TYPE LOCALITY: Indonesia, Molucca Isls, Ternate Isl.
DISTRIBUTION: Maldive Isls; New Guinea through Indonesia to Vietnam and Thailand, and adjacent islands; Solomon Isls; Philippines.
STATUS: CITES - Appendix II.
SYNONYMS: *annectens, cagayanus, canus, condorensis, enganus, fretensis, geminorum, lepidus, luteus, macassaricus, maris, robinsoni, simalurus, tomesi, tricolor, vulcanius.*
COMMENTS: Formerly included *brunneus*; see Ride (1970:180); but see Koopman (1984*c*).

Pteropus insularis Hombron and Jacquinot, 1842. *In* d'Urville, Voy. Pole Sud. Mammifères, p. 24.
TYPE LOCALITY: Caroline Isls, Truk Isl, Hogoleu (Pac. Isls Trust Terr., USA).

DISTRIBUTION: Truk Isls (C Caroline Isls).
STATUS: CITES - Appendix I; IUCN - Endangered.
SYNONYMS: *laniger.*

Pteropus leucopterus Temminck, 1853. Esquisses Zool. sur la Côte de Guine, p. 60.
TYPE LOCALITY: Philippines.
DISTRIBUTION: Philippines (known only from Luzon and Dinagat).
STATUS: CITES - Appendix II.
SYNONYMS: *chinensis.*

Pteropus livingstonii Gray, 1866. Proc. Zool. Soc. Lond., 1866:66.
TYPE LOCALITY: Comoro Isls, Anjouan Isl.
DISTRIBUTION: Comoro Isls.
STATUS: CITES - Appendix II; IUCN - Endangered.
COMMENTS: See Bergmans (1990).

Pteropus lombocensis Dobson, 1878. Cat. Chiroptera Brit. Mus., p. 34.
TYPE LOCALITY: Indonesia, Lesser Sunda Isls, Lombok Isl.
DISTRIBUTION: Alor Isl (near Timor), Lombok, and Flores (Indonesia).
STATUS: CITES - Appendix II.
SYNONYMS: *heudei, solitarius.*

Pteropus lylei K. Andersen, 1908. Ann. Mag. Nat. Hist., ser. 8, 2:367.
TYPE LOCALITY: Thailand, Bangkok.
DISTRIBUTION: Thailand, Vietnam.
STATUS: CITES - Appendix II.

Pteropus macrotis Peters, 1867. Monatsb. K. Preuss. Akad. Wiss. Berlin, 1867:327.
TYPE LOCALITY: Indonesia, Aru Isls, Wokam Isl.
DISTRIBUTION: New Guinea; Aru Isls (Indonesia).
STATUS: CITES - Appendix II; IUCN - Indeterminate.
SYNONYMS: *epularius, insignis.*

Pteropus mahaganus Sanborn, 1931. Field Mus. Nat. Hist. Publ., Zool. Ser., 2:19.
TYPE LOCALITY: Solomon Isls, Ysabel Isl, Tunnibul.
DISTRIBUTION: Bougainville and Ysabel Isls (Solomon Isls).
STATUS: CITES - Appendix II.

Pteropus mariannus Desmarest, 1822. Mammalogie, *in* Encycl. Méth., 2(Suppl.):547.
TYPE LOCALITY: Marianna Isls, Guam (USA).
DISTRIBUTION: Mariana Isls and Caroline Isls (including Palau Isls), Ryukyu Isls (Japan).
STATUS: CITES - Appendix I; U.S. ESA - Endangered in Guam as *P. m. mariannus;* IUCN - Endangered.
SYNONYMS: *keraudren, loochooensis, paganensis, pelewensis, ualanus, ulthiensis, vanikorensis, yapensis.*
COMMENTS: Includes *loochooensis, pelewensis, ualanus,* and *yapensis;* see Kuroda (1938). Corbet and Hill (1980:37-38) listed these forms as distinct species without comment; but see Yoshiyuki (1989), who treated *loochooensis* as a separate species. Probably includes *vanikorensis,* see Troughton (1930).

Pteropus mearnsi Hollister, 1913. Proc. Biol. Soc. Washington, 26:112.
TYPE LOCALITY: Philippines, Basilan Isl, Isabella.
DISTRIBUTION: Mindanao and Basilan Isl (Philippines).
STATUS: CITES - Appendix II.
COMMENTS: Probably a synonym of *P. speciosus;* see Heaney et al. (1987).

Pteropus melanopogon Peters, 1867. Monatsb. K. Preuss. Akad. Wiss. Berlin, 1867:330.
TYPE LOCALITY: Indonesia, Molucca Isls, Amboina.
DISTRIBUTION: Aru Isls, Kei Isls, Amboina, Buru, Seram, Banda Isls, Timor Laut, and adjacent islands (Indonesia). A Sangihe Isl record is erroneous; see Bergmans and Rozendaal (1988:65).
STATUS: CITES - Appendix II.
SYNONYMS: *aruensis, chrysargyrus, fumigatus, keyensis, rubiginosus.*

COMMENTS: Does not include *sepikensis*; see Koopman (1979:5). See comment under *neohibernicus*.

Pteropus melanotus Blyth, 1863. Cat. Mamm. Mus. Asiat. Soc. Calcutta, p. 20.
TYPE LOCALITY: India, Nicobar Isls.
DISTRIBUTION: Nicobar and Andaman Isls (India); Engano Isl and Nias Isl (Indonesia); Christmas Isl.
STATUS: CITES - Appendix II.
SYNONYMS: *modiglianii, natalis, niadicus, nicobaricus, satyrus, tytleri.*
COMMENTS: Includes *satyrus*; see Hill (1971*c*:6, 7).

Pteropus molossinus Temminck, 1853. Esquisses Zool. sur la Côte de Guine, p. 62.
TYPE LOCALITY: Caroline Isls, Ponape (Pac. Isls Trust Terr., USA).
DISTRIBUTION: Mortlock and Ponape Isls (Caroline Isls).
STATUS: CITES - Appendix I; IUCN - Endangered.
SYNONYMS: *breviceps.*

Pteropus neohibernicus Peters, 1876. Monatsb. K. Preuss. Akad. Wiss. Berlin, 1876:317.
TYPE LOCALITY: Papua New Guinea, Bismarck Arch., New Ireland Isl.
DISTRIBUTION: Bismarck Arch. and Admiralty Isls (Papua New Guinea), New Guinea, Mysol and Ghebi Isl.
STATUS: CITES - Appendix II.
SYNONYMS: *coronatus, degener, hilli, papuanus, sepikensis.*
COMMENTS: Includes *sepikensis*; see Koopman (1979:5).

Pteropus niger (Kerr, 1792). *In* Linnaeus, Anim. Kingdom, 1:90.
TYPE LOCALITY: Mascarene Isls, Reunion Isl (France).
DISTRIBUTION: Mascarene Isls (Reunion Isl, Mauritius Isl, subfossil on Rodrigues Isl).
STATUS: CITES - Appendix II; IUCN - Vulnerable. Extinct on Reunion Isl, see Cheke and Dahl (1981).
SYNONYMS: *fuscus, mauritianus, vulgaris.*
COMMENTS: See Bergmans (1990).

Pteropus nitendiensis Sanborn, 1930. Am. Mus. Novit., 435:2.
TYPE LOCALITY: Solomon Isls, Santa Cruz Isls, Ndeni Isl.
DISTRIBUTION: Ndeni Isl (Santa Cruz Isls).
STATUS: CITES - Appendix II.

Pteropus ocularis Peters, 1867. Monatsb. K. Preuss. Akad. Wiss. Berlin, 1867:326.
TYPE LOCALITY: Indonesia, Molucca Isls, Seram Isl.
DISTRIBUTION: Seram and Buru (Indonesia).
STATUS: CITES - Appendix II.
SYNONYMS: *ceramensis.*

Pteropus ornatus Gray, 1870. Cat. Monkeys, Lemurs, Fruit-eating Bats Brit. Mus., p. 105.
TYPE LOCALITY: New Caledonia, Noumea (France).
DISTRIBUTION: Loyalty and New Caledonia Isls.
STATUS: CITES - Appendix II.
SYNONYMS: *auratus.*
COMMENTS: Includes *auratus*; see Felten (1964*b*).

Pteropus personatus Temminck, 1825. Monogr. Mamm., 1:189.
TYPE LOCALITY: Indonesia, Molucca Isls, Ternate.
DISTRIBUTION: Halmahera Isls (Indonesia). The Sulawesi record is erroneous; see Bergmans and Rozendaal (1988:65).
STATUS: CITES - Appendix II.

Pteropus phaeocephalus Thomas, 1882. Proc. Zool. Soc. Lond., 1881:756 [1882].
TYPE LOCALITY: Caroline Isls, Mortlock Isl (Pac. Isls Trust Terr., USA).
DISTRIBUTION: Mortlock Isl (C Caroline Isls).
STATUS: CITES - Appendix I; IUCN - Endangered.
COMMENTS: Probably a subspecies of *P. insularis.*

Pteropus pilosus K. Andersen, 1908. Ann. Mag. Nat. Hist., ser. 8, 2:369.
TYPE LOCALITY: Caroline Isls, Palau Isls (Pac. Isls Trust Terr., USA).

DISTRIBUTION: Palau Isls (Caroline Isls).
STATUS: CITES - Appendix I; IUCN - Extinct; probably extinct.

Pteropus pohlei Stein, 1933. Z. Säugetierk., 8:93.
TYPE LOCALITY: Indonesia, Irian Jaya, Tjenderawasih Div., Japen Isl.
DISTRIBUTION: Japen Isl (off NW New Guinea).
STATUS: CITES - Appendix II.

Pteropus poliocephalus Temminck, 1825. Monogr. Mamm., 1:179.
TYPE LOCALITY: Australia.
DISTRIBUTION: E Australia, from S Queensland to Victoria.
STATUS: CITES - Appendix II.

Pteropus pselaphon Lay, 1829. Zool. J., 4:457.
TYPE LOCALITY: Japan, Bonin Isls.
DISTRIBUTION: Bonin and Volcano Isls (Japan).
STATUS: CITES - Appendix II.
SYNONYMS: *ursinus*.

Pteropus pumilus Miller, 1911. Proc. U.S. Natl. Mus., 38:394.
TYPE LOCALITY: Indonesia, Miangas Isl between Talaud Isls and Mindanao.
DISTRIBUTION: Philippines.
STATUS: CITES - Appendix II.
SYNONYMS: *balutus, tablasi*.
COMMENTS: Includes *balutus* and *tablasi*; see Klingener and Creighton (1984).

Pteropus rayneri Gray, 1870. Cat. Monkeys, Lemurs, Fruit-eating Bats Brit. Mus., p. 108.
TYPE LOCALITY: Solomon Isls, Guadalcanal Isl.
DISTRIBUTION: Solomon Isls.
STATUS: CITES - Appendix II.
SYNONYMS: *cognatus, grandis, lavellanus, monoensis, rennelli, rubianus*.
COMMENTS: Includes *cognatus*; see Hill (1962*a*). Corbet and Hill (1980:36) listed *cognatus* as a distinct species without comment.

Pteropus rodricensis Dobson, 1878. Cat. Chiroptera Brit. Mus., p. 36.
TYPE LOCALITY: Mascarene Isls, Rodrigues.
DISTRIBUTION: Rodrigues Isl, Round Isl near Mauritus Isl (Mascarene Isls.).
STATUS: CITES - Appendix II; U.S. ESA - Endangered; IUCN - Endangered; extinct on Round Isl.
SYNONYMS: *mascarinus*.
COMMENTS: See Bergmans (1990).

Pteropus rufus Tiedemann, 1808. Zool., v. 1, Allgemeine Zool., Mensch Säugthiere, Landshut, p. 535.
TYPE LOCALITY: Madagascar.
DISTRIBUTION: Madagascar.
STATUS: CITES - Appendix II.
SYNONYMS: *edwardsi, phaiops, princeps*.
COMMENTS: See Bergmans (1990). Commonly cited from "E. Geoffroy, 1803. Cat. Mamm. Mus. Nat. Hist. Nat. Paris, p. 47.", but this work was never published, see Appendix I.

Pteropus samoensis Peale, 1848. Mammalia *in* Repts. U.S. Expl. Surv., 8:20.
TYPE LOCALITY: Samoan Isls, Tutuila Isl (American Samoa).
DISTRIBUTION: Fiji Isls, Samoan Isls.
STATUS: CITES - Appendix I; IUCN - Endangered.
SYNONYMS: *nawaiensis, vitiensis, whitmeei*.
COMMENTS: Includes *nawaiensis*; see Hill and Beckon (1978:65).

Pteropus sanctacrucis Troughton, 1930. Rec. Aust. Mus., 18:3.
TYPE LOCALITY: Solomon Isls, Santa Cruz Isls, Ndeni Isl.
DISTRIBUTION: Santa Cruz Isls.
STATUS: CITES - Appendix II.

Pteropus scapulatus Peters, 1862. Monatsb. K. Preuss. Akad. Wiss. Berlin, 1862:574.
TYPE LOCALITY: Australia, Queensland, Cape York.

DISTRIBUTION: Australia, S New Guinea, accidental on New Zealand.
STATUS: CITES - Appendix II.
SYNONYMS: *elseyi*.

Pteropus seychellensis Milne-Edwards, 1877. Bull. Sci. Soc. Philom. Paris, ser. 7, 2:221.
TYPE LOCALITY: Seychelle Isls, Mahe Isl.
DISTRIBUTION: Seychelle Isls, Aldabra Isl, Comoro Isls, Mafia Isl (off Tanzania).
STATUS: CITES - Appendix II.
SYNONYMS: *comorensis*.
COMMENTS: Includes *comorensis*; see Hill (1971*b*:574), who also included *aldabrensis*; but see Bergmans (1990) who kept *aldabrensis* as a separate species.

Pteropus speciosus K. Andersen, 1908. Ann. Mag. Nat. Hist., ser. 8, 2:364.
TYPE LOCALITY: Philippines, Malanipa Isl (off west end of Mindanao).
DISTRIBUTION: Sulu Arch., Basilan and Mindanao (Philippines); Solombo Besar and Mata Siri (Java Sea); Talaut Isls.
STATUS: CITES - Appendix II.

Pteropus subniger (Kerr, 1792). *In* Linnaeus, Anim. Kingdom, 1:91.
TYPE LOCALITY: Mascarene Isls, Reunion Isl (France).
DISTRIBUTION: Reunion and Mauritius Isls (Mascarene Isls).
STATUS: CITES - Appendix II; IUCN - Extinct; probably extinct, see Cheke and Dahl (1981).
SYNONYMS: *collaris, ruber, rubidum, rubricollis, torquatus*.
COMMENTS: See Bergmans (1990).

Pteropus temmincki Peters, 1867. Monatsb. K. Preuss. Akad. Wiss. Berlin, 1867:331.
TYPE LOCALITY: Indonesia, Molucca Isls, Amboina Isl; see Andersen (1912:318) for clarification.
DISTRIBUTION: Buru, Amboina, Seram (Indonesia); Bismarck Arch. (Papua New Guinea); nearby small islands; perhaps Timor Isl (Indonesia).
STATUS: CITES - Appendix II.
SYNONYMS: *capistratus, liops, petersi*.

Pteropus tokudae Tate, 1934. Am. Mus. Novit., 713:1.
TYPE LOCALITY: Mariana Isls, Guam (USA).
DISTRIBUTION: Known only from Guam.
STATUS: CITES - Appendix II; U.S. ESA - Endangered; IUCN - Extinct; probably extinct.

Pteropus tonganus Quoy and Gaimard, 1830. *In* d'Urville, Voy...de Astrolabe, Zool., 1(L'Homme, Mamm., Oiseaux):74.
TYPE LOCALITY: Tonga Isls, Tongatapu Isl.
DISTRIBUTION: Karkar Isl (off NE New Guinea) and Rennell Isl (Solomon Isls), south to New Caledonia, east to Cook Isls.
STATUS: CITES - Appendix I; IUCN - Indeterminate.
SYNONYMS: *basiliscus, flavicollis, geddiei, heffernani*.
COMMENTS: Includes *geddiei*; see Sanborn (1931:14) and Felten and Kock (1972:180-181).

Pteropus tuberculatus Peters, 1869. Monatsb. K. Preuss. Akad. Wiss. Berlin, 1869:393.
TYPE LOCALITY: Solomon Isls, Santa Cruz Isls, Vanikoro Isl.
DISTRIBUTION: Vanikoro Isl (Santa Cruz Isls).
STATUS: CITES - Appendix II.
COMMENTS: Reviewed by Troughton (1927).

Pteropus vampyrus (Linnaeus, 1758). Syst. Nat., 10th ed., 1:31.
TYPE LOCALITY: Indonesia, Java.
DISTRIBUTION: Indochina, Malay Peninsula, Borneo, Philippines, Sumatra, Java, and Lesser Sunda Isls, adjacent small islands.
STATUS: CITES - Appendix II.
SYNONYMS: *celaeno, edulis, funereus, intermedius, javanicus, kopangi, lanensis, malaccensis, natunae, pluton, pteronotus, sumatrensis*.
COMMENTS: Includes *intermedius*; see Lekagul and McNeely (1977:77). Corbet and Hill (1980:37) listed *intermedius* as a distinct species without comment.

Pteropus vetulus Jouan, 1863. Mem. Soc. Imp. Sci. Nat. Cherbourg, 9:90.
TYPE LOCALITY: New Caledonia (France).
DISTRIBUTION: New Caledonia.
STATUS: CITES - Appendix II.
SYNONYMS: *germaini, macmillani.*
COMMENTS: Includes *macmillani;* see Felten (1964*b*:671).

Pteropus voeltzkowi Matschie, 1909. Sitzb. Ges. Naturf. Fr. Berlin, p. 486.
TYPE LOCALITY: Tanzania, Pemba Isl, Fufuni.
DISTRIBUTION: Pemba Isl (off coast of Tanzania).
STATUS: CITES - Appendix II; IUCN - Endangered.
COMMENTS: See Bergmans (1990).

Pteropus woodfordi Thomas, 1888. Ann. Mag. Nat. Hist., ser. 6, 1:156.
TYPE LOCALITY: Solomon Isls, Guadalcanal Isl, Aola.
DISTRIBUTION: Fauro Isl to Guadalcanal Isl (Solomon Isls).
STATUS: CITES - Appendix II.
SYNONYMS: *austini.*

Rousettus Gray, 1821. London Med. Repos., 15:299.
TYPE SPECIES: *Pteropus aegyptiacus* E. Geoffroy, 1810.
SYNONYMS: *Lissonycteris, Stenonycteris.*
COMMENTS: Includes *Lissonycteris;* see Koopman (1975:361-362). Corbet and Hill (1980:36) listed *Lissonycteris* as a distinct genus without comment.

Rousettus amplexicaudatus (E. Geoffroy, 1810). Ann. Mus. Hist. Nat. Paris, 15:96.
TYPE LOCALITY: Indonesia, Lesser Sunda Isls, Timor Isl.
DISTRIBUTION: Cambodia; Thailand; Malay Peninsula through Indonesia to New Guinea, Bismarck Arch., and Solomon Isls; Philippines.
SYNONYMS: *bocagei, brachyotis, hedigeri, infumatus, minor, philippinensis, stresemanni.*
COMMENTS: Subgenus *Rousettus.* Revised by Rookmaaker and Bergmans (1981).

Rousettus angolensis (Bocage, 1898). J. Sci. Math. Phys. Nat. Lisboa, ser. 2, 5:133.
TYPE LOCALITY: Angola, Quibula, Cahata, Pungo Andongo.
DISTRIBUTION: Senegal and Angola to Ethiopia and Mozambique; Bioko.
SYNONYMS: *crypticola, ruwenzorii, smithi.*
COMMENTS: Subgenus *Lissonycteris.* Formerly included in genus *Lissonycteris;* see Koopman (1975:361-362).

Rousettus celebensis K. Andersen, 1907. Ann. Mag. Nat. Hist., ser. 7, 19:503, 509.
TYPE LOCALITY: Indonesia, Sulawesi, Mt. Masarang, 3,500 ft. (1,067 m).
DISTRIBUTION: Sulawesi and Sangihe Isls (Indonesia).
COMMENTS: Subgenus *Rousettus.*

Rousettus egyptiacus (E. Geoffroy, 1810). Ann. Mus. Hist. Nat. Paris, 15:96.
TYPE LOCALITY: Egypt, Giza.
DISTRIBUTION: Senegal to Egypt, Cyprus, and Turkey, south to South Africa; Pakistan to Yemen; adjacent small islands.
SYNONYMS: *arabicus, geoffroyi, hottentotus, leachi, occidentalis, sjostedti, unicolor.*
COMMENTS: Subgenus *Rousettus.* Includes *leachi* and *arabicus;* see Hayman and Hill (1971:11) and Corbet (1978*c*:38).

Rousettus lanosus Thomas, 1906. Ann. Mag. Nat. Hist., ser. 7, 18:137.
TYPE LOCALITY: Uganda, Ruwenzori East, Mubuku Valley, 13,000 ft. (3,962 m).
DISTRIBUTION: Uganda, Kenya, Tanzania, S Ethiopia, E Zaire, S Sudan.
SYNONYMS: *kempi.*
COMMENTS: Subgenus *Stenonycteris.*

Rousettus leschenaulti (Desmarest, 1820). Mammalogie, *in* Encyclop. Méthod., 1:110.
TYPE LOCALITY: India, Pondicherry.
DISTRIBUTION: Sri Lanka; Pakistan to Vietnam and S China; Sumatra, Java, Bali, and Mentawai Isls (Indonesia).
SYNONYMS: *affinis, fuliginosa, fusca, infuscata, pyrivorus, seminudus, shortridgei.*
COMMENTS: Subgenus *Rousettus.* Includes *seminudus;* see Sinha (1970:82).

Rousettus madagascariensis G. Grandidier, 1928. Bull. Acad. Malgache, 11:91.
TYPE LOCALITY: Madagascar, Beforona (between Tananarive and Andevoranto).
DISTRIBUTION: Madagascar.
COMMENTS: Subgenus *Rousettus*. Considered a subspecies of *lanosus* by Hayman and Hill (1971:12); but see Bergmans (1977).

Rousettus obliviosus Kock, 1978. Proc. 4th Int. Bat Res. Conf. Nairobi, p. 208.
TYPE LOCALITY: Comoro Isls, Grand Comoro, near Boboni, 640m.
DISTRIBUTION: Comoro Isls.
COMMENTS: Subgenus *Rousettus*.

Rousettus spinalatus Bergmans and Hill, 1980. Bull. Brit. Mus. (Nat. Hist.) Zool., 38:95.
TYPE LOCALITY: Indonesia, N Sumatra, near Madan or near Prapat.
DISTRIBUTION: Sumatra, Borneo.
COMMENTS: Subgenus *Rousettus*.

Scotonycteris Matschie, 1894. Sitzb. Ges. Naturf. Fr. Berlin, p. 200.
TYPE SPECIES: *Scotonycteris zenkeri* Matschie, 1894.
COMMENTS: Reviewed by Bergmans (1990).

Scotonycteris ophiodon Pohle, 1943. Sitzb. Ges. Naturf. Fr. Berlin, p. 76.
TYPE LOCALITY: Cameroon, Bipindi.
DISTRIBUTION: Liberia to Congo Republic.
SYNONYMS: *cansdalei*.

Scotonycteris zenkeri Matschie, 1894. Sitzb. Ges. Naturf. Fr. Berlin, p. 202.
TYPE LOCALITY: Cameroon, Yaunde.
DISTRIBUTION: Liberia to Congo Republic and E Zaire; Bioko.
SYNONYMS: *occidentalis*.

Sphaerias Miller, 1906. Proc. Biol. Soc. Washington, 19:83.
TYPE SPECIES: *Cynopterus blanfordi* Thomas, 1891.

Sphaerias blanfordi (Thomas, 1891). Ann. Mus. Civ. Stor. Nat. Genova, ser. 2, 10:884, 921, 922.
TYPE LOCALITY: Burma, Karin Hills.
DISTRIBUTION: N India, Bhutan, Tibet, Burma, N Thailand, SW China.

Styloctenium Matschie, 1899. Megachiroptera Berlin Mus., p. 33.
TYPE SPECIES: *Pteropus wallacei* Gray, 1866.

Styloctenium wallacei (Gray, 1866). Proc. Zool. Soc. Lond., 1866:65.
TYPE LOCALITY: Indonesia, Sulawesi, Macassar.
DISTRIBUTION: Sulawesi.

Thoopterus Matschie, 1899. Megachiroptera Berlin Mus., p. 72, 73, 77.
TYPE SPECIES: *Cynopterus marginatus* var. *nigrescens* Gray, 1870.

Thoopterus nigrescens (Gray, 1870). Cat. Monkeys, Lemurs, Fruit-eating Bats Brit. Mus., p. 123.
TYPE LOCALITY: Indonesia, Molucca Isls, Morotai.
DISTRIBUTION: Molucca Isls, Sulawesi and Sangihe Isls (Indonesia).
SYNONYMS: *latidens*.

Subfamily Macroglossinae Gray, 1866. Proc. Zool. Soc. Lond., 1866:64.

Eonycteris Dobson, 1873. Proc. Asiat. Soc. Bengal, p. 148.
TYPE SPECIES: *Macroglossus spelaeus* Dobson, 1871.

Eonycteris major K. Andersen, 1910. Ann. Mag. Nat. Hist., ser. 8, 6:625.
TYPE LOCALITY: Malaysia, Sarawak, Mt. Dulit.
DISTRIBUTION: Borneo, Philippines, perhaps Mentawai Isls (Indonesia).
SYNONYMS: *longicauda*, *robusta*.

COMMENTS: Includes *robusta* and *longicauda*; see Tate (1942b:344); but see Heaney et al. (1987).

Eonycteris spelaea (Dobson, 1871). Proc. Asiat. Soc. Bengal, p. 105, 106.
TYPE LOCALITY: Burma, Tenasserim, Moulmein.
DISTRIBUTION: N India, Burma, S China, Thailand, W Malaysia, Borneo; Sumatra, Java, Sumba, Timor and Sulawesi (Indonesia); Philippines; Andaman Isls (India).
SYNONYMS: *glandifera, rosenbergi.*
COMMENTS: Includes *rosenbergi*; see Bergmans and Rosendaal (1988:57-61).

Macroglossus F. Cuvier, 1824. Dentes des Mammiferes, p. 248.
TYPE SPECIES: *Pteropus minimus* E Geoffroy, 1810.

Macroglossus minimus (E. Geoffroy, 1810). Ann. Mus. Hist. Nat. Paris, 15:97.
TYPE LOCALITY: Indonesia, Java.
DISTRIBUTION: Thailand to Philippines, New Guinea, Bismarck Arch. (Papua New Guinea), Solomon Isls, and N Australia.
SYNONYMS: *fructivorus, horsfieldi, kiodotes, lagochilus, meyeri, microtus, nanus, pygmaeus, rostratus.*
COMMENTS: Includes *lagochilus*; see Hill (1983:134-137). Includes *fructivorus*; see Heaney and Rabor (1982).

Macroglossus sobrinus K. Andersen, 1911. Ann. Mag. Nat. Hist., ser. 8, 3:642.
TYPE LOCALITY: Malaysia, Perak, Gunong lgari, 2,000 ft. (610 m).
DISTRIBUTION: SE Asia, Sumatra and Java, adjacent small islands.
SYNONYMS: *fraternus.*
COMMENTS: This is the species previously called *minimus*; see Hill (1983).

Megaloglossus Pagenstecher, 1885. Zool. Anz., 8:245.
TYPE SPECIES: *Megaloglossus woermanni* Pagenstecher, 1885.

Megaloglossus woermanni Pagenstecher, 1885. Zool. Anz., 8:245.
TYPE LOCALITY: Gabon, Sibange farm.
DISTRIBUTION: Liberia to Uganda, S Zaire, and N Angola; Bioko.
SYNONYMS: *prigoginii.*
COMMENTS: Reviewed by Bergmans and Van Bree (1972).

Melonycteris Dobson, 1877. Proc. Zool. Soc. Lond., 1877:119.
TYPE SPECIES: *Melonycteris melanops* Dobson, 1877.
SYNONYMS: *Nesonycteris.*
COMMENTS: Includes *Nesonycteris*; see Phillips (1968:814).

Melonycteris aurantius Phillips, 1966. J. Mammal., 47:24.
TYPE LOCALITY: Solomon Isls, Florida Isl, Haleta, 10 m.
DISTRIBUTION: Florida and Choiseul Isls (Solomon Isls).
COMMENTS: Subgenus *Nesonycteris.*

Melonycteris melanops Dobson, 1877. Proc. Zool. Soc. Lond., 1877:119.
TYPE LOCALITY: Given by Anderson (1912:790) as "New Ireland, coast adjacent to Duke of York Isl." (Papua New Guinea, Bismarck Arch.).
DISTRIBUTION: Bismarck Arch.; a New Guinea record is highly questionable.
SYNONYMS: *alboscapulatus.*

Melonycteris woodfordi (Thomas, 1887). Ann. Mag. Nat. Hist., ser. 5, 19:147.
TYPE LOCALITY: Solomon Isls, Alu Isl (near Shortland Isl).
DISTRIBUTION: Solomon Isls.
COMMENTS: Subgenus *Nesonycteris.*

Notopteris Gray, 1859. Proc. Zool. Soc. Lond., 1859:36.
TYPE SPECIES: *Notopteris macdonaldi* Gray, 1859.

Notopteris macdonaldi Gray, 1859. Proc. Zool. Soc. Lond., 1859:38.
TYPE LOCALITY: Fiji Isls., Viti Levu.
DISTRIBUTION: Vanuatu (= New Hebrides), New Caledonia, Fiji Isls, Caroline Isls.

SYNONYMS: *neocaledonica.*
COMMENTS: Includes *neocaledonica;* see Sanborn (1950:329, 330).

Syconycteris Matschie, 1899. Megachiroptera Berlin Mus., p. 94, 95, 98.
TYPE SPECIES: *Macroglossus minimus* var. *australis* Peters, 1867.
COMMENTS: Reviewed by Ziegler (1982*a*).

Syconycteris australis (Peters, 1867). Monatsb. K. Preuss. Akad. Wiss. Berlin, 1867:13.
TYPE LOCALITY: Australia, Queensland, Rockhampton.
DISTRIBUTION: E Queensland and New South Wales (Australia); New Guinea; D'Entrecasteaux Isls, Trobriand Isls, Louisiade Arch., and Bismarck Arch. (Papua New Guinea); Molucca Isls; various adjacent small islands.
SYNONYMS: *crassa, finschi, keyensis, major, naias, papuana.*
COMMENTS: Includes *naias* and *crassa;* see Lidicker and Ziegler (1968:34) and Koopman (1982:8-10).

Syconycteris carolinae Rozendaal, 1984. Zoologische Mededelingen, 58(13):200.
TYPE LOCALITY: Indonesia, Moluccas, Halmahera Isl, Gunung Gamkunora.
DISTRIBUTION: Halmahera Isl (Moluccas).

Syconycteris hobbit Ziegler, 1982. Occas. Pap. Bernice P. Bishop Mus., 25(5):1-22.
TYPE LOCALITY: Papua New Guinea, Morobe Prov., Mt. Kaindi.
DISTRIBUTION: Mountains of C New Guinea.

Family Rhinopomatidae Bonaparte, 1838. Syn. Vert. Syst., *in* Nuovi Ann. Sci. Nat., Bologna, 2:111.

Rhinopoma E. Geoffroy, 1818. Descrip. de L'Egypte, 2:113.
TYPE SPECIES: *Vespertilio microphyllus* Brünnich, 1782.
COMMENTS: Revised by Hill (1977*b*).

Rhinopoma hardwickei Gray, 1831. Zool. Misc., 1:37.
TYPE LOCALITY: India.
DISTRIBUTION: Burma to Morocco, south to Mauritania, Nigeria, and Kenya; Socotra Isl (Yemen).
SYNONYMS: *arabium, cystops, macinnesi.*
COMMENTS: See Qumsiyeh and Jones (1986, Mammalian Species, 263).

Rhinopoma microphyllum (Brünnich, 1782). Dyrenes Historie, 1:50.
TYPE LOCALITY: Egypt, Giza.
DISTRIBUTION: Morocco and Senegal to Thailand; Sumatra.
SYNONYMS: *asirensis, cordofanicum, harrisoni, kinneari, lepsianum, sumatrae, tropicalis.*

Rhinopoma muscatellum Thomas, 1903. Ann. Mag. Nat. Hist., ser. 7, 11:498.
TYPE LOCALITY: Oman, Muscat, Wadi Bani Ruha.
DISTRIBUTION: Oman, W Iran, S Afghanistan, perhaps Ethiopia.
SYNONYMS: *pusillum, seianum.*
COMMENTS: The Ethiopian record may be misidentified *R. hardwickei macinnesi.* See Qumsiyeh and Jones (1986, Mammalian Species, 263).

Family Craseonycteridae Hill, 1974. Bull. Brit. Mus. (Nat. Hist.) Zool., 27:303.

Craseonycteris Hill, 1974. Bull. Brit. Mus. (Nat. Hist.) Zool., 27:304.
TYPE SPECIES: *Craseonycteris thonglongyai* Hill, 1974.

Craseonycteris thonglongyai Hill, 1974. Bull. Brit. Mus. (Nat. Hist.) Zool., 27:305.
TYPE LOCALITY: Thailand, Kanchanaburi, Ban Sai Yoke.
DISTRIBUTION: Thailand: known only from a limited area near the type locality.
STATUS: U.S. ESA - Endangered; IUCN - Rare.
COMMENTS: See Hill and Smith (1981, Mammalian Species, 160).

Family Emballonuridae Gervais, 1856. *In* F. Comte de Castelnau, Exped. Partes Cen. Am. Sud., Zool.(Sec. 7), Vol. 1, pt. 2(Mammifères), p. 62 footnote.

Balantiopteryx Peters, 1867. Monatsb. K. Preuss. Akad. Wiss. Berlin, 1867:476.
TYPE SPECIES: *Balantiopteryx plicata* Peters, 1867.
COMMENTS: Revised by Hill (1987). A key for this genus was presented in Arroyo-Cabrales and Jones (1988*a*).

Balantiopteryx infusca (Thomas, 1897). Ann. Mag. Nat. Hist., ser. 6, 20:546.
TYPE LOCALITY: Ecuador, Esmeraldas, Cachabi.
DISTRIBUTION: W Ecuador.
COMMENTS: See Hill (1987) for information on this species. See Arroyo-Cabrales and Jones (1988*b*, Mammalian Species, 313).

Balantiopteryx io Thomas, 1904. Ann. Mag. Nat. Hist., ser. 7, 13:252.
TYPE LOCALITY: Guatemala, Alta Verapaz, Rio Dolores (near Coban).
DISTRIBUTION: S Veracruz and Oaxaca (Mexico) to EC Guatemala and Belize.
COMMENTS: See Arroyo-Cabrales and Jones (1988*b*, Mammalian Species, 313).

Balantiopteryx plicata Peters, 1867. Monatsb. K. Preuss. Akad. Wiss. Berlin, 1867:476.
TYPE LOCALITY: Costa Rica, Puntarenas.
DISTRIBUTION: Costa Rica to C Sonora and S Baja California (Mexico); N Colombia.
SYNONYMS: *ochoterenai, pallida*.
COMMENTS: See Arroyo-Cabrales and Jones (1988*a*, Mammalian Species, 301).

Centronycteris Gray, 1838. Mag. Zool. Bot., 2:499.
TYPE SPECIES: *Vespertilio calcaratus* Wied-Neuwied, 1821 (preoccupied; = *Vespertilio maximiliani* J. Fischer, 1829).

Centronycteris maximiliani (J. Fischer, 1829). Synopsis Mamm., p. 122.
TYPE LOCALITY: Brazil, Espirito Santo, Rio Jucy, Fazenda do Coroaba.
DISTRIBUTION: S Veracruz (Mexico) to Peru, Brazil, and Guianas.
SYNONYMS: *centralis, wiedi*.

Coleura Peters, 1867. Monatsb. K. Preuss. Akad. Wiss. Berlin, 1867:479.
TYPE SPECIES: *Emballonura afra* Peters, 1852.

Coleura afra (Peters, 1852). Reise nach Mossambique, Säugethiere, p. 51.
TYPE LOCALITY: Mozambique, Tete.
DISTRIBUTION: Guinea-Bissau to Somalia, south to Angola, Zaire, and Mozambique; Yemen.
SYNONYMS: *gallarum, kummeri, nilosa*.

Coleura seychellensis Peters, 1868. Monatsb. K. Preuss. Akad. Wiss. Berlin, 1868:367.
TYPE LOCALITY: Seychelle Isls, Mahe Isl.
DISTRIBUTION: Seychelle Isls; possibly Zanzibar. The Zanzibar record is extremely dubious.
STATUS: IUCN - Endangered.
SYNONYMS: *silhouettae*.

Cormura Peters, 1867. Monatsb. K. Preuss. Akad. Wiss. Berlin, 1867:475.
TYPE SPECIES: *Emballonura brevirostris* Wagner, 1843.
SYNONYMS: *Myropteryx*.

Cormura brevirostris (Wagner, 1843). Arch. Naturgesch., ser. 9, 1:367.
TYPE LOCALITY: Brazil, Amazonas, Rio Negro, Marabitanas.
DISTRIBUTION: Nicaragua to Peru and Brazil.
SYNONYMS: *pullus*.

Cyttarops Thomas, 1913. Ann. Mag. Nat. Hist., ser. 8, 11:134.
TYPE SPECIES: *Cyttarops alecto* Thomas, 1913.

Cyttarops alecto Thomas, 1913. Ann. Mag. Nat. Hist., ser. 8, 11:135.
TYPE LOCALITY: Brazil, Para, Mocajatuba.
DISTRIBUTION: Nicaragua, Costa Rica, Guyana, Amazonian Brazil.

COMMENTS: See Starrett (1972, Mammalian Species, 13).

Diclidurus Wied-Neuwied, 1820. Isis von Oken, 1819:1629 [1820].
TYPE SPECIES: *Diclidurus albus* Wied-Neuwied, 1820.
SYNONYMS: *Depanycteris*.
COMMENTS: Includes *Depanycteris*; but see Corbet and Hill (1980:46) who listed *Depanycteris* as a distinct genus without comment. A key to this genus was presented by Ceballos and Medellin (1988).

Diclidurus albus Wied-Neuwied, 1820. Isis von Oken, 1819:1630 [1820].
TYPE LOCALITY: Brazil, Bahia, Rio Pardo, Canavieiras.
DISTRIBUTION: Nayarit (Mexico) to E Brazil and Trinidad.
SYNONYMS: *freyreisii, virgo*.
COMMENTS: Subgenus *Diclidurus*. Includes *virgo*; see Goodwin (1969:48, 49), but see also Ojasti and Linares (1971). Corbet and Hill (1980:46) listed *virgo* as a distinct species without comment. See Ceballos and Medellin (1988, Mammalian Species, 316).

Diclidurus ingens Hernandez-Camacho, 1955. Caldasia, 7:87.
TYPE LOCALITY: Colombia, Caqueta, Rio Putumayo, Puerto Leguizamo.
DISTRIBUTION: Venezuela, SE Colombia, Guyana, NW Brazil.
COMMENTS: Subgenus *Diclidurus*.

Diclidurus isabellus (Thomas, 1920). Ann. Mag. Nat. Hist., ser. 9, 6:271.
TYPE LOCALITY: Brazil, Amazonas, Manacapuru (lower Solimões River).
DISTRIBUTION: NW Brazil, Venezuela.
COMMENTS: Subgenus *Depanycteris*. Formerly included in genus *Depanycteris*; see Ojasti and Linares (1971).

Diclidurus scutatus Peters, 1869. Monatsb. K. Preuss. Akad. Wiss. Berlin, 1869:400.
TYPE LOCALITY: Brazil, Para, Belem.
DISTRIBUTION: Amazonian Brazil, Venezuela, Peru, Guyana, Surinam.
COMMENTS: Subgenus *Diclidurus*.

Emballonura Temminck, 1838. Tijdschr. Nat. Gesch. Physiol., 5:22.
TYPE SPECIES: *Emballonura monticola* Temminck, 1838.
COMMENTS: Does not include *nigrescens*; see Griffiths et al. (1991).

Emballonura alecto (Eydoux and Gervais, 1836). Mag. Zool. Paris, 6:7.
TYPE LOCALITY: Philippines, Luzon, Manila.
DISTRIBUTION: Philippines, Borneo, Sulawesi and Tanimbar (Indonesia), and adjacent small islands.
SYNONYMS: *discolor, palawanensis, rivalis*.
COMMENTS: Includes *rivalis*; see Medway (1977:44).

Emballonura atrata Peters, 1874. Monatsb. K. Preuss. Akad. Wiss. Berlin, 1874:693.
TYPE LOCALITY: Madagascar.
DISTRIBUTION: E and C Madagascar.

Emballonura beccarii Peters and Doria, 1881. Ann. Mus. Civ. Stor. Nat. Genova, 16:693.
TYPE LOCALITY: Indonesia, Irian Jaya, Tjenderawasih Div., Japen Isl, Ansus.
DISTRIBUTION: New Guinea, Kei Isls, Trobriand Isls.
SYNONYMS: *clavium, locusta, meeki*.

Emballonura dianae Hill, 1956. *In* Wolff, Nat. Hist. Rennell Isl, Brit. Solomon Isls, 1:74.
TYPE LOCALITY: Solomon Isls, Rennell Isl, near Tigoa, Te-Abagua Cave, about 35 m.
DISTRIBUTION: Rennell and Malaita Isls (Solomon Isls), New Ireland (Bismarck Arch.), New Guinea.

Emballonura furax Thomas, 1911. Ann. Mag. Nat. Hist., ser. 8, 7:384.
TYPE LOCALITY: Indonesia, Irian Jaya, Kapare River, Whitewater Camp., 400 ft. (122 m).
DISTRIBUTION: New Guinea, Bismarck Arch.
STATUS: IUCN - Insufficiently known.

Emballonura monticola Temminck, 1838. Tijdschr. Nat. Gesch. Physiol., 5:25.
TYPE LOCALITY: Indonesia, Java, Mt. Munara.

DISTRIBUTION: Thailand to W Malaysia; Borneo; Sumatra, Rhio Arch., Banka, Billiton, Engano, Babi Isls, Anambas Isls, Batu Isls, Nias Isl, Mentawai Isls, Java, Sulawesi, and Karimata Isl (Indonesia).
SYNONYMS: *anambensis, peninsularis, pusilla.*

Emballonura raffrayana Dobson, 1879. Proc. Zool. Soc. Lond., 1878:876 [1879].
TYPE LOCALITY: Indonesia, Irian Jaya, Tjenderawasih Div., Numfoor Isl; for clarification see Thomas (1914*b*:442).
DISTRIBUTION: Seram Isl, Kei Isls, and Sulawesi (Indonesia); New Guinea; Bismarck Arch.; Choiseul, Ysabel and Malaita Isls (Solomon Isls).
STATUS: IUCN - Insufficiently known.
SYNONYMS: *cor, stresemanni.*

Emballonura semicaudata (Peale, 1848). Mammalia *in* Repts. U.S. Expl. Surv., 8:23.
TYPE LOCALITY: Samoa.
DISTRIBUTION: Mariana Isls and Caroline Isls (including Palau Isls), Vanuatu (= New Hebrides), Fiji Isls, Samoa.
SYNONYMS: *palauensis, rotensis, sulcata.*
COMMENTS: Includes *sulcata,* see Griffiths et al. (1991).

Mosia Gray, 1843. Ann. Mag. Nat. Hist., [ser. 1], 11:117.
TYPE SPECIES: *Mosia nigrescens* Gray, 1843.
COMMENTS: Formerly included in *Emballonura,* but see Griffiths et al. (1991).

Mosia nigrescens (Gray, 1843). Ann. Mag. Nat. Hist., [ser. 1], 11:117.
TYPE LOCALITY: Indonesia, Molucca Isls, Amboina Isl.
DISTRIBUTION: New Guinea; Kei Isls, Halmahera Isls, Schouten Isls, Sulawesi, Amboina, and Buru (Indonesia); Bismarck Arch. (Papua New Guinea); Solomon Isls; adjacent small islands.
SYNONYMS: *papuana, solomonis.*
COMMENTS: Subgenus *Mosia.* Includes *papuana;* see Laurie and Hill (1954:49). Includes *solomonis,* considered a distinct species by McKean (1972:35).

Peropteryx Peters, 1867. Monatsb. K. Preuss. Akad. Wiss. Berlin, 1867:472.
TYPE SPECIES: *Vespertilio caninus* Wied-Neuwied, 1821 (preoccupied; = *Emballonura macrotis* Wagner, 1843).
SYNONYMS: *Peronymus.*
COMMENTS: Includes *Peronymus;* see Cabrera (1958:52). Corbet and Hill (1980:45) listed *Peronymus* as a distinct genus following Sanborn (1937).

Peropteryx kappleri Peters, 1867. Monatsb. K. Preuss. Akad. Wiss. Berlin, 1867:473.
TYPE LOCALITY: Surinam.
DISTRIBUTION: S Veracruz (Mexico) to E Brazil, and Peru.
SYNONYMS: *intermedia.*
COMMENTS: Subgenus *Peropteryx.*

Peropteryx leucoptera Peters, 1867. Monatsb. K. Preuss. Akad. Wiss. Berlin, 1867:474.
TYPE LOCALITY: Surinam.
DISTRIBUTION: Peru, Colombia, N and E Brazil, Venezuela, French Guiana, Guyana, and Surinam.
SYNONYMS: *cyclops.*
COMMENTS: Subgenus *Peronymus.* Formerly included in genus *Peronymus;* see Cabrera (1958:52).

Peropteryx macrotis (Wagner, 1843). Arch. Naturgesch., ser. 9, 1:367.
TYPE LOCALITY: Brazil, Mato Grosso.
DISTRIBUTION: Guerrero and Yucatan (Mexico) to Peru, Paraguay, and S and E Brazil; Trinidad and Tobago; Margarita Isl (Venezuela); Aruba Isl (Netherlands Antilles); Grenada.
SYNONYMS: *brunnea, caninus, phaea, trinitatis.*
COMMENTS: Subgenus *Peropteryx.* Includes *trinitatis;* see Goodwin and Greenhall (1961:216). Corbet and Hill (1980:45) listed *trinitatis* as a distinct species without comment.

Rhynchonycteris Peters, 1867. Monatsb. K. Preuss. Akad. Wiss. Berlin, 1867:477.
TYPE SPECIES: *Vespertilio naso* Wied-Neuwied, 1820.

Rhynchonycteris naso (Wied-Neuwied, 1820). Reise nach Brasilien, 1:251.
TYPE LOCALITY: Brazil, Bahia, Rio Mucuri, near Morro d'Arara; for clarification see Avila-Pires (1965:9).
DISTRIBUTION: E Oaxaca and C Veracruz (Mexico) to C and E Brazil, Peru, Bolivia, French Guiana, Guyana, and Surinam; Trinidad.
SYNONYMS: *lineata, priscus, rivalis, saxatilis, villosa.*

Saccolaimus Temminck, 1838. Tijdschr. Nat. Gesch. Physiol., 5:14.
TYPE SPECIES: *Taphozous saccolaimus* Temminck, 1838.
COMMENTS: Considered a subgenus of *Taphozous* by Ellerman and Morrison-Scott (1951:104) and Corbet and Hill (1980:45), but see Barghorn (1977:5).

Saccolaimus flaviventris Peters, 1867. Proc. Zool. Soc. Lond., 1866:430 [1867].
TYPE LOCALITY: Australia.
DISTRIBUTION: Australia (except Tasmania), SE New Guinea.
SYNONYMS: *hargravei, insignis.*

Saccolaimus mixtus Troughton, 1925. Rec. Aust. Mus., 14:322.
TYPE LOCALITY: Papua New Guinea, Central Prov., Port Moresby.
DISTRIBUTION: SE New Guinea, NE Queensland (Australia).

Saccolaimus peli (Temminck, 1853). Esquisses Zool. sur la Côte de Guine, p. 82.
TYPE LOCALITY: Ghana, Boutry River.
DISTRIBUTION: Liberia to W Kenya south to Angola.

Saccolaimus pluto (Miller, 1911). Proc. U.S. Natl. Mus., 38:396.
TYPE LOCALITY: Philippines, Mindanao, near Zamboanga.
DISTRIBUTION: Philippines.
SYNONYMS: *capito.*
COMMENTS: Includes *capito*; see Lawrence (1939:42). Corbet and Hill (1980:45) listed *capito* as a distinct species without comment. Almost certainly a subspecies of *saccolaimus.*

Saccolaimus saccolaimus (Temminck, 1838). Tijdschr. Nat. Gesch. Physiol., 5:14.
TYPE LOCALITY: Indonesia, Java.
DISTRIBUTION: India and Sri Lanka through SE Asia to Borneo, Sumatra, Java, and Timor (Indonesia); New Guinea; NE Queensland (Australia); Guadalcanal Isl (Solomon Isls).
SYNONYMS: *affinis, crassus, flavomaculatus, granti, nudicluniatus, pulcher* (see Medway, 1977:45, and Goodwin, 1979:102).
COMMENTS: Corbet and Hill (1980:45) listed *nudicluniatus* as a distinct species without comment.

Saccopteryx Illiger, 1811. Prodr. Syst. Mamm. Avium., p. 121.
TYPE SPECIES: *Vespertilio lepturus* Schreber, 1774.

Saccopteryx bilineata (Temminck, 1838). Tijdschr. Nat. Gesch. Physiol., 5:33.
TYPE LOCALITY: Surinam.
DISTRIBUTION: Jalisco and Veracruz (Mexico) to Bolivia, Guianas, and E Brazil south to Rio de Janiero; Trinidad and Tobago.
SYNONYMS: *centralis, insignis, perspicillifer.*

Saccopteryx canescens Thomas, 1901. Ann. Mag. Nat. Hist., ser. 7, 7:366.
TYPE LOCALITY: Brazil, Para, Obidos.
DISTRIBUTION: Colombia, Venezuela, Guianas, N Brazil, Peru.
SYNONYMS: *pumila.*
COMMENTS: Includes *pumila*; see Husson (1962:46).

Saccopteryx gymnura Thomas, 1901. Ann. Mag. Nat. Hist., ser. 7, 7:367.
TYPE LOCALITY: Brazil, Para, Santarem.
DISTRIBUTION: Amazonian Brazil, perhaps Venezuela.

Saccopteryx leptura (Schreber, 1774). Die Säugethiere, 1(8):57.
TYPE LOCALITY: Surinam.
DISTRIBUTION: Chiapas and Tabasco (Mexico) to E Brazil and Peru; Guianas; Margarita Isl (Venezuela); Trinidad and Tobago.

Taphozous E. Geoffroy, 1818. Descrip. de L'Egypte, 2:113.
TYPE SPECIES: *Taphozous perforatus* E. Geoffroy, 1818.
SYNONYMS: *Liponycteris*.
COMMENTS: Includes *Liponycteris* but not *Saccolaimus*; see Hayman and Hill (1971:15) and Barghorn (1977:5).

Taphozous australis Gould, 1854. Mamm. Aust., p. 3.
TYPE LOCALITY: Australia, Queensland, Albany Isl (off Cape York).
DISTRIBUTION: N Queensland (Australia), Torres Strait Isls, SE New Guinea.
SYNONYMS: *fumosus*.
COMMENTS: Subgenus *Taphozous*. Includes *fumosus*; see Troughton (1925:332). Tate (1952:607) included *georgianus* in this species, but also see McKean and Price (1967).

Taphozous georgianus Thomas, 1915. J. Bombay Nat. Hist. Soc., 24:62.
TYPE LOCALITY: Australia, Western Australia, King Georges Sound.
DISTRIBUTION: Australia.
SYNONYMS: *troughtoni*.
COMMENTS: Subgenus *Taphozous*. Includes *troughtoni*; see McKean and Price (1967); but see also Chiminbu and Kitchener (1991) who recognized *troughtoni* as a distinct species.

Taphozous hamiltoni Thomas, 1920. Ann. Mag. Nat. Hist., ser. 9, 5:142.
TYPE LOCALITY: Sudan, Equatoria, Mongalla.
DISTRIBUTION: S Sudan, Chad, Kenya.
COMMENTS: Subgenus *Liponycteris*.

Taphozous hildegardeae Thomas, 1909. Ann. Mag. Nat. Hist., ser. 8, 4:98.
TYPE LOCALITY: Kenya, Coast Province, Rabai (near Mombasa).
DISTRIBUTION: Kenya, NE Tanzania, Zanzibar.
COMMENTS: Subgenus *Taphozous*.

Taphozous hilli Kitchener, 1980. Rec. W. Aust. Mus., 8:162.
TYPE LOCALITY: Australia, Western Australia, Hamersley range, near Mt. Bruce.
DISTRIBUTION: Western Australia, South Australia, and Northern Territory.
COMMENTS: Subgenus *Taphozous*.

Taphozous kapalgensis McKean and Friend, 1979. Vict. Nat., 96:239.
TYPE LOCALITY: Australia, Northern Territory, S Alligator River, near Rookery Point, Kapalga.
DISTRIBUTION: Northern Territory (Australia).
COMMENTS: Subgenus *Taphozous*.

Taphozous longimanus Hardwicke, 1825. Trans. Linn. Soc. Lond., 14:525.
TYPE LOCALITY: India, Bengal, Calcutta.
DISTRIBUTION: Sri Lanka; India to Cambodia; Malay Peninsula; Sumatra, Borneo, Java, Bali, and Flores (Indonesia).
SYNONYMS: *albipinnis, brevicaudus, cantorii, fulvidus, kampenii, leucopleura*.
COMMENTS: Subgenus *Taphozous*.

Taphozous mauritianus E. Geoffroy, 1818. Descrip. de L'Egypte, 2:127.
TYPE LOCALITY: Mauritius.
DISTRIBUTION: South Africa to Sudan and Somalia to Senegal; Mauritius and Reunion Isls (Mascarene Isls); Madagascar; Assumption Isl and Aldabra Isl.
SYNONYMS: *cinerascens, dobsoni, leucopterus*.
COMMENTS: Subgenus *Taphozous*.

Taphozous melanopogon Temminck, 1841. Monogr. Mamm., 2:287.
TYPE LOCALITY: Indonesia, W Java, Bantam.

DISTRIBUTION: Sri Lanka; India; Burma; Thailand; Laos; Vietnam; S China; Malay Peninsula and adjacent islands; Borneo; Sumatra, Java, Savu Isl, Lombok, Sumbawa, Timor, Kei Isls, and Sulawesi (Indonesia).
SYNONYMS: *achates, bicolor, cavaticus, fretensis.*
COMMENTS: Subgenus *Taphozous.*

Taphozous nudiventris Cretzschmar, 1830. *In* Rüppell, Atlas Reise Nordl. Afr., Zool., Säugeth., p. 70.
TYPE LOCALITY: Egypt, Giza.
DISTRIBUTION: Mauritania, Senegal, and Guinea-Bissau to Egypt, south to Tanzania and east to Burma.
SYNONYMS: *assabensis, babylonicus, kachhensis, magnus, nudaster, serratus, ziyidi.*
COMMENTS: Subgenus *Liponycteris.* Includes *kachhensis;* see Felten (1962:175). Formerly included in genus *Liponycteris;* see Hayman and Hill (1971:15).

Taphozous perforatus E. Geoffroy, 1818. Descrip. de L'Egypte, 2:126.
TYPE LOCALITY: Egypt, Kom Ombo.
DISTRIBUTION: Senegal to Botswana, Mozambique, Somalia and Egypt; S Arabia; S Iran; Pakistan; NW India.
SYNONYMS: *haedinus, maritimus, rhodesiae, senegalensis, sudani, swirae.*
COMMENTS: Subgenus *Taphozous.* Includes *senegalensis* and *sudani;* see Hayman and Hill (1977:16).

Taphozous philippinensis Waterhouse, 1845. Proc. Zool. Soc. Lond., 1845:9.
TYPE LOCALITY: Philippines.
DISTRIBUTION: Philippines.
SYNONYMS: *solifer.*
COMMENTS: Subgenus *Taphozous.* Includes *solifer;* see Ellerman and Morrison-Scott (1951:105). Probably a subspecies of *melanopogon;* see Heaney et al. (1987:43). Corbet and Hill (1991:51) listed *solifer* as a distinct species without comment.

Taphozous theobaldi Dobson, 1872. Proc. Asiat. Soc. Bengal, p. 152.
TYPE LOCALITY: Burma, Tenasserim.
DISTRIBUTION: C India to Vietnam; Java, Borneo and Sulawesi. A record from Maylaya appears to be in error; see Medway (1969:8).
SYNONYMS: *secatus.*
COMMENTS: Subgenus *Taphozous.*

Family Nycteridae Van der Hoeven, 1855. Handb. Dierkunde, 2nd ed., 2:1028.

Nycteris G. Cuvier and E. Geoffroy, 1795. Mag. Encyclop., 2:186.
TYPE SPECIES: *Vespertilio hispidus* Schreber, 1775.
COMMENTS: Some African species revised by Cakenberghe and de Vree (1985).

Nycteris arge Thomas, 1903. Ann. Mag. Nat. Hist., ser. 7, 12:633.
TYPE LOCALITY: Cameroon, Efulen.
DISTRIBUTION: Sierra Leone to S and E Zaire; W Kenya; SW Sudan; NE Angola; Bioko.
COMMENTS: Formerly included *intermedia;* see Hayman and Hill (1971:19), but see Cakenberghe and de Vree (1985).

Nycteris gambiensis (K. Andersen, 1912). Ann. Mag. Nat. Hist., ser. 8, 10:548.
TYPE LOCALITY: Senegal, Dialakoto.
DISTRIBUTION: Senegal, Gambia, Guinea, Sierra Leone, Ghana, Togo, Benin, Burkina Faso.

Nycteris grandis Peters, 1865. Monatsb. K. Preuss. Akad. Wiss. Berlin, 1865:358.
TYPE LOCALITY: "Guinea".
DISTRIBUTION: Senegal to Kenya, Zimbabwe, and Mozambique; Zanzibar and Pemba.
SYNONYMS: *baikii, marica, proxima.*

Nycteris hispida (Schreber, 1775). Die Säugethiere, 1:169, 188.
TYPE LOCALITY: Senegal.
DISTRIBUTION: Senegal to Somalia and south to Angola and South Africa; Zanzibar; Bioko.
SYNONYMS: *aurita, daubentoni, martini, pallida, pilosa, villosa.*

COMMENTS: Includes *aurita* and *pallida*; see Koopman (1975:377, 378).

Nycteris intermedia Aellen, 1959. Arch. Sci. Phys. Nat. Geneve, 12:218.
TYPE LOCALITY: Ivory Coast, Adiopodoume.
DISTRIBUTION: Liberia to W Tanzania and south to Angola.
COMMENTS: Formerly included in *N. arge* but see Cakenberghe and de Vree (1985).

Nycteris javanica E. Geoffroy, 1813. Ann. Mus. Hist. Nat. Paris, 20:20.
TYPE LOCALITY: Indonesia, Java.
DISTRIBUTION: Java, Bali, and Kangean Isl (Indonesia).
SYNONYMS: *bastiani*.
COMMENTS: Does not include *tragata*; see Ellerman and Morrison-Scott (1955:9).

Nycteris macrotis Dobson, 1876. Monogr. Asiat. Chiroptera, p. 80.
TYPE LOCALITY: Sierra Leone.
DISTRIBUTION: Senegal to Ethiopia, south to Zimbabwe, Malawi and Mozambique; Zanzibar; Madagascar.
SYNONYMS: *aethiopica, guineensis, luteola, madagascariensis, major* (of J. A. Allen, 1917, not K. Anderson, 1912), *oriana, vinsoni* (see Koopman, 1975:378, 1992; Cakenberghe and Vree, 1985; and Kock, 1969*a*:94-97).

Nycteris major (K. Andersen, 1912). Ann. Mag. Nat. Hist., ser. 8, 10:547.
TYPE LOCALITY: Cameroon, Ja River.
DISTRIBUTION: Liberia to Zambia.
SYNONYMS: *avakubia*.
COMMENTS: Includes *avakubia*; see Koopman (1965:6).

Nycteris nana (K. Andersen, 1912). Ann. Mag. Nat. Hist., ser. 8, 10:547.
TYPE LOCALITY: Equatorial Guinea, Rio Muni, Benito River.
DISTRIBUTION: Ivory Coast to NE Angola, W Tanzania, W Kenya, and SW Sudan.
SYNONYMS: *tristis*.
COMMENTS: Includes *tristis*; see Koopman (1975:376).

Nycteris thebaica E. Geoffroy, 1818. Descrip. de L'Egypte, 2:119.
TYPE LOCALITY: Egypt, Thebes (near Luxor).
DISTRIBUTION: Central Arabia; Israel; Sinai; Egypt to Morocco, Senegal, Benin, Somalia and Kenya, thence south to South Africa in open country; Zanzibar and Pemba.
SYNONYMS: *adana, affinis, albiventer, angolensis, aurantiaca, brockmani, capensis, damarensis, discolor, fuliginosa, labiata, media, najdiya, revoilii*.

Nycteris tragata (K. Andersen, 1912). Ann. Mag. Nat. Hist., ser. 8, 10:546.
TYPE LOCALITY: Malaysia, Sarawak, Bidi caves.
DISTRIBUTION: Burma, Thailand, W Malaysia, Sumatra, Borneo.
COMMENTS: Clearly distinct from *javanica*; see Ellerman and Morrison-Scott (1955:9).

Nycteris woodi K. Andersen, 1914. Ann. Mag. Nat. Hist., ser. 8, 13:563.
TYPE LOCALITY: Zambia, Chilanga.
DISTRIBUTION: Zambia and South Africa to SW Tanzania; Ethiopia; Somalia; Cameroon.
SYNONYMS: *benuensis, parisii, sabiensis*.
COMMENTS: For synonyms see Cakenberghe and Vree (1985).

Family Megadermatidae H. Allen, 1864. Monogr. Bats N. Am., pp. xxiii, 1.

Cardioderma Peters, 1873. Monatsb. K. Preuss. Akad. Wiss. Berlin, 1873:488.
TYPE SPECIES: *Megaderma cor* Peters, 1872.

Cardioderma cor (Peters, 1872). Monatsb. K. Preuss. Akad. Wiss. Berlin, 1872:194.
TYPE LOCALITY: Ethiopia.
DISTRIBUTION: Ethiopia, Somalia, Kenya, Uganda, E Sudan, Tanzania, Zanzibar.

Lavia Gray, 1838. Mag. Zool. Bot., 2:490.
TYPE SPECIES: *Megaderma frons* E. Geoffroy, 1810.

Lavia frons (E. Geoffroy, 1810). Ann. Mus. Hist. Nat. Paris, 15:192.
TYPE LOCALITY: Senegal.
DISTRIBUTION: Senegal to Somalia, south to Zambia and Malawi; Zanzibar.
SYNONYMS: *affinis, rex.*

Macroderma Miller, 1906. Proc. Biol. Soc. Washington, 19:84.
TYPE SPECIES: *Megaderma gigas* Dobson, 1880.

Macroderma gigas (Dobson, 1880). Proc. Zool. Soc. Lond., 1880:461.
TYPE LOCALITY: Australia, Queensland, Wilson's River, Mt. Margaret.
DISTRIBUTION: N and C Australia.
STATUS: IUCN - Vulnerable.
SYNONYMS: *saturata.*
COMMENTS: See Hudson and Wilson (1986, Mammalian Species, 260).

Megaderma E. Geoffroy, 1810. Ann. Mus. Hist. Nat. Paris, 15:197.
TYPE SPECIES: *Vespertilio spasma* Linnaeus, 1758.
SYNONYMS: *Lyroderma.*
COMMENTS: Includes *Lyroderma*, but see Hand (1985).

Megaderma lyra E. Geoffroy, 1810. Ann. Mus. Hist. Nat. Paris, 15:190.
TYPE LOCALITY: India, Madras.
DISTRIBUTION: Afghanistan to S China, south to Sri Lanka and W Malaysia.
SYNONYMS: *carnatica, carina, schistacea, sinensis, spectrum.*

Megaderma spasma (Linnaeus, 1758). Syst. Nat., 10th ed., 1:32.
TYPE LOCALITY: Indonesia, Molucca Islands, Ternate.
DISTRIBUTION: Sri Lanka and India through SE Asia to Lesser Sundas, the Philippines and Molucca Isls, various adjacent islands.
SYNONYMS: *abditum, carimatae, celebensis, ceylonense, horsfieldi, kinabalu, lasiae, majus, medium, minus, naisense, natunae, pangandarana, siumatis, trifolium.*

Family Rhinolophidae Gray, 1825. Zool. J., 2(6):242.
SYNONYMS: Hipposideridae.
COMMENTS: Includes Hipposideridae (here treated at a subfamily), see Vaughan (1978:39) and Koopman and Jones (1970); but also see Swanepoel et al. (1980:157). Hill (1982) listed Hipposideridae as a distinct family, without comment. Hill and Smith (1984) retained this separation and discussed the question.

Subfamily Rhinolophinae Gray, 1825. Zool. J., 2(6):242.

Rhinolophus Lacépède, 1799. Tabl. Div. Subd. Orders Genres Mammifères, p. 15.
TYPE SPECIES: *Vespertilio ferrum-equinum* Schreber, 1774.
SYNONYMS: *Rhinomegalophus.*
COMMENTS: Includes *Rhinomegalophus*; see Thonglongya (1973:587).

Rhinolophus acuminatus Peters, 1871. Monatsb. K. Preuss. Akad. Wiss. Berlin, 1871:308.
TYPE LOCALITY: Indonesia, Java.
DISTRIBUTION: Borneo, Sumatra (including Nias and Engano Isls), Java, Lombok, and Bali (Indonesia), Palawan (Philippines), Thailand, Laos, Cambodia.
SYNONYMS: *audax, calypso, circe, sumatranus.*

Rhinolophus adami Aellen and Brosset, 1968. Rev. Suisse Zool., 75:443.
TYPE LOCALITY: Congo Republic, Kouilou.
DISTRIBUTION: Congo Republic.

Rhinolophus affinis Horsfield, 1823. Zool. Res. Java, part 6, p. 6(unno.) of *Rhinolophus larvatus* acct. and pl. figs a,b.
TYPE LOCALITY: Indonesia, Java.
DISTRIBUTION: India to S China through Malaysia to Borneo and Lesser Sunda Isls; Andaman Isls (India); perhaps Sri Lanka.
SYNONYMS: *andamanensis, hainanus, himalayanus, macrurus, nesites, princeps, superans, tener.*
COMMENTS: Includes *andamanensis*; see Sinha (1973:612-613).

Rhinolophus alcyone Temminck, 1852. Esquisses Zool. sur la Côte de Guine, p. 80.
TYPE LOCALITY: Ghana, Boutry River.
DISTRIBUTION: Senegal to Uganda, SW Sudan, N Zaire, and Gabon; Bioko.

Rhinolophus anderseni Cabrera, 1909. Bol. Real. Soc. Esp. Hist. Nat., p. 305.
TYPE LOCALITY: Philippines, Luzon (uncertain).
DISTRIBUTION: Palawan and Luzon (Philippines).
SYNONYMS: *aequalis*.

Rhinolophus arcuatus Peters, 1871. Monatsb. K. Preuss. Akad. Wiss. Berlin, 1871:305.
TYPE LOCALITY: Philippines, Luzon.
DISTRIBUTION: Sumatra to Philippines, New Guinea, and Lesser Sundas.
SYNONYMS: *angustifolius, beccarii, exiguus, mcintyrei, proconsularis, toxopeusi, typica*.
COMMENTS: Includes *toxopeusi*; see Hill and Schlitter (1982).

Rhinolophus blasii Peters, 1866. Monatsb. K. Preuss. Akad. Wiss. Berlin, 1866:17.
TYPE LOCALITY: SE Europe; restricted to Italy by Ellerman et al. (1953:59).
DISTRIBUTION: Transvaal (South Africa) to S Zaire; Ethiopia; Somalia; Morocco; Algeria; Tunisia; Turkey; Yemen; Israel; Jordan; Syria; Iran; Yugoslavia; Albania; Bulgaria; Rumania; Transcaucasia and Turkmenistan; Afghanistan; Pakistan; Italy; Greece; Cyprus.
SYNONYMS: *andreinii, brockmani, empusa, meyeroehmi*.
COMMENTS: Includes *brockmani*; see Koopman (1975:383).

Rhinolophus borneensis Peters, 1861. Monatsb. K. Preuss. Akad. Wiss. Berlin, 1861:709.
TYPE LOCALITY: Malaysia, Sabah, Labuan Isl.
DISTRIBUTION: Borneo; Labuan and Banguey Isls (Malaysia); Java, Karimata Isls and South Natuna Isls (Indonesia); Cambodia; Vietnam.
SYNONYMS: *chaseni, importunus, spadix*.
COMMENTS: Includes *chaseni* and *importunus*; see Hill (1983). Formerly included *javanicus, celebensis, madurensis*, and *parvus*; see Goodwin (1979:104) and Hill and Thonglongya (1972:187); but see Hill (1983).

Rhinolophus canuti Thomas and Wroughton, 1909. Abstr. Proc. Zool. Soc. Lond., 1909(68):18.
TYPE LOCALITY: Indonesia, South Java.
DISTRIBUTION: Java, Timor (Indonesia).
SYNONYMS: *timoriensis*.
COMMENTS: Formerly included in *creaghi*; see Hill and Schlitter (1982).

Rhinolophus capensis Lichtenstein, 1823. Verz. Doblet. Mus. Univ. Berlin, p. 4.
TYPE LOCALITY: South Africa, Cape Prov., Cape of Good Hope.
DISTRIBUTION: Cape Prov., Natal (South Africa); Zimbabwe; Mozambique. Occurence outside Cape Prov. is doubtful; records from Zambia and Malawi are definitely erroneous.
SYNONYMS: *auritus*.

Rhinolophus celebensis K. Andersen, 1905. Proc. Zool. Soc. Lond., 1905(2):83.
TYPE LOCALITY: Sulawesi, Macassar (= Ujung Pandang).
DISTRIBUTION: Java, Madura, Bali, Timor, Sulawesi, Sangihe and Talaud Isls (Indonesia).
SYNONYMS: *javanicus, madurensis, parvus*.
COMMENTS: For synonyms see Hill (1983), but see also Bergmans and Van Bree (1986:335-337).

Rhinolophus clivosus Cretzschmar, 1828. *In* Rüppell, Atlas Reise Nordl. Afr., Zool., Säugeth., p. 47.
TYPE LOCALITY: Saudi Arabia, Muwaylih (= Mohila).
DISTRIBUTION: Turkmenistan to Afghanistan; Arabia to Algeria; subsaharan Africa to Liberia, Cameroon and South Africa.
SYNONYMS: *acrotis, andersoni, angur, bocharicus, brachygnathus, hillorum, keniensis, schwarzi, zuluensis*.
COMMENTS: Includes *bocharicus*; see Aellen (1959:362-366). *R. bocharicus* is considered a species by Hanak (1969), DeBlase (1980:94-97), Gromov and Baranova (1981), and Pavlinov and Rossolimo (1987). This species does not include *deckenii* or *silvestris*; see Koopman (1975:386).

Rhinolophus coelophyllus Peters, 1867. Proc. Zool. Soc. Lond., 1866:426 [1867].
TYPE LOCALITY: Burma, Salaween River.
DISTRIBUTION: W Malaysia, Thailand, Burma.
COMMENTS: Does not include *shameli*; see Hill and Thonglongya (1972:183-186).

Rhinolophus cognatus K. Andersen, 1906. Ann. Mus. Civ. Stor. Nat. Genova, ser. 3, 2:181.
TYPE LOCALITY: India, Andaman Isls, S Andaman Isl, Port Blair.
DISTRIBUTION: Andaman Isls (India).
SYNONYMS: *famulus*.

Rhinolophus cornutus Temminck, 1835. Monogr. Mamm., 2:37.
TYPE LOCALITY: Japan.
DISTRIBUTION: Japan (including Ryukyu Isls), perhaps SE China.
SYNONYMS: *orii, miyakonis, perditus, pumilus*.
COMMENTS: Does not include *blythi*; see Hill and Yoshiyuki (1980:186); but also see Corbet (1978*c*:43). See also comments under *pusillus*. Includes *pumilus* and *perditus*, but see Yoshiyuki (1989).

Rhinolophus creaghi Thomas, 1896. Ann. Mag. Nat. Hist., ser. 6, 18:244.
TYPE LOCALITY: Malaysia, Sabah, Sandakan.
DISTRIBUTION: Borneo; Madura Isl, Java, and Timor (Indonesia).
SYNONYMS: *pilosus*.
COMMENTS: Includes *pilosus* but not *canuti*; see Hill and Schlitter (1982:458).

Rhinolophus darlingi K. Andersen, 1905. Ann. Mag. Nat. Hist., ser. 7, 15:70.
TYPE LOCALITY: Zimbabwe, Mazoe.
DISTRIBUTION: Transvaal (South Africa), Namibia, S Angola, N and W Botswana, Zimbabwe, Malawi, Mozambique, Tanzania.
SYNONYMS: *barbertonensis, damarensis*.
COMMENTS: Includes *barbertonensis*; see Hayman and Hill (1971:23).

Rhinolophus deckenii Peters, 1867. Monatsb. K. Preuss. Akad. Wiss. Berlin, 1867:705.
TYPE LOCALITY: Tanzania, "Zanzibar coast" (mainland opposite Zanzibar).
DISTRIBUTION: Uganda, Kenya, Tanzania, Zanzibar and Pemba.
COMMENTS: Treated as a subspecies of *clivosus* by Hayman and Hill (1971:23); but see Koopman (1975:386).

Rhinolophus denti Thomas, 1904. Ann. Mag. Nat. Hist., ser. 7, 13:386.
TYPE LOCALITY: South Africa, Cape Province, Kuruman.
DISTRIBUTION: N Cape Prov. (South Africa), Namibia, Botswana, Zimbabwe, Mozambique, Guinea, Ivory Coast, Ghana.
SYNONYMS: *knorri*.

Rhinolophus eloquens K. Andersen, 1905. Ann. Mag. Nat. Hist., ser. 7, 15:74.
TYPE LOCALITY: Uganda, Entebbe.
DISTRIBUTION: Uganda, S Somalia, S Sudan, NE Zaire, Kenya, Rwanda, N Tanzania, Zanzibar and Pemba.
SYNONYMS: *perauritus*.
COMMENTS: Includes *perauritus*; see Koopman (1975:389).

Rhinolophus euryale Blasius, 1853. Arch. Naturgesch., 19(1):49.
TYPE LOCALITY: Italy, Milan.
DISTRIBUTION: Transcaucasia to Israel and S Europe; Turkmenistan; Iran; Algeria; Morocco; Tunisia; various Mediterranean islands; perhaps Egypt.
SYNONYMS: *algirus, atlanticus, barbarus, cabrerae, judaicus, meridionalis, nordmanni, toscanus*.
COMMENTS: Revised by DeBlase (1972).

Rhinolophus euryotis Temminck, 1835. Monogr. Mamm., 2:26.
TYPE LOCALITY: Indonesia, Molucca Isls, Amboina Isl.
DISTRIBUTION: Aru Isls, Buru, Amboina, Seram, and Timor Laut Isls, Kei Isls, Halmahera, and Sulawesi (Indonesia); New Guinea; Bismarck Arch.; adjacent small islands.
SYNONYMS: *aruensis, burius, praestens, tatar, timidus*.

Rhinolophus ferrumequinum (Schreber, 1774). Die Säugethiere, 1:174, pl. 62.
TYPE LOCALITY: France.

DISTRIBUTION: S England to Caucasus Mtns south to Morocco and Tunesia (but not Egypt) through Iran and Himalayas to China and Japan; adjacent small islands.
SYNONYMS: *brevitarsus, colchicus, creticum, equinus, fudisanus, germanicus, hippocrepis, homodorensis, homorodalmasiensis, insulanus, irani, italicus, korai, kosidanus, martinoi, mikadoi, nippon, norikuranus, obscurus, ogasimanus, proximus, quelpartis, regulus, rubiginosus, typicus, tragatus, ungula, unihastatus.*
COMMENTS: Revised by Strelkov et al. (1978).

Rhinolophus fumigatus Rüppell, 1842. Mus. Senckenbergianum, 3:132, 155.
TYPE LOCALITY: Ethiopia, Shoa.
DISTRIBUTION: Somalia, Ethiopia, Sudan, Kenya, Tanzania, Rwanda, Burundi, Zaire, Nigeria, Sierra Leone, Togo, Benin, Senegal, Gambia, Guinea, Burkina Faso, Ghana, Niger, Nigeria, Cameroon, Central African Republic, Zambia, Malawi, Zimbabwe, Mozambique, Angola, Namibia, South Africa.
SYNONYMS: *abae, aethiops, antinorii, diversus, exsul, foxi, macrocephalus.*
COMMENTS: Does not include *eloquens* or *perauritus*, but does include *aethiops*; see Koopman (1975:389-390).

Rhinolophus guineensis Eisentraut, 1960. Stygg. Beitr. Naturk., 39:1.
TYPE LOCALITY: Guinea, Tahiré (foot of Kelesi Plateau).
DISTRIBUTION: Guinea, Sierra Leone, Liberia.
COMMENTS: Originally described as a subspecies of *R. landeri*, but see Böhme and Hutterer (1979) who demonstrated that it was a separate species.

Rhinolophus hildebrandti Peters, 1878. Monatsb. K. Preuss. Akad. Wiss. Berlin, 1878:195.
TYPE LOCALITY: Kenya, Taita, Ndi.
DISTRIBUTION: Transvaal (South Africa) and Mozambique to Ethiopia, S Sudan, and NE Zaire.

Rhinolophus hipposideros (Bechstein, 1800). *In* Pennant, Allgemeine Ueber. Vierfuss. Thiere, 2:629.
TYPE LOCALITY: France.
DISTRIBUTION: Ireland, Iberia and Morocco through S Europe and N Africa to Kirghizia and Kashmir; Arabia; Sudan; Ethiopia.
SYNONYMS: *anomalus, bihastatus, bifer, escalerae, helvetica, intermedius, kashyiriensis, majori, midas, minimus, minutus, moravicus, pallidus, phasma, trogophilus, typus, vespa.*
COMMENTS: Revised by Felten et al. (1977).

Rhinolophus imaizumii Hill and Yoshiyuki, 1980. Bull. Natl. Sci. Mus. Tokyo, ser. A (Zool.), 6:180.
TYPE LOCALITY: Japan, Ryukyu Isls, Yayeyama Isls, Iriomote Isl, Otomi-do cave.
DISTRIBUTION: Iriomote Isl (Ryukyu Isls).

Rhinolophus inops K. Andersen, 1905. Ann. Mag. Nat. Hist., ser. 7, 16:284, 651.
TYPE LOCALITY: Philippines, Mindanao, Davao, Mt. Apo, Todaya, 1,325 m.
DISTRIBUTION: Mindanao (Philippines).

Rhinolophus keyensis Peters, 1871. Monatsb. K. Preuss. Akad. Wiss. Berlin, 1871:371.
TYPE LOCALITY: Indonesia, Molucca Isls, Kei Isl.
DISTRIBUTION: Batchian Isl (Halmahera Isls), Seram, Goram Isl (SE of Seram), Kei Isls, Wetter Isl (Flores Sea) (Indonesia).
SYNONYMS: *annectens, nanus, truncatus.*

Rhinolophus landeri Martin, 1838. Proc. Zool. Soc. Lond., 1837:101 [1838].
TYPE LOCALITY: Equatorial Guinea, Bioko.
DISTRIBUTION: Senegal to Ethiopia and Somalia, south to South Africa and Namibia; Bioko; Zanzibar.
SYNONYMS: *angolensis, axillaris, dobsoni, lobatus.*
COMMENTS: Includes *angolensis, dobsoni,* and *guineensis*, but not *brockmani* according to Hayman and Hill (1971) and Koopman (1975:388); but see Böhme and Hutterer (1979:306-307) who correctly treated *guineensis* as a separate species.

Rhinolophus lepidus Blyth, 1844. J. Asiat. Soc. Bengal, 13:486.
TYPE LOCALITY: India, Bengal, Calcutta (uncertain).

DISTRIBUTION: Afghanistan, N India, Burma, Thailand, Szechwan and Yunnan (China), W Malaysia, Sumatra.
SYNONYMS: *cuneatus, feae, monticola, refulgens, shortridgei.*
COMMENTS: Includes *feae, monticola,* and *refulgens;* see Hill and Yoshiyuki (1980:180, 186); but also see Sinha (1973:620-621) who considered *monticola* a distinct species.

Rhinolophus luctus Temminck, 1835. Monogr. Mamm., 2:24.
TYPE LOCALITY: Indonesia, Java.
DISTRIBUTION: India (including Sikkim), Nepal, Burma, Sri Lanka, S China, Taiwan, Vietnam, Laos, Thailand, Malay Peninsula, Borneo, Sumatra, Java, and Bali.
SYNONYMS: *beddomei, foetidus, formosae, lanosus, morio, perniger, sobrinus, spurcus.*
COMMENTS: Includes *lanosus;* see Ellerman and Morrison-Scott (1951:121).

Rhinolophus maclaudi Pousargues, 1897. Bull. Mus. Hist. Nat. Paris, 3:358.
TYPE LOCALITY: Guinea, Conakry.
DISTRIBUTION: Guinea, Liberia, E Zaire, W Uganda, Rwanda.
SYNONYMS: *hilli, ruwenzorii* (see Smith and Hood, 1980:170).

Rhinolophus macrotis Blyth, 1844. J. Asiat. Soc. Bengal, 13:485.
TYPE LOCALITY: Nepal.
DISTRIBUTION: N India to S China, Vietnam, and W Malaysia; Sumatra; Philippines.
SYNONYMS: *caldwelli, dohrni, episcopus, hirsutus, siamensis.*
COMMENTS: Includes *episcopus* and *hirsutus;* see Ellerman and Morrison-Scott (1951:122) and Tate (1943:2). Corbet and Hill (1980:48) listed *hirsutus* as a distinct species without comment.

Rhinolophus malayanus Bonhote, 1903. *In* N. Annandale, Fasciculi Malayenses, Zool., 1:15.
TYPE LOCALITY: Thailand, Patani, Biserat.
DISTRIBUTION: W Malaysia, Thailand, Laos, Vietnam.

Rhinolophus marshalli Thonglongya, 1973. Mammalia, 37:590.
TYPE LOCALITY: Thailand, Chantaburi, Amphoe Pong Nam Ron, foothills of Khao Soi Dao Thai.
DISTRIBUTION: Thailand.

Rhinolophus megaphyllus Gray, 1834. Proc. Zool. Soc. Lond., 1834:52.
TYPE LOCALITY: Australia, New South Wales, Murrumbidgee River.
DISTRIBUTION: E New Guinea; Misima Isl (Louisiade Arch.), Goodenough Isl (D'Entrecasteaux Isls), and Bismark Arch. (Papua New Guinea); E Queensland, E New South Wales, and E Victoria (Australia).
SYNONYMS: *fallax, ignifer, monachus, vandeuseni.*

Rhinolophus mehelyi Matschie, 1901. Sitzb. Ges. Naturf. Fr. Berlin, p. 225.
TYPE LOCALITY: Rumania, Bucharest.
DISTRIBUTION: Portugal, Spain, France, Rumania, Yugoslavia, Bulgaria, Greece, Transcaucasia; Morocco to Cyrenaica (Libya); Mediterranean islands, Iran, Afghanistan, Asia Minor, Israel, Egypt.
SYNONYMS: *carpentanus, tunetae.*
COMMENTS: Revised by DeBlase (1972).

Rhinolophus mitratus Blyth, 1844. J. Asiat. Soc. Bengal, 13:483.
TYPE LOCALITY: India, Orissa, Chaibassa.
DISTRIBUTION: N India.

Rhinolophus monoceros K. Andersen, 1905. Proc. Zool. Soc. Lond., 1905:131.
TYPE LOCALITY: Taiwan, Baksa.
DISTRIBUTION: Taiwan.
COMMENTS: Probably a subspecies of *cornutus.*

Rhinolophus nereis K. Andersen, 1905. Proc. Zool. Soc. Lond., 1905:90.
TYPE LOCALITY: Indonesia, Anamba Isls, Siantan Isl.
DISTRIBUTION: Anamba and North Natuna Isls (Indonesia).

Rhinolophus osgoodi Sanborn, 1939. Field Mus. Nat. Hist. Publ., Zool. Ser., 24:40.
TYPE LOCALITY: China, Yunnan, N of Likiang, Nguluko.

DISTRIBUTION: Yunnan (China).

Rhinolophus paradoxolophus (Bourret, 1951). Bull. Mus. Hist. Nat. Paris, ser. 2, 33:607.
TYPE LOCALITY: Vietnam, Tonkin, Lao Key, near Chapa, 1,700 m.
DISTRIBUTION: Vietnam, Thailand.

Rhinolophus pearsonii Horsfield, 1851. Cat. Mamm. Mus. E. India Co., p. 33.
TYPE LOCALITY: India, W Bengal, Darjeeling.
DISTRIBUTION: N India; Burma; Szechwan, Anhwei, and Fukien (China) to Vietnam; Thailand; W Malaysia.
SYNONYMS: *chinensis*.

Rhinolophus philippinensis Waterhouse, 1843. Proc. Zool. Soc. Lond., 1843:68.
TYPE LOCALITY: Philippines, Luzon.
DISTRIBUTION: Mindoro, Luzon, Mindanao and Negros (Phillipines); Kei Isls, Sulawesi and Timor (Indonesia); Borneo; New Guinea; NE Queensland (Australia).
SYNONYMS: *achilles, alleni, maros, montanus, robertsi, sanborni*.

Rhinolophus pusillus Temminck, 1834. Tijdschr. Nat. Gesch. Physiol., 1:29.
TYPE LOCALITY: Indonesia, Java.
DISTRIBUTION: India; Thailand; W Malaysia; Mentawai Isls, Java and Lesser Sunda Isls (Indonesia), small adjacent islands.
SYNONYMS: *blythi, calidus, gracilis, minutillus, pagi, parcus, szechwanus*.
COMMENTS: Includes *blythi, minutillus*, and *pagi*; see Hill and Yoshiyuki (1980:186). Corbet and Hill (1980:49) listed *minutillus* as a distinct species without comment.

Rhinolophus rex G. M. Allen, 1923. Am. Mus. Novit., 85:3.
TYPE LOCALITY: China, Szechwan, Wanhsien.
DISTRIBUTION: SW China.

Rhinolophus robinsoni K. Andersen, 1918. Ann. Mag. Nat. Hist., ser. 9, 2:375.
TYPE LOCALITY: Thailand, Surat Thani, Bandon.
DISTRIBUTION: W Malaysia, Thailand, adjacent small islands.
SYNONYMS: *klossi*.
COMMENTS: Includes *klossi*; see Medway (1969:24).

Rhinolophus rouxii Temminck, 1835. Monogr. Mamm., 2:306.
TYPE LOCALITY: India, Pondicherry and Calcutta.
DISTRIBUTION: Sri Lanka and India to S China and Vietnam.
SYNONYMS: *cinerascens, fulvidus, petersi, rammanika, rubidus, sinicus*.
COMMENTS: Includes *petersi*; see Sinha (1973:614, 615).

Rhinolophus rufus Eydoux and Gervais, 1836. *In* Laplace, Voy. autour du monde par les mers de l'Inde...la Favorite, 5(Zoologie), pt. 2:9.
TYPE LOCALITY: Philippines, Luzon, Manila.
DISTRIBUTION: Philippines.
COMMENTS: Name revived by Lawrence (1939:47-50).

Rhinolophus sedulus K. Andersen, 1905. Ann. Mag. Nat. Hist., ser. 7, 16:247.
TYPE LOCALITY: Malaysia, Sarawak.
DISTRIBUTION: W Malaysia, Borneo.
COMMENTS: Does not include *edax*; see Tate (1943:3, 4), but also see Chasen (1940:40).

Rhinolophus shameli Tate, 1943. Am. Mus. Novit., 1219:3.
TYPE LOCALITY: Thailand, Koh Chang Isl.
DISTRIBUTION: Burma, Thailand, Cambodia, W Malaysia.
COMMENTS: Described as a subspecies of *coelophyllus*, but see Hill and Thonglongya (1972:183-186).

Rhinolophus silvestris Aellen, 1959. Arch. Sci. Phys. Nat. Geneve, 12:228.
TYPE LOCALITY: Gabon, Latoursville, N'Dumbu Cave.
DISTRIBUTION: Gabon, Congo Republic.
COMMENTS: Considered a subspecies of *clivosus* by Hayman and Hill (1971:23), but see Koopman (1975:386).

Rhinolophus simplex K. Andersen, 1905. Proc. Zool. Soc. Lond., 1905:76.
TYPE LOCALITY: Indonesia, Lesser Sunda Isls, Lombok, 2,500 ft. (762 m).
DISTRIBUTION: Lombok, Sumbawa, and Komodo Isl (Lesser Sunda Isls).

Rhinolophus simulator K. Andersen, 1904. Ann. Mag. Nat. Hist., ser. 7, 14:384.
TYPE LOCALITY: Zimbabwe, Mazoe.
DISTRIBUTION: South Africa to S Sudan and Ethiopia; Cameroon; Nigeria; Guinea.
SYNONYMS: *alticolus, bembanicus.*
COMMENTS: Includes *alticolus* and *bembanicus;* see Koopman (1975:387) and Hayman and Hill (1971:25).

Rhinolophus stheno K. Andersen, 1905. Proc. Zool. Soc. Lond., 1905:91.
TYPE LOCALITY: Malaysia, Selangor.
DISTRIBUTION: W Malaysia; Thailand; Sumatra and Java (Indonesia).

Rhinolophus subbadius Blyth, 1844. J. Asiat. Soc. Bengal, 13:486.
TYPE LOCALITY: Nepal.
DISTRIBUTION: Assam (India), Nepal, Vietnam, Burma.

Rhinolophus subrufus K. Andersen, 1905. Ann. Mag. Nat. Hist., ser. 7, 16:283.
TYPE LOCALITY: Philippines, Luzon, Manila.
DISTRIBUTION: Philippines.
SYNONYMS: *bunkeri.*
COMMENTS: Includes *bunkeri;* see Lawrence (1939:52, 53).

Rhinolophus swinnyi Gough, 1908. Ann. Transvaal Mus., 1:72.
TYPE LOCALITY: South Africa, Cape Prov., Pondoland, Ngqeleni Dist.
DISTRIBUTION: South Africa to S Zaire and Zanzibar.
SYNONYMS: *piriensis, rhodesiae.*
COMMENTS: Probably a subspecies of *denti.*

Rhinolophus thomasi K. Andersen, 1905. Proc. Zool. Soc. Lond., 1905:100.
TYPE LOCALITY: Burma, Karin Hills.
DISTRIBUTION: Burma, Vietnam, Thailand, Yunnan (China).
SYNONYMS: *latifolius, septentrionalis.*

Rhinolophus trifoliatus Temminck, 1834. Tijdschr. Nat. Gesch. Physiol., 1:24.
TYPE LOCALITY: Indonesia, Java.
DISTRIBUTION: Malay Peninsula, SW Thailand, Burma, NE India, Borneo, Sumatra, Riau Archipelago, Banguey Isl, Java, Banka Isl and Nias Isl.
SYNONYMS: *edax, niasensis, solitarius.*
COMMENTS: Includes *edax,* see Tate (1943:3); but also see Chasen (1940:40, 41).

Rhinolophus virgo K. Andersen, 1905. Proc. Zool. Soc. Lond., 1905:88.
TYPE LOCALITY: Philippines, Luzon, Camarines Sur, Pasacao.
DISTRIBUTION: Philippines.

Rhinolophus yunanensis Dobson, 1872. J. Asiat. Soc. Bengal, 41:336.
TYPE LOCALITY: China, Yunnan, Hotha.
DISTRIBUTION: Yunnan (China), Thailand, NE India.
COMMENTS: See Lekagul and McNeely (1977:152, 154) for distinction of this species from *pearsonii.*

Subfamily Hipposiderinae Lydekker, 1891. *In* Flower and Lydekker, Mamm., Living and Extinct, p. 657.

Anthops Thomas, 1888. Ann. Mag. Nat. Hist., ser. 6, 1:156.
TYPE SPECIES: *Anthops ornatus* Thomas, 1888.

Anthops ornatus Thomas, 1888. Ann. Mag. Nat. Hist., ser. 6, 1:156.
TYPE LOCALITY: Solomon Isls, Guadalcanal Isl, Aola.
DISTRIBUTION: Solomon Isls.

Asellia Gray, 1838. Mag. Zool. Bot., 2:493.
TYPE SPECIES: *Rhinolophus tridens* E. Geoffroy, 1813.

Asellia patrizii DeBeaux, 1931. Ann. Mus. Civ. Stor. Nat. Genova, 55:186.
TYPE LOCALITY: Ethiopia, Dancalia, Gaare.
DISTRIBUTION: N Ethiopia and islands in the Red Sea.

Asellia tridens (E. Geoffroy, 1813). Ann. Mus. Hist. Nat. Paris, 20:265.
TYPE LOCALITY: Egypt, Qena, near Luxor.
DISTRIBUTION: Pakistan to Arabia, Sinai peninsula (NE Egypt) and Israel; Egypt to Morocco, Senegal, Chad, Sudan and S Somalia; Socotra (Yemen); perhaps Zanzibar.
SYNONYMS: *diluta, italosomalica, murraiana, pallida.*

Aselliscus Tate, 1941. Am. Mus. Novit., 1140:2.
TYPE SPECIES: *Rhinolophus tricuspidatus* Temminck, 1835.

Aselliscus stoliczkanus (Dobson, 1871). Proc. Asiat. Soc. Bengal, p. 106.
TYPE LOCALITY: Malaysia, West, Penang.
DISTRIBUTION: Burma, S China, Thailand, Laos, Vietnam, W Malaysia.
SYNONYMS: *trifidus, wheeleri* (see Sanborn, 1952*b*:3).

Aselliscus tricuspidatus (Temminck, 1835). Monogr. Mamm., 2:20.
TYPE LOCALITY: Indonesia, Molucca Isls, Amboina.
DISTRIBUTION: Molucca Isls, New Guinea, Bismarck Arch., Solomon Isls (including Santa Cruz Isls), Vanuatu (= New Hebrides), adjacent small islands.
SYNONYMS: *koopmani, novaeguinae, novaehebridensis.*
COMMENTS: Revised by Schlitter et al. (1983).

Cloeotis Thomas, 1901. Ann. Mag. Nat. Hist., ser. 7, 8:28.
TYPE SPECIES: *Cloeotis percivali* Thomas, 1901.
COMMENTS: Reviewed by Hill (1982).

Cloeotis percivali Thomas, 1901. Ann. Mag. Nat. Hist., ser. 7, 8:28.
TYPE LOCALITY: Kenya, Coast Prov., Takaungu.
DISTRIBUTION: Kenya, Tanzania, S Zaire, Mozambique, Zambia, Zimbabwe, SE Botswana, Swaziland, Transvaal (South Africa).
SYNONYMS: *australis.*

Coelops Blyth, 1848. J. Asiat. Soc. Bengal, 17:251.
TYPE SPECIES: *Coelops frithii* Blyth, 1848.
SYNONYMS: *Chilophylla.*
COMMENTS: Includes *Chilophylla*; see Ellerman and Morrison-Scott (1951:131).

Coelops frithi Blyth, 1848. J. Asiat. Soc. Bengal, 17:251.
TYPE LOCALITY: India, Bengal, Sunderbans.
DISTRIBUTION: NE India to S China and Vietnam, south to W Malaysia, Taiwan, and Java and Bali.
SYNONYMS: *bernsteini, formosanus, inflatus, sinicus.*

Coelops hirsutus (Miller, 1911). Proc. U.S. Natl. Mus., 38:395.
TYPE LOCALITY: Philippines, Mindoro Isl, Alag River.
DISTRIBUTION: Mindoro (Philippines).
COMMENTS: Probably a subspecies of *robinsoni*; see Hill (1972:30).

Coelops robinsoni Bonhote, 1908. J. Fed. Malay St. Mus., 3:4.
TYPE LOCALITY: Malaya, Pahang, foot of Mt. Tahan.
DISTRIBUTION: W Malaysia, Borneo. The record from Thailand is in error; see Hill (1983:152).

Hipposideros Gray, 1831. Zool. Misc., 1:37.
TYPE SPECIES: *Vespertilio speoris* Schneider, 1800.
COMMENTS: Revised by Hill (1963*b*).

Hipposideros abae J. A. Allen, 1917. Bull. Am. Mus. Nat. Hist., 37:432.
TYPE LOCALITY: Zaire, Oriental, Aba.
DISTRIBUTION: Guinea-Bissau to SW Sudan and Uganda.

Hipposideros armiger (Hodgson, 1835). J. Asiat. Soc. Bengal, 4:699.
TYPE LOCALITY: Nepal.
DISTRIBUTION: N India, Nepal, Burma, S China, Vietnam, Laos, Thailand, Malay Peninsula, Taiwan.
SYNONYMS: *debilis, swinhoei, terasensis, tranninhensis.*
COMMENTS: Includes *terasensis*, but see Yoshiyuki (1991).

Hipposideros ater Templeton, 1848. J. Asiat. Soc. Bengal, 17:252.
TYPE LOCALITY: Sri Lanka, Western Prov., Colombo.
DISTRIBUTION: Sri Lanka; India to W Malaysia, through Philippines, Indonesia, and New Guinea to N Queensland, N Northern Territory, and N Western Australia (Australia).
SYNONYMS: *albaniensis, amboiensis, antricola, aruensis, atratus, gilberti, nicobarulae, saevus, toala.*
COMMENTS: Formerly included in *bicolor*, but see Hill (1963*b*:30).

Hipposideros beatus K. Andersen, 1906. Ann. Mag. Nat. Hist., ser. 7, 17:279.
TYPE LOCALITY: Equatorial Guinea, Rio Muni, 15 mi. (24 km) from Benito River.
DISTRIBUTION: Guinea-Bissau, Sierra Leone, Liberia, Ghana, Ivory Coast, Nigeria, Cameroon, Rio Muni (Equatorial Guinea), Gabon, N Zaire.
SYNONYMS: *maximus.*

Hipposideros bicolor (Temminck, 1834). Tijdschr. Nat. Gesch. Physiol., 1:19.
TYPE LOCALITY: Indonesia, Java, Anger coast.
DISTRIBUTION: Malaysia to the Philippines, Timor (Indonesia), and adjacent small islands.
SYNONYMS: *atrox, erigens, javanicus, major.*
COMMENTS: Includes *erigens*; see Hill (1963*b*:27). Does not include *pomona, gentilis*, or *macrobullatus*; see Hill et al. (1986).

Hipposideros breviceps Tate, 1941. Bull. Am. Mus. Nat. Hist., 78:358.
TYPE LOCALITY: Indonesia, Sumatra, Mentawai Isls, N Pagi Isl.
DISTRIBUTION: Mentawai Isls (Indonesia).

Hipposideros caffer (Sundevall, 1846). Ofv. K. Svenska Vet.-Akad. Forhandl. Stockholm, 3(4):118.
TYPE LOCALITY: South Africa, Natal, near Durban.
DISTRIBUTION: Most of subsaharan Africa except the central forested region; Morocco; Yemen; Zanzibar and Pemba.
SYNONYMS: *angolensis, aurantiaca, bicornis, braima, gracilis, nanus, tephrus* (see Hayman and Hill, 1971:29).

Hipposideros calcaratus (Dobson, 1877). Proc. Zool. Soc. Lond., 1877:122.
TYPE LOCALITY: Papua New Guinea, Bismarck Archipelago, Duke of York Isl.
DISTRIBUTION: New Guinea, Bismarck Arch., Solomon Isls, adjacent small islands.
SYNONYMS: *cupidus.*
COMMENTS: Includes *cupidus*; see Smith and Hill (1981:8).

Hipposideros camerunensis Eisentraut, 1956. Zool. Jahrb. Abt. Syst. Oekol. Geogr. Tiere, 84:526.
TYPE LOCALITY: Cameroon, near Buea.
DISTRIBUTION: Cameroon, E Zaire, W Kenya.

Hipposideros cervinus (Gould, 1863). Mamm. Aust., 3: pl. 34.
TYPE LOCALITY: Australia, Queensland, Cape York.
DISTRIBUTION: W Malaysia, Sumatra, and the Philippines to Vanuatu (= New Hebrides) and NE Australia.
SYNONYMS: *batchianensis, celebensis, labuanensis, misoriensis, schneideri.*
COMMENTS: Considered distinct from *galeritus* (in which it was formerly included) by Jenkins and Hill (1981).

Hipposideros cineraceus Blyth, 1853. J. Asiat. Soc. Bengal, 22:410.
TYPE LOCALITY: Pakistan, Punjab, Salt Range, near Pind Dadan Khan.

DISTRIBUTION: Pakistan to Vietnam and Borneo; adjacent small islands; probably the Philippines.
SYNONYMS: *durgadasi, micropus, wrighti?.*
COMMENTS: Includes *durgadasi*; but see Khajuria (1982:288). Probably includes *wrighti*, see Hill and Francis (1984:308).

Hipposideros commersoni (E. Geoffroy, 1813). Ann. Mus. Hist. Nat. Paris, 20:263.
TYPE LOCALITY: Madagascar, Fort Dauphin.
DISTRIBUTION: Gambia to Ethiopia, south to Namibia, Botswana, Transvaal (South Africa) and Mozambique; Madagascar; adjacent small islands, including Saõ Thomé.
SYNONYMS: *gambiensis, gigas, marungensis, mostellum, niangarae, thomensis, viegasi, vittata.*

Hipposideros coronatus (Peters, 1871). Monatsb. K. Preuss. Akad. Wiss. Berlin, 1871:327.
TYPE LOCALITY: Philippines, Mindanao, Surigao, Mainit.
DISTRIBUTION: NE Mindanao (Philippines).

Hipposideros corynophyllus Hill, 1985. Mammalia, 49:527.
TYPE LOCALITY: Papua New Guinea, W Sepik, 3 km ENE Telephomin, 1800 m.
DISTRIBUTION: C New Guinea.

Hipposideros coxi Shelford, 1901. Ann. Mag. Nat. Hist., ser. 7, 8:113.
TYPE LOCALITY: Malaysia, Sarawak, Mt. Penrisen, 4,200 ft. (1,280 m).
DISTRIBUTION: Sarawak (Borneo, Malaysian part).

Hipposideros crumeniferus (Lesueur and Petit, 1807). *In* Péron, Voy. Decouv. Terres Austral., Atlas, pl. 35.
TYPE LOCALITY: Indonesia, Timor.
DISTRIBUTION: Timor (Indonesia).
COMMENTS: Based on plate only; not certainly determinable; see Laurie and Hill (1954:56) and Hill (1963*b*:23).

Hipposideros curtus G. M. Allen, 1921. Rev. Zool. Afr., 9:194.
TYPE LOCALITY: Cameroon, Sakbayeme.
DISTRIBUTION: Cameroon, Bioko.
SYNONYMS: *sandersoni.*
COMMENTS: Includes *sandersoni*; see Hill (1963*b*:60).

Hipposideros cyclops (Temminck, 1853). Esquisses Zool. sur la Côte de Guine, p. 75.
TYPE LOCALITY: Ghana, Boutry River.
DISTRIBUTION: Kenya and S Sudan to Senegal and Guinea-Bissau; Bioko.
SYNONYMS: *langi, micaceus.*

Hipposideros diadema (E. Geoffroy, 1813). Ann. Mus. Hist. Nat. Paris, 20:263.
TYPE LOCALITY: Indonesia, Lesser Sunda Isls, Timor Isl.
DISTRIBUTION: Burma and Vietnam through Thailand, W Malaysia and Indonesia to New Guinea, Bismarck Arch., Solomon Isls and NE and NC Australia; Philippines; Nicobar Isls.
SYNONYMS: *andersoni, ceramensis, custos, demissus, enganus, euotis, griseus, inornatus, malaitensis, masoni, mirandus, natunensis, nicobarensis, nobilis, oceanitis, ornatus, pullatus, reginae, speculator, trobrius, vicarius.*

Hipposideros dinops K. Andersen, 1905. Ann. Mag. Nat. Hist., ser. 7, 16:502.
TYPE LOCALITY: Solomon Isls, New Georgia Group, Rubiana Isl.
DISTRIBUTION: Rubiana, Ysabel, Malaita and Bougainville Isls (Solomon Isls); Peleng Isl and Sulawesi (Indonesia).
SYNONYMS: *pelingensis.*
COMMENTS: Includes *pelingensis*; see Hill (1963*b*:113).

Hipposideros doriae (Peters, 1871). Monatsb. K. Preuss. Akad. Wiss. Berlin, 1871:326.
TYPE LOCALITY: Malaysia, Sarawak.
DISTRIBUTION: Borneo.
COMMENTS: May be an earlier name for *sabanus*, which it antedates; see Hill (1963*b*:24, 46, 47).

Hipposideros dyacorum Thomas, 1902. Ann. Mag. Nat. Hist., ser. 7, 9:271.
TYPE LOCALITY: Malaysia, Sarawak, Baram, Mt. Mulu.
DISTRIBUTION: Borneo, Malaya.

Hipposideros fuliginosus (Temminck, 1853). Esquisses Zool. sur la Côte de Guine, p. 77.
TYPE LOCALITY: Ghana.
DISTRIBUTION: Liberia to Zaire and Ethiopia.

Hipposideros fulvus Gray, 1838. Mag. Zool. Bot., 2:492.
TYPE LOCALITY: India, Karnatika, Dharwar.
DISTRIBUTION: Pakistan to Vietnam, south to Sri Lanka.
SYNONYMS: *aurita, fulgens, murinus, pallidus.*

Hipposideros galeritus Cantor, 1846. J. Asiat. Soc. Bengal, 15:183.
TYPE LOCALITY: Malaysia, Penang.
DISTRIBUTION: Sri Lanka and India through SE Asia to Java and Borneo.
SYNONYMS: *brachyotis, insolens, longicauda.*
COMMENTS: Includes *longicauda;* see Hill (1963*b*:56). Formerly included *cervinus;* but see Jenkins and Hill (1981).

Hipposideros halophyllus Hill and Yenbutra, 1984. Bull. Brit. Mus. (Nat. Hist.) Zool., 47:77.
TYPE LOCALITY: Thailand, Lop Buri, Tha Woong, Khao Sa Moa Khan.
DISTRIBUTION: Thailand.

Hipposideros inexpectatus Laurie and Hill, 1954. List of land mammals of New Guinea, Celebes, and adjacent islands, p. 60.
TYPE LOCALITY: Indonesia, Sulawesi, Poso (= Posso).
DISTRIBUTION: N Sulawesi (Indonesia).

Hipposideros jonesi Hayman, 1947. Ann. Mag. Nat. Hist., ser. 11, 14:71.
TYPE LOCALITY: Sierra Leone, Makeni.
DISTRIBUTION: Sierra Leone and Guinea to Mali, Burkina Faso and Nigeria.

Hipposideros lamottei Brosset, 1984. Mammalia, 48:548.
TYPE LOCALITY: Guinea, Mt. Nimba, Pierre Richaud.
DISTRIBUTION: Mt. Nimba on Guinea-Liberia border, but probably more widespread.
COMMENTS: Distinction from *H. ruber* is not entirely clear.

Hipposideros lankadiva Kelaart, 1850. J. Sri Lanka Branch Asiat. Soc., 2(2):216.
TYPE LOCALITY: Sri Lanka, Kandy.
DISTRIBUTION: Sri Lanka, S and C India.
SYNONYMS: *indus, mixtus, unitus.*

Hipposideros larvatus (Horsfield, 1823). Zool. Res. Java, part 6:*Rhinolophus larvatus*, pl. and 10 unno. pp.
TYPE LOCALITY: Indonesia, Java.
DISTRIBUTION: Bangladesh to Vietnam; Yunnan, Kwangsi and Hainan (China); and through W Malaysia to Sumatra, Java, Borneo, and Sumba (Indonesia), and adjacent small islands.
SYNONYMS: *alongensis, barbensis, deformis, grandis, insignis, leptophyllus, neglectus, poutensis, sumbae, vulgaris.*

Hipposideros lekaguli Thonglongya and Hill, 1974. Mammalia, 38:286.
TYPE LOCALITY: Thailand, Saraburi, Kaeng Khoi, Phu Nam Tok Tak Kwang.
DISTRIBUTION: Thailand.

Hipposideros lylei Thomas, 1913. Ann. Mag. Nat. Hist., ser. 8, 12:88.
TYPE LOCALITY: Thailand, 50 mi. (80 km) N Chiengmai, Chiengdao Cave.
DISTRIBUTION: Burma, Thailand, W Malaysia.

Hipposideros macrobullatus Tate, 1941. Bull. Amer. Mus. Nat. Hist., 78:357.
TYPE LOCALITY: Indonesia, Sulawesi, Talassa (Maros).
DISTRIBUTION: Sulawesi, Ceram (Molucca Isls) and Kangean (Java Sea).
COMMENTS: Formerly included in *H. bicolor*, but see Hill et al. (1986).

Hipposideros maggietaylorae Smith and Hill, 1981. Los Angeles Cty. Mus. Contrib. Sci., 331:9.
TYPE LOCALITY: Papua New Guinea, Bismarck Arch., New Ireland, 1.3 km S, 3 km E, Lakuramau Plantation.
DISTRIBUTION: New Guinea, Bismarck Arch.
SYNONYMS: *erroris*.
COMMENTS: Formerly confused with *H. calcaratus*; see Smith and Hill (1981).

Hipposideros marisae Aellen, 1954. Rev. Suisse Zool., 61:474.
TYPE LOCALITY: Ivory Coast, Duekoue, White Leopard Rock.
DISTRIBUTION: Ivory Coast, Liberia, Guinea.

Hipposideros megalotis (Heuglin, 1862). Nova Acta Acad. Caes. Leop.-Carol., Halle, 29(8):4, 8.
TYPE LOCALITY: Ethiopia, Eritrea, Bogos Land, Keren.
DISTRIBUTION: Ethiopia, Djibouti, Somalia, Kenya, Saudi Arabia.

Hipposideros muscinus (Thomas and Doria, 1886). Ann. Mus. Civ. Stor. Nat. Genova, 4:201.
TYPE LOCALITY: Papua New Guinea, Western Prov., Fly River.
DISTRIBUTION: New Guinea.

Hipposideros nequam K. Andersen, 1918. Ann. Mag. Nat. Hist., ser. 9, 2:380.
TYPE LOCALITY: Malaysia, Selangor, Klang.
DISTRIBUTION: Known only from the type locality.
COMMENTS: See Hill (1963*b*:36).

Hipposideros obscurus (Peters, 1861). Monatsb. K. Preuss. Akad. Wiss. Berlin, 1861:707.
TYPE LOCALITY: Philippines, Luzon, Camarines, Paracale.
DISTRIBUTION: Philippines.

Hipposideros papua (Thomas and Doria, 1886). Ann. Mus. Civ. Stor. Nat. Genova, 4:204.
TYPE LOCALITY: Indonesia, Irian Jaya, Tjenderawasih Div., Misori Isl (= Biak Isl), Korido.
DISTRIBUTION: Biak Isl, W New Guinea, and N Molucca Isls; see Hill and Rozendaal (1989:103-104) for range.
STATUS: IUCN - Rare.

Hipposideros pomona K. Andersen, 1918. Ann. Mag. Nat. Hist., ser. 9, 2:380, 381.
TYPE LOCALITY: India, Mysore, Coorg, Haleri.
DISTRIBUTION: India to S China and W Malaysia.
SYNONYMS: *gentilis, sinensis*.
COMMENTS: Formerly included in *H. bicolor* but see Hill et al. (1986).

Hipposideros pratti Thomas, 1891. Ann. Mag. Nat. Hist., ser. 6, 7:527.
TYPE LOCALITY: China, Szechwan, Kiatingfu.
DISTRIBUTION: S China, Burma, Thailand, Vietnam, W Malaysia.

Hipposideros pygmaeus (Waterhouse, 1843). Proc. Zool. Soc. Lond., 1843:67.
TYPE LOCALITY: Philippines.
DISTRIBUTION: Philippines.

Hipposideros ridleyi Robinson and Kloss, 1911. J. Fed. Malay St. Mus., 4:241.
TYPE LOCALITY: Malaysia, Singapore, Botanic Gardens.
DISTRIBUTION: W Malaysia, Borneo.
STATUS: U.S. ESA - Endangered; IUCN - Indeterminate.

Hipposideros ruber (Noack, 1893). Zool. Jahrb. Abt. Syst. Oekol. Geogr. Tiere, 7:586.
TYPE LOCALITY: Tanzania, Eastern Province, Ngerengere River.
DISTRIBUTION: Senegal to Ethiopia, south to Angola, Zambia, Malawi, and Mozambique; Bioko; São Tomé and Principe.
SYNONYMS: *centralis, guineensis, niapu*.

Hipposideros sabanus Thomas, 1898. Ann. Mag. Nat. Hist., ser. 7, 1:243.
TYPE LOCALITY: Malaysia, Sarawak, Lawas.
DISTRIBUTION: Borneo, Sumatra, W Malaysia.
COMMENTS: Possibly a synonym of *doriae*; see Hill (1963*b*:47).

Hipposideros schistaceus K. Andersen, 1918. Ann. Mag. Nat. Hist., ser. 9, 2:382.
TYPE LOCALITY: India, Karnatika, Bellary.

DISTRIBUTION: S India.

Hipposideros semoni Matschie, 1903. Denks. Med. Nat. Ges. Jena (Semon Zool. Forsch. Aust.), 8:774 (Heft 6:132).
TYPE LOCALITY: Australia, Queensland, Cooktown.
DISTRIBUTION: N Queensland (Australia), E New Guinea.

Hipposideros speoris (Schneider, 1800). *In* Schreber, Die Säugethiere, pl. 59b.
TYPE LOCALITY: India, Madras, Tranquebar.
DISTRIBUTION: Peninsular India, Sri Lanka.
SYNONYMS: *apiculatus, aureus, blythi, dukhunensis, marsupialis, pulchellus, templetonii.*

Hipposideros stenotis Thomas, 1913. Ann. Mag. Nat. Hist., ser. 8, 12:206.
TYPE LOCALITY: Australia, Northern Territory, Mary River.
DISTRIBUTION: Northern Territory, N Western Australia and N Queensland (Australia). A New Guinea record is probably erroneous, see Hill (1963*b*:87).

Hipposideros turpis Bangs, 1901. Am. Nat., 35:561.
TYPE LOCALITY: Japan, Ryukyu Isls, Sakishima Isls, Ishigaki Isl.
DISTRIBUTION: Peninsular Thailand; Ishigaki Isl, Yonakuni Isl and Iriomote Isl (S Ryukyu Isls, Japan).
SYNONYMS: *pendleburyi.*

Hipposideros wollastoni Thomas, 1913. Ann. Mag. Nat. Hist., ser. 8, 12:205.
TYPE LOCALITY: Indonesia, Irian Jaya, Utakwa River, 2,500 ft. (762 m).
DISTRIBUTION: W and C New Guinea.

Paracoelops Dorst, 1947. Bull. Mus. Hist. Nat. Paris, ser. 2, 19:436.
TYPE SPECIES: *Paracoelops megalotis* Dorst, 1947.

Paracoelops megalotis Dorst, 1947. Bull. Mus. Hist. Nat. Paris, ser. 2, 19:436.
TYPE LOCALITY: Vietnam, Annam, Vinh.
DISTRIBUTION: C Vietnam (known only from the type specimen).

Rhinonicteris Gray, 1847. Proc. Zool. Soc. Lond., 1847:16.
TYPE SPECIES: *Rhinolophus aurantius* Gray, 1845.
COMMENTS: *Rhinonicteris* is the correct spelling if original orthography is adhered to. *Rhinonycteris* Gray, 1866, Proc. Zool. Soc. Lond., 1866:81, is sometimes used. Reviewed by Hill (1982) who spelled it *Rhinonycteris.*

Rhinonicteris aurantia (Gray, 1845). *In* Eyre, Central Australia, 1:405.
TYPE LOCALITY: Australia, Northern Territory, Port Essington.
DISTRIBUTION: N Western Australia, Northern Territory and NW Queensland (Australia).
STATUS: IUCN - Insufficiently known.

Triaenops Dobson, 1871. J. Asiat. Soc. Bengal, 40:455.
TYPE SPECIES: *Triaenops persicus* Dobson, 1871.
COMMENTS: Reviewed by Hill (1982).

Triaenops furculus Trouessart, 1906. Bull. Mus. Hist. Nat. Paris, 1906, 7:446.
TYPE LOCALITY: Madagascar, near Tulear, St. Augustine Bay.
DISTRIBUTION: N and W Madagascar, Aldabra Isl.
SYNONYMS: *aurita.*
COMMENTS: Includes *aurita;* see Hayman and Hill (1971:30).

Triaenops persicus Dobson, 1871. J. Asiat. Soc. Bengal, 40:455.
TYPE LOCALITY: Iran, Shiraz, 4,750 ft. (1,448 m).
DISTRIBUTION: Somalia, Ethiopia, Kenya, Tanzania, Uganda, Angola, Zanzibar, Mozambique, Yemen, Oman, Congo Republic, Iran, perhaps Egypt.
SYNONYMS: *afer, humbloti, macdonaldi, majusculus, rufus* (see Hayman and Hill, 1971:30, and Hill, 1982).

Family Noctilionidae Gray, 1821. London Med. Reposit., 15:299.

Noctilio Linnaeus, 1766. Syst. Nat., 12th ed., 1:88.
TYPE SPECIES: *Noctilio americanus* Linnaeus, 1766 (= *Vespertilio leporinus* Linnaeus, 1758).
SYNONYMS: *Dirias*.

Noctilio albiventris Desmarest, 1818. Nouv. Dict. Hist. Nat., Nouv. ed., 23:15.
TYPE LOCALITY: Brazil, Bahia, Rio Sao Francisco.
DISTRIBUTION: S Mexico to Guianas, E Brazil, N Argentina, and Peru.
SYNONYMS: *affinis, cabrerai, irex; minor* Osgood; *zaparo*.
COMMENTS: Subgenus *Dirias*. Formerly referred to as *labialis*; see Davis (1976). See Hood and Pitocchelli (1983, Mammalian Species, 197).

Noctilio leporinus (Linnaeus, 1758). Syst. Nat., 10th ed., 1:32.
TYPE LOCALITY: Surinam.
DISTRIBUTION: Sinaloa (Mexico) to Guianas, S Brazil, N Argentina, and Peru; Trinidad; Greater and Lesser Antilles; S Bahamas.
SYNONYMS: *americanus, brooksiana, dorsatus; labialis* Kerr; *mastivus, mexicanus; minor* Fermin; *rufescens, rufipes, rufus, unicolor, vittatus*.
COMMENTS: Subgenus *Dirias*. See Hood and Jones (1984, Mammalian Species, 216).

Family Mormoopidae Koch, 1862-63. Jahrb. Ver. Naturk. in Nassau, Wiesbaden, heft 17-18:358.
COMMENTS: Revised by Smith (1972).

Mormoops Leach, 1821. Trans. Linn. Soc. Lond., 13:76.
TYPE SPECIES: *Mormoops blainvillii* Leach, 1821.
SYNONYMS: *Aello*.
COMMENTS: This name is used instead of *Aello* following Opinion 462 of the International Commission on Zoological Nomenclature (1957*c*).

Mormoops blainvillii Leach, 1821. Trans. Linn. Soc. Lond., 13:77.
TYPE LOCALITY: Jamaica.
DISTRIBUTION: Greater Antilles, adjacent small islands.
SYNONYMS: *cinnamomeum, cuvieri*.

Mormoops megalophylla (Peters, 1864). Monatsb. K. Preuss. Akad. Wiss. Berlin, 1864:381.
TYPE LOCALITY: Mexico, Coahuila, Parras.
DISTRIBUTION: S Texas, S Arizona (USA), and Baja California (Mexico) to NW Peru and N Venezuela; Aruba, Curacao, and Bonaire (Netherlands Antilles); Trinidad; Margarita Isl (Venezuela).
SYNONYMS: *carteri, intermedia, rufescens, senicula, tumidiceps*.

Pteronotus Gray, 1838. Mag. Zool. Bot., 2:500.
TYPE SPECIES: *Pteronotus davyi* Gray, 1838.
SYNONYMS: *Chilonycteris, Phyllodia*.
COMMENTS: Includes *Chilonycteris* and *Phyllodia*; see Smith (1972:55). A key to this genus was presented by Herd (1983) and by Rodríguez-Durán and Kunz (1992).

Pteronotus davyi Gray, 1838. Mag. Zool. Bot., 2:500.
TYPE LOCALITY: Trinidad and Tobago, Trinidad.
DISTRIBUTION: NW Peru and N Venezuela to S Baja California, S Sonora, and Nuevo Leon (Mexico); Trinidad; S Lesser Antilles. A Brazilian record is erroneous, see Willig and Mares (1989).
SYNONYMS: *calvus, fulvus, incae*.
COMMENTS: Subgenus *Pteronotus*. See Adams (1989, Mammalian Species, 346).

Pteronotus gymnonotus Natterer, 1843. *In* Wagner, Arch. Naturgesch., 9:367.
TYPE LOCALITY: Brazil, Mato Grosso, Cuiaba.
DISTRIBUTION: S Veracruz (Mexico) to Peru, NE Brazil, and Guyana.
SYNONYMS: *centralis, suapurensis*.
COMMENTS: Subgenus *Pteronotus*. Includes *suapurensis*; see Smith (1977:245).

Pteronotus macleayii (Gray, 1839). Ann. Nat. Hist., 4:5.
TYPE LOCALITY: Cuba, Habana, Guanabacoa.
DISTRIBUTION: Cuba, Jamaica.
SYNONYMS: *griseus*.
COMMENTS: Subgenus *Chilonycteris*.

Pteronotus parnellii (Gray, 1843). Proc. Zool. Soc. Lond., 1843:50.
TYPE LOCALITY: Jamaica.
DISTRIBUTION: Peru, Brazil, Guianas, and Venezuela to S Sonora and S Tamaulipas (Mexico); Cuba; Jamaica; Puerto Rico; Hispaniola; Trinidad and Tobago; Margarita Isl (Venezuela); La Gonave Isl (Haiti).
SYNONYMS: *boothi, fuscus, gonavensis, mesoamericanus, mexicanus, osburni, paraguanensis, portoricensis, pusillus, rubiginosus*.
COMMENTS: Subgenus *Phyllodia*. See Herd (1983, Mammalian Species, 209).

Pteronotus personatus (Wagner, 1843). Arch. Naturgesch., 9:367.
TYPE LOCALITY: Brazil, Mato Grosso, São Vicente.
DISTRIBUTION: Colombia, Peru, Brazil, and Surinam to S Sonora and S Tamaulipas (Mexico); Trinidad.
SYNONYMS: *continentis, psilotis*.
COMMENTS: Includes *psilotis*; see Smith (1972:92).

Pteronotus quadridens (Gundlach, 1840). Arch. Naturgesch., 6:357.
TYPE LOCALITY: Cuba, Matanzas, Canimar.
DISTRIBUTION: Cuba, Jamaica, Hispaniola, Puerto Rico.
SYNONYMS: *fuliginosus, inflata, torrei*.
COMMENTS: Subgenus *Chilonycteris*. Includes *torrei*; For use of *quadridens* in place of *fuliginosus*; see Silva-Taboada (1976:7). See Rodríguez-Durán and Kunz (1992, Mammalian Species, 395).

Family Phyllostomidae Gray, 1825. Zool. J., 2(6):242.
COMMENTS: Includes Desmodontidae; see Jones and Carter (1976:7). For use of this familial name rather than Phyllostomatidae, see Handley (1980:10).

Subfamily Phyllostominae Gray, 1825. Zool. J., 2(6):242.

Chrotopterus Peters, 1865. Monatsb. K. Preuss. Akad. Wiss. Berlin, 1865:505.
TYPE SPECIES: *Vampyrus auritus* Peters, 1856.

Chrotopterus auritus (Peters, 1856). Monatsb. K. Preuss. Akad. Wiss. Berlin, 1865:5.
TYPE LOCALITY: Mexico, corrected to Brazil, Santa Catarina by Carter and Dolan (1978:37).
DISTRIBUTION: Veracruz (Mexico) to the Guianas, S Brazil, and N Argentina.
SYNONYMS: *australis, guianae*.
COMMENTS: See Medellin (1989, Mammalian Species, 343).

Lonchorhina Tomes, 1863. Proc. Zool. Soc. Lond., 1863:81.
TYPE SPECIES: *Lonchorhina aurita* Tomes, 1863.
COMMENTS: Reviewed by Hernandez-Camacho and Cadena-G. (1978). A key to the genus was presented by Lassieur and Wilson (1989).

Lonchorhina aurita Tomes, 1863. Proc. Zool. Soc. Lond., 1863:83.
TYPE LOCALITY: Trinidad and Tobago, Trinidad.
DISTRIBUTION: Oaxaca (Mexico) to SE Brazil, Peru, and Ecuador; Trinidad; perhaps New Providence Isl (Bahama Isls), see Jones and Carter (1976:10).
SYNONYMS: *occidentalis*.
COMMENTS: Includes *occidentalis*; see Jones and Carter (1976:10). See Lassieur and Wilson (1989, Mammalian Species, 347).

Lonchorhina fernandezi Ochoa and Ibanez, 1982. Mem. Soc. Cienc. Nat. La Salle, 42:147.
TYPE LOCALITY: Venezuela, Amazonas, 40-50 km (by road) NE Puerto Ayacucho.
DISTRIBUTION: S Venezuela.

Lonchorhina marinkellei Hernandez-Camacho and Cadena-G., 1978. Caldasia, 12:229.
TYPE LOCALITY: Colombia, Vaupes, near Mitu, Durania.
DISTRIBUTION: E Colombia to French Guiana.

Lonchorhina orinocensis Linares and Ojasti, 1971. Novid. Cient. Contrib. Occas. Mus. Hist. Nat. La Salle, Ser. Zool., 36:2.
TYPE LOCALITY: Venezuela, Bolivar, 50 km NE Puerto Paez, Boca de Villacoa.
DISTRIBUTION: Venezuela, Colombia.

Macrophyllum Gray, 1838. Mag. Zool. Bot., 2:489.
TYPE SPECIES: *Macrophyllum nieuwiedii* Gray, 1838 (= *Phyllostoma macrophyllum* Schinz, 1821).

Macrophyllum macrophyllum (Schinz, 1821). Das Thierreich, 1:163.
TYPE LOCALITY: Brazil, Bahia, Rio Mucuri.
DISTRIBUTION: Tabasco (Mexico) to Peru, Bolivia, SE Brazil, and NE Argentina.
SYNONYMS: *nieuwiedii*.
COMMENTS: See Harrison (1975, Mammalian Species, 62).

Macrotus Gray, 1843. Proc. Zool. Soc. Lond., 1843:21.
TYPE SPECIES: *Macrotus waterhousii* Gray, 1843.
COMMENTS: Revised by Anderson and Nelson (1965).

Macrotus californicus Baird, 1858. Proc. Acad. Nat. Sci. Philadelphia, 10:116.
TYPE LOCALITY: USA, California, Imperial Co., Old Fort Yuma.
DISTRIBUTION: N Sinaloa and SW Chihuahua (Mexico) to S Nevada, S California (USA); Baja California and Tamaulipas (Mexico).
COMMENTS: For a comparison with *waterhousii*, see Davis and Baker (1974:26, 34) and Greenbaum and Baker (1976). Reviewed as a subspecies of *waterhousii* by Anderson (1969*a*).

Macrotus waterhousii Gray, 1843. Proc. Zool. Soc. Lond., 1843:21.
TYPE LOCALITY: Haiti.
DISTRIBUTION: Sonora and Hidalgo (Mexico) to Guatemala; Bahama Isls; Jamaica; Cuba; Cayman Isls (NW of Jamaica); Hispaniola and Beata Isls.
SYNONYMS: *bocourtianus, bulleri, compressus, heberfolium, jamaicensis, mexicanus, minor*.
COMMENTS: Includes *mexicanus*; see Anderson and Nelson (1965:25). See Anderson (1969*a*, Mammalian Species, 1).

Micronycteris Gray, 1866. Proc. Zool. Soc. Lond., 1866:113.
TYPE SPECIES: *Phyllophora megalotis* Gray, 1842.
SYNONYMS: *Barticonycteris, Glyphonycteris, Lampronycteris, Neonycteris, Trinycteris, Xenoctenes* (see Koopman, 1978*b*:4).
COMMENTS: Corbet and Hill (1980:54) listed *Barticonycteris* as a distinct genus without comment. Revised by Sanborn (1949). A key to the genus was presented by Medellin et al. (1985).

Micronycteris behnii (Peters, 1865). Monatsb. K. Preuss. Akad. Wiss. Berlin, 1865:505.
TYPE LOCALITY: Brazil, Mato Grosso, Cuiaba (= Cuyaba).
DISTRIBUTION: C Brazil, S Peru.
COMMENTS: Subgenus *Glyphonycteris*.

Micronycteris brachyotis (Dobson, 1879). Proc. Zool. Soc. Lond., 1878:880 [1879].
TYPE LOCALITY: French Guiana, Cayenne.
DISTRIBUTION: Oaxaca (Mexico) to French Guiana and Brazil; Trinidad.
SYNONYMS: *platyceps*.
COMMENTS: Subgenus *Lampronycteris*. Includes *platyceps*; see Jones and Carter (1976:9). See Medellin et al. (1985, Mammalian Species, 251).

Micronycteris daviesi (Hill, 1964). Mammalia, 28:557.
TYPE LOCALITY: Guyana, Essequibo Prov., Potaro road, 24 mi. (39 km) from Bartica.
DISTRIBUTION: Costa Rica to Peru and French Guiana.
COMMENTS: Subgenus *Barticonycteris*. Formerly included in genus *Barticonycteris*; see Koopman (1978*b*:4).

Micronycteris hirsuta (Peters, 1869). Monatsb. K. Preuss. Akad. Wiss. Berlin, 1869:397.
TYPE LOCALITY: Costa Rica, Guanacaste, Pozo Azul.
DISTRIBUTION: Honduras to French Guiana, Trinidad, Amazonian Brazil, Peru, and Ecuador.
COMMENTS: Subgenus *Xenoctenes*.

Micronycteris megalotis (Gray, 1842). Ann. Mag. Nat. Hist., [ser. 1], 10:257.
TYPE LOCALITY: Brazil, São Paulo, Pereque.
DISTRIBUTION: Tamaulipas and Jalisco (Mexico) to Peru, Bolivia, and Brazil; Trinidad and Tobago; Margarita Isl (Venezuela); Grenada.
SYNONYMS: *elongatum, homezi, megalotes, mexicana, microtis, pygmaeus, scrobiculatum, typica.*
COMMENTS: Subgenus *Micronycteris*. Includes *microtis;* see Gardner et al. (1970:715) and Jones et al. (1977:6). See Alonso-Mejia and Medellin (1991, Mammalian Species, 376).

Micronycteris minuta (Gervais, 1856). *In* F. Comte de Castelnau, Exped. Partes Cen. Am. Sud., Zool. (Sec. 7), Vol. 1, pt. 2(Mammifères):50.
TYPE LOCALITY: Brazil, Bahia, Capela Nova.
DISTRIBUTION: Honduras to S Brazil; Peru; Guianas; Trinidad; Bolivia.
SYNONYMS: *hypoleuca.*
COMMENTS: Subgenus *Micronycteris*.

Micronycteris nicefori Sanborn, 1949. Fieldiana Zool., 31:230.
TYPE LOCALITY: Colombia, Norte de Santander, Cucuta.
DISTRIBUTION: Belize to N Colombia, Venezuela, Guianas, Amazonian Brazil, and Peru; Trinidad.
COMMENTS: Subgenus *Trinycteris*.

Micronycteris pusilla Sanborn, 1949. Fieldiana Zool., 31:228.
TYPE LOCALITY: Brazil, Amazonas, Tahuapunta (Vaupes River).
DISTRIBUTION: NW Brazil, E Colombia.
COMMENTS: Subgenus *Neonycteris*.

Micronycteris schmidtorum Sanborn, 1935. Field Mus. Nat. Hist. Publ., Zool. Ser., 20:81.
TYPE LOCALITY: Guatemala, Izabal, Bobos.
DISTRIBUTION: S Mexico to Venezuela; NE Peru; NE Brazil.
COMMENTS: Subgenus *Micronycteris*.

Micronycteris sylvestris (Thomas, 1896). Ann. Mag. Nat. Hist., ser. 6, 18:302.
TYPE LOCALITY: Costa Rica, Guanacaste, Hda. Miravalles, between 1,400 and 2,000 ft. (427-610 m).
DISTRIBUTION: Peru and SE Brazil to Nayarit and Veracruz (Mexico); Trinidad.
COMMENTS: Subgenus *Glyphonycteris*.

Mimon Gray, 1847. Proc. Zool. Soc. Lond., 1847:14.
TYPE SPECIES: *Phyllostoma bennettii* Gray, 1838.
SYNONYMS: *Anthorhina.*
COMMENTS: Includes *Anthorhina;* see Handley (1960). Gardner and Ferrell (1990:503-504) argued that *Anthorhina* is a junior objective synonym of *Tonatia*, and unavailable as a subgenus of *Mimon*. This was based on a type species designation of the preoccupied name, *Tylostoma*, however, it was clearly Lydekker's intention in renaming *Tylostoma* as *Anthorhina* to restrict it to *Tylostoma sensu* Peters and Dobson.

Mimon bennettii (Gray, 1838). Mag. Zool. Bot., 2:483.
TYPE LOCALITY: Brazil, São Paulo, Ipanema.
DISTRIBUTION: S Mexico to Colombia; Guianas; SE Brazil.
SYNONYMS: *cozumelae.*
COMMENTS: Subgenus *Mimon*. Includes *cozumelae;* see Schaldach (1965:132), Villa-R. (1966:216), and Hall (1981:112); but also see Jones and Carter (1976:12).

Mimon crenulatum (E. Geoffroy, 1810). Ann. Mus. Hist. Nat. Paris, 15:193.
TYPE LOCALITY: Brazil, Bahia; see Handley (1960).
DISTRIBUTION: Chiapas and Campeche (Mexico) to Guianas, E Brazil, Bolivia, Ecuador and E Peru; Trinidad.
SYNONYMS: *keenani, koepckeae, longifolium, peruanum, picatum* (see Jones and Carter, 1979:8; Koopman, 1978*b*:5; and Handley, 1960:462, 463).

COMMENTS: Subgenus *Anthorhina*, but see Gardner and Ferrell (1990).

Phylloderma Peters, 1865. Monatsb. K. Preuss. Akad. Wiss. Berlin, 1865:513.
TYPE SPECIES: *Phyllostoma stenops* Peters, 1865.

Phylloderma stenops Peters, 1865. Monatsb. K. Preuss. Akad. Wiss. Berlin, 1865:513.
TYPE LOCALITY: French Guiana, Cayenne.
DISTRIBUTION: S Mexico to SE Brazil, Bolivia and Peru.
SYNONYMS: *boliviensis, cayanensis, septentrionalis.*
COMMENTS: Includes *septentrionalis*; see Jones and Carter (1976:13). Bolivian form reviewed by Bárquez and Ojeda (1979).

Phyllostomus Lacépède, 1799. Tabl. Div. Subd. Order Genres Mammifères, p. 16.
TYPE SPECIES: *Vespertilio hastatus* Pallas, 1767.
COMMENTS: Does not include *Phylloderma*, but see Baker et al. (1988*b*).

Phyllostomus discolor Wagner, 1843. Arch. Naturgesch., 9(1):366.
TYPE LOCALITY: Brazil, Mato Grosso, Cuiaba.
DISTRIBUTION: Oaxaca and Veracruz (Mexico) to Guianas, SE Brazil, Paraguay, N Argentina and Peru; Trinidad; Margarita Isl (Venezeula).
SYNONYMS: *angusticeps, innominatum, verrucosus.*

Phyllostomus elongatus (E. Geoffroy, 1810). Ann. Mus. Hist. Nat. Paris, 15:182.
TYPE LOCALITY: Brazil, Mato Grosso, Rio Branco.
DISTRIBUTION: Bolivia, E Peru, Ecuador, and Colombia to Guianas and E Brazil.
SYNONYMS: *ater.*

Phyllostomus hastatus (Pallas, 1767). Spicil. Zool., 3:7.
TYPE LOCALITY: Surinam.
DISTRIBUTION: Honduras to Guianas, E Brazil, Paraguay, N Argentina, and Peru; Trinidad and Tobago; Margarita Isl (Venezuela); Bolivia.
SYNONYMS: *aruma, caucae, caurae, curaca, maximus, paeze, panamensis.*

Phyllostomus latifolius (Thomas, 1901). Ann. Mag. Nat. Hist., ser. 7, 8:142.
TYPE LOCALITY: Guyana, Essequibo Prov., Mt. Kanuku.
DISTRIBUTION: Guianas, SE Colombia.
COMMENTS: Distinction from *P. elongatus* in uncertain.

Tonatia Gray, 1827. *In* Griffith, Anim. Kingdom, Mamm., 5:71.
TYPE SPECIES: *Vampyrus bidens* Spix, 1823.
COMMENTS: A key to the species was presented by Medellin and Arita (1989).

Tonatia bidens (Spix, 1823). Sim. Vespert. Brasil., p. 65.
TYPE LOCALITY: Brazil, Bahia, Rio Sao Francisco.
DISTRIBUTION: Chiapas (Mexico) and Belize to N Argentina, Paraguay, and Brazil; Trinidad.
SYNONYMS: *childreni.*

Tonatia brasiliense (Peters, 1867). Monatsb. K. Preuss. Akad. Wiss. Berlin, 1866:674 [1867].
TYPE LOCALITY: Brazil, Bahia.
DISTRIBUTION: Veracruz (Mexico) to Bolivia, NE Brazil and Trinidad.
SYNONYMS: *minuta, nicaraguae, venezuelae.*
COMMENTS: For synonyms see Jones and Carter (1979:7); but also see Gardner (1976:3).

Tonatia carrikeri (J. A. Allen, 1910). Bull. Am. Mus. Nat. Hist., 28:147.
TYPE LOCALITY: Venezuela, Bolivar, Rio Mocho.
DISTRIBUTION: Colombia, Venezuela, Surinam, N Brazil, Bolivia, Peru.

Tonatia evotis Davis and Carter, 1978. Occas. Pap. Mus. Texas Tech Univ., 53:8.
TYPE LOCALITY: Guatemala, Izabal, 25 km S.S.W. Puerto Barrios.
DISTRIBUTION: S Mexico, Belize, Guatemala, Honduras.
COMMENTS: Formerly included in *T. silvicola*. See Medellin and Arita (1989, Mammalian Species, 334).

Tonatia schulzi Genoways and Williams, 1980. Ann. Carnegie Mus., 49:205.
TYPE LOCALITY: Surinam, Brokopondo, 3 km SW Rudi Koppelvliegveld.

DISTRIBUTION: Guianas, N Brazil.

Tonatia silvicola (d'Orbigny, 1836). Voy. Amer. Merid. Atlas Zool., 4:11, pl. 7.
TYPE LOCALITY: Bolivia, Yungas between Secure and Isiboro rivers.
DISTRIBUTION: Honduras to Bolivia, NE Argentina, Guianas, and E Brazil.
SYNONYMS: *amblyotis, centralis, colombianus, laephotis, occidentalis.*
COMMENTS: Includes *laephotis* and *amblyotis*; see Davis and Carter (1978). See Medellin and Arita (1989, Mammalian Species, 334).

Trachops Gray, 1847. Proc. Zool. Soc. Lond., 1847:14.
TYPE SPECIES: *Trachops fuliginosus* Gray, 1865 (= *Vampyrus cirrhosus* Spix, 1823).

Trachops cirrhosus (Spix, 1823). Sim. Vespert. Brasil., p. 64.
TYPE LOCALITY: Brazil, Pernambuco.
DISTRIBUTION: Oaxaca (Mexico) to Guianas, SE Brazil, Bolivia and Ecuador; Trinidad.
SYNONYMS: *coffini, ehrhardti, fuliginosus.*

Vampyrum Rafinesque, 1815. Analyse de la Nature, p. 54.
TYPE SPECIES: *Vespertilio spectrum* Linnaeus, 1758.

Vampyrum spectrum (Linnaeus, 1758). Syst. Nat., 10th ed., 1:31.
TYPE LOCALITY: Surinam.
DISTRIBUTION: Veracruz (Mexico) to Ecuador and Peru, N and SW Brazil, and Guianas; Trinidad; perhaps Jamaica.
SYNONYMS: *nelsoni.*
COMMENTS: See Navarro and Wilson (1982, Mammalian Species, 184).

Subfamily Lonchophyllinae Griffiths, 1982. Am. Mus. Novit., 2742:40.

Lionycteris Thomas, 1913. Ann. Mag. Nat. Hist., ser. 8, 12:270.
TYPE SPECIES: *Lionycteris spurrelli* Thomas, 1913.

Lionycteris spurrelli Thomas, 1913. Ann. Mag. Nat. Hist., ser. 8, 12:271.
TYPE LOCALITY: Colombia, Choco, Condoto.
DISTRIBUTION: E Panama, Colombia, Venezuela, Guianas, Amazonian Peru and Brazil.

Lonchophylla Thomas, 1903. Ann. Mag. Nat. Hist., ser. 7, 12:458.
TYPE SPECIES: *Lonchophylla mordax* Thomas, 1903.
COMMENTS: Taddei et al. (1983) gave a key to all the species.

Lonchophylla bokermanni Sazima et al., 1978. Rev. Brasil. Biol., 38:82.
TYPE LOCALITY: Brazil, Minas Gerais, Jaboticatubas, Serra do Cipo.
DISTRIBUTION: SE Brazil.

Lonchophylla dekeyseri Taddei et al., 1983. Ciencia e Cultura, 35:626.
TYPE LOCALITY: Brazil, D. F., 8 km N Brasilia.
DISTRIBUTION: E Brazil.

Lonchophylla handleyi Hill, 1980. Bull. Brit. Mus. (Nat. Hist.) Zool., 38:233.
TYPE LOCALITY: Ecuador, Morona, Santiago, Los Tayos (03°07'S, 18°12'W).
DISTRIBUTION: Ecuador, Peru, S Colombia.
COMMENTS: Formerly confused with *robusta*; see Hill (1980*a*).

Lonchophylla hesperia G. M. Allen, 1908. Bull. Mus. Comp. Zool., 52:35.
TYPE LOCALITY: Peru, Tumbes, Zorritos.
DISTRIBUTION: N Peru, Ecuador.
COMMENTS: Known only from five specimens; see Gardner (1976:5).

Lonchophylla mordax Thomas, 1903. Ann. Mag. Nat. Hist., ser. 7, 12:459.
TYPE LOCALITY: Brazil, Bahia, Lamarao.
DISTRIBUTION: Costa Rica to Ecuador; E Brazil; perhaps Peru and Bolivia.
SYNONYMS: *concava.*
COMMENTS: Includes *concava*; see Handley (1966*a*:763); but also see Jones and Carter (1976:16), who provisionally recognized it as a species.

Lonchophylla robusta Miller, 1912. Proc. U.S. Natl. Mus., 42:23.
TYPE LOCALITY: Panama, Canal Zone, Rio Chilibrillo, near Alahuela.
DISTRIBUTION: Nicaragua to Venezuela and Ecuador.
COMMENTS: Although specimens from Panama are larger than those from Costa Rica, the species is still regarded as monotypic; see Walton (1963).

Lonchophylla thomasi J. A. Allen, 1904. Bull. Am. Mus. Nat. Hist., 20:230.
TYPE LOCALITY: Venezuela, Bolivar, Ciudad Bolivar.
DISTRIBUTION: E Panama, Colombia, Venezuela, Guianas, Amazonian Brazil, Peru, Bolivia.
COMMENTS: Specimens of this species have frequently been confused with *concava*, *mordax*, and *Lionycteris spurrelli*; see Taddei et al. (1978:1, 2) and Koopman (1978*b*).

Platalina Thomas, 1928. Ann. Mag. Nat. Hist., ser. 10, 8:120.
TYPE SPECIES: *Platalina genovensium* Thomas, 1928.

Platalina genovensium Thomas, 1928. Ann. Mag. Nat. Hist., ser. 10, 8:121.
TYPE LOCALITY: Peru, near Lima.
DISTRIBUTION: Peru.

Subfamily Brachyphyllinae Gray, 1866. Proc. Zool. Soc. Lond., 1866:115.

Brachyphylla Gray, 1834. Proc. Zool. Soc. Lond., 1833:122 [1834].
TYPE SPECIES: *Brachyphylla cavernarum* Gray, 1834.
COMMENTS: Revised by Swanepoel and Genoways (1978). A key to this genus was presented by Swanepoel and Genoways (1983*a*).

Brachyphylla cavernarum Gray, 1834. Proc. Zool. Soc. Lond., 1833:123 [1834].
TYPE LOCALITY: St. Vincent (Lesser Antilles, UK).
DISTRIBUTION: Puerto Rico, Virgin Isls and throughout Lesser Antilles south to St. Vincent and Barbados.
SYNONYMS: *intermedia*, *minor*.
COMMENTS: Includes *minor*; see Swanepoel and Genoways (1978:39) and Varona (1974:27). Reviewed by Swanepoel and Genoways (1983*a*, Mammalian Species, 205).

Brachyphylla nana Miller, 1902. Proc. Acad. Nat. Sci. Philadelphia, 54:409.
TYPE LOCALITY: Cuba, Pinar del Rio, El Guama.
DISTRIBUTION: Cuba, Hispaniola, Jamaica (extinct), Grand Cayman (Cayman Isls, UK), Middle Caicos (SE Bahamas).
SYNONYMS: *pumila*.
COMMENTS: Includes *pumila*; see Jones and Carter (1976:30) and Swanepoel and Genoways (1978). Considered a subspecies of *cavernarum* by Buden (1977). Reviewed by Swanepoel and Genoways (1983*b*, Mammalian Species, 206).

Subfamily Phyllonycterinae Miller, 1907. Bull. U.S. Natl. Mus., 57:171.

Erophylla Miller, 1906. Proc. Biol. Soc. Washington, 19:84.
TYPE SPECIES: *Phyllonycteris bombifrons* Miller, 1899 (= *Phyllonycteris sezekorni* Gundlach, 1861).
COMMENTS: Revised by Buden (1976. Included as a subgenus of *Phyllonycteris* by Varona (1974:29).

Erophylla sezekorni (Gundlach, 1860). Monatsb. K. Preuss. Akad. Wiss. Berlin, 1860:818.
TYPE LOCALITY: Cuba, Pinar del Rio, Santa Cruz de los Pinos, Rangel.
DISTRIBUTION: Cuba, Jamaica, Hispaniola, Puerto Rico, Bahamas, and Cayman Isls.
SYNONYMS: *bombifrons*, *mariguanensis*, *planifrons*, *santacristobalensis*, *syops*.
COMMENTS: Includes *bombifrons*; see Buden (1976:14). Reviewed by Baker et al. (1978, Mammalian Species, 115). Based on differences in size of ears, shape of rostrum, inflation of braincase, and certain dental characters, *E. bombifrons* (Hispaniola and Puerto Rico) should probably be regarded as a distinct species. Although recognition of *E. bombifrons* represents my unpublished opinion, this species has been universally recognized up until the last several years; see Varona (1974:29) and Hall (1981:170).

Phyllonycteris Gundlach, 1860. Monatsb. K. Preuss. Akad. Wiss. Berlin, 1860:817.
TYPE SPECIES: *Phyllonycteris poeyi* Gundlach, 1860.
SYNONYMS: *Reithronycteris.*
COMMENTS: Includes the fossil species *major* from Puerto Rico and the N Lesser Antilles.

Phyllonycteris aphylla (Miller, 1898). Proc. Acad. Nat. Sci. Philadelphia, 50:334.
TYPE LOCALITY: Jamaica.
DISTRIBUTION: Jamaica.
COMMENTS: Subgenus *Reithronycteris.*

Phyllonycteris poeyi Gundlach, 1860. Monatsb. K. Preuss. Akad. Wiss. Berlin, 1860:817.
TYPE LOCALITY: Cuba, Matanzas, Canimar (cafetal "San Antonio el Fundador").
DISTRIBUTION: Cuba, Isle of Pines, Hispaniola.
SYNONYMS: *obtusa.*
COMMENTS: Subgenus *Phyllonycteris.* Includes *obtusa;* see Jones and Carter (1976:30) from Hispaniola, this subspecies was described from fossil remains and was only recently discovered as a living animal; see Klingener et al. (1978:90-92). Corbet and Hill (1980:61) listed *obtusa* as a distinct species without comment.

Subfamily Glossophaginae Bonaparte, 1845. Cat. Met. Mamm. Europe, p. 5.

Anoura Gray, 1838. Mag. Zool. Bot., 2:490.
TYPE SPECIES: *Anoura geoffroyi* Gray, 1838.
SYNONYMS: *Lonchoglossa.*
COMMENTS: Includes *Lonchoglossa;* see Cabrera (1958:74). A key to this genus was presented by Tamsitt and Nagorsen (1982).

Anoura caudifera (E. Geoffroy, 1818). Mem. Mus. Hist. Nat. Paris, 4:418.
TYPE LOCALITY: Brazil, Rio de Janeiro.
DISTRIBUTION: Colombia, Venezuela, Guianas, Brazil, Ecuador, Peru, Bolivia, NW Argentina.
SYNONYMS: *aequatoris, ecaudata, wiedii.*

Anoura cultrata Handley, 1960. Proc. U.S. Natl. Mus., 112:463.
TYPE LOCALITY: Panama, Darien, Rio Pucro, Tacarcuna Village, 3,200 ft. (975 m).
DISTRIBUTION: Costa Rica, Panama, Venezuela, Colombia, Ecuador, Peru, Bolivia.
SYNONYMS: *brevirostrum, werckleae.*
COMMENTS: Includes *brevirostrum* and *werckleae;* see Nagorsen and Tamsitt (1981). See Tamsitt and Nagorsen (1982, Mammalian Species, 179).

Anoura geoffroyi Gray, 1838. Mag. Zool. Bot., 2:490.
TYPE LOCALITY: Brazil, Rio de Janeiro.
DISTRIBUTION: Peru, Bolivia, SE Brazil, French Guiana and Ecuador to Tamaulipas and Sinaloa (Mexico); Trinidad; Grenada (Lesser Antilles).
SYNONYMS: *antricola, apolinari, lasiopyga, peruana.*
COMMENTS: Includes *apolinari;* see Sanborn (1933:26).

Anoura latidens Handley, 1984. Proc. Biol. Soc. Washington, 97:503.
TYPE LOCALITY: Venezuela, D.F., Pico Avila.
DISTRIBUTION: Venezuela, Colombia, Peru.

Choeroniscus Thomas, 1928. Ann. Mag. Nat. Hist., ser. 10, 1:122.
TYPE SPECIES: *Choeronycteris minor* Peters, 1868.

Choeroniscus godmani (Thomas, 1903). Ann. Mag. Nat. Hist., ser. 7, 11:288.
TYPE LOCALITY: Guatemala.
DISTRIBUTION: Sinaloa (Mexico) to Colombia, Venezuela, Guyana and Surinam.

Choeroniscus intermedius (J. A. Allen and Chapman, 1893). Bull. Am. Mus. Nat. Hist., 5:207.
TYPE LOCALITY: Trinidad and Tobago, Trinidad, Princestown.
DISTRIBUTION: Trinidad, Peru, Guianas, Amazonian Brazil; distribution is poorly known.
COMMENTS: Closely related to *minor,* with which it might be conspecific (Williams and Genoways, 1980*a*).

Choeroniscus minor (Peters, 1868). Monatsb. K. Preuss. Akad. Wiss. Berlin, 1868:366.
TYPE LOCALITY: Surinam.
DISTRIBUTION: Guianas, Venezuela, N Brazil, C Colombia, Ecuador, Peru, Bolivia.
SYNONYMS: *inca*.
COMMENTS: Includes *inca*; see Koopman (1978*b*:8) and Jones and Carter (1979:8).

Choeroniscus periosus Handley, 1966. Proc. Biol. Soc. Washington, 79:84.
TYPE LOCALITY: Colombia, Valle, 27 km S Buenaventura, Rio Raposo.
DISTRIBUTION: W Colombia, W Ecuador.

Choeronycteris Tschudi, 1844. Fauna Peruana, 1:70.
TYPE SPECIES: *Choeronycteris mexicana* Tschudi, 1844.

Choeronycteris mexicana Tschudi, 1844. Fauna Peruana, 1:72.
TYPE LOCALITY: Mexico.
DISTRIBUTION: Honduras and El Salvador to S California, Nevada, Arizona, and New Mexico (USA); a single record from S Texas; perhaps Venezuela.
COMMENTS: It is doubtful that *ponsi* Pirlot, 1967 (described from NW Venezuela) is referrable to this genus, and the description is inadequate to allocate it; see Jones and Carter (1976:18). See Arroyo-Cabrales et al. (1987, Mammalian Species, 291).

Glossophaga E. Geoffroy, 1818. Mem. Mus. Hist. Nat. Paris, 4:418.
TYPE SPECIES: *Vespertilio soricinus* Pallas, 1766.
COMMENTS: Revised by Miller (1913). A key to the genus was presented by Webster and Jones (1984).

Glossophaga commissarisi Gardner, 1962. Los Angeles Cty. Mus. Contrib. Sci., 54:1.
TYPE LOCALITY: Mexico, Chiapas, 10 km SE Tonala.
DISTRIBUTION: Sinaloa (Mexico) to Panama; SE Colombia; E Ecuador; E Peru; NW Brazil.
SYNONYMS: *bakeri, hespera*.

Glossophaga leachii Gray, 1844. Mammalia, *in* Zool. Voy. "Sulfur," 1:18.
TYPE LOCALITY: Nicaragua, Chinandega, Realejo.
DISTRIBUTION: Costa Rica to Guerrero, Morelos, and Tlaxcala (Mexico); Colima and Jalisco (Mexico).
SYNONYMS: *alticola, caudifer*.
COMMENTS: Originally considered a subspecies of *soricina*; see Jones and Carter (1976:14). Includes *alticola* Davis, 1944, by synonymy; see Webster and Jones (1980:4). See Webster and Jones (1984, Mammalian Species, 226).

Glossophaga longirostris Miller, 1898. Proc. Acad. Nat. Sci. Philadelphia, 50:330.
TYPE LOCALITY: Colombia, Magdalena, Sierra Nevada de Santa Marta.
DISTRIBUTION: N Ecuador; Colombia; Venezuela (including Margarita Isl); N Brazil; Guyana; Trinidad and Tobago; Grenada, St. Vincent, Curacao, Bonaire, and Aruba (Lesser Antilles). The record from Dominica is erroneous.
SYNONYMS: *campestris, elongata, major, reclusa, rostrata*.
COMMENTS: Includes *elongata*; see Jones and Carter (1976:14) and Koopman (1958:437). Revised by Webster and Handley (1986).

Glossophaga morenoi Martinez and Villa, 1938. Anal. Inst. Biol. Univ. Nac. Auto. Mexico, 9:347.
TYPE LOCALITY: Mexico, Oaxaca, Rio Guamol, 34 mi. (55 km) S (by Hwy.190) La Ventosa Jct.
DISTRIBUTION: Chiapas to Tlaxcala (Mexico).
SYNONYMS: *brevirostris, mexicana*.
COMMENTS: Includes *mexicana*; see Gardner (1986). See Webster and Jones (1985, Mammalian Species, 245).

Glossophaga soricina (Pallas, 1766). Misc. Zool., p. 48.
TYPE LOCALITY: Surinam.
DISTRIBUTION: Tamaulipas, Sonora and Tres Marias Isls (Mexico) to Guianas, SE Brazil, N Argentina and Peru; Margarita Isl (Venezuela); Trinidad; Grenada (Lesser Antilles); Jamaica; perhaps Bahama Isls.
SYNONYMS: *amplexicaudata, antillarum, handleyi, microtis, mutica, truei, valens, villosa*.

COMMENTS: Reviewed by Alvarez et al. (1991, Mammalian Species, 379).

Hylonycteris Thomas, 1903. Ann. Mag. Nat. Hist., ser. 7, 11:286.
TYPE SPECIES: *Hylonycteris underwoodi* Thomas, 1903.

Hylonycteris underwoodi Thomas, 1903. Ann. Mag. Nat. Hist., ser. 7, 11:287.
TYPE LOCALITY: Costa Rica, San Jose, Rancho Redondo.
DISTRIBUTION: W Panama to Nayarit and Veracruz (Mexico).
SYNONYMS: *minor.*
COMMENTS: Includes *minor,* but see Alvarez and Alvarez-Castañeda (1991). See Jones and Homan (1974, Mammalian Species, 32).

Leptonycteris Lydekker, 1891. *In* Flower and Lydekker, Mamm., Living and Extinct, p. 674.
TYPE SPECIES: *Ischnoglossa nivalis* Saussure, 1860.
COMMENTS: Revised by Arita and Humphrey (1988). A key for the genus was presented by Hensley and Wilkins (1988).

Leptonycteris curasoae Miller, 1900. Proc. Biol. Soc. Washington, 13:126.
TYPE LOCALITY: Curacao, Willemstad (Netherlands).
DISTRIBUTION: Curacao, Bonaire and Aruba (Netherlands Antilles); NE Colombia; N Venezuela (incl. Margarita Isl); El Salvador to Arizona and New Mexico (USA).
STATUS: U.S. ESA - Endangered as *L. sanborni* (= *yerbabuenae).*
SYNONYMS: *sanborni, tarlosti, yerbabuenae.*
COMMENTS: Includes *yerbabuenae* (= *sanborni*), but see Watkins et al. (1972).

Leptonycteris nivalis (Saussure, 1860). Rev. Mag. Zool. Paris, ser. 2, 12:492.
TYPE LOCALITY: Mexico, Veracruz, Mt. Orizaba.
DISTRIBUTION: SE Arizona and W Texas (USA) to S Mexico, and Guatemala.
STATUS: U.S. ESA - Endangered.
SYNONYMS: *longala.*
COMMENTS: Does not include *yerbabuenae,* but see Watkins et al. (1972:16). See Hensley and Wilkins (1988, Mammalian Species, 307).

Lichonycteris Thomas, 1895. Ann. Mag. Nat. Hist., ser. 6, 16:55.
TYPE SPECIES: *Lichonycteris obscura* Thomas, 1895.

Lichonycteris obscura Thomas, 1895. Ann. Mag. Nat. Hist., ser. 6, 16:55.
TYPE LOCALITY: Nicaragua, Managua, Managua.
DISTRIBUTION: Guatemala and Belize to Bolivia and Amazonian Brazil.
SYNONYMS: *degener.*
COMMENTS: Includes *degener;* see Hill (1985).

Monophyllus Leach, 1821. Trans. Linn. Soc. Lond., 13:75.
TYPE SPECIES: *Monophyllus redmani* Leach, 1821.
COMMENTS: Reviewed by Schwartz and Jones (1967). A key to the genus was published by Homan and Jones (1975*a*).

Monophyllus plethodon Miller, 1900. Proc. Washington Acad. Sci., 2:35.
TYPE LOCALITY: Barbados (Lesser Antilles), St. Michael Parish.
DISTRIBUTION: Lesser Antilles from Anguilla to St. Vincent and Barbados. Fossils known from Puerto Rico.
SYNONYMS: *frater, luciae.*
COMMENTS: Includes *luciae* and *frater;* see Schwartz and Jones (1967:13). See Homan and Jones (1975*b*, Mammalian Species, 58).

Monophyllus redmani Leach, 1821. Trans. Linn. Soc. Lond., 13:76.
TYPE LOCALITY: Jamaica.
DISTRIBUTION: Cuba, Hispaniola, Puerto Rico, Jamaica, S Bahama Isls.
SYNONYMS: *clinedaphus, cubanus, ferreus, portoricensis.*
COMMENTS: For synonyms see Schwartz and Jones (1967:13). Reviewed by Homan and Jones (1975*a*, Mammalian Species, 57).

Musonycteris Schaldach and McLaughlin, 1960. Los Angeles Cty. Mus. Contrib. Sci., 37:2.
TYPE SPECIES: *Musonycteris harrisoni* Schaldach and McLaughlin, 1960.
COMMENTS: Included in *Choeronycteris* by Handley (1966*b*:85, 86); but see Phillips (1971:99) and Webster et al. (1982).

Musonycteris harrisoni Schaldach and McLaughlin, 1960. Los Angeles Cty. Mus. Contrib. Sci., 37:3.
TYPE LOCALITY: Mexico, Colima, 2 km SE Pueblo Juarez.
DISTRIBUTION: Jalisco, Colima, Michoacan and Guerrero (Mexico).

Scleronycteris Thomas, 1912. Ann. Mag. Nat. Hist., ser. 8, 10:404.
TYPE SPECIES: *Scleronycteris ega* Thomas, 1912.

Scleronycteris ega Thomas, 1912. Ann. Mag. Nat. Hist., ser. 8, 10:405.
TYPE LOCALITY: Brazil, Amazonas, Ega.
DISTRIBUTION: Amazonian Brazil, S Venezuela.

Subfamily Carolliinae Miller, 1924. Bull. U.S. Natl. Mus., 128:53.

Carollia Gray, 1838. Mag. Zool. Bot., 2:488.
TYPE SPECIES: *Carollia braziliensis* Gray, 1838 (= *Vespertilio perspicillata* Linnaeus, 1758).
SYNONYMS: *Hemiderma*.
COMMENTS: Revised by Pine (1972).

Carollia brevicauda (Schinz, 1821). Das Thierreich, 1:164.
TYPE LOCALITY: Brazil, Espirito Santo, Jucu River, Fazenda de Coroaba.
DISTRIBUTION: San Luis Potosi (Mexico) to Peru, Bolivia, and E Brazil.
SYNONYMS: *bicolor, grayi, lanceolatum, minor*.
COMMENTS: Long confused with *perspicillata* or *subrufa*; see Pine (1972).

Carollia castanea H. Allen, 1890. Proc. Am. Philos. Soc., 28:19.
TYPE LOCALITY: Costa Rica, Angostura.
DISTRIBUTION: Honduras to Peru, Bolivia, W Brazil and Venezuela.

Carollia perspicillata (Linnaeus, 1758). Syst. Nat., 10th ed., 1:31.
TYPE LOCALITY: Surinam.
DISTRIBUTION: Oaxaca, Veracruz and Yucatan Peninsula (Mexico) to Peru, Bolivia, Paraguay, SE Brazil and Guianas; Trinidad and Tobago; Grenada (Lesser Antilles); perhaps Jamaica, N Lesser Antilles.
SYNONYMS: *amplexicaudata, azteca, brachyotus, braziliensis, calcaratum, tricolor, verrucata*.
COMMENTS: Includes *tricolor*; see Pine (1972).

Carollia subrufa (Hahn, 1905). Proc. Biol. Soc. Washington, 18:247.
TYPE LOCALITY: Mexico, Oaxaca, NW Tapanatepec, Sta. Efigenia.
DISTRIBUTION: Jalisco (Mexico) to NW Nicaragua; Guyana.
COMMENTS: Not a subspecies of *castanea*; see Pine (1972).

Rhinophylla Peters, 1865. Monatsb. K. Preuss. Akad. Wiss. Berlin, 1865:355.
TYPE SPECIES: *Rhinophylla pumilio* Peters, 1865.

Rhinophylla alethina Handley, 1966. Proc. Biol. Soc. Washington, 79:86.
TYPE LOCALITY: Colombia, Valle, 27 km S Buenaventura, Raposo River.
DISTRIBUTION: W Colombia, W Ecuador.

Rhinophylla fischerae Carter, 1966. Proc. Biol. Soc. Washington, 79:235.
TYPE LOCALITY: Peru, Loreto, 61 mi. (98 km) SE Pucallpa.
DISTRIBUTION: Peru, Ecuador, SE Colombia, Amazonian Brazil.

Rhinophylla pumilio Peters, 1865. Monatsb. K. Preuss. Akad. Wiss. Berlin, 1865:355.
TYPE LOCALITY: Brazil, Bahia.
DISTRIBUTION: Colombia, Ecuador, Peru and Bolivia to Guianas and E Brazil.

Subfamily Stenodermatinae Gervais, 1856. *In* Comte de Castelnau, Exped. Partes Cen. Am. Sud., Zool.(Sec. 7), Vol. 1, pt. 2(Mammifères):32 footnote.

Ametrida Gray, 1847. Proc. Zool. Soc. Lond., 1847:15.
TYPE SPECIES: *Ametrida centurio* Gray, 1847.
COMMENTS: Revised by Peterson (1965*b*).

Ametrida centurio Gray, 1847. Proc. Zool. Soc. Lond., 1847:15.
TYPE LOCALITY: Brazil, Para, Belem.
DISTRIBUTION: Amazonian Brazil, Guianas, Panama, Venezuela, Trinidad, Bonaire Isl (Netherlands Antilles).
SYNONYMS: *minor*.
COMMENTS: Includes *minor*; see Jones and Carter (1976:29).

Ardops Miller, 1906. Proc. Biol. Soc. Washington, 19:84.
TYPE SPECIES: *Stenoderma nichollsi* Thomas, 1891.
COMMENTS: Revised by Jones and Schwartz (1967). Included under *Stenoderma* by Varona (1974:24-26) and by Simpson (1945:58); but see Jones and Carter (1976:28).

Ardops nichollsi (Thomas, 1891). Ann. Mag. Nat. Hist., ser. 6, 7:529.
TYPE LOCALITY: Dominica (Lesser Antilles).
DISTRIBUTION: Lesser Antilles, from St. Eustatius to St. Vincent.
SYNONYMS: *annectens, koopmani, luciae, montserratensis*.
COMMENTS: For synonyms see Jones and Schwartz (1967). See Jones and Genoways (1973, Mammalian Species, 24).

Ariteus Gray, 1838. Mag. Zool. Bot., 2:491.
TYPE SPECIES: *Istiophorus flavescens* Gray, 1831.
COMMENTS: Included as a subgenus of *Stenoderma* by Varona (1974:24) and Simpson (1945:58); but see Jones and Carter (1976:29).

Ariteus flavescens (Gray, 1831). Zool. Misc., 1:37.
TYPE LOCALITY: Not designated in original publication.
DISTRIBUTION: Jamaica.
SYNONYMS: *achradophilus*.

Artibeus Leach, 1821. Trans. Linn. Soc. Lond., 13:75.
TYPE SPECIES: *Artibeus jamaicensis* Leach, 1821.
SYNONYMS: *Dermanura, Enchisthenes, Koopmania*.
COMMENTS: For synonyms see Jones and Carter (1979:8); but see also Hall (1981:163) and Owen (1991).

Artibeus amplus Handley, 1987. Fieldiana, Zool., n.s., 39:164.
TYPE LOCALITY: Venezuela, Zulia, Kasmera.
DISTRIBUTION: Venezuela, N Colombia.
COMMENTS: Subgenus *Artibeus*.

Artibeus anderseni Osgood, 1916. Field Mus. Nat. Hist. Publ., Zool. Ser., 10:212.
TYPE LOCALITY: Brazil, Rondonia, Porto Velho.
DISTRIBUTION: W Brazil, Bolivia, Ecuador, Peru.
COMMENTS: Subgenus *Dermanura*. Previously considered a subspecies of *cinereus*; but see Koopman (1978*b*:14).

Artibeus aztecus K. Andersen, 1906. Ann. Mag. Nat. Hist., ser. 7, 18:422.
TYPE LOCALITY: Mexico, Morelos, Tetela del Volcan.
DISTRIBUTION: W Panama; Costa Rica; Honduras, Guatemala; Chiapas, Oaxaca to Nuevo Leon and Sinaloa (Mexico).
SYNONYMS: *minor, major*.
COMMENTS: Subgenus *Dermanura*. Not a subspecies of *cinereus*; see Jones and Carter (1976:27). Revised by Davis (1969). See Webster and Jones (1982*b*, Mammalian Species, 177).

Artibeus cinereus (Gervais, 1856). *In* Comte de Castelnau, Exped. Partes Cen. Am. Sud., Zool.(Sec. 7), Vol. 1, pt. 2(Mammifères):36.
TYPE LOCALITY: Brazil, Para, Belem.
DISTRIBUTION: Guianas, Venezuela, N Brazil.
COMMENTS: Subgenus *Dermanura*. Does not include *gnomus*.

Artibeus concolor Peters, 1865. Monatsb. K. Preuss. Akad. Wiss. Berlin, 1865:357.
TYPE LOCALITY: Surinam, Paramaribo.
DISTRIBUTION: Guianas, Venezuela, Colombia, N Brazil, Peru.
COMMENTS: Subgenus *Koopmania*.

Artibeus fimbriatus Gray, 1838. Mag. Zool. Bot., 2:487.
TYPE LOCALITY: Brazil, Paraná, Serra do Mar, Morretes.
DISTRIBUTION: S Brazil, Paraguay.
COMMENTS: Subgenus *Artibeus*. See Handley (1989) for status of this species.

Artibeus fraterculus Anthony, 1924. Am. Mus. Novit., 114:5.
TYPE LOCALITY: Ecuador, El Oro, Portovelo, 2,000 ft.(610 m).
DISTRIBUTION: Ecuador, Peru.
COMMENTS: Subgenus *Artibeus*. Considered a subspecies of *jamaicensis* by Jones and Carter (1976:2); but see Koopman (1978*b*:14).

Artibeus glaucus Thomas, 1893. Proc. Zool. Soc. Lond., 1893:336.
TYPE LOCALITY: Peru, Junin, Chauchamayo.
DISTRIBUTION: S Mexico to Bolivia and S Brazil; Trinidad and Tobago; Grenada (Lesser Antilles).
SYNONYMS: *bogotensis, gnomus, jucundum, pumilio, rosenbergii, watsoni.*
COMMENTS: Includes *gnomus, pumilio, rosenbergii,* and *watsoni,* but see Handley (1987) who regarded *gnomus* and *watsoni* to be closely related to *A. glaucus,* but different species.

Artibeus hartii Thomas, 1892. Ann. Mag. Nat. Hist., ser. 6, 10:409.
TYPE LOCALITY: Trinidad and Tobago, Trinidad, Port of Spain.
DISTRIBUTION: Bolivia and Venezuela to Jalisco and Tamaulipas (Mexico); Arizona (USA); Trinidad.
COMMENTS: Subgenus *Enchisthenes*. Formerly included in genus *Enchisthenes;* see Jones and Carter (1979:8).

Artibeus hirsutus K. Andersen, 1906. Ann. Mag. Nat. Hist., ser. 7, 18:420.
TYPE LOCALITY: Mexico, Michoacan, La Salada.
DISTRIBUTION: Sonora to Guerrero (Mexico).
COMMENTS: Subgenus *Artibeus*. See Webster and Jones (1983, Mammalian Species, 199).

Artibeus inopinatus Davis and Carter, 1964. Proc. Biol. Soc. Washington, 77:119.
TYPE LOCALITY: Honduras, Choluteca, Choluteca.
DISTRIBUTION: El Salvador, Honduras, Nicaragua.
COMMENTS: Subgenus *Artibeus*. Possibly a subspecies of *hirsutus*. See Webster and Jones (1983, Mammalian Species, 199).

Artibeus jamaicensis Leach, 1821. Trans. Linn. Soc. Lond., 13:75.
TYPE LOCALITY: Jamaica.
DISTRIBUTION: Sinaloa and Tamaulipas (Mexico) to Ecuador, and Venezuela; Trinidad and Tobago; Greater and Lesser Antilles; Amazonian Brazil (only if *obscurus* or *planirostris* is included in this species). Perhaps Florida Keys; see Lazell and Koopman (1985); but see also Humphrey and Brown (1986).
SYNONYMS: *aequatorialis, carpolegus, coryi, eva, grenadensis, insularis, lewisi, parvipes, paulus, praeceps, richardsoni, schwartzi, trinitatis, triomylus, yucatanicus.*
COMMENTS: Subgenus *Artibeus*. Bats often treated as *jamaicensis* most likely belong to several additional species; which are *fraterculus* and *planirostris* (Koopman, 1978*b*:14-16) and *obscurus* (Handley, 1989), and are so treated here.

Artibeus lituratus (Olfers, 1818). *In* Eschwege, J. Brasilien, Neue Bibliothek. Reisenb., 15:224.
TYPE LOCALITY: Paraguay, Asuncion.
DISTRIBUTION: Sinaloa and Tamaulipas (Mexico) to S Brazil, N Argentina, and Bolivia; Trinidad and Tobago; S Lesser Antilles; Tres Marías Isls.

SYNONYMS: *femurvillosum, intermedius, koopmani, palmarum, rusbyi, superciliosum.*
COMMENTS: Subgenus *Artibeus*. Includes *palmarum* but not *fallax, hercules,* or *praeceps* (Koopman, 1978*b*:15). Includes *intermedius;* see Jones and Carter (1976:28; but see also Davis (1984). See also comment under *planirostris.*

Artibeus obscurus Schinz, 1821. *In* G. Cuvier, Das Tierreich, 1:164.
TYPE LOCALITY: Brazil, Bahia, Rio Peruhype, Villa Vicosa.
DISTRIBUTION: Colombia, Venezuela, Guianas, Ecuador, Peru, Bolivia, Brazil.
SYNONYMS: *fuliginosus.*
COMMENTS: Subgenus *Artibeus*. See Handley (1989) for status of this species.

Artibeus phaeotis (Miller, 1902). Proc. Acad. Nat. Sci. Philadelphia, 54:405.
TYPE LOCALITY: Mexico, Yucatan, Chichen-Itza.
DISTRIBUTION: Veracruz and Sinaloa (Mexico) to Ecuador and Guyana.
SYNONYMS: *nanus, palatinus, ravus, turpis.*
COMMENTS: Subgenus *Dermanura*. Includes *nanus* and *turpis;* see Jones and Lawlor (1965:412) and Davis (1970). For including *ravus* and other synomyms see Timm (1985, Mammalian Species, 235).

Artibeus planirostris (Spix, 1823). Sim. Vespert. Brasil., p. 66.
TYPE LOCALITY: Brazil, Bahia, Salvador.
DISTRIBUTION: Colombia and Venezuela, south to N Argentina and east to E Brazil.
SYNONYMS: *fallax, hercules, validum* (see Koopman, 1978*b*:15).
COMMENTS: Subgenus *Artibeus*. Handley (1987) would include this species in *A. jamaicensis.* See also comments under *lituratus*. See Handley (1991) for clarification of the holotype.

Artibeus toltecus (Saussure, 1860). Rev. Mag. Zool. Paris, ser. 2, 12:427.
TYPE LOCALITY: Mexico, Veracruz, Mirador.
DISTRIBUTION: Panama to Nuevo Leon and Sinaloa (Mexico).
SYNONYMS: *hesperus.*
COMMENTS: Subgenus *Dermanura*. Not a subspecies of *cinereus;* see Jones and Carter (1976:27). Revised by Davis (1969). Does not include *ravus,* see Handley (1987). See Webster and Jones (1982*c*, Mammalian Species, 178).

Centurio Gray, 1842. Ann. Mag. Nat. Hist., [ser. 1], 10:259.
TYPE SPECIES: *Centurio senex* Gray, 1842.

Centurio senex Gray, 1842. Ann. Mag. Nat. Hist., [ser. 1], 10:259.
TYPE LOCALITY: Nicaragua, Chinandega, Realejo.
DISTRIBUTION: Venezuela to Tamaulipas and Sinaloa (Mexico); Trinidad and Tobago.
SYNONYMS: *flavogularis, greenhalli, mexicanus, mcmurtrii, minor.*
COMMENTS: Reviewed by Paradiso (1967). See Snow et al. (1980, Mammalian Species, 138).

Chiroderma Peters, 1860. Monatsb. K. Preuss. Akad. Wiss. Berlin, 1860:747.
TYPE SPECIES: *Chiroderma villosum* Peters, 1860.
COMMENTS: Reviewed by Goodwin (1958).

Chiroderma doriae Thomas, 1891. Ann. Mus. Civ. Stor. Nat. Genova, ser. 2, 10:881.
TYPE LOCALITY: Brazil, Minas Gerais.
DISTRIBUTION: Minas Gerais and São Paulo (SE Brazil).

Chiroderma improvisum Baker and Genoways, 1976. Occas. Pap. Mus. Texas Tech Univ., 39:2.
TYPE LOCALITY: Guadeloupe (Lesser Antilles), Basse Terre, 2 km S and 2 km E Baie-Mahault (France).
DISTRIBUTION: Guadeloupe and Montserrat (Lesser Antilles).
COMMENTS: See Jones and Baker (1980, Mammalian Species, 134).

Chiroderma salvini Dobson, 1878. Cat. Chiroptera Brit. Mus., p. 532.
TYPE LOCALITY: Costa Rica.
DISTRIBUTION: Bolivia and Venezuela to Hidalgo and Chihuahua (Mexico).
SYNONYMS: *scopaeum.*

Chiroderma trinitatum Goodwin, 1958. Am. Mus. Novit., 1877:1.
TYPE LOCALITY: Trinidad and Tobago, Trinidad, Cumaca, 1,000 ft. (305 m).
DISTRIBUTION: Panama to Amazonian Brazil, Bolivia and Peru; Trinidad.
SYNONYMS: *gorgasi*.
COMMENTS: Includes *gorgasi*; see Jones and Carter (1976:25).

Chiroderma villosum Peters, 1860. Monatsb. K. Preuss. Akad. Wiss. Berlin, 1860:748.
TYPE LOCALITY: Brazil; see Carter and Dolan (1978:59).
DISTRIBUTION: Hidalgo (Mexico) to S Brazil, Bolivia and Peru; Trinidad and Tobago.
SYNONYMS: *isthmicum, jesupi* (see Handley, 1960:466).

Ectophylla H. Allen, 1892. Proc. U.S. Natl. Mus., 15:441.
TYPE SPECIES: *Ectophylla alba* H. Allen, 1892.
COMMENTS: Does not include *Mesophylla*; see Goodwin and Greenhall (1962) and Jones and Carter (1976:25); but see also Owen (1987).

Ectophylla alba H. Allen, 1892. Proc. U.S. Natl. Mus., 15:442.
TYPE LOCALITY: Honduras (= Rio Segovia) (T. McCarthy, in litt.).
DISTRIBUTION: Honduras to W Panama; also a record from W Colombia.
COMMENTS: See Timm (1982, Mammalian Species, 166).

Mesophylla Thomas, 1901. Ann. Mag. Nat. Hist., ser. 7, 8:143.
TYPE SPECIES: *Mesophylla macconnelli* Thomas, 1901.
COMMENTS: Included in *Vampyressa* by Owen (1987).

Mesophylla macconnelli Thomas, 1901. Ann. Mag. Nat. Hist., ser. 7, 8:145.
TYPE LOCALITY: Guyana, Essequibo Dist., Kunuku Mtns.
DISTRIBUTION: Nicaragua to Peru, Bolivia and Amazonian Brazil; Trinidad.
SYNONYMS: *flavescens*.

Phyllops Peters, 1865. Monatsb. K. Preuss. Akad. Wiss. Berlin, 1865:356.
TYPE SPECIES: *Phyllostoma albomaculatum* Gundlach, 1861 (= *Arctibeus falcatus* Gray, 1839).
COMMENTS: Included in *Stenoderma* by Varona (1974:24), Simpson (1945:58), and Silva-Taboada (1979:199); but see Jones and Carter (1976:28) and Corbet and Hill (1980:60).

Phyllops falcatus (Gray, 1839). Ann. Nat. Hist., 4:1.
TYPE LOCALITY: Cuba, Habana, Guanabacoa.
DISTRIBUTION: Cuba; Hispaniola; as fossil, Isle of Pines (Cuba).
SYNONYMS: *albomaculatum, haitiensis*.
COMMENTS: Includes *haitiensis*, see Koopman (1989*c*:636).

Platyrrhinus Saussure, 1860. Rev. Mag. Zool., Paris, ser. 2, 12:429.
TYPE SPECIES: *Phyllostoma lineatum* E. Geoffroy, 1810.
COMMENTS: For a history of the nomeclature of this genus and reasons for using *Platyrrhinus* in place of *Vampyrops*, see Gardner and Ferrell (1990) and Alberico and Velasco (1991).

Platyrrhinus aurarius (Handley and Ferris, 1972). Proc. Biol. Soc. Washington, 84:522.
TYPE LOCALITY: Venezuela, Bolivar, 85 km SSE El Dorado, 1,000 m.
DISTRIBUTION: S Venezuela, Colombia, Surinam.
COMMENTS: May be a synonym of *dorsalis*; see Jones and Carter (1976:23).

Platyrrhinus brachycephalus (Rouk and Carter, 1972). Occas. Pap. Mus. Texas Tech Univ., 1:1.
TYPE LOCALITY: Peru, Huanuco, 3 mi. (5 km) S Tingo Maria, 2,400 ft. (732 m).
DISTRIBUTION: N Brazil; Colombia to Guianas; Ecuador; Peru; Bolivia.
SYNONYMS: *latus, saccharus*.
COMMENTS: Includes *latus*; see Jones and Carter (1976:23).

Platyrrhinus chocoensis Alberico and Velasco, 1991. Bonn. zool. Beitr., 42:238.
TYPE LOCALITY: "Quebrada El Platinero, 12 km W Istmina (by road), 5°00' N, 76°45' W, 100 m, Departamento del Chocó, Colombia."
DISTRIBUTION: W Colombia, lowlands between the Western Cordillera of the Andes and the Pacific coast.

Platyrrhinus dorsalis (Thomas, 1900). Ann. Mag. Nat. Hist., ser. 7, 5:269.
TYPE LOCALITY: Ecuador, Paramba, 1,100 m.
DISTRIBUTION: Panama to Peru and Bolivia.
COMMENTS: Although the named forms *umbratus, oratus,* and *aquilus* were regarded as synonyms of *dorsalis* by Carter and Rouk (1973:976), Handley (1976:28) considered them distinct. See also comments under *umbratus.*

Platyrrhinus helleri (Peters, 1866). Monatsb. K. Preuss. Akad. Wiss. Berlin, 1866:392.
TYPE LOCALITY: Mexico.
DISTRIBUTION: Oaxaca and Veracruz (Mexico) to Peru, Bolivia, and Amazonian Brazil; Trinidad. A Paraguay record is erroneous.
SYNONYMS: *incarum, zarhinus.*
COMMENTS: Includes *zarhinus;* see Jones and Carter (1976:23) and Gardner and Carter (1972:78). See Ferrell and Wilson (1991, Mammalian Species, 373).

Platyrrhinus infuscus (Peters, 1880). Monatsb. K. Preuss. Akad. Wiss. Berlin, 1880:259.
TYPE LOCALITY: Peru, Cajamarca, Hualgayoc, Hac. Ninabamba.
DISTRIBUTION: Colombia to Peru, Bolivia, and NW Brazil.
SYNONYMS: *fumosus, intermedius.*
COMMENTS: Includes *intermedius* and *fumosus;* see Gardner and Carter (1972:72) who designated a neotype.

Platyrrhinus lineatus (E. Geoffroy, 1810). Ann. Mus. Hist. Nat. Paris, 15:180.
TYPE LOCALITY: Paraguay, Asuncion.
DISTRIBUTION: Colombia to Peru, Bolivia, Uruguay, N Argentina, and S and E Brazil; French Guyana; Surinam.
STATUS: CITES - Appendix III (Uruguay).
SYNONYMS: *nigellus.*
COMMENTS: Includes *nigellus;* see Jones and Carter (1979:8). See Willig and Hollander (1987, Mammalian Species, 275, as *Vampyrops lineatus*).

Platyrrhinus recifinus (Thomas, 1901). Ann. Mag. Nat. Hist., ser. 7, 8:192.
TYPE LOCALITY: Brazil, Pernambuco, Recife.
DISTRIBUTION: E Brazil. A Guyana record is erroneous, since the specimen was refered to *V. latus* (= *brachycephalus*) by Handley and Ferris (1972:521).

Platyrrhinus umbratus (Lyon, 1902). Proc. Biol. Soc. Washington, 15:151.
TYPE LOCALITY: Colombia, Magdalena, San Miguel (Macotama River).
DISTRIBUTION: Panama, N and W Colombia, N Venezuela.
SYNONYMS: *aquilus, oratus.*
COMMENTS: Formerly included in *dorsalis* by Carter and Rouk (1973:976); but see Handley (1976:28).

Platyrrhinus vittatus (Peters, 1860). Monatsb. K. Preuss. Akad. Wiss. Berlin, 1860:225.
TYPE LOCALITY: Venezuela, Carabobo, Puerto Cabello.
DISTRIBUTION: Costa Rica to Venezuela, Peru and Bolivia.

Pygoderma Peters, 1863. Monatsb. K. Preuss. Akad. Wiss. Berlin, 1863:83.
TYPE SPECIES: *Stenoderma microdon* Peters, 1863 (= *Phyllostoma bilabiatum* Wagner, 1843).

Pygoderma bilabiatum (Wagner, 1843). Arch. Naturgesch., 1:366.
TYPE LOCALITY: Brazil, São Paulo, Ipanema.
DISTRIBUTION: Surinam, Bolivia, S Brazil, Paraguay, N Argentina; the reported North American occurence is erroneous, see Jones and Carter (1976:29).
SYNONYMS: *leucomus, magna, microdon.*
COMMENTS: See Webster and Owen (1984, Mammalian Species, 220).

Sphaeronycteris Peters, 1882. Sitzb. Preuss. Akad. Wiss., 45:988.
TYPE SPECIES: *Sphaeronycteris toxophyllum* Peters, 1882.

Sphaeronycteris toxophyllum Peters, 1882. Sitzb. Preuss. Akad. Wiss., 45:989.
TYPE LOCALITY: Peru, Loreto, Pebas.
DISTRIBUTION: Colombia to Venezuela, Peru and Bolivia; Amazonian Brazil.

Stenoderma E. Geoffroy, 1818. Descrip. de L'Egypte, 2:114.
TYPE SPECIES: *Stenoderma rufa* Desmarest, 1820.
COMMENTS: Some authors include *Ardops, Phyllops,* and *Ariteus* in *Stenoderma;* see Varona (1974:23-26) and Simpson (1945:58), but most mammalogists follow the arrangement presented here; see also Jones and Carter (1976:28-29).

Stenoderma rufum Desmarest, 1820. Mammalogie, *in* Encycl. Méth., p. 117.
TYPE LOCALITY: Not designated in original publication (probably Virgin Isls).
DISTRIBUTION: Puerto Rico and Virgin Isls (St. John and St. Thomas).
SYNONYMS: *anthonyi, darioi, undatus.*
COMMENTS: See Genoways and Baker (1972, Mammalian Species, 18).

Sturnira Gray, 1842. Ann. Mag. Nat. Hist., [ser. 1], 10:257.
TYPE SPECIES: *Sturnira spectrum* Gray, 1842 (= *Phyllostoma lilium* E. Geoffroy, 1810).
SYNONYMS: *Corvira, Sturnirops* (see Jones and Carter, 1976:21).
COMMENTS: Davis (1980) gave a key to all but one of the species recognized here.

Sturnira aratathomasi Peterson and Tamsitt, 1968. R. Ontario Mus. Life Sci. Occas. Pap., 12:1.
TYPE LOCALITY: Colombia, Valle, 2 km S Pance (ca. 20 km SW Cali), 1,650 m.
DISTRIBUTION: Colombia, Ecuador, NW Venezuela, Peru.
COMMENTS: Subgenus *Sturnira.* See Soriano and Molinari (1987, Mammalian Species, 284).

Sturnira bidens Thomas, 1915. Ann. Mag. Nat. Hist., ser. 8, 16:310.
TYPE LOCALITY: Ecuador, Napo, Baeza, Upper Coca River, 6,500 ft. (1,981 m).
DISTRIBUTION: Peru, Ecuador, Colombia, Venezuela, perhaps Amazonian Brazil.
COMMENTS: Subgenus *Corvira.* Formerly included in genus *Corvira;* see Gardner and O'Neill (1969) and Jones and Carter (1976:21). See Molinari and Soriano (1987, Mammalian Species, 276).

Sturnira bogotensis Shamel, 1927. Proc. Biol. Soc. Washington, 40:129.
TYPE LOCALITY: Colombia, Cundinamarca, Bogota.
DISTRIBUTION: Colombia, Ecuador, and Peru.
COMMENTS: Subgenus *Sturnira.* Usually confused with *ludovici,* but recognized by Handley (1976:25). See also comment under *ludovici.* The correct name of this species is probably *oporaphilum;* see Anderson et al. (1982:6).

Sturnira erythromos (Tschudi, 1844). Fauna Peruana, p. 64.
TYPE LOCALITY: Peru.
DISTRIBUTION: Venezuela to NW Argentina.
COMMENTS: Subgenus *Sturnira.*

Sturnira lilium (E. Geoffroy, 1810). Ann. Mus. Hist. Nat. Paris, 15:181.
TYPE LOCALITY: Paraguay, Asuncion.
DISTRIBUTION: Lesser Antilles; Sonora and Tamaulipas (Mexico) to N Argentina, Uruguay, and E Brazil; Trinidad and Tobago; perhaps Jamaica.
SYNONYMS: *albescens, angeli, excisum, fumarium, luciae, parvidens, paulsoni, spectrum, zygomaticus.*
COMMENTS: Subgenus *Sturnira.* Includes *angeli* and *paulsoni;* see Jones and Carter (1976:20). See Gannon et al. (1989, Mammalian Species, 333).

Sturnira ludovici Anthony, 1924. Am. Mus. Novit., 139:8.
TYPE LOCALITY: Ecuador, Pichincha, near Gualea.
DISTRIBUTION: Ecuador and Guyana to Sonora and Tamaulipas (Mexico).
SYNONYMS: *hondurensis, occidentalis.*
COMMENTS: Subgenus *Sturnira.* Includes *hondurensis;* see Jones and Carter (1976:22). Bolivian records probably pertain to *bogotensis;* Peruvian ones definitely do; see Anderson et al. (1982:6).

Sturnira luisi Davis, 1980. Occas. Pap. Mus. Texas Tech Univ., 70:1.
TYPE LOCALITY: Costa Rica, Alajuela, 11 mi. (18 km) NE Naranjo, Cariblanco, 3,000 ft. (914 m).
DISTRIBUTION: Costa Rica to Ecuador and NW Peru.
COMMENTS: Subgenus *Sturnira.* The presence of this species in Colombia has not been verified; previously confused with *Sturnira ludovici* (Tamsitt, *in* Honacki et al., 1982).

Sturnira magna de la Torre, 1966. Proc. Biol. Soc. Washington, 79:267.
TYPE LOCALITY: Peru, Loreto, Iquitos, Rio Maniti, Santa Cecilia.
DISTRIBUTION: Colombia, Ecuador, Peru, Bolivia.
COMMENTS: Subgenus *Sturnira*. See Tamsitt and Häuser (1985, Mammalian Species, 240).

Sturnira mordax (Goodwin, 1938). Am. Mus. Novit., 976:1.
TYPE LOCALITY: Costa Rica, Cartago, El Sauce Peralta.
DISTRIBUTION: Costa Rica, Panama.
COMMENTS: Subgenus *Sturnira*, Formerly included in *Sturnirops*; see Jones and Carter (1976:21).

Sturnira nana Gardner and O'Neill, 1971. Occas. Pap. Mus. Zool. La. St. Univ., 42:1.
TYPE LOCALITY: Peru, Ayacucho, Huanhuachayo, 1,660 m.
DISTRIBUTION: S Peru.
COMMENTS: Subgenus *Corvira*.

Sturnira thomasi de la Torre and Schwartz, 1966. Proc. Biol. Soc. Washington, 79:299.
TYPE LOCALITY: Guadeloupe (Lesser Antilles), Sofaia, 1,200 ft. (366 m) (France).
DISTRIBUTION: Guadeloupe (Lesser Antilles).
COMMENTS: Subgenus *Sturnira*. See Jones and Genoways (1975*c*, Mammalian Species, 68).

Sturnira tildae de la Torre, 1959. Chicago Acad. Sci. Nat. Hist. Misc., 166:1.
TYPE LOCALITY: Trinidad and Tobago, Trinidad, Arima Valley.
DISTRIBUTION: Brazil, Guianas, Venezuela, Trinidad, Colombia, Ecuador, Peru, Bolivia.
COMMENTS: Subgenus *Sturnira*.

Uroderma Peters, 1866. Monatsb. K. Preuss. Akad. Wiss. Berlin, 1865:587 [1866].
TYPE SPECIES: *Phyllostoma personatum* Peters, 1865 (preoccupied; = *Uroderma bilobatum* Peters, 1866).
COMMENTS: Revised by Davis (1968).

Uroderma bilobatum Peters, 1866. Monatsb. K. Preuss. Akad. Wiss. Berlin, 1866:392.
TYPE LOCALITY: Brazil, São Paulo.
DISTRIBUTION: Veracruz and Oaxaca (Mexico) to Peru, Bolivia and Brazil; Trinidad.
SYNONYMS: *convexum, davisi, molaris, personatum, thomasi, trinitatum*.
COMMENTS: See Baker and Clark (1987, Mammalian Species, 279).

Uroderma magnirostrum Davis, 1968. J. Mammal., 49:679.
TYPE LOCALITY: Honduras, Valle, 10 km E San Lorenzo.
DISTRIBUTION: Michoacan (Mexico) to Venezuela, Peru, Bolivia and Brazil.

Vampyressa Thomas, 1900. Ann. Mag. Nat. Hist., ser. 7, 5:270.
TYPE SPECIES: *Phyllostoma pusillum* Wagner, 1843.
SYNONYMS: *Metavampyressa, Vampyriscus* (see Jones and Carter, 1976:25).
COMMENTS: Does not include *Mesophylla*, but see Owen (1987). A key for this genus is presented in Lewis and Wilson (1987).

Vampyressa bidens (Dobson, 1878). Cat. Chiroptera Brit. Mus., p. 535.
TYPE LOCALITY: Peru, Loreto, Santa Cruz (Rio Huallaga).
DISTRIBUTION: Guianas to Colombia to Peru; N Bolivia; Amazonian Brazil.
COMMENTS: Subgenus *Vampyriscus*. Formerly included in genus *Vampyriscus*; see Jones and Carter (1976:25).

Vampyressa brocki Peterson, 1968. R. Ontario Mus. Life Sci. Contrib., 73:1.
TYPE LOCALITY: Guyana, Rupununi, ca. 40 mi. (64 km) E Dadanawa, at Ow-wi-dy-wau (Oshi Wau head, near Marara Waunowa), Kuitaro River.
DISTRIBUTION: Surinam, Guyana, Amazonian Brazil, SE Colombia.
COMMENTS: Subgenus *Metavampyressa*.

Vampyressa melissa Thomas, 1926. Ann. Mag. Nat. Hist., ser. 9, 18:157.
TYPE LOCALITY: Peru, Amazonas, Chachapoyas, Puca Tambo, 1,480 m.
DISTRIBUTION: Peru, S Colombia, French Guiana.
COMMENTS: Subgenus *Vampyressa*.

Vampyressa nymphaea Thomas, 1909. Ann. Mag. Nat. Hist., ser. 8, 4:230.
TYPE LOCALITY: Colombia, Choco, Novita (San Juan River).
DISTRIBUTION: W Ecuador to Nicaragua. A record from SE Peru is suspect.
COMMENTS: Subgenus *Metavampyressa*.

Vampyressa pusilla (Wagner, 1843). Abh. Akad. Wiss., München, 5:173.
TYPE LOCALITY: Brazil, Rio de Janiero, Sapitiba.
DISTRIBUTION: Oaxaca and Veracruz (Mexico) to Bolivia and Guianas; Paraguay and SE Brazil.
SYNONYMS: *minuta, nattereri, thyone, venilla* (see Jones and Carter, 1976:24).
COMMENTS: Subgenus *Vampyressa*. See Lewis and Wilson (1987, Mammalian Species, 292).

Vampyrodes Thomas, 1900. Ann. Mag. Nat. Hist., ser. 7, 5:270.
TYPE SPECIES: *Vampyrops caracciolae* Thomas, 1889.

Vampyrodes caraccioli (Thomas, 1889). Ann. Mag. Nat. Hist., ser. 6, 4:167.
TYPE LOCALITY: Trinidad and Tobago, Trinidad.
DISTRIBUTION: Oaxaca (Mexico) to Peru, Bolivia and N Brazil; Trinidad and Tobago.
SYNONYMS: *major, ornatus*.
COMMENTS: Includes *major*; see Jones and Carter (1976:24). See Willis et al. (1990, Mammalian Species, 359).

Subfamily Desmodontinae Bonaparte, 1845. Cat. Met. Mamm. Europe, p. 5.
COMMENTS: Formerly treated as a separate family; see Jones and Carter (1976:31).

Desmodus Wied-Neuwied, 1826. Beitr. Naturgesch. Brasil, 2:231.
TYPE SPECIES: *Desmodus rufus* Wied-Neuwied, 1824 (= *Phyllostoma rotundus* E. Geoffroy, 1810).

Desmodus rotundus (E. Geoffroy, 1810). Ann. Mus. Hist. Nat. Paris, 15:181.
TYPE LOCALITY: Paraguay, Asuncion.
DISTRIBUTION: Uruguay, N Argentina, and N Chile to Sonora, Nuevo Leon and Tamaulipas (Mexico); Margarita Isl (Venezuela); Trinidad.
SYNONYMS: *cinerea, d'orbigny, ecaudatus, fuscus, mordax, murinus, rufus*.
COMMENTS: See Greenhall et al. (1983, Mammalian Species, 202).

Diaemus Miller, 1906. Proc. Biol. Soc. Washington, 19:84.
TYPE SPECIES: *Desmodus youngi* Jentinck, 1893.
COMMENTS: Included in *Desmodus* by Handley (1976:35); but see Hall (1981:174).

Diaemus youngi (Jentink, 1893). Notes Leyden Mus., 15:282.
TYPE LOCALITY: Guyana, Berbice River, upper Canje Creek.
DISTRIBUTION: Tamaulipas (Mexico) to N Argentina and E Brazil; Trinidad; Margarita Isl (Venezuela).
SYNONYMS: *cypselinus*.

Diphylla Spix, 1823. Sim. Vespert. Brasil., p. 68.
TYPE SPECIES: *Diphylla ecaudata* Spix, 1823.

Diphylla ecaudata Spix, 1823. Sim. Vespert. Brasil., p. 68.
TYPE LOCALITY: Brazil, Bahia, San Francisco River.
DISTRIBUTION: S Texas (USA) to Venezuela, Peru, Bolivia, and E Brazil.
SYNONYMS: *centralis, diphylla*.
COMMENTS: See Greenhall et al. (1984, Mammalian Species, 227).

Family Natalidae Gray, 1866. Ann. Mag. Nat. Hist., ser. 3, 17:90.

Natalus Gray, 1838. Mag. Zool. Bot., 2:496.
TYPE SPECIES: *Natalus stramineus* Gray, 1838.
SYNONYMS: *Chilonatalus, Nyctiellus*.

Natalus lepidus (Gervais, 1837). L'Inst. Paris, 5(218):253.
TYPE LOCALITY: Cuba.
DISTRIBUTION: Cuba, Bahama Isls.
COMMENTS: Subgenus *Nyctiellus*.

Natalus micropus Dobson, 1880. Proc. Zool. Soc. Lond., 1880:443.
TYPE LOCALITY: Jamaica, Kingston.
DISTRIBUTION: Cuba, Jamaica, Hispaniola, Providencia Isl (Colombia).
SYNONYMS: *brevimanus, macer*.
COMMENTS: Subgenus *Chilonatalus*. Includes *brevimanus* and *macer*; see Varona (1974:32). Formerly included *tumidifrons*; but see Ottenwalder and Genoways (1982) who revised both species. Kerridge and Baker (1978, Mammalian Species, 114) treated only the nominate subspecies.

Natalus stramineus Gray, 1838. Mag. Zool. Bot., 2:496.
TYPE LOCALITY: Unknown, given by Goodwin (1959*b*:2), who discussed the problem, as probably Antigua, Lesser Antilles.
DISTRIBUTION: S Baja California, Nuevo Leon, and Sonora (Mexico) to Brazil; Lesser Antilles; Hispaniola; Jamaica; Cuba (subfossil).
SYNONYMS: *dominicensis, jamaicensis, major, mexicanus, natalensis, primus, saturatus, tronchonii*.
COMMENTS: Subgenus *Natalus*. For synonyms see Varona (1974:32) and Goodwin (1959*b*:6). See Hoyt and Baker (1980, Mammalian Species, 130, including only the Greater Antillean subspecies). See Handley and Gardner (1990) for clarification of the holotype.

Natalus tumidifrons (Miller, 1903). Proc. Biol. Soc. Washington, 16:119.
TYPE LOCALITY: Bahamas, Watling Island.
DISTRIBUTION: Islands of the Bahamas.
COMMENTS: Subgenus *Chilonatalus*. Formerly included in *micropus*; but see Ottenwalder and Genoways (1982) who revised both species.

Natalus tumidirostris Miller, 1900. Proc. Biol. Soc. Washington, 13:160.
TYPE LOCALITY: Curacao, Hatto (Netherlands).
DISTRIBUTION: Surinam, Venezuela, Colombia, Trinidad and Tobago, Curacao and Bonaire (Netherlands Antilles).
SYNONYMS: *continentis, haymani*.
COMMENTS: Subgenus *Natalus*. Revised by Goodwin (1959*b*).

Family Furipteridae Gray, 1866. Ann. Mag. Nat. Hist., ser. 3, 17:91.

Amorphochilus Peters, 1877. Monatsb. K. Preuss. Akad. Wiss. Berlin, 1877:185.
TYPE SPECIES: *Amorphochilus schnablii* Peters, 1877.

Amorphochilus schnablii Peters, 1877. Monatsb. K. Preuss. Akad. Wiss. Berlin, 1877:185.
TYPE LOCALITY: Peru, Tumbes, Tumbes.
DISTRIBUTION: W Peru, W Ecuador, Puna Isl (Ecuador), N Chile.
SYNONYMS: *osgoodi*.

Furipterus Bonaparte, 1837. Iconogr. Fauna Ital., 1, fasc. 21.
TYPE SPECIES: *Furia horrens* F. Cuvier, 1828.

Furipterus horrens (F. Cuvier, 1828). Mem. Mus. Hist. Nat. Paris, 16:150.
TYPE LOCALITY: French Guiana, Mana River.
DISTRIBUTION: Costa Rica to Peru and E Brazil; Trinidad.
SYNONYMS: *coerulescens*.

Family Thyropteridae Miller, 1907. Bull. U.S. Natl. Mus., 57: 84, 186.

Thyroptera Spix, 1823. Sim. Vespert. Brasil., p. 61.
TYPE SPECIES: *Thyroptera tricolor* Spix, 1823.

Thyroptera discifera (Lichtenstein and Peters, 1855). Monatsb. K. Preuss. Akad. Wiss. Berlin, 1855:335.
TYPE LOCALITY: Venezuela, Carabobo, Puerto Cabello.
DISTRIBUTION: Nicaragua; Panama and Colombia to Guianas, Amazonian Brazil, Peru, and Bolivia.
SYNONYMS: *abdita, major.*
COMMENTS: See Wilson (1978, Mammalian Species, 104).

Thyroptera tricolor Spix, 1823. Sim. Vespert. Brasil., p. 61.
TYPE LOCALITY: Brazil, Amazon River.
DISTRIBUTION: Veracruz (Mexico) to Guianas, E Brazil, Bolivia, and Peru; Trinidad.
SYNONYMS: *albigula, albiventer, bicolor, juquiaensis, thyropterus.*
COMMENTS: See Wilson and Findley (1977, Mammalian Species, 71).

Family Myzopodidae Thomas, 1904. Proc. Zool. Soc. Lond., 1904(2):5.

Myzopoda Milne-Edwards and A. Grandidier, 1878. Bull. Sci. Soc. Philom. Paris, ser. 7, 2:220.
TYPE SPECIES: *Myzopoda aurita* Milne-Edwards and A. Grandidier, 1878.

Myzopoda aurita Milne-Edwards and A. Grandidier, 1878. Bull. Sci. Soc. Philom. Paris, ser. 7, 2:220.
TYPE LOCALITY: Madagascar.
DISTRIBUTION: Madagascar.
COMMENTS: See Schliemann and Maas (1978, Mammalian Species, 116).

Family Vespertilionidae Gray, 1821. London Med. Repos., 15:299.

Subfamily Kerivoulinae Miller, 1907. Bull. U.S. Natl. Mus., 57:232.

Kerivoula Gray, 1842. Ann. Mag. Nat. Hist., [ser. 1], 10:258.
TYPE SPECIES: *Vespertilio pictus* Pallas, 1767.
SYNONYMS: *Phoniscus.*
COMMENTS: Includes *Phoniscus*; see Koopman (1982:22) and Ryan (1965:518); but also see Hill (1965:524-528) and Corbet and Hill (1980:77).

Kerivoula aerosa (Tomes, 1858). Proc. Zool. Soc. Lond., 1858:333.
TYPE LOCALITY: "Eastern coast of South Africa."
DISTRIBUTION: Possibly South Africa, but more likely somewhere in SE Asia.
COMMENTS: Subgenus *Phoniscus.* See Hill (1965:552-554) for placement of this species.

Kerivoula africana Dobson, 1878. Cat. Chiroptera Brit. Mus., p. 335.
TYPE LOCALITY: Tanzania, coast opposite Zanzibar Isl.
DISTRIBUTION: Tanzania.
COMMENTS: Subgenus *Kerivoula.*

Kerivoula agnella Thomas, 1908. Ann. Mag. Nat. Hist., ser. 8, 2:372.
TYPE LOCALITY: Papua New Guinea, Louisiade Archipelago, Misima Isl.
DISTRIBUTION: Louisiade Arch. and D'Entrecasteaux Isls (Papua New Guinea).
COMMENTS: Subgenus *Kerivoula.*

Kerivoula argentata Tomes, 1861. Proc. Zool. Soc. Lond., 1861:32.
TYPE LOCALITY: Namibia, Otjoro.
DISTRIBUTION: Uganda and S Kenya to Angola, Namibia and Natal (South Africa).
SYNONYMS: *nidicola, zuluensis.*
COMMENTS: Subgenus *Kerivoula.*

Kerivoula atrox Miller, 1905. Proc. Biol. Soc. Washington, 18:230.
TYPE LOCALITY: Indonesia, Sumatra, near Kateman River.
DISTRIBUTION: S Thailand, W Malaysia, Sumatra, Borneo.
COMMENTS: Subgenus *Phoniscus.*

Kerivoula cuprosa Thomas, 1912. Ann. Mag. Nat. Hist., ser. 8, 10:41.
TYPE LOCALITY: Cameroon, Ja River, Bitye.

DISTRIBUTION: Kenya, N Zaire, S Cameroon.
COMMENTS: Subgenus *Kerivoula.*

Kerivoula eriophora (Heuglin, 1877). Reise Nordost-Afrika, 2:34.
TYPE LOCALITY: Ethiopia, Belegaz Valley, between Semian and Wogara.
DISTRIBUTION: Ethiopia.
COMMENTS: Subgenus *Kerivoula.* Very poorly known; may be conspecific with *africana* which it antedates; see Hayman and Hill (1971:53).

Kerivoula flora Thomas, 1914. Ann. Mag. Nat. Hist., ser. 8, 13:441.
TYPE LOCALITY: Indonesia, Lesser Sundas, S Flores.
DISTRIBUTION: Lesser Sunda Isls (Indonesia), Borneo.
COMMENTS: Subgenus *Kerivoula.* See Hill and Rozendaal (1989).

Kerivoula hardwickii (Horsfield, 1824). Zool. Res. Java, part 8, p. 4(unno.) of *Vespertilio Temminckii* acct.
TYPE LOCALITY: Indonesia, Java.
DISTRIBUTION: India; Sri Lanka; Burma; Thailand; China; W Malaysia; Borneo; Java, Sumatra, Mentawai Isls, Sulawesi, Bali, Lesser Sundas, Kangean Isl and Talaud Isl (Indonesia); Philippines.
SYNONYMS: *crypta, depressa, engana, fusca, malpasi.*
COMMENTS: Subgenus *Kerivoula.* Does not include *flora;* see Hill and Rozendaal (1989).

Kerivoula intermedia Hill and Francis, 1984. Bull. Brit. Mus. Nat. Hist. (Zool.), 47:323.
TYPE LOCALITY: Borneo, Sabah, Lumerao.
DISTRIBUTION: Borneo, W Malaysia.
COMMENTS: Subgenus *Kerivoula.*

Kerivoula jagorii (Peters, 1866). Monatsb. K. Preuss. Akad. Wiss. Berlin, 1866:399.
TYPE LOCALITY: Philippines, Samar.
DISTRIBUTION: Borneo, Java, Bali, Sulawesi, and Lesser Sunda Isls, Samar Isl (Philippines).
SYNONYMS: *javana, rapax* (see Hill, 1965:549-550).
COMMENTS: Subgenus *Phoniscus.*

Kerivoula lanosa (A. Smith, 1847). Illustr. Zool. S. Afr. Mamm., pl. 50.
TYPE LOCALITY: South Africa, Cape Province, 200 mi.(322 km) E Capetown.
DISTRIBUTION: Liberia to Ethiopia, south to South Africa.
SYNONYMS: *bellula, brunnea, harrisoni, lucia, muscilla.*
COMMENTS: Subgenus *Kerivoula.* Includes *harrisoni* and *muscilla;* see Hill (1977*a*).

Kerivoula minuta Miller, 1898. Proc. Acad. Nat. Sci. Philadelphia, 50:321.
TYPE LOCALITY: Thailand, Trang Province.
DISTRIBUTION: W Malaysia, S Thailand, Borneo.
COMMENTS: Subgenus *Kerivoula.*

Kerivoula muscina Tate, 1941. Bull. Am. Mus. Nat. Hist., 78:586.
TYPE LOCALITY: Papua New Guinea, Western Province, Lake Daviumbu, ca. 20 m.
DISTRIBUTION: C New Guinea.
COMMENTS: Subgenus *Kerivoula.*

Kerivoula myrella Thomas, 1914. Ann. Mag. Nat. Hist., ser. 8, 13:438.
TYPE LOCALITY: Papua New Guinea, Bismarck Archipelago, Admiralty Isls, Manus Isl.
DISTRIBUTION: Bismarck Arch.; Wetar Isl (Lesser Sunda Isls).
COMMENTS: Subgenus *Kerivoula.*

Kerivoula papillosa (Temminck, 1840). Monogr. Mamm., 2:220.
TYPE LOCALITY: Indonesia, Java, Bantam.
DISTRIBUTION: NE India, Vietnam, W Malaysia, Sumatra, Java, Sulawesi, Borneo.
SYNONYMS: *lenis, malayana.*
COMMENTS: Subgenus *Kerivoula.*

Kerivoula papuensis (Dobson, 1878). Cat. Chiroptera Brit. Mus., p. 339.
TYPE LOCALITY: Papua New Guinea, Central Prov., Port Moresby.
DISTRIBUTION: SE New Guinea; Queensland, New South Wales (Australia).
COMMENTS: Subgenus *Phoniscus.*

Kerivoula pellucida (Waterhouse, 1845). Proc. Zool. Soc. Lond., 1845:6.
TYPE LOCALITY: Philippines.
DISTRIBUTION: Borneo, Philippines, Java and Sumatra, W Malaysia.
SYNONYMS: *bombifrons*.
COMMENTS: Subgenus *Kerivoula*. Includes *bombifrons*; see Hill (1965:539).

Kerivoula phalaena Thomas, 1912. Ann. Mag. Nat. Hist., ser. 8, 10:281.
TYPE LOCALITY: Ghana, Bibianaha.
DISTRIBUTION: Liberia, Ghana, Cameroon, Congo Republic, Zaire.
COMMENTS: Subgenus *Kerivoula*.

Kerivoula picta (Pallas, 1767). Spicil. Zool., 3:7.
TYPE LOCALITY: Indonesia, Molucca Isls, Ternate Isl.
DISTRIBUTION: Sri Lanka; India to Vietnam, W Malaysia, and S China; Borneo; Sumatra, Java, Bali, Lombok, and Molucca Isls.
SYNONYMS: *bellissima*.
COMMENTS: Subgenus *Kerivoula*.

Kerivoula smithii Thomas, 1880. Ann. Mag. Nat. Hist., ser. 5, 6:166.
TYPE LOCALITY: Nigeria, Calabar.
DISTRIBUTION: Nigeria, Cameroon, N and E Zaire, Kenya, Ivory Coast, Liberia.
COMMENTS: Subgenus *Kerivoula*.

Kerivoula whiteheadi Thomas, 1894. Ann. Mag. Nat. Hist., ser. 6, 14:460.
TYPE LOCALITY: Philippines, Luzon, Isabella.
DISTRIBUTION: Philippines, Borneo, S Thailand, W Malaysia.
SYNONYMS: *bicolor, pusilla* (see Hill, 1965:532-533).
COMMENTS: Subgenus *Kerivoula*.

Subfamily Vespertilioninae Gray, 1821. London Med. Repos., 15:299.
COMMENTS: Includes Nyctophilinae, see Koopman (1984*a*).

Antrozous H. Allen, 1862. Proc. Acad. Nat. Sci. Philadelphia, 14:248.
TYPE SPECIES: *Vespertilio pallidus* Le Conte, 1856.
SYNONYMS: *Bauerus*.
COMMENTS: Includes *Bauerus*; see Pine et al. (1967); but see also Engstrom and Wilson (1981).

Antrozous dubiaquercus Van Gelder, 1959. Am. Mus. Novit., 1973:2.
TYPE LOCALITY: Mexico, Nayarit, Tres Marias Isls, Maria Magdalena Isl.
DISTRIBUTION: Tres Marias Isls, Jalisco, Veracruz, and Chiapas (Mexico); Belize; Honduras, Costa Rica.
SYNONYMS: *meyeri*.
COMMENTS: Subgenus *Bauerus*. Includes *meyeri*; see Pine (1967). Jones et al. (1977). Engstrom and Wilson (1981) placed this species in the monotypic genus *Bauerus*. See Engstrom et al. (1987*b*, Mammalian Species, 282, as *Bauerus dubiaquercus*).

Antrozous pallidus (Le Conte, 1856). Proc. Acad. Nat. Sci. Philadelphia, 7:437.
TYPE LOCALITY: USA, Texas, El Paso Co., El Paso.
DISTRIBUTION: Queretaro and Baja California (Mexico) to Kansas (USA) and British Columbia (Canada); Cuba.
SYNONYMS: *bunkeri, koopmani, minor, pacificus, packardi*.
COMMENTS: Subgenus *Antrozous*. Includes *bunkeri*; see Morse and Glass (1960). Includes *koopmani*; see Martin and Schmidly (1982). See Hermanson and O'Shea (1983, Mammalian Species, 213).

Barbastella Gray, 1821. London Med. Repos., 15:300.
TYPE SPECIES: *Vespertilio barbastellus* Schreber, 1774.

Barbastella barbastellus (Schreber, 1774). Die Säugethiere, 1:168.
TYPE LOCALITY: France, Burgundy.
DISTRIBUTION: England and W Europe to Caucasus; Turkey; Crimea (Ukraine); Morocco; larger Mediterranean islands; Canary Isls; perhaps Senegal.

SYNONYMS: *communis, daubentonii.*
COMMENTS: Does not include *leucomelas,* but see Qumsiyeh (1985).

Barbastella leucomelas (Cretzschmar, 1830). *In* Rüppell, Atlas Reise Nordl. Afr., Zool., Säugeth., p. 73.
TYPE LOCALITY: Egypt, Sinai.
DISTRIBUTION: Caucasus to The Pamirs, N Iran, Afganistan, India, and W China; Honshu, Hokkaido (Japan); Sinai (Egypt); N Ethiopia; perhaps Indo-China.
SYNONYMS: *blanfordi, caspica, darjelingensis, walteri.*

Chalinolobus Peters, 1867. Monatsb. K. Preuss. Akad. Wiss. Berlin, 1866:679 [1867].
TYPE SPECIES: *Vespertilio tuberculatus* Forster, 1844.
SYNONYMS: *Glauconycteris.*
COMMENTS: Includes *Glauconycteris;* see Koopman (1971); but also see Hill and Harrison (1987), who retained *Glauconycteris* as a distinct genus. Reviewed by Tate (1942*a*:221-227) and Ryan (1966).

Chalinolobus alboguttatus (J. A. Allen, 1917). Bull. Am. Mus. Nat. Hist., 37:449.
TYPE LOCALITY: Zaire, Oriental, Medje.
DISTRIBUTION: Zaire, Cameroon.
COMMENTS: Subgenus *Glauconycteris.*

Chalinolobus argentatus (Dobson, 1875). Proc. Zool. Soc. Lond., 1875:385.
TYPE LOCALITY: Cameroon, Western Province, Mt. Cameroon.
DISTRIBUTION: Cameroon to Kenya, south to Angola and Tanzania.
COMMENTS: Subgenus *Glauconycteris.*

Chalinolobus beatrix (Thomas, 1901). Ann. Mag. Nat. Hist., ser. 7, 8:256.
TYPE LOCALITY: Equatorial Guinea, Rio Muni, Benito River, 15 mi. (24 km) from mouth.
DISTRIBUTION: Ivory Coast, Kenya, Congo Republic, perhaps Guinea (Bissau).
SYNONYMS: *humeralis.*
COMMENTS: Subgenus *Glauconycteris.* Includes *humeralis;* see Koopman (1971:7, 8).

Chalinolobus dwyeri Ryan, 1966. J. Mammal., 47:89.
TYPE LOCALITY: Australia, New South Wales, 14 mi. (23 km) S Inverell, Copeton.
DISTRIBUTION: New South Wales and adjacent part of Queensland (Australia).
COMMENTS: Subgenus *Chalinolobus.*

Chalinolobus egeria (Thomas, 1913). Ann. Mag. Nat. Hist., ser. 8, 11:144.
TYPE LOCALITY: Cameroon, Western Province, Bibundi.
DISTRIBUTION: Cameroon, Uganda.
COMMENTS: Subgenus *Glauconycteris.*

Chalinolobus gleni (Peterson and Smith, 1973). R. Ontario Mus. Life Sci. Occas. Pap., 22:3.
TYPE LOCALITY: Cameroon, near Lomie.
DISTRIBUTION: Cameroon, Uganda.
COMMENTS: Subgenus *Glauconycteris.*

Chalinolobus gouldii (Gray, 1841). Appendix C *in* J. Two Exped. Aust., 2:401, 405.
TYPE LOCALITY: Australia, Tasmania, Launceston.
DISTRIBUTION: Australia but not Cape York Peninsula N of Cardwell; Tasmania, Norfolk Isl (Australia); New Caledonia.
SYNONYMS: *neocaledonicus, venatoris.*
COMMENTS: Subgenus *Chalinolobus.* Includes *neocaledonicus;* see Koopman (1971:5).

Chalinolobus kenyacola (Peterson, 1982.) Canadian J. Zool. 60:2521.
TYPE LOCALITY: Kenya, Coast Prov., 8.5 km N Garsen.
DISTRIBUTION: Kenya.
COMMENTS: Subgenus *Glauconycteris.*

Chalinolobus morio (Gray, 1841). Appendix C *In* J. Two Exped. Aust., 2:400, 405.
TYPE LOCALITY: Australia, Tasmania.
DISTRIBUTION: Southern Australia, Tasmania.
SYNONYMS: *australis, microdon, signifer* (see Tate, 1942*a*).
COMMENTS: Subgenus *Chalinolobus.*

Chalinolobus nigrogriseus (Gould, 1852). Mamm. Aust., pt. 4, vol. 3, pl. 43.
TYPE LOCALITY: Australia, Queensland, vic. of Moreton Bay.
DISTRIBUTION: N and E Australia; SE New Guinea and adjacent small islands.
SYNONYMS: *rogersi*.
COMMENTS: Subgenus *Chalinolobus*. Includes *rogersi*; see Van Deusen and Koopman (1971), who revised the species.

Chalinolobus picatus (Gould, 1852). Mamm. Aust., pt. 4, vol. 3, pl. 43.
TYPE LOCALITY: Australia, New South Wales, Capt. Sturt's Depot.
DISTRIBUTION: NW New South Wales, C and S Queensland, and South Australia (Australia).
COMMENTS: Subgenus *Chalinolobus*.

Chalinolobus poensis (Gray, 1842). Ann. Mag. Nat. Hist., [ser. 1], 10:258.
TYPE LOCALITY: Nigeria, Abo (lower Niger River).
DISTRIBUTION: Senegal to Uganda; Bioko.
SYNONYMS: *kraussi*.
COMMENTS: Subgenus *Glauconycteris*.

Chalinolobus superbus (Hayman, 1939). Ann. Mag. Nat. Hist., ser. 11, 3:219.
TYPE LOCALITY: Zaire, Oriental, Ituri Dist., Pawa.
DISTRIBUTION: Ivory Coast, Ghana, NE Zaire.
SYNONYMS: *sheila*.
COMMENTS: Subgenus *Glauconycteris*.

Chalinolobus tuberculatus (Forster, 1844). Descrip. Animal. Itinere Maris Aust. Terras, 1772-74:62.
TYPE LOCALITY: New Zealand.
DISTRIBUTION: New Zealand and adjacent small islands.
COMMENTS: Subgenus *Chalinolobus*.

Chalinolobus variegatus (Tomes, 1861). Proc. Zool. Soc. Lond., 1861:36.
TYPE LOCALITY: Namibia, Otjoro.
DISTRIBUTION: Senegal to Somalia, south to South Africa.
SYNONYMS: *machadoi, papilio, phalaena*.
COMMENTS: Subgenus *Glauconycteris*. Includes *machadoi*; see Koopman (1971:6).

Eptesicus Rafinesque, 1820. Ann. Nature, p. 2.
TYPE SPECIES: *Eptesicus melanops* Rafinesque, 1820 (= *Vespertilio fuscus* Beauvois, 1796).
SYNONYMS: *Neoromicia, Rhinopterus, Vespadelus*.
COMMENTS: Middle and South American species reviewed by W. B. Davis (1965) and Davis (1966). Australian species reviewed by Kitchener et al. (1987). Hill and Harrison (1987) would transfer the Australian and most of the African species (subgenera *Vespadelus* and *Neoromicia*) to *Pipistrellus*. Volleth and Tidemann (1991) would raise *Vespadelus* to full generic rank.

Eptesicus baverstocki Kitchener, Jones, and Caputi, 1987. Rec. W. Aust. Mus., 13:481.
TYPE LOCALITY: Australia, Western Australia, Yuinmery area.
DISTRIBUTION: C and S Australia.
COMMENTS: Subgenus *Vespadelus*.

Eptesicus bobrinskoi Kuzyakin, 1935. Bull. Soc. Nat. Moscow, 44:435.
TYPE LOCALITY: Kazakhstan, 65 km E Aralsk, Tyulek Wells in Aral-Kara-Kum desert.
DISTRIBUTION: North Caucasus, Kazakhstan, Uzbekistan, and Turkmenistan; an Iranian listing is apparently erroneous.
COMMENTS: Subgenus *Eptesicus*. Revised by Hanak and Gaisler (1971); see also Hanak and Horáček (1986).

Eptesicus bottae (Peters, 1869). Monatsb. K. Preuss. Akad. Wiss. Berlin, 1869:406.
TYPE LOCALITY: Yemen.
DISTRIBUTION: Turkey, Egypt, and Yemen, east to Mongolia and Pakistan.
SYNONYMS: *anatolicus, hingstoni, innesi, ognevi, omanensis, taftanimontis* (see Corbet, 1978*c*:57, and DeBlase, 1971).
COMMENTS: Subgenus *Eptesicus*. Revised by Nader and Kock (1990).

Eptesicus brasiliensis (Desmarest, 1819). Nouv. Dict. Hist. Nat., Nouv. ed., 35:478.
TYPE LOCALITY: Brazil, Goias.
DISTRIBUTION: Veracruz (Mexico) south to N Argentina and Uruguay; Trinidad and Tobago.
SYNONYMS: *andinus, arctoideus, arge, argentinus, chiriquinus, derasus, ferrugineus, hilarii, inca; melanopterus* Jentink; *nitens, thomasi.*
COMMENTS: Subgenus *Eptesicus*. Includes *chiriquinus, melanopterus*, and *andinus*; see Koopman (1978*b*:19); but see also Davis (1966).

Eptesicus brunneus (Thomas, 1880). Ann. Mag. Nat. Hist., ser. 5, 6:165.
TYPE LOCALITY: Nigeria, Eastern region, Calabar.
DISTRIBUTION: Liberia to Zaire.
COMMENTS: Subgenus *Neoromicia.*

Eptesicus capensis (A. Smith, 1829). Zool. J., 4:435.
TYPE LOCALITY: South Africa, Cape Province, Grahamstown.
DISTRIBUTION: Guinea to Ethiopia, south to South Africa; Madagascar.
SYNONYMS: *damarensis, garambae, gracilior, grandidieri, matroka, nkatiensis, notius.*
COMMENTS: Subgenus *Neoromicia*. Includes *notius*; see Koopman (1975*b*:405).

Eptesicus demissus Thomas, 1916. J. Fed. Malay St. Mus., 7:1.
TYPE LOCALITY: Thailand, Surat Thani, Khao Nong.
DISTRIBUTION: Peninsular Thailand.
COMMENTS: Subgenus *Eptesicus.*

Eptesicus diminutus Osgood, 1915. Field Mus. Nat. Hist. Publ., Zool. Ser., 10:197.
TYPE LOCALITY: Brazil, Bahia, Rio Preto, São Marcello.
DISTRIBUTION: Venezuela, E Brazil, Paraguay, Uruguay, N Argentina.
SYNONYMS: *fidelis.*
COMMENTS: Subgenus *Eptesicus*. Includes *fidelis*; see Williams (1978*c*:380-382), and also for the use of *diminutus* instead of *dorianus.*

Eptesicus douglasorum Kitchener, 1976. Rec. W. Aust. Mus., 4:295, 296.
TYPE LOCALITY: Australia, Western Australia, Kimberley, Napier Ranges, Tunnel Creek.
DISTRIBUTION: Kimberley (N Western Australia).
COMMENTS: Subgenus *Vespadelus*. Originally described as *douglasi* but emended by Kitchener et al. (1987).

Eptesicus flavescens (Seabra, 1900). J. Sci. Math. Phys. Nat. Lisboa, ser. 2, 6:23.
TYPE LOCALITY: Angola, Galanga.
DISTRIBUTION: Angola, Burundi.
SYNONYMS: *angolensis.*
COMMENTS: Subgenus *Neoromicia.*

Eptesicus floweri (de Winton, 1901). Ann. Mag. Nat. Hist., ser. 7, 7:46.
TYPE LOCALITY: Sudan, Khartoum, Wad Marium.
DISTRIBUTION: Sudan, Mali.
SYNONYMS: *lowei.*
COMMENTS: Subgenus *Rhinopterus*. Includes *lowei*; see Braestrup (1935); but also see Hayman and Hill (1971:45).

Eptesicus furinalis (d'Orbigny, 1847). Voy. Am. Merid., Atlas Zool., 4:13.
TYPE LOCALITY: Argentina, Corrientes.
DISTRIBUTION: N Argentina, Brazil, and Guianas to Jalisco and Tamaulipas (Mexico).
SYNONYMS: *carteri, chapmani, chiralensis, dorianus, findleyi, gaumeri, montosus.*
COMMENTS: Subgenus *Eptesicus*. Includes *montosus*; see Koopman (1978*b*:19), but see also Davis (1966).

Eptesicus fuscus (Beauvois, 1796). Cat. Raisonne Mus. Peale Philadelphia, p. 18.
TYPE LOCALITY: USA, Pennsylvania, Philadelphia.
DISTRIBUTION: S Canada to Colombia and N Brazil; Greater Antilles; Bahamas; Dominica and Barbados (Lesser Antilles); perhaps Alaska.
SYNONYMS: *arquatus, bahamensis, bernardinus, carolinensis, cubensis, dutertreus, hispaniolae, lynni, melanops; melanopterus* Rehn; *miradorensis, osceola, pallidus, pelliceus, peninsulae, petersoni, phaiops, ursinus, wetmorei.*

COMMENTS: Subgenus *Eptesicus*. Closely similar to *serotinus* with which it may be conspecific. Includes *lynni*, see Koopman (1989*c*). See Kurta and Baker (1990, Mammalian Species, 356).

Eptesicus guadeloupensis Genoways and Baker, 1975. Occas. Pap. Mus. Texas Tech Univ., 34:1.
TYPE LOCALITY: Guadeloupe (Lesser Antilles), Basse Terre, 2 km S and 2 km E Baiae-Mahault (France).
DISTRIBUTION: Guadeloupe (Lesser Antilles).
COMMENTS: Subgenus *Eptesicus*.

Eptesicus guineensis (Bocage, 1889). J. Sci. Math. Phys. Nat. Lisboa, ser. 2, 1:6.
TYPE LOCALITY: Guinea-Bissau, Bissau.
DISTRIBUTION: Senegal and Guinea to Ethiopia and NE Zaire; perhaps Tanzania.
SYNONYMS: *rectitragus*.
COMMENTS: Subgenus *Neoromicia*. This species is called *pusillus* by Hayman and Hill (1971); but see Koopman (1975:406, 407).

Eptesicus hottentotus (A. Smith, 1833). S. Afr. Quart. J., 2:59.
TYPE LOCALITY: South Africa, Cape Province, Uitenhage.
DISTRIBUTION: South Africa to Angola and Kenya.
SYNONYMS: *angusticeps, bensoni, megalurus, pallidior, portavernus, smithi*.
COMMENTS: Subgenus *Eptesicus*. Revised by Schlitter and Aggundey (1986).

Eptesicus innoxius (Gervais, 1841). *In* Vaillant, Voy. autour du monde...la Bonite, Zool.(Eydoux and Souleyet), 1:pl. 2.
TYPE LOCALITY: Peru, Piura, Amotape.
DISTRIBUTION: NW Peru, W Ecuador, Puna Isl (Ecuador).
SYNONYMS: *espadae, punicus*.
COMMENTS: Subgenus *Eptesicus*.

Eptesicus kobayashii Mori, 1928. Zool. Mag. (Tokyo), 40:292.
TYPE LOCALITY: Korea, Nando, Heian, Heijo.
DISTRIBUTION: Korea.
COMMENTS: Subgenus *Eptesicus*. Status uncertain; see Corbet (1978*c*:58), but probably a representative of *E. bottae*.

Eptesicus melckorum Roberts, 1919. Ann. Transvaal Mus., 6:113.
TYPE LOCALITY: South Africa, Cape Province, Berg River, Kersfontein.
DISTRIBUTION: Cape Prov. (South Africa), Zambia, Mozambique, Tanzania.
COMMENTS: Subgenus *Neoromicia*. This species has not been clearly distinguished from *capensis*.

Eptesicus nasutus (Dobson, 1877). J. Asiat. Soc. Bengal, 46:311.
TYPE LOCALITY: Pakistan, Sind, Shikarpur.
DISTRIBUTION: Arabia, Iraq, Iran, Afghanistan, Pakistan.
SYNONYMS: *batinensis, matschei, pellucens, walli*.
COMMENTS: Subgenus *Eptesicus*. Does not include *bobrinskoi*; see Corbet (1978*c*:57). Includes *walli*; see DeBlase (1980:182-188). Revised by Gaisler (1970).

Eptesicus nilssoni (Keyserling and Blasius, 1839). Arch. Naturgesch., 5(1):315.
TYPE LOCALITY: Sweden.
DISTRIBUTION: W and E Europe to E Siberia and NW China; north beyond Arctic Circle in Scandinavia, south to Bulgaria, Iraq, the Elburz Mtns (N Iran), The Pamirs and W China (not Tibet); Nepal; Honshu, Hokkaido (Japan); Sakhalin Isl (Russia).
SYNONYMS: *atratus, centralasiaticus, gobiensis, japonensis, kashgaricus, parvus, propinquus*.
COMMENTS: Subgenus *Eptesicus*. Includes *propinquus*; see W. B. Davis (1965:230). Includes *japonensis*; see Corbet (1978*c*:57), but also see Yoshiyuki (1989). Includes *gobiensis*; see Corbet (1978*c*); but see also Strelkov (1986), who discussed status, relations, and distribution, and Pavlinov and Rossolimo (1987). Revised by Wallin (1969).

Eptesicus pachyotis (Dobson, 1871). Proc. Asiat. Soc. Bengal, p. 211.
TYPE LOCALITY: India, Assam, Khasi Hills.
DISTRIBUTION: Assam (India), N Burma, N Thailand.

COMMENTS: Subgenus *Eptesicus.*

Eptesicus platyops (Thomas, 1901). Ann. Mag. Nat. Hist., ser. 7, 8:31.
TYPE LOCALITY: Nigeria, Western Region, Lagos.
DISTRIBUTION: Nigeria, Senegal, Bioko.
COMMENTS: Subgenus *Eptesicus.* Considered a subspecies of *serotinus* by Ibáñez and Valverde (1985), but no comparison with *E. bottae* was made.

Eptesicus pumilus Gray, 1841. Appendix C *in* J. Two Exped. Aust., 2:406.
TYPE LOCALITY: Australia, New South Wales, Yarrundi.
DISTRIBUTION: N Western Australia, Northern Territory, Queensland, New South Wales, and South Australia (Australia).
SYNONYMS: *caurinus, darlingtoni, finlaysoni, troughtoni.*
COMMENTS: Subgenus *Vespadelus.* Includes *darlingtoni* and *caurinus;* see McKean et al. (1978:533); but see also Kitchener et al. (1987), who described *finlaysoni* and *troughtoni,* which are included here, as separate species, and transferred *darlingtoni* to *sagittula.*

Eptesicus regulus (Thomas, 1906). Proc. Zool. Soc. Lond., 1906:470, 471.
TYPE LOCALITY: Australia, Western Australia, King Georges Sound, King River (near Albany).
DISTRIBUTION: SW and SE Australia, including Tasmania.
COMMENTS: Subgenus *Vespadelus.* See McKean et al. (1978:532, 534).

Eptesicus rendalli (Thomas, 1889). Ann. Mag. Nat. Hist., ser. 6, 3:362.
TYPE LOCALITY: Gambia, Bathurst.
DISTRIBUTION: Gambia to Somalia, south to Botswana, Malawi, and Mozambique.
SYNONYMS: *faradjius, phasma.*
COMMENTS: Subgenus *Neoromicia.*

Eptesicus sagittula McKean, et al., 1978. Aust. J. Zool., 26:535.
TYPE LOCALITY: Australia, New South Wales, 13 km N.W. Braidwood.
DISTRIBUTION: SE Australia, including Tasmania and Lord Howe Isl.
COMMENTS: Subgenus *Vespadelus.* Kitchener et al. (1987) used the name *darlingtoni* for this species (see also comments under *pumilus*).

Eptesicus serotinus (Schreber, 1774). Die Säugethiere, 1:167.
TYPE LOCALITY: France.
DISTRIBUTION: W Europe through S Asiatic Russia to Himalayas, Thailand and China, north to Korea; Taiwan; S England; N Africa; most islands in Mediterranean; perhaps Subsaharan Africa.
SYNONYMS: *albescens, andersoni, boscai, brachydigitatis, horikawai, incisivus, insularis, intermedius, isabellinus, meridionalis, mirza, okenii, pachyomus, pallens, pashtonus, rufescens, shiraziensis, sodalis, transylvanicus, turcomanicus, typus, wiedii.*
COMMENTS: Subgenus *Eptesicus.* Includes *sodalis;* see Corbet (1978*c*:57). Includes *horikawai;* see Jones (1975:189). Revised by Gaisler (1970). See comments under *fuscus* and *platyops.*

Eptesicus somalicus (Thomas, 1901). Ann. Mag. Nat. Hist., ser. 7, 8:32.
TYPE LOCALITY: Somalia, Northwest Province, Hargeisa.
DISTRIBUTION: Guinea-Bissau to Somalia, south to Namibia and South Africa; Madagascar.
SYNONYMS: *humbloti, ugandae, vansoni, zuluensis.*
COMMENTS: Subgenus *Neoromicia.* Includes *zuluensis;* see Koopman (1975:404, 405).

Eptesicus tatei Ellerman and Morrison-Scott, 1951. Checklist Palaearctic Indian Mammals, p. 158.
TYPE LOCALITY: India, Darjeeling.
DISTRIBUTION: NE India.
COMMENTS: Subgenus *Eptesicus.*

Eptesicus tenuipinnis (Peters, 1872). Monatsb. K. Preuss. Akad. Wiss. Berlin, 1872:263.
TYPE LOCALITY: "Guinea".
DISTRIBUTION: Senegal to Kenya, south to Angola and Zaire.
SYNONYMS: *ater, bicolor.*

COMMENTS: Subgenus *Neoromicia. E. bicolor* was tentatively included here by Hayman and Hill (1971:43), but may be an older name for *Pipistrellus anchietai*; see Koopman (1975:404).

Eptesicus vulturnus Thomas, 1914. Ann. Mag. Nat. Hist., ser. 8, 13:440.
TYPE LOCALITY: Australia, Tasmania.
DISTRIBUTION: SE Australia including Tasmania.
SYNONYMS: *pygmaeus*.
COMMENTS: Subgenus *Vespadelus*. Includes *pygmaeus*; see McKean et al. (1978:532).

Euderma H. Allen, 1892. Proc. Acad. Nat. Sci. Philadelphia, 43:467.
TYPE SPECIES: *Histiotus maculatus* J.A. Allen, 1891.

Euderma maculatum (J. A. Allen, 1891). Bull. Am. Mus. Nat. Hist., 3:195.
TYPE LOCALITY: USA, California, Los Angeles Co., Santa Clara Valley, Castac Creek mouth.
DISTRIBUTION: SW Canada and Montana (USA) to Queretaro (Mexico).
COMMENTS: See Watkins (1977, Mammalian Species, 77).

Eudiscopus Conisbee, 1953. Last names proposed genera subgenera Recent Mamm., p. 30.
TYPE SPECIES: *Discopus denticulus* Osgood, 1932.
COMMENTS: *Eudiscopus* is the replacement name for *Discopus* Osgood, 1932, preoccupied by *Discopus* Thompson, 1864 (a beetle).

Eudiscopus denticulus (Osgood, 1932). Field Mus. Nat. Hist. Publ., Zool. Ser., 18:236.
TYPE LOCALITY: Laos, Phong Saly, 4,000 ft. (1,219 m).
DISTRIBUTION: Laos, C Burma.
COMMENTS: See Koopman (1972, Mammalian Species, 19).

Glischropus Dobson, 1875. Proc. Zool. Soc. Lond., 1875:472.
TYPE SPECIES: *Vesperugo tylopus* Dobson, 1875.

Glischropus javanus Chasen, 1939. Treubia, 17:189.
TYPE LOCALITY: Indonesia, Java, West Java, Mt. Pangeango.
DISTRIBUTION: Java.

Glischropus tylopus (Dobson, 1875). Proc. Zool. Soc. Lond., 1875:473.
TYPE LOCALITY: Malaysia, Sabah.
DISTRIBUTION: Burma, Thailand, W Malaysia, Borneo, SW Philippines, Sumatra and N Molucca Isls.
SYNONYMS: *batjanus*.

Hesperoptenus Peters, 1868. Monatsb. K. Preuss. Akad. Wiss. Berlin, 1868:626.
TYPE SPECIES: *Vesperus (Hesperoptenus) doriae* Peters, 1868.
SYNONYMS: *Milithronycteris*.
COMMENTS: Revised by Hill (1976).

Hesperoptenus blanfordi (Dobson, 1877). J. Asiat. Soc. Bengal, 46:312.
TYPE LOCALITY: Burma, Tenasserim.
DISTRIBUTION: Burma, Thailand, Malay Peninsula, Borneo.
COMMENTS: Subgenus *Milithronycteris*.

Hesperoptenus doriae (Peters, 1868). Monatsb. K. Preuss. Akad. Wiss. Berlin, 1868:626.
TYPE LOCALITY: Malaysia, Sarawak.
DISTRIBUTION: Borneo, Malay Peninsula.
COMMENTS: Subgenus *Hesperoptenus*.

Hesperoptenus gaskelli Hill, 1983. Bull. Brit. Mus. (Nat. Hist.) Zool., 45:169.
TYPE LOCALITY: Sulawesi, Central R. Ranu.
DISTRIBUTION: Sulawesi.
COMMENTS: Subgenus *Milithronycteris*.

Hesperoptenus tickelli (Blyth, 1851). J. Asiat. Soc. Bengal, 20:157.
TYPE LOCALITY: India, Bihar, Chaibasa.

DISTRIBUTION: India (including Andaman Isls), Sri Lanka, Nepal, Bhutan, Burma, Thailand, perhaps SW China.
SYNONYMS: *isabellinus*.
COMMENTS: Subgenus *Milithronycteris*.

Hesperoptenus tomesi Thomas, 1905. Ann. Mag. Nat. Hist., ser. 7, 16:575.
TYPE LOCALITY: Malaysia, Malacca.
DISTRIBUTION: Borneo, Malay Peninsula.
COMMENTS: Subgenus *Milithronycteris*.

Histiotus Gervais, 1856. *In* F. Comte de Castelnau, Exped. Partes Cen. Am. Sud.(Sec. 7), Vol, 1, pt. 2(Mammifères):77.
TYPE SPECIES: *Plecotus velatus* I. Geoffroy, 1824.

Histiotus alienus Thomas, 1916. Ann. Mag. Nat. Hist., ser. 8, 17:276.
TYPE LOCALITY: Brazil, Santa Catarina, Joinville.
DISTRIBUTION: SE Brazil, Uruguay.

Histiotus macrotus (Poeppig, 1835). Reise Chile Peru Amaz., 1:451.
TYPE LOCALITY: Chile, Bio-Bio, Antuco.
DISTRIBUTION: Chile, NW Argentina, S Bolivia, S Peru.
SYNONYMS: *chilensis, laephotis, poeppigii*.
COMMENTS: Includes *laephotis*; see Cabrera (1958:108).

Histiotus montanus (Philippi and Landbeck, 1861). Arch. Naturgesch., p. 289.
TYPE LOCALITY: Chile, Santiago Cordillera.
DISTRIBUTION: Chile, Argentina, Uruguay, W Bolivia, S Peru, Ecuador, Colombia, Venezuela, perhaps N Peru and S Brazil.
SYNONYMS: *capucinus, colombiae, inambarus, magellanicus, segethii*.

Histiotus velatus (I. Geoffroy, 1824) Ann. Sci. Nat. Zool., 3:446.
TYPE LOCALITY: Brazil, Parana, Curitiba.
DISTRIBUTION: E Brazil, Paraguay.
SYNONYMS: *miotis*.

Ia Thomas, 1902. Ann. Mag. Nat. Hist., ser. 7, 10:163.
TYPE SPECIES: *Ia io* Thomas, 1902.
COMMENTS: Considered a subgenus of *Pipistrellus* by Ellerman and Morrison-Scott (1951:162); but see Topal (1970*a*:344, 345).

Ia io Thomas, 1902. Ann. Mag. Nat. Hist., ser. 7, 10:164.
TYPE LOCALITY: China, Hupeh, Chungyang.
DISTRIBUTION: S China, Laos, Vietnam, Thailand, NE India.
SYNONYMS: *beaulieui, longimana* (see Topal, 1970*a*:342, 343).

Idionycteris Anthony, 1923. Am. Mus. Novit., 54:1.
TYPE SPECIES: *Idionycteris mexicanus* Anthony, 1923 (= *Corynorhinus phyllotis* G. M. Allen, 1916).
COMMENTS: *Idionycteris* is considered a separate genus according to Williams et al. (1970); but see Handley (1959*b*), who retained it in *Plecotus*.

Idionycteris phyllotis (G. M. Allen, 1916). Bull. Mus. Comp. Zool., 60:352.
TYPE LOCALITY: Mexico, San Luis Potosi.
DISTRIBUTION: Distrito Federal (Mexico) to S Utah and S Nevada (USA).
SYNONYMS: *mexicanus*.
COMMENTS: Formerly included in *Plecotus*; see Williams et al. (1970). See Czaplewski (1983, Mammalian Species, 208).

Laephotis Thomas, 1901. Ann. Mag. Nat. Hist., ser. 7, 7:460.
TYPE SPECIES: *Laephotis wintoni* Thomas, 1901.
COMMENTS: Considered monotypic by Hayman and Hill (1971:49); but see J. E. Hill (1974*a*), who revised the genus.

Laephotis angolensis Monard, 1935. Arch. Mus. Bocage, 6:45.
TYPE LOCALITY: Angola, Tyihumbwe, 15 km W. Dala.
DISTRIBUTION: Angola, Zaire.

Laephotis botswanae Setzer, 1971. Proc. Biol. Soc. Washington, 84:260, 263.
TYPE LOCALITY: Botswana, 50 mi. (80 km) W and 12 mi. (19 km) S Shakawe.
DISTRIBUTION: Zaire, Zambia, Malawi, Botswana, Zimbabwe, Transvaal (South Africa).

Laephotis namibensis Setzer, 1971. Proc. Biol. Soc. Washington, 84:259.
TYPE LOCALITY: Namibia, Gobabeb, Kuiseb River.
DISTRIBUTION: Namibia.

Laephotis wintoni Thomas, 1901. Ann. Mag. Nat. Hist., ser. 7, 7:460.
TYPE LOCALITY: Kenya, Kitui, 1,150 m.
DISTRIBUTION: Ethiopia, Kenya, SW Cape Province (South Africa).

Lasionycteris Peters, 1866. Monatsb. K. Preuss. Akad. Wiss. Berlin, 1866:8.
TYPE SPECIES: *Vespertilio noctivagans* Le Conte, 1831.

Lasionycteris noctivagans (Le Conte, 1831). *In* McMurtie, Anim. Kingdom, 1(App.):431.
TYPE LOCALITY: "Eastern United States".
DISTRIBUTION: S Canada, USA (including SE Alaska, and except extreme southern parts), NE Mexico, Bermuda.
COMMENTS: See Kunz (1982, Mammalian Species, 172).

Lasiurus Gray, 1831. Zool. Misc., 1:38.
TYPE SPECIES: *Vespertilio borealis* Müller, 1776.
SYNONYMS: *Dasypterus; Nycteris* Borkhausen, 1797 (not Cuvier and Geoffroy, 1795).
COMMENTS: Treated under the name *Nycteris* by Hall (1981:219). In Opinion 111 of the International Commission on Zoological Nomenclature (1929*b*), *Lasiurus* was adopted, rather than *Nycteris*. Includes *Dasypterus*; see Hall and Jones (1961). A key to this genus was presented by Shump and Shump (1982*a*).

Lasiurus borealis (Müller, 1776). Linné's Vollständ. Natursystem, Suppl., p. 20.
TYPE LOCALITY: USA, New York.
DISTRIBUTION: Chile, Argentina, Uruguay and Brazil to C Canada; Jamaica; Cuba; Hispaniola; Puerto Rico; Bermuda; Bahamas; Trinidad and Tobago; Galapagos (Ecuador).
SYNONYMS: *blossevillii, bonariensis, brachyotis, degelidus, enslenii, frantzii, funebris, minor, lasiurus, monachus, noveboracensis, ornatus, pfeifferi, quebecensis, rubellus, rubra, rufus, salinae, teliotis, tesselatus, varius.*
COMMENTS: Subgenus *Lasiurus*. Includes *degelidus, minor, pfeifferi, blossevillii*; see Varona (1974:36), but see also Baker et al. (1988*a*). Includes *brachyotis*; see Niethammer (1964:595). See Shump and Shump (1982*a*, Mammalian Species, 183). While there is evidence that more than one species is involved in this complex, the true picture is still far from clear. Until it is better resolved, keeping all the named forms together seems preferable to a premature split.

Lasiurus castaneus Handley, 1960. Proc. U.S. Natl. Mus., 112:468.
TYPE LOCALITY: Panama, Darien, Rio Pucro, Tacarcuna Village, 3,200 ft. (975 m).
DISTRIBUTION: Panama, Costa Rica.
COMMENTS: Subgenus *Lasiurus*.

Lasiurus cinereus (Beauvois, 1796). Cat. Raisonné Mus. Peale Philadelphia, p. 18.
TYPE LOCALITY: USA, Pennsylvania, Philadelphia.
DISTRIBUTION: Colombia and Venezuela to C Chile, Uruguay, and C Argentina; Hawaii (USA); Guatemala and Mexico throughout the USA to S British Columbia, SE Mackensie, Hudson Bay and S Quebec (Canada); Galapagos; Bermuda; accidental on Cuba, Hispaniola, Iceland, and the Orkney Isls (Scotland).
STATUS: U.S. ESA - Endangered as *L. c. semotus*.
SYNONYMS: *brasiliensis, grayi, mexicana, pallescens, pruinosus, semotus, villosissimus.*

COMMENTS: Subgenus *Lasiurus*. Includes *villosissimus*, and *semotus*; see Sanborn and Crespo (1957:12), who revised the species. See Shump and Shump (1982*b*, Mammalian Species, 185).

Lasiurus ega (Gervais, 1856). *In* F. Comte de Castelnau, Exped. Partes Cen. Am. Sud.(Sec. 7), Vol. 1, pt. 2(Mammifères):73.
TYPE LOCALITY: Brazil, Amazonas, Ega.
DISTRIBUTION: S California, Arizona and S Texas (USA) to Argentina, Uruguay, and Brazil; Trinidad.
SYNONYMS: *argentinus, caudatus, fuscatus, panamensis, punensis, xanthinus.*
COMMENTS: Subgenus *Dasypterus*. Includes *xanthinus*, but see Baker et al. (1988*a*).

Lasiurus egregius (Peters, 1870). Monatsb. K. Preuss. Akad. Wiss. Berlin, 1870:275.
TYPE LOCALITY: Brazil, Santa Catarina.
DISTRIBUTION: Brazil, French Guiana, Panama.
COMMENTS: Subgenus *Lasiurus*.

Lasiurus intermedius H. Allen, 1862. Proc. Acad. Nat. Sci. Philadelphia, 14:246.
TYPE LOCALITY: Mexico, Tamaulipas, Matamoros.
DISTRIBUTION: Honduras to Sinaloa (Mexico) and through Texas to Florida and New Jersey (USA); Cuba.
SYNONYMS: *floridanus, insularis.*
COMMENTS: Subgenus *Dasypterus*. Includes *floridanus* and *insularis*; see Hall and Jones (1961:84-87); but see also Silva-Taboada (1976). See Webster et al. (1980, Mammalian Species, 132).

Lasiurus seminolus (Rhoads, 1895). Proc. Acad. Nat. Sci. Philadelphia, 47:32.
TYPE LOCALITY: USA, Florida, Pinellas Co., Tarpon Springs.
DISTRIBUTION: Florida and Texas to Oklahoma and Virginia; Pennsylvania and New York (USA); Bermuda. N Veracruz (Mexico) record unverified.
COMMENTS: Subgenus *Lasiurus*. Probably only a subspecies of *borealis* since the characters do not hold; see Koopman et al. (1957:168); but also see Baker et al. (1988*a*). See Wilkins (1987*a*, Mammalian Species, 280).

Mimetillus Thomas, 1904. Abstr. Proc. Zool. Soc. Lond., 1904(10):12.
TYPE SPECIES: *Vesperugo (Vesperus) moloneyi* Thomas, 1891.

Mimetillus moloneyi (Thomas, 1891). Ann. Mag. Nat. Hist., ser. 6, 7:528.
TYPE LOCALITY: Nigeria, Western Region, Lagos.
DISTRIBUTION: Sierra Leone to Ethiopia, south to Tanzania, Zambia, and Angola; Bioko.
SYNONYMS: *berneri, thomasi.*

Myotis Kaup, 1829. Skizz. Entwickel.-Gesch. Nat. Syst. Europ. Thierwelt, 1:106.
TYPE SPECIES: *Vespertilio myotis* Borkhausen, 1797.
SYNONYMS: *Anamygdon, Cistugo, Leuconoe, Pizonyx, Selysius.*
COMMENTS: For synonyms see Findley (1972:43), Hayman and Hill (1971:33), and Phillips and Birney (1968:495). Neotropical species revised by LaVal (1973*a*).

Myotis abei Yoshikura, 1944. Zool. Mag. (Tokyo), 56:6.
TYPE LOCALITY: Russia, Sakhalin, Shirutoru.
DISTRIBUTION: Known only from the type locality.
COMMENTS: Subgenus *Leuconoe*.

Myotis adversus (Horsfield, 1824). Zool. Res. Java, part 8, p. 3(unno.) of *Vespertilio Temminckii* acct.
TYPE LOCALITY: Indonesia, Java.
DISTRIBUTION: Taiwan and W Malaysia, south and east to New Guinea, Bismarck Arch., Solomon Isls, Vanuatu (= New Hebrides); N and E coastal Australia; perhaps Tibet (China).
SYNONYMS: *carimatae, macropus, moluccarum, orientis, solomonis, taiwanensis.*
COMMENTS: Subgenus *Leuconoe*. Includes *taiwanensis*; see Ellerman and Morrison-Scott (1951:149); but see also Findley (1972:43). Includes *carimatae*; see Hill (1983). Includes *Anamygdon solomonis*; see Phillips and Birney (1968:495).

Myotis aelleni Baud, 1979. Rev. Suisse Zool., 86:268.
TYPE LOCALITY: Argentina, Chubut, El Hoyo de Epuyen.
DISTRIBUTION: SW Argentina.
COMMENTS: Subgenus *Leuconoe*.

Myotis albescens (E. Geoffroy, 1806). Ann. Mus. Hist. Nat. Paris, 8:204.
TYPE LOCALITY: Paraguay, Paraguari, Yaguaron (of neotype).
DISTRIBUTION: S Veracruz (Mexico) to Uruguay and N Argentina.
SYNONYMS: *aenobarbus, argentatus, arsinoe, isidori, leucogaster, mundus*.
COMMENTS: Subgenus *Leuconoe*. Includes *argentatus*; see LaVal (1973*a*:25).

Myotis altarium Thomas, 1911. Abstr. Proc. Zool. Soc. Lond., 1911(90):3.
TYPE LOCALITY: China, Szechwan, Omi San.
DISTRIBUTION: Szechwan, Kweichow (China), Thailand.
COMMENTS: Subgenus *Selysius*.

Myotis annectans (Dobson, 1871). Proc. Asiat. Soc. Bengal, p. 213.
TYPE LOCALITY: India, Assam, Naga Hills.
DISTRIBUTION: NE India to Thailand.
SYNONYMS: *primula*.
COMMENTS: Subgenus *Myotis*. Includes *primula*; see Topal (1970*b*:373-375), who transfered the species from *Pipistrellus*.

Myotis atacamensis (Lataste, 1892). Actes Soc. Sci. Chile, 1:80.
TYPE LOCALITY: Chile, Antofogasta, San Pedro de Atacama.
DISTRIBUTION: S Peru, N Chile.
SYNONYMS: *nicholsoni*.
COMMENTS: Subgenus *Selysius*. Listed as a subspecies of *chiloensis* by Cabrera (1958:99). Includes *nicholsoni*; see LaVal (1973*a*:18-20).

Myotis auriculus Baker and Stains, 1955. Univ. Kansas Publ. Mus. Nat. Hist., 9:83.
TYPE LOCALITY: Mexico, Tamaulipas, Sierra de Tamaulipas, 10 mi. (16 km) W, 2 mi. (3 km) S Piedra, 1,200 ft. (366 m).
DISTRIBUTION: Arizona and New Mexico (USA) to Jalisco and Veracruz (Mexico); Guatemala.
SYNONYMS: *apache*.
COMMENTS: Subgenus *Myotis*. Listed as a subspecies of *evotis* by Hall and Kelson (1959:169); but also see Genoways and Jones (1969*b*) and Hall (1981:205). See Warner (1982, Mammalian Species, 191).

Myotis australis (Dobson, 1878). Cat. Chiroptera Brit. Mus., p. 317.
TYPE LOCALITY: Australia, New South Wales.
DISTRIBUTION: New South Wales, possibly Western Australia (Australia).
COMMENTS: Subgenus *Selysius*. Poorly known, the type and only certain specimen possibly being incorrectly labelled, or a vagrant individual of *muricola*. See also Aust. Bat Res. News, 9. A specimen from NW Australia may belong in this species (Koopman, 1984c:11-12). Hill (1983) considered *australis* a subspecies of *ater* (here included in *M. muricola*).

Myotis austroriparius (Rhoads, 1897). Proc. Acad. Nat. Sci. Philadelphia, 49:227.
TYPE LOCALITY: USA, Florida, Pinellas Co., Tarpon Springs.
DISTRIBUTION: SE USA, north to Indiana and North Carolina.
SYNONYMS: *gatesi, mumfordi*.
COMMENTS: Subgenus *Leuconoe*. Reviewed by LaVal (1970). See Jones and Manning (1989, Mammalian Species, 332).

Myotis bechsteini (Kuhl, 1817). Die Deutschen Fledermause. Hanau, p. 14, 30.
TYPE LOCALITY: Germany, Hessen, Hanau.
DISTRIBUTION: Europe to Caucasus and Iran; England; S Sweden.
SYNONYMS: *favonicus, ghidinii*.
COMMENTS: Subgenus *Myotis*.

Myotis blythii (Tomes, 1857). Proc. Zool. Soc. Lond., 1857:53.
TYPE LOCALITY: India, Rajasthan, Nasirabad.

DISTRIBUTION: Mediterranean zone of Europe and NW Africa; Crimea and Caucasus Mtns, Asia Minor, Israel to Kirgizia, Afghanistan, and Himalayas; NW Altai Mtns; Inner Mongolia and Shensi (China).

SYNONYMS: *africanus, ancilla, dobsoni, lesviacus, omari, oxygnathus, punicus, risorius* (see Corbet, 1978*c*:30; Strelkov, 1972; Felten et al., 1977; and Bogan et al., 1978).

COMMENTS: Subgenus *Myotis*.

Myotis bocagei (Peters, 1870). J. Sci. Math. Phys. Nat. Lisboa, ser. 1, 3:125.

TYPE LOCALITY: Angola, Duque de Braganca.

DISTRIBUTION: Senegal to S Yemen, south to Angola, Zambia, and Transvaal (South Africa).

SYNONYMS: *cupreolus, dogalensis, hildegardeae*.

COMMENTS: Subgenus *Leuconoe*. Includes *dogalensis*; see Corbet (1978*c*:49).

Myotis bombinus Thomas, 1906. Proc. Zool. Soc. Lond., 1905(2):337 [1906].

TYPE LOCALITY: Japan, Kiushiu, Miyasaki Ken, Tano.

DISTRIBUTION: Japan, Korea, SE Siberia, NE China.

SYNONYMS: *amurensis*.

COMMENTS: Subgenus *Myotis*. Formerly included in *M. nattereri*, but see Horáček and Hanak (1984).

Myotis brandti (Eversmann, 1845). Bull. Soc. Nat. Moscow, 18:505.

TYPE LOCALITY: Russia, Orenburgsk. Obl., S. Ural, Bolshoi-Ik River, Spasskoie. Foothills of the Ural Mountains.

DISTRIBUTION: Britain to Kazakhstan, E Siberia including Sakhalin Isls, Kamchatka Peninsula and Kurile Isls (Russia), Mongolia south to Spain, Greece, Korea, and Ussuri region (Russia); Japan.

SYNONYMS: *aureus, fujiensis, gracilis, sibiricus*.

COMMENTS: Subgenus *Selysius*. Listed as a subspecies of *mystacinus* by Ellerman and Morrison-Scott (1951:139); but also see review by Strelkov and Buntova (1982). Includes *fujiensis* and *gracilis*, but see Yoshiyuki (1989).

Myotis californicus (Audubon and Bachman, 1842). J. Acad. Nat. Sci. Philadelphia, ser. 1, 8:285.

TYPE LOCALITY: USA, California, Monterey.

DISTRIBUTION: S Alaska Panhandle (USA) to Baja California (Mexico) and Guatemala.

SYNONYMS: *agilis, caurinus, exilis, mexicanus, nitidus, oregonensis, quercinus, stephensi, tenuidorsalis*.

COMMENTS: Subgenus *Selysius*. See Miller and Allen (1928:153) for discussion of the holotype.

Myotis capaccinii (Bonaparte, 1837). Fauna Ital., 1, fasc. 20.

TYPE LOCALITY: Italy, Sicily.

DISTRIBUTION: Mediterranean zone and islands of Europe and NW Africa; Turkey; Israel; Iraq; Iran; Uzbekistan.

SYNONYMS: *blasii, bureschi, dasypus, major, megapodius, pellucens*.

COMMENTS: Subgenus *Leuconoe*. See comment under *macrodactylus*.

Myotis chiloensis (Waterhouse, 1840). Zool. Voy. H.M.S. "Beagle," Mammalia, p. 5.

TYPE LOCALITY: Chile, Chiloe Isl, Islets on eastern side.

DISTRIBUTION: C and S Chile.

SYNONYMS: *arescens, gayi*.

COMMENTS: Subgenus *Leuconoe*. See LaVal (1973*a*:43, 44) for restriction of the scope of this species.

Myotis chinensis (Tomes, 1857). Proc. Zool. Soc. Lond., 1857:52.

TYPE LOCALITY: "Southern China".

DISTRIBUTION: Szechwan and Yunnan to Kiangsu (China); Hong Kong; N Thailand.

SYNONYMS: *luctuosus*.

COMMENTS: Subgenus *Myotis*. Included in species *myotis* by Ellerman and Morrison-Scott (1951:144); but see Lekagul and McNeely (1977:206).

Myotis cobanensis Goodwin, 1955. Am. Mus. Novit., 1744:2.

TYPE LOCALITY: Guatemala, Alta Verapaz, Coban, 1,305 m.

DISTRIBUTION: C Guatemala.

COMMENTS: Subgenus *Leuconoe*. Listed as a subspecies of *velifer* by Goodwin (1955:2); but see de la Torre (1958) and Hall (1981:197).

Myotis dasycneme (Boie, 1825). Isis Jena, p. 1200.
TYPE LOCALITY: Denmark, Jutland, Dagbieg (near Wiborg).
DISTRIBUTION: France and Sweden east to Yenisei River (Russia), south to Ukraine and NW Kazakhstan; a single record from Manchuria (China).
SYNONYMS: *limnophilus, major*.
COMMENTS: Subgenus *Leuconoe*. Probably includes *surinamensis*; see Carter and Dolan (1978:73).

Myotis daubentoni (Kuhl, 1817). Die Deutschen Fledermause. Hanau, p. 14.
TYPE LOCALITY: Germany, Hessen, Hanau.
DISTRIBUTION: Europe east to Kamtschatka, Vladivostok, Sakhalin and Kurile Isls (Russia); Japan; Korea; Manchuria, E and S China; Britain and Ireland; Scandinavia; Assam (India).
SYNONYMS: *aedilus, albus, capucinellus, lanatus, laniger, loukashkini, minutellus, nathalinae, petax, staufferi, ussuriensis, volgensis*.
COMMENTS: Subgenus *Leuconoe*. Includes *laniger*; see Ellerman and Morrison-Scott (1951:147). Includes *nathalinae*; see Horáček and Hanak (1984).

Myotis dominicensis Miller, 1902. Proc. Biol. Soc. Washington, 15:243.
TYPE LOCALITY: Dominica (Lesser Antilles).
DISTRIBUTION: N Lesser Antilles.
COMMENTS: Subgenus *Selysius*. Listed as a subspecies of *nigricans* by Hall and Kelson (1959:176), but see LaVal (1973*a*:16, 17) and Hall (1981:200).

Myotis elegans Hall, 1962. Univ. Kansas Publ. Mus. Nat. Hist., 14:163-164.
TYPE LOCALITY: Mexico, Veracruz, 12.5 mi. (20 km) N Tihuatlan.
DISTRIBUTION: San Luis Potosi (Mexico) to Costa Rica.
COMMENTS: Subgenus *Selysius*.

Myotis emarginatus (E. Geoffroy, 1806). Ann. Mus. Hist. Nat. Paris, 8:198.
TYPE LOCALITY: France, Ardennes, Givet, Charlemont.
DISTRIBUTION: S Europe, north to Netherlands and S Poland, Crimea, Caucasus and Kopet Dag Mtns, east to Uzbekistan and E Iran; Israel; Morocco; Algeria; Tunisia; Lebanon; Afghanistan.
SYNONYMS: *budapestiensis, ciliatus, desertorum, lanaceus, neglectus, saturatus, turcomanicus*.
COMMENTS: Subgenus *Myotis*.

Myotis evotis (H. Allen, 1864). Smithson. Misc. Coll., 7:48.
TYPE LOCALITY: USA, California, Monterey.
DISTRIBUTION: S British Columbia, S Alberta, S Saskatchewan (Canada) to New Mexico (USA) and Baja California (Mexico).
SYNONYMS: *chrysonotus, micronyx, pacificus*.
COMMENTS: Subgenus *Myotis*. See Genoways and Jones (1969*b*:1). See Manning and Jones (1989, Mammalian Species, 329).

Myotis findleyi Bogan, 1978. J. Mammal., 59:524.
TYPE LOCALITY: Mexico, Nayarit, Tres Marías Isls, Maria Magdalena Isl.
DISTRIBUTION: Tres Marías Isls (Mexico).
COMMENTS: Subgenus *Selysius*.

Myotis formosus (Hodgson, 1835). J. Asiat. Soc. Bengal, 4:700.
TYPE LOCALITY: Nepal.
DISTRIBUTION: Afghanistan to Kweichow, Kwangsi, Kiangsu and Fukien (China); Taiwan; Korea; Tsushima Isl (Japan); Philippines; Sumatra, Java, Sulawesi, and Bali.
SYNONYMS: *andersoni, auratus, bartelsi, chofukusei, flavus, hermani, pallida, rufoniger, rufopictus, tsuensis, watasei, weberi* (see Findley, 1972:42).
COMMENTS: Subgenus *Myotis*.

Myotis fortidens Miller and Allen, 1928. Bull. U.S. Natl. Mus., 144:54.
TYPE LOCALITY: Mexico, Tabasco, Teapa.
DISTRIBUTION: Sonora and Veracruz (Mexico) to Guatemala.

SYNONYMS: *sonoriensis.*
COMMENTS: Subgenus *Leuconoe.*

Myotis frater G. M. Allen, 1923. Am. Mus. Novit., 85:6.
TYPE LOCALITY: China, Fukien, Yenping.
DISTRIBUTION: Afghanistan, Uzbekistan and S Siberia to Korea, Heilungkiang (China), SE China, and republics of the SE former USSR; Japan.
SYNONYMS: *bucharensis, kaguyae, longicaudatus.*
COMMENTS: Subgenus *Selysius.* Includes *longicaudatus;* see Corbet (1978*c*:49); but also see Wang Sung(1959).

Myotis goudoti (A. Smith, 1834). S. Afr. Quart. J., 2:244.
TYPE LOCALITY: Madagascar.
DISTRIBUTION: Madagascar, Anjouan Isl (Comoro Isls).
SYNONYMS: *anjouanensis.*
COMMENTS: Subgenus *Myotis.*

Myotis grisescens A. H. Howell, 1909. Proc. Biol. Soc. Washington, 22:46.
TYPE LOCALITY: USA, Tennessee, Marion Co., Nickajack Cave, near Shellmound.
DISTRIBUTION: Florida Panhandle to Kentucky, Indiana, Illinois, E Kansas and NE Oklahoma (USA).
STATUS: U.S. ESA - Endangered.
COMMENTS: Subgenus *Leuconoe.*

Myotis hasseltii (Temminck, 1840). Monogr. Mamm., 2:225.
TYPE LOCALITY: Indonesia, Java.
DISTRIBUTION: Burma, Thailand, Cambodia, Vietnam, W Malaysia, Sumatra, Mentawai Isls, Riau Arch., Java, Borneo, Sri Lanka.
SYNONYMS: *abboti, continentis, macellus* (see Hill, 1983).
COMMENTS: Subgenus *Leuconoe.*

Myotis horsfieldii (Temminck, 1840). Monogr. Mamm., 2:226.
TYPE LOCALITY: Indonesia, Java.
DISTRIBUTION: SE China, Thailand, India (including Andaman Isls), perhaps Sri Lanka, W Malaysi, Java, Bali, and Sulawesi, Borneo, Philippines.
SYNONYMS: *deignani, dryas, jeannei, lepidus, peshwa* (see Hill, 1983).
COMMENTS: Subgenus *Leuconoe.*

Myotis hosonoi Imaizumi, 1954. Bull. Natl. Sci. Mus. Tokyo, 1:44.
TYPE LOCALITY: Japan, Honshu, Nagano Pref., Tokiwa-Mura, Koumito.
DISTRIBUTION: Honshu (Japan).
COMMENTS: Subgenus *Selysius.*

Myotis ikonnikovi Ognev, 1912. Ann. Mus. Zool. Acad. Imp. Sci. St. Petersbourg, 16:477.
TYPE LOCALITY: Russia, Primorsk. Krai (= Ussuri Region), Dalnerechen Dist., Euseevka.
DISTRIBUTION: Ussuri region and N Korea to Lake Baikal (Russia), the Altai Mtns, and Mongolia, NE China; Sakhalin Isl (Russia) and Hokkaido Isl (Japan).
COMMENTS: Subgenus *Selysius.* Probably a subspecies of *muricola;* see Corbet (1978*c*:48), who rejected European records.

Myotis insularum (Dobson, 1878). Cat. Chiroptera Brit. Mus., p. 313.
TYPE LOCALITY: Samoa.
DISTRIBUTION: Samoa.
COMMENTS: Subgenus *Selysius.* Poorly known, the type and only specimen possibly being incorrectly labelled; see Koopman (1984*c*:11).

Myotis keaysi J. A. Allen, 1914. Bull. Am. Mus. Nat. Hist., 33:383.
TYPE LOCALITY: Peru, Puno, Inca Mines.
DISTRIBUTION: Tamaulipas (Mexico) to N Argentina and Venezuela; Trinidad.
SYNONYMS: *pilosatibialis.*
COMMENTS: Subgenus *Selysius.* Revised by LaVal (1973*a*).

Myotis keenii (Merriam, 1895). Am. Nat., 29:860.
TYPE LOCALITY: Canada, British Columbia, Queen Charlotte Isls, Graham Isl, Massett.

DISTRIBUTION: Alaska Panhandle to W Washington (USA); Mackenzie to Prince Edward Isl (Canada), south to Arkansas and Florida Panhandle (USA).
SYNONYMS: *septentrionalis*.
COMMENTS: Subgenus *Myotis*. Includes *septentrionalis*, possibly a separate species; see Van Zyll de Jong (1979). See Fitch and Shump (1979, Mammalian Species, 121).

Myotis leibii (Audubon and Bachman, 1842). J. Acad. Nat. Sci. Philadelphia, ser. 1, 8:284.
TYPE LOCALITY: USA, Pennsylvania, Erie Co.
DISTRIBUTION: N Baja California, Michoacan, and Nuevo Leon (Mexico) to S British Columbia (Canada), east to Maine (USA) and S Quebec (Canada).
SYNONYMS: *ciliolabrum*, *melanorhinus*.
COMMENTS: Subgenus *Selysius*. Called *subulatus* in Hall (1981:187); but see Glass and Baker (1968:259) who showed that *subulatus* was probalby an older name for *yumanensis*. Includes *ciliolabrum*, and *melanorhinus*, but see van Zyll de Jong (1984).

Myotis lesueuri Roberts, 1919. Ann. Transvaal Mus., 6:112.
TYPE LOCALITY: South Africa, Cape Province, Paarl Dist., Lormarins.
DISTRIBUTION: SW Cape Province (South Africa).
COMMENTS: Subgenus *Cistugo*.

Myotis levis (I. Geoffroy, 1824). Ann. Sci. Nat. (Paris), ser. 1, 3:444-445.
TYPE LOCALITY: "Southern Brazil."
DISTRIBUTION: Bolivia, Argentina, SE Brazil, Uruguay.
SYNONYMS: *alter*, *dinelli*, *nubilus*, *polythrix*.
COMMENTS: Subgenus *Leuconoe*. Included in *ruber* by Cabrera (1958:102), but see LaVal (1973*a*:36-40).

Myotis longipes (Dobson, 1873). Proc. Asiat. Soc. Bengal, p. 110.
TYPE LOCALITY: India, Kashmir, Bhima Devi Caves, 6,000 ft. (1,829 m).
DISTRIBUTION: Afghanistan, Kashmir (India).
SYNONYMS: *megalopus*.
COMMENTS: Subgenus *Leuconoe*. Included in *capaccinii* by Ellerman and Morrison-Scott (1951:148); but considered a distinct species by Hanak and Gaisler (1969).

Myotis lucifugus (Le Conte, 1831). *In* McMurtie, Anim. Kingdom, 1(App.):431.
TYPE LOCALITY: USA, Georgia, Liberty Co., near Riceboro.
DISTRIBUTION: Alaska (USA) to Labrador and Newfoundland (Canada), south to Distrito Federal (Mexico).
SYNONYMS: *affinis*, *alascensis*, *albicinctus*, *altipetens*, *baileyi*, *brevirostris*, *carissima*, *carolii*, *crassus*, *domesticus*, *gryphus*, *lanceolatus*, *occultus*, *pernox*, *relictus*, *salarii*, *virginianus*.
COMMENTS: Subgenus *Leuconoe*. Includes *occultus*; see Findley and Jones (1967). Hybridizes with *yumanensis* in some areas; see Parkinson (1979), but see Herd and Fenton (1983). See Fenton and Barclay (1980, Mammalian Species, 142).

Myotis macrodactylus (Temminck, 1840). Monogr. Mamm., 2:231.
TYPE LOCALITY: Japan.
DISTRIBUTION: Japan, Kurile Isls (Russia), S China, republics of the former SE USSR.
SYNONYMS: *fimbriatus*, *hirsutus*.
COMMENTS: Subgenus *Leuconoe*. Includes *fimbriatus*, probably conspecific with *capaccinii*, see Wallin (1969:294-296); but see also Corbet (1978*c*:51).

Myotis macrotarsus (Waterhouse, 1845). Proc. Zool. Soc. Lond., 1845:5.
TYPE LOCALITY: Philippines.
DISTRIBUTION: Philippines, N Borneo.
SYNONYMS: *saba*.
COMMENTS: Subgenus *Leuconoe*.

Myotis martiniquensis LaVal, 1973. Bull. Los Angeles Cty. Mus. Nat. Hist. Sci. Soc., 15:35.
TYPE LOCALITY: Martinique (Lesser Antilles), Tartane, 6 km E La Trinite (France).
DISTRIBUTION: Martinique, Barbados (Lesser Antilles).
SYNONYMS: *nyctor*.
COMMENTS: Subgenus *Selysius*.

Myotis milleri Elliot, 1903. Publ. Field Columb. Mus., Zool. Ser., 3:172.
TYPE LOCALITY: Mexico, Baja California, Sierra San Pedro Martir, La Grulla.
DISTRIBUTION: N Baja California (Mexico).
COMMENTS: Subgenus *Myotis*. Revised by Miller and Allen (1928:118).

Myotis montivagus (Dobson, 1874). J. Asiat. Soc. Bengal, 43:237.
TYPE LOCALITY: China, Yunnan, Hotha.
DISTRIBUTION: Yunnan to Fukien and Chihli (China); India; Burma; W Malaysia; Borneo.
SYNONYMS: *borneoensis, federatus, peytoni.*
COMMENTS: Subgenus *Leuconoe*. Includes *peytoni*; see Hill (1962*b*:126-130); but see also Findley (1972:43).

Myotis morrisi Hill, 1971. Bull. Brit. Mus. (Nat. Hist.) Zool., 21:43.
TYPE LOCALITY: Ethiopia, Walaga, Didessa River mouth.
DISTRIBUTION: Ethiopia, Nigeria.
COMMENTS: Subgenus *Myotis*.

Myotis muricola (Gray, 1846). Cat. Hodgson Coll. Brit. Mus., p. 4.
TYPE LOCALITY: Nepal.
DISTRIBUTION: Afghanistan to Taiwan and New Guinea.
SYNONYMS: *amboiensis, ater, blanfordi, browni, caliginosus, herrei, latirostris, lobipes, moupinensis, niasensis, nugax, orii, patriciae.*
COMMENTS: Subgenus *Selysius*. Includes *caliginosus, moupinensis, latirostris,* and *ater*; see Findley (1972:42, 43); but see Hill (1983), who separated *ater* (including *nugax* and *australis*). Includes *browni*, see Hill and Rozendaal (1989), and *herrei* and *patriciae*, see Heaney et al. (1987).

Myotis myotis (Borkhausen, 1797). Deutsche Fauna, 1:80.
TYPE LOCALITY: Germany, Thuringia.
DISTRIBUTION: C and S Europe, east to Ukraine; S England; most Mediterranean islands; Azores (Portugal); Asia Minor; Lebanon; Israel.
SYNONYMS: *alpinus, latipennis, macrocephalicus, myosotis, submurinus, typus.*
COMMENTS: Subgenus *Myotis*. See Corbet (1978*c*:50) for content of this species.

Myotis mystacinus (Kuhl, 1817). Die Deutschen Fledermause. Hanau, p. 15.
TYPE LOCALITY: Germany.
DISTRIBUTION: Ireland and Scandinavia to N China, south to Morocco, Iran, NW Himalayas, and S China.
SYNONYMS: *aurascens, aureus, bulgaricus, collaris, davidi, hajastanicus, humeralis, kukunoriensis, lugubris, meinertzhageni, nigrofuscus, nipalensis, pamirensis, przewalskii, rufofuscus, schinzii, schrankii, sogdianus, transcaspicus.*
COMMENTS: Subgenus *Selysius*. Reviewed by Strelkov (1983).

Myotis nattereri (Kuhl, 1817). Die Deutschen Fledermause. Hanau, p. 14, 33.
TYPE LOCALITY: Germany, Hessen, Hanau.
DISTRIBUTION: Europe (except Scandinavia); NW Africa; Turkey; Israel; Iraq; Crimea and Caucasus to Turkmenistan.
SYNONYMS: *escalerae, spelaeus, tschuliensis, typus.*
COMMENTS: Subgenus *Myotis*. Does not include *araxenus* or *bombinus*; see Horáček and Hanak (1984).

Myotis nesopolus Miller, 1900. Proc. Biol. Soc. Washington, 13:123.
TYPE LOCALITY: Curaçao, Willemstad (Netherlands).
DISTRIBUTION: NE Venezuela; Curaçao and Bonaire (Netherlands Antilles).
SYNONYMS: *larensis.*
COMMENTS: Subgenus *Selysius*. Includes *larensis*; see Genoways and Williams (1979).

Myotis nigricans (Schinz, 1821). Das Thierreich, 1:179.
TYPE LOCALITY: Brazil, Espirito Santo, between Itapemirin and Iconha Rivers.
DISTRIBUTION: Nayarit and Tamaulipas (Mexico) to Peru, N Argentina and S Brazil; Trinidad and Tobago; Grenada (Lesser Antilles).
SYNONYMS: *bondae, carteri, caucensis, chiriquensis, concinnus, dalquesti, esmeraldae, exiguus, extremus, maripensis, mundus, parvulus, punensis.*

COMMENTS: Subgenus *Selysius*. Includes *carteri*; see Corbet and Hill (1980:65), but see Bogan (1978). Neotype designated by LaVal (1973*a*). See Wilson and LaVal (1974, Mammalian Species, 39).

Myotis oreias (Temminck, 1840). Monogr. Mamm., 2:270.
TYPE LOCALITY: Singapore.
DISTRIBUTION: Known only from the type locality.
COMMENTS: Subgenus *Selysius*. Known only from the holotype.

Myotis oxyotus (Peters, 1867). Monatsb. K. Preuss. Akad. Wiss. Berlin, 1867:19.
TYPE LOCALITY: Ecuador, Mount Chimborazo, between 2,743 and 3,048 m.
DISTRIBUTION: Venezuela to Bolivia; Panama; Costa Rica.
SYNONYMS: *gardneri, thomasi.*
COMMENTS: Subgenus *Leuconoe*. Revised by LaVal (1973*a*).

Myotis ozensis Imaizumi, 1954. Bull. Natl. Sci. Mus. Tokyo, 1:49.
TYPE LOCALITY: Japan, Honshu, Gunma Pref, Ozegahara, 1,400m.
DISTRIBUTION: Honshu (Japan).
COMMENTS: Subgenus *Selysius*.

Myotis peninsularis Miller, 1898. Ann. Mag. Nat. Hist., ser. 7, 2:124.
TYPE LOCALITY: Mexico, Baja California, San Jose del Cabo.
DISTRIBUTION: S Baja California (Mexico).
COMMENTS: Subgenus *Myotis*. Listed as a subspecies of *velifer* by Hall and Kelson (1959:166), but see Hayward (1970:5); and Hall (1981:196).

Myotis pequinius Thomas, 1908. Proc. Zool. Soc. Lond., 1908:637.
TYPE LOCALITY: China, Hopeh, 30 mi. (48 km) W Peking.
DISTRIBUTION: Hopeh, Shantung, Honan and Kiangsu (China).
COMMENTS: Subgenus *Myotis*.

Myotis planiceps Baker, 1955. Proc. Biol. Soc. Washington, 68:165.
TYPE LOCALITY: Mexico, Coahuila, 7 mi. (11 km) S and 4 mi. (6 km) E Bella Union, 7,200 ft. (2,195 M).
DISTRIBUTION: Coahuila, Nuevo Leon, and Zacatecas (Mexico).
COMMENTS: Subgenus *Selysius*. See Matson (1975, Mammalian Species, 60).

Myotis pruinosus Yoshiyuki, 1971. Bull. Natl. Sci. Mus. Tokyo, 14:305.
TYPE LOCALITY: Japan, Honshu, Iwate Pref., Waga-Gun, Waga-Machi.
DISTRIBUTION: Honshu and Shikoku (Japan).
COMMENTS: Subgenus *Leuconoe*.

Myotis ricketti (Thomas, 1894). Ann. Mag. Nat. Hist., ser. 6, 14:300.
TYPE LOCALITY: China, Fukien, Foochow.
DISTRIBUTION: Fukien, Anhwei, Kiangsu, Shantung, Yunnan (China); Hong Kong.
COMMENTS: Subgenus *Leuconoe*. *M. pilosus* may be the oldest name for this species; see Ellerman and Morrison-Scott (1951:150).

Myotis ridleyi Thomas, 1898. Ann. Mag. Nat. Hist., ser. 7, 1:361.
TYPE LOCALITY: Malaysia, Selangor.
DISTRIBUTION: W Malaysia, Sumatra, Borneo.
COMMENTS: Subgenus *Selysius*. Transferred from *Pipistrellus*; see Medway (1978:35).

Myotis riparius Handley, 1960. Proc. U.S. Natl. Mus., 112:466-468.
TYPE LOCALITY: Panama, Darien, Rio Puero, Tacarcuna Village.
DISTRIBUTION: Honduras to Uruguay and E Brazil; Trinidad.
COMMENTS: Subgenus *Leuconoe*. Originally described as a subspecies of *simus*. *M. guaycuru* may be the oldest name for this species; see LaVal (1973*a*:32-35).

Myotis rosseti (Oey, 1951). Beaufortia, 1(8):4.
TYPE LOCALITY: Cambodia.
DISTRIBUTION: Cambodia, Thailand.
COMMENTS: Subgenus *Selysius*. Originally described as a species of *Glischropus*; see Hill and Topal (1973).

Myotis ruber (E. Geoffroy, 1806). Ann. Mus. Hist. Nat. Paris, 8:204.
TYPE LOCALITY: Paraguay, Neembucu, Sapucay (neotype locality).
DISTRIBUTION: SE Brazil, Paraguay, NE Argentina.
SYNONYMS: *cinnamomeus*.
COMMENTS: Subgenus *Leuconoe*. Does not include *levis*; revised by LaVal (1973*a*), who with Miller and Allen (1928:199) discussed the type.

Myotis schaubi Kormos, 1934. Foldt Kozl., 64:310.
TYPE LOCALITY: Hungary (Pliocene).
DISTRIBUTION: In the Holocene, only Transcaucasia and W Iran.
SYNONYMS: *araxenus*.
COMMENTS: Subgenus *Myotis*. The living subspecies, *araxenus*, was formerly included in *M. nattereri*; see Horáček and Hanak (1984).

Myotis scotti Thomas, 1927. Ann. Mag. Nat. Hist., ser. 9, 19:554.
TYPE LOCALITY: Ethiopia, Shoa, Djem-djem Forest (ca. 40 mi. (64 km) W Addis Ababa), 8,000 ft. (2,438 m).
DISTRIBUTION: Ethiopia.
COMMENTS: Subgenus *Selysius*.

Myotis seabrai Thomas, 1912. Ann. Mag. Nat. Hist., ser. 8, 10:205.
TYPE LOCALITY: Angola, Mossamedes.
DISTRIBUTION: NW Cape Prov. (South Africa), Namibia, SW Angola.
COMMENTS: Subgenus *Cistugo*.

Myotis sicarius Thomas, 1915. J. Bombay Nat. Hist. Soc., 23:608.
TYPE LOCALITY: India, N Sikkim.
DISTRIBUTION: Sikkim (NE India).
COMMENTS: Subgenus *Myotis*.

Myotis siligorensis (Horsfield, 1855). Ann. Mag. Nat. Hist., ser. 2, 16:102.
TYPE LOCALITY: Nepal, Siligori.
DISTRIBUTION: N India to S China and Vietnam, south to W Malaysia; Borneo.
SYNONYMS: *alticraniatus, darjilingensis, sowerbyi, thaianus*.
COMMENTS: Subgenus *Selysius*.

Myotis simus Thomas, 1901. Ann. Mag. Nat. Hist., ser. 7, 7:541.
TYPE LOCALITY: Peru, Loreto, Sarayacu (Ucayali River).
DISTRIBUTION: Colombia, Ecuador, Peru, N Brazil, Bolivia, Paraguay, NE Argentina.
COMMENTS: Subgenus *Leuconoe*. Revised by LaVal (1973*a*).

Myotis sodalis Miller and Allen, 1928. Bull. U.S. Natl. Mus., 144:130.
TYPE LOCALITY: USA, Indiana, Crawford Co., Wyandotte Cave.
DISTRIBUTION: New Hampshire to Florida Panhandle, west to Wisconsin and Oklahoma (USA).
STATUS: U.S. ESA - Endangered.
COMMENTS: Subgenus *Selysius*. See Thomson (1982, Mammalian Species, 163).

Myotis stalkeri Thomas, 1910. Ann. Mag. Nat. Hist., ser. 8, 5:384.
TYPE LOCALITY: Indonesia, Molucca Isls, Kei Isl, Ara.
DISTRIBUTION: Kei Isls (Molucca Isls).
COMMENTS: Subgenus *Leuconoe*.

Myotis thysanodes Miller, 1897. N. Am. Fauna, 13:80.
TYPE LOCALITY: USA, California, Kern Co., Tehachapi Mountains, Old Fort Tejon.
DISTRIBUTION: Chiapas (Mexico) to SW South Dakota (USA) and SC British Columbia (Canada).
SYNONYMS: *aztecus, pahasapensis, vespertinus*.
COMMENTS: Subgenus *Myotis*. Revised by Miller and Allen (1928:122-129). See O'Farrell and Studier (1980, Mammalian Species, 137).

Myotis tricolor (Temminck, 1832). *In* Smuts, Enumer. Mamm. Capensium, p. 106.
TYPE LOCALITY: South Africa, Cape Province, Capetown.
DISTRIBUTION: Ethiopia and Zaire, south to South Africa.
SYNONYMS: *loveni*.

COMMENTS: Subgenus *Myotis*. Originally *Eptesicus loveni*, see Schlitter and Aggundey (1986).

Myotis velifer (J. A. Allen, 1890). Bull. Am. Mus. Nat. Hist., 3:177.
TYPE LOCALITY: Mexico, Jalisco, Guadalajara, Santa Cruz del Valle.
DISTRIBUTION: Honduras to Kansas and SE California (USA).
SYNONYMS: *brevis, grandis, incautus, jaliscensis, magnamolaris.*
COMMENTS: Subgenus *Leuconoe*. See Hayward (1970:4-5) for scope of this species. See Fitch et al. (1981, Mammalian Species, 149).

Myotis vivesi Menegaux, 1901. Bull. Mus. Hist. Nat. Paris, 7:323.
TYPE LOCALITY: Mexico, Baja California, Partida Isl.
DISTRIBUTION: Coast of Sonora and Baja California (Mexico), chiefly on small islands.
COMMENTS: Subgenus *Leuconoe*.

Myotis volans (H. Allen, 1866). Proc. Acad. Nat. Sci. Philadelphia, 18:282.
TYPE LOCALITY: Mexico, Baja California, Cabo San Lucas.
DISTRIBUTION: Jalisco to Veracruz (Mexico); Alaska Panhandle (USA) to Baja California (Mexico), east to N Nuevo Leon (Mexico), South Dakota (USA), and C Alberta (Canada).
SYNONYMS: *altifrons, amotus, capitaneus, interior, longicrus, ruddi.*
COMMENTS: Subgenus *Leuconoe*. Revised by Miller and Allen (1928:135-147). See Warner and Czaplewski (1984, Mammalian Species, 224).

Myotis welwitschii (Gray, 1866). Proc. Zool. Soc. Lond., 1866:211.
TYPE LOCALITY: NE Angola.
DISTRIBUTION: South Africa to Ethiopia.
SYNONYMS: *venustus.*
COMMENTS: Subgenus *Myotis*.

Myotis yesoensis Yoshiyuki, 1984. Bull. Natl. Sci. Mus. Tokyo, ser. A(Zool.), 10:153.
TYPE LOCALITY: Japan, Hokkaido, Hiddaka, Mt. Petegari.
DISTRIBUTION: Hokkaido (Japan).
COMMENTS: Subgenus *Selysius*. Closest to *M. hosonoi*.

Myotis yumanensis (H. Allen, 1864). Smithson. Misc. Coll., 7:58.
TYPE LOCALITY: USA, California, Imperial Co., Old Fort Yuma.
DISTRIBUTION: Hidalgo, Morelos and Baja California (Mexico) north to British Columbia (Canada), east to Montana and W Texas (USA).
SYNONYMS: *durangae, lambi, lutosus, obscurus, oxalis, phasma, saturatus, sociabilis.*
COMMENTS: Subgenus *Leuconoe*. See comments under *leibii* and *lucifugus*.

Nyctalus Bowdich, 1825. Excursions in Madeira and Porto Santo, p. 36.
TYPE SPECIES: *Nyctalus verrucosus* Bowdich, 1825 (= *Vespertilio leisleri* Kuhl, 1817).

Nyctalus aviator (Thomas, 1911). Ann. Mag. Nat. Hist., ser. 8, 8:380.
TYPE LOCALITY: Japan, Honshu, Tokyo.
DISTRIBUTION: Hokkaido, Shikoku, Kyushu, Tsushima, Iki (Japan); Korea; E China.
COMMENTS: Listed as a subspecies of *lasiopterus* by Ellerman and Morrison-Scott (1951:161); but see Corbet (1978*c*:56).

Nyctalus azoreum (Thomas, 1901). Ann. Mag. Nat. Hist., ser. 7, 8:34.
TYPE LOCALITY: Portugal, Azores, St. Michael.
DISTRIBUTION: Azores (Portugal).
COMMENTS: Listed as a subspecies of *N. leisleri* by Corbet (1978*c*), but see Palmeirim (1991).

Nyctalus lasiopterus (Schreber, 1780). *In* Zimmermann, Geogr. Gesch. Mensch. Vierf. Thiere, 2:412.
TYPE LOCALITY: Northern Italy (uncertain).
DISTRIBUTION: W Europe to Urals and Caucasus, Asia Minor, Iran and Ust-Urt Plateau (Kazakhstan); Morocco; Libya.
SYNONYMS: *ferrugineus, maxima, sicula.*
COMMENTS: See Corbet (1978*c*:55-56) for content of this species.

Nyctalus leisleri (Kuhl, 1817). Die Deutschen Fledermause. Hanau, p. 14, 46.
TYPE LOCALITY: Germany, Hessen, Hanau.

DISTRIBUTION: W Europe to Urals and Caucasus; Britain; Ireland; Madeira Isl and Azores (Portugal); W Himalayas; E Afghanistan; NW Africa.
SYNONYMS: *dasykarpos, madeirae, pachygnathus, verrucosus.*
COMMENTS: Includes *verrucosus*, see Corbet (1978c:55), who also included *azoreum*; but see Palmeirim (1991).

Nyctalus montanus (Barrett-Hamilton, 1906). Ann. Mag. Nat. Hist., ser. 7, 17:99.
TYPE LOCALITY: India, Uttar Pradesh, Dehra Dun, Mussooree.
DISTRIBUTION: Afghanistan, Pakistan, N India, Nepal.
COMMENTS: Listed as a subspecies of *leisleri* by Ellerman and Morrison-Scott (1951:159); but see Corbet (1978c:55).

Nyctalus noctula (Schreber, 1774). Die Säugethiere, 1:166.
TYPE LOCALITY: France.
DISTRIBUTION: Europe to Urals and Caucasus; Algeria; SE Asia Minor to Israel; Western Turkmenistan to SW Siberia, Himalayas, and China, south to W Malaysia; Taiwan; Honshu (Japan); a dubious record from Mozambique.
SYNONYMS: *altivolans, furvus, labiatus, lardarius, lebanoticus, macuanus, magnus, mecklenburzevi, minima, palustris, plancei, princeps, proterus, rufescens, sinensis, velutinus.*
COMMENTS: Includes *furvus* and *velutinus*; see Corbet (1978c:55); but see also Yoshiyuki (1989).

Nycticeius Rafinesque, 1819. J. Phys. Chim. Hist. Nat. Arts Paris, 88:417.
TYPE SPECIES: *Vespertilio humeralis* Rafinesque, 1818.
SYNONYMS: *Nycticeinops, Scoteanax, Scotorepens.*
COMMENTS: Formerly included *Scotoecus*; see J. E. Hill (1974b:169-171). For content of this genus; see Koopman (1978a); but see also Kitchener and Caputi (1985), who revised the Australian species. Includes *Nycticeinops*, but see Hill and Harrison (1987).

Nycticeius balstoni (Thomas, 1906). Abstr. Proc. Zool. Soc. Lond., 1906(31):2.
TYPE LOCALITY: Australia, Western Australia, Laverton, North Pool, 503 m.
DISTRIBUTION: Mainland Australia.
SYNONYMS: *influatus, orion.*
COMMENTS: Subgenus *Scotorepens*. Includes *orion*, but not *caprenus*; see Koopman (1978a:168, 170, 171); but also see Hall and Richards (1979). Includes *influatus*; see Kitchener and Caputi (1985), who excluded *orion* and *caprenus*.

Nycticeius greyii (Gray, 1842). Zool. Voy. H.M.S. "Erebus" and "Terror," pl. 20.
TYPE LOCALITY: Australia, Northern Territory, Port Essington.
DISTRIBUTION: Western Australia (excluding the south), Northern Territory, South Australia, New South Wales, and Queensland (Australia). Records from Victoria refer to *balstoni*.
SYNONYMS: *aquilo, caprenus.*
COMMENTS: Subgenus *Scotorepens*. See Kitchener and Caputi (1985) for content.

Nycticeius humeralis (Rafinesque, 1818). Am. Mon. Mag., 3(6):445.
TYPE LOCALITY: USA, Kentucky.
DISTRIBUTION: N Veracruz (Mexico) to Nebraska, the Great Lakes, and Pennsylvania, south to Florida and the Gulf coast (USA); Cuba.
SYNONYMS: *crepuscularis, cubanus, mexicanus, subtropicalis.*
COMMENTS: Subgenus *Nycticeius*. Includes *cubanus*; see Varona (1974:37); but also see Hall (1981:227). Reviewed by Watkins (1972, Mammalian Species, 23), who did not include *cubanus*.

Nycticeius rueppellii (Peters, 1866). Monatsb. K. Preuss. Akad. Wiss. Berlin, 1866:21.
TYPE LOCALITY: Australia, New South Wales, Sydney.
DISTRIBUTION: E Queensland and E New South Wales (Australia).
COMMENTS: Subgenus *Scoteanax*. Reviewed by Kitchener and Caputi (1985).

Nycticeius sanborni (Troughton, 1937). Aust. Zool., 8:280.
TYPE LOCALITY: New Guinea, Papua, Milne Bay prov., East Cape.

DISTRIBUTION: SE New Guinea; NE Queensland, Northern Territory, and N Western Australia (Australia).

COMMENTS: Subgenus *Scotorepens*. Included in *N. balstoni* by Koopman (1978*a*), but see Kitchener and Caputi (1985).

Nycticeius schlieffeni (Peters, 1859). Monatsb. K. Preuss. Akad. Wiss. Berlin, 1859:223.

TYPE LOCALITY: Egypt, Cairo.

DISTRIBUTION: SW Arabia; Egypt to Somalia, Mozambique, Botswana, South Africa, and Namibia; Mauritania and Ghana to Sudan and Tanzania.

SYNONYMS: *adovanus, africanus, albiventer, australis, bedouin, cinnamomeus, fitzsimonsi, minimus.*

COMMENTS: Subgenus *Nycticeinops*. Includes *cinnamomeus*; see Koopman (1975:412-413).

Nyctophilus Leach, 1821. Trans. Linn. Soc. Lond., 13:78.

TYPE SPECIES: *Nyctophilus geoffroyi* Leach, 1821.

SYNONYMS: *Lamingtona.*

COMMENTS: Includes *Lamingtona*, see Hill and Koopman (1981). Australian species reviewed by Hall and Richards (1979).

Nyctophilus arnhemensis Johnson, 1959. Proc. Biol. Soc. Washington, 72:184.

TYPE LOCALITY: Australia, Northern Territory, Cape Arnhem Peninsula, S of Yirkala, Rocky Bay. (12°13'S and 136°47'E).

DISTRIBUTION: N Australia.

Nyctophilus geoffroyi Leach, 1821. Trans. Linn. Soc. Lond., 13:78.

TYPE LOCALITY: Australia, Western Australia, King George Sound.

DISTRIBUTION: Australia (except NE) including Tasmania.

SYNONYMS: *australis, geayi, leachii, novaehollandiae, pacificus, pallescens, unicolor.*

Nyctophilus gouldi Tomes, 1858. Proc. Zool. Soc. Lond., 1858:31.

TYPE LOCALITY: Australia, Queensland, Moreton Bay.

DISTRIBUTION: N Northern Territory, N and E Queensland, N and W Western Australia, E New South Wales, and Victoria (Australia); New Guinea; the Tasmanian record appears to be erroneous.

SYNONYMS: *bifax, daedalus.*

COMMENTS: Includes *bifax*; see Koopman (1984*c*:25-27); but see also Parnaby (1987).

Nyctophilus heran Kitchener, How, and Maharadatunkamsi, 1991. Rec. W. Aust. Mus., 15:100.

TYPE LOCALITY: Indonesia, Lesser Sundas, Lembata (= Lomblen).

DISTRIBUTION: Known only from the type locality.

Nyctophilus microdon Laurie and Hill, 1954. List of Land Mammals of New Guinea, Celebes, and adjacent Islands, p. 78.

TYPE LOCALITY: Papua New Guinea, Western Highlands (?) Prov., Welya (W of Hagen Range, 7,000 ft. (2,134 m)).

DISTRIBUTION: EC New Guinea.

Nyctophilus microtis Thomas, 1888. Ann. Mag. Nat. Hist., ser. 6, 2:226.

TYPE LOCALITY: Papua New Guinea, Central Prov., Astrolabe Range, Sogeri.

DISTRIBUTION: E New Guinea.

SYNONYMS: *bicolor, lophorhina.*

COMMENTS: Includes *lophorhina* as a synonym; see Hill and Koopman (1981).

Nyctophilus timoriensis (E. Geoffroy, 1806). Ann. Mus. Hist. Nat. Paris, 8:200.

TYPE LOCALITY: Indonesia, Timor (uncertain).

DISTRIBUTION: All of Australia including Tasmania; E New Guinea; probably Timor (Indonesia).

SYNONYMS: *major, sherrini.*

COMMENTS: This bat has been confused with the smaller *gouldi* in coastal SE Queensland. See Hall and Richards (1979).

Nyctophilus walkeri Thomas, 1892. Ann. Mag. Nat. Hist., ser. 6, 9:405.

TYPE LOCALITY: Australia, Northern Territory, Adelaide River.

DISTRIBUTION: Northern Territory and N Western Australia (Australia).

Otonycteris Peters, 1859. Monatsb. K. Preuss. Akad. Wiss. Berlin, 1859:223.
TYPE SPECIES: *Otonycteris hemprichii* Peters, 1859.

Otonycteris hemprichii Peters, 1859. Monatsb. K. Preuss. Akad. Wiss. Berlin, 1859:223.
TYPE LOCALITY: Egypt, Nile Valley south of Assuan (Aswan).
DISTRIBUTION: The desert zone from Morocco and Niger through Egypt and Arabia to Tadzhikistan, Afghanistan, and Kashmir.
SYNONYMS: *cinerea, jin, leucophaeus, petersi, saharae.*

Pharotis Thomas, 1914. Ann. Mag. Nat. Hist., ser. 8, 14:381.
TYPE SPECIES: *Pharotis imogene* Thomas, 1914.

Pharotis imogene Thomas, 1914. Ann. Mag. Nat. Hist., ser. 8, 14:382.
TYPE LOCALITY: Papua New Guinea, Central Prov., Lower Kemp Welch River, Kamali.
DISTRIBUTION: SE New Guinea.

Philetor Thomas, 1902. Ann. Mag. Nat. Hist., ser. 7, 9:220.
TYPE SPECIES: *Philetor rohui* Thomas, 1902 (= *Vespertilio brachypterus* Temminck, 1840).

Philetor brachypterus (Temminck, 1840). Monogr. Mamm., 2:215.
TYPE LOCALITY: Indonesia, Sumatra, Padang Dist.
DISTRIBUTION: Nepal, W Malaysia, Sumatra, Java, Borneo, Philippines, Sulawesi, New Guinea, New Britain Isl (Bismarck Arch.), perhaps Banka Isl (Indonesia).
SYNONYMS: *rohui, veraecundus* (see Hill, 1971*d*).

Pipistrellus Kaup, 1829. Skizz. Entwickel.-Gesch. Nat. Syst. Europ. Thierwelt, 1:98.
TYPE SPECIES: *Vespertilio pipistrellus* Schreber, 1774.
SYNONYMS: *Arielulus, Falsistrellus, Hypsugo, Perimyotis, Scotozous, Vansonia.*
COMMENTS: For synonyms see Ellerman and Morrison-Scott (1951:162-163); but also see Hill (1976:25); Kitchener et al. (1986); and Menu (1984). *Hypsugo* was considered a full genus by Tiunov (1986). Hill and Harrison (1987) would transfer *Neomicia* and *Vespadelus* to here from *Eptesicus*. Otherwise the subgenera they use are followed here.

Pipistrellus aegyptius (J. Fischer, 1829). Synopsis Mamm., p. 105.
TYPE LOCALITY: Egypt, Luxor.
DISTRIBUTION: Egypt, N Sudan, Lybia, Algeria, Burkina Fasso.
SYNONYMS: *deserti.*
COMMENTS: Subgenus *Pipistrellus*. For use of this name in place of *deserti* see Qumsiyeh (1985).

Pipistrellus aero Heller, 1912. Smithson. Misc. Coll., 60(12):3.
TYPE LOCALITY: Kenya, Mathews Range, Mt. Gargues.
DISTRIBUTION: NW Kenya, perhaps Ethiopia.
COMMENTS: Subgenus *Pipistrellus*. The Ethiopian specimens in the British Museum are clearly *P. kuhlii*; see Hayman and Hill (1971:41).

Pipistrellus affinis (Dobson, 1871). Proc. Asiat. Soc. Bengal, p. 213.
TYPE LOCALITY: Burma, Bhamo.
DISTRIBUTION: NE Burma, Yunnan (China), N India.
COMMENTS: Subgenus *Falsistrellus*.

Pipistrellus anchietai (Seabra, 1900). J. Sci. Math. Phys. Nat. Lisboa, ser. 2, 6:26, 120.
TYPE LOCALITY: Angola, Cahata.
DISTRIBUTION: Angola, S Zaire, Zambia.
COMMENTS: Subgenus *Hypsugo*. The oldest name for this species may be *bicolor*; see Koopman (1975:404).

Pipistrellus anthonyi Tate, 1942. Bull. Am. Mus. Nat. Hist., 80:252.
TYPE LOCALITY: Burma, Changyinku, 7,000 ft. (2,134 m).
DISTRIBUTION: Known only from the type locality.

COMMENTS: Subgenus *Hypsugo*. May be referrable to *Nyctalus* or even *Philetor*. Known only by the holotype.

Pipistrellus arabicus Harrison, 1979. Mammalia, 43:575.
TYPE LOCALITY: Oman, Wadi Sahtan (23°22'N, 57°18'E).
DISTRIBUTION: Oman.
COMMENTS: Subgenus *Hypsugo*.

Pipistrellus ariel Thomas, 1904. Ann. Mag. Nat. Hist., ser. 7, 14:157.
TYPE LOCALITY: Sudan, Kassala Province, Wadi Alagi (22°N, 35°E), 2,000 ft. (610 m).
DISTRIBUTION: Egypt, N Sudan.
COMMENTS: Subgenus *Hypsugo*.

Pipistrellus babu Thomas, 1915. J. Bombay Nat. Hist. Soc., 24:30.
TYPE LOCALITY: Pakistan, Punjab, Rawalpindi, Murree, 8,000 ft. (2,438 m).
DISTRIBUTION: Afghanistan, Pakistan, N India, Nepal, Bhutan, Burma, SW China.
COMMENTS: Subgenus *Pipistrellus*.

Pipistrellus bodenheimeri Harrison, 1960. Durban Mus. Novit., 5:261.
TYPE LOCALITY: Israel, 40 km N Eilat, Wadi Araba, Yotwata.
DISTRIBUTION: Israel, S Yemen, Oman, perhaps Socotra Isl (Yemen).
COMMENTS: Subgenus *Hypsugo*.

Pipistrellus cadornae Thomas, 1916. J. Bombay Nat. Hist. Soc., 24:416.
TYPE LOCALITY: India, Darjeeling, Pashok, 3,500 ft (1,067m).
DISTRIBUTION: NE India, Burma, Thailand.
COMMENTS: Subgenus *Hypsugo*. Listed as a subspecies of *savii* by Ellerman and Morrison-Scott (1951:170), but see Hill (1962*b*:133).

Pipistrellus ceylonicus (Kelaart, 1852). Prodr. Faun. Zeylanica, p. 22.
TYPE LOCALITY: Sri Lanka, Trincomalee.
DISTRIBUTION: Pakistan; India; Sri Lanka; Burma; Kwangsi and Hainan (China); Vietnam; Borneo.
SYNONYMS: *borneanus, chrysothrix, indicus, raptor, shanorum, subcanus, tonfangensis*.
COMMENTS: Subgenus *Pipistrellus*.

Pipistrellus circumdatus (Temminck, 1840). Monogr. Mamm., 2:214.
TYPE LOCALITY: Indonesia, Java, Tapos.
DISTRIBUTION: Java, W Malaysia, Burma, NE India, SW China.
SYNONYMS: *drungicus*.
COMMENTS: Subgenus *Arielulus*.

Pipistrellus coromandra (Gray, 1838). Mag. Zool. Bot., 2:498.
TYPE LOCALITY: India, Coromandel Coast, Pondicherry.
DISTRIBUTION: Afghanistan, Pakistan, India (including Nicobar Isls), Sri Lanka, Nepal, Bhutan, Burma, S China, Thailand, Vietnam.
SYNONYMS: *afghanus, blythii, coromandelicus, micropus, parvipes, portensis, tramatus*.
COMMENTS: Subgenus *Pipistrellus*. Does not include *aladdin*; see Corbet (1978*c*:53). See comment under *P. pipistrellus*.

Pipistrellus crassulus Thomas, 1904. Ann. Mag. Nat. Hist., ser. 7, 13:206.
TYPE LOCALITY: Cameroon, Efulen.
DISTRIBUTION: Cameroon, Zaire, S Sudan, N Angola.
COMMENTS: Subgenus *Pipistrellus*.

Pipistrellus cuprosus Hill and Francis, 1984. Bull. Brit. Mus. (Nat. Hist.) Zool., 47:312.
TYPE LOCALITY: Borneo, Sabah, Sepilok.
DISTRIBUTION: Borneo.
COMMENTS: Subgenus *Arielulus*.

Pipistrellus dormeri (Dobson, 1875). Proc. Zool. Soc. Lond., 1875:373.
TYPE LOCALITY: India, Mysore, Bellary Hills.
DISTRIBUTION: NW, S and E India, Pakistan, perhaps Taiwan.
SYNONYMS: *caurinus*.

COMMENTS: Subgenus *Scotozous*. Sometimes placed in a separate genus, *Scotozous*; see Tate (1942*a*:259) and Corbet and Hill (1980:68); but also see Ellerman and Morrison-Scott (1951:162-163) and Corbet (1978*c*:51).

Pipistrellus eisentrauti Hill, 1968. Bonn. Zool. Beitr., 19:45.
TYPE LOCALITY: Cameroon, Western Province, Rumpi Highlands, Dikume-Balue.
DISTRIBUTION: Liberia to Kenya and Somalia.
SYNONYMS: *bellieri*.
COMMENTS: Subgenus *Hypsugo*.

Pipistrellus endoi Imaizumi, 1959. Bull. Natl. Sci. Mus. Tokyo, 4:363.
TYPE LOCALITY: Japan, Honshu, Iwate Pref., Ninohe-Gun, Ashiro-cho, Horobe.
DISTRIBUTION: Honshu (Japan).
COMMENTS: Subgenus *Pipistrellus*. Very similar to *javanicus*, but see Yoshiyuki (1989).

Pipistrellus hesperus (H. Allen, 1864). Smithson. Misc. Coll., 7:43.
TYPE LOCALITY: USA, California, Imperial Co., Old Fort Yuma.
DISTRIBUTION: Washington to SW Oklahoma (USA), and Baja California, south to Hidalgo and Guerrero (Mexico).
SYNONYMS: *apus, australis, maximus, merriami, oklahomae, potosinus, santarosae*.
COMMENTS: Subgenus *Hypsugo*.

Pipistrellus imbricatus (Horsfield, 1824). Zool. Res. Java, part 8, p. 5(unno.) of *Vespertilio Temminckii* acct.
TYPE LOCALITY: Indonesia, Java.
DISTRIBUTION: Java, Kangean Isl, Bali, and Lesser Sunda Isls; Borneo; records from the Philippines are dubious.
COMMENTS: Subgenus *Hypsugo*.

Pipistrellus inexspectatus Aellen, 1959. Arch. Sci. Phys. Nat. Geneve, 12:226.
TYPE LOCALITY: Cameroon, Upper Benoue Valley, Ngaaouyanga.
DISTRIBUTION: Benin, Cameroon, Zaire, Uganda, Kenya, perhaps Sudan.
COMMENTS: Subgenus *Hypsugo*.

Pipistrellus javanicus (Gray, 1838). Mag. Zool. Bot., 2:498.
TYPE LOCALITY: Indonesia, Java.
DISTRIBUTION: S and C Japan; S Ussuri region (Russia and China); Korea; China, through SE Asia to Lesser Sunda Isls and the Philippines; perhaps Australia.
SYNONYMS: *abramus, akokomuli, bancanus, camortae, irretitus, meyeni, pumiloides; "tralatitius"* Thomas (not Horsfield).
COMMENTS: Subgenus *Pipistrellus*. Includes *abramus, "tralatitius,"* and *camortae*; but see Soota and Chaturvedi (1980) and Hill and Harrison (1987). Includes *meyeni* and *irretitus*; see Laurie and Hill (1954:67), Ellerman and Morrison-Scott (1951:165), Hill (1967:7), and Koopman (1973:115). Does not include *paterculus*, see Hill and Harrsion (1987).

Pipistrellus joffrei (Thomas, 1915). Ann. Mag. Nat. Hist., ser. 8, 15:225.
TYPE LOCALITY: Burma, Kachin Hills.
DISTRIBUTION: N Burma.
COMMENTS: Subgenus *Hypsugo*. Transferred from *Nyctalus*; see Hill (1966:383). Probably best retained in *Nyctalus*, see Koopman (1989*a*:8).

Pipistrellus kitcheneri Thomas, 1916. Ann. Mag. Nat. Hist., ser. 8, 15:229.
TYPE LOCALITY: Borneo, Kalimantan Tengah, Barito River.
DISTRIBUTION: Borneo.
COMMENTS: Subgenus *Hypsugo*. Listed as a subspecies of *imbricatus* by Chasen (1940); but see Tate (1942*a*:221) and Medway (1977:55).

Pipistrellus kuhlii (Kuhl, 1817). Die Deutschen Fledermause, Hanau, p. 14.
TYPE LOCALITY: Italy, Friuli-Venezia Giulia, Trieste.
DISTRIBUTION: S Europe through the Caucasus to Kazakhstan and Pakistan; SW Asia; most of Africa; Canary Isls (Spain).
SYNONYMS: *albicans, albolimbatus, alcythoe, broomi, calcarata, canus, fuscatus, ikhwanius, lepidus, leucotis, lobatus, marginatus, minuta, pullatus, subtilus, ursula, vispistrellus*.
COMMENTS: Subgenus *Pipistrellus*.

Pipistrellus lophurus Thomas, 1915. J. Bombay Nat. Hist. Soc., 23:413.
TYPE LOCALITY: Burma, Tenasserim, Victoria Province, Maliwun.
DISTRIBUTION: Peninsular Burma.
COMMENTS: Subgenus *Hypsugo*.

Pipistrellus macrotis (Temminck, 1840). Monogr. Mamm., 2:218.
TYPE LOCALITY: Indonesia, Sumatra, Padang.
DISTRIBUTION: W Malaysia, Sumatra, Borneo, Bali, adjacent small islands.
SYNONYMS: *curtatus, vordermanni.*
COMMENTS: Subgenus *Pipistrellus*. Listed as a subspecies of *imbricatus* by Medway (1969:39); but see Tate (1942*a*:249) and Corbet and Hill (1980:67). Includes *curtatus* and *vordermanni*, but see Hill (1983).

Pipistrellus maderensis (Dobson, 1878). Cat. Chiroptera Brit. Mus., p. 231.
TYPE LOCALITY: Madeira Isls, Madeira Isl (Portugal).
DISTRIBUTION: Madeira Isl (Portugal); Canary Isls (Spain).
COMMENTS: Subgenus *Pipistrellus*.

Pipistrellus mimus Wroughton, 1899. J. Bombay Nat. Hist. Soc., 12:722.
TYPE LOCALITY: India, Gujarat, Surat Dist., Dangs, Mheskatri.
DISTRIBUTION: Afghanistan, Pakistan, Sri Lanka, India (including Sikkim), Nepal, Bhutan, Burma, Vietnam, Thailand.
SYNONYMS: *glaucillus, principulus.*
COMMENTS: Subgenus *Pipistrellus*.

Pipistrellus minahassae (A. Meyer, 1899). Abh. Zool. Anthrop.-Ethnology. Mus. Dresden, 7(7):14.
TYPE LOCALITY: Indonesia, Sulawesi, Minahassa, Tomohon.
DISTRIBUTION: Sulawesi.
COMMENTS: Subgenus *Pipistrellus*.

Pipistrellus mordax (Peters, 1866). Monatsb. K. Preuss. Akad. Wiss. Berlin, 1866:402.
TYPE LOCALITY: Indonesia, Java.
DISTRIBUTION: Java; records from India and Sri Lanka are erroneous, based on misidentified *P. affinis*, see Hill and Harrison (1987:248).
COMMENTS: Subgenus *Falsistrellus*.

Pipistrellus musciculus Thomas, 1913. Ann. Mag. Nat. Hist., ser. 8, 11:316.
TYPE LOCALITY: Cameroon, Ja River, Bitye, 2,000 ft. (610 m).
DISTRIBUTION: Cameroon, Zaire, Gabon.
COMMENTS: Subgenus *Hypsugo*.

Pipistrellus nanulus Thomas, 1904. Ann. Mag. Nat. Hist., ser. 7, 14:198.
TYPE LOCALITY: Cameroon, Efulen.
DISTRIBUTION: Sierra Leone to Kenya; Bioko.
COMMENTS: Subgenus *Pipistrellus*.

Pipistrellus nanus (Peters, 1852). Reise nach Mossambique, Säugethiere, p. 63.
TYPE LOCALITY: Mozambique, Inhambane.
DISTRIBUTION: South Africa to Ethiopia, Sudan, Niger, Mali, and Senegal; Madagascar; Pemba and Zanzibar.
SYNONYMS: *abaensis, africanus, culex, fouriei, helios, minusculus, pagenstecheri, pusillulus, stampflii.*
COMMENTS: Subgenus *Pipistrellus*. The oldest name for this species is *africanus;* see Koopman (1975:400).

Pipistrellus nathusii (Keyserling and Blasius, 1839). Arch. Naturgesch., 5(1):320.
TYPE LOCALITY: Germany, Berlin.
DISTRIBUTION: W Europe to Urals and Caucasus, and W Asia Minor; S England.
SYNONYMS: *unicolor.*
COMMENTS: Subgenus *Pipistrellus*.

Pipistrellus paterculus Thomas, 1915. J. Bombay Nat. Hist. Soc., 24:32.
TYPE LOCALITY: Burma, Mt. Popa.
DISTRIBUTION: N India, Burma, Thailand, SW China.
SYNONYMS: *yunnanensis.*

COMMENTS: Subgenus *Pipistrellus*. Included in *abramus* by Ellerman and Morrison-Scott (1951), but see Hill and Harrison (1987).

Pipistrellus peguensis Sinha, 1969. Proc. Zool. Soc. Calcutta, 22:83.
TYPE LOCALITY: Burma, Pegu.
DISTRIBUTION: Burma.
COMMENTS: Subgenus *Pipistrellus*.

Pipistrellus permixtus Aellen, 1957. Rev. Suisse Zool., 64:200.
TYPE LOCALITY: Tanzania, Dar-es-Salaam.
DISTRIBUTION: Tanzania.
COMMENTS: Subgenus *Pipistrellus*.

Pipistrellus petersi (A. Meyer, 1899). Abh. Zool. Anthrop.-Ethnology. Mus. Dresden, 7(7):13.
TYPE LOCALITY: Indonesia, Sulawesi, North Sulawesi, Minahassa.
DISTRIBUTION: Borneo; Sulawesi; Buru and Amboina (Molucca Isls); Philippines.
COMMENTS: Subgenus *Falsistrellus*.

Pipistrellus pipistrellus (Schreber, 1774). Die Säugethiere, 1:167.
TYPE LOCALITY: France.
DISTRIBUTION: British Isles, S Scandinavia, and W Europe to the Volga and Caucasus; Morocco; Asia Minor and Israel to Kashmir, Kazakhstan, and Sinkiang (China). Perhaps Korea, Japan and Taiwan.
SYNONYMS: *aladdin, bactrianus, brachyotos, flavescens, genei, griseus, limbatus, macropterus, mediterraneus, melanopterus, minutissimus, murinus, nigricans, pusillus, pygmaeus, stenotus, typus*.
COMMENTS: Subgenus *Pipistrellus*. Includes *aladdin*; see Corbet (1978*c*:53).

Pipistrellus pulveratus (Peters, 1871). *In* Swinhoe, Proc. Zool. Soc. Lond., 1870:618 [1871].
TYPE LOCALITY: China, Fukien, Amoy.
DISTRIBUTION: Szechwan, Yunnan, Hunan, Kiangsu, Fukien (China), Hong Kong; Thailand.
COMMENTS: Subgenus *Hypsugo*.

Pipistrellus rueppelli (J. Fischer, 1829). Synopsis Mamm., p. 109.
TYPE LOCALITY: Sudan, Northern Province, Dongola.
DISTRIBUTION: Senegal, Algeria, Egypt, and Iraq, south to Botswana and Transvaal (South Africa); Zanzibar.
SYNONYMS: *coxi, fuscipes, leucomelas, pulcher, senegalensis, vernayi*.
COMMENTS: Subgenus *Pipistrellus*.

Pipistrellus rusticus (Tomes, 1861). Proc. Zool. Soc. Lond., 1861:35.
TYPE LOCALITY: Namibia, Damaraland, Olifants Vlei.
DISTRIBUTION: Liberia and Ethiopia, south to South Africa.
SYNONYMS: *marrensis*.
COMMENTS: Subgenus *Pipistrellus*.

Pipistrellus savii (Bonaparte, 1837). Fauna Ital., 1, fasc. 20.
TYPE LOCALITY: Italy, Pisa.
DISTRIBUTION: Iberia, Morocco, and the Canary (Spain) and Cape Verde Isls through the Caucasus and Mongolia to Korea, NE China and Japan, southeastward through Iran and Afghanistan to N India and Burma.
SYNONYMS: *agilis, aristippe, austenianus, alaschanicus, bonapartei, caucasicus, coreensis, darwini, leucippe, maurus, nigrans, ochromixtus, pallescens, tamerlani, tauricus, velox*.
COMMENTS: Subgenus *Hypsugo*. Includes *coreensis*, but see Yoshiyuki (1989).

Pipistrellus societatis Hill, 1972. Bull. Brit. Mus. (Nat. Hist.) Zool., 23:34.
TYPE LOCALITY: Malaysia, Pahang, Gunong Benom.
DISTRIBUTION: W Malaysia.
COMMENTS: Subgenus *Arielulus*. Synonymized with *P. circumdatus* by Heller and Volleth (1984:66-68), but see Hill and Francis (1984:314-315).

Pipistrellus stenopterus (Dobson, 1875). Proc. Zool. Soc. Lond., 1875:470.
TYPE LOCALITY: Malaysia, Sarawak.
DISTRIBUTION: W Malaysia, Sumatra, Riau Arch., N Borneo, Mindanao (Philippines).

COMMENTS: Subgenus *Hypsugo*. Transferred from *Nyctalus*, see Medway (1977:56); but probably best retained in *Nyctalus*, see Koopman (1989*a*:8).

Pipistrellus sturdeei Thomas, 1915. Ann. Mag. Nat. Hist., ser. 8, 15:230.
TYPE LOCALITY: Japan, Bonin Isls, Hillsboro (= Hahajima) Isl.
DISTRIBUTION: Bonin Isls (Japan).
COMMENTS: Subgenus *Pipistrellus*.

Pipistrellus subflavus (F. Cuvier, 1832). Nouv. Ann. Mus. Hist. Nat. Paris, 1:17.
TYPE LOCALITY: USA, Georgia.
DISTRIBUTION: Nova Scotia, S Quebec (Canada), and Minnesota (USA), south to Florida (USA) and Honduras.
SYNONYMS: *clarus, erythrodactylus, floridanus, monticola, obscurus, veraecrucis*.
COMMENTS: Subgenus *Perimyotis*. Transferred by Menu (1984) to a new genus *Perimyotis*, but comparisons are clearly inadequate; see Hill and Harrison (1987). See Fujita and Kunz (1984, Mammalian Species, 228).

Pipistrellus tasmaniensis (Gould, 1858). Mamm. Aust., 3, pl. 48.
TYPE LOCALITY: Australia, Tasmania.
DISTRIBUTION: Southern Australia except South Australia; Tasmania (Australia).
SYNONYMS: *krefftii, mackenziei*.
COMMENTS: Subgenus *Falsistrellus*. Includes *mackenziei*, but see Kitchener et al. (1986).

Pipistrellus tenuis (Temminck, 1840). Monogr. Mamm., 2:229.
TYPE LOCALITY: Indonesia, Sumatra.
DISTRIBUTION: Thailand to New Guinea, Bismarck Arch., Solomon Isls, Vanuatu (= New Hebrides) and N Australia; Cocos Keeling Isl and Christmas Isl (Indian Ocean).
SYNONYMS: *adamsi, angulatus, collinus, murrayi, nitidus, orientalis, papuanus, ponceleti, sewelanus, subulidens, wattsi, westralis*.
COMMENTS: Subgenus *Pipistrellus*. For synonyms see Koopman (1973:115, 1984*c*). See also McKean and Price (1978:346) and Kitchener et al. (1986).

Plecotus E. Geoffroy, 1818. Descrip. de L'Egypte, 2:112.
TYPE SPECIES: *Vespertilio auritus* Linnaeus, 1758.
SYNONYMS: *Corynorhinus*.
COMMENTS: Includes *Corynorhinus*; see Anderson (1972:255). Does not include *Idionycteris phyllotis*; see Williams et al. (1970); but also see Handley (1959*b*). A key to the genus was published by C. Jones (1977).

Plecotus auritus (Linnaeus, 1758). Syst. Nat., 10th ed., 1:32.
TYPE LOCALITY: Sweden.
DISTRIBUTION: Norway, Ireland, and Spain to Sakhalin Isl (Russia), Japan, N China and Nepal.
SYNONYMS: *brevimanus, communis, cornutus, megalotos, homochrous, montanus, ognevi, otus, peronii, puck, sacrimontis, typus, uenoi, velatus, vulgaris*.
COMMENTS: Subgenus *Plecotus*.

Plecotus austriacus (J. Fischer, 1829). Synopsis Mamm., p. 117.
TYPE LOCALITY: Austria, Vienna.
DISTRIBUTION: England, Spain, and Senegal to Mongolia and W China; Canary (Spain) and Cape Verde Isls.
SYNONYMS: *ariel, brevipes, christiei, hispanicus, kirschbaumii, kolombatovici, kozlovi, macrobullaris, meridionalis, mordax, wardi*.
COMMENTS: Subgenus *Plecotus*. Included in *auritus* by Ellerman and Morrison-Scott (1951:181); but see Corbet (1978*c*:61) who included *teneriffae*, but see Ibáñez and Fernández (1985).

Plecotus mexicanus (G. M. Allen, 1916). Bull. Mus. Comp. Zool., 60:347.
TYPE LOCALITY: Mexico, Chihuahua, Pacheco.
DISTRIBUTION: Sonora and Coahuila to Yucatan (Mexico); Cozumel Isl (Mexico).
COMMENTS: Subgenus *Corynorhinus*. Listed as a subspecies of *townsendii* by Hall and Kelson (1959:200); but see Handley (1959*b*) and Hall (1981:233). Formerly in genus

Corynorhinus; see Anderson (1972:255). See Tumlison (1992, Mammalian Species, 401).

Plecotus rafinesquii Lesson, 1827. Manual de Mammalogie, p. 96.
TYPE LOCALITY: USA, Illinois, Wabash Co., Mt. Carmel.
DISTRIBUTION: SE USA from Virginia to Missouri, south to E Texas and Florida.
SYNONYMS: *leconteii, macrotis.*
COMMENTS: Subgenus *Corynorhinus*. Formerly in genus *Corynorhinus*; see Anderson (1972:255). See C. Jones (1977, Mammalian Species, 69).

Plecotus taivanus Yoshiyuki, 1991. Bull. Natl. Sci. Mus. Tokyo, ser. A(Zool.), 17:189.
TYPE LOCALITY: Taiwan, Taichung Hsien, Hoping Hsiang, Mt. Anma Shan, 2250 m.
DISTRIBUTION: Taiwan.
COMMENTS: Most like *homochrous* and *puck*, here included in *P. auritus*.

Plecotus teneriffae Barrett-Hamilton, 1907. Ann. Mag. Nat. Hist., ser. 7, 20:520.
TYPE LOCALITY: Spain, Canary Isls, Teneriffe Isl.
DISTRIBUTION: Canary Isls (Spain).
COMMENTS: Synonymizied with *A. austriacus* by Corbet (1978*c*), but see Ibáñez and Fernández (1985).

Plecotus townsendii Cooper, 1837. Ann. Lyc. Nat. Hist. N.Y., 4:73.
TYPE LOCALITY: USA, Washington, Clark Co., Fort Vancouver.
DISTRIBUTION: S British Columbia (Canada) through W USA to Oaxaca (Mexico), east to Virginia.
STATUS: U.S. ESA - Endangered as *P. t. ingens* and *P. t. virginianus.*
SYNONYMS: *australis, ingens, intermedius, pallescens, virginianus.*
COMMENTS: Subgenus *Corynorhinus*. Formerly in genus *Corynorhinus*; see Anderson (1972:255). See Kunz and Martin (1982, Mammalian Species, 175).

Rhogeessa H. Allen, 1866. Proc. Acad. Nat. Sci. Philadelphia, 18:285.
TYPE SPECIES: *Rhogessa tumida* H. Allen, 1866.
SYNONYMS: *Baeodon.*
COMMENTS: Includes *Baeodon*; see Jones et al. (1977:26). Revised by LaVal (1973*b*).

Rhogeessa alleni Thomas, 1892. Ann. Mag. Nat. Hist., ser. 6, 10:477.
TYPE LOCALITY: Mexico, Jalisco, near Autlan, Santa Rosalia.
DISTRIBUTION: Oaxaca to Zacatecas (Mexico).
COMMENTS: Subgenus *Baeodon.*

Rhogeessa genowaysi Baker, 1984. Syst. Zool., 33:178.
TYPE LOCALITY: Mexico, Chiapas, 23.6 mi. NW Huixtla.
DISTRIBUTION: Pacific lowlands of S Chiapas (Mexico).
COMMENTS: Subgenus *Rhogeessa*. Morphologically inseparable from *tumida*, but with distinctive karyotype, see Baker (1984).

Rhogeessa gracilis Miller, 1897. N. Am. Fauna, 13:126.
TYPE LOCALITY: Mexico, Puebla, Piaxtla, 1,100 m.
DISTRIBUTION: Jalisco and Zacatecas to Oaxaca (Mexico).
COMMENTS: Subgenus *Rhogeessa*. See J. K. Jones, Jr. (1977, Mammalian Species, 76).

Rhogeessa minutilla Miller, 1897. Proc. Biol. Soc. Washington, 11:139.
TYPE LOCALITY: Venezuela, Margarita Isl.
DISTRIBUTION: Colombia, Venezuela (including Margarita Isl).
COMMENTS: Subgenus *Rhogeessa*. Listed as a subspecies of *parvula* by Cabrera (1958:111); but see LaVal (1973*b*:37-38).

Rhogeessa mira LaVal, 1973. Occas. Pap. Mus. Nat. Hist. Univ. Kansas, 19:26.
TYPE LOCALITY: Mexico, Michoacan, 20 km N El Infernillo.
DISTRIBUTION: S Michoacan (Mexico).
COMMENTS: Subgenus *Rhogeessa.*

Rhogeessa parvula H. Allen, 1866. Proc. Acad. Nat. Sci. Philadelphia, 18:285.
TYPE LOCALITY: Mexico, Nayarit, Tres Marías Isls.
DISTRIBUTION: Oaxaca to Sonora (Mexico); Tres Marías Isls (Mexico).

SYNONYMS: *major*.
COMMENTS: Subgenus *Rhogeessa*. For scope of this species, see LaVal (1973*b*).

Rhogeessa tumida H. Allen, 1866. Proc. Acad. Nat. Sci. Philadelphia, 18:286.
TYPE LOCALITY: Mexico, Veracruz, Mirador.
DISTRIBUTION: Tamaulipas (Mexico) to Ecuador, Bolivia, and NE Brazil; Trinidad and Tobago.
SYNONYMS: *aeneus, bombyx, io, riparia, velilla*.
COMMENTS: Subgenus *Rhogeessa*. Listed as a subspecies of *parvula* by Hall and Kelson (1959:196); but also see LaVal (1973*b*) and Hall (1981:228), who included *bombyx* and *velilla*.

Scotoecus Thomas, 1901. Ann. Mag. Nat. Hist., ser. 7, 7:263.
TYPE SPECIES: *Scotophilus albofuscus* Thomas, 1890.
COMMENTS: Considered a subgenus of *Nycticeius* by Hayman and Hill (1971:36); but see J. E. Hill (1974*b*), who revised the genus.

Scotoecus albofuscus (Thomas, 1890). Ann. Mus. Civ. Stor. Nat. Genova, 29:84.
TYPE LOCALITY: Gambia, Bathurst.
DISTRIBUTION: Gambia to Kenya and Mozambique.
SYNONYMS: *woodi*.

Scotoecus hirundo (de Winton, 1899). Ann. Mag. Nat. Hist., ser. 7, 4:355.
TYPE LOCALITY: Ghana, Gambaga.
DISTRIBUTION: Senegal to Ethiopia, south to Angola, Zambia and Malawi.
SYNONYMS: *albigula, artinii, falabae, hindei*.
COMMENTS: Includes *hindei*; see Hayman and Hill (1971:36) and Robbins (1980:84); but see also J. E. Hill (1974*b*).

Scotoecus pallidus (Dobson, 1876). Monogr. Asiat. Chiroptera, App. D:186.
TYPE LOCALITY: Pakistan, Punjab, Lahore, Mian Mir.
DISTRIBUTION: Pakistan, N India.
COMMENTS: Included in *Nycticeius* by Ellerman and Morrison-Scott (1951:177); but see J. E. Hill (1974*b*).

Scotomanes Dobson, 1875. Proc. Zool. Soc. Lond., 1875:371.
TYPE SPECIES: *Nycticejus ornatus* Blyth, 1851.
SYNONYMS: *Scoteinus*.
COMMENTS: Includes *Scoteinus*; see Sinha and Chakraborty (1971).

Scotomanes emarginatus (Dobson, 1871). Proc. Asiat. Soc. Bengal, p. 211.
TYPE LOCALITY: "India."
DISTRIBUTION: India.
COMMENTS: Included in *Nycticeius* by Ellerman and Morrison-Scott (1951:177); but see J. E. Hill (1974*b*). Probably a synonym of *S. ornatus*.

Scotomanes ornatus (Blyth, 1851). J. Asiat. Soc. Bengal, 20:511.
TYPE LOCALITY: India, Assam, Khasi Hills, Cherrapunji.
DISTRIBUTION: NE India (including Sikkim), Burma, S China, Thailand, Vietnam.
SYNONYMS: *imbrensis, nivicolus, sinensis*.

Scotophilus Leach, 1821. Trans. Linn. Soc. Lond., 13:69, 71.
TYPE SPECIES: *Scotophilus kuhlii* Leach, 1821.
SYNONYMS: *Pachyotus*.
COMMENTS: Includes *Pachyotus*; see Walker et al. (1975:359). African species revised by Koopman (1984*b*) and by C. B. Robbins et al. (1985).

Scotophilus borbonicus (E. Geoffroy, 1803). Ann. Mus. Hist. Nat. Paris, 8:201.
TYPE LOCALITY: Reunion Isl (France).
DISTRIBUTION: Madagascar, Reunion Isl (Mascarene Isls). Records from Mauritius (Mascarene Isls) are erroneous, see Cheke and Dahl (1981).
STATUS: May be extinct; see Cheke and Dahl (1981:217).

COMMENTS: Included equivocally in *leucogaster* by Hayman and Hill (1971:50-51); but see Koopman (1975:414-416). Hill (1980*b*) considered African *viridis* and *damarensis*, and possibly *leucogaster* to be conspecific with *borbonicus*, but C. B. Robbins et al. (1985) rejected any affinity with African mainland species. Also see comment under *leucogaster*. Koopman (1986), included *viridis*, *damarensis*, and *nigritellus* (but not *leucogaster*) in this species.

Scotophilus celebensis Sody, 1928. Natuurk. Tijdschr. Ned.-Ind., 88:90.
TYPE LOCALITY: Indonesia, Sulawesi, Toli Toli.
DISTRIBUTION: Sulawesi.
COMMENTS: Probably a subspecies of *S. heathi*, see Sinha (1980).

Scotophilus dinganii (A. Smith, 1833). S. Afr. Quart. J., 2:59.
TYPE LOCALITY: South Africa, Port Natal (= Durban).
DISTRIBUTION: Senegal and Sierra Leone east to Somalia and S Yemen and south to South Africa and Namibia.
SYNONYMS: *colias, herero, planirostris, pondoensis* (see C. B. Robbins et al., 1985).
COMMENTS: Placed in *leucogaster* by Koopman (1975:414-416, as *nigrita*; 1986) and Koopman et al. (1978:4-5); but see also Schlitter et al. (1980).

Scotophilus heathi (Horsfield, 1831). Proc. Zool. Soc. Lond., 1831:113.
TYPE LOCALITY: India, Madras.
DISTRIBUTION: Afghanistan to S China, including Hainan Isl, south to Sri Lanka and Vietnam.
SYNONYMS: *belangeri, flaveolus, insularis, luteus, watkinsi.*

Scotophilus kuhlii Leach, 1821. Trans. Linn. Soc. Lond., 13:71.
TYPE LOCALITY: "India".
DISTRIBUTION: Pakistan to Taiwan, south to Sri Lanka and W Malaysia, southeast to Philippines and Aru Isls (Indonesia).
SYNONYMS: *castaneus, collinus, consobrinus, gairdneri, panayensis, solutatus, swinhoei, temmincki, wroughtoni.*
COMMENTS: Generally called *temmincki*; but see Hill and Thonglongya (1972:191-193).

Scotophilus leucogaster (Cretzschmar, 1830). *In* Rüppell, Atlas Reise Nordl. Afr., Zool., Säugeth., p. 71.
TYPE LOCALITY: Sudan, Kordofan, Brunnen Nedger (Nedger Well).
DISTRIBUTION: Mauritania and Senegal to N Kenya and Ethiopia.
SYNONYMS: *altilis, flavigaster, murinoflavus, nucella.*
COMMENTS: Does not include *damarensis*, see Koopman (1986); but see C. B. Robbins et al. (1985). Includes *nucella*; but see C. B. Robbins et al. (1985). For scope of this species see Koopman (1975:414-416; as *nigrita*) and Koopman et al. (1978:4-5). However, Robbins (1978:212-213) has shown that the name *nigrita* was misapplied to another species, for which the next available name was *dinganii*, and which would be included in *leucogaster* if Koopman's (1975) reasoning were followed. However, Schlitter et al. (1980) listed *dinganii* as distinct from *leucogaster*, as did C. B. Robbins et al. (1985). See Koopman (1986).

Scotophilus nigrita (Schreber, 1774). Die Säugethiere, 1:171.
TYPE LOCALITY: Senegal.
DISTRIBUTION: Senegal to Sudan and Kenya to Mozambique.
SYNONYMS: *alvenslebeni, gigas.*
COMMENTS: For a review of this species see Robbins (1978). The identity of this species is clear and *nigrita* is the senior synonym of *gigas*. Specimens previously called *nigrita* should be called *dinganii* (see Robbins, 1978:212-213).

Scotophilus nux Thomas, 1904. Ann. Mag. Nat. Hist., ser. 7, 13:208.
TYPE LOCALITY: Cameroon, Efulen.
DISTRIBUTION: Sierra Leone to Kenya.
COMMENTS: *S. nux* has been most often recognized as a subspecies of *nigrita*, when *nigrita* was used for the species now called *dinganii*; see Allen (1939), Rosevear (1965), and Hayman and Hill (1971:50). Koopman et al. (1978:4-5) listed it as a subspecies of *leucogaster*.

Scotophilus robustus Milne-Edwards, 1881. C. R. Acad. Sci. Paris, 91:1035.
TYPE LOCALITY: Madagascar.
DISTRIBUTION: Madagascar.
COMMENTS: Recognized as a subspecies of *nigrita* (when *nigrita* was used for the species now called *dinganii*) by Hayman and Hill (1971:50). However, C. B. Robbins et al. (1985) considered it specifically distinct from *dinganii* and *borbonicus*.

Scotophilus viridis (Peters, 1852). Reise nach Mossambique, Säugethiere, p. 67.
TYPE LOCALITY: Mozambique, Mozambique Isl, 15°S.
DISTRIBUTION: Senegal to Ethiopia south to Namibia and South Africa.
SYNONYMS: *damarensis, nigritellus.*
COMMENTS: Included in *leucogaster* by Hayman and Hill (1971:50), but see Koopman (1975:414-416) and Schlitter et al. (1980). Includes *nigritellus*, see C. B. Robbins et al. (1985). Includes *damarensis*, but see C. B. Robbins et al. (1985).

Tylonycteris Peters, 1872. Monatsb. K. Preuss. Akad. Wiss. Berlin, 1872:703.
TYPE SPECIES: *Vespertilio pachypus* Temminck, 1840.

Tylonycteris pachypus (Temminck, 1840). Monogr. Mamm., 2:217.
TYPE LOCALITY: Indonesia, Java, Bantam.
DISTRIBUTION: India and S China to Philippines and Lesser Sunda Isls; Andaman Isls (India).
SYNONYMS: *aurex, bhaktii, fulvidus, meyeri.*

Tylonycteris robustula Thomas, 1915. Ann. Mag. Nat. Hist., ser. 8, 15:227.
TYPE LOCALITY: Malaysia, Sarawak, Upper Sarawak.
DISTRIBUTION: S China to Philippines, Sulawesi and Lesser Sunda Isls.
SYNONYMS: *malayana.*
COMMENTS: Includes *malayana*; see Medway (1969:37).

Vespertilio Linnaeus, 1758. Syst. Nat., 10th ed., 1:31.
TYPE SPECIES: *Vespertilio murinus* Linnaeus, 1758.

Vespertilio murinus Linnaeus, 1758. Syst. Nat., 10th ed., 1:32.
TYPE LOCALITY: Sweden.
DISTRIBUTION: Norway and Britain to Ussuri region (Russia), China, and Afghanistan.
SYNONYMS: *albogularis, discolor, kraschenninnikovi, luteus, michnoi, siculus, ussuriensis.*

Vespertilio superans Thomas, 1899. Proc. Zool. Soc. Lond., 1898:770 [1899].
TYPE LOCALITY: China, Hupeh, Ichang, Sesalin.
DISTRIBUTION: China, Ussuri region (Russia), Korea, Japan, Taiwan.
SYNONYMS: *anderssoni, aurijunctus, motoyoshii, namiyei, orientalis* (see Corbet, 1978*c*:58, and Yoshiyuki, 1989).

Subfamily Murininae Miller, 1907. Bull. U.S. Natl. Mus., 57:229.

Harpiocephalus Gray, 1842. Ann. Mag. Nat. Hist., [ser. 1], 10:259.
TYPE SPECIES: *Harpiocephalus rufus* Gray, 1842 (= *Vespertilio harpia* Temminck, 1840).

Harpiocephalus harpia (Temminck, 1840). Monogr. Mamm., 2:219.
TYPE LOCALITY: Indonesia, Java, Mt. Gede.
DISTRIBUTION: India to Taiwan and Vietnam, south to Molucca Isls, Java, and Lesser Sunda Isls.
SYNONYMS: *lasyurus, madrassius, mordax, pearsonii, rufulus, rufus.*
COMMENTS: Includes *mordax*, which may be a separate species; see Hill and Francis (1984).

Murina Gray, 1842. Ann. Mag. Nat. Hist., [ser. 1], 10:258.
TYPE SPECIES: *Vespertilio suillus* Temminck, 1840.
SYNONYMS: *Harpiola.*
COMMENTS: Includes *Harpiola*; see Corbet and Hill (1980:75).

Murina aenea Hill, 1964. Fed. Mus. J., Kuala Lumpur, N.S., 8:57.
TYPE LOCALITY: Malaysia, Pahang, Bentong Dist., near Janda Baik, Ulu Chemperoh.
DISTRIBUTION: W Malaysia, Borneo.

COMMENTS: Subgenus *Murina*.

Murina aurata Milne-Edwards, 1872. Rech. Hist. Nat. Mammifères, p. 250.
TYPE LOCALITY: China, Szechwan, Moupin.
DISTRIBUTION: Nepal to SW China and Burma.
SYNONYMS: *feae*.
COMMENTS: Subgenus *Murina*. Formerly included *ussuriensis*; see Maeda (1980:547); see also comments under *silvatica*.

Murina cyclotis Dobson, 1872. Proc. Asiat. Soc. Bengal, p. 210.
TYPE LOCALITY: India, Darjeeling.
DISTRIBUTION: Sri Lanka and India to Kwangtung and Hainan (China); Vietnam, south to W Malaysia; Borneo; Philippines; Lesser Sunda Isls.
SYNONYMS: *eileenae, peninsularis*.
COMMENTS: Subgenus *Murina*.

Murina florium Thomas, 1908. Ann. Mag. Nat. Hist., ser. 8, 2:371.
TYPE LOCALITY: Indonesia, Lesser Sunda Isls, Flores.
DISTRIBUTION: Lesser Sunda Isls, Sulawesi, New Guinea, NE Australia.
SYNONYMS: *lanosa, toxopei*.
COMMENTS: Subgenus *Murina*.

Murina fusca Sowerby, 1922. J. Mammal., 3:46.
TYPE LOCALITY: China, Manchuria, Kirin, Imienpo area.
DISTRIBUTION: Manchuria (China).
COMMENTS: Subgenus *Murina*. Listed as a subspecies of *leucogaster* by Ellerman and Morrison-Scott (1951:185) and Corbet (1978c:62); but see Wallin (1969:355).

Murina grisea Peters, 1872. Monatsb. K. Preuss. Akad. Wiss. Berlin, 1872:258.
TYPE LOCALITY: India, Uttar Pradesh, Dehra Dun, Mussooree, Jeripanee, 5,500 ft. (1,676 m).
DISTRIBUTION: NW Himalayas.
COMMENTS: Subgenus *Harpiola*.

Murina huttoni (Peters, 1872). Monatsb. K. Preuss. Akad. Wiss. Berlin, 1872:257.
TYPE LOCALITY: India, Uttar Pradesh, Dehra Dun.
DISTRIBUTION: Tibet, China (record may be doubtful); NW India to Vietnam; Fukien (China); Thailand; W Malaysia.
SYNONYMS: *rubella*.
COMMENTS: Subgenus *Murina*.

Murina leucogaster Milne-Edwards, 1872. Rech. Hist. Nat. Mammifères, p. 252.
TYPE LOCALITY: China, Szechwan, Moupin.
DISTRIBUTION: E Himalayas; Shensi, Szechwan, Fukien, and Inner Mongolia (China); Upper Yenisei River (Russia); Altai Mtns (Russia, Kazakhstan and Mongolia); Korea; Ussuri region (Russia); Sakhalin Isl (Russia); Honshu, Kyushu, and Shikiku (Japan).
SYNONYMS: *hilgendorfi, intermedia, ognevi, rubex, sibirica*.
COMMENTS: Subgenus *Murina*. Includes *hilgendorfi*, but see Yoshiyuki (1989).

Murina puta Kishida, 1924. Zool. Mag. (Tokyo), 36:130-133.
TYPE LOCALITY: Taiwan, Chang Hua, Erh-Shui.
DISTRIBUTION: Taiwan.
COMMENTS: Subgenus *Murina*. Closely related to and possibly conspecific with *M. huttoni*, see Yoshiyuki (1989:215).

Murina rozendaali Hill and Francis, 1984. Bull. Brit. Mus. Nat. Hist. (Zool.), 47:319.
TYPE LOCALITY: Borneo, Sabah, Gomantong.
DISTRIBUTION: Borneo.
COMMENTS: Subgenus *Murina*.

Murina silvatica Yoshiyuki, 1983. Bull. Natl. Sci. Mus.(Tokyo), ser. A(Zool.), 9:141.
TYPE LOCALITY: Japan, Honshu, Fukushima, Minamiaiau-Gug, Hinoemata-Mura, Ozenuma Lake.
DISTRIBUTION: Japan, including Tsushima Isls.
COMMENTS: Subgenus *Murina*. Includes specimens formerly included in *M. aurata* or *M. ussuriensis*.

Murina suilla (Temminck, 1840). Monogr. Mamm., 2:224.
TYPE LOCALITY: Indonesia, Java.
DISTRIBUTION: Java, Sumatra, Borneo, W Malaysia, nearby small islands.
SYNONYMS: *balstoni, canescens.*
COMMENTS: Subgenus *Murina*. Includes *balstoni* and *canescens*, see Koopman (1989*a*).

Murina tenebrosa Yoshiyuki, 1970. Bull. Natl. Sci. Mus. Tokyo, 13:195.
TYPE LOCALITY: Japan, Tsushima Isls, Kamishima Isl, Sago.
DISTRIBUTION: Tsushima Isls (Japan), perhaps Yakushima (Ryukyu Isls, Japan).
COMMENTS: Subgenus *Murina*.

Murina tubinaris (Scully, 1881). Proc. Zool. Soc. Lond., 1881:200.
TYPE LOCALITY: Pakistan, Gilgit.
DISTRIBUTION: Pakistan, N India, Burma, Thailand, Laos, Vietnam.
COMMENTS: Subgenus *Murina*. Listed as a subspecies of *huttoni* by Ellerman and Morrison-Scott (1951:186); but see Hill (1963*a*:49, 50).

Murina ussuriensis Ognev, 1913. Ann. Mus. Zool. Acad. Imp. Sci. St. Petersbourg, 18:402.
TYPE LOCALITY: Russia, Primorsky Krai, Kreis Imansky.
DISTRIBUTION: Ussuri region, Kurile Isls, and Sakhalin (Russia); Korea.
COMMENTS: Subgenus *Murina*. Formerly included in *aurata* see Maeda (1980:547) and Corbet (1978*c*). Japanese populations have been separated as *M. silvatica*.

Subfamily Miniopterinae Dobson, 1875. Ann. Mag. Nat. Hist., ser. 4, 16:349.

Miniopterus Bonaparte, 1837. Fauna Ital., 1, fasc. 20.
TYPE SPECIES: *Vespertilio schreibersii* Kuhl, 1817.
COMMENTS: Reviewed by (in part) Peterson (1981). See Goodwin (1979:119), Maeda (1982), and Hill (1983). All have come to very different conclusions.

Miniopterus australis Tomes, 1858. Proc. Zool. Soc. Lond., 1858:125.
TYPE LOCALITY: New Caledonia, Loyalty Isls, Lifu (21°S, 167.3°E) (France).
DISTRIBUTION: Philippines, Borneo, and Java, southeast to Vanuatu (= New Hebrides) and E Australia.
SYNONYMS: *paululus, shortridgei, solomonensis, tibialis, witkampi* (These come from a variety of sources, some of which are probably erroneous).
COMMENTS: Two species may be included in this complex; see Koopman (1989*a*:9).

Miniopterus fraterculus Thomas and Schwann, 1906. Proc. Zool. Soc. Lond., 1906:162.
TYPE LOCALITY: South Africa, Cape Province, Knysna.
DISTRIBUTION: Cape Province, Natal, and Transvaal (South Africa), Malawi, Zambia, Angola, Mozambique, Madagascar.

Miniopterus fuscus Bonhote, 1902. Novit. Zool., 9:626.
TYPE LOCALITY: Ryukyu Isls, Okinawa.
DISTRIBUTION: Ryukyu Isls (Japan), SE China, Thailand, W Malaysia, Borneo, Java, Sulawesi, Philippines, New Guinea.
SYNONYMS: *medius, yayeyamae.*
COMMENTS: Called *medius* by Hill (1983), but *fuscus* is the older name.

Miniopterus inflatus Thomas, 1903. Ann. Mag. Nat. Hist., ser. 7, 12:634.
TYPE LOCALITY: Cameroon, Efulen.
DISTRIBUTION: Ethiopia, Somalia, Kenya, Uganda, Burundi, E and S Zaire, Cameroon, Gabon, Mozambique, Liberia, Madagascar, perhaps Nigeria. W African distribution uncertain because of confusion with *schreibersi*.
SYNONYMS: *africanus, rufus.*

Miniopterus magnater Sanborn, 1931. Field Mus. Nat. Hist. Publ., Zool. Ser., 18:26.
TYPE LOCALITY: Papua New Guinea, E Sepik, Marienberg.
DISTRIBUTION: NE India and SE China to Timor (Indonesia) and New Guinea.
SYNONYMS: *macrodens.*

Miniopterus minor Peters, 1867. Monatsb. K. Preuss. Akad. Wiss. Berlin, 1866:885 [1867].
TYPE LOCALITY: Tanzania, coast opposite Zanzibar Isl.

DISTRIBUTION: Kenya, Tanzania, Zaire, Congo Republic, Madagascar, São Tomé Isl, Comoro Isls.
SYNONYMS: *grivaudi, manavi, newtoni.*

Miniopterus pusillus Dobson, 1876. Monogr. Asiat. Chiroptera, p. 162.
TYPE LOCALITY: India, Nicobar Islands (NW of Sumatra).
DISTRIBUTION: India to the Philippines and Moluccas; Solomons to New Caledonia; perhaps New Guinea.
SYNONYMS: *macrocneme.*
COMMENTS: See Hill (1983) for content.

Miniopterus robustior Revilliod, 1914. *In* Sarasin and Roux, Nova Caledonia, A. Zool., 1:359.
TYPE LOCALITY: New Caledonia, Loyalty Isls, Lifu Isl, Quepenee (France).
DISTRIBUTION: Loyalty Isls (E of New Caledonia).
COMMENTS: See Hill (1971*a*:579) for the status of this species.

Miniopterus schreibersi (Kuhl, 1817). Die Deutschen Fledermause, Hanau, p. 14.
TYPE LOCALITY: Rumania, Banat, near Coronini, Kolumbacs Cave.
DISTRIBUTION: S Europe and Morocco through the Caucasus and Iran to most of China and Japan; most of Indo-Malayan region; New Guinea; Solomon Isls (including Bougainville Isl); Australia; subsaharan Africa; Madagascar; Bismarck Arch.
SYNONYMS: *arenarius, baussencis, blepotis, breyeri, chinensis, dasythrix, eschscholtzii, fuliginosus, harardai, inexpectatus, italicus, japoniae, majori, natalensis, oceanensis, orianae, orsinii, pallidus, parvipes, pulcher, ravus, scotinus, smitianus, ursinii, vicinior, villiersi.*
COMMENTS: Formerly included *magnater.* Reviewed by Crucitti (1976).

Miniopterus tristis (Waterhouse, 1845). Proc. Zool. Soc. Lond., 1845:3.
TYPE LOCALITY: Philippine Isls.
DISTRIBUTION: Philippines, Sulawesi, New Guinea; Bismarck Arch., Solomon Isls, Vanuatu (= New Hebrides).
SYNONYMS: *bismarckensis, celebensis, grandis, insularis, melanesiensis, propritristis.*
COMMENTS: Includes *propritristis;* see Koopman (1984*c*).

Subfamily Tomopeatinae Miller, 1907. Bull. U.S. Natl. Mus., 57:237.

Tomopeas Miller, 1900. Ann. Mag. Nat. Hist., ser. 7, 6:570.
TYPE SPECIES: *Tomopeas ravus* Miller, 1900.

Tomopeas ravus Miller, 1900. Ann. Mag. Nat. Hist., ser. 7, 6:571.
TYPE LOCALITY: Peru, Cajamarca, Yayan, 1,000 m.
DISTRIBUTION: W Peru.

Family Mystacinidae Dobson, 1875. Ann. Mag. Nat. Hist., ser. 4, 16:349.

Mystacina Gray, 1843. *In* Dieffenbach, Travels in New Zealand, 2:296.
TYPE SPECIES: *Mystacina tuberculata* Gray, 1843.
COMMENTS: Revised by Hill and Daniel (1985).

Mystacina robusta Dwyer, 1962 Zool. Publ. Victoria Univ., Wellington, 28:3.
TYPE LOCALITY: New Zealand, Big South Cape Isl.
DISTRIBUTION: Vicinity of type locality only, though originally widely distributed in New Zealand.
STATUS: Probably extinct.

Mystacina tuberculata Gray, 1843. Mammalia, *in* Voy. "Sulphur," Zool., p. 23.
TYPE LOCALITY: New Zealand.
DISTRIBUTION: New Zealand.
SYNONYMS: *aupourica, rhyacobia, velutina.*

Family Molossidae Gervais, 1856. *In* Comte de Castelnau, Exped. Partes Cen. Am. Sud., Zool.(Sec. 7), Vol. 1, pt. 2(Mammifères):53 footnote.
COMMENTS: Revised by Freeman (1981). For a very different arrangement see Legendre (1984).

Chaerephon Dobson, 1874. J. Asiat. Soc. Bengal, 43:144.
TYPE SPECIES: *Molossus (Nyctinomus) johorensis* Dobson, 1873.
COMMENTS: Formerly included in *Tadarida*; see Freeman (1981:60, 133, 150); see also Legendre (1984).

Chaerephon aloysiisabaudiae (Festa, 1907). Bol. Mus. Zool. Anat. Comp. Univ. Torino, 22(546):1.
TYPE LOCALITY: Uganda, Toro.
DISTRIBUTION: Ghana, Gabon, Zaire, Uganda, perhaps Ethiopia.
SYNONYMS: *cyclotis*.

Chaerephon ansorgei (Thomas, 1913). Ann. Mag. Nat. Hist., ser. 8, 11:318.
TYPE LOCALITY: Angola, Malange.
DISTRIBUTION: Cameroon to Ethiopia, south to Angola and Natal (South Africa).
SYNONYMS: *rhodesiae*.
COMMENTS: Distinct from *bivittata*; see Eger and Peterson (1979:1889), who revised it.

Chaerephon bemmeleni (Jentink, 1879). Notes Leyden Mus., 1:125.
TYPE LOCALITY: "Liberia", but see Kuhn (1965).
DISTRIBUTION: Liberia, Cameroon, Sudan, Zaire, Uganda, Kenya, Tanzania.
SYNONYMS: *cistura*.
COMMENTS: Includes *cistura*; see Koopman (1975:425).

Chaerephon bivittata (Heuglin, 1861). Nova Acta Acad. Caes. Leop.-Carol., Halle, 29(8):413.
TYPE LOCALITY: Ethiopia, Eritrea, Keren.
DISTRIBUTION: Sudan, Ethiopia, Uganda, Kenya, Tanzania, Zambia, Zimbabwe, Mozambique.
COMMENTS: Revised by Eger and Peterson (1979).

Chaerephon chapini J. A. Allen, 1917. Bull. Am. Mus. Nat. Hist., 37:461.
TYPE LOCALITY: Zaire, Oriental, Faradje.
DISTRIBUTION: Ethiopia, Zaire, Uganda, Angola, Namibia, Botswana, Zimbabwe, Zambia.
SYNONYMS: *lancasteri, shortridgei*.

Chaerephon gallagheri (Harrison, 1975). Mammalia, 39:313.
TYPE LOCALITY: Zaire, Kivu, 30 km SW Kindu, Scierie Forest (3°10'S and 25°49'E).
DISTRIBUTION: Zaire.

Chaerephon jobensis (Miller, 1902). Proc. Biol. Soc. Washington, 15:246.
TYPE LOCALITY: Indonesia, Irian Jaya, Tjenderawasih Div., Japen Isl, Ansus.
DISTRIBUTION: New Guinea, N and C Australia, Solomon Isls, Vanuatu (= New Hebrides), Fiji, perhaps Bismarck Arch.
SYNONYMS: *bregullae, colonicus, solomonis*.
COMMENTS: Listed as a subspecies of *plicata* by Laurie and Hill (1954:63); but see Hill (1961*b*:54-55), who included *solomonis* in this species. Revised by Felten (1964*a*).

Chaerephon johorensis (Dobson, 1873). Proc. Asiat. Soc. Bengal, p. 22.
TYPE LOCALITY: Malaysia, Johore.
DISTRIBUTION: W Malaysia, Sumatra.

Chaerephon major (Trouessart, 1897). Cat. Mamm. Viv. Foss., 1:146.
TYPE LOCALITY: N Sudan, 5th Cataract of the Nile.
DISTRIBUTION: Liberia, Mali, Burkina Faso, Ghana, Togo, Nigeria, Niger, Sudan, NE Zaire, Uganda, Tanzania.
SYNONYMS: *abae, emini*.

Chaerephon nigeriae Thomas, 1913. Ann. Mag. Nat. Hist., ser. 8, 11:319.
TYPE LOCALITY: Nigeria, Northern Region, Zaria Province.

DISTRIBUTION: Ghana and Niger to Saudi Arabia (record by Nader and Kock, 1979) and Ethiopia, south to Namibia, Botswana and Zimbabwe.
SYNONYMS: *spillmani*.

Chaerephon plicata (Buchanan, 1800). Trans. Linn. Soc. Lond., 5:261.
TYPE LOCALITY: India, Bengal.
DISTRIBUTION: India and Sri Lanka to S China and Vietnam, southeast to Philippines, Borneo and Lesser Sunda Isls; Hainan (China); Cocos Keeling Isl (Indian Ocean).
SYNONYMS: *adustus, bengalensis, dilatatus, insularis, luzonus, murinus, tenuis*.
COMMENTS: Includes *luzonus*; see Hill (1961*b*:52).

Chaerephon pumila (Cretzschmar, 1830). *In* Rüppell, Atlas Reise Nordl. Afr., Zool., Säugeth., p. 69.
TYPE LOCALITY: Eritrea, Massawa.
DISTRIBUTION: Senegal to Yemen, south to South Africa; Bioko; Pemba and Zanzibar; Comoro Isls; Aldabra and Amirante Isls (Seychelles); Madagascar.
SYNONYMS: *cristatus, elphicki, faini, frater, gambianus, hindei, langi, leucogaster, limbata, naivashae, nigri, pusillus, websteri*.
COMMENTS: Includes *pusillus*; see Hayman and Hill (1971:64).

Chaerephon russata J. A. Allen, 1917. Bull. Am. Mus. Nat. Hist., 37:458.
TYPE LOCALITY: Zaire, Oriental, Medje.
DISTRIBUTION: Ghana, Cameroon, Zaire, Kenya.

Cheiromeles Horsfield, 1824. Zool. Res. Java, part 8:*Cheiromeles torquatus*, pl. and 10 unno. pp.
TYPE SPECIES: *Cheiromeles torquatus* Horsfield, 1824.

Cheiromeles torquatus Horsfield, 1824. Zool. Res. Java, part 8:*Cheiromeles torquatus*, pl. and 10 unno. pp.
TYPE LOCALITY: Malaysia, Penang.
DISTRIBUTION: W Malaysia, Sumatra and Java, Borneo, SW Philippines, Sulawesi and nearby small islands.
SYNONYMS: *caudatus, cheiropus, jacobsoni, parvidens*.
COMMENTS: Includes *parvidens*; see Koopman (1989*a*).

Eumops Miller, 1906. Proc. Biol. Soc. Washington, 19:85.
TYPE SPECIES: *Molossus californicus* Merriam, 1890 (= *Molossus perotis* Schinz, 1821).
COMMENTS: Revised by Eger (1977).

Eumops auripendulus (Shaw, 1800). Gen. Zool. Syst. Nat. Hist., 1(1):137.
TYPE LOCALITY: French Guiana.
DISTRIBUTION: Oaxaca and Yucatan (Mexico) to Peru, N Argentina, E Brazil, and Trinidad.
SYNONYMS: *abrasus* Allen (not Temminck), *amplexicaudatus, barbatus, leucopleura, longimanus, major, milleri, oaxacensis*.
COMMENTS: Called *abrasus* in Hall and Kelson (1959:211); but see Husson (1962:243-246) and Hall (1981:248).

Eumops bonariensis (Peters, 1874). Monatsb. K. Preuss. Akad. Wiss. Berlin, 1874:232.
TYPE LOCALITY: Argentina, Buenos Aires.
DISTRIBUTION: Veracruz (Mexico) to NW Peru, N Argentina, Uruguay, and Brazil. The Patagonian record is probably accidental or erroneous, see Cabrera (1958:125).
SYNONYMS: *beckeri, delticus, nanus, patagonicus*.

Eumops dabbenei Thomas, 1914. Ann. Mag. Nat. Hist., ser. 8, 13:481.
TYPE LOCALITY: Argentina, Chaco.
DISTRIBUTION: Colombia, Venezuela, Paraguay, N Argentina.
SYNONYMS: *mederai*.

Eumops glaucinus (Wagner, 1843). Arch. Naturgesch., 9(1):368.
TYPE LOCALITY: Brazil, Mato Grosso, Cuiaba.
DISTRIBUTION: Jalisco (Mexico) to Peru, N Argentina and Brazil; Jamaica; Cuba; Florida (USA).
SYNONYMS: *ferox, floridanus, orthotis*.
COMMENTS: Includes *floridanus*; see Eger (1977:43).

Eumops hansae Sanborn, 1932. J. Mammal., 13:356.
TYPE LOCALITY: Brazil, Santa Catarina, Joinville, Colonia Hansa.
DISTRIBUTION: Costa Rica, Panama, Venezuela, Guyana, French Guiana, Peru, Bolivia, Brazil.
SYNONYMS: *amazonicus.*
COMMENTS: Includes *amazonicus;* see Gardner et al. (1970:727) and Eger (1977:44).

Eumops maurus (Thomas, 1901). Ann. Mag. Nat. Hist., ser. 7, 7:141.
TYPE LOCALITY: Guyana, Kanuku Mtns.
DISTRIBUTION: Guyana, Surinam.
SYNONYMS: *geijskesi.*
COMMENTS: Includes *geijskesi;* see Eger (1977:46-48).

Eumops perotis (Schinz, 1821). *In* Cuvier, Das Thierreich, 1:870.
TYPE LOCALITY: Brazil, Rio de Janiero, Campos do Goita Cazes, Villa São Salvador.
DISTRIBUTION: California and Texas (USA) to Zacatecas and Hidalgo (Mexico); Colombia to N Argentina and E Brazil; Cuba.
SYNONYMS: *californicus, gigas, renatae, trumbulli.*
COMMENTS: Includes *trumbulli;* see Koopman (1978*b*:22); but also see Eger (1977:53).

Eumops underwoodi Goodwin, 1940. Am. Mus. Novit., 1075:2.
TYPE LOCALITY: Honduras, La Paz, 6 km N Chinacla.
DISTRIBUTION: Arizona (USA) to Nicaragua.
SYNONYMS: *sonoriensis.*

Molossops Peters, 1866. Monatsb. K. Preuss. Akad. Wiss. Berlin, 1865:575 [1866].
TYPE SPECIES: *Dysopes temminckii* Burmeister, 1854.
SYNONYMS: *Cabreramops, Cynomops, Neoplatymops.*
COMMENTS: Includes *Cabreramops, Cynomops,* and *Neoplatymops* as subgenera; see Jones et al. (1977:28) and Freeman (1981:133) respectively.

Molossops abrasus (Temminck, 1827). Monogr. Mamm., 1:232.
TYPE LOCALITY: "Brazil."
DISTRIBUTION: Venezuela, Guyana, Surinam, Peru, Brazil, Bolivia, Paraguay, N Argentina.
SYNONYMS: *brachymeles, cerastes, mastivus.*
COMMENTS: Subgenus *Cynomops.* Called *brachymeles* by Cabrera (1958:118-119) and Freeman (1981:155; who included it in the subgenus *Cynomops*); but see Carter and Dolan (1978:84).

Molossops aequatorianus Cabrera, 1917. Trab. Mus. Nac. Cienc. Nat. Zool., 31:20.
TYPE LOCALITY: Ecuador, Los Rios, Babahoyo.
DISTRIBUTION: Ecuador.
COMMENTS: Subgenus *Cabreramops.* See Ibanez (1980:105), who separated *Cabreramops* generically.

Molossops greenhalli (Goodwin, 1958). Am. Mus. Novit., 1877:3.
TYPE LOCALITY: Trinidad and Tobago, Trinidad, Port of Spain, Botanic Gardens.
DISTRIBUTION: Nayarit (Mexico) to Ecuador and NE Brazil; Trinidad.
SYNONYMS: *mexicanus.*
COMMENTS: Subgenus *Cynomops;* see Freeman (1981:155).

Molossops mattogrossensis Vieira, 1942. Argent. Zool. Sao Paulo, 3:430.
TYPE LOCALITY: Brazil, Mato Grosso, Juruena River, São Simao.
DISTRIBUTION: Venezuela, Guyana, C and NE Brazil.
COMMENTS: Subgenus *Neoplatymops;* see Freeman (1981:155). Listed as a subspecies of *Molossops temminckii* by Cabrera (1958:117); but see Peterson (1965*a*:3-5), who considered *Neoplatymops* a distinct genus. See Willig and Jones (1985, Mammalian Species, 244, as *Neoplatymops mattogrossensis*).

Molossops neglectus Williams and Genoways, 1980. Ann. Carnegie Mus., 49(25):489.
TYPE LOCALITY: Surinam, Surinam, 1 km S, 2 km E Powaka (5°25'N, 55°3'W).
DISTRIBUTION: Surinam, Amazonian Brazil and Peru.
COMMENTS: Subgenus *Molossops.*

Molossops planirostris (Peters, 1866). Monatsb. K. Preuss. Akad. Wiss. Berlin, 1865:575 [1866].
TYPE LOCALITY: French Guiana, Cayenne.
DISTRIBUTION: Panama to Peru, N Argentina, Paraguay, Brazil and Surinam.
SYNONYMS: *espiritosantensis, milleri, paranus.*
COMMENTS: Subgenus *Cynomops;* see Freeman (1981:155). Includes *milleri* and *paranus;* see Koopman (1978*b*:20); but also see Handley (1976) and Williams and Genoways (1980*c*), who regarded *paranus* as a distinct species.

Molossops temminckii (Burmeister, 1854). Syst. Uebers. Thiere Bras., p. 72.
TYPE LOCALITY: Brazil, Minas Gerais, Lagoa Santa.
DISTRIBUTION: Venezuela, Colombia, Peru, Bolivia, S Brazil, Paraguay, N Argentina, Uruguay.
SYNONYMS: *griseiventer, hirtipes, sylvia.*
COMMENTS: Subgenus *Molossops.*

Molossus E. Geoffroy, 1805. Ann. Mus. Hist. Nat. Paris, 6:151.
TYPE SPECIES: *Vespertilio molossus* Pallas, 1766.
COMMENTS: Middle American species reviewed by Dolan (1989).

Molossus ater E. Geoffroy, 1805. Bull. Sci. Soc. Philom. Paris, 3(96):279.
TYPE LOCALITY: French Guiana, Cayenne.
DISTRIBUTION: Tamaulipas and Sinaloa (Mexico) to Peru, N Argentina, Brazil and Guianas; Trinidad.
SYNONYMS: *albus, alecto, castaneus, fluminensis, holosericeus, macdougalli, malagai, myosurus, nigricans, rufus, ursinus.*
COMMENTS: Called *rufus* by Dolan (1989), but see Husson (1962) and Hall (1981:252). Includes *malagai;* see Jones (1965). Includes *macdougalli;* see Jones et al. (1977), but also see Hall (1981:252), who placed it in *pretiosus.*

Molossus bondae J. A. Allen, 1904. Bull. Am. Mus. Nat. Hist., 20:228.
TYPE LOCALITY: Colombia, Magdalena, Bonda.
DISTRIBUTION: Honduras to Ecuador and Venezuela; Cozumel Isl (Mexico).
COMMENTS: May include *coibensis,* according to Freeman (1981:157), but not according to Dolan (1989).

Molossus molossus (Pallas, 1766). Misc. Zool., p. 49-50.
TYPE LOCALITY: France, Martinique (Lesser Antilles).
DISTRIBUTION: Sinaloa and Coahuila (Mexico) to Peru, N Argentina, Uruguay, Brazil and Guianas; Greater and Lesser Antilles; Margarita Isl (Venezuela); Curacao and Bonaire (Netherlands Antilles); Trinidad and Tobago.
SYNONYMS: *acuticaudatus, amplexicaudus, aztecus, barnesi, cherriei, crassicaudatus, coibensis, currentium, daulensis, debilis, fortis, fusciventer, lambi, longicaudatus, major, milleri, moxensis, obscurus, olivaceofuscus, pygmaeus, tropidorhynchus, velox, verrilli.*
COMMENTS: Includes *fortis, milleri, debilis,* and *tropidorhynchus;* see Varona (1974:42). Called *major* by Hall and Kelson (1959:216) and Cabrera (1958:130) but see Husson (1962:251-259). Hall (1981:255) included *coibensis* and *aztecus* in *molossus,* but see Dolan (1989). Antillean populations reviewed by Genoways et al. (1981). Includes *daulensis,* but see Albuja (1982).

Molossus pretiosus Miller, 1902. Proc. Acad. Nat. Sci. Philadelphia, p. 396.
TYPE LOCALITY: Venezuela, Caracas, LaGuaira.
DISTRIBUTION: Guerrero, Oaxaca (Mexico); Nicaragua to Colombia, Venezuela and Guyana.
COMMENTS: Listed as a synonym of *rufus* by Cabrera (1958:132); but see Jones et al. (1977).

Molossus sinaloae J. A. Allen, 1906. Bull. Am. Mus. Nat. Hist., 22:236.
TYPE LOCALITY: Mexico, Sinaloa, Esquinapa.
DISTRIBUTION: Sinaloa (Mexico) to Colombia and Surinam; Trinidad.
SYNONYMS: *trinitatus.*
COMMENTS: Includes *trinitatus,* but see Freeman (1981:158).

Mops Lesson, 1842. Nouv. Tabl. Regn. Anim. Mammifères, p. 18.
TYPE SPECIES: *Mops indicus* Lesson, 1842 (= *Molossus mops* Blainville, 1840).
SYNONYMS: *Philippinopterus, Xiphonycteris.*

COMMENTS: Formerly included in *Tadarida;* for synonyms see Freeman (1981:158). See also Legendre (1984).

Mops brachypterus (Peters, 1852). Reise nach Mossambique, Säugethiere, p. 59.
TYPE LOCALITY: Mozambique, Mozambique Isl (15°S, 40°42'E).
DISTRIBUTION: Gambia to Kenya; Tanzania (including Zanzibar); Mozambique.
SYNONYMS: *leonis, ochraceus.*
COMMENTS: Subgenus *Xiphonycteris.* Includes *leonis;* see El-Rayah (1981:6).

Mops condylurus (A. Smith, 1833). S. Afr. Quart. J., 1:54.
TYPE LOCALITY: South Africa, Natal, Durban.
DISTRIBUTION: Senegal to Somalia, south to Angola, Botswana, and Natal (South Africa); Madagascar.
SYNONYMS: *angolensis, fulva, leucostigma, occidentalis, orientis, osborni, wonderi.*
COMMENTS: Subgenus *Mops.*

Mops congicus J. A. Allen, 1917. Bull. Am. Mus. Nat. Hist., 37:467.
TYPE LOCALITY: Zaire, Oriental, Medje.
DISTRIBUTION: Ghana, Nigeria, Cameroon, Zaire, Uganda, perhaps Gambia.
COMMENTS: Subgenus *Mops.* Does not include *trevori;* see Peterson (1972).

Mops demonstrator (Thomas, 1903). Ann. Mag. Nat. Hist., ser. 7, 12:504.
TYPE LOCALITY: Sudan, Equatoria, Mongalla.
DISTRIBUTION: Sudan, Zaire, Uganda, Burkina Faso.
SYNONYMS: *faradjius.*
COMMENTS: Subgenus *Mops.*

Mops midas (Sundevall, 1843). K. Svenska Vet.-Akad. Handl. Stockholm, 1842:207 [1843].
TYPE LOCALITY: Sudan, Blue Nile, White Nile River, West bank, Jebel el Funj.
DISTRIBUTION: Senegal to Saudi Arabia, south to Botswana and Transvaal (South Africa); Madagascar.
SYNONYMS: *miarensis, unicolor.*
COMMENTS: Subgenus *Mops.*

Mops mops (Blainville, 1840). Osteogr. Mamm., pt. 5(Vespertilio), p. 101.
TYPE LOCALITY: Indonesia, Sumatra.
DISTRIBUTION: W Malaysia, Sumatra, Borneo, perhaps Java.
SYNONYMS: *indicus.*
COMMENTS: Subgenus *Mops.*

Mops nanulus J. A. Allen, 1917. Bull. Am. Mus. Nat. Hist., 37:477.
TYPE LOCALITY: Zaire, Oriental, Niangara.
DISTRIBUTION: Gambia to Ethiopia and Kenya.
SYNONYMS: *calabarensis.*
COMMENTS: Subgenus *Xiphonycteris.* Distinction from *spurrelli* not certain; see Koopman (1989*b*).

Mops niangarae J. A. Allen, 1917. Bull. Am. Mus. Nat. Hist., 37:468.
TYPE LOCALITY: Zaire, Niangara.
DISTRIBUTION: Known only from the holotype.
COMMENTS: Subgenus *Mops.* Peterson (1972) included this species in *trevori;* Hayman and Hill (1971) listed it as a subspecies of *Tadarida congica* (= *Mops congicus*). Freeman (1981:111, 159) considered it a distinct species pending collection of additional specimens.

Mops niveiventer Cabrera and Ruxton, 1926. Ann. Mag. Nat. Hist., ser. 9, 17:594.
TYPE LOCALITY: Zaire, Kasai Occidental, Luluabourg.
DISTRIBUTION: Zaire, Rwanda, Burundi, Tanzania, Angola, Zambia, Mozambique.
SYNONYMS: *chitauensis.*
COMMENTS: Subgenus *Mops.* Probably a subspecies of *demonstrator.* Records from Botswana and Madagascar definitely *condylurus,* see Meester et al. (1986:74) and Hayman and Hill (1971:61).

Mops petersoni (El Rayah, 1981). R. Ontario Mus. Life Sci. Occas. Pap., 36:3.
TYPE LOCALITY: Cameroon, 15 km S Kumba (4°39'N, 9°26'E).

DISTRIBUTION: Cameroon and Ghana, perhaps Sierra Leone.
COMMENTS: Subgenus *Xiphonycteris*. Described in *Tadarida* (*Xiphonycteris*), but see comments under *Tadarida* and *Mops*.

Mops sarasinorum (A. Meyer, 1899). Abh. Zool. Anthrop.-Ethnology. Mus. Dresden, 7(7):15.
TYPE LOCALITY: Indonesia, Sulawesi, Batulappa (North of Lake Tempe).
DISTRIBUTION: Sulawesi (Indonesia) and adjacent small islands; Philippines.
SYNONYMS: *lanei*.
COMMENTS: Subgenus *Mops*. Includes *lanei* (formerly included in *Philippinopterus*); see Freeman (1981:160).

Mops spurrelli (Dollman, 1911). Ann. Mag. Nat. Hist., ser. 8, 7:211.
TYPE LOCALITY: Ghana, Bibianaha.
DISTRIBUTION: Liberia, Ivory Coast, Ghana, Togo, Benin, Rio Muni and Bioko (Equatorial Guinea), Zaire.
COMMENTS: Subgenus *Xiphonycteris*.

Mops thersites (Thomas, 1903). Ann. Mag. Nat. Hist., ser. 7, 12:634.
TYPE LOCALITY: Cameroon, Efulen.
DISTRIBUTION: Sierra Leone to Rwanda; Bioko; perhaps Mozambique and Zanzibar.
SYNONYMS: *occipitalis*.
COMMENTS: Subgenus *Xiphonycteris*.

Mops trevori J. A. Allen, 1917. Bull. Am. Mus. Nat. Hist., 37:468.
TYPE LOCALITY: Zaire, Oriental, Faradje.
DISTRIBUTION: NW Zaire, Uganda.
COMMENTS: Subgenus *Mops*. Formerly included *niangarae*; see Freeman (1981:111, 159).

Mormopterus Peters, 1865. Monatsb. K. Preuss. Akad. Wiss. Berlin, 1865:258.
TYPE SPECIES: *Nyctinomus* (*Mormopterus*) *jugularis* Peters, 1865.
SYNONYMS: *Micronomus*, *Platymops*, *Sauromys*.
COMMENTS: Formerly included in *Tadarida*; see Koopman (1975:419-421); but also see Freeman (1981:133, 160), who included *Sauromys* and *Platymops* as subgenera, and also *Micronomus*.

Mormopterus acetabulosus (Hermann, 1804). Observ. Zool., p. 19.
TYPE LOCALITY: Mauritius, Port Louis.
DISTRIBUTION: Reunion and Mauritius (Mascarene Isls), Madagascar, South Africa, Ethiopia.
SYNONYMS: *natalensis*.
COMMENTS: Subgenus *Mormopterus*.

Mormopterus beccarii Peters, 1881. Monatsb. K. Preuss. Akad. Wiss. Berlin, 1881:484.
TYPE LOCALITY: Indonesia, Molucca Isls, Amboina Isl.
DISTRIBUTION: Molucca Isls, New Guinea, adjacent small islands, N Australia.
SYNONYMS: *astrolabiensis*.
COMMENTS: Subgenus *Mormopterus*. Includes *astrolabiensis*; see Freeman (1981:160).

Mormopterus doriae K. Andersen, 1907. Ann. Mus. Civ. Stor. Nat. Genova, 3(38):42.
TYPE LOCALITY: Indonesia, Sumatra, Deli, Soekaranda.
DISTRIBUTION: Sumatra.
COMMENTS: Subgenus *Mormopterus*.

Mormopterus jugularis (Peters, 1865). *In* Sclater, Proc. Zool. Soc. Lond., 1865:468.
TYPE LOCALITY: Madagascar, Antananarivo.
DISTRIBUTION: Madagascar.
SYNONYMS: *albiventer*.
COMMENTS: Subgenus *Mormopterus*.

Mormopterus kalinowskii (Thomas, 1893). Proc. Zool. Soc. Lond., 1893:334.
TYPE LOCALITY: "Central Peru."
DISTRIBUTION: Peru, N Chile.
COMMENTS: Subgenus *Mormopterus*.

Mormopterus minutus (Miller, 1899). Bull. Am. Mus. Nat. Hist., 12:173.
TYPE LOCALITY: Cuba, Las Villas, Trinidad, San Pablo.

DISTRIBUTION: Cuba.
COMMENTS: Subgenus *Mormopterus*.

Mormopterus norfolkensis (Gray, 1840). Ann. Nat. Hist., 4:7.
TYPE LOCALITY: Australia, Norfolk Isl (S Pacific Ocean); uncertain.
DISTRIBUTION: Norfolk Isl?, SE Queensland, E New South Wales (Australia).
SYNONYMS: *wilcoxii*.
COMMENTS: Subgenus *Mormopterus*. There is considerable doubt as to the status of this species; see Hill (1961*b*:44) and Koopman (1984*c*:29-31).

Mormopterus petrophilus (Roberts, 1917). Ann. Transvaal Mus., 6:4.
TYPE LOCALITY: South Africa, Transvaal, near Rustenburg, Bleskap.
DISTRIBUTION: South Africa, Namibia, Botswana, Zimbabwe, Mozambique, perhaps Ghana.
SYNONYMS: *erongensis, fitzsimonsi, haagneri, umbratus*.
COMMENTS: Subgenus *Sauromys*; see Freeman (1981:133, 161). Formerly included in genus *Sauromys* by Peterson (1965*a*:12).

Mormopterus phrudus (Handley, 1956). Proc. Biol. Soc. Washington, 69:197.
TYPE LOCALITY: Peru, Cuzco, Machu Picchu, Urubamba River, San Miguel Bridge.
DISTRIBUTION: Peru.
COMMENTS: Subgenus *Mormopterus*.

Mormopterus planiceps (Peters, 1866). Monatsb. K. Preuss. Akad. Wiss. Berlin, 1866:23.
TYPE LOCALITY: Australia. Probably New South Wales, Sydney; see Iredale and Troughton (1934) for discussion.
DISTRIBUTION: Australia, New Guinea.
SYNONYMS: *cobourgiana, loriae, petersi, ridei*.
COMMENTS: Subgenus *Mormopterus*. Includes *loriae*; see Hill (1961*b*:45-46) and see Koopman (1984*c*:29-30).

Mormopterus setiger Peters, 1878. Monatsb. K. Preuss. Akad. Wiss. Berlin, 1878:196.
TYPE LOCALITY: Kenya, Taita.
DISTRIBUTION: S Sudan, Ethiopia, Kenya.
SYNONYMS: *barbatogularis, macmillani, parkeri*.
COMMENTS: Subgenus *Platymops*; see Freeman (1981:133, 161). Formerly included in *Platymops* by Harrison and Fleetwood (1960:277-278).

Myopterus E. Geoffroy, 1818. Descrip. de L'Egypte, 2:113.
TYPE SPECIES: *Myopterus senegalensis* Oken, 1816 (not available) (= *Myopterus daubentonii* Desmarest, 1820).
SYNONYMS: *Eomops*.
COMMENTS: Includes *Eomops*; see Hayman and Hill (1971:56).

Myopterus daubentonii Desmarest, 1820. Mammalogie, *in* Encyclop. Méth., 1:132.
TYPE LOCALITY: Senegal.
DISTRIBUTION: Senegal, Ivory Coast, NE Zaire.
SYNONYMS: *albatus*.
COMMENTS: Holotype lost. Includes *albatus*; see Koopman (1989*b*).

Myopterus whitleyi (Scharff, 1900). Ann. Mag. Nat. Hist., ser. 7, 6:569.
TYPE LOCALITY: Nigeria, Mid-Western Region, Benin City.
DISTRIBUTION: Ghana, Nigeria, Cameroon, Zaire, Uganda.

Nyctinomops Miller, 1902. Proc. Acad. Nat. Sci. Philadelphia, 54:393.
TYPE SPECIES: *Nyctinomus femorosaccus* Merriam, 1889.
COMMENTS: Formerly included in *Tadarida*; see Hall (1981:243); but also see Freeman (1981:124). A key to the species in this genus was presented by Kumirai and Jones (1990).

Nyctinomops aurispinosus (Peale, 1848). Mammalia, *in* Repts. U.S. Expl. Surv., 8:21.
TYPE LOCALITY: Brazil, Rio Grande do Norte, 100 mi. (161 km) off Cape Sao Roque.
DISTRIBUTION: Sonora and Tamaulipas (Mexico) to Peru, Bolivia, and Brazil.
SYNONYMS: *similis*.

COMMENTS: Includes *similis;* see Jones and Arroyo-Cabrales (1990, Mammalian Species, 350).

Nyctinomops femorosaccus (Merriam, 1889). N. Am. Fauna, 2:23.
TYPE LOCALITY: USA, California, Riverside Co., Palm Springs.
DISTRIBUTION: Guerrero (Mexico) to New Mexico, Arizona, California (USA) and Baja California (Mexico).
COMMENTS: See Kumirai and Jones (1990, Mammalian Species, 349).

Nyctinomops laticaudatus (E. Geoffroy, 1805). Ann. Mus. Hist. Nat. Paris, 6:156.
TYPE LOCALITY: Paraguay, Asuncion.
DISTRIBUTION: Tamaulipas and Jalisco (Mexico) to NW Peru, N Argentina, and Brazil; Trinidad; Cuba.
SYNONYMS: *caecus, europs, ferruginea, gracilis, macarenensis, yucatanica.*
COMMENTS: Includes *yucatanicus, europs,* and *gracilis;* see Silva-Taboada and Koopman (1964:3, 4) and Freeman (1981:162).

Nyctinomops macrotis (Gray, 1840). Ann. Nat. Hist., 4:5.
TYPE LOCALITY: Cuba.
DISTRIBUTION: SW British Columbia and Iowa (USA) to Peru, N Argentina and Uruguay; Cuba; Jamaica; Hispaniola.
SYNONYMS: *aequatoralis, affinis, auritus, depressus, megalotis; molossa* Hershkovitz (not Pallas); *nevadensis.*
COMMENTS: Called *Tadarida molossa* by Hall and Kelson (1959:208); but see Husson (1962:256-259). See Milner et al. (1990, Mammalian Species, 351).

Otomops Thomas, 1913. J. Bombay Nat. Hist. Soc., 22:91.
TYPE SPECIES: *Nyctinomus wroughtoni* Thomas, 1913.

Otomops formosus Chasen, 1939. Treubia, 17:186.
TYPE LOCALITY: Indonesia, Java, Tjibadak.
DISTRIBUTION: Java.

Otomops martiensseni (Matschie, 1897). Arch. Naturgesch., 63(1):84.
TYPE LOCALITY: Tanzania, Tanga, Magrotto Plantation.
DISTRIBUTION: Djibouti and Central African Republic to Angola and Natal (South Africa); Madagascar.
SYNONYMS: *icarus, madagascariensis.*

Otomops papuensis Lawrence, 1948. J. Mammal., 29:413.
TYPE LOCALITY: Papua New Guinea, Gulf Prov., Vailala River.
DISTRIBUTION: SE New Guinea.

Otomops secundus Hayman, 1952. *In* Laurie, Bull. Brit. Mus. (Nat. Hist.), Zool., 1:314.
TYPE LOCALITY: Papua New Guinea, Madang Prov., Tapu.
DISTRIBUTION: NE New Guinea.
COMMENTS: May be a subspecies of *papuensis,* according to the describer; see Hill (1983:197-198).

Otomops wroughtoni (Thomas, 1913). J. Bombay Nat. Hist. Soc., 22:87.
TYPE LOCALITY: India, Mysore, Kanara, near Talewadi.
DISTRIBUTION: S India.

Promops Gervais, 1856. *In* Comte de Castelnau, Exped. Partes Cen. Am. Sud., Zool.(Sec. 7), Vol 1, pt. 2(Mammifères):58.
TYPE SPECIES: *Promops ursinus* Gervais, 1856 (= *Molossus nasutus* Spix, 1823).

Promops centralis Thomas, 1915. Ann. Mag. Nat. Hist., ser. 8, 16:62.
TYPE LOCALITY: Mexico, Yucatan.
DISTRIBUTION: Jalisco and Yucatan (Mexico) to Peru, N Argentina, and Surinam; Trinidad.
SYNONYMS: *davisoni, occultus* (see Ojasti and Linares, 1971:433-434).

Promops nasutus (Spix, 1823). Sim. Vespert. Brasil., p. 58.
TYPE LOCALITY: Brazil, Bahia, Sao Francisco River.

DISTRIBUTION: Venezuela, Trinidad, Surinam, Brazil, Ecuador, Peru, Bolivia, Paraguay, N Argentina.
SYNONYMS: *ancilla, downsi, fosteri, fumarius, pamana, rufocastaneus, ursinus.*
COMMENTS: Includes *pamana*; see Goodwin and Greenhall (1962).

Tadarida Rafinesque, 1814. Precis Som., p. 55.
TYPE SPECIES: *Cephalotes teniotis* Rafinesque, 1814.
SYNONYMS: *Austronomus, Nyctinomus, Rhizomops.*
COMMENTS: Formerly included *Chaerephon, Mops, Mormopterus,* and *Nyctinomops*; see Freeman (1981:133). Includes *Rhizomops,* but see Legendre (1984); see also Owen et al. (1990). Mahoney and Walton (1988) regarded *Nyctinomus* as the prior name for this genus.

Tadarida aegyptiaca (E. Geoffroy, 1818). Descrip. de L'Egypte, 2:128.
TYPE LOCALITY: Egypt, Giza.
DISTRIBUTION: South Africa to Nigeria, Algeria, and Egypt to Yemen and Oman, east to India and Sri Lanka. Records from Saudi Arabia and Oman are by Jennings (1979).
SYNONYMS: *anchietae, bocagei, brunneus, geoffroyi, gossei, sindica, talpinus, thomasi, tongaensis, tragata.*
COMMENTS: Includes *tragata*; see Corbet (1978c:63) and Freeman (1981:165).

Tadarida australis (Gray, 1839). Mag. Zool. Bot., 2:501.
TYPE LOCALITY: Australia, New South Wales.
DISTRIBUTION: S and C Australia, New Guinea.
SYNONYMS: *albidus, atratus, kuboriensis.*
COMMENTS: Includes *kuboriensis*; see Koopman (1982:23-24); but also see McKean and Calaby (1968:377).

Tadarida brasiliensis (I. Geoffroy, 1824). Ann. Sci. Nat. (Paris), ser. 1, 1:343.
TYPE LOCALITY: Brazil, Parana, Curitiba.
DISTRIBUTION: S Brazil, Argentina, and Chile to Oregon, S Nebraska and Ohio (USA); Greater and Lesser Antilles.
SYNONYMS: *antillularum, bahamensis, californicus, constanzae, cynocephala, fuliginosus, intermedia, mexicana, mohavensis, multispinosus, murina, muscula, naso, nasutus, peruanus, rugosus, texana.*
COMMENTS: Placed in distinct genus (*Rhizomops*) by Legendre (1984), but see Freeman (1981:68) and Owen et al. (1990). See Wilkins (1989, Mammalian Species, 331).

Tadarida espiritosantensis (Ruschi, 1951). Bol. Mus. Biol. Prof. Mello-Leitao Zool., Santa Teresa, Espírito Santo, 7:19.
TYPE LOCALITY: Brazil, Espírito Santo, Santa Teresa, Treis Barras.
DISTRIBUTION: Brazil.
COMMENTS: Mentioned in Pine and Ruschi (1976); listed as a species by Freeman (1981:166). Probably a synonym of *Nyctinomops laticaudatus.*

Tadarida fulminans (Thomas, 1903). Ann. Mag. Nat. Hist., ser. 7, 12:501.
TYPE LOCALITY: Madagascar, Betsilio, Fianarantsoa.
DISTRIBUTION: E Zaire, Rwanda, Kenya, Tanzania, Zambia, Malawi, Zimbabwe, Transvaal (South Africa), Madagascar.
SYNONYMS: *mastersoni.*

Tadarida lobata (Thomas, 1891). Ann. Mag. Nat. Hist., ser. 6, 7:303.
TYPE LOCALITY: Kenya, West Pokot, Turkwell Gorge.
DISTRIBUTION: Kenya, Zimbabwe.

Tadarida teniotis (Rafinesque, 1814). Precis Som., p. 12.
TYPE LOCALITY: Italy, Sicily.
DISTRIBUTION: France, Portugal and Morocco to Japan, S China, and Taiwan; Madeira (Portugal) and Canary Isls (Spain).
SYNONYMS: *cestoni, cinerea, coecata, insignis, latouchei, nigrogriseus, rueppellii, savii.*
COMMENTS: Revised by Aellen (1966) and Kock and Nader (1984). Includes *insignis,* but see Yoshiyuki (1989) and Yoshiyuki et al. (1989).

Tadarida ventralis (Heuglin, 1861). Nouv. Act. Acad. Caes. Leop.-Carol., 29(8):4,11.
TYPE LOCALITY: Ethiopia, Eritrea, Keren.
DISTRIBUTION: Ethiopia to South Africa.
SYNONYMS: *africana*.
COMMENTS: For use of this name in place of *africana*, see Kock (1975).

ORDER PRIMATES
by Colin P. Groves

ORDER PRIMATES

Family Cheirogaleidae Gray, 1873. Proc. Zool. Soc. Lond. 1872:849 [1873].
COMMENTS: Formerly included in Lemuridae. For status of this taxon, see Rumpler (1975).

Subfamily Cheirogaleinae Gray, 1873. Proc. Zool. Soc. Lond., 1872:849 [1873].

Allocebus Petter-Rousseaux and Petter, 1967. Mammalia, 31:574.
TYPE SPECIES: *Cheirogaleus trichotis* Günther, 1875.
COMMENTS: Previously included in *Cheirogaleus*, but very distinct.

Allocebus trichotis (Günther, 1875). Proc. Zool. Soc. Lond., 1875:78.
TYPE LOCALITY: Madagascar, between Tamatave and Morondava.
DISTRIBUTION: E Madagascar, vicinity of Morondava Bay.
STATUS: CITES - Appendix I; U.S. ESA and IUCN - Endangered.

Cheirogaleus É. Geoffroy, 1812. Ann. Mus. Hist. Nat. Paris, 19:172.
TYPE SPECIES: *Cheirogaleus major* É. Geoffroy, 1812; fixed by Elliot (1907*b*:548).
SYNONYMS: *Altililemur, Cebugale, Mioxocebus, Myspithecus, Opolemur.*
COMMENTS: Revised by Petter et al. (1977:27-79).

Cheirogaleus major É. Geoffroy, 1812. Ann. Mus. Hist. Nat. Paris, 19:172.
TYPE LOCALITY: Madagascar, Fort Dauphin.
DISTRIBUTION: E Madagascar.
STATUS: CITES - Appendix I; U.S. ESA - Endangered.
SYNONYMS: *adipicaudatus, commersonii, crossleyi, griseus, melanotis, milii, sibreei, typicus, typus.*
COMMENTS: *Cheirogaleus crossleyi* Grandidier, 1870, may be a full species.

Cheirogaleus medius É. Geoffroy, 1812. Ann. Mus. Hist. Nat. Paris, 19:172.
TYPE LOCALITY: Madagascar, Fort Dauphin.
DISTRIBUTION: W and S Madagascar.
STATUS: CITES - Appendix I; U.S. ESA - Endangered.
SYNONYMS: *samati, thomasi.*

Microcebus É. Geoffroy, 1834. Cours Hist. Nat. Mamm., lecon 11, 1828:24.
TYPE SPECIES: *Lemur pusillus* É. Geoffroy, 1795 (= *Lemur murinus* J. F. Miller, 1777).
SYNONYMS: *Azema, Gliscebus, Mirza, Murilemur, Myocebus, Myscebus, Scartes.*
COMMENTS: *Mirza* Gray, 1870, may be a separate genus. Revised by Petter et al. (1977:27-79).

Microcebus coquereli (A. Grandidier, 1867). Rev. Mag. Zool. Paris, ser. 2, 19:85.
TYPE LOCALITY: Madagascar, Morondava.
DISTRIBUTION: W Madagascar.
STATUS: CITES - Appendix I; U.S. ESA - Endangered; IUCN - Vulnerable.
COMMENTS: Placed in a separate genus *Mirza*, Gray, 1870, by Schwartz and Tattersall (1985).

Microcebus murinus (J. F. Miller, 1777). Cimelia Physica, p. 25.
TYPE LOCALITY: Madagascar.
DISTRIBUTION: W and S Madagascar.
STATUS: CITES - Appendix I; U.S. ESA - Endangered.
SYNONYMS: *gliroides, griseorufus, madagascarensis, minima, minor, myoxinus, palmarum, prehensilis, pusillus.*

Microcebus rufus É. Geoffroy, 1834. Cours Hist. Nat. Mamm., lecon 11, 1828:24.
TYPE LOCALITY: Madagascar.
DISTRIBUTION: E and N Madagascar.

STATUS: CITES - Appendix I; U.S. ESA - Endangered.
SYNONYMS: *smithii*.
COMMENTS: Separated from *murinus* by Petter et al. (1977:30).

Subfamily Phanerinae Rumpler, 1974. *In* Martin et al. (eds.), Prosimian Biology, p. 867.

Phaner Gray, 1870. Cat. Monkeys, Lemurs, Fruit-eating Bats Brit. Mus., p. 135.
TYPE SPECIES: *Lemur furcifer* Blainville, 1839.

Phaner furcifer (Blainville, 1839). Osteogr. Mamm., Primates, p. 35.
TYPE LOCALITY: Madagascar, Morondava.
DISTRIBUTION: W, NE, and extreme SE Madagascar.
STATUS: CITES - Appendix I; U.S. ESA - Endangered; IUCN - Rare.
SYNONYMS: *electromontis, pallescens, parienti*.
COMMENTS: May consist of more than one species, see Groves and Tattersall (1991).

Family Lemuridae Gray, 1821. London Med. Repos., 15:296.
COMMENTS: Reviewed by Petter et al. (1977).

Eulemur Simons and Rumpler, 1988. C. R. Acad. Sci. Paris, III, 307:547.
TYPE SPECIES: *Lemur mongoz* Linnaeus, 1766.
SYNONYMS: *Petterus* Groves and Eaglen, 1988.

Eulemur coronatus (Gray, 1842). Ann. Mag. Nat. Hist., [ser. 1], 10:257.
TYPE LOCALITY: Madagascar.
DISTRIBUTION: Mt. Ambre (N Madagascar).
STATUS: CITES - Appendix I; U.S. ESA and IUCN - Endangered.
SYNONYMS: *chrysampyx*.
COMMENTS: Separated from *mongoz* by Petter et al. (1977:151).

Eulemur fulvus (É. Geoffroy, 1796). Mag. Encyclop., 1:47.
TYPE LOCALITY: Madagascar.
DISTRIBUTION: Coastal Madagascar, except extreme south; Mayotte (Comoro Isls).
STATUS: CITES - Appendix I; U.S. ESA - Endangered; IUCN - Rare as *E. f. albifrons, E. f. fulvus* and *E. f. rufus*, Vulnerable as *E. f. albocollaris, E. f. collaris, E. f. mayottensis* and *E. f. sanfordi*.
SYNONYMS: *albifrons, albocollaris, bruneus, cinereiceps, collaris, frederici, macromangoz, mayottensis, melanocephala, rufifrons, rufus, sanfordi, xanthomystax*.
COMMENTS: Sympatric with *macaco*; see Tattersall (1976). Includes *rufus, albifrons, collaris,* and *sanfordi*; see Petter and Petter (1977:6) and Groves (1978*a*, 1989:87-92); some of these nominal forms may prove to be distinct species.

Eulemur macaco (Linnaeus, 1766). Syst. Nat., 12th ed., 1:34.
TYPE LOCALITY: Madagascar.
DISTRIBUTION: Nosi Be and NW Madagascar.
STATUS: CITES - Appendix I; U.S. ESA - Endangered; IUCN - Endangered as *E. m. flavifrons*, otherwise Vulnerable.
SYNONYMS: *flavifrons, leucomystax, niger, nigerrimus*.

Eulemur mongoz (Linnaeus, 1766). Syst. Nat., 12th ed., 1:44.
TYPE LOCALITY: Comoros, Anjouan Isl.
DISTRIBUTION: NW Madagascar, between Majunga and Betsiboka; Anjouan, Moheli (Comoro Isls).
STATUS: CITES - Appendix I; U.S. ESA and IUCN - Endangered.
SYNONYMS: *albimanus, anjuanensis, brissonii, bugi, cuvieri, dubius, johannae, micromongoz, ocularis, nigrifrons, roussardii*.

Eulemur rubriventer (I. Geoffroy, 1850). C. R. Acad. Sci. Paris, 31:876.
TYPE LOCALITY: Madagascar, Tamatave.
DISTRIBUTION: E Madagascar.
STATUS: CITES - Appendix I; U.S. ESA - Endangered; IUCN - Vulnerable.
SYNONYMS: *flaviventer, rufipes, rufiventer*.

Hapalemur I. Geoffroy, 1851. L'Inst. Paris, 19(929):341.
TYPE SPECIES: *Lemur griseus* É. Geoffroy, 1812 (= *Lemur griseus* Link, 1795).
SYNONYMS: *Prohapalemur* Lamberton, 1936; *Prolemur* Gray, 1871.
COMMENTS: Placed in Lepilemuridae (= Megaladapidae) by Tattersall (1982).

Hapalemur aureus Meier, Albignac, Peyriéras, Rumpler, and Wright, 1987. Folia Primatol., 48:211.
TYPE LOCALITY: Madagascar, 6.25 km from the village of Ranomafana, 21°16'38"S, 47°23'50"E.
DISTRIBUTION: Between Namorona River and Bevoahazo Village, SE Madagascar.
STATUS: CITES - Appendix I; U.S. ESA and IUCN - Endangered.

Hapalemur griseus (Link, 1795). Beytr. Naturg., 1:65.
TYPE LOCALITY: Madagascar.
DISTRIBUTION: E, NW, and Antsalova Dist. of W Madagascar.
STATUS: CITES - Appendix I; U.S. ESA - Endangered; IUCN - Endangered as *H. g. alaotrensis*, Insufficiently known as *H. g. griseus*, Vulnerable as *H. g. occidentalis*.
SYNONYMS: *alaotrensis, cinereus, meridionalis, occidentalis, olivaceus, schlegeli*.
COMMENTS: Groves (1989:87) suggested that *H. g. alaotrensis* may be a full species.

Hapalemur simus Gray, 1870. Cat. Monkeys, Lemurs, Fruit-eating Bats Brit. Mus., p. 133.
TYPE LOCALITY: Madagascar.
DISTRIBUTION: E Madagascar, inland from Mananjary.
STATUS: CITES - Appendix I; U.S. ESA and IUCN - Endangered.
SYNONYMS: *gallieni*.
COMMENTS: Includes *H. gallieni* Standing, 1905 (see Vuillaume-Randriamanantena et al. (1985).

Lemur Linnaeus, 1758. Syst. Nat., 10th ed., 1:24.
TYPE SPECIES: *Lemur catta* Linnaeus, 1758.
SYNONYMS: *Catta, Maki, Mococo, Odorlemur, Procebus, Prosimia*.
COMMENTS: Revised by Petter et al. (1977:128-213). Restricted to *L. catta* by Groves and Eaglen (1988:533) and Simons and Rumpler (1988:547) .

Lemur catta Linnaeus, 1758. Syst. Nat., 10th ed., 1:30.
TYPE LOCALITY: Madagascar.
DISTRIBUTION: S Madagascar.
STATUS: CITES - Appendix I; U.S. ESA - Endangered; IUCN - Vulnerable.
SYNONYMS: *mococo*.

Varecia Gray, 1863. Proc. Zool. Soc. Lond., 1863:135.
TYPE SPECIES: *Lemur varius* É. Geoffroy (= *Lemur macaco variegatus* Kerr, 1792).
COMMENTS: Separated from *Lemur* by J.-J. Petter (1962).

Varecia variegata (Kerr, 1792). *In* Linnaeus, Anim. Kingdom, 1:85.
TYPE LOCALITY: Madagascar.
DISTRIBUTION: East coast of Madagascar, to 19°S latitude.
STATUS: CITES - Appendix I; U.S. ESA and IUCN - Endangered.
SYNONYMS: *editorum, erythromela, ruber, subcinctus, vari, varius*.

Family Megaladapidae Major, 1893. Proc. Roy. Soc., 54:176.
SYNONYMS: Lepilemuridae.
COMMENTS: Includes many extinct (subfossil) genera, but also *Lepilemur* according to Schwartz and Tattersall (1985:20) and Groves (1989:92).

Lepilemur I. Geoffroy, 1851. Cat. Meth. Coll. Mamm. Ois. (Mus. Hist. Nat. Paris), Primates, p. 75.
TYPE SPECIES: *Lepilemur mustelinus* I. Geoffroy, 1851.
SYNONYMS: *Galeocebus, Lepidilemur; Mixocebus*, Peters, 1874.
COMMENTS: Revised by Petter et al. (1977:274-318). Six of the seven species are known to be karyotypically distinct. Rumpler (1975) and Corbet and Hill (1980:83) placed this

genus in a subfamily of Lemuridae. Petter and Petter (1977:6) placed it in its own family Lepilemuridae.

Lepilemur dorsalis Gray, 1870. Cat. Monkeys, Lemurs, Fruit-eating Bats Brit. Mus., p. 135.
TYPE LOCALITY: NW Madagascar.
DISTRIBUTION: Nosi Be and Ambanja Region (NW Madagascar).
STATUS: CITES - Appendix I; U.S. ESA - Endangered; IUCN - Vulnerable.
SYNONYMS: *grandidieri*.

Lepilemur edwardsi (Forbes, 1894). Handbook of Primates, 1:87.
TYPE LOCALITY: Madagascar, Betsaka, 12 mi inland from Majunga.
DISTRIBUTION: E Madagascar, between Ankarafantsika and Antsalova.
STATUS: CITES - Appendix I; U.S. ESA - Endangered; IUCN - Rare.
SYNONYMS: *rufescens*.
COMMENTS: Includes *rufescens*, which Petter and Petter (1977:7) considered a distinct species.

Lepilemur leucopus (Major, 1894). Ann. Mag. Nat. Hist., ser. 6, 13:211.
TYPE LOCALITY: Madagascar, Fort Dauphin (Bevilany).
DISTRIBUTION: Arid zone of S Madagascar.
STATUS: CITES - Appendix I; U.S. ESA - Endangered; IUCN - Rare.

Lepilemur microdon (Forbes, 1894). Handbook of Primates, 1:88.
TYPE LOCALITY: Madagascar, east of Betsileo.
DISTRIBUTION: E Madagascar, between Perinet and Ft. Dauphin.
STATUS: CITES - Appendix I; U.S. ESA - Endangered; IUCN - Rare.

Lepilemur mustelinus I. Geoffroy, 1851. Cat. Meth. Coll. Mamm.Ois. (Mus. Hist. Nat. Paris), Primates, p. 76.
TYPE LOCALITY: Madagascar, north of Tamatave.
DISTRIBUTION: E Madagascar, between Tamatave and Antalana.
STATUS: CITES - Appendix I; U.S. ESA - Endangered; IUCN - Rare.
SYNONYMS: *caniceps*.

Lepilemur ruficaudatus A. Grandidier, 1867. Rev. Mag. Zool. Paris, ser. 2, 19:256.
TYPE LOCALITY: Madagascar, Morondava.
DISTRIBUTION: SW Madagascar, between Morondava and Sakarana.
STATUS: CITES - Appendix I; U.S. ESA - Endangered; IUCN - Rare.
SYNONYMS: *globiceps*, *pallidicauda*.

Lepilemur septentrionalis Rumpler and Albignac, 1975. Am. J. Phys. Anthropol., 42:425.
TYPE LOCALITY: Madagascar, Sahafary Forest.
DISTRIBUTION: Extreme N Madagascar.
STATUS: CITES - Appendix I; U.S. ESA - Endangered; IUCN - Vulnerable.
SYNONYMS: *andrafiamensis*, *ankaranensis*, *sahafarensis*.

Family Indridae Burnett, 1828. Quart. J. Sci. Arts, 2:306-307.
COMMENTS: Reviewed by Petter et al. (1977). Usually called Indriidae, but see Jenkins (1987:43).

Avahi Jourdan, 1834. L'Inst., Paris, 2:231.
TYPE SPECIES: *Lemur laniger* Gmelin, 1788.
SYNONYMS: *Habrocebus*, *Iropocus*, *Microrhynchus*, *Semnocebus*.
COMMENTS: *Lichanotus* has commonly been used for this genus, but see Jenkins (1987:55).

Avahi laniger (Gmelin, 1788). Syst. Nat., 13th ed., 1:44.
TYPE LOCALITY: Madagascar.
DISTRIBUTION: E coast and Ankarafantsika Dist. in NW Madagascar.
STATUS: CITES - Appendix I; U.S. ESA - Endangered; IUCN - Vulnerable.
SYNONYMS: *awahi*, *lanatus*, *longicaudatus*, *occidentalis*, *orientalis*.
COMMENTS: Rumpler et al. (1990) suggested that *occidentalis* may be a distinct species.

Indri É. Geoffroy and G. Cuvier, 1796. Mag. Encyclop., 1:46.
TYPE SPECIES: *Lemur indri* Gmelin, 1788.
SYNONYMS: *Lichanotus, Pithelemur.*

Indri indri (Gmelin, 1788). Syst. Nat., 13th ed., 1:42.
TYPE LOCALITY: Madagascar.
DISTRIBUTION: NE to EC Madagascar.
STATUS: CITES - Appendix I; U.S. ESA and IUCN - Endangered.
SYNONYMS: *ater, brevicaudatus, mitratus, niger, variegatus.*

Propithecus Bennett, 1832. Proc. Zool. Soc. Lond., 1832:20.
TYPE SPECIES: *Propithecus diadema* Bennett, 1832.
SYNONYMS: *Macromerus.*

Propithecus diadema Bennett, 1832. Proc. Zool. Soc. Lond., 1832:20.
TYPE LOCALITY: Madagascar.
DISTRIBUTION: N and E Madagascar.
STATUS: CITES - Appendix I; U.S. ESA and IUCN - Endangered.
SYNONYMS: *albus, bicolor, candidus, edwardsi, holomelas, perrieri, sericeus, typicus.*

Propithecus tattersalli Simons, 1988. Folia Primatol., 50:146.
TYPE LOCALITY: Madagascar, 6-7 km NE of Daraina, Antseranana Prov., 13°9'S, 49°41'E.
DISTRIBUTION: Ampandrana and Daraina districts, Madagascar.
STATUS: CITES - Appendix I; U.S. ESA and IUCN - Endangered.

Propithecus verreauxi A. Grandidier, 1867. Rev. Mag. Zool. Paris, ser. 2, 19:84.
TYPE LOCALITY: Madagascar, Tsifanihy (N of Cape Ste.-Marie).
DISTRIBUTION: NC to SW Madagascar.
STATUS: CITES - Appendix I; U.S. ESA - Endangered; IUCN - Vulnerable.
SYNONYMS: *coquereli, coronatus, damanus, damoni, deckenii, majori.*

Family Daubentoniidae Gray, 1863. Proc. Zool. Soc. Lond., 1863:151.
SYNONYMS: Chiromyidae.
COMMENTS: Groves (1989:65, 74-78) proposed separating this family to its own infraorder, Chiromyiformes.

Daubentonia É. Geoffroy, 1795. Decad. Philos. Litt., 28:195.
TYPE SPECIES: *Sciurus madagascarensis* Gmelen, 1788.
SYNONYMS: *Aye-aye, Cheiromys; Chiromys* Illiger, 1811; *Myslemur, Myspithecus, Psilodactylus, Scolecophagus.*

Daubentonia madagascariensis (Gmelin, 1788). Syst. Nat., 13th ed., 1:152.
TYPE LOCALITY: NW Madagascar.
DISTRIBUTION: NE and NW Madagascar (discontinuous).
STATUS: CITES - Appendix I; U.S. ESA and IUCN - Endangered.
SYNONYMS: *daubentonii, laniger, psilodactylus, robusta.*
COMMENTS: The form *robusta* may be a distinct species.

Family Loridae Gray, 1821. London Med. Repos., 15:298.
SYNONYMS: Lorisidae.
COMMENTS: Previously known as Lorisidae; for correct form of name see Jenkins (1987:1).

Arctocebus Gray, 1863. Proc. Zool. Soc. Lond., 1863:150.
TYPE SPECIES: *Perodicticus calabarensis* J. A. Smith, 1860.

Arctocebus aureus de Winton, 1902. Ann. Mag. Nat. Hist., ser. 7, 9:48.
TYPE LOCALITY: Equatorial Guinea, 50 miles up Benito River.
DISTRIBUTION: C Africa, S of Sanaga River, W and N of Zaire/Oubungui River system.
STATUS: CITES - Appendix II.
SYNONYMS: *ruficeps.*
COMMENTS: Formerly classified as a subspecies of *Arctocebus calabarensis*, but as a full species by Maier (1980:567) and Groves (1989:100-101).

Arctocebus calabarensis (J. A. Smith, 1860). Proc. Roy. Phys. Soc. Edinburgh, 2:177.
TYPE LOCALITY: Nigeria, Old Calabar.
DISTRIBUTION: C Africa, between Niger and Sanaga Rivers.
STATUS: CITES - Appendix II; IUCN - Insufficiently known.

Loris É. Geoffroy, 1796. Mag. Encyclop., 1:48.
TYPE SPECIES: *Lemur tardigradus* Linnaeus, 1758.
SYNONYMS: *Stenops, Tardigradus.*

Loris tardigradus (Linnaeus, 1758). Syst. Nat., 10th ed., 1:29.
TYPE LOCALITY: Sri Lanka.
DISTRIBUTION: Sri Lanka, S India.
STATUS: CITES - Appendix II.
SYNONYMS: *ceylonicus, gracilis, grandis, lydekkerianus, malabaricus, nordicus, nycticeboides, zeylanicus.*
COMMENTS: More than one species may be concealed under this name.

Nycticebus É. Geoffroy, 1812. Ann. Mus. Hist. Nat. Paris, 19:163.
TYPE SPECIES: *Tardigradus coucang* Boddart, 1785.

Nycticebus coucang (Boddaert, 1785). Elench. Anim., p. 67.
TYPE LOCALITY: Malaysia, Malacca.
DISTRIBUTION: Sulu Arch. (S Philippines); Assam (India) to Vietnam and Malay Peninsula; W Indonesia; Yunnan, perhaps Kwangsi (China).
STATUS: CITES - Appendix II.
SYNONYMS: *bancanus, beugalensis, borneanus, cinereus, hilleri, incanus, insularis, javanicus, natunae, malaianus, menagensis, ornatus, sumatrensis, tardigradus, tenasserimensis.*

Nycticebus pygmaeus Bonhote, 1907. Abstr. Proc. Zool. Soc. Lond., 1907(38):2.
TYPE LOCALITY: Vietnam, Nhatrang.
DISTRIBUTION: Laos; Cambodia; Vietnam, east of Mekong River; S Yunnan (China).
STATUS: CITES - Appendix II; U.S. ESA - Threatened; IUCN - Vulnerable.
SYNONYMS: *intermedius.*
COMMENTS: Includes *intermedius* see Groves (1971*c*) and Lekagul and McNeely (1977:270).

Perodicticus Bennett, 1831. Proc. Zool. Soc. Lond., 1830:109 [1831].
TYPE SPECIES: *Lemur potto* Müller, 1766.
SYNONYMS: *Potto.*

Perodicticus potto (Müller, 1766). *In* Linnaeus, Vollstand Natursyst. Suppl., p. 12.
TYPE LOCALITY: Ghana, Elmina.
DISTRIBUTION: Cameroon to Guinea; Congo Republic; Gabon; Zaire to W Kenya.
STATUS: CITES - Appendix II.
SYNONYMS: *arrhenii, batesi, bosmannii, edwardsi, faustus, geoffroyi, guineensis, ibeanus, ju-ju, nebulosus.*

Family Galagonidae Gray, 1825. Ann. Philos., n.s., 10:338.
SYNONYMS: Galagidae.
COMMENTS: Formerly considered a subfamily of Lorisidae; see Hill and Meester (1977:2) and Jenkins (1987:85). Synonymous with Galagidae but for the correct form of the name see Jenkins (1987:1).

Euoticus Gray, 1863. Proc. Zool. Soc. Lond., 1863:140.
TYPE SPECIES: *Otogale pallida* Gray, 1863.
COMMENTS: Recognized as a genus by Groves (1989:103).

Euoticus elegantulus (Le Conte, 1857). Proc. Acad. Nat. Sci. Philadelphia, 9:10.
TYPE LOCALITY: W Africa.
DISTRIBUTION: Gabon, Congo Republic, Cameroon south of Sanaga River.
STATUS: CITES - Appendix II.
SYNONYMS: *apicalus, tonsor.*

Euoticus pallidus (Gray, 1863). Proc. Zool. Soc. Lond., 1863:140.
TYPE LOCALITY: Equatorial Guinea, Bioko.
DISTRIBUTION: Bioko (Equatorial Guinea); between Sanaga River (Cameroon) and Niger River.
STATUS: CITES - Appendix II.
SYNONYMS: *talboti*.
COMMENTS: Accepted as a species by Groves (1989:104).

Galago É. Geoffroy, 1796. Mag. Encyclop., 1:49.
TYPE SPECIES: *Galago senegalensis* É. Geoffroy, 1796.
SYNONYMS: *Chirosciurus, Macropus, Otolicnus, Sciurocheirus*.

Galago alleni Waterhouse, 1838. Proc. Zool. Soc. Lond., 1837:87 [1838].
TYPE LOCALITY: Equatorial Guinea, Bioko.
DISTRIBUTION: Bioko (Equatorial Guinea), Gabon, Cameroon, Congo Republic.
STATUS: CITES - Appendix II.
SYNONYMS: *batesi, cameronensis, gabonensis*.

Galago gallarum Thomas, 1901. Ann. Mag. Nat. Hist., ser. 7, 8:27.
TYPE LOCALITY: Ethiopia, Webi Dau, Boran County.
DISTRIBUTION: Between Tana River (Kenya) and Shebele River (Somalia), west of Ethiopian Rift.
STATUS: CITES - Appendix II.
COMMENTS: Recognized as a species by Nash et al. (1989) and by Groves (1989).

Galago matschiei Lorenz, 1917. Ann. K. K. Naturhist. Hofmus, Wien, 31:237.
TYPE LOCALITY: Zaire: Moera, Ituri River.
DISTRIBUTION: E Zaire; perhaps Uganda.
STATUS: CITES - Appendix II.
SYNONYMS: *inustus*.
COMMENTS: Placed in the subgenus *Euoticus* by Petter and Petter-Rousseaux (1979). This name replaces the more commonly used *inustus* (Groves, 1989:102; Nash et al., 1989:69-70).

Galago moholi A. Smith, 1836. Rept. Exped. Exploring Central Africa, 1834:42 [1836].
TYPE LOCALITY: South Africa, W Transvaal, Marico-Limpopo confluence.
DISTRIBUTION: Southern Africa, to SW Tanzania and E Zaire.
STATUS: CITES - Appendix II.
SYNONYMS: *australis, bradfieldi, conspicillatus, intontoi, mossambicus, nyassae, tumbolensis*.
COMMENTS: Separated from *senegalensis* by Jenkins (1987), Nash et al. (1989), and Groves (1989).

Galago senegalensis É. Geoffroy, 1796. Mag. Encyclop., 1:38.
TYPE LOCALITY: Senegal.
DISTRIBUTION: Senegal to Ethiopia, south to South Africa and Angola; perhaps Namibia.
STATUS: CITES - Appendix II.
SYNONYMS: *acaciarum, albipes, braccatus, calago, dunni, galago, geoffroyi, intontoi, pupulus, sennariensis, sotikae, teng*.
COMMENTS: Hill and Meester (1977:2) included *granti* in this species; but see Smithers and Wilson (1979).

Galagoides A. Smith, 1833. S. African Quart. J., (2):32.
TYPE SPECIES: *Galago demidoff* G. Fischer, 1806.
SYNONYMS: *Hemigalago*.

Galagoides demidoff (G. Fischer, 1806). Mem. Soc. Imp. Nat. Moscow, 1:24.
TYPE LOCALITY: Senegal.
DISTRIBUTION: Senegal to Uganda; Bioko (Equatorial Guinea); isolated forests of Kenya and Tanzania south to Malawi.
STATUS: CITES - Appendix II; IUCN - Insufficiently known as *G. thomasi*.
SYNONYMS: *anomurus, demidovii, medius, murinus, orinus, peli, phasma, poensis, pusillus, thomasi*.
COMMENTS: Nash et al. (1989) recognized *G. thomasi*, Elliot, 1907, as a full species, partially sympatric with *demidoff*. For use of *demidoff* in place of *demidovii*, see Jenkins (1987:98).

Galagoides zanzibaricus (Matschie, 1893). Sitzb. Ges. Naturf. Fr. Berlin, p. 111.
TYPE LOCALITY: Tanzania, Zanzibar, Yambiani.
DISTRIBUTION: E African coast from Tana River, south to S Mozambique; Zanzibar.
STATUS: CITES - Appendix II; IUCN - Vulnerable.
SYNONYMS: *cocos*; *granti* Thomas and Wroughton, 1907; *mertensi*.
COMMENTS: Separated from *senegalensis* by Kingdon (1971:309), Groves (1974*b*:463, 1989:103), Jenkins (1987:118), and Nash et al. (1989).

Otolemur Coquerel, 1859. Rev. Mag. Zool. Paris, ser. 2, 11:458.
TYPE SPECIES: *Otolemur agyisymbanus* Coquerel, 1859 (= *Otolicnus garnetti* Ogilby, 1838).
SYNONYMS: *Callotus*, *Otogale*.
COMMENTS: Removed from *Galago* by Groves (1974*b*:461-463, 1989) and Jenkins (1987:122).

Otolemur crassicaudatus (É. Geoffroy, 1812). Ann. Mus. Hist. Nat. Paris, 19:166.
TYPE LOCALITY: Mozambique, Quelimane (see Thomas, 1917*b*).
DISTRIBUTION: Kenya, Tanzania and Rwanda to Natal (South Africa) and Angola.
STATUS: CITES - Appendix II.
SYNONYMS: *argentatus*, *badius*, *kirkii*, *lestradei*, *lonnbergi*, *monteiri*, *umbrosus*, *zuluensis*.

Otolemur garnettii (Ogilby, 1838). Proc. Zool. Soc. Lond., 1838:6.
TYPE LOCALITY: Zanzibar (designated by Thomas, 1917*b*:48).
DISTRIBUTION: S Somalia to SE Tanzania (including Zanzibar, Pemba and Mafia Isls).
STATUS: CITES - Appendix II.
SYNONYMS: *agyisymbanus*, *hindei*, *hindsi*, *kikuyuensis*, *lasiotis*, *lasiotus*, *panganiensis*.
COMMENTS: Olson (1979, in litt.[Unpubl. Ph.D. Dissertation, Univ. London]) treated *garnettii* as a distinct species. Partly sympatric with *O. crassicaudatus*.

Family Tarsiidae Gray, 1825. Ann. Philos., n.s., 10:338.

Tarsius Storr, 1780. Prodr. Meth. Mamm., p. 33.
TYPE SPECIES: *Simia syrichta* Linnaeus, 1758.

Tarsius bancanus Horsfield, 1821. Zool. Res. Java, part 2:*Tarsius Bancanus*, pls.(col. and fig. g.) and 4 unno. pp.
TYPE LOCALITY: Indonesia, SE Sumatra, Bangka Isl.
DISTRIBUTION: Indonesia: Bangka Isl, Sumatra, Karimata Isl, Billiton Isl, and Sirhassen Isl (South Natuna Isls); Borneo.
STATUS: CITES - Appendix II.
SYNONYMS: *borneanus*, *natunensis*, *saltator*.

Tarsius dianae Niemitz, Nietsch, Warter, and Rumpler, 1991. Folia Primatol., 56:105.
TYPE LOCALITY: C Sulawesi, 2 km SE of Kamarora, 700m, 01°10'S, 120°09'E, near northern boundary of Lore Lindu National Park.
DISTRIBUTION: C Sulawesi (Indonesia).
STATUS: CITES - Appendix II. Status unknown.

Tarsius pumilus Miller and Hollister, 1921. Proc. Biol. Soc. Washington, 34:103.
TYPE LOCALITY: C Sulawesi, Rano Rano.
DISTRIBUTION: Known only from type locality and Latimojong Mtns, C Sulawesi (Indonesia).
STATUS: CITES - Appendix II; IUCN - Indeterminate.
COMMENTS: Separated from *T. spectrum* by Musser and Dagosto (1987).

Tarsius spectrum (Pallas, 1779). Nova Spec. Quadruped, Glir. Ord., p. 275.
TYPE LOCALITY: Indonesia, Sulawesi Selatan, Ujung Pandang (= Makasar).
DISTRIBUTION: Sulawesi lowlands, Peleng Isl, Sangir Isls, Savu Isl, and Selayar Isl (Indonesia).
STATUS: CITES - Appendix II.
SYNONYMS: *dentatus*, *fischerii*, *fuscomanus*, *fuscus*, *pallassii*, *pelengensis*, *podje*, *sangirensis*.
COMMENTS: Sangir form may be a separate species (Feiler, 1990:85). Niemitz et al. (1991) implied that this species may be restricted to N Sulawesi.

Tarsius syrichta (Linnaeus, 1758). Syst. Nat., 10th ed., 1:29.
TYPE LOCALITY: Philippine Isls, Samar Isl.
DISTRIBUTION: Mindanao, Bohol Isl, Samar Isl, Leyte Isl (Philippines).
STATUS: CITES - Appendix II; U.S. ESA - Threatened; IUCN - Endangered.
SYNONYMS: *buffonnii, carbonarius, daubentonii, fraterculus, macauco, macrotarsos, minimus, philippensis, tarsier, tarsius.*

Family Callitrichidae Gray, 1821. London Med. Repos., 15:298.
SYNONYMS: Hapalidae.
COMMENTS: Generally spelt Callithricidae, but see Napier and Napier (1967).

Callimico Miranda-Ribeiro, 1912. Brasil. Rundsch., p. 21.
TYPE SPECIES: *Callimico snethlageri* Miranda-Ribeiro, 1912 (= *Callimico goeldii* Thomas, 1904).
COMMENTS: Placed in a separate family, Callimiconidae, by Hershkovitz (1977), but recognized as a Callitrichid by Pocock (1920*a*), Napier (1976), and Groves (1989).

Callimico goeldii (Thomas, 1904). Ann. Mag. Nat. Hist., ser. 7, 14:189.
TYPE LOCALITY: Brazil, Acre, Rio Yaco.
DISTRIBUTION: W Brazil, N Bolivia, E Peru, Colombia: Upper Amazon Rainforests.
STATUS: CITES - Appendix I; U.S. ESA - Endangered; IUCN - Rare.
SYNONYMS: *snethlageri.*

Callithrix Erxleben, 1777. Syst. Regni Anim., p. 55.
TYPE SPECIES: *Simia jacchus* Linnaeus, 1758.
SYNONYMS: *Anthopithecus, Arctopithecus, Cebuella, Hapale, Jacchus, Liocephalus, Mico, Micoella, Middas, Ouistitis.*
COMMENTS: Includes *Mico,* see Cabrera (1958:185); and *Cebuella,* see Groves (1989:110-115).

Callithrix argentata (Linnaeus, 1771). Mantissa Plantarum, 2, Appendix:521.
TYPE LOCALITY: Brazil, Pará, Cametá, on banks of Rio Tocantins.
DISTRIBUTION: N and C Brazil, E Bolivia.
STATUS: CITES - Appendix II; IUCN - Vulnerable as *C. a. leucippe* and *C. humeralifer intermedius.*
SYNONYMS: *emiliae, intermedius, leucippe, leucomerus, leukeurin, melanura.*
COMMENTS: Includes *emiliae, melanura* and *leucippe* (Hershkovitz, 1977:436), and *intermedius* (Ávila-Pires, 1985). De Vivo (1985) recognized *emiliae* (Thomas, 1920) as a full species. Coimbra-Filho (1990) recognized *emiliae* and *intermedius* as distinct species.

Callithrix aurita (É. Geoffroy, 1812). Ann. Mus. Hist. Nat. Paris, 19:119.
TYPE LOCALITY: Brazil, Rio de Janeiro.
DISTRIBUTION: SE Brazilian coast.
STATUS: CITES - Appendix I; U.S. ESA and IUCN - Endangered.
SYNONYMS: *chrysopyga, coelestis, itatiayae, petronius.*
COMMENTS: Accepted as a species by Mittermeier et al. (1988) and Groves (1989).

Callithrix flaviceps (Thomas, 1903). Ann. Mag. Nat. Hist., ser. 7, 12:240.
TYPE LOCALITY: Brazil, Rive, Espírito Santo.
DISTRIBUTION: S Espírito Santo.
STATUS: CITES - Appendix I; U.S. ESA and IUCN - Endangered.
SYNONYMS: *flavescente.*
COMMENTS: Accepted as a species by Mittermeier et al. (1988) and Groves (1989). Regarded as a subspecies of *aurita* by Coimbra-Filho (1990).

Callithrix geoffroyi (Humboldt, 1812). Rec. Obs. Zool. Anat. Comp., 360.
TYPE LOCALITY: Brazil, Victoria.
DISTRIBUTION: EC Brazil (coast of Bahia).
STATUS: CITES - Appendix II.
SYNONYMS: *albifrons, leucocephala, leucogenys, maximiliani.*
COMMENTS: Accepted as a species by Mittermeier et al. (1988) and Groves (1989).

Callithrix humeralifera (É. Geoffroy, 1812). Ann. Mus. Hist. Nat. Paris, 19:120.
TYPE LOCALITY: Brazil, left bank of Rio Tapajós, Paricatuba.

DISTRIBUTION: Brazil, between Madeira and Tapajós rivers, south of the Amazon.
STATUS: CITES - Appendix II; IUCN - Insufficiently known as *C. h. chrysoleuca*.
SYNONYMS: *chrysoleuca, melanoleucus, santaremensis, sericeus*.
COMMENTS: For synonyms see Hershkovitz (1977:595-598); *intermedius* was transfered to *C. argentata* by Ávila-Pires (1985). Coimbra-Filho (1990), who called this species *humeralifera*, listed *chrysoleuca* and *intermedia* as distinct species.

Callithrix jacchus (Linnaeus, 1758). Syst. Nat., 10th ed., 1:27.
TYPE LOCALITY: Brazil, Pernambuco.
DISTRIBUTION: Brazilian coast: Piauí, Ceará, and Pernambuco Provinces.
STATUS: CITES - Appendix II.
SYNONYMS: *albicollis, communis, hapale, leucotis, tertius, vulgaris*.
COMMENTS: Includes *aurita, flaviceps, geoffroyi*, and *penicillata* according to Hershkovitz (1977:489-527), but refuted by Mittermeier et al. (1988) who recognized all of these species.

Callithrix kuhlii (Wied-Neuwied, 1826). Beitr. Naturgesch. Brasil, 2:142.
TYPE LOCALITY: Brazil, north of Rio Belmonte.
DISTRIBUTION: Between Rio de Contas and Rio Jequitinhonha, SW Brazil.
STATUS: CITES - Appendix II.
COMMENTS: Accepted as a species by Mittermeier et al. (1988) and Groves (1989).

Callithrix penicillata (É. Geoffroy, 1812). Ann. Mus. Hist. Nat., Paris, 19:119.
TYPE LOCALITY: Brazil, Lamarão, Bahia.
DISTRIBUTION: Brazilian coast: Bahia to São Paulo, inland to Goiás.
STATUS: CITES - Appendix II.
SYNONYMS: *jordani, melanotis, trigonifer*.
COMMENTS: Accepted as a species by Mittermeier et al. (1988) and Groves (1989).

Callithrix pygmaea (Spix, 1823). Sim. Vespert. Brasil., p. 32.
TYPE LOCALITY: Brazil, Amazonas, Solimões River, Tabatinga.
DISTRIBUTION: N and W Brazil, N Peru, Ecuador.
STATUS: CITES - Appendix II.
SYNONYMS: *nigra, niveiventris*.
COMMENTS: Revised by Hershkovitz (1977:462-464), placed in *Callithrix* by Groves (1989).

Leontopithecus Lesson, 1840. Spec. Mamm. Bim. Quadrum., 1844:184, 200.
TYPE SPECIES: *Leontopithecus marikina* Lesson, 1840 (= *Simia rosalia* Linnaeus, 1766).
SYNONYMS: *Leontideus, Leontocebus*.
COMMENTS: For synonyms see Hershkovitz (1977:807-808).

Leontopithecus caissara Lorini and Persson, 1990. Bol. Mus. Nac., Rio de Janeiro, N.S., 338:2.
TYPE LOCALITY: Brazil, Superagui Isl (south of São Paulo), Guaraquecaba, 25°18'S, 48°11'W.
DISTRIBUTION: Superagui Isl.
STATUS: CITES - Appendix I; U.S. ESA and IUCN - Endangered.
COMMENTS: Considered a subspecies of *chrysopygus* by Coimbra-Filho (1990).

Leontopithecus chrysomelas (Kuhl, 1820). Beitr. Zool. Vergl. Aust., 51.
TYPE LOCALITY: Brazil, Ribeirao das Minhocas, S Bahia.
DISTRIBUTION: Brazil, coastal Bahia.
STATUS: CITES - Appendix I; U.S. ESA and IUCN - Endangered.
SYNONYMS: *chrysurus*.
COMMENTS: Recognized as a full species by Rosenberger and Coimbra-Filho (1984).

Leontopithecus chrysopygus (Mikan, 1823). Delectus florae et faunae Brasiliensis, Vienna, 3, plate.
TYPE LOCALITY: Brazil, Ipanema, São Paulo.
DISTRIBUTION: Brazil, São Paulo region.
STATUS: CITES - Appendix I; U.S. ESA and IUCN - Endangered.
SYNONYMS: *ater*.
COMMENTS: Recognized as a full species by Rosenberger and Coimbra-Filho (1984).

Leontopithecus rosalia (Linnaeus, 1766). Syst. Nat., 12th ed., 1:41.
TYPE LOCALITY: Brazil, coast, Rio São João (right bank), between 22° and 23°S. Locality is for *L. r. rosalia*, see Wied-Neuwied (1826) and Carvalho (1965).
DISTRIBUTION: SE Brazil: Rio Doce (Espírito Santo) south into Rio de Janeiro and Guanabara.
STATUS: CITES - Appendix I; U.S. ESA and IUCN - Endangered.
SYNONYMS: *aurora, brasiliensis, guyannensis, leoninus, marikina.*
COMMENTS: See Kleiman (1981, Mammalian Species, 148).

Saguinus Hoffmannsegg, 1807. Mag. Ges. Naturf. Fr., 1:101.
TYPE SPECIES: *Saguinus ursula* Hoffmannsegg, 1807 (= *Simia midas* Linnaeus, 1758).
SYNONYMS: *Hapanella, Leontocebus, Marikina, Midas, Mystax, Oedipomidus, Oedipus, Seniocebus, Tamarin, Tamarinus.*
COMMENTS: See Hershkovitz (1977:601-603).

Saguinus bicolor (Spix, 1823). Sim. Vespert. Brasil., p. 30.
TYPE LOCALITY: Brazil, Manaus, Barra de Rio Negro.
DISTRIBUTION: N Brazil; perhaps NE Peru.
STATUS: CITES - Appendix I; U.S. ESA and IUCN - Endangered.
SYNONYMS: *martinsi, ochraceus.*
COMMENTS: See Hershkovitz (1977:744).

Saguinus fuscicollis (Spix, 1823). Sim. Vespert. Brasil., p. 27.
TYPE LOCALITY: Brazil, between Solimões River and Iça River, São Paulo de Olivença.
DISTRIBUTION: N and W Brazil, N Bolivia, E Peru, E Ecuador, SW Colombia.
STATUS: CITES - Appendix II.
SYNONYMS: *acrensis, apiculatus, avilapiresi, bluntschlii, crandalli, cruzlimai, devillei, elegans, flavifrons, fuscus, hololeucus, illigeri, imberbis, lagonotus, leonina, leucogenys, melanoleucus, micans, mounseyi, nigrifrons, pacator, pebilis, primitivus, purillus, weddelli.*
COMMENTS: See Hershkovitz (1977:640-642). Coimbra-Filho (1990) regarded *melanoleucus* as a distinct species, with subspecies *acrensis* and *crandalli.*

Saguinus geoffroyi (Pucheran, 1845). Rev. Mag. Zool. Paris, 8:336.
TYPE LOCALITY: Panama, Canal Zone.
DISTRIBUTION: SE Costa Rica to NW Colombia.
STATUS: CITES - Appendix I.
SYNONYMS: *salaguiensis, spixii.*
COMMENTS: Distinct from *S. oedipus*; see Natori (1988).

Saguinus imperator (Goeldi, 1907). Proc. Zool. Soc. Lond., 1907:93.
TYPE LOCALITY: Brazil, Rio Acre.
DISTRIBUTION: W Brazil, E Peru.
STATUS: CITES - Appendix II; IUCN - Indeterminate.
SYNONYMS: *subgrisescens.*

Saguinus inustus (Schwartz, 1951). Am. Mus. Novit., 1508:1.
TYPE LOCALITY: Brazil, Amazonas, Tabocal.
DISTRIBUTION: NW Brazil, SW Colombia.
STATUS: CITES - Appendix II.

Saguinus labiatus (É. Geoffroy, 1812). Rec. Observ. Zool., 1:361.
TYPE LOCALITY: Brazil, Amazonas, Lake Joanacan.
DISTRIBUTION: W Brazil, E Peru.
STATUS: CITES - Appendix II.
SYNONYMS: *elagantulus, erythrogaster, griseovertex, rufiventer, thomasi.*

Saguinus leucopus (Günther, 1877). Proc. Zool. Soc. Lond., 1876:746 [1877].
TYPE LOCALITY: Colombia, Antioquia, Medellin.
DISTRIBUTION: N Colombia.
STATUS: CITES - Appendix I; U.S. ESA - Threatened; IUCN - Endangered.
SYNONYMS: *pegasis.*

Saguinus midas (Linnaeus, 1758). Syst. Nat., 10th ed., 1:28.
TYPE LOCALITY: Surinam.
DISTRIBUTION: N Brazil, Guyana, French Guiana, Surinam.

STATUS: CITES - Appendix II.
SYNONYMS: *egens, gracilis, lacepedii, negre, niger, rufimanus, tamarin, umbratus, ursula, ursulus.*
COMMENTS: Includes *tamarin*; see Hershkovitz (1977:711).

Saguinus mystax (Spix, 1823). Sim. Vespert. Brasil., p. 29.
TYPE LOCALITY: Brazil, Amazonas, between Solimões River and Içа River.
DISTRIBUTION: W Brazil, Peru.
STATUS: CITES - Appendix II.
SYNONYMS: *juruanus, pileatus, pluto.*
COMMENTS: See Hershkovitz (1977:700).

Saguinus nigricollis (Spix, 1823). Sim. Vespert. Brasil., p. 28.
TYPE LOCALITY: Brazil, Amazonas, São Paulo de Olivença.
DISTRIBUTION: W Brazil, E Peru, E Ecuador.
STATUS: CITES - Appendix II.
SYNONYMS: *graellsi, hernandezi, rufoniger.*
COMMENTS: Includes *graellsi*; see Hershkovitz (1977:628).

Saguinus oedipus (Linnaeus, 1758). Syst. Nat., 10th ed., 1:28.
TYPE LOCALITY: Colombia, Bolivar, lower Rio Sinu.
DISTRIBUTION: N Colombia, Panama.
STATUS: CITES - Appendix I; U.S. ESA - Endangered; IUCN - Endangered as *S. o. oedipus.*
SYNONYMS: *doguin, meticulosus, titi.*
COMMENTS: Does not include *geoffroyi*; see Natori (1988).

Saguinus tripartitus (Milne-Edwards, 1878). Arch. Mus. Hist. Nat. Paris, (2)1:161.
TYPE LOCALITY: Equador, Rio Napo.
DISTRIBUTION: East of Rio Curaray, Brazil-Colombia border; sympatric with *S. fuscicollis* around Curaray-Napo Confluence.
STATUS: CITES - Appendix II.
COMMENTS: Given specific rank by Thorington (1988) on the evidence of marginal sympatry with *S. fuscicollis.*

Family Cebidae Bonaparte, 1831. Saggio Dist. Metod. Anim. Vert., p. 6.
SYNONYMS: Aotidae, Atelidae, Callicebidae, Pitheciidae.
COMMENTS: Groves (1989) divided this into several different families; Martin (1990) recommended the conservative approach.

Subfamily Alouattinae Trouessart, 1897. Cat. Mamm., new ed., fasc. I, p. 32.

Alouatta Lacépède, 1799. Tabl. Div. Subd. Orders Genres Mammifères, p. 4.
TYPE SPECIES: *Simia belzebul* Linnaeus, 1766.
SYNONYMS: *Mycetes* Illiger, 1811; *Stentor* É. Geoffroy, 1812.

Alouatta belzebul (Linnaeus, 1766). Syst. Nat., 12th ed., 1:37.
TYPE LOCALITY: Brazil, Pará, Rio Capim.
DISTRIBUTION: N Brazil (mainly south of Lower Amazon, east of Rio Madeira); Mexiana Isl (Brazil); in Pará Prov. (Brazil), N of Amazon.
STATUS: CITES - Appendix II.
SYNONYMS: *beelzebub; discolor* Spix, 1823; *flavimanus, mexianae; nigerrima* Lönnberg, 1961; *rufimanus* Kuhl, 1820; *tapojozensis, ululata, villosus.*

Alouatta caraya (Humboldt, 1812). Rec. Observ. Zool., 1:355.
TYPE LOCALITY: Paraguay.
DISTRIBUTION: N Argentina to Mato Grosso (Brazil).
STATUS: CITES - Appendix II.
SYNONYMS: *barbatus, niger, nigra.*

Alouatta coibensis Thomas, 1902. Novit. Zool., 9:135.
TYPE LOCALITY: Panama, Coiba Isl.
DISTRIBUTION: Coiba Isl and Azuero Penninsula, Panama.
STATUS: Not in CITES but very threatened.
SYNONYMS: *trabeata.*

COMMENTS: Recognized as a full species by Froehlich and Froehlich (1987).

Alouatta fusca (É. Geoffroy, 1812). Ann. Mus. Hist. Nat. Paris, 19:108.
TYPE LOCALITY: Brazil.
DISTRIBUTION: N Bolivia (?- *beniensis*); SE and EC Brazil.
STATUS: CITES - Appendix II; IUCN - Vulnerable.
SYNONYMS: *beniensis*?, *bicolor*, *clamitans*; *guariba*, Humboldt, 1812; *iheringi*; *ursinus* É. Geoffroy, 1812.
COMMENTS: *A. guariba* (Humboldt, 1812) may be the correct name for this species. *A. "fusca" beniensis* may actually be *A. seniculus* (Mittermeier et al., 1988:13-75).

Alouatta palliata (Gray, 1849). Proc. Zool. Soc. Lond., 1848:138 [1849].
TYPE LOCALITY: Nicaragua, Lake Nicaragua.
DISTRIBUTION: W Ecuador to Veracruz and Oaxaca (Mexico).
STATUS: CITES - Appendix I; U.S. ESA - Endangered.
SYNONYMS: *aequatorialis*, *inclamax*, *inconsonans*, *matagalpae*, *mexicana*, *quichua*.

Alouatta pigra Lawrence, 1933. Bull. Mus. Comp. Zool., 75:333.
TYPE LOCALITY: Guatemala, Petén, Uaxactun.
DISTRIBUTION: Yucatan and Chiapas (Mexico) to Belize and Guatemala.
STATUS: CITES - Appendix II; U.S. ESA - Threatened; IUCN - Insufficiently known as *A. villosa*.
SYNONYMS: *luctuosa*, *villosa*?.
COMMENTS: The name *villosa* has been applied to this species (see Napier, 1976:76) but Lawrence (1933) regarded it as a *nomen dubium*; see Smith (1970:366) and Hall (1981:260, 263).

Alouatta sara Elliot, 1910. Ann. Mag. Nat. Hist., ser. 8, 5:283.
TYPE LOCALITY: Bolivia, Sara Prov., Santa Cruz.
DISTRIBUTION: Environs of Rio Paray, Santa Cruz, Bolivia.
STATUS: CITES - Appendix II.
COMMENTS: Separated from *A. seniculus* by Minezawa et al. (1985).

Alouatta seniculus (Linnaeus, 1766). Syst. Nat., 12th ed., 1:37.
TYPE LOCALITY: Colombia, Bolivar, Rio Magdalena, Cartagena.
DISTRIBUTION: Trinidad; Bolivia to Ecuador; Colombia to Guyana, French Guiana, Surinam, and NC Brazil.
STATUS: CITES - Appendix II.
SYNONYMS: *amazonica*, *arctoidea*, *auratus*, *bogotensis*, *caucensis*, *caquetensis*, *chrysurus*, *insularis*, *juara*, *juruana*, *laniger*, *macconnelli*, *puruensis*, *rubicunda*, *rubiginosa*, *straminea*; *ursina* Humboldt, 1815.

Subfamily Aotinae Poche, 1908. Zool. Annalen, 2, 4:269.
COMMENTS: Groves (1989) suggested that this is a full family.

Aotus Illiger, 1811. Prodr. Syst. Mamm. Avium., p. 71.
TYPE SPECIES: *Simia trivirgata* Humboldt, 1811.
SYNONYMS: *Nyctipithecus*.

Aotus azarai (Humboldt, 1811). Rec. Observ. Zool., 1:359.
TYPE LOCALITY: Argentina, right bank of Rio Paraguay.
DISTRIBUTION: Bolivia south of Rio Madre de Dios, south to Paraguay and N Argentina.
STATUS: CITES - Appendix II.
SYNONYMS: *bidentatus*; *boliviensis* Elliot, 1907; *miriquouina*, *rufipes*, *stenorrhina*.
COMMENTS: The form of this name is changed in accordance with Art. 31 (a(ii)) of the International Code of Zoological Nomenclature (International Commission on Zoological Nomenclature, 1985)

Aotus brumbacki Hershkovitz, 1983. Ann. J. Primatol., 4;217.
TYPE LOCALITY: Colombia, Dept. of Meta, Villavicencio region.
DISTRIBUTION: E Colombia, from E Boyaca east to Dept. of Meta, perhaps to Venezuelan border.
STATUS: CITES - Appendix II. Status unknown.

Aotus hershkovitzi Ramirez-Cerquera, 1983. IX Cong. Latinoamer. Zool. [abstracts], Arequipa, Peru, p. 148.
TYPE LOCALITY: Colombia, Dept. of Meta, east side of Cordillera Oriental.
DISTRIBUTION: Known from the type locality only.
STATUS: CITES - Appendix II. Status unknown.

Aotus infulatus (Kuhl, 1820). Beitr. Zool. Vergl. Anat., 1820:30.
TYPE LOCALITY: "Brazil".
DISTRIBUTION: Brazil, south of Lower Amazon, east of the Tapajoz-Juruena.
STATUS: CITES - Appendix II.
SYNONYMS: *commersoni, felina, noctivaga; roberti* Dollman, 1909.
COMMENTS: May be conspecific with *A. azarai*; see Pieczarka and Nagamuchi (1988).

Aotus lemurinus (I. Geoffroy, 1843). C. R. Acad. Sci. Paris, 16:1151.
TYPE LOCALITY: Colombia, Dept. of Caldas, Quindio.
DISTRIBUTION: Panama, Equador and Colombia west of Cordillera Oriental.
STATUS: CITES - Appendix II.
SYNONYMS: *aversus, bipunctatus; griseimembra* Elliot, 1912; *hirsutus, lanius, pervigilis, villosus; zonalis* Goldman, 1914.

Aotus miconax Thomas, 1927. Ann. Mag. Nat. Hist., ser. 9, 19:365.
TYPE LOCALITY: Peru, Amazonas, San Nicolas, 4500 ft.
DISTRIBUTION: A small area in Peru between Rio Ucayali and the Andes, south of Rio Marañon.
STATUS: CITES - Appendix II.

Aotus nancymaae Hershkovitz, 1983. Am. J. Primatol., 4:223.
TYPE LOCALITY: Peru, Loreto, right bank of Rio Samiria, above Estacion Pithecia, 130 m.
DISTRIBUTION: Loreto Dept. (Peru) to Rio Jandiatuba, south of Rio Solimões (Brazil); and enclave between Rios Tigre and Pastaza (Peru).
STATUS: CITES - Appendix II.
COMMENTS: The form of the name is changed in accordance with the International Code of Zoological Nomenclature, Art. 31 (a(ii)) (International Commission on Zoological Nomenclature, 1985).

Aotus nigriceps Dollman, 1909. Ann. Mag. Nat. Hist., ser. 8, 4:200.
TYPE LOCALITY: Peru, Chanchamayo, 1000 m.
DISTRIBUTION: Brazil, south of Rio Solimões, west of Rio Tapajós Juruena, west into Peru.
STATUS: CITES - Appendix II.
SYNONYMS: *senex.*

Aotus trivirgatus (Humboldt, 1811). Rec. Observ. Zool., 1:306.
TYPE LOCALITY: Venezuela, Duida Range, Rio Casiquiare.
DISTRIBUTION: Venezuela, south of Rio Orinoco, south to Brazil north of Rios Negro and Amazon.
STATUS: CITES - Appendix II.
SYNONYMS: *duruculi; felinus* Spix, 1823; *humboldtii* Schultz, 1823; *rufus.*
COMMENTS: Divided into 9 species by Hershkovitz (1983).

Aotus vociferans (Spix, 1823). Sim. Vespert. Brasil. 25.
TYPE LOCALITY: Brazil, upper Marañon, Tabatinga.
DISTRIBUTION: Colombia, east of Cordillera Oriental, west of Rio Negro, south to Brazil (north of Amazon-Solimões Rivers).
STATUS: CITES - Appendix II.
SYNONYMS: *gularis* Dollman, 1909; *microdon* Dollman, 1909; *oseryi, spixii.*

Subfamily Atelinae Gray, 1825. Ann. Philos., n.s., 10:338.
COMMENTS: Combined with Alouattinae as family Atelidae by Rosenberger (1977).

Ateles É. Geoffroy, 1806. Ann. Mus. Hist. Nat. Paris, 1:262.
TYPE SPECIES: *Simia paniscus* Linnaeus, 1758.
SYNONYMS: *Ameranthropoides* Montandon; *Atelocheirus; Montaneia* Ameghino; *Paniscus, Sapaja.*

COMMENTS: All *Ateles* are considered conspecific by Hernandez-Camacho and Cooper (1976:66).

Ateles belzebuth É. Geoffroy, 1806. Ann. Mus. Hist. Nat. Paris, 7:27.
TYPE LOCALITY: Venezuela, Esmeralda.
DISTRIBUTION: Cordillera Oriental, Colombia to Venezuela and N Peru.
STATUS: CITES - Appendix II; IUCN - Vulnerable.
SYNONYMS: *bartlettii, braccatus, brissonii, brunneus, chuva, fuliginosus; hybridus* I. Geoffroy; *loysi, problema, variegatus.*

Ateles chamek (Humboldt, 1812). Rec. Observ. Zool., 1:353.
TYPE LOCALITY: Peru, Cuzco, Rio Comberciato.
DISTRIBUTION: NE Peru, E Bolivia to Brazil west of Rio Juruá and south of Rio Solimões.
STATUS: CITES - Appendix II.
SYNONYMS: *longimembris, peruvianus.*
COMMENTS: Separated from *paniscus* by Groves (1989).

Ateles fusciceps Gray, 1866. Proc. Zool. Soc. Lond., 1865:733 [1866].
TYPE LOCALITY: NW Ecuador, Imbabura Prov., Hacienda Chinipamba, 1500 m.
DISTRIBUTION: SE Panama to Ecuador, Colombia to W Cordillera (Paraguay).
STATUS: CITES - Appendix II; IUCN - Vulnerable.
SYNONYMS: *dariensis* Goldman, 1915; *robustus* J. A. Allen, 1914.

Ateles geoffroyi Kuhl, 1820. Beitr. Zool. Vergl. Anat., 1:26.
TYPE LOCALITY: Nicaragua, San Juan del Norte.
DISTRIBUTION: S Mexico to Panama.
STATUS: CITES - Appendix I and U.S. ESA - Endangered as *A. g. frontatus* and *A. g. panamensis* only, otherwise Appendix II; IUCN - Vulnerable.
SYNONYMS: *azuerensis, cucullatus, frontatus, grisescens, melanocercus, melanochir, neglectus, ornatus, pan, panamensis, rufiventris, trianguligera, tricolor, vellerosus, yucatanensis.*

Ateles marginatus É. Geoffroy, 1809. Ann. Mus. Hist. Nat. Paris, 13:97.
TYPE LOCALITY: Brazil, Rio Tocantins, Cametá.
DISTRIBUTION: South of Lower Amazon, R Tapajós to Rio Tocantins.
STATUS: CITES - Appendix II.
SYNONYMS: *albifrons, frontalis.*
COMMENTS: Separated from *belzebuth* by Groves (1989).

Ateles paniscus (Linnaeus, 1758). Syst. Nat., 10th ed., 1:26.
TYPE LOCALITY: French Guyana.
DISTRIBUTION: North of the Amazon (east of Rio Negro).
STATUS: CITES - Appendix II; IUCN - Vulnerable.
SYNONYMS: *ater* F. Cuvier; *cayennensis, pentadactylus, subpentadactylus, surinamensis.*

Brachyteles Spix, 1823. Sim. Vespert. Brasil, 36.
TYPE SPECIES: *Brachyteles macrotarsus* Spix, 1823 (= *Ateles arachnoides* É. Geoffroy, 1806).
SYNONYMS: *Brachyteleus, Eriodes.*

Brachyteles arachnoides (É. Geoffroy, 1806). Ann. Mus. Hist. Nat. Paris, 7:271.
TYPE LOCALITY: Brazil, Rio de Janeiro, Brazil.
DISTRIBUTION: SE Brazil.
STATUS: CITES - Appendix I; U.S. ESA and IUCN - Endangered.
SYNONYMS: *eriodes, hemidactylus, hypoxanthus, macrotarsus, tuberifer.*

Lagothrix É. Geoffroy, 1812. *In* Humboldt, Rec. Observ. Zool., 1:356.
TYPE SPECIES: *Lagothrix humboldtii* É. Geoffroy, 1812 (= *Simia lagothricha* Humboldt, 1812).
SYNONYMS: *Gastrimargus; Oreonax* Thomas, 1927.

Lagothrix flavicauda Humboldt, 1812. Rec. Observ. Zool., 1:363.
TYPE LOCALITY: Peru, San Martin, Puca Tambo, 5100 ft.
DISTRIBUTION: E Andes in San Martin (Peru) and Amazonas (Brazil).
STATUS: CITES - Appendix I; U.S. ESA and IUCN - Endangered.
SYNONYMS: *hendeei* Thomas.

Lagothrix lagotricha (Humboldt, 1812). Rec. Observ. Zool., 1:322.
TYPE LOCALITY: Colombia, Úapes, Rio Guaviare.
DISTRIBUTION: Middle and Upper Amazon.
STATUS: CITES - Appendix II; IUCN - Vulnerable.
SYNONYMS: *barrigo, cana* E. Geoffroy; *caparro, caroarensis, castelnaui, geoffroyi; humboldtii* E. Geoffroy; *infumata, lugens, olivaceus, poeppigii, puruensis, thomasi, tschudii, ubericola.*

Subfamily Callicebinae Pocock, 1925. Proc. Zool. Soc. Lond., 1925:45.
COMMENTS: Regarded as a full family by Groves (1989).

Callicebus Thomas, 1903. Ann. Mag. Nat. Hist., ser. 7, 12:456.
TYPE SPECIES: *Simia personatus* É. Geoffroy, 1812.
COMMENTS: Revised by Hershkovitz (1990*b*).

Callicebus brunneus (Wagner, 1842). Arch. Naturgesch., 8:357.
TYPE LOCALITY: Brazil, Rondônia, upper Rio Madeira, Cachoeira da Bananeira.
DISTRIBUTION: Middle to upper Madeira basin in Peru and Brazil, to upper Rio Purús (Brazil) and Ucayali (Peru).
STATUS: CITES - Appendix II.
COMMENTS: Perhaps a subspecies of *moloch,* see Groves (1992).

Callicebus caligatus (Wagner, 1842). Arch. Naturgesch., 8:357.
TYPE LOCALITY: Brazil, Rio Madeira, Borba.
DISTRIBUTION: W Brazil, south of Rio Solimões, from Rio Ucayali-Tapiche in Peru.
STATUS: CITES - Appendix II.
SYNONYMS: *castaneoventris, usto-fuscus.*
COMMENTS: Perhaps a synonym of *cupreus,* see Groves (1992).

Callicebus cinerascens (Spix, 1823). Sim. Vespert. Brasil., p. 20.
TYPE LOCALITY: "Río Putumayo or Iça, Peruvian border of Amazonas, Brazil"; Hershkovitz (1990*b*) doubted its accuracy.
DISTRIBUTION: Rio Madeira basin (SW Brazil).
STATUS: CITES - Appendix II.

Callicebus cupreus (Spix, 1823). Sim. Vespert. Brasil., p. 23.
TYPE LOCALITY: Brazil, Rio Solimões, Tabatinga.
DISTRIBUTION: South of the Amazon from Rio Purús to Rio Ucayali, Brazil and Peru.
STATUS: CITES - Appendix II.
SYNONYMS: *acreanus, discolor, egeria, leucometopa, napoleon, ornatus, paenulatus, rutteri, subrufus, toppini.*

Callicebus donacophilus (d'Orbigny, 1836). Voy. Am. Merid., Atlas Zool., pl. 5.
TYPE LOCALITY: Bolivia, Moxos Prov., Río Marmoré.
DISTRIBUTION: WC Bolivia, El Beni and Santa Cruz Provs.; W Paraguay; Mato Grosso (Brazil).
STATUS: CITES - Appendix II.
SYNONYMS: *pallescens.*

Callicebus dubius Hershkovitz, 1988. Proc. Acad. Nat. Sci. Philadelphia, 140:264.
TYPE LOCALITY: Brazil, probably lower Rio Purús, opposite Lago di Aiapuá.
DISTRIBUTION: Brazil, between Rio Madeira and Rio Purús.
STATUS: CITES - Appendix II.
COMMENTS: Perhaps a synonym of *cupreus,* see Groves (1992).

Callicebus hoffmannsi Thomas, 1908. Ann. Mag. Nat. Hist., ser. 8, 2:89.
TYPE LOCALITY: Brazil, Pará, Urucurituba, Rio Tapajós.
DISTRIBUTION: C Brazil, south of Amazon, between Rios Madeira and Tapajós.
STATUS: CITES - Appendix II.
SYNONYMS: *baptista.*
COMMENTS: Perhaps a subspecies of *moloch,* see Groves (1992).

Callicebus modestus Lönnberg, 1939. Ark. f. Zool., 31A:17.
TYPE LOCALITY: Bolivia, Beni, El Consuelo.
DISTRIBUTION: Upper Río Beni basin (Bolivia).

STATUS: CITES - Appendix II.

Callicebus moloch (Hoffmannsegg, 1807). Magas. Ges. Nativf. Fr., 9:97.
TYPE LOCALITY: Brazil, Pará, or near Belém.
DISTRIBUTION: C Brazil, between Rios Tapajós and Araguaia.
STATUS: CITES - Appendix II.
SYNONYMS: *emiliae, geoffroyi, hypokantha, remulus, sakir.*
COMMENTS: See Jones and Anderson (1978, Mammalian Species, 112).

Callicebus oenanthe Thomas, 1924. Ann. Mag. Nat. Hist., ser. 9, 14:286.
TYPE LOCALITY: Peru, San Martín Dept., Moyobamba, 840 m.
DISTRIBUTION: Rio Mayo valley (N Peru).
STATUS: CITES - Appendix II.

Callicebus olallae Lönnberg, 1939. Ark. f. Zool., 31A:16.
TYPE LOCALITY: Bolivia, El Beni Prov., La Laguna (5 km from Santa Rosa).
DISTRIBUTION: Known only from the type locality.
STATUS: CITES - Appendix II.

Callicebus personatus (É. Geoffroy, 1812). *In* Humboldt, Rec. Observ. Zool., 1:357.
TYPE LOCALITY: Brazil, Espírito Santo, Rio Doce (see Hershkovitz, 1990*b*).
DISTRIBUTION: SE Brazil, between Rio Itapicuru and Rio Tietê, inland to Rio Paraná-Paraíba.
STATUS: CITES - Appendix II; IUCN - Endangered.
SYNONYMS: *barbarabrownae, brunello, chloroenemis, crinicaudus, gigot, grandis, incanescens, melanochir, melanops, migrufus, nigrifrons.*

Callicebus torquatus (Hoffmannsegg, 1807). Magas. Ges. Nativf. Fr., 10:86.
TYPE LOCALITY: Brazil, Codajás, north bank of Rio Solimões.
DISTRIBUTION: S Columbia, Venezuela to Upper Amazon, 7°S, east to Río Negro and Río Purus.
STATUS: CITES - Appendix II.
SYNONYMS: *amictus, duida, ignitus, lucifer, lugens, medemi, purinus, regulus, vidua.*

Subfamily Cebinae Bonaparte, 1831. Saggio. Dist. Metod. Anim. Vert., p. 6.
SYNONYMS: Saimirinae.

Cebus Erxleben, 1777. Syst. Regni Anim., p. 44.
TYPE SPECIES: *Simia capucina* Linnaeus, 1758.
SYNONYMS: *Calyptrocebus, Eucebus, Otocebus, Pseudocebus, Sapajus.*

Cebus albifrons (Humboldt, 1812). Rec. Observ. Zool., 1:324.
TYPE LOCALITY: Venezuela, Orinoco River.
DISTRIBUTION: Venezuela, Colombia, Ecuador, N Peru, NW Brazil, Trinidad.
STATUS: CITES - Appendix II.
SYNONYMS: *adustus, aequatorialis, castaneus, cesarae, chrysopus, cuscinus, flavescens; flavus* Elliot; *gracilis; hypoleuca* Humboldt, 1812; *leucocephalus, malitiosus, pleei; unicolor* Spix, 1823; *versicolor, yuracus.*
COMMENTS: May be conspecific with *C. capucinus;* intermediates occur in Columbia (middle San Jorge Valley; Lower Cauca River), see Hernandez-Camacho and Cooper (1976:58).

Cebus apella (Linnaeus, 1758). Syst. Nat., 10th ed., 1:28.
TYPE LOCALITY: French Guiana.
DISTRIBUTION: N and C South America.
STATUS: CITES - Appendix II.
SYNONYMS: *avus, azarae, barbatua, buffonii, caliginosus, capillatus, cay, chacoensis, cirrifer, crassiceps, cristatus, cucullatus, elegans, fallax; fatuellus* Linnaeus, 1766; *fistulato; flavus* Wied; *frontalus, fulvus, hypomelas, juruanus, leucogenys; libidinosus* Spix, 1823; *lunatus; macrocephalus* Spix, 1823; *magnus, maranonis, margaritae, monachus, morrulus, niger; nigritus,* Goldfuss, 1809; *pallidus, paraguayansis, peruanus, robustus, sagitta, subcristatus, tocantinus, trepida, trinitatis, variegatus, vellerosus, versutus, xanthocephalus, xanthosternos.*
COMMENTS: Mittermeier et al. (1988:13-75) suggested that *C. a. xanthosternos* is a distinct species.

Cebus capucinus (Linnaeus, 1758). Syst. Nat., 10th ed., 1:29.
TYPE LOCALITY: N Colombia.
DISTRIBUTION: W Ecuador to Honduras.
STATUS: CITES - Appendix II.
SYNONYMS: *albulus, curtus; hypoleucus* of authors (not Humboldt, 1812); *imitator* Thomas, 1903; *limitaneus, nigripectus.*

Cebus olivaceus Schomburgk, 1848. Reise Brit. Guiana, 2:247.
TYPE LOCALITY: Venezuela, Bolivar, southern base of Mt. Roraima, 930 m.
DISTRIBUTION: Guyana, French Guiana, Surinam, N Brazil, Venezuela, N Colombia.
STATUS: CITES - Appendix II.
SYNONYMS: *annellatus, apiculatus, barbatus, brunneus, castaneus, griseus, leporinus, nigrivittatus, pucheranii.*
COMMENTS: Replaces *nigrivittatus;* see Husson (1978:223). Mittermeier and Coimbra-Filho (1981) queried the distinction of this species from *C. capucinus.*

Saimiri Voigt, 1831. *In* Cuvier, Das Thierreich geord. nach. seiner Org., 1:95.
TYPE SPECIES: *Simia sciurea* Linnaeus, 1758.
SYNONYMS: *Chrysothrix, Pithesciurus.*
COMMENTS: Placed in a separate subfamily Saimirinae by Hershkovitz (1970*a*) and Napier (1976).

Saimiri boliviensis (I. Geoffroy and Blainville, 1834). Nouv. Ann. Mus. Hist. Nat. Paris, 3:89.
TYPE LOCALITY: Bolivia, Santa Cruz, Rio San Miguel, Guarayos Mission.
DISTRIBUTION: Upper Amazon in Peru; SW Brazil; Bolivia.
STATUS: CITES - Appendix II.
SYNONYMS: *entomophagus, jaburuensis, peruviensis, pluvialis.*
COMMENTS: Separated from *S. sciureus* by Hershkovitz (1984).

Saimiri oerstedii (Reinhardt, 1872). Vidensk. Medd. Nat. Hist. Kjobenhaven, p. 157.
TYPE LOCALITY: Panama, Chiriquí, David.
DISTRIBUTION: Panama, Costa Rica.
STATUS: CITES - Appendix I; U.S. ESA and IUCN - Endangered.
SYNONYMS: *citrinellus.*
COMMENTS: Hershkovitz (1972*a*) considered *oerstedii* a subspecies of *sciureus;* but see Hershkovitz (1984).

Saimiri sciureus (Linnaeus, 1758). Syst. Nat., 10th ed., 1:29.
TYPE LOCALITY: French Guyana, Cayenne.
DISTRIBUTION: N Brazil, Marajo Isl (Brazil), Guyana, French Guiana, Surinam, Venezuela, Colombia, E Ecuador, NE Peru.
STATUS: CITES - Appendix II.
SYNONYMS: *albigena, apedia, caquetensis, cassiquiarensis, codajazensis, collinsi, juruana, leucopsis, lunulatus, macrodon, morta, nigriceps, nigrivittata, petrina.*

Saimiri ustus (I. Geoffroy, 1843). C. R. Acad. Sci. 16(21):1157.
TYPE LOCALITY: Brazil, Amazonas, Rio Madeira, Humaitu.
DISTRIBUTION: S Brazil: Amazonas, Pará, Rondonia, probably Mato Grosso del Norte.
STATUS: CITES - Appendix II. Status uncertain.
SYNONYMS: *madeirae* Thomas, 1908.

Saimiri vanzolinii Ayres, 1985. Papeis Avulsos de Zoologia, São Paulo, 36:148.
TYPE LOCALITY: Brazil, Amazonas, mouth of Rio Japura, left bank of Lago Mamirauá.
DISTRIBUTION: Between Rios Japura, Solimões and (probably) Paranado Jaraua (Brazil); Tarara and Capucho Isls.
STATUS: CITES - Appendix II. Endangered.
COMMENTS: A subspecies of *S. boliviensis* according to Hershkovitz (1987*c*:22, fin.).

Subfamily Pitheciinae Mivart, 1865. Proc. Zool. Soc. Lond., 1865:547.

Cacajao Lesson, 1840. Spec. Mamm. Bim. et Quadrum., p. 181.
TYPE SPECIES: *Simia melanocephalus* Humboldt, 1812.

SYNONYMS: *Brachyurus* Spix, 1823; *Cercoptochus, Cothurus, Neocothurus, Ouakaria.*
COMMENTS: Revised by Hershkovitz (1987*c*).

Cacajao calvus (I. Geoffroy, 1847). C. R. Acad. Sci. Paris, 24:576.
TYPE LOCALITY: Brazil, Amazonas, Fonte Boa.
DISTRIBUTION: NW Brazil, E Peru.
STATUS: CITES - Appendix I; U.S. ESA - Endangered; IUCN - Vulnerable.
SYNONYMS: *alba, novaesi, rubicundus, ucayalii.*
COMMENTS: Includes *rubicundus;* see Hershkovitz (1972*a*), but also see Szalay and Delson (1979:290) who listed it as a distinct species.

Cacajao melanocephalus (Humboldt, 1812). Rec. Observ. Zool., 1:317.
TYPE LOCALITY: Venezuela, Casiquiare Forests, Mision de San Francisco Solano.
DISTRIBUTION: SW Venezuela, NW Brazil.
STATUS: CITES - Appendix I; U.S. ESA - Endangered; IUCN - Vulnerable.
SYNONYMS: *ouakary, spixii.*

Chiropotes Lesson, 1840. Spec. Mamm. Bim. et Quadrum., p. 178.
TYPE SPECIES: *Pithecia (Chiropotes) couxio* Lesson, 1840 (= *Cebus satanas* Hoffmannsegg, 1807).
SYNONYMS: *Saki.*
COMMENTS: Reviewed by Hershkovitz (1985).

Chiropotes albinasus (I. Geoffroy and Deville, 1848). C. R. Acad. Sci. Paris, 27:498.
TYPE LOCALITY: Brazil, Pará, Santarém.
DISTRIBUTION: NC Brazil.
STATUS: CITES - Appendix I; U.S. ESA - Endangered; IUCN - Vulnerable.
SYNONYMS: *roosevelti* J. A. Allen, 1914.

Chiropotes satanas (Hoffmannsegg, 1807). Mag. Ges. Naturf. Fr., 10:93.
TYPE LOCALITY: Brazil, Pará, lower Rio Tocantins, Cametá.
DISTRIBUTION: N Brazil, Guyana, French Guiana, Surinam, S Venezuela.
STATUS: CITES - Appendix II; U.S. ESA - Endangered as *C. s. satanas;* IUCN - Endangered as *C. s. satanas.*
SYNONYMS: *ater, chiropotes, couxio, fulvofusca, israelita, nigra, sagulata, utahicki.*

Pithecia Desmarest, 1804. Tabl. Méth. Hist Nat., *in* Nouv. Dict. Hist. Nat., 24:8.
TYPE SPECIES: *Simia pithecia* Linnaeus, 1766.
SYNONYMS: *Callitrix; Yarkea* Lesson, 1840.
COMMENTS: Revised by Hershkovitz (1987*d*).

Pithecia aequatorialis Hershkovitz, 1987. Am. J. Primatol., 12:429.
TYPE LOCALITY: Peru, Loreto, lower Rio Nanay, Santa Luisa.
DISTRIBUTION: Napo (Ecuador) to Loreto (Peru).
STATUS: CITES - Appendix II.
COMMENTS: Based upon specimens included by Hershkovitz (1979) in *P. monachus.*

Pithecia albicans Gray, 1860. Proc. Zool. Soc. Lond., 1860:231.
TYPE LOCALITY: Brazil, Amazonas, Tefé (south bank of Solimões River).
DISTRIBUTION: South bank of Amazon, between lower Jurua and lower Purús Rivers.
STATUS: CITES - Appendix II.
COMMENTS: Separated from *monachus* by Hershkovitz (1979, 1987*d*).

Pithecia irrorata Gray, 1842. Ann. Mag. Nat. Hist., [ser. 1], 10:256.
TYPE LOCALITY: Brazil, Pará, West bank of Rio Tapajós, Parque Nacional de Amazônia.
DISTRIBUTION: South of the Amazon in SW Brazil; SW Peru; E Bolivia.
STATUS: CITES - Appendix II.
SYNONYMS: *vanzolinii.*
COMMENTS: Includes some specimens identified by Hershkovitz (1979) as *P. hirsuta.* May prove to be a subspecies of *P. monachus* (Hershkovitz, 1987*d*).

Pithecia monachus (É. Geoffroy, 1812). Rec. Observ. Zool. 1:359.
TYPE LOCALITY: Brazil, left bank of Rio Solimões between Tabatinga and Rio Tocantins.

DISTRIBUTION: West of Rio Jurua and Rio Japura-Caqueta (in Brazil), Colombia, Ecuador and Peru.
STATUS: CITES - Appendix II.
SYNONYMS: *guapo; hirsuta* Spix, 1823; *inusta, milleri, napensis.*
COMMENTS: Does not include *albicans*, see Hershkovitz (1979). Includes *hirsuta*, see Hershkovitz (1987*d*).

Pithecia pithecia (Linnaeus, 1766). Syst. Nat., 12th ed., 1:40.
TYPE LOCALITY: French Guiana, Cayenne.
DISTRIBUTION: Guyana; French Guiana; Surinam; N Amazon, east of Río Negro and Rio Orinoco (N Brazil, S Venezuela).
STATUS: CITES - Appendix II.
SYNONYMS: *adusta, capillamentosa, chrysocephala, leucocephala, lotichiusi, ochrocephala, pongonias, rufibarbata, rufiventer, saki.*

Family Cercopithecidae Gray, 1821. London Med. Repos., 15:297.
COMMENTS: Hill (*in* Honacki et al., 1982:230) and Groves (1989) would divide this family into the Colobidae and Cercopithecidae.

Subfamily Cercopithecinae Gray, 1821. London Med. Repos., 15:297.
SYNONYMS: Cercocebini, Papinae, Papioninae.

Allenopithecus Lang, 1923. Am. Mus. Novit., 87:1.
TYPE SPECIES: *Cercopithecus nigroviridis* Pocock, 1907.
COMMENTS: Separated from *Cercopithecus* by Thorington and Groves (1970:638), Szalay and Delson (1979), and Groves (1989).

Allenopithecus nigroviridis (Pocock, 1907). Proc. Zool. Soc. Lond., 1907:739.
TYPE LOCALITY: Zaire, upper Congo River.
DISTRIBUTION: NW Zaire, NE Angola.
STATUS: CITES - Appendix II; IUCN - Insufficiently known.

Cercocebus É. Geoffroy, 1812. Ann. Mus. Hist. Nat. Paris, 19:97.
TYPE SPECIES: *Cercocebus fuliginosus* É. Geoffroy, 1812 (= *Simia* (*Cercopithecus*) *aethiops torquatus* Kerr, 1792)
SYNONYMS: *Aethiops* Martin, 1841.
COMMENTS: Van Gelder (1977*b*:8) and Groves (1978*b*) included *Cercocebus* in *Cercopithecus.*

Cercocebus agilis Milne-Edwards, 1886. Rev. Scient., 12:15.
TYPE LOCALITY: Zaire, Republic Poste des Ouaddas (junction of Oubangui and Congo Rivers).
DISTRIBUTION: Equatorial Guinea, Cameroon, NE Gabon, Central African Republic, N Congo Republic, Zaire.
STATUS: CITES - Appendix II.
SYNONYMS: *chrysogaster* Lydekker, 1900; *fumosus* Matschie, 1914; *hagenbecki* Lydekker, 1900; *oberlaenderi* Lorenz, 1915.
COMMENTS: Includes *chrysogaster*; separated from *galeritus* by Groves (1978*b*). A mangabey, probably of this species, was recently discovered in Uzungwa Mtns, E Tanzania. Wasser (1985) used the name *sanje*, and IUCN listed *Cercocebus galeritus sanje* as Endangered in Tanzania, but I have been unable to locate a valid description; no museum specimens exist.

Cercocebus galeritus Peters, 1879. Monatsb. K. Preuss. Akad. Wiss. Berlin, 1879:830.
TYPE LOCALITY: Kenya, Tana River, Mitole (2°10'S, 40°10'E).
DISTRIBUTION: Lower Tana River (Kenya).
STATUS: CITES - Appendix I *C. g. galeritus*, otherwise Appendix II; U.S. ESA - Endangered; IUCN - Endangered as *C. g. galeritus.*
COMMENTS: Formerly included *agilis*; but see Groves (1978*b*).

Cercocebus torquatus (Kerr, 1792). *In* Linnaeus, Anim. Kingdom, 1:67.
TYPE LOCALITY: West Africa.
DISTRIBUTION: Guinea to Gabon.

STATUS: CITES - Appendix II; U.S. ESA - Endangered; IUCN - Vulnerable.
SYNONYMS: *aethiopicus* F. Cuvier, 1821; *atys* Andebert, 1797; *collaris* Gray, 1843; *crossi* Gray, 1843; *fuliginosus* E. Geoffroy, 1812; *lunulatus* Temminck, 1853.
COMMENTS: Includes *atys*; see Dandelot (1974:12) and Groves (1978*b*).

Cercopithecus Linnaeus, 1758. Syst. Nat., 10th ed., 1:26.
TYPE SPECIES: *Simia diana* Linnaeus, 1758.
SYNONYMS: *Allochrocebus; Callithrix* Erxleben; *Cercocephalus, Diademia, Diana, Insignicebus; Lasiopyga* Reichenbach; *Melanocebus, Mona; Monichus* Reichenbach; *Neocebus, Otopithecus, Petaurista, Pogonocebus, Rhinosticteus, Rhinostigma.*
COMMENTS: Dandelot (1974:14) included *Allenopithecus, Erythrocebus,* and *Miopithecus* in this genus; but see Groves (1978*b*) and Corbet and Hill (1980:89) who considered them to be distinct genera; Szalay and Delson (1979) considered *Miopithecus* a subgenus of *Cercopithecus.* Van Gelder (1977*b*:8) included *Cercocebus, Papio,* and *Theropithecus* in this genus, but see Groves (1978*b*), Ansell (1978:33), and Cronin and Meikle (1979:259). Designated as a subgroup of *Simia* by Linnaeus; type species *S. diana* designated by Stiles and Orleman (1926:52). *Simia* was suppressed by Opinion 114 of the International Commision on Zoological Nomenclature (1929*c*).

Cercopithecus ascanius (Audebert, 1799). Hist. Nat. Singes Makis, 4(2):13.
TYPE LOCALITY: Angola (NW, by lower Congo River). See Machado (1969).
DISTRIBUTION: Uganda, Zaire, Zambia, Angola, marginally in Central African Republic; W Kenya.
STATUS: CITES - Appendix II.
SYNONYMS: *atrinasus, cirrhorhinus, enkamer, histrio, ituriensis, kaimosae, kassaicus, katangae, melanogenys, montanus, mpangae, omissus, orientalis, pelorhinus, picturatus, rutschuricus, sassae, schmidti, whitesidei.*

Cercopithecus campbelli Waterhouse, 1838. Proc. Zool. Soc. Lond., 1838:61.
TYPE LOCALITY: Sierra Leone.
DISTRIBUTION: Gambia to Ghana.
STATUS: CITES - Appendix II.
SYNONYMS: *burnettii, lowei, temminickii.*
COMMENTS: McAllan and Bruce (1989) argued that the original publication of this species should be: The Analyst, 24:298-299 [publ. 2 July 1838].

Cercopithecus cephus (Linnaeus, 1758). Syst. Nat., 10th ed., 1:27.
TYPE LOCALITY: Africa.
DISTRIBUTION: Gabon, Congo Republic, S Cameroon, Equatorial Guinea, SW Central African Republic, NW Angola.
STATUS: CITES - Appendix II.
SYNONYMS: *buccalis, cephodes, inobservatus, pulcher.*
COMMENTS: May include *erythrotis*; see Struhsaker (1970:374-376); but also see Dandelot (1974:23); may include *sclateri*, but see Kingdon (1980:461).

Cercopithecus diana (Linnaeus, 1758). Syst. Nat., 10th ed., 1:26.
TYPE LOCALITY: Liberia.
DISTRIBUTION: Sierra Leone to Ghana.
STATUS: CITES - Appendix I; U.S. ESA - Endangered; IUCN - Vulnerable.
SYNONYMS: *faunus* Linnaeus; *ignita, palatinus, roloway.*
COMMENTS: Type species; see comment under *Cercopithecus.* Includes *roloway*; see Dandelot (1974:25).

Cercopithecus dryas Schwartz, 1932. Rev. Zool. Bot. Afr., 21:251.
TYPE LOCALITY: Zaire, Ikela Zone, Yapatsi. (See Thys van den Audenaerde, 1977:1006).
DISTRIBUTION: Known only from a few localities in C Zaire (Wamba Dist., 22°31'-33°E, and 0°01'N-0°01'S); see Kuroda et al. (1985, as *Cercopithecus salongo*).
STATUS: CITES - Appendix II; IUCN - Vulnerable as *C. salongo.*
SYNONYMS: *salongo* Thys van den Audenaerde, 1977.
COMMENTS: Not a subspecies of *diana*; possibly related to *Chlorocebus aethiops*; see Thys van den Audenaerde (1977:1007). *C. salongo* is an age-variant of this species, see Colyn et al. (1991).

Cercopithecus erythrogaster Gray, 1866. Proc. Zool. Soc. Lond., 1866:169.
TYPE LOCALITY: Nigeria, Lagos.
DISTRIBUTION: S Nigeria.
STATUS: CITES - Appendix II; U.S. ESA and IUCN - Endangered.

Cercopithecus erythrotis Waterhouse, 1838. Proc. Zool. Soc. Lond., 1838:59.
TYPE LOCALITY: Equatorial Guinea, Bioko.
DISTRIBUTION: S and E Nigeria, Cameroon, Bioko.
STATUS: CITES - Appendix II; U.S. ESA - Endangered; IUCN - Vulnerable.
COMMENTS: Considered a subspecies of *cephus* by Struhsaker (1970:374-376); but also see Dandelot (1974:23). McAllan and Bruce (1989) argued that the original publication of this species should be: The Analyst, 24:298-299 [publ. 2 July 1838].

Cercopithecus hamlyni Pocock, 1907. Ann. Mag. Nat. Hist., ser. 7, 20:521.
TYPE LOCALITY: Zaire, Ituri Forest.
DISTRIBUTION: E Zaire, Rwanda.
STATUS: CITES - Appendix II; IUCN - Vulnerable.
SYNONYMS: *aurora* Thomas and Wroughton, 1910; *kahuziensis* Colyn and Verheyen, 1988.

Cercopithecus lhoesti P. Sclater, 1899. Proc. Zool. Soc. Lond., 1898:586 [1899].
TYPE LOCALITY: Zaire, Tschepo River, near Stanleyville.
DISTRIBUTION: E Zaire, W Uganda, Rwanda, Burundi.
STATUS: CITES - Appendix II; U.S. ESA - Endangered; IUCN - Vulnerable.
SYNONYMS: *rutschuricus, thomasi.*
COMMENTS: Does not include *preussi*; see Harrison (1988:562).

Cercopithecus mitis Wolf, 1822. Abbild. Beschreib. Merkw. Naturgesch. Gegenstandes, 2:145.
TYPE LOCALITY: Angola.
DISTRIBUTION: Ethiopia to South Africa, S and E Zaire, NW Angola.
STATUS: CITES - Appendix II.
SYNONYMS: *albogularis, albotorquatus, beirensis, boutourlinii, brindei, carruthersi, diadematus, dilophos, doggetti, elgonis, erythrarchus, francescae, heymansi, hindei, insignis, kandti, kibonotensis, kima, kolbi, leucampyx, labiatus, maesi, maritima, mauae, moloneyi, monoides, mossambicus, neumanni, nigrigenis, nubilus, nyasae, omensis, opisthostictus, otoleucus, phylax, pluto, princeps, rufilatus, samango, schubotzi, schoutedeni, schwartzi, sibatoi, stairsi, stuhlmanni, zammaranoi.*
COMMENTS: Includes *albogularis*; see Booth (1968); but also see Dandelot (1974:19).

Cercopithecus mona (Schreber, 1774). Die Säugethiere, 1:103.
TYPE LOCALITY: "Guinea".
DISTRIBUTION: Ghana to Cameroon; introduced into Lesser Antilles (Caribbean).
STATUS: CITES - Appendix II.
SYNONYMS: *monacha, monella.*

Cercopithecus neglectus Schlegel, 1876. Mus. Hist. Nat. Pays-Bas. Simiae, p. 70.
TYPE LOCALITY: Sudan, "White Nile".
DISTRIBUTION: SE Cameroon to Uganda and N Angola, W Kenya, SW Ethiopia, and S Sudan.
STATUS: CITES - Appendix II.
SYNONYMS: *brazzae, brazziformis, ezrae, uelensis.*

Cercopithecus nictitans (Linnaeus, 1766). Syst. Nat., 12th ed., 1:40.
TYPE LOCALITY: Equatorial Guinea, Benito River.
DISTRIBUTION: Liberia; Ivory Coast; Nigeria to NW Zaire, Central African Republic; Rio Muni and Bioko (Equatorial Guinea).
STATUS: CITES - Appendix II.
SYNONYMS: *insolitus, ludio, laglaizei, martini, stampflii, sticticeps.*

Cercopithecus petaurista (Schreber, 1774). Die Säugethiere, 1:97, 185.
TYPE LOCALITY: "Guinea".
DISTRIBUTION: Gambia to Benin.
STATUS: CITES - Appendix II.
SYNONYMS: *albinasus, büttikoferi, fantiensis, pygrius.*

Cercopithecus pogonias Bennett, 1833. Proc. Zool. Soc. Lond., 1833:67.
TYPE LOCALITY: Equatorial Guinea, Bioko.
DISTRIBUTION: SE Nigeria, Cameroon, Bioko and Rio Muni (Equatorial Guinea), N and W Gabon, W Zaire, Congo Republic.
STATUS: CITES - Appendix II.
SYNONYMS: *erxlebeni, grayi, nigripes, pallidus, petronellae.*

Cercopithecus preussi Matschie, 1898. Sitzber. Ges. Naturf. Fr. Berlin, 76.
TYPE LOCALITY: Cameroon, Victoria.
DISTRIBUTION: Region of Mt. Cameroon; Bioko (Equatorial Guinea).
STATUS: CITES - Appendix II; IUCN - Endangered.
SYNONYMS: *crossi, insularis.*
COMMENTS: Not a subspecies of *C. lhoesti.*

Cercopithecus sclateri Pocock, 1904. Abstr. Proc. Zool. Soc. Lond., 1904(5):18.
TYPE LOCALITY: Nigeria, Benin City.
DISTRIBUTION: SE Nigeria.
STATUS: CITES - Appendix II; IUCN - Endangered.
COMMENTS: Recognized as a full species by Kingdon (1980:661).

Cercopithecus solatus M. J. S. Harrison, 1988. J. Zool. (London), 215:562.
TYPE LOCALITY: C Gabon, SE of Booue, Faret des Abeilles, River Bali, 0°14'S, 12°15'E.
DISTRIBUTION: C Gabon.
STATUS: CITES - Appendix II; IUCN - Vulnerable; endangered.

Cercopithecus wolfi A. Meyer, 1891. Notes Leyden Mus., 13:63.
TYPE LOCALITY: "Central West Africa."
DISTRIBUTION: Zaire, NE Angola, W Uganda, Central African Republic.
STATUS: CITES - Appendix II.
SYNONYMS: *denti, elegans, liebrechtsi, pyrogaster.*
COMMENTS: Includes *denti*; see Dandelot (1974:25).

Chlorocebus Gray, 1870. Cat. of Monkeys, Lemurs Fruit-eating Bats Brit. Mus., p. 5.
TYPE SPECIES: *Simia sabaea* Linnaeus, 1766 (= *Simia aethiops* Linnaeus, 1758).
SYNONYMS: *Cynocebus.*
COMMENTS: Recognized as a full genus distinct from *Cercopithecus* by Groves (1989).

Chlorocebus aethiops (Linnaeus, 1758). Syst. Nat., 10th ed., 1:28.
TYPE LOCALITY: Sudan, Sennaar.
DISTRIBUTION: Senegal to Ethiopia, south to South Africa; Zanzibar, Pemba, and Mafia (Tanzania); introduced into Lesser Antilles (Caribbean).
STATUS: CITES - Appendix II.
SYNONYMS: *alexandri, arenarius, beniana, budgetti, calliaudi, callidus, callitrichus, cano-viridis, centralis, chrysurus, cinereo-viridis, circumcinctus, cloetei, contigua, cynosuros, djamdjamensis, ellenbecki, excubutor, flavidus, graueri, griseistictus, griseoviridis, helvescens, hilgerti, itimbiriensis, johnstoni, katangensis, lalandii, lukonzolwae, luteus, marjoriae, marrensis, matschiei, nesiotes, ngamiensis, nifoviridis, passargei, pembae, pousarguei, pusillus, pygerythrus, rubellus, rufoniger, rufoviridis, sabaeus, silaceus, tantalus, tephrops, tholloni, toldti, tumbili, viridis, voeltzkowi, weidholzi, werneri, weynsi, whytei.*
COMMENTS: Includes *pygerythrus, sabaeus,* and *tantalus*; see Struhsaker (1970:376).

Erythrocebus Trouessart, 1897. Cat. Mamm. Viv. Foss., 1:19.
TYPE SPECIES: *Simia patas* Schreber, 1775.
COMMENTS: Recognized as a distinct genus by Thorington and Groves (1970:638-639) and Szalay and Delson (1979).

Erythrocebus patas (Schreber, 1775). Die Säugethiere, 1:98.
TYPE LOCALITY: Senegal.
DISTRIBUTION: Savannahs, from W Africa to Ethiopia, Kenya, and Tanzania.
STATUS: CITES - Appendix II.
SYNONYMS: *albigenus, albosignatus, baumstarki, circumcinctus, formosus, kerstingi, langheldi, poliomystax, poliophaeus, pyrrhonotus, rubra, rufa, sannio, whitei, zechi.*

Lophocebus Palmer, 1903. Science, N.S., 17:873.
TYPE SPECIES: *Presbytis albigena* Gray, 1850.
SYNONYMS: *Cercolophocebus* Matschie, 1914; *Semnocebus* Gray, 1870.
COMMENTS: Formerly included in *Cercocebus*; a subgenus of *Cercocebus*, according to Szalay and Delson (1979), but see Groves (1979).

Lophocebus albigena (Gray, 1850). Proc. Zool. Soc. Lond., 1850:77.
TYPE LOCALITY: Zaire, Mayombe.
DISTRIBUTION: SE Nigeria, Cameroon, Congo Republic, Gabon, Equatorial Guinea, NE Angola, Central African Republic, Zaire, W Uganda, Burundi, W Kenya, W Tanzania.
STATUS: CITES - Appendix II; IUCN - Insufficiently known as *Cercocebus aterrimus*.
SYNONYMS: *aterrimus*; *coelognathus* Matschie, 1914; *congicus* Sclater, 1900; *hamlyni* Pocock, 1906; *ituricus* Matschie, 1913; *jamrachi* Pocoe, 1906; *johnstoni* Lydekker, 1900; *mawambicus* Lorenz, 1917; *opdenboschi*; *osmani* Groves, 1978; *rothschildi* Lydekker, 1900; *ugandae* Matschie, 1913; *weynsi* Matschie, 1913; *zenkeri* Schwarz, 1910.
COMMENTS: Includes *aterrimus* and *opdenboschi*; see Groves (1978*b*), but see also Szalay and Delson (1979).

Macaca Lacépède, 1799. Tabl. Div. Subd. Orders Genres Mammifères, p. 4.
TYPE SPECIES: *Simia inuus* Linnaeus, 1766 (= *Simia sylvanus* Linnaeus, 1758).
SYNONYMS: *Aulaxinus, Cynamolgus, Cynomacaca, Cynopithecus, Gymnopyga, Inuus, Lyssodes, Magotus, Magus, Maimon, Nemestrinus, Pithes, Rhesus, Salmacis, Silenus, Silvanus, Vetulus, Zati.*

Macaca arctoides (I. Geoffroy, 1831). *In* Belanger (ed.), Voy. Indes Orient., Mamm., 3(Zool.):61.
TYPE LOCALITY: "Cochin-China" (Indochina).
DISTRIBUTION: Assam (India) to S China and Malay Peninsula.
STATUS: CITES - Appendix II; U.S. ESA - Threatened.
SYNONYMS: *brunneus, harmandi, melanotus, melli, pullus, rufescens*; *speciosa* Blyth, 1875; *spenoca, ursinus.*
COMMENTS: Reviewed by Fooden et al. (1985). *M. speciosa* Blyth, 1875 (not *speciosa* I. Geoffroy, 1826, which is a synonym of *M. fuscata*) is a junior synonym; see Fooden (1969, 1976).

Macaca assamensis (M'Clelland, 1840). Proc. Zool. Soc. Lond., 1839:148 [1840].
TYPE LOCALITY: India, Assam.
DISTRIBUTION: Nepal to N Vietnam, S China.
STATUS: CITES - Appendix II.
SYNONYMS: *coolidgei, pelops, problematicus, rhesosimilis.*
COMMENTS: Reviewed by Fooden (1982).

Macaca cyclopis (Swinhoe, 1863). Proc. Zool. Soc. Lond., 1862:350 [1863].
TYPE LOCALITY: Taiwan, Jusan, Takao Pref.
DISTRIBUTION: Taiwan.
STATUS: CITES - Appendix II; U.S. ESA - Threatened; IUCN - Vulnerable.
SYNONYMS: *affinis.*
COMMENTS: Probably conspecific with *mulatta*; but see Fooden (1980).

Macaca fascicularis (Raffles, 1821). Trans. Linn. Soc. Lond., 13:246.
TYPE LOCALITY: Indonesia, Sumatra, Bengkulen.
DISTRIBUTION: Indochina and Burma to Borneo and Timor (Indonesia); Philippine Isls; Nicobar Isls (India).
STATUS: CITES - Appendix II.
SYNONYMS: *agnatus, alacer, apoensis, argentimembris, atriceps, aureus, baweanus, bintangensis, cagayanus, capitalis, carbonarius, carimatae, cupida, cynomolgus, dollmani, fuscus, impudens, irus, karimondjawae, karimoni, laetus, lapsus, lasiae, limitis, lingae, lingungensis, mandibularis, mansalaris, mindanensis, mindorus, mordax, phaeura, philippensis, pumilus, resina, sublimitus, submordax, suluensis, umbrosus, validus, vitiis.*
COMMENTS: Includes *irus*; see Medway (1977:70-71). Includes *cynomolgus*; see W. C. O. Hill (1974:476-477). Intergrades, and possibly conspecific, with *mulatta*; see Fooden (1964, 1971:24-32); but also see Fooden (1980).

Macaca fuscata (Blyth, 1875). J. Asiat. Soc. Bengal, 44:6.
TYPE LOCALITY: Japan.
DISTRIBUTION: Honshu, Shikoku, Kyushu, and adjacent small islands (Japan); Yaku Isl (Ryukyu Isls, Japan).
STATUS: CITES - Appendix II; U.S. ESA - Threatened; IUCN - Endangered as *M. f. yakui*.
SYNONYMS: *japanensis, speciosa, yakui*.
COMMENTS: Includes *speciosa* I. Geoffroy, 1826 (not *speciosa* Blyth, 1875) which was suppressed by Opinion 920 of the International Commision on Zoological Nomenclature (1970); see Fooden (1976).

Macaca maura (Schinz, 1825). *In* Cuvier, Das Thierreich, p. 257.
TYPE LOCALITY: Indonesia, Sulawesi Selatan.
DISTRIBUTION: S Sulawesi, south of Tempe Depression (Indonesia).
STATUS: CITES - Appendix II; IUCN - Vulnerable.
SYNONYMS: *brachyurus, hypomelas, inornatus, majuscula*.
COMMENTS: Type species of *Gymnopyga*; see Fooden (1969:79). Included in *nigra* by Corbet and Hill (1980:87).

Macaca mulatta (Zimmermann, 1780). Geogr. Gesch. Mensch. Vierf. Thiere, 2:195.
TYPE LOCALITY: India, Nepal Terai.
DISTRIBUTION: Afghanistan and India to N Thailand, China, and Hainan Isl (China).
STATUS: CITES - Appendix II.
SYNONYMS: *brachyurus, brevicaudatus, erythraea, fulvus, lasiotus, littoralis, mcmahoni, nipalensis, orinops, rhesus, sanctijohannis, siamica, tcheliensis, vestita, villosa*.
COMMENTS: See comments under *fascicularis*.

Macaca nemestrina (Linnaeus, 1766). Syst. Nat., 12th ed., 1:35.
TYPE LOCALITY: Indonesia, Sumatra.
DISTRIBUTION: Malay Peninsula, Borneo, Sumatra and Bangka Isl (Indonesia), Burma (including Mergui Arch.), Thailand (including Phuket), Yunnan (China), Laos.
STATUS: CITES - Appendix II; IUCN - Endangered as *M. pagensis*.
SYNONYMS: *adusta, andamanensis, blythii, broca, carpolegus, insulana, indochinensis, leonina, nucifera, pagensis*.
COMMENTS: Includes *pagensis*; see Fooden (1975:67, 1980:7) and Szalay and Delson (1979). Wilson and Wilson (1977:216) considered *pagensis* a distinct species.

Macaca nigra (Desmarest, 1822). Mammalogie, *in* Encyclop. Meth., 2(Suppl.):534.
TYPE LOCALITY: Indonesia, Sulawesi, Maluku, Bacan Isl.
DISTRIBUTION: Sulawesi, northeast of Gorontalo, Bacan, and adjacent islands (Indonesia).
STATUS: CITES - Appendix II.
SYNONYMS: *lembicus, nigrescens*.
COMMENTS: Type species of *Cynopithecus*; see Fooden (1969). Includes *nigrescens*; see Groves (1980*c*). Corbet and Hill (1980:88) also included *brunnescens* and *maura*; but see Fooden (1969:1-9).

Macaca ochreata (Ogilby, 1841). Proc. Zool. Soc. Lond., 1841:56.
TYPE LOCALITY: Unknown.
DISTRIBUTION: SE Sulawesi, Muna, and Butung (Indonesia).
STATUS: CITES - Appendix II; IUCN - Insufficiently known as *M. brunnescens*.
SYNONYMS: *brunnescens, fusco-ater*.
COMMENTS: Perhaps conspecific with *maura*. Fooden (1969) recognized *brunnescens*; but Groves (1980*c*:1-9) included it in *ochreata*.

Macaca radiata (É. Geoffroy, 1812). Ann. Mus. Hist. Nat. Paris, 19:98.
TYPE LOCALITY: India; see W. C. O. Hill (1974).
DISTRIBUTION: S India.
STATUS: CITES - Appendix II.
SYNONYMS: *diluta*.
COMMENTS: Probably conspecific with *sinica*; but see Fooden (1980) and Fooden et al. (1981:463). Revised by Fooden (1981).

Macaca silenus (Linnaeus, 1758). Syst. Nat., 10th ed., 1:26.
TYPE LOCALITY: "Ceylon" India, Western Ghats; see W. C. O. Hill (1974:652).

DISTRIBUTION: SW India, Western Ghats.
STATUS: CITES - Appendix I; U.S. ESA and IUCN - Endangered.
SYNONYMS: *ferox, veter.*

Macaca sinica (Linnaeus, 1771). Mantissa Plantarum, 2, Appendix:521.
TYPE LOCALITY: Probably Sri Lanka; see Fooden (1979).
DISTRIBUTION: Sri Lanka.
STATUS: CITES - Appendix II; U.S. ESA - Threatened.
SYNONYMS: *audeberti, aurifrons, inaureus, longicaudata, opisthomelas, pileatus.*
COMMENTS: See Fooden (1979).

Macaca sylvanus (Linnaeus, 1758). Syst. Nat., 10th ed., 1:25.
TYPE LOCALITY: North Africa, "Barbary coast".
DISTRIBUTION: Morocco, Algeria, Gibraltar (introduced).
STATUS: CITES - Appendix II; IUCN - Vulnerable.
SYNONYMS: *ecaudatus; inuus* Linnaeus; *pithecus, pygmaeus.*
COMMENTS: See Fooden (1976:226) for the use of this name.

Macaca thibetana (Milne-Edwards, 1870). C. R. Acad. Sci. Paris, 70:341.
TYPE LOCALITY: China, Szechwan, Moupin.
DISTRIBUTION: E Tibet; Szechwan to Kwangtung (China).
STATUS: CITES - Appendix II; IUCN - Insufficiently known.
SYNONYMS: *esau.*
COMMENTS: Includes *esau*; see Fooden (1967:160). Reviewed by Fooden (1983).

Macaca tonkeana (Meyer, 1899). Abh. Zool. Anthrop.-Ethnology. Mus. Dresden, 7(7):3.
TYPE LOCALITY: Indonesia, Sulawesi Tengah, Tonkean.
DISTRIBUTION: C Sulawesi, south to Latimojong, northeast to Gorontalo (Indonesia); Togian Isls (Indonesia).
STATUS: CITES - Appendix II; IUCN - Vulnerable as *M. hecki.*
SYNONYMS: *hecki, tonsus, togeanus.*
COMMENTS: Includes *hecki*; see Groves (1980*c*); but see also Fooden (1969:1-9). Formerly included in *Cynopithecus*; see Fooden (1969:106-115).

Mandrillus Ritgen, 1824. Natureichen Eintheilung der Säugethiere, p. 33.
TYPE SPECIES: *Simia maimon* Linnaeus, 1766; *Simia mormon* Alstromer, 1766 (= *Simia sphinx* Linnaeus, 1758).
SYNONYMS: *Drill, Maimon, Mandril, Mormon.*
COMMENTS: Not a synonym of *Papio* (see Groves, 1989).

Mandrillus leucophaeus (F. Cuvier, 1807). Ann. Mus. Hist. Nat. Paris, 9:477.
TYPE LOCALITY: Africa.
DISTRIBUTION: SE Nigeria; Cameroon, north of the Sanaga River and just south of it; Bioko (Equatorial Guinea). See Grubb (1973) for details.
STATUS: CITES - Appendix I; U.S. ESA and IUCN - Endangered.
SYNONYMS: *cinerea, drill, hagenbecki, mundamensis, poensis, sylvicola.*
COMMENTS: Delson and Napier (1976:46) considered this species in genus *Papio*, subgenus *Papio*. Placed in subgenus *Mandrillus* by Dandelot (1974:9). *Mandrillus* considered a full genus by Groves (1989).

Mandrillus sphinx (Linnaeus, 1758). Syst. Nat., 10th ed., 1:25.
TYPE LOCALITY: Cameroon, Ja River, Bitye.
DISTRIBUTION: Cameroon, south of the Sanaga River; Rio Muni (Equatorial Guinea); Gabon; Congo Republic. See Grubb (1973) for details.
STATUS: CITES - Appendix I; U.S. ESA - Endangered; IUCN - Vulnerable.
SYNONYMS: *burlacei, ebolowae, escherichi, insularis, madarogaster, maimon, mormon, planirostris, schreberi, suilla, tessmanni, zenkeri.*
COMMENTS: Delson and Napier (1976:46) considered *sphinx* in genus *Papio*, subgenus *Papio*. Placed in subgenus *Mandrillus* by Dandelot (1974:9). *Mandrillus* considered a full genus by Groves (1989).

Miopithecus I. Geoffroy, 1862 C. R. Acad. Sci. Paris, 15:720.
TYPE SPECIES: *Simia talapoin* Schreber, 1774.

COMMENTS: Is a subgenus of *Cercopithecus* according to Szalay and Delson (1979), but also see Groves (1978*b*, 1989). A second species of *Miopithecus*, described by Machado (1969) has not yet been named; its distribution is Gabon, Equatorial Guinea and Cameroon.

Miopithecus talapoin (Schreber, 1774). Die Säugethiere, 1:101, 186, pl. 17.
TYPE LOCALITY: Angola.
DISTRIBUTION: Angola, SW Zaire.
STATUS: CITES - Appendix II.
SYNONYMS: *ansorgei, capillatus, melarhinus, pileatus, pilettei.*

Papio Erxleben, 1777. Systema Regni Animalis, 1, Mammalia:xxx, 15.
TYPE SPECIES: *Cynocephalus papio* Desmarest, 1820 (= *Simia hamadryas* Linnaeus, 1758).
SYNONYMS: *Chaeropitheus* Gervais, 1839; *Comopithecus* J. A. Allen, 1925; *Cynocephalus* J. A. Allen, 1925; *Hamadryas* Lesson, 1840.
COMMENTS: Opinion 1199 of the International Commision on Zoological Nomenclature (1982) fixed this as the first available name, and fixed the type species.

Papio hamadryas (Linnaeus, 1758). Syst. Nat., 10th ed., 1:27.
TYPE LOCALITY: Egypt.
DISTRIBUTION: Senegal to Somalia and S Arabia, south to South Africa.
STATUS: CITES - Appendix II; IUCN - Rare.
SYNONYMS: *aegyptiaca, anubis, antiquorum, brockmani, chaeropitheus, chobiensis, choras, comatus, cynocephalus, doguera, furax, graueri, griseipes, heuglini, ibeanus, jubilaeus, kindae, langheldi, lestes, lydekkeri, neumanni, ngamiensis, nigeriae, nigripes, occidentalis, ochraceus, olivaceus, orientalis, papio, porcarius, pruinosus, rhodesiae, rubescens, strepitus, tesselatum, thoth, tibestianus, transvaalensis, ursinus, variegata, vigilis, werneri, yokoensis.*
COMMENTS: Includes *anubis, cynocephalus, papio,* and *ursinus*; see Szalay and Delson (1979:336) and Jolly and Brett (1973:85). Dandelot (1974:9), Corbet and Hill (1980:88,) and others recognized these as distinct species. Groves (1989) suggested some of these might be distinct species. Maples and McKern (1967:2) discussed integradation in these forms.

Theropithecus I. Geoffroy, 1843. Arch. Mus. Hist. Nat. Paris, 1841, 2:576.
TYPE SPECIES: *Macacus gelada* Rüppell, 1835.
SYNONYMS: *Gelada.*
COMMENTS: Considered a distinct genus by Cronin and Meikle (1979:259). Van Gelder (1977*b*:8) included this genus in *Cercopithecus.*

Theropithecus gelada (Rüppell, 1835). Neue Wirbelt. Fauna Abyssin. Gehörig. Säugeth., p. 5.
TYPE LOCALITY: Ethiopia, Semyen (Simien).
DISTRIBUTION: N Ethiopia, highlands.
STATUS: CITES - Appendix II; U.S. ESA - Threatened; IUCN - Rare.
SYNONYMS: *nedjo, obscurus, ruppelli, senex.*

Subfamily Colobinae Jerdon, 1867. Mammals of India, p. 3.
SYNONYMS: Presbytinae, Semnopithecinae.
COMMENTS: Separated provisionally as a full family (Colobidae) by Groves (1989). On the name of this subfamily, see Delson (1976), and Brandon-Jones (1978).

Colobus Illiger, 1811. Prodr. Syst. Mamm. Avium., p. 69.
TYPE SPECIES: *Simia polycomos* Schreber, 1800 (= *Cebus polykomos* Zimmerman, 1780).
SYNONYMS: *Colobolus, Guereza, Pterycolobus, Stachycolobus.*
COMMENTS: Does not include *Procolobus*, see Groves (1989).

Colobus angolensis P. Sclater, 1860. Proc. Zool. Soc. Lond., 1860:245.
TYPE LOCALITY: Angola, 300 mi. (483 km) inland from Bembe.
DISTRIBUTION: NE Angola, S and E Zaire, Rwanda, Burundi, NE Zambia, SE Kenya, E Tanzania.
STATUS: CITES - Appendix II.

SYNONYMS: *adolfi-friederici, benamakimae, cordieri, cottoni, langheldi, maniemae, mawambicus, nahani, palliatus, prigoginei, ruwenzorii, sandbergi, sharpei, weynsi.*
COMMENTS: Thorington and Groves (1970:629-647), Dandelot (1974:37), and Corbet and Hill (1980:89) listed *angolensis* as a distinct species.

Colobus guereza Rüppell, 1835. Neue Wirbelt. Fauna Abyssin. Gehörig. Säugeth., p. 1.
TYPE LOCALITY: Ethiopia, Gojjam and Kulla.
DISTRIBUTION: Nigeria to Ethiopia; Kenya; Uganda; Tanzania.
STATUS: CITES - Appendix II.
SYNONYMS: *abyssinicus*, Oken (unavailable); *albocaudatus, brachychaites, caudatus, dianae, dodingae, elgonis, escherichi, gallarum, ituricus, kikuyuensis, laticeps, managaschae, matschiei, occidentalis, poliurus, percivali, roosevelti, ruppelli, rutschiricus, terrestris, thikae, uellensis.*

Colobus polykomos (Zimmermann, 1780). Geogr. Gesch. Mensch. Vierf. Thiere, 2:202.
TYPE LOCALITY: Sierra Leone.
DISTRIBUTION: Gambia to Benin; Nigeria.
STATUS: CITES - Appendix II.
SYNONYMS: *bicolor, comosa, dollmani, leucomeros, regalis, tetradactyla, ursinus, vellerosus.*
COMMENTS: Oates and Trocco (1983) classified *vellerosus* as a distinct species.

Colobus satanas Waterhouse, 1838. Proc. Zool. Soc. Lond., 1837:87 [1838].
TYPE LOCALITY: Equatorial Guinea, Bioko.
DISTRIBUTION: SW Gabon, Rio Muni and Bioko (Equatorial Guinea), SW Cameroon.
STATUS: CITES - Appendix II; U.S. ESA and IUCN - Endangered.
SYNONYMS: *anthracinus, limbarenicus, municus, zenkeri.*
COMMENTS: McAllen and Bruce (1989) argued that the original publication of this species should be: The Analyst, 24:298-299 [publ. 2 July 1838].

Nasalis É. Geoffroy, 1812. Ann. Mus. Hist. Nat. Paris, 19:89.
TYPE SPECIES: *Cercopithecus larvatus* Wurmb, 1787.
SYNONYMS: *Simias.*
COMMENTS: Includes *Simias*; see Groves (1970:639). Szalay and Delson (1979) and Delson (1975:217) considered *Simias* a subgenus; but also see Krumbiegel (1978).

Nasalis concolor (Miller, 1903). Smithson. Misc. Coll., 45:67.
TYPE LOCALITY: Indonesia, W Sumatra, S Pagai Isl.
DISTRIBUTION: Mentawai Isls (Indonesia).
STATUS: CITES - Appendix I; U.S. ESA and IUCN - Endangered.
SYNONYMS: *siberu.*
COMMENTS: Type species of *Simias* Miller, 1903, which was considered a subgenus by Delson (1975:217).

Nasalis larvatus (Wurmb, 1787). Verh. Batav. Genootsch., 3:353.
TYPE LOCALITY: Indonesia, W Kalimantan, Pontianak.
DISTRIBUTION: Borneo.
STATUS: CITES - Appendix I; U.S. ESA - Endangered; IUCN - Vulnerable.
SYNONYMS: *capistratus, nasica, orientalis, recurvus.*
COMMENTS: Subgenus *Nasalis*; see Delson (1975:217).

Presbytis Eschscholtz, 1821. Reise (Kotzebue), 3:196.
TYPE SPECIES: *Presbytis mitrata* Eschscholtz, 1821 (= *Simia melalophos* Raffles, 1821).
SYNONYMS: *Corypithecus, Lophopitheus, Presbypitheus.*
COMMENTS: Does not include *Semnopithecus* and *Trachypithecus* (Groves, 1989; Hooijer, 1962:20-24); but see Thorington and Groves (1970:629-647). Szalay and Delson (1979:402) included these and *Kasi* as subgenera.

Presbytis comata (Desmarest, 1822). Mammalogie, *in* Encycl. Meth., 2(Suppl.):533.
TYPE LOCALITY: Indonesia, W Java.
DISTRIBUTION: W and C Java.
STATUS: CITES - Appendix II; IUCN - Endangered.
SYNONYMS: *aygula*, various authors, *fredericae.*

COMMENTS: Formerly called *P. aygula*, but see Napier and Groves (1983) who showed that *aygula* is a *nomen oblitum* for *Macaca fascicularis*.

Presbytis femoralis (Martin, 1838). Mag. Nat. Hist. [Charlesworth's], 2:436.
TYPE LOCALITY: Singapore.
DISTRIBUTION: Malay Penninsula, Singapore, NE Sumatra, Natuna Isl (Indonesia), Sarawak (Borneo, Malaysia).
STATUS: CITES - Appendix II.
SYNONYMS: *arwasca, australis, cana, catemana, chrysomelas, cruciger, dilecta, keatii, natunae, neglectus, nigrimanis, nubigena, paenulata, percura, rhionis, robinsoni, siamensis.*
COMMENTS: Separated from *P. melalophos* by Wilson and Wilson (1977:217-222); recognized as a species by Aimi et al. (1986).

Presbytis frontata (Müller, 1838). Tijdschr. Nat. Gesch. Physiol., 5:136.
TYPE LOCALITY: Indonesia, SE Kalimantan: Murung and "Pulu Lampy", near Banjarmasin, Pematang, Kuala (Medway, 1965:82).
DISTRIBUTION: NW and E Borneo, except southwest.
STATUS: CITES - Appendix II.
SYNONYMS: *nudifrons.*

Presbytis hosei (Thomas, 1889). Proc. Zool. Soc. Lond., 1889:159.
TYPE LOCALITY: Malaysia, Sarawak, Barman District.
DISTRIBUTION: N and E Borneo.
STATUS: CITES - Appendix II.
SYNONYMS: *canicrus, everetti, sabana.*
COMMENTS: Separated from "*aygula*" (= *comata*) by Medway (1970:544).

Presbytis melalophos (Raffles, 1821). Trans. Linn. Soc. Lond., 13:245.
TYPE LOCALITY: Indonesia, Sumatra, Bengkulen.
DISTRIBUTION: Sumatra (Indonesia).
STATUS: CITES - Appendix II.
SYNONYMS: *aurata, batuana, ferruginea, flavimanus, fluviatilis, fuscomurina, margae, mitrata, nobilis, sumatrana.*
COMMENTS: Type species of *Presbytis* (as *P. m. mitrata*). Does not include *P. femoralis*, which was regarded as a separate species by Wilson and Wilson (1977:217-222).

Presbytis potenziani (Bonaparte, 1856). C. R. Acad. Sci. Paris, 43:412.
TYPE LOCALITY: Indonesia, W Sumatra, Sipora Isl.
DISTRIBUTION: Mentawai Isls (Indonesia).
STATUS: CITES - Appendix I; U.S. ESA - Threatened; IUCN - Endangered.
SYNONYMS: *chrysogaster, siberu.*

Presbytis rubicunda (Müller, 1838). Tijdschr. Nat. Gesch. Physiol., 5:137.
TYPE LOCALITY: Indonesia, S Kalimantan, Mt. Sekumbang (SE of Banjermassin).
DISTRIBUTION: Borneo; Karimata Isl (Indonesia).
STATUS: CITES - Appendix II.
SYNONYMS: *chrysea, ignita, karimatae, rubida.*

Presbytis thomasi (Collett, 1893). Proc. Zool. Soc. Lond., 1892:613 [1893].
TYPE LOCALITY: Indonesia, Sumatra, Aceh, Langkat.
DISTRIBUTION: Sumatra: Aceh, south to about 3°50'N.
STATUS: CITES - Appendix II.
SYNONYMS: *nubilus.*
COMMENTS: Separated from "*aygula*" (= *comata*) by Medway (1970:544).

Procolobus Rochebrune, 1877. Faune de Sénégambie, Suppl. Vert., Mamm., 1:95.
TYPE SPECIES: *Colobus verus* Van Beneden, 1838.
SYNONYMS: *Lophocolobus, Piliocolobus, Tropicolobus.*
COMMENTS: Separate from *Colobus*, see Corbet and Hill (1980:90) and Groves (1989).

Procolobus badius (Kerr, 1792). *In* Linnaeus, Anim. Kingdom, 1:74.
TYPE LOCALITY: Sierra Leone.
DISTRIBUTION: Senegal to Ghana.

STATUS: CITES - Appendix II; IUCN - Rare as *P. b. temminckii*, Endangered as *P. b. waldroni*, otherwise Vulnerable.
SYNONYMS: *ferriginea, fuliginosus, rufo-fuliginus, rufoniger, temminckii, waldroni*.
COMMENTS: Subgenus *Piliocolobus*; includes *waldroni* and *temminckii*; see Dandelot (1974:33); but also see Rahm (1970).

Procolobus pennantii (Waterhouse, 1838). Proc. Zool. Soc. Lond., 1838:57.
TYPE LOCALITY: Equatorial Guinea, Bioko.
DISTRIBUTION: Congo Republic; Bioko (Equatorial Guinea); Zaire; W Uganda; W and SE Tanzania; Zanzibar (Tanzania).
STATUS: CITES - Appendix I as *C. p. kirki*, otherwise Appendix II; U.S. ESA - Endangered as *C. kirki*; IUCN - Vulnerable as subspecies *tephrosceles*, Insufficiently known as subspecies *ellioti, foae, oustaleti* and *tholloni*, otherwise Endangered.
SYNONYMS: *anzeliusi, bouvieri, brunneus, ellioti, foai, gordonorum, graueri, gudoviusi, kabambarei, kirki, langi, lulidicus, lovizettii, melanochir, multicolor, nigrimanus, oustaleti, parmentieri, powelli, schubotzi, tephrosceles, tholloni, umbrinus, variabilis*.
COMMENTS: Subgenus *Piliocolobus*; includes *bouvieri, kirki*, and *tholloni*; but also see Rahm (1970) who considered this species a subspecies of *badius*, and Dandelot (1974:35), who considered *kirki* and *tholloni* distinct species.

Procolobus preussi (Matschie, 1900). Sitzb. Ges. Naturf. Fr. Berlin, 1900:183.
TYPE LOCALITY: Cameroon, Barombi (on Elephant Lake).
DISTRIBUTION: Yabassi Dist. (Cameroon).
STATUS: CITES - Appendix II; U.S. ESA and IUCN - Endangered.
COMMENTS: Subgenus *Piliocolobus*. Considered by Rahm (1970) to be a subspecies of *badius*, but see Dandelot (1974:37).

Procolobus rufomitratus (Peters, 1879). Monatsb. K. Preuss. Akad. Wiss. Berlin, 1879:829.
TYPE LOCALITY: Kenya, Tana River, Muniuni.
DISTRIBUTION: Lower Tana River (Kenya).
STATUS: CITES - Appendix I; U.S. ESA - Endangered as *C. r. rufomitratus*; IUCN - Endangered as *P. r. rufomitratus*, otherwise Vulnerable.
COMMENTS: Subgenus *Piliocolobus*.

Procolobus verus (Van Beneden, 1838). Bull. Acad. Sci. Belles-Letters Bruxelles, 5:347.
TYPE LOCALITY: Africa.
DISTRIBUTION: Sierra Leone to Togo; Idah Dist. (E Nigeria, see Menzies, 1970, for comments).
STATUS: CITES - Appendix II; IUCN - Rare.
SYNONYMS: *chrysurus, cristatus, olivaceus, verus*.
COMMENTS: Subgenus *Procolobus*; see Dandelot (1974:37) and Corbet and Hill (1980:90).

Pygathrix É. Geoffroy, 1812. Ann. Mus. Hist. Nat. Paris, 19:90.
TYPE SPECIES: *Simia nemaeus* Linnaeus, 1771.
SYNONYMS: *Presbytiscus, Rhinopithecus*.
COMMENTS: Includes *Rhinopithecus*; see Groves (1970) and Szalay and Delson (1979:404).

Pygathrix avunculus (Dollman, 1912). Abstr. Proc. Zool. Soc. Lond., 1912(106):18.
TYPE LOCALITY: Vietnam, Songkoi River, Yen Bay.
DISTRIBUTION: N Vietnam.
STATUS: CITES - Appendix I; U.S. ESA and IUCN - Endangered.
COMMENTS: Subgenus *Rhinopithecus*; see Szalay and Delson (1979:404).

Pygathrix bieti Milne-Edwards, 1897. Bull. Mus. Hist. Nat. Paris, 3:157.
TYPE LOCALITY: China, Yunnan, left bank of upper Mekong, Kiape, 28°25'N, 98°55'E, a day's journey south of Atentse.
DISTRIBUTION: Mountains of left bank of upper Mekong (China).
STATUS: CITES - Appendix I; U.S. ESA and IUCN - Endangered.
COMMENTS: Regarded by Groves (1970:569) as a subspecies of *roxellana*, but regarded as a full species by Peng et al. (1988).

Pygathrix brelichi (Thomas, 1903). Proc. Zool. Soc. Lond., 1903(1):224.
TYPE LOCALITY: China, N Guizhou, Van Gin Shan Range (= Fanjinshan).

DISTRIBUTION: Fanjinshan (Guizhou, China).
STATUS: CITES - Appendix I; U.S. ESA and IUCN - Endangered.
COMMENTS: Subgenus *Rhinopithecus*. Considered a valid species by Groves (1970:569).

Pygathrix nemaeus (Linnaeus, 1771). Mantissa Plantarum, p. 521.
TYPE LOCALITY: Cochin-China (Indo-China).
DISTRIBUTION: S and C Vietnam, E Laos, E Cambodia, ? Hainan Isl (China).
STATUS: CITES - Appendix I; U.S. ESA and IUCN - Endangered.
SYNONYMS: *moi, nigripes*.
COMMENTS: Type species of subgenus *Pygathrix*.

Pygathrix roxellana (Milne-Edwards, 1870). C. R. Acad. Sci. Paris, 70:341.
TYPE LOCALITY: China, Sichuan, Moupin (= Baoxing, 30°26′N, 102°50′E).
DISTRIBUTION: NW Yunnan, Sichuan, S Ganssu (China).
STATUS: CITES - Appendix I; U.S. ESA - Endangered; IUCN - Vulnerable.
COMMENTS: Type species of subgenus *Rhinopithecus*; see Groves (1970:569) and Szalay and Delson (1979:404).

Semnopithecus Desmarest, 1822. Mammalogie, *in* Encycl. Meth, 2(Suppl.):532.
TYPE SPECIES: *Simia entellus* Dufresne, 1797.
COMMENTS: Considered a subgenus of *Presbytis* by Szalay and Delson (1979); separated from *Presbytis* by Groves (1989).

Semnopithecus entellus (Dufresne, 1797). Bull. Sci. Soc. Philom. Paris, ser. 1, 7:49.
TYPE LOCALITY: India, Bengal.
DISTRIBUTION: India, Nepal, S Tibet, Sri Lanka, Pakistan and Kashmir.
STATUS: CITES - Appendix I; U.S. ESA - Endangered.
SYNONYMS: *achates, achilles, aeneas, ajax, albipes, anchises, dussumieri, elissa, hector, hypoleucos, iulus, lania, nipalensis, pallipes, petrophilus, priam, priamelus, schistaceus, thersites*.

Trachypithecus Reichenbach, 1862. Vollständ. Nat. Affen, p. 89.
TYPE SPECIES: *Semnopithecus pyrrhus* Horsfield, 1823 (= *Cercopithecus auratus* É. Geoffroy, 1812).
SYNONYMS: *Kasi*.
COMMENTS: Separated from *Presbytis* by Hooijer (1962) and Groves (1989).

Trachypithecus auratus (É. Geoffroy, 1812). Ann. Mus. Hist. Nat. Paris, 19:93.
TYPE LOCALITY: Java, Semarang (Müller, 1840:16).
DISTRIBUTION: Java, Bali, and Lombok (Indonesia).
STATUS: CITES - Appendix II.
SYNONYMS: *kohlbruggei, pyrrhus, sondaicus*.
COMMENTS: Subgenus *Trachypithecus*. Separated from *T. cristatus* by Weitzel and Groves (1985).

Trachypithecus cristatus (Raffles, 1821). Trans. Linn. Soc. Lond., 13:244.
TYPE LOCALITY: Indonesia, Sumatra, Bengkulen (Bengkulu).
DISTRIBUTION: Burma and Indochina to Borneo.
STATUS: CITES - Appendix II.
SYNONYMS: *atrior, barbei, caudalis, germaini, koratensis, mandibularis, margarita, pruinosus, pullata, ultima, vigilans*.
COMMENTS: Subgenus *Trachypithecus*.

Trachypithecus francoisi (Pousargues, 1898). Bull. Mus. Hist. Nat. Paris, 4:319.
TYPE LOCALITY: China, Kwangsi, Lungchow.
DISTRIBUTION: N Vietnam, C Laos, Kwangsi (China).
STATUS: CITES - Appendix II; U.S. ESA and IUCN - Endangered.
SYNONYMS: *delacouri, hatinhensis, laotum, leucocephalus, poliocephalus*.
COMMENTS: Subgenus *Trachypithecus*. Some or all of the subspecies may be distinct species.

Trachypithecus geei Khajuria, 1956. Ann. Mag. Nat. Hist., ser. 12, 9:86.
TYPE LOCALITY: India, Assam, Goalpara Dist., Jamduar Forest Rest House, E bank of Sankosh River.

DISTRIBUTION: Between Sankosh and Manas Rivers, Indo-Bhutan border (on both sides).
STATUS: CITES - Appendix I; U.S. ESA - Endangered; IUCN - Rare.
COMMENTS: Subgenus *Trachypithecus*. For authorship of the name, see Biswas (1967). May prove to be a subspecies of *pileatus*

Trachypithecus johnii (J. Fischer, 1829). Synopsis Mamm., p. 25.
TYPE LOCALITY: India, Tellicherry.
DISTRIBUTION: S India.
STATUS: CITES - Appendix II; IUCN - Endangered.
SYNONYMS: *cucullatus, jubatus*.
COMMENTS: Subgenus *Kasi*; see Szalay and Delson (1979:402). Probably conspecific with *vetulus*.

Trachypithecus obscurus (Reid, 1837). Proc. Zool. Soc. Lond., 1837:14.
TYPE LOCALITY: Malaysia, Malacca.
DISTRIBUTION: S Thailand and Malay Peninsula, and small adjacent islands.
STATUS: CITES - Appendix II.
SYNONYMS: *carbo, corax, corvus, flavicauda, halonifer, leucomystax, sanctorum, seimundi, smithi, styx*.
COMMENTS: Subgenus *Trachypithecus*.

Trachypithecus phayrei (Blyth, 1847). J. Asiat. Soc. Bengal, 16:733.
TYPE LOCALITY: Burma, Arakan.
DISTRIBUTION: Laos, Burma, C Vietnam, C and N Thailand, Yunnan (China).
STATUS: CITES - Appendix II.
SYNONYMS: *argenteus, crepuscula, holotephreus, melamera, shanicus, wroughtoni*.
COMMENTS: Subgenus *Trachypithecus*.

Trachypithecus pileatus (Blyth, 1843). J. Asiat. Soc. Bengal, 12:174.
TYPE LOCALITY: India, Assam.
DISTRIBUTION: Assam, NW Burma, E Bangladesh, S Yunnan (China).
STATUS: CITES - Appendix I; U.S. ESA - Endangered.
SYNONYMS: *argentatus, belliger, brahma, durga, saturatus, shortridgei, tenebrica*.
COMMENTS: Subgenus *Trachypithecus*.

Trachypithecus vetulus (Erxleben, 1777). Syst. Regni Anim., p. 24.
TYPE LOCALITY: Sri Lanka, Hill country of South.
DISTRIBUTION: Sri Lanka.
STATUS: CITES - Appendix II; U.S. ESA - Threatened as *Presbytis senex*.
SYNONYMS: *albinus, harti, kelaartii, kephalopterus, latibarba, latibarbatus, leucoprymnus, monticola, nestor, philbricki, phillipsi, porphyrops, purpuratus, senex, ursinus*.
COMMENTS: Type of subgenus *Kasi*; see Szalay and Delson (1979:402). Referred to as *vetulus* by Corbet and Hill (1980:90) without comment. *Presbytis vetulus* was named in the same paper as *senex* on page 25 (following *senex*). On the correct usage of *vetulus* for the species, see Napier (1985:72).

Family Hylobatidae Gray, 1871. Cat. Monkeys, Lemurs, Fruit-eating Bats Brit. Mus., p. 4.
COMMENTS: Vaughan (1978:39-40) included this family in Pongidae (which is here considered a part of Hominidae); but see Delson and Andrews (1975:441) and Thenius (1981). Szalay and Delson (1979:461) included Hylobatidae in Hominidae.

Hylobates Illiger, 1811. Prodr. Syst. Mamm. Avium., p. 67.
TYPE SPECIES: *Homo lar* Linnaeus, 1771.
SYNONYMS: *Brachiopithecus, Brachitanytes, Bunopithecus, Cheiron, Hoolock, Loratus, Methylobates, Nomascus, Siamanga, Symphalangus*.
COMMENTS: Includes *Symphalangus*, see Anderson (1967:175); and *Nomascus*, see Corbet and Hill (1980:91). Revised by Groves (1972*b*). Reviewed by Marshall and Marshall (1976).

Hylobates agilis F. Cuvier, 1821. *In* É. Geoffroy and F. Cuvier, Hist. Nat. Mammifères, pt. 2, 3(32):1-3, "Wouwou".
TYPE LOCALITY: Indonesia, W Sumatra.

DISTRIBUTION: Malay Peninsula from the Mudah and Thepha rivers on the north to the Perak and Kelanton rivers on the south; Sumatra (Indonesia), SE of Lake Toba and the Singkil River; Kalimantan (Indonesian Borneo) between the Kapuas and Barito rivers.
STATUS: CITES - Appendix I; U.S. ESA - Endangered.
SYNONYMS: *albibarbis, ablogriseus, albonigrescens, rafflei, unko.*
COMMENTS: Subgenus *Hylobates*. Not a subspecies of *lar*. May include *muelleri* as a subspecies; or *albibarbis* (here recognized as a subspecies of *agilis*) may be a subspecies of *muelleri*.

Hylobates concolor (Harlan, 1826). J. Acad. Nat. Sci. Philadelphia, ser. 5, 4:231.
TYPE LOCALITY: Vietnam, Tonkin.
DISTRIBUTION: Hainan Isl (China); E of the Mekong River in S Yunnan (China), Laos, and Vietnam to Red River in Vietnam.
STATUS: CITES - Appendix I; U.S. ESA - Endangered; IUCN - Vulnerable.
SYNONYMS: *furvogaster, hainanus, harlani, jingdongensis, lu, nasutus, niger.*
COMMENTS: Type species of subgenus *Nomascus* which was recognized as a separate genus by Lekagul and McNeely (1977:308); but see Corbet and Hill (1980:91) and Szalay and Delson (1979:461).

Hylobates gabriellae Thomas, 1909. Ann. Mag. Nat. Hist., ser. 8, 4:112.
TYPE LOCALITY: Vietnam, Langbian.
DISTRIBUTION: S Laos, S Vietnam, Cambodia.
STATUS: CITES - Appendix I; U.S. ESA - Endangered.
SYNONYMS: *siki.*
COMMENTS: Subgenus *Nomascus*. Separated from *leucogenys* by Groves and Wang (1989).

Hylobates hoolock (Harlan, 1834). Trans. Am. Philos. Soc., 4:52.
TYPE LOCALITY: India, Assam, Garo Hills.
DISTRIBUTION: Between the Brahmaputra and Salween rivers in Assam (India), Burma, and Yunnan (China).
STATUS: CITES - Appendix I; U.S. ESA - Endangered; IUCN - Vulnerable.
SYNONYMS: *choromandus, fuscus, golock, hulock, leuconedys, scyritus.*
COMMENTS: Subgenus *Bunopithecus*.

Hylobates klossii (Miller, 1903). Smithson. Misc. Coll., 45:70.
TYPE LOCALITY: Indonesia, West Sumatra, S Pagai Isl.
DISTRIBUTION: Mentawai Isls (Indonesia).
STATUS: CITES - Appendix I; U.S. ESA and IUCN - Endangered.
COMMENTS: Subgenus *Hylobates*.

Hylobates lar (Linnaeus, 1771). Mantissa Plantarum, p. 521.
TYPE LOCALITY: Malaysia, Malacca (restricted by Kloss, 1929).
DISTRIBUTION: Between the Salween and Mekong rivers from S Yunnan (China) south to the Mun R. (Thailand) and the Mudah and Thepha rivers on the Malay Peninsula; S Malay Peninsula south of the Perak and Kelantan rivers; Sumatra (Indonesia) NW of Lake Toba and the Singkil R.; E and S Burma.
STATUS: CITES - Appendix I; U.S. ESA - Endangered.
SYNONYMS: *albimana, carpenteri, entelloides, longimana, varius, variegatus, vestitus, yunnanensis.*
COMMENTS: Subgenus *Hylobates*.

Hylobates leucogenys Ogilby, 1840. Proc. Zool. Soc. Lond., 1840:20.
TYPE LOCALITY: Laos, Muang Khi (Fooden, 1987).
DISTRIBUTION: SW Yunnan (China) to 19°N in Vietnam.
STATUS: CITES - Appendix I; U.S. ESA - Endangered.
SYNONYMS: *henrici.*
COMMENTS: Subgenus *Nomascus*. Separated from *concolor* by Dao (1983) and Ma and Wang (1986).

Hylobates moloch (Audebert, 1798). Hist. Nat. Singes Makis, 1st fasc., sect. 2, pl. 2.
TYPE LOCALITY: Indonesia, W Java, Mt. Salak (restricted by Sody, 1949*b*).
DISTRIBUTION: Java (Indonesia).
STATUS: CITES - Appendix I; U.S. ESA and IUCN - Endangered.

SYNONYMS: *cinerea, javanicus, leucisca, pongoalsoni.*
COMMENTS: Subgenus *Hylobates.*

Hylobates muelleri Martin, 1841. Nat. Hist. Mamm. Anim., p. 444.
TYPE LOCALITY: Indonesia, Kalimantan, "Southeast Borneo"; restricted by Lyon (1911:142).
DISTRIBUTION: Borneo from the N bank of the Kapuas River clockwise around the island to the E bank of the Barito River.
STATUS: CITES - Appendix I; U.S. ESA - Endangered.
SYNONYMS: *abbotti, funereus.*
COMMENTS: Subgenus *Hylobates.* May be conspecific with *H. agilis;* a wide intergrade zone with *H. agilis albibarbis* in Borneo (Marshall and Sugardjito, 1986).

Hylobates pileatus (Gray, 1861). Proc. Zool. Soc. Lond., 1861:136.
TYPE LOCALITY: Cambodia.
DISTRIBUTION: SE Thailand and Cambodia south of the Mun and Takhrong rivers and west of the Mekong River.
STATUS: CITES - Appendix I; U.S. ESA and IUCN - Endangered.
COMMENTS: Subgenus *Hylobates.*

Hylobates syndactylus (Raffles, 1821). Trans. Linn. Soc. Lond., 13:241.
TYPE LOCALITY: Indonesia, W Sumatra, Bengkeulen.
DISTRIBUTION: Barisan Mountains of Sumatra (Indonesia); mountains of Malay Peninsula south of Perak River.
STATUS: CITES - Appendix I; U.S. ESA - Endangered.
SYNONYMS: *continentis, gibbon, subfossilis, volzi.*
COMMENTS: Type species of subgenus *Symphalangus;* see Groves (1972*b*).

Family Hominidae Gray, 1825. Ann. Philos., n.s., 10:344.
SYNONYMS: Pongidae.
COMMENTS: For combining all genera in one family, see Groves (1989).

Gorilla (I. Geoffroy, 1852). C. R. Acad. Sci. Paris, 36:933.
TYPE SPECIES: *Troglodytes gorilla* Savage and Wyman, 1847.
SYNONYMS: *Pseudogorilla.*
COMMENTS: A subgenus of *Pan* according to Tuttle (1967); but see Groves (1989) who stated that they are not closely related.

Gorilla gorilla (Savage and Wyman, 1847). Boston J. Nat. Hist., 5:417.
TYPE LOCALITY: Gabon, Gabon Estuary, Mpongwe country.
DISTRIBUTION: SE Nigeria, Cameroon, Rio Muni (Equatorial Guinea), Congo Republic, SW Central African Republic, Gabon, N and E Zaire, SW Uganda, N Rwanda.
STATUS: CITES - Appendix I; U.S. ESA - Endangered; IUCN - Endangered as *G. g. berengei* and *G. g. graueri,* otherwise Vulnerable.
SYNONYMS: *adrotes, beringei, castaneiceps, diehli, ellioti, gigas, gina, graueri, halli, hansmeyeri, jacobi, manyema, matschiei, mayema, mikenensis, rex-pygmaeorum, savagei, schwartzi, uellensis, zenkeri.*

Homo Linnaeus, 1758. Syst. Nat., 10th ed., 1:20.
TYPE SPECIES: *Homo sapiens* Linnaeus, 1758.
SYNONYMS: *Africanthropus, Cyphanthropus, Palaeanthropus, Pithecanthropus, Sinanthropus, Telanthropus.*

Homo sapiens Linnaeus, 1758. Syst. Nat., 10th ed., 1:20.
TYPE LOCALITY: Sweden, Uppsala.
DISTRIBUTION: Cosmopolitan.
STATUS: CITES - Appendix II as Order Primates; absolutely not endangered.
SYNONYMS: *heidelbergensis, neanderthalensis, palestinus, rhodesiensis,* hundreds of others.

Pan Oken, 1816. Lehrb. Naturgesch., ser. 3, 2:xi.
TYPE SPECIES: *Simia troglodytes* Blumenbach, 1775.
SYNONYMS: *Anthropopithecus, Bonobo, Chimpansee, Engeco, Fsihego, Hylanthropus, Mimetes, Pseudanthropos, Theranthropus, Troglodytes.*

COMMENTS: In accordance with the provisions of Art. 80 of the International Code of Zoological Nomenclature (International Commission on Zoological Nomenclature, 1985), *Pan* is used instead of *Chimpansee*. Reviewed by Hill (1969).

Pan paniscus Schwartz, 1929. Rev. Zool. Bot. Afr., 16:4.
TYPE LOCALITY: Zaire, S of the upper Maringa River, 30 km S of Befale.
DISTRIBUTION: Congo Basin of Zaire, on south side of Congo River.
STATUS: CITES - Appendix I; U.S. ESA - Endangered; IUCN - Vulnerable.

Pan troglodytes (Blumenbach, 1775). De generis humani varietate nativa, p. 37.
TYPE LOCALITY: Gabon, Mayoumba.
DISTRIBUTION: S Cameroon; Gabon; S Congo Republic; Uganda; W Tanzania; E and N Zaire; W Central African Republic; Guinea to W Nigeria, south to Congo River in W Africa.
STATUS: CITES - Appendix I; U.S. ESA - Endangered in the wild, Threatened in captivity; IUCN - Endangered as *P. t. verus*, otherwise Vulnerable.
SYNONYMS: *adolfifriederici, angustimanus, aubryi, calvescens, calvus, castanomale, chimpanse, cottoni, ellioti, fuliginosus, fuscus, graueri, hecki, ituriensis, koolookamba, lagaros, leucoprymnus, livingstonii, mafuca, marungensis, nahani, niger, ochroleucus, oertzeni, papio, pfeifferi, purschei, pusillus, raripilosus, reuteri, satyrus, schneideri, schubotzi, schweinfurthii, steindachneri, tschego, vellerosus, verus, yambuyae.*
COMMENTS: For authorship of this name, see the International Commission on Zoological Nomenclature (1988).

Pongo Lacépède, 1799. Tabl. Div. Subd. Orders Genres Mammifères, p. 4.
TYPE SPECIES: *Simia pygmaeus* Linnaeus, 1760.
SYNONYMS: *Lophotus, Macrobates, Satyrus.*

Pongo pygmaeus (Linnaeus, 1760). Amoenit. Acad., 6:68.
TYPE LOCALITY: "Borneo".
DISTRIBUTION: Sumatra, NW of Lake Toba (Indonesia); discontinuous in Borneo.
STATUS: CITES - Appendix I as *P. p. abelii* and *P. p. pygmaeus*, otherwise Appendix II; U.S. ESA and IUCN - Endangered.
SYNONYMS: *abelii, abongensis, agrias, batangtuensis, bicolor, bornaensis, brookei, curtus, dadappensis, deliensis, genepaiensis, gigantica, landakkensis, langkatensis, morio, owenii, rantarensis, rufus, satyrus, skalauensis, sumatranus, tuakensis, wallichii, wurmbii.*
COMMENTS: On the nomenclature, see Groves and Holthuis (1985). Reviewed by Groves (1971*a*, Mammalian Species, 4).

ORDER CARNIVORA

by W. Christopher Wozencraft

ORDER CARNIVORA

COMMENTS: Recent studies on family level relationships have agreed in several key areas, among which are: (1) The sister group to the Carnivora is the †Creodonta. (2) The Recent Carnivora are organized into two monophyletic groups; Suborder Feliformia: felids, herpestids, hyaenids and viverrids; and Suborder Caniformia: canids, ursids, mustelids, odobenids, otariids, phocids, and procyonids. (3) The pinnipeds (otariids, odobenids, and phocids) are included within the suborder Caniformia; placing them in a separate Order would make the Carnivora paraphyletic (Flynn et al., 1988; Tedford, 1976; Wozencraft, 1989*a*; Berta, 1991).

Family Canidae G. Fischer, 1817. Mém. Soc. Imp. Nat. Moscow, 5:372.

COMMENTS: Conservation status and distribution reviewed by Ginsberg and Macdonald (1990). Reviewed by Langguth (1975) and Stains (1975). Revisions by Langguth (1969), Clutton-Brock et al. (1976), Van Gelder (1978), Berta (1987, 1988), Wayne and O'Brien (1987), and Wayne et al. (1987*a*, *b*, 1989) gave little support to the subfamilies recognized by Simpson (1945); therefore, no subfamilies are recognized here. Van Gelder's (1977*b*) hybridization criteria for generic classification resulted in the recognition of only a few genera, including some paraphyletic groups.

Alopex Kaup, 1829. Skizz. Entwickel. Gesch. Nat. Syst. Europ. Theirwelt, 1:85.

TYPE SPECIES: *Canis lagopus* Linnaeus, 1758, by monotypy (International Commission on Zoological Nomenclature, 1956*a*; Melville and Smith, 1987).

COMMENTS: Bobrinskii et al. (1965) considered *Alopex* a subgenus of *Vulpes*; Van Gelder (1978) considered it a subgenus of *Canis*.

Alopex lagopus (Linnaeus, 1758). Syst. Nat., 10th ed., 1:40.

TYPE LOCALITY: "alpibus Lapponicis, Sibiria," restricted by Thomas (1911*a*) to "Sweden (Lapland)."

DISTRIBUTION: Circumpolar, entire tundra zone of the Holarctic, including most of the Arctic islands.

SYNONYMS: *arctica* Oken, 1816; *argenteus* Billberg, 1827; *beringensis* Merriam, 1902; *beringianus* Cherski, 1920; *caerulea* Nilsson, 1820; *fuliginosus* Bechstein, 1799; *groenlandicus* Bechstein, 1799; *hallensis* Merriam, 1900; *innuitus* Merriam, 1902; *kenaiensis* Brass, 1911; *pribilofensis* Merriam, 1902; *spitzbergenensis* Barrett-Hamilton and Bonhote, 1898; *typicus* Barrett-Hamilton and Bonhote, 1898; *ungava* Merriam, 1902.

COMMENTS: Viable hybrids have been recorded between *Alopex lagopus* and *Vulpes vulpes* (Chiarelli, 1975).

Atelocynus Cabrera, 1940. Notas Mus. La Plata 5:14.

TYPE SPECIES: *Canis microtis* Sclater, 1883, by original designation.

COMMENTS: See comments under *Dusicyon*. Placed in *Atelocynus* by Cabrera (1931, 1958), Langguth (1975), Stains (1975), and Berta (1985, 1986, 1988). Van Gelder (1978) considered *Atelocynus* a subgenus of *Canis*.

Atelocynus microtis (Sclater, 1883). Proc. Zool. Soc. Lond., 1882:631 [1883].

TYPE LOCALITY: "Amazons," restricted by Hershkovitz (1957*a*) to "south bank of the Rio Amazonas, Pará, Brazil."

DISTRIBUTION: Amazonian Brazil, Peru, Ecuador, and Colombia; perhaps Bolivia, and a second, allopatric population in S Brazil, Paraguay and N Argentina (Ginsberg and Macdonald, 1990).

STATUS: IUCN - Insufficiently known.

SYNONYMS: *sclateri* J. A. Allen, 1905.

COMMENTS: Reviewed by Hershkovitz (1961*a*) and Berta (1986, Mammalian Species, 256).

Canis Linnaeus, 1758. Syst. Nat., 10th ed., 1:38.

TYPE SPECIES: *Canis familiaris* Linnaeus, 1758 (= *Canis lupus* Linnaeus, 1758) by Linnean tautonomy (Melville and Smith, 1987).

COMMENTS: Van Gelder (1978) included *Alopex, Atelocynus, Cerdocyon, Pseudalopex, Lycalopex, Dusicyon,* and *Vulpes* as subgenera. However, this arrangement is not currently employed by most mammalogists (Berta, 1987, 1988; Corbet, 1978*c*; Corbet and Hill, 1980; Gromov and Baranova, 1981; Hall, 1981; Wozencraft, 1989*b*).

Canis adustus Sundevall, 1847. Ofv. K. Svenska Vet.-Akad. Forhandl. Stockholm, 1846, 3:121 [1847].

TYPE LOCALITY: "Caffraria Interiore"; listed as "Magaliesberg" [South Africa] by Sclater (1900).

DISTRIBUTION: Open woodland and semi-arid grassland from Senegal to Ethiopia, south to N Namibia, N Botswana, Zimbabwe, Mozambique, and N South Africa. The W African population is largely "unknown" (Ginsberg and Macdonald, 1990).

SYNONYMS: *bweha* Heller, 1914; *centralis* Schwarz, 1915; *holubi* Lorenz, 1895; *kaffensis* Neumann, 1902; *lateralis* Sclater, 1870; *notatus* Heller, 1914; *studeri* Hilzheimer, 1906; *wunderlichi* Noack, 1897.

Canis aureus Linnaeus, 1758. Syst. Nat., 10th ed., 1:40.

TYPE LOCALITY: "oriente", restricted by Thomas (1911*a*) to "Benná Mts., Laristan, S. Persia" [Iran].

DISTRIBUTION: N and E Africa, south to Senegal, Nigeria, and Tanzania; SW Asia; SE Europe; Transcaucasia; C Asia; Iran; Afghanistan; S Asia to Thailand, including Sri Lanka.

STATUS: CITES - Appendix III (India).

SYNONYMS: *algirensis* Wagner, 1841; *anthus* Cuvier, 1820; *balcanicus* Brusina, 1892; *barbarus* H[amilton]. Smith, 1839; *bea* Heller, 1914; *caucasica* Kolenati, 1858; *cruesemanni* Matschie, 1900; *dalmatinus* Wagner, 1841; *doederleini* Hilzheimer, 1906; *ecsedensis* Kretzoi, 1947; *gallaensis* Lorenz, 1906; *graecus* Wagner, 1841; *grayi* Hilzheimer, 1906; *hadramauticus* Noack, 1896; *hungaricus* Ehik, 1938; *indicus* Hodgson, 1833; *kola* Wroughton, 1916; *lanka* Wroughton, 1916; *lupaster* Hemprich and Ehrenberg, 1833; *maroccanus* Cabrera, 1921; *mengesi* Noack, 1897; *minor* Mojsisovico, 1897; *moreotica* I. Geoffroy Saint-Hilaire, 1835; *naria* Wroughton, 1916; *nubianus* Cabrera, 1921; *riparius* Hemprich and Ehrenberg, 1832; *sacer* Hemprich and Ehrenberg, 1833; *senegalensis* H[amilton]. Smith, 1839; *somalicus* Lorenz, 1906; *soudanicus* Thomas, 1903; *studeri* Hilzheimer, 1906; *syriacus* Hemprich and Ehrenberg, 1833; *thooides* Hilzheimer, 1906; *tripolitanus* Wagner, 1841; *typicus* Kolenati, 1858; *variegatus* Cretzschmar, 1826; *vulgaris* Wagner, 1841.

Canis latrans Say, 1823. *In* James, Account Exped. Pittsburgh to Rocky Mtns, 1:168.

TYPE LOCALITY: "Engineer cantonment" reported at "latitude 41°25'N, and longitude...95°47'30'W" (p. XVIII, vol. 2). Reported in Honacki et al. (1982) as "U.S.A., Nebraska, Washington Co., Engineer Cantonment, about 12 mi. (19.2 km) S. E. Blair".

DISTRIBUTION: Originally may have occurred west of the Mississippi R., north to about 55°N, and south to Mexico City. Has extended its range north to N Alaska (USA), Northwest Territories, and Hudson Bay (Canada), approximately to 70°N, south to Costa Rica to approximately 10°N, and east to the Atlantic coast. Introduced in Florida and Georgia (Bekoff, 1977).

SYNONYMS: *cagottis* H[amilton]. Smith, 1839; *clepticus* Elliot, 1903; *dickeyi* Nelson, 1932; *estor* Merriam, 1897; *frustror* Woodhouse, 1850; *goldmani* Merriam, 1904; *hondurensis* Goldman, 1936; *impavidus* Allen, 1903; *jamesi* Townsend, 1912; *lestes* Merriam, 1897; *mearnsi* Merriam, 1897; *microdon* Merriam, 1897; *ochropus* Eschscholtz, 1829; *peninsulae* Merriam, 1897; *texensis* Bailey, 1905; *thamnos* Jackson, 1949; *umpquensis* Jackson, 1949; *vigilis* Merriam, 1897.

COMMENTS: Revised by Young (1951), and reviewed by Bekoff (1977, Mammalian Species, 79).

Canis lupus Linnaeus, 1758. Syst. Nat., 10th ed., 1:39.

TYPE LOCALITY: "Europæ sylvis, etjam frigidioribus", restricted by Thomas (1911*a*) to "Sweden".

DISTRIBUTION: Throughout the N hemisphere: North America south to 20°N in Oaxaca (Mexico); Europe; Asia, including the Arabian Peninsula and Japan, excluding Indochina and S India. Extirpated from most of the continental USA, Europe, and SE China and Indochina (Ginsberg and Macdonald, 1990).

STATUS: CITES - Appendix I (Indian, Pakistan, Bhutan, and Nepal populations); otherwise Appendix II. U.S. ESA - Endangered in the USA (48 conterminous States, except Minnesota) and Mexico; Threatened in Minnesota (USA). IUCN - Vulnerable.

SYNONYMS: *albus* Kerr, 1792; *alces* Goldman, 1941; *altaicus* Noak, 1911; *arabs* Pocock, 1934; *arctos* Pocock, 1935; *argunensis* Dybowski, 1922; *ater* Richardson, 1839; *baileyi* Nelson and Goldman, 1929; *banksianus* Anderson, 1943; *beothucus* G. M. Allen and Barbour, 1937; *bernardi* Anderson, 1943; *campestris* Dwigubski, 1804; *canadensis* de Blainville, 1843; *canus* de Sélys Longchamps, 1839; *chanco* Gray, 1863; *columbianus* Goldman, 1941; *communis* Dwigubski, 1804; *coreanus* Abe, 1923; *crassodon* Halla, 1932; *cubanensis* Ognev, 1923; *deitanus* Cabrera, 1907; *desertorum* Bogdanov, 1992; *dybowskii* Kerr, 1792; *ekloni* Przewalski, 1883; *familiaris* Linnaeus, 1758; *filchneri* Matschie, 1908; *flavus* Kerr, 1792; *fulvus* de Sélys Longchamps, 1839; *fusca* Richardson, 1839; *gigas* Townsend, 1850; *griseoalbus* Baird, 1858; *hattai* Kishida, 1931; *hodophilax* Temminck, 1839; *hudsonicus* Goldman, 1941; *irremotus* Goldman, 1937; *italicus* Altobello, 1921; *japonicus* Nehring, 1883; *kamtschaticus* Dybowski, 1922; *karanorensis* Matschie, 1908; *knightii* Anderson, 1947; *kurjak* Bolkay, 1925; *labradorius* Goldman, 1937; *laniger* Hodgson, 1847; *ligoni* Goldman, 1937; *lupus-griseus* Sabine, 1823; *lycaon* Schreberrt, 1775; *mackenzii* Anderson, 1943; *major* Ogérien, 1863; *manningi* Anderson, 1943; *minor* Ogérien, 1863; *mogollonensis* Goldman, 1937; *monstrabilis* Goldman, 1937; *nubilus* Say, 1823; *occidentalis* Richardson, 1829; *orientalis* Wagner, 1841; *orion* Pocock, 1935; *pallipes* Sykes, 1831; *pambasileus* Elliot, 1905; *rex* Pocock, 1935; *signatus* Cabrera, 1907; *sticte* Richardson, 1839; *tschiliensis* Matschie, 1908; *tundrarum* Miller, 1912; *turuchanensis* Ognev, 1923; *ungavensis* Comeau, 1940; *variabilis* Say, 1823; *youngi* Goldman, 1937.

COMMENTS: Reviewed by Mech (1974, Mammalian Species, 37). *C. familiaris* has page priority over *C. lupus* in Linnaeus (1758), but both were published simultaneously, and *C. lupus* has been universally used for this species.

Canis mesomelas Schreber, 1775. Die Säugethiere, 2(14):pl. 95[1775]; text, 3(21):370[1776], 586[1777].

TYPE LOCALITY: "Vorgebirge der guten Hofnung" [South Africa, Cape Prov., Cape of Good Hope].

DISTRIBUTION: Africa, south of the tropical rain-forest in the west and as far north as Ethiopia and Sudan in the east; two populations, (1) Angola across to mid-Mozambique and south; and (2) Ethiopia to Tanzania and northern most Zimbabwe east of Zaire (Ginsberg and Macdonald, 1990).

SYNONYMS: *achrotes* Thomas, 1926; *arenarum* Thomas, 1926; *elgonae* Heller, 1914; *mcmillani* Heller, 1914; *schmidti* Noack, 1897; *variegatoides* Smith, 1834.

Canis rufus Audubon and Bachman, 1851. Viviparous Quadrupeds of North America, 2:240.

TYPE LOCALITY: Not given. Restricted by Goldman (1937) to "15 miles west of Austin, Texas" [USA], based on accounts from Audubon and Bachman (1851).

DISTRIBUTION: SE and SC USA, from Florida to C Texas and north to S Indiana and Missouri.

STATUS: U.S. ESA - Endangered (except Dare, Tyrrell, Hyde, and Washington Counties, North Carolina, which have a nonessential experimental population); IUCN - Endangered.

SYNONYMS: *floridanus* Miller, 1912; *gregoryi* Goldman, 1937; *niger* Bartram, 1791.

COMMENTS: Reviewed and recognized by Paradiso and Nowak (1972, Mammalian Species, 22); also recognized by Paradiso (1968), Atkins and Dillion (1971), and Nowak (1979). The widely used name *C. niger* is invalid (International Commission on Zoological Nomenclature, 1957*a*). The validity of *rufus* as a full species was questioned by Clutton-Brock et al. (1976), due to the existence of natural hybrids with *lupus* and *latrans*. Natural hybridization may be a consequence of habitat disruption by man (Paradiso and Nowak, 1971). Nowak (1979) provided evidence for specific distinctness. All specimens examined by Wayne and Jenks (1991) had either a *lupus* or *latrans* mtDNA genotype.

Canis simensis Rüppell, 1840. Neue Wirbelt. Fauna Abyssin. Gehörig. Säugeth., 1:39, pl. 14.
TYPE LOCALITY: "Wir beobachteten diesen wolfsartigen Hund in den Bergen von Simen..." [Ethiopia, mountains of Simen].
DISTRIBUTION: C Ethiopia.
STATUS: U.S. ESA and IUCN - Endangered.
SYNONYMS: *sinus* Gervais, 1855; *walgie* Heuglin, 1863.
COMMENTS: Sometimes placed in subgenus *Simenia* Gray, 1868.

Cerdocyon H[amilton]. Smith, 1839. Jardine's Natur. Libr., 9:259.
TYPE SPECIES: *Canis azarae* Wied, 1824 (= *Canis Thous* Linnaeus, 1766) by subsequent designation (Thomas, 1914*a*).

Cerdocyon thous (Linnaeus, 1766). Syst. Nat., 12th ed., 1:60.
TYPE LOCALITY: "Surinamo" [Surinam].
DISTRIBUTION: Uruguay and N Argentina to lowlands of Bolivia, Venezuela; Colombia and Guyanas; Brazil except Amazonia.
STATUS: CITES - Appendix II.
SYNONYMS: *affinis* Marelli, 1931; *angulensis* Thomas, 1903; *apollinaris* Thomas, 1914; *aquilus* Bangs, 1898; *azarae* Wied, 1824; *brachyteles* de Blainville, 1843; *brasiliensis* Wied, 1824; *cancrivora* Brongniart, 1792; *fronto* Lönnberg, 1919; *fulvogriscus* Zukowsky, 1950; *germanus* G. M. Allen, 1923; *guaraxa* H[amilton]. Smith, 1829; *jucundus* Thomas, 1921; *lunaris* Thomas, 1914; *melampus* Wagner, 1841; *melanostomus* Wagner, 1843; *mimax* Thomas, 1914; *riograndensis* Ihering, 1911; *robustior* Lund, 1843; *rudis* Günther, 1879; *savannarum* Osgood, 1912; *tucumanus* Thomas, 1921; *vetulus* Studer, 1905.
COMMENTS: Reviewed by Berta (1982, Mammalian Species, 186). Placed in *Cerdocyon* by Langguth (1975), Stains (1975), and Berta (1982); placed in subgenus *Canis* (*Cerdocyon*) by Van Gelder (1978).

Chrysocyon H[amilton]. Smith, 1839. Jardine's Natur. Libr., 9:241.
TYPE SPECIES: *Canis jubatus* Desmarest, 1820 (= *Canis brachyurus* Illiger, 1815).
COMMENTS: Recognized by Langguth (1975), Stains (1975), Van Gelder (1978), and Berta (1988).

Chrysocyon brachyurus (Illiger, 1815). Abh. Phys. Klasse K. Pruess. Akad. Wiss., 1804-1811:121.
TYPE LOCALITY: Listed by Cabrera (1958) as "los esteros del Paraguay".
DISTRIBUTION: NE Argentina; Paraguay; lowlands of Bolivia; Brazil from Rio Grande do Sul to Minas Gerais, Goiás and Mato Grosso.
STATUS: CITES - Appendix II; U.S. ESA - Endangered; IUCN - Vulnerable.
SYNONYMS: *campestris* Wied[-Neuwied], 1826; *cancrosa* Oken, 1816; *isodactylus* Ameghino, 1906; *jubatus* Desmarest, 1820; *vulpes* Larranaga, 1923.
COMMENTS: See Dietz (1985, Mammalian Species, 234).

Cuon Hodgson, 1838. Ann. Mag. Nat. Hist., [ser. 1], 1:152.
TYPE SPECIES: *Canis primaevus* Hodgson, 1838, by monotypy (Melville and Smith, 1987).
COMMENTS: Placed in subfamily Simocyoninae Dawkins, 1868, by Simpson (1945) and Stains (1975).

Cuon alpinus (Pallas, 1811). Zoogr. Rosso-Asiat., 1:34.
TYPE LOCALITY: "Udskoi Ostrog"; reported in Honacki et al. (1982) as "U.S.S.R., Amurskaya Obl., Udskii-Ostrog."
DISTRIBUTION: Java; Sumatra; Malaysia; montane forest areas of the Indian peninsula and N Pakistan through Tibet and Xinjiang (Tian Shan and Altai—extinct); Indochina and China to Korea and Ussuri region (Russia), and mountains of S Siberia and N Mongolia.
STATUS: CITES - Appendix II; U.S. ESA - Endangered; IUCN - Vulnerable.
SYNONYMS: *clamitans* Heude, 1892; *dukhunensis* Sykes, 1831; *fumosus* Pocock, 1936; *grayiformes* Hodgson, 1863; *hesperius* Afanasiev and Zolotarev, 1935; *infuscus* Pocock, 1936; *javanicus* Desmarest, 1820; *laniger* Pocock, 1936; *lepturus* Heude, 1892; *primaevus* Hodgson, 1833; *rutilans* Blanford, 1839; *sumatrensis* Hardwicke, 1822.
COMMENTS: Reviewed by Cohen (1978, Mammalian Species, 100).

Dusicyon H[amilton]. Smith, 1839. Jardine's Natur. Libr., 9:248.
TYPE SPECIES: *Canis antarcticus* Bechstein, 1799 (= *C. australis* Kerr, 1792), by subsequent designation by Cabrera (1931).
COMMENTS: There has been general disagreement as to generic classification of the South American canids, with most of the disagreement centered on the species *australis, culpaeus, griseus, gymnocercus, microtis, sechurae, thous,* and *vetulus*. Van Gelder (1978) proposed placing these taxa into *Canis* and giving only subgeneric recognition. The other extreme is best represented by Cabrera (1931) who recognized 5 genera for this group. Langguth (1969) first followed Cabrera's classification, but later (1975) decided to group most taxa into *Canis*, because he felt differences were not sufficient to warrant generic distinctions. The phenetic approaches of Clutton-Brock et al. (1976) and Wayne and O'Brien (1987) confirmed the close similarities of these taxa. Berta's (1987, 1988) phylogenetic hypothesis is followed here.

Dusicyon australis (Kerr, 1792). *In* Linnaeus, Anim. Kingdom, p. 144.
TYPE LOCALITY: "America and Falkland islands."
DISTRIBUTION: Falkland Islands.
STATUS: IUCN - Extinct.
SYNONYMS: *antarcticus* Bechstein, 1799; *darwini* Thomas, 1914.
COMMENTS: Placed in *Dusicyon* by Cabrera (1931) and Berta (1987, 1988), and considered as a subgenus separate from other "foxes" (i.e., *culpaeus, griseus, gymnocercus,* and *sechurae*) by Langguth (1975) and Van Gelder (1978).

Lycaon Brookes, 1827. *In* Griffith et al., Anim. Kingdom, 5:151.
TYPE SPECIES: *Lycaon tricolor* Brookes, 1827 (= *Hyaena picta* Temminck, 1820) by monotypy (Melville and Smith, 1987).
COMMENTS: Placed in Simocyoninae Dawkins, 1868, by Simpson (1945) and Stains (1975).

Lycaon pictus (Temminck, 1820). Ann. Gen. Sci. Phys., 3:54, pl. 35.
TYPE LOCALITY: "à la côte de Mosambique" [Mozambique].
DISTRIBUTION: Subsaharan Africa, except in the desert area of the north and the tropical rain-forest of the west. See Ginsberg and Macdonald (1990, fig. 6) for present status of population.
STATUS: U.S. ESA and IUCN - Endangered.
SYNONYMS: *cacondae* Matschie, 1915; *dieseneri* Matschie, 1915; *ebermaieri* Matschie, 1915; *fuchsi* Matschie, 1915; *gansseri* Matschie, 1915; *gobabis* Matschie, 1915; *hennigi* Matschie, 1915; *huebneri* Matschie, 1915; *kondoae* Matschie, 1915; *krebsi* Matschie, 1915; *lademanni* Matschie, 1915; *lalandei* Matschie, 1915; *langheldi* Matschie, 1915; *luchsingeri* Matschie, 1915; *lupinus* Thomas, 1902; *manguensis* Matschie, 1915; *mischlichi* Matschie, 1915; *prageri* Matschie, 1912; *richteri* Matschie, 1915; *ruppelli* Matschie, 1915; *ruwanae* Matschie, 1915; *sharicus* Thomas, 1907; *ssongeae* Matschie, 1915; *stierlingi* Matschie, 1915; *styxi* Matschie, 1915; *taborae* Matschie, 1915; *takanus* Matschie, 1915; *tricolor* Brookes, 1827; *typicus* A. Smith, 1833; *venatica* Burchell, 1822; *windhorni* Matschie, 1915; *wintgensi* Matschie, 1915; *zedlitzi* Matschie, 1915; *zuluensis* Thomas, 1904.

Nyctereutes Temminck, 1839. Tijdschr. Nat. Gesch. Physiol., 5:285.
TYPE SPECIES: *Canis viverrinus* Temminck, 1839 (= *Canis procyonoides* Gray, 1834).

Nyctereutes procyonoides (Gray, 1834). Illustr. Indian Zool., 2:pl. 1.
TYPE LOCALITY: Unknown; restricted to "vicinity of Canton, China" by Allen (1938).
DISTRIBUTION: Ussuri region (Russia), Korea, China, Japan, and N Indochina.
SYNONYMS: *albus* Hornaday, 1904; *amurensis* Matschie, 1908; *koreensis* Mori, 1922; *orestes* Thomas, 1923; *sinensis* Brass, 1904; *stegmanni* Matschie, 1908; *ussuriensis* Matschie, 1908; *viverrinus* Temminck, 1839.
COMMENTS: See Ward and Wurster-Hill (1990, Mammalian Species, 358).

Otocyon Müller, 1836. Arch. Anat. Physiol., Jahresber. Fortschr. Wiss., 1835:1 [1836].
TYPE SPECIES: *Otocyon caffer* Müller, 1836 (= *Canis megalotis* Desmarest, 1822), by monotypy (Melville and Smith, 1987).

Otocyon megalotis (Desmarest, 1822). Mammalogie, *in* Encyclop. Meth., 2(Suppl.):538.
TYPE LOCALITY: "le Cap de Bonne-Espérance" [South Africa, Cape Prov., Cape of Good Hope].
DISTRIBUTION: Allopatric south (Botswana, Namibia, South Africa) and east (Kenya, Ethiopia, Tanzania) African popultions (Ginsberg and Macdonald, 1990).
SYNONYMS: *auritus* H[amilton]. Smith, 1840; *caffer* Müller, 1836; *canescens* Cabrera, 1910; *lalandi* Desmoulins, 1823; *steinhardti* Zukowsky, 1924; *virgatus* Miller, 1909.

Pseudalopex Burmeister, 1856. Erläut Fauna Brasiliens, p. 24.
TYPE SPECIES: *Canis magellanicus* Gray, 1836 (= *Canis culpaeus* Molina, 1782), by subsequent designation by Cabrera (1931).
COMMENTS: See comments under *Dusicyon*. Although combining taxa included here with *Dusicyon* would not be in conflict with Berta (1987, 1988), her analyses suggested that other genera, now extinct, are more closely related to *Dusicyon*. Berta (1987, 1988) presented derived features that would support a single origin for those taxa recognized here in *Pseudalopex* (which would also agree with Cabrera, 1958 and Stains, 1975). A detailed comparative morphological study by Langguth (1969) caused him to conclude (1975:193) that *Pseudalopex* merited generic rank.

Pseudalopex culpaeus (Molina, 1782). Sagg. Stor. Nat. Chile, p. 293.
TYPE LOCALITY: "Chili" restricted by Cabrera (1931) to "the Santiago Province."
DISTRIBUTION: From Tierra del Fuego through the Andes of Chile and Argentina to the highlands of Bolivia, Peru, Ecuador, and Colombia.
STATUS: CITES - Appendix II.
SYNONYMS: *albigula* Philippi, 1903; *amblyodon* Philippi, 1903; *andina* Thomas, 1914; *chilensis* Kerr, 1792; *culpaeolus* Thomas, 1914; *ferrugineus* Huber, 1925; *inca* Thomas, 1914; *lycoides* Philippi, 1896; *magellanicus* Gray, 1836; *montanus* Prichard, 1902; *prichardi* Thomas, 1914; *reissii* Hilzheimer, 1906; *riveti* Trouessart, 1906; *smithersi* Thomas, 1914.
COMMENTS: Placed in *Pseudalopex* by Berta (1987, 1988); and in *Dusicyon* by Cabrera (1958). Considered in *Canis* (*Pseudalopex*) by Langguth (1975), Clutton-Brock et al. (1976), and Van Gelder (1978). Includes *culpaeolus* and *inca*, see Langguth (1967).

Pseudalopex griseus (Gray, 1837). Mag. Nat. Hist. [Charlesworth's], 1:578.
TYPE LOCALITY: "Magellan", listed in Cabrera (1958) as "Costa del Estrecho de Magallanes" [Chile].
DISTRIBUTION: Atacama (Chile) and Santiago del Estero (Argentina) south to Tierra del Fuego.
STATUS: CITES - Appendix II; IUCN - Vulnerable.
SYNONYMS: *domeykoanus* Philippi, 1901; *fulvipes* Martin, 1837; *gracilis* Burmeister, 1861; *lagopus* Molina, 1782; *maullinicus* Philippi, 1903; *patagonicus* Philippi, 1866; *rufipes* Philippi, 1901; *torquatus* Philippi, 1903; *zorrula* Thomas, 1921.
COMMENTS: Placed in *Pseudalopex* by Berta (1988) and *Dusicyon* by Cabrera (1958). Considered in *Canis* (*Pseudalopex*) by Langguth (1975, and Van Gelder (1978). Cabrera (1958) and Osgood (1943) considered *Vulpes fulvipes* distinct, however, Langguth (1969) presented data to support its inclusion.

Pseudalopex gymnocercus (G. Fischer, 1814). Zoognosia, 3:xi, 178.
TYPE LOCALITY: "Paraguay", restricted by Cabrera (1958) to "a los alrededores de Asunción."
DISTRIBUTION: Argentina, north of Rio Negro; Paraguay; Uruguay; S Brazil; E Bolivia.
STATUS: CITES - Appendix II.
SYNONYMS: *antiquus* Ameghino, 1889; *brasiliensis* Schinz, 1821; *cinereoargenteus* Larranaga, 1923; *entrerianus* Burmeister, 1861; *fossilis* H. Gervais and Ameghino, 1889; *protalopex* Lund, 1839.
COMMENTS: Placed in *Pseudalopex* by Berta (1988) and in *Dusicyon* by Cabrera (1958). Considered in *Canis* (*Pseudalopex*) by Langguth (1975) and Van Gelder (1978).

Pseudalopex sechurae (Thomas, 1900). Ann. Mag. Nat. Hist., ser. 7, 5:148.
TYPE LOCALITY: "Desert of Sechura, N.W. Peru. . . Sullana".
DISTRIBUTION: NW Peru and SW Ecuador.
STATUS: IUCN - Insufficiently known.

COMMENTS: Placed in *Pseudalopex* by Berta (1988) and in *Dusicyon* by Cabrera (1958). Considered in *Canis* (*Pseudalopex*) by Langguth (1975) and Van Gelder (1978).

Pseudalopex vetulus (Lund, 1842). K. Dansk. Vid. Selsk. Naturv. Math. Afhandl., 9:4.
TYPE LOCALITY: Listed by Cabrera (1958) as "Lagoa Santa, Minas Gerais" [Brazil].
DISTRIBUTION: C Brazilian Highlands in the States of Mato Grosso, Goiás, Minas Gerais, Bahia, and São Paulo.
STATUS: IUCN - Insufficiently known.
SYNONYMS: *azarae* Lund, 1839; *chilensis* Gray, 1868; *fulvicaudus* Lund, 1843; *parvidens* Mivart, 1890; *sladeni* Thomas, 1903; *urostictus* Mivart, 1890.
COMMENTS: Placed in *Pseudalopex* by Berta (1987, 1988); in *Dusicyon* (*Lycalopex*) by Cabrera (1958) (and implied by Stains, 1975); in *Lycalopex* by Langguth (1975); and considered in *Canis* (*Lycalopex*) by Van Gelder (1978).

Speothos Lund, 1839. Ann. Sci. Nat. Zool. (Paris), ser. 2, 11:224.
TYPE SPECIES: *Speothos pacivorus* Lund, 1839 (extinct).
COMMENTS: Berta and Marshall (1978) included *Icticyon*. Placed in Simocyoninae Dawkins, 1868, by Simpson (1945) and Stains (1975).

Speothos venaticus (Lund, 1842). K. Dansk. Vid. Selsk. Naturv. Math. Afhandl., 9:67.
TYPE LOCALITY: "Lagoa Santa" [Minas Gerais, Brazil].
DISTRIBUTION: Forested areas of Bolivia, Paraguay and Brazil (except the semiarid NE); E Peru; Ecuador; Colombia; Venezuela; Guyana; French Guiana; Surinam; Panama.
STATUS: CITES - Appendix I; IUCN - Vulnerable.
SYNONYMS: *baskii* Schinz, 1849; *melanogaster* Gray, 1846; *wingei* Ihering, 1911.

Urocyon Baird, 1857. Mammalia, *in* Repts. U.S. Expl. Surv., 8(1):121, 138.
TYPE SPECIES: *Canis virginianus* Schreber, 1775 (= *Canis cinereo argenteus* Schreber, 1775) by subsequent designation (Elliot, 1901; Melville and Smith, 1987).
COMMENTS: Considered a subgenus of *Vulpes* by Clutton-Brock et al. (1976).

Urocyon cinereoargenteus (Schreber, 1775). Die Säugethiere, 2(13):pl. 92[1775]; text: 3(21):361[1776].
TYPE LOCALITY: "Sein Vaterland ist Carolina und die Wärmeren Gegenden von Nordamerica, vielleicht auch Surinam."
DISTRIBUTION: North America from Oregon, Nevada, Utah, and Colorado in the West and the USA-Canadian border in the East through Central America to N Colombia and Venezuela.
SYNONYMS: *borealis* Merriam, 1903; *californicus* Mearns, 1897; *colimensis* Goldman, 1938; *costaricensis* Goodwin, 1938; *floridanus* Rhoads, 1895; *fraterculus* Elliot, 1896; *furvus* Allen and Barbour, 1923; *guatemalae* Miller, 1899; *inyoensis* Elliot, 1904; *madrensis* Burt and Hooper, 1941; *nigrirostris* Lichtenstein, 1850; *ocythous* Bangs, 1899; *orinomus* Goldman, 1938; *parvidens* Miller, 1899; *peninsularis* Huey, 1928; *pensylvanicus* Boddaert, 1784; *scottii* Mearns, 1891; *sequoiensis* Dixon, 1910; *texensis* Mearns, 1897; *townsendi* Merriam, 1899; *virginianus* Schreber, 1775.
COMMENTS: Reviewed by Fritzell and Haroldson (1982, Mammalian Species, 189). Placed in *Canis* (*Vulpes*) by Van Gelder (1978); Clutton-Brock et al. (1976) placed it in *Vulpes* (*Urocyon*).

Urocyon littoralis (Baird, 1857). Mammalia, *in* Repts. U.S. Expl. Surv., 8(1):143.
TYPE LOCALITY: "island of San Miguel, on the coast of California."
DISTRIBUTION: Islands off the Pacific coast of S California (USA).
STATUS: IUCN - Rare.
SYNONYMS: *catalinae* Merriam, 1903; *clementae* Merriam, 1903; *dickeyi* Grinnell and Linsdale, 1930; *santacruzae* Merriam, 1903; *santarosae* Grinnell and Linsdale, 1930.
COMMENTS: Placed in *Canis* (*Vulpes*) by Van Gelder (1978); Clutton-Brock et al. (1976) placed it in *Vulpes* (*Urocyon*). May be conspecific with *U. cinereoargenteus* (Stains, 1975; Van Gelder, 1978).

Vulpes Frisch, 1775. Das Natur-System der Vierfüssigen Thiere, p. 15.
TYPE SPECIES: *Canis vulpes* Linnaeus, 1758, by designation under the plenary powers (Melville and Smith, 1978).

SYNONYMS: *Fennecus* Desmarest, 1804.

COMMENTS: Although Frisch (1774[1775]) has been ruled a rejected work for nomenclatural purposes, *Vulpes* has been retained (International Commission on Zoological Nomenclature, 1979). Considered a subgenus of *Canis* by Van Gelder (1978); however, this arrangement is not currently employed by most mammalogists (Corbet, 1978*c*; Corbet and Hill, 1980; Gromov and Baranova, 1981; Hall, 1981; Wozencraft, 1989*b*). Includes *Fennecus*; see comments under *V. zerda* and *Urocyon*.

Vulpes bengalensis (Shaw, 1800). Gen. Zool. Syst. Nat. Hist., 1(2), Mammalia, p. 330.

TYPE LOCALITY: "Bengal."

DISTRIBUTION: India, Pakistan, S Nepal.

STATUS: CITES - Appendix III (India); IUCN - Indeterminate.

SYNONYMS: *chrysurus* Gray, 1837; *hodgsonii* Gray, 1837; *indicus* Hodgson, 1833; *kokree* Sykes, 1831; *rufescens* Gray, 1833; *xanthura* Gray, 1837.

Vulpes cana Blanford, 1877. J. Asiat. Soc. Bengal, 2:321.

TYPE LOCALITY: "Gwadar, Baluchistan", [Pakistan].

DISTRIBUTION: Turkmenistan, Afghanistan, NE Iran, Pakistan.

STATUS: CITES - Appendix II; IUCN - Insufficiently known.

Vulpes chama (A. Smith, 1833). S. Afr. Quart. J., 2:89.

TYPE LOCALITY: "Namaqualand and the country on both sides of the Orange river" [Namibia]; fixed by Shortridge (1942) as "Port Nolloth, Little Namaqualand."

DISTRIBUTION: Southern Africa, S Angola, and S of Zambia.

SYNONYMS: *caama* Smith, 1839; *hodgsoni* Noack, 1910; *variegatoides* Layard, 1861.

Vulpes corsac (Linnaeus, 1768). Syst. Nat., 12th ed., 3: appendix 223.

TYPE LOCALITY: "in campis magi deserti ab Jaco fluvio verus Irtim"; listed by Honacki et al. (1982) as "U.S.S.R., N. Kazakhstan, steppes between Ural and Irtysh rivers, near Petropavlovsk."

DISTRIBUTION: Kazakhstan, Russia, C Asia, Mongolia, Transbaikalia, NE China, N Afghanistan.

STATUS: IUCN - Insufficiently known.

SYNONYMS: *kalmykorum* Ognev, 1935; *nigra* Kastschenko, 1912; *scorodumovi* Bobrinskii, 1944; *turkmenica* Ognev, 1935.

Vulpes ferrilata Hodgson, 1842. J. Asiat. Soc. Bengal, 11:278.

TYPE LOCALITY: "brought from Lassa" [Tibet, China].

DISTRIBUTION: China: Tibet, Tsinghai, Kansu, and Yunnan; Nepal.

Vulpes pallida (Cretzschmar, 1827). *In* Rüppell, Atlas Reise Nordl. Afr., Zool., Säugeth., p. 33, pl. 11.

TYPE LOCALITY: "Kordofan" [Sudan].

DISTRIBUTION: Semiarid sahelian region of Africa from Senegal through Nigeria, Cameroon, and Sudan to Somalia.

STATUS: IUCN - Insufficiently known.

SYNONYMS: *edwardsi* Rochebrune, 1883; *harterti* Thomas and Hinton, 1921; *oertzeni* Matschie, 1910; *sabbar* Hemprich and Ehrenberg, 1832.

Vulpes rueppellii (Schinz, 1825). *In* G. Cuvier, Das Thierreich, 4:508.

TYPE LOCALITY: "Vatherland Dongola".

DISTRIBUTION: Arid areas of N Africa from Morocco to Somalia; Egypt; Sinai; Arabia; Iran; parts of Pakistan and Afghanistan.

STATUS: IUCN - Insufficiently known.

SYNONYMS: *caesia* Thomas, 1921; *cyrenaica* Festa, 1921; *famelicus* Cretzschmar, 1826; *sabaea* Pocock, 1934; *somaliae* Thomas, 1918; *zarudnyi* Birula, 1912.

Vulpes velox (Say, 1823). *In* James, Account of an Exped. from Pittsburgh to the Rocky Mtns, 1:487.

TYPE LOCALITY: "camp on the river Platte, at the fording place of the Pawnee Indians, twenty-seven miles below the confluence of the North and South, or Paduca Forks." [Camp on 20 June 1820 reported to be at 40°59'15'N (vol. 2)].

DISTRIBUTION: C North America from SE British Columbia, SC Alberta and SW Saskatchewan (Canada) to NW Texas (panhandle) and E New Mexico, east of Rockies (USA).

STATUS: U.S. ESA - Endangered as *V. v. hebes* (Canada populations) and as *V. macrotis mutica*.

SYNONYMS: *arizonensis* Goldman, 1931; *arsipus* Elliot, 1903; *devia* Nelson and Goldman, 1909; *hebes* Merriam, 1902; *macrotis* Merriam, 1888; *muticus* Merriam, 1902; *neomexicana* Merriam, 1902; *nevadensis* Goldman, 1931; *tenuirostris* Nelson and Goldman, 1931; *zinseri* Benson, 1938.

COMMENTS: Reviewed by Egoscue (1979, Mammalian Species, 122) and McGrew (1979, Mammalian Species, 123, as *V. macrotis*). Revised by Waithman and Roest (1977) and Dragoo et al. (1990). Blair et al. (1968), Lechleitner (1969), Bueler (1973), and Dragoo et al. (1990) considered *macrotis* and *velox* conspecific. Packard and Bowers (1970), Rohwer and Kilgore (1973), and Thornton and Creel (1975) (who found hybrids between *velox* and *macrotis* but concluded they were of reduced viability) retained both as species.

Vulpes vulpes (Linnaeus, 1758). Syst. Nat., 10th ed., 1:40.

TYPE LOCALITY: "Europa, Asia, Africa, antrafodiens," restricted by Thomas (1911*a*), to "Sweden (Upsala)."

DISTRIBUTION: Europe and continental Asia except the tundra; N India and peninsular Indochina; Japan; Palearctic Africa; N America as far south as Texas and New Mexico (USA), but absent in part of the Central Plains and the Arctic. Introduced to Australia; see Corbet and Hill (1980:93).

STATUS: CITES - Appendix III (India) as *V. v. griffithi*, *V. v. montana*, and *V. v. pusilla* (= *leucopus*).

SYNONYMS: *abietorum* Merriam, 1900; *acaab* Cabrera, 1916; *aegyptiacus* Sonnini, 1816; *alascensis* Merriam, 1900; *alba* Borkhausen, 1797; *algeriensis* Loche, 1858; *alopex* Blanford, 1881; *alpherakyi* Satunin, 1906; *alticola* Ognev, 1926; *anadyrensis* J. A. Allen, 1903; *anatolica* Thomas, 1920; *anubis* Hemprich and Ehrenberg, 1833; *arabica* Thomas, 1902; *atlantica* Wagner, 1841; *aurantioluteus* Matschie, 1907; *bangsi* Merriam, 1900; *barbarus* Shaw, 1800; *beringiana* Middendorff, 1875; *cascadensis* Merriam, 1900; *caucasica* Dinnik, 1914; *cinera* Bechstein, 1801; *communis* Burnett, 1830; *crucigera* Bechstein, 1789; *daurica* Ognev, 1931; *deletrix* Bangs, 1898; *dolichocrania* Ognev, 1926; *dorsalis* Gray, 1837; *eckloni* Jacobi, 1923; *flavescens* Gray, 1843; *fulvus* Desmarest, 1820; *griffithii* Blyth, 1854; *harrimani* Merriam, 1900; *himalaicus* Ogilby, 1837; *hoole* Swinhoe, 1870; *huli* Sowerby, 1923; *hypomelas* Wagner, 1841; *ichnusae* Miller, 1907; *indutus* Miller, 1907; *jakutensis* Ognev, 1923; *japonica* Gray, 1868; *kamtschadensis* Brass, 1911; *karagan* Erxleben, 1777; *kenaiensis* Merriam, 1900; *kiyomasai* Kishida and Mori, 1929; *krimeamontana* Brauner, 1914; *kurdistanica* Satunin, 1906; *ladacensis* Matschie, 1907; *leucopus* Blyth, 1854; *lineatus* Billberg, 1827; *lineiventer* Swinhoe, 1871; *lutea* Bechstein, 1801; *macrourus* Baird, 1852; *melanogaster* Bonaparte, 1832; *melanotus* Pallas, 1811; *meridionalis* Fitzinger, 1855; *montana* Pearson, 1836; *necator* Merriam, 1900; *nepalensis* Gray, 1837; *nigra* Borkhausen, 1797; *nigro-argenteus* Nillson, 1820; *nigrocaudatus* Billberg, 1827; *niloticus* Desmarest, 1820; *ochroxantha* Ognev, 1926; *palaestina* Thomas, 1920; *peculiosa* Kishida, 1924; *pennsylvanicus* Rhoads, 1894; *persicus* Blanford, 1875; *pusilla* Blyth, 1854; *regalis* Merriam, 1900; *rubricosa* Bangs, 1898; *schrencki* Kishida, 1924; *silaceus* Miller, 1907; *sitkaensis* Brass, 1911; *splendens* Thomas, 1902; *splendidissima* Kishida, 1924; *stepensis* Brauner, 1914; *tobolica* Ognev, 1926; *tschiliensis* Matschie, 1907; *ussuriensis* Dybowski, 1922; *vafra* Bangs, 1897; *variagatus* Billberg, 1827; *vulgaris* Oken, 1816; *vulpecula* Hemprich and Ehrenberg, 1833; *waddelli* Bonhote, 1906.

Vulpes zerda (Zimmermann, 1780). Geogr. Gesch. Mensch. Vierf. Thiere, 2:247.

TYPE LOCALITY: "Es bewohnt die Soara und andere Theile von Nordafrika hinter den Atlas, der Ritter Bruce behauptet, man fände es auch in tripolitanischen."

DISTRIBUTION: Morocco to Arabia.

STATUS: CITES - Appendix II; IUCN - Insufficiently known.

SYNONYMS: *arabicus* Desmarest, 1804; *aurita* Meyer, 1793; *brucei* Desmarest, 1820; *cerdo* Gmelin, 1788; *denhamii* Boitard, 1842; *fennecus* Lesson, 1827; *saarensis* Skjöldebrand, 1777; *zaarensis* Gray, 1843.

COMMENTS: Placed in *Fennecus* by Stains (1975).

Family Felidae G. Fischer, 1817. Mém. Soc. Imp. Nat. Moscow, 5:372.

COMMENTS: Revised by Pocock (1917, 1951), Weigel (1961), Hemmer (1978), Král and Zima (1980), Kratochvíl (1982*c*), Groves (1982*a*), and Collier and O'Brien (1985). Some (Honacki et al., 1982; Van Gelder, 1977*b*) have followed Simpson (1945) and placed the majority of taxa in *Felis*, except for the large cats (*Panthera* and *Acinonyx*); however, this is not well supported by primary systematic studies and only poorly represents relationships below the family level. Although taxonomic disagreements appear at first to be considerable, most revolve around the recognition of categories at the generic versus the subgeneric level and interpretations of degrees of similarity.

Subfamily Acinonychinae Pocock, 1917. Ann. Mag. Nat. Hist., ser. 8, 20:332.

COMMENTS: Type genus: *Acinonyx* Brookes, 1828.

Acinonyx Brookes, 1828. Cat. Anat. Zool. Mus. J. Brookes, London, p. 16, 33.

TYPE SPECIES: *Acinonyx Venator* Brookes, 1828 (= *Felis jubata* Schreber, 1775), by monotypy (International Commission on Zoological Nomenclature, 1956*a*; Melville and Smith, 1987).

SYNONYMS: *Cynaelurus* Gloger, 1841; *Cynailurus* Wagler, 1830; *Cynofelis* Lesson, 1842; *Guepar* Boitard, 1842; *Guepardus* Duvernoy, 1834.

Acinonyx jubatus (Schreber, 1775). Die Säugethiere, 2(15):pl. 105[1775]; text 3(22):392[1777].

TYPE LOCALITY: "südliche Afrika; man bekömmt die Felle vom Vorgebirge der guten Hofnung" [South Africa, Cape Province, Cape of Good Hope].

DISTRIBUTION: Afghanistan, Angola, Benin, Botswana, Burkina Faso, Cameroon, Central African Republic, Chad, Egypt, Ethiopia, India, Iran, Iraq, Kenya, Libya, Malawi, Mali, Mauritania, Morocco, Mozambique, Namibia, Niger, Nigeria, Pakistan, Saudia Arabia, Senegal, Somalia, South Africa, Sudan, Tadzhikistan, Tanzania, Turkmenistan, Uganda, Western Sahara, Zaire, Zambia, Zimbabwe. Now restricted to N Iran and Subsaharan Africa, except rain forest and some areas in the south.

STATUS: CITES - Appendix I; U.S. ESA - Endangered; IUCN - Endangered as *A. j. venaticus*, otherwise Vulnerable.

SYNONYMS: *fearonii* Smith, 1834; *fearonis* Fitzinger, 1869; *guttata* Hermann, 1804; *hecki* Hilzheimer, 1913; *lanea* Sclater, 1877; *megabalica* Heuglin, 1863; *ngorongorensis* Hilzheimer, 1913; *obergi* Hilzheimer, 1913; *raddei* Hilzheimer, 1913; *raineyi* Heller, 1913; *rex* Pocock, 1927; *senegalensis* Blainville, 1843; *soemmeringii* Fitzinger, 1855; *velox* Heller, 1913; *venatica* Griffith, 1821; *venator* Brookes, 1828; *wagneri* Hilzheimer, 1913.

COMMENTS: Placed in *Acinonyx* by Pocock (1917), Weigel (1961), Hemmer (1978), Král and Zima (1980), Kratochvíl (1982*c*), and Groves (1982*a*).

Subfamily Felinae G. Fischer, 1817. Mém. Soc. Imp. Nat. Moscow, 5:372.

COMMENTS: Type genus: *Felis* Linnaeus, 1758.

Caracal Gray, 1843. List. Specimens Mamm. Coll. Brit. Mus. p. 46.

TYPE SPECIES: *Caracal melanotis* Gray, 1843 (= *Felis caracal* Schreber, 1776), by monotypy.

SYNONYMS: *Urolynchus* Severtzov, 1858.

Caracal caracal (Schreber, 1776). Die Säugethiere, 3(16):pl. 110[1776]; text 3(24):413, 587[1777].

TYPE LOCALITY: "Vorgebirge der guten Hofnung", restricted by Allen (1924:281) to "Table Mountain, near Cape Town, South Africa".

DISTRIBUTION: Aden, Afghanistan, Algeria, Angola, Arabia, Botswana, Egypt, Ethiopia, Gabon, India, Iran, Iraq, Israel, Kenya, Kuwait, Libya, Malawi, Mauritania, Morocco, Mozambique, Namibia, Niger, Pakistan, Senegal, Somalia, South Africa, Sudan, Syria, Tanzania, Turkey, Turkmenistan, Uganda, Zaire, Zambia, Zimbabwe. See Stuart (1984) for distribution and status information.

STATUS: CITES - Appendix I as *Felis caracal* (Asian population); otherwise, Appendix II; IUCN - Rare as *F. c. michaelis*.

SYNONYMS: *aharonii* Matschie, 1912; *algira* Wagner, 1841; *bengalensis* Fischer, 1829; *berberorum* Matschie, 1892; *coloniae* Thomas, 1926; *corylinus* Matschie, 1912;

damarensis Roberts, 1926; *limpopoensis* Roberts, 1926; *lucani* Rochebrune, 1885; *medjerdae* Matschie, 1912; *melanotis* Gray, 1843; *michaelis* Heptner, 1945; *nubicus* Fischer, 1829; *poecilotis* Thomas and Hinton, 1921; *roothi* Roberts, 1926; *schmitzi* Matschie, 1912; *spatzi* Matschie, 1912.

COMMENTS: Allen (1924) discussed the authority and the type locality. Several recent studies (Groves, 1982*a*; Král and Zima, 1980) emphasized the closeness of this taxon to *Felis* and the separation of it from *Lynx* (*sensu* Simpson, 1945). Weigel (1961), Hemmer (1978), and Werdelin (1981), placed *caracal* in the monotypic *Caracal* (= *Urolynchus* of Kratochvíl, 1982*c*) followed here.

Catopuma Severtzov, 1858. Rev. Mag. Zool. Paris, ser. 2, 10:387.

TYPE SPECIES: *Felis moormensis* Hodgson, 1831 (= *Felis temminckii* Vigors and Horsfield, 1827), by monotypy.

SYNONYMS: *Badiofelis* Pocock, 1932.

Catopuma badia (Gray, 1874). Proc. Zool. Soc. Lond., 1874:322.

TYPE LOCALITY: "Borneo, Sarawak" [Malaysia].

DISTRIBUTION: Borneo.

STATUS: CITES - Appendix II; IUCN - Rare.

COMMENTS: This poorly known species was placed in *Catopuma* by Hemmer (1978) and Groves (1982*a*). Placed in the monotypic *Badiofelis* by Pocock (1932*d*:749), followed by Weigel (1961).

Catopuma temminckii (Vigors and Horsfield, 1827). Zool. J., 3:451.

TYPE LOCALITY: "Sumatra" [Indonesia].

DISTRIBUTION: Bangladesh, Burma, China (from Shaanxi and Jiangxi to the south), Indonesia (Sumatra), India, Cambodia, Laos, Malaysia, Nepal, Thailand, Vietnam.

STATUS: CITES - Appendix I; U.S. ESA - Endangered; IUCN - Indeterminate.

SYNONYMS: *aurata* Blyth, 1863; *badiodorsalis* Howell, 1926; *bainsei* Sowerby, 1924; *dominicanorum* Sclater, 1898; *mitchelli* Lydekker, 1908; *moormensis* Hodgson, 1831; *nigrescens* Gray, 1863; *semenovi* Satunin, 1904; *tristis* Milne-Edwards, 1872.

COMMENTS: Placed in *Catopuma* by Hemmer (1978) and Groves (1982*a*). Placed in *Profelis* by Pocock (1932*d*), followed by Weigel (1961), Král and Zima (1980), and Kratochvíl (1982*c*).

Felis Linnaeus, 1758. Syst. Nat., 10th ed., 1:41.

TYPE SPECIES: *Felis catus* Linnaeus, 1758 (= *Felis silvestris* Schreber, 1775), by Linnean tautonymy (Melville and Smith, 1987).

SYNONYMS: *Catolynx* Severtzov, 1858; *Chaus* Gray, 1843; *Eremaelurus* Ognev, 1927; *Microfelis* Roberts, 1926; *Otailurus* Severtzov, 1858.

COMMENTS: Revised by Schwangart (1943), Pocock (1951), and Haltenorth (1953).

Felis bieti Milne-Edwards, 1892. Rev. Gen. Sci. Pures Appl., 3:671.

TYPE LOCALITY: "Batang á Tatsien-Lou", restricted by Pousargues (1898:358) to "environ de Tongolo et de Ta-tsien-lou" [China, Sichuan].

DISTRIBUTION: Endemic to the dry steppe from the eastern flanks of the Tibetan plateau in C China north to Nei Monggol; perhaps to Mongolia.

STATUS: CITES - Appendix II.

SYNONYMS: *chutuchta* Birula, 1917; *pallida* Büchner, 1894; *subpallida* Jacobi, 1922; *vellerosa* Pocock, 1943.

COMMENTS: Haltenorth (1953) suggested *chutuchta* and *vellerosa* belonged in *sylvestris*.

Felis chaus Schreber, 1777. Die Säugethiere, 2(13):pl. 110.B[1777]; text, 3(24):414[1777].

TYPE LOCALITY: "wohnt in den sumpfigen mit Schilf bewachsenen oder bewaldeten Gegenden der Steppen um das kaspische Meer, und die in selbiges fallenden Flüse. Auf der Nordseite des Terekflusses und der Festung Kislar . . . desto Hünfiger aber bey der Mündung der Kur . . .". Listed in Honacki et al. (1982) as "U.S.S.R., Dagestan, Terek River, N. of the Caucasus".

DISTRIBUTION: Afghanistan, Algeria, Arabia, Benin, Burma, China, Egypt, India, Iran, Iraq, Israel, Kenya, Malawi, Morocco, Mozambique, Nepal, Pakistan, Sri Lanka, Syria, Thailand, republics of the former USSR, Vietnam, Yemen, Zambia, Zimbabwe.

STATUS: CITES - Appendix II.

SYNONYMS: *affinis* Gray, 1830; *catolynx* Pallas, 1811; *chrysomelanotis* Nehring, 1902; *erythrotus* Hodgson, 1836; *fulvidina* Thomas, 1929; *furax* de Winton, 1898; *jacquemonti* Geoffroy, 1844; *kelaarti* Pocock, 1939; *kutas* Pearson, 1832; *libycus* Olivier, 1804; *nilotica* de Winton, 1898; *prateri* Pocock, 1939; *ruppelii* Brandt, 1832; *shawiana* Blanford, 1876; *typica* de Winton, 1898.
COMMENTS: *F. chaus* Güldenstädt, 1776, is invalid (Allen, 1920).

Felis margarita Loche, 1858. Rev. Mag. Zool. Paris, ser. 2, 10:49.
TYPE LOCALITY: "environs de Négonca (Sahara)" [Algeria].
DISTRIBUTION: Deserts in Algeria, Egypt, Iran, Libya, Morocco, Niger, Oman, Pakistan, Qatar, Saudia Arabia, Sudan, Tunisia, Turkmenistan, Uzbekistan, Yemen.
STATUS: CITES - Appendix II; U.S. ESA and IUCN - Endangered as *F. margarita scheffeli*.
SYNONYMS: *airensis* Pocock, 1938; *harrisoni* Hemmer, Grubb, and Groves, 1976; *margaritae* Trouessart, 1897; *margueritei* Trouessart, 1904; *meinertzhageni* Pocock, 1938; *scheffeli* Hemmer, 1974; *thinobius* Ognev, 1927.
COMMENTS: Revised by Schauenberg (1974) and Hemmer et al. (1976). *F. marginata* Gray, 1867, is an incorrect subsequent spelling. Pocock (1951) and Schauenberg (1974) included *Eremaelurus thinobia*, which was recognized as separate by Haltenorth (1953) and Weigel (1961), but see discussion by Hemmer et al. (1976). Král and Zima (1980) suggested this species was closely related to *O. manul*.

Felis nigripes Burchell, 1824. Travels in Interior of Southern Africa, 2:592.
TYPE LOCALITY: Burchell (1824:509) implied the country of the "Bachapins", presumably in the capital, "the town of Litákun (Letárkoon)...27°.6'.44".[S]...24°.39'.27"[E]" [South Africa].
DISTRIBUTION: South Africa, Namibia, Botswana.
STATUS: CITES - Appendix I; U.S. ESA - Endangered.
SYNONYMS: *thomasi* Shortridge, 1931.
COMMENTS: Král and Zima (1980) noted a distinctly different karotype from other *Felis*.

Felis silvestris Schreber, 1775. Die Säugethiere, 2(15):pl. 107[1775]; text 3(23):397[1777].
TYPE LOCALITY: Not given. Fixed by Haltenorth (1953) as "vielleicht Nordfrankreich". Listed by Pocock (1951) as "Germany".
DISTRIBUTION: Afghanistan, Algeria, Angola, Arabia, Botswana, Chad, China, Egypt, Ethiopia, France, Germany, Guinea, India, Iran, Iraq, Israel, Italy, Kazakhstan, Libya, Malawi, Mali, Mauritania, Morocco, Mozambique, Namibia, Pakistan, Poland, Senegal, South Africa, Spain, Sudan, Syria, Tanzania, Tunisia, Turkistan, United Kingdom, Zambia, Zimbabwe. Introduced to: Australia, Brazil, Canada, and Madagascar.
STATUS: CITES - Appendix II (the domestic cat, *Felis catus*, is specifically excluded from protection).
SYNONYMS: *agrius* Bate, 1906; *algiricus* Fischer, 1829; *angorensis* Gmelin, 1788; *antiquorum* Fischer, 1829; *aureus* Kerr, 1792; *bouvieri* Rochebrune, 1883; *brevicaudata* Schinz, 1844; *brockmani* Pocock, 1944; *bubastis* Hemprich and Ehrenberg, 1832; *caeruleus* Erxleben, 1777; *caffra* A. Smith, 1826; *cafra* Desmarest, 1822; *caligata* Temminck, 1824; *catus* Linnaeus, 1758; *caucasicus* Satunin, 1905; *caudatus* Gray, 1874; *cretensis* Haltenorth, 1953; *cristata* Lataste, 1885; *cumana* Schinz, 1844; *cyrenarum* Ghigi, 1920; *daemon* Satunin, 1904; *domesticus* Erxleben, 1777; *euxina* Pocock, 1943; *ferox* Martorelli, 1896; *ferus* Erxleben, 1777; *foxi* Pocock, 1944; *grampia* Miller, 1907; *griselda* Thomas, 1926; *griseoflava* Zukowski, 1914; *gulata* Herman, 1804; *haussa* Thomas and Hinton, 1921; *hispanicus* Erxleben, 1777; *huttoni* Blyth, 1846; *iraki* Cheesman, 1920; *issikulensis* Ognev, 1930; *japonica* Fischer, 1829; *jordansi* Schwarz, 1930; *kozlovi* Satunin, 1905; *longiceps* Bechstein, 1800; *longipilis* Zukowsky, 1915; *lowei* Pocock, 1944; *lybica* Forster, 1780; *lybiensis* Kerr, 1792; *lynesi* Pocock, 1944; *macrothrix* Zukowsky, 1915; *madagascariensis* Kerr, 1792; *maniculata* Temminck, 1824; *matschiei* Zukowsky, 1914; *mauritana* Cabrera, 1906; *mediterranea* Martorelli, 1896; *megalotis* Müller, 1839; *mellandi* Schwann, 1904; *molisana* Altobello, 1921; *morea* Trouessart, 1904; *murgabensis* Zukowsky, 1914; *namaquana* Thomas, 1926; *nandae* Heller, 1913; *nesterovi* Birula, 1916; *nubiensis* Kerr, 1792; *obscura* Desmarest, 1822; *ocreata* Gmelin, 1791; *ornata* Gray, 1830; *pulchella* Gray, 1837; *pyrrhus* Pocock, 1944; *reyi* Lavauden, 1929; *ruber* Gmelin, 1788; *rubida* Schwann, 1904; *ruppelii* Schinz, 1825; *rusticana* Thomas, 1928; *salvanicola*

Dekeyser, 1950; *sarda* Lataste, 1885; *schnitnikovi* Birula, 1914; *servalina* Jardine, 1834; *shawiana* Blanford, 1876; *siamensis* Trouessart, 1904; *sinensis* Kerr, 1792; *striatus* Bechstein, 1800; *syriaca* Fischer, 1829; *taitae* Heller, 1913; *tartessia* Miller, 1907; *torquata* Blyth, 1863; *tralatitia* Fischer, 1829; *trapezia* Blackler, 1916; *tristrami* Pocock, 1944; *ugandae* Schwann, 1904; *vernayi* Roberts, 1932; *vulgaris* Fischer, 1829; *xanthella* Thomas, 1926.

COMMENTS: Revised by Ragni and Randi (1986), who included *libyca*, and by Haltenorth (1953), who included *chutuchta*, *libyca*, and *vellerosa*; however, *chutuchta* and *vellerosa* were placed in *bieti* in Pocock's (1951) revision, provisionally followed here. Includes *F. catus* (worldwide), which was domesticated from this species (Corbet, 1978*c*:181). Ellerman and Morrison-Scott (1951) argued that *lybica* Foster, 1780, was a *lapsus* for *libyca*; however, there is no clear internal evidence that the name was misspelled (Meester et al., 1986). Rosevear (1974), Ansell (1978), Smithers (1983), and Meester et al. (1986) retained *lybica* as separate from *silvestris*.

Herpailurus Severtzov, 1858. Rev. Mag. Zool. Paris, ser. 2, 10:390.

TYPE SPECIES: *Felis yaguarondi* Lacépède, 1809, by monotypy.

Herpailurus yaguarondi (Lacépède, 1809). *In* Azara, Voy. Am. Mérid., Atlas, 1809: pl. 10.

TYPE LOCALITY: "Paraguay", restricted by Hershkovitz (1951) to "Cayenne, French Guiana".

DISTRIBUTION: Argentina, Belize, Bolivia, Brazil, Colombia, Costa Rica, El Salvador, French Guiana, Guatemala, Guyana, Honduras, Nicaragua, Panama, Paraguay, Peru, Mexico, Surinam, USA (Arizona and Texas), Venezuela.

STATUS: CITES - Appendix I (North and Central American populations); otherwise Appendix II. U.S. ESA - Endangered as *F. y. cacomitli*, *F. y. fossata*, *F. y. panamensis*, and *F. y. tolteca*; IUCN - Indeterminate.

SYNONYMS: *ameghinoi* Holmberg, 1898; *apache* Mearns, 1901; *cacomitli* Berlandier, 1859; *darwini* Martin, 1837; *eira* Desmarest, 1816; *eyra* Fischer, 1814; *fossata* Mearns, 1901; *melantho* Thomas, 1914; *panamensis* Allen, 1904; *tolteca* Thomas, 1898; *unicolor* Traill, 1819.

COMMENTS: Others have used *yagouaroundi* É. Geoffroy Saint-Hilaire, 1803, or *yaguarondi* Desmarest, 1816, however, the former was never officially published (see Appendix I) and the latter is a junior synonym. Placed in *Herpailurus* by Weigel (1961), Hemmer (1978), and Kratochvíl (1982*c*).

Leopardus Gray, 1842. Ann. Mag. Nat. Hist., [ser. 1], 10:260.

TYPE SPECIES: *Leopardus griseus* Gray, 1842 (= *Felis pardalis* Linnaeus, 1758), by subsequent designation by Pocock (1916*b*:316).

SYNONYMS: *Margay* Gray, 1867; *Oncilla* Allen, 1919; *Oncoides* Severtzov, 1858; *Pardalis* Gray, 1867.

COMMENTS: Revised by Allen (1919*a*) and Pocock (1941*b*). Considered at the subgeneric level by Cabrera (1958), who also included *Oncifelis geoffroyi* and *O. guigna* (but not *O. colocolo*). Hemmer (1978) placed *tigrinus* in *Oncifelis*.

Leopardus pardalis (Linnaeus, 1758). Syst. Nat., 10th ed., 1:42.

TYPE LOCALITY: "America", restricted to "Mexico", by Thomas (1911*a*:136), further restricted by Allen (1919*a*:345) to "State of Vera Cruz".

DISTRIBUTION: N Argentinia, Belize, E Bolivia, Brazil, Colombia, Costa Rica, Ecuador, El Salvador, French Guiana, Guatemala, Guyana, Honduras, Mexico, Nicaragua, Panama, Paraguay, E Peru, Surinam, Trinidad, USA (Arizona and Texas), Uruguay, Venezuela.

STATUS: CITES - Appendix I; U.S. ESA - Endangered; IUCN - Vulnerable.

SYNONYMS: *aequatorialis* Mearns, 1902; *albescens* Pucheran, 1855; *armillatus* F. Cuvier, 1832; *brasiliensis* Schinz, 1844; *buffoni* Brass, 1911; *canescens* Swainson, 1838; *chati* Gray, 1827; *chibigouazou* Gray, 1827; *chibiguazu* Fischer, 1830; *costaricensis* Mearns, 1902; *griffithii* Fischer, 1830; *griseus* Gray, 1842; *hamiltoni* Fischer, 1830; *limitis* Mearns, 1901; *ludoviciana* Brass, 1911; *maracaya* Wagner, 1841; *maripensis* J. A. Allen, 1904; *mearnsi* J. A. Allen, 1904; *melanura* Ball, 1844; *mexicana* Kerr, 1792; *minimus* Wilson, 1860; *mitis* F. Cuvier, 1820; *nelsoni* Goldman, 1925; *ocelot* Link, 1795; *pseudopardalis* Boitard, 1842; *pusaea* Thomas, 1914; *sanctaemartae* J. A. Allen, 1904; *smithii* Swainson, 1838; *sonoriensis* Goldman, 1925; *steinbachi* Pocock, 1941; *tumatumari* J. A. Allen, 1915.

COMMENTS: Placed in *Leopardus* by Allen (1919*a*:345), Weigel (1961), Hemmer (1978), and Kratochvíl (1982*c*).

Leopardus tigrinus (Schreber, 1775). Die Säugethiere, 2(15):pl. 106[1775]; text, 3(23):396[1777].

TYPE LOCALITY: "südlichen Amerika", restricted by Allen (1919*a*:356), to "Cayenne" [French Guiana].

DISTRIBUTION: N Argentina, Brazil, Colombia, Costa Rica, Ecuador, French Guiana, Guyana, Paraguay, Peru, Surinam, Venezuela.

STATUS: CITES - Appendix I; U.S. ESA - Endangered; IUCN - Vulnerable.

SYNONYMS: *andina* Thomas, 1903; *carrikeri* J. A. Allen, 1904; *caucensis* J. A. Allen, 1915; *elenae* J. A. Allen, 1915; *emerita* Thomas, 1912; *emiliae* Thomas, 1914; *geoffroyi* Elliot, 1872; *guttula* Hensel, 1872; *margay* Müller, 1776; *oncilla* Thomas, 1903; *pardinoides* Gray, 1867.

COMMENTS: Placed in *Leopardus* by Allen (1919*a*:345), Weigel (1961), and Kratochvíl (1982*c*); placed in *Oncifelis* (with *O. guigna* and *O. geoffroyi*) by Hemmer (1978). *L. tigrinus* shares a derived chromosomal number with *pardalis* and *wiedii* (Wurster-Hill, 1973). Includes *Felis pardinoides* (Cabrera, 1958:286); Allen (1919*a*:345) and Weigel (1961) considered *pardinoides* as distinct, but closely related to *tigrinus*.

Leopardus wiedii (Schinz, 1821). *In* G. Cuvier, Das Thierreich, 1:235.

TYPE LOCALITY: "Brasilien", restricted by Allen (1919*a*:357) to "northern Espirito Santo, Brazil", and further restricted by Cabrera (1958:290), to "Brasil, restringida al Morro de Arará, sobre el rio Mucurí, estado de Baía".

DISTRIBUTION: N Argentina, Belize, Bolivia, Brazil, Colombia, Costa Rica, Ecuador, El Salvador, Guatemala, Guyana, Honduras, Mexico, Nicaragua, Panama, Paraguay, Peru, Surinam, USA (Texas), Uruguay, Venezuela.

STATUS: CITES - Appendix I; U.S. ESA - Endangered (from Mexico southward); IUCN - Vulnerable.

SYNONYMS: *amazonica* Cabrera, 1917; *andina* Allen, 1916; *boliviae* Pocock, 1941; *catenata* H. Smith, 1827; *cooperi* Goldman, 1943; *elegans* Lesson, 1830; *geoffroyi* Rochebrune, 1895; *glaucula* Thomas, 1903; *ludovici* Lönnberg, 1925; *macroura* Wied, 1823; *macrura* Hensel, 1872; *mexicana* Saussure, 1860; *nicaraguae* J. A. Allen, 1919; *oaxacensis* Nelson and Goldman, 1931; *pardictis* Pocock, 1941; *pirrensis* Goldman, 1914; *salvinia* Pocock, 1941; *sanctaemartae* Allen, 1906; *tigrinoides* Gray, 1842; *vigens* Thomas, 1904; *yucatanica* Nelson and Goldman, 1931.

COMMENTS: Placed in *Leopardus* by Weigel (1961), Hemmer (1978), and Kratochvíl (1982*c*). Allen (1919*a*) and Weigel (1961) suggested that *wiedii* (in part) may be conspecific with *tigrinus*; however, Hemmer (1978) considered differences between *wiedii* and *tigrinus* to warrant generic distinction.

Leptailurus Severtzov, 1858. Rev. Mag. Zool. Paris, ser. 2, 10:389.

TYPE SPECIES: *Felis serval* Schreber, 1776, by monotypy.

Leptailurus serval (Schreber, 1776). Die Säugethiere, 3(16):pl. 108[1776]; text 3(23):407[1777].

TYPE LOCALITY: "Ostindien und Tibet in gebirgegen Gegenden, vielleicht auch am Vorgebirge der guten Hofnung und dem heissern Afrika"; restricted by Allen (1924) to the "Cape region of South Africa".

DISTRIBUTION: Algeria, Angola, Benin, Botswana, Burundi, Cameroon, Central African Republic, Chad, Equatorial Guinea, Ethiopia, Gabon, Gambia, Ghana, Guinea, Ivory Coast, Kenya, Liberia, Malawi, Mali, Mauritania, Mozambique, Namibia, Niger, Nigeria, Rwanda, Senegal, N Sierra Leone, Somalia, South Africa, Sudan, Tanzania, Togo, Tunisia, Uganda, Zaire, Zambia, Zimbabwe; formerly in Morocco.

STATUS: CITES - Appendix II; U.S. ESA - Endangered as *F. s. constantina*.

SYNONYMS: *algiricus* Fischer, 1829; *beirae* Wroughton, 1910; *brachyura* Wagner, 1841; *capensis* Forster, 1781; *constantina* Forster, 1780; *faradjius* Allen, 1924; *ferrarii* de Beaux, 1924; *galeopardus* Desmarest, 1820; *hamiltoni* Roberts, 1931; *hindei* Wroughton, 1910; *ingridi* Lundholm, 1955; *kempi* Wroughton, 1910; *kivuensis* Lönnberg, 1919; *larseni* Thomas, 1913; *limpopoensis* Roberts, 1926; *liposticta* Pocock, 1907; *lonnbergi* Cabrera, 1910; *mababiensis* Roberts, 1932; *niger* Lönnberg, 1897; *ogilbyi* Schinz, 1844; *pantasticta* Pocock, 1907; *phillipsi* Allen, 1914; *pococki* Cabrera, 1910; *poliotricha* Pocock, 1907;

senegalensis Lesson, 1839; *servalina* Ogilby, 1839; *tanae* Pocock, 1944; *togoensis* Matschie, 1893.

COMMENTS: Placed in *Leptailurus* by Weigel (1961), Hemmer (1978), and Kratochvíl (1982*c*); Groves (1982*a*) considered *Leptailurus* a subgenus of *Felis*.

Lynx Kerr, 1792. *In* Linnaeus, Anim. Kingdom, 1:155.

TYPE SPECIES: *Felis lynx* Linnaeus, 1758; type species by absolute tautonymy (Melville and Smith, 1987).

SYNONYMS: *Cervaria* Gray, 1867; *Eucervaria* Palmer, 1903; *Lynchus* Jardine, 1834; *Lynceus* Gray, 1821; *Lyncus* Gray, 1825; *Pardina*; Kaup, 1829.

COMMENTS: Revised by Matyushkin (1979), Werdelin (1981), and García-Perea (1992), who recognized the generic status of *Lynx*; Groves (1982*a*) and Hemmer (1978) considered *Lynx* a subgenus of *Felis*.

Lynx canadensis Kerr, 1792. *In* Linnaeus, Anim. Kingdom, 1:157.

TYPE LOCALITY: "Canada"; listed in Miller (1912*b*) as "Eastern Canada".

DISTRIBUTION: Taiga zone of North America, south to C Utah and SW Colorado, NE Nebraska, S Indiana, and West Virginia (USA).

STATUS: CITES - Appendix II.

SYNONYMS: *mollipilosus* Stone, 1900; *subsolanus* Bangs, 1897.

COMMENTS: Considered distinct from *L. lynx* by Kurtén and Anderson (1980), Matyushkin (1979), Werdelin (1981), and García-Perea (1992). Weigel (1961) and Tumlison (1987) considered these taxa conspecific.

Lynx lynx (Linnaeus, 1758). Syst. Nat., 10th ed., 1:43.

TYPE LOCALITY: "Europæ sylvis and desertis", subsequently restricted by Thomas (1911*a*:136) to "Wennersborg, S. Sweden".

DISTRIBUTION: Taiga forests from Scandinavia through E Siberia and Sakhalin; from China (Gansu, Qinghai, Shaanxi, and Sichuan) through montane Europe (formerly widespread, now restricted to Balkans, Carpathians, Pyreneans [French side], and Alps; reintroduced to French Vosges and Jura Mt., Swiss Alps, Austria, and Yugoslavia).

STATUS: CITES - Appendix II.

SYNONYMS: *borealis* Thunberg, 1798; *carpathicus* Kratochvíl and Stollmann, 1963; *cervaria* Temminck, 1824; *dinniki* Satunin, 1915; *isabellina* Blyth, 1847; *kamensis* Satunin, 1904; *kozlovi* Fetisov, 1950; *lupulinus* Thunberg, 1825; *lyncula* Nilsson, 1820; *martinoi* Miríc, 1978; *neglectus* Stroganov, 1962; *orientalis* Satunin, 1905; *stroganovi* Heptner, 1969; *tibetanus* Gray, 1863; *vulgaris* Kerr, 1792; *vulpinus* Thunberg, 1825; *wardi* Lydekker, 1904; *wrangeli* Ognev, 1928.

COMMENTS: Does not include *L. canadensis* or *L. pardina*, following Matyushkin (1979), García-Perea (1992), and Werdelin (1981). Includes *isabellina* (Gao, *in* Gao, 1987:336). Reviewed by Tumlison (1987, Mammalian Species, 269, as *Felis lynx*).

Lynx pardinus (Temminck, 1827). Monogr. Mamm., 1:116.

TYPE LOCALITY: "Portugal, puisque le commerce reçoit des peaux préparées de Lisbonne, et que M. le baron de Vionénil tua, en 1818, sur les bords du Tage, à dix lieues de Lisbonne".

DISTRIBUTION: SW Spain and Portugal.

STATUS: CITES - Appendix I; U.S. ESA and IUCN - Endangered.

SYNONYMS: *pardella* Miller, 1907.

COMMENTS: Given specific status by Matyushkin (1979), García-Perea (1992), and Werdelin (1981); however Weigel (1961) and Tumlison (1987) considered *pardinus* conspecific with *L. lynx*.

Lynx rufus (Schreber, 1777). Säugethiere, 3(25):pl. 109.B[1777]; text 3(24):412[1777].

TYPE LOCALITY: "Provinz New York in Amerika".

DISTRIBUTION: S British Columbia to Nova Scotia (Canada), south to Oaxaca (Mexico).

STATUS: CITES - Appendix II. U.S. ESA - Endangered as *Felis rufus escuinapae*.

SYNONYMS: *baileyi* Merriam, 1890; *californicus* Mearns, 1897; *eremicus* Mearns, 1897; *escuinapae* Allen, 1903; *fasciatus* Rafinesque, 1817; *floridanus* Rafinesque, 1817; *gigas* Bangs, 1897; *maculata* Horsfield and Vigors, 1829; *montanus* Rafinesque, 1817; *oaxacensis* Goodwin, 1963; *oculeus* Bangs, 1899; *pallescens* Merriam, 1899; *peninsularis*

Thomas, 1898; *superiorensis* Peterson and Downing, 1952; *texensis* Allen, 1895; *uinta* Merriam, 1902.

Oncifelis Severtzov, 1858. Rev. Mag. Zool. Paris, ser. 2, 10:386.

TYPE SPECIES: *Felis geoffroyi* d'Orbigny and Gervais, 1842, by monotypy.

SYNONYMS: *Dendrailurus* Severtzov, 1858; *Lynchailurus* Severtzov, 1858; *Noctifelis* Severtzov, 1858; *Pajeros* Gray, 1867.

COMMENTS: Although there is general agreement as to the monophyletic nature of this genus, some have chosen to emphasize the distinctiveness of *O. colocolo* and placed it in monotypic *Lynchailurus*. Cabrera (1958) suggested that *O. geoffroyi* and *O. guigna* were more closely related to taxa considered here under *Leopardus* rather than to *O. colocolo*, which he placed in a separate subgenus. Hemmer (1978) also included *L. tigrinus* in this genus.

Oncifelis colocolo (Molina, 1782). Sagg. Stor. Nat. Chile, p. 295.

TYPE LOCALITY: "che abitano i boschi del Chili", restricted by Osgood (1943) to "Province of Valparaiso" [Chile].

DISTRIBUTION: Argentina, Bolivia, Brazil, Chile, Ecuador, Paraguay, Peru, Uruguay.

SYNONYMS: *albescens* Fitzinger, 1869; *braccata* Cope, 1889; *budini* Pocock, 1941; *crespoi* Cabrera, 1957; *crucina* Thomas, 1901; *garleppi* Matschie, 1912; *huina* Pocock, 1941; *munoai* Ximénez, 1961; *neumayeri* Matschie, 1912; *pajeros* Desmarest, 1816; *pampa* Schinz, 1831; *pampanus* Gray, 1867; *passerum* Sclater, 1871; *steinbachi* Pocock, 1941; *thomasi* Lönnberg, 1913.

COMMENTS: The validity of *colocolo* was questioned by Osgood (1943), with the next available name that of *pajeros* Desmarest, 1816 (used by Weigel, 1961; Hemmer, 1978; Král and Zima, 1980; and Kratochvíl, 1982*c*). However, Wolffsohn (1908) and Cabrera (1940, 1958) defended the original description.

Oncifelis geoffroyi (d'Orbigny and Gervais, 1844). Bull. Sci. Soc. Philom. Paris, 1844:40.

TYPE LOCALITY: "des rives du Rio Negro, en Patagonie".

DISTRIBUTION: Argentina, Andean Bolivia, S Paraguay, S Brazil, Uruguay.

STATUS: CITES - Appendix I.

SYNONYMS: *euxanthus* Pocock, 1940; *himalayanus* Gray, 1843; *leucobaptus* Pocock, 1940; *macdonaldi* Marelli, 1932; *melas* Bertoni, 1914; *paraguae* Pocock, 1940; *pardoides* Gray, 1867; *salinarum* Thomas, 1903; *warwickii* Gray, 1867.

COMMENTS: Revised by Pocock (1940*b*) and reviewed by Ximenez (1975, Mammalian Species, 54, as *Felis geoffroyi*). Placed in *Oncifelis* by Allen (1919*a*), Weigel (1961), Hemmer (1978), Král and Zima (1980), and Kratochvíl (1982*c*). Ximenez (1975) followed Cabrera (1958), and placed *geoffroyi* in subgenus *Leopardus* of *Felis*. Includes *F. pardoides* (Cabrera, 1958:280).

Oncifelis guigna (Molina, 1782). Sagg. Stor. Nat. Chile, p. 295.

TYPE LOCALITY: "Chili", restricted by Thomas (1903:240) to "Valdivia" [Chile].

DISTRIBUTION: Argentina and Chile.

STATUS: CITES - Appendix II.

SYNONYMS: *guina* Philippi, 1870; *molinae* Osgood, 1943; *santacrucensis* Artayeta, 1950; *tigrillo* Schinz, 1844.

COMMENTS: Placed in *Oncifelis* by Weigel (1961) and Hemmer (1978). Placed in subgenus *Leopardus* of *Felis* by Cabrera (1958).

Oreailurus Cabrera, 1940. Notas Mus. La Plata 5(29):16.

TYPE SPECIES: *Felis jacobita* Cornalia, 1865, by original designation.

Oreailurus jacobita (Cornalia, 1865). Mem. Soc. Ital. Sci. Nat., 1:3.

TYPE LOCALITY: "Bolivia, circa Potosi et Humacuaca in montibus sat elevatis"; further clarified by Cabrera (1958:297) as "Sur del departamento boliviano de Potosi, cerca de la frontera argentina, entre Potosi y Humahuaca".

DISTRIBUTION: NW Argentina, SE Bolivia, NE Chile, S Peru.

STATUS: CITES - Appendix I; U.S. ESA - Endangered; IUCN - Rare.

COMMENTS: Placed in *Oreailurus* by Cabrera (1940), Weigel (1961), and Hemmer (1978). Later, Cabrera (1958:297) reconsidered *Oreailurus* as a subgenus of *Felis*.

Otocolobus Brandt, 1842. Bull. Sci. Acad. Imp. Sci. St. Petersbourg, 9:col. 38.

TYPE SPECIES: *Felis manul* Pallas, 1776, by original designation.

SYNONYMS: *Trichaelurus* Satunin, 1905.

COMMENTS: Palmer (1904:487) and Allen (1938) believed that *Otocolobus* Severtzov, 1858, was a junior homonym of *Otocolobus* Brandt, 1844, a subgenus of Sciuridae, and suggested using *Trichaelurus* Satunin, 1905. Brandt apparently reassigned the generic name in 1844 after his initial use for a felid in 1842.

Otocolobus manul (Pallas, 1776). Reise Prov. Russ. Reichs., 3:692.

TYPE LOCALITY: "Frequens in rupestribus, apricis totius Tatariae Mongoliaeque desertae". Listed in Honacki et al. (1982) as "U.S.S.R., S.W. Transbaikalia, Buryat. Mongolsk. A.S.S.R.;S. of Lake Baikal, Kulusutai".

DISTRIBUTION: Steppe and semidesert from Caspian Sea to Kashmir, Transbaikalia, Mongolia, and C China, north to Inner Mongolia and east to Hopei.

STATUS: CITES - Appendix II.

SYNONYMS: *ferrugineus* Ognev, 1928; *mongolicus* Satunin, 1905; *nigripectus* Hodgson, 1842; *satuni* Lydekker, 1907.

COMMENTS: Revised by Pocock (1907), Birula (1913, 1916), Ognev (1935), and Schwangart (1936).

Prionailurus Severtzov, 1858. Rev. Mag. Zool. Paris, ser. 2, 10:387.

TYPE SPECIES: *Felis pardochrous* Hodgson, 1844 (= *Felis bengalensis* Kerr, 1792), by original designation.

SYNONYMS: *Ictailurus* Severtzov, 1858; *Mayailurus* Imaizumi, 1967; *Viverriceps* Gray, 1867; *Zibethailurus* Severtzov, 1858.

COMMENTS: Although there is little dispute that these four taxa are distinct and represent a single monophyletic group, some have chosen to emphasize their distinctiveness by placing them in monotypic genera.

Prionailurus bengalensis (Kerr, 1792). *In* Linnaeus, Anim. Kingdom, 1:151.

TYPE LOCALITY: "Bengal" [India].

DISTRIBUTION: Afghanistan, Bangladesh, Burma, Cambodia, China, India, Indonesia, Japan (Tsushima and Iriomote Isls), Korea, Laos, Malaysia, Nepal, Pakistan, Philippine Isls, Taiwan, Thailand, republics of the former USSR, and Vietnam.

STATUS: CITES - Appendix I as *Felis bengalensis bengalensis* (except for Chinese population); otherwise Appendix II. U.S. ESA - Endangered; IUCN - Endangered as *F. iriomotensis*.

SYNONYMS: *alleni* Sody, 1949; *anastasiae* Satunin, 1905; *borneoensis* Brongersma, 1935; *chinensis* Gray, 1837; *ellioti* Gray, 1842; *euptilura* Elliot, 1871; *horsfieldi* Gray, 1842; *ingrami* Bonhote, 1903; *iriomotensis* Imaizumi, 1967; *javanensis* Desmarest, 1816; *manchurica* Mori, 1922; *microtis* Milne-Edwards, 1872; *minuta* Temminck, 1827; *nipalensis* Horsfield and Vigors, 1829; *pardochrous* Hodgson, 1844; *raddei* Trouessart, 1904; *reevesii* Gray, 1843; *ricketti* Bonhote, 1903; *scripta* Milne-Edwards, 1872; *sumatrana* Horsfield, 1821; *tenasserimensis* Gray, 1867; *tingia* Lyon, 1908; *trevelyani* Pocock, 1939; *undata* Radde, 1862; *wagati* Gray, 1867.

COMMENTS: Placed in *Prionailurus* by Pocock (1917), Weigel (1961), Hemmer (1978), Groves (1982*a*), and Kratochvíl (1982*c*). Includes *euptilura* following Allen (1939) and Gao (*in* Gao, 1987). Heptner (1971) and Gromov and Baranova (1981), considered *euptilura* a distinct species; however, Gao (*in* Gao, 1987) pointed out that Heptner compared Russian specimens with those from Southeast Asia, whereas when intervening Chinese populations are included, his distinctions do not hold. Includes *minuta* following Chasen (1940). Imaizumi (1967) described *Mayailurus iriomotensis*, but his key characters are polymorphic in *bengalensis* (Glass and Todd, 1977; Petzsch, 1970). Groves (1982*a*) believed that *iriomotensis* warranted species level recognition.

Prionailurus planiceps (Vigors and Horsfield, 1827). Zool. J., 3:449.

TYPE LOCALITY: "Sumatra" [Indonesia].

DISTRIBUTION: Peninsular Thailand and Malaysia; Indonesia (Sumatra, Borneo).

STATUS: CITES - Appendix I; U.S. ESA - Endangered; IUCN - Indeterminate.

COMMENTS: Placed in *Prionailurus* by Weigel (1961), Hemmer (1978), Kratochvíl (1982*c*), and Groves (1982*a*).

Prionailurus rubiginosus (I. Geoffroy Saint-Hilaire, 1831). *In* Bélanger (ed.), Voy. Indes Orient., Mamm., 3(Zoologie):140.

TYPE LOCALITY: "bois de lataniers qui couvrent une hauteur voisine de Pondichéry" [India, Pondicherry].

DISTRIBUTION: India and Sri Lanka (see Chakraborty, 1978).

STATUS: CITES - Appendix I (Indian population), otherwise Appendix II; IUCN - Insufficiently known.

SYNONYMS: *koladivinus* Deraniyagala, 1956; *phillipsi* Pocock, 1939.

COMMENTS: Placed in *Prionailurus* by Weigel (1961), Hemmer (1978), Kratochvíl (1982*c*), and Groves (1982*a*).

Prionailurus viverrinus (Bennett, 1833). Proc. Zool. Soc. Lond., 1833:68.

TYPE LOCALITY: "from the continent of India".

DISTRIBUTION: Bangladesh, Burma, S China, India, Indonesia, Malaysia, Nepal, Pakistan, Sri Lanka, Taiwan, Thailand, and Vietnam.

STATUS: CITES - Appendix II.

SYNONYMS: *bennettii* Gray, 1867; *himalayanus* Jardine, 1834; *rizophoreus* Sody, 1936; *viverriceps* Hodgson, 1836.

COMMENTS: Placed in *Prionailurus* by Weigel (1961), Hemmer (1978), Kratochvíl (1982*c*), and Groves (1982*a*).

Profelis Severtzov, 1858. Revue Mag. Zool. Paris, ser. 2, 10:386.

TYPE SPECIES: *Felis celidogaster* Temminck, 1827 (= *Felis aurata* Temminck, 1827), by monotypy.

Profelis aurata (Temminck, 1827). Monogr. Mamm., 1:120.

TYPE LOCALITY: "Nous ne savons pas au juste dans quelle partie du globe a été trouvé"; fixed by Van Mensch and Van Bree (1969) to "probably the coastal region of Lower Guinea (Between Cross River and River Congo. . .)".

DISTRIBUTION: N Angola, Burundi, Cameroon, Central African Republic, Gabon, Gambia, Ghana, Kenya, Liberia, Nigeria, Rwanda, Sierra Leone, Uganda, S Zaire.

STATUS: CITES - Appendix II.

SYNONYMS: *celidogaster* Temminck, 1827; *chalybeata* H[amilton]. Smith, 1827; *chrysothrix* Temminck, 1827; *cottoni* Lydekker, 1906; *maka* Van Saceghem, 1942; *neglecta* Gray, 1838; *rutilus* Waterhouse, 1843.

COMMENTS: Revised by Van Mensch and Van Bree (1969). Placed in *Profelis* by Pocock (1917), Weigel (1961), Hemmer (1978), Kratochvíl (1982*c*), and Groves (1982*a*). Král and Zima (1980) placed in *Felis*.

Puma Jardine, 1834. Natur. Libr., 2:266.

TYPE SPECIES: *Felis concolor* Jardine, 1834, by original designation.

Puma concolor (Linnaeus, 1771). Mantissa Plantarum, 2:522.

TYPE LOCALITY: "Brassilia", restricted by Goldman (*in* Young and Goldman, 1946:200); to "Cayenne region, French Guiana".

DISTRIBUTION: Argentina, Belize, Bolivia, Brazil, Canada, Chile, Colombia, Costa Rica, Ecuador, El Salvador, Honduras, Guatemala, Guyana, Mexico, Nicaragua, Panama, Paraguay, Peru, Surinam, USA, Venezuela.

STATUS: CITES - Appendix I as *F. c. coryi*, *F. c. costaricensis*, and *F. c. couguar*; otherwise Appendix II. U.S. ESA - Endangered as *F. c. coryi*, *F. c. costaricensis*, and *F. c. couguar*. IUCN - Endangered as *F. c. coryi* and *F. c. cougar*.

SYNONYMS: *acrocodia* Goldman, 1943; *anthonyi* Nelson and Goldman, 1931; *araucanus* Osgood, 1943; *arundivaga* Hollister, 1911; *aztecus* Merriam, 1901; *bangsi* Merriam, 1901; *borbensis* Nelson and Goldman, 1933; *browni* Merriam, 1903; *cabrerae* Pocock, 1940; *californica* May, 1896; *capricornensis* Goldman and Young, 1946; *coryi* Bangs, 1899; *costaricensis* Merriam, 1901; *cougar* Kerr, 1792; *floridana* Cory, 1896; *greeni* Nelson and Goldman, 1931; *hippolestes* Merriam, 1897; *hudsoni* Cabrera, 1957; *improcera* Philipps, 1912; *incarum* Nelson and Goldman, 1929; *kaibabensis* Nelson and Goldman, 1931; *mayensis* Nelson and Goldman, 1929; *missoulensis* Goldman, 1943; *nigra* Jardine, 1834; *olympus* Merriam, 1897; *oregonensis* Rafinesque, 1832; *osgoodi* Nelson and Goldman, 1943; *patagonica* Merriam, 1901; *pearsoni* Thomas, 1901; *puma*

Molina, 1782; *punensis* Housse, 1950; *schorgeri* Jackson, 1955; *soasoaranna* Lesson, 1842; *soderstromii* Lönnberg, 1913; *stanleyana* Goldman, 1938; *sucuacuara* Liais, 1872; *vancouverensis* Nelson and Goldman, 1932; *wavula* Lesson, 1842; *youngi* Goldman, 1936.

COMMENTS: Reviewed by Currier, 1983 (Mammalian Species, 200, as *Felis concolor*). Placed in *Puma* by Pocock (1917), Weigel (1961), Hemmer (1978), and Kratochvíl (1982*c*).

Subfamily Pantherinae Pocock, 1917. Ann. Mag. Nat. Hist. ser. 8, 20:332.

COMMENTS: Type genus: *Panthera* Oken, 1816. Pocock's (1917) original classification for this subfamily placed *Neofelis* in the Felinae.

Neofelis Gray, 1867. Proc. Zool. Soc. Lond., 1867:265.

TYPE SPECIES: *Felis macrocelis* Horsfield, 1825 (= *Felis nebulosa* Griffith, 1821), by subsequent designation by Pocock (1917:343).

COMMENTS: Placed in Pantherinae by Hemmer (1978) and Weigel (1961). Placed in Neofelinae by Kratochvíl (1982*c*).

Neofelis nebulosa (Griffith, 1821). Gen. Particular Descrip. Vert. Anim. (Carn.), p. 37, pl.

TYPE LOCALITY: "brought from Canton" [China, Guangdong: Guangzhou].

DISTRIBUTION: Burma, Cambodia, China, India, Indonesia, Malaysia, Nepal, Taiwan, Thailand, and Vietnam.

STATUS: CITES - Appendix I; U.S. ESA - Endangered; IUCN - Vulnerable.

SYNONYMS: *brachyurus* Swinhoe, 1862; *diardi* Cuvier, 1823; *macrocelis* Horsfield, 1825; *macrosceloides* Hodgson, 1853.

COMMENTS: Placed in *Neofelis* by Pocock (1917), Weigel (1961), Hemmer (1978), and Kratochvíl (1982*c*). Groves (1982*a*) placed in *Panthera*.

Panthera Oken, 1816. Lehrb. Naturgesch., ser. 3, 2:1052.

TYPE SPECIES: *Felis pardus* Linnaeus, 1758, by subsequent designation by Allen (1902:378).

SYNONYMS: *Jaguarius* Severtzov, 1858; *Leo* Oken, 1816; *Leonina* Grevé, 1894; *Pardotigris* Kretzoi, 1929; *Pardus* Fitzinger, 1868; *Tigris* Oken, 1816.

COMMENTS: Revised by Hemmer (1966, 1968, 1974). *Panthera* Oken, 1816, has been ruled available (International Commission on Zoological Nomenclature, 1985*c*). Includes *Tigris* following Pocock (1916*b*). Van Gelder (1977*b*:13) included *Panthera* as a synonym of *Felis*.

Panthera leo (Linnaeus, 1758). Syst. Nat., 10th ed., 1:41.

TYPE LOCALITY: "Africa", restricted by Allen (1924:222) to "the Barbary coast region of Africa, or, more explicity, Constantine, Algeria".

DISTRIBUTION: Present (except in tropical rain forests) in Botswana, Ethiopia, India, Kenya, Malawi, Mali, Mozambique, Namibia, Senegal, Somalia, South Africa, Sudan, Uganda, Zambia, and Zimbabwe. Formerly present but now extinct in Algeria, Arabia, Egypt, Greece, Iran, Iraq, Israel, Libya, Morocco, Pakistan, and Tunisia.

STATUS: CITES - Appendix I as *P. l. persica*; otherwise Appendix II. U.S. ESA and IUCN - Endangered as *P. l. persica*.

SYNONYMS: *adusta* Pocock, 1927; *africanus* Brehm, 1829; *asiaticus* Jardine, 1834; *azandicus* Allen, 1924; *barbaricus* Meyer, 1826; *barbarus* Fischer, 1829; *bengalensis* Bennett, 1829; *bleyenberghi* Lönnberg, 1914; *capensis* Fischer, 1829; *gambianus* Gray, 1843; *goojratensis* Smee, 1833; *hollisteri* Allen, 1924; *indicus* de Blainville, 1843; *kamptzi* Matschie, 1900; *krugeri* Roberts, 1929; *maculatus* Huevelmans, 1955; *massaicus* Neumann, 1900; *melanochaitus* H. Smith, 1842; *nigra* Loche, 1858; *nobilis* Gray, 1867; *nubicus* Blainville, 1843; *nyanzae* Heller, 1913; *persicus* Meyer, 1826; *roosevelti* Heller, 1913; *sabakiensis* Lönnberg, 1905; *senegalensis* Meyer, 1826; *somaliensis* Noack, 1891; *suahelicus* Neumann, 1900; *vernayi* Roberts, 1948; *webbiensis* Zukowsky, 1964.

COMMENTS: Revised by Pocock (1930*c*). Placed in *Panthera* by Pocock (1930*c*), Weigel (1961), Kratochvíl (1982*c*), Hemmer (1978), and Groves (1982*a*).

Panthera onca (Linnaeus, 1758). Syst. Nat., 10th ed., 1:42.

TYPE LOCALITY: "America meridionali", fixed by Thomas (1911*a*:136) to "Pernambuco" [Brazil].

DISTRIBUTION: N Argentina, Belize, Bolivia, Brazil, Colombia, Costa Rica, El Salvador, French Guiana, Guatemala, Guyana, Honduras, S Mexico, Nicaragua, Panama, Paraguay, Peru, Surinam, Venezuela; formerly in USA (Arizona, California, New Mexico, Texas).

STATUS: CITES - Appendix I; U.S. ESA - Endangered (from Mexico southward); IUCN - Vulnerable.

SYNONYMS: *alba* Fitzinger, 1869; *antiqua* Ameghino, 1889; *arizonensis* Goldman, 1932; *boliviensis* Nelson and Goldman, 1933; *centralis* Mearns, 1901; *coxi* Nelson and Goldman, 1933; *fossilis* Ameghino, 1889; *goldmani* Mearns, 1901; *hernandesii* Gray, 1858; *jaguapara* Liais, 1872; *jaguar* Link, 1795; *jaguarete* Liais, 1872; *jaguatyrica* Liais, 1872; *madeirae* Nelson and Goldman, 1933; *major* Fischer, 1830; *mexianae* Hagmann, 1908; *milleri* Nelson and Goldman, 1933; *minor* Fischer, 1830; *nigra* Erxleben, 1777; *notialis* Hollister, 1914; *onssa* Ihering, 1911; *onza* Brehm, 1876; *palustris* Ameghino, 1888; *paraguensis* Hollister, 1914; *paulensis* Nelson and Goldman, 1933; *peruviana* de Blainville, 1843; *proplatensis* Ameghino, 1904; *ramsayi* Miller, 1930; *ucayalae* Nelson and Goldman, 1933; *veraecrucis* Nelson and Goldman, 1933.

COMMENTS: Revised by Nelson and Goldman (1933) and Pocock (1939*b*). Placed in *Panthera* by Pocock (1939*b*), Weigel (1961), Hemmer (1978), Kratochvíl (1982*c*), and Groves (1982*a*). Reviewed by Seymour (1989, Mammalian Species, 340).

Panthera pardus (Linnaeus, 1758). Syst. Nat., 10th ed., 1:41.

TYPE LOCALITY: "Indiis", fixed by Thomas (1911*a*:135), as "Egypt"; see discussion by Pocock (1930*a*).

DISTRIBUTION: Afghanistan, Algeria, Angola, Arabia, Botswana, Burma, Cameroon, Central African Republic, Chad, China, Congo, Egypt, Ethiopia, Gabon, Guinea-Bissau, India, Indonesia (Java), Iran, Iraq, Kenya, Korea, Liberia, Laos, Malawi, Malaysia, Mauritania, Morocco, Mozambique, Namibia, Nepal, Niger, Nigeria, Pakistan, Senegal, Sierra Leone, Somalia, South Africa, Sri Lanka, Sudan, Tanzania, Thailand, Tunisia, Turkey, Uganda, republics of the former USSR, Vietnam, Zaire, Zambia, and Zimbabwe.

STATUS: CITES - Appendix I; U.S. ESA - Endangered (except in Africa, in the wild, south of, and including Gabon, Congo, Zaire, Uganda, and Kenya, where this species is Threatened). IUCN - Threatened.

SYNONYMS: *adersi* Pocock, 1932; *adusta* Pocock, 1927; *antinorii* de Beaux, 1923; *antiquorum* Fitzinger, 1868; *barbarus* de Blainville, 1839; *bedfordi* Pocock, 1930; *brockmani* Pocock, 1932; *centralis* Lönnberg, 1917; *chinensis* Gray, 1867; *chui* Heller, 1913; *ciscaucasicus* Satunin, 1914; *delacouri* Pocock, 1930; *fontanierii* A. M. Edwards, 1867; *fortis* Heller, 1913; *fusca* Meyer, 1794; *grayi* Trouessart, 1904; *hanensis* Matschie, 1907; *iturensis* J. A. Allen, 1924; *japonensis* Gray, 1862; *jarvisi* Pocock, 1932; *kotiya* Deraniyagala, 1956; *leopardus* Schreber, 1777; *longicaudata* Valenciennes, 1856; *melanotica* Günther, 1885; *melas* G. Cuvier, 1809; *millardi* Pocock, 1930; *minor* Matschie, 1895; *nanopardus* Thomas, 1904; *niger* Fitzinger, 1868; *nimr* Hemprich and Ehrenberg, 1833; *orientalis* Schlegel, 1857; *palearia* Cuvier, 1832; *perniger* Hodgson, 1863; *poecilura* Valenciennes, 1856; *poliopardus* Brehm, 1863; *puella* Pocock, 1932; *reichenowi* Cabrera, 1918; *ruwenzorii* Camerano, 1906; *saxicolor* Pocock, 1927; *shortridgei* Pocock, 1932; *sindica* Pocock, 1930; *suahelicus* Neumann, 1900; *tulliana* Valenciennes, 1856; *variegata* Wagner, 1841; *varius* Gray, 1843; *villosa* Bonhote, 1903; *vulgaris* Oken, 1816.

COMMENTS: Revised by Pocock (1930*a*, *b*, 1932*c*). Placed in *Panthera* by Pocock (1930*a*), Weigel (1961), Hemmer (1978), Kratochvíl (1982*c*), and Groves (1982*a*).

Panthera tigris (Linnaeus, 1758). Syst. Nat., 10th ed., 1:41.

TYPE LOCALITY: "Asia", fixed by Thomas (1911*a*:135) as "Bengal" [India].

DISTRIBUTION: Bangladesh, Bhutan, Burma, China, Laos, India, Indonesia (Sumatra), Korea, Malaysia, Nepal, Thailand, Vietnam, and republics of the former USSR. Formerly found in Afghanistan, Pakistan, Iran, and Indonesia (Java and Bali).

STATUS: CITES - Appendix I; U.S. ESA and IUCN - Endangered.

SYNONYMS: *altaica* Temminck, 1844; *amoyensis* Hilzheimer, 1905; *amurensis* Dode, 1871; *balica* Schwarz, 1912; *corbetti* Mazak, 1968; *coreensis* Brass, 1904; *lecoqi* Schwarz, 1916; *longipilis* Fitzinger, 1868; *mandshurica* Baykov, 1925; *mikadoi* Satunin, 1915; *mongolica* Lesson, 1842; *nigra* Lesson, 1842; *regalis* Gray, 1842; *septentrionalis* Satunin, 1904; *sondaica* Temminck, 1844; *striatus* Severtzov, 1858; *styani* Pocock, 1929; *sumatrae*

Pocock, 1929; *sumatrana* de Blainville, 1839; *trabata* Schwarz, 1916; *virgata* Illiger, 1815.

COMMENTS: Revised by Pocock (1929), and Mazák (1979, 1981 [Mammalian Species, 152]). Placed in *Panthera* by Pocock (1929), Weigel (1961), Hemmer (1978), Kratochvíl (1982*c*), and Groves (1982*a*).

Pardofelis Severtzov, 1858. Rev. Mag. Zool. Paris, ser. 2, 10:387.

TYPE SPECIES: *Felis marmorata* Martin, 1837, by monotypy.

COMMENTS: There is considerable cotroversy over the correct placement of this genus. Hemmer (1978), Král and Zima (1980), Groves (1982*a*), and Kratochvíl (1982*c*) suggested a close relationship with *Panthera*. Pocock (1932*d*) and Weigel (1961) suggested a relationship with felines while recognizing similarities with *Panthera*. It is perhaps best considered *incertae sedis*.

Pardofelis marmorata (Martin, 1837). Proc. Zool. Soc. Lond., 1836:108 [1837].

TYPE LOCALITY: "Java or Sumatra" [Indonesia], restricted by Robinson and Kloss (1919*a*:261), to "Sumatra".

DISTRIBUTION: N India and Nepal to Vietnam, Thailand, Malaysia, Sumatra, and Borneo; S China.

STATUS: CITES - Appendix I; U.S. ESA - Endangered; IUCN - Indeterminate.

SYNONYMS: *charltoni* Gray, 1846; *dosul* Hodgson, 1863; *duvaucelli* Hodgson, 1863; *longicaudata* Blainville, 1843; *ogilbii* Hodgson, 1847.

COMMENTS: Revised by Pocock (1932*d*). Placed in *Pardofelis* by Pocock (1932*d*), Weigel (1961), Král and Zima (1980), Kratochvíl (1982*c*), Hemmer (1978), and Groves (1982*a*).

Uncia Gray, 1854. Ann. Mag. Nat. Hist., ser. 2, 14:394.

TYPE SPECIES: *Felis irbis* Ehrenberg, 1830 (= *Felis uncia* Schreber, 1775), by subsequent designation (Palmer, 1904:710).

COMMENTS: Revised by Pocock (1916*b*).

Uncia uncia (Schreber, 1775). Die Säugethiere, 2(14):pl. 100[1775]; text, 3(22):386-7[1777].

TYPE LOCALITY: "Barbarey, Persien, Ostindien, und China", restricted by Pocock (1930*b*:332) to "Altai Mountains". Ognev (1962*c*:221) disagreed and restricted the locality to "Kopet-Dagh Mountains...the southern slopes of these mountains adjacent to Iran".

DISTRIBUTION: Afghanistan, Bhutan, China, India, Mongolia, Nepal, Pakistan, republics of the former USSR.

STATUS: CITES - Appendix I; U.S. ESA and IUCN - Endangered.

SYNONYMS: *irbis* Ehrenberg, 1830; *schneideri* Zukowsky, 1950; *uncioides* Horsfield, 1855.

COMMENTS: Revised by Pocock (1930*b*). Placed in *Uncia* by Pocock (1930*b*), Weigel (1961), Kratochvíl (1980*c*), and Heptner and Naumov (1967). Placed in *Uncia* and reviewed by Hemmer (1972, Mammalian Species, 20).

Family Herpestidae Bonaparte, 1845. Cat. Meth. Mamm. Europe, p. 3.

SYNONYMS: Mungotidae Pocock, 1916.

COMMENTS: Type genus: *Herpestes* Illiger, 1811 (Melville and Smith, 1987). For separation of taxa included here from Viverridae (after Simpson, 1945), see comments under Viverridae.

Subfamily Galidiinae Gray, 1865. Proc. Zool. Soc. London, 1864:508 [1865].

Galidia I. Geoffroy Saint-Hilaire, 1837. C. R. Acad. Sci. Paris, 5:580.

TYPE SPECIES: *Galidia elegans* I. Geoffroy Saint-Hilaire, 1837, by monotypy.

Galidia elegans I. Geoffroy Saint-Hilaire, 1837. C. R. Acad. Sci. Paris, 5:581.

TYPE LOCALITY: "Madagascar".

DISTRIBUTION: Endemic to Madagascar.

SYNONYMS: *afra* Kerr, 1792.

Galidictis I. Geoffroy Saint-Hilaire, 1839. Mag. Zool., Mamm. Art. No. 5, p. 33, footnote, 37.
TYPE SPECIES: *Mustela striata* I. Geoffroy Saint-Hilaire, 1837 (= *Viverra fasciata* Gmelin, 1788) by original designation.
COMMENTS: Gregory and Hellman (1939) separated *Galidictis* from other galidiines and placed it in the Viverridae.

Galidictis fasciata (Gmelin, 1788). *In* Linnaeus, Syst. Nat., 13th ed., 1:92.
TYPE LOCALITY: Erroneously listed by Gmelin as "in India".
DISTRIBUTION: Endemic to Madagascar.
STATUS: IUCN - Indeterminate.
SYNONYMS: *eximius* Pocock, 1915; *ornatus* Pocock, 1915; *rufa* Grandidier, 1869; *striata* I. Geoffroy Saint-Hilaire, 1839; *vittatus* Schinz, 1844.

Galidictis grandidieri Wozencraft, 1986. J. Mammal., 67:561.
TYPE LOCALITY: "Madagascar".
DISTRIBUTION: Known only from the spiny desert of SW Madagascar.
STATUS: IUCN - Insufficiently known.
COMMENTS: *G. grandidiensis* Wozencraft, 1986 was emended to *G. grandidieri* by Wozencraft (1987).

Mungotictis Pocock, 1915. Ann. Mag. Nat. Hist., ser. 8, 16:120.
TYPE SPECIES: *Galidictis vittatus* Gray, 1848 (= *Galidia decemlineata* Grandidier, 1867).

Mungotictis decemlineata (A. Grandidier, 1867). Rev. Mag. Zool. Paris, ser. 2, 19:85.
TYPE LOCALITY: "à la côte ouest de Madagascar" (pg. 84).
DISTRIBUTION: Endemic to Madagascar.
STATUS: IUCN - Vulnerable.
SYNONYMS: *lineatus* Pocock, 1915; *substriatus* Pocock, 1915; *vittatus* Gray, 1848.

Salanoia Gray, 1865. Proc. Zool. Soc. Lond., 1864:523 [1865].
TYPE SPECIES: *Galidia concolor* I. Geoffroy Saint-Hilaire, 1837.

Salanoia concolor (I. Geoffroy Saint-Hilaire, 1837). C. R. Acad. Sci. Paris, 5:581.
TYPE LOCALITY: "Madagascar".
DISTRIBUTION: Endemic to Madagascar.
STATUS: IUCN - Insufficiently known.
SYNONYMS: *olivacea* I. Geoffroy Saint-Hilaire, 1839; *unicolor* I. Geoffroy Saint-Hilaire, 1837.
COMMENTS: Geoffroy Saint-Hilaire (1839) noted that his first listed species name, *unicolor*, was a typographical error and should have been *concolor* (Coetzee, 1977*b*:35).

Subfamily Herpestinae Bonaparte, 1845. Cat. Meth. Mamm. Europe, p. 3.
COMMENTS: Type genus: *Herpestes* Illiger, 1811 (Melville and Smith, 1987). Thomas (1882) proposed a separate monotypic subfamily (Surcatinae) for *Suricata*. Wozencraft (1989*b*) placed *Crossarchus, Cynictis, Dologale, Helogale, Liberiictis, Mungos, Paracynictis,* and *Suricata* in the Mungotinae but gave no supporting rationale. Fredga's (1972) analysis of chromosomes would support Wozencraft's Mungotinae (with the inclusion of *Bdeogale* and *Ichneumia*). The phylogenetic analysis of allozyme data by Taylor et al. (1991) also supported *Cynictis, Suricata,* and *Helogale* as a monophyletic group.

Atilax F. G. Cuvier, 1826. *In* E. Geoffroy Saint-Hilaire and F. G. Cuvier, Hist. Nat. Mammifères, pt. 3, 5(54), "Vansire," 2 pp.
TYPE SPECIES: *Herpestes paludinosus* G. [Baron] Cuvier, 1829, by original designation (Melville and Smith, 1987).
COMMENTS: Fredga's (1972) comparative chromosome study of mongooses suggested that recognition of *Atilax* as distinct from *Herpestes* would make *Herpestes* paraphyletic. However, allozyme data support *Atilax* as the first early offshoot of the main herpestine branch (Taylor et al., 1991).

Atilax paludinosus (G.[Baron] Cuvier, 1829). Regn. Anim., Nouv. ed., 1:158.
TYPE LOCALITY: "une grand des marais du Cap" [South Africa, Cape Prov., Cape of Good Hope].

DISTRIBUTION: Algeria, Angola, Botswana, Cameroon, Equatorial Guinea, Ethiopia, Gabon, Ivory Coast, Liberia, Malawi, Mozambique, Niger, Ruwanda, Senegal, Sierra Leone, Somalia, South Africa, Sudan, Tanzania, Uganda, Zaire, and Zambia.
SYNONYMS: *atilax* Wagner, 1841; *galera* Schreber, 1777; *macrodon* Allen, 1924; *mitis* Thomas, 1902; *mordax* Thomas, 1912; *nigerianus* Thomas, 1912; *paludosus* Gray, 1865; *pluto* Temminck, 1853; *robustus* Gray, 1864; *rubellus* Thomas and Wroughton, 1908; *rubescens* Hollister, 1912; *spadiceus* Cabrera, 1921; *transvaalensis* Roberts, 1933; *urinatrix* Smith, 1829; *vansire* F. Cuvier, 1826; *voangshire* Zimmermann, 1777.

Bdeogale Peters, 1850. Spenersche Z., 25 June, 1850 (unpaginated).
TYPE SPECIES: *Bdeogale crassicauda* Peters, 1852; by subsequent designation by Thomas (1882) (Melville and Smith, 1987).
COMMENTS: Matschie (1895), Pocock (1916*a*), Coetzee (1977*b*), Kingdon (1977), and Meester et al. (1986) included *Galeriscus* Thomas, 1894. Rosevear (1974) believed that no one had advanced any "reasoned argument" for combining *Galeriscus* with *Bdeogale*, and followed Schouteden (1945) and Hill and Carter (1941) who considered them distinct; all have agreed that *jacksoni* and *nigripes* are sister groups.

Bdeogale crassicauda Peters, 1852. Monatsb. K. Preuss. Akad. Wiss., Berlin, 1852:81.
TYPE LOCALITY: "Africa orient., Tette, Boror, 17-18° Lat. austr". (pg. 82). Restricted by Moreau et al. (1946:410) to "Tette" [Mozambique].
DISTRIBUTION: Kenya, Malawi, C Mozambique, Tanzania (incl. Zanzibar), S and E Zambia, NE Zimbabwe.
STATUS: IUCN - Endangered as *B. c. omnivora*.
SYNONYMS: *omnivora* Heller, 1913; *puisa* Peters, 1852; *tenuis* Thomas and Wroughton, 1908.
COMMENTS: Reviewed by Taylor (1987, Mammalian Species, 294) and revised by Sale and Taylor (1970).

Bdeogale jacksoni (Thomas, 1894). Ann. Mag. Nat. Hist., ser. 6, 13:522.
TYPE LOCALITY: "Mianzini, Masailand, 8000 feet" (pg. 523). Restricted by Moreau et al. (1946:410) to "Mianzini. . . a few miles E.S.E. of Naivasha and on the southern end of the Kinangop Plateau. . . 9000 ft" [Kenya].
DISTRIBUTION: C Kenya, SE Uganda.
STATUS: IUCN - Insufficiently known.
COMMENTS: Rosevear (1974) placed *jacksoni* in *Galeriscus*. Kingdon (1977) considered *jacksoni* conspecific with *nigripes*; however, Rosevear (1974) and Coetzee (1977*b*) noted skull and skin differences.

Bdeogale nigripes Pucheran, 1855. Rev. Mag. Zool. Paris, 7(2):111.
TYPE LOCALITY: "Gubon" [Gabon].
DISTRIBUTION: Nigeria to N Angola.
COMMENTS: Rosevear (1974) placed *nigripes* in *Galeriscus*. Kingdon (1977) considered *jacksoni* conspecific with *nigripes*; however, Rosevear (1974) and Coetzee (1977*b*) noted skull and skin differences.

Crossarchus F. G. Cuvier, 1825. *In* E. Geoffroy Saint-Hilaire and F. G. Cuvier, Hist. Nat. Mammifères, pt. 3, 5(47), "le Mangue" 3 pp., 1 pl.
TYPE SPECIES: *Crossarchus obscurus* F. G. Cuvier, 1825, by original designation (Melville and Smith, 1987).
COMMENTS: Revised by Goldman (1984). Placed in *Mungos* by Hill and Carter (1941). Van Rompaey and Colyn (1992) presented a key to the species.

Crossarchus alexandri Thomas and Wroughton, 1907. Ann. Mag. Nat. Hist., ser. 7, 19:373.
TYPE LOCALITY: "from Banzyville, Ubanghi" [= Mobayi, Zaire, 4°N, 21°11'E (Goldman, 1984)].
DISTRIBUTION: Central African Republic, Congo, Uganda, Zaire.
SYNONYMS: *minor* Goldman, 1984.

Crossarchus ansorgei Thomas, 1910. Ann. Mag. Nat. Hist., ser. 8, 5:195.
TYPE LOCALITY: "Dalla Tando" [= Angola, Ndala Tando, 9°18'S, 14°54'E (Goldman, 1984)].
DISTRIBUTION: N Angola, SE Zaire.
SYNONYMS: *nigricolor*.

COMMENTS: Reviewed by Van Rompaey and Colyn (1992, Mammalian Species, 402).

Crossarchus obscurus F. G. Cuvier, 1825. *In* E. Geoffroy Saint-Hilaire and F. G. Cuvier, Hist. Nat. Mammifères, pt. 3, 5(47), "Mangue" 3 pp.
TYPE LOCALITY: "côtes occidentales de l'Afrique, et vraisemblablement des parties qui sont au midi de la Gambie" restricted by Cuvier (1829:158) to "Sierra Leone".
DISTRIBUTION: Benin, Cameroon, Central African Republic, Ghana, Ivory Coast, Liberia, Malawi, Nigeria, Sierra Leone.
SYNONYMS: *platycephalus* Goldman, 1984; *somalicus* Thomas, 1895.
COMMENTS: Goldman (1984, 1987) separated central (*C. platycephalus*) from western (*C. obscurus*) African populations based on phenetic differences in skull proportions. Wozencraft (1989*b*) argued for consideration of these populations as conspecific. Reviewed by Goldman (1987, Mammalian Species, 290).

Cynictis Ogilby, 1833. Proc. Zool. Soc. Lond., 1833:48.
TYPE SPECIES: *Cynictis steedmanni* Ogilby, 1833 (= *Herpestes penicillatus* G. Cuvier, 1829).

Cynictis penicillata (G. [Baron] Cuvier, 1829). Regn. Anim., Nouv. ed., 1:158.
TYPE LOCALITY: "du Cap", restricted by Roberts (1951:151) to "Uitenhage, C.P." [South Africa].
DISTRIBUTION: S Angola, Botswana, Namibia, South Africa, SW Zimbabwe.
SYNONYMS: *bechuanae* Roberts, 1932; *brachyura* Roberts, 1924; *bradfieldi* Roberts, 1924; *cinderella* Thomas, 1927; *coombsi* Roberts, 1929; *intensa* Schwann, 1906; *kalaharica* Roberts, 1932; *karasensis* Roberts, 1938; *lepturus* Smith, 1839; *levaillantii* Smith, 1829; *ogilbyi* Smith, 1834; *pallidior* Thomas and Schwann, 1904; *steedmanni* Ogilby, 1833; *typicus* Smith, 1834.
COMMENTS: Revised by Lundholm (1955*b*).

Dologale Thomas, 1926. Ann. Mag. Nat. Hist., ser. 9, 17:183.
TYPE SPECIES: *Crossarchus dybowskii* Pousargues, 1893, by original designation.
COMMENTS: Revised by Hayman (1936). Although originally placed in *Crossarchus*, most since Hayman (1936) believed this genus to be the sister group to *Helogale*; Allen (1924) identified some specimens of this taxon as *Helogale hirtula robusta*.

Dologale dybowskii (Pousargues, 1893). Bull. Soc. Zool. Fr., 18:51.
TYPE LOCALITY: "Ubangi, Congo Belge", restricted by Moreau et al. (1945:410) to "on the Upper Kemo, a tributary to the north of the Ubangui, about 6°17'N, 19°12'E" [Central African Republic].
DISTRIBUTION: Central African Republic, S Sudan, W Uganda, NE Zaire.
SYNONYMS: *nigripes* Kershaw, 1924; *robusta* Allen, 1924.
COMMENTS: Pousargues (1894) later redescribed the species in detail.

Galerella Gray, 1865. Proc. Zool. Soc. Lond., 1864:564 [1865].
TYPE SPECIES: *Herpestes ochraceus* Gray, 1849 (= *Herpestes sanguineus* Rüppell, 1836) by original designation.
COMMENTS: Revised by Lynch (1981), Watson and Dippenaar (1987), Watson (1990), and Taylor et al. (1991) who considered these taxa a monophyletic group. Crawford-Cabral (1989*a*:2) regarded these taxa as a "superspecies with several allospecies." These taxa are provisionally separated from *Herpestes* (*sensu latu*) following these revisions and reviews by Rosevear (1974), Ansell (1978), Smithers (1983), and Meester et al. (1986) (see discussion under *Herpestes*).

Galerella flavescens (Bocage, 1889). J. Sci. Math. Phys. Nat. Lisboa, ser 12, 1:179.
TYPE LOCALITY: "Benguella", [Angola].
DISTRIBUTION: S Angola, C and N Namibia.
SYNONYMS: *annulatus* Lundholm, 1955; *nigratus* Thomas, 1928; *shortridgei* Roberts, 1932.
COMMENTS: Included in *sanguinea* by Taylor (1975). The form *flavescens* was not mentioned in Meester et al. (1986) or Watson and Dippenaar's (1987) revision; Crawford-Cabral (1989*a*) considered *nigratus* conspecific with *flavescens*, the senior synonym. Meester et al. (1986) listed *nigratus* as a synonym of *G. pulverulenta*.

Galerella pulverulenta (Wagner, 1839). Gelehrte. Anz. I.K. Bayer. Akad. Wiss., München., 9:426.

TYPE LOCALITY: "Kap" [Cape of Good Hope, South Africa].

DISTRIBUTION: South Africa, south of 27°S latitude (Bronner, 1990).

SYNONYMS: *apiculatus* Gray, 1865; *basuticus* Roberts, 1936; *caffra* Smith, 1826; *lasti* Wroughton, 1907; *maritimus* Roberts, 1919; *ratlamuchi* Thomas, 1929; *ruddi* Rhomas, 1903; *rufescens* Lorenz, 1898.

COMMENTS: Revised by Lynch (1981) and Watson and Dippenaar (1987), who removed *annulata* and *shortridgei* (considered here as *incertae sedis*) and *nigrata* (placed here in *flavescens*) from *pulverulenta*; although Meester et al. (1986) did not. Crawford-Cabral (1989*a*) included these taxa with *flavescens*.

Galerella sanguinea (Rüppell, 1836). Neue Wirbelt. Fauna Abyssin. Gehörig. Säugeth., 1:27.

TYPE LOCALITY: "Kordofan" [Sudan].

DISTRIBUTION: Angola, Benin, Botswana, Burkina Faso, Cameroon, Cape Verde Isls, Central African Republic, Congo, Equatorial Guinea, Ethiopia, Ghana, Ivory Coast, Kenya, Liberia, Malawi, Mauritana, Mozambique, Namibia, Niger, Rwanda, Senegal, Sierra Leone, Somalia, South Africa, Sudan, Tanzania, Togo, Uganda, Zaire, Zambia, Zimbabwe.

SYNONYMS: *auratus* Thomas and Wroughton, 1908; *badius* Smith, 1838; *bocagei* Thomas and Wroughton, 1905; *bradfieldi* Roberts, 1932; *bruneoochracea* Matschie, 1914; *caldatus* Thomas, 1927; *canus* Wroughton, 1907; *caurinus* Thomas, 1926; *cauui* Smith, 1836; *condradsi* Matschie, 1914; *dasilvai* Roberts, 1938; *dentifer* Heller, 1913; *elegans* Matschie, 1914; *emini* Matschie, 1914; *ererensis* Matschie, 1914; *erlangeri* Matschie, 1914; *erongensis* Roberts, 1946; *flaviventris* Matschie, 1914; *fulvidior* Thomas, 1904; *fuscus* Rüppell, 1835; *galbus* Wroughton, 1909; *galinieri* Guérin and Ferret, 1847; *gracilis* Rüppell, 1835; *granti* Gray, 1864; *ignitoides* Roberts, 1932; *ignitus* Roberts, 1913; *iodoprymnus* Heuglin, 1861; *kalaharicus* Roberts, 1932; *kaokoensis* Roberts, 1932; *khanensis* Roberts, 1932; *lancasteri* Roberts, 1932; *lefebvrei* Desmurs and Prévost, 1850; *lundensis* Monard, 1935; *marae* Matschie, 1914; *melanura* Martin, 1836; *mossambica* Matschie, 1914; *mustela* Schwarz, 1935; *mutgigella* Rüppell, 1835; *mutscheltschela* Heuglin, 1877; *neumanni* Matschie, 1894; *ngamiensis* Roberts, 1932; *nigricaudatus* I. Geoffroy Saint-Hilaire, 1839; *ochraceus* Gray, 1849; *ochromelas* Pucheran, 1855; *okavangensis* Roberts, 1932; *orestes* Heller, 1911; *ornatus* Peters, 1852; *parvipes* Hollister, 1916; *perfulvidus* Thomas, 1904; *phoenicurus* Thomas, 1912; *proteus* Thomas, 1907; *punctulatus* Gray, 1849; *ratlamuchi* Smith, 1836; *rendilis* Lönnberg, 1912; *ruasae* Matschie, 1914; *ruficauda* Heuglin, 1877; *saharae* Thomas, 1925; *schimperi* Matschie, 1914; *swinnyi* Roberts, 1913; *talboti* Thomas and Wroughton, 1907; *turstigi* Matschie, 1914; *ugandae* Wroughton, 1909; *upingtoni* Shortridge, 1934; *venatica* Gray, 1865; *zombae* Wroughton, 1907.

COMMENTS: This engimatic group, reviewed by Taylor (1975), is represented by several allopatric populations (A situation similar to the *Genetta genetta* complex where they are recognized as conspecific). Watson and Dippenaar (1987), in their revision, argued for the separation of *nigratus* Thomas, 1928 (= *flavescens*), and *swalius* Thomas, 1926, and considered *swinnyi* Roberts, 1913, as *incertae sedis* (included here), although their study did not include representative samples from NE and W Africa. It is believed that *swinnyi* has been extirpated from the type locality (Watson, in litt.). Taylor (1989) suggested that the allopatric *ochraceus* from Somalia warrants full specific status. Certainly, these studies suggests that a thorough revision, inclusive of all of the African forms of *sanguinea* is badly needed.

Galerella swalius (Thomas, 1926). Proc. Zool. Soc. Lond., 1926:292.

TYPE LOCALITY: "Great Brukaros Mountain, 3500'" [Namibia].

DISTRIBUTION: S and C Namibia.

COMMENTS: Included in *sanguinea* by Meester et al. (1989) and Taylor (1975).

Helogale Gray, 1862. Proc. Zool. Soc. Lond., 1861:308 [1862].

TYPE SPECIES: *Herpestes parvulus* Sundevall, 1846, by subsequent designation by Thomas (1882) (Melville and Smith, 1987).

COMMENTS: The number of taxa ascribed to this genus is provisional. Ellerman et al. (1953) recognized 7 species, Allen (1939) listed eleven. The acceptance here of two, follows

Coetzee (1977*b*). The range of *H. hirtula* is included within that of *H. parvula*.

Helogale hirtula Thomas, 1904. Ann. Mag. Nat. Hist., ser. 7, 14:97.

TYPE LOCALITY: "Gabridehari, 60 mi West of Gerlogobi", restricted by Moreau et al. (1946:410) to "south-east Ethiopia (Ogaden) at about 7°0'N, 45°20'E". Further restricted by Yalden et al. (1980) to "Gabridehari (=Gabredarre, Kebridar) 6°45'N, 44°17'E".

DISTRIBUTION: S Ethiopia, S and C Somalia, N and C Kenya.

SYNONYMS: *ahlselli* Lönnberg, 1912; *annulata* Drake-Brockman, 1912; *lutescens* Thomas, 1911; *powelli* Drake-Brockman, 1912.

Helogale parvula (Sundevall, 1847). Ofv. K. Svenska Vet.-Akad. Forhandl. Stockholm, 1846, 3(4):121 [1847].

TYPE LOCALITY: "Caffraria superiore, juxta tropicum", restricted by Roberts (1951) to "Zoutpansberg" [South Africa].

DISTRIBUTION: Angola, Botswana, Ethiopia, Gambia, Kenya, Malawi, Mozambique, Namibia, Somalia, South Africa, Sudan, Tanzania, Uganda, Zaire, Zambia.

SYNONYMS: *affinis* Hollister, 1916; *atkinsoni* Thomas, 1897; *bradfieldi* Roberts, 1928; *brunetta* Thomas, 1926; *brunnula* Thomas and Schwann, 1906; *ivori* Thomas, 1919; *macmillani* Thomas, 1906; *mimetra* Thomas, 1926; *nero* Thomas, 1928; *ochracea* Thomas, 1910; *parvus* Hollister, 1912; *ruficeps* Kershaw, 1922; *rufula* Thomas, 1910; *undulatus* Peters, 1852; *varia* Thomas, 1902; *vetula* Thomas, 1911; *victorina* Thomas, 1902.

Herpestes Illiger, 1811. Prodr. Syst. Mamm. Avium., p. 135.

TYPE SPECIES: *Viverra ichneumon* Linnaeus, 1758, by absolute tautonomy, through the replaced name *Ichneumon* Lacépède, 1799 (Melville and Smith, 1987).

COMMENTS: Revised by Pocock (1919, 1937, 1941*a*) and Bechthold (1939). Coetzee (1977*b*), and Hayman (*in* Sanderson, 1940) included *Xenogale* (see discussion under *naso*). Allen (1924), considered only *ichneumon* in this genus and separated *sanguineus* and *pulverulentus* into *Galerella*; for support, he contrasted the large *ichneumon* with the smaller *sanguineus-pulverulentus* complex and reported proportional differences in measurements of skeleton and skull. His rationale has been repeated, in some cases verbatum, by Rosevear (1974), Ansell (1978), Smithers (1983), Meester et al. (1986), and Watson and Dippenaar (1987). Taylor et al. (1991) presented an allozyme analysis and argued for generic recognition, however, they did not include Asiatic *Herpestes*, and their consensus tree made the placement of the *sanguineus/pulverulentus* clade equivocal. Fredga's (1972) comparative chromosome analysis looked at variation including Asiatic and African *Herpestes*, and based on this, recognition of *Galerella* would make *Herpestes* paraphyletic. Comparison of measurements from Allen (1924), Rosevear (1974), and Smithers (1983) for African forms, and Bechthold (1939) and Pocock (1941*a*) for Asiatic forms reveals that the large morphological gaps originally identified by Allen (1924), dissolve when Asiatic species are included. Ellerman et al. (1953) and Wozencraft (1989*b*) suggested that differences between these taxa and other *Herpestes* are less than those found within *Herpestes*. Morphological criteria similar to that used by Allen (1924) have mostly been used at the specific level in other carnivores. Although the case is not strong, many recent authors have recognized *Galerella* and this is provisionally followed here (see comments under *Galerella*).

Herpestes brachyurus Gray, 1837. Proc. Zool. Soc. Lond., 1836:88 [1837].

TYPE LOCALITY: "Indian Islands", restricted by Kloss (1917) to "Borneo", however, Thomas (1921*c*) believed it to be from "Malacca". Pocock, *in* Chasen (1940), believed the type to be a "Malayan Race".

DISTRIBUTION: S India, Indonesia (Borneo, Sumatra), Malaysia, Philippine Isls, Singapore, Sri Lanka, Vietnam.

SYNONYMS: *ceylanicus* Nevill, 1887; *ceylonicus* Thomas, 1924; *dyacorum* Thomas, 1921; *flavidens* Kelaart, 1852; *fulvescens* Kelaart, 1851; *fusca* Waterhouse, 1838; *hosei* Jentink, 1903; *javanensis* Bechthold, 1936; *maccarthiae* Gray, 1851; *palawanus* Allen, 1910; *parvus* Jentink, 1895; *phillipsi* Thomas, 1924; *rafflesii* Anderson, 1875; *rajah* Thomas, 1921; *rubidior* Pocock, 1937; *siccatus* Thomas, 1924; *sumatrius* Thomas, 1921.

COMMENTS: Bechthold (1939), followed here, included *hosei* and *fusca*, and listed characteristics suggesting that in some respects, *semitorquatus* was intermediate between *brachyurus* and *urva*; this was followed by Medway (1977). Bechthold (1939) believed that *fusca* (*sensu stricto*) is most closely related to far-eastern *brachyurus* forms and considered them conspecific (both forms are short tailed mongooses); however, he gave features of the skull and pelage (used elsewhere at the specific level, i.e., *edwardsii* vs. *javanicus*) that distinguished the S India/Sri Lankan populations from those of SE Asia. Here they are provisionally treated as allopatric subspecies. Schwarz (1947) believed *semitorquatus* to be a red color morph of the dark *brachyurus*, although he did not address the most distinguishing feature of the collared mongoose - the collar - present in *semitorquatus* and absent in *brachyurus*. Medway (1977), followed by Payne et al. (1985), recognized *hosei* based on differences in the shape of the coronoid process of the mandible.

Herpestes edwardsii (E. Geoffroy Saint-Hilaire, 1818). Descrip. de L'Egypte, 2:139.

TYPE LOCALITY: "Indes orientales".

DISTRIBUTION: Afghanistan, Bahrain, India, Indonesia, Iran, Japan, Kuwait, Malaya (introduced; Wells, 1989), Nepal, Pakistan, Saudia Arabia, Sri Lanka. Populations believed to be introductions on Ryukyu Isls, Mauritus, and Reunion Isl (Corbet and Hill, 1980).

STATUS: CITES - Appendix III (India).

SYNONYMS: *andersoni* Murray, 1884; *carnaticus* Wroughton, 1921; *ellioti* Wroughton; *ferrugineus* Blanford, 1874; *fimbriatus* Temminck, 1853; *frederici* Desmarest, 1823; *griseus* E. Geoffroy Saint-Hilaire 1818; *lanka* Wroughton, 1915; *malaccensis* Fischer, 1829; *moerens* Wroughton, 1915; *montanus* Bechthold, 1936; *mungo* Blanford, 1888; *nyula* Hodgson, 1836; *pallens* Ryley, 1914; *pallidus* Wagner, 1841; *pondiceriana* Gervais, 1841; *ruddi* Thomas, 1903.

Herpestes ichneumon (Linnaeus, 1758). Syst. Nat., 10th ed., 1:43.

TYPE LOCALITY: "in Ægypto ad ripas Nili,...in India primario; mansuescit", restricted by Thomas (1911*a*) to "Egypt".

DISTRIBUTION: Algeria, Angola, Botswana, Cameroon, Chad, Egypt, Ethiopia, Gambia, Ghana, Gibralter, Guinea, Israel, Italy, Ivory Coast, Jordan, Kenya, Lebanon, Liberia, Lybia, Malawi, Morocco, Mozambique, Niger, Portugal, Rwanda, Senegal, Sierra Leone, South Africa, Spain, Sudan, Syria, Tanzania, Togo, Tunisia, Turkey, Uganda, Zaire, Zambia.

SYNONYMS: *aegyptiae* Tiedemann, 1808; *angolensis* Bocage, 1890; *bennettii* Gray, 1837; *cafra* Gmelin, 1788; *centralis* Lönnberg, 1917; *dorsalis* Gray, 1864; *egypti* Tiedemann, 1808; *ferruginea* Seabra, 1909; *funestus* Osgood, 1910; *grandis* Thomas, 1890; *griseus* Smuts, 1832; *lademanni* Matschie, 1914; *mababiensis* Roberts, 1932; *madagascarensis* Smith, 1834; *major* Geoffroy Saint-Hilaire 1818; *nems* Kerr, 1792; *numidianus* Gray, 1865; *numidicus* Cuvier, 1834; *parvidens* Lönnberg, 1908; *pharaon* Lacépède, 1799; *sabiensis* Roberts, 1926; *sangronizi* Cabrera, 1924; *widdringtonii* Gray, 1842.

Herpestes javanicus (E. Geoffroy Saint-Hilaire, 1818). Descrip. de L'Egypte, 2:138.

TYPE LOCALITY: "Java".

DISTRIBUTION: Afghanistan, Bangladesh, Bhutan, Burma, Cambodia, China, India, Indonesia, Malaysia, Nepal, Pakistan, Thailand, Vietnam. Introduced to Cuba, Dominican Republic, Fiji Isls, Hawaiian Isls, Jamacia, Japan, Puerto Rico, Surinam, West Indies, and many other tropical regions.

STATUS: CITES - Appendix III (India) as *H. auropunctatus*.

SYNONYMS: *auropunctata* Hodgson, 1836; *birmanicus* Thomas, 1886; *exilis* Gervais, 1841; *helvus* Ryley, 1914; *incertus* Kloss, 1917; *nepalensis* Gray, 1837; *pallipes* Blyth, 1845; *peninsulae* Schwarz, 1910; *perakensis* Kloss, 1917; *persicus* Gray, 1865; *rubrifrons* Allen, 1909; *rutilus* Gray, 1861; *siamensis* Kloss, 1917.

COMMENTS: Bechthold (1939), Pocock (1941*a*), and Lekagul and McNeeley (1977) included *Mangusta auropunctata*. Wells (1989) discussed the situation for the morphotypes in Indochina. Reviewed by Nellis (1989, Mammalian Species, 342, as *H. auropunctatus*).

Herpestes naso de Winton, 1901. Bull. Liverpool Mus., 3:35.

TYPE LOCALITY: "Cameroon River, West Africa" [Cameroon].

DISTRIBUTION: Cameroon, Congo, Equatorial Guinea, Gabon, Kenya, Niger, Tanzania, Zaire.

SYNONYMS: *microdon* Allen, 1919.

COMMENTS: Placed in *Xenogale* by Allen (1919*b*), and followed by Rosevear (1974) and Ansell (1978). This taxon, and *ichneumon*, which is generally recognized as its sister taxon (Allen, 1919*b*; Rosevear, 1974; Hayman, *in* Sanderson, 1940), can be distinguished principally by proportional differences of the interorbital region (Rosevear, 1974). Hayman (*in* Sanderson, 1940), Wenzel and Haltenorth (1972), and Coetzee (1977*b*) did not feel these differences were sufficient to warrant generic distinction, as did Allen (1919*b*), and Rosevear (1974). Recognition of *Xenogale* would make *Herpestes* paraphyletic.

Herpestes palustris Ghose, 1965. Proc. Zool. Soc. Calcutta, 18:174.

TYPE LOCALITY: "Nalbani (*c* 8 km east of Calcutta), North Salt Lake, 24-Parganas district, West Bengal, India".

DISTRIBUTION: India: West Bengal.

COMMENTS: Wenzel and Haltenorth (1972) speculated that this taxon may be conspecific with *auropunctatus* (= *javanicus*); whereas Ewer (1973), based on Ghose's observation that *palustris* can emit scent similiar to *urva* speculated on such a relationship to *urva*.

Herpestes semitorquatus Gray, 1846. Ann. Mag. Nat. Hist., [ser. 1], 18:211.

TYPE LOCALITY: "Borneo" [Mainland opposite Labuan (= Brunei)].

DISTRIBUTION: Indonesia (Borneo and Sumatra).

SYNONYMS: *uniformis* Robinson and Kloss, 1919.

COMMENTS: Schwarz (1947) concluded that *semitorquatus* was a red color morph of the dark *brachyurus*, see comments under *brachyurus*. Placed in subgenus *Urva* by Bechthold (1939).

Herpestes smithii Gray, 1837. Mag. Nat. Hist. [Charlesworth's], 1:578.

TYPE LOCALITY: Not given. Thomas (1923) suggested that it was from the "Bombay Region" but this was questioned by Pocock (1937).

DISTRIBUTION: India, Sri Lanka.

STATUS: CITES - Appendix III (India).

SYNONYMS: *canens* Thomas, 1921; *ellioti* Blyth, 1851; *jerdonii* Gray, 1865; *monticolus* Jerdon, 1867; *rubiginosus* Kelaart, 1852; *rusanus* Thomas, 1921; *thysanurus* Wagner, 1841; *torquatus* Kelaart, 1852; *zeylanius* Thomas, 1921.

Herpestes urva (Hodgson, 1836). J. Asiat. Soc. Bengal, 5:238.

TYPE LOCALITY: "Central and Northern Regions" [Nepal].

DISTRIBUTION: Burma, China, India, Laos, Malaysia (Wells and Francis, 1988), Nepal, Taiwan, Thailand, Vietnam.

STATUS: CITES - Appendix III (India).

SYNONYMS: *annamensis* Bechthold, 1936; *cancrivora* Hodgson, 1837; *formosanus* Bechthold, 1936; *fusca* Gray, 1830; *hanensis* Matschie, 1907; *sinensis* Bechthold, 1936.

Herpestes vitticollis Bennett, 1835. Proc. Zool. Soc. Lond., 1835:67.

TYPE LOCALITY: "in forests about twenty miles inland from Kolun or Quilon, in the Travancore country" [India].

DISTRIBUTION: S India, Sri Lanka.

STATUS: CITES - Appendix III (India).

SYNONYMS: *inornatus* Pocock, 1941; *rubiginosus* Wagner, 1841.

Ichneumia I. Geoffroy Saint-Hilaire, 1837. Ann. Sci. Nat. Zool. (Paris), 8(2):251.

TYPE SPECIES: *Herpestes albicaudus* G. [Baron] Cuvier, 1829, by designation of Geoffroy Saint-Hilaire (1839) (Melville and Smith, 1987).

Ichneumia albicauda (G. [Baron] Cuvier, 1829). Regn. Anim., Nouv. ed., 21:158.

TYPE LOCALITY: "l'Afrique australe et le Sénégal".

DISTRIBUTION: Angola, Botswana, Burkina Faso, Central African Republic, Ghana, Ivory Coast, Kenya, Kenya, Mozambique, Namibia, Niger, Nigeria, Oman, Senegal, Sierra Leone, Somalia, South Africa, Sudan, Yemen, Zaire, Zambia, Zimbabwe.

SYNONYMS: *abuwudan* Fitzinger and Heuglin, 1866; *albescens* I. Geoffroy Saint-Hilaire 1839; *almodovari* Cabrerra, 1902; *dialeucos* Hollister, 1916; *ferox* Heller, 1913; *grandis* Thomas, 1890; *haagneri* Roberts, 1924; *ibeanus* Thomas, 1904; *leucurus* Hemprich and

Ehrenberg, 1833; *loandae* Thomas, 1904; *loempo* Temminck, 1853; *nigricauda* Pucheran, 1855.
COMMENTS: Reviewed by Taylor (1972, Mammalian Species, 12).

Liberiictis Hayman, 1958. Ann. Mag. Nat. Hist., ser. 13, 1:449.
TYPE SPECIES: *Liberiictis kuhni* Hayman, 1958, by original designation.

Liberiictis kuhni Hayman, 1958. Ann. Mag. Nat. Hist., ser. 13, 1:449.
TYPE LOCALITY: "Kpeaplay, north-east Liberia, about 6°36'N, 8°30'W".
DISTRIBUTION: Liberia and Ivory Coast (Taylor, in litt.).
STATUS: IUCN - Endangered.
COMMENTS: Reviewed by Goldman and Taylor (1990, Mammalian Species, 348).

Mungos E. Geoffroy Saint-Hilaire and F. G. Cuvier, 1795. Mag. Encyclop., 2:184, 187.
TYPE SPECIES: Not given; *Viverra mungo* Gmelin, 1788, designated by Muirhead (1819) (Melville and Smith, 1987).
COMMENTS: Allen (1919*b*) discussed the nomenclatural history of this name.

Mungos gambianus (Ogilby, 1835). Proc. Zool. Soc. Lond., 1835:102.
TYPE LOCALITY: "Gambia".
DISTRIBUTION: Gambia, Ghana, Ivory Coast, Niger, Nigeria, Senegal, Sierra Leone, Togo.

Mungos mungo (Gmelin, 1788). *In* Linnaeus, Syst. Nat., 13th ed., 1:84.
TYPE LOCALITY: "Bengala, Persia, aliisque asiae", restricted by Ogilby (1835:101) to "Gambia". However, Thomas (1882) believed it to be in the eastern part of South Africa, Cape Prov., as did Roberts (1929).
DISTRIBUTION: Angola, Botswana, Burundi, Cameroon, Central African Republic, Chad, Ethiopia, Guinea-Bissau, Kenya, Malawi, Mozambique, Namibia, Niger, Nigeria, Rwanda, Senegal, Somalia, South Africa, Sudan, Tanzania, Uganda, Zaire, Zambia, Zimbabwe.
SYNONYMS: *adailensis* Heuglin, 1861; *bororensis* Roberts, 1929; *colonus* Heller, 1911; *damarensis* Zukowsky, 1956; *fasciatus* Desmarest, 1823; *gothneh* Heuglin and Fitzinger, 1866; *grisonax* Thomas, 1926; *leucostethicus* Heuglin and Fitzinger, 1866; *macrosus* Lydekker, 1907; *macrurus* Thomas, 1907; *mandjarum* Schwarz, 1915; *ngamiensis* Roberts, 1932; *pallidipes* Roberts, 1929; *rossi* Roberts, 1929; *senescens* Thomas and Wroughton, 1907; *somalicus* Thomas, 1895; *taenianotus* Smith, 1834; *talboti* Thomas and Wroughton, 1908; *zebra* Rüppell, 1835; *zebroides* Lönnberg, 1908.

Paracynictis Pocock, 1916. Ann. Mag. Nat. Hist., ser. 8, 17:177.
TYPE SPECIES: *Cynictis selousi* de Winton, 1896, by original designation (Melville and Smith, 1987).

Paracynictis selousi (de Winton, 1896). Ann. Mag. Nat. Hist., ser. 6, 18:469.
TYPE LOCALITY: "found on a grassy heap under a tree, EssexVale, Matabeleland. . . near Bulawayo" [Zimbabwe].
DISTRIBUTION: Angola, Botswana, Malawi, Mozambique, Namibia, South Africa, Zambia, Zimbabwe.
SYNONYMS: *bechuanae* Roberts, 1932; *ngamiensis* Roberts, 1932; *sengaani* Roberts, 1931.

Rhynchogale Thomas, 1894. Proc. Zool. Soc. Lond., 1894:139.
TYPE SPECIES: *Rhinogale melleri* Gray, 1865, by monotypy through the replaced name *Rhinogale* Gray, 1865.

Rhynchogale melleri (Gray, 1865). Proc. Zool. Soc. Lond., 1864:575 [1865].)
TYPE LOCALITY: "from a ravine on the outskirts of the Otto Estate, near Mbweni, about 2½ miles west of Kilosa, Tanganyika Territory" [Tanzania] (pg. 79).
DISTRIBUTION: Malawi, Mozambique, South Africa, Tanzania, Zaire, Zambia, Zimbabwe.
SYNONYMS: *caniceps* Kershaw, 1924; *langi* Roberts, 1938.

Suricata Desmarest, 1804. Tabl. Méth. Hist. Nat., *in* Nouv. Dict. Hist. Nat., 24:15.
TYPE SPECIES: *Suricata capensis* Desmarest, 1804 (= *Viverra suricatta* Erxleben, 1777), by monotypy (Melville and Smith, 1987).

Suricata suricatta (Schreber, 1776). Die Säugethiere, pl. 117 [1776].
TYPE LOCALITY: Listed as "Cape of Good Hope" by Meester et al. (1986), restricted by Thomas and Schwann (1905:133) to "Deelfontein" [South Africa].
DISTRIBUTION: Angola, Namibia, South Africa, S Botswana.
SYNONYMS: *capensis* Desmarest, 1804; *hahni* Thomas, 1927; *hamiltoni* Thomas and Schwann, 1905; *lophurus* Thomas and Schwann, 1905; *marjoriae* Bradfield, 1936; *namaquensis* Thomas and Schwann, 1905; *surakatta* Smith, 1826; *tetradactyla* Pallas 1778; *typicus* Smith, 1834; *viverrina* Desmarest, 1819; *zenik* Scopoli, 1786.

Family Hyaenidae Gray, 1821. London Med. Repos., 15:302.
COMMENTS: Type genus: *Hyaena* Brünnich, 1771, by original designation. Reviewed by Ronnefeld (1969) and revised by Werdelin and Solounias (1991).

Subfamily Hyaeninae Gray, 1821. London Med. Repos., 15:302.
COMMENTS: Type genus: *Hyaena* Brünnich, 1771, by original designation. Mivart (1882) was first to restrict the subfamily to taxa considered here.

Crocuta Kaup, 1828. Oken's Isis. Encyclop. Zeit, 21(11), column 1145.
TYPE SPECIES: *Canis crocuta* Erxleben, 1777, by original designation.
COMMENTS: Antedated by *Crocuta* Meigen, 1800 (an insect), but this name has been suppressed (International Commission on Zoological Nomenclature, 1962).

Crocuta crocuta (Erxleben, 1777). Syst. Regni Anim., 1:578.
TYPE LOCALITY: "Guinea, Aethiopia, ad caput bonae spei in terrae rupiumque caueis", restricted by Cabrera (1911:95) to "Senegambia".
DISTRIBUTION: Angola, Botswana, Cameroon, Ethiopia, Gabon, Gambia, Guinea, Kenya, Malawi, Mauritania, Mozambique, Namibia, Nigeria, Senegal, Sierra Leone, Somalia, South Africa, Sudan, Tanzania, Togo, Uganda, Zaire, Zambia, Zimbabwe.
SYNONYMS: *capensis* Desmarest, 1817; *colvini* Lydekker, 1844; *croacuta* A. Smith, 1826; *cuvieri* Boitard, 1842; *encrita* Smith, 1827; *felina* Lydekker, 1844; *fisi* Heller, 1914; *fortis* Allen, 1924; *gariepensis* Matschie, 1900; *germinans* Matschie, 1900; *habessynica* Blainville, 1844; *kibonotensis* Lönnberg, 1905; *leontiewi* Satunin, 1905; *maculata* Thunberg, 1811; *noltei* Matschie, 1900; *nyasae* Cabrera, 1911; *nzoyae* Cabrera, 1911; *panganensis* Lönnberg, 1905; *rufa* Desmarest, 1817; *rufopicta* Cabrera, 1911; *sivalensis* Falconer and Cautley, 1868; *spelaea* Goldfuss, 1823; *thierryi* Matschie, 1900; *thomasi* Cabrera, 1911; *togoensis* Matschie, 1900; *ultima* Matsumoto, 1915; *venustula* Ewer, 1954; *weissmanni* Trouessart, 1904; *wissmanni* Matschie, 1900.
COMMENTS: Revised by Matthews (1939).

Hyaena Brünnich, 1771. Zool. Fundamenta, p. 34, 42, 43.
TYPE SPECIES: *Canis hyaena* Linnaeus, 1758, by original designation.
COMMENTS: Revised by Pocock (1934c). *Hyaena* Brisson, 1762, is unavailable (International Commission on Zoological Nomenclature, 1955). See comments under *Parahyaena* for exclusion of *brunnea* here.

Hyaena hyaena (Linnaeus, 1758). Syst. Nat. 10th ed., 1:40.
TYPE LOCALITY: "India", restricted by Thomas (1911*a*:134) to "Benna Mts., Laristan, S. Persia".
DISTRIBUTION: Afghanistan, Algeria, Egypt, Ethiopia, India, Iran, Iraq, Israel, Kenya, Libya, Mali, Morocco, Nepal, Nigeria, Pakistan, Saudia Arabia, Sierra Leone, Somalia, South Africa, Sudan, Tanzania, republics of the former USSR, Yemen.
STATUS: U.S. ESA and IUCN - Endangered as *H. h. barbara*.
SYNONYMS: *antiquorum* Temminck, 1820; *barbara* de Blainville, 1844; *bergeri* Matschie, 1910; *bilkiewiczi* Satunin, 1905; *bokcharensis* Satunin, 1905; *dubbah* Meyer, 1793; *dubia* Schinz, 1821; *fasciata* Thunberg, 1820; *hienomelas* Matschie, 1900; *hyaenomelas* Desmarest, 1820; *indica* de Blainville, 1844; *orientalis* Tiedemann, 1808; *rendilis* Lönnberg, 1912; *satunini* Matschie, 1910; *schillingsi* Matschie, 1900; *striata* Zimmerman, 1777; *suilla* Filippi, 1853; *sultana* Pocock, 1934; *syriaca* Matschie, 1900; *virgata* Ogilby, 1840; *vulgaris* Desmarest, 1820; *zarudnyi* Satunin, 1905.
COMMENTS: Reviewed by Rieger (1981, Mammalian Species, 150).

Parahyaena Hendey, 1974. Ann. S. Afr. Mus., 63:149.
TYPE SPECIES: *Hyaena brunnea* Thunberg, 1820.
COMMENTS: Some have considered *Hyaena* to include *brunnea*, however, this would be a paraphyletic group among extant (Werdelin and Solounias, 1991) and/or fossil taxa (Galiano and Frailey, 1977; Hendey, 1974).

Parahyaena brunnea (Thunberg, 1820). K. Svenska Vet.-Acad. Handl. Stockholm, p. 59.
TYPE LOCALITY: "Goda Hopps Udden; Södra Afrika" [South Africa, Cape Prov., Cape of Good Hope].
DISTRIBUTION: Botswana, Mozambique, Namibia, Orange Free State, South Africa, Zimbabwe.
STATUS: CITES - Appendix I; U.S. ESA - Endangered; IUCN - Vulnerable.
SYNONYMS: *fusca* Geoffroy Saint-Hilaire, 1825; *makapani* Toerien, 1952; *melampus* Pocock, 1935; *striata* Smith, 1826; *villosa* Smith, 1827.
COMMENTS: Reviewed by Mills (1982, Mammalian Species, 194, as *Hyaena brunnea*).

Subfamily Protelinae I. Geoffroy Saint-Hilaire, 1851. Cat. Méth. Coll. Mamm. Ois. (Mus. Hist. Nat. Paris):xiv.
COMMENTS: Type genus: *Proteles* I. Geoffroy Saint-Hilaire, 1824, by original designation (Melville and Smith, 1987). Ronnefeld (1969) and Meester et al. (1986) placed *Proteles* in a separate family.

Proteles I. Geoffroy Saint-Hilaire, 1824. Bull. Sci. Soc. Philom. Paris, 1824:139.
TYPE SPECIES: *Proteles lalandii* I. Geoffroy Saint-Hilaire, 1824 (= *Viverra cristata* Sparrman, 1783), by original designation (Melville and Smith, 1987).

Proteles cristatus (Sparrman, 1783). Resa Goda-Hopps-Udden. I., 1783:581.
TYPE LOCALITY: English translation (Sparrman, 1786) of original locality: "Agter-Bruntjes hoogte...which takes in the upper part of Kleine Visch-rivier, and is separated from Camdebo by Bruntjes hoogtens..."; listed in Allen (1939) as "Near Little Fish River, Somerset East, Cape Colony" [South Africa].
DISTRIBUTION: Angola, Botswana, Central African Republic, Egypt, Ethiopia, Kenya, Mozambique, Namibia, Somalia, South Africa, Sudan, Tanzania, Uganda, Zambia, Zimbabwe.
STATUS: CITES - Appendix III (Botswana).
SYNONYMS: *canescens* Shortridge and Carter, 1938; *harrisoni* Rothschild, 1902; *hyenoides* Desmarest, 1822; *lalandii* I. Geoffroy Saint-Hilaire, 1824; *pallidior* Cabrera, 1910; *septentrionalis* Rothschild, 1902; *termes* Heller, 1913; *transvaalensis* Roberts, 1932; *typicus* A. Smith, 1833.
COMMENTS: Reviewed by Koehler and Richardson (1990, Mammalian Species, 363).

Family Mustelidae G. Fischer, 1817. Mém. Soc. Imp. Nat. Moscow, 5:372.
COMMENTS: For review of subfamily relationships, see Muizon (1982*b*) and Wozencraft (1989*a*). Conservation status of family reviewed by Schreiber et al. (1989).

Subfamily Lutrinae Bonaparte, 1838. Nuovi Ann Sci. Nat., 2:111.
COMMENTS: Revised by Pocock (1941*a*), van Zyll de Jong (1972, 1987, 1991*a*), and Pohle (1920). Reviewed by Harris (1968) and I. I. Sokolov (1973). Foster-Turley et al. (1990) reviewed the conservation status and distribution of otters.

Amblonyx Rafinesque, 1832. Atlantic Journal and Friend of Knowledge, 1(2):62.
TYPE SPECIES: *Lutra concolor* Rafinesque, 1832 (= *Lutra cinerea* Illiger, 1815).
SYNONYMS: *Leptonyx* Lesson, 1842; *Micraonyx* J.A. Allen, 1919.
COMMENTS: Osgood (1932), Ellerman and Morrison-Scott (1951), Coetzee (1977*b*), and Meester et al. (1986) considered *Amblonyx* Rafinesque, 1832, as congeneric with *Aonyx*; here separated based on Harris (1968), van Zyll de Jong (1972, 1987), and Medway (1977).

Amblonyx cinereus (Illiger, 1815). Abh. Phys. Klasse K. Preuss. Akad. Wiss., 1804-1811:99 [1815].
TYPE LOCALITY: "Batavia" [Indonesia, Java, Jakarta].

DISTRIBUTION: India, Bangladesh, Burma, S China and Hainan Isl; Taiwan; Indochina to Indonesia (Sumatra, Java, Borneo); Philippines (Palawan Isl).
STATUS: CITES - Appendix II; IUCN - Insufficiently known.
SYNONYMS: *concolor* Rafinesque, 1832; *indigitatus* Hodgson, 1839; *leptonyx* Horsfield, 1823; *nirnai* Pocock, 1940; *sikimensis* Horsfield, 1855; *swinhoei* Gray, 1867.

Aonyx Lesson, 1827. Manual de Mammalogie, p. 157.
TYPE SPECIES: *Aonyx Delalandi* Lesson, 1827 (= *Lutra capensis* Schinz, 1821) by subsequent designation (Palmer, 1904).
SYNONYMS: *Anahyster* Murray, 1860; *Paraonyx* Hinton, 1921.
COMMENTS: Osgood (1932), Ellerman and Morrison-Scott (1951), and Coetzee (1977*b*) considered *Amblonyx* Rafinesque, 1832, as congeneric; here separated based on Harris (1968), Medway (1977), and van Zyll de Jong (1972, 1987). There is little question that *capensis* and *congicus* are sister species.

Aonyx capensis (Schinz, 1821). *In* G. Cuvier, Das Thierreich, 1:211.
TYPE LOCALITY: "Capischer otter. . . Afrika" [South Africa, Cape Province].
DISTRIBUTION: Angola, Benin, Botswana, Burkina Faso, Cameroon, Central African Republic, Chad, Ethiopia, Ghana, Guinea-Bissau, Ivory Coast, Kenya, Lesotho, Liberia, Malawi, Mozambique, Namibia, Niger, Nigeria, Rwanda, Senegal, Sierra Leone, South Africa, Sudan, Swaziland, Tanzania, Togo, Uganda, Zaire, Zambia, Zimbabwe.
STATUS: CITES - Appendix II.
SYNONYMS: *angolae* Thomas, 1908; *calaboricus* Murray, 1860; *coombsi* Roberts, 1926; *delalandi* Lesson, 1827; *gambianus* Gray, 1865; *helios* Heller, 1913; *hindei* Thomas, 1905; *inunguis* F. G. Cuvier, 1823; *lenoiri* Rochebrune, 1888; *meneleki* Thomas, 1902; *poensis* Waterhouse, 1838.

Aonyx congicus Lönnberg, 1910. Ark. Zool., 7(9):1.
TYPE LOCALITY: "Lower Congo." [Zaire]
DISTRIBUTION: Angola, Cameroon, Central African Republic, Gabon, Kenya, Nigeria, Rwanda, Uganda, Zaire.
STATUS: CITES - Appendix I; U.S. ESA - Endangered as *A.* (*Paraonyx*) *congica microdon*.
SYNONYMS: *microdon* Pohle, 1920; *philippsi* Hinton, 1921.
COMMENTS: Considered a distinct species by Pohle (1920) and van Zyll de Jong (1987). Perret and Aellen (1956) included *Paraonyx philippsi* Hinton, 1921, and *Aonyx microdon* Pohle, 1920; however, Harris (1968) listed these as distinct.

Enhydra Fleming, 1822. Philos. Zool., 2:187.
TYPE SPECIES: *Mustela lutris* Linnaeus, 1758, by monotypy (Melville and Smith, 1987).
SYNONYMS: *Enydris* Lichtenstein, 1827; *Latax* Gloger, 1827; *Sutra* Elliot, 1874.

Enhydra lutris (Linnaeus, 1758). Syst. Nat., 10th ed., 1:45.
TYPE LOCALITY: "Asia et America septentrionali," restricted by Thomas (1911*a*), to "Kamtchatka", then by Roest (1973), to "the east central coast of Kamchatka, opposite the Commander Islands." [Russia].
DISTRIBUTION: Russia (Sakhalin Isl, Kurile Isls, Commander Isls, Kamchatka), Canada, USA (Aleutian Isls, and S Alaska to California). Formerly in Mexico (Baja California) and Japan (coastal Hokkaido).
STATUS: CITES - Appendix I as *E. l. nereis*; otherwise Appedix II. U.S. ESA - Threatened as *E. l. nereis*.
SYNONYMS: *aterrima* Pallas, 1811; *gracilis* Bechstein, 1800; *kamtschatica* Dybowski, 1922; *kenyoni* Wilson, 1990; *marina* Steller, 1751; *nereis* Merriam, 1904; *orientalis* Oken, 1816; *stelleri* Lesson, 1827.
COMMENTS: Reviewed by Roest (1973), Davis and Lidicker (1975), Wilson et al. (1991), and Estes (1980, Mammalian Species, 133).

Lontra Gray, 1843. Ann. Mag. Nat. Hist., [ser. 1], 11:118.
TYPE SPECIES: *Lutra canadensis* Gray, 1843.
SYNONYMS: *Latax* Gray, 1843; *Lataxina* Gray, 1843; *Nutria* Gray, 1865.
COMMENTS: Van Zyll de Jong (1972, 1987, 1991*a*) argued that the New World otters represent a single radiation and questioned whether *Lutra* (*sensu stricto*) or *Aonyx*

was the closest sister group. There has been no published work to refute his hypothesis, although it has not received general acceptance. Hall (1981) chose not to question the monophyletic nature of the group, but to lower it to the subgeneric rank, feeling that the characters where not sufficient enough to warrant generic distinction. Regardless of the "morphological gap" between the monophyletic New World otters and the Old World otters, if *Lutra* (*sensu stricto*) is the closest sister group, then inclusion within *Lutra* could be maintained. However, if, as van Zyll de Jong (1987) suggested, *Aonyx* is the closest outgroup, then recognition at the generic level is necessary.

Lontra canadensis (Schreber, 1777). Die Säugethiere, 3(18):pl. 126.B[1776], text: 3(26):457, 588(index)[1777]) (First occurance of name on pg. 588)

TYPE LOCALITY: "Der Fischotter ...Europa überall gemein, ... Einwohner des nordlichen Theils von Asien, bis nach Kamatschatka hinaus, und ...Persian hinunter, und von Nordamerika." Miller (1912*b*:113) listed the type locality as "Eastern Canada".

DISTRIBUTION: Most of North America south to Arizona, Texas, and Florida.

STATUS: CITES - Appendix II.

SYNONYMS: *americana* Wyman, 1847; *atterima* Elliot, 1901; *brevipilosus* Grinnell, 1914; *californica* Baird, 1857; *chimo* Anderson, 1945; *degener* Bangs, 1898; *destructor* Barnston, 1863; *evexa* Goldman, 1935; *hudsonica* Merriam, 1899; *interior* Swenk, 1920; *kodiacensis* Goldman, 1935; *lataxina* F. G. Cuvier, 1823; *mira* Goldman, 1935; *mollis* Gray, 1843; *nexa* Goldman, 1935; *optiva* Goldman, 1935; *pacifica* Grinnell, 1933; *paranensis* Elliot, 1901; *periclyzomae* Elliot, 1905; *preblei* Goldman, 1935; *rhoadsi* Cope, 1897; *sonora* Rhoads, 1898; *texensis* Goldman, 1935; *vaga* Bangs, 1898; *vancouverensis* Goldman, 1935; *yukonensis* Goldman, 1935.

Lontra felina (Molina, 1782). Sagg. Stor. Nat. Chile, p. 284.

TYPE LOCALITY: "Chili" [Chile].

DISTRIBUTION: West coast of South America from N Peru to Straits of Magellan.

STATUS: CITES - Appendix I; U.S. ESA - Endangered; IUCN - Vulnerable.

SYNONYMS: *brachydactyla* Wagner, 1841; *californica* Gray, 1837; *chilensis* Bennet, 1832; *cinerea* Thomas, 1908; *peruensis* Pohle, 1920; *peruviensis* Gervais, 1841.

COMMENTS: Placed in *Lontra* by van Zyll de Jong (1972, 1987).

Lontra longicaudis (Olfers, 1818). *In* Eschwege, J. Brasilien, Neue Bibliothek Reisenb., 15(2):233.

TYPE LOCALITY: "Brasilien."

DISTRIBUTION: Mexico, Central America, W South America south to Peru, E South America south to Uruguay.

STATUS: CITES - Appendix I; U.S. ESA - Endangered; IUCN - Vulnerable as *L. l. longicaudis*.

SYNONYMS: *annectens* Major, 1897; *colombiana* J. A. Allen, 1904; *emerita* Thomas, 1908; *enudris* Cuvier, 1823; *fusco-rufa* Gray, 1865; *incarum* Thomas, 1908; *insularis* Cuvier, 1823; *latidens* J. A. Allen, 1908; *latifrons* Nehring, 1887; *lutris* Larrañaga, 1923; *mesopetes* Cabrera, 1924; *mitis* Thomas, 1908; *paraensis* Burmeister, 1861; *parilina* Thomas, 1914; *platensis* Waterhouse, 1838; *pratensis* Gerrard, 1862; *repanda* Goldman, 1914; *solitaria* Wagner, 1842.

COMMENTS: Van Zyll de Jong (1972) included *annectens*, *enudris*, *incarum*, *mesopetes*, and *platensis*; however, Pohle (1920), Cabrera (1958), and Harris (1968) recognized these taxa as distinct species. These taxa, often referred to as the *annectens*-group were chiefly distinguished by variation in the shape of the rhinarium; van Zyll de Jong's (1972) analysis suggested that these should be considered conspecific.

Lontra provocax (Thomas, 1908). Ann. Mag. Nat. Hist., ser. 8, 1:391.

TYPE LOCALITY: "south of Lake Nahuel Huapi, Patagonia." [Argentina]

DISTRIBUTION: Patagonia (C and S Chile, W Argentina).

STATUS: CITES - Appendix I; U.S. ESA - Endangered; IUCN - Vulnerable.

SYNONYMS: *huidobria* Gray, 1847.

COMMENTS: Placed in *Lontra* by van Zyll de Jong (1987).

Lutra Brünnich, 1771. Zool. Fundamenta, p. 34, 42.

TYPE SPECIES: *Mustela lutra* Linnaeus, 1758.

SYNONYMS: *Barangia* Gray, 1865; *Hydrictis* Pocock, 1921; *Lutronectes* Gray, 1867.

COMMENTS: Some have used *Lutra* Brisson, 1762, which was ruled unavailable (International Commission on Zoological Nomenclature, 1955). Pocock (1921*c*, 1941*a*) recognized *Lutrogale* Gray, 1865, and *Hydrictis* Pocock, 1921. Harris (1968) and van Zyll de Jong (1987; 1991*a*) considered *lutra*, *maculicollis*, and *sumatrana* to represent a single monophyletic group; furthermore, van Zyll de Jong's analysis supported separation of *perspicillata* from other *Lutra*, which is followed here. Van Zyll de Jong (1972) referred New World otters to *Lontra* (see comment therein), and although this has not been followed by some recent checklists (Corbet and Hill, 1991; Hall, 1981; Honacki, et al., 1982), van Zyll de Jong's hypothesis has not been refuted.

Lutra lutra (Linnaeus, 1758). Syst. Nat., 10th ed., 1:45.

TYPE LOCALITY: "Europæ aquis dulcibus, fluviis, flagnis, piscinis," subsequently restricted by Thomas (1911*a*) to "Upsala" [Sweden].

DISTRIBUTION: Eurasia (excl. tundra and desert): Afghanistan, Albania, Algeria, Austria, Bangladesh, Belgium, Bulgaria, China, Czechoslovakia, Denmark, England, Estonia, Finland, France, Germany, Greece, Hungary, India, Indonesia, Iraq, Ireland, Israel, Italy, Japan, Jordan, Korea, Laos, Lithuania, Malaysia, Mongolia, Morocco, Norway, Pakistan, Poland, Portugal, Scotland, Spain, Sri Lanka, Sweden, Switzerland, Taiwan, Tunisia, Turkey, republics of the former USSR, Vietnam, and Yugoslavia.

STATUS: CITES - Appendix I; IUCN - Vulnerable as *L. l. lutra*.

SYNONYMS: *amurensis* Dybowski, 1922; *angustifrons* Lataste, 1885; *aureventer* Blyth, 1863; *aurobrunneus* Hodgson, 1839; *baicalensis* Dybowski, 1922; *barang* F. G. Cuvier, 1823; *ceylonica* Pohle, 1920; *chinensis* Gray, 1837; *fluviatilis* Leach, 1816; *hanensis* Matschie, 1907; *indica* Gray, 1837; *intermedia* Pohle, 1920; *japonica* Nehring, 1887; *kamtschatica* Dybowski, 1922; *kutab* Schinz, 1844; *marinus* Billberg, 1827; *meridonalis* Ognev, 1931; *monticolus* Hodgson, 1839; *nair* F. G. Cuvier, 1823; *nepalensis* Gray, 1865; *nippon* Imaizumi and Yoshiyuki, 1989; *nudipes* Melchior, 1834; *oxiana* Birula, 1915; *piscatoria* Kerr, 1792; *roensis* Ogilby, 1834; *seistanica* Birula, 1912; *sinensis* Trouessart, 1897; *splendida* Cabrera, 1906; *stejnegeri* Goldman, 1936; *vulgaris* Erxleben, 1777; *whiteleyi* Gray, 1867.

COMMENTS: Imaizumi and Yoshiyuki (1989) considered Japanese otters a distinct species (*L. nippon*).

Lutra maculicollis Lichtenstein, 1835. Arch. Naturgesch., 1:89.

TYPE LOCALITY: "Kafferlandes am östlichen Abhange der Bambusberge." Listed in Honacki et al. (1982) as "Namibia, Orange River, Bambusbergen" [According to US Board of Geographic Names gazetteer, Bamboesberg Mts are in South Africa, at 31°30'S 26°20'E].

DISTRIBUTION: Subsaharan Africa, except east coast and SW African deserts, including: Angola, Benin, Botswana, Burkina Faso, Cameroon, Central African Republic, Chad, Ethiopia, Gabon, Ivory Coast, Kenya, Liberia, Malawi, Mozambique, Namibia, Nigeria, Rwanda, Sierra Leone, South Africa, Sudan, Swaziland, Tanzania, Togo, Uganda, Zaire, Zambia.

STATUS: CITES - Appendix II.

SYNONYMS: *chobiensis* Roberts, 1932; *grayii* Gerrard, 1862; *kivuana* Pohle, 1920; *matschiei* Cabrera, 1903; *mutandae* Hinton, 1921; *nilotica* Thomas, 1911.

COMMENTS: Pocock (1921*c*) and Cabrera (1929) placed *maculicollis* in the monotypic *Hydrictis*; however, Ansell (1978) and Harris (1968) considered *Hydrictis* a subgenus (see comments under *Lutra*). Van Zyll de Jong (1987) placed as sister groups *L. perspicillata* and (*L. lutra*+*L. sumatrana*). Pohle (1920) placed *maculicollis* in subgenus *Lutra*.

Lutra sumatrana (Gray, 1865). Proc. Zool. Soc. Lond., 1865:123.)

TYPE LOCALITY: "Sumatra (Raffles); Malacca (B.M.)," restricted by Pocock (1941*a*) to "Sumatra."

DISTRIBUTION: Indonesia (Sumatra, Borneo), Cambodia, Malaysia, Thailand, Vietnam.

STATUS: CITES - Appendix II; IUCN - Insufficiently known.

SYNONYMS: *lovii* Günther, 1877.

Lutrogale Gray, 1865. Proc. Zool. Soc. Lond., 1865:127.
TYPE SPECIES: Not designated.
COMMENTS: See comments under *Lutra*. Consideration here as a separate genus is consistent with Pohle (1920), Pocock (1941*a*), and Van Zyll de Jong (1987).

Lutrogale perspicillata (I. Geoffroy Saint-Hilaire, 1826). *In* Bory de Saint-Vincent, Dict. Class. Hist. Nat. Paris, 9:519.
TYPE LOCALITY: "Sumatra" [Indonesia].
DISTRIBUTION: Afghanistan, Bangladesh, China, India, Indonesia (Sumatra, Java, and Borneo), Iraq, Malaysia, Nepal, Pakistan, Thailand, Vietnam.
STATUS: CITES - Appendix II; IUCN - Insufficiently known.
SYNONYMS: *ellioti* Anderson, 1879; *macrodus* Gray, 1865; *simung* Lesson, 1827; *sindica* Pocock, 1940; *tarayensis* Hodgson, 1839.
COMMENTS: Pocock (1941*a*), Davis (1978), and van Zyll de Jong (1972) placed *perspicillata* in the monotypic *Lutrogale*, considered a subgenus by Pohle (1920); see comments under *Lutra*. Van Zyll de Jong's (1987) analysis placed as sister groups *L. maculicollis* and (*L. lutra*+*L. sumatrana*).

Pteronura Gray, 1837. Mag. Nat. Hist. [Charlesworth's], 1:580.
TYPE SPECIES: *Pteronura sambachii* Gray, 1837 (= *Mustela brasilienesis* Gmelin, 1788), by monotypy. (Melville and Smith, 1987).
SYNONYMS: *Pterura* Wiegmann, 1839; *Saricovia* Zimmermann, 1777.

Pteronura brasiliensis (Gmelin, 1788). *In* Linnaeus, Syst. Nat., 13th ed., 1:93.
TYPE LOCALITY: "in fluviis americae meridionalis"; Cabrera (1958:274) restricted to "rió São Francisco, en la orilla correspondiente al estado de Alagoas", Brazil.
DISTRIBUTION: Argentina, Bolivia, Brazil, Colombia, Ecuador, Guyana, Peru, Suriname, Venezuela.
STATUS: CITES - Appendix I; U.S. ESA - Endangered; IUCN - Vulnerable.
SYNONYMS: *lupina* Thomas, 1889; *nitens* Olfers, 1818; *paraguensis* Schinz, 1821; *paranensis* Rengger, 1830; *paroensis* Lesson, 1842; *sambachii* Schinz, 1844; *sanbachii* Wiegmann, 1838.
COMMENTS: See lengthly comments by Harris (1968) concerning the correct identity of the type, the confusion in published synonomies, and the type locality.

Subfamily Melinae Bonaparte, 1838. Nuovi Ann Sci. Nat., 2:111.
COMMENTS: Reviewed by Long (1978) and Long and Killingley (1983). Revised by Pocock (1921*b*, *d*, 1941*a*), Petter (1971), and Wozencraft (1989*a*), who suggested that this may not be a monophyletic group. Pocock recognized as subfamilies, and Petter as tribes, four groups: (1) *Meles*, *Arctonyx*; (2) *Melogale*; (3) *Mydaus*; and (4) *Taxidea*. Pocock (1920*b*), Long (1981), and Wozencraft (1989*b*) argued for the separation of *Taxidea* into a monotypic subfamily, Taxidiinae, which would be consistent with the phylogenetic hypotheses refered to above and which will be followed here. Petter (1971), Radinsky (1973), and Schmidt-Kittler (1981) suggested that *Mydaus* may be the sister group to Mephitinae. Its placement in this subfamily is provisional.

Arctonyx F. G. Cuvier, 1825. *In* E. Geoffroy Saint-Hilaire and F. G. Cuvier, Hist. Nat. Mammifères, pt. 3, 5(51), "Bali-saur", 2 pp.
TYPE SPECIES: *Arctonyx collaris* F. G. Cuvier, 1825.
SYNONYMS: *Trichomanis* Hubrecht, 1891.
COMMENTS: Revised by Pocock (1940*a*).

Arctonyx collaris F. G. Cuvier, 1825. *In* E. Geoffroy Saint-Hilaire and F. G. Cuvier, Hist. Nat. Mammifères, pt. 3, 5(51), "Bali-saur", 2 pp.
TYPE LOCALITY: "dans les montagnes qui séparent le Boutan de l'Indoustan."
DISTRIBUTION: Widespread in China; ranges from Assam (India) and Burma to Indochina, Thailand, Sumatra, and probably Perak in Malaya.
SYNONYMS: *albogularis* Blyth, 1853; *consul* Pocock, 1940; *dictator* Thomas, 1910; *isonyx* Horsfield, 1856; *leucolaemus* Milne-Edwards, 1867; *obscurus* Milne-Edwards, 1871; *taraiyensis* Hodgson, 1863; *taxoides* Blyth, 1853.

Meles Boddaert, 1785. Elench. Anim., 1:45.
TYPE SPECIES: [*Ursus*] *Meles* Linnaeus, 1758.

Meles meles (Linnaeus, 1758). Syst. Nat., 10th ed., 1:48.
TYPE LOCALITY: "Europa inter rimas rupium et lapidum," restricted by Thomas (1911*a*) to "Upsala" [Sweden].
DISTRIBUTION: Scandinavia to S Siberia, south to Israel; Iraq; China, Korea, and Japan; and on Ireland, Britain, Crete, and Rhodes.
SYNONYMS: *aberrans* Stroganov, 1962; *alba* Gmelin, 1788; *altaicus* Kastschenko, 1901; *amurensis* Schrenk, 1858; *anakuma* Temminck, 1844; *arcalus* Miller, 1907; *arenarius* Satunin, 1895; *blanfordi* Matschie, 1902; *britannicus* Satunin, 1906; *canescens* Blanford, 1875; *caninus* Billberg, 1827; *caucasicus* Ognev, 1926; *chinensis* Gray, 1868; *communis* Billberg, 1827; *danicus* Degerbøl, 1933; *europaeus* Desmarest, 1816; *hanensis* Matschie, 1907; *heptneri* Ognev, 1931; *leptorhynchus* Milne-Edwards, 1867; *leucurus* Hodgson, 1847; *maculata* Gmelin, 1788; *marianensis* Graells, 1897; *melanogenys* J. A. Allen and Andrews, 1913; *minor* Satunin, 1905; *raddei* Kastschenko, 1901; *rhodius* Festa, 1914; *schrenkii* Nehring, 1891; *severzovi* Heptner, 1940; *sibiricus* Kastschenko, 1900; *siningensis* Matschie, 1907; *talassicus* Ognev, 1931; *tauricus* Ognev, 1926; *taxus* Boddaert, 1785; *tianschanensis* Hayninger-Huene, 1910; *tsingtanensis* Matschie, 1907; *typicus* Barrett-Hamilton, 1899; *vulgaris* Tiedemann, 1808.
COMMENTS: Although Ognev (1931) considered *leptorhynchus* (including *amurensis*) a distinct species; most have considered *meles* and *leptorhynchus* conspecific (Allen, 1938; Ellerman and Morrison-Scott, 1951; Novikov, 1956; Stroganov, 1962; Long and Killingley, 1983). Baryshnikov and Potapova (1990) suggested that *M. meles* and *M. anakuma* are not conspecific.

Melogale I. Geoffroy Saint-Hilaire, 1831. *In* Bélanger (ed.), Voy. Indes Orient., 3(Zoologie):129, pl. 5.[issued 13 March 1831].
TYPE SPECIES: *Melogale personata* I. Geoffroy Saint-Hilaire, 1831.
SYNONYMS: *Helictis* Gray, 1831; *Nesictis* Thomas, 1922.
COMMENTS: There is uncertainty as to the the total number of species in this genus. Most recent authors tend to consider *everetii* and/or *orientalis* as conspecific with *personata*; however, Pocock (1941*a*), Everts (1968), Long (1978, 1981), and Long and Killingley (1983) supported the recognition of these populations as distinct. As Long and Killingley pointed out, there is no published information to refute Pocock's (1941*a*) revision.

Melogale everetti (Thomas, 1895). Ann. Mag. Nat. Hist., ser. 6, 15:331-332.
TYPE LOCALITY: "Mount Kina Balu, N. Borneo, about 4000 ft."
DISTRIBUTION: Borneo.
STATUS: IUCN - Insufficiently known.
COMMENTS: Medway (1977) included *everetti* in *orientalis*; however, see comments under genus.

Melogale moschata (Gray, 1831). Proc. Zool. Soc. Lond., 1831:94.
TYPE LOCALITY: "China," restricted by Allen (1929) to "Canton, Kwangtung Province, South China, where the original specimen was secured by John Reeves."
DISTRIBUTION: India (Naga Hills near Manipur, Assam), to Taiwan; Hainan Isl, C and SE China, N Laos, and N Vietnam.
SYNONYMS: *ferreo-griseus* Hilzheimer, 1905; *millsi* Thomas, 1922; *modesta* Thomas, 1922; *sorella* G. M. Allen, 1929; *subaurantiaca* Swinhoe, 1862; *taxilla* Thomas, 1925.

Melogale orientalis (Horsfield, 1821). Zool. Res. Java., part 2:*Gulo orientalis*, pl. and 4 unno. pp.
TYPE LOCALITY: "limited...to...south of Mountain Prahu, between the two prinicpal cones of the central part of Java, the Mountain Sumbing, and...Teggal,...Baggulen and Banyumas...to Gowong in the east." [Indonesia, Java].
DISTRIBUTION: Java (Indonesia).
STATUS: IUCN - Insufficiently known.
SYNONYMS: *fusca* Gúerin, 1835; *maccourus* Temminck, 1824.

Melogale personata I. Geoffroy Saint-Hilaire, 1831. *In* Bélanger (ed.) Voy. Indes Orient., 3(Zoologie):137, pl. 5.[issued 13 March 1831].
TYPE LOCALITY: "environs de Rangoun" [Burma].
DISTRIBUTION: Nepal through India (Assam) and most of Indochina, including peninsular Thailand and Malaysia.
SYNONYMS: *laotum* Thomas, 1922; *nipalensis* Hodgson, 1836; *orientalis* Blanford, 1888; *pierrei* Bonhote, 1903; *tonquinia* Thomas, 1922.

Mydaus F. G. Cuvier, 1821. *In* E. Geoffroy Saint-Hilaire and F. G. Cuvier, Hist. Nat. Mammifères, pt. 2, 3(27), "Telagon", 2 pp.
TYPE SPECIES: *Mydaus meliceps* F. G. Cuvier, 1821 (= *Mephitis javanensis* Desmarest, 1820).
SYNONYMS: *Suillotaxus* Lawrence, 1939.
COMMENTS: Lawrence (1939) believed that differences in dentition and pelage warranted separation of these taxa into separate, monotypic genera (i.e., *Suillotaxus marchei, Mydaus javanensis*). These differences parallel those found within the genus *Melogale* (Long and Killingley, 1983). This genus is provisionally placed in Melinae (see comments therein), its position may be better considered as *incertae sedis*.

Mydaus javanensis (Desmarest, 1820). Mammalogie, *in* Encycl. Meth., I:187
TYPE LOCALITY: "l'île de Java." [Indonesia, Java].
DISTRIBUTION: Indonesia (Java, Borneo, Sumatra and the Natuna Isls) and Malaysia (Borneo).
SYNONYMS: *lucifer* Thomas, 1902; *luciferoides* Lönnberg and Mjöberg, 1925; *meliceps* Cuvier, 1821.

Mydaus marchei (Huet, 1887). Le Naturaliste, ser. 2, 9(13):149-151.
TYPE LOCALITY: "l'ile Palaouan" [Philippine Islands, Palawan].
DISTRIBUTION: Philippine Isls (Palawan and Calamian Isls).
SYNONYMS: *schadenbergii* Jentink, 1895.
COMMENTS: Referred to the genus *Suillotaxus* by Lawrence (1939). *Suillotaxus* was considered a subgenus of *Mydaus* by Long (1978, 1981).

Subfamily Mellivorinae Gray, 1865. Proc. Zool. Soc. Lond., 1865:103.
COMMENTS: Type genus *Mellivora* Storr, 1780. Wozencraft (1989*b*) placed *Mellivora* in the Mustelinae.

Mellivora Storr, 1780. Prodr. Meth. Mamm., p. 34, Tabl. A.
TYPE SPECIES: *Viverra Ratel* Sparrmann, 1778 (= *Viverra capensis* Schreber, 1776), by designation (Sclater, 1900); (Melville and Smith, 1987).
SYNONYMS: *Lipotus* Sundevall, 1843; *Melitoryx* Gloger, 1841; *Ratellus* Gray, 1827; *Ratelus* Bennett, 1830; *Ursitaxus* Hodgson, 1835.

Mellivora capensis (Schreber, 1776). Die Säugethiere, 3(18):pl. 125[1776]; text, 3(26):450[1777].
TYPE LOCALITY: "Vorgebirge der guten Hofnung" [South Africa, Cape Prov., Cape of Good Hope].
DISTRIBUTION: Savanna and steppe from Nepal, E India and Turkmenistan west to Lebanon, south of the Mediterranean to South Africa.
STATUS: CITES - Appendix III (Ghana and Botswana).
SYNONYMS: *abyssinica* Hollister, 1910; *brockmani* Wroughton and Cheesman, 1920; *buchanani* Thomas, 1925; *concisa* Thomas and Wroughton, 1907; *cottoni* Lydekker, 1906; *inauritus* Hodgson, 1836; *indicus* Kerr, 1792; *leuconota* Sclater, 1867; *maxwelli* Thomas, 1923; *mellivorus* Cuvier, 1798; *pumilio* Pocock, 1946; *ratel* Sparrmann, 1778; *ratelus* Fraser, 1862; *sagulata* Hollister, 1910; *signata* Pocock, 1909; *typicus* Smith, 1833; *vernayi* Roberts, 1932; *wilsoni* Cheesman, 1920.

Subfamily Mephitinae Bonaparte, 1845. Cat. Meth. Mamm. Europe, p. 1.

Conepatus Gray, 1837. Mag. Nat. Hist. [Charlesworth's], 1:581.
TYPE SPECIES: *Conepàtus humbóldtii* Gray, 1837, by monotypy (Melville and Smith, 1987).

SYNONYMS: *Marputius* Gray, 1837; *Oryctogale* Merriam, 1902; *Ozolictis* Gloger; *Thiosmus* Lichtenstein, 1838.
COMMENTS: Revised by Kipp (1965), who studied an extensive series of southern South American specimens and could not recognize distinctive groups among them based on skull morphology, and only two groups based on pelage coloration.

Conepatus chinga (Molina, 1782). Sagg. Stor. Nat. Chile, p. 288.
TYPE LOCALITY: "Chili," restricted by Cabrera (1958) to "alrededores de Valparaíso." [Chile].
DISTRIBUTION: Chile, Peru, N Argentina, Bolivia, Uruguay, S Brazil.
SYNONYMS: *ajax* Thomas, 1913; *arequipae* Thomas, 1898; *budini* Thomas, 1919; *calurus* Thomas, 1919; *chilensis* Link, 1795; *chinghe* Bechstein, 1800; *chorensis* Thomas, 1902; *dimidiata* Fischer, 1814; *enuchus* Thomas, 1927; *feuillei* Gervais, 1841; *furcata* Wagner, 1841; *gibsoni* Thomas, 1910; *hunti* Thomas, 1903; *inca* Thomas, 1900; *mapurito* Tschudi, 1844; *mendosus* Yepes, 1939; *monzoni* Aplin, 1894; *pampanus* Thomas, 1921; *porcinus* Thomas, 1902; *rex* Thomas, 1898; *suffocans* Burmeister, 1879; *vittata* Larrañaga, 1923.
COMMENTS: Kipp (1965) considered *rex* as conspecific; however, it was listed as separate by Osgood (1943) and Cabrera (1958).

Conepatus humboldtii Gray, 1837. Mag. Nat. Hist. [Charlesworth's], 1:581.
TYPE LOCALITY: "Magellan Straits." [Chile].
DISTRIBUTION: NE Argentina and Paraguay south to the Straits of Magellan.
STATUS: CITES - Appendix II.
SYNONYMS: *castaneus* d'Orbigny and Gervais, 1847; *gaucho* Thomas, 1927; *patachonica* Burmeister, 1859; *patagonica* Lichtenstein, 1836; *proteus* Thomas, 1902.
COMMENTS: Kipp (1965) considered *castaneus* as conspecific; Cabrera (1958) considered it separate.

Conepatus leuconotus (Lichtenstein, 1832). Darst. Säugeth., text: "*Mephitis leuconota*" [not paginated], pl. 44. fig. 1.
TYPE LOCALITY: "oberen Lauf des Rio Alvarado" [Mexico, Veracruz, Rio Alvarado].
DISTRIBUTION: S Gulf coast of Texas (USA), south along coast to Vera Cruz (Mexico).
SYNONYMS: *texensis* Merriam, 1902.
COMMENTS: May be only subspecifically distinct from *mesoleucus* (Hall, 1981).

Conepatus mesoleucus (Lichtenstein, 1832) Darst. Säugeth., text: "*Mephitis mesoleuca*" [not paginated] pl. 44, fig. 2.
TYPE LOCALITY: "Gegend von Chico in Mexico" [Mexico, Hidalgo, near El Chico].
DISTRIBUTION: Arizona, Colorado, Texas (USA), south to Nicaragua.
STATUS: IUCN - Indeterminate as *C. m. telmalestes*.
SYNONYMS: *figginsi* Miller, 1925; *filipensis* Merriam, 1902; *fremonti* Miller, 1933; *mearnsi* Merriam, 1902; *nelsoni* Goldman, 1922; *nicaraguae* J. A. Allen, 1910; *pediculus* Merriam, 1902; *sonoriensis* Merriam, 1902; *telmalestes* Bailey, 1905; *venaticus* Goldman, 1922.
COMMENTS: May be conspecific with *leuconotus* (Hall, 1981).

Conepatus semistriatus (Boddaert, 1785). Elench. Anim., 1:84.
TYPE LOCALITY: "Mexico"; Cabrera (1958) listed the type locality as "Minas de Montuosa, cerca de Pamplona, departamento del norte de Santander, Colombia".
DISTRIBUTION: Veracruz, Tabasco, and Yucatan (Mexico) to Peru and E Brazil.
SYNONYMS: *amazonica* Lichtenstein, 1838; *bahiensis* Ihering, 1911; *chilensis* Gray, 1865; *conepatl* Gmelin, 1788; *gumillae* Lichtenstein, 1836; *mapurito* Gmelin, 1788; *putorius* Mutis, 1770; *quitensis* Humboldt, 1812; *taxinus* Thomas, 1924; *trichurus* Thomas, 1905; *tropicalis* Merriam, 1902; *westermanni* Reinhardt, 1856; *yucatanicus* Goldman, 1943; *zorilla* Fischer, 1829; *zorrino* Thomas, 1901.

Mephitis E. Geoffroy Saint-Hilaire and G. [Baron] Cuvier, 1795. Mag. Encyclop., 2:187.
TYPE SPECIES: *Viverra mephitis* Schreber, 1776.
SYNONYMS: *Chincha* Lesson, 1842; *Leucomitra* Howell, 1901.

Mephitis macroura Lichtenstein, 1832. Darst. Säugeth., text: "*Mephitis macroura*," [not paginated], pl. 46.

TYPE LOCALITY: "Gebirgs-Gegenden nordwestlich von der Stadt Mexico." [Mexico, mountains NW of Mexico City].

DISTRIBUTION: S Arizona, S New Mexico, and W Texas (USA), through Mexico to Costa Rica.

SYNONYMS: *concolor* Gray, 1865; *edulis* Coues, 1877; *eximius* Hall and Dalquest, 1950; *mexicana* Gray, 1837; *milleri* Mearns, 1897; *richardsoni* Goodwin, 1957; *vittata* Lichtenstein, 1832.

Mephitis mephitis (Schreber, 1776). Die Säugethiere, 3(17):pl. 121[1776], text, 3(26):444, 588 (index)[1777].

TYPE LOCALITY: "Amerika".

DISTRIBUTION: SW Northwest Territories to Hudson Bay and S Quebec (Canada), south to Florida (USA), N Tamaulipas, N Durango, and N Baja California (Mexico).

SYNONYMS: *americana* Desmarest, 1818; *avia* Bangs, 1898; *bivirgata* H[amilton]-Smith, 1842; *dentata* Brass, 1911; *elongata* Bangs, 1895; *estor* Merriam, 1890; *fetidissima* Boitard, 1842; *foetulenta* Elliot, 1899; *frontata* Coues, 1875; *holzneri* Mearns, 1897; *hudsonica* Richardson, 1829; *major* Howell, 1901; *mesomelas* Lichtenstein, 1832; *minnesotoe* Brass, 1911; *newtonensis* Brown, 1908; *nigra* Peale and Palisot de Beauvois, 1796; *notata* Howell, 1901; *occidentalis* Baird, 1858; *olida* Boitard, 1842; *platyrhina* Howell, 1901; *putida* Boitard, 1842; *scrutator* Bangs, 1896; *spissigrada* Bangs, 1898; *varians* Gray, 1837.

COMMENTS: Reviewed by Wade-Smith and Verts (1982, Mammalian Species, 173).

Spilogale Gray, 1865. Proc. Zool. Soc. Lond., 1865:150.

TYPE SPECIES: *Mephitis interrupta* Rafinesque, 1820 (= [*Viverra*] *putorius* Linnaeus, 1758) by monotypy (Melville and Smith, 1987).

COMMENTS: Revised by Van Gelder (1959).

Spilogale putorius (Linnaeus, 1758). Syst. Nat., 10th ed., 1:44.

TYPE LOCALITY: "America septentrionali" restricted by Thomas (1911*a*), to "South Carolina." [USA].

DISTRIBUTION: SW Canada east to Minnesota and S Pennsylvania (USA); south to Costa Rica.

SYNONYMS: *ambigua* Mearns, 1897; *angustifrons* Howell, 1902; *bicolor* Gray, 1837; *gracilis* Merriam, 1890; *interrupta* Rafinesque, 1820; *larvatus* Hodgson, 1849; *leucoparia* Merriam, 1890; *lucasona* Merriam, 1890; *microdon* Howell, 1906; *mupurita* Müller, 1776; *olympica* Elliot, 1899; *phenax* Merriam, 1890; *putida* Cuvier, 1798; *quaterlinearis* Winans, 1859; *ringens* Merriams, 1890; *saxatalis* Merriam, 1890; *striata* Shaw, 1800; *tenuis* Howell, 1902; *texensis* Merriam, 1890; *tibetanus* Horsfield, 1851; *zorilla* Schreber, 1776.

COMMENTS: Mead (1968) argued that *S. p. gracilis* and "possibly" *leucoparia* are reproductively isolated from eastern populations and therefore should be considered distinct species; both were included by Van Gelder (1959).

Spilogale pygmaea Thomas, 1898. Proc. Zool. Soc. Lond., 1897:898 [1898].

TYPE LOCALITY: "Rosario, Sinaloa, W. Mexico."

DISTRIBUTION: Sinaloa to Oaxaca (Mexico).

COMMENTS: Ewer (1973) argued that *pygmaea* is conspecific with *putorius*.

Subfamily Mustelinae G. Fischer, 1817. Mém. Soc. Imp. Nat. Moscow, 5:372.

Eira H[amilton]. Smith, 1842. Jardine's Natur. Libr., 35:201.

TYPE SPECIES: *Mustela barbara* Linnaeus, 1758.

SYNONYMS: *Galera* Gray, 1843 (preoccupied by *Galera* Browne, 1789); *Gulo* Lichtenstein, 1825; *Mustela* Schinz, 1821; *Tayra* Oken, 1816 (invalid, see International Commission on Zoological Nomenclature, 1956*b*); *Viverra* Traill, 1821.

Eira barbara (Linnaeus, 1758). Syst. Nat., 10th ed., 1:46.

TYPE LOCALITY: "Brasilia," restricted by Lönnberg (1913) to "Pernambuco."

DISTRIBUTION: Sinaloa and Tamaulipas (Mexico), south to Argentina, including: Brazil, Colombia, Costa Rica, Ecuador, Guatemala, Guyana, Honduras, Mexico, Nicaragua, Panama, Peru, Venezuela, Argentina, Belize, Bolivia, Surinam, and Trinidad.
STATUS: CITES - Appendix III (Honduras).
SYNONYMS: *bimaculata* Martinez, 1873; *biologiae* Thomas, 1900; *brunnea* Thomas, 1907; *canescens* Lichtenstein, 1825; *gulina* Schinz, 1821; *ilya* H[amilton]. Smith, 1842; *irara* J. A. Allen, 1904; *kriegi* Krumbiegal, 1942; *leira* F. Cuvier, 1849; *madeirensis* Lönnberg, 1913; *peruana* Nehring, 1886; *poliocephalus* Traill, 1821; *senex* Thomas, 1900; *senilis* J.A. Allen, 1913; *sinuensis* Humboldt, 1812; *tucumana* Lönnberg, 1913.
COMMENTS: Reviewed by Thomas (1900) and Lönnberg (1913).

Galictis Bell, 1826. Zool. J., 2:552.
TYPE SPECIES: *Viverra vittata* Schreber, 1776, by original designation.
SYNONYMS: *Grison* Oken, 1816 (invalid, see International Commission on Zoological Nomenclature, 1956*b*); *Grisonella* Thomas, 1912; *Grisonia* Gray, 1865; *Gulo* Desmarest, 1820; *Mustela* Bechstein, 1800.

Galictis cuja (Molina, 1782). Sagg. Stor. Nat. Chile, p. 291.
TYPE LOCALITY: "Chili" restricted by Thomas (1912) to "S. Chili (Temuco)"; Cabrera (1958) restricted the locality to "alrededores de Santiago" [Chile]. Honacki et al. (1982) listed the type locality for *G. c. furax* Thomas, 1912.
DISTRIBUTION: Argentina, Bolivia, Peru, Brazil, Chile, Paraguay.
SYNONYMS: *albifrons* Larrañaga, 1923; *barbara* Hudson, 1903; *brasiliensis* D'Orbigny, 1838; *chilensis* Ihering, 1886; *furax* Thomas, 1907; *huronax* Thomas, 1921; *luteolus* Thomas, 1907; *melinus* Thomas, 1912; *quiqui* Molina, 1782; *ratellina* Thomas, 1921; *shiptoni* Thomas, 1926.

Galictis vittata (Schreber, 1776). Säugethiere, 3(18):pl. 124[1776], text, 3(26):418, 447[1777].
TYPE LOCALITY: "Surinam".
DISTRIBUTION: San Luis Potosi and Veracruz (Mexico) to Peru and S Brazil, at lower elevations, including: Argentina, Bolivia, Colombia, Costa Rica, Guatemala, Guyana, Panama, and Venezuela.
STATUS: CITES - Appendix III (Costa Rica).
SYNONYMS: *allamandi* Bell, 1841; *andina* Thomas, 1903; *brasiliensis* Thunberg, 1820; *canaster* Nelson, 1901; *crassidens* Nehring, 1885; *gujanensis* Bechstein, 1800; *intermedia* Lund, 1845.
COMMENTS: Krumbiegel (1942) included *allamandi*.

Gulo Pallas, 1780. Spicil. Zool., 14:25.
TYPE SPECIES: *Gulo sibiricus* Pallas, 1780 (= [*Mustela*] *gulo* Linnaeus, 1758), by absolute tautonymy (Melville and Smith, 1987).
COMMENTS: Corbet (1978*c*) attributed *Gulo* to Storr, 1780. *Gulo* Frisch, 1775, is invalid (International Commission on Zoological Nomenclature, 1954*b*).

Gulo gulo (Linnaeus, 1758). Syst. Nat., 10th ed., 1:45.
TYPE LOCALITY: "alpibus Lapponiæ, Ruffiae, Sibiriae, sylvis vastissimis", restricted by Thomas (1911*a*) to "Lapland".
DISTRIBUTION: Holarctic taiga and S tundra, south to 50°N (Eurasia) and 37°N (North America).
STATUS: IUCN - Vulnerable.
SYNONYMS: *arcticus* Desmarest, 1820; *arctos* Kaup, 1829; *auduboni* Matschie, 1918; *bairdi* Matschie, 1918; *biedermanni* Matschie, 1918; *borealis* Nilsson, 1820; *hylaeus* Elliot, 1905; *kamtschaticus* Dybowsky, 1922; *katschenakensis* Matschie, 1918; *luscus* Linnaeus, 1758; *luteus* Elliot, 1904; *niediecki* Matschie, 1918; *sibirica* Pallas, 1780; *vancouverensis* Goldman, 1935; *vulgaris* Oken, 1816; *wachei* Matschie, 1918.
COMMENTS: Degerbøl (1935) and Kurtén and Rausch (1959) demonstrated that *gulo* and *luscus* are conspecific.

Ictonyx Kaup, 1835. Das Thierreich in Seinen Hauptformen, 1:352.
TYPE SPECIES: *Ictonyx capensis* Kaup, 1835 (= *Bradypus striatus* Perry, 1810) (Melville and Smith, 1987).

SYNONYMS: *Ictidonyx* Agassiz, 1846; *Ictomys* Roberts, 1936; *Poecilictis* Thomas and Hinton, 1920; *Rhabdogale* Wiegmann, 1838; *Zorilla* I. Geoffroy Saint-Hilaire, 1826.

COMMENTS: There is considerable controversy over the correct name for this genus (Herschkovitz, 1955*b*; Van Gelder, 1966). *Zorilla* Oken, 1816, is invalid (International Commission on Zoological Nomenclature, 1956*b*). *Zorilla* Geoffroy Saint-Hilaire, 1826, was suppressed under the plenary powers for the purposes of the Principle of Priority (International Commission on Zoological Nomenclature, 1967). Rosevear (1974) strongly suggested, and Dekeyser (1955) and Niethammer (1987*a*) argued that *Poecilictis* Thomas and Hinton, 1920, and *Ictonyx* are congeneric. The principal skull features used to erect the new genus by Thomas and Hinton (1920) could not be supported when a more extensive series of specimens was measured (Rosevear, 1974).

Ictonyx libyca (Hemprich and Ehrenberg, 1833). Symb. Phys. Mamm., vol. 1, pt. 2, sig. K, verso.

TYPE LOCALITY: "Libyae" [Libya].

DISTRIBUTION: Fringes of the Sahara Desert from Morocco and Egypt on the north, and from Mauretania and N Nigeria to Sudan on the south.

SYNONYMS: *frenata* Sundevall, 1843; *multivittata* Wagner, 1841; *oralis* Thomas and Hinton, 1920; *rothschildi* Thomas and Hinton, 1920; *vaillantii* Loche, 1856.

Ictonyx striatus (Perry, 1810). Arcana, Mus. Nat. Hist., Signature Y, Fig. [41][1810].

TYPE LOCALITY: "South America". This is clearly in error and Hollister (1915) fixed the type locality as "Cape of Good Hope". [South Africa].

DISTRIBUTION: Senegal to SE Egypt and Ethiopia, south to South Africa, including: Botswana, Kenya, Malawi, Namibia, Nigeria, South Africa, Sudan, Tanzania, Uganda, and Zimbabwe.

SYNONYMS: *africana* Lichtenstein, 1838; *albescens* Heller, 1913; *arenarius* Roberts, 1924; *capensis* A. Smith, 1826; *elgonis* Granvik, 1924; *erythraea* de Winton, 1898; *ghansiensis* Roberts, 1932; *giganteus* Roberts, 1932; *intermedia* Anderson and de Winton, 1902; *kalaharicus* Roberts, 1932; *lancasteri* Roberts, 1932; *limpopoensis* Roberts, 1917; *maximus* Roberts, 1924; *mustelina* Wagner, 1841; *nigricaudus* Roberts, 1932; *orangiae* Roberts, 1924; *ovamboensis* Roberts, 1951; *pondoensis* Roberts, 1924; *pretoriae* Roberts, 1924; *senegalensis* Fischer, 1829; *shoae* Thomas, 1906; *shortridgei* Roberts, 1932; *sudanicus* Thomas and Hinton, 1923; *variegata* Lesson, 1842; *zorilla* Smuts, 1832.

Lyncodon Gervais, 1845. *In* d'Orbigny, Dict. Univ. Hist. Nat., 4:685.

TYPE SPECIES: *Mustela patagonica* Blainville, 1842.

Lyncodon patagonicus (Blainville, 1842). Osteogr. Mamm., pt. 10(Viverra):1.

TYPE LOCALITY: Listed in Cabrera (1958) as "cercanías del río Negro." [Argentina].

DISTRIBUTION: Argentina and S Chile.

SYNONYMS: *anticola* Burmeister, 1869; *lujanensis* Ameghino, 1889; *quiqui* Burmeister, 1861.

Martes Pinel, 1792. Actes Soc. Hist. Nat. Paris, 1:55.

TYPE SPECIES: *Martes domestica* Pinel, 1792 (= *Mustela foina* Erxleben, 1777).

SYNONYMS: *Charronia* Gray, 1865; *Lamprogale* Ognev, 1928; *Mustela* Blasius, 1857; *Pekania* Gray, 1865; *Zibellina* Kaup, 1829.

Martes americana (Turton, 1806). *In* Linnaeus, Gen. Syst. Nat., 1:60.

TYPE LOCALITY: "North America".

DISTRIBUTION: Alaska and Canada south to N California, south in the Sierra Nevada and Rocky Mtns to 35°N.

SYNONYMS: *abieticola* Preble, 1902; *abietinoides* Gray, 1865; *actuosa* Osgood, 1900; *atrata* Bangs, 1897; *boria* Elliot, 1905; *brumalis* Bangs, 1898; *caurina* Merriam, 1890; *humboldtensis* Grinnell and Dixon, 1926; *huro* F. Cuvier, 1823; *kenaiensis* Elliot, 1903; *leucopus* Kuhl, 1820; *martinus* Ames, 1874; *nesophila* Osgood, 1901; *origenes* Rhoads, 1902; *sierrae* Grinnell and Storer, 1916; *vancouverensis* Grinnell and Dixon, 1926; *vulpina* Rafinesque, 1819.

COMMENTS: May be conspecific with *martes*, *melampus*, and *zibellina* (Anderson, 1970; Hagmeier, 1961). See Clark et al. (1987, Mammalian Species, 289).

Martes flavigula (Boddaert, 1785). Elench. Anim., 1:88.
TYPE LOCALITY: Not given; fixed by Pocock (1941*a*) as "Nepal".
DISTRIBUTION: Primorski Krai (Russia); Korea; China, west along Himalayan foothills to NW Pakistan; isolates in S India, SE Asia, Taiwan, Sumatra, Java, and Borneo.
STATUS: CITES - Appendix III (India); U.S. ESA - Endangered as *M. f. chrysospila*; IUCN - Indeterminate as *M. f. chrysospila*.
SYNONYMS: *aterrima* Pallas, 1811; *chrysogaster* Hamilton-Smith, 1842; *chrysospila* Swinhoe, 1866; *guadricolor* Shaw, 1800; *hardwickei* Horsfield, 1828; *henricii* Westermann, 1851; *indochinensis* Kloss, 1916; *lasiotis* Gray, 1850; *leucocephalus* Gray, 1865; *leucotis* Bechstein, 1800; *melina* Kerr, 1792; *melli* Matschie, 1922; *peninsularis* Bonhote, 1901; *quadricolor* Shaw, 1800; *typica* Bonhote, 1901; *yuenshanensis* Shih, 1930.

Martes foina (Erxleben, 1777). Syst. Regni Anim., 1:458.
TYPE LOCALITY: "Europa inque Persia", listed by Miller (1912*b*) as "Germany."
DISTRIBUTION: From Spain to S and C Europe, through Caucasus Mtns, to the Altai (Russia, Kazakhstan), Mongolia, and Himalayas; adjacent China; islands of Corfu, Crete and Rhodes.
STATUS: CITES - Appendix III (India) as *M. f. intermedia*.
SYNONYMS: *domestica* Pinel, 1792; *intermedia* Severtzov, 1873; *leucolachnaea* Blanford, 1879; *mediterranea* Barrett-Hamilton, 1898; *rosanowi* Martino, 1917; *toufoeus* Wroughton, 1919.

Martes gwatkinsii Horsfield, 1851. Cat. Mamm. Mus. E. India Co. p. 90.
TYPE LOCALITY: "Madras" [India].
DISTRIBUTION: S India.
STATUS: CITES - Appendix III (India); IUCN - Indeterminate.
COMMENTS: Included in *Martes flavigula* by Honacki et al. (1982) and Corbet (1978*c*); however, separated by Bonhote (1901*b*), Pocock (1936*a*, 1941*a*), Ellerman and Morrison-Scott (1951), and Anderson (1970).

Martes martes (Linnaeus, 1758). Syst. Nat., 10th ed., 1:46.
TYPE LOCALITY: "sylvis antiquis", restricted by Thomas (1911*a*) to "Upsala" [Sweden].
DISTRIBUTION: Britain and Ireland; N and W Europe to W Siberia, south to Sicily, Sardinia, Corsica, the Elburz Mtns (Iran), and the Caucasus Mtns.
SYNONYMS: *abietum* Gray, 1865; *henricii* Westermann, 1851; *sylvatica* Nilsson, 1820; *sylvestris* Oken, 1816; *vulgaris* Griffith, 1827.
COMMENTS: May be conspecific with *americana*, *melampus*, and *zibellina* (Anderson, 1970; Hagmeier, 1961).

Martes melampus (Wagner, 1841). *In* Schreber, Die Säugethiere., Suppl., 2:229.
TYPE LOCALITY: "Japan".
DISTRIBUTION: Japan (Honshu, Kyushu, Shikoku, Tsushima, introduced on Sado Isl); Korea.
STATUS: IUCN - Indeterminate as *M. m. tsuensis*.
SYNONYMS: *japonica* Gray, 1865; *melanopus* Gray, 1865.
COMMENTS: May be conspecific with *americana*, *martes* and *zibellina* (Anderson, 1970; Hagmeier, 1961). Heptner and Naumov (1967) included Japanese and Korean *melampus* in *zibellina*.

Martes pennanti (Erxleben, 1777). Syst. Regni Anim., 1:470.
TYPE LOCALITY: "in America boreali, vulgaris victitans quadrupedibus minoribus." Listed by Miller and Rehm (1901) as "Eastern Canada."
DISTRIBUTION: Yukon to E Quebec (Canada), NW USA, Sierra Nevadas, N Rocky Mtns, Great Lakes Region, New England, to North Carolina (USA).
SYNONYMS: *canadensis* Schreber, 1778; *columbiana* Goldman, 1935; *melanorhyncha* Boddaert, 1784; *nigra* Turton, 1802; *pacifica* Rhoads, 1898; *piscator* Shaw, 1800; *varietas* Richardson, 1829.
COMMENTS: Reviewed by Powell (1981, Mammalian Species, 156).

Martes zibellina (Linnaeus, 1758). Syst. Nat., 10th ed., 1:46.
TYPE LOCALITY: "asia septentrionali," restricted by Thomas (1911*a*) to "N. Asia." Restricted by Ognev (1935) to "Tobol'skuyu gub. v ee severnoi chasti" ["northern part of Tobol'sk Province" (1962*c* translation)] [Russia].

DISTRIBUTION: Ural Mtns to Siberia, Kamchatka, Sakhalin (Russia); Mongolia; Sinkiang and NE China; N Korea; Hokkaido (Japan); originally west to N Scandanavia and W Poland.

SYNONYMS: *brachyura* Temminck, 1844.

COMMENTS: Reviewed by Pavlinin (1966). May be conspecific with *americana, martes,* and *melampus* (Anderson, 1970; Hagmeier, 1961). Heptner and Naumov (1967) included Japanese and Korean *melampus* in *zibellina*.

Mustela Linnaeus, 1758. Syst. Nat., 10th ed., 1:45.

TYPE SPECIES: [*Mustela*] *Putorius* Linnaeus, 1758.

SYNONYMS: *Grammogale* Cabrera, 1940; *Lutreola* Wagner, 1841; *Putorius* Cuvier, 1817; *Vison* Gray, 1843.

COMMENTS: Revised by Hall (1951) and Youngman (1982). Youngman (1982) recognized 5 subgenera: *Putorius* (*putorius, eversmannii, nigripes*); *Lutreola* (*lutreola, sibirica, nudipes, lutreolina,* and perhaps *strigidorsa*); *Mustela* (*erminea, nivalis, altaica, frenata*); *Vison* (*vison*); and *Grammogale* (*africana, felipei*). Some have chosen to elevate these groupings to generic status; because of the lack of any comprehensive phylogenetic approach to this problem, they are provisionally recognized here as valid subgenera.

Mustela africana Desmarest, 1818. Nouv. Dict. Hist. Nat., Nouv. ed., 9:376.

TYPE LOCALITY: "Africa", type locality is in error, fixed by Cabrera (1958) as "arrabales de Belem, la antigua Pará." [Brazil]

DISTRIBUTION: Amazon Basin in Brazil, Ecuador, and Peru.

SYNONYMS: *paraensis* Goeldi, 1897; *stolzmanni* Taczanowski, 1881.

COMMENTS: Youngman (1982) placed *africana* in subgenus *Grammogale*. Izor and de la Torre (1978) suggested that *africana* and *felipei* form a monophyletic group. Cabrera (1958) considered *Grammogale* a valid genus.

Mustela altaica Pallas, 1811. Zoogr. Rosso-Asiat., I:98.

TYPE LOCALITY: "qui alpes altaicas adibunt" [Altai Mtns].

DISTRIBUTION: E Kazakhstan, S and SE Siberia, Primorski Krai (Russia); Mongolia; Tibet, W and N China; Korea.

STATUS: CITES - Appendix III (India).

SYNONYMS: *alpina* Gebler, 1823; *astutus* Milne-Edwards, 1870; *birulai* Ognev, 1928; *longstaffi* Wroughton, 1911; *raddei* Ognev, 1928; *sacana* Thomas, 1914; *temon* Hodgson, 1857.

COMMENTS: Youngman (1982) placed *altaica* in the subgenus *Mustela*; however, Ognev (1935) considered *sibirica* and *altaica* closely related.

Mustela erminea Linnaeus, 1758. Syst. Nat., 10th ed., 1:46.

TYPE LOCALITY: "Europa and Asia frigidiore; hyeme praefertim in alpinis regionibus nivea".

DISTRIBUTION: Circumboreal, tundra and forested regions of Palearctic, south to the Pyrenees, alpine Slovenia, Alps, Caucasus, W Himalayas, and Xinjiang (China); N Mongolia, N China; and C Honshu (Japan); and, in the Nearctic south to C California, N New Mexico, N Iowa and Maryland (USA). Introduced to New Zealand.

STATUS: CITES - Appendix III (India).

SYNONYMS: *alascensis* Merriam, 1896; *algiricus* Thomas, 1895; *anguinae* Hall, 1932; *angustidens* Brown, 1908; *arcticus* Merriam, 1896; *audax* Barrett-Hamilton, 1904; *bangsi* Hall, 1945; *celenda* Hall, 1944; *cigognanii* Bonaparte, 1838; *fallenda* Hall, 1945; *ferghanae* Thomas, 1895; *gulosa* Hall, 1945; *haidarum* Preble, 1898; *herminea* Oken, 1816; *hibernicus* Thomas and Barrett-Hamilton, 1895; *imperii* Barrett-Hamilton, 1904; *initis* Hall, 1944; *invicta* Hall, 1945; *kadiacensis* Merriam, 1896; *kanei* G. M. Allen, 1914; *labiata* Degerbøl, 1935; *leptus* Merriam, 1903; *lymani* Hollister, 1912; *microtis* J. A. Allen, 1903; *mortigena* Bangs, 1913; *muricus* Bangs, 1899; *nippon* Cabrera, 1913; *olympica* Hall, 1945; *polaris* Barrett-Hamilton, 1904; *pusilla* DeKay, 1842; *richardsonii* Bonaparte, 1838; *rixosa* Svihla and Svihla, 1932; *salva* Hall, 1944; *seclusa* Hall, 1944; *semplei* Sutton and Hamilton, 1932; *streatori* Merriam, 1896; *vulgaris* Griffith, 1827; *whiteheadi* Wroughton, 1908.

COMMENTS: Revised by Eger (1990). Reviewed by C. M. King (1983, Mammalian Species, 195). Youngman (1982) placed *erminea* in the subgenus *Mustela*.

Mustela eversmannii Lesson, 1827. Manuel de Mammalogie, p. 144.

TYPE LOCALITY: "trouvé...entre Orembourg et Bukkara," restricted by Stroganov (1962:338) to "basseinu sredneyo techeniya r. Ileka, ...r. Bol'shoi Khobdy" [Russia, Orenburg Obl., S of Orenburg, mouth of Khobda River, a tributary of Ilek River.]

DISTRIBUTION: Steppes and subdeserts of E Europe, and republics of the former USSR; Mongolia; W, C and NE China.

SYNONYMS: *larvatus* Hodgson, 1849; *lineiventer* Hollister, 1913; *tiarata* Hollister, 1913.

COMMENTS: Reviewed by Kostro (1948), Heptner (*in* Heptner and Naumov, 1967), and Anderson (1977). Youngman (1982) placed *eversmannii* in the subgenus *Putorius*. Anderson (1977) and Kurtén and Anderson (1980) suggested that *nigripes* and *eversmannii* may be conspecific. Pocock (1936*b*) and Ellerman and Morrison-Scott (1966) considered *eversmannii* and *putorius* conspecific; however, Ognev (1931), Stroganov (1962), and Heptner (*in* Heptner and Naumov, 1967), recognized them as distinct species.

Mustela felipei Izor and de la Torre, 1978. J. Mammal., 59:92.

TYPE LOCALITY: "Santa Marta, elevation 2,700 m, near San Agustin, Huila, Colombia".

DISTRIBUTION: The type locality and Cauca, Colombia (two localities separated by 70 km, on opposite sides of Cordillera Central.

COMMENTS: Izor and de la Torre (1978) suggested that *africana* and *felipei* form a monophyletic group. Youngman (1982) placed *felipei* in subgenus *Grammogale*.

Mustela frenata Lichtenstein, 1831. Darst. Säugeth., text:"Das gezäumte Wiesel" [not paginated], and plate 42.

TYPE LOCALITY: "der Nähe von Mexico".

DISTRIBUTION: S Canada to Venezuela and Bolivia, excluding the SW deserts of the USA.

SYNONYMS: *aequatorialis* Coues, 1877; *affinis* Gray, 1874; *agilis* Tschudi, 1844; *alleni* Merriam, 1896; *altifrontalis* Hall, 1936; *arizonensis* Mearns, 1891; *arthuri* Hall, 1927; *aureoventris* Gray, 1864; *boliviensis* Hall, 1938; *brasiliensis* Sevastianoff, 1813; *cicognanii* Henninger, 1921; *costaricensis* Goldman, 1912; *effera* Hall, 1936; *fusca* DeKay, 1842; *goldmani* Merriam, 1896; *gracilis* Brown, 1908; *helleri* Hall, 1935; *inyoensis* Hall, 1936; *jelskii* Taczanowski, 1881; *latirostra* Hall, 1936; *leucoparia* Merriam, 1896; *longicauda* Bonaparte, 1838; *macrophonius* Elliot, 1905; *macrura* Taczanowski, 1874; *meridana* Hollister, 1914; *mexicanus* Coues, 1877; *munda* Grinnell, 1933; *mundus* Bangs, 1899; *neomexicanus* Barber and Cockerell, 1898; *nevadensis* Hall, 1936; *nicaraguae* Allen, 1916; *nigriauris* Hall, 1936; *notius* Bangs, 1899; *noveboracensis* Emmons, 1840; *occisor* Bangs, 1899; *olivacea* Howell, 1913; *oregonensis* Merriam, 1896; *oribasus* Bangs, 1899; *panamensis* Hall, 1932; *peninsulae* Rhoads, 1894; *perdus* Merriam, 1902; *perotae* Hall, 1936; *primulina* Jackson, 1913; *pulchra* Hall, 1936; *richardsonii* Baird, 1858; *saturatus* Merriam, 1896; *spadix* Bangs, 1896; *texensis* Hall, 1936; *tropicalis* Merriam, 1896; *washingtoni* Merriam, 1896; *xanthogenys* Gray, 1843.

COMMENTS: Youngman (1982) placed *frenata* in the subgenus *Mustela*.

Mustela kathiah Hodgson, 1835. J. Asiat. Soc. Bengal, 4:702.

TYPE LOCALITY: "Kachar region" [Nepal].

DISTRIBUTION: Himalayas from N Pakistan through Nepal to Burma; S and E China; Indochinese Peninsula.

STATUS: CITES - Appendix III (India).

SYNONYMS: *auriventer* Hodgson, 1837; *caporiaccoi* de Beaux, 1935; *dorsalis* Trouessart, 1895; *tsaidamensis* Hilzheimer, 1910;

Mustela lutreola (Linnaeus, 1761). Fauna Suecica, 2nd ed., p. 5.

TYPE LOCALITY: "Finlandiae aquolis", restricted by Matschie (1912) to "Südwest-Finnland."

DISTRIBUTION: NE Spain, France; disjunct populations throughout Europe to the Irtysh and Ob Rivers (Russia, Kazakhstan).

STATUS: IUCN - Vulnerable.

SYNONYMS: *alba* de Sélys Longchamps, 1839; *albica* Matschie, 1912; *alpinus* Ogérien, 1863; *armorica* Matschie, 1912; *biedermanni* Matschie, 1912; *binominata* Ellerman and Morrison-Scott, 1951; *borealis* Novikov, 1939; *budina* Matschie, 1912; *caucasica* Novikovi, 1939; *cylipena* Matschie, 1912; *europeae* Homeyer, 1879; *fulva* Kerr, 1792; *glogeri* Matschie, 1912; *hungarica* Ehik, 1932; *minor* Erxleben, 1977; *novikovi* Ellerman

and Morrison-Scott, 1951; *taivana* Thomas, 1913; *transsylvanica* Ehik, 1932; *turovi* Kuznetzov and Novikov, 1939; *varina* Matschie, 1912; *wyborgensis* Matschie, 1912.
COMMENTS: Youngman (1982) placed *lutreola* in subgenus *Lutreola*. Revised by Matschie (1912), Novikov (1939), and Youngman (1982; 1990, Mammalian Species, 362). Occasional hybrids occur between *lutreola* and *putorius* (Youngman, 1982).

Mustela lutreolina Robinson and Thomas, 1917. Ann. Mag. Nat. Hist., ser. 8, 20:261-262.
TYPE LOCALITY: "Tjibodas, West Java, 5500'"; identified by Van Bree and Boeadi (1978) as "6°44'S 107°00'E".
DISTRIBUTION: Indonesia (Java, Sumatra).
STATUS: IUCN - Insufficiently known.
COMMENTS: Revised by Van Bree and Boeadi (1978). Ellerman and Morrison-Scott (1966) implied, and Novikov (1956) and Corbet (1978*c*) believed *lutreolina* and *sibirica* conspecific; however this has not been supported by primary studies (Van Bree and Boeadi, 1978). Youngman (1982) placed *lutreolina* in the subgenus *Lutreola*.

Mustela nigripes (Audubon and Bachman, 1851). Viviparous Quadrupeds of North America, 2:297.
TYPE LOCALITY: "lower waters of the Platte River", restricted by Hayden (1863:138) to "Fort Laramie" [Wyoming, USA].
DISTRIBUTION: Formerly, S Alberta and Saskatchewan (Canada) south to Arizona, Oklahoma, and NW Texas (USA). Viable populations now only in captivity.
STATUS: CITES - Appendix I; U.S. ESA and IUCN - Endangered.
COMMENTS: Reviewed by Hillman and Clark (1980, Mammalian Species, 126) and Anderson (1977). Youngman (1982) placed *nigripes* in the subgenus *Putorius*. Anderson (1977) and Kurtén and Anderson (1980) suggested that *nigripes* and *eversmannii* may be conspecific.

Mustela nivalis Linnaeus, 1766. Syst. Nat., 12th ed., 1:69.
TYPE LOCALITY: "Westrobothnia" [Sweden].
DISTRIBUTION: The Palearctic (excl. Ireland, the Arabian Peninsula, and Arctic Isls); Japan; the Nearctic in Alaska (USA), Canada, and USA, south to Wyoming and North Carolina. Introduced to New Zealand; see Corbet and Hill (1980).
SYNONYMS: *africana* Gray, 1865; *albipes* Mina Palumbo, 1868; *allegheniensis* Rhoads, 1901; *alpinus* Burg, 1920; *atlas* Barrett-Hamilton, 1904; *boccamela* Bechstein, 1800; *campestris* Jackson, 1913; *caraftensis* Kishida, 1936; *caucasicus* Barrett-Hamilton, 1900; *corsicanus* Cavazza, 1908; *dinniki* Satunin, 1907; *dombrowskii* Matschie, 1901; *eskimo* Stone, 1900; *fulva* Mina Palumbo, 1868; *gale* Pallas, 1811; *galinthias* Bate, 1906; *italicus* Barrett-Hamilton, 1900; *kamtschatica* Dybowski, 1922; *major* Fatio, 1905; *meridionalis* Costa, 1869; *minor* Nilsson, 1820; *minutus* Pomel, 1853; *monticola* Cavazza, 1908; *mosanensis* Mori, 1927; *namiyei* Kuroda, 1921; *nikolskii* Smirnov, 1899; *numidicus* Pucheran, 1855; *pallidus* Barrett-Hamilton, 1900; *punctata* Domaniewski, 1926; *pusillus* Fatio, 1869; *pygmaeus* J. A. Allen, 1903; *rixosus* Bangs, 1896; *russelliana* Thomas, 1911; *siculus* Barrett-Hamilton, 1900; *stoliczkana* Blanford, 1877; *subpalmata* Hemprich and Ehrenberg, 1833; *tonkinensis* Björkegren, 1942; *trettaui* Kleinschmidt, 1937; *typicus* Barrett-Hamilton, 1900; *vulgaris* Erxleben, 1777; *yesoidsuna* Kishida, 1936.
COMMENTS: Reviewed by Reichstein (1957) and van Zyll de Jong (1992), who considered *Putorius rixosus* and *Mustela nivalis* conspecific. Youngman (1982) placed *nivalis* in the subgenus *Mustela*.

Mustela nudipes Desmarest, 1822. Mammalogie, *in* Encycl. Meth., 2(Suppl.):537.
TYPE LOCALITY: "L'île de Java". Locality is in error, fixed by Robinson and Kloss (1919*b*) as "West Sumatra" [Indonesia].
DISTRIBUTION: Thailand, Malaysia, Indonesia (Sumatra, Java, Borneo).
SYNONYMS: *leucocephalus* Gray, 1865.
COMMENTS: Pocock (1941*a*) believed *strigidorsa* and *nudipes* to be closely related. Youngman (1982) placed *nudipes* in the subgenus *Lutreola*.

Mustela putorius Linnaeus, 1758. Syst. Nat., 10th ed., 1:46.
TYPE LOCALITY: "inter Europae rupes et lapidum acervos", restricted by Thomas (1911*a*) to "Scania, S. Sweden."
DISTRIBUTION: Europe (excluding Ireland and most of Scandinavia), east to the Ural Mtns.

SYNONYMS: *ambigua* Mearns, 1897; *angustifrons* Howell, 1902; *bicolor* Gray, 1837; *gracilis* Merriam, 1890; *interrupta* Rafinesque, 1820; *larvatus* Hodgson, 1849; *leucoparia* Merriam, 1890; *lucasona* Merriam, 1890; *microdon* Howell, 1906; *mupurita* Müller, 1776; *olympica* Elliot, 1899; *phenax* Merriam, 1890; *putida* Cuvier, 1798; *quaterlinearis* Winans, 1859; *ringens* Merriams, 1890; *saxatalis* Merriam, 1890; *striata* Shaw, 1800; *tenuis* Howell, 1902; *texensis* Merriam, 1890; *tibetanus* Horsfield, 1851; *zorilla* Schreber, 1776.

COMMENTS: Reviewed by Heptner (*in* Heptner and Naumov, 1967). Probable ancestor of domestic ferret, *furo* (Rempe, 1970; Volobuev, 1971). Youngman (1982) placed *putorius* in the subgenus *Putorius*. Pocock (1936*b*) and Ellerman and Morrison-Scott (1966) considered *eversmannii* and *putorius* conspecific; however, Ognev (1931), Stroganov (1962), and Heptner (*in* Heptner and Naumov, 1967), recognized them as distinct species.

Mustela sibirica Pallas, 1773. Reise Prov. Russ. Reichs., 2:701.

TYPE LOCALITY: "Sibiriae montanis, sylvis densissimis", restricted by Pocock (1941*a*) to "Vorposten Tigerazkoi, near Usstkomengorsk, W. Altai," based on Pallas (1773:570). Listed in Honaki et al. (1982) as "U.S.S.R., E. Kazakhstan, vic. of Ust-Kamenogorsk, Tigeretskoie."

DISTRIBUTION: Russia (Tataria, Dalnevostochny Rayon), Ural Mtns, Siberia; Pakistan east to N Burma; N Thailand; Taiwan; China; Korea; Japan; introduced to Sakhalin, Russia.

STATUS: CITES - Appendix III (India).

SYNONYMS: *asaii* Kuroda, 1943; *australis* Satunin, 1911; *canigula* Hodgson, 1842; *charbinensis* Lowkashkin, 1934; *coreanus* Domaniewski, 1926; *davidianus* Milne-Edwards, 1872; *fontainierii* Milne-Edwards, 1871; *hamptoni* Thomas, 1921; *hodgsoni* Gray, 1843; *horsfieldii* Blyth, 1843; *humeralis* Blyth, 1842; *itatsi* Temminck, 1844; *katsurai* Kishida, 1931; *major* Hilzheimer, 1910; *manchurica* Brass, 1911; *melli* Matschie, 1922; *miles* Barrett-Hamilton, 1904; *moupinensis* Milne-Edwards, 1868; *natsi* Temminck, 1844; *noctis* Barrett-Hamilton, 1904; *peninsulae* Kishida, 1931; *quelpartis* Thomas, 1908; *sho* Kuroda, 1924; *sibirica* Pallas, 1773; *stegmanni* Matschie, 1907; *subhemachalanus* Hodgson, 1837; *tafeli* Hilzheimer, 1910; *taivana* Thomas, 1913.

COMMENTS: Youngman (1982) placed *sibirica* in the subgenus *Lutreola*. Ognev (1935) considered *altaica* and *sibirica* closely related. Ellerman and Morrison-Scott (1966) implied, and Novikov (1956) and Corbet (1978*c*) believed *lutreolina* and *sibirica* conspecific; however this has not been supported by primary studies (Van Bree and Boeadi, 1978).

Mustela strigidorsa Gray, 1855. Proc. Zool. Soc. Lond., 1853:191 [1855].

TYPE LOCALITY: Not given. Gray (1853) based the type description on a manuscript given to him by Hodgson. Horsfield (1855) later fixed the type locality as, "the Sikim Hills of Tarai." [India, Sikkim].

DISTRIBUTION: Nepal east through Burma, Yunnan (China) and Thailand to Laos.

COMMENTS: Youngman (1982) suggested that *strigidorsa* belonged in subgenus *Lutreola*.

Mustela vison Schreber, 1777. Die Säugethiere, 3(19):pl. 127.B[1777]; text, 3(26):463[1777].

TYPE LOCALITY: "Man findet das Vison in Canada un Pensilvanien".

DISTRIBUTION: North America from Alaska and Canada through all of USA except SW deserts. Introduced to Iceland, NC Europe, British Isls, Norway, Belarussia, Baltic States, Spain, and Siberia.

SYNONYMS: *antiquus* Loomis, 1911; *borealis* Brass, 1911; *energumenos* Bangs, 1896; *evagor* Hall, 1932; *evergladensis* Hamilton, 1948; *ingens* Osgood, 1900; *lacustris* Preble, 1902; *letifera* Hollister, 1915; *lowii* Anderson, 1945; *lutensis* Bangs, 1898; *lutreocephala* Harlan, 1825; *macrodon* Prentiss, 1903; *melampeplus* Elliot, 1903; *mink* Peale and Palisot de Beauvois, 1796; *minx* Turton, 1800; *nesolestes* Heller, 1909; *nigrescens* Audubon and Bachman, 1854; *rufa* Smith, 1858; *vulgivagus* Bangs, 1895; *winingus* Baird, 1857.

COMMENTS: Manville (1966) demonstrated that *macrodon* is conspecific, although Kurtén and Anderson (1980) recognized it as a distinct species. Hall (1951), Heptner and Yurgenson (1967), and Hollister (1913*a*) considered *vison* closely related to *lutreola*, however analyses by Graphodatskii et al. (1976) and Youngman (1982) supported *vison* to be one of the earliest offshoots of the *Mustela* lineage. Youngman (1982) placed *vison* in the subgenus *Vison*.

Poecilogale Thomas, 1883. Ann. Mag. Nat. Hist., ser. 5, 11:370.
TYPE SPECIES: *Zorilla albinucha* Gray, 1864, by monotypy (Melville and Smith, 1987).

Poecilogale albinucha (Gray, 1864). Proc. Zool. Soc. Lond., 1864:69, plate X.
TYPE LOCALITY: "it was without any habitat". Fixed by Coetzee (1977*b*) as "Cape Colony".
DISTRIBUTION: South Africa, Zimbabwe, Mozambique, Botswana, Namibia, Malawi, Zambia, Angola, Zaire, Rwanda, Burundi, Uganda, Tanzania, Kenya.
SYNONYMS: *africana* Peters, 1865; *bechuanae* Roberts, 1931; *doggetti* Thomas and Schwann, 1904; *flavistriata* Bocage, 1865; *lebombo* Roberts, 1931; *transvaalensis* Roberts, 1926.

Vormela Blasius, 1884. Ber. Naturforsch Ges. Bemberg, 13:9.
TYPE SPECIES: *Mustela sarmatica* Pallas, 1771 (= *Mustela peregusna* Güldenstädt, 1770), by original designation (Melville and Smith, 1987).

Vormela peregusna (Güldenstädt, 1770). Nova Comm. Imp. Acad. Sci. Petropoli, 14(1):441.
TYPE LOCALITY: "habitat in campis apricis desertis Tanaicensibus". Listed in Honacki et al. (1982) as "U.S.S.R., Rostov Obl., steppes at lower Don River."
DISTRIBUTION: Steppes and deserts of SE Europe, Caucasus, Kazakhstan, Middle Asia; SW Asia (excl. Arabia); N China and S Mongolia.
STATUS: IUCN - Vulnerable.
SYNONYMS: *alpherakii* Birula, 1910; *euxina* Pocock, 1936; *koshewnikowi* Satunin, 1910; *negans* Miller, 1910; *ornata* Pocock, 1936; *peregusna* Güldenstädt, 1770; *sarmatica* Pallas, 1771; *syriaca* Pocock, 1936; *tedschenika* Satunin, 1910;

Subfamily Taxidiinae Pocock, 1920. Proc. Zool. Soc. London, 1920:424.
COMMENTS: Type genus: *Taxidea* Waterhouse (1838). See comments under Melinae for rationale in separating this taxon from old world "badgers."

Taxidea Waterhouse, 1839. Proc. Zool. Soc. Lond., 1838:154 [1839].
TYPE SPECIES: *Ursus Meles labradorius* Gmelin, 1788 (= *Ursus taxus* Schreber, 1777), by original designation. (Melville and Smith, 1987).
SYNONYMS: *Taxus* Say, 1823.

Taxidea taxus (Schreber, 1777). Säugethiere, 3(26):pl. 142[1778], text, 3(26):520[1777].
TYPE LOCALITY: "Er wohnt in Labrador und um die Hudsonsbay," restricted by Long (1972), to "Carman, Manitoba." [Canada].
DISTRIBUTION: SW Canada, British Columbia, Alberta, Saskatschewan, Manitoba, Ontario, to Baja California and Puebla (Mexico), east to Lake Ontario.
SYNONYMS: *americanus* Boddaert, 1784; *apache* Schantz, 1948; *berlandieri* Baird, 1858; *californica* Gray, 1865; *dacotensis* Schantz, 1946; *halli* Schantz, 1951; *hallorani* Schantz, 1949; *infusca* Thomas, 1898; *iowae* Schantz, 1947; *jacksoni* Schantz, 1945; *jeffersonii* Harlan, 1825; *kansensis* Schantz, 1950; *labradorius* Gmelin, 1788; *littoralis* Schantz, 1949; *marylandica* Gidley and Gaxin, 1933; *merriami* Schantz, 1950; *montanus* Richardson, 1829; *neglecta* Mearns, 1891; *nevadensis* Schantz, 1949; *obscurata* de Beaux, 1924; *papagoensis* Skinner, 1943; *phippsi* Figgins, 1918; *robusta* Hay, 1921; *sonoriensis* Goldman, 1939; *sulcata* Cope, 1878.
COMMENTS: Reviewed by Long (1972; 1973, Mammalian Species, 26).

Family Odobenidae Allen, 1880. U.S. Geol. and Geog. Surv. Territ., 12:ix, 5.
COMMENTS: Trichecidae Gray, 1821 and Rosmaridae Gill, 1866 are invalid (International Commission on Zoological Nomenclature, 1959). Considered as a subfamily of Otariidae by Mitchell and Tedford (1973), Tedford (1976), Hall (1981), Berta and Deméré (1986), Barnes (1989), and Wozencraft (1989*a*, *b*); however, Wyss (1987) and Berta (1991) contended that this would make the otariids paraphyletic.

Odobenus Brisson, 1762. Regne Anim., 2nd ed., p. 30.
TYPE SPECIES: *Odobenus odobenus* Brisson, 1762 (= *Phoca rosmarus* Linnaeus, 1758).

COMMENTS: Although the names in Brisson (1762) are invalid, *Odobenus* has been retained (International Commission on Zoological Nomenclature, 1957*e*).

Odobenus rosmarus (Linnaeus, 1758). Syst. Nat., 10th ed., 1:38.
TYPE LOCALITY: "intra Zonam arcticam Europae, Asiae, Americae".
DISTRIBUTION: Arctic seas: south as far as New England (USA), Great Britain, Scandinavia, Pribilof Isls, and Honshu (Japan) at least occasionally.
STATUS: CITES - Appendix III (Canada); IUCN - Insufficiently known as *O. r. laptevi*.
SYNONYMS: *arcticus* Pallas, 1811; *cookii* Fremerij, 1831; *divergens* Illiger, 1815; *obesus* Illiger, 1815; *orientalis* Dybowski, 1922.
COMMENTS: Reviewed by Fay (1985, Mammalian Species, 238).

Family Otariidae Gray, 1825. Ann. Philos., n.s., 10:340.
COMMENTS: Reviewed by Allen (1880), Repenning et al. (1971), Mitchell and Tedford (1973), J. E. King (1983), Berta and Deméré (1986), and Barnes (1989). Does not include *Odobenus*, which was included in a monotypic subfamily (Odobeninae) by Mitchell and Tedford (1973), Tedford (1976), Hall (1981), Barnes (1989), and Wozencraft (1989*a*, *b*); however, see Wyss (1987) and Berta (1991) who are followed here. Berta and Deméré (1986) separated *Arctocephalus* and *Callorhinus* into the Arctocephalinae. Repenning et al. (1971), Repenning and Tedford (1977), and others argued against the recognition of subfamilies.

Arctocephalus E. Geoffroy Saint-Hilaire and F. Cuvier, 1826. *In* F. Cuvier, Dict. Sci. Nat., 39:553 [1826].
TYPE SPECIES: "*Phoca ursina* " (= *Phoca pusilla* Schreber, 1777; not *Phoca ursina* Linnaeus, 1758; Allen, 1905 discussed confusion in designation of type species).
COMMENTS: Reviewed by King (1954) and Repenning et al. (1971) who included *Arctophoca* Peters, 1867. Van Gelder (1977*b*) considered *Zalophus* and *Arctocephalus* congeneric.

Arctocephalus australis (Zimmermann, 1783). Geogr. Gesch. Mensch. Vierf. Thiere, 3:276.
TYPE LOCALITY: Zimmermann (1783) based the name on the "Falkland Isle Seal" of Pennant (1781), however, he added that it "Wohnt um Juan Fernandez, und über haupt in dortigen Meeren." [Falkland Isls, UK].
DISTRIBUTION: Coasts of South America from Lima (Peru) to Rio de Janeiro (Brazil); Falkland Isls.
STATUS: CITES - Appendix II.
SYNONYMS: *brachydactyla* Philippi, 1892; *falclandicus* Nahring, 1887; *falklandica* Desmarest, 1817; *gracilis* Nehring, 1887; *grayi* Scott, 1873; *hauvillii* Lesson, 1827; *latirostris* Gray, 1872; *leucostoma* Philippi, 1892; *lupina* Molina, 1782; *nigrescens* Gray, 1850; *philippi* Philippi, 1892; *porcina* Molina, 1782; *shawii* Lesson, 1828; *ursinus* Gray, 1843.
COMMENTS: Scheffer (1958) included *galapagoensis* Heller, 1904; but this was not followed by Repenning et al. (1971) or J. E. King (1983).

Arctocephalus forsteri (Lesson, 1828). *In* Bory de Saint-Vincet (ed.), Dict. Class. Hist. Nat. Paris, 13:421.
TYPE LOCALITY: Scheffer (1958) restricted the type locality to "Dusky Sound, New Zealand."
DISTRIBUTION: New Zealand and nearby subantarctic isls; S and W Australia.
STATUS: CITES - Appendix II.

Arctocephalus galapagoensis Heller, 1904. Proc. Calif. Acad. Sci., ser. 3, 3(7):245.
TYPE LOCALITY: "Wenman Island" [Ecuador, Galapagos Isls].
DISTRIBUTION: Specimens recorded only from Galapagos Isls [Ecuador].
STATUS: CITES - Appendix II.
COMMENTS: Repenning et al. (1971) supported recognition at the specific level, followed by J. E. King (1983); however, Scheffer (1958) considered *galapagoensis* conspecific with *australis*, which would be the most closely related taxon. Reviewed by Clark (1975, Mammalian Species, 64).

Arctocephalus gazella (Peters, 1875). Monatsb. K. Preuss. Akad. Wiss. Berlin, 1875:393, 396.
TYPE LOCALITY: "von Seehunden aus Kerguelenland". Restriced by Scheffer (1958) to "Anse Betsy (49°09'S, 70°11'E)."

DISTRIBUTION: Islands south of Antarctic convergence (Kerguelen, S Sandwich, S Orkney, Heard, Bouver, S Georgia, S Shetland Isls).
STATUS: CITES - Appendix II.
COMMENTS: Reviewed by King (1959*a*, *b*). Peters' (1875) original description, based on two specimens, was considered by Allen (1880) as invalid (*nomen nudum*). A year later Peters (1876) learned the second specimen came from St. Paul or Amsterdam Island and gave it the name *Otaria* (*Arctophoca*) *elegans*. Allen (1892) and King (1959*b*) considered these as subspecific populations. Placed in *Arctocephalus* by Repenning et al. (1971).

Arctocephalus philippii (Peters, 1866). Monatsb. K. Preuss. Akad. Wiss. Berlin, 1866:276, pl. 2a, b, c.
TYPE LOCALITY: "Insel Juan Fernandez". Listed by Scheffer (1958) as "Isla Más a Tierra, Islas Juan Fernández, Chile".
DISTRIBUTION: Specimens recorded from Juan Fernandez and San Felix Isls (Chile).
STATUS: CITES - Appendix II; IUCN - Vulnerable.
SYNONYMS: *argentata* Philippi, 1871; *aurita* Tschudi, 1844.

Arctocephalus pusillus (Schreber, 1775). Die Säugethiere, 2(13):pl. 85[1775]; text, 3(17):314 [1776].
TYPE LOCALITY: Unknown. "Diese Gattung findet sich in den levantischen, und nach dem Herrn Grafen von Büffon, im indischen Meere"; discussion of type locality in Allen (1880).
DISTRIBUTION: SW Africa from Angola to Algoa Bay; SE Australia; Tasmania.
STATUS: CITES - Appendix II.
SYNONYMS: *antarctica* Thunberg, 1811; *compressa* Gray, 1874; *delalandii* Gray, 1859; *doriferus* Wood Jones, 1925; *nivosus* Gray, 1868; *parva* Boddaert, 1785; *peronii* Desmarest, 1820; *schisthyperoes* Turner, 1868; *tasmanicus* Scott and Lord, 1926.
COMMENTS: Repenning et al. (1971) and J. E. King (1983) included *doriferus* Wood-Jones, 1925; however, Scheffer (1958) considered it a distinct species.

Arctocephalus townsendi Merriam, 1897. Proc. Biol. Soc. Wash., 11:175.
TYPE LOCALITY: "Guadalupe Island, off Lower California. . . collected on the beach on west side of Guadalupe." [Mexico]
DISTRIBUTION: Guadalupe Isl (Mexico), Channel Isls (California, USA).
STATUS: CITES - Appendix I; U.S. ESA - Threatened; IUCN - Vulnerable.
COMMENTS: Formerly included in *Arctophoca*; see Repenning et al. (1971). Considered conspecific with *philippii* by Scheffer (1958).

Arctocephalus tropicalis (Gray, 1872). Proc. Zool. Soc. Lond., 1872:653, 659.
TYPE LOCALITY: "North coast of Australia." This is in error, fixed by King (1959*b*) to "'Australasian sea'...to include the islands of St. Paul and Amsterdam as these are the islands nearest to Australia...".
DISTRIBUTION: Islands north of Antarctic Convergence (Tristan, Gough, Marion, Crozet, Amsterdam, Macquarie Isls).
STATUS: CITES - Appendix II.

Callorhinus Gray, 1859. Proc. Zool. Soc. Lond., 1859:359.
TYPE SPECIES: *Arctocephalus ursinus* Gray, 1859 (= *Phoca ursina* Linnaeus, 1758), by original designation.

Callorhinus ursinus (Linnaeus, 1758). Syst. Nat., 10th ed., 1:37.
TYPE LOCALITY: "in Camschatcæ maritimus inter Asiam and Americam proximam, primario in infula Beringri," restricted by Thomas (1911*a*) to "Bering Island."
DISTRIBUTION: North Pacific, in Okhotsk and Bering Seas; Commander and Pribilof Isls (Russia), south to Japan, Shantung (China), and S California (USA).
SYNONYMS: *alascanus* Jordan and Clark, 1898; *californianus* Gray, 1866; *curilensis* Jordan and Clark, 1898; *cynocephala* Walbaum, 1792; *mimica* Tilesius, 1835.

Eumetopias Gill, 1866. Proc. Essex Inst. Salem, 5:7.
TYPE SPECIES: "*Otaria californiana* Lesson., = *Arctocephalus monterienis* Gray." (= *Phoca jubata* Schreber, 1776).

COMMENTS: For a discussion of the type, see Scheffer (1958).

Eumetopias jubatus (Schreber, 1776). Die Säugethiere, text, 3(17):300[1776]; 3(17):pl. 83.B[1776].

TYPE LOCALITY: "...Aufenthalt in dem nördlichen Theil des stillen Meeres...westlichen Küste von Amerika... östlichen von Kamtschatka...Inseln...Küsten unter dem 56ten Grade der Breite liegen." [N part of the Pacific. Russia, Commander and Bering Isls].

DISTRIBUTION: N Pacific, from Hokkaido (Japan) to S California (USA).

STATUS: U.S. ESA - Threatened.

SYNONYMS: *marinus* Steller, 1751; *monteriensis* Gray, 1859; *stellerii* Lesson, 1828.

COMMENTS: Scheffer (1958) pointed out that *jubata* Forster, 1775, is invalid. Reviewed by Loughlin et al. (1987, Mammalian Species, 283).

Neophoca Gray, 1866. Ann. Mag. Nat. Hist., ser. 3, 18:231.

TYPE SPECIES: *Arctocephalus lobatus* Gray, 1828 (= *Otaria cinerea* Péron, 1816).

COMMENTS: Sivertsen (1954) and Scheffer (1958) considered *Neophoca* congeneric with *Phocarctos*; however, it was retained as separate by J. E. King (1960, 1983), Rice (1977), and Barnes (1989).

Neophoca cinerea (Péron, 1816). Voy. Decouv. Terres. Austral., 2:54.

TYPE LOCALITY: "L'ile Decrès" [Australia, South Australia, Kangaroo Isl].

DISTRIBUTION: Houtmans Abrolhos (Western Australia) to Kangaroo Isl (South Australia).

SYNONYMS: *albicollis* Péron, 1816; *australis* Quoy and Gaimard, 1830; *forsteri* Wood Jones, 1992 (not Lesson, 1828); *hookeri* Gray, 1844; *lobatus* Gray, 1828; *williamsi* McCoy, 1877.

COMMENTS: Allen (1880:203-204) questioned the validity of the type description. Reviewed by Ling (1992, Mammalian Species, 392).

Otaria Péron, 1816. Voy. Decouv. Terres. Austral., 2:37.

TYPE SPECIES: *Otaria leonina* Péron, 1816 (= *Phoca byroni* de Blainville, 1820).

Otaria byronia (Blainville, 1820). J. Phys. Chim. Hist. Nat. Arts Paris, 91:300.

TYPE LOCALITY: "the island of Tinian...située à l'est des Philippines ou par le 15° de latitude méridionale et le 215° de long. méridionale de Greenwich." This is an error, as there are no sea lions on these islands, fixed by Scheffer (1958) as "it probably came from the Strait of Magellan or Islas Juan Fernández" based on Allen (1905) and Hamilton (1934).

DISTRIBUTION: South American coasts from Peru to Uruguay; Falkland Isl; occasionally north to coast of Brazil.

SYNONYMS: *chilensis* Müller, 1841; *chonotica* Philippi, 1892; *flavescens* Shaw, 1800; *fluva* Philippi, 1892; *godeffroyi* Peters, 1866; *hookeri* Sclater, 1866; *leonina* Molina, 1782; *minor* Gray, 1874; *molossina* Lesson and Garnot, 1826; *pernettyi* Lesson, 1828; *pygmaea* Gray, 1874; *rufa* Philippi, 1892; *scont* Boddaert, 1784; *ulloae* Tschudi, 1844; *uraniae* Lesson, 1827; *velutina* Philippi, 1892.

COMMENTS: *Phoca flavescens* Shaw, 1800, was described based on a juvenile specimen in the Leverian Museum from the "Magellanic Straits" [Chile]. His description, as stated, does not fit any know otariid from this region (Allen, 1880, 1905). Cabrera (1940) believed that the description was sufficient to distinguish it from *Arctocephalus*, the only other possibility; based on the molting of young pups. King (1978) who reviewed the controversy, pointed out that at molting stage, the size of the pup would exclude it from Shaw's (1800) description.

Phocarctos Peters, 1866. Monatsb. K. Preuss. Akad. Wiss. Berlin, 1866:269.

TYPE SPECIES: *Otaria hookeri* (*Arctocephalus Hookeri* Gray, 1844).

COMMENTS: Sivertsen (1954) and Scheffer (1958) considered *Phocarctos* congeneric with *Neophoca*, however, it was retained as separate by J. E. King (1960, 1983), Rice (1977), and Barnes (1989).

Phocarctos hookeri (Gray, 1844). Zool. Voy. H.M.S. "Erebus" and "Terror," 1:4.

TYPE LOCALITY: "Falkland Islands and Cape Horn." Locality in error; fixed by Clark (1873) as "Auckland Islands. . . between 800 and 900 miles S. of Tasmania, in lat. 50°48'S., long. 166°42'E." [New Zealand]

DISTRIBUTION: Subantarctic islands of New Zealand.

Zalophus Gill, 1866. Proc. Essex Inst. Salem, 5:7.
TYPE SPECIES: "*Otaria Gilliespii* Macbain", 1858 (= *Otaria californiana* Lesson, 1828).
COMMENTS: Included in *Arctocephalus* by Van Gelder (1977*b*). Mohr (1952) reported successful matings between *Arctocephalus pusillus* and *Z. californianus*.

Zalophus californianus (Lesson, 1828). *In* Bory de Saint-Vincet (ed.), Dict. Class. Hist. Nat. Paris, 13:420.
TYPE LOCALITY: "les rochers dans le voisinage de la baie San-Francisco sont ordinairement couverts de lion marins." [USA, California, San Francisco Bay].
DISTRIBUTION: N Pacific from near the Mexico-Guatemala border to British Columbia (Canada); Japan; Galapagos Isls (Gallo-Reynosa and Solorzano-Velasco, 1991).
STATUS: IUCN - Extinct? as *Z. c. japonicus*.
SYNONYMS: *gillespii* M'Bain, 1858; *japonica* Peters, 1866; *lobatus* Jentink, 1892; *philippii* Allen, 1905; *wollebaeki* Sivertsen, 1953.

Family Phocidae Gray, 1821. London Med. Repos., 15:297.
COMMENTS: Reviewed by Chapskii (1955), Scheffer (1958), J. E. King (1966, 1983), Hendey (1972), Muizon (1982*a*), and Wyss (1988). Muizon (1982*a*), and Wyss's (1988) phylogenetic analyses agreed on three areas: 1) The monophyletic nature of two groups they refer to as the Lobodontini (*Hydrurga, Leptonychotes, Lobodon*, and *Ommatophoca*), and the Phocinae (*Erignathus, Cystophora, Halichoerus*, and *Phoca*), 2) The lobodonts, along with *Monachus* and *Mirounga* have been traditionally refered to as the Monachinae (kept by Muizon), however, they both suggested that this group may be paraphyletic, 3) Because of the "unsettled" nature of these taxa, no subfamilies are recognized at this time.

Cystophora Nilsson, 1820. Skand. Faun. Dagg. Djur., 1:382.
TYPE SPECIES: *Cystophora borealis* Nilsson, 1820 (= *Phoca cristata* Erxleben, 1777).
COMMENTS: Revised by King (1966).

Cystophora cristata (Erxleben, 1777). Syst. Regni Anim., 1:590.
TYPE LOCALITY: "Habitat in Groenlandia australiori et Newfoundland". [S Greenland and Newfoundland].
DISTRIBUTION: N Atlantic and Arctic oceans, from Newfoundland (Canada) to S Greenland, Svalbard and Novaya Zemlya (Russia); occasionally south to Portugal and Florida (USA).
SYNONYMS: *borealis* Nilsson, 1820; *dimidiata* Rüppell, 1842.
COMMENTS: Reviewed by Kovacs and Lavigne (1986, Mammalian Species, 258).

Erignathus Gill, 1866. Proc. Essex Inst. Salem, 5:5.
TYPE SPECIES: *Phoca barbata* Fabricius, 1791 (= *Phoca barbata* Erxleben, 1777).

Erignathus barbatus (Erxleben, 1777). Syst. Regni Anim., 1:590.
TYPE LOCALITY: "ad Scotiam atque Groelandiam australiorem, vulgaris circa Islandiam" [North Atlantic, S Greenland].
DISTRIBUTION: Circumpolar Arctic seas (Canada, Greenland, Japan, Norway, USA, republics of the former USSR), south to Hokkaido (Japan) and Newfoundland.
SYNONYMS: *albigena* Pallas, 1811; *lepechenii* Lesson, 1828; *leporina* Lepechin, 1778; *nautica* Pallas, 1811; *parsonii* Lesson, 1828.

Halichoerus Nilsson, 1820. Skand. Faun. Dagg. Djur., 1:376.
TYPE SPECIES: *Halichoerus griseus* Nilsson, 1820 (= *Phoca grypus* Fabricius, 1791).
COMMENTS: Mohr (1952) described successful mating in captivity between *Phoca hispida* and *Halichoerus grypus*.

Halichoerus grypus (Fabricius, 1791). Skr. Nat. Selsk. Copenhagen, 1(2):167.
TYPE LOCALITY: Listed by Scheffer (1958) as "Greenland".
DISTRIBUTION: Temperate and subarctic waters around Britain, Canada (Newfoundland area), Iceland, Greenland, Norway, Russia (Kola Peninsula), and Baltic Sea.

SYNONYMS: *atlantica* Nehring, 1886; *baltica* Nehring, 1886; *griseus* Nilsson, 1820; *halichoerus* Thienemann, 1824; *macrorhynchus* Hornschuch and Schilling, 1851; *pachyrhynchus* Hornschuch and Schilling, 1851.

Hydrurga Gistel, 1848. Naturgesch. des Thierreichs, p. xi.
TYPE SPECIES: *Phoca leptonyx* Blainville, 1820.
COMMENTS: Replacement name for preoccupied *Stenorhinchus* E. Geoffroy Saint-Hilaire and F. Cuvier, 1826.

Hydrurga leptonyx (Blainville, 1820). J. Phys. Chim. Hist. Nat. Arts Paris, 91:298.
TYPE LOCALITY: "des environs des îles Falckland ou Malouines" [Falkland Isls (UK)].
DISTRIBUTION: Circumpolar, in southern oceans (Antarctica, Australia, Chile, Kerguelen Islands, South Sandwich Isls, Falkland Isls, southern Africa, New Zealand).
SYNONYMS: *homei* Lesson, 1828.

Leptonychotes Gill, 1872. Smithson. Misc. Coll., 11:70.
TYPE SPECIES: *Otaria weddellii* Lesson, 1826, by monotypy.
COMMENTS: Replacement name for *Leptonyx* Gray, 1837, which is preoccupied by *Leptonyx* Swainson, 1821.

Leptonychotes weddellii (Lesson, 1826). Bull. Sci. Nat. Geol., 7:437.
TYPE LOCALITY: "sur les côtes des Orcades australes, situées sour 60 degrés 37 minutes de lat" [South Orkney Isl (Br. Antarct. Trust Terr.)].
DISTRIBUTION: Coastal fast ice areas of Antarctic continent and adjacent islands; occasionally north to South America, Australia, and New Zealand.
SYNONYMS: *leopardina* Hamilton, 1839.
COMMENTS: Reviewed by Stirling (1971, Mammalian Species, 6) and Kooyman (1981).

Lobodon Gray, 1844. Zool. Voy. H.M.S. "Erebus" and "Terror," 1:2.
TYPE SPECIES: *Phoca carcinophaga* Hombron and Jacquinot, 1842.

Lobodon carcinophagus (Hombron and Jacquinot, 1842). *In* Dumont d'Uville, Voy. Pole Sud., Zool., Altas: Mammifères, pl. 10 [1842], vol. 3:Mammifères et Oiseaux, p. 27 [1853].
TYPE LOCALITY: "capturé sur les glaces du Pole Sud, entre les îles Sandwich et les îles Powels, à 150 lieues de distance de chacune de ces îles." [Scotia Sea (midway between South Orkney and South Sandwich Isls) (Br. Antarct. Trust Terr.)].
DISTRIBUTION: Antarctic seas, frequently on pack ice around Antarctic Continent, limited records from New Zealand, Australia, Tasmania, and Argentina.
SYNONYMS: *serridens* Owen, 1843.

Mirounga Gray, 1827. *In* Griffith et al., Anim. Kingdom, 5:179.
TYPE SPECIES: *Phoca proboscidea* Péron, 1816 (= *Phoca leonina* Linnaeus, 1758).
COMMENTS: Revised by King (1966). Ling and Bryden (1992) provided a key to the species.

Mirounga angustirostris (Gill, 1866). Proc. Essex Inst. Salem, 5:13.
TYPE LOCALITY: "California" restricted by Poole and Schantz (1942) to "St. Bartholomews Bay, lower California, Mexico." Clarified by Scheffer (1958) as "Bahía Tórtola (= Bahía San Bartolomé) 27°39'N, 114°51'W, Baja California, Mexico".
DISTRIBUTION: SE Alaska to California (USA) and Baja California (Mexico).

Mirounga leonina (Linnaeus, 1758). Syst. Nat., 10th ed., 1:37.
TYPE LOCALITY: "ad polum Antarcticum" restricted by Thomas (1911*a*) to "Juan Fernandez", further restricted by Hamilton (1940) as "Isla Mas a Tierra" [Chile].
DISTRIBUTION: Subantarctic islands, including Macquarie, Kerguelen, S Georgia Isls.
STATUS: CITES - Appendix II.
SYNONYMS: *ansonii* de Blainville, 1820; *ansonina* de Blainville, 1820; *coxii* Desmarest, 1817; *crosetensis* Lydekker, 1909; *dubia* Fischer, 1829; *elephantina* Gray, 1844; *falclandicus* Lydekker, 1909; *falklandica* Peters, 1875; *kerguelensis* Peters, 1875; *macquariensis* Lydekker, 1909; *patagonica* Gray, 1827; *proboscidea* Peron, 1816; *resima* Peron, 1816.
COMMENTS: Reviewed by Ling and Bryden (1992, Mammalian Species, 391).

Monachus Fleming, 1822. Philos. Zool., 2:187.
TYPE SPECIES: *Phoca monachus* Hermann, 1779.
COMMENTS: Revised by King (1956). Wyss (1988) suggested that this may be a paraphyletic group.

Monachus monachus (Hermann, 1779). Beschaft. Berlin Ges. Naturforsch. Fr., 4:501.
TYPE LOCALITY: King (1956) listed the locality as "captured in the autumn of 1777 in the Dalmation Sea at Ossero." [Yugoslavia].
DISTRIBUTION: Mediterranean and Black Seas; NW Africa to Cape Blanc.
STATUS: CITES - Appendix I; U.S. ESA and IUCN - Endangered.
SYNONYMS: *albiventer* Boddaert, 1785; *altantica* Gray, 1854; *bicolor* Shaw, 1800; *byronii* Desmarest, 1820; *hermannii* Lesson, 1828; *leucogaster* Peron, 1817; *mediterraneus* Nilsson, 1838.

Monachus schauinslandi Matschie, 1905. Sitzb. Ges. Naturf. Fr. Berlin, 1905:258.
TYPE LOCALITY: "Laysan ist eine kleine Koralleinsel, nordwestlich der Sandwich-Inseln" [USA, Laysan Isl, 25°50'N, 171°50'W].
DISTRIBUTION: NW Hawaiian Isls, from Nihoa to Kure.
STATUS: CITES - Appendix I; U.S. ESA and IUCN - Endangered.

Monachus tropicalis (Gray, 1850). Cat. Spec. Mamm. Coll. Br. Mus., Part 2(Seals), p. 28.
TYPE LOCALITY: "Jamaica" restricted by King (1956) to "Pedro Cays, 80 km. south of Jamaica".
DISTRIBUTION: Caribbean Sea and Yucatan.
STATUS: CITES - Appendix I; U.S. ESA - Endangered; IUCN - Extinct?.
SYNONYMS: *antillarum* Gray, 1849.
COMMENTS: "Extinct since the early 1950's" (Kenyon, 1977).

Ommatophoca Gray, 1844. Zool. Voy. H.M.S. "Erebus" and "Terror," 1:3.
TYPE SPECIES: *Ommatophoca Rossii* Gray, 1844, by monotypy.

Ommatophoca rossii Gray, 1844. Zool. Voy. H.M.S. "Erebus" and "Terror," 1:3.
TYPE LOCALITY: "Antarctic ocean", restricted by Barrett-Hamilton (1902) to "pack ice, north of Ross Sea 68°S, 176°E".
DISTRIBUTION: Circumpolar, Antarctic pack ice, particularly King Haakon VII Sea.

Phoca Linnaeus, 1758. Syst. Nat., 10th ed., 1:37.
TYPE SPECIES: *Phoca vitulina* Linnaeus, 1758, by tautonomy.
COMMENTS: Burns and Fay (1970), Rice (1977), McDermid and Bonner (1975), Gromov and Baranova (1981), J. E. King (1983), and Wyss (1988) considered *Phoca, Pusa, Histriophoca*, and *Pagophilus* a monophyletic group. However, Muizon (1982*a*) suggested that *Histricophoca* and *Pagophilus* may be more closely related to *Cystophora* than to *Phoca* (*sensu stricto*). Burns and Fay (1970) and McDermid and Bonner (1975) argued that these differences should be recognized only at the subgeneric level which is followed here.

Phoca caspica Gmelin, 1788. *In* Linnaeus, Syst. Nat., 13th ed., 1:64.
TYPE LOCALITY: "in mari, praesertim septentrionali, etiam Pacifico et Caspico" [Caspian Sea].
DISTRIBUTION: Caspian Sea.
COMMENTS: Placed in *Pusa* Scopoli, 1777, followed by Scheffer (1958).

Phoca fasciata Zimmermann, 1783. Geogr. Gesch. Mensch. Vierf. Thiere, 3:277.
TYPE LOCALITY: "Wohnt um die Kurilischen Inseln" [Russia, Kurile Isls].
DISTRIBUTION: Okhotsk, W Bering, Chukchiand Japan Seas.
SYNONYMS: *equestris* Pallas, 1831.
COMMENTS: Reviewed by Burns and Fay (1970). Placed in *Histriophoca* Gill, 1873 by Scheffer (1958) and Muizon (1982*a*)

Phoca groenlandica Erxleben, 1777. Syst. Regni Anim., 1:588.
TYPE LOCALITY: "in Groenlandia et Newfoundland."
DISTRIBUTION: N Atlantic and Arctic oceans from E Canada to the White Sea (Russia).

SYNONYMS: *albicauda* Desmarest, 1822; *albini* Alessandrini, 1851; *dorsata* Pallas, 1811; *leucopla* Thienemann, 1824; *oceanica* Lepechin, 1778; *semilunaris* Boddaert, 1785.
COMMENTS: Placed in *Pagophilus* Gray, 1844, followed by Scheffer (1958) and Muizon (1982*a*).

Phoca hispida Schreber, 1775. Die Säugethiere, 2(13):pl. 86[1775]; text, 3(17):312[1776].
TYPE LOCALITY: "Man fängt ihn auf den Küsten von Grönland und Labrader".
DISTRIBUTION: Arctic Ocean, Okhotsk, Bering, and Baltic Seas; Finland (Saimaa Lake); Russia (Ladoga Lake); Canada (Nettilling Lake, Baffin Isl).
STATUS: IUCN - Endangered as *P. h. saimensis*.
SYNONYMS: *annellata* Nilsson, 1820; *beaufortiana* Anderson, 1943; *birulai* Smirnov, 1929; *botnica* Gmelin, 1788; *foetida* Fabricius, 1776; *gichigensis* Allen, 1902; *krascheninikovi* Naumov and Smirnov, 1936; *ladogensis* Nordquist, 1899; *octonata* Kutorga, 1839; *pomororum* Smirnov, 1929; *pygmaea* Zukowsky, 1914; *saimensis* Nordquist, 1899; *soperi* Anderson, 1943.
COMMENTS: Placed in *Pusa* Scopoli, 1777, followed by Scheffer (1958). Mohr (1952) described successful mating in captivity between *Phoca hispida* and *Halichoerus grypus*.

Phoca largha Pallas, 1811. Zoogr. Rosso-Asiat., 1:113.
TYPE LOCALITY: "quam quod observetur tantum ad orientale littus Camtschatcae", Shaughnessy and Fay (1977) listed as "Eastern coast of Kamchatka" [Russia].
DISTRIBUTION: Associated with pack ice in coastal N Pacific, Bering and Okhotsk Seas, Aleutian Isls; south to Kiangsu (China) and Japan.
SYNONYMS: *chorisi* Lesson, 1828; *macrodens* Allen, 1902; *nummularis* Temminck, 1847; *ochotensis* Allen, 1902; *petersi* Mohr, 1941; *pribilofensis* Allen, 1902; *tigrina* Lesson, 1827.
COMMENTS: Scheffer (1958) considered *largha* as conspecific with *vitulina*; however, Shaughnessy and Fay (1977) and J. E. King (1983) separated the two.

Phoca sibirica Gmelin, 1788. *In* Linnaeus, Syst. Nat., 13th ed., 1:64.
TYPE LOCALITY: "Baikal et Orom" [Lake Baikal and Lake Oron (=Ozero Oron), U.S.S.R.].
DISTRIBUTION: Endemic to Lake Baikal (Russia).
SYNONYMS: *baicalensis* Dybowski, 1873; *oronensis* Dybowski, 1922.
COMMENTS: Placed in *Pusa* Scopoli, 1777, followed by Scheffer (1958). Reviewed by Thomas et al. (1982, Mammalian Species, 188).

Phoca vitulina Linnaeus, 1758. Syst. Nat., 10th ed., 1:38.
TYPE LOCALITY: "in mari Europæo" restricted by Thomas (1911*a*) to "Mari Bothnico et Baltico", however, presently it does not occur in the Gulf of Bothnia (Bobrinski et al., 1944).
DISTRIBUTION: Kurile Isls and Kamchatka (Russia); south to Kiangsu (China); Alaska to Mexico; Greenland and E Canada to NE USA; Iceland; Britain and Europe; Seal Lakes, Ungava Peninsula (Canada); Iliamna Lake, Alaska (USA).
STATUS: IUCN - Vulnerable as *P. v. stejnegeri*.
SYNONYMS: *antarcticus* Peale, 1848; *californica* Gray, 1866; *canina* Pallas, 1811; *concolor* DeKay, 1842; *geronimensis* Allen, 1902; *insularis* Belkin, 1964; *kurilensis* McLaren, 1966; *linnaei* Lesson, 1828; *littorea* Thienemann, 1824; *mellonae* Doutt, 1942; *pealii* Gill, 1866; *pribilofensis* Allen, 1902; *richardii* Gray, 1873; *scopulicola* Thienemann, 1824; *thienemannii* Lesson, 1828; *variegata* Nilsson, 1820.
COMMENTS: The position of *stejnegeri* Allen, 1902, remains uncertain; Scheffer (1958) placed in *largha* Pallas, 1811; J. E. King (1983) placed in *vitulina* Linnaeus, 1758; and Shaughnessy and Fay (1977) suggested *incertae sedis*.

Family Procyonidae Gray, 1825. Ann. Philos., n.s., 10:339.
COMMENTS: Revised by Hollister (1915), Pocock (1921*a*), Baskin (1982, 1989), and Decker and Wozencraft (1991). Does not include *Ailurus* or *Ailuropoda*, following information presented in Davis (1964), Todd and Pressman (1968), Sarich (1976), Ginsburg (1982), Wozencraft (1989*b*), and Decker and Wozencraft (1991); however, Hollister (1915), Gregory (1936), Thenius (1979), and Flynn et al. (1988) considered *Ailurus* in the Procyonidae.

Subfamily Potosinae Trouessart, 1904. Cat. Mamm. Viv. Foss. 1:183.
COMMENTS: Type genus: *Potos* E. Geoffroy Saint-Hilaire and G. [Baron] Cuvier, 1795, by original designation.

Bassaricyon J. A. Allen, 1876. Proc. Acad. Nat. Sci. Philadelphia, 28:20, pl. 1.
TYPE SPECIES: *Bassaricyon Gabbii* J. A. Allen, 1876, by original designation.
COMMENTS: Reviewed by Poglayen-Neuwall (1965). Several workers have suggested that the several named forms of *Bassaricyon* are conspecific (Decker and Wozencraft, 1991; Ewer, 1973; Hall and Kelson, 1959; Stains, 1967; Wozencraft, 1989*b*); but supporting systematic work is lacking.

Bassaricyon alleni Thomas, 1880. Proc. Zool. Soc. Lond., 1880:397.
TYPE LOCALITY: "Sarayacu, on the Bobonasa river, Upper Pastasa river", [Ecuador].
DISTRIBUTION: Ecuador east of the Andes, and Peru to Cuzco Prov.; Bolivia; possibly into Venezuela.

Bassaricyon beddardi Pocock, 1921. Ann. Mag. Nat. Hist., ser. 9, 7:231.
TYPE LOCALITY: "Bastrica woods, Essequibo River, British Guiana".
DISTRIBUTION: Guyana, and possibly adjacent Venezuela and Brasil.
COMMENTS: A renaming of *alleni* Sclater, 1895 (not Thomas, 1880). Cabrera (1958) erroneously listed *Bassaricyon beddardi* as *Bassariscus beddardi*.

Bassaricyon gabbii J. A. Allen, 1876. Proc. Acad. Nat. Sci. Philadelphia, 28:21.
TYPE LOCALITY: "Costa Rica," restricted by Allen (1908) to "Talamanca".
DISTRIBUTION: C Nicaragua, Costa Rica, Panama, W Colombia, W Ecuador.
STATUS: CITES - Appendix III (Costa Rica).
SYNONYMS: *medius* Thomas, 1909; *orinomus* Goldman, 1912; *richardsoni* J. A. Allen, 1908; *siccatus* Thomas, 1927.

Bassaricyon lasius Harris, 1932. Occas. Pap. Mus. Zool., Univ. Michigan, 248:3.
TYPE LOCALITY: "Estrella de Cartago, Costa Rica. This locality is six to eight miles south of Cartago near the source of the Rio Estrella, at an altitude of about 4,500 feet."
DISTRIBUTION: Known only from the type locality.

Bassaricyon pauli Enders, 1936. Proc. Acad. Nat. Sci. Philadelphia, 88:365.
TYPE LOCALITY: "Between Rio Chiriqui Viejo and Rio Colorado, on a hill known locally as Cerro Pando, elevation 4800 feet, about ten miles from El Volcan, Province de Chiriqui, R. de Panama."
DISTRIBUTION: Known only from the type locality.

Potos E. Geoffroy Saint-Hilaire and F. G. Cuvier, 1795. Mag. Encyclop., 2:187.
TYPE SPECIES: *Viverra caudivolvula* Schreber, 1777 (= *Lemur flavus* Schreber, 1774), by original designation.
COMMENTS: Hernandez-Camacho (1977) placed *Potos* in Cercoleptidae Bonaparte, 1838.

Potos flavus (Schreber, 1774). Die Säugethiere, 1(9):pl. 42[1774]; text, p. 187[189](index)[1774].
TYPE LOCALITY: "Er ist, der Sage nach, auf den Gebirgen in Jamaica einheimisch"; restricted by Thomas (1902*b*), to "Surinam". Ford and Hoffmann (1988) discussed the confusion over the name and type locality.
DISTRIBUTION: S Tamaulipas and Guerrero (possibly Michoacan), Mexico through Guatemala, Belize, Nicaragua, Costa Rica, and Panama, to the Mato Grosso of Brazil and N South American (Colombia, Venezuela, Guyana, Surinam, Ecuador, Peru, and Bolivia).
STATUS: CITES - Appendix III (Hondurus).
SYNONYMS: *arborensis* Goodwin, 1938; *aztecus* Thomas, 1902; *boothi* Goodwin, 1957; *brachyotos* Schinz, 1844; *brachyotus* Martin, 1836; *brasiliensis* Ihering, 1911; *campechensis* Nelson and Goldman, 1931; *caucensis* J. A. Allen, 1904; *caudivolvula* Schreber, 1777; *chapadensis* J. A. Allen, 1904; *chiriquensis* Allen, 1904; *dugesii* Villa, 1944; *guerrerensis* Goldman, 1915; *isthmicus* Goldman, 1913; *lepida* Illiger, 1815; *mansuetus* Thomas, 1914; *megalotus* Martin, 1836; *meridensis* Thomas, 1902; *modestus*

Lönnberg, 1921; *nocturna* Wied[-Neuwied], 1826; *potto* Müller, 1776; *prehensilis* Kerr, 1792; *simiasciurus* Schreber, 1774; *tolimensis* J. A. Allen, 1913.
COMMENTS: Revised by Kortlucke (1973) and Hernandez-Camacho (1977). Reviewed by Ford and Hoffmann (1988, Mammalian Species, 321), Cabrera (1958), and Husson (1978).

Subfamily Procyoninae Gray, 1825. Ann. Philos., n.s., 10:339.
COMMENTS: Type genus: *Procyon* Storr, 1780.

Bassariscus Coues, 1887. Science, 9:516.
TYPE SPECIES: *Bassariscus astutus*, by monotypy through the replaced name *Bassaris astuta* Lichtenstein, 1830 (Melville and Smith, 1987).
COMMENTS: Hollister (1915) placed in monotypic Bassariscidae.

Bassariscus astutus (Lichtenstein, 1830). Abh. König. Akad. Wiss., Berlin, 1827:119 [1830].
TYPE LOCALITY: "Mexico" [near city of Mexico].
DISTRIBUTION: SW Oregon, N Nevada, Utah, SW Wyoming and W Colorado, south through California, Arizona, New Mexico and Texas (USA), to the Isthmus of Tehuantepec (Mexico); Tiburon Isl and several other islands in the Gulf of California.
SYNONYMS: *albipes* Elliot, 1904; *arizonensis* Goldman, 1932; *bolei* Goldman, 1945; *consitus* Nelson and Goldman, 1932; *flavus* Rhoads, 1894; *insulicola* Nelson and Goldman, 1909; *macdougalli* Goodwin, 1956; *nevadensis* Miller, 1913; *octavus* Hall, 1926; *oregonus* Rhoads, 1894; *palmarius* Nelson and Goldman, 1909; *raptor* Baird, 1859; *saxicola* Merriam, 1897; *willetti* Stager, 1950; *yumanensis* Huey, 1937.
COMMENTS: Revised by Rhoads (1893). See Poglayen-Neuwall and Toweill (1988, Mammalian Species, 327).

Bassariscus sumichrasti (Saussure, 1860). Rev. Mag. Zool. Paris, ser. 2, 12:7.
TYPE LOCALITY: "Cet animal habite les greniers dans la région chaude du Mexique." Hall and Kelson (1959) listed "Mexico, Veracruz, Mirador."
DISTRIBUTION: Guerrero and S Veracruz (Mexico) to W Panama including Guatemala, Belize, Nicaragua, El Salvador, and Costa Rica.
STATUS: CITES - Appendix III (Costa Rica).
SYNONYMS: *campechensis* Nelson and Goldman, 1932; *monticola* Cordero, 1875; *notinus* Thomas, 1903; *oaxacensis* Goodwin, 1956; *variabilis* Peters, 1874.

Nasua Storr, 1780. Prodr. Meth. Mamm., p. 35, tabl. A.
TYPE SPECIES: *Viverra nasua* Linnaeus, 1766, by absolute tautomy (Melville and Smith, 1987).
COMMENTS: Revised by Allen (1879) and Decker (1991). Reviewed by Cabrera (1958).

Nasua narica (Linnaeus, 1766). Syst. Nat., 12th ed., 1:64.
TYPE LOCALITY: "America", restricted by Allen (1879) to "Veracruz, Mexico"; Herkovitz (1951) further restricted it to "Achotal, Isthmus of Techuantpec, Vera Cruz."
DISTRIBUTION: USA (S Arizona and SW New Mexico), south through Mexico (except Baja California), Central America (Belize, Costa Rica, El Salvador, Guatemala, Honduras, Nicaragua, and Panama), to northernmost Colombia (Gulf of Uraba).
STATUS: CITES - Appendix III (Honduras) as *Nasua nasua* (= *narica*); IUCN - Insufficiently known as *N. nelsoni*.
SYNONYMS: *bullata* Allen, 1904; *isthmica* Goldman, 1942; *mexicana* Weinland, 1860; *molaris* Merriam, 1902; *nelsoni* Merriam, 1901; *pallida* Allen, 1904; *panamensis* Allen, 1904; *richmondi* Goldman, 1932; *tamaulipensis* Goldman, 1942; *thersites* Thomas, 1901; *vulpecula* Erxleben, 1777; *yucatanica* Allen, 1904.
COMMENTS: Includes *nelsoni* (Decker, 1991).

Nasua nasua (Linnaeus 1766). Syst. Nat., 12th ed., 1:64.
TYPE LOCALITY: "America."; listed by Cabrera (1958) as "Pernambuco".
DISTRIBUTION: Argentina, Bolivia, Brazil, Colombia, Guyana, Peru, Venezuela, Paraguay, Surinam, Uruguay.
STATUS: CITES - Appendix III as *N. n. solitaria* (Uruguay) and *Nasua nasua* (Honduras).
SYNONYMS: *annulata* Desmarest, 1820; *aricana* Vieira, 1945; *boliviensis* Cabrera, 1956; *candace* Thomas, 1912; *cinerascens* Lönnberg, 1921; *dichromatica* Tate, 1939; *dorsalis* Gray, 1866;

fusca Schonburgk, 1839; *gualeae* Lönnberg, 1921; *gualeae* Lönnberg, 1921; *henseli* Lönnberg, 1921; *jivaro* Thomas, 1914; *judex* Thomas, 1914; *jurvana* Ihering, 1911; *leucorhynchus* Tschudi, 1845; *manium* Thomas, 1912; *mephisto* Thomas, 1927; *mexianae* Vieira, 1945; *mondie* Olfers, 1818; *montana* Tschudi, 1844; *monticola* Schinz, 1844; *nasica* Winge, 1859; *nasuta* Tschudi, 1844; *obfuscata* Olfers, 1818; *phaeocephala* Allen, 1904; *quasje* Gmelin, 1788; *quichua* Thomas, 1901; *rufa* Fischer [von Waldheim], 1814; *rufina* Tschudi, 1844; *rusca* Fischer [von Waldheim], 1814; *sociabilis* Schinz, 1821; *soederstroemmi* Lönnberg, 1921; *solitaria* Schinz, 1821; *vittata* Tschudi, 1845.

Nasuella Hollister, 1915. Proc. U.S. Natl. Mus., 49:148.
TYPE SPECIES: *Nasua olivacea meridensis* Thomas, 1901, by original designation.

Nasuella olivacea (Gray, 1865). Proc. Zool. Soc. Lond., 1864:703 [1865].
TYPE LOCALITY: "Santa Fé de Bogota" [Colombia], subsequently restricted by Cabrera (1958:249) to "Bogotá, lo que debe interpretarse como las montañas próximas a esta capital."
DISTRIBUTION: Andes of Colombia, W Venezuela, and Ecuador.

Procyon Storr, 1780. Prodr. Meth. Mamm., p. 35.
TYPE SPECIES: *Ursus lotor* Linnaeus, 1758, by designation by Elliot (1901).
COMMENTS: Reviewed by Goldman (1950) and Lotze and Anderson (1979); the later suggested *gloveralleni*, *maynardi*, and *minor* may be conspecific with *lotor*; Corbet and Hill (1986) considered *gloveralleni*, *insularis*, *minor*, and *pygmaeus* as conspecific with *lotor*; however, primary systematic works supporting these views are lacking.

Procyon cancrivorus (G. Cuvier, 1798). Tabl. Elem. Hist. Nat. Anim., p. 113.
TYPE LOCALITY: "se trouve à Cayenne" [French Guiana, Cayenne].
DISTRIBUTION: Argentina, Bolivia, Brazil, Colombia, Costa Rica, Guyana, Panama, Peru, Surinam, Trinidad and Tobago, and Venezuela.
SYNONYMS: *aequatorialis* J. A. Allen, 1915; *brasiliensis* Ihering, 1911; *nigripes* Mivart, 1886; *panamensis* Goldman, 1913; *proteus* J. A. Allen, 1904.

Procyon gloveralleni Nelson and Goldman, 1930. J. Mammal., 11:453.
TYPE LOCALITY: "Island of Barbados, Lesser Antilles, West Indies."
DISTRIBUTION: Known only from the type locality.
STATUS: IUCN - Extinct? (see Hall, 1981:973).

Procyon insularis Merriam, 1898. Proc. Biol. Soc. Washington, 12:17.
TYPE LOCALITY: "Maria Madre Island, Tres Marias Ids., Mexico."
DISTRIBUTION: María Madre Isl (*P. i. insularis*), and María Magdalena Isl (*P. i. vicinus*).
SYNONYMS: *vicinus* Nelson and Goldman, 1931.

Procyon lotor (Linnaeus, 1758). Syst. Nat., 10th ed., 1:48.
TYPE LOCALITY: "Americæ maritimis," restricted by Thomas (1911*a*) to "Pennsylvania" [USA].
DISTRIBUTION: S Canada throughout USA (except parts of the Rocky Mtns), Mexico, and Central America to C Panama. Introductions into France, Germany and republics of the former USSR.
SYNONYMS: *annulatus* Fischer, 1814; *auspicatus* Nelson, 1930; *brachyurus* Wiegmann, 1837; *californicus* Mearns, 1914; *castaneus* de Beaux, 1910; *crassidens* Hollister, 1914; *dickeyi* Nelson and Goldman, 1931; *elucus* Bangs, 1898; *excelsus* Nelson and Goldman, 1930; *flavidus* de Beaux, 1910; *fusca* Burmeister, 1850; *fuscipes* Mearns, 1914; *grinnelli* Nelson and Goldman, 1930; *gularis* H. Smith, 1848; *hernandezii* Wagler, 1831; *hirtus* Nelson and Goldman, 1930; *hudsonicus* Brass, 1911; *incautus* Nelson, 1930; *inesperatus* Nelson, 1930; *litoreus* Nelson and Goldman, 1930; *marinus* Nelson, 1930; *maritimus* Dozier, 1948; *megalodous* Lowery, 1943; *melanus* Gray, 1864; *mexicana* Baird, 1858; *nivea* Gray, 1837; *obscurus* Wiegmann, 1837; *ochraceus* Mearns, 1914; *pacifica* Merriam, 1899; *pallidus* Merriam, 1900; *proteus* Brass, 1911; *psora* Gray, 1842; *pumilus* Miller, 1911; *rufescens* de Beaux, 1910; *shufeldti* Nelson and Goldman, 1931; *simus* Gidley, 1906; *solutus* Nelson and Goldman, 1931; *vancouverensis* Nelson and Goldman, 1930; *varius* Nelson and Goldman, 1930; *vulgaris* Tiedemann, 1808.
COMMENTS: Reviewed by Lotze and Anderson (1979, Mammalian Species, 119).

Procyon maynardi Bangs, 1898. Proc. Biol. Soc. Washington, 12:92.
TYPE LOCALITY: "Nassau Island, Bahamas".
DISTRIBUTION: Known only from the type locality.
COMMENTS: Koopman et al. (1957) examined the type series and believed it to be conspecific with *P. lotor.*

Procyon minor Miller, 1911. Proc. Biol. Soc. Washington, 24:4.
TYPE LOCALITY: "Ponite-à-Pitre, Guadeloupe, Lesser Antilles".
DISTRIBUTION: Guadeloupe Isl (Lesser Antilles).

Procyon pygmaeus Merriam, 1901. Proc. Biol. Soc. Washington, 14:101.
TYPE LOCALITY: "Cozumel Island, Yucatan" [Mexico].
DISTRIBUTION: Known only from the type locality.
STATUS: IUCN - Insufficiently known.

Family Ursidae G. Fischer, 1817. Mém. Soc. Imp. Nat. Moscow, 5:372.
COMMENTS: *Ailurus* and *Ailuropoda* have been placed in a separate family by some; however, morphological and molecular evidence strongly supports the placement of *Ailuropoda* in this family (Chorn and Hoffmann, 1978; Davis, 1964; Goldman et al., 1989; Hendy, 1980*a*; O'Brien et al., 1985; Wozencraft, 1989*a*). Thenius (1979) placed *Ailuropoda* in the monotypic family Ailuropodidae, and *Ailurus* in the Procyonidae. The placement of *Ailurus* is certainly more controversial, with most phenetic studies placing it in the procyonids and most phylogenetic studies placing it near or in the ursids. See comments under *Ailurus* concerning its placement here. Morphological studies have supported the monophyly of three subfamilies (Hendy, 1980*a*; Kurtén, 1966; Thenius, 1979), although this has not been collaborated by a recent molecular approach (Goldman et al., 1989).

Subfamily Ailurinae Gray, 1843. List Specimens Mamm. Coll. Brit. Mus., p. xxi.
COMMENTS: Hendy (1980*a*, *b*) placed *Ailuropoda* and *Ailurus* in tribe Ailuropodini, subfamily Agriotheriinae (Ailurinae has priority).

Ailuropoda Milne-Edwards, 1870. Ann. Sci. Nat. Zool. (Paris), ser. 5, 13(10):1.
TYPE SPECIES: *Ursus melanoleucus* David, 1869, by monotypy.
COMMENTS: Revised by Davis (1964) and Hendey (1980*a*). Reviewed by Chorn and Hoffmann (1978).

Ailuropoda melanoleuca (David, 1869). Nouv. Arch. Mus. Hist. Nat. Paris, Bull., 5:12-13.
TYPE LOCALITY: "Mou-pin" [China, Sichuan Sheng, Baoxing (=Moupin) 30°23'N, 102°50'E].
DISTRIBUTION: China: Sichuan, Shensi, Gansu; perhaps Qinghai, on E edge of Tibetan plateau.
STATUS: CITES - Appendix I; U.S. ESA and IUCN - Endangered.
COMMENTS: Regarded by Hendey (1980*a*, *b*) as the only surviving species in the subfamily Agriotheriinae. Placed in the monotypic family Ailuropodidae by Thenius (1979). Reviewed by Chorn and Hoffmann (1978, Mammalian Species, 110).

Ailurus F. G. Cuvier, 1825. *In* E. Geoffroy Saint-Hilaire and F. G. Cuvier, Hist. Nat. Mammifèeres, pt. 3, 5(50), "Panda" 3 pp.
TYPE SPECIES: *Ailurus fulgens* F. G. Cuvier, 1825, by monotypy (Melville and Smith, 1987).
COMMENTS: Biochemical and molecular evidence has suggested that the enigmatic *Ailurus* is intermediate between the procyonids and the ursids (O'Brien et al., 1985; Sarich, 1973; Tagle et al., 1986; Wayne et al., 1989; Wurster and Benirschke, 1968); more closely related to ursids than to procyonids (Todd and Pressmann, 1968; Zhang and Shi, 1991); or more closely related to the procyonids than to ursids (Goldman et al., 1989). Morphological studies have pointed out the lack of any shared derived features with the procyonids (Bugge, 1978; Decker and Wozencraft, 1991; Ginsburg, 1982; Hunt, 1974; Mayr, 1986; Schmidt-Kittler, 1981; Wozencraft, 1989*a*, *b*). Flynn et al. (1988) could not find any unambiguous features to place *Ailurus* with the procyonids, and only two characters to unite *Ailurus* with some procyonids, the loss of M/3 (shared also with mustelids) and the presence of a cusp on P4 on some, but not all procyonids. Wozencraft (1989*a*) found 9 shared derived features with the

Ursidae, and Decker and Wozencraft (1991) identified 6 unambiguous shared derived features of the procyonids, all of which are lacking in *Ailurus*. Only phenetic studies have indicated any kind of relationship to the procyonids (Gregory, 1936; Thenius, 1979) and only when compared to *Procyon*, a derived procyonid (Baskin, 1982, 1989; Decker and Wozencraft, 1991).

Ailurus fulgens F. G. Cuvier, 1825. *In* E. Geoffroy Saint-Hilaire and F. G. Cuvier, Hist. Nat. Mammifères, pt. 3, 5(50), "Panda" 3 pp.

TYPE LOCALITY: "Indes orientales."

DISTRIBUTION: Yunnan and Szechwan (China), N Burma, Sikkim (India), Nepal.

STATUS: CITES - Appendix II; IUCN - Insufficiently known.

SYNONYMS: *ochraceus* Hodgson, 1847; *refulgens* Milne-Edwards, 1868; *styani* Thomas, 1902.

COMMENTS: Although this species is sometimes included in the Procyonidae because of its ringed tail, and superficial resemblance of teeth and rounded skull to *Procyon*, this species does not have the shared derived morphological characters that would place it there (Decker and Wozencraft, 1991). See comments under the genus. Nearly all of the derived features that it shares with *Ailuropoda* (and not with other bears) relate directly to feeding on bamboo (Davis, 1964; Gregory, 1936). Reviewed by Roberts and Gittleman (1984, Mammalian Species, 222).

Subfamily Ursinae G. Fischer, 1817. Mém. Soc. Imp. Nat. Moscow, 5:372.

COMMENTS: Revised by Erdbrink (1953). Although there is general agreement as to the monophyletic origin of this subfamily, there is considerable disagreement as to the relationships within (Goldman et al., 1989; Hendy, 1980*a*; Thenius, 1979). Erdbrink (1953) collapsed all taxa into a broadly encompassing "*Ursus*"; also suggested by the hybridization criteria of Van Gelder (1977*b*). A consensus tree, not shown, but derived from hypotheses presented in Goldman et al. (1989) would not elucidate relationships below this level. Simpson (1945) followed Pocock (1941*a*) and recognized these taxa as monotypic genera.

Helarctos Horsfield, 1825. Zool. J., 2(6):221.

TYPE SPECIES: *Helarctos euryspilus* Horsfield, 1825 (= *Ursus malayanus* Raffles, 1821), by original designation (Melville and Smith, 1987).

COMMENTS: Revised by Pocock (1932*b*). Van Gelder (1977) placed *Helarctos* in *Melursus*. Pocock (1941*a*) first suggested the close relationship between *M. ursinus* and *H. malayanus*; however, this was not well supported by Goldman et al. (1989).

Helarctos malayanus (Raffles, 1821). Trans. Linn. Soc. Lond., 13:254.

TYPE LOCALITY: "Sumatra" [Indonesia].

DISTRIBUTION: Burma, China (Yunnan and Szechwan), India, Indonesia (Sumatra, Borneo), Laos, Taiwan, Malaysia, Thailand, Vietnam.

STATUS: CITES - Appendix I; IUCN - Vulnerable.

SYNONYMS: *anmamiticus* Heude, 1901; *euryspilus* Horsfield, 1825; *wardi* Lydekker, 1906.

Melursus Meyer, 1793. Zool. Entdeck., p. 155.

TYPE SPECIES: *Melursus lybius* Meyer, 1793 (= *Bradypus ursinus* Shaw, 1791), by monotypy (Melville and Smith, 1987).

COMMENTS: Revised by Pocock (1932*b*). See comments under *Helarctos* concerning the relationship between these taxa.

Melursus ursinus (Shaw, 1791). Nat. Misc., 2 (unpaged) pl. 58.

TYPE LOCALITY: "Abinteriore Bengala"; restricted by Pocock (1941*a*) as "Patna, north of the Ganges, Bengal" [India].

DISTRIBUTION: Sri Lanka; India, north to the Indian desert and to the foothills of the Himalayas.

STATUS: CITES - Appendix I; IUCN - Vulnerable.

SYNONYMS: *inornatus* Pucheran, 1855; *labiatus* de Blainville, 1817; *longirostris* Tiedemann, 1820; *lybius* Meyer, 1793; *niger* Goldfuss, 1804.

Tremarctos Gervais, 1855. Hist. Nat. Mammifères, 2:20.

TYPE SPECIES: *Ursus ornatus* F. G. Cuvier, 1825.

COMMENTS: Thenius (1976) argued that differences between *Tremarctos* and other bears warranted separation at the subfamily level, however, Kurtén (1966) and Hendey (1980*a*) supported its inclusion here.

Tremarctos ornatus (F. G. Cuvier, 1825). *In* E. Geoffroy Saint-Hilaire and F. G. Cuvier, Hist. Nat. Mammifèeres, pt. 3, 5(50) "Ours des cordiliéres du Chili," 2 pp.

TYPE LOCALITY: "cordiliéres du Chili," restricted by Cabrera (1958) to "los montañas al este de Trujillo, departamento de la Libertad, Perú."

DISTRIBUTION: Mountainous regions of W Venezuela, Colombia, Ecuador, Peru, and W Bolivia; perhaps Panama.

STATUS: CITES - Appendix I; IUCN - Vulnerable.

SYNONYMS: *frugilegus* Tschudi, 1844; *lasallei* Maria, 1924; *majori* Thomas, 1902; *nasutus* Sclater, 1868; *thomasi* Hornaday, 1911.

Ursus Linnaeus, 1758. Syst. Nat., 10th ed., 1:47.

TYPE SPECIES: *Ursus arctos* Linnaeus, 1758, by tautonymy (Melville and Smith, 1987).

SYNONYMS: *Thalarctos*.

COMMENTS: The close relationship of these four species has been generally recognized by morphological and molecular studies (Goldman et al., 1989; Hendey, 1980*a*; Kurtén and Anderson, 1980; Shields and Kocher, 1990). Allen (1938) proposed a close relationship between *thibetanus* and *americanus*. Thenius (1953), Goldman et al. (1989), and Shields and Kocher (1991) gave support to the monophyly of *arctos* with *maritimus*.

Ursus americanus Pallas, 1780. Spicil. Zool., 14:5.

TYPE LOCALITY: Not given. In Pallas' (1780) description, he refered to Brickell (1737) who implied North Carolina (USA) by stating they "are very common in this province." Palmer (1904) listed the locality as "eastern North America".

DISTRIBUTION: NC Alaska to Labrador and Newfoundland (Canada), south to C California, N Nevada (USA), N Nayarit and S Tamaulipas (Mexico), and Florida (USA).

STATUS: CITES - Appendix II.

SYNONYMS: *altifrontalis* Elliot, 1903; *amblyceps* Baird, 1859; *californiensis* Miller, 1900; *carlottae* Osgood, 1901; *cinnamomum* Audubon and Bachman, 1854; *emmonsii* Dall, 1895; *eremicus* Merriam, 1904; *floridanus* Merriam, 1896; *glacilis* Kells, 1897; *hamiltoni* Cameron, 1957; *hunteri* Anderson, 1944; *hylodromus* Elliot, 1904; *kenaiensis* Allen, 1910; *kermodei* Hornaday, 1905; *luteolus* Griffith, 1821; *machetes* Elliot, 1903; *perniger* Allen, 1910; *pugnax* Swarth, 1911; *randi* Anderson, 1944; *sornborgeri* Bangs, 1898; *vancouveri* Hall, 1928.

Ursus arctos Linnaeus, 1758. Syst. Nat., 10th ed., 1:47.

TYPE LOCALITY: "sylvis Europæ frigidæ" restricted by Thomas (1911*a*) to, "Northern Sweden."

DISTRIBUTION: Formerly, NW Africa, all of the Palearctic from W Europe, Near and Middle East through N Himalayas to W and N China and Chukot (Russia); Hokkaido (Japan); W North America, north from N Mexico.

STATUS: CITES - Appendix I as *U. arctos* (Mexico, Bhutan, China, and Mongolia populations), *U. a. isabellinus*, and *U. a. pruinosus*; otherwise Appendix II. U.S. ESA - Endangered as *U. a. arctos* in Italy, *U. a. nelsoni* in Mexico, and *U. a. pruinosus*; Threatened as *U. a. horribilis* in the USA (48 conterminous states); IUCN - Extinct as *U. a. nelsoni*.

SYNONYMS: *absarokus* Merriam, 1914; *alascensis* Merriam, 1896; *albus* Gmelin, 1788; *alexandrae* Merriam, 1914; *alpinus* Fischer, 1814; *andersoni* Merriam, 1918; *annulatus* Billberg, 1827; *apache* Merriam, 1916; *argenteus* Billberg, 1827; *arizonae* Merriam, 1916; *atnarko* Merriam, 1918; *aureus* Fitzinger, 1855; *badius* Schrank, 1798; *baikalensis* Ognev, 1924; *bairdi* Merriam, 1918; *beringiana* Middendorff, 1853; *bisonophagus* Merriam, 1918; *bosniensis* Bolkay, 1925; *brunneus* Billberg, 1827; *cadaverinus* Eversmann, 1840; *californicus* Merriam, 1896; *canadensis* Merriam, 1914; *candescens* Hamilton-Smith, 1827; *caucasicus* Smirnov, 1919; *caurinus* Merriam, 1914; *cavifrons* Heude, 1901; *chelan* Merriam, 1916; *chelidonias* Merriam, 1918; *cinereus* Desmarest, 1820; *collaris* Cuvier and Geoffroy Saint-Hilarie, 1824; *colusus* Merriam, 1914; *crassodon* Merriam, 1918; *crassus* Merriam, 1918; *cressonus* Merriam, 1916; *crowtheri* Schinz, 1844; *dalli* Merriam,

1896; *dusorgus* Merriam, 1916; *eltonclarki* Merriam, 1914; *ereunetes* Merriam, 1918; *eulophus* Merriam, 1904; *euryrhinus* Nilsson, 1847; *eversmanni* Gray, 1864; *eximius* Merriam, 1916; *falciger* Reichenbach, 1836; *ferox* Temminck, 1844; *formicarius* Billberg, 1828; *fuscus* Gmelin, 1788; *grandis* Gray, 1864; *griseus* Kerr, 1792; *gyas* Merriam, 1902; *henshawi* Merriam, 1914; *holzworthi* Merriam, 1929; *hoots* Merriam, 1916; *horriaeus* Baird, 1858; *horribilis* Ord, 1815; *hylodromus* Elliot, 1904; *idahoensis* Meriam, 1918; *imperator* Merriam, 1914; *impiger* Merriam, 1918; *innuitus* Merriam, 1914; *inopinatus* Merriam, 1918; *insularis* Merriam, 1916; *internationalis* Merriam, 1914; *isabellinus* Horsfield, 1826; *jeniseensis* Ognev, 1924; *kadiaki* Kleinschmidt, 1911; *kenaiensis* Merriam, 1904; *kennerleyi* Merriam, 1914; *kidderi* Merriam, 1902; *klamathensis* Meriam, 1914; *kluane* Meriam, 1916; *kodiaki* Kleinschmidt, 1911; *kolymensis* Ognev, 1924; *kwakiutl* Merriam, 1916; *lagomyiarius* Przewalski, 1883; *lasiotus* Gray, 1867; *lasistanicus* Satunin, 1913; *latifrons* Merriam, 1914; *leuconyx* Severtzov, 1873; *macfarlani* Merriam, 1918; *machetes* Elliot, 1903; *macrodon* Merriam, 1918; *magister* Merriam, 1914; *major* Nilsson, 1820; *mandchuricus* Heude, 1898; *marsicanus* Altobello, 1921; *melanarctos* Heude, 1898; *mendocinensis* Merriam, 1916; *meridionalis* Middendorff, 1851; *merriamii* J. A. Allen, 1902; *middendorffi* Merriam, 1896; *minor* Nilsson, 1820; *mirabilis* Merriam, 1916; *mirus* Meriam, 1918; *myrmephagus* Billberg, 1827; *navaho* Merriam, 1914; *neglectus* Merriam, 1916; *nelsoni* Merriam, 1914; *niger* Gmelin, 1788; *normalis* Gray, 1864; *nortoni* Merriam, 1914; *norvegicus* Fischer, 1829; *nuchek* Merriam, 1916; *ophrus* Merriam, 1916; *orgiloides* Merriam, 1918; *orgilos* Merriam, 1914; *oribasus* Merriam, 1918; *pallasi* Merriam, 1916; *pamirensis* Ognev, 1924; *pellyensis* Meriam, 1918; *persicus* Lönnberg, 1925; *perturbans* Merriam, 1918; *pervagor* Merriam, 1914; *phaeonyx* Merriam, 1904; *piscator* Pucheran, 1855; *planiceps* Merriam, 1918; *polonicus* Gray, 1864; *pruinosus* Blyth, 1854; *pulchellus* Merriam, 1918; *pyrenaicus* Fischer, 1829; *richardsoni* Swainson, 1838; *rogersi* Merriam, 1918; *rossicus* Gray, 1864; *rufus* Borkhausen, 1797; *rungiusi* Merriam, 1918; *russelli* Merriam, 1914; *sagittalis* Merriam, 1918; *scandinavicus* Gray, 1864; *schmitzi* Matschie, 1917; *selkirki* Merriam, 1916; *shanorum* Thomas, 1906; *sheldoni* Merriam, 1910; *shirasi* Merriam, 1914; *shoshone* Merriam, 1914; *sibiricus* Gray, 1864; *sitkeenensis* Merriam, 1914; *sitkensis* Merriam, 1896; *smirnovi* Lönnberg, 1925; *stenorostris* Gray, 1864; *syriacus* Hemprich and Ehrenberg, 1828; *tahltanicus* Merriam, 1914; *texensis* Merriam, 1914; *toklat* Merriam, 1914; *townsendi* Merriam, 1916; *tularensis* Merriam, 1918; *tundrensis* Merriam, 1914; *ursus* Boddaert, 1772; *utahensis* Merriam, 1914; *warburtoni* Merriam, 1916; *washake* Merriam, 1916; *yesoensis* Lydekker, 1897.

COMMENTS: Reviewed by Couturier (1954), Rausch (1963*a*), Kurtén (1973), and Hall (1984). Ognev (1931) and Allen (1938) recognized *U. pruinosus* as distinct; not followed by Ellerman and Morrison-Scott (1951), Gao (1987), and Strognov (1962). Lönnberg (1923*b*) believed that differences between *pruinosus* and *arctos* warranted subgeneric distinction as (*Mylarctos*) *pruinosus*; however, this was not supported by Pocock's (1932*a*) thorough revision.

Ursus maritimus Phipps, 1774. Voyage Towards North Pole, p. 185.

TYPE LOCALITY: "on the main land of Spitsbergen" [Norway].

DISTRIBUTION: Circumpolar in the Arctic, S limits determined by ice pack.

STATUS: CITES - Appendix II; IUCN - Vulnerable.

SYNONYMS: *eogroenlandicus* Knotterus-Meyer, 1908; *groenlandicus* Birula, 1932; *jenaensis* Knotterus-Meyer, 1908; *labradorensis* Knotterus-Meyer, 1908; *marinus* Pallas, 1776; *polaris* Shaw, 1792; *spitzbergensis* Knotterus-Meyer, 1908; *ungavensis* Knottnerus-Meyer, 1908.

COMMENTS: Revised by Wilson (1976). Reviewed by DeMaster and Stirling (1981, Mammalian Species, 145). Placed in subgenus *Thalarctos* by Gromov and Baranova (1981). Sister species to *arctos* (Goldman et al., 1989; Shields and Kocher, 1991).

Ursus thibetanus G.[Baron] Cuvier, 1823. Rech. Oss. Foss., Nouv. ed., 4:325.

TYPE LOCALITY: "Cet ours a été trouvé d'abord par M. Wallich dans les montagnes du Napaul, et je l'ai rencontré également dans celles du Sylhet" [India, Assam, Sylhet].

DISTRIBUTION: Afghanistan, China, India, Indochina, Japan, Korea, Laos, Nepal, Pakistan, Taiwan, Thailand, Russia (SE Primorski Krai), Vietnam.

STATUS: CITES - Appendix I; U.S. ESA - Endangered as *U. t. gedrosianus*; IUCN - Endangered as *U. t. gedrosianus*, otherwise Vulnerable.

SYNONYMS: *arboreus* Gray, 1864; *clarki* Sowerby, 1920; *formosanus* Swinhoe, 1864; *gedrosianus* Blanford, 1877; *japonicus* Schlegel, 1857; *laniger* Pocock, 1932; *leuconyx* Heude, 1901; *macneilli* Lydekker, 1909; *melli* Matschie, 1922; *mupinensis* Heude, 1901; *rexi* Matschie, 1897; *torquatus* Wagner, 1841; *ussuricus* Heude, 1901; *wulsini* Howell, 1928.
COMMENTS: Placed in subgenus *Selenarctos* by Gromov and Baranova (1981); and in subgenus *Euarctos* by Thenius (1979). Allen (1938) suggested a close relationship to *U. americanus*; Pocock (1932*b*) retained in a separate genus, there is molecular support for both positions (Goldman et al., 1989).

Family Viverridae Gray, 1821. London Med. Repos., 1821:301.
COMMENTS: Type genus: *Viverra* Linnaeus, 1758, by original designation (Melville and Smith, 1987). Does not include subfamilies Herpestinae Bonaparte, 1845, or Galidiinae Gray, 1865; see Pocock (1916*c*, 1919), Gregory and Hellman (1939), Wurster and Benirschke (1968), Thenius (1972), Radinsky (1975), Hunt (1987), Flynn et al. (1988), and Wozencraft (1989*a*, *b*).

Subfamily Cryptoproctinae Gray, 1865. Proc. Zool. Soc. Lond., 1864:508 [1865].
COMMENTS: Type genus: *Cryptoprocta* Bennett, 1833, by monotypy (Melville and Smith, 1987). Considered a subfamily of felids by Gregory and Hellman (1939) and Beaumont (1964). There is disagreement as to the proper subfamilies for the Malagasy viverrids; see Pocock (1915*a*), Petter (1974), and Wozencraft (1989*a*, *b*).

Cryptoprocta Bennett, 1833. Proc. Zool. Soc. Lond., 1833:46.
TYPE SPECIES: *Cryptoprocta ferox* Bennett, 1833, by original designation (Melville and Smith, 1987).
COMMENTS: Beaumont (1964) placed *Cryptoprocta* with the Felidae because of similarities in dental characters. Albignac (1970), Thenius (1972), Radinsky (1975), Coetzee (1977*b*), Flynn et al. (1988), and Wozencraft (1989*a*, *b*) placed it in the Viverridae; Hemmer (1978) suggested an intermediate position.

Cryptoprocta ferox Bennett, 1833. Proc. Zool. Soc. Lond., 1833:46.
TYPE LOCALITY: "Madagascar".
DISTRIBUTION: Endemic to Madagascar.
STATUS: CITES - Appendix II; IUCN - Insufficiently known.
SYNONYMS: *typicus* A. Smith, 1834.
COMMENTS: Reviewed by Köhncke and Leonhardt (1986, Mammmalian Species, 254).

Subfamily Euplerinae Chenu, 1852. Ency. Hist. Nat., 21:165.
COMMENTS: Type genus: *Eupleres* Doyère, 1835, by monotypy (Melville and Smith, 1987) for those who consider *Eupleres* and *Hemigalus* in different family groups. There is disagreement as to the subfamilies for Malagasy viverrids; Pocock (1915*a*) placed *Fossa* in the SE Asian Hemigalinae. The evidence for this is weak (Albignac, 1973; Wozencraft, 1989*a*). Gregory and Hellman (1939), Albignac (1974), and Petter (1974), placed it in Fossinae; however, Eupleridae Chenu, 1852, is the senior synonym.

Eupleres Doyère, 1835. Bull. Soc. Sci. Nat., 3:45.
TYPE SPECIES: *Eupleres goudotii* Doyère, 1835, by monotypy (Melville and Smith, 1987).
COMMENTS: Gregory and Hellman (1939) followed Chenu (1852), and suggested placing *Eupleres* in a separate family; however, Albignac (1973, 1974), Petter (1974), Coetzee (1977*b*), and Wozencraft (1989*a*) included it in Viverridae.

Eupleres goudotii Doyère, 1835. Bull. Soc. Sci. Nat., 3:45.
TYPE LOCALITY: "Tamatave" [Madagascar, 18°10'S, 49°23'E].
DISTRIBUTION: Endemic to Madagascar.
STATUS: CITES - Appendix II. IUCN - Vulnerable; considered by IUCN to have high conservation priority status for viverrids (Schreiber et al., 1989).
SYNONYMS: *major* Lavauden, 1929.
COMMENTS: Includes *E. major* (Albignac, 1973; Coetzee, 1977*b*).

Fossa Gray, 1865. Proc. Zool. Soc. Lond., 1864:518 [1865].
TYPE SPECIES: *Fossa d'aubentonii* Gray, 1865 (= *Viverra Fossana* Müller, 1776).

Fossa fossana (Müller, 1776). Linné's Vollstand, Natursyst., Suppl., p. 32.
TYPE LOCALITY: "Madagascar".
DISTRIBUTION: Endemic to Madagascar.
STATUS: CITES - Appendix II; IUCN - Vulnerable.
SYNONYMS: *d'aubentonii* Gray, 1865; *fossa* Schreber, 1777; *majori* Dollman, 1909.
COMMENTS: The commonly used *Viverra fossa* Schreber, 1777, is a junior synonym (G. Petter, 1962, 1974).

Subfamily Hemigalinae Gray, 1865. Proc. Zool. Soc. Lond., 1864:508 [1865].
COMMENTS: Simpson (1945) also included *Eupleres* and *Fossa,* following Pocock (1915*a*). Pocock (1933*d*) and Gregory and Hellman (1939) placed *Cynogale* in the monotypic Cynogalinae, although both recognized the close relationship of *Cynogale* to other hemigalines. Placed in Hemigalinae by Simpson (1945) and Ellerman and Morrison-Scott (1951).

Chrotogale Thomas, 1912. Abstr. Proc. Zool. Soc. Lond., 1912(106):17.
TYPE SPECIES: *Chrotogale owstoni* Thomas, 1912, by monotypy.

Chrotogale owstoni Thomas, 1912. Abstr. Proc. Zool. Soc. Lond., 1912(106):17.
TYPE LOCALITY: "Yen-bay, on the Song-koi River, Tonkin" [Vietnam: Yen Bay on the Songhoi River; 21°43'N 104°54'E].
DISTRIBUTION: China (Yunnan, Guangxi), Laos, Vietnam.
STATUS: IUCN - Insufficiently known; considered by IUCN to have high conservation priority status for viverrids (Schreiber et al., 1989).
COMMENTS: Revised by Pocock (1933*d*).

Cynogale Gray, 1837. Proc. Zool. Soc. Lond., 1836:88 [1837].
TYPE SPECIES: *Cyongale bennettii* Gray, 1837, by monotypy (Melville and Smith, 1987).
SYNONYMS: *Lamictis* de Blainville, 1837; *Potamophilus* Müller, 1838.

Cynogale bennettii Gray, 1837. Proc. Zool. Soc. Lond., 1836:88 [1837].
TYPE LOCALITY: "Sumatra" [Indonesia].
DISTRIBUTION: Brunei, Indonesia (Sumatra, Borneo), Malaysia, Thailand, Vietnam.
STATUS: CITES - Appendix II. IUCN - Insufficiently known; *Cynogale* populations from Vietnam (*C. lowei*) considered by IUCN to have high conservation priority status for viverrids (Schreiber et al., 1989).
SYNONYMS: *barbatus* Müller, 1838; *carcharias* de Blainville, 1837; *lowei* Pocock, 1933.
COMMENTS: Revised by Pocock (1933*d*). Includes *C. lowei,* which is only known from the type, a poorly preserved juvenile skin from N Vietnam (Ellerman and Morrison-Scott, 1951).

Diplogale Thomas, 1912. Abstr. Proc. Zool. Soc. Lond., 1912(106):18.
TYPE SPECIES: *Hemigale hosei* Thomas, 1892, by original designation.

Diplogale hosei (Thomas, 1892) Ann. Mag. Nat. Hist. ser. 6, 9:250.
TYPE LOCALITY: "Mount Dulit, N. Borneo, 4000 ft" [Malaysia, Sarawak, Gunung Dulit, 3°15'N, 114°15'E].
DISTRIBUTION: Malaysia (Sabah, Sarawak).
COMMENTS: Thomas published two accounts of the type description in 1892, one printed in August (Thomas, 1892*a*), and another printed in October (Thomas, 1892*b*). Although Pocock (1933*d*) supported Thomas (1912) in separating this species into *Diplogale;* Chasen (1940), Medway (1977), and Payne et al. (1985) did not. Similiar differences in pelage and dentition have been used elsewhere (e.g., *Osbornictis* vs. *Genetta*) at the generic level.

Hemigalus Jourdan, 1837. C. R. Acad. Sci. Paris, 5:442.
TYPE SPECIES: *Hemigalus zebra* Gervais, 1841 (= *Paradoxùrus Derbyànus* Gray, 1837), by monotypy.

SYNONYMS: *Hemigale* Gray, 1865.

Hemigalus derbyanus (Gray, 1837). Mag. Nat. Hist. [Charlesworth's], 1:579.
TYPE LOCALITY: Not given. Fixed by Gray (1837) as "in Peninsulâ Malayanâ".
DISTRIBUTION: Peninsular Burma, Indonesia (Sipora Isl, South Pagi Isl, Borneo, Sumatra), Malaysia, Thailand.
STATUS: CITES - Appendix II.
SYNONYMS: *boiei* Müller, 1838; *derbianus* Thomas, 1915; *derbyi* Temminck, 1841; *incursor* Thomas, 1915; *invisus* Gray, 1830; *minor* Miller, 1903; *zebra* Gray, 1837.
COMMENTS: Gervais (1841) was the first to place in *Hemigalus*. Gray (1849) later considered *Paradoxurus derbyanus* a junior synonym of *Viverra hardwicki* Gray, 1830, and placed it also in the genus *Hemigalea* after Jourdan (1837). However, *V. hardwicki* is a junior synonym of *Prionodon linsang* Raffles, 1821.

Subfamily Nandiniinae Pocock, 1929. Ency. Brit. (ed. 14), 3:898.
COMMENTS: Type genus: *Nandinia* Gray, 1843, by monotypy. Listed in Nandiniinae by Gregory and Hellman (1939), and Coetzee (1977*b*); separated at the family level by Pocock (1929), and Hunt (1987). Although *Nandinia* shares many derived features of the scent gland and external morphology with the viverrids (especially the Paradoxurinae), Hunt (1987) felt that the retention of the primitive features of the auditory bullae were sufficient to warrant separation at the family level.

Nandinia Gray, 1843. List Specimens Mamm. Coll. Brit. Mus., p. 54.
TYPE SPECIES: *Viverra binotata* Gray, 1830, by monotypy (Melville and Smith, 1987).

Nandinia binotata (Gray, 1830). Spicil. Zool., 2:9.
TYPE LOCALITY: "Africa, Ashantee" [Ghana; Ashanti Region; aproximately at 6°55'N 0°32'E].
DISTRIBUTION: Angola, Benin, Cameroon, Central African Republic, Congo, Equatorial Guinea, Gabon, Ghana, Guinea, Guinea-Bissau, Ivory Coast, Kenya, Liberia, Malawi, Mozambique, Nigeria, Rwanda, Senegal, Sierra Leone, Sudan, Tanzania, Togo, Uganda, Zaire, Zambia, Zimbabwe.
SYNONYMS: *arborea* Heller, 1913; *gerrardi* Thomas, 1893; *hamiltonii* Gray, 1832.

Subfamily Paradoxurinae Gray, 1865. Proc. Zool. Soc. Lond., 1864:508 [1865].
COMMENTS: Type genus: *Paradoxurus* F. G. Cuvier, 1821.

Arctictis Temminck, 1824. Prospectus de Monographies de Mammiferes, p. xxi. [issued March, 1824].
TYPE SPECIES: *Viverra ? binturong* Raffles, 1821, by monotypy (Melville and Smith, 1987).
SYNONYMS: *Ictides* Valenciennes, 1825.
COMMENTS: First placed in Paradoxurinae by Gray (1869).

Arctictis binturong (Raffles, 1821). Trans. Linn. Soc. Lond., 13:253.
TYPE LOCALITY: "Malacca".
DISTRIBUTION: Bangladesh, Bhutan, Burma, China (Yunnan), India (incl. Sikkim), Indonesia (Borneo, Java, Sumatra), Laos, Malaysia, Nepal, Philippine Isls (Palawan), Thailand, Vietnam.
STATUS: CITES - Appendix III (India).
SYNONYMS: *albifrons* F. G. Cuvier, 1822; *ater* F. G. Cuvier and E. Geoffroy Saint-Hilaire, 1824; *gairdneri* Thomas, 1916; *niasensis* Lyon, 1916; *pageli* Schwarz, 1911; *penicillatus* Temminck, 1835; *whitei* Allen, 1910.
COMMENTS: Revised by Pocock (1933*d*).

Arctogalidia Merriam, 1897. Science, 5:302.
TYPE SPECIES: *Paradoxurus trivirgatus* Gray, 1832, by monotypy through the replaced name *Arctogale* Gray, 1865 (Melville and Smith, 1987).
SYNONYMS: *Arctogale* Gray, 1865.
COMMENTS: Gray's (1864[1865]) generic name stood until Merriam (1897) pointed out that the name *Arctogale* was preoccupied (= *Arctogale erminea* Kaup, 1829). First placed in the Paradoxurinae by Gray (1869); later separated into the monotypic Arctogalidinae

by Pocock (1933*d*), and followed by Gregory and Hellman (1939); however, Simpson (1945), and Ellerman and Morrison-Scott (1951) retained it in Paradoxurinae.

Arctogalidia trivirgata (Gray, 1832). Proc. Zool. Soc. Lond., 1832:68.

TYPE LOCALITY: "from a specimen in the Leyden Museum, sent from the Molúccas", restricted by Jentink (1887) to "Java, Buitenzorg" [= Indonesia, Java, Bogor] but see comments.

DISTRIBUTION: Bangladesh, Burma, China (Yunnan), India, Indonesia, Laos, Malaysia, Thailand, Vietnam.

STATUS: IUCN - Indeterminate as *A. t. trilineata*.

SYNONYMS: *bancana* Schwarz, 1913; *bicolor* Miller, 1913; *depressa* Miller, 1913; *fusca* Miller, 1906; *inornata* Miller, 1901; *leucotis* Horsfield, 1851; *macra* Miller, 1913; *major* Miller, 1906; *melli* Matschie, 1922; *millsi* Wroughton, 1921; *mima* Miller, 1913; *minor* Lyon, 1907; *simplex* Miller, 1902; *stigmaticus* Temminck, 1853; *sumatrana* Lyon, 1908; *tingia* Lyon, 1908; *trilineatus* Wagner, 1841.

COMMENTS: Revised by Pocock (1933*d*) and Van Bemmel (1952). Gray (1832) originally described the type from the "Moluccas"; later Temminck (1841) refered to the same specimen as being from "Java". Gray (1843), then corrected the presumed geographic error and listed the same type as from "Malacca". Jentink (1887) listed the same type from "Buitenzorg". However, Van Bemmel (1952) stated that the collector, Reinwardt, was in the eastern part of the Indo-Australian Archipelago in 1821 and the type did not match other specimens from Java.

Macrogalidia Schwarz, 1910. Ann. Mag. Nat. Hist., ser. 5, 8:423.

TYPE SPECIES: *Paradoxurus musschenbroekii* Schlegel, 1877, by monotypy.

COMMENTS: First placed in the Paradoxurinae by Pocock (1933*d*).

Macrogalidia musschenbroekii (Schlegel, 1877). Prosp. mus. publ., 1877.

TYPE LOCALITY: "in the Northern parts of the isle of Celebes", restriced by Jentink (1887), to "Celebes, Menado-Kinilo" [Indonesia, Kinilou, 1°22'N, 124°51'E].

DISTRIBUTION: Indonesia (Sulawesi).

STATUS: IUCN - Rare.

COMMENTS: Schlegel (1877) circulated an unnumbered "Prospectus," for the "Annals of the Royal Zoological Museum of the Netherlands at Leyden", which contained the first mention of the new species. The prospectus was to preceed the first issue of the new "Annals" which was never printed. He republished the type description in the Notes of the Leyden Museum, (1879).

Paguma Gray, 1831. Proc. Comm. Sci. and Corres. Zool. Soc. Lond., 1831:94.

TYPE SPECIES: *Gulo larvatus* C. E. H[amilton] Smith, 1827.

COMMENTS: Reviewed by Pocock (1933*c*). First placed in the Paradoxurinae by Gray (1864).

Paguma larvata (H[amilton]. Smith, 1827). *In* Griffith et al., Anim. Kingdom, 2:281.

TYPE LOCALITY: Not given. Fixed by Temminck (1841) as "Nepal". Gray (1864) discounted this because he knew of no specimens from Nepal, and reassigned the name to two specimens from Canton, China collected by J. R. Reeve (Pocock, 1934*a*).

DISTRIBUTION: Bangladesh, Burma, Cambodia, China (Hainan Dao north to Hopei, Shanxi and the vicinity of Beijing), India (and S Andaman Isls), Indonesia (N Borneo, Sumatra), Japan (introduced), Laos, Malaysia, Nepal, Pakistan, Singapore, Taiwan, Thailand, Vietnam.

STATUS: CITES - Appendix III (India).

SYNONYMS: *annectens* Robinson and Kloss, 1918; *aurata* Blainville, 1842; *grayi* Bennett, 1835; *hainana* Thomas, 1909; *intrudens* Wroughton, 1910; *janetta* Thomas, 1928; *jourdanii* Gray, 1837; *laniger* Hodgson, 1841; *lanigerus* Hodgson, 1836; *leucocephala* Gray,, 1850; *leucomystax* Gray, 1837; *neglecta* Pocock, 1934; *nigriceps* Pocock, 1939; *nipalensis* Hodgson, 1836; *ogilbyi* Fraser, 1849; *pallasii* Otto, 1835; *reevesi* Matschie, 1908; *rivalis* Thomas, 1921; *robustus* Miller, 1906; *rubidus* Blyth, 1858; *taivana* Swinhoe, 1862; *tytlerii* Tytler, 1864; *vagans* Kloss, 1919; *wroughtoni* Schwarz, 1913; *yunalis* Thomas, 1921.

COMMENTS: Pocock (1934*b*) included *Paradoxurus tytlerii*. The "imperfect, no doubt immature skin, without skull (B.M. no. 43.1.12.103)" (Pocock, 1941*a*:416) provisionally recognized by Pocock as *P. lanigera* does not contain diagnostic features that would definitively align the specimen with *Paguma* (Ellerman and Morrison-Scott, 1951).

Paradoxurus F. G. Cuvier, 1821. *In* E. Geoffroy Saint-Hilaire and F. G. Cuvier, Hist. Nat. Mammifères, pt. 2, 3(24):15 "Martre des Palmiers" and pl. 186.
TYPE SPECIES: *Paradoxurus typus* F. G. Cuvier, 1821 (= *Viverra hermaphrodita* Pallas, 1777), by indication (Melville and Smith, 1987).
COMMENTS: Revised by Pocock (1933*c*).

Paradoxurus hermaphroditus (Pallas, 1777). *In* Schreber, Die Säugethiere, 3(25):426, [1777].
TYPE LOCALITY: Uncertain. "Das Vaterland des beschreibenen Thieres ist die Barbarey".
DISTRIBUTION: Bhutan, Burma, Cambodia, China, India, Indonesia, Japan, Laos, Malaysia, Nepal, New Guinea, Philippine Isls, Singapore, Sri Lanka, Thailand, Vietnam; scattered records in Sulawesi, Moluccas, and Aru Isls, probably resulting from introductions.
STATUS: CITES - Appendix III (India); IUCN - Endangered as *P. lignicolor*.
SYNONYMS: *baritensis* Lönnberg, 1925; *birmanicus* Wroughton, 1917; *bondar* Desmarest, 1820; *brunneipes* Miller, 1906; *canescens* Lyon, 1907; *canus* Miller, 1913; *celebensis* Schwarz, 1911; *cochinensis* Schwarz, 1911; *crassiceps* Pucheran, 1855; *crossi* Gray, 1832; *dubius* Gray, 1832; *exitus* Schwarz, 1911; *felinus* Wagner, 1841; *fossa* Marsden, 1811; *fuliginosus* Gray, 1832; *fuscus* Miller, 1913; *hamiltonii* Gray, 1832; *hanieli* Schwarz, 1912; *hirsutus* Hodgson, 1836; *javanica* Horsfield, 1824; *kangeanus* Thomas, 1910; *kutensis* Chasen and Kloss, 1916; *laneus* Pocock, 1934; *laotum* Gyldenstolpe, 1917; *lehmanni* Mertens, 1929; *leucopus* Ogilby, 1829; *lignicolor* Miller, 1903; *macrodus* Gray, 1865; *milleri* Kloss, 1908; *minax* Thomas, 1909; *minor* Bonhote, 1903; *musanga* Raffles, 1822; *musangoides* Gray, 1837; *nictitatans* Taylor, 1891; *niger* Blanford, 1885; *nigra* Desmarest, 1820; *nigrifrons* Gray, 1843; *padangus* Lyon, 1908; *pallasii* Gray, 1832; *pallens* Miller, 1913; *parvus* Miller, 1913; *pennantii* Gray, 1832; *philippinensis* F. G. Cuvier, 1837; *prehensilis* Desmarest, 1820; *pugnax* Miller, 1913; *pulcher* Miller, 1913; *quadriscriptus* Horsfield, 1855; *ravus* Miller, 1913; *rindjanicus* Mertens, 1929; *robustus* Miller, 1906; *rubidus* Blyth, 1858; *sabanus* Thomas, 1909; *sacer* Miller, 1913; *scindiae* Pocock, 1934; *senex* Miller, 1913; *setosus* Jacquinot and Pucheran, 1853; *siberu* Chasen and Kloss, 1928; *simplex* Miller, 1913; *strictis* Horsfield, 1855; *sumatrensis* Fischer, 1829; *sumbanus* Schwarz, 1910; *torvus* Thomas, 1909; *typus* F. G. Cuvier and E. Geoffroy Saint-Hilaire, 1821; *vellerosus* Pocock, 1934; *vicinus* Schwarz, 1910.

Paradoxurus jerdoni Blanford, 1885. Proc. Zool. Soc. Lond., 1885:613.
TYPE LOCALITY: "Kodaikanal, on the Palni (or Pulney) hills in the Madura district, Madras Presidency" [India, Tamil Nadu Province, Palni Hills, Kodaikanal; 10°15'N, 77°31'E].
DISTRIBUTION: S India.
STATUS: CITES - Appendix III (India); IUCN - Indeterminate.
SYNONYMS: *caniscus* Pocock, 1933.

Paradoxurus zeylonensis (Pallas, 1777). *In* Schreber, Die Säugethiere, 3(26):451[1777].
TYPE LOCALITY: "Cenlon" [= Sri Lanka].
DISTRIBUTION: Endemic to Sri Lanka.
SYNONYMS: *aureus* F. G. Cuvier, 1822; *montanus* Kelaart, 1852.

Subfamily Viverrinae Gray, 1821. London Med. Repos., 15:301.
COMMENTS: Type genus: *Viverra* Linnaeus, 1758, by original designation (Melville and Smith, 1987). Pocock (1933*d*) and Gregory and Hellman (1939) placed *Poiana* and *Prionodon* in the Prionodontinae, considered a sister group to the remaining viverrines. This was not followed by Gill (1872), Simpson (1945), Ellerman and Morrison-Scott (1951), Rosevear (1974), and Wozencraft (1989*b*).

Civettictis Pocock, 1915. Proc. Zool. Soc. Lond., 1915:134.
TYPE SPECIES: *Viverra civetta* Schreber, 1776, by monotypy (Melville and Smith, 1987).
COMMENTS: Included in *Viverra* by Coetzee (1977*b*); recognized as *Civettictis* by Rosevear (1974), Kingdon (1977), Ansell (1978), Smithers (1983), and Wozencraft (1989*b*).

Civettictis civetta (Schreber, 1776). Die Säugethiere, 3(16):pl. 111[1776]; text 3(24):418, 3:index, p. 587[1777].

TYPE LOCALITY: "Guinea, Kongo, das Vorgebirge der guten Hofnung und Aethiopien", restricted by Allen (1924:117) to "Guinea".

DISTRIBUTION: Angola, Benin, Botswana, Cameroonn, Central African Republic, Congo, Equatorial Guinea, Ethiopia, Gabon, Gambia, Guinea, Ivory Coast, Kenya, Liberia, Malawi, Mozambique, Namibia, Niger, Nigeria, Rwanda, Senegal, Sierra Leone, South Africa, Sudan, Tanzania, Thailand, Uganda, Zaire, Zambia, Zimbabwe.

STATUS: CITES - Appendix III (Botswana).

SYNONYMS: *australis* Lundholm, 1955; *congica* Cabrera, 1929; *matschiei* Pocock, 1933; *megaspila* Noack, 1891; *orientalis* Matschie, 1891; *poortmanni* Pucheran, 1855; *schwarzi* Cabrera, 1929; *volkmanni* Lundholm, 1955.

COMMENTS: Some authors have placed this species in *Viverra*, see Coetzee (1977*b*); most have followed Pocock (1915*b*), who placed this species in *Civettictis*. Rosevear (1974) noted differences in scent glands; G. Petter (1969) discussed dental differences.

Genetta G. Cuvier, 1816. Regne Anim., 1:156.

TYPE SPECIES: *Viverra genetta* Linnaeus, 1758, by original designation.

SYNONYMS: *Paragenetta*, *Pseudogenetta*.

COMMENTS: For reviews, see Crawford-Cabral (1966, 1969, 1970, 1973, 1981), Rosevear (1974), Coetzee (1977*b*), Schlawe (1980, 1981), and Wozencraft (1984, 1989*b*). Includes *Paragenetta* and *Pseudogenetta* as subgenera.

Genetta abyssinica (Rüppell, 1836). Neue Wirbelt. Fauna Abyssin. Gehörig. Säugeth., 1:33.

TYPE LOCALITY: "In Abyssinien, wo es sehr häufig vorkömmt, führt es beiden Landeseingebornen zu Gondar"; Ethiopia, Gondar (12°36'N, 37°28'E).

DISTRIBUTION: Egypt, Ethiopia, Somalia, Sudan.

STATUS: IUCN - Insufficiently known.

COMMENTS: Subgenus *Pseudogenetta*.

Genetta angolensis Bocage, 1882. J. Sci. Math. Phys. Nat. Lisboa, ser. 1, 9:29.

TYPE LOCALITY: "Calcuimba" [= Angola, Caconda (13°47'S, 15°08'E)].

DISTRIBUTION: Angola, Malawi, Mozambique, Tanzania, Zaire, Zambia, Zimbabwe.

SYNONYMS: *hintoni* Schwarz, 1929; *mossambica* Matschie, 1902.

COMMENTS: Subgenus *Genetta*.

Genetta genetta (Linnaeus, 1758). Syst. Nat., 10th ed., 1:45.

TYPE LOCALITY: "oriente juxta rivos", restricted by Linnaeus (1766) to "oriente juxta rivos, Hispania", later listed by Thomas (1911*a*) as "Spain". Cabrera (1914), synonymizing *G. peninsulæ*, further restricted the type locality to "El Pardo, cerca de Madrid" [Spain, El Pardo, near Madrid (40°32'N, 3°46'W)].

DISTRIBUTION: Algeria, Angola, Arabia, Belgium, Botswana, Burkina Faso, Egypt, Ethiopia, France, Germany, Holland, Kenya, Liberia, Libya, Mauritania, Morocco, Mozambique, Namibia, Nigeria, Orange Free State, Portugal, Senegal, Spain, Somalia, South Africa, Sudan, Tanzania, Tunisia, Uganda, Western Sahara, Yemen, Zambia, Zimbabwe.

STATUS: IUCN - Rare as *G. g. isabelae*.

SYNONYMS: *afra* F. G. Cuvier, 1825; *albipes* Trouessart, 1904; *balearica* Thomas, 1902; *barbar* Matschie, 1902; *barbara* Hamilton-Smith, 1842; *bella* Matschie, 1902; *bonapartei* Loche, 1857; *communis* Burnett, 1830; *dongolana* Hemprich and Ehrenberg, 1833; *felina* Thunberg, 1811; *gallica* Oken, 1816; *grantii* Thomas, 1902; *guardafuensis* Neumann, 1902; *hararensis* Neumann, 1902; *hispanica* Oken, 1816; *leptura* Reichenbach, 1836; *ludia* Thomas and Schwann, 1906; *macrura* Jentink, 1892; *melas* de la Paz Graells, 1897; *neumanni* Matschie, 1902; *peninsulæ* Cabrera, 1905; *pulchra* Matschie, 1902; *rhodanica* Matschie, 1902; *senegalensis* Fischer, 1829; *tedescoi* de Beaux, 1924; *terraesanctae* Neumann, 1902; *vulgaris* Lesson, 1827.

COMMENTS: Subgenus *Genetta*. Schlawe (1981) included *afra*, *bonapartei*, *barbar*, *barbara*, *balearica*, *lusitanica*, *melas*, *peninsulae*, *pyrenaica*, *terraesanctae*, *rhodanica*, and *isabelae*; and provisionally separated into *G. felina* the following: *guardafuensis*, *hararensis*, *leptura*, *senegalensis*, *dongolana*, *granti*, *neumanni*, *bella*, *pulchra*, and *ludia*, which are included here, following Crawford-Cabral (1966, 1969; 1981), Coetzee (1977*b*),

Smithers (1983), and Wozencraft (1984, 1989*b*). Rosevear (1974) separated *senegalensis* from *genetta*; however, this was not followed by Coetzee (1977*b*), Kingdon (1977), Ansell (1978), Crawford-Cabral (1981), or Wozencraft (1989*b*). Wozencraft (1984) recognized four allopatric populations: *genetta* (Meditrerrean), *dongolana* (East African), *felina* (South African), and *senegalensis* (West African).

Genetta johnstoni Pocock, 1908. Proc. Zool. Soc. Lond., 1907:1041 [1908].
TYPE LOCALITY: "in a district from fifteen to twenty miles west of the Putu Mountains, which lie west of the Duobe and Cavally Rivers". [Liberia]
DISTRIBUTION: Ghana, Guinea, Ivory Coast, Liberia.
STATUS: IUCN - Insufficiently known.
SYNONYMS: *lehmanni* Kuhn, 1960.
COMMENTS: Subgenus *Paragenetta*.

Genetta maculata (Gray, 1830). Spicil. Zool., 2:9.
TYPE LOCALITY: "in Africa Boreali".
DISTRIBUTION: Angola, Botswana, Burkina Faso, Cameroon, Central African Republic, Chad, Equatorial Guinea, Ethiopia, French Equatorial Africa, Gabon, Gambia, Ghana, Guinea, Ivory Coast, Kenya, Liberia, Malawi, Mozambique, Namibia, Niger, Nigeria, Senegal, Sierra Leone, South Africa, Sudan, Tanzania, Togo, Uganda, Zaire, Zambia, Zimbabwe.
SYNONYMS: *aequatorialis* Heuglin and Fitzinger, 1866; *albiventris* Schwarz, 1930; *amer* Gray, 1843; *bettoni* Thomas, 1902; *bini* Rosevear, 1974; *deorum* Funaiolo and Simonetta, 1960; *dubia* Matschie, 1902; *erlangeri* Matschie, 1902; *fieldiana* Du Chaillu, 1860; *genettoides* Temminck, 1853; *gleimi* Matschie, 1902; *insularis* Cabrera, 1921; *letabae* Thomas and Schwann, 1906; *loandae* Matschie, 1930; *matschiei* Neumann, 1902; *pantherina* Hamilton-Smith, 1842; *pardina* I. Geoffroy Saint Hilaire, 1832; *poensis* Waterhouse, 1838; *pumila* Hollister, 1916; *rubiginosa* Pucheran, 1855; *schraderi* Matschie, 1902; *soror* Schwarz, 1929; *stuhlmanni* Matschie, 1902; *suahelica* Matschie, 1902; *zambesiana* Matschie, 1902; *zuluensis* Roberts, 1924.
COMMENTS: Schlawe (1981) included *aequatorialis, amer* (*sensu* Schwartz, 1930), *bini* (not a synonym of *servalina*), *deorum, dubia, erlangeri, fieldiana, genettoides, gleimi, insularis, letabae, matschiei, pantherina, pardina, poensis, pumila, rubiginosa, schraderi, stuhlmanni, suahelica*, and *zambesiana*. May be conspecific with *tigrina*; see Cotezee (1977*b*), Pringle (1977), and Schlawe (1981); but also see Ansell (1978) and Wozencraft (1984). Crawford-Cabral (1981) and Ansell (1978) placed genets west of the Dahomey Gap in *pardina* and southern and eastern populations in *rubiginosa* (except for the extreme southern *tigrina*).

Genetta servalina Pucheran, 1855. Rev. Mag. Zool. Paris, 7(2):154.
TYPE LOCALITY: "Gabon".
DISTRIBUTION: Cameroon, Central African Republic, Congo, Equatorial Guinea, Gabon, Kenya, Niger, Tanzania, Uganda, Zaire.
SYNONYMS: *aubryana* Pucheran, 1855; *bettoni* Thomas, 1902; *cristata* Hayman, 1940; *intensa* Lönnberg, 1917.
COMMENTS: Includes *cristata, bettoni*, and *aubryana*, but not *bini*, which is here considered a junior synonym of *maculata*; see Crawford-Cabral (1970:323), Wenzel and Haltenorth (1972), Coetzee (1977*b*), and Schlawe (1981); but also see Rosevear (1974) and Crawford-Cabral (1981), who recognized *cristata* as a separate species.

Genetta thierryi Matschie, 1902. Verh. V. Internat. Zool. Congr., 1901:1142.
TYPE LOCALITY: "Hinterland von Togo von 9° n. Br. ab", restricted to "Borogu = Borgou, (10.78 N., 0.65 E)?" by Schlawe (1981:159).
DISTRIBUTION: Benin, Burkina Faso, Cameroon, Gambia, Ghana, Ivory Coast, Mali, Nigeria, Niger, Sierra Leone, Senegal, Togo.
SYNONYMS: *villersi* Dekeyser, 1949.
COMMENTS: Subgenus *Pseudogenetta*. Includes *Pseudogenetta villiersi*, see Kuhn (1960), Crawford-Cabral (1969, 1981), Rosevear (1974), and Schlawe (1981).

Genetta tigrina (Schreber, 1776). Die Säugethiere, 3(17):pl. 115[1776]; text, 3(25):425 [1777].
TYPE LOCALITY: "von dem Vorgebirge der guten Hofnug" [South Africa, Cape Prov., Cape of Good Hope].

DISTRIBUTION: From the Cape of Good Hope to S Natal (South Africa), Lesotho, and Orange Free State.

SYNONYMS: *methi* Roberts, 1948.

COMMENTS: Subgenus *Genetta*. May be conspecific with *maculata*; see Crawford-Cabral (1966), Pringle (1977), Ansell (1978), Schlawe (1981), and Meester et al. (1986).

Genetta victoriae Thomas, 1901. Proc. Zool. Soc. Lond., 1901(2):87.

TYPE LOCALITY: "Entebbe, Uganda". Subsequently restricted by Moreau et al. (1945:410) to "Near Lupanzula's, ten miles west of Beni, Ituri Forest, Congo Belge". See Allen (1924) for discussion.

DISTRIBUTION: E Zaire, mountainous regions.

COMMENTS: Subgenus *Genetta*.

Osbornictis J. A. Allen, 1919. J. Mammal., 1:25.

TYPE SPECIES: *Osbornictis piscivora* J. A. Allen, 1919, by original designation.

COMMENTS: Sister taxon to *Genetta*.

Osbornictis piscivora J. A. Allen, 1919. J. Mammal., 1:25.

TYPE LOCALITY: "Niapu, Belgian Congo" [Zaire, Niapu, 2°25'N, 26°28'E].

DISTRIBUTION: Zaire.

COMMENTS: Reviewed by Van Rompaey (1988, Mammalian Species, 309).

Poiana Gray, 1865. Proc. Zool. Soc. Lond., 1864:507, 520 [1865].

TYPE SPECIES: *Genetta richardsonii* Thomson, 1842, by monotypy (Melville and Smith, 1987).

COMMENTS: Pocock (1933*d*) believed it very likely that *richardsonii* pocessed scent glands and therefore separated it from *Prionodon*, although it is the loss of scent glands in *Prionodon* which is derived (Wozencraft, 1984).

Poiana richardsonii (Thomson, 1842). Ann. Mag. Nat. Hist., ser. 1, 10:204.

TYPE LOCALITY: "Fernando Po" [Equatorial Guinea: Bioko].

DISTRIBUTION: Cameroon, Congo, Equatorial Guinea, Gabon, Ivory Coast, Liberia, Zaire.

STATUS: IUCN - Indeterminate as *P. r. liberiensis*.

SYNONYMS: *leightoni* Pocock, 1908; *poensis* Mivart, 1882.

COMMENTS: Rosevear (1974) raised *leightoni* to the specific level. *Poiana* records are few and scattered and recent studies placed *leightoni* as a subspecies; see de Beaufort (1965), Michaelis (1972), and Kingdon (1977).

Prionodon Horsfield, 1823. Zool. Res. Java, part 5, p. 13(unno.) of *Mangusta Javanica* acct.

TYPE SPECIES: *Felis gracilis* Horsfield, 1823 (= *Viverra ? linsang* Hardwicke, 1821).

SYNONYMS: *Pardictis* Thomas, 1925 (see Pocock, 1933*d*).

Prionodon linsang (Hardwicke, 1821). Trans. Linn. Soc. Lond., 13:236, pl. 24.

TYPE LOCALITY: "Malaysia, Malacca", restricted by Robinson and Kloss (1920:264) to "Malacca".

DISTRIBUTION: Bankga Isl; Java; Borneo; Billiton Isl; peninsular Burma, Malaysia (West) to Sumatra.

STATUS: CITES - Appendix II.

SYNONYMS: *gracilis* Horsfield, 1823; *maculosus* Blanford, 1878.

Prionodon pardicolor Hodgson, 1842. Calcutta J. Nat. Hist., 2:57.

TYPE LOCALITY: "Sikim. . . Sub-Hemalayan mountains". [India].

DISTRIBUTION: Burma, China (Yunnan), India, Indonesia, Laos, Western Malaysia, Nepal, Thailand, Vietnam.

STATUS: CITES - Appendix I; U.S. ESA - Endangered.

SYNONYMS: *pardochrous* Gray, 1863; *perdicator* Schinz, 1844; *presina* Thomas, 1925.

Viverra Linnaeus, 1758. Syst. Nat., 10th ed., 1:43.

TYPE SPECIES: *Viverra zibetha* Linnaeus, 1758, by subsequent designation (Sclater, 1900; Melville and Smith, 1987).

COMMENTS: Pocock (1933*b*) placed *civettina* and *megaspila* in *Moschothera* which was not recognized by Ellerman and Morrison-Scott (1951), Wang (*in* Gao, 1987), or

Wozencraft (1989*b*). Does not include *Civettictis*; see Kingdon (1977) and Ansell (1978); but also see Coetzee (1977*b*).

Viverra civettina Blyth, 1862. J. Asiatic Soc. Bengal, 31:332.
TYPE LOCALITY: "Southern Malabar" restricted by Pocock (1933*a*:446) to "Travancore" [India].
DISTRIBUTION: Endemic to S India.
STATUS: U.S. ESA and IUCN - Endangered.
COMMENTS: Considered a subspecies of *V. megaspila* by Ellerman and Morrison-Scott (1951); however, considered at the specific level by Lindsay (1928), Pocock (1941*a*), and Wozencraft (1984, 1989*b*).

Viverra megaspila Blyth, 1862. J. Asiatic Soc. Bengal, 31:331.
TYPE LOCALITY: "vicinity of Prome" [Burma, Prome (= Pye) 18°49'N, 95°13'E].
DISTRIBUTION: Burma, Malaysia, Thailand, Vietnam.
STATUS: CITES - Appendix III (India).
COMMENTS: Does not include *V. civettina*; reviewed by Lindsay (1928), Pocock (1941*a*), and Wozencraft (1989*b*).

Viverra tangalunga Gray, 1832. Proc. Zool. Soc. Lond., 1832:63.
TYPE LOCALITY: Not given. Fixed by Gray (1843:48) as "Sumatra" [Indonesia].
DISTRIBUTION: China, Indonesia (Sumatra, Rhio-Lingga Arch., Bangka Isl, Borneo, Karimata Isl, Sulawesi, Buru, Amboina), Cambodia, Malaysia, Philippines, Thailand. Introduced to many SE Asian Islands.
SYNONYMS: *lankavensis* Robinson and Kloss, 1920.

Viverra zibetha Linnaeus, 1758. Syst. Nat., 10th ed., 1:44.
TYPE LOCALITY: "Indiis", subsequently restricted by Thomas (1911*a*:137) to "Bengal".
DISTRIBUTION: Burma, Cambodia, China (Anhui, Shaanxi, Zhejiang and Jiangsu), India, Indonesia, Laos, Western Malaysia, Nepal, Thailand, Vietnam.
STATUS: CITES - Appendix III (India).
SYNONYMS: *ashtoni* Swinhoe, 1865; *civettoides* Hodgson, 1842; *filchneri* Matschie, 1908; *melanura* Hodgson, 1842; *orientalis* Hodgson, 1842; *picta* Wroughton, 1915; *piscator* Shaw, 1800; *pruinosa* Wroughton, 1917; *sigillata* Robinson and Kloss, 1920; *surdaster* Thomas, 1927; *undulata* Gray, 1830.

Viverricula Hodgson, 1838. Ann. Mag. Nat. Hist., [ser. 1], 1:152.
TYPE SPECIES: *Civetta indica* Desmarest, 1804, by subsequent designation (Sclater, 1891; Melville and Smith, 1987).

Viverricula indica (Desmarest, 1804). Tabl. Méth. Hist. Nat., *in*, Nouv. Dict. Hist. Nat., 24:9, 17.
TYPE LOCALITY: "l'Inde" [India].
DISTRIBUTION: Bangladesh, Burma, Cambodia, China, Hong Kong, India, Indonesia (Sumatra, Java, Kangean Isl, Sumbawa, Bali), Laos, Malaysia, Pakistan, Sri Lanka, Taiwan, Thailand, Vietnam. Introduced to Yemen, Zanzibar, Socotra, Madagascar, the Comoro Isls, and the Philippines; scattered distribution on many SE Asian islands due to introductions.
STATUS: CITES - Appendix III (India).
SYNONYMS: *baptistae* Pocock, 1933; *bengalensis* Hardwicke, 1833; *deserti* Bonhote, 1898; *hanensis* Matschie, 1908; *malaccensis* Gmelin, 1788; *mayori* Pocock, 1933; *pallida* Gray, 1832; *rasse* Horsfield, 1824; *schlegelii* Pollen, 1868; *taivana* Schwarz, 1911; *thai* Kloss, 1919; *wellsi* Pocock, 1933.
COMMENTS: The correct identity of *V. malaccensis* is uncertain (Pocock, 1933*b*).

ORDER CETACEA

by James G. Mead and Robert L. Brownell, Jr.

ORDER CETACEA

COMMENTS: The definition of oceanic water masses follows Briggs (1974). Includes as suborders Mysticeti (Balaenidae, Balaenopteridae, Eschrichtiidae, and Neobalaenidae) and Odontoceti (Delphinidae, Monodontidae, Phocoenidae, Physeteridae, Platanistidae, and Ziphiidae).

Family Balaenidae Gray, 1821. Lond. Med. Repos., 15:310.

COMMENTS: Commonly included *Caperea*, which is here put in a separate family, Neobalaenidae, following Barnes and McLeod (1984).

Balaena Linnaeus, 1758. Syst. Nat., 10th ed., 1:75.

TYPE SPECIES: *Balaena mysticetus* Linnaeus, 1758.

SYNONYMS: *Leiobalaena*.

Balaena mysticetus Linnaeus, 1758. Syst. Nat., 10th ed., 1:75.

TYPE LOCALITY: "Habitat in Oceano Groenlandico" (= Greenland Sea).

DISTRIBUTION: Northern Hemisphere: arctic waters. Strays have occured in Japan, Gulf of St. Lawrence, and Massachusetts.

STATUS: CITES - Appendix I; U.S. ESA - Endangered; IUCN - Vulnerable.

COMMENTS: Reviewed by Reeves and Leatherwood (1985).

Eubalaena Gray, 1864. Proc. Zool. Soc. Lond., 1864(2):201.

TYPE SPECIES: *Balaena australis* Desmoulins, 1822.

SYNONYMS: *Halibalaena, Hunterius*.

COMMENTS: Corbet and Hill (1980) used this genus. Formerly considered to include three separate species, *glacialis, australis* and *japonicus*; see Hershkovitz (1961*b*).

Eubalaena australis (Desmoulins, 1822). *In* Bory de Saint-Vincent (ed.), Dict. Class. Hist. Nat. Paris, 2:161, pl.

TYPE LOCALITY: Algoa Bay, Cape of Good Hope, South Africa.

DISTRIBUTION: Southern Hemisphere: antarctic to temperate waters; occasionally along the northern part of the Antarctic Peninsula.

STATUS: CITES - Appendix I; U.S. ESA - Endangered (included with *E. glacialis*); IUCN - Vulnerable.

SYNONYMS: *antarctica, antipodarum, temminckii*.

COMMENTS: Reviewed by Cummings (1985*b*). Included in *glacialis* by some recent authors.

Eubalaena glacialis (Müller, 1776). Zool. Danicae Prodr., p. 7.

TYPE LOCALITY: None given, listed as Norway, Finnmark, Nord Kapp (vicinity of North Cape) by Eschricht and Reinhardt (1861).

DISTRIBUTION: Northern Hemisphere: temperate to tropical waters; one stray record from Hawaii (Scarff, 1986).

STATUS: CITES - Appendix I; U.S. ESA - Endangered; IUCN - Endangered.

SYNONYMS: *biscayensis, japonica, nordcaper, sieboldi*.

COMMENTS: Reviewed by Cummings (1985*b*); see Hershkovitz (1961*b*).

Family Balaenopteridae Gray, 1864. Proc. Zool. Soc. London, 1864:203.

Balaenoptera Lacépède, 1804. Hist. Nat. Cetacees, p. 114.

TYPE SPECIES: *Balaenoptera gibbar* Lacépède, 1804 (= *Balaena physalus* Linnaeus, 1758).

SYNONYMS: *Catoptera, Cuvierius, Physalus, Pterobalaena, Rorqualus, Sibbaldius*.

Balaenoptera acutorostrata Lacépède, 1804. Hist. Nat. Cetacees, p. 134.

TYPE LOCALITY: France, "pris aux environs de la rade de Cherbourg", Manche.

DISTRIBUTION: Worldwide: arctic to tropical waters.

STATUS: CITES - Appendix I.

SYNONYMS: *bonaerensis, davidsoni, huttoni, minimus, rostrata*.

COMMENTS: Reviewed by Stewart and Leatherwood (1985). Two forms have been described from SW Pacific waters (Arnold et al., 1987). May represent two or three species (Wada and Numachi, 1991).

Balaenoptera borealis Lesson, 1828. Hist. Nat. Gen. Part. Mamm. Oiseaux, 1:342.
TYPE LOCALITY: Germany, Schleswig-Holstein, Lubeck Bay, near Gromitz (see Rudolphi, 1822).
DISTRIBUTION: Worldwide: cold-temperate to tropical waters. Distributional records sometimes confused with *B. edeni*.
STATUS: CITES - Appendix I; U.S. ESA - Endangered.
SYNONYMS: *rostrata, schlegellii*.
COMMENTS: Reviewed by Gambell (1985*a*).

Balaenoptera edeni Anderson, 1879. Anat. Zool. Res., Yunnan, p. 551, pl. 44.
TYPE LOCALITY: Burma, "found its way into the Thaybyoo Choung, which runs into the Gulf of Martaban between the Sittang and Beeling Rivers, and about equidistant from each".
DISTRIBUTION: Worldwide: warm-temperate to tropical waters. Distributional records sometimes confused with *B. borealis*.
STATUS: CITES - Appendix I.
SYNONYMS: *brydei*.
COMMENTS: Reviewed by Cummings (1985*a*). May represent more than one species (Wada and Numachi, 1991).

Balaenoptera musculus (Linnaeus, 1758). Syst. Nat., 10th ed., 1:76.
TYPE LOCALITY: UK, Scotland, Firth of Forth ("Habitat in mari Scotico").
DISTRIBUTION: Worldwide: arctic to tropical waters.
STATUS: CITES - Appendix I; U.S. ESA - Endangered; IUCN - Endangered.
SYNONYMS: *brevicauda, gigas, indica, intermedia, major, sibbaldii, sibbaldius, sulfureus*.
COMMENTS: Reviewed by Yochem and Leatherwood (1985). Includes subspecies *B. m. brevicauda* Ichihara, 1966 (not Zemsky and Boronin, 1964, which is a *nomen nudum* (Rice, 1977:6).

Balaenoptera physalus (Linnaeus, 1758). Syst. Nat., 10th ed., 1:75.
TYPE LOCALITY: "Habitat in Oceano Europeao", restricted to Norway, near Svalbard, Spitsbergen Sea by Thomas (1911*a*).
DISTRIBUTION: Worldwide: arctic to tropical waters.
STATUS: CITES - Appendix I; U.S. ESA - Endangered; IUCN - Vulnerable.
SYNONYMS: *antiquorum, boops, gibbar, patachonica, velifera*.
COMMENTS: Reviewed by Gambell (1985*b*).

Megaptera Gray, 1846. Ann. Mag. Nat. Hist., [ser. 1], 17:83.
TYPE SPECIES: *Megaptera longipinna* Gray, 1846 (= *Balaena novaeangliae* Borowski, 1781).
SYNONYMS: *Cyphobalaena, Kyphobalaena, Perqualus, Poescopia*.

Megaptera novaeangliae (Borowski, 1781). Gemein. Naturgesch. Thier., 2(1):21.
TYPE LOCALITY: USA, "de la nouvelle Angleterre" (= coast of New England).
DISTRIBUTION: Worldwide: cold-temperate to tropical waters.
STATUS: CITES - Appendix I; U.S. ESA - Endangered; IUCN - Vulnerable.
SYNONYMS: *braziliensis, burmeisteri, lalandii, longimana, longipinna, nodosa, versabilis*.
COMMENTS: Reviewed by Winn and Reichley (1985).

Family Eschrichtiidae Ellerman and Morrison-Scott, 1951. Checklist of Palearctic Indian Mammals, p. 713.
SYNONYMS: Rhachianectidae.

Eschrichtius Gray, 1864. Ann. Mag. Nat. Hist., ser. 3, 14:350.
TYPE SPECIES: *Balaenoptera robusta* Lilljeborg, 1861.
SYNONYMS: *Cyphonotus, Rhachianectes*.

Eschrichtius robustus (Lilljeborg, 1861). Forh. Skand. Naturf. Ottende Mode, Kopenhagen, 1860, 8:602 [1861].

TYPE LOCALITY: Sweden, "på Gräsön i Roslagen"; "Benen lägo 840 fot frän hafsstranden, ungefär 12 à 15 fot öfver hafvets yta" (= Uppland, Graso Isl).

DISTRIBUTION: North Pacific: warm temperate to arctic waters. Formerly present in the North Atlantic. Sometimes enters tropical water at the southern boundaries of its distribution; see Henderson (1990) for further details.

STATUS: CITES - Appendix I; U.S. ESA - Endangered.

SYNONYMS: *gibbosus, glaucus.*

COMMENTS: See Rice and Wolman (1971), Jones et al. (1984), and Wollman (1985).

Family Neobalaenidae Gray, 1873. Ann. Mag. Nat. Hist., ser. 4, 11:108.

COMMENTS: See Barnes and McLeod (1984) for comments. Gray, 1874 (Trans. Proc. N. Z. Inst. 6(18):93-97) is cited by Barnes and McLeod for Neobalaenidae.

Caperea Gray, 1864. Proc. Zool. Soc. Lond. 1864(2):202.

TYPE SPECIES: *Balaena (Caperea) antipodarum* Gray, 1846 (= *Balaena marginata* Gray, 1846).

SYNONYMS: *Neobalaena.*

Caperea marginata (Gray, 1846). Zool. Voy. H.M.S. "Erebus" and "Terror", 1:48.

TYPE LOCALITY: "Inhab. W. Australia" (= Southern Hemisphere, temperate waters; see Baker, 1985).

DISTRIBUTION: Southern Hemisphere: cold-temperate waters.

STATUS: CITES - Appendix I.

SYNONYMS: *antipodarum.*

COMMENTS: Reviewed by Baker (1985).

Family Delphinidae Gray, 1821. London Med. Repos., 15(1):310.

COMMENTS: Includes Globicephalidae, Grampidelphidae, Stenidae, Orcinae, Lissodelphinae, Cephalorhynchinae and Delphininae (Fraser and Purves, 1960); see Kasuya (1973), Mead (1975), Barnes (1978). Also includes *Orcaella* (see Heyning, 1989*a* and Lint et al., 1990), sometimes put in the family Monodontidae.

Cephalorhynchus Gray, 1846. Zool. Voy. H.M.S. "Erebus" and "Terror", 1:36.

TYPE SPECIES: *Delphinus heavisidii* Gray, 1828.

SYNONYMS: *Eutropia.*

COMMENTS: Revised by Harmer (1922).

Cephalorhynchus commersonii (Lacépède, 1804). Hist. Nat. Cetacees, p. 317.

TYPE LOCALITY: Chile, "de la terre de Feu et dans le détroit de Magellan" (= Tierra del Fuego, Straits of Magellan).

DISTRIBUTION: Argentina to Chile: Gulf of San Matias, Argentina, to the Chilean side of the Straits of Magellan; South Shetland, Falkland and Kerguelen Isls. See Brownell and Praderi (1985) for further discussion.

STATUS: CITES - Appendix II; IUCN - Insufficiently known.

SYNONYMS: *floweri.*

COMMENTS: Reviewed by Goodall et al. (1988).

Cephalorhynchus eutropia Gray, 1846. Zool. Voy. H.M.S. "Erebus" and "Terror", 1:pl. 34.

TYPE LOCALITY: None given, listed as Pacific Ocean, off the coast of Chile by Gray (1850:112).

DISTRIBUTION: Chile: coastal waters between Valparaiso and Navarino Island, Tierra del Fuego.

STATUS: CITES - Appendix II; IUCN - Insufficiently known.

SYNONYMS: *albiventris, obtusata.*

COMMENTS: Reviewed by Goodall et al. (1988). *Tursio? panope* is not a synonym (see *Lagenorhynchus obscurus*).

Cephalorhynchus heavisidii (Gray, 1828). Spicil. Zool., 1:2.

TYPE LOCALITY: South Africa, Cape Prov., "Inhab. Cape of Good Hope".

DISTRIBUTION: South Africa to perhaps S Angola: coastal waters from Cape Town to 17°09'S (Namibia).
STATUS: CITES - Appendix II; IUCN - Insufficiently known.
SYNONYMS: *hastatus*.

Cephalorhynchus hectori (Van Beneden, 1881). Bull. R. Acad. Belg., ser. 3, 4:877, pl. 11.
TYPE LOCALITY: "capturé sur la côte nord-est de la Nouvelle-Zélande." (= New Zealand, North coast).
DISTRIBUTION: New Zealand: coastal waters. Harrison's (1960) reference to the occurence of this species around Sarawak is undocumented by specimens or photos.
STATUS: CITES - Appendix II; IUCN - Vulnerable.
SYNONYMS: *albifrons*.

Delphinus Linnaeus, 1758. Syst. Nat., 10th ed., 1:77.
TYPE SPECIES: *Delphinus delphis* Linnaeus, 1758.
SYNONYMS: *Eudelphinus, Rhinodelphis*.

Delphinus delphis Linnaeus, 1758. Syst. Nat., 10th ed., 1:77.
TYPE LOCALITY: E North Atlantic ("Oceano Europaeo").
DISTRIBUTION: Worldwide: temperate and tropical waters, including the Black Sea.
STATUS: CITES - Appendix II.
SYNONYMS: *bairdii, capensis, tropicalis*.
COMMENTS: Includes *bairdii* (see van Bree and Purves, 1972) and *tropicalis* (see van Bree and Gallagher, 1978 and Casinos, 1984).

Feresa Gray, 1870. Proc. Zool. Soc. Lond., 1870(1):77.
TYPE SPECIES: *Delphinus intermedius* Gray, 1827 (= *Feresa attenuata* Gray, 1875).
COMMENTS: *Delphinus intermedius* Gray, 1827 was preoccupied by *Delphinus intermedius* Harlan, 1827 (= *Globicephala melas*). Gray subsequentally changed generic designations of that nominal taxon (*Grampus intermedius* Gray, 1843; *Orca intermedia* Gray, 1846).

Feresa attenuata Gray, 1875. J. Mus. Godeffroy (Hamburg), 8:184.
TYPE LOCALITY: "South Seas."
DISTRIBUTION: Worldwide: tropical to warm-temperate waters.
STATUS: CITES - Appendix II.
SYNONYMS: *intermedius, occulta*.

Globicephala Lesson, 1828. Compl. Oeuvres Buffon Hist. Nat., 1:441.
TYPE SPECIES: *Delphinus globiceps* Cuvier, 1812 (= *Delphinus melas* Traill, 1809).
SYNONYMS: *Cetus, Globiceps, Sphaerocephalus*.
COMMENTS: Reviewed by van Bree (1971).

Globicephala macrorhynchus Gray, 1846. Zool. Voy. H.M.S. "Erebus" and "Terror", 1:33.
TYPE LOCALITY: "South Seas".
DISTRIBUTION: Worldwide: tropical and warm-temperate waters; cold-temperate waters of the N Pacific, where it appears to stray as far north as the Gulf of Alaska (Pike and MacAskie, 1969).
STATUS: CITES - Appendix II.
SYNONYMS: *brachypterus, scammonii, sieboldii*.
COMMENTS: See Van Bree (1971).

Globicephala melas (Traill, 1809). Nicholson's J. Nat. Philos. Chem. Arts, 22:81.
TYPE LOCALITY: UK, Scotland, "in Scapay Bay, in Pomona, one of the Orkneys".
DISTRIBUTION: North Atlantic and southern Oceans: cold-temperate waters. Kasuya (1975) described the historic distribution in the NW Pacific.
STATUS: CITES - Appendix II.
SYNONYMS: *edwardii, globiceps, leucosagmaphora, svineval*.
COMMENTS: See Van Bree (1971). Formerly called *G. melaena* but Article 31b of the third edition of the International of Zoological Nomenclature (1985) specifically gave *melas* as an example of a Greek adjective that does not change its ending when transferred to a genus of another gender (see Schevill, 1990*a*, *b*; Rice, 1990).

Grampus Gray, 1828. Spicil. Zool., 1:2.
TYPE SPECIES: *Delphinus griseus* Cuvier, 1812.
SYNONYMS: *Grampidelphis, Grayius.*

Grampus griseus (G. Cuvier, 1812). Ann. Mus. Hist. Nat. Paris, 19:13.
TYPE LOCALITY: France, Finistere, "envoyé de Brest".
DISTRIBUTION: Worldwide: temperate to tropical waters.
STATUS: CITES - Appendix II.
SYNONYMS: *rissoanus, stearnsii.*
COMMENTS: Corbet and Hill (1980:110) included *rectipinna* in this species but it belongs in *Orcinus orca.*

Lagenodelphis Fraser, 1956. Sarawak Mus. J., n.s., 8(7):496.
TYPE SPECIES: *Lagenodelphis hosei* Fraser, 1956.

Lagenodelphis hosei Fraser, 1956. Sarawak Mus. J., n.s., 8(7):496.
TYPE LOCALITY: "Collected at the mouth of Lutong River, Baram, Borneo."
DISTRIBUTION: Worldwide: warm-temperate to tropical waters.
STATUS: CITES - Appendix II.

Lagenorhynchus Gray, 1846. Ann. Mag. Nat. Hist., [ser 1.], 17:84.
TYPE SPECIES: *Delphinus albirostris* Gray, 1846.
SYNONYMS: *Electra, Leucopleurus, Sagmatius.*
COMMENTS: Reviewed by Fraser (1966).

Lagenorhynchus acutus (Gray, 1828). Spicil. Zool., 1:2.
TYPE LOCALITY: None given, listed as North Sea, Faeroe Isls. (Denmark) (uncertain) by Gray (1846:36).
DISTRIBUTION: North Atlantic: cold temperate waters; *L. acutus* tends to be distributed to the south of *L. albirostris.*
STATUS: CITES - Appendix II.
SYNONYMS: *gubernator, leucopleurus, perspicillatus.*

Lagenorhynchus albirostris (Gray, 1846). Ann. Mag. Nat. Hist., [ser. 1], 17:84.
TYPE LOCALITY: None given in original description, given by Gray (1846:35) as UK, England, "North Sea, coast of Norfolk.", and by Gray (1850) as Great Yarmouth.
DISTRIBUTION: North Atlantic: cold-temperate waters; *L. albirostris* tends to be distributed to the north of *L. acutus.*
STATUS: CITES - Appendix II.
SYNONYMS: *pseudotursio.*

Lagenorhynchus australis (Peale, 1848). Mammalia *in* Repts. U.S. Expl. Surv., 8:33, pl. 6.
TYPE LOCALITY: "South Atlantic Ocean, off the coast of Patagonia", Argentina, 1 days sail north of the Straits of LeMaire.
DISTRIBUTION: Chile to Argentina: Valparaiso to Commodoro Rivadavia and Falkland Isls: Cold-temperate waters. One published (photograph) sighting in the tropical waters of the South Pacific, Cook Isls (Leatherwood et al. (1991).
STATUS: CITES - Appendix II.
SYNONYMS: *amblodon, chilöensis.*
COMMENTS: Included in *cruciger* by Bierman and Slijper (1947) and Hershkovitz (1966*a*:67), but considered a distinct species by Fraser (1966), Rice (1977), Brownell (1974), and Mitchell (1975).

Lagenorhynchus cruciger (Quoy and Gaimard, 1824). Voy. autour du Monde...l'Uranie et la Physicienne, Zool., p. 87, pl. 2.
TYPE LOCALITY: Pacific Ocean, "entre la Nouvelle-Hollande et le cap Horn [= between Australia and Cape Horn]...par 49° [S] de latitude".
DISTRIBUTION: Southern Hemisphere: antarctic and cold-temperate waters.
STATUS: CITES - Appendix II.
SYNONYMS: *albigena, bivattus, clanculus, wilsoni.*
COMMENTS: Formerly included *australis* and *obscurus,* see Hershkovitz (1966*a*) and comments under *australis* and *obscurus.*

Lagenorhynchus obliquidens Gill, 1865. Proc. Acad. Nat. Sci. Philadelphia, 17:177.
TYPE LOCALITY: USA, "obtained at San Francisco, California".
DISTRIBUTION: North Pacific: cold-temperate waters except warm-temperate waters of the ends of its range. Undocumented sighting from Hong Kong (Hammond and Leatherwood, 1984:495).
STATUS: CITES - Appendix II.
SYNONYMS: *longidens, ognevi.*
COMMENTS: May be a Northern Hemisphere form of *L. obscurus. Lagenorhynchus thicolea* is not synonymous with *L. obliquidens* (see *Lissodelphis*).

Lagenorhynchus obscurus (Gray, 1828). Spicil. Zool., 1:2.
TYPE LOCALITY: South Africa, Cape Prov., "Inhab. Cape of Good Hope".
DISTRIBUTION: Southern Hemisphere: cold-temperate continental waters.
STATUS: CITES - Appendix II.
SYNONYMS: *breviceps, fitzroyi, Tursio? panope, similis, supercilliosus.*
COMMENTS: Included in *cruciger* by Hershkovitz (1966*a*:65), but considered a distinct species by Rice (1977), Brownell (1974), and Mitchell (1975). Previously reported from Kerguelen Isls, reidentified as young specimen of *Cephalorhynchus commersonii* (Robineau, 1989).

Lissodelphis Gloger, 1841. Gemein. Naturgesch. Thier., 1:169.
TYPE SPECIES: *Delphinus peronii* Lacépède, 1804.
SYNONYMS: *Delphinapterus* (part), *Leucorhamphus, Tursio.*
COMMENTS: This may be a monotypic genus. The holotype of *Lagenorhynchus thicolea,* previously associated with *Lagenorhynchus,* is a specimen of *Lissodelphis* spp.

Lissodelphis borealis (Peale, 1848). Mammalia *in* Repts. U.S. Expl. Surv., 8:35, pl. 8.
TYPE LOCALITY: "North Pacific Ocean, latitude 46° 6' 50" N., 134° 5' W. from Greenwich.", 10°W of Astoria, Oregon, USA.
DISTRIBUTION: North Pacific: cold-temperate waters.
STATUS: CITES - Appendix II.

Lissodelphis peronii (Lacépède, 1804). Hist. Nat. Cetacees, p. 316.
TYPE LOCALITY: Indian Ocean, "dans les environs du cap sud de la terre de Diémen, et par conséquent vers le quarante-quatrime degré de latitude australe." (= about 44°S, 141°E, south of Tasmania).
DISTRIBUTION: Southern Hemisphere: cold-temperate waters, occasionally Antarctic waters south of Argentina.
STATUS: CITES - Appendix II.
SYNONYMS: *leucorhamphus.*

Orcaella Gray, 1866. Cat. Seals Whales Brit. Mus., p. 285.
TYPE SPECIES: *Orca* (*Orcaella*) *brevirostris* Gray, 1866.
COMMENTS: We follow Fordyce (1989), Heyning (1989*a*), and Lint et al. (1990) in including *Orcaella* in the Delphinidae, not in the Monodontidae as was recently proposed (Kasuya, 1973; Barnes et al., 1985).

Orcaella brevirostris (Gray, 1866). Cat. Seals Whales Brit. Mus., p. 285, fig. 57.
TYPE LOCALITY: "Inhab. East coast of India, the harbour of Vizagapatam" (= Vishakhapatnam Harbor, in Bay of Bengal).
DISTRIBUTION: SE Asia, N Australia and Papua New Guinea: tropical coastal waters and large rivers.
STATUS: CITES - Appendix II; IUCN - Insufficiently known.
SYNONYMS: *fluminalis.*
COMMENTS: Reviewed by Marsh et al. (1989).

Orcinus Fitzinger, 1860. Wiss.-Pop. Naturgesch. Säugeth., 6:204.
TYPE SPECIES: *Delphinus orca* Linnaeus, 1758.
SYNONYMS: *Gladiator, Grampus, Orca.*

Orcinus orca (Linnaeus, 1758). Syst. Nat., 10th ed., 1:77.
TYPE LOCALITY: E North Atlantic ("Oceano Europaeo").

DISTRIBUTION: Worldwide: all seas and oceans.
STATUS: CITES - Appendix II.
SYNONYMS: *ater, capensis, gladiator, rectipinna.*
COMMENTS: Reviewed by Heyning and Dahlheim (1988, Mammalian Species, 304).

Peponocephala Nishiwaki and Norris, 1966. Sci. Rep. Whales Res. Inst., 20:95.
TYPE SPECIES: *Lagenorhynchus electra* Gray, 1846.
SYNONYMS: *Electra.*
COMMENTS: Formerly included in *Lagenorhynchus.*

Peponocephala electra (Gray, 1846). Zool. Voy. H.M.S. "Erebus" and "Terror", 1:35.
TYPE LOCALITY: None given, unknown.
DISTRIBUTION: Worldwide: tropical to warm-temperate waters.
STATUS: CITES - Appendix II.
SYNONYMS: *asia, fusiformis, pectoralis.*
COMMENTS: Historically this species was included in the genus *Lagenorhynchus.*

Pseudorca Reinhardt, 1862. Overs. Danske Vidensk. Selsk. Forh., 1862:151.
TYPE SPECIES: *Phocaena crassidens* Owen, 1846.
SYNONYMS: *Neoorca.*

Pseudorca crassidens (Owen, 1846). Hist. Brit. Foss. Mamm. Birds, p. 516, fig. 213.
TYPE LOCALITY: UK, England, "in the great fen of Lincolnshire beneath the turf, in the neighborhood of the ancient town of Stamford". (subfossil).
DISTRIBUTION: Worldwide: temperate to tropical waters.
STATUS: CITES - Appendix II.
SYNONYMS: *destructor, meridionalis.*

Sotalia Gray, 1866. Cat. Seals Whales Brit. Mus., p. 401.
TYPE SPECIES: *Delphinus guianensis* Van Beneden, 1864 (= *Delphinus fluviatilis* Gervais and Deville, 1853).
SYNONYMS: *Steno, Tucuxa.*

Sotalia fluviatilis (Gervais and Deville, 1853). *In* Gervais, Bull. Soc. Agric. Herault, p. 148.
TYPE LOCALITY: Peru, Loreto, Rio Maranon above Pebas.
DISTRIBUTION: Western Atlantic: coastal waters from Panama to Santos, São Paulo, Brazil: Amazon and Orinoco river systems. See Vidal (1990) and Borobia et al. (1991).
STATUS: CITES - Appendix I.
SYNONYMS: *guianensis, pallida, tucuxi.*
COMMENTS: Due to the difficulty in finding the original work, the full citation is included here: Gervais, F. L. P. [and Deville]. 1853. Sur les mammiféres marins qui fréquentent les côtes de la France et plus particulìerement sur une novelle espéce de dauphins propre a la Méditerranés. Bulletin Sociéte Centrale d'Agriculture et des Comices Agricoles du Départment de l'Herault, Montpellier, 40me année, pp. 140-155, 1 pl.

Sousa Gray, 1866. Proc. Zool. Soc. Lond., 1866(2):213.
TYPE SPECIES: *Steno lentiginosus* Gray, 1866 (= *Delphinus chinensis* Osbeck, 1765).
SYNONYMS: *Sotalia, Steno, Stenopontistes.*
COMMENTS: Formerly included in *Sotalia* (Hershkovitz 1966*a*:18).

Sousa chinensis (Osbeck, 1765). Reise nach Ostind. China Rostock, 1:7.
TYPE LOCALITY: China, Guangdong Prov., Zhujiang Kou (mouth of Canton River).
DISTRIBUTION: Indian Ocean: coastal waters and rivers from False Bay, South Africa, east to S China and Moreton Bay, Queensland (Australia, see Corkeron, 1990).
STATUS: CITES - Appendix I.
SYNONYMS: *borneensis, lentiginosa, plumbea, zambezicus.*
COMMENTS: See Perrin (1975) who placed *Delphinus malayanus* in *Stenella attenuata*, as is done here; Pilleri and Gihr (1973-74) considered *borneensis, plumbea* and *lentiginosa* to be distinct species. Mitchell (1975) combined those species into *S. chinensis*; Brownell (1975*b*) included *Stenopontistes zambezicus* as a synonym of *S. plumbea.*

Sousa teuszii (Kükenthal, 1892). Zool. Jahrb. Syst., 6:442, pl. 21.
TYPE LOCALITY: "aus Kamerun" (= Cameroon), Cameroun Oriental, Bay of Warships, near Douala.
DISTRIBUTION: E South Atlantic: coastal waters in river mouths from S Morocco (W Sahara; see Beaubrun, 1990) to Cameroon.
STATUS: CITES - Appendix I.
COMMENTS: Reviewed by Pilleri and Gihr (1972).

Stenella Gray, 1866. Proc. Zool. Soc. Lond., 1866:213.
TYPE SPECIES: *Steno attenuatus* Gray, 1846.
SYNONYMS: *Clymene, Euphrosyne, Fretidelphis, Micropia, Prodelphinus.*
COMMENTS: Reviewed, in part, by Perrin (l975) and Perrin et al. (1981, 1987). The International Commission on Zoological Nomenclature (1991) conserved *Stenella* Gray, 1846.

Stenella attenuata (Gray, 1846). Zool. Voy. H.M.S. "Erebus" and "Terror", 1:44.
TYPE LOCALITY: None given, unknown (possibly India, see Gray, 1843).
DISTRIBUTION: Worldwide: temperate to tropical waters.
STATUS: CITES - Appendix II.
SYNONYMS: *albirostratus, brevimanus, capensis, consimilis, graffmani, malayanus, pseudodelphis, punctata, velox.*
COMMENTS: Perrin et al. (1987) revised this species. *D. dubius* is a *nomen nudum* [sic *dubium*] (Perrin et al., 1987). Opinion 1660 of the International Commission on Zoological Nomenclature (1991) conserved *attenuata* Gray, 1846 and suppressed *velox* Cuvier, 1839, *pseudodelphis* Schlegel, 1841, and *brevimanus* Wagner, 1846.

Stenella clymene (Gray, 1846). Zool. Voy. H.M.S. "Erebus" and "Terror", 1:39.
TYPE LOCALITY: None given, unknown.
DISTRIBUTION: Atlantic Ocean including the Gulf of Mexico: warm-temperate to tropical waters.
STATUS: CITES - Appendix II.
SYNONYMS: *metis, normalis.*
COMMENTS: Recognized by Hershkovitz (1966*a*), but not by Mitchell (1975) who included it in *longirostris*. See Perrin et al. (1981) for redescription.

Stenella coeruleoalba (Meyen, 1833). Nova Acta Acad. Caes. Nat. Curios., 16(2):609, pl. 43.
TYPE LOCALITY: "an der östlichen Küste von Südamerika; wir karpunirten ihn in der Gegend des Rio de la Plata." (= South Atlantic Ocean near Rio de la Plata, off coast of Argentina and Uruguay).
DISTRIBUTION: Worldwide: cold-temperate to tropical waters.
STATUS: CITES - Appendix II.
SYNONYMS: *asthenops, crotaphiscus, euphrosyne, styx, tethyos.*
COMMENTS: See Mitchell (1970:720). Perrin et al. (1981, 1987) gave a revised synonymy of this species.

Stenella frontalis (G. Cuvier, 1829). Règne Anim., Nouv. ed., 1:288.
TYPE LOCALITY: "découvert un aux îles du Cap-Vert". (= off Cape Verde Islands, West Africa).
DISTRIBUTION: Atlantic Ocean including the Gulf of Mexico: warm-temperate to tropical waters.
STATUS: CITES - Appendix II.
SYNONYMS: *doris, froenatus, plagiodon.*
COMMENTS: Perrin et al. (1987) revised this species. The International Commission on Zoological Nomenclature (1977*a*) suppressed *D. pernettensis* de Blainville, 1817 and *D. pernettyi* Desmarest, 1820, which Hershkovitz (1966*a*) used as a senior synonym for *S. plagiodon.*

Stenella longirostris (Gray, 1828). Spicil. Zool., 1:1.
TYPE LOCALITY: None given, unknown.
DISTRIBUTION: Worldwide: warm-temperate to tropical waters.
STATUS: CITES - Appendix II.
SYNONYMS: *alope, centroamericana, longirostris, microps, orientalis, roseiventris.*

COMMENTS: See Perrin (1975:206). Perrin (1990) established three subspecies (*centroamericana*, *longirostris*, and *orientalis*).

Steno Gray, 1846. Zool. Voy. H.M.S. "Erebus" and "Terror", 1:43.
TYPE SPECIES: *Delphinus rostratus* Cuvier, 1833 (= *Delphinus bredanensis* Lesson, 1828).
SYNONYMS: *Glyphidelphis*.

Steno bredanensis (Lesson, 1828). Hist. Nat. Gen. Part. Mamm. Oiseaux, 1:206.
TYPE LOCALITY: Coast of France.
DISTRIBUTION: Worldwide: warm-temperate to tropical waters.
STATUS: CITES - Appendix II.
SYNONYMS: *compressus, frontatus, perspicillatus, rostratus*.
COMMENTS: *Stenopontistes zambezicus* is not a synonym, see comment under *Sousa chinensis*. See Schevill (1987*a*) for further taxonomic notes.

Tursiops Gervais, 1855. Hist. Nat. Mammifères, 2:323.
TYPE SPECIES: *Delphinus truncatus* Montagu, 1821.
SYNONYMS: *Gadamu, Tursio*.
COMMENTS: This highly polymorphic genus is currently considered to be monotypic.

Tursiops truncatus (Montagu, 1821). Mem. Wernerian Nat. Hist. Soc., 3:75, pl. 3.
TYPE LOCALITY: UK, England, Devonshire, "in Duncannon Pool, near Stoke Gabriel, about five miles up the River Dart".
DISTRIBUTION: Worldwide: temperate to tropical waters, including the Black Sea.
STATUS: CITES - Appendix II.
SYNONYMS: *aduncus, gephyreus, gillii, nesarnack, nuuanu*.
COMMENTS: See Leatherwood and Reeves (1990). Ross and Cockroft (1990:124) considered *aduncus* to be synonymous with *truncatus*. Hall (1981:885-887) considered *nesarnack* and *gillii* distinct species, and synonymized *truncatus* with *nesarnack*. Opinion 1413 of the International Commision on Zoological Nomenclature (1986) conserved *truncatus* Montagu, 1821 and suppressed *nesarnack* Lacépède, 1804.

Family Monodontidae Gray, 1821. London Med. Repos., 15(1):310.
COMMENTS: Does not include *Orcaella*, a delphinid.

Delphinapterus Lacépède, 1804. Hist. Nat. Cetacees, p. 241.
TYPE SPECIES: *Delphinapterus beluga* Lacépède 1804 (= *Delphinus leucas* Pallas, 1776).
SYNONYMS: *Argocetus, Beluga*.
COMMENTS: Reviewed by Kleinenberg et al. (1969) and T. G. Smith et al. (1990).

Delphinapterus leucas (Pallas, 1776). Reise Prov. Russ. Reichs, 3(1):85 [footnote].
TYPE LOCALITY: NE Siberia, "die im Obischen Meerbusen" (= mouth of Ob River).
DISTRIBUTION: Circumpolar in Arctic seas; Okhotsk and Bering Seas; northern Gulf of Alaska (Cook Inlet); Gulf of St. Lawrence: arctic to cold-temperate waters; occasionally strays south to Honshu, Japan; France; and Massachusetts, USA.
STATUS: CITES - Appendix II; IUCN - Insufficiently known.
SYNONYMS: *albicans, beluga, catodon, dorofeevi, marisalbi*.
COMMENTS: Reviewed by Kleinenberg et al. (1969), T. G. Smith et al. (1990), Stewart and Stewart (1989, Mammalian Species, 336) and Brodie (1989).

Monodon Linnaeus, 1758. Syst. Nat., 10th ed., 1:75.
TYPE SPECIES: *Monodon monoceros* Linnaeus, 1758.
SYNONYMS: *Ceratodon, Diodon, Narwalus, Tachynices*.
COMMENTS: Reviewed by Reeves and Tracey (1980).

Monodon monoceros Linnaeus, 1758. Syst. Nat., 10th ed., 1:75.
TYPE LOCALITY: "Habitat in Oceano Septentrionali Americae, Europae." (= northern seas of Europe and America).
DISTRIBUTION: Arctic Ocean; rarely in Beaufort, Chuckchi and East Siberian Seas; occasional strays as far south as the Newfoundland, the Netherlands, British Isles and Japan.
STATUS: CITES - Appendix II; IUCN - Insufficiently known.

SYNONYMS: *microcephalus, monodon, narhval, vulgaris.*
COMMENTS: Reviewed by Reeves and Tracey (1980, Mammalian Species, 127) and Hay and Mansfield (1989).

Family Phocoenidae Gray, 1825. Ann. Philos., n.s., 10:340.
COMMENTS: Formerly considered a subfamily of Delphinidae; see Gromov and Baranova (1981:222).

Australophocaena Barnes, 1985. Mar. Mammal. Sci. 1(2):149-165.
TYPE SPECIES: *Phocoena dioptrica* Lahille, 1912.
COMMENTS: Reviewed by Brownell (1975*a*).

Australophocaena dioptrica (Lahille, 1912). Ann. Mus. Nat. Hist., Buenos Aires, 23:269.
TYPE LOCALITY: Argentina, Buenos Aires, "capturado en Punta Colares, cerca de Quilmes".
DISTRIBUTION: Southern Hemisphere: cold-temperate waters; Uruguay, Argentina; Falkland, South Georgia, Heard, Macquarie and the Auckland Isls, perhaps Kerguelen Isls. Perhaps circumpolar, see Baker (1977).
STATUS: CITES - Appendix II.
SYNONYMS: *stornii.*
COMMENTS: Reviewed by Brownell (1975*a*, Mammalian Species, 66, as *Phocoena dioptrica*). Barnes (1985) proposed *Australophocaena* to house this species. *Phocaena obtusata* is synonymous with *Cephalorhynchus eutropia*. See Goodall et al. (1988).

Neophocaena Palmer, 1899. Proc. Biol. Soc. Washington, 13:23.
TYPE SPECIES: *Delphinus phocaenoides* Cuvier, 1829.
SYNONYMS: *Meomeris, Neomeris.*
COMMENTS: Includes *Neomeris*; see Rice (1977) and Pilleri and Chen (1980).

Neophocaena phocaenoides (G. Cuvier, 1829). Règne Anim., Nouv. ed., 1:291.
TYPE LOCALITY: South Africa, Cape Prov., Cape of Good Hope ("à découvert au Cap,". Almost certainly erroneous; unknown today from coast of Africa.
DISTRIBUTION: Indo-Pacific: warm-temperate to tropical waters; Persian Gulf to Malaysia, north coast of Java (Tas'an and Leatherwood, 1984), China, and Japan: coastal waters and some rivers.
STATUS: CITES - Appendix I.
SYNONYMS: *asiaeorientalis; melas* Temminck (not Traill), *sunameri.*
COMMENTS: Includes as subspecies *asiaeorientalis* and *sunameri*. Reviewed by Pilleri and Gihr (1972, 1975:657, 673, 1980*b*). Van Bree (1973) considered *asiaeorientalis* to be of subspecific rank and *sunameri* to be synonymous with *phocaenoides*.

Phocoena G. Cuvier, 1817. Règne Anim., Nouv. ed., 1:279.
TYPE SPECIES: *Delphinus phocoena* Linnaeus, 1758.
SYNONYMS: *Acanthodelphis.*
COMMENTS: *Phocaena* and *Phocena* are later spellings.

Phocoena phocoena (Linnaeus, 1758). Syst. Nat., 10th ed., 1:77.
TYPE LOCALITY: "Habitat in Oceono Europaeo, & Balthico." (= Baltic Sea, "Swedish Seas").
DISTRIBUTION: N Pacific and N Atlantic: arctic to cold-temperate waters, isolated population in Black Sea; extends south to Senegal in the E Atlantic.
STATUS: CITES - Appendix II; IUCN - Insufficiently known.
SYNONYMS: *americana, communis, lineata, relicta, vomerina.*
COMMENTS: Reviewed by Gaskin et al. (1974, Mammalian Species, 42).

Phocoena sinus Norris and McFarland, 1958. J. Mammal., 39:22, pl. 1-4.
TYPE LOCALITY: "from the northeast shore of Punta San Felipe, Baja California Norte, Gulf of California, Mexico".
DISTRIBUTION: North Pacific: warm-temperate waters; northern Gulf of California (Mexico); erroneously reported from the S Gulf of California, including Tres Marías Isls and N Jalisco (Brownell, 1986).
STATUS: CITES - Appendix I; U.S. ESA and IUCN - Endangered.
COMMENTS: Reviewed by Brownell (1983, Mammalian Species, 198).

Phocoena spinipinnis Burmeister, 1865. Proc. Zool. Soc. Lond., 1865:228, figs 1-5.
TYPE LOCALITY: Argentina, Buenos Aires, "captured in the mouth of the River Plata".
DISTRIBUTION: Southern Hemisphere: coastal temperate waters of South America, from Rio Urucanga, Santa Catarina, Brazil to Tierra del Fuego to Paita, Peru.
STATUS: CITES - Appendix II.
SYNONYMS: *philippii*.
COMMENTS: Reviewed by Brownell and Praderi (1984, Mammalian Species, 217). A recent specimen referred to this species from Heard Island has been reidentified as *Australophocaena dioptrica* (Brownell et al., 1989).

Phocoenoides Andrews, 1911. Bull. Am. Mus. Nat. Hist., 30:31.
TYPE SPECIES: *Phocoenoides truei* Andrews, 1911 (= *Phocaena dalli* True, 1885).

Phocoenoides dalli (True, 1885). Proc. U.S. Natl. Mus., 8:95, pls. 2-5.
TYPE LOCALITY: USA, Alaska, "in the strait west of Adakh [sic] Island, one of the Aleutian group".
DISTRIBUTION: North Pacific: cold-temperate waters.
STATUS: CITES - Appendix II.
SYNONYMS: *truei*.
COMMENTS: Reviewed by Jefferson (1988, Mammalian Species 319).

Family Physeteridae Gray, 1821. London Med. Repos., 15(1):310.
COMMENTS: *Kogia* is sometimes put in a separate family, Kogiidae.

Kogia Gray, 1846. Zool. Voy. H.M.S. "Erebus" and "Terror", 1:22.
TYPE SPECIES: *Physeter breviceps* Blainville, 1838.
SYNONYMS: *Callignathus, Cogia, Euphysetes*.
COMMENTS: Reviewed by Handley (1966*c*).

Kogia breviceps (Blainville, 1838). Ann. Franc. Etr. Anat. Phys., 2:337.
TYPE LOCALITY: South Africa, Cape Prov., "rapportée des mers du cap de Bonne-Espérance" (= Cape of Good Hope).
DISTRIBUTION: Worldwide: temperate to tropical waters.
STATUS: CITES - Appendix II.
SYNONYMS: *floweri, goodei, grayii*.
COMMENTS: Reviewed by Caldwell and Caldwell (1989).

Kogia simus (Owen, 1866). Trans. Zool. Soc. Lond., 6(1):30, pls. 10-14.
TYPE LOCALITY: India, Andhra Pradesh (= Madras Presidency), "taken at Waltair".
DISTRIBUTION: Worldwide: warm-temperate to tropical waters, occasionally strands in cold-temperate areas.
STATUS: CITES - Appendix II.
COMMENTS: Reviewed by Nagorsen (1985, Mammalian Species, 239) and Caldwell and Caldwell (1989).

Physeter Linnaeus, 1758. Syst. Nat., 10th ed., 1:76.
TYPE SPECIES: *Physeter macrocephalus* Linnaeus, 1758 (= *Physeter catodon* Linnaeus, 1758) by subsequent selection (Palmer, 1904:5).
SYNONYMS: *Catodon, Cetus, Meganeuron, Megistosaurus, Physalus*.

Physeter catodon Linnaeus, 1758. Syst. Nat., 10th ed., 1:76.
TYPE LOCALITY: "Habitat in Oceano Septentrionali.", restricted to Netherlands, Middenpiat by Husson and Holthuis (1974).
DISTRIBUTION: Worldwide: antarctic and cold-temperate waters (northern hemisphere) to tropical waters.
STATUS: CITES - Appendix I; U.S. ESA - Endangered.
SYNONYMS: *australasianus, australis, macrocephalus*.
COMMENTS: Neotype designated by Husson and Holthuis (1974:212). Linnaeus used both *catodon* and *macrocephalus* in the 10th edition. *P. catodon* has line priority. See Hershkovitz (1966*a*:121), Schevill (1986, 1987*b*), Holthuis (1987), and Rice (1989, who also reviewed the species).

Family Platanistidae Gray, 1846. Zool. Voy. H.M.S. "Erebus" and "Terror", 1:25.
SYNONYMS: Iniidae, Lipotidae, Pontoporiidae, Stenodelphinidae, Susuidae.
COMMENTS: See Barnes et al. (1985) and Heyning (1989*a*) for alternative classifications. We use one family for river dophins due to the lack of consensus regarding the composition of the individual family group names.

Inia d'Orbigny, 1834. Nouv. Ann. Mus. Hist. Nat. Paris, 3:31.
TYPE SPECIES: *Inia boliviensis* d'Orbigny 1834 (= *Delphinus geoffrensis* Blainville, 1817).
COMMENTS: Reviewed by Pilleri and Gihr (1980*a*).

Inia geoffrensis (Blainville, 1817). Nouv. Dict. Hist. Nat., Nouv. ed., 9:151.
TYPE LOCALITY: "sur la côte du Brésil.", probably upper Amazon River.
DISTRIBUTION: Peru, Ecuador, Brazil, Bolivia, Venezuela, Colombia: Amazon, Negro, Mamore (Bolivia), and Orinoco River systems.
STATUS: CITES - Appendix II; IUCN - Vulnerable.
SYNONYMS: *boliviensis*.
COMMENTS: Reviewed by Best and da Silva (1989). Includes *boliviensis*, see Casinos and Ocaña (1979); but also see Pilleri and Gihr (1977), who considered it a distinct species.

Lipotes Miller, 1918. Smithson. Misc. Coll., 68(9):1.
TYPE SPECIES: *Lipotes vexillifer* Miller, 1918.

Lipotes vexillifer Miller, 1918. Smithson. Misc. Coll., 68(9):1.
TYPE LOCALITY: "Tung Ting Lake, about 600 miles up the Yangtze River, [Hunan] China".
DISTRIBUTION: China: Chang Jiang (Yangtze) and Qiantang Jiang (mouth of Fuchun Jiang) river systems.
STATUS: CITES - Appendix I; U.S. ESA and IUCN - Endangered.
COMMENTS: Reviewed by Chen (1989), Zhou et al. (1978, 1979). Reviewed by Brownell and Herald (1972, Mammalian Species, 10).

Platanista Wagler, 1830. Naturliches Syst. Amphibien, p. 35.
TYPE SPECIES: *Delphinus gangetica* Roxburgh, 1801.
SYNONYMS: *Susu*.
COMMENTS: Authorship reviewed by Pilleri (1978). The International Commission on Zoological Nomenclature (1989) conserved *Platanista* Wagler, 1830 and *gangeticus* Roxburgh, 1801 and suppressed *Susu* Lesson, 1828.

Platanista gangetica (Roxburgh, 1801). Asiat. Res. Trans. Soc. (Calcutta ed.), 7:170, pl. 5.
TYPE LOCALITY: India, West Bengal, "in the Ganges. . . rivers, and creeks, which intersect in the delta of that river to the South, S. E. and east of Calcutta." (= Hooghly River, Ganges River delta).
DISTRIBUTION: India, Nepal, Bhutan, and Bangladesh: Ganges, Bramaputra, Meghna, Karnaphuli, and Hooghly river systems.
STATUS: CITES - Appendix I; IUCN - Vulnerable.
COMMENTS: Reviewed by Reeves and Brownell (1989). Formerly included *minor* (= *indi*), see van Bree (1976), Pilleri and Gihr (1971), and Pilleri (1978).

Platanista minor Owen, 1853. Descrip. Cat. Osteol. R. Mus. Coll. Surgeons, 2:448.
TYPE LOCALITY: Pakistan, "from the Indus" River.
DISTRIBUTION: Pakistan, Indus River system.
STATUS: CITES - Appendix I; U.S. ESA and IUCN - Endangered.
SYNONYMS: *indi*.
COMMENTS: Reviewed by Reeves and Brownell (1989). See van Bree (1976). Formerly included in *gangetica*; see Pilleri and Gihr (1971) and Pilleri and Gihr (1976*a*, *b*).

Pontoporia Gray, 1846. Zool. Voy. H.M.S. "Erebus" and "Terror", 1:46.
TYPE SPECIES: *Delphinus blainvillei* Gervais and d'Orbigny, 1844.
SYNONYMS: *Stenodelphis*.

Pontoporia blainvillei (Gervais and d'Orbigny, 1844). Bull. Sci. Soc. Philom. Paris, 1844:39.
TYPE LOCALITY: Uruguay, "qui a été pris à Montevideo" = mouth of the Rio de La Plata near Montevideo.
DISTRIBUTION: Brazil to Argentina: coastal waters from Doce River, Regencia, Espírito Santo to Peninsula Valdez.
STATUS: CITES - Appendix II; IUCN - Insufficiently known.
SYNONYMS: *tenuirostris*.
COMMENTS: Reviewed by Brownell (1989).

Family Ziphiidae Gray, 1865. Proc. Zool. Soc. Lond., 1865:528.
SYNONYMS: Hyperoodontidae.
COMMENTS: Although Hyperoodontidae Gray, 1846 has priority over Ziphiidae, we have chosen to use the latter name following Article 23(b) of the International Code of Zoological Nomenclature (1985) because Ziphiidae has been the name of choice for more than 100 years. Family reviewed by Moore (1968).

Berardius Duvernoy, 1851. Ann. Sci. Nat. Zool. (Paris), ser. 3, 15:41.
TYPE SPECIES: *Berardius arnuxii* Duvernoy, 1851.
COMMENTS: This may be a monotypic genus.

Berardius arnuxii Duvernoy, 1851. Ann. Sci. Nat. Zool. (Paris), ser. 3, 15:52, fig. 1.
TYPE LOCALITY: "échoué sur la côte, dans le port d'Akaroa, presqu'île de Bancks, dands la Nouvelle-Zélande." (= New Zealand, Canterbury Prov., Akaroa).
DISTRIBUTION: Southern Hemisphere: circumpolar, temperate waters.
STATUS: CITES - Appendix I.
COMMENTS: Reviewed by Balcomb (1989).

Berardius bairdii Stejneger, 1883. Proc. U.S. Natl. Mus., 6:75.
TYPE LOCALITY: Russia, Commander Isls, "found stranded in Stare Gavan, on the eastern shore of Bering Island".
DISTRIBUTION: North Pacific: temperate waters.
STATUS: CITES - Appendix I.
SYNONYMS: *vegae*.
COMMENTS: Reviewed by Balcomb (1989); possibly a subspecies of *arnuxii*, see Davies (1963) and McLachlan et al. (1966).

Hyperoodon Lacépède, 1804. Hist. Nat. Cetacees, xliv, 319.
TYPE SPECIES: *Hyperoodon butskopf* Lacépède, 1804 (= *Balaena ampullata* Forster, 1770).
SYNONYMS: *Anodon, Chaenodelphinus, Frasercetus, Heterodon, Lagocetus, Uranodon*.
COMMENTS: Includes *Frasercetus* Moore, 1968 as a subgenus.

Hyperoodon ampullatus (Forster, 1770). *In* Kalm, Travels into N. Am., 1:18.
TYPE LOCALITY: "See Mr. Pennant's [1769] British Zoology Vol. 3, p. 43, where it is called the beaked whale, and very well described;" Pennant (1769:43) gave Maldon (England) as the locality and 1717 as the date stranded.
DISTRIBUTION: North Atlantic: arctic to cold-temperate waters. The Mediterranean record represents a stray (J. G. Mead, 1989*b*).
STATUS: CITES - Appendix I; IUCN - Vulnerable.
SYNONYMS: *butskopf, latifrons, rostratus*.
COMMENTS: Reviewed by J. G. Mead (1989*b*).

Hyperoodon planifrons Flower, 1882. Proc. Zool. Soc. Lond., 1882:392, figs. 1, 2.
TYPE LOCALITY: "found upon the sea-beach of Lewis Island in the Dampier Archipelago, North-western Australia."
DISTRIBUTION: Southern Hemisphere: circumpolar, antarctic to temperate waters, occasionally into tropical waters. May occur in the W North Pacific.
STATUS: CITES - Appendix I.
SYNONYMS: *burmeisterei*.
COMMENTS: Reviewed by J. G. Mead (1989*b*). Moore (1968) erected the subgenus *Frasercetus* for this species.

Indopacetus Moore, 1968. Fieldiana Zool., 53(4):254.
TYPE SPECIES: *Mesoplodon pacificus* Longman, 1926.
COMMENTS: Considered by many authors to be included in *Mesoplodon*. Known only from two specimens.

Indopacetus pacificus (Longman, 1926). Mem. Queensl. Mus., 8(3):269, pl. 43.
TYPE LOCALITY: Australia, Queensland, "found at Mackay".
DISTRIBUTION: Indian Ocean and W South Pacific: tropical waters.
STATUS: CITES - Appendix II.
COMMENTS: Reviewed by J. G. Mead (1989*c*); known only from two skulls, the second from Somalia. Commonly included in *Mesoplodon* (Heyning, 1989*a*; J. G. Mead, 1989*c*).

Mesoplodon Gervais, 1850. Ann. Sci. Nat. Zool. (Paris), ser. 3, 14:16.
TYPE SPECIES: *Delphinus sowerbensis* Blainville, 1817 (= *Physeter bidens* Sowerby, 1804).
SYNONYMS: *Aodon, Dioplodon, Dolichodon, Micropterus, Oulodon, Nodus, Paikea.*
COMMENTS: *Mesoplodon* Gervais, 1850 and *Physeter bidens* Sowerby were conserved; *Nodus, Micropteron*, and *Mikropteron* were suppressed by the International Commission on Zoological Nomenclature (1985*b*).

Mesoplodon bidens (Sowerby, 1804). Trans. Linn. Soc. Lond., 7:310.
TYPE LOCALITY: UK, Scotland, "stranded on the estate of James Brodie, Esq. F. L. S., in the county of Elgin."
DISTRIBUTION: North Atlantic and Baltic Sea: temperate waters. Occurence in the Mediterranean Sea was discussed by van Bree (1975), who considered the evidence unconvincing; however, Casinos and Filella (1981) supported a report from the Italian coast (Brunelli and Fasella, 1929). There is one report from the Gulf of Mexico (Bonde and O'Shea, 1989) that is also considered a stray.
STATUS: CITES - Appendix II.
SYNONYMS: *dalei, micropterus, sowerbensis, sowerbyi.*
COMMENTS: Reviewed by J. G. Mead (1989*c*).

Mesoplodon bowdoini Andrews, 1908. Bull. Am. Mus. Nat. Hist., 24:203, figs. 1-5, pl. 13.
TYPE LOCALITY: "collected at New Brighton Beach, Canterbury Province, New Zealand".
DISTRIBUTION: Southern Hemisphere, South Pacific and Indian oceans, cold-temperate waters of Australia and New Zealand. The record from Kerguelen Isls (Robineau, 1973) is erroneus (J. G. Mead, 1989*c*).
STATUS: CITES - Appendix II.
COMMENTS: Reviewed by J. G. Mead (1989*c*). McCann (see Mead et al., 1982) felt that *M. bowdoini* was synonymous with *M. stejnegeri*.

Mesoplodon carlhubbsi Moore, 1963. Am. Midl. Nat., 70:396, figs. 1-3, 7, 8, 13-15.
TYPE LOCALITY: "La Jolla, California, 32° 51' 41" N. Lat., 117° 15' 19" W. Long."
DISTRIBUTION: North Pacific: temperate waters.
STATUS: CITES - Appendix II.
COMMENTS: Reviewed by J. G. Mead (1989*c*). Very closely related to *bowdoini*. Orr believed that this species was synonymous with *M. stejnegeri* (see Mead et al., 1982). Hubbs (1946) first identified the holotype of this species as *M. bowdoini*.

Mesoplodon densirostris (Blainville, 1817). Nouv. Dict. Hist. Nat., Nouv. ed., 9:178.
TYPE LOCALITY: None given, unknown.
DISTRIBUTION: World-wide: temperate to tropical waters.
STATUS: CITES - Appendix II.
SYNONYMS: *seychellensis.*
COMMENTS: Reviewed by J. G. Mead (1989*c*).

Mesoplodon europaeus (Gervais, 1855). Hist. Nat. Mammifères, 2:320.
TYPE LOCALITY: English Channel, "qui provient d'un individu harponné dans la Manche."
DISTRIBUTION: Aside from the type, one specimen from Ireland, one specimen from Guinea-Bissau, and three records from Ascension Isl, it is only known from the W North Atlantic: temperate to tropical waters.
STATUS: CITES - Appendix II.
SYNONYMS: *gervaisi.*

COMMENTS: Reviewed by J. G. Mead (1989*c*). The type was not harpooned, as stated by Gervais, but was found as a "cadavre" (Deslongschamps, 1866:177).

Mesoplodon ginkgodens Nishiwaki and Kamiya, 1958. Sci. Rep. Whales Res. Inst. (Tokyo), 13:53, 13 figs., 17 pls.
TYPE LOCALITY: Japan, "Oiso Beach, Sagami Bay, near Tokyo."
DISTRIBUTION: North Pacific and Indian Oceans: warm-temperate to tropical waters; Japan, Taiwan, Baja California, Sri Lanka, Indonesia and Australia.
STATUS: CITES - Appendix II.
SYNONYMS: *hotaula*.
COMMENTS: Reviewed by J. G. Mead (1989*c*)

Mesoplodon grayi Von Haast, 1876. Proc. Zool. Soc. Lond., 1876:9.
TYPE LOCALITY: New Zealand, "the Chatham Islands. . . from specimens stranded. . . on the Waitangi beach of the main island of that group."
DISTRIBUTION: Southern Hemisphere: cold-temperate waters; one specimen found in the Netherlands (Boschma, 1950:779).
STATUS: CITES - Appendix II.
SYNONYMS: *australis, haasti*.
COMMENTS: Reviewed by J. G. Mead (1989*c*).

Mesoplodon hectori (Gray, 1871). Ann. Mag. Nat. Hist., ser. 4, 8:116.
TYPE LOCALITY: New Zealand, Wellington, "killed in Tatai [sic] Bay, Cook's Straits" (= Titai Bay).
DISTRIBUTION: Southern Hemisphere, North Pacific: temperate waters.
STATUS: CITES - Appendix II.
SYNONYMS: *knoxi*.
COMMENTS: Reviewed by J. G. Mead (1989*c*).

Mesoplodon layardii (Gray, 1865). Proc. Zool. Soc. Lond., 1865:357, fig.
TYPE LOCALITY: None given, probably South Africa.
DISTRIBUTION: Southern Hemisphere: temperate waters.
STATUS: CITES - Appendix II.
SYNONYMS: *floweri, guntheri, longirostris, thomsoni, traversii*.
COMMENTS: Reviewed by J. G. Mead (1989*c*).

Mesoplodon mirus True, 1913. Smithson. Misc. Coll., 60(25):1.
TYPE LOCALITY: USA, "stranded in the outer bank of Bird Island Shoal in the harbor of Beaufort, North Carolina".
DISTRIBUTION: North Atlantic, South Atlantic coast of South Africa, Australia: temperate waters.
STATUS: CITES - Appendix II.
COMMENTS: Reviewed by J. G. Mead (1989*c*).

Mesoplodon peruvianus Reyes, Mead, and Van Waerebeek, 1991. Marine Mammal Sci., 7(1):1, 6 figs.
TYPE LOCALITY: "Playa Paraiso (11° 12' S), Huacho, Lima, Peru."
DISTRIBUTION: E South Pacific, E North Pacific: cold-temperate to tropical waters. Known from the coast of Peru between Playa Paraiso (11°S) and San Juan de Marcona (15°S). Two specimens are known from near La Paz, Baja California, Mexico (Urban-Ramirez and Aurioles-Gamboa, in press).
STATUS: CITES - Appendix II.

Mesoplodon stejnegeri True, 1885. Proc. U.S. Natl. Mus., 8:584, pl. 25.
TYPE LOCALITY: Russia, Commander Isls, "Bering Island".
DISTRIBUTION: North Pacific: cold-temperate waters.
STATUS: CITES - Appendix II.
COMMENTS: Reviewed by Loughlin and Perez (1985, Mammalian Species, 250) and J. G. Mead (1989*c*).

Tasmacetus Oliver, 1937. Proc. Zool. Soc. Lond., 107:371.
TYPE SPECIES: *Tasmacetus shepherdi* Oliver, 1937.

Tasmacetus shepherdi Oliver, 1937. Proc. Zool. Soc. Lond., 107:371, pls. 1-5.
TYPE LOCALITY: New Zealand, North Island, "cast upon the beach at Ohawe, in the province of Taranaki."
DISTRIBUTION: Southern Hemisphere: cold-temperate waters, particularly off New Zealand, Chile, Argentina and Tristan de Cunha.
STATUS: CITES - Appendix II.
COMMENTS: Reviewed by J. G. Mead (1989*a*).

Ziphius G. Cuvier, 1823. Rech. Oss. Foss., Nouv. ed., 5:350.
TYPE SPECIES: *Ziphius cavirostris* G. Cuvier, 1823.
SYNONYMS: *Diodon, Hypodon, Petrorhynchus, Ziphiorhynchus.*

Ziphius cavirostris G. Cuvier, 1823. Rech. Oss. Foss., Nouv. ed., 5(1):350.
TYPE LOCALITY: France, "dans le département des Bouches-du-Rhône, entre de Fos et l'embouchure du Galégeon" (= between Fos and the mouth of the Galégeon River).
DISTRIBUTION: Worldwide: cold-temperate to tropical waters.
STATUS: CITES - Appendix II.
SYNONYMS: *australis, capensis, chathamensis, indicus.*
COMMENTS: Reviewed by Heyning (1989*b*).

ORDER SIRENIA
by Don E. Wilson

ORDER SIRENIA

Family Dugongidae Gray, 1821. London Med. Repos., 15:309.
SYNONYMS: Halicoridae, Halitheriidae, Hydrodamalidae.

Dugong Lacépède, 1799. Tab. Div. Subd. Orders Genres Mammiféres, 14:17.
TYPE SPECIES: *Dugong indicus* Lacépède, 1799 (= *Trichechus dugon* Müller, 1776).
SYNONYMS: *Dugungus, Halicore, Platystomus.*
COMMENTS: Subfamily Dugonginae.

Dugong dugon (Müller, 1776). Linne's Vollstand. Natursyst. Suppl., p. 21.
TYPE LOCALITY: Cape of Good Hope to the Philippines.
DISTRIBUTION: Tropical coastal waters of Indian and W Pacific Oceans.
STATUS: CITES - Appendix I, except Australian population which is Appendix II; U.S. ESA - Endangered; IUCN - Vulnerable.
SYNONYMS: *australis, cetacea, dugung, hemprichii, indicus, lottum, tabernaculi.*
COMMENTS: Reviewed by Husar (1978*a*, Mammalian Species, 88).

Hydrodamalis Retzius, 1794. K. Svenska Vet.-Akad. Handl. Stockholm, 15, p. 292.
TYPE SPECIES: *Hydrodamalis Stelleri* Retzius, 1794 (= *Manati gigas* Zimmermann, 1780).
SYNONYMS: *Nepus, Rytina, Stellerus.*
COMMENTS: Subfamily Hydrodamalinae; see Domning (1978).

Hydrodamalis gigas (Zimmermann, 1780). Geogr. Gesch. Mensch. Vierf. Thiere, 2:426.
TYPE LOCALITY: Bering Sea, Commander Isls, Bering Isl.
DISTRIBUTION: Known only from the Commander Islands, Bering Sea.
STATUS: IUCN - Extinct.
SYNONYMS: *balaenurus, borealis, stelleri.*
COMMENTS: See Forsten and Youngman (1982, Mammalian Species, 165).

Family Trichechidae Gill, 1872. Smithson. Misc. Coll., 11(1):14.
SYNONYMS: Manatidae.

Trichechus Linnaeus, 1758. Syst. Nat., 10th ed., 1:34.
TYPE SPECIES: *Trichechus manatus* Linnaeus, 1758.
SYNONYMS: *Halipaedisca, Manatus, Oxystomus.*
COMMENTS: Revised by Hatt (1934*a*); evolutionary history summarized by Domning (1982).

Trichechus inunguis (Natterer, 1883). *In* Pelzeln, Verh. Zool.-Bot. Ges. Wien, 33:89.
TYPE LOCALITY: Brazil, Amazonas, Rio Madeira, Borba.
DISTRIBUTION: Amazon basin of Brazil, Colombia, Ecuador, Guyana, and Peru.
STATUS: CITES - Appendix I; U.S. ESA - Endangered; IUCN - Vulnerable.
COMMENTS: Reviewed by Husar (1977, Mammalian Species, 72).

Trichechus manatus Linnaeus, 1758. Syst. Nat., 10th ed., 1:34.
TYPE LOCALITY: "Mari Americano"; restricted by Thomas (1911*a*) to "West Indies."
DISTRIBUTION: Caribbean coastal areas and river systems from Virginia, USA to Espírito Santo, Brazil, including Belize, Colombia, Costa Rica, French Guiana, Guyana, Honduras, Mexico, Nicaragua, Panama, Suriname, Venezuela, and the West Indies including the Bahamas, Cuba, Dominican Republic, Haiti, Jamaica, Puerto Rico, and formerly the Virgin Isls.
STATUS: CITES - Appendix I; U.S. ESA - Endangered; IUCN - Vulnerable.
SYNONYMS: *amazonius, americanus, antillarum, atlanticus, clusii, guyannensis, koellikeri, latirostris, minor, oronocensis, trichechus.*

COMMENTS: See Domning (1981); reviewed by Husar (1978*c*, Mammalian Species, 93).

Trichechus senegalensis Link, 1795. Beitr. Naturgesch., 1(2):209.
TYPE LOCALITY: Senegal.
DISTRIBUTION: Coastal W Africa including river systems from Angola to Senegal.
STATUS: CITES - Appendix II; U.S. ESA - Threatened; IUCN - Vulnerable.
SYNONYMS: *africanus, australis, nasutus, owenii, stroggylonurus, vogelii.*
COMMENTS: Reviewed by Husar (1978*b*, Mammalian Species, 89).

ORDER PROBOSCIDEA
by Don E. Wilson

ORDER PROBOSCIDEA

Family Elephantidae Gray, 1821. London Med. Repos., 15:305.
COMMENTS: Revised by Maglio (1973).

Elephas Linnaeus, 1758. Syst. Nat., 10th ed., 1:33.
TYPE SPECIES: *Elephas maximus* Linnaeus, 1758.
SYNONYMS: *Elephantus, Hesperoloxodon, Hypselephas, Leith-adamsia, Omoloxodon, Paleoloxodon, Pilgrimia, Platelephas, Sivalikia, Stegoloxodon.*

Elephas maximus Linnaeus, 1758. Syst. Nat., 10th ed., 1:33.
TYPE LOCALITY: "Zeylonae" [Sri Lanka].
DISTRIBUTION: Bangladesh, Burma, China, Cambodia, India, Indonesia, Laos, Malaysia, Thailand, Vietnam.
STATUS: CITES - Appendix I; U.S. ESA and IUCN - Endangered.
SYNONYMS: *asiaticus, asurus, bengalensis, birmanicus, borneensis, ceylanicus, dakhunensis, dauntela, gigas, heterodactylus, hirsutus, indicus, isodactylus, mukna, rubridens, sinhaleyus, sondaicus, sumatranus, vilaliya, zeylanicus.*
COMMENTS: See Shoshani and Eisenberg (1982, Mammalian Species, 182).

Loxodonta Cuvier, 1825. *In* E. Geoffroy St.-Hilaire and F. G. Cuvier, Hist. Nat. Mammifères, 3(52):2.
TYPE SPECIES: *Elephas africanus* Blumenbach, 1797.

Loxodonta africana (Blumenbach, 1797). Handb. Naturgesch., 5th ed., p. 125.
TYPE LOCALITY: Restricted to the Orange River, South Africa by Pohle (1926; see Allen, 1939).
DISTRIBUTION: From S Mauritania, Mali, Chad, and Sudan south to N South Africa, Botswana, and Namibia.
STATUS: CITES - Appendix I; U.S. ESA - Threatened; IUCN - Vulnerable.
SYNONYMS: *albertensis, angammensis, angolensis, berbericus, capensis, cavendishi, cottoni, cyclotis, fransseni, hannibaldi, knochenhaueri, mocambicus, orleansi, oxyotis, peeli, pharaohensis, prima, priscus, pumilio, rothschildi, selousi, toxotis, typicus, zukowskyi.*
COMMENTS: See Laursen and Bekoff (1978, Mammalian Species, 92).

ORDER PERISSODACTYLA

by Peter Grubb

ORDER PERISSODACTYLA

Family Equidae Gray, 1821. London Med. Repos., 15:307.

Equus Linnaeus, 1758. Syst. Nat., 10th ed., 1:73.

TYPE SPECIES: *Equus caballus* Linnaeus, 1758.

SYNONYMS: *Asinohippus, Asinus, Dolichohippus, Grevya, Hemionus, Hemippus, Hippotigris, Ludolphozecora, Microhippus, Megacephalon, Megacephalonella, Onager, Pseudoquagga, Quagga, Quaggoides, Zebra* (see Bennett, 1980; Groves and Willoughby, 1981).

Equus asinus Linnaeus, 1758. Syst. Nat., 10th ed., 1:73.

TYPE LOCALITY: "Habitat in oriente" (= Middle East?).

DISTRIBUTION: NE Sudan (now extinct); NE Ethiopia; N Somalia; up until the third century A.D. in N Algeria, Morocco and Tunisia; domesticated worldwide; feral or possibly wild in Hoggar (S Algeria) and Tibesti (N Chad); feral in Sudan, Saudi Arabia, Socotra Isl (Yemen), Sri Lanka, Australia, USA (including Hawaiian Isls), Galapagos Isls, Chagos Isls and probably other oceanic islands.

STATUS: CITES - Appendix I as *E. africanus*; U.S. ESA and IUCN - Endangered as *E. africanus* (= *asinus*).

SYNONYMS: *aethiopicus, africanus, atlanticus, dianae, somalicus, taeniopus, vulgaris*.

COMMENTS: Ansell (1974*a*:6) recommended use of *africanus* as specific name, since the name *asinus* was based upon domestic populations.

Equus burchellii (Gray, 1824). Zool. J., 1:247.

TYPE LOCALITY: South Africa, N Cape Province, Kuruman, Little Klibbolikhonni Fontein.

DISTRIBUTION: SE Sudan, SW Ethiopia and S Somalia south and southwest to SE Zaire, S and E Angola, N Namibia, N and E Botswana; Natal and Transvaal (South Africa); formerly more widespread in southern Africa, south to Orange River.

SYNONYMS: *annectens, antiquorum, boehmi, borensis, campestris, chapmanni, crawshaii, cuninghamei, festivus, foai, goldfinchi, granti, isabella, jallae, kaufmanni, kaokensis, mariae, markhami, muansae, paucistriatus, pococki, selousii, tigrinus, transvaalensis, wahlbergi, zambeziensis, zebroides*.

COMMENTS: A species separate from *E. quagga*; see Gentry (1975), Eisenmann and Turlot (1978), and Bennett (1980). Many previous workers regarded *quagga* and *burchellii* as conspecific; see Rau (1978). Recently regarded as conspecific by Groves (1985*b*). Reviewed by Grubb (1981, Mammalian Species, 157).

Equus caballus Linnaeus, 1758. Syst. Nat., 10th ed., 1:73.

TYPE LOCALITY: "Habitat in Europa" (= Sweden?); based on domestic horses.

DISTRIBUTION: In classical antiquity, wild horses said to have ranged as far west as Spain; into the late 18th Century, from Poland and Russian Steppes east to Turkestan and Mongolia; wild population survived (at least until recently) in SW Mongolia and adjacent Kansu, Sinkiang, and Inner Mongolia (China). Domesticated worldwide; feral in Portugal, Spain, France, Greece, Iran, Sri Lanka, Australia, New Zealand, Colombia, Hispaniola, Canada, USA (incl. Hawaiian Isls), Galapagos and probably other oceanic islands.

STATUS: CITES - Appendix I and U.S. ESA - Endangered as *Equus przewalskii*; IUCN - Extinct? as *E. przewalskii*.

SYNONYMS: *ferus, gmelini, gutsenensis, hagenbecki, przewalskii, silvatica, silvestris*.

COMMENTS: Horses have been assigned to two different species, *E. caballus* (including *ferus* and *gmelini*) and *E. przewalskii*, but recent authors include *przewalskii* in *caballus*; see Corbet (1978*c*:194), Groves (1974*a*), and Bennett (1980). Groves (1971*b*) and Corbet (1978*c*:194) proposed that *ferus* replace *caballus*, objecting to the use of specific names based on domestic animals. Gromov and Baranova (1981:333-334) continued to recognize two species, *gmelini* and *przewalskii*.

Equus grevyi Oustalet, 1882. La Nature (Paris), 10(2):12.
TYPE LOCALITY: Ethiopia, Galla Country.
DISTRIBUTION: Dry desert regions of N Kenya, S Somalia, and S and E Ethiopia.
STATUS: CITES - Appendix I; U.S. ESA - Threatened; IUCN - Endangered.
SYNONYMS: *berberensis*.

Equus hemionus Pallas, 1775. Nova Comm. Imp. Acad. Sci. Petrop., 19:394.
TYPE LOCALITY: Russia, Transbaikalia, S Chitinsk. Obl., Tarei-Nor, 50°N, 115°E.
DISTRIBUTION: Formerly much of Mongolia, north to Transbaikalia (Russia); east to NE Inner Mongolia (China) and possibly W Manchuria (China); and west to Dzungarian Gate. Survives in SW and SC Mongolia and adjacent China; see Sokolov and Orlov (1980:248).
STATUS: CITES - Appendix I as *E. h. hemionus*, otherwise Appendix II; U.S. ESA - Endangered; IUCN - Vulnerable.
SYNONYMS: *bedfordi, castaneus, finschi, luteus*.
COMMENTS: Revised by Groves and Mazak (1967) and Schlawe (1986).

Equus kiang Moorcroft, 1841. Travels in the Himalayan Provinces, 1:312.
TYPE LOCALITY: India, Kashmir, Ladak.
DISTRIBUTION: Ladak (India), Tibet, Tsinghai and Szechwan (China), adjacent Nepal and Sikkim (India).
SYNONYMS: *equioides, holdereri, nepalensis, polyodon, tafeli*.
COMMENTS: Revised by Groves and Mazak (1967), who with Bennett (1980) separated *kiang* from *hemionus*; but Schlawe (1986) regarded *kiang* as a subspecies of *hemionus*.

Equus onager Boddaert, 1785. Elench. Anim., p. 160.
TYPE LOCALITY: NW Persia (= Iran), Kasbin, near Caspian.
DISTRIBUTION: Formerly Kazakhstan north to upper Irtysh and Ural Rs. (Russia); westward north of the Caucasus and Black Sea at least to Dniestr River (Ukraine); and SE of Caspian Sea, Anatolia, N Iraq, Iran, Afghanistan, and Pakistan to Thar Desert of NW India; survives as isolated populations in Rann of Kutch (India), Badkhys Preserve, Turkmenia, and C Iran; also reestablished on Barsa-khelmes Isl (Aral Sea, Uzbekistan).
STATUS: CITES - Appendix I *E. hemionus khur*; IUCN - Endangered as *E. h. khur*, Extinct as *E. h. hemippus*.
SYNONYMS: *bahram, blanfordi, dzigguetai, hamar, hemippus, indicus, khur, kulan, syriacus*.
COMMENTS: Revised by Groves and Mazak (1967), who with Groves (1986) and Schlawe (1986) included *onager* in *hemionus*, but Bennett (1980) considered *onager* a distinct species.

Equus quagga Boddaert, 1785. Elench. Anim., p. 160.
TYPE LOCALITY: South Africa.
DISTRIBUTION: South Africa, south of the Vaal River.
STATUS: IUCN - Extinct as *E. q. quagga*; extinct, last specimen, a captive, died in 1872.
SYNONYMS: *danielli, greyi, isabellinus, lorenzi, trouessarti*.
COMMENTS: See comments under *burchellii*.

Equus zebra Linnaeus, 1758. Syst. Nat., 10th ed., 1:74.
TYPE LOCALITY: South Africa, SW Cape Prov., Paardeburg, near Malmesbury.
DISTRIBUTION: S Angola, Namibia, SW and SC Cape Prov. (South Africa). Now much reduced in numbers and, in South Africa, confined to a few nature reserves.
STATUS: CITES - Appendix I as *E. z. zebra*, Appendix II as *E. z. hartmannae*; U.S. ESA - Endangered as *E. z. zebra*, Threatened as *E. z. hartmannae*; IUCN - Endangered as *E. z. zebra*, Vulnerable as *E. z. hartmannae*.
SYNONYMS: *frederici, greatheadi, hartmannae, indica, matschiei, montanus, penricei*.
COMMENTS: Reviewed by Penzhorn (1988, Mammalian Species, 314).

Family Tapiridae Gray, 1821. London Med. Repos., 15:306.

Tapirus Brünnich, 1771. Zool. Fundamenta, pp. 44, 45.
TYPE SPECIES: *Hippopotamus terrestris* Linnaeus, 1758.
SYNONYMS: *Acrocodia, Cinchacus, Elasmognathus, Rhinochoerus, Syspotamus, Tapirella*.

Tapirus bairdii (Gill, 1865). Proc. Acad. Nat. Sci. Philadelphia, 17:183.
TYPE LOCALITY: Panama, Isthmus of Panama, restricted to Canal Zone by Hershkovitz (1954).
DISTRIBUTION: S Veracruz and S Oaxaca (Mexico) east of Isthmus of Tehuantepec to Colombia west of the Rio Cauca and Ecuador west of the Andes to the Gulf of Guayaquil.
STATUS: CITES - Appendix I; U.S. ESA - Endangered; IUCN - Vulnerable.
SYNONYMS: *dowii*.
COMMENTS: Revised by Hershkovitz (1954).

Tapirus indicus Desmarest, 1819. Nouv. Dict. Hist. Nat., Nouv. ed., 32:458,
TYPE LOCALITY: Malaysia, Malay Peninsula.
DISTRIBUTION: Burma and Thailand south of 18°N, south through Peninsular Malaysia; Sumatra. Said to have survived into this Century in S Vietnam; see Harper (1945). Recorded approximately 3,000-1,500 years BP from Uttar Pradesh (N India) and Henan (China); see Banerjee and Ghosh (1981) and Teilhard de Chardin and Young (1936).
STATUS: CITES - Appendix I; U.S. ESA and IUCN - Endangered.
SYNONYMS: *bicolor, malayanus, sumatranus*.
COMMENTS: Separated as genus *Acrocodia* by Eisenberg et al. (1987).

Tapirus pinchaque (Roulin, 1829). Ann. Sci. Nat. Zool., 18:46.
TYPE LOCALITY: Colombia, Cundinamarca, Paramo de Sumapaz.
DISTRIBUTION: Andes of Colombia and Ecuador; perhaps W Venezuela and N Peru.
STATUS: CITES - Appendix I; U.S. ESA - Endangered; IUCN - Vulnerable.
SYNONYMS: *andicola, leucogenys, pinchacus, roulinii, villosus*.
COMMENTS: Revised by Hershkovitz (1954).

Tapirus terrestris (Linnaeus, 1758). Syst. Nat., 10th ed., 1:74.
TYPE LOCALITY: "Brasilia", i.e., Brazil, Pernambuco.
DISTRIBUTION: Venezuela and Colombia south to S Brazil, N Argentina and Paraguay, east of the Andes.
STATUS: CITES - Appendix II; U.S. ESA - Endangered.
SYNONYMS: *aenigmaticus, americanus, anta, anulipes, brasiliensis, colombianus, ecuadorensis, guianae, laurillardi, maypuri, mexianae, peruvianus, rufus, sabatyra, spegazzinii, suillus, tapir, tapirus*.
COMMENTS: Revised by Hershkovitz (1954).

Family Rhinocerotidae Gray, 1821. London Med. Repos., 15:306.

Ceratotherium Gray, 1868. Proc. Zool. Soc. Lond., 1867:1027 [1868].
TYPE SPECIES: *Rhinoceros simus* Burchell, 1817.

Ceratotherium simum (Burchell, 1817). Bull. Sci. Soc. Philom. Paris, p. 97.
TYPE LOCALITY: South Africa, Cape Prov., Makuba Range, Chue Spring (= Heuningvlei; about 26°15'S, 23°10'E).
DISTRIBUTION: Formerly S Chad, Central African Republic, S Sudan, NE Zaire, Uganda, S Zambia and from Zambezi River in Zimbabwe and S Mozambique to Vaal River in South Africa. Now much restricted in distribution; in south of range, extinct except in E Natal (South Africa), but reintroduced into other parts of South Africa (Natal, Transvaal, Orange Free State), Namibia, Swaziland, Mozambique, Zimbabwe and Botswana. Introduced into Zambia and Kenya. In north of range, now confined to NE Zaire.
STATUS: CITES - Appendix I; U.S. ESA and IUCN - Endangered as *C. s. cottoni*.
SYNONYMS: *burchellii, camptoceros, camus, crossii, cottoni, kaiboaba, kulamanae, kulamane, oswellii, prostheceros*.
COMMENTS: Reviewed by Groves (1972*a*, Mammalian Species, 8). Revised by Groves (1975*b*).

Dicerorhinus Gloger, 1841. Gemein Hand.-Hilfsbuch. Nat., p. 125.
TYPE SPECIES: *Rhinoceros sumatrensis* Fischer, 1814.
SYNONYMS: *Ceratorhinus, Didermocerus*.

COMMENTS: *Didermocerus* Brookes, 1828, has been rejected, and *Dicerorhinus* validated (International Commission on Zoological Nomenclature, 1977*b*).

Dicerorhinus sumatrensis (G. Fischer, 1814). Zoognosia, 3:301.
TYPE LOCALITY: Indonesia, Sumatra, Bencoolen (= Bintuhan) Dist., Fort Marlborough.
DISTRIBUTION: Formerly Assam (India), Chittagong Hills (Bangladesh), Burma, Thailand, and Vietnam south through Peninsular Malaysia to Sumatra; probably also S China, Laos, and Cambodia; Borneo, and Mergui Isl. Survives in Tenasserim Range (Thailand-Burma), Petchabun Range (Thailand), and other scattered localities in Burma, Peninsular Malaysia, Sumatra, and Borneo.
STATUS: CITES - Appendix I; U.S. ESA and IUCN - Endangered.
SYNONYMS: *blythii, harrissoni, lasiotis, niger.*
COMMENTS: Reviewed by Groves and Kurt (1972, Mammalian Species, 21). Revised by Groves (1967*c*).

Diceros Gray, 1821. London Med. Repos., 15:306.
TYPE SPECIES: *Rhinoceros bicornis* Linnaeus, 1758.

Diceros bicornis (Linnaeus, 1758). Syst. Nat., 10th ed., 1:56.
TYPE LOCALITY: South Africa, Cape Prov.
DISTRIBUTION: Formerly in suitable open habitats in Africa south of about 10°N from N Nigeria, Chad, S Sudan and N Somalia, and from Angola, south to Cape Province (South Africa). Very much reduced in numbers this century, particularly in recent decades, and probably now extinct in many countries which it formerly occupied. Survives in reserves in Kenya, Tanzania, Namibia, Zambia, Zimbabwe and Zululand (South Africa), and possibly still in Cameroon, Chad, Central African Republic, Sudan, Rwanda, Malawi, Mozambique, Angola and Botswana; widely reintroduced into parts of South Africa (Cumming et al., 1990).
STATUS: CITES - Appendix I; U.S. ESA and IUCN - Endangered.
SYNONYMS: *africanus, angolensis, atbarensis, brucii, camperi, capensis, chobiensis, gordoni, holmwoodi, keitloa, ladoensis, longipes, major, michaeli, minor, niger, occidentalis, palustris, platyceros, plesioceros, punyana, rendilis, somaliensis.*
COMMENTS: Revised by Groves (1967*b*).

Rhinoceros Linnaeus, 1758. Syst. Nat., 10th ed., 1:56.
TYPE SPECIES: *Rhinoceros unicornis* Linnaeus, 1758.
SYNONYMS: *Eurhinoceros, Monocerorhinus.*

Rhinoceros sondaicus Desmarest, 1822. Mammalogie, *in* Encycl. Meth., 2:399.
TYPE LOCALITY: Java (Indonesia).
DISTRIBUTION: Formerly Bangladesh, Burma, Thailand, Laos, Cambodia, Vietnam, and probably S China through Peninsular Malaysia to Sumatra and Java. Survives in Ujung Kulon (W Java) and in Vietnam; perhaps in small areas of Burma, Thailand, Laos, and Cambodia.
STATUS: CITES - Appendix I; U.S. ESA and IUCN - Endangered.
SYNONYMS: *annamiticus, floweri, inermis, javanicus, nasalis.*
COMMENTS: Revised by Groves (1967*c*).

Rhinoceros unicornis Linnaeus, 1758. Syst. Nat., 10th ed., 1:56.
TYPE LOCALITY: India, Assam Terai.
DISTRIBUTION: Within the present millenium, Indus Valley (Pakistan) east in N India to Assam and perhaps N Burma. Survives in India (Assam, West Bengal), Nepal and possibly N Burma.
STATUS: CITES - Appendix I; U.S. ESA and IUCN - Endangered.
SYNONYMS: *asiaticus, indicus, jamrachii, stenocephalus.*
COMMENTS: Reviewed by Laurie et al. (1983, Mammalian Species, 211).

ORDER HYRACOIDEA

by Duane A. Schlitter

ORDER HYRACOIDEA

Family Procaviidae Thomas, 1892. Proc. Zool. Soc. Lond., 1892:51.

COMMENTS: Hyracidae Gray, 1821, is a group name based on *Hyrax* Hermann, 1783. The number of valid species is uncertain; see Bothma (1971) and Corbet (1979). Revised by Hahn (1934:207). Also see Allen (1939), Meyer (1978), and Roberts (1951). The generic definitions are also controversial; Roche (1972) retained only *Procavia* and *Dendrohyrax* but Hoeck (1978) and Meester et al. (1986:178) retained *Procavia, Heterohyrax,* and *Dendrohyrax* as separate genera. A modern key to the genera was developed by Meester et al. (1986).

Dendrohyrax Gray, 1868. Ann. Mag. Nat. Hist., ser. 4, 1:48.

TYPE SPECIES: *Hyrax arboreus* A. Smith, 1827.

COMMENTS: A key to the species was published by Jones (1978).

Dendrohyrax arboreus (A. Smith, 1827). Trans. Linn. Soc. Lond., 15:468.

TYPE LOCALITY: South Africa, Cape Prov., forests of Cape of Good Hope.

DISTRIBUTION: Cape Prov. and Natal (South Africa); Mozambique; Zambia; Malawi; Zaire; Tanzania to Kenya and the Sudan.

SYNONYMS: *adolfi-friederici, bettoni, braueri, crawshayi, helgei, mimus, ruwenzorii, scheelei, scheffleri, schubotzi, stuhlmanni, vilhelmi.*

Dendrohyrax dorsalis (Fraser, 1855). Proc. Zool. Soc. Lond., 1854:99 [1855].

TYPE LOCALITY: Equatorial Guinea, Bioko.

DISTRIBUTION: Gambia to N Angola; Bioko (Equatorial Guinea); C and NE Zaire; N Uganda.

SYNONYMS: *adametzi, aschantiensis, beniensis, brevimaculatus, congoensis, emini, latrator, marmota, nigricans, rubriventer, stampflii, sylvestris, tessmanni, zenkeri.*

COMMENTS: See Jones (1978, Mammalian Species, 113).

Dendrohyrax validus True, 1890. Proc. U.S. Natl. Mus., 13:228.

TYPE LOCALITY: Tanzania (= Tanganyika), Mt. Kilimanjaro.

DISTRIBUTION: E Tanzania (including Zanzibar and Pemba Isls); S Kenya; Tumbatu Isl.

STATUS: IUCN - Insufficiently known.

SYNONYMS: *adersi, neumanni, schusteri, terricola, validus, vosseleri.*

COMMENTS: Bothma (1971:1) questioned the distinctness of this species from *arboreus.*

Heterohyrax Gray, 1868. Ann. Mag. Nat. Hist., ser. 4, 1:50.

TYPE SPECIES: *Dendrohyrax blainvillii* Gray, 1868 (= *Hyrax brucei* Gray, 1868).

COMMENTS: Included as a subgenus of *Dendrohyrax* by Roche (1972).

Heterohyrax antineae (Heim de Balsac and Begouen, 1932). Bull. Mus. Hist. Nat. Paris, ser. 2, 4:479.

TYPE LOCALITY: Algeria, C Sahara, Ahaggar.

DISTRIBUTION: Ahaggar Mtns (S Algeria).

COMMENTS: Bothma (1971), Hatt (1936), and Schwarz (1933*b*) stated that this species may be conspecific with *brucei,* and was so placed by Roche (1972:41).

Heterohyrax brucei (Gray, 1868). Ann. Mag. Nat. Hist., ser. 4, 1:44.

TYPE LOCALITY: Ethiopia (= Abyssinia).

DISTRIBUTION: Egypt to Somalia to N South Africa to WC Angola.

SYNONYMS: *albipes, arboricola, bakeri, blainvillii, bocagei, borana, chapini, dieseneri, frommi, granti, grayi, hararensis, hindei, hoogstraali, irroratus, kempi, lademanni, maculata, manningi, mossambicus, münzneri, princeps, prittwitzi, pumilus, rhodesiae, ruckwaensis, ruddi, rudolfi, somalicus, ssongaea, thomasi, victoria-njansae, webensis.*

COMMENTS: Allen (1939) considered *Hyrax syriacus* a prior name for this species; Bothma (1971) referred *syriacus* to *Procavia;* see comment under *P. capensis.* Bothma (1971) and Hatt (1936:132) stated that *H. chapini* may be conspecific with *brucei,* and was so placed by Roche (1972:41).

Procavia Storr, 1780. Prodr. Meth. Mamm., p. 40.
TYPE SPECIES: *Cavia capensis* Pallas, 1766.
SYNONYMS: *Euhyrax*, *Hyrax*.
COMMENTS: See Bothma (1971).

Procavia capensis (Pallas, 1766). Misc. Zool., p. 30.
TYPE LOCALITY: South Africa, Cape Prov., Cape of Good Hope.
DISTRIBUTION: Syria; Lebanon; Turkey; Israel; Saudi Arabia; Yemen; NE Africa; Senegal to Somalia to N Tanzania; S Malawi to S Angola, Namibia and South Africa; isolated mountains in Algeria and Libya.
SYNONYMS: *abbyssinicus, alpini, antineae, bamendae, bounhioli, buchanani, burtonii, butleri, capillosa, chiversi, comata, coombsi, daemon, dongolanus, ebneri, ehrenbergi, elberti, erlangeri, ferrugineus, flavimaculata, goslingi, habessinicus, ituriensis, jacksoni, jayakari, johnstoni, kamerunensis, kerstingi, latastei, letabae, lopesi, luteogaster, mackinderi, marlothi, marrensis, matschiei, melfica, meneliki, minor, natalensis, naumanni, orangiae, oweni, pallida, ruficeps, reuningi, schmitzi, schultzei, scioanus, semicircularis, sharica, sinaiticus, slatini, syriaca, varians, volkmanni, waterbergensis, welwitschii, windhuki, zelotes*.
COMMENTS: Includes *habessinica, johnstoni, ruficeps, syriaca*, and *welwitschii*; see Corbet (1979), Roche (1972), and Skinner and Smithers (1990:558); but see also Bothma (1971). See Olds and Shoshani (1982, Mammalian Species, 171).

ORDER TUBULIDENTATA
by Duane A. Schlitter

ORDER TUBULIDENTATA

Family Orycteropodidae Gray, 1821. London Med. Repos., 15:305.

Orycteropus G. Cuvier, 1798. Tabl. Elem. Hist. Nat. Anim., 1798:144.
TYPE SPECIES: *Myrmecophaga capensis* Gmelin, 1788 (= *Myrmecophaga afra* Pallas, 1766).
COMMENTS: Sherborn (1902:701) gave *Orycteropus* "Geoffroy, Decad. Phil. et Litt. XXVIII. 1795", but it is untracable. Meester et al. (1986:182) reviewed and assigned the correct generic name.

Orycteropus afer (Pallas, 1766). Misc. Zool., p. 64.
TYPE LOCALITY: South Africa, Cape Prov., Cape of Good Hope.
DISTRIBUTION: Savannah zones of West Africa to E Sudan and Ethiopia; Kenya; S Somalia; N and W Uganda to Tanzania; Rwanda; N, E, and C Zaire; W Angola; Namibia; Botswana; Zimbabwe; Zambia; Mozambique; South Africa.
SYNONYMS: *adametzi, aethiopicus, albicaudus, angolensis, erikssoni, faradjius, haussanus, kordofanicus, lademanni, leptodon, matschiei, observandus, ruvanensis, senegalensis, somalicus, wardi, wertheri.*
COMMENTS: Reviewed by Melton (1976), Pocock (1924), and Shoshani et al. (1988, Mammalian Species, 300).

ORDER ARTIODACTYLA
by Peter Grubb

ORDER ARTIODACTYLA

COMMENTS: Sequence of non-ruminant families follows Simpson (1945); sequence of ruminant families based on Janis and Scott (1987).

Family Suidae Gray, 1821. London Med. Repos., 15:306.

Subfamily Babyrousinae Thenius, 1970. Z. Säugetierk., 35:334.

Babyrousa Perry, 1811. Arcana, Mus. Nat. Hist. (plate and 2 pages, unno.).
TYPE SPECIES: *Babyrousa quadricornua* Perry, 1811 (= *Sus babyrussa* Linnaeus, 1758).
COMMENTS: Revised by Groves (1980*b*).

Babyrousa babyrussa (Linnaeus, 1758). Syst. Nat., 10th ed., 1:50.
TYPE LOCALITY: "Borneo" (= Buru Isl, Indonesia).
DISTRIBUTION: N and C Sulawesi, Lembeh Isl, Buru (N Molucca Isls), Sula Isls, Malengi Isl (Togean Isls).
STATUS: CITES - Appendix I; U.S. ESA - Endangered; IUCN - Vulnerable.
SYNONYMS: *alfurus, beruensis, bolabatuensis, celebensis, frosti, quadricornua, togeanensis.*

Subfamily Phacochoerinae Gray, 1868. Proc. Zool. Soc. Lond., 1868:21, 45.

Phacochoerus F. Cuvier, 1826. Dict. Sci. Nat., 39:383.
TYPE SPECIES: *Aper aethiopicus* Pallas, 1766.
SYNONYMS: *Aper, Dinochoerus, Eureodon, Macrocephalus.*

Phacochoerus aethiopicus (Pallas, 1766). Misc. Zool., p. 16.
TYPE LOCALITY: South Africa, between Kaffraria and Great Namaqualand, two hundred leagues from the Cape of Good Hope. See Vosmaer (1766).
DISTRIBUTION: Cape Province, South Africa (extinct); N Kenya and Somalia. See Grubb (in press).
SYNONYMS: *delamerei, edentatus, pallasii, typicus* (see Grubb, in press).
COMMENTS: For distinctions from *P. africanus*, see Ewer (1957).

Phacochoerus africanus (Gmelin, 1788). *In* Linnaeus, Syst. Nat., 13th ed., 1:220.
TYPE LOCALITY: Senegal, Cape Verde.
DISTRIBUTION: Outside forest zone of Africa from Senegal to Somalia, south to S Africa, Botswana and Nambia.
SYNONYMS: *aeliani, barkeri, bufo, centralis, fossor, haroia, incisivus, massaicus, sclateri, shortridgei, sundevallii.*
COMMENTS: Specifically distinct from *P. aethiopicus*, see Cooke and Wilkinson (1978), Ewer (1957), and Grubb (in press).

Subfamily Suinae Gray, 1821. London Med. Repos., 15:306.

Hylochoerus Thomas, 1904. Nature, 70:577.
TYPE SPECIES: *Hylochoerus meinertzhageni* Thomas, 1904.

Hylochoerus meinertzhageni Thomas, 1904. Nature, 70:577.
TYPE LOCALITY: Kenya, Nandi Forest, near Kaimosi, 7,000 ft. See Allen and Lawrence (1936).
DISTRIBUTION: Guinea to Ghana; E Nigeria to Kenya and N Tanzania; SW Ethiopia.
SYNONYMS: *gigliolii, ituriensis, ivoriensis, rimator, schulzi.*
COMMENTS: See Thomas (1904 [1905]) for designation of the type specimen.

Potamochoerus Gray, 1854. Proc. Zool. Soc. Lond., 1852:129 [1854].
TYPE SPECIES: *Choiropotamus pictus* Gray, 1852 (= *Sus porcus* Linnaeus, 1758).
SYNONYMS: *Choiropotamus, Koiropotamus, Nyctochoerus.*
COMMENTS: Revised by de Beaux (1924).

Potamochoerus larvatus (F. Cuvier, 1822). Mem. Mus. Hist. Nat. Paris, 8:447.
TYPE LOCALITY: Madagascar (no precise locality).
DISTRIBUTION: Ethiopia, S Sudan and E Zaire south to E and S South Africa, west to N Botswana and Angola; Madagascar and Comoro Isls (introduced?).
SYNONYMS: *africanus, arrhenii, choeropotamus, congicus, cottoni, daemonis, edwardsi, hassama, hova, intermedius, johnstoni, keniae, koiropotamus, madagascariensis, maschona, nyasae, somaliensis.*
COMMENTS: Specifically distinct from *P. porcus*, see de Beaux (1924) and Grubb (in press).

Potamochoerus porcus (Linnaeus, 1758). Syst. Nat., 10th ed., 1:50.
TYPE LOCALITY: "Guinea" (= West Africa).
DISTRIBUTION: Rainforest zone of Africa from Senegal to Zaire.
SYNONYMS: *albifrons, albinuchalis, mawambicus, penicillatus, pictus, ubangensis.*

Sus Linnaeus, 1758. Syst. Nat., 10th ed., 1:49.
TYPE SPECIES: *Sus scrofa* Linnaeus, 1758.
SYNONYMS: *Aulacochoerus, Centuriosus, Dasychoerus, Euhys, Microsus, Porcula, Scrofa, Sinisus.*
COMMENTS: Revised by Groves (1981*a*).

Sus barbatus Müller, 1838. Tijdschr. Nat. Gesch. Physiol., 5:149.
TYPE LOCALITY: Indonesia, Kalimantan, Banjarmasin.
DISTRIBUTION: Palawan, Balabac Isl, and Calamian Isls (Philippines); Sumatra, Banka Isl, Rhio Arch., Peninsular Malaysia, Borneo.
STATUS: IUCN - Vulnerable as *S. b. oi.*
SYNONYMS: *ahoenobarbus, balabacensis, branti, calamianensis, edmondi, gargantua, longirostris, oi, palavensis, sumatranus.*

Sus bucculentus Heude, 1892. Mem. Hist. Nat. Emp. Chin., 2:XXB, fig. 7.
TYPE LOCALITY: Viet Nam, Cochin China, "les bords du Donnai" (Dong Nai River).
DISTRIBUTION: South Vietnam.
COMMENTS: Known only from two skulls; "closely related" to *S. verrucosus* but treated as a separate species (Groves, 1981*a*).

Sus cebifrons Heude, 1888. Mem. Hist. Nat. Emp. Chin., 2, pl. 17, fig. 5.
TYPE LOCALITY: Philippines, Cebu Isl.
DISTRIBUTION: Philippines: Cebu and Negros Isls.
STATUS: IUCN - Vulnerable.
SYNONYMS: *negrinus.*
COMMENTS: Specifically distinct from *S. barbatus* and *S. philippensis* (Groves and Grubb, in press; Sanborn, 1952*a*).

Sus celebensis Müller and Schlegel, 1843. *In* Temminck, Verh. Nat. Gesch. Nederland. Overz. Bezitt., Zool.,p. 177[1845]; pl. 28[1843].
TYPE LOCALITY: Indonesia, Sulawesi, Manado.
DISTRIBUTION: Indonesia: Sulawesi and neighboring small islands; feral on Halmahera and Simaleue Isls.
SYNONYMS: *amboinensis, macassaricus, maritimus, mimus, nehringii, niadensis, weberi.*
COMMENTS: Although this species is commonly cited from "Müller, 1840. *In* Temminck, Verh. Nat. Gesch. Nederland. Overz. Bezitt., Zool., Zoogd. Indisch. Archipel, p. 42", it is not found on that page. A species distinct from *S. verrucosus* (Groves, 1981*a*).

Sus heureni Hardjasasmita, 1987. Scripta Geol., 85:46.
TYPE LOCALITY: Indonesia, W Flores Isl, Manggarai Plain near Pota.
DISTRIBUTION: Flores Isl (Indonesia).
COMMENTS: A prior name for this species may be *Microsus floresianus* Heude, 1898; see Groves (1981*a*) who regarded warty pigs from Flores as a feral population of *S. celebensis.*

Sus philippensis Nehring, 1886. Sber. Ges. Naturf. Fr., Berlin, 1886:83.
TYPE LOCALITY: Philippines, Luzon Isl.
DISTRIBUTION: Philippines: Luzon, Mainit, Mindanao, Jolo, Mindoro, Catanduanis and Samar Isls.

SYNONYMS: *arietinus, crassidens, effrenus, frenatus, inconstans, joloensis, mainitensis, marchei, megalodontus, microtis, mindanensis, minutus.*
COMMENTS: Regarded as a species distinct from *S. barbatus* (Groves and Grubb, in press).

Sus salvanius (Hodgson, 1847). J. Asiat. Soc. Bengal, 16:423.
TYPE LOCALITY: India, Sikkim Terai.
DISTRIBUTION: Bhutan, S Nepal, N India (incl. Sikkim).
STATUS: CITES - Appendix I; U.S. ESA and IUCN - Endangered.

Sus scrofa Linnaeus, 1758. Syst. Nat., 10th ed., 1:49.
TYPE LOCALITY: Germany.
DISTRIBUTION: N Africa; Europe, S Russia and China south to Middle East, India, Sri Lanka, and Indonesia (Sumatra, Java east to Bali and Sumbawa Isls). Extinct in British Isles and Scandinavia. Populations of Corsica and Sardinia and formerly in Egypt and N Sudan are or were of old feral origin. Widespread as feral populations in Norway, Sweden, South Africa, Lesser Sunda Isls, Australia, USA, West Indies, Central and South America and numerous oceanic islands, including Andaman Isls and Mauritius (Indian Ocean); Hawaiian, Galapagos and Fiji Isls (Pacific Ocean). Feral and domestic populations of Molucca Isls, New Guinea and Solomon Isls thought to originate from *scrofa* X *celebensis* hybrids.
STATUS: IUCN - Vulnerable as *S. s. riukiuanus.*
SYNONYMS: *acrocranius, affinis, aipomus, algira, andamanensis, andersoni, aper, *aruensis, attila, babi, baeticus, barbarus, bengalensis, canescens, castilianus, celtica, *ceramensis, chirodontus, collinus, continentalis, coreanus, cristatus, curtidens, davidi, dicrurus, enganus, europaeus, falzfeini, ferus, flavescens; floresianus* Jentink, 1905; *frontosus, gigas, *goramensis, indicus, isonotus, japonica, jubatulus, jubatus, laticeps, leucomystax, leucorhinus, libycus, majori, mandchuricus, mediterraneus, melas, meridionalis, microdontus, milleri, moupinensis, natunensis, nicobaricus, *niger, nigripes, nipponicus, oxyodontus, paludosus, palustris, *papuensis, peninsularis, planiceps, raddeanus, reiseri, rhionis, riukiuanus, sahariensis, sardous, scrofoides, sennaarensis, setosus, sibiricus, songaricus, spatharius, taininensis, taivanus, *ternatensis, tuancus, ussuricus, vittatus, zeylonensis* (* may be based on descendants of *scrofa* X *celebensis* hybrids).
COMMENTS: For systematics, origin, and distribution of feral populations see Groves (1981*a*), Lever (1985), Uerpmann (1987), and Vigne (1988).

Sus timoriensis Müller, 1840. *In* Temminck, Verh. Nat. Gesch. Nederland. Overz. Bezitt., Zool., Zoogd. Indisch. Archipel, p. 42[1840].
TYPE LOCALITY: Indonesia, "Timor" Isl, Pritti near Kupang.
DISTRIBUTION: Timor Isl (Indonesia).
COMMENTS: This species was further described by Müller and Schlegel, 1845, *in* Temminck, Verh. Nat. Gesch. Nederland. Overz. Bezitt., Zool., Mammalia, p. 173. A feral population of *S. celebensis* according to Groves (1981*a*), but a valid species according to Hardjasasmita (1987).

Sus verrucosus Müller, 1840. *In* Temminck, Verh. Nat. Gesch. Nederland. Overz. Bezitt., Zool., Zoogd. Indisch. Archipel., p. 42[1840].
TYPE LOCALITY: Indonesia, "Java".
DISTRIBUTION: Java, Madoera Isl, Bawean Isl.
STATUS: IUCN - Vulnerable.
SYNONYMS: *blouchi, borneensis, ceramica, mystaceus, olivieri.*
COMMENTS: This species was further described by Müller and Schlegel, *in* Temminck, Verh. Nat. Gesch. Nederland. Overz. Bezitt., Zool., Mammalia, p. 175[1845], pl. 28[1843]. Synonyms apparently from Borneo and Seram were based on wrongly located specimens (Groves, 1981*a*).

Family Tayassuidae Palmer, 1897. Proc. Biol. Soc. Washington, 11:174.

Catagonus Ameghino, 1904. An. Mus. Soc. Cient. Argent., 58:188.
TYPE SPECIES: *Catagonus metropolitanus* Ameghino, 1904 (extinct).

Catagonus wagneri (Rusconi, 1930). An. Mus. Nac. Hist. Nat. "Bernardino Rivadavia," 36:231.
TYPE LOCALITY: Argentina, Santiagodel Estero, Llajta Manca.

DISTRIBUTION: Gran Chaco of Paraguay, Argentina, and Bolivia.
STATUS: CITES - Appendix I; IUCN - Vulnerable.
COMMENTS: Originally described from pre-Hispanic and subfossil remains; subsequently discovered alive; see Wetzel et al. (1975), and Wetzel (1977, 1981). Reviewed by Mayer and Wetzel (1986, Mammalian Species, 259).

Pecari Reichenbach, 1835. Bildergalerie der Thierwelt, part 6, p. 1.
TYPE SPECIES: *Dicotyles torquatus* Cuvier, 1816 (= *Sus tajacu* Linnaeus, 1758).
SYNONYMS: *Adenonotus, Notophorus.*
COMMENTS: The Collared Peccary should be assigned to a separate genus from the White-lipped species according to Woodburne (1968), Husson (1978:347-348), and Wright (1989). Use of appropriate generic names for these taxa is controversial. Genotypes of *Tayassu* and *Dicotyles* by subsequent designation are White-lipped Peccaries, see Miller and Rehn (1901:12), and Miller (1912*b*:384), so the correct generic name for Collared Peccaries appears to be *Pecari*, with type by monotypy *Dicotyles torquatus.*

Pecari tajacu (Linnaeus, 1758). Syst. Nat., 10th ed., 1:50.
TYPE LOCALITY: Mexico, designated by Thomas (1911*a*:140); however, Linnaeus's name *Sus tajacu* is evidently based on the tajacu of Marcgraf, from Pernambuco, Brazil; see Cabrera (1961:319) and Hershkovitz (1987*b*).
DISTRIBUTION: N Argentina and NW Peru to NC Texas, SW New Mexico and Arizona (USA). Introduced to Cuba.
STATUS: CITES - Appendix II as *Tayassu* spp.
SYNONYMS: *angulatus, bangsi, crassus, crusnigrum, humeralis, macrocephalus, minor, modestus, nanus, nelsoni, niger, nigrescens, patira, sonoriensis, tajassu, torquatus, torvus, yucatanensis.*

Tayassu G. Fischer, 1814. Zoognosia, 3:284.
TYPE SPECIES: *Tayassu pecari* Fischer, 1814 (= *Sus pecari* Link, 1795).
SYNONYMS: *Dicotyles, Olidosus.*
COMMENTS: By subsequent designation, the type of *Tayassu* is *T. pecari* Fischer, 1814 (= *Sus pecari* Link, 1795); see Miller and Rehn (1901:12). By subsequent designation, the type of *Dicotyles* is *D. labiatus* G. Cuvier; see Miller (1912*b*:384). *Sus pecari* and *Dicotyles labiatus* are synonyms of *Tayassu pecari*; see Hershkovitz (1963). Therefore, *Dicotyles* is a synonym of *Tayassu*, but see Husson (1978:347-348) and Woodburne (1968) for contrary views.

Tayassu pecari (Link, 1795). Beitr. Naturgesch., 2:104.
TYPE LOCALITY: French Guiana, Cayenne.
DISTRIBUTION: Oaxaca and Veracruz (Mexico) to W Ecuador, Paraguay, Brazil, Uruguay, and NE Argentina. Introduced to Cuba.
STATUS: CITES - Appendix II.
SYNONYMS: *aequatoris, albirostris, beebei, labiatus, ringens, spiradens.*
COMMENTS: Includes *albirostris*; see Husson (1978:353). Reviewed by Mayer and Wetzel (1987, Mammalian Species, 293).

Family Hippopotamidae Gray, 1821. London Med. Repos., 15:306.

Hexaprotodon Falconer and Cautley, 1836. Asia. Res. Calcutta, 19:51.
TYPE SPECIES: *Hippopotamus sivalensis* Falconer and Cautley, 1836 (extinct fossil Asiatic species).
SYNONYMS: *Choerodes, Choeropsis, Diprotodon.*
COMMENTS: Includes *Choeropsis* Leidy, 1853, following Coryndon (1977).

Hexaprotodon liberiensis (Morton, 1849). J. Acad. Nat. Sci. Phila., (2)1:232.
TYPE LOCALITY: Liberia, St. Paul's River.
DISTRIBUTION: Sierra Leone to Ivory Coast; SC Nigeria (extinct?).
STATUS: CITES - Appendix II; IUCN - Vulnerable.
SYNONYMS: *heslopi, minor.*

Hexaprotodon madagascariensis (Guldberg, 1883). Videnskabs-Selsk. Christiania. Forh., 6:1-24.
TYPE LOCALITY: Madagascar, near Antsirabé.

DISTRIBUTION: Known from subfossil material from the central highlands of Madagascar; apparently survived into present millenium (Stuenes, 1989).
COMMENTS: Transfered to *Hexaprotodon* from *Hippopotamus*; see J.M. Harrris (1991).

Hippopotamus Linnaeus, 1758. Syst. Nat., 10th ed., 1:74.
TYPE SPECIES: *Hippopotamus amphibius* Linnaeus, 1758.

Hippopotamus amphibius Linnaeus, 1758. Syst. Nat., 10th ed., 1:74.
TYPE LOCALITY: Egypt, Nile River.
DISTRIBUTION: Rivers of savanna zone of Africa, and main rivers of forest zone in Zaire, south to rivers of S Cape Province (South Africa), and north along White Nile to Nile Delta (Egypt); now extinct along most of the White Nile and in most of South Africa, except in N and E Transvaal and N Natal.
STATUS: CITES - Appendix III (Ghana).
SYNONYMS: *abyssinicus, australis, capensis, constrictus, kiboko, senegalensis, tschadensis, typus.*

Hippopotamus lemerlei Grandidier, 1868. *In* Milne Edwards, C. R. Acad. Sci. (Paris), 67:1165.
TYPE LOCALITY: Madagascar, Ambolisatra.
DISTRIBUTION: Known from subfossil material from coastal S Madagascar; survived into present millenium, 9th-13th Century (Stuenes, 1989).
SYNONYMS: *leptorhynchus, standini.*

Family Camelidae Gray, 1821. London Med. Repos., 15:307.

Camelus Linnaeus, 1758. Syst. Nat., 10th ed., 1:65.
TYPE SPECIES: *Camelus bactrianus* Linnaeus, 1758.

Camelus bactrianus Linnaeus, 1758. Syst. Nat., 10th ed., 1:65.
TYPE LOCALITY: "Bactria" (= Uzbekistan, Bokhara) (domesticated stock).
DISTRIBUTION: Exists in the wild only in SW Mongolia, Kansu, Tsinghai, and Sinkiang (China); domesticated in Iran, Afghanistan, and Pakistan, north to Kazakhstan, Mongolia, and China.
STATUS: U.S. ESA - Endangered; IUCN - Vulnerable.
SYNONYMS: *ferus.*
COMMENTS: Includes *ferus* Przewalski, 1883, based on wild specimen; *bactrianus* Linnaeus, 1758, has priority. Produces fertile hybrids with *dromedarius*. Though their distributions merge, breeding is regulated (as all individuals are domesticated in the zone of contact). Corbet (1978*c*:197), citing A.P. Gray (1954), stated that male hybrids are sterile.

Camelus dromedarius Linnaeus, 1758. Syst. Nat., 10th ed., 1:65.
TYPE LOCALITY: "Africa," deserts of Libya and Arabia (domesticated stock).
DISTRIBUTION: Extinct in the wild; first domesticated about 4,000 yr BP from wild populations which had become restricted to the S Arabian Peninsula; domesticated from Senegal and Mauritania to Somalia and Kenya, throughout N Africa, the Middle East, Arabia, and Iran to NW India; feral populations in Australia.
COMMENTS: Produces fertile hybrids with *bactrianus* (see comments therein). Bohlken (1961) considered *dromedarius* a synonym of *bactrianus*. Reviewed by Köhler-Rollefson (1991, Mammalian Species, 375). Biology reviewed by Gauthier-Pilters and Innis Dagg (1981). For history of domestication, see R. T. Wilson (1984).

Lama G. Cuvier, 1800. Lecon Anat. Comp., I, tab. 1.
TYPE SPECIES: *Camelus glama* Linnaeus, 1758.
SYNONYMS: *Auchenia, Neoauchenia, Pacos.*

Lama glama (Linnaeus, 1758). Syst. Nat., 10th ed., 1:65.
TYPE LOCALITY: Peru, Andes (domesticated stock).
DISTRIBUTION: Domesticated in S Peru, W Bolivia, NW Argentina.
SYNONYMS: *ameghiniana, araucanus, castelnaudi, chilihueque, cordubensis, crequii, domestica, ensenadensis, intermedia, lama, llacma; mesolithica* Ameghino, 1889; *moromoro, peruana.*

Lama guanicoe (Müller, 1776). Linné's Vollstand. Natursyst. Suppl., p. 50.
TYPE LOCALITY: Chile, Andes of Patagonia.
DISTRIBUTION: Cordilleras of the Andes: S Peru, Bolivia, Argentina, and Chile; Patagonia and Tierra del Fuego (Chile and Argentina); Navarino Isl (Chile).
STATUS: CITES - Appendix II.
SYNONYMS: *cacsilensis, fera, huanacus, loennbergi; mesolithica* H. Gervais and Ameghino, 1880; *molinaei, peruviana, voglii.*
COMMENTS: Has previously been included with the Llama, *L. glama*, of which it may be the wild ancestor, see Lydekker (1915). See also Hemmer (1990).

Lama pacos (Linnaeus, 1758). Syst. Nat., 10th ed., 1:65.
TYPE LOCALITY: Peru (domesticated stock).
DISTRIBUTION: Domesticated in S Peru, W Bolivia.
SYNONYMS: *lujanensis.*
COMMENTS: Often regarded as a synonym for *glama*; see Corbet and Hill (1991:126). Probably originated from hybrids between *Lama glama* and *Vicugna vicugna*; see Hemmer (1990).

Vicugna Lesson, 1842. Nouv. Tabl. Regn. Anim. Mammifères, p. 167.
TYPE SPECIES: *Camelus vicugna* Molina, 1782.

Vicugna vicugna (Molina, 1782). Sagg. Stor. Nat. Chile, p. 313.
TYPE LOCALITY: Chile, cordilleras of Coquimbo and Copiapo.
DISTRIBUTION: S Peru, W Bolivia, NW Argentina, N Chile.
STATUS: CITES - Appendix I, some Chilean and Peruvian populations Appendix II; U.S. ESA - Endangered; IUCN - Vulnerable.
SYNONYMS: *cristina, elfridae, frontosa, gracilis, mensalis, minuta, pristina, provicugna.*

Family Tragulidae Milne-Edwards, 1864. Ann. Sci. Nat. Zool. Paris, ser. 5, 2:157.

Hyemoschus Gray, 1845. Ann. Mag. Nat. Hist., [ser. 1], 16:350.
TYPE SPECIES: *Moschus aquaticus* Ogilby, 1841.

Hyemoschus aquaticus (Ogilby, 1841). Proc. Zool. Soc. Lond., 1840:35 [1841].
TYPE LOCALITY: Sierra Leone, Bulham Creek.
DISTRIBUTION: Sierra Leone to Ghana; Nigeria to Zaire, marginally entering Uganda.
STATUS: CITES - Appendix III (Ghana).
SYNONYMS: *batesi, cottoni.*

Moschiola Hodgson, 1843. Calcutta J. Nat. Hist., 4:292.
TYPE SPECIES: *Tragulus mimenoides* Hodgson, 1842 (= *Moschus meminna* Erxleben, 1777).
COMMENTS: Treated as a full genus by Groves and Grubb (1987), following Flerov (1931).

Moschiola meminna (Erxleben, 1777). Syst. Regn. Anim., 1:322.
TYPE LOCALITY: Sri Lanka.
DISTRIBUTION: Sri Lanka, peninsular India, Nepal.
SYNONYMS: *indica, malaccensis, mimenoides.*

Tragulus Pallas, 1779. Spicil. Zool., 13:27.
TYPE SPECIES: *Cervus javanicus* Osbeck, 1765.
COMMENTS: Original citation, Brisson, 1762, Regn. Anim., 2nd ed., 12:65-68, is not available according to Hopwood (1947).

Tragulus javanicus (Osbeck, 1765). Reise nach Ostindien und China, p. 357.
TYPE LOCALITY: Indonesia, W Java, Udjon Kulon Peninsula.
DISTRIBUTION: Indochina, Thailand, Yunnan (China), Malaysia, Sumatra, Borneo, Java, many adjacent islands.
SYNONYMS: *affinis, angustiae, brevipes, carimatae, everetti, focalinus, fulvicollis, fulviventer, fuscatus, hosei; indicus* Brisson, 1765 (not available); *insularis, kanchil, klossi, lampensis, lancavensis, longipes, luteicollis, masae, mergatus, natunae, pallidus, pelandoc, penangensis, pidonis, pierrei, pinius, pumilus, ravulus, ravus, rubeus, russeus, russulus, subrufus, virgicollis, williamsoni.*
COMMENTS: In older literature, known as *T. kanchil*; see Van Bemmel (1949*b*).

Tragulus napu (F. Cuvier, 1822). *In* E. Geoffroy and F. Cuvier, Hist. Nat. Mammifères, pt. 2, 4(37):4 pp. "Chevrotain napu".

TYPE LOCALITY: Indonesia, S Sumatra.

DISTRIBUTION: Indochina, Thailand, Malaysia, Borneo, Balabac Isl (Philippines), Sumatra, many adjacent islands.

SYNONYMS: *abjectus, abruptus, amoenus, anambensis, annae, bancanus, banguei, batuanus, billitonus, borneanus, bunguranensis, canescens, flavicollis, formosus, hendersoni, jugularis, lutescens, neubronneri, niasis, nigricans, nigricollis, nigrocinctus, parallelus, perflavus, pretiellus, pretiosus, rufulus, sebucus, siantanicus, stanleyanus, terutus, umbrinus, versicolor.*

COMMENTS: In older literature, mistakenly given the name *T. javanicus*; see Van Bemmel (1949*b*).

Family Giraffidae Gray, 1821. London Med. Repos., 15:307.

COMMENTS: Placement of this family follows Janis and Scott (1987).

Giraffa Brünnich, 1771. Zool. Fundamenta, p. 36.

TYPE SPECIES: *Cervus camelopardalis* Linnaeus, 1758.

SYNONYMS: *Camelopardalis, Orasius, Trachelotherium.*

Giraffa camelopardalis (Linnaeus, 1758). Syst. Nat., 10th ed., 1:66.

TYPE LOCALITY: "Æthiopia et Sennar", restricted to Sudan, Sennar, by Harper (1940:322).

DISTRIBUTION: Formerly, Gambia and Senegal to Ethiopia and Somalia, south to Central African Republic, NE Zaire, Uganda and Tanzania; E and SW Zambia; S Angola, Zimbabwe and S Mozambique to South Africa, mostly north of the Orange River. Distribution now much restricted; in W Africa still present at least until recently in Mali, Burkina Faso, Niger, NE Nigeria and N Cameroon; in southern Africa, now ranging no further south than N Namibia, Botswana and E Transvaal (South Africa).

SYNONYMS: *aethiopica, angolensis, antiquorum, australis, biturigum, capensis, congoensis, cottoni, giraffa, hagenbecki, infumata, maculata, nigrescens, peralta, renatae, reticulata, rothschildi, schillingsi, senaariensis, thornicrofti, tippelskirchi, wardi.*

COMMENTS: Reviewed by Dagg (1971, Mammalian Species, 5).

Okapia Lankester, 1901. Nature, 64:24.

TYPE SPECIES: *Equus johnstoni* P. L. Sclater, 1901.

Okapia johnstoni (P. L. Sclater, 1901). Proc. Zool. Soc. Lond., 1901(1):50.

TYPE LOCALITY: Zaire, Semliki Forest, Mundala.

DISTRIBUTION: N and E Zaire; perhaps adjacent areas.

SYNONYMS: *erikssoni, kibalensis, liebrechtsi, tigrinum.*

Family Moschidae Gray, 1821 London Med. Repos., 15:307.

COMMENTS: A family separate from the Cervidae; see Flerov (1960), Webb and Taylor (1980), Groves and Grubb (1987), and Janis and Scott (1987).

Moschus Linnaeus, 1758. Syst. Nat., 10th ed., 1:66.

TYPE SPECIES: *Moschus moschiferus* Linnaeus, 1758.

SYNONYMS: *Odontodorcus.*

COMMENTS: Species limits in E Himalayas are still uncertain; see Cai and Feng (1981), Groves (1976, 1980*a*), Groves and Grubb (1987), and Grubb (1982*a*).

Moschus berezovskii Flerov, 1929. C. R. Acad. Sci. U.S.S.R., 1928A:519 [1929].

TYPE LOCALITY: China, Szechwan, near Lungan, Ho-tsi-how Pass.

DISTRIBUTION: S and C China including Anhwei, Tibet?; N Vietnam.

STATUS: CITES - Appendix II; U.S. ESA - Endangered in Tibet and Yunnan (China).

SYNONYMS: *anhuiensis, caobangis.*

COMMENTS: Includes *anhuiensis*; see Groves and Feng (1986). A well-defined species; see Gao (1963), Groves (1976), and Grubb (1982*a*).

Moschus chrysogaster (Hodgson, 1839). J. Asiat. Soc. Bengal, 8:203.

TYPE LOCALITY: "Nepal" (Probably Tibetan Plateau).

DISTRIBUTION: Himalayas of N Afghanistan, N Pakistan, N India (incl. Sikkim), and Nepal; C Tibet to C China.
STATUS: CITES - Appendix I in Afghanistan, Bhutan, India, Nepal, and Pakistan; otherwise Appendix II; U.S. ESA - Endangered in Afghanistan, Bhutan, China (Yunnan and Tibet), India, Nepal, Pakistan, and Sikkim.
SYNONYMS: *cupreus, leucogaster, sifanicus.*
COMMENTS: A well defined species; see both Groves (1976) and Gao (1963), under the name *sifanicus*. The name *sifanicus* should be included in *M. chrysogaster*, but "*chrysogaster*" of Cai and Feng (1981) is subspecifically or specifically distinct. Available name for this taxon may be *leucogaster* Hodgson, 1839; see Grubb (1982*a*). Groves and Grubb (1987) provisionally treated *leucogaster* as a species; Grubb (1990) listed it as a Himalayan subspecies-group of *M. chrysogaster.*

Moschus fuscus Li, 1981. Zool. Res., 2:159.
TYPE LOCALITY: China, Yunnan, Gongshan-Xian, Bapo, 3,500 m.
DISTRIBUTION: W Yunnan and SE Tibet (China); N Burma; Assam and Sikkim (India); Bhutan; Nepal.
STATUS: CITES - Appendix I in Bhutan, Burma, India, and Nepal; otherwise Appendix II; U.S. ESA - Endangered in Bhutan, Burma, China, Sikkim, India, and Nepal.
COMMENTS: Gao (1985) treated *fuscus* as a subspecies of *chrysogaster.*

Moschus moschiferus Linnaeus, 1758. Syst. Nat., 10th ed., 1:66.
TYPE LOCALITY: "Tataria versus Chinam"; restricted to Russia, SW Siberia, Altai Mtns by Heptner et al. (1961).
DISTRIBUTION: Forests of E Siberia, N Mongolia, N China west to Kansu, Korea, Sakhalin Isl.
STATUS: CITES - Appendix II.
SYNONYMS: *altaicus, arcticus, parvipes, sachalinensis, sibiricus, turowi.*
COMMENTS: Includes *sibiricus*; see Corbet (1978*c*:198).

Family Cervidae Goldfuss, 1820. Handb. Zool., 2:xx, 374.
COMMENTS: Reviewed by Whitehead (1972) and by Groves and Grubb (1987). For introduced populations, see Lever (1985).

Subfamily Cervinae Goldfuss, 1820. Handb. Zool., 2:xx, 374.

Axis H. Smith, 1827. *In* Griffith et al., Anim. Kingdom, 5:312.
TYPE SPECIES: *Cervus axis* Erxleben, 1777.
SYNONYMS: *Hyelaphus.*
COMMENTS: Treated as a full genus, not a subgenus of *Cervus*, by Groves and Grubb (1987).

Axis axis (Erxleben, 1777). Syst. Regn. Anim., 1:312.
TYPE LOCALITY: India, Bihar, "banks of the Ganges."
DISTRIBUTION: India (incl. Sikkim); Sri Lanka; Nepal; introduced to the former Yugoslavia, republics of the former W USSR (still extant?), Andaman Isls, Australia, Hawaiian Isls and Texas (USA), Brazil, Argentina and Uruguay.
SYNONYMS: *ceylonensis, indicus, maculatus, major, minor, nudipalpebra.*

Axis calamianensis (Heude, 1888). Mem. Hist. Nat. Emp. Chin., 2:49.
TYPE LOCALITY: Philippines, Calamian Isls, Culion Isl.
DISTRIBUTION: Philippines, Calamian Isls.
STATUS: CITES - Appendix I; U.S. ESA - Endangered; IUCN - Vulnerable.
SYNONYMS: *culionensis.*
COMMENTS: Included in *A. porcinus* by Haltenorth (1963), but treated as a full species by Groves and Grubb (1987).

Axis kuhlii (Müller, 1840). *In* Temminck, Verh. Nat. Gesch. Nederland. Bezitt., Zool., Zoogd. Indisch. Archipel., p. 45[1840].
TYPE LOCALITY: "Java en Borneo", but is found only on Bawean Isl, Indonesia.
DISTRIBUTION: Indonesia, Bawean Isl.
STATUS: CITES - Appendix I; U.S. ESA - Endangered; IUCN - Rare.
COMMENTS: This species was further described by Müller and Schlegel, *in* Temminck, Verh. Nat. Gesch. Nederland. Overz. Bezitt., Zool., Mammalia, p. 223[1845], pl. 44[1842].

Included in *A. porcinus* by Haltenorth (1963), but treated as a full species by Groves and Grubb (1987).

Axis porcinus (Zimmermann, 1780). Geogr. Gesch. Mensch. Vierf. Thiere, 2:131.
TYPE LOCALITY: India, Bengal.
DISTRIBUTION: Pakistan and N India to Vietnam; Yunnan (China); Sri Lanka (introduced?); introduced to S Australia.
STATUS: CITES - Appendix I as *Cervus (= Axis) porcinus annamiticus;* U.S. ESA - Endangered as *Axis (= Cervus) porcinus annamiticus.*
SYNONYMS: *annamiticus, hecki, oryzus, pumilio.*
COMMENTS: *Cervus porcinus* Zimmermann, 1777, is not an available name as it was published in an unavailable work (Spec. Zool. Geogr., p. 532): see Hemming (1950:547).

Cervus Linnaeus, 1758. Syst. Nat., 10th ed., 1:66.
TYPE SPECIES: *Cervus elaphus* Linnaeus, 1758.
SYNONYMS: *Elaphoceros, Elaphus, Harana, Hippelaphus, Melanaxis, Panolia, Procervus, Przewalskium, Pseudaxis, Pseudocervus, Rucervus, Rusa, Sambur, Sika, Strongyloceros, Sikaillus, Thaocervus, Ussa.*
COMMENTS: Includes *Rusa, Rucervus,* and *Przewalskium* as subgenera, with *Panolia, Sika* and *Thaocervus* as synonyms of subgenera, see Groves and Grubb (1987). Van Gelder (1977*b*) also included *Elaphurus, Axis, Dama* and *Hyelaphus.*

Cervus albirostris Przewalski, 1883. Third Journey in Central Asia, p. 124.
TYPE LOCALITY: China, Kansu, 3 km above mouth of Kokusu River, Humboldt Mtns, Nan Shan.
DISTRIBUTION: Tibet, Tsinghai, Kansu, and Szechwan (China).
STATUS: IUCN - Vulnerable.
SYNONYMS: *dybowskii* Sclater, 1889; *sellatus, thoroldi.*

Cervus alfredi Sclater, 1870. Proc. Zool. Soc. Lond., 1870:381.
TYPE LOCALITY: Philippines.
DISTRIBUTION: Philippines: Masbate, Panay and Negros Isls; formerly also Seguinjor, Guimares, Cebu, Bohol and perhaps other islands.
STATUS: U.S. ESA and IUCN - Endangered.
SYNONYMS: *breviceps, masbatensis.*
COMMENTS: Included in *C. mariannus* by Haltenorth (1963). Revised by Grubb and Groves (1983), where treated as a full species.

Cervus duvaucelii G. Cuvier, 1823. Rech. Oss. Foss., Nouv. ed., 4:505.
TYPE LOCALITY: N India.
DISTRIBUTION: N and C India, SW Nepal; extinct in Pakistan.
STATUS: CITES - Appendix I; U.S. ESA and IUCN - Endangered.
SYNONYMS: *bahrainja, branderi, dimorphe, elaphoides, euceros, eucladoceros, ranjitsinhi, smithii.*
COMMENTS: Revised by Groves (1982*b*).

Cervus elaphus Linnaeus, 1758. Syst. Nat., 10th ed., 1:67.
TYPE LOCALITY: S Sweden.
DISTRIBUTION: Tunisia; NE Algeria; Europe east to Crimea and Caucasus; Turkey; N Iran; C Asia from N Afghanistan, Kashmir (India) and Russian Turkestan east to Siberia, Mongolia, W and N China and Ussuri region (Russia); in Corsica and Sardinia only since Neolithic; North America, where now restricted to western areas and reserves. Red Deer (*C. e. elaphus* and related subspecies) introduced to Morocco, USA, Argentina, Chile, Australia, and New Zealand; Wapiti (*C. e. canadensis* and related subspecies) introduced to Ural Mtns and Volga Steppe (Russia) and New Zealand.
STATUS: CITES - Appendix I as *C. e. hanglu;* Appendix II as *C. e. bactrianus;* Appendix III (Tunisia) as *C. e. barbarus.* U.S. ESA - Endangered as *C. e. bactrianus, C. e. barbarus, C. e. corsicanus, C. e. hanglu, C. e. macneilli, C. e. wallichi,* and *C. e. yarkandensis;* IUCN - Endangered as *C. e. bactrianus, C. e. corsicanus, C. e. hanglu, C. e. wallichi,* and *C. e. yarkandensis;* Vulnerable as *C. e. barbarus;* Indeterminate as *C. e. macneilli.*
SYNONYMS: *affinis, alashanicus, albicus, albifrons, albus, asiaticus, atlanticus, bactrianus, baicalensis, bajovaricus, balticus, barbarus, bedfordianus, biedermanni, bolivari, brauneri,*

campestris, canadensis, carpathicus, cashmeriensis, casperianus, caspius, caucasicus, corsicanus, debilis, eustephanus, germanicus, hanglu, hagenbecki, hippelaphus, hispanicus, isubra, kansuensis, luedorfi, macneilli, major, manitobensis, maral, mediterraneus, merriami, minor, montanus, nannodes, naryanus, neglectus, nelsoni, occidentalis, rhenanus, roosevelti, saxonicus, scoticus, sibiricus, songaricus, tauricus, tibetanus, ussuricus, varius, visurgensis, vulgaris, wachei, wallichi, wapiti, wardi, xanthopygus, yarkandensis.

COMMENTS: For synonyms see Corbet (1978*c*:201), McCullough (1969), and Hall (1981:1084-1087). Reviewed by J.M. Dolan (1988).

Cervus eldii M'Clelland, 1842. Calcutta J. Nat. Hist., 2:417.

TYPE LOCALITY: India, Assam, Manipur.

DISTRIBUTION: Manipur (N India), Burma, Thailand, Laos, Cambodia, Vietnam, Hainan Isl (China); now much reduced in numbers or extinct in several of these countries.

STATUS: CITES - Appendix I; U.S. ESA - Endangered; IUCN - Endangered as *C. e. eldii* and *C. e. siamensis*, otherwise Vulnerable.

SYNONYMS: *acuticauda, acuticornis, brucei, cornipes, frontalis, hainanus, lyratus, platyceros, siamensis, thamin.*

Cervus mariannus Desmarest, 1822. Mammalogie, *in* Encycl. Meth., 2:436.

TYPE LOCALITY: Mariana Isls, Guam (introduced).

DISTRIBUTION: Philippines: Luzon, Mindoro, Mindanao and Basilan Isls. Introduced to Mariana, Caroline and Bonin Isls (W Pacific Ocean).

SYNONYMS: *ambrosianus, apoensis, atheneensis, barandanus, baryceros, basilanensis, boninensis, brachyceros, chrysotrichos, cinereus, corteanus, crassicornis, dailliardianus, elegans, elorzanus, francianus, garcianus, gonzalinus, gorrichanus, guevaranus, guidoteanus, hippolitianus, longicuspis, macarianus, maraisianus, marzaninus, michaelinus, microdontus, nigellus, nigricans, nublanus, philippinus, ramosianus, rosarianus, roxasianus, rubiginosa, spatharius, steerii, telesforianus, tuasoninus, verzosianus, vidalinus, villemerianus.*

COMMENTS: Treated as a separate species from *C. unicolor* by Haltenorth (1963), and by Grubb and Groves (1983), who revised this taxon.

Cervus nippon Temminck, 1838. Coup d'oeil sur la faune des iles de la Sonde et de l'empire du Japon, p. xxii.

TYPE LOCALITY: Japan, probably Kyushu; see Groves and Smeenk (1978).

DISTRIBUTION: Taiwan, E China, Manchuria, Korea (incl. Cheju Isl), adjacent E Siberia, Japan (incl Tsushima Isls), Vietnam. Introduced to British Isles, mainland Europe (incl. Lithuania and Ukraine), Caucasus region, New Zealand, USA, Solo Isl (Philippines, still extant?), Kerama Isls (Ryukyu Isls) and small islands off Japan.

STATUS: U.S. ESA - Endangered as *C. n. grassianus, C. n. keramae, C. n. kopschi, C. n. mandarinus*, and *C. n. taiouanus*; IUCN - Extinct? as *C. n. grassianus*, Extinct as *C. n. mandarinus*, Endangered as *C. n. keramae, C. n. pseudaxis*, and *C. n. taiouanus*, Indeterminate as *C. n. sichuanicus*.

SYNONYMS: *aceros, andreanus, aplodonticus, aplodontus, arietinus, blakistoninus, brachypus, brachyrhinus, centralis, consobrinus, cycloceros, cyclorhinus, daimius, dejardinus, devilleanus, dolichorhinus, dominicanus, dugennianus; dybowskii* Taczanowski, 1876; *elegans, ellipticus, euopis, frinianus, fuscus, gracilis, granulosus, grassianus, grilloanus, hollandianus, hortulorum, hyemalis, ignotus, imperialis, infelix, japonicus, joretianus, kematoceros, keramae, kopschi, lacrymosus, latidens, legrandianus, mageshimae, mandarinus, mantchuricus, marmandianus, matsumotei, microdontus, microspilus, minoensis, minor, minutus, mitratus, modestus, morrisianus, novioninus, orthopodicus, orthopus, oxycephalus, paschalis, pouvrelianus, pseudaxis, pulchellus, regulus, rex, riverianus, rutilus, schizodonticus, schlegeli, schulzianus, sendaiensis, sica, sicarius, sichuanicus, sika, soloensis, surdescens; swinhoei* Glover, 1956; *sylvanus, tai-oranus, taiouanus, taivanus, typicus, xendaiensis, yakushimae, yesoensis, yuanus.*

COMMENTS: Includes *hortulorum, taiouanus* and *pulchellus*, which were considered species by Imaizumi (1970*a*). Includes *soloensis*, see Grubb and Groves (1983). Reviewed by Feldhamer (1980, Mammalian Species, 128). Revised in part by Groves and Smeenk (1978).

Cervus schomburgki Blyth, 1863. Proc. Zool. Soc. Lond., 1863:155.

TYPE LOCALITY: Thailand.

DISTRIBUTION: Thailand.

COMMENTS: Included in *duvaucelii* by Haltenorth (1963:58) and Groves (1982*b*); but treated as a full species by Lekagul and McNeely (1977). Extinct; last specimen killed in 1932; see Harper (1945).

Cervus timorensis Blainville, 1822. J. Phys. Chim. Hist. Nat. Arts Paris, 94:267.

TYPE LOCALITY: Indonesia, Lesser Sunda Isls, Timor Isl.

DISTRIBUTION: Java; Bali and probably introduced in antiquity to Lesser Sunda Isls, Molucca Isls (including Buru and Seram), Sulawesi and Timor (Indonesia). Since 17th century, introduced to Kalimantan (Borneo, extinct?), New Guinea, New Britain, Aru Isls (Indonesia), Mauritius, Comoro Isls, Madagascar (extinct?), Australia, New Zealand, New Caledonia and small islands in Indonesia and off the coast of Australia.

SYNONYMS: *buruensis, djonga, floresiensis, hippelaphus, hoevellianus, javanicus, laronesiotes, lepidus, macassaricus, menadensis, moluccensis, peronii, renschi, russa, sumbavanus, tavistocki, tunjuc.*

COMMENTS: Revised by Van Bemmel (1949*a*). Includes *tavistocki*; see Grubb and Groves (1983).

Cervus unicolor Kerr, 1792. *In* Linnaeus, Anim. Kingdom, p. 300.

TYPE LOCALITY: Sri Lanka.

DISTRIBUTION: India and Sri Lanka east to S China, Hainan Isl and Taiwan; south to Peninsular Malaysia, Sumatra, Borneo, Siberut, Sipora, and Pagi and Nias Isls; introduced to Australia and New Zealand.

SYNONYMS: *albicornis, aristotelis, brachyrhinus, brookei, cambojensis, colombertinus, combalbertinus, curvicornis, dejeani, equinus, errardianus, hainana, hamiltonianus, heterocerus, jarai, joubertianus, latidens, lemeanus, leschenaulti, lignarius, longicornis, major, malaccensis, nepalensis, niger, oceanus, officialis, outreyanus, pennantii, planiceps, planidens, simoninus; swinhoii* Sclater, 1862; *verutus.*

Dama Frisch, 1775. Das Natur-System der Vierfüssigen Thiere, 3.

TYPE SPECIES: *Cervus dama* Linnaeus, 1758.

COMMENTS: *Dama* as generic name for the Fallow Deer was validated by Opinion 581, International Commission on Zoological Nomenclature (1960). Treated as a full genus, not a subgenus of *Cervus* by Groves and Grubb (1987).

Dama dama (Linnaeus, 1758). Syst. Nat., 10th ed., 1:67.

TYPE LOCALITY: Sweden (introduced).

DISTRIBUTION: S Turkey; introduced to Europe (incl. Lithuania and Ukraine), South Africa, Australia, New Zealand, USA, Argentina, Chile, Peru, Uruguay and islands in Fijian group, Lesser Antilles and off W Canadian Coast. For present distribution, see Chapman and Chapman (1980); for natural recent distribution see Uerpmann (1987).

SYNONYMS: *albus, leucaethiops, maura, mauricus, niger, platyceros, plinii, varius, vulgaris.*

COMMENTS: Reviewed by Feldhamer et al. (1988, Mammalian Species, 317).

Dama mesopotamica (Brooke, 1875). Proc. Zool. Soc. Lond., 1875:264.

TYPE LOCALITY: Iran, Luristan Province.

DISTRIBUTION: Formerly Lebanon, Syria, Jordan, Israel and Iraq; survives in W Iran.

STATUS: CITES - Appendix I; U.S. ESA and IUCN - Endangered.

COMMENTS: Regarded as a separate species from *D. dama* by Haltenorth (1959), Ferguson et al. (1985), Uerpmann (1987), and Harrison and Bates (1991).

Elaphurus Milne-Edwards, 1866. Ann. Sci. Nat. Zool. (Paris), 5:382.

TYPE SPECIES: *Elaphurus davidianus* Milne-Edwards, 1866.

Elaphurus davidianus Milne-Edwards, 1866. Ann. Sci. Nat. Zool., 5:382.

TYPE LOCALITY: China, Chihli, Pekin, Imperial Hunting Park.

DISTRIBUTION: Formerly NE China; extinct in wild since 3rd or 4th Century; now reintroduced to its former range, near Peking and near Shanghai.

STATUS: IUCN - Endangered.

SYNONYMS: *menziesianus, tarandoides.*

COMMENTS: Included in *Elaphurus* by Corbet (1978*c*:201); but see Van Gelder (1977*b*).

Subfamily Hydropotinae Trouessart, 1898. Cat. Mamm. Viv. Foss., new ed., fasc. 4:865.

Hydropotes Swinhoe, 1870. Athenaeum, 2208:264.
TYPE SPECIES: *Hydropotes inermis* Swinhoe, 1870.
SYNONYMS: *Hydrelaphus*.
COMMENTS: Original description usually given as Proc. Zool. Soc. Lond., 1870:90 [publ. June, 1870], but McAllan and Bruce (1989) have shown that publication in The Athenaeum was earlier (19 Feb. 1870).

Hydropotes inermis Swinhoe, 1870. Athenaeum, 2208:264.
TYPE LOCALITY: China, Kiangsu, Chingkiang, Yangtze River, Deer Isl.
DISTRIBUTION: Lower Yangtze Basin, south to Guangxi (China); Korea; introduced in England and France.
STATUS: IUCN - Rare.
SYNONYMS: *affinis, argyropus, kreyenbergi*.
COMMENTS: Original description usually given as Proc. Zool. Soc. Lond., 1870:89 [publ. June, 1870], but McAllan and Bruce (1989) have shown that publication in The Athenaeum was earlier (19 Feb. 1870).

Subfamily Muntiacinae Knottnerus-Meyer, 1907. Arch. Naturgesch., 73:14, 97.
COMMENTS: Relegated to tribal status, in Cervinae, by Groves and Grubb (1987); then treated as a full family by Groves and Grubb (1990). Assignment to Cervinae now supported by evidence in Kraus and Miyamoto (1991); yet generally regarded as a subfamily (Haltenorth, 1963).

Elaphodus Milne-Edwards, 1872. Nouv. Arch. Mus. Hist. Nat. Paris, Bull., 7:93.
TYPE SPECIES: *Elaphodus cephalophus* Milne-Edwards, 1872.
SYNONYMS: *Lophotragus*.
COMMENTS: For year of publication, see Ellerman and Morrison-Scott (1953).

Elaphodus cephalophus Milne-Edwards, 1872. Nouv. Arch. Mus. Hist. Nat. Paris, Bull., 7:93.
TYPE LOCALITY: China, Sichuan, Baoxing (= Szechwan, Moupin).
DISTRIBUTION: S and C China; N Burma.
SYNONYMS: *fociensis, ichangensis, michianus*.

Muntiacus Rafinesque, 1815. Analyse de la Nature, p. 56.
TYPE SPECIES: *Cervus muntjak* Zimmerman, 1780.
SYNONYMS: *Cervulus, Procops, Prox, Stylocerus*.
COMMENTS: *Muntiacus* Rafinesque is a *nomen nudum*, but was conserved by Opinion 460 of the International Commission on Zoological Nomenclature (1957*b*). Revised by Groves and Grubb (1990).

Muntiacus atherodes Groves and Grubb, 1982. Zool. Meded. Leiden, 56:210.
TYPE LOCALITY: Malaysia, Sabah, Tawau, Coconut Research Station, Forest Camp 1.
DISTRIBUTION: Borneo.
COMMENTS: Formerly included in *M. muntjak*, or in a separate species, *M. pleiharicus*; see Chasen (1940:203); however *pleiharicus* is a synonym of *muntjak*; see Groves and Grubb (1982).

Muntiacus crinifrons (Sclater, 1885). Proc. Zool. Soc. Lond., 1885:1, pl. 1.
TYPE LOCALITY: China, Chekiang (= Zhejiang), near Ningpo.
DISTRIBUTION: E China: Zhejiang, N Fujian and S Anhui; formerly from Yunnan and Guangdong to Jaingsu (China) according to Shou (1962:454).
STATUS: CITES - Appendix I; IUCN - Vulnerable.
COMMENTS: Included in *muntjak* by Haltenorth (1963:42).

Muntiacus feae (Thomas and Doria, 1889). Ann. Mus. Civ. Stor. Nat. Genova, 7:92.
TYPE LOCALITY: Burma, Tenasserim, Thagata Juva, southeast of Mt. Mulaiyit.
DISTRIBUTION: Thailand; Laos; peninsular Burma; SE Yunnan (China).
STATUS: U.S. ESA and IUCN - Endangered.
SYNONYMS: *rooseveltorum*.

COMMENTS: Included in *muntjak* by Haltenorth (1963:42). Includes *rooseveltorum*; see Groves and Grubb (1990). For spelling of the species name as *feai*, see Grubb (1977); but see International Code of Zoological Nomenclature (International Commission on Zoological Nomenclature, 1985) for retention of original spelling *feae*.

Muntiacus gongshanensis Ma, 1990. *In* Ma et al., Zool. Res. Kunming, 11:47.
TYPE LOCALITY: China, Yunnan, Gongshan county, E slope of N section of Gaoligong Mtns, Puladi, Mijiao.
DISTRIBUTION: N Burma; SE Tibet and W Yunnan (China).

Muntiacus muntjak (Zimmermann, 1780). Geogr. Gesch. Mensch. Vierf. Thiere, 2:131.
TYPE LOCALITY: Indonesia, Java.
DISTRIBUTION: Sri Lanka; India; NE Pakistan; Nepal; Bhutan; Bangladesh; S China (Yunnan, Guangxi, Hainan Isl) south through Indochina to Peninsular Malaysia, Sumatra, Borneo, Java, Bali, Lombok and many smaller Indonesian islands.
SYNONYMS: *albipes, annamensis, aureus, bancanus, curvostylis, grandicornis, malabaricus, melas, menglalis, montanus, moschatus, nainggolani, nigripes, peninsulae, pleiharicus, ratwa, robinsoni, rubidus, styloceros, subcornutus, tamulicus, vaginalis, yunnanensis.*
COMMENTS: Includes *pleiharicus*, listed as a distinct species by Chasen (1940:203), and *vaginalis*. Made to include *reevesi, feae, rooseveltorum* and *crinifrons* by Haltenorth (1963:42).

Muntiacus reevesi (Ogilby, 1839). Proc. Zool. Soc. Lond., 1838:105 [1839].
TYPE LOCALITY: China, Kwangtung (= Guangdong), near Canton.
DISTRIBUTION: Shensi, Kansu and S China; Taiwan; introduced to England and France.
SYNONYMS: *bridgemani, lachrymans, micrurus, pingshiangicus, sclateri, sinensis, teesdalei.*
COMMENTS: Included in *muntjak* by Haltenorth (1963:42); but see Corbet (1978*c*:199).

Subfamily Capreolinae Brookes, 1828. Cat. Mus. J. Brookes, p. 62.

Alces Gray, 1821. London Med. Repos., 15:307.
TYPE SPECIES: *Cervus alces* Linnaeus, 1758.
SYNONYMS: *Alcelaphus, Paralces.*

Alces alces (Linnaeus, 1758). Syst. Nat., 10th ed., 1:66.
TYPE LOCALITY: Sweden.
DISTRIBUTION: N Eurasia from Scandinavia and Poland, east to Anadyr region (E Siberia); south to Ukraine and S Siberia, N Mongolia and N China (Xingjiang, Inner Mongolia, Heilongjiang); Alaska (USA); Canada; N USA; extinct in Caucasus region since last century; introduced to New Zealand.
SYNONYMS: *americanus, andersoni, angusticephalus, antiquorum, bedfordiae, buturlini, cameloides, caucasicus, columbae, coronatus, europaeus, gigas, jubata, machlis, meridionalis, palmatus, pfizenmayeri, shirasi, tymensis, uralensis, yakutskensis.*
COMMENTS: Revised by R.L. Peterson (1952); reviewed by Franzmann (1981, Mammalian Species, 154).

Blastocerus Gray, 1850. Gleanings, Knowsley Menagerie, p. 68.
TYPE SPECIES: *Cervus paludosus* Desmarest, 1822 (= *Cervus dichotomus* Illiger, 1815).
SYNONYMS: *Antifer, Elaphalces, Epieuryceros, Paraceros.*
COMMENTS: Included in *Odocoileus* by Haltenorth (1963:44-45); but see Groves and Grubb (1987). For first valid use of generic name see Hershkovitz (1958) who argued that it was Gray, 1850.

Blastocerus dichotomus (Illiger, 1815). Abh. Phys. Klasse K.-Preuss. Akad. Wiss., 1804-1811:117 [1815].
TYPE LOCALITY: Paraguay, Lake Ypoa, south of Asuncion.
DISTRIBUTION: C Brazil to Paraguay and N Argentina.
STATUS: CITES - Appendix I; U.S. ESA - Endangered; IUCN - Vulnerable.
SYNONYMS: *azpatianus, ensenadensis, furcata, melanopus, paludosus, palustris.*
COMMENTS: Reviewed by Pinder and Grosse (1991, Mammalian Species, 380).

Capreolus Gray, 1821. London Med. Repos., 15:307.
TYPE SPECIES: *Cervus capreolus* Linnaeus, 1758.

Capreolus capreolus (Linnaeus, 1758). Syst. Nat., 10th ed., 1:68.
TYPE LOCALITY: Sweden.
DISTRIBUTION: Europe to W Russia; Turkey; Caucasus region; NW Syria; N Iraq; N Iran; extinct in Lebanon and Israel.
SYNONYMS: *albus, albicus, armenius, baleni, balticus, canus, capraea, cistaunicus, coxi, decorus, dorcas, europaeus, grandis, italicus, joffrei, niger, plumbeus, rhenanus, thotti, transsylvanicus, transvosagicus, varius, vulgaris, warthae, whittalli, zedlitzi.*

Capreolus pygargus (Pallas, 1771). Reise Prov. Russ. Reichs, 1:453.
TYPE LOCALITY: Russia, Volga, Samara district, River Sok. Given by Rossolimo (in litt.) as Orenburgskaia Obl., Bulgulma-Belebei uplands, upper flow of River Sok.
DISTRIBUTION: S Ural and N Caucasus Mtns (Russia), SE and E Kazakhstan, Tien Shan and S Siberia (Russia) eastward to Pacific coast, south into N and C China, N Mongolia and Korea.
SYNONYMS: *bedfordi, caucasica, ferghanicus, mantschuricus, melanotis, ochracea, tianschanicus.*
COMMENTS: Now regarded by most Russian authors as a species distinct from *C. capreolus*; see Sokolov and Gromov (1990). Rossolimo (in litt.) gave the original year of publication as 1773.

Hippocamelus Leuckart, 1816. Diss. Inaug. de *Equo bisulco* Molinae, p. 23.
TYPE SPECIES: *Hippocamelus dubius* Leuckart, 1816 (= *Equus bisulcus* Molina, 1792).
SYNONYMS: *Anomalocera, Cervequus, Craegoceros, Furcifer, Huamela, Xenelaphus.*
COMMENTS: Included in *Odocoileus* by Haltenorth (1963:44, 46).

Hippocamelus antisensis (d'Orbigny, 1834). Ann. Mus. Hist. Nat. Paris, 3:91.
TYPE LOCALITY: Bolivian Andes, near La Paz.
DISTRIBUTION: Andes from Ecuador to NW Argentina.
STATUS: CITES - Appendix I; U.S. ESA - Endangered; IUCN - Vulnerable.
SYNONYMS: *anomalocera, huamel.*

Hippocamelus bisulcus (Molina, 1782). Sagg. Stor. Nat. Chile, p. 320.
TYPE LOCALITY: Chilean Andes, Colchagua Prov.
DISTRIBUTION: Andes of S Chile and S Argentina.
STATUS: CITES - Appendix I; U.S. ESA and IUCN - Endangered.
SYNONYMS: *andicus, cerasina, chilensis, dubius, equinus, huemel, leucotis.*

Mazama Rafinesque, 1817. Am. Mon. Mag., 1(5):363.
TYPE SPECIES: *Mazama pita* Rafinesque, 1817 (= *Moschus americanus* Erxleben, 1777).
SYNONYMS: *Coassus, Doratoceros, Homelaphus, Nanelaphus, Passalites, Subulo.*
COMMENTS: Reviewed by Czernay (1987).

Mazama americana (Erxleben, 1777). Syst. Regni Anim., 1:324.
TYPE LOCALITY: French Guiana, Cayenne.
DISTRIBUTION: S Tamaulipas and Yucatan (Mexico) to S Brazil, N Argentina, S Bolivia, and Paraguay; Trinidad and Tobago.
STATUS: CITES - Appendix III (Guatemala) as *M. a. cerasina.*
SYNONYMS: *auritus, carrikeri, cerasina, dolichurus, fuscata, gualea, inornatus, jucunda, juruana, pandora, pita, reperticia, rosii, rufa, sarae, sartorii, sheila, tema, temama, toba, trinitatis, tumatumari, whitelyi, zamora, zetta.*

Mazama bricenii Thomas, 1908. Ann. Mag. Nat. Hist., ser. 8, 1:349.
TYPE LOCALITY: Venezuela, Merida, Paramo de la Culata.
DISTRIBUTION: W Venezuela.
COMMENTS: A species distinct form *M. rufina* according to Czernay (1987).

Mazama chunyi Hershkovitz, 1959. Proc. Biol. Soc. Washington, 72:45.
TYPE LOCALITY: Bolivia, La Paz, Cocopunco, 3200 m.
DISTRIBUTION: Bolivian Andes, S Peru.
COMMENTS: Prior to 1959 this species was confused with *Pudu mephistophiles*; see comment under that species.

Mazama gouazoupira (G. Fischer, 1814). Zoognosia, 3:465.
TYPE LOCALITY: Paraguay, Asuncion region.
DISTRIBUTION: San Jose Isl (Panama); Peru, Ecuador and Colombia east to Brazil and south to Bolivia, Paraguay, N Argentina and Uruguay.
SYNONYMS: *argentina, bira, cita, fusca, gouazoubira, kozeritzi, mexianae, murelia, namby, nemorivaga, permira, rondoni, sanctaemartae, simplicicornis, superciliaris, tschudii.*
COMMENTS: Although the specific name is based on the gouazoubira of Azara, the original spelling was "*gouazoupira*" not "*gouazoubira*".

Mazama nana (Hensel, 1872). Abhandl. Pruess. Akad. Wiss., 1872:99.
TYPE LOCALITY: Brazil, Rio Grande do Sul.
DISTRIBUTION: C and SE Brazil, E Paraguay, N Argentina.
COMMENTS: A species distinct form *M. rufina* according to Czernay (1987).

Mazama rufina (Bourcier and Pucheran, 1852). Rev. Zool. Paris, p. 561.
TYPE LOCALITY: Ecuador, Pichincha, Pichincha Mtns, Lloa valley.
DISTRIBUTION: Ecuador, S Colombia.

Odocoileus Rafinesque, 1832. Atlantic Journal and Friend of Knowledge, 1:109.
TYPE SPECIES: *Odocoileus spelaeus* Rafinesque, 1832 (= *Dama virginianus* Zimmermann, 1780).
SYNONYMS: *Aplacerus, Cariacus, Dorcelaphus, Eucervus, Gymnotis, Macrotis, Oplacerus, Otelaphus, Palaeodocoileus, Protomazama, Reduncina.*
COMMENTS: Hall (1981:1087) employed *Dama* Zimmerman, 1780, of which *Dama virginiana* (= *Odocoileus virginianus*) is the type, for this genus, but *Dama* Frisch, 1775, with *Cervus dama* (= *Dama dama*) as type has priority and thus preoccupies *Dama* Zimmerman, 1780 (International Commission on Zoological Nomenclature, 1960).

Odocoileus hemionus (Rafinesque, 1817). Am. Mon. Mag., 1:436.
TYPE LOCALITY: USA, South Dakota, mouth of Big Sioux River.
DISTRIBUTION: Baja California and Sonora to N Tamaulipas (Mexico); W USA (to Minnesota); W Canada; Alaskan Panhandle (USA). Introduced to Kauai (Hawaiian Isls) and Argentina.
STATUS: U.S. ESA and IUCN - Endangered as *O. h. cerrosensis.*
SYNONYMS: *auritus, californicus, canus, cerrosensis, columbianus, crooki, eremicus, fuliginatus, inyoensis, lewisii, macrotis, montanus, peninsulae, punctulatus, pusilla, richardsoni, scaphiotus, sheldoni, sitkensis, virgultus.*
COMMENTS: Revised by Cowan (1936); reviewed by A.E. Anderson and Wallmo (1984, Mammalian Species, 219).

Odocoileus virginianus (Zimmermann, 1780). Geogr. Gesch. Mensch. Vierf. Thiere, 2:129.
TYPE LOCALITY: "Bewohnt in grossen Heerben Carolina v), Virginien, Louisiana w), und geht vielleicht bis Panama x) hinunter"; restricted by Hershkovitz (1948*c*:43) to USA, Virginia.
DISTRIBUTION: W and S Canada; NW, SW, C and E USA to Bolivia, Guianas and N Brazil. Introduced to Czechoslovakia, Finland, New Zealand and West Indies, where these deer may survive on Cuba, Curacao, St. Croix and St. Thomas Isls.
STATUS: CITES - Appendix III (Guatemala) as *O. v. mayensis* [listing in the process of being withdrawn]; U.S. ESA - Endangered as *O. v. clavium* and *O. v. leucurus*; IUCN - Rare as *O. v. clavium.*
SYNONYMS: *abeli, acapulcensis, aequatorialis, antonii, battyi, borealis, brachyceros, campestris, cariacou, carminis, chiriquensis, clavatus, clavium, columbicus, consul, costaricensis, couesi, curassavicus, dacotensis, fraterculus, goudotii, gracilis, gymnotis, hiltonensis, lasiotis, leucurus, lichtensteini, louisianae, macrourus, margaritae, mcilhennyi, mexicanus, miquihuanensis, nelsoni, nemoralis, nigribarbis, oaxacensis, ochrourus, osceola, peruvianus, philippii, rothschildi, savannarum, seminolus, sinaloae, spelaeus, spinosus, suacuapara, sylvaticus, taurinsulae, texanus, thomasi, toltecus, tropicalis, truei, ustus, venatorius, veraecrucis, wiegmanni, wisconsinensis, yucatanensis.*
COMMENTS: Reviewed by W.P. Smith (1991, Mammalian Species, 388).

Ozotoceros Ameghino, 1891. Rev. Argent. Hist. Nat., 1:243.
TYPE SPECIES: *Cervus campestris* F. Cuvier, 1817 (= *Cervus bezoarticus* Linnaeus, 1758).
SYNONYMS: *Blastoceros, Ozelaphus.*

COMMENTS: *Ozotoceros* is the name to be used for *Blastoceros*, Fitzinger, 1860, if *Blastoceros* is regarded as an invalid emendation of *Blastocerus*; see Hershkovitz (1958). Included in *Odocoileus* by Haltenorth (1963:46) and Bianchini and Delupi (1979); but see Groves and Grubb (1987).

Ozotoceros bezoarticus (Linnaeus, 1758). Syst. Nat., 10th ed., 1:67.
TYPE LOCALITY: Brazil, Pernambuco; identified by Thomas (1911*a*:151).
DISTRIBUTION: Brazil, N Argentina, Paraguay, Uruguay, S Bolivia.
STATUS: CITES - Appendix I; U.S. ESA - Endangered; IUCN - Endangered as *O. b. celer*.
SYNONYMS: *albus, azarae, caenosus, campestris, celer, comosus, cuguapara, dickii, fossilis, leucogaster, pampaeus, sylvestris*.
COMMENTS: Reviewed by Jackson (1987, Mammalian Species, 295).

Pudu Gray, 1852. Proc. Zool. Soc. Lond., 1850:242 [1852].
TYPE SPECIES: *Capra puda* Molina, 1782.
SYNONYMS: *Pudella*.
COMMENTS: Included in *Mazama* by Haltenorth (1963:48). Includes *Pudella*; revised by Hershkovitz (1982).

Pudu mephistophiles (de Winton, 1896). Proc. Zool. Soc. Lond., 1896:508.
TYPE LOCALITY: Ecuador, Napo-Pastaza Prov., Paramo de Papallacta.
DISTRIBUTION: Andes of Colombia, Ecuador, and Peru.
STATUS: CITES - Appendix II; IUCN - Indeterminate.
SYNONYMS: *fusca, wetmorei*.
COMMENTS: *P. mephistophiles* of Matschie and Sanborn (not de Winton) is *Mazama chunyi*.

Pudu puda (Molina, 1782). Sagg. Stor. Nat. Chile, p. 310
TYPE LOCALITY: Chile, Chiloe Prov.
DISTRIBUTION: S Chile, SW Argentina.
STATUS: CITES - Appendix; I U.S. ESA - Endangered.
SYNONYMS: *chilensis, humilis*.
COMMENTS: For original spelling of specific name, see Hershkovitz (1982).

Rangifer H. Smith, 1827. *In* Griffith et al., Anim. Kingdom, 5:304.
TYPE SPECIES: *Cervus tarandus* Linnaeus, 1758.
SYNONYMS: *Achlis, Tarandus*.
COMMENTS: Revised by Banfield (1961).

Rangifer tarandus (Linnaeus, 1758). Syst. Nat., 10th ed., 1:67.
TYPE LOCALITY: Sweden, Alpine Lapland (domesticated stock).
DISTRIBUTION: Circumboreal in tundra and taiga from Svalbard, Norway, Finland, Russia, Alaska (USA) and Canada including most arctic islands, and Greenland, south to N Mongolia; Inner Mongolia and Heilungkiang, China (now feral?); Sakhalin Isl; N Idaho and Great Lakes region (USA). Introduced to, and feral in, Iceland, Kerguelen Isls, South Georgia Isl, Pribilof Isls, St. Matthew Isl. Extinct in Sweden.
STATUS: U.S. ESA - Endangered as *R. t. caribou* in Canada (SE British Columbia at the Canadian-USA border, Columbia R., Kootenay, R. Kootenay Lake, and Kootenai R.) and USA (Idaho, Washington).
SYNONYMS: *angustirostris, arcticus, asiaticus, borealis, buskensis, caboti, caribou, chukchensis, cilindricornis, dawsoni, dichotomus, eogroenlandicus, excelsifrons, fennicus, fortidens, furcifer, granti, groenlandicus, keewatinensis, labradorensis, lapponum, lenensis, montanus, mcguirei, ogilvyensis, osborni, pearsoni, pearyi, phylarchus, platyrhynchus, rangifer, selousi, setoni, sibiricus, silvicola, spetsbergensis, stonei, sylvestris, taimyrensis, terraenovae, transuralensis, valentinae, yakutskensis*.

Family Antilocapridae Gray, 1866. Ann. Mag. Nat. Hist., ser 3, 18:325-326, 468.

Antilocapra Ord, 1818. J. Phys. Chim. Hist. Nat. Arts Paris, 87:149.
TYPE SPECIES: *Antilope americana* Ord, 1815.
COMMENTS: Included in Bovidae by O'Gara and Matson (1975); but restored to separate family status by Janis and Scott (1987) and Soulounias (1988).

Antilocapra americana (Ord, 1815). *In* Guthrie, New Geogr., Hist. Coml. Grammar., Philadelphia, 2nd ed., 2:292.
TYPE LOCALITY: USA, Plains and Highlands of the Missouri River.
DISTRIBUTION: S Alberta and S Saskatchewan (Canada) south through W USA to Hidalgo, Baja California, W Sonora (Mexico). Introduced to Lanai Isl (Hawaiian Isls).
STATUS: CITES - Appendix I (Mexican populations); U.S. ESA and IUCN - Endangered as *A. a. peninsularis* and *A. a. sonoriensis*.
SYNONYMS: *anteflexa, mexicana, oregona, peninsularis, sonoriensis*.
COMMENTS: Reviewed by O'Gara (1978, Mammalian Species, 90).

Family Bovidae Gray, 1821. London Med. Repos., 15:308.
COMMENTS: Distribution and status of introduced populations reviewed by Lever (1985). Distribution and status of African species reviewed by East (1988, 1989, 1990). Systematics of African species reviewed by Ansell (1972) and Gentry (1972).

Subfamily Aepycerotinae Gray, 1872. Cat. Ruminant Mamm. Brit. Mus., p. 4, 42.

Aepyceros Sundevall, 1847. Kongl. Svenska Vet.-Akad. Handl. Stockholm, 1845:271 [1847].
TYPE SPECIES: *Antilope melampus* Lichtenstein, 1812.

Aepyceros melampus (Lichtenstein, 1812). Reisen Sudl. Africa, 2, pl. 4 opp. p. 544.
TYPE LOCALITY: South Africa, Cape Prov., Kuruman, Khosis.
DISTRIBUTION: NE South Africa to Angola, S Zaire, Rwanda, Uganda, and Kenya.
STATUS: U.S. ESA and IUCN - Endangered as *A. m. petersi*.
SYNONYMS: *holubi, johnstoni, katangae, pallah, petersi, rendilis, suara*.
COMMENTS: Includes *petersi*; see Ansell (1972:57).

Subfamily Alcelaphinae Brooke, 1876. *In* Wallace, Geog. Dist. Animals, 2:224.

Alcelaphus Blainville, 1816. Bull. Sci. Soc. Philom. Paris, 1816:75.
TYPE SPECIES: *Antilope bubalis* Pallas, 1767 (= *Antilope buselaphus* Pallas, 1766).
SYNONYMS: *Acronotus, Bubalis, Damalis*.
COMMENTS: Van Gelder (1977*b*) included *Sigmoceros* and *Damaliscus* in this genus, but has not been followed by recent authors; see Swanepoel et al. (1980:187). Formerly included *Sigmoceros lichtensteinii*, see comment under *Sigmoceros*.

Alcelaphus buselaphus (Pallas, 1766). Misc. Zool., p. 7.
TYPE LOCALITY: Morocco.
DISTRIBUTION: Senegal to Ethiopia, south to E Zaire, Uganda, Kenya and N Tanzania; S Angola, W Zimbabwe, Botswana, Namibia and South Africa. Formerly N Morocco, N Algeria, and, up to Dynastic times, east to Egypt. Extinct in N Africa, Somalia, and much of its former South African range.
STATUS: U.S. ESA and IUCN - Endangered as *A. b. swaynei* and *A. b. tora*.
SYNONYMS: *bubalis, bubastis, caama, cokii, deckeni, digglei, evalensis, heuglini, insignis, invadens, jacksoni, keniae, kongoni, lelwel, luzarchei, major, matschiei, mauretanicus, modestus, nakurae, neumanni, niediecki, noacki, obscurus, oscari, rahatensis, ritchiei, roosevelti, rothschildi, sabakiensis, schillingsi, schulzi, selbornei, senegalensis, swaynei, tanae, tora, tschadensis, tunisianus, wembaerensis*.
COMMENTS: Includes *caama*; see Ellerman et al. (1953:202).

Connochaetes Lichtenstein, 1812. Mag. Ges. Naturf. Fr. Berlin, 6:152.
TYPE SPECIES: *Antilope gnou* Gmelin, 1788.
SYNONYMS: *Butragus, Catablepas, Cemas, Gorgon*.

Connochaetes gnou (Zimmermann, 1780). Geogr. Gesch. Mensch. Vierf. Thiere, 2:102.
TYPE LOCALITY: South Africa, Cape Prov., Colesberg (Harper, 1940:329).
DISTRIBUTION: Originally from Transvaal and Natal south to Cape Prov. (South Africa); now only in captivity, or as reintroduced populations in Lesotho; Swaziland; Orange Free State, Cape Prov., Transvaal and Natal (South Africa).
SYNONYMS: *capensis, connochaetes, operculatus*.
COMMENTS: Reviewed by Von Richter (1974, Mammalian Species, 50).

Connochaetes taurinus (Burchell, 1823). Travels in Interior of Southern Africa, 2:278(footnote) [1824].

TYPE LOCALITY: South Africa, Cape Prov., Kuruman, Khosis.

DISTRIBUTION: S Kenya, Tanzania and Zambia south to Angola, Namibia, Botswana and NE South Africa. Extinct in Malawi.

SYNONYMS: *albojubatus, borlei, cooksoni, corniculatus, fasciatus, gorgon, hecki, henrici, johnstoni, lorenzi, mattosi, mearnsi, reichei, rufijianus, schulzi.*

COMMENTS: For date of publication, see Ellerman et al. (1953:205).

Damaliscus P.L. Sclater and Thomas, 1894. Book of Antelopes, 1(part 1):3, 51.

TYPE SPECIES: *Antilope pygargus* Pallas, 1767.

SYNONYMS: *Beatragus, Damalis.*

COMMENTS: Includes *Beatragus;* see Ansell (1972:54). Placed in *Alcelaphus* by Van Gelder (1977*b*:18); but see also Vrba (1979) and Gentry (1990).

Damaliscus hunteri (Sclater, 1889). Proc. Zool. Soc. Lond., 1889:58.

TYPE LOCALITY: Kenya, E bank of Tana River.

DISTRIBUTION: S Somalia to N Kenya.

STATUS: IUCN - Vulnerable.

COMMENTS: Included in *D. lunatus* by Haltenorth (1963:100). Formerly in *Beatragus;* see Ansell (1972:54); retained in *Beatragus* by Gentry and Gentry (1978) and Gentry (1990).

Damaliscus lunatus (Burchell, 1823). Travels in Interior of Southern Africa, 2:334 [1824].

TYPE LOCALITY: South Africa, Cape Prov., Matlhawareng River, NE of Kuruman.

DISTRIBUTION: Formerly Mauritania and Senegal east to W Ethiopia and S Somalia, and south to Tanzania; also Zambia to South Africa. Now extinct in Mauritana, Mali, Senegal, Gambia, Guinea, Ghana, Mozambique and most of South African Range; survives in Niger, Chad, Burkina Faso, Benin, Togo, N Nigeria, Cameroon, Central African Republic, Sudan, W Ethiopia, S Somalia, Uganda, Kenya, Tanzania, Rwanda, E and S Zaire, NE Namibia, N Botswana, E Angola, W and C Zambia, Zimbabwe and Transvaal (South Africa).

STATUS: CITES - Appendix III (Ghana); IUCN - Vulnerable as *D. l. korrigum.*

SYNONYMS: *eurus, floweri, jimela, jonesi, korrigum, lyra, phalius, purpurescens, reclinis, selousi, senegalensis, tiang, topi, ugandae.*

COMMENTS: Includes *korrigum;* see Ansell (1972:56). For date of publication, see Ellerman et al. (1953:201).

Damaliscus pygargus (Pallas, 1767). Spicil. Zool., 1:10.

TYPE LOCALITY: South Africa, Cape Prov., Swart River.

DISTRIBUTION: Formerly from SW Cape Prov. to E Transvaal (South Africa); now only in captivity, or as reintroduced populations in Lesotho, Swaziland, and South Africa.

STATUS: CITES - Appendix II and U.S. ESA - Endangered, as *D. dorcas dorcas;* IUCN - Vulnerable as *D. d. dorcas,* otherwise Rare as *D. dorcas.*

SYNONYMS: *albifrons, dorcas, grisea, maculata, personata, phillipsi, scripta.*

COMMENTS: Includes *phillipsi* and *albifrons;* see Ansell (1972:55). Includes *dorcas, pygargus* being the valid name; see Rookmaaker (1991). The type locality of *Damaliscus dorcas* was restricted to South Africa, Cape Prov., Riversdale District, Kafferkiuls River by Harper (1940). But Bigalke (1948), in making an alternative restriction of the type locality to Swart River, used the name *Damaliscus pygargus.* Since this is now the valid name for the species, Bigalke's restriction must be accepted.

Sigmoceros Heller, 1912. Smithson. Misc. Coll., 60(8):4.

TYPE SPECIES: *Bubalis lichtensteinii* Peters, 1852 (= *Antilope lichtensteinii* Peters, 1849.

COMMENTS: Formerly included in *Alcelaphus;* see Vrba (1979:223).

Sigmoceros lichtensteinii (Peters, 1849). Spenerschen Zeitung, 23 December, 1849, p. unknown; reprinted in 1912 in Gesellschaft Natuurforschender Freunde zu Berlin for 1839-59.

TYPE LOCALITY: Mozambique, Tette.

DISTRIBUTION: Tanzania, SE Zaire and NE Angola to NE Zimbabwe and C Mozambique.

SYNONYMS: *bangae, basengae, dieseneri, frommi, gendagendae, godonga, godowiusi, gombensis, gorongozae, grotei, hennigi, heuferi, inkulanondo, janenschi, kangosa, konzi, lacrymalis, lademanni, leucoprymnus, leupolti, lindicus, munzneri, niediecki, petersi, prittwitzi, rendalli, rowumae, rukwai, saadanicus, schmitti, schusteri, senganus, shirensis, stierlingi, tendagurucus, ufipae, ugalae, ulanagae, ungonicus, ungoniensis, uwendensis, wiesei, wintgensis.*

COMMENTS: Included in *Alcelaphus buselaphus* by Haltenorth (1963:102), but placed in separate genus, *Sigmoceros* by Vrba (1979). Included in *Alcelaphus* by Gentry (1990).

Subfamily Antilopinae Gray, 1821. London Med. Repos., 15:307.

Ammodorcas Thomas, 1891. Proc. Zool. Soc. Lond., 1891:207, pl. 21,22.
TYPE SPECIES: *Cervicapra clarkei* Thomas, 1891.

Ammodorcas clarkei (Thomas, 1891). Ann. Mag. Nat. Hist., ser. 6, 7:304.
TYPE LOCALITY: Somalia, Habergerhagi's Country, Buroa Wells.
DISTRIBUTION: E Ethiopia, N Somalia.
STATUS: U.S. ESA - Endangered; IUCN - Vulnerable.
COMMENTS: Reviewed by Schomber (1964).

Antidorcas Sundevall, 1847. Kongl. Svenska Vet.-Akad. Handl. Stockholm, 1845:271 [1847].
TYPE SPECIES: *Antilope euchore* Forster, 1790 (= *Antilope marsupialis* Zimmermann, 1780).

Antidorcas marsupialis (Zimmermann, 1780). Geogr. Gesch. Mensch. Vierf. Thiere, 2:427.
TYPE LOCALITY: South Africa, Cape of Good Hope.
DISTRIBUTION: Namibia, SW Angola, Botswana; South Africa (range here now much reduced).
SYNONYMS: *angolensis, centralis, dorsata, euchore, hofmeyri, pygargus, saccata, saliens, saltans.*
COMMENTS: Revised by Groves (1981*b*).

Antilope Pallas, 1766. Misc. Zool., p. 1.
TYPE SPECIES: *Capra cervicapra* Linnaeus, 1758.
SYNONYMS: *Cervicapra.*

Antilope cervicapra (Linnaeus, 1758). Syst. Nat., 10th ed., 1:69.
TYPE LOCALITY: India, Travancore, inland of Trivandrum.
DISTRIBUTION: E Pakistan (extinct but reintroduced); India from Punjab south to Madras and east to Bihar (formerly up to Assam); extinct in Bangladesh and now localized in India; introduced to Nepal, Texas (USA), and Argentina.
STATUS: CITES - Appendix III (Nepal).
SYNONYMS: *bezoartica, bilineata, centralis, hagenbecki, rajputanae, rupicapra, strepsiceros.*
COMMENTS: Revised by Groves (1982*c*).

Dorcatragus Noack, 1894. Zool. Anz., 17:202.
TYPE SPECIES: *Oreotragus megalotis* Menges, 1894.

Dorcatragus megalotis (Menges, 1894). Zool. Anz., 17:130.
TYPE LOCALITY: Somalia, Hekebo Plateau, 35 mi (56 km) SW of Berbera.
DISTRIBUTION: N Somalia, Djibouti, and marginally in adjacent parts of Ethiopia.
STATUS: IUCN - Insufficiently known.

Gazella Blainville, 1816. Bull. Sci. Soc. Philom. Paris, 1816:75.
TYPE SPECIES: *Capra dorcas* Linnaeus, 1758.
SYNONYMS: *Dorcas, Eudorcas, Korin, Leptoceros, Matscheia, Nanger, Trachelocele, Tragops, Tragopsis.*
COMMENTS: Revised in part by Groves (1969*a*).

Gazella arabica (Lichenstein, 1827). Darst. Säugeth., pl. 6.
TYPE LOCALITY: Saudi Arabia, Farasan Isls.
DISTRIBUTION: Not certainly known.

COMMENTS: Treated as a separate species from *G. gazella* by Groves (1985*a*). Only known from two specimens; see Groves (1983). Even if formerly present on Farasan Isls, now replaced there by *G. gazella farasani*; see Thouless and Al Bassri (1991).

Gazella bennettii (Sykes, 1831). Proc. Zool. Soc. Lond., 1830-1831:104 [1831].
TYPE LOCALITY: India, "Deccan".
DISTRIBUTION: Iran to NW India.
SYNONYMS: *christii, fuscifrons, hayi, hazenna, kennioni.*
COMMENTS: A species distinct from *G. gazella* according to Furley et al. (1988) and Groves (1985*a*, 1988).

Gazella bilkis Groves and Lay, 1985. Mammalia, 49:29.
TYPE LOCALITY: Yemen, Wadi Maleh 5 mi E. of Ta'izz, El Hauban.
DISTRIBUTION: N Yemen.

Gazella cuvieri (Ogilby, 1841). Proc. Zool. Soc. Lond., 1840:35 [1841].
TYPE LOCALITY: Morocco, Mogador.
DISTRIBUTION: Morocco, N Algeria, C Tunisia.
STATUS: CITES - Appendix III (Tunisia); U.S. ESA and IUCN - Endangered.
SYNONYMS: *cinerascens, corinna, vera.*
COMMENTS: Assigned to *G. gazella* by Haltenorth (1963:111); but a distinct species according to Groves (1969*a*).

Gazella dama (Pallas, 1766). Misc. Zool., p. 5.
TYPE LOCALITY: Africa, near Lake Chad.
DISTRIBUTION: Formerly from Morocco, Western Sahara, Mauritania and Senegal east to Egypt and Sudan. Now extinct in Mauritania, Senegal, Morocco, Algeria and Egypt; survives at least in Mali, Niger, Chad, Burkina Faso and Sudan.
STATUS: CITES - Appendix I; U.S. ESA - Endangered as *G. d. lozanoi* and *G. d. mhorr*; IUCN - Endangered.
SYNONYMS: *addra, damergouensis, doria, lozanoi, mhorr, nanguer, occidentalis, orientalis, permista, reducta, ruficollis, weidholzi.*

Gazella dorcas (Linnaeus, 1758). Syst. Nat., 10th ed., 1:69.
TYPE LOCALITY: Lower Egypt.
DISTRIBUTION: Morocco south to Mauritania (and formerly to Senegal) east to S Israel and Egypt and from there south to Sudan, NE Ethiopia and N Somalia.
STATUS: CITES - Appendix III (Tunisia); U.S. ESA - Endangered as *G. d. massaesyla* and *G. d. pelzelni* (sic); IUCN - Vulnerable.
SYNONYMS: *beccarii, cabrerai, corinna, isabella, isidis, kevella, littoralis, maculata, massaesyla, neglecta, osiris, pelzelnii, rueppelli, sundevalli.*
COMMENTS: Includes *pelzelnii*; see Gentry (1972:89); but also see Haltenorth (1963:112). Reviewed by Ferguson (1981) and Groves (1981*c*).

Gazella gazella (Pallas, 1766). Misc. Zool., p. 7.
TYPE LOCALITY: Syria.
DISTRIBUTION: Syria, Israel, Lebanon, Arabian Peninsula.
STATUS: U.S. ESA - Endangered; IUCN - Vulnerable.
SYNONYMS: *cora, erlangeri, farasani, hanishi, merilli, muscatensis.*

Gazella granti Brooke, 1872. Proc. Zool. Soc. Lond., 1872:602.
TYPE LOCALITY: Tanzania, W Kinyenye, Ugogo.
DISTRIBUTION: SE Sudan, NE Uganda and S Ethiopia south to S Somalia, Kenya, and N Tanzania.
SYNONYMS: *brighti, gelidjiensis, lacuum, notata, petersii, raineyi, robertsi, roosevelti, serengetae.*

Gazella leptoceros (F. Cuvier, 1842). *In* E. Geoffroy and F. Cuvier, Hist. Nat. Mammifères, pt. 4, 7(72):1-2, "Antilope aux longues cornes".
TYPE LOCALITY: Sudan, Sennar, but probably Egypt, between Giza and Wadi Natron.
DISTRIBUTION: Algeria, Tunisia, Libya, W Egypt, Niger, N Chad and apparently Mali and Sudan.
STATUS: CITES - Appendix III (Tunisia); U.S. ESA and IUCN - Endangered.
SYNONYMS: *abuharab, cuvieri, loderi.*

Gazella rufifrons Gray, 1846. Ann. Mag. Nat. Hist., [ser. 1], 18:214.
TYPE LOCALITY: Senegal.
DISTRIBUTION: Senegal to NE Ethiopia, with southern limits of distribution extending from N Togo to N Central African Republic; extinct in N Ghana.
STATUS: IUCN - Vulnerable.
SYNONYMS: *centralis, hasleri, kanuri, laevipes, salmi, senegalensis, tilonura.*
COMMENTS: Includes *tilonura*; see Gentry (1972:90); but also see Haltenorth (1963:112). Groves (1969*a*) included *rufifrons* in *cuvieri*, but subsequently (1975*a*) separated them.

Gazella rufina Thomas, 1894. Proc. Zool. Soc. Lond., 1894:467.
TYPE LOCALITY: Interior of Algeria.
DISTRIBUTION: Algeria.
SYNONYMS: *pallaryi.*
COMMENTS: Thought to be extinct; see Corbet (1978*c*:210).

Gazella saudiya Carruthers and Schwarz, 1935. Proc. Zool. Soc. Lond., 1935:155.
TYPE LOCALITY: Saudi Arabia, 150 mi NE of Mecca, Dhalm.
DISTRIBUTION: Formerly Saudi Arabia, Kuwait, S Iraq. Extinct in the wild.
STATUS: U.S. ESA - Endangered.
COMMENTS: A species distinct from *G. dorcas* according to Groves (1988).

Gazella soemmerringii (Cretzschmar, 1828). *In* Rüppell, Atlas Reise Nordl. Afr., Zool., Säugeth., p. 49, pl. 19.
TYPE LOCALITY: E Ethiopia.
DISTRIBUTION: N Somalia, Ethiopia, EC Sudan.
STATUS: IUCN - Vulnerable.
SYNONYMS: *berberana, butteri, casanovae, erlangeri, sibyllae.*

Gazella spekei Blyth, 1863. Cat. Mamm. Mus. Asiat. Soc. Calcutta, p. 172.
TYPE LOCALITY: N Somalia.
DISTRIBUTION: Somalia, E Ethiopia.
STATUS: IUCN - Vulnerable.

Gazella subgutturosa (Güldenstaedt, 1780). Acta Acad. Sci. Petropoli, for 1778, 1:251 [1780].
TYPE LOCALITY: Azerbaijan, Steppes of E Transcaucasia.
DISTRIBUTION: Israel; Jordan, C Arabia and E Caucasus through Iran; Afghanistan; WC Pakistan; Kazakhstan; Turmenistan; Uzbekistan; Mongolia; W China.
STATUS: U.S. ESA and IUCN - Endangered as *G. s. marica.*
SYNONYMS: *gracilicornis, hilleriana, marica, mongolica, persica, reginae, sairensis, seistanica, yarkandensis.*

Gazella thomsonii Günther, 1884. Ann. Mag. Nat. Hist., ser. 5, 14:427.
TYPE LOCALITY: Kenya, Kilimanjaro Dist.
DISTRIBUTION: SE Sudan, SW Ethiopia, S and C Kenya, N Tanzania.
SYNONYMS: *albonotatus, baringoensis, behni, bergeri, bergerinae, biedermanni, dieseneri, dongilanensis, langheldi, macrocephala, manharae, marwitzi, mundorosica, nakuroensis, nasalis, ndjiriensis, ruwanae, sabakiensis, schillingsi, seringetica, wembaerensis.*
COMMENTS: Groves (1969*a*) included *thomsonii* in *G. cuvieri*; but see also Gentry (1972:88, 90-91). Groves (1985*a*, 1988) included *thomsonii* in *G. rufifrons.*

Litocranius Kohl, 1886. Ann. K. K. Naturhist. Hofmus. Wien, 1:79.
TYPE SPECIES: *Gazella walleri* Brooke, 1879.
SYNONYMS: *Lithocranius.*
COMMENTS: Revised by Schomber (1963).

Litocranius walleri (Brooke, 1879). Proc. Zool. Soc. Lond., 1878:929, pl. 56 [1879].
TYPE LOCALITY: "mainland of Africa, north of the island of Zanzibar, about lat. 30°S and long. 38°E" and therefore apparently in Kenya, but shown to be correctly "Somalia, coast near Juba River" by Sclater and Thomas (1898) and Moreau et al. (1946).
DISTRIBUTION: E Ethiopia, Somalia, Kenya, NE Tanzania.
SYNONYMS: *sclateri.*

Madoqua Ogilby, 1837. Proc. Zool. Soc. Lond., 1836:137 [1837].
TYPE SPECIES: *Antilope saltiana* Desmarest, 1816.
SYNONYMS: *Rhynchotragus*.
COMMENTS: Includes *Rhynchotragus*; see Ansell (1972:61). Revised in part by Yalden (1978).

Madoqua guentheri Thomas, 1894. Proc. Zool. Soc. Lond., 1894:324.
TYPE LOCALITY: Ethiopia, Ogaden (6°30'N, 42°30'E), 3,000 ft. (914 m).
DISTRIBUTION: S and C Somalia, S Ethiopia, SE Sudan, NE Uganda, N Kenya.
SYNONYMS: *hodsoni, nasoguttata, smithii, wroughtoni*.

Madoqua kirkii (Günther, 1880). Proc. Zool. Soc. Lond., 1880:17.
TYPE LOCALITY: Somalia, Brava.
DISTRIBUTION: Kenya, N and C Tanzania, S Somalia, SW Angola, Namibia.
SYNONYMS: *cavendishi, damarensis, hindei, langi, minor, nyikae, thomasi, variani*.
COMMENTS: Includes *cavendishi, damarensis* and *thomasi*; see Ansell (1972:64). May really constitute two species; see Ryder et al. (1989).

Madoqua piacentinii Drake-Brockman, 1911. Proc. Zool. Soc. Lond., 1911:981.
TYPE LOCALITY: Somalia, Gharabwein, near Obbia (5°25'N, 48°25'E).
DISTRIBUTION: E Somalia.
COMMENTS: Included in *swaynei* by Ansell (1972:62); but see Yalden (1978:262). See also comment under *saltiana*.

Madoqua saltiana (Desmarest, 1816). Nouv. Dict. Hist. Nat., Nouv. ed., 2:192.
TYPE LOCALITY: Ethiopia (= Abyssinia).
DISTRIBUTION: NE Sudan, N Ethiopia, Somalia, Djibouti.
SYNONYMS: *citernii, cordeauxi, erlangeri, gubanensis, hararensis, hemprichiana, lawrancei, madoka, phillipsi, swaynei*.
COMMENTS: Includes *cordeauxi*; see Ansell (1972:63). Includes *erlangeri, phillipsi*, and *swaynei*; see Yalden (1978:262); but see also Ansell (1972:62).

Neotragus H. Smith, 1827. *In* Griffith et al., Anim. Kingdom, 5:349.
TYPE SPECIES: *Capra pygmaea* Linnaeus, 1758.
SYNONYMS: *Hylarnus, Memina, Nesotragus*.
COMMENTS: Includes *Nesotragus*; see Ansell (1972:68).

Neotragus batesi de Winton, 1903. Proc. Zool. Soc. Lond., 1903(1):192.
TYPE LOCALITY: Cameroon, Bulu Country, Efulen.
DISTRIBUTION: SE Nigeria, SE Cameroon, NE Gabon, N Congo Republic, E Zaire, W Uganda.
SYNONYMS: *harrisoni*.

Neotragus moschatus (Von Dueben, 1846). *In* Sundevall, Ofv. K. Svenska Vet.-Akad. Forhandl., Stockholm, 3(7):221.
TYPE LOCALITY: Tanzania, Chapani Isl., 2 mi. (3 km) from Zanzibar.
DISTRIBUTION: SE Kenya south to Malawi, Mozambique, and Natal and E Transvaal (South Africa); Zanzibar and Mafia Isls (Tanzania).
STATUS: U.S. ESA - Endangered as *N. m. moschatus*.
SYNONYMS: *akeleyi, deserticola, kirchenpaueri, livingstonianus, zanzibaricus, zuluensis*.
COMMENTS: Includes *livingstonianus*; see Ellerman et al. (1953).

Neotragus pygmeus (Linnaeus, 1758). Syst. Nat., 10th ed., 1:69.
TYPE LOCALITY: "Guinea, India" (= west coast of Africa).
DISTRIBUTION: Sierra Leone to Ghana.
SYNONYMS: *perpusillus, regia, spinigera*.

Oreotragus A. Smith, 1834. S. Afr. Quart. J., 2:212.
TYPE SPECIES: *Antilope oreotragus* Zimmermann, 1783.

Oreotragus oreotragus (Zimmermann, 1783). Geogr. Gesch. Mensch. Vierf. Thiere, 3:269.
TYPE LOCALITY: South Africa, Cape of Good Hope.
DISTRIBUTION: C Nigeria, N Central African Republic, E Sudan and Ethiopia south to South Africa and west to S Zaire, Zambia and E Botswana; Namibia.
STATUS: IUCN - Endangered as *O. o. porteousi*.

SYNONYMS: *aceratos, aureus, centralis, cunenensis, hyatti, klippspringer, porteusi, saltator, saltatrixoides, schillingsi, somalicus, steinhardti, stevensoni, transvaalensis, tyleri.*

Ourebia Laurillard, 1842. *In* d'Orbigny, Dict. Univ. D'Hist. Nat., 1:622.
TYPE SPECIES: *Antilope scoparia* Schreber, 1799 (= *Antilope ourebi* Zimmermann, 1783).
SYNONYMS: *Quadriscopa, Scopophorus.*

Ourebia ourebi (Zimmermann, 1783). Geogr. Gesch. Mensch. Vierf. Thiere, 3:268.
TYPE LOCALITY: South Africa, Cape of Good Hope (= Uitenhage Dist., see Roberts, 1951).
DISTRIBUTION: Senegal to W and C Ethiopia and S Somalia, south and southwest to E South Africa, N Botswana, N Namibia, Angola, and S Zaire.
SYNONYMS: *aequatoria, brevicaudata, cottoni, dorcas, gallarum, goslingi, grayi, haggardi, hastata, kenyae, leucopus, masakensis, melanura, microdon, montana, nigricaudata, pitmani, quadriscopa, rutila, scoparia, smithii, splendida, ugandae.*
COMMENTS: For synonyms see Ansell (1972:66).

Pantholops Hodgson, 1834. Proc. Zool. Soc. Lond., 1834:81.
TYPE SPECIES: *Antilope hodgsonii* Aber, 1826.

Pantholops hodgsonii (Abel, 1826). Calcutta Gov't Gazette., see Phil. Mag., 1826, 68:234.
TYPE LOCALITY: China, Tibet, Kooti Pass in Arrun Valley, Tingri Maiden.
DISTRIBUTION: Tibet, Tsinghai, Szechwan (China); Ladak (N India).
STATUS: CITES - Appendix I.
SYNONYMS: *chiru, kemas.*

Procapra Hodgson, 1846. J. Asiat. Soc. Bengal, 15:334.
TYPE SPECIES: *Procapra picticaudata* Hodgson, 1846.
SYNONYMS: *Prodorcas.*
COMMENTS: Revised by Groves (1967*a*). Gromov and Baranova (1981:393) considered *Procapra* a subgenus of *Gazella*; but Groves (1985*a*) maintained its status as a genus.

Procapra gutturosa (Pallas, 1777). Spicil. Zool., 12:46.
TYPE LOCALITY: Russia, SE Transbaikalia, Chitinsk. Obl., upper Onon River.
DISTRIBUTION: Formerly, Mongolia except mountains and SW desert, N China, and Transbaikalia (Russia); now restricted to E Mongolia and Inner Mongolia (China); see Sokolov and Orlov (1980:280).
SYNONYMS: *altaica, orientalis.*

Procapra picticaudata Hodgson, 1846. J. Asiat. Soc. Bengal, 15:334, pl. 2.
TYPE LOCALITY: China, Tibet, Hundes Dist; but according to Groves (1967*a*), the locality is more likely the district N of Sikkim.
DISTRIBUTION: Szechwan, Tsinghai, Tibet (China); adjacent Indian Himalayas; possibly Sinkiang (China).

Procapra przewalskii (Büchner, 1891). Melanges Biol. Soc. St. Petersb., 13:161.
TYPE LOCALITY: Mongolia, Chagrin-Gol (= Steppe); erroneously given as China, S Ordos Desert by Ellerman and Morrison-Scott (1951), which was followed by Gromov and Baranova (1981).
DISTRIBUTION: Tsinghai to Inner Mongolia (NC China).
SYNONYMS: *diversicornis.*
COMMENTS: Considered a subspecies of *picticaudata* by Ellerman and Morrison-Scott (1951:388).

Raphicerus H. Smith, 1827. *In* Griffith et al., Animal Kingdom, 5:342.
TYPE SPECIES: *Cerophorus acuticornis* Blainville, 1816 (= *Antilope campestris* Thunberg, 1811).
SYNONYMS: *Calotragus, Grysbock, Nototragus, Pediotragus.*

Raphicerus campestris (Thunberg, 1811). Mem. Acad. Imp. Sci. St. Petersbourg, 3:313.
TYPE LOCALITY: South Africa, Cape of Good Hope.
DISTRIBUTION: N and C Tanzania; S Kenya; Angola, W Zambia, Zimbabwe, and S Mozambique south to South Africa.
SYNONYMS: *acuticornis, bourquii, capensis, capricornis, cunenensis, fulvorubescens, grayi, hoamibensis, horstockii, ibex, kelleni, natalensis, neumanni, pallida, pediotragus, rupestris,*

steinhardti, stigmatus, subulata, tragulus, ugabensis, zukowskyi, zuluensis.

Raphicerus melanotis (Thunberg, 1811). Mem. Acad. Imp. Sci. St. Petersbourg, 3:312.
TYPE LOCALITY: South Africa, Cape of Good Hope.
DISTRIBUTION: S Cape Prov. (South Africa).
SYNONYMS: *grisea, rubroalbescens, rufescens.*

Raphicerus sharpei Thomas, 1897. Proc. Zool. Soc. Lond., 1896:796, pl. 9 [1867].
TYPE LOCALITY: Malawi (= Nyasaland), S Angoniland.
DISTRIBUTION: Tanzania, SE Zaire, Zambia, Malawi, Mozambique, Zimbabwe, N Botswana, Swaziland, NE Transvaal (South Africa).
SYNONYMS: *colonicus.*
COMMENTS: Included in *melanotis* by Haltenorth (1963:78) but see Ansell (1972:67).

Saiga Gray, 1843. List Specimens Mamm. Coll. Brit. Mus., p. xxvi.
TYPE SPECIES: *Capra tatarica* Linnaeus, 1766.
SYNONYMS: *Colus.*
COMMENTS: This generic name is spelt "*Saiga*" on p. xxvi and "*Siaga*" on p. 160 of the original citation.

Saiga tatarica (Linnaeus, 1766). Syst. Nat., 12th ed., 1:97.
TYPE LOCALITY: W Kazakhstan, "Ural Steppes."
DISTRIBUTION: N Caucasus (Kalmyk Steppe, Russia), Kazakhstan, N Uzbekistan, SW Mongolia, Sinkiang (China). Formerly west to Poland.
STATUS: U.S. ESA - Endangered as *S. t. mongolica.*
SYNONYMS: *colus, mongolica, saiga, scythica.*
COMMENTS: Reviewed by Sokolov (1974, Mammalian Species, 38).

Subfamily Bovinae Gray, 1821. London Med. Repos., 15:308.
COMMENTS: Tribe Bovini reviewed by Groves (1981*d*).

Bison H. Smith, 1827. *In* Griffith et al., Animal Kingdom, 5:373.
TYPE SPECIES: *Bos bison* Linnaeus, 1758.
COMMENTS: Revised by Bohlken (1967), McDonald (1981), and Van Zyll de Jong (1986). A synonym of *Bos* according to Groves (1981*d*).

Bison bison (Linnaeus, 1758). Syst. Nat., 10th ed., 1:72.
TYPE LOCALITY: "Mexico" (= C Kansas, "Quivira"), see Hershkovitz (1957*b*); redesignated as Canadian River valley, E New Mexico (USA), see McDonald (1981:62).
DISTRIBUTION: Formerly NW and C Canada, south through USA, to Chihuahua, Coahuila (Mexico). Exterminated in the wild except in Yellowstone Park, Wyoming (USA) and Wood Buffalo Park, Northwest Territory (Canada). Reintroduced widely within native range and in C Alaska.
STATUS: CITES - Appendix I and U.S. ESA - Endangered as *B. b. athabascae.*
SYNONYMS: *americanus, athabascae, haningtoni, montanae, oregonus, pennsylvanicus, septemtrionalis* (sic).
COMMENTS: Reviewed by Meagher (1986, Mammalian Species, 266).

Bison bonasus (Linnaeus, 1758). Syst. Nat., 10th ed., 1:71.
TYPE LOCALITY: Bialowieza Forest, Poland.
DISTRIBUTION: Europe, originally from S Sweden south to the Pyrenees Mtns and Balkans east to R. Don and W Caucasus Mtns (Russia), surviving in France until about 6th century, in Germany, Rumania and W Russia into 18th century and in Caucasus Mtns and Poland until early part of present century; extinct except where now reintroduced to E Poland, W Russia, and Caucasus Mtns.
STATUS: IUCN - Vulnerable.
SYNONYMS: *arbustotundrarum, caucasicus, europaeus, hungarorum, nostras; urus* Boddaert, 1785.
COMMENTS: Considered conspecific with *bison* by Bohlken (1967) and Van Zyll de Jong (1986); but not included in *B. bison* by Meagher (1986). Reviewed by Flerov (1979).

Bos Linnaeus, 1758. Syst. Nat., 10th ed., 1:71.
TYPE SPECIES: *Bos taurus* Linnaeus, 1758.

SYNONYMS: *Bibos, Bubalibos, Gauribos, Gaveus, Novibos, Poephagus, Taurus, Uribos, Urus.*
COMMENTS: Includes *Bibos, Novibos,* and *Poephagus;* see Ellerman and Morrison-Scott (1951:380).

Bos frontalis Lambert, 1804. Trans. Linn. Soc. Lond., 7:57.
TYPE LOCALITY: Bangladesh, NE Chittagong (domesticated stock).
DISTRIBUTION: India, Nepal, Burma, Thailand, S Tibet and Yunnan (China), S Vietnam, Cambodia, Peninsular Malaysia.
STATUS: CITES - Appendix I, U.S. ESA - Endangered, and IUCN - Vulnerable as *Bos gaurus.*
SYNONYMS: *annamiticus, asseel, brachyrhinus, cavifrons, fuscicornis, gaurus, guavera, hubbacki, laosiensis, platyceros, readei, subhemachalus, sylhetanus, sylvanus.*
COMMENTS: Includes *gaurus;* but see Corbet and Hill (1991:130). Formerly placed in *Bibos.*

Bos grunniens Linnaeus, 1766. Syst. Nat., 12th ed., 1:99.
TYPE LOCALITY: Boreal Asia (domesticated stock).
DISTRIBUTION: Sinkiang, Tibet, Tsinghai, Szechwan (China); Ladak (N India); Nepal; domesticated in C Asia.
STATUS: CITES - Appendix I as *B.* (=[sic] *grunniens*) *mutus;* U.S. ESA - Endangered as *B. g. mutus;* IUCN - Endangered.
SYNONYMS: *mutus.*
COMMENTS: Includes *mutus;* but see Corbet (1978*c*:206). Formerly placed in *Poephagus.* Reviewed by Olsen (1990).

Bos javanicus d'Alton, 1823. Die Skelete der Wiederkauer, abgebildt und verglichen, p. 7.
TYPE LOCALITY: Indonesia, Java.
DISTRIBUTION: Burma, Thailand, and Indochina south to N Peninsular Malaysia; Java; Borneo; introduced to Australia, Bali Isl, Sangihe and Enggano Isls; domesticated in SE Asia.
STATUS: U.S. ESA - Endangered; IUCN - Vulnerable.
SYNONYMS: *banteng, birmanicus, butleri, discolor, domesticus, leucoprymnus, longicornis, lowi, porteri, sondaicus.*
COMMENTS: For use of *javanicus* instead of *banteng,* see Hooijer (1956). Formerly placed in *Bibos.*

Bos sauveli Urbain, 1937. Bull. Soc. Zool. Fr., 62:307.
TYPE LOCALITY: Cambodia, near Tchep Village.
DISTRIBUTION: Cambodia, SE Thailand, S Laos, W Vietnam.
STATUS: CITES - Appendix I; U.S. ESA and IUCN - Endangered.
COMMENTS: Included in *Novibos* by Coolidge (1940); but see Ellerman and Morrison-Scott (1951:380). Reviewed by MacKinnon and Stuart (1989).

Bos taurus Linnaeus, 1758. Syst. Nat., 10th ed., 1:71.
TYPE LOCALITY: Sweden, Upsala according to Thomas (1911*a*); but Linnaeus (1758) cited "Habitat in Poloniæ", possibly refering to the Aurochs, which he recognized as conspecific with domestic cattle.
DISTRIBUTION: Europe and W Russia to the Middle East, surviving at least into the Iron Age in the Middle East; extinct in the wild, except in Poland, by commencement of 15th century; last wild individual died in 1627. Distributed worldwide under domestication; feral populations in Spain, France, Australia, New Guinea, USA, Colombia, Argentina and many islands, including Hawaiian, Galapagos, Dominican Republic/Haiti, Tristan da Cunha, New Amsterdam and Juan Fernandez Isls.
SYNONYMS: *bunnelli, indicus, primigenius; urus* Erxleben, 1777.
COMMENTS: Includes *primigenius* (extinct wild ancestor) and *indicus;* but see Corbet (1978*c*:206).

Boselaphus Blainville, 1816. Bull. Sci. Soc. Philom. Paris, 1816:75.
TYPE SPECIES: *Antilope tragocamelus* Pallas, 1766.
SYNONYMS: *Portax.*

Boselaphus tragocamelus (Pallas, 1766). Misc. Zool., p. 5.
TYPE LOCALITY: India, no exact locality ("Plains of Peninsular India").
DISTRIBUTION: E Pakistan and N India south to Bombay and Mysore; introduced into Texas (USA).

SYNONYMS: *albipes, hippelaphus, picta, risia.*

Bubalus H. Smith, 1827. *In* Griffith et al., Animal Kingdom, 5:371.
TYPE SPECIES: *Bos bubalis* Linnaeus, 1758.
SYNONYMS: *Anoa.*
COMMENTS: Includes *Anoa*; see Groves (1969*b*).

Bubalus bubalis (Linnaeus, 1758). Syst. Nat., 10th ed., 1:72.
TYPE LOCALITY: "Asia"?; restricted by Thomas (1911*a*) to Italy, Rome, but Linnaeus (1758) stated "Habitat in Asia, cultus in Italia".
DISTRIBUTION: Formerly India to Indochina; true wild populations survive in Assam and Orissa (India), Nepal, N Thailand, and possibly in Bangladesh, Cambodia and Vietnam; domesticated in N Africa and S Europe east to Indonesia and in E South America; feral populations in Sri Lanka, Borneo and other parts of SE Asia, New Britain and New Ireland (Bismarck Arch., Papua New Guinea), and Australia.
STATUS: CITES - Appendix III (Nepal) as *B. arnee* "formerly listed as *B. bubalis*, a nonprotected, domesticated form"- but see comments below; IUCN - Endangered.
SYNONYMS: *arnee, fulvus, hosei, indicus, kerabau, macroceros, mainitensis, moellendorfii, septentrionalis.*
COMMENTS: Includes *arnee*, the name used for the species by those workers who do not employ specific names based on domestic animals: *bubalis* is the senior synonym; see Ellerman and Morrison-Scott (1951:383); but see also Corbet and Hill (1991:130).

Bubalus depressicornis (H. Smith, 1827). *In* Griffith et al., Animal Kingdom, 4:293.
TYPE LOCALITY: Indonesia, Sulawesi (= Celebes).
DISTRIBUTION: Sulawesi.
STATUS: CITES - Appendix I; U.S. ESA and IUCN - Endangered.
SYNONYMS: *anoa, celebensis, fergusoni, platycerus.*
COMMENTS: Includes *anoa*; see Groves (1982*e*). Formerly included in *Anoa* but placed in genus *Bubalus*, subgenus *Anoa* by Groves (1969*b*:3).

Bubalus mephistopheles Hopwood, 1925. Ann. Mag. Nat. Hist., ser. 9, 16:238.
TYPE LOCALITY: China, Honan, Chang-Te-Ho on the Yellow River near Chang-Te-Fu.
DISTRIBUTION: Extinct; formerly in NE China, survived until the Shang Dynasty, 18-12th Century BC; see Teilhard de Chardin and Young (1936).

Bubalus mindorensis (Heude, 1888). Mem. Hist. Nat. Emp. Chin., 2:4.
TYPE LOCALITY: Philippines, Mindoro.
DISTRIBUTION: Mindoro (Philippines).
STATUS: CITES - Appendix I; U.S. ESA and IUCN - Endangered.
COMMENTS: A subspecies of *B. bubalis* according to Bohlken (1958), but restored to specific status, in subgenus *Bubalus*, by Groves (1969*b*:10).

Bubalus quarlesi (Ouwens, 1910). Bull. Dépt. Agric. Indes Néerl., 38:7.
TYPE LOCALITY: Indonesia, Sulawesi, mountains of C Toradja Dist.
DISTRIBUTION: Mountains of Sulawesi.
STATUS: CITES - Appendix I; U.S. ESA and IUCN - Endangered.
COMMENTS: Subgenus *Anoa*; see Groves (1969*b*). Formerly included in *A. depressicornis*; see Haltenorth (1963:131).

Syncerus Hodgson, 1847. J. Asiat. Soc. Bengal, ser. 2, 16:709.
TYPE SPECIES: *Bos brachyceros* Gray, 1837 (= *Bos caffer* Sparrman, 1779).
SYNONYMS: *Planiceros.*
COMMENTS: Considered a subgenus of *Bubalus* by Haltenorth (1963:133). Reviewed by Grubb (1972).

Syncerus caffer (Sparrman, 1779). K. Svenska Vet.-Akad. Handl. Stockholm, 40:79.
TYPE LOCALITY: South Africa, Cape of Good Hope, Sunday River, Algoa Bay.
DISTRIBUTION: Senegal to S Ethiopia to S Africa.
SYNONYMS: *adamauae, adametzi, adolfifriederici, aequinoctialis, athiensis, azrakensis, beddingtoni, bornouensis, brachyceros, bubuensis, centralis, corniculatus, cottoni, cubangensis, cunenensis, diehli, gariepensis, gazae, geoffroyi, houyi, hunti, hylaeus, lomamiensis, limpopoensis, massaicus, matthewsi, mayi, nanus, neumanni, niediecki, nuni,*

pegasus, planiceros, pumilus, pungwensis, radcliffei, reclinis, ruahaensis, rufuensis, sankurrensis, schillingsi, simpsoni, solvayi, tanae, thierryi, urundicus, ussanguensis, wembarensis, wiesei, wintgensi.
COMMENTS: For synonyms see Ansell (1972:18).

Taurotragus Wagner, 1855. *In* Schreber Die Säugethiere, Suppl., 5:438.
TYPE SPECIES: *Antilope oreas* Pallas, 1777 (= *Antilope oryx* Pallas, 1766).
SYNONYMS: *Doratoceros, Oreas.*
COMMENTS: This genus has been included in *Tragelaphus*; see Van Gelder (1977*a, b*) and Ansell (1978:53). Generic rank was restored by Smithers (1983:679), Meester et al. (1986:216), and Ansell and Dowsett (1988:87).

Taurotragus derbianus (Gray, 1847). Ann. Mag. Nat. Hist., [ser. 1], 20:286.
TYPE LOCALITY: Gambia.
DISTRIBUTION: Formerly, Gambia to S Sudan and NW Uganda; now in Senegal to Guinea and Cameroon to S Sudan; may survive in S Mali and NE Nigeria.
STATUS: U.S. ESA and IUCN - Endangered as *T. d. derbianus*; IUCN - Vulnerable as *Tragelaphus derbianus gigas.*
SYNONYMS: *cameroonensis, colini, congolanus, gigas.*
COMMENTS: Regarded as conspecific with *T. oryx* by Haltenorth (1963:86), but usually treated as a full species; see Ansell (1972:26).

Taurotragus oryx (Pallas, 1766). Misc. Zool., p. 9.
TYPE LOCALITY: South Africa, Cape of Good Hope.
DISTRIBUTION: Angola to S Africa; thence north through Botswana, S Zaire, Malawi, Zambia, and Tanzania to SE Sudan and SW Ethiopia.
SYNONYMS: *alces, canna, barbatus, billingae, kaufmanni, livingstonii, niediecki, oreas, pattersonianus, selousi, triangularis.*

Tetracerus Leach, 1825. Trans. Linn. Soc. Lond., 14:524.
TYPE SPECIES: *Antilope chickara* Hardwicke, 1825 (= *Cerophorus quadricornis* Blainville, 1816).

Tetracerus quadricornis (Blainville, 1816). Bull. Sci. Soc. Philom. Paris, 1816:75.
TYPE LOCALITY: India, no exact locality ("Plains of Peninsular India").
DISTRIBUTION: Peninsular India, north to Nepal.
STATUS: CITES - Appendix III (Nepal).
SYNONYMS: *chickara, iodes, paccerois, striatocornis, subquadricornis, tetracornis.*

Tragelaphus Blainville, 1816. Bull. Sci. Soc. Philom. Paris, 1816:75.
TYPE SPECIES: *Antilope sylvatica* Sparrman, 1780 (= *Antilope scripta* Pallas, 1766).
SYNONYMS: *Ammelaphus, Boocercus, Calliope, Euryceros, Limnotragus, Nyala, Strepsicerastes, Strepsicerella, Strepsiceros.*
COMMENTS: Includes *Boocercus, Limnotragus, Nyala* and *Strepsiceros*; see Ansell (1972:20) and Van Gelder (1977*a, b*).

Tragelaphus angasii Gray, 1849. *In* Angas, Proc. Zool. Soc. Lond., 1848:89 [1849].
TYPE LOCALITY: South Africa, Natal, Zululand, St. Lucia Bay.
DISTRIBUTION: S Malawi; Mozambique; N and S Zimbabwe; Swaziland; Natal and E Transvaal (South Africa).

Tragelaphus buxtoni (Lydekker, 1910). Nature (London), 84:397.
TYPE LOCALITY: Ethiopia, Bak Prov., Sahatu Mtns, southeast of Lake Zwai, 9,000 ft. (2,743 m).
DISTRIBUTION: Ethiopia, east of Rift Valley.
STATUS: IUCN - Endangered.

Tragelaphus eurycerus (Ogilby, 1837). Proc. Zool. Soc. Lond., 1836:120 [1837].
TYPE LOCALITY: West Africa.
DISTRIBUTION: Sierra Leone to Benin; S Cameroon and N Gabon to Central African Republic, S Sudan and Zaire; S Kenya.
STATUS: CITES - Appendix III (Ghana).
SYNONYMS: *albovirgatus, cooperi, isaaci, katanganus.*

COMMENTS: *T. euryceros* is a later spelling; see Haltenorth (1963:86). Formerly placed in *Boocercus*. Reviewed by Ralls (1978, Mammalian Species, 111).

Tragelaphus imberbis (Blyth, 1869). Proc. Zool. Soc. Lond., 1869:55.
TYPE LOCALITY: "Abyssinia" (= Ethiopia).
DISTRIBUTION: Kenya, E Tanzania, NE Uganda, Somalia, SE Ethiopia, SE Sudan, Yemen, SW Saudi Arabia.
SYNONYMS: *australis*.
COMMENTS: Arabian records are based on only two specimens (Harrison and Bates, 1991:192).

Tragelaphus scriptus (Pallas, 1766). Misc. Zool., p. 8.
TYPE LOCALITY: Senegal.
DISTRIBUTION: S Mauritania to Ethiopia and S Somalia, and south to N Namibia and South Africa.
SYNONYMS: *bor, barkeri, brunneus; cottoni* Matschie, 1912; *dama, decula, delamerei, dianae, dodingae, eldomae, fasciatus, fulvoochraceus, haywoodi, heterochrous, insularis, johannae, knutsoni, laticeps, locorinae, makalae, massaicus, meneliki, meridionalis, meruensis, multicolor, nigrinotatus, obscurus, olivaceus, ornatus, phaleratus, pictus, powelli, punctatus, roualeynei, reidae, sassae, signatus, simplex, sylvaticus, tjaderi, uellensis.*

Tragelaphus spekii Sclater, 1863. *In* Speke, Journal of the Discovery of the Source of the Nile, p. 223.
TYPE LOCALITY: Tanzania, Karagwe, W of Lake Victoria, restricted to Bukoba district, Lake Lwelo, 2°S, 30°57'E by Moreau et al. (1946:441).
DISTRIBUTION: Gambia and Senegal; Togo to S Sudan, south to Angola, Caprivi Strip (NE Namibia), N Botswana and extreme NW of Zimbabwe, and east to W Kenya, W Tanzania and Zambia (and C Mozambique?).
STATUS: CITES - Appendix III (Ghana).
SYNONYMS: *albonotatus, gratus, inornatus, larkenii, selousi, sylvestris, ugallae, wilhelmi.*

Tragelaphus strepsiceros (Pallas, 1766). Misc. Zool., p. 9.
TYPE LOCALITY: South Africa, Cape of Good Hope; or Namibia, Gammafluss (= Lowen River). See Meester et al. (1986:217).
DISTRIBUTION: S Chad, N Central African Republic, W and E Sudan, NE Uganda, Ethiopia and Somalia south and southwest to South Africa, Namibia, Angola and SE Zaire.
SYNONYMS: *abyssinicus, bea, burlacei, capensis, chora; cottoni* Dollman and Burlace, 1928; *excelsus, frommi, hamiltoni, koodoo, torticornis, zambesiensis.*

Subfamily Caprinae Gray, 1821. London Med. Repos., 15:307.

Ammotragus Blyth, 1840. Proc. Zool. Soc. Lond., 1840:13.
TYPE SPECIES: *Antilope lervia* Pallas, 1777.
COMMENTS: Ansell (1972:70) included *Ammotragus* in *Capra*; but see comment under *Capra*.

Ammotragus lervia (Pallas, 1777). Spicil. Zool., 12:12.
TYPE LOCALITY: Algeria, Department of Oran.
DISTRIBUTION: Western Sahara to W Egypt; Mali to Sudan; introduced to USA, N Mexico and Spain.
STATUS: CITES - Appendix II; IUCN - Vulnerable.
SYNONYMS: *angusi, blainei, fassini, ornata, sahariensis, tragelaphus.*
COMMENTS: Reviewed by G.G. Gray and Simpson (1980, Mammalian Species, 144).

Budorcas Hodgson, 1850. J. Asiat. Soc. Bengal, 19:65.
TYPE SPECIES: *Budorcas taxicolor* Hodgson, 1850.

Budorcas taxicolor Hodgson, 1850. J. Asiat. Soc. Bengal, 19:65.
TYPE LOCALITY: India, Assam, Mishmi Hills.
DISTRIBUTION: Sikkim and Mishmi Hills (NE India); Bhutan; N Burma; SE Tibet, Sichuan, Gansu, Shaanxi and Yunnan (China).
STATUS: CITES - Appendix II; IUCN - Rare as *B. t. bedfordi*, Indeterminate as *B. t. tibetana*.
SYNONYMS: *bedfordi, mitchelli, sinensis, tibetana, whitei.*

COMMENTS: Reviewed by Neas and Hoffmann (1987, Mammalian Species, 277).

Capra Linnaeus, 1758. Syst. Nat., 10th ed., 1:68.
TYPE SPECIES: *Capra hircus* Linnaeus, 1758.
SYNONYMS: *Aegoceros, Aries, Eucapra, Euibex, Hircus, Ibex, Orthaegoceros, Tragus, Turocapra, Turus.*
COMMENTS: Includes *Orthaegoceros*; see Heptner et al. (1961:593). Various authors have included *Ammotragus* and *Ovis*; see Ansell (1972:70) and Van Gelder (1977*b*); probably should also include *Pseudois*. However, most authors have not followed this arrangement; see G.G. Gray and Simpson (1980), Gromov and Baranova (1981), Hall (1981), and Corbet and Hill (1991). There is no consensus concerning the number of species to be recognized in this genus; some would recognize only two (*hircus* and *falconeri*); see Haltenorth (1963), while others would recognize up to nine. Heptner et al. (1961) are followed here.

Capra caucasica Güldenstaedt and Pallas, 1783. Acta Acad. Sci. Petropoli, for 1779, 2:273 [1783].
TYPE LOCALITY: Russia, Caucasus Mtns, between Malka and Baksan Rivers, east of Mt. Elbrus.
DISTRIBUTION: W Caucasus Mtns (Russia).
SYNONYMS: *dinniki, raddei, severtzovi.*
COMMENTS: Possibly *caucasica* is the prior name for *cylindricornis*, in which case this species should be termed *severtzovi*; see Ellerman and Morrison-Scott (1951:407). See also comment under *cylindricornis*.

Capra cylindricornis (Blyth, 1841). Proc. Zool. Soc. Lond., 1840:68 [1841].
TYPE LOCALITY: Georgia, E Caucasus Mtns, Mt. Kasbek.
DISTRIBUTION: E Caucasus Mtns (Azerbaijan, E Georgia, Russia).
SYNONYMS: *pallasii.*
COMMENTS: Possibly a junior synonym of *caucasica*; see Gromov and Baranova (1981:404). See also comment under *caucasica*.

Capra falconeri (Wagner, 1839). Gelehrt. Anz. I. K. Bayer Akad. Wiss., München, 9:430.
TYPE LOCALITY: India, Kashmir, Astor.
DISTRIBUTION: Afghanistan; N and C Pakistan; N India; Kashmir; S Uzbekistan and Tadzhikistan.
STATUS: CITES - Appendix I *C. f. chialtanensis, C. f. falconeri, C. f. heptneri, C. f. jerdoni*, and *C. f. megaceros*; otherwise Appendix II; U.S. ESA - Endangered as *C. f. jerdoni, C. f. megaceros*, and in Chiltan Range of WC Pakistan as *C. aegagrus* (= *C. f. chiltanensis* [sic]); IUCN - Endangered as *C. f. megaceros*, otherwise Vulnerable.
SYNONYMS: *cashmiriensis, chialtanensis, chitralensis, gilgitensis, heptneri, jerdoni, megaceros, ognevi.*

Capra hircus Linnaeus, 1758. Syst. Nat., 10th ed., 1:68.
TYPE LOCALITY: Sweden (domesticated stock).
DISTRIBUTION: Turkey, Caucasus region, Turkmenistan, Iraq, Iran, SW Afghanistan, S Pakistan, Oman (feral?), extinct in Lebanon and Syria; domesticated worldwide; feral populations in British Isles, islands in the Mediterranean, USA, Canada, Chile, Argentina, Venezuela, Australia, New Zealand and many oceanic islands including Bonin, Hawaiian, Galapagos, Seychelles, and Juan Fernandez Isls.
SYNONYMS: *aegagrus, blythi, cilicica, cretica, dorcas, florstedti, gazella, jourensis, neglecta, persica, picta, turcmenica.*
COMMENTS: Includes *aegagrus*, but see Corbet (1978*c*:214).

Capra ibex Linnaeus, 1758. Syst. Nat., 10th ed., 1:68.
TYPE LOCALITY: Switzerland, Valais.
DISTRIBUTION: Formerly the mountains of France, Switzerland, Austria, Germany, and N Italy; extinct except in Italy but now reintroduced into much of its former range.
SYNONYMS: *alpina, europea, graicus.*

Capra nubiana F. Cuvier, 1825. *In* É. Geoffroy and F. Cuvier, Hist. Nat. Mammifères, pt. 3, 6(50):2 pp. "Bouc sauvage de la Haute-Egypte".
TYPE LOCALITY: Egypt, "Upper Egypt".

DISTRIBUTION: Egypt east of the Nile, NE Sudan, and Syria south to Yemen and east to Oman.
SYNONYMS: *arabica, beden, mengesi, sinaitica.*
COMMENTS: Treated as a species distinct from *C. ibex* by Uerpmann (1987).

Capra pyrenaica Schinz, 1838. N. Denkschr. Schweiz. Ges. Natur. Wiss., 2:9.
TYPE LOCALITY: Spain, Pyrenees Mtns, Huesca, near Maladetta Pass.
DISTRIBUTION: Iberian Peninsula.
STATUS: U.S. ESA and IUCN - Endangered as *C. p. pyrenaica.*
SYNONYMS: *cameranoi, hispanica, lusitanica, victoriae.*

Capra sibirica (Pallas, 1776). Spicil. Zool., 11:52.
TYPE LOCALITY: Russia, Siberia, Sayan Mtns, near Munku Sardyx.
DISTRIBUTION: Mountain ranges of N Afghanistan, N Pakistan, and N India through Tadzhikistan, Kirgizia, E Kazakhstan and S Siberia (Russia) to Sinkiang (China) and S and W Mongolia.
SYNONYMS: *alaiana, almasyi, altaica, dauvergnii, dementievi, fasciata, filippii, formosovi, hagenbecki, hemalayanus, lorenzi, lydekkeri, merzbacheri, pallasii, pedri, sakeen, transalaiana, wardi.*
COMMENTS: Treated as a species distinct from *C. ibex* by Heptner et al. (1961).

Capra walie Rüppell, 1835. Neue Wirbelt. Fauna Abyssin. Gehörig, Säugeth., 1:16.
TYPE LOCALITY: Ethiopia, mountains of Simien.
DISTRIBUTION: N Ethiopia.
STATUS: U.S. ESA and IUCN - Endangered.
COMMENTS: Treated as a species distinct from *C. ibex* by Ansell (1972:70) and Yalden et al. (1984).

Hemitragus Hodgson, 1841. Calcutta J. Nat. Hist., 2:218.
TYPE SPECIES: *Capra jharal* Hodgson, 1833 (= *Capra jemlahica* H. Smith, 1826).

Hemitragus hylocrius (Ogilby, 1838). Proc. Zool. Soc. Lond., 1837:81 [1838].
TYPE LOCALITY: India, Nilgiri Hills.
DISTRIBUTION: SW and S India.
STATUS: IUCN - Vulnerable.
SYNONYMS: *warryato.*
COMMENTS: Included in *jemlahicus* by Haltenorth (1963:125) but generally regarded as a full species, for example by Corbet and Hill (1991).

Hemitragus jayakari Thomas, 1894. Ann. Mag. Nat. Hist., ser. 6, 13:365.
TYPE LOCALITY: Oman, Jebel Akhdar Range, Jebel Taw.
DISTRIBUTION: Oman, United Arab Emirates.
STATUS: U.S. ESA and IUCN - Endangered.
COMMENTS: Included in *jemlahicus* by Haltenorth (1963:125) but see Harrison (1968:324).

Hemitragus jemlahicus (H. Smith, 1826). *In* Griffith et al., Animal Kingdom, Vol. 4, plate [1826] opp. p. 308 [1827].
TYPE LOCALITY: Nepal, Jemla Hills.
DISTRIBUTION: Himalayas, from Kashmir through N India to Nepal and Sikkim; S Tibet (China). Introduced in New Zealand and SW Cape Prov. (South Africa).
SYNONYMS: *jharal, quadrimammis, schaeferi, tubericornis.*
COMMENTS: Specific name is spelt "*jemlanica*" on p. 308 and "*jemlahica*" in legend to the plate on the opposite unnumbered page, dated 1826, in the original description which was published in 1827.

Naemorhedus H. Smith, 1827. *In* Griffith et al., Animal Kingdom, 5:352.
TYPE SPECIES: *Antilope goral* Hardwicke, 1825.
SYNONYMS: *Austritragus, Capricornis, Capricornulus, Caprina, Kemas, Lithotragus, Urotragus.*
COMMENTS: The original spelling is "*Naemorhedus*". *Nemorhaedus, Naemorhaedus, Nemorhedus,* and *Nemorrhedus* are later spellings. Reviewed by J.M. Dolan (1963); and also by Groves and Grubb (1985) who included *Capricornis* in this genus.

Naemorhedus baileyi Pocock, 1914. J. Bombay Nat. Hist. Soc., 23:32.
TYPE LOCALITY: China, Tibet, Bomi, Dre on banks of Yigron Tso.
DISTRIBUTION: SE Tibet and Yunnan (China); N Burma; Assam (NE India).
STATUS: IUCN - Vulnerable.
SYNONYMS: *cranbrooki.*
COMMENTS: Regarded as a valid species by Groves and Grubb (1985); and by Zhang Cizu (1987, under the name *cranbrooki*).

Naemorhedus caudatus (Milne-Edwards, 1867). Ann. Sci. Nat. Zool. (Paris), 7:377.
TYPE LOCALITY: Russia, Amurland, Bureja Mtns.
DISTRIBUTION: Soviet Far East (Russia), E China, E Burma, W Thailand.
SYNONYMS: *aldridgeanus, arnouxianus, cinerea, curvicornis, evansi, fantozatianus; fargesianus* Heude, 1894; *galeanus, griseus, henryanus, initialis, iodinus, niger, pinchonianus, raddeanus, versicolor; vidianus* Heude, 1894; *xanthodeiros.*
COMMENTS: Regarded as a species distinct from *N. goral* by Groves and Grubb (1985).

Naemorhedus crispus (Temminck, 1845). Fauna Japonica, 1(Mamm.), p. 55, pl. 18,19.
TYPE LOCALITY: Japan, Honshu.
DISTRIBUTION: Honshu, Shikoku and Kyushu (Japan).
SYNONYMS: *pryerianus, saxicola.*
COMMENTS: Included in *sumatraensis* by Haltenorth (1963:119); but a valid species according to Dolan (1963).

Naemorhedus goral (Hardwicke, 1825). Trans. Linn. Soc. Lond., 14:518.
TYPE LOCALITY: Nepal, in the Himalayas.
DISTRIBUTION: N Pakistan, N India, Nepal, Bhutan, Sikkim (NE India).
STATUS: CITES - Appendix I; U.S. ESA - Endangered.
SYNONYMS: *bedfordi, duvaucelii, hodgsoni.*
COMMENTS: Reviewed by J.I. Mead (1989, Mammalian Species, 335, as *Nemorhaedus goral*).

Naemorhedus sumatraensis (Bechstein, 1799). *In* Pennant, Allgemeine, Ueber. Vierfuss. Thiere, 1:98.
TYPE LOCALITY: Indonesia, Sumatra.
DISTRIBUTION: N India; Nepal; Sikkim (NE India); Burma; China north to Gansu and Anhui; Thailand; Indochina; Malay Peninsula; Sumatra.
STATUS: CITES - Appendix I; U.S. ESA - Endangered; IUCN - Endangered as *Capricornis sumatraensis sumatraensis.*
SYNONYMS: *annectens, argyrochaetes, benetianus, brachyrhinus, bubalina, chrysochaetes, collasinus, cornutus, edwardsii, erythropygius; fargesianus* Heude, 1894; *gendrelianus, humei, jamrachi, longicornis, marcolinus, maritimus, maxillaris, microdontus, milneedwardsii, montinus, nasutus, osborni, platyrhinus, pugnax, robinsoni, rocherianus, rodoni, rubidus, swettenhami, thar, ungulosus; vidianus* Heude, 1894.

Naemorhedus swinhoei (Gray, 1862). Ann. Mag. Nat. Hist., ser. 3, 10:320.
TYPE LOCALITY: Taiwan.
DISTRIBUTION: Taiwan.
STATUS: IUCN - Vulnerable.
COMMENTS: Regarded as a species distinct form *N. crispus* by Groves and Grubb (1985).

Oreamnos Rafinesque, 1817. Am. Mon. Mag., 2:44.
TYPE SPECIES: *Mazama dorsata* Rafinesque, 1817 (= *R[upicapra] americanus* Blainville, 1816).

Oreamnos americanus (de Blainville, 1816). Bull. Sci. Soc. Philom. Paris, 1816:80.
TYPE LOCALITY: USA, Washington, Mt. Adams.
DISTRIBUTION: SE Alaska (USA), S Yukon and SW Mackenzie (Canada) to NC Oregon, C Idaho, and Montana (USA). Introduced to Kodiak, Chichagof, and Baranof Isls (Alaska), Olympic Peninsula (Washington), C Montana, Black Hills (South Dakota), and Colorado (USA).
SYNONYMS: *columbiae, columbiana, dorsata, kennedyi, lanigera, missoulae, montanus, sericea.*
COMMENTS: Reviewed by Rideout and Hoffmann (1975, Mammalian Species, 63).

Ovibos Blainville, 1816. Bull. Sci. Soc. Philom. Paris, 1816:76.
TYPE SPECIES: *Bos moschatus* Zimmerman, 1780.

Ovibos moschatus (Zimmermann, 1780). Geogr. Gesch. Mensch. Vierf. Thiere, 2:86.
TYPE LOCALITY: Canada, Manitoba, between Seal and Churchill Rs.
DISTRIBUTION: Formerly Point Barrow, Alaska (USA) east to NE Greenland, south to NE Manitoba (Canada). Range now much reduced. Introduced to Seward Peninsula and Nunivak Isl, Alaska (USA); Taimyr Peninsula and Wrangel Isl (Russia); Svalbard (Norway).
SYNONYMS: *mackenzianus, melvillensis, niphoecus, pearyi, wardi.*
COMMENTS: Reviewed by Lent (1988, Mammalian Species, 302).

Ovis Linnaeus, 1758. Syst. Nat., 10th ed., 1:70.
TYPE SPECIES: *Ovis aries* Linnaeus, 1758.
SYNONYMS: *Ammon, Argali, Caprovis, Musimon, Pachyceros.*
COMMENTS: Placed in *Capra* by Van Gelder (1977*b*); see comments under *Capra*. There is no consensus concerning the number of species to be recognized in this genus; some would recognize only one (*ammon*; see Haltenorth, 1963:126-128); others two (*ammon, canadensis*; see Corbet, 1978*c*:218); while others recognize up to seven, as do the most recent reviews (Korobitsyna et al., 1974; Nadler et al., 1973). Six species are listed here.

Ovis ammon (Linnaeus, 1758). Syst. Nat., 10th ed., 1:70.
TYPE LOCALITY: Kazakhstan, Vostochno-Kazakhstansk. Obl., Altai Mtns, Bukhtarma; near Ust-Kamenogorsk.
DISTRIBUTION: S Siberia, E Kazakhstan, Kirgizia, and Tadzhikistan, south to Pamir Range in NE Afghanistan and N Pakistan; Ladak (N India); Nepal; Sikkim (NE India); Sinkiang and Tibet to Inner Mongolia (China); Mongolia.
STATUS: CITES - Appendix I *O. a. hodgsoni*, otherwise Appendix II; U.S. ESA - Endangered as *O. a. hodgsoni.*
SYNONYMS: *adametzi, altaica, ammonoides, argali, asiaticus, bambhera, blythi, brookei, collium, comosa, dalailamae, darwini, dauricus, heinsii, henrii, hodgsonii, humei, intermedia; jubata* Peters, 1876; *karelini, kozlovi, littledalei; mongolica* Severtzov, 1873; *nigrimontana, polii, przevalskii, sairensis.*
COMMENTS: Haltenorth (1963:121) and Corbet (1978*c*:218) included *orientalis* (= *aries*), *musimon* and *vignei*; but see review by Nadler et al. (1973) and Corbet and Hill (1991:136). Subspecies reviewed by Sopin (1982).

Ovis aries Linnaeus, 1758. Syst. Nat., 10th ed., 1:70.
TYPE LOCALITY: Sweden (domesticated stock).
DISTRIBUTION: S and E Turkey, Armenia, S Azerbaijan, N Iraq, W Iran. Domesticated worldwide; primitive domestic populations (mouflon) feral on Corsica and Sardinia, introduced from there to Europe, Crimea (USSR), USA (incl. Hawaiian Isls), Chile, Kerguelen Isls, Tenerife (Canary Isls); on Cyprus; and on St. Kilda and other small islands off the British Isles; improved domestic stock feral in Norway, Sweden, USA, islands off coasts of British Isles and New Zealand, Kerguelen Isls, and probably other oceanic islands.
STATUS: CITES - Appendix I as *O. orientalis ophion*; U.S. ESA - Endangered as *O. musimon ophion*; IUCN - Vulnerable as *O. orientalis musimon* and *O. orientalis ophion.*
SYNONYMS: *anatolica, armeniana, corsicosardinensis, cyprius, erskinei, gmelini, laristanica, isphaganica; jubata* Kerr, 1792; *matschiei; mongolica* Fitzinger, 1860; *musimon, occidentalis, occidentosardinensis, ophion, orientalis, sinesella, urmiana.*
COMMENTS: Includes *orientalis*; see Nadler et al. (1973). Also includes *musimon* and *ophion*, primitive domestic sheep, now feral; see S. Payne (1968), Vigne (1988), and Hemmer (1990). Hybridizes with *vignei* in C Iran; see Nadler et al. (1971*b*) and Valdez et al. (1978).

Ovis canadensis Shaw, 1804. Nat. Misc., 51, text to pl. 610.
TYPE LOCALITY: Canada, Alberta, Mountains on Bow River, near Exshaw.
DISTRIBUTION: S British Columbia and SW Alberta (Canada) to Coahuila, Chihuahua, Sonora and Baja California (Mexico).
STATUS: CITES - Appendix II for Mexican population.
SYNONYMS: *auduboni, californiana, cervina, cremnobates, gaillardi, mexicana, montana, nelsoni, palmeri, pygargus, samilkameenensis, sheldoni, sierrae, texianus, weemsi.*

COMMENTS: Corbet (1978*c*:218) included *nivicola*; but see also Korobitsyna et al. (1974) and Corbet and Hill (1991:135). Revised by Cowan (1940). Reviewed by Shackleton (1985, Mammalian Species, 230).

Ovis dalli Nelson, 1884. Proc. U.S. Natl. Mus., 7:13.
TYPE LOCALITY: Alaska, west bank of Yukon River, Tanana Hills.
DISTRIBUTION: Alaska to N British Columbia and W Mackenzie (Canada).
SYNONYMS: *cowani, fannini, kenaiensis, liardensis, niger, stonei.*
COMMENTS: Revised by Cowan (1940). Reviewed by Bowyer and Leslie (1992, Mammalian Species, 393).

Ovis nivicola Eschscholtz, 1829. Zool. Atlas, Part 1, p. 1, pl. 1.
TYPE LOCALITY: Russia, E Kamchatka.
DISTRIBUTION: Putorana Mtns, NC Siberia; NE Siberia from Lena River east to Chukotka and Kamchatka (Russia).
SYNONYMS: *alleni, borealis, koriakorum, lydekkeri, middendorfi, potanini, storcki.*
COMMENTS: Corbet (1978*c*:218) and others included *nivicola* in *canadensis*; but see Korobitsyna et al. (1974) and Gromov and Baranova (1981:407).

Ovis vignei Blyth, 1841. Proc. Zool. Soc. Lond., 1840:70 [1841].
TYPE LOCALITY: India, Kashmir, Astor.
DISTRIBUTION: Uzbekistan, Tadzhikistan, and NE Iran to Afghanistan, Pakistan, and NW India; Oman (introduced?).
STATUS: CITES - Appendix I; U.S. ESA - Endangered as *O. v. vignei*.
SYNONYMS: *arabica, arkal, blanfordi, bochariensis, cycloceros, dolgopolovi, punjabiensis, severtzovi, varentsowi.*
COMMENTS: Hybridizes with *orientalis* (= *aries*) in C Iran; see Nadler et al. (1971*b*) and Valdez et al. (1978).

Pseudois Hodgson, 1846. J. Asiat. Soc. Bengal, 15:343.
TYPE SPECIES: *Ovis nayaur* Hodgson, 1833.
COMMENTS: Revised by Groves (1978*c*). Reviewed by Wang Xiao-ming and Hoffmann (1987).

Pseudois nayaur (Hodgson, 1833). Asiat. Res., 18(2):135.
TYPE LOCALITY: Nepal, Tibetan frontier.
DISTRIBUTION: Pamir Range in Tadzhikistan, N Pakistan, Ladak (N India) and Tibet (China) east to Yunnan, Sichuan, Gansu and Shaanxi (China).
SYNONYMS: *burrhel, caesia, nahoor, szechuanensis.*
COMMENTS: See Wang Xiao-ming and Hoffmann (1987, Mammalian Species, 278).

Pseudois schaeferi Haltenorth, 1963. Handb. Zool., 8(32):126.
TYPE LOCALITY: China, upper Yangtze Gorge, Drupalong, south of Batang.
DISTRIBUTION: Upper Yangtze Gorge (W China).
COMMENTS: A separate species according to Groves (1978*c*:183). See Wang Xiao-ming and Hoffmann (1987, Mammalian Species, 278).

Rupicapra Blainville, 1816. Bull. Sci. Soc. Philom. Paris, 1816:75.
TYPE SPECIES: *Capra rupicapra* Linnaeus, 1768.
SYNONYMS: *Capella.*
COMMENTS: Revised by Lovari and Scala (1980, 1984), Scala and Lovari (1984), and Nascetti et al. (1985).

Rupicapra pyrenaica Bonaparte, 1845. Cat. Meth. Mamm. Europe, p. 17.
TYPE LOCALITY: Spain, Pyrenees.
DISTRIBUTION: N Spain; Appenine Mtns (Italy).
STATUS: CITES - Appendix I and U.S. ESA - Endangered as *R. rupicapra ornata*; IUCN - Vulnerable as *R. p. ornata*.
SYNONYMS: *ornata, parva.*
COMMENTS: Regarded as a species distinct from *R. rupicapra* by Lovari (1985, 1987).

Rupicapra rupicapra (Linnaeus, 1758). Syst. Nat., 10th ed., 1:68.
TYPE LOCALITY: Switzerland.

DISTRIBUTION: Alps of France and Switzerland east to Carpathians (Romania), Tatra Mtns (Czechoslovakia), Caucasus Mtns, and Turkey; introduced to New Zealand.
STATUS: IUCN - Endangered as *R. r. cartusiana*, Rare as *R. r. tatrica*.
SYNONYMS: *alpina, asiatica, balcanica, capella, carpatica, cartusiana, caucasica, dorcas, europea, faesula, hamulicornis, olympica, sylvatica, tatrica, tragus.*

Subfamily Cephalophinae Gray, 1871. Proc. Zool. Soc. Lond., 1871:588.

Cephalophus H. Smith, 1827. *In* Griffith et al., Animal Kingdom, 5:344.
TYPE SPECIES: *Antilope silvicultrix* Afzelius, 1815.
SYNONYMS: *Cephalophella, Cephalophia, Cephalophidium, Cephalophops, Cephalophorus, Cephalophula, Guevei, Philantomba, Potamotragus, Terpone.*
COMMENTS: Van Gelder (1977*b*) included *Sylvicapra*. However, recent authors have not followed this arrangement; see Swanepoel et al. (1980:188) and Meester et al. (1986).

Cephalophus adersi Thomas, 1918. Ann. Mag. Nat. Hist., ser. 9, 2:151.
TYPE LOCALITY: Tanzania, Zanzibar.
DISTRIBUTION: Zanzibar (Tanzania), Sokoke Forest (Kenya).
STATUS: IUCN - Vulnerable.
COMMENTS: Possibly conspecific with *natalensis* and/or *callipygus*; see Ansell (1972:33).

Cephalophus callipygus Peters, 1876. Monatsb. K. Preuss. Akad. Wiss. Berlin, 1876:483.
TYPE LOCALITY: Gabon, Gabon River.
DISTRIBUTION: West of Congo and Ubangi Rivers in Congo Republic; S Central African Republic; Gabon and S Cameroon.
COMMENTS: Possibly conspecific with *natalensis* and/or *adersi*; see Ansell (1972:33). According to Groves and Grubb (1974), *callipygus* is not related to *natalensis*.

Cephalophus dorsalis Gray, 1846. Ann. Mag. Nat. Hist., [ser. 1], 18:165.
TYPE LOCALITY: Sierra Leone.
DISTRIBUTION: Sierra Leone to Togo; SE Nigeria to Central African Republic and Zaire, south to N Angola.
STATUS: CITES - Appendix II.
SYNONYMS: *arrhenii, badius, breviceps, castaneus, kuha, leucochilus, orientalis.*

Cephalophus harveyi (Thomas, 1893). Ann. Mag. Nat. Hist., ser. 6, 11:48.
TYPE LOCALITY: Tanzania, Kilimanjaro Dist., Kahe Forest west of Taveta.
DISTRIBUTION: S Somalia, E Kenya, E and S Tanzania, N Malawi, E. Zambia, sight records from E Ethiopia.
SYNONYMS: *bottegoi, kenniae.*
COMMENTS: Treated as a species separate from C. *natalensis* by Kingdon (1982:297).

Cephalophus jentinki Thomas, 1892. Proc. Zool. Soc. Lond., 1892:417.
TYPE LOCALITY: Liberia.
DISTRIBUTION: Sierra Leone, Liberia, W Ivory Coast.
STATUS: CITES - Appendix I; U.S. ESA and IUCN - Endangered.
COMMENTS: Reviewed by Kuhn (1968).

Cephalophus leucogaster Gray, 1873. Ann. Mag. Nat. Hist., ser. 4, 12:43.
TYPE LOCALITY: Gabon.
DISTRIBUTION: Cameroon, Gabon, Congo Republic, SW and E Zaire.
SYNONYMS: *seke.*

Cephalophus maxwellii (H. Smith, 1827). *In* Griffith et al., Animal Kingdom, 4:267.
TYPE LOCALITY: Sierra Leone.
DISTRIBUTION: Senegal and Gambia to SW Nigeria.
SYNONYMS: *danei, frederici, liberiensis, lowei, philantomba, whitfieldi.*
COMMENTS: Included in *monticola* by Haltenorth and Diller (1977:43). Reviewed by Ralls (1973, Mammalian Species, 31).

Cephalophus monticola (Thunberg, 1789). Resa uti Europa, Africa, Asia..., 2:66.
TYPE LOCALITY: South Africa, Cape Province, Knysna District, Langkloof (= "Lange Kloof"), 33°40'S, 23°40'E; see Meester et al. (1986:196).

DISTRIBUTION: E Nigeria to Kenya and Tanzania, south to Angola, Zambia, Malawi, E Zimbabwe and Mozambique; Natal and E Cape Province (South Africa); Zanzibar; Bioko; Pemba Isl.
STATUS: CITES - Appendix II.
SYNONYMS: *aequatorialis, anchietae, bakeri, bicolor, caerula, caffer, congicus, defriesi, fuscicolor, hecki, ludlami, lugens, melanorheus, musculoides, minuta, nyasae, pembae, perpusilla, ruddi, schultzei, schusteri, simpsoni, sundevalli.*
COMMENTS: May include *maxwellii*; see Haltenorth and Diller (1977:43).

Cephalophus natalensis A. Smith, 1834. S. Afr. Quart. J., 2:217.
TYPE LOCALITY: South Africa, Natal, Durban ("Port Natal").
DISTRIBUTION: S Tanzania, S Malawi, Mozambique, Natal (South Africa).
SYNONYMS: *amoenus, bradshawi, lebombo, robertsi, vassei.*
COMMENTS: Possibly includes *adersi* and/or *callipygus*; see Ansell (1972:34), who also included *weynsi*.

Cephalophus niger Gray, 1846. Ann. Mag. Nat. Hist., [ser. 1], 18:165.
TYPE LOCALITY: "Guinea Coast" = Ghana.
DISTRIBUTION: Guinea to Nigeria, west of lower Niger River.
SYNONYMS: *pluto.*

Cephalophus nigrifrons Gray, 1871. Proc. Zool. Soc. Lond., 1871:598.
TYPE LOCALITY: Gabon.
DISTRIBUTION: Cameroon to Zaire, Rwanda and W Uganda, south to N Angola; Mt. Elgon (Uganda/Kenya); Aberdare Range and Mt. Kenya (Kenya).
SYNONYMS: *apanbanga, aureus, claudi, emini, fosteri, hooki, kivuensis, lusumbi, mixtus.*

Cephalophus ogilbyi (Waterhouse, 1838). Proc. Zool. Soc. Lond., 1838:60.
TYPE LOCALITY: Equatorial Guinea, Fernando Po (= Bioko).
DISTRIBUTION: Sierra Leone, Liberia, W Ivory Coast, W Ghana, SE Nigeria, S Cameroon, Bioko, S Gabon.
STATUS: CITES - Appendix II; IUCN - Vulnerable.
SYNONYMS: *brookei, crusalbum.*

Cephalophus rubidus Thomas, 1901. Proc. Zool. Soc. Lond., 1901(2):89.
TYPE LOCALITY: Uganda, Ruwenzori District.
DISTRIBUTION: Ruwenzori Mtns in E Zaire and W Uganda.
STATUS: IUCN - Endangered.
COMMENTS: Regarded as a species separate from *C. nigrifrons* by Kingdon (1982:292).

Cephalophus rufilatus Gray, 1846. Ann. Mag. Nat. Hist., [ser. 1], 18:166.
TYPE LOCALITY: Sierra Leone, Waterloo Village.
DISTRIBUTION: Senegal to SW Sudan and NE Uganda south to Cameroon and N Zaire.
SYNONYMS: *cuvieri, rubidior.*

Cephalophus silvicultor (Afzelius, 1815). Nova Acta Reg. Soc. Sci. Upsala, 7:265, pl. 8, fig. 1.
TYPE LOCALITY: "Sierra Leone and region of Pongas and Quia Rivers [Guinea]."
DISTRIBUTION: Senegal to SW Sudan, W Uganda and Rwanda, south to Angola and Zambia; W Kenya.
STATUS: CITES - Appendix II.
SYNONYMS: *coxi, ituriensis, longiceps, melanoprymnus, punctulatus, ruficrista, sclateri, thomasi.*
COMMENTS: Reviewed by Lumpkin and Kranz (1984, Mammalian Species, 225, as *Cephalophus sylvicultor*).

Cephalophus spadix True, 1890. Proc. U.S. Natl. Mus., 13:227.
TYPE LOCALITY: Tanzania, Mt. Kilimanjaro.
DISTRIBUTION: Highlands of NE and C Tanzania.
STATUS: IUCN - Vulnerable.
COMMENTS: Possibly a subspecies of *silvicultor* (Haltenorth, 1963:71).

Cephalophus weynsi Thomas, 1901. Ann. Mus. Congo Zool., 2(1):15.
TYPE LOCALITY: Zaire, near Stanley Falls.
DISTRIBUTION: Zaire, Uganda, Rwanda, W Kenya.
SYNONYMS: *barbertoni, centralis, ignifer, johnstoni, leopoldi, lestradei, rutshuricus.*

COMMENTS: Formerly included in *callipygus*, but separate according to Groves and Grubb (1974).

Cephalophus zebra Gray, 1838. Ann. Nat. Hist., 1:27.
TYPE LOCALITY: Sierra Leone.
DISTRIBUTION: E Sierra Leone, Liberia, W Ivory Coast.
STATUS: CITES - Appendix II; IUCN - Vulnerable.
SYNONYMS: *doria, zebrata.*
COMMENTS: For synonyms see Ansell (1980) and the International Commission on Zoological Nomenclature (1985*a*). Reviewed by Kuhn (1966).

Sylvicapra Ogilby, 1837. Proc. Zool. Soc. Lond., 1836:138 [1837].
TYPE SPECIES: *Antilope mergens* Desmarest, 1816 (= *Capra grimmia* Linnaeus, 1758).
SYNONYMS: *Cephalophora, Grimmia.*
COMMENTS: Included in *Cephalophus* by Haltenorth (1963:71) and Van Gelder (1977*b*:18); but see Ansell (1978:57), Swanepool et al. (1980:188), and Meester et al. (1986).

Sylvicapra grimmia (Linnaeus, 1758). Syst. Nat., 10th ed., 1:70.
TYPE LOCALITY: South Africa, Cape Prov., Capetown.
DISTRIBUTION: Senegal to Ethiopia and S Somalia, south to N and E Zaire, Rwanda and Tanzania; S Gabon, S Congo Republic and S Zaire south to South Africa.
SYNONYMS: *abyssinica, altifrons, altivallis, bradfieldi, burchellii, caffra, campbelliae, cana, coronata, cunenensis, deserti, flavescens, hindei, irrorata, leucoprosopus, lobeliarum, lutea, madoqua, mergens, nictitans, noomei, nyansae, ocularis, omurambae, pallidior, platous, ptoox, roosevelti, shirensis, splendidula, steinhardti, transvaalensis, ugabensis, uvirensis, vernayi, walkeri.*

Subfamily Hippotraginae Brooke, 1876. *In* Wallace, Geog. Dist. Animals, 2:223.

Addax Rafinesque, 1815. Analyse de la Nature, p. 56.
TYPE SPECIES: *Cerophorus nasomaculata* Blainville, 1816.

Addax nasomaculatus (Blainville, 1816). Bull. Sci. Soc. Philom. Paris, 1816:75.
TYPE LOCALITY: Probably Senegambia.
DISTRIBUTION: Mauritania to Sudan; formerly Egypt to Tunisia.
STATUS: CITES - Appendix I; IUCN - Endangered.
SYNONYMS: *addax, gibbosa, mytilopes, suturosa.*
COMMENTS: Nearly extinct in wild (East, 1990).

Hippotragus Sundevall, 1846. K. Svenska Vet.-Akad. Handl. Stockholm, for 1844, p. 196 [1846].
TYPE SPECIES: *Antilope leucophaea* Pallas, 1766.
SYNONYMS: *Aigererus, Aigocerus, Egocerus, Ozanna.*

Hippotragus equinus (Desmarest, 1804). Dict. Class. Hist. Nat. Paris, p. 24.
TYPE LOCALITY: N South Africa, Lataku (= Takoon); see Harper (1940).
DISTRIBUTION: Senegal to W Ethiopia; south to N South Africa, N Botswana and Namibia.
SYNONYMS: *aethiopica, aurita, bakeri, barbata, cottoni, docoi, dogetti, gambianus, jubata, koba, langheldi, rufopallidus, scharicus, truteri.*

Hippotragus leucophaeus (Pallas, 1766). Misc. Zool., p. 4.
TYPE LOCALITY: South Africa, Cape of Good Hope, Swellendam Dist.
DISTRIBUTION: S Cape Province (South Africa); extirpated about 1799.
STATUS: IUCN - Extinct.
SYNONYMS: *capensis, glauca.*
COMMENTS: Reviewed by Mohr (1967).

Hippotragus niger (Harris, 1838). Athenaeum, 535:71.
TYPE LOCALITY: South Africa, Transvaal, Cashan Range (= Magaliesberge), near Pretoria.
DISTRIBUTION: SE Kenya to N South Africa, west to N Botswana; Angola, between Cuanza and Loando Rs.
STATUS: CITES - Appendix I, and U.S. ESA and IUCN - Endangered, as *H. n. variani.*

SYNONYMS: *anselli, harrisi, kaufmanni, kirkii, roosevelti, variani.*
COMMENTS: Includes *variani*; see Ansell (1972:47). Original publication usually assumed to be Proc. Zool. Soc. Lond., 1838:2 (publ. July, 1838), but McAllan and Bruce (1989) have shown that an earlier publication is The Athenaeum (publ. 27 Jan., 1838).

Oryx Blainville, 1816. Bull. Sci. Soc. Philom. Paris, 1816:75.
TYPE SPECIES: *Antilope oryx* Pallas, 1777 (= *Capra gazella* Linnaeus, 1758).
SYNONYMS: *Aegoryx.*

Oryx dammah (Cretzschmar, 1827). *In* Rüppell, Atlas Reise Nordl. Afr., Zool., Säugheth., p. 22.
TYPE LOCALITY: Sudan, Haraza, "probably Kordofan."
DISTRIBUTION: Formerly Western Sahara and Tunisia to Egypt; Mauritania to Sudan; survives only in Chad.
STATUS: CITES - Appendix I; IUCN - Endangered.
SYNONYMS: *algazel, ensicornis, nubica, senegalensis, tao.*
COMMENTS: Includes *tao*; see Ansell (1972:48). The name *algazel* Oken, 1816, was declared invalid by Opinion 417 of the International Commission on Zoological Nomenclature (1956*b*).

Oryx gazella (Linnaeus, 1758). Syst. Nat., 10th ed., 1:69.
TYPE LOCALITY: South Africa.
DISTRIBUTION: NE Ethiopia and SE Sudan to Somalia, NE Uganda and N Tanzania; SW Angola, Botswana and W Zimbabwe to N South Africa.
SYNONYMS: *annectens, aschenborni, beisa, bezoartica, blainei, callotis, capensis, gallarum, pasan, recticornis, subcallotis.*
COMMENTS: Includes *beisa*; see Ansell (1972:49).

Oryx leucoryx (Pallas, 1777). Spicil. Zool., 12:17.
TYPE LOCALITY: Arabia.
DISTRIBUTION: SE Arabian Peninsula; formerly Iraq. Probably became extinct in the wild, but maintained in captivity, and recently reintroduced into Arabia.
STATUS: CITES - Appendix I; U.S. ESA and IUCN - Endangered.
SYNONYMS: *asiatica, beatrix, latipes, pallasii.*
COMMENTS: Included in *O. gazella* by Haltenorth (1963:88).

Subfamily Peleinae Gray, 1872. Cat. Ruminant Mamm. Brit. Mus., p. 3, 29.

Pelea Gray, 1851. Proc. Zool. Soc. Lond., 1850:126 [1851].
TYPE SPECIES: *Antelope* (sic) *capreolus* Forster, 1790.

Pelea capreolus (Forster, 1790). *In* Levaillant, Erste Reise Afrika, p. 71.
TYPE LOCALITY: South Africa, Cape Prov., Caledon, Houhoek (= "Ouwe-hoeck"); see Skead (1973:79).
DISTRIBUTION: Cape Prov. to Natal and Transvaal (South Africa); Lesotho and Swaziland.
SYNONYMS: *lanata, villosa.*

Subfamily Reduncinae Knottnerus-Meyer, 1907. Arch. Naturgesch., 73:39.

Kobus A. Smith, 1840. Illustr. Zool. S. Afr. Mamm., Part 12, pl. 28 plus text.
TYPE SPECIES: *Antilope ellipsiprymnus* Ogilby, 1833.
SYNONYMS: *Adenota, Hydrotragus, Onotragus, Pseudokobus.*
COMMENTS: Includes *Adenota* and *Onotragus*; see Ansell (1972:40).

Kobus ellipsiprymnus (Ogilby, 1833). Proc. Zool. Soc. Lond., 1833:47.
TYPE LOCALITY: South Africa, between Lataku (near Kuruman) and W coast of Africa, N of Orange River, on Molopo River.
DISTRIBUTION: Senegal to Somalia, Kenya and Tanzania; S Gabon, S Congo Republic and S Zaire, south to Caprivi Strip (Namibia), N and E Botswana and NE South Africa.
SYNONYMS: *abyssinica; adolfifriderici* Matschie, 1910; *albertensis, angusticeps, annectens, avellanifrons, breviceps, canescens, cottoni, crawshayi, defassa, dianae, frommi, fulvifrons, griseotinctus, harnieri, hawashensi, kondensis, kulu, kuru, ladoensis, lipuwa, matschiei,*

muenzneri, nzoiae, pallidus, penricei, powelli, raineyi, schubotzi, senegalensis, singsing, thikae, tjaederi, togoensis, tschadensis, ugandae, unctuosus, uwendensis.

COMMENTS: Includes *defassa*; see Ansell (1972:42).

Kobus kob (Erxleben, 1777). Syst. Regni Anim., 1:293.

TYPE LOCALITY: Upper Guinea, towards Senegal.

DISTRIBUTION: Senegal to W Ethiopia and Sudan; south to N Zaire, Uganda, W Kenya and NW Tanzania. Now extinct in Tanzania.

SYNONYMS: *adansoni, adenota, adolfi; adolfifriderici* Schwarz, 1913; *alurae, annulipes, bahrkeetae, buffonii, forfex, fraseri, kul, leucotis, loderi, neumanni, nigricans, nigroscapulatus; notatus* Rothschild, 1913; *pousarguesi, riparia, thomasi, ubangiensis, vaughani.*

Kobus leche Gray, 1850. Gleanings, Knowsley Menagerie, 2:23.

TYPE LOCALITY: Botswana (= Bechuanaland), Botletle (= Zoaga) River, near Lake Ngami.

DISTRIBUTION: N Botswana, NE Namibia, SE Angola, SE Zaire and Zambia.

STATUS: CITES - Appendix II; U.S. ESA - Threatened; IUCN - Vulnerable.

SYNONYMS: *amboellensis, grandicornis, kafuensis; notatus* Matschie, 1912; *robertsi, smithemani.*

Kobus megaceros (Fitzinger, 1855). Sitzb. K. Akad. Wiss. Wien, 17:247.

TYPE LOCALITY: Sudan, Bahr-el-Ghazal, Sobat River.

DISTRIBUTION: S Sudan, W Ethiopia.

SYNONYMS: *maria.*

Kobus vardonii (Livingstone, 1857). Missionary Travels and Researches in South Africa, p. 256.

TYPE LOCALITY: Zambia, Barotseland, Chobe Valley, near Libonta (40°30'S, 23°15'E).

DISTRIBUTION: S Zaire, Zambia, Malawi and S Tanzania south to N Angola, NE Namibia, N Botswana; vagrant in N Zimbabwe.

SYNONYMS: *senganus.*

COMMENTS: Included in *kob* by Haltenorth (1963:92) but see Ansell (1972:44).

Redunca H. Smith, 1827. *In* Griffith et al., Animal Kingdom, 5:337.

TYPE SPECIES: *Antilope redunca* Pallas, 1767.

SYNONYMS: *Cervicapra, Eleotragus, Nagor, Oreodorcas.*

Redunca arundinum (Boddaert, 1785). Elench. Anim., 1:141.

TYPE LOCALITY: South Africa, Cape of Good Hope, Bathurst Division. See A. Roberts (1951:292).

DISTRIBUTION: S Gabon, S Congo Republic, S Zaire and Tanzania, south to N Namibia, N and E Botswana, Zimbabwe, Mozambique and E South Africa.

SYNONYMS: *algoensis, caffra, cinerea, eleotragus, isabellina, multiannulata, occidentalis, penricei, thomasinae.*

Redunca fulvorufula (Afzelius, 1815). Nova Acta Reg. Soc. Sci. Upsala, 7:250.

TYPE LOCALITY: South Africa, E Cape Prov.

DISTRIBUTION: E Nigeria and W Cameroon; SE Sudan; NE Uganda; C Ethiopia; Kenya; N Tanzania; Lesotho, E South Africa, and Swaziland marginally extending into S Mozambique and SE Botswana.

STATUS: IUCN - Endangered as *R. f. adamauae.*

SYNONYMS: *adamauae, chanleri, lalandia, schoana, subalpina.*

Redunca redunca (Pallas, 1767). Spicil. Zool., 1:8.

TYPE LOCALITY: Senegal, Goree (= Gori) Isl.

DISTRIBUTION: Senegal to C Ethiopia; south to N Zaire, Rwanda, Burundi and Tanzania.

SYNONYMS: *bayoni, bohor, cottoni, dianae, donaldsoni, nagor, nigeriensis, odrob, reversa, rufa, tohi, ugandae, wardi.*

ORDER PHOLIDOTA

by Duane A. Schlitter

ORDER PHOLIDOTA

Family Manidae Gray, 1821. London Med. Repos., 15:305.

Manis Linnaeus, 1758. Syst. Nat., 10th ed., 1:36.
TYPE SPECIES: *Manis pentadactyla* Linnaeus, 1758.
SYNONYMS: *Paramanis, Phataginus, Smutsia, Uromanis* (see Corbet and Hill, 1980:127).
COMMENTS: Morphological evidence suggests a subdivision of the genus; see Patterson (1978).

Manis crassicaudata Gray, 1827. *In* Griffith et al., Anim. Kingdom, 5:282.
TYPE LOCALITY: India.
DISTRIBUTION: Pakistan, east to W Bengal (India) and Yunnan (China) south to Sri Lanka.
STATUS: CITES - Appendix II.
COMMENTS: Formerly erroneously called *pentadactyla*; see Emry (1970:460) and Ellerman and Morrison-Scott (1951).

Manis gigantea Illiger, 1815. Abh. Phys. Klasse K. Pruess Konigl. Akad. Wiss., p. 84.
TYPE LOCALITY: Not found.
DISTRIBUTION: Senegal to W Kenya, south to Rwanda, C Zaire and SW Angola.
STATUS: CITES - Appendix III (Ghana).

Manis javanica Desmarest, 1822. Mammalogie, *in* Encycl. Méth., 2:377.
TYPE LOCALITY: Indonesia, Java.
DISTRIBUTION: Burma; Thailand; Indochina; Sumatra and Java (Indonesia); Borneo; SW Philippines; adjacent islands.
STATUS: CITES - Appendix II.
SYNONYMS: *culionensis*.
COMMENTS: Reviewed by Ellerman and Morrison-Scott (1951); includes *culionensis*, considered a separate species by Sanborn (1952*a*:114).

Manis pentadactyla Linnaeus, 1758. Syst. Nat., 10th ed., 1:36.
TYPE LOCALITY: Taiwan (= Formosa).
DISTRIBUTION: Nepal to S China, Hainan Isl (China), and N Indochina; Taiwan.
STATUS: CITES - Appendix II.
SYNONYMS: *aurita*.
COMMENTS: Includes *aurita*; see Emry (1970:460) and Ellerman and Morrison-Scott (1951).

Manis temminckii Smuts, 1832. Enumer. Mamm. Capensium, p. 54.
TYPE LOCALITY: South Africa, N Cape Prov., Latakou (= Litakun), near Kuruman.
DISTRIBUTION: N South Africa, N and E Namibia, Zimbabwe, Mozambique, Botswana, Angola, Kenya, S Zaire, S Sudan, Chad.
STATUS: CITES - Appendix I; U.S. ESA - Endangered.
COMMENTS: Reviewed by Stuart (1980).

Manis tetradactyla Linnaeus, 1766. Syst. Nat., 12th ed., 1:53.
TYPE LOCALITY: West Africa.
DISTRIBUTION: Senegal and Gambia to W Uganda, south to SW Angola.
STATUS: CITES - Appendix III (Ghana).
SYNONYMS: *longicaudata*.
COMMENTS: Includes *longicaudata* see Meester (1972:2).

Manis tricuspis Rafinesque, 1821. Ann. Sci. Phys. Brux., 7:215.
TYPE LOCALITY: West Africa, "Guinee."
DISTRIBUTION: Senegal to W Kenya, south to NE Zambia and SW Angola; Bioko (Equatorial Guinea).
STATUS: CITES - Appendix III (Ghana).
SYNONYMS: *tridentata*.
COMMENTS: Includes *tridentata*; see Ansell (1982).

ORDER RODENTIA
SUBORDER SCIUROGNATHI

FAMILY APLODONTIDAE
by Don E. Wilson

Family Aplodontidae Brandt, 1855. Mem. Acad. Imp. Sci. St. Petersbourg, Ser. 6, VII; Sci. Nat., p. 151.
SYNONYMS: Allomyidae, Haplodontidae.

Aplodontia Richardson, 1829. Zool. J., 4:334.
TYPE SPECIES: *Anisonyx rufa* Rafinesque, 1817.
SYNONYMS: *Haplodon*.
COMMENTS: Revised by Taylor (1918).

Aplodontia rufa (Rafinesque, 1817). Am. Mon. Mag., 2:45.
TYPE LOCALITY: USA, Oregon, neighborhood Columbia River.
DISTRIBUTION: SW British Columbia (Canada) to C California (USA).
STATUS: IUCN - Indeterminate as *A. r. nigra* and *A. r. phaea*.
SYNONYMS: *californica, chryseola, columbiana, grisea, humboldtiana, leporinus, major, nigra, olympica, pacifica, phaea, rainieri*.
COMMENTS: See Taylor (1918).

FAMILY SCIURIDAE

by Robert S. Hoffmann, Charles G. Anderson,
Richard W. Thorington, Jr., and Lawrence R. Heaney

Family Sciuridae Hemprich, 1820. Grundriss Naturgesch., p. 32.

COMMENTS: Ellerman (1940) reviewed the history of sciurid classification. The first modern classification (Pocock, 1923) recognized six subfamilies: Sciurinae, Tamiasciurinae, Funambulinae, Callosciurinae, Xerinae, and Marmotinae. Simpson (1945) recognized the same taxa, but all at the tribal level. Ellerman (1940) avoided formal designation, but recognized seven "sections" in the Sciurus "group," which often do not conform with the above. Moore (1959) recognized Simpson's six tribes and, in addition, Ratufini and Paraxerini for certain genera that had previously been included in Funambulini. Black (1963) elevated Tamiini to tribal level (previously in Marmotini). Gromov et al. (1965) elevated the ground squirrels to subfamily rank, Marmotinae, which included the tribes Tamiini Black, Otospermophilini Gromov, Citellini Gromov, Marmotini Simpson (part), and Cynomyini Gromov. He also recognized the subfamilies Xerinae and Sciurinae, but did not treat any other groups. Emry and Thorington (1984) submerged Tamiasciurini in Sciurini and Tamiini in Marmotini. They also (1982) described the oldest known sciurid, *Protosciurus*, which they interpreted as an arboreal form similar to *Sciurus* from which both flying squirrels and ground squirrels have been derived. These authors treated flying squirrels as a monophyletic sister group to other squirrels, either as a family (Pocock, 1923) or subfamily (other authors). The monophyly of flying squirrels has been both supported (Thorington, 1984) and challenged (Hight et al., 1974). Heaney (1985) recommended that the subtribe Hyosciurina of Moore (1959) be discarded because it lacks defining characters, and that the component genera (*Hyosciurus, Prosciurillus, Rubisciurus,* and *Exilisciurus*) be retained in tribe Callosciurini Simpson, 1945. A catolog of Indian sciurids was presented by Agrawal and Chakraborty (1979).

Subfamily Sciurinae Hemprich, 1820. Grundriss Naturgesch., p. 32.

Ammospermophilus Merriam, 1892. Proc. Biol. Soc. Washington, 7:27.

TYPE SPECIES: *Tamias leucurus* Merriam, 1889.

COMMENTS: Tribe Otospermophilini (Gromov et al., 1965) or Marmotini. Formerly included in *Spermophilus* (Hershkovitz, 1949*b*); Bryant (1945) considered *Ammospermophilus* a distinct genus. Recent biochemical and chromosomal data support Bryant's recognition of generic status.

Ammospermophilus harrisii (Audubon and Bachman, 1854). Viviparous Quadrupeds of North America, 3:267.

TYPE LOCALITY: Unknown. Restricted by Mearns (1896:444) to Santa Cruz Valley at the Mexican boundary, Santa Cruz Co., Arizona [USA].

DISTRIBUTION: Arizona to SW New Mexico (USA) and adjoining Sonora, Mexico.

SYNONYMS: *kinoensis* Huey, 1937; *saxicolus* (Mearns, 1896).

COMMENTS: See *A. insularis*; reviewed by Best et al. (1990*c*, Mammalian Species, 366).

Ammospermophilus insularis Nelson and Goldman, 1909. Proc. Biol. Soc. Washington, 22:24.

TYPE LOCALITY: "Espiritu Santo Island, Lower California [=Baja California Sur], Mexico."

DISTRIBUTION: Known only from the type locality.

COMMENTS: Considered a distinct species (Hall, 1981:381); may be most closely related to *A. harrisii* (Mascarello and Bolles, 1980). Reviewed by Best et al. (1990*a*, Mammalian Species, 364).

Ammospermophilus interpres (Merriam, 1890). N. Am. Fauna, 4:21.

TYPE LOCALITY: "El Paso, [El Paso Co.], Texas [USA]."

DISTRIBUTION: New Mexico and W Texas (USA) to Coahuila, Chihuahua, and Durango (Mexico).

COMMENTS: Most divergent species of the genus (Bolles, 1981), and probable primitive sister-species to remainder (Hafner, 1981). Reviewed by Best et al. (1990*b*, Mammalian Species, 365).

Ammospermophilus leucurus (Merriam, 1889). N. Am. Fauna, 2:19.
TYPE LOCALITY: "San Gorgonio Pass, [Riverside Co.], California [USA]."
DISTRIBUTION: E California and SE Oregon to Colorado and New Mexico (USA), south to Baja California Sur (Mexico).
SYNONYMS: *canfieldiae* Huey, 1929; *cinamomeus* (Merriam, 1890); *escalante* (Hansen, 1955); *extimus* Nelson and Goldman, 1929; *notom* (Hansen, 1955); *peninsulae* (J. Allen, 1893); *pennipes* A. Howell, 1931; *tersus* Goldman, 1929; *vinnulus* (Elliot, 1904).
COMMENTS: Reviewed by Belk and Smith (1990, Mammalian Species, 368).

Ammospermophilus nelsoni (Merriam, 1893). Proc. Biol. Soc. Washington, 8:129.
TYPE LOCALITY: "Tipton, San Joachin Valley, [Tulare Co.], California [USA]."
DISTRIBUTION: San Joaquin Valley (S California, USA).
STATUS: May now be restricted to southern half of its former range (Hafner, 1981).
SYNONYMS: *amplus* Taylor, 1916.
COMMENTS: Most closely related to *interpres* (Hafner, 1981). Reviewed by Best et al. (1990*d*, Mammalian Species, 367).

Atlantoxerus Forsyth Major, 1893. Proc. Zool. Soc. Lond., 1893:189.
TYPE SPECIES: *Sciurus getulus* Linnaeus, 1758.
COMMENTS: Subfamily Xerinae (Gromov et al., 1965:70) or tribe Xerini (Moore, 1959). *Atlantoxerus* was originally a subgenus of *Xerus*, and later raised to full generic rank by Thomas (1909*a*). Closely related to *Xerus* (Corbet, 1978*c*:79).

Atlantoxerus getulus (Linnaeus, 1758). Syst. Nat., 10th ed., 1:64.
TYPE LOCALITY: "in Africa." Restricted by Thomas (1911*a*:149) to "Barbary;" and by Cabrera (1932:217) to Agadir, Morocco.
DISTRIBUTION: Grand and Middle Atlas south to Agadir and N edge of Sahara (Morocco), NW Algeria.
SYNONYMS: *praetextus* Wagner, 1842; *trivittatus* (Gray, 1842); see Corbet (1978*c*:79).

Callosciurus Gray, 1867. Ann. Mag. Nat. Hist., ser. 3, 20:277.
TYPE SPECIES: *Sciurus rafflesii* Vigors and Horsfield, 1828 (= *Sciurus prevostii* Desmarest, 1822).
SYNONYMS: *Baginia* Gray, 1867; *Erythrosciurus* Gray, 1867; *Hessonoglyphotes* Moore, 1959; *Heterosciurus* Trouessart, 1880; *Tomeutes* Thomas, 1915.
COMMENTS: Tribe Callosciurini (Moore, 1959). Formerly included *Sundasciurus* and *Prosciurillus* (in part; see Moore, 1958) and *Tamiops* (Moore and Tate, 1965). See also Corbet and Hill (1992). Reviewed in part by Chakraborty (1985).

Callosciurus adamsi (Kloss, 1921). J. Str. Br. Roy. Asiat. Soc., 83:151.
TYPE LOCALITY: Malaysia, Sarawak, Baram River, Long Mujan, 150 mi. (241 km) up Baram River, 700-900 ft. (213-274 m).
DISTRIBUTION: Lowlands and hills in Sabah and Sarawak, Borneo, below the range of *orestes*.
COMMENTS: Distinct from *albescens*; see Medway (1977:93). Member of the *notatus* species group, closely related to *orestes*. Chromosomes of "*C. albescens*" from Sabah by Harada and Kobayashi (1980) refer to either this species or *C. orestes*.

Callosciurus albescens (Bonhote, 1901). Ann. Mag. Nat. Hist., ser. 7, 7:446.
TYPE LOCALITY: Indonesia, Sumatra, Acheen (= Atjeh).
DISTRIBUTION: Sumatra.
SYNONYMS: *albiculus* G. Miller, 1942.
COMMENTS: Considered a distinct species by Medway (1977:93). Member of the *notatus* species group. Corbet and Hill (1992) considered *albescens* and *albiculus* assignable to *C. notatus*.

Callosciurus baluensis (Bonhote, 1901). Ann. Mag. Nat. Hist., ser. 7, 7:174.
TYPE LOCALITY: Malaysia, Sabah, Mount Kinabalu, Borneo, alt. 1000 ft.

DISTRIBUTION: Above 300 m elevation in Sabah and Sarawak, Malaysia.

SYNONYMS: *baramensis* (Chasen, 1940); *medialis* Allen and Coolidge, 1940.

COMMENTS: Sometimes considered a subspecies of *prevostii* (see Medway, 1977:86), but often sympatric with that species (Payne et al., 1985). Corbet and Hill (1992) included *prevostii erythromelas* in this species, with *schlegeli* a synonym.

Callosciurus caniceps (Gray, 1842). Ann. Mag. Nat. Hist., [ser. 1], 10:263.

TYPE LOCALITY: "Bhotan." Restricted by Robinson and Kloss (1918*a*:206) to N Tenasserim, Burma.

DISTRIBUTION: Thailand, peninsular Burma, peninsular Malaysia, and adjacent islands.

SYNONYMS: *adangensis* (Miller, 1903), *altinsularis* (Miller, 1903); *bentincanus* (Miller, 1903); *bimaculatus* (Temminck, 1853); *casensis* (Miller, 1903); *chrysonotus* (Blyth, 1847); *concolor* (Blyth, 1856); *davisoni* (Bonhote, 1901); *domelicus* (Miller, 1903); *epomophorus* (Bonhote, 1901); *erubescens* Cabrera, 1917; *fallax* (Robinson and Kloss, 1914); *fluminalis* (Robinson and Wroughton, 1911); *hastilis* Thomas, 1923; *helgei* (Gyldenstolpe, 1917); *helvus* (Shamel, 1930); *inexpectatus* (Kloss, 1916); *lancavensis* (Miller, 1903); *lucas* (Miller, 1903); *mapravis* Thomas and Robinson, 1921; *matthaeus* (Miller, 1903); *milleri* (Robinson and Wroughton, 1911); *moheius* Thomas and Robinson, 1921; *mohillius* Thomas and Robinson, 1921; *nakanus* Thomas and Robinson, 1921; *panjioli* Thomas and Robinson, 1921; *panjius* Thomas and Robinson, 1921; *pipidonis* Thomas and Robinson, 1921; *samuiensis* (Robinson and Kloss, 1914); *sullivanus* (Miller, 1903); *tabaudius* Thomas, 1922; *tacopius* Thomas and Robinson, 1921; *telibius* Thomas and Robinson, 1921; *terutavensis* (Thomas and Wroughton, 1909).

COMMENTS: Revised by Moore and Tate (1965). Member of the *caniceps* species group.

Callosciurus erythraeus (Pallas, 1779). Nova Spec. Quad. Glir. Ord., p. 377.

TYPE LOCALITY: Not known; restricted to Assam, India by Bonhote (1901*a*); further restricted to the Garo Hills of Assam by Moore and Tate (1965).

DISTRIBUTION: West of Irrawaddy River in India, Burma, and SE China. East of Irrawaddy River in Burma, Thailand, peninsular Malaysia, Indochina, S China, and Taiwan.

SYNONYMS: *albifer* (Hilzheimer, 1906); *aquilo* Wroughton, 1921; *atrodorsalis* (Gray, 1842); *bartoni* (Thomas, 1914); *bhutanensis* (Bonhote, 1901); *bolovensis* Osgood, 1932; *bonhotei* (Robinson and Kloss, 1911); *canigenus* Howell, 1927; *careyi* Thomas and Wroughton, 1916; *castaneoventris* (Gray, 1842); *centralis* (Bonhote, 1901); *cinnamomeiventris* (Swinhoe, 1862); *contumax* Thomas, 1927; *crotalius* Thomas and Wroughton, 1916; *crumpi* Wroughton, 1916; *cucphuongis* Dao Van Tien, 1965; *dabshanensis* Xu and Chen, 1989; *dactylinus* Thomas, 1927; *erythrogaster* (Blyth, 1842); *flavimanus* (I. Geoffroy,1831); *fryanus* Thomas and Wroughton, 1916; *fumigatus* (Bonhote, 1907); *gloveri* Thomas, 1921; *gongshanensis* Wang, 1981; *gordoni* (Anderson, 1871); *griseimanus* (Milne-Edwards, 1867); *griseopectus* (Blyth, 1847); *griseopectus* (Milne-Edwards, 1874); *haemobaphes* (G. M. Allen, 1912); *haringtoni* (Thomas, 1905); *hendeei* Osgood, 1932; *hyperythrus* (Blyth, 1856); *insularis* (J. A. Allen, 1906); *intermedius* (Anderson, 1879); *kemmisi* (Wroughton, 1908); *kinneari* Thomas and Wroughton, 1916; *leucopus* (Gray, 1867); *leucurus* (Hilzheimer, 1905); *michianus* (Robinson and Wroughton, 1911); *midas* (Thomas, 1914); *millardi* Thomas and Wroughton, 1916; *nagarum* Thomas and Wroughton, 1916; *nigridorsalis* Kuroda, 1935; *ningpoensis* (Bonhote, 1901); *phanrangis* Robinson and Kloss, 1922; *pirata* Thomas, 1929; *pranis* (Kloss, 1916); *primus* Allen and Coolidge, 1940; *punctatissimus* (Gray, 1867); *quantulus* Thomas, 1927; *quinlingensis* Xu and Chen, 1989; *roberti* (Bonhote, 1901); *rubeculus* (Miller, 1903); *rubex* (Thomas, 1914); *shanicus* (Ryley, 1914); *shortridgei* Thomas and Wroughton, 1916; *siamensis* (Gray, 1860); *sladeni* (Anderson, 1871); *solutus* (Thomas, 1914); *styani* (Thomas, 1894); *tachin* (Kloss, 1916); *thai* (Kloss, 1917); *thaiwanensis* (Bonhote, 1901); *tsingtanensis* (Hilzheimer, 1905); *tsingtauensis* (Hilzheimer, 1906); *vassali* (Bonhote, 1907); *vernayi* Carter, 1942; *wellsi* Wroughton, 1921; *woodi* Harris, 1931; *wuliangshanensis* Li and Wang, 1981; *youngi* (Robinson and Kloss, 1914); *zimmeensis* Robinson and Wroughton, 1916.

COMMENTS: Includes *sladeni*; see Moore and Tate (1965). Includes *flavimanus* (Corbet and Hill, 1992), which was formerly considered distinct (Moore and Tate, 1965). Chromosomes described (as *flavimanus*) by Nadler et al. (1975*b*). Indian populations reviewed by Agrawal and Chakraborty (1979), and Vietnamese populations by Dao (1965).

Callosciurus finlaysonii (Horsfield, 1824). Zool. Res. Java, part 7, p. 7(unno.) of *Sciurus Plantnai* acct.
TYPE LOCALITY: "the Islands called Sichang, in the Gulf of Siam", Koh Si Chang (Gulf of Thailand).
DISTRIBUTION: SC Burma, Thailand, Cambodia, Laos, Vietnam.
SYNONYMS: *albivexilli* (Kloss, 1916); *annellatus* Thomas, 1929; *bocourti* (Milne-Edwards, 1867); *bonhotei* Robinson and Wroughton, 1911; *boonsongi* Moore and Tate, 1965; *cinnamomeus* (Temminck, 1853); *cockerelli* Thomas, 1928; *dextralis* (Wroughton, 1908); *ferrugineus* (F. Cuvier, 1829); *floweri* (Bonhote, 1901); *folletti* (Kloss, 1915); *frandseni* (Kloss, 1916); *germaini* (Milne-Edwards, 1867); *grutei* (Gyldenstolpe, 1917); *harmandi* (Milne-Edwards, 1877); *herberti* Robinson and Kloss, 1922; *keraudrenii* (Lesson, 1830); *leucocephalus* (Bonhote, 1901); *leucogaster* (Milne-Edwards, 1867); *lylei* (Wroughton, 1908); *menamicus* Thomas, 1929; *nox* (Wroughton, 1908); *pierrei* Robinson and Kloss, 1922; *portus* (Kloss, 1915); *prachin* Kloss, 1921; *rajasima* Kloss, 1921; *sinistralis* (Wroughton, 1908); *splendens* (Gray, 1861); *tachardi* Robinson, 1916; *trotteri* (Kloss, 1916); *williamsoni* Robinson and Kloss, 1922.
COMMENTS: Revised by Moore and Tate (1965), expanded to include *ferrugeneus* by Corbet and Hill (1992). Chromosomes described by Nadler et al. (1975*b*).

Callosciurus inornatus (Gray, 1867). Ann. Mag. Nat. Hist., ser. 3, 20:282.
TYPE LOCALITY: "Loo Mountains." Restricted by Moore and Tate (1965:209) to "Mountains in Laos."
DISTRIBUTION: Laos, N Vietnam, S Yunnan (China).
SYNONYMS: *imitator* Thomas, 1925.
COMMENTS: Revised by Moore and Tate (1965); member of the *caniceps* species group.

Callosciurus melanogaster (Thomas, 1895). Ann. Mus. Civ. Stor. Nat. Genova, 14:668.
TYPE LOCALITY: Indonesia, Mentawi Isls, Sipora Isl, "Si Oban".
DISTRIBUTION: Mentawi Isls (Indonesia).
SYNONYMS: *atratus* (Miller, 1903); *mentawi* (Chasen and Kloss, 1927).
COMMENTS: Member of the *notatus* species group. Part of the rodent fauna endemic to the Mentawi Archipelago (see comments under *Leopoldamys siporanus*).

Callosciurus nigrovittatus (Horsfield, 1824). Zool. Res. Java, part 7, p. 5(unno.) of *Sciurus Plantani* acct.
TYPE LOCALITY: Indonesia, "The island of Java." Restricted by Robinson and Kloss (1918*a*:222) to "Java (probably East and central parts)."
DISTRIBUTION: S Vietnam, Malaysia, peninsular Thailand, Sumatra, Borneo, Java, adjacent small islands.
SYNONYMS: *acraeus* Miller, 1942; *bantamensis* Sody, 1949; *besuki* (Kloss, 1921); *bilimitatus* (Miller, 1903); *bocki* (Robinson and Wroughton, 1911); *johorensis* (Robinson and Wroughton, 1911); *klossi* (Miller, 1900); *madsoedi* Sody, 1929; *microrhynchus* (Kloss, 1908); *phoenicurus* Sody, 1949; *salakensis* Sody, 1949; *tengerensis* Sody, 1949.
COMMENTS: Member of the *notatus* species group.

Callosciurus notatus (Boddaert, 1785). Elench. Anim., p. 119.
TYPE LOCALITY: Indonesia, West Java.
DISTRIBUTION: From peninsular Malaysia and Thailand to Java, Bali, and Borneo; Lombok Isl; Salayer Isl (south of Sulawesi; Musser, 1987 considered this population an introduction, possibly from Java); widespread on smaller islands.
SYNONYMS: *abbottii* (Miller, 1900); *anambensis* (Miller, 1900); *andrewsii* (Bonhote, 1901); *aoris* (Miller, 1903); *arendsis* (Lyon, 1911); *atristriatus* (Miller, 1913); *badjing* (Kerr, 1792); *balstoni* (Robinson and Wroughton, 1911); *bilineatus* (E. Geoffroy, 1817); *billitonus* (Lyon, 1906); *conipus* (Lyon, 1911); *datus* (Lyon, 1911); *dilutus* (Miller, 1913); *director* (Lyon, 1909); *dulitensis* (Bonhote, 1901); *famulus* (Robinson, 1912); *guillemardi* (Kloss, 1926); *ictericus* (Miller, 1903); *kalianda* Sody, 1949; *lamucotanus* (Lyon, 1911); *lautensis* (Miller, 1901); *lighti* (Chasen and Kloss, 1924); *lunaris* (Chasen and Kloss, 1924); *lutescens* (Miller, 1901); *madurae* (Thomas, 1910); *magnificus* Sody, 1949; *malawali* (Chasen and Kloss, 1932); *maporensis* (Robinson, 1916); *marinsularis* (Lyon, 1911); *microtis* (Jentink, 1879); *miniatus* (Miller, 1900); *nesiotes* (Thomas and Wroughton, 1909); *nicotianae* Sody, 1936; *pannovianus* (Miller, 1903); *pemangilensis* (Miller, 1903);

peninsularis (Miller, 1903); *percommodus* Chasen, 1940; *perhentiani* (Kloss, 1911); *plantani* (Ljungh, 1801); *plasticus* (Kloss, 1911); *prinsulae* Sody, 1949; *poliopus* (Lyon, 1911); *pretiosus* (Miller, 1903); *proteus* (Kloss, 1911); *raptor* Hill, 1960; *rubidiventris* (Miller, 1901); *rupatius* (Lyon, 1908); *rutiliventris* (Miller, 1901); *saturatus* (Miller, 1903); *scottii* (Kloss, 1911); *seraiae* (Miller, 1901); *serutus* (Miller, 1906); *singapurensis* Robinson, 1916; *siriensis* (Lyon, 1911); *stellaris* (Chasen and Kloss, 1924); *stresemanni* (Thomas, 1913); *subluteus* (Thomas and Wroughton, 1909); *tamansari* (Kloss, 1921); *tapanulius* (Lyon, 1907); *tarussanus* Lyon, 1907; *tedongus* (Lyon, 1906); *tenuirostris* (Miller, 1900); *tinggius* Hill, 1960; *toupai* Lesson, 1827; *ubericolor* (Miller, 1903); *vanheurni* Sody, 1929; *verbeeki* Sody, 1929; *vinocastaneus* Sody, 1949; *vittatus* (Raffles, 1821); *watsoni* (Kloss, 1911).

COMMENTS: Member of the *notatus* species group. Corbet and Hill (1992) tentatively placed *dschinschicus* (Gmelin, 1788)(= *ginginianus* Shaw, 1801) here. Chromosomes described by Nadler et al. (1975*b*).

Callosciurus orestes (Thomas, 1895). Ann. Mag. Nat. Hist., ser. 6, 15:529.

TYPE LOCALITY: Malaysia, Sarawak, Mount Dulit, 4000 ft.

DISTRIBUTION: Sabah and Sarawak [Borneo], Malaysia, at middle elevations (Payne et al., 1985).

SYNONYMS: *canalvus* (Moore, 1959); *venetus* (Chasen, 1940).

COMMENTS: Formerly considered a subspecies of *C. nigrovittatus* (e.g., Medway, 1977:92), but clearly distinct (Payne et al., 1985). Includes *Glyphotes canalvus*, which Moore (1959) placed in subgenus *Hessonoglyphotes*. Member of the *notatus* species group, closely related to *adamsi* (see comments therein).

Callosciurus phayrei (Blyth, 1856). J. Asiat. Soc. Bengal, 24:472.

TYPE LOCALITY: Burma, Martaban, "Sent from Moulmein" (Robinson and Kloss, 1918*a*:225).

DISTRIBUTION: Upper Irrawaddy R. and Sittang R. eastward to Salween R., S Burma.

SYNONYMS: *blanfordii* (Blyth, 1862); *heinrichi* Tate, 1954.

COMMENTS: Considered a distinct species by Moore and Tate (1965). A member of the *caniceps* species group.

Callosciurus prevostii (Desmarest, 1822). Mammalogie, *in* Encyclop. Méth., p. 335.

TYPE LOCALITY: Malaysia, Malacca Prov., "Settlement of Malacca."

DISTRIBUTION: Malaysian subregion except Java; N Sulawesi. Musser (1987) considered the population in Sulawesi as introduced.

SYNONYMS: *armalis* (Lyon, 1911); *atricapillus* (Schlegel, 1863); *atrox* (Miller, 1913); *bangkanus* (Schlegel, 1863); *banksi* (Chasen, 1933); *borneoensis* (Müller and Schlegel, 1842); *caedis* (Chasen and Kloss, 1932); *carimatae* (Miller, 1906); *carimonensis* (Miller, 1906); *caroli* (Bonhote, 1901); *condurensis* (Miller, 1906); *coomansi* Sody, 1949; *erebus* (Miller, 1903); *erythromelas* (Temminck, 1853); *griseicauda* (Bonhote, 1901); *harrisoni* (Stone and Rehn, 1902); *humei* (Bonhote, 1901); *indica* (Müller and Schlegel, 1842, not *indicus* Erxleben, 1777); *kuchingensis* (Bonhote, 1901); *melanops* (Miller, 1902); *mendanauus* (Lyon, 1906); *mimellus* (Miller, 1900); *mimiculus* (Miller, 1900); *navigator* (Bonhote, 1901); *nyx* (Lyon, 1908); *palustris* (Lyon, 1907); *pelapius* (Lyon, 1911); *penialius* (Lyon, 1908); *piceus* (Peters, 1866); *pluto* (Gray, 1867); *proserpinae* (Lyon, 1907); *rafflesii* (Vigores and Horsfield, 1828); *redimitus* (Boon Mesch, 1829); *rufogularis* (Gray, 1842); *rufoniger* (Motley and Dillwyn, 1855); *rufonigra* (Gray, 1842); *sanggaus* (Lyon, 1907); *sarawakensis* (Gray, 1867); *schlegeli* (Gray, 1867); *suffusus* (Bonhote, 1901); *sumatranus* (Schlegel, 1863); *waringensis* Sody, 1949; *wrayi* (Kloss, 1910).

COMMENTS: Formerly included *baluensis*; but the two are often sympatric (Payne et al., 1985). Reviewed by Heaney (1978).

Callosciurus pygerythrus (I. Geoffroy Saint Hilaire, 1831). Mag. Zool. Paris, p. 5, pl. 4-6.

TYPE LOCALITY: "from forest of Syriam, near Pegu, Burma" (Moore and Tate, 1965:217).

DISTRIBUTION: Nepal and NE India to Burma, N Vietnam, and Yunnan (China).

SYNONYMS: *assamensis* (Gray, 1843); *bellona* (Thomas and Wroughton, 1916); *blythii* (Tytler, 1854); *janetta* (Thomas, 1914); *lokroides* (Hodgson, 1836); *mearsi* (Bonhote, 1906); *owensi* (Thomas and Wroughton, 1916); *similis* (Gray, 1867); *stevensi* (Thomas, 1908); *virgo* (Thomas and Wroughton, 1916).

COMMENTS: Revised by Moore and Tate (1965:209).

Callosciurus quinquestriatus (Anderson, 1871). Proc. Zool. Soc. Lond., 1871:142.
TYPE LOCALITY: "common at Ponsee, on the Kakhyen range of hills, east of Bhamo, at an elevation of from 2000 to 3000 ft" [Burma].
DISTRIBUTION: NE Burma, Yunnan (China).
SYNONYMS: *beebei* (J. Allen, 1911); *imarius* Thomas, 1926; *sylvester* Thomas, 1926.
COMMENTS: Included in *flavimanus* by Moore and Tate (1965:209); but see Corbet and Hill (1992).

Cynomys Rafinesque, 1817. Am. Mon. Mag., 2:43.
TYPE SPECIES: *Cynomys socialis* Rafinesque, 1817 (= *Arctomys ludoviciana* Ord, 1815).
SYNONYMS: *Arctomys* Ord, 1815; *Cynomomus* Osborn, 1894; *Leucocrossuromys* Hollister, 1916; *Mamcynomiscus* Herrera, 1899; *Monax* Warden, 1819.
COMMENTS: Tribe Cynomyini (Gromov et al., 1965) or Marmotini. Revised by Pizzimenti (1975). Clark et al. (1971) published a key to the genus. Includes *Cynomys* and *Leucocrossuromys* as subgenera. Relationships of *Cynomys* to other ground squirrels are unclear. While generally regarded as monophyletic, and a sister-group to North American *Spermophilus*, recent evidence suggests that *Cynomys* may be most closely related to subgenus *Spermophilus*, making genus *Spermophilus* paraphyletic (Hafner, 1984:17).

Cynomys gunnisoni (Baird, 1855). Proc. Acad. Nat. Sci. Philadelphia, 7:334.
TYPE LOCALITY: "Cochitope [Cochetopa] Pass of Rocky Mountains." [Saguache Co., Colorado, USA].
DISTRIBUTION: SE Utah, SW Colorado, NE Arizona, and NW New Mexico (USA).
SYNONYMS: *zuniensis* Hollister, 1916.
COMMENTS: Subgenus *Leucocrossuromys* (Hall, 1981). Reviewed by Pizzimenti and Hoffmann (1973, Mammalian Species, 25) and Pizzimenti (1976).

Cynomys leucurus Merriam, 1890. N. Am. Fauna, 3:59.
TYPE LOCALITY: "Fort Bridger, [Uinta Co.] Wyoming" (Merriam, 1890:33).
DISTRIBUTION: SC Montana, W and C Wyoming, NE Utah, and NW Colorado (USA).
COMMENTS: Subgenus *Leucocrossuromys* (Hall, 1981). Reviewed by Clark et al. (1971, Mammalian Species, 7) and Pizzimenti (1976).

Cynomys ludovicianus (Ord, 1815). *In* Guthrie, New Geogr., Hist. Coml. Grammar, Philadelphia, 2nd ed., 2:292.
TYPE LOCALITY: "vicinity of the Missouri." Restricted by Hollister (1916*a*:14) to "Upper Missouri River."
DISTRIBUTION: Saskatchewan (Canada); Montana to E Nebraska, W Texas, New Mexico, and SE Arizona (USA); NE Sonora, and N Chihuahua (Mexico).
SYNONYMS: *arizonensis* Mearns, 1890; *cinereus* Richardson, 1829; *grisea* Rafinesque, 1817; *latrans* (Harlan, 1815); *missouriensis* (Warden, 1819); *pyrrotrichus* Elliot, 1905; *socialis* Rafinesque, 1817.
COMMENTS: Subgenus *Cynomys*. Intraspecific variation reviewed by Chesser (1983).

Cynomys mexicanus Merriam, 1892. Proc. Biol. Soc. Washington, 7:157.
TYPE LOCALITY: "La Ventura, Coahuila, Mexico."
DISTRIBUTION: Coahuila, and San Luis Potosi; perhaps Nuevo Leon, and Zacatecas (NC Mexico).
STATUS: CITES - Appendix I; U.S. ESA and IUCN - Endangered.
COMMENTS: Subgenus *Cynomys* (Hall, 1981:412); reviewed by Ceballos-G. and Wilson (1985, Mammalian Species, 248) and Treviño-V. (1991).

Cynomys parvidens J. A. Allen, 1905. Mus. Brooklyn Inst. Arts and Sci., Sci. Bull., 1:119.
TYPE LOCALITY: USA, "Buckskin Valley, Iron County, Utah."
DISTRIBUTION: SC Utah (USA).
STATUS: U.S. ESA - Threatened; IUCN - Vulnerable.
COMMENTS: Subgenus *Leucocrossuromys* (Hall, 1981). Reviewed by Pizzimenti and Collier (1975, Mammalian Species, 52) and Pizzimenti and Nadler (1972).

Dremomys Heude, 1898. Mem. Hist. Nat. Emp. Chin., 4(2):54.
TYPE SPECIES: *Sciurus pernyi* Milne-Edwards, 1867.

SYNONYMS: *Zetis* Thomas, 1908.
COMMENTS: Tribe Callosciurini (Moore, 1959). Reviewed by Moore and Tate (1965).

Dremomys everetti (Thomas, 1890). Ann. Mag. Nat. Hist., ser. 6, 6:171.
TYPE LOCALITY: "Mount Penrisen, West Sarawak," [Malaysia].
DISTRIBUTION: Mountains of N and W Borneo (Kalimantan, Sarawak, Sabah), above 3,200 ft.

Dremomys lokriah (Hodgson, 1836). J. Asiat. Soc. Bengal, 5:232.
TYPE LOCALITY: "central and Northern regions of Nipál" [Nepal].
DISTRIBUTION: C Nepal east to Salween River; Tibet (China); N Burma; mountains in E India; Bhutan.
SYNONYMS: *bhotia* Thomas and Wrougton, 1916; *garonum* Thomas, 1922; *macmillani* Thomas and Wroughton, 1916; *motuoensis* Cai and Zhang, 1980; *pagus* Moore, 1956; *subflaviventris* Thomas, 1922.
COMMENTS: Revised by Moore and Tate (1965). See also Agrawal and Chakraborty (1979).

Dremomys pernyi (Milne-Edwards, 1867). Rev. Mag. Zool. (Paris), ser. 2, 19:19.
TYPE LOCALITY: "les montagnes de la principaute de Moupin [Muping]" [= Baoxing, Sichuan, China].
DISTRIBUTION: NE India; N Burma; N Vietnam; Tibet; Sichuan, Yunnan, Guizhou, Hunan, Hubei, Jiangxi, Fujian, Anhui (China); Taiwan.
SYNONYMS: *calidior* Thomas, 1916; *chintalis* Thomas, 1916; *flavior* G. Allen, 1912; *griselda* Thomas, 1916; *howelli* Thomas, 1922; *imus* Thomas, 1922; *lentus* A.B. Howell, 1927; *lichiensis* Thomas, 1922; *mentosus* Thomas, 1922; *modestus* Thomas, 1916; *owstoni* (Thomas, 1908); *senex* G. Allen, 1912.

Dremomys pyrrhomerus (Thomas, 1895). Ann. Mag. Nat. Hist., ser. 6, 16:242.
TYPE LOCALITY: "Ichang, Yang-tse-kiang [river] [Hupei, China]."
DISTRIBUTION: C and S China, extreme N Vietnam, Hainan Island (China).
SYNONYMS: *gularis* Osgood, 1932; *melli* Matschie, 1922; *riudonensis* (J. Allen, 1906).
COMMENTS: Moore and Tate (1965) treated *pyrrhomerus* as a separate species based on evidence of geographic parapatry, *D. p. gularis* with *D. r. rufigenis* (Osgood, 1932). Corbet and Hill (1992) listed *gularis* as a separate species, but considered other forms of *pyrrhomerus* to be conspecific with *rufigenis*.

Dremomys rufigenis (Blanford, 1878). J. Asiat. Soc. Bengal, 47(2):156.
TYPE LOCALITY: Burma, Tenasserim, Mt. Mooleyit.
DISTRIBUTION: NE India, N and C Burma, Yunnan (China), Laos, south through Vietnam, Thailand, peninsular Malaysia.
SYNONYMS: *adamsoni* Thomas, 1914; *belfieldi* (Bonhote, 1908); *fuscus* (Bonhote, 1907); *laomache* Thomas, 1921; *opimus* Thomas and Wroughton, 1916.
COMMENTS: Formerly included *pyrrhomerus*; but see Moore and Tate (1965). Chromosomes described by Nadler and Hoffmann (1970).

Epixerus Thomas, 1909. Ann. Mag. Nat. Hist., ser. 8, 3:472.
TYPE SPECIES: *Sciurus wilsoni* Du Chaillu, 1860.
COMMENTS: Tribe Protoxerini (Moore, 1959).

Epixerus ebii (Temminck, 1853). Esquisses Zool. sur la Côte de Guine, p. 129.
TYPE LOCALITY: Ghana. "...les grandes forêts de la Guiné, et se trouve dans les mêmes localités que l'espèce précédente, [...les confins du pays des Fantes...] mais parait être moins abondante dans les parties boisées de Dabocrom."
DISTRIBUTION: Sierra Leone, Liberia, Ivory Coast, Ghana.
STATUS: CITES - Appendix III (Ghana).
SYNONYMS: *jonesi* Hayman, 1954.
COMMENTS: *E. ebii* and *E. wilsoni* considered distinct species on the basis of cranial differences cited by Perret and Aellen (1956), Verheyen (1959), and Rosevear (1969).

Epixerus wilsoni (Du Chaillu, 1860). Proc. Boston Soc. Nat. Hist., 7:364.
TYPE LOCALITY: Gabon. "Gaboon,...the headwaters of the Ovenga River."
DISTRIBUTION: Cameroon, Rio Muni, Gabon.
SYNONYMS: *mayumbicus* Verheyen, 1959.

COMMENTS: *E. ebii* and *E. wilsoni* considered distinct species on the basis of cranial differences cited by Perret and Aellen (1956), Verheyen (1959) and Rosevear (1969). Amtmann (1975) included *E. wilsoni* with *E. ebii*.

Exilisciurus Moore, 1958. Am. Mus. Novit., 1914:4.

TYPE SPECIES: *Sciurus exilis* Müller, 1838.

COMMENTS: Tribe Callosciurini (Moore, 1959). The species included in *Exilisciurus* were formerly included in *Nannosciurus*; see Moore (1958, 1959) and Heaney (1985).

Exilisciurus concinnus (Thomas, 1888). Ann. Mag. Nat. Hist., ser. 6, 2:407.

TYPE LOCALITY: Philippines, Zamboanga Prov., Basilan Island, Isabela.

DISTRIBUTION: Mindanao faunal region (Heaney et al., 1987) including Mindanao, Basilan, Biliran, Bohol, Dinagat, Leyte, Samar, and Siargao islands (Heaney, 1985).

SYNONYMS: *luncefordi* Taylor, 1934; *samaricus* Thomas, 1897; *surrutilus* Hollister, 1913.

COMMENTS: Revised by Heaney (1985).

Exilisciurus exilis (Müller, 1838). Tijdschr. Nat. Gesch. Physiol., 5:138.

TYPE LOCALITY: Indonesia, Kalimantan, Kapuas River Basin, Tanah Laut (see Medway, 1977, and Heaney, 1985).

DISTRIBUTION: Borneo and Banggi Island.

SYNONYMS: *retectus* Thomas, 1910; *sordidus* Chasen and Kloss, 1928.

Exilisciurus whiteheadi (Thomas, 1887). Ann. Mag. Nat. Hist., ser. 5, 20:127.

TYPE LOCALITY: Malaysia, Sabah, Mt. Kinabalu.

DISTRIBUTION: Mountains of Sabah and Sarawak (Malaysia), above 900 m, and adjacent parts of West Kalimantan, Indonesia (Medway, 1977).

Funambulus Lesson, 1835. Illustr. Zool., pl. 43.

TYPE SPECIES: *Sciurus indicus* Lesson, 1835 (= *Sciurus palmarum* Linnaeus, 1776).

SYNONYMS: *Palmista* Gray, 1867; *Prasadsciurus* Moore and Tate, 1965; *Tamiodes* Pocock, 1923.

COMMENTS: Tribe Funambulini (Moore, 1959:170). Reviewed by Moore (1960) and Moore and Tate (1965). Includes *Funambulus* and *Prasadsciurus* as subgenera. Prasad (1957) proposed a separate subfamily for the genus, based on the distinctive anatomy of the male reproductive tract. Trends in evolution of karyotypes analyzed by Aswathanarayana (1987).

Funambulus layardi (Blyth, 1849). J. Asiat. Soc. Bengal, 18:602.

TYPE LOCALITY: "Upland districts" [Sri Lanka]. Restricted by Thomas (1924:241) to "Ambigamoa Hills...Central Province (7° N., 80° 3'E.)...."

DISTRIBUTION: S and C Sri Lanka, and mountains of S India.

SYNONYMS: *dravidianus* Robinson, 1917; *signatus* Thomas, 1924.

COMMENTS: Subgenus *Funambulus*. The presence of this species in S India is based on very few specimens.

Funambulus palmarum (Linnaeus, 1766). Syst. Nat., 12th ed., 1:86.

TYPE LOCALITY: "*in* America, Asia, Africa." Restricted by Wroughton (1905*a*:409) to E coast of Madras, India.

DISTRIBUTION: C and S India, Sri Lanka.

SYNONYMS: *bellaricus* Wroughton, 1916; *bengalensis* Wroughton, 1916; *brodiei* (Blyth, 1849); *comorinus* Wroughton, 1905; *favonicus* Thomas and Wroughton, 1915; *gossei* Wroughton and Davidson, 1919; *indicus* (Lesson, 1835, not Erxleben, 1777); *kelaarti* (Layard, 1851); *matugamensis* Lindsay, 1926; *olympius* Thomas and Wroughton, 1915; *penicillatus* (Leach, 1814); *robertsoni* Wroughton, 1916.

COMMENTS: Subgenus *Funambulus*. Distribution largely allopatric with that of *F. pennantii*, but limited area of sympatry along W coast of India, south of 16° N lat. (Moore, 1960).

Funambulus pennantii Wroughton, 1905. J. Bombay Nat. Hist. Soc., 16:411.

TYPE LOCALITY: "Mandvi Taluka of Surat District," Guzerath (= Gudjerat), India.

DISTRIBUTION: SE Iran through Pakistan to Nepal and N and C India. Perhaps adjacent Afghanistan.

SYNONYMS: *argentescens* Wroughton, 1905; *lutescens* Wroughton, 1916.

COMMENTS: Subgenus *Prasadsciurus* (Moore and Tate, 1965:71). Reviewed by Agrawal and Chakraborty (1979).

Funambulus sublineatus (Waterhouse, 1838). Proc. Zool. Soc. Lond., 1838:19.
TYPE LOCALITY: India, Madras, Nilgiri Hills.
DISTRIBUTION: SW India, C Sri Lanka.
SYNONYMS: *delesserti* (Gervais, 1841); *kathleenae* Thomas and Wroughton, 1915; *obscura* (Pelzeln and Kohl, 1886); *trilineatus* (Blyth, 1849).
COMMENTS: Subgenus *Funambulus.*

Funambulus tristriatus (Waterhouse, 1837). Mag. Nat. Hist. [Charlesworth's], 1:499.
TYPE LOCALITY: "...the more southern parts of Hindostan." Restricted by Wroughton (1905*a*:411) to Travancore; further restricted by Moore and Tate (1965:89) to Western Ghats, south of 12 deg. N. lat.
DISTRIBUTION: West coast of India, from below 20° N to southern tip.
SYNONYMS: *annandalei* Robinson, 1917; *dussumieri* (Milne-Edwards, 1867); *numarius* Wroughton, 1916; *thomasi* Wroughton and Davidson, 1919; *wroughtoni* Ryley, 1913.
COMMENTS: Subgenus *Funambulus.*

Funisciurus Trouessart, 1880. Le Naturaliste, 2(37):293.
TYPE SPECIES: *Sciurus isabella* Gray, 1862.
COMMENTS: Tribe Funambulini (Moore, 1959). Raised to full generic rank by Thomas (1897*b*). This treatment follows Amtmann (1975).

Funisciurus anerythrus (Thomas, 1890). Proc. Zool. Soc. Lond., 1890:447.
TYPE LOCALITY: Uganda, "Buguera", S of Lake Albert.
DISTRIBUTION: SW Nigeria, Cameroon, Central African Republic, NE Zaire, Uganda; SW Zaire and N Shaba Prov. (Zaire).
SYNONYMS: *bandarum* Thomas, 1915; *mystax* De Winton, 1898; *niapu* Allen, 1922; *ochrogaster* Cabrera and Ruxton, 1926; *raptorum* Thomas, 1903.

Funisciurus bayonii (Bocage, 1890). J. Sci. Math. Phys. Nat. Lisboa, (2)2:3.
TYPE LOCALITY: N Angola, "...du Duque de Bragança..."
DISTRIBUTION: NE Angola, SW Zaire.

Funisciurus carruthersi Thomas, 1906. Ann. Mag. Nat. Hist., ser. 7, 18:140.
TYPE LOCALITY: Uganda, "Ruwenzori East, 6500' (1900 m)."
DISTRIBUTION: Ruwenzori (S Uganda), Rwanda, Burundi.
SYNONYMS: *birungensis* Gyldenstolpe, 1927; *chrysippus* Thomas, 1923; *tanganyikae* Thomas, 1909.

Funisciurus congicus (Kuhl, 1820). Beitr. Zool. Vergl. Anat. Abt., 2:66.
TYPE LOCALITY: "Congo". No specific locale given, probably Angola (Hill and Carter, 1941:71).
DISTRIBUTION: Zaire, Angola, Namibia.
SYNONYMS: *damarensis* (Roberts, 1938); *flavinus* Thomas, 1904; *interior* Thomas, 1916; *oenone* Thomas, 1926; *olivellus* Thomas, 1904; *poolii* (Jentink, 1906); *praetextus* Wagner, 1843.

Funisciurus isabella (Gray, 1862). Proc. Zool. Soc. Lond., 1862:180.
TYPE LOCALITY: Cameroon, "...from the Camaroon Mountains, 7000 feet (2100 m) above the level of the sea..."
DISTRIBUTION: Cameroon, Central African Republic, Congo.
SYNONYMS: *dubosti* Eisentraut, 1969; *duchaillui* Sanborn, 1953.

Funisciurus lemniscatus (Le Conte, 1857). Proc. Acad. Nat. Sci. Philadelphia, p. 11.
TYPE LOCALITY: Equatorial Guinea, Rio Muni. "...from Western Africa."
DISTRIBUTION: S of Sanaga River (Cameroon), Central African Republic, Zaire.
SYNONYMS: *mayumbicus* Kershaw, 1923; *sharpei* (Gray, 1873).

Funisciurus leucogenys (Waterhouse, 1842). Ann. Mag. Nat. Hist., [ser. 1], 10: 202.
TYPE LOCALITY: Equatorial Guinea, Bioko "...brought from Fernando Po".
DISTRIBUTION: Ghana, Togo, Benin, Nigeria, Cameroon, Central African Republic, Rio Muni, Bioko.

SYNONYMS: *auriculatus* (Matschie, 1891); *beatus* (Thomas, 1910); *boydi* (Thomas, 1910); *erythrogenys* (Waterhouse, 1843); *oliviae* (Dollman, 1911).
COMMENTS: Waterhouse (1842[1843]) renamed this species *erythrogenys*, "red-cheeked" in an attempt to replace the inappropriate name *leucogenys*, "white-cheeked". This is an unjustified emendation.

Funisciurus pyrropus (F. Cuvier, 1833). *In* E. Geoffroy and F. Cuvier, Hist. Nat. Mammifères, No. 66, 2 unno. pp and pl.[1833]; Tab 4:240 [1842].
TYPE LOCALITY: Gabon. "...et elle venoit de l'île Fernandopô, dans le gulfe de Guinée..."
DISTRIBUTION: Gambia, S Senegal, Guinea Bissau, W Guinea, Sierra Leone, Liberia, S Ivory Coast, SW Ghana, W Nigeria, W Cameroon, Rio Muni (Equatorial Guinea), W Congo, Uganda, Rwanda, Burundi, Zaire, NW Angola.
SYNONYMS: *akka* De Winton, 1895; *emini* (De Winton, 1895; not Stuhlman, 1894); *erythrops* (Gray, 1867); *leonis* Thomas, 1905; *leucostigma* (Temminck, 1853); *mandingo* Thomas, 1903; *nigrensis* Thomas, 1909; *niveatus* Thomas, 1923; *pembertoni* Thomas, 1904; *rubripes* (Du Chaillu, 1860); *talboti* Thomas, 1909; *victoriae* Allen and Loveridge, 1942; *wintoni* Neumann, 1900.
COMMENTS: Cuvier wrote, "Je donnerai à cet Écuriel le nom de *Pyrropus*, á cause de la couleur rousse de ses pieds." Schinz (1845) spelled the species name *pyrrhopus* with no mention of the previous spelling. This constitutes an unjustified emendation. Not found on Bioko, Equatorial Guinea (= "Fernandopô", see type locality), and since the animal was a pet, it probably was captured on the mainland (Thomas, 1890:447).

Funisciurus substriatus De Winton, 1899. Ann. Mag. Nat. Hist., ser. 7, 4:357.
TYPE LOCALITY: Ghana. "...near Kintampo, Gold Coast hinterland, 800 feet (240 m)."
DISTRIBUTION: Ivory Coast, S Ghana, Togo, Benin, SE Nigeria.

Glyphotes Thomas, 1898. Ann. Mag. Nat. Hist., ser. 7, 2:251.
TYPE SPECIES: *Glyphotes simus* Thomas, 1898.
COMMENTS: Tribe Callosciurini (Moore, 1959). Corbet and Hill (1992) placed this genus in *Callosciurus*. Both Payne et al. (1985) and Corbet and Hill (1992) considered *Hessonoglyphotes* Moore, 1959, to be a junior synonym of *Callosciurus*.

Glyphotes simus Thomas, 1898. Ann. Mag. Nat. Hist., ser. 7, 2:251.
TYPE LOCALITY: "Mount Kina Balu, N. Bórneo" [Sabah, Malaysia].
DISTRIBUTION: Mountains of Borneo; Sabah and Sarawak (Malaysia), 1000 to 1700 m.
COMMENTS: Reviewed by Hill (1959).

Heliosciurus Trouessart, 1880. Le Naturaliste, 2nd year, 1:292.
TYPE SPECIES: *Sciurus gambianus* Ogilby, 1835 as designated by Opinion 464 of the International Commission on Zoological Nomenclature (1957*d*) in which *Sciurus annulatus* Desmarest, 1822 was suppressed.
COMMENTS: Tribe Protoxerini (Moore, 1959). Includes *Aethosciurus ruwenzorii*. Reviewed in part by Grubb (1982*b*).

Heliosciurus gambianus (Ogilby, 1835). Proc. Zool. Soc. Lond., 1835:103.
TYPE LOCALITY: Gambia, possibly near Ft. St. Mary. "...brought from the Gambia..." "Through... Mr. Rendall, who has lately arrived from the Gambia, where his brother is lieutenent-governor of Fort St. Mary and the other British possessions in that neighbourhood..."
DISTRIBUTION: Senegal, Gambia, Guinea Bissau, Guinea, Sierra Leone, Liberia, Ivory Coast, Burkina Faso, Ghana, Togo, Benin, Nigeria, Chad, Central African Republic, Sudan, Ethiopia, Uganda, Kenya, Burundi, Tanzania, Zaire, Angola, Zimbabwe, Zambia.
SYNONYMS: *abassensis* (Neumann, 1902); *albina* (Gray, 1867); *annularis* (Schinz, 1845); *annulatus* (Desmarest, 1822); *bongensis* Heuglin, 1877; *canaster* Thomas and Hinton, 1923; *dysoni* (St. Leger, 1937); *elegans* Thomas, 1909; *hoogstraali* Setzer, 1954; *kaffensis* (Neumann, 1902); *lateris* (Thomas, 1909); *limbatus* Schwarz, 1915; *loandicus* Thomas, 1923; *madogae* (Heller, 1911); *multicolor* (Rüppel 1835); *omensis* (Thomas, 1904); *rhodesiae* (Wroughton, 1907); *senescens* Thomas, 1909; *simplex* (Lesson, 1838).

Heliosciurus mutabilis (Peters, 1852) Bericht Verhandl. K. Preuss Akad. Wiss., Berlin, 17:273.
TYPE LOCALITY: Mozambique, Boror, 19 km NW of Quelimane, 17°S . "Africa orientalis, Boror, 17° Lat. Austr."
DISTRIBUTION: Malawi; S and SW highlands, Tanzania; NW of the Zambezi R near Beira (Mozambique); Chirinda Forest, Melsetter Dist., Sabi/Lundi River confluence, Vumba, Umtali (SE Zimbabwe).
SYNONYMS: *beirae* Roberts, 1913; *chirindensis* Roberts, 1913; *shirensis* (Gray, 1867); *smithersi* (Lundholm, 1955); *vumbae* Roberts, 1937.
COMMENTS: Allen (1939) considered *mutabilis* a distinct species. Ellerman (1940) treated it as a subspecies of *H. gambianus* following Ingoldby (1927). Rosevear (1963), followed by Amtmann (1975), treated it as a subspecies of *H. rufobrachium*. Grubb (1982*b*) again treated *H. mutabilis* as a distinct species.

Heliosciurus punctatus (Temminck, 1853) Esquisses Zool. sur la Côte de Guiné, p. 138.
TYPE LOCALITY: Guinea coast, no exact locality given. "...dans toutes les forêts de la Guiné..." Given by Ingoldby (1927) as Ghana: "Secondi and Bibiani, Gold Coast."
DISTRIBUTION: E Liberia, S Ivory Coast, S Ghana (E to Lake Volta).
SYNONYMS: *savannius* Thomas, 1923.
COMMENTS: Treated as a subspecies of *H. gambianus* by Ingoldby (1927), Ellerman (1940), Rosevear (1969), and Amtmann (1975). Considered a distinct species by Allen (1939), and Roth and Thorington (1982).

Heliosciurus rufobrachium (Waterhouse, 1842). Ann. Mag. Nat. Hist., [ser. 1], 10:202.
TYPE LOCALITY: Equatorial Guinea, Bioko "...brought from Fernando Po...".
DISTRIBUTION: Senegal, W Gambia, W Guinea Bissau, W Guinea, Sierra Leone, Liberia, S Ivory Coast, S Ghana, S Togo, Benin, S Nigeria, Cameroon, Bioko and Rio Muni (Equatorial Guinea), SW Central African Republic, SE Sudan, Uganda, Rwanda, Burundi, SW and SE Kenya, E and NW Tanzania, Malawi, Mozambique, E Zimbabwe, Zaire.
SYNONYMS: *acticola* Thomas, 1923; *arrhenii* (Lönnberg, 1917); *aschantiensis* (Neumann, 1902); *aubryi* (Milne-Edwards, 1867); *benga* Cabrera, 1917; *brauni* St. Leger, 1935; *caurinus* Thomas, 1923; *coenosus* (Thomas, 1909); *emissus* Thomas, 1923; *hardyi* Thomas, 1923; *isabellinus* (Gray, 1867); *keniae* (Neumann, 1902); *leonensis* Thomas, 1923; *libericus* (Miller, 1900); *lualabae* Thomas, 1923; *maculatus* (Temminck, 1853); *medjianus* Allen, 1922; *nyansae* (Neuman, 1902); *obfuscatus* Thomas, 1923; *occidentalis* (Monard, 1941); *pasha* (Schwann, 1904); *rubricatus* Allen, 1922; *rufo-brachiatus* (Waterhouse, 1843); *semlikii* Thomas, 1907; *waterhousii* (Gray, 1867); *leakyi* Toschi, 1946.
COMMENTS: Inclusion of *aubryi* and *emissus* questioned (Amtmann, 1975; Grubb, 1978:158).

Heliosciurus ruwenzorii (Schwann, 1904). Ann. Mag. Nat. Hist., ser. 7, 13:71.
TYPE LOCALITY: E Zaire, Ruwenzori, Wimi Valley.
DISTRIBUTION: Ruwenzori Mtns in E Zaire; Rwanda; Burundi; SW Uganda.
SYNONYMS: *ituriensis* (Prigogone, 1954); *schoutedeni* (Prigogine, 1954); *vulcanius* Thomas, 1909.
COMMENTS: Formerly included in *Aethosciurus*, which is here included in *Paraxerus* following Moore (1959).

Heliosciurus undulatus (True, 1892) Proc. U.S. Natl. Mus., 15:465.
TYPE LOCALITY: Tanzania, Mt. Kilimanjaro. "Male. Mount Kilima-Njaro, June 12, 1888. 6,000 feet (1800 m). Female. Kahé, south of Mount Kilima-Njaro, September 6, 1888."
DISTRIBUTION: SE Kenya; NE Tanzania, including Mafia and Zanzibar Isls.
SYNONYMS: *daucinus* Thomas, 1909; *dolosus* Thomas, 1909; *marwitzi* Müller, 1911; *shindi* (Heller, 1914).

Hyosciurus Archbold and Tate, 1935. Am. Mus. Novit., 801:2.
TYPE SPECIES: *Hyosciurus heinrichi* Tate and Archbold, 1935.
COMMENTS: Tribe Callosciurini (Moore, 1959:174).

Hyosciurus heinrichi Archbold and Tate, 1935. Am. Mus. Novit., 801: 2.
TYPE LOCALITY: "Latimodjong Mtns., central Celebes, 2300 meters" [C Sulawesi, Indonesia].
DISTRIBUTION: Restricted to mountains of C Sulawesi (Musser and Dagosto, 1987:44).
COMMENTS: Formerly included *ileile* (Tate and Archbold, 1936:1); see comments therein.

Hyosciurus ileile Tate and Archbold, 1936. Am. Mus. Novit., 846:1.
TYPE LOCALITY: "Ile-ile, North Celebes, 1700 meters" [N Sulawesi, Indonesia].
DISTRIBUTION: Mountains of N Sulawesi.
COMMENTS: Formerly considered a subspecies of *heinrichi* (Tate and Archbold, 1936), but see Musser (1987).

Lariscus Thomas and Wroughton, 1909. Proc. Zool. Soc. Lond., 1909:389.
TYPE SPECIES: *Sciurus insignis* F. Cuvier, 1821.
SYNONYMS: *Laria* Gray, 1867 (not Scopoli, 1763); *Paralariscus* Ellerman, 1947.
COMMENTS: Tribe Callosciurini (Moore, 1959:173). Includes *Paralariscus*, treated as a genus by Ellerman (1947:259) and Moore (1959:173).

Lariscus hosei (Thomas, 1892). Ann. Mag. Nat. Hist., ser. 6, 10:215.
TYPE LOCALITY: "Batu Sang Mount [Mt. Batu Song], Baram River, N. Borneo (5000 feet)" [Baram Dist., Sarawak, Malaysia].
DISTRIBUTION: Mountains of Sarawak and Sabah (Malaysia).
COMMENTS: Sometimes put in the genus *Paralariscus*; see Ellerman (1947:259); but also see Medway (1977:97).

Lariscus insignis (F. Cuvier, 1821). *In* E. Geoffroy St. Hilaire and F. Cuvier, Hist. Nat. Mammifères, Pt. 2, 4(34):2 pp
TYPE LOCALITY: "Sumatra," [Indonesia]. Restricted by Sody (1949*a*:113) to Lampongs, S Sumatra.
DISTRIBUTION: Peninsular Malaysia and Thailand, Sumatra, Java, Borneo, adjacent isls.
SYNONYMS: *atchinensis* Sody, 1949; *castaneus* (Miller, 1900); *diversoides* Sody, 1949; *diversus* (Thomas, 1898); *fornicatus* Robinson, 1917; *jalorensis* (Bonhote, 1903); *javanus* (Thomas and Wroughton, 1909); *meridionalis* Robinson and Kloss, 1911; *murianus* Sody, 1937; *peninsulae* (Miller, 1903); *rostratus* (Miller, 1903); *saturatus* Chasen, 1935; *vulcanus* Kloss, 1921.
COMMENTS: Formerly included *niobe* and *obscurus*; see Chasen (1940:145-146); but see also Corbet and Hill (1991:141); see comments under *niobe*.

Lariscus niobe (Thomas, 1898) Ann. Mag. Nat. Hist., ser. 7, 2:249.
TYPE LOCALITY: "Pajo, Sumatra" [Sumatra, Indonesia]. Restricted by Robinson and Kloss (1918*b*:37) to Pajokombo, Padang Highlands, West Sumatra.
DISTRIBUTION: Mountains of Sumatra, Mentawi Isls, and Java (Idjen Mtns).

Lariscus obscurus (Miller, 1903). Smithson. Misc. Coll., 45:23.
TYPE LOCALITY: Indonesia, Mentawi Isls, "South Pagi Island, Sumatra".
DISTRIBUTION: Siberut Isl and S Pagi Isl (Sumatra, Indonesia).
SYNONYMS: *auroreus* Sody, 1949; *siberu* Chasen and Kloss, 1928.
COMMENTS: Formerly included in *insignis* (Chasen, 1940) or *niobe* (Chasen and Kloss (1927). Part of the rodent fauna endemic to the Mentawi Archipelago (see comments under *Leopoldamys siporanus*).

Marmota Blumenbach, 1779. Hand. Hilfsb. Nat., 1:79.
TYPE SPECIES: *Mus marmota* Linnaeus, 1758.
SYNONYMS: *Arctomys* Schreber, 1780; *Glis* Erxleben, 1777; *Lagomys* Storr, 1780; *Lipura* Storr, 1780; *Marmotops* Pocock, 1922.
COMMENTS: Tribe Marmotini (Moore, 1959). North American species reviewed by Howell (1915); Eurasian species revised by Gromov et al. (1965); amphiberingian species reviewed by Hoffmann et al. (1979). Frase and Hoffmann (1980) provided a key to North American species.

Marmota baibacina Kastschenko, 1899. Rezul't. Altaisk. Zool. Exp. 1898, p. 62.
TYPE LOCALITY: "...Multa River, near Nizhne-Uimon in the Altai Mountains" [Altaisk. Krai, Russia] (Ognev, 1963*a*:252). Alternatively, Aktol' River near Cherga, Gorno-Altaisk. A.O. (Kuznetsov, *in* Ellerman and Morrison-Scott, 1951:514).
DISTRIBUTION: Altai Mtns, SW Siberia (Russia), SE Kazakhstan, Kirgizistan; Mongolia; Xinjiang (China). Introduced into Caucasus Mtns (Dagestan, Russia; Gromov et al., 1965:360).

SYNONYMS: *aphanasievi* Kuznetsov, 1965; *centralis* (Thomas, 1909); *kastschenkoi* Stroganov and Yudin, 1956; *lewisi* (Audubon and Bachman, 1854); *ognevi* Scalon, 1950.

COMMENTS: Placed by Ellerman and Morrison-Scott (1951:514) in *marmota*, and by Corbet (1978*c*:81) in *bobak*; Kapitonov (1966) analyzed purported hybridization between *baibacina* and *bobak*, while Nikol'skii (1974) and Nikol'skii et al. (1983) found species-specific vocalizations. Most Russian authors retain both as distinct species (Gromov et al., 1965:337-387; Zimina, 1978; Zholnerovskaya et al., 1990) and include *centralis* in this species. Kapitonov (1966) indicated that the population called *aphanasievi* is included in this species; but also see Corbet (1978*c*:81). Includes *lewisi*, a *nomen oblitum* (Hoffmann, 1977); *baibacina* (Brandt, 1843) is a *nomen nudum*. See also *bobak*, *sibirica*.

Marmota bobak (Müller, 1776). Linné's Vollstand. Natursyst. Suppl., p. 40.

TYPE LOCALITY: "Poland." Restricted by Ognev (1963*a*:221) to "right [W] bank of the Dnepr" [River], Ukraine.

DISTRIBUTION: Steppes of E Europe, east through Belarus, Ukraine, and Russia to N and C Kazakhstan.

STATUS: Regaining parts of former range (Bibikov, 1991).

SYNONYMS: *arctomys* (Pallas, 1779); *baibac* (Pallas, 1811); *bobac* (Schreber, 1780); *kozlovi* Fokanov, 1966; *tschaganensis* Bazhanov, 1930.

COMMENTS: See comments under *baibacina*, *himalayana*, and *sibirica*. Includes *kozlovi* (see Fokanov, 1966) and *tschaganensis* (see Gromov et al., 1965).

Marmota broweri Hall and Gilmore, 1934. Canadian Field Nat., 48:57.

TYPE LOCALITY: USA, "Point Lay, Arctic Coast of Alaska" Restricted by Rausch (1953:117) to head of Kukpowruk River, Alaska.

DISTRIBUTION: Brooks Range of N Alaska (USA) from near coast of Chukchi Sea to Alaska-Yukon border; perhaps also N Yukon (Canada).

COMMENTS: Regarded as a synonym of *caligata* (Hall, 1981) but Rausch and Rausch (1965, 1971) and Hoffmann et al. (1979) considered *broweri* a distinct species.

Marmota caligata (Eschscholtz, 1829). Zool. Atlas, Part 2, p. 1, pl. 6.

TYPE LOCALITY: Not specified (?). Restricted by Allen (1877:927) to near Bristol Bay, Alaska, USA.

DISTRIBUTION: C Alaska (USA), Yukon and Northwest Territories (Canada) south to W and NE Washington, C Idaho, and W Montana (USA).

SYNONYMS: *cascadensis* A. Howell, 1914; *nivaria* A. Howell, 1914; *okanagana* (King, 1836); *oxytona* Hollister, 1914; *raceyi* Anderson, 1932; *sheldoni* A. Howell, 1914; *sibila* Hollister, 1912; *vigilis* Heller, 1909.

Marmota camtschatica (Pallas, 1811). Zoogr. Rosso-Asiat., p. 156.

TYPE LOCALITY: "Kamchatka" [Kamchatsk. Obl., Russia].

DISTRIBUTION: E Siberia from Transbaikalia to Chukotka and Kamchatka (Russia), in several geographically isolated populations (Nikol'skii et al., 1991).

SYNONYMS: *bungei* (Kastschenko, 1901); *cliftoni* (Thomas, 1902); *doppelmayeri* Birula, 1922.

COMMENTS: Regarded as a synonym of *marmota* (Ellerman and Morrison-Scott, 1951; Rausch, 1953). Hoffmann et al. (1979) reviewed this and related species, and affirmed its specific status. Kapitonov (1978) concluded that morphological differences justified independent specific status for *doppelmayeri*, but Nikol'skii et al. (1991) showed similarity of vocalization between it and *bungei*, while the nominate form differed, and recommended that *doppelmayeri* be retained provisionally in this species.

Marmota caudata (Geoffroy, 1844). *In* Jacquemont, Voy. dans l'Inde, 4, Zool., p. 66.

TYPE LOCALITY: "Hombur [Ghombur] area, upper reaches of the Indus in Kashmir [India]" (Ognev, 1963*a*:284).

DISTRIBUTION: W Tien Shan through the Pamirs (Kirgizistan, Tadzhikistan) to Hindu Kush (Afghanistan), Pakistan, Kashmir (India), and mtns of extreme W Xinjiang and Xizang (China).

STATUS: CITES - Appendix III (India).

SYNONYMS: *aurea* (Blanford, 1875); *dichrous* (Anderson, 1875); *flavina* (Thomas, 1909); *littledalei* (Thomas, 1909); *stirlingi* Thomas, 1916.

COMMENTS: Includes *dichrous* (Corbet, 1978c:82); but also see Gromov et al. (1965:440) who listed it as a distinct species.

Marmota flaviventris (Audubon and Bachman, 1841). Proc. Acad. Nat. Sci. Philadelphia, 1:99.
TYPE LOCALITY: "Mountains between Texas and California" Restricted by Howell (1915) to Mt. Hood [Oregon, USA].
DISTRIBUTION: SC British Columbia and S Alberta (Canada) south to N New Mexico, S Utah, Nevada, and California (USA).
SYNONYMS: *avara* (Bangs, 1899); *campioni* Figgins, 1915; *dacota* (Merriam, 1889); *engelhardti* J. Allen, 1905; *fortirostris* Grinnell, 1921; *luteola* A. Howell, 1914; *nosophora* A. Howell, 1914; *notioros* Warren, 1934; *obscura* A. Howell, 1914; *parvula* A. Howell, 1915; *sierrae* A. Howell, 1915; *warreni* A. Howell, 1914.
COMMENTS: Reviewed by Frase and Hoffmann (1980, Mammalian Species, 135).

Marmota himalayana (Hodgson, 1841). J. Asiat. Soc. Bengal, 10:777.
TYPE LOCALITY: "Himalaya...and sandy plains of Tibet"; "potius Tibetensis" (Hodgson, 1843). Restricted by Blanford (1875) to "the Kachar of Nepal."
DISTRIBUTION: Montane regions of W China, Nepal, and N India to Ladak.
STATUS: CITES - Appendix III (India).
SYNONYMS: *hemachalana* (Hodgson, 1843); *hodgsoni* (Blanford, 1879); *robusta* (Milne-Edwards, 1872); *tataricus* (Jameson, 1847); *tibetanus* (Gray, 1847).
COMMENTS: Placed in *bobak* (Ellerman and Morrison-Scott, 1951:515; Corbet, 1978c:81), but geographically separated from that species; evidence for specific status in Gromov et al. (1965); see also comment under *baibacina*.

Marmota marmota (Linnaeus, 1758). Syst. Nat., 10th ed., 1:60.
TYPE LOCALITY: "in alpibus Helveticis" Restricted by Thomas (1911a:147) to Swiss Alps.
DISTRIBUTION: Swiss, Italian, and French Alps; W Austria; S Germany; Carpathian (Romania) and Tatra Mtns (Czechoslovakia, Poland); introduced into French Pyrennes, E Austria and N Yugoslavia.
SYNONYMS: *alba* (Bechstein, 1801); *alpina* Blumenbach, 1779; *latirostris* Kratochvil, 1961; *marmotta* Trouessart, 1904; *nigra* (Bechstein, 1801); *tigrina* (Bechstein, 1801).
COMMENTS: Formerly included *baibacina*, *broweri*, *caligata*, *camtschatica*, and *menzbieri* (Ellerman and Morrison-Scott, 1951; Rausch, 1953).

Marmota menzbieri (Kashkarov, 1925). Trans. Turk. Sci. Soc., 2:47.
TYPE LOCALITY: "Chigyr-Tash, in the headwaters of the Ugam River, Talass Ala Tau" [Yuzhno-Kazakhstansk. Obl., Kazakhstan].
DISTRIBUTION: W Tien Shan Mtns, in S Kazakhstan and NW Kirgizia.
STATUS: IUCN - Vulnerable.
SYNONYMS: *zachidovi* Petrov, 1963.
COMMENTS: Regarded by Ellerman and Morrison-Scott (1951:514) as a probable synonym of *marmota*; but see Corbet (1978c:81).

Marmota monax (Linnaeus, 1758). Syst. Nat., 10th ed., 1:60.
TYPE LOCALITY: "in America septentrionalis." Restricted by Thomas (1911a:147) to Maryland [USA].
DISTRIBUTION: Alaska (USA) through S Canada to S Labrador to NE and SC USA; south in Rocky Mtns, possibly to N Idaho.
SYNONYMS: *bunkeri* Black, 1935; *canadensis* (Erxleben, 1777); *empetra* (Pallas, 1778); *ignava* (Bangs, 1899); *johnsoni* Anderson, 1943; *melanopus* (Kuhl, 1820); *ochracea* Swarth, 1911; *petrensis* A. Howell, 1915; *preblorum* A. Howell, 1914; *rufescens* A. Howell, 1914; *sibila* (Wolf, 1808).

Marmota olympus (Merriam, 1898). Proc. Acad. Nat. Sci. Philadelphia, 50:352.
TYPE LOCALITY: "From Timberline at head of Soleduc River, Olympic Mountains, [Olympic Nat. Park] Washington [USA]."
DISTRIBUTION: Olympic Mtns of W Washington (USA).
COMMENTS: Considered a subspecies of *marmota* (Rausch, 1953). Reviewed by Hoffmann et al. (1979) who confirmed its specific status.

Marmota sibirica (Radde, 1862). Reise in den Suden von Ost-Sibierien, p. 159.
TYPE LOCALITY: "Kulusutai, near Lake Torei-Nor, southeast Transbaikal" [Chitinsk Obl., Russia].
DISTRIBUTION: SW Siberia, Tuva, Transbaikalia (Russia); N and W Mongolia; Heilungjiang and Inner Mongolia (China).
SYNONYMS: *caliginosus* Bannikov and Skalon, 1949; *dahurica* (Dybowski, 1922).
COMMENTS: Placed (with *baibacina*) in *bobak* (Ellerman and Morrison-Scott, 1951:515; Corbet, 1978*c*:81). Gromov et al. (1965) and Zimina (1978) provided evidence of specific distinctness and included *caliginosus* in this species; see comment under *baibacina*. Nikol'skii (1974) and Smirin et al. (1985) analyzed contact between *baibacina* and *sibirica* in Tuva and the Mongolian Altai; Sokolov and Orlov (1980:329) also indicated sympatry in NW Mongolia; limited hybridization is possible.

Marmota vancouverensis Swarth, 1911. Univ. California Publ. Zool., 7:201.
TYPE LOCALITY: "Mt. Douglas (altitude 4,200 feet), twenty miles south of Alberni, Vancouver Island, British Columbia" [Canada].
DISTRIBUTION: Mountains of Vancouver Isl (British Columbia, Canada).
STATUS: U.S. ESA and IUCN - Endangered.
COMMENTS: Considered a subspecies of *marmota* by Rausch (1953). Reviewed by Hoffmann et al. (1979) who confirmed its specific status, and by Nagorsen (1987, Mammalian Species, 270).

Menetes Thomas, 1908. J. Bombay Nat. Hist. Soc., 18:244.
TYPE SPECIES: *Sciurus berdmorei* Blyth, 1849.
COMMENTS: Tribe Callosciurini (Moore, 1959:173). Reviewed by Moore and Tate (1965).

Menetes berdmorei (Blyth, 1849). J. Asiat. Soc. Bengal, 18(30):603.
TYPE LOCALITY: "Thougyeen district" [Tenasserim, Burma].
DISTRIBUTION: S Vietnam, Cambodia, S Laos, Thailand, S Yunnan (China) to C Burma.
SYNONYMS: *amotus* (Miller, 1913); *consularis* Thomas, 1914; *decoratus* Thomas, 1914; *koratensis* (Gyldenstolpe, 1917); *moerescens* Thomas, 1914; *mouhotei* (Gray, 1861); *peninsularis* Kloss, 1919; *pyrrocephalus* (Milne-Edwards, 1867); *rufescens* Kloss, 1916; *umbrosus* Kloss, 1916.
COMMENTS: For taxonomic history, see Moore and Tate (1965:294). Chromosomes described by Nadler and Hoffmann (1970).

Microsciurus J. A. Allen, 1895. Bull. Am. Mus. Nat. Hist., 7:332.
TYPE SPECIES: *Sciurus alfari* J. A. Allen, 1895.
COMMENTS: Tribe Microsciurini (Moore, 1959). See Emmons and Feer (1990).

Microsciurus alfari (J. A. Allen, 1895). Bull. Am. Mus. Nat. Hist., 7:333.
TYPE LOCALITY: Costa Rica, Jimenez.
DISTRIBUTION: Colombia, Costa Rica, Nicaragua, Panama.
SYNONYMS: *alticola* Goodwin, 1943; *browni* (Bangs, 1902); *fusculus* (Thomas, 1910); *septentrionalis* Anthony, 1920; *venustulus* Goldman, 1912.
COMMENTS: See Hall (1981:439-440).

Microsciurus flaviventer (Gray, 1867). Ann. Mag. Nat. Hist., ser. 3, 20:432.
TYPE LOCALITY: Brazil. Cabrera (1961) suggested the type locality could be restricted to Pebas, based on Thomas (1928*b*).
DISTRIBUTION: Amazon basin of Colombia, Ecuador, Peru, and Brazil west of the Rios Negro and Jurua.
SYNONYMS: *avunculus* Thomas, 1914; *brevirostris*; *florenciae* J. A. Allen, 1914; *manarius* Thomas, 1920; *napi* (Thomas, 1900); *otinus* (Thomas, 1901); *peruanus* (J. A. Allen, 1897); *rubicollis* Thomas, 1914 (lapsus for *rubrirostris*); *rubrirostris* J. A. Allen, 1914; *sabanillae* Anthony, 1922; *similis* (Nelson, 1899); *simonsi* (Thomas, 1900); *vivatus* Goldman, 1912.

Microsciurus mimulus (Thomas, 1898). Ann. Mag. Nat. Hist., ser. 7, 2:266.
TYPE LOCALITY: Ecuador, Esmeraldas, Cachavi, 665 ft. (203 m).
DISTRIBUTION: NW Ecuador, N Colombia, and Panama.
SYNONYMS: *boquetensis* (Nelson, 1903); *isthmius* (Nelson, 1899); *palmeri* (Thomas, 1909).

COMMENTS: See Hall (1981:440) and Handley (1966*a*).

Microsciurus santanderensis (Hernandez-Camacho, 1957). *In* Borerro and Hernandez-Camacho, Anal. Soc. Biol. Bogota, 7:219.
TYPE LOCALITY: Colombia, Santander Dept., Meseta de los Caballeros, NE of La Albania.
DISTRIBUTION: Colombia, between the Magdalena River and the Cordillera Oriental.
COMMENTS: See Eisenberg (1989) and Hernandez-Camacho (1960).

Myosciurus Thomas, 1909. Ann. Mag. Nat. Hist., ser. 8, 3:474.
TYPE SPECIES: *Sciurus minutus* Du Chaillu, 1860 (= *Sciurus pumilio* Le Conte, 1857).
COMMENTS: Tribe Funambulini (Moore, 1959).

Myosciurus pumilio (Le Conte, 1857). Proc. Acad. Nat. Sci. Philadelphia, 9:11.
TYPE LOCALITY: Gabon. "...the head waters of the Ovenga River...".
DISTRIBUTION: SE Nigeria, Cameroon, Gabon, Bioko (Equatorial Guinea).
SYNONYMS: *minutulus* Hollister, 1921; *minutus* (Du Chaillu, 1860).

Nannosciurus Trouessart, 1880. Le Naturaliste, p. 292.
TYPE SPECIES: *Sciurus melanotis* Müller, 1840.
COMMENTS: Tribe Callosciurini (Moore, 1959). Formerly included the species now placed in *Exilisciurus*; see Moore (1958) and Heaney (1985).

Nannosciurus melanotis (Müller, 1840). *In* Temminck, Verh. Nat. Gesch. Nederland. Overz. Bezitt., Zool., Zoogd. Indisch. Archipel., p. 35[1840], see comments.
TYPE LOCALITY: "Borneo".
DISTRIBUTION: Sumatra, Java, Borneo, adjacent small islands.
SYNONYMS: *bancanus* Lyon, 1906; *borneanus* Lyon, 1906; *pallidus* Chasen and Kloss, 1928; *pulcher* Miller, 1902; *soricinus* (Waterhouse, 1838); *sumatranus* Lyon, 1906.
COMMENTS: This species was further described by Müller and Schlegel, *in* Temminck, Verh. Nat. Gesch. Nederland. Overz. Bezitt., Zool., Mammalia, pp. 87, 88, 98[1844], pl. 14, fig 4, 5[1841] (from which this species is often cited). See Appendix I. Reviewed by Heaney (1985).

Paraxerus Forsyth Major, 1893. Proc. Zool. Soc. Lond., 1893:189.
TYPE SPECIES: *Sciurus cepapi* Smith, 1836.
SYNONYMS: *Aethosciurus* Thomas, 1916; *Montisciurus* Eisentraut, 1976; *Tamiscus* Thomas, 1918.
COMMENTS: Tribe Funambulini (Moore, 1959). Originally a subgenus of *Xerus*; includes *Aethosciurus* (with the exception of *ruwenzorii*, here included in *Heliosciurus*), *Montisciurus*, and *Tamiscus*. This treatment follows Moore (1959).

Paraxerus alexandri (Thomas and Wroughton, 1907). Ann. Mag. Nat. Hist., ser. 7, 19:376.
TYPE LOCALITY: Zaire, Gudima, River Iri, Upper Welle.
DISTRIBUTION: NE Zaire, Uganda.

Paraxerus boehmi (Reichenow, 1886). Zool. Anz., 9:315.
TYPE LOCALITY: SE Zaire, Marungu. "Marungu (Inner-Afrika)."
DISTRIBUTION: S Sudan, N, S, and E Zaire, Uganda, W Kenya, NW Tanzania, N Zambia.
SYNONYMS: *antoniae* Thomas and Wroughton, 1907; *emini* (Stuhlman, 1894, not De Winton, 1895); *gazellae* (Thomas, 1918); *lunaris* (Thomas, 1918); *tanganyikae* (Thomas, 1918); *ugandae* (Neumann, 1902); *vulcanorum* (Thomas, 1918).

Paraxerus cepapi (A. Smith, 1836). Rept. Exped. Exploring Central Africa, p. 43.
TYPE LOCALITY: South Africa, W Transvaal, Rustenberg Dist., Marico River.
DISTRIBUTION: S Angola, Zambia, SE Zaire, Malawi, SW Tanzania, Mozambique, N Namibia, N Botswana, Zimbabwe, Transvaal (South Africa).
SYNONYMS: *bororensis* Roberts, 1946; *carpi* Lundholm, 1955; *cepate* (lapsis for *cepapi*; Gray, 1843); *cepapoides* Roberts, 1946; *chobiensis* Roberts, 1932; *kalaharicus* Roberts, 1932; *maunensis* Roberts, 1932; *phalaena* Thomas, 1926; *quotus* Wroughton, 1909; *sindi* Thomas and Wroughton, 1908; *soccatus* Wroughton, 1909; *yulei* (Thomas, 1902).

Paraxerus cooperi Hayman, 1950. Ann. Mag. Nat. Hist., ser. 12, 3:262.
TYPE LOCALITY: Cameroon, Kumba Div., Rumpi Hills, 5°N, 9°15′E.

DISTRIBUTION: Cameroon.
COMMENTS: Eisentraut (1976) put *cooperi* in a separate genus, *Montisciurus*.

Paraxerus flavovittis (Peters, 1852). Bericht Verhandl. K. Preuss. Akad. Wiss. Berlin, 17:274.
TYPE LOCALITY: NE Mozambique, Mocímboa, 11°S on the coast. "Africa orientalis, Mossimboa, Quitangonha, a 11° ad 15° Lat. Austr."
DISTRIBUTION: S Kenya, Tanzania, N Mozambique.
SYNONYMS: *exgeanus* (Hinton, 1920); *ibeanus* (Hinton, 1920); *mossambicus* (Thomas, 1919).
COMMENTS: Commonly spelled *flavivittis* but this was an unjustified emendation by Peters (1852).

Paraxerus lucifer (Thomas, 1897). Proc. Zool. Soc. Lond., 1897:430.
TYPE LOCALITY: Malawi, Kombe Forest, Misuku Mtns, 9°43′S, 33°31′E.
DISTRIBUTION: N Malawi, SW Tanzania, E Zambia.

Paraxerus ochraceus (Huet, 1880). Nouv. Arch. Mus. Hist. Nat. Paris, (2)3:154.
TYPE LOCALITY: Tanzania, Bagamoyo, (06°25′S, 38°54′E). "Cette petite espèce provient de Bagamoyo, station de nos missionnaires, sur la côte de Zanguebar,..."
DISTRIBUTION: S Sudan, Kenya, Tanzania.
SYNONYMS: *affinis* (Trouessart, 1897); *animosus* Dollman, 1911; *aruscensis* (Pagenstecher, 1885); *augustus* Dollman, 1911; *capitis* (Thomas, 1909); *electus* Thomas, 1909; *ganana* (Rhoads, 1896); *jacksoni* (De Winton, 1897); *kahari* Heller, 1911; *pauli* Matschie, 1894; *percivali* Dollman, 1911; *salutans* Thomas, 1909.

Paraxerus palliatus (Peters, 1852). Bericht Verhandl. K. Preuss. Akad. Wiss., Berlin, 17:273.
TYPE LOCALITY: Mozambique, mainland near Mocambique Isl. "Africa orientalis, Quintangonha, 15° Lat. Austr."
DISTRIBUTION: S Somalia, E Kenya, E Tanzania, Malawi, Mozambique, Zimbabwe, Natal (South Africa).
SYNONYMS: *auriventris* Roberts, 1926; *barawensis* (Neumann, 1902); *bridgemani* Dollman, 1914; *frerei* (Gray, 1873); *lastii* Thomas, 1906; *ornatus* (Gray, 1864); *sponsus* Thomas and Wroughton, 1907; *suahelicus* (Neumann, 1902); *swynnertoni* Wroughton, 1908; *tanae* (Neumann, 1902); *tongensis* Roberts, 1931.
COMMENTS: Reviewed by Viljoen (1989).

Paraxerus poensis (A. Smith, 1830). S. Afr. Quart. J., 2:128.
TYPE LOCALITY: Equatorial Guinea, Bioko "Fernando Póo."
DISTRIBUTION: Sierra Leone, SE Guinea, Liberia, Ivory Coast, Ghana, Benin, S Nigeria, Cameroon, Bioko (Equatorial Guinea), Congo, W Zaire.
SYNONYMS: *affinis* (Rhoads, 1896); *musculinus* (Temminck, 1853); *olivaceus* (Milne-Edwards, 1867); *subviridescens* (Le Conte, 1857).
COMMENTS: Subgenus *Aethosciurus* according to Moore (1959).

Paraxerus vexillarius (Kershaw, 1923). Ann. Mag. Nat. Hist., ser. 9, 11:591.
TYPE LOCALITY: Tanzania, Usambara, Lushoto, Wilhelmsthal.
DISTRIBUTION: C and E Tanzania.
SYNONYMS: *byatti* (Kershaw, 1923); *laetus* (Allen and Loveridge, 1933).
COMMENTS: It is possible that *vexillarius* and *byatti* are separate species, as treated by Allen (1939) and Ellerman (1940). Amtmann (1975) combined them but stated that they may be distinct species.

Paraxerus vincenti Hayman, 1950. Ann. Mag. Nat. Hist., ser. 12, 3:263.
TYPE LOCALITY: Mozambique, Namuli Mtn, N of the Zambezi River, "...collected at Namuli Mountain, Portuguese East Africa (15°21′S, 37°4′E), at 5000 ft (1500 m)..."
DISTRIBUTION: N Mozambique.

Prosciurillus Ellerman, 1947. Proc. Zool. Soc. Lond., 117:259.
TYPE SPECIES: *Sciurus murinus* Müller and Schlegel, 1844.
COMMENTS: Tribe Callosciurini; closely related to *Sundasciurus* (see Heaney, 1985). Reviewed by Moore (1958), who transferred *leucomus* from *Callosciurus* to this genus.

Prosciurillus abstrusus Moore, 1958. Am. Mus. Novit., 1890:3.

TYPE LOCALITY: "...1500 meters elevation, Gunong Tanke Salokko, 'Mengkoka Geb.' [Mekongqa Gebirgte], in the southeastern peninsula of Celebes, latitude 3° 40' S., longitude 121° 13' E.," [Sulawesi, Indonesia].

DISTRIBUTION: Known only from the type locality in SE Sulawesi.

SYNONYMS: *obscurus* Moore, 1959.

COMMENTS: Includes *obscurus*, probably a *nomen nudum*; see Moore (1959:203).

Prosciurillus leucomus (Müller and Schlegel, 1844). *In* Temminck, Verh. Nat. Gesch. Nederland. Overz. Bezitt., Zool., Mammalia, p. 87.

TYPE LOCALITY: "Celebes." Restricted by Meyer (1898) to Minahassa, NE Celebes [Sulawesi, Indonesia].

DISTRIBUTION: Sulawesi; Buton and Sangihe (= Sanghir) Isls.

SYNONYMS: *elbertae* (Schwarz, 1911); *hirsutus* (Hayman, 1946); *mowewensis* (Roux, 1910); *occidentalis* (Meyer, 1898); *rosenbergii* (Jentink, 1879); *sarasinorum* (Meyer, 1898); *tingahi* (Meyer, 1896); *tonkeanus* (Meyer, 1896); *topapuensis* (Roux, 1910).

COMMENTS: Formerly included in *Callosciurus*; see Moore (1958). Feiler (1990) raised the insular *rosenbergii* (including *tingahi*) to full specific status. For authority and date, see comments under *Rubrisciurus rubriventer.*

Prosciurillus murinus (Müller and Schlegel, 1844). *In* Temminck, Verh. Nat. Gesch. Nederland. Overz. Bezitt., Zool., Mammalia, p. 87.

TYPE LOCALITY: "Celebes." Restricted by Sody (1949*a*) to NE Celebes [Sulawesi, Indonesia].

DISTRIBUTION: NE and C Sulawesi.

SYNONYMS: *evidens* (Miller and Hollister, 1921); *griseus* (Sody, 1949); *necopinus* (Miller and Hollister, 1921).

COMMENTS: For authority and date, see comments under *Rubrisciurus rubriventer.*

Prosciurillus weberi (Jentink, 1890). Weber's Zool. Ergebn. 1:115.

TYPE LOCALITY: Indonesia, C Celebes, Luwu (near Palopo).

DISTRIBUTION: C Sulawesi (Indonesia).

COMMENTS: Elevated to specific status by Corbet and Hill (1992).

Protoxerus Forsyth Major, 1893. Proc. Zool. Soc. Lond., 1893:189.

TYPE SPECIES: *Sciurus stangeri* Waterhouse, 1842.

SYNONYMS: *Allosciurus* Conisbee, 1953.

COMMENTS: Tribe Protoxerini (Moore, 1959). Includes *Allosciurus* (replaced *Myrsilas* Thomas, 1909, which was preoccupied by *Myrsilas* Stål, 1865 [Hemiptera]) and *Protoxerus* as subgenera. Originally a subgenus of *Xerus*, raised to full generic rank by Thomas (1897*b*).

Protoxerus aubinnii (Gray, 1873). Ann. Mag. Nat. Hist., ser. 4, 12:65.

TYPE LOCALITY: Ghana, Ashanti Prov., Fanti. "Fantee".

DISTRIBUTION: Liberia, Ivory Coast, Ghana.

SYNONYMS: *salae* (Jentink, 1881).

COMMENTS: Subgenus *Allosciurus*.

Protoxerus stangeri (Waterhouse, 1842). Ann. Mag. Nat. Hist., [ser. 1], 10:202 (footnote).

TYPE LOCALITY: Equatorial Guinea, Bioko. "...brought from Fernando Po...".

DISTRIBUTION: Sierra Leone, Liberia, Ivory Coast, W Ghana, Togo, S Niger, Nigeria, Cameroon, Rio Muni, Bioko (Equatorial Guinea), Gabon, E Congo, N Angola, S Central African Republic, S Sudan, Zaire, Uganda, Rwanda, Burundi, W Kenya, N Tanzania.

SYNONYMS: *bea* Heller, 1912; *calliurus* (Peters, 1874); *caniceps* (Temminck, 1853); *centricola* (Thomas, 1906); *cooperi* Kingdon, 1971; *dissonus* Thomas, 1923; *eborivorus* (Du Chaillu, 1860); *kabobo* Verheyen, 1960; *kwango* Verheyen, 1960; *loandae* (Thomas, 1906); *moerens* Thomas, 1923; *nigeriae* (Thomas, 1906); *nordhoffi* (Du Chaillu, 1860); *notabilis* Thomas, 1923; *personatus* Kershaw, 1923; *signatus* Thomas, 1910; *subalbidus* (Du Chaillu, 1860); *temminckii* (Anderson, 1879); *torrentium* Thomas, 1923.

COMMENTS: Subgenus *Protoxerus*.

Ratufa Gray, 1867. Ann. Mag. Nat. Hist., ser. 3, 20:273.
TYPE SPECIES: *Sciurus indicus* Erxleben, 1777.
SYNONYMS: *Eosciurus* Trouessart, 1880; *Rukaia* Gray, 1867.
COMMENTS: Tribe Ratufini (Moore, 1959). Reviewed in part by Moore and Tate (1965).

Ratufa affinis (Raffles, 1821). Trans. Linn. Soc. Lond., 13:259.
TYPE LOCALITY: "Singapore Island".
DISTRIBUTION: S Vietnam, Malaysian region (except Java and SW Philippines).
STATUS: CITES - Appendix II.
SYNONYMS: *albiceps* (Jentinck, 1897); *arusinus* Lyon, 1907; *aureiventer* (I. Geoffroy, 1831); *balae* Miller, 1903; *bancana* Lyon, 1906; *banguei* Chasen and Kloss, 1932; *baramensis* Bonhote, 1900; *bulana* Lyon, 1909; *bunguranensis* (Thomas and Hartert, 1894); *carimonensis* Miller, 1906; *catemana* Lyon, 1907; *condurensis* Miller, 1906; *confinis* Miller, 1906; *conspicua* Miller, 1903; *cothurnata* Lyon, 1911; *dulitensis* Lönnberg and Mjöberg, 1925; *ephippium* (Müller, 1838); *femoralis* Miller, 1903; *frontalis* Kloss, 1932; *griseicollis* Lyon, 1911; *hypoleucos* (Horsfield, 1823); *insignis* Miller, 1903; *interposita* Kloss, 1932; *johorensis* Robinson and Kloss, 1911; *klossi* J. E. Hill, 1960; *lumholzi* Lönnberg, 1925; *masae* Miller, 1903; *nanogigas* (Thomas and Hartert, 1895); *nigrescens* Miller, 1903; *notabilis* Miller, 1902; *piniensis* Miller, 1903; *polia* Lyon, 1906; *pyrsonota* Miller, 1900; *sandakanensis* Bonhote, 1900; *sirhassenensis* Bonhote, 1900; *vittata* Lyon, 1911; *vittatula* Lyon, 1911.

Ratufa bicolor (Sparrman, 1778). Samhelle Hand. (Wet. Afd.), 1:70.
TYPE LOCALITY: Indonesia, W Java, Anjer.
DISTRIBUTION: E Nepal; SE Tibet to S Yunnan and Hainan (China); Assam (India), Burma, Thailand, Laos, Cambodia, and Vietnam, south through the Malay Peninsula to Java and Bali.
STATUS: CITES - Appendix II.
SYNONYMS: *albiceps* (Desmarest, 1817); *anambae* Miller, 1900; *angusticeps* Miller, 1901; *baliensis* Thomas, 1913; *batuana* Lyon, 1916; *celaenopepla* Miller, 1913; *condorensis* Kloss, 1920; *dicolorata* Robinson and Kloss, 1914; *felli* Thomas and Wroughton, 1916; *fretensis* Thomas and Wroughton, 1909; *gigantea* (McClelland, 1839); *hainana* J. Allen, 1906; *humeralis* (Coulon, 1836); *javensis* (Zimmerman, 1780); *laenata* Miller, 1903; *leschnaultii* (Desmarest, 1822); *leucogenys* Kloss, 1916; *lutrina* Thomas and Wroughton, 1916; *macruroides* (Hodgson, 1849); *major* Miller, 1911; *marana* Thomas and Wroughton, 1916; *melanopepla* Miller, 1900; *palliata* Miller, 1902; *penangensis* Robinson and Kloss, 1911; *peninsulae* (Miller, 1913); *phaeopepla* Miller, 1913; *sinus* Kloss, 1916; *sondaica* (Müller and Schlegel, 1844); *smithi* Robinson and Kloss, 1922; *stigmosa* Thomas, 1923; *tiomanensis* Miller, 1900.

Ratufa indica (Erxleben, 1777). Syst. Regn. Anim., 1:420.
TYPE LOCALITY: "in India orientali" [Bombay, India]. Based on "Pennant's Bombay squirrel" (Moore and Tate, 1965:43).
DISTRIBUTION: C and S India, excluding central lowlands.
STATUS: CITES - Appendix II; IUCN - Endangered as *R. i. dealbata* and *R. i. elphinstonei*.
SYNONYMS: *bengalensis* (Blanford, 1897); *bombaya* (Boddaert, 1785); *centralis* Ryley, 1913; *dealbata* (Blanford, 1897); *elphinstoni* (Sykes, 1831); *malabarica* (Scopoli, 1786); *maxima* (Schreber, 1784); *purpureus* (Zimmermann, 1777); *superans* Ryley, 1913.
COMMENTS: Pennant (1771:281) described, but did not name this species, and said it "Inhabits Bombay."

Ratufa macroura (Pennant, 1769). Indian Zool., 1, pl. 1.
TYPE LOCALITY: "Ceylon and Malabar....Malacca...Goa and Amboina." Restricted by Phillips (1933) to highlands of Central and Uva Provs., Sri Lanka.
DISTRIBUTION: Sri Lanka and S India.
STATUS: CITES - Appendix II.
SYNONYMS: *albipes* (Blyth, 1859); *ceilonensis* (Boddaert, 1785); *ceylonica* (Erxleben, 1777); *dandolena* Thomas and Wroughton, 1915; *macrura* Blanford, 1891; *melanochra* Thomas and Wroughton, 1915; *montana* (Kalaart, 1852); *sinhala* Phillips, 1931; *tennentii* (Layard, 1849); *zeylanicus* (Ray, 1693).

Rheithrosciurus Gray, 1867. Ann. Mag. Nat. Hist., ser. 3, 20:273.
TYPE SPECIES: *Sciurus macrotis* Gray, 1857.
SYNONYMS: *Rhithrosciurus* Hose, 1893.
COMMENTS: Tribe Sciurini (Moore, 1959:177), or *incertae sedis* (Simpson, 1945:78).

Rheithrosciurus macrotis (Gray, 1857). Proc. Zool. Soc. Lond., 1856:341 [1857].
TYPE LOCALITY: "Sarawak," [Malaysia].
DISTRIBUTION: Borneo.

Rhinosciurus Blyth, 1856. J. Asiat. Soc. Bengal, 14:477.
TYPE SPECIES: *Sciurus laticaudatus* Müller, 1840.
COMMENTS: Tribe Callosciurini (Moore, 1959:173).

Rhinosciurus laticaudatus (Müller, 1840). *In* Temminck, Verh. Nat. Gesch. Nederland. Overz. Bezitt., Zool., Zoogd. Indisch. Archipel., p. 34[1840], see comments.
TYPE LOCALITY: "Pontianak" [W Kalimantan, Indonesia].
DISTRIBUTION: Malaysian region except Java and SW Philippines.
SYNONYMS: *incultus* Lyon, 1916; *leo* Thomas and Wroughton, 1909; *peracer* Thomas and Wroughton, 1909; *rhionis* Thomas and Wroughton, 1909; *robinsoni* Thomas, 1908; *saturatus* Robinson and Kloss, 1919; *tupaioides* Blyth, 1855.
COMMENTS: This species was further described by Müller and Schlegel, *in* Temminck, Verh. Nat. Gesch. Nederland. Overz. Bezitt., Zool., Mammalia, pp. 87, 100[1844], pl. 15[1841] (from which this species is often cited). See Appendix I.

Rubrisciurus Ellerman, 1954. *In* Laurie and Hill, Land of Mammals of New Guinea, Celebes, and Adjacent Islands, p. 94.
TYPE SPECIES: *Sciurus rubriventer* Müller and Schlegel, 1844.
SYNONYMS: *Tomeutes* Thomas, 1915.
COMMENTS: Tribe Callosciurini (Moore, 1959:174). Sometimes considered a subgenus of *Callosciurus*; see Laurie and Hill (1954:93) and Corbet and Hill (1992); but see also Moore (1959) and McLaughlin (1984:273), who considered it a distinct genus.

Rubrisciurus rubriventer (Müller and Schlegel, 1844). *In* Temminck, Verh. Nat. Gesch. Nederland. Overz. Bezitt., Zool., Mammalia, p. 86[1844].
TYPE LOCALITY: "Celebes" [Minahassa, NE Sulawesi, Indonesia].
DISTRIBUTION: N, C, and SE Sulawesi.
COMMENTS: Ellerman (1940:375) dated the species description as 1839, and ascribed it to Forsten, who was listed by Müller and Schlegel as the author, but Forsten was the collector, not the author. See Appendix I for the correct publication date.

Sciurillus Thomas, 1914. Proc. Zool. Soc. Lond., 1914:416.
TYPE SPECIES: *Sciurus pusillus* Desmarest, 1817.
COMMENTS: Tribe Sciurillini (Moore, 1959).

Sciurillus pusillus (Desmarest, 1817). Nouv. Dict. Hist. Nat., Nouv. ed., 10:109.
TYPE LOCALITY: French Guiana, Cayenne.
DISTRIBUTION: Brazil, French Guiana, Peru, Suriname.
SYNONYMS: *glaucinus* Thomas, 1914; *hoehnei* Miranda Ribeiro, 1941; *kuhlii* (Gray, 1867).
COMMENTS: Husson (1978) dated the name from Geoffroy's (1803) catalog, a work considered unpublished (see Appendix I) by other authors, and designated a lectotype.

Sciurotamias Miller, 1901. Proc. Biol. Soc. Washington, 14:23.
TYPE SPECIES: *Sciurus davidianus* Milne-Edwards, 1867.
SYNONYMS: *Rupestes* Thomas, 1922.
COMMENTS: Includes *Rupestes* and *Sciurotamias* as subgenera; reviewed by Moore and Tate (1965). *S. davidianus* has a penile duct and Cowper's glands, and its glans and baculum are similar to those of *Ratufa*; therefore, Callahan and Davis (1982) removed this taxon from the Tamiasciurini (Moore, 1959:182) or Tamiini (Gromov et al., 1965:124) and tentatively referred it to Ratufini.

Sciurotamias davidianus (Milne-Edwards, 1867). Rev. Mag. Zool. Paris, ser. 2, 19:196.
TYPE LOCALITY: "Mountains of Peking," Hebei Prov., China.
DISTRIBUTION: Hebei to Sichuan and Hubei; Guizhou (China).
SYNONYMS: *collaris* (Heude, 1898); *consobrinus* (Milne-Edwards, 1874); *latro* (Heude, 1898); *owstoni* J. Allen, 1909; *saltitans* (Heude, 1898); *thayeri* G. Allen, 1912.
COMMENTS: Subgenus *Sciurotamias*.

Sciurotamias forresti (Thomas, 1922). Ann. Mag. Nat. Hist., ser. 9, 10:399.
TYPE LOCALITY: "Mekong-Yangtze Divide on 27° 20' N 7000-9000'." Yunnan Prov., China.
DISTRIBUTION: Yunnan and Sichuan Provs. (China).
COMMENTS: Subgenus *Rupestes* (Moore and Tate, 1965).

Sciurus Linnaeus, 1758. Syst. Nat., 10th ed., 1:63.
TYPE SPECIES: *Sciurus vulgaris* Linnaeus, 1758.
SYNONYMS: *Aphrontis* Schultze, 1893; *Araeosciurus* Nelson, 1899; *Baiosciurus* Nelson, 1899; *Echinosciurus* Trouessart, 1880; *Guerlinguetus* Gray, 1821; *Hadrosciurus* J. Allen, 1915; *Hesperosciurus* Nelson, 1899; *Histriosciurus* J. Allen, 1915; *Leptosciurus* J. Allen, 1915; *Macroxus* F. Cuvier, 1823; *Mesosciurus* J. Allen, 1915; *Neosciurus* Trouessart, 1880; *Oreosciurus* Ognev, 1935; *Otosciurus* Nelson, 1899; *Parasciurus* Trouessart, 1880; *Simosciurus* J. Allen, 1915; *Tenes* Thomas, 1909; *Urosciurus* J. Allen, 1915.
COMMENTS: Tribe Sciurini (Moore, 1959:177); includes *Guerlinguetus*, *Hesperosciurus*, *Otosciurus*, *Sciurus* (Hall, 1981:417-436); *Tenes* (Corbet, 1978c:76) and *Hadrosciurus* (Cabrera, 1961:374) as subgenera. Moore (1959) considered *Guerlinguetus* a distinct genus (*Hadrosciurus*, *Urosciurus* as subgenera) and included *Otosciurus* and *Hesperosciurus* in subgenus *Sciurus*.

Sciurus aberti Woodhouse, 1853. Proc. Acad. Nat. Sci. Philadelphia, 1852, 6:110, 220 [1853].
TYPE LOCALITY: "...in the San Francisco Mountains, New Mexico" [= Coconino Co., Arizona, USA].
DISTRIBUTION: SE Utah, S and W Colorado, extreme SE Wyoming, W and C New Mexico, and Arizona (USA); Chihuahua, Durango, and Sonora (NW Mexico).
SYNONYMS: *barberi* J. Allen, 1904; *castanonotus* Baird, 1858; *castanotus* Baird, 1855; *chuscensis* Goldman, 1931; *concolor* True, 1894; *dorsalis* Woodhouse, 1853; *durangi* Thomas, 1893; *ferreus* True, 1900; *kaibabensis* (Merriam, 1904); *mimus* Merriam, 1904; *navajo* Durrant and Kelson, 1947; *phaeurus* J. Allen, 1904.
COMMENTS: Subgenus *Otosciurus* (Hall, 1981:434). Includes *kaibabensis* (Hoffmeister and Diersing, 1978). Reviewed by Nash and Seaman (1977, Mammalian Species, 80). Post-Pleistocene dispersal analyzed by Davis and Brown (1989).

Sciurus aestuans Linnaeus, 1766. Syst. Nat., 12th ed., 1:88.
TYPE LOCALITY: Surinam.
DISTRIBUTION: Brazil, French Guiana, Guyana, Suriname, and Venezuela.
SYNONYMS: *alphonsei* Thomas, 1903; *bancrofti* Kerr, 1792; *garbei* (Pinto, 1931); *georgihernandezi* Barriga-Bonilla, 1966; *guajanensis* Kerr, 1792; *guianensis* Peters, 1863; *guerlingus* (Shaw, 1801); *henseli* Miranda Ribeiro, 1941; *ingrami* Thomas, 1901; *macconnelli* Thomas, 1901; *olivascens* Illiger, 1815; *poaiae* (Moojen, 1942); *quelchii* Thomas, 1901; *roberti* Thomas, 1903; *venustus* (J. A. Allen, 1940).
COMMENTS: Subgenus *Guerlinguetus*; see Cabrera (1961:359). Formerly included *gilvigularis*; see Avila-Pires (1964).

Sciurus alleni Nelson, 1898. Proc. Biol. Soc. Washington, 12:147.
TYPE LOCALITY: "Monterey, Tamaulipas [= Nuevo Leon]," [Mexico].
DISTRIBUTION: SE Coahuila through C Nuevo Leon, south through W Tamaulipas to extreme N San Luis Potosi (Mexico).
COMMENTS: Subgenus *Sciurus* (Hall, 1981:430).

Sciurus anomalus Güldenstaedt,1785. *In* Schreber, Die Säugeth., 4:781.
TYPE LOCALITY: "without exact indication of locality" (Ognev, 1966:368). Restricted by Güldenstadt (1785, see Ognev, 1940:423) to Sabeka, 25 km SW of Kutais, Georgia.
DISTRIBUTION: Turkey, Transcaucasia (Armenia, Azerbaidzhan, Georgia), N and W Iran, Syria, Lebanon, Israel.

SYNONYMS: *caucasicus* Pallas, 1811; *fulvus* Blanford, 1875; *historicus* Gray, 1867; *pallescens* (Gray, 1867); *persicus* Erxleben, 1777 (may be based on *Glis glis*); *russatus* Wagner, 1842; *syriacus* Ehrenberg, 1828.

COMMENTS: Subgenus *Tenes* (Corbet, 1978*c*:76). Russian authors ascribe this species to Gmelin, 1778. Syst. Nat., 13th ed., 1:148

Sciurus arizonensis Coues, 1867. Am. Nat., 1:357.

TYPE LOCALITY: "Fort Whipple," [Yavapai Co., Arizona, USA].

DISTRIBUTION: C and SE Arizona and WC New Mexico (USA); NE Sonora (Mexico).

SYNONYMS: *catalinae* Doutt, 1931; *huachuca* J. Allen, 1894.

COMMENTS: Subgenus *Sciurus* (Hall, 1981:432).

Sciurus aureogaster F. Cuvier, 1829. *In* E. Geoffroy and F. Cuvier, Hist. Nat. Mammifères, pt. 3, 6(59):1-2 "Ecureuil de la Californie".

TYPE LOCALITY: "California" Restricted by Nelson (1899*b*:38) to Alta Mira, Tamaulipas, Mexico.

DISTRIBUTION: SW and C Guatemala to Guanajuato to Nayarit and Nuevo Leon (Mexico).

SYNONYMS: *affinis* Alston, 1878; *albipes* Wagner, 1837; *cervicalis* J. Allen, 1890; *chiapensis* Nelson, 1899; *chrysogaster* Giebel, 1855; *cocos* Nelson, 1898; *colimensis* Nelson, 1898; *effugius* Nelson, 1898; *ferruginiventris* Audubon and Bachman, 1841; *frumentor* Nelson, 1898; *griseoflavus* Gray, 1867; *hernandezi* Nelson, 1898; *hirtus* Nelson, 1898; *hypopyrrhus* Wagler, 1831; *hypoxanthus* I. Geoffroy, 1855; *leucogaster* F. Cuvier, 1831; *leucops* Gray, 1867; *littoralis* Nelson, 1907; *maurus* Gray, 1867; *morio* Gray, 1867; *mustelinus* Audubon and Bachman, 1841; *nelsoni* Merriam, 1893; *nemoralis* Nelson, 1898; *nigrescens* Bennett, 1833; *perigrinator* Nelson, 1904; *poliopus* Fitzinger, 1867; *quercinus* Nelson, 1898; *raviventer* Lichtenstein, 1830; *rufipes* Fitzinger, 1867; *rufiventris* Rovirosa, 1887; *senex* Nelson, 1904; *socialis* Wagner, 1837; *tepicanus* J. Allen, 1906; *varius* Wagner, 1843; *wagneri* J. Allen, 1898.

COMMENTS: Subgenus *Sciurus* (Hall, 1981:418). Revised by Musser (1968), who also designated a lectotype and confirmed the type locality restriction (Musser, 1970*c*). Includes *griseoflavus*, *nelsoni*, *poliopus*, and *socialis* (Musser, 1968). Introduced to Elliot Key, Dade County, Florida (USA) (Brown and McGuire, 1975).

Sciurus carolinensis Gmelin, 1788. *In* Linnaeus, Syst. Nat., 13th ed., 1:148.

TYPE LOCALITY: "Carolina."

DISTRIBUTION: E Texas (USA) to Saskatchewan (Canada) and east to Atlantic Coast. Introduced into Britain, South Africa, and various localities in W North America.

SYNONYMS: *extimus* Bangs, 1896; *fuliginosus* Bachman, 1839; *hiemalis* Ord, 1815; *hypophaeus* Merriam, 1886; *leucotis* Gapper, 1830; *matecumbei* Bailey, 1937; *migratorius* Audubon and Bachman, 1849; *minutus* Bailey, 1937; *pennsylvanicus* Ord, 1815.

COMMENTS: Subgenus *Sciurus* (Hall, 1981:417).

Sciurus colliaei Richardson, 1839. Zool. Capt. Beechey's Voy., p. 8.

TYPE LOCALITY: "San Blas, Tepic, [Nayarit,] Mexico."

DISTRIBUTION: Mexico: WC coast including Sonora, Chihuahua, Sinaloa, Durango, Nayarit, Jalisco, and Colima.

SYNONYMS: *nuchalis* Nelson, 1899; *sinaloensis* Nelson, 1899; *truei* Nelson, 1899.

COMMENTS: Subgenus *Sciurus* (Hall, 1981:421). Includes *sinaloensis* and *truei* (Anderson, 1962).

Sciurus deppei Peters, 1863. Monatsb. K. Preuss. Akad. Wiss. Berlin, 1863:654.

TYPE LOCALITY: "Papantla, Veracruz, Mexico." (Nelson, 1899*b*:101).

DISTRIBUTION: Tamaulipas (Mexico) to Costa Rica.

STATUS: CITES - Appenidx III (Costa Rica).

SYNONYMS: *matagalpae* J. Allen, 1908; *miravallensis* Harris, 1931; *negligens* Nelson, 1898; *taeniurus* Gray, 1867; *tephrogaster* Gray, 1867; *vivax* Nelson, 1901.

COMMENTS: Subgenus *Sciurus* (Hall, 1981:426).

Sciurus flammifer Thomas, 1904. Ann. Mag. Nat. Hist., ser. 7, 14:33.

TYPE LOCALITY: Venezuela, Bolivar, Caura Valley, La Union.

DISTRIBUTION: Venezuela south of Orinoco River from the Colombian border to Cuidad Bolivar.

COMMENTS: Subgenus *Hadrosciurus*; see Cabrera (1961:374).

Sciurus gilvigularis Wagner, 1842. Arch. Naturgesch., 2:43.
TYPE LOCALITY: Brazil, Borba, Rio Madeira.
DISTRIBUTION: N Brazil, Guyana, Venezuela.
SYNONYMS: *gilviventris* Pelzeln, 1883; *paraensis* Goeldi and Hagmann, 1904.
COMMENTS: Subgenus *Guerlinguetus;* see Avila-Pires (1964). Cabrera (1961:359) included *gilvigularis* in *aestuans.*

Sciurus granatensis Humboldt, 1811. Rec. Observ. Zool., 1(1805):8.
TYPE LOCALITY: Colombia, Dept. Bolivar, Cartagena.
DISTRIBUTION: Costa Rica, Colombia, Ecuador, Margarita Isl, Panama, Trinidad, Tobago, Venezuela.
SYNONYMS: *agricolae* Hershkovitz, 1947; *baudensis* J. A. Allen, 1915; *bondae* J. A. Allen, 1899; *candelensis* (J. A. Allen, 1914); *carchensis* Harris and Hershkovitz, 1938; *chapmani* J. A. Allen, 1899; *chiriquensis* Bangs, 1902; *choco* Goldman, 1915; *chrysuros* Pucheran, 1845; *cucutae* J. A. Allen, 1914; *ferminae* (Cabrera, 1917); *gerrardi* Gray, 1861; *griseimembra* (J. A. Allen, 1914); *griseogena* (Gray, 1867); *hoffmanni* Peters, 1863; *hyporrhodus* Gray, 1867; *imbaburae* Harris and Hershkovitz, 1938; *inconstans* Osgood, 1921; *klagesi* Thomas, 1914; *leonis* Lawrence, 1933; *llanensis* Mondolfi and Boher, 1984; *magdalenae* J. A. Allen; *manavi* (J. A. Allen, 1914); *maracaibensis* Hershkovitz, 1947; *meridensis* Thomas, 1901; *milleri* J. A. Allen; *morulus* Bangs, 1900; *nesaeus* G. M. Allen, 1902; *norosiensis* Hershkovitz, 1947; *perijae* Hershkovitz, 1947; *quebradensis* J. A. Allen, 1899; *quindianus* (J. A. Allen, 1914); *rhoadsi* (J. A. Allen, 1914); *rufoniger* Pucheran, 1845; *salaquensis* J. A. Allen, 1914; *saltuensis* Bangs, 1898; *söderströmi* Stone, 1914; *splendidus* Gray, 1842; *sumaco* (Cabrera, 1917); *tamae* Osgood, 1912; *tarrae* Hershkovitz, 1947; *tephrogaster* (Gray, 1867); *tobagensis* Ogsood, 1910; *valdiviae* (J. A. Allen, 1915); *variabilis* I. Geoffroy, 1832; *versicolor* Thomas, 1900; *xanthotus* (Gray, 1867); *zuliae* Osgood, 1910.
COMMENTS: Subgenus *Guerlinguetus;* see Hall (1981:436) and Nitikman (1985, Mammalian Species, 246).

Sciurus griseus Ord, 1818. J. Phys. Chim. Hist. Nat. Arts Paris, 87:152.
TYPE LOCALITY: "The Dalles of the Columbia" [River, Wasco Co., Oregon, USA].
DISTRIBUTION: C Washington, W Oregon, and California (USA) to Baja California Norte (Mexico).
SYNONYMS: *anthonyi* Mearns, 1897; *fossor* Peale, 1848; *heermanni* Le Conte, 1852; *leporinus* Audubon and Bachman, 1841; *nigripes* Bryant, 1889.
COMMENTS: Subgenus *Hesperosciurus* (Hall, 1981:433).

Sciurus ignitus (Gray, 1867). Ann. Mag. Nat. Hist., ser. 3, 20:429.
TYPE LOCALITY: Bolivia, near Yungas, upper Rio Beni.
DISTRIBUTION: Argentina, Bolivia, Brazil, and Peru.
SYNONYMS: *argentinius* Thomas, 1921; *boliviensis* Osgood, 1921; *cabrerai* Moojen, 1958; *cuscinus* Thomas, 1899; *irroratus* (Gray, 1867); *leucogaster* (J. A. Allen, 1915); *ochrescens* Thomas, 1914.
COMMENTS: Subgenus *Guerlinguetus;* see Cabrera (1961:370-371).

Sciurus igniventris Wagner, 1842. Arch. Naturg., 1:360.
TYPE LOCALITY: Brazil, Amazonas, north of the Rio Negro, Marabitanos.
DISTRIBUTION: Brazil, Colombia, Ecuador, Peru, Venezuela.
SYNONYMS: *cocalis* Thomas, 1900; *duida* J. A. Allen, 1914; *fulminatus* Thomas, 1926; *manhanensis* (Moojen, 1942); *taedifer* Thomas, 1900; *zamorae* J. A. Allen, 1914.
COMMENTS: Subgenus *Urosciurus;* see Patton (1984). Lawrence (1988:1) restricted *duida* to this species although it has also been referred to *S. spadiceus;* the holotype is a composite.

Sciurus lis Temminck, 1844. Fauna Japonica, 1(Mamm.), p. 45.
TYPE LOCALITY: "Japan". Restricted by Corbet (1978*c*:78) to Honshu, [Japan].
DISTRIBUTION: Honshu, Shikoku, and Kyushu (Japan).
COMMENTS: Subgenus *Sciurus* (Corbet, 1978*c*:77).

Sciurus nayaritensis J. A. Allen, 1890. Bull. Am. Mus. Nat. Hist., 2:7, footnote.
TYPE LOCALITY: "Sierra Valparaiso, Zacatecas," [Mexico.]
DISTRIBUTION: Jalisco (Mexico) north to SE Arizona (USA).

SYNONYMS: *alstoni* J. Allen, 1889; *apache* J. Allen, 1893; *chiricahuae* Goldman, 1933.
COMMENTS: Subgenus *Sciurus* (Hall, 1981:431). Includes *chiricahuae* and *apache* (Hall, 1981:431; Lee and Hoffmeister, 1963).

Sciurus niger Linnaeus, 1758. Syst. Nat., 10th ed., 1:64.
TYPE LOCALITY: "in America septentrionalis." Restricted by Thomas (1911*a*:149) to S South Carolina [USA].
DISTRIBUTION: Texas (USA) and adjacent Mexico, north to Manitoba (Canada) east to the Atlantic Coast.
STATUS: U.S. ESA - Endangered as *S. n. cinereus*; nonessential experimental population in Sussex Co., Delaware (USA); IUCN - Endangered as *S. n. cinereus*.
SYNONYMS: *avicinnia* A. Howell, 1919; *bachmani* Lowery and Davis, 1942; *bryanti* Bailey, 1920; *capistratus* Bosc, 1802; *cinereus* Linnaeus, 1758; *limitis* Baird, 1855; *ludovicianus* Custis, 1806; *macroura* Say, 1823; *magnificaudatus* Harlan, 1825; *neglectus* (Gray, 1867); *ruber* Rafinesque, 1820; *rubicaudatus* Audubon and Bachman, 1851; *rufiventer* E. Geoffroy, 1803; *sayii* Audubon and Bachman, 1851; *shermani* Moore, 1956; *subauratus* Bachman, 1839; *texianus* Bachman, 1839; *vulpinus* Gmelin, 1788.
COMMENTS: Subgenus *Sciurus* (Hall, 1981:427).

Sciurus oculatus Peters, 1863. Monatsb. K. Preuss. Akad. Wiss. Berlin, 1863:653.
TYPE LOCALITY: "Eastern Mexico." Restricted by Nelson (1899*b*:88) to "near Las Vigas [Veracruz] Mexico."
DISTRIBUTION: Mexico: San Luis Potosi, Hidalgo, Veracruz, Puebla, Mexico, Queretaro and Guanajuato.
SYNONYMS: *capistratus* Lichtenstein, 1830; *melanonotus* Thomas, 1890; *shawi* Dalquest, 1950; *tolucae* Nelson, 1898.
COMMENTS: Subgenus *Sciurus* (Hall, 1981:430).

Sciurus pucheranii (Fitzinger, 1867). Sitzb. Math. Naturw. Cl., 55, Abth., 1:487.
TYPE LOCALITY: Colombia, near Bogota.
DISTRIBUTION: Colombian Andes.
SYNONYMS: *caucensis* Nelson, 1899; *medellinensis* (Gray, 1867); *minor* (Alston, 1878); *salentensis* (J. A. Allen, 1914).
COMMENTS: Subgenus *Guerlinguetus*; see Cabrera (1961:372) and Eisenberg (1989). Moore (1959) placed *pucheranii* in the genus *Microsciurus*.

Sciurus pyrrhinus Thomas, 1898. Ann. Mag. Nat. Hist., ser. 7, 2:265.
TYPE LOCALITY: Peru, Junin Dept., Vitoc, Garita del Sol.
DISTRIBUTION: E slopes of the Andes of Peru.
SYNONYMS: *variabilis* Tschudi, 1844.
COMMENTS: Subgenus *Hadrosciurus*; see Cabrera (1961:378). J. A. Allen (1915*b*) placed *pyrrhinus* in the genus *Mesosciurus*.

Sciurus richmondi Nelson, 1898. Proc. Biol. Soc. Washington, 12:146.
TYPE LOCALITY: "Escondido River (50 mi. above Bluefields) Nicaragua" (Nelson, 1899*b*:100).
DISTRIBUTION: Nicaragua.
COMMENTS: Subgenus *Guerlinguetus* (Hall, 1981:436). Reviewed by Jones and Genoways (1975*b*, Mammalian Species, 53).

Sciurus sanborni Osgood, 1944. Field Mus. Nat. Hist. Publ., Zool. Ser., 29:191.
TYPE LOCALITY: Peru, Madre de Dios Dept., La Pampa, between the Rio Inambari and Rio Tambopata, 33 km N of Santo Domingo, 570 m.
DISTRIBUTION: Madre de Dios Dept., Peru.
COMMENTS: Subgenus *Guerlinguetus*; see Cabrera (1961:373).

Sciurus spadiceus Olfers, 1818. *In* Eschwege, J. Brasilien Neue Bibliothek. Reisenb., 15(2):208.
TYPE LOCALITY: Brazil, restricted by Hershkovitz (1959*a*) to Cuyabá, Matto Grosso.
DISTRIBUTION: Bolivia, Brazil, Colombia, Ecuador, Peru.
SYNONYMS: *brunneo-niger* (Gray, 1867); *castus* Thomas, 1903; *fumigatus* (Gray, 1867); *juralis* Thomas, 1926; *langsdorffi* Brandt, 1835; *morio* Wagner, 1848; *nigratus* (Pinto, 1931); *purusianus* (Moojen, 1942); *pyrrhonotus* Wagner, 1842; *rondoniae* (Moojen, 1942); *steinbachi* J. A. Allen, 1914; *taparius* Thomas, 1926; *tricolor* Tschudi, 1844; *urucumus* J. A. Allen, 1914.

COMMENTS: Subgenus *Urosciurus*; see Patton (1984).

Sciurus stramineus Eydoux and Souleyet, 1841. *In* Vaillant, Voy. autour du monde...la Bonite, Zool., 1:73.

TYPE LOCALITY: Peru, Piura Dept., Omatope.

DISTRIBUTION: Extreme NW Peru and SW Ecuador in the area surrounding the Gulf of Guayaquil.

SYNONYMS: *fraseri* (Gray, 1867); *guayanus* Thomas, 1900; *nebouxii* I. Geoffroy, 1855; *zarumae* J. A. Allen, 1914.

COMMENTS: Subgenus *Guerlinguetus*; see Cabrera (1961:373).

Sciurus variegatoides Ogilby, 1839. Proc. Zool. Soc. Lond., 1839:117.

TYPE LOCALITY: "...west coast of South America." Restricted by Nelson (1899*b*:80) to "[El] Salvador."

DISTRIBUTION: S Chiapas (Mexico), through Central America to Panama.

SYNONYMS: *adolphei* Lesson, 1842; *annalium* Thomas, 1905; *atrirufus* Harris, 1930; *austini* Harris, 1933; *bangsi* Dickey, 1928; *belti* Nelson, 1899; *boothiae* Gray, 1843; *dorsalis* Gray, 1849; *fuscovariegatus* Schinz, 1845; *goldmani* Nelson, 1898; *griseocaudatus* Gray, 1843; *helveolus* Goldman, 1912; *intermedius* Gray, 1867; *loweryi* McPherson, 1972; *managuensis* Nelson, 1898; *melania* (Gray, 1867); *nicoyana* Gray, 1867; *pyladei* Lesson, 1842; *richardsoni* Gray, 1842; *rigidus* Peters, 1863; *thomasi* Nelson, 1899; *underwoodi* Goldman, 1932.

COMMENTS: Subgenus *Sciurus*; includes *goldmani* (Hall, 1981:424). Revised by Harris (1937).

Sciurus vulgaris Linnaeus, 1758. Syst. Nat., 10th ed., 1:63.

TYPE LOCALITY: "in Europae arboribus." Restricted by Thomas (1911*a*:148) to Uppsala, Sweden.

DISTRIBUTION: Forested regions of Palearctic, from Iberia and Great Britain east to Kamchatka Peninsula and Sakhalin Isl (Russia); south to Mediterranean and Black Seas, N Mongolia, Korea, and NE China.

SYNONYMS: *albonotatus* Billberg, 1827; *albus* Billberg, 1827; *alpinus* Desmarest, 1822; *altaicus* Serebrennikov, 1928; *ameliae* Cabrera, 1924; *anadyrensis* Ognev, 1929; *arcticus* Trouessart, 1906; *argenteus* Kerr, 1792; *baeticus* Cabrera, 1905; *balcanicus* Heinrich, 1936; *bashkiricus* Ognev, 1935; *brunnea* Altum, 1876; *chiliensis* Sowerby, 1921; *cinerea* Hermann, 1804; *coreae* Sowerby, 1921; *coreanus* Kishida, 1924; *croaticus* Wettstein, 1927; *dulkeiti* Ognev, 1929; *europaeus* Gray, 1843; *exalbidus* Pallas, 1778; *fedjushini* Ognev, 1935; *formosovi* Ognev, 1935; *fuscoater* Altum, 1876; *fusconigricans* Dwigubski, 1804; *fuscorubens* Dwigubski, 1804; *golzmajeri* Smirnov, 1960; *gotthardi* Fatio, 1905; *graeca* Altum, 1876; *hoffmanni* Valverde, 1968; *infuscatus* Cabrera, 1905; *istrandjae* Heinrich, 1936; *italicus* Bonaparte, 1838; *jacutensis* Ognev, 1929; *jenissejensis* Ognev, 1935; *kalbinensis* Selevin, 1924; *kessleri* Migulin, 1928; *leucourus* Kerr, 1792; *lilaeus* Miller, 1907; *mantchuricus* Thomas, 1909; *martensi* Matschie, 1901; *meridionalis* Lucifero, 1907; *nadymensis* Serebrennikov, 1928; *niger* Billberg, 1827; *nigrescens* Altum, 1876; *numantius* Miller, 1907; *ognevi* Migulin, 1928; *orientis* Thomas, 1906; *rhodopensis* Heinrich, 1936; *rufus* Kerr, 1792; *rupestris* Thomas, 1907; *russus* Miller, 1907; *rutilans* Miller, 1907; *segurae* Miller, 1909; *silanus* Hecht, 1931; *subalpinus* Burg, 1920; *talahutky* Brass, 1911; *typicus* Barrett-Hamilton, 1899; *ukrainicus* Migulin, 1928; *uralensis* Ognev, 1935; *varius* Gmelin, 1789.

COMMENTS: Subgenus *Sciurus*. For discussion of taxonomy, see Sidorowicz (1971) and Corbet (1978*c*).

Sciurus yucatanensis J. A. Allen, 1877. *In* Coues and Allen, Mongr. N Am. Rodentia (U.S. Geol. Geograph. Survey Terr., Rep., 11:705).

TYPE LOCALITY: "Merida, Yucatan," [Mexico].

DISTRIBUTION: Yucatan Peninsula (Mexico); N and SW Belize; N Guatemala.

SYNONYMS: *baliolus* Nelson, 1901; *phaeopus* Goodwin, 1932.

COMMENTS: Subgenus *Sciurus* (Hall, 1981:422).

Spermophilopsis Blasius, 1884. Tageblatt. Versamml. Deutsch. Naturf. Magdeburg, 57:325.

TYPE SPECIES: *Arctomys leptodactylus* Lichtenstein, 1823.

COMMENTS: Subfamily Xerinae (Gromov et al., 1965:70) or tribe Xerini (Moore, 1959).

Spermophilopsis leptodactylus (Lichtenstein, 1823). Naturh. Abh. Eversmann's Reise, p. 119.
TYPE LOCALITY: "Vicinity of Kara Ata, 140 km northwest of the Old Town of Bukhara" [Uzbekistan] (Ognev, 1966:394).
DISTRIBUTION: SE Kazakstan, Turkmenistan, Uzbekistan, W Tadzhikistan, NE Iran, NW Afghanistan.
SYNONYMS: *bactrianus* (Scully, 1888); *heptopotamicus* Heptner and Ismagilov, 1952; *schumakovi* (Satunin, 1908); *turcomanus* (Eichwald, 1834).
COMMENTS: Related to African xerine squirrels (*Atlantoxerus* and *Xerus*) (Nadler et al., 1969; Nadler and Hoffmann, 1974).

Spermophilus F. Cuvier, 1825. Dentes des Mammiferes, p. 255.
TYPE SPECIES: *Mus citellus* Linnaeus, 1766.
SYNONYMS: *Anisonyx* Rafinesque, 1817; *Arctomys* Schreber, 1780; *Callospermophilus* Merriam, 1897; *Citellus* Oken, 1816; *Citillus* Lichtenstein, 1830; *Colobates* Milne-Edwards, 1874; *Colobotis* Brandt, 1844; *Ictidomoides* Mearns, 1907; *Ictidomys* J. Allen, 1877; *Notocitellus* A. Howell, 1938; *Otocolobus* Brandt, 1844; *Otospermophilus* Brandt, 1844; *Poliocitellus* A. Howell, 1938; *Spermatophilus* Wagler, 1830; *Spermophila* Richardson, 1825; *Spermophilis* Richardson, 1839; *Urocitellus* Obolenskij, 1927; *Xerospermophilus* Merriam, 1892.
COMMENTS: Tribe Marmotini (Moore, 1959). *Citellus* Oken, 1816 has been widely used, but is invalid (Corbet, 1978*c*:82; Hershkovitz, 1949*b*). Includes *Callospermophilus, Ictidomys, Otospermophilus, Poliocitellus, Spermophilus,* and *Xerospermophilus* as subgenera (Hall, 1981:382); Gromov et al. (1965) gave the first three taxa generic rank. North American species revised by Howell (1938). Eurasian species revised by Gromov et al. (1965) who also recognized *Colobotis* and *Urocitellus* as subgenera; but see Hall (1981:382), who included them in subgenus *Spermophilus*. Holarctic species reviewed by Nadler et al. (1982, 1984). A key to the genus was given by Rickart and Yensen (1991).

Spermophilus adocetus (Merriam, 1903). Proc. Biol. Soc. Washington, 16:79.
TYPE LOCALITY: "La Salada, 40 miles [64 km] south of Uruapan, Michoacan, Mexico."
DISTRIBUTION: E Jalisco, Michoacan, and N Guerrero (WC Mexico).
SYNONYMS: *arceliae* Villa-R., 1942; *infernatus* Alvarez and Ramírez-P., 1968.
COMMENTS: Subgenus *Otospermophilus* (Hall, 1981:399); but see also Birney and Genoways (1973) who suggested it is closer to subgenus *Ictidomys*.

Spermophilus alashanicus Büchner, 1888. Wiss. Res. Przewalski Cent. Asien Zool. I:(Säugeth.):11.
TYPE LOCALITY: "Southern Ala Shan" [Desert, China]. (Ognev, 1963*a*:150).
DISTRIBUTION: SC Mongolia; Ala Shan and E Nan Shan (N China).
SYNONYMS: *dilutus* (Formozov, 1929); *obscurus* Büchner, 1888; *siccus* (G. Allen, 1925).
COMMENTS: Subgenus *Spermophilus* (Gromov et al. 1965:208). Placed by Corbet (1978*c*) in *dauricus;* Orlov and Davaa (1975) provided evidence of specific distinctness.

Spermophilus annulatus Audubon and Bachman, 1842. J. Acad. Nat. Sci. Philadelphia, 8:319.
TYPE LOCALITY: "Western prairies." Restricted by Howell (1938:163) to Manzanillo, Colima, Mexico.
DISTRIBUTION: Nayarit to N Guerrero (Mexico).
SYNONYMS: *goldmani* Merriam, 1902.
COMMENTS: Subgenus *Notocitellus* (Howell, 1938:162) or *Otospermophilus* (Hall, 1981:403).

Spermophilus armatus Kennicott, 1863. Proc. Acad. Nat. Sci. Philadelphia, 15:158.
TYPE LOCALITY: "In the foothills of the Uinta Mountains, near Fort Bridger, [Uinta Co.] Wyo[ming]." [USA].
DISTRIBUTION: SC Utah to S Montana, SE Idaho to W Wyoming (USA).
COMMENTS: Subgenus *Spermophilus* (Hall, 1981:386).

Spermophilus atricapillus W. Bryant, 1889. Proc. California Acad. Sci., ser. 2, 2:26.
TYPE LOCALITY: "Comondu, Lower California" [Baja California Sur, Mexico].
DISTRIBUTION: Baja California (Mexico).
COMMENTS: Subgenus *Otospermophilus* (Hall, 1981:402).

Spermophilus beecheyi (Richardson, 1829). Fauna Boreali-Americana, 1:170.
TYPE LOCALITY: "neighborhood of San Francisco and Monterey, in Calfornia." Restricted by Grinnell (1933) to Monterey, Monterey Co., California, USA.
DISTRIBUTION: W Washington (USA) to Baja California Norte (Mexico).
SYNONYMS: *douglasii* (Richardson, 1829); *fisheri* Merriam, 1893; *nesioticus* (Elliot, 1904); *nudipes* (Huey, 1931); *parvulus* (A. Howell, 1931); *rupinarum* (Huey, 1931); *sierrae* (A. Howell, 1931).
COMMENTS: Subgenus *Otospermophilus* (Hall, 1981:401).

Spermophilus beldingi Merriam, 1888. Ann. N.Y. Acad. Sci., 4:317.
TYPE LOCALITY: "Donner, [Placer Co.,] California [USA]."
DISTRIBUTION: E Oregon, SW Idaho, NE California, N Nevada, and NW Utah (USA).
SYNONYMS: *creber* (Hall, 1940); *oregonus* Merriam, 1898.
COMMENTS: Subgenus *Spermophilus* (Hall, 1981:387). Reviewed by Jenkins and Eshelman (1984, Mammalian Species, 221).

Spermophilus brunneus (A. H. Howell, 1928). Proc. Biol. Soc. Washington, 41:211.
TYPE LOCALITY: "New Meadows, Adams County, Idaho [USA]."
DISTRIBUTION: WC Idaho (USA), in three isolated areas; north of Payette R. to Hitt and Cuddy Mtns; between Cuddy and Seven Devils Mtns, and east of West Mtns.
STATUS: "...limited ranges and small breeding populations...vulnerable..." (Yensen, 1991:597).
SYNONYMS: *endemicus* Yensen, 1991.
COMMENTS: Subgenus *Spermophilus* (Hall, 1981:385). Includes *endemicus*, which "may be reaching species-level separation" (Yensen, 1991:597).

Spermophilus canus Merriam, 1898. Proc. Biol. Soc. Washington, 12:70.
TYPE LOCALITY: "Antelope, Wasco County, Oregon." [USA].
DISTRIBUTION: USA: E Oregon, except NE and SE corners; extreme NW Nevada; W side of Snake River in WC Idaho.
SYNONYMS: *vigilis* (Merriam, 1913).
COMMENTS: Subgenus *Spermophilus* (Hall, 1981:383). Includes *vigilis* (see Vorontsov and Lyapunova, 1970; Nadler et al., 1984, wherein that junior synonym was employed as the species name). Formerly considered a subspecies of *townsendii* (Hall, 1981:383-384), but differs in diploid chromosome number (Nadler et al., 1984). No hybridization between *canus* (2n=46) and adjacent *mollis* (2n=38) or *townsendii* (2n=36) has been reported (Rickart et al., 1985). Reviewed (in part) by Rickart (1987), as *S. townsendii*.

Spermophilus citellus (Linnaeus, 1766). Syst. Nat., 12th ed., 1:80.
TYPE LOCALITY: "Austria"; restricted by Martino and Martino (1940) to "Wagram, Niederosterrich" (Bauer, 1960:254).
DISTRIBUTION: SE Germany, Czechoslovakia, SW Poland through SE Europe to European Turkey, Moldova and W Ukraine.
SYNONYMS: *balcanicus* (Markov, 1957); *citillus* (Pallas, 1779); *gradojevici* (Martino, 1929); *istricus* (Calinescu, 1934); *karamani* (Martino, 1940); *laskarevi* (Martino, 1940); *macedonicus* Fraguedakis-Tsolis and Ondrias, 1977; *martinoi* (Peshchev, 1955); *thracius* (Mursaloglu, 1964).
COMMENTS: Subgenus *Spermophilus*. Formerly included *dauricus* and *xanthoprymnus* as in Ellerman and Morrison-Scott (1951:506); but see Gromov et al. (1965:208, 237), Vorontsov and Lyapunova (1970), and Orlov and Davaa (1975). Cytogenetics described by Belcheva and Peshev (1985) and Soldatovic et al. (1984).

Spermophilus columbianus (Ord, 1815). *In* Guthrie, New Geogr., Hist., Coml. Grammar, Philadelphia, 2nd ed., 2:292.
TYPE LOCALITY: "Between the forks of the Clearwater and Kooskooskie rivers," [Idaho Co., Idaho, USA].
DISTRIBUTION: SE British Columbia and W Alberta (Canada) to NE Oregon, C Idaho, and C Montana (USA).
SYNONYMS: *albertae* (J. Allen, 1903); *brachiura* (Rafinesque, 1817); *erythrogluteia* (Richardson, 1829); *ruficaudus* (A. Howell, 1928).

COMMENTS: Subgenus *Urocitellus* according to Gromov et al. (1965:196), but Hall (1981:381) included *Urocitellus* in subgenus *Spermophilus*. Chromosomes described by Nadler et al. (1975*a*). Reviewed by Elliot and Flinders (1991, Mammalian Species, 372).

Spermophilus dauricus Brandt, 1843. Bull. Phys. Math. Acad. Sci. St. Petersbourg, 2:379.

TYPE LOCALITY: "...circa Torei lacum exiccatum Dauuriae et ad Onon Bursa rivum." Torei-Nor (Lake), Chitinsk. Obl., Russia.

DISTRIBUTION: Transbaikalia (Russia), Mongolia, N China.

SYNONYMS: *mongolicus* (Milne-Edwards, 1867); *ramosus* (Thomas, 1909); *umbratus* (Thomas, 1908); *yamashinae* (Kuroda, 1939).

COMMENTS: Subgenus *Spermophilus* (Gromov et al., 1965:244). Corbet (1978*c*:83) tentatively included *alashanicus* in this species, but see Orlov and Davaa (1975) who provided evidence of specific distinctness. See comment under *alashanicus*. Ellerman and Morrison-Scott (1951:506) included *dauricus* in *citellus;* but see Gromov et al. (1965:244) who considered *dauricus* a distinct species.

Spermophilus elegans Kennicott, 1863. Acad. Nat. Sci. Philadelphia, p. 158.

TYPE LOCALITY: "Fort Bridger," [Uinta Co., Wyoming, USA].

DISTRIBUTION: NE Nevada, SE Oregon, S Idaho, and SW Montana to C Colorado and W Nebraska (USA).

SYNONYMS: *aureus* (Davis, 1939); *nevadensis* (A. Howell, 1928).

COMMENTS: Subgenus *Spermophilus* (Hall, 1981:385). Regarded by Howell (1938) and Hall (1981:385) as a subspecies of *richardsonii;* but Nadler et al. (1971*a*), Robinson and Hoffmann (1975), Koeppl et al. (1978) and Fagerstone (1982) provided evidence of specific distinctness and included *aureus* and *nevadensis* in *elegans*. Reviewed by Zegers (1984, Mammalian Species, 214).

Spermophilus erythrogenys Brandt, 1841. Bull. Acad. Sci. St. Petersbourg, p. 43.

TYPE LOCALITY: "...vicinity of Barnaul" [Altaisk Krai, Russia] (Ognev, 1963*a*:60).

DISTRIBUTION: E Kazakhstan, SW Siberia (Russia), Xinjiang (China). Isolated population (*pallidicaudus*) in Mongolia and Inner Mongolia (China).

SYNONYMS: *brevicauda* Brandt, 1843; *brunnescens* (Beljaev, 1943); *carruthersi* (Thomas, 1912); *iliensis* (Beljaev, 1943); *intermedius* Brandt, 1843; *pallidicauda* (Satunin, 1903); *saryarka* Selevin, 1937; *selevini* (Vinogradov and Argyropulo, 1941).

COMMENTS: Subgenus *Colobotis* according to Gromov et al. (1965:315), but see Hall (1981:381) who included *Colobotis* in subgenus *Spermophilus*. Includes *brevicauda* (= *intermedius*), *carruthersi*, and *pallidicauda;* Ellerman and Morrison-Scott (1951:508, 511) regarded *brevicauda*, *intermedius* and *carruthersi* as synonyms of *pygmaeus*, and *pallidicauda* as a full species. Sludskii et al. (1969) considered *intermedius* (= *brevicauda*) a full species. Provisionally included in *major* by Corbet (1978*c*:84); but see Gromov et al. (1965:315), Vorontsov and Lyapunova (1970), and Nikol'skii (1984) for evidence of specific distinctness. See also *major*.

Spermophilus franklinii (Sabine, 1822). Trans. Linn. Soc. Lond., 13:587.

TYPE LOCALITY: None specified. Restricted by Preble (1908:165) to Carlton House, Saskatchewan, Canada.

DISTRIBUTION: N Great Plains; Alberta, Saskatchewan, and Manitoba (Canada), south to Kansas, Illinois, and Indiana (USA).

COMMENTS: Subgenus *Poliocitellus* (Hall, 1981:397). Gromov et al. (1965:155) considered *Poliocitellus* a subgenus of the genus *Ictidomys*.

Spermophilus fulvus (Lichtenstein, 1823). Naturh. Abh. Eversmann's Reise, p. 119.

TYPE LOCALITY: "near the Kuvandzhur River, east of Mugodzhary Mountains, north of Aral Sea" [Kazakhstan] (Ognev, 1963*a*:29).

DISTRIBUTION: Kazakhstan, from the Caspian Sea and the Volga River to Lake Balkash; south through Uzbekistan, W Tadzhikistan and Turkmenistan to NE Iran, and N Afghanistan; W Xinjiang (China).

SYNONYMS: *concolor* (Fischer, 1829); *concolor* I. Geoffroy, 1831; *giganteus* (Fischer, 1829); *hypoleucos* (Satunin, 1909); *maximus* (Pallas, 1778); *nanus* (Fischer, 1829); *nigrimontanus* (Antipin, 1942); *orlovi* (Ognev, 1937); *oxianus* (Thomas, 1915); *parthianus* (Thomas, 1915).

COMMENTS: Subgenus *Colobotis* according to Gromov et al. (1965:276), but see Hall (1981:381) who included *Colobotis* in subgenus *Spermophilus*. See also *major*.

Spermophilus lateralis (Say, 1823). *In* Long, Account Exped. Pittsburgh to Rocky Mtns, 2:46.
TYPE LOCALITY: "near Cañon City." Restricted by Merriam (1905:163) to Arkansas River, about 26 mi. [42 km] below Canyon City, Fremont Co., Colorado [USA].
DISTRIBUTION: Montane W North America, from C British Columbia to S New Mexico in the Rocky Mtns, and the Columbia River south to S California and Nevada.
SYNONYMS: *arizonensis* (V. Bailey, 1913); *bernardinus* Merriam, 1898; *brevicaudus* Merriam, 1893; *caryi* (A. Howell, 1917); *castanurus* (Merriam, 1890); *certus* (Goldman, 1921); *chrysodeirus* (Merriam, 1890); *cinerascens* (Merriam, 1890); *connectens* (A. Howell, 1931); *mitratus* (A. Howell, 1931); *tescorum* (Hollister, 1911); *trepidus* (Taylor, 1910); *trinitatus* (Merriam, 1901); *wortmani* (J. Allen, 1895).
COMMENTS: Subgenus *Callospermophilus* (Hall, 1981:406). Gromov et al. (1965:150) considered *Callospermophilus* a subgenus of the genus *Otospermophilus*.

Spermophilus madrensis (Merriam, 1901). Proc. Washington Acad. Sci., 3:563.
TYPE LOCALITY: "from Sierra Madre, near Guadalupe y Calvo, Chihuahua, Mexico (7,000 feet altitude)."
DISTRIBUTION: SW Chihuahua (Mexico).
COMMENTS: Subgenus *Callospermophilus* (Hall, 1981:410). Gromov et al. (1965:150) considered *Callospermophilus* a subgenus of the genus *Otospermophilus*. Reviewed by Best and Thomas (1991*b*, Mammalian Species, 378).

Spermophilus major (Pallas, 1778). Nova Spec. Quad. Glir. Ord., p. 125.
TYPE LOCALITY: "Steppe near Samara," [Kuibyshev, Kuibyshevsk. Obl., Russia] (Ognev, 1963*a*:34).
DISTRIBUTION: Steppe between Volga and Irtysh rivers (Russia; N Kazakhstan). Formerly, steppe between Don and Volga rivers (Russia; Gromov et al., 1965:291).
SYNONYMS: *argyropuloi* (Bazhanov, 1947); *heptneri* (Vasil'eva, 1964); *rufescens* (Keyserling and Blasius, 1840); *ungae* (Martino, 1923) .
COMMENTS: Subgenus *Colobotis* according to Gromov et al. (1965:290), but see Hall (1981:381) who included *Colobotis* in subgenus *Spermophilus*. Occasionally hybridizes with *erythrogenys* and *fulvus* (Denisov, 1963; Ognev, 1947). Corbet (1978*c*:84) provisionally included *erythrogenys* and *brevicauda* in this species, but Gromov et al. (1965:290) and Vorontsov and Lyapunova (1970) considered *erythrogenys* a distinct species, and Gromov et al. (1965:315) included *brevicauda* in *erythrogenys*; see comment under *erythrogenys*.

Spermophilus mexicanus (Erxleben, 1777). Syst. Regn. Anim., 1:428.
TYPE LOCALITY: "in nova Hispania?" Restricted by Mearns (1896:443) to "Toluca, [Mexico,] Mexico."
DISTRIBUTION: S New Mexico and W Texas (USA) to Jalisco and S Puebla (C Mexico).
SYNONYMS: *parvidens* Mearns, 1896.
COMMENTS: Subgenus *Ictidomys* (Hall, 1981:394). Known to hybridize at several localities with *tridecemlineatus* (Cothran and Honeycutt, 1984; Cothran et al., 1977). Reviewed by Young and Jones (1982, Mammalian Species, 164).

Spermophilus mohavensis Merriam, 1889. N. Am. Fauna, 2:15.
TYPE LOCALITY: "Mohave River, California [USA]." Restricted by Grinnell and Dixon (1918) to near Rabbit Springs, about 15 mi. (24 km) E Hesperia, San Bernardino Co.
DISTRIBUTION: NW Mohave Desert and Owens Valley (S California, USA).
COMMENTS: Subgenus *Xerospermophilus*; (Hall, 1981:405). Most closely related to *S. tereticaudus* (Hafner and Yates, 1983); hybridizes at one locality.

Spermophilus mollis Kennicott, 1863. Proc. Acad. Nat. Sci. Philadelphia, 15:157.
TYPE LOCALITY: "Camp Floyd, near Fairfield, [Utah Co.,] Utah [USA]."
DISTRIBUTION: Two disjunct populations: in Washington, N of Yakima, W of Columbia, rivers (*nancyae* only); and SE corner of Oregon, Snake River valley (Idaho) southward through Nevada (except extreme S), extreme EC California, and W Utah.
SYNONYMS: *artemesiae* (Merriam, 1913); *idahoensis* (Merriam, 1913); *leurodon* (Merriam, 1913); *nancyae* Nadler, 1968; *pessimus* (Merriam, 1913); *stephensi* Merriam, 1898; *washoensis* (Merriam, 1913).

COMMENTS: Subgenus *Spermophilus* (Hall, 1981:383). Formerly considered a subspecies of *townsendii* (Hall, 1981:383), but differs chromosomally (Nadler et al., 1984). No hybridization between *mollis* (2n=38) and adjacent *canus* (incl. *vigilis*) (2n=46) has been reported (Rickart et al., 1985:97-98). The taxon *nancyae* (2n=38) is disjunct, but provisionally included here. Reviewed (in part) by Rickart (1987) as *S. townsendii*.

Spermophilus musicus Ménétries, 1832. Cat. Raisonne des Objets de Zoologie, St. Petersbourg, p. 21.

TYPE LOCALITY: "Il habite le Caucase sur les montagnes le plus élevées et pas loin des nieges éternalles." Restricted by Sviridenko (1927, see Ognev, 1963*a*:114) to "Ush-Kulan," [Georgia].

DISTRIBUTION: N Caucasus Mtns (Georgia).

SYNONYMS: *boehmii* (Krassovskii, 1932); *saturatus* (Ognev, 1947).

COMMENTS: Subgenus *Spermophilus* (Gromov et al., 1965:249). Regarded by Ellerman and Morrison-Scott (1951:508) and Corbet (1978*c*:83) as a subspecies of *pygmaeus*. Gromov et al. (1965:249) and Vorontsov and Lyapunova (1970) provided evidence of specific distinctness.

Spermophilus parryii (Richardson, 1825). *In* Parry, Voy. discovery Northwest Passage, Vol. 6-app. second voy., p. 316.

TYPE LOCALITY: Restricted by Preble (1902:46) to "Five Hawser Bay, Lyon Inlet, Melville Peninsula, [Hudson Bay, Keewatin District, Northwest Territories], Canada."

DISTRIBUTION: NW Canada; Alaska (USA); NE Yakutia, Anadyrsk. Krai, and Chukotka (Russia).

SYNONYMS: *ablusus* (Osgood, 1903); *barrowensis* Merriam, 1900; *beringensis* Merriam, 1900; *buxtoni* (J. Allen, 1903); *coriacorum* (Portenko, 1963); *janensis* (Ognev, 1937); *kennicottii* (Ross, 1861); *kodiacensis* J. Allen, 1874; *leucostictus* Brandt, 1844; *lyratus* (Hall and Gilmore, 1932); *nebulicola* (Osgood, 1903); *osgoodi* Merriam, 1900; *phaeognatha* (Richardson, 1829); *plesius* Osgood, 1900; *stejnegeri* (J. Allen, 1903); *stonei* (J. Allen, 1903); *tschuktschorum* Chernyavskii, 1972.

COMMENTS: Subgenus *Urocitellus* according to Gromov et al. (1965:184), but see Hall (1981:381) who included *Urocitellus* in subgenus *Spermophilus*. Regarded by Ellerman and Morrison-Scott (1951:511) and Hall and Kelson (1959:343) as a synonym of *undulatus*. Gromov et al. (1965:184) and Nadler et al. (1974) provided evidence of specific distinctness. Reviewed by Chernyavskii (1972), Nadler and Hoffmann (1977), Serdyuk (1979) (Palearctic), and Pearson (1981) (Nearctic). Nikol'skii and Wallschläger (1982) noted differences in alarm calls between Siberian and Alaskan populations.

Spermophilus perotensis Merriam, 1893. Proc. Biol. Soc. Washington, 8:131.

TYPE LOCALITY: "Perote, Veracruz, Mexico."

DISTRIBUTION: Veracruz and Puebla (EC Mexico).

COMMENTS: Subgenus *Ictidomys* (Hall, 1981:397).

Spermophilus pygmaeus (Pallas, 1778). Nova Spec. Quad. Glir. Ord., p. 122.

TYPE LOCALITY: "Maximos et paene dixerim monstrosos Citillós passim ad inferiorum laikum in campis squalidis." Restricted by Ognev (1963*a*:102) to "lower reaches of the Ural River" "Indersk" [Kazakhstan].

DISTRIBUTION: SW Ukraine; S Ural Mtns to Crimea (Russia); Kazakhstan; NW Uzbekistan; Dagestan (Georgia).

SYNONYMS: *atricapilla* (Orlov, 1927); *binominatus* (Ellerman, 1940); *brauneri* (Martino, 1914); *ellermani* (Harris, 1944); *flavescens* (Pallas, 1779); *herbicola* (Martino, 1914); *kalabuchovi* (Ognev, 1937); *kazakstanicus* (Goodwin, 1935); *mugosaricus* (Lichtenstein, 1823); *nikolskii* (Heptner, 1934); *orlovi* (Ellerman, 1940); *pallidus* (Orlov and Fenyuk, 1927); *planicola* (Satunin, 1909); *satunini* (Sveridenko, 1922); *septentrionalis* (Obolenskii, 1927).

COMMENTS: Subgenus *Spermophilus*, see Gromov et al. (1965:257) and comment under *musicus*. Hybridizes rarely with *erythrogenys*, *fulvus*, and *major* (Bazhanov 1944; Denisov, 1964); hybridizes extensively with *suslicus* in zone of contact SW of Saratov (Russia) (Denisov and Smirnova, 1976).

Spermophilus relictus (Kashkarov, 1923). Trans. Turk. Sci. Soc., 1:185.

TYPE LOCALITY: "Kara-Bura Gorge and Kumysh-Tagh Gorge in the Talus Ala Tau" [Talassk. Obl., Kirgizistan] (Ognev, 1963*a*:70).

DISTRIBUTION: Tien Shan Mtns in Kirgizistan and SE Kazakhstan.

SYNONYMS: *arenicola* (Rall, 1935); *ralli* (Kuznetsov, 1948).

COMMENTS: Subgenus *Spermophilus* (Gromov et al., 1965:198).

Spermophilus richardsonii (Sabine, 1822). Trans. Linn. Soc. Lond., 13:589.

TYPE LOCALITY: "Carleton-House," [Saskatchewan, Canada].

DISTRIBUTION: N Great Plains in S Alberta, S Saskatchewan, S Manitoba (Canada), Montana (see Swenson, 1981), North Dakota, NE South Dakota, W Minnesota, and NW Iowa (USA).

COMMENTS: Subgenus *Spermophilus* (Hall, 1981:385). Formerly included *elegans*; see comment under that species. Reviewed by Michener and Koeppl (1985, Mammalian Species, 243).

Spermophilus saturatus (Rhoads, 1895). Proc. Acad. Nat. Sci. Philadelphia, 47:43.

TYPE LOCALITY: "Lake Kichelos [= Keechelus], Kittitas Co., Wash[ingto]n, (elevation 8,000 feet [2,438 m])."

DISTRIBUTION: Cascade Mtns of W Washington (USA) and SW British Columbia (Canada).

COMMENTS: Subgenus *Callospermophilus* (Hall, 1981:409). Gromov et al. (1965:150) considered *Callospermophilus* a subgenus of genus *Otospermophilus*. Reviewed by Trombulak (1988, Mammalian Species, 322).

Spermophilus spilosoma Bennett, 1833. Proc. Zool. Soc. Lond., 1833:40.

TYPE LOCALITY: "that part of California which adjoins to Mexico." Restricted by Howell (1938:122) to Durango [City], Durango, Mexico.

DISTRIBUTION: C Mexico to S Texas, SW South Dakota, and NW Arizona (USA).

SYNONYMS: *altiplanensis* Anderson, 1972; *ammophilus* Hoffmeister, 1959; *annectens* Merriam, 1893; *arens* V. Bailey, 1902; *bavicorensis* Anderson, 1972; *cabrerai* (Dalquest, 1951); *canescens* Merriam, 1890; *cryptospilotus* Merriam, 1890; *macrospilotus* Merriam, 1890; *major* Merriam, 1890; *marginatus* V. Bailey, 1902; *microspilotus* Elliot, 1901; *obsidianus* Merriam, 1890; *obsoletus* Kennicott, 1863; *oricolus* Alvarez, 1962; *pallescens* (A. Howell, 1928); *pratensis* Merriam, 1890.

COMMENTS: Subgenus *Ictidomys* (Hall, 1981:395). Reviewed by Streubel and Fitzgerald (1978*a*, Mammalian Species, 101).

Spermophilus suslicus (Güldenstaedt, 1770). Nova Comm. Acad. Sci. Petropoli, 14:389.

TYPE LOCALITY: "...in campis vastissimus tanaicensibus precipue urbes et Tambov" [Voronezh area, Voronezhsk. Obl., Russia].

DISTRIBUTION: Steppes of E and S Europe, including Poland, E Rumania, Ukraine north to Oka River and east to the Volga River (Russia).

SYNONYMS: *averini* (Migulin, 1927); *boristhenicus* (Pusanov, 1958); *guttatus* (Pallas, 1770); *guttulatus* Schinz, 1845; *leucopictus* (Dondorff, 1792); *meridioccidentalis* (Migulin, 1927); *odessana* Nordmann, 1842; *ognevi* (Reshetnik, 1946); *volhynensis* (Reshetnik, 1946).

COMMENTS: Subgenus *Spermophilus* Gromov et al. (1965:212). See also comment under *pygmaeus*.

Spermophilus tereticaudus Baird, 1858. Mammalia, *in* Repts. U.S. Expl. Surv., 8(1):315.

TYPE LOCALITY: [Old] "Fort Yuma" [Imperial Co., California, USA].

DISTRIBUTION: Deserts of SE California, S Nevada, W Arizona (USA), NE Baja California and Sonora (Mexico).

SYNONYMS: *apricus* (Huey, 1927); *arizonae* (Grinnell, 1918); *chlorus* (Elliot, 1904); *eremonomus* (Elliot, 1904); *neglectus* Merriam, 1889; *sonoriensis* Ward, 1891; *vociferans* (Huey, 1926).

COMMENTS: Subgenus *Xerospermophilus* (Hall, 1981:405). Reviewed by Ernest and Mares (1987, Mammalian Species, 274). See also *mohavensis*.

Spermophilus townsendii Bachman, 1839. J. Acad. Nat. Sci. Philadelphia, 8:61.

TYPE LOCALITY: "On the Columbia River, about 300 miles above its mouth." Restricted by Howell (1938:60, 62) to west bank of Walla Walla River near confluence with Columbia River. [near Wallula, Walla Walla Co., Washington, USA].

DISTRIBUTION: SE Washington (USA), S of Yakima River and W and N of Columbia River.
SYNONYMS: [*mollis*] *yakimensis* Merriam, 1898.
COMMENTS: Subgenus *Spermophilus* (Hall, 1981:382). Formerly included two cytotypes (*mollis, canus*) now considered distinct species (Nadler et al., 1984; Vorontsov and Lyapunova, 1970). Reviewed by Rickart (1987, Mammalian Species, 268).

Spermophilus tridecemlineatus (Mitchill, 1821). Med. Repos. (NY), (n.s.), 6(21):248.
TYPE LOCALITY: "...region bordering the sources of the river Mississippi..."; restricted by Allen (1895*b*:338) to C Minnesota [USA].
DISTRIBUTION: Great Plains, from C Texas to E Utah, Ohio (USA) and SC Canada.
SYNONYMS: *alleni* Merriam, 1898; *arenicola* (A. Howell, 1928); *badius* Bangs, 1899; *blanca* Armstrong, 1971; *hollisteri* (V. Bailey, 1913); *hoodii* (Sabine, 1822); *monticola* (A. Howell, 1928); *olivaceous* J. Allen, 1895; *pallidus* J. Allen, 1874; *parvus* J. Allen, 1895; *texensis* Merriam, 1898.
COMMENTS: Subgenus *Ictidomys* (Hall, 1981:391). Reviewed by Streubel and Fitzgerald (1978*b*, Mammalian Species, 103). See also comment under *mexicanus*.

Spermophilus undulatus (Pallas, 1778). Nova Spec. Quad. Glir. Ord., p. 122.
TYPE LOCALITY: "Selenga River valley," [Buryat ASSR, Russia].
DISTRIBUTION: E Kazakhstan; S Siberia, Transbaikalia (Russia); N Mongolia; Heilungjiang and Xinjiang (China).
SYNONYMS: *altaicus* (Brandt, 1841); *eversmanni* (Brandt, 1841); *intercedens* (Ognev, 1937); *jacutensis* (Brandt, 1844); *menzbieri* (Ognev, 1937); *stramineus* (Obolenskii, 1927); *transbaikalicus* (Obolenskii, 1927).
COMMENTS: Subgenus *Urocitellus* according to Gromov et al. (1965:162), but see Hall (1981:381) who included *Urocitellus* in subgenus *Spermophilus*. Formerly included *parryii* (Hall and Kelson, 1959:343); but see Nadler et al. (1974) and comments under *parryii*. Chromosomes described by Nadler et al. (1975*a*).

Spermophilus variegatus (Erxleben, 1777). Syst. Regn. Anim., 1:421.
TYPE LOCALITY: "in Mexico." Restricted by Nelson (1898:898) to "Valley of Mexico, near City of Mexico," [Distrito Federal, Mexico].
DISTRIBUTION: S Nevada to SW Texas and Utah (USA) to Puebla (C Mexico).
SYNONYMS: *buccatus* (Lichtenstein, 1830); *buckleyi* Slack, 1861; *couchii* Baird, 1855; *grammurus* (Say, 1823); *juglans* (V. Bailey, 1913); *macrourus* Bennett, 1833; *robustus* (Durrant and Hansen, 1954); *rupestris* (J. Allen, 1903); *tiburonensis* Jones and Manning, 1989; *tularosae* (Benson, 1932); *utah* (Merriam, 1903).
COMMENTS: Subgenus *Otospermophilus* (Hall, 1981:399). Reviewed by Oaks et al. (1987, Mammalian Species, 272).

Spermophilus washingtoni (A. H. Howell, 1938). N. Am. Fauna, 56:69.
TYPE LOCALITY: "Touchet, Walla Walla Co., Wash[ington]." [USA].
DISTRIBUTION: SE Washington, NE Oregon (USA).
SYNONYMS: *loringi* A. Howell, 1938.
COMMENTS: Subgenus *Spermophilus* Hall (1981:384). Reviewed by Rickart and Yensen (1991, Mammalian Species, 371).

Spermophilus xanthoprymnus (Bennett, 1835). Proc. Zool. Soc. Lond., 1835:90.
TYPE LOCALITY: "Erzurum" [Turkey].
DISTRIBUTION: Transcaucasia (Armenia, possibly Azerbaidzhan), Turkey, Syria, and Israel.
SYNONYMS: *schmidti* (Satunin, 1908).
COMMENTS: Subgenus *Spermophilus* (Gromov et al. 1965:237). Regarded by Ellerman and Morrison-Scott (1951:506) and Corbet (1978*c*:83) as a synonym of *citellus*; but see Vasil'eva (1961), Gromov et al. (1965:237), and Vorontsov and Lyapunova (1970), who provided evidence of specific distinctness.

Sundasciurus Moore, 1958. Am. Mus. Novit., 1914:2.
TYPE SPECIES: *Sciurus lowii* Thomas, 1892.
SYNONYMS: *Aletesciurus* Moore, 1958.
COMMENTS: Reviewed in part by Heaney (1979). Includes *Aletesciurus* and *Sundasciurus* as subgenera (Moore, 1958:3).

Sundasciurus brookei (Thomas, 1892). Ann. Mag. Nat. Hist., ser. 6, 9:253.
TYPE LOCALITY: "Mount Dulit, N. Borneo" [Sarawak, Malaysia].
DISTRIBUTION: Mountains of Borneo, from 600 to 1500 m.
COMMENTS: Subgenus *Sundasciurus*.

Sundasciurus davensis (Sanborn, 1952). Fieldiana Zool., 33:117.
TYPE LOCALITY: "Madaum, 25 feet altitude, Tagum Municipality, Davao Province, Mindanao Island, Philippines Islands."
DISTRIBUTION: Known only from the type locality.
SYNONYMS: *cagsi* (Meyer, 1890).
COMMENTS: Subgenus *Aletesciurus*. May be conspecific with *mindanensis*, *philippinensis*, and *samarensis* (Heaney et al., 1987; Corbet and Hill, 1992).

Sundasciurus fraterculus (Thomas, 1895). Ann. Mus. Civ. Storia Nat. Genova, Ser. 2a, 14:669.
TYPE LOCALITY: Indonesia, Mentawi Isls, Sipora Isl, "Sereinu".
DISTRIBUTION: Sipora Isl, Siberut Isl, and N Pagi Isl (Sumatra, Indonesia).
SYNONYMS: *pumilus* (Miller, 1903); *siberu* (Chasen and Kloss, 1928).
COMMENTS: Formerly included in *lowii* (Chasen, 1940:144); but see Moore (1959). Geographic variation reviewed by Jenkins and Hill (1982). Part of the rodent fauna endemic to the Mentawi Archipelago (see comments under *Leopoldamys siporanus*).

Sundasciurus hippurus (I. Geoffroy, 1831). *In* Bélanger (ed.), Voy. Indes Orient., Mamm., 3(Zoologie):149.
TYPE LOCALITY: "Java." Restricted by Robinson and Kloss (1918*a*:226) to Malacca, Malaysia.
DISTRIBUTION: S Vietnam, Malaysian subregion except Java and SW Philippines.
SYNONYMS: *borneensis* (Gray, 1867); *grayi* (Bonhote, 1901); *hippurellus* (Lyon, 1907); *hippurosus* (Lyon, 1907); *inquinatus* (Thomas, 1908); *ornatus* Dao and Cao, 1990; *pryeri* (Thomas, 1892); *rufogaster* (Gray, 1842).
COMMENTS: Subgenus *Aletesciurus*.

Sundasciurus hoogstraali (Sanborn, 1952). Fieldiana Zool., 33:115.
TYPE LOCALITY: "Dimaniang, Busuanga Island, Calamianes group, Philippine Islands."
DISTRIBUTION: Busuanga Isl (Philippines).
COMMENTS: Subgenus *Aletesciurus*. Member of the *steerii* group; see comment under *steerii*. Placed in *moellendorffi* by Corbet and Hill (1991:140), without comment.

Sundasciurus jentinki (Thomas, 1887). Ann. Mag. Nat. Hist., ser. 5, 20:128.
TYPE LOCALITY: "Mount Kina-Balu," [Sabah, Malaysia].
DISTRIBUTION: Mountains of N Borneo.
SYNONYMS: *subsignanus* (Chasen, 1937).
COMMENTS: Subgenus *Sundasciurus*. Chromosomes described by Harada and Kobayashi (1980).

Sundasciurus juvencus (Thomas, 1908). Ann. Mag. Nat. Hist., ser. 8, 2:498.
TYPE LOCALITY: "the islands of Palawan....Puerto Princesa [Philippines]".
DISTRIBUTION: N Palawan Isl.
COMMENTS: Subgenus *Aletesciurus*. Member of the *steerii* group; see comment under *steerii*.

Sundasciurus lowii (Thomas, 1892). Ann. Mag. Nat. Hist., ser. 6, 2:253.
TYPE LOCALITY: "Lumbidan, on the mainland opposite Labuan" [Sarawak, Malaysia].
DISTRIBUTION: Malaysian subregion, except Java and SW Philippines.
SYNONYMS: *alacris* (Thomas, 1908); *balae* (Miller, 1903); *bangueyae* (Thomas, 1910); *humilis* (Miller, 1913); *lingungensis* (Miller, 1901); *natunensis* (Thomas, 1895); *piniensis* (Miller, 1903); *robinsoni* (Bonhote, 1903); *seimundi* (Thomas and Wroughton, 1909); *vanakeni* (Robinson and Kloss, 1916).
COMMENTS: Subgenus *Sundasciurus*. Formerly included *fraterculus* and *pumilus* (Chasen, 1940:144).

Sundasciurus mindanensis (Steere, 1890). List of the Birds and Mammals collected by the Steere Expedition to the Philippines. Ann Arbor, Mich., p. 29.
TYPE LOCALITY: "Mindanao," [Philippines].
DISTRIBUTION: Mindanao and adjacent small islands (Philippines).
COMMENTS: Subgenus *Aletesciurus*. May be conspecific with *davensis*, *philippinensis*, and *samarensis* (Heaney et al., 1987; Corbet and Hill, 1992).

Sundasciurus moellendorffi (Matschie, 1898). Sitzb. Gesell. Naturf. Fr., Berlin, 5:41.
TYPE LOCALITY: "Calamianes....von Culion" [Calamian Isls, Philippines].
DISTRIBUTION: Known only from the type locality.
SYNONYMS: *albicauda* (Matschie, 1898).
COMMENTS: Subgenus *Aletesciurus*. The form *albicauda* was listed as a distinct species by Corbet and Hill (1980:132) without comment, but included in *moellendorffi* by Corbet and Hill (1986:149), as was *hoogstraali* (in Corbet and Hill, 1991:140) without comment. Spelling of name changed in accordance with Art. 32d(i) of the International Code of Zoological Nomenclature (1985).

Sundasciurus philippinensis (Waterhouse, 1839). Proc. Zool. Soc. Lond., 1839:117.
TYPE LOCALITY: "Mindanado" [Mindanao Isl, Philippines].
DISTRIBUTION: S and W Mindanao, and Basilan (Philippines).
COMMENTS: Subgenus *Aletesciurus*. May be conspecific with *davensis*, *mindanensis*, and *samarensis* (Heaney et al., 1987; Corbet and Hill, 1992).

Sundasciurus rabori Heaney, 1979. Proc. Biol. Soc. Washington, 92:281.
TYPE LOCALITY: "Magtaguimbong, Mt. Mantalingajan, Palawan Isl, Republic of the Philippines, Island between 3,600 and 4,350 feet elevation...approximately 8° 48'N, 117° 40'E."
DISTRIBUTION: Above 800 m in mountains on Palawan (Philippines).
COMMENTS: Subgenus *Aletesciurus* (see Heaney, 1979).

Sundasciurus samarensis (Steere, 1890). List of the Birds and Mammals Collected by the Steere Expedition to the Philippines. Ann Arbor, Mich., p. 30.
TYPE LOCALITY: "Samar and Leyte," [Philippines].
DISTRIBUTION: Samar and Leyte Isls (Philippines).
COMMENTS: Subgenus *Aletesciurus*. May be conspecific with *davensis*, *mindanensis*, and *philippinensis* (Heaney et al., 1987; Corbet and Hill, 1992).

Sundasciurus steerii (Günther, 1877). Proc. Zool. Soc. Lond., 1876:735 [1877].
TYPE LOCALITY: "Balabac" [Isl, Philippines].
DISTRIBUTION: Balabac and S Palawan Isls in lowlands (Philippines).
COMMENTS: Subgenus *Aletesciurus*. Member of the *steerii* group, which contains 1-3 species; see Heaney (1979).

Sundasciurus tenuis (Horsfield, 1824). Zool. Res. Java, part 7, p. 9(unno.) of *Sciurus Plantani* acct.
TYPE LOCALITY: Singapore.
DISTRIBUTION: Malaysian subregion, except Java and SW Philippines.
SYNONYMS: *altitudinus* (Robinson and Kloss, 1916); *bancarus* (Miller, 1903); *batus* (Lyon, 1916); *gunong* (Robinson and Kloss, 1914); *mansalaris* (Miller, 1903); *modestus* (Müller, 1840); *parvus* (Miller, 1901); *procerus* (Miller, 1901); *siantanicus* (Chasen and Kloss, 1928); *sordidus* (Kloss, 1911); *surdus* (Miller, 1900); *tahan* (Bonhote, 1908); *tiomanicus* (Robinson, 1917).
COMMENTS: Subgenus *Sundasciurus*. The name *pumilus* (see synonym in *fraterculus*) was previously assigned to *tenuis* (Robinson and Kloss, 1918*a*:229); but see Chasen (1940:144).

Syntheosciurus Bangs, 1902. Bull. Mus. Comp. Zool., 39:25.
TYPE SPECIES: *Syntheosciurus brochus* Bangs, 1902.
COMMENTS: Tribe Microsciurini (Moore, 1959). Goodwin (1946) suggested that *Syntheosciurus* might be a subgenus of *Sciurus*; but see Heaney and Hoffmann (1978), Enders (1980), and Hall (1981:438). Moore (1959:179) included *Mesosciurus* (*granatensis*, *pyrrhinus*) as a subgenus, but see *Sciurus*.

Syntheosciurus brochus Bangs, 1902. Bull. Mus. Comp. Zool., 39:25.
TYPE LOCALITY: "Boquete, 7,000 ft." [2,134 m, Chiriqui, Panama].
DISTRIBUTION: Costa Rica to N Panama.
STATUS: IUCN - Insufficiently known.
SYNONYMS: *poasensis* (Goodwin, 1942).
COMMENTS: Reviewed by Enders (1953:509), and Wells and Giacalone (1985, Mammalian Species, 249). Includes *poasensis* (Hall, 1981:438; Heaney and Hoffmann, 1978; Enders, 1980).

Tamias Illiger, 1811. Prodr. Syst. Mamm. Avium., p. 83.
TYPE SPECIES: *Sciurus striatus* Linnaeus, 1758.
SYNONYMS: *Eutamias* Trouessart, 1880; *Neotamias* A. Howell, 1929.
COMMENTS: Tribe Marmotini (Moore, 1959). Nearctic forms revised by Howell (1929). Sutton (1992) provided a key to the species. Includes *Eutamias* (*sibiricus*), *Tamias* (*striatus*), and *Neotamias* as subgenera (Corbet, 1978*c*:85; Ellerman, 1940:428; Levenson et al., 1985; Nadler et al., 1977). Disagreement exists regarding the status of *Eutamias* and *Neotamias*; see White (1953), Ellis and Maxson (1979), Hall (1981:337), and Patterson and Heaney (1987). Levenson et al. (1985) found that *T.* (*T.*) *striatus* and *T.* (*E.*) *sibiricus* together formed the primitive sister group to *Neotamias* species; if this is confirmed, it renders *Eutamias* in the generic sense paraphyletic.

Tamias alpinus Merriam, 1893. Proc. Biol. Soc. Washington, 8:137.
TYPE LOCALITY: "Big Cottonwood Meadows, ... just south of Mount Whitney, altitude 3,050 meters or 10,000 feet" [Tulare Co., California, USA].
DISTRIBUTION: Alpine zone in Sierra Nevada, from Tuolumne to Tulare Counties (EC California, USA).
COMMENTS: Subgenus *Neotamias*.

Tamias amoenus J. A. Allen, 1890. Bull. Am. Mus. Nat. Hist., 3:90.
TYPE LOCALITY: "Fort Klamath, [Klamath Co.,] Oregon." [USA].
DISTRIBUTION: C British Columbia (Canada) south to C California east to C Montana and W Wyoming (USA).
SYNONYMS: *affinis* J. Allen, 1890; *albiventris* (Booth, 1947); *canicaudus* (Merriam, 1903); *caurinus* (Merriam, 1898); *celeris* (Hall and Johnson, 1940); *cratericus* (Blossom, 1937); *felix* Rhoads, 1895; *ludibundus* (Hollister, 1911); *luteiventris* J. Allen, 1890; *monoensis* (Grinnell and Storer, 1916); *ochraceus* (A. Howell, 1925); *propinquus* (Anthony, 1913); *septentrionalis* (Cowan, 1946); *vallicola* (A. Howell, 1922).
COMMENTS: Subgenus *Neotamias*. Reviewed by Sutton (1992, Mammalian Species, 390).

Tamias bulleri J. A. Allen, 1889. Bull. Am. Mus. Nat. Hist., 2:173.
TYPE LOCALITY: "Sierra de Valparaiso, Zacatecas," [Mexico].
DISTRIBUTION: Sierra Madre, in S Durango, W Zacatecas, and N Jalisco (Mexico).
COMMENTS: Subgenus *Neotamias*. Formerly included *durangae* and *solivagus*, which were considered *incertae sedis* by Callahan (1980); see *durangae*.

Tamias canipes (V. Bailey, 1902). Proc. Biol. Soc. Washington, 15:117.
TYPE LOCALITY: "Guadalupe Mts" [Culberson Co., Texas, USA]. Restricted by A. Howell (1929:101) to head of Dog Canyon, 7,000 ft [2,130 m].
DISTRIBUTION: Mountains of SE New Mexico and W Texas (USA).
SYNONYMS: *sacramentoensis* Fleharty, 1960.
COMMENTS: Subgenus *Neotamias*. Elevated from subspecies of *cinereicollis* by Fleharty (1960). Reviewed by Findley et al. (1975:103-112).

Tamias cinereicollis J. A. Allen, 1890. Bull. Am. Mus. Nat. Hist., 3:94.
TYPE LOCALITY: "San Francisco Mountain, [Coconino Co.,] Arizona." [USA].
DISTRIBUTION: Mountains of C and E Arizona and C and SW New Mexico (USA).
SYNONYMS: *cinereus* (V. Bailey, 1911).
COMMENTS: Subgenus *Neotamias*.

Tamias dorsalis Baird, 1855. Proc. Acad. Nat. Sci. Philadelphia, 7:332.
TYPE LOCALITY: "Fort Webster, Coppermines of the Mimbres" Restricted by Howell (1929:131) to near present site of Santa Rita, Grant Co., New Mexico.
DISTRIBUTION: E Nevada, S Idaho, Utah, SW Wyoming, and NW Colorado south through Arizona and W New Mexico (USA) to NW Durango, W Coahuila, and coastal Sonora (Mexico).
SYNONYMS: *canescens* (J. Allen, 1904); *carminis* (Goldman, 1938); *grinnelli* (Burt, 1931); *nidoensis* (Lidicker, 1960); *sonoriensis* (Callahan and Davis, 1977); *utahensis* (Merriam, 1897).
COMMENTS: Subgenus *Neotamias*. Reviewed by Callahan and Davis (1977) who included *sonoriensis* in this species and by Hart (1992, Mammalian Species, 399).

Tamias durangae (J. A. Allen, 1903). Bull. Am. Mus. Nat. Hist., 19:594.
TYPE LOCALITY: "Arroyo de Bucy, Sierra de Candella, ...about 7,500 feet [2,134 m], northwestern Durango, Mexico."
DISTRIBUTION: SW Chihuahua to WC Durango; SE Coahuila (Mexico).
SYNONYMS: *nexus* Elliot, 1905; *solivagus* (A. Howell, 1922).
COMMENTS: Subgenus *Neotamias*. Formerly included in *bulleri*; includes *solivagus*; perhaps conspecific with *canipes*; see Callahan (1980), who considered both *durangae* and *solivagus*, *incertae sedis*.

Tamias merriami J. A. Allen, 1889. Bull. Am. Mus. Nat. Hist., 2:176.
TYPE LOCALITY: "San Bernardino Mts" [N of San Bernardino, 4,500 feet (1,372 m), San Bernardino Co., California, USA].
DISTRIBUTION: San Francisco Bay southward in the Coast Range, and south of Columbia (California, USA) in the Sierra Nevada, to extreme N Baja California (Mexico).
SYNONYMS: *kernensis* (Grinnell and Storer, 1916); *mariposae* (Grinnell and Storer, 1916); *pricei* J. Allen, 1895.
COMMENTS: Subgenus *Neotamias*. Formerly included *meridionalis* and *obscurus*; see Callahan (1977). See also comment under *obscurus*.

Tamias minimus Bachman, 1839. J. Acad. Nat. Sci. Philadelphia, 8:71.
TYPE LOCALITY: "Green River, near mouth of Big Sandy Creek, [Sweetwater Co.,] Wyo[ming]." [USA].
DISTRIBUTION: C Yukon (Canada) south through Sierra Nevada and S New Mexico, east to Michigan (USA) and W Quebec (Canada).
STATUS: IUCN - Endangered as *T. m. atristriatus*.
SYNONYMS: *arizonensis* (A. Howell, 1922); *atristriatus* (V. Bailey, 1913); *borealis* J. Allen, 1877; *cacodemus* (Cary, 1906); *caniceps* (Osgood, 1900); *caryi* (Merriam, 1908); *clarus* (V. Bailey, 1918); *confinis* (A. Howell, 1925); *consobrinus* J. Allen, 1890; *grisescens* (A. Howell, 1925); *hudsonius* (Anderson and Rand, 1944); *jacksoni* (A. Howell, 1925); *lectus* (J. Allen, 1905); *melanurus* Merriam, 1890; *neglectus* J. Allen, 1890; *operarius* (Merriam, 1905); *oreocetes* (Merriam, 1897); *pallidus* J. Allen, 1874; *pictus* J. Allen, 1890; *scrutator* (Hall and Hatfield, 1934); *selkirki* (Cowan, 1946); *silvaticus* (White, 1952).
COMMENTS: Subgenus *Neotamias*. Southern Rocky Mtn populations reviewed by Sullivan (1985).

Tamias obscurus J. A. Allen, 1890. Bull. Am. Mus. Nat. Hist., 3:70.
TYPE LOCALITY: "San Pedro [Martir] Mountains" [near Vallecitos, Baja California Norte, Mexico].
DISTRIBUTION: S California (San Bernardino Co., USA) to C Baja California (Mexico).
SYNONYMS: *davisi* (Callahan, 1977); *meridionalis* (Nelson and Goldman, 1909).
COMMENTS: Subgenus *Neotamias*. Regarded by Howell (1929) as a synonym of *merriami*; but see Callahan (1977) who provided evidence of specific distinctness and included *davisi* and *meridionalis* in this species.

Tamias ochrogenys (Merriam, 1897). Proc. Biol. Soc. Washington, 11:195, 206.
TYPE LOCALITY: "Mendocino, [Mendocino Co.,] California." [USA].
DISTRIBUTION: Coast of N California from Van Duzen River south to S Sonoma Co. (USA).
COMMENTS: Subgenus *Neotamias*. Elevated from subspecies of *townsendii* by Sutton and Nadler (1974); supported by Kain (1985), Sutton (1987), and Gannon and Lawlor (1989).

Tamias palmeri (Merriam, 1897). Proc. Biol. Soc. Washington, 11:208.
TYPE LOCALITY: "Charleston Peak, [Clark Co.,] Nevada [USA] (altitude about 2450 meters or 8000 feet)."
DISTRIBUTION: Charleston Mtns (S Nevada, USA).
COMMENTS: Subgenus *Neotamias*.

Tamias panamintinus Merriam, 1893. Proc. Biol. Soc. Washington, 8:134.
TYPE LOCALITY: "Johnson Cañon...Panamint Mountains, [Inyo Co.,] California." [USA] Restricted by Grinnell (1933:128) to vicinity of Hungry Bill's Ranch, about 5,000 ft. [1,524 m].
DISTRIBUTION: Mountains of SE California and SW Nevada (USA).

SYNONYMS: *acrus* (Johnson, 1943); *juniperus* (Burt, 1931).
COMMENTS: Subgenus *Neotamias*.

Tamias quadrimaculatus Gray, 1867. Ann. Mag. Nat. Hist., ser. 3, 20:435.
TYPE LOCALITY: "California, Michigan Bluff (Gruber)" [Placer Co., USA].
DISTRIBUTION: Sierra Nevada of EC California (Plumas to Mariposa cos.); C and adjacent WC Nevada (USA).
SYNONYMS: *macrorhabdotes* Merriam, 1886.
COMMENTS: Subgenus *Neotamias*.

Tamias quadrivittatus (Say, 1823). *In* James, Account Exped. Pittsburgh to Rocky Mtns, 2:45.
TYPE LOCALITY: "[Arkansas River]...the place where the river leaves the mountains," Restricted by Merriam (1905:163) to about 26 mi. [42 km] below Cañon City, Fremont Co., Colorado [USA].
DISTRIBUTION: Mountains of Colorado and E Utah south to NE Arizona and S New Mexico (USA).
STATUS: IUCN - Endangered as *T. q. australis*.
SYNONYMS: *animosus* (Warren, 1909); *australis* (Patterson, 1980); *gracilis* J. Allen, 1890.
COMMENTS: Subgenus *Neotamias*. Includes *australis* (Patterson, 1980). Formerly included *hopiensis*, but see Patterson (1984:452), who regarded it a *nomen dubium*, and *rufus*, now considered distinct (Patterson, 1984); but see Hoffmeister and Ellis (1979). See also *umbrinus*.

Tamias ruficaudus (A. H. Howell, 1920). Proc. Biol. Soc. Washington, 33:91.
TYPE LOCALITY: "Upper St. Mary's Lake, [Glacier Co.,] Montana." [USA].
DISTRIBUTION: NE Washington to W Montana (USA), and SE British Columbia (Canada).
SYNONYMS: *simulans* (A. Howell, 1922).
COMMENTS: Subgenus *Neotamias*. Patterson and Heaney (1987) suggested that *simulans* may be specifically distinct from *ruficaudus*, but the nature of contact between the two forms is not known.

Tamias rufus (Hoffmeister and Ellis, 1979). Southwest. Nat., 24:656.
TYPE LOCALITY: USA, "10 mi SW Page, Coconino County, Arizona".
DISTRIBUTION: E and S Utah, extreme W Colorado, and NE Arizona, USA.
COMMENTS: Formerly included in *quadrivittatus*, but see Patterson (1984).

Tamias senex J. A. Allen, 1890. Bull. Am. Mus. Nat. Hist., 3:83.
TYPE LOCALITY: "Summit of Donner Pass, Placer Co., Cal[ifornia, USA]."
DISTRIBUTION: Sierra Nevada of EC California and WC Nevada to N coast of California, and NC Oregon (USA).
COMMENTS: Subgenus *Neotamias*. Elevated from a subspecies of *townsendii* by Sutton and Nadler (1974); distinction also suggested by Kain (1985), Sutton (1987), and Gannon and Lawlor (1989).

Tamias sibiricus (Laxmann, 1769). Sibirische Briefe, Gottingen, p. 69.
TYPE LOCALITY: "Vicinity of Barnaul" [Altaisk Krai, Russia] (Chaworth-Musters, 1937).
DISTRIBUTION: N European and Siberian Russia to Sakhalin; S Kurile Isls (Russia); extreme E Kazakhstan to N Mongolia, China, Korea, and Hokkaido (Japan).
SYNONYMS: *albogularis* (J. Allen, 1909); *altaicus* (Hollister, 1912); *asiaticus* (Gmelin, 1788); *barberi* (Johnson and Jones, 1955); *intercessor* (Thomas, 1908); *jacutensis* (Ognev, 1935); *lineatus* (Siebold, 1824); *okadae* (Kuroda, 1932); *ordinalis* (Thomas, 1908); *orientalis* Bonhote, 1899; *pallasi* Baird, 1856; *senescens* (Miller, 1898); *striatus* (Pallas, 1778); *umbrosus* (A. Howell, 1927); *uthensis* (Pallas, 1811).
COMMENTS: Subgenus *Eutamias*; see Gromov et al. (1965:125).

Tamias siskiyou (A. H. Howell, 1922). J. Mammal., 3:180.
TYPE LOCALITY: "Near summit of White Mountain, Siskiyou Mountains, altitude 6,000 feet [1,829 m], [Siskiyou Co.,] California" [USA].
DISTRIBUTION: Siskiyou Mtns and coast of N California to C Oregon (USA).
COMMENTS: Subgenus *Neotamias*. Elevated from a subspecies of *townsendii* by Sutton and Nadler (1974); distinction supported by Kain (1985), Sutton (1987), and Gannon and Lawlor (1989).

Tamias sonomae (Grinnell, 1915). Univ. California Publ. Zool., 12:321.
TYPE LOCALITY: "One mile [1.6 km] west of Guerneville, Sonoma County, California." [USA].
DISTRIBUTION: NW California, from San Francisco Bay north to Siskiyou Co. (USA).
SYNONYMS: *alleni* (A. Howell, 1922).
COMMENTS: Subgenus *Neotamias*.

Tamias speciosus Merriam, 1890. *In* J. A. Allen, Bull. Am. Mus. Nat. Hist., 3:86.
TYPE LOCALITY: "San Bernardino Mts., [San Bernardino Co.,] Cal[ifornia, USA]." Restricted by Howell (1929:89) to Whitewater Creek, 7,500 ft. [2,255 m].
DISTRIBUTION: USA: Sierra Nevada from Mt. Lassen to San Bernardino Mtns (California); W Nevada.
SYNONYMS: *callipeplus* Merriam, 1893; *frater* J. Allen, 1890; *sequoiensis* (A. Howell, 1922).
COMMENTS: Subgenus *Neotamias*.

Tamias striatus (Linnaeus, 1758). Syst. Nat., 10th ed., 1:64.
TYPE LOCALITY: Restricted by Howell (1929:14) to "upper Savannah River, S[outh] C[arolina]." [USA].
DISTRIBUTION: S Manitoba and Nova Scotia (Canada) to Louisiana, Alabama, and Georgia, east to Atlantic Coast (USA).
SYNONYMS: *americanus* Gmelin, 1788; *doorsiensis* Long, 1971; *fisheri* A. Howell, 1925; *griseus* Mearns, 1891; *lysteri* (Richardson, 1829); *ohioensis* Bole and Moulthrop, 1942; *peninsulae* Hooper, 1942; *pipilans* Lowery, 1943; *quebecensis* Cameron, 1950; *rufescens* Bole and Moulthrop, 1942; *venustus* Bangs, 1896.
COMMENTS: Subgenus *Tamias*; see Gromov et al. (1965:134). Reviewed by Snyder (1982, Mammalian Species, 168).

Tamias townsendii Bachman, 1839. J. Acad. Nat. Sci. Philadelphia, 8(1):68.
TYPE LOCALITY: "lower Columbia River, near lower mouth of Willamette River, [Multnomah Co.,] Oreg[on]." [USA].
DISTRIBUTION: SW British Columbia (Canada), W Washington and Oregon to the Rogue River (USA).
SYNONYMS: *cooperi* Baird, 1855; *hindei* Gray, 1842; *littoralis* Elliot, 1903.
COMMENTS: Subgenus *Neotamias*. Formerly included *ochrogenys, siskiyou* and *senex*; see comments under those species.

Tamias umbrinus J. A. Allen, 1890. Bull. Am. Mus. Nat. Hist., 3:96.
TYPE LOCALITY: "Uintah Mountains, south of Ft. Bridger" Restricted by Howell (1929:94) to Blacks Fork, about 8,000 ft. [2,438 m], Summit Co., Utah [USA].
DISTRIBUTION: E California and N Arizona to N Colorado, SE and NW Wyoming, and extreme SW Montana (USA).
SYNONYMS: *adsitus* (J. Allen, 1905); *fremonti* (White, 1953); *inyoensis* (Merriam, 1897); *montanus* (White, 1953); *nevadensis* (Burt, 1931); *sedulus* (White, 1953).
COMMENTS: Subgenus *Neotamias*. Bergstrom and Hoffmann (1991) found species-specific vocalizations, habitats, and bacular characters in sympatric *umbrinus* and *quadrivittatus*, but convergence in electromorphs.

Tamiasciurus Trouessart, 1880. Le Naturaliste, 2(37):292.
TYPE SPECIES: *[Sciurus vulgaris] hudsonicus* Erxleben, 1777.
COMMENTS: Tribe Tamiasciurini (Moore, 1959).

Tamiasciurus douglasii (Bachman, 1839). Proc. Zool. Soc. Lond., 1838:99 [1839].
TYPE LOCALITY: "Shores of Columbia River" Restricted by Allen (1898:284) to mouth of Columbia River, [Clatsop Co., Oregon, USA].
DISTRIBUTION: Coast and Cascade ranges and Sierra Nevada of SW British Columbia (not Vancouver Isl) (Canada) to S California (USA).
SYNONYMS: *belcheri* (Gray, 1842); *cascadensis* (J. Allen, 1898); *mollipilosus* (Audubon and Bachman, 1841); *orarius* (Bangs, 1897); *suckleyi* (Baird, 1855).
COMMENTS: Formerly included *mearnsi*; see Lindsay (1981). Hall (1981:466) suggested that *douglasii* might be conspecific with *hudsonicus*, but Lindsay (1982) showed that apparent hybrids were probably due to character convergence.

Tamiasciurus hudsonicus (Erxleben, 1777). Syst. Regn. Anim., 1:416.

TYPE LOCALITY: "ad fretum Hudsonis" Restricted by Howell (1936:134) to the mouth of Severn River, Hudson Bay, Ontario, Canada.

DISTRIBUTION: Alaska (USA), throughout Canada (south of tundra); including Vancouver Isl, W USA in mountain states; NE USA, south to NW South Carolina.

STATUS: U.S. ESA and IUCN - Endangered as *T. h. grahamensis*.

SYNONYMS: *abieticola* (A. Howell, 1929); *baileyi* (J. Allen, 1898); *columbianus* A. Howell, 1936; *dakotensis* (J. Allen, 1894); *dixiensis* Hardy, 1942; *fremonti* (Audubon and Bachman, 1853); *grahamensis* (J. Allen, 1894); *gymnicus* (Bangs, 1899); *kenaiensis* A. Howell, 1936; *lanuginosus* (Bachman, 1839); *laurentianus* Anderson, 1942; *loquax* (Bangs, 1896); *lychnuchus* (Stone and Rehn, 1903); *minnesota* (J. Allen, 1899); *mogollonensis* (Mearns, 1890); *murii* A. Howell, 1943; *neomexicanus* (J. Allen, 1898); *pallescens* A. Howell, 1942; *petulans* (Osgood, 1900); *picatus* (Swarth, 1921); *preblei* A. Howell, 1936; *regalis* A. Howell, 1936; *richardsoni* (Bachman, 1839); *rubrolineatus* (Desmarest, 1822); *streatori* (J. Allen, 1898); *ungavensis* Anderson, 1942; *vancouverensis* (J. Allen, 1890); *ventorum* (J. Allen, 1898); *wasatchensis* Hardy, 1950.

COMMENTS: Includes *fremonti*; see Hardy (1950). See also *douglasii*.

Tamiasciurus mearnsi (Townsend, 1897). Proc. Biol. Soc. Washington, 11:146.

TYPE LOCALITY: "San Pedro Martir Mountains, Lower California (altitude about 7,000 feet)" [Baja California Norte, Mexico].

DISTRIBUTION: Sierra San Pedro Martir Mtns (Baja California Norte, Mexico).

COMMENTS: Formerly included in *douglasii*; see Lindsay (1981).

Tamiops J. A. Allen, 1906. Bull. Am. Mus. Nat. Hist., 22: 475.

TYPE SPECIES: *Tamiops macclellandi hainanus* J. A. Allen, 1906 (= *T. maritimus hainanus*).

COMMENTS: Tribe Callosciurini (Moore, 1959:173). See Osgood (1932), and Moore and Tate (1965), who revised *Tamiops*.

Tamiops macclellandi (Horsfield, 1840). Proc. Zool. Soc. Lond., 1839:152 [1840].

TYPE LOCALITY: "Bengal as well as Assam" [India]. Restricted by Ellerman (1940:354) to Assam.

DISTRIBUTION: E Nepal through Assam (India); N and C Burma, and Yunnan (China), and south through Thailand, Vietnam, Laos, and Cambodia to the S Malay Peninsula.

SYNONYMS: *barbei* (Blyth, 1847); *collinus* Moore, 1958; *inconstans* Thomas, 1920; *kongensis* (Bonhote, 1901); *leucotis* (Temminck, 1853); *lylei* Thomas, 1920; *manipurensis* (Bonhote, 1900); *novemlineatus* (Miller, 1903); *pembertoni* (Blyth, 1843).

Tamiops maritimus (Bonhote, 1900). Ann. Mag. Nat. Hist., ser. 7, 5:51.

TYPE LOCALITY: "Foochow, province of Fokien" [Fukien, China].

DISTRIBUTION: Hubei, Anhui and Zhejiang, south through Guangxi and Guangdong (China), to S Vietnam and Laos; isls of Hainan (China); Taiwan.

SYNONYMS: *formosanus* (Bonhote, 1900); *hainanus* J. Allen, 1906; *laotum* Robinson and Kloss, 1922; *moi* Robinson and Kloss, 1922; *monticolus* (Bonhote, 1900); *riudoni* J. Allen, 1906; *sauteri* J. Allen, 1911.

Tamiops rodolphei (Milne-Edwards, 1867). Rev. Mag. Zool. Paris, ser. 2, 19:227.

TYPE LOCALITY: "Cochin China near Saigon," [Vietnam].

DISTRIBUTION: E Thailand, Cambodia, S Laos, S Vietnam.

SYNONYMS: *dolphoides* Kloss, 1921; *elbeli* Moore, 1958; *holti* (Ellerman, 1940); *liantis* Kloss, 1919; *lylei* Thomas, 1920.

Tamiops swinhoei (Milne-Edwards, 1874). Rech. Hist. Nat. Mammifères, 1:308.

TYPE LOCALITY: "Moupin [Muping]," [= Baoxing, Sichuan, China]. Restricted by Allen (1940:673) to "Hongchantin...about 6000 feet."

DISTRIBUTION: Extreme SW Gansu south through Tibet, Sichuan and Yunnan (China) to N Burma and N Vietnam; isolated population in Hebei (China).

SYNONYMS: *chingpingensis* Lu and Qyan, 1965; *clarkei* Thomas, 1920; *forresti* Thomas, 1920; *holti* (Ellerman, 1940); *olivaceus* Osgood, 1932; *russeolus* Jacobi, 1923; *spencei* Thomas, 1921; *vestitus* Miller, 1915.

COMMENTS: The form *forresti* was originally named as a subspecies of *T. maritimus*, but assigned to *swinhoei* by Moore and Tate (1965:248).

Xerus Hemprich and Ehrenberg, 1833. Symb. Phys. Mamm., vol. 1, sig. ee.
TYPE SPECIES: *Sciurus (Xerus) brachyotis* Hemprich and Ehrenberg, 1833 (= *Sciurus rutilus* Cretzchmar, 1828).
SYNONYMS: *Euxerus* Thomas, 1909; *Geosciurus* Smith, 1834.
COMMENTS: Includes *Euxerus*, *Geosciurus*, and *Xerus* as subgenera (Amtmann, 1975; Ellerman, 1940; Moore, 1959).

Xerus erythropus (Desmarest, 1817). *In* Nouv. Dict. Hist. Nat., Nouv. ed., 10:110.
TYPE LOCALITY: Senegal (neotype). Origin of original type unknown. "Inconnue."
DISTRIBUTION: SE Morocco, S Mauritania, Senegal, Gambia, Guinea Bissau, Guinea, Sierra Leone, Ivory Coast, S Mali, Burkina Faso, Ghana, Togo, Benin, SE Niger, NE Nigeria, Cameroon, NE Congo, SE Chad, NE Central African Republic, Sudan, Zaire, NW Uganda, Rwanda, W Ethiopia, W Kenya, N Tanzania.
SYNONYMS: *agadius* (Thomas and Hinton, 1921); *albovittatus* (Desmarest, 1817); *chadensis* (Thomas, 1905); *fulvior* (Thomas, 1905); *lacustris* (Thomas, 1905); *lessonii* (Fitzinger, 1867, new name for *marabutus* Lesson); *leucoumbrinus* (Rüppell, 1835); *limitaneus* (Thomas and Hinton, 1923); *maestus* (Thomas, 1910); *marabutus* (Lesson, 1838); *microdon* Thomas, 1905; *prestigiator* (Lesson, 1838).
COMMENTS: Placed in *Euxerus* which is considered a subgenus of *Xerus* by Ellerman (1940), Moore (1959), and Amtmann (1975). The name *erythopus* or *erythropus* is commonly attributed to E. Geoffroy, 1803. Cat. Mamm. Mus. Hist. Nat., Paris, p. 178, who gave the name as *erythopus*; however, this work was never published and is therefore unavailable (see Appendix I). The next available use of the name *erythopus* was by Desmarest, 1817. Shinz, 1845, used the spelling *erythropus*. Opinion 945 of the International Commission on Zoological Nomenclature (1971) ruled that *erythopus* Geoffroy, 1803 be changed to *erythropus* as an incorrect original spelling. The proper latin root is "erythro". In the last 100 years nearly all authors have used the spelling *erythropus*. In the interest of orthographic stability we advocate that the specific name be spelled *erythropus*. It is not desirable to perpetuate the lapsus in spelling by early workers.

Xerus inauris (Zimmermann, 1780). Geogr. Gesch. Mensch. Vierf. Thiere, 2:344.
TYPE LOCALITY: South Africa, Kaffirland, 100 mi. (160 km) N of Cape of Good Hope. "Bewohnt die Cafferen, über hundert Meilen nordwärts des Vorgebürges der guten Hofnung...".
DISTRIBUTION: S Angola, Namibia, Botswana, W Zimbabwe, South Africa.
SYNONYMS: *africanus* (Shaw, 1801); *capensis* (Kerr, 1792); *dschinshicus* (Gmelin, 1788); *ginginianus* (Shaw, 1801); *levaillantii* (Kuhl, 1820); *namaquensis* (Lichtenstein, 1793); *setosus* (Smuts, 1832).
COMMENTS: Subgenus *Geosciurus*. Distinction from *X. princeps* supported by minor chromosomal differences reported by Robinson et al. (1986).

Xerus princeps (Thomas, 1929). Proc. Zool. Soc. Lond., 1929:106.
TYPE LOCALITY: N Namibia, C Koakoveld, Otjitundua.
DISTRIBUTION: W Namibia, S Angola, restricted to the Kaokoland escarpment of Namibia and Angola as far north as 14°10'S 16°0'E.
COMMENTS: Subgenus *Geosciurus*. Distinction from *X. inauris* supported by minor chromosomal differences reported by Robinson et al. (1986).

Xerus rutilus (Cretzschmar, 1828). *In* Rüppell, Atlas Reise Nordl. Afr., Zool., Säugeth., p. 59.
TYPE LOCALITY: Ethiopia, eastern slope of Abyssynia. "Der östliche Abhang Abyssiniens, wo es häufig vorkommmt." Probably Massawa, according to Mertens (1925:26) "...wahrsheinlich Massaua".
DISTRIBUTION: SE Sudan, E and S Ethiopia, Djibouti, Somalia, Kenya, NE Uganda, NE Tanzania.
SYNONYMS: *abessinicus* (Gmelin, 1788); *brachyotis* (Hemprich and Ehrenberg, 1833); *dabagala* Heuglin, 1861; *dorsalis* Dollman, 1911; *fuscus* (Huet, 1880); *intensus* Thomas, 1904; *massaicus* Toschi, 1945; *rufifrons* Dollman, 1911; *saturatus* (Neumann, 1900); *stephanicus* Thomas, 1906.
COMMENTS: Subgenus *Xerus*. Chromosomes described by Nadler and Hoffmann (1974). Reviewed by O'Shea (1991, Mammalian Species, 370).

Subfamily Pteromyinae Brandt, 1855. Kaiserlich. Akad. Wiss. Méth. Math, Phys. Nat., 7:157.
SYNONYMS: Petauristidae Miller, 1912.
COMMENTS: Controversy continues over whether flying squirrels constitute a monophyletic group; see Johnson-Murray (1977), Hight et al. (1974), and Thorington (1984). Bibliography by Lin et al. (1985). For validity of subfamilial name, see comment under *Pteromys*.

Aeretes G. M. Allen, 1940. Nat. Hist. Cent. Asia, II, 2:745.
TYPE SPECIES: *Pteromys melanopterus* Milne-Edwards, 1867.
COMMENTS: *Aeretes* is often dated as G. M. Allen, "p.vii, September 2, 1938," but this is a *nomen nudum*.

Aeretes melanopterus (Milne-Edwards, 1867). Ann. Sci. Nat. Zool., 8:375.
TYPE LOCALITY: "Les forêts qui couvrent la chaine montagneuse du Tscheli" [= Chihli; old name for Hebei Prov., China].
DISTRIBUTION: Hebei and Sichuan (China).
SYNONYMS: *sulcatus* (Howell, 1927); *szechuanensis* Wang, Tu, and Wang, 1966.
COMMENTS: Known only from two widely separated areas.

Aeromys Robinson and Kloss, 1915. J. Fed. Malay St. Mus., 6:23.
TYPE SPECIES: *Pteromys tephromelas* Günther, 1873.

Aeromys tephromelas (Günther, 1873). Proc. Zool. Soc. Lond., 1873:413.
TYPE LOCALITY: "Pinang" [Wellesley, Penang Isl., Malaysia].
DISTRIBUTION: Malaysian region, except Java and SW Philippines.
SYNONYMS: *bartelsi* (Sody, 1936); *phaeomelas* (Günther, 1873).
COMMENTS: Includes *phaeomelas*; see Medway (1977:101).

Aeromys thomasi (Hose, 1900). Ann. Mag. Nat. Hist., ser. 7, 5:215.
TYPE LOCALITY: "Silat River, about 70 miles south of Claudetown, Eastern Sarawak" [Baram, Sarawak, Malaysia].
DISTRIBUTION: Borneo, except SE.
SYNONYMS: *nitidus* Jentink, 1897.

Belomys Thomas, 1908. Ann. Mag. Nat. Hist., ser. 8, 1:2.
TYPE SPECIES: *Sciuropterus pearsonii* Gray, 1842.

Belomys pearsonii (Gray, 1842). Ann. Mag. Nat. Hist., [ser. 1], 10:263.
TYPE LOCALITY: "India, [Assam,] Dargellan" (= Darjeeling).
DISTRIBUTION: Sikkim and Assam (India) to Hunan, Sichuan, Yunnan, Guizhou, Guangxi, Hainan, (China); Bhutan; Taiwan, Indochina, and N Burma (see Agrawal and Chakraborty, 1979).
SYNONYMS: *blandus* Osgood, 1932; *kaleensis* (Swinhoe, 1863); *trichotis* Thomas, 1908; *villosus* (Blyth, 1847).
COMMENTS: Corbet and Hill (1992) synonomyzed this monotypic genus with *Trogopterus*.

Biswamoyopterus Saha, 1981. Bull. Zool. Surv. India, 4:331.
TYPE SPECIES: *Biswamoyopterus biswasi* Saha, 1981.

Biswamoyopterus biswasi Saha, 1981. Bull. Zool. Surv. India, 4:333.
TYPE LOCALITY: "Namdapha, Tirap District, Arunachal Pradesh, India."
DISTRIBUTION: Known only from type locality, western slope, Patkai Range.
COMMENTS: "Close to *Aeromys*" (Saha, 1981:333).

Eupetaurus Thomas, 1888. J. Asiat. Soc. Bengal, 57:256.
TYPE SPECIES: *Eupetaurus cinereus* Thomas, 1888.

Eupetaurus cinereus Thomas, 1888. J. Asiat. Soc. Bengal, 57:258.
TYPE LOCALITY: "Gilgit [Valley],...about 6000 feet" [Pakistan].
DISTRIBUTION: High elevations from N Pakistan and Kashmir to Sikkim (India; Agrawal and Chakraborty, 1970) to Tibet and possibly Yunnan (China).
COMMENTS: Reviewed by McKenna (1962).

Glaucomys Thomas, 1908. Ann. Mag. Nat. Hist., ser. 8, 1:5.
TYPE SPECIES: *Mus volans* Linnaeus, 1758.
COMMENTS: Revised by Howell (1918); karyotypes and evolution evaluated by Rausch and Rausch (1982).

Glaucomys sabrinus (Shaw, 1801). Gen. Zool., 2:157.
TYPE LOCALITY: Not specified. Restricted by Howell (1918:33) to the mouth of Severn River, Ontario, Canada.
DISTRIBUTION: Alaska and Canada, NW USA to S California and W South Dakota (Black Hills), NE USA to S Appalachian Mtns.
STATUS: U.S. ESA and IUCN - Endangered as *G. s. coloratus* and *G. s. fuscus*.
SYNONYMS: *alpinus* (Richardson, 1828); *bangsi* (Rhoads, 1897); *bullatus* A. Howell, 1915; *californicus* (Rhoads, 1897); *canadensis* (E. Geoffroy, 1803); *canescens* A. Howell, 1915; *coloratus* Handley, 1953; *columbiensis* A. Howell, 1915; *flaviventris* A. Howell, 1915; *fuliginosus* (Rhoads, 1897); *fuscus* Miller, 1936; *goodwini* Anderson, 1943; *gouldi* Anderson, 1943; *griseifrons* A. Howell, 1934; *hudsonicus* Gmelin, 1788; *klamathensis* (Merriam, 1897); *lascivus* (Bangs, 1899); *latipes* A. Howell, 1915; *lucifugus* Hall, 1934; *macrotis* (Mearns, 1898); *makkovikensis* (Sornborger, 1900); *murinauralis* Musser, 1961; *olympicus* (Elliot, 1899); *oregonensis* (Bachman, 1839); *reductus* Cowan, 1937; *stephensi* (Merriam, 1900); *yukonensis* (Osgood, 1900); *zaphaeus* (Osgood, 1905).
COMMENTS: Reviewed by Wells-Gosling and Heaney (1984, Mammalian Species, 229).

Glaucomys volans (Linnaeus, 1758). Syst. Nat., 10th ed., 1:63.
TYPE LOCALITY: "Virginia, Mexico"; restricted by Elliot (1901:109) to "Virginia (U.S.A.)".
DISTRIBUTION: Texas, Kansas, and Minnesota (USA) to Nova Scotia (Canada) and E USA; montane populations scattered from NW Mexico to Honduras.
STATUS: IUCN - Rare as *G. v. goldmani*, *G. v. guerreroensis*, and *G. v. oaxacensis*.
SYNONYMS: *americana* (Oken, 1816); *chontali* Goodwin, 1961; *cucullatus* (Fischer, 1829); *goldmani* (Nelson, 1904); *guerreroensis* Diersing, 1980; *herreranus* Goldman, 1936; *madrensis* Goldman, 1936; *nebrascensis* (Swenk, 1915); *oaxacensis* Goodwin, 1961; *querceti* (Bangs, 1896); *saturatus* A. Howell, 1915; *silus* (Bangs, 1896); *texensis* A. Howell, 1915; *underwoodi* Goodwin, 1936; *virginianus* (Tiedemann, 1808); *volucella* (Pallas, 1778);
COMMENTS: Reviewed by Dolan and Carter (1973, Mammalian Species, 78). For systematics and distribution of mesoamerican populations, see Diersing (1980*a*) and Braun (1988).

Hylopetes Thomas, 1908. Ann. Mag. Nat. Hist., ser. 8, 1:6.
TYPE SPECIES: *Sciuropterus everetti* Thomas, 1908 (= *Hylopetes spadiceus everetti*).
SYNONYMS: *Eoglaucomys* Howell, 1915.
COMMENTS: Includes *Eoglaucomys*; see Ellerman and Morrison-Scott (1951), McLaughlin (1967), and Corbet and Hill (1992), but also see McKenna (1962).

Hylopetes alboniger (Hodgson, 1836). J. Asiat. Soc. Bengal, 5:231.
TYPE LOCALITY: "central and Northern regions of Nipál" [Nepal].
DISTRIBUTION: Nepal and Assam (India) to Sichuan, Yunnan, and Hainan (China) and Indochina.
SYNONYMS: *chianfengensis* Wang and Lu, 1966; *leachii* (Gray, 1837); *orinus* (G. Allen, 1940); *turnbulli* (Gray, 1838).

Hylopetes baberi (Blyth, 1847). J. As. Soc. Bengal, 16:866.
TYPE LOCALITY: "Mountain districts of Nijrow," [Kohistan, Afghanistan].
DISTRIBUTION: Mountains of EC and NW Afghanistan, between 1,600 and 3,500 m; Kashmir.
COMMENTS: Formerly considered a subspecies of *fimbriatus* (Ellerman and Morrison-Scott, 1951:468), but elevated to specific status by Chakraborty (1981).

Hylopetes bartelsi (Chasen, 1939). Treubia, 17:185.
TYPE LOCALITY: "Tjilondong, Mt. Pangrango, West Java, about 900 metres" [Indonesia].
DISTRIBUTION: Java.
COMMENTS: Formerly included in *Petinomys*; but see Corbet and Hill (1992).

Hylopetes fimbriatus (Gray, 1837). Ann. Mag. Nat. Hist., 1:584.
TYPE LOCALITY: "India." Restricted by Robinson and Kloss (1918*a*:184) to "Western Himalayas," and by Ellerman and Morrison Scott (1955:468) to Simla, Punjab.

DISTRIBUTION: Kashmir and Punjab (India) from 1800 to 3600 m.
COMMENTS: Placed in monotypic genus *Eoglaucomys* by McKenna (1962) and others, but included in *Hylopetes* by McLaughlin (1967:215) and Corbet and Hill (1992).

Hylopetes lepidus (Horsfield, 1823). Zool. Res. Java, part 5:*Pteromys lepidus*, pl. and 2 unno. pp.
TYPE LOCALITY: "..only found in the closest forests of Java," [Indonesia].
DISTRIBUTION: S Vietnam, Thailand to Java; Borneo.
SYNONYMS: *aurantiacus* Wagner, 1841; *platyurus* (Jentink, 1890); *sagitta* (Linnaeus, 1766, part).
COMMENTS: Formerly called *sagitta*; see Medway (1977:104). Includes *platyurus* (Hill, 1961*c*; Medway, 1977:104; Corbet and Hill, 1992).

Hylopetes nigripes (Thomas, 1893). Ann. Mag. Nat. Hist., ser. 6, 12:30.
TYPE LOCALITY: "Puerta Princesa, Palawan" [Philippines].
DISTRIBUTION: Palawan and Bancalan Isls (Philippines).
SYNONYMS: *elassodontus* (Osgood, 1918).

Hylopetes phayrei (Blyth, 1859). J. Asiat. Soc. Bengal, 28:278.
TYPE LOCALITY: "Rangoon, Merqui" [Burma].
DISTRIBUTION: Burma; Thailand; Laos; S Vietnam; Fukien and Hainan (China).
SYNONYMS: *anchises* Allen and Coolidge, 1940; *electilis* (G. Allen, 1925); *laotum* (Thomas, 1914); *probus* (Thomas, 1914).
COMMENTS : Includes *electilis*, which was in *Petinomys*; see Corbet and Hill (1992).

Hylopetes sipora Chasen, 1940. Bull. Raffles Mus., 15:117.
TYPE LOCALITY: "Sipora Island, Mentawi Islands, West Sumatra" [Indonesia].
DISTRIBUTION: Sipora Isl (Indonesia).
COMMENTS: Formerly included in *spadiceus*; see Hill (1961*c*) who noted that an adult specimen is needed to clarify the status of this taxon. Part of the rodent fauna endemic to the Mentawi Archipelago (see comments under *Leopoldamys siporanus*).

Hylopetes spadiceus (Blyth, 1847). J. Asiat. Soc. Bengal, 16:867.
TYPE LOCALITY: "Arracan" [Arakan, Burma].
DISTRIBUTION: Burma, Thailand, S Vietnam, Sumatra, and Malaysia.
SYNONYMS: *amoenus* (Miller, 1906); *belone* (Thomas, 1908); *caroli* Gyldenstolpe, 1920; *everetti* (Thomas, 1895); *harrisoni* (Stone, 1900); *sumatrae* Sody, 1949.
COMMENTS: Includes *harrisoni*; see Medway (1977:105). Formerly included *sipora*; see Hill (1961*c*). Corbet and Hill (1980:137) listed *spadiceus* in *lepidus*, without comment.

Hylopetes winstoni (Sody, 1949). Treubia Buitenzorg, 20:75.
TYPE LOCALITY: "Baleq, E. Atjeh, N. Sumatra, 1200 m," [Indonesia].
DISTRIBUTION: Known only from the type locality.
COMMENTS: Originally described as a species of *Iomys*, but placed in *Hylopetes* by Corbet and Hill (1992).

Iomys Thomas, 1908. Ann. Mag. Nat. Hist., ser. 8, 1:1.
TYPE SPECIES: *Pteromys horsfieldii* Waterhouse, 1838.
COMMENTS: Formerly included *winstoni*, here placed in *Hylopetes*.

Iomys horsfieldii (Waterhouse, 1838) Proc. Zool. Soc. Lond., 1837:87 [1838].
TYPE LOCALITY: "Either from Java or Sumatra." Restricted by Chasen (1940:114) to Sumatra.
DISTRIBUTION: Malay Peninsula to Java; Borneo.
SYNONYMS: *davisoni* (Thomas, 1886); *everetti* (G. Allen and Coolidge, 1940, not Thomas, 1895); *lepidus* Lyon, 1911 (not Horsfield, 1824); *penangensis* Chasen, 1940; *thomsoni* Thomas, 1900.
COMMENTS: Formerly included *sipora* and *winstoni* (Corbet and Hill, 1980, 1986).

Iomys sipora Chasen and Kloss, 1928. Proc. Zool. Soc. Lond., 1927(4):819 [1928].
TYPE LOCALITY: "Sipora Island, West Sumatra." [Indonesia].
DISTRIBUTION: Mentawi Isls [Indonesia].
COMMENTS: Part of the rodent fauna endemic to the Mentawi Archipelago (see comments under *Leopoldamys siporanus*).

Petaurillus Thomas, 1908. Ann. Mag. Nat. Hist., ser. 8, 1:3.
TYPE SPECIES: *Sciuropterus hosei* Thomas, 1900.

Petaurillus emiliae Thomas, 1908. Ann. Mag. Nat. Hist., ser. 8, 1:8.
TYPE LOCALITY: "Baram, E. Sarawak," [Malaysia].
DISTRIBUTION: Sarawak.

Petaurillus hosei (Thomas, 1900). Ann. Mag. Nat. Hist., ser. 7, 5:275.
TYPE LOCALITY: "Baram District, Eastern Sarawak...Toyut River" [Malaysia].
DISTRIBUTION: Sarawak.

Petaurillus kinlochii (Robinson and Kloss, 1911). J. Fed. Malay St. Mus., 4:171.
TYPE LOCALITY: Kapar, Selangor, Malaysia.
DISTRIBUTION: Selangor (Malay Peninsula).
COMMENTS: Corbet and Hill (1992) included this form in *hosei*.

Petaurista Link, 1795. Zool. Beytr., 1(2):52, 78.
TYPE SPECIES: *Sciurus petaurista* Pallas, 1766.
SYNONYMS: *Galeolemur* Lesson, 1840.

Petaurista alborufus (Milne-Edwards, 1870). C. R. Acad. Sci. (Paris), 70:342.
TYPE LOCALITY: Moupin [= Baoxing, Sichuan, China].
DISTRIBUTION: Taiwan, S and C China.
SYNONYMS: *alborusus* (Hilzheimer, 1906); *castaneus* Thomas, 1923; *leucocephalus* (Hilzheimer, 1905); *lena* Thomas, 1907; *ochraspis* Thomas, 1923; *pectoralis* (Swinhoe, 1871).
COMMENTS: Includes *lena* which was treated as a separate species by Kuntz and Ming (1970); see Jones (1975). Provisionally includes *pectoralis* (Corbet and Hill, 1992). Reviewed by Day (1988).

Petaurista elegans (Müller, 1840). *In* Temminck, Verhandl. Nat. Gesch. Nederland. Overz. Bezitt., Zool., Zoogd. Indisch. Archipel, pp. 35, 56[1840], see comments.
TYPE LOCALITY: "Java" [Indonesia].
DISTRIBUTION: Nepal, Sikkim (India), Sichuan and Yunnan (China), N and W Burma, Laos, Tonkin (Vietnam), Malay Peninsula, Sumatra, Java (Indonesia), Borneo.
SYNONYMS: *banksi* Chasen, 1933; *caniceps* (Gray, 1842); *clarkei* Thomas, 1922; *gorkhali* (Lindsay, 1929); *marica* Thomas, 1912; *punctatus* (Gray, 1846); *senex* (Hodgson, 1844); *slamatensis* Sody, 1949; *sumatrana* Kloss, 1921; *sybilla* Thomas and Wroughton, 1916.
COMMENTS: Includes *clarkei* and *marica*; see Ellerman and Morrison-Scott (1966:460-461). This species was described in greater detail by Schlegel and Müller, *in* Temminck, Verh. Nat. Gesch. Nederland. Overz. Bezitt., Zool., Mammalia, pp. 107, 112[1845]. See Appendix I. Corbet and Hill (1992) considered *caniceps* from Nepal, Sikkim, N Burma, and W China, and *sybilla* from a few localities in Burma and W China, as distinct species, sympatric with *elegans* in W Yunnan (China).

Petaurista leucogenys (Temminck, 1827). Monogr. Mamm., 1:27.
TYPE LOCALITY: "...de forêts dans les provinces de Figo." Restricted by Kuroda (1938:50) to "Higo, Kiusiu" [Kyushu, Japan].
DISTRIBUTION: Japan, except Hokkaido; Gansu, Sichuan, Yunnan (China).
SYNONYMS: *filchnerinae* (Matschie, 1908); *hintoni* Mori, 1923; *nikkonis* Thomas, 1905; *oreas* Thomas, 1905; *osiui* Kuroda, 1938; *pectoralis* (Swinhoe, 1870); *thomasi* Kuroda and Mori, 1923 (not Hose, 1900); *tosae* Thomas, 1905; *watasei* Mori, 1927.
COMMENTS: Formerly included *xanthotis*, see Corbet (1978*c*:86). McKenna (1962) and Corbet and Hill (1991:145) considered it distinct.

Petaurista magnificus (Hodgson, 1836). J. Asiat. Soc. Bengal, 5:231.
TYPE LOCALITY: "central and Northern regions of Nipál" [Nepal].
DISTRIBUTION: Tibet (China), Nepal, Bhutan, Sikkim (India).
SYNONYMS: *hodgsoni* Ghose and Saha, 1981.
COMMENTS: Formerly included *nobilis* (Ellerman and Morrison-Scott, 1966:464); but also see Ghose and Saha (1981:95).

Petaurista nobilis (Gray, 1842). Ann. Mag. Nat. Hist., 10:263.
TYPE LOCALITY: "India, Dargellan" [Darjeeling].
DISTRIBUTION: C Nepal, Sikkim (India), Bhutan.
SYNONYMS: *chrysotrix* (Hodgson, 1844); *singhei* Saha, 1977.

COMMENTS: Formerly included in *magnificus;* elevated to specific rank by Ghose and Saha (1981:95) and Corbet and Hill (1992).

Petaurista petaurista (Pallas, 1766). Misc. Zool., p. 54.
TYPE LOCALITY: Not stated. Restricted by Robinson and Kloss (1918*a*:172, 221) to Preanger Regencies, Western Java, Indonesia.
DISTRIBUTION: E Afghanistan, Kashmir and Punjab east to Assam (India); Yunnan, Sichuan, Fukien (?) (China); Burma; Thailand; Indochina; Malaysia; Sumatra, Java (Indonesia); Borneo.
SYNONYMS: *albiventer* (Gray, 1834); *barroni* Kloss, 1916; *batuana* Miller, 1903; *birrelli* Wroughton, 1911; *candidula* Wroughton, 1911; *cicur* Robinson and Kloss, 1914; *fulvinus* Wroughton, 1911; *inornatus* (Geoffroy, 1844); *interceptio* Sody, 1949; *lumholtzi* Gyldenstolpe, 1920; *marchio* Thomas, 1908; *melanotis* (Gray, 1837); *mimicus* Miller, 1913; *nigrescens* Medway, 1965; *nigricaudatus* Robinson and Kloss, 1918; *nitida* (Desmarest, 1818); *nitidula* Thomas, 1900; *penangensis* Robinson and Kloss, 1918; *rajah* Thomas, 1908; *rufipes* Sody, 1949; *sodyi* Harris, 1951; *stellaris* Chasen, 1940; *taylori* Thomas, 1914; *terutaus* Lyon, 1907.
COMMENTS: Reviewed by Corbet and Hill (1992), who recognized *philippensis* as distinct, and allocated many of the forms that were formerly assigned to *petaurista* to it; the two species are widely sympatric. Also reviewed by Day (1988); karyotypic variation reported by Yong and Dhaliwal (1976).

Petaurista philippensis (Elliot, 1839). Madras J. Litt. Sci., 10:217.
TYPE LOCALITY: Near Madras [India].
DISTRIBUTION: Sri Lanka; India, north to Bombay and Rajastan, S Bihar; Burma, Thailand; S China, including Hainan and Taiwan; Indonesia (Agrawal and Chakraborty, 1979).
SYNONYMS: *annamensis* Thomas, 1914; *badiatus* Thomas, 1925; *cineraceus* (Blyth, 1847); *cinderella* Wroughton, 1911; *grandis* (Swinhoe, 1863); *griseiventer* (Gray, 1843); *hainana* G. Allen, 1925; *lanka* Wroughton, 1911; *lylei* Bonhote, 1900; *mergulus* Thomas, 1922; *miloni* Bourret, 1942; *nigra* Wang, 1981; *oral* (Tickell, 1842); *primrosei* Thomas, 1926; *reguli* Thomas, 1926; *rubicundus* Howell, 1927; *rufipes* G. Allen, 1925; *stockleyi* Carter, 1933; *venningi* Thomas, 1914; *yunanensis* (Anderson, 1875).
COMMENTS: Formerly included in *petaurista;* but see Corbet and Hill (1992), who reviewed the species.

Petaurista xanthotis (Milne-Edwards, 1872). Rech. Hist. Nat. Mammifères, p. 301.
TYPE LOCALITY: Moupin [= Baoxing, Sichuan, China].
DISTRIBUTION: Mountains of W China (Sichuan, Yunnan, E Tibet, Gansu).
SYNONYMS: *buechneri* (Matschie, 1907); *filchnerinae* (Matschie, 1907).
COMMENTS: Formerly included in *leucogenys;* but see Corbet and Hill (1992).

Petinomys Thomas, 1908. Ann. Mag. Nat. Hist., ser. 8, 1:6.
TYPE SPECIES: *Sciuropterus lugens* Thomas, 1908 (= *Petinomys hageni lugens*).
SYNONYMS: *Olisthomys* Carter, 1942.
COMMENTS: Formerly included *bartelsi* and *electilis*, here included in *Hylopetes;* see McKenna (1962:35) and Corbet and Hill (1992).

Petinomys crinitus (Hollister, 1911). Proc. Biol. Soc. Washington, 24:185.
TYPE LOCALITY: "Basilan Island, Philippines."
DISTRIBUTION: Basilan, Dinagat, Siargao, and Mindanao Isls (Philippines; Heaney and Rabor, 1982).
SYNONYMS: *mindanensis* (Rabor, 1939); *nigricaudus* Sanborn, 1953.
COMMENTS: Holotype of *Hylopetes mindanensis* destroyed during World War II; paratype in National Museum of Natural History is probably *P. crinitus.*

Petinomys fuscocapillus (Jerdon, 1847). J. Asiat. Soc. Bengal, 16:867.
TYPE LOCALITY: "S. India" Travancore.
DISTRIBUTION: S India, Sri Lanka.
SYNONYMS: *layardi* (Kelaart, 1850).

Petinomys genibarbis (Horsfield, 1822). Zool. Res. Java, part 4:*Pteromys genibarbis*, pls.(col. and fig. w) and 6 unno. pp.
TYPE LOCALITY: "...the forests of Pugar...Eastern portion of Java." [Indonesia].
DISTRIBUTION: Malaya to Sumatra, Java, and Borneo.
SYNONYMS: *borneoensis* (Thomas, 1908); *malaccanus* (Thomas, 1908).

COMMENTS: Perhaps conspecific with *sagitta*; see comment therein, and Medway (1977:102).

Petinomys hageni (Jentink, 1888). Notes Leyden Mus., 11:26.
TYPE LOCALITY: "Deli" [North East Sumatra, Indonesia].
DISTRIBUTION: Borneo, Sumatra.
SYNONYMS: *ouwensi* Sody, 1949.
COMMENTS: Formerly included *lugens* (Chasen, 1940:119); see comments therein. Medway (1977:112) considered the status of *ouwensi* doubtful.

Petinomys lugens (Thomas, 1895). Ann. Mus. Civ. Storia Nat. Geneva, Ser. 2a, 14:666.
TYPE LOCALITY: Indonesia, Mentawi Isls, Sipora Isl, "Si Oban".
DISTRIBUTION: Siberut and Sipora Isls (Sumatra, Indonesia).
SYNONYMS: *maerens* (Miller, 1903).
COMMENTS: Formerly included in *hageni*; but see Chasen and Kloss (1927:819) and Jenkins and Hill (1982). Part of the rodent fauna endemic to the Mentawi Archipelago (see comments under *Leopoldamys siporanus*).

Petinomys sagitta (Linnaeus, 1766). Syst. Nat., 12th ed., 1:88.
TYPE LOCALITY: Indonesia, Java.
DISTRIBUTION: Java.
COMMENTS: Formerly included in *Hylopetes*; see Medway (1977:102). May be conspecific with *genibarbis*; see Ellerman and Morrison-Scott (1955:31). Medway (1977:102) considered it inadvisable to synonymize *sagitta* and *genibarbis* until the relationship has been certainly established. Corbet and Hill (1992) considered *sagitta* as *incertae sedis*.

Petinomys setosus (Temminck, 1844). Fauna Japonica, 1(Mamm.), p. 49.
TYPE LOCALITY: "...Padang, êle de Sumatra" [Indonesia].
DISTRIBUTION: Burma, Malaysia, Sumatra, Borneo [Indonesia].
SYNONYMS: *morrisi* (Carter, 1942).
COMMENTS: Includes *morrisi*; see Muul and Thonglongya (1971) and Corbet and Hill (1980:137). McKenna (1962) considered *morrisi* assignable to a distinct genus *Olisthomys*.

Petinomys vordermanni (Jentink, 1890). Notes Leyden Mus., 12:150.
TYPE LOCALITY: "Billiton" [Belitung Isl, Indonesia].
DISTRIBUTION: S Burma, Malaya, Borneo.
SYNONYMS: *phipsoni* (Thomas, 1916).
COMMENTS: McKenna (1962) considered *vordermanni* representative of an undescribed genus, but Muul and Thonglongya (1971) and Hill (1961*c*) retained it in *Petinomys*.

Pteromys G. Cuvier, 1800. Lecon's Anat. Comp., I, tab. 1.
TYPE SPECIES: *Sciurus volans* Linnaeus, 1758.
SYNONYMS: *Sciuropterus* F. Cuvier, 1825.
COMMENTS: *Sciuropterus* was previously employed for this genus by Simpson (1945:80); he believed *Pteromys* to be a synonym of *Petaurista*, but Ellerman and Morrison-Scott (1951:466) presented evidence for the validity of *Pteromys*.

Pteromys momonga Temminck, 1844. Fauna Japonica, 1(Mamm.), p. 47.
TYPE LOCALITY: "...les forêts de l'interieur," Restricted by Kishida (in Kuroda, 1938:51) to "Kiusiu" [Kyushu, Japan].
DISTRIBUTION: Kyushu and Honshu (Japan).
SYNONYMS: *amygdali* (Thomas, 1906); *interventus* (Kuroda, 1941); *momoga* Temminck, 1845.

Pteromys volans (Linnaeus, 1758). Syst. Nat., 10th ed., 1:64.
TYPE LOCALITY: "in borealibus Europae, Asiae, et Americae." Restricted by Thomas (1911*a*:149) to "Finland." Ognev (1966:268) proposed restriction to "central Sweden," but the species does not occur there (Sulkava, 1978:76).
DISTRIBUTION: Palearctic taiga, from N Finland east to Chukotka (Russia); south to E Baltic shore; S Ural Mtns, Altai Mtns (Russia), Mongolia; N China; Korea; Sakhalin Isl (Russia), and Hokkaido (Japan); perhaps mtns of W China.
SYNONYMS: *aluco* (Thomas, 1907); *anadyrensis* Ognev, 1940; *arsenjevi* Ognev, 1935; *athene* (Thomas, 1907); *betulinus* Serebrennikov, 1930; *buechneri* Satunin, 1903; *gubari* Ognev, 1935; *incanus* Miller, 1918; *ognevi* Stroganov, 1936; *orii* (Kuroda, 1921); *russicus*

Tiedemann, 1808; *sibiricus* Desmarest, 1922; *turovi* Ognev, 1919; *vulgaris* Wagner, 1842; *wulungshanensis* (Mori, 1939).

COMMENTS: Chromosomes described by Rausch and Rausch (1982).

Pteromyscus Thomas, 1908. Ann. Mag. Nat. Hist., ser. 8, 1:3.

TYPE SPECIES: *Sciuropterus pulverulentus* Günther, 1873.

Pteromyscus pulverulentus (Günther, 1873). Proc. Zool. Soc. Lond., 1873:413.

TYPE LOCALITY: "Pinang" [Penang, Malaysia].

DISTRIBUTION: S Thailand to Sumatra; Borneo.

SYNONYMS: *borneanus* Thomas, 1908.

Trogopterus Heude, 1898. Mem. Hist. Nat. Emp. Chin., 4(1): 46-47.

TYPE SPECIES: *Pteromys xanthipes* Milne-Edwards, 1867.

Trogopterus xanthipes (Milne-Edwards, 1867). Ann. Sci. Nat. Zool., 8: 376.

TYPE LOCALITY: "Les forêts Qui couvrent la chaine montagneuse du Tscheli" [= Chihli, old name for Hebei Prov., China].

DISTRIBUTION: Montane forests, from Yunnan to C and E China.

SYNONYMS: *edithae* Thomas, 1923; *himalaicus* Thomas, 1914; *minax* Thomas, 1923; *mordax* Thomas, 1914.

COMMENTS: Includes *edithae* (Yunnan), *himalaicus* (S Tibet), *minax* (Sichuan), and *mordax* (Hubei); closely related to *Belomys* from SE Asia, which Corbet and Hill (1992) include in *Trogopterus*.

FAMILY CASTORIDAE

by Don E. Wilson

Family Castoridae Hemprich, 1820. Grundriss Naturgesch., p. 33.

Castor Linnaeus, 1758. Syst. Nat., 10th ed., 1:58.
TYPE SPECIES: *Castor fiber* Linnaeus, 1758.

Castor canadensis Kuhl, 1820. Beitr. Zool. Vergl. Anat., 1:64.
TYPE LOCALITY: Canada, Hudson Bay.
DISTRIBUTION: Brooks Range in Alaska (USA) to Labrador (Canada), Tamaulipas (Mexico), and N Florida (USA). Introduced in Europe and Asia.
STATUS: IUCN - Endangered as *C. c. frondator* and *C. c. mexicanus*.
SYNONYMS: *acadicus, baileyi, belugae, caecator, carolinensis, concisor, duchesnei, frondator, idoneus, labradorensis, leucodontus, mexicanus, michiganensis, missouriensis, pallidus, phaeus, repentinus, rostralis, sagittatus, shastensis, subauratus, taylori, texensis.*
COMMENTS: Lavrov and Orlov (1973) considered this species distinct from *fiber*. Reviewed by Jenkins and Busher (1979, Mammalian Species, 120).

Castor fiber Linnaeus, 1758. Syst. Nat., 10th ed., 1:58.
TYPE LOCALITY: Sweden.
DISTRIBUTION: NW and NC Eurasia, east to Lake Baikal, and south to France and Mongolia.
STATUS: U.S. ESA - Endangered as *C. f. birulai*; IUCN - Endangered as *C. f. bindai*.
SYNONYMS: *albicus, albus, balticus, bindai, birulai, flavus, fulvus, galliae, gallicus, niger, pohlei, propiius, solitarius, variegatus, varius, vistulanus.*
COMMENTS: Lavrov and Orlov (1973) clarified the distinction between this species and *canadensis*. Lavrov (1979) considered *albicus* a distinct species, but also see Corbet and Hill (1991) and Freye (1978).

FAMILY GEOMYIDAE

by James L. Patton

Family Geomyidae Bonaparte, 1845. Cat. Meth. Mamm. Europe, p. 5.

COMMENTS: Revised by Russell (1968*a*), who placed all Recent genera in one of two tribes in the subfamily Geomyinae: Geomyini (*Geomys, Pappogeomys, Orthogeomys,* and *Zygogeomys*) and Thomomyini (*Thomomys*); he included *Macrogeomys* and *Heterogeomys* as subgenera of *Orthogeomys,* an action followed by more recent authors (Hall, 1981; Hafner, 1982, 1991), and *Cratogeomys* as a subgenus of *Pappogeomys* (see also Russell, 1968*b*). The latter action was accepted by Hall (1981) but not by Jones et al. (1986), Lee and Baker (1987), Davidow-Henry et al. (1989), and Hollander (1990), who considered *Cratogeomys* as a separate genus. *Platygeomys* Merriam was synonymized with *Pappogeomys* by Hooper (1946). Phallic morphology has been reviewed by Williams (1982*b*); Hafner (1982) reported on phylogenetic relationships based on biochemical characters; and Wahlert (1985) provided a classification of the superfamily Geomyoidea, including Geomyidae. Russell (1968*a*) and Hall (1981) provided keys to Recent genera. Specific boundaries within all pocket gopher genera are often difficult to define because of an apparent inability to subdivide the subterranean niche and thus resultant allopatric distribution patterns (see Patton, 1990).

Geomys Rafinesque, 1817. Am. Mon. Mag., 2(1):45.

TYPE SPECIES: *Geomys pinetis* Rafinesque, 1817.

SYNONYMS: *Ascomys* Lichtenstein, *Diplostoma* Rafinesque, *Parageomys* Hibbard, *Pseudostoma* Say, *Saccophorus* Kuhl.

COMMENTS: Revised by Merriam (1895*a*). Specific boundaries in this genus are poorly defined. Complex relationships have been described for several definable geographic units based on a variety of contact zone analyses that have employed morphological, karyological, allozyme, and/or mitochondrial and nuclear DNA biochemical analyses. These have indicated varying degrees of hybridization between geographically differentiated forms, and authors vary in their recognition of these entities at the specific or subspecific levels. While the rather traditional species (*arenarius, bursarius, personatus, pinetis,* and *tropicalis*; Hall, 1981) are clearly an inadequate representation of specific diversity in the genus (see Block and Zimmerman, 1991), this arrangement is followed here due to lack of a thorough geographic analysis throughout the range of the genus and a concensus among workers. Keys to the species of *Geomys* were published by Baker and Williams (1974) and Hall (1981).

Geomys arenarius Merriam, 1895. N. Am. Fauna, 8:139.

TYPE LOCALITY: USA, Texas, El Paso Co., El Paso.

DISTRIBUTION: Extreme W Texas, SW and SC New Mexico (USA); N Chihuahua (Mexico).

SYNONYMS: *brevirostris* Hall.

COMMENTS: Reviewed by Williams and Baker (1974, Mammalian Species, 36) and Williams and Genoways (1978). Considered as a subspecies of *bursarius* by Hafner and Geluso (1983). For subspecies, see Hall (1981).

Geomys bursarius (Shaw, 1800). Philosophical Magazine, 6:215.

TYPE LOCALITY: USA, upper Mississippi Valley (restricted to Minnesota, Sherburne Co., Elk River by Swenk, 1939).

DISTRIBUTION: SC Manitoba (Canada) to NW Indiana, SW Louisiana, SC Texas, and EC New Mexico (USA).

SYNONYMS: *alba* Rafinesque, *ammophilus* Davis, *attwateri* Merriam; *brazensis* Davis, *breviceps* Baird, *canadensis* Lichtenstein, *dutcheri* Davis, *fusca* Rafinesque, *hylaeus* Blossom, *illinoensis* Komarek and Spencer, *industrius* Villa and Hall, *jugossicularis* Hooper, *knoxjonesi* Baker and Genoways, *levisagittalis* Swenk, *llanensis* Bailey, *ludemani* Davis, *lutescens* Merriam, *major* Davis, *majusculus* Swenk, *missouriensis* McLaughlin, *pratincola* Davis, *saccatus* Mitchell, *sagittalis* Merriam, *terricolus* Davis, *texensis* Merriam, *vinaceus* Swenk, *wisconsinensis* Jackson.

COMMENTS: For subspecies, see Hall (1981:499-500). Revised by Merriam (1895*a*) and, in part, by Honeycutt and Schmidly (1979). The form *lutescens* was considered as a valid species by Heaney and Timm (1983*a*, 1985) but as a subspecies by Burns et al. (1985) and Sudman et al. (1987). Tucker and Schmidly (1981) and Williams and Cameron (1991) considered *attwateri* distinct from *bursarius*, as did Bradley et al. (1991) for *knoxjonesi*, Sulentich et al. (1991) for *breviceps*, and Block and Zimmerman (1991) for *attwateri*, *breviceps*, *knoxjonesi*, and *texensis*.

Geomys personatus True, 1889. Proc. U.S. Natl. Mus., 11:159.
TYPE LOCALITY: USA, Texas, Cameron Co., Padre Island.
DISTRIBUTION: S Texas, south of San Antonio and Del Rio, including Padre and Mustang Islands (USA); barrier beaches of extreme NE Tamaulipas (Mexico).
SYNONYMS: *fallax* Merriam, *fuscus* Davis, *maritimus* Davis, *megapotamus* Davis, *streckeri* Davis.
COMMENTS: Reviewed by Davis (1940), Williams and Genoways (1981), and Williams (1982*a*, Mammalian Species, 170).

Geomys pinetis Rafinesque, 1817. Amer. Monthly Mag., 2:45.
TYPE LOCALITY: USA, Georgia, in the region of the pines (restricted to Screven Co. by Harper, 1952).
DISTRIBUTION: C Florida to S Georgia and S Alabama (USA).
SYNONYMS: *austrinus* Bangs, *colonus* Bangs, *cumberlandius* Bangs, *floridanus* Audubon and Bachman, *fontanelus* Sherman, *goffi* Sherman, *mobilensis* Merriam, *tuza* Barton.
COMMENTS: *Mus tuza* Barton, 1806, a senior synomym of *pinetis* according to Merriam (1895*a*), was considered of uncertain application and not available by Harper (1952). Laerm (1981) supported the synonymy of *cumberlandius*. Reviewed by Pembleton and Williams (1978, Mammalian Species, 86), Williams and Genoways (1980*b*), and Wilkins (1987*b*).

Geomys tropicalis Goldman, 1915. Proc. Biol. Soc. Washington, 28:134.
TYPE LOCALITY: Mexico, Tamaulipas, Altamira.
DISTRIBUTION: Vicinity of Altamira and Tampico in SE Tamaulipas (Mexico).
COMMENTS: Elevated to specific status by Alvarez (1963*a*). Reviewed by Baker and Williams (1974, Mammalian Species, 35) and Williams and Genoways (1977).

Orthogeomys Merriam, 1895. N. Am. Fauna, 8:172.
TYPE SPECIES: *Geomys scalops* Thomas, 1894.
SYNONYMS: *Heterogeomys* Merriam, *Macrogeomys* Merriam.
COMMENTS: Russell (1968*a*) revised the genus, and included *Heterogeomys* and *Macrogeomys* as valid subgenera. Biochemical systematics of species of subgenus *Macrogeomys* was presented by Hafner (1991). Species limits are poorly defined, as most are known from single localities or otherwise geogeographically restricted areas.

Orthogeomys cavator (Bangs, 1902). Bull. Mus. Comp. Zool., 39:42.
TYPE LOCALITY: Panama, Chiriqui Prov., Boquete, 4,800 ft. (1,463 m).
DISTRIBUTION: NW Panama to C Costa Rica.
SYNONYMS: *nigrescens* Goodwin, *pansa* Bangs.
COMMENTS: Subgenus *Macrogeomys*. Reviewed by Goodwin (1946).

Orthogeomys cherriei (J. A. Allen, 1893). Bull. Am. Mus. Nat. Hist., 5:337.
TYPE LOCALITY: Costa Rica, Limón Prov., Santa Clara.
DISTRIBUTION: NC Costa Rica (see Hafner and Hafner, 1987; Hafner, 1991).
SYNONYMS: *carlosensis* Goodwin, *costaricensis* Merriam.
COMMENTS: Subgenus *Macrogeomys*. Revised by Goodwin (1946).

Orthogeomys cuniculus Elliot, 1905. Proc. Biol. Soc. Washington, 18:224.
TYPE LOCALITY: Mexico, Oaxaca, Zanatepec (corrected from Yautepec, Oaxaca by Elliot, 1907*a*).
DISTRIBUTION: Known only from the type locality.
COMMENTS: Subgenus *Orthogeomys*. Reviewed by Nelson and Goldman (1930).

Orthogeomys dariensis (Goldman, 1912). Smithson. Misc. Coll., 60(2):8.
TYPE LOCALITY: Panama, Darien Prov., Cana, upper Rio Tuyra, 2,000 ft. (610 m).

DISTRIBUTION: E Panama.
COMMENTS: Subgenus *Macrogeomys*. May include *thaeleri*.

Orthogeomys grandis (Thomas, 1893). Ann. Mag. Nat. Hist., ser. 6, 12:270.
TYPE LOCALITY: Guatemala, Sacatepequez Prov., Dueñas.
DISTRIBUTION: Honduras to Jalisco (Mexico).
SYNONYMS: *alleni* Nelson and Goldman, *alvarezi* Schaldach, *annexus* Nelson and Goldman, *carbo* Goodwin, *engelhardi* Felten, *felipensis* Nelson and Goldman, *guerrerensis* Nelson and Goldman, *huixtlae* Villa, *latifrons* Merriam, *nelsoni* Merriam, *pluto* Lawrence, *pygacanthus* Dickey, *scalops* Thomas, *soconuscensis* Villa, *vulcani* Nelson and Goldman.
COMMENTS: Subgenus *Orthogeomys*. Burt and Stirton (1961) and Hall (1981) included *pygacanthus* in *grandis*; Russell (1968*a*) considered it a distinct species.

Orthogeomys heterodus (Peters, 1865). Monatsb. K. Preuss. Akad. Wiss. Berlin, p. 177.
TYPE LOCALITY: Costa Rica, exact locality unknown.
DISTRIBUTION: C Costa Rica.
SYNONYMS: *cartagoensis* Goodwin, *dolichocephalus* Merriam.
COMMENTS: Subgenus *Macrogeomys*. Reviewed by Goodwin (1946). Phylogenetic relationships given by Hafner (1991).

Orthogeomys hispidus (Le Conte, 1852). Proc. Acad. Nat. Sci., Philadelphia, 6:158.
TYPE LOCALITY: Mexico (restricted to Veracruz, near Jalapa, by Merriam, 1895*a*).
DISTRIBUTION: Yucatan Peninsula, Belize, Guatemala, and NW Honduras, to S Tamaulipas (Mexico).
SYNONYMS: *cayoensis* Burt, *chiapensis* Nelson and Goldman, *concavus* Nelson and Goldman, *hondurensis* Davis, *isthmicus* Nelson and Goldman, *latirostris* Hall and Alvarez, *negatus* Goodwin, *teapensis* Goldman, *tehuantepecus* Goldman, *torridus* Merriam, *yucantanensis* Nelson and Goldman.
COMMENTS: Subgenus *Heterogeomys*. May include *lanius* (see Hall, 1981:511-512). Revised by Nelson and Goldman (1929).

Orthogeomys lanius (Elliot, 1905). Proc. Biol. Soc. Washington, 18:235.
TYPE LOCALITY: Mexico, Veracruz, Xuchil, SE side of Mt. Orizaba.
DISTRIBUTION: Known only from the type locality.
COMMENTS: Subgenus *Heterogeomys*. May be conspecific with *hispidus* (Hall, 1981:512).

Orthogeomys matagalpae (J. A. Allen, 1910). Bull. Amer. Mus. Nat. Hist., 28:97.
TYPE LOCALITY: Nicaragua, Matagalpa, Peña Blanca.
DISTRIBUTION: NC Nicaragua to SC Honduras.
COMMENTS: Subgenus *Macrogeomys*.

Orthogeomys thaeleri Alberico, 1990. *In* Peters and Hutterer (eds.), Vertebrates in the Tropics, Mus. Alex. Koenig, Bonn, p. 104.
TYPE LOCALITY: *ca.* 7 km S. Bahía Solano, Municipio Bahía Solano, Depto. Chocó, Colombia, 100 m.
DISTRIBUTION: Serranía de Baudó (extreme NW Colombia).
COMMENTS: Subgenus *Macrogeomys*. May be conspecific with *O. dariensis*.

Orthogeomys underwoodi (Osgood, 1931). Field Mus. Nat. Hist. Publ., Zool. Ser., 295, 18:143.
TYPE LOCALITY: Costa Rica, San Jose Prov., Alto de Jabillo Pirris, between San Geronimo and Pozo Azul.
DISTRIBUTION: Central Pacific coast of Costa Rica.
COMMENTS: Subgenus *Macrogeomys*.

Pappogeomys Merriam, 1895. N. Am. Fauna, 8:145.
TYPE SPECIES: *Geomys bulleri* Thomas, 1892.
SYNONYMS: *Cratogeomys* Merriam, *Platygeomys* Merriam.
COMMENTS: Hooper (1946) included *Platygeomys* within *Pappogeomys*. *Cratogeomys* was recognized by Russell (1968*a*, *b*) as a valid subgenus, a position followed by Hall (1981). Honeycutt and Williams (1982), based on a phylogenetic analysis of allozyme data, retained *alcorni* and *bulleri* in *Pappogeomys*, but considered *Cratogeomys* generically distinct, containing the remaining species listed by Russell. Lee and Baker (1987), Davidow-Henry et al. (1989), and Hollander (1990) followed Honeycutt and

Williams in considering *Cratogeomys* a distinct genus. Available data (Russell, 1968*a*, *b*; Honeycutt and Williams, 1982) support a sister-taxon relationship between *Pappogeomys* and *Cratogeomys*, but it remains a matter of personal opinion as to recognition status at the generic or subgeneric levels. Revision by Nelson and Goldman (1934; *Cratogeomys* only) and Russell (1968*b*).

Pappogeomys alcorni Russell, 1957. Univ. Kansas Publ., Mus. Nat. Hist., 9:359.
TYPE LOCALITY: Mexico, Jalisco, 4 mi. (6 km) W. of Mazamitla, 6,600 ft. (2,012 m).
DISTRIBUTION: S Jalisco (Mexico), in Sierra del Tigre.
COMMENTS: Subgenus *Pappogeomys*.

Pappogeomys bulleri (Thomas, 1892). Ann. Mag. Nat. Hist., ser. 6, 10:196.
TYPE LOCALITY: Mexico, Jalisco, Talpa, W slope Sierra de Mascota, 8,500 ft. (2,591 m) (probably about 5000 ft., 1500 m).
DISTRIBUTION: Nayarit, Jalisco, and Colima (Mexico).
SYNONYMS: *albinasus* Merriam, *amecensis* Goldman, *burti* Goldman, *flammeus* Goldman, *infuscus* Russell, *lagunensis* Goldman, *lutulentus* Russell, *melanurus* Genoways and Jones, *nayaritensis* Goldman, *nelsoni* Merriam.
COMMENTS: Subgenus *Pappogeomys*. Reviewed by Genoways and Jones (1969*a*). Polytypic; subspecies reviewed by Russell (1968*b*) and are listed in Hall (1981).

Pappogeomys castanops (Baird, 1852). *In* Stansbury, Expl. Surv. Valley Great Salt Lake, Utah, App. C(Zool.), p. 313.
TYPE LOCALITY: USA, Colorado, Bent Co., along prairie road to Bent's Fort, near present town of Las Animas.
DISTRIBUTION: SE Colorado and SW Kansas (USA) to C San Luis Potosi (Mexico).
SYNONYMS: *angusticeps* Nelson and Goldman, *bullatus* Russell and Baker, *clarkii* Baird, *consitus* Nelson and Goldman, *elibatus* Russell, *excelsus* Nelson and Goldman, *goldmani* Merriam, *hirtus* Nelson and Goldman, *jucundus* Russell and Baker, *lacrimalis* Nelson and Goldman, *parviceps* Russell, *perexiguus* Russell, *peridoneus* Nelson and Goldman, *perplanus* Nelson and Goldman, *planifrons* Nelson and Goldman, *pratensis* Russell, *rubellus* Nelson and Goldman, *simulans* Russell, *sordidulus* Russell and Baker, *subnubilus* Nelson and Goldman, *subsimus* Nelson and Goldman, *surculus* Russell, *tamaulipensis* Nelson and Goldman, *torridus*, Russell, *ustulatus* Russell and Baker.
COMMENTS: Subgenus *Cratogeomys*. Russell (1968*b*) delineated two subspecies groups, which are karyotypically distinct and may represent separate species (see Berry and Baker, 1972, and Lee and Baker, 1987). Reviewed by Davidow-Henry et al. (1989, Mammalian Species, 338, as *Cratogeomys castanops*); subspecies in the US were revised by Hollander (1990).

Pappogeomys fumosus (Merriam, 1892). Proc. Biol. Soc. Washington, 7:165.
TYPE LOCALITY: Mexico, Colima, 3 mi. (5 km) W of Colima, 1,700 ft. (518 m).
DISTRIBUTION: Plain of E Colima (Mexico).
COMMENTS: Subgenus *Cratogeomys*.

Pappogeomys gymnurus (Merriam, 1892). Proc. Biol. Soc. Washington, 7:166.
TYPE LOCALITY: Mexico, Jalisco, Zapotlán (= Ciudad Guzmán), 4,000 ft. (1,219 m).
DISTRIBUTION: S and C Jalisco and NE Michoacan (Mexico).
SYNONYMS: *imparilis* Goldman, *russelli* Genoways and Jones, *tellus* Russell.
COMMENTS: Subgenus *Cratogeomys*. Polytypic; subspecies reviewed by Russell (1968*b*).

Pappogeomys merriami (Thomas, 1893). Ann. Mag. Nat. Hist., ser. 6, 12:271.
TYPE LOCALITY: Mexico, "Southern Mexico," probably Valley of Mexico.
DISTRIBUTION: WC Veracruz to Distrito Federal, Morelos, and surrounding areas, including SE Central Plateau and S Sierra Madre Oriental (Mexico).
SYNONYMS: *estor* Merriam, *fulvescens* Merriam, *irolonis* Nelson and Goldman, *oreocetes* Merriam, *peraltus* Goldman, *peregrinus* Merriam, *perotensis* Merriam, *saccharalis* Nelson and Goldman.
COMMENTS: Subgenus *Cratogeomys*. Polytypic; subspecies reviewed by Russell (1968*b*).

Pappogeomys neglectus (Merriam, 1902). Proc. Biol. Soc. Washington, 15:68.
TYPE LOCALITY: Mexico, Querétaro, Cerro de la Calentura, about 8 mi (13 km) NW of Pinal de Amoles, 9,500 ft. (2,896 m).

DISTRIBUTION: Known only from the type locality.
COMMENTS: Subgenus *Cratogeomys*.

Pappogeomys tylorhinus (Merriam, 1895). N. Am. Fauna, 8:167.
TYPE LOCALITY: Mexico, Hidalgo, Tula, 6,800 ft. (2,073 m).
DISTRIBUTION: Distrito Federal and Hidalgo to C Jalisco (Mexico).
SYNONYMS: *angustirostris* Merriam, *arvalis* Hooper, *atratus* Russell, *brevinasus* Russell, *planiceps* Merriam, *varius* Goldman, *zodius* Russell.
COMMENTS: Subgenus *Cratogeomys*. Polytypic; subspecies reviewed by Russell (1968*b*).

Pappogeomys zinseri (Goldman, 1939). J. Mammal., 20:91.
TYPE LOCALITY: Mexico, Jalisco, Lagos, 6,150 ft. (1,875 m).
DISTRIBUTION: NE Jalisco (Mexico).
COMMENTS: Subgenus *Cratogeomys*.

Thomomys Wied-Neuwied, 1839. Nova Acta Phys.-Med. Acad. Caes. Leop.-Carol., 19(1):377.
TYPE SPECIES: *Thomomys rufescens* Wied-Neuwied, 1939 (= *T. talpoides rufescens*).
SYNONYMS: *Megascapheus* Elliot, 1903.
COMMENTS: Recent species allocated to two subgenera (Thaeler, 1980): *Thomomys*, which includes the species *talpoides*, *clusius*, *idahoensis*, *mazama*, and *monticola*; and *Megascapheus* including *bottae*, *bulbivorus*, *townsendii*, and *umbrinus*. Phylogenetic relationships among members of the subgenus *Megascapheus* were presented by Patton and Smith (1981), based on molecular data. Specific boundaries for some forms are poorly defined, but Patton and Smith (1989) provided an operational definition that can be applied to all pocket gophers.

Thomomys bottae (Eydoux and Gervais, 1836). Mag. Zool., Paris, 6:23.
TYPE LOCALITY: USA, coast of California (restricted to vicinity of Monterey, Monterey Co., by Baird, 1857).
DISTRIBUTION: SW and W USA, north to Oregon, east to Colorado, and south to Sinaloa and Nuevo Leon (Mexico).
SYNONYMS: *abbotti* Huey, *absonus* Goldman, *abstrusus* Hall and Davis, *acrirostratus* Grinnell, *actuosus* Kelson, *adderrans* Huey, *affinis* Huey, *agricolaris* Grinnell, **albatus* Grinnell, *albicaudatus* Hall, **alexandrae* Goldman, *alienus* Goldman, **alpinus* Merriam, *alticolus* J. A. Allen, **altivallis* Rhoads, *amargosae* Grinnell, *analogus* Goldman, **angularis* Merriam, *angustidens* Baker, *anitae* J. A. Allen, **apache* Bailey, **aphrastus* Elliot, *ardicola* Huey, **argusensis* Huey, *aureiventris* Hall, **aureus* J. A. Allen, *awahnee* Merriam, **baileyi* Merriam, *basilicae* Benson and Tillotson, *birdseyei* Goldman, *bonnevillei* Durant, *boregoensis* Huey, *boreorarius* Durham, *borjasensis* Huey, *brazierhowelli* Huey, *brevidens* Hall, **cabezonae* Merriam, *cactophilus* Huey, *camargensis* Anderson, *camoae* Burt, *caneloensis* Lange, **canus* Bailey, *carri* Lange, *catalinae* Goldman, *catavinensis* Huey, *cedrinus* Huey, *centralis* Hall, **cervinus* J. A. Allen, *chiricahuae* Nelson and Goldman, **chrysonotus* Grinnell, *cinereus* Hall, *collinus* Goldman, *collis* Hooper, *comobabiensis* Huey, *concisor* Hall and Davis, *confinalis* Goldman, *connectens* Hall, *contractus* Durrant, *convergens* Nelson and Goldman, *convexus* Durrant, *crassus* Chattin, *cultellus* Kelson, *cunicularis* Huey, *curtatus* Hall, *depauperatus* Grinnell and Hill, *depressus* Hall, **desertorum* Merriam, *desitus* Goldman, *detumidus* Grinnell, **diaboli* Grinnell, *dissimilis* Goldman, *divergens* Nelson and Goldman, *estanciae* Benson and Tillotson, *extenuatus* Goldman, *flavidus* Goldman, **fulvus* Woodhouse, *fumosus* Hall, *grahamensis* Goldman, *growlerensis* Huey, *guadalupensis* Goldman, **harquahalae* Grinnell, *homorus* Huey, *howelli* Goldman, *hualpaiensis* Goldman, *hueyi* Goldman, *humilis* Baker, *imitabilis* Goldman, *incomptus* Goldman, **infrapallidus* Grinnell, *ingens* Grinnell, *internatus* Goldman, **jacinteus* Grinnell and Swarth, *jojobae* Huey, *juarezensis* Huey, **lachuguilla* Bailey, *lacrymalis* Hall, **laticeps* Baird, **latirostris* Merriam, *latus* Hall and Davis, *lenis* Goldman, **leucodon* Merriam, *levidensis* Goldman, *limitaris* Goldman, *limpiae* Blair, *litoris* Burt, *lorenzi* Huey, *lucidus* Hall, *lucrificus* Hall and Durham, **magdalenae* Nelson and Goldman, *martirensis* J. A. Allen, **mearnsi* Bailey, **melanotis* Grinnell, **mewa* Merriam, *minimus* Durrant, *minor* Bailey, *modicus* Goldman, *mohavensis* Grinnell, *morulus* Hooper, **muralis* Goldman, *mutabilis* Goldman, *nanus* Hall, *nasutus* Hall, *navus* Merriam, **neglectus* Bailey, *nesophilus* Durrant, *nicholi* Goldman, **nigricans* Rhoads,

occipitalis Benson and Tillotson, **operarius* Merriam, *operosus* Hatfield, *optabilis* Goldman, *opulentus* Goldman, **oreoecus* Burt, *osgoodi* Goldman, *paguatae* Hooper, *pallescens* Rhoads, *parvulus* Goldman, *pascalis* Merriam, *patulus* Goldman, **pectoralis* Goldman, *peramplus* Goldman, **perpallidus* Merriam, *perpes* Merriam, *pervagus* Merriam, *pervarius* Goldman, *phasma* Goldman, **phelleoecus* Burt, *pinalensis* Goldman, *piutensis* Grinnell and Hill, *planirostris* Burt, *planorum* Hooper, *powelli* Durrant, **providentialis* Grinnell, *proximarinus* Huey, *proximus* Burt and Campbell, *puertae* Grinnell, *pusillus* Goldman, *retractus* Baker, *rhizophagus* Huey, *riparius* Grinnell and Hill, *robustus* Durrant, *rubidus* Youngman, *rufidulus* Hoffmeister, *ruidosae* Hall, *rupestris* Chattin, *ruricola* Huey, *russeolus* Nelson and Goldman, *sanctidiegi* Huey, *saxatilis* Grinnell, **scapterus* Elliot, *scotophilus* Davis, *sevieri* Durrant, *siccovallis* Huey, *silvifugus* Grinnell, **simulus* Nelson and Goldman, **sinaloae* Merriam, **solitarius* Grinnell, *spatiosus* Goldman, *stansburyi* Durrant, **sturgisi* Goldman, *suboles* Goldman, *subsimilis* Goldman, *texensis* Bailey, *tivius* Durrant, **toltecus* J.A. Allen, *trumbullensis* Hall and Davis, *tularosae* Hall, *vanrosseni* Huey, *varus* Hall and Long, *vescus* Hall and Davis, *villai* Baker, *virgineus* Goldman, *wahwahensis* Durrant, *winthropi* Nelson and Goldman, *xerophilus* Huey (*those taxa originally described as distinct species or treated as such in the literature).

COMMENTS: Subgenus *Megascapheus*. The taxonomic history of this, and related species has been contentious. Hall and Kelson (1959) included *bottae* within *umbrinus* but considered *baileyi* and *townsendii* as separate species, but Hall (1981) included all four of these taxa within his concept of *umbrinus*. Anderson (1966, 1972), Hoffmeister (1969, 1986), and Patton and co-workers (Patton and Dingman, 1968; Patton, 1973; Patton and Smith, 1981) considered *bottae* (including *baileyi*) separate from *umbrinus*. Thaeler (1968*b*), Patton et al. (1984), Patton and Smith (1989), and Rogers (1991*a*, *b*) supported the specific separation of *townsendii* from Hall's (1981) *umbrinus*. As envisioned here, *bottae* includes a number of taxa considered by earlier workers to be full species but which are each now relegated to subspecific status or are recognized as synonyms of other subspecies. Revisions of those subspecies occurring in Arizona and California have been presented by Hoffmeister (1986) and Patton and Smith (1990), respectively. Hoffmeister (1969), Patton and Dingman (1968), and Patton (1973) reported on hybridization between *bottae* and *umbrinus* in S Arizona; Thaeler (1968*b*) and Patton et al. (1984) examined hybridization between *bottae* and *townsendii* in NE California. For subspecies, see Hall (1981), as modified for Arizona by Hoffmeister (1986), and for California by Patton and Smith (1990).

Thomomys bulbivorus (Richardson, 1829). Quadrupeds, *in* Fauna Boreali-Americana, 1:206.

TYPE LOCALITY: USA, banks of Columbia River (probably Portland, Multnomah Co., Oregon; see Bailey, 1915).

DISTRIBUTION: Willamette Valley (NW Oregon, USA).

COMMENTS: Subgenus *Megascapheus*. Revised by Bailey (1915).

Thomomys clusius Coues, 1875. Proc. Acad. Nat. Sci., Philadelphia, 27:138.

TYPE LOCALITY: USA, Wyoming, Carbon Co., Bridger Pass, 18 mi (29 km) SW Rawlins.

DISTRIBUTION: Carbon and Sweetwater Cos., SC Wyoming (USA).

COMMENTS: Subgenus *Thomomys*. Included in *talpoides* by Hall (1981:457); revised by Thaeler and Hinesley (1979).

Thomomys idahoensis Merriam, 1901. Proc. Biol. Soc. Washington, 14:114.

TYPE LOCALITY: USA, Idaho, Clark Co., Birch Creek (10 mi [16 km] S Nicholia [Lemhi Co.], about 6,400 ft. [1940 m]).

DISTRIBUTION: EC Idaho, adjacent Montana and W Wyoming, and N Utah (USA).

SYNONYMS: *confinus* Davis, *pygmaeus* Merriam.

COMMENTS: Subgenus *Thomomys*. Thaeler (1972, 1977) revised this species. Formerly included in *talpoides* by Hall and Kelson (1959:441) and Hall (1981:457, 463; but see Hall, 1981:1179).

Thomomys mazama Merriam, 1897. Proc. Biol. Soc. Washington, 11:214.

TYPE LOCALITY: USA, Oregon, Klamath Co., Anna Creek near Crater Lake, Mt. Mazama, 6,000 ft. (1,829 m).

DISTRIBUTION: NW Washington through C Oregon to N California (USA).

SYNONYMS: *couchi* Goldman, *glacialis* Dalquest and Scheffer, *helleri* Elliot, *hesperus* Merriam, *louiei* Gardner, *melanops* Merriam, *nasicus* Merriam, *niger* Merriam, *oregonus* Merriam, *premaxillaris* Grinnell, *pugetensis* Dalquest and Scheffer, *tacomensis* Taylor, *tumuli* Dalquest and Scheffer, *yelmensis* Merriam.

COMMENTS: Subgenus *Thomomys*. Revised by Johnson and Benson (1960); considered a subspecies of *monticola* by Bailey (1915) and Hall and Kelson (1959) but geographic sympatry with *monticola* in N California documented by Thaeler (1968*a*). Subspecies listed in Hall (1981:465-467).

Thomomys monticola J. A. Allen, 1893. Bull. Am. Mus. Nat. Hist., 5:48.

TYPE LOCALITY: USA, California, El Dorado Co., Mt. Tallac, 7,500 ft. (2,286 m).

DISTRIBUTION: Sierra Nevada Mtns of C and N California and extreme WC Nevada (USA).

SYNONYMS: *pinetorum* Merriam.

COMMENTS: Subgenus *Thomomys*. Revised by Johnson and Benson (1960); Thaeler (1968*a*) provided details of distribution in N California.

Thomomys talpoides (Richardson, 1828). Zool. J., 3:518.

TYPE LOCALITY: Restricted to Canada, Saskatchewan, North Saskatchewan River, Carlton House near Fort Carlton by Bailey (1915:97).

DISTRIBUTION: S British Columbia to C Alberta and SW Manitoba (Canada), south to C South Dakota and N New Mexico, N Arizona, N Nevada, and NE California (USA).

SYNONYMS: *aequalidens* Dalquest, *andersoni* Goldman, *attenuatus* Hall and Montague, *badius* Goldman, *borealis* Richardson, *bridgeri* Merriam, *bullatus* Bailey, *caryi* Bailey, *cheyennensis* Swenk, *cognatus* Johnstone, *columbianus* Bailey, *devexus* Hall and Dalquest, *douglasii* Richardson, *duranti* Kelson, *falcifer* Grinnell, *fisheri* Merriam, *fossor* J. A. Allen, *fuscus* Merriam, *gracilis* Durrant, *immunis* Hall and Dalquest, *incensus* Goldman, *kaibabensis* Goldman, *kelloggi* Goldman, *levis* Goldman, *limosus* Merriam, *loringi* Bailey, *macrotis* Miller, *medius* Goldman, *meritus* Hall, *monoensis* Huey, *moorei* Goldman, *myops* Merriam, *nebulosus* Bailey, *ocius* Merriam, *oquirrhensis* Durrant, *parowanensis* Goldman, *pierreicolus* Swenk, *pryori* Bailey, *quadratus* Merriam, *ravus* Durrant, *relicinus* Goldman, *retrorsus* Hall, *rostralis* Hall and Montague, *rufescens* Wied-Neuwied, *saturatus* Bailey, *segregatus* Johnstone, *shawi* Taylor, *taylori* Hooper, *tenellus* Goldman, *trivialis* Goldman, *uinta* Merriam, *unisulcatus* Gray, *wallowa* Hall and Orr, *wasatchensis* Durrant, *whitmani* Drake and Booth, *yakimensis* Hall and Dalquest.

COMMENTS: Formerly included *idahoensis* and *clusius*; see Thaeler (1972) and Thaeler and Hinesley (1979). Partial revision by Bailey (1915) and Thaeler (1985). The considerable degree of chromosomal differentiation among geographic representatives of this form (Thaeler, 1985) suggests that more than one biological species is currently included under the name *talpoides*.

Thomomys townsendii (Bachman, 1839). J. Acad. Nat. Sci., Philadelphia, 8:105.

TYPE LOCALITY: USA, erroneously given as "Columbia River", but restricted to Idaho, Canyon Co., near Nampa, by Bailey (1915).

DISTRIBUTION: Snake River Valley of Idaho south and west to SE Oregon, NE California, and N Nevada (USA).

SYNONYMS: *atrogriseus* Bailey, *bachmani* Davis, *elkoensis* Davis, *nevadensis* Merriam, *owyhensis* Davis, *relictus* Grinnell, *similis* Davis.

COMMENTS: Revised by Davis (1937) and Rogers (1991*a*, *b*). Considered a distinct species by Thaeler (1968*b*), Patton et al. (1984), Patton and Smith (1989), and Rogers (1991*a*, *b*), despite limited hybridization with *bottae* in NE California. Hall (1981:469, 495) reviewed Thaeler's evidence and included *townsendii* in *umbrinus* (*sensu* Hall).

Thomomys umbrinus (Richardson, 1829). Quadrupeds, *in* Fauna Boreali-Americana, 1:202.

TYPE LOCALITY: Originally given as Cadadaguios, southwestern Louisiana, but restricted to Mexico, Veracruz (probably Puebla), vicinity of Boca del Norte, by Bailey (1906).

DISTRIBUTION: SC Arizona and SW New Mexico (USA) south to Puebla and Veracruz (Mexico).

STATUS: IUCN - Rare.

SYNONYMS: *albigularis* Nelson and Goldman, *arriagensis* Dalquest, *atrodorsalis* Nelson and Goldman, *atrovarius* J. A. Allen, *burti* Huey, *caliginosus* Nelson and Goldman, *chihuahuae* Nelson and Goldman, *crassidens* Nelson and Goldman, *durangi* Nelson and

Goldman, *emotus* Goldman, *enixus* Nelson and Goldman, *evexus* Nelson and Goldman, *eximius* Nelson and Goldman, *extimus* Nelson and Goldman, *goldmani* Merriam, *intermedius* Mearns, *juntae* Anderson, *madrensis* Nelson and Goldman, *martinensis* Nelson and Goldman, *musculus* Nelson and Goldman, *nelsoni* Merriam, *newmani* Dalquest, *orizabae* Merriam, *parviceps* Nelson and Goldman, *perditus* Nelson and Goldman, *peregrinus* Merriam, *potosinus* Nelson and Goldman, *pullus* Hall and Villa, *quercinus* Burt and Campbell, *sheldoni* Nelson and Goldman, *sonoriensis* Nelson and Goldman, *supernus* Nelson and Goldman, *tolucae* Nelson and Goldman, *vulcanius* Nelson and Goldman, *zacatecae* Nelson and Goldman.

COMMENTS: Hybridization with *bottae* reported in Patagonia Mtns of S Arizona by Patton and Dingman (1968), Patton (1973), and Hoffmeister (1969). Considered a species separate from *bottae* by these authors and by Anderson (1966, 1972). Hall (1981:469) included *baileyi, bottae,* and *townsendii* in this species based on a his view of the hybridization data in Hoffmeister (1969), Patton (1973), and Thaeler (1968*b*); he reported intergradation between *umbrinus* and *bottae* in Nuevo Leon (Mexico). Phylogenetic relationships among geographic units of *umbrinus* were examined by Hafner et al. (1987).

Zygogeomys Merriam, 1895. N. Am. Fauna, 8:195.

TYPE SPECIES: *Zygogeomys trichopus* Merriam, 1895.

Zygogeomys trichopus Merriam, 1895. N. Am. Fauna, 8:196.

TYPE LOCALITY: Mexico, Michoacan, Nahuatzen, 8,000-8,500 ft. (2,438-2,591 m).

DISTRIBUTION: Known only from small areas west and south of Lago Pátzcuaro, NC Michoacan (Mexico).

STATUS: IUCN - Indeterminate. Local temporal extirpation documented by Hafner and Barkley (1984).

FAMILY HETEROMYIDAE

by James L. Patton

Family Heteromyidae Gray, 1868. Proc. Zool. Soc. Lond., 1868:201.

COMMENTS: Currently divided into three subfamilies: Dipodomyinae containing the genera *Dipodomys* and *Microdipodops*, Heteromyinae with *Heteromys* and *Liomys*, and Perognathinae comprised of *Chaetodipus* and *Perognathus*. Content defined by Wood (1935), Hafner and Hafner (1983), and Wahlert (1985).

Subfamily Dipodomyinae Gervais, 1853. Ann. Sci. Nat., Paris, ser. 3., 20:245.

COMMENTS: Wood (1935) allocated only the Recent genus *Dipodomys* to the subfamily; Hafner and Hafner (1983; see also Hafner, 1982) included *Microdipodops*, as did Wahlert (1985) and Ryan (1989*a*).

Dipodomys Gray, 1841. Ann. Mag. Nat. Hist., [ser. 1], 7:521.

TYPE SPECIES: *Dipodomys phillipsii* Gray 1841.

SYNONYMS: *Dipodops* Merriam, *Perodipus* Fitzinger.

COMMENTS: Interspecific relationships summarized by Setzer (1949), Lidicker (1960), Johnson and Selander (1971), Stock (1974), Best and Schnell (1974), and Schnell et al. (1978). Key to species given by Hall (1981:563-564) and Best (1991).

Dipodomys agilis Gambel, 1848. Proc. Acad. Nat. Sci. Philadelphia, 4:77.

TYPE LOCALITY: USA, California, Los Angeles Co., Los Angeles (see Grinnell, 1922).

DISTRIBUTION: SW and SC California (USA); Baja California (Mexico).

SYNONYMS: *antiquarius* Huey, *australis* Huey, *cabezonae* Merriam, *eremoecus* Huey, *fuscus* Boulware, *latimaxillaris* Huey, *martirensis* Huey, *paralius* Huey, *pedionomus* Huey, *peninsularis* Merriam, *perplexus* Merriam, *plectilis* Huey, *simulans* Merriam, *wagneri* LeConte.

COMMENTS: Reviewed by Best (1978, 1983*a*), Best et al. (1986), and Lackey (1967); see also Hall (1981:1179).

Dipodomys californicus Merriam, 1890. N. Am. Fauna, 4:49.

TYPE LOCALITY: USA, California, Mendocino Co., Ukiah.

DISTRIBUTION: SC Oregon and N California to north of San Francisco Bay (USA).

STATUS: Subspecies *eximius* is listed as of Special Concern by the State of California Dept. of Fish and Game.

SYNONYMS: *eximius* Grinnell, *saxatilis* Grinnell and Linsdale.

COMMENTS: Considered distinct from *heermanni* based on chromosomal (Fashing, 1973) and biochemical data (Patton et al., 1976). Hall (1981:578) listed *californicus* as a subspecies of *heermanni*, without discussion of Patton et al. (1976). Reviewed by Kelt (1988*b*, Mammalian Species, 324).

Dipodomys compactus True, 1889. Proc. U.S. Natl. Mus., 11:160.

TYPE LOCALITY: USA, Texas, Cameron Co., Padre Island.

DISTRIBUTION: Mainland, Padre and Mustang Isls of S Texas (USA) and barrier islands of N Tamaulipas (Mexico).

SYNONYMS: *largus* Hall, *parvabullatus* Hall, *sennetti* J. A. Allen.

COMMENTS: Considered distinct from *ordii* by Johnson and Selander (1971), Schmidly and Hendricks (1976), and Baumgardner and Schmidly (1981). Hall (1981:565) provisionally listed this species as a subspecies of *ordii*. Reviewed by Baumgardner (1991, Mammalian Species, 369).

Dipodomys deserti Stephens, 1887. Am. Nat., 21:42.

TYPE LOCALITY: USA, California, San Bernardino, Mohave River (3 to 4 mi [5-7 km] from, and opposite, Hesperia; see Hall, 1981:588).

DISTRIBUTION: Deserts of E California, to S and W Nevada, SW Utah, W and SC Arizona (USA), NW Sonora and NE Baja California Norte (Mexico).

SYNONYMS: *aquilus* Nader, *arizonae* Huey, *sonoriensis* Goldman.

COMMENTS: Revised by Nader (1978). Reviewed by Best et al. (1989, Mammalian Species, 339).

Dipodomys elator Merriam, 1894. Proc. Biol. Soc. Washington, 9:109.
TYPE LOCALITY: USA, Texas, Clay Co., Henrietta.
DISTRIBUTION: SW Oklahoma and NC Texas (USA).
STATUS: IUCN - Rare.
COMMENTS: Probably no longer occurs in Oklahoma (Caire et al., 1989). Reviewed by Carter et al. (1985, Mammalian Species, 232).

Dipodomys elephantinus (Grinnell, 1919). Univ. California Publ. Zool., 21:43.
TYPE LOCALITY: USA, California, San Benito Co., Bear Valley, 1 mi (1.6 km) N Cook P.O., 1,300 ft. (396 m).
DISTRIBUTION: WC California in San Benito and Monterey Cos. (USA).
COMMENTS: May be conspecific with *venustus*; see Stock (1974) and Schnell et al. (1978). Reviewed by Best (1986, Mammalian Species, 255).

Dipodomys gravipes Huey, 1925. Proc. Biol. Soc. Washington, 38:83.
TYPE LOCALITY: Mexico, Baja California Norte, 2 mi (3 km) W Santo Domingo Mission, 30°45'N, 115°58'W.
DISTRIBUTION: NW Baja California Norte (Mexico).
STATUS: IUCN - Endangered.
COMMENTS: Considered distinct from *agilis* by Best (1978). Reviewed by Best (1983*b*) and Best and Lackey (1985, Mammalian Species, 236).

Dipodomys heermanni Le Conte, 1853. Proc. Acad. Nat. Sci. Philadelphia, 6:224.
TYPE LOCALITY: USA, California, Sierra Nevada; restricted to Calaveras River, Calaveras Co. by Grinnell (1922:47).
DISTRIBUTION: C California (USA).
STATUS: U.S. ESA, IUCN and the State of California Dept. of Fish and Game - Endangered as *D. h. morroensis*.
SYNONYMS: *arenae* Boulware, *berkeleyensis* Grinnell, *dixoni* Grinnell, *goldmani* Merriam, *jolonensis* Grinnell, *morroensis* Merriam, *streatori* Merriam, *swarthi* Grinnell, *tularensis* Merriam.
COMMENTS: Revised by Grinnell (1922). Does not include *californicus*, see Patton et al. (1976) and comment under that species. Reviewed by Kelt (1988*a*, Mammalian Species, 323).

Dipodomys ingens (Merriam, 1904). Proc. Biol. Soc. Washington, 17:141.
TYPE LOCALITY: USA, California, San Luis Obispo Co., Carrizo Plain, Painted Rock, 20 [=25.5] mi (32 km) SE of Simmler.
DISTRIBUTION: Western edge of Joaquin Valley, adjacent Carrizo and Elkhorn plains and upper Cuyama Valley of WC California (USA).
STATUS: U.S. ESA and State of California Dept. of Fish and Game - Endangered; IUCN - Indeterminate. Extirpated over much of its original range.
COMMENTS: Reviewed by Williams and Kilburn (1991, Mammalian Species, 377).

Dipodomys insularis Merriam, 1907. Proc. Biol. Soc. Washington, 20:77.
TYPE LOCALITY: Mexico, Baja California Sur, Gulf of California, San José Island.
DISTRIBUTION: Known only from the type locality.
COMMENTS: Possibly a subspecies of *merriami* (see Lidicker, 1960). Reviewed by Best and Thomas (1991*a*, Mammalian Species, 374) who considered it a full species.

Dipodomys margaritae Merriam, 1907. Proc. Biol. Soc. Washington, 20:76.
TYPE LOCALITY: Mexico, Baja California Sur, Santa Margarita Island.
DISTRIBUTION: Known only from the type locality.
COMMENTS: Lidicker (1960) included *margaritae* in *merriami*, but Huey (1964) and Hall (1981:587) considered it a distinct species. Reviewed by Best (1992, Mammalian Species, 400).

Dipodomys merriami Mearns, 1890. Bull. Am. Mus. Nat. Hist., 2:290.
TYPE LOCALITY: USA, Arizona, Maricopa Co., New River, between Phoenix and Prescott (see Lidicker, 1960:165).
DISTRIBUTION: NW Nevada and NE California to Texas (USA), south to Baja California Sur, N Sinaloa, and Mexican Plateau to San Luis Potosi (Mexico).

SYNONYMS: *ambiguus* Merriam, *annulus* Huey, *arenivagus* Elliot, *atronasus* Merriam, *brunensis* Huey, *collinus* Lidicker, *frenatus* Bole, *kernensis* Merriam, *llanoensis* Huey, *mayensis* Goldman, *melanurus* Merriam, *mitchelli* Mearns, *mortivallis* Elliot, *nevadensis* Merriam, *olivaceus* Swarth, *parvus* Rhoads, *platycephalus* Merriam, *quintinensis* Huey, *regillus* Goldman, *semipallidus* Huey, *similis* Rhoads, *simiolus* Rhoads, *trinidadensis* Huey, *vulcani* Benson.

COMMENTS: Revised by Lidicker (1960) who included *margaritae*; but see Huey (1964).

Dipodomys microps (Merriam, 1904). Proc. Biol. Soc. Washington, 17:145.

TYPE LOCALITY: USA, California, Inyo Co., Owens Valley, Lone Pine.

DISTRIBUTION: SE Oregon and SW Idaho, south through NW and SE Californa, Nevada, and W Utah, to NW Arizona (USA).

STATUS: IUCN - Insufficiently known as *D. m. leucotis*.

SYNONYMS: *alfredi* Goldman, *aquilonius* Willett, *bonnevillei* Goldman, *celsus* Goldman, *centralis* Hall and Dale, *idahoensis* Hall and Dale, *leucotis* Goldman, *levipes*, Merriam, *occidentalis* Hall and Dale, *panamintinus* Elliot (in part, not *panamintinus* Merriam), *preblei* Goldman, *russeolus* Goldman, *subtenuis* Goldman.

COMMENTS: Reviewed by Csuti (1979) and Hayssen (1991, Mammalian Species, 389).

Dipodomys nelsoni Merriam, 1907. Proc. Biol. Soc. Washington, 20:75.

TYPE LOCALITY: Mexico, Coahuila, La Ventura.

DISTRIBUTION: Mexican Plateau from N Coahuila and S Chihuahua to N San Luis Potosi and S Nuevo Leon (Mexico).

COMMENTS: Nader (1978) included *nelsoni* in *spectabilis*; Anderson (1972) and Matson (1980) presented evidence of specific distinctness. Reviewed by Best (1988*b*, Mammalian Species, 326).

Dipodomys nitratoides Merriam, 1894. Proc. Biol. Soc. Washington, 9:112.

TYPE LOCALITY: USA, California, Tulare Co., San Joaquin Valley, Tipton.

DISTRIBUTION: S San Joaquin Valley, WC California (USA).

STATUS: U.S. ESA and IUCN - Endangered as *D. n. nitratoides* and *D. n. exilis*. California Dept. of Fish and Game - Endangered as *D. n. exilis*; *D. n. nitratoides* is of Special Concern; and *D. n. brevinasus* is California Fully Protected.

SYNONYMS: *brevinasus* Grinnell, *exilis* Merriam.

COMMENTS: Very similar to *merriami* except in bacular morphology (Best and Schnell, 1974). Revised by Grinnell (1922) and reviewed by Best (1991, Mammalian Species, 381).

Dipodomys ordii Woodhouse, 1853. Proc. Acad. Nat. Sci. Philadelphia, 6:224.

TYPE LOCALITY: USA, Texas, El Paso Co., El Paso.

DISTRIBUTION: SW Saskatchewan and SE Alberta (Canada) and SE Washington south through Great Plains and intermontane basins of western USA, to Mexican Plateau as far south as Hidalgo (Mexico).

SYNONYMS: *attenuatus* Bryant, *celeripes* Durrant and Hall, *chapmani* Mearns, *cinderensis* Hardy, *cineraceus* Goldman, *columbianus* Merriam, *cupidineus* Goldman, *durranti* Setzer, *evexus* Goldman, *extractus* Setzer, *fetosus* Durrant and Hall, *fremonti* Durrant and Setzer, *idoneus* Setzer, *inaquosus* Hall, *longipes* Merriam, *luteolus* Goldman, *marshalli* Goldman, *medius* Setzer, *monoensis* Grinnell, *montanus* Baird, *nexilis* Goldman, *obscurus* J. A. Allen, *oklahomae* Trowbridge and Whitaker, *pallidus* Durrant and Setzer, *palmeri* J. A. Allen, *panguitchensis* Hardy, *priscus* Hoffmeister, *pullus* Anderson, *richardsoni* J. A. Allen, *sanrafaeli* Durrant and Setzer, *terrosus* Hoffmeister, *uintensis* Durrant and Setzer, *utahensis* Merriam.

COMMENTS: Revised by Setzer (1949) and reviewed by Garrison and Best (1990, Mammalian Species, 353). Does not include *compactus*, see Schmidly and Hendricks (1976), Baumgardner and Schmidly (1981), and comment under that species.

Dipodomys panamintinus (Merriam, 1894). Proc. Biol. Soc. Washington, 9:114.

TYPE LOCALITY: USA, California, Inyo Co., Panamint Mtns, head of Willow Creek.

DISTRIBUTION: Deserts of E California and W Nevada (USA).

SYNONYMS: *argusensis* Huey, *caudatus* Hall, *leucogenys* Grinnell, *mohavensis* Grinnell.

COMMENTS: Reviewed by Intress and Best (1990, Mammalian Species, 354).

Dipodomys phillipsii Gray, 1841. Ann. Mag. Nat. Hist., [ser. 1], 7:522.
TYPE LOCALITY: Mexico, Hidalgo, Valley of Mexico, near Real del Monte (= Mineral de Monte).
DISTRIBUTION: C Durango south to N Oaxaca (Mexico).
SYNONYMS: *ornatus* Merriam, *oaxacae* Hooper, *perotensis* Merriam.
COMMENTS: Reviewed by Genoways and Jones (1971) and Jones and Genoways (1975*a*, Mammalian Species, 51).

Dipodomys spectabilis Merriam, 1890. N. Am. Fauna, 4:46.
TYPE LOCALITY: USA, Arizona, Cochise Co., Dos Cabezos.
DISTRIBUTION: SC Arizona, New Mexico, W Texas (USA) south to N Sonora, Chihuahua and San Luis Potosi (Mexico).
SYNONYMS: *baileyi* Goldman, *clarencei* Goldman, *cratodon* Merriam, *intermedius* Nader, *perblandus* Goldman, *zygomaticus* Goldman.
COMMENTS: Revised by Nader (1978) who included *nelsoni*; but also see Anderson (1972), Matson (1980), and Hall (1981:581), who presented evidence of specific distinctness. Reviewed by Best (1988*a*, Mammalian Species, 311).

Dipodomys stephensi (Merriam, 1907). Proc. Biol. Soc. Washington, 20:78.
TYPE LOCALITY: USA, California, Riverside Co., San Jacinto Valley, W. of Winchester.
DISTRIBUTION: Riverside, San Bernardino, and San Diego Cos. of S California (USA).
STATUS: U.S. ESA and IUCN - Endangered; State of California Dept. of Fish and Game - Threatened.
SYNONYMS: *cascus* Huey.
COMMENTS: Relationships to other species of the *heermanni* group studied by Lackey (1967). Reviewed by Bleich (1977, Mammalian Species, 73).

Dipodomys venustus (Merriam, 1904). Proc. Biol. Soc. Washington, 17:142.
TYPE LOCALITY: USA, California, Santa Cruz Co., Santa Cruz.
DISTRIBUTION: From San Francisco Bay to Estero Bay (WC California, USA).
SYNONYMS: *sanctiluciae* Grinnell.
COMMENTS: Revised by Grinnell (1922). May be conspecific with *elephantinus* (see comment under that species); also see Hall (1981:576).

Microdipodops Merriam, 1891. N. Am. Fauna, 5:115.
TYPE SPECIES: *Microdipodops megacephalus* Merriam, 1891.
COMMENTS: Revised by Hall (1941); also see Hafner et al. (1979). Wood (1935), Hafner (1978), and Hall (1981) considered *Microdipodops* a member of the subfamily Perognathinae. Hafner (1982), Hafner and Hafner (1983), Wahlert (1985), and Ryan (1989*a*) summarized evidence for referring the genus to the subfamily Dipodomyinae. A key to species is given by Hall (1981:560).

Microdipodops megacephalus Merriam, 1891. N. Am. Fauna, 5:116.
TYPE LOCALITY: USA, Nevada, Elko Co., Halleck.
DISTRIBUTION: SE Oregon, S Idaho, NE and EC California, N and C Nevada, and WC Utah (USA).
SYNONYMS: *albiventer* Hall and Durrant, *ambiguus* Hall, *atrirelictus* Hafner, *californicus* Merriam, *leucotis* Hall and Durrant, *medius* Hall, *nasutus* Hall, *nexus* Hall, *oregonus* Merriam, *paululus* Hall and Durrant, *polionotus* Grinnell, *sabulonis* Hall.
COMMENTS: Reviewed by O'Farrell and Blaustein (1974*a*, Mammalian Species, 46). The suggestion by Hall (1941:380-382) of hybridization between *megacephalus* and *pallidus* was discounted by Hafner et al. (1979). Also, Hall's suggestion (1981:560) that *leucotis* may warrant specific status is not supported (see Hafner and Hafner, 1983).

Microdipodops pallidus Merriam, 1901. Proc. Biol. Soc. Washington, 14:127.
TYPE LOCALITY: USA, Nevada, Churchill Co., Mountain Well.
DISTRIBUTION: EC California, W and SC Nevada (USA).
SYNONYMS: *ammophilus* Hall, *dickeyi* Goldman, *lucidus* Goldman, *purus* Hall, *restrictus* Hafner, *ruficollaris* Hall.
COMMENTS: Reviewed by O'Farrell and Blaustein (1974*b*, Mammalian Species, 47).

Subfamily Heteromyinae Gray, 1868. Proc. Zool. Soc. Lond., 1868:201.
COMMENTS: Contains the Recent genera *Heteromys* and *Liomys*, following Wood (1935), Wahlert (1985), and Ryan (1989*a*). However, a review of the generic limits is warranted since biochemical data (Rogers, 1990) suggested that *Heteromys* as currently defined is paraphyletic relative to *Liomys*.

Heteromys Desmarest, 1817. Nouv. Dict. Hist. Nat., Nouv. ed., 14:181.
TYPE SPECIES: *Mus anomalus* Thompson, 1815.
SYNONYMS: *Xylomys* Merriam, 1902.
COMMENTS: Includes *Xylomys* as a subgenus (see also Hall, 1981). Revised by Goldman (1911). Rogers and Schmidly (1982) revised the *desmarestianus* group. Hall and Kelson (1959:543) and Hall (1981:596) commented on the need for revision of this genus. Rogers (1989, 1990) discussed phylogenetic relationships among species, and remarked on at least one undescribed species from Costa Rica, as did Handley (1976) from Venezuela. Key to the species given in Schmidt et al. (1989).

Heteromys anomalus (Thompson, 1815). Trans. Linn. Soc. Lond., 11:161.
TYPE LOCALITY: Trinidad and Tobago, Trinidad.
DISTRIBUTION: W and N Colombia east to N Venezuela, Trinidad, Tobago, and Margarita Island.
SYNONYMS: *brachialis* Osgood, *hershkovitzi* Hernández-Camacho, *jesupi* J. A. Allen, *melanoleucus* Gray, *thompsoni* Lesson.
COMMENTS: Subgenus *Heteromys*. South American subspecies listed by Cabrera (1961) and distribution mapped by Eisenberg (1989).

Heteromys australis Thomas, 1901. Ann. Mag. Nat. Hist., ser. 7, 7:174.
TYPE LOCALITY: Ecuador, Esmeraldas Prov., Cachabi River, San Javier, below Cachabi.
DISTRIBUTION: E Panama south to SW Colombia and NW Ecuador.
SYNONYMS: *conscius* Goldman, *lomitensis* J. A. Allen, *pacificus* Pearson.
COMMENTS: Subgenus *Heteromys*. North American subspecies listed by Hall (1981); South American ones by Cabrera (1961).

Heteromys desmarestianus Gray, 1868. Proc. Zool. Soc. Lond., 1868:204.
TYPE LOCALITY: Guatemala, Coban.
DISTRIBUTION: SE Tobasco (Mexico) south to NW Colombia.
SYNONYMS: *chiriquensis* Enders, *crassirostris* Goldman, *fuscatus* J. A. Allen, *griseus* Merriam, *lepturus* Merriam, *longicaudatus* Gray, *nigricaudatus* Goodwin, *panamensis* Goldman, *planifrons* Goldman, *psakastus* Dickey, *repens* Bangs, *subaffinis* Goldman, *temporalis* Goldman, *underwoodi* Goodwin, *zonalis* Goldman.
COMMENTS: Subgenus *Heteromys*. Partial revisions by Goodwin (1969) and Rogers and Schmidly (1982); subspecies listed by Hall (1981).

Heteromys gaumeri J. A. Allen and Chapman, 1897. Bull. Am. Mus. Nat. Hist., 9:9.
TYPE LOCALITY: Mexico, Yucatan, Chichen-Itza.
DISTRIBUTION: Yucatan Peninsula (Mexico) and N Guatemala.
COMMENTS: Subgenus *Heteromys*. Reviewed by Engstrom et al. (1987*a*) and Schmidt et al. (1989, Mammalian Species, 345).

Heteromys goldmani Merriam, 1902. Proc. Biol. Soc. Washington, 15:41.
TYPE LOCALITY: Mexico, Chiapas, Chicharras.
DISTRIBUTION: S Chiapas (Mexico) and adjacent W Guatemala.
COMMENTS: Subgenus *Heteromys*.

Heteromys nelsoni Merriam, 1902. Proc. Biol. Soc. Washington, 15:43.
TYPE LOCALITY: Mexico, Chiapas, Pinabete, 8,200 ft. (2,499 m).
DISTRIBUTION: Known only from the type locality.
COMMENTS: Type species of subgenus *Xylomys* Merriam. Reviewed by Rogers and Rogers (1992*b*, Mammalian Species, 397).

Heteromys oresterus Harris, 1932. Occas. Pap. Mus. Zool., Univ. Michigan, 248:4.
TYPE LOCALITY: Costa Rica, Cordillera de Talamanca, El Copey de Dota, 6,000 ft. (1,829 m).
DISTRIBUTION: Talamanca Range of Costa Rica.

COMMENTS: Hall (1981) placed *oresterus* in the subgenus *Xylomys* but Rogers (1989, 1990) presented evidence that *oresterus* was not closely related to *nelsoni*, the type species of *Xylomys*. Reviewed by Rogers and Rogers (1992*a*, Mammalian Species, 396).

Liomys Merriam, 1902. Proc. Biol. Soc. Washington, 15:44.
TYPE SPECIES: *Heteromys alleni* Coues, 1881 (= *L. irroratus alleni*, see Genoways, 1973).
COMMENTS: Revised by Genoways (1973). Phylogenetic relationships among species and relative to *Heteromys* presented by Rogers (1989, 1990). Keys to the species of *Liomys* were presented by Genoways (1973:44-45) and Dowler and Genoways (1978).

Liomys adspersus (Peters, 1874). Monatsb. K. Preuss. Akad. Wiss. Berlin, p. 357.
TYPE LOCALITY: Panama, City of Panama (restricted by Goldman, 1920).
DISTRIBUTION: C Panama.

Liomys irroratus (Gray, 1868). Proc. Zool. Soc. Lond., 1868:205.
TYPE LOCALITY: Mexico, Oaxaca, Oaxaca (restricted by Genoways, 1973:111).
DISTRIBUTION: S Texas (USA), and SC Chihuahua to Oaxaca (Mexico).
SYNONYMS: *acutus* Hall and Villa-R., *albolimbatus* Gray, *alleni* Coues, *bulleri* Thomas, *canus* Merriam, *exiguus* Elliot, *guerrerensis* Goldman, *jaliscensis* J. A. Allen, *minor* Merriam, *pretiosus* Goldman, *pullus* Hooper, *texensis* Merriam, *torridus* Merriam, *yautepecus* Goodwin.
COMMENTS: Reviewed by Dowler and Genoways (1978, Mammalian Species, 82).

Liomys pictus (Thomas, 1893). Ann. Mag. Nat. Hist., ser. 6, 12:233.
TYPE LOCALITY: Mexico, Jalisco, San Sebastian, 4,300 ft. (1,311 m).
DISTRIBUTION: West coast of Mexico from Sonora to Chiapas, and east coast in Veracruz south to extreme NW Guatemala.
SYNONYMS: *annectens* Merriam, *escuinapae* J. A. Allen, *hispidus* J. A. Allen, *isthmius* Merriam, *obscurus* Merriam, *orbitalis* Merriam, *paralius* Elliot, *parviceps* Goldman, *phaeura* Merriam, *pinetorum* Goodwin, *plantinarensis* Merriam, *rostratus* Merriam, *sonorana* Merriam, *veraecrucis* Merriam.
COMMENTS: Reviewed by McGhee and Genoways (1978, Mammalian Species, 83). It is likely that more than one species is represented by *pictus* as currently defined (see Morales and Engstrom, 1989, and Rogers, 1990).

Liomys salvini (Thomas, 1893). Ann. Mag. Nat. Hist., ser. 6, 11:331.
TYPE LOCALITY: Guatemala, Sacatepequez, Duenas.
DISTRIBUTION: E Oaxaca (Mexico) south to C Costa Rica.
SYNONYMS: *anthonyi* Goodwin, *aterrimus* Goodwin, *crispus* Merriam, *heterothrix* Merriam, *nigrescens* Thomas, *setosus* Merriam, *vulcani* J. A. Allen.
COMMENTS: Reviewed by Carter and Genoways (1978, Mammalian Species, 84).

Liomys spectabilis Genoways, 1971. Occas. Papers Mus. Nat. Hist., Univ. Kansas, 5:1.
TYPE LOCALITY: Mexico, Jalisco, 2.2 mi (3.5 km) NE Contla, 3,850 ft. (1,173 m).
DISTRIBUTION: SE Jalisco (Mexico).

Subfamily Perognathinae Coues, 1875. Proc. Acad. Nat. Sci. Philadelphia, 27:277.
COMMENTS: Subfamily name emended by Wood (1935); originally given by Coues as Perognathidinae. Wood (1935:89), Hafner (1978), and Hall (1981) included *Microdipodops* in the subfamily, a course not followed by Hafner and Hafner (1983), Wahlert (1985), and Ryan (1989*a*).

Chaetodipus Merriam, 1889. N. Am. Fauna, 1:5.
TYPE SPECIES: *Perognathus spinatus* Merriam, 1889.
SYNONYMS: *Burtognathus* Hoffmeister.
COMMENTS: Species revised by Merriam (1889) and Osgood (1900) under the generic name *Perognathus*; both authors considered *Chaetodipus* as a valid subgenus (see also Hall, 1981). Raised to generic status by Hafner and Hafner (1983), an action followed by most subsequent authors. Includes *Burtognathus* (see Hoffmeister, 1986), defined as a subgenus to contain the single species *C. hispidus*. Chromosomal and biochemical systematics provided by Patton (1967*a*) and J. L. Patton et al. (1981).

Chaetodipus arenarius Merriam, 1894. Proc. California Acad. Sci., ser. 2, 4:461.
TYPE LOCALITY: Mexico, Baja California Sur, San Jorge, near Comondu.
DISTRIBUTION: Baja California (Mexico).
SYNONYMS: *albescens* Huey, *albulus* Nelson and Goldman, *ambiguus* Nelson and Goldman, *ammophilus* Osgood, *dalquesti* Roth, *helleri* Elliot, *mexicalis* Huey, *paralios* Huey, *sabulosus* Huey, *siccus* Osgood, *sublucidus* Nelson and Goldman.
COMMENTS: Reviewed by Huey (1964) and Lackey (1991*a*, Mammalian Species, 384, as *Perognathus arenarius*). Includes *dalquesti*, considered a full species by Hall (1981).

Chaetodipus artus Osgood, 1900. N. Am. Fauna, 18:55.
TYPE LOCALITY: Mexico, Chihuahua, Batopilas.
DISTRIBUTION: S Sonora, SW Chihuahua, W Durango, and Sinaloa (Mexico).
COMMENTS: Revised by Anderson (1964).

Chaetodipus baileyi Merriam, 1894. Proc. Acad. Nat. Sci. Philadelphia, 46:262.
TYPE LOCALITY: Mexico, Sonora, Magdalena.
DISTRIBUTION: S California, S Arizona, SW New Mexico (USA), south to N Sinaloa and Baja California (Mexico).
SYNONYMS: *domensis* Goldman, *extimus* Nelson and Goldman, *fornicatus* Burt, *hueyi* Nelson and Goldman, *insularis* Townsend, *knekus* Elliot, *mesidios* Huey, *rudinoris* Elliot.
COMMENTS: Reviewed by Paulson (1988*a*, Mammalian Species, 297).

Chaetodipus californicus Merriam, 1889. N. Am. Fauna, 1:26.
TYPE LOCALITY: USA, California, Alameda Co., Berkeley.
DISTRIBUTION: C California (USA) to N Baja California Norte (Mexico).
SYNONYMS: *armatus* Merriam, *bensoni* von Bloeker, *bernardinus* Benson, *dispar* Osgood, *femoralis* J. A. Allen, *marinensis* von Bloeker, *mesopolius* Elliot, *ochrus* Osgood.
COMMENTS: Subspecies listed by Hall (1981).

Chaetodipus fallax Merriam, 1889. N. Am. Fauna, 1:19.
TYPE LOCALITY: USA, California, San Bernardino Co., Reche Canyon (3 mi [5 km] SE Colton), 1,250 ft. (381 m).
DISTRIBUTION: SW California (USA) to W Baja California Norte (Mexico).
SYNONYMS: *anthonyi* Osgood, *inopinus* Nelson and Goldman, *majusculus* Huey, *pallidus* Mearns, *xerotrophicus* Huey.
COMMENTS: Reviewed by Huey (1964). Includes *anthonyi*, considered as a full species by Hall (1981).

Chaetodipus formosus Merriam, 1889. N. Am. Fauna, 1:17.
TYPE LOCALITY: USA, Utah, Washington Co., St. George.
DISTRIBUTION: W Utah, Nevada, E California, and NW Arizona (USA), and E coast of Baja California to Bahía Concepcion (Baja California Sur, Mexico).
SYNONYMS: *cinerascens* Nelson and Goldman, *domisaxensis* Cockrum, *incolatus* Hall, *infolatus* Huey, *melanocaudus* Cockrum, *melanurus* Hall, *mesembrinus* Elliot, *mohavensis* Huey.
COMMENTS: Reviewed by Huey (1964). Included in *Perognathus* by Hall (1981:542) and earlier workers, but allocated to *Chaetodipus* by J. L. Patton et al. (1981) and Hafner and Hafner (1983).

Chaetodipus goldmani Osgood, 1900. N. Am. Fauna, 18:54.
TYPE LOCALITY: Mexico, Sinaloa, Sinaloa.
DISTRIBUTION: NE to S Sonora, SW Chihuahua, and N Sinaloa (Mexico); see Straney and Patton (1981).
COMMENTS: Revised by Anderson (1964).

Chaetodipus hispidus (Baird, 1858). Mammalia *in* Repts. U.S. Expl. Surv., 8(1):421.
TYPE LOCALITY: Mexico, Tamaulipas, Charco Escondido.
DISTRIBUTION: Great Plains from S North Dakota to SE Arizona and W Louisiana (USA), south to Tamaulipas and Hidalgo (Mexico).
SYNONYMS: *conditi* J. A. Allen, *latirostris* Rhoads, *maximus* Elliot, *paradoxus* Merriam, *spilotus* Merriam, *zacatecae* Osgood.
COMMENTS: Revised by Glass (1947); subspecies listed by Hall (1981). Reviewed by Paulson (1988*b*, Mammalian Species, 320). Type species of monotypic subgenus *Burtognathus*.

Chaetodipus intermedius Merriam, 1889. N. Am. Fauna, 1:18.
TYPE LOCALITY: USA, Arizona, Mohave Co., Mud Spring.
DISTRIBUTION: SC Utah and Arizona to W Texas (USA), south to C Sonora and C Chihuahua (Mexico).
SYNONYMS: *ater* Dice, *crinitus* Benson, *lithophilus* Huey, *minimus* Burt, *nigrimontis* Blossom, *obscurus* Merriam, *phasma* Goldman, *pinacate* Blossom, *rupestris* Benson, *umbrosus* Benson.
COMMENTS: Subspecies listed by Hoffmeister (1974) and Hall (1981).

Chaetodipus lineatus Dalquest, 1951. J. Washington Acad. Sci, 41:362.
TYPE LOCALITY: Mexico, San Luis Potosi, 1 km S of Arriaga.
DISTRIBUTION: San Luis Potosi (Mexico).
COMMENTS: Probably conspecific with *penicillatus*.

Chaetodipus nelsoni Merriam, 1894. Proc. Acad. Nat. Sci. Philadelphia, 46:266.
TYPE LOCALITY: Mexico, San Luis Potosi, Hacienda La Parada, about 25 mi (40km) NW Ciudad San Luis Potosi.
DISTRIBUTION: Chihuahuan desert plateau from SE New Mexico and W Texas (USA) to Jalisco and San Luis Potosi (Mexico).
SYNONYMS: *canescens* Merriam, *collis* Blair, *popei* Blair.
COMMENTS: Subspecies listed by Hall (1981).

Chaetodipus penicillatus (Woodhouse, 1852). Proc. Acad. Nat. Sci. Philadelphia, 6:200.
TYPE LOCALITY: USA, San Francisco Mountains, New Mexico (fixed to Arizona, Yuma Co., 1 mi [1.6 km] SW Parker by Hoffmeister and Lee, 1967).
DISTRIBUTION: SE California and S Nevada to S Arizona, New Mexico, and W Texas (USA) to NE Baja California, Sonora, and Central Plateau to San Luis Potosi (Mexico).
SYNONYMS: *angustirostris* Osgood, *atrodorsalis* Dalquest, *eremicus* Mearns, *goldmani* Townsend (not *goldmani* Osgood), *pricei* J. A. Allen, *seri* Nelson, *seorsus* Goldman (not *seorsus* Burt), *sobrinus* Goldman, *stephensi* Merriam.
COMMENTS: Revised by Hoffmeister and Lee (1967). Chromosomal (Patton, 1969) and biochemical (J. L. Patton et al., 1981) data suggest that the Chihuahan Desert subspecies *atrodorsalis* and *eremicus* are a separate species. May include *lineatus*.

Chaetodipus pernix J. A. Allen, 1898. Bull. Am. Mus. Nat. Hist., 10:149.
TYPE LOCALITY: Mexico, Sinaloa, Rosario.
DISTRIBUTION: Coastal lowlands from S Sonora to N Nayarit (Mexico).
SYNONYMS: *rostratus* Osgood.
COMMENTS: Chromosomal and biochemical evidence suggests that the northern subspecies *rostratus* is specifically distinct from *pernix* (J. L. Patton et al., 1981).

Chaetodipus spinatus Merriam, 1889. N. Am. Fauna, 1:21.
TYPE LOCALITY: USA, California, San Bernardino Co., 25 mi (40km) below The Needles, Colorado River.
DISTRIBUTION: S Nevada, SE California (USA); Baja California Norte and Baja California Sur (Mexico).
SYNONYMS: *broccus* Huey, *bryanti* Merriam, *evermanni* Nelson and Goldman, *guardiae* Burt, *lambi* Benson, *latijularis* Burt, *lorenzi* Banks, *magdalenae* Osgood, *macrosensis* Burt, *margaritae* Merriam, *occultus* Nelson, *oribates* Huey, *peninsulae* Merriam, *prietae* Huey, *pullus* Burt, *refescens* Huey, *seorsus* Burt.
COMMENTS: Reviewed by Lackey (1991*b*, Mammalian Species, 385).

Perognathus Wied-Neuwied, 1839. Nova Acta Phys.-Med. Acad. Caes. Leop.-Carol., 19(1):368.
TYPE SPECIES: *Perognathus fasciatus* Wied-Neuwied, 1839.
SYNONYMS: *Abromys* Gray, *Cricetodipus* Peale, *Otognosis* Coues.
COMMENTS: Revised by Merriam (1889) and Osgood (1900), who also included those species listed here for *Chaetodipus*. Chromosomal relationships reviewed by Patton (1967*b*) and Williams (1978*a*).

Perognathus alticola Rhoads, 1894. Proc. Acad. Nat. Sci. Philadelphia, 45:412.
TYPE LOCALITY: USA, California, San Bernardino Co., San Bernardino Mtns, Squirrel Inn, Little Bear Valley, 5,500 ft. (1,576 m).
DISTRIBUTION: SC California (USA).

STATUS: IUCN - Vulnerable. Listed as a Species of Special Concern by the State of California Dept. of Fish and Game.

SYNONYMS: *inexpectatus* Huey.

COMMENTS: Subspecies listed by Hall (1981); probably only subspecifically distinct from *parvus*.

Perognathus amplus Osgood, 1900. N. Am. Fauna, 18:32.

TYPE LOCALITY: USA, Arizona, Yavapai Co., Fort Verde.

DISTRIBUTION: W and C Arizona (USA) to NW Sonora (Mexico).

SYNONYMS: *ammodytes* Benson, *cineris* Benson, *jacksoni* Goldman, *pergracilis* Goldman, *rotundus* Goldman, *taylori* Goldman.

COMMENTS: Subspecies listed in Hall (1981).

Perognathus fasciatus Wied-Neuwied, 1839. Nova Acta Phys.-Med. Acad. Caes. Leop.-Carol., 19(1):369.

TYPE LOCALITY: USA, North Dakota, Williams Co., upper Missouri River near jct. with the Yellowstone, near Buford.

DISTRIBUTION: Great Plains from SE Alberta, Saskatchewan, and SW Manitoba (Canada) to NE Utah, S Colorado, and E South Dakota (USA).

SYNONYMS: *callistus* Osgood, *infraluteus* Thomas, *litus* Cary, *olivaceogriseus* Swenk.

COMMENTS: Revised by Williams and Genoways (1979). Reviewed by Manning and Jones (1988, Mammalian Species, 303).

Perognathus flavescens Merriam, 1889. N. Am. Fauna, 1:11.

TYPE LOCALITY: USA, Nebraska, Cherry Co., Kennedy.

DISTRIBUTION: Great Plains and intermountain basins from Minnesota and N Utah (USA) to N Chihuahua (Mexico).

SYNONYMS: *apache* Merriam, *caryi* Goldman, *cleomophila* Goldman, *cockrumi* Hall, *copei* Rhoads, *gypsi* Dice, *melanotis* Osgood, *perniger* Osgood, *relictus* Goldman.

COMMENTS: Reviewed by Williams (1978*b*).

Perognathus flavus Baird, 1855. Proc. Acad. Nat. Sci. Philadelphia, 7:332.

TYPE LOCALITY: USA, Texas, El Paso Co., El Paso.

DISTRIBUTION: SW Great Plains and intermountain plateaus from South Dakota, E Wyoming, and SE Utah (USA) south to Sonora and Puebla (Mexico).

SYNONYMS: *bimaculatus* Merriam, *bunkeri* Cockrum, *fuliginosus* Merriam, *fuscus* Anderson, *goodpasteri* Hoffmeister, *hopiensis* Goldman, *medius* Baker, *mexicanus* Merriam, *pallescens* Baker, *parviceps* Baker, *piperi* Goldman, *sanluisi* Hill, *sonoriensis* Nelson and Goldman.

COMMENTS: Revised by Baker (1954); subspecies listed by Hall (1981). Wilson (1973) considered *merriami* conspecific, but Lee and Engstrom (1991) presented evidence of sympatry with limited hybridization.

Perognathus inornatus Merriam, 1889. N. Am. Fauna, 1:15.

TYPE LOCALITY: USA, California, Fresno Co., Fresno.

DISTRIBUTION: Central and Salinas valleys of California (USA).

STATUS: Listed as a Species of Special Concern by the State of California Dept. of Fish and Game.

SYNONYMS: *neglectus* Taylor, *sillimani* von Bloeker.

COMMENTS: Revised by Osgood (1918); subspecies listed by Hall (1981).

Perognathus longimembris (Coues, 1875). Proc. Acad. Nat. Sci. Philadelphia, 27:305.

TYPE LOCALITY: USA, California, Kern Co., Tehachapi Mtns, Old Fort Tejon.

DISTRIBUTION: SE Oregon and W Utah (USA) south to N Sonora and Baja California Norte (Mexico).

STATUS: IUCN - Endangered as subspecies *brevinasus* and *psammophilus*. Two California subspecies (*brevinasus* and *pacificus*) listed as Species of Special Concern by the State of California Dept. of Fish and Game.

SYNONYMS: *aestivus* Huey, *arcus* Benson, *arenicola* Stephens, *arizonensis* Goldman, *bangsi* Mearns, *bombycinus* Osgood, *brevinasus* Osgood, *cantwelli*, von Bloeker, *elibatus* Elliot, *gulosus* Hall, *internationalis* Huey, *kinoensis* Huey, *nevadensis* Merriam, *pacificus* Merans, *panamintinus* Merriam, *pericalles* Elliot, *pimensis* Huey, *psammophilus* von Bloeker, *salinensis* Bole, *tularensis* Richardson, *venustus* Huey, *virginis* Huey.

COMMENTS: Revised by Osgood (1918); subspecies listed by Hall (1981).

Perognathus merriami J. A. Allen, 1892. Bull. Am. Mus. Nat. Hist., 4:45.

TYPE LOCALITY: USA, Texas, Cameron Co., Brownsville.

DISTRIBUTION: SE New Mexico east to S Texas (USA), east from N Chihuahua to Tamaulipas (Mexico).

SYNONYMS: *gilvus* Osgood, *mearnsi* J. A. Allen.

COMMENTS: Synonymized with *flavus* by Wilson (1973), but Lee and Engstrom (1991) considered *merriami* a separate species based on biochemical genetics.

Perognathus parvus (Peale, 1848). Mammalia *in* Repts. U.S. Expl. Surv., 8:53.

TYPE LOCALITY: USA, Oregon, Wasco Co., probably near The Dalles.

DISTRIBUTION: Great Basin from S British Columbia (Canada), south to E California and east to SE Wyoming and NW Arizona (USA).

SYNONYMS: *amoenus* Merriam, *bullatus* Durrant and Lee, *clarus* Goldman, *columbianus* Merriam, *idahoensis* Goldman, *laingi* Anderson, *magruderensis* Osgood, *lordi* Gray, *mollipilosus* Coues, *monticola* Baird, *olivaceus* Merriam, *plerus* Goldman, *trumbullensis* Benson, *yakimensis* Broadbrooks.

COMMENTS: Reviewed by Verts and Kirkland (1988, Mammalian Species, 318). Probably includes *xanthonotus* Grinnell.

Perognathus xanthanotus Grinnell, 1912. Proc. Biol. Soc. Washington, 25:128.

TYPE LOCALITY: USA, California, Kern Co., E Slope Walker Pass, Freeman Canyon, 4900 ft.

DISTRIBUTION: Known only from the vicinity of the type locality.

COMMENTS: Probably only a subspecies of *parvus*.

FAMILY DIPODIDAE

by Mary Ellen Holden

Family Dipodidae G. Fischer, 1817. Mem. Soc. Imp. Nat., Moscow, 5:372.

SYNONYMS: Allactagidae, Dipodes, Dipodum, Dipsidae, Jaculidae, Sicistidae, Sminthidae, Zapodidae.

COMMENTS: The monophyly of dipodids seems well established (Ellerman, 1940; Klingener, 1984; Shenbrot, 1992; Stein, 1990; Vinogradov, 1930). Authors have frequently recognized either a single family, Dipodidae, or two families, Zapodidae (including sicistines and zapodines), and Dipodidae; recently Shenbrot (1992) recognized four families including Allactagidae, Dipodidae, Sminthidae (Sicistidae), and Zapodidae. In a phylogenetic study based on limb myology, Stein (1990) proposed to recognize Sicistidae and Dipodidae (including zapodines and all other dipodids). Sicistines are often regarded as representing the most primitive dipodids morphologically; however, Sokolov et al. (1987*b*) and Vorontsov (1969) hypothesized that zapodines possess the most primitive karyotype of dipodids, from which those of sicistines and other dipodids are derived.

Shenbrot (1992) incorporated internal and external phallic morphology, coronal structure of the molars, and bullar morphology in a phylogenetic analysis. His final classification is eclectic, based on the results of cladistic analysis and degree of morphological divergence (phenetic distance). His arrangement differed significantly from that of Stein (1990) in that *Sicista* was not shown to be the most primitive dipodoid (all four families are sister taxa), and Euchoreutinae was hypothesized to be most closely related to Sicistinae, and was united with Sicistinae in Sminthidae (Sicistidae); Stein (1990) proposed Euchoreutinae to be a sister group of zapodines and allactagines, and placed all groups except *Sicista* in Dipodidae.

The true sister group of dipodids has not been established, and comparison with other outgroups such as sciurids, myoxids, heteromyids, and muroids that retain more primitive characters than *Phodopus* (used in the phylogenetic study by Stein, 1990) and other Cricetinae (as defined by Carleton and Musser, 1984; implied as an outgroup by Shenbrot, 1992), is needed for confirmation of assigned character polarities. Reviews of the relationship of dipodids to other rodents were given by Klingener (1964, 1984).

An integrative phylogenetic analysis including appropriate outgroups, and incorporating the characters listed above, and/or molecular data sets, should be undertaken to help elucidate dipodid relationships. Vinogradov (1930, 1937) proposed a classification, modified slightly by Ellerman (1940), that provided the foundation for systematic research of dipodids. Phylogenetic studies have supported much of this original classification, though the proposed relationships among the higher taxa have changed significantly. Pending further study, a single family is recognized here, but hypotheses supported by Stein (1990) and Shenbrot (1992) are incorporated at the subfamilial level.

Reviews of dipodid research and classification were contributed by Gambaryan et al. (1980), Heptner (1984), Klingener (1984), Shenbrot (1986, 1992), and Stein (1990). Cranial and dental characters were investigated by Vinogradov (1930). Comparative myology studied by Klingener (1964); facial myology by Gambaryan et al. (1980); myology of postcrania by Fokin (1971) and Stein (1990). Review of distribution and habitat of dipodids (excluding sicistines and zapodines) given by Kulik (1980). Chromosome numbers of members of each subfamily provided by Vorontsov (1969). Male genitalia studied by Vinogradov (1925) and Pavlinov and Shenbrot (1983). Comparative behavior and its taxonomic significance studied by Rogovin (1984). The distributions of species occuring in the former USSR, and taxonomic problems were verified and much enhanced by personal communication with G. I. Shenbrot; those of China were likewise greatly improved by the efforts of Lin Yonglie.

Subfamily Allactaginae Vinogradov, 1925. Proc. Zool. Soc. Lond., 1925(1):578.

COMMENTS: Dental morphology and evolution studied by Shenbrot (1984).

Allactaga (F. Cuvier, 1837). Proc. Zool. Soc. Lond., 1836:141 [1837].
TYPE SPECIES: *Mus jaculus* Pallas, 1778 (= *Dipus sibericus major* Kerr, 1792); see comments below.
SYNONYMS: *Mesoallactaga, Microallactaga, Orientallactaga, Paralactaga, Scarturus, Scirteta, Scirtetes, Scirtomys* (see Ellerman and Morrison-Scott, 1951; and Pavlinov and Rossolimo, 1987).
COMMENTS: Does not include *Allactodipus* (see comment under *Allactodipus bobrinskii*). Shenbrot's (1984) subgeneric classification is followed below except where noted. Subspecific revision of species occuring in the independant countries of the former USSR provided by Shenbrot (1991*d*). *Mus jaculus* Linnaeus, 1758 is the type species of *Jaculus* (see comments therein).

Allactaga balikunica Hsia and Fang, 1964. Acta Zootaxon. Sin., 6:16.
TYPE LOCALITY: China, Xinjiang, Balikun.
DISTRIBUTION: Mongolia, from Altai Sumon east to Bordzon-Gobi (Sokolov et al., 1981*a*), and NE Xinjiang, China (Ma et al., 1987).
SYNONYMS: *nataliae*.
COMMENTS: Subgenus *Orientallactaga*. Closely related to *A. bullata*, with which it is sympatric in S Mongolia. *A. nataliae* Sokolov, 1981 is a junior synonym of *A. balikunica* (see Sokolov and Shenbrot, 1987*a*).

Allactaga bullata Allen, 1925. Am. Mus. Novit., 161:2.
TYPE LOCALITY: Mongolia, Altai Gobi, Tsagan-Nur (Tsagaan Nuur).
DISTRIBUTION: Deserts of S and W Mongolia (Bannikov, 1954; Sokolov et al., 1981*a*; adjacent Chinese provinces of Nei Mongol, E Xinjiang, Ningxia (Ma et al., 1987), Gansu (Chen and Wang, 1985; Zheng and Zhang, 1990) and N Shaanxi (Wang, 1990).
COMMENTS: Subgenus *Orientallactaga*. Closely related to *A. balikunica* (see comments therein).

Allactaga elater (Lichtenstein, 1828). Abh. König. Akad. Wiss., Berlin, 1825:155 [1828].
TYPE LOCALITY: W Kazakhskaya, Kirgiz Steppe. The type locality given by Lichtenstein (1828) is in W Kazakhstan, according to Vinogradov (1937), Kuznetsov (1944, 1965), and Ognev (1963*b*), not E Kazakhstan as reported by Ellerman and Morrison-Scott (1951:529) and Corbet (1978*c*:154).
DISTRIBUTION: W Pakistan (Roberts, 1977); Afghanistan (Hassinger, 1973); Iran (Lay, 1967); NE Turkey; in former USSR (see Kuznetsov, 1965; Sludskii, 1977; and Shenbrot, 1991*d*), Armenia, Azerbaydzhan, Georgia, N Caucasus, north along W Caspian Sea to Lower Volga south to Turkmenistan, east through Kazakhstan to NE Xinjiang, Nei Mongol, and N Gansu, China (Ma et al., 1987), in desert and semi-desert zones.
SYNONYMS: *aralychensis, bactriana, caucasicus, dzungariae, heptneri, indica, kizljaricus, strandi, turkmeni, zaisanicus* (see Corbet, 1978*c*, 1984; Shenbrot, 1991*d*).
COMMENTS: Subgenus *Allactaga*. Does not include *vinogradovi* (see comment therein). Detailed review provided by Ognev (1963*b*). Karyotype contributed by Vorontsov et al. (1969*c*). Subspecific revision and additional distributional data provided by Shenbrot (1991*d*).

Allactaga euphratica Thomas, 1881. Ann. Mag. Nat. Hist., ser. 5, 8:15.
TYPE LOCALITY: Iraq.
DISTRIBUTION: Steppe and semi-desert from Turkey, Syria, E Jordan, east to N Saudi Arabia and Kuwait; north through Iraq to the Caucasus; N Iran; E Afghanistan. Figured by Harrison and Bates (1991).
SYNONYMS: *caprimulga, laticeps, schmidti, williamsi* (see Corbet, 1978*c*).
COMMENTS: Subgenus *Paralactaga*. Many authors incorrectly employ the name *williamsi* for this species. Following Corbet (1978*c*:155) *A. euphratica* is considered a distinct species from *A. hotsoni*. Reviewed by Ognev (1963*b*). Detailed habitat data provided by Naumov and Lobachev (1975).

Allactaga firouzi Womochel, 1978. Fieldiana Zool., 72(5):65.
TYPE LOCALITY: Iran, Isfahan Prov., 18 mi S Shah Reza (Qomisheh), 2253 m.
DISTRIBUTION: Known only from the type locality, a flat plain with a gravel substrate and sparse, mountain steppe vegetation (Womochel, 1978).

COMMENTS: Not allocated to subgenus by Shenbrot (1984). *A. firouzi* appears to be morphologically distinct from *A. euphratica* and *A. hotsoni* (Womochel, 1978), but its relationship with these and other species of allactagines needs further study. Shenbrot (1991*d*) tentatively synonymized *firouzi* with *A. elater turkmeni*, but later (in litt.) examined the type specimen and considered *firouzi* synonymous with *hotsoni*.

Allactaga hotsoni Thomas, 1920. J. Bombay Nat. Hist. Soc., 26(4):936.
TYPE LOCALITY: Iran, Persian Baluchistan, 20 mi. SW Sib, Kant (Kont), 3950ft.
DISTRIBUTION: Iranian Baluchistan (Lay, 1967); W Pakistan (Roberts, 1977); and S Afghanistan (Hassinger, 1973).
COMMENTS: Subgenus *Allactaga* (see Pavlinov and Shenbrot, 1985). Following Corbet (1978*c*:155) *A. hotsoni* is considered a distinct species from *A. elater* and *A. euphratica*; see comment under *A. firouzi*.

Allactaga major (Kerr, 1792). *In* Linnaeus, Anim. Kingdom, p. 274.
TYPE LOCALITY: Kazakhstan, between Caspian Sea and Irtysh River.
DISTRIBUTION: Steppes and deserts from Caucasus N to Moscow and Kiev E to Ob River (W Siberia), Kazakhstan, Turkmenistan, and Uzbekistan; figured by Kuznetsov (1965) and Sludskii (1977).
SYNONYMS: *brachyotis, chachlovi, decumanus, djetysuensis, flavescens, fuscus, hochlovi, intermedius, jaculus, macrotis, nigricans, spiculum, vexillarius* (see Ellerman and Morrison-Scott, 1951; Corbet, 1978*c*; and Shenbrot, 1991*d*).
COMMENTS: Subgenus *Allactaga*. *A. jaculus* Pallas is a junior synonym (see Ognev, 1963*b*:94; Pavlinov and Rossolimo, 1987:154). Subspecific revision and additional distributional data provided by Shenbrot (1991*d*). Karyotype contributed by Vorontsov et al. (1969*c*). Reviewed by Ognev (1963*b*). Detailed habitat data provided by Naumov and Lobachev (1975).

Allactaga severtzovi Vinogradov, 1925. Proc. Zool. Soc. Lond., 1925:583.
TYPE LOCALITY: SE Kazakhstan, Taldy-Kurgan (Kopal) dist., Tamar-Utkul.
DISTRIBUTION: Kazakhstan, Uzbekistan, NE Turkmenistan, and SW Tadzhikistan; figured by Kuznetsov (1965) and Sludskii (1977).
SYNONYMS: *chorezmi*.
COMMENTS: Subgenus *Allactaga*. Taxonomic study provided by Shenbrot (1991*d*), who described *chorezmi* as a subspecies of *A. severtzovi*. Reviewed by Ognev (1963*b*). Karyotype provided by Vorontsov et al. (1969*c*). Detailed habitat data provided by Naumov and Lobachev (1975).

Allactaga sibirica (Forster, 1778). K. Svenska Vet.-Akad. Handl. Stockholm, 39:112.
TYPE LOCALITY: SE Transbaikalia, Chitinskaya Oblast, near Lake Tarei-Nur.
DISTRIBUTION: From lower Ural River (Kazakhstan) and Caspian Sea east to Chitinskaya Oblast south to N Turkmenistan (Kuznetsov, 1965; Shenbrot, 1991*d*; Sludskii, 1977); Mongolia (Bannikov, 1954); China: Nei Mongol, Xinjiang, Qinghai, Gansu, Ningxia, Shaanxi, Shanxi, N Hebei, W Liaoning, W Jilin, and W Heilongjiang (in China see Ho et al., 1986; Laing and Zhang, 1985; Liu et al., 1990; Ma et al., 1987; Mi et al., 1990; Qin, 1991; Shou, 1962; Zhang and Wang, 1963; and Zheng and Zhang, 1990); no valid record from Korea (see Corbet, 1978*c*:154).
SYNONYMS: *alactaga, altorum, annulata, brachyurus, bulganensis, dementiewi, grisescens, halticus, longior, media, mongolica, ognevi, ruckbeili, salicus, saliens, saltator, semideserta, suschkini* (see Ellerman and Morrison-Scott, 1951; Corbet, 1978*c*, 1984; Pavlinov and Rossolimo, 1987; and Shenbrot, 1991*d*).
COMMENTS: Subgenus *Orientallactaga*. Reviewed by Ognev (1963*b*). Subspecific revision and additional distributional data provided by Shenbrot (1991*d*). Geographic variation studied by Varshavsky (1991). Karyotype provided by Vorontsov et al. (1969*c*). Detailed habitat data provided by Naumov and Lobachev (1975).

Allactaga tetradactyla (Lichtenstein, 1823). Verz. Doublet. Zool. Mus. Univ. Berlin, p. 2.
TYPE LOCALITY: Libyan Desert between Siwa and Alexandria.
DISTRIBUTION: Coastal gravel plains of Egypt and E Libya, from near Alexandria to the Gulf of Sirte (see Ranck, 1968, and Osborn and Helmy, 1980).
SYNONYMS: *brucii* (see Ellerman and Morrison-Scott, 1951).
COMMENTS: Subgenus *Scarturus*.

Allactaga vinogradovi Argyropulo, 1941. Fauna SSSR, Mlekopitaiushchiy, Opredelitel grizunov, Izdatelstvo Akademii Nauk, SSSR, p. 138.
TYPE LOCALITY: Kazakhstan, Dzhambul region, Burnoye and Rovnoye.
DISTRIBUTION: S Kazakhstan, E Uzbekistan, Kirghizia, and Tadzhikistan (see Shenbrot, 1991*d*).
COMMENTS: Subgenus *Allactaga*. Shenbrot (1991*d*) showed that *A. vinogradovi* differs from *A. elater* in toothrow length and phallic morphology, and that the two are sympatric along the Talas River and in SE Betpak-Dala (Kazakhstan).

Allactodipus Kolesnikov, 1937. Bull. Sredne-Az. Gos. Univ., 22(29):255.
TYPE SPECIES: *Allactodipus bobrinskii* Kolesnikov, 1937.
COMMENTS: Reviewed by Shenbrot (1974, 1984), who showed that in postcranial and dental characters, particularly the height of the molars and alveolar pattern, *A. bobrinskii* falls outside the range of variation of the genus *Allactaga*. He restored *bobrinskii* to its original genus, *Allactodipus*, a hypothesis followed by Pavlinov and Rossolimo (1987). A critical phylogenetic study incorporating these and other morphological and/or molecular differences between *A. bobrinskii* and other allactagines and dipodines is needed.

Allactodipus bobrinskii Kolesnikov, 1937. Bull. Sredne-Az. Gos. Univ., 22(29):255.
TYPE LOCALITY: Uzbekistan, Kizil-kum (Kyzylkum) Desert, 140 km NW of Bukhara, Khala-Ata.
DISTRIBUTION: W and N Turkmenistan and C and W Uzbekistan, in the Kyzylkum and Karakumy deserts; figured by Kuznetsov (1965) and Sludskii (1977).
COMMENTS: Karyotype provided by Vorontsov et al. (1969*c*). Detailed habitat data provided by Naumov and Lobachev (1975).

Pygeretmus Gloger, 1841. Gemeinn. Hand. Hilfsbuch. Nat., 1:106.
TYPE SPECIES: *Dipus platurus* Lichtenstein, 1823 (emended to *D. platyurus* by Lichtenstein, 1828).
SYNONYMS: *Alactagulus*, *Platycercomys*, *Pygerethmus* (see Ellerman and Morrison-Scott, 1951).
COMMENTS: Includes *Alactagulus* (see Heptner, 1984; Pavlinov and Rossolimo, 1987; Shenbrot, 1984). Vorontsov et al. (1969*b*) found no distinguishing karyological characters between subgenera *Alactagulus* and *Pygeretmus*; in the same study karyotypes and morphometric comparisons of subgenus *Pygeretmus* were provided. Shenbrot's (1984) subgeneric classification is followed below.

Pygeretmus platyurus (Lichtenstein, 1823). Naturh. Abh. Eversmann's Reise, p. 121.
TYPE LOCALITY: Kazakhstan, E shore of Aral Sea, Kuwan-Darya River.
DISTRIBUTION: W, C, and E Kazakhstan.
SYNONYMS: *platurus*, *vinogradovi* (see Pavlinov and Rossolimo, 1987; and Shenbrot, 1988).
COMMENTS: Subgenus *Pygeretmus*. Corbet (1978*c*) and Gromov and Baranova (1981) synonymized *P. vinogradovi* with *P. platyurus* without comment. Heptner (1984) agreed because he felt the differences between the two forms were not sharp enough to warrant specific recognition, a view tentatively followed by Pavlinov and Rossolimo (1987). Shenbrot's (1988) results indicated that *P. vinogradovi* falls within the range of variation of *P. platyurus*, and that *vinogradovi* should not even be recognized at the subspecific level. Review and distribution map provided by Silverstov et al. (1969) and Sludskii (1977); detailed habitat data provided by Naumov and Lobachev (1975).

Pygeretmus pumilio (Kerr, 1792). *In* Linnaeus, Anim. Kingdom, p. 275.
TYPE LOCALITY: Kazakhstan, between Caspian Sea and Irtysh River. Pavlinov and Rossolimo (1987) incorrectly listed Kirghiz Steppe as the type locality for this species.
DISTRIBUTION: From the Don River (Russia) through Kazakhstan to the Irtysh River (Kuznetsov, 1965; Sludskii, 1977), south to NE Iran (Lay, 1967); E to S Mongolia (Bannikov, 1954); China: W Nei Mongol (Ma et al., 1987), N Xinjiang (Ma et al., 1987; Chen and Wang, 1985), and probably Ningxia.
SYNONYMS: *acontion*, *aralensis*, *dinniki*, *minor*, *minutus*, *pallidus*, *potanini*, *pygmaeus*, *sibericus*, *tanaiticus*, *turcomanus* (see Ellerman and Morrison-Scott, 1951; and Corbet, 1978*c*).

COMMENTS: Subgenus *Alactagulus* (see Pavlinov and Rossolimo, 1987). The name *pygmaeus* is preoccupied and is an invalid junior synonym of *pumilio*, (see Ellerman and Morrison-Scott, 1951; and Corbet, 1978*c*). Reviewed by Ognev (1963*b*). Detailed habitat data provided by Naumov and Lobachev (1975).

Pygeretmus shitkovi (Kuznetsov, 1930). Doklady Acad. Nauk S.S.S.R., Leningrad, 1930A:623.

TYPE LOCALITY: SE Kazakhstan, Taldy-Kurgan district, on NW shore of Ala-Kul (Alakol) Lake, Rybalnoje.

DISTRIBUTION: E Kazakhstan, in region of Lake Balkhash (Corbet, 1978*c*; Kuznetsov, 1965; Sludskii, 1977).

SYNONYMS: *schitkovi*, *zhitkovi* (see Ellerman and Morrison-Scott, 1951; and Pavlinov and Rossolimo, 1987).

COMMENTS: Subgenus *Pygeretmus*. Corbet (1978*c*) regarded the emendation of *shitkovi* to *zhitkovi* invalid, but other workers disagree; there has been no ruling by the International Commission on Zoological Nomenclature. Reviewed by Ognev (1963*b*).

Subfamily Cardiocraniinae Vinogradov, 1925. Proc. Zool. Zoc. Lond., 1925(1):578.

SYNONYMS: Salpingotini.

COMMENTS: Following Pavlinov (1980*b*) two tribes, Cardiocraniini and Salpingotini, are recognized.

Cardiocranius Satunin, 1903. Ann. Mus. Zool. Acad. Imp. Sci. St. Petersbourg, 7:582.

TYPE SPECIES: *Cardiocranius paradoxus* Satunin, 1903.

COMMENTS: Tribe Cardiocraniini. Detailed review provided by Ognev (1963*b*).

Cardiocranius paradoxus Satunin, 1903. Ann. Mus. Zool. Acad. Imp. Sci. St. Petersbourg, 7:584.

TYPE LOCALITY: China, NW Gansu, Nan Shan, Shargol-Dzhin.

DISTRIBUTION: China: N Xinjiang, W Nei Mongol, N Ningxia, and Gansu (see Chen and Wang, 1985; Ma et al., 1987; Mi et al., 1990; Qin, 1991; and Zhou et al., 1985); Mongolia; S Tuvinskaya Oblast, and E Kazakhstan (see Sokolov and Shenbrot,1988).

COMMENTS: A morphometric study by Sokolov and Shenbrot (1988) indicated that the Kazakhstan population falls within the range of intraspecific variation for *C. paradoxus*, and is not a separate species as was suggested by Gromov and Baranova (1981). For detailed habitat data see Naumov and Lobachev (1975).

Salpingotus Vinogradov, 1922. *In* Kozlov, Mongolia and Amdo, p. 540.

TYPE SPECIES: *Salpingotus kozlovi* Vinogradov, 1922.

SYNONYMS: *Anguistodontus*, *Prosalpingotus*, *Salpingotulus* (see Pavlinov and Rossolimo, 1987).

COMMENTS: Tribe Salpingotini. The subgeneric arrangement hypothesized by Vorontsov and Shenbrot (1984), based on morphometric analyses and skull, dental, and genital morphology, is followed here, except that *Salpingotulus* is recognized at the subgeneric, rather than generic, level (see comment under *S. michaelis*). Comparative myology of pelvic girdle studied by Fokin (1971). Distribution of species of *Salpingotus* shown in Vorontsov and Shenbrot (1984).

Salpingotus crassicauda Vinogradov, 1924. Zool. Anz., 61:150.

TYPE LOCALITY: China, N Xinjiang, Altai Gobi, near Schara-sumé (Sharasume), approx. 160km S Russia-Mongolian border. Most authors list the type locality as being located in W Mongolia, but it is actually in N Xinjiang, China (Bannikov, 1954).

DISTRIBUTION: Steppes and deserts of NW China (Chen and Wang, 1985; Ma et al., 1987; and Zheng and Zhang, 1990), S and SW Mongolia (Sokolov and Shenbrot, 1988), and adjacent E Kazakhstan in Lake Zaysan basin (Naumov and Lobachev, 1975; Vorontsov and Shenbrot, 1984; Vorontsov et al., 1969*a*).

SYNONYMS: *gobicus* Sokolov and Shenbrot, 1988.

COMMENTS: Subgenus *Anguistodontus*. Populations S of Lake Balkhash and N of Aral Sea previously included in *S. crassicauda* now recognized as a distinct species, *S. pallidus*. Detailed habitat data given by Naumov and Lobachev (1975). Reviewed by Ognev (1963*b*). Taxonomic study of Mongolia and Zaysan Basin populations provided by Sokolov and Shenbrot (1988). Karyotype provided by Vorontsov et al. (1969*d*).

Salpingotus heptneri Vorontsov and Smirnov, 1969. *In* Vorontsov (ed.) [The mammals: evolution, karyology, taxonomy, fauna.], Novosibirsk, p. 60.

TYPE LOCALITY: Uzbekistan, NW Kizil-Kum (Kyzylkum) Desert, 80 km NE Takhta-Kupir, 8 km E Gori Kok-Tobe.

DISTRIBUTION: Uzbekistan and S Kazakhstan, NW and N Kyzylkum desert (Vorontsov and Shenbrot, 1984).

COMMENTS: Subgenus *Prosalpingotus*.

Salpingotus kozlovi Vinogradov, 1922. *In* Kozlov, Mongolia and Amdo, p. 542.

TYPE LOCALITY: Mongolia: Gobi desert, Khara-Khoto.

DISTRIBUTION: Deserts of S and SE Mongolia; and China: Nei Mongol, Xinjiang, Gansu, N Shaanxi, and Ningxia (see Chen and Wang, 1985; Ma et al., 1987; Mi et al., 1990; Qian et al., 1965; Qin, 1991; Wang, 1990; and Zheng and Zhang, 1990).

COMMENTS: Subgenus *Salpingotus*. See Corbet (1978*c*) for comment regarding allocation of specimen from Irtysh river on Kazakhstan-Chinese border to *S. crassicauda*. Reviewed by Ognev (1963*b*). Study of geographic variation in Mongolian samples provided by Sokolov and Shenbrot (1988).

Salpingotus michaelis Fitzgibbon, 1966. Mammalia, 30(3):431.

TYPE LOCALITY: Pakistan, NW Baluchistan, Nushki Plateau, appox. 29°N, 66°E, altitude 3,500 ft.

DISTRIBUTION: Pakistan, NW Baluchistan (Roberts, 1977).

COMMENTS: Subgenus *Salpingotulus*. Pavlinov (1980*b*) placed *S. michaelis* in a separate genus, *Salpingotulus*, based on genital morphology, tooth characters, and shape of the condylar process; this arrangement was followed by Vorontsov and Shenbrot (1984). The results of a morphometric and qualitative study of *Salpingotus* by Vorontsov and Shenbrot (1984) showed substructure within the genus based on overall similarity, with *S. michaelis* joining the clusters of other *Salpingotus* at a high level of dissimilarity, followed by *S. kozlovi*. Their results indicated that generic separation of *michaelis* was not supported, particularly if *kozlovi* was retained in *Salpingotus*. In addition, no phylogenetic study has been done. Pavlinov and Rossolimo (1987:162) suggested that *Salpingotus* included *Salpingotulus*, a hypothesis followed here.

For discussion of *S. michaelis* versus *S. thomasi*, and the possible occurance of *S. michaelis* in Afghanistan, see Hassinger (1973) and Roberts (1977). See also comment under *S. thomasi*.

Salpingotus pallidus Vorontsov and Shenbrot, 1984. Zool. Zh., 63(5):740.

TYPE LOCALITY: Kazakhstan, Aktyubinskaya, Chelkarskii, Peski Bol'shiye Barsuki.

DISTRIBUTION: Deserts of N Aral and S Balkhash regions.

SYNONYMS: *sludskii*.

COMMENTS: Subgenus *Prosalpingotus*. Shenbrot and Mazin (1989) named *sludskii* as a subspecies of *S. pallidus*.

Salpingotus thomasi Vinogradov, 1928. Ann. Mag. Nat. Hist., ser. 10, 1:373.

TYPE LOCALITY: Afghanistan or S Tibet (Ognev, 1963*b*; Vinogradov, 1928).

DISTRIBUTION: Known only from the type specimen, of which the country of origin, "Afghanistan", is questionable (see discussion in Hassinger, 1973, and Roberts, 1977).

COMMENTS: Subgenus *Prosalpingotus* (see Pavlinov and Shenbrot, 1985). Reviewed by Ognev (1963*b*).

Subfamily Dipodinae G. Fischer, 1817. Mem. Soc. Imp. Nat. Moscow, 5:372.

SYNONYMS: Dipsidae, Dipina, Dipodes, Dipodum, Jaculinae.

COMMENTS: *Paradipus*, traditionally included in Dipodinae, is placed in its own subfamily (see comment under Paradipodinae).

Dipus Zimmermann, 1780. Geogr. Gesch. Mensch. Vierf. Thiere, 2:354.

TYPE SPECIES: *Mus sagitta* Pallas, 1773.

SYNONYMS: *Dipodipus* (see Ellerman and Morrison-Scott, 1951).

Dipus sagitta (Pallas, 1773). Reise Prov. Russ. Reichs., 2:706.

TYPE LOCALITY: N Kazakhstan, Pavlodarskaya Oblast, right bank of Irtysh River near Yamyshevskaya at Podpusknoi (see Ellerman and Morrison-Scott, 1951:535).

DISTRIBUTION: Desert, steppe and dry woodland from Don River (Russia), NW coast of Caspian Sea, and N Iran, through Turkmenistan, Uzbekistan, and Kazakhstan to S Tuva, Russia; Mongolia (Bannikov, 1954); China: Nei Mongol, Xinjiang, Qinghai, Gansu, Ningxia, N Shaanxi, N Shanxi, Liaoning, and Jilin (see Chen and Wang, 1985; Liu et al., 1990; Ma et al., 1987; Mi et al., 1990; Qian et al., 1965; Qin, 1991; Shou, 1962; Wang, 1990; Zhang and Wang, 1963; Zheng and Zhang, 1990; and Zhou et al., 1985).

SYNONYMS: *aksuensis, austrouralensis, bulganensis, deasyi, fuscocanus, halli, innae, kalmikensis, lagopus, megacranius, nogai, sowerbyi, turanicus, ubsanensis, usuni, zaissanensis* (see Corbet, 1984; Ellerman and Morrison-Scott, 1951; Pavlinov and Rossolimo, 1987; Shenbrot, 1991*a*, *c*; and Wang, 1964).

COMMENTS: Karyotype provided by Vorontsov et al. (1969*d*). Reviewed by Ognev (1963*b*). Taxonomic study, analysis of geographic variation, and distribution throughout most of range provided by Shenbrot (1991*a*, *c*). Detailed habitat data provided by Naumov and Lobachev (1975).

Eremodipus Vinogradov, 1930. Izv. Akad. Nauk S.S.S.R., Leningrad, Otdel. Phyz.-Math, p. 334.

TYPE SPECIES: *Scirtopoda lichtensteini* Vinogradov, 1927.

COMMENTS: Following Vinogradov (1930), Heptner (1975), Gromov and Baranova (1981), Pavlinov and Rossolimo (1987) and Shenbrot (1990*b*), *Eremodipus* is recognized as a genus distinct from *Jaculus*.

Eremodipus lichtensteini (Vinogradov, 1927). Z. Säugetierk., 2(1):92.

TYPE LOCALITY: Turkmenistan, vicinity of Merv.

DISTRIBUTION: Kazakhstan, Turkmenistan, and Uzbekistan, from Caspian Sea to Aral Sea, and south of Lake Balkhash.

SYNONYMS: *balkashensis, jaxartensis*.

COMMENTS: Reviewed by Ognev (1963*b*). Taxonomic study, distribution, and review contributed by Shenbrot (1990*b*), who described *balkashensis* and *jaxartensis* as subspecies. Detailed habitat data provided by Naumov and Lobachev (1975). Karyotype given by Vorontsov et al. (1969*d*).

Jaculus Erxleben, 1777. Syst. Regni Anim., 1:404.

TYPE SPECIES: *Mus jaculus* Linnaeus, 1758, as fixed by Opinion 730 of the International Commission on Zoological Nomenclature (1965); not *Jaculus orientalis* Erxleben, 1777.

SYNONYMS: *Scirtopoda* (incl. *Haltomys*) (see Ellerman and Morrison-Scott, 1951; and Corbet, 1978*c*).

COMMENTS: Myology, in context of adaptive and phylogenetic significance, studied by Klingener (1964).

Jaculus blanfordi (Murray, 1884). Ann. Mag. Nat. Hist., ser. 5, 14:98.

TYPE LOCALITY: Iran, Bushire.

DISTRIBUTION: E and S Iran (Lay, 1967), S and W Afghanistan (Hassinger, 1973), and W Pakistan (Roberts, 1977).

COMMENTS: See comment under *J. turcmenicus*.

Jaculus jaculus (Linnaeus, 1758). Syst. Nat., 10th ed., 1:63.

TYPE LOCALITY: Egypt, Giza Pyramids.

DISTRIBUTION: Africa, NE Nigeria (Happold, 1987) and Niger, from SW Mauritania to Morocco, E through Algeria (Kowalski and Rzebik-Kowalska, 1991), Tunisia (Vesmanis, 1984) and Libya (Ranck, 1968) to Egypt (Osborn and Helmy, 1980), Sudan, and Somalia; throughout Arabia (Harrsion and Bates, 1991) to SW Iran (Lay, 1967).

SYNONYMS: *aegyptius, airensis, arenaceous, butleri, centralis, collinsi, cufrensis, darricarrerei, deserti, elbaensis, favillus, favonicus, florentiae, fuscipes, gordoni, hirtipes?, loftusi, macromystax, macrotarsus?, oralis, rarus, schlueteri, sefrius, syrius, tripolitanicus, vastus, vocator, vulturnus, whitchurchi* (see Ellerman and Morrison-Scott, 1951; and Corbet, 1978*c*).

COMMENTS: Ranck (1968) recognized two species in this complex, *J. jaculus* and *J. deserti*, but Harrison (1978) showed that they are conspecific based on Ranck's criteria (see also discussion in Corbet, 1978*c*:152). Karyotype given by Al Saleh and Khan (1984). See comment under *J. turcmenicus*.

Jaculus orientalis Erxleben, 1777. Syst. Regn. Anim., 1:404.
TYPE LOCALITY: Egypt, in the "mountains separating Egypt from Arabia" (Allen, 1939:424).
DISTRIBUTION: Deserts of N Africa and Arabia (Harrison and Bates, 1991), from Morocco E through Algeria (Kowalski and Rzebik-Kowalska, 1991), Tunisia (Vesmanis, 1984), and Libya (Ranck, 1968) to Egypt (Osborn and Helmy, 1980), Sinai and S Israel.
SYNONYMS: *bipes, gerboa, locusta, mauritanicus* (see Ellerman and Morrison-Scott, 1951).
COMMENTS: *J. orientalis* has been identified from the late Pliocene in Ethiopia (Wesselman, 1984) and Plio-Pleistocene in Kenya (Black and Krishtalka, 1986). See comment under *J. turcmenicus*.

Jaculus turcmenicus Vinogradov and Bondar, 1949. Doklady Akad. Nauk S.S.S.R., Leningrad, 65:559.
TYPE LOCALITY: Turkmenskaya, Nebitdagskom Region, Tchagil sands, near S coast of Kara-Bogaz-Gol.
DISTRIBUTION: SE coast of Caspian Sea through Turkmenistan to the Kyzylkum Desert, C Uzbekistan; see Kuznetsov (1965).
SYNONYMS: *margianus* (see Shenbrot, 1990*a*).
COMMENTS: Heptner (1975) concluded that *J. turcmenicus* was conspecific with *J. blanfordi*, but the measurements given for the two species differ considerably, no tests of significance were performed, and only two specimens of *J. blanfordi* were included in the study. The two species are retained here, as in Shenbrot (1990*a*), pending critical revision. The presence of spines on the bacula of *J. blanfordi* and *J. orientalis* (figured by Didier and Petter, 1960), is shared by *J. turcmenicus* (figured in Heptner, 1975), suggesting that these three species may form a monophyletic group that excludes *J. jaculus*; see also comment in Corbet (1978*c*) under *J. blanfordi*. Taxonomic study, distribution, and review provided by Shenbrot (1990*a*). Detailed habitat data provided by Naumov and Lobachev (1975). Morphology and habitat discussed in Stalmakova (1957). Karyotype provided by Vorontsov et al. (1969*d*).

Stylodipus Allen, 1925. Am. Mus. Novit., 161:4.
TYPE SPECIES: *Stylodipus andrewsi* Allen, 1925.
SYNONYMS: *Halticus* (see Ellerman and Morrison-Scott, 1951).
COMMENTS: *Scirtopoda* is often incorrectly used for this genus, but is a junior synonym of *Jaculus* (see Ellerman and Morrison-Scott, 1951:536; and Corbet, 1978*c*:153).

Stylodipus andrewsi Allen, 1925. Am. Mus. Novit., 161:4.
TYPE LOCALITY: S Mongolia, near Mt. Uskuk (Ussuk), Camp Ondai Sair (Andrews, 1932:101).
DISTRIBUTION: NW, S, and C Mongolia east of Barun Khurai (Baruun Huuray) Valley; and adjacent China: Nei Mongol, Xinjiang, Gansu, and Ningxia (see Ma et al., 1987; Qin, 1991; and Zheng and Zhang, 1990).
COMMENTS: Ellerman and Morrison-Scott (1951) and Corbet (1978*c*) included *andrewsi* as a subspecies of *S. telum*, but Sokolov and Orlov (1980) recognized it as a distinct species. Sokolov and Shenbrot (1987*b*) showed that in addition to the retention of a rudimentary P4, *S. andrewsi* is differentiated from *S. sungorus* and *S. telum* in dental and phallic characters, and in greater bullar inflation. The ranges of *S. andrewsi* and *S. sungorus* are adjacent but do not overlap (Sokolov and Shenbrot, 1987*b*). See comments under *S. sungorus* and *S. telum*.

Stylodipus sungorus Sokolov and Shenbrot, 1987. Zool. Zh., 66(4):580.
TYPE LOCALITY: SW Mongolia, Altai Gobi, north slope Takhin-Shara-Nuru (Tahiyn-Shar-Nuruu) range, 15 km E Tsargin.
DISTRIBUTION: SW Mongolia, possibly Xinjiang, China (see Sokolov and Shenbrot, 1987*b*:585).
COMMENTS: On the basis of the length and breadth of the molar row, size of auditory bullae, and phallic characters (Sokolov and Shenbrot, 1987*b*), *S. sungorus* appears to be distinct from *S. telum*. Relationship and distribution relative to *S. andrewsi* and *S. telum* is unclear and needs further study (see Sokolov and Shenbrot, 1987*b*). See comment under *S. andrewsi*.

Stylodipus telum (Lichtenstein, 1823). Naturhist. Anhang (or Eversmann's Reise Orenburg), p. 120.
TYPE LOCALITY: Kazakhstan, steppe along NE shore of Aral Sea (Ognev, 1963*b*:303).

DISTRIBUTION: E Ukraine, N Caucasus, N and W Turkmenistan, W Uzbekistan and Kazakhstan (Kuznetsov, 1965; Shenbrot, 1991*b*); E to N Xinjiang, N Gansu, and Nei Mongol, China (see Chen and Wang, 1985; Ma et al., 1987; Mi et al., 1990; Qian et al., 1965; Shou, 1962; and Zheng and Zhang, 1990; but see comment below).
SYNONYMS: *amankaragai, birulae, falzfeini, halticus, karelini, nastjukovi, turovi, proximus* (see Ellerman and Morrison-Scott, 1951; and Shenbrot, 1991*b*).
COMMENTS: Reviewed by Ognev (1963*b*). Subspecific revision contributed by Shenbrot (1991*b*). Detailed habitat data provided by Naumov and Lobachev (1975). Karyotype provided by Vorontsov et al. (1969*d*). Because *S. andrewsi* is considered a synonym of *S. telum* by some workers, some of the records of *S. telum* from China may represent *S. andrewsi*. See comments under *S. andrewsi* and *S. sungorus*.

Subfamily Euchoreutinae Lyon, 1901. Proc. U. S. Natl. Mus., 23 (1228):666.
COMMENTS: Based on crown structure of the molars, and characters of the mastoid region, Shenbrot (1992) proposed that Euchoreutinae is most closely related to Sicistinae. The results of Stein's (1990) study of limb musculature placed Euchoreutinae as a sister group of Zapodinae and Allactaginae. See comment under Dipodidae.

Euchoreutes Sclater, 1891. Proc. Zool. Soc. Lond., 1890:610 [1891].
TYPE SPECIES: *Euchoreutes naso* Sclater, 1891.
COMMENTS: Detailed review provided by Ognev (1963*b*).

Euchoreutes naso Sclater, 1891. Proc. Zool. Soc. Lond., 1890:610 [1891].
TYPE LOCALITY: NW China; W Xinjiang, W of Taklimakan Shamo (Takla-Makan Desert), near Shache (Yarkand).
DISTRIBUTION: S Mongolia; China: Nei Mongol, Xinjiang, Qinghai, Gansu, and Ningxia (see Chen and Wang, 1985; Liu et al., 1990; Ma et al., 1987; Qian et al., 1965; Shou, 1962; Zhang and Wang, 1963; and Zheng and Zhang, 1990).
SYNONYMS: *alashanicus, yiwuensis* (see Corbet, 1978*c*, 1984).

Subfamily Paradipodinae Pavlinov and Shenbrot, 1983. Trudy Zoolog. Inst. Akad. Nauk SSSR, Lenin, 119:87.
COMMENTS: Shenbrot (1992) showed that *Paradipus* is highly differentiated from Dipodinae (in which it has traditionally been placed), and that Paradipodinae appears to be most closely related to Cardiocraniinae, based on molar and mastoid characters. *Paradipus* was not included in Stein's (1990) study of limb myology.

Paradipus Vinogradov, 1930. Izv. Acad. Sci. U.S.S.R., p. 333.
TYPE SPECIES: *Scirtopoda ctenodactyla* Vinogradov, 1929.

Paradipus ctenodactylus (Vinogradov, 1929). Doklady Akad. Nauk S.S.S.R., Leningrad, 1929:248.
TYPE LOCALITY: E Turkmenistan, near Repetek.
DISTRIBUTION: Sand deserts of Turkmenistan, Uzbekistan, and E Aral region of Kazakhstan; see Kuznetsov (1965) and Sludskii (1977).
COMMENTS: Reviewed by Ognev (1963*b*). Detailed habitat data provided by Naumov and Lobachev (1975). Karyotype provided by Vorontsov et al. (1969*d*). Os penis described and figured by Shenbrot (1992).

Subfamily Sicistinae Allen, 1901. Proc. Biol. Soc. Washington, 14:185.
SYNONYMS: Sminthi, Sminthinae.
COMMENTS: Pavlinov and Rossolimo (1987), Shenbrot (1992), and others use Sminthidae for this group, because Brandt's (1855) Sminthi predates Sicistinae Allen, 1901*a*. However, article 40b of the International Code of Zoological Nomenclature (1985:81) states that if a family-group name has been replaced before 1961 because of synonymy of the type genus, and the replacement name has won general acceptance, it is to be maintained. Allen (1901*a*) replaced Sminthinae with Sicistinae because he synonymized *Sminthus* Nordmann, 1840 under *Sicista* Gray, 1827. As Sicistinae is also most widely adopted, it is the valid family-group name. The separation of Sicistinae

from Zapodinae, suggested by Ellerman (1940), is supported by the results of Shenbrot (1986, 1992), Sokolov et al. (1987*b*), Stein (1990), and Vorontsov (1969).

Though the modern distribution of *Sicista* is Palaearctic, Sicistinae has been recorded from Pleistocene deposits in North America (Martin, 1989*a*).

Sicista Gray, 1827. *In* Griffith et al., Anim. Kingdom, 5:228.

TYPE SPECIES: *Mus subtilis* Pallas, 1773.

SYNONYMS: *Clonomys* Thilesius, 1850, *Sminthus* (see Ellerman and Morrison-Scott, 1951).

COMMENTS: Reviewed by Ognev (1963*b*). Karyological research and systematic problems in the genus reviewed by Sokolov et al. (1987*b*). Several species in the *S. concolor* complex (*S. armenica, S. caucasica, S. caudata, S. kazbegica, S. kluchorica, S. tianshanica*) have been distinguished by Sokolov and colleagues primarily by karyotypic and spermatozoal differences. These species are provisionally recognized here, but need further documentation and corroborative data sets to firmly establish their specific status.

Myology, in context of adaptive and phylogenetic significance, studied by Klingener (1964). Comparative myology of pelvic girdle studied by Fokin (1971).

Sicista armenica Sokolov and Baskevich, 1988. Zool. Zh., 67(2):301.

TYPE LOCALITY: NW Armyanskaya, Malyy Kavkaz, Pazdanskiy Region, Pambakskiy Range, near Ankavan, head of Marmarik River, subalpine zone, 2200 meters.

DISTRIBUTION: Known only from the type locality.

COMMENTS: Sokolov and Baskevich (1988) presented karyological and spermatozoal characters that distinguished *S. armenica* from *S. caucasica, S. kluchorica,* and *S. kazbegica*. See also comment under *Sicista*.

Sicista betulina (Pallas, 1779). Nova Spec. Quad. Glir. Ord., p. 332.

TYPE LOCALITY: SW Siberia, birch plain on bank of Ishim River, and Barabinskaya Step.

DISTRIBUTION: Boreal and montane forests from Norway, Denmark; east to Ozero Baykal, and possibly the Ussuri region of China and SE Siberia, north to the Arctic Circle at the White Sea and Usa River, south to Austria, Carpathian and Sayan Mtns (Corbet, 1978*c*). In the former USSR see Kuznetsov (1965) and Sludskii (1977). Sokolov et al. (1989) considered the Ussuri region records questionable.

SYNONYMS: *montana, norvegica, taigica, tatricus* (see Ellerman and Morrison-Scott, 1951; and Corbet, 1978*c*).

COMMENTS: Sokolov et al. (1982, 1987*b*) gave karyological, and spermatozoal characters that distinguished this species from *S. napaea* and *S. pseudonapaea*. Pallas' type specimen was probably not preserved (Ognev, 1963*b*:33). Review and distribution in Europe provided by Pucek (1982). See comment under *S. strandi*.

Sicista caucasica Vinogradov, 1925. Proc. Zool. Soc. Lond., 1925:584.

TYPE LOCALITY: Russia, N Caucasus, Krasnodarskiy Kray (Kuban Prov.), Maykop (Maikop) District, 7000-9000 feet.

DISTRIBUTION: NW Caucasus; distribution figured by Sokolov et al. (1987*a*).

COMMENTS: Sokolov et al. (1981*b*, 1987*b*) gave karyological and phallic characters that distinguished this species from *S. kluchorica* and *S. concolor, S. armenica* (Sokolov and Baskevich, 1988), and *S. kazbegica* (Sokolov et al., 1986*b*). See also comment under *Sicista*, and in Corbet (1984:25).

Sicista caudata Thomas, 1907. Proc. Zool. Soc. Lond., 1907:413.

TYPE LOCALITY: Russia, Sakhalin Oblast, Sakhalin Isl, 17 miles NW Korsakov.

DISTRIBUTION: Ussuri region of NE China and Primorski Kray, Sikhote-Alin range, and Sakhalin Isl, Russia.

COMMENTS: Sokolov et al. (1982, 1987*b*) and Sokolov and Kovalskaya (1990) gave karyological and spermatozoal characters that distinguished this species from *S. tianshanica*, and from *S. concolor* (Sokolov et al., 1980). See also comment under *Sicista* and *S. concolor*.

Sicista concolor (Büchner, 1892). Bull. Sci. Acad. Imp. Sci. St. Petersbourg, 35(3):107.

TYPE LOCALITY: China, Gansu, N slope of the mountains of Xining, Guiduisha.

DISTRIBUTION: China: Xinjiang, Qinghai, Gansu, Shaanxi, and W Sichuan (Ma et al., 1987; Wang, 1990; and Zheng and Zhang, 1990); India; Kashmir; and N Pakistan.

SYNONYMS: *flavus, leathemi, weigoldi* (see Ellerman and Morrison-Scott, 1951).
COMMENTS: *S. concolor* has been reported from the Chinese provinces of Heilongjinag and Jilin (Yang et al., 1991); the relationship of these populations to *S. caudata* needs further study. Does not include *S. armenica, S. caucasica, S. caudata, S. kazbegica, S. kluchorica*, or *S. tianshanica*; see comments under respective species and under *Sicista*. See also comment in Corbet (1978*c*:149, 1984:25).

Sicista kazbegica Sokolov, Baskevich, and Kovalskaya, 1986. Zool. Zh., 65(6):949.
TYPE LOCALITY: Georgia, Kazbegi District, 14 km NW Kobi, Suatisi Gap, upper reaches Terek River, subalpine zone, 2200 meters.
DISTRIBUTION: Georgia, Kazbegi District (Sokolov et al., 1986*b*, 1987*a*).
COMMENTS: Sokolov et al. (1986*b*) gave karyological and spermatozoal characters that distinguished this species from *S. caucasica* and *S. kluchorica*, and *S. armenica* (Sokolov and Baskevich, 1988). See also comment under *Sicista*.

Sicista kluchorica Sokolov, Kovalskaya, and Baskevich, 1980. Gryzuny Severnovo Kavkaza., p.38.
TYPE LOCALITY: Russia, N Caucasus, Karachayevo-Cherkess Autonomous Region, upper North Klukhor River at Klukhor Pass, 2100 meters.
DISTRIBUTION: NW Caucasus; see Sokolov et al. (1987*a*).
COMMENTS: Sokolov et al. (1981*b*, 1987*b*) gave karyological and phallic characters that distinguished this species from *S. caucasica* and *S. concolor, S. armenica* (Sokolov and Baskevich, 1988), and *S. kazbegica* (Sokolov et al., 1986*b*). See also comment under *Sicista* and in Corbet (1984:25).

Sicista napaea Hollister, 1912. Smithson. Misc. Coll., 60(14):2.
TYPE LOCALITY: Russia, Altai Krai, Altai Mtns, Seminsk Ridge, Tapuchii (Tapucha).
DISTRIBUTION: E Kazakhskaya and Russia, Altai Krai, NW Altai Mtns; see Kuznetsov (1965) and Sludskii (1977).
COMMENTS: Sokolov et al. (1982, 1987*b*) gave karyological and spermatozoal characters that distinguished this species from *S. pseudonapaea* and *S. betulina*.

Sicista pseudonapaea Strautman, 1949. Vestn. Akad. Nauk Kazakh. SSR, 5:109.
TYPE LOCALITY: E Kazakhstan, Altai Mtns, N slope of Narymskiy Range, Katon-Karagay.
DISTRIBUTION: E Kazakhstan, Taiga of S Altai Mtns; see Kuznetsov (1965) and Sludskii (1977).
COMMENTS: Sokolov et al. (1982, 1987*b*) gave karyological, spermatozoal, and phallic characters that distinguished this species from *S. napaea* and *S. betulina*. *S. pseudonapaea* is provisionally recognized here, but requires further documentation, and the inclusion of other data sets, to establish its specific status.

Sicista severtzovi Ognev, 1935. Byullet. Nauchno-issled. Inst. Zool., Mosk., 2:54.
TYPE LOCALITY: Russia, Voronezh Oblast, Bobrov District, Kamennaya Steppe Experimental Station.
DISTRIBUTION: S Russia.
COMMENTS: Sokolov et al. (1986*a*) separated this species from *S. subtilis* by its distinctive karyotype. It is provisionally recognized here, but further documentation, and the inclusion of other data sets, are needed to establish its specific status.

Sicista strandi (Formozov, 1931). Folia Zool. Hydrob. Riga, 3:79.
TYPE LOCALITY: Russia, Caucasus, Stavropol Krai, Karachayevo-Cherkess Region, Karachayevsk District, Uchkulan (Utschkulak), Igera, 2100 meters; shown in Sokolov et al. (1989).
DISTRIBUTION: N Caucasus, north to Kursk District of S Russia; see Sokolov et al. (1989).
COMMENTS: Sokolov et al. (1989) distinguished this species from *S. betulina* primarily by its different karyotype. Pavlinov and Rossolimo (1987) tentatively listed *S. strandi* as a separate species, a hypothesis provisionally followed here. Further documentation and incorporation of other character sets is essential in order to assess whether or not *strandi* should be included in *S. betulina*.

Sicista subtilis (Pallas, 1773). Reise Prov. Russ. Reichs., 1(2):705.
TYPE LOCALITY: Russia, Kurgan Oblast, on Tobol River near Kaminskaya Kur'ya (suburb), on road from Zverinogolovskoye to Kurgan (Ognev, 1963*b*:27).

DISTRIBUTION: Steppes from E Austria, Hungary, Yugoslavia, and Rumania through S Russia, N Kazakhstan, and SW Sibiria to the Altai Range, Lake Balkhash, Lake Baikal, and NW Xinjiang, China (see Li and Wang, 1981; and Ma et al., 1987). In former USSR see Kuznetsov (1965) and Sludskii (1977).

SYNONYMS: *interstriatus, interzonus, lineatus, loriger, nordmanni, pallida, siberica, tripartitus, tristriatus, trizona, vaga, virgulosus* (see Ellerman and Morrison-Scott, 1951).

COMMENTS: Type specimen was probably not preserved (Ognev, 1963*b*:27). Karyology studied by Sokolov et al. (1986*a*). Review and distribution in Europe provided by Pucek (1982). See comment under *S. severtzovi*.

Sicista tianshanica (Salensky, 1903). Ezheg. Zool. Muz. Akad. Nauk, 8:17.

TYPE LOCALITY: China, Xinjiang, S slope Tien Shan Mtns, between Kapchagay (Chapzagai-gol) and Tsaima (Zanma) Rivers.

DISTRIBUTION: Tien Shan Mtns of Kazakhstan (see Sludskii, 1977); Tien Shan Mtns and E Tarbagatay Mtns of Xinjiang, China (see Ma et al., 1987).

COMMENTS: Sokolov et al. (1982, 1987*b*) and Sokolov and Kovalskaya (1990) gave karyological and spermatozoal characters that distinguished this species from *S. caudata*, and from *S. concolor* (Sokolov et al., 1980). See also comment under *Sicista*.

Subfamily Zapodinae Coues, 1875. Bull. U.S. Geol. Geog. Surv. Terr., ser. 2, 5(3):253.

COMMENTS: The recognition of three distinct genera of zapodines has been challenged (Corbet, 1978*c*). Higher level relationships among zapodines have been addressed by Preble (1899), who described *Eozapus* and *Napaeozapus* as new subgenera of *Zapus*, and by Krutzsch (1954), who supported generic separation based on differences in tooth number and occlusal pattern, bacula, and ear ossicles. Klingener (1964:75) found no consistant differences in the myology of *Zapus* versus *Napaeozapus* (*Eozapus* was not included in his study), but favored generic separation of the two based on dental morphology. The dental differences documented in Preble (1899) and Krutzsch (1954) are substantial and phylogenetically significant; following Ellerman (1940) and Krutzsch (1954), three genera are retained here.

A critical study documenting other morphological characters, and/or molecular differentiation among zapodines, as well as the level of differentiation of zapodines relative to other dipodids, is needed.

Eozapus Preble, 1899. N. Am. Fauna, 15:37.

TYPE SPECIES: *Zapus setchuanus* Pousargues, 1896.

Eozapus setchuanus (Pousargues, 1896). Bull. Mus. Hist. Nat. Paris, 2:13.

TYPE LOCALITY: China, Sichuan Prov., Tatsienlu (Kangding).

DISTRIBUTION: China: Qinghai, Gansu, Ningxia, Shaanxi, Sichuan, and NW Yunnan (Qin, 1991; Wang, 1990; Zhang and Wang, 1963; Zheng and Zhang, 1990).

SYNONYMS: *vicinus* (see Corbet, 1978*c*).

Napaeozapus Preble, 1899. N. Am. Fauna, 15:33.

TYPE SPECIES: *Zapus insignis* Miller, 1891.

COMMENTS: For verification of the absence of cheek pouches in *Napaeozapus* see Klingener (1971).

Napaeozapus insignis (Miller, 1891). Am. Nat., 25:742.

TYPE LOCALITY: Canada, New Brunswick, Restigouche River.

DISTRIBUTION: Canada: SE Manitoba, SW and E Ontario, S and E Quebec north to S Labrador. USA: E Minnesota, N and C Wisconsin, upper peninsular and N lower peninsular Michigan, E Ohio, Pennsylvania; north and east to NW New Jersey, New York, Connecticut, Rhode Island, W Massachusetts (isolated population in Martha's Vineyard), Vermont, New Hampshire, and Maine; south to West Virginia, W Virginia, E Kentucky, E Tennessee, W North Carolina, NW South Carolina, and NE Georgia.

SYNONYMS: *abietorum, algonquinensis, frutectanus, gaspensis, roanensis, saguenayensis* (see Hall, 1981).

COMMENTS: A specimen of *Napaeozapus* was collected in Park County, Indiana (Lyon, 1942), though subsequent trapping failed to yield further examples (Mumford, 1969). The identity of the specimen was verified by Klingener (1965:645) and Wrigley (1972:42).

Systematic revision and biology provided by Wrigley (1972). Myology, in context of adaptive and phylogenetic significance, studied by Klingener (1964). Diagnosis, range map, and records provided by Hall (1981). Reviewed by Whitaker and Wrigley (1972, Mammalian Species, 14).

Zapus Coues, 1875. Bull. U.S. Geol. Geog. Surv. Terr., ser. 2, 5(3):253.

TYPE SPECIES: *Dipus hudsonius* Zimmermann, 1780.

COMMENTS: Revised by Preble (1899) and Krutzsch (1954). Myology, in context of adaptive and phylogenetic significance, studied by Klingener (1964). Dental evolution investigated by Klingener (1963). For verification of the absence of cheek pouches in *Zapus* see Klingener (1971). Phallic morphology described by Shenbrot (1992).

Zapus hudsonius (Zimmermann, 1780). Geogr. Gesch. Mensch. Vierf. Thiere, 2:358.

TYPE LOCALITY: Canada, Ontario, Hudson Bay, Fort Severn (see Anderson, 1942*b*).

DISTRIBUTION: USA and Canada: S Alaska to S Coast Hudson Bay to Labrador, south to E North Carolina and NW South Carolina, southwest to NW Alabama, north to NE Mississippi and Tennessee, west to NE Oklahoma, northwest to SE Montana, northeast to SE Saskatchewan, northwest to C and S British Columbia. Isolated populations in S Wyoming, NC Colorado, N and C New Mexico, and EC Arizona.

STATUS: IUCN - Endangered as *Z. h. luteus*.

SYNONYMS: *acadicus, alascensis, americanus, australis, brevipes, campestris, canadensis, hardyi, labradorius, intermedius, ladas, luteus, microcephalus, ontarioensis, pallidus, preblei, rafinesquei, tenellus* (see Hall, 1981).

COMMENTS: The S Rocky Mtn subspecies *luteus*, formerly assigned to *princeps*, was shown to represent *hudsonius* by Hafner et al. (1981). Diagnosis, records and range map (excluding Mississippi), provided by Hall (1981). Mississippi record given by Kennedy et al. (1982). Reviewed by Whitaker (1972, Mammalian Species, 11).

Zapus princeps Allen, 1893. Bull. Am. Mus. Nat. Hist., 5:71.

TYPE LOCALITY: USA, Colorado, La Plata Co., Florida.

DISTRIBUTION: Canada and USA: S Yukon southeast to NE South Dakota, west to C Montana, southeast to SE Wyoming, S to NC New Mexico; northwest to N and C Utah (isolated population in E Utah), N and C Nevada, EC California north to SW, C and E Oregon, SE Washington northwest to S Yukon.

SYNONYMS: *alleni, chrysogenys, cinereus, curtatus, idahoensis, kootenayensis, major, minor, nevadensis, oregonus, pacificus, palatinus, saltator, utahensis* (see Hall, 1981).

COMMENTS: Formerly included *luteus*; see comment under *hudsonius*. Diagnosis, records, and range map provided by Hall (1981).

Zapus trinotatus Rhoads, 1895. Proc. Acad. Nat. Sci. Philadelphia, 1894, 47:421 [1895].

TYPE LOCALITY: Canada, British Columbia, mouth of the Frazer River, Lulu Isl.

DISTRIBUTION: Canada and USA: SW British Columbia, W Washington, coastal and WC Oregon, along the N California coast south to the Marin Peninsula.

STATUS: IUCN - Indeterminate as *Z. t. orarius*.

SYNONYMS: *eureka, montanus, orarius* (see Hall, 1981).

COMMENTS: Diagnosis, records, and range map provided by Hall (1981). Reviewed by Gannon (1988, Mammalian Species, 315).

FAMILY MURIDAE

by Guy G. Musser and Michael D. Carleton

Family Muridae Illiger, 1815. Abhandl. K. Akad. Wiss., Berlin for 1804-11, p. 46, 129.

COMMENTS: The scope of the Family Muridae employed herein is practically equivalent to the Superfamily Muroidea as classically arranged by Miller and Gidley (1918), Simpson (1945), and others (see Carleton, 1980; Carleton and Musser, 1984; and Hooper and Musser, 1964*a*; for reviews of muroid classification). The chapter's organization generally follows that of Carleton and Musser (1984), which serves as a convenient framework to discuss the generic-and specific-level diversity of muroids and their attendant taxonomic problems and uncertainties. It is not a higher-order classification of the family or superfamily. Family-group synonyms are listed under the various subfamilies.

We have sought to present the 1326 species and 281 genera recognized within the dialogue of the original descriptive and primary revisionary literature, especially where differences of opinion exist over a taxon's rank and validity, rather than secondary and tertiary sources. In many instances, we have gone back to examine collections and type material in order to gain a better grasp of a taxonomic issue or to illuminate distributions; the holdings of the American Museum of Natural History, British Museum of Natural History, Field Museum of Natural History, Museum of Comparative Zoology, Museum National d'Histoire Naturelle, National Museum of Natural History, and Senckenberg Museum were especially important in this regard. In reading the various generic and specific accounts, a clear message is repeated: much alpha taxonomic uncertainty and confusion persist within the Muridae, a situation which warrants very basic, museum-based revisionary attention.

Contents (listed alphabetically):
Arvicolinae Gray, 1821 (26 genera, 143 species)
Calomyscinae Vorontsov and Potapova, 1979 (1 genus, 6 species)
Cricetinae G. Fischer, 1817 (7 genera, 18 species)
Cricetomyinae Roberts, 1951 (3 genera, 6 species)
Dendromurinae Alston, 1876 (8 genera, 23 species)
Gerbillinae Gray, 1825 (14 genera, 110 species)
Lophiomyinae Milne Edwards, 1867 (1 genus, 1 species)
Murinae Illiger, 1815 (122 genera, 529 species)
Myospalacinae Lilljeborg, 1866 (1 genus, 7 species)
Mystromyinae Vorontsov, 1966 (1 genus, 1 species)
Nesomyinae Major, 1897 (7 genera, 14 species)
Otomyinae Thomas, 1897 (2 genera, 14 species)
Petromyscinae Roberts, 1951 (2 genera, 5 species)
Platacanthomyinae Alston, 1876 (2 genera, 3 species)
Rhizomyinae Winge, 1887 (3 genera, 15 species)
Sigmodontinae Wagner, 1843 (79 genera, 423 species)
Spalacinae Gray, 1821 (2 genera, 8 species)

Subfamily Arvicolinae Gray, 1821. London Med. Repos., 15:303.

SYNONYMS: Alticoli, Bramini, Clethrionomyini, Dicrostonychinae, Ellobiini, Fibrini, Lagurini, Lemminae, Microtinae, Microtoscoptini, Myodini, Neofibrini, Ondatrini, Phenacomyini, Pitymyini, Pliomyini, Pliophenacomyini, Prometheomyinae, Synaptomyini.

COMMENTS: See Kretzoi (1962, 1969) for use of Arvicolinae Gray, 1821, instead of Microtinae Miller, 1896; a group concept of arvicoline rodents actually had emerged long prior to Miller's (1896) seminal monograph (e.g., Murray, 1866; Alston, 1876). Carleton and Musser (1984) provided a general diagnosis and review of the limits and contents of the subfamily. Synthetic regional taxonomic treatments include: Ellerman and Morrison-Scott (1951), Corbet (1978*c*, 1984), and Agadzhanyan and Yatsenko (1984) for Palaearctic species; Ognev (1963*b*, 1964), Gromov and Polyakov (1977), and Pavlinov and Rossolimo (1987) for Asian forms; Niethammer and Krapp (1982*a*) for European species; and Hall and Cockrum (1953) and Hall (1981) for North

American voles and lemmings. Biochronology of arvicolines in the northern hemisphere was comprehensively reviewed by Repenning et al. (1990) and Repenning (1990, and references therein).

Broad, multispecies surveys have been undertaken on morphological and biochemical systems of arvicolines that bear on issues of their phenetic divergence and phylogenetic relationships. For example, comparative and functional studies of the dentition (Hinton, 1926; Koenigswald, 1980, 1982; Miller, 1896); of the cranium (Gromov, 1990; Kratochvil, 1982*a*; Pietsch, 1980); of middle ear anatomy (Hooper, 1968*a*; Pavlinov, 1984); of cutaneous and subcutaneous glands (Quay, 1954*a*, 1968; Sokolov and Dzhemukhadze, 1991); of myology (Kesner, 1980, 1986; Repenning, 1968; Stein, 1986, 1987); of the digestive tract (Carleton, 1981; Quay, 1954*b*; Vorontsov, 1979); and of reproductive structures (Anderson, 1960; Hooper and Hart, 1962; Niethammer, 1972). Molecular studies that address phylogenetic questions have assessed allozymic variation (Chaline and Graf, 1988; Gill et al., 1987; Graf, 1982; Moore and Janecek, 1990), DNA-DNA hybridization (Catzeflis, 1990; Catzeflis et al., 1987), and chromosomal morphology and banding patterns (Burgos et al., 1989; Modi, 1987; Radosavlievic et al., 1990; Zagorodnyuk, 1990, 1991*c*; Zima and Král, 1984*a*).

The proliferation of family-group names, practically a one-to-one correspondence with recognized genera, does not necessarily connote unambiguous delineation of higher-order relationships—e.g., compare the tribal contents of Ognev (1963*b*), Hooper and Hart (1962), Kretzoi (1969), Gromov and Polyakov (1977), and Repenning et al. (1990). See Kretzoi (1969), Chaline (1972), and Gromov and Polyakov (1977) for authorship of synonyms, for the most part used at the tribal level.

Alticola Blanford, 1881. J. Asiat. Soc. Bengal, 50:96.

TYPE SPECIES: *Arvicola stoliczkanus* Blanford, 1875.

SYNONYMS: *Aschizomys, Platycranius.*

COMMENTS: Hinton's (1926) monograph remains one of the best overall reviews of the genus, as supplemented by the morphological studies of Rossolimo et al. (1988) and Rossolimo (1989) and the cytogenetic results of Hielscher et al. (1992). Schwarz (1939) also reviewed the Himalayan species. We follow Gromov and Polyakov (1977), along with Pavlinov and Rossolimo (1987), in recognizing the subgenera *Alticola*, *Aschizomys*, and *Platycranius*. DNA-DNA hybridization studies indicated that *Alticola macrotis* is a sister species of *Clethrionomys rufocanus* and that *A. argentatus* is phylogenetically closer to *C. rutilus* and *C. glareolus* than to *A. macrotis* (Gileva et al., 1989). *Alticola* is broadly related to *Clethrionomys*, an affinity Hooper and Hart (1962) acknowledged by placing *Alticola* with *Clethrionomys, Eothenomys, Hyperacrius, Dolomys*, and *Phenacomys* in the tribe Clethrionomyini, a division also recognized by Gromov and Polyakov (1977). Comparative study of appendicular myology and osteology reinforces the monophyly of *Alticola* and its close association with *Clethrionomys* and *Eothenomys* (Stein, 1987).

Alticola albicauda (True, 1894). Proc. U. S. Natl. Mus., 17:12.

TYPE LOCALITY: India, Baltistan, Braldu Valley.

DISTRIBUTION: Recorded only from the Himalayan portions of Baltistan and Kashmir, NW India.

SYNONYMS: *acmaeus*.

COMMENTS: Subgenus *Alticola*. Although usually included in *A. roylei* (Corbet, 1978*c*; Ellerman and Morrison-Scott, 1951; Gromov and Polyakov, 1977), the diagnostic traits of *albicauda* that set it apart as a distinct species were earlier pointed out by Hinton (1926). Schwarz's (1939) *acmaeus* identifies a geographic variant of the same species.

Alticola argentatus (Severtzov, 1879). Izv. Soc. Nat. Anthrop. Etnogr., 8, 2:82.

TYPE LOCALITY: Kazakhstan, Chimkentskaia Obl., Karatau Mtns, Mashat.

DISTRIBUTION: Mountainous region from Xinjiang, NW China (Ma et al., 1987), southwest through the Dzhungarski Alatau, the Talasskiy Alatau, the Tien Shan, Pamirs, and into the mountains of N Afghanistan, N Pakistan, and N India.

SYNONYMS: *alaica, argurus, blanfordi, gracilis, lahulis, leucura, longicauda, longicaudata, parvidens, phasma, rosanvi, saurica, shnitnikovi, severtzovi, subluteus, villosa, worthingtoni.*

COMMENTS: Subgenus *Alticola*. Included in *A. roylei* by Ellerman and Morrison-Scott (1951) and Corbet (1978*c*), but separated as a species by Rossolimo (1989), who recognized *blanfordi, phasma, severtzovi, subluteus,* and *worthingtoni* as subspecies. Attribution of synonyms follows Ellerman and Morrison-Scott (1951) and Pavlinov and Rossolimo (1987).

Alticola barakshin Bannikov, 1947. Bull. Moscow Soc. Nat., Biol., 52, 4:217.

TYPE LOCALITY: Mongolia, Gobi Altai Mtns, Gurvan Saikhan Ridge, Dzun Saikhan.

DISTRIBUTION: From Tuva region in Russia east through W and S Mongolia; probably occurs in adjoining Chinese regions (see Rossolimo et al., 1988).

COMMENTS: Subgenus *Alticola*. Included in *A. stoliczkanus* by Corbet (1978*c*) and Pavlinov and Rossolimo (1987), but revised as a separate species by Rossolimo et al. (1988). Chromosomal data provided by Yatsenko (1980).

Alticola lemminus (Miller, 1898). Proc. Acad. Nat. Sci. Philadelphia, 1898:369.

TYPE LOCALITY: Siberia, Bering Strait, Plover Bay, Kelsey Station.

DISTRIBUTION: NE Siberia from Chukotka area west through the Anadyr region to mouth of River Lena, and south throughout the Lena basin.

COMMENTS: Subgenus *Aschizomys*. Corbet (1978*c*) transferred the species to *Eothenomys*, but Russian workers refered *lemminus* to *Alticola* (Gromov and Polyakov, 1977; Ognev, 1964; Pavlinov and Rossolimo, 1987). Earlier, Hinton (1926:279) recognized *Aschizomys* but speculated that the species "will prove to be a member of the *E. rufocanus* group." Miller (1940*a*:94) reexamined the holotype and identified it as "nothing more than an alcohol-discolored specimen of the extreme East Asian representative of *Clethrionomys rufocanus*." Based on our inspection of the holotype of *lemminus*, we endorse Ognev's (1964) allocation to *Alticola*, subgenus *Aschizomys*, and are impressed by its diagnostic features. Significant chromosomal and morphological differences between samples of *A. lemminus* from Chukotka and Yakutia regions led Bykova et al. (1978) to speculate that "*lemminus*" may be a composite of two species.

Alticola macrotis (Radde, 1862). Reise in den Suden von Ost-Sibierien, 1:196.

TYPE LOCALITY: Siberia, S Krasnoyarsk Krai, E Sayan Mtns.

DISTRIBUTION: From Altai (Siberia and Xinjiang) and Sayan Mtns east and northeast through N Mongolia and Lake Baikal region to Region of Olekminski in Yakutsk, Russia.

SYNONYMS: *altaica, fetisovi, vicina, vinogradovi.*

COMMENTS: Subgenus *Alticola*. Ognev (1964) placed this species in the subgenus *Alticola*, but Gromov and Polyakov (1977) and Pavlinov and Rossolimo (1987) listed it under the subgenus *Aschizomys*. We follow Ognev until the phylogenetic position of *lemminus*, the type-species of *Aschizomys*, is defined relative to species of *Alticola*.

Alticola montosa (True, 1894). Proc. U. S. Natl. Mus., 17:11.

TYPE LOCALITY: India, central Kashmir, 11,000 ft.

DISTRIBUTION: Known only from Kashmir, 8000 to 13,000 ft (Hinton, 1926).

SYNONYMS: *imitator.*

COMMENTS: Subgenus *Alticola*. Usually incorporated in *A. roylei* (Corbet, 1978*c*; Gromov and Polyakov, 1977), but the diagnostic traits of *montosa* clearly distinguish it from the geographically adjacent *roylei*, as Hinton (1926) long ago noted.

Alticola roylei (Gray, 1842). Ann. Mag. Nat. Hist., [ser. 1], 10:265.

TYPE LOCALITY: India, Kumaon.

DISTRIBUTION: W Himalayas; recorded only from N Kumaon and N Himachal Pradesh (Lahul region) of N India.

SYNONYMS: *cautus.*

COMMENTS: Subgenus *Alticola*. Once considered the broadest-ranging species of *Alticola* in central Asia (Corbet, 1978*c*), but with the removal of *argentatus* and its synonyms (Rossolimo, 1989), the geographic and morphological definition of *roylei* conforms to that presented by Hinton (1926).

Alticola semicanus (Allen, 1924). Am. Mus. Novit., 133:6.
TYPE LOCALITY: Mongolia, Khangai Mtns, upper flow of Ongyin Gol River "Sain Noin Khan."
DISTRIBUTION: From the Tuva region of Russia throughout N Mongolia (see Rossolimo et al., 1988).
SYNONYMS: *alleni*.
COMMENTS: Subgenus *Alticola*. Originally described as a subspecies of *Microtus worthingtoni*, later synonymized with *Alticola roylei* (Corbet, 1978*c*), eventually listed as a separate species (Pavlinov and Rossolimo, 1987), and finally taxonomically revised (Rossolimo et al., 1988).

Alticola stoliczkanus (Blanford, 1875). J. Asiat. Soc. Bengal, 44:107.
TYPE LOCALITY: India, N Ladakh, Kuenlun Mtns.
DISTRIBUTION: From N Ladakh and Nepal through W and N Xizang, Tibet, to Gansu in N China; range limits unknown.
SYNONYMS: *acrophilus, lama, nanschanicus*.
COMMENTS: Subgenus *Alticola*. Geographic range is allopatric to that of *A. stracheyi* (Feng et al., 1986), with which it was once united.

Alticola stracheyi (Thomas, 1880). Ann. Mag. Nat. Hist., ser. 5, 6:332.
TYPE LOCALITY: India, Kashmir, Ladakh (as amended by Hinton, 1926:322).
DISTRIBUTION: Himalayas from E Kashmir in N India east through S Xizang (Tibet) and N Nepal to N Sikkim.
SYNONYMS: *cricetulus, bhatnagari*.
COMMENTS: Subgenus *Alticola*. Included in *A. stoliczkanus* by Schwarz (1939), Gromov and Polyakov (1977) and Corbet (1978*c*), but reinstated as a species by Feng et al. (1986), which reflects Hinton's (1926) earlier arrangement.

Alticola strelzowi (Kastschenko, 1899). Izv. Imp. Tomsk. Univ., 16:50.
TYPE LOCALITY: Russia, Altai Krai, Altai Mtns, near Lake Teniga.
DISTRIBUTION: From the Altai Mtns of NW Mongolia, Siberia, and Xinjiang in NW China (Ma et al., 1987) west through Kazakhstan to Karaganda region.
SYNONYMS: *depressus, desertorum*.
COMMENTS: Subgenus *Platycranius*. Citations and synonyms are discussed by Pavlinov and Rossolimo (1987).

Alticola tuvinicus Ognev, 1950. [Mammals of USSR and Adjacent Countries], 7:520.
TYPE LOCALITY: Russia, Tuvinskaya (Tuva), Kyzyl Mozhalyk.
DISTRIBUTION: The Altai Mtns and Tuva region, N Khubsugul Lake Valley, S Bailkal Lake Valley, and and nearby regions in NW Mongolia (see Rossolimo et al., 1988).
SYNONYMS: *baicalensis, khubsugulensis, kosogol, olchonensis*.
COMMENTS: Subgenus *Alticola*. Originally described as a species and then synonymized under the ubiquitous *A. roylei*. Listed as a separate species by Pavlinov and Rossolimo (1987) and revised by Rossolimo et al. (1988), who recognized *kosogol* and *olchonensis* as subspecies. Gromov and Polyakov (1977) considered *olchonensis* closely related to *A. macrotis*; its status needs to be reassessed.

Arborimus Taylor, 1915. Proc. California Acad. Sci., ser. 4, 5:119.
TYPE SPECIES: *Phenacomys longicaudus* True, 1890.
SYNONYMS: *Paraphenacomys*.
COMMENTS: Described as a subgenus of *Phenacomys* and conventionally recognized as such or as a complete synonym (Carleton and Musser, 1984; Hall, 1981; Howell, 1926*b*). Evidence for generic stature marshalled by Johnson (1968, 1973). Nonetheless, whether *Arborimus* is most closely related to *Phenacomys* or to some other arvicoline has not been substantiated with broad taxonomic sampling that includes critical species like *albipes*. Specific and subspecific classification basically set forth by Howell (1926*b*) and Hall and Cockrum (1953), as part of *Phenacomys*, and by Johnson and George (1991).

Arborimus albipes (Merriam, 1901). Proc. Biol. Soc. Washington, 14:125.
TYPE LOCALITY: USA, California, Humboldt Co., Humboldt Bay, redwood forest near Arcata.

DISTRIBUTION: Pacific coastal zone south of Columbia River, from W Oregon to extreme NW California, USA.

COMMENTS: More generalized terrestrial form and habits have suggested a closer relationship to *Phenacomys intermedius* (e.g., Hall, 1981, placed *albipes* and *intermedius* together in subgenus *Phenacomys*). Johnson and Maser (1982) enumerated character states that instead support closer congruence of *albipes* with species of *Arborimus*. However, see Repenning and Grady (1988), who diagnosed the subgenus *Paraphenacomys* of *Phenacomys* to contain *albipes*, which they viewed as more distantly related to the sister species *intermedius* and *longicaudus*.

Arborimus longicaudus (True, 1890). Proc. U. S. Natl. Mus., 13:303.

TYPE LOCALITY: USA, Oregon, Coos Co., Marshfield.

DISTRIBUTION: Pacific coast of W Oregon, north of Klamath Mtns, USA.

SYNONYMS: *silvicola*.

COMMENTS: Although Howell (1926*b*) recognized *silvicola* as a nominal species, subsequent research has favored its synonymy under *A. longicaudus* (Johnson, 1968), where it has been maintained as a subspecies (Hall, 1981; Johnson and George, 1991). Formerly included populations in California assigned to the new species *A. pomo* (see next account).

Arborimus pomo Johnson and George, 1991. Los Angeles Co. Nat. Hist. Mus., Contr. Sci., 429:12.

TYPE LOCALITY: USA, California, Sonoma Co., 0.8 km N Jenner, Jenner Ridge; 38°27'N, 123°06'W.

DISTRIBUTION: Coastal coniferous forest of NW California, south of Klamath Mtns as far as Sonoma Co., USA.

COMMENTS: Closely related to *A. longicaudus* (see Johnson and George, 1991).

Arvicola Lacepede, 1799. Tab. Div. Subd. Orders Genres Mammifères, p. 10.

TYPE SPECIES: *Mus amphibius* Linnaeus, 1758 (= *Mus terrestris* Linnaeus, 1758).

SYNONYMS: *Alviceola, Hemiotomys, Ochetomys, Paludicola, Praticola*.

COMMENTS: Excludes the North American *Microtus richardsoni* (see account of that species), which was placed in *Arvicola* by Hooper and Hart (1962). Generic reviews were provided by Corbet (1978*c*, 1984) and Gromov and Polyakov (1977), who recognized only two extant species, *A. sapidus* and *A. terrestris*. Earlier, Miller (1912*a*) defined seven species (*amphibius, illyricus, italicus, musignani, sapidus, scherman*, and *terrestris*) and Hinton (1926) recognized four (*amphibius, sapidus, scherman*, and *terrestris*). We follow Corbet's (1978*c*) unsatisfactory arrangement pending careful systematic revision of the genus, which will likely reflect a classification more similar to that of Hinton (1926). *Arvicola* is phylogenetically close to *Microtus* (Burgos et al., 1989; Chaline and Graf, 1988; Graf, 1982). Heinrich (1990) hypothesized that it evolved from the extinct *Mimomys*, a view already presented by Hinton (1926), and Rekovets (1990) summarized the fossil history leading to the modern groups of *Arvicola*.

Arvicola sapidus Miller, 1908. Ann. Mag. Nat. Hist., ser. 8, 1:195.

TYPE LOCALITY: Spain, Burgos Prov., Santo Domingo de Silos.

DISTRIBUTION: Portugal, Spain, and W France (see map in Reichstein, 1982*a*:214).

SYNONYMS: *musiniani, tenebricus*.

COMMENTS: Reviewed by Hinton (1926), Corbet (1978*c*), and Reichstein (1982*a*), with new morphological and ecological information supplied by Ventura and Gosalbez (1990) and Ventura et al. (1989).

Arvicola terrestris (Linnaeus, 1758). Syst. Nat., 10th ed., 1:61.

TYPE LOCALITY: Sweden, Uppsala.

DISTRIBUTION: Europe (except C and S Spain but including N Spain and N Portugal, W France, and SW Italy), from mountains of Mediterranean region to Arctic Sea, east through Siberia almost to Pacific coast, south to Israel, Iran, Lake Baikal and N Tien Shan Mtns of NW China (Corbet, 1978*c*; European range mapped by Reichstein, 1982*b*; former USSR portion outlined in Kuznetsov, 1965; Portugal record from Ramalhinho and Mathias, 1988).

SYNONYMS: *abrukensis, albus, americana, amphibius, aquaticus, argentoratensis, argyropus, armenius, ater, barabensis, brigantium, buffonii, cantabriae, canus, castaneus, caucasicus, cernjavskii, cubanensis, destructor, djukovi, exitus, ferrugineus, fuliginosus, hintoni, hyperryphaeus, illyricus, italicus, jacutensis, jenissijensis, karatshaicus, korabensis, kuruschi, kuznetzovi, littoralis, martinoi, meridionalis, minor, monticola, musignani, niger, nigricans, obensis, ognevi, pallasii, paludosus, persicus, pertinax, reta, rufescens, scherman, schermous, seythius, stankovici, tanaiticus, tataricus, tauricus, turovi, uralensis, variabilis, volgensis.*

COMMENTS: European populations are reviewed by Reichstein (1982*b*) and northern Spanish samples by Ventura and Gosalbez (1989). Morphometric analyses contrasting two subspecies in Netherlands reported by Warmerdam (1982). Morphological variability in context of taxonomic and distributional studies provided by Kratochvíl (1980, 1983), Krystufek and Tvrtkovic (1984), Nikolaeva (1982), and Ventura (1991). More than one species is represented in this complex (see generic comments).

Blanfordimys Argyropulo, 1933. Z. Säugetierk., 8:182.

TYPE SPECIES: *Microtus bucharicus* Vinogradov, 1930.

COMMENTS: Originally proposed as a subgenus of *Microtus*, a ranking traditionally acknowledged by Russian authors (Golenishchev and Sablina, 1991; Gromov and Polyakov, 1977; Ognev, 1964; Pavlinov and Rossolimo, 1987). In other taxonomic variations, Corbet (1978*c*) assigned *afghanus*, the type-species of *Blanfordimys*, to the genus *Pitymys*, and Chaline (1974) placed it in *Neodon*, subgenus *Microtus*. Ellerman (1941) considered the diagnostic traits of *afghanus* so impressive that he recognized the genus, as did Ellerman and Morrison-Scott (1951) and recently Zagorodnyuk (1990). We concur. *Blanfordimys* is defined by derived features that set it apart from other voles. Furthermore, the significance of these features has never been adequately assessed in the context of a careful phylogenetic study involving the many subgenera now included in *Microtus* and other genera considered to be closely related to it. Isolating *Blanfordimys* from *Microtus* defines an explicit hypothesis of relationships that should be critically tested with morphological, chromosomal, and biochemical data.

Blanfordimys afghanus (Thomas, 1912). Ann. Mag. Nat. Hist., ser. 8, 9:349.

TYPE LOCALITY: Afghanistan, Badkhiz, Gulran.

DISTRIBUTION: Recorded from high steppes and semi-desert in S Turkmenistan, Uzbekistan, Tadzhikistan, and Afghanistan; isolated population in Great Balkhan Mtns on E coast of Caspian Sea (see Golenishchev and Sablina, 1991; Hassinger, 1973; Niethammer, 1970).

SYNONYMS: *balchanensis, dangarinensis.*

COMMENTS: Taxonomy and distribution of Afghanistan populations reported by Niethammer (1970) and Hassinger (1973); morphometric and karyological analyses provided by Golenishchev and Sablina (1991), who recognized three subspecies.

Blanfordimys bucharicus (Vinogradov, 1930). Rukovodstvok opredeleniyu gryzunov Srednei Azii [Key to Determine Rodents of Central Asia], p. 45.

TYPE LOCALITY: Tadzhikistan, Zeravshan Range, 8 km S Pendzhikent, near village of Zivan, 2200 m.

DISTRIBUTION: Mtns of SW Tadzhikistan, possibly N Afghanistan; limits unresolved.

SYNONYMS: *davydovi.*

COMMENTS: Usually included in *afghanus* (Corbet, 1978*c*; Ellerman and Morrison-Scott, 1951; Ognev, 1964; Pavlinov and Rossolimo, 1987), but morphometric and karyologic analyses indicate that *bucharicus* is a separate species with two distinct geographic components, one newly described as the subspecies *davydovi* (Golenishchev and Sablina, 1991).

Chionomys Miller, 1908. Ann. Mag. Nat. Hist., ser. 8, 1:97.

TYPE SPECIES: *Arvicola nivalis* Martins, 1842.

COMMENTS: Tribe Arvicolini. Although described as a genus, Miller (1912*a*) later employed *Chionomys* as a subgenus, a status which became entrenched in the literature (Corbet, 1978*c*, Krapp, 1982*a*) with rare dissenters (e.g., Gromov and Polyakov, 1977; Lehmann, 1969). Recent analyses reveal that *Chionomys* is not part of the monophyletic group containing *Microtus* (Chaline and Graf, 1988; Graf, 1982; Nadachowski, 1990*a*; Pavlinov and Rossolimo, 1987; Zagorodnyuk, 1990). Van der Meulen (1978) considered *Suranomys* to be a junior synonym of *Chionomys*, but the type species of *Suranomys* (= *Microtus malei*) is regarded as a *Microtus* related to the *oeconomus* group, not to *Chionomys* (see Nadachowski, 1990*a*). New World *Microtus longicaudus* was referred to *Chionomys* by Anderson (1960), but a variety of data sources allies the former with *Microtus* proper (Chaline and Graf, 1988; Graf, 1982).

Discussing the origin and phylogeny of *Chionomys*, Nadachowski (1990*a*, 1991) suggested that two branches developed in Europe, one leading to *C. nivalis*, the other to *C. roberti* and *C. gud*. All three species are sympatric in the Caucasus (Nadachowski, 1990*a*). Karyotypic variation among the three species is reported by Sablina (1988) and Zima and Král (1984*a*).

Chionomys gud (Satunin, 1909). Izv. Kavkas. Mus., 4:272.

TYPE LOCALITY: Georgia, Caucasus, Gudauri, S of Krestovyi Pass.

DISTRIBUTION: Recorded only from Caucasus Mtns, and NE Turkey.

SYNONYMS: *gotschobi, lghesicus, lasistanius, lucidus, neujukovi, oseticus*.

COMMENTS: Reviewed by Corbet (1978*c*). Ellerman and Morrison-Scott (1951) and Corbet (1978*c*) listed *gotschobi* and *lghesicus* as synonyms of *nivalis*, but Pavlinov and Rossolimo (1987) included them in *C. gud*.

Chionomys nivalis (Martins, 1842). Rev. Zool. Paris, p. 331.

TYPE LOCALITY: Switzerland, Berner Oberland, Faulhorn.

DISTRIBUTION: Mountains of Europe from Spain through the Alps to Tatra, the Carpathians, Balkans, Mt Olympus and Pindus Range, east to W Caucasus, Turkey, Israel, Lebanon, Syria, Transcaucasia, Kopet Dag, and Zagros Mtns of Iran (see Krapp, 1982*a*, for European range and Harrison and Bates, 1991, for Middle East). Records in Greece were provided by Niethammer (1987*b*).

SYNONYMS: *abulensis, aleco, alpinus, aquitanius, cedrorum, dementievi, hermonis, lebrunii, leucurus* (of Gerbe, 1852, not Blyth, 1863), *loginovi, malyi, mirhanreini, nivicola, olympius, petrophilus, pontius, radnensis, satunini, spitzenbergerae, trialeticus, ulpius, wagneri*.

COMMENTS: European populations reviewed by Krapp (1982*a*). Intraspecific morphological variation among Carpathian samples were analyzed by Kratochvíl (1981) along with a review of European and Turkish subspecies. The subspecies *spitzenbergerae* denotes a population from S Turkey that was previously identified as *C. gud* (Nadachowski, 1990*b*), a species which is known only from NE Turkey where it is sympatric with *C. nivalis*. Analyses of vertical distribution of *C. nivalis* in Yugoslavia (Krystufek and Kovacic, 1989) and geographic variation among samples from Austria and Yugoslavia (Krystufek, 1990) amplify knowledge of morphological variation within the species. Chromosomal variation in Bulgarian populations was reported by Peshev and Belcheva (1979). Allozyme variation and differentiation among samples of *C. nivalis* from N Italy and Israel were reported by Filippucci et al. (1991), who also noted that the Israeli population from Mt. Hermon (*hermonis*) might represent a separate species.

Chionomys roberti (Thomas, 1906). Ann. Mag. Nat. Hist., ser. 7, 17:418.

TYPE LOCALITY: Turkey, Pontus Prov., Sumila, 30 mi S Trebizond (= Trabzon).

DISTRIBUTION: Recorded only from forests of W Caucasus Mtns, and NE Turkey.

SYNONYMS: *circassicus, occidentalis personatus, pshavus, turovi*.

COMMENTS: Reviewed by Corbet (1978*c*).

Clethrionomys Tilesius, 1850. Isis, 2:28.

TYPE SPECIES: *Mus rutilus* Pallas, 1779.

SYNONYMS: *Craseomys, Glareomys, Eotomys, Euotomys, Evotomys, Neoaschizomys*.

COMMENTS: Placed in tribe Clethrionomyini, usually with voles such as *Eothenomys, Alticola*, and *Hyperacrius* (Koenigswald, 1980; Kretzoi, 1969). *Clethrionomys* is perhaps most closely related to *Eothenomys*, if not congeneric with it as suggested by some authors (Corbet, 1978*c*, 1984; Hooper and Hart, 1962). Large segments of the genus were taxonomically reviewed by Aimi (1980), Corbet (1978*c*, 1984), and Gromov and Polyakov (1977); chromosomal data summarized and interspecific relationships discussed by Gamperl (1982), Modi and Gamperl (1989), Nadler et al. (1976); Sokolov et al. (1990), and Vorontsov et al. (1978); electrophoretic data presented by Nadler et al. (1978) and Chaline and Graf (1988). The relationship among European and Japanese species based on living and Pleistocene samples was monographed by Kawamura (1988).

Species in this genus were commonly listed under *Evotomys* (e.g., Hinton, 1926) until Palmer (1928) established the priority of *Clethrionomys*. Recently, Pavlinov and Rossolimo (1987) have questioned whether *Myodes* is a senior synonym of *Clethrionomys* and Zagorodnyuk (1990) employed the former name. This issue warrants prompt settlement before the literature on red-backed voles is diffused under yet another generic name.

Clethrionomys californicus (Merriam, 1890). N. Am. Fauna, 4:26.

TYPE LOCALITY: USA, California, Humboldt Co., Eureka.

DISTRIBUTION: Pacific-coast coniferous forest from the Columbia River south through W Oregon to NW California, USA.

SYNONYMS: *mazama, obscurus*.

COMMENTS: The name *occidentalis* was formerly applied to this species (e.g., Hall and Cockrum, 1953), but populations north of the Columbia River, which include *occidentalis* and *caurinus*, have been reassigned to *C. gapperi* (Cowan and Guiguet, 1965; Johnson and Ostenson, 1959).

Clethrionomys centralis Miller, 1906. Ann. Mag. Nat. Hist., ser. 7, 17:373.

TYPE LOCALITY: Kazakhstan, W Tien Shan Mtns, Koksu valley, 9000 ft.

DISTRIBUTION: Known only from Tien Shan Mtns, Kazakhstan and Kirghizia; and adjacent Xinjiang, China.

SYNONYMS: *frater*.

COMMENTS: A distinctive species recognized as such by Hinton (1926), included in *C. glareolus* by Corbet (1978*c*), and recognized again as separate (Corbet, 1984; Pavlinov and Rossolimo, 1987). Some Russian (Gromov and Polyakov, 1977) and Chinese (Ma et al., 1987) workers still use *frater* for the species. Chromosomal data recorded by Vorontsov et al. (1978) and Sokolov et al. (1990).

Clethrionomys gapperi (Vigors, 1830). Zool. J., 5:204.

TYPE LOCALITY: Canada, Ontario, between York (= Toronto) and Lake Simcoe.

DISTRIBUTION: Most of Canada from N British Columbia to Labrador, excluding Newfoundland; south in the Appalachians to N Georgia, in the Great Plains to N Iowa, and in the Rockies to C New Mexico and EC Arizona, USA.

SYNONYMS: *arizonensis, athabascae, brevicaudus, carolinensis, cascadensis, caurinus, fuscodorsalis, galei, gaspeanus, gauti, hudsonius, idahoensis, limitis, loringi, maurus, nivarius, occidentalis, ochraceus, pallescens, paludicola, phaeus, proteus, pygmaeus, rhoadsii, rufescens, rupicola, saturatus, solus, stikinensis, uintaensis, ungava, wrangeli*.

COMMENTS: Information on specific relationships somewhat contradictory. Interspecific hybrids of reduced fertility produced from laboratory crosses with Eurasian *C. glareolus*, which led Grant (1974) to view the two as semispecies of recent divergence. Others (Bee and Hall, 1956; Youngman, 1975) have suggested, without presentation of data, that *gapperi* and *rutilus* are conspecific. Based on biochemical data, Nadler et al. (1978) viewed Old World *C. rufocanus* as closely related to the *gapperi-rutilus* complex. Most highly variable in gastric morphology among species of *Clethrionomys* studied by Carleton (1981). See account of *C. californicus* for allocation of *occidentalis* and *caurinus* to *C. gapperi*. See Merritt (1981, Mammalian Species, 146).

Clethrionomys glareolus (Schreber, 1780). Die Säugethiere, 4:680.

TYPE LOCALITY: Denmark, Lolland Island.

DISTRIBUTION: Forests of W Palaearctic from France and Scandinavia to Lake Baikal, south to N Spain, N Italy (isolated montane populations farther south), the Balkans (but not most of Greece), W Turkey, N Kazakhstan and the Altai and Sayan Mtns; also occurs on Britain and SW Ireland (see Corbet, 1978*c*, and Viro and Niethammer, 1982:117).

SYNONYMS: *alstoni, bernisi, bosniensis, britannicus, caesarius, cantueli, curcio, devius, erica, fulvus, garganicus, gorka, hallucalis, helveticus, hercynicus, insulaebellae, intermedius, istericus, italicus, jurassicus, makedonicus, minor, nageri, norvegicus, ognevi, petrovi, pirinus, ponticus, pratensis, reinwaldti, riparia, rubidus, rufescens, ruttneri, saianicus, sibericus, skomerensis, sobrus, suecicus, tomensis, variscicus, vasconiae, vesanus, wasjuganensis.*

COMMENTS: European populations reviewed by Viro and Niethammer (1982). Analysis of evolutionary relationships among British samples was reported by Steven (1953), biochemical differentiation among populations over short geographic distances was presented by Leitner and Hartl (1988). Evolutionary significance in morphology of third upper molar in extant and fossil samples analyzed by Bauchau and Chaline (1987).

Clethrionomys rufocanus (Sundevall, 1846). Ofv. K. Svenska Vet.-Akad. Forhandl. Stockholm, 3:122.

TYPE LOCALITY: Sweden, Lappmark.

DISTRIBUTION: N Palearctic from Scandinavia through Siberia to Kamchatka, Russia, south to S Ural Mtns, the Altai Mtns, Mongolia, Transbaikal, N China (Xinjiang and Heilongjiang), Korea, and N Japan (Hokkaido and Rishiri Isls) (see Aimi, 1980; Corbet, 1978*c*; Henttonen and Viitala, 1982; Kaneko, 1992; and Ma et al., 1987).

SYNONYMS: *akkeshii, arsenjevi, bargusinensis, bedfordiae, bromleyi, irkutensis, kamtschaticus, kolymensis, kurilensis, latastei, montanus, rex, siberica, wosnessenskii, yesomontanus.*

COMMENTS: European populations reviewed by Henttonen and Viitala (1982). Morphological discrimination of *C. rufocanus* and *Eothenomys regulus* and their geographic distributions in the former USSR, NE China, and Korea assessed by Kaneko (1990). Variation in morphology of upper third molar in context of systematic, age, and seasonal significance was reported by Abe (1982). Chromosomal data suggest the need to re-evaluate the specific status of *montanus* and *bedfordiae* (see Kashiwabara and Onoyama, 1988). Allocation of synonyms follows Aimi (1980), Ellerman and Morrison-Scott (1951), and Pavlinov and Rossolimo (1987).

Clethrionomys rutilus (Pallas, 1779). Nova Spec. Quadr. Glir. Ord., p. 246.

TYPE LOCALITY: Siberia, center of Ob River delta.

DISTRIBUTION: Holarctic: in Old World, from N Scandinavia east to Chukotski Peninsula, and south to N Kazakhstan, Mongolia, Transbaikalia, NE China, Korea, and islands of Sakhalin and Hokkaido (see Corbet, 1978*c*; Henttonen and Peiponen, 1982); St. Lawrence Isl, Bering Sea; in New World, from Alaska east to Hudson Bay, and south to N British Columbia and extreme NE Manitoba, Canada.

SYNONYMS: *alascensis, albiventer, amurensis, baikalensis, dawsoni, dorogostaiskii, glacialis, hintoni, insularis, jacutensis, jochelsoni, laticeps, latigriseus, lenaensis, mikado, mollessonae, narymensis, orca, otus, parvidens, platycephalus, rjabovi, rossicus, russatus, salairicus, tugarinovi, tundrensis, uralensis, vinogradovi, volgensis, washburni, watsoni.*

COMMENTS: Conspecificity of Old and New World populations advanced by Rausch (1953) and corroborated by subsequent studies (e.g., Nadler et al., 1976, 1978; Rausch and Rausch, 1975*a*). European populations reviewed by Henttonen and Peiponen (1982). Variation in pattern of third upper molar and its systematic implications reported by Nakatsu (1982) for Japanese populations. North American populations revised, as *C. dawsoni*, by Orr (1945), and, as *C. rutilus* by Manning (1956). Also see account of *C. gapperi*.

Clethrionomys sikotanensis (Tokuda, 1935). Mem. Coll. Sci. Kyoto Imp. Univ., ser. B, 10:241.

TYPE LOCALITY: Russia, Kuril Islands, Sikotan Island.

DISTRIBUTION: Islands of Sikotan, Daikoku, and Rishiri (Imaizumi, 1971); limits of insular distribution unresolved.

SYNONYMS: *microtinus.*

COMMENTS: Allocated by Corbet (1978*c*) and Aimi (1980) to *C. rufocanus* but recognized as a distinct species by Gromov and Polyakov (1977), Imaizumi (1971, 1972), and Pavlinov and Rossolimo (1987). The status of *microtinus* was discussed by Gromov and Polyakov (1977) and Pavlinov and Rossolimo (1987).

Dicrostonyx Gloger, 1841. Gemein Hand.- Hilfsbuch. Nat., 1:97.

TYPE SPECIES: *Mus hudsonius* Pallas, 1778.

SYNONYMS: *Borioikon, Cuniculus, Misothermus, Tylonyx.*

COMMENTS: At first *Dicrostonyx* was grouped with other lemmings following Miller's (1896) classic Lemmi-Microti division (e.g., Ellerman, 1941; Hinton, 1926; Ognev, 1963*b*;

Simpson, 1945). An impressive variety of data, however, requires its tribal separation (Dicrostonychini) from the true lemmings (Lemmini) and suggests that the origin of *Dicrostonyx* dates to the earliest radiation of arvicolines (Carleton, 1981; Chaline and Graf, 1988; Gromov and Polyakov, 1977; Hinton, 1926; Hooper and Hart, 1962; Kretzoi, 1969; Modi, 1987).

The simple viewpoint of a single circumpolar species, *D. torquatus*, as advanced by Ognev (1963*b*) and Rausch (1953, 1963*b*), has been unsettled by karyotypic reports of the past two decades (Chernyavskii and Kozlovskii, 1980; Krohne, 1982; Rausch, 1977; Rausch and Rausch, 1972). The occurrence of varying lemmings in quite different tundra biotopes (e.g., see Youngman, 1975) alone might have questioned the existence of only one species, but the karyological and breeding results of Rausch and Rausch (1972) first drew attention to the possibility of a superspecies complex among North American *Dicrostonyx*, an interpretation reiterated by Rausch (1977; see summary of chromosomal variation in Krohne, 1982). Later authorities either listed the North American karyotypic morphs as species (Corbet and Hill, 1991; Honacki et al., 1982; Jones et al., 1986) or continued to recognize most as subspecies of *D. groenlandicus*, together with *D. exsul* on St. Lawrence Island and *D. hudsonius* on the Ungava Peninsula (Hall, 1981).

After examining museum specimens, we readily appreciate the specific distinctiveness of *groenlandicus*, *hudsonius*, *richardsoni*, and *unalascensis*. The morphological discrimination of others (*kilangmiutak*, *nelsoni*, and *rubricatus*) is more subtle but may yield to careful description and analysis; we have not seen examples of *nunatakensis*, an isolated taxon described as a subspecies of *torquatus* by Youngman (1967). Unfortunately, the provocative findings of Rausch and others have not been explored and substantiated using other taxonomic information; in this regard, Youngman's (1975) narrative overview of species and racial distributions warrants attention in future efforts. We herein continue to list the species of *Dicrostonyx* as provisional and stress that their repetition in checklists like this enhances neither our confidence in the number of biological entities nor our understanding of their biogeographic significance.

Dicrostonyx exsul G. M. Allen, 1919. Bull. Mus. Comp. Zool., 62:532.
TYPE LOCALITY: USA, Alaska, Bering Sea, St. Lawrence Island.
DISTRIBUTION: Known only from the type locality.

Dicrostonyx groenlandicus (Traill, 1823). *In* Scoresby, J. Voy. to Northern Whale-Fishery..., p. 416.
TYPE LOCALITY: Greenland, Jamesons Land.
DISTRIBUTION: N Greenland and Queen Elizabeth Islands, south to Baffin and Southampton islands and NE District of Keewatin, Canada; limits uncertain.
SYNONYMS: *clarus*, *lentus*.

Dicrostonyx hudsonius (Pallas, 1778). Nova Spec. Quad. Glir. Ord., p. 208.
TYPE LOCALITY: Canada, Labrador.
DISTRIBUTION: Labrador and N Quebec, Canada.
COMMENTS: Unbanded karyotype resembles that of *D. richardsoni* (see Krohne, 1982).

Dicrostonyx kilangmiutak Anderson and Rand, 1945. J. Mammal., 26:305.
TYPE LOCALITY: Canada, Northwest Territories, SE Victoria Island on Victoria Strait, DeHaven Point.
DISTRIBUTION: Victoria and Banks islands and the adjacent Canadian mainland; poorly known.

Dicrostonyx nelsoni Merriam, 1900. Proc. Washington Acad. Sci., 2:25.
TYPE LOCALITY: USA, Alaska, Norton Sound, St. Michael.
DISTRIBUTION: W Alaska and Alaskan Peninsula, USA.
SYNONYMS: *peninsulae*.

Dicrostonyx nunatakensis Youngman, 1967. Proc. Biol. Soc. Washington, 80:31.
TYPE LOCALITY: Canada, Yukon Territory, Ogilvie Mtns, 20 mi S Chapman Lake, 5500 ft; 64°35'N, 138°13'W.
DISTRIBUTION: Known only from the Ogilvie Mtns, NC Yukon Territory, Canada.

COMMENTS: As remarked by Youngman (1967), this form contrasts markedly with nearby *rubricatus* and *kilangmiutak*; tentatively retained as a species by Honacki et al. (1982).

Dicrostonyx richardsoni Merriam, 1900. Proc. Washington Acad. Sci., 2:26.
TYPE LOCALITY: Canada, Manitoba, Fort Churchill.
DISTRIBUTION: W coast of Hudson Bay west to vicinity of Great Slave Lake, District of MacKenzie, Canada; extent of westward distribution unknown.

Dicrostonyx rubricatus (Richardson, 1889). Zool. Capt. Beechey's Voy., p. 7.
TYPE LOCALITY: USA, Alaska, Bering Strait.
DISTRIBUTION: N Alaska, USA.
SYNONYMS: *alascensis*.

Dicrostonyx torquatus (Pallas, 1778). Nova Spec. Quad. Glir. Ord., p. 206.
TYPE LOCALITY: Siberia, mouth of River Ob.
DISTRIBUTION: Palearctic tundra from White Sea, W Russia, to Chukotski Peninsula, NE Siberia, and Kamchatka; including Novaya Zemlya and New Siberian islands, Arctic Ocean (see Corbet, 1978*c*).
SYNONYMS: *chionopaes, lenae, lenensis, pallida, ungulatus*.
COMMENTS: Chromosomal traits of populations from the Polar Urals (*D. t. torquatus*), the Laptev Sea coast and Rautan Island off the coast of the Chukotka Peninsula (*D. t. chionopaes*) are similar, and progeny of crosses between these two subspecies are fertile (Gileva, 1980). Unusual sex-chromosome constitution and other chromosomal information were summarized by Gileva et al. (1980), Gileva (1983), and Zima and Král (1984*a*). Once believed to encompass most or all New World populations (e.g., Rausch, 1953, 1963*b*) but their level of relationship to *D. torquatus* proper is now unclear (see remarks under genus).

Dicrostonyx unalascensis Merriam, 1900. Proc. Washington Acad. Sci., 2:25.
TYPE LOCALITY: USA, Alaska, Umnak Island.
DISTRIBUTION: Umnak and Unalaska islands of Aleutian Archipelago, Alaska, USA.
SYNONYMS: *stevensoni*.
COMMENTS: Only species of *Dicrostonyx* to lack molt to white winter pelage and acquistion of snow claws (see Rausch and Rausch, 1972).

Dicrostonyx vinogradovi Ognev, 1948. [Mammals of the U.S.S.R. and adjacent countries], 6:509.
TYPE LOCALITY: SE Siberia, Wrangel Island (Os. Vrangelya), off coast of Anadyr region.
DISTRIBUTION: Known only from the type locality.
COMMENTS: Included in *D. torquatus* by Corbet (1978*c*), but chromosomal data, morphological traits, and breeding results indicate *vinogradovi* is a separate species (Chernyavskii and Kozlovskii, 1980).

Dinaromys Kretzoi, 1955. Acta Geol. Acad. Sci. Hung., 3:347-353.
TYPE SPECIES: *Microtus* (*Chionomys*) *marakovici* Bolkay, 1924 (= *Microtus bogdanovi* Martino, 1922).
COMMENTS: Often referenced as *Dolomys* until Corbet (1978*c*) explained the correct usage of *Dinaromys* for *bogdanovi*. The genus has been allocated to the Ondatrini (see Corbet, 1978*c*) or Clethrionomyini (Gromov and Polyakov, 1977; Hooper and Hart, 1962), but Kretzoi's (1969) referral to the Pliomyini perhaps reflects a stronger hypothesis of its phylogenetic affinities (also see Zagorodnyuk, 1990). Based on molar enamel microstructure, Koenigswald (1980) discovered no close affinity with any living arvicoline and suggested a relationship with an extinct species of *Propliomys*, a late Pliocene genus also placed in Pliomyini by Kretzoi (1969).

Dinaromys bogdanovi (Martino, 1922). Ann. Mag. Nat. Hist., ser. 9, 9:413.
TYPE LOCALITY: Yugoslavia, Rijeka Prov., Montenegro, Cetinje.
DISTRIBUTION: Recorded only from mountains of Yugoslavia (see map in Petrov and Todorovic, 1982); limits unresolved.
SYNONYMS: *coeruleus, grebenscikovi, korabensis, longipedis, marakovici, preniensis, trebevicensis*.
COMMENTS: The extant species is most closely related to two Pleistocene species: *D. dalmatinus*, recorded from N Italy, Yugoslavia, and S Greece (see Petrov and

Todorovic, 1982); and *D. topachevskii* from Uzbekistan (see Nesin and Skorik, 1989). Zoogeographic aspects of *D. bogdanovi* were discussed by Petrov (1979) and chromosomal data by Zima and Král (1984*a*).

Ellobius G. Fischer, 1814. Zoognosia, 3:72.

TYPE SPECIES: *Mus talpinus* Pallas, 1770.

SYNONYMS: *Afganomys, Afghanomys, Chthonergus, Lemmomys, Myospalax* (of Blyth, 1846, not Laxmann, 1769, or Hermann, 1783).

COMMENTS: Gromov and Polyakov (1977) excluded *Ellobius* from arvicolines, and Pavlinov and Rossolimo (1987) viewed the Ellobiini as Cricetidae *incertae sedis,* questioning whether it belonged in Arvicolinae or Cricetinae. Most workers, however, have recognized *Ellobius,* albeit highly specialized and unusual, as the only extant member of the tribe Ellobiini of Arvicolinae (Corbet, 1978*c*; Hooper and Hart, 1962; Kretzoi, 1969; Topachevskii and Rekovets, 1982). Topachevskii and Rekovets (1982) interpreted relationships and morhological trends in diversification within the genus from the late Pliocene to Recent. The present range of *Ellobius* is only part of a wider distribution that once embraced Israel and North Africa in the middle to late Pleistocene (Jaeger, 1988). Ultrastructure, meiotic behavior, and evolution of sex chromosomes are discussed by Kolomiets et al. (1991). Two subgenera are recognized, *Ellobius* and *Afganomys;* Zagorodnyuk (1990) listed the latter as a genus.

Ellobius alaicus Vorontsov et al., 1969. *In* Vorontsov (ed.), [The Mammals: Evolution, karyology, taxonomy, fauna], Novosibirsk, p. 127.

TYPE LOCALITY: Kirghizia, Alai Valley, between Sary-Tashem and Bardabo, 3300 m.

DISTRIBUTION: Recorded only from the Alai Mtns, S Kirghizia.

COMMENTS: Subgenus *Ellobius.* Included, with question, in *E. talpinus* by Corbet (1978*c*). Chromosomal data and breeding results reveal that *alaicus* does not belong in *E. talpinus* and is reproductively isolated from *E. tancrei,* with which it is parapatric (Corbet, 1984; Lyapunova et al., 1990, and references therein). Closely related to *E. tancrei* and provisionally placed in that species by Pavlinov and Rossolimo (1987).

Ellobius fuscocapillus Blyth, 1843. J. Asiat. Soc. Bengal, 11:887.

TYPE LOCALITY: Pakistan, Baluchistan Region, Quetta Div., Quetta.

DISTRIBUTION: E Iran, Afghanistan, W Pakistan, and S Turkmenistan in the Kopet Dag Mtns.

SYNONYMS: *farsistani, intermedius.*

COMMENTS: Subgenus *Afganomys.* Chromosomal data presented by Vorontsov et al. (1980) and Lyapunova et al. (1980) in context of assessing specific differences within the genus and variation in sex chromosomes. Morphology described in detail by Hinton (1926), in context of surveying interrelationships among arvicoline genera.

Ellobius lutescens Thomas, 1897. Ann. Mag. Nat. Hist., ser. 6, 20:308.

TYPE LOCALITY: Turkey, Kurdistan, Van.

DISTRIBUTION: From S Caucasus Mtns south through E Turkey and NW Iran.

SYNONYMS: *legendrei, woosnami.*

COMMENTS: Subgenus *Afganomys.* Treated by Corbet (1978*c*) as a synonym of *E. fuscocapillus* but now shown to be a distinct species (Corbet, 1984; Vorontsov et al., 1980, and references therein), as earlier listed by Ellerman and Morrison-Scott (1951). This is a species with a low diploid number (17) and an unusual mode of sex determination, which has been the subject of the many reports reviewed by Zima and Král (1984*a*), as well as current inquiry (Vogel et al., 1988, and references therein).

Ellobius talpinus (Pallas, 1770). Nova Comm. Acad. Sci. Petropoli, 14, I:568.

TYPE LOCALITY: Russia, W Bank of Volga River, between Kuibyshev (= Samara) and Kostychi.

DISTRIBUTION: Steppes from S Ukraine and Crimea east through Kazakhstan to N of Balkhash Lake, and in Turkmenistan (see Pavlinov and Rossolimo, 1987; Yakimenko and Lyapunova, 1986).

SYNONYMS: *ciscaucasicus, murinus, rufescens, tanaiticus, transcaspiae.*

COMMENTS: Subgenus *Ellobius.* Chromosomal polymorphism among samples from the Pamir-Alai Mtns was analyzed by Lyapunova et al. (1980); other chromosomal data summarized and reviewed by Zima and Král (1984*a*). Cytological identification of

Turkmenian populations as *E. talpinus* and comparison with *T. tancrei* provided by Yakimenko and Lyapunova (1986).

Ellobius tancrei Blasius, 1884. Zool. Anz., 7:197.
TYPE LOCALITY: Kazakhstan, Zaissan Lake Valley, Kendyrlik (= Przevalskoie).
DISTRIBUTION: From NE Turkmenistan (Yakimenko and Lyapunova, 1986) and Uzbekistan east through E Kazakhstan to E China (Xinjiang and Nei Mongolia) and Mongol.
SYNONYMS: *coenosus, fusciceps, fuscipes, kastschenkoi, larvatus, ognevi, orientalis, ursulus.*
COMMENTS: Included in *E. talpinus* by Corbet (1978*c*) but now regarded as a distinct species whose geographic range is allopatric to that of *E. talpinus* (see references in Corbet, 1984; Pavlinov and Rossolimo, 1987; Yakimenko and Lyapunova, 1986). Chromosomal contrasts with *E. talpinus* recorded by Yakimenko and Lyapunova (1986), comparisons with *E. alaicus* reported by Lyapunova et al. (1990).

Eolagurus Argyropulo, 1946. Vestn. Akad. Nauk Kazakh. SSR, 7-8:44.
TYPE SPECIES: *Georychus luteus* Eversmann, 1840.
COMMENTS: Corbet (1978*c*) viewed this taxon as part of *Lagurus*, but subsequent authorities have considered the two separate genera (Corbet and Hill, 1991; Gromov and Polyakov, 1977; Pavlinov and Rossolimo, 1987; Zagorodnyuk, 1990). Pavlinov and Rossolimo (1987) clarified the taxonomic decisions behind origin of the generic name, and pointed out that Corbet's (1978*c*) statement that it was meant as a subgenus for *luteus* may be the first validation of *Eolagurus*. A member of the tribe Lagurini (see account of *Lagurus*).

Eolagurus luteus (Eversmann, 1840). Bull. Soc. Nat. Moscow, p. 25.
TYPE LOCALITY: Kazakhstan, NW of Aral Sea.
DISTRIBUTION: Formerly Kazakhstan region, but now extinct (see Corbet, 1978*c*, and reference therein); Nxingiang (Ma et al., 1987), and W Mongolia.
COMMENTS: Reviewed by Corbet (1978*c*) and Gromov and Polyakov (1977). Cranial and dental morphology described by Hinton (1926) in context of surveying interrelationships among arvicoline genera.

Eolagurus przewalskii (Büchner, 1889). Wiss. Res. Przewalski Cent.-Asien, Reisen, Zool., I:(Säugeth.), p. 127.
TYPE LOCALITY: China, Qinghai (Tsinghai), Tsaidam region, shore of Iche-zaidemin Nor.
DISTRIBUTION: From S Xinjiang and N Xizang in W China, east through Quinghai and N Gansu to S Mongolia and Nei Mongol; limits unknown.
COMMENTS: This very distinctive species was incorrectly united with *E. luteus* by Corbet (1978*c*) but treated as separate by Allen (1940), Corbet and Hill (1991), and Gromov and Polyakov (1977).

Eothenomys Miller, 1896. N. Am. Fauna, 12:45.
TYPE SPECIES: *Arvicola melanogaster* Milne-Edwards, 1871.
SYNONYMS: *Anteliomys, Caryomys.*
COMMENTS: Revised by Corbet (1978*c*), who also included *Phaulomys* (a genus we separate from *Eothenomys*) and *Aschizomys* (which we include in *Alticola*). *Eothenomys* is closely related to *Hyperacrius, Alticola,* and *Clethrionomys*, all of which are usually placed in the Clethrionomyini (Gromov and Polyakov, 1977; Hooper and Hart, 1962; Koenigswald, 1980).

Eothenomys chinensis (Thomas, 1891). Ann. Mag. Nat. Hist., ser. 6, 8:117.
TYPE LOCALITY: China, W Sichuan, Kiatingfu.
DISTRIBUTION: Recorded only from Sichuan and Yunnan between 2000-4000 m (Corbet, 1978*c*).
SYNONYMS: *tarquinus, wardi.*
COMMENTS: The type-species of *Anteliomys* (see Hinton, 1926).

Eothenomys custos (Thomas, 1912). Ann. Mag. Nat. Hist., ser. 8, 9:517.
TYPE LOCALITY: China, Yunnan, Atuntsi, 11,500-12,500 ft.
DISTRIBUTION: Recorded only from mtns of Sichuan and Yunnan, where it occurs up to 4200 m (Corbet, 1978*c*).
SYNONYMS: *hintoni, rubelius, rubellus.*

COMMENTS: Listed as a species of *Anteliomys* by Hinton (1926).

Eothenomys eva (Thomas, 1911). Abstr. Proc. Zool. Soc. London, 1911(90):4.
TYPE LOCALITY: China, Gansu (Kansu), SE of Tauchow, 10,000 ft.
DISTRIBUTION: China, mtns of S Gansu and adjoining regions of Shaanxi, Sichuan, and Hubei (Allen, 1940; Corbet, 1978*c*; Kaneko, 1992).
SYNONYMS: *alcinous, aquilus.*
COMMENTS: Listed as a member of *Evotomys* or *Clethrionomys* by Hinton (1926), Ellerman and Morrison-Scott (1951), and Gromov and Polyakov (1977), but correctly described and revised as a distinct species of *Eothenomys* by Allen (1940), Corbet (1978*c*), and Kaneko (1992).

Eothenomys inez (Thomas, 1908). Abstr. Proc. Zool. Soc. London, 1908(63):45.
TYPE LOCALITY: China, Shanxi (Shansi), mtns 12 mi NW Kolanchow, 7000 ft.
DISTRIBUTION: Shaanxi and Shanxi provinces, China.
SYNONYMS: *nux.*
COMMENTS: Following Hinton (1926), *inez* was synonymized with *Clethrionomys rufocanus* (Ellerman and Morrison-Scott, 1951; Gromov and Polyakov, 1977), but its status as a distinct species of *Eothenomys* is well documented (Allen, 1940; Corbet, 1978*c*; Kaneko, 1992).

Eothenomys melanogaster (Milne-Edwards, 1871). Nouv. Arch. Mus. Hist. Nat. Paris, Bull., 7:93.
TYPE LOCALITY: China, W Sichuan, Moupin.
DISTRIBUTION: W and S China, north to S Gansu and Ningxia, south to N Thailand and N Burma; also on Taiwan.
SYNONYMS: *aurora, bonzo, cachinus, colurnus, confinii, eleusis, fidelis, kanoi, libonotus, miletus, mucronatus.*
COMMENTS: Allen (1940) treated *miletus* and *eleusis* as distinct species, and Corbet (1978*c*) acknowledged that more than one species may be represented in what he identified as *E. melanogaster.* After checking large museum series, we share this reservation and only provisionally ally these synonyms with *melanogaster* pending a systematic revision of the group. Conventional karyotype of Taiwan populations reported by Harada et al. (1991).

Eothenomys olitor (Thomas, 1911). Abstr. Proc. Zool. Soc. London, 1911(100):50.
TYPE LOCALITY: China, Yunnan, Chaotungfu, 6700 ft.
DISTRIBUTION: Recorded only from Yunnan between 6000-7000 ft (see Allen, 1940).
COMMENTS: Hinton (1926) recognized this species as a member of *Eothenomys,* but Gromov and Polyakov (1977) placed it in *Anteliomys.*

Eothenomys proditor Hinton, 1923. Ann. Mag. Nat. Hist., ser. 9, 11:152.
TYPE LOCALITY: China, Yunnan, Likiang Range (27°30'N), 13,000 ft.
DISTRIBUTION: Known only from the mtns of Yunnan and Sichuan, China (Allen, 1940; Corbet, 1978*c*).
COMMENTS: Treated as a species of *Eothenomys* by Hinton (1926), but placed in *Anteliomys* by Gromov and Polyakov (1977).

Eothenomys regulus (Thomas, 1907). Proc. Zool. Soc. London, 1906:863 [1907].
TYPE LOCALITY: Korea, Mingyong, 110 mi SE Seoul.
DISTRIBUTION: Korea.
COMMENTS: Usually included in *Clethrionomys rufocanus* (Allen, 1940; Ellerman and Morrison-Scott, 1951; Gromov and Polyakov, 1977) but treated as a separate species by Corbet (1978*c*). In a careful morphological study of *C. rufocanus* and *E. regulus* from the former USSR, NE China, and Korea, Kaneko (1990) discovered *E. regulus* to be a Korean endemic and suggested (p. 129) "that the true geographical demarcation line between the two species lies on the western and southern boundary of the Kaima Plateau, North Korea."

Eothenomys shanseius (Thomas, 1908). Proc. Zool. Soc. London, 1908:643.
TYPE LOCALITY: China, Shanxi (Shansi), Chao Cheng Shan (= Mt. Nanyan Shan), 8000 ft; 37°54'N, 111°30'E (as restricted by Kaneko, 1992:93).
DISTRIBUTION: Shanxi and Hebei provinces, China (see Kaneko, 1992).

SYNONYMS: *jeholicus*.

COMMENTS: Regarded by Allen (1940), Ellerman and Morrison-Scott (1951), Gromov and Polyakov (1977), and Hinton (1926) as a member of *Evotomys* or *Clethrionomys rufocanus*, but Corbet (1978*c*) and Kaneko (1992) reassociated the species with *Eothenomys*.

Hyperacrius Miller, 1896. N. Am. Fauna, 12:54.

TYPE SPECIES: *Arvicola fertilis* True, 1894.

COMMENTS: Taxonomy, geographic distribution, and ecology reviewed by Phillips (1969). According to Corbet (1978*c*), phylogenetically near *Alticola* but more fossorial, and assigned to Clethrionomyini by Hooper and Hart (1962) and Gromov and Polyakov (1977).

Hyperacrius fertilis (True, 1894). Proc. U. S. Natl. Mus., 17:10.

TYPE LOCALITY: India, Kashmir, Pir Panjal Mtns, 8500 ft.

DISTRIBUTION: Recorded only from Kashmir region and N Pakistan (Corbet, 1978*c*; Phillips, 1969).

SYNONYMS: *aitchisoni, brachelix, zygomaticus*.

Hyperacrius wynnei (Blanford, 1881). J. Asiat. Soc. Bengal,49:244.

TYPE LOCALITY: Pakistan, Murree, 7000 ft (locality of lectotype, as selected by Phillips, 1969).

DISTRIBUTION: N Pakistan, Murree Hills in the lower Kahgan Valley E of Indus River, and W of the Indus in Swat (see Phillips, 1969).

SYNONYMS: *traubi*.

Lagurus Gloger, 1841. Gemein Hand.-Hilfsbuch. Nat., 1:97.

TYPE SPECIES: *Mus lagurus* Pallas, 1773.

SYNONYMS: *Eremiomys*.

COMMENTS: Closely related to *Eolagurus*, both of which are the only extant members of the Tribe Lagurini (Gromov and Polyakov, 1977; Hooper and Hart, 1962; Zagorodnyuk, 1990). Pavlinov and Rossolimo (1987) clarified the status of *Mus lagurus* as the type species instead of *L. migratorius* as indicated by Corbet (1978*c*). Excludes North American *Lemmiscus curtatus* (see that account).

Lagurus lagurus (Pallas, 1773). Reise Prov. Russ. Reichs., 2:704.

TYPE LOCALITY: Kazakhstan, mouth of Ural River.

DISTRIBUTION: Steppes from Ukraine through N Kazakhstan to W Mongolia and NW China (Xinjiang) (see Corbet, 1978*c*).

SYNONYMS: *abacanicus, agressus, altorum, migratorius, occidentalis, saturatus*.

COMMENTS: Chromosomal data summarized and reviewed by Zima and Král (1984*a*).

Lasiopodomys Lataste, 1887. Ann. Mus. Civ. Stor. Nat. Genova, 2a, 4:268.

TYPE SPECIES: *Arvicola brandtii* Radde, 1861.

COMMENTS: Although systematists agree that *Lasiopodomys* belongs in Arvicolini, they have disagreed over its generic status. Some have relegated it to a subgenus of *Microtus* (Allen, 1940; Corbet, 1978*c*; Corbet and Hill, 1991; Ellerman and Morrison-Scott, 1951). However, Hinton (1926) noted the diagnostic features that sets *Lasiopodomys* apart as a genus, a ranking broadly acknowledged by both neontologists and paleontologists (Gromov and Polyakov, 1977; Pavlinov and Rossolimo, 1987; Repenning et al., 1990; Smorkacheva et al., 1990; Zagorodnyuk, 1990; Zheng and Li, 1990). The allocation to *Microtus* has not issued from careful phylogenetic study; until data is obtained suggesting otherwise, *Lasiopodomys* should be retained as a genus.

Lasiopodomys brandtii (Radde, 1861). Melanges Biol. Acad. St. Petersbourg, 3:683.

TYPE LOCALITY: Russia, NE Mongolia, near Tarei-Nor.

DISTRIBUTION: Mongolia and adjacent Transbaikalia, Russia, and the Chinese provinces of E Nei Mongol, Heilungkiang, Jilin, and Hebei.

SYNONYMS: *aga, hangaicus, warringtoni*.

Lasiopodomys fuscus (Büchner, 1889). Wiss. Res. Przewalski Cent.-Asien Reisen, Zool., I:(Säuget.), p. 125.
TYPE LOCALITY: China, Qinghai Prov., "Dy-Tschju River (upper reaches of Yellow and Blue Rivers), approximately 34°N, 93°E" (as given by Ellerman and Morrison-Scott, 1951:682).
DISTRIBUTION: Recorded from Qinghai Prov.; extent of range unresolved.
COMMENTS: Included in *Pitymys leucurus* by Ellerman and Morrison-Scott (1951) and Corbet (1978*c*), but separated as a species and placed in *Lasiopodomys* by Zheng and Wang (1980).

Lasiopodomys mandarinus (Milne-Edwards, 1871). Rech. Hist. Nat. Mammifères, p. 129.
TYPE LOCALITY: China, Shanxi (Shansi), probably near Saratsi.
DISTRIBUTION: C and NE China (Nei Mongolia, Hebei, Shaanxi, Shanxi, Jiangsu, Anhui, Heilongjiang); N Mongolia; Transbaikal region and E and SE Siberia of Russia; Korea; limits of range uncertain.
SYNONYMS: *faeceus, jeholensis, johannes, kishidai, pullus, vinogradovi.*

Lemmiscus Thomas, 1912. Ann. Mag. Nat. Hist., ser. 8, 9:401.
TYPE SPECIES: *Arvicola curtata* Cope, 1868.
COMMENTS: Named as a subgenus of *Lagurus* to segregate New World sagebrush voles from Old World steppe voles. Davis (1939) underscored the morphological separation between New and Old World forms and raised *Lemmiscus* to a genus, a view supported by Carleton's (1981) study of gastric anatomy. Subsequent faunal studies and checklists have variously listed *Lemmiscus* as a genus (Carleton and Musser, 1984; Gromov and Polyakov, 1977) or as a subgenus of *Lagurus* (Hall, 1981; Honacki et al., 1982). Certain morphological traits associate *Lemmiscus* with *Microtus* (Carleton, 1981; Davis, 1939), but chromosomal banding patterns provide little resolution of its phylogenetic affinity (Modi, 1987).

Lemmiscus curtatus (Cope, 1868). Proc. Acad. Nat. Sci. Philadelphia, 20:2.
TYPE LOCALITY: USA, Nevada, Esmeralda Co., Mt. Magruder, Pigeon Spring.
DISTRIBUTION: Sagebrush steppe and desert from S Alberta and SE Saskatchewan, Canada, south to NW Colorado and EC California, including the Columbia Basin of interior Oregon and Washington, USA.
SYNONYMS: *artemisiae, intermedius, levidensis, orbitus, pallidus, pauperrimus.*
COMMENTS: See Carroll and Genoways, 1980 (Mammalian Species, 124, as *Lagurus curtatus*).

Lemmus Link, 1795. Beitr. Naturgesch., 1(2):75.
TYPE SPECIES: *Mus lemmus* Linnaeus, 1758.
SYNONYMS: *Brachyurus, Hypudaeus, Lemnus, Myodes.*
COMMENTS: Nominative genus of Miller's (1896) classic tribe Lemmi, then including *Dicrostonyx* (see that account). Distinctiveness still recognized as the tribe Lemmini, including *Myopus* and *Synaptomys*, a clade believed to represent an early line of arvicoline evolution (Carleton, 1981; Chaline and Graf, 1988; Graf, 1982; Gromov and Polyakov, 1977; Hinton, 1926; Hooper and Hart, 1962; Koenigswald, 1980). Fossil history reviewed by Koenigswald and Martin (1984) and zoogeography discussed by Rausch and Rausch (1975*b*). Old World taxa reviewed by Corbet (1978*c*, 1984), Gromov and Polyakov (1977), and Pavlinov and Rossolimo (1987). Chromosomal information summarized and reviewed by Zima and Král (1984*a*). Also see account of *Myopus*.

Lemmus amurensis Vinogradov, 1924. Ann. Mag. Nat. Hist., ser. 9, 14:186.
TYPE LOCALITY: Siberia, Pikan, on River Zeya, a tributary of the Amur River.
DISTRIBUTION: Larch taiga of E Siberia, from east of Lake Baikal through the upper Amur River basin, and north in the Verkhoyansk and Cherskogo Mtns to the River Omolon (see Chernyavskii et al., 1980).
SYNONYMS: *ognevi.*
COMMENTS: A distinctive species revised by Chernyavskii et al. (1980). Additional chromosomal data analyzed by Gileva et al. (1984) in context of distinguishing species of *Lemmus*.

Lemmus lemmus (Linnaeus, 1758). Syst. Nat., 10th ed., 1:59.

TYPE LOCALITY: Sweden, Lappmark.

DISTRIBUTION: Mountains of Scandinavia and tundra from Lapland to the White Sea (Corbet, 1978*c*; see Tast, 1982*a*, for European range).

SYNONYMS: *borealis, iretator, migratorius, norvegicus.*

COMMENTS: European populations reviewed by Tast (1982*a*).

Lemmus sibiricus (Kerr, 1792). *In* Linnaeus, Anim. Kingdom, p. 241.

TYPE LOCALITY: Russia, Yamalo-Nenetskaya Nats. Okr., between Polar Ural Mtns and lower course of Ob River.

DISTRIBUTION: Holarctic tundra landscapes: in Palearctic, from White Sea, W Russia, to Chukotski Peninsula, NE Siberia, and Kamchatka; including Nunivak and St. George islands in the Bering Sea; in Nearctic, from W Alaska east to Baffin Island and Hudson Bay, and south in the Rocky Mtns to C British Columbia, Canada.

SYNONYMS: *alascensis, bungei, chrysogaster, flavescens, harroldi, helvolus, iterator, kittlitzi, minor, minusculus, nigripes, novosibiricus, obensis, paulus, phaiocephalus, portenkoi, subarticus, trimucronatus, xanthotrichus, yukonensis.*

COMMENTS: North American races revised, as *L. trimucronatus*, by Davis (1944) and retained as such by Hall and Cockrum (1953) and Hall and Kelson (1959). Rausch (1953) proposed the synonymy of *trimucronatus* and *nigripes* under Old World *L. sibiricus*, a taxonomic arrangement elaborated by Rausch and Rausch (1975*b*) and maintained in subsequent faunal works (Banfield, 1974; Hall, 1981; Jones et al., 1986). Gileva (1983) and Gileva et al. (1984) believed that the cytogenetic peculiarities of *chrysogaster* confirm its independence as a species relative to *L. amurensis, L. lemmus*, and *L. sibiricus*, and suggested it may be conspecific with North American *L. trimucronatus*. The sample they identified as *chrysogaster*, however, comes from the Chukotski Peninsula on the coast of the East Siberian Sea, not from the west coast of the Okhotsk Sea, the type locality of *chrysogaster*. Pavlinov and Rossolimo (1987) retained, with reservation, *chrysogaster* in the synonymy of *L. sibiricus* pending further study and accurate identification of lemmings from the Chukotski Peninsula. Corbet and Hill (1991) continued to recognize the St. George Island form *nigripes* as a species.

Microtus Schrank, 1798. Fauna Boica, 1(1):72.

TYPE SPECIES: *Microtus terrestris* Schrank, 1798 (= *Mus arvalis* Pallas, 1778).

SYNONYMS: *Agricola, Alexandromys, Ammomys, Arbusticola, Arvalomys, Aulacomys, Campicola, Chilotus, Euarvicola, Hemiotomys, Herpetomys, Iberomys, Lemmimicrotus, Meridiopitymys, Micrurus, Mynomes, Neodon, Orthriomys, Pallasiinus, Parapitymys, Pedomys, Phaiomys, Pinemys, Pitymys, Psammomys* (of Le Conte, 1830, not Cretzschmar, 1828), *Stenocranius, Sumeriomys, Suranomys, Sylvicola, Terricola, Tetramerodon.*

COMMENTS: Nowhere are the explosiveness and recency of arvicoline evolution more dramatically highlighted than by the inconsistency of systematic treatment of genus-group taxa to be subsumed by *Microtus*. No consensus exists concerning the morphological limits or monophyly of many of these taxa, a situation which in part reflects the narrow reliance of our classifications on dental characters undergoing rapid change (see Guthrie, 1971; Koenigswald, 1980). Such variability and inconsistency of systematic opinion are epitomized by the taxon *Pitymys*. As noted by Carleton and Musser (1984:321), "Generally, paleontologists and European mammalogists accord *Pitymys* separate generic status (Corbet, 1978*c*; Koenigswald, 1980; Repenning, 1983), whereas North American workers view it as a subgenus of *Microtus* (Hall, 1981; Jones et al., 1975)." The geographic split among systematists is by itself instructive, as is the nature of the character base consulted by paleontologists versus neontologists. Similar disputes have surrounded the taxonomic history of other genus-group taxa associated with *Microtus*, such as *Arvicola, Blanfordimys, Chionomys, Lasiopodomys, Neodon, Phaiomys*, and *Proedromys*.

Zagorodnyuk (1990) developed an interesting reclassification of *Microtus* and its kin. Although probably wrong on details, we believe that it offers an important philosophical alternative in the continuing examination of relationships among *Microtus*-like forms. In his arrangement of Arvicolini, Zagorodnyuk emphasized hypotheses of intracontinental origin and regional diversification of major clades of *Microtus*. Thus, pitymyine forms of the Old World (*Terricola*) are segregated from those of the New World (*Pitymys*); New World common voles (*Mynomes*) are separated from Old World voles such as *Microtus* proper, *Alexandromys*, and *Agricola*; and the invasion of the New World semiaquatic niche is recognized (*Aulacomys*) as independent of the Old World water-vole radiation (*Arvicola*). If we interpret them correctly, such a viewpoint is consistent with the preliminary results emerging from electrophoretic studies (Chaline and Graf, 1988; Graf, 1982; Moore and Janecek, 1990) and with some paleontological perspectives (e.g., Chaline, 1974). The more traditional notion of intercontinental dispersal and broad transcontinental distributions still receives stong support in the publications of Martin (1974, 1987), Repenning (1980, 1983), Repenning et al. (1990), and van der Muelen (1978). Although we question Zagorodnyuk's cardinal reliance on chomosomal traits and do not wholly embrace his ranking of taxa (e.g., *Neodon* and *Terricola* as genera), his classification of *Microtus* and related forms deserves serious attention in future studies.

See accounts of *Blanfordimys*, *Chionomys*, *Lasiopodomys*, and *Proedromys*, often included in *Microtus* but which are here treated as genera. North American forms revised by Bailey (1900) and taxonomy updated by Hall and Cockrum (1953) and Hall (1981); many aspects of paleontology, taxonomy, zoogeography, and anatomy covered in Tamarin (1985). For synoptic coverage of the more diverse Palearctic *Microtus* fauna, see Corbet (1978*c*), Gromov and Polyakov (1977), Niethammer and Krapp (1982*a*), and Ognev (1963*b*, 1964).

Microtus abbreviatus Miller, 1899. Proc. Biol. Soc. Washington, 13:13.

TYPE LOCALITY: USA, Alaska, Bering Sea, Hall Island.

DISTRIBUTION: Hall and St. Matthew Islands, Bering Sea (Alaska, USA).

SYNONYMS: *fisheri*.

COMMENTS: An insular relative of *M. miurus* of the Alaskan mainland, generally retained as a species (Fedyk, 1970; Rausch and Rausch, 1968). Karyotype reported by Rausch and Rausch (1968) and chromosomal affinities within subgenus *Stenocranius* discussed by Fedyk (1970). Also see accounts of *M. gregalis* and *M. miurus*.

Microtus agrestis (Linnaeus, 1761). Fauna Suecica, 2nd ed., p. 11.

TYPE LOCALITY: Sweden, Uppsala.

DISTRIBUTION: Britain and nearby small islands, Scandinavia, and France east through Europe and Siberia to Lena River; south to Pyrennes of France and Spain, and to N Portugal; east to N Yugoslavia, S Urals, Altai Mtns, NW China (Xinjiang), and Lake Baikal region (Corbet, 1978*c*; Krapp and Niethammer, 1982).

SYNONYMS: *angustifrons, arcturus, argyropoli, argyropuli, argyropuloi, armoricanus, bailloni, campestris, britannicus, carinthiacus, enez-groezi, estiae, exsul, fiona, gregarius, hirta, insularis, intermedia, latifrons, levenedii, luch, macgillivrayi, mial, mongol, neglectus, nigra, nigricans, ognevi, orioecus, pannonicus, punctus, rozianus, rufa, scaloni, tridentinus, wettsteini*.

COMMENTS: Subgenus *Agricola*, *agrestis* species group *sensu* Zagorodnyuk (1990). Although regarded as conspecific with North American *M. pennsylvanicus* on morphological grounds (Klimkiewicz, 1970), chomosomal differences led Vorontsov and Lyapunova (1986) to conclude that their similarities represented convergence, not phylogenetic alliance; instead they remarked upon possible closer relationship of *M. agrestis* to North American *M. chrotorrhinus*. Zagorodnyuk (1990) emphasized this evolutionary distance by placing *M. pennsylvanicus* in the subgenus *Mynomes*. Chromosomal data summarized and reviewed by Zima and Král (1984*a*); European populations reviewed by Krapp and Niethammer (1982).

Microtus arvalis (Pallas, 1778). Nova Spec. Quadr. Glir. Ord., p. 78.

TYPE LOCALITY: Germany; neotype from Leningrad Oblast.

DISTRIBUTION: From C and N Spain throughout Europe (including Denmark) to western margin of Black Sea in the south and northeast to Kirov region (west of the Urals) in Russia; also populations on the Orkney Islands, Guernsey (Channel Islands), and Yeu (France) (see Niethammer and Krapp, 1982*b*, and Zagorodnyuk, 1991*a*).

SYNONYMS: *albus, arvensis, angularis, assimilis, asturianus, brauneri, calypsus, caucasicus, cimbricus, contigua, cunicularius, depressa, duplicatus, flava, fulva, galliardi, grandis, heptneri, howelkae, igmanensis, incertus, incognitus, levis, meldensis, meridianus, orcadensis, oyaensis, principalis, ronaldshaiensis, rousiensis, rufescentefuscus, ruthenus, sandayensis, sarnius, simplex, terrestris* (of Schrank, 1798, not Linnaeus, 1758), *variabilis, vulgaris, westrae.*

COMMENTS: Subgenus *Microtus*, *arvalis* species group *sensu* Zagorodnyuk (1990). Taxonomy and distribution generally reviewed by Corbet (1978*c*) and European populations by Niethammer and Krapp (1982). Morphological variation among samples from NE Spain was documented by Gosalbez and Sans-Coma (1977). Chromosomal data and comparisons with other species summarized by Zima and Král (1984*a*) and Burgos et al. (1989). Some of the synonyms listed and much of the southern and eastern distribution outlined for *M. arvalis* by Corbet (1978*c*) actually refer to either *M. obscurus* or *M. rossiaemeridionalis* (see those accounts).

Microtus bavaricus König, 1962. Senckenberg. Biol., 43:2.

TYPE LOCALITY: Germany, Bavarian Alps, Garmisch-Partenkirchen, 730 m.

DISTRIBUTION: Germany, Bavarian Alps (see König, 1982).

STATUS: IUCN - Extinct.

COMMENTS: Subgenus *Terricola*, *subterraneus* species group *sensu* Chaline et al. (1988).

Microtus breweri (Baird, 1858). Mammalia, *in* Repts. U.S. Expl. Surv., 8(1):525.

TYPE LOCALITY: USA, Massachusetts, Muskeget Island, off Nantucket.

DISTRIBUTION: Known only from the type locality.

STATUS: IUCN - Rare.

COMMENTS: Subgenus *Mynomes*, *pennsylvanicus* species group *sensu* Zagorodnyuk (1990). An insular vicariant of *M. pennsylvanicus*, the two are inseparable karyotypically (Fivush et al., 1975; Modi, 1986), marginally distinct electrophoretically (Kohn and Tamarin, 1978), but morphologically sharply discrete (Bailey, 1900; Miller, 1896; Moyer et al., 1988). Although posited as conspecific with *M. pennsylvanicus* (e.g., Corbet and Hill, 1991; Jones et al., 1986; Modi, 1986), Moyer et al. (1988) mustered convincing evidence for the retention of *breweri* as a species. See Tamarin and Kunz (1974, Mammalian Species, 45).

Microtus cabrerae Thomas, 1906. Ann. Mag. Nat. Hist., ser. 7, 17:576.

TYPE LOCALITY: Spain, Madrid Prov., Sierra de Guadarrama, near Rascafria.

DISTRIBUTION: Recorded only from Spain, Portugal, and French side of the Pyrenees (see Niethammer, 1982*f*, and Corbet, 1984).

SYNONYMS: *dentatus.*

COMMENTS: Subgenus *Agricola*, *agrestis* species group *sensu* Zagorodnyuk (1990). Reviewed by Niethammer (1982*f*). Close relative is the Pleistocene *M. brecciensis* from Spain and S France (see Niethammer, 1982*f*). Chaline (1974) used the subgenus *Iberomys*, a name superceded by *Agricola*, for *dentatus* and *brecciensis*. Biochemical and chromosomal data reported by Millet et al. (1982), Burgos et al. (1989), and Jimenez et al. (1991); other chromosomal data summarized and reviewed by Zima and Král (1984*a*).

Microtus californicus (Peale, 1848). Mammalia *in* Repts. U.S. Expl. Surv., 8:46.

TYPE LOCALITY: USA, California, Santa Clara Co., vicinity of San Francisco Bay, San Francisquito Creek near Palo Alto (as fixed by Kellogg, 1918:5).

DISTRIBUTION: Oak woodlands and grasslands of Pacific coast, from SW Oregon through California, USA, to N Baja California, Mexico.

STATUS: U.S. ESA and IUCN - Endangered as *M. c. scirpensis.*

SYNONYMS: *aequivocatus, aestuarinus, constrictus, edax, eximius, grinnelli, halophilus, huperuthrus, kernensis, mariposae, mohavensis, neglectus, paludicola, perplexibilis, sanctidiegi, sanpabloensis, scirpensis, stephensi, trowbridgii, vallicola.*

COMMENTS: Broadly affiliated with other North American species of *Microtus*, but evidence for nearest specific relative contradictory (compare assessments of Anderson, 1959; Hooper and Hart, 1962; Moore and Janecek, 1990). Zagorodnyuk (1990) acknowledged enigmatic phyletic stature as sole member of *californicus* species group, subgenus *Mynomes*. Geographic races delineated by Kellogg (1918); Gill (1980) recorded instances of sterility in hybrids between *M. c. californicus* and *M. c. stephensi.*

Microtus canicaudus Miller, 1897. Proc. Biol. Soc. Washington, 11:67.

TYPE LOCALITY: USA, Oregon, Polk Co., Willamette Valley, McCoy.

DISTRIBUTION: Willamette Valley of NW Oregon and adjacent Washington, USA.

COMMENTS: Subgenus *Mynomes, montanus* species group *sensu* Zagorodnyuk (1990). Reduced to a subspecies of *M. montanus* by Hall and Kelson (1951); resurrected to specific status based on karyotypic and electrophoretic evidence (Hsu and Johnson, 1970; Johnson, 1968; Modi, 1986). Viewed as sibling species to *M. montanus* (see Hoffmann and Koeppl, 1985; Modi, 1986, 1987). See Verts and Carraway (1987, Mammalian Species, 267).

Microtus chrotorrhinus (Miller, 1894). Proc. Boston Soc. Nat. Hist., 26:190.

TYPE LOCALITY: USA, New Hampshire, Coos Co., Mount Washington, head of Tuckerman's Ravine, 5300 ft.

DISTRIBUTION: S Labrador southwest through S Quebec and Ontario, Canada, to NE Minnesota, USA; south in Appalachian Mtns to E Tennessee and W North Carolina, USA.

SYNONYMS: *carolinensis, ravus.*

COMMENTS: Subgenus *Aulacomys,* sole member of *chrotorrhinus* species group *sensu* Zagorodnyuk (1990). Although conventionally viewed as closely related to (Anderson, 1960), if not conspecific with (Hall and Kelson, 1959) *M. xanthognathus,* morphological and chromosomal traits reveal their more distant kinship (Bailey, 1900; Guilday, 1982; R. L. Martin, 1973, 1979; Rausch and Rausch, 1974). Genic variation evaluated by Kilpatrick and Crowell (1985). See Kirkland and Jannett (1982, Mammalian Species, 180).

Microtus daghestanicus (Shidlovskii, 1919). Raboty Zemskoi Opytnoi Stantsi, 2:12.

TYPE LOCALITY: Russia, Daghestan, Caucasus Mtns, Karda.

DISTRIBUTION: Caucasus Mtns, S Daghestan.

SYNONYMS: *intermedius, suramensis.*

COMMENTS: Subgenus *Terricola, subterraneus* species group *sensu* Zagorodnyuk (1990). Included in *Pitymys subterraneus* by Ellerman and Morrison-Scott (1951) and in *P. majori* by Corbet (1978*c*). However, Kratochvíl and Král (1974) and Baskevich et al. (1984) provided evidence supporting *daghestanicus* as a separate species, a view endorsed by Pavlinov and Rossolimo (1987). Zima and Král (1984*a*) and Achverdjan et al. (1992) reviewed chromosomal information. Corbet (1978*c*) discussed the status of the replacement name *suramensis.*

Microtus duodecimcostatus de Selys-Longchamps, 1839. Rev. Zool. Paris, p. 8.

TYPE LOCALITY: France, Gard, Montpellier.

DISTRIBUTION: SE France, E and S Spain, and Portugal.

SYNONYMS: *centralis, flavescens, fuscus, ibericus, pascuus, provincialis, regulus.*

COMMENTS: Subgenus *Terricola, duodecimcostatus* species group *sensu* Chaline et al. (1988). Craniometric analyses contrasting this species with *M. gerbei* and *M. lusitanicus* documented by Spitz (1978). Taxonomy and biology reviewed by Niethammer (1982*i*); chromosomal data summarized by Zima and Král (1984*a*); meiotic behavior of sex chromosomes reported by Carnero et al. (1991).

Microtus evoronensis Kovalskaya and Sokolov, 1980. Zool. Zh., 59:1410.

TYPE LOCALITY: Russia, Khabarovsk Krai, Lake Evoron basin, Devyatka River.

DISTRIBUTION: Known only from the type locality.

COMMENTS: Subgenus *Alexandromys, maximowiczii* species group *sensu* Zagorodnyuk (1990). Considered by Meier (1983) to be closely related to *M. mujanensis* and *M. maximowiczii.*

Microtus felteni Malec and Storch, 1963. Senckenberg. Biol., 44:171.

TYPE LOCALITY: Yugoslavia, Rep. Makedonija (Macedonia), Pelister Mtns, near Trnovo-Magarevo.

DISTRIBUTION: S Yugoslavia and Greece (see Niethammer, 1982*h*, 1987*b*).

COMMENTS: Subgenus *Terricola, savii* species group *sensu* Chaline et al. (1988). Reviewed by Niethammer (1982*h*). Chromosomal data summarized and reviewed by Zima and Král (1984*a*). Biochemical comparisons among *M. felteni, M. subterraneus,* and other voles are documented by Gill et al. (1987).

Microtus fortis Büchner, 1889. Wiss. Res. Przewalski's Cent.- Asien. Reisen, Zool., I:(Säugeth.), p. 99.

TYPE LOCALITY: China, Nei Mongol, Ordos Desert, Huang Ho Valley, Sujan.

DISTRIBUTION: Transbaikalia and Amur region south through Nei Mongol and E China to lower Yangtse Valley and Fujian (see Kovalskaya et al., 1988).

SYNONYMS: *calamarum, dolichocephalus, fujianensis, michnoi, pelliceus, uliginosus.*

COMMENTS: Subgenus *Alexandromys*. Geographic variation in chromosomal traits and its significance was reported by Kovalskaya et al. (1988; 1991). Most researchers view this species as a close phylogenetic relative of *M. mongolicus* (Meier, 1983; Radjabli et al., 1984). Zagorodnyuk (1990) listed *M. fortis* as the only member of its species group and related to the *M. middendorffi* group, which contains *M. middendorffi, M. miurus, M. mongolicus,* and *M. sachalinensis*. Reviewed by Corbet (1978*c*) and Gromov and Polyakov (1977).

Microtus gerbei (Gerbe, 1879). Le Naturaliste, 1:51.

TYPE LOCALITY: France, Loire-Inferieure, Dreneuf.

DISTRIBUTION: SW France, and the Pyrenees Mtns of France and Spain (see Krapp, 1982*d*).

SYNONYMS: *brunneus, pyrenaicus.*

COMMENTS: Subgenus *Terricola*, *savii* species group *sensu* both Chaline et al. (1988) and Zagorodnyuk (1990). This species was reviewed by Krapp (1982*d*) under the name *pyrenaicus*, but according to Spitz (1978) the latter is a *nomen dubium* and *gerbei* assumes priority. The species was included in *savii* by Corbet (1978*c*), but chromosomal evidence and craniometric analysis underscore its specific integrity (Kratochvíl and Král, 1974; Spitz, 1978). Chromosomal data (as *pyrenaicus*) summarized by Zima and Král (1984*a*).

Microtus gregalis (Pallas, 1779). Nova Spec. Quad. Glir. Ord., p. 238.

TYPE LOCALITY: Siberia, E of Chulym River.

DISTRIBUTION: Palearctic tundra of Siberia from White Sea to far northeast, and wooded steppe from S Urals east to Amur and NE China (Heilongjiang), south to Aral Sea, Pamirs, Tien Shan and Altai Mtns, N Mongolia, and NW China (Xinjiang) (see Corbet, 1978*c*; Rausch, 1964).

SYNONYMS: *angustus, brevicauda, buturlini, castaneus, dolguschini, dukelskiae, eversmanni, kossogolicus, major, montosus, nordenskioldi, pallasii, raddei, ravidulus, sirtalaensis, slowzovi, talassicus, tarbagataicus, tianschanicus, tundrae, unguiculatus, zachvatkini.*

COMMENTS: Subgenus *Stenocranius*. The only Asian species in the subgenus (Zagorodnyuk, 1990). Intraspecific variation in chromosomal traits among Mongolian samples documented by Kovalskaya (1989) and earlier chromosomal data summarized and reviewed by Zima and Král (1984*a*). Based on morphological and zoogeographic criteria, Rausch (1964) considered the North American *miurus* to be conspecific with Asian *gregalis*, a connection refuted by Fedyk (1970) with chromosomal data and by Vorontsov and Lyapunova (1986). Zagorodnyuk (1990) emphasized the karyotypic divergence between *gregalis* and *miurus* by listing the latter in the *M. middendorffi* species group along with *M. mongolicus* and *M. sachalinensis*, subgenus *Alexandromys*.

Microtus guatemalensis Merriam, 1898. Proc. Biol. Soc. Washington, 12:108.

TYPE LOCALITY: Guatemala, Huehuetenango Dept., Todos Santos, 10,000 ft.

DISTRIBUTION: Highland meadows of C Chiapas, Mexico, and C Guatemala.

COMMENTS: Type species of *Herpetomys*, usually placed in *Microtus*, with (Bailey, 1900) or without (Hall, 1981) subgeneric ranking; or placed in *Pitymys* when used as a genus (Honacki et al., 1982; Martin, 1974).

Microtus guentheri (Danford and Alston, 1880). Proc. Zool. Soc. London, 1880:62.

TYPE LOCALITY: Turkey, Maras Prov., Taurus Mtns, near Maras (= Marash).

DISTRIBUTION: S Bulgaria, S Yugoslavia, E Greece, and W Turkey (see Niethammer, 1982*e*).

SYNONYMS: *hartingi, lydius, macedonicus, martinoi, strandzensis.*

COMMENTS: Subgenus *Microtus*, *socialis* species group *sensu* Zagorodnyuk (1990). Gromov and Polyakov (1977) and Pavlinov and Rossolimo (1987) placed this species in subgenus *Sumeriomys*, but Zagorodnyuk (1990) left it in the subgenus *Microtus*. Corbet (1978*c*) initially included *guentheri* in *M. socialis* but later recognized its specific distinctness (Corbet, 1984), which was earlier demonstrated by Felten et al. (1971) and Morlok (1978). Harrison and Bates (1991) continued to unite *guentheri* with *M. socialis*. Reviewed by Niethammer (1982*e*); chromosomal data summarized and reviewed by Zima and Král (1984*a*).

Microtus hyperboreus Vinogradov, 1934. Trav. L'Inst. Zool. Acad. Sci., 1933:1.
TYPE LOCALITY: Siberia, Verhoiansk Mtns.
DISTRIBUTION: NE Siberia, basin of Yana River, Berhoiansk Range, and Taimyr Peninsula (see Ellerman and Morrison-Scott, 1951).
SYNONYMS: *swerevi*.
COMMENTS: Subgenus *Alexandromys*. Reviewed by Gromov and Polyakov (1977). Corbet (1978*c*) included *hyperboreus* in *M. middendorffi* because Gileva (1972) had demonstrated their complete interfertility. Pavlinov and Rossolimo (1987), however, recognized *M. hyperboreus*, noting that the two are clearly morphologically distinct and that material in published studies probably omitted true *hyperboreus*. Meier (1983) recognized the two species as closely related and noted that no study of variability within *middendorffi* had been made, and more importantly, that insufficient material from the type locality had been analyzed. Vorontsov and Lyapunova (1976) considered *M. hyperboreus* and *M. middendorffi* to be chromosomally closely related to North American *M. miurus*.

Microtus irani Thomas, 1921. J. Bombay Nat. Hist. Soc., 27:581.
TYPE LOCALITY: Iran, Fars Prov., Shiraz, Bagh-i-Rezi.
DISTRIBUTION: E Turkey, N Syria, Lebanon, Israel, W Jordan, Cyrenaica in Libya, N Iraq, W and N Iran, and Kopet Dag Mtns in Turkmenistan.
SYNONYMS: *mustersi, paradoxus, mystacinus*.
COMMENTS: Subgenus *Microtus*, *socialis* species group *sensu* Zagorodnyuk (1990). Pavlinov and Rossolimo (1987) listed *M. irani* in the subgenus *Sumeriomys*, a name not recognized by Zagorodnyuk (1990). Corbet (1978*c*) and Harrison and Bates (1991) included *irani* in *M. socialis*, but the two are distinct species as documented by Morlok (1978) and Kock and Nader (1983), who also noted that most records of *M. guentheri* from Syria, Lebanon, and Israel are *M. irani* and that *philistinus* may be an older name for *irani*. The isolated Lybian population was reviewed, under *mustersi*, by Ranck (1968). The status of *paradoxus* (from Kopet Dag Mtns) deserves re-evaluation: Pavlinov and Rossolimo (1987) treated it as a synonym of *irani* but Zagorodnyuk (1990) maintained it as distinct.

Microtus irene (Thomas, 1911). Abstr. Proc. Zool. Soc. London, 1911(90):5.
TYPE LOCALITY: China, Sichuan, Tatsienlu.
DISTRIBUTION: High mtns in N Burma and the Chinese provinces of E Xizang (Feng et al., 1986), Yunnan, Sichuan, and S Gansu (see Niethammer, 1970).
SYNONYMS: *forresti, oniscus*.
COMMENTS: Subgenus *Neodon*. Ellerman and Morrison-Scott (1951) listed *irene* as a valid species, but Corbet (1978*c*) allocated it to *Pitymys sikimensis*, citing Gruber's (1969) "detailed study" as substantiation. However, the union of *irene* and *sikimensis* is insufficiently demonstated by the data of Gruber, who even summarized the trenchant differences in body size and molar occlusal patterns that clearly distinguish the two. Furthermore, Feng et al. (1986) recorded *irene* from E Tibet and *sikimensis* from just west of that locality, without evidence of overlap in the diagnostic features of each species.

Microtus juldaschi (Severtzov, 1879). Sap. Turk. Otd. Obsh.Lubit. Estestv., 1:63.
TYPE LOCALITY: Kirghizia, Kara Kul Lake basin, near Aksu.
DISTRIBUTION: Tien Shan and Pamir Mtns in Kirghizia and Tadzhikistan west to about Samarkand, south to NE Afghanistan and N Pakistan (Roberts, 1977), and east to NW Xizang (Tibet) (see Feng et al., 1986; Niethammer, 1970).
SYNONYMS: *carruthersi, pamirensis, talassensis*.
COMMENTS: Subgenus *Neodon*, a member of its own species group (Zagorodnyuk, 1990). The status of the distinctive *carruthers* is unresolved: treated as a separate species (Ellerman and Morrison-Scott, 1951; Ognev, 1964); then synonymized with *juldaschi* (Corbet, 1978*c*); and listed as a synonym of the latter but with question (Gromov and Polyakov, 1977; Pavlinov and Rossolimo, 1987). Chromosomal analyses by Gileva et al. (1982) revealed two distinctive groups, which they considered to be two species

but with probable different boundaries than usually described for *juldaschi* and *carruthersi*.

Microtus kermanensis Roguin, 1988. Rev. Suisse Zool., 95:601.
TYPE LOCALITY: Iran, Kerman Prov., Zahrud-e Bala, 70 km S Kerman, 2700 m.
DISTRIBUTION: Recorded from two isolated mountain populations in SE Iran: Kuh-e Laleh-Zar and Kuh-e Hazar S of Kerman (see Roguin, 1988).
COMMENTS: Subgenus *Microtus*, *arvalis* species group *sensu* Zagorodnyuk (1990).

Microtus kirgisorum Ognev, 1950. [Mammals of USSR and Adjacent Countries], 7:219.
TYPE LOCALITY: Kazakhstan, Tuyouk, Khirgiski Mtns.
DISTRIBUTION: Recorded from S Kazakhstan, Kirghizia, Tadzhikistan, and SE Turkmenistan; may also occur in N Afghanistan.
COMMENTS: Subgenus *Microtus*, *arvalis* species group *sensu* Zagorodnyuk (1990). Originally described as a subspecies of *M. arvalis* but later elevated to a separate species (Meier, 1983; Meier et al., 1981), as observed by Corbet (1984) and Pavlinov and Rossolimo (1987). Chromosomal data presented by Orlov et al. (1983) in context of phylogenetic study among members of the *M. arvalis* group.

Microtus leucurus (Blyth, 1863). J. Asiat. Soc. Bengal, 32:89.
TYPE LOCALITY: India, Ladakh, near Lake Chomoriri (= Tsomoriri).
DISTRIBUTION: Chinese provinces of S Qinghai (Zheng and Wang, 1980) and Xizang (Feng et al., 1986) on the Tibetan Plateau, and high altitudes in the Himalayas west to Kashmir (see Niethammer, 1970).
SYNONYMS: *everesti, petulans, strauchi, tsaidamensis, waltoni, zadoensis*.
COMMENTS: Classified by Zagorodnyuk (1990) under the subgenus *Phaiomys* of the genus *Neodon*; others have recognized *Phaiomys* as a genus (Repenning et al., 1990). Whatever its ranking, *leucurus* may date from the early evolution of the genus, as suggested by its association with the late Pliocene *Allophaiomys*, a taxon regarded by most as ancestral to *Microtus* (see Martin, 1989*b*; Repenning et al., 1990). As classified in the genus *Microtus*, this species can still be called *leucurus*; the older *leucurus* Gerbe, 1852, is a synonym of *nivalis*, a species which has been reallocated from *Microtus* to *Chionomys* (see that account).

Microtus limnophilus Büchner, 1889. Wiss. Res. Przewalski Cent.-Asien. Reis. Zool., I:(Säugeth.), p. 110.
TYPE LOCALITY: China, Qinghai.
DISTRIBUTION: From Qinghai Prov. to W Mongolia.
COMMENTS: Subgenus *Pallasiinus*, *oeconomus* species group *sensu* Zagorodnyuk (1990). Formerly included in *M. oeconomus* but separated as a species parapatric with *M. oeconomus* in W Mongolia (Malygin et al., 1990, and references therein).

Microtus longicaudus (Merriam, 1888). Am. Nat., 22:934.
TYPE LOCALITY: USA, South Dakota, Custer Co., Black Hills, Custer, 5500 ft.
DISTRIBUTION: Rocky Mountains and adjacent foothills, from E Alaska and N Yukon, south through British Columbia and SW Alberta, Canada, to E California and W Colorado; including Pacific coastal taiga to N California; disjunct southern pockets in S California, Arizona, and New Mexico, USA.
SYNONYMS: *abditus, alticola, angusticeps, angustus, baileyi, bernardinus, cautus, coronarius, halli, incanus, latus, leucophaeus, littoralis, macrurus, mordax, sierrae, vellerosus*.
COMMENTS: Sometimes viewed as a Nearctic member of *Chionomys* (Anderson, 1959), or allocated to subgenus *Microtus* (Chaline, 1974; Hall, 1981), or to subgenus *Aulacomys* (Zagorodnyuk, 1990). Although strongly differentiated relative to other North American *Microtus* (e.g., Hooper and Hart, 1962; Modi, 1987; Moore and Janecek, 1990), the phyletic affinity of *M. longicaudus* lies with this complex and not Old World *Chionomys* (Chaline and Graf, 1988; Gromov and Polyakov, 1977; Zagorodnyuk, 1990). The extensive karyotypic variation reported (Judd and Cross, 1980) suggests the need for taxonomic revision. Jones et al. (1986) viewed *coronarius* as an insular derivative and subspecies of *M. longicaudus*. See Smolen and Keller (1987, Mammalian Species, 271).

Microtus lusitanicus (Gerbe, 1879). Rev. Mag. Zool. Paris, Ser. 3, 7:44.
TYPE LOCALITY: Portugal.

DISTRIBUTION: Portugal, NW Spain, and SW France (see Niethammer, 1982*j*).
SYNONYMS: *depressus, gerritmilleri, hurdanensis, mariae, pelandonius, planiceps.*
COMMENTS: Subgenus *Terricola, duodecimcostatus* species group *sensu* Chaline et al. (1988). Spitz (1978) provided craniometric analyses in context of contrasting *M. lusitanicus* with *M. duodecimcostatus* and *M. gerbei.* Reviewed by Niethammer (1982*j*) and Corbet (1984); chromosomal data reviewed by Zima and Král (1984*a*). Both Graf (1982) and Chaline and Graf (1988) apparently used *mariae* for this species, which reflects Winking's (1976) usage in separating the latter from *M. duodecimcostatus;* however, the proper name is *lusitanicus* (Niethammer (1982*j*).

Microtus majori Thomas, 1906. Ann. Mag. Nat. Hist., ser. 7, 17:419.
TYPE LOCALITY: Turkey, Trabzon Prov., Sumela (= Meryemana), 30 mi S Trebizond (= Trabzon).
DISTRIBUTION: Mt Olympus in Greece (Niethammer, 1987*b*), S Yugoslavia, N and W Turkey, and W and N Caucasus (see Storch, 1982).
SYNONYMS: *ciscaucasicus, colchicus, dinniki, fingeri, labensis, rubelianus, transcaucasicus, vinogradovi* (see Pavlinov and Rossolimo, 1987).
COMMENTS: Subgenus *Terricola.* Included in the *subterraneus* species group by Chaline et al. (1988); considered the sole member of its own species group by Zagorodnyuk (1990). Reviewed by Kratochvíl and Král (1974), Storch (1982) and Corbet (1978*c*, 1984). The European Turkish population was discussed by Kivanc (1986). Zima and Král (1984*a*) and Achverdjan et al. (1992) summarized comparative chromosomal data.

Microtus maximowiczii (Schrenk, 1859). Reisen und Forsch. *in* Säugeth. Amurlande St. Petersburg., 1:140.
TYPE LOCALITY: Russia, Chita Oblast, upper Amur region, mouth of Omutnaya River.
DISTRIBUTION: From E shore of Lake Baikal to upper Amur region, E Mongolia, and NE China (Heilongjiang).
SYNONYMS: *ungurensis.*
COMMENTS: Subgenus *Alexandromys, maximowiczii* species group *sensu* Zagorodnyuk (1990). Reviewed by Gromov and Polyakov (1977), Corbet (1978*c*), Orlov and Kovalskaya (1978), and Meier (1983). Additional intraspecific analyses of chromosomal polymorphism provided by Kovalskaya (1977).

Microtus mexicanus (Saussure, 1861). Rev. Mag. Zool. Paris, Ser. 2, 13:3.
TYPE LOCALITY: Mexico, Puebla, Volcan de Orizaba.
DISTRIBUTION: Highly dissected in mountains from extreme S Utah and SW Colorado, USA, south in Sierra Madres through interior Mexico to C Oaxaca.
STATUS: U.S. ESA and IUCN - Endangered as *M. m. hualpaiensis;* IUCN - Indeterminate as *M. m. navaho.*
SYNONYMS: *fulviventer, fundatus, guadalupensis, hualpaiensis, madrensis, mogollonensis, navaho, neveriae, phaeus, salvus, subsimus.*
COMMENTS: Interspecific affinities unclear: affiliated with *Microtus sensu stricto,* in particular *M. californicus,* by Anderson (1959, 1960); associated with members of subgenus *Pitymys* by Hooper and Hart (1962); genic similarity to *M. montanus* and *M. pennsylvanicus* disclosed by Moore and Janecek (1990); also see Hoffmann and Koeppl, 1985); chromosomal banding data uninformative (Modi, 1987). Morphometric, electrophoretic, and karyotypic variation among isolated populations in southwestern USA studied by Wilhelm (1982). Musser (1964) reduced *fulviventer* to a subspecies of *M. mexicanus.*

Microtus middendorffi (Poliakov, 1881). Mem. Acad. Imp. Sci. St. Petersbourg, 39:70.
TYPE LOCALITY: Russia, Krasnoyarsk Krai, Taimyr Peninsula.
DISTRIBUTION: NC Siberia, from the Polar Ural Mtns to the N Lena River region, south in Urals to 62°N, and in the Lena Valley near Yakutsk.
SYNONYMS: *obscurus, ryphaeus, swerevi, tasensis, uralensis.*
COMMENTS: Subgenus *Alexandromys, middendorffi* species group *sensu* Zagorodnyuk (1990). Meier (1983) considered *M. middendorffi* to be phylogenetically distant from other northern and central Asian species of *Microtus.*

Microtus miurus Osgood, 1901. N. Am. Fauna, 21:64.

TYPE LOCALITY: USA, Alaska, Cook Inlet, Turnagain Arm, head of Bear Creek in mtns near Hope City.

DISTRIBUTION: Wet tundra and streambanks of N Alaska, N Yukon, and westernmost Northwest Territories; allopatric segment in SE Alaska and SW Yukon.

SYNONYMS: *andersoni, cantator, muriei, oreas, paneaki.*

COMMENTS: Classically treated as a member of the subgenus *Stenocranius*, related to Old World *M. gregalis* and synonymized with same by Rausch (1964) and Rausch and Rausch (1968). Other data clearly support their specific distinctiveness (see Anderson, 1960; Fedyk, 1970; Vorontsov and Lyapunova, 1986). Zagorodnyuk (1990) further emphasized the distant kinship of *M. gregalis* and *M. miurus* by placing them in different subgenera, allying *miurus* with certain Old World species of the subgenus *Alexandromys.*

Microtus mongolicus (Radde, 1861). Melange Biol. Acad. St. Petersbourg, 3:681.

TYPE LOCALITY: Russia, Transbaikalia (Chitinskaya Oblast), Omutnaya River, a tributary to the Amur River.

DISTRIBUTION: Transbaikalia, Mongolia, and NE China.

SYNONYMS: *baicalensis, poljakovi, xerophilus.*

COMMENTS: Subgenus *Alexandromys, middendorffi* species group *sensu* Zagorodnyuk (1990). Taxonomy reviewed by Allen (1940), Gromov and Polyakov (1977), Corbet (1978*c*, 1984), and Meier (1983). Karyotypic analyses recorded by Orlov et al. (1983), Yatsenko et al. (1980), and Radjabli et al. (1984) in context of phylogenetic comparisons among Asian species of *Microtus.*

Microtus montanus (Peale, 1848). Mammalia *in* Repts. U.S. Expl. Surv., 8:44.

TYPE LOCALITY: USA, California, Siskiyou Co., headwaters of Sacramento River, near Mount Shasta.

DISTRIBUTION: Cascade, Sierra Nevada, and Rocky Mountain ranges: SC British Columbia, Canada, south to EC California, S Utah, and NC New Mexico, USA; disjunct populations in S Nevada, EC Arizona, and NE New Mexico.

SYNONYMS: *amosus, arizonensis, canescens, caryi, codiensis, dutcheri, fucosus, fusus, longirostris, micropus, nanus, nevadensis, nexus, pratincola, rivularis, undosus, yosemite, zygomaticus.*

COMMENTS: Subgenus *Mynomes, montanus* species group *sensu* Zagorodnyuk (1990). Geographic variation and subspecific classification assessed by Anderson (1959), who viewed *M. oeconomus* as its sister species. Hooper and Hart (1962) arranged *M. montanus* with *M. pennsylvanicus* and *M. townsendii*, a general relationship corroborated by karyotypic and genic analyses (Modi, 1987; Moore and Janecek, 1990). Two karyotypic morphs reported by Judd et al. (1980), who raised the question of their specific distinction.

Microtus montebelli (Milne-Edwards, 1872). Rech. Hist. Nat. Mammifères, p. 285.

TYPE LOCALITY: Japan, Honshu Island, Fusiyama.

DISTRIBUTION: Japanese islands of Honshu, Sado, and Kyushu; and Sikotan Island in the Kuriles (Corbet, 1978*c*).

SYNONYMS: *brevicorpus, hatanezumi.*

COMMENTS: Subgenus *Pallasiinus, oeconomus* species group *sensu* Zagorodnyuk (1990). Chromosomal data summarized by Tsuchiya (1981). Variation in molar patterns from Holocene and Pleistocene samples exhaustively summarized by Kawamura (1988) in context of defining Pleistocene species endemic to Japan.

Microtus mujanensis Orlov and Kovalskaya, 1978. Zool. Zh., 57:1224.

TYPE LOCALITY: Russia, Buryat, Bauntovski Oblast, Vitim River Basin, Muya Valley.

DISTRIBUTION: Known only from the type locality.

COMMENTS: Subgenus *Alexandromys, maximowiczii* species group *sensu* Zagorodnyuk (1990). The species group also embraces *M. evoronensis*, which reflects Meier's (1983) interpretation of propinquity among the three species.

Microtus multiplex (Fatio, 1905). Arch. Sci. Phys. Nat. Geneve, Ser. 4, 19:193.

TYPE LOCALITY: Switzerland, Ticino Canton, near Lugano.

DISTRIBUTION: S Alps and N Apennines in Switzerland, Austria, Italy, and France; also NW and C Yugoslavia (see map in Krapp, 1982*b*).

SYNONYMS: *fatoii, druentius, leponticus, liechtensteini, orientalis, petrovi*.

COMMENTS: Subgenus *Terricola*. Placed in *subterraneus* species group by Chaline et al. (1988) but in the *multiplex* group, along with *M. tatricus*, by Zagorodnyuk (1990). Reviewed by Krapp (1982*b*). Corbet (1978*c*, and references therein) viewed *liechtensteini* as a separate species, but Krapp (1982*b*) included it in *M. multiplex*. Two distinctive karyotypes exist, one with 2N=48 (*multiplex*) and the other with 2N=46 (*liechtensteini*), but there is a record of hybridization (Zima and Král, 1984*a*). Tarsal glands compared (as *liechtensteini*) with those of *M. subterraneus* by Hrabe (1977).

Microtus nasarovi (Shidlovsky, 1938). Work of the Zemstvo Experimental Station (Tifh's), 2:12.

TYPE LOCALITY: Russia, Daghestan, Gunib, Karda.

DISTRIBUTION: NE Caucasus.

COMMENTS: Subgenus *Terricola*, *subterraneus* species group *sensu* Zagorodnyuk (1990). Included in *majori* by Corbet (1978*c*) but retained as a species by Pavlinov and Rossolimo (1987).

Microtus oaxacensis Goodwin, 1966. Am. Mus. Novit., 2243:1-4.

TYPE LOCALITY: Mexico, Oaxaca, Ixtlan Dist., near Vista Hermosa, Tarabundi Ranch, 5000 ft.

DISTRIBUTION: Sierra de Juarez, NC Oaxaca, Mexico.

COMMENTS: Relationships obscure: differentially diagnosed with respect to *M. mexicanus* and *M. umbrosus* by Goodwin (1966); description amplified and resemblance to *M. guatemalensis* noted by Jones and Genoways (1967); grouped with other relictual "pitymyine" species found at southern edge of Nearctic-*Microtus* distribution by Hoffmann and Koeppl (1985).

Microtus obscurus (Eversmann, 1841). Mem. Univ. Kazan, 156.

TYPE LOCALITY: Siberia, Altai Mtns.

DISTRIBUTION: Russia from the Crimea east through Siberia to the upper Yenesei River, south through NW Mongolia, NW China (Xinjiang), the Altai Mtns, Lake Balkash, to the Caucusus and N Iran (see Zagorodnyuk, 1991*a*).

SYNONYMS: *brevirostris, gudauricus, ilaeus, ileos, innae, iphigeniae, macrocranius, transcaucasicus, transuralensis*.

COMMENTS: Subgenus *Microtus*, *arvalis* species group *sensu* Zagorodnyuk (1990). Formerly included in *M. arvalis* (e.g., Corbet, 1978*c*) but shown to be a distinct species by Zagorodnyuk (1991*a*, *b*) based on morphological and chromosomal features. The species appears to be parapatric with *M. arvalis* (Zagorodnyuk, 1991*a*).

Microtus ochrogaster (Wagner, 1842). *In* Schreber, Die Säugethiere, Suppl. 3:592.

TYPE LOCALITY: USA, Indiana, Posey Co., New Harmony (as fixed by Bole and Moulthrop, 1942:157).

DISTRIBUTION: Northern and Central Great Plains: EC Alberta to S Manitoba, Canada, south to N Oklahoma and Arkansas, eastwards to C Tennessee and westernmost West Virginia, USA; relictual populations in C Colorado, N New Mexico, and coastal prairies of SW Louisiana and adjacent Texas, USA.

SYNONYMS: *austerus, cinnamonea, haydenii, ludovicianus, minor, ohioensis, similis, taylori*.

COMMENTS: A member of *Pedomys*, a taxon used as a subgenus of *Microtus* (Bailey, 1900; van der Meulen, 1978), as a full synonym of *Microtus* (Hall, 1981), or as a synonym of *Pitymys*, whether the latter is used as a genus (Repenning, 1983) or subgenus (Hooper and Hart, 1962; Zagorodnyuk, 1990). The purported close affinity of *M. ochrogaster* with North American species of *Pitymys* has received little support from biochemical studies (Chaline and Graf, 1988; Moore and Janacek, 1990). Geographic variation over the Central Great Plains studied by Choate and Williams (1978). The strong morphometric segregation of *minor* from other *M. ochrogaster* advises renewed scrutiny of its status (Severinghaus, 1977). Includes *ludovicianus*, an isolated form of the coastal prairies formerly consided a separate species and now apparently extinct (see Lowery, 1974). See Stalling (1990, Mammalian Species, 355).

Microtus oeconomus (Pallas, 1776). Reise Prov. Russ. Reichs., 3:693.

TYPE LOCALITY: Siberia, Ishim Valley.

DISTRIBUTION: Tundra and northern taiga of Holarctic: in Palearctic, from Scandinavia and the Netherlands across to borderlands of Bering Sea, including Sakhalin and Kurile islands, and south to E Germany, Ukraine, S Kazakhstan, Mongolia, and the Ussuri region; St. Lawrence Isl in Bering Sea; in Nearctic, from Alaska through Yukon Territory to W Northwest Territories and extreme NW British Columbia, Canada.

SYNONYMS: *altaicus, amakensis, anikini, arenicola, dauricus, elymocetes, endoecus, finmarchicus, flaviventris, gilmorei, hahlovi, innuitus, kamtschatica, karaginensis, kjusjurensis, kodiacensis, koreni, macfarlani, malcolmi, medius, mehelyi, montium-caelestinum, naumovi, operarius, ouralensis, petshorae, popofensis, punukensis, ratticeps, shantaricus, sitkensis, stimmingi, suntaricus, tschuktschorum, uchidae, unalascensis, uralensis, yakutatensis.*

COMMENTS: Subgenus *Pallasiinus*, *oeconomus* species group, including *M. montebelli* and *M. limnophilus* (Zagorodnyuk, 1990). Conspecific stature of Old and New World populations averred by Zimmermann (1942) and sustained by subsequent studies (Nadler et al., 1976, 1978; Ognev, 1964; Rausch, 1953). Geographic variation and subspecies of Nearctic populations reviewed by Paradiso and Manville (1961); Russian populations by Gromov and Polyakov (1977); European populations by Tast (1982*b*). Chromosomal data is summarized and evaluated by Zima and Král (1984*a*). Ecological and distributional details of the species at its southern limit in Poland are described by Salata-Pilacinska (1990). Angermann (1984) analysed intraspecific molar patterns in the context of assessing the range of variation against which patterns in extinct species could be tested.

Ognev (1964), but not Corbet (1978*c*), identified *ratticeps* Keyserling and Blasius, 1841, as the proper name applicable to this species (also see discussions in Ellerman and Morrison-Scott, 1951:705; and Hall, 1981:805). Although systematists continue to utilize *oeconomus* Pallas, 1776, for reasons of familiarity, this nomenclatural uncertainty needs formal resolution.

Microtus oregoni (Bachman, 1839). J. Acad. Nat. Sci. Philadelphia, 8:60.

TYPE LOCALITY: USA, Oregon, Clatsop Co., Astoria.

DISTRIBUTION: Moist coniferous forest seres of Pacific Northwest, from SW British Columbia, Canada, south to NW California, USA.

SYNONYMS: *adocetus, bairdi, cantwelli, morosus, serpens.*

COMMENTS: Subgenus *Mynomes*, sole member of *oregoni* species group *sensu* Zagorodnyuk (1990). Type species of *Chilotus*, conventionally recognized as a subgenus of *Microtus* (see Anderson, 1959, 1960; Bailey, 1900), and occasionally including Old World *M. socialis* as a comember (Chaline, 1974; Ognev, 1964) or not (Anderson, 1959). Hooper and Hart (1962), however, found the morphological evidence insufficient to warrant subgeneric segregation of *oregoni* from North American species of *Microtus*, a viewpoint sustained by genetic distance comparisons (Moore and Janecek, 1990) and reflected in the classification of Zagorodnyuk (1990). Diploid number and sex-determining mechanism unique among North American *Microtus* studied (Ohno et al., 1963). See Carraway and Verts, 1985 (Mammalian Species, 233).

Microtus pennsylvanicus (Ord, 1815). *In* Guthrie, New Geogr., Hist., Comml., Grammar, Philadelphia, 2nd ed., 2:292.

TYPE LOCALITY: USA, Pennsylvania, "meadows below Philadelphia."

DISTRIBUTION: Meadowlands interspersed across boreal and mixed coniferous-deciduous biomes of North America: C Alaska to Labrador, including Newfoundland and Prince Edwards Island, Canada; south in Rocky Mountains to N New Mexico, in Great Plains to N Kansas, and in Appalachians to N Georgia, USA; isolated populations in W New Mexico and Florida, USA, and in N Chihuahua, Mexico.

STATUS: U.S. ESA - Endangered as *M. p. dukecampbelli*; IUCN - Endangered as *M. p. chihuahuensis* and *M. p. dukecampbelli*.

SYNONYMS: *acadicus, admiraltiae, alborufescens, alcorni, aphorodemus, arcticus, aztecus, chihuahuensis, copelandi, dekayi, drummondii, dukecampbelli, enixus, finitus, fontigenus, fulva, funebris, hirsutus, insperatus, kincaidi, labradorius, longipilis, magdalenensis, microcephalus, modestus, nasuta, nesophilus, nigrans, noveboracensis, oneida, palustris, pratensis, provectus, pullatus, rubidus, rufescens, rufidorsum, shattucki, stonei, tananaensis, terraenovae, uligocola, wahema.*

COMMENTS: Subgenus *Mynomes*, *pennsylvanicus* species group *sensu* Zagorodnyuk (1990). Proposed as conspecific with Old World *M. agrestis* by Klimkiewicz (1970), but G-banded chromosomal differences support their recognition as distinct species (Modi, 1987; Vorontsov and Lyapunova, 1986). Aside from the probable insular derivative *M. breweri* (see above account), *pennsylvanicus* may be closely related to *M. montanus* and *M. townsendii* among New World species (see Hooper and Hart, 1962; Modi, 1987; Moore and Janecek, 1990). Regional studies of variation undertaken (e.g., Anderson, 1956; Anderson and Hubbard, 1971; Weddle and Choate, 1983), but comprehensive review of entire species warranted. Insular form *provectus* relegated to subspecific status by Chamberlain (1954) and Moyer et al. (1988) and *nesophilus* by Jones et al. (1986). See Reich (1981, Mammalian Species, 159).

Microtus pinetorum (Le Conte, 1830). Ann. Lyc. Nat. Hist., 3:133.

TYPE LOCALITY: USA, Georgia, Liberty Co., old LeConte plantation near Riceboro (as interpreted by Bailey, 1900:63).

DISTRIBUTION: Temperate deciduous forest zone of E USA: eastern shoreline from S Maine to N Florida, west to C Wisconsin and E Texas.

SYNONYMS: *auricularis, carbonarius, kennicottii, nemoralis, parvulus, scalopsoides, schmidti.*

COMMENTS: Type species of *Pitymys* (see remarks under genus for variable treatment of this taxon). Van der Meulen (1978) regarded *nemoralis* and *parvulus* as species distinct from *pinetorum* in the genus *Pitymys*, as did Repenning (1983) for *nemoralis*. Their conclusions have yet to be critically examined with a broader array of information bases, although limited chromosomal and genetic surveys promise future insight (Moore and Janecek, 1990; J. W. Wilson, 1984). See Smolen (1981, Mammalian Species, 147).

Microtus quasiater (Coues, 1874). Proc. Acad. Nat. Sci. Philadelphia, 26:191.

TYPE LOCALITY: Mexico, Veracruz, Jalapa.

DISTRIBUTION: SE San Luis Potosi to N Oaxaca, Mexico.

COMMENTS: Subgenus *Pitymys*. Named as a variety of *M. pinetorum* but specific status affirmed thereafter (Bailey, 1900; Hall and Cockrum, 1953; Repenning, 1983). Typically viewed as most closely related to *pinetorum* in subgenus or genus *Pitymys* (Anderson, 1960; Hall and Cockrum, 1953; van der Meulen, 1978), a convention challenged by Repenning (1983) and Moore and Janecek (1990), who supported sister-group kinship between *ochrogaster* and *quasiater.*

Microtus richardsoni (DeKay, 1842). Zoology of New York, Part I, Mammals, p. 91.

TYPE LOCALITY: Canada, Alberta Prov., vicinity of Jasper House (as interpreted by Bailey, 1900:60).

DISTRIBUTION: Subalpine and alpine meadows of Rocky Mountains, from S British Columbia and Alberta, Canada, to W Wyoming and C Utah, USA; and of Cascade Mountains, from SW British Columbia south through WC Oregon.

SYNONYMS: *arvicoloides, macropus, myllodontus, principalis.*

COMMENTS: Senior synonym of the type (= *arvicoloides*) of *Aulacomys*. In early classifications (Bailey, 1900; Miller, 1896), *richardsoni* was placed with Old World water voles of the subgenus *Arvicola* (including *Aulacomys*) of *Microtus*, a relationship reaffirmed by Hooper and Hart (1962) and followed by others, with *Arvicola* employed either at the subgeneric (Hall, 1981) or generic level (Jones et al., 1975). Other evidence favors the retention of *richardsoni* within *Microtus* and the restriction of *Arvicola* to Old World forms (Carleton, 1981; Gromov and Polyakov, 1977; Hinton, 1926; Koenigswald, 1980; Repenning, 1980; Zagorodnyuk, 1990). Hoffmann and Koeppel (1985) further suggested the *de novo* origin of *M. richardsoni* in North America from an early stock that also gave rise to *M. xanthognathus*, an idea consistent with the biochemical similarity of *richardsoni* to certain New World *Microtus* (Nadler et al., 1978) and with Zagorodnyuk's (1990) expanded concept of the subgenus *Aulacomys* (including *M. chrotorrhinus, M. longicaudus,* and *M. xanthognathus*). The phylogenetic placement of *richardsoni* in *Arvicola* versus *Microtus* bears critically on interpretations of historical zoogeography, but the appropriate sampling of New and Old World species needed to resolve the problem has yet to be achieved within a single study. See Ludwig (1984, Mammalian Species, 223).

Microtus rossiaemeridionalis Ognev, 1924. Gryzuny Servernogo Kavkaza, Rostov-on-Don, [Rodentia N. Caucasus] p. 27.

TYPE LOCALITY: Russia, Bobrov subdistrict of Voronej Govt., Novii Kurlak (after Ellerman and Morrison-Scott, 1951:698).

DISTRIBUTION: From Finland east through Russia to Urals, south to region just north of the Caucusus and through the Ukraine to Rumania, Bulgaria, S Yugoslavia, N Greece, and NW Turkey (see Petrov and Ruzic, 1982; Zagorodnyuk, 1991*b*; Zima et al, 1991); introduced to Svalbard (see Fredga et al., 1990).

SYNONYMS: *caspicus, epiroticus, ghalgai, muhlisi, relictus, rhodopensis, subarvalis.*

COMMENTS: Subgenus *Microtus*, *arvalis* species group. This is the species (with 2N=54, FN=56) that was listed as *M. subarvalis* or *epiroticus* by Corbet (1978*c*, 1984) and reviewed under the latter name by Petrov and Ruzic (1982). Cytological and morphological comparisons to *M. arvalis* and *M. obscurus* underscore the specific distinctness of *M. rossiaemeridionalis* (Gavrila et al., 1986; Král et al., 1981; Kratochvíl, 1982*b*; Naumova et al., 1990; Zagorodnyuk, 1991*a*, *b*; Zima et al., 1991). The geographic ranges of *M. arvalis* and *M. rossiaemeridionalis* broadly overlap (see Zagorodnyuk, 1991*b*:30). The record (as *epiroticus*) from Svalbard is documented by Fredga et al. (1990), who suggested that the voles were introduced.

Microtus sachalinensis Vasin, 1955. Zool. Zh., 34:427.

TYPE LOCALITY: Russia, Sakhalin Island, Poronaysk region, Olen River.

DISTRIBUTION: An endemic of Sakhalin Island.

COMMENTS: Subgenus *Alexandromys*, *middendorffi* species group *sensu* Zagorodnyuk (1990). Reviewed by Corbet (1978*c*) and Gromov and Polyakov (1977).

Microtus savii (de Selys-Longchamps, 1838). Rev. Zool. Paris, p. 248.

TYPE LOCALITY: Italy, near Pisa.

DISTRIBUTION: SE France, Italy, Sicily, and Elba (see map in Krapp, 1982*c*).

SYNONYMS: *brachycercus, nebrodensis, selysii.*

COMMENTS: Subgenus *Terricola*. Both Chaline et al. (1988) and Zagorodnyuk (1990) used this species as the core of a *savii* species group. Reviewed by Corbet (1978*c*, 1984) and Krapp (1982*c*); chromosomal data summarized and reviewed by Zima and Král (1984*a*); karyology of Italian samples documented by Galleni et al. (1992). Closely related to the fossil *M. henseli* from Sardinia and Corsica (Brunet-Lecomte and Chaline, 1991; Krapp, 1982*c*).

Microtus schelkovnikovi (Satunin, 1907). Izv. Kavkas. Mus., 3, 1:242.

TYPE LOCALITY: Azerbaijan, Talysk Mtns, near village of Dzhi.

DISTRIBUTION: S Azerbaijan, Talysk and Elburz Mtns; limits of range unknown.

SYNONYMS: *dorothea.*

COMMENTS: Subgenus *Terricola*, sole member of its own species group *sensu* Zagorodnyuk (1990). Lumped under *Pitymys subterraneus* by Ellerman Morrison-Scott (1951), but its validity as a separate species has been documented by Kratochvil (1970) and Kratochvíl and Král (1974) and is so listed in faunal catalogues (Corbet, 1978*c*; Pavlinov and Rossolimo, 1987). The name *kaznakovi* may be valid for this species (Corbet, 1984), but Pavlinov and Rossolimo (1987) avoided its use and cited Gromov and Polyakov (1977) for the opinion that it belongs with *Alticola*. Chromosomal data summarized by Zima and Král (1984*a*) and Achverdjan et al. (1992).

Microtus sikimensis (Hodgson, 1849). Ann. Mag. Nat. Hist., ser. 2, 3:203.

TYPE LOCALITY: India, Sikkim.

DISTRIBUTION: Himalayas from W Nepal east through Sikkim, Bhutan, to S and E Xizang (Tibet; Feng et al., 1986).

SYNONYMS: *thricolis.*

COMMENTS: Subgenus *Neodon*, a member, along with *irene*, of the *sikimensis* species group *sensu* Zagorodnyuk (1990). Geographic distribution is parapatric on E Tibetan Plateau with *M. irene* (see that account).

Microtus socialis (Pallas, 1773). Reise Prov. Russ. Reichs., 2:705.

TYPE LOCALITY: Kazakhstan, probably Gur'evsk Oblast between Volga and Ural rivers.

DISTRIBUTION: Palearctic steppe from Dneper River and Crimea east to Lake Balkhash and NW Xinjiang in China, south through E Turkey to N Syria and NE Iran.

SYNONYMS: *astrachanensis, binominatus, bogdoensis, colchicus, goriensis, gravesi, hyrcania, nikolajevi, parvus, satunini, schidlovskii, shevketi, syriacus.*

COMMENTS: Subgenus *Microtus, socialis* species group *sensu* Zagorodnyuk (1990), along with *M. guentheri* and *M. irani.* Chromosomal data summarized and reviewed by Zima and Král (1984*a*). Harrison and Bates (1991) included *guentheri* and *irani* under *M. socialis* because they could not discriminate the three in their samples but provided no substantiating data or analyses (see accounts of *guentheri* and *irani*).

Microtus subterraneus (de Selys-Longchamps, 1836). Essai Monogr. sur les Campagnols des Env. de Liege., 10.

TYPE LOCALITY: Belgium, Liege, Waremme.

DISTRIBUTION: Bulk of range from N and C France through C Europe to Ukraine and the Don River, south through Yugoslavia and the Balkans into N Greece (European range mapped in Niethammer, 1982*g*); isolated populations in NE Russia (see Zagorodnyuk, 1989).

SYNONYMS: *atratus, capucinus, dacius, dinaricus, ehiki, fusca, hungaricus, incertoides, incertus, klozeli, kupelwieseri, martinoi, matrensis, mustersi* (of Martino, 1937, not Hinton, 1926), *neuhauseri, nyirensis, rufofuscus, transsylvanicus, transvolgensis, ukrainicus, volaensis, wettsteini, zimmermanni.*

COMMENTS: Subgenus *Terricola, subterraneus* species group *sensu* Chaline et al. (1988) and Zagorodnyuk (1990). Reviewed by Corbet (1978*c*, 1984) and Niethammer (1982*g*). Taxonomy, geographic distribution, and morphological variation in context of defining species in subgenus *Terricola* reported by Zagorodnyuk (1989, 1992); karyotypic information presented by Sablina et al. (1989), Zima (1986), and Zima and Král (1984*a*); biochemical comparisons among *M. subterraneus, M. felteni,* and other voles documented by Gill et al. (1987). Synonyms referenced according to Zagorodnyuk (1989), who speculated that *dacius* is a separate species.

Microtus tatricus Kratochvil, 1952. Acta Acad. Sci. Nat. Moravo-Siles., 24:155-194.

TYPE LOCALITY: Czechoslovakia, Poprad Dist., Velka Studena Dolina valley, High Tatra Mtns.

DISTRIBUTION: W and E Carpathian Mtns; montane spruce forest of Tatra Mtns between Czechoslovakia and Poland, Pilsko Mtn, and Beslksid Ziwiecki Mtns; also W Ukraine (see Zagorodnyuk, 1988; Zagorodnyuk et al., 1992).

SYNONYMS: *zykovi.*

COMMENTS: Subgenus *Terricola, subterraneus* species group *sensu* Chaline et al. (1988). Reviewed by Corbet (1984) and Niethammer (1982*l*). Chromosomal data evaluated by Zima and Král (1984*a*). Taxonomy, geographic distribution, and morphological variation in context of discriminating species of the subgenus *Terricola* were covered by Zagorodnyuk (1989), who also named the subspecies *zykovi,* and Zagorodnyuk et al. (1992).

Microtus thomasi (Barrett-Hamilton, 1903). Ann. Mag. Nat. Hist., ser. 7, 11:306.

TYPE LOCALITY: Yugoslavia, Montenegro, Vranici.

DISTRIBUTION: From S coastal Yugoslavia into Greece (including Evvoia Isl); see map in Niethammer (1982*k*).

SYNONYMS: *atticus.*

COMMENTS: Subgenus *Terricola, duodecimcostatus* species group *sensu* Chaline et al. (1988) or a member of its own species group *sensu* Zagorodnyuk (1990). Reviewed by Corbet (1978*c*) and Niethammer (1982*k*); chromosomal information summarized by Zima and Král (1984*a*).

Microtus townsendii (Bachman, 1839). J. Acad. Nat. Sci. Philadelphia, 8:60.

TYPE LOCALITY: USA, Oregon, Multnomah Co., Wappatoo (Sauvie) Island, lower Columbia River near mouth of Willamette River (as located by Bailey, 1900:46).

DISTRIBUTION: Pacific northwest, from extreme SW British Columbia, Canada, to NW California, USA, including Vancouver and neighboring islands.

SYNONYMS: *cowani, cummingi, laingi, occidentalis, pugeti, tetramerus.*

COMMENTS: Subgenus *Mynomes, pennsylvanicus* species group *sensu* Zagorodnyuk (1990). Broadly allied with *M. pennsylvanicus* and *M. montanus* (Anderson, 1959; Hooper and Hart, 1962; Modi, 1987), though precise cladistic picture unresolved. See Cornely and Verts (1988, Mammalian Species, 325).

Microtus transcaspicus Satunin, 1905. Izv. Kavkas. Mus., 2:57.
TYPE LOCALITY: Turkmenistan, Kopet Dag Mtns, Chuli Valley, near Ashkhabad.
DISTRIBUTION: S Turkmenistan, N Afghanistan, and N Iran.
SYNONYMS: *khorkoutensis*.
COMMENTS: Subgenus *Microtus*, *arvalis* species group *sensu* Zagorodnyuk (1990). Taxonomy reviewed by Gromov and Polyakov (1977), Corbet (1978*c*), Malygin (1978), Meier et al. (1981), and Meier (1983).

Microtus umbrosus Merriam, 1898. Proc. Biol. Soc. Washington, 12:107.
TYPE LOCALITY: Mexico, Oaxaca, Mt. Zempoaltepec, 8200 ft.
DISTRIBUTION: Known only from Mt. Zempoaltepec.
COMMENTS: Type species and sole member of *Orthriomys*, a taxon typically identified as a subgenus of *Microtus* (Anderson, 1959; Bailey, 1900; Hall and Cockrum, 1953; Zagorodnyuk, 1990). Martin (1974:61) suggested that this species "may eventually" be included in *Neodon*.

Microtus xanthognathus (Leach, 1815). Zool. Misc., 1:60.
TYPE LOCALITY: Canada, Manitoba, Hudson Bay.
DISTRIBUTION: Western boreal taiga zone: EC Alaska to W Northwest Territories, southeastwards to C Alberta and western coast of Hudson Bay, Canada.
COMMENTS: Relationships obscure: probably not a sister species to *M. chrotorrhinus* as once believed (see that account and Lidicker and Yang, 1986); placed by Zagorodnyuk (1990) in the subgenus *Aulacomys*, *richardsoni* species group. Karyotype reported and affinities discussed by Rausch and Rausch (1974).

Myopus Miller, 1910. Smithson. Misc. Coll., 52:497.
TYPE SPECIES: *Myodes schisticolor* Lilljeborg, 1844.
COMMENTS: Tribe Lemmini. Conventionally treated as a genus until Chaline (1972) regarded the differences in molar pattern between *schisticolor* and *Lemmus* to only reflect specific-level distinctions (also see Chaline and Mein, 1979; Chaline et al., 1989). Koenigswald and Martin (1984) also cited molar-pattern similarity for their allocation of *Myopus* to a subgenus of *Lemmus*. Niethammer and Henttonen (1982) and Pavlinov and Rossolimo (1987), however, maintained the generic segregation of *Myopus*, as we do here. *Lemmus* and *Myopus* share certain dental resemblances, but they are readily distinguished by other features that have not been addressed in the context of assessing the interrelationship of forms comprising these genus-group taxa. In our opinion, the phenotypically restricted character set mustered from a paleontological perspective is insufficient to falsify the hypothesis that *schisticolor* represents a monophyletic group separate from species of *Lemmus*.

Myopus schisticolor (Lilljeborg, 1844). Ofv. K. Svenska Vet.-Akad. Forhandl. Stockholm, I, p. 33.
TYPE LOCALITY: Norway, Gulbrandsdal, N end of Mjosen, near Lillehammer.
DISTRIBUTION: Coniferous forest from Norway and Sweden through Siberia to Kolyma River and Kamchatka, south to the Altai Mtns, N Mongolia and NE China (Heilungkiang), and the Sikhote Alin Range (Corbet, 1978*c*); also a southern isolate in the Ural Mtns, near source of Ural River, about 450 km south of previously recorded limit at 58°N (see Corbet, 1984).
SYNONYMS: *middendorfi*, *morulus*, *saianicus*, *thayeri*, *vinogradovi*.
COMMENTS: Described in detail by Miller (1912*a*) and Hinton (1926); reviewed by Gromov and Polyakov (1977) and Corbet (1978*c*, 1984). European populations reviewed by Niethammer and Henttonen (1982). Morphological variation among Norwegian populations analyzed by Kratochvíl (1979) in context of assessing age and geographic variation. Intraspecific chromosomal variation and extraordinary sex-chromosome traits are documented by Gropp et al. (1976), Kozlovskii (1986), Lau et al. (1992), and Gileva and Fedorov (1991).

Neofiber True, 1884. Science, 4:34.
TYPE SPECIES: *Neofiber alleni* True, 1884.
COMMENTS: A relictual form once more widespread in the eastern USA (see Frazier, 1977, and Hibbard and Dalquest, 1973). Although sometimes placed with *Ondatra* in the tribe Ondatrini (Chaline and Mein, 1979; Kretzoi, 1969; Repenning et al., 1990), a

variety of morphological features conveys their very distant relationship (Carleton, 1981; Hooper and Hart, 1962; Koenigswald, 1980), as does fossil evidence (Martin, 1975; Zakrzewski, 1974). Alternatively, *Neofiber* either has been cited as Arvicolinae *incertae sedis* (Gromov and Polyakov, 1977; Koenigswald, 1980) or affiliated with *Arvicola* and *Microtus* in a broadly-defined tribe Arvicolini (Zagorodnyuk, 1990).

Neofiber alleni True, 1884. Science, 4:34.
TYPE LOCALITY: USA, Florida, Brevard Co., Georgiana.
DISTRIBUTION: Most of peninsular Florida to extreme SE Georgia, USA.
SYNONYMS: *apalachicolae, exoristus, nigrescens, struix.*
COMMENTS: Geographic races delimited by Schwartz (1953); and see Birkenholz (1972, Mammalian Species, 15).

Ondatra Link, 1795. Beytrage zur Naturgeschichte, 1(2):76.
TYPE SPECIES: *Castor zibethicus* Linnaeus, 1766.
SYNONYMS: *Fiber, Moschomys, Simotes.*
COMMENTS: Tribe Ondatrini, either as sole member (Hooper and Hart, 1962; Koenigswald, 1980) or with *Neofiber* (Chaline and Mein, 1979; Kretzoi, 1969; Repenning et al., 1990). Evidence from fossil history and morphology suggests that the two genera of "muskrats" are more distantly related than their inclusion in the same tribe would connote (Carleton, 1981; Hooper and Hart, 1962; Koenigswald, 1980; L. D. Martin, 1975, 1979). Believed to have originated from Pliocene form *Pliopotamys* (Eshelman, 1975; Zakrzewski, 1974).

Ondatra zibethicus (Linnaeus, 1766). Syst. Nat., 12th ed., 1:79.
TYPE LOCALITY: E Canada.
DISTRIBUTION: North America, north to the treeline, including Newfoundland; south to the Gulf of Mexico, Rio Grande and lower Colorado River valleys. Introduced to Czechoslovakia in 1905 and now widespread in the Palearctic, including C and N Europe, most of Ukraine, Russia, and Siberia, adjacent parts of China and Mongolia, and Honshu Isl, Japan (see Corbet, 1978*c*); also into southernmost Argentina (see Olrog and Lucero, 1981).
SYNONYMS: *albus, americana, aquilonius, bernardi, cinnamominus, goldmani, hudsonius, macrodon, maculosa, mergens, niger, nigra, obscurus, occipitalis, osoyoosensis, pallidus, ripensis, rivalicius, spatulatus, varius, zalophus.*
COMMENTS: Revised, under the name *Fiber*, by Hollister (1911). Comprehensive summaries of ecology, population biology, and economic status provided by Pietsch (1982) and Perry (1982); history of introductions, mostly for fur-farms, and population spread in Eurasia and USA reviewed by Storer (1937). See Willner et al. (1980, Mammalian Species, 141).

Phaulomys Thomas, 1905. Ann. Mag. Nat. Hist., ser. 7, 15:493.
TYPE SPECIES: *Evotomys smithii* Thomas, 1905.
COMMENTS: Originally described as a subgenus of *Evotomys* (= *Clethrionomys*) and included in *Eothenomys* by most workers (Aimi, 1980; Corbet, 1978*c*). Kawamura (1988), however, reinstated *Phaulomys* as a genus containing the two species listed below. Although the extant species have rootless molars, they can be derived through a transitional series of late Pleistocene samples from the middle Pleistocene *Clethrionomys japonicus*, which has molars with definite roots. Study of the variation in molar and external traits among species of *Eothenomys* and *Anteliomys* also led Tanaka (1971) to recognize *Phaulomys* as a genus distinct from *Eothenomys*. Chromosomal analyses of *smithii* suggest that *Clethrionomys* and *Phaulomys* have been derived from a common ancestor (Ando et al., 1988). The removal of Japanese *Phaulomys* from *Eothenomys*, whose species-diversity centers in the mountains of China, and its proposed association with *Clethrionomys* is zoogeographically plausible and outlines a precise hypothesis that can be tested by analyses of other data sets. Aimi (1980) retraced the complex taxonomic history of species allocated to either *Evotomys, Clethrionomys, Anteliomys, Phaulomys,* or *Eothenomys*.

Phaulomys andersoni (Thomas, 1905). Abstr. Proc. Zool. Soc. London, 1905(23):18.
TYPE LOCALITY: Japan, Honshu, Iwate Prefecture, Tsunagi, near Morioka.

DISTRIBUTION: Known only from Honshu.
SYNONYMS: *imaizumii, niigatae.*
COMMENTS: Revised by Aimi (1980) as a species of *Eothenomys.*

Phaulomys smithii (Thomas, 1905). Ann. Mag. Nat. Hist., ser. 7, 15:493.
TYPE LOCALITY: Japan, Honshu, Hyogo Prefecture, Kobe.
DISTRIBUTION: Recorded from the Japanese islands of Dogo, Honshu, Shikoku, and Kyushu.
SYNONYMS: *kageus, okiensis.*
COMMENTS: Revised by Aimi (1980) as a species of *Eothenomys;* evidence for including *kageus* was presented by Kaneko (1985) and Ando et al. (1988). Intraspecific chromosomal variation was documented by Ando et al. (1991).

Phenacomys Merriam, 1889. N. Am. Fauna, 2:32.
TYPE SPECIES: *Phenacomys intermedius* Merriam, 1889.
COMMENTS: Placed by Gromov and Polyakov (1977) as Arvicolinae *incertae sedis;* by Zagorodnyuk (1990) as tribe Phenacomyini, including *Arborimus;* by Repenning et al. (1990) as Arvicolini, including *Arvicola* and certain extinct genera. Revised by Howell (1926*b*) and Hall and Cockrum (1953), then including species assigned to *Arborimus* (see that account).

Phenacomys intermedius Merriam, 1889. N. Am. Fauna, 2:32.
TYPE LOCALITY: Canada, British Columbia Prov., 20 mi NNW Kamloops.
DISTRIBUTION: SW British Columbia and adjacent Alberta, Canada, south to N New Mexico, C Utah, and N California, USA; disjunct populations in EC California and W Nevada, USA.
SYNONYMS: *celsus, constablei, laingi, levis, olympicus, oramontis, orophilus, preblei, pumilus, truei.*
COMMENTS: Howell (1926*b*) originally recognized three species (*intermedius, mackenzii,* and *ungava*) of heather voles, later reduced to two (*intermedius* and *ungava* including *mackenzii*) by Anderson (1942*a*, 1947). Crowe (1943) further lumped all under *intermedius* based on suspected intergrades from SW Alberta, and the recognition of a single species has been generally followed (e.g., Banfield, 1974; Corbet and Hill, 1991; Hall and Cockrum, 1953) but not exclusively (Cowan and Guiguet, 1965; Peterson, 1966*a*). Foster and Peterson (1961) questioned Crowe's appreciation of age-effects in his identification of the alledged intergrades between *intermedius* and *ungava*. As remarked by Cowan and Guiguet (1965), the matter of their synonymy "requires more detailed examination before a decision can be reached," an appraisal which stands equally valid today. See McAllister and Hoffmann (1988, Mammalian Species, 305, including *ungava*).

Phenacomys ungava Merriam, 1889. N. Am. Fauna, 2:35.
TYPE LOCALITY: Canada, Quebec Prov., Fort Chimo, near Ungava Bay.
DISTRIBUTION: S Yukon across much of Canada to E Labrador; as far south as S Alberta, along the northern Great Lakes and St. Lawrence River.
SYNONYMS: *celatus, crassus, latimanus, mackenzii, soperi.*
COMMENTS: See remarks under *P. intermedius* and McAllister and Hoffmann (1988) on proper usage of *ungava* for this form.

Proedromys Thomas, 1911. Abstr. Proc. Zool. Soc. London, 1911(90):4.
TYPE SPECIES: *Proedromys bedfordi* Thomas, 1911.
COMMENTS: Tribe Arvicolini. Ellerman and Morrison-Scott (1951) and Corbet (1978*c*) included this taxon in *Microtus*, but it is kept separate by Allen (1940), Wang et al. (1966), and Gromov and Polyakov (1977).

Proedromys bedfordi Thomas, 1911. Abstr. Proc. Zool. Soc. London, 1911(90):4.
TYPE LOCALITY: China, Gansu (Kansu), 60 mi SE Minchow.
DISTRIBUTION: Recorded from Gansu and Sichuan, China.
COMMENTS: This is the only extant species in the genus. The Sichuan record is reported by Wang et al (1966). Pleistocene fragments from Shanxi, Hebei, and Shandong have been referred to *P.* cf. *bedfordi* (Zheng and Li, 1990).

Prometheomys Satunin, 1901. Zool. Anz., 24:572.

TYPE SPECIES: *Prometheomys schaposchnikowi* Satunin, 1901.

COMMENTS: The only extant member of an archaic line of the arvicoline radiation, which most specialists isolate as the tribe Prometheomyini within Arvicolinae (Gromov and Polyakov, 1977; Hooper and Hart, 1962; Kretzoi, 1969; Pavlinov and Rossolimo, 1987; Zagorodnyuk, 1990). Repenning et al. (1990), however, aligned *Prometheomys* with another presumed relic, *Ellobius*, in the subfamily Prometheomyinae.

Prometheomys schaposchnikowi Satunin, 1901. Zool. Anz., 24:572.

TYPE LOCALITY: Georgia (Gruzinskaya), Caucasus Mtns, Gudaur, S of Krestovyi Pass, 6500 ft (as given by Ognev, 1963*b*).

DISTRIBUTION: Alpine zone of Caucasus Mtns, Georgia, and extreme NE Turkey.

COMMENTS: Reviewed by Gromov and Polyakov (1977) and Corbet (1978*c*); cranial and dental morphology detailed by Hinton (1926); chromosomal data summarized by Zima and Král (1984*a*).

Synaptomys Baird, 1858. Mammalia *in* Repts. U.S. Expl. Surv., 8(1):558.

TYPE SPECIES: *Synaptomys cooperi* Baird, 1858.

SYNONYMS: *Mictomys, Plioctomys.*

COMMENTS: Many taxonomic characters link *Synaptomys* with the true lemmings (*Lemmus* and *Myopus*) in a clade, usually treated as the tribe Lemmini, believed to represent an early line of arvicoline evolution (Carleton, 1981; Chaline and Graf, 1988; Graf, 1982; Gromov and Polyakov, 1977; Hinton, 1926; Hooper and Hart, 1962; Koenigswald, 1980). Fossil history reviewed by Koenigswald and Martin (1984). Although described as a genus, Miller (1896) arranged *Mictomys* as a subgenus of *Synaptomys*, as conventionally recognized (Hall, 1981; Howell, 1927). Paleontologists, however, have emphasized the dental divergence of *Synaptomys* and *Mictomys* at the generic level (Koenigswald and Martin, 1984; Repenning and Grady, 1988), although both are presumed to have separated from a common ancestor (*Plioctomys*) in the late Pliocene. Current specific and subspecific classification essentially derives from Howell (1927).

Synaptomys borealis (Richardson, 1828). Zool. J., 3:517.

TYPE LOCALITY: Canada, District of Mackenzie, Great Bear Lake, Fort Franklin.

DISTRIBUTION: Alaska to N Washington, USA, eastwards across much of interior Canada to Labrador; disjunct range segment from Gaspe Peninsula, Quebec, to C New Hampshire, USA.

SYNONYMS: *andersoni, artemisiae, bullatus, chapmani, dalli, innuitus, medioximus, smithi, sphagnicola, truei, wrangeli.*

COMMENTS: Subgenus *Mictomys*. Pleistocene records establish the species in the Great Basin, far to the south of its current range (see Mead et al., 1992).

Synaptomys cooperi Baird, 1858. Mammalia *in* Repts. U.S. Expl. Surv., 8(1):558.

TYPE LOCALITY: USA, New Hampshire, Carroll Co., Jackson (as fixed by Bole and Moulthrop, 1942:146).

DISTRIBUTION: Midwestern and E USA through SE Canada, including Nova Scotia and Cape Breton Island; as far south as W North Carolina and NE Arkansas; outlying populations in SW Kansas and W Nebraska.

SYNONYMS: *fatuus, gossii, helaletes, jesseni, kentucki, paludis, relictus, saturatus, stonei.*

COMMENTS: Subgenus *Synaptomys*. Geographic diversification evaluated by Wetzel (1955). See Linzey (1983, Mammalian Species, 210).

Volemys Zagorodnyuk, 1990. Vestn. Zool., 2:28.

TYPE SPECIES: *Microtus musseri* Lawrence, 1982.

COMMENTS: Tribe Arvicolini. A small group of alpine or subalpine species isolated on high mountains in SW China, N Burma and on Taiwan. All have traditionally been placed in *Microtus*, although their phylogenetic relationships with other species in that genus were regarded as obscure or equivocal (see the discussion in Zagorodnyuk, 1990).

Volemys clarkei (Hinton, 1923). Ann. Mag. Nat. Hist., ser. 9, 11:158.

TYPE LOCALITY: China, Yunnan, divide between the Kuikiang and Salween rivers, 11,000 ft; 28 N.

DISTRIBUTION: Recorded from mountains of NW Yunnan and N Burma, 3400-4000 m (Allen, 1940; Corbet, 1978*c*); range limits unresolved.
COMMENTS: Regarded as the only member of the *V. clarkei* species group by Zagorodnyuk (1990).

Volemys kikuchii (Kuroda, 1920). Dobuts. Zasshi, 32:40-41.
TYPE LOCALITY: Taiwan, Mt. Morrison, 10,000 ft.
DISTRIBUTION: Endemic to highlands of Taiwan; see Lin et al. (1987).
COMMENTS: Subgenus *Volemys*. Zagorodnyuk (1990) listed this as the only member of the *kikuchii* species group in *Volemys*. Morphometric analyses of crania and dentitions are provided by Kaneko (1987); conventional karyotype reported by Harada et al. (1991); information about the holotype and type locality was summarized by Jones (1975).

Volemys millicens (Thomas, 1911). Abstr. Proc. Zool. Soc. London, 1911(100):49.
TYPE LOCALITY: China, Sichuan, Weichoe, Si-ho River Valley, 12,000 ft.
DISTRIBUTION: Recorded from mountains of E Xizang (Tibet; see Feng et al., 1986) and Sichuan, China.
COMMENTS: Closely related to *V. musseri* according to Lawrence (1982) and Zagorodnyuk (1990).

Volemys musseri (Lawrence, 1982). Am. Mus. Novit., 2745:6.
TYPE LOCALITY: China, Sichuan, Qionglai Shan, 30 mi W Wenquan, 9000 ft.
DISTRIBUTION: Recorded from the Qionglai Shan in W Sichuan, China, 7000-12,000 ft.
COMMENTS: Zagorodnyuk (1990) placed this species with *V. millicens* in the same species group. This cluster may be part of an old fauna that is peculiar to the western highlands of Sichuan (see Lawrence, 1982).

Subfamily Calomyscinae Vorontsov and Potapova, 1979. Zool. Zh., 58:1393.
COMMENTS: Although earlier authors associated it with New World sigmodontines, Vorontsov and Potapova (1979) recognized *Calomyscus*'s combination of distinctive features by setting it apart as a tribe in the subfamily Cricetinae of the family Cricetidae. Pavlinov (1980*c*) kept it with sigmodontines, and Agusti (1989:387) summarily regarded it as the only living member of the Myocricetodontinae. Carleton and Musser (1984:313) discussed the different taxonomic assignments of *Calomyscus*, its characteristics that did not fit with Old World hamsters (Cricetinae), and the dental similarities between *Calomyscus* and Miocene *Democricetodon*, and remarked that "*Calomyscus* could be justifiably classified among the cricetodontines, a group hitherto supposed extinct."

Calomyscus Thomas, 1905. Abstr. Proc. Zool. Soc. Lond., 1905(24):23.
TYPE SPECIES: *Calomyscus bailwardi* Thomas, 1905.
COMMENTS: *Calomyscus* has been considered monotypic (Ellerman and Morrison-Scott, 1951; Peshev, 1991). However, Vorontsov et al. (1979), in a comprehensive revision of the genus, treated most of the subspecies of *C. bailwardi* as distinct species, an arrangement that should be tested with additional data.
The population from Syria was discovered only recently (see Peshev, 1991), indicating that the geographic ranges of living species have yet to be defined. The modern range of the genus is a remnant of a broader distribution that in the Tertiary extended as far west as Spain (see Agusti, 1989, and references therein).

Calomyscus bailwardi Thomas, 1905. Abstr. Proc. Zool. Soc. Lond., 1905(24):23.
TYPE LOCALITY: Iran, Khuzistan, 120 km SE Ahwaz, Mala-i-Mir (= Jzeh).
DISTRIBUTION: Iran.
SYNONYMS: *grandis*.
COMMENTS: True *bailwardi* has been recorded only from Iran; its range as usually described in the literature (Corbet, 1978*c*) represents other species of *Calomyscus* (Vorontsov et al., 1979). Our distributional sketch is documented by specimens from C Iran (in the Field Museum of Natural History) and records from SW Iran (Vorontsov et al., 1979). Schlitter and Setzer (1973) described *grandis* as a subspecies of *C. bailwardi*.

Calomyscus baluchi Thomas, 1920. J. Bombay Nat. Hist. Soc., 26:939.
TYPE LOCALITY: Pakistan, Baluchistan, Kelat Dist.

DISTRIBUTION: W Pakistan (from NW Frontier and Swat region in the northwest south to Baluchistan) and NC and E Afghanistan.

SYNONYMS: *mustersi*.

COMMENTS: See Vorontsov et al. (1979) for status of *mustersi*. Most of our knowledge about the distribution is documented by specimens in the Field Museum of Natural History.

Calomyscus hotsoni Thomas, 1920. J. Bombay Nat. Hist. Soc., 26:938.

TYPE LOCALITY: Pakistan, Baluchistan, Gwambuk Kaul, 50 km SW Panjgur (26°30'N, 63°50'E).

DISTRIBUTION: Known only from vicinity of type locality.

COMMENTS: In body size, the smallest of all the species of *Calomyscus* (Vorontsov et al., 1979).

Calomyscus mystax Kashkarov, 1925. Trans. Turkestansk. Nauch. ob-va pri Sredniaziatsk. Univ. (Tashkent), 2:43.

TYPE LOCALITY: Turkmenistan, Great Balkhan Mtns, Bashi-Mugur.

DISTRIBUTION: Great and Little Balkhan and Kopet Dag Mtns and Badkhiz desert, S Turkmenia, NC (Mazanderan Prov.) and NE (Khorassan Prov.) Iran, and NW Afghanistan.

SYNONYMS: *elburzensis*.

COMMENTS: See Vorontsov et al. (1979) for status of *elburzensis*. See comments under *C. urartensis*. The Afghanistan distribution is based upon specimens in the Field Museum of Natural History; we relied on other series from that Institution and records in Vorontsov et al. (1979) for the Iranian segment.

Calomyscus tsolovi Peshev, 1991. Mammalia, 55:112.

TYPE LOCALITY: Syria, Thafas, 18 km NW Derra (= Der'a).

DISTRIBUTION: SW Syria; limits unknown.

COMMENTS: Described by Peshev (1991) as a subspecies of *C. bailwardi*, but its small size and short tail clearly distinguish it from any other described form now recognized as a species; such an option was even suggested by Peshev (1991:112).

Calomyscus urartensis Vorontsov and Kartavseva, 1979. Zool. Zh., 58:1218.

TYPE LOCALITY: Azerbaidzhan, Nakhichevansk., Alindzhachai River, 7 km N Dzhul'ta.

DISTRIBUTION: Extreme S Transcaucasus (Azerbaidzhan), far NW Iran (Azarbaijan Prov.; series in the Field Museum of Natural History).

COMMENTS: *C. urartensis* is chromosomally and morphologically closely similar to *C. mystax* (Vorontsov et al., 1979, and our study of material in the Field Museum).

Subfamily Cricetinae G. Fischer, 1817. Mém. Soc. Imp. Nat. Moscow, 5:372.

COMMENTS: Reviewed by Corbet (1978*c*, 1984), who (1978*c*:88) noted that "generic divisions within the group are rather unstable and a fresh, comprehensive classification is required," a conclusion we echo after studying specimens and literature. One of these divisions is *Cricetulus*, and Corbet (1978*c*:90) acknowledged that some species he listed there are frequently placed in *Allocricetulus* and *Tscherskia*, but "pending a review of generic classification in the subfamily as a whole," he preferred to include those genera in *Cricetulus*. We prefer the opposite: to recognize the separate genera until the subfamily is systematically revised. In cranial morphology, *Allocricetulus* closely resembles *Cricetus*, not *Cricetulus*, and the latter is more like *Phodopus* than either *Tscherskia* or *Allocricetulus*. Pavlinov and Rossolimo (1987) also treated *Cricetulus*, *Allocricetulus*, and *Tscherskia* as genera. Some of the Chinese hamsters were reviewed by Wang and Zheng (1973). Keys to the European species were provided by Niethammer (1982*a*). Chromosomal information for European species was recorded by Zima and Král (1984*a*).

Carleton and Musser (1984) diagnosised the subfamily and reviewed general characters, major fossil groups, and other topics.

Allocricetulus Argyropulo, 1933. Z. Säugetierk., 8:133.

TYPE SPECIES: *Cricetus eversmanni* Brandt, 1859.

Allocricetulus curtatus (Allen, 1925). Am. Mus. Novit., 179:3.

TYPE LOCALITY: China, W Nei Mongol (Inner Mongolia), Iren Dabasu (= Ehrlien).

DISTRIBUTION: China; "Steppes of Mongolia north of the Altai and eastwards to Inner Mongolia" (Corbet, 1978*c*:92); also recorded from N Xinjiang Prov. in NW China (Ma et al., 1987), Ningxia (Qin, 1991), and Anhui (Liu et al., 1985).

COMMENTS: Ma et al. (1987) treated *curtatus* as a subspecies of *A. eversmanni*, but Pavlinov and Rossolimo (1987) and Corbet and Hill (1991) continued to recognize it as a separate species.

Allocricetulus eversmanni (Brandt, 1859). Melanges. Biol. Acad. St. Petersbourg, p. 210.

TYPE LOCALITY: Russia, Orenburg Oblast, near Orenburg.

DISTRIBUTION: N Kazakhstan steppes from Volga River to the upper Irtysh at Zaysan (Corbet, 1978*c*).

SYNONYMS: *beljawi, belajevi, beljaevi, microdon, pseudocurtatus.*

COMMENTS: See Pavlinov and Rossolimo (1987) for allocating *pseudocurtatus* to *A. eversmanni.*

Cansumys Allen, 1928. J. Mammal., 9:244.

TYPE SPECIES: *Cansumys canus* Allen, 1928.

COMMENTS: Listed as a synonym of *Cricetulus* by Ellerman (1941), Ellerman and Morrison-Scott (1951), and Corbet (1978*c*) and of *Tscherskia* by Carleton and Musser (1984) and Pavlinov and Rossolimo (1987), but *Cansumys* does not belong with either of these. Its combination of pelage pattern, large body size, long tail, and high-crowned selenodont-like molars make it unique among cricetines (Ross, 1988).

Cansumys canus Allen, 1928. J. Mammal., 9:245.

TYPE LOCALITY: China, S Gansu (Kansu) Prov, Choni.

DISTRIBUTION: Known from vicinity of the type locality and Shaanxi Prov., China (Wang, 1990).

COMMENTS: Represented by the holotype and young animal described by Allen (1940) a specimen in the Field Museum of Natural History (36067), and Wang's (1990) record. Ellerman and Morrison-Scott (1951), and Corbet (1978*c*) included *canus* as a form of *Cricetulus triton*, but Corbet and Hill (1991) listed it as a species of *Cansumys*.

Cricetulus Milne-Edwards, 1867. Ann. Sci. Nat. (Paris), 7:376.

TYPE SPECIES: *Cricetulus griseus* Milne-Edwards, 1867 (= *Mus barabensis* Pallas, 1773).

SYNONYMS: *Urocricetus.*

COMMENTS: See comments under subfamily.

Cricetulus alticola Thomas, 1917. Ann. Mag. Nat. Hist., ser. 8, 19:455.

TYPE LOCALITY: India, Ladak, Shushul, 13,500 ft.

DISTRIBUTION: Known only from W Nepal (Lim and Ross, 1992), N India in Kashmir and Ladak, and western part of Tibetan Plateau in China.

COMMENTS: Feng et al. (1986) considered *alticola* to be a subspecies of *C. kamensis*, occuring in W Xizang Prov, but Corbet and Hill (1991) recognized it as a separate species.

Cricetulus barabensis (Pallas, 1773). Reise Prov. Russ. Reichs., 2:704.

TYPE LOCALITY: Russia, W Siberia, banks of River Ob.

DISTRIBUTION: Steppes of S Siberia from River Irtysh to Ussuri region, and south to Mongolia, N China (Xinjiang through Nei Mongol), and Korea.

SYNONYMS: *ferrugineus, fumatus, furunculus, griseus* (Milne-Edwards, 1867, not Kashkarov, 1923), *manchuricus, mongolicus, obscurus, pseudogriseus, tuvinicus, xinganensis.*

COMMENTS: Pavlinov and Rossolimo (1987) listed *pseudogriseus* as a species, and Corbet and Hill (1991) treated it and *griseus* as species. However, these were originally diagnosed by chromosomal traits, and in a study that included *C. barabensis*, Král et al. (1984) found it difficult to karyotypically characterize the three species because of the extensive homology among chromosome arms. Our treatment reflects that of Corbet (1978*c*), who discussed the problem and included *pseudogriseus* and *griseus* in *C. barabensis*. See Malygin et al. (1992) for an opposite treatment.

Cricetulus kamensis (Satunin, 1903). Ann. Zool. Mus. St. Petersbourg, 7:574.

TYPE LOCALITY: China, NE Tibet (Xizang Prov), Mekong Dist, River Moktschjun.

DISTRIBUTION: China, Tibetan Plateau.

SYNONYMS: *kozlovi, lama, tibetanus.*

COMMENTS: See discussion under *C. alticola*.

Cricetulus longicaudatus (Milne-Edwards, 1867). Rech. Hist. Nat. Mammifères, p. 13.
TYPE LOCALITY: China, N Shanxi (Shansi), near Saratsi.
DISTRIBUTION: Altai and Tuva regions of Russia and Kazakhstan, NW China (Xinjiang), Mongolia, and adjacent regions of China south to N Tibet (Feng et al., 1986; Qin, 1991).
SYNONYMS: *andersoni, chiumalaiensis, dichrootis, griseiventris* (Satunin, 1903, not Thomas, 1917), *kozhantschikovi, nigrescens.*

Cricetulus migratorius (Pallas, 1773). Reise Prov. Russ. Reichs., 2:703.
TYPE LOCALITY: W Kazakhstan, lower Ural River.
DISTRIBUTION: S European Russia and SE Europe (Greece, Rumania, Bulgaria) through Kazakhstan to S Mongolia and N China (Xinjiang, Ningxia; Qin, 1991), north nearly to Moscow, south to Israel, Jordan, Lebanon, Iraq, Iran, Pakistan, Afghanistan, and Turkey.
SYNONYMS: *accedula, arenarius, atticus, bellicosus, caesius, cinerascens, cinereus, coerulescens, elisarjewi, falzfeini, fulvus, griseiventris* (Thomas, 1917, not Satunin, 1903); *griseus* (Kashkarov, 1923, not Milne-Edwards, 1867), *isabellinus, murinus, myosurus, neglectus, ognevi, pamirensis, phaeus, pulcher, sviridenkoi, tauricus, vernula, zvierezombi.*
COMMENTS: Taxonomy, morphology, distribution, fossils, and biology summarized for European segment by Niethammer (1982*d*).

Cricetulus sokolovi Orlov and Malygin, 1988. Zool. Zh., 67:305.
TYPE LOCALITY: W Mongolia.
DISTRIBUTION: W and S Mongolia, C Nei Mongol of N China.
COMMENTS: A distinctive species defined by diagnostic chromosomal and pelage traits (Orlov and Malygin, 1988). Samples from the Mongolian segment of the range had been identified as *C. obscurus* (Král et al., 1984, and references therein; Orlov and Malygin, 1988); real *obscurus* is a form of *C. barabensis.*

Cricetus Leske, 1779. Anfansgr. Naturg., 1:168.
TYPE SPECIES: *Mus cricetus* Linnaeus, 1758.
SYNONYMS: *Hamster, Heliomys.*

Cricetus cricetus (Linnaeus, 1758). Syst. Nat., 10th ed., 1:60.
TYPE LOCALITY: Germany.
DISTRIBUTION: From Belgium across C Europe, W Siberia and N Kazakhstan to the upper Yenesei and Altai region and NW China (Xinjiang).
SYNONYMS: *albus, babylonicus, canescens, frumentarius, fulvus, fuscidorsis, germanicus, jeudii, latycranius, nehringi, niger, nigricans* (Lacépède, 1799, not Brandt, 1832), *polychroma, rufescens, stavropolicus, tauricus, tomensis, varius, vulgaris.*
COMMENTS: Taxonomy, morphology, distribution, karyotype, and biology of European populations reviewed by Niethammer (1982*b*), who also noted that closest relatives are extinct species described from Pleistocene samples. More recent analyses of morphological variability among European samples were reported by Grulich (1987, 1991, and references therein). Significance of geographic range in Netherlands was discussed by Lenders and Pelzers (1982).

Mesocricetus Nehring, 1898. Zool. Anz., 21:49.
TYPE SPECIES: *Cricetus nigricans* Brandt, 1832 (= *Cricetus raddei* Nehring, 1894).
SYNONYMS: *Mediocricetus, Semicricetus.*
COMMENTS: Ellerman and Morrison-Scott (1951) recognized only one extant species in the genus, but current checklists recognize at least three (Corbet, 1978*c*; Corbet and Hill, 1991). Pieper (1984) described *M. rathgeberi* based on Holocene fossils from the Greek island of Armathia (off coast of Kasos Isl between Kriti and Rodhos).

Mesocricetus auratus (Waterhouse, 1839). Proc. Zool. Soc. Lond., 1839:57.
TYPE LOCALITY: Syria, Aleppo.
DISTRIBUTION: Vicinity of type locality.
COMMENTS: Lyman and O'Brien (1977) discussed the geographic range of *auratus* and the evidence for separating it from *brandti.*

Mesocricetus brandti (Nehring, 1898). Zool. Anz., 21:331.
TYPE LOCALITY: Georgia, near Tbilisi.
DISTRIBUTION: Anatolian Turkey, south into Israel, Lebanon, Syria, N Iraq, NW Iran, N Transcaucasia, and Kurdistan.
SYNONYMS: *koenigi*.
COMMENTS: Corbet (1978*c*) included *brandti* in *M. auratus* but acknowledged that it might be a separate species, as it is treated by Niethammer (1982*a*, *c*) and Pavlinov and Rossolimo (1987). However, *M. brandti* was earlier revised by Lyman and O'Brien (1977), who presented morphological, chromosomal, and breeding evidence supporting its specific status, and also provided distributional, ecological, and husbandry information. Additional chromosomal data were documented by Fang and Jagiello (In Press).

Mesocricetus newtoni (Nehring, 1898). Zool. Anz., 21:329.
TYPE LOCALITY: Bulgaria, Kolarovgrad (= Schumla or Shumen).
DISTRIBUTION: E Rumania and Bulgaria.
COMMENTS: A distinct species (Corbet, 1978*c*; Corbet and Hill, 1991). Niethammer (1982*c*) summarized taxonomy, morphology, chromosomal data, distribution, paleontology, biology, and provided comparisons with other species of *Mesocricetus*.

Mesocricetus raddei (Nehring, 1894). Zool. Anz., 18:148.
TYPE LOCALITY: Russia, N Caucasus, Daghestan, Samur River.
DISTRIBUTION: Russia, steppes along N slopes of Caucasus to Don River and Sea of Azov (Corbet, 1978*c*).
SYNONYMS: *avaricus*, *nigricans* (Brandt, 1832, not Lacépède, 1799), *nigriculus*.
COMMENTS: Recognized as a species by Pavlinov and Rossolimo (1987).

Phodopus Miller, 1910. Smithson. Misc. Coll., 52:498.
TYPE SPECIES: *Cricetulus bedfordiae* Thomas, 1908 (= *Cricetulus roborovskii* Satunin, 1903).
SYNONYMS: *Cricetiscus*.
COMMENTS: Chromosomal data reported by Spyropoulos et al. (1982) and Schmid et al. (1986).

Phodopus campbelli (Thomas, 1905). Ann. Mag. Nat. Hist., ser. 7, 15:322.
TYPE LOCALITY: NE Mongolia, Shaborte (see Pavlinov and Rossolimo, 1987, for additional data).
DISTRIBUTION: Transbaikalia in Russia; Mongolia, and adjacent China from Heilungkiang through Nei Mongol to Xinjiang.
SYNONYMS: *crepidatus*, *tuvinicus*.
COMMENTS: Included in *P. sungorus* by Corbet (1978*c*), but regarded as separate species by Pavlinov and Rossolimo (1987).

Phodopus roborovskii (Satunin, 1903). Ann. Zool. Mus. St. Petersbourg, 7:571.
TYPE LOCALITY: China, Nan Shan Mtns, upper part of Shargol Dzhin River.
DISTRIBUTION: Tuva (Russia) and E Kazakhstan, W and S Mongolia, adjacent parts of China from Heilongjiang west to N Xinjiang (Ma et al., 1987; Qin, 1991).
SYNONYMS: *bedfordiae*, *praedilectus*, *przewalskii*.
COMMENTS: The status of *przewalskii* was discussed by Corbet (1978*c*) and Pavlinov and Rossolimo (1987).

Phodopus sungorus (Pallas, 1773). Reise Prov. Russ. Reichs., 2:703.
TYPE LOCALITY: E Kazakhstan, 100 km west of Semipalatinsk, near Grachevsk.
DISTRIBUTION: E Kazakhstan and SW Siberia.

Tscherskia Ognev, 1914. Moskva Dnev. Zool. otd. obsc. liub. jest., 2:102.
TYPE SPECIES: *Tscherskia albipes* Ognev, 1914 (= *Cricetulus triton* de Winton, 1899).
SYNONYMS: *Asiocricetus*.
COMMENTS: See comments under subfamily. The genus is considered to be monotypic, but Storch (1974) described *T. rusa* from Holocene subfossils found in Iran.

Tscherskia triton (de Winton, 1899). Proc. Zool. Soc. Lond., 1899:575.
TYPE LOCALITY: China, N Shantung.

DISTRIBUTION: NE China from Shaanxi to SE Manchuria (Heilongjiang) and south to Anhui (Liu et al., 1985), Korea, and north to upper Ussuri in Russia (Corbet, 1978*c*; Qin, 1991).

SYNONYMS: *albipes, arenosus, bampensis, collinus, fuscipes, incanus, meihsienensis, nestor, ningshaanensis, yamashinai.*

COMMENTS: Whether one or more species is represented by the named forms remains to be resolved by systematic revision. Song (1985) proposed *ningshaanensis* for a sample of *T. triton* from Shaanxi.

Subfamily Cricetomyinae Roberts, 1951. Mammals S Africa, p. 434.

SYNONYMS: Saccostomurinae.

COMMENTS: Although initially classified within Murinae (e.g., Ellerman, 1941; Simpson, 1945; Thomas, 1897*a*), systematists acknowledged a closer affinity of African pouched rats to one another than to other murines, a distinction later formalized by Roberts (1951). Subsequent systematic arrangements have followed Petter (1966*a*) in allying cricetomyines with cricetids (Reig, 1980; Rosevear, 1969). Anatomy of internal cheek pouches supports the monophyly of the subfamily (see Ryan, 1989*b*); other morphological traits and phylogenetic interpretations summarized by Carleton and Musser (1984). Roberts (1951) segregated *Saccostomus*, as the lone member of Saccostomurinae, from other African pouched rats (Cricetomyinae), a division not recognized by later systematists (Petter, 1966*a*; Ryan, 1989*b*).

Beamys Thomas, 1909. Ann. Mag. Nat. Hist., ser. 8, 4:107.

TYPE SPECIES: *Beamys hindei* Thomas, 1909.

COMMENTS: Two species have been generally recognized as originally described (e.g., Ellerman, 1941; Ellerman et al., 1953; Hanney, 1965; Misonne, 1974), until Ansell and Ansell (1973) suggested their synonymy, a classification adopted by others (e.g., Ansell, 1978; Corbet and Hill, 1991; Honacki et al., 1982). Two species appear evident among USNM series, but the matter deserves the attention of a serious revision.

Beamys hindei Thomas, 1909. Ann. Mag. Nat. Hist., ser. 8, 4:108.

TYPE LOCALITY: Kenya, Coast Prov., Taveta.

DISTRIBUTION: S Kenya and neighboring NE Tanzania.

Beamys major Dollman, 1914. Ann. Mag. Nat. Hist., ser. 8, 14:428.

TYPE LOCALITY: Malawi, Southern Region, Mulanje.

DISTRIBUTION: Malawi, NE Zambia, and S Tanzania.

Cricetomys Waterhouse, 1840. Proc. Zool. Soc. Lond., 1840:2.

TYPE SPECIES: *Cricetomys gambianus* Waterhouse, 1840.

COMMENTS: Six nominal species were recognized (e.g., Allen, 1939) before Ellerman (1941) reduced all to subspecies of *C. gambianus*. Genest-Villard's (1967) revision provided evidence of two kinds, a predominantly savannah-dwelling species (*C. gambianus*) and a lowland forest form (*C. emini*). Her character states for discriminating the two species are generally workable in West Africa but break down in parts of eastern Africa, which suggests the need for additional alpha-level study.

Cricetomys emini Wroughton, 1910. Ann. Mag. Nat. Hist., ser. 8, 5:269.

TYPE LOCALITY: Zaire, Monbuttu, Gadda.

DISTRIBUTION: Lowland rain forest: in West Africa, from Sierra Leone to S Nigeria; in Central Africa, from Cameroon and Gabon, through Congo and N Zaire, to S Uganda; including Bioko (= Fernando Poo).

SYNONYMS: *dolichops, kivuensis, liberiae, luteus, poensis, proparator, raineyi, sanctus.*

Cricetomys gambianus Waterhouse, 1840. Proc. Zool. Soc. Lond., 1840:2.

TYPE LOCALITY: Gambia, River Gambia.

DISTRIBUTION: Savannah and forest edges from Senegal and Sierra Leone east to S Sudan and N Uganda; southwards, exclusive of Congo forest block, to S Angola, S Zambia, and E Natal, South Africa.

SYNONYMS: *adventor, ansorgei, buchanani, cunctator, cosensi, dichrurus, dissimilis, elgonis, enguvi, goliath, grahami, haagneri, kenyensis, langi, microtis, oliviae, osgoodi, selindensis, servorum, vaughanjonesi, viator.*

Saccostomus Peters, 1846. Bericht Verhandl. K. Preuss. Akad. Wiss., Berlin, 11:258.
TYPE SPECIES: *Saccostomus campestris* Peters, 1846.
SYNONYMS: *Eosaccomys.*
COMMENTS: Ellerman (1941) thought that all nominal forms would prove to be races of the single species *S. campestris*, a speculation widely observed in later classifications and faunal reports (e.g., Delany, 1975; Ellerman et al., 1953; Kingdon, 1974; Misonne, 1974). Hubert (1978*a*), however, presented karyotypic and morphological evidence that differentiates an eastern African species (*S. mearnsi*) from a southern African one (*S. campestris*), as earlier intimated by Allen and Lawrence (1936).

Saccostomus campestris Peters, 1846. Bericht Verhandl. K. Preuss. Akad. Wiss., Berlin, 11:258.
TYPE LOCALITY: Mozambique, Tete Dist., Zambezi River, Tete.
DISTRIBUTION: Broadly occurring in arid to mesic southern savannahs and grasslands: from SW Tanzania across to C Angola; south through most of Malawi, Zambia, Zimbabwe, Botswana, and Namibia; to S Mozambique and Cape Province, South Africa.
SYNONYMS: *anderssoni, angolae, elegans, fuscus, hildae, lapidarius, limpopoensis, mashonae, pagei, streeteri.*
COMMENTS: A composite of two or more species as suggested by extraordinary chromosomal variation (Gordon, 1986; Gordon and Rautenbach, 1980); the status of *anderssoni* and *mashonae* especially deserves attention in regard to *campestris* proper.

Saccostomus mearnsi Heller, 1910. Smithson. Misc. Coll., 54:3.
TYPE LOCALITY: Kenya, Coast Prov., Changamwe.
DISTRIBUTION: Extreme S Ethiopia and S Somalia, through E Uganda and Kenya, to NE Tanzania.
SYNONYMS: *cricetulus, isiolae, umbriventer.*
COMMENTS: See remarks under generic account.

Subfamily Dendromurinae Alston, 1876. Proc. Zool. Soc. Lond., 1876:82.
SYNONYMS: Deomyinae, Dendromyinae.
COMMENTS: Carleton and Musser (1984) provided diagnosis of the subfamily, general characters, habits, habitats, and other information; summarized historical judgments about relationships of the group, which by 1984 had disassociated dendromurines from murines and either aligned them to "cricetids" or treated them as a separate family (Chaline et al., 1977); and cautioned that more research was required to determine if the dendromurines as currently defined are a natural group or instead a polyphyletic assemblage of specialized relicts evolved from a primitive African muroid stock. The association of dendromurines as a subfamily of Cricetidae was retained by Lindsay (1988).

Extant members of Dendromurinae are found only in subsaharan Africa (Dieterlen, 1971), and the group is represented in parts of that region by Pleistocene, Pliocene, and Miocene fossils (Carleton and Musser, 1984; Conroy et al., 1992; Lavocat, 1978; Senut et al., 1992), and in North Africa by Miocene samples (Lindsay, 1988). But at one time the geographic range of dendromurines extended beyond Africa: examples of *Dendromus* are recorded from the late Miocene deposits of S Spain, and extinct genera mark the first appearance of dendromurines in the middle Miocene of Pakistan and Thailand (Lindsay, 1988, and references therein).

Dendromus A. Smith, 1829. Zool. J. London, 4:438.
TYPE SPECIES: *Dendromus typus* A. Smith, 1829 (= *Mus mesomelas* Brants, 1827).
SYNONYMS: *Chortomys, Dendromys, Poemys.*
COMMENTS: In his checklist of African mammals, Allen (1939) listed 28 species of *Dendromus*, which was reduced to four by Bohmann's (1942) revision. Dieterlen (1971) focused on reporting morphological, ecological, and other traits of the central African forms and recognized five species from that region. We recognize eleven. Definitions of *D. kahuziensis, D. lovati, D. insignis, D. vernayi*, and *D. nyikae* are clear,

even though the first two are known by few specimens. Four (*D. melanotis, D. mesomelas, D. messorius,* and *D. mystacalis*) need to be systematically revised—each may actually be a complex of species; some synonyms associated with each of them may be incorrectly allocated. Two others, *D. kivu* and *D. oreas* require critical comparison with samples of true *D. mesomelas* to resolve their affinities. Chromosomal information for several species was reported by Matthey (1967, 1970).

Dendromus insignis (Thomas, 1903). Ann. Mag. Nat. Hist., ser. 7, 12:341.

TYPE LOCALITY: Kenya, Nandi.

DISTRIBUTION: Discontinuous highland range from Ethiopia (Yalden et al., 1976, recorded under *mesomelas*), E Kenya (Hollister, 1919), Western Rift mountains from W Uganda south to Rwanda, and Mt. Kilimanjaro in NE Tanzania; extent of montane distribution unknown.

SYNONYMS: *abyssinicus, kilimandjari, percivali.*

COMMENTS: Although *insignis* is now included in *D. mesomelas* (Bohmann, 1942; Misonne, 1974), Thomas (1916*b*) noted that the "lumping of *insignis* with the southern *D. mesomelas*" appeared to be "unfounded," an evaluation matching our conclusion based on study of specimens. *D. insignis* is easily distinguished from *D. mesomelas* by its very large body size and cranial traits. Furthermore, it occurs at the same localities in the Western Rift mountains as *D. kivu*, which is the species most like *D. mesomelas*. Bohmann (1939) described *kilimandjari* as a subspecies of *D. mesomelas*.

Dendromus kahuziensis Dieterlen, 1969. Z. Säugetierk., 34:348-353.

TYPE LOCALITY: Zaire, Kivu, Mt Kahuzi, 2100 m.

DISTRIBUTION: Zaire, Kivu region.

COMMENTS: A very distinct long-tailed species known only by few specimens from moss forest-bamboo habitat (Dieterlen, 1969*a*, 1976*c*).

Dendromus kivu Thomas, 1916. Ann. Mag. Nat. Hist., ser. 8, 18:242.

TYPE LOCALITY: Zaire, near Lake Kivu, Buhamba, 2000 m.

DISTRIBUTION: A possible montane Western Rift endemic; recorded from the west and east slopes of the Ruwenzoris in E Zaire and W Uganda (Osgood, 1936, and specimens in the Field Museum of Natural History) and the Kivu region in E Zaire (samples in the American Museum of Natural History and the British Museum of Natural History).

SYNONYMS: *lunaris.*

COMMENTS: Osgood (1936:236) discussed two series of *Dendromus* collected at the same time and place on the western slope of Mt. Ruwenzori. He described one as *D. lunaris*, based on two examples, and identified the other as *D. insignis*, a conclustion we verified from study of his specimens. The two species are also sympatric in the Kivu Region (series in American Museum of Natural History). Thomas originally described *kivu* as a subspecies of *D. insignis*, but our study of the holotype reveals it to be the same as Osgood's *lunaris*, which it predates. Both *kivu* and *insignis* have been included in *D. mesomelas* (Bohmann, 1942; Misonne, 1974). It is the smaller-bodied *kivu* that is morphologically similar to *D. mesomelas* and the nature of the relationship between the two will have to be revealed by systematic revision of the *mesomelas* group.

Dendromus lovati de Winton, 1900. Proc. Zool. Soc. Lond., 1899:986 [1900].

TYPE LOCALITY: Ethiopia, Managasha, near Addis Ababa.

DISTRIBUTION: Endemic to Ethiopian Plateau where it is recorded from 2500-3550 m (Rupp, 1980; Yalden et al., 1976).

COMMENTS: A grassland species differing from other members of *Dendromus* in its apparently strictly terrestrial habits (Yalden et al., 1976; Yalden, 1988).

Dendromus melanotis Smith, 1834. S. Afr. Quart. J., 2:158.

TYPE LOCALITY: South Africa, near Port Natal (Durban).

DISTRIBUTION: From South Africa (see map in Skinner and Smithers, 1990:306) north to Uganda, and west through Nigeria to Mt-Nimba in Guinea; also from Ethiopia (#8119 in American Museum of Natural History), but published records might be *D. mystacalis* (Yalden et al., 1976). Limits of range unresolved.

SYNONYMS: *arenarius, basuticus, capensis, carteri, chiversi, concinnus, exoneratus, insignis* (Shortridge and Carter, 1938, not Thomas, 1903), *leucostomus, nigrifrons, pallidus, pretoriae, shortridgei, spectabilis, subtilis, thorntoni, vulturnus.*

Dendromus mesomelas (Brants, 1827). Het. Geslacht der Muizen, p. 122.

TYPE LOCALITY: South Africa, E Cape Prov., Sunday's River, east of Port Elizabeth.

DISTRIBUTION: Discontinuous ranges in South Africa (SW Cape Prov. east to Natal and E Transvaal), C Mozambique, extreme NE Zambia, N Botswana (see Map in Skinner and Smithers, 1990:307), extreme NW and NE Zaire (Ansell, 1978), highlands of N and S Malawi (Ansell and Dowsett, 1988), mountains of SW (Rungwe) and EC (Uluguru) Tanzania, and SE Zaire (Marungu); extent of range unresolved.

SYNONYMS: *ayresi, hintoni, major, nyasae, pumilio, typicus, typus.*

COMMENTS: We treat some of the names usually included in *D. mesomelas* (see Misonne, 1974) as distinct species (see accounts of *insignis, kivu, oreas,* and *vernayi*). Judged by our study of specimens, samples from SE Zaire and SW Tanzania are morphologically similar to series of *D. kivu*, and the relationship among these and samples of true *mesomelas* has to be resolved by critical systematic revision. We include *major*, which was described as a subspecis of *D. mesomelas* (see St. Leger, 1930), but its relationships should be reexamined because of its large body size.

Dendromus messorius Thomas, 1903. Ann. Mag. Nat. Hist., ser. 7, 12:340.

TYPE LOCALITY: Cameroon, Efulen.

DISTRIBUTION: Benin, Nigeria to Zaire, Uganda, Kenya, and Sudan; limits not yet resolved.

SYNONYMS: *haymani, kumasi, ruddi.*

COMMENTS: Treated as part of *D. mystacalis* by Misonne (1974), but regarded as a separate species by Hatt (1940*a*) and Dieterlen (1971), who pointed out its sympatric distribution with *D. mystacalis.*

Dendromus mystacalis Heuglin, 1863. Nova Acta Acad. Caes. Leop.-Carol., Halle, 30:2, suppl. 5.

TYPE LOCALITY: Ethiopia, Baschlo region (Ellerman et al., 1953, offered additional comments).

DISTRIBUTION: From South Africa (see map in Skinner and Smithers, 1990:308) north through most of the continent to Ethiopia, S Sudan, Zaire, and N Angola; extent of range unknown.

SYNONYMS: *acraeus, ansorgei, capitis, jamesoni, lineatus, nairobae, ochropus, pallescens, pongolensis, uthmoelleri, whytei.*

Dendromus nyikae Wroughton, 1909. Ann. Mag. Nat. Hist., ser. 8, 3:248.

TYPE LOCALITY: N Malawi, Nyika Plateau.

DISTRIBUTION: SC Zaire, N and C Angola, South Africa (E Transvaal), E Zimbabwe, Zambia, Malawi, SW Tanzania, and Mozambique; limits unknown.

SYNONYMS: *angolensis, bernardi, longicaudatus, pecilei.*

COMMENTS: The record from Zaire is documented by specimens in the American Museum of Natural History collected at Kananga (Luluabourg) and labelled *pecilei.*

Dendromus oreas Osgood, 1936. Field Mus. Nat. Hist., Zool. Ser., 20:236.

TYPE LOCALITY: Cameroon, Southwest Prov., southwest side of Mt. Cameroon, 9000 ft.

DISTRIBUTION: Possibly endemic to mountains of extreme E Cameroon in the Southest Prov.; recorded from Mt. Cameroon and Mt. Kupe and Mt. Manenguba in the massif about 70 mi northeast of Mt. Cameroon (Osgood, 1936; Rosevear, 1969).

COMMENTS: Although originally described by Osgood (1936) as a distinct species, *oreas* was later included in *D. mesomelas* (Bohmann, 1942; Rosevear, 1969; Misonne, 1974). Our study of the holotype supports Osgood's (1936) view that *oreas* may be related to *D. lunaris* (= *kivu*), but is not a geographic sample of *D. insignis*. Judged from the wide range in length of upper molar row that Rosevear (1969:472) recorded for *oreas*, his Cameroon sample is either extremely variable, which is unusual in this group, or consists of two species.

Dendromus vernayi Hill and Carter, 1937. Am. Mus. Novit., 913:4.

TYPE LOCALITY: EC Angola, Chitau, 4930 ft.

DISTRIBUTION: Known only from the type locality (Hill and Carter, 1941; Crawford-Cabral, 1986).

COMMENTS: Hill and Carter described *vernayi* as a subspecies of *D. mesomelas*, an association which has not been questioned (Bohmann, 1942; Misonne, 1974; Crawford-Cabral, 1986). Our study of the holotype and original series revealed *vernayi* to be distinguished from *mesomelas* by its smaller size, bright buffy venter, much shorter tail, and shorter and stubby rostrum. No data supports its affiliation with *D. mesomelas* as a subspecies and it is clearly distinct from the much larger *D. insignis*.

Dendroprionomys Petter, 1966. Mammalia, 30:129.
TYPE SPECIES: *Dendroprionomys rousseloti* Petter, 1966.

Dendroprionomys rousseloti Petter, 1966. Mammalia, 30:129.
TYPE LOCALITY: SE Congo, Brazzaville, Zoological Gardens.
DISTRIBUTION: Congo, recorded only from vicinity of Brazzaville.
COMMENTS: A species combining features of *Dendromus* with those of *Prionomys* that is still known only by a few specimens (Petter, 1966*b*).

Deomys Thomas, 1888. Proc. Zool. Soc. Lond., 1888:130.
TYPE SPECIES: *Deomys furrugineus* Thomas, 1888.
COMMENTS: Although currently included in Dendromurinae, the morphological traits of *Deomys* do not fit well in that group (Rosevear, 1969), and influenced Ellerman (1941) to erect the Deomyinae to contain the genus (see the discussion in Carleton and Musser, 1984).

Deomys ferrugineus Thomas, 1888. Proc. Zool. Soc. Lond., 1888:130.
TYPE LOCALITY: Lower Congo.
DISTRIBUTION: Uganda, Rwanda, Zaire, SW Central African Republic, S Cameroon, Gabon, Congo, Equatorial Guinea, Bioko.
SYNONYMS: *christyi, poensis, vandenberghei.*
COMMENTS: Hatt (1940*a*) and Rosevear (1969) provided good reviews. Myological and skeletal interaction in mastication was reported by Lemire (1966).

Leimacomys Matschie, 1893. Sitzb. Ges. Naturf. Fr. Berlin, p. 107.
TYPE SPECIES: *Leimacomys buettneri* Matschie, 1893.
SYNONYMS: *Limacomys*.
COMMENTS: Apparently more closely related to *Steatomys* than to other dendromurines (Dieterlen, 1976*a*). Rosevear (1969) provided a particularly good summary of past taxonomic allocations of *Leimacomys* and the nature of the type specimens.

Leimacomys buettneri Matschie, 1893. Sitzb. Ges. Naturf. Fr. Berlin, p. 109.
TYPE LOCALITY: Togo, Bismarckburg, 710 m (see Misonne, 1966, for additional data).
DISTRIBUTION: Known only from the type locality.
COMMENTS: Still known only by two specimens (Dieterlen, 1976*a*; Misonne, 1966; Rosevear, 1969).

Malacothrix Wagner, 1843. *In* Schreber, Die Säugethiere, Suppl., 3:496.
TYPE SPECIES: *Otomys typicus* A. Smith, 1834.
SYNONYMS: *Otomys* (Smith, 1834, not Cuvier, 1823).
COMMENTS: Generally considered monotypic (Meester et al., 1986), but this view needs testing by systematic revision. See Matthey (1967) for reference to chromosomal information and karyological contrasts with other dendromurine genera.

Malacothrix typica (A. Smith, 1834). S. Afr. Quart. J., 2:148.
TYPE LOCALITY: South Africa, E Cape Prov., Graaff Reinet Dist.
DISTRIBUTION: South Africa (Cape Prov., Orange Free State, SW Transvaal), S Botswana, Namibia, and SW Angola (see map in Skinner and Smithers, 1990:303).
SYNONYMS: *damarensis, egeria, fryi, harveyi, kalaharicus, molopensis*.
COMMENTS: Species is now absent from Natal in South Africa but was present up to about 60,000 years ago (Avery, 1991).

Megadendromus Dieterlen and Rupp, 1978. Z. Säugetierk., 43:129.
TYPE SPECIES: *Megadendromus nikolausi* Dieterlen and Rupp, 1978.

COMMENTS: A spectacular dendromurine with derived molar traits that were interpreted by Dieterlen and Rupp (1978) to indicate a closer relationship between Dendromurinae and Murinae than was previously suspected.

Megadendromus nikolausi Deiterlen and Rupp, 1978. Z. Säugetierk., 43:131.
TYPE LOCALITY: Ethiopia, S Goba, Bale Mtns.
DISTRIBUTION: Apparently endemic to the Bale Mtns of Ethiopia.
COMMENTS: Known only by two complete specimens and two fragmented skulls extracted from owl pellets (Demeter and Topal, 1982; Dieterlen and Rupp, 1978). Morphology and ecology were reported by Dieterlen and Rupp (1978); Demeter and Topal (1982) provided additional habitat information.

Prionomys Dollman, 1910. Ann. Mag. Nat. Hist., ser. 8, 6:226.
TYPE SPECIES: *Prionomys batesi* Dollman, 1910.

Prionomys batesi Dollman, 1910. Ann. Mag. Nat. Hist., ser. 8, 6:228.
TYPE LOCALITY: Cameroon, Ja River, Bitye, 2000 ft.
DISTRIBUTION: Recorded only from W and S Cameroon and S Central African Republic; limits unknown.
COMMENTS: Petter (1966*b*) discussed *P. batesi* from Central African Republic and contrasted its morphology with that of *Dendroprionomys*. The specimen from W Cameroon (55 km NE Obala) is in the American Museum of Natural History (241344).

Steatomys Peters, 1846. Bericht Verhandl. K. Preuss. Akad. Wiss. Berlin, 11:258.
TYPE SPECIES: *Steatomys pratensis* Peters, 1846.
COMMENTS: Reviewed by Coetzee (1977*a*). Rosevear (1969) provided an excellent review of the West African species, and Swanepoel and Schlitter (1978) produced a more comprehensive revision of them. Still, the widely distributed species such as *S. pratensis* and *S. parvus* need careful systematic revision to better resolve actual diversity of species represented in samples and their geographic ranges.

Steatomys caurinus Thomas, 1912. Ann. Mag. Nat. Hist., ser. 9, 9:271.
TYPE LOCALITY: Nigeria, Panyam, 4000 ft.
DISTRIBUTION: Endemic to W Africa from Senegal west to C Nigeria (see map in Swanepoel and Schlitter, 1978); limits unknown.
SYNONYMS: *roseveari*.
COMMENTS: Listed as a subspecies of *S. pratensis* by Coetzee (1977*a*), but documented as a separate species in the revision by Swanepoel and Schlitter (1978).

Steatomys cuppedius Thomas and Hinton, 1920. Novit. Zool., 27:318.
TYPE LOCALITY: Nigeria, Farniso (= Panisau), near Kano, 1700 ft.
DISTRIBUTION: A W African endemic occurring from Senegal to NC Nigeria and SC Niger (see map in Swanepoel and Schlitter, 1978); actual range unresolved.
COMMENTS: Revised by Swanepoel and Schlitter (1978), who regarded *cuppedius* as a distinct species and not a member of *S. parvus*, as Coetzee (1977*a*) had indicated.

Steatomys jacksoni Hayman, 1936. Proc. Zool. Soc. Lond., 1935:930 [1936].
TYPE LOCALITY: Ghana, Ashanti, Wenchi.
DISTRIBUTION: Known only from type locality and SW Nigeria.
COMMENTS: This species was known only by the holotype, but a few specimens have been recorded from SW Nigeria (Anadu, 1979), and earlier Rosevear (1969) had suggested that two young animals from Wulehe, Togo, might be examples of *S. jacksoni*. Included in *S. pratensis* by Coetzee (1977*a*), but retained as a species by Swanepoel and Schlitter (1978), who preferred to recognize it until significance of the diagnostic character (size and shape of interparietal) could be assessed by additional specimens.

Steatomys krebsii Peters, 1852. Reise nach Mossambique, Säugethiere, p. 165.
TYPE LOCALITY: South Africa, Kaffraria.
DISTRIBUTION: S Angola, W Zambia, NE and SE Botswana, Caprivi Strip, and South Africa (W, N, E Cape Province, NW Natal, N Orange Free State, S and SW Transvaal); see map in Skinner and Smithers (1990).
SYNONYMS: *angolensis, bensoni, bradleyi, chiversi, leucorhynchus, mariae, orangiae, pentonyx, transvaalensis*.

COMMENTS: Reviewed by Ansell (1978).

Steatomys parvus Rhoads, 1896. Proc. Acad. Nat. Sci. Philadelphia, p. 529.
TYPE LOCALITY: Ethiopia, Lake Rudolf, Rusia.
DISTRIBUTION: South Africa (N Natal, Zululand), N Botswana, NW Zimbabwe, N Namibia, Angola, Zambia, Tanzania, Kenya, Uganda, Somalia, S Sudan, and S Ethiopia.
SYNONYMS: *aquilo, athi, kalaharicus, loveridgei, minutus, swalius, thomasi, tongensis, umbratus.*
COMMENTS: The single Ethiopian record was listed as *S. pratensis* by Yalden et al. (1976).

Steatomys pratensis Peters, 1846. Bericht Verhandl. K. Preuss. Akad. Wiss. Berlin, 11:258.
TYPE LOCALITY: Mozambique, Zambezi River, Tete.
DISTRIBUTION: South Africa (Natal, Zululand, Transvaal), Swaziland, Mozambique, Zimbabwe, N Botswana, NE and N Namibia, N Malawi, Zambia, Angola, Zaire, SW Sudan and west to Cameroon.
SYNONYMS: *bocagei, edulis, gazellae, kasaicus, maunensis, natalensis, nyasae, opimus.*

Subfamily Gerbillinae Gray, 1825. Ann. Philos., n.s., 10:342.
SYNONYMS: Ammodillini, Desmodilliscini, Gerbillurini, Merionidae, Pachyuromyini, Rhombomyini, Taterillinae.
COMMENTS: Whether viewed as a subfamily or family, species of gerbils form a distinct group defined by a suite of derived traits (see Carleton and Musser, 1984). In a series of reports, Pavlinov (1980*a*, 1981, 1982, 1985, 1987) presented analyses of skeletal and dental characters of gerbils and developed a hypothesis of their phylogeny and a classification. The monograph by Pavlinov et al. (1990) represents a culmination of these efforts, in which the phylogeny, species classification, morphology, ecology, and geographical distribution of Gerbillidae are comprehensively reviewed.

Cytogenetic data for the subfamily summarized by Qumsiyeh and Schlitter (1991) and Viegas-Pequignot et al. (1986). Rates of protein, chromosomal, and morphological evolution in four genera reported by Qumsiyeh and Chesser (1988). Chromosomal and biochemical results of several genera documented and discussed in a phylogenetic context by Benazzou et al. (1984). Anatomy, physiology, adaptive significance, and evolution of the middle and inner ear of gerbillines documented by Lay (1972) and Pavlinov (1988); significance of acoustic emissions and morphology of cochlea in context of adaptation and systematics reported by Bridelance (1987) and Plassmann et al. (1987); and the relationship of acoustical adaptations in relation to steppe and desert environments summarized by Petter et al. (1984).

The origin and evolution of North African gerbils discussed by Tong (1989) whose results were presented in a phylogenetic classification of genera and compared with other classifications based on morphological, chromosomal, and biochemical data. Palearctic species reviewed by Corbet (1978*c*, 1984); Russian species checklisted by Pavlinov and Rossolimo (1987). See Pavlinov et al. (1990) for authors and publication dates of family-group names.

Ammodillus Thomas, 1904. Ann. Mag. Nat. Hist., ser. 7, 14:102.
TYPE SPECIES: *Gerbillus imbellis* de Winton, 1898.
COMMENTS: Reviewed by Pavlinov (1981), who was so impressed with its unique morphological traits that he proposed the monotypic tribe Ammodillini.

Ammodillus imbellis (de Winton, 1898). Ann. Mag. Nat. Hist., ser. 7, 1:249.
TYPE LOCALITY: Somalia, Goodar.
DISTRIBUTION: Somalia and E Ethiopia.
COMMENTS: Roche and Petter (1968) reviewed the species and mapped its distribution.

Brachiones Thomas, 1925. Ann. Mag. Nat. Hist., ser. 9, 16:548.
TYPE SPECIES: *Gerbillus przewalskii* Büchner, 1889.
COMMENTS: Placed in the tribe Rhombomyini by Pavlinov et al. (1990); more closely related to *Meriones* and *Sekeetamys* than to any other genera.

Brachiones przewalskii (Büchner, 1889). Wiss. Res. Przewalski Cent.-Asian Reisen., Zool., I:(Säugeth.), p. 51.
TYPE LOCALITY: China, Xinjiang (Sinkiang), Lob Nor.

DISTRIBUTION: China, deserts from Xinjiang to Gansu.
SYNONYMS: *arenicolor, callichrous.*
COMMENTS: Reviewed by Corbet (1978*c*).

Desmodilliscus Wettstein, 1916. Anz. Akad. Wiss. Wien, 53:153.
TYPE SPECIES: *Desmodilliscus braueri* Wettstein, 1916.
COMMENTS: The only member of subtribe Desmodilliscina, tribe Gerbillini, as arranged by Pavlinov et al. (1990).

Desmodilliscus braueri Wettstein, 1916. Anz. Akad. Wiss. Wien, 53:153.
TYPE LOCALITY: Sudan, S of El Obeid.
DISTRIBUTION: Recorded from Sahel savanna in N and C Sudan, N Cameroon, W Niger, N Nigeria, C Mali, N Burkina Faso (Upper Volta), Senegal, and W Mauritania (see Good, 1947; Hutterer and Dieterlen, 1986).
SYNONYMS: *buchanani, fuscus.*
COMMENTS: Review of distribution and geographic variation provided by Hutterer and Dieterlen (1986). A sample from Burkina Faso was reported by Gautun et al. (1985).

Desmodillus Thomas and Schwann, 1904. Abstr. Proc. Zool. Soc. London, 1904(2):6.
TYPE SPECIES: *Gerbillus auricularis* Smith, 1834.
COMMENTS: A South African endemic that is a member of the tribe Gerbillurini, along with *Gerbillurus*, in the classification scheme of Pavlinov et al. (1990).

Desmodillus auricularis (Smith, 1834). S. Afr. Quart. J., Ser. 2, 2:160.
TYPE LOCALITY: South Africa, Little Namaqualand, Kamiesberg (Meester et al., 1986).
DISTRIBUTION: South Africa (Cape Prov., SW Orange Free State, SW Transvaal), S Botswana, and Namibia (see Skinner and Smithers, 1990).
SYNONYMS: *brevicaudatus, caffer, hoeschi, pudicus, robertsi, shortridgei, wolfi.*
COMMENTS: Taxonomy and geographic distribution summarized by Meester et al. (1986). Ecology and distribution reviewed by Griffin (1990).

Gerbillurus Shortridge, 1942. Ann. S. Afr. Mus., 36:27-1.
TYPE SPECIES: *Gerbillus vallinus* Thomas, 1918.
SYNONYMS: *Paratatera, Progerbillurus.*
COMMENTS: Concordance of morphological, allozymic, and chromosomal data supports the monophyly of *Gerbillurus* and indicates its close relationship to the *robusta* group of *Tatera* (Qumsiyeh et al., 1987, and references therein). Pavlinov (1987) and Pavlinov et al. (1990), however, considered *Gerbillurus* to be a sister-species of *Desmodillus* and to form the monophyletic tribe Gerbillurini. Comparisons in thermal parameters, macro-and micro-environments, and interspecific aggression among four sympatric species of *Gerbillurus* were documented by Downs and Perrin (1989, 1990) and Dempster and Perrin (1990) in context of adaptive significance and phylogenetic relationships. Standard karyotypic data reported by Schlitter et al. (1984), and an hypothesis of phylogenetic relationships among the four species, derived from chromosomal data was advanced by Qumsiyeh et al. (1991).

Gerbillurus paeba (A. Smith, 1836). Rept. Exped. Exploring Central Africa, app., p. 43.
TYPE LOCALITY: South Africa, Cape Province, Vryberg (as restricted by Roberts, 1951).
DISTRIBUTION: South Africa (C, W, and N Cape Prov., SW Orange Free State, W and N Transvaal), W Mozambique, W Zimbabwe, Botswana, Namibia, and SW Angola.
SYNONYMS: *broomi, calidus, coombsi, exilis, infernus, kalaharicus, leucanthus, mulleri, oralis, swakopensis, swalius, tenuis.*
COMMENTS: Subgenus *Progerbillurus*. Meester et al. (1986) and Skinner and Smithers (1990) summarized South African taxonomy and distribution; Griffin (1990) reviewed Namibian populations. Based on analyses of chromosomal data, Qumsiyeh (1986) hypothesized that *G. paeba* and *G. vallinus* evolved from a common ancestor.

Gerbillurus setzeri (Schlitter, 1973). Bull. S. California Acad. Sci., 72:13.
TYPE LOCALITY: Namibia, Gobabeb, 1 mi E Namib Desert Research Station.
DISTRIBUTION: Namib Desert, from the Namib-Naukluft Nat. Park north through Namibia to extreme SW Angola (Meester et al., 1986; Skinner and Smithers, 1990).

COMMENTS: Subgenus *Gerbillurus*. Interpretation of chromosomal data indicated *G. setzeri* and *G. tytonis* are closely related (Qumsiyeh et al., 1991). Reviewed by Griffin (1990), who also refered to an undescribed form (*G.* cf. *setzeri*) coexisting with *G. setzeri*.

Gerbillurus tytonis (Bauer and Niethammer, 1960). Bonn. Zool. Beitr., (1959), 10:255 [1960].
TYPE LOCALITY: Namibia, Namib Desert, Sossusvlei.
DISTRIBUTION: Namibia (Sossusvlei, Sandwich Harbour and Bobabeb, Namib Desert, and Farm Canaan near diamond region of Namibia).
COMMENTS: Subgenus *Paratatera*. Taxonomic history summarized by Meester et al. (1986); distribution and habitat reviewed by Griffin (1990) and Skinner and Smithers (1990).

Gerbillurus vallinus (Thomas, 1918). Ann. Mag. Nat. Hist., ser. 9, 2:148.
TYPE LOCALITY: South Africa, Cape Province, Bushmanland, Kenhart, Tuin (see Meester et al., 1986).
DISTRIBUTION: From South Africa (NW Cape Prov.) northwest through Namibia towards Brukaros-Karas Mtns and central Namib Desert.
SYNONYMS: *seeheimi*.
COMMENTS: Subgenus *Gerbillurus*. Taxonomy summarized by Meester et al. (1986); ecology and range reviewed by Griffin (1990) and Skinner and Smithers (1990).

Gerbillus Desmarest, 1804. Tabl. Méth. Hist Nat., *in* Nouv. Dict. Hist. Nat., 24:22.
TYPE SPECIES: *Gerbillus aegyptius* Desmarest, 1804 (= *Dipus gerbillus* Olivier, 1801).
SYNONYMS: *Dipodillus, Endecapleura, Hendecapleura, Monodia, Petteromys*.
COMMENTS: This genus has never been adequately revised. Lay (1983) summarized taxonomic difficulties of the genus, discussed significant character complexes that would bear on any scheme to separate species into subgenera (nature of plantar surfaces, size of auditory bulla, relative tail length, dental traits, accessory tympanum, and karyotype), and provided an annotated checklist of the species he considered valid. Lack of concordance among the suites of characters discouraged Lay from allocating species to higher categories conventionally recognized as either subgenera or genera (*Dipodillus, Hendecapleura, Monodia, Petteromys*; see Pavlinov et al., 1990), and Lay felt compelled to recognize a single genus without subgenera until the group was systematically re-evaluated. No such comprehensive revision has emerged since 1983, and Lay's taxonomic overview remains the best working hypothesis of specific diversity in *Gerbillus*. With a few departures, we follow his treatment. Pavlinov et al. (1990) recognized subgenera of *Gerbillus* but admitted the weakness of these categories.

An excellent account of species occurring east of the Euphrates River was provided by Lay and Nadler (1975). The distribution of six species endemic to North Africa was mapped and discussed by Cheylan (1990) in the context of assessing endemism and speciation of Mediterranean mammals. Tranier and Julien-Laferriere (1990) commented on suggested identities of several species from North Africa. A contribution to allozymic variation within and between species was offered by Nevo (1982). Evolutionary tendencies within *Gerbillus*, as reflected in dental traits, were discussed by Petter (1973*c*). And the significance of variation in incisor microstructure within the genus was reported by Flynn (1982). Species of *Gerbillus* were included by Bonhomme et al. (1985) and Pascale et al. (1990) in their assessment of phylogenetic relationships among muroid rodents with electrophoretic and DNA-sequence data.

Gerbillus acticola Thomas, 1918. Ann. Mag. Nat. Hist., ser. 9, 2:147.
TYPE LOCALITY: Somalia, Berbera.
DISTRIBUTION: Somalia.
COMMENTS: Lay (1983) explained why Petter's (1975*b*) synonymy of *G. acticola* with *G. pyramidum* should not be followed.

Gerbillus agag Thomas, 1903. Proc. Zool. Soc. London, 1903:296.
TYPE LOCALITY: Sudan, W Kordofan, Agageh Wells.
DISTRIBUTION: N Nigeria, Mali, and Niger to Chad, Sudan, and Kenya.
COMMENTS: The forms *cosensi, maradius, nigeriae*, and *sudanensis* have been associated with *G. agag*, but Lay (1983) found the limits of this species impossible to define based on published analyses and regarded it as monotypic pending systematic revision.

Gerbillus allenbyi Thomas, 1918. Ann. Mag. Nat. Hist., ser. 9, 2:146.
TYPE LOCALITY: Israel, Rehoboth.
DISTRIBUTION: Endemic to Israel, coastal dunes from N of Gaza to Haifa.
COMMENTS: Lay (1983) mentioned the synonymy of this form with *G. andersoni*, as accepted by Harrison and Bates (1991), but kept it separate pending a revision of the genus.

Gerbillus amoenus (de Winton, 1902). Ann. Mag. Nat. Hist., ser. 7, 9:46.
TYPE LOCALITY: Egypt, Giza Prov.
DISTRIBUTION: Recorded only from Egypt and Libya.
COMMENTS: Lay (1983) speculated that the species may range across Tunisia and Algeria to Mauritania, noted its past associations with *dasyurus* and *campestris*, and advised future comparison with *G. nanus*. Ranck (1968) reviewed the Libyan populations and recorded significant geographic variation.

Gerbillus andersoni de Winton, 1902. Ann. Mag. Nat. Hist., ser. 7, 9:45.
TYPE LOCALITY: Egypt, E Alexandria, Mandara.
DISTRIBUTION: Egypt, Nile Delta south to El Faiyum (as mapped for *G. a. andersoni* by Osborn and Helmy, 1980:120).
COMMENTS: The forms *allenbyi*, *inflatus*, and *bonhotei* have all been listed as synonyms of *G. andersoni* (e.g., Harrison and Bates, 1991; Osborn and Helmy, 1980), but Lay (1983) argued that current evidence does not support the union of these forms. He also suggested that *blanci* and *eatoni* be tentatively associated with *G. andersoni*. The latter species should be considered monotypic pending a revision. With the range as described above, *G. andersoni* is allopatric with *G. eatoni* to the west and *G. bonhotei* to the east (see those accounts).

Gerbillus aquilus Schlitter and Setzer, 1972. Proc. Biol. Soc. Washington, 86:167.
TYPE LOCALITY: Iran, 60 km W Kerman.
DISTRIBUTION: SE Iran, W Pakistan, S Afghanistan (see Lay and Nadler, 1975).
SYNONYMS: *subsolanus*.
COMMENTS: Reviewed by Lay and Nadler (1975).

Gerbillus bilensis Frick, 1914. Ann. Carnegie Mus., 9:12.
TYPE LOCALITY: Ethiopia, near Bilen.
DISTRIBUTION: Known only from the type locality.
COMMENTS: Lay (1983) questioned the synonymy of this species with *pulvinatus* as allocated by Petter (1975*b*).

Gerbillus bonhotei Thomas, 1919. Ann. Mag. Nat. Hist., ser. 9, 3:560.
TYPE LOCALITY: Egypt, Sinai, Khubra Abu Guzoar.
DISTRIBUTION: NE Sinai Peninsula.
COMMENTS: Usually included in *G. andersoni* (Harrison and Bates, 1991; Osborn and Helmy, 1980) but maintained as distinct pending systematic review of the genus (Lay, 1983). Harrison and Bates (1991) recorded the form from Jordan, an occurrence which requires corroboration according to Lay (1983).

Gerbillus bottai Lataste, 1882. Le Naturaliste, 4:36.
TYPE LOCALITY: Sudan, Sennar.
DISTRIBUTION: Recorded only from Sudan and Kenya.
COMMENTS: Although *luteolus* and *harwoodi* have been included in *bottai*, Lay (1983) related reasons, based on presence or absence of accessory tympanum, for provisionally regarding the species as monotypic.

Gerbillus brockmani (Thomas, 1910). Ann. Mag. Nat. Hist., ser. 8, 5:420.
TYPE LOCALITY: Somalia, Burao, 85 mi S Berbera.
DISTRIBUTION: Somalia.
COMMENTS: Petter (1975*b*) placed *brockmani* in synonymy with *G. nanus*, but no evidence indicates that the species occurs anywhere remotely near Somalia (see Lay, 1983).

Gerbillus burtoni F. Cuvier, 1838. Trans. Zool. Soc. London, 2:145, pl. 25.
TYPE LOCALITY: Sudan, Dharfur.
DISTRIBUTION: Known only from the type locality.
COMMENTS: Usually treated as a synonym of *G. pyramidum*, but Lay (1983) mustered diagnostic cranial traits that appear to distinguish them.

Gerbillus campestris (Loche, 1867). Expl. Sci. Alg. Zool. Mamm., p. 106.
TYPE LOCALITY: Algeria, Constantine Prov., Philipeville.
DISTRIBUTION: N Africa, from Morocco to Egypt and Sudan.
SYNONYMS: *brunnescens, cinnamomeus, dodsoni, haymani, patrizii, riparius, rozsikae, somalicus, venustus, wassifi.*
COMMENTS: Nineteen species-group names have been associated with *G. campestris* by different authors in various combinations, as summarized by Lay (1983). He also noted that most opinions lacked supportive evidence and that some of the synonyms are unidentifiable or *nomina nuda*. Those names listed here are probably correctly associated with *G. campestris*, but the species requires refined definition through careful systematic revision. Different geographical populations are reviewed by Kowalski and Rzebik-Kowalska (1991, Algeria), Ranck (1968, Libya), and Osborn and Helmy (1980, Egypt). Benazzou and Zyadi (1990) conducted a biometric study analyzing variation among Moroccan populations. Cockrum and Setzer (1976:643) clarified the author and date of publication of *campestris*.

Gerbillus cheesmani Thomas, 1919. J. Bombay Nat. Hist. Soc., 26:748.
TYPE LOCALITY: Iraq, Lower Euphrates, near Basra.
DISTRIBUTION: SW Iran, C and S Iraq, Saudi Arabia, Oman, North Yemen, South Yemen, and Kuwait (see Lay and Nadler, 1975; Harrison and Bates, 1991).
SYNONYMS: *arduus, maritimus.*
COMMENTS: Reviewed by Lay and Nadler (1975), Lay (1983), and Harrison and Bates (1991). Chromosomal polymorphism among samples from Kuwait was analyzed and reported by Badr and Asker (1980). Detailed comparisons between Qatarian *G. cheesmani* and *G. nanus* in cranial morphology were reported by Madkour (1984).

Gerbillus cosensi Dollman, 1914. Abstr. Proc. Zool. Soc. London, 1914(131):25.
TYPE LOCALITY: Kenya, Kozibiri River, Ngamatak, Turkwel River.
DISTRIBUTION: Known only from the type locality.
COMMENTS: Petter (1975*b*) synonymized this form with *G. agag*, but Lay (1983) regarded it as separate pending revision.

Gerbillus dalloni Heim de Balsac, 1936. Mem. Acad. Sci. de l'Inst. de France, ser. 2, 62:43 (Mem. no. 1), text F. 1.
TYPE LOCALITY: Chad, Tibesti region.
DISTRIBUTION: Known only from the type locality.
COMMENTS: Another species united with *G. agag* by Petter (1975*b*) without supporting documentation, "but the type localities are separated by more than 2300 km" (Lay, 1983:340), and should "be regarded as a valid species pending revision."

Gerbillus dasyurus (Wagner, 1842). Arch. Naturgesch., 8:20.
TYPE LOCALITY: Sinai.
DISTRIBUTION: Arabian Peninsula, Iraq, Syria, Lebanon, Israel, Sinai (see Harrison and Bates, 1991).
SYNONYMS: *dasyroides, gallagheri, leosollicitus, palmyrae.*
COMMENTS: Review of morphology and distribution, along with citations for synonyms, provided by Harrison and Bates (1991). Those authors also listed *lixa* as a synonym of *G. dasyurus*, but the holotype, a young animal, has an accessory tympanum and bare soled hind feet, which is uncharacteristic of *G. dasyurus* but does suggest alliance with *G. nanus* (Lay, 1983).

Gerbillus diminutus (Dollmann, 1911). Ann. Mag. Nat. Hist., ser. 8, 7:520.
TYPE LOCALITY: Kenya, Nyama Nyango, Guaso Nyiro.
DISTRIBUTION: Kenya.
COMMENTS: United with *G. pusillus* by Petter (1975*b*) without substantiation; should be kept separate until its status can be clarified by systematic revision (Lay, 1983).

Gerbillus dongolanus (Heuglin, 1877). Reise in Nordost-Afrika, 2:79.
TYPE LOCALITY: Sudan, Dongola.
DISTRIBUTION: Known only from the type locality.
COMMENTS: Lay (1983) noted that several authors have synonymized this taxon with *G. pyramidum* but always without confirmatory documentation. Until that data is available, the species should be considered valid.

Gerbillus dunni Thomas, 1904. Ann. Mag. Nat. Hist., ser. 7, 14:101.
TYPE LOCALITY: Somalia, Gerlogobi.
DISTRIBUTION: Ethiopia, Somalia, Djibouti.
COMMENTS: The conspecificity of *dunni* with *G. latastei* has been suggested, but the latter's range lies more than 4000 km from the distribution of *G. dunni* (Lay, 1983).

Gerbillus famulus Yerbury and Thomas, 1895. Proc. Zool. Soc. London, 1895:551.
TYPE LOCALITY: South Yemen, Aden, Lehej.
DISTRIBUTION: Endemic to South Yemen and North Yemen (see Harrison and Bates, 1991:272).
COMMENTS: A large and elegant gerbil whose morphology, geographic range, and ecology was reviewed by Harrison and Bates (1991).

Gerbillus floweri Thomas, 1919. Ann. Mag. Nat. Hist., ser. 9, 3:559.
TYPE LOCALITY: Egypt, Sinai, S of El Arish.
DISTRIBUTION: Known only from the type locality.
COMMENTS: Usually considered a synonym of *G. pyramidum*, but Lay (1983) explained why it should not be united with that species.

Gerbillus garamantis Lataste, 1881. Le Naturaliste, 3:507.
TYPE LOCALITY: Algeria, Ouargla, Sidi Roueld.
DISTRIBUTION: Algeria.
COMMENTS: Most workers have included this taxon in *G. nanus* (e.g., Kowalski and Rzebik-Kowalska, 1991), but never with supporting evidence, and Lay elaborated why it should be considered valid pending revision of the *G. nanus* group. The species (listed under *G. nanus*) was reviewed by Kowalski and Rzebik-Kowalska (1991).

Gerbillus gerbillus (Olivier, 1801). Bull. Sci. Soc. Philom. Paris, 2:121.
TYPE LOCALITY: Egypt, Giza Prov.
DISTRIBUTION: From Israel through Egypt and N Sudan to Morocco; also N Mali, N Niger, and N Chad (see Corbet, 1978*c*; Harrison and Bates, 1991:283; Osborn and Helmy, 1980:131).
SYNONYMS: *aegyptius, aeruginosus, asyutensis, discolor, foleyi, hirtipes, longicaudatus, psammophilous, sudanensis.*
COMMENTS: Geographic portions reviewed by Ranck (1968), Corbet (1978*c*), Lay (1983), Kowalski and Rzebik-Kowalska (1991), and Harrison and Bates (1991). In 1983, Lay drew attention to the lack of inquiry into variation in this species which has such an extensive range; that complaint stands today and the species needs careful taxonomic review. The form *hirtipes* was synonymized with *G. gerbillus* by Cockrum (1976), but because of his inadequate documentation, Lay (1983) was reluctant to accept this union. Cockrum's evidence was scanty, but we are swayed by Kowalski and Rzebik-Kowalska's (1991) argument for merging *hirtipes* with *G. gerbillus*. The origin of multiple sex chromosomes in this species was discussed by Wahrman et al. (1983).

Gerbillus gleadowi Murray, 1886. Ann. Mag. Nat. Hist., ser. 5, 17:246.
TYPE LOCALITY: Pakistan, Upper Sind, Rohri Dist, Mirpur-Drahrki Taluka, 15 mi SW Rehti, Beruto.
DISTRIBUTION: NW India, sand dunes along Indus Valley of Pakistan (see Lay and Nadler, 1975).
COMMENTS: A distinctive species defined by diagnostic morphological and chromosomal data (Lay and Nadler, 1975).

Gerbillus grobbeni Klaptocz, 1909. Zool. Jahrb., Syst., 27:252.
TYPE LOCALITY: Libya, Cyrenaica, Dernah.
DISTRIBUTION: Known only from the type locality.
COMMENTS: Petter (1975*b*) and Corbet (1978*c*) included *grobbeni* in *G. nanus*, but Lay (1983) agreed with Ranck (1968) who maintained it as valid.

Gerbillus harwoodi Thomas, 1901. Ann. Mag. Nat. Hist., ser. 7, 8:275.
TYPE LOCALITY: Kenya, Lake Naivasha.
DISTRIBUTION: Kenya.
SYNONYMS: *luteus.*
COMMENTS: See comments under account of *G. bottai.*

Gerbillus henleyi (de Winton, 1903). Novit. Zool., 10:284.
TYPE LOCALITY: Egypt, Wadi Natron, Zaghig.
DISTRIBUTION: From Algeria through N Africa to Israel and Jordan, scattered records in W Saudia Arabia, N Yemen, and Oman; also recorded from Burkina Faso (Maddalena et al., 1988) and N Senegal (Duplantier et al. 1991*a*).
SYNONYMS: *jordani, makrami, mariae.*
COMMENTS: Broad segments of the species reviewed by Ranck (1968, Libya), Osborn and Helmy (1980, Egypt), Harrison and Bates (1991, Arabian Peninsula), and Kowalski and Rzebik-Kowalska (1991, Algeria). The occurrences in Burkina Faso and Senegal were postulated to reflect the southward expansion of Saharan environments (Duplantier et al., 1991*a*).

Gerbillus hesperinus Cabrera, 1936. Bol. Real. Soc. Esp. Hist. Nat., p. 365.
TYPE LOCALITY: Morocco, Mogador (= Essouira).
DISTRIBUTION: Coastal Morocco N of Middle Atlas Mtns.
COMMENTS: See the references cited by Lay (1983) for characters defining this distinctive species.

Gerbillus hoogstraali Lay, 1975. Fieldiana Zool., 65:90.
TYPE LOCALITY: Morocco, 7 km S Taroudannt.
DISTRIBUTION: Known only from the type locality.
COMMENTS: Comparisons and diagnostic features presented by Lay (1975).

Gerbillus jamesi Harrison, 1967. Mammalia, 31:383.
TYPE LOCALITY: Tunisia, between Bou Ficha and Enfidaville.
DISTRIBUTION: Tunisia.
COMMENTS: Recognized as valid by Petter (1975*b*) and Lay (1983).

Gerbillus juliani (St. Leger, 1935). Ann. Mag. Nat. Hist., ser. 10, 15:669.
TYPE LOCALITY: Somalia, Bulhar.
DISTRIBUTION: Somalia.
COMMENTS: Roche and Petter (1968) reviewed this species under *Monodia*, but Petter (1975*b*) later synonymized it with *G. watersi* without supporting evidence. Lay (1983) recognized *juliani* as valid pending revision of the genus.

Gerbillus latastei Thomas and Trouessart, 1903. Bull. Soc. Zool. France, 28:172.
TYPE LOCALITY: Tunisia, Kebili.
DISTRIBUTION: Tunisia and Libya.
SYNONYMS: *aureus, favillus, nalutensis.*
COMMENTS: The three synonyms listed here along with *bonhotei, dunni, perpallidus, riggenbachi,* and *rosalinda* were united with *G. latastei* by Cockrum (1977) when he reported on the specific identity of that species. Lay (1983) discussed the basis for considering each of these five as separate species. The Libyan populations (under *G. aureus*) were reviewed by Ranck (1968).

Gerbillus lowei (Thomas and Hinton, 1923). Proc. Zool. Soc. London, 1923:261.
TYPE LOCALITY: Sudan, Jebel Marra.
DISTRIBUTION: Known only from the type locality.
COMMENTS: Synonymized with *G. campestris* by Petter (1975*b*) without supporting evidence, but should be kept separate pending revision of the genus (Lay, 1983).

Gerbillus mackillingini (Thomas, 1904). Ann. Mag. Nat. Hist., ser. 7, 14:158.
TYPE LOCALITY: Egypt, Wadi Alagai, eastern desert of Nubia.
DISTRIBUTION: E desert of S Egypt (see Osborn and Helmy, 1980) and probably adjacent Sudan.
COMMENTS: Although some authors have placed this species with *G. nanus* (see references in Lay, 1983), Osborn and Helmy (1980) demonstrated its specific distinction.

Gerbillus maghrebi Schlitter and Setzer, 1972. Proc. Biol. Soc. Washington, 84:387.
TYPE LOCALITY: Morocco, Fes Prov., 15 km WSW Taounate (see Lay, 1983).
DISTRIBUTION: Known only from the type locality.
COMMENTS: A distinctive species related to *G. campestris* according to Lay (1983).

Gerbillus mauritaniae (Heim de Balsac, 1943). Bull. Mus. Hist. Nat. Paris, 15:287.
TYPE LOCALITY: Mauritania, Aouker Region, S of Archane Titarek.
DISTRIBUTION: Known only from the type locality.
COMMENTS: Type species of *Monodia*, but according to Lay (1983) the diagnostic traits do not warrant generic distinction.

Gerbillus mesopotamiae Harrison, 1956. J. Mammal, 37:417.
TYPE LOCALITY: Iraq, SW of Faluja, W bank of Euphrates River, near Amiriya.
DISTRIBUTION: Iraq and SW Iran in valleys of the Tigris, Euphrates, and Karun Rivers (see Lay and Nadler, 1975).
COMMENTS: A distinctive species defined and reviewed by Lay and Nadler (1975) and Harrison and Bates (1991).

Gerbillus muriculus (Thomas and Hinton, 1923). Proc. Zool. Soc. London, 1923:263.
TYPE LOCALITY: Sudan, Darfur, Madu, 80 mi NE El Fasher.
DISTRIBUTION: Sudan.
COMMENTS: Lay (1983) regarded this species as valid pending revision.

Gerbillus nancillus Thomas and Hinton, 1923. Proc. Zool. Soc. London, 1923:260.
TYPE LOCALITY: Sudan, Plains of Darfur, 45 mi N El Fasher.
DISTRIBUTION: Vicinity of El Fasher.
COMMENTS: Possibly a distinct species (Lay, 1983).

Gerbillus nanus Blanford, 1875. Ann. Mag. Nat. Hist., ser. 4, 16:312.
TYPE LOCALITY: Pakistan, Gedrosia (see Lay, 1983).
DISTRIBUTION: An extensive range from the Baluchistan region of NW India, Pakistan, S Afghanistan, and Iran through the Arabian Peninsula, Iraq, Jordan, Israel, and North Africa to Morocco.
SYNONYMS: *arabium, hilda, indus, lixa, mimulus, setonbrownei.*
COMMENTS: Regional reviews of the species provided by Lay and Nadler (1975), Harrison and Bates (1991), Osborn and Helmy (1980), Ranck (1968), and Kowalski and Rzebik-Kowalska (1991). Lay (1983) remarked that *G. nanus* and *G. amoenus* share several morphological and chromosomal traits and that the nature of their relationship should be explored by careful revision. Cranial morphology of Qatarian *G. nanus* and *G. cheesmani* contrasted by Madkour (1984).

Gerbillus nigeriae Thomas and Hinton, 1920. Novit. Zool., 27:317.
TYPE LOCALITY: Nigeria, Farniso near Kano.
DISTRIBUTION: N Nigeria and Burkina Faso.
COMMENTS: Lay (1983) maintained this species as valid even though it has been synonymized with *G. agag* by some workers. The sample from Burkina Faso was documented and karyotyped by Gautun et al. (1985).

Gerbillus occiduus Lay, 1975. Fieldiana Zool., 65:94.
TYPE LOCALITY: Morocco, Aoreora, 80 km WSW Goulimine.
DISTRIBUTION: Known only from the type locality.
COMMENTS: See Lay (1983) for a review.

Gerbillus percivali (Dollman, 1914). Ann. Mag. Nat. Hist., ser. 8, 14:488.
TYPE LOCALITY: Kenya, Voi.
DISTRIBUTION: Kenya.
COMMENTS: Although this species was synonymized with *G. pusillus* by Petter (1975*b*), it should be kept separate pending a revision of the genus.

Gerbillus perpallidus Setzer, 1958. J. Egypt Publ. Health Assoc., 33:221.
TYPE LOCALITY: Egypt, Bir Victoria.
DISTRIBUTION: N Egypt, W of the Nile River.
COMMENTS: See Lay (1983) for a review of this distinctive species.

Gerbillus poecilops Yerbury and Thomas, 1895. Proc. Zool. Soc. London, 1895:549.
TYPE LOCALITY: South Yemen, Aden, Lahej.
DISTRIBUTION: South Yemen, North Yemen, and SW Saudi Arabia.
COMMENTS: A valid species reviewed by Harrison and Bates (1991).

Gerbillus principulus (Thomas and Hinton, 1923). Proc. Zool. Soc. London, 1923:262.
TYPE LOCALITY: Sudan, Jebel Meidob, El Malha.
DISTRIBUTION: Known only from the type locality.
COMMENTS: This species has been associated with *G. nanus* or *G. watersi*, but Lay (1983) regarded it as valid pending systematic revision.

Gerbillus pulvinatus Rhoads, 1896. Proc. Acad. Nat. Sci. Philadelphia, p. 537.
TYPE LOCALITY: Ethiopia, Lake Rudolf, Rusia.
DISTRIBUTION: Ethiopia.
COMMENTS: Lay (1983) explained why this species should be considered valid and monotypic, although Petter (1975*b*) placed *bilensis* in it.

Gerbillus pusillus Peters, 1878. Monatsb. K. Preuss. Akad. Wiss. Berlin, 1879:201 [1878].
TYPE LOCALITY: Kenya, Ndi and Kitui.
DISTRIBUTION: Kenya, Ethiopia, and S Sudan (Dieterlen and Nikolaus, 1985).
COMMENTS: Petter (1975*b*) and Lay (1983) recognized this species.

Gerbillus pyramidum Geoffroy, 1825. Dict. Class. Hist. Nat. Paris, 7:321.
TYPE LOCALITY: Egypt, Giza Prov.
DISTRIBUTION: Egypt, Nile delta and valley south to N Sudan, oases of Western Desert and SE Eastern Desert (see Osborn and Helmy, 1980:97); possibly Khartoum region in EC Sudan.
SYNONYMS: *elbaensis, gedeedus.*
COMMENTS: Many authors have viewed this species as ranging extensively from the Sinai throughout North Africa, and containing many synonyms. Lay (1983), however, restricted the distribution of *G. pyramidum* proper to the region mapped by Osborn and Helmy (1980). Tawill and Niethammer (1989) discussed the morphological and chromosomal identification of a sample from Khartoum as most probably *G. pyramidum.*

Gerbillus quadrimaculatus Lataste, 1882. Le Naturaliste, 2:27.
TYPE LOCALITY: Sudan, Nubia.
DISTRIBUTION: Known only from the type locality, NE Sudan.
COMMENTS: Lay (1983) remarked that although most authors list this form equivocably under *G. nanus*, it should be kept separate until a revision is available.

Gerbillus riggenbachi Thomas, 1903. Novit. Zool., 10:301.
TYPE LOCALITY: Western Sahara, Rio de Oro.
DISTRIBUTION: Recorded from the type locality and N Senegal.
COMMENTS: This form has been included in either *G. pyramidum* (Petter, 1975*b*) or *G. latastei* (Cockrum, 1977), but Lay (1983:347) contended that it "should be regarded as distinct pending comprehensive revision." The Senegal record was reported by Duplantier et al. (1991*a*).

Gerbillus rosalinda St. Leger, 1929. Ann. Mag. Nat. Hist., ser. 10, 4:295.
TYPE LOCALITY: Sudan, Kordofan, Abu Zabad, 145 km SW El Obeid.
DISTRIBUTION: Sudan.
COMMENTS: Both Petter (1975*b*) and Lay (1983) listed this species as distinct pending revision of the genus.

Gerbillus ruberrimus Rhoads, 1896. Proc. Acad. Nat. Sci. Philadelphia, p. 538.
TYPE LOCALITY: Ethiopia, Finik, near Webi Shebeli.
DISTRIBUTION: E Ethiopia, Somalia and Kenya.
COMMENTS: Another form that should be considered distinct until revisionary work demonstrates otherwise (Lay, 1983).

Gerbillus simoni Lataste, 1881. Le Naturaliste, 3:499.
TYPE LOCALITY: Algeria, Oued Magra.
DISTRIBUTION: Egypt, W of Nile Delta, Libya, Tunisia, and Algeria.
SYNONYMS: *kaiseri, zakariai.*
COMMENTS: A distinctive species that is the type species of *Dipodillus* (Lay, 1983). Regional reviews are available for Egypt (Osborn and Helmy, 1980), Libya (Ranck, 1968), and Algeria (Kowalski and Rzebik-Kowalska, 1991).

Gerbillus somalicus (Thomas, 1910). Ann. Mag. Nat. Hist., ser. 8, 5:197.
TYPE LOCALITY: Somalia, Upper Sheikh.
DISTRIBUTION: Somalia.
COMMENTS: Tentatively regarded as valid following Lay (1983).

Gerbillus stigmonyx Heuglin, 1877. Reise in Nordost-Afrika, 2:78.
TYPE LOCALITY: Sudan, Khartoum.
DISTRIBUTION: Sudan.
SYNONYMS: *luteolus*.
COMMENTS: Listed as a synonym of *G. campestris* by Petter (1975*b*); retained as separate by Lay (1983).

Gerbillus syrticus Misonne, 1974. Bull. Inst. R. Sci. Nat. Belg., 50:1.
TYPE LOCALITY: Libya, 12 km N Nofilia.
DISTRIBUTION: Known only from the type locality.
COMMENTS: Relationships obscure (see Lay, 1983).

Gerbillus tarabuli Thomas, 1902. Proc. Zool. Soc. London, 1902:5.
TYPE LOCALITY: Libya, Sebha.
DISTRIBUTION: Libya.
SYNONYMS: *hamadensis*.
COMMENTS: Usually listed as a synonym of *G. pyramidum*, *G. tarabuli* can be distinguished by morphological traits (Lay et al., 1975). Future inquiry, according to Lay (1983:347) "should examine the possibility that the 2N = 40...forms reported from Tunisia, Algeria, Morocco and Senegal...are conspecific and may be referable to *G. tarabuli*."

Gerbillus vivax (Thomas, 1902). Proc. Zool. Soc. London, 1902:8.
TYPE LOCALITY: Libya, Sebha.
DISTRIBUTION: Libya.
COMMENTS: Variously placed in either *G. dasyurus*, *G. amoenus*, or *G. nanus*, but Lay (1983) disputed its association with *G. dasyurus* and urged that its level of relationship to *G. amoenus* and *G. nanus* be assessed by systematic revision.

Gerbillus watersi de Winton, 1901. Novit. Zool., 8:399.
TYPE LOCALITY: Sudan, Upper Nile, Shendi.
DISTRIBUTION: Somalia, Sudan.
COMMENTS: Listed both as a subspecies of *G. nanus* or as a valid species by Petter (1975*b*) in the same report. The species should be considered distinct until revisionary studies advise otherwise (Lay, 1983).

Meriones Illiger, 1811. Prodr. Syst. Mamm. Avium., p. 82.
TYPE SPECIES: *Mus tamariscinus* Pallas, 1773.
SYNONYMS: *Cheliones*, *Idomeneus*, *Meraeus*, *Pallasiomys*, *Parameriones*.
COMMENTS: A member of the tribe Rhombomyini in the scheme of Pavlinov et al. (1990). No modern systematic revision is available for *Meriones*. The early revision by Chaworth-Musters and Ellerman (1947), as updated and modified by Ellerman and Morrison-Scott (1951) and Corbet (1978*c*), represents the most current review of species in the genus. Additional taxonomic, distributional, and evolutionary views are found in the taxonomic reports and regional faunal studies cited throughout the accounts below. Most workers agree on definitions of the species we list here, but results of careful systematic revision will probably uncover a greater number of species. All chromosomal data concerning *Meriones* up to 1967 was summarized by Nadler and Lay (1967) in the context of assessing relationships among the species. Additional chromosomal data and its significance to understanding phylogenetic relationships among four species of *Meriones* reported by Benazzou et al. (1982). Lay and Nadler (1969) summarized laboratory hybridization attempts among several species of *Meriones*. Records of species from Iran and Pakistan reported by Lay et al. (1970).

Meriones arimalius Cheesman and Hinton, 1924. Ann. Mag. Nat. Hist., ser. 9, 14:554.
TYPE LOCALITY: Saudi Arabia, Yabrin (Jabrin), Djebel Agoula.
DISTRIBUTION: "Northern sands of the Rub al Khali in Saudia Arabia and Oman" (Harrison and Bates, 1991:297).

COMMENTS: Subgenus *Pallasiomys*. Ellerman and Morrison-Scott (1951) listed this form as a valid species, but it was later included in *M. libycus* (Corbet, 1978c; Harrison and Bates, 1991). Pavlinov et al. (1990:294) reinstated *arimalius* as a separate species and reviewed its salient characters. Even from the terse description of its diagnostic traits provided by Harrison and Bates (1991:297), who recognized the form as a subspecies of *M. libycus*, it is evident that *arimalius* is quite morphologically different from populations of *lybicus* north of it in Saudia Arabia. The species was also considered distinct by Nadler and Lay (1967).

Meriones chengi Wang, 1964. Acta Zootaxon. Sinica, 1:9.
TYPE LOCALITY: China, N Xinjiang (Sinkiang), Da-Ho-Yien, Turfan.
DISTRIBUTION: Known only from the type locality.
COMMENTS: Subgenus *Pallasiomys*. Wang (1964) considered this species, based on a series of adult and immature specimens, to be most closely related to *M. meridianus*, which also occurs in Xinjiang Prov (Ma et al., 1987). The relationship of *chengi* to the latter species needs to be assessed by revision of *Meriones*, especially the *M. meridianus* complex.

Meriones crassus Sundevall, 1842. K. Svenska Vet. Akad., Ser. 3, p. 233.
TYPE LOCALITY: Egypt, Sinai, Fount of Moses (Ain Musa).
DISTRIBUTION: Across North Africa from Morocco through Niger, Sudan, and Egypt to Israel, Jordan, Syria, Saudi Arabia, Iraq, Iran, and Afghanistan.
SYNONYMS: *asyutensis, charon, ismahelis, longifrons, pallidus, pelerinus, perpallidus, swinhoei.*
COMMENTS: Subgenus *Pallasiomys*. Reviewed by Corbet (1978c). Regional reviews of the species are available for Algeria (Kowalski and Rzebik-Kowalska, 1991), Libya (Ranck, 1968), Egypt (Osborn and Helmy, 1980), the Arabian Peninsula (Harrison and Bates, 1991), Iran (Lay, 1967), and Afghanistan (Hassinger, 1973). See Koffler (1972, Mammalian Species, 9).

Meriones dahli Shidlovsky, 1962. Definition Rodents Zakabkaziya, p. 115.
TYPE LOCALITY: Armenia, Sadarak steppe, foothills of Vardanis (Saraibulak) Ridge.
DISTRIBUTION: Local occurrence in sandy habitats in Armenia.
COMMENTS: Subgenus *Pallasiomys*. Included in *M. meridianus* by Corbet (1978c) but shown to be a separate species by Dyatlov and Avanyan (1987), whose results were based on morphological, biochemical, and chromosomal traits, as well as interbreeding experiments. Reviewed by Pavlinov et al. (1990).

Meriones hurrianae Jordon, 1867. Mamm. India, p. 186.
TYPE LOCALITY: India, Hurriana Dist.
DISTRIBUTION: Primarily in Thar Desert in SE Iran, Pakistan, and NW India.
SYNONYMS: *collinus.*
COMMENTS: Subgenus *Cheliones*. The Pakistan population reviewed by Roberts (1977). Hassinger (1973) discussed old records of the species from Afghanistan as probably erroneous.

Meriones libycus Lichtenstein, 1823. Verz. Doublet. Zool. Mus. Univ. Berlin, p. 5.
TYPE LOCALITY: Egypt, near Alexandria.
DISTRIBUTION: North Africa from Western Sahara (Rio de Oro) to Egypt, through Saudi Arabia, Jordan, Iraq, Syria, Iran, Afghanistan, and into S Turkestan to W China (Xinjiang).
SYNONYMS: *afghanus, aquilo, arimalius, auratus, azizi, caucasius, caudatus, collium, edithae, erythrourus, evelynae, eversmanni, farsi, gaetulus, guyonii, heptneri, iranensis, marginae, mariae, maxeratis, melanurus, oxianus, renaultii, schousboeii, schwarzovi, sogdianus, syrius, turfanen.*
COMMENTS: Subgenus *Pallasiomys*. Reviewed by Corbet (1978c). Regional studies cover populations in Algeria (Kowalski and Rzebik-Kowalska, 1991), Libya (Ranck, 1968, as *caudatus*), Egypt (Osborn and Helmy, 1980), Arabian Penninsula (Harrison and Bates, 1991), Iran (Lay, 1967), and Afghanistan (Hassinger, 1973). Results of comparative craniometric analyses between Moroccan samples of *M. libycus* and *M. shawi* obtained in sympatry were reported by Zaime and Pascal (1988). Morphological and karyotypic contrasts between these same two species as well as laboratory hybridization experiments, were recorded by Lay and Nadler (1969). In North Africa,

M. libycus inhabits the Sahara desert, but does extend to the Mediterranean in Morocco, Algeria, and Libya where it overlaps the distribution of *M. shawi*, which is primarily Mediterranean littoral (Lay and Nadler, 1969; Zaime and Pascal, 1988). Citations for synonyms among Russian samples were supplied by Pavlinov and Rossolimo (1987); *afghanus* is proposed as a new subspecies in that checklist. The Xinjiang population was reviewed by Ma et al. (1987).

Lay and Nadler (1969) also clarified why *caudatus*, used by Ranck (1968) as a species name for Libyan samples, simply refers to *M. libycus*. Corbet (1978*c*:127), apparently unaware of the report by Lay and Nadler (1969), followed Ranck and listed *caudatus* as a species, but cautioned that *caudatus* may be "conspecific with *M. libycus* and that the Libyan forms assigned by Ranck to *M. libycus* should really be allocated to *M. shawi*." This is correct and restates the past confusion that has "clouded the taxonomy of *M. shawi* and *M. libycus* because of uncertainty concerning the number of species in this complex and their nomenclature" (Lay and Nadler, 1969:44).

Meriones meridianus (Pallas, 1773). Reise Prov. Russ. Reichs., p. 702.

TYPE LOCALITY: Astrakhanskaya Oblast, Dosang (as restricted by Heptner, *in* Vinogradov et al., 1936, not Chaworth-Musters and Ellerman, 1947—see Pavlinov and Rossolimo, 1987).

DISTRIBUTION: From Lower Don River and N of the Caucasus to Mongolia and the Chinese provinces of Xinjiang, Qinghai, Shanxi and Hebei, south to E Iran and N Afghanistan. The isolated segment in Armenia mentioned by Corbet (1978*c*) refers to *M. dahli* (see that account).

SYNONYMS: *auceps, brevicaudatus, buechneri, cryptorhinus, fulvus, heptneri, jei, karelini, lepturus, littoralis, massagetes, muleiensis, nogaiorum, penicilliger, psammophilus, roborowskii, shitkovi, tropini, urianchaicus, uschtaganicus, zhitkovi.*

COMMENTS: Subgenus *Pallasiomys*. The significance of intraspecific chromosomal variation among Russian samples was reported by Korobitsyna and Kartavtseva (1988). Utilizing several sets of data, including results from hybridization studies, Dyatlov and Avanyan (1987) tested conspecificity of the subspecies *meridianus, nogaiorum,* and *dahli*, and concluded that *nogaiorum* should be considered a semispecies and *dahli* a species (see that account). We retain *nogaiorum* in *M. meridianus* pending unequivocable results demonstrating its evolutionary status. The Xinjiang population was discussed by Ma et al. (1987).

Meriones persicus (Blanford, 1875). Ann. Mag. Nat. Hist., ser. 4, 16:312.

TYPE LOCALITY: Iran, Kohrud, N of Isfahan.

DISTRIBUTION: From Iran, adjacent regions of Transcaucasia, Turkey, Iraq, Turkmenistan, Afghanistan and Pakistan (W of Indus River).

SYNONYMS: *ambrosius, baptistae, gurganensis, rossicus, suschkini.*

COMMENTS: Subgenus *Parameriones*. Regional studies available for the Middle East (Harrison and Bates, 1991), Pakistan (Roberts, 1977), Afghanistan (Hassinger, 1973), and Iran (Lay, 1967).

Meriones rex Yerbury and Thomas, 1895. Proc. Zool. Soc. London, 1895:552.

TYPE LOCALITY: Yemen, Lahej, near Aden.

DISTRIBUTION: SW Arabia, from Mecca to Aden (see Harrison and Bates, 1991:289).

SYNONYMS: *buryi, philbyi.*

COMMENTS: Subgenus *Parameriones*, according to Nadler and Lay (1967) and Harrison and Bates (1991), but not allocated to a subgenus by Pavlinov et al. (1990).

Meriones sacramenti Thomas, 1922. Ann. Mag. Nat. Hist., ser. 9, 10:552.

TYPE LOCALITY: Israel, 10 mi S Beersheba.

DISTRIBUTION: A small range in Israel, on coastal plain S of the River Yarqon and in the northern Negev.

COMMENTS: Subgenus *Pallasiomys*. A distinctive Israeli endemic, reviewed by Harrison and Bates (1991).

Meriones shawi (Duvernoy, 1842). Mem. Soc. Sci. Nancy, 3:22.

TYPE LOCALITY: Algeria, Oman.

DISTRIBUTION: Mediterranean littoral from Morocco to N Sinai, never found more than about 150 mi inland (Lay and Nadler, 1969, and references therein).

SYNONYMS: *albipes, auziensis, crassibulla, grandis, isis, laticeps, longiceps, richardii, savii, sellysii, trouessarti.*

COMMENTS: Subgenus *Pallasiomys*. A distinctive species that ranges mostly north of *M. libycus* but is sympatric with it in several regions (Lay and Nadler, 1969; Zaime and Pascal, 1988). The distribution maps of Algerian *M. shawi* and *M. libycus* in Kowalski and Rzebik-Kowalska (1991) illustrate the geographic relationships of these species—mostly parapatric, but sympatric near the coast. *M. shawi* and *M. libycus* are often confused in museum collections and published reports (see the review by Lay and Nadler, 1969).

Meriones tamariscinus (Pallas, 1773). Reise Prov. Russ. Reichs., 2:702.

TYPE LOCALITY: Kazakhstan, Saraitschikowski (= Saraichik).

DISTRIBUTION: N Caucasus and Kazakhstan to the Altai Mtns, and through N Xinjiang and W Gansu of China.

SYNONYMS: *ciscaucasicus, collium, jaxartensis, kokandicus, montanus, satschouensis, tamaricinus.*

COMMENTS: Subgenus *Meriones*. Reviewed by Corbet (1978c); regional reviews offered by Allen (1940, China), and Ma et al. (1987, Xinjiang).

Meriones tristrami Thomas, 1892. Ann. Mag. Nat. Hist., ser. 6, 9:148.

TYPE LOCALITY: Israel, Dead Sea region.

DISTRIBUTION: From Israel, Lebanon, and Jordan to E Turkey, Syria, N Iraq, NW Iran, and Transcaucasia (see Harrison and Bates, 1991:294).

SYNONYMS: *blackleri, bodenheimeri, bogdanovi, intraponticus, karieteni, lycaon.*

COMMENTS: Subgenus *Pallasiomys*. The species was generally reviewed by Corbet (1978c) and regionally reviewed by Harrison and Bates (1991) and Lay (1967). Chromosomal polymorphism and its significance among Transcaucasian samples was reported by Korobitsyna and Korablev (1980).

Meriones unguiculatus (Milne-Edwards, 1867). Ann. Sci. Nat. Zool., Ser. 7, 5:377.

TYPE LOCALITY: China, N Shanxi, 10 mi NE of Tschang-Kur, Eul-che san hao (= Ershi san hao).

DISTRIBUTION: Mongolia, adjacent regions of Siberia (Transbaikalia), and of China from N Gansu through Nei Mongol to Heilongjiang.

SYNONYMS: *chihfengensis, koslovi, kurauchii, selenginus.*

COMMENTS: Subgenus *Pallasiomys*. Reviewed by Allen (1940) and Corbet (1978c). Corbet also included Xinjiang in the distribution of the species, but Ma et al. (1987) did not record it there. See Gulotta, 1971 (Mammalian Species, 3).

Meriones vinogradovi Heptner, 1931. Zool. Anz., 94:122.

TYPE LOCALITY: Iran, Persian Azarbaidjan.

DISTRIBUTION: SE Turkey, N Syria, N Iran, and Armenia and Azerbaijan (see Harrison and Bates, 1991).

COMMENTS: Subgenus *Pallasiomys*. A distinctive species (Pavlinov and Rossolimo, 1987) that was reviewed by Harrison and Bates (1991). The Iranian population was reviewed by Lay (1967).

Meriones zarudnyi Heptner, 1937. Byull. Moscow Ova. Ispyt. Prir. Otd. Biol., 46:19.

TYPE LOCALITY: Turkmenistan, Kushka (Afghanistan-Turkmenistan border).

DISTRIBUTION: NE Iran, N Afghanistan, and S Turkmenistan.

COMMENTS: Subgenus *Pallasiomys*. Reviewed by Corbet (1978c) and listed by Pavlinov and Rossolimo (1987) as a distinctive species. The Afghanistan population was reviewed by Hassinger (1973).

Microdillus Thomas, 1910. Ann. Mag. Nat. Hist., ser. 8, 5:197.

TYPE SPECIES: *Gerbillus peeli* de Winton, 1898.

COMMENTS: A member of the Tribe Gerbillini in the classification of Pavlinov et al. (1990).

Microdillus peeli (de Winton, 1898). Ann. Mag. Nat. Hist., ser. 7, 1:250.

TYPE LOCALITY: Somalia, Eyk.

DISTRIBUTION: Recorded only from Somalia.

COMMENTS: Reviewed by Roche and Petter (1968), who provided a distribution map.

Pachyuromys Lataste, 1880. Le Naturaliste, 2(40):313.
TYPE SPECIES: *Pachyuromys duprasi* Lataste, 1880.
COMMENTS: The sole member of the tribe Pachyuromyini in Pavlinov et al.'s (1990) classification.

Pachyuromys duprasi Lataste, 1880. Le Naturaliste, 2(40):313.
TYPE LOCALITY: Algeria, Laghouat.
DISTRIBUTION: N Sahara desert from W Morocco to N Egypt.
SYNONYMS: *faroulti, natronensis*.
COMMENTS: The Algerian population was reviewed by Kowalski and Rzebik-Kowalska (1991), the Libyan segment by Ranck (1968), and the Egyptian by Osborn and Helmy (1980).

Psammomys Cretzschmar, 1828. *In* Rüppell, Atlas Reise Nordl. Afr., Zool., Säugeth., p. 56.
TYPE SPECIES: *Psammomys obesus* Cretzschmar, 1828.
COMMENTS: Reviewed by Corbet (1978*c*, 1984). A member of the tribe Rhombomyini in the classification scheme of Pavlinov et al. (1990).

Psammomys obesus Cretzschmar, 1828. *In* Rüppell, Atlas Reise Nordl. Afr., Zool., Säugeth., p. 58, pl. 22.
TYPE LOCALITY: Egypt, Alexandria.
DISTRIBUTION: In North Africa from Algeria (Kowalski and Rzebik-Kowalska, 1991) through Tunisia and coastal region of Egypt (Osborn and Helmy, 1980) into Syria, Jordan, Israel, and parts of Arabia (Harrison and Bates, 1991); also on coast of Sudan (Corbet, 1978*c*).
SYNONYMS: *algiricus, dianae, edusa, nicolli, roudairei, terraesanctae, tripolitanus*.
COMMENTS: Electrophoretic, chromosomal, and morphological traits were analyzed by Qumsiyeh and Chesser (1988) in the context of assessing evolutionary change among four genera of gerbils.

Psammomys vexillaris Thomas, 1925. Ann. Mag. Nat. Hist., ser. 9, 16:198.
TYPE LOCALITY: Libya, Tripolitania Prov., Bu Ngem (Bondjem).
DISTRIBUTION: Algeria, Tunisia, and Libya.
COMMENTS: Sometimes included in *P. obesus* but a separate species as pointed out by Ranck (1968) and Cockrum et al. (1977). Kowalski and Rzebik-Kowalska (1991) did not recognize this species in Algeria and discussed only *P. obesus*.

Rhombomys Wagner, 1841. Gelehrte Anz. I. K. Bayer Akad. Wiss., München, 12, 52:421.
TYPE SPECIES: *Rhombomys pallidus* Wagner, 1841 (= *Meriones opimus* Lichtenstein, 1823).
COMMENTS: Reviewed by Corbet (1978*c*). The tribe Rhombomyini contains this genus along with *Sekeetamys, Meriones, Brachiones*, and *Psammomys* (Pavlinov et al., 1990).

Rhombomys opimus (Lichtenstein, 1823). Naturh. Abh. Eversmann's Reise, p. 122.
TYPE LOCALITY: Kazakhstan, Kzyl-Ordinskaya, KaraKumy Desert (see Pavlinov and Rossolimo, 1987).
DISTRIBUTION: From S Mongolia through Ningxia, Gansu, and Xinjiang in China to Kazakhstan, Iran, Afghanistan, and SW Pakistan (Corbet, 1978*c*; Ma et al., 1987).
SYNONYMS: *alaschanicus, dalversinicus, funicolor, giganteus, nigrescens, pallidus, pevzovi, sargadensis, sodalis*.
COMMENTS: Citations for synonyms were referenced by Pavlinov and Rossolimo (1987). Reviewed by Pavlinov et al. (1990). Regional reviews cover Pakistan (Roberts, 1977), Iran (Lay, 1967), Afghanistan (Hassinger, 1973), and Xinjiang (Ma et al., 1987).

Sekeetamys Ellerman, 1947. Proc. Zool. Soc. London, 117:271.
TYPE SPECIES: *Gerbillus calurus* Thomas, 1892.
COMMENTS: Reviewed by Corbet (1978*c*), who also discussed the past allocation of *calurus* to either *Meriones* or *Gerbillus*. One of four gerbil genera in which electrophoretic, chromosomal, and morphological traits were examined to assess rates of evolutionary change (Qumsiyeh and Chesser, 1988). Placed in the tribe Rhombomyini by Pavlinov et al. (1990).

Sekeetamys calurus (Thomas, 1892). Ann. Mag. Nat. Hist., ser. 6, 9:76.
TYPE LOCALITY: Egypt, Sinai, near Tor.
DISTRIBUTION: From E Egypt through Sinai, S Israel and Jordan into C Saudi Arabia (see Osborn and Helmy, 1980; Harrison and Bates, 1991).
SYNONYMS: *makrami*.
COMMENTS: Morphology, taxonomy, and ecology summarized by Harrison and Bates (1991).

Tatera Lataste, 1882. Le Naturaliste, Paris 2:126.
TYPE SPECIES: *Dipus indicus* Hardwicke, 1807.
SYNONYMS: *Gerbilliscus, Taterona*.
COMMENTS: A member of the tribe Taterillini according to Pavlinov et al. (1990). Taxonomic revisions of various inclusiveness were provided by Pirlot (1955) and Bates (1985, 1988). Davis (1975*a*) presented a review of the genus, in which, aside from *T. boehmi*, he considered all the species to cluster either in an *afra* group or *robusta* group. Chromosomal information for some species was reported by Matthey and Petter (1970). Results of craniometric studies of Angolan *Tatera* were presented by Crawford-Cabral (1988) and Crawford-Cabral and Pacheco (1991). Pavlinov et al. (1990) regarded true *Tatera* to consist only of the Asian species *T. indica*, placed all the African species in the genus *Gerbilliscus* (subgenera *Gerbilliscus* and *Taterona*), and identified *Taterillus* as its closest relative. There are distinctive features separating the Asian from all the African species, but we are unconvinced they are not part of the same monophyletic group and treat them that way by allocating the species among the subgenera *Tatera, Taterona*, and *Gerbilliscus*.

Tatera afra (Gray, 1830). Spicil. Zool., p. 10.
TYPE LOCALITY: South Africa, Cape of Good Hope, vicinity of Cape Town (as restricted by Meester et al., 1986).
DISTRIBUTION: SW Cape Province, South Africa (see Skinner and Smithers, 1990).
SYNONYMS: *africanus, gilli, schlegelii*.
COMMENTS: Subgenus *Taterona*. Taxonomy and distribution summarized by Meester et al. (1986), who listed the species in the *T. afra* group.

Tatera boehmi (Noack, 1887). Zool. Jahrb. Syst., 2:241.
TYPE LOCALITY: Zaire, Marunga, Qua Mpala.
DISTRIBUTION: Angola, S Zaire, Zambia, Malawi, Tanzania, Kenya and Uganda.
SYNONYMS: *bohmi, fallax, fraterculus, varia*.
COMMENTS: Subgenus *Gerbilliscus*. Revised by Bates (1988). The type locality is in S Zaire (Ansell, 1978), although it has been also identified as N Zambia (Allen, 1939; Bates, 1988). The double-grooved incisors and fringed, white-tipped tail in this species are unique among species of *Tatera*, thus its allocation to a separate subgenus.

Tatera brantsii (Smith, 1836). Rept. Exped. Exploring Central Africa, p. 43.
TYPE LOCALITY: South Africa, Ladybrand, E Orange Free State, near Lesotho border (see Meester et al., 1986, for details).
DISTRIBUTION: South Africa (most of Cape Prov., N and W Natal, Zululand, Orange Free State, Transvaal; see Skinner and Smithers, 1990), W Zimbabwe, Botswana, C and E Namibia, S Angola, and SW Zambia.
SYNONYMS: *breyeri, draco, griquae, joanae, maccalinus, maputa, miliaria, montanus, namaquensis, natalensis, perpallida, ruddi, tongensis*.
COMMENTS: Subgenus *Taterona*. Taxonomy and distribution summarized by Meester et al. (1986), who assigned the species to the *T. afra* group. Geographic variation in protein and enzyme markers among samples from Lesotho was reported by Maurer et al. (1976).

Tatera guineae Thomas, 1910. Ann. Mag. Nat. Hist., ser. 8, 5:351.
TYPE LOCALITY: Guinea-Bissau, Gunnal.
DISTRIBUTION: Gambia, Senegal, to Ghana and Burkina Faso.
SYNONYMS: *picta*.
COMMENTS: Subgenus *Taterona*. Reviewed by Rosevear (1969). Shown to be morphologically distinct from *T. robusta* by Bates (1985). Gautun et al. (1985) provided chromosomal information.

Tatera inclusa Thomas and Wroughton, 1908. Proc. Zool. Soc. London, 1908:169.
TYPE LOCALITY: Mozambique, Gorongoza Dist., Tambarara.
DISTRIBUTION: E Zimbabwe, Mozambique to NE Tanzania.
COMMENTS: Subgenus *Taterona*. Taxonomy and geographic range summarized by Meester et al. (1986), who listed the species in *T. afra* group.

Tatera indica (Hardwicke, 1807). Trans. Linn. Soc. Lond., 8:279.
TYPE LOCALITY: India, United Prov., between Benares and Hardwar.
DISTRIBUTION: An extensive range from Syria, Iraq, and Kuwait through Iran, Afghanistan, and Pakistan into most of Indian Peninsula north to the Terai region of S Nepal; also Sri Lanka (see Bates, 1988).
SYNONYMS: *bailwardi, ceylonica, cuvieri, dunni, hardwickei, monticola, otarius, persica, pitmani, scansa, sherrini, taeniurus.*
COMMENTS: Subgenus *Tatera*. Revised by Bates (1988), who recognized three distinctive subspecies. Regional reviews of the species include the segments from Arabian Peninsula (Harrison and Bates, 1991), Iran (Lay, 1967), Afghanistan (Hassinger, 1973), and Pakistan (Roberts, 1977). This is the only species that Pavlinov et al. (1990) allocated to the genus *Tatera*, redefining all African species to the genus *Gerbilliscus*.

Tatera kempi Wroughton, 1906. Ann. Mag. Nat. Hist., ser. 7, 17:375.
TYPE LOCALITY: Nigeria, Aguleri.
DISTRIBUTION: Senegal and Guinea (Mt Nimba) to Cameroon.
SYNONYMS: *gambiana, giffardi, hopkinsoni, welmanni.*
COMMENTS: Bates (1988) listed *kempi* as a synonym of *T. valida*. With the exception of Nigeria, no specimens were examined from the range of *T. kempi* (including the forms synonymized here). Davis (1975*a*) also recognized *kempi* as a subspecies of *T. valida*, including the same synonyms listed below. Although Rosevear (1969) recognized *kempi, hopkinsoni*, and *welmanni* as separate species within West Africa, he commented that *hopkinsoni* and *welmanni* were probably only racially distinct. We separate *T. kempi* from *T. valida* and provisionally treat *hopkinsoni* and *welmanni* as synonyms, pending a critical revision. Chromosomal data were reported from samples collected in Burkina Faso (Gautun et al., 1985, as *hopkinsoni*) and Mt Nimba in Guinea (Gautun et al., 1986, as either *kempi* or *hopkinsoni*).

Tatera leucogaster (Peters, 1852). Bericht Verhandl. K. Preuss., Akad. Wiss., Berlin, 17:274.
TYPE LOCALITY: Mozambique, north of Zambezi River, Mesuril, (as restricted by Davis, 1949:1004).
DISTRIBUTION: South Africa (W Orange Free State, Cape Prov. north of Orange River, NE Natal and Zululand, most of Transvaal; see Skinner and Smithers, 1990), Mozambique, Zimbabwe, Botswana, Namibia, Malawi, Zambia, S Angola, SW Tanzania, and S Zaire.
SYNONYMS: *angolae, bechuanae, beirae, beirensis, kaokoensis, limpopoensis, littoralis, lobengulae, mashonae, mitchelli, ndolae, nigrotibialis, nyasae, panja, pestis, pretoriae, salsa, schinzi, shirensis, stellae, tenuis, tzaneenensis, waterbergensis, zuluensis.*
COMMENTS: Subgenus *Taterona*. Taxonomy and distribution summarized by Meester et al. (1986), who included the species in the *T. robusta* group. Namibian populations reviewed by Griffin (1990). Nongeographic variation in Botswana sample analyzed by Swanepoel et al. (1979).

Tatera nigricauda (Peters, 1878). Monatsb. K. Preuss. Akad. Wiss., Berlin, 1879:200 [1878].
TYPE LOCALITY: Kenya, Taita, Ndi.
DISTRIBUTION: Kenya, Somalia, and Tanzania.
SYNONYMS: *nyama, percivali.*
COMMENTS: Subgenus *Taterona*. Revised by Bates (1988).

Tatera phillipsi (de Winton, 1898). Ann. Mag. Nat. Hist., ser. 7, 1:253.
TYPE LOCALITY: Somalia, Hanka Dadi.
DISTRIBUTION: Somalia and the Rift Valley in Ethiopia and Kenya (see Bates, 1988).
SYNONYMS: *bodessana, umbrosa.*
COMMENTS: Subgenus *Taterona*. Revised by Bates (1988). The form *minusculus* is most probably synonymous with either *T. phillipsi* or *T. robusta* (Bates, 1988).

Tatera robusta (Cretzschmar, 1830). *In* Rüppell, Atlas Reise Nordl. Afr., Zool., Säugeth., p. 75.
TYPE LOCALITY: Sudan, Ambukol.
DISTRIBUTION: Burkina Faso, Chad, Sudan, Ethiopia, Somalia, Uganda, Kenya, and Tanzania (see Bates, 1985, 1988).
SYNONYMS: *bayeri, bodessae, iconica, loveridgei, macropus, mombasae, muansae, pothae, shoana, swaythlingi, taylori, vicina.*
COMMENTS: Subgenus *Taterona*. Revised by Bates (1988).

Tatera valida (Bocage, 1890). J. Sci. Math. Phys. Nat. Lisboa, 2(5):6.
TYPE LOCALITY: Angola, Rio Cuando, Ambaca, Quissange, Caconda.
DISTRIBUTION: Central African Republic, Chad, Sudan, Ethiopia, Uganda and Kenya to SW Tanzania, Zaire, Zambia, and Angola.
SYNONYMS: *beniensis, benvenuta, dichrura, dundasi, liodon, lucia, neavei, nigrita, ruwenzorii, smithi, soror, taborae.*
COMMENTS: Subgenus *Taterona*. Revised by Bates (1988). Allopatric with and closely related to the West African *T. kempi* (see that account).

Taterillus Thomas, 1910. Ann. Mag. Nat. Hist., ser. 8, 6:222.
TYPE SPECIES: *Gerbillus emini* Thomas, 1892.
SYNONYMS: *Taterina.*
COMMENTS: A member of the tribe Taterillini in the arrangement of Pavlinov et al. (1990). Robbins (1971) proposed a new dental terminology for the genus based on a large sample of *T. gracilis*, analysed (1973) nongeographic variation drawn from the same sample, and summarized (1977) morphometric and chomosomal differentiation among the seven species that he considered valid (only *T. petteri* has been added). Chromosomal data were reported for some species by Matthey and Petter (1970). Like many genera of African muroid rodents, *Taterillus* requires critical systematic revision to determine species definitions and their distributional limits. The species now recognized are morphologically similar to one another, prompting many workers to consider them "sibling" or cryptic species (Sicard et al., 1988). Whether or not such described entities represent species or allopatric segments of a larger interbreeding unit is impossible to assess without further refinement of our present taxonomic understanding of the genus.

Taterillus arenarius Robbins, 1974. Proc. Biol. Soc. Washington, 87:399.
TYPE LOCALITY: Mauritania, Trarza Region, Tiguent.
DISTRIBUTION: N Sahel savanna and subdesert from Mauritania and through Mali to Niger (see Robbins, 1974); eastern limits unknown although Sicard et al. (1988) believed the species to be confined to the left bank of the Niger River.
COMMENTS: Robbins (1974) compared his new species with samples of *T. gracilis* and *T. pygarus*. The latter and *T. arenarius* are sympatric in S Mauritania.

Taterillus congicus Thomas, 1915. Ann. Mag. Nat. Hist., ser. 8, 16:147.
TYPE LOCALITY: Zaire, Upper Uele (Welle), Poko.
DISTRIBUTION: Cameroon, Chad, Central African Republic, Zaire, Sudan, Uganda.
SYNONYMS: *clivosus.*
COMMENTS: Assignment of *clivosus* follows Robbins (1977).

Taterillus emini (Thomas, 1892). Ann. Mag. Nat. Hist., ser. 6, 9:78.
TYPE LOCALITY: Uganda, Wadelai.
DISTRIBUTION: Sudan, W Ethiopia, Uganda, NW Kenya, NE Zaire.
SYNONYMS: *anthonyi, butleri, gyas.*
COMMENTS: Allocation of the synonyms to this species follows Robbins (1977).

Taterillus gracilis (Thomas, 1892). Ann. Mag. Nat. Hist., ser. 6, 9:77.
TYPE LOCALITY: Gambia (see Robbins, 1974).
DISTRIBUTION: N Nigeria, Niger, and Burkina Faso to Gambia and Senegal.
SYNONYMS: *angelus, nigeriae.*
COMMENTS: Chromosomal data were reported by Gautun et al. (1985) for the sample from Burkina Faso. Rosevear (1969) included *angelus* in this species, but Sicard et al. (1988) discussed its possible specific status. The form *nigeriae* was provisionally considered

valid by Rosevear (1969) but was referred to a subspecies of *T. gracilis* by Robbins (1974).

Taterillus harringtoni (Thomas, 1906). Ann. Mag. Nat. Hist., ser. 7, 18:303.
TYPE LOCALITY: Ethiopia, east of Lake Turkana (Rudolf), near Mutti Galeb.
DISTRIBUTION: Central African Republic, Sudan, Ethiopia, Somalia, E Uganda, Kenya, Tanzania.
SYNONYMS: *illustris, kadugliensis, lorenzi, lowei, melanops, meneghetti, nubilus, osgoodi, perluteus, rufus, tenebricus, zammarani.*
COMMENTS: Allocation of synonyms to this species follows Robbins (1977).

Taterillus lacustris (Thomas and Wroughton, 1907). Ann. Mag. Nat. Hist., ser. 7, 19:37.
TYPE LOCALITY: Nigeria, Lake Chad (= Kaddai).
DISTRIBUTION: NE Nigeria and Cameroon.
COMMENTS: Rosevear (1969) synonymized this with *T. gracilis*, but both Robbins (1974, 1977) and Petter (1975*b*) treated it as a distinct species.

Taterillus petteri Gautun, Tranier, and Sicard, 1985. Mammalia, 49:538.
TYPE LOCALITY: Burkina Faso (Upper Volta), Oudalan Prov., near Oursi Pond; 14°38′N, 00°26′W (as refined by Sicard et al., 1988:188).
DISTRIBUTION: Sahel savannah of E Burkina Faso and W Niger west of Niger River.
COMMENTS: As explained by Ansell (1989*a*), the availability of *petteri* should properly date from its first checklist citation (Gautun et al., 1985), not from its later formal description (Sicard et al., 1988). According to Sicard et al. (1988), *T. petteri* is confined to the loop of the Niger River and is parapatric with *T. gracilis*, from which it differs in morphology, ecology, biochemistry, and physiology. In the same paper, they stated that *petteri* may actually be conspecific with *angelus* (from Gambia), although they acknowledged not examining the holotype of the latter, and they speculated that *angelus* and *lacustris* may prove to be the same, but, because the type of the former is young and that of the latter is old, the relationship is difficult to demonstrate. The contribution of *petteri* to the systematic literature of African gerbils hardly improves the picture of relationships among populations of *Taterillus*.

Taterillus pygargus (F. Cuvier, 1838). Trans. Zool. Soc. London, 2:142.
TYPE LOCALITY: Senegal, probably St. Louis (as suggested by Robbins, 1977:191).
DISTRIBUTION: Gambia, Senegal, S Mauritania, and W Mali.
COMMENTS: Cuvier's *pygargus* was included in *Gerbillus pyramidum*, but the holotype is an example of *Taterillus* (Petter et al., 1972; Petter, 1975*b*, and references therein). Apparently unaware of Petter's observation, Lay (1983) listed *pygargus* as a species of *Gerbillus* known only from the type locality, which he thought was Egypt.

Subfamily Lophiomyinae Milne-Edwards, 1867. Arch. Mus. Hist. Nat. Paris. Memoires C, III, p. 81-116.
COMMENTS: See Carleton and Musser (1984) for diagnosis and general characterisitics, as well as other discussion. They also summarized past estimates of relationships, which were generally reflected in arrangement of *Lophiomys* in its own family or in a subfamily of either Cricetidae or Nesomyidae. Wahlert (1984) allocated *Lophiomys* and *Cricetops* to Lophiomyinae and considered this cluster closely related to Cricetinae.

Lophiomys Milne-Edwards, 1867. L'Institut, Paris, 35:46.
TYPE SPECIES: *Lophiomys imhausii* Milne-Edwards, 1867.
SYNONYMS: *Phractomys, Phragmomys.*
COMMENTS: Closest relatives are *Microlophiomys vorontsovi* from late Miocene of Ukraine (Topacevski and Skorik, 1984), and *Cricetops dormitor* from middle Oligocene of Mongolia and Kazakhstan (Wahlert, 1984).

Lophiomys imhausi Milne-Edwards, 1867. L'Institut, Paris, 35:46.
TYPE LOCALITY: Somalia ("Probably from African coast opposite Aden, where it was purchased," Allen, 1939:315; also see Thomas, 1910*c*:222).
DISTRIBUTION: E Sudan, Ethiopia, Somalia, Kenya, Uganda, and Tanzania; from sea level up to 3300 m in Ethiopia (Yalden et al., 1976), but apparently restricted to mountain

forest in Kenya and Uganda (Hollister, 1919; Delany, 1975). Known from Israel by subfossils and may occur in Arabia (Harrison and Bates, 1991).

SYNONYMS: *aethiopicus, bozasi, hindei, ibeanus, smithi, testudo, thomasi.*

COMMENTS: Thomas (1910*c*) recognized four species of *Lophiomys*, but Ellerman (1940:636) noted that "Thomas evidently came to the conclusion that all the East African 'species' were one, as there is a note in his tracts to this effect. I am inclined to go further and think that until more material comes to hand all forms must be treated as races of the earliest name *imhausi*." Ellerman's view prevails today and has yet to be tested by careful taxonomic revision.

Subfamily Murinae Illiger, 1815. Abh. Phys. Klasse K.-Preuss Akad. Wiss. Berlin, for 1804-11, p. 46, 129 [1815].

SYNONYMS: Anisomyini, Conilurinae, Hydromyinae, Murina, Phloeomyinae, Pseudomyinae, Rhynchomyinae.

COMMENTS: Diagnosis, characteristics, and contents of subfamily generally as presented by Carleton and Musser (1984). No general tribal arrangement of genera is available except for provincial groupings (Conilurini and Hydromyini for Australian species [summarized in Watts and Aslin, 1981] and Anisomyini for some New Guinea genera [Lidicker and Brylski, 1987]). *Acomys* and *Uranomys* would be excluded from the subfamily by some (see those accounts). Comparative chromosomal data provided in phylogenetic framework and other contexts for European species by Zima and Král (1984*a*), for some Asian groups by Markvong et al. (1973), Raman and Sharma (1977), Gadi and Sharma (1983), and Cao and Tran (1984), for African species by Robbins and Baker (1978), for Australian forms by Baverstock et al. (1977*c-e*, 1983*a*, *b*), for New Guinea species by Donnellan (1987), and for murines in general by Viegas-Pequignot et al. (1983, 1985, 1986). Phylogenetic relationships among Australian species based on biochemical results reported by Baverstock et al. (1977*a*, *b*, 1980) and Watts et al. (1992), those among other species reported by Iskandar and Bonhomme (1984) and Bonhomme et al. (1985). DNA-DNA hybridization results reported in phylogenetic context by Catzeflis (1990) and Catzeflis et al. (1987). Amplification of DNA (L1) in relationship to murine divergence from other muroids reported by Pascale et al. (1990). Comparative data on hair morphology (Keogh, 1985), soft palate topography (Eisentraut, 1969*a*; Fülling, 1992), and digestive system anatomy (Perrin and Curtis, 1980) provided results for phylogenetic analyses. The discrepency between the relativley rapid divergence of murine taxa as revealed by fossils and the much slower rates indicated by molecular data is reconciled by Jaeger et al. (1986) by postulating accelerated rates of evolution for certain proteins and a higher rate of nucleotide substitution in murines than ordinarily seen in other eutherians.

Abditomys Musser, 1982. Am. Mus. Novit., 2730:3.

TYPE SPECIES: *Rattus latidens* Sanborn, 1952.

COMMENTS: Belongs to the Philippine New Endemic cluster (Musser and Heaney, 1992).

Abditomys latidens (Sanborn, 1952). Fieldiana Zool., 33:125.

TYPE LOCALITY: Philippines, Luzon Isl, Mountain Prov, Mt Data, 7500 ft.

DISTRIBUTION: Known only by two specimens from Luzon.

COMMENTS: Taxonomic, morphological, and ecological data provided by Musser (1982*a*). Shown by Musser and Heaney (1992) to be close phylogenetic relative of *Tryphomys adustus*, another Luzon endemic.

Acomys I. Geoffroy, 1838. Ann. Sci. Nat. Zool. (Paris), ser. 2, 10:126.

TYPE SPECIES: *Mus cahirinus* Desmarest, 1819.

SYNONYMS: *Acanthomys, Peracomys.*

COMMENTS: Except for a brief list of species and subspecies made by Setzer (1975), the partial reviews by Matthey (1965*a*, *b*, 1968) based on chromosomal data, by Petter (1983) using morphology, by Janecek et al. (1991) that incorporated genic data, and the regional systematic revision by Dippenaar and Rautenbach (1986), no systematic revision of *Acomys* is available. Even the inclusion of *Acomys* within Murinae is questioned. Morphological evidence has been used to support a close relationship to

Mus (see Jacobs, 1978) and *Uranomys* (see Hinton, 1921; Misonne, 1969), but reproductive biology of *Acomys* is special among murines (Dieterlen, 1961, 1962, 1963), and biochemical data suggested *Acomys* is not closely related to *Mus* but to *Uranomys*, and is either distantly related to murines or not even a member of the subfamily (Bonhomme et al., 1985; Pascale et al., 1990; Sarich, 1985; Wilson et al., 1987). Dental evidence links *Acomys, Uranomys*, and *Lophuromys* to the exclusion of all other extant African muroids (Denys and Michaux, 1992). According to Denys (1990), *Acomys* is at least 4.5 million years old and not recently evolved, and careful comparisons between *Acomys* and primitive murines as well as muroids in general are necessary to determine phylogenetic position of genus.

Extant species are sorted into two subgenera (*Acomys* and *Peracomys*), but these groupings require reassesment by systematic revision of genus. Chromosomal data were summarized by Volobouev et al. (1991) and Sokolov et al. (1992).

Acomys cahirinus (Desmarest, 1819). Nouv. Dict. Hist. Nat., Nouv. ed., 29:70.

TYPE LOCALITY: Egypt, Cairo.

DISTRIBUTION: W Sahara to Egypt (including Sinai), N Nigeria, N Ethiopia, N Sudan, Jordan, Israel, Lebanon, Syria, Yemen, Oman, Saudi Arabia, S Iraq, Iran, and Pakistan.

SYNONYMS: *airensis, albigena, chudeaui, dimidiatus, flavidus, helmyi, hispidus, homericus, hunteri, megalodus, megalotis, nubicus, sabryi, seurati, viator, whitei.*

COMMENTS: Subgenus *Acomys*. Synonyms have either been treated as separate species, or as subspecies or synonyms of either *A. cahirinus* or *A. dimidiatus* (Ellerman, 1941; Harrison and Bates, 1991; Petter, 1983; Setzer, 1975). The *cahirinus-dimidiatus* complex needs critical systematic revision and may consist of several morphologically similar species (Corbet and Hill, 1991; Petter, 1983). Benazzou (1983), for example, recognized *chudeaui* as a species, and Le Berre and Le Guelte (1990) listed *airensis* and *chudeaui* as separate species. Chromosomal data reported by Tranier (1975, under *A. airensis*), Al-Saleh (1988), Volobouev et al. (1991, under *A. dimidiatus*), and Harrison and Bates (1991).

Acomys cilicicus Spitzenberger, 1978. Ann. Nat. Hist. Mus. Wien, 81:444.

TYPE LOCALITY: Asiatic Turkey, Vil Mersin, 17 km E Silifke.

DISTRIBUTION: Known only from the type locality.

COMMENTS: Subgenus *Acomys*. Corbet (1984) commented on this species.

Acomys cineraceus Fitzinger and Heuglin, 1866. Sitzb. K. Akad. Wiss. Wien, 54:573.

TYPE LOCALITY: WC Sudan, Doka, E Sennaar (Allen, 1939:364).

DISTRIBUTION: From N Ghana and Burkina Faso through N Togo, N Benin, S Niger, and N Nigeria to C and S Sudan, N Uganda, and C and S Ethiopia (specimens in the National Museum of Natural History).

SYNONYMS: *cinerascens, hawashensis, hystrella, intermedius, johannis, lowei, witherbyi.*

COMMENTS: Subgenus *Acomys*. Dieterlen (in litt.) noted that *A. cineraceus* is a distinct species and one of four (*A. wilsoni, A. percivali*, and *A. cahirinus*) occurring in Sudan. Petter (1983) recognized *witherbyi* as a species, and reported that it coexists with a member of the *cahirinus-dimidiatus* complex in Sudan. In morphology, *A. cineraceus* is closely similar to *A. kempi*; systematic revision would reveal whether each is a species, or simply represents a population of one species.

Acomys ignitus Dollman, 1910. Ann. Mag. Nat. Hist., ser. 8, 6:229.

TYPE LOCALITY: Kenya, Voi.

DISTRIBUTION: Usambara Mtns in NE Tanzania (specimens in Field Museum of Natural History), and Kenya; limits unknown.

COMMENTS: Subgenus *Acomys*. Petter (1983) recognized *ignitus* as a valid species, clashing with Setzer (1975), who regarded it as part of *A. dimidiatus*. Petter's action reflects reality, reinforcing Hollister (1919) and Ellerman (1941), who recognized *ignitus* as a distinct species, but associated the names *pulchellus, kempi*, and *montanus* with it either as subspecies or direct synonyms. Janecek et al. (1991) considered *ignitus* distinct and phylogenetically closely related to *A. cahirinus*, based on genic data.

Acomys kempi Dollman, 1911. Ann. Mag. Nat. Hist., ser. 8, 8:125.

TYPE LOCALITY: Kenya, Chanler Falls, N Guaso Nyiro.

DISTRIBUTION: S Somalia, Kenya and NE Tanzania (samples in the Field Museum of Natural History and the National Museum of Natural History); limits unknown.
SYNONYMS: *montanus*, *pulchellus*.
COMMENTS: Subgenus *Acomys*. Originally described by Dollman as a subspecies of *A. ignitus* and listed that way by Ellerman (1941) and Hollister (1919), but considered a subspecies of *A. cahirinus* by Setzer (1975). Treated as a species by Janecek et al. (1991) with closest evolutionary ties to *A. cahirinus*. The morphological characteristics and geographic range of *kempi* may represent the estern segment of *A. cineraceus*. Hollister (1919) correctly explained why *pulchellus* is a synonym of *A. kempi*; we include *montanus* based on our studies.

Acomys louisae Thomas, 1896. Ann. Mag. Nat. Hist., ser. 6, 18:269.
TYPE LOCALITY: Somalia, 40 mi south of Berbera.
DISTRIBUTION: Somalia; limits unknown.
SYNONYMS: *umbratus*.
COMMENTS: Subgenus *Peracomys*. Considered distinct, and the type-species of subgenus *Peracomys*, by Petter and Roche (1981). Petter (1983) included *umbratus* in the species.

Acomys minous Bate, 1906. Proc. Zool. Soc. Lond., 1905(2):321 [1906].
TYPE LOCALITY: Greece, Crete Isl, Kanea.
DISTRIBUTION: Crete (Greece).
COMMENTS: Subgenus *Acomys*. Treated as a species by Dieterlen (1978*a*) and Corbet and Hill (1991).

Acomys mullah Thomas, 1904. Ann. Mag. Nat. Hist., ser. 7, 14:103.
TYPE LOCALITY: Ethiopia.
DISTRIBUTION: Ethiopia and Somalia; limits unknown.
SYNONYMS: *brockmani*.
COMMENTS: Subgenus *Acomys*. Petter (1983) recognized *brockmani* as a valid species, suggesting it might be referable to *mullah*, which is the older name. These two forms are characterized by large molar rows, also diagnostic of *lowei*, which Petter implicitly associated with both *mullah* and *brockmani* as species. Ellerman (1941) also recognized *mullah* and *brockmani* as species, but Setzer (1975) arranged *lowei* as a subspecies of *A. cahirinus*, Dieterlen (in litt.) treated it as a synonym of *A. cineraceus*, and Setzer (1975) treated *mullah* and *brockmani* as subspecies of *A. dimidiatus*. Yalden et al. (1976) listed *mullah* as a synonym of *A. cahirinus*.

Acomys nesiotes Bate, 1903. Ann. Mag. Nat. Hist., ser. 7, 11:565.
TYPE LOCALITY: Cyrus, Kernyia Hills, near Dikomo.
DISTRIBUTION: Cyprus.
COMMENTS: Subgenus *Acomys*. Listed as a subspecies of *A. dimidiatus* by Ellerman (1941) and included in *A. cahirinus* by Corbet (1978*c*), but treated as a distinct species by Spitzenberger (1978).

Acomys percivali Dollman, 1911. Ann. Mag. Nat. Hist., ser. 8, 8:126.
TYPE LOCALITY: Kenya, Chanler Falls, Nyiro.
DISTRIBUTION: S Sudan (E of White Nile), Uganda, Kenya, Ethiopia, and S Somalia.
COMMENTS: Subgenus *Acomys*. Treated as a synonym of *kempi*, which in turn was included within *A. cahirinus* by Setzer (1975). However, Hollister (1919) treated *percivali* as a species based on many specimens, as did Ellerman (1941). Both Matthey (1968) and Hubert (1978*b*) identified specimens from Ethiopia as *A. percivali*. Petter (1983) acknowledged the specific status of *percivali* as did Neal (1983). Janecek et al. (1991) regarded *percivali* as the species genically most closely related to *A. wilsoni*.

Acomys russatus (Wagner, 1840). Abh. Akad. Wiss. Münchin, 3:195.
TYPE LOCALITY: Egypt, Sinai.
DISTRIBUTION: E Egypt, Sinai, Jordan, Israel, and Saudi Arabia.
SYNONYMS: *aegyptiacus*, *affinis*, *harrisoni*, *lewisi*.
COMMENTS: Subgenus *Acomys*. Qumsiyeh et al. (1986) retained *lewisi* as a species because although the karyotype of a sample from Jordan was indistinguishable from that of *A. russatus*, fur color and bacular morphology of *lewisi* was distinctive (Atallah, 1967). However, based on morphological evidence, *lewisi* was included in *A. russatus*

by Corbet (1978*c*), Osborn and Helmy (1980), and Harrison and Bates (1991). This allocation was also supported by genic data (Janecek et al., 1991).

Acomys spinosissimus Peters, 1852. Reise nach Mossambique, Säugeth., p. 160.

TYPE LOCALITY: Mozambique, Tette and Buio.

DISTRIBUTION: NE Tanzania (Amani; series in the National Museum of Natural History) and EC Tanzania, (Kilosa and Morogoro regions; series in the Field Museum of Natural History), SE Zaire, Zambia, Malawi, Zimbabwe, E and SE Botswana, C Mozambique, and N and NW South Africa.

SYNONYMS: *selousi, transvaalensis.*

COMMENTS: Subgenus *Acomys*. Revised by Dippenaar and Rautenbach (1986). Interpretation of genic data indicated *A. spinosissimus* should be placed in a species-group separate from other *Acomys* (Janecek et al., 1991).

Acomys subspinosus (Waterhouse, 1838). Proc. Zool. Soc. Lond., 1837:104 [1838].

TYPE LOCALITY: South Africa, Cape Prov, Cape of Good Hope.

DISTRIBUTION: South Africa; restricted to S and SW Cape Prov. (Dippenaar and Rautenbach, 1986; Skinner and Smithers, 1990).

COMMENTS: Subgenus *Acomys*. Revised by Dippenaar and Rautenbach (1986). Should be in a species-group by itself, according to the results of genic analyses by Janecek et al. (1991).

Acomys wilsoni Thomas, 1892. Ann. Mag. Nat. Hist., ser. 6, 10:22.

TYPE LOCALITY: Kenya, Mombasa.

DISTRIBUTION: S Sudan, S Ethiopia, S Somalia, Kenya, and south to EC Tanzania (Kondoa; specimens in the American Museum of Natural History); limits unknown.

SYNONYMS: *ablutus, argillaceus, bovonei, enid, nubilus.*

COMMENTS: Subgenus *Acomys*. Formerly included in *subspinosus* by Setzer (1975) but considered a species by Hollister (1919), Ellerman (1941), Matthey (1968), Yalden et al. (1976), Rupp (1980), Petter and Roche (1981), Petter (1983), and Corbet and Hill (1991). Genically most similar to *A. percivali* (Janecek et al., 1991). Specimens of *nubilus* are larger and longer-tailed than *wilsoni* and may represent a separate species; de Beaux's (1934) description of *bovonei* recalls *nubilus*.

Aethomys Thomas, 1915. Ann. Mag. Nat. Hist., ser. 8, 16:477.

TYPE SPECIES: *Epimys hindei* Thomas, 1902.

SYNONYMS: *Micaelamys*.

COMMENTS: Genus uncritically reviewed by Ellerman (1941), more fully by Davis (1975*b*), and still requires careful systematic revision to understand specific diversity, geographic distributions, and phylogenetic relationships of the genus with other African murines. *Micaelamys* is traditionally used as a subgenus for *A. granti* and *A. namaquensis*, but the characters separating them from other species in *Aethomys* are no more significant than the traits distinguishing those other species. Chromosomal data for various species provided by Matthey (1964) and Baker et al. (1988*c*).

Aethomys bocagei (Thomas, 1904). Ann. Mag. Nat. Hist., ser. 7, 13:416.

TYPE LOCALITY: Angola, Pungo Andongo.

DISTRIBUTION: Recorded only from C and W Angola; limits unknown.

COMMENTS: Morphology is very similar to *A. silindensis* (see that account).

Aethomys chrysophilus (de Winton, 1897). Proc. Zool. Soc. Lond., 1896:801 [1897].

TYPE LOCALITY: S Zimbabwe, Mashonaland, Mazoe.

DISTRIBUTION: From SE Kenya south through Tanzania, Malawi, Zambia, Mozambique, Zimbabwe, and E Botswana into South Africa, Namibia, and north into S Angola.

SYNONYMS: *alticola, capricornis, fouriei, harei, imago, ineptus, magalakuini, pretoriae, singidae, tongensis, tzaneenensis, voi.*

COMMENTS: Data from chromosomes (Gordon and Rautenbach, 1980; Visser and Robinson, 1986) and spermatozoal morphology (Gordon and Watson, 1986; Visser and Robinson, 1987) indicated that populations now identified as *A. chrysophilus* consist of two species: *chrysophilus* itself and another species not yet identified by scientific name. The segment occuring in southern African subregion reviewed by Skinner and Smithers (1990).

Aethomys granti (Wroughton, 1908). Ann. Mag. Nat. Hist., ser. 8, 1:257.
TYPE LOCALITY: South Africa, Deelfontein.
DISTRIBUTION: South Africa, known only from SC Cape Prov. (see map in Skinner and Smithers, 1990:279).
COMMENTS: Meester et al. (1986:292) provided historical taxonomic allocations of *granti*, which ranged from *Myomys*, through *Rattus* and *Mastomys* to *Aethomys*. Reviewed and compared with *A. namaquensis*, its closest relative, by Skinner and Smithers (1990).

Aethomys hindei (Thomas, 1902). Ann. Mag. Nat. Hist., ser. 7, 9:218.
TYPE LOCALITY: Kenya, Machakos.
DISTRIBUTION: N Camaroon, N and NE Zaire, S Sudan, SW Ethiopia, Uganda, Kenya, and Tanzania (no farther south than Muanza); southern limits unresolved.
SYNONYMS: *alghazal, centralis, helleri, medicatus, norae.*
COMMENTS: Originally described as a species but later incorrectly arranged as a subspecies of *A. kaiseri* (e.g., Hollister, 1919; Swynnerton and Hayman, 1951) with which it is sympatric. Actual geographic range of *A. hindei* is unresolved because many series in museum collections and reported in the literature are misidentified as *A. kaiseri*. See Bekele and Schlitter (1989) for discussion of the Ethiopian record. Davis (1975*b*) recognized two discrete populations, one to the east of the Rift Valley, and one to the west. The complex requires revisionary study to determine significance of appreciable geographic variation in morphological traits among samples.

Aethomys kaiseri (Noack, 1887). Zool. Jahrb. Syst.,2:228.
TYPE LOCALITY: Zaire, Marungu.
DISTRIBUTION: SW Uganda, S Kenya, Rwanda, S and E Zaire, Tanzania, Malawi, Zambia, and E Angola.
SYNONYMS: *amalae, hintoni, manteufeli, pedester, turneri, vernayi, walambae.*
COMMENTS: Significance of the morphological variation among samples from different regions needs to be assessed by a systematic revision.

Aethomys namaquensis (A. Smith, 1834). S. Afr. Quart. J., 2:160.
TYPE LOCALITY: South Africa, Cape Prov., Namaqualand, Cape of Good Hope.
DISTRIBUTION: S Angola, South Africa (except parts of Cape Prov., coastal Natal, and Namib Desert), Botswana, Zimbabwe, S and C Mozambique, S Malawi, and SE Zambia (see map in Skinner and Smithers, 1990:278).
SYNONYMS: *arborarius* (Peters, 1852, not True, 1892), *auricomis, avarillus, avunculus, calarius, capensis, centralis, drakensbergi, epupae, grahami, klaverensis, lechochloides, lehocla, longicaudatus, monticularis, namibensis, phippsi, siccatus, waterbergensis.*
COMMENTS: Significance of the appreciable variation in body size and pelage coloration among geographic samples needs to be assessed by systematic revision to determine whether that variation reflects one or more species. Reviewed by Meester et al. (1986) and Skinner and Smithers (1990).

Aethomys nyikae (Thomas, 1897). Proc. Zool. Soc. Lond., 1897:431.
TYPE LOCALITY: N Malawi, Nyika Plateau.
DISTRIBUTION: N Zambia, Malawi, N Angola, and S Zaire; limits unknown.
SYNONYMS: *dollmani.*
COMMENTS: Delany (1975) considered *nyikae* to be a synonym of *kaiseri*, but Davis (1975*b*) and Ansell (1978) correctly treated it as a separate species. Ansell (1978) provided documentation for the geographic range and discussed the identity of *dollmani*. The published record from Eastern Ngorima Reserve in E Zimbabwe is possibly based on a misidentification as inferred by Skinner and Smithers (1990:280).

Aethomys silindensis Roberts, 1938. Ann. Transvaal Mus., 19:245.
TYPE LOCALITY: E Zimbabwe, Mt Silinda.
DISTRIBUTION: E Zimbabwe (see map in Skinner and Smithers, 1990:277).
COMMENTS: For more than 30 years known only by two specimens from the type locality, but samples are now recorded from the nearby Ngorima Reserve in Melsetter Dist. and Stapleford in Umtali Dist, and the species may also occur in adjacent parts of Mozambique (Skinner and Smithers, 1990). Particular derived external, cranial, and dental traits phylogenetically tie *A. silindensis* to the Angolan *A. bocagei*.

Aethomys stannarius Thomas, 1913. Ann. Mag. Nat. Hist., ser. 8, 11:482.
TYPE LOCALITY: N Nigeria, Bauchi Prov., Kabwir.
DISTRIBUTION: N Nigeria to W Cameroon.
COMMENTS: A West African endemic that was included in *A. hindei* by Davis (1975*b*), but was correctly treated as a distinct species by Ellerman (1941:145), Rosevear (1969:389), and Hutterer and Joger (1982:126).

Aethomys thomasi (de Winton, 1897). Ann. Mag. Nat. Hist., ser. 6, 20:327.
TYPE LOCALITY: Angola, Galanga.
DISTRIBUTION: W and C Angola.
COMMENTS: A distinctive species related to *A. kaiseri*. Distribution of *A. thomasi* is mostly parapatric with *A. kaiseri* although Crawford-Cabral (1986) suspected it overlaps in the central highlands.

Anisomys Thomas, 1904. Proc. Zool. Soc. Lond., 1903(2):199 [1904].
TYPE SPECIES: *Anisomys imitator* Thomas, 1904.
COMMENTS: Lidicker (1968) documented phallic morphology of *Anisomys* and other New Guinea endemic murines, and concluded that the phallic morphology of *Anisomys* retained a high proportion of ancestral states. Used as the type genus of Tribe Anisomyini by Lidicker and Brylski (1987). Member of the New Guinea Old Endemics (Musser, 1981*c*).

Anisomys imitator Thomas, 1904. Proc. Zool. Soc. Lond., 1903:200 [1904].
TYPE LOCALITY: Papua New Guinea, Central Prov., Aroa River, Avera.
DISTRIBUTION: Forested mountain backbone of mainland New Guinea.
COMMENTS: Flannery (1990*b*) provided a description of the species and a distribution map. Chromosomal data reported by Donnellan (1987).

Anonymomys Musser, 1981. Bull. Am. Mus. Nat. Hist., 168:300.
TYPE SPECIES: *Anonymomys mindorensis* Musser, 1981.

Anonymomys mindorensis Musser, 1981. Bull. Am. Mus. Nat. Hist., 168:300.
TYPE LOCALITY: Philippines, Mindoro Isl, Halcon Range, Ilong Peak, 4500 ft.
DISTRIBUTION: Known only from the type locality.
COMMENTS: An endemic of Mindoro Island phylogenetically more closely related to native Sundaic murines than to any Philippine endemic (Musser and Heaney, 1992; Musser and Newcomb, 1983).

Apodemus Kaup, 1829. Skizz. Entwickel.-Gesch. Nat. Syst. Europ. Thierwelt, 1:154.
TYPE SPECIES: *Mus agrarius* Pallas, 1771.
SYNONYMS: *Alsomys, Karstomys, Nemomys, Petromys, Sylvaemus.*
COMMENTS: Palaearctic species reviewed by Corbet (1978*c*, 1984) and Kobayashi (1985), Chinese species by Xia (l984, 1985). Recognized species have been allocated among the subgenera *Apodemus, Sylvaemus, Alsomys,* and *Karstomys* (Corbet, 1978*c*; Zimmermann, 1962) but whether these names designate monophyletic clusters and should be retained as subgenera or instead raised to generic rank remains to be answered by critical systematic revision of the entire group, which is currently unavailable. Most taxonomists recognize at least the subgeneric validity of *Apodemus* and *Sylvaemus*; some suggest *Sylvaemus* should be raised to generic rank because of its great morphological and genic divergence from *Apodemus* (Britton-Davidian et al., 1991; Mezhzherin and Zykov, 1991); others treat *Sylvaemus* as a separate genus (Bonhomme et al., 1985; Mezhzherin and Lashkova, 1992). Phallic morphological comparisons among five European species provided by Williams et al. (1980) and among Chinese species documented by Yang and Fang (1988). Taxonomic differences in testes size among European species documented by Kratochvil (1971). Comparative chromosomal studies among European species provided by Soldatovic et al. (1975), Bekasova et al. (1980), and Vujosevic et al. (1984); chromosomal contrasts among Japanese species presented by Tsuchiya (1981). Electrophoretic variations of enzymes among species of *Apodemus* documented by Darviche et al. (1979, and references cited therein), Gemmeke (1980), Gill et al. (1987), and Fraguedakis-Tsolis (1983) in systematic context; electrophoretic, karyological, and

morphological distinctions among several species reported by Vorontsov et al. (1989) and Britton-Davidian et al. (1991). Differences in restriction endonuclease of nuclear DNA from three Austrian species reported by Csaikl et al. (1990). DNA-DNA hybridization results from analyses of three species presented by Catzeflis et al. (1987) and Catzeflis (1990). Numerous taxonomic provincial studies described morphological and other distinctions among sympatric species of *Apodemus*; examples are the study of five species from Bulgaria by Popov (1981), three species from Poland by Ruprecht (1979), and two species from Korea by Koh (1988) and Park et al. (1990). Vorontsov et al. (1989) described the presence of five species in the Caucasus, some of which were distinguished only by biochemical traits; Vorontsov et al. (1992) recently identified and defined four species from there. Tchernov (1979) reported on polymorphism, size trends and Pleistocene paleoclimatic responses of three Israeli species in an evolutionary context.

Apodemus agrarius (Pallas, 1771). Reise Prov. Russ. Reichs., 1:454.

TYPE LOCALITY: Russia, Ulianovsk Obl, middle Volga River, Ulianovsk (formerly Simbirsk).

DISTRIBUTION: C Europe to Lake Baikal, south to Thrace, Caucasus, and Tien Shan Mtns; Amur River through Korea to E Xizang and E Yunnan, W Sichuan, Fujiau, and Taiwan (China); Quelpart Isl (Korea).

SYNONYMS: *albostriatus, caucasicus, chejuensis, coreae, gloveri, harti, henrici, insulaemus, istrianus, kahmanni, karelicus, maculatus, mantchuricus, nikolskii, ningpoensis, ognevi, pallescens, pallidior, rubens, septentrionalis, tianschanicus, volgensis.*

COMMENTS: Subgenus *Apodemus*. Karyological data reported by Kang and Koh (1976), Koh (1982), and Lungeanu et al. (1986). Chromatic, morphological, and biochemical information presented by Wang (1985*b*), Zhao and Lu (1986), and Liu et al. (1991) in context of subspecific relationships. Age and Geographic variation in Korean, Polish, and Yugolavian populations as reflected by results of morphometric analyses documented by Sikorski (1982), Koh (1983, 1991), and Krystufek (1985*b*). European and Palaearctic populations reviewed by Böhme (1978*b*) and Karaseva et al. (1992).

Apodemus alpicola Heinrich, 1952. J. Mammal., 33:260.

TYPE LOCALITY: Allgäu, Osterachtal, S Germany.

DISTRIBUTION: NW parts of the Alps: S Germany, Austria, Liechtenstein, Switzerland, and N Italy.

SYNONYMS: *alpinus.*

COMMENTS: Subgenus *Sylvaemus*. Reviewed by Storch and Lütt (1989) who noted that *S. alpicola* occured syntopically with *A. sylvaticus* and *A. flavicollis*. The specific integrity of *alpicola* was biochemically confirmed by Vogel et al. (1991).

Apodemus argenteus (Temminck, 1844). *In* Siebold, Temminck, and Schlegel, Fauna Japonica, Arnz et Socii, Lugduni Batavorum, p. 51.

TYPE LOCALITY: Japan.

DISTRIBUTION: Japan; the four main islands along with some of the smaller ones (Corbet, 1978*c*:136).

SYNONYMS: *celatus, geisha, hokkaidi, sagax, tanei, yakui.*

COMMENTS: Subgenus *Alsomys*, but subgeneric allocation was questioned by Corbet (1978*c*:136). Because the type-series of *Mus argenteus* Temminck, 1844 is composite, a lectotype was chosen by Smeenk et al. (1982), which stabilizes the name of this species. Year of publication of *argenteus* is usually listed as 1845 but description appeared in 1844 (Smeenk et al., 1982). Electrophoretic analyses of 17 enzymes reported by Saitoh et al. (1989) in context of biochemical systematics of Japanese *Apodemus*. Chromosomal and morphometric comparisons between *A. argenteus* and other Japanese *Apodemus* reviewed by Vorontsov et al. (1977*a*).

Apodemus arianus (Blanford, 1881). Ann. Mag. Nat. Hist., ser. 5, 7:162.

TYPE LOCALITY: N Persia (Iran), Kohrud.

DISTRIBUTION: From N Iran west through Iraq to Lebanon and N Israel (see Harrison and Bates, 1991); limits unknown.

SYNONYMS: *erythronotus, witherbyi.*

COMMENTS: Subgenus *Sylvaemus*. Usually listed as a subspecies of *A. sylvaticus* (Ellerman and Morrison-Scott, 1951: Harrison and Bates, 1991), we separate it on the basis of its distinctive pelage and smaller body size. Both *arianus* and a larger, darker form (possibly *wardi*) occur in N Iran, although Lay (1967:186) considered them ecological subspecies. The relationship between *arianus, wardi*, and the small *A. uralensis* requires taxonomic resolution. *A. arianus* may be the species represented by the sample from Qazvin, N Iraq, that Darviche et al. (1979) separated electrophoretically from *A. sylvaticus* and *A. flavicollis*, but still considered closer to the latter. Vorontsov et al. (1992) included *arianus* in *A. uralensis*.

Apodemus chevrieri (Milne-Edwards, 1868). Rech. Hist. Nat. Mammifères, p. 288.

TYPE LOCALITY: China, Sichuan, Moupin.

DISTRIBUTION: W China, from W Hubei and S Gansu south through Sichuan into N and C Yunnan.

COMMENTS: Subgenus *Apodemus*. Included in *A. agrarius* by Allen (1940), Ellerman and Morrison-Scott (1951), and Corbet (1978*c*), but was earlier correctly listed as a species by Ellerman (1941:102). Currently recognized as a species distinct from *A. agrarius* by Xia (1984), Wang (1985*b*), and Yang and Fang (1988). Sympatric with *A. agrarius* in Sichuan and Ghizhou provinces (Wang, 1985*b*; Xia, 1985).

Apodemus draco (Barrett-Hamilton, 1900). Proc. Zool. Soc. Lond., 1900:418.

TYPE LOCALITY: S China, NW Fujian, Kuatun.

DISTRIBUTION: China (Hebei, SW Fujian, Henan, Shanxi, Shaanxi, Ninxia, Gansu, Yunnan, Sichuan, Guizhou, Xixang), Burma, and India (Assam).

SYNONYMS: *argenteus* (Swinhoe, not Temminck), *bodius, ilex, orestes*.

COMMENTS: Subgenus *Alsomys*. Originally described as a subspecies of *sylvaticus*, and listed that way by Allen (1940:945) and Ellerman and Morrison-Scott (1951:571), but correctly listed as a separate species by Ellerman (1941:101), Corbet (1978*c*:137), Corbet and Hill (1991), and Xia (1985), who also noted the close relationship between *A. draco* and *A. argenteus*.

Apodemus flavicollis (Melchior, 1834). Dansk. Staat. Norg. Pattedyr, p. 99.

TYPE LOCALITY: Denmark, Sieland Isl.

DISTRIBUTION: England and Wales, from NW Spain, France, Denmark, S Scandinavia through European Russia to Urals, S Italy, the Balkans, Syria, Lebanon, and Israel (Corbet, 1978*c*:134); also Netherlands (Van der Straeten, 1977).

SYNONYMS: *brauneri, cellarius, dietzi, fennicus, geminae, princeps, saturatus, wintoni*.

COMMENTS: Subgenus *Sylvaemus*. B chromosome polymorphism among populations reported by Vujosevic et al. (1991). European populations reviewed by Niethammer (1978*b*). Syrian, Lebanese, and Israeli populations should be re-examined to see if they represent *A. fulvipectus*, or another species, rather than *A. flavicollis*.

Apodemus fulvipectus Ognev, 1924. Rodentia of N Caucasus, Rostov-on-Don, p. 47.

TYPE LOCALITY: Georgia, N Caucasus, Dusheti Dist., near Kobi.

DISTRIBUTION: From E of the Dnepr River in the Ukraine, the Crimea, through Russia into the N Caucasus, Trancaucasus (Georgia, Armenia, and Azervaijan) and possibly in N Turkey, N Iran, and east to Kopet-Dag Mts (Mezhzherin and Zagorodnyuk, 1989).

SYNONYMS: *chorassanicus, falzfeini*.

COMMENTS: Subgenus *Sylvaemus*. Mezhzherin and Zagorodnyuk (1989) described *falzfeini* as a species, which O. Rossolimo (in litt.) considered to be the same as *A. fulvipectus*. Subsequent results from genetic studies led Mezhzherin and Zykov (1991) to treat *falzfeini* as identical with *chorassanicus*, which was originally described as a subspecies of *A. sylvaticus*. Vorontsov et al. (1992) recognized *fulvipectus* as the oldest name for this species; it was one of the electrophoretic siblings in the Caucasus revealed by Vorontsov et al. (1989).

Apodemus gurkha Thomas, 1924. J. Bombay Nat. Hist. Soc., 29:888.

TYPE LOCALITY: Nepal, Gorkha, Laprak.

DISTRIBUTION: Nepal.

COMMENTS: Subgenus *Alsomys*. Reviewed and contrasted with *A. sylvaticus* by Martens and Niethammer (1972). Chromosomal data reported and contrasted with other *Apodemus* by Gemmeke and Niethammer (1983). Also listed as a distinct species by Corbet (1978*c*) and Corbet and Hill (1991).

Apodemus hermonensis Filippucci, Simson, and Nevo, 1989. Boll. Zool., 56:374.

TYPE LOCALITY: Mt Hermon, Israel.

DISTRIBUTION: Alpine "tragacanthic" belt at about 2000 m on Mt Hermon, but may also occur in Lebanon (Filippucci et al., 1989).

COMMENTS: Subgenus *Sylvaemus*. Morphologically and genetically closely related to *A. flavicollis*, which it displaces on Mt Hermon (and possibly also Lebanon and Antilebanon mountain ranges) at elevations above 1900 m (Filippucci et al., 1989).

Apodemus hyrcanicus Vorontsov, Boyeskorov, and Mezhzherin, 1992. Zool. Zh., 71:127.

TYPE LOCALITY: Azerbaijan, Caucasus, Astarinski (R-NO), Hirkauski Preserve, "Piavolil" Dist., 450 m.

DISTRIBUTION: E Caucasus (see map in Vorontsov et al., 1992:124) where it is apparently endemic in the low mountain broadleaf forests of the Talysh region.

COMMENTS: *Apodemus hyrcanicus*, along with *A. uralensis*, *A. fulvipectus*, and *A. ponticus*, are all found in the Caucasus and have recently been revised by Vorontsov et al. (1992) using chromosomal, biochemical, and morphological characters.

Apodemus latronum Thomas, 1911. Abstr. Proc. Zool. Soc. Lond., 100:49.

TYPE LOCALITY: China, W Szechwan, Tatsienlu.

DISTRIBUTION: China (E Xizang, Sichuan, Yunnan) and N Burma.

COMMENTS: Subgenus *Alsomys*. Treated as a subspecies of *A. draco* by Feng et al. (1986), but *latronum* is a separate species that is sympatric with *draco* in Sichuan (Corbet, 1978*c*).

Apodemus mystacinus (Danford and Alston, 1877). Proc. Zool. Soc. Lond., 1877:279.

TYPE LOCALITY: Turkey, Adana Prov., Bulgar Dagh Mt, Zebil.

DISTRIBUTION: SE Europe, Israel, Lebanon, Jordan, Iraq, Iraq, NW Iran, S Georgia in Caucasus, Rhodes, Crete, and other inshore Aegean isls, and N Arabia (Corbet, 1978*c*; Niethammer, 1978*a*).

SYNONYMS: *epimelas, euxinus, pohlei, rhodius, smyrnensis*.

COMMENTS: Subgenus *Karstomys*. Reviewed by Storch (1977) and Niethammer (1978*a*), who included *krkensis* as a subspecies; that form, however, is a color phase of *A. sylvaticus* (see that account). The Arabian portion of its range includes the form named *pohlei* (Harrison and Bates, 1991). Vorontsov et al. (1989) provided chromosomal data in context of defining species in Caucasia.

Apodemus peninsulae (Thomas, 1907). Proc. Zool. Soc. Lond., 1906:862 [1907].

TYPE LOCALITY: Korea, 110 m SE of Seoul, Mingyoung.

DISTRIBUTION: SE Siberia from NE China (Xinjiang) and Altai Mtns to Ussuri, south through NE China and Korea, and E Mongolia to SW China (Sichuan and E Xizang); also on N Japanese islands of Sakhalin and Hokkaido.

SYNONYMS: *giliacus, major, majusculus, nigritalus, praetor, qinghaiensis, rufulus, sowerbyi*.

COMMENTS: Subgenus *Alsomys*. The Chinese *qinghaiensis* was described as a subspecies (Feng et al., 1983). Results of B-chromosomal analyses and references to chromosomal studies of *A. peninsulae* provided by Kolomiets et al. (1988) and Borisov and Malygin (1991). Biochemical systematics in reference to *A. speciosus* and the Hokkaido *giliacus* reported by Saitoh et al. (1989), who also reviewed the contrasting treatment of *giliacus* as either a subspecies of *A. peninsulae* or a separate species, and its bigeographical implications. Corbet (1978*c*) listed *nigritalus* as a synonym of *A. sylvaticus*, but the holotype is *A. peninsulae* (also see Pavlinov and Rossolimo, 1987:221). *A. peninsulae nigritalus* is sympatric with the smaller-bodied *A. uralensis tscherga* in the Altai region, which Hollister had noted (but under different names) in 1913*b* (see Ellerman and Morrison-Scott, 1951:567). Placed in subgenus *Apodemus* by Pavlinov and Rossolimo (1987) and Mezhzherin and Zykov (1991).

Apodemus ponticus Sviridenko, 1936. Abs. Works Zool. Inst. Moscow State Univ., 3:103.

TYPE LOCALITY: N Caucasus, Chernomorski Dist., (Black Sea) Olgino Village.

DISTRIBUTION: From shore of Azov Sea through Ciscaucasia, south through Caucasus into Armenia, E Turkey, Iraq, and possibly NW Iran; limits unknown. Parts of the Russian distribution were mapped by Vereshchagin (1967; as *fulvipectus*), Mezhzherin (1991), and Vorontsov et al. (1992).

SYNONYMS: *argyropuli, argyropuloi, brevicauda, parvus, persicus, planicola, samaricus, samariensis, saxatilis*.

COMMENTS: Subgenus *Sylvaemus*. The names above have been associated with *A. flavicollis* (Ellerman and Morrison-Scott, 1951; Harrison and Bates, 1991), but they identify samples of *A. ponticus* which is a distinct species (Bobrinskii et al., 1944; Mezhzherin, 1991; O. Rossolimo, in litt.; Vereshchagin, 1967:510; Vorontsov et al., 1992). True *flavicollis* does not occur in the Caucasus and parts of the M East. The names *argyropuli* Ellerman and Morrison-Scott, 1951, and *argyropuloi* Heptner, 1948 (see Harrison and Bates, 1991) were proposed to replace *parvus*.

Apodemus rusiges Miller, 1913. Proc. Biol. Soc. Washington, 26:81.

TYPE LOCALITY: N India, C Kashmir.

DISTRIBUTION: N India (Kashmir, Himachal Pradesh, and Kumaun); range limits unresolved.

SYNONYMS: *griseus*.

COMMENTS: Subgenus *Sylvaemus*. Miller's *rusiges* was a renaming of True's *griseus* (see Ellerman, 1941:100). Listed as a subspecies of *A. flavicollis* by Ellerman (1941, 1961) and Ellerman and Morrison-Scott (1951), and as a subspecies of *A. sylvaticus* by Corbet (1978*c*), the long-tailed and distinctively patterned *rusiges* appears to be a separate species endemic to N India.

Apodemus semotus Thomas, 1908. Ann. Mag. Nat. Hist., ser. 8, 1:44

TYPE LOCALITY: Taiwan.

DISTRIBUTION: Endemic to Taiwan.

COMMENTS: Subgenus *Alsomys*. Closely related in morphology to mainland *A. draco* (Corbet, 1978*c*).

Apodemus speciosus (Temminck, 1844). *In* Siebold, Temminck and Schlegel, Fauna Japonica, Arnz et Socii, Lugduni Batavorum, p. 52.

TYPE LOCALITY: Japan.

DISTRIBUTION: Endemic to four primary islands and smaller islands of Japan (see map in Tsuchiya, 1974).

SYNONYMS: *ainu, dorsalis, insperatus, miyakensis, navigator, sadoensis, tusimaensis*.

COMMENTS: Subgenus *Alsomys*. Sympatric with *A. peninsulae* on Hokkaido. The citation is usually cited as 1845, but was published in 1844; see Holthuis and Sakai (1970). Some of the names (*ainu, navigator,* and *miyakensis*) have been listed as species (Vorontsov et al., 1977*a*; also discussion in Corbet, 1978*c*), but we follow Corbet (1978*c*) in recognizing only one species. This action was also reflected by Tsuchiya (1974), as he divided the names into two groups of *A. speciosus* based on biochemical and chromosomal evidence. Saitoh et al. (1989) provided additional biochemical information in context of systematic comparisons among Japanese *Apodemus*. Placed in subgenus *Apodemus* by Pavlinov and Rossolimo (1987) and Mezhzherin and Zykov (1991).

Apodemus sylvaticus (Linnaeus, 1758). Syst. Nat., 10th ed., 1:62.

TYPE LOCALITY: Sweden, Uppsala.

DISTRIBUTION: Europe north to Scandinavia and east to NW Ukraine and N Byelorussia, and on many islands (Iceland, Britain, Ireland, numerous nearby isls, most Mediterranean isls); see map in Niethammer (1978*c*:341). Also mountains of N Africa from Atlas Mtns in Morocco east across Algiers to Tunisia (see map in Kock and Felton, 1980).

SYNONYMS: *albus, algirus, alpinus, bergensis, butei, callipides, candidus, celticus, chamaeropsis clanceyi, creticus, cumbrae, dichruroides, dichrurus, eivissensis, fiolagan, flaviventris, flavobrunneus, fridariensis, frumentariae, ghia, grandiculus, granti, hamiltoni, hayi, hebridensis, hermani, hessei, hirtensis, iconicus, ifranensis, ilvanus, intermedius, isabellinus, islandicus, kilikiae, krkensis, larus, leucocephalus, maclean, maximus, milleri, nesiticus, niger, parvus* (Bechstein, 1793, not Argyropulo, 1941), *pecchioli, rufescens, spadix, stankovici, tauricus, thuleo, tirae, tural, varius*.

COMMENTS: Subgenus *Sylvaemus*. The geographic range as outlined here is primarily European and N African. Its occurrence as mapped by Corbet (1978*c*) east of Byelorussia and W Ukraine reflects ranges of other species (*uralensis, fulvipectus, arianus, wardi, rusiges*) once included within *A. sylvaticus*. Contrary to published records, *A. sylvaticus* is not part of the modern Israeli fauna (Filippucci et al., 1989), but is represented there by fossils from between 40,000 and 10,000 B.C. (Tchernov, 1979). The Arabian Peninsula record at Qatar is based on *Mus* (Kock and Nader, 1990). Biochemical (Byrne et al., 1990; Fernandes et al., 1991; Gemmeke, 1981) and morphometric (Alcantara, 1991; Murback, 1979) intrapopulational analyses provided results of differentiation within species and change in its evolutionary history (see

Berry, 1973, and references therein), as well as insular gigantism (Libois and Fons, 1990). Morphometric contrasts between *A. sylvaticus* and *A. flavicollis* in context of evolutionary divergence reported by Hedges (1969), Mezhzherin and Lashkova (1992) and Van der Straeten and Van der Straeten-Harrie (1977); chromosomal and biochemical (mtDNA) contrasts by Debrot and Mermod (1977), Hirning et al. (1989) and Tegelstrom and Jaarola (1989). European populations (virtually entire species) reviewed by Niethammer (1978*c*). North African populations reviewed by Kock and Felten (1980) and Kowalski and Rzebik-Kowalska (1991); the latter also point out that *algirus* Pomel, 1856 and *chamaeropsis* Loche, 1867 may refer to species of *Mus* or *Gerbillus*. Conspecificity of *krkensis* with *A. sylvaticus* was documented by Williams et al. (1980) and Dolan and Yates (1981).

Apodemus uralensis (Pallas, 1811). Zoogr. Rosso-Asiat., 1:168.

TYPE LOCALITY: Russia, S Ural Mts.

DISTRIBUTION: From E Europe and Turkey (see map in Steiner, 1978, under *A. microps*) east to the Altai Mts and NW China (Xinjiang), south into the Caucasus; S and E limits unknown.

SYNONYMS: *baessleri, balchaschensis, charkovensis, ciscaucasicus, microps, microtis, mosquensis, nankiangensis, pallidus, pallipes, tokmak, tscherga, vohlynensis.*

COMMENTS: Subgenus *Sylvaemus*. Listed as a subspecies of *A. sylvaticus* by Ellerman and Morrison-Scott (1951), who also noted that Kuznetzov (*in* Bobrinskii et al., 1944) had already suspected that the Altai *tscherga* and *uralensis* refered to the same species. This identity, including those of most of the synonyms, is documented by biochemical and morphological results (Mezhzherin and Mikhailenko, 1991; Mezhzherin and Zykov, 1991). Gemmeke (1983) and Vorontsov et al. (1989, 1992) also provided chromosomal and biochemical information (under *microps*), and Tvrtkovic and Đzukic (1977) provided morphological data in context of distinguishing species of *Apodemus*. Ma et al. (1987) and Pavlinov and Rossolimo (1987) treated *tscherga* as a subspecies of *A. sylvaticus*, but Corbet (1978*c*) listed it under *A. peninsulae*. The NW Chinese *nankiangensis* was described as a subspecies of *A. sylvaticus* (see Corbet, 1984; Ma et al., 1987); we provisionally include it and three of the other synonyms (*pallidus, balchaschensis*, and *pallipes*) pending a systematic revision of E Russian, Middle Eastern, and W Chinese *Apodemus*. E European populations reviewed by Steiner (1978, under *microps*).

Apodemus wardi (Wroughton, 1908). J. Bombay Nat. Hist. Soc., 18:282.

TYPE LOCALITY: N India, Kashmir, Ladakh, Saspul.

DISTRIBUTION: NC Nepal (Martens and Niethammer, 1972) through Kashmir, N Pakistan, and Afghanistan (Ellerman and Morrison-Scott, 1951) to NW Iran; limits unknown.

SYNONYMS: *bushengensis, pentax.*

COMMENTS: Subgenus *Sylvaemus*. Originally described as a subspecies of *sylvaticus*, then listed as a subspecies of *A. flavicollis* by Ellerman and Morrison-Scott (1951), arranged as a subspecies of *A. sylvaticus* by Corbet (1978*c*), and suspected to be a different species than the latter by Gemmeke and Niethammer (1982) based on biochemical analyses of Nepalese and Iranian samples (which were documented by Darviche et al, 1979). The form *bushengensis* from SW Xizang (Tibet) was described as a subspecies of *A. sylvaticus* (Feng et al., 1986), and we allocate it to *wardi* pending a revision of the W Chinese, N Indian, and E Russian *Apodemus*. The relationship between this species, *A. peninsulae, A. uralensis*, and European *A. sylvaticus* requires better definition. Marshall (in press) identifies the holotype of *sublimis* Blanford, 1879, from Ladakh, as an example of *wardi*; if correct, *sublimis* may be the oldest name for this entity.

Apomys Mearns, 1905. Proc. U.S. Natl. Mus., 28:455.

TYPE SPECIES: *Apomys hylocoetes* Mearns, 1905.

COMMENTS: At one time included in *Rattus*, but is a distinct genus and forms a monophyletic group of its own within the assemblage of Philippine Old Endemics (Musser and Heaney, 1992). Taxonomic history of the genus and preliminary systematic revision provided by Musser (1982*b*); additional taxonomic notes and phylogenetic relationships outlined by Musser and Heaney (1992). Undescribed species recorded from Negros and Sibuyan islands (Musser and Heaney, 1992); actual distribution in archipelago and number of species still unknown.

Apomys abrae (Sanborn, 1952). Fieldiana Zool., 33(2):133.
TYPE LOCALITY: Philippines, Luzon Isl, Abra Prov, Abra, 3500 ft.
DISTRIBUTION: Known only from middle and high elevations on Luzon.
COMMENTS: Included within the "*Apomys abrae-hylocetes* Group" by Musser (1982*b*).

Apomys datae (Meyer, 1899). Abh. Mus. Dresden, ser. 7, 7:25.
TYPE LOCALITY: Philippines, N Luzon Isl, Lepanto.
DISTRIBUTION: Known only from W highlands in N Luzon.
SYNONYMS: *major*.
COMMENTS: The sole member of "*Apomys datae* Group" (Musser, 1982*b*).

Apomys hylocoetes Mearns, 1905. Proc. U.S. Natl. Mus., 28:456.
TYPE LOCALITY: Philippines, Mindanao Isl, Mt Apo, 6000 ft.
DISTRIBUTION: Known only from 6000-7600 ft on Mindanao.
SYNONYMS: *petraeus*.
COMMENTS: Included within "*Apomys abrae-hylocetes* Group" by Musser (1982*b*).

Apomys insignis Mearns, 1905. Proc. U.S. Natl. Mus., 28:459.
TYPE LOCALITY: Philippines, S Mindanao Isl, Mt Apo, 6000 ft.
DISTRIBUTION: Known only from middle to high elevations on Mindanao (Musser and Heaney, 1992).
SYNONYMS: *bardus*.
COMMENTS: Included within "*Apomys abrae-hylocetes* Group" by Musser (1982*b*).

Apomys littoralis (Sanborn, 1952). Fieldiana Zool., 33(2):134.
TYPE LOCALITY: Philippines, Mindanao Isl, Bugasan, Cotabato, 50 ft.
DISTRIBUTION: Formerly recorded from Negros and Mindanao (Musser, 1982*b*), but revised view excludes it from Negros, and records it from lowlands of Mindanao, Bohol, Biliran, Dinagat, and Leyte (Musser and Heaney, 1992).
COMMENTS: Included within "*Apomys abrae-hylocetes* Group" by Musser (1982*b*).

Apomys microdon Hollister, 1913. Proc. U.S. Natl. Mus., 46:327.
TYPE LOCALITY: Philippines, Cataduanes Isl, Biga.
DISTRIBUTION: Cataduanes and lowlands of S Luzon (Musser and Heaney, 1992). Formerly and incorrectly thought to also occur on Leyte and Dinagat (Musser, 1982*b*).
SYNONYMS: *hollisteri*.
COMMENTS: Included within "*Apomys abrae-hylocetes* Group" by Musser (1982*b*).

Apomys musculus Miller, 1911. Proc. U.S. Natl. Mus., 38:403.
TYPE LOCALITY: Philippines, Luzon Isl, Benguet, Baguio, Camp John Hay, 5000 ft.
DISTRIBUTION: Luzon and Mindoro, Philippines.
COMMENTS: Included within "*Apomys abrae-hylocetes* Group" by Musser (1982*b*).

Apomys sacobianus Johnson, 1962. Proc. Biol. Soc. Washington, 75:318.
TYPE LOCALITY: Philippines, Luzon Isl, Pampanga Prov, Sacobia River, Clark Air Base.
DISTRIBUTION: Known only from the type locality.
COMMENTS: Included within "*Apomys abrae-hylocetes* Group" by Musser (1982*b*).

Archboldomys Musser, 1982. Bull. Am. Mus. Nat. Hist., 174:30.
TYPE SPECIES: *Archboldomys luzonensis* Musser, 1982.
COMMENTS: Member of the Philippine Old Endemics most closely related to *Crunomys*; nature of relationship of this monophyletic group formed by *Archboldomys* and *Crunomys* to other Philippine shrew rats presently unresolvable (Musser and Heaney, 1992).

Archboldomys luzonensis Musser, 1982. Bull. Am. Mus. Nat. Hist., 174:30.
TYPE LOCALITY: Philippines, SE Luzon Isl, Camarines Sur Prov, Mt Isarog, 6560 ft.
DISTRIBUTION: Known only from montane habitat on Mt Isarog in SE peninsula of Luzon (Rickart et al., 1991).

COMMENTS: Morphological descriptions and comparisons with *Crunomys* and Sulawesi shrew rats provided by Musser (1982*c*). Elevational and ecologial notes in Rickart et al., 1991). Most closely related to Philippine *Crunomys* (Musser and Heaney, 1992).

Arvicanthis Lesson, 1842. Nouv. Tabl. Regn. Anim. Mammalifères, p. 147.

TYPE SPECIES: *Lemmus niloticus* Desmarest, 1822.

SYNONYMS: *Isomys*.

COMMENTS: Opinions on the number of species in the genus have varied from one (Misonne, 1974) to several (Allen, 1939; Corbet and Hill, 1991; Dollman, 1911; Ellerman, 1941; Rousseau, 1983). The first, and in many ways still most useful, review of *Arvicanthis* was made by Dollman (1911). Several recent attempts to assess variability in the genus using breeding studies (Petter et al., 1969), morphometric analyses (Rousseau, 1983), chromosomal evidence (Capanna and Civitelli, 1988; Orlov et al., 1992; Volobouev et al., 1987; and references therein), and electrophoretic patterns of blood proteins (Kaminski et al., 1984) have provided partial insights into species-diversity within *Arvicanthis*, as have results obtained based on examination of museum specimens (focusing on qualitative traits as well as external and craniodental measurements), and study of the discussions by Dollman (1911), Hollister (1919), Osgood (1936), and Corbet and Yalden (1972). Of the five species we list, three (*abyssinicus*, *blicki*, and *neumanni*) are easily diagnosed and recognizable, but two (*niloticus* and *nairobae*) are not and the geographic ranges and synonyms we provide for each of them are provisional until a sound systematic revision becomes available.

Arvicanthis abyssinicus (Rüppell, 1842). Mus. Senckenberg., 3:104.

TYPE LOCALITY: Ethiopia, Simien Prov., Simien Mtns, Entschetqab (see Osgood, 1936:252).

DISTRIBUTION: Ethiopia, between 1300-3400 m (Yalden et al., 1976).

SYNONYMS: *fluvicinctus*, *rufodorsalis*, *saturatus*.

COMMENTS: An Ethiopian endemic and morphologically distinctive species whose closest relative is *A. blicki*, another Ethiopian endemic. Historically the species was perceived to embrace forms occurring from Ethiopia south to Zambia (Allen, 1939; Dollman, 1911; Ellerman, 1941), but Osgood (1936:252) discussed several of *abyssinicus*'s diagnostic features, noting that it "is not unlikely that it is confined to Ethiopia and at least some of the forms of Kenya and Uganda which have been associated with it will need other allocation." This view has been reinforced by morphometric analyses (Rousseau, 1983), and our examination of specimens.

Arvicanthis abyssinicus and *A. blicki* are the only species of *Arvicanthis* endemic to Ethiopia, both are part of middle and high altitude grassland and moorland endemic mammal faunas of Ethiopia (Demeter and Topal, 1982; Rupp, 1980; Yalden, 1988), and both are more closely related to each other than to any other species of the genus. *Arvicanthis niloticus* and *A. neumanni* also occur in Ethiopia (see those accounts), but their ranges also extend far beyond that country's boundaries.

Rousseau (1983) regarded *mearnsi*, described as a subspecies of *A. abyssinicus* by Frick (1914), to also be a subspecies of *A. abyssinicus*, but its qualitative dental and chromatic characteristics fall within the range of variation seen in *A. niloticus*, as Osgood (1936) and Yalden et al. (1976) noted (discussed by them under either *lacernatus* or *dembeensis*).

Chromosomal data reported by Orlov et al. (1992) within the context of studying chromosomal variation among Ethiopian samples of *Arvicanthis*.

Arvicanthis blicki Frick, 1914. Ann. Carnegie Mus., 9:20.

TYPE LOCALITY: Ethiopia, South Chilalo Mtns, Hora Mt base camp, 9000 ft.

DISTRIBUTION: Ethiopia; between 2750 and 3500 m from plateau on E side of Ethiopian Rift Valley (Yalden et al., 1976).

COMMENTS: An Ethiopian endemic that is a characteristic diurnal member of Afro-Alpine moorland zone above 3500 m (Demeter and Topal, 1982; Rupp, 1980; Yalden, 1988; Yalden et al., 1976). In morphology, ecological and geographic distributions, and habits, *A. blicki* is a very distinctive species, as recognized by Dorst (1972) that, judged by similarities in molar topography and dorsal fur patterning, is more closely related to *A. abyssinicus* than to any other species of *Arvicanthis*.

Arvicanthis nairobae J. A. Allen, 1909. Bull. Am. Mus. Nat. Hist., 26:168.

TYPE LOCALITY: Kenya, Nairobi.

DISTRIBUTION: Recorded in and east of the Rift Valley from Mount Lololokwi ("an isolated mountain east of the Mathews Range, about midway between Mount Kenia and Mount Marsabit," Hollister, 1919) in C Kenya south to the Dodoma region in EC Tanzania; N and S limits unknown.

SYNONYMS: *chanleri, pallescens, praeceps, virescens.*

COMMENTS: The name *nairobae* is the oldest applicable to samples from east of the Rift Valley in Kenya and Tanzania containing animals smaller in body size and generally brighter and buffier in pelage than those larger and darker specimens we have identified as *A. niloticus* from west of the Rift. Some specimens of *A. nairobae* closely resemble those of *A. neumanni* in pelage coloration, and might be mistaken for it, but are larger in body size. Corbet and Yalden (1972) even suggested that *chanleri* might represent *somalicus* (= *neumanni*), but the holotype is larger and fits within the range of variation seen among samples of *nairobae*. Records of sympatry between *nairobae* and *neumanni* are documented by series from Mount Lololokwi (specimens in the National Museum of Natural History), and the Dodoma region of Tanzania (samples in the American Museum of Natural History), and the two are probably broadly sympatric throughout their ranges in Kenya and Tanzania.

The morphological and geographic relationships between *A. nairobae* and *A. niloticus* require resolution, which could be accomplished by systematic review of the *niloticus-nairobae* complex. For example, *nairobae* is found with *A. neumanni* in the Dodoma region and they are the only species of *Arvicanthis* represented by specimens from E Tanzania. In contrast, samples from the Tabora area, to the west of Dodoma, contain *A. neumanni* and a species significantly larger in body size with darker pelage than samples of *A. nairobae* from Dodoma. It is this larger species that ranges south into Zambia and north through W Tanzania, W Kenya, and Uganda. Studying the distribution of characters from samples collected along a transect between the Dodoma and Tabora regions would elucidate character and geographic relationships between the two kinds.

Arvicanthis neumanni (Matschie, 1894). Sitz.-Ber. Ges. Naturf. Fr. Berlin, 1894:204.

TYPE LOCALITY: Central Tanzania, Kondoa District, Barungi.

DISTRIBUTION: N and E Rift Valleys of Ethiopia (Yalden et al., 1976), Somalia, extreme SE Sudan (Dieterlen and Nikolaus, 1985), and south through Kenya in and east of the Rift Valley (Dollman, 1911; Hollister, 1919) to C and EC Tanzania on both sides of the Rift; limits have yet to be resolved.

SYNONYMS: *reptans, rumruti, somalicus.*

COMMENTS: Under the name *somalicus* Thomas, 1903*a*, *A. neumanni* has been treated as a species (Allen, 1939; Dollman, 1911; Ellerman, 1941; Hollister, 1919), and with an exception or two (e.g., Misonne, 1974), retains that status (Rousseau, 1983; Yalden et al., 1976). Both reliable published records (see above) and specimens we examined document the northern and central range of *A. neumanni*. Known southern limits of the species are defined by samples from the Mawele region south of Tabora (Mwanasomano's, 31 mi S Tabora) in C Tanzania and southeast of there at Kilosa in EC Tanzania (specimens in Museum of Comparative Zoology, Harvard). The species is sympatric with both *A. niloticus* and *A. nairobae* (see those accounts). Allen and Loveridge (1933:117) recorded three specimens from Mwanza on the southern margin of Lake Victoria in Tanzania under the name *muansae* thinking they represented topotypes of that form, but the holotype of *muansae* from Mwanza is very large with dark pelage (Matschie, 1911) and is an example of *A. niloticus*, while their sample is clearly the much smaller and paler *A. neumanni*; Mwanza is probably another site of sympatry.

Arvicanthis niloticus (Desmarest, 1822). Mammalogie *in* Encyclop. Méth., 2:281.

TYPE LOCALITY: Egypt.

DISTRIBUTION: Specimens we examined were from S Mauritania, Senegal, and Gambia east through Sierra Leone, Ivory Coast, Ghana, Burkina Faso, Togo, Benin, Nigeria, S Niger, S Chad, Sudan, and Egypt to the western half of Ethiopia (to and in the Rift Valley); south through NE Zaire, Uganda, S Burundi, Kenya and Tanzania west of the Rift Valley, into E Zamibia, where the population is isolated from the nearest one, which is in SW Tanzania (Ansell, 1978); there is a record from N Malawi (1 skull only, living animals never found; Ansell and Dowsett, 1988); and the species occurs in SW Arabia.

SYNONYMS: *ansorgei, centralis, centrosus, dembeensis, discolor, jebelae, kordofanensis, luctosus, variegatus* var. *major, mearnsi, variegatus* var. *minor, mordax, muansae, naso, nubilans, occidentalis, ochropus, pelliceus, raffertyi, reichardi, rhodesiae, rossii, rubescens, rufinus, setosus, solatus, tenebrosus, testicularis, variegatus, zaphiri.*

COMMENTS: We include *dembeensis,* which has been recognized as a distinct species occurring in Ethiopia and nearby regions (Yalden et al., 1976; Demeter, 1983). Thomas (1910*b*) had once associated *dembeensis* with *Desmomys,* noting that it had already been referred to *Mus, Arvicanthis, Golunda, Pelomys,* and *Oenomys.* Later, however, Thomas (1928*a*) referred *dembeensis* to *Arvicanthis,* a view confirmed by Osgood (1936) and reinforced by Dieterlen (1974). The name *lacernatus* was once used in place of *dembeensis* for the Ethiopian samples of *Arvicanthis* (Allen, 1939; Corbet and Yalden, 1972; Osgood, 1936) but cranium of the holotype is an example of *Meriones* (Yalden et al., 1976). *Arvicanthis dembeensis* was thought to be a separate species because Ethiopian samples were originally contrasted by Corbet and Yalden (1972, under the name *lacernatus*) with *abyssinicus,* which they treated as a subspecies of *A. niloticus.* However, *abyssinicus* is a distinctive species related to *A. blicki* and not to *A. niloticus.* We observed that the diagnostic features attributed to *dembeensis* also described samples of *Arvicanthis* from throughout Africa west and south of Ethiopia.

Sympatry with other species of *Arvicanthis* has been recorded at several places in Uganda (Delany, 1975; Dollman, 1911; Hollister, 1919), but we cannot distinguish two kinds at the localities. The only verifiable sympatry is between the large-bodied *A. niloticus* and the small-bodied *A. neumanni* in the Rift Valley of Ethiopia (*niloticus* was recorded as either *dembeensis* or *lacernatus, neumanni* as *somalicus;* Corbet and Yalden, 1972; Demeter, 1983) and in the Mawele region (Mwanasomano's, 30 mi S Tabora) In C Tanzania (documented by specimens in the Museum of Comparative Zoology, Harvard). Allopatry or parapatry, however, is usual. In Ethiopia, *A. niloticus* occurs up to about 2000 ft and is replaced at higher elevations by *A. abyssinicus* and *A. blicki* (Rupp, 1980; Yalden et al., 1976); in Kenya and Tanzania, *A. niloticus* ranges west of the E Rift Valley and is allopatric with *A. nairobae,* found in and east of the Rift.

Our definition of *A. niloticus* is unsatisfactory. There is appreciable intra- and interpopulational variation in body size, pelage coloration, and cytological and biochemical traits, and the significance of this variation should be assessed by a critical systematic revision; more than one species may be present in the complex, as suggested by morphometric (Rousseau, 1983), chromosomal (Viegas-Pequignot et al., 1983; Volobouev et al., 1987) and electrophoretic (Kaminski et al., 1984) analyses. Furthermore, Corbet (1984) questioned including the Arabian *naso* with *A. niloticus.* Capanna and Civitelli (1988) recorded a karyotype with 2N=44 from a sample they identified as *A. niloticus* from Somalia; this contrasts with the range of 56-62 usually reported among samples of the species.

Bandicota Gray, 1873. Ann. Mag. Nat. Hist., ser. 4, 12:418.

TYPE SPECIES: *Mus giganteus* Hardwicke, 1804 (= *Mus indicus* Bechstein, 1800).

SYNONYMS: *Gunomys.*

COMMENTS: *Nesokia* has been considered the closest relative of *Bandicota* (Misonne, 1969: Niethammer, 1977; Radtke and Niethammer, 1984/85; Wroughton, 1908), but outside of that alliance, its phylogenetic position among murines is debatable. Misonne (1969) was unsure where to place *Bandicota* except that it was not related to *Rattus;* but Niethammer (1977) and Gemmeke and Niethammer (1984) concluded from morphological, chromosomal, and biochemical data that *Rattus* is close to *Bandicota.*

Bandicota bengalensis (Gray and Hardwicke, 1833). Illustr. Indian Zool., pl. 21.

TYPE LOCALITY: India, Bengal.

DISTRIBUTION: Approximate natural range: Sri Lanka, peninsular India to Pakistan, Kashmir, Nepal, NE India (Assam), Bangladesh, and Burma. Introduced to Penang Isl off the W coast of Malay Peninsula (Chasen, 1936), Sumatra and Java (Musser and Newcomb, 1983), and Saudi Arabia (Kock et al., 1990).

SYNONYMS: *barclayanus, blythianus, daccaensis, dubius, gracilis, insularis, kok, lordi, morungensis, plurimammis, providens, sindicus, sundavensis, tarayensis, varillus, varius, wardi.*

COMMENTS: The most morphologically divergent of the species now placed in *Bandicota;* so impressive are the differences that *bengalensis* has been placed in its own genus, *Gunomys* (see Wroughton, 1908). Geographic variation and one view of subspecies presented by Agrawal and Chakraborty (1976). Chromosomal data reported by Sharma and Raman (1971, 1973) and Dubey and Raman (1992). Lekagul and Felten (1989) recognized *varius* as a distinct species in Thailand, but that record was based on specimens of *B. savilei* (in Senckenberg Museum).

Bandicota indica (Bechstein, 1800). *In* Pennant, Allgemeine Ueber Vierfuss. Thiere, 2:497.

TYPE LOCALITY: India, Pondicherry.

DISTRIBUTION: Sri Lanka, peninsular India north to Nepal, NE India (Assam), Burma, S China (Yunnan and Hong Kong Isl), Taiwan, Thailand, Laos, and Vietnam. Introduced into Kedah and Perlis regions of Malay Peninsula (Harrison, 1956; J. T. Marshall, Jr., 1977*a*) as well as Java (Musser and Newcomb, 1983). Its spotty distribution may reflect other geographic introductions (Taiwan for example); "since it is commensal, large, and delicious to eat, this bandicoot may have been spread by man in comparatively recent times" (J. T. Marshall, Jr., 1977*a*:428).

SYNONYMS: *bandicota* (of Bechstein, 1800; see Ellerman, 1941), *elliotanus, eloquens, gigantea, jabouillei, kagii, macropus, malabarica, mordax, nemorivaga, perchal, setifera, siamensis, sonlaensis, taiwanus.*

COMMENTS: Morphologically more closely related to *B. savilei* than to *B. bengalensis*, but based on gel electrophoretic comparisons closer to *Nesokia* than to other species of *Bandicota* (Radtke and Niethammer, 1984/85). Chromosomal data for Thai samples provided by Markvong et al. (1973). A careful systematic revision is necessary to assess the significance of morphological and biochemical variation among samples. Pradhan et al. (1989), for example, argued that *B. gigantea* is specifically distinct from *B. indica,* and indicated they are pursuing a taxonomic revision of the genus.

Bandicota savilei Thomas, 1916. J. Bombay Nat. Hist. Soc., 24:641.

TYPE LOCALITY: Burma, Mt Popa, about 2500 ft.

DISTRIBUTION: Recorded from E Burma, Thailand, and Vietnam; probably also occurs in S Laos and Cambodia.

SYNONYMS: *bangchakensis, curtata, giaraiensis, hichensis.*

COMMENTS: Occurs sympatrically with *B. indica* in C and S Thailand and S Vietnam; found with *B. bengalensis* in Burma where *B. savilei* lives in fields and *B. bengalensis* in village houses. *Bandicota savilei* is a very distinctive species as Thomas (1916*d*) pointed out when he described it; unfortunately it was later incorrectly treated as a subspecies of *B. indica* (Ellerman, 1961; Ellerman and Morrison-Scott, 1951). Specimens are usually misidentified as *B. bengalensis*, but in external, cranial, and dental morphology, *B. savilei* is linked to *B. indica*. Electrophoretic data, however, indicated a closer relationship with *B. bengalensis* than with *B. indica* (Radtke and Niethammer, 1984/85). Chromosomal information reported by Markvong et al. (1973). Lekagul and Felten (1989) described *bangchakensis* as a species; *hichensis* and *giaraiensis* were proposed as subspecies of *B. bengalensis* (Dao, 1961; Dao and Cao, 1990).

Batomys Thomas, 1895. Ann. Mag. Nat. Hist., ser. 6, 16:162.

TYPE SPECIES: *Batomys granti* Thomas, 1895.

SYNONYMS: *Mindanaomys.*

COMMENTS: A Philippine Old Endemic in the same monophyletic group as *Crateromys* and *Carpomys* (Musser and Heaney, 1992). An undescribed species is found on Dinagat Isl. Actual distribution in archipelago and number of species in genus yet to be determined.

Batomys dentatus Miller, 1911. Proc. U.S. Natl. Mus., 38:400

TYPE LOCALITY: Philippines, N Luzon Isl, Benguet, Haights-in-the-Oaks, 7000 ft.

DISTRIBUTION: Known only from the type locality.

COMMENTS: Known only by the holotype.

Batomys granti Thomas, 1895. Ann. Mag. Nat. Hist., ser. 6, 16:162.
TYPE LOCALITY: Philippines, N Luzon Isl, Mountain Prov, Mt Data, 7000 ft.
DISTRIBUTION: Luzon Isl, Mt Data in north and Mt Isarog in SE peninsula.
STATUS: IUCN - Indeterminate.
COMMENTS: Probably most closely related to *B. salomonseni* than to other species of *Batomys*.

Batomys salomonseni (Sanborn, 1953). Vidensk. Medd. Nat. Foren. Kjobenhavn, 115:287.
TYPE LOCALITY: Philippines, Mindanao Isl, Bukidnon Prov, Mt. Katanglad, 1600 m.
DISTRIBUTION: Islands of Mindanao, Biliran, and Leyte at low and middle elevations.
COMMENTS: Originally described as the only species in *Mindanaomys* by Sanborn (1953), but that genus was considered inseparable from *Batomys* by Misonne (1969) and Musser and Heaney (1992). Probably more closely related to *B. granti* than to *B. dentatus*.

Berylmys Ellerman, 1947. Proc. Zool. Soc. Lond., 117:261.
TYPE SPECIES: *Epimys manipulus* Thomas, 1916.
COMMENTS: Originally proposed as a subgenus of *Rattus* by Ellerman (1947-1948), but elevated to generic rank in a revision by Musser and Newcomb (1983) which recorded taxonomic history of names and groups associated with *Berylmys*, and reported past evolutionary histories of species (which have been centered in Indochina), finding that *Berylmys* was dentally similar to *Niviventer, Maxomys*, and *Leopoldamys*, but shared some derived cranial characters with *Rattus* and that phylogenetic relationships were still unknown. Sperm morphology united *Berylmys* with *Sundamys, Rattus*, and *Leopoldamys* (Breed and Yong, 1986), but that union was based on a shared spermatozoal form which is probably primitive.

Berylmys berdmorei (Blyth, 1851). J. Asiat. Soc. Bengal, 20:173.
TYPE LOCALITY: S Burma (S Tenasserim), Mergui.
DISTRIBUTION: S Burma, N and SE Thailand, Cambodia, N Laos, and S Vietnam.
SYNONYMS: *magnus, mullulus*.
COMMENTS: *B. berdmorei* is the only species of *Berylmys* recorded from a small island (Con Son, S Vietnam) off continental Indochina.

Berylmys bowersi (Anderson, 1879). Anat. Zool. Res., Yunnan, p. 304.
TYPE LOCALITY: China, Yunnan Prov., Kakhyen Hills, Hotha, 4500 ft.
DISTRIBUTION: NE India, N and C Burma, S China (Yunnan, Guangxi, Fujiau, and S Anhui; Liu et al., 1985), N and peninsular Thailand, N Laos, N Vietnam, Malay Peninsula, and NW Sumatra (Medan).
SYNONYMS: *ferreocanus, kennethi, lactiventer, latouchei, totipes, wellsi*.
COMMENTS: Not known to occur on small islands off continental margin; *bowersi* is the only species of *Berylmys* found on Malay Peninsula and an island of the Sunda Shelf. Spermatozoal morphology and its significance documented by Breed and Yong (1986).

Berylmys mackenziei (Thomas, 1916). J. Bombay Nat. Hist. Soc., 24:40.
TYPE LOCALITY: Burma, Chin Hills, 50 mi west of Kindat.
DISTRIBUTION: NE India (Assam), C and S Burma, China (Szechwan), and S Vietnam.
SYNONYMS: *feae*.

Berylmys manipulus (Thomas, 1916). J. Bombay Nat. Hist. Soc., 24, 3:413.
TYPE LOCALITY: C Burma, Kabaw Valley, Kampat, 20 mi west of Kindat.
DISTRIBUTION: India (Assam), N and C Burma, and China (Yunan).
SYNONYMS: *kekrimus*.

Bullimus Mearns, 1905. Proc. U.S. Natl. Mus., 28:450.
TYPE SPECIES: *Bullimus bagobus* Mearns, 1905.
COMMENTS: Although described as a distinctive genus by Mearns, *Bullimus* has been treated as a subgenus of *Rattus* in most taxonomic works (e.g., Ellerman, 1941; Misonne, 1969; Simpson, 1945) until generically reinstated by Musser (1982*c*) and Musser and Newcomb (1983) and rediagnosed by Musser and Heaney (1992), who also documented taxonomic history and past associations with other murine genera. Belongs to the group of Philippine New Endemics, but its closest phylogenetic relative has yet to be resolved (Musser and Heaney, 1992). Genus needs critical

systematic revision, number of actual species is uncertain, those listed below are provisional (Musser and Heaney, 1992; Musser and Newcomb, 1983).

Bullimus bagobus Mearns, 1905. Proc. U.S. Natl. Mus., 28:450.

TYPE LOCALITY: Philippines, Mindanao Isl, Mt Apo, Todaya, 4000 ft.

DISTRIBUTION: Philippines (known from islands of Samar, Calicoan, Leyte, Dinagat, Siargao, Mindanao, Bohol, and Maripipi, but probably occurs on other islands in the archipelago).

SYNONYMS: *barkeri, rabori*.

COMMENTS: See comments under *luzonicus*.

Bullimus luzonicus (Thomas, 1895). Ann. Mag. Nat. Hist., ser. 6, 16:163.

TYPE LOCALITY: Philippines, Luzon Isl, Lepanto, Mt Data, 8000 ft.

DISTRIBUTION: Known only from Luzon.

COMMENTS: Whether *luzonicus* and the insular samples of what are now called *bagobus* are really different entities or represent the same species remains to be determined.

Bunomys Thomas, 1910. Ann. Mag. Nat. Hist., ser. 8, 6:508.

TYPE SPECIES: *Mus coelestis* Thomas, 1896.

SYNONYMS: *Bunomys* (Grandidier, 1905, not Thomas, 1910), *Frateromys*.

COMMENTS: An endemic of Sulawesi. Recognized as a genus by Tate (1936), but included in *Rattus* by Ellerman (1941), Simpson (1945), Laurie and Hill (1954), and Misonne (1969); treated as separate genus by Musser and Newcomb (1983), Musser (1984, 1987, 1991), and Corbet and Hill (1991). Phylogenetic relationships unresolved, but part of Sulawesi New Endemics; spermatozoal morphology distinctive although some species resemble that in species of *Rattus*, but resemblance reflects shared primitive features (Breed and Musser, 1991). Kitchener et al. (1991*a*) considered *Bunomys* a very close relative of *Paulamys*, a Flores endemic; both are probably ecological if not morphological equivalents (Musser et al., 1986). Geographic and altitudinal distributions of species, along with forest types, summarized by Musser and Dagosto (1987), Musser (1991), and Musser and Holden (1991). Grandidier (1905:50) proposed *Bunomys* for a Madagascar rat, and Sody (1941:260) used *Frateromys*, but both are *nomina nuda*.

Bunomys andrewsi (J. A. Allen, 1911). Bull. Am. Mus. Nat. Hist., 30:366.

TYPE LOCALITY: Indonesia, NE Sulawesi, Buton Isl.

DISTRIBUTION: Sulawesi; central core, SE peninsula, possibly lowlands of SW peninsula; limits unresolved.

SYNONYMS: *adspersus, inferior*.

COMMENTS: Recognized as a distinct species in the "*Rattus chrysocomus* Group" by Tate (1936); placed in the "*chrysocomus* group" of *Rattus* by Ellerman (1941) then in the "*coelestis* group" of *Rattus* by Ellerman (1949*a*); treated as a species in subgenus *Rattus* by Laurie and Hill (1954); but included in *Bunomys* by Musser (1981*c*) and Musser and Newcomb (1983). In morphology and pelage features, *B. andrewsi* is most closely related to *B. fratrorum* of the NW peninsula and *B. heinrichi* in the SW arm of Sulawesi.

Bunomys chrysocomus (Hoffmann, 1887). Abh. Zool. Anthrop.-Ethnology Mus. Dresden, 3:17

TYPE LOCALITY: Indonesia, N Sulawesi, Minahassa.

DISTRIBUTION: Sulawesi; throughout island generally between 200 and 1500 m, but reaches 2200 m in some places.

SYNONYMS: *brevimolaris, koka, nigellus, rallus*.

COMMENTS: Placed in the "*Rattus chrysocomus* Group" by Tate (1936), in subgenus *Rattus* by Ellerman (1941:190), then subgenus *Maxomys* of *Rattus* by Ellerman (1949*a*:70) and Laurie and Hill (1954:118); but treated as a species of *Bunomys* by Musser (1981*c*, 1991) and Musser and Newcomb (1983).

Bunomys coelestis (Thomas, 1896). Ann. Mag. Nat. Hist., ser. 6, 18:248.

TYPE LOCALITY: Indonesia, SW Sulawesi, Bonthain Peak (Gunung Lampobatang), 6000 ft.

DISTRIBUTION: Sulawesi; known only from the slopes of Gunung Lampobatang.

COMMENTS: Listed as a species of *Bunomys* by Tate (1936), as a species of *Rattus* in the "*coelestis* group" by Ellerman (1941, 1949*a*), but as a species in subgenus *Rattus* by

Laurie and Hill (1954). Musser (1981*c*) and Musser and Newcomb (1983) included *coelestis* in *B. chrysocomus*, but Musser (1991) and Musser and Holden (1991) treated it as a species restricted to montane forest formations on south end of SW peninsula.

Bunomys fratrorum (Thomas, 1896). Ann. Mag. Nat. Hist., ser. 6, 18:246.
TYPE LOCALITY: Indonesia, NE Sulawesi, Rurukan, 3500 ft.
DISTRIBUTION: Sulawesi; NE peninsula only.
COMMENTS: Despite Thomas' clear declaration that *fratrorum* was distinct from *chrysocomus*, an evaluation also supported by Tate (1936) and Sody (1941), *fratrorum* was listed as a synonym of *chrysocomus* by Ellerman (1949*a*) and Laurie and Hill (1954). The two are sympatric on the NE peninsula (Musser, 1970*d*, 1981*c*, 1991; Musser and Newcomb, 1983). Sody (1941) made *fratrorum* the type species of *Frateromys*.

Bunomys heinrichi (Tate and Archbold, 1935). Am. Mus. Novit., 802:6.
TYPE LOCALITY: Indonesia, SW Sulawesi, Gunung Lampobatang, Lambasang, 1100 m.
DISTRIBUTION: Sulawesi; known only from montane forest formations on the upper slopes of Gunung Lampobatang.
COMMENTS: Originally described as a subspecies of *Rattus penitus* and listed that way by Ellerman in 1941, but as a subspecies of *Rattus adspersus* in 1949*a*, which was followed by Laurie and Hill (1954). Included within *Bunomys penitus* by Musser (1981*c*) and Musser and Newcomb, 1983), and within *Frateromys penitus* by Sody (1941), but treated as a species by Musser (1991) and Musser and Holden (1991).

Bunomys penitus (Miller and Hollister, 1921). Proc. Biol. Soc. Washington, 34:72.
TYPE LOCALITY: Indonesia, C Sulawesi, Gunung Lehio, above 6000 ft.
DISTRIBUTION: Sulawesi; central core and SE peninsula only.
SYNONYMS: *sericatus*.
COMMENTS: Described as a species of *Rattus*, then listed as a species in the *chrysocomus* group of *Rattus* by Tate (1936) and Ellerman (1941), treated as a subspecies of *Rattus adspersus* by Laurie and Hill (1954), and included in *Frateromys* by Sody (1941). However, *Bunomys penitus* is distinctive in morphology and confined to montane forest formations (Musser, 1987, 1991; Musser and Dagosto, 1987; Musser and Holden, 1991; Musser and Newcomb, 1983).

Bunomys prolatus Musser, 1991. Am. Mus. Novit., 3001:4.
TYPE LOCALITY: Indonesia, C Sulawesi, Gunung Tambusisi, 6000 ft.
DISTRIBUTION: Known only from 6000 ft on Gunung Tambusisi, Sulawesi.
COMMENTS: Morphologically most like *Bunomys chrysocomus*, with which it is nearly sympatric (both were caught at the same elevation but in different habitats; Musser, 1991).

Canariomys Crusafont-Pairo and Petter, 1964. Mammalia, 28, Suppl.:607.
TYPE SPECIES: *Canariomys bravoi* Crusafont-Pairo and Petter, 1964 (fossil).
COMMENTS: *Canariomys*, along with *Malpaisomys*, represent the known murine fauna that was endemic to the Canary Islands, and is apparently now extinct (Boye et al., 1992; Hutterer et al., 1988; Lopez-Martinez and Lopez-Jurado, 1987).

Canariomys tamarani Lopez-Martinez and Lopez-Jurado, 1987. Donana, Publ. ocas., 2:10.
TYPE LOCALITY: Canary Islands, Gran Canaria Isl, "La Aldea de San Nicolas de Tolentino" (Lopez-Martinez and Lopez-Jurado (1987:12).
DISTRIBUTION: Canary Islands, Gran Canaria Isl.
COMMENTS: An endemic of Gran Canaria Isl known only by specimens from the Holocene to pre-hispanic epoch (about 2000 year before present). Its congener, *C. bravoi* is documented by Pliocene-Pleistocene fossils from Tenerife Isl in the Canary group (Crusafont-Pairo and Petter, 1964). A cladistic analysis by Lopez-Martinez and Lopez-Jurado (1987) suggested that *C. bravoi* and *C. tamarani* were in the same monophyletic group, one related to the African *Pelomys-Arvicanthis* assemblage. However, a cladistic analysis by Hutterer et al. (1988) indicated that each species of *Canariomys* was in a different monophyletic cluster. Careful re-examination of the material will be necessary to resolve these contrasting views.

Carpomys Thomas, 1895. Ann. Mag. Nat. Hist., ser. 6, 16:161.
TYPE SPECIES: *Carpomys melanurus* Thomas, 1895.
COMMENTS: A Philippine Old Endemic in the same monophyletic group as *Crateromys* and *Batomys* (Musser and Heaney, 1992).

Carpomys melanurus Thomas, 1895. Ann. Mag. Nat. Hist., ser. 6, 16:162.
TYPE LOCALITY: Philippines, N Luzon Isl, Mt Data, 7000-8000 ft.
DISTRIBUTION: Known only from Mt. Data, but probably occurs in other highlands of N Luzon.

Carpomys phaeurus Thomas, 1895. Ann. Mag. Nat. Hist., ser. 6, 16:162.
TYPE LOCALITY: Philippines, N Luzon Isl, Mt Data, 7000-8000 ft.
DISTRIBUTION: Known only from Mt Data and Mt Kapilingan (Sanborn, 1952*a*), but likely occurs elsewhere in mountains of N Luzon.

Celaenomys Thomas, 1898. Trans. Zool. Soc. London, 14(6):390.
TYPE SPECIES: *Xeromys silaceus* Thomas, 1895.
COMMENTS: A Philippine Old Endemic in the same monophyletic group as *Chrotomys* (Musser and Heaney, 1992).

Celaenomys silaceus (Thomas, 1895). Ann. Mag. Nat. Hist., ser. 6, 16:161.
TYPE LOCALITY: Philippines, N Luzon Isl, Mt Data, 8000 ft.
DISTRIBUTION: Known only from N Luzon on Mt Data and Haights-in-the-Oaks (Sanborn, 1952*a*), but probably lives in montane forest formations elsewhere on Luzon.

Chiromyscus Thomas, 1925. Proc. Zool. Soc. Lond., 1925:503.
TYPE SPECIES: *Mus chiropus* Thomas, 1891.
COMMENTS: Reviewed by Musser (1981*b*), who regarded *Chiromyscus* as closely related to *Niviventer.*

Chiromyscus chiropus (Thomas, 1891). Ann. Mus. Civ. Stor. Nat. Genova, ser. 2, 10:884.
TYPE LOCALITY: E Burma, Karin Hills.
DISTRIBUTION: E Burma, N Thailand, Vietnam, and C Laos (Musser, 1981*b*).

Chiropodomys Peters, 1869. Monatsb. K. Preuss. Akad. Wiss. Berlin, p. 448.
TYPE SPECIES: *Mus gliroides* Blyth, 1856.
SYNONYMS: *Insulaemus.*
COMMENTS: Revised by Musser (1979). Among Asian murines, phylogenetic position of *Chiropodomys* is ambiguous. Musser and Newcomb (1983) suggested a phylogenetic link between *Chiropodomys* and *Hapalomys,* and a weaker tie to *Haeromys.* Spermatozoal morphology of *Chiropodomys* resembles that described for species of *Hapalomys, Maxomys,* and *Haeromys,* and is also similar to *Mus* (Breed and Musser, 1991; Breed and Yong, 1986). Musser (1979) and Musser and Newcomb (1983) recorded other published opinions about relationships of *Chiropodomys.* No data supports the inclusion of *Chiropodomys* within the Phloeomyinae—along with *Coryphomys, Lenomys, Pogonomys, Mallomys, Phloeomys,* and *Crateromys*—which is where Simpson (1945) listed it.

Chiropodomys calamianensis (Taylor, 1934). Monogr. Bur. Sci. Manila, 30:470.
TYPE LOCALITY: Philippines, Calamian Isls, Busuanga Isl, Minuit, sea level.
DISTRIBUTION: Busuanga, Balabac, and Palawan Isls.
COMMENTS: Originally described as the type species of *Insulaemus* (Taylor, 1934). In morphology and geographic continuity, *C. calamianensis* is most closely related to *C. major* of mainland Borneo (Musser, 1979).

Chiropodomys gliroides (Blyth, 1856). J. Asiat. Soc. Bengal, 24:721.
TYPE LOCALITY: India (Assam), Khasi Hills, Cherrapunji.
DISTRIBUTION: Documented from China (Guangxi and Yunnan), India (Assam), Burma, Thailand, Laos, Vietnam, Malay Peninsula, S Sumatra, Pulau Nias, Kepulauan Tujuh, Kepulauan Natuna, Java, and Bali (Musser, 1979; Wu and Deng, 1984), but probably occurs on other small islands of the Sunda Shelf, in Cambodia, and in other provinces of S China.
SYNONYMS: *ana, jingdongensis, niadis, peguensis, penicillatus.*

COMMENTS: Variation among some samples in certain morphological features is significantly correlated with geography both on mainland Indochina and islands of the Sunda Shelf (Musser, 1979). The geographic variation in Indochina is reflected in Wu and Deng's (1984) description of *C. jingdongensis*, based on a small sample from Yunnan, in which a few cranial dimensions average slightly larger than most samples of *C. gliroides* from elsewhere in its range, but are not otherwise significantly different. Chromosomal data reported by Yong (1973, 1983), J. T. Marshall, Jr. (1977*a*), and Tsuchiya et al. (1979).

Chiropodomys karlkoopmani Musser, 1979. Bull. Am. Mus. Nat. Hist., 162(6):389.

TYPE LOCALITY: Indonesia, Kepulauan Mentawai, Pulau Pagai Utara.

DISTRIBUTION: Known only from Pagai (Musser, 1979), and Siberut (Jenkins and Hill, 1982) in the Mentawai Isls.

COMMENTS: The largest in body size of any described *Chiropodomys*, and possibly most closely related to *C. major*. Part of the rodent fauna endemic to the Mentawi Archipelago (see account of *Leopoldamys siporanus*).

Chiropodomys major Thomas, 1893. Ann. Mag. Nat. Hist., ser. 6, 11:344.

TYPE LOCALITY: Malaysia (Borneo), Sarawak, Sadong.

DISTRIBUTION: Borneo: recorded only from Sarawak and Sabah (Musser, 1979), but probably also occurs in Kalimantan.

SYNONYMS: *legatus, pictor.*

COMMENTS: Most closely related to *C. calamianensis* in morphology and geography.

Chiropodomys muroides Medway, 1965. J. Malay. Branch R. Asiat. Soc., 36(3):133.

TYPE LOCALITY: Malaysia (Borneo), Sabah, Gunung Kinabalu, Bundu Tuhan, 4000 ft.

DISTRIBUTION: Known only by a few specimens from Gunung Kinabalu and N Kalimantan, but probably occurs elsewhere on Borneo.

COMMENTS: The smallest in body size of any known species of *Chiropodomys;* its closest phylogenetic allies may be *C. gliroides* and *C. pusillus*.

Chiropodomys pusillus Thomas, 1893. Ann. Mag. Nat. Hist., ser. 6, 11:345.

TYPE LOCALITY: Malaysia (Borneo), Sabah, Gunung Kinabalu, 1000 ft.

DISTRIBUTION: Recorded only from Sabah, Sarawak, and S Kalimantan (Musser, 1979), but probably occurs throughout Borneo.

COMMENTS: Thomas described *pusillus* as a species, but Musser (1979) treated it as a distinctive subspecies of *C. gliroides*. However, the few known specimens of *pusillus* represent a population of small-bodied mice in which the range of variation of most dimensions are outside of range recorded for all other samples of *C. gliroides*.

Chiruromys Thomas, 1888. Proc. Zool. Soc. Lond., 1888:237.

TYPE SPECIES: *Chiruromys forbesi* Thomas, 1888.

COMMENTS: Member of the New Guinea Old Endemics. Chromosomal and morphometric study, but not a systematic revision, produced by Dennis and Menzies (1979). Closest phylogenetic relative was thought to be *Pogonomys* (e.g., Tate, 1951), but is not that genus, and remains to be discovered. Flannery (1990*b*) provided photographs of the species and summarized distributional and biological information.

Chiruromys forbesi Thomas, 1888. Proc. Zool. Soc. Lond., 1888:239.

TYPE LOCALITY: Papua New Guinea, Central Prov., Astrolabe Range, Sogeri, 1500 ft.

DISTRIBUTION: New Guinea; endemic to mainland of SE Papua New Guinea (sea level to 700 m; on N side not known west of Oomsis in valley of Markham River [see map in Brass, 1959], and on S side not recorded west of type locality) and d'Entrecasteaux Isls (Goodenoough, Fergusson, and Normanby). Dennis and Menzies (1979) included the Louisiade Isls, but samples in the American Museum of Natural History of *Chiruromys* from there are not *forbesi*.

SYNONYMS: *major, mambatus, pulcher, satisfactus, shawmayeri, vulturnus.*

COMMENTS: Significance of the appreciable geographic variation in body size and other traits needs to be assessed in a critical systematic review of the species.

Chiruromys lamia (Thomas, 1897). Ann. Mus. Civ. Stor. Nat. Genova, 18:615.

TYPE LOCALITY: Papua New Guinea, Central Prov., lower Kemp Welch River, Ighibirei (see Thomas, 1897*a*:4).

DISTRIBUTION: New Guinea; known only from mainland of SE Papua New Guinea, sea level to 2300 m (see maps in Dennis and Menzies, 1979; Flannery, 1990*b*).
SYNONYMS: *kagi*.

Chiruromys vates (Thomas, 1908). Ann. Mag. Nat. Hist., ser. 8, 2:495.
TYPE LOCALITY: Papua New Guinea, Central Prov., Upper St. Joseph's River (= Angabunga River), Madeu, 50 m NE Hall Sound, 2000-3000 ft.
DISTRIBUTION: Papua New Guinea; S side of Central Cordillera, from Lake Murray area of Western Prov. east to region of type locality, sea level to 1500 m (see map in Flannery, 1990*b*).

Chrotomys Thomas, 1895. Ann. Mag. Nat. Hist., ser. 6, 16:161.
TYPE SPECIES: *Chrotomys whiteheadi* Thomas, 1895.
COMMENTS: A Philippine Old Endemic most closely related to *Celaenomys*, and unrelated to shrew rats and shrew mice endemic to Australia and New Guinea (Musser and Heaney, 1992). Actual distribution in archipelago and number of species in genus still unknown.

Chrotomys gonzalesi Rickart and Heaney, 1991. Proc. Biol. Soc. Washington 104:389.
TYPE LOCALITY: Philippines, SE Luzon Isl, Camarines Sur Prov., W slope Mt Isarog, 4 km N, 21 km E Naga, 1350 m.
DISTRIBUTION: Known only from W slope Mt Isarog, Luzon, between 1350 and 1750 m.
COMMENTS: Morphology, ecology, and comparisons with other species of *Chrotomys* provided by Rickart and Heaney (1991) and Rickart et al. (1991).

Chrotomys mindorensis Kellogg, 1945. Proc. Biol. Soc. Washington 58:123.
TYPE LOCALITY: Philipines, Mindoro Isl, 3 mi SSE of San Jose (Central), 200 ft.
DISTRIBUTION: Lowlands of Luzon and Mindoro.
COMMENTS: Considered specifically distinct from *C. whiteheadi* by Musser et al. (1982*b*).

Chrotomys whiteheadi Thomas, 1895. Ann. Mag. Nat. Hist., ser. 6, 16:161.
TYPE LOCALITY: Philippines, N Luzon Isl, Mt Data, 8000 ft.
DISTRIBUTION: Luzon; one specimen is recorded from Irisan, Benguet (see Hollister, 1913*b*); all other examples are from the type locality.

Coccymys Menzies, 1990. Science in New Guinea, 16:132.
TYPE SPECIES: *Pogonomelomys ruemmleri* Tate and Archbold, 1941.
COMMENTS: A New Guinea Old Endemic with uncertain relationships to *Pogonomelomys*, *Melomys*, and *Uromys*.

Coccymys albidens (Tate, 1951). Bull. Am. Mus. Nat. Hist., 97:286.
TYPE LOCALITY: New Guinea, Irian Jaya, 15 mi N Mt Wilhelmina, Lake Habbema, 3225 m.
DISTRIBUTION: New Guinea, Irian Jaya; known only from the type locality and 9 km NE Lake Habbema, 2800 m.
STATUS: IUCN - Insufficiently known.
COMMENTS: Still represented only by the six specimens in the type series collected in 1938. Originally described as a species of *Melomys* (Tate, 1951), but its diagnostic traits are unlike those defining species within that genus, and similar to those characterizing *ruemmleri*. This conclusion reached independently by Flannery (1990*b*) and Menzies (1990), although neither of them formally allocated *albidens* and *ruemmleri* to the same genus.

Coccymys ruemmleri (Tate and Archbold, 1941). Am. Mus. Novit., 1101:6.
TYPE LOCALITY: New Guinea, Irian Jaya, N slope Mt Wilhelmina, Lake Habbema, 3225 m.
DISTRIBUTION: New Guinea; Central Cordillera from Mt Wilhelmina in Irian Jaya to Mt Kaindi in Papua New Guinea, above 1900 m.
SYNONYMS: *shawmayeri*.
COMMENTS: The species *ruemmleri* was originally described as a *Pogonomelomys* (Tate and Archbold, 1941), but its morphology was so different from the other species in the genus that Tate (1951) placed it in a group within *Pogonomelomys* separate from *mayeri* and *bruijnii*, the species considered typical of the genus. The distinctivness of *ruemmleri* was reinforced by Lidicker's (1968) study of phallic morphology; others have noted that *ruemmleri* was not part of the same monophyletic group containing

the other species of *Pogonomelomys* (e.g., Flannery, 1990*b*). Finally, Menzies (1990) made *ruemmleri* the type species of *Coccymys*. Before the reports of Lidicker, Flannery, and Menzies, the unique character of *ruemmleri* had been ascertained by Jack Mahoney, who died before he could finish his revision of the group. The form *shawmayeri* was described by Hinton (1943) as a remarkable species of *Rattus* unlike any of the New Guinea species and possibly closely related to Nepal and Sikkim endemics.

Colomys Thomas and Wroughton, 1907. Ann. Mag. Nat. Hist., ser. 7, 19:379.
TYPE SPECIES: *Colomys goslingi* Thomas and Wroughton, 1907.
SYNONYMS: *Nilopegamys*.
COMMENTS: Hayman (1966) discussed the identity of *Nilopegamys* with *Colomys*.

Colomys goslingi Thomas and Wroughton, 1907. Ann. Mag. Nat. Hist., ser. 7, 19:380.
TYPE LOCALITY: Zaire, Uele River, Gambi.
DISTRIBUTION: Recorded from Liberia (Lofa), Cameroon, NE Angola, NW Zambia, Zaire, Ruanda, Uganda, E Kenya, S Sudan, and W Ethiopia; limits unknown.
SYNONYMS: *bicolor, denti, eisentrauti, plumbeus, ruandensis*.
COMMENTS: Study of external, cranial, and dental variation in context of taxonomic revision was provided by Dieterlen (1983). *Colomys goslingi* is a carnivore preying on limnetic macroinvertebrates and small vertebrates, and possesses special morphological and neurological adaptations as well as behavioral repitoire that are correlated with extracting such a diet from streams (Dieterlen and Statzner, 1981; Stephan and Dieterlen, 1982). The species is apparently restricted to banks of small flowing streams in tropical rain forest (Dieterlen, 1983), although it has been taken along streams in grassland far from forest (Hayman, 1966). Comparisons with Neotropical ichthyomyines provided by Voss (1988).

Conilurus Ogilby, 1838. Trans. Linn. Soc. Lond., 18:124.
TYPE SPECIES: *Conylurus constructor* Ogilby, 1838 (= *Hapalotis albipes* Lichtenstein, 1829).
SYNONYMS: *Conylurus, Hapalotis*.
COMMENTS: Member of the Australian Old Endemics (Musser, 1981*c*:167) that is phylogenetically closely allied to *Mesembriomys* and *Leporillus* (see Watts et al., 1992, and references therein); part of the group is also designated as the Conilurini where *Conilurus* is placed (Baverstock, 1984). Mahoney and Richardson (1988:154) cataloged taxonomic, distributional, and biological references.

Conilurus albipes (Lichtenstein, 1829). Darst. Säugeth., 6, 2 unno. text pages and pl. 29.
TYPE LOCALITY: SE Australia (Neuholland); see Mahoney and Richardson (1988:155).
DISTRIBUTION: Australia; once occurred in a strip from SE Queensland through coastal New South Wales and Victoria into SE South Australia (see Watts and Aslin, 1981:130).
STATUS: IUCN - Extinct.
SYNONYMS: *constructor, destructor*.
COMMENTS: For complete citation see Mahoney and Richardson (1988:155). Living *C. albipes* have not been encountered by naturalists for over a century and the species is apparently extinct (Mahoney and Richardson, 1988; Watts and Aslin, 1981).

Conilurus penicillatus (Gould, 1842). Proc. Zool. Soc. Lond., 1842:12.
TYPE LOCALITY: Australia, Northern Territory, Port Essington, sea shore.
DISTRIBUTION: Australia; "northern coastal area of the Northern Territory, adjacent islands [including Melville Island, Groote Eylandt, and the Pellew Group], and the extreme north-east of Western Australia" (Watts and Aslin, 1981:133). SC Papua New Guinea in Morehead region (Waithman, 1979).
SYNONYMS: *hemileucurus, melanura, melibius, randi*.
COMMENTS: Analyses of external morphology of glans penis and spermatozoal structure and form provided by Breed (1984), Breed and Sarafis (1978), and Morrissey and Breed (1982). Results from studies of chromosomal features (Baverstock et al., 1977*c*, 1983*b*), electrophoretic data (Baverstock et al., 1981), phallic morphology (Lidicker and Brylski, 1987), and dental traits (Misonne, 1969) supported the hypothesis that *C. penicillatus* is phylogenetically closely related to *Mesembriomys gouldii*, and microcomplement fixation data placed it in the same clade close to species of

Leporillus and *Mesembriomys* (see Watts et al., 1992). *Conilurus penicillatus* is one of the few Old Endemic Australian murines that also occurs in SC New Guinea where it is still represented by only two specimens (Flannery, 1990*b*:192).

Coryphomys Schaub, 1937. Verh. Naturf. Ges. Basel, 48:2.
TYPE SPECIES: *Coryphomys buhleri* Schaub, 1937.
COMMENTS: Simpson (1945) listed *Coryphomys*, along with *Lenomys*, *Pogonomys*, *Chiropodomys*, *Mallomys*, *Phloeomys*, and *Crateromys* in the Phloeomyinae, but no data supports such an allocation.

Coryphomys buehleri Schaub, 1937. Verh. Naturf. Ges. Basel, 48:2.
TYPE LOCALITY: Indonesia, Nusa Tenggara, W Timor, cave deposit near Nikiniki.
DISTRIBUTION: Known only from Timor.
COMMENTS: Known only by subfossil fragments collected at the type locality and from sites in E Timor (Glover, 1986). Related to endemic New Guinea murines and not to species of *Papagomys*, *Hooijeromys*, *Komodomys*, or *Paulamys* on Flores. *Coryphomys buehleri* is one of four species, each in its own genus, of giant rats (three have yet to be named and described) endemic to Timor; all are known only by subfossils (Glover, 1986).

Crateromys Thomas, 1895. Ann. Mag. Nat. Hist., ser. 6, 16:163.
TYPE SPECIES: *Phloeomys schadenbergi* Meyer, 1895.
COMMENTS: Reviewed by Musser and Gordon (1981) and Musser et al. (1985). A Philippine Old Endemic in the same monophyletic group as *Batomys* and *Carpomys* (Musser and Heaney, 1992). Except for a suggested but distant phylogenetic link to *Phloeomys* that has to be further tested (Musser and Heaney, 1992), no data supports inclusion of *Crateromys* in the Phloeomyinae along with *Coryphomys*, *Lenomys*, *Pogonomys*, *Chiropodomys*, and *Mallomys*, as Simpson (1945) indicated. An undescribed species related to *C. schadenbergi* was recorded from Panay Isl (Musser and Heaney, 1992); real distribution in the archipelago and actual number of species in the genus have yet to be determined.

Crateromys australis Musser, Heaney, and Rabor, 1985. Am. Mus. Novit., 2821:3.
TYPE LOCALITY: Philippines, Surigao del Norte Prov., Dinagat Isl, Loreto Municipality, Balitbiton.
DISTRIBUTION: Known only from the type locality.
COMMENTS: Known only by the holotype.

Crateromys paulus Musser and Gordon, 1981. J. Mammal., 62:515.
TYPE LOCALITY: Philippines, Mindoro Occidental Prov., Ilin Isl.
DISTRIBUTION: Known only from the type locality.
COMMENTS: Known only by the holotype.

Crateromys schadenbergi (Meyer, 1895). Abh. Mus. Dresden, 6:1.
TYPE LOCALITY: Philippines, N Luzon Isl, Mt Data.
DISTRIBUTION: Known only from the mountains of N Luzon.

Cremnomys Wroughton, 1912. J. Bombay Nat. Hist. Soc., 21:340.
TYPE SPECIES: *Cremnomys cutchicus* Wroughton, 1912.
SYNONYMS: *Madromys*.
COMMENTS: An Indian endemic that was incorporated into subgenus *Rattus* (Ellerman, 1941), then arranged as a valid subgenus within *Rattus* (Ellerman, 1961; Ellerman and Morrison-Scott, 1951), and finally reinstated as a very distinctive genus related to *Millardia* (Gadi and Sharma, 1983; Misonne, 1969; Raman and Sharma, 1977).

Cremnomys blanfordi (Thomas, 1881). Ann. Mag. Nat. Hist., ser. 5, 7:24.
TYPE LOCALITY: India, Madras, Kadapa.
DISTRIBUTION: Endemic to Sri Lanka and Peninsular India, from S provinces north to Behar and C Provinces.
COMMENTS: Once allocated to *Rattus* (Ellerman, 1941, 1961), *blanfordi* is phylogenetically distant from any species in that genus and is related to species in *Cremnomys*, a conclusion based on dental morphology (Misonne, 1969) and chromosomal evidence

(Gadi and Sharma, 1983; Raman and Sharma, 1977; Rao and Lakhotia, 1972). Chromosomal number and configuration of C. *blanfordi* are very similar to that recorded for C. *cutchicus* and C. *elvira* (Raman and Sharma, 1977), but differ in amount of C-band-positive constitutive heterochromatin (Sharma and Gadi, 1977). Sody (1941) proposed *Madromys*, a *nomen nudum*, for this species.

Cremnomys cutchicus Wroughton, 1912. J. Bombay Nat. Hist. Soc., 21:340.
TYPE LOCALITY: India, Kutch, Dhonsa.
DISTRIBUTION: An Indian endemic: Kutch, Kathiawar, S Rajputana, and Bihar in NW India; Mysore, Bellary, and Eastern Ghats in S Peninsula.
SYNONYMS: *australis, caenosus, leechi, medius, siva, rajput*.
COMMENTS: Ellerman (1961) suggested that *australis* was possibly a valid species. The complex needs taxonomic revision. Chromosomal information reported by Raman and Sharna (1977), Sharma and Gadi (1977), Rishi and Puri (1984), and Sobti and Gill (1984).

Cremnomys elvira (Ellerman, 1946). Ann. Mag. Nat. Hist., ser. 11, 13:207.
TYPE LOCALITY: India, E Ghats, Salem Dist, Kurumbapatti.
DISTRIBUTION: Known only from SE India.
COMMENTS: Still represented by few specimens from the region of the type locality.

Crossomys Thomas, 1907. Ann. Mag. Nat. Hist., ser. 7, 20:70.
TYPE SPECIES: *Crossomys moncktoni* Thomas, 1907.
COMMENTS: Member of New Guinea Old Endemics (Musser, 1981*c*).

Crossomys moncktoni Thomas, 1907. Ann. Mag. Nat. Hist., ser. 7, 20:71.
TYPE LOCALITY: New Guinea, NE Papua New Guinea, Central Prov., Brown River, Serigina, 4500 ft.
DISTRIBUTION: Papua New Guinea; known only from Papuan Central Cordillera and mountains on Huon Peninsula (see map in Flannery, 1990*b*:191).
COMMENTS: Based on phallic morphology, Lidicker (1968) speculated that C. *moncktoni* is not closely related to *Hydromys* and its relatives. Flannery (1990*b*) provided photographs and summaries of distributional and biological information. Chromosomal data were reported by Donnellan (1987). Comparisons with Neotropical ichthyomyines made by Voss (1988).

Crunomys Thomas, 1897. Trans. Zool. Soc. London, 14(6):393.
TYPE SPECIES: *Crunomys fallax* Thomas, 1897.
COMMENTS: Revised by Musser (1982*c*). An Old Endemic of the Philippines (Musser and Heaney, 1992) and Sulawesi (Musser, 1981*c*). Phylogenetically closely related to Philippine *Archboldomys* (Musser and Heaney, 1992).

Crunomys celebensis Musser, 1982. Bull. Am. Mus. Nat. Hist., 174:16.
TYPE LOCALITY: Indonesia, C Sulawesi, near Tomado, 3500 ft.
DISTRIBUTION: C Sulawesi; known only from mountain valley of Danau Lindu and upper drainage of Sungai Miu in Kulawi region.
COMMENTS: Known only by three specimens. Morphologically very distinct from Philippine species of *Crunomys*, and without close relatives on Sulawesi (Musser, 1982*c*). Spermatozoal morphology distinctive (Breed and Musser, 1991), but unrevealing in assessing phlogenetic relationships.

Crunomys fallax Thomas, 1897. Trans. Zool. Soc. London, 14:394.
TYPE LOCALITY: Philippines, NC Luzon Isl, Isabella Prov, 1000 ft.
DISTRIBUTION: Known only from the type locality.
COMMENTS: Known only by the holotype. Morphologically more similar to the Philippine C. *melanius* and C. *rabori* than to the Sulawesian C. *celebensis* (Musser, 1982*c*).

Crunomys melanius Thomas, 1907. Proc. Zool. Soc. Lond., 1907:141.
TYPE LOCALITY: Philippines, Mindanao Isl, Davao Prov., Mt Apo, 3000 ft.
DISTRIBUTION: Known only from Mindanao Isl.
COMMENTS: Represented only by five specimens, and closely related to C. *rabori* (Musser, 1982*c*).

Crunomys rabori Musser, 1982. Bull. Am. Mus. Nat. Hist., 174:14.
TYPE LOCALITY: Philippines, Leyte Isl, Leyte Prov., Mt Lobi Range, Barrio Buri.
DISTRIBUTION: Known only from the type locality.
COMMENTS: Known only by the holotype.

Dacnomys Thomas, 1916. J. Bombay Nat. Hist. Soc., 24(3):404.
TYPE SPECIES: *Dacnomys millardi* Thomas, 1916.
COMMENTS: Reviewed by Musser (1981*b*). Closest phylogenetic alliance may be with *Niviventer*, particularly with members of the *N. andersoni* Division of that genus (Musser, 1981*b*).

Dacnomys millardi Thomas, 1916. J. Bombay Nat. Hist. Soc., 24(3):404.
TYPE LOCALITY: India, near Darjeeling, Gopaldhara, 3440 ft.
DISTRIBUTION: E Nepal, NE India (Bengal Presidency, Assam), N Laos, and S China (S Yunnan); probably occurs over a wider range (see map in Musser, 1981*b*).
SYNONYMS: *ingens, wroughtoni*.
COMMENTS: Known only by a few specimens (Li et al., 1987; Musser, 1981*b*).

Dasymys Peters, 1875. Monatsb. K. Preuss. Akad. Wiss., Berlin, p. 12.
TYPE SPECIES: *Dasymys gueinzii* Peters, 1875 (= *Mus incomtus* Sundevall, 1847; see Allen, 1939).
COMMENTS: Chromosomal data summarized by Carleton and Martinez (1991). Closest phylogenetic allies yet to be determined, but many pelage, cranial, and dental traits of *Dasymys* suggest alliance with *Aethomys*.

Dasymys foxi Thomas, 1912. Ann. Mag. Nat. Hist., ser. 8, 9:685.
TYPE LOCALITY: Nigeria, Panyam, 4000 ft.
DISTRIBUTION: Known only from Jos Plateau in Nigeria.
COMMENTS: Reviewed by Carleton and Martinez (1991), who contrasted its distinctive morphological and distributional traits with *D. rufulus*, the common species of *Dasymys* in West Africa.

Dasymys incomtus (Sundevall, 1847). Ofv. K. Svenska Vet.-Akad. Forhandl, Stockholm, 1846, 3:120 [1847].
TYPE LOCALITY: South Africa, "Caffraria prope Portum Natal," (= Durban, Natal).
DISTRIBUTION: S Africa, Malawi, Zambia, Zimbabwe, Angola, Zaire, Uganda, Kenya, Tanzania, Ethiopia, S Sudan; limits unknown.
SYNONYMS: *alleni, bentleyae, capensis, edsoni, fuscus, griseifrons, gueinzii, helukus, medius, nigridius, orthos, palustris, savannus, shawi*.
COMMENTS: Possibly a complex of several species. Chromosomal variation from South African specimens reported by Gordon (1991:413) who noted it "is equivocal whether the different chromosomal forms represent distinct species characterized by fixed rearrangements or a chromosomally polymorphic species."

Dasymys montanus Thomas, 1906. Ann. Mag. Nat. Hist., ser. 7, 18:143.
TYPE LOCALITY: Uganda, Ruwenzori East, Mubuku Valley, 12,500 ft.
DISTRIBUTION: Known only from Ruwenzori Mtns, Uganda, at very high altitudes; a montane Western Rift endemic.
COMMENTS: Usually included in *D. incomtus* (Delany, 1975), but distinguished from that species by its altitudinal distribution, long and fine fur, very short tail, and diagnostic cranial traits (Thomas, 1906*a*).

Dasymys nudipes (Peters, 1870). Jorn. Sci. Math., Phys. Nat., Lisboa, ser. 1, 3:126.
TYPE LOCALITY: Angola, Huilla.
DISTRIBUTION: S Angola, SW Zambia, NE Namibia, N Botswana; limits unknown, possibly also occurs in NW Zimbabwe.
COMMENTS: Treated by Allen (1939), Hill and Carter (1941), and Roberts (1951) as a species, but included in *D. incomtus* by Ellerman (1941) and most later writers of lists (e.g., Meester et al., 1986; Misonne, 1974). Crawford-Cabral (1983) recorded sympatry between *D. nudipes* and *D. incomtus* in Angola, and our survey of series from Chitau identified as *D. nudipes* by Hill and Carter (1941:98) revealed it consists of both *nudipes* and *incomtus*, qualitative observations supported by morphometric analyses

(Crawford-Cabral and Pacheco, 1989). Lukolela and Luluabourg, Zaire, have been included within range of *D. nudipes* (Crawford-Cabral, 1983), but specimens from there (in the American Museum of Natural History) are examples of *D. incomtus*. The holotype and only specimen of *edsoni*, described as subspecies of *D. nudipes* from Lukolela, middle Zaire (Hatt, 1934*b*), is an *incomtus*, not a *nudipes*.

Dasymys rufulus Miller, 1900. Proc. Washington Acad. Sci., 2:639.

TYPE LOCALITY: Liberia, Mount Coffee.

DISTRIBUTION: Specimen records are from Sierra Leone, Liberia, Ivory Coast, Ghana, Togo, Benin, E Nigeria, and Cameroon.

SYNONYMS: *longipilosus*.

COMMENTS: Morphometric and geographic contrasts with *D. foxi* presented by Carleton and Martinez (1991). Relationships with populations from Zaire and elsewhere now treated as *D. incomtus bentleyae* need to be clarified by critical systematic revision.

Dephomys Thomas, 1926. Ann. Mag. Nat. Hist., ser. 9, 17:177.

TYPE SPECIES: *Mus defua* Miller, 1900.

COMMENTS: Another genus founded by Thomas and either relegated to *Rattus* as a subgenus or combined with *Stochomys*, which was then treated as a subgenus of either *Rattus* or *Aethomys* (see discussions in D. H. S. Davis, 1965; Rosevear, 1969; and Van der Straeten, 1984). The generic integrity of *Dephomys* was recognized by Misonne (1969) and Rosevear (1969), who also suggested it was not especially closely related to *Stochomys*, which was the conventional view. Molar characters and morphometric data clearly point to *Hybomys* as the closest phylogenetic ally of *Dephomys* (Misonne, 1969; Van der Straeten, 1984).

Dephomys defua (Miller, 1900). Proc. Washington Acad. Sci., 2:635.

TYPE LOCALITY: Liberia, Mt Coffee.

DISTRIBUTION: Specimens are from Sierra Leone, Guinea (Mt Nimbo), Liberia, Ivory Coast, and Ghana.

COMMENTS: Part of the murine fauna endemic to W Africa (see account of *Grammomys buntingi*). Taxonomy, morphology, range, and habits reviewed by Rosevear (1969) and Van der Straeten (1984).

Dephomys eburnea (Heim de Balsac et Bellier, 1967). Mammalia, 31:157.

TYPE LOCALITY: Ivory Coast, Lamto.

DISTRIBUTION: Records are from Ivory Coast and Liberia.

COMMENTS: Originally described as a subspecies of *defua* (see discussion and references in Rosevear, 1969), but shown to be a separate species by Van der Straeten (1984). Part of the murine fauna endemic to W Africa (see account of *Grammomys buntingi*). Chromosomal data reported by Tranier and Dosso (1979).

Desmomys Thomas, 1910. Ann. Mag. Nat. Hist., ser. 8, 5:284.

TYPE SPECIES: *Pelomys harringtoni* Thomas, 1902.

COMMENTS: Listed as a genus by Allen (1939), but usually treated as a subgenus of *Pelomys* (Corbet and Hill, 1991; Ellerman, 1941; Rupp, 1980; Yalden et al., 1976). Most of the diagnostic traits described by Thomas (1910*b*) are outside the range of morphological variation seen among species of *Pelomys*. Our study of specimens revealed that general external traits and cranial conformation of *D. harringtoni*, the only species in the genus, resemble species of *Mylomys* and *Pelomys*, but that *Desmomys* has its own derived dental patterns (ridge-like cusp t9 connecting central cusp t8 with labial cusp t6 on first and second molars, ridge-like cusp t7 on second upper molar). The phylogenetic relationships among species of *Desmomys*, *Pelomys*, and *Mylomys* has to be assessed by revisionary study. See the account of *Mylomys* for the status of *rex*, which is usually placed in *Desmomys* (Yalden et al., 1976).

Desmomys harringtoni (Thomas, 1902). Proc. Zool. Soc. Lond., 1902:313.

TYPE LOCALITY: Ethiopia, W Shoa, Kutai, Katchisa.

DISTRIBUTION: Ethiopian plateau between 1500 and 3000 m (Rupp, 1980; Yalden et al., 1976).

COMMENTS: Another Ethiopian endemic, and one that is common, judged from the large series in museum collections.

Diomys Thomas, 1917. J. Bombay Nat. Hist. Soc., 25:203.
TYPE SPECIES: *Diomys crumpi* Thomas, 1917.
COMMENTS: Misonne (1969) suggested that *Diomys* is related to *Chiromyscus, Dacnomys,* and *Niviventer* (Misonne used *Maxomys* for this group), but Musser and Newcomb (1983) hypothesized that *Millardia, Cremnomys,* and other Indian genera may be more closely allied to *Diomys;* both views require testing.

Diomys crumpi Thomas, 1917. J. Bombay Nat. Hist. Soc., 25:203.
TYPE LOCALITY: India, Bihar, Hazaribagh, Mt Paresnath.
DISTRIBUTION: Recorded from NE India and W Nepal; limits unknown.
COMMENTS: Reviewed by Ingles et al. (1980) and Musser and Newcomb (1983).

Diplothrix Thomas, 1916. J. Bombay Nat. Hist. Soc., 24:404.
TYPE SPECIES: *Lenothrix legata* Thomas, 1906.

Diplothrix legatus (Thomas, 1906). Ann. Mag. Nat. Hist., ser. 7, 17:88.
TYPE LOCALITY: Japan, Ryukyu Isls, Amami-oshima Isl.
DISTRIBUTION: Japan, Ryukyu Isls of Amami-oshima, Tokun-oshima, and Okinawa (known by modern specimens only in north, but by Quaternary fossils from farther south on island, and from Miyako Isl about 250 km S W of Okinawa) (see Kawamura, 1989).
SYNONYMS: *bowersii* var. *okinavensis.*
COMMENTS: Historical and incorrect allocation of *legatus* with *Lenothrix* and *Rattus* is reviewed by Kawamura (1989). Phylogenetic relationship, discerned from molar occlusal patterns, close to *Rattus* and far from *Lenothrix;* "phylogeny of this unique genus will be sufficiently understood, when the fossil murids from China, India and Southeast Asia will be investigated in detail" (Kawmura, 1989:110). Chromosomal data reviewed by Tsuchiya (1981).

Echiothrix Gray, 1867. Proc. Zool. Soc. Lond., 1867:599.
TYPE SPECIES: *Echiothrix leucura* Gray, 1867.
SYNONYMS: *Craurothrix.*
COMMENTS: Thomas (1898*b*:397) explained why *Echiothrix* is the proper name and not *Craurothrix,* which he proposed thinking *Echiothrix* was preoccupied. Based upon shared cranial features, Thomas (1898*b*) thought *Echiothrix* was related to the Philippine *Rhynchomys* and placed them in the Rhynchomyinae. Later workers disagreed (see summary in Musser, 1990), and, except to recognize that *Echiothrix* is an Old Endemic of Sulawesi (Musser, 1981*c*), no one has discovered its closest phylogenetic ally. Spermatozoal morphology resembles that of *Margaretamys, Maxomys,* and in some aspects even *Rattus,* but is ambiguous in illuminating possible phylogenetic alliances (Breed and Musser, 1991). Revision of the genus required to determine if N and C Sulawesian samples represent one or two species.

Echiothrix leucura Gray, 1867. Proc. Zool. Soc. Lond., 1867:600.
TYPE LOCALITY: Indonesia, N Sulawesi. Gray thought the holotype came from Australia, but Jentink (1883) thought the species would prove to be found only on Sulawesi, and Laurie and Hill (1954) listed the type locality as probably N Sulawesi.
DISTRIBUTION: Sulawesi: N arm and C core in tropical lowland evergreen rain forest (Musser, 1990; Musser and Holden, 1991).
SYNONYMS: *brevicula, centrosa.*
COMMENTS: External, cranial, and dental morphology reviewed by Musser (1969*b*, 1990), karyotype of male documented by Musser (1990).

Eropeplus Miller and Hollister, 1921. Proc. Biol. Soc. Washington, 34:94.
TYPE SPECIES: *Eropeplus canus* Miller and Hollister, 1921.
COMMENTS: Ellerman (1941) allied *Eropeplus* closely with *Rattus,* but its nearest phylogenetic relative is Sulawesian *Lenomys,* an affinity supported by external, cranial, and spermatozoal characters (Breed and Musser, 1991; Musser, 1981*c*; Tate, 1936). Spermatozoal morphology of *Eropeplus* and *Lenomys* was unique among Sulawesian taxa sampled (Breed and Musser, 1991).

Eropeplus canus Miller and Hollister, 1921. Proc. Biol. Soc. Washington, 34:94.
TYPE LOCALITY: Indonesia, C Sulawesi, Gunung Lehio, above 6000 ft.

DISTRIBUTION: C Sulawesi; known only by small samples from a few places in montane tropical rainforest formations (Musser, 1970*d*; Musser and Holden, 1991), but probably occurs throughout at least the central core of the island in suitable montane habitat.

Golunda Gray, 1837. Mag. Nat. Hist. [Charlesworth's], 1:586.
TYPE SPECIES: *Golunda ellioti* Gray, 1837.
COMMENTS: Reviewed and compared with *Hadromys* and *Mylomys* by Musser (1987). Pliocene fragments identified as a species of *Golunda* have been recorded from Ethiopia, but Musser (1987) explained why they do not represent this genus.

Golunda ellioti Gray, 1837. Mag. Nat. Hist. [Charlesworth's], 1:586.
TYPE LOCALITY: India, Dharwar.
DISTRIBUTION: SE Iran (Misonne, 1990), Pakistan, Nepal, N and NE India south through Indian peninsula to Sri Lanka.
SYNONYMS: *bombax, coenosa, coffaeus, coraginis, gujerati, hirsutus, limitaris, myothrix, newara, nuwara, paupera, watsoni.*

Grammomys Thomas, 1915. Ann. Mag. Nat. Hist., ser. 8, 16:150.
TYPE SPECIES: *Mus dolichurus* Smuts, 1832.
COMMENTS: A distinctive genus as asserted by Ellerman (1941) and other workers (e.g., Hutterer and Dieterlen, 1984; Rosevear, 1969), and unrelated to *Thamnomys* with which it has often been united as a subgenus (Allen, 1939; Hatt, 1940*b*; Hollister, 1919; Misonne, 1974; Petter and Tranier, 1975). Partly reviewed by Petter and Tranier (1975) and Hutterer and Dieterlen (1984), who provided morphological, distributional, and chromosomal comparisons. Morphological and chromosomal similarities exist with *Thallomys* (Olert et al., 1978).

Grammomys aridulus Thomas and Hinton, 1923. Proc. Zool. Soc. Lond., 1923:268.
TYPE LOCALITY: Sudan, Darfur, Wadi Aribo, Kulme.
DISTRIBUTION: WC Sudan (Dieterlen and Nikolaus, 1985).
COMMENTS: Usually either listed as a subspecies of *G. macmillani* (Allen, 1939; Ellerman, 1941; Setzer, 1956) or included in *G. dolichurus* (Misonne, 1974), but considered a distinct species by Hutterer and Dieterlen (1984).

Grammomys buntingi (Thomas, 1911). Ann. Mag. Nat. Hist., ser. 8, 7:381.
TYPE LOCALITY: Liberia, Bassa, Gonyon.
DISTRIBUTION: Zone of high forest, coastal scrub or Guinea woodland in West Africa from Sierra Leone and Guinea to Ivory Coast and Liberia.
COMMENTS: Except for Misonne (1974), who included it in *G. dolichurus*, *buntingi* has always been listed or discussed as a species of *Grammomys* (Allen, 1939; Ellerman, 1941; Hutterer and Dieterlen, 1984; Petter and Trainer, 1975; Rosevear, 1969). Apparently *G. buntingi* is part of a suite of species endemic to W Africa: *Dephomys defua, D. eburnea, Hybomys planifrons, H. trivirgatus, Hylomyscus baeri, Lemniscomys bellieri, Malacomys cansdalei, M. edwardsi, Myomys daltoni, M. derooi*, and *Praomys rostratus* (Carleton and Robbins, 1985; Hutterer and Dieterlen, 1984).

Grammomys caniceps Hutterer and Dieterlen, 1984. Stuttg. Beitr. Naturk., A, 374:12.
TYPE LOCALITY: Kenya, Malindi.
DISTRIBUTION: N Kenya and S Somalia.
COMMENTS: Karyotype of Kenyan sample described and discussed by Hutterer and Dieterlen (1984), and chromosomal variation among individuals from S Somalia described and discussed in detail by Roche et al. (1984) who assigned their sample to the *dolichurus* group, but who also, and independant of Hutterer and Dieterlen (1984), noted the distinctive quality of the Somalian species as indicated by chromosomal evidence.

Grammomys cometes (Thomas and Wroughton, 1908). Proc. Zool. Soc. Lond., 1908:549.
TYPE LOCALITY: Mozambique, Inhambane.
DISTRIBUTION: From Pirie Forest (northwest of King William's Town) in SE Cape Prov. of South Africa north through Natal and Transvaal into E Zimbabwe (Melsetter and

Umtali districts) and Mozambique south of the Zambezi River (see Meester et al., 1986; Smither and Tello, 1976; Skinner and Smithers, 1990).

SYNONYMS: *silindensis*.

COMMENTS: The geographic range of *G. cometes* has been outlined as extending north from South Africa through East Africa to S Sudan (Hutterer and Dieterlen, 1984) but pending revisionary study of the genus we restrict it to the eastern segment of the Southern African Subregion south of Zambezi River (similar to the rang mapped by Skinner and Smithers, 1990:225), and consider samples north of that river to be *G. ibeanus* (see that account). We studied the holotype of *cometes* and the other specimens in the type series noted by Thomas and Wroughton (1908); these animals are on average larger and have more highly inflated bullae than do those from north of the Zambesi River. Ansell (1978) and Ansell and Dowsett (1988) assigned samples from Zambia and Malawi to *cometes*, but were also impressed with the chromatic and morphological contrast between them and the holotype from Inhambane. The specimen from the Pirie Forest (in the American Museum of Natural History) represents a range extension south of Natal.

Grammomys dolichurus (Smuts, 1832). Enumer. Mamm. Capensium, p. 38.

TYPE LOCALITY: South Africa, near Cape Town.

DISTRIBUTION: From Nigeria east to SW Ethiopia, south through N Zaire, Uganda, Kenya, Tanzania, and Malawi, and to South Africa (Natal and Cape Prov.), and west through Zimbabwe and Zambia to Angola; limits of geographic range unresolved.

SYNONYMS: *angolensis*, *arborarius* (of True, 1892, not Peters, 1852), *baliolus*, *discolor*, *elgonis*, *insignis*, *littoralis*, *polionops*, *surdaster*, *tongensis*.

COMMENTS: The number of scientific names reflects morphological and chromosomal variation correlated with geography that suggests more than one species is represented (Hutterer and Dieterlen, 1984; Meester et al., 1986); the complex requires careful revision. For example, specimens of true *dolichurus* from South Africa have duller pelage and more inflated bullae than animals from East and West Africa; should these prove to be diagnostic specific differences, the northern populations should be identified as *G. surdaster*. The Ethiopian locality is based on a specimen from Kefa (in the National Museum of Natural History) and not on those recorded by Yalden et al. (1976), which represent other species (see Hutterer and Dieterlen, 1984).

Grammomys dryas (Thomas, 1907). Ann. Mag. Nat. Hist., ser. 7, 19:123.

TYPE LOCALITY: "Kenya Colony" (Uganda), Ruwenzori East, 6-7000 ft.

DISTRIBUTION: A montane Western Rift endemic: Ruwenzoris and Kivu region in Uganda and Zaire, NW Burundi (specimens in the Field Museum of Natural History).

COMMENTS: Originally described as a species of *Thamnomys* by Thomas (1907*a*) and listed that way by Allen (1939). Ellerman (1941), however, treated it as valid species of *Grammomys*, which, in the absence of a critical systematic revision of the genus, best expresses current knowledge. Thomas (1907*a*) noted the diagnostic mammary count in *dryas*, which, in combination with cranial traits, set it apart from other described forms.

Grammomys gigas (Dollman, 1911). Ann. Mag. Nat. Hist., ser. 8, 7:527.

TYPE LOCALITY: Kenya, Mt Kenya, Solai, 9000 ft.

DISTRIBUTION: Known only from the vicinity of Mt Kenya.

COMMENTS: Recorded only by the holotype. Recognized as a species in most lists (Allen, 1939; Ellerman, 1941). Hutterer and Dieterlen (1984) insisted the species has to be recognized because of its large teeth, an opinion we share based on our study of the holotype; but the possibility that it may simply be a large individual of *G. ibeanus* is a hypothesis that deserves testing.

Grammomys ibeanus (Osgood, 1910). Field Mus. Nat. Hist. Publ., Zool. Ser., 10:8.

TYPE LOCALITY: Kenya, Molo.

DISTRIBUTION: From extreme NE Zambia (Ansell, 1978) and Malawi (Ansell and Dowsett, 1988) north through SC Tanzania (specimens in Museum of Comparative Zoology, Harvard) and Kenya to S Sudan (Hollister, 1919; Hutterer and Dieterlen, 1984).

COMMENTS: Morphological and geographic definition of *G. ibeanus* is unsatisfactory, particularly the extent of its distribution in Tanzania. Hutterer and Dieterlen (1984) treated *ibeanus* as a form of *G. cometes*, but the striking morphological distinctions between samples of *ibeanus* and the type series of *cometes* prompted our specific ranking of *ibeanus* (see also account of *G. cometes*).

Grammomys macmillani (Wroughton, 1907). Ann. Mag. Nat. Hist., ser. 7, 20:504.

TYPE LOCALITY: Ethiopia, north of Lake Rudolf, Wouida.

DISTRIBUTION: Sierra Leone, Liberia, Central African Republic, S Sudan, S Ethiopia, N Zaire, Kenya, Uganda (including Bugala Isl in Lake Victoria), Tanzania, Malawi, Mozambique, and E Zimbabwe.

SYNONYMS: *callithrix, erythropygus, gazellae, oblitus, ochraceus, usambarae, vumbaensis, vumbensis.*

COMMENTS: Hutterer and Dieterlen (1984) provided historical association of the name with other taxa. Chromosomal variation (under name of *gazellae*) documented by Civitelli et al. (1989). The records form Sierra Leone, Tanzania, Malawi, and Mozambique are based on series in the American Museum of Natural History, the Museum of Comparative Zoology at Harvard, and the National Museum of Natural History and represent significant range extensions beyond that outlined by Hutterer and Dieterlen (1984). Judged by Roberts's (1938) description and our study of one of the specimens in his original series (24046 in Museum of Comparative Zoology, Harvard), *vumbaensis* from E Zimbabwe (Vunba and Mount Selinda) clearly belongs in the synonymy of *G. macmillani* rather than *dolichurus* where Meester et al. (1986) listed it. Matschie's (1915) *usambarae* from N Tanzania is also likely an example of *macmillani* because of its small size and short molar row.

Grammomys minnae Hutterer and Dieterlen, 1984. Stuttg. Beitr. Naturk., A, 374:10.

TYPE LOCALITY: S Ethiopia, Sidamo Prov., edge of Bulcha Forest, 1800 m.

DISTRIBUTION: S Ethiopia; limits unknown.

Grammomys rutilans (Peters, 1876). Monatsb. K. Preuss. Akad. Wiss. Berlin, p. 478.

TYPE LOCALITY: Gabon, Limbareni.

DISTRIBUTION: Tropical rain forest block and outlying patches: from Guinea (Mt Nimba) eastward through Ivory Coast, Ghana, Togo, S Nigeria to Central African Republic and Zaire, then south through Cameroon, Equatorial Guinea, and Gabon to N Angola. Also found on Bioko (= Fernando Poo) and in forest patches in W Uganda.

SYNONYMS: *centralis, kuru, poensis.*

COMMENTS: Usually considered a member of *Thamnomys* (Allen, 1939; Ellerman, 1941; Hutterer and Dieterlen, 1984), but external, cranial, and dental morphology is more similar to that characterizing species of *Grammomys* (where it was listed by D. H. S. Davis, 1965, and Misonne, 1974), and we include it within *Grammomys* pending critical systematic evaluation of samples now identified as *rutilans*. The form *kuru*, treated here and by D. H. S. Davis (1965) as a synonym of *G. rutilans*, is sometimes listed as a species (e.g., Ellerman, 1941), but with provisions (Hatt, 1940*a*; Hutterer and Dieterlen, 1984; Thomas, 1915); our study of the holotype revealed it is a very young adult *rutilans*. Tranier and Dosso (1979) reported 2N=36 for an individual captured in Ivory Coast; this contrasts with 2N=50 usually recorded for *G. rutilans* (Matthey, 1963) and either signals exceptional chromosomal variation within *rutilans* or the presence of a morphologically undetected species.

Hadromys Thomas, 1911. J. Bombay Nat. Hist. Soc., 20:999.

TYPE SPECIES: *Mus humei* Thomas, 1886.

COMMENTS: Usually considered closely allied to *Arvicanthis* and its relatives, especially *Golunda*, but a combination of primitive and derived cranial and dental traits divorces *Hadromys* from that group: "best hypothesis now available on phylogenetic affinities is to consider the species of *Hadromys* to have been derived from some late Miocene ancestor, a species of *Karnimata*, for example" (Musser, 1987:19).

Hadromys humei (Thomas, 1886). Proc. Zool. Soc. Lond., 1886:63.

TYPE LOCALITY: India, Manipur, Moirang.

DISTRIBUTION: NE India (Manipur to NW Assam) and S China (W Yunnan).

SYNONYMS: *yunnanensis.*

COMMENTS: Reviewed by Musser (1987). Closest relative is *H. loujacobsi*, known by fossil fragments from early Pleistocene sediments in Punjab region of N Pakistan (Musser,

1987). Yang and Wang (1987) described *yunnanensis* as a very distinctinve subspecies of *H. humei*.

Haeromys Thomas, 1911. Ann. Mag. Nat. Hist., ser. 8, 7:207.
TYPE SPECIES: *Mus margarettae* Thomas, 1893.
COMMENTS: A Sundaic and Sulawesian endemic. Cranial, dental, and spermatozoal morphology suggested a distant phylogenetic link to *Chiropodomys* (Breed and Musser, 1991; Musser and Newcomb, 1983); chromosomal data was ambiguous in assessing its closest relatives (Musser, 1990).

Haeromys margarettae (Thomas, 1893). Ann. Mag. Nat. Hist., ser. 6, 11:346.
TYPE LOCALITY: Malaysia, Sarawak, Penrisen Hills.
DISTRIBUTION: Borneo; known only from the type locality and Sabah (Chasen and Kloss, 1932).
COMMENTS: Known only by the holotype and a specimen from Sabah; record from East Kalimantan by Medway (1977) is a nestling of *Sundamys muelleri*.

Haeromys minahassae (Thomas, 1896). Ann. Mag. Nat. Hist., ser. 6, 18:247.
TYPE LOCALITY: Indonesia, Sulawesi, NE Sulawesi, Minahassa, Rurukan.
DISTRIBUTION: Sulawesi; recorded only from the NE peninsula and central core (Musser, 1990).
COMMENTS: External, cranial, and dental features clearly distinguish Sulawesi species from those on Borneo and Palawan. Two species occur in C Sulawesi, one restricted to tropical lowland evergreen rain forest, the other (yet to be named and described) found only in montane rain forest (Musser, 1990; Musser and Holden, 1991). Karyotype is primitive, spermatozoal morphology is distinctive (Musser, 1990; Breed and Musser, 1991).

Haeromys pusillus (Thomas, 1893). Ann. Mag. Nat. Hist., ser. 6, 11:232.
TYPE LOCALITY: Malaysia, Sabah, Mt Kinabalu.
DISTRIBUTION: Borneo (Sarawak, Sabah, and East Kalimantan) and Palawan Isl (Philippines).
COMMENTS: Represented by few specimens. A close phylogenetic relative of *H. margarettae*.

Hapalomys Blyth, 1859. J. Asiat. Soc. Bengal, 28:296.
TYPE SPECIES: *Hapalomys longicaudatus* Blyth, 1859.
COMMENTS: Reviewed by Musser (1972). One of the few genera with representatives in both Indochina and on the Sunda Shelf (Musser and Newcomb, 1983).

Hapalomys delacouri Thomas, 1927. Proc. Zool. Soc. Lond., 1927:55.
TYPE LOCALITY: S Vietnam, Dakto.
DISTRIBUTION: S China (Hainan Isl), N Laos, and S Vietnam; limits unknown.
SYNONYMS: *marmosa, pasquieri*.
COMMENTS: Known only by few specimens.

Hapalomys longicaudatus Blyth, 1859. J. Asiat. Soc. Bengal, 28:296.
TYPE LOCALITY: Burma, Tenasserim, Sitang River Valley.
DISTRIBUTION: SE Burma, SW and peninsular Thailand, and Malaya Peninsula; limits unknown.
COMMENTS: Karyotype uninformative about phylogenetic relationships (Yong, et al., 1982). Spermatozoal morphology similar to that of *Chiropodomys* and *Maxomys*, which also resembled the basic structure found in species of *Mus* and *Apodemus* (Breed and Yong (1986). Among Sundaic genera, *H. longicaudatus* may be most closely related to *Chiropodomys*, based on external and cranial traits (Musser and Newcomb, 1983).

Heimyscus Misonne, 1969. Mus. Roy. l'Afrique Cent., Tervuren, Zool., no. 172:125.
TYPE SPECIES: *Hylomyscus fumosus* Brosset, Dubost, and Heim de Balsac, 1965.
COMMENTS: A very distinctive genus whose closest phylogenetic affinites have yet to be resolved.

Heimyscus fumosus (Brosset, Dubost, and Heim de Balsac, 1965). Biologia Gabonica, 1:154.
TYPE LOCALITY: Gabon, Makokou.
DISTRIBUTION: Recorded only from Gabon, S Cameroon (Robbins et al., 1980), and Central African Republic (Petter and Genest, 1970).

COMMENTS: Originally described as a species of *Hylomyscus*, but differs significantly from any species in that genus in morphological and chromosomal traits (Misonne, 1969; Robbins et al., 1980).

Hybomys Thomas, 1910. Ann. Mag. Nat. Hist., ser. 8, 5:85.

TYPE SPECIES: *Mus univittatus* Peters, 1876.

SYNONYMS: *Typomys*.

COMMENTS: Carleton and Robbins (1985) provided a general review of morphometric traits, qualitative features, chromosomal data, geographic distributions, taxonomic assessments, and ecological and zoogeographical discussions of several species, with a focus on West African forms. The species have traditionally been arranged into two groups and reflected taxonomically as either subgenera (*Hybomys* and *Typomys*) or genera (see Rosevear's, 1969, excellent exposition). Results of biometric analyses led Van der Straeten (1984) to argue strongly for recognizing two genera. Carleton and Robbins (1985:983) corroborated that dichotomy with qualitative features and chromosomal data, but noted that whether "*Typomys* and *Hybomys* merit generic segregation must await an evaluation of character variation among *Hybomys* and its near relatives."

Hybomys basilii Eisentraut, 1965. Zool. Jahrb. Syst., 92:20.

TYPE LOCALITY: Fernando Poo (Bioko), Mocatal, 1200 m.

DISTRIBUTION: Known only from Bioko.

COMMENTS: Subgenus *Hybomys*. Originally described as a subspecies of *H. univittatus*, but raised to specific rank by Van der Straeten (1985).

Hybomys eisentrauti Van der Straeten and Hutterer, 1986. Mammalia, 50:36.

TYPE LOCALITY: W Cameroon, Mt Lefo, Bambului, 1800 m.

DISTRIBUTION: Recorded only from Mt Lefo and Mt Oku in W Cameroon (Bamenda-Banso highlands).

COMMENTS: Subgenus *Hybomys*. Definition of this species and its comparisons with others is discussed by Van der Straeten and Hutterer (1986).

Hybomys lunaris (Thomas, 1906). Ann. Mag. Nat. Hist., ser. 7, 18:145.

TYPE LOCALITY: Uganda, Ruwenzori East, Mubuku Valley, 6000 ft.

DISTRIBUTION: NE and E Zaire, W Uganda, and Rwanda; limits unknown.

COMMENTS: Subgenus *Hybomys*. Originally described as a subspecies of *H. univittatus* by Thomas, *lunaris* is a separate species according to the biometrical analyses of Van der Straeten et al. (1986).

Hybomys planifrons (Miller, 1900). Proc. Washington Acad. Sci., 2:641.

TYPE LOCALITY: Liberia, Mt Coffee.

DISTRIBUTION: NE Sierra Leone, Liberia, SE Guinea, and W Ivory Coast, west of the Sassandra River (see map in Carleton and Robbins, 1985:990).

COMMENTS: Subgenus *Typomys*. Part of the murine fauna endemic to W Africa (see account of *Grammomys buntingi*). Taxonomic status, phylogenetic relationship, and significance of distribution in Liberian forest refuge reviewed by Carleton and Robbins (1985). Additional data reported by Gautun et al. (1986).

Hybomys trivirgatus (Temminck, 1853). Esquisses Zool. sur la Côte de Guine, p. 159.

TYPE LOCALITY: Ghana (= Gold Coast), Dabocrom.

DISTRIBUTION: From E Sierra Leone west to S Nigeria west of the Niger River (see map in Carleton and Robbins, 1985).

SYNONYMS: *pearsei*.

COMMENTS: Subgenus *Typomys*. Part of the murine fauna endemic to W Africa (see account of *Grammomys buntingi*). Morphology, phylogenetic affinities, and geographic range reviewed by Carleton and Robbins (1985).

Hybomys univittatus (Peters, 1876). Monatsb. K. Preuss. Akad. Wiss., Berlin, p. 479.

TYPE LOCALITY: Gabon, Dongila.

DISTRIBUTION: From SE Nigeria (on E side of Cross River), through Cameroon, Equatorial Guinea, Gabon, Congo, S Central African Republic, Zaire and extreme NW Zambia to S Uganda and W Rwanda (see map in Carleton and Robbins, 1985:990).

SYNONYMS: *badius*, *rufocanus*.

COMMENTS: Subgenus *Hybomys*. Review of species and comparisons with *H. trivirgatus* and *H. planifrons* reported by Carleton and Robbins (1985).

Hydromys E. Geoffroy, 1804. Bull. Sci. Soc. Philom. Paris, 3(93):353.
TYPE SPECIES: *Hydromys chrysogaster* E. Geoffroy, 1804.
SYNONYMS: *Baiyankamys*.
COMMENTS: Member of the Australian and New Guinea Old Endemics (Musser, 1981*c*:167). Comparisons with Neotropical ichthyomyines made by Voss (1988). Flannery (1990*b*) provided photographs and distributional and biological summaries of species. Phallic morphology of *H. chrysogaster* and *H. habbema* described by Lidicker (1968). Mahoney (1968) explained why *Baiyankamys* is a synonym of *Hydromys*.

Hydromys chrysogaster E. Geoffroy, 1804. Bull. Sci. Soc. Philom. Paris, 93:354.
TYPE LOCALITY: Australia, Tasmania, Bruny Isl (see Mahoney and Richardson, 1988:157).
DISTRIBUTION: Australia: freshwater lakes and rivers as well as swamp, salt marsh, and supralittoral habitats (absent from central Australian region); also found on Tasmania and numerous smaller islands off the coast of Austalia (Friend and Thomas, 1990; Watts and Aslin, 1981:67); Kei Isls and Aru Isls; New Guinea: throughout most of the island from sea level to 1900 m (see map in Flannery, 1990*b*:188). Also on Obi Isl in the Moluccas (Flannery, in litt.).
SYNONYMS: *beccarii, caurinus, esox, fuliginosus, fulvogaster, fulvolavatus, fulvoventer, grootensis, illuteus, lawnensis, leucogaster, longmani, lutrilla, melicertes, moae, nauticus, oriens, reginae*.
COMMENTS: Considered an Australian member of Hydromyini (Baverstock, 1984). Chromosomal data presented by Baverstock et al. (1977*c*, 1983*b*). Biochemical evidence supports weak link between *H. chrysogaster* and *Xeromys* (Baverstock et al., 1981). Morphology of spermatozoa and male reproductive tract discussed in context of comparative study of Australian murines (Breed, 1984, 1986; Breed and Sarafis, 1978; Morrissey and Breed, 1982). References to distributional, taxonomic and biological literature cataloged by Mahoney and Richardson (1988:156). Significance of variation in body size and pelage color needs to be assessed in context of critical systematic revision of the species.

Hydromys habbema Tate and Archbold, 1941. Am. Mus. Novit., 1101:3.
TYPE LOCALITY: New Guinea, Irian Jaya, 15 km N Mt Wilhelmina, Lake Habbema, 3225 m.
DISTRIBUTION: New Guinea, Irian Jaya, known only from the type locality and the NE slope of Mt Wilhelmina between 3560 and 3600 m (see Tate, 1951:227); limits unknown.

Hydromys hussoni Musser and Piik, 1982. Zool. Meded. Leiden, 56:157.
TYPE LOCALITY: New Guinea, Irian Jaya, Wissel Lakes, Lake Paniai, Enarotali, 1765 m (see Musser and Piik, 1982, for other details).
DISTRIBUTION: New Guinea; known only from the Wissel Lakes region in Irian Jaya and from the Maprik area in East Sepik Prov. of Papua New Guinea (see maps in Flannery, 1990*b*:186; Musser and Piik, 1982:156); limits unknown.
COMMENTS: Overall morphology and anatomical propotions of *H. hussoni* are like those of *H. chrysogaster*, although it is much smaller than that species (Musser and Piik, 1982).

Hydromys neobrittanicus Tate and Archbold, 1935. Am. Mus. Novit., 803:8.
TYPE LOCALITY: Bismarck Archipelago, New Britain Isl, Wide Bay, Balayang, Bainings.
DISTRIBUTION: Known only from New Britain Isl.
COMMENTS: Included in *H. chrysogaster* by Ziegler (1982*b*:883), but it was recognized as a New Britain endemic by Flannery and White (1991) and should retain this status until the significance of its diagnostic traits can be assessed in a systematic revision of the large-bodied forms of *Hydromys*, as Tate (1951:236) already noted.

Hydromys shawmayeri (Hinton, 1943). Ann. Mag. Nat. Hist., ser. 11, 10:552.
TYPE LOCALITY: Papua New Guinea, SE Bismarck Range, Purari-Ramu Divide, Baiyanka, 6500 ft.
DISTRIBUTION: New Guinea; recorded only along Papuan Central Cordillera from Hagen Range in the west to Mt Kaindi area in the east (localities 1-4 that were mapped as "*H. habbema*" in Musser and Piik, 1982:156).

COMMENTS: Although identified as *H. habbema*, specimens from Papua New Guinea were noted to exhibit significant morphological differences from true *habbema* in Irian Jaya, not only in body size but also in certain cranial proportions (Mahoney, 1968; Musser and Piik, 1982; Tate, 1951). These and other contrasts support the hypothesis of an eastern (*shawmayeri*) and a western (*habbema*) species, a pattern that is common to other New Guinea murines (*Pseudohydromys murinus* and *P. occidentalis*, for example). Mahoney (1968) discussed the problems associated with the original holotype of *shawmayeri*, which was described as a species of *Baiyankamys*.

Hylomyscus Thomas, 1926. Ann. Mag. Nat. Hist., ser. 9, 17:174.

TYPE SPECIES: *Epimys aeta* Thomas, 1911.

COMMENTS: Taxonomic, distributional, and biological summaries of West African forms provided by Rosevear (1969). Morphometric, distributional, and chromosomal data for some species in Camaroon reported by Robbins et al. (1980). Species from Ivory Coast reviewed by Heim de Balsac and Aellen (1965). Distributions of species listed below based primarily on study of museum specimens. See Rosevear (1969) and Robbins et al. (1980) for taxonomic history of the alternating use of *Hylomyscus* as a genus or subgenus.

Hylomyscus aeta (Thomas, 1911). Ann. Mag. Nat. Hist., ser. 8, 7:591.

TYPE LOCALITY: Cameroon, Bitye, Ja River.

DISTRIBUTION: The central forest block from Equatorial Guinea (Bioko = Fernando Poo) to Cameroon, Gabon, Central African Republic, Republic of the Congo, Zaire, W Uganda, and NW Burundi; limits unknown.

SYNONYMS: *grandis, laticeps, shoutedeni, weileri*.

COMMENTS: *H. aeta* is easily distinguished from all other species of *Hylomyscus* by its distinct supraorbital shelves. Hatt (1940*a*) tenatively included *aeta, shoutedeni,* and *weileri* in *H. carillus* but our study of samples and holotypes does not support his arrangement. Variation in body size exists among the samples but its significance has not been assessed by careful study. The range in Burundi is based on specimens in the Field Museum of Natural History. Swynnerton and Hayman (1951:316) listed a record from the Uluguru Mtns in EC Tanzania but we have not been able to verify it. Chromosomal data were reported by Robbins et al. (1980). Eisentraut's (1969*b*) *grandis* from Mt. Oku is a separate species (Hutterer, 1992).

Hylomyscus alleni (Waterhouse, 1838). Proc. Zool. Soc. Lond., 1837:77 [1838].

TYPE LOCALITY: Equatorial Guinea, Fernando Poo (= Bioko), Gulf of Guinea.

DISTRIBUTION: Equatorial Guinea (Bioko = Fernando Poo); and W Africa from Guinea (Mt. Nimba) to Gabon and Cameroon; limits unknown.

SYNONYMS: *canus, simus*.

COMMENTS: We do not follow Rosevear (1969) in restricting *H. alleni* to Fernando Poo, and agree with Eisentraut (1969*b*) and Robbins et al. (1980) in recognizing the species in W Africa (based on our study of Eisentraut's series and other specimens). Robbins et al. (1980) reported sympatry between *alleni* and *stella* in S Cameroon, as well as cranial and chromosomal distinctions. However, the morphological differences they noted are slight and variable even with a single sample, the karyotypes have the same 2N of 46 and only differ slightly in fundamental numbers (68 versus 70), and they did not demonstrate that the contrasts did not represent intrapopulational variation. The chromosomal contrasts are not impressive considering that a sample of *H. stella* from Burundi had 2N=48 and NF=86 (Maddalena et al., 1989). Some authors (Heim de Balsac and Aellen, 1965; Brosset et al., 1965) regard *simus* as the species distributed throughout W Africa and into Angola, and *stella* to be the form in C Africa, but we follow Rosevear (1969) in not being able to distinguish any sample that could be called *simus*. The definitions of both *alleni* and *stella* on continental Africa need clarification, as Rosevear (1969) so cogently discussed.

Hylomyscus baeri Heim de Balsac and Aellen, 1965. Biologia Gabonica, 1:175.

TYPE LOCALITY: Ivory Coast, Adiopodoume.

DISTRIBUTION: Recorded only from Ivory Coast and Ghana.

COMMENTS: Part of the murine fauna endemic to W Africa (see account of *Grammomys buntingi*). Distributional and morphometric data summarized by Robbins and Setzer (1979).

Hylomyscus carillus (Thomas, 1904). Ann. Mag. Nat. Hist., ser. 7, 13:418.
TYPE LOCALITY: Angola, Andongo, Pungo, 1200 m.
DISTRIBUTION: N Angola.
COMMENTS: Associated with either *H. aeta* (see Hatt, 1940*a*) or *H. alleni* (see Allen, 1939), the affinities of *carillus* are closer to *H. stella* (based on our study of the holotype and series from Pungo in the British Museum of Natural History).

Hylomyscus denniae (Thomas, 1906). Ann. Mag. Nat. Hist., ser. 7, 18:144.
TYPE LOCALITY: Uganda, Mubuku Valley, Ruwenzori East, 7000 ft.
DISTRIBUTION: Montane forest islands from WC Angola, extreme E Zaire, Uganda W Rwanda, and Kenya through Tanzania to NE Zambia.
SYNONYMS: *anselli, endorobae, vulcanorum*.
COMMENTS: The species was superficially reviewed by Hatt (1940*a*) and more thoroughly by Bishop (1979). The records of *H. carillus* from Chitau and Hanha in WC Angola (Hill and Carter, 1941:98) are based on examples of *H. denniae*. additional samples (in the Field Museum of Natural History) were collected in the same region on Mt. Moco and Mt. Soque. Although a long way west from the nearest records of *H. denniae*, most morphological characteristics of the Angolan series fall within the range of variation among samples now defined as *H. denniae*. That variation, however, is appreciable, especially in body size, and its significance in determining whether one or more species is present in what is now regarded as *denniae* has to be assessed by systematic revision.

Hylomyscus parvus Brosset, Dubost, and Heim de Balsac, 1965. Biologia Gabonica, 1:149.
TYPE LOCALITY: Gabon, Belinga, 800 m.
DISTRIBUTION: N and E Zaire, S Central African Republic, N Gabon, and S Cameroon (see map in Dudu et al., 1989).
COMMENTS: A distinctive species discussed by Dudu et al. (1989).

Hylomyscus stella (Thomas, 1911). Ann. Mag. Nat. Hist., ser. 8, 7:590.
TYPE LOCALITY: E Zaire, Ituri Forest, between Mawambi and Avakubi.
DISTRIBUTION: From Gabon, Cameroon, S Nigeria, Central African Republic, S Sudan to Zaire, N Angola, Uganda, and W Kenya to EC Tanzania, Burundi, and Rwanda; limits unknown.
SYNONYMS: *kaimosae*.
COMMENTS: The new records from Sudan and N Angola are documented by specimens in the Field Museum of Natural History; those from Tanzania are in the British Museum of Natural History. Some east African samples were reviewed by Bishop (1979). Using electrophoretic traits, Iskandar et al. (1988) documented two species occurring together in NW Gabon. One is *stella* but they could not place a name on the other. Recognizable morphological variation exists among the samples of *H. stella* and its significance has to be assessed by critical systematic revision. The variation in 2N and FN among samples identified as *H. stella* was documented by Robbins et al. (1980) and Maddalena et al. (1989).

Hyomys Thomas, 1904. Proc. Zool. Soc. Lond., 1903(2):198 [1904].
TYPE SPECIES: *Hyomys meeki* Thomas, 1904 (= *Mus goliath* Milne-Edwards, 1900).
COMMENTS: Member of the New Guinea Old Endemics (Musser, 1981*c*). Phallic morphology documented by Lidicker (1968). Photograph and distributional and biological data summarized by Flannery (1990*b*). Whether only one or more species are present in this genus has never been satisfactorily resolved (Flannery, 1990*b*; Rümmler, 1938; Tate, 1951), but examination of museum specimens revealed the two species listed below.

Hyomys dammermani Stein, 1933. Z. Säugetierk., 8:95.
TYPE LOCALITY: New Guinea, Irian Jaya, Weyland Range, Kunupi Mt.

DISTRIBUTION: New Guinea; from Weyland Range and Snow Mts in Irian Jaya east along Central Cordillera to Schrader Range and the Mt Hagen and Nondugl region in Papua New Guinea; limits unknown.

COMMENTS: Originally described as a subspecies of *H. meeki* (Rümmler, 1938; Stein, 1933), *dammermani* is a small-bodied species with only traces of white wisps about the ears.

Hyomys goliath (Milne-Edwards, 1900). Bull. Mus. Hist. Nat., Paris, 6:165.

TYPE LOCALITY: Papua New Guinea, Central Prov., highlands of Aroa River basin.

DISTRIBUTION: Papua New Guinea; Central Cordillera from Mt Dayman in the east to Kratke Mts in the west, and mountains of Huon Peninsula; limits unknown.

SYNONYMS: *meeki*, *strobilurus* (see Laurie and Hill, 1954).

COMMENTS: This is the large-bodied species with prominent white auricular tufts that is identified as *meeki* in the older literature.

Kadarsanomys Musser, 1981. Zool. Verhandelingen, 189:5.

TYPE SPECIES: *Rattus canus sodyi* Bartels, 1937.

COMMENTS: The only murine genus endemic to Java (Musser and Newcomb, 1983).

Kadarsanomys sodyi (Bartels, 1937). Treubia, 16:45.

TYPE LOCALITY: Indonesia, W Java, Gunung Pangrango-Gede, 1000 m.

DISTRIBUTION: Java.

COMMENTS: Morphology, natural history, and comparisons with *Rattus* and *Lenothrix* reported by Musser (1981*a*). Represented only by modern series collected in W Java during 1933-1935, and subfossil fragments from C and E Java (Musser and Newcomb, 1983). Among Sundaic murines, *Kadarsanomys* has no close phylogenetic allies, but some cranial and dental traits suggest a distant relationship with *Rattus* (Musser and Newcomb, 1983).

Komodomys Musser and Boeadi, 1980. J. Mammal., 61:397.

TYPE SPECIES: *Rattus rintjanus* Sody, 1941.

COMMENTS: An endemic of Nusa Tenggara, Indonesia.

Komodomys rintjanus (Sody, 1941). Treubia, 18:310.

TYPE LOCALITY: Indonesia, Nusa Tenggara, Pulau Rintja, Lohoboeaja.

DISTRIBUTION: Nusa Tenggara: islands of Rintja, Padar, and Flores; probably occurs on other islands in the Lesser Sunda chain (e.g., Komodo).

COMMENTS: Dental morphology tied *Komodomys* to *Papagomys*, another endemic of Nusa Tenggara found only on Flores Isl (Musser, 1981*c*; Musser and Boeadi, 1980).

Lamottemys Petter, 1986. Cimbebasia, Ser. A, 8:98.

TYPE SPECIES: *Lamottemys okuensis* Petter, 1986.

COMMENTS: A distinctive genus whose closest phylogenetic relative is probably *Oenomys* (Dieterlen and Van der Straeten, 1988; Petter, 1986).

Lamottemys okuensis Petter, 1986. Cimbebasia, Ser. A, 8:98.

TYPE LOCALITY: W Cameroon, Mt Oku.

DISTRIBUTION: Known only by a small sample from Mt Oku (Dieterlen and Van der Straeten, 1988; Fülling, 1992; Petter, 1986).

Leggadina Thomas, 1910. Ann. Mag. Nat. Hist., ser. 8, 6:606.

TYPE SPECIES: *Mus forresti* Thomas, 1906.

COMMENTS: Sometimes included in *Pseudomys*, but a distinctive genus that is a member of the Australian Old Endemics (Musser, 1981*c*:167), which includes the Conilurini, which is where *Leggadina* is usually placed (Baverstock, 1984). Some species of *Pseudomys*, namely *delicatulus* and *hermannsburgensis*, are often included in *Leggadina* (Lidicker and Brylski, 1987) but only *forresti* and *lakedownensis* belong there (Mahoney and Posamentier, 1975; Mahoney and Richardson, 1988; Watts and Aslin, 1981). Data from microcomplement fixation of albumin indicates that *L. forresti* is so distinct from *Pseudomys* that it forms a separate clade that is also isolated from all the other Australian endemics (Watts et al., 1992).

Morphology of the male reproductive tract, external anatomy of the glans penis, and spermatozoal structure documented in context of comparative study of

Australian murines by Breed (1980, 1984, 1986), Morrissey and Breed (1982), and Lidicker and Brylski (1987). Biochemical and chromosomal data discussed by Baverstock et al. (1976*a*, 1981, 1983*b*). Taxonomic, distributional, and biological references for species catalogued by Mahoney and Richardson (1988).

Leggadina forresti (Thomas, 1906). Abstr. Proc. Zool. Soc. Lond., 1906(32):6.
TYPE LOCALITY: Australia, Northern Territory, Alexandria (for additional information, see Mahoney and Richardson, 1988).
DISTRIBUTION: Inland Australia; W half of Queensland, NW New South Wales, N South Australia, S Northern Territory, and scattered in Western Australia, incl. Thevenard Isl (see map in Watts and Aslin, 1981).
SYNONYMS: *berneyi, messorius, waitei.*

Leggadina lakedownensis Watts, 1976. Trans. R. Soc. S. Aust., 100:105.
TYPE LOCALITY: Australia, Queensland, Lakeland Downs, 110 km north of Cooktown.
DISTRIBUTION: Australia; N Queensland, where it has been recorded only from region of Princess Charlotte Bay and Lakeland Downs (see Watts and Aslin, 1981:211), and Western Australia at Kimberley (specimens in the Western Australian Museum; Watts, in litt.).
STATUS: IUCN - Insufficiently known.
COMMENTS: A distinctive species distinguished from its close relative *L. forresti* by a suite of morphological, biochemical, and chromosomal traits (Baverstock et al., 1976*a*; Watts, 1976).

Lemniscomys Trouessart, 1881. Bull. Soc. Etudes Sci. Angers, 10:124.
TYPE SPECIES: *Mus barbarus* Linnaeus, 1766.
COMMENTS: Chromosomal data is summarized for several species by Gautun et al. (1985) and Filippucci et al. (1986). Karyological and morphological comparisons between *L. striatus* and *L. bellieri* are reported by Van der Straeten and Verheyen (1978*a*).

Lemniscomys barbarus (Linnaeus, 1766). Syst. Nat., 12th ed., 1:addenda.
TYPE LOCALITY: Morocco (= "Barbaria," see Allen, 1939).
DISTRIBUTION: Recorded from Tunisia, Algeria, Morocco, W Sahara, Senegal, Gambia, Ivory Coast, Ghana, Togo, Benin, Burkina Faso, Nigeria, Cameroon, Sudan, Ethiopia, Kenya N Uganda, Tanzania, and E Zaire.
SYNONYMS: *albolineatus, convictus, dunni, ifniensis, manteufeli, nigeriae, nubalis, olga, orientalis* (Hatt, 1935, not Desmarest, 1819), *oweni, spekei, zebra.*
COMMENTS: Significance of the considerable variation in coat color and pattern as well as body size among geographic samples will have to be assessed by a careful systematic revision. Closest relative is *L. hoogstraali.*

Lemniscomys bellieri Van der Straeten, 1975. Rev. Zool. Afr., 89:906.
TYPE LOCALITY: Ivory Coast, Ayeremou (= Lamto).
DISTRIBUTION: Guinea and Doka woodland of Ivory Coast.
COMMENTS: Part of the murine fauna endemic to W Africa (see account of *Grammomys buntingi*). A relative of *L. macculus*, according to Van der Straeten (1975), who also recorded chromosomal information.

Lemniscomys griselda (Thomas, 1904). Ann. Mag. Nat. Hist., ser. 7, 13:414.
TYPE LOCALITY: Angola, Jinga country, Muene Coshi.
DISTRIBUTION: Known only from Angola.
COMMENTS: Morphometrically related to *L. rosalia* (Van der Straeten, 1980*b*).

Lemniscomys hoogstraali Dieterlen, 1991. Bonn. Zool. Beitr., 42:11.
TYPE LOCALITY: Sudan, Upper Nile Prov, Paloich, 12 mi N Niayok.
DISTRIBUTION: Known only from the type locality.
COMMENTS: Known only by the holotype. A member of the *Lemniscomys barbarus* group. Dieterlen (1991) summarized current knowledge of this species.

Lemniscomys linulus (Thomas, 1910). Ann. Mag. Nat. Hist., ser. 8, 6:429.
TYPE LOCALITY: Senegal (= French Gambia), Gamon.
DISTRIBUTION: Sudan savanna and forest clearings in Senegal and Ivory Coast (see map in Van der Straeten, 1980*a*).

COMMENTS: Once treated as a subspecies of *L. griselda* (Allen, 1939), but now considered a separate species related to *L. griselda* and *L. rosalia*.

Lemniscomys macculus (Thomas and Wroughton, 1910). Trans. Zool. Soc. London, 19:515.
TYPE LOCALITY: Uganda, SE Ruwenzori, Mokia.
DISTRIBUTION: Recorded from Savannahs of NE Zaire, S Sudan, Ethiopia, Uganda, and Kenya; limits unknown.
SYNONYMS: *akka*.
COMMENTS: Reviewed by Van der Straeten and Verheyen (1979*a*). The species is often confused with *L. striatus*, but occurs sympatric with it (Hollister, 1919).

Lemniscomys mittendorfi Eisentraut, 1968. Bonn. Zool. Beitr., 19:7.
TYPE LOCALITY: Cameroon, Lake Oku.
DISTRIBUTION: Known only from the type locality.
COMMENTS: Included in *striatus* by Misonne (1974), but treated as a separate species by Van der Straeten and Verheyen (1980), who also suggested it has morphometric affinities with *L. macculus* and *L. bellieri*.

Lemniscomys rosalia (Thomas, 1904). Ann. Mag. Nat. Hist., ser. 7, 13:414.
TYPE LOCALITY: Tanzania, Nguru Mtns, Monda.
DISTRIBUTION: N Namibia, South Africa (E Natal and Zululand, C and N Transvaal), E Swaziland, Zimbabwe, C and N Botswana, Mozambique, Zambia, Malawi, Tanzania, and S Kenya.
SYNONYMS: *calidior; dorsalis* (Smith, 1845, not Fischer, 1814), *fitzsimonsi, maculosus, mearnsi phaeotis, sabiensis, sabulatus, spinalis, zuluensis*.
COMMENTS: Once included in *L. griselda* (Allen, 1939), but now considered a distinct species (Van der Straeten, 1980*b*).

Lemniscomys roseveari Van der Straeten, 1980. Ann. Cape Prov. Mus. Nat. Hist., 13(5):55.
TYPE LOCALITY: Zambia, Zambezi (= Balovale), 1015 m (see Van der Straeten, 1980*b*, for additional data).
DISTRIBUTION: Known only from the type locality and Solwezi in Zambia.
COMMENTS: Related to the *L. rosalia* complex (Van der Straeten, 1980*b*).

Lemniscomys striatus (Linnaeus, 1758). Syst. Nat., 10th ed., 1:62.
TYPE LOCALITY: "India" (= Sierra Leone; see Allen, 1939:394).
DISTRIBUTION: From Burkina Faso and Sierra Leone west to Ethiopia, and south into NW Angola and through Kenya, Uganda, Rwanda, Zaire, and Tanzania into NE Zambia and N Malawi.
SYNONYMS: *ardens, dieterleni, fasciatus, lulae, lynesi, massaicus, micropus, orientalis* (Desmarest, 1819, not Hatt, 1935), *pulchella, pulcher, spermophilus, venustus, versustos, wroughtoni*.
COMMENTS: Reviewed by Van der Straeten and Verheyen (1980).

Lenomys Thomas, 1898. Trans. Zool. Soc. London, 14:409.
TYPE SPECIES: *Mus meyeri* Jentink, 1879.
COMMENTS: A Sulawesi endemic reviewed by Musser (1970*d*, 1981*c*, 1984). Simpson (1945) listed *Lenomys* as a member of the Phloeomyinae along with *Coryphomys, Pogonomys, Mallomys, Phloeomys, Chiropodomys*, and *Crateromys*, but no evidence supports this arrangement. External, cranial, and spermatozoal traits tied *Lenomys* phylogenetically close to *Eropeplus*, another Sulawesian endemic (Breed and Musser, 1991; Musser, 1981*c*).

Lenomys meyeri (Jentink, 1879). Notes Leyden Mus., 1:12.
TYPE LOCALITY: Indonesia, Sulawesi, NE peninsula, Menado.
DISTRIBUTION: Recorded only from N, C, and SW Sulawesi.
SYNONYMS: *lampo, longicaudus*.
COMMENTS: Represented by a small extant series from the N peninsula and central core, and by both modern specimens and subfossils from the SW peninsula of Sulawesi (Musser, 1970*d*; 1984). A second species (yet to be named and described) is known only from subfossil fragment collected in SW arm of Sulawesi (Musser and Holden, 1991).

Lenothrix Miller, 1903. Proc. U.S. Natl. Mus., 26(1317):466.

TYPE SPECIES: *Lenothrix canus* Miller, 1903.

COMMENTS: Chromosomal and biochemical data suggested a close phylogenetic link between *Lenothrix* and *Niviventer* (Chan et al., 1979), but derived dental morphology is shared with *Pithecheir*, and spermatozoal conformation is highly divergent from any Sundaic murine. In sum, *Lenothrix* is a Sundaic endemic characterized by many primitive external, cranial, dental, and chromosomal features and a few derived dental and spermatozoal traits; despite several claims, its phylogenetic relationships still require illumination (see discussions in Breed and Yong, 1986; Musser, 1981*b*; Musser and Newcomb, 1983).

Lenothrix canus Miller, 1903. Proc. U.S. Natl. Mus., 26(1317):466.

TYPE LOCALITY: Indonesia, Tuangku Isl (west of Sumatra).

DISTRIBUTION: Malay Peninsula, Penang Isl, Tuangku Isl, and Borneo (Sarawak and Sabah).

SYNONYMS: *malaisia*.

COMMENTS: Historical allocations of *canus* to either *Rattus* or *Lenothrix*, as well as comparisons between *L. canus* and other Sundaic endemics documented by Musser (1981*a-c*) and Musser and Newcomb (1983).

Leopoldamys Ellerman, 1947. Proc. Zool. Soc. Lond., 117:267.

TYPE SPECIES: *Mus sabanus* Thomas, 1887.

COMMENTS: Definition and contrasts with *Rattus* and *Niviventer* provided by Musser (1981*b*), who also reviewed morphological, chromosomal, and distributional characteristics. *Leopoldamys* is dentally similar to *Berylmys, Maxomys*, and *Niviventer*. Sperm morphology united *Leopoldamys* with *Berylmys, Sundamys*, and *Rattus* (Breed and Yong, 1986), but alliance is based on shared spermatozoal form that is likely primitive. Analyses of chromosomal traits suggested *Leopoldamys* is more closely related to *Bandicota, Berylmys, Nesokia, Rattus*, and *Sundamys*, than to *Lenothrix, Maxomys*, or *Niviventer* (Gadi and Sharma, 1983); biochemical data separated *Leopoldamys* far from *Rattus* (Chan et al., 1979).

Leopoldamys edwardsi (Thomas, 1882). Proc. Zool. Soc. Lond., 1882:587.

TYPE LOCALITY: China, mountains of W Fujian (probably Kuatun).

DISTRIBUTION: India (W Bengal, Sikkim, N Assam), N Burma, S and C China (to S Anhui; Liu et al., 1985), N Thailand, Laos, Vietnam, Malay Peninsula, and W Sumatra.

SYNONYMS: *ciliatus, garonum, gigas, hainanensis, listeri, melli, milleti, setiger*.

COMMENTS: Requires taxonomic revision. Samples from Indochina may represent a different species than the one sampled from Malay Peninsula and Sumatra (Musser, 1981*b*). In additon to synonyms assembled by Musser (1981*b*), Xu and Yu (1985) recently described *hainanensis*. Phallic morphology described by Yang and Fang (1988) in context of assessing phylogenetic relationships among Chinese murines.

Leopoldamys neilli (Marshall, 1976). Family Muridae: rats and mice, p. 485. Privately printed by Government Printing Office, Bangkok.

TYPE LOCALITY: Thailand, Saraburi Prov, Kaengkhoi Dist, "outside the entrance to the bat cave, half-way up the face of a wooded limestone cliff, 200 meters altitude."

DISTRIBUTION: C and W Thailand.

COMMENTS: A Thai endemic. See Musser (1981*b*:237) for discussion of original publication.

Leopoldamys sabanus (Thomas, 1887). Ann. Mag. Nat. Hist., ser. 5, 20:269.

TYPE LOCALITY: Malaysia, Sabah (N Borneo), Gunung Kinabalu.

DISTRIBUTION: Bangladesh, Thailand, Vietnam, Cambodia, Laos, Malay Peninsula, Sumatra, Java, Borneo, and smaller islands on the Sunda Shelf.

SYNONYMS: *balae, bunguranensis, clarae, dictatorius, fremens, heptneri, herberti, insularum, lancavensis, lucas, luta, macrourus, mansalaris, masae, matthaeus, mayapahit, nasutus, revertens, salanga, stentor, strepitans, stridens, stridulus, tapanulius, tersus, tuancus, ululans, vociferans*.

COMMENTS: Appreciable morphological variation exists between samples from north and south of Isthmus of Kra, and among insular samples from the Sunda Shelf; systematic revision is required to assess whether variation is characteristic of one or more than one species (Musser, 1981*b*). The name *macrourus* has priority over *sabanus*, but is based on a specimen of uncertain origin (Musser, 1981*b*).

Leopoldamys siporanus (Thomas, 1895). Ann. Mus. Civ. Stor. Nat. Genova, 34:11.

TYPE LOCALITY: Indonesia, Kepulauan Mentawai, Pulau Sipora.

DISTRIBUTION: Endemic to Mentawai Archipelago; islands of Siberut, Sipora, North Pagai, and South Pagai.

SYNONYMS: *soccatus*.

COMMENTS: *L. siporanus* joins *Maxomys pagensis, Chiropodomys karlkoopmani, Rattus lugens, Iomys sipora, Hylopetes sipora, Petinomys lugens, Callosciurus melanogaster, Sundasciurus fraterculus*, and *Lariscus obscurus* as part of the rodent fauna endemic to the Mentawi Archipelago.

Leporillus Thomas, 1906. Ann. Mag. Nat. Hist., ser. 7, 17:83.

TYPE SPECIES: *Hapalotis apicalis* Gould, 1853.

COMMENTS: Member of the Australian Old Endemics (Musser, 1981*c*:167), which includes the Conilurini where Baverstock (1984) listed *Leporillus*. Mahoney and Richardson (1988) cataloged taxonomic, distributional, and biological references.

Leporillus apicalis (Gould, 1853). Proc. Zool. Soc. Lond., 1853:126.

TYPE LOCALITY: Australia, South Australia.

DISTRIBUTION: Australia: once found in S Northern Territory, C and SE Western Australia, South Australia, and western parts of New South Wales and Victoria, but now presumed to be extinct; extent of former range indicated by specimens caught in late 1800's and early 1900's and distribution of empty stick nests (Mahoney and Richardson, 1988:159; Watts and Aslin, 1981:152).

STATUS: IUCN - Extinct.

Leporillus conditor (Sturt, 1848). Narr. Exped. C. Aust., 1:120.

TYPE LOCALITY: Australia, New South Wales, Polia area, about 45 miles from Laidley Ponds.

DISTRIBUTION: Australia; once ranged on mainland from lower Darling River to Nullarbor Plain in New South Wales, South Australia, and SE corner of Western Australia, and presumed to be extinct; living population on Franklin Isl in the Nuyt's Archipelago of W South Australia (Mahoney and Richardson, 1988:160; Watts and Aslin, 1981:147).

STATUS: CITES - Appendix I; U.S. ESA - Endangered; IUCN - Rare.

SYNONYMS: *jonesi*.

COMMENTS: Phylogenetic significance of spermatozoal morphology reported by Breed and Sarafis (1978); chromosomal morphology described by Baverstock et al. (1977*c*). Electrophoretic data indicated *L. conditor* is phylogenetically closely allied to *Pseudomys* (Baverstock et al., 1981), but information from analyses of phallic and dental morphology placed *L. conditor* in same monophyletic group as *Conilurus* and *Mesembriomys*, to the exclusion of *Pseudomys* (Lidicker and Brylski, 1987; Misonne, 1969), which is also supported by albumin data (Watts et al., 1992).

Leptomys Thomas, 1897. Ann. Mus. Civ. Stor. Nat. Genova, 18:610.

TYPE SPECIES: *Leptomys elegans* Thomas, 1897.

COMMENTS: Member of the New Guinea Old Endemics (Musser, 1981*c*). Most lists and faunal studies published since 1951 recognized only one species in *Leptomys* (Flannery, 1990*b*; Laurie and Hill, 1954; Menzies and Dennis, 1979; Tate, 1951), but Rummler's (1938) revision in which he identified two species (*elegans* and *ernstmayri*), and Tate and Archbold's (1938) description of a third (*signatus*) accurately reflects the known diversity. A derived cephalic arterial pattern, along with other morphological features, is shared by *Leptomys, Mayrmys, Neohydromys, Pseudohydromys*, and *Lorentzimys* (Musser and Heaney, 1992). Certain phallic traits (Lidicker and Brylski, 1987), also united *Leptomys* and *Lorentzimys*. Chromosomal data were provided by Donnellan (1987), but whether the sampled species was *L. elegans* or *L. ernstmayri* is unclear.

Leptomys elegans Thomas, 1897. Ann. Mus. Civ. Stor. Nat. Genova, 18:610.

TYPE LOCALITY: Papua New Guinea (= "British New Guinea"), Central Prov., Astrolabe Range behind Port Moresby (according to Tate, 1951:223, but Laurie and Hill, 1954:132 noted that no exact locality was published).

DISTRIBUTION: Papua New Guinea; known only from Owen Stanley Range below 1250 m from region near and northwest of Port Morseby east to Mt Dayman; limits unknown.

COMMENTS: Poorly represented in museum collections.

Leptomys ernstmayri Rümmler, 1932. Das Aquarium, 6:134.
TYPE LOCALITY: Papua New Guinea, Morobe Prov., Huon Peninsula, Saruwaged Mts, Ogeramnang, 1785 m.
DISTRIBUTION: Papua New Guinea (above 1250 m in Owen Stanley Range from Mt Dayman west to Kratke Mts and Purosa region in Eastern Highlands Prov., then northeast to mountains on Huon Peninsula; probably occurs farther west in Papuan Central Cordillera, but limits unknown); Irian Jaya (Arfak Mts only; limits unknown).
COMMENTS: A distinctive species distinguished from *L. elegans* by the diagnostic and distributional traits enumerated by Rümmler (1932). It is endemic to mountain rainforest and in Owen Stanley Range replaces *L. elegans* at elevations above 1000 m.

Leptomys signatus Tate and Archbold, 1938. Am. Mus. Novit., 982:2.
TYPE LOCALITY: Papua New Guinea, Western Prov., Sturt Isl Camp, Fly River (N bank), near Fairfax Isls, sea level.
DISTRIBUTION: Known only from the type locality along lower Fly River; limits unknown.
COMMENTS: The diagnostic traits Tate and Archbold (1938) used to separate this species from the other two identify a lowland population of *Leptomys* that shows no morphological intergradation with samples of either *L. elegans* or *L. ernstmayri*. This species is still known only by the four examples in the type series.

Limnomys Mearns, 1905. Proc. U.S. Natl. Mus., 28:451.
TYPE SPECIES: *Limnomys sibuanus* Mearns, 1905.
COMMENTS: Taxonomic history and past erroneous association with *Rattus* reviewed by Musser (1977*b*), and Musser and Heaney (1992). Revised by Musser and Heaney (1992), who considered *Limnomys* a member of the Philippine New Endemics.

Limnomys sibuanus Mearns, 1905. Proc. U.S. Natl. Mus., 28:452.
TYPE LOCALITY: Philippines, SE Mindanao, Mt Apo, 6600 ft.
DISTRIBUTION: Between 6200 and 9000 ft on Mt Apo and Mt Malindang, Mindanao.
SYNONYMS: *mearnsi*.
COMMENTS: Known only by seven specimens.

Lophuromys Peters, 1874. Monatsb. K. Preuss. Akad. Wiss. Berlin, p. 234.
TYPE SPECIES: *Lasiomys afer* Peters, 1866 (= *Mus sikapusi* Temminck, 1853).
SYNONYMS: *Kivumys*, *Lasiomys* (Peters, 1866, not Burmeister, 1854), *Neanthomys*.
COMMENTS: Revised by Dieterlen (1976*b*, 1987), who recognized two subgenera, *Lophuromys* and *Kivumys* (1987). Related to *Acomys* and *Uranomys* (our studies; Watts, in litt.; Denys and Michaux, 1992).

Lophuromys cinereus Dieterlen and Gelmroth, 1974. Z. Säugetierk., 39:338.
TYPE LOCALITY: Zaire, Marais Mukaba, Parc National du Kahuzi-Biega.
DISTRIBUTION: Known only from the type locality.
COMMENTS: Subgenus *Lophuromys*. Apparently occurs in swampy areas above 2000 m in montane regions of central African Rift Valley. The specific status of *cinereus* was questioned by Dieterlen (1987).

Lophuromys flavopunctatus Thomas, 1888. Proc. Zool. Soc. Lond., 1888:14.
TYPE LOCALITY: Ethiopia, Shoa, probably Ankober (see explanation in Allen, 1939).
DISTRIBUTION: From NE Angola throughout Zaire, Uganda, Kenya, and south through Tanzania, to Malawi, N Zambia, and N Mozambique; an isolated segment in Ethiopia. See map in Dieterlen (1976*b*).
SYNONYMS: *aquilus*, *brevicaudus*, *brunneus*, *chrysopus*, *laticeps*, *major*, *margarettae*, *rita*, *rubecula*, *simensis*, *zaphiri*, *zena*.
COMMENTS: Subgenus *Lophuromys*. Chromosomal data for sample from Burundi was reported by Maddalena et al. (1989). The appreciable character variation among samples probably reflects more than one species.

Lophuromys luteogaster Hatt, 1934. Am. Mus. Novit., 708:4.
TYPE LOCALITY: Zaire, Ituri Dist, Medje.
DISTRIBUTION: Recorded from NE and E Zaire (Medje, Irangi, Bafwasende, and Tungula (see map in Dieterlen, 1987); limits unknown.

COMMENTS: Subgenus *Kivumys*. References to this poorly known species were summarized by Dieterlen (1975, 1987).

Lophuromys medicaudatus Dieterlen, 1975. Bonn. Zool. Beitr., 26:295.
TYPE LOCALITY: Zaire, Lemera-Nyabutera.
DISTRIBUTION: Montane forests of E Zaire and Rwanda above 2000 m (see map in Dieterlen, 1987); a montane Western Rift endemic.
COMMENTS: Subgenus *Kivumys*. Information summarized by Dieterlen (1975, 1987).

Lophuromys melanonyx Petter, 1972. Mammalia, 36:177.
TYPE LOCALITY: Ethiopia, Bale Dist, Dinshu.
DISTRIBUTION: S and C Ethiopia, west of the Ethiopian Rift Valley (Dieterlen, 1987) and east of the Rift Valley in the Bale Mtns, 3100-4050 m.
COMMENTS: Subgenus *Lophuromys*. Endemic to Afro-alpine moorland of Ethiopia where it shares habitat with the other moorland specialists *Stenocephalemys albocaudata, Arvicanthis blicki, Tachyoryctes macrocephalus*, and *Otomys typus* (Demeter and Topal, 1982; Rupp, 1980; Yalden, 1988; Yalden et al., 1976).

Lophuromys nudicaudus Heller, 1911. Smithson. Misc. Coll., 56(17):11.
TYPE LOCALITY: Cameroon, Bula country, Efulen.
DISTRIBUTION: Recorded from W Cameroon, Equatorial Guinea (incl. Bioko = Fernando Poo), and Gabon (see map in Dieterlen, 1978*b*).
SYNONYMS: *naso*.
COMMENTS: Subgenus *Lophuromys*. Included in *L. sikapusi* by Misonne (1974), but is a distinct species (see discussion in Rosevear, 1969). Dieterlen (1978*b*) explained why *naso* is a synonym. The karyotype of specimens from Cameroon was described by Verheyen and Van der Straeten (1980).

Lophuromys rahmi Verheyen, 1964. Rev. Zool. Bot. Afr., 69:206.
TYPE LOCALITY: Zaire, Kivu, Bogamanda near Lemera.
DISTRIBUTION: Recorded from montane forest in E Zaire (Kivu) and Rwanda above 1800 m; limits unknown, possibly a montane Western Rift endemic.
COMMENTS: Subgenus *Lophuromys*. Distributional and biological information summarized by Verheyen (1964*a*), Van der Straeten and Verheyen (1983), and Dieterlen (1976*b*, 1987).

Lophuromys sikapusi (Temminck, 1853). Esquisses Zool. sur la Côte de Guine, p. 160.
TYPE LOCALITY: Ghana, Dabacrom.
DISTRIBUTION: From Sierra Leone through West Africa to Zaire, Uganda, and W Kenya; also in N Angola (see map in Dieterlen, 1976*b*:10).
SYNONYMS: *afer, ansorgei, eisentrauti, manteufeli, pyrrhus, tullbergi*.
COMMENTS: Subgenus *Lophuromys*. West African samples were reviewed by Rosevear (1969). Dieterlen (1978*b*) described *eisentrauti* as a subspecies in the Bamenda highlands of W Camaroon, but Hutterer et al. (1992) raised it to species rank. Chromosomal information summarized by Gautun et al. (1986).

Lophuromys woosnami Thomas, 1906. Ann. Mag. Nat. Hist., ser. 7, 18:146.
TYPE LOCALITY: Uganda, Ruwenzori East, Mubuku Valley, 6000 ft.
DISTRIBUTION: E Zaire, W Uganda, Rwanda, and Burundi; a montane Western Rift endemic.
SYNONYMS: *prittiei*.
COMMENTS: Subgenus *Kivumys*. The impressive genic differences between this species and *L. flavopunctatus* prompted Verheyen et al. (1986) to suggest that the two may belong in different subgenera, reflecting Dieterlen's (1976*b*) earlier conclusion based on morphology. Chromosomal data for sample from Burundi were reported by Maddalena et al. (1989).

Lorentzimys Jentink, 1911. Nova Guinea, 9:166.
TYPE SPECIES: *Lorentzimys nouhuysi* Jentink, 1911.
COMMENTS: Member of the New Guinea Old Endemics (Musser, 1981*c*). Photograph, distributional and biological information summarized by Flannery (1990*b*:197). Phallic morphology described by Lidicker (1968). Analyses of phallic traits (Lidicker and Brylski, 1987) indicated *Lorentzimys* and *Leptomys* are related, a hypothesis also supported by dental features and shared derived cephalic arterial pattern.

Lorentzimys nouhuysi Jentink, 1911. Nova Guinea, 9:166.
TYPE LOCALITY: New Guinea, Irian Jaya, Noord (= Lorentz River), Bivak 2400 m.
DISTRIBUTION: New Guinea; Records are from S lowlands and middle altitudes, throughout Central Cordillera from Mt Wilhelm region in the west to Mt Dayman in the east, and the Torricelli Mtns (see map in Flannery, 1990*b*:197); limits unknown.
SYNONYMS: *alticola*.
COMMENTS: Some authors recognized *alticola* as a species (Lidicker, 1968; Lidicker and Brylski, 1987; Tate, 1951). Chromosomal data reported by Donnellan (1987).

Macruromys Stein, 1933. Z. Säugetierk., 8:94.
TYPE SPECIES: *Macruromys elegans* Stein, 1933.
COMMENTS: Member of the New Guinea Old Endemics (Musser, 1981*c*). Distributional and biological information summarized by Flannery (1990*b*).

Macruromys elegans Stein, 1933. Z. Säugetierk., 8:95.
TYPE LOCALITY: New Guinea, Irian Jaya, Weyland Range, Mt Kunupi, 1400-1800 m.
DISTRIBUTION: Known only from the type locality.
STATUS: IUCN - Insufficiently known.
COMMENTS: Possibly endemic to Weyland Range. Still represented by very few specimens.

Macruromys major Rümmler, 1935. Z. Säugetierk., 10:105.
TYPE LOCALITY: Papua New Guinea, Eastern Highlands Prov., Kratke Mtns, Buntibasa Dist, 4000-5000 ft m.
DISTRIBUTION: New Guinea; known from a few mid-altitude localities on N slopes of Central Cordillera in Irian Jaya and Papua New Guinea, and on Huon Peninsula (Mt Rawlinson); see map in Flannery (1990*b*); limits unknown.
COMMENTS: Phallic morphology and its systematic significance described by Lidicker (1968).

Malacomys Milne-Edwards, 1877. Bull. Sci. Soc. Philom. Paris, ser. 6, 12:10.
TYPE SPECIES: *Malacomys longipes* Milne-Edwards, 1877.
COMMENTS: Study of dental features led Misonne (1969:106) to write that "*Malacomys* stands apart from all the other African genera by its unusual dental characters. This is an advanced genus which cannot be clearly related to any other one." However, external, cranial, and dental traits indicates *Malacomys* to be phylogenetically closely allied with *Praomys*, particularly through the morphologically annectant *M. lukolelae*, a species usually included in *Praomys* and relegated to the *P. tullbergi* complex (Van der Straeten and Dudu, 1990). Two revisions (Rautenbach and Schlitter, 1978; Van der Straeten and Verheyen, 1979*b*) provided somewhat different evaluations of morphological variation and its interpretation in context of specific diversity.

Malacomys cansdalei Ansell, 1958. Ann. Mag. Nat. Hist., ser. 13, 1:342.
TYPE LOCALITY: Ghana, Oda.
DISTRIBUTION: Forest zone of S Ghana, S Ivory Coast, and E Liberia.
SYNONYMS: *giganteus*.
COMMENTS: Part of the murine fauna endemic to West Africa (see account of *Grammomys buntingi*). Rautenbach and Schlitter (1978) considered *cansdalei* a subspecies of *M. longipes*.

Malacomys edwardsi Rochebrune, 1885. Bull. Sci. Soc. Philom. Paris, ser. 7, 9:87.
TYPE LOCALITY: Liberia (= Rep. of Guinea), Mellacoree (= Melikhoure) River.
DISTRIBUTION: Sierra Leone, Guinea, Liberia, and southern regions of Ivory Coast, Ghana, and Nigeria.
COMMENTS: Part of the murine fauna endemic to West Africa (see account of *Grammomys buntingi*). Chromosomal data reported by Matthey (1958) and Van der Straeten and Verheyen (1979*b*).

Malacomys longipes Milne-Edwards, 1877. Bull. Sci. Soc. Philom. Paris, ser. 6, 12:10.
TYPE LOCALITY: Gabon, Gaboon River (vicinity of Ogooue, Gabon; see Rautenbach and Schlitter, 1978:414).
DISTRIBUTION: Guinea (Mt Nimba) eastward to S Sudan, Uganda and Rwanda, south to NW Zambia and NE Angola.

SYNONYMS: *australis, centralis, giganteus, wilsoni.*
COMMENTS: Chromosomal data reported by Viegas-Pequignot et al. (1983).

Malacomys lukolelae (Hatt, 1934). Am. Mus. Novit., 708:13.
TYPE LOCALITY: Zaire, Lukolela.
DISTRIBUTION: Known only from the type locality.
COMMENTS: Petter (1975*c*) recorded *lukolelae* from Central African Republic, but the specimens we have seen from that series represent an undescribed species of *Praomys; M. lukolelae* is still only represented by the three specimens in the type series (Van der Straeten and Dudu, 1990). Originally described by Hatt (1934*b*) as a subspecies of *Praomys tullbergi*, but *lukolelae*'s long and very slim hind feet, tip of short fifth digit extending only to base of digital pad of fourth digit, very large ears, cranial conformation, and molar dental patterns associate it with *Malacomys*, especially *M. verschureni.*

Malacomys verschureni Verheyen and Van der Straeten, 1977. Rev. Zool. Afr., 91:739.
TYPE LOCALITY: Zaire, Mamiki.
DISTRIBUTION: Known only from NE Zaire (eastern edge of Central African high forest block).
COMMENTS: Original description was based on one specimen (Verheyen and Van der Straeten, 1977), and the species is still represented by very few examples (Dieterlen and Van der Straeten, 1984; Robbins and Van der Straeten, 1982). Chromosomal information reported by Robbins and Van der Straeten (1982).

Mallomys Thomas, 1898. Novit. Zool., 5:1.
TYPE SPECIES: *Mallomys rothschildi* Thomas, 1898.
SYNONYMS: *Dendrosminthus.*
COMMENTS: Listed as a member of the Phloeomyinae by Simpson (1945), but no data supports such an allocation. Member of the New Guinea Old Endemics (Musser, 1981*c*). Revised by Flannery et al. (1989).

Mallomys aroaensis (De Vis, 1907). Ann. Queensl. Mus., 7:11.
TYPE LOCALITY: Papua New Guinea, Central Prov., head of Aroa River (see Flannery et al., 1989, for other data).
DISTRIBUTION: New Guinea; Papuan Central Cordillera from Mt Sisa in the west to Mt Simpson in the east; also highlands of Huon Peninsula (see map in Flannery et al., 1989:95).
SYNONYMS: *hercules.*

Mallomys gunung Flannery, Aplin, Groves, and Adams, 1989. Rec. Aust. Mus., 41:101.
TYPE LOCALITY: New Guinea, Irian Jaya, 2 km E Mt Wilhelmina, 3800 m (see Flannery et al., 1989, for details).
DISTRIBUTION: New Guinea, Irian Jaya, Snow Mts; known only from vicinity of the type locality and Mt Carstenz (see map in Flannery et al., 1989:97).
COMMENTS: A very distinctive species known only by a few specimens collected between 2400 and 4000 m.

Mallomys istapantap Flannery, Aplin, Groves, and Adams, 1989. Rec. Aust. Mus., 41:96.
TYPE LOCALITY: Papua New Guinea, Mt Hagen Dist., Korelum (Flannery et al., 1989, provided additional data).
DISTRIBUTION: New Guinea; high altitudes along Central Cordillera from Mt Victoria in Owen Stanley Range of E Papua New Guinea to Bele River region in Irian Jaya.
COMMENTS: Published records are only from the Central Cordillera in Papua New Guinea (see map in Flannery et al., 1989:97), but one specimen (in the American Museum of Natural History, 151342) comes from 18 km N of Lake Habbema in Irian Jaya.

Mallomys rothschildi Thomas, 1898. Novit. Zool., 5:2.
TYPE LOCALITY: Papua New Guinea, Central Prov., Owen Stanley Range, between Mt Musgrave and Mt Scratchley, 5000-6000 ft (Flannery et al., 1989 provided additional data).
DISTRIBUTION: New Guinea; Central Cordillera, from Weyland Range in W Irian Jaya to Owen Stanley Range in Papua New Guinea (see map in Flannery et al., 1989:93).
SYNONYMS: *argentata, weylandi.*

COMMENTS: *M. rothschildi* and *M. aroaensis* are thought to be sympatric at a few localities (Flannery et al., 1989), but the morphological variation present among samples requires re-examination to test whether it reflects presence of one or two species.

Malpaisomys Hutterer, Lopez-Martinez, and Michaux, 1988. Palaeovertebrata, 18:246.
TYPE SPECIES: *Malpaisomys insularis* Hutterer, Lopez-Martinez, and Michaux, 1988.
COMMENTS: Represented only by Late Pleistocene and Holocene samples.

Malpaisomys insularis Hutterer, Lopez-Martinez, and Michaux, 1988. Palaeovertebrata, 18:246.
TYPE LOCALITY: Canary Islands, Fuerteventura Isl, Cueva Villaverde near La Oliva, at stratigraphic level dated about 1070 before present (see Hutterer et al., 1988 for additional information).
DISTRIBUTION: Recorded from the islands of Fuerteventura, Lanzorote, and Graciosa in the Canary Isls (see map in Hutterer et al., 1988).
COMMENTS: *M. insularis*, the lava mouse, has been recorded from sediments dated between 25,000 and 32,000 years before present up to historical times when the species became extinct, sometime between 800 years before present and now (Hutterer et al., 1988; Michaux et al., 1991). Reconstruction and study of the postcranial skeleton suggest the species was adapted to living in lava fields (Boye et al., 1992). Boye et al. (1992) reported that about 2000 years before present, *Mus musculus* was apparently casually imported to the islands by humans who also arrived then, and from that time to the historical period, populations of *M. insularis* declined and were progressively replaced by house mice. This interaction between lava mice and house mice is the hypothesized causual reason for extinction of the lava mice.
Malpaisomys insularis is part of a mammalian fauna endemic to the eastern Canary Islands that includes the shrew *Crocidura canariensis* (Hutterer et al., 1987*a*; Michaux et al., 1991; Boye et al., 1992).

Margaretamys Musser, 1981. Bull. Am. Mus. Nat. Hist., 168(3):275.
TYPE SPECIES: *Mus beccarii* Jentink, 1880.
COMMENTS: Morphological, chromosomal, and distributional features monographed by Musser (1981*b*). Spermatozoal morphology distinctive and similar among the three species (Breed and Musser, 1991). Member of the Sulawesi Old Endemics (Musser, 1981*c*).

Margaretamys beccarii (Jentink, 1880). Notes Leyden Mus., 2:11.
TYPE LOCALITY: Indonesia, Sulawesi, NE peninsula, Menado-Langowah.
DISTRIBUTION: NE and C Sulawesi in lowland tropical evergreen rain forest.
SYNONYMS: *thysanurus*.
COMMENTS: History of incorrect taxonomic allocations of *beccarii* recorded by Musser (1971*e*, 1981*b*).

Margaretamys elegans Musser, 1981. Bull. Am. Mus. Nat. Hist., 168:286.
TYPE LOCALITY: Indonesia, C Sulawesi, Gunung Nokilalaki, 6500 ft.
DISTRIBUTION: Known only from 5300-7500 ft in montane rain forest on Gunung Nokilalaki, but probably occurs on mountains elsewhere in central core of Sulawesi.

Margaretamys parvus Musser, 1981. Bull. Am. Mus. Nat. Hist., 168:294.
TYPE LOCALITY: Indonesia, C Sulawesi, Gunung Nokilalaki, 7400 ft.
DISTRIBUTION: Known only from 6000-7500 ft in montane rain forest on Gunung Nokilalaki, but probably occurs elsewhere in mountainous central part of Sulawesi.

Mastomys Thomas, 1915. Ann. Mag. Nat. Hist., ser. 8, 16:477.
TYPE SPECIES: *Mus coucha* Smith, 1834.
COMMENTS: The historical taxonomic use of *Mastomys* as a genus or subgenus of either *Rattus* or *Praomys* is summarized by Meester et al. (1986). Important to studies of specific diversity, biogeography, and medicine, samples of *Mastomys* have been examined in several contexts (see reviews by Keogh and Price, 1981, and Skinner and Smithers, 1990). Definition of species by chromosomal and biochemical traits from some regions (Duplantier et al., 1990*a*; Green et al., 1980; Hubert et al., 1983) has proceeded faster than definitions based on morphology. The result is a new view of

species-diversity in the genus, but also an ignorance of morphological limits of those species and their real geographic distributions. Robbins and Van der Straeten (1989) analyzed all the taxa associated with *Mastomys* but did not allocate any of them to species. We can tie only those associated primarily with South African samples to species following Meester et al. (1986) but otherwise list no synonyms and refer readers to Robbins and Van der Straeten (1989). The genus requires careful taxonomic revision.

Mastomys angolensis (Bocage, 1890). Jorn. Sci. Math., Phys. Nat., Lisboa, ser. 2, 2:12.

TYPE LOCALITY: Angola, Capangombe, interior of Mossamedes (additional information provided by Crawford-Cabral, 1989*b*).

DISTRIBUTION: Angola and S Zaire.

SYNONYMS: *angolae*.

COMMENTS: Sometimes placed in *Myomys* or *Myomyscus* (Allen, 1939; D. H. S. Davis, 1965; Hill and Carter, 1941), *angolensis* is more closely related to species of *Mastomys* (Crawford-Cabral, 1989*b*; Misonne, 1974). It is morphologically similar to *M. shortridgei* and some authors treat the latter as a subspecies of *M. angolensis* (Ansell, 1978; Ellerman et al., 1953). *M. angolensis* is either sympatric (Hill and Carter, 1941; Crawford-Cabral, 1983) or altitudinally parapatric (specimens in the Field Museum of Natural History from Mt. Soque) with what is probably *M. natalensis*. Crawford-Cabral (1989*b*) proposed *Praomys angolae*, which is apparently nothing more than a renaming of *angolensis*.

Mastomys coucha (Smith, 1834). Rept. Exped. Exploring Central Africa, p. 43.

TYPE LOCALITY: South Africa, N Cape Prov., between Orange River and Tropic of Capricorn (see Meester et al., 1986:286).

DISTRIBUTION: South Africa (E and N Cape Prov., Zululand, Lesotho, Orange Free State, S and W Transvaal), S and W Zimbabwe, C Namibia (see map in Skinner and Smithers, 1990:270); extent of range beyond this region unresolved.

SYNONYMS: *bradfieldi, breyeri, limpopoensis, marikquensis, sicialis, socialis*.

COMMENTS: Characterized by 2N=36 (FN=56) and a distinctive hemoglobin electromorph, *coucha* occurs sympatrically with *M. natalensis*, which is distinguished by different hemoglobin pattern and 2N=32, FN=54 (Green et al., 1980). The two species also differ in cranial, phallic, and spermatozoal morphology as well as reproductive behavior, ultrasonic vocalizations, and phermones (see references in Skinner and Smithers, 1990). Synonyms listed are only those pertaining to samples from South Africa for reasons explained by Meester et al. (1986); of these, Robbins and Van der Straeten (1989) regarded *marikquensis* to be a *Myomys*, and Robbins (in Meester et al., 1986) claimed it may be a species distinct from *Myomys verreauxii*.

Mastomys erythroleucus (Temminck, 1853). Esquisses Zool. sur la Côte de Guine, p. 160.

TYPE LOCALITY: Guinea.

DISTRIBUTION: Morocco, occurs also from Gambia and Senegal eastward through W Africa to S Ethiopia and Somalia, and south through E Africa to E Zaire (Kivu Dist) and Burundi; southern limits in E Africa unresolved.

COMMENTS: Chromosomal (2N=38, FN=52) and electrophoretic data clearly set this species apart from other *Mastomys* with which it occurs (see Duplantier et al. 1990*a*, *b*, and references therein); it is sympatric with *M. hildebrandtii* (called *huberti* in the reference) and what might be *M. natalensis* in Senegal (Duplantier et al., 1990*a*).

Mastomys hildebrandtii (Peters, 1878). Monatsb. K. Preuss. Akad. Wiss., Berlin, p. 200.

TYPE LOCALITY: Kenya, Taita, Ndi.

DISTRIBUTION: From Senegal and Gambia east through West Africa to Central African Republic and N Zaire, extending to Djibuti and Somalia, then south to Burundi and Kenya; southern limits in E and W Africa unresolved.

COMMENTS: According to Qumsiyeh et al. (1990), *hildebrandtii* is an older name for *huberti*, the *Mastomys* with 2N=32, FN=44. Geographic range is abstracted from the map in Hubert et al. (1983). This is not the same species as *M. natalensis* from South Africa which is also characterized by 2N=32, but a different FN (54; Duplantier et al., 1990*a*), hemoglobin pattern (C. B. Robbins et al., 1983), and apparently serum proteins (Robbins and Van der Straeten, 1989).

Mastomys natalensis (Smith, 1834). S. Afr. Quart. J., ser. 2, 2:156.

TYPE LOCALITY: South Africa, Natal, Port Natal (= Durban).

DISTRIBUTION: Recorded from South Africa (S and E Cape Prov., Transkei, Natal, E Transvaal), Zimbabwe, C and NE Namibia (see map in Skinner and Smithers, 1990:268); apparently also EC Tanzania (Morogoro, see Leirs et al., 1989) and Senegal in W Africa.

SYNONYMS: *caffer, illovoensis, komatiensis, microdon, muscardinus, ovamboensis, zuluensis.*

COMMENTS: Characterized by 2N=32, FN=54 and a distinctive hemoglobin electromorph (Green et al., 1980). Samples with these chromosomal features have also been found in Senegal (Duplantier et al., 1990*a*) and EC Tanzania (Leirs et al., 1989). This species probably occurs in Angola, S Zaire, Zambia, Malawi, Mozambique, and farther north in Tanzania and perhaps may even range more extensively in W Africa, but at this time samples from those regions have not been identified by linking chromosomal and biochemical data to morphology. Realizing this problem, some regional faunal accounts provisionally list specimens under *M. natalensis* (e.g., Ansell and Dowsett, 1988, for Malawi mammals). Listed synonyms are only those pertaining to South Africa for reasons explained by Meester et al. (1986).

Mastomys pernanus (Kershaw, 1921). Ann. Mag. Nat. Hist., ser. 9, 8:568.

TYPE LOCALITY: SW Kenya, Amala (Mara) River; see Allen (1939:406) and Misonne and Verschuren (1964:655) for details.

DISTRIBUTION: Published records are from SW Kenya, NW Tanzania, and Rwanda (Misonne and Verschuren, 1964); limits are unknown.

COMMENTS: Ellerman (1941) listed *pernanus* as a distinctive species of *Rattus* in the subgenus *Mastomys*, but earlier Allen (1939) had treated it as a species of *Myomys*. Robbins and Van der Straeten (1989) claimed this a possibility. Misonne and Verschuren (1964, 1966*b*), however, discussed the problem and concluded that *pernanus* should be associated with *Mastomys*. Either postulated generic association is unsupported by critical analyses of comparative data so the allocation of *pernanus* remains clouded. The species is known by very few specimens (Misonne and Verschuren, 1964; 114439 in the American Museum of Natural History is from NW Tanzania).

Mastomys shortridgei (St. Leger, 1933). Proc. Zool. Soc. Lond., 1933:411.

TYPE LOCALITY: N Namibia, Grootfontein Dist., Okavango-Omatako junction.

DISTRIBUTION: Extreme NW Botswana and NE Namibia in the region of the confluence of Okavango and Kwito rivers (see map in Skinner and Smithers, 1990:270).

SYNONYMS: *legerae.*

COMMENTS: Taxonomy reviewed by Meester et al. (1986:285). Some workers considered *shortridgei* closely related to or the same as *M. angolensis* (Meester et al., 1986). Occurs sympatrically with other species of *Mastomys* (Skinner and Smithers, 1990).

Mastomys verheyeni Robbins and Van der Straeten, 1989. Senckenberg. Biol., 69:4.

TYPE LOCALITY: Nigeria, Bornu Region, Mudu (see Robbins and Van der Straeten, 1989, for more information).

DISTRIBUTION: "Known only from Nigeria and Cameroon savanna area immediately surrounding the southern part of Lake Chad" (Robbins and Van der Straeten, 1989:9).

Maxomys Sody, 1936. Natuurk. Tidjschr. Ned.-Ind., 96:55.

TYPE SPECIES: *Mus bartelsii* Jentink, 1910.

COMMENTS: Definition and contents of *Maxomys* provided by Musser et al. (1979), who discussed historical allocations of species to groups of *Rattus*, and summarized chromosomal information. The exclusion of *Maxomys* from *Rattus* was also supported by biochemical data (Chan et al., 1979). Morphological analyses indicated that *Maxomys* shares all its derived molar traits with *Berylmys, Mus,* and *Niviventer,* and some of its external features with *Mus* (Musser and Newcomb, 1983). Hypothetical phylogenetic estimate of relationships based on karyotypic data placed *Maxomys* close to *Niviventer* and *Lenothrix* and far from *Rattus* (Gadi and Sharma, 1983). Spermatozoal morphology of Malayan *Maxomys* and one Sulawesian species was similar to *Chiropodomys* and *Hapalomys*, a configuration also like that seen in *Mus* and *Apodemus* (Breed and Musser, 1991; Breed and Yong, 1986). Sperm conformation in the other species of Sulawesian *Maxomys* were unlike the Malayan species and resembled that of the Sulawesian endemic *Margaretamys* (Breed and Musser, 1991).

Maxomys alticola (Thomas, 1888). Ann. Mag. Nat. Hist., ser. 6, 2:408.
TYPE LOCALITY: Malaysia, Sabah (N Borneo), Gunung Kinabalu ("from a high level").
DISTRIBUTION: Known only from two mountains in Sabah, N Borneo (from 3500 to 11,000 ft on Gunung Kinabalu, and also on Gunung Trus Madi).
SYNONYMS: *kinabaluensis*.
COMMENTS: The name *alticola* was once applied to populations from lower elevations on Gunung Kinabalu, the Malay Peninsula, and Sumatra (Chasen, 1940), but was shown to represent a distinct species endemic to mountains of Sabah (Medway, 1964, 1977).

Maxomys baeodon (Thomas, 1894). Ann. Mag. Nat. Hist., ser. 6, 14:452.
TYPE LOCALITY: Malaysia, Sabah (N Borneo), Gunung Kinabalu.
DISTRIBUTION: Known only from a few scattered localities in Sabah and Sarawak (N Borneo).
SYNONYMS: *trachynotus*.
COMMENTS: A very distinctive species represented in collections by few specimens (Musser et al., 1979).

Maxomys bartelsii (Jentink, 1910). Notes Leyden Mus., 33:69.
TYPE LOCALITY: Indonesia, W Java, Gunung Pangerango-Gede, 6000 ft.
DISTRIBUTION: Endemic to the mountains of W and C Java (Van Peenen et al., 1974).
SYNONYMS: *obscuratus*, *tjibunensis*.
COMMENTS: Morphology, chromosomes, and geographic distribution reviewed by Van Peenen et al. (1974); skull and dentition illustrated in Musser and Newcomb (1983).

Maxomys dollmani (Ellerman, 1941). Families Genera Living Rodents, 2:218.
TYPE LOCALITY: Indonesia, C Sulawesi, Quarles Mtns, Rantekaroa, 6000 ft.
DISTRIBUTION: Type locality and Gunung Tanke Salokko in the SE peninsula of Sulawesi; known only from high elevations in montane forests.
COMMENTS: Originally described by Ellerman as a subspecies of *Rattus hellwaldii*, but shown to be a distinct species by Musser (1969*c*), and morphologically allied to *M. hellwaldii* (Musser, 1991).

Maxomys hellwaldii (Jentink, 1878). Notes Leyden Mus., 1:11.
TYPE LOCALITY: Indonesia, Sulawesi, NE peninsula, Menado.
DISTRIBUTION: Throughout Sulawesi in tropical lowland evergreen rain forest (Musser and Holden, 1991).
SYNONYMS: *cereus*, *griseogenys*, *localis*.
COMMENTS: Among Sulawesian representatives of *Maxomys*, *M. hellwaldii* is most closely related to *M. dollmani* (Musser, 1991). Spermatozoal morphology was unlike species of Malayan *Maxomys* studied, and more similar to species of *Margaretamys*, another Sulawesian endemic (Breed and Musser, 1991).

Maxomys hylomyoides (Robinson and Kloss, 1916). J. Str. Br. Roy. Asiat. Soc., 73:273.
TYPE LOCALITY: Indonesia, W Sumatra, Korinchi Peak, 7300 ft.
DISTRIBUTION: Endemic to montane forests in mountainous backbone of W Sumatra.
COMMENTS: At one time listed as a subspecies of *alticola* (Chasen, 1940), but later reinstated as a distinctive species (Medway, 1964; Musser et al., 1979).

Maxomys inas (Bonhote, 1906). Proc. Zool. Soc. Lond., 1906:9.
TYPE LOCALITY: Malaysia, Perak (Malay Peninsula), Gunung Inas.
DISTRIBUTION: Endemic of Malay Peninsula (and possibly Peninsular Thailand south of Isthmus of Kra) in montane forests, rarely occurring below 3000 ft (Medway, 1969).
COMMENTS: Once treated as a subspecies of *alticola* (Chasen, 1940), but identified as a separate species by Medway (1964). A close phylogenetic ally to *M. whiteheadi*, as indicated by morphological, biochemical, chromosomal, and spermatozoal traits (Breed and Yong, 1986; Chan et al., 1978, 1979; Yong, 1969).

Maxomys inflatus (Robinson and Kloss, 1916). J. Str. Br. Roy. Asiat. Soc., 73:273.
TYPE LOCALITY: Indonesia, W Sumatra, Korinchi Peak, Sungei Kumbang, 4700 ft.
DISTRIBUTION: Endemic to mountain forests of W Sumatra.
COMMENTS: One of the most distinctive species of *Maxomys*, as reflected by its wide rostrum and inflated nasolacrimal capsules (Musser et al., 1979).

Maxomys moi (Robinson and Kloss, 1922). Ann. Mag. Nat. Hist., ser. 9, 9:95.
TYPE LOCALITY: Vietnam, Lang Bian Mtns, Arbre Broye, 5400 ft.
DISTRIBUTION: Endemic to S Vietnam and S Laos.
COMMENTS: Originally described as a species by Robinson and Kloss (1922), later listed as a subspecies of either *surifer* or *coxingi*, but finally shown to be a distinct species (Musser et al., 1979; Van Peenen et al., 1969). Closest phylogenetic relative is *M. surifer*, an estimate based on cranial, dental, and chromosomal traits shared by both species (Duncan and Van Peenen, 1971; Musser et al., 1979).

Maxomys musschenbroekii (Jentink, 1878). Notes Leyden Mus., 1:10.
TYPE LOCALITY: Indonesia, Sulawesi, NE peninsula, Menado.
DISTRIBUTION: Throughout Sulawesi (Musser, 1991; Musser and Holden, 1991).
SYNONYMS: *aspinatus, lalawora, tetricus.*
COMMENTS: Considered conspecific with Sundaic *M. whiteheadi* by Ellerman and Morrison-Scott (1951), but *musschenbroekii* is a distinct species endemic to Sulawesi (Medway, 1977; Musser, 1991). Spermatozoal morphology more similar to that described for Malayan species of *Maxomys* than to sperm of other sampled Sulawesian species (Breed and Musser, 1991).

Maxomys ochraceiventer (Thomas, 1894). Ann. Mag. Nat. Hist., ser. 6, 14:451.
TYPE LOCALITY: Malaysia, Sabah (N Borneo), Gunung Kinabalu, below 3000 ft.
DISTRIBUTION: Sabah and Sarawak (N Borneo), and East Kalimantan (E Borneo); apparently restricted to hills (Medway, 1964).
SYNONYMS: *perasper.*
COMMENTS: Once treated as a lower altitudinal subspecies of the higher montane *alticola* (Chasen, 1940), but the two species are sympatric at 3500 ft on slopes of Gunung Kinabalu (Medway, 1977).

Maxomys pagensis (Miller, 1903). Smithson. Misc. Coll., 45:39.
TYPE LOCALITY: Indonesia, Kepulauan Mentawai, Pulau Pagai Selattan (S Pagai Isl), off coast W Sumatra.
DISTRIBUTION: Endemic to islands of South Pagai, North Pagai, Sipora, and Siberut in Mentawai Archipelago.
COMMENTS: Usually listed as a subspecies of *M. surifer* (Chasen, 1940), but treated as a species by Musser et al. (1979). Closest phylogenetic relative is probably *M. surifer.* Part of the rodent fauna endemic to Mentawi Archipelago (see account of *Leopoldamys siporanus*).

Maxomys panglima (Robinson, 1921). Ann. Mag. Nat. Hist., ser. 9, 7:235.
TYPE LOCALITY: Philippines, Palawan Isl.
DISTRIBUTION: Endemic to Balabac, Palawan, Busuanga, and Culion Isls; politically part of Philippines, but faunistically an extension of the Sunda Shelf.
SYNONYMS: *palawanensis.*
COMMENTS: Treated in the past as a subspecies of *M. surifer*, but morphological features supported its independence as a species; uncertain whether it is more closely related to *M. surifer* or *M. rajah* (Musser et al., 1979).

Maxomys rajah (Thomas, 1894). Ann. Mag. Nat. Hist., ser. 6, 14:451.
TYPE LOCALITY: Malaysia, Sarawak (N Borneo), Gunung Batu Song.
DISTRIBUTION: Endemic to the Sunda Shelf; Peninsular Thailand south of Isthmus of Kra, Malay Peninsula, Riau Archipelago, Sumatra, amd Borneo; absent from Java and Bali.
SYNONYMS: *hidongis, lingensis, pellax, similis.*
COMMENTS: Distribution similar in broad outline to that of *M. whiteheadi*. Occurs sympatrically with *M. surifer*, and although samples of each are often misidentified, the two differ in a suite of morphological, ecological, behavioral, and biochemical features (Chan et al., 1979; see references in Musser et al., 1979).

Maxomys surifer (Miller, 1900). Proc. Biol. Soc. Washington, 13:148.
TYPE LOCALITY: Peninsular Thailand, Trang.
DISTRIBUTION: Indochina (S Burma, Thailand, Laos, Cambodia, and Vietnam) and the Sunda Shelf (Peninsular Thailand, Malay Peninsula, Borneo, Sumatra, Java, and many smaller islands).

SYNONYMS: *anambae, antucus, aoris, banacus, bandahara, bentincanus, binominatus, butangensis, casensis, carimatae, catellifer, changensis, connectens, domelicus, eclipsis, flavidulus, flavigrandis, finis, grandis, koratis, kramis, kutensis, leonis, luteolus, mabalus, manicalis, microdon, muntia, natunae, pelagius, pemangilis, perflavus, pidonis, pinacus, puket, ravus, saturatus, serutus, siarma, solaris, spurcus, telibon, tenebrosus, ubecus, umbridorsum, verbeeki.*

COMMENTS: The only species of *Maxomys* with a range encompassing Indochinese and Sundaic faunal regions. The two groups of samples differ in morphological features, and the significance of this variation should be determined in a careful systematic revision of the genus. *Berylmys bowersi, Chiropodomys gliroides, Leopoldamys sabanus,* and *L. edwardsi* have roughly concordant geographic ranges and demonstrate similar geographic variation (Musser et al., 1979; Musser and Newcomb, 1983). Comparative spermatozoal morphology documented by Breed and Yong (1986), chromosomal and biochemical data summarized by Chan et al. (1979) in phylogenetic context.

Maxomys wattsi Musser, 1991. Am. Mus. Novit., 3001:20.

TYPE LOCALITY: Indonesia, C Sulawesi, Gunung Tambusisi, Tambusisi Damar, 4700 ft.

DISTRIBUTION: Known only from 4700-6000 ft on Gunung Tambusisi, C Sulawesi.

COMMENTS: A morphologically distinctive species; assessing its phylogenetic relations to other species will require systematic revision of *Maxomys* (Musser, 1991).

Maxomys whiteheadi (Thomas, 1894). Ann. Mag. Nat. Hist., ser. 6, 14:452.

TYPE LOCALITY: Malaysia, Sabah (N Borneo), Gunung Kinabalu.

DISTRIBUTION: Peninsular Thailand south of Isthmus of Kra, Malay Peninsula, Sumatra, Borneo, and various adjacent islands; absent from Java and Bali.

SYNONYMS: *asper, batamanus, batus, coritzae, klossi, mandus, melanurus, melinogaster, perlutus, piratae, subitus.*

COMMENTS: A Sundaic endemic. Ellerman and Morrison-Scott (1951) included *whiteheadi* in the Sulawesi *musschenbroekii*, but the two are separate species (Chasen, 1940; Medway, 1977; Musser, 1991; Musser et al., 1979; Tate, 1936). Spermatozoal morphology (Breed and Yong, 1986), and data from biochemical, morphological, and cytological studies (Chan et al., 1978, 1979; Yong, 1969) pointed to a close relationship between *M. whiteheadi* and the Malayan *M. inas*.

Mayermys Laurie and Hill, 1954. List of Land Mammals of New Guinea, Celebes and Adjacent Islands, p. 133.

TYPE SPECIES: *Mayermys ellermani* Laurie and Hill, 1954.

COMMENTS: Member of the New Guinea Old Endemics (Musser, 1981*c*:167). Phallic morphology similar to *Neohydromys* (Lidicker, 1968). *Mayermys, Leptomys, Neohydromys,* and *Pseudohydromys* share a derived configuration of cephalic arterial pattern (Musser and Heaney, 1992).

Mayermys ellermani Laurie and Hill, 1954. List of Land Mammals of New Guinea, Celebes and Adjacent Islands, 1952, p. 134.

TYPE LOCALITY: Papua New Guinea, Chimbu Prov., Bismarck Range, N slopes of Mt Wilhelm, 8000 ft.

DISTRIBUTION: Papua New Guinea; known only from scattered localities along Central Cordillera from Telefomin region in west to Wau area in the east (see map in Flannery 1990*b*:183).

COMMENTS: Flannery (1990*b*:183) provided a photograph of the animal and a summary of distributional and biological data.

Melasmothrix Miller and Hollister, 1921. Proc. Biol. Soc. Washington, 34:93.

TYPE SPECIES: *Melasmothrix naso* Miller and Hollister, 1921.

COMMENTS: Reviewed by Musser (1982*c*). Closely related to *Tateomys* as judged by morphological, ecological, and spermatozoal characters (Breed and Musser, 1991; Musser, 1982*c*). Placed in the group of Sulawesian Old Endemics by Musser (1981*c*).

Melasmothrix naso Miller and Hollister, 1921. Proc. Biol. Soc. Washington, 34:93.

TYPE LOCALITY: Indonesia, C Sulawesi, Rano Rano, 6000 ft.

DISTRIBUTION: Sulawesi; known only from upper montane rain forest at Rano Rano and on Gunung Nokilalaki, but probably occurs on other mountains in central core of island.

Melomys Thomas, 1922. Ann. Mag. Nat. Hist., ser. 9, 9:261.
TYPE SPECIES: *Uromys rufescens* Alston, 1877.
SYNONYMS: *Paramelomys*.
COMMENTS: Member of the Australian and New Guinea region Old Endemics (Musser, 1981*c*, 1982*b*). Morphological and distributional limits need revision. Currently, three primary groups can be recognized: (1) species endemic or indigenous to New Guinea region (*fellowsi, gracilis, leucogaster, levipes, lorentzii, mollis, moncktoni, platyops*, and *rufescens*); (2) Australian species (*capensis, cervinipes, rubicola*), one occurring on Australia and New Guinea (*burtoni*), and some Moluccan species (*aerosus, fraterculus*, and *obiensis*); (3) New Guinea *lanosus* and *rattoides*. Data from microcomplement fixation of albumin indicated Australian *Melomys* was closely related to *Uromys* and in the same monophyletic group with *Mesembriomys, Leporillus, Conilurus*, and *Zyzomys* (Watts et al., 1992).
Mahoney and Richardson (1988) cataloged taxonomic, distributional, and biological references to Australian species; Flannery (1990*b*) summarized the same for many of the New Guinea *Melomys*.

Melomys aerosus (Thomas, 1920). Ann. Mag. Nat. Hist., ser. 9, 6:428.
TYPE LOCALITY: Indonesia, Pulau Seram, Mt Manusela, 6000 ft.
DISTRIBUTION: Endemic to Seram Isl.
COMMENTS: Tate (1951:292) suggested a relationship between *M. aerosus* and the New Guinea *M. levipes* group, but cranial morphology indicates *M. aerosus* to be more closely related to Australian species of *Melomys*, especially *M. cervinipes*, than to the endemic *Melomys* of New Guinea (data from specimens in the British Museum of Natural History).

Melomys bougainville Troughton, 1936. Rec. Aust. Mus., 19:344.
TYPE LOCALITY: Solomon Isls, Bougainville Isl, Bouin.
DISTRIBUTION: Endemic to islands of Buka and Bougainville in Solomon Arch. (Flannery and Wickler, 1990).
COMMENTS: Although historically treated as a subspecies of *M. rufescens, M. bougainville* is a separate species known by small samples of extant and archaeological specimens (Flannery and Wickler, 1990).

Melomys burtoni (Ramsay, 1887). Proc. Linn. Soc. N.S.W., ser. 2, 2:531.
TYPE LOCALITY: Australia, Western Australia, near Derby.
DISTRIBUTION: Australia; along the coast "from just south of the New South Wales-Queensland border, north to the tip of Cape York, and in coastal areas of the Northern Territory and north-eastern Western Australia" (Watts and Aslin, 1981:84); also found on many offshore islands, including those in Torres Strait. New Guinea; patchy distribution (grassland habitats) on both sides of Central Cordillera, near sea level to 2200m (see map in Flannery, 1990*b*:231); also on Woodlark Isl in d'Entrecasteaux Arch., and Misima Isl in Louisiade Arch.
SYNONYMS: *albiventer, australius, callopes, frigicola, froggatti, hintoni, insulae, littoralis, lutillus, melicus, mixtus, murinus, muscalis*.
COMMENTS: Results of chromosomal and electrophoretic studies reported by Baverstock et al. (1977*c*, 1980, 1981, 1983*b*). Anatomy of male reproductive tract and spermatozoal morphology presented by Breed and Sarafis (1978), Morrissey and Breed (1982), and Breed (1984, 1986). We follow Tate (1951) and Mahoney and Richardson (1988) in treating populations from Australian and New Guinea region as a single species, but the complex needs revision using data from morphological, chromosomal, and molecular sets to assess significance of the variation apparent both among samples from New Guinea and between those from New Guinea and Australia.

Melomys capensis Tate, 1951. Bull. Am. Mus. Nat. Hist., 97:295.
TYPE LOCALITY: Australia, Queensland, Nesbit River, Rocky Scrub (E of Coen), 1500 ft.
DISTRIBUTION: Australia, Queensland, Iron and McIlwraith Ranges of Cape York (north of Cooktown).
COMMENTS: Originally described as a subspecies of *M. cervinipes* by Tate (1951:295), but separated as a distinct species by Baverstock et al. (1980), who based their conclusions on differences in blood proteins; otherwise, the two species are similar in

morphological traits and body size (Watts and Aslin, 1981:82). Other electrophoretic results presented by Baverstock et al. (1981).

Melomys cervinipes (Gould, 1852). Mamm. Aust., pt. 4, 3:pl. 14.
TYPE LOCALITY: Australia, Queensland, Stradbrook Isl.
DISTRIBUTION: Australia; extant range is closed forest and more open habitat along the E Australian coast from Cooktown region of Cape York in Queensland south to Gosford area of New South Wales (Watts and Aslin, 1981:79). Late Pleistocene specimens indicated distribution once extended farther south to the Pyramids Cave region in Victoria (Wakefield, 1972*a*).
SYNONYMS: *banfieldi, bunya, eboreus, limicauda, pallidus.*
COMMENTS: Anatomy of male reproductive tract and spermatozoa reported by Breed and Sarafis (1978), Morrissey and Breed (1982), and Breed (1984, 1986). Chromosomal morphology, G-banding homologies, and results of electrophoretic analyses presented by Baverstock et al. (1977*c*, 1980, 1981, 1983*b*), who (1980) reported that *M. cervinipes* was phylogenetically close to *M. capensis*, but electrophoretically distant, having experienced a rapid rate of electrophoretic evolution relative to that found in *M. capensis* and *M. burtoni*.

Melomys fellowsi Hinton, 1943. Ann. Mag. Nat. Hist., ser. 11, 10:554.
TYPE LOCALITY: Papua New Guinea, Purari-Ramu Divide, Baiyanka, 8000 ft.
DISTRIBUTION: Papua New Guinea; known only from a few places at high elevations in the central highlands; see map in Flannery (1990*b*:222).
COMMENTS: A distinctive species that resembles *M. rufescens* and *M. leucogaster* in some morphometric features (Menzies, 1990).

Melomys fraterculus (Thomas, 1920). Ann. Mag. Nat. Hist., ser. 9, 6:428.
TYPE LOCALITY: Indonesia, Pulau Seram, Mt Manusela, 6000 ft.
DISTRIBUTION: Endemic to Seram Isl.
COMMENTS: Placed in *Pogonomelomys* by Rümmler (1938), kept there by Tate (1951), but returned to *Melomys* by Laurie and Hill (1954). Based on our study of specimens, *Melomys fraterculus* shares many derived cranial features with the Australian *M. cervinipes* complex and is probably more closely related to the indigenous Australian *Melomys* than to those on New Guinea.

Melomys gracilis (Thomas, 1906). Ann. Mag. Nat. Hist., ser. 7, 17:328.
TYPE LOCALITY: E Papua New Guinea, Northern Prov., Angabunga (St Joseph's) River, Owgarra, 1800 m.
DISTRIBUTION: Papua New Guinea; specimens from high altitudes (above 1200 m) along Central Cordillera from Owen Stanley Range (type locality, head of Aroa River, Mt Tafa, Mafulu,) through the Eastern Highlands (Garaina area, Wau region, Kratke Mts, Upper Ramu River Plateau, Okapa area) to the slopes of Mt Hagen (Tomba).
SYNONYMS: *dollmani.*
COMMENTS: Although usually listed as a subspecies or synonym of *M. rufescens* (Flannery, 1990*b*; Laurie and Hill, 1954; Tate, 1951), *M. gracilis* is a separate species. It is sympatric with *M. rufescens* in the Kratke Mts, on the Upper Ramu River Plateau, in the Okapa area, and at Tomba (specimens in the American Museum of Natural History and the British Museum of Natural History).

Melomys lanosus Thomas, 1922. Ann. Mag. Nat. Hist. ser. 9, 9:263.
TYPE LOCALITY: New Guinea, Irian Jaya, Doormanpad-bivak (3°30'S, 138°30'E), 2400 m.
DISTRIBUTION: Mountain forests (not recorded below 1500 m) of N New Guinea; from the type locality in Irian Jaya eastward along Central Cordillera through the Telefomin region, Schrader Range, Kratke Mts, and Mt Hagen region to the Wau area of Papua New Guinea; also from Cyclops Range in the N coastal mts; not known from Huon Peninsula.
SYNONYMS: *shawmayeri.*
COMMENTS: Once treated as a subspecies of *M. levipes, lanosus*, as Flannery (1990*b*:230) noted, is a separate species. It is unrelated to *M. levipes*, but instead is phylogenetically allied with *M. rattoides*, a larger-bodied species that displaces it in lower altitudes. This close association was also reflected in Menzies' (1990) similarity coefficient analysis. Our study of specimens in the American Museum of Natural

History revealed that *Melomys lanosus* and *M. rattoides* share derived traits (only one pair of teats, and the derived cephalic arterial circulation and its osseous reflection in the cranium; see figures in Musser and Heaney, 1992) not found in any other species of *Melomys*. Any systematic revision of *Melomys* will have to determine whether these two species are members of that genus, which seems unlikely. Flannery listed *shawmayeri* as a synonym of *M. rattoides*, but holotype is example of *M. lanosus*.

Melomys leucogaster (Jentink, 1908). Nova Guinea, 9:3.

TYPE LOCALITY: New Guinea, S Irian Jaya, Lorentz River, Alkmaar, 300 m.

DISTRIBUTION: New Guinea; S side of Central Cordillera from the type locality in Irian Jaya east to E Papua New Guinea on the mainland (see map in Flannery, 1990*b*:234); also Yule Isl off the coast of E Papua New Guinea in Central Prov., Conflict Isls, and Rossel Isl in Louisiade Arch., Moluccas, Seram and Talaud Isls.

SYNONYMS: *arcium, caurinus, fulgens, latipes, talaudium*.

COMMENTS: A close morphological relative of *M. rufescens*. *Melomys leucogaster* is one of the few New Guinea species that also occurs on archipelagos to the SE and W of that island continent. The population on Rossel Isl was listed as *M. arcium* by Laurie and Hill (1954), and that on Seram and the Talauds as *M. fulgens*.

Melomys levipes (Thomas, 1897). Ann. Mus. Civ. Stor. Nat. Genova, 18:617.

TYPE LOCALITY: New Guinea, Papua, Central Prov., Sogeri Plateau, Haveri, 700 m (additional information provided by Laurie and Hill, 1954:121, and Menzies, 1989).

DISTRIBUTION: Papua New Guinea, S Central Prov.; Sogeri Plateau and Atrolabe Range near Port Moresby, below about 700 m; limits unknown.

COMMENTS: A lectotype was designated by Menzies (1989), but Rümmler (1938) had already indicated which of Thomas' two cotypes should be considered the holotype. Our examination of series indicated that true *levipes* is documented by the holotype and Tate's (1951:291) series from Baruari and Itiki in the Astrolabe Range (other material Tate lists as *M. levipes* is either *M. platyops* or *M. mollis*). All the other records usually associated with *M. levipes* (Laurie and Hill, 1954; Tate, 1951) represent *M. lanosus*, *M. lorentzii*, *M. mollis*, and *M. rattoides*. In morphology and altitudinal distribution, *M. levipes* is very similar to *M. lorentzii* and may represent the SE Papuan form of that species, an alliance also discerned by Thomas (1913*b*). A specimen from New Britain in the American Museum of Natural History (194397) is cranially and dentally similar to *M. levipes*.

Melomys lorentzii (Jentink, 1908). Nova Guinea, 9:3.

TYPE LOCALITY: New Guinea, Irian Jaya, Lorentz River, Resi Camp, 900 m.

DISTRIBUTION: New Guinea; specimens are from Vogelkop (Oransbari), Weyland Range (midaltitudes), and along the south side of Central Cordillera, from Kapare River in SW Irian Jaya to middle altitudes and lowlands of the Fly River drainage in Papua New Guinea; limits unknown. Altitudinal range from sea level to about 1500 m.

SYNONYMS: *naso, sturti, weylandi*.

COMMENTS: The form *naso* is usually listed as a subspecies of *M. levipes* (Flannery, 1990*b*; Laurie and Hill, 1954), but Rümmler (1938) correctly identified the holotype as an *M. lorentzii*. Tate's (1951:296) *sturti* was described as a subspecies of *M. moncktoni*, but the holotype and type series are examples of *M. lorentzii*. The form *weylandi*, from the Weyland Range in SE Irian Jaya, is usually listed as a subspecies of *M. levipes*, but specimens in the type series are examples of *M. lorentzii*.

Usually listed as a subspecies of *M. levipes* (Laurie and Hill, 1954; Tate, 1951), *lorentzii* is a distinct species that resembles *M. levipes* in many morphological traits; however, *M. moncktoni* may be closer to *M. lorentzii* judged by Menzies's (1990) morphometric similarity coefficient analysis. Significance of the appreciable variation in body size among samples needs to be assessed in a revision of *M. lorentzii*, and a phylogenetic inquiry into its closest relatives.

Melomys mollis (Thomas, 1913). Ann. Mag. Nat. Hist., ser. 8, 12:210.

TYPE LOCALITY: New Guinea, Irian Jaya, Nassau Range, upper Utakwa River, S slope Mount Carstenz, Camp Padang, 6c, 5500 ft.

DISTRIBUTION: New Guinea; scattered in montane forest throughout Central Cordillera from Arfak Mtns in Vogelkop of Irian Jaya to Mt Dayman in extreme E Papua New Guinea.

SYNONYMS: *arfakianus, clarae, meeki, stevensi*.

COMMENTS: Specimens representing *M. mollis* have been identified as *M. levipes* in the literature, and the synonyms listed here have also been associated with that species. However, judged by our study of specimens in the American Museum of Natural History and the British Museum of Natural History, *Melomys mollis* is a distinct species tied to montane forest formations and differs from *M. levipes* in habitat and morphology. Specimens from W Irian Jaya are more darkly pigmented than those from E Papua, but series from Papuan Central and Eastern highlands bridge this chromatic gap.

Melomys moncktoni (Thomas, 1904). Ann. Mag. Nat. Hist., ser. 7, 14:399.

TYPE LOCALITY: Papua New Guinea, Northern Prov., NE coast, 8°30'S, 148°20'E (Kumusi River).

DISTRIBUTION: Papua New Guinea; reliable records are from coastal plains and foothills (not exceeding 700 m) of extreme SE Papua New Guinea, from the type locality on the NE coast eastward to southern lowlands where westernmost record is from foothills northeast of Port Morseby area; limits unknown.

COMMENTS: Horizontal and altitudinal distributions of this species have been misunderstood due to incorrect identifications of specimens. *Melomys moncktoni* was thought to have a primarily southern distribution from Irian Jaya to Papua, and to extend up into moss forest (see map and discussion in Flannery, 1990*b*:225), but all samples from moss forest represent other species. Examples of *M. moncktoni* come only from the restricted range described above (based on series in the American Museum of Natural History and the British Museum of Natural History).

The forms *intermedius*, *shawi*, and *sturti* are usually associated with *M. moncktoni* (Flannery, 1990*b*; Laurie and Hill, 1954; Tate, 1951), but the holotype of *shawi* is a *M. rubex*, the type series of *intermedius* belongs to *M. platyops*, and the type series of *sturti* represents *M. lorentzii*.

Melomys obiensis (Thomas, 1911). Ann. Mag. Nat. Hist., ser. 8, 7:208

TYPE LOCALITY: Indonesia, Malukus, Pulau Obi.

DISTRIBUTION: Endemic to Obi Isl, S of Halmahera.

COMMENTS: Judged by morphological traits, a close relative of *M. fraterculus* from Seram, and more closely related to Australian *Melomys cervinipes* than to any New Guinea species, an observation gleaned from our study of specimens and earlier recorded by Tate (1951:297).

Melomys platyops (Thomas, 1906). Ann. Mag. Nat. Hist., ser. 7, 17:327.

TYPE LOCALITY: SE Papua New Guinea, Central Prov., head of Aroa River.

DISTRIBUTION: New Guinea, lowlands and midmountain altitudes both north and south of the Central Cordillera, and on islands; in the north from Nabire (coastal lowlands) in NW Irian Jaya to the east end of Papua New Guinea, then west through S New Guinea to the Utakwa River in SW Irian Jaya. Also occurs on islands of Yapen (Japen) and Biak in Irian Jaya; New Britain in the Bismarck Arch.; and the isls of Normanby, Fergusson, and Goodenough in the d'Entrecasteaux Arch. Altitudinal range from sea level to 900 m.

SYNONYMS: *fuscus, intermedius, jobiensis, mamberanus*.

COMMENTS: This species was thought to range primarily throughout N New Guinea (see map in Flannery, 1990*b*:224) but our reidentification of museum specimens and holotypes reveals otherwise. The form *intermedius* was originally described as a subspecies of *M. moncktoni* (see Rümmler, 1938), but type series from Utakwa River (type locality) in SW Irian Jaya have only one hair per scale (all *M. moncktoni* have three hairs per scale), as does *M. platyops*, and their cranial, dental, and other external traits are also characteristic of *platyops*, not *moncktoni*. Most of the 25 topotypes Tate (1951:297) identified as *M. moncktoni sturti* from the lower Fly River are also examples of *M. platyops*. Geographic variation in body size exists among samples of *M. platyops*, and a careful systematic revision of the species is required to assess its significance.

Melomys rattoides Thomas, 1922. Ann. Mag. Nat. Hist., ser. 9, 9:263.

TYPE LOCALITY: New Guinea, NW Irian Jaya, Mamberano River, Pionier-bivak (2°20'S, 138°0' E), 200 ft.

DISTRIBUTION: New Guinea; N slopes and lowlands from the type locality in Irian Jaya through slopes N of Idenberg River to the Telefomin region in West Sepik Prov. of Papua New Guinea (Flannery and Seri, 1990); also on Pulau Yapen (Japen Isl); limits of range unknown. No records from above 1200 m.

COMMENTS: The distribution is based on the few reliable identifications of samples. Tate's (1951:290) records of *M. rattoides* from Cyclops Mts, for example, were based on *M. lanosus*, as were the easternmost dots on Flannery's (1990*b*:229) map. Closest relative is *M. lanosus*, which occurs in moss forests at altitudes higher than the distribution of *M. rattoides* (see account of *M. lanosus*).

Melomys rubex Thomas, 1922. Ann. Mag. Nat. Hist., ser. 9, 9:263.

TYPE LOCALITY: New Guinea, Irian Jaya, Mamberano River, Doormanpad-bivak, 1410 m.

DISTRIBUTION: New Guinea; above 800 m in the Central Cordillera from the Arfak Mtns in Irian Jaya to Mt Dayman at the end of the Owen Stanley Range in SE Papua New Guinea; also in the Torricelli Mtns and Huon Peninsula.

SYNONYMS: *alleni, arfakiensis, clarus, pohlei, rutilus, shawi, steini, stressmani, tafa.*

COMMENTS: The many names applied to *M. rubex* reflect morphological variation among some samples that is concordant with interrupted highland distributions. Taxonomic status of some names associated with *rubex* briefly reviewed by Menzies (1974). The form *shawi* is usually listed as a subspecies of *M. moncktoni* (Laurie and Hill, 1954:123), but the holotype is an example of *M. rubex*.

Melomys rubicola Thomas, 1924. Ann. Mag. Nat. Hist., ser. 9, 13:298.

TYPE LOCALITY: Australia, Queensland, Torres Strait, Bramble Cay, about 9°S, 144°E.

DISTRIBUTION: Australia; endemic to Bramble Cay at the extreme northern end of the Great Barrier Reef of Queensland (Limpus et al., 1983).

COMMENTS: Studies of blood proteins and morphology suggested *M. rubicola* was closely related to *M. capensis*, which is endemic to Cape York in N Queensland (Limpus et al., 1983). Structure of sperm head described by Breed (1984).

Melomys rufescens (Alston, 1877). Proc. Zool. Soc. Lond., 1877:124.

TYPE LOCALITY: "Duke of York Isl. or adjacent parts of New Britain or New Ireland" (Tate, 1951:304).

DISTRIBUTION: New Guinea; throughout the island continent, from the Vogelkop in Irian Jaya to the east end of Papua New Guinea, coastal lowlands to high altitudes in mountains. Also on New Britain, Duke of York Isl, and New Ireland (see Flannery and White, 1991) in the Bismarck Arch.

SYNONYMS: *calidior, hageni, musavora, niviventer, sexplicatus, stalkeri.*

COMMENTS: The significant geographic variation in morphological traits present among samples allows recognition of four distinct groups; *rufescens* from N and W New Guinea and the Bismarck Arch., *niviventer* from Fly River drainage, *stalkeri* from E Papua New Guinea, and *hageni* from the Eastern Highlands (based on our study of specimens in the American Museum of Natural History and the British Museum of Natural History).

Melomys spechti Flannery and Wickler, 1990. Aust. Mammal., 13:130.

TYPE LOCALITY: Solomon Isls, Buka Isl, Kilu rockshelter.

DISTRIBUTION: Recorded only from Buka Isl.

COMMENTS: A distinctive species known only by archaeological fragments (Flannery and Wickler, 1990).

Mesembriomys Palmer, 1906. Proc. Biol. Soc. Washington, 19:97.

TYPE SPECIES: *Mus hirsutus* Gould, 1842 (= *Hapalotis gouldii* Gray, 1843).

SYNONYMS: *Ammomys.*

COMMENTS: Member of the Australian Old Endemics (Musser, 1981*c*:167), which includes the Conilurini where Baverstock (1984) placed *Mesembriomys*, and closely related to *Leporillus* and *Conilurus* (Watts et al., 1992). Mahoney and Richardson (1988:164) cataloged taxonomic, distributional, and biological references.

Mesembriomys gouldii (Gray, 1843). List Specimens Mamm. Coll. Br. Mus., p. 116.

TYPE LOCALITY: Australia, Northern Territory, Port Essington.

DISTRIBUTION: Australia; N Western Australia, N Northern Territory, N Queensland, Melville Isl, and Bathurst Isl (Watts and Aslin, 1981; Friend, 1991).
SYNONYMS: *hirsutus* (Gould, 1842, not Elliot, 1839), *melvillensis*, *rattoides*.
COMMENTS: Analyses of chromosomal and electrophoretic data (Baverstock et al., 1977*a*, *c*, 1981, 1983*b*) as well as phallic and dental morphology (Lidicker and Brylski, 1987; Misonne, 1969) indicated *M. gouldii* is phylogenetically most closely related to species of *Conilurus*.

Mesembriomys macrurus (Peters, 1876). Monatsb. K. Preuss. Akad. Wiss. Berlin, p. 355.
TYPE LOCALITY: Australia, Western Australia, "at a small mainland creek, Mermaid Strait" (Mahoney and Richardson, 1988:164).
DISTRIBUTION: N Western Australia and N Northern Territory (see map and discussion in Watts and Aslin, 1981:128); probably extinct in NW central region of Western Australia (Mahoney and Richardson, 1988:164).
SYNONYMS: *boweri*.

Microhydromys Tate and Archbold, 1941. Am. Mus. Novit., 1101:2.
TYPE SPECIES: *Microhydromys richardsoni* Tate and Archbold, 1941.
COMMENTS: Member of the New Guinea Old Endemics (Musser, 1981*c*). Reviewed by Flannery (1989).

Microhydromys musseri Flannery, 1989. Proc. Linn. Soc. N.S.W, 111:216.
TYPE LOCALITY: Papua New Guinea, West Sepik Prov., Torricelli Mtns, Mt Somoro, 1350 m (see Flannery, 1989, for details).
DISTRIBUTION: Known only from the type locality.
COMMENTS: Represented only by the holotype. A very distinct *Microhydromys* that is part of a highland fauna endemic to the N Coast ranges of Papua New Guinea (Flannery, 1989).

Microhydromys richardsoni Tate and Archbold, 1941. Am. Mus. Novit., 1101:2.
TYPE LOCALITY: New Guinea, Irian Jaya, Idenburg River, 4 km SW Bernhard Camp, 850 m.
DISTRIBUTION: New Guinea; scattered localities in hill forest from type locality east to Sogeri in Port Moresby region (see map in Flannery, 1990*b*:184).
COMMENTS: Known only by four specimens (Flannery, 1989). Distributional and biological data summarized by Flannery (1990*b*:184). *Microhydromys richardsoni* possesses the primitive pattern of the cephalic arterial circulation, a conformation shared with species of *Crossomys*, *Hydromys*, and *Parahydromys*, but not with *Leptomys* or the other genera of shrew mice (*Mayermys*, *Neohydromys*, and *Pseudohydromys*).

Micromys Dehne, 1841. *Micromys agilis*, Ein Neues Säugetier der Fauna von Dresden, p. 1.
TYPE SPECIES: *Micromys agilis* Dehne, 1841 (= *Mus minitus* Pallas, 1771).
COMMENTS: Only one extant species is recognized by most workers, but at least five others are documented by Miocene (Storch, 1987) and Pliocene (Weerd, 1979) fossils. Whether the extant samples represent one or more species has yet to be resolved by critical systematic revision.

Micromys minutus (Pallas, 1771). Reise Prov. Russ. Reichs., 1:454.
TYPE LOCALITY: Russia, Ulyanovsk. Obl. Middle Volga River, Simbirsk (now Ulyanovsk).
DISTRIBUTION: From NW Spain through most of Europe, across Siberia to Ussuri region and Korea, north to about 65° in European Russia and Yakutia, south to N edge of Caucasus and N Mongolia; isolated ranges in S China west through Yunnan to SE Tibet and NE India (Assam). Island distributions include Britain, Japan (Honshu, Shikoku, Kyushu, and Tsushima), Quelpart Isl (Korea), and Taiwan; see Corbet (1978*c*) for details.
SYNONYMS: *agilis*, *aokii*, *arundinaceus*, *avenarius*, *batarovi*, *berezowskii*, *campestris*, *danubialis*, *erythrotis*, *fenniae*, *flavus*, *hertigi*, *hondonis*, *japonicus*, *kytmanovi*, *meridionalis*, *messorius*, *minatus*, *minimus*, *oryzivorus*, *parvulus*, *pendulinus*, *pianmaensis*, *pratensis*, *pumilus*, *sareptae*, *soricinus*, *subobscurus*, *takasagoensis*, *triticeus*, *ussuricus*.
COMMENTS: Reviewed by Corbet (1978*c*, 1984). Chromosomal information reported by Jüdes (1981), Zima (1983), Lungeanu et al. (1984), Solleder et al. (1984), and Schmid et al. (1987). European populations reviewed by Böhme (1978*a*). Phallic morphology of

Chinese samples described by Yang and Fang (1988) in context of assessing relationships among Chinese murines.

Millardia Thomas, 1911. J. Bombay Nat. Hist. Soc., 20:998.

TYPE SPECIES: *Golunda meltada* Gray, 1837.

SYNONYMS: *Grypomys, Guyia, Millardomys.*

COMMENTS: Listed as a distinct genus by Ellerman in 1941, but later as a subgenus of *Rattus* (Ellerman, 1961). By 1969, *Millardia* was again treated as a genus and thought to be closely related to the Indian *Cremnomys* (Misonne, 1969). Subsequent analyses of morphological features and particularly chromosomal traits has demonstrated the great phylogenetic distance between *Rattus* and *Millardia* (Gadi and Sharma, 1983; Mishra and Dhandra, 1975; Raman and Sharma, 1977). Cytogenetic analyses resulted in a phylogenetic hypothesis isolating *Millardia* and *Cremnomys* from other Asian genera (Gadi and Sharma, 1983). Two species of *Millardia*, until 1982 thought to be strictly an Asian genus, were recorded from the Pliocene of Ethiopia (Sabatier, 1982). One of the species is apparently a *Millardia* but the other was reidentified as an *Acomys* (Denys, 1990).

Millardia gleadowi (Murray, 1886). Proc. Zool. Soc. Lond., 1885:809 [1886].

TYPE LOCALITY: Pakistan, Sind, Karachi.

DISTRIBUTION: Pakistan, adjacent India and Afghanistan.

COMMENTS: Raman and Sharna (1977) reported essentially no similarity between karyotypes of *M. meltada* and *M. gleadowi.*

Millardia kathleenae Thomas, 1914. J. Bombay Nat. Hist. Soc., 23:29.

TYPE LOCALITY: Burma, Pagan.

DISTRIBUTION: C Burma.

COMMENTS: Sody (1941) proposed the genus *Millardomys*, which is a *nomen nudum*, for this species.

Millardia kondana Mishra and Dhanda, 1975. J. Mammal., 56:76.

TYPE LOCALITY: India, Maharashtra State, Poona Dist, Sinhgarh (18°23'N, 73°42'E).

DISTRIBUTION: India; known only from the Sinhgarh Plateau in the Maharashtra region.

COMMENTS: Morphological comparisons between this distinctive species and *M. gleadowi*, *M. kathleenae*, and *M. meltada* were reported by Mishra and Dhanda (1975).

Millardia meltada (Gray, 1837). Ann. Mag. Nat. Hist., [ser. 1], 1:586.

TYPE LOCALITY: India, S Mahratta, Dharwar.

DISTRIBUTION: Sri Lanka; Indian Peninsula west to Gujarat and Rajasthan, north to Himachal Pradesh, and east to West Bengal; E Pakistan; and Terai region of Nepal (see Rana, 1985).

SYNONYMS: *comberi, dunni, lanuginosus, listoni, pallidior, singuri.*

COMMENTS: Cytogenetics of this species is the subject of a growing body of literature (Nanda and Raman, 1981; Raman and Sharma, 1977; Sobti and Gill, 1984; Yosida, 1978). Mandal and Ghosh (1981) described *singuri* as a subspecies of *M. meltada*.

Muriculus Thomas, 1903. Proc. Zool. Soc. Lond., 1902(2):314 [1903].

TYPE SPECIES: *Mus imberbis* Rüppell, 1842.

COMMENTS: When Thomas proposed *Muriculus* he suggested it might be related to *Lophuromys*, but Osgood (1936) found little affinity between the two genera and instead noted a closer morphological relationship with *Mus* and *Zelotomys*. The close tie to *Mus* is real, and Misonne (1974) suggested that *Muriculus* might be merged with *Mus*; however, the middorsal stripe and morphological specializations of rostrum, mandible, incisors, and increased expanse of incisor enamel associated with pronounced proodonty, are not part of the character suite defining *Mus*, and set *Muriculus* apart as a very distinctive genus.

Muriculus imberbis (Rüppell, 1842). Mus. Senckenberg., 3:110.

TYPE LOCALITY: N Ethiopia, Simien, 3300 m.

DISTRIBUTION: Mountains of Ethiopia on both sides of the Rift Valley between 1900 and 3400 m (Rupp, 1980).

SYNONYMS: *chilaloensis.*

COMMENTS: Member of a unique rodent fauna endemic to high mountains of Ethiopia. Osgood (1936) described *chilaloensis* as a subspecies of *M. imberbis*.

Mus Linnaeus, 1758. Syst. Nat., 10th ed., 1:59.

TYPE SPECIES: *Mus musculus* Linnaeus, 1758.

SYNONYMS: *Budamys, Coelomys, Drymomys, Gatamiya, Hylenomys, Leggada, Leggadilla, Musculus, Mycteromys, Nannomys, Oromys, Pseudoconomys, Pyromys, Tautatus.*

COMMENTS: Extant species of *Mus* are contained in four distinct subgenera (*Coelomys, Mus, Nannomys,* and *Pyromys*), each diagnosed by a suite of discrete morphological traits (J. T. Marshall, Jr., 1977*b*, 1986) and biochemical features (Bonhomme, 1986; She et al., 1990). *Nannomys, Pyromys,* and *Coelomys* were treated as genera by Bonhomme (1986), a view also adopted by She et al. (1990). The Asian species were reviewed by J. T. Marshall, Jr. (1977*b*) and the European forms by Marshall (1981, 1986), Marshall and Sage (1981), Bonhomme et al. (1984), Gerasimov et al. (1990), and She et al. (1990). Most of the African species need careful systematic review (see accounts below of species in subgenus *Nannomys*) and Ansell (in Meester et al., 1986:280) believed the African segment of the genus to be "over-split." Relevance of metrical, chromosomal, and allozyme variation to systematics of *Mus* was discussed by Corbet (1990). See Carleton and Musser (1984) for authors and publication dates of synonyms.

Mus baoulei (Vermeiren and Verheyen, 1980). Rev. Zool. Afr., 94:573.

TYPE LOCALITY: Ivory Coast, Lamto.

DISTRIBUTION: Known only from Ivory Coast and E Guinea.

COMMENTS: Subgenus *Nannomys*. A species occurring sympatrically with *M. musculoides* and *M. setulosus*, and whose morphology is closely similar to those in the *M. sorella* group (Vermeiren and Verheyen, 1980). The diagnostic traits reported for *M. baoulei* by Vermeiren and Verheyen (1980) are those that set *M. sorella* apart from other species of *Mus* (Verheyen, 1965*a*). Judged by their description, *baoulei* is distinguished from *M. sorella* by smaller size, a contrast that also exists between *M. sorella* and *M. neavei* (see that account). The relationship of *baoulei* to other members of the *M. sorella* group requires fresh assessment.

Mus booduga (Gray, 1837). Mag. Nat. Hist. [Charlesworth's], 1:586.

TYPE LOCALITY: India, S Mahratta.

DISTRIBUTION: Sri Lanka, Peninsular India (north to Jammu and Kashmir), S Nepal, and C Burma.

COMMENTS: Subgenus *Mus*. Results from chromosomal analyses reported by Sen and Sharma (1983) and Sharma et al. (1986) in context of evolutionary divergence relative to other species of *Mus*.

Mus bufo (Thomas, 1906). Ann. Mag. Nat. Hist., ser. 7, 18:145.

TYPE LOCALITY: Uganda, Ruwenzori East, 6000 ft.

DISTRIBUTION: E Zaire (Kivu region), adjacent Uganda, Rwanda, and Burundi; a montane Western Rift endemic.

SYNONYMS: *ablutus, wambutti.*

COMMENTS: Subgenus *Nannomys*. In body size and morphology, *M. bufo* superficially resembles the large-bodied *M. triton*, but the two are distinguished by dental features and tail length (Petter and Matthey, 1975) as well as karyotypes (Robbins and Baker, 1978), and both occur together in the Kivu region of E Zaire (specimens in the American Museum of Natural History). Chromosomal data for Burundi samples reported by Maddalena et al. (1989).

Mus callewaerti (Thomas, 1925). Ann. Mag. Nat. Hist., ser. 9, 15:668.

TYPE LOCALITY: S Zaire, Lualaba, Luluabourg, 610 m.

DISTRIBUTION: Recorded only from N and C Angola, S and W Zaire; limits unknown.

COMMENTS: Subgenus *Nannomys*. Thomas described this species as a member of the genus *Hylenomys*, which is now united with *Mus* (Hill and Carter, 1941; Misonne, 1965). It is the largest-bodied of any of the African *Mus* (Petter and Matthey, 1975), has ivory incisors and very large auditory bullae, and in external features resembles *M. triton*, which misled Hatt (1940*a*) into treating *callewaerti* as a subspecies of *triton*. Misonne (1965) summarized distributional and other information.

Mus caroli Bonhote, 1902. Novit. Zool., 9:627.
TYPE LOCALITY: Japan, Ryukyu (= Liukiu) Isls, Okinawa Isl.
DISTRIBUTION: Natural range probably from Ryukyu Isls to Taiwan, S China (Fujian and Yunnan Provs., Hainan Isl, Hong Kong), Vietnam, Laos, Cambodia, and Thailand (N of Isthmus of Kra; see map in J. T. Marshall, Jr., 1977*a*). Also recorded from Malay Peninsula (S Kedah State), Sumatra, Java, Madura, and Flores Isls in Nusa Tenggara, all places where it was likely inadvertently introduced (Musser and Newcomb, 1983).
SYNONYMS: *boninensis, formosanus, kakhyenensis, kukilensis, ouwensi.*
COMMENTS: Subgenus *Mus.*

Mus cervicolor Hodgson, 1845. Ann. Mag. Nat. Hist., [ser. 1], 15:268.
TYPE LOCALITY: Nepal.
DISTRIBUTION: Indigenous range from Nepal east through Sikkim into NE India (Assam), Burma, Thailand, Laos, Cambodia, and Vietnam (see map in J. T. Marshall, Jr., 1977*a*). Also recorded from Sumatra and Java where it has likely been inadvertently introduced (Musser and Newcomb, 1983).
SYNONYMS: *annamensis, cunicularis, imphalensis, nitidulus, popaeus, strophiatus.*
COMMENTS: Subgenus *Mus.*

Mus cookii Ryley, 1914. J. Bombay Nat. Hist. Soc., 22:663.
TYPE LOCALITY: N Burma, Shan States, Gokteik, 2133 ft.
DISTRIBUTION: Peninsular India, Nepal through NE India (Assam) to Burma, S China (SW Yunnan), N and C Thailand, Laos, and N Vietnam (see map in J. T. Marshall, Jr., 1977*a*).
SYNONYMS: *darjilingensis, nagarum, palnica.*
COMMENTS: Subgenus *Mus.*

Mus crociduroides (Robinson and Kloss, 1916). J. Str. Br. Roy. Asiat. Soc., 73:271
TYPE LOCALITY: Indonesia, W Sumatra, Korinchi Peak, 10,000 ft.
DISTRIBUTION: Upper montane rain forest in mountain chain along W Sumatra.
COMMENTS: Subgenus *Coelomys.* A distinct montane species endemic to the mountains of W Sumatra (Musser, 1986; Musser and Newcomb, 1983). Listed by Chasen (1940) as a species of *Mycteromys,* which also contained the Javan *M. vulcani,* the closest relative of *M. crociduroides.*

Mus famulus Bonhote, 1898. J. Bombay Nat. Hist. Soc., 12:99.
TYPE LOCALITY: S India, Nilgiri Hills, Coonoor, 5000 ft.
DISTRIBUTION: Recorded only from the Nilgiri Hills in S India.
COMMENTS: Subgenus *Coelomys.*

Mus fernandoni (Phillips, 1932). Spolia Zeylan., 16:325.
TYPE LOCALITY: Sri Lanka, Mulhalkelle Dist., Kubalgamuwa, 3000 ft.
DISTRIBUTION: Endemic to Sri Lanka.
COMMENTS: Subgenus *Pyromys.* Phillips (1980) summarized distributional and biological information.

Mus goundae Petter and Genest, 1970. Mammalia, 34:455.
TYPE LOCALITY: Central African Republic, vicinity of Gounda River (Petter, 1981*b*, provided coordinates).
DISTRIBUTION: Recorded only from the vicinity of the type locality (see map in Jotterand, 1972); limits unknown.
COMMENTS: Subgenus *Nannomys.* Petter (1981*b*) treated *goundae* as a species related to others in the *M. sorella* group, but the nature of that relationship remains unresolved (see account of *M. sorella*). Chromosomal data were reported by Jotterand (1972).

Mus haussa (Thomas and Hinton, 1920). Novit. Zool., 27:319.
TYPE LOCALITY: Nigeria, Farniso.
DISTRIBUTION: Records are from Senegal and S Mauritania through Mali, Ivory Coast, Burkina Faso, Ghana, S Niger, and Benin to N Nigeria; limits unknown.
COMMENTS: Subgenus *Nannomys.* Body size, pelage color and pattern, and other morphological traits of *M. haussa* are very similar to those of *M. tenellus* (Petter, 1963*c*, 1972*a*; Rosevear, 1969), and F. Petter (1969) came close to combining them. The

relationship between *M. haussa* and *M. tenellus* needs to be assessed by systematic revision of the group.

Mus indutus (Thomas, 1910). Ann. Mag. Nat. Hist., ser. 8, 5:89.
TYPE LOCALITY: South Africa, N Cape Prov., Molopo River, west of Morokwen.
DISTRIBUTION: South Africa (N Orange Free State, Transvaal, N Cape Prov.), W Zimbabwe, Botswana, C and N Namibia, (see map in Skinner and Smithers, 1990:264).
SYNONYMS: *deserti, pretoriae, valschensis.*
COMMENTS: Subgenus *Nannomys*. Skinner and Smithers (1990) and Meester et al. (1986) discussed the morphological and chromosomal distinctions between *M. indutus* and *M. minutoides*. Skinner and Smithers (1990) also summarized biological data, and Meester et al. (1986) provided citations for synonyms. The definition of this species is ambiguous (Meester et al., 1986; Skinner and Smithers, 1990). Meester et al. (1986) included the Angolan *sybilla* in *M. indutus*, but our study of the holotype revealed that *sybilla* belongs with *M. musculoides*.

Mus kasaicus (Cabrera, 1924). Bol. Real. Soc. Espanola Hist. Nat., Madrid, 24:222.
TYPE LOCALITY: Zaire, Kasai, St. Joseph de Kananga (Luluabourg).
DISTRIBUTION: Recorded only from the vicinity of the type locality.
COMMENTS: Subgenus *Nannomys*. Now treated as a distinct species belonging to the *M. sorella* group (Petter, 1981*b*), but its relationship to that species needs to be assessed in a careful revision of the *M. sorella* complex (see account of the latter).

Mus macedonicus Petrov and Ruzic, 1983. Proc. Fauna SR Serbia, Serbian Acad. Sci. and Arts, Belgrade, 2:177.
TYPE LOCALITY: Yugoslavia, Macedonia, near Valandovo.
DISTRIBUTION: Yugoslavia, Bulgaria, Turkey, Iran, Syria, Jordan, and Israel.
SYNONYMS: *spretoides.*
COMMENTS: Subgenus *Mus*. Originally desribed as a subspecies of *M. hortulanus* (Petrov and Ruzic, 1985), but Bonhomme (1986) and Marshall (in litt.) treated it as a distinct species occurring sympatrically with *M. musculus* and *M. spicilegus* (also see map in Petrov and Ruzic, 1985:234). Results of biochemical and morphometric analyses of *M. macedonicus* (under *spretoides*) in context of phylogenetic studies provided by Bonhomme et al. (1984), Auffray et al. (1990*b*), Gerasimov et al. (1990), and She et al. (1990). Fossil history of Israeli *M. macedonicus* (referred to as *spretoides*) relative to colonization and origin of commensalism in *M. musculus* reported by Auffray et al. (1988).

Mus mahomet Rhoads, 1896. Proc. Acad. Nat. Sci. Philadelphia, p. 532.
TYPE LOCALITY: SC Ethiopia, Sheikh Mahomet.
DISTRIBUTION: Ethiopian highlands (1500-3400 m; Rupp, 1980; specimens in the Field Museum of Natural History; see map in Yalden et al., 1976), SW Uganda and SW Kenya (specimens in the National Museum of Natural History); limits unknown.
SYNONYMS: *emesi.*
COMMENTS: Subgenus *Nannomys*. Heller (1911) described *emesi* as a subspecies of *M. musculoides*, but Hollister (1919:96) and Hatt (1940*a*) treated it a a distinct species. Our study of Hatt's series from NE Zaire in the American Museum of Natural History revealed they consisted of *M. musculoides* and *M. sorella*. The holotype of *emesi* and most of Hollister's other examples from Uganda do represent a species distinct from *M. musculoides*; in morphology and chromatic traits, we cannot distinguish the series of *emesi* from the large samples of *mahomet* collected by Osgood in Ethiopia. *M. mahomet* is sympatric with *M. musculoides* in Uganda and Kenya and narrowly sympatric or closely parapatric with *M. setulosus* in Ethiopia (Yalden et al., 1976; specimens in the Field Museum of Natural History). Yalden et al. (1976:30) suspected *kerensis* might be the correct name for *M. mahomet*.

Mus mattheyi Petter, 1969. Mammalia, 33:118.
TYPE LOCALITY: Ghana, Accra.
DISTRIBUTION: Recognized only from the type locality.
COMMENTS: Subgenus *Nannomys*. This form has been recorded from Senegal, Ivory Coast, Burkina Faso, and Ghana (F. Petter, 1969; Petter et al., 1971; Petter and Matthey, 1975), but after studying large series (in the Museum National d'Histoire Naturelle

and the National Museum of Natural History) of the *haussa* and *musculoides* complexes from west Africa, we are unable to assign anything to *mattheyi*. The morphological traits used to distinguish *mattheyi* from *haussa* by F. Petter (1969) and Petter and Matthey (1975) vary in a continuous fashion from typical *haussa* morphology to that considered diagnostic for *mattheyi*. The status of *mattheyi* needs to be reassessed in a taxonomic revision of the group. A chromosomal complement of 2N=36, FN=36 characterizes samples identified as *M. mattheyi* and is considered primitive for African *Mus* (F. Petter, 1969; Jotterand-Bellomo, 1986).

Mus mayori (Thomas, 1915). J. Bombay Nat. Hist. Soc., 23:415.
TYPE LOCALITY: Sri Lanka, Central Mountains, Pattipola, 6200 ft.
DISTRIBUTION: Endemic to forested regions of Sri Lanka.
SYNONYMS: *bicolor* (Thomas, 1915, not Tichomirow and Kortchagin, 1889), *pococki*.
COMMENTS: Subgenus *Coelomys*. Phillips (1980) recognized highland (*mayori*) and lowland (*pococki*) as subspecies, and summarized distributional and biological information for both; also see J. T. Marshall, Jr. (1977*b*) for discussion of chromatic variation within each form.

Mus minutoides Smith, 1834. S. Afr. Quart. J., ser. 2, 2:157.
TYPE LOCALITY: South Africa, S Cape Prov., Cape Town.
DISTRIBUTION: South Africa (W, S, and E Cape Province, Natal, Zululand, Lesotho, Orange Free State, C and E Transvaal), Swaziland; northern limits unknown (see Skinner and Smithers, 1990:264; Meester et al., 1986:283).
SYNONYMS: *minimus* (Peters, 1852, not White, 1789), *umbratus*.
COMMENTS: Subgenus *Nannomys*. Relationship of this species to *M. musculoides* has to be assessed by careful systematic revision of the *minutoides-musculoides* complex (see following account).

Mus musculoides Temminck, 1853. Esquisses Zool. sur la Côte de Guine, p. 161.
TYPE LOCALITY: West Africa, "Côte de Guine."
DISTRIBUTION: Subsaharan Africa (including Ethiopia and Somalia) southward to contact with *M. minutoides*.
SYNONYMS: *bella, enclavae, gallarum, gondokorae, grata, marica, paulina, petila, soricoides, sungarae, sybilla, vicina*.
COMMENTS: Subgenus *Nannomys*. Whether the samples reflect only one or a complex of species is unresolved. For example, *grata* (or *gratus*) is often listed as a separate species (Hatt, 1940*a*; Hollister, 1919); Petter and Matthey, 1975). Unresolved also is the geographic distribution of *M. musculoides*, and the nature of the biological relationship between it and *M. minutoides*. The considerable chromosoal variation among samples from West Africa was documented by Jotterand (1972) and Jotterand-Bellomo (1984, 1986) under the identification of *minutoides/musculoides*, a label that reflects current understanding of specific limits in this complex.
Neither Yalden et al. (1976) nor Rupp (1980) recorded *M. musculoides* from Ethiopia, but we have seen many specimens from that country (in the British Museum and the Field Museum of Natural History).

Mus musculus Linnaeus, 1758. Syst. Nat., 10th ed., 1:62.
TYPE LOCALITY: Sweden, Uppsala County, Uppsala.
DISTRIBUTION: Spread throughout most of world through its close association with humans (Ellerman and Morrison-Scott, 1951); in some areas restricted to human dwellings and habitats maintained by human activity; sometimes feral where introduced.
SYNONYMS: *abbotti, adelaidensis, airolensis, albertisii, albicans; albidiventris* (Burg, 1923, not Blyth, 1852), *musculus* var. *albinus, albus, tomensis* morph *amurensis, ater, azoricus, bactrianus, bateae, bicolor, bieni, borealis, brevirostris, canacorum, candidus* (Laurent, 1937, not Bechstein, 1796), *caoccii, castaneus, caudatus, cinereo-maculatus, commissarius, decolor, domesticus, dubius* (Hodgson, 1845, not Fischer, 1829), *dunckeri, faeroensis, far, flavescens* (Fischer, 1872, not Barrett-Hamilton, 1896, Elliot, 1839, or Waterhouse, 1837), *flavus* (Bechstein, 1801, not Kerr, 1792), *formosovi, fredericae, funereus, gansuensis, gentilis, gentilulus, gerbillinus, germanicus, gilvus, hanuma, hapsaliensis, helgolandicus, helviticus, helvolus, heroldii, homourus, hortulanus, hydrophilus, indianus, jalapae, jamesoni, kaleh-peninsularis, kambei, kuro, lundii, maculatus, major* (Severtzov, 1873, not Brants, 1827, or Pallas, 1779), *manchu, manei* (Gray, 1843), *manei* (Kelaart,

1852), *musculus* var. *melanogaster, minotaurus, mohri, mollissimus, molossinus, momiyamai, mongolium, muralis, mykinessiensis, mystacinus* (Mohr, 1923, not Danford and Alston, 1867), *nattereri, niger, nipalensis, niveus, nogaiorum, nordmanni, musculus* var. *nudoplicatus, orientalis* (Cretzschmar, 1826, not Desmarest, 1819), *orii, oxyrrhinus, pachycercus, pallescens, percnonotus, peruvianus, polonicus, poschiavinus, praetextus, pygmaeus* (Biswas and Khajuria, 1955, not Milne-Edwards, 1874), *raddei, rama, reboudi, rotans, musculus* var. *rubicundus, tomensis* morph *rufiventris, sareptanicus, severtzovi, simsoni, sinicus, striatus, subcaeruleus, subterraneus, taitensis, taiwanus, takagii, molossinus* var. *takayamai, tantillus, tataricus, theobaldi, tomensis, tytleri, urbanus, utsuryonis, variabilis, varius, viculorum, vignaudii, vinogradovi, wagneri, yamashinai, yesonis, yonakuni.* (see Ellerman, 1941; Ellerman and Morrison-Scott, 1951; Mahoney and Richardson, 1988; and Marshall, in press).

COMMENTS: Subgenus *Mus*. Schwarz and Schwarz (1943) provided a revision that was followed with minor changes by Ellerman and Morrison-Scott (1951). This arrangement was criticized by Jones and Johnson (1965:394) who found that specimens from Asia they studied "bear little or no relation to this idealized classification." Subsequent treatments of this group were presented by J. T. Marshall, Jr. (1977*b*, 1981, 1986) and Marshall and Sage (1981). The most recent classification combined biochemical analyses of European, Asian, and African mice (Bonhomme et al., 1984), and the translation of these results, as well as the incorporation of morphological data, into a new view of *Mus musculus* and its allies (Marshall, in press), that in allocation of the many names to *M. musculus* is suprisingly concordant with the treatment of Ellerman and Morrison-Scott (1951). The scientific names listed under *M. musculus* by those authors, but now placed elsewhere, are here associated with *M. spicilegus, M. spretus*, and *M. macedonicus* (see those accounts).

Samples of *M. musculus* have been the focus of numerous morphometric, chromosomal, and biochemical studies undertaken in context of phylogenetic inquiries; some of the latest contributions that also summarized and cited earlier reports are Evans (1981), Foster et al. (1981), Bonhomme et al. (1984), Potter et al. (1986), Giagia et al. (1987), Hubner (1988), Nishioka (1987), Winking et al., (1988), Britton-Davidian (1990), Corti and Ciabatti (1990), Gerasimov et al. (1990), She et al. (1990), Searle (1991), Viroux and Bauchau (1992), Karn and Dlouhy (1991), Bush and Paigen (1992), and Scriven and Bauchau (1992). Analyses, using palaeontological and archaeozoological approaches, of the colonization process of W Eurasia by house mice and their origin of commensalism were presented by Auffray et al. (1988, 1990*a*, *c*) and Auffray and Britton-Davidian (1992). Analyses of mitochondrial DNA variation among samples from Japanese islands reveal a polyphyletic origin of Japanese *M. musculus* derived from *musculus, castaneus*, and *domesticus* strains (Bonhomme et al., 1989).

Mus neavei (Thomas, 1910). Ann. Mag. Nat. Hist., ser. 8, 5:90.

TYPE LOCALITY: E Zambia, E Loangwe Dist., Petauke, 2400 ft.

DISTRIBUTION: S Zaire, E Zambia, S Zimbabwe, Transvaal of South Africa, W Mozambique, and S Tanzania; limits unknown.

COMMENTS: Subgenus *Nannomys*. Originally described as a species, *neavei* was later treated as a subspecies of *M. sorella* (Verheyen, 1965*a*), an arrangement accepted by Ansell (1978), Meester et al. (1986), and Skinner and Smithers (1990). Petter (1981*b*), however, pointed out that while a member of the *M. sorella* group, *neavei* should be treated as a separate species; in morphology and body size it appears to be close to *M. oubanguii* (Petter, 1981*b*). Our study (series in the American Museum of Natural History, the British Museum of Natural History, and the National Museum of Natural History) corroborates Petter's view. *M. neavei* is a distinct species and easily distinguished from *M. sorella* by its richer tawny fur, much smaller size, more delicate cranium, and shorter molar rows (3.0-3.2 mm in 7 examples of *neavei*, 3.2-3.7 mm in nine *sorella*). How *M. neavei* is related to *M. oubanguii* and the small-bodied *M. baoulei* is unresolved. Meester et al. (1986:282) summarized published distributional information. Supposed records of *M. neavei* from Malawi represent other species (Ansell and Dowsett, 1988).

Mus orangiae (Roberts, 1926). Ann. Transvaal Mus., 11:251.
TYPE LOCALITY: South Africa, N Orange Free State, Kruisementifontein, Viljoensdrift, near Vereeniging.
DISTRIBUTION: South Africa; Orange Free State (see map in Vermeiren and Verheyen, 1983).
COMMENTS: Subgenus *Nannomys*. Treated as a species possibly allied to *M. setzeri* by Vermeiren and Verheyen (1983), but listed as a subspecies of *M. minutoides* by Meester et al. (1986), an arrangement followed by Skinner and Smithers (1990).

Mus oubanguii Petter and Genest, 1970. Mammalia, 34:454.
TYPE LOCALITY: Central African Republic, La Maboke, Ippy, Bangassou (Petter, 1981*b*, provided coordinates).
DISTRIBUTION: Recorded only from Central African Republic (savanna north of Oubangui River); see map in Jotterand (1972:332).
COMMENTS: Subgenus *Nannomys*. Sympatric with *M. setulosus* and *M. musculoides* (Petter and Genest, 1970), but a phylogenetic member of the *M. sorella* group, according to Petter (1981*b*), who also noted that its morphology, except for a dental trait, is similar to that of *M. neavei* (see that account). Chromosomal information, in context of understanding chromosomal evolution among species of African *Mus*, was documented by Jotterand (1972) and Jotterand-Bellomo (1984, 1986).

Mus pahari Thomas, 1916. J. Bombay Nat. Hist. Soc., 24:414.
TYPE LOCALITY: India, Sikkim, Batasia, 6000 ft.
DISTRIBUTION: From NE India (Sikkim and Assam) through Burma, S China (Yunnan), Thailand, Laos, and Vietnam (see map in J. T. Marshall, Jr., 1977*a*).
SYNONYMS: *gairdneri, jacksoniae, meator, mocchauensis.*
COMMENTS: Subgenus *Coelomys*. Dao (1978) described *mocchauensis* as a subspecies of *M. pahari*.

Mus phillipsi Wroughton, 1912. J. Bombay Nat. Hist. Soc., 21:772.
TYPE LOCALITY: India, Central Prov., Nimur Dist, Asirgarh, 1500 ft.
DISTRIBUTION: Peninsular India.
SYNONYMS: *siva, surkha.*
COMMENTS: Subgenus *Pyromys*.

Mus platythrix Bennett, 1832. Proc. Zool. Soc. Lond., 1832:121.
TYPE LOCALITY: Peninsular India, Dukhun.
DISTRIBUTION: Peninsular India.
SYNONYMS: *bahadur, grahami, hannyngtoni.*
COMMENTS: Subgenus *Pyromys*.

Mus saxicola Elliot, 1839. Madras J. Litt. Sci., 10:215.
TYPE LOCALITY: India, Madras.
DISTRIBUTION: Peninsular India, S Pakistan, and S Nepal.
SYNONYMS: *cinderella, gurkha, priestlyi, ramnadensis, sadhu.*
COMMENTS: Subgenus *Pyromys*.

Mus setulosus Peters, 1876. Monatsb. K. Preuss. Akad. Wiss. Berlin, p. 480.
TYPE LOCALITY: Cameroon, Victoria.
DISTRIBUTION: From Guinea (Mt Nimba) and Sierra Leone eastward through Liberia, Ivory Coast, Ghana, Togo, Benin, Nigeria, Cameroon, Gabon, Central African Republic, N Zaire (Haut-Zaïre), S Sudan, WC and S Ethiopia to N Uganda and W Kenya (documented by Rosevear, 1969; Petter and Genest, 1970; and our study of samples in the American Museum of Natural History, British Museum, Field Museum of Natural History, and National Museum of Natural History).
SYNONYMS: *pasha, proconodon.*
COMMENTS: Subgenus *Nannomys*. A distinct species sometimes confused with *M. musculoides*, which occurs over approximately the same region (Rosevear, 1969). Both *pasha* (Thomas, 1910*a*) and *proconodon* (Rhoads, 1896) were originally described as species. Osgood (1936) associated *pasha* with *M. proconodon*, and we agree with his identification. Petter and Matthey (1975) regarded *pasha* as a species, noting that it might be referable to *M. setulosus*. Both Osgood (1936) and Yalden et al. (1976) recognized *proconodon* as a distinct species endemic to Ethiopia. Our study of Osgood's specimens, some of which are near-topotypes, revealed that their

morphological traits fell within the range of variation typical of *M. setulosus*. Our identification was forshadowed by Petter and Matthey (1975) who cited the range of *M. setulosus* to include Ethiopia, based on a letter from J. Prevost. Chromosomal data for samples from W Africa were documented by Jotterand (1972), Jotterand-Bellomo (1981, 1986), and Matthey (1964).

Mus setzeri Petter, 1978. Mammalia, 42:377.

TYPE LOCALITY: Botswana, 82 km W of Mohembo (near Nambian border).

DISTRIBUTION: NE Namibia, NW and S Botswana, and W Zambia (see map in Vermeiren and Verheyen, 1983).

COMMENTS: Subgenus *Nannomys*. A unique desert species reviewed by Vermeiren and Verheyen (1983) and Skinner and Smithers (1990).

Mus shortridgei (Thomas, 1914). J. Bombay Nat. Hist. Soc., 23:30.

TYPE LOCALITY: Burma, Mt Popa, 4961 ft.

DISTRIBUTION: Burma, Thailand, Cambodia (see map in J. T. Marshall, Jr., 1977*a*:431), and NW Vietnam (Dao, 1978).

SYNONYMS: *nghialoensis*.

COMMENTS: Subgenus *Pyromys*. Dao (1966) described *nghialoensis* as a subspecies of *M. platythrix*, which at the time embraced the Indochinese *shortridgei*.

Mus sorella (Thomas, 1909). Ann. Mag. Nat. Hist., ser. 8, 4:548.

TYPE LOCALITY: W Kenya, Mt Elgon, Kirui, 6000 ft.

DISTRIBUTION: Documented by specimens from E Cameroon, EC Angola, NE and SE Zaire, Uganda, Kenya, and N Tanzania (Petter, 1981*b*; Verheyen, 1965*a*; specimens in the American Museum of Natural History, and the National Museum of Natural Hitory); limits unknown

SYNONYMS: *acholi*, *wamae*.

COMMENTS: Subgenus *Nannomys*. Closest relatives are *M. baoulei*, *M. goundae*, *M. kasaicus*, *M. neavei*, and *M. oubanguii*; Petter (1981*b*) placed these (except *baoulei*) together in the *M. sorella* group. Petter (1981*b*) also recognized *wamae* and *acholi* as species in the *sorella* complex, but after examining holotypes and other specimens we agree with Verheyen (1965*a*), who united them with *M. sorella*. There are at least two distinct species in the group, *M. sorella* and *M. neavei* (see later account), but the nature of their phylogenetic relationship to other forms in this complex needs to be assessed by critical systematic review. Reidentification of museum specimens might also help resolve geographic ranges. The specimens from Angola, for example, were originally identified by Hill and Carter (1941) as *M. bella*.

Mus spicilegus Petenyi, 1882. Termeszetrajzi Fuzetek, Budapest, 5:114.

TYPE LOCALITY: Hungary, Budapest, Rakos Plains.

DISTRIBUTION: Hungary, Rumania, Yugoslavia, Bulgaria, and steppes of Crimea and S Ukraine.

SYNONYMS: *mehelyi*, *sergii*.

COMMENTS: Subgenus *Mus*. This is the mouse that constructs soil-covered storage mounds of grain, and was formally known as *M. hortulanus* (see Corbet, 1984); however, the holotype of *hortulanus* is actually a *M. musculus*, so the earliest name for the species is *spicilegus* (Gerasimov et al., 1990). Results of morphometric and biochemical analyses were reported by Bonhomme et al. (1984), Petrov and Ruzic (1985), Gerasimov et al. (1990), She et al. (1990), and Lyalyukhina et al. (1991). Other cytogenetic and biochemical contrasts between the species (reported as *hortulanus*) and *M. musculus* were recorded by Bulatova and Kotenkova (1990), and Yakimenko et al. (1990).

Mus spretus Lataste, 1883. Acta Linn. Soc. Bordeaux, ser. 7, 4:27.

TYPE LOCALITY: Algeria, Oued Magra, between M'sila and Barika, north of Hodna.

DISTRIBUTION: S France, Spain (incl. Balearic Isls), Portugal, Morocco, Algeria, Tunisia, and Libya (see map in Marshall, 1981).

SYNONYMS: *hispanicus*, *lusitanicus*, *lynesi*, *mogrebinus*, *parvus*, *rifensis*.

COMMENTS: Subgenus *Mus*. This species was the subject of biometrical and morphological analyses (Darviche and Orsini, 1982; Engels, 1980, 1983*b*; Gerasimov et al., 1990; Palomo, 1988; Palomo et al., 1983; Vargas, et al., 1984), as well as chromosomal and electrophoretic studies (Cano et al., 1984; Engels, 1983*a*; Matsuda and Chapman, 1992; Traut et al., 1992). Analysis of variability in mitochrondrial DNA revealed two genetically distinct phylogenetic groups within *spretus* (Boursot, et al., 1985). Differences between *M. spretus* and *M. musculus* in thermoregulatory capabilities reported by Gorecki et al. (1990). Alcover et al. (1985) described *parvus* as a subspecies of *M. spretus*.

Mus tenellus (Thomas, 1903). Proc. Zool. Soc. Lond., 1903(1):298.

TYPE LOCALITY: Sudan, Blue Nile, Roseires.

DISTRIBUTION: Sudan, S Ethiopia (below 2000 m; Rupp, 1980), S Somalia, and south through Kenya to C Tanzania (Dodoma); limits unknown.

SYNONYMS: *aequatorius, delamensis, gerbillus, suahelica*.

COMMENTS: Subgenus *Nannomys*. Reviewed by Petter (1972*a*) and Yalden et al. (1976). Morphologically and ecologically closely similar to *M. haussa* (see that account). The southernmost record is based on the holotype of *gerbillus* (Allen and Loveridge, 1933), which is an example of *M. tenellus*. Yalden et al. (1976) listed *gallarum* as a synonym of *M. tenellus*, but the holotype is an example of *M. musculoides*. Judged by our studies of museum specimens, most published Ethiopian records of *M. tenellus* are actually *M. musculoides*.

Mus terricolor Blyth, 1851. J. Asiat. Soc. Bengal, 20:172.

TYPE LOCALITY: S India, Bengal, neighborhood of Calcutta.

DISTRIBUTION: Indigenous to peninsular India, Nepal, and Pakistan; occurs also in Medan region of N Sumatra (Indonesia) where it was probably inadvertently introduced (Musser and Newcomb, 1983, discussed under *dunni*).

SYNONYMS: *beavanii, dunni*.

COMMENTS: Subgenus *Mus*. Formerly referred to as *M. dunni* (J. T. Marshall, Jr., 1977*b*, 1986), but *terricolor* is the older name. Chromosomal results presented by Sharma et al. (1986, under *dunni*) in context of evolutionary divergence from other species of *Mus*.

Mus triton (Thomas, 1909). Ann. Mag. Nat. Hist., ser. 8, 4:548.

TYPE LOCALITY: Kenya, Mt Elgon, Kirui, 6000 ft.

DISTRIBUTION: N Zaire and E Zaire (Kivu region), Uganda, Kenya, Tanzania, Malawi, Tete Dist. of Mozambique, Zambia, and Angola.

SYNONYMS: *birungensis, fors, imatongensis, murilla, naivashae*.

COMMENTS: Subgenus *Nannomys*. Listed as a questionable synonym of *M. mahomet* by Yalden et al. (1976:30) who were unsure about the equivalence of *mahomet* and *triton* and merely noted that Ethiopian samples previously identified as *triton* were really *mahomet*. The description of *Mus birungensis* (Lonnberg and Gyldenstolpe, 1925) mirrors the range of variation of *M. triton* in series (in the American Museum of Natural History) we have examinied from the Kivu region of E Zaire. Considerable chromosomal polymorphism has been reported in samples identified as *M. triton* (Robbins and Baker, 1978). Extant southern limit of species is Zambia and Tete Dist. of Mozambique (about 17°S), but it was present in Natal, South Africa up to about 60,000 years ago (Avery, 1991).

Mus vulcani (Robinson and Kloss, 1919). Ann. Mag. Nat. Hist., ser. 9, 4:378.

TYPE LOCALITY: Indonesia, W Java, Gunung Gede, Gandang Badak, 7900 ft.

DISTRIBUTION: Endemic to moss forests in mountains of W Java.

COMMENTS: Subgenus *Coelomys*. Originally described as a subspecies of *Mycteromys crociduroides* (see Chasen, 1940), but is a separate species. Both *M. vulcani* and *M. crociduroides* are the only native *Mus* known from islands on the Sunda Shelf (Musser, 1986; Musser and Newcomb, 1983).

Mylomys Thomas, 1906. Ann. Mag. Nat. Hist., ser. 7, 18:224.

TYPE SPECIES: *Mylomys cuninghamei* Thomas, 1906 (= *Golunda dybowskii* Pousargues, 1893).

COMMENTS: The Indian *Golunda* and *Mylomys* are usually considered close relatives of each other, but Musser (1987) discussed traits that indicated a distant relationship between the two genera and a closer phylogenetic alliance between *Mylomys* and *Pelomys*.

Mylomys dybowskii (Pousargues, 1893). Bull. Soc. Zool. Fr., 18:163.

TYPE LOCALITY: Central African Republic (= French Congo), Kemo River.

DISTRIBUTION: Guinea (Mt Nimba), Ivory Coast, Ghana, S Cameroon, Congo, Central African Republic, W, N and E Zaire, Rwanda, Tanzania, Kenya, Uganda, and S Sudan.

SYNONYMS: *alberti, christyi, cuninghamei, lowei, massaicus, rex, richardi, roosevelti.*

COMMENTS: Hatt (1940*a*) noted that the cotypes of *dybowskii* are examples of *Mylomys* and not *Pelomys*, under which the name had been listed (Ellerman, 1941), and selected a lectotype. The identity was verified by F. Petter (1962*b*). Significance of geographic variation in chromatic and morphological traits has yet to be assessed by critical systematic revision; whether the genus is monotypic or contains more than one species is unresolved. Chromosomal data for sample from Central African Republic reported by Matthey (1970), and those from Mt Nimba (Guinea) provided by Gautun et al. (1986).

The taxon *rex*, represented only by the holotype, a skin without skull from Kaffa in C Ethiopia, was described by Thomas (1906*a*) as a species of *Arvicanthis*, but later "provisionally considered as a giant member of *Desmomys*" (Thomas, 1916*a*:68). Dieterlen (1974) challenged the validity of *rex*, but Yalden et al. (1976) pointed out the features distinguishing the holotype from samples of *D. harringtoni*, and treated *rex* as another distinctive species endemic to Ethiopia. Our study of the holotype skin reveals it to be a large and probably old adult of *Mylomys* that is not as brightly pigmented as most samples of that genus. Whether the holotype actually came from Ethiopia, or represents a separate species of *Mylomys* are unknown; we provisionally list *rex* in the synonymy of *M. dybowskii*.

Myomys Thomas, 1915. Ann. Mag. Nat. Hist., ser. 8, 16:477.

TYPE SPECIES: *Epimys colonus* (see Allen, 1939; = *Mus verreauxii* A. Smith, 1834).

SYNONYMS: *Myomyscus.*

COMMENTS: Because some workers claimed *colonus* was unidentifiable and therefore *Myomys* is invalid, *Myomyscus* was proposed by Shortridge (1942) to replace *Myomys*. The problem and historical opinions were reviewed by Roberts (1951), Ellerman et al. (1953), Rosevear (1969), Van der Straeten and Verheyen (1978*b*) and Meester et al. (1986). Should *Myomys* really prove to have no nomenclatural status, *Myomyscus* is the name to use for this group.

In aspects of their morphology, the species seem to be arboreal and scansorial counterparts of those in *Mastomys*, all of which are primarily terrestrial. Some species of *Myomys* may be closely related to the *Mastomys* complex. Chromosomal and immunological data related *Myomys fumatus* to species of *Mastomys* (Qumsiyeh et al., 1990), and morphometric analyses placed *M. verreauxii* closer to *Mastomys* than to other species of *Myomys* (Van der Straeten and Dieterlen, 1983). Some workers (e.g., Misonne, 1969, 1974; Qumsiyeh et al., 1990) included *Myomys* and *Myomyscus* within *Praomys*. We retain *Myomys* separate from *Mastomys* and *Praomys* until the species within each of these groups are defined, and the systematic relationships among the groups are assessed by careful revision.

Myomys albipes (Rüppell, 1842). Mus. Senckenberg., 3:107.

TYPE LOCALITY: Ethiopia, Massawa.

DISTRIBUTION: Ethiopia; endemic to Ethiopian Plateau between 1500-3300 m (Yalden et al., 1976; Van der Straeten and Dieterlen, 1983).

SYNONYMS: *rufidorsalis alettensis, rufidorsalis ankoberensis, leucopus, albipes* var. *minor* (see Yalden et al., 1976).

COMMENTS: Usually listed as either a species of *Myomys* (Allen, 1939) or *Praomys* (e.g., Yalden et al., 1976). Morphometric traits related *albipes* closely to species of *Stenocephalemys* and what was described as *Praomys ruppi* (Van der Straeten and Dieterlen, 1983). Furthermore, qualitative external, cranial (see figure 5 in Rupp, 1980), and molar (see figures in Misonne, 1969) traits of *albipes* are more similar to species of *Stenocephalemys* than to most species in *Myomys*. The phylogenetic significance of this morphologically annectant relationship of *M. albipes* between other *Myomys* and *Stenocephalemys* needs to be assessed by taxonomic revision of both groups. The Sudanese specimens reported by Setzer (1956) as *P. albipes fuscirostris* are not *M. albipes*.

Myomys daltoni (Thomas, 1892). Ann. Mag. Nat. Hist., ser. 6, 10:181.
TYPE LOCALITY: West Africa (see discussion in Rosevear, 1969:412).
DISTRIBUTION: From Gambia and Senegal through Sierra Leone, N Ivory Coast, S Mali, Burkina Faso, Ghana, Togo, Benin, Nigeria, S Chad and Central African Republic to SW Sudan; eastern limits unresolved.
SYNONYMS: *butleri, ingoldbyi, saturatus* (Ingoldby, 1929, not Lyon, 1911), *tuareg*.
COMMENTS: Reviewed by Rosevear (1969) and Van der Straeten and Verheyen (1978*b*). Although its range is allopatrically complementary to the distribution of *M. fumatus*, *M. daltoni* is probably not conspecific with that E African species. Chromosomal morphology is presented by Matthey (1964). The name *tuareg* was described as a subspecies of *Grammomys macmillani*, but listed by Rosevear (1969) as a species of *Grammomys* of doubtful validity, and finally identified by Braestrup and Hutterer (1985) as a possibly distinct subspecies of *M. daltoni*. Setzer (1956) retained *butleri*, known only by the holotype collected in SW Sudan, as a species, but its morphological traits, judged by Setzer's description, are those of *M. daltoni*.

Myomys derooi Van der Straeten and Verheyen, 1978. Z. Säugetierk., 43:33.
TYPE LOCALITY: Togo, Borgou, 160 m.
DISTRIBUTION: Recorded from Ghana, Togo, Benin, and W Nigeria (see map in Van der Straeten and Verheyen, 1978*b*).
COMMENTS: A form found living in and around human dwellings. Part of the murine fauna endemic to W Africa (see account of *Grammomys buntingi*).

Myomys fumatus (Peters, 1878). Monatsb. K. Preuss. Akad. Wiss., p. 200.
TYPE LOCALITY: Kenya, Ukamba.
DISTRIBUTION: Africa; from EC Tanzania north through Kenya and N Uganda into Somalia, Ethiopia, and S Sudan; W and S limits unknown.
SYNONYMS: *allisoni, brockmani, niveiventris, oweni, subfuscus, ulae*.
COMMENTS: Kenya samples were discussed by Hollister (1919), the Uganda series by Delany (1975), and the Ethiopian segment by Yalden et al. (1976). Chromosomal and immunological information provided by Qumsiyeh et al. (1990).

Myomys ruppi (Van der Straeten and Dieterlen, 1983). Annls. Mus. R. Afr. C., 237:121.
TYPE LOCALITY: Ethiopia, Bonke, north of Bulta, 2800-3200 m.
DISTRIBUTION: Ethiopia; known only from Bonke and Bulta in the Gamo Gofa region of SW Ethiopia, 2700-3200 m.
COMMENTS: Originally described as a species of *Praomys* (see Van der Straeten and Dieterlen, 1983) based on material collected by Rupp (1980), who illustrated the skull as "*Praomys albipes*, stenocephaler Typ" (p. 92). *M. ruppi* combines morphological features of both *M. albipes* and *Stenocephalemys*, an observation reinforced by morphometric analyses (Van der Straeten and Dieterlen, 1983). Were it not for the long tail of *M. ruppi* (a trait shared with *M. albipes*), the species could as easily be included within *Stenocephalemys*.

Myomys verreauxii (Smith, 1834). S. Afr. Quart. J., 2:156.
TYPE LOCALITY: South Africa, Cape of Good Hope, near Cape Town.
DISTRIBUTION: South Africa, SW Cape Prov., from Olifants River in the west to Nature's Valley, Plettenberg Bay in the east (see map in Skinner and Smithers, 1990:271).
SYNONYMS: *colonus, veroxii*.
COMMENTS: A South African endemic. Taxonomy reviewed by Meester et al. (1986); distributional and biological information provided by Skinner and Smithers (1990).

Myomys yemeni Sandborn and Hoogstraal, 1953. Fieldiana, Zoology, 34:241.
TYPE LOCALITY: Yemen, Kariet Wadi Dhahr, six miles northwest of San'a, 6400 ft.
DISTRIBUTION: Recorded only from N Yemen and SW Saudia Arabia (see map in Harrision and Bates, 1991:249).
COMMENTS: Originally described by Sanborn and Hoogstraal (1953) as a subspecies of *M. fumatus*, the diagnostic traits of *yemeni* are outside the range of variation recorded for any sample of *fumatus*. Our study of holotype and specimens of *yemeni* and *fumatus* at the Field Museum of Natural History revealed that *yemeni* is much larger than *fumatus* (no overlap in length of molar rows, for example), has paler pelage and significantly larger ears and auditory bullae (both absolutely and relative to body

size). The morphological attributes of *yemeni* define a distinctive species of *Myomys*; its phylogenetic relationships to other species in the genus have yet to be resolved. The species (under *fumatus*) was reviewed by Harrison and Bates (1991).

Neohydromys Laurie, 1952. Bull. Br. Mus. (Nat. Hist.) Zool., 1:311.
TYPE SPECIES: *Neohydromys fuscus* Laurie, 1952.
COMMENTS: Member of the New Guinea Old Endemics (Musser, 1981*c*).

Neohydromys fuscus Laurie, 1952. Bull. Br. Mus. (Nat. Hist.) Zool., 1:311.
TYPE LOCALITY: Papua New Guinea, Chimbu Prov., N slopes Mt Wilhelm, 9000-10,000 ft.
DISTRIBUTION: Papua New Guinea; scattered localities from Mt Wilhelm east to highlands of Wau region (see map in Flannery, 1990*b*:180).
COMMENTS: Phallic morphology similar to that characterizing *Mayermys* (Lidicker, 1968). Certain derived cranial and dental features unite *Neohydromys*, *Mayermys*, and *Pseudohydromys* as close relatives (e.g., all share the derived pattern of cephalic arterial circulation; Musser and Heaney, 1992).

Nesokia Gray, 1842. Ann. Mag. Nat. Hist., ser. 7, 10:264.
TYPE SPECIES: *Arvicola indica* Gray and Hardwicke, 1830.
SYNONYMS: *Erythronesokia*, *Spalacomys*.

Nesokia bunnii (Khajuria, 1981). Bull. Nat. Hist. Res. Centre, 7:162.
TYPE LOCALITY: Iraq, Basra Province, Al-Qurna.
DISTRIBUTION: Marshes at the confluence of Tigris and Euphrates rivers in SE Iraq; probably also occurs in bordering portion of Iran.
COMMENTS: Originally described under genus *Erythronesokia* by Khajuria (1981}), but shown to be a very distinctive species of *Nesokia* by Al-Robaae and Felten (1990).

Nesokia indica (Gray and Hardwicke, 1830). Illustr. Indian Zool., 1:pl. 11.
TYPE LOCALITY: India (uncertain).
DISTRIBUTION: Modern range covers Bangladesh, NE India (Bihar), NW India (Kumaon and Rajputana), Pakistan, Afghanistan, Iran, Iraq, Syria, Saudi Arabia, Israel, NE Egypt, NW China (Xinjiang), Turkmenistan, Uzbekistan, and Tadzhikistan. Late Pleistocene sites are beyond modern range in Egypt and in N Sudan (Osborn and Helmy, 1980).
SYNONYMS: *bacheri*, *bailwardi*, *beaba*, *boettgeri*, *brachyura*, *buxtoni*, *chitralensis*, *dukelskiana*, *griffithi*, *hardwickei*, *huttoni*, *insularis*, *legendrei*, *myosura*, *satunini*, *scullyi*, *suilla*.
COMMENTS: Chromosomal data in different contexts reported by Thelma and Rao (1982), Rao et al. (1983), Juyal et al. (1989), and Dubey and Raman (1992). External, cranial, and dental morphology supported a close phylogenetic relationship with *Bandicota* (Misonne, 1969; Niethammer, 1977; Wroughton, 1908), and electrophoretic comparisons of eight loci indicated a sister-species alliance with *B. indica* (Radtke and Niethammer, 1984/85). Substantial morphological variation is present among geographic samples of *N. indica*, and careful systematic revision is required to determine whether this variation represents one or more species.

Niviventer Marshall, 1976. Family Muridae: rats and mice. Government Printing Office, Bangkok, p. 402.
TYPE SPECIES: *Mus niviventer* Hodgeson, 1836.
COMMENTS: Diagnosed and contrasted with other Indo-Sundaic genera by Musser (1981*b*), who also reviewed morphological, chromosomal, and distributional information. Morphological and geographic aspects of seven species (*andersoni*, *brahma*, *cremoriventer*, *eha*, *excelsior*, *hinpoon*, and *langbianis*) are defined; limits of the others (*confucianus*, *coxingi*, *culturatus*, *fulvescens*, *niviventer*, and *rapit*) require resolution by taxonomic revision.

Closest phylogenetic relatives are Indochinese *Chiromyscus* and *Dacnomys*; among Sundaic genera, *Niviventer* shares dental derivations with *Berylmys*, *Leopoldamys*, and *Maxomys* (Musser, 1981*b*; Musser and Newcomb, 1983). Analyses of chromosomal data postulated chromosomal similarities among *Niviventer*, *Lenothrix*, and possibly *Maxomys*, and an origin from a common ancestor (Gadi and Sharma, 1983). Analyses of biochemical and morphological data for Malayan Peninsula species documented by Chan et al. (1979), who demonstrated substantial separation from *Rattus* and alliance with *Lenothrix* in protein variation ,but equivocal affinities in morphological context. Spermatozoal morphology equivocal in assessing phylogenetic relationships (Breed and Yong, 1986). Phallic morphology of three Chinese taxa described by Yang and Fang (1988) in context of assessing phylogenetic relationships among Chinese murines. See Musser (1981*b*) for discussion of the original citation.

Niviventer andersoni (Thomas, 1911). Abstr. Proc. Zool. Soc. Lond., 1911(90):4.
TYPE LOCALITY: China, Sichuan, Omi San, 6000 ft.
DISTRIBUTION: China (SE Tibet, Yunnan, Sichuan, and Shaanxi).
COMMENTS: Reviewed and contrasted with *N. excelsior* and *N. confucianus* by Musser and Chiu (1979). Closest relative is *Niviventer excelsior;* the two species are set apart from other species of *Niviventer* by primitive traits they share, and both are isolated in the high mountains of W China (Musser, 1981*b*).

Niviventer brahma (Thomas, 1914). J. Bombay Nat. Hist. Soc., 23:232.
TYPE LOCALITY: India, N Assam, Anzong Valley in Mishmi Hills, 6000 ft.
DISTRIBUTION: N Assam (India) and N Burma.
COMMENTS: Morphological and geographic limits outlined by Musser (1970*b*, 1973*a*, 1981*b*), who also reported that the species is represented by few specimens, and is most closely related to *N. eha*.

Niviventer confucianus (Milne-Edwards, 1871). Nouv. Arch. Mus. Hist. Nat. Paris, 7, Bull.:93.
TYPE LOCALITY: China, Szechwan, Moupin.
DISTRIBUTION: N Burma, N Thailand, and highlands of China north to Jilin Prov.; limits uncertain.
SYNONYMS: *canorus, chihliensis, elegans, littoreus, luticolor, mentosus, naoniuensis, sacer, sinianus, yaoshanensis, yushuensis, zappeyi*.
COMMENTS: Wang and Zheng (1981) reported results from a systematic study under the name of *Rattus niviventer*, and described *yushuensis* as a subspecies. Zhang and Zhao (1984) proposed *naoniuensis* as a subspecies. Usually included in *N. niviventer*, but that allocation is not supported by present evidence (Abe, 1983; Musser, 1981*b*). Geographic and elevational relationships between northern *N. confucianus* and southern *N. fulvescens*, especially in W and S China, needs to be resolved by careful taxonomic revision.

Niviventer coxingi (Swinhoe, 1864). Proc. Zool. Soc. Lond., 1864:185.
TYPE LOCALITY: Taiwan.
DISTRIBUTION: Endemic to Taiwan.
SYNONYMS: *coninga*.
COMMENTS: Musser (1981*b*) included a population from N Burma, but its relationship to other groups of large-bodied *Niviventer*, especially, *N. coxingi*, is unclear. Whether *N. coxingi* is an insular form of populations now found in the mountains of Indochina or is simply a large-bodied insular derivative of mainland *N. fulvescens* will have to be evaluated in context of critical systematic revision (Musser, 1981*b*).

Niviventer cremoriventer (Miller, 1900). Proc. Biol. Soc. Washington, 13:144.
TYPE LOCALITY: Thailand, Trang Prov.
DISTRIBUTION: Peninsular Thailand, Malay Peninsula and offshore islands, Mergui Archipelago, Anambas Islands, Sumatra, Nias, Billiton and Banka islands, Borneo and offshore islands, Java, and Bali.
SYNONYMS: *barussanus, cretaceiventer, flaviventer, gilbiventer, kina, malawali, mengurus, solus, spatulatus, sumatrae*.
COMMENTS: Revised and discussed by Musser (1973*c*, 1981*b*), who also described the species as a Sundaic endemic whose closest phylogenetic relative is the Indochinese endemic, *N. langbianis*.

Niviventer culturatus (Thomas, 1917). Ann. Mag. Nat. Hist., ser. 8, 20:198.
TYPE LOCALITY: Taiwan, Mt Arizan, 8000 ft.
DISTRIBUTION: Endemic to mountains of Taiwan.
COMMENTS: Either listed as a species (Ellerman, 1941), a subspecies of *N. niviventer* (Ellerman and Morrison-Scott, 1951; Wang and Zheng, 1981), or included in *N. confucianus* (Musser, 1981*b*). It is a distinctive insular form that in morphology

resembles mainland *N. confucianus*, but differs sufficiently that it should be treated as a species until relationships can be assessed by systematic revision of the genus.

Niviventer eha (Wroughton, 1916). J. Bombay Nat. Hist. Soc., 24:428.
TYPE LOCALITY: India, Sikkim, Lachen, 8800 ft.
DISTRIBUTION: Recorded from Nepal, India (Darjeeling, Sikkim, and N Assam), N Burma, and China (N Yunnan).
SYNONYMS: *ninus*.
COMMENTS: Reviewed by Musser (1970*b*). Shared morphological traits and proportions support the hypothesis of close phylogenetic relationship between *N. eha* and *N. brahma*.

Niviventer excelsior (Thomas, 1911). Abstr. Proc. Zool. Soc. Lond., 1911(90):4.
TYPE LOCALITY: China, W Sichuan, Tatsienlu, 9000 ft.
DISTRIBUTION: China (Sichuan).
COMMENTS: A Sichuan endemic related to and occurring sympatrically with *N. andersoni* (Musser, 1981*b*; Musser and Chiu, 1979).

Niviventer fulvescens (Gray, 1847). Cat. Hodgson Coll. Br. Mus., p. 18.
TYPE LOCALITY: Nepal.
DISTRIBUTION: From S Himalayas (Nepal and N India) through Bangladesh, S China (incl. Hainan Isl), and Indochina (incl. Con Son Isl off Vietnam) to Peninsular Thailand, Malay Peninsula, Sumatra, Java, and Bali.
SYNONYMS: *baturus, besuki, blythi, bukit, caudatior, cinnamomeus, condorensis, flavipilis, gracilis, huang, jacobsoni, jerdoni, lepidus, lepturoides, lieftincki, ling, lotipes, marinus, mekongis, mentosus, minor, octomammis, orbus, pan, temmincki, treubii, vulpicolor, wongi.*
COMMENTS: Some authors have referred to populations on the Sunda Shelf and S Indochina as *bukit*, and those occurring farther north as *fulvescens* (Chasen, 1940; J. T. Marshall, Jr., 1977*a*; Musser, 1981*b*), pending a taxonomic revision of the group. Recent study now supports the hypothesis that samples of *bukit* represent *N. fulvescens* (Abe, 1983), an arrangement reflecting the earlier view of Osgood (1932:305): "The relationship of *fulvescens* to southern forms is obvious in several instances, especially in that of *R. f. bukit* which can at most be no more than a subspecies." This hypothesis will require testing by carful systematic revision of the *fulvescens-bukit* complex.
Niviventer fulvescens is the only member of the genus with a geographic distribution encompassing SE Asian mainland and some islands of the Sunda Shelf; other murines with roughly equivalent ranges are *Berylmys bowersi, Chiropodomys gliroides, Leopoldamys sabanus, L. edwardsi*, and *Maxomys surifer* (Musser and Newcomb, 1983). Spermatozoal morphology of Malayan *bukit* described by Breed and Yong (1986) in comparative context.

Niviventer hinpoon (Marshall, 1976). Family Muridae: rats and mice, p. 459. Privately printed by Government Printing Office, Bangkok.
TYPE LOCALITY: Thailand, Saraburi Prov., Kaengkhoi Dist, "outside the entrance to the bat cave, half-way up the face of a wooded limestone cliff, 200 meters altitude."
DISTRIBUTION: Endemic to Korat Plateau in Thailand.
COMMENTS: See Musser (1981*b*) for discussion of original citation.

Niviventer langbianis (Robinson and Kloss, 1922). Ann. Mag. Nat. Hist., ser. 9, 9:96.
TYPE LOCALITY: S Vietnam, Langbian Peak, 1800-2300 m.
DISTRIBUTION: Recorded from India (Assam), Burma, Thailand north of Isthmus of Kra, Laos, and Vietnam.
SYNONYMS: *indosinicus, quangninhensis, vientianensis*
COMMENTS: Morphological limits and comparisons with *N. cremoriventer*, its closest relative, were reported by Musser (1973*c*, 1981*b*). Dao and Cao (1990) described *quangninhensis* as a subspecies of *Rattus cremoriventer.*

Niviventer lepturus (Jentink, 1879). Notes Leyden Mus., 2:17.
TYPE LOCALITY: Indonesia, W Java, Gunung Gede.
DISTRIBUTION: Endemic to montane forest in W and C Java.
SYNONYMS: *fredericae, maculipectus*.
COMMENTS: Reviewed and compared with other species of *Niviventer* by Musser (1981*b*). A distinctive member of the endemic Javan murine fauna (Musser, 1986; Musser and Newcomb, 1983).

Niviventer niviventer (Hodgson, 1836). J. Asiat. Soc. Bengal, 5:234.
TYPE LOCALITY: Nepal, Katmandu.
DISTRIBUTION: NE Pakistan, Nepal, and N India (Punjab, Kumaon, Darjeeling, Sikkim).
SYNONYMS: *lepcha, monticola, niveiventer.*
COMMENTS: Nature of relationship of *N. niviventer* with *N. confucianus* to the east requires resolution in context of systematic revision of the genus, but to date "there is yet no convincing evidence that the Nepalese populations are the same as those from areas farther east in northern Burma and China" (Musser, 1981*b*:253) or from N Thailand (Abe, 1983:160).

Niviventer rapit (Bonhote, 1903). Ann. Mag. Nat. Hist., ser. 7, 11:123.
TYPE LOCALITY: Malaysia, Sabah (N Borneo), Gunung Kinabalu.
DISTRIBUTION: Cameron Highlands of Malay Peninsula, mountains of Sumatra, and highlands of N Borneo.
SYNONYMS: *atchinensis, cameroni, fraternus.*
COMMENTS: Reviewed and contrasted with other species of *Niviventer* by Musser (1981*b*). The hypothesis that only one species is involved in such a disjunct insular distribution requires evaluation in context of systematic revision of *Niviventer.*

Niviventer tenaster (Thomas, 1916). Ann. Mag. Nat. Hist., ser. 8, 17:425.
TYPE LOCALITY: Burma (Tenasserim), Mt Mulaiyit, 5000-6000 ft.
DISTRIBUTION: Mountains of Assam (India), S Burma (also possibly N Burma), and Vietnam.
SYNONYMS: *champa.*
COMMENTS: A large-bodied species that is sympatric with what has been identified as either *N. bukit* or *N. fulvescens* (Musser, 1981*b*). Samples from N Burma, which Musser provisionally referred to as *N. coxingi*, are not that species and are similar in morphology to *N. tenaster*, but larger in body size and darker in fur coloration. To determine whether the N Burma population is a separate species or a geographic variant of *N. tenaster*, and the nature of the relationship of all these highland Indochinese populations with the large-bodied *N. coxingi* of Taiwan, will have to be resolved by careful systematic revision (Musser, 1981*b*). Robinson and Kloss (1922) described *champa* as a subspecies of *Rattus bukit.*

Notomys Lesson, 1842. Nouv. Tabl. Regne Anim. Mammifères, p. 129.
TYPE SPECIES: *Dipus mitchellii* Ogilby, 1838.
SYNONYMS: *Ascopharynx, Podanomalus, Thylacomys.*
COMMENTS: Member of the Australian Old Endemics (Musser, 1981*c*:167), which includes the Conilurini where Baverstock (1984) placed *Notomys*. Gross and microscopical anatomy of neck glands described by Watts (1975); morphological variation in female reproductive tract documented by Breed (1985); morphology of male reproductive tract, glans penis, and spermatozoa described by Breed (1980, 1984, 1986), Breed and Sarafis (1978), and Morrissey and Breed (1982); results of electrophoretic studies presented by Baverstock et al. (1977*b*, 1981); chromosomal evolution and G-banding homologies addressed by Baverstock et al. (1977*c*, *e*, 1983*b*). Mahoney and Richardson (1988) cataloged references to taxonomy, distribution, and biology of the species.
Members of *Notomys* form a monophyletic group diagnosed by a suite of distinctive morphological and genic traits; closest phylogenetic relatives are species of *Pseudomys* (see Lidicker and Brylski, 1987, and references therein; Watts et al., 1992).

Notomys alexis Thomas, 1922. Ann. Mag. Nat. Hist., ser. 9, 9:316.
TYPE LOCALITY: Australia, Northern Territory, 35 miles SW of Alroy, 800 ft (see Mahoney and Richardson, 1988:166).
DISTRIBUTION: Australia; Western Australia, Northern Territory, South Australia, and W Queensland (see map in Watts and Aslin, 1981:109).
SYNONYMS: *everardensis, reginae.*
COMMENTS: Of all the species of *Notomys*, *N. alexis* has the most extensive geographic range (Watts and Aslin, 1981). Variation in sperm head morphology documented by Breed and Sarafis (1983).

Notomys amplus Brazenor, 1936. Mem. Nat. Mus. Melb., 9:7.
TYPE LOCALITY: Australia, Northern Territory, Charlotte Waters.
DISTRIBUTION: Australia; S Northern Territory and N South Australia (see map in Watts and Aslin, 1981:104).
STATUS: IUCN - Extinct.
COMMENTS: Known by only two extant specimens from the type locality (Watts and Aslin, 1981) and a skin collected during the last cenruty from Burt Plain near Alice Springs (in the Australian Museum, Flannery, in litt.), but also represented by owl pellet deposits from Flinders Ranges of South Australia; apparently extinct (Watts and Aslin, 1981; Mahoney and Richardson, 1988).

Notomys aquilo Thomas, 1921. Ann. Mag. Nat. Hist., ser. 9, 8:540.
TYPE LOCALITY: Australia, Queensland, Cape York.
DISTRIBUTION: Australia; N Queensland and N Northern Territory (Groot Eylandt and N Arnhem Land); see map in Watts and Aslin (1981:113).
STATUS: U.S. ESA - Endangered; IUCN - Insufficiently known.
SYNONYMS: *carpentarius*.
COMMENTS: Apparently found only in coastal sand ridges around Gulf of Carpentaria.

Notomys cervinus (Gould, 1853). Proc. Zool. Soc. Lond., 1851:127 [1853].
TYPE LOCALITY: Australia, "Interior of South Australia."
DISTRIBUTION: Australia; SW Queensland, South Australia, and S Northern Territory (see map in Watts and Aslin, 1981:99).
SYNONYMS: *aistoni*.
COMMENTS: For date of publication see Mahoney and Richardson (1988:167).

Notomys fuscus (Jones, 1925). Rec. S. Aust. Mus., 3:3.
TYPE LOCALITY: Australia, South Australia, Ooldea Dist.
DISTRIBUTION: Australia; SE Western Australia, S Northern Territory, South Australia, and SW Queensland (see map in Watts and Aslin, 1981:115). Also in W New South Wales (Watts, in litt.).
STATUS: IUCN - Vulnerable.
SYNONYMS: *eyreius*, *filmeri*.

Notomys longicaudatus (Gould, 1844). Proc. Zool. Soc. Lond., 1844:104.
TYPE LOCALITY: Australia, Western Australia, Moore River.
DISTRIBUTION: Australia; Western Australia and Northern Territory (see Watts and Aslin, 1981:107).
STATUS: IUCN - Extinct.
SYNONYMS: *sturti*.
COMMENTS: No living animals have either been seen or trapped since 1901, and the species is apparently extinct (Watts and Aslin, 1981). Male reproductive anatomy and spermatozoal morphology is described by Breed (1990).

Notomys macrotis Thomas, 1921. Ann. Mag. Nat. Hist., ser. 9, 8:538.
TYPE LOCALITY: Australia, Western Australia, Moore River.
DISTRIBUTION: Known only from the type locality.
STATUS: IUCN - Extinct.
SYNONYMS: *megalotis*.
COMMENTS: Represented only by the holotype and paratype from "Australia" (Mahoney, 1975). Apparently extinct. Closest phylogenetic relative is probably *Notomys cervinus* (Mahoney, 1975).

Notomys mitchellii (Ogilby, 1838). Lond. Edinb. Philos. Mag. J. Sci., 12:96.
TYPE LOCALITY: Australia, Victoria, about 12 km southeast of Lake Boga (see Mahoney and Richardson, 1988:169).
DISTRIBUTION: Australia; S Western Australia, S South Australia, and W Victoria (see map and discussion in Watts and Aslin, 1981:118); once occurred in SW New South Wales, but is now apparently extinct there (Mahoney and Richardson, 1988:170).
SYNONYMS: *alutacea*, *gouldi*, *macropus*, *richardsonii*.

Notomys mordax Thomas, 1922. Ann. Mag. Nat. Hist., ser. 9, 9:317.
TYPE LOCALITY: Australia, Queensland, Darling Downs.

DISTRIBUTION: Known only from the type locality.
STATUS: IUCN - Extinct.
COMMENTS: Still represented only by the the skull of the holotype (Mahoney, 1977). Apparently extinct (Watts and Aslin, 1981; Mahoney and Richardson, 1988). Mahoney (1977) considered *N. mordax* to be closely related to *N. mitchellii*; Watts and Aslin (1981:121) claimed that "it is not possible to be sure that this one skull really represents a distinct species, or whether it is simply that of a large specimen of Mitchell's hopping-mouse."

Oenomys Thomas, 1904. Ann. Mag. Nat. Hist., ser. 7, 13:416.
TYPE SPECIES: *Mus hypoxanthus* Pucheran, 1855.
SYNONYMS: *Aenomys*.

Oenomys hypoxanthus (Pucheran, 1855). Revue Mag. Zool. Paris, ser. 2, 7:206.
TYPE LOCALITY: Gabon.
DISTRIBUTION: Tropical forest block from S Nigeria south to N Angola, and east across Zaire (incl. islands of Zaire river between Kisangani and Kinshasa; Colyn and Dudu, 1986) to Rwanda, Burundi, and Uganda; isolated forest patches in S Sudan, SW Ethiopia, Kenya and W Tanzania (see section of map east of Ghana in Dieterlen and Rupp, 1976).
SYNONYMS: *albiventris, anchietae, bacchante, editus, marungensis, moerens, oris; rufinus* (Matschie, 1895, not Temminck, 1855), *talangae, unyori, vallicola*.
COMMENTS: Another African species showing appreciable geographic variation in fur color and body size (Dieterlen and Rupp, 1976; Thomas, 1915). Chromosomal data reported by Matthey (1963, 1967) and Maddalena et al. (1989). Closest phylogenetic relative is probably *Thamnomys* (Hatt, 1940*a*).

Oenomys ornatus Thomas, 1911. Ann. Mag. Nat. Hist., ser. 8, 7:378.
TYPE LOCALITY: Ghana, Bibianaha, near Dunkwa.
DISTRIBUTION: SE Guinea (Mt Nimba) to Ghana.
COMMENTS: Originally described by Thomas (1911*b*) as a distinct species, and subsequently either listed that way (Allen, 1939; Ellerman, 1941) or included in *hypoxanthus* (Misonne, 1974). Rosevear (1969), however, recognized *ornatus* as a very distinctive subspecies of *O. hypoxanthus*, and Tranier and Gautun (1979) reinstated its specific uniqueness diagnosed by chromosomal, morphological, and distributional attributes.

Palawanomys Musser and Newcomb, 1983. Bull. Am. Mus. Nat. Hist., 174:335.
TYPE SPECIES: *Palawanomys furvus* Musser and Newcomb, 1983.
COMMENTS: Phylogenetic affinities unclear, but among murines native to the Sunda Shelf, *Palawanomys* appears most closely allied to the cluster of genera that would include *Rattus*; broader regional comparisons required before affinities can be determined (Musser and Newcomb, 1983).

Palawanomys furvus Musser and Newcomb, 1983. Bull. Am. Mus. Nat. Hist., 174:335.
TYPE LOCALITY: Philippines, Palawan Isl, Brooke's Point Municipality, Mt Mantalingajan, 4500 ft.
DISTRIBUTION: Known only from the type locality.
COMMENTS: Known only by four melanistic examples.

Papagomys Sody, 1941. Treubia, 18:322.
TYPE SPECIES: *Mus armandvillei* Jentink, 1892.
COMMENTS: Formerly thought closely related to *Mallomys*, but has no affinity to that New Guinea Old Endemic, and instead is phylogenetically related to *Komodomys* and Pleistocene *Hooijeromys*, both endemics of Nusa Tenggara (Musser, 1981*c*). Species of *Papagomys* reviewed by Musser (1981*c*).

Papagomys armandvillei (Jentink, 1892). Weber's Zool. Ergebn., 3:79, pl. 5.
TYPE LOCALITY: Indonesia, Nusa Tenggara (Lesser Sunda Isls), Pulau Flores.
DISTRIBUTION: Known only from Flores Isl.
SYNONYMS: *besar, verhoeveni*.
COMMENTS: Known by extant specimens as well as subfossil fragments (3000-4000 years old), and still living on Flores.

Papagomys theodorverhoeveni Musser, 1981. Bull. Am. Mus. Nat. Hist., 169:95.
TYPE LOCALITY: Indonesia, Nusa Tenggara (Lesser Sunda Isls), Pulau Flores, Menggarai Prov., Liang Toge, a cave near Warukia, 1 km south of Lepa.
DISTRIBUTION: Known only from Flores Isl.
COMMENTS: Known only by subfossil fragments (3000-4000 years old), but possibly still living on Flores.

Parahydromys Poche, 1906. Zool. Anz., 30:326.
TYPE SPECIES: *Limnomys asper* Thomas, 1906.
SYNONYMS: *Drosomys, Limnomys* (Thomas, 1906, not Mearns, 1905).
COMMENTS: Member of the New Guinea Old Endemics (Musser, 1981*c*). Phallic morphology of incomplete specimen described by Lidicker (1968).

Parahydromys asper (Thomas, 1906). Ann. Mag. Nat. Hist., ser. 7, 17:326.
TYPE LOCALITY: Papua, New Guinea, Central Prov., Owen Stanley Range, Richardson Range, Mt Gayata, 2000-4000 m.
DISTRIBUTION: New Guinea; Central Cordillera from Weyland Range in Irian Jaya to Owen Stanley Range in Papua New Guinea, as well as Huon Peninsula (see map in Flannery, 1990*b*:185).
COMMENTS: Thought to be closely related to *Hydromys* (Flannery, 1989; Lidicker, 1968; Tate, 1951), but no critical assessment of that hypothesis has been made. Flannery (1990*b*) provided distributional and biological notes. Donnellan (1987) provided chromosomal data.

Paraleptomys Tate and Archbold, 1941. Am. Mus. Novit., 1101:1.
TYPE SPECIES: *Paraleptomys wilhelmina* Tate and Archbold, 1941.
COMMENTS: Member of the New Guinea Old Endemics (Musser, 1981*c*). Based on cranial and phallic traits (Lidicker, 1968; Tate, 1951), *Paraleptomys* is traditionally considered closely related to *Leptomys*, but this hypothesis requires testing.

Paraleptomys rufilatus Osgood, 1945. Fieldiana Zool., 31:1.
TYPE LOCALITY: New Guinea, Irian Jaya, Cyclops Mtns, Mt Dafonsero, 4700 ft.
DISTRIBUTION: NC New Guinea; known only from N Coast Ranges in Irian Jaya (Cyclops Mtns) and adjacent Papua New Guinea (Torricelli Mtns); see map in Flannery (1990*b*:179).
COMMENTS: A very distinct species (Flannery, 1990*b*; Osgood, 1945) that along with *Microhydromys musseri* and *Petaurus abidi* is endemic to the N Coastal Ranges (Flannery, 1990*b*).

Paraleptomys wilhelmina Tate and Archbold, 1941. Am. Mus. Novit., 1101:1.
TYPE LOCALITY: New Guinea, Irian Jaya, near Mt Wilhelmina, 9 km NE Lake Habbema, 2800 m.
DISTRIBUTION: C New Guinea; known only from N slopes of Snow Mtns between Idenburg River and Mt Wilhelmina in Irian Jaya (Tate and Archbold, 1941) and the Tifalmin Valley in W Papua New Guinea (Flannery and Seri, 1990:189).

Paruromys Ellerman, 1954. *In* Laurie and Hill, List of land mammals of New Guinea, Celebes and adjacent islands, p. 117. [1954]
TYPE SPECIES: *Rattus dominator* Thomas, 1921.
COMMENTS: Described by Ellerman as a subgenus of *Rattus*, but now recognized as distinct genus (Musser and Newcomb, 1983).

Paruromys dominator (Thomas, 1921). Ann. Mag. Nat. Hist., ser. 9, 7:244.
TYPE LOCALITY: Indonesia, N Sulawesi, Minahassa, Mt Masarang, 4000 ft.
DISTRIBUTION: Sulawesi; throughout the island except upper slopes of Gunung Lampobatang in SW peninsula (Musser and Holden, 1991).
SYNONYMS: *frosti* (see Musser, 1971*b*).
COMMENTS: Taxonomic allocations of *dominator* from the time it was originally described as a species of *Rattus* by Thomas (1921*a*), through its use as type-species of subgenus *Paruromys* by Ellerman (*in* Laurie and Hill, 1954), up to its inclusion in subgenus *Bullimus* by Misonne (1969) were reviewed by Musser and Newcomb (1983). Sody (1941) listed *dominator* as a species of *Taeromys*. Spermatozoal morphology of

dominator is unlike species of *Rattus* or any other species for which data from spermatozoal morphology are available (Breed and Musser, 1991).

Paruromys ursinus (Sody, 1941). Treubia, 18:312.
TYPE LOCALITY: Indonesia, Sulawesi, SE peninsula, Gunung Lampobatang, Wawokaraeng, 2200 m.
DISTRIBUTION: Known only from the upper slopes of Gunung Lampobatang, Sulawesi.
COMMENTS: Originally described as a subspecies of *Taeromys dominator* by Sody (1941:312), and usually listed as a subspecies of *Paruromys dominator* (Musser, 1984), but treated as distinct species by Musser and Holden (1991).

Paulamys Musser, 1986. *In* Musser et al., Am. Mus. Novit., 2850:2.
TYPE SPECIES: *Floresomys naso* Musser, 1981.
SYNONYMS: *Floresomys* (Musser, 1981, not Fries et al., 1955; see reference in Musser et al., 1986).

Paulamys naso (Musser, 1981). Bull. Am. Mus. Nat. Hist., 169:112.
TYPE LOCALITY: Indonesia, Nusa Tenggara (Lesser Sunda Isls), Pulau Flores, Menggarai Prov., Liang Toge, cave near Warukia, 1 km south of Lepa.
DISTRIBUTION: Known only from Flores Isl.
COMMENTS: Originally described and known only from subfossil fragments (Musser, 1981*c*; Musser et al., 1986), but one extant specimen has been referred to this species by Kitchener et al. (1991*a*), who suggested it is closely related to Sulawesian *Bunomys*. This assessment needs testing by study of more extant specimens of *P. naso* in a phylogenetic context that would also compare the sample to species of native New Guinea and Australian *Rattus*.

Pelomys Peters, 1852. Bericht Verhandl. K. Preuss. Akad. Wiss. Berlin, 17:275.
TYPE SPECIES: *Mus (Pelomys) fallax* Peters, 1852.
SYNONYMS: *Komemys*.
COMMENTS: Definition and phylogenetic position of *Pelomys* needs to be reassessed in context of a systematic revision of arvicanthine murines. In overall morphology, the genus is most closely related to *Mylomys* and *Desmomys*. *Komemys* is usually treated as a subgenus for the species *hopkinsi* and *isseli* (e.g., Delany, 1975), but we recognize it here as a synonym of *Pelomys*.

Pelomys campanae (Huet, 1888). Le Naturaliste, ser. 2, 10 (31):143.
TYPE LOCALITY: W Angola, Landana.
DISTRIBUTION: WC and N Angola and W Zaire.
COMMENTS: A distinctive species that occurs either sympatrically or parapatrically with *P. fallax* in parts of its range (Crawford-Cabral, 1983).

Pelomys fallax (Peters, 1852). Bericht Verhandl. K. Preuss. Akad. Wiss. Berlin, 17:275.
TYPE LOCALITY: Mozambique, Caya Dist., Zambesi River and Boror, Licuare River.
DISTRIBUTION: S Kenya, SW Uganda, Tanzania, Zaire, Angola, Zambia, Malawi, Mozambique, E and NW Zimbabwe, and N Botswana.
SYNONYMS: *australis, concolor, frater, insignatus, iridescens, luluae, rhodesiae, vumbae* (see Allen, 1939; Meester et al., 1986).
COMMENTS: Appreciable variation in size and fur color exists between samples from Angola and Zambia and those from the rest of the geographic range of *P. fallax*, suggesting more than one species may be present in this complex.
No extant records are from South Africa but the species occurred in Natal more than 17,000 years before present when the region was covered with deciduous woodland instead of thornveld (Avery, 1991).

Pelomys hopkinsi Hayman, 1955. Rev. Zool. Bot. Afr., 52:323.
TYPE LOCALITY: SW Uganda, Kigezi, Rwamachuchu.
DISTRIBUTION: Rwanda, Uganda, and SW Kenya (see Bekele and Schlitter, 1989).
COMMENTS: Morphologically similar to *P. isseli*, but significant distinguishing traits suggested that *hopkinsi* and *isseli* should be viewed as separate species (Bekele and Schlitter, 1989).

Pelomys isseli (de Beaux, 1924). Ann. Mus. Civ. Stor. Nat. Genova, 51:207.
TYPE LOCALITY: Uganda, Lake Victoria, Kome Isl.
DISTRIBUTION: Uganda; endemic to islands of Kome, Bugala, and Bunyama in Lake Victoria (Bekele and Schlitter, 1989; Delany, 1975).
COMMENTS: Closely related to *P. hopkinsi*.

Pelomys minor Cabrera and Ruxton, 1926. Ann. Mag. Nat. Hist., ser. 9, 17:601.
TYPE LOCALITY: Zaire, Luluabourg.
DISTRIBUTION: N Angola, NW Zambia, S and E Zaire, and W Tanzania.
COMMENTS: A very distinct species, but aspects of its morphology resemble some species of *Lemniscomys* (e.g., *L. griselda*).

Phloeomys Waterhouse, 1839. Proc. Zool. Soc. Lond., 1839:108.
TYPE SPECIES: *Mus (Phloeomys) cumingi* Waterhouse, 1839.
COMMENTS: Part of the Philippine Old Endemics, but actual phylogenetic relationships relative to other members of that group and to genera in other areas of Indo-Australian region unclear (Musser and Heaney, 1992). Listed as a member of Phloeomyinae by Simpson (1945), along with *Chiropodomys, Coryphomys, Crateromys, Lenomys, Mallomys,* and *Pongonomys*, but no data supports such an allocation.

Phloeomys cumingi (Waterhouse, 1839). Proc. Zool. Soc. Lond., 1839:108.
TYPE LOCALITY: Philippines, Luzon Isl.
DISTRIBUTION: S Luzon, Marinduque, and Catanduanes Isls (Heaney et al., 1991; Musser and Heaney, 1992).
SYNONYMS: *albayensis, nomen nudum* (see Ellerman, 1941:293), *elegans*.
COMMENTS: Sometimes considered conspecific with *P. pallidus*, but *cumingi* is a distinct species (Thomas, 1898*b*).

Phloeomys pallidus Nehring, 1890. Sitzb. Ges. Naturf. Fr. Berlin, p. 106.
TYPE LOCALITY: Philippines, Luzon Isl.
DISTRIBUTION: N Luzon; limits unknown.

Pithecheir F. G. Cuvier, 1833. *In* E. Geoffroy and F. G. Cuvier, Hist. Nat. Mammifères, pt. 4, 7(66):1-2 "Pithéchéir Mélanure".
TYPE SPECIES: *Pithecheir melanurus* Cuvier, 1838.
COMMENTS: An endemc of the Sunda Shelf that Musser and Newcomb (1983) suggested, on the basis of cranial and dental traits, was distantly related to *Lenothrix*, an estimate of phylogenetic affinites unsupported by chromosomal data (Yong et al., 1982) and spermatozoal morphology (Breed and Yong, 1986).

Pithecheir melanurus F. G. Cuvier, 1833. *In* E. Geoffroy and F. G. Cuvier, Hist. Nat. Mammifères, pt. 4, 7(66):1-2 "Pithéchéir Mélanure".
TYPE LOCALITY: Indonesia, Java.
DISTRIBUTION: Java only (see Musser, 1982*d*:76).
COMMENTS: Represented by few specimens.

Pithecheir parvus Kloss, 1916. J. Fed. Malay St. Mus., 6:250.
TYPE LOCALITY: Malay Peninsula, Selangor, Bukit Kutu, near Kuala Kubu, 3,400 ft.
DISTRIBUTION: Malay Peninsula (Pahang and Selangor).
COMMENTS: Originally described as a subspecies of *P. melanurus*, but is a distinctive species endemic to Malay Peninsula (Musser and Newcomb, 1983; Muul and Lim, 1971). Karyotype has same diploid number as *Hapalomys longicaudus*, but is uninformative about inferring phylogenetic relationships (Yong, et al., 1982). Spermatozoal morphology very distinctive (no apical hook), unlike that of any other Sundaic endemic (Breed and Yong, 1986), and resembles sperm of Sulawesian *Lenomys* and *Eropeplus* (Breed and Musser, 1991).

Pogonomelomys Rümmler, 1936. Z. Säugetierk., 11:248.
TYPE SPECIES: *Melomys mayeri* Rothschild and Dollman, 1932.
SYNONYMS: *Abeomelomys*.
COMMENTS: Member of the New Guinea Old Endemics (Musser, 1981*c*). Revised by Menzies (1990). Originally described as a subgenus of *Melomys*, but now regarded as a distinct

genus (Laurie and Hill, 1954; Menzies, 1990; Tate, 1951); its phylogenetic relationships to *Melomys* and other related genera still requires elucidation.

Pogonomelomys bruijni (Peters and Doria, 1876). Ann. Mus. Civ. Stor. Nat. Genova, 8:336.
TYPE LOCALITY: W New Guinea, Irian Jaya, Pulau Salawati, off W coast of Vogelkop.
DISTRIBUTION: New Guinea; known only by 11 specimens from type locality, Vogelkop mainland, lower Fly River, and low altitudes on Mt Bosavi and Mt Sisa in Papua New Guinea (Menzies, 1990).
SYNONYMS: *brassi*.
COMMENTS: The lowland morphological and phylogenetic counterpart of the highland *P. mayeri*.

Pogonomelomys mayeri (Rothschild and Dollman, 1932). Abstr. Proc. Zool. Soc. Lond., 1932(353):14.
TYPE LOCALITY: New Guinea, Irian Jaya, Weyland Range, Gebroeders Mtns, 5000 ft.
DISTRIBUTION: New Guinea; mountains from Weyland Range in Irian Jaya to Wau region in Morobe Prov. of Papua New Guinea; not known from farther east in Owen Stanley Range.
COMMENTS: Phallic morphology described by Lidicker (1968).

Pogonomelomys sevia (Tate and Archbold, 1935). Am. Mus. Novit., 803:3.
TYPE LOCALITY: Papua New Guinea, Morobe Prov., Huon Peninsula, Cromwell Range, Sevia, 1400 m.
DISTRIBUTION: New Guinea; Central Cordillera of Papua New Guinea from Star Mtns in the west to the Wau region of Morobe Prov. in the east, including mountains of Huon Peninsula (Flannery, 1990*b*; Menzies, 1990).
SYNONYMS: *tatei*.
COMMENTS: Originally described as a species of *Melomys* by Tate and Archbold (1935) and subsequently transferred to *Pogonomelomys* by Rümmler (1938) where it remained (Laurie and Hill, 1954; Tate, 1951) until Menzies (1990) made *sevia* the type species of *Abeomelomys*. The species has always been considered distinctive compared to *mayeri* and *bruijni* (Flannery, 1990*b*; Tate, 1951), but the traits used by Menzies to diagnose *Abeomelomys* do not distinguish that genus from *Pogonomelomys*.

Pogonomys Milne-Edwards, 1877. C. R. Acad. Sci. Paris, 85:1081.
TYPE SPECIES: *Mus* (*Pogonomys*) *macrourus* Milne-Edwards, 1877.
COMMENTS: Member of the New Guinea and Australian Old Endemics (Musser, 1981*c*). A chromosomal and morphometric study, representing an incomplete review of genus, was offered by Dennis and Menzies (1979). Additional chromosomal data were reported by Donnellan (1987).

Pogonomys championi Flannery, 1988. Rec. Aust. Mus., 40:333.
TYPE LOCALITY: Papua New Guinea, West Sepik Prov., Telefomin Valley, Ofektaman, 1400 m (see Flannery, 1988, for additional information).
DISTRIBUTION: Papua New Guinea; known only from Telefomin and Tifalmin valleys between 1400 and 2300 m (Flannery, 1988).
COMMENTS: Morphologically similar to *P. sylvestris*.

Pogonomys loriae Thomas, 1897. Ann. Mus. Civ. Stor. Nat. Genova, 18:613.
TYPE LOCALITY: Papua New Guinea, Central Prov., mountains behind Astrolabe Range, near Mt Wori, Haveri, 700 m (see Laurie and Hill, 1954:96, for details).
DISTRIBUTION: New Guinea; throughout highland habitats from Vogelkop in the west to the Owen Stanley Range in the east; also a small sample from Fly River drainage in SC Papua New Guinea (in the American Museum of Natural History); and on Goodenough and Fergusson Isls in the the d'Entrecasteaux Arch. Australia, NE coastal Queensland (see discussion and map in Watts and Aslin, 1981:64).
SYNONYMS: *dryas*, *fergussoniensis*.
COMMENTS: The morphological and geographic definition of this species is unsatisfactory. Significance of the appreciable geographic variation in external, cranial, and dental dimensions among samples need to be assessed by a careful revision; more than one species may be represented. Mahoney and Richardson (1988:170) catalogued taxonomic, distributional, and biological references for the Australian sample (which

was identiifed as *P. mollipilosus* by Watts and Aslin, 1981); we provisionally include it under *P. loriae*.

Pogonomys macrourus (Milne-Edwards, 1877). C. R. Acad. Sci. Paris, 85:1081.
TYPE LOCALITY: New Guinea, Irian Jaya, Vogelkop, Arfak Mtns, Amberbaki.
DISTRIBUTION: New Guinea; throughout lowland and midmontane forests from sea level to 1500 m (see map in Flannery, 1990*b*:203); also recorded from Yapen (Japen Isl) and New Britain Isl in the Bismarck Arch (Rümmler, 1938).
SYNONYMS: *derimapa, huon, lepidus, mollipilosus* (see Tate, 1951; Dennis and Menzies, 1979).
COMMENTS: Phallic morphology described by Lidicker (1968).

Pogonomys sylvestris Thomas, 1920. Ann. Mag. Nat. Hist., ser. 9, 6:534.
TYPE LOCALITY: Papua New Guinea, Morobe Prov., Rawlinson Mtns, 1500 m.
DISTRIBUTION: Papua New Guinea; mountains above 1300 m.
COMMENTS: True *sylvestris* has been recorded only from Papua New Guinea. Previous records from Irian Jaya represent other species or morphologically distinctive montane isolates related to *sylvestris* at either the specific or subspecific level.

Praomys Thomas, 1915. Ann. Mag. Nat. Hist., ser. 8, 15:4
TYPE SPECIES: *Epimys tullbergi* Thomas, 1894.
COMMENTS: *Myomys* (or *Myomyscus*), *Mastomys*, and *Hylomyscus* have been united with *Praomys* as subgenera (D. H. S. Davis, 1965; Misonne, 1974), but are treated as separate genera here and by other workers such as Rosevear (1969), who also reviewed the taxonomic history of some species in *Praomys* as well as its generic status relative to the other genera allied with it. Van der Straeten and Dieterlen (1987) and Van der Straeten and Dudu (1990) provided brief reviews of the historical and current allocations of forms to the *P. tullbergi, P. jacksoni*, and *P. delectorum* complexes. Chromosomal and biochemical traits reviewed or referenced by Qumsiyeh et al. (1990), who combined *Myomys* and *Mastomys* with *Praomys*. Additional chromosomal information recorded by Maddalena et al. (1989). Not only do the contents of *Praomys* require careful systematic revision, but its phylogenetic relationships relative to *Mastomys, Myomys*, and *Hylomyscus* also needs resolution through revisionary studies.

Praomys delectorum (Thomas, 1910). Ann. Mag. Nat. Hist., ser. 8, 6:430.
TYPE LOCALITY: S Malawi, Mlanji Plateau, 5500 ft.
DISTRIBUTION: High plateaus and isolated mountains from NE Zambia (Nyika Plateau, Makutus, and Mafingas; see map in Ansell, 1978) and Malawi (see map in Ansell and Dowsett, 1988), through Tanzania to SE Kenya.
SYNONYMS: *melanotus, octomastis, taitae.*
COMMENTS: Of the synonyms belonging here, *taitae* was described as a species (Heller, 1912) and recognized as such by Hollister (1919) and Swynnerton and Hayman (1951), *melanotus* was described as a form of *P. tullbergi* (Allen and Loveridge, 1933), and *octomastis* was presented as a subspecies of *P. jacksoni* (Hatt, 1940*b*). Demeter and Hutterer (1986) suggested that *taitae* was synonymous with *Hylomyscus denniae*, but it is not, judging from our study of specimens and holotypes.

Praomys hartwigi Eisentraut, 1968. Bonn. Zool. Beitr., 19:8-11.
TYPE LOCALITY: Cameroon, Lake Oku.
DISTRIBUTION: Known only from the type locality and Gotel Mtns in Nigeria (Hutterer et al., 1992; Nikolaus and Dowsett, 1989).
SYNONYMS: *obscurus.*
COMMENTS: Still recorded only by a few specimens (Misonne, 1974; Nikolaus and Dowsett, 1989).

Praomys jacksoni (de Winton, 1897). Ann. Mag. Nat. Hist., ser. 6, 20:318.
TYPE LOCALITY: Uganda, Entebbe.
DISTRIBUTION: From C Nigeria through Cameroon and Central African Republic to S Sudan, Zaire, N Angola, Uganda, Rwanda, Kenya, and southward through E Tanzania to N and E Zambia.
SYNONYMS: *montis, peromyscus, sudanensis, viator.*

COMMENTS: At one time listed as a subspecies of *P. tullbergi* (e.g., Hollister, 1919), *jacksoni* is a distinct species (Allen, 1939; Ansell, 1978; Van der Straeten and Dieterlen, 1987; Van der Straeten and Dudu, 1990) occurring sympatrically with *P. tullbergi*. The problem of identifying the holotype of *jacksoni* as well as current names associated with the species was reviewed and discussed by Van der Straeten and Dieterlen (1987) and Van der Straeten and Dudu (1990). The latter authors recognized *montis* and *peromyscus* as species; the *P. jacksoni* complex is probably composite, but until the forms representing potential species (*viator, montis, peromyscus*) have been diagnosed and their geographic ranges described we include them here under *P. jacksoni*. Van der Straeten and Dieterlen (1992) reported results from craniometrical camparisons among four samples of *P. jacksoni* collected at different altitudes (850-3300 m).

Praomys minor Hatt, 1934. Am. Mus. Novit., 708:11.
TYPE LOCALITY: C Zaire, Lukolela.
DISTRIBUTION: Known only from the type locality.
COMMENTS: Originally described as a subspecies of *P. jacksoni* (Hatt, 1934*b*), and then treated as a subspecies of *P. tullbergi* (Petter, 1975*c*), *minor* is a distinct species in the *P. jacksoni* complex (Van der Straeten and Dudu, 1990).

Praomys misonnei Van der Straeten and Dieterlen, 1987. Stuttg. Beitr. Naturk. ser. A, 402:3.
TYPE LOCALITY: E Zaire, Kivu region, Irangi.
DISTRIBUTION: N and E Zaire.
COMMENTS: Sympatric with *P. jacksoni* at the type locality and with *P. jacksoni* and *P. mutoni* in Haut-Zaïre (Van der Straeten and Dudu, 1990). Regarded as related to *P. tullbergi* by Van der Straeten and Dieterlen (1987). The type series and examples of *misonnei* we examined from the Ituri Forest in E Zaire and Gamangui in Haut-Zaïre, where *P. jacksoni* was also trapped, are morphologically very similar to *P. tullbergi*; the possibility that *misonnei* simply represents populations of *P. tullbergi* at the eastern margins of its geographic range needs to be considered in any systematic revision of the complex. Qumsiyeh et al. (1990) identified Kenyan samples as *P. misonnei*, and their chromosomal and electrophoretic characteristics relate the samples to *Mastomys hildebrandti*; identity of the Kenyan material should be reassessed.

Praomys morio (Trouessart, 1881). Bull. Soc. Etudes Sci. Angers, 10:121.
TYPE LOCALITY: Cameroon, Mt. Cameroon, 7000 ft; see Rosevear (1969:399).
DISTRIBUTION: Mt. Cameroon.
SYNONYMS: *maurus* (of Gray, 1862, not Waterhouse, 1839).
COMMENTS: A member of the *P. tullbergi* complex. We restrict *P. morio* to Mt. Cameroon, although Eisentraut (1970) recorded it from Bioko, and Petter (1965) discussed samples from the Central African Republic. The species requires definition; alledged distinctions between it and *P. tullbergi* may not reflect specific differences (Hutterer, in litt.). Our study revealed that series from outside of Mt. Cameroon identified as *morio* are either *tullbergi* or an undescribed species of *Praomys* (the series from Central African Republic, for example).

Praomys mutoni Van der Straeten and Dudu (1990). *In* Peters and Hutterer (eds.), Vertebrates in the tropics, Museum Alexander Koenig, Bonn, p. 75.
TYPE LOCALITY: N Zaire (Haut-Zaïre), Batiabongena (Masako Forest Reserve), 00°36'N, 25°13'E.
DISTRIBUTION: Recorded only from the type locality.
COMMENTS: A distinct forest species related to *P. jacksoni* and occurring sympatrically with it (Van der Straeten and Dudu, 1990).

Praomys rostratus (Miller, 1900). Proc. Washington Acad. Sci., 2:637.
TYPE LOCALITY: Liberia, Mt Coffee.
DISTRIBUTION: Recorded only from forest in Liberia, Mt Nimba region of Guinea, and Ivory Coast; limits unresolved.
COMMENTS: Originally described as a subspecies of *tullbergi*, but Van der Straeten and Verheyen (1981) distinguished *rostratus* from *tullbergi* by its greater body size, noted that both kinds were sympatric, and raised *rostratus* to specific rank. A similar distribution of body sizes, as well as different ecologies, were recorded by Gautun et al. (1986) from Mt Nimba, and they also separated their samples into either *P.*

tullbergi or *P. rostratus*. *P. rostratus* is part of the murine fauna endemic to W Africa (see account of *Grammomys buntingi*).

Praomys tullbergi (Thomas, 1894). Ann. Mag. Nat. Hist., ser. 6, 13:205.

TYPE LOCALITY: Ghana, Ashanti, Wasa, Ankober River.

DISTRIBUTION: Forest and Guinea woodland from Gambia River in the west through Cameroon to N and E Zaire; Bioko; also NW Angola (specimens in the Field Museum of Natural History); limits unknown.

SYNONYMS: *burtoni* (of Thomas, 1892, not Ramsay, 1887).

COMMENTS: Reviewed by Rosevear (1969), Van der Straeten and Verheyen (1981), and more recent reports cited in the above species accounts. Closely related to *P. misonnei*, *P. morio*, and *P. rostratus*.

Pseudohydromys Rümmler, 1934. Z. Säugetierk., 9:47

TYPE SPECIES: *Pseudohydromys murinus* Rümmler, 1934.

COMMENTS: Member of the New Guinea Old Endemics (Musser, 1981*c*). Closely related to *Mayermys* and *Neohydromys* in phallic morphology (Lidicker, 1968) as well as external, cranial, and dental traits.

Pseudohydromys murinus Rummler, 1934. Z. Säugetierk., 9:48.

TYPE LOCALITY: Papua New Guinea, Morobe Prov., Mt Missim, 7000 ft.

DISTRIBUTION: Papua New Guinea; known only from the type locality and high slopes of Mt Wilhelm (Laurie, 1952).

Pseudohydromys occidentalis Tate, 1951. Bull. Am. Mus. Nat. Hist., 97:224.

TYPE LOCALITY: New Guinea, Irian Jaya, north of Lake Wilhelmina, Lake Habbema, 3225 m.

DISTRIBUTION: New Guinea; known only from the area around Lake Habbema and slopes of Mt Wilhelmina in Irian Jaya (Tate, 1951:225) and the Star Mtns and Victor Emmanuel Range in W Papua New Guinea (Flannery, 1990*b*:181).

Pseudomys Gray, 1832. Proc. Zool. Soc. Lond., 1832:39.

TYPE SPECIES: *Pseudomys australis* Gray, 1832.

SYNONYMS: *Gyomys*, *Mastacomys*, *Paraleporillus*, *Thetomys*.

COMMENTS: Taxonomic, distributional, and biological references to all species cataloged by Mahoney and Richardson (1988). Member of the Australian Old Endemics (Musser, 1981*c*), part of which is the Conilurini where Lee et al. (1981) and Baverstock (1984) placed *Pseudomys*.

Data from several character suites have been used to estimate relationships among species of *Pseudomys*: anatomy of male and female reproductive tracts (Breed, 1980, 1985, 1986); phallic morphology (Lidicker and Brylski, 1987; Morrissey and Breed, 1982); spermatozoal morphology (Breed, 1983, 1984; Breed and Sarafis, 1978); electrophoretic (Baverstock et al., 1977*a*, 1981); and chromosomal (Baverstock et al., 1977*c*, 1983*b*). Despite such different approaches, no study estimating phylogenetic relationships is available that integrates all these data with information from skins, skulls, and dentitions. So of all the distinctive groups of species comprising what is now called *Pseudomys*, opinion still ranges from including them all in one genus (*Pseudomys*) to allocating them to three genera (*Pseudomys*, *Gyomys*, and *Leggadina*) to merging *Mastacomys* and *Leporillus* with *Pseudomys* (see discussions in Baverstock et al., 1981; Watts and Aslin, 1981; Breed, 1983; Lidicker and Brylski, 1987; and Watts et al., 1992). Watts et al. (1992) discussed the futility of attempting to split *Pseudomys* and they regarded it as a single but complex genus in which results from microcomplement fixation confirmed its monophyly relative to the other Australian murines except *Mastacomys*, which they merge with *Pseudomys*. We follow that taxonomic decision here.

Pseudomys albocinereus (Gould, 1845). Proc. Zool. Soc. Lond., 1845:78.

TYPE LOCALITY: Australia, Western Australia, "scrubby plains near Perth" (see Mahoney and Richardson, 1988:171).

DISTRIBUTION: Australia, SW Western Australia (from Shark Bay area southeast to Israelite Bay); also found on islands of Bernier, Dorre, Shark Bay, and Woody (see map in Watts and Aslin, 1981:196).

SYNONYMS: *squalorum*.

COMMENTS: Analysis of phallic morphology suggested *P. albocinereus* belongs in group with *P. fumeus* and *P. shortridgei* (Lidicker and Brylski, 1987), but electrophoretic data placed it in a cluster containing *P. apodemoides* and seven other species, excluding *P. fumeus* and *P. shortridgei* (Baverstock et al., 1981). Dental traits suggested *P. albocinereus* is closely related to the Pliocene *P. vandycki*, and if resemblance reflects monophyly, the two species form a distinct group within *Pseudomys* (Godthelp, 1989). But Watts (in litt.) wrote that "virtually all data supports close relationships between *P. apodemoides* and *P. albocinereus*. Relationships beyond this are any ones guess." See also Watts et al. (1992).

Pseudomys apodemoides Finlayson, 1932. Trans. R. Proc. Soc. S. Aust., 56:170.
TYPE LOCALITY: Australia, Southern Australia, Coombe.
DISTRIBUTION: Australia; SE South Australia and W Victoria (Murray-Darling Basin).
COMMENTS: Phylogenetic relationships are equivocal to some (see discussion in Lidicker and Brylski, 1987:635, and references therein), but not to other workers (see preceeding account).

Pseudomys australis Gray, 1832. Proc. Zool. Soc. Lond., 1832:39.
TYPE LOCALITY: Australia, New South Wales, SW side of Liverpool Plains.
DISTRIBUTION: Australia; New South Wales, S Queensland, South Australia, and S Northern Territory; Late Pleistocene to Recent remains from W Victoria (Watts and Aslin, 1981); probably extinct in New South Wales (Mahoney and Richardson, 1988:172).
SYNONYMS: *auritus, flavescens, lineolatus, minnie, murinus, stirtoni.*
COMMENTS: Microscopic structure of hooks on sperm head reported by Flaherty and Breed (1982). Phallic information suggested *P. australis* is related to *P. gouldii, P. higginsi,* and *P. nanus*, to the exclusion of other species of *Pseudomys* (Lidicker and Brylski, 1987:635); electrophoretic data (Baverstock et al., 1981) and spermatozoal morphology (Breed, 1983) are not discordant with this association.

Pseudomys bolami Troughton, 1932. Rec. Aust. Mus., 18:292.
TYPE LOCALITY: Australia, South Australia, Ooldea.
DISTRIBUTION: Australia, S South Australia and S Western Australia (see map in Kitchener, 1985:216).
COMMENTS: Originally described by Troughton (1932*a*) as a subspecies of *P. hermannsburgensis*, but distinguished from that species and redescribed by Kitchener et al. (1984*a*), who also reported sympatry of both species at Goongarrie, Western Australia.

Pseudomys chapmani Kitchener, 1980. Rec. West. Aust. Mus., 8:405.
TYPE LOCALITY: Australia, Western Australia, Pilbara Dist., East Hammersley Range, West Angelas Mine Site (Kitchener, 1980 provided additional information).
DISTRIBUTION: Australia, NW Western Australia; extant specimens known only from Pilbara Dist. (see map in Kitchener, 1985:216), but distribution of pebble mounds indicated range once extended through Gascoyne to Murchison Dist. with southern limit near Mileura, northern limit the Great Sandy Desert, and eastern limit the Gibson Desert (see discussion and map in Dunlop and Pound, 1981).
COMMENTS: This species builds pebble mounds and is sympatric with *P. hermannsburgensis* (Kitchener, 1980), which does not construct pebble mounds (Dunlop and Pound, 1981), but phylogenetically most closely allied to *P. johnsoni*, another species that constructs pebble mounds (Kitchener, 1985).

Pseudomys delicatulus (Gould, 1842). Proc. Zool. Soc. Lond., 1842:13.
TYPE LOCALITY: Australia, Northern Territory, Port Essington.
DISTRIBUTION: Australia; N coastal region from near Port Hedland in Western Australia to Bundaberg area in Queensland (see map in Watts and Aslin, 1981:188). SC Papu New Guinea, Morehead region on the trans-Fly plains (Waithman, 1979).
SYNONYMS: *mimulus, pumilus.*
COMMENTS: Electrophoretic data and spermatozoal morphology supported a close relationship between *P. delicatulus, P. novaehollandiae,* and *P. pilligaensis* (Breed, 1983; Briscoe et al., 1981).

Pseudomys desertor Troughton, 1932. Rec. Aust. Mus., 18:293.
TYPE LOCALITY: Australia, "Central Australia" (see Mahoney and Richardson, 1988:174).

DISTRIBUTION: Arid regions of Australia; Western Australia, South Australia, Northern Territory, Queensland, New South Wales, and Victoria (Mahoney and Richardson, 1988:174).

SYNONYMS: *murrayensis, subrufus*.

COMMENTS: Spermatozoal morphology similar to *P. australis* and many other species of *Pseudomys* (Breed, 1983). Both names listed above are unused senior synonyms of *desertor* and should not be used (Mahoney and Richardson, 1988:174).

Pseudomys fieldi (Waite, 1896). Rept. Horn Sci. Exped. Cent. Aust., Zool., 2:403.

TYPE LOCALITY: Australia, S Northern Territory, Alice Springs.

DISTRIBUTION: Known only from the type locality.

STATUS: U.S. ESA - Endangered; IUCN - Extinct.

COMMENTS: Still represented only by the holotype. "It is difficult to determine whether or not this represents a distinct species or a rather aberrant specimen of some other species" (Watts and Aslin, 1981:171).

Pseudomys fumeus Brazenor, 1934. Mem. Nat. Mus. Melb., 8:158.

TYPE LOCALITY: Australia, Victoria, Turton's Pass, Otway Forest.

DISTRIBUTION: Australia, Victoria (see map in Watts and Aslin, 1981:202). Range during Late Pleistocene extended into E New South Wales (Wakefield, 1972*a*).

STATUS: U.S. ESA - Endangered; IUCN - Rare.

COMMENTS: Electrophoretic data separated *P. fumeus* from other species of *Pseudomys* (Baverstock et al., 1981); phallic morphology suggested a close tie between *P. fumeus*, *P. albocinereus*, and *P. shortridgei*, a group for which the generic name *Gyomys* is available (Lidicker and Brylski, 1987:635); but spermatozoal structure tied *P. fumeus* with most other species of *Pseudomys* (Breed, 1983).

Pseudomys fuscus (Thomas, 1882). Ann. Mag. Nat. Hist., ser. 5, 9:413.

TYPE LOCALITY: Australia, Tasmania.

DISTRIBUTION: Australia; modern records from E New South Wales, S Victoria, and Tasmania, but Late Pleistocene-Holocene fragments indicated range once included Kangaroo Island, Carrieton, and Naracoorte in South Australia (Archer et al., 1984; Pledge, 1990).

SYNONYMS: *brazenori, mordicus, wombeyensis*.

COMMENTS: The form *fuscus* and the other taxa listed above were all described and revised under *Mastacomys* (Ride, 1956; Wakefield, 1972*b*), and this genus has always been recognized as part of the Australian fauna (Watts and Aslin, 1981; Mahoney and Richardson, 1988). However, chromosomal morphology (Baverstock et al., 1977*c*), G-banding homologies (Baverstock et al., 1983*b*), electrophoretic data (Baverstock et al., 1981), and phallic morphology (Lidicker and Brylski, 1987) linked *fuscus* with some species of *Pseudomys*, and Watts et al. (1992) united *fuscus* with *Pseudomys*. Sperm head structure reported by Breed (1984), and variation in external morphology of glans penis documented by Morrissey and Breed (1982). Taxonomic, distributional, and biological references cataloged by Mahoney and Richardson (1988:160).

Pseudomys glaucus Thomas, 1910. Ann. Mag. Nat. Hist., ser. 8, 6:609.

TYPE LOCALITY: Australia, S Queensland.

DISTRIBUTION: Australia, Murray-Darling basin in New South Wales and S Queensland (Mahoney and Richardson, 1988:175).

COMMENTS: Possibly extinct (Mahoney and Richardson, 1988:175).

Pseudomys gouldii (Waterhouse, 1839). Zool. Voy. H.M.S. "Beagle," Mammalia, 2:67.

TYPE LOCALITY: Australia, New South Wales, north of Hunter River (see Mahoney and Richardson, 1988:175).

DISTRIBUTION: Australia; range based on Recent and subfossil specimens includes W Western Australia and Murray-Darling basin in E South Australia, New South Wales, and N Victoria.

STATUS: U.S. ESA - Endangered; IUCN - Extinct.

SYNONYMS: *rawlinnae*.

COMMENTS: For full citation and other information see Mahoney and Richardson (1988:175). Apparently extinct (Mahoney and Richardson, 1988:175); no live animals seen or collected since the middle 1850's (Watts and Aslin, 1981:169). Phallic morphology indicated *gouldii* is clustered with *P. australis, P. higginsi*, and *P. nanus* (Lidicker and Brylski, 1987).

Pseudomys gracilicaudatus (Gould, 1845). Proc. Zool. Soc. Lond., 1845:77.
TYPE LOCALITY: Australia, Queensland, Darling Downs, Oakey Creek.
DISTRIBUTION: Australia; modern range along the eastern coast from Townsville in N Queensland to Sydney area in New South Wales; subfossil specimens from farther south in New South Wales (Mahoney and Posamentier, 1975) and from S Victoria (see map in Watts and Aslin, 1981:180).
SYNONYMS: *ultra*.
COMMENTS: Phylogenetically closely related to *P. nanus*, an estimate based on spermatozoal morphology (Breed, 1983), electrophoretic data (Baverstock et al., 1977*a*, 1981), and morphology of skin, skull, and teeth (Watts and Aslin, 1981).

Pseudomys hermannsburgensis (Waite, 1896). Rept. Horn Sci. Exped. Cent. Aust., Zool., 2:405.
TYPE LOCALITY: Australia, Northern Territory, George Gill Range (see Mahoney and Richardson, 1988:176).
DISTRIBUTION: Australia; arid parts of Western Australia, South Australia, Northern Territory, W Queensland, W New South Wales, and NW Victoria (see map in Watts and Aslin, 1981:190).
SYNONYMS: *brazenori*.
COMMENTS: Electrophoretic data (Baverstock et al., 1981) and spermatozoal morphology (Breed, 1983) indicated *P. hermannsburgensis* is clustered with most other species in *Pseudomys*, but phallic morphology interpreted by Lidicker and Brylski (1987) as indicating closer affinity to species they place in *Leggadina*.

Pseudomys higginsi (Trouessart, 1897). Cat. Mamm. Viv. Foss., 1:473.
TYPE LOCALITY: Australia, Tasmania, Kentishbury.
DISTRIBUTION: Australia; extant population known only from Tasmania; represented on mainland in Victoria and E New South Wales by Late Pleistocene samples (Wakefield, 1972*b*).
SYNONYMS: *australiensis, leucopus* (of Higgins and Petterd, 1881, not Rafinesque, 1818).
COMMENTS: Spermatozoal morphology (Breed, 1983) and electrophoretic data (Baverstock et al., 1981) placed *P. higginsi* with most other species of *Pseudomys*, but phallic anatomy (Lidicker and Brylski, 1987) clustered *P. higginsi* with *P. australis, P. gouldii*, and *P. nanus*. Wakefield (1972*b*) described *australiensis* as a subspecies based on Late Pleistocene fossils.

Pseudomys johnsoni Kitchener, 1985. Rec. West. Aust. Mus., 12:208.
TYPE LOCALITY: Australia, C Northern Territory, Kurundi Station, Kurinelli Mine, 150 m (Kitchener, 1985, provided additional information).
DISTRIBUTION: Australia; known only from small area in arid C Northern Territory (see map in Kitchener, 1985:216).
COMMENTS: Closest phylogenetic relative is apparently *P. chapmani*, which occurs in NW Western Australia (Kitchener, 1985).

Pseudomys laborifex Kitchener and Humphreys, 1986. Rec. West. Aust. Mus., 12:420.
TYPE LOCALITY: Australia, N Western Australia, Kimberley Region, Mitchell Plateau, adjacent to Camp Creek, 270 m (Kitchener and Humphreys, 1986:421, provided more information, including description of habitat at type locality).
DISTRIBUTION: Australia, N Western Australia and N Northern Territory (see maps in Kitchener and Humphreys, 1986:430, 1987:292).
SYNONYMS: *calabyi*.
COMMENTS: Kitchener and Humphreys (1987) proposed *calabyi* as a subspecies.

Pseudomys nanus (Gould, 1858). Proc. Zool. Soc. Lond., 1858:242.
TYPE LOCALITY: Australia, Western Australia, Victoria Plains (see Mahoney and Richardson, 1988:177).
DISTRIBUTION: Australia; W coast of Western Australia (also Barrow Isl), between Port Hedland and the Barkly Tableland in NE Western Australia, N Northern Territory, and NW Queensland (also South-West Isl in Gulf of Carpentaria); see map in Watts and Aslin (1981:177).
SYNONYMS: *ferculinus*.

COMMENTS: A close relative of *P. gracilicaudatus* to the exclusion of other species of *Pseudomys*, judged by analyses of electrophoretic data (Baverstock et al., 1977*a*, 1981); but a member of the group that includes only *P. australis*, *P. gouldii*, and *P. higginsi*, a conclusion based on phallic morphology (Lidicker and Brylski, 1987); but not a distinctive relative to most other species in the genus as judged by spermatozoal form (Breed, 1983).

Pseudomys novaehollandiae (Waterhouse, 1843). Proc. Zool. Soc. Lond., 1842:146 [1843].
TYPE LOCALITY: Australia, New South Wales, upper Hunter River, Yarrundi.
DISTRIBUTION: Australia; Coastal region of E New South Wales, S Victoria, and N Tasmania (see map in Watts and Aslin, 1981:193).
STATUS: U.S. ESA - Endangered.
COMMENTS: External, cranial, and dental morphology, along with electrophoretic data and spermatozoal anatomy, pointed to a close relationship between *P. novaehollandiae* and *P. pilligaensis* (Briscoe et al., 1981; Breed, 1983).

Pseudomys occidentalis Tate, 1951. Bull. Am. Mus. Nat. Hist., 97:246.
TYPE LOCALITY: Australia, Western Australia, Tambellup.
DISTRIBUTION: Australia; extant range in SW Western Australia, subfossil specimens indicate species extended along S coastline to Kangaroo Isl off coast of South Australia (see map and discussion in Watts and Aslin, 1981:205).
STATUS: U.S. ESA - Endangered; IUCN - Rare.
COMMENTS: Considered "rare and likely to become extinct," but range was contracting before arrival of Europeans (Watts and Aslin, 1981:205). Electrophoretic data clustered *P. occidentalis* with all other *Pseudomys* analyzed except *P. fumeus*, *P. gracilicaudatus*, *P. nanus*, and *P. shortridgei* (Baverstock et al., 1981).

Pseudomys oralis Thomas, 1921. Ann. Mag. Nat. Hist., ser. 9, 8:621.
TYPE LOCALITY: Australia; exact place unknown, but likely located in NE New South Wales or SE Queensland (Mahoney and Richardson, 1988:179).
DISTRIBUTION: Australia; extant specimens from NE New South Wales and SE Queensland, but Late Pleistocene fossils are from farther south in New South Wales and E Victoria (see map in Watts and Aslin, 1981:170).
STATUS: IUCN - Rare.
COMMENTS: Originally described as a subspecies of *P. australis* (see Mahoney and Richardson, 1988:179). One of the least known and rarest of *Pseudomys* (Watts and Aslin, 1981).

Pseudomys patrius (Thomas and Dollman, 1909). Proc. Zool. Soc. Lond., 1908:791 [1909].
TYPE LOCALITY: Australia, Queensland, Mt Inkerman.
DISTRIBUTION: Known only from the region of the type locality.
COMMENTS: Treated as a distinct species by Fox and Briscoe (1980) and Mahoney and Richardson (1988:179), but earlier judged by Mahoney to be a synonym of *P. delicatulus* (see Kitchener, 1985:218).

Pseudomys pilligaensis Fox and Briscoe, 1980. Aust. Mamm., 3:112.
TYPE LOCALITY: Australia, New South Wales, Merriwindi State Forest, 3 km west of Pilliga-Baradine Road., Cumberdeen Road (Fox and Briscoe, 1980, provided additional information).
DISTRIBUTION: Australia, N New South Wales, collected from a few localities within the Pilliga Scrub (see map in Fox and Briscoe, 1980:119).
STATUS: IUCN - Indeterminate.
COMMENTS: Chromosomal morphology described by Fox and Briscoe (1980). Morphological and electrophoretic data supported a close phylogenetic relationship of *P. pilligaensis* to *P. delicatulus* and *P. novaehollandiae* (Briscoe et al., 1981), which was reinforced by spermatozoal morphology (Breed, 1983).

Pseudomys praeconis Thomas, 1910. Ann. Mag. Nat. Hist., ser. 8, 6:608.
TYPE LOCALITY: Australia, Western Australia, Peron Peninsula.
DISTRIBUTION: Australia, Western Australia Shark Bay region and Bernier Isl (see map in Watts and Aslin, 1981:166).
STATUS: CITES - Appendix I; U.S. ESA - Endangered; IUCN - Rare.

COMMENTS: Clustered with most other species of *Pseudomys*, judged by electrophoretic data (Baverstock et al., 1981).

Pseudomys shortridgei (Thomas, 1907). Proc. Zool. Soc. Lond., 1906:765 [1907].

TYPE LOCALITY: Australia, SW Western Australia, neighborhood of Woyerling Reserve, 973 ft (additional information in Mahoney and Richardson, 1988:180).

DISTRIBUTION: Australia; SW Western Australia and SW Victoria (Grampian Mtns and Portland areas); see map in Watts and Aslin (1981:185).

STATUS: U.S. ESA - Endangered; IUCN - Vulnerable.

COMMENTS: Thought to be extinct in Western Australia (Watts and Aslin, 1981), but recently rediscovered there (Baynes et al., 1987). Electrophoretic data suggested *P. shortridgei* is phylogenetically isolated from all other species of *Pseudomys* (Baverstock et al., 1981); spermatozoal morphology unlike most other *Pseudomys* but similar to that of *P. delicatulus*, *P. novaehollandiae*, and *P. pilligaensis* (Breed, 1983); phallic anatomy linked *P. shortridgei* to *P. albocinereus* and *P. fumeus*, a group that could be generically recognized by calling it *Gyomys* (Lidicker and Brylski, 1987:635).

Rattus G. Fischer, 1803. Natl. Mus. Nat. Paris, 2:128.

TYPE SPECIES: *Mus decumanus* Pallas, 1778 (see Hollister, 1916*b*; = *Mus norvegicus* Berkenhout, 1769).

SYNONYMS: *Acanthomys* (Gray, 1867, not Lesson, 1842, or Tokuda, 1941), *Christomys*, *Cironomys*, *Epimys*, *Geromys*, *Mollicomys*; *Octomys* (Sody, 1941, not Thomas, 1922), *Pullomys*, *Togomys*.

COMMENTS: *Rattus* Frisch, 1775, is unavailable. Sody (1941) proposed the genera *Christomys*, *Cironomys*, *Geromys*, *Mollicomys*, *Octomys*, and *Pullomys* for various species we list in *Rattus*; all are *nomina nuda*. *Togomys* is based on *R. exulans* (Dieterlen, in Ansell, 1989*a*). Taxonomic changes altering the definition of *Rattus* as understood by Tate (1936), Ellerman (1941, 1949*a*, 1961), and Simpson (1945), were described and summarized by Misonne (1969), Musser (1981*b*), Musser and Newcomb (1983), Musser and Holden (1991), and Musser and Heaney (1992). Species we list here can be sorted into the following groups.

1. The *norvegicus* group. *Rattus norvegicus*, the type species of the genus, is divergent from *Rattus rattus* in morphology as well as electrophoretic and immunological traits (Chan, 1977; Chan et al., 1979; Baverstock et al., 1983*a*, *c*, 1986; Watts, in litt.)

2. The *rattus* group (*adustus*, *argentiventer*, *baluensis*, *burrus*, *everetti*, *hoffmanni*, *koopmani*, *losea*, *lugens*, *mindorensis*, *mollicomulus*, *nitidus*, *osgoodi*, *palmarum*, *rattus*, *tanezumi*, *sikkimensis*, *simalurensis*, *tawitawiensis*, *tiomanicus*, and *turkestanicus*). Morphology of these species generally reflects the conception of what is usually called subgenus *Rattus* (Musser and Holden, 1991). Whether or not this cluster, along with the *norvegicus* group, will eventually form the only contents of *Rattus* is a speculation that has to be assessed by systematic revision of the genus (Musser and Heaney, 1985, 1992; Musser and Holden, 1991). Comparative chromosomal data for many of these species were summarized by Musser and Holden (1991). Schwarz and Schwarz (1967) offered a most peculiar and idiosyncratic revision of the group.

3. Native Australian species (*colletii*, *fuscipes*, *lutreolus*, *sordidus*, *tunneyi*, and *villosissimus*), which were revised by Taylor and Horner (1973) and are reviewed by Watts and Aslin (1981); one (*sordidus*) also occurs in New Guinea. Results from biochemical and chromosomal studies indicated the species form a monophyletic cluster to the exclusion of either *R. rattus* or *R. norvegicus* (Baverstock et al., 1977*d*, 1983*a*, 1986). Baverstock et al. (1983*a*) proposed a hypothesis of phylogenetic relationships based on cladistic analyses of electrophoretic, immunologic, and chromosomal data, and another proposed set of relationships assessed by isozyme elecrophoresis (Baverstock et al., 1986). Mahoney and Richardson (1988) cataloged taxonomic, distributional, and biological references. This group may also include the endemics from Timor (*timorensis*) and Flores (*hainaldi*). Furthermore, such endemic Nusa Tenggara (Lesser Sunda Isls) genera as *Komodomys* are morphologically similar to some members of the Australian *Rattus*.

4. Native New Guinea species (*jobiensis*, *leucopus*, *mordax*, *novaeguineae*, *praetor*, *sanila*, *steini*, and probably *giluwensis*), which are indigenous to New Guinea and

adjacent archipelagos; one (*leucopus*) also occurs in NE Australia. All have been the subject of a systematic revision by Taylor et al. (1982, 1983). Known endemics of the Moluccas (*elaphinus, feliceus,* and probably *morotaiensis*) are also related to this group. Nature of the relationship among this group and members of *Stenomys*, members of the Australian cluster, and those genera of New Endemics native to the Philippine Isls (Musser and Heaney, 1992) remains to be resolved.

5. The *xanthurus* group (*bontanus, foramineus, marmosurus, pelurus,* and *xanthurus*), which occurs on Sulawesi and adjacent Peleng Isl, and may eventually be removed from *Rattus* (Musser and Holden, 1991).

Phylogenetic affinities of the remaining species listed below (*annandalei, enganus, exulans, hoogerwerfi, macleari, montanus, nativitatis, ranjiniae, stoicus,* and *korinchi*) are unresolved; some may eventually be excluded from the genus.

Rattus adustus Sody, 1940. Treubia, 17:397.

TYPE LOCALITY: Indonesia, Pulau Enggano, off the coast of W Sumatra (and off the continental shelf), Kiojoh, sea level.

DISTRIBUTION: Known only from Enggano Isl.

COMMENTS: Known only by the holotype. Although Sody (1940) described *adustus* as a species of *Rattus*, he later listed it as a subspecies of *R. rattus* in a section also containing *lugens* and *mentawi*, populations endemic to the Mentawai islands (Sody, 1941). The Enggano rat is distinctive; in morphology and geographic proximity it is related to *R. lugens* (Musser and Heaney, 1985).

Rattus annandalei (Bonhote, 1903). *In* Annandale, Fasciculi Malayenses, Zool., pt. I:30.

TYPE LOCALITY: Malaysia (Malay Peninsula), S Perak, Sungkei.

DISTRIBUTION: Malay Peninsula, Singapore, E Sumatra, and islands of Padang and Rupat off the coast of E Sumatra (see Musser and Newcomb, 1983:515, and references therein).

SYNONYMS: *bullatus, villosus.*

COMMENTS: A Sundaic endemic; superficially resembles *Sundamys muelleri* and some species of *Rattus* in primitive external, cranial, and dental features, but in other specialized traits *R. annandalei* is unlike any species of *Sundamys* and may even eventually be removed from *Rattus* (Musser and Newcomb, 1983).

Rattus argentiventer (Robinson and Kloss, 1916). J. Strs. Br. Roy. Asiat. Soc., 73:274.

TYPE LOCALITY: Indonesia, west coast of Sumatra, Pasir Ganting.

DISTRIBUTION: Recorded from Thailand, Koh Samui off the east coast of peninsular Thailand, Cambodia, and S Vietnam in Indochina; the Malay Peninsula, Sumatra, Java, Borneo, Kangean Isl, and Bali on the Sunda Shelf; islands of Lombok, Sumbawa, Komodo, Rintja, Flores, Sumba, and Timor in Nusa Tenggara (Lesser Sunda Isls); Mindoro and Mindanao Isls in the Philippines; Sulawesi; and one place and date of collection in New Guinea.

SYNONYMS: *bali, brevicaudatus, chaseni, hoxaensis, pesticulus, saturnus, umbriventer.*

COMMENTS: The incorrect historical association of *argentiventer* as a subspecies of *Rattus rattus* was summarized by Musser (1973*b*). Judged by its close morphological alliance with species Ellerman (1941) placed in subgenus *Rattus*, which are mostly mainland Asian in origins, and its peculiar geographic distribution that is discordant with ranges of endemic species, *R. argentiventer* seems clearly an element that is native to Indochina and was inadvertently introduced into the endemic and highly distinctive murine faunas of the Sunda Shelf, Philippines, Sulawesi, Nusa Tenggara, and New Guinea (Musser, 1973*b*; Musser and Holden, 1991; Musser and Newcomb, 1983; Taylor et al., 1982), possibly with the spread of rice culture. *Rattus hoxaensis* from C Vietnam, described by Dao (1960), probably represents *R. argentiventer.*

Rattus baluensis (Thomas, 1894). Ann. Mag. Nat. Hist., ser. 6, 14:454.

TYPE LOCALITY: Malaysia, Sabah (N Borneo), Gunung Kinabalu, 8000 ft.

DISTRIBUTION: Known only from 7000-12,500 ft on slopes of Gunung Kinabalu, N Borneo (Musser, 1986).

COMMENTS: Taxonomic history of past, and incorrect, association of *baluensis* with *Rattus rattus* was documented by Musser (1986), who also noted that its closest relative is probably *R. tiomanicus*, which occurs in lowlands of Borneo and on many islands of the Sunda Shelf.

Rattus bontanus Thomas, 1921. Ann. Mag. Nat. Hist., ser. 9, 7:246.
TYPE LOCALITY: Indonesia, SW Sulawesi, Mt Bonthain (Gunung Lampobatang), 2000 ft.
DISTRIBUTION: Known only from 600-2500 m on the slopes of Gunung Lampobatang, SW Sulawesi.
COMMENTS: Sody (1941) questionably included this species in *Taeromys*, Laurie and Hill (1954), and Musser (1984) treated it as a subspecies of *R. xanthurus*, but Musser and Holden (1991) contended that it is a distinct species most closely related to *R. foramineus*, which occurs in coastal lowlands of the southern end of the SW peninsula of Sulawesi.

Rattus burrus (Miller, 1902). Proc. U. S. Nat. Mus., 24:768.
TYPE LOCALITY: India, Nicobar Isls, Trinkat Island.
DISTRIBUTION: Islands of Trinkat, Little Nicobar, and Great Nicobar in the Nicobar Archipelago.
SYNONYMS: *burrescens*.
COMMENTS: Except for larger body size, morphology closely resembles that of most samples of *R. tiomanicus* from the Sunda Shelf. Whether *burrus* is an endemic of the Nicobars or whether samples from the three islands are insular variants of *R. tiomanicus* needs to be tested by a systematic revision of the *R. tiomanicus* complex (Musser, 1986; Musser and Califia, 1982; Musser and Heaney, 1985).

Rattus colletti (Thomas, 1904). Novit. Zool., 11:599.
TYPE LOCALITY: Australia, Northern Territory, South Alligator River.
DISTRIBUTION: Australia; known only from the coastal floodplains of the Northern Territory, the most restricted range of all the native Australian *Rattus*. (see map in Watts and Aslin, 1981).
COMMENTS: Arranged as a subspecies of *R. sordidus* by Taylor and Horner (1973), but subsequently treated as a separate species (Mahoney and Richardson, 1988; Watts and Aslin, 1981). Although diploid counts are very different, *Rattus colletti* (2N=42) is genically close to *R. villosissimus* (2N=50); the two hybridize in the laboratory, but the offspring exhibit severely reduced fertility (Baverstock et al., 1983*a*, 1986). Fertile hybrids have been obtained between laboratory crosses of *R. colletti* and *R. tunneyi*, but the two are sympatric in the wild where no hybrids have been found (Baverstock et al., 1983*a*).

Rattus elaphinus Sody, 1941. Treubia, 18:307.
TYPE LOCALITY: Indonesia, Kepulauan Sula, Pulau Taliabu (east of Pulau Peleng and Sulawesi).
DISTRIBUTION: Known only from Palau (Isl) Taliabu, Indonesia (Musser and Holden, 1991) and adjacent Pulau Manggole (Flannery, in litt.).
COMMENTS: A morphologically distinctive species that is most closely related to native *Rattus* on the Moluccas and New Guinea. Amplified description and comparisons with Sulawesian *R. hoffmanni* and *R. koopmani* from Peleng Isl provided by Musser and Holden (1991).

Rattus enganus (Miller, 1906). Proc. U.S. Natl. Mus., 30:821.
TYPE LOCALITY: Indonesia, Pulau Enggano (west of Sumatra and off the continental margin).
DISTRIBUTION: Known only from Enggano Isl.
COMMENTS: Represented only by the holotype. Some cranial, dental, and external features resemble those in *R. macleari* from Christmas Island and *R. xanthurus* from Sulawesi, but the holotype of *R. enganus* is morphologically distinctive; determining its phylogenetic affinities within *Rattus* will require more specimens from Enggano Isl and revisionary study of the genus (Musser and Newcomb, 1983).

Rattus everetti (Günther, 1879). Proc. Zool. Soc. Lond., 1879:75.
TYPE LOCALITY: Philippines, N Mindanao Isl (Heaney and Rabor, 1982).
DISTRIBUTION: Islands of Luzon, Catanduanes, Mindoro, Sibuyan, Ticao, Camiguin, Samar, Calicoan, Leyte, Dinagat, Siargao, Mindanao, Basilan, Bohol, Biliran, and Maripipi; probably occurs on other islands in the Philippine archipelago (Musser and Heaney, 1992).
SYNONYMS: *albigularis*, *gala*, *tagulayensis*, *tyrannus*.

COMMENTS: May be more than one species represented in insular samples (*tyrannus*, for example), and the *everetti* complex needs critical systematic revision. Similar to the core group of *Rattus* in cranial, dental, and spermatozoal characters (Breed and Musser, 1991; Musser and Heaney, 1992), but not closely related to other species of *Rattus* endemic to the Philipines. Member of the Philippine New Endemics, but distantly related to other genera in that group (Musser and Heaney, 1992).

Rattus exulans (Peale, 1848). Mammalia *in* Repts. U.S. Expl. Surv., 8:47.

TYPE LOCALITY: Society Isls, Tahiti Isl (France).

DISTRIBUTION: E Bangladesh, Burma, Thailand, Laos, Cambodia, Vietnam, Sundaic region (incl. Mentawai isls, and islands of Enggano, Nias, and Simeulule), Christmas Isl, Sulawesi, Philippines, Moluccas, Nusa Tenggara (Lesser Sunda Isls), New Guinea Region, two islands off the coast of N and NE Australia (not on mainland), Micronesia, New Zealand, and Polynesia, including Hawaii and Easter Isl. Not recorded from Andaman or Nicobar isls, despite assertion of Wodzicki and Taylor (1984).

SYNONYMS: *aemuli, aitape, apicus, basilanus, bocourti, buruensis, calcis, clabatus, concolor, browni echimyoides, ephippium, equile, eurous, gawae, hawaiiensis, huegeli, jessook, lassacquerei, leucophaetus, luteiventris, malengiensis, manoquarius, maorium, mayonicus, Togomys melanoderma, meringgit, micronesiensis, negrinus, obscurus, ornatulus, otteni, pantarensis, praecelsus, pullus, querceti, raveni, rennelli, schuitemakeri, solatus, stragulum, suffectus, surdus, tibicen, todayensis, vigoratus, vitiensis, vulcani, wichmanni* (see Dieterlen, *in* Ansell, 1989*a*; Ellerman, 1961; Laurie and Hill, 1954; Musser, 1970*e*, 1977*a*; Musser and Newcomb, 1983; and Taylor et al., 1982).

COMMENTS: Inadvertant human introduction responsible for most of the Pacific insular occurrences (see Roberts, 1991), and possibly for distribution outside of mainland SE Asia where species may have originated (Musser and Newcomb, 1983).

Rattus exulans has traditionally been included within subgenus *Rattus* along with *R. argentiventer*, *R. nitidus*, *R. norvegicus*, *R. rattus*, and others which have, in the past, formed the core of that group (e.g., Ellerman, 1941; Misonne, 1969). Some electrophoretic and chromosomal data supported this allocation (Raman and Sharma, 1977; Chan, 1977; Chan et al., 1979), but morphological and other biochemical data suggested the species is distant from *R. rattus* and its close relatives (Gemmeke and Niethammer, 1984; Medway and Yong, 1976; Pasteur et al., 1982).

Rattus feliceus Thomas, 1920. Ann. Mag. Nat. Hist., ser. 9, 6:423.

TYPE LOCALITY: Indonesia, Kepulauan Maluku, Pulau Seram, Gunung Manusela, 6000 ft.

DISTRIBUTION: Known only from Seram Isl.

COMMENTS: An amplified description and comparison with *R. koopmani* from Peleng Isl were provided by Musser and Holden (1991), who also discussed past, and incorrect, subspecific allocations of *feliceus* to other species of *Rattus*. Phylogenetic affinities of *R. feliceus* are with species of *Rattus* endemic to New Guinea region (Musser and Holden, 1991).

Rattus foramineus Sody, 1941. Treubia, 18:308.

TYPE LOCALITY: Indonesia, SW Sulawesi, Bulukumba.

DISTRIBUTION: Known only from a few localities on coastal lowlands near the end of the SW peninsula of Sulawesi.

COMMENTS: Sody described *pelurus* from Pulau Peleng as a subspecies of *R. foramineus*, but the Peleng Isl rat is a distinct species (Musser and Holden, 1991). *Rattus foramineus* is represented only by four modern specimens (Sody, 1941) and subfossil fragments (Musser, 1984; Musser and Holden, 1991). Laurie and Hill (1954) listed *R. foramineus* as *incertae sedis*, but Musser (1984) treated it as a subspecies of *R. xanthurus*, and Musser and Holden (1991) considered it to be a species most closely related to *R. bontanus*, which replaces it on slopes of Gunung Lampobatang, Sulawesi.

Rattus fuscipes (Waterhouse, 1839). Zool. Voy. H.M.S. "Beagle," Mammalia, p. 66.

TYPE LOCALITY: Neotype from Australia, Western Australia, Albany, "Little Grove" on Princess Royal Harbor, 4 mi S Mt Melville; holotype was lost (Mahoney and Richardson, 1988; Taylor and Horner, 1973).

DISTRIBUTION: Coastal, subcoastal, and offshore islands of SW Western Australia; S coast from Eyre Peninsula in South Australia to W Victoria; coastal and subcoastal Victoria from Otway Peninsula north to near Rockhampton in Queensland; coastal Queensland from Townsville to Cooktown (see map in Taylor and Horner, 1973:15).

SYNONYMS: *assimilis, brazenori, coracius, glauerti, greyii, manicatus, mondraineus, murrayi, peccatus, pelori, ravus.*

COMMENTS: Taylor and Horner (1973) suggested, on morphological grounds, that the Queensland population of *R. fuscipes* (*coracius*) has a common ancestry with Queensland *R. leucopus*, a hypothesis reasserted by Taylor et al. (1982, 1983). This relationship, however, is not supported by either chromosomal (Dennis and Menzies, 1978) or biochemical data (Baverstock et al., 1983*a*, 1986). See Taylor and Calaby (1988*a*, Mammalian Species, 298).

Rattus giluwensis Hill, 1960. J. Mammal., 41:277.

TYPE LOCALITY: NE New Guinea, Papua, Mt Giluwe, 11,000-12,000 ft.

DISTRIBUTION: Recorded only from Mt Giluwe, Papua New Guinea and adjoining highlands between 2195 and 3660 m (see map in Taylor et al., 1982).

STATUS: IUCN - Rare.

COMMENTS: Originally described as *R. ruber melanurus* by Laurie and Hill (1954:112), but the name is preoccupied by *Rattus melanurus* Shamel, 1940, which refers to a sample of *Maxomys whiteheadi*. A morphologically distinctive species whose phylogenetic affinities seem equivocal (Taylor et al., 1982, 1983).

Rattus hainaldi Kitchener, How, and Maharadatunkamsi, 1991. Rec. West. Aust. Mus., 15:557.

TYPE LOCALITY: Indonesia, Nusa Tenggara, Pulau Flores, Gunung Ranakah, 1300 m, above Kampong Robo, Desa Longko, 8 km SSE Ruteng.

DISTRIBUTION: Known only from Flores Isl.

COMMENTS: Represented by two specimens (Kitchener et al., 1991*c*). Phylogenetic affinities uncertain, possibly allied to native Australian and New Guinea species of *Rattus*.

Rattus hoffmanni (Matschie, 1901). Abh. Senckenb. Naturforsch. Ges., 25:284.

TYPE LOCALITY: Indonesia, NE Sulawesi, Minahassa.

DISTRIBUTION: Sulawesi; throughout island except upper slopes of Gunung Lampobatang at the end of the SW peninsula (distribution is concordant with that of *Paruromys dominator*); also on Pulau Malenge in Kepulauan Togian (Musser and Holden, 1991).

SYNONYMS: *biformatus, celebensis* (of Hoffmann, 1887, not Gray, 1867), *linduensis, mengkoka, mollicomus, tatei* (see Musser, 1971*b*).

COMMENTS: Morphological, chromosomal, distributional, and ecological boundaries of species elaborated by Musser and Holden (1991), who also documented taxonomic history. Closest relative of *R. hoffmanni* is *R. mollicomulus* from Gunung Lampobatang, Sulawesi; both form the monophyletic *Rattus hoffmanni* group. Phylogenetic affinities of this group to other species of *Rattus* is unresolved.

Rattus hoogerwerfi Chasen, 1939. Treubia, 17(3):496.

TYPE LOCALITY: Indonesia, W Sumatra, Aceh, Gunung Leuser, Blang Kedjeren, 2900 ft.

DISTRIBUTION: Known only from between 2900 and 9300 ft in foothills and upper slopes of Gunung Leuser, Sumatra (see map in Musser, 1986).

COMMENTS: Very distinctive and possibly not a member of *Rattus*. Morphology and comparisons with other species documented by Musser and Newcomb (1983) and Musser (1986). Although phylogenetic affinities with species of *Rattus* from Sulawesi, the Philippines, and Christmas Island have been claimed by Miller (1942), and with *R. korinchi* by Chasen (1939), no compelling evidence links *R. hoogerwerfi* with any other species (Musser, 1986).

Rattus jobiensis Rümmler, 1935. Z. Säugetierk., 10:116.

TYPE LOCALITY: New Guinea, Irian Jaya, Pulay Yapen (Japen Isl) in Teluk Cendrawasih (Geelvinck Bay).

DISTRIBUTION: New Guinea, Irian Jaya, recorded only from islands of Yapen, Owi, and Biak in Geelvinck Bay (see map in Taylor et al., 1982:262).

SYNONYMS: *biakensis, owiensis.*

COMMENTS: Originally described as a subspecies of *R. leucopus*, *jobiensis* is a distinctive species that may be more closely related to Moluccan endemics than to native mainland New Guinea species of *Rattus* (Taylor et al., 1982).

Rattus koopmani Musser and Holden, 1991. Bull. Am. Mus. Nat. Hist., 206:389.

TYPE LOCALITY: Indonesia, Kepulauan Banggai, Pulau Peleng (1°23'S, 123°14'E), which is separated from mainland Sulawesi by deepwater Selat Peleng.

DISTRIBUTION: Known only from Peleng Isl.

COMMENTS: Represented only by the holotype. In some characters, *R. koopmani* resembles the Sulawesian *R. hoffmanni*, but it also shares other traits with *R. elaphinus* from Taliabu Isl to the east of Peleng Isl. More specimens are needed to assess morphological variation (Musser and Holden, 1991) and better estimate phylogenetic affinity.

Rattus korinchi (Robinson and Kloss, 1916). J. Str. Br. Roy. Asiatic Soc., 73:275.

TYPE LOCALITY: Indonesia, W Sumatra, Propinsi Jambi, Gunung Kerinci, Sungai Kring, 7300 ft.

DISTRIBUTION: Recorded only from Gunung Kerinci and Gunung Talakmau in W Sumatra (see map in Musser, 1986:4).

COMMENTS: Revised by Musser (1986). Known by very few specimens, and morphologically unlike any other described species of *Rattus*.

Rattus leucopus (Gray, 1867). Proc. Zool. Soc. Lond., 1867:598.

TYPE LOCALITY: Australia, Queensland, Cape York = Locality of lectotype (see Mahoney and Richardson, 1988:183).

DISTRIBUTION: Australia, Queensland; one population ranges from the tip of Cape York south down the E side of the Peninsula to the vicinity of Coen, another from the region of Cooktown south along the coast to Tully; all records are east of the Great Dividing Range (see maps in Taylor and Horner, 1973:43; Watts and Aslin, 1981). New Guinea; widespread in lowlands south of Central Cordillera, and in N and S lowland regions fringing the Owen Stanley Range (see map in Taylor et al., 1982:232).

SYNONYMS: *cooktownensis, dobodurae, mcilwraithi, personata, ratticolor, ringens, terra-reginae,*

COMMENTS: This species and *R. sordidus* are the only two native *Rattus* occurring on both New Guinea and the NE coastal region of Australia (Taylor et al., 1982). Morphologically related to other species of *Rattus* native to New Guinea (Taylor et al., 1982). Morphological data interpreted by Taylor and Horner (1973) to indicate close affiliation between *R. leucopus* and *R. fuscipes* from coastal Queensland; genic data discordant with this view (Baverstock et al., 1983*a*, 1986).

Rattus losea (Swinhoe, 1871). Proc. Zool. Soc. Lond., 1870:637 [1871].

TYPE LOCALITY: Taiwan.

DISTRIBUTION: Recorded from Taiwan, Pescadores Isls, S China (Fujian, Guangdong, Jiangxi, Hainan Isl), C Vietnam, S Laos, and Thailand (excluding peninsular Thailand); probably will also be found in Cambodia, N Laos and Vietnam, and parts of Burma borderingThailand (Musser and Newcomb, 1985).

SYNONYMS: *exiguus, sakeratensis.*

COMMENTS: Known morphological, geographical, and altitudinal boundaries; correctly and incorrectly assoicated scientific names; and geographic variation reviewed by Musser and Newcomb (1985). They also suggested that *R. losea* is morphologically and probably phylogenetically nearer *R. osgoodi* from the highlands of S Vietnam than any other species of *Rattus*, and speculated that *R. losea* is also phylogenetically linked to *R. argentiventer.*

Rattus lugens (Miller, 1903). Smithson. Misc. Coll., 45:33.

TYPE LOCALITY: Indonesia, Kepulauan Mentawai, Pulau Utara Pagai (North Pagai Isl).

DISTRIBUTION: Islands of Siberut, Sipora, Pagai Utara, and Pagai Selattan in the Mentawai Archipelago off coast of SW Sumatra; these islands lie on a slender peninsula of the continental shelf.

SYNONYMS: *mentawi.*

COMMENTS: Chasen (1940) listed *lugens* and *mentawi* each as a subspecies of *R. rattus*, but the Mentawai endemic is a morphologically distinctive species closely related to *R. adustus* from Enggano Isl, and to the *R. tiomanicus* complex (Musser, 1986; Musser and Califia, 1982; Musser and Heaney, 1985). Part of the rodent fauna endemic to Mentawi Archipelago (see account of *Leopoldamys siporanus*).

Rattus lutreolus (J. E. Gray, 1841). *In* G. Gray, App. C *in* Jour. Two Exped. Aust., II:409.
TYPE LOCALITY: Australia, New South Wales, Hunter River, Moscheto Isl (locality of lectotype; see Mahoney and Richardson, 1988).
DISTRIBUTION: Tasmania; coastal and subcoastal habitats from vicinity of Adelaide in SE South Australia east and north through Victoria, New South Wales to SE Queensland; isolated populations occur in N Queensland (see maps in Taylor and Horner, 1973:53, and Watts and Aslin, 1981:231).
SYNONYMS: *cambricus, imbil, lacus, pachyurus, petterdi, tetragonurus, vellerosus, velutinus.*
COMMENTS: Watts and Aslin (1981) provided a comprehensive discussion of the species. See Taylor and Calaby (1988*b*, Mammalian Species, 299).

Rattus macleari (Thomas, 1887). Proc. Zool. Soc. Lond., 1887:513.
TYPE LOCALITY: Christmas Isl (Australia).
DISTRIBUTION: Was endemic to Christmas Isl, 320 km south of Java in the Indian Ocean, but was thought to be extinct by 1908 (Andrews, 1909) and is now considered extinct (Flannery, 1990*c*).
COMMENTS: Ellerman (1941) first listed the species as the only member of "*macleari*" group in subgenus *Rattus*, then placed it and *R. nativitatis* in same group within subgenus *Stenomys* of *Rattus* (Ellerman, 1949*a*). Chasen (1940) thought *macleari* to be nearest *Sundamys muelleri*, but in their comparisons, Musser and Newcomb (1983) found no support for this alliance. Misonne (1969) included *macleari* in subgenus *Rattus*. Sody (1941) proposed genus *Christomys* for *macleari*. In the original description, Thomas (1887*c*) indicated *macleari* to belong to a group that included *celebensis, everetti, meyeri*, and *xanthurus*; of these, only *xanthurus* resembled *macleari* (Musser and Newcomb, 1983). Phylogenetic relationships remain unresolved, but Musser (1986) suggested *macleari* should be compared with a group of species that includes *annandalei, enganus, korinchi, montanus, nativitatis*, and *xanthurus*, all not part of subgenus *Rattus*, and distantly related to it.

Rattus marmosurus Thomas, 1921. Ann. Mag. Nat. Hist., ser. 9, 7:246.
TYPE LOCALITY: Indonesia, NE Sulawesi, Minahassa, Gunung Masarang, 2000 ft.
DISTRIBUTION: NE Sulawesi only, from vicinity of Teluk Kuandang (0°50'N, 122°52'E) east to Gunung Klabat (1°28'N, 125°02'E).
SYNONYMS: *tondanus.*
COMMENTS: Although *marmosurus* has been listed as a subspecies of *R. xanthurus* (Ellerman, 1941; Laurie and Hill, 1954), most researchers besides Thomas have recognized its specific uniqueness (Misonne, 1969; Musser, 1971*e*, 1984; Musser and Holden, 1991; Sody, 1941; Tate, 1936). Furthermore, the two are sympatric in NE Sulawesi (Musser, 1971*e*). Sody (1941) questionably included *marmosurus* in *Taeromys*, but it is a member of the *Rattus xanthurus* group. Musser (1971*e*) explained why *tondanus* is a synonym.

Rattus mindorensis (Thomas, 1898). Trans. Zool. Soc. London, 14:402.
TYPE LOCALITY: Philippines, Mindoro Isl, Mt Dulangan, 5000 ft.
DISTRIBUTION: Highlands of Mindoro (Philippines).
SYNONYMS: *picinus.*
COMMENTS: Phylogenetically distant from other species of *Rattus* native to the Philippines (Musser and Heaney, 1992). Morphologically closely related to *Rattus tiomanicus*, which is native to Malay Peninsula and islands on the Sunda Shelf, and possibly only an insular variant of that species (Musser and Califia, 1982; Musser and Heaney, 1985). Musser (1977*b*) explained why *picinus* is a synonym.

Rattus mollicomulus Tate and Archbold, 1935. Am. Mus. Novit., 802:4.
TYPE LOCALITY: Indonesia, Sulawesi, SW Sulawesi, Gunung Lampobatang, Wawokaraeng, 1500 m.
DISTRIBUTION: Known only from the upper slopes of Gunung Lampobatang, Sulawesi; distribution is concordant with *Paruromys ursinus* (Musser and Holden, 1991).
COMMENTS: Morphological and distributional limits outlined by Musser and Holden (1991), who also provided a history of past allocations of the name. Closest relative is *R. hoffmanni*, which occurs in lowlands of the SW peninsula and throughout the rest of Sulawesi.

Rattus montanus Phillips, 1932. Ceylon J. Sci., Sec. B, 16:323.

TYPE LOCALITY: Sri Lanka (Ceylon), West Haputale, Ohiya, 5200-6000 ft.

DISTRIBUTION: Known only from the type locality, Horton Plains at 7000 ft, and Nuwara Eliya at 6000 ft in primary montane forests of Central and Uva Provinces of Sri Lanka (Phillips, 1980).

COMMENTS: A montane endemic on Sri Lanka and morphologically so unlike most other species of *Rattus* it may not belong in same genus, despite assertion of one author that it is nothing more than a large form of *R. rattus* (see Musser, 1986:22). Like *R. annandalei, R. hoogerwerfi, R. korinchi, R. macleari, R. nativitatis,* and members of the *R. xanthurus* group, the Ceylon endemic seems isolated within present morphological boundaries of *Rattus* (Musser, 1986).

Rattus mordax (Thomas, 1904). Ann. Mag. Nat. Hist., ser. 7, 14:398.

TYPE LOCALITY: Papua New Guinea, Kumusi River (see Tate, 1951:333, for comments).

DISTRIBUTION: Papua New Guinea; Huon Peninsula and both sides of Owen Stanley Range on mainland. Also found in the d'Entrecasteaux Isl, the Louisiade Arch., and Woodlark Isl in the Trobriand Isls (see map in Taylor et al., 1982:225).

SYNONYMS: *fergussoniensis.*

COMMENTS: Originally described by Thomas as a species, *mordax* has been historically treated as a subspecies of either *leucopus, ringens,* or *ruber* until it was reinstated by Taylor et al. (1982).

Rattus morotaiensis Kellogg, 1945. Proc. Biol. Soc. Washington, 58:66.

TYPE LOCALITY: Indonesia, Moluccas, Pulau Morotai, off N coast of Pulau Halmahera.

DISTRIBUTION: Recorded only from Morotai Isl but probably occurs on Halmahara and nearby islands.

COMMENTS: A distinctive species probably related to native *Rattus* of New Guinea.

Rattus nativitatis (Thomas, 1889). Proc. Zool. Soc. Lond., 1888:533 [1889].

TYPE LOCALITY: Christmas Isl (Australia).

DISTRIBUTION: Endemic to Christmas Isl, 320 km south of Java in the Indian Ocean; suspected to be extinct by 1908 (Andrews, 1909) and is now considered extinct (Flannery, 1990*c*).

COMMENTS: For Thomas (1888*b*), the morphology of *R. nativitatis* distanced it from any other described species of *Rattus*. Ellerman (1941) first listed the species as the only member of the "*nativitatis*" group in subgenus *Rattus*, then placed it and *R. macleari* in same group within subgenus *Stenomys* of *Rattus* (Ellerman, 1949*a*). Chasen (1940) thought *R. nativitatis* to be without close relatives in Malaysia, but Misonne (1969) placed it close to *rajah* in the subgenus *Leopoldamys* of *Rattus*, an allocation rejected by Musser (1981*b*) and Musser and Newcomb (1983). Three hypotheses about phylogenetic position of *R. nativitatis* require testing: 1. It is more closely related to *R. macleari*, the other endemic on Christmas Island than to any other species of *Rattus*; 2. It is not related to *R. macleari* but to other species in the genus or is phylogenetically isolated; 3. It is not even a member of *Rattus*.

Rattus nitidus (Hodgson, 1845). Ann. Mag. Nat. Hist., [ser. 1], 15:267.

TYPE LOCALITY: Nepal.

DISTRIBUTION: Records on mainland Southeast Asia are from S China (including Hainan Isl), Vietnam, Laos, N Thailand, Burma, India (Assam, Bhutan, Sikkim, and Kumaun), Bangladesh, and Nepal; these probably represent the indigenous range. Records east of the continental shelf are from Sulawesi, Luzon Isl in the Philippines, Seram Isl in the Moluccas, the Vogelkop Peninsula of Irian Jaya, and the Palau isls; this range likely represents introductions mediated by human agency (Musser and Holden, 1991).

SYNONYMS: *aequicaudalus, guhai, horeites, manuselae, obsoletus, rahengis, ruber, rubricosa, subditivus, vanheurni* (see Ellerman, 1941; Khajuria et al., 1977; J. T. Marshall, Jr., 1977*a, b*; Musser, 1981*c*; Musser and Holden, 1991; and Taylor et al., 1982).

COMMENTS: Ellerman (1941) listed *pyctoris* as a synonym of *R. nitidus*, but that name was based on a specimen of *R. turkestanicus*. Phallic morphology described by Yang and Fang (1988) in context of assessing phylogenetic relationships among Chinese murines.

Rattus norvegicus (Berkenhout, 1769). Outlines of the Natural History of Great Britain and Ireland, 1:5.

TYPE LOCALITY: Great Britain.

DISTRIBUTION: Original distribution assumed to be SE Siberia and N China (Heilongjiang), but introduced worldwide where it is more common in colder climates of high latitudes (Kucheruk, 1990); in warmer regions and tropics restricted to habitats highly modified by humans—sewers, buildings, wharves, and breakwaters, for example (Johnson, 1962*a*).

SYNONYMS: *norvegicus* var. *albus, caraco, caspius, cauquenensis, decaryi, decumanoides* (*nomen nudum*), *decumanus, discolor, griseipectus, hibernicus, hoffmanni* (Trouessart, 1904, not Matschie, 1901), *humiliatus, hybridus, insolatus, javanus, leucosternum, lutescens, magnirostris, decumanus* var. *major, maniculatus, norvegicus* var. *orii, norvegicus* var. *otomoi, plumbeus, praestans, primarius, simpsoni* (Philippi, 1900, not Ellerman, 1949), *socer, sowerbyi, surmulottus, tamarensis* (see Ellerman and Morrison-Scott, 1951; Jones and Johnson, 1965; Laurie and Hill, 1954; Mahoney and Richardson, 1988; Musser, 1977*a*; Musser and Newcomb, 1985; Sody, 1941; and Osgood, 1943).

COMMENTS: Geographic variation among presumably native Chinese populations reported by Wu (1982). Samples from native Asian, free-living introduced, and laboratory populations were the subjects of numerous morphological (e.g., Bugge, 1970; Greene, 1935), physiological, chromosomal and molecular (many are summarized in Yosida, 1980, and Levan et al., 1990) studies, which have produced among the mass of data a gene map of *R. norvegicus* (Levan et al., 1990), and results of attempted hybridizations between *R. norvegicus* and different forms of *R. rattus* (summarized by Yosida, 1980). Review of European populations provided by Becker (1978*b*); overall review of systematics reported by Milyutin (1990). Phylogenetic relationships to other members of subgenus *Rattus* equivocal (e.g., contrast Chan et al., 1979, with Pasteur et al., 1982), but morphological and biochemical data indicate significant phylogenetic divergence between *R. norvegicus* and *R. rattus* (Baverstock et al., 1983*a*, *b*, 1986; Chan, 1977). Phallic morphology of Chinese samples described by Yang and Fang (1988) in context of assessing phylogenetic relationships among Chinese murines.

Rattus novaeguineae Taylor and Calaby, 1982. *In* Taylor et al., Bull. Am. Mus. Nat. Hist., 173:259

TYPE LOCALITY: Papua New Guinea, Kalolo Creek, 1070 m.

DISTRIBUTION: Recorded only from Papua New Guinea "from Kassam westward to Karimui in the north, and southward to Koranga, at altitudes ranging from 740 to 1525 m" (Taylor et al., 1982:259; see map on p. 258).

Rattus osgoodi Musser and Newcomb, 1985. Am. Mus. Nat. Hist., 2814:18.

TYPE LOCALITY: S Vietnam, Tuyen Duc Prov., Langbian Peak, 5000 ft.

DISTRIBUTION: Highlands (3000-6000 ft) of southern Vietnam.

COMMENTS: Morphologically and probably phylogenetically related to *R. losea*.

Rattus palmarum (Zelebor, 1869). Reise Oesterr. Fregatte Novara. Zool., (Wirbelthiere), Säugeth., p. 26.

TYPE LOCALITY: India, Nicobar Isls, probably Carr Nicobar (Dr. K. Bauer, Naturhistorisches Museum Wien, in litt.).

DISTRIBUTION: Nicobar Isls, probably Carr Nicobar.

SYNONYMS: *novarae, palmarum*.

COMMENTS: Known only by the few specimens in the original series. A distinctive species (Musser and Heaney, 1985; Musser and Newcomb, 1983) that may be related to the *R. tiomanicus* complex (Musser, 1986). Both synonyms are Fitzinger names dating from 1861 and are *nomina nuda* (see Miller, 1902:759).

Rattus pelurus Sody, 1941. Treubia, 18:308.

TYPE LOCALITY: Indonesia, Kepulauan Banggai, Pulau Peleng (1°23'S, 123°14'E), east of Sulawesi.

DISTRIBUTION: Known only from Peleng Isl.

COMMENTS: Originally described by Sody as a subspecies of *R. foramineus*, but provisionally treated as an insular population of *R. xanthurus* by Musser (1984), and finally reviewed as a distinctive species by Musser and Holden (1991), who also placed it in the *R. xanthurus* group.

Rattus praetor (Thomas, 1888). Ann. Mag. Nat. Hist., [ser. 1], 2:158.

TYPE LOCALITY: Solomon Isls, Guadalcanal Isl, Aola.

DISTRIBUTION: New Guinea; on the mainland found from Vogelkop throughout both sides of Central Cordillera in Irian Jaya and lowlands of N New Guinea to about the Sepik-Ramu drainage. Also occurs on Admiralty and Solomon Isls and in the Bismarck Arch. on Umboi, New Britain, and New Ireland (Flannery and White, 1991; Taylor et al., 1982).

SYNONYMS: *bandiculus, coenorum, mediocris, purdiensis, sansapor, tramitius, utakwa.*

COMMENTS: In Tate's (1951) monograph, this species was listed as a subspecies of *R. ruber*; the holotype of *ruber* is an example of the introduced *R. nitidus* (Taylor et al., 1983, summarized the taxonomic history).

Rattus ranjiniae Agrawal and Ghosal, 1969. Proc. Zool. Soc. Calcutta, 22:41.

TYPE LOCALITY: SW Indian Peninsula, India, Kerala State, Trivandrum.

DISTRIBUTION: Known only from the type locality and Trichur, also in Kerala.

COMMENTS: Represented by the four specimens taken at the type locality and two others collected at Trichur. Most information about the species is meager and contained in the original description. Our study of two paratypes kindly loaned to us by Dr. S. Chakraborty revealed that *R. ranjiniae* is characterized by large claws relative to body size, very long and slender hind feet, large body size, long molar rows, small bullae, narrow incisive foramina, and a short bony palate that does not extend past the third molars. These traits combine in a morphology that is uniquely distinct compared with all other species now placed in *Rattus*. Phylogenetic relationships of *ranjiniae* are unknown; possibly this species should be removed from *Rattus*.

Rattus rattus (Linnaeus, 1758). Syst. Nat., 10th ed., 1:61.

TYPE LOCALITY: Sweden, Uppsala County, Uppsala.

DISTRIBUTION: Native to Indian Peninsula, and introduced worldwide in the tropics and temperate zone (Becker, 1978*a*; de Roguin, 1991; Dieterlen, 1979; Duplantier et al., 1991*b*; Johnson, 1962*a*, *b*; Niethammer, 1975; Taylor and Horner, 1973; Taylor et al., 1982; Twigg, 1992; Yosida, 1980; Yosida et al., 1985).

SYNONYMS: *aethiops, albiventer, albus, alexandrino-rattus, alexandrinus, arboreus, arboricola, asiaticus, ater, atratus, atridorsum, auratus, beccarii, brookei, brunneusculus, caeruleus, ceylonus, chionogaster, coquimbensis, crassipes, cyaneus, doboensis, domesticus, doriae, erythronotus, flavescens, flaviventris, frugivorus, fuliginosus, fulvaster, fuscus, gangutrianus, girensis, griseocaeruleus, indicus* (Desmarest, 1822, not Bechstein, 1800), *infralineatus* (*nomen nudum*), *intermedius, jujensis, jurassicus, kandianus, kandiyanus, kelaarti, kijabius, latipes, leucogaster, muansae, narbadae, nemoralis, nericola, osorninus, picteti, rattiformis, rattoides, rufescens, ruthenus, saltuum, samharensis, satarae, siculae, subcaeruleus, subrufus, sueirensis, sylvestris, tectorum, tetragonurus, tettensis, tompsoni, variabilis, varius* (see Allen, 1939; Ellerman and Morrison-Scott, 1951; Mahoney and Richardson, 1988; Osgood, 1943; Schlitter and Thonglongya, 1971; Taylor et al., 1982).

COMMENTS: Numerous cytogenetic studies focusing on the *R. rattus* complex, summarized by Baverstock et al. (1983*c*), Bekasova and Mezhova (1983), Niethammer (1975), and Yosida (1980), have revealed the complex to consist of two basic groups of populations. The Oceanian or European type has 2N=38 (40 in some), the Asian type is characterized by 2N=42; the two are also distinguished by biochemical features (Baverstock et al., 1983*c*) as well as morphological traits (Schwabe, 1979). Where the Asian type is indigenous, the Oceanian form is restricted to ports or on ships in harbor. Both chromosomal kinds apparently occur together without evidence of interbreeding on the Polynesian island of Fiji (Yosida et al., 1985), but do hybridize in the laboratory (usually producing sterile offspring) and on the South Pacific islands of Chichijima and Eniwetok (with apparent introgression). The biological status of the two kinds were best summarized by Baverstock et al. (1983*c*:978), who noted that if "the chromosomal, electrophoretic and laboratory hybridization data are considered together, it seems that the 2n=38 and 2n=42 forms are best considered as incipient species. Where they meet, they may introgress, become sympatric without interbreeding or one may replace the other depending upon the prevailing biological conditions," a view earlier espoused by Capanna (1974). *Rattus rattus* is the name for the 2N=38/40 group, and we list it as a species separate from the 2N=42 form, for which the oldest name is *R. tanezumi* (see that account).

Rattus sanila Flannery and White, 1991. Nat. Geog. Res. Explor., 7:102.
TYPE LOCALITY: Bismarck Arch., New Ireland, Balof, site 2.
DISTRIBUTION: Apparently endemic to New Ireland.
COMMENTS: Represented only by subfossil fragments dated at 3000 before present and older, but may still occur in primary forest, which has not been adequately sampled. Flannery and White (1991) described *sanila* as a subspecies of *R. mordax*, but most of the dental measurements of *sanila* exceed and do not overlap those of even the largest known *mordax*, suggesting the former to be a separate species, which even Flannery and White acknowledged.

Rattus sikkimensis Hinton, 1919. J. Bombay Nat. Hist. Soc., 26:394.
TYPE LOCALITY: India, Sikkim, Pashok, 3500 ft.
DISTRIBUTION: S China (incl. islands of Hong Kong and Hainan), Vietnam, Laos, Cambodia, Thailand (incl. Koh Klum off SE Thailand in the Gulf of Siam), C and N Burma, NE India (Sikkim, Darjeeling, Bhutan), and possibly E Nepal. Not recorded on the mainland of peninsular Thailand south of Isthmus of Kra (10°30'N), but occurs on four islands (Koh Tau, Koh Pangan, Koh Samui, and Koh Kra) off the coast well south of the Isthmus (see map in Musser and Heaney, 1985).
SYNONYMS: *hainanicus, klumensis, koratensis, kraensis, remotus, yaoshanensis*.
COMMENTS: Originally described as a subspecies of *R. rattus* (Hinton, 1919), then listed as a synonym of *R. r. brunneusculus* (Ellerman, 1961). South Vietnamese samples discussed under *R. sladeni* (Van Peenen et al., 1969), N Vietnamese series under *R. koratensis* (Dao, 1985), and Thai samples under *R. koratensis* and *R. remotus* (J. T. Marshall, Jr., 1977*a*).

Rattus simalurensis (Miller, 1903). Proc. U.S. Natl. Mus., 26:458.
TYPE LOCALITY: Indonesia, Sumatra, Pulau Simeulue (Simalur).
DISTRIBUTION: Simalur Isl and nearby islands of Siumat, Lasia, and Babi.
SYNONYMS: *babi, lasiae* (see Musser and Heaney, 1985).
COMMENTS: The form *simalurensis* and its two synonyms were each listed as a separate subspecies of *R. rattus* by Chasen (1940), but *simalurensis* is distinct from *R. rattus* and in morphology represents slightly larger-bodied island variants of *R. tiomanicus* (Musser, 1986; Musser and Califia, 1982; Musser and Heaney, 1985). Whether populations on Simalur islands are distinct species or insular representatives of *R. tiomanicus* occurring off the margin of the continental shelf will have to be determined by systematic revision of the *R. tiomanicus* complex.

Rattus sordidus (Gould, 1858). Proc. Zool. Soc. Lond., 1857:242 [1858].
TYPE LOCALITY: Australia, Queensland, open plains Darling Downs = locality of lectotype (see Mahoney and Richardson, 1988).
DISTRIBUTION: Australia; E coast from the tip of Cape York to NE New South Wales, and some off-shore islands (see Watts and Aslin, 1981:239). New Guinea; lowlands south of Central Cordillera from Dobodura in E Papua New Guinea west and north to Koembe in Irian Jaya (see map in Taylor et al., 1983:265).
SYNONYMS: *aramia, brachyrhinus, bunae, conatus, gestri, gestroi, youngi*.
COMMENTS: One of the two species of native *Rattus* in the New Guinea-Australian region that occurs on both land masses. On morphological evidence, Taylor and Horner (1973) arranged *villosissimus* and *colletti* as subspecies of *R. sordidus*. Later evaluations, however, based on chromosomal, biochemical, and hybridization data, suggested the three should be viewed as separate species in the same monophyletic cluster (Baverstock et al., 1977*d*, 1983*a*, 1986), which is the way they are treated in current faunal accounts and catalogs (Mahoney and Richardson, 1988; Watts and Aslin, 1981).

Rattus steini Rümmler, 1935. Z. Säugetierk., 10:115.
TYPE LOCALITY: New Guinea, Irian Jaya, Weyland Range, Kunupi Mt.

DISTRIBUTION: Mid-montane elevations along Central Cordillera of Irian Jaya and Papua New Guinea, as well as highlands along the N coast and on Huon Peninsula (see map in Taylor et al., 1982:243).

SYNONYMS: *baliemensis, foesteri, hageni, rosalinda*.

COMMENTS: Until Taylor et al.'s (1982) revision, the distinctness of *steini* was obscured by its historical allocation as a subspecies of either *leucopus, mordax, ringens, ruber,* or *verecundus*.

Rattus stoicus (Miller, 1902). Proc. U.S. Natl. Mus., 24:759.

TYPE LOCALITY: India, Andaman Isls, Henry Lawrence Isl.

DISTRIBUTION: The islands of Henry Lawrence, Little Andaman, and South Andaman in the Andaman Arch. (Musser and Newcomb, 1983).

SYNONYMS: *rogersi, taciturnus* (see Musser and Heaney, 1985).

COMMENTS: Despite past confusion with *Sundamys muelleri, R. stoicus* is defined by a unique set of derived and primitive features, and is endemic to the Andaman Isls. (Musser and Newcomb, 1983). Its closest phylogenetic ally has yet to be determined (Musser, 1986).

Rattus tanezumi Temminck, 1844 *In* Siebold, Temminck, and Schlegel, Fauna Japonica, Arnz et Socii, Lugduni Batavorum, p. 51.

TYPE LOCALITY: Japan, possibly from near Nagasaki on Kyushu Isl (see Jones and Johnson, 1965).

DISTRIBUTION: Apparently indigenous to SE Asia, from E Afghanistan through highlands of Nepal and N India into S and C China (incl. Hainan Isl), Korea, and mainland Indochina (incl. offshore islands) south to Isthmus of Kra; also probably native to Mergui Arch., Andaman Isls, and some of the Nicobar Isls; also in SW peninsular India. Whether native or introduced to Taiwan and Japan is unknown (but see Yosida and Harada, 1985). Most likely introduced to the Malay Peninsula and islands on the Sunda Shelf (Medway and Yong, 1976) and nearby archipelagos just off of the Shelf, including the Mentawais (Musser and Califia, 1982; Musser and Newcomb, 1983). Certainly introduced to the Cocos-Keeling Isls (Musser and Califia, 1982), the Philippines (Musser, 1977*a*), Sulawesi (Musser and Holden, 1991), and numerous islands east through the Moluccas and Nusa Tenggara (Musser, 1970*a*, 1972, 1981*c*) to W New Guinea (Sody, 1941), and farther east through Micronesia to islands of Eniwetok and Fiji (Johnson, 1962*a*, *b*), but not to the Samoas where *R. rattus* occurs (Yosida et al., 1985).

SYNONYMS: *alangensis, amboinensis, andamanensis, argyraceus, auroreus, barussanoides, benguetensis, bhotia, brevicaudus* Chakraborty, 1975, *brevicaudus* Kuroda, 1952; *brunneus, bullocki, burrulus, coloratus, dammermani, dentatus, diardii, exsul, flavipectus, flebilis, fortunatus, germaini, griseiventer, holchu, insulanus, kadanus, keelingensis, kelleri, khyensis, kramensis, lalolis, lanensis, longicaudus, lontaris, macmillani, makassarius, makensis, mansorius, masaretes, mesanis, mindanensis, moheius, molliculus, moluccarius, neglectus, nezumi, obiensis, ouangthomae, palelae, palembang, panjius, pannellus, pannosus, pelengensis, pipidonis, poenitentiarii, portus, povolnyi, pulliventer, rangensis, robiginosus, robonsoni, robustulus, samati, santalum, sapoensis, septicus, sladeni, sumbae, tablasi, talaudensis, tatkonensis, thai, tikos, tistae, toxi, turbidus, yeni, yunnanensis, wroughtoni, zamboangae* (see Ellerman, 1941; Ellerman and Morrison-Scott, 1951; Johnson, 1962*b*; Jones and Johnson, 1965; Laurie and Hill, 1954; Musser, 1970*a*, 1972, 1977*a*; Musser and Califia, 1982; Musser and Holden, 1991; Musser and Newcomb, 1983, 1985; Sody, 1941).

COMMENTS: The authority is usually cited as 1845, but was published in 1844; see Holthuis and Sakai (1970). The name *tanezumi* is the oldest for the 2N=42 group of Asian houserats that is distinguished from the 2N=38/40 *R. rattus* not only by chromosomal characters but also by morphological and biochemical traits (see account of *R. rattus*). The indigenous range is generally north and east of peninsular India, but in Mysore State of SW India, the two chromosomal types (*wroughtoni*, 2N=42, and *rufescens*, 2N=38) occur together (Lakhotia et al., 1973; Niethammer, 1975). Nature of the specific relationship between *R. rattus* and *R. tanezumi* on the Indian subcontinent where indigenous ranges are either parapatric or overlap still requires resolution. Phallic morphology of Chinese *flavipectus* described by Yang and Fang (1988) in context of assessing phylogenetic relationships among Chinese murines.

Rattus tawitawiensis Musser and Heaney, 1985. Am. Mus. Novit., 2818:5.
TYPE LOCALITY: Philippines, Sulu Arch., Tawitawi Isl, Batu Batu.
DISTRIBUTION: Tawitawi Island in southern part of Sulu Arch.
COMMENTS: Known only by three specimens. Phylogenetic affinities unclear; in some traits resembles Sulawesian *R. hoffmanni* more closely than any other described species of *Rattus* from the Indo-Australian region (Musser and Heaney, 1985).

Rattus timorensis Kitchener, Aplin, and Boeadi, 1991. Rec. West. Aust. Mus., 15:446.
TYPE LOCALITY: Indonesia, Nusa Tenggara, Timor, Gunung Mutis, 1900 m, 7 km E Desa Nenas.
DISTRIBUTION: Known only from the type locality.
COMMENTS: Represented only by the holotype. Some of the large series of subfossil fragments collected in E Timor by Glover (1986) may be this species. Phylogenetic affinities unknown, and possibly not even a member of *Rattus* (Kitchener et al., 1991*b*).

Rattus tiomanicus (Miller, 1900). Proc. Washington Acad. Sci., 2:212.
TYPE LOCALITY: Malaysia, Pahang, Tioman Isl, off the east coast Malay Peninsula.
DISTRIBUTION: Endemic to the Sunda Shelf and some offshore islands. Records on the Shelf are from peninsular Thailand south of Isthmus of Kra (10°30'N), the Malay Peninsula, Sumatra, Java, Bali, Borneo, Palawan, and many smaller islands. Off the Sunda Shelf, *R. tiomanicus* is documented from Enggano Isl, southwest of Sumatra, and Maratua Arch., east of Borneo (Musser and Calafia, 1982; Musser and Heaney, 1985).
SYNONYMS: *ambersoni, banguei, batin, blangorum, delirius, ducis, generatius, jalorensis, jarak, jemuris, julianus, kabanicus, kunduris, lamucotanus, larusius, luxuriosus, maerens, mangalumis, mara, pauper, payanus, pemanggis, perhentianus, pharus, piperis, rhionis, roa, roquei, rumpia, sabae, sebasianus, siantanicus, sribuatensis, tambelanicus, tenggolensis, terutavensis, tingius, tua, vernalus, viclana.*
COMMENTS: Reviewed by Musser and Califia (1982), who also summarized and provided references documenting the incorrect historical association of *tiomanicus* and the other synonyms listed here as subspecies of *R. rattus*. They also pointed out that a careful study of interisland variation among named forms of the *R. tiomanicus* complex is necessary before relationships among the insular populations can be discerned; more than one species, for example, may be represented in what is now viewed as *R. tiomanicus*.

Rattus mindorensis from Mindoro Island in the Philippines, *R. simalurensis* from the islands of Babi, Lasia, Siumat, and Simalule, off the northwest coast of Sumatra; and *R. burrus* from some of the Nicobar islands are also morphologically very similar to the *R. tiomanicus* complex and should be considered part of it (Musser, 1986; Musser and Heaney, 1985). Whether they are species or island forms of *R. tiomanicus* has yet to be determined.

Rattus tunneyi (Thomas, 1904). Novit. Zool., 11:223.
TYPE LOCALITY: Australia, Northern Territory, Mary River.
DISTRIBUTION: Australia; NE and SW Western Australia, Northern Territory, E Queensland, and NE New South Wales. Also recorded from offshore islands. Extant range vastly reduced from former distribution (see map in Taylor and Horner, 1973:89, and discussion in Watts and Aslin, 1981).
SYNONYMS: *apex, austrinus, culmorum, dispar, melvilleus, vallesius, woodwardi.*
COMMENTS: This species hybridized in the laboratory with *R. colletti* (Baverstock et al., 1983*a*, 1986).

Rattus turkestanicus (Satunin, 1903). Ann. Mus. Zool. Acad. Imp. Sci. St. Petersbourg, 7:588.
TYPE LOCALITY: Kirghizia, Oshskaya Obl., Lenniskii p-h, Arslanbob (= "Assam-bob"; see Pavlinov and Rossolimo, 1987).
DISTRIBUTION: Records are from Kirghizia, NE Iran, N and E Afghanistan, N Pakistan, N India (Kashmir, Sikkim), Nepal, and S China (Yunnan and Guangdong) (see Musser and Newcomb, 1985:15).
SYNONYMS: *celsus, gilgitianus, khumbuensis, rattoides* (Hodgson, 1845, not Pictet and Pictet, 1844), *shigarus, vicerex* (see Corbet, 1978*c*; Musser and Newcomb, 1985).

COMMENTS: Despite continued use of *rattoides* for this complex (Caldarini et al., 1989; Corbet, 1978*c*), it is a synonym of *R. rattus* (Schlitter and Thonglongya, 1971). Three distinctive morphological, chromosomal, and geographic forms are included under *turkestanicus* (Caldarini et al., 1989; Niethammer and Martens, 1975) and were first recognized by Hinton (1922) who treated all three as species (*R. turkestanicus*, *R. vicerex*, and *R. rattoides*) but also suggested that each may instead be a well-differentiated subspecies. *Rattus turkestanicus* and *R. vicerex* were reported to occur sympatrically in Kashmir (Chakraborty, 1983), but those identifications have to be verified. A careful systematic treatment is needed to determine whether the three groups represent species or geographic variants.

Oldest name for the complex is *pyctoris* (Hodgson, 1845; incorrectly listed as a synonym of *R. nitidus* by Ellerman, 1961) and would either replace *turkestanicus* if all samples represent a single species, or would identify Nepal and Sikkim populations.

Rattus villosissimus (Waite, 1898). Proc. R. Soc. Victoria, 10:125.

TYPE LOCALITY: Australia, Queensland, "from probably the vicinity of Goonhaghooheeny Billabong, Cooper Creek" (see Mahoney and Richardson, 1988).

DISTRIBUTION: Australia; broad inland range from NW Western Australia through Northern Territory into most of Queensland and N South Australia and N New South Wales (see map in Watts and Aslin, 1981:245).

SYNONYMS: *longipilis* (Gould, 1854, not Waterhouse, 1837), *profusus*.

COMMENTS: Geographic range is allopatric to the coastal *R. sordidus* in Queensland and *R. colletti* in Northern Territory (see map in Taylor and Horner, 1973:72). The three species are closely related; *villosissimus* was treated as a subspecies of *R. sordidus* by Taylor and Horner (1973), but is considered genically closer to *colletti* by Baverstock et al. (1983*a*, 1986). See accounts of *sordidus* and *colletti*. Analyses of electrophoretic data by Gemmeke and Niethammer (1984) indicated *R. villosissimus* to be greatly separated from *R. argentiventer*, *R. exulans*, *R. norvegicus*, and *R. tiomanicus*, and closer to species of *Bandicota* and *Maxomys*.

Rattus xanthurus (Gray, 1867). Proc. Zool. Soc. Lond., 1867:598.

TYPE LOCALITY: Indonesia, NE Sulawesi, Tondano, 3600 ft.

DISTRIBUTION: Sulawesi; northern arm, central core, and SE peninsula. Absent from the SW peninsula where it is replaced by *R. foramineus* in coastal lowlands and *R. bontanus* on Gunung Lampobatang. Occurs sympatrically with *R. marmosurus* on the NE Peninsula.

SYNONYMS: *faberi*, *facetus*, *orientalis*, *paraxanthus*, *salocco* (see Musser, 1971*c-e*, 1984).

COMMENTS: Member of the group which includes *R. bontanus*, *R. foramineus*, *R. marmosurus*, and *R. pelurus*. Sody (1941) included *xanthurus* in *Taeromys*. Although the *R. xanthurus* group may eventually be removed from *Rattus* (Musser and Holden, 1991), it does not belong in *Taeromys*.

Rhabdomys Thomas, 1916. Ann. Mag. Nat. Hist., ser. 8, 18:69.

TYPE SPECIES: *Mus pumilio* Sparrman, 1784.

Rhabdomys pumilio (Sparrman, 1784). K. Svenska Vet.-Akad. Handl. Stockholm, p. 236.

TYPE LOCALITY: S Africa, S Cape Prov., east of Knysna, Tsitsikamma Forest, Slangrivier.

DISTRIBUTION: S and C Angola, Namibia, Botswana, South Africa, E Zimbabwe, WC Mozambique, Malawi (Nyika Plateau and Mulanje massif), NE Zambia (Nyika plateau), and highlands in Tanzania, Kenya, Uganda, and SE Zaire.

SYNONYMS: *angolae*, *bechuanae*, *bethuliensis*, *chakae*, *cinereus*, *cradockensis*, *deserti*, *dilectus*, *diminutus*, *donavani*, *fouriei*, *griquae*, *griquoides*, *intermedius*, *lineatus*, *pumilio* var. *major*, *meridionalis*, *moshesh*, *namaquensis*, *namibensis*, *nyasae*, *orangiae*, *prieskae*, *septemvittatus*, *typicus*, *vaalensis*, *vittatus*.

COMMENTS: When Wroughton (1905*b*) reviewed *pumilio*, he distinguished four groups, each with different forms, and although unsure about the taxonomic status to give the forms, thought each group represented a separate species. Distribution of character variation, however, forced him to conclude that (p. 630) "in view of the absolute identity of pattern, the variability of coloration, and the difficulty of deciding the inter-relationship of the different forms, the simpler and safer way is to call them all subspecies of the original species *pumilio*." Wroughton's view prevails today.

Checklists (Allen, 1939; Ellerman, 1941; Ellerman et al., 1953), faunal studies (e.g., Ansell, 1978; Ansell and Dowsett, 1988; Roberts, 1951; Skinner and Smithers, 1990; Smithers, 1971), a study of possible influences of climate on length of tail (Coetzee, 1970), and preliminary studies on geographic variation (see Meester et al., 1986:275) have yet to critically analyze patterns of chromatic and morphological variation in context of assessing whether only one or several species exist. Hill and Carter (1941:101) recognized two species in Angola; *R. pumilio* from central and southern regions, and *R. bechuanae* form the arid southwestern portion north of Namibia. They noted the lack of intergradation between the two kinds and we have not found any evidence that such intergradation exists. The significance of this observation can only be assessed by careful sysematic review of *Rhabdomys*.

Morphology of digestive system in relation to diet and evolution described by Perrin and Curtis (1980).

Rhynchomys Thomas, 1895. Ann. Mag. Nat. Hist., ser. 6, 16:160.

TYPE SPECIES: *Rhynchomys soricoides* Thomas, 1895.

COMMENTS: Member of the Philippine Old Endemics, phylogenetic relationships with other genera of shrew rats are ambiguous (Musser and Heaney, 1992).

Rhynchomys isarogensis Musser and Freeman, 1981. J. Mammal., 62:154.

TYPE LOCALITY: Philippines, Camarines Sur Prov., SE Peninsula of Luzon Isl, Mt Isarog, 5000ft.

DISTRIBUTION: Known only from Mt. Isarog, Luzon, in montane forest formations at 1125 m and above (Rickart et al., 1991).

Rhynchomys soricoides Thomas, 1895. Ann. Mag. Nat. Hist., ser. 6, 16:160.

TYPE LOCALITY: Philippines, N Luzon Isl, Mountain Prov., Mt Data, 8000 ft.

DISTRIBUTION: Known only from Mount Data but probably occurs elsewhere in montane forest formations in N Luzon.

Solomys Thomas, 1922. Ann. Mag. Nat. Hist., ser. 9, 9:261.

TYPE SPECIES: *Uromys sapientis* Thomas, 1902.

SYNONYMS: *Unicomys*.

COMMENTS: Member of the New Guinea region Old Endemics (Musser, 1981*c*). Included in *Melomys* by Ellerman (1941:226), then transferred to *Uromys* by Tate (1951:312), but finally recognized again as a distinct genus (Flannery and Wickler, 1990; Laurie and Hill, 1954:128). Presumably related to *Uromys* and *Melomys*, but no adequate diagnosis and definition of the genus and assessment of its phylogenetic relationships is yet available.

Solomys ponceleti (Troughton, 1935). Rec. Aust. Mus., 19:260.

TYPE LOCALITY: Solomon Isls, Bougainville Isl, about 10 mi inland from Buin.

DISTRIBUTION: Solomon Isls; endemic to islands of Buka, Bougainville, and Choiseul (Flannery and Wickler, 1990; Flannery, in litt.).

STATUS: IUCN - Endangered.

COMMENTS: Known by a few extant specimens and archaeological fragments (Flannery et al., 1988; Flannery and Wickler, 1990).

Solomys salamonis (Ramsay, 1883). Proc. Linn. Soc. N.S.W., 7:43.

TYPE LOCALITY: Solomon Isls, Florida Isl in the Nggela Group (see Flannery and Wickler, 1990:13).

DISTRIBUTION: Endemic to Florida Isl in the Solomon Arch. (Flannery and Wickler, 1990).

COMMENTS: Originally described as a *Mus*, the species was transferred to *Uromys* (Rümmler, 1938; Tate, 1951), but then correctly placed in *Solomys* by Troughton (1936). Known by a few extant specimens. *S. salamonis* is also recorded from Nggela Island (Flannery and Wickler, 1990), but Flannery (in litt.) claimed the species is known only from the skull of the holotype.

Solomys salebrosus Troughton, 1936. Rec. Aust. Mus., 19:436

TYPE LOCALITY: Solomon Isls, Bougainville Isl.

DISTRIBUTION: Endemic to islands of Buka, Bougainville, and Choliseul in Solomon Arch.

COMMENTS: Known by extant specimens and archaeological fragments (Flannery and Wickler, 1990).

Solomys sapientis (Thomas, 1902). Ann. Mag. Nat. Hist., ser. 7, 9:44.
TYPE LOCALITY: Solomon Isls, Santa Ysabel Isl.
DISTRIBUTION: Endemic to Ysabel Isl (Flannery and Wickler, 1990).
COMMENTS: Represented by extant specimens.

Solomys spriggsarum Flannery and Wickler, 1990. Aust. Mamm., 13:133.
TYPE LOCALITY: Solomon Isls, Buka Isl, Kilu rockshelter.
DISTRIBUTION: Endemic to Buka Isl.
COMMENTS: Known only by subfossil archaeological material (Flannery and Wickler, 1990).

Spelaeomys Hooijer, 1957. Zool. Meded. Leiden, 35:306.
TYPE SPECIES: *Spelaeomys florensis* Hooijer, 1957.
COMMENTS: Reviewed by Musser (1981*c*), who also recorded past opinions about phylogenetic affinities of *Spelaeomys*, noted that it is not closely related to *Hooijeromys*, *Komodomys*, *Papagomys*, or *Paulamys*—the other Nusa Tenggara endemics—and hypothesized that "*Spelaeomys* belongs with the old native genera of New Guinea, possibly Australia, and likely Timor." Additional data is required to test this notion.

Spelaeomys florensis. Hooijer, 1957. Zool. Meded. Leiden, 35:306.
TYPE LOCALITY: Indonesia, Nusa Tenggara, Palau Flores, Manggarai Prov., Liang Toge, a cave near Warukia, 1 km south of Lepa.
DISTRIBUTION: Known only from Flores Isl.
COMMENTS: Represented only by subfossil fragments (3000-4000 years old), but may still live on Flores, and possibly other nearby islands.

Srilankamys Musser, 1981. Bull. Am. Mus. Nat. Hist., 168:268.
TYPE SPECIES: *Rattus ohiensis* Phillips, 1929.
COMMENTS: A montane insular relict with possible phylogenetic ties to *Chiromyscus*, *Maxomys*, and *Niviventer*, but not to *Rattus* (Musser, 1981*b*).

Srilankamys ohiensis (Phillips, 1929). Ceylon J. Sci., Sec. B, 15:167.
TYPE LOCALITY: Sri Lanka (Ceylon), Ohiya, W Haputale, 6000 ft.
DISTRIBUTION: Primary lowland evergreen tropical and montane rainforest formations on mountains in Uva and Central Provinces of Sri Lanka between 3000 and 7000 ft (Phillips, 1980).
COMMENTS: Musser (1981*b*) described the history of past and incorrect allocations of *ohiensis* to groups that were eventually recognized as the genera *Apomys*, *Lenothrix*, *Leopoldamys*, *Maxomys*, *Niviventer*, and *Rattus*.

Stenocephalemys Frick, 1914. Ann. Carnegie Mus., 9:7.
TYPE SPECIES: *Stenocephalemys albocaudata* Frick, 1914.
COMMENTS: An Ethiopian endemic that is phylogenetically closely related to the Ethiopian endemics *Myomys albipes* and *M. ruppi* (Van der Straeten and Dieterlen, 1983; see those accounts), and through them to the other species of *Myomys*.

Stenocephalemys albocaudata Frick, 1914. Ann. Carnegie Mus., 9:8.
TYPE LOCALITY: S Ethiopia, Chilalo Mtns, Inyala Camp.
DISTRIBUTION: Ethiopia; endemic to the E plateau between 3000 and 4050 m.
COMMENTS: Inhabits the Afro-alpine moorland where it occurs with *Arvicanthis blicki*, *Lophuromys melanonyx*, and *Tachyoryctes macrocephalus*, which are also moorland specialists (Rupp, 1980; Yalden, 1988).

Stenocephalemys griseicauda Petter, 1972. Mammalia, 36:171.
TYPE LOCALITY: Ethiopia, Bale Mtns, Dinsho.
DISTRIBUTION: Ethiopia; S and N highlands, 2400-3900 m (Demeter and Topal, 1982; Yalden, 1988).
COMMENTS: Another Ethiopian mountain endemic that overlaps altitudinally with *S. albocaudata* but occupies bushy areas rather than moorland (Yalden, 1988).

Stenomys Thomas, 1910. Ann. Mag. Nat. Hist., ser. 8, 6:507.
TYPE SPECIES: *Mus verecundus* Thomas, 1904.
SYNONYMS: *Nesoromys*.
COMMENTS: Flannery (1990*b*) provided photographs of animals along with distributional and biological summaries of New Guinea species. Rümmler (1938) united *Nesoromys* with *Stenomys*.

Stenomys ceramicus (Thomas, 1920). Ann. Mag. Nat. Hist., ser. 9, 6:425.
TYPE LOCALITY: Indonesia, Moluccas, Pulau Seram, Gunung Manusela, 6000 ft.
DISTRIBUTION: Endemic to montane forests of Seram Isl.
COMMENTS: Apparently morphologically related to the *S. niobe* complex of New Guinea. Originally described as a species of *Nesoromys* (Rümmler, 1938).

Stenomys niobe (Thomas, 1906). Ann. Mag. Nat. Hist., ser. 7, 17:327.
TYPE LOCALITY: Papua New Guinea, Angabunga River, Owgarra, 2750 m.
DISTRIBUTION: New Guinea, nearly all montane regions, including the Huon Peninsula (see map in Taylor et al., 1982:191).
SYNONYMS: *arfakiensis, arrogans, clarae, haymani, klossi, pococki, rufulus, stevensi*.
COMMENTS: Revised by Taylor et al. (1982), who recognized two distinct subspecies, *arrogans* and *niobe*. Flannery (1990*b*:241), however, noted that unpublished results from biochemical studies indicated that two species are present in what Taylor et al. (1982) defined as a single species, one at high altitudes, the other at lower altitudes. Chromosomal morphology presented by Dennis and Menzies (1978). Phallic anatomy discussed by Lidicker (1968).

Stenomys richardsoni (Tate, 1949). Am. Mus. Novit., 1421:1.
TYPE LOCALITY: New Guinea, Irian Jaya, near Lake Habbema, 3225 m.
DISTRIBUTION: New Guinea, Irian Jaya; high mountain slopes in W portion of Central Cordillera (see map and discussion in Taylor et al., 1982).
SYNONYMS: *omichlodes*.
COMMENTS: Inhabits grassland and tundra-like cold and wet habitats at or above limits of montane forest (Taylor et al., 1982).

Stenomys vandeuseni (Taylor and Calaby, 1982). *In* Taylor et al., Bull. Am. Mus. Nat. Hist., 173:211.
TYPE LOCALITY: Papua New Guinea, Maneau Range, N slope Mt Dayman, Middle Camp, 1540 m.
DISTRIBUTION: Known only from the type locality.
COMMENTS: Taylor and Calaby described *vandeuseni* as a subspecies of *verecundus* (Taylor et al., 1982), but its distinctive morphology and habitat indicate otherwise, as the describers even suggested, and as Flannery (1990*b*:245) speculated. *Stenomys verecundus* occurs on Mt Dayman in rain forest at 700 m and is replaced at 1540 m in montane oak forest by *S. vandeuseni*, a relationship that is similar to the distributions of *Leptomys elegans* (rain forest) and *L. ernstmayri* (oak forest) on Mt Dayman.

Stenomys verecundus (Thomas, 1904). Novit. Zool., 11:598.
TYPE LOCALITY: Papua New Guinea, Central Prov., Aroa River, Avera, 200 m.
DISTRIBUTION: New Guinea; Vogelkop and Weyland Range of Irian Jaya, Central Cordillera east of 141° to easternmost Papua New Guinea; 150 to 2750 m (see map in Taylor et al., 1982:204).
SYNONYMS: *mollis, tomba, unicolor*.
COMMENTS: Revised by Taylor et al. (1982), who recognized three distinct subspecies—*mollis, unicolor*, and *verecundus*. Chromosomal morphology discussed by Dennis and Menzies (1978).

Stochomys Thomas, 1926. Ann. Mag. Nat. Hist., ser. 9, 17:176.
TYPE SPECIES: *Dasymys longicaudatus* Tullberg, 1893.
COMMENTS: Listed as a genus by Allen (1939), but allocated to *Rattus* as a subgenus by Ellerman (1941). D. H. S. Davis (1965) placed *Stochomys* in *Aethomys* as a subgenus, which reflected Thomas's (1915) early allocation of *longicaudatus* to subgenus *Aethomys*. Misonne (1969) and Rosevear (1969), however, reinstated the generic position of *Stochomys*, correctly noting that it was distinctive and not closely related

to *Rattus*. It has been phylogenetically associated with *Dephomys* (D. H. S. Davis, 1965; Misonne, 1969), but is unrelated to that genus, as reflected by morphometric analyses (Van der Straeten, 1984), and analyses of qualitative traits. Features of pelage, skull, and dentition characterizing *Stochomys* are most similar to those found in species of *Aethomys*, and suggest the two to be members of the same monophyletic clade in which *Stochomys* is the high forest partner of Savanna *Aethomys*.

Stochomys longicaudatus (Tullberg, 1893). Nova Acta Reg. Soc. Sci. Upsala, ser. 3, 16:36.
TYPE LOCALITY: Cameroon.
DISTRIBUTION: Tropical evergreen forest; recorded from Togo, S Nigeria, Central African Republic, Cameroon, Gabon, Zaire, and Uganda.
SYNONYMS: *hypoleucus* (Pucheran, 1855, not Sundevall), *ituricus*, *sebastianus*.
COMMENTS: Rosevear (1969) described the historical allocations of *longicaudatus* to *Aethomys*, *Dasymys*, *Epimys*, *Mus*, *Rattus*, and *Stochomys*. Morphometric analyses, which resulted in distinguishing two distinct subspecies, reported by Van der Straeten (1984).

Sundamys Musser and Newcomb, 1983. Bull. Am. Mus. Nat. Hist., 174:401.
TYPE SPECIES: *Mus mülleri* Jentink, 1879.
COMMENTS: Endemic to the Sunda Shelf. Revised by Musser and Newcomb (1983).

Sundamys infraluteus (Thomas, 1888). Ann. Mag. Nat. Hist., ser. 6, 2:409.
TYPE LOCALITY: Malaysia, Sabah (N Borneo), Kinabalu.
DISTRIBUTION: Gunung Kinabalu and Gunung Trus Madi in Sabah; W mountain chain of Sumatra.
SYNONYMS: *atchinus*.
COMMENTS: A montane species related to *S. muelleri*, which occurs at middle elevations and lowlands.

Sundamys maxi (Sody, 1932). Natuurh. Maandbl. Maastricht., 21:157.
TYPE LOCALITY: Indonesia, Java, Bandung, Cibuni.
DISTRIBUTION: Known only from W Java between 900 and 1350 m.
COMMENTS: A distinctive species and member of the suite of endemic Javanese murines (Musser, 1986). Still only represented by specimens collected between 1932 and 1935 by Max Bartels Jr. Phylogenetic relationship may be closer to *S. muelleri* than to *S. infraluteus*.

Sundamys muelleri (Jentink, 1879). Notes Leyden Mus., 2:16.
TYPE LOCALITY: Indonesia, W Sumatra, Batang Singgalang, Padang Highlands.
DISTRIBUTION: Sunda Shelf only: SW peninsular Burma, peninsular Thailand, Malay Peninsula, Sumatra, Borneo, Palawan, and many smaller islands on the Shelf; records from islands off of Sunda Shelf (Nicobars, for example) proved to represent other species (see maps in Musser and Newcomb, 1983).
SYNONYMS: *balabagensis*, *balmasus*, *borneanus*, *campus*, *chombolis*, *crassus*, *credulus*, *culionensis*, *domitor*, *firmus*, *foederis*, *integer*, *otiosus*, *pinatus*, *pollens*, *potens*, *sebucus*, *terempa*, *valens*, *validus*, *victor*, *virtus*, *waringensis*.
COMMENTS: Morphological, chromosomal, and spermatozoal data suggested *S. muelleri* is a distant relative of *Rattus* but in the same monophyletic clade (Breed and Yong, 1986; Musser and Newcomb, 1983). Populations from SW Burma, Peninsular Thailand, and the Malay Peninsula are significantly larger in body size than those from elsewhere on the Sunda Shelf and may be different species (Musser and Newcomb, 1983).

Taeromys Sody, 1941. Treubia, 18:260.
TYPE SPECIES: *Mus* (*Gymnomys*) *celebensis* Gray, 1867.
SYNONYMS: *Arcuomys*.
COMMENTS: A Sulawesi endemic. Formerly included in *Rattus* by Ellerman (1949*a*), and in subgenus *Bullimus* of genus *Rattus* by Misonne (1969), but considered a distinct genus by Musser (1984, 1987) and Musser and Newcomb (1983). Partially reviewed and contrasted with Sundaic *Sundamys* by Musser and Newcomb (1983). Marked interspecific contrasts in spermatozoal morphology evident among species and reflected differences in cranial and dental traits (Breed and Musser, 1991).

Taeromys arcuatus (Tate and Archbold, 1935). Am. Mus. Novit., 802:9.
TYPE LOCALITY: Indonesia, SE Sulawesi, Pegunnungan Mekongga, Tanke Salokko, 1500 m.
DISTRIBUTION: Known only from the type locality.
COMMENTS: Closest relative is the population (as yet unnamed and undescribed) in C Sulawesi (Musser and Holden, 1991). Sody (1941) proposed *Arcuomys*, a *nomen nudum*, for this species.

Taeromys callitrichus (Jentink, 1878). Notes Leyden Mus., p. 12.
TYPE LOCALITY: Indonesia, NE Sulawesi, Kakas.
DISTRIBUTION: Sulawesi: recorded from a few places on the NE peninsula, in the central core, and on the SE peninsula (Musser, 1970*d*; Musser and Holden, 1991).
SYNONYMS: *jentinki, maculipilis, microbullatus*.
COMMENTS: Reviewed by Musser (1970*d*). Most closely related to *T. arcuatus*.

Taeromys celebensis (Gray, 1867). Proc. Zool. Soc. Lond., 1867:598.
TYPE LOCALITY: Indonesia, NE Peninsula, Manado.
DISTRIBUTION: Sulawesi: recorded from throughout island (Musser and Holden, 1991).
COMMENTS: Recorded from SW peninsula only by subfossil fragment (Musser, 1984), but from elsewhere by extant specimens. Morphologically distant from other species in the genus (Breed and Musser, 1991; Musser and Newcomb, 1983).

Taeromys hamatus (Miller and Hollister, 1921). Proc. Biol. Soc. Washington, 34:96.
TYPE LOCALITY: Indonesia, C Sulawesi, Gunung Lehio, 6000 ft.
DISTRIBUTION: Sulawesi: known only from a few montane localities in the central core of the island.
COMMENTS: Closest relative is *T. taerae*.

Taeromys punicans (Miller and Hollister, 1921). Proc. Biol. Soc. Washington, 34:98.
TYPE LOCALITY: Indonesia, C Pinedapa, 100 ft.
DISTRIBUTION: Sulawesi: known only from the central part and the SW peninsula.
COMMENTS: A lowland species represented by two modern specimens from the type locality and a few subfossil fragments from the SW peninsula (Musser, 1984).

Taeromys taerae (Sody, 1932). Natuurh. Maandbl. Maastricht., 21:158.
TYPE LOCALITY: Indonesia, NE Sulawesi, Lembean, near Tondano.
DISTRIBUTION: Sulawesi: recorded only from highlands of the NE peninsula.
SYNONYMS: *simpsoni* (Ellerman, 1949, not Philippi, 1900); *tatei* (Sody, 1941, not Ellerman, 1941).
COMMENTS: Reviewed by Musser (1971*d*).

Tarsomys Mearns, 1905. Proc. U.S. Natl. Mus., 28:453.
TYPE SPECIES: *Tarsomys apoensis* Mearns, 1905.
COMMENTS: Revised by Musser and Heaney (1992) who recorded the history of the incorrect inclusion of *Tarsomys* in *Rattus*, and placed the genus in the group of Philippine New Endemics, although its closest relative has yet to be determined.

Tarsomys apoensis Mearns, 1905. Proc. U.S. Natl. Mus., 28:453.
TYPE LOCALITY: Philippines, S Mindanao Isl, Davao City Prov., Mt Apo, 6750 ft.
DISTRIBUTION: Known from several mountains on Mindanao above 5200 ft (Musser and Heaney, 1992).

Tarsomys echinatus Musser and Heaney, 1992. Bull. Am. Mus. Nat. Hist., 138:33.
TYPE LOCALITY: Philippines, S Mindanao, South Cotabato Prov., Mt Matutum, Tupi, Balisong, 2700-3700 ft.
DISTRIBUTION: Known only from the vicinity of type locality, but probably occurs elsewhere on Mindanao.

Tateomys Musser, 1969. Am. Mus. Novit., 2384:2.
TYPE SPECIES: *Tateomys rhinogradoides* Musser, 1969.
COMMENTS: Originally, and incorrectly, linked to the "*Rattus chrysocomus* group" by Musser (1969*b*), but morphological, spermatozoal, and ecological characteristics clearly united *Tateomys* and the other Sulawesian shrew rat *Melasmothrix* in the same monophyletic group to the exclusion of either *Rattus*, *Bunomys*, or any other

Sulawesian endemic (Breed and Musser, 1991; Musser, 1982*c*). Listed as part of the Sulawesi Old Endemics by Musser (1981*c*).

Tateomys macrocercus Musser, 1982. Bull. Am. Mus. Nat. Hist., 174:64.
TYPE LOCALITY: Indonesia, C Sulawesi, Gunung Nokilalaki, 7500 ft.
DISTRIBUTION: Known only between 6500 and 7500 ft in the upper montane rain forest on Gunung Nokilalaki, but probably occurs in other mountainous regions of at least C Sulawesi.

Tateomys rhinogradoides Musser, 1969. Am. Mus. Novit., 2384:3.
TYPE LOCALITY: Indonesia, C Sulawesi, Gunung Latimodjong, 2200 m.
DISTRIBUTION: Known from the type locality, Gunung Tokala, and Gunung Nokilalaki at high elevations in upper montane rain forest in central core of Sulawesi.

Thallomys Thomas, 1920. Ann. Mag. Nat. Hist., ser. 9, 5:141.
TYPE SPECIES: *Mus nigricauda* Thomas, 1882.
COMMENTS: After being proposed as a genus by Thomas (1920), *Thallomys* was used in checklists (e.g., Allen, 1939; Ellerman, 1941) until Ellerman et al. (1953) united it with subgenus *Aethomys* of *Rattus*. *Thallomys* was reinstated by Lundholm (1955*c*), who also suggested it was closely related to *Thamnomys*. Misonne (1969) pointed out that *Thallomys* has nothing to do with *Rattus*, and is most closely related to Tertiary european *Parapodemus*, an evaluation based on molar occlusal patterns. Although clearly unrelated to *Rattus*, *Thallomy*'s phylogenetic position within the diversity of African murines is still unresolved.

Early checklists and faunal accounts recognized several species (Ellerman, 1941; Roberts, 1951), but for last 20-30 years, *Thallomys* has been either treated as monotypic (D. H. S. Davis, 1965; Meester et al., 1986; Misonne, 1974), or as containing at least two species (Petter, 1973*a*; Skinner and Smithers, 1990). That considerable and significant morphological diversity exists in the genus has been acknowedged. D. H. S. Davis (1965), for example, recognized five groups of subspecies. Petter (1973*a*) clearly demonstrated that at least two species could be diagnosed. Regional faunal reports of southern African mammals discussed presence of two distinct kinds (Skinner and Smithers, 1990; Smithers and Wilson, 1979). Meester et al. (1986) announced the need for revision of *Thallomys*, and that both morphological and chromosomal data suggested the presence of two species (see Gordon, 1987). We list four species. Our review is based on original descriptions of taxa, Roberts' (1951) useful monograph and other regional faunal reports, large collections of specimens, and holotypes. Gastric anatomy reported by Perrin (1986).

Thallomys loringi (Heller, 1909). Smithson. Misc. Coll., 52:471.
TYPE LOCALITY: E Kenya, Lake Naivasha.
DISTRIBUTION: Known only by specimens from E Kenya, and N and E Tanzania; limits unknown.
COMMENTS: Originally described as a species of *Thamnomys* (Heller, 1909), then listed as a subspecies of *Thallomys nigricauda* (Allen, 1939; Ellerman, 1941), and finally allocated to *T. damarensis* (here included in *nigricauda*) as a subspecies (Petter, 1973*a*). *T. loringi* is morphologically and probably phylogenetically allied to *T. nigricauda*, but no critical documentation of its conspecificity with that species has ever been presented. Because *T. loringi* can be diagnosed by pelage and other distinctions, we list it as a species, a hypothesis that can be tested by careful systematic revision of the genus.

Thallomys nigricauda (Thomas, 1882). Proc. Zool. Soc. Lond., 1882:266.
TYPE LOCALITY: Namibia, Great Namaqualand, Hountop (= Hudup or Hutop) River, west of Gibeon (Meester et al., 1986).
DISTRIBUTION: Angola, Namibia, N Cape Prov. of South Africa, Botswana, and SE Zamibia; north and eastern limits unknown.
SYNONYMS: *bradfieldi, damarensis, davisi, herero, kalaharicus, leuconoe, molopensis, nitela, quissamae, robertsi.*
COMMENTS: *T. nigricauda* and *T. paedulcus* are now recognized as occurring in the Southern African Subregion by Skinner and Smithers (1990), who also summarized some of the chromosomal, morphological, and ecological distinctions between the two species. They also suggested that *herero* and *leuconoe* may represent samples of *T. paedulcus*,

but the holotypes are examples of *T. nigricauda*. Morphological and geographic definitions of *T. nigricauda* are unsatisfactory. Appreciable geographic variation in body size, length of molar row, pelage coloration, and tail pilosity exists among samples and its significance will have to be assessed by critical systematic revision.

Thallomys paedulcus (Sundevall, 1846). Ofv. K. Svenska Vet.-Akad. Forhandl. Stockholm, 3:120.

TYPE LOCALITY: South Africa, "In Caffraria interiore, prope tropicum," according to Ellerman et al. (1953). D. H. S. Davis (1965:127) reported that "type locality was in the Magaliesberg area and has been provisionally fixed as Crocodile Drift, Brits, Transvaal."

DISTRIBUTION: From S Africa (N Natal, Transvaal, Zululand), Swaziland, and S Botswana north through Zimbabwe, S Zambia, Mozambique, Malawi, Tanzania, Kenya to S Ethiopia and S Somalia; limits unknown.

SYNONYMS: *acaciae, lebomboensis, moggi, rhodesiae, ruddi, scotti, somaliensis, stevensoni, zambeziana*.

COMMENTS: In body size, the smallest of all the species. Identification of *paedulcus* as a separate species compared with the larger *T. damarensis*, and association of *scotti* with it, was correctly perceived and documented by Petter (1973*a*). Thomas and Wroughton's (1908) *ruddi*, which was described as a species of *Thamnomys*, is a *Thallomys* (Lawrence and Loveridge, 1953) and another example of *T. paedulcus*. Identification of the holotype of *paedulcus*, along with critical measurements, was recorded by Ellerman et al. (1953), and verified by Petter (1973*a*). The names *acaciae, lebomboensis*, and *stevensoni* were listed by Roberts (1951) as subspecies of *T. moggi*; *rhodesiae* was described by Osgood (1910) as a subspecies of *Mus damarensis* (which is here synonymized in *T. nigricauda*); and *zambeziana* was proposed by Lundholm (1955*b*) as a subspecies of *T. nigricauda*. Roche (1964) described *somaliensis* as a subspecies of *T. paedulcus*.

Thallomys shortridgei Thomas and Hinton, 1923. Proc. Zool. Soc. Lond., 1923:492.

TYPE LOCALITY: S Africa, NW Cape Prov., Louisvale, S bank of Orange River.

DISTRIBUTION: South Africa; known only from south bank of Orange River from about Upington west to Goodhouse, Little Namaqualand, in NW Cape Prov.; limits unknown.

COMMENTS: Thomas and Hinton's (1923) acute description pointed to a distinctive species distinguished by a diagnostic combination of chromatic and cranial (especially the small bullae) traits. Ellerman (1941) treated *shortridgei* as a species, its distinction has been recognized by others (Roberts, 1951; Van Rooyen *in* De Graff, 1978), and despite recent checklists where it is listed as a subspecies of either *T. paedulcus* (Meester et al., 1986) or *T. nigricauda* (Skinner and Smithers, 1990), the name identifies a valid species.

Thamnomys Thomas, 1907. Ann. Mag. Nat. Hist., ser. 7, 19:121.

TYPE SPECIES: *Thamnomys venustus* Thomas, 1907.

COMMENTS: Despite assertions that species of *Thamnomys* and *Grammomys* are in the same monophyletic group but separable at the subgeneric level (Allen, 1939; Hatt, 1940*a*; Hollister, 1919; Misonne, 1974; Petter and Tranier, 1975; and others—see references in Meester et al., 1986), *Thamnomys* is a distinct genus, as explained by Ellerman (1941), Hutterer and Dieterlen (1984), Misonne (1969), Rosevear (1969), and other workers (see references in Meester et al., 1986). Those systematists, however, also included *rutilans* in *Thamnomys*, but we place it in *Grammomys* pending systematic revision of *rutilans*.

Oenomys is the closest phylogenetic relative, a view presented more than fifty years ago by Hatt (1940*a*:522): "*Oenomys* and *Thamnomys* bear much resemblance to each other, and I am inclined to believe that they represent respectively semi-arboreal and arboreal descendants of a common stock."

Thamnomys kempi Dollman, 1911. Ann. Mag. Nat. Hist., ser. 8, 8:658.

TYPE LOCALITY: E Zaire, Buhamba, near Lake Kivu, 6000 ft.

DISTRIBUTION: E Zaire, W Uganda, and NW Burundi; a montane Western Rift Valley endemic.

SYNONYMS: *major.*
COMMENTS: A distinctive species as listed by Allen (1939) and Ellerman (1941), not a subspecies of *T. venustus*. Both species are sympatric at Kibati in E Zaire (documented by specimens in the Museum of Comparative Zoology, Harvard).

Thamnomys venustus Thomas, 1907. Ann. Mag. Nat. Hist., ser. 7, 19:122.
TYPE LOCALITY: Uganda, Ruwenzori East, 7000 ft.
DISTRIBUTION: E Zaire, Uganda, and Rwanda: a montane Western Rift Valley endemic.
SYNONYMS: *kivuensis, schoutedeni.*
COMMENTS: Allen and Loveridge (1942) described *kivuensis* as a subspecies of *T. venustus*.

Tokudaia Kuroda, 1943. Biogeographica, 139:61.
TYPE SPECIES: *Rattus jerdoni osimensis* Abe, 1934.
SYNONYMS: *Acanthomys* (Tokuda, 1941, not Lesson, 1842), *Tokudamys*.
COMMENTS: In addition to two species listed below, a third, yet to be named and described, occurs on Tokun-oshima Isl in the Ryukyu group (Tsuchiya, 1981; Tsuchiya et al., 1989). Best estimate of phylogenetic relationships, based on dental morphology, placed *Tokudaia* in a group containing *Apodemus*, Pliocene *Rhagapodemus*, and Quaternary *Rhagamys*; and suggested that *Tokudaia* evolved from Miocene *Parapodemus-Apodemus* ancestral stock; its evolutionary link to primitive species of *Apodemus* may be found in Miocene or Pliocene sediments of China (see Kawamura, 1989).

Tokudaia muenninki Johnson, 1946. Proc. Biol. Soc. Washington, 59:170.
TYPE LOCALITY: Japan, Ryukyu Isls, N Okinawa Isl, Hentona (western coast).
DISTRIBUTION: Known by modern specimens from Okinawa, and Late Pleistocene and Holocene samples from Okinawa and adjacent island of Le-jima (Kawamura, 1989).
STATUS: IUCN - Endangered.
COMMENTS: Originally described as a subspecies of *T. osimensis* (Johnson, 1946), chromosomal evidence, as well as external and cranial morphology, clearly distinguished *muenninki* as a separate species (Tsuchiya, 1981).

Tokudaia osimensis (Abe, 1934). J. Sci. Hiroshima Univ., ser. B, div. 1, 3:107.
TYPE LOCALITY: Japan, Ryukyu Isls, Amami-oshima Isl, village of Sumiyo.
DISTRIBUTION: Known only by modern samples from Amami-oshima Isl.
COMMENTS: Impressive chromosomal distinctions between *osimensis*, *muenninki*, and a third unnamed species from Tokun-oshima Isl reported by Tsuchiya (1981), and Tsuchiya et al. (1989).

Tryphomys Miller, 1910. Proc. U.S. Natl. Mus., 38:399.
TYPE SPECIES: *Tryphomys adustus* Miller, 1910.
COMMENTS: Musser and Newcomb (1983) and Musser and Heaney (1992) revised the genus and documented the history of incorrect association of *Tryphomys* with *Rattus*. Treated as a Philippine New Endemic by Musser and Heaney (1992). *Tryphomys* and *Abditomys*, another Luzon endemic, form a monophyletic cluster; the nature of their phylogenetic alliances to other genera of Philippine New Endemics, or to genera native to other regions in the Indo-Australian region, remains unresolved.

Tryphomys adustus Miller, 1910. Proc. U.S. Natl. Mus., 38:399.
TYPE LOCALITY: Philippines, Luzon Isl, Benguet, Haights-in-the-Oaks.
DISTRIBUTION: Luzon; known only from very few localities in highlands and lowlands (Barbehenn et al., 1972-1973; Sanborn, 1952*a*).

Uranomys Dollman, 1909. Ann. Mag. Nat. Hist., ser. 8, 4:551.
TYPE SPECIES: *Uranomys ruddi* Dollman, 1909.
COMMENTS: Allen (1939) listed seven species of *Uranomys* but noted that six of them would probably prove to be subspecies of *U. ruddi*, which is how the genus is currently treated (Verheyen, 1964*b*). In external, cranial, and dental traits, *Uranomys* is closely related to *Acomys* (Hinton, 1921; Misonne, 1969), a phylogenetic alliance also supported by DNA (LINE) sequence analyses (Pascale et al., 1990), and *Lophuromys* (our studies; Watts, in litt.; Denys and Michaux, 1992).

Uranomys ruddi Dollman, 1909. Ann. Mag. Nat. Hist., ser. 8, 4:552.
TYPE LOCALITY: Kenya, Mt Elgon, Kirui, 6000 ft.
DISTRIBUTION: Senegal, Guinea, Ivory Coast, Togo, N Nigeria, N Camaroon, NE Zaire, Uganda, Kenya, C Mozambique, Malawi, and SE Zimbabwe; limits unknown; a savanna species.
SYNONYMS: *acomyoides, foxi, oweni, shortridgei, tenebrosus, ugandae, woodi.*
COMMENTS: Chromosomal data was presented by Viegas-Pequignot et al. (1983) in context of chromosomal phylogeny of selected murines; protein electrophoretic data analysed by Iskandar and Bonhomme (1984). More than one species may be represented among samples. Specimens of *shortridgei*, for example, are darker than the others and have much larger molars. The significance of geographic variation in fur color and craniodental traits has to be assessed in a systematic revision of the genus.

Uromys Peters, 1867. Monatsb. K. Preuss. Akad. Wiss. Berlin, p. 343.
TYPE SPECIES: *Mus macropus* Gray, 1866 (= *Hapalotis caudimaculatus* Krefft, 1867).
SYNONYMS: *Cyromys, Gymnomys, Melanomys* (Winter, 1983, not Thomas, 1902).
COMMENTS: Member of the New Guinea and Australian Old Endemics (Musser, 1981*c*). The genus has been revised by Groves and Flannery (in litt.), who arrange the species in two subgenera, *Uromys* and *Cyromys*).

Uromys anak Thomas, 1907. Ann. Mag. Nat. Hist., ser. 7, 20:72.
TYPE LOCALITY: Papua New Guinea, Central Prov., Brown River, Efogi, "not less than" 4000 ft.
DISTRIBUTION: New Guinea; throughout the Central Cordillera from Weyland Range in the west to Mt Dayman in the east as well as the Huon Peninsula (see map in Flannery, 1990*b*:219).
SYNONYMS: *rothschildi.*
COMMENTS: Subgenus *Uromys*. Closely related to *U. neobritanicus*.

Uromys caudimaculatus (Krefft, 1867). Proc. Zool. Soc. Lond., 1867:316.
TYPE LOCALITY: Australia, Queensland, Cape York.
DISTRIBUTION: Australia; NE coastal Queensland in tropical forests from Townsville area north to tip of Cape York, and a few islands off the coast of N Queensland (Watts and Aslin, 1981:91). New Guinea; widespread throughout lowland and midmontane regions on the mainland, also on Aru Isls, Kei Isls, Waigeo Isl, and Normanby and Fergusson in the d'Entrecasteaux Arch.
SYNONYMS: *aruensis, ductor, exilis, lamington, macropus, multiplicatus, nero, papuanus, prolixus, scaphax, sherrini, siebersi, validus, waigeuensis.*
COMMENTS: Subgenus *Uromys*. The Australian population has been studied from viewpoints of chromosomal morphology (Baverstock et al., 1977*c*), heterochromatin variation (Baverstock et al., 1976*b*, 1982), electrophoretic data (Baverstock et al., 1981), G-banding homologies (Baverstock et al., 1983*b*), morphology of male reproductive tract (Breed, 1986), and spermatozoal morphology (Breed, 1984; Breed and Sarafis, 1978). Donnellan (1987) provided chromosomal data for samples from New Guinea. Mahoney and Richardson (1988:189) cataloged taxonomic, distributional, and biological references covering Australian populations.

Two different chromosomal forms of Australian *U. caudimaculatus* exist, one extending from McIlwraith Ranges northward, the other from Cooktown southward; each is separated by a 200 km break in rainforest (Baverstock et al., 1976*b*, 1977*c*).

Phallic morphology of New Guinea samples reported by Lidicker (1968). Significance of the morphological variation among samples from New Guinea and adjacent islands needs to be assessed in context of systematic revision to determine how many species actually exist, a study that has now been completed (Groves and Flannery, in litt.).

Uromys hadrourus (Winter, 1983). *In* The Australian Museum Complete Book of Australian Mammals (R. Strahan, ed.), p. 379.
TYPE LOCALITY: Australia, NE Queensland, Thornton Peak summit area, 1200 m (see Winter, 1984, for additional data).
DISTRIBUTION: Known only by five specimens (Groves and Flannery, in litt.) from rainforest on Mareeba Franite of the Thornton Peak massif (Winter, 1983, 1984).

COMMENTS: Subgenus *Uromys*. Originally described as a species of *Melomys*, Groves and Flannery (in litt.) have shown that *hadrourus* belongs in *Uromys* and is closely related to *U. caudimaculatus*. *U. hadrourus* is a member of a group of species that are endemic to "the Townsville to Cooktown region, and considered to be relicts of a wet- and cool-adapted fauna which may have orinigated in Australia from a common pre-Pleisocene stock of Australia and New Guinea" (Winter, 1984:525). Chromosomal morphology reported by Baverstock et al. (1977*c*). McAllan and Bruce (1989) explained why 1983 is the publication date for *hadrourus* instead of 1984 as is usually accepted, and pointed out that the species was first given the name *Melanomys hadrourus*. That generic name is preoccupied by *Melanomys* Thomas, 1902.

Uromys imperator (Thomas, 1888). Ann. Mag. Nat. Hist., ser. 6, 1:157.
TYPE LOCALITY: Solomon Isls, Guadalcanal Isl, Aola.
DISTRIBUTION: Endemic to Guadalcanal Isl.
COMMENTS: Subgenus *Cyromys*. Historically, this species has been assigned to either *Uromys* or *Cyromys*, but is now regarded as a member of the former (Flannery and Wickler, 1990; Groves and Flannery, in litt.). Known only by a few specimens (Flannery, 1991).

Uromys neobritanicus Tate and Archbold, 1935. Am. Mus. Novit., 803:4.
TYPE LOCALITY: Papua New Guinea, Bismarck Arch., New Britain Isl.
DISTRIBUTION: Endemic to New Britain Isl.
COMMENTS: Subgenus *Uromys*. Included in *U. anak* by Ziegler (1982*b*:882) without explanation, but *neobritanicus* is diagnosed by distinctive external and cranial traits that are outside the range of morphological variation characteristic of *U. anak* (Groves and Flannery, in litt.). Chromosomal data reported by Donnellan (1987).

Uromys porculus Thomas, 1904. Ann. Mag. Nat. Hist., ser. 7, 14:400.
TYPE LOCALITY: Solomon Isls, Guadalcanal Isl, Aola.
DISTRIBUTION: Endemic to Guadalcanal Isl.
COMMENTS: Subgenus *Cyromys*. Although originally described as a species of *Uromys*, it was transferred to *Melomys* (Ellerman, 1941; Rümmler, 1938), placed back in *Uromys* (Tate, 1951), again put in *Melomys* (Laurie and Hill, 1954), and has currently been returned to *Uromys* (Flannery and Wickler, 1990; Groves and Flannery, in litt.).

Uromys rex (Thomas, 1888). Ann. Mag. Nat. Hist., ser. 6, 1:157.
TYPE LOCALITY: Solomon Isls, Guadalcanal Isl, Aola.
DISTRIBUTION: Found only on Guadalcanal Isl.
COMMENTS: Subgenus *Cyromys*. Another Guadalcanal species that has historically been placed in either *Uromys* or *Cyromys*, but is currently allocated to the former (Flannery and Wickler, 1990; Groves and Flannery, in litt.). Still represented by less than a dozen specimens (Flannery, 1991).

Vandeleuria Gray, 1842. Ann. Mag. Nat. Hist., [ser. 1], 10:265
TYPE SPECIES: *Mus oleraceus* Bennett, 1832.
COMMENTS: Allied to *Chiropodomys* and *Micromys*, a hypothesis based on dental morphology (Misonne, 1969) and one that requires testing with other character sets.

Vandeleuria nolthenii Phillips, 1929. Ceylon J. Sci., Sec. B, 15:165.
TYPE LOCALITY: Sri Lanka, Uva, Ohiya, West Haputale, 6000 ft.
DISTRIBUTION: Highlands (above 3800 ft) of Central and Uva provinces of Sri Lanka.
COMMENTS: Usually considered a subspecies of *V. oleracea* (Agrawal and Chakraborty, 1980; Ellerman, 1961), *nolthenii* is distinct in its montane distribution, pelage coloration, and external and cranial traits from *V. oleracea* occuring at lower elevations, and should be treated as a separate species (Musser, 1979).

Vandeleuria oleracea (Bennett, 1832). Proc. Zool. Soc. Lond., 1832:121.
TYPE LOCALITY: India, Madras, Deccan region.
DISTRIBUTION: Peninsular India, Sri Lanka (lowlands), Nepal to Burma, S China (Yunnan), Thailand (except peninsula south of Isthmus of Kra, 10°30'N), and Vietnam; probably occurs throughout Indochina.
SYNONYMS: *badius, domecolus, dumeticola, marica, modesta, nilagirica, povensis, rubida, sibylla, spadicea, wroughtoni*.

COMMENTS: Chromosomal features vary geographically (see Winking et al, 1979, and references therein), as do chromatic and cranial traits; *oleracea* is possibly a composite of species, and despite Agrawal and Chakraborty's (1980) review of geographic variation, needs careful systematic revision.

Vernaya Anthony, 1941. Field Mus. Nat. Hist. Publ. Zool. Ser., 27:110.
TYPE SPECIES: *Chiropodomys fulvus* G. M. Allen, 1927.
SYNONYMS: *Octopodomys*.
COMMENTS: Dental morphology interpreted by Misonne (1969) to indicate close relationship with *Chiropodomys*; hypothesis requires testing with other characters. About the same time Anthony proposed *Vernaya*, Sody (1941) proposed *Octopodomys*.

Vernaya fulva (G. M. Allen, 1927). Am. Mus. Novit., 270.
TYPE LOCALITY: China, Yunnan, Yinpankai, Mekong River.
DISTRIBUTION: S China (Sichuan and Yunnan) and N Burma; known only from above 7000 ft.
SYNONYMS: *foramena*.
COMMENTS: Originally described by Allen as a species of *Chiropodomys*, but later reidentified by him as the only example of *Vandeleuria dumeticola* known from Yunnan (Allen, 1940). Anthony (1941) correctly pointed out the morphological uniqueness of *fulva* by erecting a new genus to contain it. Still known only by very few specimens. Wang et al. (1980) described *foramena* as a species of *Vernaya*, but diagnostic traits simply represent individual variation found in *V. fulva*.

Xenuromys Tate and Archbold, 1941. Am. Mus. Novit., 1101:3.
TYPE SPECIES: *Mus barbatus* Milne-Edwards, 1900.
COMMENTS: Member of the New Guinea Old Endemics (Musser, 1981*c*).

Xenuromys barbatus (Milne-Edwards, 1900). Bull. Mus. Hist. Nat. Paris, 6:167.
TYPE LOCALITY: Papua New Guinea, "British New Guinea."
DISTRIBUTION: New Guinea; Represented by only a few specimens collected along the Central Cordillera from Mt Dayman in E Papua New Guinea to the Idenberg River in Irian Jaya (Flannery, 1990*b*; Flannery et al., 1985).
STATUS: IUCN - Rare.
SYNONYMS: *guba*.
COMMENTS: Closest phylogenetic relatives appear to be species of *Uromys* (Tate, 1951).

Xeromys Thomas, 1889. Proc. Zool. Soc. Lond., 1889:248.
TYPE SPECIES: *Xeromys myoides* Thomas, 1889.
COMMENTS: Member of the Australian Old Endemics (Musser, 1981*c*:166) or Hydromyini of Baverstock (1984), and Lee et al. (1981). *Xeromys* is phylogenetically distantly related to *Hydromys*, although the two are still sister genera (Watts et al., 1992).

Xeromys myoides Thomas, 1889. Proc. Zool. Soc. Lond., 1889:248.
TYPE LOCALITY: Australia, Queensland, Mackay.
DISTRIBUTION: Australia; C and S Queensland, North Stradbroke Island off the coast of SE Queensland (Van Dyck et al., 1979), Northern Territory, and Melville Island off the coast of Northern Territory; probably has a wider range (Watts and Aslin, 1981).
STATUS: CITES - Appendix I; U.S. ESA - Endangered; IUCN - Rare.
COMMENTS: Closely associated with tidal mangrove swamps and still known only by small samples, most of which have been collected since 1970 (Dwyer et al., 1979; Van Dyck et al., 1979; Watts and Aslin, 1981). Morphology of sperm head structure and male reproductive tract studied by Breed (1984, 1986) in context of comparative study among Australian murines. Chromosomal complement similar to that of *Hydromys chrysogaster* (Baverstock et al., 1977*c*), as is primitive pattern of cephalic arterial configuration. Mahoney and Richardson (1988:190) cataloged references to taxonomy and natural history.

Zelotomys Osgood, 1910. Field Mus. Nat. Hist. Publ., Zool. Ser., 10:7.
TYPE SPECIES: *Mus hildegardeae* Thomas, 1902.
SYNONYMS: *Ochromys*.

COMMENTS: Similar to *Lophuromys* in molar occlusal patterns (Misonne, 1969) and other features, but strength of phylogenetic link to either that genus or other murines requires assessment by further study.

Zelotomys hildegardeae (Thomas, 1902). Ann. Mag. Nat. Hist., ser. 7, 9:219.
TYPE LOCALITY: Kenya (Kenya Colony), Machakos.
DISTRIBUTION: Angola, Zambia, Malawi (Nyika Plateau), Tanzania, Kenya, W Uganda, Rwanda, Burundi, Zaire, S Sudan, and Central African Republic.
SYNONYMS: *instans, kuvelaiensis, lillyana, shortridgei, vinaceus.*

Zelotomys woosnami (Schwann, 1906). Proc. Zool. Soc. Lond., 1906:108.
TYPE LOCALITY: S Botswana, Molopo River.
DISTRIBUTION: S Africa (N Cape Province), SW and N Botswana, and Namibia (see map in Skinner and Smithers, 1990).
COMMENTS: Past allocations have been with *Aethomys, Rattus,* and *Thallomys;* see D. H. S. Davis (1965) and references in Meester et al. (1986).

Zyzomys Thomas, 1909. Ann. Mag. Nat. Hist., ser. 8, 3:372.
TYPE SPECIES: *Mus argurus* Thomas, 1889.
SYNONYMS: *Laomys.*
COMMENTS: Member of the Australian Old Endemics (Musser, 1981*c*:167) or Conilurini of Baverstock (1984) and Lee et al. (1981). Taxonomy of species appraised by Kitchener (1989). Chromosomal morphology distinct from other Australian murines (Baverstock et al., 1977*c*), but this result was not supported by electrophoretic data (Baverstock et al., 1981), and G-banding homologies suggested karyotype of *Zyzomys* "to be derived from the ancestral karyotype characterizing *Conilurus, Mesembriomys,* and *Leggadina*" (Baverstock et al., 1983*b*). Molar traits resembled those characterizing *Conilurus* (Misonne, 1969), but phallic characters linked *Zyzomys* to *Pseudomys* and its close relatives (Lidicker and Brylski, 1987). Data from microcomplement fixation of albumin indicated *Zyzomys* belongs in a clade with *Mesembriomys, Conilurus, Leporillus,* and *Melomys* (Watts et al., 1992). External morphology of glans penis and spermatozoal structure reported by Breed, 1984), Breed and Sarafis, 1978; and Morrissey and Breed (1982). Mahoney and Richardson (1988) cataloged taxonomic, distributional, and biological references to some of the species.

Zyzomys argurus (Thomas, 1889). Ann. Mag. Nat. Hist., ser. 6, 3:433.
TYPE LOCALITY: Australia, probably Northern Territory (see Mahoney and Richardson (1988:191).
DISTRIBUTION: Australia; always in rocky outcrops in Pilbara region of Western Australia through Kimberleys to N coastal Queensland between Cooktown and Townsville; also on offshore islands (see Watts and Aslin, 1981:138).
SYNONYMS: *indutus.*
COMMENTS: Electrophoretic comparisons between populations discordant with chromosomal differences (Baverstock et al., 1977*d*).

Zyzomys maini Kitchener, 1989. Rec. West. Aust. Mus., 14:357.
TYPE LOCALITY: Australia, Northern Territory, Djawamba Massif, 1.5 km east of Ja Ja Billabong, 150 m; habitat described by Kitchener (1989:357).
DISTRIBUTION: Australia, Northern Territory in region of East and South Alligator Rivers on outliers of stony Arnhem Land escarpment (Kitchener, 1989).

Zyzomys palatilis Kitchener, 1989. Rec. West. Aust. Mus., 14:361.
TYPE LOCALITY: Australia, Northern Territory, Echo Gorge, Wollogorang Station, 180 m; habitat described by Kitchener (1989:361).
DISTRIBUTION: Australia; recorded only from Echo Gorge (in Gulf Country near Queensland border).

Zyzomys pedunculatus (Waite, 1896). Rept. Horn Sci. Exped. Cent. Aust., Zool., 2:395.
TYPE LOCALITY: Australia, Northern Territory, Alice Springs, but this locality is suspect (Kitchener, 1989).
DISTRIBUTION: C Australia; Northern Territory (see map and discussion in Watts and Aslin, 1981:143).
STATUS: CITES - Appendix I; U.S. ESA and IUCN - Endangered.

SYNONYMS: *brachyotis*.

COMMENTS: Rare (now even believed extinct, Flannery, in litt.) and restricted to rocks. Distribution of fossils indicated species "may once have extended across the rocky ranges of central Western Australia, through the Hamersleys to the coast" (Watts and Aslin, 1981:143).

Zyzomys woodwardi (Thomas, 1909). Ann. Mag. Nat. Hist., ser. 8, 3:373.

TYPE LOCALITY: Australia, Western Australia, E Kimberly, Parrys Creek, near Windham, 100 ft.

DISTRIBUTION: Australia; Kimberleys of Western Australia and Arnhem Land (western escarpments) of Northern Territory (see map in Watts and Aslin, 1981:141).

COMMENTS: Restricted to rocky regions, especially boulders at the base of cliffs; limited in distribution but common at some localities (Watts and Aslin, 1981).

Subfamily Myospalacinae Lilljeborg, 1866. Syst. Öfversigt. Gnag. Däggdjuren, p. 25.

SYNONYMS: Myotalpinae, Siphneinae.

COMMENTS: Diagnosis, morphological and chromosomal characteristics, and general remarks on habits, habitat, distribution, geologic range, fossil groups, and phylogenetic relationships within the subfamily and with other muroid groups were provided by Carleton and Musser (1984). Results of phylogenetic analysis of living and some extinct species were reported by Lawrence (1991).

Myospalax Laxmann, 1769. Sibirische Briefe, Gottingen, p. 75.

TYPE SPECIES: *Mus myospalax* Laxmann, 1773.

SYNONYMS: *Aspalomys, Eospalax, Myotalpa, Prosiphneus, Siphneus, Zokor.*

COMMENTS: Phylogenetic relationships reviewed by Lawrence (1991), who provided references to systematic and genetic studies of living forms in the former USSR. Chinese taxa have been reviewed, with different results, by Fan and Shi (1982) and Li and Chen (1989).

Myospalax aspalax (Pallas, 1776). Reise Prov. Russ. Reichs., 3:692.

TYPE LOCALITY: Russia, Transbaikalia, Dauuria ("Doldogo, on Onon River, below Atchinsk," Ellerman and Morrison-Scott, 1951:652).

DISTRIBUTION: Russia (Upper Amur basin) and China (Nei Mongol).

SYNONYMS: *armandii, dybowskii, hangaicus, talpinus, zokor.*

COMMENTS: Listed, with a question mark, as a subspecies of *M. myospalax* by Ellerman and Morrison-Scott (1951), and unequivically as *M. m. aspalax* by Corbet (1978*c*), but separated on morphological grounds by Lawrence (1991) and others cited in her report. Closest phylogenetic relative is the Lower Pleistocene *M. pseudarmandi*, and both are in same monophyletic species-group containing the modern *M. epsilanus, M. myospalax*, and Pliocene *M. youngi* (Lawrence, 1991).

Myospalax epsilanus Thomas, 1912. Ann. Mag. Nat. Hist., ser. 8, 9:94.

TYPE LOCALITY: China, Heilongjiang Prov. (Manchuria), Khingan Mtns, 3400 ft.

DISTRIBUTION: NE China (valleys of Great Khiaghau in Kirin Province) and Russia (E Transbaikalia between Amur and S Ussuri Territory); limits unknown.

COMMENTS: The most primitive member of the monophyletic *myospalax* species-group, which includes species represented by modern specimens as well as Pliocene and Pleistocene fossils (Lawrence, 1991).

Myospalax fontanierii (Milne-Edwards, 1867). Ann. Sci. Nat. (Paris), 7:376.

TYPE LOCALITY: China, Kansu.

DISTRIBUTION: Dry grasslands from Gansu, Ningxia (Qin, 1991) through Hupeh and Anhui to Sichuan (China).

SYNONYMS: *baileyi, cansus, fontanus, kukunoriensis, rufescens, shenseius.*

COMMENTS: Phylogenetically in same monophyletic group as *M. smithii* and *M. rothschildi*, which is the most derived group within the subfamily (Lawrence, 1991). Fan and Shi (1982) recognized *cansus* and *baileyi* as species, but Li and Chen (1989) argued for their recognition as subspecies of *M. fontanierii*.

Myospalax myospalax (Laxmann, 1773). Kongl. Svenska. Vet.-Akad. Handl. Stockholm, 34:134.

TYPE LOCALITY: Russia, Altai Krai, 100 km SE of Barnaul, Sommaren, near Paniusheva on Alei River.

DISTRIBUTION: Russia and Kazakhstan (entire N region of Cisaltai plain and foothills as well as the W and C Altai).

SYNONYMS: *incertus, komurai, laxmanni, tarbagataicus.*

COMMENTS: Member of monophyletic *myospalax* species-group, which includes living *M. epsilanus, M. aspalax,* and the Pliocene and Pleistocene *M. youngi* and *M. pseudarmandi* (Lawrence, 1991).

Myospalax psilurus (Milne-Edwards, 1874). Rech. Hist. Nat. Mammifères, p. 126.

TYPE LOCALITY: China, Chihli, south of Peking.

DISTRIBUTION: Transbaikalia and Ussuri region of Russia to E Mongolia, NE and C China; limits unknown.

SYNONYMS: *spilurus.*

COMMENTS: The sole member of the *psilurus* species-group, which is defined, except for rootless hypsodonty, by retention of the greatest number of primitive traits of any species in the genus (Lawrence, 1991).

Myospalax rothschildi Thomas, 1911. Ann. Mag. Nat. Hist., ser. 8, 8:722.

TYPE LOCALITY: China, Gansu, 40 mi SE Tao-chou.

DISTRIBUTION: China; Kansu and Hubei.

SYNONYMS: *hubeinensis, minor.*

COMMENTS: The sister-species of *M. smithii,* and in same monophyletic cluster as *M. fontainieri,* as determined by cladistic analysis of external, cranial, and dental traits (Lawrence, 1991). Recently, *hubeinensis* was described as subspecies of *M. rothschildi* by Li and Chen (1989).

Myospalax smithii Thomas, 1911. Ann. Mag. Nat. Hist., ser. 8, 8:720.

TYPE LOCALITY: China, Gansu, 30 mi SE Tao-chou.

DISTRIBUTION: China, Gansu and Ningxia (Qin, 1991).

COMMENTS: Phylogenetically closely allied to *M. rothschildi,* and in same monophyletic group with *M. fontanierii,* a cluster defined by a suite of highly derived morphological features (Lawrence, 1991).

Subfamily Mystromyinae Vorontsov, 1966. Zool. Zh., 45:437.

COMMENTS: Phylogenetic allocation of *Mystromys* so puzzled Ellerman (1941:445) that he wrote "I am entirely at a loss to suggest the relationships of this genus, which seems not only isolated from the Palaearctic and Neotropical genera, but to have no marked generic characters..." Based on study of external, skeletal, dentition, gastrointestinal, genital and other anatomical characteristics of *Mystromys albicaudatus,* Vorontsov (1966) concluded the species was not closely related to Palaearctic hamsters and placed *Mystromys* in the monotypic tribe Mystromyini. Carleton and Musser (1984:313) remarked that *Mystromys* is possibly a survivor of an ancient phyletic line, pointing out that Lavocat (1973, 1978) "raised such a novel possibility for *Mystromys* by noting its probable derivation from afrocricetodontine rodents and by recognizing the subfamily Mystromyinae in the Nesomyidae, a family composed of archaic African cricetids that he derived from the afrocricetodontines, the African counterpart to European and Asian cricetodontines." Pocock (1987) and Skinner and Smithers (1990) retained *Mystromys* in the Cricetinae of family Cricetidae.

Mystromys Wagner, 1841. Gelehrte Anz. I. K. Bayer. Akad. Wiss., München, 12(54), col. 434.

TYPE SPECIES: *Mystromys albipes* Wagner, 1841 (= *Otomys albicaudatus* A. Smith, 1834).

COMMENTS: Gross morphology of male accessory glands described by Voss and Linzey (1981) in context of survey assessing systematic implications of these structures among muroid rodents. Gross and histological stomach anatomy, as well as developmental and physiological aspects of gastric structure and processes, reported by Perrin and Curtis (1980), Maddock and Perrin (1981, 1983), and Perrin and Maddock (1983), and presented in comparative context to discern evolutionary relationships among Southern African muroid rodents, and in context of function and diet.

Mystromys albicaudatus (A. Smith, 1834). S Afr. Quart. J., 2:148.
TYPE LOCALITY: S Africa, E Cape Province, Albany Dist.
DISTRIBUTION: Endemic to southern Africa: Cape Prov., W and N Natal, Orange Free State, and SW and S Transvaal (South Africa), and Swaziland; see map in Skinner and Smithers (1990:297).
SYNONYMS: *albipes, fumosus, lanuginosus* (see Meester et al., 1986).
COMMENTS: Closest phylogenetic relatives are *Mystromys hausleitneri, M. pocockei*, and *Proodontomys cookei*, all extinct and represented by fossils from Pliocene-Pleistocene australopithecine sites in the Transvaal of South Africa (Denys, 1991; Pocock, 1987).

Subfamily Nesomyinae Major, 1897. Proc. Zool. Soc. Lond., 1897:718.
SYNONYMS: Brachytarsomyes, Brachyuromyes, Eliuri, Gymnuromyinae.
COMMENTS: Group exceedingly diverse morphologically, defying an unambiguous diagnosis and questioning their monophyletic origin (see discussion in Carleton and Musser, 1984). Proponents of a single ancestral origin have usually arranged nesomyines as a subfamily of Cricetidae (e.g., Miller and Gidley, 1918; Simpson, 1945) or as a subfamily within a broadly defined family Nesomyidae, which includes other archaic groups like cricetomyines, tachyoryctines, and *Mystromys* (Chaline et al., 1977; Lavocat, 1978). Ellerman (1941, 1949*a*) argued that nesomyines are polyphyletic and dispersed the seven genera among four subfamilies of Muridae *sensu lato*, as reflected in the family-group synonyms all named by Ellerman (1941). Lavocat (1978) viewed *Protarsomys*, lower Miocene of Kenya, as close to the ancestry of Malagasy Nesomyinae, and Chaline et al. (1977) placed the Miocene fossil in synonymy under extant *Macrotarsomys*. Carleton and Schmidt (1990) disputed this relationship and generic equivalence; nesomyines otherwise known only from the Holocene of Madagascar.
Ellerman (1949*a*) provided the most valuable synopsis of nesomyine taxa and set forth the basic species-level classification currently recognized; taxonomy updated by Petter (1972*c*, 1975*a*). Locality data, geographic ranges, and type localities of named forms summarized by Carleton and Schmidt (1990). Revisionary and ecological studies required for all genera.

Brachytarsomys Günther, 1875. Proc. Zool. Soc. Lond., 1875:79.
TYPE SPECIES: *Brachytarsomys albicauda* Günther, 1875.
COMMENTS: The incipient prismatic condition of the dentition persuaded Ellerman (1941) to classify *Brachytarsomys* as a primitive microtine.

Brachytarsomys albicauda Günther, 1875. Proc. Zool. Soc. Lond., 1875:80.
TYPE LOCALITY: Madagascar, between "Tamatave and Morondava."
DISTRIBUTION: E Madagascar rain forest at middle elevations.
SYNONYMS: *villosa*.
COMMENTS: The indeterminate nature of Günther's (1875) type locality might be restricted with archival research since the species occurs only in the east. Carleton and Schmidt (1990) suggested that *villosa*, named as a subspecies by F. Petter (1962*a*) based on a zoo specimen, is a distinct species; its geographic occurrence is unknown.

Brachyuromys Major, 1896. Ann. Mag. Nat. Hist., ser. 6, 18:322.
TYPE SPECIES: *Brachyuromys ramirohitra* Major, 1896.
COMMENTS: Ellerman (1941) allied *Brachyuromys* as a tribe within Tachoryctinae, an affinity earlier considered plausible by Major (1897).

Brachyuromys betsileoensis (Bartlett, 1880). Proc. Zool. Soc. Lond., 1879:770 [1880].
TYPE LOCALITY: Madagascar, "S.E. Betsileo."
DISTRIBUTION: Madagascar: Central plateau and its eastern fringes.
COMMENTS: Described as a species of *Nesomys* but referred to *Brachyuromys* by Major (1896); occurs sympatrically with *B. ramirohitra*.

Brachyuromys ramirohitra Major, 1896. Ann. Mag. Nat. Hist., ser. 6, 18:323.
TYPE LOCALITY: Madagascar, Fianarantsoa Prov., 6 hours SE Fandriana, Ampitambe forest.
DISTRIBUTION: Madagascar: southern part of central highlands; extent poorly documented.

Eliurus Milne-Edwards, 1885. Ann. Sci. Nat., Zool. Paleontol. (Paris), 20: Art. 1 bis.
TYPE SPECIES: *Eliurus myoxinus* Milne-Edwards, 1885.
COMMENTS: Genus arrayed as a monotypic tribe within Murinae by Ellerman (1941). Interpreted by Ellerman (1949*a*) as consisting only of a large polytypic species (*E. myoxinus*) and a small monotypic one (*E. minor*). Revised by Carleton (1993), who recognized 6 species and summarized their morphological identification and distributions.

Eliurus majori Thomas, 1895. Ann. Mag. Nat. Hist., ser. 6, 16:164.
TYPE LOCALITY: Madagascar, Fianarantsoa Prov., Ambolimitambo forest, 4500 ft.
DISTRIBUTION: Madagascar: known from three widely isolated localities in N, C, and S highlands.
COMMENTS: Closely related to *E. penicillatus*.

Eliurus minor Major, 1896. Ann. Mag. Nat. Hist., ser. 6, 18:462.
TYPE LOCALITY: Madagascar, Fianarantsoa Prov., Ampitambe.
DISTRIBUTION: Madagascar: broadly distributed in E forest, from near sea level to 1500 m.
COMMENTS: Co-occurs with most other species of *Eliurus*.

Eliurus myoxinus Milne-Edwards, 1885. Ann. Sci. Nat., Zool. Paleontol. (Paris), 20: Art. 1 bis.
TYPE LOCALITY: Madagascar, Toliara Prov., Tsilambana.
DISTRIBUTION: Dry deciduous forest and xerophilous habitats in SW and S Madagascar.
COMMENTS: *Eliurus myoxinus sensu stricto* is actually endemic to western biotopes but its range approaches that of eastern species in the extreme south.

Eliurus penicillatus Thomas, 1908. Ann. Mag. Nat. Hist., ser. 8, 2:453.
TYPE LOCALITY: Madagascar, Fianarantsoa Prov., Ampitambe.
DISTRIBUTION: Known only from the type locality.
COMMENTS: Tentatively retained as a species by Carleton (1993) but its level of differentiation from *E. majori* requires further study.

Eliurus tanala Major, 1896. Ann. Mag. Nat. Hist., ser. 6, 18:462.
TYPE LOCALITY: Madagascar, Fianarantsoa Prov., 30 mi S Fianarantsoa, near Vinanitelo.
DISTRIBUTION: Madagascar: middle to upper elevation E rain forest.
COMMENTS: Distribution overlaps that of its morphologically similar congener *E. webbi* at lower elevations.

Eliurus webbi Ellerman, 1949. Families Genera Living Rodents, 3(App. II):163.
TYPE LOCALITY: Madagascar, Fianarantsoa Prov., 20 mi S Farafangana.
DISTRIBUTION: Madagascar: elongate belt of low to middle elevation E rain forest.
COMMENTS: Substantial variation exhibited by populations allocated to this species deserves additional scrutiny.

Gymnuromys Major, 1896. Ann. Mag. Nat. Hist., ser. 6, 18:324.
TYPE SPECIES: *Gymnuromys roberti* Major, 1896.
COMMENTS: Ellerman (1941) created the new subfamily Gymnuromyinae in recognition of the distinctive features of this species, which he believed to be derived from a *Nesomys*-like ancestor.

Gymnuromys roberti Major, 1896. Ann. Mag. Nat. Hist., ser. 6, 18:324.
TYPE LOCALITY: Madagascar, Fianarantsoa Prov., Ampitambe forest.
DISTRIBUTION: E Madagascar rain forest, 500 to 950 m.
COMMENTS: Some have voiced concern that populations of *G. roberti* are being supplanted by introduced *Rattus* (summary in Carleton and Schmidt, 1990).

Hypogeomys A. Grandidier, 1869. Rev. Mag. Zool. Paris, ser. 2, 21:338.
TYPE SPECIES: *Hypogeomys antimena* A. Grandidier, 1869.
COMMENTS: Retained by Ellerman (1941) within Cricetinae as a genus of obscure relationships.

Hypogeomys antimena A. Grandidier, 1869. Rev. Mag. Zool. Paris, ser. 2, 21:339.
TYPE LOCALITY: Madagascar, Toliara Prov., between the banks of the Tsiribihina and Andranomena rivers.
DISTRIBUTION: Narrow coastal zone of sandy soils in WC Madagascar.
COMMENTS: Ecology and conservation status of extant populations discussed by Cook et al. (1991).

Macrotarsomys Milne-Edwards and G. Grandidier, 1898. Bull. Mus. Hist. Nat. Paris, ser. 1, 4:179.
TYPE SPECIES: *Macrotarsomys bastardi* Milne Edwards and G. Grandidier, 1898.
COMMENTS: Retained within Cricetinae by Ellerman (1941). Lavocat (1978) and Chaline et al. (1977) aligned *Macrotarsomys* with the Miocene Kenyan fossil *Protarsomys*, a relationship questioned by Carleton and Schmidt (1990).

Macrotarsomys bastardi Milne-Edwards and G. Grandidier, 1898. Bull. Mus. Hist. Nat. Paris, ser. 1, 4:179.
TYPE LOCALITY: Madagascar, Fianarantsoa Prov., east of Ihosy River, near Ravori; see Carleton and Schmidt (1990:14).
DISTRIBUTION: Dryer regions of W and S Madagascar, including deciduous forest and open savannah.
SYNONYMS: *occidentalis*.

Macrotarsomys ingens Petter, 1959. Mammalia, 23:140.
TYPE LOCALITY: Madagascar, Mahajanga Prov., Ankarafantsika Reserve, near Ampijoroa.
DISTRIBUTION: Known only from the type locality.
COMMENTS: Occurs sympatrically with *M. bastardi*.

Nesomys Peters, 1870. Sitzb. Ges. Naturf. Fr. Berlin, p. 54.
TYPE SPECIES: *Nesomys rufus* Peters, 1870.
SYNONYMS: *Hallomys*.
COMMENTS: Major (1897) synonymized Jentink's (1879) *Hallomys* (type species = *H. audeberti*) under *Nesomys* and allocated the genus to the Cricetinae.

Nesomys rufus Peters, 1870. Sitb. Ges. Naturf. Fr. Berlin, p. 55.
TYPE LOCALITY: Madagascar, Antsiranana Prov., Vohima.
DISTRIBUTION: Broadly distributed in N and E forest (*audeberti* and *rufus*), and one locality in WC Madagascar (*lambertoni*).
SYNONYMS: *audeberti*, *lambertoni*.
COMMENTS: Ellerman (1941, 1949*a*) listed *audeberti* and *lambertoni* as species as described, but Petter (1972*c*, 1975*a*) arranged them as subspecies of *N. rufus*. Carleton and Schmidt (1990) indicated that each will stand as a species upon revision.

Subfamily Otomyinae Thomas, 1897. Proc. Zool. Soc. Lond., 1896:1017 [1897].
COMMENTS: Morphologically, a strongly circumscribed group of species indigenous to subsaharan Africa; early on ranked as a subfamily of Muridae *sensu stricto* (Miller and Gidley, 1918; Simpson, 1945), later as a subfamily of Cricetidae (Misonne, 1974) or Nesomyidae (Chaline et al., 1977; Lavocat, 1978). Paleontological evidence and anatomical considerations have reopened the question of their phyletic origin from African murines, especially arvicanthine forms (Carleton and Musser, 1984; Pocock, 1976), although the phylogenetic significance of the annectant fossil genus (*Euryotomys* Pocock, 1976), as a murid or cricetid, has been recently disputed (see Denys et al., 1987).

Number of genera recognized has varied from five (Roberts, 1951) to three (Thomas, 1918*b*; Pocock, 1976) to one (Bohmann, 1952), usually just the two listed here (Ellerman, 1941; Ellerman et al., 1953; Misonne, 1974; De Graaff, 1981; Meester et al., 1986; Smithers, 1983). The diverse generic arrangements principally reflect the emphasis on dentition versus bullar development as a diagnostic key. Integration of a more heterogeneous information base, such as the allozymic survey initiated by Taylor et al. (1989), would better illuminate phylogenetic relationships and perhaps stabilize our generic classification. Morphological features described by Bernard et al. (1990), Bohmann (1952), Perrin and Curtis (1980), and Tullberg (1899). Few multispecies surveys of chromosomes and proteins undertaken to date; available information covered by Robinson and Elder (1987) and Taylor et al. (1989).

Otomys F. Cuvier, 1824. Dents des Mammifères, p. 255.

TYPE SPECIES: *Euryotis irrorata* Brants, 1827.

SYNONYMS: *Anchotomys, Euryotis, Lamotomys, Metotomys, Myotomys, Oreinomys, Oreomys, Palaeotomys.*

COMMENTS: Bohmann (1952) included all otomyine species in *Otomys*, but most classifications have accorded the large-bullar forms seperate generic status as *Parotomys* (see below). Other synonyms have been infrequently treated as genera, namely *Myotomys*, including *unisulcatus* and *sloggetti* (see Thomas, 1918*b*; Pocock, 1976), and *Lamotomys*, including *laminatus* (see Roberts, 1951). Oldest known fossil *Otomys* species date from late Pliocene (2-3 mya) in South Africa and from early Pleistocene (1-2 mya) in East Africa (see Denys, 1989).

Otomys anchietae Bocage, 1882. J. Sci. Acad. Lisbon, 9:26.

TYPE LOCALITY: Angola, Huila, Caconda.

DISTRIBUTION: Isolated segments in C Angola, SW Tanzania, N Malawi, and W Kenya.

SYNONYMS: *barbouri, lacustris.*

COMMENTS: Type species of *Anchotomys*, employed as a subgenus of *Otomys* by Thomas (1918*b*). Believed to be closely related to *O. irroratus* (Bohmann, 1952), if not conspecific with it (Petter, 1982). Dieterlen and Van der Straeten (1992) treated *barbouri* and *lacustris* as separate species.

Otomys angoniensis Wroughton, 1906. Ann. Mag. Nat. Hist., ser. 7, 18:274.

TYPE LOCALITY: Malawi, Misuku Range, Matipa Forest, 7000 ft (as refined by Ansell and Dowsett, 1991).

DISTRIBUTION: SE savannah and grasslands, from S Kenya to NE Cape Province, South Africa.

SYNONYMS: *canescens, divinorum, elassodon, mashona, nyikae, pretoriae, rowleyi, sabiensis, tugelensis.*

COMMENTS: Confused under *O. irroratus* by Bohmann (1952); morphological and karyotypic differences support their separate specific status (Davis, 1962; Matthey, 1964; Misonne, 1974). The form *maximus* (see account below) is viewed by some as a subspecies of *O. angoniensis*. See Bronner and Meester (1988, Mammalian Species, 306).

Otomys denti Thomas, 1906. Ann. Mag. Nat. Hist., ser. 7, 18:142.

TYPE LOCALITY: Uganda, east slope of Mount Ruwenzori, Mubuku Valley, 6000 ft (as restricted by Moreau et al., 1946:420).

DISTRIBUTION: Intermittently found in EC Africa, from Mt. Ruwenzori, Uganda, through the Virunga volcanoes, to the Nyika Plateau of N Malawi and Zambia and to the Usambara and Uluguru mountains, EC Tanzania.

SYNONYMS: *kempi, sungae.*

COMMENTS: The form *kempi* has been treated as a distinct species (e.g., Ellerman, 1941; Thomas, 1918*b*); reduced to subspecific rank by Bohmann (1952) and so observed currently (Delany, 1975; Misonne, 1974).

Otomys irroratus (Brants, 1827). Het Geslacht der Muizen, p. 94.

TYPE LOCALITY: South Africa, Cape Province, Cape Town district, near Constantia (as fixed by A. Smith, 1834:149).

DISTRIBUTION: Mesic savannah and grasslands of southern Africa: S Cape Province to C Transvaal, South Africa; disjunct populations in W South Africa and in E Zimbabwe and contiguous Mozambique.

SYNONYMS: *auratus, bisulcatus, capensis, coenosus, cupreoides, cupreus, natalensis, obscura, orientalis, randensis, typicus.*

COMMENTS: Bohmann (1952) established a broad, highly polymorphic definition of *O. irroratus*, then including among his 23 subspecies *angoniensis* and *maximus* (see those accounts). Presumably excepting *angoniensis* and *maximus*, Dieterlen (1968) and Petter (1982) further enlarged Bohmann's concept of *irroratus* to subsume the following forms here (and elsewhere) treated as separate species (see individual accounts): *anchietae, laminatus, tropicalis,* and *typus*. Although followed to a greater or lesser extent (e.g., Delany, 1975; Kingdon, 1974), such an inclusive species construct is disputed by others who restrict *O. irroratus* proper to southern Africa (e.g., Meester et al., 1986; Misonne, 1974). G-banded comparisons reveal a karyotype that is highly derived (Robinson and Elder, 1987). Cytogenetic variation extensive among South African population samples (Contrafatto et al., 1992*a*). See Bronner et al. (1988, Mammalian Species, 308).

Otomys laminatus Thomas and Schwann, 1905. Abst. Proc. Zool. Soc. Lond., 1905(18):23.
TYPE LOCALITY: South Africa, Natal, Zululand, Nkandhla, Sibudeni, 1050 m.
DISTRIBUTION: Discontinuous in South Africa, from SE Transvaal, parts of Natal and Transkei, to SW Cape Province; also Swaziland.
SYNONYMS: *fannini, mariepsi, pondoensis, silberbaueri.*
COMMENTS: A species with a highly derived molar pattern, believed to intergrade with *O. irroratus* by Petter (1982) but retained as a species in recent faunal studies (De Graaff, 1981; Meester et al., 1986; Smithers, 1983). Designated as the type of *Lamotomys*, a taxon employed as a subgenus of *Otomys* by Thomas (1918*b*) and as a genus by Roberts (1951).

Otomys maximus Roberts, 1924. Ann. Transvaal Mus., 10:70.
TYPE LOCALITY: Zambia, Machile River, a northern tributary of the Zambesi.
DISTRIBUTION: S Angola, SW Zambia, Okavango region of Botswana, and Caprivi Strip of Namibia.
SYNONYMS: *cuanzensis, davisi.*
COMMENTS: Described as a subspecies of *irroratus* and later elevated by Roberts (1951) to specific level. Viewed as a subspecies of *O. angoniensis* by Davis (1974), and so ranked in many faunal treatises (e.g., Meester et al., 1986; Misonne, 1974), or continued as a distinct species (Smithers, 1983; Swanepoel et al., 1980), which leaves the matter of their synonymy as inconclusive to date.

Otomys occidentalis Dieterlen and Van der Straeten, 1992. Boon. Zool. Beitr., 43:386.
TYPE LOCALITY: SE Nigeria, Gotel Mtns, Chappal Waddi.
DISTRIBUTION: Recorded only from the type locality and Mt. Oku in W Cameroon.
COMMENTS: Closely related to populations in mountains of E Africa (Dieterlen and Van der Straeten, 1992).

Otomys saundersiae Roberts, 1929. Ann. Transvaal Mus., 13:115.
TYPE LOCALITY: South Africa, Cape Province, Grahamstown.
DISTRIBUTION: Isolated populations in SW Cape Province and in E Cape to S Orange Free State and Lesotho.
SYNONYMS: *karoensis.*
COMMENTS: Named as a subspecies of *tugelensis* (= *O. angoniensis*) by Roberts and later (1951) raised by him to full species.

Otomys sloggetti Thomas, 1902. Ann. Mag. Nat. Hist., ser. 7, 10:311.
TYPE LOCALITY: South Africa, Cape Province, Deelfontein, north of Richmond.
DISTRIBUTION: E Cape Province, Lesotho, and NW Natal, South Africa (Lynch and Watson, 1992).
SYNONYMS: *basuticus, jeppei, robertsi, turneri.*
COMMENTS: A species with many primitive dental traits, viewed as closely related to *O. unisulcatus* (Bohmann, 1952; Roberts, 1951; Thomas, 1918*b*). Karyotypic and genetic data supplied by Contrafatto et al. (1992*b*).

Otomys tropicalis Thomas, 1902. Ann. Mag. Nat. Hist., ser. 7, 10:314.
TYPE LOCALITY: Kenya, west slope of Mt. Kenya, 3000 m.
DISTRIBUTION: Irregular highland distribution in W Kenya, Uganda, Rwanda, Burundi, and bordering Zaire; outlying populations on Mount Cameroon.
SYNONYMS: *burtoni, dollmani, elgonis, faradjius, ghigii, nubilus, ruberculus, vivax, vulcanicus.*
COMMENTS: Regarded as conspecific with *O. irroratus* by Bohmann (1952) and accordingly recognized in regional treatments (Delany, 1975; Kingdon, 1974); however, others have maintained the southern African form *irroratus* as specifically distinct from eastern African *tropicalis* (De Graaff, 1981; Meester et al., 1986; Misonne, 1974). Even so, the taxa herein assembled under *tropicalis* (after Misonne, 1974) remain a composite of two or more species, whose definition and range will require careful

revisionary work, involving simultaneous attention to possible relationships with species-group taxa now lumped under *O. typus* (see next account).

Otomys typus Heuglin, 1877. Reise in Nordost-Afrika, 2:77.

TYPE LOCALITY: Abyssinia (= Ethiopia), Shoa.

DISTRIBUTION: Disjunct occurrence at high elevations in E Africa, from Ethiopian highlands south through Uganda and W Kenya, to the Nyika Plateau of N Malawi and adjacent Zambia and to the Uzungwe Mnts of C Tanzania.

SYNONYMS: *dartmouthi, degeni, fortior, giloensis, helleri, jacksoni, malkensis, malleus, orestes, percivali, thomasi, squalus, uzungwensis, zinki.*

COMMENTS: Dieterlen (1968) and Petter (1982) viewed *typus* as another variant of a highly polymorphic *O. irroratus*, a conclusion which seems at odds with their morphological discrimination as presented elsewhere (e.g., Ansell, 1978; Bohmann, 1952; Kingdon, 1974; Misonne, 1974). The list of synonyms observes the classifications of Bohmann (1952) and Misonne (1974). Their conspecific stature is highly suspect and invites rigorous specimen-based corroboration that also addresses possible relationships to certain forms masquerading under *tropicalis*. The occurrence of greater species endemism thoughout the isolated East African highlands and volcanoes deserves more serious consideration than it has received to date.

Otomys unisulcatus F. Cuvier, 1829. *In* E. Geoffroy and F. Cuvier, Hist. Nat. Mammifères, pt. 3, 6(60):1-2 "Otomys cafre".

TYPE LOCALITY: South Africa, Cape Province, SW Karroo, Matjiesfontein, SW of Laingsburg (as designated by Roberts, 1946:318).

DISTRIBUTION: NW Cape Province, through the Great and Little Karroo, to E Cape Province, South Africa.

SYNONYMS: *albaniensis, bergensis, broomi, grantii.*

COMMENTS: A species having a relictual distribution and exhibiting many traits interpreted as plesiomorphic for the subfamily (e.g., Bohmann, 1952). Genetic distance data reported by Taylor et al. (1989) suggest the inclusion of *unisulcatus* with species of *Parotomys*, an intriguing possibility that deserves substantiation with a broader taxonomic sampling. Allozymic variation unappreciable over species range and questions validity of subspecific divisions (see Van Dyk et al., 1991).

Parotomys Thomas, 1918. Ann. Mag. Nat. Hist., ser. 9, 2:204.

TYPE SPECIES: *Euryotis brantsii* A. Smith, 1834.

SYNONYMS: *Liotomys.*

COMMENTS: Bohmann (1952) allocated *Parotomys* as another synonym of an all-inclusive genus *Otomys*. Maintained as a separate genus by Ellerman (1941) and in most later systematic accounts (Ellerman et al., 1953; Misonne, 1974; De Graaff, 1981; Meester et al., 1986). Roberts (1951) also elevated *Liotomys* to generic rank. See comments under subfamily.

Parotomys brantsii (A. Smith, 1834). S. Afr. Quart. J., Ser. 2, 2: 150.

TYPE LOCALITY: South Africa, Cape Province, Little Namaqualand, "toward the mouth of the Orange River."

DISTRIBUTION: C and NW Cape Province, South Africa, to SW Botswana and SE Namibia.

SYNONYMS: *deserti, luteolus, pallida, rufifrons.*

COMMENTS: Although Port Nolloth, as restricted by Thomas and Schwann (1904:178), is usually cited as the type locality of *P. brantsii* (e.g., Ellerman et al., 1953; Meester et al., 1986), a careful reading of Thomas and Schwann suggests that they had actually associated one of Smith's cotypes with a series collected at Klipfontein, a place some 50 miles inland from Port Nolloth. The notion that the type locality was "restricted" to Port Nolloth apparently stems from an inadvertent indication in Roberts (1951). A future revisor of the species should clarify this matter.

Parotomys littledalei Thomas, 1918. Ann. Mag. Nat. Hist., ser. 9, 2:205.

TYPE LOCALITY: South Africa, Cape Province, Bushmanland, Kenhardt, Tuin.

DISTRIBUTION: C and NW Cape Province, South Africa, to S and W Namibia.

SYNONYMS: *molopensis, namibensis.*

COMMENTS: Type species of *Liotomys*, named as a subgenus of *Parotomys* by Thomas (1918*b*) and viewed as a genus by Roberts (1951).

Subfamily Petromyscinae Roberts, 1951. Mammals S Africa, p. 434.

COMMENTS: Of the two genera in this subfamily, *Delanymys* and *Petromyscus*, Lavocat (1964) wrote that their molar cusp patterns showed structural links between *Mystromys* and the dendromurines, a view shared by Verheyen (1965*b*). Petter (1967*b*), however, interpreted the derived molar traits shared by *Delanymys* and *Petromyscus* to indicate close phylogenetic alliance and he united them in the subfamily Petromyscinae, also pointing out that they had no close relationship to dendromurines. Carleton and Musser (1984) provided a diagnosis, characterization, and other remarks on contents and definition of the Petromyscinae. They also noted that aside from the shared molar peculiarities, the two genera are very different from one another and may not be part of the same monophyletic group. The subfamily is represented in the Plio-Pleistocene by the South African *Stenodontomys darti* (Pocock, 1987).

Delanymys Hayman, 1962. Rev. Zool. Bot. Afr., 65:1-2.

TYPE SPECIES: *Delanymys brooksi* Hayman, 1962.

Delanymys brooksi Hayman, 1962. Rev. Zool. Bot. Afr., 65:1-2.

TYPE LOCALITY: SW Uganda, Kigezi, near Kanaba, Echuya (or Muchuya) Swamp, 7500 ft (see Hayman, 1962).

DISTRIBUTION: SW Uganda, Zaire (Kivu), and Rwanda (Volcanos and Nyungwe forest).

COMMENTS: Available information on morphology, geographic distribution, and ecology can be found in Hayman (1962, 1963*a*), Verheyen (1965*b*), Dieterlen (1969*b*), and Van der Straeten and Verheyen (1983).

Petromyscus Thomas, 1926. Ann. Mag. Nat. Hist., ser. 9, 17:179.

TYPE SPECIES: *Praomys collinus* Thomas and Hinton, 1925.

COMMENTS: Most faunal lists, either without question or with reservations, recognized only two species (Ellerman et al., 1953; Meester et al., 1986), but our study of museum specimens and original descriptions of taxa revealed the presence of four distinctive species, a conclusion also recorded by Skinner and Smithers (1990).

Petromyscus barbouri Shortridge and Carter, 1938. Ann. S. Afr. Mus., 32:288.

TYPE LOCALITY: South Africa, NW Cape Province, Little Namaqualand, Witwater, Kamiesberg, 3500-3800 ft.

DISTRIBUTION: Known only from Springbok and Kamiesberg regions and Loeriesfontein area in Little Namaqualand, NW Cape Province, South Africa (see map in Skinner and Smithers, 1990:315); limits unknown.

COMMENTS: Although listed as a subspecies of *P. collinus* by Meester et al. (1986), *P. barbouri* is separated from *P. collinus* not only by the diagnostic short and bicolored tail noted by Shortridge and Carter (1938), but by its smaller skull, relatively shorter rostrum, much shorter molar rows, and lack of postaxillary teats.

Petromyscus collinus (Thomas and Hinton, 1925). Proc. Zool. Soc. Lond., 1925:237.

TYPE LOCALITY: Namibia, Damaraland, Karibib, northwest of Windhoek, 3142 ft.

DISTRIBUTION: From Kaokoveld region in N Namibia south through Namibia to extreme NW Cape Prov. of South Africa south of the Orange River in Goodhouse and Pella areas; also recorded from C and W Cape Prov. and SW Angola by Skinner and Smithers (1990:314).

SYNONYMS: *bruchus, capensis, kaokoensis, kurzi, namibensis, rufus, variabilis.*

COMMENTS: Roberts (1951) listed *capensis*, known only from Goodhouse, as a species, and Meester et al. (1986) treated it as a synonym of *barbouri*. In describing it as a subspecies of *P. collinus*, however, Shortridge and Carter (1938) reflected its true affinities because its morphology is unlike *P. barbouri*, which Shortridge and Carter described in the same paper and knew well.

Roberts (1951) also treated *bruchus* (type locality is Great Brukkaros Mtn in S Namibia) as a separate species, allocating the northern *shortridgei* to it as a subspecies. The latter is clearly a species separate from *P. collinus* (see comments under *shortridgei*), but the status of *bruchus* will have to be illuminated in a revision of *Petromyscus*; the few specimens we studied from Great Brukkaros Mtn were examples of *P. collinus*.

Petromyscus monticularis (Thomas and Hinton, 1925). Proc. Zool. Soc. Lond., 1925:238.
TYPE LOCALITY: Namibia, Great Namaqualand, Great Brukkaros Mtn, near Berseba.
DISTRIBUTION: S Namibia: vicinity of type locality and south of there between Aus region in the west and South African border near Rietfontein area in the east. South Africa: in extreme N Cape Prov. on south bank of the Orange River at Augrabies Falls (USNM 452333).
COMMENTS: Geographic limits of this distinctive species have yet to be determined.

Petromyscus shortridgei Thomas, 1926. Proc. Zool. Soc. Lond., 1926:302.
TYPE LOCALITY: Extreme S Angola, Rua Cana Falls, 3350 ft.
DISTRIBUTION: W and S Angola and N Namibia (S to Erongo Mtns and Okahandja region); limits unknown; see map in Skinner and Smithers (1990:316).
COMMENTS: Roberts (1951) treated *shortridgei* as a subspecies of *P. bruchus*, and Meester et al. (1986) listed it as a subspecies of *P. collinus*. However, Thomas correctly expressed the distinctness of the animal by describing it as a species, which was also Schlitter's evaluation (*in* Meester et al., 1986), and that of Skinner and Smithers (1990). The larger size of *shortridgei*, darker fur with a less silky texture, and lack of postaxillary teats clearly separate it from *P. collinus*, a judgement based upon specimens in the American Museum of Natural History and the Field Museum of Natural History. Our estimate of its geographic range, which overlaps that of *P. collinus*, is also derived from those series.

Subfamily Platacanthomyinae Alston, 1876. Proc. Zool. Soc. Lond., 1876:81.
COMMENTS: Diagnosis, general characteristics, and natural history provided by Carleton and Musser (1984), who also referenced changes in allocation of subfamily from Myoxidae to Muridae, and explained why platacanthomyines are not dormice. Closest relatives of *Platacanthomys* and *Typhlomys* are two species of *Neocometes* from European Miocene (Carleton and Musser, 1984), and a third species from Lower Miocene in N Thailand (Mein et al., 1990). "*Platacanthomys* and *Typhlomys*, although chracterized by many specialized features, appear to be relicts of an assemblage that is recognizable as platacanthomyine as far back as the early Miocene, a group that may have had its origins in some primitive and as yet unknown Eocene or Oligocene [muroid] stock, probably in Asia" (Carleton and Musser, 1984:368).

Platacanthomys Blyth, 1859. J. Asiat. Soc. Bengal, 28:288.
TYPE SPECIES: *Platacanthomys lasiurus* Blyth, 1859.

Platacanthomys lasiurus Blyth, 1859. J. Asiat. Soc. Bengal, 28:289.
TYPE LOCALITY: India, Malabar, Alipi.
DISTRIBUTION: Forests below 3000 ft in S India.
COMMENTS: Known by very few specimens.

Typhlomys Milne-Edwards, 1877. Bull. Sci. Soc. Philom. Paris, ser. 6, 12:9 [1877].
TYPE SPECIES: *Typhlomys cinereus* Milne-Edwards, 1877.
COMMENTS: The Miocene *Neocometes* is morphologically closely allied to *Typhlomys* (see review in Carleton and Musser, 1984).

Typhlomys chapensis Osgood, 1932. Field Mus. Nat. Hist., Zool. Ser., 18:298.
TYPE LOCALITY: N Vietnam, Chapa.
DISTRIBUTION: Known only from the type locality.
COMMENTS: Originally described as subspecies of *T. cinereus*, but its far greater size, pointed out by Osgood, clearly distinguishes it from the Chinese species. Still only represented by the 14 specimens "obtained by Delacour and Lowe at Chapa, all received from native collectors" (Osgood, 1932:298).

Typhlomys cinereus Milne-Edwards, 1877. Bull. Sci. Soc. Philom. Paris, ser. 6, 12:9 [1877].
TYPE LOCALITY: China, W Fujian.
DISTRIBUTION: S China (Yunnan, Fujian, Guangxi, and S Anhui); records are from montane forest (Allen, 1940; Wu and Wang, 1984; Liu et al., 1985).
SYNONYMS: *jindongensis*.
COMMENTS: Wu and Wang (1984) described *jindongensis* as a subspecies of *T. cinereus*.

Subfamily Rhizomyinae Winge, 1887. E Museo Lundii, 1:109.

SYNONYMS: Tachyoryctinae.

COMMENTS: Diagnosed and reviewed by Carleton and Musser (1984) and Flynn (1990), who also summarized relationships of extant and extinct genera in form of a cladogram. Flynn treated the group as a separate family and utilized Tachyoryctinae for *Tachyoryctes* and Rhizomyinae for *Rhizomys* and *Cannomys;* in our scheme these two groups would be treated as tribes.

Cannomys Thomas, 1915. Ann. Mag. Nat. Hist., ser. 8, 16:57.

TYPE SPECIES: *Rhizomys badius* Hodgson, 1841.

COMMENTS: The sister-genus of *Rhizomys,* a relationship based on cladistic analyses of skeletal and dental traits (Flynn, 1990).

Cannomys badius (Hodgson, 1841). Calcutta J. Nat. Hist., 2:60.

TYPE LOCALITY: Nepal.

DISTRIBUTION: E Nepal, through N and NE India (Bhutan, Sikkim, Assam), SE Bangladesh, Burma, S China (Yunnan), Thailand, and Cambodia.

SYNONYMS: *castaneus, lonnbergi, minor, pater, plumbescens.*

COMMENTS: Kock and Posamentier (1983) and Lekagul and McNeely (1977) provided detailed and general distribution maps.

Rhizomys Gray, 1831. Proc. Zool. Soc. Lond., 1831:95.

TYPE SPECIES: *Rhizomys sinensis* Gray, 1831.

SYNONYMS: *Nyctocleptes.*

Rhizomys pruinosus Blyth, 1851. J. Asiat. Soc. Bengal, 20:519.

TYPE LOCALITY: India, Assam, Khasi Hills, Cherrapunji.

DISTRIBUTION: S China (Yunnan, Guangxi, Guangdong), NE India (Assam), E Burma, Thailand, Laos, Cambodia, Vietnam, south to Perak on Malay Peninsula.

SYNONYMS: *latouchei, pannosus, prusianus, senex, umbriceps* (see Ellerman and Morrison-Scott, 1951; Medway, 1969).

Rhizomys sinensis Gray, 1831. Proc. Zool. Soc. Lond., 1831:95.

TYPE LOCALITY: China, Guangdon, near Canton.

DISTRIBUTION: S and C China (Sichuan, north to S Gansu and S Shaanxi, east and south through Hubei, Anhui, and Fujian; see Allen, 1940; Ellerman and Morrison-Scott, 1951; Wang, 1990; Zheng and Zhang, 1990), N Burma, and Vietnam.

SYNONYMS: *chinensis, davidi, reductus, troglodytes, vestitus, wardi* (see Ellerman and Morrison-Scott, 1951).

COMMENTS: Dao and Cao (1990) described *reductus* as a subspecies of *R. sinensis* from S Vietnam.

Rhizomys sumatrensis (Raffles, 1821). Trans. Linn. Soc. Lond., 13:258.

TYPE LOCALITY: Malaysia, Malacca.

DISTRIBUTION: Sumatra, Malay Peninsula, Thailand, Laos, Cambodia, Vietnam, S China (Yunnan), and Burma.

SYNONYMS: *cinereus, dekan, erythrogenys, insularis, javanus, padangensis.*

Tachyoryctes Rüppell, 1835. Neue wirbelth. Fauna Abyssin. Gehörig., Säugeth., 1:35, footnote.

TYPE SPECIES: *Bathyergus splendens* Rüppell, 1835.

SYNONYMS: *Chrysomys.*

COMMENTS: Hollister (1919:40) listed eight species of *Tachyoryctes,* noting that "all have constant characters of differentiation, and intergradation between any two of them is not indicated by this material," and speculated that the "numerous forms will doubtlessly be connected by complete chains of intergrades and the final monographer of the genus will be obliged to reduce many of the named forms to the rank of subspecies." Allen (1939) and Ellerman (1941) listed 14 species of *Tachyoryctes,* but these were reduced to *T. macrocephalus* and *T. splendens* by Misonne (1974) who was followed by Corbet and Hill (1991). Of the twenty forms described, Rahm (1980) considered about 16 of them to be valid subspecies of *T. splendens.*

This change from 14 to 2 species was not based on careful analyses of morphological variation characterizing the named forms. Aside from Bekele's (1986) inconclusive univariate analyses of craniometric data from samples of *Tachyoryctes*, no assessment of morphological variation in the genus is available in context of a systematic revision. Until such a study provides documentation and definition of specific limits and their geographic distributions, we return to a modifed treatment of the arrangement by Allen (1939) and Ellerman (1941), which reflects the diagnostic information now available.

Tachyoryctes ankoliae Thomas, 1909. Ann. Mag. Nat. Hist., ser. 8, 4:545.
TYPE LOCALITY: S Uganda, Burumba, Ankole.
DISTRIBUTION: S Uganda; limits unknown.
COMMENTS: Recorded as a species by Swynnerton and Hayman (1951) but as a subspecies of *T. splendens* by Delany (1975), who nevertheless noted the cranial traits by which *ankoliae* could be distinguished from the adjacent *ruddi*.

Tachyoryctes annectens Thomas, 1891. Ann. Mag. Nat. Hist., ser. 6, 7:304.
TYPE LOCALITY: Kenya, Mianzini, E of Lake Naivasha (Hollister, 1919).
DISTRIBUTION: Known only from vicinity of type locality.
COMMENTS: Hollister (1919) commented on the distinctive large size of the holotype.

Tachyoryctes audax Thomas, 1910. Ann. Mag. Nat. Hist., ser. 8, 5:421.
TYPE LOCALITY: Kenya, summit of Aberdare Range, 10,000 ft.
DISTRIBUTION: Recorded from Aberdare Mtns, Kenya (Hollister, 1919).

Tachyoryctes daemon Thomas, 1909. Ann. Mag. Nat. Hist., ser. 8, 4:545.
TYPE LOCALITY: N Tanzania, Mt Kilimanjaro, 5000 ft.
DISTRIBUTION: Recorded from N Tanzania (Mt Kilimanjaro, Mt Meru and other highlands between Kilimanjaro and Lake Victoria and between the lake and Rwanda); limits unknown.
COMMENTS: Records are from Hollister (1919) and Swynnerton and Hayman (1951), who also listed *daemon* as a species.

Tachyoryctes macrocephalus Rüppell, 1842. Mus. Senckenberg., 3:97.
TYPE LOCALITY: Ethiopia, Shoa.
DISTRIBUTION: Ethiopia; endemic to high southern plateau (Rupp, 1980).
SYNONYMS: *hecki*.
COMMENTS: Inhabitant of Afro-alpine moorland and grassland, which it shares with the other Ethiopian endemics *Stenocephalemys albocaudata*, *Lophuromys melanonyx*, and *Arvicanthis blicki* (Yalden, 1988). *Tachyoryctes splendens* ranges from below 1000 m to the high moorland (4000 m) where it occurs together with *T. macrocephalus* (Rupp, 1980; Yalden et al., 1976). Yalden (1975) provided ecological observations for *T. macrocephalus* and morphological contrasts between it and *T. splendens*. See Yalden (1985, Mammalian Species, 237).

Tachyoryctes naivashae Thomas, 1909. Ann. Mag. Nat. Hist., ser. 8, 4:547.
TYPE LOCALITY: Kenya, Lake Naivasha, 6350 ft.
DISTRIBUTION: Kenya; recorded from W and S of Lake Naivasha (Hollister, 1919).
COMMENTS: Considered by Thomas (1909*b*:547) to be the smallest of the East African *Tachyoryctes* and occurring "quite close" to the two largest (*annectens* and *storeyi*).

Tachyoryctes rex Heller, 1910. Smithson. Misc. Coll., 56:4.
TYPE LOCALITY: Kenya, W slopes of Mt. Kenya, 10,000 ft.
DISTRIBUTION: Recorded only from 9000 to 10,700 ft on slopes of Mt. Kenya (Hollister, 1919).
COMMENTS: A large-bodied and distinctive species (Hollister, 1919).

Tachyoryctes ruandae Lönnberg and Gyldenstolpe, 1925. Ark. Zool., 17B, no. 5:6.
TYPE LOCALITY: Rwanda, Mt. Muhavura.
DISTRIBUTION: E Zaire (Kivu), Rwanda, and Burundi.
COMMENTS: See Elbl et al. (1966) and Rahm (1967) for distributional and ecological information. Stomach morphology was described by Rahm (1976) and chromosomal data was reported by Matthey (1967).

Tachyoryctes ruddi Thomas, 1909. Ann. Mag. Nat. Hist., ser. 8, 4:547.
TYPE LOCALITY: Kenya, Mt. Elgon, Kirui, 6000 ft.
DISTRIBUTION: SW Kenya (vicinity of Mt. Elgon) and SE Uganda.
SYNONYMS: *badius*.
COMMENTS: Hollister (1919) commented on pelage traits.

Tachyoryctes spalacinus Thomas, 1909. Ann. Mag. Nat. Hist., ser. 8, 4:547.
TYPE LOCALITY: Kenya, Embi, near Mt. Kenya, 5400 ft.
DISTRIBUTION: Recorded from the plains in vicinity of Mt. Kenya (Hollister, 1919).

Tachyoryctes splendens (Rüppell, 1835). Neue wirbelt. Fauna Abyssin. Gehörig., Säugeth., 1:36.
TYPE LOCALITY: Ethiopia, Dembea Prov, Gondar.
DISTRIBUTION: Ethiopia (500-3900 m; Rupp, 1980), Somalia, and NW Kenya; limits unknown.
SYNONYMS: *canicaudus, cheesmani, gallarum, ibeanus, omensis, pontifex, somalicus*.
COMMENTS: Yalden et al. (1976) claimed that the synonyms (except for *ibeanus*) listed here clearly apply to one species. Osgood (1936), however, noted that two species, *splendens* and *cheesmani*, could be distinguished among the Ethiopian samples he examined. We include *ibeanus* because it is geographically close to the range of *T. splendens* and was originally described as a subspecies of that species (see Allen, 1939). Hollister (1919) identified series from SE Kenya as *T. ibeanus*.

Subfamily Sigmodontinae Wagner, 1843. *In* Schreber, Die Säugethiere, Suppl., 3:398.
SYNONYMS: Akodontini, Hesperomyinae, Ichthyomyini, Neotominae, Onychomyini, Oryzomyini, Peromyscini, Phyllotini, Reithrodontini, Reithrodontomyini, Scapteromyini, Thomasomyini, Tylomyinae, Wiedomyini.
COMMENTS: Second only to Murinae in generic and specific diversity. Priority of family-group name Sigmodontinae set forth by Hershkovitz (1966*b*) and Reig (1980). Taxonomic and nomenclatural histories of many forms compiled by Tate (1932*a-h*). State-of-the-art alpha-level classifications presented by Miller (1924), Gyldenstolpe (1932), Ellerman (1941), Hall and Kelson (1959), Cabrera (1961), and Hall (1981). The studies of Carleton (1980), Gardner and Patton (1976), Hershkovitz (1962, 1966*c*), Hooper and Musser (1964*a*), and Reig (1980, 1984, 1987) contain information on higher-level relationships and classificatory arrangements. For an overview of phylogenetic diversification and biogeography of Sigmodontinae, see Hershkovitz (1966*b*) and Reig (1984, 1986); for a paleontological background, see Baskin (1986), Martin (1980), Marshall (1979), Reig (1978), and Slaughter and Ubelaker (1984). Sigmodontine genera have been informally or formally grouped into tribes (see Carleton and Musser, 1989; Hershkovitz, 1966*c*; and Reig, 1980, 1984); however, convincing evidence of monophyly has been mustered for only one of these tribal constructs (see Voss, 1988). The tribal affiliations, cited here in their informal adjectival construction, basically conform to Reig (1980, 1984), except that we maintain the thomasomyines (= *Aepeomys, Delomys, Phaenomys, Rhagomys, Rhipidomys, Thomasomys*, and *Wilfredomys*) as distinct from the oryzomyines (see Thomas, 1906*d*, 1917*c*; Hershkovitz, 1966*c*; Carleton and Musser, 1989).

Abrawayaomys Cunha and Cruz, 1979. Bol. Mus. Biol. Prof. Mello-Leitao Zool., 96:2.
TYPE SPECIES: *Abrawayaomys ruschii* Cunha and Cruz, 1979.
COMMENTS: Diagnostic traits of the new genus seem to combine aspects of *Neacomys, Oryzomys*, and *Akodon*, and Reig (1987) acknowledged the enigmatic affinities of *Abrawayaomys* as Sigmodontinae *incertae sedis*. Certain cranial features of *Abrawayaomys* suggest an archaic thomasomyine, perhaps distantly related to other thomasomyine genera of SE Brazil (e.g., see comment under *Delomys*).

Abrawayaomys ruschii Cunha and Cruz, 1979. Bol. Mus. Biol. Prof. Mello-Leitao Zool., 96:2.
TYPE LOCALITY: Brazil, Espírito Santo, Forno Grande, Castelo.
DISTRIBUTION: Known only from the states of Espírito Santo and Minas Gerais, Brazil, and Misiones province, Argentina (after Massoia et al., 1991).

Aepeomys Thomas, 1898. Ann. Mag. Nat. Hist., ser. 7, 1:452.
TYPE SPECIES: *Oryzomys lugens* Thomas, 1896.

COMMENTS: Thomasomyine. Synonymized under *Thomasomys* by Osgood (1933*c*) and so followed by Ellerman (1941) and Cabrera (1961); generic status maintained by Gyldenstolpe (1932) and Gardner and Patton (1976). Level of recognition of *Aepeomys* requires phylogentic investigation involving other thomasomyines.

Aepeomys fuscatus J. A. Allen, 1912. Bull. Am. Mus. Nat. Hist., 31:89.
TYPE LOCALITY: Colombia, Valle del Cauca Dept., San Antonio, 2040 m.
DISTRIBUTION: W and C Andes of Colombia.
COMMENTS: Formerly ranked as a subspecies of *A. lugens* (Cabrera, 1961). Chromosomal morphology highly divergent among thomasomyines reported by Gardner and Patton (1976), who listed it as a species without explanation.

Aepeomys lugens (Thomas, 1896). Ann. Mag. Nat. Hist., ser. 6, 18:306.
TYPE LOCALITY: Venezuela, Merida, La Loma del Morro, 3000 m.
DISTRIBUTION: Merida Andes of Venezuela to Andean Ecuador.
SYNONYMS: *ottleyi, vulcani*.

Akodon Meyen, 1833. Verhandl. Kais. Leop.-Carol. Akad. Wiss., 16(2):599.
TYPE SPECIES: *Akodon boliviensis* Meyen, 1833.
SYNONYMS: *Abrothrix, Chalcomys, Deltamys, Hypsimys, Microxus, Thaptomys*.
COMMENTS: The morphotypical akodontine genus stands at the nexus of a host of specific- and generic-level taxonomic problems. To review all classificatory variations here would be more confusing than enlightening; suffice it to say that the nomenclatural history of the following genus-group taxa has been variously intertwined with that of *Akodon*: *Abrothrix, Bolomys, Chalcomys, Chroeomys, Deltamys, Hypsimys, Microxus, Thalpomys*, and *Thaptomys* (especially see Cabrera, 1961; Ellerman, 1941; Gyldenstolpe, 1932; Reig, 1984, 1987; Tate, 1932*g*; Thomas, 1916*c*). Comprehension of akodont systematics has been hindered by lack of explicit specimen documentation, careful character analyses, and a phylogenetic context, which in turn erodes decisions about the ranking of genus-group taxa. Fortunately, renewed interest in basic revisionary study will render the alpha systematics of akodonts less imposing (see Apfelbaum and Reig, 1989; Hershkovitz, 1990*a, c*; Myers, 1989; Myers et al., 1990; Patton et al., 1989; Reig, 1987). See accounts of *Bolomys, Chroeomys*, and *Thalpomys* for history and arguments on their generic status.

Akodon aerosus Thomas, 1913. Ann. Mag. Nat. Hist., ser. 8, 11:406.
TYPE LOCALITY: Ecuador, Tungurahua Prov., upper Rio Pastaza, Mirador, 1500 m.
DISTRIBUTION: SE Ecuador, E Peru, and NW Bolivia.
SYNONYMS: *baliolus*.
COMMENTS: Type species of *Chalcomys*, usually placed as a synonym of the subgenus *Akodon* (Cabrera, 1961; Reig, 1987). Formerly included as a subspecies of *A. urichi* by Cabrera (1961), but Gardner and Patton (1976), noting the pronounced difference in diploid number, listed the two as separate species. Probably more than one species yet masquerades under the name of *A. aerosus* (see Patton et al., 1990).

Akodon affinis (J. A. Allen, 1912). Bull. Am. Mus. Nat. Hist., 31:89.
TYPE LOCALITY: Colombia, Valle del Cauca Dept., San Antonio, near Cali, 8000 ft.
DISTRIBUTION: Cordillera Occidental of W Colombia.
SYNONYMS: *tolimae*.
COMMENTS: Subgenus *Akodon*. Tentatively affiliated with *Microxus* by Gyldenstolpe (1932) but Cabrera (1961) placed it with *Akodon* proper.

Akodon albiventer Thomas, 1897. Ann. Mag. Nat. Hist., ser. 6, 20:217.
TYPE LOCALITY: Argentina, Salta Prov., Bajo Rio Cachi.
DISTRIBUTION: SE Peru, through WC Bolivia, to N Argentina and Chile.
SYNONYMS: *berlepschii*.
COMMENTS: Subgenus *Akodon*. Sometimes referenced as a member of *Bolomys* (Bianchi et al., 1971; Gardner and Patton, 1976), but its inclusion, together with *berlepschii*, within *Akodon* proper is more strongly supported (Pine et al., 1979; Reig, 1987).

Akodon azarae (J. Fischer, 1829). Synopsis Mamm., p. 325.
TYPE LOCALITY: Argentina, Entre Rios Prov., about 30°30'S latitude between the Uruguay and Parana Rivers.

DISTRIBUTION: NE Argentina, southernmost Bolivia, Paraguay, Uruguay, and extreme S Brazil.

SYNONYMS: *agreste, arenicola, bibianae, hunteri.*

COMMENTS: Subgenus *Akodon*. Olrog and Lucero (1981) maintained *arenicola* as a species distinct from *A. azarae*, but karyotypic and other data support their union (Vitullo et al., 1986; Ximenez et al., 1972).

Akodon bogotensis Thomas, 1895. Ann. Mag. Nat. Hist., ser. 6, 16:369.

TYPE LOCALITY: Colombia, Cundinamarca, Dept., Bogota region, 8750 ft.

DISTRIBUTION: Andes of W Venezuela, E and C Colombia.

COMMENTS: Subgenus *Microxus*. As noted by Patton et al. (1989), generic-level representation of the divergence of *Microxus* from *Akodon* is more often based on traits of the species *bogotensis*, not the type species of *Microxus* (= *Oxymycterus mimus*). The status of *bogotensis*, together with that of *latebricola*, deserves reconsideration with respect to *A. mimus* (see comments therein) and other *Akodon*. Also see Reig (1987) and Voss and Linzey (1981).

Akodon boliviensis Meyen, 1833. Verhandl. Kais. Leop.-Carol. Akad. Wiss., 16(2):600, pl. 43, fig. 1.

TYPE LOCALITY: Peru, Puno Dept., Pichu-Pichun, 14,000 ft (location clarified by Myers et al., 1990:49).

DISTRIBUTION: Altiplano of SE Peru and NC Bolivia.

SYNONYMS: *pacificus.*

COMMENTS: Subgenus *Akodon*. Karyology and geographic variation evaluated by Myers et al. (1990), who more narrowly defined the morphological and distributional boundaries of *A. boliviensis*, excluding *spegazzinii* and *subfuscus*, forms which had been arranged as subspecies (e.g., Cabrera, 1961). Southern range limits poorly understood (see Myers et al., 1990).

Akodon budini (Thomas, 1918). Ann. Mag. Nat. Hist., ser. 9, 1:191.

TYPE LOCALITY: Argentina, Jujuy Prov., Leon, 1500 m.

DISTRIBUTION: Mountains of NW Argentina.

SYNONYMS: *deceptor.*

COMMENTS: Type species of *Hypsimys*, which Cabrera (1961) and Reig (1987) recognized as a subgenus of *Akodon*. Karyotype reported by Vitullo et al. (1986) and biochemical divergence by Apfelbaum and Reig (1989).

Akodon cursor (Winge, 1887). E Museo Lundii, 1(3):25.

TYPE LOCALITY: Brazil, Minas Gerais, Lagoa Santa, Rio das Velhas.

DISTRIBUTION: C and SE Brazil, Uruguay, E Paraguay, and NE Argentina.

SYNONYMS: *montensis.*

COMMENTS: Subgenus *Akodon*. Formerly included in *arviculoides* (e.g., Cabrera, 1961; Gyldenstolpe, 1932), which Reig (1978, 1987) reallocated to *Bolomys lasiurus*.

Akodon dayi Osgood, 1916. Field Mus. Nat. Hist. Publ., Zool. Ser., 10:208.

TYPE LOCALITY: Bolivia, Cochabamba Dept., Todos Santos, Chapare River.

DISTRIBUTION: C to SC Bolivia, 250-700 m.

COMMENTS: Subgenus *Akodon*, *varius* group *sensu* Myers (1989). Interpreted as a subspecies of *A. tapirapoanus* (= *Bolomys lasiurus*) by Cabrera (1961); considered a species closely related to *A. toba* by Myers (1989).

Akodon dolores Thomas, 1916. Ann. Mag. Nat. Hist., ser. 8, 18:324.

TYPE LOCALITY: Argentina, Cordoba Prov., near Villa Dolores, Yacanto, 900 m.

DISTRIBUTION: Sierra de Cordoba, C Argentina.

COMMENTS: Subgenus *Akodon*, *varius* group *sensu* Myers (1989). Considered closely related to *A. molinae* (Bianchi et al., 1979; Myers, 1989) if not conspecific with it (Hershkovitz, 1990*c*).

Akodon fumeus Thomas, 1902. Ann. Mag. Nat. Hist., ser. 7, 9:137.

TYPE LOCALITY: Bolivia, Cochabamba Dept., Rio Secure, Choro, 3500 m.

DISTRIBUTION: E Andean slopes of SE Peru and W Bolivia.

COMMENTS: Subgenus *Akodon*. Treated as a subspecies of *A. mollis* (Cabrera, 1961; Gyldenstolpe, 1932; Hershkovitz, 1990*c*); viewed as a distinct species more closely related to *A. kofordi* by Myers and Patton (1989*b*).

Akodon hershkovitzi Patterson, Gallardo, and Freas, 1984. Fieldiana Zool., New Ser., 23:8.
TYPE LOCALITY: Chile, Magallanes Prov., Isla Capitan Aracena, head of Bahia Morris, 60 m; 54°14'S, 71°30'W.
DISTRIBUTION: Outer islands of the Chilean Archipelago.
COMMENTS: Subgenus *Akodon* after Patterson et al. (1984); subgenus *Abrothrix sensu* Reig (1987). A yellow-nosed species related to *A. xanthorhinus*.

Akodon illuteus (Thomas, 1925). Ann. Mag. Nat. Hist., ser. 9, 15:582.
TYPE LOCALITY: Argentina, Tucuman Prov., Sierra de Aconquija, 3000-4000 m.
DISTRIBUTION: NW Argentina.
COMMENTS: Variably placed with *Abrothrix*, usually as a subgenus (Cabrera, 1961; Reig, 1987), or with *Akodon sensu stricto* (Gardner and Patton, 1976).

Akodon iniscatus Thomas, 1919. Ann. Mag. Nat. Hist., ser. 9, 3:205.
TYPE LOCALITY: Argentina, Chubut Prov., Valle del Lago Blanco, Koslowsky region, 100 m.
DISTRIBUTION: WC to S Argentina.
SYNONYMS: *collinus*, *nucus*.
COMMENTS: Subgenus *Akodon*. The form *nucus* has been recently listed as a species (Hershkovitz, 1990*c*; Reig, 1987).

Akodon juninensis Myers, Patton, and Smith, 1990. Misc. Publ. Mus. Zool., Univ. Michigan, 177:41.
TYPE LOCALITY: Peru, Junín Dept., 22 km (by road) N La Oroya (at junction of Hwy 3 to Junín and Hwy 20 to Tarma), 4040 m.
DISTRIBUTION: E and W Andean slopes, above 2700 m, of C Peru, south along western slopes to Dept. Ayacucho.
COMMENTS: Subgenus *Akodon*. Considered a member of the *A. boliviensis* species group (see Myers et al., 1990).

Akodon kempi (Thomas, 1917). Ann. Mag. Nat. Hist., ser. 8, 20:99.
TYPE LOCALITY: Argentina, Buenos Aires Prov., Rio Parana, Isla Ella, 1 m.
DISTRIBUTION: EC Argentina and adjacent Uruguay.
SYNONYMS: *langguthi*.
COMMENTS: Type species of *Deltamys*, usually treated as a subgenus of *Akodon* (e.g., Cabrera, 1961; Reig, 1987), although Massoia (1980*b*) accorded the taxon generic status.

Akodon kofordi Myers and Patton, 1989. Occas. Pap. Mus. Zool., Univ. Michigan, 721:14.
TYPE LOCALITY: Peru, Puno Dept., 9 km (by road) N Limbani, Agualani, 2840 m.
DISTRIBUTION: Depts. Cusco and Puno, SE Peru, 2750 to 2900 m.
COMMENTS: Subgenus *Akodon*. Myers and Patton (1989*b*) informally united their new species with *A. fumeus* as the *A. fumeus* species group.

Akodon lanosus (Thomas, 1897). Ann. Mag. Nat. Hist., ser. 6, 20:218.
TYPE LOCALITY: Argentina, Tierra del Fuego Prov., Bahia Monteith, Straits of Magellan.
DISTRIBUTION: Southernmost Chile and Argentina.
COMMENTS: Subgenus *Abrothrix*. Relegated to a subspecies of *A. longipilis* by Mann (1978) but others have affirmed its specific status (Osgood, 1943; Yañez et al., 1978). Earlier associated with *Microxus* (e.g., Gyldenstolpe, 1932) but later allocated to subgenus *Abrothrix* of *Akodon* (Osgood, 1943; Reig, 1987).

Akodon latebricola (Anthony, 1924). Am. Mus. Novit., 139:3.
TYPE LOCALITY: Ecuador, Tungurahua Prov., Rio Cusutagua, Hacienda San Francisco, east of Ambato, 8000 ft.
DISTRIBUTION: Andean Ecuador.
COMMENTS: Subgenus *Microxus* (see remarks under *A. bogotensis* and *A. mimus*).

Akodon lindberghi Hershkovitz, 1990. Fieldiana Zool., New Ser., 57:16.
TYPE LOCALITY: Brazil, Distrito Federal, 20 km NW Brasilia, Parque Nacional de Brasilia, Matosa, 1100 m.

DISTRIBUTION: Cerrado habitat of the Parque Nacional de Brasilia.
COMMENTS: Subgenus *Akodon*. Morphology, karyotype, and habitat described by Hershkovitz (1990*c*), who assigned the species to the *A. boliviensis* species group. Hershkovitz (1990*c*) also discussed the *nomen nudum Plectomys paludicola* and its possible equivalence to the species *A. lindberghi*.

Akodon longipilis (Waterhouse, 1837). Proc. Zool. Soc. Lond., 1837:16.
TYPE LOCALITY: Chile, Coquimbo Prov., Coquimbo.
DISTRIBUTION: C to S Chile and Argentina.
SYNONYMS: *angustus, apta, brachytarsus, castanea, francei, fusco-ater, hirta, melampus, modestior, moerens, nubila, porcinus, suffusa.*
COMMENTS: Type species of *Abrothrix*, typically ranked as a subgenus of *Akodon* (Cabrera, 1961; Ellerman, 1941; Reig, 1978, 1987), occasionally as a genus (Gyldenstolpe, 1932), or with *Microxus* as a synonym (Hershkovitz, 1966*c*). See Reig (1987) for a review of treatment of *Abrothrix* and its specific contents. Species closely related to, perhaps conspecific with *sanborni* (see Osgood, 1943; Pine et al., 1979); also see *A. lanosus*. Pearson (1984) assigned *Chelemys angustus* to synonymy with *A. longipilis*; other synonyms follow Osgood (1943).

Akodon mansoensis De Santis and Justo, 1980. Neotropica, 26(75):121.
TYPE LOCALITY: Argentina, Rio Negro Prov., Bariloche Dept., Estacion Aforo, Rio Manso Superior.
DISTRIBUTION: Andes of Rio Negro Prov., Argentina.
COMMENTS: Subgenus *Abrothrix*. Reig (1987) questioned the subgeneric assignment as well as the validity of this form.

Akodon markhami Pine, 1973. Ann. Inst. Patagonia, 4(1-3):423-426.
TYPE LOCALITY: Chile, Magallanes Prov., Isla Wellington, about 1.2 km NW Puerto Eden.
DISTRIBUTION: Isla Wellington, S Chile.
COMMENTS: Subgenus *Akodon*.

Akodon mimus (Thomas, 1901). Ann. Mag. Nat. Hist., ser. 7, 7:183.
TYPE LOCALITY: Peru, Puno Dept., Limbane, 2600 m.
DISTRIBUTION: E Andean slopes of SE Peru to WC Bolivia.
COMMENTS: Type species of *Microxus*, a taxon arranged either as a subgenus of *Akodon* (Cabrera, 1961), as a distinct akodontine genus (Ellerman, 1941; Gyldenstolpe, 1932; Reig, 1987; Thomas, 1916*c*), or as a synonym of the oxymycterine genus *Abrothrix* (Hershkovitz, 1966*c*). Preliminary electrophoretic evidence indicates that *mimus* is phyletically closer to species of *Akodon sensu stricto* than to those of *Bolomys* or *Chroeomys* (see Patton et al., 1989; Smith and Patton, 1991). Also see remarks under *A. bogotensis*.

Akodon molinae Contreras, 1968. Zool. Platense, 1(2):9-12.
TYPE LOCALITY: Argentina, Buenos Aires Prov., Partido de Villarino, Laguna Chasico.
DISTRIBUTION: EC Argentina.
COMMENTS: Subgenus *Akodon*, *varius* group *sensu* Myers (1989). See remarks under *A. dolores*.

Akodon mollis Thomas, 1894. Ann. Mag. Nat. Hist., ser. 6, 14:363.
TYPE LOCALITY: Peru, Piura Dept., Tumbes.
DISTRIBUTION: High Andes of Ecuador to NC Peru.
SYNONYMS: *altorum, fulvescens.*
COMMENTS: Subgenus *Akodon*. See remarks under *A. fumeus*.

Akodon neocenus Thomas, 1919. Ann. Mag. Nat. Hist., ser. 9, 3:213.
TYPE LOCALITY: Argentina, Neuquen Prov., upper Rio Negro, Rio Limay.
DISTRIBUTION: C and S Argentina.
COMMENTS: Subgenus *Akodon*, *varius* group *sensu* Myers (1989). A subspecies of *A. varius* according to Cabrera (1961); a species more closely related to *A. dolores* or *A. toba* according to Myers (1989).

Akodon nigrita (Lichtenstein, 1829). Darst. Säugeth., 7:pl. 35, fig. 1.
TYPE LOCALITY: Brazil, vicinity of Rio de Janeiro.
DISTRIBUTION: SE Brazil, E Paraguay, and NE Argentina.

SYNONYMS: *fuliginosus, henseli, orycter, subterraneus.*

COMMENTS: Member of *Thaptomys*, usually segregated as a subgenus of *Akodon* (Cabrera, 1961; Massoia, 1963*a*) though Reig (1978, 1987) considered *Thaptomys* a full synonym of *Akodon sensu stricto*. Includes *subterraneus*, the type of *Thaptomys* (see Massoia, 1963*a*).

Akodon olivaceus (Waterhouse, 1837). Proc. Zool. Soc. Lond., 1837:16.

TYPE LOCALITY: Chile, Aconcagua Prov., Valparaiso.

DISTRIBUTION: N through midsouthern Chile, bordering area of westernmost Argentina, about to 48°S latitude.

SYNONYMS: *atratus, beatus, brachiotis, brevicaudatus, chonoticus, foncki, germaini, landbecki, lepturus, macronychos, mochae, nasica, nemoralis, pencanus, psilurus, renggeri, ruficaudus, senilis, trichotis, vinealis, xanthopus.*

COMMENTS: Subgenus *Akodon*. Yañez et al. (1979) arranged *xanthorhinus* and *canescens* as subspecies of *A. olivaceus*; Patterson et al. (1984) presented distributional and morphological arguments that sustain their specific distinction. Listing of Phillipi's numerous epithets follows Osgood (1943).

Akodon orophilus Osgood, 1913. Field Mus. Nat. Hist. Publ., Zool. Ser., 10:98.

TYPE LOCALITY: Peru, Amazonas Dept., Leimabamba, Alto Utcubamba, 2400 m.

DISTRIBUTION: Mountains of N Peru.

SYNONYMS: *orientalis.*

COMMENTS: Subgenus *Akodon*. Although initially described as a subspecies of *A. mollis*, Osgood (1943:197) later removed *orophilus* as a distinct species, at which rank it has since remained (Cabrera, 1961; Reig, 1987). Level of differentiation from *A. torques*, which Cabrera (1961) arranged as a subspecies of *A. orophilus*, uncertain.

Akodon puer Thomas, 1902. Ann. Mag. Nat. Hist., ser. 7, 9:136.

TYPE LOCALITY: Bolivia, Cochabamba Dept., upper Rio Secure, Choquecamate, 4000 m.

DISTRIBUTION: High altiplano of C Peru (Puno), through W Bolivia, to NW Argentina.

SYNONYMS: *caenosus, lutescens, polius.*

COMMENTS: Subgenus *Akodon*. Karyologic and morphometric variation presented, as *A. caenosus*, by Barquez et al. (1980). In their revision of the *boliviensis* group, Myers et al. (1990) referred *caenosus*, maintained as a species by Cabrera (1961), and *lutescens*, classified as a subspecies of *A. andinus* by Cabrera (1961), to subspecies of *A. puer* (also see Vitullo et al., 1986). Hershkovitz (1990*c*) listed both *caenosus* and *lutescens* as species affiliated with *A. boliviensis*.

Akodon sanborni Osgood, 1943. Field Mus. Nat. Hist. Publ., Zool. Ser., 30:194.

TYPE LOCALITY: Chile, Isla Chiloe, mouth of Rio Inio.

DISTRIBUTION: S Chile and adjacent Argentina.

COMMENTS: Subgenus *Abrothrix* (see Osgood, 1943, and Reig, 1987). Pine et al. (1979) urged further study of its differentiation from *A. longipilis*.

Akodon sanctipaulensis Hershkovitz, 1990. Fieldiana Zool., New Ser., 57:23.

TYPE LOCALITY: Brazil, São Paulo, Primeiro Morro.

DISTRIBUTION: Serra do Mar, SE Brazil.

COMMENTS: Subgenus *Akodon*. Considered a member of the *A. boliviensis* species group.

Akodon serrensis Thomas, 1902. Ann. Mag. Nat. Hist., ser. 7, 9:61.

TYPE LOCALITY: Brazil, Paraná, Serra do Mar, Roca Nova, 1000 m.

DISTRIBUTION: SE Brazil.

SYNONYMS: *leucogula.*

COMMENTS: Subgenus *Akodon*. Hershkovitz (1990*c*) listed *leucogula* as a species.

Akodon siberiae Myers and Patton, 1989. Occas. Pap. Mus. Zool., Univ. Michigan, 720:4.

TYPE LOCALITY: Bolivia, Cochabamba Dept., 28 km (by road) W Comarapa, 2800 m; 17°51'S, 64°40'W.

DISTRIBUTION: Known only from the vicinity of the type locality.

COMMENTS: Subgenus *Hypsimys*. Morphology and chromosomes suggest that *A. siberiae* is closely related to *A. budini*, the type species of *Hypsimys* (see Myers and Patton, 1989*a*).

Akodon simulator Thomas, 1916. Ann. Mag. Nat. Hist., ser. 8, 18:335.
TYPE LOCALITY: Argentina, Tucuman Prov., San Pablo, Villa Nouges, 1200 m.
DISTRIBUTION: E Andean foothills from SC Bolivia to NW Argentina.
SYNONYMS: *glaucinus, tartareus.*
COMMENTS: Subgenus *Akodon, varius* group. Placed as a subspecies of *A. varius* (Cabrera, 1961; Thomas, 1926*b*); returned to species rank by Myers (1989), who recognized *glaucinus* and *tartareus* as subspecies. Karyology and morphometrics reviewed, as *A. varius*, by Barquez et al. (1980).

Akodon spegazzinii Thomas, 1897. Ann. Mag. Nat. Hist., ser. 6, 20:216.
TYPE LOCALITY: Argentina, Salta Prov., lower Rio Cachi.
DISTRIBUTION: E Andean slopes, 400 to 1000 m, of NW Argentina.
SYNONYMS: *alterus, tucumanensis.*
COMMENTS: Subgenus *Akodon.* Karyology and morphometrics reviewed, as *A. boliviensis tucumanensis*, by Barquez et al. (1980). Myers et al. (1990) segregated *spegazzinii*, with *tucumanensis* as a subspecies, as a species distinct from *A. boliviensis.* Significance of marginal genetic and morphologic divergence of *alterus* evaluated by Blaustein et al. (1992).

Akodon subfuscus Osgood, 1944. Field Mus. Nat. Hist. Publ., Zool. Ser., 29:195.
TYPE LOCALITY: Peru, Puno Dept., drainage of Rio Inambari, Limbani, 9000 ft.
DISTRIBUTION: W and E Andean slopes of SC Peru to NW Bolivia (La Paz).
SYNONYMS: *arequipae.*
COMMENTS: Subgenus *Akodon.* Described as a subspecies of *A. boliviensis;* raised to specific level and another subspecies named by Myers et al. (1990).

Akodon surdus Thomas, 1917. Smithson. Misc. Coll., 68(4):2.
TYPE LOCALITY: Peru, Cuzco Dept., Huadquina, 5000 ft.
DISTRIBUTION: Andes of SE Peru.
COMMENTS: Subgenus *Akodon.* Distribution and affinities little known; retained as a species by Reig (1987), Myers (1989), and Hershkovitz (1990*c*).

Akodon sylvanus Thomas, 1921. Ann. Mag. Nat. Hist., ser. 9, 7:184.
TYPE LOCALITY: Argentina, Jujuy Prov., Sierra de Santa Barbara, Sunchal, 1200 m.
DISTRIBUTION: NW Argentina.
SYNONYMS: *pervalens.*
COMMENTS: Subgenus *Akodon.* Retained as a species until Cabrera (1961) synonymized it as a race of *A. azarae;* Myers (1989) tentatively listed *sylvanus* and *azarae* as separate species. The status and relationship of *pervalens* deserves resolution: retained as a nominal species similar to *A. cursor* by Myers (1989); placed as a synonym of *A. serrensis* by Hershkovitz (1990*c*). Arrangment here follows the describer, Thomas (1925).

Akodon toba Thomas, 1921. Ann. Mag. Nat. Hist., ser. 9, 7:178.
TYPE LOCALITY: Paraguay, Presidente Hayes Dept., northern Chaco, Jesematathla, 100 m.
DISTRIBUTION: Chaco of W Paraguay, E Boliva, and N Argentina.
COMMENTS: Subgenus *Akodon, varius* group *sensu* Myers (1989). Relegated to a subspecies of *A. varius* by Cabrera (1961); karyotypic and morphological discrimination from *varius* proper summarized by Myers (1989).

Akodon torques (Thomas, 1917). Smithson. Misc. Coll., 68:3.
TYPE LOCALITY: Peru, Cuzco Dept., Machu Picchu, 10,000 ft.
DISTRIBUTION: E Andean cloud forest of SE Peru.
COMMENTS: Subgenus *Akodon.* Relegated to a subspecies of *A. orophilus* (e.g., Cabrera, 1961), but recent studies have revealed that *torques* is a genetically distinctive form related to *A. mollis* (Patton et al., 1989, 1990). Described as a species of *Microxus* but considered a member of *Akodon* proper by later authors (Ellerman, 1941; Patton et al., 1989; Thomas, 1927).

Akodon urichi J. A. Allen and Chapman, 1897. Bull. Am. Mus. Nat. Hist., 9:19.
TYPE LOCALITY: Trinidad, Caparo.
DISTRIBUTION: Trinidad, Tobago, Venezuela, E Colombia, N Brazil.
SYNONYMS: *chapmani, meridensis, saturatus, venezuelensis.*

COMMENTS: Subgenus *Akodon*. Gardner and Patton (1976) considered *A. aerosus* distinct from *A. urichi*.

Akodon varius Thomas, 1902. Ann. Mag. Nat. Hist., ser. 7, 9:134.
TYPE LOCALITY: Bolivia, Cochabamba Dept., Cochabamba, 2400 m.
DISTRIBUTION: E Andean slopes, 2000-3000 m, of W Bolivia.
COMMENTS: Subgenus *Akodon*, *varius* group. Cabrera (1961) included as subspecies *neocenus*, *simulator*, and *toba*, forms here arranged as species according to the preliminary revision of Myers (1989).

Akodon xanthorhinus (Waterhouse, 1837). Proc. Zool. Soc. Lond., 1837:17.
TYPE LOCALITY: Chile, Magallanes Prov., Isla Hoste, Hardy Peninsula.
DISTRIBUTION: Extreme S Chile and Argentina, including Tierra del Fuego.
SYNONYMS: *canescens*, *llanoi*.
COMMENTS: Subgeneric allocation questionable: a member of *Akodon* following Patterson et al. (1984) or *Abrothrix* according to Reig (1987). Type locality, synonymy, distribution, and karyotype discussed by Patterson et al. (1984), who judged the form *infans*, synonymized under *A. xanthorhinus* by Osgood (1943), as a *nomen dubium* and considered *llanoi*, described by Pine (1976), as a subjective synonym. Genetically highly differentiated from typical *Akodon* (see Apfelbaum and Reig, 1989).

Andalgalomys Williams and Mares, 1978. Ann. Carnegie Mus., 47:197.
TYPE SPECIES: *Andalgalomys olrogi* Williams and Mares, 1978.
COMMENTS: A phyllotine that includes forms formerly associated with *Graomys*. Morphological and karyotypic traits summarized by Olds et al. (1987), who emended the generic diagnosis.

Andalgalomys olrogi Williams and Mares, 1978. Ann. Carnegie Mus., 47:203.
TYPE LOCALITY: Argentina, Catamarca Prov., 15 km (on route 62) W Andalgala, west bank Rio Amanao.
DISTRIBUTION: Known only from the vicinity of the type locality.

Andalgalomys pearsoni (Myers, 1977). Occas. Pap. Mus. Zool., Univ. Michigan, 676:1.
TYPE LOCALITY: Paraguay, Boqueron Dept., 410 km (by road) NW Villa Hayes.
DISTRIBUTION: Chaco of W Paraguay and SE Bolivia (Santa Cruz).
SYNONYMS: *dorbignyi*.
COMMENTS: Diploid count differs by one pair (76, 78) in the two subspecies (Myers, 1977; Olds et al., 1987).

Andinomys Thomas, 1902. Proc. Zool. Soc. Lond., 1902(1):116.
TYPE SPECIES: *Andinomys edax* Thomas, 1902.
COMMENTS: Phyllotine. Revised by Hershkovitz (1962). Standard karyotype reported and compared to other phyllotines by Pearson and Patton (1976) and Simonetti and Spotorno (1980).

Andinomys edax Thomas, 1902. Proc. Zool. Soc. Lond., 1902(1):116.
TYPE LOCALITY: Bolivia, Potosi Dept., between Potosi and Sucre, El Cabrado, 3700 m.
DISTRIBUTION: Altiplano of extreme S Peru (Puno) and N Chile (in Chile see Spotorno, 1976; Pine et al., 1979), through W Bolivia, to NW Argentina (Jujuy and Catamarca).
SYNONYMS: *lineicaudatus*.

Anotomys Thomas, 1906. Ann. Mag. Nat. Hist., ser. 7, 17:86.
TYPE SPECIES: *Anotomys leander* Thomas, 1906.
COMMENTS: Ichthyomyine. Formerly included *trichotis* (Handley, 1976), which Voss (1988) removed to the new genus *Chibchanomys*. Phylogenetic relationships studied by Voss (1988).

Anotomys leander Thomas, 1906. Ann. Mag. Nat. Hist., ser. 7, 17:87.
TYPE LOCALITY: Ecuador, Prov. Pichincha, Volcán Pichincha, 11,500 ft.
DISTRIBUTION: N Ecuador at high elevations.
COMMENTS: Highest diploid number (2n=92) yet reported for Mammalia (Gardner, 1971). Taxonomy and distribution reviewed by Voss (1988).

Auliscomys Osgood, 1915. Field Mus. Nat. Hist. Publ., Zool. Ser., 10:190.
TYPE SPECIES: *Reithrodon pictus* Thomas, 1884.
SYNONYMS: *Loxodontomys*.
COMMENTS: Phyllotine. Alpha systematics revised by Pearson (1958) and Hershkovitz (1962) as part of *Phyllotis*. Diagnosed as a subgenus of *Phyllotis*; variably recognized afterwards at that rank (Osgood, 1947; Pearson, 1958) or as a genus (Gyldenstolpe, 1932; Thomas, 1926*a*). Recent evidence supports the monophyly and probable earlier divergence of *Auliscomys* (Pearson and Patton, 1976; Simonetti and Spotorno, 1980).Chromosomal evolution of three of the four extant species presented by Walker and Spotorno (1992).

Auliscomys boliviensis (Waterhouse, 1846). Proc. Zool. Soc. Lond., 1846:9.
TYPE LOCALITY: Bolivia, Potosi Dept., "a few leagues" south of Potosi, 12,000 ft.
DISTRIBUTION: Altiplano from S Peru (Arequipa Dept.) to extreme N Chile and W Boliva (Potosi Dept.).
SYNONYMS: *flavidior*.

Auliscomys micropus (Waterhouse, 1837). Proc. Zool. Soc. Lond., 1837:17.
TYPE LOCALITY: Argentina, Santa Cruz Prov., "interior plains of Patagonia" at 50°S latitude, near the Rio Santa Cruz (probably near La Argentina *fide* Hershkovitz, 1962:392).
DISTRIBUTION: S Andes of Chile and Argentina, from about 38°S latitude to Straits of Magellan.
SYNONYMS: *alsus, fumipes*.
COMMENTS: Type species of Osgood's (1947) subgenus *Loxodontomys* and revised in this context by Pearson (1958). Banded karyotype published by Spotorno and Walker (1979); transferred to *Auliscomys* by Simonetti and Spotorno (1980).

Auliscomys pictus (Thomas, 1884). Proc. Zool. Soc. Lond., 1884:457.
TYPE LOCALITY: Peru, Junín Dept., Junín, 13,700 ft.
DISTRIBUTION: High Andes from C Peru (Ancash Dept.) to NW Bolivia (La Paz Dept.).
SYNONYMS: *decoloratus*.

Auliscomys sublimis (Thomas, 1900). Ann. Mag. Nat. Hist., ser. 7, 6:467.
TYPE LOCALITY: Peru, Arequipa Dept., Rinconado Malo Pass, between Caylloma and Calalla, 18,000 ft.
DISTRIBUTION: Altiplano from S Peru (Ayacucho Dept.), through SW Bolivia and adjacent Chile, to NW Argentina.
SYNONYMS: *leucurus*.

Baiomys True, 1894. Proc. U. S. Natl. Mus. (1893), 16:758 [1894].
TYPE SPECIES: *Hesperomys taylori* Thomas, 1887.
COMMENTS: Peromyscine. Revised by Packard (1960). Relegated to a subgenus of *Peromyscus* by Osgood (1909) but reinstated as a separate genus by Miller (1912*b*). Relationship to South American forms (e.g., *Calomys*) advanced by Packard (1960) but other studies have disclosed cognate affinity to *Scotinomys* (Carleton, 1980; Carleton et al., 1975; Hooper, 1960; Hooper and Musser, 1964*a*). Karyology evaluated by Yates et al. (1979) and biochemical variation by Calhoun et al. (1989). Hershkovitz (1962) transferred the form *hummelincki*, erroneously described under *Baiomys*, to the phyllotine genus *Calomys*.

Baiomys musculus (Merriam, 1892). Proc. Biol. Soc. Washington, 7:170.
TYPE LOCALITY: Mexico, Colima, Colima.
DISTRIBUTION: S Nayarit and C Veracruz, Mexico, to NW Nicaragua, excluding Yucatan Peninsula and Caribbean tropical lowlands.
SYNONYMS: *brunneus, grisescens, handleyi, infernatis, nebulosus, nigrescens, pallidus, pullus*.
COMMENTS: See Packard and Montgomery (1978, Mammalian Species, 102).

Baiomys taylori (Thomas, 1887). Ann. Mag. Nat. Hist., ser. 5, 19:66.
TYPE LOCALITY: USA, Texas, Duval Co., San Diego.
DISTRIBUTION: SE Arizona and SW New Mexico, E Texas, USA, south to Michoacan, C Hidalgo, and C Veracruz, Mexico.
SYNONYMS: *allex, analogus, ater, canutus, fuliginatus, paulus, subater*.
COMMENTS: See Eshelman and Cameron (1987, Mammalian Species, 285).

Bibimys Massoia, 1979. Physis, sec. C., 38(95):2.
TYPE SPECIES: *Bibimys torresi* Massoia, 1979.
COMMENTS: Diagnosed as a third genus of scapteromyines, also including *Kunsia* and *Scapteromys* (see Hershkovitz, 1966*b*), and so maintained by Reig (1984, 1986).

Bibimys chacoensis (Shamel, 1931). J. Washington Acad. Sci., 21:247.
TYPE LOCALITY: Argentina, Chaco Prov., Las Palmas.
DISTRIBUTION: NE Argentina.
COMMENTS: Described as a species of *Akodon*; transferred to *Bibimys* by Massoia (1980*a*), who noted the need to clarify the differentiation of this form from *B. torresi*.

Bibimys labiosus (Winge, 1887). E Museo Lundii, 1(3):25.
TYPE LOCALITY: Brazil, Minas Gerais, Pleistocene cave deposits near Lagoa Santa.
DISTRIBUTION: Minas Gerais, Brazil.
COMMENTS: Described as a species of *Scapteromys*, referred to *Akodon* by Hershkovitz (1966*c*), and reidentified as a species of *Bibimys* by Massoia (1980*a*). As with other sigmodontine species described by Winge (1887) from the Lagoa Santa caves, the relationship of *B. labiosus* to species named later warrants critical investigation (see Voss and Myers, 1991).

Bibimys torresi Massoia, 1979. Physis, sec. C., 38(95):3.
TYPE LOCALITY: Argentina, Buenos Aires Prov., Paraná River delta at the confluence of Arroyo Las Piedras with Arroyo Cucarachas, Campana.
DISTRIBUTION: EC Argentina.

Blarinomys Thomas, 1896. Ann. Mag. Nat. Hist., ser. 6, 18:310.
TYPE SPECIES: *Oxymycterus breviceps* Winge, 1887.
COMMENTS: Akodontine (see Reig, 1987).

Blarinomys breviceps (Winge, 1887). E Museo Lundii, 1(3):34.
TYPE LOCALITY: Brazil, Minas Gerais, Rio das Velhas, Pleistocene cave deposits near Lagoa Santa.
DISTRIBUTION: Bahia to Minas Gerais and Rio de Janeiro, SE Brazil.
COMMENTS: See Matson and Abravaya (1977, Mammalian Species, 74).

Bolomys Thomas, 1916. Ann. Mag. Nat. Hist., ser. 8, 18:339.
TYPE SPECIES: *Akodon amoenus* Thomas, 1900.
SYNONYMS: *Cabreramys*.
COMMENTS: Akodontine. After diagnosis as a genus, taxon retained as a subgenus of *Akodon* by Ellerman (1941) and Cabrera (1961). Emendation of diagnostic traits, synonymy of *Cabreramys*, and provisional list of species presented by Reig (1987). Complex taxonomic history and diagnosis further refined based on Bolivian species by Anderson and Olds (1989). Chromosomal traits and inappropriate karyotypic references reviewed by Maia and Langguth (1981) and Reig (1987).

Bolomys amoenus (Thomas, 1900). Ann. Mag. Nat. Hist., ser. 7, 6:468.
TYPE LOCALITY: Peru, Arequipa Dept., Calalla, Rio Colca, near Sumbay, 3500 m.
DISTRIBUTION: Highlands of SE Peru and WC Bolivia (after Anderson and Olds, 1989).
COMMENTS: Morphological definition as type species of genus clarified by Reig (1987).

Bolomys lactens (Thomas, 1918). Ann. Mag. Nat. Hist., ser. 9, 1:188.
TYPE LOCALITY: Argentina, Jujuy Prov., Leon, about 1500 m.
DISTRIBUTION: Highlands of SC Bolivia (see Anderson and Olds, 1989) and NW Argentina.
SYNONYMS: *leucolimnaeus, negrito, orbus*.

Bolomys lasiurus (Lund, 1841). K. Dansk. Vid. Selsk. Naturv. Math. Afhandl., 8:50.
TYPE LOCALITY: Brazil, Minas Gerais, Rio das Velhas, Lagoa Santa.
DISTRIBUTION: E Bolivia, Paraguay, N Argentina, and Brazil south of the Amazon River.
SYNONYMS: *arviculoides, brachyurus, fuscinus, lasiotus* (of Lund, 1838), *lenguarum, orobinus, pixuna, tapirapoanus*.
COMMENTS: Revised as part of *Zygodontomys* (Hershkovitz, 1962). Inclusion of *lasiurus* within *Bolomys* (or *Akodon*), not *Zygodontomys*, supported by karyological and morphological information (see Gardner and Patton, 1976; Maia and Langguth, 1981;

Voss and Linzey, 1981). Geographic variation assessed by Macêdo and Mares (1987), who retained only *l. fuscinus* and *l. lasiurus* as subspecies. They further placed *lenguarum*, a form tentatively retained as a species by Reig (1987) and Anderson and Olds (1989), in synonymy under *B. l. lasiurus*.

Bolomys obscurus (Waterhouse, 1837). Proc. Zool. Soc. Lond., 1837:16.
TYPE LOCALITY: Uruguay, Maldonado.
DISTRIBUTION: S Uruguay and EC Argentina.
SYNONYMS: *benefactus*.
COMMENTS: Type species of *Cabreramys* (Massoia and Fornes, 1967), considered a junior synonym of *Bolomys* by Reig (1987).

Bolomys punctulatus (Thomas, 1894). Ann. Mag. Nat. Hist., ser 6, 14:361.
TYPE LOCALITY: Ecuador.
DISTRIBUTION: Indeterminate area of Ecuador and perhaps Colombia.
COMMENTS: Described as a species of *Akodon* but referred to a subspecies of *Zygodontomys brevicauda* by Hershkovitz (1962). Voss (1991*b*) recognized the *Bolomys*-like traits of *punctulatus* and provisionally retained it as a species while noting its similarity to *B. lasiurus*. The enigmatic distribution of the few fragmentary specimens assignable to *punctulatus*, which originate from a region outside of the currently-known geographic range of *Bolomys*, was discussed by Voss (1991*b*).

Bolomys temchuki (Massoia, 1980). Hist. Nat., 1(25):179.
TYPE LOCALITY: Argentina, Misiones Prov., Depto. Capital, en terrenos del INTA, Arroyo Zaiman.
DISTRIBUTION: Region of Misiones, Argentina.
SYNONYMS: *elioi, liciae*.

Calomys Waterhouse, 1837. Proc. Zool. Soc. Lond., 1837:21.
TYPE SPECIES: *Mus (Calomys) bimaculatus* Waterhouse, 1837 (= *Mus laucha* Fischer, 1814).
SYNONYMS: *Hesperomys*.
COMMENTS: Phyllotine. Cladistic position unclear, but generally viewed as a primitive clade relative to other phyllotine genera (Hershkovitz, 1962; Pearson and Patton, 1976). Possible relationship of *Calomys* to North American Pliocene form *Bensonomys* debated by Baskin (1978, 1986) and Reig (1980). Generic revision by Hershkovitz (1962), who reduced the previously-recognized 10-15 species (e.g., Cabrera, 1961; Ellerman, 1941) to only four, two of which, *laucha* and *callosus*, have been later shown to be species complexes (Corti et al., 1987; Massoia et al., 1968; Pearson and Patton, 1976; Reig, 1986; Williams and Mares, 1978). Regional names, however, are sometimes employed in contradictory or inconsistent fashion, which leaves unsettled the question of relationship to extraregional forms and the identification of older synonyms to be employed. The species listed here, their distributional extent, and the allocation of synonyms must be viewed cautiously pending another generic revision.

Calomys boliviae (Thomas, 1901). Ann. Mag. Nat. Hist., ser. 7, 8:253.
TYPE LOCALITY: Bolivia, La Paz Dept., Rio Solocame, 1200 m.
DISTRIBUTION: W Bolivia.
SYNONYMS: *fecundus*.
COMMENTS: Conventionally arranged as a subspecies or synonym of *C. callosus* (Cabrera, 1961; Hershkovitz, 1962). Use here follows informal listing of Reig (1986), who included *fecundus* as a subjective synonym. Karyotype reported, as *C. fecundus*, by Pearson and Patton (1976).

Calomys callidus (Thomas, 1916). Ann. Mag. Nat. Hist., ser. 8, 17:182.
TYPE LOCALITY: Argentina, Corrientes Prov., Goya, 600 ft.
DISTRIBUTION: EC Argentina and E Paraguay.
COMMENTS: Diagnosed as a subspecies of *C. venustus* and recognized as such (e.g., Cabrera, 1961) until Hershkovitz (1962) placed it in synonymy with *C. callosus callosus*. Specific status of *callidus* affirmed by Corti et al. (1987), who summarized its distribution and karyotypic discrimination.

Calomys callosus (Rengger, 1830). Naturgesch. Säugeth. Paraguay, p. 231.

TYPE LOCALITY: Paraguay, Neembucu Dept., banks of Rio Paraguay opposite mouth of Rio Bermejo (as interpreted by Hershkovitz, 1962:172).

DISTRIBUTION: N Argentina, E Bolivia, W Paraguay, WC to EC Brazil (after Mares et al., 1981*b*, 1989*a*).

SYNONYMS: *expulsus, muriculus, venustus.*

COMMENTS: Homogeneity of populations unified here should be viewed skeptically. According to Hershkovitz (1962), *callosus* also encompassed *boliviae* (see species account), *callidus* (see species account), and *fecundus* (see *C. boliviae*), together with species-group taxa enumerated above. Others have referenced, without explanation, *muriculus* (Williams and Mares, 1978) and *venustus* (Corti et al., 1987) as separate species, although Reig (1986) did not. Relationship and possible priority of *venustus* with respect to *boliviae* and *fecundus* especially merits study. Karyotype of nominate form published by Pearson and Patton (1976).

Calomys hummelincki (Husson, 1960). Stud. Faun. Curacao Carib. Isl., 43:34.

TYPE LOCALITY: Netherlands West Indies, Curaçao, Klein Santa Martha.

DISTRIBUTION: Llanos of NE Colombia (La Guajira), N Venezuela, and continental-shelf islands Curaçao and Aruba.

COMMENTS: Originally named as a species of *Baiomys*; included in *C. laucha* by Hershkovitz (1962), who considered the Venezuelan populations to be introductions. Indigenous status reasserted by Handley (1976).

Calomys laucha (G. Fischer, 1814). *In* Eschwege, J. Brasilien, Neue Bibliothek. Reisenb., 15(2):209.

TYPE LOCALITY: Paraguay, vicinity of Asunci6n (as restricted by Hershkovitz, 1962:153).

DISTRIBUTION: N Argentina and Uruguay, SE Bolivia, W Paraguay, and WC Brazil.

SYNONYMS: *bimaculatus, bonariensis, dubius, gracilipes, pusillus.*

COMMENTS: Formerly included *hummelincki* (See Handley, 1976), *musculinus* (see Massoia et al., 1968), and *tener* (see species account). Attribution of *pusillus* follows Hershkovitz (1962; also see Osgood, 1943:239).

Calomys lepidus (Thomas, 1884). Proc. Zool. Soc. Lond., 1884:454.

TYPE LOCALITY: Peru, Junín Dept., Junín.

DISTRIBUTION: Altiplano of C Peru through W Bolivia, to NE Chile and NW Argentina.

SYNONYMS: *argurus, carillus, ducillus, marcarum, montanus.*

COMMENTS: Karyotype reported by Pearson and Patton (1976).

Calomys musculinus (Thomas, 1913). Ann. Mag. Nat. Hist., ser. 8, 11:138.

TYPE LOCALITY: Argentina, Jujuy Prov., Maimara, 2230 m.

DISTRIBUTION: N Argentina, E Paraguay.

SYNONYMS: *cordovensis, cortensis, murillus.*

COMMENTS: Included in *C. laucha* by Hershkovitz (1962) but morphological and karyotypic evidence supports its specific status (Massoia et al., 1968; Pearson and Patton, 1976). Synonymy follows Contreras and Rosi (1980); Reig (1986) listed *murillus* as a species.

Calomys sorellus (Thomas, 1900). Ann. Mag. Nat. Hist., ser. 7, 6:297.

TYPE LOCALITY: Peru, Libertad Dept., 8 mi S Huamachuco, 3500 m.

DISTRIBUTION: Peruvian Andes, above 2000 m, from Dept. Libertad to Puno.

SYNONYMS: *frida, miurus.*

COMMENTS: Classified as a subspecies of *C. lepidus* (Cabrera, 1961); discrimination from and sympatry with *C. lepidus* documented by Hershkovitz (1962), who relegated *frida* and *miurus* to synonymy under *C. sorellus* (also see Pearson and Patton, 1976).

Calomys tener (Winge, 1887). E Museo Lundii, 1(3):15.

TYPE LOCALITY: Brazil, Minas Gerais, Rio das Velhas, Lagoa Santa.

DISTRIBUTION: EC Brazil.

COMMENTS: Arranged as a subspecies of *C. laucha* by Hershkovitz (1962), but others have retained *tener*, along with *C. laucha*, as distinct species (Cabrera, 1961; Mares et al., 1989*a*; Moojen, 1952), which suggests the need for further study of their level of differentiation.

Chelemys (Thomas, 1903). Ann. Mag. Nat. Hist., ser. 7, 12:242.
TYPE SPECIES: *Akodon megalonyx* Thomas, 1903.
COMMENTS: Akodontine. Named as a subgenus of *Akodon*, later ranked as a genus (Gyldenstolpe, 1932; Reig, 1987; Thomas, 1927) or consolidated under *Notiomys*, together with *Geoxus* (Cabrera, 1961; Ellerman, 1941; Osgood, 1925, 1943). Pearson (1984) enumerated diagnostic traits of these three long-clawed, semi-fossorial taxa, and Reig (1987) amplified their morphological definition. *Chelemys angustus* Thomas, 1927, is based on a specimen of *Akodon longipilis* (see Pearson, 1984).

Chelemys macronyx (Thomas, 1894). Ann. Mag. Nat. Hist., ser. 6, 14:362.
TYPE LOCALITY: Argentina, Mendoza Prov., Fort San Rafael.
DISTRIBUTION: E and W flanks of S Andes along Chile-Argentina boundary, about 34°S latitude south to Straits of Magellan. C Chilean distribution augmented by Pine et al. (1979).
SYNONYMS: *alleni, connectens, fumosus, vestitus*.
COMMENTS: Treated as a subspecies of *C. megalonyx* by Mann (1978) and Tamayo and Frassinetti (1980); recognized as species by Osgood (1943) and Pearson (1984). Osgood (1943) designated the skin of *Notiomys connectens* as the holotype of this composite specimen and synonymized it with [*C.*] *macronyx vestitus*.

Chelemys megalonyx (Waterhouse, 1845). Proc. Zool. Soc. Lond., 1844:154 [1845].
TYPE LOCALITY: Chile, Valparaiso Prov., Lake Quintero.
DISTRIBUTION: C Chile, Coquimbo Prov. south to Magallanes.
SYNONYMS: *delfini, microtis, niger, scalops*.
COMMENTS: The association of *delfini*—retained by Osgood (1943) as a species related to *C. macronyx* but listed as a subspecies of *C. megalonyx* by Tamayo and Frassinetti (1980)—needs resolution.

Chibchanomys Voss, 1988. Bull. Am. Mus. Nat. Hist., 188:321.
TYPE SPECIES: *Ichthyomys trichotis* Thomas, 1897.
COMMENTS: Ichthyomyine. Described as a species of *Ichthyomys*, *trichotis* was thereafter associated first with *Rheomys* (Cabrera, 1961; Tate, 1932*h*) and then *Anotomys* (Handley, 1976). Generic distinctiveness and phylogenetic relationships substantiated by Voss (1988).

Chibchanomys trichotis (Thomas, 1897). Ann. Mag. Nat. Hist., ser. 6, 20:220.
TYPE LOCALITY: Colombia, "W. Cundinamarca."
DISTRIBUTION: Highlands in W Venezuela, Colombia, and Peru.
COMMENTS: Provenience of type, taxonomy, and distribution reviewed by Voss (1988).

Chilomys Thomas, 1897. Ann. Mag. Nat. Hist., ser. 6, 19:501.
TYPE SPECIES: *Oryzomys instans* Thomas, 1895.
COMMENTS: Arranged with oryzomyines (Reig, 1984) but tribal affinity obscure; perhaps more closely related to thomasomyines. Aspects of morphology reported by Carleton (1973) and Voss and Linzey (1981).

Chilomys instans (Thomas, 1895). Ann. Mag. Nat. Hist., ser. 6, 16:368.
TYPE LOCALITY: Colombia, Cundinamarca Dept., Bogota region, Hacienda de La Selva, 1380 m.
DISTRIBUTION: N Andes, from N Ecuador through C and N Colombia, to W Venezuela. Venezuelan distribution amplified by Handley (1976).
SYNONYMS: *fumeus*.

Chinchillula Thomas, 1898. Ann. Mag. Nat. Hist., ser. 7, 1:280.
TYPE SPECIES: *Chinchillula sahamae* Thomas, 1898.
COMMENTS: Phyllotine. Revised by Hershkovitz (1962); karyology reviewed by Pearson and Patton (1976).

Chinchillula sahamae Thomas, 1898. Ann. Mag. Nat. Hist., ser. 1, 7:280.
TYPE LOCALITY: Bolivia, La Paz Dept., Esperanza, 50 km N Mount Sajama, Pacajes, 4200 m.
DISTRIBUTION: Altiplano region of S Peru, W Bolivia, extreme NW Argentina, and N Chile.

Chroeomys Thomas, 1916. Ann. Mag. Nat. Hist., ser. 8, 18:340.

TYPE SPECIES: *Akodon pulcherrimus* Thomas, 1897 (= *Akodon jelskii* Thomas, 1894).

COMMENTS: Akodontine. Described as a genus (Thomas, 1916*c*) but afterwards conventionally viewed as a subgenus of *Akodon* (Cabrera, 1961; Ellerman, 1941; Gardner and Patton, 1976; Reig, 1987). Among the most genetically divergent akodonts surveyed by Patton et al. (1989) and Smith and Patton (1991), including species of typical *Akodon, Bolomys,* and *Microxus,* and recognized by those authors as a separate genus.

Chroeomys andinus (Philippi, 1858). Arch. Naturgesh., 23(1):77.

TYPE LOCALITY: Chile, Santiago Prov., Altos Andes.

DISTRIBUTION: Peru, Bolivia, N Chile, N Argentina.

SYNONYMS: *cinnamomea, dolichonyx, gossei, jucundus.*

COMMENTS: Although not colorfully marked like *C. jelskii,* genetic data support the transfer of *andinus* to *Chroeomys* (see Smith and Patton, 1991).

Chroeomys jelskii (Thomas, 1894). Ann. Mag. Nat. Hist., ser. 6, 14:360.

TYPE LOCALITY: Peru, Junín Dept., Junín.

DISTRIBUTION: Altiplano of S Peru to WC Bolivia and NW Argentina.

SYNONYMS: *bacchante, cayllomae, cruceri, inambarii, inornatus, ochrotis, pulcherrimus, pyrrhotis, scalops, sodalis.*

COMMENTS: Revised by Sanborn (1947*b*).

Delomys Thomas, 1917. Ann. Mag. Nat. Hist., ser. 8, 20:196.

TYPE SPECIES: *Hesperomys dorsalis* Hensel, 1872.

COMMENTS: Thomasomyine. *Delomys* has variously stood as a genus (Avila-Pires, 1960*b*; Gyldenstolpe, 1932; Tate, 1932*f*) or placed within *Thomasomys,* with or without formal subgeneric division (Ellerman, 1941; Moojen, 1952; Osgood, 1933*d*). As remarked by Osgood (1933*d*), Thomas in part created *Delomys* to taxonomically underscore the geographic disjunction between thomasomyines in SE Brazil and the diverse radiation of typical *Thomasomys* in the N Andes (e.g., see Reig, 1986). There are features that unite the forms of *Delomys,* but decision on its classificatory rank—and that of other SE Brazilian endemics like *Wilfredomys, Rhagomys,* and *Phaenomys*—must await broader-based phylogenetic inquiry involving Andean thomasomyines.

Number of recognized species listed as four (Moojen, 1952), three (Reig, 1986), two (Ellerman, 1941; Gyldenstolpe, 1932), or one (Corbet and Hill, 1991; Cabrera, 1961), a situation which advises that future list-making would profit from critical specimen examination. Here we recognize two, *D. dorsalis* and *D. sublineatus,* following the karyotypic study of Zanchin et al. (1992*b*). Avila-Pires (1960*b*) associated *Calomys plebejus* Winge, 1887, as another form of *Delomys;* its status with regard to *dorsalis* and *sublineatus* warrents clarification.

Delomys dorsalis (Hensel, 1872). Abh. König. Akad. Wiss. Berlin, p. 42.

TYPE LOCALITY: Brazil, Rio Grande do Sul.

DISTRIBUTION: Atlantic coastal region of SE Brazil and extreme NE Argentina.

SYNONYMS: *collinus, lechei, obscura, plebejus.*

Delomys sublineatus (Thomas, 1903). Ann. Mag. Nat. Hist., ser. 7, 12:240.

TYPE LOCALITY: Brazil, Espírito Santo, Engenheiro Reeve, 500 m.

DISTRIBUTION: Atlantic coastal forest of SE Brazil, Espírito Santo to Paraná.

Eligmodontia F. Cuvier, 1837. Ann. Sci. Nat. (Paris), ser. 2, 7:169.

TYPE SPECIES: *Eligmodontia typus* F. Cuvier, 1837.

SYNONYMS: *Eligmodon, Heligmodontia.*

COMMENTS: A phyllotine member closely allied to *Calomys* (Hershkovitz, 1962; Williams and Mares, 1978). Revised by Hershkovitz (1962), who synonymized all forms as *E. typus*. Morphological, chromosomal, and distributional evidence, however, supports a view of greater species diversity (e.g., Kelt et al., 1991; Mann, 1978; Mares et al., 1989*b*; Osgood, 1943). For example, Kelt et al. (1991) identified three chomosomal morphs to which they applied the names *typus* (2n=44), *puerulus* (2n=50), and *morgani* (2n=32). The association of names, particularly for *typus* and *morgani,* is based on the nearness of their sample localities to type localities. This is not the same as karyotyping topotypic animals or comparing karyotyped specimens to holotypes. In view of the complex interdigitation of specific ranges among the ridges and valleys of the southern Andes, the use of these chromosomal formulae as a diagnostic key to species identifications should be employed cautiously pending a full revision of the genus. Williams and Mares (1978) referred *Graomys hypogaeus* to synonymy under *E. typus sensu lato* and questioned whether the type is a composite specimen; resolution of its status and proper synonymy requires restudy of the holotype as part of a revision of *Eligmodontia*.

Eligmodontia moreni Thomas, 1896. Ann. Mag. Nat. Hist., ser. 6, 18:307.

TYPE LOCALITY: Argentina, La Rioja Prov., Chilecito, 1200 m.

DISTRIBUTION: Atlantic-facing Andean slopes at intermediate elevations, El Salta to Neuquen Prov., NW Argentina; limits poorly documented.

COMMENTS: Mares et al. (1981*b*, 1989*b*) noted the morphological, distributional, and ecological distinction of *E. moreni* and *E. puerulus* in El Salta Prov., Argentina. The differentiation of *moreni* from *morgani* in the S Andes and from *typus* in the Pampas warrants further study.

Eligmodontia morgani Allen, 1901. Bull. Am. Mus. Nat. Hist., 14:409.

TYPE LOCALITY: Argentina, Santa Cruz Prov., basaltic canyons 50 mi SE Lake Buenos Aires.

DISTRIBUTION: W Patagonian region of S Argentina and adjacent Chile; limits of distribution unknown, perhaps as far north as Neuquen Prov., Argentina.

COMMENTS: This species epithet was used by Kelt et al. (1991) for the 2n=32 chromosomal form but whether this karyotype characterizes animals from the type locality needs verification. Other forms of *Eligmodontia* occur in the area.

Eligmodontia puerulus (Philippi, 1896). Anal. Mus. Nac. Chile, Zool. Ent., 13a:20.

TYPE LOCALITY: Chile, Antofagasta Prov., San Pedro de Atacama, 3223 m.

DISTRIBUTION: Altiplano of extreme S Peru, through NE Chile and W Bolivia, to NW Argentina.

SYNONYMS: *hirtipes, jucunda, marica, tarapacensis*.

COMMENTS: Reduced to a subspecies of *E. typus* by Hershkovitz (1962) but morphological (Mann, 1978; Osgood, 1943) and karyotypic (Kelt et al., 1991; Ortells et al., 1989) data sustain the specific recognition of *E. puerulus*.

Eligmodontia typus F. Cuvier, 1837. Ann. Sci. Nat. (Paris), ser. 2, 7:169.

TYPE LOCALITY: Argentina, "Buenos Aires."

DISTRIBUTION: C Argentina and contiguous parts of Chile; range uncertain.

SYNONYMS: *elegans, hypogaeus, pamparum*.

COMMENTS: Hershkovitz (1962:185) discussed the uncertainty of the type locality (also see Ortells et al., 1989:138), which should be formally restricted in order to settle the morphological identification of *E. typus* and promote a critical definition of the species within the genus.

Euneomys Coues, 1874. Proc. Acad. Nat. Sci. Philadelphia, 26:185.

TYPE SPECIES: *Reithrodon chinchilloides* Waterhouse, 1839.

SYNONYMS: *Bothriomys, Chelemyscus*.

COMMENTS: Phyllotine. Few specimens exist to substantiate the specific taxonomy and distribution of the genus. Revised by Hershkovitz (1962), who acknowledged several nominal species but indicated their probable synonymy under *E. chinchilloides*, as Mann (1978) so formalized for Chilean populations. Yañez et al. (1987) summarized extant specimen data for *Euneomys* and concluded that one species is represented. Others have recognized two or more (e.g., Osgood, 1943; Pearson and Christie, 1991; Pine et al., 1979). The latter arrangement is more nearly correct but must be sustained within the context of rigorous revisions. Holotype of *Reithrodon fossor* Thomas, 1899, type species of *Chelemyscus* Thomas, 1925, is a composite, the designated type skull referrable to *Euneomys* (Osgood, 1943:164; Pearson, 1984:231).

Euneomys chinchilloides (Waterhouse, 1839). Zool. Voy. H. M. S. "Beagle", Mammalia, p. 72.

TYPE LOCALITY: Chile, Tierra del Fuego, Straits of Magellan, south shore near eastern entrance.

DISTRIBUTION: Isla Grande de Tierra del Fuego, neighboring islands, and southernmost Chile; limits uncertain.
SYNONYMS: *ultimus*.
COMMENTS: Hershkovitz (1962) had suggested that other nominal species would prove to be synonyms of *E. chinchilloides*. This taxonomy has been followed (Mann, 1978; Yañez et al., 1987) but misrepresents species diversity in the genus (see Pearson and Christie, 1991).

Euneomys fossor (Thomas, 1899). Ann. Mag. Nat. Hist., ser. 7, 4:280.
TYPE LOCALITY: Argentina, "Prov. Salta."
DISTRIBUTION: Known only from the type locality, which is questionable as *Euneomys* is otherwise unrecorded this far north in Argentina (see Mares et al., 1989*b*).
COMMENTS: Status doubtful—"the name *Chelemyscus fossor* should be attached to an appropriate species of *Euneomys* by some future revisor" (Pearson, 1984:231).

Euneomys mordax Thomas, 1912. Ann. Mag. Nat. Hist., ser. 8, 10:410.
TYPE LOCALITY: Argentina, Fort San Rafael; uncertain, probably not the San Rafael in Mendoza Prov. (see Pearson and Christie, 1991).
DISTRIBUTION: WC Argentina and adjacent region of Chile (Santiago and Malleco Provs.); range limits unknown.
SYNONYMS: *noei*.
COMMENTS: Pearson and Christie (1991) treated *noei* as conspecific with *E. mordax*.

Euneomys petersoni J. A. Allen, 1903. Bull. Am. Mus. Nat. Hist., 19:192.
TYPE LOCALITY: Argentina, Santa Cruz Prov., upper Rio Chico, near the Cordilleras.
DISTRIBUTION: Extreme S Argentina and adjacent Chile, excluding Tierra del Fuego; limits uncertain.
SYNONYMS: *dabbenei*.
COMMENTS: Relegated to a subspecies of *E. chinchilloides* by Hershkovitz (1962) and Pearson and Christie (1991), but Osgood (1943) had noted the nearby occurrence of *E. petersoni* and *E. chinchilloides* without evidence of intergradation.

Galenomys Thomas, 1916. Ann. Mag. Nat. Hist., ser. 8, 17:143.
TYPE SPECIES: *Phyllotis garleppi* Thomas, 1898.
COMMENTS: Phyllotine. Named as a subgenus of *Euneomys*, later elevated to genus (Thomas, 1926*a*), and reassociated as a subgenus of *Phyllotis* (Ellerman, 1941; Osgood, 1947). Revised by Hershkovitz (1962), who considered *Galenomys* a distinct genus, as did Pearson (1958).

Galenomys garleppi (Thomas, 1898). Ann. Mag. Nat. Hist., ser. 7, 1:279.
TYPE LOCALITY: Bolivia, La Paz Dept., Esperanza, northeast of Mount Sajama, 4140 m.
DISTRIBUTION: Altiplano of S Peru, N Chile, and adjacent Bolivia.

Geoxus Thomas, 1919. Ann. Mag. Nat. Hist., ser. 9, 3:209.
TYPE SPECIES: *Geoxus fossor* Thomas, 1919 (= *Oxymycterus valdivianus* Philippi, 1858).
COMMENTS: Akodontine. Revised by Osgood (1925, 1943) as part of *Notiomys*—see remarks under *Chelemys*.

Geoxus valdivianus (Philippi, 1858). Arch. Naturgesch., 24(1):303.
TYPE LOCALITY: Chile, Valdivia Prov.
DISTRIBUTION: C and S Chile, including Mocha and Chiloe islands, to Straits of Magellan, and S Argentina (Neuquen Prov. to Chubut).
SYNONYMS: *araucanus, bicolor, bullocki, chiloensis, fossor, michaelseni, microtis*.
COMMENTS: Pearson (1984) noted that the generic association and specific status of *michaelseni*, arranged by Osgood (1943) as a subspecies, are uncertain.

Graomys Thomas, 1916. Ann. Mag. Nat. Hist., ser. 8, 17:141.
TYPE SPECIES: *Mus griseoflavus* Waterhouse, 1837.
COMMENTS: Phyllotine. Previously included in *Phyllotis* as a formal subgeneric division (Osgood, 1947; Pearson, 1958) or not (Hershkovitz, 1962). Returned to generic stature based on karyological data (Pearson and Patton, 1976) and other traits (see Reig, 1978; Olds and Anderson, 1989). Williams and Mares (1978) transferred *Graomys*

pearsoni Myers, 1977, to *Andalgalomys* and synonymized *G. hypogaeus* in *Eligmodontia typus*.

Graomys domorum (Thomas, 1902). Ann. Mag. Nat. Hist., ser. 7, 9:132.
TYPE LOCALITY: Bolivia, Cochabamba Dept., Tapacari, 3000 m.
DISTRIBUTION: E slopes of Andes of SC Bolivia and NW Argentina.
SYNONYMS: *taterona*.
COMMENTS: Ranked as a subspecies of *G. griseoflavus* by Hershkovitz (1962). Chromosomal divergence reported by Pearson and Patton (1976), who reinstated *domorum* as a species (also see Olds et al., 1987). Reig (1978) followed Cabrera (1961) in allocating *taterona* to *G. domorum*.

Graomys edithae Thomas, 1919. Ann. Mag. Nat. Hist., ser. 9, 3:495.
TYPE LOCALITY: Argentina, La Rioja Prov., Otro Cerro, about 18 km NNW Chumbicha, 3000 m.
DISTRIBUTION: Known only from the type locality.
COMMENTS: Status uncertain—placed as a synonym of *G. griseoflavus* by Cabrera (1961) or retained as a nominal species of *Phyllotis* (Hershkovitz, 1962) or of *Graomys* (Myers, 1977; Williams and Mares, 1978). Reinvestigation of the original type series, reportedly acquired in near sympatry with *cachinus* and *medius* (Thomas, 1919), would shed much light.

Graomys griseoflavus (Waterhouse, 1837). Proc. Zool. Soc. Lond., 1837:28.
TYPE LOCALITY: Argentina, Rio Negro Prov., mouth of Rio Negro.
DISTRIBUTION: S Bolivia, W Paraguay, and nearby Brazil, south to S Chubut Prov., Argentina.
SYNONYMS: *cachinus*, *centralis*, *chacoensis*, *lockwoodi*, *medius*.
COMMENTS: Even with removal of *G. domorum*, specific homogeneity of included populations and list of species-group synonyms remain highly suspect.

Habromys (Hooper and Musser, 1964). Occas. Pap. Mus. Zool., Univ. Michigan, 635:12.
TYPE SPECIES: *Peromyscus lepturus* Merriam, 1898.
COMMENTS: Peromyscine. Circumscribes species originally placed with the *Peromyscus mexicanus* complex, subgenus *Peromyscus* (Osgood, 1909). Hooper and Musser (1964*b*) acknowledged their distinctive morphology as the subgenus *Habromys* of *Peromyscus* (and Hooper, 1968*b*), a clade which Carleton (1980) viewed at the generic level. Systematic evidence weakly supports common ancestry of *Habromys* with *Neotomodon* and/or *Podomys* (Carleton, 1980; Hooper and Musser, 1964*b*; Stangl and Baker, 1984*b*). Male reproductive tract examined by Hooper (1958), Hooper and Musser (1964*b*), and Linzey and Layne (1969, 1974).

Habromys chinanteco (Robertson and Musser, 1976). Occas. Pap. Mus. Nat. Hist., Univ. Kansas, 47:1.
TYPE LOCALITY: Mexico, Oaxaca, north slope Cerro Pelon, 31.6 km S Vista Hermosa, 2650 m.
DISTRIBUTION: Vicinity of the type locality in the Sierra de Juarez, NC Oaxaca, Mexico.
COMMENTS: A small-bodied species morphologically similar to *H. simulatus* and sympatric with *H. lepturus* (see Robertson and Musser, 1976).

Habromys lepturus (Merriam, 1898). Proc. Biol. Soc. Washington, 12:118.
TYPE LOCALITY: Mexico, Oaxaca, Cerro Zempoaltepec, 8200 ft.
DISTRIBUTION: Cloud forests on Cerro Zempoaltepec and Sierra de Juarez, NC Oaxaca, Mexico.
SYNONYMS: *ixtlani*.
COMMENTS: Musser (1969*a*) arranged the distinctive form *ixtlani* as a subspecies.

Habromys lophurus (Osgood, 1904). Proc. Biol. Soc. Washington, 17:72.
TYPE LOCALITY: Guatemala, Huehuetenango, Todos Santos, 10,000 ft.
DISTRIBUTION: Highlands of Chiapas, Mexico, C Guatemala, and NW El Salvador.
COMMENTS: Robertson and Musser (1976) noted much diversity among their samples of *H. lophurus*.

Habromys simulatus (Osgood, 1904). Proc. Biol. Soc. Washington, 17:72.
TYPE LOCALITY: Mexico, Veracruz, near Jico, 6000 ft.
DISTRIBUTION: Restricted area on E slopes of Sierra Madre Oriental, C Veracruz, Mexico.

COMMENTS: Known by only three specimens—the type, paratype, and one from near Zacualpan, Veracruz (see Robertson and Musser, 1976); distributional limits unknown.

Hodomys Merriam, 1894. Proc. Acad. Nat. Sci. Philadelphia, 46:232.

TYPE SPECIES: *Neotoma alleni* Merriam, 1892.

COMMENTS: Neotomine. Maintained as a genus (e.g., Goldman, 1910; Ellerman, 1941) until arranged as a subgenus of *Neotoma* by Burt and Barkalow (1942). Carleton (1980) expressed relationships and differentiation of *Hodomys* at the generic rank, perhaps closely related to *Xenomys* (also see Carleton, 1973; Hooper, 1960; and Schaldach, 1960).

Hodomys alleni (Merriam, 1892). Proc. Biol. Soc. Washington, 7:168.

TYPE LOCALITY: Mexico, Colima, Manzanillo.

DISTRIBUTION: Southernmost Sinaloa to Oaxaca; interior Mexico along basin of Rio Balsas to central Puebla.

SYNONYMS: *elattura, guerrerensis, vetulus.*

COMMENTS: Kelson (1952) placed *vetulus* as a subspecies of *N. alleni*. See Genoways and Birney, 1974 (Mammalian Species, 41, as *Neotoma alleni*), who first reported the karyotype (2n=48).

Holochilus Brandt, 1835. Mem. Acad. Imp. Sci. St. Petersbourg, ser. 6, 3(2):428.

TYPE SPECIES: *Mus (Holochilus) leucogaster* Brandt, 1835 (= *Mus brasiliensis* Desmarest, 1819).

COMMENTS: Arranged by Hershkovitz (1955*a*) as one of four genera of sigmodont rodents. Based on reproductive anatomy, Hooper and Musser (1964*a*) remarked that *Holochilus* may represent a "well differentiated oryzomyine rather than a sigmodont." Standard karyotypic data have provided little resolution of generic affinities (Gardner and Patton, 1976) but g-banded evaluations supported an oryzomyine affiliation (R. J. Baker et al., 1983*a*). Retained, with *Sigmodon* proper, in the tribe Sigmodontini by Reig (1984, 1986).

Genus revised by Hershkovitz (1955*a*), who consolidated 13 nominal species (e.g., Ellerman, 1941) under *H. brasiliensis* and diagnosed a new one, *H. magnus*. Subsequent studies have revealed that *brasiliensis* of Hershkovitz (1955*a*) is a composite of three or more species (Aguilera and Perez-Zapata, 1989; Gardner and Patton, 1976; Massoia, 1980*a*, 1981; Reig, 1986). In general, evidence for the species acknowledged below issues from more localized studies that lack the perspective of a full generic review; these names, distributions, and synonyms must therefore be accepted as tentative pending such a synoptic revision. Association of synonyms basically observes the species classification of Massoia (1980*a*).

Holochilus brasiliensis (Desmarest, 1819). Nouv. Dict. Hist. Nat., Nouv. ed., 29:62.

TYPE LOCALITY: Brazil, Minas Gerais, Lagoa Santa (as restricted by Hershkovitz, 1955*a*:662).

DISTRIBUTION: SE Brazil, Uruguay, and EC Argentina.

SYNONYMS: *canellinus, darwini, leucogaster, russatus, vulpinus.*

COMMENTS: Morphological definition and geographic range reduced in scope by Massoia (1980*a*, 1981), who segregated *H. sciureus* as a species distinct from *H. brasiliensis*.

Holochilus chacarius Thomas, 1906. Ann. Mag. Nat. Hist., ser. 7, 18:446.

TYPE LOCALITY: Paraguay, eastern Chaco, one league NW Concepcion, 500 ft.

DISTRIBUTION: Paraguay and NE Argentina.

SYNONYMS: *balnearum.*

COMMENTS: Considered distinct by Massoia (1980*a*), who treated *balnearum* as a synonym; Reig (1986) mentioned *balnearum* as another good species. Highly variable in diploid and fundamental numbers (Vidal et al., 1976; Nachman and Myers, 1989).

Holochilus magnus Hershkovitz, 1955. Fieldiana Zool., 37:657.

TYPE LOCALITY: Uruguay, about 40 km south Treinta y Tres, Rio Cebollati, Paso de Averias.

DISTRIBUTION: Uruguay and southernmost Brazil; not recorded from Argentina (Olrog and Lucero, 1981).

COMMENTS: Massoia (1980*a*) referred Winge's (1887) *Hesperomys molitor* to *Holochilus* and noted its close resemblance to *H. magnus*; their relationship and status warrant additional study (also see Voss and Myers, 1991). Banded karyotypic comparisons of *H. magnus* and *H. brasiliensis* presented by Freitas et al. (1983).

Holochilus sciureus Wagner, 1842. Arch. Naturgesch., ser. 8, 1:16.
TYPE LOCALITY: Brazil, Minas Gerais, Rio São Francisco.
DISTRIBUTION: Broad reaches of Orinoco and Amazon river basins: E and S Venezuela, Guianas, N and C Brazil, and Amazonian regions of Colombia, Ecuador, Peru, and Bolivia.
SYNONYMS: *amazonicus, berbericensis, guianae, incarum, nanus, venezuelae.*
COMMENTS: Discrimination from *H. brasiliensis* documented by Massoia (1980*a*, 1981), who raised *sciureus* to specific rank. Many forms swept under Hershkovitz's (1955*a*) concept of *brasiliensis* actually belong to this "species," which itself may be a composite. Reig (1986), for instance, listed *amazonicus, guianae,* and *venezuelae* as probable valid species, a taxonomy which should be given empirical testing (also see Aguilera and Perez-Zapata, 1989).

Ichthyomys Thomas, 1893. Proc. Zool. Soc. Lond., 1893:337.
TYPE SPECIES: *Ichthyomys stolzmanni* Thomas, 1893.
COMMENTS: Ichthyomyine. Revised by Voss (1988), who defined four species among the eight described taxa, and studied phylogenetic relationships.

Ichthyomys hydrobates (Winge, 1891). Vidensk. Medd. Nat. Foren. Kjobenhavn, ser. 5, 3:20.
TYPE LOCALITY: Venezuela, Mérida, Sierra de Mérida.
DISTRIBUTION: W Venezuela, Colombia, and Ecuador.
SYNONYMS: *nicefori, soderstromi.*
COMMENTS: The holotype may have originated from the vicinity of Mérida, 1600-1700 m (Voss, 1988). Taxonomy and distribution reviewed by Voss (1988), who retained *nicefori* and *soderstromi* as subspecies.

Ichthyomys pittieri Handley and Mondolfi, 1963. Acta Biol. Venezuela, 3:417.
TYPE LOCALITY: Venezuela, Aragua, Rancho Grande, near headwaters of Río Limon.
DISTRIBUTION: N Venezuela.
COMMENTS: Known by four specimens from the Carribean Coastal Range of Venezuela. Taxonomy reviewed by Voss (1988).

Ichthyomys stolzmanni Thomas, 1893. Proc. Zool. Soc. Lond., 1893:339.
TYPE LOCALITY: Peru, Junín Dept., Chanchamayo, near Tarma, 923 m.
DISTRIBUTION: Ecuador and Peru.
SYNONYMS: *orientalis.*
COMMENTS: Taxonomy and distribution reviewed by Voss (1988), who retained *orientalis* as a subspecies.

Ichthyomys tweedii Anthony, 1921. Am. Mus. Novit., 20:1.
TYPE LOCALITY: Ecuador, Prov. El Oro, Portovelo, 615 m.
DISTRIBUTION: W Ecuador and C Panama.
SYNONYMS: *caurinus.*
COMMENTS: Taxonomy and distribution reviewed by Voss (1988), who placed *caurinus* in full synonymy.

Irenomys Thomas, 1919. Ann. Mag. Nat. Hist., ser. 9, 3:201.
TYPE SPECIES: *Reithrodon longicaudatus* Philippi, 1900 (= *Mus tarsalis* Philippi, 1900).
COMMENTS: Closest generic relatives obscure; usually grouped with phyllotines (Olds and Anderson, 1989; Vorontsov, 1959).

Irenomys tarsalis (Philippi, 1900). Ann. Mus. Nac. Chile Zool., 14:10.
TYPE LOCALITY: Chile, Valdivia Prov., Fundo San Juan, near La Union.
DISTRIBUTION: C and S Chile, including Chiloe and Guaitecas Islands, and adjacent Argentina (Neuquen Prov.).
SYNONYMS: *longicaudatus.*

Isthmomys Hooper and Musser, 1964. Occas. Pap. Mus. Zool., Univ. Michigan, 635:12.
TYPE SPECIES: *Megadontomys flavidus* Bangs, 1902.

COMMENTS: Peromyscine. Species associated here were originally classified in *Megadontomys*, used either as a genus or as a subgenus of *Peromyscus* (Osgood, 1909). *Isthmomys* was later diagnosed as a subgenus of *Peromyscus* (Hooper and Musser, 1964*b*) and maintained at this rank by Hooper (1968*b*); accorded generic status by Carleton (1980, 1989). Sister-group relationship with *Megadontomys* proposed by Carleton (1980) but unsupported by chromosomal banding data (Stangl and Baker, 1984*b*). Aspects of morphology considered by Carleton (1973, 1980), Hooper and Musser (1964*b*), and Linzey and Layne (1969, 1974); karyology by Stangl and Baker (1984*b*).

Isthmomys flavidus (Bangs, 1902). Bull. Mus. Comp. Zool., 9:27.
TYPE LOCALITY: Panama, Chiriqui Prov., Volcan de Chiriqui, Boquete.
DISTRIBUTION: Intermediate elevations in W Panama (Chiriqui region) and on Azuero Peninsula (see Handley, 1966*a*).
COMMENTS: Whether *I. pirrensis* is a junior synonym of *I. flavidus*, as opined by Hooper (1968*b*), has yet to be assessed.

Isthmomys pirrensis (Goldman, 1912). Smithson. Misc. Coll., 60(2):5.
TYPE LOCALITY: Panama, Darien Prov., Mount Pirri, headwaters of Rio Limon, 4500 ft.
DISTRIBUTION: Easternmost Panama and perhaps adjacent Colombia.

Juscelinomys Moojen, 1965. Rev. Brasil. Biol., 25:281.
TYPE SPECIES: *Juscelinomys candango* Moojen, 1965.
COMMENTS: Little information available beyond original description; believed to be "an akodontine closely related to *Oxymycterus* and *Lenoxus*" (Reig, 1987:361).

Juscelinomys candango Moojen, 1965. Rev. Brasil. Biol., 25:281.
TYPE LOCALITY: Brazil, Federal District, Brasilia, Parque Zoobotanico, 1030 m.
DISTRIBUTION: C Brazil.

Juscelinomys talpinus (Winge, 1887). E Museo Lundii, 1(3):36.
TYPE LOCALITY: Brazil, Minas Gerais, Pleistocene cave deposits near Lagoa Santa.
DISTRIBUTION: Known only from the type locality.
COMMENTS: Moojen (1965) provisionally referred Winge's *Oxymycterus talpinus* to *Juscelinomys*. The present existence of *J. talpinus* in the region and its status with respect to *J. candango* invite attention (see Voss and Myers, 1991).

Kunsia Hershkovitz, 1966. Z. Säugetierk., 31(2):112.
TYPE SPECIES: *Mus tomentosus* Lichtenstein, 1830.
COMMENTS: Species formerly included in *Scapteromys* until set apart in *Kunsia* by Hershkovitz (1966*c*), who arranged both genera in the scapteromyine "group," which he viewed as closely related to oxymycterines. Formal tribal segregation of the two genera affirmed by Reig (1980). Karyology summarized by Gardner and Patton (1976).

Kunsia fronto (Winge, 1887). E Museo Lundii, 1(3):44.
TYPE LOCALITY: Brazil, Minas Gerais, Rio das Velhas, Pleistocene cave deposits near Lagoa Santa.
DISTRIBUTION: Isolated localities in NE Argentina and EC Brazil; range poorly documented.
SYNONYMS: *chacoensis*, *planaltensis*.
COMMENTS: Taxonomy and distribution reviewed by Avila-Pires (1972).

Kunsia tomentosus (Lichtenstein, 1830). Darst. Säugeth., 7(15):33.
TYPE LOCALITY: Brazil, southeastern area along the Rio Uruguay (as restricted by Hershkovitz, 1966*c*:120).
DISTRIBUTION: NE Bolivia (Beni Prov.) and WC Brazil (Mato Grosso); Pleistocene cave samples in Minas Gerais, Brazil.
SYNONYMS: *gnambiquarae*, *principalis*.
COMMENTS: Rare in collections; range inadequately known. See Massoia and Fornes (1965) and Hershkovitz (1966*c*) for justification of synonymy.

Lenoxus Thomas, 1909. Ann. Mag. Nat. Hist., ser. 8, 4:236.
TYPE SPECIES: *Oxymycterus apicalis* J. A. Allen, 1900.

COMMENTS: Akodontine. Viewed as closely related to *Oxymycterus* by Reig (1987), but electrophoretic data reveal *Lenoxus* as the most highly differentiated sister taxon to all other akodonts surveyed, including *Oxymycterus* (Patton et al., 1989).

Lenoxus apicalis (J. A. Allen, 1900). Bull. Am. Mus. Nat. Hist., 13:224.
TYPE LOCALITY: Peru, Puno Dept., Valley of Rio Inambari, Santo Domingo Mine, 6000 ft.
DISTRIBUTION: Cloud forest of E Andean slopes in SE Peru and W Bolivia.
SYNONYMS: *boliviae*.

Megadontomys Merriam, 1898. Proc. Biol. Soc. Washington, 12:115.
TYPE SPECIES: *Peromyscus thomasi* Merriam, 1898.
COMMENTS: Peromyscine. Used variably as a genus until Osgood (1909) stabilized its taxonomic ranking as a subgenus of *Peromyscus*, and so followed by Hooper and Musser (1964*b*) and Hooper (1968*b*). Carleton (1980, 1989) viewed the relationships and differentiation of *Megadontomys* at the generic level (but see Rogers, 1983). Aspects of morphology studied by Carleton (1973, 1980), Hooper and Musser (1964*b*), and Linzey and Layne (1969, 1974). Karyological affinities evaluated by Rogers (1983), Rogers et al. (1984), and Stangl and Baker (1984*b*). Werbitsky and Kilpatrick (1987) reported relatively low levels of genetic similarity between the nominal forms.

Megadontomys cryophilus (Musser, 1964). Occas. Pap. Mus. Zool., Univ. Michigan, 636:13.
TYPE LOCALITY: Mexico, Oaxaca, Distrito de Ixtlan, 13 mi NE Llano de las Flores, south slope Cerro Pelon, 9200 ft.
DISTRIBUTION: Sierra de Juarez, NC Oaxaca, Mexico.
COMMENTS: Named as a subspecies of *Peromyscus thomasi*. Citing morphological traits and genetic differentiation reported by Werbitsky and Kilpatrick (1987), Carleton (1989) arranged *cryophilus* as a species.

Megadontomys nelsoni (Merriam, 1898). Proc. Biol. Soc. Washington, 12:116.
TYPE LOCALITY: Mexico, Veracruz, Jico, 6000 ft.
DISTRIBUTION: E slopes of Sierra Madre Oriental, from SE Hidalgo to C Veracruz, Mexico.
COMMENTS: Relegated to a subspecies of *Peromyscus thomasi* by Musser (1964); reinstated to species rank by Carleton (1989).

Megadontomys thomasi (Merriam, 1898). Proc. Biol. Soc. Washington, 12:116.
TYPE LOCALITY: Mexico, Guerrero, mountains near Chilpancingo, 9700 ft.
DISTRIBUTION: High Sierra Madre del Sur of Guerrero, Mexico.
COMMENTS: Formerly encompassed *cryophilus* and *nelsoni* as subspecies (Musser, 1964); see above accounts.

Megalomys Trouessart, 1881. Le Naturaliste, 1:357.
TYPE SPECIES: *Mus desmarestii* Fischer, 1829.
SYNONYMS: *Moschomys* (of Trouessart, 1903), *Moschophoromys*.
COMMENTS: Named as a subgenus of the all-inclusive *Hesperomys* but later associated with *Oryzomys* by Major (1901); typically recognized as an oryzomyine genus endemic to the Lesser Antilles (Ellerman, 1941; Hall, 1981; Tate, 1932*c*). Closest mainland relative may be the poorly known *Oryzomys hammondi* of Ecuador (Ray, 1962). Both species are extinct but persisted perhaps until the late 1800s (Ray, 1962; Woods, 1989*a*).

Megalomys desmarestii (J. Fischer, 1829). Synopsis Mamm., p. 316.
TYPE LOCALITY: Lesser Antilles, Martinique.
DISTRIBUTION: Known only from the type locality.
STATUS: Extinct.
SYNONYMS: *pilorides* (of Desmarest, 1826).

Megalomys luciae (Forsyth Major, 1901). Ann. Mag. Nat. Hist., ser. 7, 7:206.
TYPE LOCALITY: Lesser Antilles, Santa Lucia.
DISTRIBUTION: Known only from the type locality.
STATUS: Extinct.

Melanomys Thomas, 1902. Ann. Mag. Nat. Hist., ser. 7, 10:248.
TYPE SPECIES: *Oryzomys phaeopus* Thomas, 1894 (= *Hesperomys caliginosus* Tomes, 1860).

COMMENTS: Oryzomyine. A morphologically distinctive group usually arranged as a subgenus of *Oryzomys* since Goldman's (1918) revision (e.g., Tate, 1932*e*; Ellerman, 1941; Cabrera, 1961; Hall, 1981). Allen (1913), however, provided morphological criteria, as contrasted to the type species (= *Mus palustris*) of *Oryzomys*, defending his retention of *Melanomys* as a genus; its synonymy in *Oryzomys* proper deserves more rigorous, character-based, phylogenetic substantiation. The taxon has not been revised; the nominal species listed stem from Cabrera (1961), who acknowledged his solely literature-based interpretation as provisional.

Melanomys caliginosus (Tomes, 1860). Proc. Zool. Soc. Lond., 1860:263.
TYPE LOCALITY: Ecuador, Esmeraldas Prov., Esmeraldas.
DISTRIBUTION: C American lowlands from easternmost Honduras through Panama; in South America, N and W Colombia to extreme NW Venezuela and to SW Ecuador.
SYNONYMS: *affinis, buenavistae, chrysomelas, columbianus, idoneus, lomitensis, monticola, obscurior, olivinus, oroensis, phaeopus, tolimensis, vallicola.*
COMMENTS: Chromosomal complement described by Gardner and Patton (1976).

Melanomys robustulus Thomas, 1914. Ann. Mag. Nat. Hist., ser. 8, 14:243.
TYPE LOCALITY: Ecuador, Morona-Zamora Prov., Gualaquiza, 2500 ft.
DISTRIBUTION: SE Ecuador.

Melanomys zunigae (Sanborn, 1949). Publ. Mus. Hist. Nat., Javier Prado, Zool., 1(3):2.
TYPE LOCALITY: Peru, Lima Dept., Lomas de Atocongo.
DISTRIBUTION: WC Peru.

Microryzomys Thomas, 1917. Smithson. Misc. Coll., 68:1.
TYPE SPECIES: *Hesperomys minutus* Tomes, 1860.
SYNONYMS: *Thallomyscus*.
COMMENTS: Oryzomyine. Named as a subgenus of *Oryzomys* and either retained as such (Cabrera, 1961; Ellerman, 1941; Osgood, 1933*a*) or placed in synonymy with *Oligoryzomys* (Gyldenstolpe, 1932; Tate, 1932*e*; Thomas, 1926*c*). Junior synonymy of *Thallomyscus* established by Osgood (1933*a*). Raised to genus by Carleton and Musser (1984) and later revised by them (1989).

Microryzomys altissimus (Osgood, 1933). Field Mus. Nat. Hist. Publ., Zool. Ser., 20:5.
TYPE LOCALITY: Peru, Pasco Dept., La Quinua, mountains north of Cerro de Pasco, 11,600 ft.
DISTRIBUTION: High Andes of Colombia, Ecuador, and Peru.
SYNONYMS: *chotanus, hylaeus.*
COMMENTS: Diagnosed as a subspecies of *Oryzomys minutus* but elevated to species by Hershkovitz (1940), who named two additional subspecies. Subspecific arrangement followed by Cabrera (1961) but none retained by Carleton and Musser (1989).

Microryzomys minutus (Tomes, 1860). Proc. Zool. Soc. Lond., 1860:215.
TYPE LOCALITY: Ecuador, Chimborazo Prov., probably near Pallatanga.
DISTRIBUTION: Montane forest from N Venezuela, through Colombia, Ecuador and Peru, to WC Bolivia.
SYNONYMS: *aurillus, dryas, fulvirostris, humilior.*
COMMENTS: Formerly included *altissimus* as a subspecies by Osgood (1933*a*), who also treated *aurillus, humilior,* and *fulvirostris* as subspecies. No races were deemed diagnosable by Carleton and Musser (1989).

Neacomys Thomas, 1900. Ann. Mag. Nat. Hist., ser. 7, 5:153.
TYPE SPECIES: *Hesperomys spinosus* Thomas, 1882.
COMMENTS: Oryzomyine. Recent revisionary standard lacking other than regional review of Lawrence (1941) and faunal checklists (e.g., Handley, 1966*a*, 1976; Husson, 1978). Traits for species recognition and distributional limits poorly delineated.

Neacomys guianae Thomas, 1905. Ann. Mag. Nat. Hist., ser. 7, 16:310.
TYPE LOCALITY: Guyana, Demerara River, 120 ft.
DISTRIBUTION: Guianas, S Venezuela, and N Brazil.

Neacomys pictus Goldman, 1912. Smithson. Misc. Coll., 60:2.
TYPE LOCALITY: Panama, Darien Prov., Cana, 1800 ft.

DISTRIBUTION: Known only from easternmost Panama.
COMMENTS: Cabrera (1961) allocated *pictus* to subspecific standing under *N. tenuipes*, and Handley (1966*a*) concurred, but the relationship and status of this gracile form bear reexamination.

Neacomys spinosus (Thomas, 1882). Proc. Zool. Soc. Lond., 1882:105.
TYPE LOCALITY: Peru, Amazonas Dept., Huambo, 3700 ft.
DISTRIBUTION: C and W Brazil to Andean foothills and lowlands of E Colombia, Ecuador, and Peru, probably including E Bolivia.
SYNONYMS: *amoenus, carceleni, typicus.*
COMMENTS: Karyotype compared to other oryzomyines by Gardner and Patton (1976).

Neacomys tenuipes Thomas, 1900. Ann. Mag. Nat. Hist., ser. 7, 5:153.
TYPE LOCALITY: Colombia, Cundinamarca Dept., Bogota region, Guaquimay.
DISTRIBUTION: W and NC Colombia, N Venezuela, E Ecuador, and N Brazil.
SYNONYMS: *pusillus.*
COMMENTS: Named and recognized (e.g., Ellerman, 1941) as a subspecies of *N. spinosus* until elevated to specific status by Lawrence (1941).

Nectomys Peters, 1861. Abh. König. Akad. Wiss. Berlin, 1860:151 [1861].
TYPE SPECIES: *Mus squamipes* Brants, 1827.
SYNONYMS: *Potamys.*
COMMENTS: Oryzomyine. R. J. Baker et al. (1983*a*) evaluated G-banded karyotypes of *Nectomys* with respect to other oryzomyines. Revised by Hershkovitz (1944), then including *Sigmodontomys* as a subgenus; Gardner and Patton (1976) urged removal of *Sigmodontomys* to *Oryzomys*.

Nectomys palmipes Allen and Chapman, 1893. Bull. Am. Mus. Nat. Hist., 5:209.
TYPE LOCALITY: Trinidad, Victoria County, Princes Town.
DISTRIBUTION: Island of Trinidad and nearby region of NE Venezuela: limits of distribution unknown.
COMMENTS: Arranged by Hershkovitz (1944) as one of many subspecies of *N. squamipes*. Barros et al. (1992) reinstated *palmipes* to species based on its inordinately low diploid number (2n=16-17) as compared to other populations of *Nectomys*, which range from 2n=38 to 2n=59 (Barros et al., 1992; Gardner and Patton, 1976).

Nectomys parvipes Petter, 1979. Mammalia, 43:507.
TYPE LOCALITY: French Guiana, Comte River, Cacao; 4°35′N, 52°28′W.
DISTRIBUTION: Known only from the type locality.

Nectomys squamipes (Brants, 1827). Het Geslacht der Muizen, p. 138.
TYPE LOCALITY: Brazil, São Paulo, São Sebastiao (as restricted by Hershkovitz, 1944:32).
DISTRIBUTION: Basin of the Rios Magdalena and Cauca, NC Colombia; east of the Andes broadly distributed from Guianas, Venezuela, and E Colombia southwards to NE Argentina, Uruguay, and SE Brazil; sea level to 2000 m.
SYNONYMS: *amazonicus, apicalis, aquaticus, brasiliensis, cephalotes, fulvinus, garleppii, grandis, magdalenae, mattensis, melanius, montanus, napensis, olivaceus, pollens, rattus, robustus, saturatus, tarrensis, tatei, vallensis.*
COMMENTS: Hershkovitz (1944) arrayed most nominal taxa of water rats as subspecies of *N. squamipes*, a view maintained by Cabrera (1961). Gardner and Patton (1976) intimated the mixed composition of *squamipes*; Reig (*in* Honacki et al., 1982; 1986) has offered enumerations of valid species and probable synonymies. Fullscale generic revision warranted. See Ernest (1986, Mammalian Species, 265).

Nelsonia Merriam, 1897. Proc. Biol. Soc. Washington, 11:277.
TYPE SPECIES: *Nelsonia neotomodon* Merriam, 1897.
COMMENTS: Neotomine. Revised by Hooper (1954) and recently by Engstrom et al. (1992), who resurrected *goldmani* as a species distinct from *N. neotomodon*. Viewed as a basal member of clade including *Neotoma* and related genera (Carleton, 1980; Hooper, 1954, 1960). Phylogenetic significance of banded karyotype discussed by Engstrom and Bickham (1983).

Nelsonia goldmani Merriam, 1903. Proc. Biol. Soc. Washington, 16:80.
TYPE LOCALITY: Mexico, Michoacan, Mount Tancitaro.
DISTRIBUTION: Transverse Volcanic Range, from Colima and S Jalisco eastward through N Michoacan to N Mexico, Mexico.
SYNONYMS: *cliftoni*.

Nelsonia neotomodon Merriam, 1897. Proc. Biol. Soc. Washington, 11:278.
TYPE LOCALITY: Mexico, Zacatecas, mountains near Plateado, 8200 ft.
DISTRIBUTION: Sierra Madre Occidental from S Durango to N Jalisco and Aguascalientes, Mexico.

Neotoma Say and Ord, 1825. J. Acad. Nat. Sci. Philadelphia, 4:345.
TYPE SPECIES: *Mus floridana* Ord, 1818.
SYNONYMS: *Homodontomys, Teanopus, Teonoma*.
COMMENTS: Neotomine. Revised by Goldman (1910), then including only *Homodontomys, Teonoma*, and the nominate subgenus. Burt and Barkalow (1942) established current subgeneric framework, also relegating *Hodomys* and *Teanopus* to subgenera. See Birney (1976) and Mascarello (1978) for modifications of species-group associations since Goldman (1910). Phylogenetic relationships of the genus considered by Hooper and Musser (1964*a*) and Carleton (1980), who reinstated *Hodomys* as a genus (see above account); interspecific relationships evaluated by Carleton (1980) and Koop et al. (1985). Anatomical systems described by Arata (1964), Burt and Barkalow (1942), Carleton (1973, 1980), Hooper (1960), and Howell (1926*a*). Karyotypic variation and evolution assessed by Mascarello and Hsu (1976) and Koop et al. (1985).

Neotoma albigula Hartley, 1894. Proc. California Acad. Sci., Ser. 2, 4:157.
TYPE LOCALITY: USA, Arizona, Pima Co., vicinity of Fort Lowell, near Tucson.
DISTRIBUTION: Extreme SE California to S Colorado to W Texas, USA, south to NE Michoacan and W Hidalgo, Mexico.
SYNONYMS: *angusticeps, brevicauda, cumulator, durangae, grandis, laplataensis, latifrons, leucodon, mearnsi, melanura, melas, montezumae, robusta, seri, sheldoni, subsolana, venusta, warreni, zacatecae*.
COMMENTS: Subgenus *Neotoma*. Subspecific taxonomy updated by Hall and Genoways (1970). Hybridization suspected with *N. micropus* in Colorado (Finley, 1958) and apparent intergradation of the two in Coahuila (Anderson, 1969*b*). Closely related to *N. floridana* and *N. micropus* (Birney, 1976), the three considered semispecies by Zimmerman and Nejtek (1977). See Macêdo and Mares (1988, Mammalian Species, 310).

Neotoma angustapalata Baker, 1951. Univ. Kansas Mus. Nat. Hist. Misc. Publ., 5:217.
TYPE LOCALITY: Mexico, Tamaulipas, 70 km (by highway) S Ciudad Victoria and 6 km W Panamerican Highway, El Carrizo.
DISTRIBUTION: SW Tamaulipas and adjacent San Luis Potosi, Mexico.
COMMENTS: Subgenus *Neotoma*. Specific status maintained by Birney (1973) but level of relationship to *N. micropus* unclear.

Neotoma anthonyi J. A. Allen, 1898. Bull. Am. Mus. Nat. Hist., 10:151.
TYPE LOCALITY: Mexico, Baja California Norte, Todos Santos Island.
DISTRIBUTION: Known only from the type locality.
STATUS: IUCN - Endangered.
COMMENTS: Subgenus *Neotoma*. Listed as nominal species of *N. lepida* group by Goldman (1932); also see Mascarello (1978).

Neotoma bryanti Merriam, 1887. Am. Nat., 21:191.
TYPE LOCALITY: Mexico, Baja California Norte, Cedros Island.
DISTRIBUTION: Known only from the type locality.
COMMENTS: Subgenus *Neotoma*. Listed as nominal species of *N. lepida* group by Goldman (1932); also see Mascarello (1978).

Neotoma bunkeri Burt, 1932. Trans. San Diego Soc. Nat. Hist., 7:181.
TYPE LOCALITY: Mexico, Baja California Sur, Coronados Island, 26°06'N, 111°18'W.
DISTRIBUTION: Known only from the type locality.
STATUS: IUCN - Endangered.

COMMENTS: Subgenus *Neotoma*. Likely conspecific with *N. lepida* (see Mascarello, 1978).

Neotoma chrysomelas J. A. Allen, 1908. Bull. Am. Mus. Nat. Hist., 24:653.
TYPE LOCALITY: Nicaragua, Matagalpa, Matagalpa.
DISTRIBUTION: NW Nicaragua, Honduras.
COMMENTS: Subgenus *Neotoma*. Hall (1981) suggested that *chrysomelas* is conspecific with *N. mexicana*, a proposal that should be considered appropos of a much-needed revision of the latter.

Neotoma cinerea (Ord, 1815). *In* Guthrie, New Geogr., Hist., Comml., Grammar, Philadelphia, 2nd ed., 2:292.
TYPE LOCALITY: USA, Montana, Cascade Co., Great Falls.
DISTRIBUTION: SE Yukon and westernmost Northwest Territories, south through British Columbia and W Alberta, Canada, to NW USA, as far south as N New Mexico and Arizona and east to W Dakotas.
SYNONYMS: *acraia, alticola, apicalis, arizonae, cinnamomea, columbiana, drummondii, fusca, grangeri, lucida, macrodon, occidentalis, orolestes, pulla, rupicola, saxamans.*
COMMENTS: Subgenus *Teonoma*. Independent evidence for sister-species relationship to *N. fuscipes* (Carleton, 1980; Koop et al., 1985) supports synonymy of *Homodontomys* under subgenus *Teonoma*, not subgenus *Neotoma* as enacted by Burt and Barkalow (1942).

Neotoma devia Goldman, 1927. Proc. Biol. Soc. Washington, 40:205.
TYPE LOCALITY: Arizona, Painted Desert, Tanner Tank, 5200 ft.
DISTRIBUTION: EC and S Utah, W Arizona, USA; NW Sonora, Mexico.
SYNONYMS: *aureotunicata, auripila, bensoni, flava, harteri, monstrabilis, sanrafaeli.*
COMMENTS: Subgenus *Neotoma*. Once ranked as a subspecies of *N. lepida*, from which several data sources support its specific-level divergence (Koop et al., 1985; Mascarello, 1978). Allocation of species-group synonyms requires further affirmation as does the delimitation of its geographic range.

Neotoma floridana (Ord, 1818). Bull. Sci. Soc. Philom. Paris, 1818:181.
TYPE LOCALITY: USA, Florida, Duval Co., St. Johns River, near Jacksonville.
DISTRIBUTION: SC and E USA from EC Colorado to E Texas, eastwards along Appalachians to W Connecticut, and along gulf-coast states to S North Carolina and C Florida.
STATUS: U.S. ESA and IUCN - Endangered as *N. f. smalli.*
SYNONYMS: *attwateri, baileyi, campestris, haematoreia, illinoensis, magister, osagensis, pennsylvanica, rubida, smalli.*
COMMENTS: Subgenus *Neotoma*. Hybridization with *N. micropus* possible but introgression along narrow contact zone judged insubstantial (Birney, 1973). Birney (1976) noted that further study may reveal *magister* as genetically isolated from *N. floridana*. See Wiley (1980, Mammalian Species, 139).

Neotoma fuscipes Baird, 1858. Mammalia *in* Repts. U.S. Expl. Surv., 8(1):495.
TYPE LOCALITY: USA, California, Sonoma Co., Petaluma.
DISTRIBUTION: W Oregon through W and C California, USA, to N Baja California, Mexico.
STATUS: IUCN - Vulnerable as *N. f. riparia.*
SYNONYMS: *affinis, annectens, bullatior, cnemophila, dispar, luciana, macrotis, martirensis, mohavensis, monochroura, perplexa, riparia, simplex, splendens, streatori.*
COMMENTS: Subgenus *Teonoma* (see account of *N. cinerea*). See Carraway and Verts (1991*a*, Mammalian Species, 386).

Neotoma goldmani Merriam, 1903. Proc. Biol. Soc. Washington, 16:48.
TYPE LOCALITY: Mexico, Coahuila, Saltillo, 5,000 ft.
DISTRIBUTION: SE Chihuahua to WC San Luis Potosi, Mexico.
COMMENTS: Subgenus *Neotoma*.

Neotoma lepida Thomas, 1893. Ann. Mag. Nat. Hist., ser. 6, 12:235.
TYPE LOCALITY: USA, "Simpson's Route" between Camp Floyd (= Fairfield), Utah and Carson City, Nevada (as restricted by Goldman, 1932:61).
DISTRIBUTION: SE Oregon and SW Idaho, south through Nevada and S California, USA, to S Baja California, Mexico.

SYNONYMS: *abbreviata, arenacea, aridicola, bella, californica, desertorum, egressa, felipensis, gilva, grinnelli, insularis, intermedia, latirostra, marcosensis, marshalli, molagrandis, nevadensis, notia, nudicauda, perpallida, petricola, pretiosa, ravida, sola, vicina.*
COMMENTS: Subgenus *Neotoma*. Aside from removal of *N. devia*, may still represent a composite of two species (see Mascarello, 1978).

Neotoma martinensis Goldman, 1905. Proc. Biol. Soc. Washington, 18:28.
TYPE LOCALITY: Mexico, Baja California Norte, San Martin Island.
DISTRIBUTION: Known only from the type locality.
STATUS: IUCN - Endangered.
COMMENTS: Subgenus *Neotoma*. Listed as nominal species of *N. lepida* group by Goldman (1932); also see Mascarello (1978).

Neotoma mexicana Baird, 1855. Proc. Acad. Nat. Sci. Philadelphia, 7:333.
TYPE LOCALITY: Mexico, Chihuahua, mountains near Chihuahua.
DISTRIBUTION: SE Utah and C Colorado, USA, southwards through W and interior Mexico, to highlands of Guatemala, El Salvador, and W Honduras.
SYNONYMS: *atrata, bullata, chamula, distincta, eremita, fallax, ferruginea, fulviventer, griseoventer, inopinata, inornata, isthmica, madrensis, navus, ochracea, orizabae, parvidens, picta, pinetorum, scopulorum, sinaloae, solitaria, tenuicauda, torquata, tropicalis, vulcani.*
COMMENTS: Subgenus *Neotoma*. Subspecies classification revised by Hall (1955). Specific homogeneity of included taxa doubtful. See Cornely and Baker (1986, Mammalian Species, 262).

Neotoma micropus Baird, 1855. Proc. Acad. Nat. Sci. Philadelphia, 7:333.
TYPE LOCALITY: Mexico, Tamaulipas, Charco Escondido.
DISTRIBUTION: SE Colorado and SW Kansas through W Texas and most of New Mexico, USA; south in Mexico to N Chihuahua, E San Luis Potosi, and S Tamaulipas.
SYNONYMS: *canescens, leucophaea, littoralis, planiceps, surberi.*
COMMENTS: Subgenus *Neotoma*. Limited hybridization documented with *N. floridana* in Oklahoma (Birney, 1973) and believed probable with *N. albigula* in Colorado (Finley, 1958) and Coahuila (Anderson, 1969*b*); also see comments under *N. albigula* and *N. floridana*. See Braun and Mares (1989, Mammalian Species, 330).

Neotoma nelsoni Goldman, 1905. Proc. Biol. Soc. Washington, 18:29.
TYPE LOCALITY: Mexico, Veracruz, Perote, 7800 ft.
DISTRIBUTION: Known only from the type locality.
COMMENTS: Subgenus *Neotoma*. Affiliated with *N. albigula* but separate stature uncertain (see Hall and Genoways, 1970).

Neotoma palatina Goldman, 1905. Proc. Biol. Soc. Washington, 18:27.
TYPE LOCALITY: Mexico, Jalisco, Bolanos, 2800 ft.
DISTRIBUTION: EC Jalisco, Mexico.
COMMENTS: Subgenus *Neotoma*. Specific distinctiveness from *N. albigula* reasserted by Hall and Genoways (1970).

Neotoma phenax (Merriam, 1903). Proc. Biol. Soc. Washington, 16:81.
TYPE LOCALITY: Mexico, Sonora, Rio Mayo, Camoa.
DISTRIBUTION: SW Sonora and NW Sinaloa, Mexico.
COMMENTS: Subgenus *Teanopus*. Type species of *Teanopus*, which was reduced to a subgenus by Burt and Barkalow (1942). Karyotype interpreted as highly derived (Koop et al., 1985). See Jones and Genoways (1978, Mammalian Species, 108).

Neotoma stephensi Goldman, 1905. Proc. Biol. Soc. Washington, 18:32.
TYPE LOCALITY: USA, Arizona, Mohave Co., Hualapai Mtns., 6300 ft.
DISTRIBUTION: Extreme SC Utah, N Arizona, and NW New Mexico, USA.
SYNONYMS: *relicta.*
COMMENTS: Subgenus *Neotoma*. Morphological separation from *N. lepida* and geographic variation reviewed by Hoffmeister and de la Torre (1960). See Jones and Hildreth (1989, Mammalian Species, 328).

Neotoma varia Burt, 1932. Trans. San Diego Soc. Nat. Hist., 7:178.
TYPE LOCALITY: Mexico, Sonora, Turner Island, 28°43'N, 112°19'W.

DISTRIBUTION: Known only from the type locality.
COMMENTS: Subgenus *Neotoma*. Affiliated with *N. albigula* (see Hall and Genoways, 1970).

Neotomodon Merriam, 1898. Proc. Biol. Soc. Washington, 12:127.
TYPE SPECIES: *Neotomodon alstoni* Merriam, 1898.
COMMENTS: Peromyscine. Conventionally ranked as a genus, later judged closely related to *Peromyscus sensu lato* (Hooper and Musser, 1964*b*). Synonymy as a subgenus of *Peromyscus* advocated by Yates et al. (1979) and J. C. Patton et al. (1981), whereas Carleton (1980, 1989) retained *Neotomodon* as a genus. Various kinds of evidence suggest the phyletic association of *Neotomodon*, *Podomys*, and perhaps *Habromys* (Carleton, 1980; Hooper and Musser, 1964*b*; Linzey and Layne, 1969; J. C. Patton et al., 1981; Stangl and Baker, 1984*b*).

Neotomodon alstoni Merriam, 1898. Proc. Biol. Soc. Washington, 12:128.
TYPE LOCALITY: Mexico, Michoacan, Nahuatzin, 8500 ft.
DISTRIBUTION: Cordillera Transvolcanica, Mexico, from WC Michoacan eastwards to C Veracruz.
SYNONYMS: *orizabae*, *perotensis*.
COMMENTS: Williams and Ramirez-Pulido's (1984) evaluation of geographic variation disclosed no basis for subspecific divisions. See Williams et al. (1985, Mammalian Species, 242, as *Peromyscus alstoni*).

Neotomys Thomas, 1894. Ann. Mag. Nat. Hist., ser. 6, 14:346.
TYPE SPECIES: *Neotomys ebriosus* Thomas, 1894.
COMMENTS: Revised by Sanborn (1947*a*); distribution augmented by Pine et al. (1979). Although grouped with sigmodont rodents by Hershkovitz (1955*a*), other studies have convincingly linked the genus with phyllotines (Olds and Anderson, 1989; Pearson and Patton, 1976).

Neotomys ebriosus Thomas, 1894. Ann. Mag. Nat. Hist., ser. 6, 14:348.
TYPE LOCALITY: Peru, Junín Dept., Vitoc Valley.
DISTRIBUTION: Altiplano of C Peru (Junín), south through northernmost Chile and W Bolivia, to NW Argentina; above 3350 m. Argentine locality records summarized by Barquez (1983).
SYNONYMS: *vulturnus*.
COMMENTS: Sanborn (1947*a*) reduced *vulturnus* to a subspecies of *N. ebriosus*.

Nesoryzomys Heller, 1904. Proc. California Acad. Sci., 3:241.
TYPE SPECIES: *Nesoryzomys narboroughi* Heller, 1904 (= *Oryzomys indefessus* Thomas, 1899).
COMMENTS: Oryzomyine. Realigned as a subgenus of *Oryzomys* by Ellerman (1941), following the comments of Goldman (1918). Morphological, genic, and karyological information, however, sustains the generic separation of *Nesoryzomys* (Beaufort, 1963; Gardner and Patton, 1976; Patton and Hafner, 1983) and intimates that the group is an old Galapagos immigrant, originating *ca.* 3-3.5 mya (see Patton and Hafner, 1983).

Nesoryzomys darwini Osgood, 1929. Field Mus. Nat. Hist. Publ., Zool. Ser., 17:23.
TYPE LOCALITY: Ecuador, Galapagos Archipelago, Santa Cruz Island, Academia Bay.
DISTRIBUTION: Santa Cruz (= Indefatigable) Island.
COMMENTS: Probably extinct, last recorded in 1930 (see Patton and Hafner, 1983).

Nesoryzomys fernandinae Hutterer and Hirsch, 1979. Bonn. Zool. Beitr., 30:278.
TYPE LOCALITY: Ecuador, Galapagos Archipelago, Fernandina Island.
DISTRIBUTION: Known only from the type locality.
COMMENTS: Type material recovered from fresh owl pellets. Believed closely related to *N. darwini*; sympatric with *N. indefessus narboroughi* on Fernandina Island (see Hutterer and Hirsch, 1979).

Nesoryzomys indefessus (Thomas, 1899). Ann. Mag. Nat. Hist., ser. 7, 4:280.
TYPE LOCALITY: Ecuador, Galapagos Archipelago, Santa Cruz Island, Academia Bay.
DISTRIBUTION: Santa Cruz, Baltra (= South Seymour), and Fernandina (= Narborough) islands.
SYNONYMS: *narboroughi*.

COMMENTS: Patton and Hafner (1983:539) recommended that *indefessus*, *narboroughi*, and *swarthi* are "best considered races of a single species, which differ primarily in pelage color." Their analyses sustain this conclusion with regard to *indefessus* and *narboroughi* but not the craniodental differentiation of *N. swarthi*. *N. i. indefessus* is probably extinct, none documented since 1934 (see Patton and Hafner, 1983). Populations of *N. i. narboroughi* on Fernandina Island, which lacks commensal *Rattus* and *Mus*, appear stable (see Patton and Hafner, 1983).

Nesoryzomys swarthi Orr, 1938. Proc. California Acad. Sci., 23(21):304.
TYPE LOCALITY: Ecuador, Galapagos Archipelago, San Salvador Island, Sullivan Bay.
DISTRIBUTION: Known only from San Salvador (= James, Santiago) Island.
COMMENTS: Status viewed as a insular race of *N. indefessus* by Patton and Hafner (1983), but Orr (1938) underscored the trenchant diagnostic traits that separate *N. swarthi* from both *N. indefessus* and *narboroughi*. Remnant populations thought to exist as of 1965 (see Peterson, 1966*b*); recent efforts have uncovered only *Rattus* and *Mus* (see Patton and Hafner, 1983).

Neusticomys Anthony, 1921. Am. Mus. Novit., 20:2.
TYPE SPECIES: *Neusticomys monticolus* Anthony, 1921.
SYNONYMS: *Daptomys*.
COMMENTS: Ichthyomyine. Phylogenetic relationships studied by Voss (1988), who allocated *Daptomys* as a full synonym.

Neusticomys monticolus Anthony, 1921. Am. Mus. Novit., 20:2.
TYPE LOCALITY: Ecuador, Prov. Pichincha, Nono Farm "San Francisco," 10,500 ft.
DISTRIBUTION: Colombia and Ecuador.
COMMENTS: Taxonomy and distribution reviewed by Voss (1988).

Neusticomys mussoi Ochoa G. and Soriano, 1991. J. Mammal., 72:97.
TYPE LOCALITY: Venezuela, Edo. Tachira, 14 km SE Pregonero, Rio Potosí, Paso Hondo, 1050 m.
DISTRIBUTION: Known only from the type locality.
COMMENTS: The two known specimens represent the least aquatically specialized ichthyomyine species described thus far.

Neusticomys oyapocki (Dubost and Petter, 1978). Mammalia, 42:436.
TYPE LOCALITY: French Guiana, Trois-Sauts, near the banks of the Oyapock River; 02°10'N, 53°11'W.
DISTRIBUTION: Known only from the type locality.
COMMENTS: The single known specimen was originally referred to *Daptomys*. Maintained as a species by Voss (1988).

Neusticomys peruviensis (Musser and Gardner, 1974). Am. Mus. Novit., 2537:7.
TYPE LOCALITY: Peru, Depto. Loreto, Balta, 300 m; 10°08'S, 17°13'W.
DISTRIBUTION: Known only from the type locality.
COMMENTS: The single known specimen was originally referred to *Daptomys*. Maintained as a species by Voss (1988).

Neusticomys venezuelae (Anthony, 1929). Am. Mus. Novit., 383:2.
TYPE LOCALITY: Venezuela, Edo. Sucre, 24 km W Cumanacoa, headwaters of Río Neverí, 2400 ft.
DISTRIBUTION: S Venezuela and Guyana.
COMMENTS: Type species of *Daptomys*, considered a junior synonym of *Neusticomys* by Voss (1988).

Notiomys Thomas, 1890. *In* Milne-Edwards, Mission Sci. Cap. Horn, 1882-3, 6, Mamm., p. 23.
TYPE SPECIES: *Notiomys edwardsii* Thomas, 1890.
COMMENTS: Akodontine. Alpha systematics revised by Osgood (1925), who viewed *Chelemys* and *Geoxus* as synonyms (see comments under those genera and in Pearson, 1984, and Reig, 1987).

Notiomys edwardsii Thomas, 1890. *In* Milne-Edwards, Mission Sci. Cap. Horn, 1882-3, 6, Mamm., p. 24.
TYPE LOCALITY: Argentina, Santa Cruz Prov., south of Santa Cruz, near 50°S latitude.

DISTRIBUTION: S Argentina, from Rio Negro Prov. to Santa Cruz Prov.
COMMENTS: Known by only six specimens; morphology and habits amplified by Pearson (1984).

Nyctomys Saussure, 1860. Rev. Mag. Zool. Paris, ser. 2, 12:106.
TYPE SPECIES: *Hesperomys sumichrasti* Saussure, 1860.
COMMENTS: Middle American endemic of enigmatic phyletic position. Placed with thomasomyine group of South American sigmodontines by Hershkovitz (1944, 1962); others have disputed the association with thomasomyines, instead suggesting distant kinship to neotomine-peromyscines or a cladistic origin prior to both North and South American sigmodontines (Arata, 1964; Carleton, 1980; Haiduk et al., 1988; Hooper and Musser, 1964*a*; Voss and Linzey, 1981).

Nyctomys sumichrasti (Saussure, 1860). Rev. Mag. Zool. Paris, ser. 2, 12:107.
TYPE LOCALITY: Mexico, Veracruz, Uvero, 20 km NW Santiago Tuxtla (as restricted by Alvarez, 1963*b*).
DISTRIBUTION: S Jalisco and S Veracruz, Mexico, south to C Panama, excluding the Yucatan Peninsula.
SYNONYMS: *colimensis, costaricensis, decolorus, florencei, nitellinus, pallidulus, salvini, venustulus.*
COMMENTS: Standard and banded karyotypes described by Lee and Elder (1977) and Haiduk et al. (1988), respectively.

Ochrotomys Osgood, 1909. N. Am. Fauna, 28:222.
TYPE SPECIES: *Arvicola nuttalli* Harlan, 1832.
COMMENTS: Although diagnosed as a subgenus of *Peromyscus*, the distant kinship and generic segregation of *Ochrotomys* have been repeatedly sustained (Blair, 1942; Carleton, 1980; Hooper and Musser, 1964*b*; Patton and Hsu, 1967). A peromyscine, but this conventional placement has been questioned (see Engstrom and Bickham, 1982; Carleton, 1989:115).

Ochrotomys nuttalli (Harlan, 1832). Mon. Am. J. Geol. Nat. Sci. Philadelphia, p. 446.
TYPE LOCALITY: USA, Virginia, Norfolk Co., Norfolk.
DISTRIBUTION: SE Missouri across to S Virginia, south to E Texas, the Gulf coast, and C Florida.
SYNONYMS: *aureolus, flammeus, lewisi, lisae.*
COMMENTS: Subspecific classification revised by Packard (1969). See Linzey and Packard (1977, Mammalian Species, 75).

Oecomys Thomas, 1906. Ann. Mag. Nat. Hist., ser. 7, 18:444.
TYPE SPECIES: *Rhipidomys benevolens* Thomas, 1901 (= *Hesperomys bicolor* Tomes, 1860).
COMMENTS: Oryzomyine. Diagnosed as a subgenus of *Oryzomys* to segregate arboreal, pencil-tailed sigmodontines with a long palate from *Rhipidomys*, under which many of the species included here were first described. Thereafter treated alternatively as a subgenus of *Oryzomys* (Ellerman, 1941; Goldman, 1918) or as full genus (Gyldenstolpe, 1932; Thomas, 1917*c*) until Hershkovitz's (1960) revision stabilized its ranking as a subgenus (e.g., Cabrera, 1961; Hall, 1981). Systematists have recently acknowledged the morphological and karyotypic distinctiveness of *Oecomys* at the generic level (Carleton and Musser, 1984; Gardner and Patton, 1976; Reig, 1984, 1986), but this recognition as yet lacks convincing substantiation from a phylogenetic perspective.

Revised by Hershkovitz (1960), who consolidated some 25 species (e.g., Ellerman, 1941) into the two species *bicolor* and *concolor*. Although this gross underestimation of the species diversity within *Oecomys* has been intimated by other authors (e.g., Gardner and Patton, 1976; Reig, 1986), it has yet to be documented within a taxonomic revision. The species identified here issue from our revision in progress; we are confident that all of these will stand as valid yet some, such as *O. trinitatis*, are undoubtedly composites even now.

Oecomys bicolor (Tomes, 1860). Proc. Zool. Soc. Lond., 1860:217.
TYPE LOCALITY: Ecuador, Morona-Santiago Prov., Gualaquiza, Rio Gualaquiza, 885 m.

DISTRIBUTION: E Panama to W Colombia and Ecuador; Venezuela, Guianas, N and C Brazil; Amazonian drainage of E Bolivia, Peru, and Colombia.
SYNONYMS: *benevolens, dryas* (of Thomas, 1900), *endersi, florenciae, milleri, nitedulus, occidentalis, phelpsi, rosilla, trabeatus.*

Oecomys cleberi Locks, 1981. Bol. Mus. Nac., Zool., 300:1.
TYPE LOCALITY: Brazil, Federal District, Universidade de Brasilia, Fazenda Agua Limpa; 45°54′W, 15°57′S.
DISTRIBUTION: Known only from the type locality.
COMMENTS: Allied to *O. bicolor,* or *O. paricola.*

Oecomys concolor (Wagner, 1845). Arch. Naturgesch., 11:147.
TYPE LOCALITY: Brazil, Amazonas, Rio Curicuriari, a tributary of the upper Rio Negro, below São Gabriel.
DISTRIBUTION: S Venezuela, NW Brazil north of Amazon, E Colombia, and N Bolivia.
SYNONYMS: *marmosurus.*

Oecomys flavicans (Thomas, 1894). Ann. Mag. Nat. Hist., ser. 6, 14:351.
TYPE LOCALITY: Venezuela, Mérida, Mérida, 1600 m.
DISTRIBUTION: Coastal Range and Cordillera de Mérida of N and W Venezuala, west to Sierra de Santa Marta of NE Columbia, perhaps including the Cordillera Oriental.
SYNONYMS: *illectus, mincae.*

Oecomys mamorae (Thomas, 1906). Ann. Mag. Nat. Hist., ser. 7, 18:445.
TYPE LOCALITY: Bolivia, Cochabamba Dept., upper Rio Mamore, Mosetenes.
DISTRIBUTION: E Bolivia, N Paraguay, and WC Brazil.

Oecomys paricola (Thomas, 1904). Ann. Mag. Nat. Hist., ser. 7, 14:194.
TYPE LOCALITY: Brazil, Pará, Igarape-Assu, 50 m.
DISTRIBUTION: SE Venezuela, Guianas, and N and C Brazil.
SYNONYMS: *auyantepui.*

Oecomys phaeotis (Thomas, 1901). Ann. Mag. Nat. Hist., ser. 7, 7:181.
TYPE LOCALITY: Peru, Puno Dept., upper Rio Inambari, Sagrario, 1000 m.
DISTRIBUTION: Eastern slopes of Peruvian Andes.

Oecomys rex Thomas, 1910. Ann. Mag. Nat. Hist., ser. 8, 6:504.
TYPE LOCALITY: Guyana, Demerara Dist., Rio Supinaam.
DISTRIBUTION: Extreme E Venezuela, Guianas, and NE Brazil north of the Amazon (Amapa and Amazonas).
SYNONYMS: *regalis.*

Oecomys roberti (Thomas, 1904). Proc. Zool. Soc. Lond., 1903(2):237 [1904].
TYPE LOCALITY: Brazil, Mato Grosso, Santa Anna de Chapada, 800 m.
DISTRIBUTION: S Venezuela, Guianas, and Amazonian region of N Brazil, E Peru, and N Bolivia.
SYNONYMS: *guianae, tapajinus.*

Oecomys rutilus Anthony, 1921. Am. Mus. Novit., 19:4.
TYPE LOCALITY: Guyana, Mazaruni-Potaro Dist., Kartabo.
DISTRIBUTION: Guyana, Surinam, and French Guiana.

Oecomys speciosus (J. A. Allen and Chapman, 1893). Bull. Am. Mus. Nat. Hist., 5:212.
TYPE LOCALITY: Trinidad, Princes Town.
DISTRIBUTION: Savannahs of NE Colombia, C and N Venezuela, and Trinidad.
SYNONYMS: *caicarae, trichurus.*

Oecomys superans Thomas, 1911. Ann. Mag. Nat. Hist., ser. 8, 8:250.
TYPE LOCALITY: Ecuador, Pastaza Prov., Rio Bobonaza, Canelos, 2100 ft.
DISTRIBUTION: Lower Andean slopes and foothills of E Colombia, Ecuador, and Peru.
SYNONYMS: *melleus, palmeri.*

Oecomys trinitatis (J. A. Allen and Chapman, 1893). Bull. Am. Mus. Nat. Hist., 5:213.
TYPE LOCALITY: Trinidad, Princes Town.
DISTRIBUTION: Neotropical rainforests from SW Costa Rica to SE Brazil, including Guianas, Trinidad and Tobago; E Andean slopes of WE Colombia to SC Peru.

SYNONYMS: *bahiensis, catherinae, cinnamomeus, frontalis, fulviventer, helvolus, klagesi, osgoodi, palmarius, splendens, subluteus, tectus, vicencianus.*

Oligoryzomys Bangs, 1900. New England Zool. Club., 1:94.
TYPE SPECIES: *Oryzomys navus* Bangs, 1900 (= *Hesperomys fulvescens* Saussure, 1860).
COMMENTS: Oryzomyine. Described as a subgenus of *Oryzomys* and usually recognized as such (Ellerman, 1941; Hall, 1981; Tate, 1932*e*) or as a genus (Contreras and Berry, 1983; Gyldenstolpe, 1932), with *Microryzomys* as a full synonym (Gyldenstolpe, 1932; Tate, 1932*e*) or not (Cabrera, 1961). Diagnosis emended at the generic level by Carleton and Musser (1989). Species-level revisions required: estimates range from one (Hershkovitz, 1966*c*) to 30 (Tate, 1932*e*), usually around 12 (Cabrera, 1961). Regional studies have recorded three or four species in sympatry or parapatry (Contreras and Berry, 1983; Massoia, 1973; Myers and Carleton, 1981; Olds and Anderson, 1987). Karyology of many species presented in Espinosa and Reig (1991), Gallardo and Patterson (1985), Gardner and Patton (1976), and Myers and Carleton (1981). The species recognized here observe the preliminary review of Carleton and Musser (1989).

Oligoryzomys andinus (Osgood, 1914). Field Mus. Nat. Hist. Publ., Zool. Ser., 10:156.
TYPE LOCALITY: Peru, La Libertad Dept., upper Rio Chicama, Hacienda Llagueda, 6000 ft.
DISTRIBUTION: W Peru and WC Bolivia; geographic and altitudinal limits uncertain.
COMMENTS: Karyotype reported by Gardner and Patton (1976).

Oligoryzomys arenalis (Thomas, 1913). Ann. Mag. Nat. Hist., ser. 8, 12:571.
TYPE LOCALITY: Peru, Lambayeque Dept., Eten, 10 m.
DISTRIBUTION: Arid and semiarid coastal plain of Peru.
COMMENTS: Level of relationship to *O. fulvescens* warrants clarification.

Oligoryzomys chacoensis (Myers and Carleton, 1981). Misc. Publ. Mus. Zool., Univ. Michigan, 161:19.
TYPE LOCALITY: Paraguay, Boqueron Dept., km 419 along Trans Chaco Hwy, northwest of Villa Hayes.
DISTRIBUTION: Dryer habitats of W Paraguay, SE Bolivia, WC Brazil, and N Argentina. May be more broadly distributed in cerrado and caatinga habitats in SE Brazil.
COMMENTS: Karyotype reported by Myers and Carleton (1981); morphometric variation by Myers and Carleton (1981) and Olds and Anderson (1987). Status with regard to *O. andinus* unresolved.

Oligoryzomys delticola (Thomas, 1917). Ann. Mag. Nat. Hist., ser. 8, 20:96.
TYPE LOCALITY: Argentina, Buenos Aires Prov., Delta del Parana, Isla Ella, 1 m.
DISTRIBUTION: EC Argentina, Uruguay, and S Brazil (Rio Grande do Sul).

Oligoryzomys destructor (Tschudi, 1844). Fauna Peruana, 1:182.
TYPE LOCALITY: E Peru.
DISTRIBUTION: Andes of S Colombia, through Ecuador and Peru, to WC Bolivia.
SYNONYMS: *maranonicus, melanostoma, spodiurus, stolzmanni.*
COMMENTS: Karyotype reported by Gardner and Patton (1976) as *Oryzomys longicaudatus* variant (4). As discussed by Hershkovitz (1940:81), Tschudi's specimens may have originated from haciendas along the Rio Chinchao, Huanuco Dept., Peru, 900-1000 m.

Oligoryzomys eliurus (Wagner, 1845). Arch. Naturgesch., 11:147.
TYPE LOCALITY: Brazil, São Paulo, Ytarare.
DISTRIBUTION: C and SE Brazil.
SYNONYMS: *pygmaeus, utiaritensis.*
COMMENTS: Perhaps conspecific with *O. nigripes* (see Myers and Carleton, 1981).

Oligoryzomys flavescens (Waterhouse, 1837). Proc. Zool. Soc. Lond., 1837:19.
TYPE LOCALITY: Uruguay, Maldonado.
DISTRIBUTION: SE Brazil, Uruguay, and Argentina (south to Chubut Prov.).
SYNONYMS: *antoniae, occidentalis.*
COMMENTS: Chromosomal variation reported and taxonomic implications discussed by Sbalqueiro et al. (1991).

Oligoryzomys fulvescens (Saussure, 1860). Revue Mag. Zool. Paris, ser. 2, 12:102.
TYPE LOCALITY: Mexico, Veracruz, Orizaba (as restricted by Merriam, 1901:295).
DISTRIBUTION: W and E versants of S Mexico, through Mesoamerica, to Ecuador, northernmost Brazil, and Guianas in South America.
SYNONYMS: *costaricensis, delicatus, engraciae, lenis, mayensis, messorius, munchiquensis, navus, nicaraguae, pacificus, tenuipes.*
COMMENTS: All Central American forms retained as subspecies (see Hall, 1981). Variable karyotypic descriptions intimate that more than one species occurs among the listed synonyms (Gardner and Patton, 1976; Haiduk et al., 1979).

Oligoryzomys griseolus (Osgood, 1912). Field Mus. Nat. Hist. Publ., Zool. Ser., 10:49.
TYPE LOCALITY: Venezuela, Tachira, upper Rio Tachira, west of Paramo de Tama, 6000-7000 ft.
DISTRIBUTION: Tachira Andes of W Venezuela and Cordillera Oriental of E Colombia.
COMMENTS: As noted by Osgood (1912), this distinctive form contrasts sharply with neighboring populations of *O. fulvescens* (e.g., *navus* and *tenuipes*) and instead resembles *O. vegetus* of W Panama.

Oligoryzomys longicaudatus (Bennett, 1832). Proc. Zool. Soc. Lond., 1832:2.
TYPE LOCALITY: Chile, Valparaiso Prov. (as suggested by Osgood, 1943:143).
DISTRIBUTION: NC to S Andes, approximately to 50°S latitude, of Chile and Argentina.
SYNONYMS: *agilis, amblyrrhynchus, araucanus, commutatus, coppingeri, diminutivus, dumetorum, exiguus, glaphyrus, macrocercus, melaenus, melanizon, mizurus, nigribarbis, pernix, peteroanus, philippii, saltator.*
COMMENTS: Formerly encompassed most Andean populations of *Oligoryzomys* (see Cabrera, 1961). Morphometric variation among Chilean populations investigated by Gallardo and Palma (1990), who placed *philippii* in full synonymy with *O. longicaudatus* and removed *magellanicus* to specific status. Relationships and level of divergence from allopatric forms like *O. destructor* and *O. nigripes* warrants study (see Carleton and Musser, 1989). Attribution of numerous Philippi (1900) epithets follows Osgood (1943).

Oligoryzomys magellanicus (Bennett, 1836). Proc. Zool. Soc. Lond., 1835:191 [1836].
TYPE LOCALITY: Chile, Magallanes Prov., Straits of Magellan, Port Famine.
DISTRIBUTION: S Patagonian region of Chile and Argentina, including Tierra del Fuego.
COMMENTS: Karyotypic, morphometric, and phallic differention from *O. longicaudatus* supports the specific recognition of *O. magellanicus* (Gallardo and Palma, 1990; Gallardo and Patterson, 1985).

Oligoryzomys microtis (Allen, 1916). Bull. Am. Mus. Nat. Hist., 35:525.
TYPE LOCALITY: Brazil, Amazonas, lower Rio Solimões.
DISTRIBUTION: C Brazil south of Rios Solimões-Amazon, and contiguous lowlands of Peru, Bolivia, Paraguay, and Argentina.
SYNONYMS: *chaparensis, fornesi, mattogrossae.*
COMMENTS: Karyotype reported by Gardner and Patton (1976) as *Oryzomys longicaudatus* variant (2) and by Myers and Carleton (1981) as *Oryzomys fornesi.*

Oligoryzomys nigripes (Olfers, 1818). *In* Eschwege, J. Brasilien, Neue Bibliothek. Reisenb., 15:209.
TYPE LOCALITY: Paraguay, Paraguari Dept., Ybycui National Park, 85 km SSE Atyra (as restricted by Myers and Carleton, 1981:14).
DISTRIBUTION: E Paraguay and N Argentina.
COMMENTS: Neotype designated, diagnosis emended, and karyotype described by Myers and Carleton (1981). Distributional extent unclear; relationship to allopatric forms such as *eliurus* and *longicaudatus* warrants investigation. Myers and Carleton (1981) recommended that *Mus longitarsus* Rengger, 1830, which could apply to either *O. microtis* or *O. nigripes*, be considered a *nomen dubium.*

Oligoryzomys vegetus (Bangs, 1902). Bull. Mus. Comp. Zool., 39:35.
TYPE LOCALITY: Panama, Chiriqui Prov., Volcan de Chiriqui, Boquete, 4000 ft.
DISTRIBUTION: Westernmost Panama; limits uncertain.
SYNONYMS: *creper, reventazoni.*
COMMENTS: Relegated to a subspecies of *O. fulvescens* by Goldman (1918) and so arranged thereafter (e.g., Hall, 1981). Bangs (1902), however, correctly recognized the sympatry of his new species with *O. fulvescens.*

Oligoryzomys victus (Thomas, 1898). Ann. Mag. Nat. Hist., ser. 7, 1:178.
TYPE LOCALITY: Lesser Antilles, Saint Vincent.
DISTRIBUTION: Known only from the type locality.
COMMENTS: Earlier classifications listed *O. victus* as an *Oryzomys* of uncertain affinity (Ellerman, 1941; Goldman, 1918), and Hall and Kelson (1959) erroneously placed it with their "*tectus* group" (= *Oecomys*). Thomas (1898a) and later Ray (1962) emphasized its alliance with species of *Oligoryzomys*. Known only by the holotype; presumably extinct (see Ray, 1962).

Onychomys Baird, 1858. Mammalia *in* Repts. U.S. Expl. Surv., 8(1):458.
TYPE SPECIES: *Hypudaeus leucogaster* Wied-Neuwied, 1841.
COMMENTS: Peromyscine. Revised by Hollister (1914). Chromosomal evolution among three species investigated by Baker et al. (1979), allozymic differentiation by Sullivan et al. (1986), and mitochondrial-DNA phylogeny by Riddle and Honeycutt (1990); historical biogeography interpreted by Riddle and Honeycutt (1990); affinity of Recent and fossil forms evaluated by Carleton and Eshelman (1979). Phyletic affinity to other sigmodontine genera assessed by Carleton (1980), Hooper and Musser (1964*b*), and Stangl and Baker (1984*b*).

Onychomys arenicola Mearns, 1896. Preliminary diagnosis of new mammals from the Mexican border of the U. S., p. 3 (preprint of Proc. U. S. Natl. Mus., 19:137-140).
TYPE LOCALITY: USA, Texas, El Paso Co., 6 mi above El Paso.
DISTRIBUTION: Chihuahuan Desert: SE Arizona, SC New Mexico, and W Texas, USA, south into C Mexico, to Aguascalientes, San Luis Potosi, and W Tamaulipas.
SYNONYMS: *canus, surrufus.*
COMMENTS: Placed in full synonymy of *O. t. torridus* by Hollister (1914). Sympatry with *O. torridus* and karyotypic discrimination reported by Hinesley (1979), who raised *O. arenicola* to species—also see Baker et al. (1979). Sullivan et al. (1986) and Riddle and Honneycutt (1990) viewed *O. leucogaster* and *O. arenicola* as sister taxa. Specimen-based documentation of geographic range and synonymy of species-group taxa provisionally associated here under *O. arenicola* highly welcomed.

Onychomys leucogaster (Wied-Neuwied, 1841). Reise in Nord-America, 2:99.
TYPE LOCALITY: USA, North Dakota, Oliver Co., Mandan village near Fort Clark.
DISTRIBUTION: S Alberta, Saskatchewan, and SW Manitoba, Canada, south through Great Plains and Great Basin region of USA, to N Tamaulipas, Mexico.
SYNONYMS: *albescens, arcticeps, breviauritus, brevicaudus, capitulatus, durranti, fuliginosus, fuscogriseus, longipes, melanophrys, missouriensis, pallescens, pallidus, ruidosae, utahensis.*
COMMENTS: Geographic variation and subspecific taxonomy reviewed for central Great Plains (Engstrom and Choate, 1979) and for Great Basin region (Riddle and Choate, 1986). Biogeographic scenario of intraspecific differentiation developed by Riddle and Choate (1986). See McCarty (1978, Mammalian Species, 87).

Onychomys torridus (Coues, 1874). Proc. Acad. Nat. Sci. Philadelphia, 26:183.
TYPE LOCALITY: USA, Arizona, Graham Co., Camp Grant.
DISTRIBUTION: C California, S Nevada, and extreme SW Utah, USA, south to N Baja California, W Sonora, and northernmost Sinaloa, Mexico.
SYNONYMS: *clarus, knoxjonesi, longicaudus, macrotis, perpallidus, pulcher, ramona, tularensis, yakiensis.*
COMMENTS: See comments under *O. arenicola*. See McCarty (1975, Mammalian Species, 59, including *arenicola*). Hollander and Willig (1992) described the Sinoloan *knoxjonesi*.

Oryzomys Baird, 1858. Mammalia *in* Repts. U.S. Expl. Surv., 8(1):458.
TYPE SPECIES: *Mus palustris* Harlan, 1837.
COMMENTS: Oryzomyine. A taxonomically complex and nomenclaturally confused group whose generic definition was successively broadened by Goldman (1918), Tate (1932*d*, *e*), and Ellerman (1941). By the time of Cabrera (1961), the genus embraced as subgenera *Melanomys, Microryzomys, Nesoryzomys, Oecomys,* and *Oligoryzomys*—an agglomeration of taxa perhaps as evolutionarily divergent from one another and from *Oryzomys sensu stricto* as *Neacomys* and *Nectomys*, forms traditionally accorded

generic status (see Carleton and Musser, 1989:52). Others have recognized all of these or various ones as genera (e.g., Thomas, 1917*c*; Gyldenstolpe, 1932; Gardner and Patton, 1976; Reig, 1986), as we do here; see appropriate generic accounts for taxonomic histories. Karyotypic information for many species supplied by R. J. Baker et al. (1983*a*), Gardner and Patton (1976), and Haiduk et al. (1979); for morphological surveys, see Carleton (1973, 1980), Hooper and Musser (1964*a*), and Voss and Linzey (1981). Much basic alpha-revision yet required.

Oryzomys albigularis (Tomes, 1860). Proc. Zool. Soc. Lond., 1860:264.
TYPE LOCALITY: Ecuador, Chimborazo Prov., Pallatanga, 4950 ft.
DISTRIBUTION: Montane forests of N and W Venezuela, easternmost Panama, Andes of Colombia and Ecuador, to N Peru.
SYNONYMS: *caracolus, childi, maculiventer, meridensis, moerex, oconnelli, pectoralis, pirrensis.*
COMMENTS: Hershkovitz's (1944) footnoted listing of specific synonyms of *O. albigularis* set the precedent for Cabrera's (1961) arrangement of the South American forms as subspecies, a viewpoint reiterated in regional studies (e.g., Handley, 1966*a*, 1976). Gardner and Patton (1976) demonstrated the composite nature of Hershkovitz's (1944) and Cabrera's (1961) concept of *albigularis*; however, the determination of priority and refinement of distributions require much museum-based research. Here we tentatively follow the taxonomy of Gardner and Patton (1976) and Patton et al. (1990) and recognize *O. auriventer, O. devius, O. keaysi,* and *O. levipes* as separate species (see those accounts). Gardner and Patton (1976) reassociated Cabrera's (1961) name-combination *O. a. boliviae* as a junior synonym of *O. nitidus*.

Oryzomys alfaroi (J. A. Allen, 1891). Bull. Am. Mus. Nat. Hist., 3:214.
TYPE LOCALITY: Costa Rica, Alajuela Prov., San Carlos.
DISTRIBUTION: Lowland to lower montane forests from S Tamaulipas and Oaxaca, Mexico, through Middle America, to W Colombia and Ecuador.
SYNONYMS: *agrestis, dariensis, gloriaensis, gracilis, incertus* (of J. A. Allen, 1908), *intagensis, palatinus, palmirae.*
COMMENTS: Goldman (1918) forged a broad definition of the species, expanded more so by Hall and Kelson (1959), which encompassed many forms previously treated as distinct (e.g., Merriam, 1901). We recognize *chapmani, rhabdops,* and *saturatior* as species (revision in progress). *Oryzomys alfaroi* proper may be more closely related to *melanotis-rostratus* than to the *chapmani-saturatior* group. Karyotype reported by Haiduk et al. (1979) and Engstrom (1984).

Oryzomys auriventer Thomas, 1890. Ann. Mag. Nat. Hist., ser. 4, 7:379.
TYPE LOCALITY: Ecuador, Tungurahua Prov., upper Rio Pastaza, Mirador, 1500 m.
DISTRIBUTION: E Ecuador and N Peru.
SYNONYMS: *nimbosus.*
COMMENTS: A subspecies of *O. albigularis sensu* Cabrera (1961); considered distinct by Gardner and Patton (1976).

Oryzomys balneator Thomas, 1900. Ann. Mag. Nat. Hist., ser. 7, 5:273.
TYPE LOCALITY: Ecuador, Napo-Pastaza Prov., upper Rio Pastaza, Mirador, 1500 m.
DISTRIBUTION: E and S Ecuador, N Peru.
SYNONYMS: *hesperus.*
COMMENTS: Relationships obscure; range extent uncertain.

Oryzomys bolivaris J. A. Allen, 1901. Bull. Am. Mus. Nat. Hist., 14:405.
TYPE LOCALITY: Ecuador, Bolivar Prov., Porvenir, 1800 m.
DISTRIBUTION: Lowland evergreen forest from E Honduras, through E Nicaragua, Costa Rica and Panama, to W Colombia and W Ecuador.
SYNONYMS: *alleni, bombycinus, castaneus, orinus, rivularis.*
COMMENTS: Names conventionally applied to this species include *bombycinus,* as reviewed by Pine (1971), and *rivularis,* as discussed by Gardner and Patton (1976). Priority of *bolivaris* presented by Musser and Gardner (in litt.), who summarized the morphological variation and geographic distribution of the species.

Oryzomys buccinatus (Olfers, 1818). *In* Eschwege, J. Brasilien, Neue Bibliothek. Reisenb., 15:209.

TYPE LOCALITY: Paraguay, Caraguatay Dept., 45 km east of Asunción, Atyra.

DISTRIBUTION: E Paraguay and NE Argentina.

SYNONYMS: *angouya* (of Desmarest, 1819), *anguya*.

COMMENTS: Level of differentiation from *O. subflavus* of E Brazil requires study. Hershkovitz (1959*a*) mistakenly included *O. ratticeps* within *O. buccinatus*, but the two species co-occur in E Paraguay (Cabrera, 1961; Myers, 1982). As with other names based on Azara's (1801) characterizations, the clear fixation of this name to a species morphology and the proper usage of *angouya* Fischer, 1814, versus *buccinatus* Olfers, 1818, should be formally established.

Oryzomys capito (Olfers, 1818). *In* Eschwege, J. Brasilien, Neue Bibliothek. Reisenb., 15:209.

TYPE LOCALITY: Paraguay, San Ignacio Guazu.

DISTRIBUTION: S Venezuela and Guianas, Brazil and Amonzonian regions of Colombia, Ecuador, Peru, and Bolivia.

SYNONYMS: *cephalotes, goeldi, laticeps* (of Lund, 1841), *modestus, perenensis, saltator, velutinus*.

COMMENTS: Taxonomic understanding of this species embroiled by the infelicitous footnote of Hershkovitz (1960:544), who suggested the synonymy of some 20 taxa under *O. capito*, an opinion expanded by Cabrera (1961) and followed by other authors (e.g., Handley, 1966*a*, 1976). The extremeness of this viewpoint was seriously challenged by the karyotypic study of Gardner and Patton (1976); as a result, the following forms considered synonyms by Hershkovitz (1960) and/or Cabrera (1961) are now acknowledged as distinct species or as synonyms of other species (see separate accounts): *O. intermedius, O. legatus, O. macconnelli, O. nitidus, O. oniscus, rivularis* (of *O. bolivaris*), *O. talamancae*, and *O. yunganus*. As with other names based on Azara's (1801) characterizations, the clear fixation of this name to a species morphology and the proper usage of *megacephalus* Fischer, 1814, versus *capito* Olfers, 1818, should be formally established.

Oryzomys chapmani Thomas, 1898. Ann. Mag. Nat. Hist., ser. 7, 1:179.

TYPE LOCALITY: Mexico, Veracruz, Jalapa, 4400 ft.

DISTRIBUTION: Cloud forest elevations of Cordillera Oriental (Tamaulipas to Veracruz), Sistema Montanosa (N Oaxaca), and Sierra Madre del Sur (S Oaxaca and Guerrero), Mexico.

SYNONYMS: *caudatus, dilutior, guerrerensis, huastecae*.

COMMENTS: Relegated to a subspecies of *O. alfaroi* by Goldman (1918). Goodwin (1969) recognized *caudatus* as distinct from *O. alfaroi* in N Oaxaca, but, as earlier arranged by Merriam (1901), *O. chapmani* has priority for this Mexican species. We view *O. saturatior* as the vicariant relative of *O. chapmani* (revision in progress). Karyotype reported, as *O. caudatus*, by Haiduk et al. (1979).

Oryzomys couesi (Alston, 1877). Proc. Zool. Soc. Lond., 1876:756 [1877].

TYPE LOCALITY: Guatemala, Alta Verapaz Dept., Coban.

DISTRIBUTION: Extreme S Texas, USA; Mexico, excluding NC plateau region, south through most of Central America, to NW Colombia (see Hershkovitz, 1987*a*); including Jamaica and Isla Cozumel.

SYNONYMS: *albiventer, antillarum, apatelius, aquaticus, aztecus, azuerensis, bulleri, cozumelae, crinitus, fulgens, gatunensis, goldmani, jalapae, lambi, mexicanus, peninsulae, peragrus, pinicola, regillus, richardsoni, richmondi, rufinus, rufus, teapensis, zygomaticus*.

COMMENTS: Retained as a species by Goldman (1918) until Hall (1960) considered it only subspecifically distinct from *O. palustris*. Benson and Gehlbach (1979) returned *O. couesi* to specific status based on morphological contrasts with *O. p. texensis* in supposed area of intergradation. Karyotype reported by Benson and Gehlbach (1979) and Haiduk et al. (1979); morphometric comparisons to *O. palustris* by Humphrey and Setzer (1989).

Following Hall's (1960) example, other insular or localized subspecies—namely, *antillarum, azuerensis, cozumelae, fulgens, gatunensis*, and *peninsulae* (see Handley, 1966*a*; Hershkovitz, 1971; Jones and Lawlor, 1965)—were swept under *O. palustris*. They are here included in *O. couesi* because of geographic proximity, but their placement, together with other Central American populations referred to *O. couesi*, should be critically reviewed.

Oryzomys devius Bangs, 1902. Bull. Mus. Comp. Zool., 39:34.
TYPE LOCALITY: Panama, Chiriqui Prov., Volcan de Chiriqui, Boquete, 5000 ft.
DISTRIBUTION: Highlands of Costa Rica and westernmost Panama.
COMMENTS: Maintained as a species until relegated to synonymy under *O. albigularis* by Handley (1966*a*), emulating the treatment of South American *albigularis*-like forms by Cabrera (1961). Gardner (1983*a*), however, continued to rank *devius* as a species. The status of *devius* must be evaluated within a revisionary context of the entire *albigularis* complex.

Oryzomys dimidiatus (Thomas, 1905). Ann. Mag. Nat. Hist., ser. 7, 15:586.
TYPE LOCALITY: Nicaragua, Zelaya Dept., Rio Escondido, 7 mi below Rama.
DISTRIBUTION: SE Nicaragua.
COMMENTS: Described as a species of *Nectomys* and included therein as *incertae sedis* by Hershkovitz (1944). Later designated as the type species of *Micronectomys*, subgenus *Oryzomys*, by Hershkovitz (1948*b*), who afterwards (1970*b*) acknowledged the taxon as an *nomen nudum* (also see Pine and Wetzel, 1975:653). Morphological features resemble those of the *O. couesi-palustris* complex (Hershkovitz, 1970*b*). Only two specimens known (see Jones and Engstrom, 1986).

Oryzomys galapagoensis (Waterhouse, 1839). Zool. Voy. H. M. S. "Beagle," Mammalia, p. 66.
TYPE LOCALITY: Ecuador, Galapagos Islands, Chatham Island.
DISTRIBUTION: San Cristobal (= Chatham) and Sante Fe (= Barrington) islands.
SYNONYMS: *bauri*.
COMMENTS: Cabrera (1961) assigned *bauri* as a synonym of *O. galapagoensis*, an action more fully documented by Patton and Hafner (1983). Related to *Oryzomys xantheolus* complex on mainland South America (Gardner and Patton, 1976; Patton and Hafner, 1983). Extirpated from San Cristobal Isl but populations (*O. g. bauri*) still inhabit Sante Fe Isl (see Patton and Hafner, 1983).

Oryzomys gorgasi Hershkovitz, 1971. J. Mammal., 52:700.
TYPE LOCALITY: Colombia, Antioquia Dept., basin of Rio Atrato, Loma Teguerre, just below and opposite Sautata (Choco), 1 m; 7°54'N, 77°W.
DISTRIBUTION: Known only from the type locality.
COMMENTS: Known only by the holotype; related to *O. couesi* and *O. palustris* (see Hershkovitz, 1971).

Oryzomys hammondi (Thomas, 1913). Ann. Mag. Nat. Hist., ser. 8, 12:570.
TYPE LOCALITY: Ecuador, Pichincha Prov., Mindo, 4213 ft.
DISTRIBUTION: NW Ecuador.
COMMENTS: Described as a species of *Nectomys* and included therein as *incertae sedis* by Hershkovitz (1944). Later designated as the type species of *Macruroryzomys*, subgenus *Oryzomys*, by Hershkovitz (1948*b*), who afterwards (1970*b*) acknowledged the taxon as a *nomen nudum* (also see Pine and Wetzel, 1975:653). Distributional limits poorly defined and affinities obscure, perhaps related to the extirpated Antillean form *Megalomys* (Ray, 1962).

Oryzomys intectus Thomas, 1921. Ann. Mag. Nat. Hist., ser. 9, 8:356.
TYPE LOCALITY: Colombia, Antioquia Dept., 8 km E Medellin, Santa Elena, 9000 ft.
DISTRIBUTION: NC Colombia.
COMMENTS: Distinctive species of uncertain affinities, variously associated with *O. balneator* or *Melanomys* (Gyldenstolpe, 1932), with *Nectomys* (Ellerman, 1941), or with subgenus *Oryzomys* (Cabrera, 1961). Such a diversity of opinion begs renewed study.

Oryzomys intermedius (Leche, 1886). Zool. Jahrb., 1:693.
TYPE LOCALITY: Brazil, Rio Grande do Sul, Taquara do Mundo Novo.
DISTRIBUTION: SE Brazil (Bahia to Rio Grande do Sul), E Paraguay, and NE Argentina (*fide* Massoia, 1975).
COMMENTS: Classified as a subspecies of *O. capito* by Cabrera (1961) although its distinction from *laticeps* (= *O. capito*) was earlier maintained by Moojen (1952). Gardner and Patton (1976) perceived the closer kinship of *intermedius* to *O. nitidus* and suggested

their synonymy. *Oryzomys intermedius, O. legatus, O. macconnelli,* and *O. nitidus* form a closely-related complex whose interrelationships, specific stature, and distributions deserve investigation.

Oryzomys keaysi J. A. Allen, 1900. Bull. Am. Mus. Nat. Hist., 13:225.
TYPE LOCALITY: Peru, Puno Dept., valley of upper Rio Inambari, Inca Mines, 6000 ft.
DISTRIBUTION: Montane rainforest of E Peruvian Andes.
COMMENTS: A subspecies of *O. albigularis sensu* Cabrera (1961). Genetically divergent from *O. albigularis* and *O. levipes;* contiguously allopatric to the latter (see Patton et al., 1990).

Oryzomys kelloggi Avila-Pires, 1959. Atas Soc. Biol. Rio de Janiero, 3(4):2.
TYPE LOCALITY: Brazil, Minas Gerais, Alem Paraiba.
DISTRIBUTION: SE Brazil.
COMMENTS: Status with regard to *O. intermedius* needs investigation.

Oryzomys lamia Thomas, 1901. Ann. Mag. Nat. Hist., ser. 7, 8:528.
TYPE LOCALITY: Brazil, Minas Gerais, along Rio Jordao, a small tributary of the Rio Paranaiba, 700-900 m.
DISTRIBUTION: SE Brazil.
COMMENTS: Status and distributional extent unknown; allied to *O. intermedius* according to Thomas (1901*b*) and Gyldenstolpe (1932).

Oryzomys legatus Thomas, 1925. Ann. Mag. Nat. Hist., ser. 9, 15:577.
TYPE LOCALITY: Bolivia, Tarija Dept., Carapari, 1000 m.
DISTRIBUTION: E Andean slopes of SC Bolivia and NW Argentina.
COMMENTS: Considered conspecific with *O. capito* by Hershkovitz (1960) and Cabrera (1961). Gardner and Patton (1976) viewed *legatus* as a probable synonym of *O. nitidus,* to which it appears closely related. Retained here as a species following the study of Massoia (1975; also Mares et al., 1989*b*) and pending determination of its status with regard to *O. nitidus* and *O. intermedius* (see comments therein).

Oryzomys levipes Thomas, 1902. Ann. Mag. Nat. Hist., ser. 7, 9:129.
TYPE LOCALITY: Peru, Puno Dept., Limbane, 2200 m.
DISTRIBUTION: Cloud forest of SE Peru to WC Bolivia.
COMMENTS: Synonymized under *O. albigularis keaysi* by Cabrera (1961). Genetically divergent from and altitudinally parapatric to *O. keaysi* in Peru (see Patton et al., 1990).

Oryzomys macconnelli Thomas, 1910. Ann. Mag. Nat. Hist., ser. 8, 6:186.
TYPE LOCALITY: Guyana, Demerara Dist., Rio Supinaam, a tributary of the Lower Essequibo.
DISTRIBUTION: Lowlands of SC Colombia, E Ecuador and Peru, east to S Venezuela, Guianas, and N Brazil.
SYNONYMS: *incertus* (of J. A. Allen, 1913), *mureliae.*
COMMENTS: Included in *O. capito* by Hershkovitz (1960) but retained as species by Cabrera (1961). Morphological and karyotypic basis for species recognition reinforced by Pine (1973*b*), Gardner and Patton (1976), and Husson (1978) (see comments under *O. intermedius*).

Oryzomys melanotis Thomas, 1893. Ann. Mag. Nat. Hist., ser. 6, 11:404.
TYPE LOCALITY: Mexico, Jalisco, Mineral San Sebastian.
DISTRIBUTION: Low to intermediate elevations of W Mexico, from S Sinaloa to SW Oaxaca.
SYNONYMS: *colimensis.*
COMMENTS: Revised by Goldman (1918), who recognized *melanotis* and *rostratus* as separate species within a *melanotis* species group. Hooper (1953) viewed the geographic complementarity of the two as a subspecific pattern, and so recognized by Hall and Kelson (1959) and Hall (1981). Engstrom (1984) returned *rostratus* to a separate species based on a robust variety of data interpreted within a zoogeographic context.

Oryzomys nelsoni Merriam, 1898. Proc. Biol. Soc. Washington, 12:15.
TYPE LOCALITY: Mexico, Nayarit, Maria Madre Island.
DISTRIBUTION: Known only from the type locality.
STATUS: IUCN - Extinct?

COMMENTS: Known only by 4 specimens of the type series. Allied to *O. couesi*, under which Hershkovitz (1971) listed it as a subspecies. Others (Goldman, 1918; Hall, 1981) have maintained it as a species in view of its unique differentiation from mainland *O. palustris*. Presumed to be extinct (see Wilson, 1991).

Oryzomys nitidus (Thomas, 1884). Proc. Zool. Soc. Lond., 1884:452.

TYPE LOCALITY: Peru, Junín Dept., valley of Rio Tulumayo, 10 km S San Ramon, Amable Maria, 2000 ft (as located by Gardner and Patton, 1976:42).

DISTRIBUTION: Lowland rain forest of E Ecuador, Peru and Bolivia, and WC Brazil (Mato Grosso).

SYNONYMS: *boliviae*.

COMMENTS: A species formerly included in *O. capito sensu* Cabrera (1961) or affiliated with *O. alfaroi* by Hershkovitz (1966*c*). Gardner and Patton (1976) provided karyotypic and morphological evidence justifying the specific distinction of *O. nitidus* from both and listed *boliviae, intermedius*, and *legatus* as likely synonyms (see remarks under *O. intermedius*).

Oryzomys oniscus Thomas, 1904. Ann. Mag. Nat. Hist., ser. 7, 13:142.

TYPE LOCALITY: Brazil, Pernambuco, São Lorenzo, 50 m.

DISTRIBUTION: E Brazil (Pernambuco, Bahia).

COMMENTS: Arranged as a subspecies of *O. capito* by Cabrera (1961). Although closely related to *O. capito*, this form is specifically distinct (Moojen, 1952).

Oryzomys palustris (Harlan, 1837). Am. J. Sci., 31:385.

TYPE LOCALITY: USA, New Jersey, Salem Co., "Fastland," near Salem.

DISTRIBUTION: SE USA: SE Kansas to E Texas, eastwards to S New Jersey and peninsular Florida.

STATUS: U.S. ESA - Endangered in the Lower Florida Keys (west of the Seven Mile Bridge) as *O. p. natator*; IUCN - Indeterminate as *O. argentatus* in Florida.

SYNONYMS: *argentatus, coloratus, natator, oryzivora, planirostris, sanibeli, texensis*.

COMMENTS: Formerly included *couesi* and related forms as subspecies (see account of *O. couesi*). Geographic variation evaluated by Humphrey and Setzer (1989), who acknowledged the nominate and one other subspecies, *O. p. natator*. The status of *argentatus* has oscillated recently: described as a species from the Florida Keys by Spitzer and Lazell (1978); synonymized under *O. palustris natator* by Humphrey and Setzer (1989); reinstated as species by Goodyear (1991). Here we follow Humphrey and Setzer (1989) based on their broader geographic sampling and more robust analyses. In light of the subspecific endemism displayed among other Florida-Key mammals, the status of *argentatus* merits further study drawing upon genetic evidence. See Wolfe (1982, Mammalian Species, 176).

Oryzomys polius Osgood, 1913. Field Mus. Nat. Hist. Publ., Zool. Ser., 10:97.

TYPE LOCALITY: Peru, Amazonas Dept., mountains east of Balsas, Tambo Carrizal, 5000 ft.

DISTRIBUTION: NC Peru.

COMMENTS: Affinities obscure; compared to *O. xantheolus* by Osgood (1913).

Oryzomys ratticeps (Hensel, 1873). Abh. König Akad. Wiss. Berlin, 1872:36 [1873].

TYPE LOCALITY: Brazil, Rio Grande do Sul.

DISTRIBUTION: E Brazil, NE Argentina, and Paraguay.

SYNONYMS: *moojeni, paraganus, rex* (of Winge, 1887), *tropicius*.

COMMENTS: Referred to *O. buccinatus* by Hershkovitz (1959*a*) but its separate specific status is well-established (Avila-Pires, 1960*a*; Moojen, 1952).

Oryzomys rhabdops Merriam, 1901. Proc. Washington Acad. Sci., 3:292.

TYPE LOCALITY: Guatemala, Quezaltenango Dept., Calel, 10,000 ft.

DISTRIBUTION: Highlands of S Chiapas, Mexico, and C Guatemala.

SYNONYMS: *angusticeps*.

COMMENTS: Allocated to subspecies of *O. alfaroi* by Goldman (1918); we reinstate it as a species (revision in progress).

Oryzomys rostratus Merriam, 1901. Proc. Washington Acad. Sci., 3:293.

TYPE LOCALITY: Mexico, Puebla, Metlaltoyuca.

DISTRIBUTION: Deciduous and evergreen tropical forests from central Tamaulipas to Oaxaca and Yucatan Peninsula, Mexico, through Guatemala, El Salvador, and Honduras, to S Nicaragua.

SYNONYMS: *carrorum, megadon, salvadorensis, yucatanensis.*

COMMENTS: Considered a subspecies of *O. melanotis* (Hall, 1981; Hooper, 1953). Engstrom (1984) effectively argued the specific distinctiveness of *O. rostratus* (also see account of *O. melanotis*). Distributional records in Nicaragua reviewed, as *O. melanotis megadon,* by Jones and Engstrom (1986). Homogeneity of populations assigned to *O. rostratus* warrants additional study.

Oryzomys saturatior Merriam, 1901. Proc. Washington Acad. Sci., 3:290.

TYPE LOCALITY: Mexico, Chiapas, Tumbala, 5000 ft.

DISTRIBUTION: Cloud forest elevations from S Oaxaca and Chiapas, Mexico, through Guatemala, Honduras, and El Salvador, to NC Nicaragua.

SYNONYMS: *hylocetes.*

COMMENTS: Described as a subspecies of *O. chapmani* (Merriam, 1901) and later retained as a subspecies of *O. alfaroi* (Goldman, 1918; Hall and Kelson, 1959). In our revision (in prep.) of the *alfaroi* complex we view the divergence of *chapmani* and *saturatior,* whose ranges are separated by the Isthmus of Tehuantepec, as sister species.

Oryzomys subflavus (Wagner, 1842). Arch. Naturgesh., 8(1):362.

TYPE LOCALITY: Brazil, Minas Gerais, probably Lagoa Santa.

DISTRIBUTION: E Brazil.

SYNONYMS: *laticeps* (of Winge, 1887), *vulpinoides.*

COMMENTS: Related to, and possibly conspecific with, populations in Paraguay and Argentina usually identified as *O. buccinatus* (see remarks therein). Hershkovitz (1960) attributed *catherinae* Thomas, and *rex* Thomas, as synonyms of *O. subflavus;* however, both are forms of *Oecomys* (which see). The supposed distribution of *O. subflavus* in the Guianas (e.g., Honacki et al., 1982) issued from this erroneus allocation of names (e.g., not recorded by Husson, 1978).

Oryzomys talamancae J. A. Allen, 1891. Proc. U. S. Natl. Mus., 14:193.

TYPE LOCALITY: Costa Rica, Limón Prov., Talamanca.

DISTRIBUTION: Forested lowlands of E Costa Rica, Panama, W and NC Colombia, N Venezuela, and W Ecuador.

SYNONYMS: *carrikeri, magdalenae, medius, mollipilosus, panamensis, sylvaticus, villosus.*

COMMENTS: Considered a junior synonym of *O. capito* (Hershkovitz, 1960), an opinion that led to its arrangement as a subspecies thereof (Hall, 1981; Handley, 1966*a*). Species status reasserted by Gardner (1983*a*); morphological variation, distributional limits, and synonymy clarified by Musser and Williams (1985).

Oryzomys xantheolus Thomas, 1894. Ann. Mag. Nat. Hist., ser. 6, 14:354.

TYPE LOCALITY: Peru, Piura Dept., Tumbez.

DISTRIBUTION: SW Ecuador to W Peru.

SYNONYMS: *baroni, ica.*

COMMENTS: Probable mainland relative of *O. galapagoensis* (Gardner and Patton, 1976; Patton and Hafner, 1983).

Oryzomys yunganus Thomas, 1902. Ann. Mag. Nat. Hist., ser. 7., 9:130.

TYPE LOCALITY: Bolivia, Cochabamba Dept., Charuplaya, Rio Secure, 1350 m.

DISTRIBUTION: Low to intermediate elevations of Guianas and S Venezuela (Ochoa et al., 1988), and along E Andean foothills from C Colombia to WC Bolivia.

COMMENTS: Considered a subspecies of *O. capito* by Cabrera (1961); specific differences illuminated by Gardner and Patton (1976).

Osgoodomys Hooper and Musser, 1964. Occas. Pap. Mus. Zool., Univ. Michigan, 635:12.

TYPE SPECIES: *Peromyscus banderanus* J. A. Allen, 1897.

COMMENTS: Peromyscine. Diagnosed as a subgenus of *Peromyscus* by Hooper and Musser (1964*b*) and retained there by Hooper (1968*b*); proposed as a distinct genus by Carleton (1980). Anatomy of male reproductive system detailed by Linzey and Layne (1969, 1974). Karyological relationships assessed by Rogers et al. (1984) and genetic similarity by Schmidly et al. (1985).

Osgoodomys banderanus (J. A. Allen, 1897). Bull. Am. Mus. Nat. Hist., 9:51.
TYPE LOCALITY: Mexico, Nayarit, Valle de Banderas.
DISTRIBUTION: Coastal plain of S Nayarit to S Guerrero, interior of Michoacan and Guerrero along the basin of the Rio Balsas, Mexico.
SYNONYMS: *vicinior.*
COMMENTS: The forms *angelensis*, *coatlanensis*, and *sloeops* have been mistaken as subspecies of *O. banderanus* and were reassigned to *Peromyscus mexicanus* by Musser (1969*a*).

Otonyctomys Anthony, 1932. Am. Mus. Novit., 586:1.
TYPE SPECIES: *Otonyctomys hatti* Anthony, 1932.
COMMENTS: Middle American endemic, presumably the sister-taxon of *Nyctomys* (e.g., see Hooper and Musser, 1964*a*), but systematic biology poorly known.

Otonyctomys hatti Anthony, 1932. Am. Mus. Novit., 586:1.
TYPE LOCALITY: Mexico, Yucatan, Chichen-Itza.
DISTRIBUTION: Yucatan Peninsula, Mexico, south to N Belize and NE Guatemala (Peten Dept.).

Ototylomys Merriam, 1901. Proc. Washington Acad. Sci., 3:561.
TYPE SPECIES: *Ototylomys phyllotis* Merriam, 1901.
COMMENTS: Closedly related to *Tylomys* (Carleton, 1980; Hooper, 1960; Hooper and Musser, 1964*b*; Lawlor, 1969). Tribal-level assignment unresolved—viewed as a member of neotomines (*sensu* Hooper and Musser, 1964*b*) or of an archaic clade that originated prior to differentiation of other neotomine-peromyscines (Carleton, 1980).

Ototylomys phyllotis Merriam, 1901. Proc. Washington Acad. Sci., 3:562.
TYPE LOCALITY: Mexico, Yucatan, Tunkas.
DISTRIBUTION: C Costa Rica north to Yucatan Peninsula, S Tabasco, and N Chiapas, Mexico; isolated record from NC Guerrero, Mexico.
SYNONYMS: *affinis*, *australis*, *brevirostris*, *connectens*, *fumeus*, *guatemalae*, *phaeus*.
COMMENTS: Distribution and geographic variation examined by Lawlor (1969), who retained *australis* and *connectens* as subspecies. See Lawlor (1982, Mammalian Species, 181).

Oxymycterus Waterhouse, 1837. Proc. Zool. Soc. Lond., 1837:21.
TYPE SPECIES: *Mus nasutus* Waterhouse, 1837.
COMMENTS: Akodontine. Specific taxonomy needs full revision; nominal species here recognized follow Cabrera (1961), Reig's (1987) provisional tally of valid forms, and Hinojosa et al. (1987). Although initial studies of electromorphic variation support the monophyly of *Oxymycterus* and its placement as an akodont (Hinojosa et al., 1987; Patton et al., 1989), cladistic relationship with *Abrothrix*, *Microxus*, and certain *Akodon* needs further exploration. Studies addressing morphological definition of *Oxymycterus* include Carleton (1973), Hooper and Musser (1964*a*), Hinojosa et al. (1987), Vorontsov (1967), and Voss and Linzey (1981). Diploid number of species thus far karyotyped have proven to be identical (Vitullo et al., 1986).

Oxymycterus akodontius Thomas, 1921. Ann. Mag. Nat. Hist., ser. 9, 8:615.
TYPE LOCALITY: Argentina, Jujuy Prov., Higuerilla, 20 km E Tilcara, 2000 m.
DISTRIBUTION: NW Argentina.
COMMENTS: Perhaps conspecific with *O. paramensis* according to Cabrera (1961) and Vitullo et al. (1986).

Oxymycterus angularis Thomas, 1909. Ann. Mag. Nat. Hist., ser. 8, 4:237.
TYPE LOCALITY: Brazil, Pernambuco, São Lourenco, 30 m.
DISTRIBUTION: E Brazil.

Oxymycterus delator Thomas, 1903. Ann. Mag. Nat. Hist., ser. 7, 11:489.
TYPE LOCALITY: Paraguay, Paraguari Dept., Sapucai.
DISTRIBUTION: E Paraguay.

Oxymycterus hiska Hinojosa, Anderson, and Patton, 1987. Am. Mus. Novit., 2898:14.
TYPE LOCALITY: Peru, Puno Dept., 14 km W Yanahuaya, 2210 m; 14°19'S, 69°21'W.
DISTRIBUTION: Known only from the type locality.

COMMENTS: Smallest species of *Oxymycterus*, morphologically resembling *O. hucucha*.

Oxymycterus hispidus Pictet, 1843. Mem. Soc. Phys. Hist. Nat. Geneve, 10:212.
TYPE LOCALITY: Brazil, Bahia.
DISTRIBUTION: NE Argentina (Misiones) to E Brazil (Bahia).
SYNONYMS: *judex, misionalis, quaestor.*

Oxymycterus hucucha Hinojosa, Anderson, and Patton, 1987. Am. Mus. Novit., 2898:15.
TYPE LOCALITY: Bolivia, Cochabamba Dept., 28 km (by road) W Comarapa, 2800 m; 17°51'S, 64°40'W.
DISTRIBUTION: Vicinity of type locality.
COMMENTS: Morphologically similar to *O. hiska*.

Oxymycterus iheringi Thomas, 1896. Ann. Mag. Nat. Hist., ser. 6, 18:308.
TYPE LOCALITY: Brazil, Rio Grande do Sul, Taquara do Mundo Novo, Rio dos Linos.
DISTRIBUTION: NE Argentina (Misiones) and SE Brazil.
COMMENTS: Formerly included in *Microxus* or *Akodon* (e.g., Cabrera, 1961; Ellerman, 1941); reallocated to *Oxymycterus* by Massoia (1963*b*) and morphological recognition amplified by Massoia and Fornes (1969).

Oxymycterus inca Thomas, 1900. Ann. Mag. Nat. Hist., ser. 7, 6:298.
TYPE LOCALITY: Peru, Junín Dept., Rio Perene, 800 m.
DISTRIBUTION: SC Peru to WC Bolivia.
SYNONYMS: *doris, iris, juliacae.*
COMMENTS: Distributional extent uncertain; identity of *Oxymycterus* populations in Amazon Basin with this species requires substantiation.

Oxymycterus nasutus (Waterhouse, 1837). Proc. Zool. Soc. Lond., 1837:16.
TYPE LOCALITY: Uruguay, Maldonado.
DISTRIBUTION: Uruguay and adjacent SE Brazil.
COMMENTS: Comparisons with *O. iheringi* provided by Massoia and Fornes (1969); they (and Vitullo et al., 1986) further noted the distinction of *O. nasutus* from *O. rufus*, under which it had been ranked as a subspecies (Cabrera, 1961). Range limits unknown.

Oxymycterus paramensis Thomas, 1902. Ann. Mag. Nat. Hist., ser. 7, 9:139.
TYPE LOCALITY: Bolivia, Cochabamba Dept., Rio Secure, Choquecamate, 4000 m.
DISTRIBUTION: Upper E Andean slopes in SE Peru, WC Bolivia, and NW Argentina.
SYNONYMS: *jacentior, nigrifrons.*
COMMENTS: Genetic distance data relative to other akodonts presented by Hinojosa et al. (1987). Placement of *jacentior* with this species merits reconsideration.

Oxymycterus roberti Thomas, 1901. Ann. Mag. Nat. Hist., ser. 7, 8:530.
TYPE LOCALITY: Brazil, Minas Gerais, Rio Jordao, Paranaiba, 700-900 m.
DISTRIBUTION: Vicinity of type locality.

Oxymycterus rufus (G. Fischer, 1814). Zoognosia, 3:71.
TYPE LOCALITY: Argentina, 32.5°S latitude along lower Rio Parana.
DISTRIBUTION: EC Argentina, Uruguay, and SE Brazil.
SYNONYMS: *dasytrichus, platensis, rostellatus, rutilans.*
COMMENTS: Priority of *Mus rufus* Fischer, 1814, over *Mus rutilans* Olfers, 1818, both based on Azara's (1801) descriptions, deserves formal stabilization as does clarification of the morphological identity of the species with Azara's "rat roux." Karyotypes of *O. rufus* and *platensis* support their union as one species (Vitullo et al., 1986). Level of relationship to other nominal species and range limits uncertain.

Cabrera (1961) cited Rengger (1830) for the placement of the type locality near Asunci6n, Paraguay, as it is usually given. Rengger's basis for this interpretation is unclear since Azara (1802) mentioned collecting the species in an arroyo at 32.5 degrees, presumably south latitude, which places the locality somewhere in southern Entre Rios province near the Rio Parana. The origin of the larger, rufescent *O. rufus* from this region seems more plausible than around Asunci6n where the smaller, dark-colored *O. delator* occurs (see Myers, 1982). A future revisor should clarify this discrepancy.

Peromyscus Gloger, 1841. Gemein Hand.-Hilfsbuch. Nat., 1:95.

TYPE SPECIES: *Peromyscus arboreus* Gloger, 1841 (= *Mus leucopus* Rafinesque, 1818).

SYNONYMS: *Haplomylomys, Sitomys, Trinodontomys, Vesperimus.*

COMMENTS: The *Drosophila* of North American mammalogy, the alpha-level classification of the genus has been revised three times (Osgood, 1909; Hooper, 1968*b*; Carleton, 1989) and its biology and evolution have been twice monographed (King, 1968; Kirkland and Layne, 1989). Multispecies surveys have broadly sampled the morphology of the genus (Carleton, 1973, 1980; Hooper, 1957, 1958; Hooper and Musser, 1964*b*; Linzey and Layne, 1969, 1974), its karyology (Robbins and Baker, 1981; L. W. Robbins et al., 1983; Rogers et al., 1984; Stangl and Baker, 1984*b*), biochemical variation (Avise et al., 1974*a*, *b*, 1979; Brownell, 1983; Fuller et al., 1984; J. C. Patton et al., 1981; Rogers and Engstrom, 1992; Schmidly et al., 1985; Zimmerman et al., 1978), and geographical ecology (Glazier, 1980).

Major subdivisions of *Peromyscus* have received added revisionary attention, especially the *eremicus* (Avise et al., 1974*b*; Lawlor, 1971*a*, *b*), *maniculatus* (Allard et al., 1987; Gunn and Greenbaum, 1986), *boylii* (Avise et al., 1974*a*; Bradley and Schmidly, 1987; Bradley et al., 1989; Carleton, 1977, 1979; Schmidly, 1973*a*; Schmidly et al., 1988; Smith, 1990), *truei* (Janecek, 1990; Modi and Lee, 1984; Schmidly, 1973*b*; Zimmerman et al., 1975), and *mexicanus* (Huckaby, 1980; Musser, 1971*a*; Rogers and Engstrom, 1992; Smith et al., 1986) species groups.

Greater emphasis on phylogenetic systematics has altered Osgood's (1909) original generic scope. *Baiomys* and *Ochrotomys* have been removed and ranked as separate genera (Carleton, 1980; Hooper, 1958; Hooper and Musser, 1964*b*). Taxa defined as subgenera by Hooper (1968*b*—*Habromys, Isthmomys, Megadontomys, Osgoodomys*, and *Podomys*) have been considered distinct genera by Carleton (1980) but not others (Rogers, 1983; Stangl and Baker, 1984*b*; Yates et al., 1979). Expansion of the generic limits of Hooper (1968*b*) has been advocated to encompass *Neotomodon* (Stangl and Baker, 1984*b*; Yates et al., 1979) and perhaps *Onychomys* (Stangl and Baker, 1984*b*). Nomenclatural resolution of these alternative proposals awaits further study. *Haplomylomys* has been used as a subgenus to contain the *californicus* and *eremicus* species groups, all others being assigned to the subgenus *Peromyscus*; Carleton (1989) did not employ subgeneric divisions.

Peromyscus attwateri J. A. Allen, 1895. Bull. Am. Mus. Nat. Hist., 7:330.

TYPE LOCALITY: USA, Texas, Kerr Co., Turtle Creek.

DISTRIBUTION: Edwards Plateau of NC Texas, north through E Oklahoma, to SE Kansas, SW Missouri, and NW Arkansas, USA.

SYNONYMS: *bellus, cansensis, laceyi.*

COMMENTS: Classified as a geographic race of *P. boylii* (Osgood, 1909) until rediagnosed as a species by Schmidly (1973*a*). Biochemical evolution studied by Kilpatrick (1984). Cognate relationship to *P. difficilis* suggested by Janecek (1990). See Schmidly (1974*a*, Mammalian Species, 48). *P. boylii* species group.

Peromyscus aztecus (Saussure, 1860). Rev. Mag. Zool. Paris, Ser. 2, 12:105.

TYPE LOCALITY: Mexico, Veracruz, vicinity of Mirador.

DISTRIBUTION: Intermediate to high altitudes from S Jalisco and C Veracruz, Mexico, through highlands of Guatemala, to Honduras and N El Salvador.

SYNONYMS: *cordillerae, evides, hondurensis, hylocetes, oaxacensis.*

COMMENTS: Traditionally viewed as a subspecies of *P. boylii* (Osgood, 1909). Species status recognized by Alvarez (1961); morphological recognition, distribution, and synonymies clarified by Musser (1969*a*) and Carleton (1979). Several of the junior synonyms (*evides, hylocetes, oaxacensis*) have been treated as species (Hooper, 1968*b*; Osgood, 1909) and their inclusion under *P. aztecus* deserves further scrutiny (e.g., see Smith et al., 1989; Sullivan and Kilpatrick, 1991). Systematic relationships examined by Bradley and Schmidly (1987), Smith (1990), and Sullivan and Kilpatrick (1991). Segregated together with *P. spicilegus* and *P. winkelmanni* as the *aztecus* species group by Carleton (1989).

Peromyscus boylii (Baird, 1855). Proc. Acad. Nat. Sci. Philadelphia, 7:335.

TYPE LOCALITY: USA, California, Eldorado Co., Middle Fork of American River, near Auburn.

DISTRIBUTION: California to westernmost Oklahoma, USA, south to Queretaro and W Hidalgo, Mexico.

SYNONYMS: *gaurus, glasselli, major, metallicola, parasiticus, pinalis, robustus, rowleyi, utahensis.*

COMMENTS: Many taxa previously arrayed as subspecies by Osgood (1909) or Hall (1981) have been elevated to species, including *P. attwateri* (see Schmidly, 1973*a*), *P. aztecus* (see Alvarez, 1961; Carleton, 1979; Hooper, 1968*b*), *P. levipes* (see Schmidly et al., 1988), *P. madrensis* (see Carleton, 1977; Carleton et al., 1982), *P. simulus* (see Carleton, 1977), and *P. spicilegus* (see Carleton, 1977). Others have been realigned with other species—*cordillerae* and *evides* under *P. aztecus* (Carleton, 1979); *ambiguus, beatae,* and *sacarensis* under *P. levipes* (Carleton, 1979; Schmidly et al., 1988); and *penicillatus* under *P. difficilis* (Diersing, 1976). Chromosomal variation of *P. boylii* and its kin investigated by Houseal et al. (1987), Lee et al. (1972), and Schmidly and Schroeter (1974); biochemical variation by Avise et al. (1974*a*), Kilpatrick and Zimmerman (1975), Rennert and Kilpatrick (1986, 1987); morphological variation by Bradley and Schmidly (1987), Carleton (1977), and Schmidly et al. (1988). *P. boylii* species group.

Peromyscus bullatus Osgood, 1904. Proc. Biol. Soc. Washington, 17:63.

TYPE LOCALITY: Mexico, Veracruz, Perote.

DISTRIBUTION: Vicinity of type locality and 3 km W Limon.

COMMENTS: Retained as a species since its discovery (Carleton, 1989; Hoffmeister, 1951; Osgood, 1909), although Hooper (1968*b*) suspected that it would prove to be a subspecies of *P. truei*, a possibility which has yet to be addressed. *P. truei* species group.

Peromyscus californicus (Gambel, 1848). Proc. Acad. Nat. Sci. Philadelphia, 4:78.

TYPE LOCALITY: USA, California, Monterey Co., Monterey.

DISTRIBUTION: C and S California, USA, excluding San Joaquin Valley, to NW Baja California Norte, Mexico.

SYNONYMS: *benitoensis, insignis, mariposae, parasiticus.*

COMMENTS: Electrophoretic and morphometric variation investigated by Smith (1979), who retained only a northern (*c. californicus*) and southern (*c. insignis*) subspecies. See Merritt (1978, Mammalian Species, 85). *P. californicus* species group.

Peromyscus caniceps Burt, 1932. Trans. San Diego Soc. Nat. Hist., 7:174.

TYPE LOCALITY: Mexico, Baja California Sur, Monserrate Island, 25°38'N, 111°02'W.

DISTRIBUTION: Known only from Monserrate Island.

COMMENTS: Relationship to *P. eva* highlighted by Lawlor (1971*a*, *b*), who suggested its possible ranking as a subspecies thereof but later (1983) maintained it as an insular species. Hall (1981) listed *P. caniceps* in the *crinitus* species group without explanation. *P. eremicus* species group.

Peromyscus crinitus (Merriam, 1891). N. Am. Fauna, 5:53.

TYPE LOCALITY: USA, Idaho, Jerome Co., Shoshone Falls, north side of Snake River.

DISTRIBUTION: E Oregon and SW Idaho, south through Nevada and parts of Utah and W Colorado, USA, to Baja California Norte and NW Sonora, Mexico.

SYNONYMS: *auripectus, delgadilli, disparilus, doutii, pallidissimus, pergracilis, peridoneus, petraius, rupicolus, scitulus, scopulorum, stephensi.*

COMMENTS: Revised by Osgood (1909); subspecific taxonomy updated by Hall and Hoffmeister (1942). Initially placed in subgenus *Haplomylomys* (Osgood, 1909); later transferred to subgenus *Peromyscus* (Hooper and Musser, 1964*b*). Karyotype viewed as primitive for the genus (Greenbaum and Baker, 1978; Stangl and Baker, 1984*b*). Specific homogeneity questionable. *P. crinitus* species group. See Johnson and Armstrong (1987, Mammalian Species, 287).

Peromyscus dickeyi Burt, 1932. Trans. San Diego Soc. Nat. Hist., 7:176.

TYPE LOCALITY: Mexico, Baja California Sur, Tortuga Island, 27°21'N, 111°54'W.

DISTRIBUTION: Known only from Tortuga Island.

COMMENTS: Genetically similar to *P. eremicus* (Avise et al., 1974*b*) and presumably originated from an *eremicus*-like progenitor (Lawlor, 1983). *P. eremicus* species group.

Peromyscus difficilis (J. A. Allen, 1891). Bull. Am. Mus. Nat. Hist., 3:298.

TYPE LOCALITY: Mexico, Zacatecas, Sierra de Valparaiso.

DISTRIBUTION: W Chihuahua and SE Coahuila, south to C Oaxaca, Mexico.

SYNONYMS: *amplus, felipensis, petricola, saxicola.*

COMMENTS: Revised by Hoffmeister and de la Torre (1961) to include *nasutus* (see species account below) and *comanche* (transferred to *P. truei*—Schmidly, 1973*b*). Karyologic and biochemical evidence has suggested that *P. difficilis* and *P. nasutus* are sibling species (Avise et al., 1979; Zimmerman et al., 1975, 1978); classified as such by Carleton (1989), but see Janecek (1990). *P. truei* species group.

Peromyscus eremicus (Baird, 1858). Mammalia *in* Repts. U.S. Expl. Surv., 8(1):479.

TYPE LOCALITY: USA, California, Imperial Co., Old Fort Yuma, Colorado River opposite Yuma, Arizona.

DISTRIBUTION: S California east to Transpecos Texas, USA, most of Baja California peninsula, south along coast to C Sinaloa and on the Mexican Plateau to N San Luis Potosi, Mexico.

SYNONYMS: *alcorni, anthonyi, arenarius, avius, cedrosensis, cinereus, collatus, fraterculus, herronii, homochroia, insulicola, nigellus, papagensis, phaeurus, polypolius, propinquus, pullus, sinaloensis, tiburonensis.*

COMMENTS: Relationships of *P. eremicus* to various insular forms in Gulf of California elucidated by Lawlor (1971*a*, *b*; 1983), who relegated *collatus* to a subspecies. Genetic variation surveyed by Avise et al. (1974*b*), who noted differentiation of *eremicus* populations east and west of the Colorado River. Once included *P. merriami* (see Hoffmeister and Lee, 1963) and *P. eva* (see Lawlor, 1971*a*). See Veal and Caire (1979, Mammalian Species, 118). *P. eremicus* species group.

Peromyscus eva Thomas, 1898. Ann. Mag. Nat. Hist., ser. 7, 1:44.

TYPE LOCALITY: Mexico, Baja California Sur, San Jose del Cabo.

DISTRIBUTION: S Baja California Sur and Carmen Island, Mexico.

SYNONYMS: *carmeni.*

COMMENTS: Ranked by Osgood (1909) as a race of *P. eremicus*, but Lawlor (1971*a*), noting morphological differences and sympatry of the two, resurrected *P. eva* as a species. *P. eremicus* species group.

Peromyscus furvus J. A. Allen and Chapman, 1897. Bull. Am. Mus. Nat. Hist., 9:201.

TYPE LOCALITY: Mexico, Veracruz, 1.5 mi E Jalapa, 4400 ft.

DISTRIBUTION: E flanks of Sierra Madre Oriental from extreme S San Luis Potosi to NW Oaxaca, Mexico.

SYNONYMS: *angustirostris, latirostris.*

COMMENTS: Morphological limits, geographic variation, and distribution reviewed by Huckaby (1980); full synonyms include *latirostris* (see Hall, 1971) and *angustirostris* (see Musser, 1964). Considered a member of the *mexicanus* group by Hooper (1968*b*); segregated as *P. furvus* species group, tentatively including *P. ochraventer* and *P. mayensis*, by Carleton (1989).

Peromyscus gossypinus (Le Conte, 1853). Proc. Acad. Nat. Sci. Philadelphia, 6:411.

TYPE LOCALITY: USA, Georgia, Liberty Co., near Riceboro, probably Le Conte Plantation.

DISTRIBUTION: SE USA, from SE Oklahoma, extreme S Illinois and SE Virginia, south to Gulf of Mexico and peninsular Florida.

STATUS: U.S. ESA and IUCN - Endangered as *P. g. allapaticola.*

SYNONYMS: *allapaticola, anastasae, cognatus, insulanus, nigriculus, megacephalus, mississippiensis, palmarius, restrictus, telmaphilus.*

COMMENTS: Hybridization in lab with *P. leucopus* (Bradshaw, 1968), but differentiation of *P. gossypinus* in field is well documented (Engstrom et al., 1982; Price and Kennedy, 1980; L. W. Robbins et al., 1985). See Wolfe and Linzey (1977, Mammalian Species, 70). *P. leucopus* species group.

Peromyscus grandis Goodwin, 1932. Am. Mus. Novit., 560:4.

TYPE LOCALITY: Guatemala, Alta Verapaz, Finca Concepcion, 3 mi S San Miguel Tucuru, 3750 ft.

DISTRIBUTION: S Alta Verapaz and NE Baja Verapaz, Guatemala; limits of distribution unknown.

COMMENTS: Maintained as a species by Huckaby (1980). The level of relationship of this species to the allopatric forms *P. guatemalensis* and *P. zarhynchus* requires investigation. *P. mexicanus* species group.

Peromyscus gratus Merriam, 1898. Proc. Biol. Soc. Washington, 12:123.

TYPE LOCALITY: Mexico, Distrito Federal, Tlalpan.

DISTRIBUTION: SW New Mexico, USA, south from W Chihuahua and SE Coahuila, through interior Mexico to C Oaxaca.

SYNONYMS: *erasmus, gentilis, pavidus, zapotecae, zelotes.*

COMMENTS: Mexican populations revised, as part of *P. truei*, by Hoffmeister (1951). Divergent karyotypes and genetic distances had questioned unity of *P. truei* (Lee et al., 1972; Zimmerman et al., 1978); sympatry documented in New Mexico by Modi and Lee (1984), who raised *P. gratus* to a species (also see Janecek, 1990). *P. truei* species group.

Peromyscus guardia Townsend, 1912. Bull. Am. Mus. Nat. Hist., 31:126.

TYPE LOCALITY: Mexico, Baja California Norte, Angel de la Guarda Island, 29°33'N, 113°35'W.

DISTRIBUTION: Angel de la Guarda Island, Granito and Mejia Islands, N Gulf of California, Mexico.

SYNONYMS: *harbisoni, mejiae.*

COMMENTS: Status and relationships with respect to *P. eremicus* and *P. interparietalis* illuminated by Brand and Ryckman (1969), Lawlor (1971*b*), and Avise et al. (1974*b*). Formerly included *P. interparietalis* (see Banks, 1967). *P. eremicus* species group.

Peromyscus guatemalensis Merriam, 1898. Proc. Biol. Soc. Washington, 12:118.

TYPE LOCALITY: Guatemala, Huehuetenango, Todos Santos, 10,000 ft.

DISTRIBUTION: Intermediate to high elevations in mountains of S Chiapas, Mexico, and SW Guatemala.

SYNONYMS: *altilaneus.*

COMMENTS: Morphological variation and distribution clarified by Huckaby (1980). Level of relationship to morphologically similar allopatric forms *P. grandis* and *P. zarhynchus* unresolved. Status of *altilaneus* problematic: synonymized under *guatemalensis* by Huckaby (1980; also see Carleton and Huckaby, 1975) but listed as a species by Hall (1981), who followed Osgood (1909). Carleton (1989) noted cranial resemblance of *altilaneus* to *P. mexicanus. P. mexicanus* species group.

Peromyscus gymnotis Thomas, 1894. Ann. Mag. Nat. Hist., ser. 6, 14:365.

TYPE LOCALITY: "Guatemala."

DISTRIBUTION: Pacific coastal plain and adjacent foothills from S Chiapas, Mexico, to S Nicaragua, west of Lake Nicaragua.

SYNONYMS: *allophylus.*

COMMENTS: Formerly classified as a subspecies of *P. mexicanus* (Osgood, 1909); Musser (1971*a*) elevated *gymnotis* to species and pointed out the junior status of *allophylus*. Differentiation from *P. mexicanus* sustained by Huckaby (1980) and Jones and Yates (1983), who also amplified the range and morphological recognition of *P. gymnotis. P. mexicanus* species group.

Peromyscus hooperi Lee and Schmidly, 1977. J. Mammal., 58:263.

TYPE LOCALITY: Mexico, Coahuila, 2.5 mi W, 21 mi S Ocampo, 3500 ft.

DISTRIBUTION: C Coahuila to NE tip of Zacatecas, Mexico.

COMMENTS: Allocated, with reservation, to the subgenus *Peromyscus* as sole member of *P. hooperi* species group (Schmidly et al., 1985).

Peromyscus interparietalis Burt, 1932. Trans. San Diego Soc. Nat. Hist., 7:175.

TYPE LOCALITY: Mexico, Baja California Norte, San Lorenzo Sur Island, 28°36'N, 112°51'W.

DISTRIBUTION: North and South San Lorenzo Islands, and Salsipuedes Island, N Gulf of California, Mexico.

SYNONYMS: *lorenzi, ryckmani.*

COMMENTS: Considered distinct from *P. guardia* by Banks (1967); also see Avise et al. (1974*b*), Brand and Ryckman (1969), and Lawlor (1971*b*). *P. eremicus* species group.

Peromyscus leucopus (Rafinesque, 1818). Am. Mon. Mag., 3:446.

TYPE LOCALITY: USA, Kentucky, "pine barrens," presumably near mouth of Ohio River.

DISTRIBUTION: C and E USA, excluding Florida; northwards to S Alberta and to S Ontario, Quebec and Nova Scotia, Canada; southwards to N Durango and along Caribbean coast to Isthmus of Tehuantepec and NW Yucatan Peninsula, Mexico.

SYNONYMS: *affinis, ammodytes, arboreus, aridulus, arizonae, brevicaudus, campestris, canus, castaneus, caudatus, cozulmelae, easti, emmonsi, flaccidus, fuscus, incensus, lachiguiriensis, mearnsii, mesomelas, michiganensis, minnisotae, myoides, musculoides, noveboracensis, ochraceus, texanus, tornillo.*

COMMENTS: Revised by Osgood (1909). Distinctive northeastern (*leucopus*) and southwestern (*texanus*) cytotypes reported (R. J. Baker et al., 1983*b*), with introgression across hybrid zone in central Oklahoma (Nelson et al., 1987; Stangl, 1986; Stangl and Baker, 1984*a*). Genetic variation investigated by Browne (1977)—also see above account of *P. gossypinus*. See Lackey et al. (1985, Mammalian Species, 247).

Peromyscus levipes Merriam, 1898. Proc. Biol. Soc. Washington, 12:123.

TYPE LOCALITY: Mexico, Tlaxcala, Mount Malinche, 8400 ft.

DISTRIBUTION: Mexico—from E Nayarit to C Nuevo Leon and Tamaulipas, south through mountains of Oaxaca and Chiapas—to highlands of Guatemala, El Salvador, and Honduras.

SYNONYMS: *ambiguus, beatae, sacarensis, sagax.*

COMMENTS: Systematic studies have confirmed sympatry of *P. boylii rowleyi* and *P. levipes* in Queretaro and Hidalgo and supported its elevation to species (Houseal et al., 1987; Rennert and Kilpatrick, 1987; Schmidly et al., 1988). The taxon *beatae* was provisionally retained by Carleton (1989) under *P. levipes*, whereas Schmidly et al. (1988) viewed it as a cryptic species distinct from *P. levipes* (including *ambiguus*—but see Bradley et al., 1989). The status and distributional extent of these taxa and other chromosomal forms deserve attention (see Bradley et al., 1989; Houseal et al., 1987; Smith et al., 1989). *P. boylii* species group.

Peromyscus madrensis Merriam, 1898. Proc. Biol. Soc. Washington, 12:16.

TYPE LOCALITY: Mexico, Nayarit, Tres Marías Islands, María Madre Island.

DISTRIBUTION: Tres Marías Islands, Mexico.

COMMENTS: Treated as a subspecies of *P. boylii* by Osgood (1909) but considered distinct by Carleton (1977); viewed as closely related to *P. simulus* by Carleton et al. (1982). Perhaps extirpated on María Magdalena Island, where *Rattus rattus* is now abundant (Carleton et al., 1982; Wilson, 1991). *P. boylii* species group.

Peromyscus maniculatus (Wagner, 1845). Arch. Naturgesch., 11, 1:148.

TYPE LOCALITY: Canada, Labrador, Moravian settlements.

DISTRIBUTION: Panhandle of Alaska and across N Canada, south through most of continental USA, excluding the SE and E seaboard, to southernmost Baja California Sur and to NC Oaxaca, Mexico.

SYNONYMS: *abietorum, akeleyi, algidus, alpinus, anacapae, angustus, anticostiensis, argentatus, artemisiae, assimilis, austerus, bairdii, balaclavae, beresfordi, blandus, borealis, canadensis, cancrivorus, carli, catalinae, cineritius, clementis, coolidgei, deserticolus, dorsalis, doylei, dubius, elusus, eremus, exiguus, exterus, fulvus, gambeli, georgiensis, geronimensis, gracilis, gunnisoni, hollisteri, hueyi, hylaeus, imperfectus, inclarus, insolatus, keeni, labecula, luteus, macrorhinus, magdalenae, margaritae, maritimus, martinensis, medius, osgoodi, nebrascensis, nubiterrae, oresterus, ozarkiarum, pallescens, perimekurus, plumbeus, pluvialis, prevostensis, rubidus, rubriventer, rufinus, sanctaerosae, santacruzae, sartinensis, saturatus, saxamans, serratus, sonoriensis, streatori, subarcticus, thurberi, triangularis, umbrinus.*

COMMENTS: Status and relationships with regard to *P. melanotis* addressed by Bowers et al. (1973), Bowers (1974), and Greenbaum and Baker (1978); to *P. oreas* by Allard et al. (1987), Allard and Greenbaum (1988), and Gunn and Greenbaum (1986); and to *P. polionotus* by Avise et al. (1979) and Robbins and Baker (1981). Distinction between short-tailed and long-tailed geographic races (e.g., Hooper, 1968*b*; Koh and Peterson, 1983) obscurely reflected in chromosomal and biochemical data (Avise et al., 1979; Bowers et al., 1973; Bradshaw and Hsu, 1972; Calhoun et al., 1988; Lansman et al., 1983). Specific homogeneity of included taxa doubtful. *P. maniculatus* species group.

Peromyscus mayensis Carleton and Huckaby, 1975. J. Mammal., 56:444.
TYPE LOCALITY: Guatemala, Depto. Huehuetenango, about 7 km NW Santa Eulalia, Yaiquich, 2950 m.
DISTRIBUTION: Known only from the type locality.
COMMENTS: Nearest specific relatives obscure; provisionally assigned to the *furvus* species group (Carleton, 1989).

Peromyscus megalops Merriam, 1898. Proc. Biol. Soc. Washington, 12:119.
TYPE LOCALITY: Mexico, Oaxaca, La Cieneguilla ranch, near Santa Maria Ozolotepec, 10,000 ft.
DISTRIBUTION: Sierra Madre del Sur of Guerrero and Oaxaca, Mexico.
SYNONYMS: *auritus, comptus.*
COMMENTS: Following Osgood (1909), Hall (1981) continued to maintain *melanurus* as a subspecies, a form which Huckaby (1980) documented as a species distinct from *P. megalops*. Distribution and ecology reviewed by Musser (1964). Carleton (1989) observed Osgood's (1909) earlier arrangement of a *megalops* species group (including *P. melanocarpus* and *P. melanurus*) apart from the *mexicanus* species group.

Peromyscus mekisturus Merriam, 1898. Proc. Biol. Soc. Washington, 12:124.
TYPE LOCALITY: Mexico, Puebla, Chalchicomula, 8400 ft.
DISTRIBUTION: SE Puebla, Mexico.
COMMENTS: A distinctive species known only by two specimens, the type and one from Tehuacan, Puebla, Mexico (Hooper, 1947). Provisionally retained with the *melanophrys* species group (Carleton, 1989).

Peromyscus melanocarpus Osgood, 1904. Proc. Biol. Soc. Washington, 17:73.
TYPE LOCALITY: Mexico, Oaxaca, Cerro Zempoaltepec, above Yacochi.
DISTRIBUTION: Cloud forest of Sierra de Juarez and Sierra de Zempoaltepec, NC Oaxaca, Mexico.
COMMENTS: Range and distinctive morphology substantiated by Huckaby (1980). See Rickart and Robertson (1985, Mammalian Species, 241). *P. megalops* species group.

Peromyscus melanophrys (Coues, 1874). Proc. Acad. Nat. Sci. Philadelphia, 26:181.
TYPE LOCALITY: Mexico, Oaxaca, Santa Efigenia.
DISTRIBUTION: S Durango and Coahuila, south through interior Mexico to Chiapas.
SYNONYMS: *coahuilensis, consobrinus, gadovii, leucurus, micropus, xenurus, zamorae.*
COMMENTS: Revised by Baker (1952) to include *xenurus*, which Osgood (1909) had retained as a species. Relationships evaluated by Schmidly et al. (1985) and Stangl and Baker (1984*b*). The level of differentiation of *micropus* from Jalisco invites study. *P. melanophrys* species group.

Peromyscus melanotis J. A. Allen and Chapman, 1897. Bull. Am. Mus. Nat. Hist., 9:203.
TYPE LOCALITY: Mexico, Veracruz, Las Vigas, 8000 ft.
DISTRIBUTION: Cordillera Transvolcanica in C Mexico (E Jalisco to C Veracruz), northwards along Sierra Madre Oriental to S Nuevo Leon and along Sierra Madre Occidental to W Chihuahua; isolated populations in SE Arizona, USA.
SYNONYMS: *cecilii, zamelas.*
COMMENTS: Genetic, karyotypic, and geographic variation studied by Bowers et al. (1973), who included populations in S Arizona previously identified as *P. maniculatus rufinus*. However, see Hoffmeister (1986) who disputed their conclusions and maintained all Arizona populations as *P. maniculatus*. Bowers (1974) discerned no basis for subspecific divisions. *P. maniculatus* species group.

Peromyscus melanurus Osgood, 1909. N. Am. Fauna, 28:215.
TYPE LOCALITY: Mexico, Oaxaca, below Pluma Hidalgo, 3000 ft.
DISTRIBUTION: Pacific slopes of the Sierra Madre del Sur of Oaxaca, Mexico.
COMMENTS: Although named as a subspecies of *P. megalops*, Huckaby (1980) substantiated the specific status and range of *melanurus*. *P. megalops* species group.

Peromyscus merriami Mearns, 1896. Preliminary diagnoses of new mammals from the Mexican border of the United States, p. 2; preprint of Proc. U. S. Nat. Mus., 19:138.
TYPE LOCALITY: Mexico, Sonora, Sonoyta, on Sonoyta River.
DISTRIBUTION: SC Arizona, USA, through Sonora to C Sinaloa, Mexico.
SYNONYMS: *goldmani.*

COMMENTS: Osgood (1909) synonymized *merriami* under *P. eremicus* but Hoffmeister and Lee (1963) documented their sympatric occurrence and morphological differentiation. Geographic variation evaluated by Hoffmeister and Diersing (1973), who retained *goldmani* as a subspecies. Bears close kinship to *P. eremicus* (Avise et al., 1974*b*; Lawlor, 1971*a*) and to *P. pembertoni* (Lawlor, 1971*a*, 1983). *P. eremicus* species group.

Peromyscus mexicanus (Saussure, 1860). Rev. Mag. Zool. Paris, Ser. 2, 12:103.

TYPE LOCALITY: Mexico, Veracruz, 10 km E Mirador (restricted by Dalquest, 1950:8).

DISTRIBUTION: In Mexico—along the Atlantic coast from S San Luis Potosi to the Isthmus of Tehuantepec, and along the Pacific coast, from the Guerrero-Oaxaca border to C Chiapas; upper foothills and middle-elevation mountains in Guatemala, through El Salvador, Honduras, and Nicaragua, to upper highlands in Costa Rica and W Panama (Chiriqui region).

SYNONYMS: *angelensis, azulensis, cacabatus, coatlanensis, hesperus, nicaraguae, nudipes, orientalis, orizabae, philombrius, putlaensis, salvadorensis, saxatilis, sloeops, teapensis, tehuantepecus, totontepecus, tropicalis.*

COMMENTS: Geographic range, variation, and taxonomic synonymy summarized by Huckaby (1980). Includes forms formerly viewed as members of *Osgoodomys banderanus* (*angelensis, coatlanensis,* and *sloeops*—see Musser, 1969*a*), *P. guatemalensis* (*tropicalis*—see Musser, 1969*a*), and *P. megalops* (*azulensis*—see Huckaby, 1980). Although these taxa were erroneously associated by their original describers, the conspecific status of populations now arranged under *P. mexicanus* direly needs investigation. In particular, southern populations identified as *nudipes,* arranged as a synonym by Huckaby (1980) and Carleton (1989), have been considered a distinct species (Osgood, 1909; Hooper, 1968*b*) and deserve additional scrutiny. Euchromatic banding patterns identical in *P. mexicanus* and related species so far examined (Smith et al., 1986). See account of *P. gymnotis,* which had been arranged as a race of *P. mexicanus. P. mexicanus* species group.

Peromyscus nasutus (J. A. Allen, 1891). Bull. Amer. Mus. Nat. Hist., 3:229.

TYPE LOCALITY: USA, Colorado, Latimer Co., Estes Park.

DISTRIBUTION: C Colorado and SE Utah, south through New Mexico and Transpecos, Texas, USA, to NW Coahuila, Mexico.

SYNONYMS: *griseus, penicillatus.*

COMMENTS: Considered a distinct species (Osgood, 1909) until synonymized under *P. difficilis* by Hoffmeister and de la Torre (1961). Recognized as a sibling species separate from *P. difficilis* by Zimmerman et al. (1975, 1978). Includes *penicillatus,* which Osgood (1909) had misclassified under *P. boylii* (see Diersing, 1976). *P. truei* species group.

Peromyscus ochraventer Baker, 1951. Univ. Kansas Mus. Nat. Hist. Misc. Publ., 5:213.

TYPE LOCALITY: Mexico, Tamaulipas, El Carrizo, 70 km (by highway) S Ciudad Victoria and 6 km W Panamerican Highway, 2800 ft.

DISTRIBUTION: Moist forests of S Tamaulipas and adjacent San Luis Potosi, Mexico.

COMMENTS: Revised by Huckaby (1980) as part of the *mexicanus* species group (*sensu* Hooper, 1968*b*). Departs from conservative karyotypic pattern exhibited by species of the *mexicanus* group (Robbins and Baker, 1981; Smith et al., 1986); provisionally assigned to the *furvus* species group by Carleton (1989).

Peromyscus oreas Bangs, 1898. Proc. Biol. Soc. Washington, 12:84.

TYPE LOCALITY: Canada, British Columbia, Mount Baker Range, 6500 ft, near boundary of Whatcom Co., Washington.

DISTRIBUTION: SW British Columbia, including Vancouver Island and neighboring islands, Canada, and W Washington, USA.

SYNONYMS: *interdictus, isolatus.*

COMMENTS: Maintained as a subspecies of *P. maniculatus* by Osgood (1909) and Hooper (1968*b*). Sheppe (1961) considered *oreas* a distinct species, a conclusion ratified by recent taxonomic research (Allard et al. 1987; Allard and Greenbaum, 1988; Calhoun and Greenbaum, 1991; Gunn and Greenbaum, 1986; Sullivan et al., 1990). Viewed as sister species of *P. sitkensis* (Gunn and Greenbaum, 1986). Further clarification of relationship to taxa in *P. maniculatus* (e.g., *keeni* and *macrorhinus*) and to *P. sitkensis* is required. *P. maniculatus* species group.

Peromyscus pectoralis Osgood, 1904. Proc. Biol. Soc. Washington, 17:59.
TYPE LOCALITY: Mexico, Queretaro, Jalpan.
DISTRIBUTION: SE New Mexico and C Texas, USA, south to N Jalisco and Hidalgo, Mexico.
SYNONYMS: *collinus, eremicoides, laceianus.*
COMMENTS: Revised by Schmidly (1972), who recognized three subspecies which correspond to patterns of genetic differentiation (Kilpatrick and Zimmerman, 1976). Consanguinity of included populations nevertheless suspect (see Avise et al., 1974*a*). See Schmidly (1974*b*, Mammalian Species, 49). *P. boylii* species group.

Peromyscus pembertoni Burt, 1932. Trans. San Diego Soc. Nat. Hist., 7:176.
TYPE LOCALITY: Mexico, Sonora, San Pedro Nolasco Island, 27°58′N, 111°24′W.
DISTRIBUTION: Known only from the type locality.
STATUS: IUCN - Extinct.
COMMENTS: Derivation from a *P. merriami*-like ancestor postulated by Lawlor (1971*a*). Probably extinct—see Lawlor (1983). *P. eremicus* species group.

Peromyscus perfulvus Osgood, 1945. J. Mammal., 26:299.
TYPE LOCALITY: Mexico, Michoacan, 10 km W Apatzingan, 1040 ft.
DISTRIBUTION: Coastal lowlands of Jalisco and Colima, along Rio Balsas to interior Michocan and northernmost Guerrero, Mexico.
SYNONYMS: *chrysopus.*
COMMENTS: Placed with *P. melanophrys* group by Hooper (1968*b*), an association supported by electrophoretic data (Schmidly et al., 1985).

Peromyscus polionotus (Wagner, 1843). Arch. Naturgesch., 9, 2:52.
TYPE LOCALITY: USA, Georgia.
DISTRIBUTION: NE Mississippi to W South Carolina, south through Alabama and Georgia, to W and most of peninsular Florida, USA.
STATUS: U.S. ESA and IUCN - Endangered as *P. p. allophrys, P. p. ammobates, P. p. phasma,* and *P. p. trissyllepsis;* U.S. ESA - Treatened and IUCN - Indeterminate as *P. p. niveiventris.*
SYNONYMS: *albifrons, allophrys, ammobates, arenarius, baliolus, colemani, decoloratus, griseobracatus, leucocephalus, lucubrans, niveiventris, peninsularis, phasma, rhoadsi, subgriseus, sumneri, trissyllepsis.*
COMMENTS: Revised by Osgood (1909). Pleisitocene origination of subspecies hypothesized by Bowen (1968); intraspecific genetic differentiation surveyed by Selander et al. (1971). Relationships assessed by Avise et al. (1979), Rogers et al. (1984), and Stangl and Baker (1984*b*). *P. maniculatus* species group.

Peromyscus polius Osgood, 1904. Proc. Biol. Soc. Washington, 17:61.
TYPE LOCALITY: Mexico, Chihuahua, Colonia Garcia.
DISTRIBUTION: WC Chihuahua, Mexico.
COMMENTS: Species-group affiliation enigmatic: initially placed in *truei* group (Osgood, 1909) but provisionally reassigned to *boylii* group (Hoffmeister, 1951). Genetic distance data weakly support the latter association (Kilpatrick and Zimmerman, 1975; Schmidly et al., 1985); also see Bradley and Schmidly (1987). *P. boylii* species group.

Peromyscus pseudocrinitus Burt, 1932. Trans. San Diego Soc. Nat. Hist., 7:173.
TYPE LOCALITY: Mexico, Baja California Sur, Coronados Island, 26°06′N, 111°18′W.
DISTRIBUTION: Known only from the type locality.
COMMENTS: Tentatively allocated to *crinitus* species group by Hooper (1968*b*); affinities to *P. eremicus* asserted by Lawlor (1971*a*) and believed derived therefrom (Lawlor, 1983). *P. eremicus* species group.

Peromyscus sejugis Burt, 1932. Trans. San Diego Soc. Nat. Hist., 7:171.
TYPE LOCALITY: Mexico, Baja California Sur, Santa Cruz Island, 25°17′N, 110°43′W.
DISTRIBUTION: Santa Cruz and San Diego islands, S Gulf of California, Mexico.
COMMENTS: Highly differentiated morphologically and genetically from mainland *P. maniculatus* (Avise et al., 1974*b*, 1979; Burt, 1932), though probably derived from a *maniculatus*-like ancestor (Lawlor, 1983). *P. maniculatus* species group.

Peromyscus simulus Osgood, 1904. Proc. Biol. Soc. Washington, 17:64.
TYPE LOCALITY: Mexico, Nayarit, San Blas.
DISTRIBUTION: Coastal plain of Nayarit and S Sinaloa, Mexico.
COMMENTS: Previously classified as a geographic race of *P. boylii* (Osgood, 1909; also see Hall, 1981); elevated to species by Carleton (1977). Relationships addressed by Bradley and Schmidly (1987) and Carleton et al. (1982). *P. boylii* species group.

Peromyscus sitkensis Merriam, 1897. Proc. Biol. Soc. Washington, 11:223.
TYPE LOCALITY: USA, Alaska, Baranof Island, Sitka.
DISTRIBUTION: Outer islands of Alexander Archipelago, off Alaskan panhandle, USA.
SYNONYMS: *oceanicus*.
COMMENTS: Banded karyotypes reported by Pengilly et al. (1983). Insular forms off southern British Columbia referred to *sitkensis* by Thomas (1973) but identified as *P. oreas* by Gunn and Greenbaum (1986). These studies underscore the need to refine relationships among *P. sitkensis, P. oreas,* and certain forms of *P. maniculatus.P. maniculatus* species group.

Peromyscus slevini Mailliard, 1924. Proc. California Acad. Sci., ser. 4, 12:1221.
TYPE LOCALITY: Mexico, Baja California Sur, Santa Catalina Island, 17 mi NE Punta San Marcial, 25°43'50"N.
DISTRIBUTION: Known only from the type locality.
COMMENTS: Proposed derivation from a *maniculatus*-like ancestor (Lawlor, 1983) needs critical reexamination; association with *maniculatus* species group judged tenuous by Carleton (1989; also see Burt, 1934).

Peromyscus spicilegus J. A. Allen, 1897. Bull. Am. Mus. Nat. Hist., 9:50.
TYPE LOCALITY: Mexico, Jalisco, Mascota, Mineral San Sebastian.
DISTRIBUTION: Intermediate elevations of Sierra Madre Occidental from S Sinaloa and SW Durango to WC Michoacan, Mexico.
COMMENTS: Arranged as a subpecies of *P. boylii* by Osgood (1909; and Hall, 1981). Morphological, chromosomal, and distributional data require its recognition as a species having closer affinity to *P. aztecus* and its allies (see Bradley and Schmidly, 1987; Carleton, 1977, 1979; Carleton et al., 1982; Hooper, 1968*b*). Includes populations in Michoacan once identified as *evides* (see Carleton, 1977; Smith et al., 1989). Variation with altitude investigated by Sanchez-Cordero and Villa-Ramirez (1988). *P. aztecus* species group.

Peromyscus stephani Townsend, 1912. Bull. Am. Mus. Nat. Hist., 31:126.
TYPE LOCALITY: Mexico, Sonora, San Esteban Island, 28°34'N, 113°21'W.
DISTRIBUTION: Known only from Isla San Esteban.
COMMENTS: Lawlor (1971*b*) demonstrated the *boylii*-like traits of *P. stephani,* which was previously classified in the subgenus *Haplomylomys* (Hooper, 1968*b*). Electrophoretic investigations support its specific status (Avise et al., 1974*a*, *b*). *P. boylii* species group.

Peromyscus stirtoni Dickey, 1928. Proc. Biol. Soc. Washington, 41:5.
TYPE LOCALITY: El Salvador, La Union, Rio Goascoran, 13 deg. 30' N, 100 ft.
DISTRIBUTION: Intermittently found in dry to semiarid lowlands from SE Guatemala, through El Salvador and Honduras, to NC Nicaragua.
COMMENTS: Although Hooper (1968*b*) questioned the specific recognition of *P. stirtoni,* others have substantiated its distinctive morphology and habitat (Huckaby, 1980; Jones and Yates, 1983). Late Holocene remains from Guanacaste Prov., Costa Rica, suggest a recent range contraction of this species (see Woodman, 1988). See Jones (1990, Mammalian Species, 361). Tentatively retained in the *mexicanus* species group (Carleton, 1989).

Peromyscus truei (Shufeldt, 1885). Proc. U. S. Natl. Mus., 8:407.
TYPE LOCALITY: USA, New Mexico, McKinley Co., Fort Wingate.
DISTRIBUTION: USA, SW Oregon to W and SE Colorado, south to Baja California (Mexico), Arizona, New Mexico, and NC Texas.
SYNONYMS: *chlorus, comanche, dyselius, gilberti, hemionotis, lagunae, lasius, martirensis, megalotis, montipinoris, nevadensis, preblei, sequoiensis.*

COMMENTS: Revised by Hoffmeister (1951), who included Mexican populations later recognized as *P. gratus* (see above species account). Schmidly (1973*b*) reallocated *comanche* as a subspecies of *P. truei* (also see Modi and Lee, 1984, and Janecek, 1990). See Hoffmeister (1981, Mammalian Species, 161). *P. truei* species group.

Peromyscus winkelmanni Carleton, 1977. Occas. Pap. Mus. Zool., Univ. Michigan, 675:2.
TYPE LOCALITY: Mexico, Michoacan, 6.3 mi (by road) WSW Dos Aguas, 8000 ft.
DISTRIBUTION: Isolated localities in the Sierra de Coalcoman, Michoacan, and in Guerrero, Mexico (after Sullivan et al., 1991).
COMMENTS: Karyotype reported by Smith et al. (1989); relationships evaluated by Bradley and Schmidly (1987) and Sullivan et al. (1991). Removed to *aztecus* species group by Carleton (1989).

Peromyscus yucatanicus J. A. Allen and Chapman, 1897. Bull. Am. Mus. Nat. Hist., 9:8.
TYPE LOCALITY: Mexico, Yucatan, Chichen-Itza.
DISTRIBUTION: N Yucatan Peninsula, Mexico.
SYNONYMS: *badius*.
COMMENTS: Huckaby (1980) followed Lawlor (1965) in treating *P. yucatanicus* as monotypic; Hall (1981) accepted Osgood's (1909) subspecific arrangement. The notion (e.g., Lawlor, 1965; Hooper, 1968*b*) that *P. yucatanicus* is most closely related to *P. mexicanus* has not received support (Carleton, 1973; Huckaby, 1980). See Young and Jones (1983, Mammalian Species, 196). *P. mexicanus* species group.

Peromyscus zarhynchus Merriam, 1898. Proc. Biol. Soc. Washington, 12:117.
TYPE LOCALITY: Mexico, Chiapas, mountains above Tumbala, 5500 ft.
DISTRIBUTION: Middle to high elevation cloud forest in mountains of NC Chiapas, Mexico.
SYNONYMS: *cristobalensis*.
COMMENTS: Distribution and morphological variation reviewed by Huckaby (1980). Relationship and taxonomic status with respect to *P. guatemalensis* and *P. grandis* unresolved. *P. mexicanus* species group.

Phaenomys Thomas, 1917. Ann. Mag. Nat. Hist., ser. 8, 20:196.
TYPE SPECIES: *Oryzomys ferrugineus* Thomas, 1894.
COMMENTS: Thomasomyine. Rare in collections; knowledge of taxonomy and distribution minimal.

Phaenomys ferrugineus (Thomas, 1894). Ann. Mag. Nat. Hist., ser. 6, 14:352.
TYPE LOCALITY: Brazil, Rio de Janeiro.
DISTRIBUTION: Vicinity of type locality.

Phyllotis Waterhouse, 1837. Proc. Zool. Soc. Lond., 1837:27.
TYPE SPECIES: *Mus darwini* Waterhouse, 1837.
SYNONYMS: *Paralomys*.
COMMENTS: Phyllotine. Full generic revisions by Pearson (1958) and Hershkovitz (1962). Pearson (1958) utilized *Auliscomys, Graomys*, and *Loxodontomys* as subgenera, whereas Hershkovitz (1962) placed all three, plus *Paralomys*, in complete synonymy—see accounts of *Auliscomys* and *Graomys* for their reinstatement as genera. New karyotypic and distributional data bearing on species status and relationships provided by Pearson (1972), Spotorno (1976), and Pearson and Patton (1976).

Phyllotis amicus Thomas, 1900. Ann. Mag. Nat. Hist., ser. 7, 5:355.
TYPE LOCALITY: Peru, Cajamarca Dept., Tolon, 100 m.
DISTRIBUTION: Coast and lower Pacific slopes of W Peru (Depts. Lambayeque to Ayacucho).
SYNONYMS: *maritimus, montanus*.

Phyllotis andium Thomas, 1912. Ann. Mag. Nat. Hist., ser. 8, 10:409.
TYPE LOCALITY: Ecuador, Canar Prov., Canar, 8500 ft.
DISTRIBUTION: E and W Andean slopes from C Ecuador (Tungurahua) to C Peru (Lima).
SYNONYMS: *fruticicolus, melanius, stenops, tamborum*.

Phyllotis bonaeriensis Crespo, 1964. Neotropica, 10:99.
TYPE LOCALITY: Argentina, Buenos Aires Prov., Sierra de la Ventana.

DISTRIBUTION: Buenos Aires Prov., Argentina.
COMMENTS: Formerly included in *P. darwini* (see Reig, 1978).

Phyllotis caprinus Pearson, 1958. Univ. California Publ. Zool., 56:435.
TYPE LOCALITY: Argentina, Jujuy Prov., Tilcara, 8000 ft.
DISTRIBUTION: Upper slopes of E Andes from S Bolivia to northernmost Argentina.
COMMENTS: Demoted to a subspecies of *P. darwini* by Hershkovitz (1962), but distributional and morphological evidence substantiates the specific recognition of *P. caprinus* (Pearson, 1958).

Phyllotis darwini (Waterhouse, 1837). Proc. Zool. Soc. Lond., 1837:28.
TYPE LOCALITY: Chile, Coquimbo Prov., Coquimbo.
DISTRIBUTION: C Peru (Junín), south through W Bolivia, to C Chile and WC Argentina.
SYNONYMS: *abrocodon, arenarius, boedeckeri, campestris, capito, chilensis, dichrous, fulvescens, glirinus, griseoflavus* (of Philippi, 1900), *illapelinus, lanatus, limatus, megalotis, melanotis, melanotus, mollis, platytarsus, posticalis, ricarulus, segethi*.
COMMENTS: See accounts of *P. bonaeriensis, P. caprinus, P. definitus, P. magister, P. osgoodi, P. wolffsohni*, and *P. xanthopygus*, previously arrayed as subspecies of *P. darwini* by Pearson (1958) and/or Hershkovitz (1962), for references addressing their specific status. The complexity of populations comprising this species urges still further revisionary attention.

Phyllotis definitus Osgood, 1915. Field Mus. Nat. Hist. Publ., Zool. Ser., 10:189.
TYPE LOCALITY: Peru, Ancash Dept., Macate, 9000 ft.
DISTRIBUTION: Andes of Ancash, Peru.
COMMENTS: Included in *P. magister* by Pearson (1958) and in *P. darwini* by Hershkovitz (1962) but distinctive karyotype is unlike either of those species (see Pearson, 1972).

Phyllotis gerbillus Thomas, 1900. Ann. Mag. Nat. Hist., ser. 7, 5:151.
TYPE LOCALITY: Peru, Piura Dept., Piura.
DISTRIBUTION: Sechura Desert of NW Peru.
COMMENTS: Type species of *Paralomys*, a genus named by Thomas (1926*a*); included within *Phyllotis* by Osgood (1947), a position supported by its derived karyotype (Pearson and Patton, 1976). The relationship of *gerbillus* to other *Phyllotis* deserves reconsideration.

Phyllotis haggardi Thomas, 1908. Ann. Mag. Nat. Hist., ser. 7, 2:270.
TYPE LOCALITY: Ecuador, Pichincha Prov., Mount Pichincha, above Quito, 3400-4000 m.
DISTRIBUTION: Andes of C Ecuador.
SYNONYMS: *elegantulus, fuscus*.

Phyllotis magister Thomas, 1912. Ann. Mag. Nat. Hist., ser. 8, 10:406.
TYPE LOCALITY: Peru, Arequipa Dept., Arequipa, 2300 m.
DISTRIBUTION: Upper Pacific slopes of Andes from C Peru to N Chile.
COMMENTS: Hershkovitz (1962) treated *magister* as a race of *P. darwini*, a position inconsistent with records of sympatry (Pearson, 1958) and karyotypic differences (Pearson, 1972; Pearson and Patton, 1976). Pearson (1958) arranged *definitus* as a subspecies of *P. magister* but later (1972) supported its specific status.

Phyllotis osgoodi Mann, 1945. Biologica, 2:81.
TYPE LOCALITY: Chile, Tarapaca Prov., Parinacota.
DISTRIBUTION: Altiplano of NE Chile.
COMMENTS: Relegated to synonymy under *P. darwini* (Hershkovitz, 1962; Pearson, 1958) but considerable evidence supports its validity as a species (Spotorno, 1976; Spotorno and Walker, 1979, 1983).

Phyllotis osilae J. A. Allen, 1901. Bull. Am. Mus. Nat. Hist., 14:44.
TYPE LOCALITY: Peru, Puno Dept., Osila (Pearson, 1958:426, considered the place-name equivalent to Asillo, 17 mi ENE Ayaviri, 13,000 ft).
DISTRIBUTION: Upper Andean slopes on Atlantic drainage, from SC Peru (Cuzco), through WC Bolivia, to N Argentina (Catamarca).
SYNONYMS: *lutescens, nogalaris, phaeus, tucumanus*.
COMMENTS: All-acrocentric karyotype (2n=68) interpreted as ancestral state of the genus (Pearson and Patton, 1976).

Phyllotis wolffsohni Thomas, 1902. Ann. Mag. Nat. Hist., ser. 7, 9:131.
TYPE LOCALITY: Bolivia, Cochabamba, Tapacari, 9900 ft.
DISTRIBUTION: Upper E Andean slopes in C Bolivia.
COMMENTS: Hershkovitz (1962) synonymized *wolffsohni* as a subspecies of *P. darwini*, a relationship at variance with morphological and karyotypic information (Pearson, 1958; Pearson and Patton, 1976).

Phyllotis xanthopygus (Waterhouse, 1837). Proc. Zool. Soc. Lond., 1837:28.
TYPE LOCALITY: Argentina, Santa Cruz Prov., coast of Santa Cruz.
DISTRIBUTION: NW Argentina (Catamarca) and C Chile (Atacama), south along both flanks of Andes to Santa Cruz Prov., Argentina, and adjacent Magallanes Prov., Chile.
SYNONYMS: *oreigenus, rupestris, vaccarum, wolffhuegeli.*
COMMENTS: Viewed as a geographic race of *P. darwini* by Pearson (1958) and Hershkovitz (1962). Morphometric, electrophoretic, and karyotypic differentiation support the specific distinction of *P. xanthopygus* (Spotorno and Walker, 1983; Walker et al., 1984).

Podomys Osgood, 1909. N. Am. Fauna, 28:226.
TYPE SPECIES: *Hesperomys floridanus* Chapman, 1889.
COMMENTS: Peromyscini. Named as a subgenus of *Peromyscus* by Osgood (1909) and maintained as such by Hooper (1968*b*). Carleton (1980, 1989) argued for generic recognition, a ranking disputed by others (Rogers et al., 1984; Stangl and Baker, 1984*b*). Various morphological features suggest relationship to *Neotomodon* and/or *Habromys* (Carleton, 1980; Hooper and Musser, 1964*b*; Linzey and Layne, 1969). Accessory reproductive gland and spermatozoan morphology described by Linzey and Layne (1969, 1974).

Podomys floridanus (Chapman, 1889). Bull. Am. Mus. Nat. Hist., 2:117.
TYPE LOCALITY: USA, Florida, Alachua Co., Gainesville.
DISTRIBUTION: Peninsular Florida, USA.
COMMENTS: Smith et al. (1973) noted low heterozygosity levels within populations and high genetic similarity between them. Status considered threatened by Florida agencies due to disappearance of scrub habitat (see Layne, 1990).

Podoxymys Anthony, 1929. Am. Mus. Novit., 383:4.
TYPE SPECIES: *Podoxymys roraimae* Anthony, 1929.
COMMENTS: Akodontine. Systematics little known aside from original comparisons, report on gastric morphology by Carleton (1973), and brief comments of Reig (1987), and new data on systematics and karyology by Pérez-Zapata et al. (1992).

Podoxymys roraimae Anthony, 1929. Am. Mus. Novit., 383:4.
TYPE LOCALITY: Guyana, summit of Mount Roraima, 8600 ft.
DISTRIBUTION: Guyana and probably adjacent portions of Venezuela and Brazil.
COMMENTS: Known only by six specimens (Pérez-Zapata et al., 1992).

Pseudoryzomys Hershkovitz, 1962. Fieldiana Zool., 31:208.
TYPE SPECIES: *Oryzomys wavrini* Thomas, 1921 (= *Hesperomys simplex* Winge, 1887).
COMMENTS: Associated with phyllotines by Hershkovitz (1962), but excluded from same by Olds and Anderson (1989). Oryzomyine characteristics illuminated by Voss and Myers (1991), who eschewed the "traditional phenetic practices of Neotropical muroid classification" and left the genus as Sigmodontinae *incertae sedis* pending phylogenetic studies of its tribal-level affinity. See Voss and Myers (1991:418) on the formal availability of the genus-group name.

Pseudoryzomys simplex (Winge, 1887). E Museo Lundii, 1(3):11.
TYPE LOCALITY: Brazil, Minas Gerais, near Lagoa Santa.
DISTRIBUTION: NE Argentina, W Paraguay, and SE Bolivia to E Brazil (Edo. Pernambuco).
SYNONYMS: *wavrini, reigi.*
COMMENTS: Formerly classified as *Oryzomys incertae sedis* (Cabrera, 1961). Lectotype designated, synonymy presented, karyotype and distribution discussed by Voss and Myers (1991).

Punomys Osgood, 1943. J. Mammal., 24:369.
TYPE SPECIES: *Punomys lemminus* Osgood, 1943.
COMMENTS: Generic-level affinities uncertain (Osgood, 1943; Reig, 1980, 1984), although formally assigned to Tribe Phyllotini by Vorontsov (1959) and Olds and Anderson (1989).

Punomys lemminus Osgood, 1943. J. Mammal., 24:369.
TYPE LOCALITY: Peru, Puno Dept., San Antonio de Esquilache, 4500 m.
DISTRIBUTION: Altiplano of S Peru.

Reithrodon Waterhouse, 1837. Proc. Zool. Soc. Lond., 1837:29.
TYPE SPECIES: *Reithrodon typicus* Waterhouse, 1837 (= *Mus auritus* G. Fischer, 1814).
COMMENTS: Arranged by Hershkovitz (1955*a*) with sigmodont rodents; other evidence points to its kinship with phyllotines (Olds and Anderson, 1989; Pearson and Patton, 1976).

Reithrodon auritus (G. Fischer, 1814). Zoognosia, 3:71.
TYPE LOCALITY: Argentina, Buenos Aires Prov., pampas south of Buenos Aires, south bank of the Rio de la Plata.
DISTRIBUTION: Argentina, adjacent Chile, and Uruguay.
SYNONYMS: *caurinus, cuniculoides, currentium, evae, flammarum, hatcheri, marinus, obscurus, pachycephalus, pampanus, physodes, typicus*.
COMMENTS: The taxa associated here follow Osgood (1943) and Hershkovitz (1959*a*), whose listings of probable synonyms require substantiation, as does the formal stabilization of the proper name as *auritus* Fischer, 1814, versus *physodes* Olfers, 1818.

Reithrodontomys Giglioli, 1874. Bull. Soc. Geogr. Ital., Roma, 11:326.
TYPE SPECIES: *Reithrodon megalotis* Baird, 1858.
SYNONYMS: *Aporodon, Ochetodon*.
COMMENTS: Peromyscine. Alpha taxonomy revised by Allen (1895*a*), Howell (1914), and Hooper (1952), the last of whom framed the currently-used subgeneric division and species groupings. Distributions and species limits of Latin American harvest mice, especially those of the subgenus *Aporodon*, require renewed systematic attention. Generic relationships investigated by Carleton (1980), Hooper and Musser (1964*a*), and J. C. Patton et al. (1981). For comparative studies of morphology, see Arata (1964), Carleton (1973, 1980), and Hooper (1952, 1959); of karyology, see Carleton and Myers (1979), Engstrom et al. (1981), Hood et al. (1984), and Robbins and Baker (1980); of genic variation, see Nelson et al. (1984). A key to the species is found in Spencer and Cameron (1982).

Reithrodontomys brevirostris Goodwin, 1943. Am. Mus. Novit., 1231:1.
TYPE LOCALITY: Costa Rica, Alajuela Prov., canyons above Villa Quesada, 5000 ft.
DISTRIBUTION: Allopatric populations in highlands of NC Nicaragua (see Jones and Genoways, 1970) and C Costa Rica.
SYNONYMS: *nicaraguae*.
COMMENTS: Subgenus *Aporodon*, *mexicanus* species group. See Jones and Baldassarre (1982, Mammalian Species, 192).

Reithrodontomys burti Benson, 1939. Proc. Biol. Soc. Washington, 52:147.
TYPE LOCALITY: Mexico, Sonora, Rio Sonora, Rancho de Costa Rica.
DISTRIBUTION: WC Sonora to C Sinaloa, Mexico.
COMMENTS: Subgenus *Reithrodontomys*, *megalotis* species group. A relict species having close affinity to *R. montanus* (Hooper, 1952).

Reithrodontomys chrysopsis Merriam, 1900. Proc. Biol. Soc. Washington, 13:152.
TYPE LOCALITY: Mexico, Mexico, Volcan Popocatepetl, 11,500 ft.
DISTRIBUTION: Transverse Volcanic Range, SE Jalisco to WC Veracruz, Mexico.
SYNONYMS: *colimae, orizabae, perotensis, tolucae*.
COMMENTS: Subgenus *Reithrodontomys*, *megalotis* species group.

Reithrodontomys creper Bangs, 1902. Bull. Mus. Comp. Zool., 39:39.
TYPE LOCALITY: Panama, Chiriqui Prov., Volcan de Chiriqui, 11,000 ft.
DISTRIBUTION: Upper elevations in Cordilleras Central and Talamanca, Costa Rica, to Chiriqui region, W Panama.

COMMENTS: Subgenus *Aporodon*, *tenuirostris* species group.

Reithrodontomys darienensis Pearson, 1939. Not. Naturae, Acad. Nat. Sci. Philadelphia, 6:1.
TYPE LOCALITY: Panama, Darien Prov., Santa Cruz de Cana, upper Rio Tuyra, 2000 ft.
DISTRIBUTION: E Panama, including Azuero Peninsula, and perhaps adjacent Colombia.
COMMENTS: Subgenus *Aporodon*, *mexicanus* species group. Hooper (1952) viewed *R. darienensis* as closely related to *R. gracilis*.

Reithrodontomys fulvescens J. A. Allen, 1894. Bull. Am. Mus. Nat. Hist., 6:319.
TYPE LOCALITY: Mexico, Sonora, Oposura, 2000 ft.
DISTRIBUTION: SC Arizona to SW Missouri to WC Mississippi, USA, south through Mexico, to W Nicaragua; excluding Yucatan Peninsula and Caribbean coastal lowlands.
SYNONYMS: *amoenus, aurantius, canus, chiapensis, chrysotis, difficilis, griseoflavus, helvolus, inexpectatus, infernatis, intermedius, laceyi, meridionalis, mustelinus, nelsoni, tenuis, toltecus, tropicalis*.
COMMENTS: Subgenus *Reithrodontomys*, *fulvescens* species group. Chromosomal complement (2n=50) interpreted as primitive for the genus (Robbins and Baker, 1980). See Spencer and Cameron (1982, Mammalian Species, 174).

Reithrodontomys gracilis J. A. Allen and Chapman, 1897. Bull. Am. Mus. Nat. Hist., 9:9.
TYPE LOCALITY: Mexico, Yucatan, Chichen-Itza.
DISTRIBUTION: Yucatan Peninsula and coastal Chiapas, Mexico, south along Pacific watershed to NW Costa Rica.
SYNONYMS: *anthonyi, harrisi, insularis, pacificus*.
COMMENTS: Subgenus *Aporodon*, *mexicanus* species group. See Young and Jones (1984, Mammalian Species, 218).

Reithrodontomys hirsutus Merriam, 1901. Proc. Washington Acad. Sci., 3:553.
TYPE LOCALITY: Mexico, Jalisco, Ameca, 4000 ft.
DISTRIBUTION: SC Nayarit and NW Jalisco, Mexico.
SYNONYMS: *levipes*.
COMMENTS: Subgenus *Reithrodontomys*, *fulvescens* species group.

Reithrodontomys humulis (Audubon and Bachman, 1941). Proc. Acad. Nat. Sci. Philadelphia, 1:97.
TYPE LOCALITY: USA, South Carolina, Charleston Co., Charleston.
DISTRIBUTION: SE USA, from SE Oklahoma and E Texas east to the Atlantic seaboard, from S Maryland to C peninsular Florida.
SYNONYMS: *carolinensis, dickinsoni, impiger, lecontii, merriami, virginianus*.
COMMENTS: Subgenus *Reithrodontomys*, *megalotis* species group. Karyotype departs markedly from other species of *megalotis* group (Carleton and Myers, 1979; Engstrom et al., 1981).

Reithrodontomys megalotis (Baird, 1858). Mammalia *in* Repts. U.S. Expl. Surv., 8(1):451.
TYPE LOCALITY: Mexico-USA boundary region, between Janos, Chihuahua, and San Luis Springs, Grant Co., New Mexico.
DISTRIBUTION: SC British Columbia and SE Alberta, Canada; W and NC USA; south to N Baja California and through interior Mexico to central Oaxaca.
SYNONYMS: *alticolus, amoles, arizonensis, aztecus, caryi, catalinae, cinereus, deserti, distichlis, dychei, hooperi, klamathensis, limicola, longicaudus, nebrascensis, nigrescens, pallidus, pectoralis, peninsulae, ravus, santacruzae, saturatus, sestinensis*.
COMMENTS: Subgenus *Reithrodontomys*, *megalotis* species group. Formerly included *R. zacatecae*, which Hood et al. (1984) identified as a separate species. The substantial chromosomal variation reported (see Engstrom et al., 1981) cautions that other species may be lumped under *R. megalotis*. See Webster and Jones (1982*a*, Mammalian Species, 167).

Reithrodontomys mexicanus (Saussure, 1860). Rev. Mag. Zool. Paris, Ser. 2, 12:109.
TYPE LOCALITY: Mexico, Veracruz, Mirador (as restricted by Hooper, 1952:140).
DISTRIBUTION: S Tamaulipas and WC Michoacan, Mexico, south through Middle American highlands to W Panama; Andes of W Colombia and N Ecuador.

SYNONYMS: *cherrii, costaricensis, eremicus, garichensis, goldmani, howelli, jalapae, lucifrons, milleri, minusculus, ocotepequensis, orinus, potrerograndei, riparius, scansor, soederstroemi.*
COMMENTS: Subgenus *Aporodon, mexicanus* species group.

Reithrodontomys microdon Merriam, 1901. Proc. Washington Acad. Sci., 3:548.
TYPE LOCALITY: Guatemala, Huehuetenango Dept., Todos Santos, 10,000 ft.
DISTRIBUTION: Isolated pockets in highlands of N Michoacan and Distrito Federal, N Oaxaca, and C Chiapas, Mexico, and WC Guatemala.
SYNONYMS: *albilabris, wagneri.*
COMMENTS: Subgenus *Aporodon, tenuirostris* species group.

Reithrodontomys montanus (Baird, 1855). Proc. Acad. Nat. Sci. Philadelphia, 7:335.
TYPE LOCALITY: USA, Colorado, Saguache Co., upper part of the San Luis Valley (as restricted by Allen, 1895*a*:125). See Armstrong's (1972:190) lucid summary of the contradictory references to the type locality.
DISTRIBUTION: High Plains of C USA, from W South Dakota and E Wyoming to EC Texas and extreme SE Arizona; NE Sonora and Chihuahua to N Durango, Mexico.
SYNONYMS: *albescens, griseus.*
COMMENTS: Subgenus *Reithrodontomys, megalotis* species group. See Wilkins (1986, Mammalian Species, 257).

Reithrodontomys paradoxus Jones and Genoways, 1970. Occas. Pap. W. Found. Vert. Zool., 2:12.
TYPE LOCALITY: Nicaragua, Carazo Dept., 3 mi NNW Diriamba, about 660 m.
DISTRIBUTION: Isolated records from SW Nicaragua and WC Costa Rica.
COMMENTS: Subgenus *Aporodon, mexicanus* species group. Morphologically closest to *R. brevirostris*; extent of distribution unknown. See Jones and Baldassarre (1982, Mammalian Species, 192).

Reithrodontomys raviventris Dixon, 1908. Proc. Biol. Soc. Washington, 21:197.
TYPE LOCALITY: USA, California, San Mateo Co., Redwood City.
DISTRIBUTION: Salt marshes around San Francisco Bay, California, USA.
STATUS: U.S. ESA and IUCN - Endangered.
SYNONYMS: *halicoetes.*
COMMENTS: Subgenus *Reithrodontomys, megalotis* species group. Morphological and chromosomal differentiation between the subspecies *raviventris* and *halicoetes* thought to represent terminal stages of speciation (Fisler, 1965; Shellhammer, 1967). Banded chromosomes and genic data reveal sister-group relationship to *R. montanus* (Hood et al., 1984; Nelson et al., 1984). See Shellhammer (1982, Mammalian Species, 169).

Reithrodontomys rodriguezi Goodwin, 1943. Am. Mus. Novit., 1231:1.
TYPE LOCALITY: Costa Rica, Cartago Prov., Volcan de Irazu, 9000 ft.
DISTRIBUTION: Reported only from Volcan de Irazu, Costa Rica.
COMMENTS: Subgenus *Aporodon, tenuirostris* species group.

Reithrodontomys spectabilis Jones and Lawlor, 1965. Univ. Kansas Publ., Mus. Nat. Hist., 16:413.
TYPE LOCALITY: Mexico, Quintana Roo, Isla Cozumel, 2.5 km N San Miguel.
DISTRIBUTION: Restricted to Cozumel Island, Mexico.
COMMENTS: Subgenus *Aporodon, mexicanus* species group. A large insular species, perhaps sharing common ancestry with *R. gracilis* (Jones and Lawlor, 1965). See Jones (1982, Mammalian Species, 193).

Reithrodontomys sumichrasti (Saussure, 1861). Rev. Mag. Zool. Paris, Ser. 2, 13:3.
TYPE LOCALITY: Mexico, Veracruz, Mirador (as restricted by Hooper, 1952:72).
DISTRIBUTION: Allopatric segments in Middle American highlands: SW Jalisco and S San Luis Potosi to C Guerrero and EC Oaxaca, Mexico; C Chiapas, Mexico, to NC Nicaragua; C Costa Rica to W Panama.
SYNONYMS: *alleni, australis, dorsalis, luteolus, modestus, nerterus, otus, rufescens, seclusus, underwoodi, vulcanius.*
COMMENTS: Subgenus *Reithrodontomys, megalotis* species group.

Reithrodontomys tenuirostris Merriam, 1901. Proc. Washington Acad. Sci., 3:547.
TYPE LOCALITY: Guatemala, Huehuetenango Dept., Todos Santos, 10,000 ft.

DISTRIBUTION: Mountains of S Chiapas, Mexico, and C Guatemala.
SYNONYMS: *aureus*.
COMMENTS: Subgenus *Aporodon*, *tenuirostris* species group.

Reithrodontomys zacatecae Merriam, 1901. Proc. Washington Acad. Sci., 3:557.
TYPE LOCALITY: Mexico, Zacatecas, Sierra de Valparaiso.
DISTRIBUTION: W Chihuahua to WC Michoacan, Mexico.
SYNONYMS: *obscurus*.
COMMENTS: Subgenus *Reithrodontomys*, *megalotis* species group. Hooper (1952) noted the phenetic distinction of *zacatecae* from *R. megalotis saturatus* in Jalisco and Michoacan but elected to retain it as a subspecies. Karyotypic divergence of *zacatecae* reported by Hood et al. (1984), who proposed its elevation to species. Distributional limits, morphological discrimination, and assignment of species-group synonyms deserve amplification.

Rhagomys Thomas, 1917. Ann. Mag. Nat. Hist., ser. 8, 20:192.
TYPE SPECIES: *Hesperomys rufescens* Thomas, 1886.
COMMENTS: Generic affinities uncertain: originally allied with *Oryzomys*-*Oecomys* (Thomas, 1917*c*), or included among oryzomyine genera (Tate, 1932*f*), or listed as Sigmodontinae *incertae sedis* (Reig, 1980, 1984).

Rhagomys rufescens (Thomas, 1886). Ann. Mag. Nat. Hist., ser. 5, 17:250.
TYPE LOCALITY: Brazil, Rio de Janeiro.
DISTRIBUTION: Known only from the type locality.
COMMENTS: Known only by two specimens.

Rheomys Thomas 1906. Ann. Mag. Nat. Hist., ser. 7, 17:421.
TYPE SPECIES: *Rheomys underwoodi* Thomas, 1906.
SYNONYMS: *Neorheomys*.
COMMENTS: Ichthyomyine. All named forms suggested as conspecific under *trichotis* by Hershkovitz (1966*b*) but Voss (1988) recognized four species. Formerly included *trichotis* (Tate, 1932*h*), a species later designated as the type of *Chibchanomys* (Voss, 1988). *Neorheomys*, diagnosed as a subgenus (Goodwin, 1959*a*), was placed in full synonymy by Voss (1988). Separation from *Ichthyomys* questioned by Ellerman (1941) and Hall (1981); relationships and generic stature illuminated by Voss (1988).

Rheomys mexicanus Goodwin, 1959. Am. Mus. Novit., 1967:4.
TYPE LOCALITY: Mexico, Oaxaca, Dist. Miahuatlan, San José Lachiguirí, 4000 ft.
DISTRIBUTION: Oaxaca, Mexico.
COMMENTS: Viewed as closely related to *R. underwoodi* (Voss, 1988). Type species of Goodwin's (1959*a*) *Neorheomys*.

Rheomys raptor Goldman, 1912. Smithson. Misc. Coll., 60:7.
TYPE LOCALITY: Panama, Darién Prov., Cerro Pirre, near headwaters of Río Limon, 4500 ft.
DISTRIBUTION: Costa Rica and Panama.
SYNONYMS: *hartmanni*.
COMMENTS: Treated as a subspecies of *Rheomys trichotis* (Cabrera, 1961), but species status affirmed by Voss (1988). Enders (1939) described *hartmanni* as a species; recognized as such (e.g., Hall, 1981) until Voss (1988) allocated the form to a subspecies of *raptor*.

Rheomys thomasi Dickey, 1928. Proc. Biol. Soc. Washington, 41:11.
TYPE LOCALITY: El Salvador, San Miguel Dept., Finca San Felipe, Cerro Cacaguatique, 3500 ft.
DISTRIBUTION: S Mexico, Guatemala, and El Salvador.
SYNONYMS: *chiapensis*, *stirtoni*.
COMMENTS: Of two named subspecies, Voss (1988) retained only *stirtoni* as a race.

Rheomys underwoodi Thomas, 1906. Ann. Mag. Nat. Hist., ser. 7, 17:422.
TYPE LOCALITY: Costa Rica, Cartago Prov., near Tres Ríos.
DISTRIBUTION: C Costa Rica and W Panama.
COMMENTS: Viewed as closely related to *R. mexicanus* (Voss, 1988).

Rhipidomys Tschudi, 1844. Arch. Naturgesch., 10, 1:252.

TYPE SPECIES: *Hesperomys leucodactylus* Tschudi, 1844.

COMMENTS: Thomasomyine. No critical synopsis of valid species exists apart from the original descriptions. Compared to the overly lumped classification of Cabrera (1961), Handley's (1976) regional listing reveals the greater species diversity within the genus, but how these Venezuelan forms relate to *Rhipidomys* populations inhabiting other parts of the Andes and Amazonia remains unknown. The following enumeration must be regarded as highly provisional and attempts to amalgamate the differing viewpoints of Cabrera (1961), Gyldenstolpe (1932), and Handley (1976). Species distributions are in general poorly documented. Preliminary karyotypic surveys provided by Gardner and Patton (1976) and Zanchin et al. (1992*a*).

Rhipidomys austrinus Thomas, 1921. Ann. Mag. Nat. Hist., ser. 9, 7:183.

TYPE LOCALITY: Argentina, Jujuy Prov., Sierra de Santa Barbara, Sunchal, 1200 m.

DISTRIBUTION: E Andean slopes of SC Bolivia and NW Argentina.

SYNONYMS: *collinus*.

COMMENTS: Whether this form intergrades with *R. leucodactylus*, as usually identified (e.g., Mares et al., 1989*b*), or with *R. couesi* deserves specimen-based corroboration.

Rhipidomys caucensis J. A. Allen, 1913. Bull. Am. Mus. Nat. Hist., 32:601.

TYPE LOCALITY: Colombia, Cauca Dept., Munchique, 8225 ft.

DISTRIBUTION: W Andes of Colombia.

Rhipidomys couesi (J. A. Allen and Chapman, 1893). Bull. Am. Mus. Nat. Hist., 5:214.

TYPE LOCALITY: Trinidad, Princes Town.

DISTRIBUTION: Trinidad, Venezuela, Colombia, Ecuador, and Peru.

SYNONYMS: *cumananus*, *goodfellowi*, *modicus*.

COMMENTS: Cabrera (1961) identified *couesi* as a subspecies of *R. sclateri*. As emphasized by its original describer Thomas (1887*b*), *sclateri* closely resembles Peruvian *R. leucodactylus*, and, in both Venezuela (see Handley, 1976) and Peru, *R. couesi* and *R. leucodactylus* are morphologically distinct.

Rhipidomys fulviventer Thomas, 1896. Ann. Mag. Nat. Hist., ser. 6, 18:304.

TYPE LOCALITY: Colombia, Cundinamarca Dept., Aqua Dulce, 2400 ft.

DISTRIBUTION: Andes of W Venezuela and Colombia.

SYNONYMS: *elatturus*, *tenuicauda*.

COMMENTS: Listed as a subspecies of *R. latimanus* by Cabrera (1961) but clearly a species separate from the *latimanus-venezuelae* complex in Venezuela (Handley, 1976). Compared with *R. wetzeli* by Gardner (1989).

Rhipidomys latimanus (Tomes, 1860). Proc. Zool. Soc. Lond., 1860:213.

TYPE LOCALITY: Ecuador, Chimborazo Prov., Pallatanga, 1485 m.

DISTRIBUTION: Montane forests of C and W Colombia and Ecuador.

SYNONYMS: *cocalensis*, *microtis*, *mollissimus*, *pictor*.

COMMENTS: This species may encompass Andean populations in E Colombia and Venezuela identified as *R. venezuelae* (Handley, 1976); its status with regard to *R. scandens* and *R. mastacalis* also needs clarification. Cabrera (1961) arranged *fulviventer* and *venustus* as subspecies, but these two forms are distinct from one another and from the *latimanus-venezuelae* complex in Venezuela (Handley, 1976). Karyotype reported by Gardner and Patton (1976).

Rhipidomys leucodactylus (Tschudi, 1844). Fauna Peruana, 1:183.

TYPE LOCALITY: Eastern Peru; Cabrera (1961) suggested the type's origin from the upper Rio Huallaga, a restriction which should be considered by a future revisor.

DISTRIBUTION: Guianas, S Venezuela, N Brazil, Ecuador, and Peru.

SYNONYMS: *aratayae*, *bovallii*, *equatoris*, *lucullus*, *rex*, *sclateri*.

COMMENTS: Thomas (1887*b*) remarked that *sclateri*, named from British Guiana, was the eastern counterpart of Peruvian *R. leucodactylus*; Guillotin and Petter (1984) described *aratayae*, a subspecies of *R. leucodactylus*, from French Guiana.

Rhipidomys macconnelli de Winton, 1900. Trans. Linn. Soc. Lond., 8:52.

TYPE LOCALITY: Venezuela, Bolivar, Mount Roraima, 2600 m.

DISTRIBUTION: S Venezuela, and perhaps northernmost Brazil and adjacent Guyana.

SYNONYMS: *subnubis.*

COMMENTS: Sometimes misallocated as a species of *Thomasomys* (e.g., Gyldenstolpe, 1932) but see Hershkovitz (1959*b*).

Rhipidomys mastacalis (Lund, 1840). K. Dansk. Vid. Selsk. Naturv. Math. Afhandl., p. 24.

TYPE LOCALITY: Brazil, Minas Gerais, Rio das Velhas, Lagoa Santa.

DISTRIBUTION: C and E Brazil, and perhaps Amazonian regions of Venezuela, Colombia, Ecuador, and Peru.

SYNONYMS: *cearanus, emiliae, macrurus, maculipes, yuruanus.*

COMMENTS: Degree of differentiation from the Andean forms *latimanus* and *venezuelae* unresolved; occurs in close proximity to a smaller species here listed as *R. nitela.* Distributional limits unclear. Zanchin et al. (1992*a*) considered *cearanus* as a species separate from *R. mastacalis.*

Rhipidomys nitela Thomas, 1901. Ann. Mag. Nat. Hist., ser. 7, 8:148.

TYPE LOCALITY: Guyana, Kanuku Mountains, Kwaimatta, 240 ft.

DISTRIBUTION: S Venezuela, Guianas, NC Brazil.

SYNONYMS: *fervidus, milleri, tobagi.*

COMMENTS: Previously listed either as a race of *R. venezuelae* (Gyldenstolpe, 1932) or *R. mastacalis* (Cabrera, 1961), but distinct from true *venezuelae* in Venezuela (see Handley, 1976, who used the species name *mastacalis*) and from *R. mastacalis* in N Brazil.

Rhipidomys ochrogaster J. A. Allen, 1901. Bull. Am. Mus. Nat. Hist., 14:43

TYPE LOCALITY: Peru, Puno Dept., valley of Rio Inambari, Inca Mines, 6000 ft.

DISTRIBUTION: SE Peru.

COMMENTS: Retained as a species until judged the same as *R. l. leucodactylus* by Cabrera (1961). Allen (1901*b*), however, specifically contrasted his new species with Peruvian *R. leucodactylus.* Geographic range and specific allies uncertain.

Rhipidomys scandens Goldman, 1913. Smithson. Misc. Coll., 60(22):8.

TYPE LOCALITY: Panama, Darien Prov., Mount Pirre, headwaters of Rio Limon, 5000 ft.

DISTRIBUTION: Known only from the Serrania del Darien and Serrania de Pirre, easternmost Panama.

COMMENTS: See remarks under *R. latimanus.*

Rhipidomys venezuelae Thomas, 1896. Ann. Mag. Nat. Hist., ser. 6, 18:303.

TYPE LOCALITY: Venezuela, Merida, Merida, 1630 m.

DISTRIBUTION: Mountains of N and W Venezuela and E Colombia.

COMMENTS: This form may represent the E Andean complement of *R. latimanus.*

Rhipidomys venustus Thomas, 1900. Ann. Mag. Nat. Hist., ser. 7, 5:152.

TYPE LOCALITY: Venezuela, Merida, Las Vegas del Chama, 1400 m.

DISTRIBUTION: Mountains of N Venezuela.

COMMENTS: Placed as a subspecies of *R. latimanus* by Cabrera (1961) but distinct from the *latimanus-venezuelae* complex in Venezuela (Handley, 1976).

Rhipidomys wetzeli Gardner, 1989. *In* Eisenberg, Advances in Neotropical Mammalogy, p. 417.

TYPE LOCALITY: Venezuela, Territorio Federal Amazonas, Cerro de la Neblina, 1800 m; 00°50′N, 65°58′W.

DISTRIBUTION: Highlands in S Venezuela.

COMMENTS: A small species related to the *R. fulviventer* group (see Gardner, 1989).

Scapteromys Waterhouse, 1837. Proc. Zool. Soc. Lond., 1837:20.

TYPE SPECIES: *Mus tumidus* Waterhouse, 1837.

COMMENTS: Scapteromyine. Hershkovitz (1966*c*) removed *tomentosus* and *fronto* to *Kunsia,* the presumed cognate genus of *Scapteromys,* and Massoia (1980*a*) transferred *labiosus* to *Bibimys.* Karyology summarized by Gardner and Patton (1976).

Scapteromys tumidus (Waterhouse, 1837). Proc. Zool. Soc. Lond., 1837:15.

TYPE LOCALITY: Uruguay, Maldonado.

DISTRIBUTION: Southernmost Brazil (Rio Grande do Sul), Uruguay and adjacent Argentina, E Paraguay.

SYNONYMS: *aquaticus.*

COMMENTS: Distribution and morphological identity reviewed by Hershkovitz (1966*c*). Massoia and Fornes (1964) realigned *aquaticus* as a subspecies of *S. tumidus*. The taxonomic significance of the reported karyotypic differences between *aquaticus* and *tumidus* warrants re-examination (see Brum et al., 1973, and Gentile de Fronza, 1970).

Scolomys Anthony, 1924. Am. Mus. Novit., 139:1.
TYPE SPECIES: *Scolomys melanops* Anthony, 1924.
COMMENTS: Rare in collections. Phylogenetic affinity not critically explored, although listed in tribe Oryzomyini (Reig, 1984).

Scolomys melanops Anthony, 1924. Am. Mus. Novit., 139:2.
TYPE LOCALITY: Ecuador, Pastaza Prov., Mera, 3800 ft.
DISTRIBUTION: Known only from the vicinity of the type locality.

Scolomys ucayalensis Pacheco, 1991. Publ. Mus. Hist. nat., Ser. A Zool., Univ. Nac. Mayor de San Marcos, 37:1.
TYPE LOCALITY: Peru, Loreto Dept., 2.8 km E Jenaro Herrera, right bank of Ucayali River, 135 m; 79°39'W, 04°52'S.
DISTRIBUTION: Known only from the type locality.
COMMENTS: Known only by two specimens recovered from dense undergrowth in disturbed lowland rain forest.

Scotinomys Thomas, 1913. Ann. Mag. Nat. Hist., ser. 8, 11:408.
TYPE SPECIES: *Hesperomys teguina* Alston, 1877.
COMMENTS: Peromyscine. Until diagnosed by Thomas, forms associated here had been affiliated with *Akodon* (e.g., Bangs, 1902). Revised by Hooper (1972); for other aspects of biology and systematics, see Hooper and Carleton (1976), Carleton et al. (1975), and Rogers and Heske (1984). Broadly affiliated with neotomine-peromyscines *sensu* Hooper and Musser (1964*a*); common ancestry with *Baiomys* proposed on the basis of morphological traits (Hooper, 1960; Hooper and Musser, 1964*b*; Carleton, 1980), a relationship weakened by karyological banding data (Rogers and Heske, 1984).

Scotinomys teguina (Alston, 1877). Proc. Zool. Soc. Lond., 1876:755 [1877].
TYPE LOCALITY: Guatemala, Alta Verapaz Dept., Coban.
DISTRIBUTION: Intermediate elevations of Middle America from E Oaxaca, Mexico, to W Panama.
SYNONYMS: *apricus, cacabatus, endersi, episcopi, escazuensis, garichensis, irazu, leridensis, rufoniger, stenopygius, subnubilis.*
COMMENTS: Hooper (1972) retained *apricus, irazu,* and *rufoniger* as subspecies.

Scotinomys xerampelinus (Bangs, 1902). Bull. Mus. Comp. Zool., 39:41.
TYPE LOCALITY: Panama, Chiriqui Prov., Volcan de Chiriqui, 10,300 ft.
DISTRIBUTION: High elevations in Cordilleras Central and Talamancae of Costa Rica to Volcan Chiriqui region in W Panama.
SYNONYMS: *harrisi, longipilosus.*
COMMENTS: Hooper (1972) synonymized *harrisi* and *longipilosus* under *S. xerampelinus*, without subspecies.

Sigmodon Say and Ord, 1825. J. Acad. Nat. Sci. Philadelphia, 4(2):352.
TYPE SPECIES: *Sigmodon hispidus* Say and Ord, 1825.
SYNONYMS: *Deilemys, Lasiomys, Sigmomys.*
COMMENTS: North American forms revised by Bailey (1902), with subsequent partial revisions by Baker (1969), Voss (1992), and Zimmerman (1970); further alpha-taxonomic study dearly needed. Standard karyology summarized by Zimmerman (1970), banded karyotypes by Elder (1980) and Elder and Lee (1985), albumin differentiation by Fuller et al. (1984). Skin-skull morphology (Bailey, 1902; Baker, 1969) and karyology (Elder and Lee, 1985; Zimmerman, 1970) have yielded conflicting pictures of species-group associations. *Sigmomys*, type species *Sigmodon alstoni*, variously treated as a distinct genus (Handley, 1976), a subgenus of *Sigmodon* (Husson, 1978), or a full synonym (Cabrera, 1961). Hershkovitz (1955*a*) arrayed *Sigmodon* with *Holochilus, Neotomys,* and *Reithrodon* as the sigmodont group, but other

evidence has questioned the close affinity of each to *Sigmodon* (see those generic accounts).

Sigmodon alleni Bailey, 1902. Proc. Biol. Soc. Washington, 15:112.
TYPE LOCALITY: Mexico, Jalisco, Mascota, Mineral San Sebastian.
DISTRIBUTION: W Mexico, from S Sinaloa to S Oaxaca.
SYNONYMS: *guerrerensis, macdougalli, macrodon, minor, planifrons, setzeri, vulcani.*
COMMENTS: See Shump and Baker (1978*a*, Mammalian Species, 95).

Sigmodon alstoni (Thomas, 1881). Proc. Zool. Soc. Lond., 1880:691 [1881].
TYPE LOCALITY: Venezuela, Sucre, Cumaná.
DISTRIBUTION: Intermittently distributed in savannas over NE Colombia, N and E Venezuela, Guyana, Surinam, and N Brazil. Range amplified by Handley (1976), Husson (1978), and Voss (1992).
SYNONYMS: *savannarum, venester.*

Sigmodon arizonae Mearns, 1890. Bull. Am. Mus. Nat. Hist., 2:287.
TYPE LOCALITY: USA, Arizona, Yavapai Co., Fort Verde.
DISTRIBUTION: Extreme SE California and SC Arizona, USA, south in W Mexico to Nayarit.
STATUS: IUCN - Endangered as *S. a. plenus.*
SYNONYMS: *cienegae, jacksoni, major, plenus.*
COMMENTS: Chromosomal and morphological differences have revealed the specific distinction of *S. arizonae* and *S. hispidus* (see Zimmerman, 1970; Severinghaus and Hoffmeister, 1978; Elder and Lee, 1985).

Sigmodon fulviventer J. A. Allen, 1889. Bull. Am. Mus. Nat. Hist., 2:180.
TYPE LOCALITY: Mexico, Zacatecas, Zacatecas.
DISTRIBUTION: SE Arizona and WC New Mexico, USA, south through interior Mexico to Guanajuato and NW Michoacan.
SYNONYMS: *dalquesti, goldmani, melanotis, minimus, woodi.*
COMMENTS: See Baker and Shump (1978*a*, Mammalian Species, 94).

Sigmodon hispidus Say and Ord, 1825. J. Acad. Nat. Sci. Philadelphia, 42:354.
TYPE LOCALITY: USA, Florida, St. Johns River.
DISTRIBUTION: SE USA, from S Nebraska to C Virginia south to SE Arizona and peninsular Florida; interior and E Mexico through Middle America to C Panama; in South America to N Colombia and N Venezuela.
SYNONYMS: *alfredi, austerulus, berlandieri, bogotensis, borucae, chiriquensis, confinis, eremicus, exsputus, fervidus, floridanus, furvus, griseus, hirsutus, insulicola, komareki, littoralis, microdon, obvelatus, pallidus, sanctaemartae, saturatus, solus, spadicipygus, texianus, toltecus, tonalensis, villae, virginianus, zanjonensis.*
COMMENTS: See comments under *S. arizonae, S. inopinatus, S. mascotensis,* and *S. peruanus,* forms formerly ranked as subspecies of *S. hispidus.* The homogeneity of species-group taxa assembled here must be viewed skeptically. The synonymy of still other forms with *S. mascotensis* is likely. See Cameron and Spencer (1981, Mammalian Species, 158).

Sigmodon inopinatus Anthony, 1924. Am. Mus. Novit., 114:3.
TYPE LOCALITY: Ecuador, Chimborazo Prov., Mt. Chimborazo, Urbina, 11,400 ft.
DISTRIBUTION: Known only from high Andes in Azuay and Chimborazo provinces, Ecuador.
COMMENTS: Lumped under *S. hispidus* by Cabrera (1961) following the footnoted opinion of Hershkovitz (1955*a*); reinstated as a species by Voss (1992).

Sigmodon leucotis Bailey, 1902. Proc. Biol. Soc. Washington, 15:115.
TYPE LOCALITY: Mexico, Zacatecas, Sierra de Valparaiso, 2653 m.
DISTRIBUTION: Interior Mexico, from SW Chihuahua and S Nuevo Leon to C Oaxaca.
SYNONYMS: *alticola, amoles.*
COMMENTS: See Shump and Baker (1978*b*, Mammalian Species, 96).

Sigmodon mascotensis J. A. Allen, 1897. Bull. Am. Mus. Nat. Hist., 9:54.
TYPE LOCALITY: Mexico, Jalisco, Mascota, Mineral San Sebastian.
DISTRIBUTION: W Mexico, from S Nayarit south to E Oaxaca.
SYNONYMS: *atratus, colimae, inexoratus, ischyrus.*

COMMENTS: Removed from *S. hispidus* and reinstated as species by Zimmerman (1970); also see Severinghaus and Hoffmeister (1978) and Elder and Lee (1985). Limits of geographic distribution need refinement.

Sigmodon ochrognathus Bailey, 1902. Proc. Biol. Soc. Washington, 15:115.
TYPE LOCALITY: USA, Texas, Brewster Co., Chisos Mountains, 8000 ft.
DISTRIBUTION: SE Arizona, extreme SW New Mexico, and Transpecos, Texas, USA, south to C Durango, Mexico. Northern outlier population may persist in Guadalupe Mtns, Transpecos, Texas (see Stangl and Dalquest, 1991).
SYNONYMS: *baileyi, madrensis, montanus.*
COMMENTS: See Baker and Shump (1978*b*, Mammalian Species, 97).

Sigmodon peruanus J. A. Allen, 1897. Bull. Am. Mus. Nat. Hist., 9:118.
TYPE LOCALITY: Peru, La Libertad Dept., Trujillo, 200 ft.
DISTRIBUTION: Pacific coastal plain and contiguous Andean foothills of W Ecuador and NW Peru.
SYNONYMS: *chonensis, lonnbergi, puna, simonsi.*
COMMENTS: Lumped under *S. hispidus* by Cabrera (1961) following the footnoted opinion of Hershkovitz (1955*a*); reinstated as a species by Voss (1992).

Sigmodontomys J. A. Allen, 1897. Bull. Am. Mus. Nat. Hist., 9:38.
TYPE SPECIES: *Sigmodontomys alfari* J. A. Allen, 1897.
COMMENTS: Oryzomyine. Named as a genus but later viewed as a synonym of *Nectomys* (Ellerman, 1941; Gyldenstolpe, 1932), usually ranked as a subgenus (Hershkovitz, 1944). Hershkovitz (1944:71) stressed the weakness of their association: "The apparent relationship of *Sigmodontomys* to *Nectomys*...is probably attributable to an independant development...from the common oryzomyine stock rather than to a divergence from a...more recent *Nectomys*-like stock," an assessment later corroborated by other data (Gardner and Patton, 1976; Hooper and Musser, 1964*a*). Gardner and Patton (1976) formally transferred *Sigmodontomys* to a subgenus of *Oryzomys*, and we here provisionally list it as genus, noting the need to refine its relationships to *Oryzomys* proper.

Sigmodontomys alfari J. A. Allen, 1897. Bull. Am. Mus. Nat. Hist., 9:39.
TYPE LOCALITY: Costa Rica, Limón Prov., Jimenez, 700 ft.
DISTRIBUTION: Lowland forest from E Honduras to Panama; C and W Colombia to NW Venezuela and NW Ecuador.
SYNONYMS: *barbacoas, efficax, esmeraldarum, ochraceus, ochrinus, russulus.*
COMMENTS: Karyotype reported by Gardner and Patton (1976).

Sigmodontomys aphrastus (Harris, 1932). Occas. Pap. Mus. Zool., Univ. Michigan, 248:5.
TYPE LOCALITY: Costa Rica, San José Prov., San Joaquin de Dota, 4000 ft.
DISTRIBUTION: Known only from the type locality and Chiriqui Prov., W Panama.
COMMENTS: Usually listed as a species of *Oryzomys* of uncertain relationship (Hall, 1981). Assignment to *Sigmodontomys* tentative following the observations of Ray (1962). Distribution meagerly documented, known by only three specimens.

Thalpomys Thomas, 1916. Ann. Mag. Nat. Hist., ser. 8, 18:339.
TYPE SPECIES: *Thalpomys lasiotis* Thomas, 1916 (not *Mus lasiotis* of Lund, 1841).
COMMENTS: Akodontine. Recognized as a genus, as initially diagnosed, by Gyldenstolpe (1932); reclassified as a subgenus of *Akodon* by Ellerman (1941) and so observed by Cabrera (1961). See Hershkovitz (1990*a*) for availability of genus-group name, its differentiating characters from typical *Akodon*, and definition of included species.

Thalpomys cerradensis Hershkovitz, 1990. J. Nat. Hist., 24:777.
TYPE LOCALITY: Brazil, Distrito Federal, Parque Nacional de Brasilia, 1100 m.
DISTRIBUTION: Cerrado of C Brazil.

Thalpomys lasiotis Thomas, 1916. Ann. Mag. Nat. Hist., ser. 8, 18:339.
TYPE LOCALITY: Brazil, Minas Gerais, Lagoa Santa.
DISTRIBUTION: Cerrado of C Brazil.
SYNONYMS: *reinhardti.*

COMMENTS: For synonymy of the replacement name *Akodon reinhardti*, see Hershkovitz (1990*a*).

Thomasomys Coues, 1884. Am. Nat., 18:1275.

TYPE SPECIES: *Hesperomys cinereus* Thomas, 1882.

SYNONYMS: *Erioryzomys, Inomys*.

COMMENTS: Thomasomyine. A complex genus whose nomenclatural history is intertwined with *Aepeomys, Delomys*, and *Wilfredomys*, taxa which have been used as subgenera (e.g., Ellerman, 1941; Cabrera, 1961) or as genera (see their accounts). Comparative anatomical studies involving some *Thomasomys* include Carleton (1973), Hooper and Musser (1964*a*), and Voss and Linzey (1981); chromosomal numbers of several species published by Gardner and Patton (1976). Revisionary studies required to critically overhaul species taxonomy and to place ranking of various genus-group taxa in a phylogenetic context. Species recognized here basically follow Cabrera (1961).

Thomasomys aureus (Tomes, 1860). Proc. Zool. Soc. Lond., 1860:219.

TYPE LOCALITY: Ecuador, Chimborazo Prov., Pallatanga, 4950 ft.

DISTRIBUTION: Andean forests of NC Venezuela, C and W Colombia, Ecuador, and N Peru.

SYNONYMS: *altorum, nicefori, popayanus, praetor, princeps*.

COMMENTS: Highly differentiated morphologically from other species of *Thomasomys* (see Carleton, 1973; Hooper and Musser, 1964*a*; Voss and Linzey, 1981). Karyotype reported by Gardner and Patton (1976).

Thomasomys baeops (Thomas, 1899). Ann. Mag Nat. Hist., ser. 7, 3:152.

TYPE LOCALITY: Ecuador, El Oro Prov., Chilla Valley, Rio Pita, 3500 m.

DISTRIBUTION: W Andes of Ecuador.

Thomasomys bombycinus Anthony, 1925. Am. Mus. Novit., 178:1.

TYPE LOCALITY: Colombia, Antioquia Dept., Paramillo, 12,500 ft.

DISTRIBUTION: Cordillera Occidental of Colombia.

Thomasomys cinereiventer J. A. Allen, 1912. Bull. Am. Mus. Nat. Hist., 31:80.

TYPE LOCALITY: Colombia, Cauca Dept., 64 km W Popayan, 3070 m.

DISTRIBUTION: Upper Andean elevations in Colombia and Ecuador.

SYNONYMS: *contradictus, dispar, erro*.

Thomasomys cinereus (Thomas, 1882). Proc. Zool. Soc. Lond., 1882:108.

TYPE LOCALITY: Peru, Cajamarca Dept., Cutervo, 9200 ft.

DISTRIBUTION: SW Ecuador and N Peru.

SYNONYMS: *caudivarius*.

Thomasomys daphne Thomas, 1917. Smithson. Misc. Coll., 68(4):2.

TYPE LOCALITY: Peru, Cuzco Dept., Ocabamba Valley, 9100 ft.

DISTRIBUTION: S Peru to C Bolivia.

SYNONYMS: *australis*.

Thomasomys eleusis Thomas, 1926. Ann. Mag. Nat. Hist., ser. 9, 17:614.

TYPE LOCALITY: Peru, Amazonas Dept., mountains east of Balsas, Tambo Jenes, 12,000 ft.

DISTRIBUTION: NC Peru.

Thomasomys gracilis Thomas, 1917. Smithson. Misc. Coll., 68(4):2.

TYPE LOCALITY: Peru, Cuzco Dept., Machu Picchu, 12,000 ft.

DISTRIBUTION: Andean Ecuador to SE Peru.

SYNONYMS: *cinnameus, hudsoni*.

Thomasomys hylophilus Osgood, 1912. Field Mus. Nat. Hist. Publ., Zool. Ser., 10(5):50.

TYPE LOCALITY: Colombia, Santander Dept., Upper Rio Tachira, Paramo de Tama, 7500 ft.

DISTRIBUTION: Cordillera Oriental of Colombia, Cordillera de Merida, W Venezuela (see Handley, 1976).

Thomasomys incanus (Thomas, 1894). Ann. Mag. Nat. Hist., ser. 6, 14:350.

TYPE LOCALITY: Peru, Junín Dept., Vitoc Valley.

DISTRIBUTION: Andes of C Peru.

SYNONYMS: *fraternus*.

COMMENTS: Type species of *Inomys*.

Thomasomys ischyurus Osgood, 1914. Field Mus. Nat. Hist. Publ., Zool. Ser., 10(12):162.
TYPE LOCALITY: Peru, Amazonas Dept., 65 km E Chachapoyas, near Uchco, Tambo Almirante, 5000 ft.
DISTRIBUTION: N to C Peru.

Thomasomys kalinowskii (Thomas, 1894). Ann. Mag. Nat. Hist., ser. 6, 14:349.
TYPE LOCALITY: Peru, Junín Dept., Vitoc Valley.
DISTRIBUTION: Andes of C Peru.
COMMENTS: Chromosomal formula reported by Gardner and Patton (1976).

Thomasomys ladewi Anthony, 1926. Am. Mus. Novit., 239:1.
TYPE LOCALITY: Bolivia, La Paz Dept., Rio Aceramarca, 10,800 ft.
DISTRIBUTION: Andes of NW Bolivia.

Thomasomys laniger (Thomas, 1895). Ann. Mag. Nat. Hist., ser. 6, 16:59.
TYPE LOCALITY: Colombia, Cundinamarca Dept., Bogota Region, 8750 ft.
DISTRIBUTION: Andes of C Colombia and adjacent W Venezuela (see Handley, 1976).
SYNONYMS: *emeritus*.

Thomasomys monochromos Bangs, 1900. Proc. New England Zool. Club, 1:97.
TYPE LOCALITY: Colombia, Magdalena Dept., Sierra Nevada de Santa Marta, Macotama, 3300 m.
DISTRIBUTION: Extreme NE Colombia.
COMMENTS: Included as a subspecies of *T. laniger* by Cabrera (1961); karyotype reported by Gardner and Patton (1976) as a species without comment.

Thomasomys niveipes (Thomas, 1896). Ann. Mag. Nat. Hist., ser. 6, 18:305.
TYPE LOCALITY: Colombia, Cundinamarca Dept., La Oya del Barro.
DISTRIBUTION: C Colombia.

Thomasomys notatus Thomas, 1917. Smithson. Misc. Coll., 68(4):2.
TYPE LOCALITY: Peru, Cuzco Dept., Torontoy, 9500 ft.
DISTRIBUTION: SE Peru.
COMMENTS: Standard karyotype reported by Gardner and Patton (1976).

Thomasomys oreas Anthony, 1926. Am. Mus. Novit., 239:2.
TYPE LOCALITY: Bolivia, La Paz Dept., Cocopunco, 10,000 ft.
DISTRIBUTION: WC Bolivia.

Thomasomys paramorum Thomas, 1898. Ann. Mag. Nat. Hist., ser. 7, 1:453.
TYPE LOCALITY: Ecuador, Chimborazo Dept., paramo south of Mount Chimborazo.
DISTRIBUTION: Andean Ecuador.

Thomasomys pyrrhonotus Thomas, 1886. Ann. Mag. Nat. Hist., ser. 5, 18:421.
TYPE LOCALITY: Peru, Cajamarca Dept., Rio Malleta, Tambillo, 5800 ft.
DISTRIBUTION: Andes of S Ecuador and NW Peru.
SYNONYMS: *auricularis*.
COMMENTS: Morphology redescribed and additional specimens reported by Pine (1980*b*).

Thomasomys rhoadsi Stone, 1914. Proc. Acad. Nat. Sci. Philadelphia, 66:12.
TYPE LOCALITY: Ecuador, Pichincha Prov., Mount Pichincha, Hacienda Garzon, 10,500 ft.
DISTRIBUTION: Andes of Ecuador.
SYNONYMS: *fumeus*.

Thomasomys rosalinda Thomas and St. Leger, 1926. Ann. Mag. Nat. Hist., ser. 9, 18:345.
TYPE LOCALITY: Peru, Amazonas Dept., Goncha, 8500 ft.
DISTRIBUTION: NC Peru.

Thomasomys silvestris Anthony, 1924. Am. Mus. Novit., 114:2.
TYPE LOCALITY: Ecuador, Pichincha Prov., Santo Domingo trail on western slope of Mt. Corazon, Las Maquinas, 7000 ft.
DISTRIBUTION: W Andes of Ecuador.

Thomasomys taczanowskii (Thomas, 1882). Proc. Zool. Soc. Lond., 1882:109.
TYPE LOCALITY: Peru, Cajamarca Dept., Rio Malleta, Tambillo, 5800 ft.

DISTRIBUTION: NW Peru.
COMMENTS: Karyology reviewed by Gardner and Patton (1976).

Thomasomys vestitus (Thomas, 1898). Ann. Mag. Nat. Hist., ser. 7, 1:454.
TYPE LOCALITY: Venezuela, Merida, Rio Milla, 1630 m.
DISTRIBUTION: Merida Andes of W Venezuela.

Tylomys Peters, 1866. Monatsb. K. Preuss. Akad. Wiss. Berlin, 1866:404.
TYPE SPECIES: *Hesperomys nudicaudus* Peters, 1866.
COMMENTS: Tribal-level membership unresolved; closely related to *Ototylomys* (see remarks under same). Basic revision much needed; forms listed follow Hall (1981:626), who remarked that "study...may show that some of the species are only subspecies."

Tylomys bullaris Merriam, 1901. Proc. Washington Acad. Sci., 3:561.
TYPE LOCALITY: Mexico, Chiapas, Tuxtla Gutierrez.
DISTRIBUTION: Known only from the type locality.

Tylomys fulviventer Anthony, 1916. Bull. Am. Mus. Nat. Hist., 35:366.
TYPE LOCALITY: Panama, Darien Prov., Tacarcuna, 4200 ft.
DISTRIBUTION: Easternmost Panama.
COMMENTS: Possibly a subspecies of *T. mirae* according to Cabrera (1961) or a synonym of *T. panamensis* according to Handley (1966*a*).

Tylomys mirae Thomas, 1899. Ann. Mag. Nat. Hist., ser. 7, 4:278.
TYPE LOCALITY: Ecuador, Imbabura Prov., Rio Mira, Paramba.
DISTRIBUTION: W Colombia and NW Ecuador.
SYNONYMS: *bogotensis*.

Tylomys nudicaudus (Peters, 1866). Monatsb. K. Preuss. Akad. Wiss. Berlin, 1866:404.
TYPE LOCALITY: Guatemala.
DISTRIBUTION: C Guerrero and C Veracruz, Mexico, south to S Nicaragua, excluding Yucatan Peninsula.
SYNONYMS: *gymnurus, microdon, villai*.
COMMENTS: Goodwin (1934) passingly noted that La Primavera (Dept. Alto Verapaz) may be the type locality, a restriction which should be formally considered by a future revisor. Goodwin (1969) relegated *gymnurus* to a subspecies of *T. nudicaudus*.

Tylomys panamensis (Gray, 1873). Ann. Mag. Nat. Hist., ser. 4, 12:417.
TYPE LOCALITY: Panama.
DISTRIBUTION: Easternmost Panama (see Goldman, 1920).
COMMENTS: Handley (1966*a*) suggested that *fulviventer* and *watsoni* may prove to be junior synonyms of *T. panamensis*.

Tylomys tumbalensis Merriam, 1901. Proc. Washington Acad. Sci., 3:560.
TYPE LOCALITY: Mexico, Chiapas, Tumbala.
DISTRIBUTION: Known only from the type locality.

Tylomys watsoni Thomas, 1899. Ann. Mag. Nat. Hist., ser. 7, 4:278.
TYPE LOCALITY: Panama, Chiriqui Prov., Volcan de Chiriqui, Bogava (= Bugaba), 800 ft.
DISTRIBUTION: Costa Rica and W Panama.

Wiedomys Hershkovitz, 1959. Proc. Biol. Soc. Washington, 72:5.
TYPE SPECIES: *Mus pyrrhorhinos* Wied-Neuwied, 1821.
COMMENTS: Wiedomyine, a tribe formally diagnosed by Reig (1980) to contain the problematic form *pyrrhorhinos*, which has been variously classified as a species of *Oryzomys* or *Thomasomys* (see Tate, 1932*f*; Osgood, 1933*d*; and Hershkovitz, 1959*b*), and a new fossil genus, *Cholomys*, recovered from E Argentina.

Wiedomys pyrrhorhinos (Wied-Neuwied, 1821). Reise nach Brasilien, 2:177.
TYPE LOCALITY: Brazil, Bahia, caatingas along the Riacho da Ressaca, between the farms Tamboril and Ilha (as clarified by Avila-Pires, 1965).
DISTRIBUTION: SE Brazil from Ceara to Rio Grande do Sul.
COMMENTS: Locality records clarified by Pine (1980*b*).

Wilfredomys Avila-Pires, 1960. Bol. Mus. Nac., Nov. Ser., Rio de Janeiro, 220:3.
TYPE SPECIES: *Thomasomys oenax* Thomas, 1928.
COMMENTS: Genus diagnosed to encompass another problematic thomasomyine species from SE Brazil; see Osgood (1933*b*) and Pine (1980*b*) for historical review of the generic affiliations of the type species. The categorical recognition of this taxon, conventionally treated as a subgenus of *Thomasomys* (e.g., Carleton and Musser, 1984; Corbet and Hill, 1991; Pine, 1980*b*), will depend on phylogenetic studies involving other thomasomyines (also see comments under *Delomys*).

Wilfredomys oenax (Thomas, 1928). Ann. Mag Nat. Hist., ser. 10, 1:154.
TYPE LOCALITY: Brazil, Rio Grande do Sul, San Lorenzo.
DISTRIBUTION: SE Brazil to C Uruguay.
COMMENTS: Morphology redescribed by Pine (1980*b*).

Wilfredomys pictipes (Osgood, 1933). Field Mus. Nat. Hist. Publ., Zool. Ser., 20(2):11.
TYPE LOCALITY: Argentina, Misiones Prov., Caraguatay, Rio Parana, 100 mi S Rio Iguazu.
DISTRIBUTION: NE Argentina and SE Brazil.
COMMENTS: Provisionally assigned to *Wilfredomys* on the assertions of Osgood (1933*b*) and Pine (1980*b*) that *pictipes* is most closely related to *oenax*.

Xenomys Merriam, 1892. Proc. Biol. Soc. Washington, 7:160.
TYPE SPECIES: *Xenomys nelsoni* Merriam, 1892.
COMMENTS: Neotomine. Knowledge of biology and systematic relationships meager; among woodrats may share common ancestry with *Hodomys* (Carleton, 1980). Banded karyotype reported by Haiduk et al. (1988), who noted its resemblance to those of *Onychomys* and *Peromyscus*.

Xenomys nelsoni Merriam, 1892. Proc. Biol. Soc. Washington, 7:161.
TYPE LOCALITY: Mexico, Colima, Hacienda Magdalena, between Ciudad Colima and Manzanillo.
DISTRIBUTION: Colima and westernmost Jalisco, Mexico.

Zygodontomys J. A. Allen, 1897. Bull. Am. Mus. Nat. Hist., 9:38.
TYPE SPECIES: *Oryzomys cherriei* J. A. Allen, 1895 (= *Oryzomys brevicauda* J. A. Allen and Chapman, 1893).
COMMENTS: First revised by Hershkovitz (1962), then thought to include southern populations now allocated to *Bolomys lasiurus* (see Maia and Langguth, 1981; Reig, 1987; Voss and Linzey, 1981). Definition of genus emended and species taxonomy revised by Voss (1991*a*); Reig (*in* Reig et al., 1990*a*) listed four species as valid. Karyology summarized by Gardner and Patton (1976) and Reig et al. (1990*a*). Hershkovitz (1962) considered *Zygodontomys* a member of the phyllotine group, an association disputed by other studies (Hooper and Musser, 1964*a*; Olds and Anderson, 1989; Pearson and Patton, 1976). Unsettled phylogenetic position expressed as Sigmodontinae *incertae sedis* (Reig, 1984; Voss, 1991*a*).

Zygodontomys brevicauda (J. A. Allen and Chapman, 1893). Bull. Am. Mus. Nat. Hist., 5:215.
TYPE LOCALITY: Trinidad, Princes Town.
DISTRIBUTION: Savannas from SE Costa Rica through Panama, Colombia, Venezuela, and the Guianas, to Brazil north of the Amazon River; including Trinidad and Tobago and smaller continental-shelf islands adjacent Panama and Venezuela.
SYNONYMS: *cherriei, fraterculus, frustrator, griseus, microtinus, reigi, sanctaemartae, seorsus, soldadoensis, stellae, thomasi, tobagi, ventriosus*.
COMMENTS: Geographic variation evaluated by Voss (1991*a*), who retained three subspecies: *brevicauda, cherriei*, and *microtinus*. Formerly included *punctulatus*, which Voss (1991*b*) reidentified as a member of *Bolomys*.

Zygodontomys brunneus Thomas, 1898. Ann. Mag. Nat. Hist., ser. 7, 2:269.
TYPE LOCALITY: Colombia, "El Saibal, W. Cundinamarca." Indeterminate nature of type locality discussed by Voss (1991*a*).
DISTRIBUTION: Intermontane valleys of N Colombia.
SYNONYMS: *borreroi*.

COMMENTS: Relegated to a subspecies of *Z. brevicauda* by Gyldenstolpe (1932) and Hershkovitz (1962), but specific status documented by Voss (1991*a*).

Subfamily Spalacinae Gray, 1821. London Med. Repos., 15:303.

SYNONYMS: Aspalacidae.

COMMENTS: Taxonomy and distribution reviewed by Ognev (1963*a*) and Corbet (1978*c*, 1984); subfamily monographed by Topachevskii (1969). For an introduction to morphology, taxonomy, and paleontology of European species, see Savic (1982*a*) and Kivanc (1988). Carleton and Musser (1984) provided a diagnosis of the subfamily and summarized general characters, distribution, taxonomy, and fossils; and Savic and Nevo (1990) gave an excellent synopsis of the evolutionary history, speciation, and population biology of mole rats. Catzeflis et al. (1989) used DNA-DNA hybridization to compare spalacines with arvicolines and murines to estimate their time of divergence. Vorontsov et al. (1977*b*) distinguished four species groups based upon biochemical data: *N. nehringi*, *N. leucodon*, *S. microphthalmus*, and the last containing *S. graecus*, *S. polonicus*, *S. arenarius*, and *S. giganteus*. The species defined biochemically are the same as those defined by Ognev (1963*a*) and Topachevskii (1969) using morphological traits and by Lyapunova et al. (1974) with chromosomal evidence. To these seven species is added *N. ehrenbergi*. Although Savic and Nevo (1990:133) acknowledged eight extant species, they concluded that the systematics is unrealistic because "it is based primarily on classical morphology, ignoring the central phenomenon of Spalacid evolution, i.e., chromosomal speciation which suggests that more than 30 living karyotypes, or species have been described and the end is not yet in sight."

Results from sequencing alpha crystalline A chain, a lens protein of the eye in *Spalax*, and its significance for molecular clock hypothesis and phenetic analysis was discussed by McKenna (1992).

Nannospalax Palmer, 1903. Science, n.s., 17:873.

TYPE SPECIES: *Spalax kirgisorum* Nehring, 1898 (= *Spalax ehrenbergi* Nehring, 1898).

SYNONYMS: *Mesospalax*, *Microspalax* (of Mehely, 1909, not Megnin and Trouessart, 1885), *Ujhelyiana*.

COMMENTS: The species were monographed under the genus *Microspalax* Mehely, 1909, by Topachevskii (1969). Mehely's generic name, however, is preoccupied and that is why *Nannospalax*, as well as *Ujhelyiana*, were proposed. *Nannospalax* is differentiated from *Spalax* by a suite of skeletal and dental traits (Topachevskii, 1969). Compared with species of *Spalax*, karyotypes of those in *Nannospalax* have low diploid and fundamental numbers and exhibit significant interpopulational differences that correlate with biological and ecological distinctions. Such properties have led investigators to suspect that the three entities listed below are superspecies, each made up of separate biological species (Savic and Nevo, 1990).

Nannospalax ehrenbergi (Nehring, 1898). Sitzb. Ges. Naturf. Fr. Berlin (for December, 1897), 178, pl. 2 [1898].

TYPE LOCALITY: Israel, Jaffa.

DISTRIBUTION: From Syria, Lebanon, Jordan, and Israel through N Egypt to N Libya (see Lay and Nadler, 1972, for North African range); possibly SW Turkey (Savic and Nevo, 1990).

SYNONYMS: *aegyptiacus*, *berytensis*, *fritschi*, *intermedius*, *kirgisorum*.

COMMENTS: Chromosomal data for North African populations provided by Lay and Nadler (1972). Libyan populations reviewed by Ranck (1968) and those in Egypt by Osborn and Helmy (1980). For more than 30 years, Israeli populations of this "superspecies" have been the subjects of intensive research concerning evolutionary theory and the processes of speciation and adaptive radiation (see Nevo, 1991).

Nannospalax leucodon (Nordmann, 1840). Demidoff Voy., 3:34.

TYPE LOCALITY: Ukraine, near Odessa.

DISTRIBUTION: From Yugoslavia through Hungary, Rumania, Bulgaria, Greece, and extreme NW Turkey, to SW Ukraine just east of Dnestr River in Odessa region (see Vorontsov, 1977*b*; Savic, 1982*b*).

SYNONYMS: *ceylanus, epiroticus, hellenicus, hercegovinensis, hungaricus, insularis, makedonicus, martinoi, montanoserbicus, montanosyrmiensis, monticola, ovchepolensis, peloponnesiacus, serbicus, strumiciensis, syrmiensis, thermaicus, thessalicus, thracius, transsylvanicus, turcicus, vasvárii.*

COMMENTS: This species was morphologically characterized by Topachevskii (1969) and is now viewed as a superspecies based on chromosomal studies (Giagia et al., 1982; Peshev, 1983; Savic, 1982*b*; Savic and Nevo, 1990; Savic and Soldatovic, 1977, 1979). Savic (1982*b*) delineated six clusters that he interpreted as chromosomal species: *leucodon* group (with *hungaricus, montanosyrmiensis, monticola,* and *transsylvanicus*), *makedonicus* group, *strumiciensis* group (with *ovchepolensis* and *serbicus*), *epiroticus* (with *hellenicus*), *turcicus* group (with *thracius*), and *montanoserbicus* group (with *hercegovinensis* and *syrmiensis*). Mikes et al. (1982) analysed the pelvis of *N. leucodon* in the context of assessing sexual dimorphism and taxonomic differences. Morphological and biometric analyses by Kivanc (1988) indicated that *N. leucodon* occurs in most of Turkey (subspecies *anatolicus, armeniacus, cilicicus, nehringi,* and *turcicus*) and that *N. ehrenbergi* extends up into SW Turkey (subspecies *intermedius* and *kirgisorum*). However, Savic and Nevo (1990) viewed these results skeptically and urged elucidation by chromosomal evidence. Available biochemical and chromosomal data clearly do not support the conspecificity of *leucodon* and *nehringi* (Vorontsov et al., 1977*b*, and references therein).

Nannospalax nehringi (Satunin, 1898). Zool. Anz., 21:314.

TYPE LOCALITY: Armenia, Kasikoporan.

DISTRIBUTION: Turkey, Armenia, and Georgia (see Topachevskii, 1969; Vorontsov et al., 1977*b*).

SYNONYMS: *anatolicus, armeniacus, captorum, cilicicus, corybantium.*

COMMENTS: Ellerman and Morrison-Scott (1951) and Corbet (1978*c*) listed *nehringi* as a synonym of *S. leucodon*, but Topachevskii (1969) treated it as a distinct species (under subgenus *Mesospalax*, genus *Microspalax*), a revision followed by Corbet (1984) and Pavlinov and Rossolimo (1987). Topachevskii (1969) included *turcicus* in the synonymy of this species, but Savic (1982*b*) placed it in *leucodon*. Morphometric analyses have been interpreted to indicate that *nehringi* and its synonyms are geographic variants of *N. leucodon* (see Kivanc, 1988; Savic and Nevo, 1990, and account of *leucodon*).

Spalax Guldenstaedt, 1770. Nova Comm. Acad. Sci. Petropoli, ser. 14, 1:410.

TYPE SPECIES: *Spalax microphthalmus* Guldenstaedt, 1770.

SYNONYMS: *Anotis, Aspalax, Macrospalax, Myospalax* (of Hermann, 1783, not Laxmann, 1769), *Ommatostergus, Talpoides.*

COMMENTS: Peshev (1989*a, b*) illuminated interspecific differences among the five species, as well as sexual dimorphism within each, using morphometric techniques. Compared with *Nannospalax*, karyotypic characteristics of *Spalax* species include high diploid and fundamental numbers, no acrocentric chromosomes, smaller interspecific differences in karyotypic structure, and lack of significant interpopulation variation (Savic and Nevo, 1990).

Spalax arenarius Reshetnik, 1939. Reports Zool. Mus. Kiev, 23:11.

TYPE LOCALITY: Ukraine, Nikolaev Region, Golaya Pristan, NW shore of Black Sea.

DISTRIBUTION: Small range in S Ukraine (see Vorontsov et al., 1977*b*).

COMMENTS: Listed as a subspecies of *S. microphthalmus* by Corbet (1978*c*), but elevated to species by Topachevskii (1969) and recognized as such by Lyapunova et al. (1974) and Pavlinov and Rossolimo (1987). Topachevskii (1969) viewed the restricted geographic range of *S. arenarius* as a relictual distribution.

Spalax giganteus Nehring, 1898. Sitzb. Ges. Naturf. Fr. Berlin, p. 169.

TYPE LOCALITY: Russia, Daghestan, W shore of Caspian Sea, near MakhachKaly.

DISTRIBUTION: Isolated populations in steppes NW of the Caspian Sea and in Kazakhstan between the Volga River and W margin of the Caspian Sea between the Dnepr and Ural rivers (see Topachevskii, 1969; Vorontsov et al., 1977*b*).

SYNONYMS: *uralensis.*

COMMENTS: Revised by Topachevskii (1969) and listed as a separate species by Pavlinov and Rossolimo (1987). Geographic range is parapatric with that of *S. microphthalmus* (Corbet, 1978*c*).

Spalax graecus Nehring, 1898. Zool. Anz., 21:228.

TYPE LOCALITY: Ukraine, Bukovina region, vicinity of Chernovtsy (as restriced by Topachevskii, 1969).

DISTRIBUTION: Romania and SW Ukraine (see Topachevskii, 1969; Savic, 1982*d*).

SYNONYMS: *antiquus, istricus, mezosegiensis.*

COMMENTS: Included in *S. microphthalmus* by Ellerman and Morrison-Scott (1951) and Corbet (1978*c*), but arranged as a distinctive species by Topachevskii (1969) and so listed by Pavlinov and Rossolimo (1987). Savic (1982*d*) reviewed in detail the European segment. Topachevskii (1969) recognized *istricus*, apparently restricted to Rumania, as a subspecies, and Murariu and Torcea (1984) separated it as a species based on cranial traits. Their sample, however, was small, and variation of the traits believed to discriminate *istricus* from *S. graecus* should be reassessed using larger samples from both Rumania and the Ukraine.

Spalax microphthalmus Guldenstaedt, 1770. Nova Comm. Acad. Sci. Petropoli, 14:1

TYPE LOCALITY: Russia, Voronezhskaya Oblast, Novokhoper Steppe.

DISTRIBUTION: Steppes in Ukraine and S Russia between Dnieper and Volga rivers, north to Orel-Kursk line, and south to Ciscaucasia (see Topachevskii, 1969; Vorontsov et al., 1977*b*).

SYNONYMS: *pallasi, typhlus.*

COMMENTS: Monographed by Topachevskii (1969) and maintained as a species by Pavlinov and Rossolimo (1987).

Spalax zemni Erxleben, 1777. Syst. Regni Anim., 1:370-371.

TYPE LOCALITY: Ukraine, Ternopolsk region (see Topachevskii, 1969).

DISTRIBUTION: SE Poland east into Ukraine between Dnestr and Dnepr rivers and south to margin of Black Sea (see Topachevskii, 1969; Savic, 1982*c*; Vorontsov et al., 1977*b*).

SYNONYMS: *podolicus, polonicus.*

COMMENTS: Included in *S. microphthalmus* by Corbet (1978*c*), but monographed as a separate species (under the name *polonicus*) by Topachevskii (1969) and listed as such by Pavlinov and Rossolimo (1987). Reviewed (as *polonicus*) by Savic (1982*c*). Topachevskii (1969) indicated that *zemni* was a *nomen nudum*, but it is properly available (Ellerman, 1949*b*) and becomes the earliest name for this species (Pavlinov and Rossolimo, 1987).

FAMILY ANOMALURIDAE

by Fritz Dieterlen

Family Anomaluridae Gervais, 1849. *In* D'Orbigny, Dict. Univ. Hist. Nat., 11:203.
COMMENTS: See Misonne (1974:3-5).

Subfamily Anomalurinae Gervais, 1849. *In* D'Orbigny, Dict. Univ. Hist. Nat., 11:203.

Anomalurus Waterhouse, 1843. Proc. Zool. Soc. Lond., 1842:124 [1843].
TYPE SPECIES: *Anomalurus fraseri* Waterhouse, 1843 (= *Pteromys derbianus* Gray, 1842).
SYNONYMS: *Anomalurella* Matschie, 1914; *Anomalurodon* Matschie, 1914; *Anomalurops* Matschie, 1914; *Aroaethrus* Waterhouse, 1843.
COMMENTS: See Misonne (1974:3-5).

Anomalurus beecrofti Fraser, 1853. Proc. Zool. Soc. Lond., 1852:17 [1853].
TYPE LOCALITY: Equatorial Guinea, "Fernando Po" (= Bioko).
DISTRIBUTION: Senegal to Uganda and Zaire (Kasai); Bioko (Equatorial Guinea).
STATUS: CITES - Appendix III (Ghana).
SYNONYMS: *argenteus* Schwann, 1904; *chapini* J. A. Allen, 1922; *citrinus* Thomas, 1916; *fulgens* Gray, 1869; *hervoi* Dekeyser and Villiers, 1951; *laniger* Temminck, 1853; *schoutedeni* Verheyen, 1968.
COMMENTS: See Misonne (1974:4). Matschie (1914) placed *beecrofti* in his subgenus *Anomalurops*. Includes *schoutedeni* which is considered a distinct species by Cabral (1971).

Anomalurus derbianus (Gray, 1842). Ann. Mag. Nat. Hist., [ser. 1], 10:262.
TYPE LOCALITY: "Sierra Leone".
DISTRIBUTION: Sierra Leone to Angola, east to Kenya, south to Mozambique and Zambia.
STATUS: CITES - Appendix III (Ghana).
SYNONYMS: *beldeni* Du Chaillu, 1860; *chrysophaenus* Dubois, 1888; *cinereus* Thomas, 1895; *erythronotus* Milne-Edwards, 1879; *fortior* Lönnberg, 1917; *fraseri* Waterhouse, 1842; *griselda* Dollman, 1914; *imperator* Dollman, 1911; *jacksoni* de Winton, 1898; *jordani* St. Leger, 1935; *laticeps* Aguilar-Amat, 1922; *neavei* Dollman, 1909; *nigrensis* Thomas, 1904; *orientalis* Peters, 1880; *perustus* Thomas, 1914; *squamicaudus* Schinz, 1845.
COMMENTS: See Misonne (1974:4-5).

Anomalurus pelii (Schlegel and Müller, 1845). *In* Temminck, Verh. Nat. Gesch. Nederland. Overz. Bezitt., Zool., Mammalia, p. 109[1845].
TYPE LOCALITY: "bÿ Daboerom, aan de Goudkust" (= Ghana, Dabacrom).
DISTRIBUTION: Sierra Leone to Ghana.
STATUS: CITES - Appendix III (Ghana).
SYNONYMS: *auzembergeri* Matschie, 1914.
COMMENTS: See Misonne (1974:4).

Anomalurus pusillus Thomas, 1887. Ann. Mag. Nat. Hist., ser. 5, 20:440.
TYPE LOCALITY: Zaire, "Bellima and Tingasi, Monbuttu".
DISTRIBUTION: S Cameroon, Gabon, Zaire.
SYNONYMS: *batesi* de Winton, 1897.
COMMENTS: See Misonne (1974:5).

Subfamily Zenkerellinae Matschie, 1898. Sitzb. Ges. Naturf. Fr. Berlin, 4:26.
SYNONYMS: Idiurinae Miller and Gidley, 1918.

Idiurus Matschie, 1894. Sitzb. Ges. Naturf. Fr. Berlin, 8:194.
TYPE SPECIES: *Idiurus zenkeri* Matschie, 1894.
COMMENTS: Revised by Verheyen (1963). See Misonne (1974:4).

Idiurus macrotis Miller, 1898. Proc. Biol. Soc. Washington, 12:73.
TYPE LOCALITY: "Efulen, Cameroon district, West Africa".
DISTRIBUTION: Sierra Leone to E Zaire.

STATUS: CITES - Appendix III (Ghana).

SYNONYMS: *cansdalei* Hayman, 1946; *kivuensis* Lönnberg, 1917; *langi* J. A. Allen, 1922; *panga* J. A. Allen, 1922.

COMMENTS: Includes *kivuensis*, originally considered (by Lönnberg) as subspecies of *zenkeri*, then considered a valid species by Hayman (1946:211); but see Verheyen (1963:169-178, 183) who included *kivuensis* in *macrotis* and revised this species.

Idiurus zenkeri Matschie, 1894. Sitzb. Ges. Naturf. Fr. Berlin, 8:197.

TYPE LOCALITY: S Cameroon, Yaunde.

DISTRIBUTION: S Cameroon to Uganda.

SYNONYMS: *haymani* Verheyen, 1963.

COMMENTS: Revised by Verheyen (1963:159-169). Does not include *kivuensis*, see comment under *macrotis*.

Zenkerella Matschie, 1898. Sitzb. Ges. Naturf. Fr. Berlin, 4:23.

TYPE SPECIES: *Zenkerella insignis* Matschie, 1898.

SYNONYMS: *Aethurus* de Winton, 1898.

COMMENTS: Includes *Aethurus*; see Walker (1968:754). See Misonne (1974:4).

Zenkerella insignis Matschie, 1898. Sitzb. Ges. Naturf. Fr. Berlin, 4:24.

TYPE LOCALITY: "Kamerun [Cameroon], Afr. occ., Yaunde".

DISTRIBUTION: SW Cameroon, Rio Muni (Equatorial Guinea), Gabon, Central African Republic.

SYNONYMS: *glirinus* de Winton, 1898.

COMMENTS: See Misonne (1974:4).

FAMILY PEDETIDAE

by Fritz Dieterlen

Family Pedetidae Gray, 1825. Ann. Philos., n.s., 10:342.
COMMENTS: See Misonne (1974:9).

Pedetes Illiger, 1811. Prodr. Syst. Mamm. Avium., p. 81.
TYPE SPECIES: *Yerbua capensis* Forster, 1778.
SYNONYMS: *Helamis* F. Cuvier, 1821; *Helamys* G. Cuvier, 1816; *Pedestes* Gray, 1843; *Yerbua* Forster, 1778.
COMMENTS: See Misonne (1974:9).

Pedetes capensis (Forster, 1778). K. Svenska Vet.-Akad. Handl. Stockholm, (1)39:109.
TYPE LOCALITY: South Africa, Cape Prov., Cape of Good Hope.
DISTRIBUTION: South Africa, Namibia, Angola, Zimbabwe, Mozambique, Zambia, S Zaire, Tanzania, Kenya.
SYNONYMS: *albaniensis* Roberts, 1946; *angolae* Hinton, 1920; *cafer* Pallas, 1778; *currax* Hollister, 1918; *damarensis* Roberts, 1926; *dentatus* Miller, 1927; *fouriei* Roberts, 1938; *larvalis* Hollister, 1918; *orangiae* Wroughton, 1907; *salinae* Wroughton, 1907; *surdaster* Thomas, 1902; *taborae* G. M. Allen and Loveridge, 1927; *typicus* A. Smith, 1834.
COMMENTS: See Misonne (1974:8).

FAMILY CTENODACTYLIDAE
by Fritz Dieterlen

Family Ctenodactylidae Gervais, 1853. Ann. Sci. Nat. (Paris), ser. 3, 20:245.
COMMENTS: Reviewed by W. George (1979).

Ctenodactylus Gray, 1830. Spicil. Zool., p. 10.
TYPE SPECIES: *Ctenodactylus massonii* Gray, 1830 (= *Mus gundi* Rothmann, 1776).
COMMENTS: Reviewed by W. George (1979).

Ctenodactylus gundi (Rothmann, 1776). *In* Schlözer, Briefwechsel, 1:339.
TYPE LOCALITY: Libya, Gharian, 80 km S of Tripoli.
DISTRIBUTION: N Morocco to NW Libya.
SYNONYMS: *arabicus* Shaw, 1801; *massonii* Gray, 1830; *typicus* A. Smith, 1834.
COMMENTS: Reviewed by W. George (1979).

Ctenodactylus vali Thomas, 1902. Proc. Zool. Soc. Lond., 1902(2):11.
TYPE LOCALITY: Libya, "Wadi Bey" (NW of Bonjem, Tripoli).
DISTRIBUTION: S Morocco, W Algeria, NW Libya.
SYNONYMS: *joleaudi* Heim de Balsac, 1936.
COMMENTS: Corbet (1978*c*:160) included *vali* in *gundi*, but George (1982) and Corbet and Hill (1991) listed both as distinct species.

Felovia Lataste, 1886. Le Naturaliste, 7(36):287.
TYPE SPECIES: *Felovia vae* Lataste, 1886.
COMMENTS: Proposed as a subgenus of *Massoutiera*; recognized as a valid genus by Thomas (1913*a*:31) and St. Leger (1931:978).

Felovia vae Lataste, 1886. Le Naturaliste, 7(36):287.
TYPE LOCALITY: Senegal, upper Senegal River, Medina Dist., Felou.
DISTRIBUTION: Senegal, Mauritania, Mali.

Massoutiera Lataste, 1885. Le Naturaliste, 7(3):21.
TYPE SPECIES: *Ctenodactylus mzabi* Lataste, 1881.

Massoutiera mzabi (Lataste, 1881). Bull. Soc. Zool. Fr., 6:314.
TYPE LOCALITY: Algeria, Ghardaia.
DISTRIBUTION: SE Algeria, SW Libya, NE Mali, N Niger, N Chad.
SYNONYMS: *harterti* Thomas, 1913; *rothschildi* Thomas and Hinton, 1921.

Pectinator Blyth, 1856. J. Asiat. Soc. Bengal for 1855, (2)24:294 [1856].
TYPE SPECIES: *Pectinator spekei* Blyth, 1855.
SYNONYMS: *Petrobates* Heuglin, 1860.

Pectinator spekei Blyth, 1856. J. Asiat. Soc. Bengal for 1855, (2)24:294 [1856].
TYPE LOCALITY: Somalia (between Goree Bunder and Nogal), *ca* 09°N, 47°E.
DISTRIBUTION: Ethiopia, Somalia, Djibouti.
SYNONYMS: *legerae* de Beaux, 1934; *meridionalis* de Beaux.

FAMILY MYOXIDAE

by Mary Ellen Holden

Family Myoxidae Gray, 1821. London Med. Repos., 15:303.

SYNONYMS: Gliridae Thomas, 1897 (excluding Platacanthomyinae); Leithiidae Lydekker, 1895; Muscardinidae Palmer, 1899 (excluding Platacanthomyinae).

COMMENTS: Simpson (1945) deemed Myoxidae Gray, 1821 invalid due to the apparent synonymy of the type genus *Myoxus* Zimmermann, 1780, with *Glis* Brisson, 1762, and used Gliridae Thomas, 1897. Hopwood (1947) argued that Brisson's names are invalid because they are not Linnaean or binomial. As noted by Hopwood (1947), *Glis* is valid in Erxleben (1777) for marmots, ground-squirrels, voles, and lemmings, rendering *Glis* Storr, 1780 (which included pedetids, dormice, and other rodents) invalid. Thus the oldest available name to replace *Glis* Brisson is *Myoxus* Zimmermann, valid in Linnaeus (1788) for dormice, and the correct family name for dormice is Myoxidae (for further discussion see Wahlert et al., In Press).

Carleton and Musser (1984) verified and discussed Alston's (1876) placement of Platacanthomyinae within the Muridae. Wahlert et al. (In Press) provided a history of classification and an introduction to morphological characters of myoxids. With the exception of the Graphiurinae, the extant genera are surviving members of distinct evolutionary lineages which were clearly differentiated by the early to medial Miocene (de Bruijn, 1967; Daams, 1981). De Bruijn (1967) and Daams (1981) each proposed classifications of fossil and extant myoxids based on dental characters. Wahlert et al. (In Press) incorporated dental characters plus forty-three osteological characters in a phylogenetic analysis of extant myoxid genera. Their resulting classification is similar to that of de Bruijn (1967) except that *Glirulus* is included in Myoxinae (not placed in its own subfamily), Leithiinae is subdivided into two tribes, and *Selevinia* is placed as a close relative of *Myomimus*. The classification of Wahlert et al. (In Press) is followed below. The distributions of the species occuring in China were verified and much enhanced by the efforts of Lin Yonglie.

Subfamily Graphiurinae Winge, 1887. E Museu Lundii, 1:109, 123.

COMMENTS: Daams (1981) suggested that the Graphiurinae are most closely related to the Leithiinae, based on tooth morphology. The results of Wahlert et al. (In Press) indicated that the Graphiurinae represent the earliest or most primitive branch of myoxids.

Graphiurus Smuts, 1832. Enumer. Mamm. Capensium, pp. 32-33.

TYPE SPECIES: *Sciurus ocularis* Smith, 1829.

SYNONYMS: *Aethoglis, Claviglis, Gliriscus* (see Ellerman, 1940; Ellerman et al., 1953).

COMMENTS: The revision of *Graphiurus* by Genest-Villard (1978), based mostly on size grades, underestimated species diversity, particularly in the *murinus* group. Species limits were defined in parts of Africa (e.g., Ansell and Dowsett, 1988; Robbins and Schlitter, 1981). A systematic revision of *Graphiurus* is in progress and the species recognized below reflect information in the literature as well as examination of museum specimens; results are provisional and assignment of synonyms incomplete.

Graphiurus christyi Dollman, 1914. Rev. Zool. Afr., 4(1):80.

TYPE LOCALITY: Zaire, Mambaka.

DISTRIBUTION: N Zaire (see Hatt, 1940*a*; Schlitter et al., 1985), S Cameroon (see Robbins and Schlitter, 1981).

Graphiurus crassicaudatus (Jentink, 1888). Notes Leyden Mus., 10:41.

TYPE LOCALITY: Liberia, Du Queah River.

DISTRIBUTION: Liberia (Kuhn, 1965), Ivory Coast (in Mt. Nimba reserve see Heim de Balsac and Lamotte, 1958), Ghana, Togo, Nigeria; Cameroon (Robbins and Schlitter, 1981; Schlitter et al., 1985); perhaps Bioko (see Rosevear, 1969).

SYNONYMS: *dorotheae* (see Allen, 1939).

Graphiurus hueti Rochebrune, 1883. Act. Soc. Linn. Bordeaux, 37, 4(7):110.
TYPE LOCALITY: Senegal, near Saint-Louis.
DISTRIBUTION: Senegal, south to Sierra Leone, Liberia (Kuhn, 1965), Ivory Coast (Aellen, 1965, and in Mt. Nimba Reserve see Heim de Balsac and Lamotte, 1958), Ghana, Nigeria (Happold, 1987), Cameroon (Robbins and Schlitter, 1981; Schlitter et al., 1985), Central African Republic, and Gabon. In W Africa see Rosevear (1969).
SYNONYMS: *argenteus, nagtglasii* (see Allen, 1939).
COMMENTS: Does not include *monardi* (see comment under *G. monardi*). Karyotype of Ivory Coast specimen given by Tranier and Dosso (1979).

Graphiurus kelleni (Reuvens, 1890). Notes Leyden Mus., 13:74.
TYPE LOCALITY: Angola: Mossamedes district, "Damara-land" (see Hill and Carter, 1941).
DISTRIBUTION: Angola (Hayman, 1963*b*; Hill and Carter, 1941), Zambia (Ansell, 1978), Malawi (Ansell and Dowsett, 1988; Ansell, 1989*b*), and Zimbabwe.
SYNONYMS: *ansorgei, johnstoni, nanus* (see Allen, 1939, for citations).
COMMENTS: See comments under *G. parvus*.

Graphiurus lorraineus Dollman, 1910. Ann. Mag. Nat. Hist., ser. 8, 5:285.
TYPE LOCALITY: Zaire, Welle (Uele) River, Molegbwe, south of Setema Rapids.
DISTRIBUTION: Sierra Leone, Liberia (Kuhn, 1965), Ivory Coast (Aellen, 1965; Mt Nimba Reserve see Heim de Balsac and Lamotte, 1958), Ghana, Togo, Benin, Nigeria (Happold, 1987), Cameroon (Robbins and Schlitter, 1981; Schlitter et al., 1985), Gabon, N Angola (Hayman, 1963*b*), Zaire (Hatt, 1940*a*; Petter, 1967*a*; Verheyen and Verschuren, 1966), Uganda, SW Tanzania (Swynnerton and Hayman, 1951); perhaps Gambia. In W Africa see Rosevear (1969).
SYNONYMS: *collaris, haedulus, spurrelli*.
COMMENTS: Discussion of Gambian specimens and possible synonymy with *G. coupeii incertae sedis* given in Schlitter et al. (1985).

Graphiurus microtis (Noack, 1887). Zool. Jahrb., 2:248.
TYPE LOCALITY: Zaire, Marungu, Mpala.
DISTRIBUTION: Zambia, Malawi, and Tanzania; limits unknown.
COMMENTS: Following Ansell (1989*a*, *b*) and Ansell and Dowsett (1988), *microtis* is recognized as a species distinct from *G. murinus*. See comment under *G. murinus* regarding synonyms for this species.

Graphiurus monardi (St. Leger, 1936). Ann. Mag. Nat. Hist., ser. 10, 17:465.
TYPE LOCALITY: Angola, Tyihumbwe (Chiumbe) River, 15 km above Dala (see Hill and Carter, 1941).
DISTRIBUTION: E Angola (Hayman, 1963*b*), NW Zambia (Ansell, 1978), and S Zaire; limits unknown.
SYNONYMS: *schoutedeni* (see Ansell, 1989*a*,).
COMMENTS: Allen (1939) listed *monardi* as a subspecies of *G. hueti*, an arrangement followed by Genest-Villard (1978). Ellerman et al. (1953) were correct in stating that *monardi* has no close affinity with *G. hueti*.

Graphiurus murinus (Desmarest, 1822). Mammalogie, *in* Encyclop. Méth., 2(Suppl.):542.
TYPE LOCALITY: South Africa, Cape of Good Hope.
DISTRIBUTION: Sudan (Setzer, 1956), Uganda (Delany, 1975), Ethiopia (Corbet and Yalden, 1972; Yalden et al., 1976), Kenya (Hollister, 1919), Tanzania (Swynnerton and Hayman, 1951), Malawi (Ansell and Dowsett, 1988; Ansell, 1989*b*), Mozambique (Smithers and Tello, 1976), E Zaire (Rahm and Christiaensen, 1963), S Angola, Zambia, Zimbabwe (Smithers and Wilson, 1979), Botswana (Smithers, 1971), E and N Namibia, southern Africa (de Graaff, 1981; Roberts, 1951; Skinner and Smithers, 1990).
SYNONYMS: *alticola, butleri, cineraceus, cinerascens, dasilvai, erythrobronchus, etoschae, griselda, isolatus, lalandianus, littoralis, marrensis, pretoriae, raptor, saturatus, schneideri, selindensis, soleatus, streeteri, sudanensis, tzaneenensis, vandami, vulcanicus, woosnami, zuluensis* (see Allen, 1939, and Setzer, 1956, for citations).
COMMENTS: The synonyms (and therefore the distribution) listed here for *G. murinus* almost certainly contain names which are actually synonyms of *G. microtis* and other species, but pending systematic revision they cannot be confidently separated here, and are

kept in their traditional listing under *murinus*. Meester et al. (1986) broke *murinus* into three subspecies: *murinus, microtis* and *griselda*. Though *microtis* is considered a valid species here and in Ansell (1989*a, b*) and Ansell and Dowsett (1988), the allocation of synonyms listed under this name in Meester et al. (1986) has yet to be documented. Three different karyotypes were found in the *G. murinus* group in southern Africa (Dippenaar et al., 1983), providing another indication of the complexity of the relationships of named forms traditionally placed in this species.

Graphiurus ocularis (A. Smith, 1829). Zool. J., 4:439.
TYPE LOCALITY: South Africa, Cape Prov., near Plettenberg Bay.
DISTRIBUTION: South Africa, in the Cape Prov. and SW Transvaal (Channing, 1984; de Graaff, 1991; Skinner and Smithers, 1990).
SYNONYMS: *capensis, elegans, typicus* (see Allen, 1939).

Graphiurus olga (Thomas, 1925). Ann. Mag. Nat. Hist. ser. 9, 16:191.
TYPE LOCALITY: Niger, Air, Tehsiderak (Tassederek), 2350 ft. (see Schlitter et al., 1985).
DISTRIBUTION: Known only from three localities in N Niger, N Nigeria and NE Cameroon; see Schlitter et al. (1985).
COMMENTS: See comments under *G. parvus*.

Graphiurus parvus (True, 1893). Proc. U.S. Natl. Mus., 16(95A):601.
TYPE LOCALITY: Kenya, Tana River, between the coast and Hameye.
DISTRIBUTION: Gambia, Ivory Coast, Ghana, Nigeria, Mali, Sudan, N Uganda and Kenya (see Hollister, 1919), Ethiopia, Somalia, Tanzania (see Swynnerton and Hayman, 1951). In W Africa see Rosevear (1969).
SYNONYMS: *brockmani, dollmani, foxi, internus, personatus* (see Allen, 1939, for citations).
COMMENTS: Schlitter et al. (1985) discussed taxonomic problems associated with the small dormice occurring in savannahs, scrub, and woodlands of west, east, and southern Africa. Three species of this grade are provisionally recognized here; *G. kelleni, G. olga,* and *G. parvus*.

Graphiurus platyops Thomas, 1897. Ann. Mag. Nat. Hist., ser. 6, 19:388.
TYPE LOCALITY: S Zimbabwe, Mashonaland, Enkeldorn.
DISTRIBUTION: Southern Africa, incl. South Africa (de Graaff, 1981; Skinners and Smithers, 1990), Zimbabwe (Smithers and Wilson, 1979), Zambia (Ansell, 1978), Botswana (Smithers, 1971), Angola (Hayman, 1963*b*; Hill and Carter, 1941), S Zaire, Malawi, (Ansell and Dowsett, 1988), and Mozambique (Smithers and Tello, 1976).
SYNONYMS: *angolensis, eastwoodae, jordani, parvulus* (see Ellerman et al., 1953).
COMMENTS: Discussion of *G. platyops* vs. *G. platyops parvulus* given by Ansell (1978).

Graphiurus rupicola (Thomas and Hinton, 1925). Proc. Zool. Soc. Lond., 1925:232.
TYPE LOCALITY: Namibia, Karibib, 3842 ft.
DISTRIBUTION: Namibia (see Thomas and Hinton, 1925) and NW South Africa (see Shortridge and Carter, 1938), in a narrow strip from Karibib and Mt. Brukaros, Namibia, south to Port Nolloth and Eenriet, NW Cape Prov., South Africa.
SYNONYMS: *australis, kaokoensis, montosus* (see Roberts, 1951).
COMMENTS: Ellerman et al. (1953) and Genest-Villard (1978) listed *rupicola* as a subspecies of *G. platyops*, but Roberts (1951) recognized it as a distinct species, a position followed here based on study of specimens.

Graphiurus surdus Dollman, 1912. Ann. Mag. Nat. Hist., ser. 8, 9:314.
TYPE LOCALITY: Equatorial Guinea, Rio Muni Prov., Benito River.
DISTRIBUTION: Rio Muni Prov. (Equatorial Guinea) and S Cameroon.
SYNONYMS: *schwabi* (see Allen, 1939, for citation).
COMMENTS: Allen (1939) and Robbins and Schlitter (1981) listed *schwabi* as a synonym of *haedulus* (= *G. lorraineus*), but the type specimen of *schwabi*, a juvenile, represents *G. surdus*.

Subfamily Leithiinae Lydekker, 1896. Proc. Zool. Soc. Lond., 1895:862 [1896].
SYNONYMS: Dryomyinae, Myomiminae, Seleviniinae.
COMMENTS: Lydekker (1895) proposed the family Leithiidae to separate the giant Pleistocene dormouse of Malta from other myoxids, and *Leithia* for the type genus. Major (1899) argued that *Leithia* was in fact a myoxid, and Leithiidae a junior

synonym of Myoxidae. De Bruijn (1967) proposed Dryomyinae, which included *Leithia, Dryomys, Eliomys,* and other genera. The International Code of Zoological Nomenclature (1985) mandates that when a nominal taxon is lowered in rank in the family group, its type genus remains the same. Because Dryomyinae de Bruijn contains *Leithia,* the correct name for the subfamily is Leithiinae. The results of the phylogenetic analysis by Wahlert et al. (In Press) indicated that the genera listed below form a monophyletic group composed of two tribes: Leithiini (*Dryomys* and *Eliomys*) and Myomimini (*Myomimus* and *Selevinia*). All genera in Leithiinae share derived dental and osteological characters; see comments in each genus.

Dryomys Thomas, 1906. Proc. Zool. Soc. Lond., 1905(2):348 [1906].

TYPE SPECIES: *Mus nitedula* Pallas, 1778.

SYNONYMS: *Chaetocauda* (see comment under *D. sichuanensis*), *Dyromys* (see Ellerman and Morrison-Scott, 1951:544, and footnote in Simpson, 1945:92), *Elius* (excluding *glis*).

COMMENTS: Tribe Leithiini. *Dryomys* seems most closely related to *Eliomys* (Wahlert et al., In Press). See also comments under Leithiinae.

Dryomys laniger Felten and Storch, 1968. Senckenberg. Biol., 49(6):429.

TYPE LOCALITY: Turkey, Antalya Prov., 20 km SSE Elmali, Bey Mtns, Ciglikara, 2000 m.

DISTRIBUTION: Turkey, W and C Toro (Taurus) Mtns, 1620-2000 m; limits unknown. Reported only from karst regions (Spitzenberger, 1976).

COMMENTS: Photographs provided by Felten and Storch (1968).

Dryomys nitedula (Pallas, 1778). Nova Spec. Quad. Glir. Ord., p. 88.

TYPE LOCALITY: Russia, lower Volga River.

DISTRIBUTION: SE Germany, Switzerland, Austria (Niethammer, 1960; Spitzenberger, 1983), Czechoslovakia (Andera, 1987; Kratochvil, 1967), Poland (Daoud, 1989); Ukraine (Bezrodny, 1991) and Byelorussia north to Lithuania and (in Russia) Vyshni-Volochek east to Moscow, S Gorkiy, and upper Volga R. to Kazan, south to lower Dnepr R., mouth of Volga R., and Caucasus Mtns (Ognev, 1947; Vereshchagin, 1959; also Kuznetsov, 1965; Pavlinov and Rossolimo, 1987), Italy (Filippucci, 1986; Paolucci et al., 1987), Hungary, Yugoslavia (Gazaryan, 1985; Krystufek, 1985*a*), Romania, Bulgaria, Greece (Ondrias, 1966), Turkey, Arabia (Harrison and Bates, 1991), E Syria (von Lehmann, 1965), N Israel (Atallah, 1978; Nevo and Amir, 1961, 1964), N Iraq (Jawdat, 1977), Iran (Lay, 1967), Afghanistan (Hassinger, 1973), N Pakistan (Roberts, 1977); Tadzhikistan (Allobergenov, 1986; Davydov, 1984), Uzbekistan, Turkmenistan, Kirghizia, and C Kazakhstan north to the S Altai Gobi (again see Kuznetsov, 1965; Ognev, 1947; Pavlinov and Rossolimo, 1987), the Tarbagaty Mtns east to the eastern limit of the Tien Shan Mtns, China (Ma et al., 1981, 1987; Wang and Yang, 1983); perhaps Lebanon (Lewis et al., 1967). In Europe see also Storch (1978).

SYNONYMS: *angelus, aspromontis, bilkjewiczi, carpathicus, caucasicus, daghestanicus, diamesus, dryas, intermedius, kurdistanicus, milleri, obolenskii, ognevi, pallidus, phrygius, pictus, ravijojla, robustus, saxatilis, tanaiticus, tichomirowi, wingei* (see Ellerman and Morrison-Scott, 1951; Corbet, 1978*c*).

COMMENTS: There has been no critical revision of this species throughout its range, and *D. nitedula* may contain two or more species. Taxonomy of subspecies occuring in the independant republics of the former USSR studied by Rossolimo (1971). Study of pelage color variation in Europe and Syria provided by Roesler and Witte (1969), in Yugoslavia by Krystufek (1985*a*), in Greece and SE Europe by Ondrias (1966), and in the independant republics of the former USSR by Ognev (1947). Chromosomal data provided by Filippucci et al. (1985), Zima (1987), and Zima and Kral (1984*a*). Phallic structure studied by Hrabe (1969). Illustrations and taxonomic implications of the os and glans penis, and stomach anatomy provided by Kratochvil (1973).

Dryomys sichuanensis (Wang, 1985). Acta Theriol. Sinica, 5(1):67.

TYPE LOCALITY: China, N Sichuan Prov., Pinwu county, Wang-lang Natural Reserve.

DISTRIBUTION: Known only from the type locality, a subalpine deciduous and coniferous forest in the Sichuan highlands; limits unknown.

COMMENTS: Wang (1985*a*) described *sichuanensis* as a species of *Chaetocauda,* a genus he placed in the Myomiminae. Based on Wang's comparative chart and photograph, the shape and details of the skull and teeth greatly resemble those of *Dryomys* , not

Myomimus; hence its placement here. Careful systematic study is required to determine whether or not *sichuanensis* is a distinct species of *Dryomys.*

Eliomys Wagner, 1840. Abh. math-phys. Cl. K. Bayer. Akad. Wiss., München, 3:176.

TYPE SPECIES: *Eliomys melanurus* Wagner, 1839.

SYNONYMS: *Bifa* (see Corbet, 1978*c*), *Eivissia* (see Alcover and Agusti, 1985), *Hypnomys, Maltamys, Tyrrhenoglis* (see Maempel and de Bruijn, 1982).

COMMENTS: Tribe Leithiini. See Nader et al. (1983) and Wahlert et al. (In Press) for clarification of the publication date for *Eliomys* and *Eliomys melanurus* Wagner (1839 vs. 1840). Corbet (1978*c*) cautioned that humans have affected the present distribution of the species. *Eliomys* appears most closely related to *Dryomys* (Wahlert et al., In Press). See also comments under Leithiinae.

Eliomys melanurus (Wagner, 1839). Gelehrte Anz. I. K. Bayer. Akad. Wiss., München, 8(37):299.

TYPE LOCALITY: Sinai (restricted to Mt. Sinai by Nader et al., 1983).

DISTRIBUTION: S Turkey (Misonne, 1957), Syria (Kahmann, 1981), Iraq (Kahmann, 1981; Nadachowski et al., 1978), Jordan (Atallah, 1978; Bodenheimer, 1958; Kahmann, 1981; Tristram, 1877), Lebanon (G. M. Allen, 1915; Lewis et al., 1967), Israel (Bodenheimer, 1958; Ilani and Shalmon, 1983; Kahmann, 1981), Saudi Arabia (Kahmann, 1981; Nader et al., 1981; Vesey-Fitzgerald, 1953; also Harrison and Bates, 1991), the Sinai penin. (Haim and Tchernov, 1974; Osborn and Helmy, 1980; Wassif and Hoogstraal, 1954; Kahmann, 1981), Egypt (Osborn and Helmy, 1980), Libya (Ranck, 1968), Tunisia (Kahmann and Thoms, 1987), Algeria (Kowalski and Rzebik-Kowalska, 1991), Morocco (Moreno and Delibes, 1981). Also, in N Africa, see Niethammer (1959).

SYNONYMS: *cyrenaicus, denticulatus, larotina, larotinus, munbyanus, occidentalis, tunetae* (see Ellerman and Morrison-Scott, 1951; Corbet, 1978*c*).

COMMENTS: Systematics of Moroccan population examined by Moreno (1989) and Moreno and Delibes (1981); comparative study of African populations by Kahmann and Thoms (1973*b*); descriptions of type specimens, karyotypes, color plates of skins, and biology of *melanurus* group provided by Kahmann and Thoms (1981); os penis figured by Didier (1953); biometric study of Tunesian populations by Kahmann and Thoms (1987); chromosomal data provided by Delibes et al. (1980), Dutrillaux (1986), Filippucci et al. (1988*a*, *b*, 1990), and Tranier and Petter (1978).

Allozyme data given by Filippucci et al. (1988*c*) indicated that N African and Middle Eastern populations form a monophyletic group, and supported the recognition of *melanurus* as a distinct species of *Eliomys.* This hypothesis is tentatively followed here, but is based on few samples and needs corroboration. A critical morphological study of *Eliomys* throughout its range is needed.

Eliomys quercinus (Linnaeus, 1766). Syst. Nat., 12th ed., 1:84.

TYPE LOCALITY: Germany.

DISTRIBUTION: Portugal, Spain (Abad, 1987; Moreno, 1989), Balearic Isls (Alcover, 1986; Kahmann and Alcover, 1974; Kahmann and Thoms, 1973*a*), Andorra (Gosalbez-Noguera et al., 1989), France, Corsica (Orsini and Cheylan, 1988), Belgium (Luyts, 1986), Germany (Feustel, 1984; Görner and Henkel, 1988; Mockel, 1986; Rehage, 1984; Schoppe, 1986), Netherlands (Foppen et al., 1989), Czechoslovakia (Andera, 1986; Kratochvil, 1967), Poland (Daoud, 1989), Finland, Ukraine (Bezrodny, 1991), in independant republics of the former USSR see Kuznetsov (1965) and Ognev (1947), Estonia (Masing and Timm, 1988), Switzerland, Austria (Spitzenberger, 1983), Italy, including Sardinia and Sicily (Cristaldi and Amori, 1988), the Dalmatia region of Yugoslavia. In Europe see Storch (1978). Roman record in England (probably introduced) reported by O'Conner (1986).

SYNONYMS: *nitela* var. *amori, cincticauda, dalmaticus, dichrurus, gotthardus, gymnesicus, hamiltoni, hortualis, jurrasicus, liparensis, nitela* var. *lusitanica, nitela, ophiusae, pallidus, raiticus, sardus, superans, valverdei* (see Ellerman and Morrison-Scott, 1951; Corbet, 1978*c*).

COMMENTS: Systematic study of Central European and Spanish populations by Moreno (1989) and Moreno et al. (1986). Chromosomal data given by Arroyo et al. (1982), Dutrillaux (1986), Filippucci et al. (1988*b*; 1990), Murariu et al. (1985), and Zima and Kral (1984*a*). Allozyme data analyzed by Filippucci et al. (1988*c*).

Myomimus Ognev, 1924. Priroda Okhota Ukraine [Nat. and Hunting in Ukraine], Kharkov, 1-2:115-116.

TYPE SPECIES: *Myomimus personatus* Ognev, 1924.

SYNONYMS: *Philistomys* (see Kowalski, 1963).

COMMENTS: Tribe Myomimini. This genus is in need of systematic revision. Distribution shown in Vorontsov (1986). The closest living relative of *Myomimus* may be *Selevinia* (Wahlert et al., In Press). See also comments under Leithiinae and *Selevinia*.

Myomimus personatus Ognev, 1924. Priroda Okhota Ukraine [Nat. and Hunting in Ukraine], Kharkov, 1-2:115.

TYPE LOCALITY: Turkmenistan, Kopet-Dag Mtns, near Kaine-Kassyr on the Sumbar R., on the Turkmenistan-Iran border.

DISTRIBUTION: Mountains in extreme NE Iran; Kopet-Dag and Malyy Balkhan Mtns, Turkmenistan (Kuznetsov, 1965; Marinina et al., 1987; Ognev, 1947; Shcherbina et al., 1988); Iskander, Uzbekistan (Zykov, 1987); limits unknown.

COMMENTS: External morphology of the glans penis and its taxonomic significance discussed by Rossolimo and Pavlinov (1985). Comparison with *M. roachi* provided by Rossolimo (1976*b*). Skull, teeth, os penis, feet, and ear ossicles figured by Ognev (1947).

Myomimus roachi (Bate, 1937). Ann. Mag. Nat. Hist., ser. 10, 20:399.

TYPE LOCALITY: Israel, Mt. Carmel, Tabun Cave, upper Pleistocene layers.

DISTRIBUTION: Pleistocene in Israel, extant in SE Bulgaria, Thrace, and W Turkey; limits unknown. New localities in Thrace figured by Kurtonur and Ozkan (1990).

SYNONYMS: *bulgaricus* (see Storch, 1978).

COMMENTS: Distribution and review provided by Storch (1978). Detailed description and figures of SE Bulgaria population given by Pechev et al. (1964). Taxonomic study of Bulgarian samples and comparison with *M. personatus* provided by Rossolimo (1976*b*).

Myomimus setzeri Rossolimo, 1976. Vestn. Zool., 4:51.

TYPE LOCALITY: Iran, Kurdistan, 4 km W of Bane.

DISTRIBUTION: W Iran; limits unknown

COMMENTS: Skull and teeth figured in Rossolimo (1976*a*).

Selevinia Belosludov and Bazhanov, 1939. Uchen. Zap. Kaz. Univ. Alma-Ata, 1(1):81.

TYPE SPECIES: *Selevinia betpakdalaensis* Belosludov and Bazhanov, 1939.

COMMENTS: Tribe Myomimini. Belosludov and Bazhanov (1939) recognized *Selevinia* as a new genus of murids, and then as a separate family of rodents seemingly allied to myoxids (Bazhanov and Belosludov, 1941). Based on a suite of morphological characters, Ognev (1947) hypothesized that *Selevinia* was a highly differentiated dormouse most closely related to *Myomimus*. Beacuse of its distinct dental formula and structure, Ognev (1947) listed *Selevinia* in a separate subfamily, a hypothesis independantly suggested by Ellerman (1949*a*). Published descriptions and figures of *Selevinia* included derived character states that allowed Wahlert et al. (In Press) to place this genus within Myomimini, supporting Ognev's (1947) hypothesis. Though the nearest extant relative of *Selevinia* is proposed to be *Myomimus*, *Plioselevinia*, a fossil genus known from Lower Pliocene breccia of Poland, may be the closest relative of this genus.

Selevinia betpakdalaensis Belosludov and Bazhanov, 1939. Uchen. Zap. Kaz. Univ. Alma-Ata, 1(1):81.

TYPE LOCALITY: S Kazakhstan, N Betpak-Dala Desert, Kyzyl-Ui.

DISTRIBUTION: SE and E Kazakhstan, deserts E, N and W of Lake Balkhash (Burdelov and Rossinskaya, 1959; Ismagilov, 1961; Kuznetsov, 1965).

SYNONYMS: *paradoxa* (see Bazhanov and Belosludov, 1941).

COMMENTS: Measurements and illustration of cranium, and biological observations provided by Bazhanov and Belosludov (1941). Skull, ear, ear ossicles, and foot figured in Ognev (1947). Photographs of live specimen, distribution, and biological data contributed by Sludskii (1977). Teeth and embryo figured by Bazhanov (1951). Mastoid portion of the bullae discussed by Pavlinov (1988).

Subfamily Myoxinae Gray, 1821. London Med. Repos., 15:303.

SYNONYMS: Glirinae (including only *Glis*), Glirulinae, Muscardininae.

COMMENTS: The results of Wahlert et al. (In Press) indicated that *Myoxus, Muscardinus,* and *Glirulus* form a monophyletic group. See also comments in each genus.

Glirulus Thomas, 1906. Proc. Zool. Soc. Lond., 1905(2):347 [1906].

TYPE SPECIES: *Myoxus javanicus* Schinz, 1845 (= *lapsus* for *japonicus;* see Thomas (1905*a*) for an explanation of the emendation of *javanicus* to *japonicus.*

SYNONYMS: *Amphidyromys, Paraglirulus* (see van der Meulen and de Bruijn, 1982).

COMMENTS: De Bruijn (1967) placed *Glirulus* and related fossil genera in their own subfamily based on dental characters. Daams (1981) included these genera in the Dryomyinae (= Leithiinae), as did van der Meulen and de Bruijn (1982). However, the phylogenetic study by Wahlert et al. (In Press) supported the inclusion of *Glirulus* within the Myoxinae (see also comment under Myoxinae). *Glirulus* appears to be the sister group of *Myoxus* and *Muscardinus* (Wahlert et al., In Press).

Glirulus japonicus (Schinz, 1845). Syst. Verzeichniss Säugeth., 2:530.

TYPE LOCALITY: Japan.

DISTRIBUTION: Japan, Isls of Honshu, Shikoku, and Kyushu.

SYNONYMS: *elegans, lasiotis* (see Ellerman and Morrison-Scott, 1951; Corbet, 1978*c*).

COMMENTS: *Myoxus elegans* Temminck, 1844, is preoccupied by *Graphiurus elegans* Ogilby, 1838, placed in the genus *Myoxus* by Wagner, 1843 (see Thomas, 1905*a*). Species of *Glirulus* are represented in Europe by early Miocene, Pliocene and early Pleistocene fossils (Hugueney and Mein, 1965; Kowalski, 1963; van der Meulen and de Bruijn, 1982). Medial and late Pleistocene representatives of *G. japonicus* on Japan were discussed by Kawamura (1989) and Kowalski and Hasegawa (1976). External genital morphology and its taxonomic significance were examined by Rossolimo and Pavlinov (1985).

Muscardinus Kaup, 1829. Skizz. Entwikel.-Gesch. Nat. Syst. Europ. Thierwelt, 1:139.

TYPE SPECIES: *Mus avellanarius* Linnaeus, 1758.

SYNONMYS: *Muscardinulus* (see Thaler, 1966).

COMMENTS: Kratochvil (1973) proposed to place *Muscardinus* in its own subfamily, Muscardininae, based on the uniquely derived stomach anatomy, molar morphology, and phallic features. De Bruijn (1967) and Daams (1981) both included *Muscardinus* in the Myoxinae based on dental morphology. This arrangement was supported by the results of Wahlert et al. (In Press). *Muscardinus* appears to be most closely related to *Myoxus* (Wahlert et al., In Press). See also comment under Myoxinae.

Muscardinus avellanarius (Linnaeus, 1758). Syst. Nat., 10th ed., 1:62.

TYPE LOCALITY: Sweden.

DISTRIBUTION: Cumbria and S England (Hurrell and McIntosh, 1984), France, Switzerland, Italy, Sicily, Austria (Spitzenberger, 1983), Germany (Görner and Henkel, 1988; Mockel, 1988; Rehage and Steinborn, 1984; Schoppe, 1986; Schulze, 1986, 1987), Netherlands (van Laar, 1984), Denmark (Jensen, 1980), S and C Sweden, Poland (Daoud, 1989; Kaluza, 1987; Wilk, 1987); Lithuania, Byelorussia, and Ukraine (Bezrodny, 1991) east to Russia (Kuznetsov, 1965; Ognev, 1947); Czechoslovakia (Andera, 1987), Hungary, Romania, Yugoslavia, Bulgaria (Belcheva et al., 1989), Greece (Ondrias, 1966), Corfu, N Turkey (Kivanc, 1983). In Europe see Storch (1978).

SYNONYMS: *abanticus, anglicus, corilinum, kroecki, muscardinus, niveus, pulcher, speciosus, trapezius, zeus* (see Ellerman and Morrison-Scott, 1951; Corbet, 1978*c*, 1984).

COMMENTS: Systematic study of W and C European subspecies provided by Witte (1962) and Roesler and Witte (1969), and of Turkish subspecies by Kivanc (1983). Chromosomal data reported by Belcheva et al. (1989), Zima (1987), and Zima and Kral (1984*a*); phallic structure examined by Hrabe (1969); illustrations and taxonomic implications of os and glans penis, and stomach anatomy, provided by Kratochvil (1973).

Myoxus von Zimmermann, 1780. Geogr. Gesch. Mensch. Vierf. Thiere, 2:351.

TYPE SPECIES: *Sciurus glis* Linnaeus, 1766.

SYNONYMS: *Elius* (excluding *dryas*), *Myorus* (see Palmer, 1904).

COMMENTS: See comments under Myoxidae and discussion in Wahlert et al. (In Press) for explanation of the validity of *Myoxus* versus *Glis*. *Myoxus* appears to be most closely related to *Muscardinus* (Wahlert et al., In Press).

Myoxus glis (Linnaeus, 1766). Syst. Nat., 12th ed., 1(1):87.

TYPE LOCALITY: Germany.

DISTRIBUTION: N Spain, France (Gautherin, 1988), Switzerland, Netherlands, Germany (Feustel, 1984; Görner and Henkel, 1988; Labes et al., 1987; Pankow, 1989; Rehage and Preywisch, 1984; Schoppe, 1986; Schulze, 1986; von Vietinghoff-Riesch, 1960), Poland (Bielecka, 1986; Daoud, 1989); Ukraine north to Byelorussia, east to Volga R., south to Saratov and Voronezh; Caucasus Mtns, south to N Iran (Lay, 1967) and SW Turkmenistan (see Bezrodny, 1991; Kuznetsov, 1965; Ognev, 1947; Ruprecht and Szwagrzak, 1986; Vereshchagin, 1959); the Mediterranean (except S and C Iberia, Balearic Isls), Corsica, Sardinia, Sicily, Elba, Italy (Cristaldi and Amori, 1988; Pilastro, 1990; Salotti, 1984; Vesmanis and Vesmanis, 1980; Witte, 1962), N Adriatic Isls, Austria, Czechoslovakia (Andera, 1986), Hungary (Becsy, 1982), Yugoslavia, Romania (Vasiliu, 1961), Bulgaria (Atanassov and Peschev, 1963), Greece (Ondrias, 1966; in Macedonia see Vohralik and Sofianidou, 1987), Crete, Corfu, Cephalonia, N Turkey; introduced to England (Corbet, 1984). In Europe see also Storch (1978) and Ondrias (1966).

SYNONYMS: *abruttii, argenteus, avellanus, caspicus, caspius, esculentus, giglis, insularis, intermedius, italicus, martinoi, melonii, minutus, orientalis, persicus, petruccii, pindicus, postus, pyrenaicus, spoliatus, subalpinus, tschetshenicus, vagneri, vulgaris* (see Ellerman and Morrison-Scott, 1951; Corbet, 1978*c*).

COMMENTS: Biometry and taxonomy of Asiago Plateau population by Franco (1988); diagnosis and distribution of subspecies in SE Europe by Ondrias (1966). Chromosomal data reported by Belcheva et al. (1989), Dutrillaux (1986), Zima (1987) and Zima and Kral (1984*a*); phallic structure examined by Hrabe (1969); illustrations and taxonomic significance of the os and glans penis, and stomach anatomy, provided by Kratochvil (1973). The significance of geographical variation within *Myoxus*, in the context of subspecific or specific level differentiation among populations, has not been investigated throughout its range.

SUBORDER HYSTRICOGNATHI
by Charles A. Woods

SUBORDER HYSTRICOGNATHI

SYNONYMS: Hystricomorpha.

COMMENTS: Originally used as a tribe (Tribus) by Tullberg under his Order Glires, Suborder Simplicidentati, the first time this natural group was united together without various hystricomorphous forms. Includes Hystricidae, Thryonomyoidea, Bathyergoidea, "Caviomorpha", and the Eocene-Oligocene Franimorpha. The term "Caviomorpha" is inappropriate since it is unlikely all New World forms are part of a single radiation, and the Caviidae are the least characteristic family of the group, so it is best to refer to these collectively as the New World hystricognaths, and discuss them in their superfamily groupings (Chinchilloidea, Cavioidea, Erethizontoidea, and Octodontoidea). See Tullberg (1899:69-71) and Wood (1985:478-495) for definitions of hystricognath characters and lists of taxa.

Family Bathyergidae Waterhouse, 1841. Ann. Mag. Nat. Hist., [ser. 1], 8:81.

COMMENTS: Reviewed by Honeycutt et al. (1991). The family has been traditionally divided into two subfamilies: Bathyerginae with grooved upper incisors (*Bathyergus*); and Georychinae with ungrooved upper incisors (*Cryptomys, Georychus, Heliophobius, Heterocephalus*). Honeycutt et al. (1991:59) did not support such groupings.

Bathyergus Illiger, 1811. Prodr. Syst. Mamm. Avium., p. 86.

TYPE SPECIES: *Mus maritimus* Gmelin, 1788 (= *Mus suillus* Schreber, 1782).

SYNONYMS: *Orycterus*.

Bathyergus janetta Thomas and Schwann, 1904. Abstr. Proc. Zool. Soc. Lond., 1904(2):6.

TYPE LOCALITY: South Africa, NW Cape Prov., coastal Little Namaqualand, Port Nolloth.

DISTRIBUTION: SW South Africa; S Namibia.

SYNONYMS: *inselbergensis, plowsi*.

COMMENTS: Ellerman et al. (1953) included *janetta* as a subspecies of *suillus*, but de Graaff (1975) regarded *janetta* and *suillus* as separate species.

Bathyergus suillus (Schreber, 1782). Die Säugethiere, 4:715.

TYPE LOCALITY: South Africa, Cape of Good Hope.

DISTRIBUTION: S South Africa.

SYNONYMS: *africana, intermedius, maritimus*.

COMMENTS: Includes *intermedius* (de Graaff, 1975:3).

Cryptomys Gray, 1864. Proc. Zool. Soc. Lond., 1864:124.

TYPE SPECIES: *Georychus holosericeus* Wagner, 1842 (= *Bathyergus hottentotus* Lesson, 1826).

SYNONYMS: *Coetomys, Typhloryctes*.

COMMENTS: Originally a subgenus of *Georychus*. See review by Nevo et al. (1987).

Cryptomys bocagei (de Winton, 1897). Ann. Mag. Nat. Hist., ser. 6, 20:323.

TYPE LOCALITY: Angola, Hanha.

DISTRIBUTION: C Angola, NW Zambia, S Zaire.

SYNONYMS: *kubangensis*.

Cryptomys damarensis (Ogilby, 1838). Proc. Zool. Soc. London, 1838:5.

TYPE LOCALITY: Namibia, Damaraland.

DISTRIBUTION: E Namibia, Botswana, W Zimbabwe, S Zambia, S Angola.

SYNONYMS: *lugardi, micklemi, ovamboensis*.

Cryptomys foxi (Thomas, 1911). Ann. Mag. Nat. Hist., ser. 8, 7:462.

TYPE LOCALITY: Nigeria, Panyam (1212 m).

DISTRIBUTION: C Nigeria.

COMMENTS: Honeycutt et al. (1991:50) recognized *foxi* as distinct.

Cryptomys hottentotus (Lesson, 1826). Zool., 1:166.

TYPE LOCALITY: South Africa, SW Cape Prov., near Paarl (east of Capetown).

DISTRIBUTION: South Africa to Tanzania, S Zaire, and Namibia.
SYNONYMS: *abberrans, albus, amatus, anomalus, arenarius, beirae, bigalkei, caecutiens, cradockensis, darlingi, exenticus, holosericeus, jamesoni, jorisseni, junodi, komatiensis, langi, ludwigii, mahali, melanoticus, montanus, natalensis, nemo, nimrodi, orangiae, palki; pallidus* Roberts, 1917; *pretoriae, rufulus, stellatus, streeteri, talpoides, transvaalensis, valschensis, vandami, vetensis, vryburgensis, whytei, zimbitiensis, zuluensis.*
COMMENTS: Includes *darlingi, holosericeus,* and *natalensis* (de Graaff, 1975:3-4; Honeycutt et al., 1991:51). Corbet and Hill (1991:208) recognized *natalensis* as a distinct species without comment.

Cryptomys mechowi (Peters, 1881). Sitzb. Ges. Naturf. Fr. Berlin, p. 133.
TYPE LOCALITY: Angola, Malange.
DISTRIBUTION: Angola, S Zaire, Malawi, Zambia, Tanzania.
COMMENTS: Includes *ansorgei, blainei,* and *mellandi* (de Graaff, 1975:3). Corbet and Hill (1991:207) spelled the name *mechowii.*

Cryptomys ochraceocinereus (Heuglin, 1864). Nouv. Acta Acad. Caes. Leop. Dresden, 31:3.
TYPE LOCALITY: Sudan, Upper Bahr-el-Ghazal.
DISTRIBUTION: E Nigeria, Central African Republic, N Zaire, S Sudan, NW Uganda.
SYNONYMS: *kummi, lechei.*
COMMENTS: Includes *kummi* and *lechei* (de Graaff, 1975:3).

Cryptomys zechi (Matschie, 1900). Sitzb. Ges. Naturf. Fr., Berlin, p. 146.
TYPE LOCALITY: Ghana, near Kete-Kradji.
DISTRIBUTION: EC Ghana, WC Togo.
COMMENTS: Elevated to species level by Honeycutt et al. (1991:52).

Georychus Illiger, 1811. Prodr. Syst. Mamm. Avium., p. 87.
TYPE SPECIES: *Mus capensis* Pallas, 1778.
COMMENTS: Honeycutt et al. (1991:53) indicated there may be two species in South Africa.

Georychus capensis (Pallas, 1778). Nova Spec. Quad. Glir. Ord., 76:172.
TYPE LOCALITY: South Africa, Cape of Good Hope.
DISTRIBUTION: South Africa.
SYNONYMS: *buffonii, canescens, leucops, yatesi.*
COMMENTS: Includes *canescens* and *yatesi* (de Graaff, 1975:3).

Heliophobius Peters, 1846. Bericht Verhandl. K. Preuss. Akad. Wiss. Berlin, 11:259.
TYPE SPECIES: *Heliophobius argenteocinereus* Peters, 1846.
SYNONYMS: *Myoscalops.*

Heliophobius argenteocinereus Peters, 1846. Bericht Verhandl. K. Preuss. Akad. Wiss. Berlin, 11:259.
TYPE LOCALITY: Mozambique, Tete (on the Zambezi River).
DISTRIBUTION: Zimbabwe, E Zambia, and N Mozambique to Zaire, Kenya, and N Tanzania.
SYNONYMS: *albifrons, angonicus, emini, kapiti, marungensis, mottoulei, pallidus* Gray, 1864, *robustus, spalax.*
COMMENTS: Honeycutt et al. (1991:54-55) concluded that the characters used to separate *H. spalax* from *H. argenteocinereus* are due to age variation, and that the genus is monotypic. See also de Graaff (1975:2). The chromosome no. is 2n=60.

Heterocephalus Rüppell, 1842. Mus. Senckenbergianum Abh., 3(2):99.
TYPE SPECIES: *Heterocephalus glaber* Rüppell, 1842.

Heterocephalus glaber Rüppell, 1842. Mus. Senckenbergianum Abh., 3(2):99.
TYPE LOCALITY: Ethiopia, Shoa.
DISTRIBUTION: C Somalia, C and E Ethiopia, C and S Kenya.
STATUS: Common.
SYNONYMS: *ansorgei, dunni, phillipsi, progrediens, scortecci, stygius.*
COMMENTS: According to Honeycutt et al. (1991:58) genetic data indicate there are two geographic groups of this species in Kenya. The chromosome no. is 2n=60.

Family Hystricidae G. Fischer, 1817. Mém. Soc. Imp. Nat., Moscow, 5:372.
COMMENTS: Reviewed by Mohr (1965) and by Van Weers (1976, 1977, 1978, 1979, 1983). Usually divided into two subfamilies (Atherurinae, Hystricinae), but Van Weers did not agree.

Atherurus F. Cuvier, 1829. Dict. Sci. Nat., 59:483.
TYPE SPECIES: *Hystrix macroura* Linnaeus, 1758.
COMMENTS: See Van Weers (1977:213).

Atherurus africanus Gray, 1842. Ann. Mag. Nat. Hist., [ser. 1], 10:261.
TYPE LOCALITY: Sierra Leone.
DISTRIBUTION: Gambia, Sierra Leone, Liberia, Ghana, Zaire, Kenya, Uganda, S Sudan.
SYNONYMS: *armata, burrowsi, centralis, turneri.*
COMMENTS: Includes *centralis* and *turneri*; see Misonne (1974:8).

Atherurus macrourus (Linnaeus, 1758). Syst. Nat., 10th ed., 1:57.
TYPE LOCALITY: "Habitat in Asia", restricted to Malaysia, Malacca by Lyon (1907:584).
DISTRIBUTION: E Assam (India), Szechwan, Yunnan, Hupei, and Hainan (China) to Malaya, Sumatra, and adjacent islands, Thailand, Laos, Vietnam, Burma.
SYNONYMS: *angustiramus, assamensis, hainanus, pemangilis, retardatus, stevensi, terutaus, tionis, zygomatica.*

Hystrix Linnaeus, 1758. Syst. Nat., 10th ed., 1:56.
TYPE SPECIES: *Hystrix cristata* Linnaeus, 1758.
SYNONYMS: *Acanthion*; *Oedocephalus* (see Allen, 1939), *Thecurus.*
COMMENTS: Divided into three subgenera: *Acanthion* Cuvier, 1823; *Hystrix* Linnaeus, 1758; and *Thecurus* Lyon, 1907 (see Van Weers, 1977, 1978, 1979).

Hystrix africaeaustralis Peters, 1852. Reise nach Mossambique, Säugethiere, p.170.
TYPE LOCALITY: Mozambique, Querimba coast and Tette, about 10°30' to 12°S, 40°30'E, sea level.
DISTRIBUTION: Mouth of the Congo River to Rwanda, Uganda, Kenya, W and S Tanzania, Mozambique, and South Africa. Sympatric with *H. cristata.*
STATUS: Common.
SYNONYMS: *capensis, prittwitzi, stegmanni, zuluensis.*
COMMENTS: Subgenus *Hystrix*. Revised by Corbet and Jones (1965); see Van Weers (1983).

Hystrix brachyura Linnaeus, 1758. Syst. Nat., 10th ed., 1:57.
TYPE LOCALITY: Malaysia, Malacca.
DISTRIBUTION: Nepal, Sikkim and Assam (India), C and S China, Burma, Thailand, Indochina, Malaya, Sumatra, Borneo, Malay Islands of Pinang, Singapore.
SYNONYMS: *alophus, bengalensis, grotei, hodgsoni, klossi, longicauda, millsi, mülleri, papae, subcristata, yunnanensis.*
COMMENTS: Subgenus *Acanthion* (Van Weers, 1979:233); also see Lekagul and McNeely (1988:492).

Hystrix crassispinis (Günther, 1877). Proc. Zool. Soc. Lond., 1876:736 [1877].
TYPE LOCALITY: Malaysia, Sabah, opposite Labuan Isl.
DISTRIBUTION: N Borneo.
SYNONYMS: *major.*
COMMENTS: Subgenus *Thecurus* (Van Weers, 1978:22).

Hystrix cristata Linnaeus, 1758. Syst. Nat., 10th ed., 1:56.
TYPE LOCALITY: "Asia", restricted to near Rome, Italy by Thomas (1911*a*:141).
DISTRIBUTION: Morocco to Egypt; Senegal to Ethiopia and N Tanzania; Sicily, Italy, Albania, and N Greece (European populations possibly introduced). Sympatric with *H. africaeaustralis* in central Africa.
STATUS: CITES - Appendix III (Ghana).
SYNONYMS: *aerula*(?), *cuvieri, daubentoni, europaea, galeata, occidanea, senegalica.*
COMMENTS: In subgenus *Hystrix*. Includes *galeata*; see Corbet (1978*c*:159) and Corbet and Jones (1965). Allen (1939:441) commented on the status of *cuvieri* in N Africa.

Hystrix indica Kerr, 1792. *In* Linnaeus, Anim. Kingdom, p. 213.

TYPE LOCALITY: India.

DISTRIBUTION: Transcaucasus; Asia Minor; Israel; Arabia to S Kazakhstan and India; Sri Lanka; Tibet (China).

STATUS: Locally common.

SYNONYMS: *aharonii, blanfordi, cuneiceps, hirsutirostris, leucurus, malabarica, mersinae, mesopotamica, narynensis, satunini, schmidtzi, zeylonensis.*

COMMENTS: Subgenus *Hystrix*. Citation based on Smellie's translation of Buffon (1781:206). Gromov and Baranova (1981:102) employed the name *leucura* for this species without reference to Kerr, 1792.

Hystrix javanica (F. Cuvier, 1823). Mem. Mus. Hist. Nat. Paris, 9:431.

TYPE LOCALITY: Indonesia, Java.

DISTRIBUTION: Java, Bali, Sumbawa, Flores, Lombok, Madura, Tonahdjampea, and S Sulawesi (Indonesia).

SYNONYMS: *brevispinosa, ecaudata, javanicum, sumbawae, torquata.*

COMMENTS: Subgenus *Acanthion*. Formerly included in *brachyura* by Chasen (1940), but see Van Weers (1979).

Hystrix pumila (Günther, 1879). Ann. Mag. Nat. Hist., ser. 5, 4:106.

TYPE LOCALITY: Philippines, Palawan, Paragua (= Puerto Princesa).

DISTRIBUTION: Palawan and Busuanga Isls (Philippines).

COMMENTS: Subgenus *Thecurus*.

Hystrix sumatrae (Lyon, 1907). Proc. U.S. Natl. Mus., 32:583.

TYPE LOCALITY: Indonesia, E coast of Sumatra, Aru Bay.

DISTRIBUTION: Sumatra.

COMMENTS: Subgenus *Thecurus*. Treated as a subspecies of *crassispinis* by Chasen (1940), but see Van Weers (1978).

Trichys Günther, 1877. Proc. Zool. Soc. Lond., 1876:739 [1877].

TYPE SPECIES: *Trichys lipura* Günther, 1877 (= *Hystrix fasciculata* Shaw, 1801).

COMMENTS: Reviewed by Van Weers (1976).

Trichys fasciculata (Shaw, 1801). Gen. Zool., 2:11.

TYPE LOCALITY: "Malacca", but no holotype. Van Weers (1976) designated a neotype from "Runuk Tanjong." Malaysia.

DISTRIBUTION: Borneo, Sumatra, Malaya.

SYNONYMS: *guentheri, lipura, macrotis.*

COMMENTS: Medway (1977:135) considered *lipura* a distinct species, but did not mention Van Weers (1976).

Family Petromuridae Tullberg, 1899. Ueber das System der Nagethiere, p. 147.

SYNONYMS: Petromyidae.

COMMENTS: Swanepoel et al. (1980:161) employed Petromuridae instead of Petromyidae for this family (see *Petromus*). The group had an elaborate Tertiary radiation composed of at least two subfamilies and seven known genera.

Petromus A. Smith, 1831. S. Afr. Quart. J., 1(5):10.

TYPE SPECIES: *Petromus typicus* A. Smith, 1831.

SYNONYMS: *Petromys.*

COMMENTS: Originally named *Petromus*, but unjustifiably emended to *Petromys* by Smith (1834), see Swanepoel et al. (1980:161).

Petromus typicus A. Smith, 1831. S. Afr. Quart. J., 1(5):11.

TYPE LOCALITY: South Africa, "Mountains towards mouth of Orange River".

DISTRIBUTION: W South Africa, Namibia, to SW Angola.

STATUS: Common.

SYNONYMS: *ausensis, barbiensis, cinnamomeus, cunealis, greeni, guinasensis, karasensis, kobosensis, marjoriae, namaquensis, pallidior, tropicalis, windhoekensis.*

COMMENTS: Numerous subspecies have been named, most of which are invalid.

Family Thryonomyidae Pocock, 1922. Proc. Zool. Soc. London, 1922:423.
COMMENTS: Several fossil forms from the Oligocene of North Africa.

Thryonomys Fitzinger, 1867. Sitzb. Akad. Wiss. Wien, 56(1):141.
TYPE SPECIES: *Aulacodus swinderianus* Temminck, 1827.
SYNONYMS: *Aulacodus, Choeromys, Triaulacodus.*

Thryonomys gregorianus (Thomas, 1894). Ann. Mag. Nat. Hist., ser. 6, 13:202.
TYPE LOCALITY: Kenya, Kiroyo, Luiji Reru River (00°35'S, 37°05'E).
DISTRIBUTION: Cameroon, Central African Republic, Zaire, S Sudan, Ethiopia, Kenya, Uganda, Tanzania, Malawi, Zambia, Zimbabwe, Mozambique.
STATUS: Common.
SYNONYMS: *camerunensis, harrisoni, logonensis, rutshuricus, sclateri.*
COMMENTS: See Misonne (1974:7).

Thryonomys swinderianus (Temminck, 1827). Monogr. Mamm., 1:248.
TYPE LOCALITY: Sierra Leone.
DISTRIBUTION: Africa south of the Sahara.
SYNONYMS: *calamophagus, raptorum, semipalmatus, variegatus.*

Family Erethizontidae Bonaparte, 1845. Cat. Meth. Mamm. Europe, p. 5.
COMMENTS: Spelled Erithizontidae by Corbet and Hill (1980:189); but see comment under *Erethizon*. Extinct forms date to the Oligocene of South America. The genus *Chaetomys* is moved to Echimyidae based on its retention of the deciduous premolars and other dental features (see Patterson and Wood, 1982:394), although retained in Erethizontidae by Corbet and Hill (1991:200).

Coendou Lacépède, 1799. Tabl. Mamm., p. 11.
TYPE SPECIES: *Hystrix prehensilis* Linnaeus, 1758.
COMMENTS: Cabrera (1961:600), Walker et al. (1975:1012), Hall (1981:853), Emmons and Feer (1990), Corbet and Hill (1991:200), and Handley and Pine (1992) included *Sphiggurus* in *Coendou*; but see Husson (1978:484-490). Woods (1984) and Nowak (1991) followed Husson and recognized *Sphiggurus* as distinct from *Coendou*. Includes C. sp. nov. (Black Dwarf Porcupine) of Emmons and Feer (1990:199) which is "presumably the same species as" *C. koopmani* (see Handley and Pine, 1992:237).

Coendou bicolor (Tschudi, 1844). Fauna Peruana, p. 186.
TYPE LOCALITY: Peru, Junín Dept., between Tulumayo and Chanchamayo Rivers.
DISTRIBUTION: Bolivia, Peru, Andean and W Ecuador, N Colombia, perhaps SW Colombia. Lives up to 2,500 m elevation.
SYNONYMS: *quichua, richardsoni, sanctamartae, simonsi.*
COMMENTS: May include *rothschildi* (Hall, 1981:854); Corbet and Hill (1991:200) listed *rothschildi* as a subspecies of *bicolor* (see comment under *rothschildi*). Emmons and Feer (1990:198) considered C. *quichua* Thomas, 1899, of the Andes of Ecuador to be a distinct species based on its smaller size and shorter tail.

Coendou koopmani Handley and Pine, 1992. Mammalia, 56:238.
TYPE LOCALITY: Brazil, Pará, Belém.
DISTRIBUTION: Amazonian lowlands south of Rio Amazonas, from Ilha de Marajó, Belém, and both banks of the Rio Tocantins south to Ilha do Tayaúna, and westward to Aurá Igarapé, ca. 15 km W Borba, on the Rio Madeira.
COMMENTS: Sympatric with *C. prehensilis* throughout its range.

Coendou prehensilis (Linnaeus, 1758). Syst. Nat., 10th ed., 1:57.
TYPE LOCALITY: Brazil, Pernambuco.
DISTRIBUTION: E Venezuela, Guyanas, C and E Brazil, Bolivia, Trinidad (see Goodwin and Greenhall, 1961:202). Lives up to 1,500 m in elevation.
SYNONYMS: *boliviensis, brandtii, centralis, cuandu, longicaudatus, platycentrotus.*

Coendou rothschildi Thomas, 1902. Ann. Mag. Nat. Hist., ser. 7, 10:169.
TYPE LOCALITY: Panama, Chiriqui, Sevilla Isl.
DISTRIBUTION: Panama.

COMMENTS: Possibly a subspecies of *bicolor* (Goldman, 1920:135; Hall, 1981:854). Corbet and Hill (1991:200) listed *rothschildi* as a subspecies of *bicolor*. Emmons and Feer (1990:198) stated that if *C. rothschildi* is valid, the *C. "bicolor"* west of the Andes are probably *C. rothschildi*.

Echinoprocta Gray, 1865. Proc. Zool. Soc. Lond., 1865:321.
TYPE SPECIES: *Erethizon (Echinoprocta) rufescens* Gray, 1865.

Echinoprocta rufescens (Gray, 1865). Proc. Zool. Soc. Lond., 1865:321.
TYPE LOCALITY: Colombia.
DISTRIBUTION: Colombia, montane areas of the eastern cordillera of Andes between 800 and 2,000 m elevation (Eisenberg, 1989).
STATUS: Unknown.
SYNONYMS: *epixanthus, sneiderni* (see comment under *Sphiggurus*).

Erethizon F. Cuvier, 1823. Mem. Mus. Hist. Nat. Paris, 9:432.
TYPE SPECIES: *Hystrix dorsata* Linnaeus, 1758.
COMMENTS: *Erithizon* Burnett, 1829, is a later spelling (Corbet and Hill 1980:189).

Erethizon dorsatum (Linnaeus, 1758). Syst. Nat., 10th ed., 1:57.
TYPE LOCALITY: E Canada (= Quebec Prov.).
DISTRIBUTION: C Alaska (USA) to S Hudson Bay and Labrador (Canada), south to E Tennessee, C Iowa, and C Texas (USA), N Coahuila, Chihuahua, and Sonora (Mexico), and S California (USA).
STATUS: Common.
SYNONYMS: *bruneri, couesi, doani, epixanthus, myops, nigrescens, picinum*.
COMMENTS: Reviewed by Miller and Kellogg (1955:631), who restricted the type locality; also see Anderson and Rand (1943). Includes *couesi*; (see review by Woods, 1973, Mammalian Species, 29). There are six subspecies.

Sphiggurus F. Cuvier, 1825. Dentes des Mammiferes, p. 256.
TYPE SPECIES: *Hystrix spinosa* F. Cuvier, 1823.
SYNONYMS: *Cercolabes, Sinoetherus*.
COMMENTS: This genus possibly dates from F. Cuvier, 1823. Mém Mus. Hist. Nat. Paris., 9:427, 433-435, where *Sphiggurus* seems only a French name "Sphiggure" except on pp. 433-434, where it is abbreviated "*S. spinosa*". Formerly included in *Coendou* by Cabrera (1961:600), Walker et al. (1975:1012), and Hall (1981:853); see also comments under *Coendou*. Considered a distinct genus by Husson (1978:484-490), Woods (1984), and Nowak (1991). Emmons and Feer (1990:201) recognized *Coendou [Sphiggurus] sneiderni*, the White-fronted Hairy Dwarf Porcupine, from Cauca Department of Colombia, which is apparently known from only one specimen, but this species is not recognized here, and is tentatively placed in the synonymy of *Echinoprocta* following Cabrera (1961).

Sphiggurus insidiosus (Lichtenstein, 1818). Zool. Mus. Univ. Berlin, p. 18-19.
TYPE LOCALITY: Brazil, Bahía, Salvador.
DISTRIBUTION: E and Amazonian Brazil, Surinam.
STATUS: Unknown.
SYNONYMS: *melanurus*.
COMMENTS: Formerly included in *Coendou*; see Husson (1978:484), and comment under *Sphiggurus*. Emmons and Feer (1990:199-200) considered *melanurus* Wagner, 1842, to be a separate species (Black-tailed Hairy Dwarf Porcupine).

Sphiggurus mexicanus (Kerr, 1792). *In* Linnaeus, Anim. Kingdom, 1:214.
TYPE LOCALITY: Mexico, in the mountains.
DISTRIBUTION: San Luis Potosi and Yucatan (Mexico) to W Panama.
STATUS: CITES - Appendix III (Honduras).
SYNONYMS: *laenatus, liebmani, yucataniae*.

Sphiggurus pallidus (Waterhouse, 1848). Nat. Hist. Mamm., 2:434.
TYPE LOCALITY: Unknown, probably West Indies.
DISTRIBUTION: West Indies.

STATUS: Extinct.
COMMENTS: Known only by the two immature specimens upon which the original description was based (Hall, 1981:854). Included in *Sphiggurus* based on the description of its pelage.

Sphiggurus spinosus (F. Cuvier, 1823). Mem. Mus. Hist. Nat. Paris, 9:433.
TYPE LOCALITY: Paraguay, along the Paraná River.
DISTRIBUTION: Paraguay, S and E Brazil, NE Argentina, Uruguay.
STATUS: CITES - Appendix III (Uruguay).
SYNONYMS: *affinis, couiy, nigricans; paraguayensis* Miranda Ribeiro, 1936; *roberti, sericeus.*
COMMENTS: Possibly restricted to the shores of the Paraná River. Formerly included in *Coendou;* see Husson (1978) and comment under *Sphiggurus.* Emmons and Feer (1990:200) considered the correct species name to be *paragayensis* Oken, 1816, which does predate Cuvier's name, but Cabrera (1961:602) noted that Oken's names are not recognized. If Oken's name is accepted, then the type species for *Sphiggurus* becomes *paragayensis* (see Tate, 1935:307).

Sphiggurus vestitus (Thomas, 1899). Ann. Mag. Nat. Hist., ser. 7, 4:284.
TYPE LOCALITY: Colombia.
DISTRIBUTION: Colombia, W Venezuela south of Lake Maracaibo at [+/-] 2,600 m elevation.
STATUS: Extremely rare.
SYNONYMS: *pruinosus.*
COMMENTS: Includes *pruinosus* Thomas, 1905 (see Cabrera, 1961:284). But also see Handley (1976:55), who listed *pruinosus* as distinct species, without comment. Emmons and Feer (1990:201) also considered *pruinosus* as a distinct species (Frosted Hairy Dwarf Porcupine), which is "apparently rare, known only from a few specimens" from the Andes near Mérida, Venezuela. Formerly included in *Coendou* (see comment under *Sphiggurus*).

Sphiggurus villosus (F. Cuvier, 1823). Mem. Mus. Hist. Nat. Paris, 9:434.
TYPE LOCALITY: Brazil, mountains near Rio de Janeiro, Corcoracto.
DISTRIBUTION: Minas Gerais to Rio Grande do Sul (SE Brazil).
COMMENTS: Considered a distinct species by Husson (1978:489). Cabrera (1961:600-601) included this species in *insidiosus;* see comment under *Sphiggurus;* formerly included in *Coendou.*

Family Chinchillidae Bennett, 1833. Proc. Zool. Soc. London, 1833:58.
SYNONYMS: Lagostomidae.

Chinchilla Bennett, 1829. Gard. Menag. Zool. Soc., 1:1.
TYPE SPECIES: *Mus laniger* Molina, 1782.

Chinchilla brevicaudata Waterhouse, 1848. Nat. Hist. Mamm., 2:241.
TYPE LOCALITY: Peru.
DISTRIBUTION: Andes of S Bolivia, S Peru, NW Argentina, and Chile.
STATUS: CITES - Appendix I (South American populations only); U.S. ESA - Endangered as *C. b. boliviana;* IUCN - Indeterminate.
SYNONYMS: *boliviana; chinchilla* (Lichtenstein, 1830), *intermedia, major.*
COMMENTS: Pine et al. (1979:362-363) included *brevicaudata* in *lanigera* without comment.

Chinchilla lanigera (Molina, 1782). Sagg. Stor. Nat. Chile, p. 301.
TYPE LOCALITY: Chile, Coquimbo Prov., Coquimbo.
DISTRIBUTION: N Chile, in foothills of the Andes and coastal mountains south to Coquimbo.
STATUS: CITES - Appendix I (South American populations only); IUCN - Indeterminate; Almost extinct in the wild.
SYNONYMS: *chinchilla* (Fischer, 1814), *velligera.*

Lagidium Meyen, 1833. Nouv. Acta Acad. Caes. Leop.-Carol., 16(2):576.
TYPE SPECIES: *Lagidium peruanum* Meyen, 1833.
COMMENTS: Revised by Osgood (1943:137). *Viscaccia* Oken, 1816 has precedence over *Lagidium,* but *Lagidium* was adopted in preference to *Viscaccia* by suspension of the Rules (Opinion 110, International Commission on Zoological Nomenclature, 1929*a*).

However, names from Oken's 1916 "Lehrbuch der Naturgeschichte" are non-Linnaean and not available (Hershkovitz, 1949*b*:289) and neither an opinion nor suspension of the Rules was required for the recognition of *Lagidium* (Hershkovitz, 1949*b*:296).

Lagidium peruanum Meyen, 1833. Nouv. Acta Acad. Caes. Leop.-Carol., 16(2):578.
TYPE LOCALITY: Peru, Puno Dept., Pisacoma.
DISTRIBUTION: C and S Peru.
STATUS: Unknown.
SYNONYMS: *arequipe, inca, punensis, saturata, subrosea.*

Lagidium viscacia (Molina, 1782). Sagg. Stor. Nat. Chile, p. 307.
TYPE LOCALITY: Chile, Santiago Prov., Cordillera de Santiago.
DISTRIBUTION: W Argentina, S and W Bolivia, N Chile, S Peru.
STATUS: Unknown.
SYNONYMS: *boxi, crassidens, criniger, crinigerum, cuscus, cuvieri, famatinae, lockwoodi, lutescens, moreni, pallipes, sarae, tontalis, tucumanum, vulcani.*

Lagidium wolffsohni (Thomas, 1907). Ann. Mag. Nat. Hist., ser. 7, 19:440.
TYPE LOCALITY: Argentina, Santa Cruz, Baguales and Vizcachas Mtns (50°50' S, 72°20' W).
DISTRIBUTION: SW Argentina and adjacent Chile.
STATUS: Common.

Lagostomus Brookes, 1828. Trans. Linn. Soc. Lond., 16:96.
TYPE SPECIES: *Lagostomus trichodactylus* Brookes, 1828 (= *Dipus maximus* Desmarest, 1817).
SYNONYMS: *Viscaccia* Schinz, 1825 (not Oken, 1816).

Lagostomus maximus (Desmarest, 1817). Nouv. Dict. Hist. Nat., Nouv. ed., 13:117.
TYPE LOCALITY: Unknown; possibly from pampas of Buenos Aires, Argentina; see Cabrera (1961:559).
DISTRIBUTION: N, C, and E Argentina, S and W Paraguay, SE Bolivia.
SYNONYMS: *americana, diana, inmollis, pamparum, petilidens, trichodactylus.*

Family Dinomyidae Troschel, 1874. Archiv Naturgesch., Bd. 2:132.
COMMENTS: Includes many very large extinct species.

Dinomys Peters, 1873. Monatsb. K. Preuss. Akad. Wiss. Berlin, p. 551.
TYPE SPECIES: *Dinomys branickii* Peters, 1873.

Dinomys branickii Peters, 1873. Monatsb. K. Preuss. Akad. Wiss. Berlin, p. 551.
TYPE LOCALITY: Peru, Junín Dept., Montaña de Vitoc, Amable Maria.
DISTRIBUTION: Venezuela, Colombia, Ecuador, Peru, Brazil, and Bolivia.
STATUS: IUCN - Endangered; rare.
SYNONYMS: *gigas, occidentalis, pacarana.*

Family Caviidae Gray, 1821. London Med. Repos., 15:304.

Subfamily Caviinae Gray, 1821. London Med. Repos., 15:304.
COMMENTS: Numerous extinct genera are present in the fossil record.

Cavia Pallas, 1766. Misc. Zool., p. 30.
TYPE SPECIES: *Cavia cobaya* Pallas, 1766 (= *Mus porcellus* Linnaeus, 1758).
SYNONYMS: *Anoema, Cobaya.*
COMMENTS: Reviewed by Hückinghaus (1961) and Cabrera (1961). See Tate (1935:343) for the reason *porcellus* is "excluded from consideration" as the type because the name was not included under the generic name at the time of its original publication.

Cavia aperea Erxleben, 1777. Syst. Regni Anim., 1:348.
TYPE LOCALITY: Brazil, Pernambuco.
DISTRIBUTION: Colombia, Ecuador, Venezuela, Guianas, Brazil, N Argentina, Uruguay, Paraguay.

SYNONYMS: *azarae, guianae, hilaria, hypoleuca; leucopyga* Brandt, 1815; *nana, pamparum, rosida.*

COMMENTS: Includes *guianae;* see Hückinghaus (1961:58) and Husson (1978:449); but also see Cabrera (1961:578) who placed *guianae* in *porcellus.* Corbet and Hill (1991:201) listed *guianae* as a distinct species without comment. Includes *pamparum* according to Massoia and Fornes (1967) and Hückinghaus (1961:57), but Cabrera (1961:577) listed *pamparum* as a distinct species. Also includes *nana* (see Hückinghaus, 1961:58).

Cavia fulgida Wagler, 1831. Isis, 24:512.

TYPE LOCALITY: "Amazonia"; probably an error (Cabrera, 1961:577).

DISTRIBUTION: E Brazil, between Minas Gerais and Santa Catarina.

STATUS: Unknown.

SYNONYMS: *nigricans, rufescens.*

Cavia magna Ximinez, 1980. Rev. Nordest. Biol., 3 (especial):148.

TYPE LOCALITY: "en las orillas del arroyo Imbé, municipio de Tramandaí, estado de Rio Grande del Sur, Brasil".

DISTRIBUTION: Dept. of Rocha, Uruguay, to Estados Rio Grande del Sur and Santa Catarina, Brazil.

Cavia porcellus (Linnaeus, 1758). Syst. Nat., 10th ed., 1:59.

TYPE LOCALITY: Brazil, Pernambuco (questionable).

DISTRIBUTION: Domesticated worldwide; possibly feral in N South America.

SYNONYMS: *anolaimae, cobaya; cutleri* Bennett, 1836; *leucopyga* Cabanis, 1848; *longipilis.*

COMMENTS: Husson (1978:451) reserved the use of *porcellus* to denote domesticated guinea pigs, which are probably derived from *tschudii* (Corbet and Hill, 1991:201), but also see Hückinghaus (1961:96), who regarded *porcellus* as a synonym of *aperea.* This species may be a domesticated animal with no established wild population. K. F. Koopman (pers. comm.) believes that N South American populations may be feral domestic guinea pigs; but see comment under *aperea.*

Cavia tschudii Fitzinger, 1857. Sitzb. Akad. Wiss. Wien, p. 154.

TYPE LOCALITY: Peru, Ica Dept., Ica.

DISTRIBUTION: Peru, S Bolivia, NW Argentina, N Chile.

STATUS: Common.

SYNONYMS: *arequipae, atahualpae; cutleri* Tschudi, 1844, *festina, osgoodi, pallidior, sodalis, stolida, umbrata.*

COMMENTS: Formerly included in *aperea* by Hückinghaus (1961:57); but see Cabrera (1961:579) and Pine et al. (1979:361), who considered *tschudii* a distinct species. Includes *stolida* (Cabrera, 1961:579), but also see Hückinghaus (1961:58), who considered *stolida* a distinct species.

Galea Meyen, 1832. Nouv. Acta Acad. Caes. Leop.-Carol., 16:597.

TYPE SPECIES: *Galea musteloides* Meyen, 1832.

COMMENTS: Reviewed by Hückinghaus (1961) and Cabrera (1961).

Galea flavidens (Brandt, 1835). Mem. Acad. Imp. Sci. St. Petersbourg, ser. 6, 3:439.

TYPE LOCALITY: Unknown; possibly Minas Gerais, Brazil.

DISTRIBUTION: Brazil.

SYNONYMS: *bilobidens.*

COMMENTS: Paula Couto (1950:232) considered *flavidens* synonymous with *spixii,* but see Cabrera (1961:573) who believed both to be distinct species and discussed the type locality.

Galea musteloides Meyen, 1832. Nouv. Acta Acad. Caes. Leop.-Carol., 16:597.

TYPE LOCALITY: Peru, Paso de Tacna, on road to Lake Titicaca.

DISTRIBUTION: S Peru, Bolivia, Argentina, N Chile.

STATUS: Common.

SYNONYMS: *auceps, boliviensis, comes, demissa, leucoblephara, littoralis, negrensis.*

COMMENTS: Reviewed by Mann (1950:2).

Galea spixii (Wagler, 1831). Isis, 24:512.

TYPE LOCALITY: Brazil, restricted to Minas Gerais, Lagoa Santa by Cabrera (1961:575).

DISTRIBUTION: Brazil, Bolivia, east of the Andes.
SYNONYMS: *campicola, palustris, saxatilis, wellsi.*
COMMENTS: Includes *wellsi* (Corbet and Hill, 1991:201); also see Cabrera (1961:575). Hückinghaus (1961:71) considered *wellsi* a distinct species.

Kerodon F. Cuvier, 1825. Dentes des Mammiferes, p. 151.
TYPE SPECIES: *Kerodon moco* Lesson, 1827 (= *Cavia rupestris* Wied, 1820).
SYNONYMS: *Cerodon.*
COMMENTS: Reviewed by Hückinghaus (1961). The name *Kerodon acrobata* was used by Mares and Ojeda (1982) for a second species in Brazil, although I can find no formal description of this species.

Kerodon rupestris (Wied-Neuwied, 1820). Isis, 6:43.
TYPE LOCALITY: Brazil, Bahia, Rio Belmonte. Designated by Moojen (1952) as Rio Grande de Belmonte, Rio Pardo, Rio San Francisco; but see Cabrera (1961:580).
DISTRIBUTION: E Brazil.
SYNONYMS: *aciurens, moco.*

Microcavia H. Gervais and Ameghino, 1880. Mamm. Fos. Am. Sud., p. 50.
TYPE SPECIES: *Microcavia typus* H. Gervais and Ameghino, 1880 (fossil).
SYNONYMS: *Caviella, Monticavia, Nanocavia.*
COMMENTS: Includes *Monticavia.* Reviewed by Hückinghaus (1961) and Cabrera (1961).

Microcavia australis (I. Geoffroy and d'Orbigny, 1833). Mag. Zool. Paris, p. 3.
TYPE LOCALITY: Argentina, Patagonia, vicinity of the lower part of the Rio Negro.
DISTRIBUTION: Argentina between Jujuy and Santa Cruz Provs.; Aisen Prov. (Chile); extreme S Bolivia.
STATUS: Common.
SYNONYMS: *joannia, kingii, maenas, nigriana, salinia.*
COMMENTS: Reviewed by Thomas (1921*b*:445).

Microcavia niata (Thomas, 1898). Ann. Mag. Nat. Hist., ser. 7, 1:282.
TYPE LOCALITY: Bolivia, La Paz Dept., Esperanza, Monte Sajama, 4,000 m.
DISTRIBUTION: SW Bolivia in the high Andes.
STATUS: Unknown.
SYNONYMS: *pallidior.*

Microcavia shiptoni (Thomas, 1925). Ann. Mag. Nat. Hist., ser. 9, 15:419.
TYPE LOCALITY: Argentina, Catamarca Prov., Laguna Blanca, 3,400 m.
DISTRIBUTION: NW Argentina in the mtns of Tucumán, Catamarca, and Salta Provinces.

Subfamily Dolichotinae Pocock, 1922. Proc. Zool. Soc. Lond., 1922:426.

Dolichotis Desmarest, 1820. Jour. Phys. Chim. Hist. Nat. Arts Paris, 88:205.
TYPE SPECIES: *Cavia patachonica* Shaw, 1801 (= *Cavia salinicola* Zimmerman, 1780).
SYNONYMS: *Mara, Pediolagus.*
COMMENTS: Includes *Pediolagus* Marelli, 1927 (Starrett, 1967:263); also see Cabrera (1961:580) who considered *Pediolagus* a distinct genus.

Dolichotis patagonum (Zimmermann, 1780). Geogr. Gesch. Mensch. Vierf. Thiere, 2:328.
TYPE LOCALITY: Argentina, Santa Cruz Prov., Puerto Deseado.
DISTRIBUTION: Argentina, approx. 28°S (Bolson de Pipanco, Catamarca Prov.) to 50°S.
SYNONYMS: *australis, centricola, magellanica, patachonicha.*

Dolichotis salinicola Burmeister, 1876. Proc. Zool. Soc. Lond., 1875:634 [1876].
TYPE LOCALITY: Argentina, SW Catamarca Prov., between Totoralejos and Recreo.
DISTRIBUTION: Chaco of Paraguay; NW Argentina as far south as Cordoba Prov.; extreme S Bolivia.
SYNONYMS: *ballivianensis, centralis, cyniclus.*
COMMENTS: Formerly included in *Pediolagus* (Starrett, 1967:263); also see comment under *Dolichotis.*

Family Hydrochaeridae Gray, 1825. Ann. Philos., n.s., 10:341.
COMMENTS: Some fossil forms extended into North America during late Pliocene and Pleistocene.

Hydrochaeris Brunnich, 1772. Zool. Fundamenta, p. 44.
TYPE SPECIES: *Sus hydrochaeris* Linnaeus, 1758.
SYNONYMS: *Capiguara, Xenohydrochoerus.*
COMMENTS: Mones and Ojasti (1986:5) recommended retention of the original name *Hydrochoerus;* Husson (1978:456-457) discussed the spelling of the generic name and concluded that *Hydrochoerus* Brisson, 1762 is not valid because Brisson's (1762) work was not consistently binominal.

Hydrochaeris hydrochaeris (Linnaeus, 1766). Syst. Nat., 12th ed., 1:103.
TYPE LOCALITY: "Habitat in Surinamo". Cabrera (1961:583) gave Pernambuco, Brazil, but see Husson (1978:451) for restriction to Suriname.
DISTRIBUTION: Panama, Colombia, Venezuela, the Guyanas and Peru, south through Brazil, Paraguay, NE Argentina, and Uruguay.
STATUS: Locally common, uncommon in Panama; farmed for meat in some areas.
SYNONYMS: *capybara, cobaya, dabbenei, irroratus, isthmius, notialis, uruguayensis.*
COMMENTS: Includes *isthmius;* see Handley (1966*a*:785). Reviewed by Mones and Ojasti (1986, Mammalian Species, 264, as *Hydrochoerus hydrochaeris*).

Family Dasyproctidae Bonaparte, 1838. Syn. Vert. Syst., *in* Nuovi Ann. Sci. Nat., Bologna, 2:112.
COMMENTS: The Agoutidae has been included in Dasyproctidae by several authors, but see comments under Agoutidae.

Dasyprocta Illiger, 1811. Prodr. Syst. Mamm. Avium., p. 93.
TYPE SPECIES: *Mus aguti* Linnaeus, 1766 (= *Mus leporinus* Linnaeus, 1758).
SYNONYMS: *Cloromis.*
COMMENTS: This group is in need of revision, and many of the species are questionable. Most are based on geographic distribution at this time. The West Indian agoutis are descendents of forms introduced to the islands. The pattern appears to be *D. leporina aguti* (from Brazil) to the Virgin Islands; *D. l. albida* on St. Vincent and Granada; *D. l. fulvus* on Martanique and St. Lucia; and *D. l. noblei* on Guadeluope, St. Kitts, Dominica, and Montserrat).

Dasyprocta azarae Lichtenstein, 1823. Verz. Doublet. Zool. Mus. Berlin, p. 3.
TYPE LOCALITY: Brazil, São Paulo.
DISTRIBUTION: EC and S Brazil, Paraguay, NE Argentina.
STATUS: Uncommon.
SYNONYMS: *acuti, aurea, catrinae, caudata, felicia, paraguayensis.*

Dasyprocta coibae Thomas, 1902. Novit. Zool., 9:136.
TYPE LOCALITY: Panama, Coiba Isl.
DISTRIBUTION: Endemic to Coiba Isl.
COMMENTS: Reviewed by Hall (1981:862).

Dasyprocta cristata (Desmarest, 1816). Nouv. Dict. Hist. Nat., Nouv. ed., 2(1):215.
TYPE LOCALITY: Suriname.
DISTRIBUTION: Guianas.
STATUS: Unknown.
COMMENTS: May be synonymous with *leporina;* see Husson (1978:464-466), and Hershkovitz (1972*b*:311-341). Husson (1978:464) designated the author as É. Geoffroy (1803:165), a work considered unavailable as it was never published (see Appendix I).

Dasyprocta fuliginosa Wagler, 1832. Isis, 25:1221.
TYPE LOCALITY: Brazil, Amazonas, Borba, on lower Rio Madeira (= "Amazon River").
DISTRIBUTION: Colombia, S Venezuela, Surinam, N Brazil, Peru.
STATUS: Common in many areas.
SYNONYMS: *caudata.*
COMMENTS: Reviewed by J. A. Allen (1915*a*:625).

Dasyprocta guamara Ojasti, 1972. Mem. Soc. Cienc. Nat. La Salle, 32:176.
TYPE LOCALITY: Venezuela, Delta Amacuro, Araguablsi, 9°13'27"N, 61°0'16"W.
DISTRIBUTION: Orinoco Delta (Venezuela).

Dasyprocta kalinowskii Thomas, 1897. Ann. Mag. Nat. Hist., ser. 6, 20:219.
TYPE LOCALITY: Peru, Cuzco Dept., Santa Ana Valley, Idma.
DISTRIBUTION: SE Peru.

Dasyprocta leporina (Linnaeus, 1758). Syst. Nat., 10th ed., 1:59.
TYPE LOCALITY: Suriname, Peninka, Peninka Creek and Cennewijne River.
DISTRIBUTION: Lesser Antilles, Venezuela, Guianas, Amazonian and E Brazil, introduced into the Virgin Islands.
STATUS: Common.
SYNONYMS: *aguti, aliba, cayana, cayennae, croconota, flavescens, fulvus, lucifer, lunaris, maraxica, noblei, rubrata.*
COMMENTS: Husson (1978:462-463) explained why *Mus leporinus* has priority, and why *aguti* should be synonymized. Recognized as *D. aguti cayana* by Cabrera (1961:585-586); may also include *cristata*. Includes *albida, antillensis,* and *noblei*; see Varona (1974:75), who included these forms in *aguti*.

Dasyprocta mexicana Saussure, 1860. Rev. Mag. Zool. Paris, ser. 2, 12:53.
TYPE LOCALITY: Mexico, probably Veracruz (= "hot zone of Mexico"); see Hall (1981:859).
DISTRIBUTION: C Veracruz and E Oaxaca (Mexico); introduced into W and E Cuba.

Dasyprocta prymnolopha Wagler, 1831. Isis, 24:619.
TYPE LOCALITY: Brazil, Pará. Original designation of Guyana was probably an error (Cabrera, 1961:588).
DISTRIBUTION: NE Brazil.
SYNONYMS: *nigriclunis.*
COMMENTS: Ojasti (1972:164) considered the status of this form uncertain.

Dasyprocta punctata Gray, 1842. Ann. Mag. Nat. Hist., [ser. 1], 10:264.
TYPE LOCALITY: Nicaragua, Chinadega, El Realejo.
DISTRIBUTION: Chiapas and Yucatan Peninsula (S Mexico) to S Bolivia, N Argentina, and SW Brazil. Introduced into W and E Cuba and the Cayman Isls.
STATUS: CITES - Appendix III (Honduras).
SYNONYMS: *bellula, boliviae, callida, chiapensis, chocoensis, columbiana, dariensis, isthmica, nuchalis, pallidiventris, pandora, richmondi, underwoodi, urucuma, variegata, yungarum, zamorae.*
COMMENTS: Includes *variegata* (Goldman, 1913:11); but also see Handley (1976:56) and Emmons and Feer (1990:207) who listed *variegata* as a distinct species.

Dasyprocta ruatanica Thomas, 1901. Ann. Mag. Nat. Hist., ser. 7, 8:272.
TYPE LOCALITY: Honduras, Roatan Isl.
DISTRIBUTION: Endemic to Roatan Isl.
STATUS: Threatened.
COMMENTS: Similar to *D. punctata* but much smaller.

Myoprocta Thomas, 1903. Ann. Mag. Nat. Hist., ser. 7, 12:464.
TYPE SPECIES: *Cavia acouchy* Erxleben, 1777.
COMMENTS: Cabrera (1961:591) and Emmons and Feer (1990:210-211) recognized Greenish (*M. acouchy*) and Reddish (*M. pratti*) Acuchis as distinct species. This group needs to be revised.

Myoprocta acouchy (Erxleben, 1777). Syst. Regn. Anim., 1:354.
TYPE LOCALITY: French Guiana, Cayenne.
DISTRIBUTION: Guianas, Amazonian Brazil, Ecuador, N Peru, S Venezuela, S Colombia (upper Rio Uaupes).
STATUS: Common.
SYNONYMS: *acoushy, archidonae, caymanum, demararae, limanus, milleri, parva, pratti, puralis.*
COMMENTS: Includes *pratti* (see Husson, 1978:471-472). Formerly included *exilis* (Husson, 1978:468).

Myoprocta exilis (Wagler, 1831). Isis, 24:621.
TYPE LOCALITY: Brazil, mouth of Rio Negro; reviewed by Allen (1916).
DISTRIBUTION: Guianas, S Venezuela and Colombia, E Ecuador, N Peru, Amazon Basin (Brazil).
STATUS: Unknown.
SYNONYMS: *leptura*.
COMMENTS: Listed by Cabrera (1961:591) as *M. acouchy exilis*; *exilis* considered a species by Husson (1978:468-472) and Tate (1939:151-229).

Family Agoutidae Gray, 1821. London Med. Repos., 15:304.
SYNONYMS: Cuniculidae.
COMMENTS: The familial or subfamilial status of this taxon is debated; Husson (1978:472) and Cabrera (1961:593) placed in Agoutidae; Hall (1981:858) placed in Agoutinae of the Dasyproctidae; Starrett (1967:269) placed in Cuniculinae of the Dasyproctidae; and Ellerman (1940:221) in the Cuniculidae. As *Cuniculus* is unavailable (see comments under *Agouti*), Agoutidae is the proper familial name.

Agouti Lacépède, 1799. Tabl. Mamm., p. 9.
TYPE SPECIES: *Mus paca* Linnaeus, 1766.
SYNONYMS: *Coelogenys, Cuniculus, Paca, Stictomys*.
COMMENTS: See Handley (1976:55) and Husson (1978:47). *Cuniculus* Brisson, 1762 is unavailable because Brisson's (1762) work was not consistently binominal (see Hopwood, 1947; International Commission on Zoological Nomenclature, 1955:197).

Agouti paca (Linnaeus, 1766). Syst. Nat., 12th ed., 1:81.
TYPE LOCALITY: French Guiana, Cayenne.
DISTRIBUTION: SE San Luis Potosi (Mexico) to Paraguay, Guianas, and S Brazil. Introduced into Cuba.
STATUS: CITES - Appendix III (Honduras).
SYNONYMS: *alba, fulvus, guanta, mexianae, sublaevis, subniger, venezuelica*.

Agouti taczanowskii (Stolzmann, 1865). Proc. Zool. Soc. Lond., 1865:161.
TYPE LOCALITY: Ecuador, Andes.
DISTRIBUTION: Mountains of Peru, Ecuador, Colombia, and NW Venezuela.
STATUS: Unknown.
SYNONYMS: *andina, sierrae*.
COMMENTS: Cabrera (1961:595) placed this species in *Stictomys*, but Handley (1976:55) and Gardner (1971:1088) included it in *Agouti*.

Family Ctenomyidae Lesson, 1842. Nouv. Tabl. Règne Animal., Mamm., p. 105.
COMMENTS: This assemblage of 56 named species is in need of revision. The species are variable in chromosome number, but fairly uniform in morphology, suggesting that the major radiation of species was in the Pleistocene (Roig and Reig, 1969:666; Reig et al. (1990*b*:89). *Ctenomys* is most closely allied to the octodontid *Octodontomys*. Whether the group should be recognized as a subfamily (Ctenomyinae, Reig, 1958), or as a family is debated. Although the designation of subfamily within the Octodontidae best reflects the evolutionary history of the group, it is more common to treat the group as a distinct family specialized for fossorial life. Glanz and Anderson (1990) gave a cladogram and list of synapomorphies for the Ctenomyidae. Here 39 species are recognized. Reig (1986:408) stated that his analysis indicated there may be 44 valid species. See Reig et al. (1990*b*) for an overview of ctenomyid taxonomy, and Cook et al. (1990:22-24) for a discussion of the possible significance of the extensive chromosomal variation in *Ctenomys* and octodontids (2n=10-102), which is nearly as great as the known variation for all mammals.

Ctenomys Blainville, 1826. Bull. Sci. Soc. Philom. Paris, p. 62.
TYPE SPECIES: *Ctenomys brasiliensis* Blainville, 1826.
SYNONYMS: *Chacomys, Haptomys, Orycteromys*.
COMMENTS: Cabrera (1961, based on Osgood, 1946) and Nowak (1991) listed two subgenera: *Ctenomys*, with all but *C. conoveri*, which is placed in subgenus *Chacomys*.

Revision of this genus may reduce the number of species recognized. The chromosomal variation within *Ctenomys* is 2n=10-70.

Ctenomys argentinus Contreras and Berry, 1982. Historia Natural (Argentina), 2(20):166.
TYPE LOCALITY: Argentina, Chaco Prov., Libertador Gen. San Martín Dept., Campo Araos (26°31'S, 59°15'W).
DISTRIBUTION: Argentina, NC Chaco Prov.
COMMENTS: The chromosome no. is 2n=44.

Ctenomys australis Rusconi, 1934. Rev. Chil. Nat. Hist., 38:108.
TYPE LOCALITY: Argentina, Buenos Aires Prov.
DISTRIBUTION: E Argentina in Buenos Aires Prov.
COMMENTS: Considered distinct from *porteousi* by Contreras and Reig (1965:62) and Roig and Reig (1969:670). The chromosome no. is 2n=48.

Ctenomys azarae Thomas, 1903. Ann. Mag. Nat. Hist., ser. 7, 11:228.
TYPE LOCALITY: Argentina, Buenos Aires Prov. (37°45'S, 65°W).
DISTRIBUTION: Only known from La Pampa Prov. (Argentina).
COMMENTS: Considered distinct from *mendocinus* by Roig and Reig (1969:670). The chromosome no. is 2n=47-48.

Ctenomys boliviensis Waterhouse, 1848. Nat. Hist. Mamm., 2:278.
TYPE LOCALITY: Bolivia, Santa Cruz Dept., Santa Cruz de la Sierra.
DISTRIBUTION: C Bolivia at the foot of the Andes, W Paraguay, and Formosa Prov., Argentina.
SYNONYMS: *goodfellowi*.
COMMENTS: The chromosome no. is 2n=34-46.

Ctenomys bonettoi Contreras and Berry, 1982 Historia Natural (Argentina), 2(14):123.
TYPE LOCALITY: Argentina, Chaco Prov., Dept. Sargento Cabral, 7.5 km SE of Capitán Solari (26°48'S, 59°33'W).
DISTRIBUTION: Argentina, Chaco Prov.

Ctenomys brasiliensis Blainville, 1826. Bull. Sci. Soc. Philom. Paris, 3:62.
TYPE LOCALITY: Brazil, Minas Gerais.
DISTRIBUTION: E Brazil.

Ctenomys colburni J. A. Allen, 1903. Bull. Am. Mus. Nat. Hist., 19:188.
TYPE LOCALITY: Argentina, Santa Cruz Prov., Arroyo Aiken, 95 km SE Lake Buenos Ayres.
DISTRIBUTION: Extreme W Santa Cruz Prov. (Argentina).

Ctenomys conoveri Osgood, 1946. Fieldiana Zool., 31:47.
TYPE LOCALITY: Paraguay, Boqueron, 16 km W of Filadelfia (22°15'S, 60°10'W).
DISTRIBUTION: Chaco of Paraguay, and adjacent Argentina.
COMMENTS: Cabrera (1961:556) based on Osgood (1946:47) placed this form in a separate subgenus *Chacomys*, as did Nowak (1991:940). The chromosome no. is 2n=48-50.

Ctenomys dorsalis Thomas, 1900. Ann. Mag. Nat. Hist., ser. 7, 6:385.
TYPE LOCALITY: Paraguay, Boqueron Prov., N Chaco.
DISTRIBUTION: Paraguay, west of the River Paraguay in the northern Chaco.

Ctenomys emilianus Thomas and St. Leger, 1926. Ann. Mag. Nat. Hist., ser. 9, 18:637.
TYPE LOCALITY: Argentina, Neuquén Prov., Chos Malal.
DISTRIBUTION: Neuquén Prov., at the base of the Andes (Argentina).

Ctenomys frater Thomas, 1902. Ann. Mag. Nat. Hist., ser. 7, 9:185.
TYPE LOCALITY: Argentina, Jujuy Prov., Sunchal.
DISTRIBUTION: Mountains south of Jujuy and in Salta Prov. (Argentina); SW Bolivia at 2,000-4,500 m.
STATUS: Uncommon.
SYNONYMS: *barbarus, budini, mordosus, sylvanus, utibilis*.
COMMENTS: See Cook et al. (1990:1-6). The chromosome no. is 2n=52; FN=78.

Ctenomys fulvus Philippi, 1860. Reise Wuste Atacama, p. 157.
TYPE LOCALITY: Chile, Antofagasta Prov., Pingo-Pingo.
DISTRIBUTION: Mountains and Monte desert of NW Argentina and N Chile.

SYNONYMS: *atracamensis, chilensis, coludo, famosus, fulvus, johannis, pallidus, pernix, robustus, tolduco.*

COMMENTS: Includes *robustus,* a Rassenkreis subspecies restricted to the oasis of Pica in Tarapaca Prov., Chile (Mann, 1978:292). Includes *coludo, famosus, johanis,* and *tulduco;* see Cabrera (1961:548-549) but also see Contreras et al. (1977), who considered these forms provisionally distinct; also see comment under *validus.* The chromosome no. is 2n=26.

Ctenomys haigi Thomas, 1917. Ann. Mag. Nat. Hist., ser. 9, 3:210.

TYPE LOCALITY: Argentina, Chubut Prov., El Maitén, 700 m.

DISTRIBUTION: Chubut and Río Negro Prov., Argentina.

SYNONYMS: *lentulus.*

COMMENTS: Was combined with *mendocinus,* but Pearson (1984:231) elevated to a species based on its unique karyotype. The chromosome no. is 2n=50.

Ctenomys knighti Thomas, 1919. Ann. Mag. Nat. Hist., ser. 9, 3:498.

TYPE LOCALITY: Argentina, Catamarca Prov., Otro Cerro.

DISTRIBUTION: Mountains between Tucumán and La Rioja (W Argentina), north to Salta.

SYNONYMS: *viperinus.*

COMMENTS: Cabrera (1961:550) believed *viperinus* synonymous with *knighti.* The chromosome no. is 2n=36.

Ctenomys latro Thomas, 1918. Ann. Mag. Nat. Hist., ser. 9, 1:38.

TYPE LOCALITY: Argentina, Tucumán Prov., Tapia, 600 m.

DISTRIBUTION: NW Argentina in Tucumán and Salta Provs.

COMMENTS: Considered distinct from *mendocinus* by Roig and Reig (1969:670). The chromosome no. is 2n=40-42.

Ctenomys leucodon Waterhouse, 1848. Nat. Hist. Mamm., 2:281.

TYPE LOCALITY: Bolivia, La Paz Dept., San Andres de Machaca.

DISTRIBUTION: W Bolivia and E Peru around Lake Titicaca.

COMMENTS: See Cook et al. (1990:2-3). The chromosome no. is 2n=36; FN=68.

Ctenomys lewisi Thomas, 1926. Ann. Mag. Nat. Hist., ser. 9, 17:323.

TYPE LOCALITY: Bolivia, Tarija Dept., Sama, 4,000 m.

DISTRIBUTION: S Bolivia.

COMMENTS: May be semi-aquatic (Anderson, pers. comm.). See Cook et al. (1990:13-16). The chromosome no. is 2n=56; FN=74.

Ctenomys magellanicus Bennett, 1836. Proc. Zool. Soc. Lond., 1835:190 [1836].

TYPE LOCALITY: Chile, Magallanes, Bahia de San Gregorio.

DISTRIBUTION: Extreme S Chile and S Argentina.

STATUS: Reduced in numbers by sheep grazing.

SYNONYMS: *dicki, fodax, fueginus, neglectus, osgoodi; robustus* J. A. Allen, 1903.

COMMENTS: The chromosome no. is 2n=34-36.

Ctenomys maulinus Philippi, 1872. Z. Ges. Naturw., N. F., 6:442.

TYPE LOCALITY: Chile, Talca Prov., Laguna de Maule.

DISTRIBUTION: Between Talca and Cautin Provs. (SC Chile), and Neuquén Prov. (Argentina).

SYNONYMS: *brunneus.*

COMMENTS: The chromosome no. is 2n=26.

Ctenomys mendocinus Philippi, 1869. Arch. Naturgesch., 1:38.

TYPE LOCALITY: Argentina, Mendoza Prov., Mendoza.

DISTRIBUTION: East of the mountains from Salta Prov. to Chubut (Argentina).

SYNONYMS: *bergi, fochi, juris, pundti, recessus.*

COMMENTS: Formerly included *azarae, latro, occultus,* and *tucumanus* (Roig and Reig, 1969:670; Reig et al., 1966). Pearson (1984) recognized *haigi* as a distinct species. The chromosome no. is 2n=47-48.

Ctenomys minutus Nehring, 1887. Sitzb. Ges. Naturf. Fr. Berlin, p. 47.

TYPE LOCALITY: Brazil, Rio Grande do Sul, Campos E of Mondo Novo.

DISTRIBUTION: Rio Grande do Sul and Mato Grosso (SW Brazil), Uruguay, NW Argentina.

SYNONYMS: *bicolor, rionegrensis.*

COMMENTS: Reviewed by Langguth and Abella (1970). Altuna and Lessa (1985) regarded *rionegrensis* as a full species. The chromosome no. is 2n=46-55.

Ctenomys nattereri Wagner, 1848. Arch. Naturgesch., 1:72.
TYPE LOCALITY: Brazil, Mato Grosso, Caicara.
DISTRIBUTION: Mato Grosso.
SYNONYMS: *brasiliensis* Pelzen, 1883; *rondoni*.

Ctenomys occultus Thomas, 1920. Ann. Mag. Nat. Hist., ser. 9, 6:243.
TYPE LOCALITY: Argentina, Monteagudo, 80 km SE Tucumán City.
DISTRIBUTION: NW Argentina in Tucumán and adjacent provs.
COMMENTS: Revised by Reig et al. (1966); also see Cabrera (1961:552) who included this species in *mendocinus*. The chromosome no. is 2n=22.

Ctenomys opimus Wagner, 1848. Arch. Naturgesch., 1:75.
TYPE LOCALITY: Bolivia, Oruro Dept., Monte Sajama.
DISTRIBUTION: NW Argentina, SW Bolivia, S Peru, N Chile between 2,000 and 5,000 m elevation on the high Andean steppe (= Puna).
STATUS: Common.
SYNONYMS: *luteolus*.
COMMENTS: May be conspecific with *fulvus*; *opimus* and *fulvus* cannot be separated on the basis of karyotypes; see Gallardo (1979:80) and Cook et al. (1990). The chromosome no. is 2n=26; FN=48.

Ctenomys pearsoni Lessa and Langguth, 1983. Res. Com. Jour. Cienc. Nat. Montevideo (Uruguay) 3:86.
TYPE LOCALITY: Uruguay, Dept. Colonia, 25 km SE of Carmelo, Arroyo Limetas.
DISTRIBUTION: Soriana, San José, and Colonia Depts. (Uruguay).
COMMENTS: The chromosome no. is 2n=56-70.

Ctenomys perrensis Thomas, 1898. Ann. Mag. Nat. Hist., ser. 6, 18:311.
TYPE LOCALITY: Argentina, Corrientes Prov., Goya.
DISTRIBUTION: Corrientes, Entre Ríos, and Misiones Provs. (NE Argentina).
COMMENTS: The chromosome no. is 2n=50-56.

Ctenomys peruanus Sanborn and Pearson, 1947. Proc. Biol. Soc. Washington, 60:13.
TYPE LOCALITY: Peru, Puno Dept., Pisacoma.
DISTRIBUTION: Altiplano of extreme S Peru.

Ctenomys pontifex Thomas, 1918. Ann. Mag. Nat. Hist., ser. 9, 1:39.
TYPE LOCALITY: Argentina, Mendoza Prov., San Rafael.
DISTRIBUTION: W Argentina, east of the Andes in San Luis and Mendoza Provs.

Ctenomys porteousi Thomas, 1916. Ann. Mag. Nat. Hist., ser. 8, 18:304.
TYPE LOCALITY: Argentina, Buenos Aires Prov., Bonifacio.
DISTRIBUTION: Buenos Aires and La Pampa Provs. (E Argentina).
COMMENTS: Cabrera (1961:555) included *australis* in this species; but see Contreras and Reig (1965:62) and Roig and Reig (1969:670) who considered *australis* as a distinct species; this is followed here. The chromosome no. is 2n=47-48.

Ctenomys saltarius Thomas, 1912. Ann. Mag. Nat. Hist., ser. 8, 10:639.
TYPE LOCALITY: Argentina, Salta Prov., Salta.
DISTRIBUTION: Salta and Jujuy Provs. (N Argentina).
STATUS: Common.

Ctenomys sericeus J. A. Allen, 1903. Bull. Am. Mus. Nat. Hist., 19:187.
TYPE LOCALITY: Argentina, Santa Cruz Prov., Rio Chico.
DISTRIBUTION: SW Argentina in Santa Cruz, Chubut, and Río Negro Provs.

Ctenomys sociabilis Pearson and Christie, 1985. Hist. Nat. (Argentina), 5:338.
TYPE LOCALITY: Argentina, Prov. Neuquén, 2 km W Cerro Puntudo, Estancia Fortín Chacabuco (1075 m).
DISTRIBUTION: In region of Reserva Nacional del Parque Nacional Nahuel Huapi, Neuquén Prov., Argentina.
COMMENTS: The chromosome no. is 2n=56.

Ctenomys steinbachi Thomas, 1907. Ann. Mag. Nat. Hist., ser. 7, 20:164.
TYPE LOCALITY: Bolivia, Santa Cruz Dept., near Santa Cruz de la Sierra.
DISTRIBUTION: Bolivia, east of the Andes.
COMMENTS: The chromosome no. is 2n=10.

Ctenomys talarum Thomas, 1898. Ann. Mag. Nat. Hist., ser. 7, 1:285.
TYPE LOCALITY: Argentina, Buenos Aires Prov., Los Talas.
DISTRIBUTION: Along the coast in Buenos Aires Prov. (E Argentina), possibly to Santa Fe Prov.
SYNONYMS: *antonii*.
COMMENTS: Possibly a subspecies of *mendocinus* (Cabrera, 1961:556). The chromosome no. is 2n=46-50.

Ctenomys torquatus Lichtenstein, 1830. Darst. Säugeth., text of pl. 31.
TYPE LOCALITY: Brazil, S provinces, and banks of Uruguay River. See Moojen (1952:188).
DISTRIBUTION: Uruguay, NE Argentina, extreme S Brazil.
COMMENTS: Considered a distinct species by Ellerman (1940:167) and Novak (1991:939); but see Cabrera (1961:547) who provisionally included it in *brasiliensis*. According to Redford and Eisenberg (1992), the chromosome no. is 2n=56-70, indicating the group is probably a superspecies, although Reig et al. (1990*b*) reported a chromosome no. of 2n=44-46.

Ctenomys tuconax Thomas, 1925. Ann. Mag. Nat. Hist., ser. 9, 15:583.
TYPE LOCALITY: Argentina, Tucumán Prov., Concepcion, 500 m.
DISTRIBUTION: East of the mountains to 3,000 m in Tucumán Prov. (NW Argentina).
COMMENTS: The chromosome no. is 2n=58-61.

Ctenomys tucumanus Thomas, 1900. Ann. Mag. Nat. Hist., ser. 7, 6:301.
TYPE LOCALITY: Argentina, Tucumán Prov., 450 m.
DISTRIBUTION: NW Argentina.
COMMENTS: The chromosome no. is 2n=28. Considered distinct from *mendocinus* by Roig and Reig (1969:670).

Ctenomys validus Contreras, Roig, and Suzarte, 1977. Physis, sec. C., 36(92):160.
TYPE LOCALITY: Argentina, Mendoza Prov., Guaymallen Dept., sand banks of El Borbollon, El Algarrobal (near city of Mendoza).
DISTRIBUTION: Mendoza Prov. (Argentina).
COMMENTS: Closely related to *johanis* (here included in *fulvus*) according to Contreras et al. (1977).

Family Octodontidae Waterhouse, 1840. Proc. Zool. Soc. Lond., 1839:172 [1840].
SYNONYMS: Spalacopidae.
COMMENTS: Includes Spalacopidae Lilljeborg, 1866. Sometimes considered the most primitive group of South American hystricognaths with numerous fossil genera from Oligocene on, but Reig (1986:418) reserved this distinction for the Echimyidae. *Ctenomys* is often placed here as subfamily, and is closely related to octodontids (see comments under Ctenomyidae and in Cook et al., 1990:22-23). Chromosomal variation is not as conservative as previously thought, and is 2n=38-102.

Aconaemys Ameghino, 1891. Rev. Argent. Hist. Nat., 1:245.
TYPE SPECIES: *Schizodon fuscus* Waterhouse, 1842.
COMMENTS: Reig (1986:408) noted that *Aconaemys* is not separable from the Pliocene/ Pleistocene form *Pithanotomys* Ameghino, 1887 at the generic level. If indeed the two are not separable, *Pithanotomys* would have priority as a name over *Acomaemys*.

Aconaemys fuscus (Waterhouse, 1842). Proc. Zool. Soc. Lond., 1841:91 [1842].
TYPE LOCALITY: Argentina, Mendoza Prov., Valle de las Cuevas; see comments.
DISTRIBUTION: High Andes of Chile and Argentina (between 33° and 41°S).
STATUS: Locally abundant.
SYNONYMS: *porteri*.
COMMENTS: Waterhouse never designated a type species, which was finally designated as a lectotype by Thomas (1917*a*). Waterhouse's original type locality was "from Chile",

but in 1848 he added that the locality was from the Valle de las Cuevas near the Volcano of Peteroa, which would make the locality in Argentina (Pearson, 1984). These complications make it difficult to evaluate the status of the following species. The chromosome no. is 2n=56.

Aconaemys sagei Pearson, 1984. Jour. Zool. (London), 202:229.
TYPE LOCALITY: Argentina, Neuquén Prov. 2 km S Cerro Quillén at Pampa de Hui Hui (1050 m).
DISTRIBUTION: Known only from near Lago Quillén and Lago Hui Hui in Neuquén Prov. of Argentina, but reports suggest it might also occur in Malleco Prov. of Chile.
STATUS: Locally abundant, but may have very limited distribution.
COMMENTS: Much smaller than *A. fuscus*.

Octodon Bennett, 1832. Proc. Zool. Soc. Lond., 1832:46.
TYPE SPECIES: *Octodon cumingi* Bennett, 1832 (= *Sciurus degus* Molina, 1782).
SYNONYMS: *Dendrobius*.

Octodon bridgesi Waterhouse, 1845. Proc. Zool. Soc. Lond., 1844:155 [1845].
TYPE LOCALITY: Chile, Curico Prov., Rio Teno.
DISTRIBUTION: Chilean Andes from 34°15' to at least 40°S (Redford and Eisenberg, 1992).
STATUS: Once common, but now quite rare (Redford and Eisenberg, 1992).
COMMENTS: Massoia (1979:36) reported *Octodon*, perhaps *bridgesi*, from S Argentina. The chromosome no. is 2n=58.

Octodon degus (Molina, 1782). Sagg. Stor. Nat. Chile, p. 303.
TYPE LOCALITY: Chile, Santiago Prov., Santiago.
DISTRIBUTION: Chile, west slope of the Andes between Vallenar and Curico, to 1,200 m.
STATUS: Common.
SYNONYMS: *alba, clivorum, cumingii, getulus, kummingii, pallidus, peruana*.
COMMENTS: Reviewed by Woods and Boraker (1975, Mammalian Species, 67). The chromosome no. is 2n=58.

Octodon lunatus Osgood, 1943. Field Mus. Nat. Hist. Publ., Zool. Ser., 30:110.
TYPE LOCALITY: Chile, Valparaiso Prov., Olmue.
DISTRIBUTION: Coastal mountains of Valparaiso, Aconcagua, and Coquimbo Provs. (Chile) between 31°30' and 35°S.
STATUS: Common.
COMMENTS: Included in *Octodon bridgesi* by Tamayo and Frassinetti (1980:354). The chromosome no. is 2n=78 (Spotorno et al., 1988).

Octodontomys Palmer, 1903. Science, n.s., 17:873.
TYPE SPECIES: *Neoctodon simonsi* Thomas, 1902 (= *Octodon gliroides* Gervais and d'Orbigny, 1844).

Octodontomys gliroides (Gervais and d'Orbigny, 1844). Bull. Sci. Soc. Philom. Paris, p. 22.
TYPE LOCALITY: Bolivia, La Paz Dept., near La Paz.
DISTRIBUTION: Andes of N Chile, SW Bolivia, and NW Argentina; occurs between 2,000 and 5,000 m in xeric habitats.
STATUS: Common.
SYNONYMS: *simonsi*.
COMMENTS: The chromosome no. is 2n=38.

Octomys Thomas, 1920. Ann. Mag. Nat. Hist., ser. 9, 6:117.
TYPE SPECIES: *Octomys mimax* Thomas, 1920.
COMMENTS: Formerly included *Octomys barrerae*, which is now separated as *Tympanoctomys*.

Octomys mimax Thomas, 1920. Ann. Mag. Nat. Hist., ser. 9, 6:118.
TYPE LOCALITY: Argentina, Catamarca Prov., La Puntilla.
DISTRIBUTION: Foothills and lower montane slopes of the Andes, and portions of the Monte desert of Catamarca, La Rioja, San Juan, and N Mendoza Provs. (Argentina).
SYNONYMS: *joannius*.
COMMENTS: Occurs at high elevations in Andes, and is similar in habits to the North American Woodrat (Redford and Eisenberg, 1992).

Spalacopus Wagler, 1832. Isis, 25:1219.
TYPE SPECIES: *Spalacopus poeppigii* Wagler, 1832 (= *Mus cyanus* Molina, 1782).

Spalacopus cyanus (Molina, 1782). Sagg. Stor. Nat. Chile, p. 300.
TYPE LOCALITY: Chile, Valparaiso Prov.
DISTRIBUTION: Chile, west of the Andes between 27° and 36°S.
STATUS: Locally common, very fossorial.
SYNONYMS: *ater, maulinus, noctivagus, poeppigii, tabanus.*
COMMENTS: Includes *tabanus* (Cabrera, 1961:517; Corbet and Hill, 1991:203). Chromosome no. is 2n = 58.

Tympanoctomys Yepes, 1941. Rev. Inst. Bact., 9, 1940 [1941]:569.
TYPE SPECIES: *Octomys barrerae* Lawrence, 1941.
COMMENTS: Widely considered a synonym of *Octomys* (see Reig, 1986:408; Corbet and Hill, 1991:204), but Cabrera (1961:516), and Redford and Eisenberg (1992) considered it a separate genus based on very enlarged tympanic bullae.

Tympanoctomys barrerae (Lawrence, 1941). Proc. New England Zool. Club, 18:43.
TYPE LOCALITY: Argentina, Mendoza Prov., La Paz.
DISTRIBUTION: Arid plains of Mendoza Province.
STATUS: IUCN - Vulnerable; rare.
COMMENTS: The chromosome no. is 2n=102 (Contreras and Tores-Mura, 1987).

Family Abrocomidae Miller and Gidley, 1918. Jour. Washington Acad. Sci., 8(13):447.

Abrocoma Waterhouse, 1837. Proc. Zool. Soc. Lond., 1837:30.
TYPE SPECIES: *Abrocoma Bennettii* Waterhouse, 1837.
SYNONYMS: *Habrocoma.*

Abrocoma bennettii Waterhouse, 1837. Proc. Zool. Soc. Lond., 1837:31.
TYPE LOCALITY: Chile, Aconcagua Prov., vicinity Aconcagua, flanks of Cordillera.
DISTRIBUTION: Chile from Copiapo to the area of Rio Biobio.
STATUS: Common.
SYNONYMS: *cuvieri, helvina, laniger, murrayi.*

Abrocoma boliviensis Glanz and Anderson, 1990. Am. Mus. Novit., 2991:23.
TYPE LOCALITY: Bolivia, Dept. Santa Cruz, Manual M. Caballero Prov., Comarapa.
DISTRIBUTION: Only known from the type locality.
STATUS: Apparently rare.

Abrocoma cinerea Thomas, 1919. Ann. Mag. Nat. Hist., ser. 9, 4:132.
TYPE LOCALITY: Argentina, Jujuy Prov., Cerro Casabindo.
DISTRIBUTION: SE Peru, W Bolivia, N Chile, and NW Argentina.
STATUS: Rare.
SYNONYMS: *budini, famatina, schistacea, vaccarum.*

Family Echimyidae Gray, 1825. Ann. Philos., n.s., 10:341.
COMMENTS: This group is complex, and in need of revision. The family includes the most primitive fossil New World hystricognaths from the Early Oligocene of Patagonia. Reig (1986:418) noted that some living taxa in this family with brachyodont and pentalophodont molars (*Mesomys* and *Lonchothrix*) are of the type expected in the ancestoral New World Hystricognathi. The family is also the most diverse of all Hystricognathi.

Subfamily Chaetomyinae Thomas, 1897. Proc. Zool. Soc. London, 1896:1026 [1897].
SYNONYMS: Cercolabidae.
COMMENTS: Originally in Erethizontidae, but removed by Patterson and Wood (1982:394) because this form retains the deciduous premolars (an echimyid derived character) unlike all known erethizontids.

Chaetomys Gray, 1843. List Specimens Mamm. Coll. Brit. Mus., p. 123.
TYPE SPECIES: *Hystrix subspinosus* Olfers, 1818.

SYNONYMS: *Plectrochoerus*.
COMMENTS: Formerly included in Erethizontidae; see Patterson and Wood (1982). Retained in Erethizontidae by Corbet and Hill (1991:200).

Chaetomys subspinosus (Olfers, 1818). Neue Bibl. Reisenb., p.211.
TYPE LOCALITY: Brazil, Pará, Cameta. Ávila-Pires (1967:178) gave Brazil, Bahia, Ilheus as the type locality.
DISTRIBUTION: Atlantic forests of S Bahía and N Espírito Santo states of Brazil (not Amazonia).
STATUS: U.S. ESA - Endangered; IUCN - Indeterminate. More common in suitable habitats than previously thought (Oliver and Santos, 1991:17), and may be quite common near Uruçuca in S Bahía.
SYNONYMS: *moricandi, rutila, tortilis, volubilis*.
COMMENTS: See Oliver and Santos (1991) for a discussion of the confusion surrounding the distribution of this species.

Subfamily Dactylomyinae Tate, 1935. Bull. Am. Mus. Nat. Hist., 68:295.
COMMENTS: Reig (1986:409) questioned placement of this subfamily in Echimyidae, and raised the possibility that dactylomyines are capromyids. If West Indian Spiny Rats are placed in Capromyidae (see Woods, 1982) then dactylomyines might be too, but here the West Indian forms are left in Echimyidae in their own subfamily (see comment under Heteropsomyinae).

Dactylomys I. Geoffroy, 1838. Ann. Sci. Nat. Zool. (Paris), ser. 2, 10:126.
TYPE SPECIES: *Dactylomys typus* I. Geoffroy, 1838 (= *Echimys dactylinus* Desmarest, 1817).
SYNONYMS: *Lachnomys*.
COMMENTS: Includes *Lachnomys* (Cabrera, 1961:543).

Dactylomys boliviensis Anthony, 1920. J. Mammal., 1:82.
TYPE LOCALITY: Bolivia, Cochabamba Dept., Misión de San Antonio, Río Chimoré, 390 m.
DISTRIBUTION: C Bolivia, SE Peru.

Dactylomys dactylinus (Desmarest, 1817). Nouv. Dict. Hist. Nat., Nouv. ed., 10:57.
TYPE LOCALITY: Upper Amazon area (no locality in original description).
DISTRIBUTION: Upper Amazon Basin in N Brazil, Peru, Ecuador, and perhaps Colombia.
STATUS: Locally common.
SYNONYMS: *canescens, modestus, typus*.

Dactylomys peruanus J. A. Allen, 1900. Bull. Am. Mus. Nat. Hist., 13:220.
TYPE LOCALITY: Peru, Juliaca; corrected by Allen (1901) to Inca Mines, "about 200 miles northeast of Jaliaca" 1,830 m, 13°30'S, 70°W.
DISTRIBUTION: SE Peru, and Bolivia between 1,000 and 3,000 m.
STATUS: Unknown.

Kannabateomys Jentink, 1891. Notes Leyden Mus., 13:109.
TYPE SPECIES: *Dactylomys amblyonyx* Wagner, 1845.

Kannabateomys amblyonyx (Wagner, 1845). Arch. Naturgesch., 1:146.
TYPE LOCALITY: Brazil, São Paulo, Ipanema.
DISTRIBUTION: E Brazil, Paraguay, NE Argentina; lives in bamboo thickets.
STATUS: Locally common.
SYNONYMS: *pallidior*.

Olallamys Emmons, 1988. J. Mammal., 69(2):421.
TYPE SPECIES: *Thrinacodus albicauda* Günther, 1879.
COMMENTS: Emmons (1988) indicated that *Thrinacodus* Günther, 1879, is preoccupied by *Thrinacodus* St. John and Worthen, 1875 (a shark).

Olallamys albicauda (Günther, 1879). Proc. Zool. Soc. Lond., 1879:144.
TYPE LOCALITY: Colombia, Antioquia Dept., Medellin.
DISTRIBUTION: NW and C Colombia, west of the Cordillera Central; occurs up to 3,000 m, often in dense bamboo thickets (Eisenberg, 1989:411).

STATUS: Unknown.
SYNONYMS: *apolinari*.

Olallamys edax (Thomas, 1916). Ann. Mag. Nat. Hist., ser. 8, 18:299.
TYPE LOCALITY: Venezuela, Merida, Sierra de Merida.
DISTRIBUTION: W Venezuela and adjacent N Colombia.
STATUS: Unknown.

Subfamily Echimyinae Gray, 1825. Ann. Philos., n.s., 10:341.
SYNONYMS: Loncherinae.

Diplomys Thomas, 1916. Ann. Mag. Nat. Hist., ser. 8, 18:240.
TYPE SPECIES: *Loncheres caniceps* Günther, 1877.

Diplomys caniceps (Günther, 1877). Proc. Zool. Soc. Lond., 1876:745 [1877].
TYPE LOCALITY: Colombia, Antioquia Dept., Medellin.
DISTRIBUTION: W Colombia, N Ecuador.
STATUS: Locally common.
COMMENTS: See description in Emmons and Feer (1990:222).

Diplomys labilis (Bangs, 1901). Am. Nat., 35:638.
TYPE LOCALITY: Panama, San Miguel Isl.
DISTRIBUTION: Panama (including San Miguel Isl), W Colombia, and (probably) N Ecuador.
STATUS: Locally common.
SYNONYMS: *darlingi*.
COMMENTS: Includes *darlingi* (Handley, 1966*a*:787; Hall, 1981:874).

Diplomys rufodorsalis (J. A. Allen, 1899). Bull. Am. Mus. Nat. Hist., 12:197.
TYPE LOCALITY: Colombia, Magdalena Dept., Onaca.
DISTRIBUTION: NE Colombia.
STATUS: Rare (very restricted range).

Echimys G. Cuvier, 1809. Bull. Sci. Soc. Philom. Paris, 24:394.
TYPE SPECIES: *Myoxus chrysurus* Zimmermann, 1780.
SYNONYMS: *Echinomys, Loncheres, Nelomys, Phyllomys*.
COMMENTS: Formerly included *armatus* which is the type species of the genus *Makalata* (see that account). *Makalata* is not recognized by Eisenberg (1989), nor Emmons and Feer (1990). Emmons and Feer separated *Nelomys* from *Echimys*, but here it is synonymized.

Echimys blainvillei (F. Cuvier, 1837). Ann. Sci. Nat. Zool. (Paris), ser. 2, 8:371.
TYPE LOCALITY: Brazil, Bahia, Isla de Deos.
DISTRIBUTION: SE Brazil, mainland and coastal islands.
STATUS: Unknown.
SYNONYMS: *medius*.
COMMENTS: Placed in *Nelomys* by Emmons and Feer (1990:220).

Echimys braziliensis (Waterhouse, 1848). Nat. Hist. Mamm., 2:330.
TYPE LOCALITY: Brazil, Minas Gerais, Lagoa Santa.
DISTRIBUTION: S Brazil in Minas Gerais and probably coastal forests from S Bahia to Rio de Janeiro.
SYNONYMS: *armatus* Winge, 1887.
COMMENTS: According to Cabrera (1961:540) *Phyllomys brasiliensis* is a *nomen nudum*. Emmons and Feer (1990:220) recognized as *Nelomys* sp., stating that the correct name for *brasiliensis* is not clear.

Echimys chrysurus (Zimmermann, 1780). Geogr. Gesch. Mensch. Vierf. Theire, 2:352.
TYPE LOCALITY: Suriname.
DISTRIBUTION: Guianas to lower Amazonian NE Brazil.
STATUS: Uncommon.
SYNONYMS: *cristatus, paleaceus*.

Echimys dasythrix (Hensel, 1872). Abh. Konigl. Akad. Wiss. Berlin, p. 49.
TYPE LOCALITY: Brazil, Río Grande do Sul.
DISTRIBUTION: SE and E Brazil.
COMMENTS: Recognized as *Nelomys dasythrix* by Emmons and Feer (1990:220).

Echimys grandis (Wagner, 1845). Arch. Naturgesch., 1:145.
TYPE LOCALITY: Brazil, Amazonas, Manaqueri.
DISTRIBUTION: Amazonian Brazil along the banks of the Amazon River from Río Negro to Ilha Caviana.
STATUS: Locally common.
COMMENTS: Formerly included *rhipidurus*, which was given specific status by Emmons and Feer (1990).

Echimys lamarum (Thomas, 1916). Ann. Mag. Nat. Hist., ser. 8, 18:297.
TYPE LOCALITY: Brazil, Bahia, Lamarao 112 km NE of Salvador.
DISTRIBUTION: E Brazil.
COMMENTS: Emmons and Feer (1990) elevated to species and placed in genus *Nelomys*.

Echimys macrurus (Wagner, 1842). Arch. Naturgesch., 1:360.
TYPE LOCALITY: Brazil, Amazonas, Borba.
DISTRIBUTION: Brazil, south of the Amazon River.
COMMENTS: Not recognized by Emmons and Feer (1990).

Echimys nigrispinus (Wagner, 1842). Arch. Naturgesch., 1:361.
TYPE LOCALITY: Brazil, São Paulo, Ipanema.
DISTRIBUTION: E Brazil.
COMMENTS: Recognized as a *Nelomys* by Emmons and Feer (1990:219).

Echimys pictus (Pictet, 1841). Notice Anim. Nouv. Mus. Geneve, p. 29.
TYPE LOCALITY: Brazil, Bahia.
DISTRIBUTION: E Brazil in S Bahia near Ilhéus.
COMMENTS: Treated as a species of *Nelomys* by Emmons and Feer (1990:218), but as *Isothrix pictus* by Patton and Emmons (1985:12) and Corbet and Hill (1991:207).

Echimys rhipidurus Thomas, 1928. Ann. Mag. Nat. Hist., ser. 10, 2:291.
TYPE LOCALITY: Peru, Dept. Loreto, Pebas.
DISTRIBUTION: C and N Amazonian Peru.
STATUS: Uncommon.
COMMENTS: Elevated to species by Emmons and Feer (1990:216).

Echimys saturnus Thomas, 1928. Ann. Mag. Nat. Hist., ser. 10, 2:409.
TYPE LOCALITY: Ecuador, Napo-Pastaza Prov., Río Napo.
DISTRIBUTION: Ecuador and N Peru, east of the Andes to ± 1,000 m.

Echimys semivillosus (I. Geoffroy, 1838). Ann. Sci. Nat. Zool. (Paris), ser. 2, 10:125.
TYPE LOCALITY: Colombia, Bolivar Dept., Cartagena.
DISTRIBUTION: N Colombia, Venezuela, Margarita Isl.
STATUS: Locally common.
SYNONYMS: *carrikeri, flavidus, punctatus.*
COMMENTS: See Cabrera (1961:542-543).

Echimys thomasi (Ihering, 1871). Rev. Mus. Paulista, 2:171.
TYPE LOCALITY: Brazil, Bahia, Isla de São Sabastião.
DISTRIBUTION: Endemic to the type locality.
STATUS: Unknown, but apparently stable.
COMMENTS: Placed in *Nelomys* by Emmons and Feer (1990:219).

Echimys unicolor (Wagner, 1842). Arch. Naturgesch., 1:361.
TYPE LOCALITY: Brazil.
DISTRIBUTION: Brazil; exact distribution unknown (Corbet and Hill, 1991:207).
COMMENTS: According to Cabrera (1961:543), Thomas believed that *unicolor* was a synonym of *braziliensis*. Not recognized by Emmons and Feer (1990).

Isothrix Wagner, 1845. Arch. Naturgesch., 1:145.
TYPE SPECIES: *Isothrix bistriata* Wagner, 1845 (selected by Goldman, 1916).

SYNONYMS: *Lasiuromys*.
COMMENTS: See Patton and Emmons (1985) for a review of *Isothrix*.

Isothrix bistriata Wagner, 1845. Arch. Naturgesch., 1:146.
TYPE LOCALITY: Brazil, Río Guaporé.
DISTRIBUTION: Bolivia, SW to NC Brazil, S Venezuela, adjacent Colombia.
STATUS: Uncommon.
SYNONYMS: *molliae, negrensis, orinoci, pagurus, picta, villosa*.
COMMENTS: Patton and Emmons (1985) recognized 2 subspecies (*I. b. bistriata* and *I. b. orinoci*). They combined *villosus* as part of subspecies *bistriata*. Listed as *bistriatus* by Corbet and Hill (1991:206).

Isothrix pagurus Wagner, 1845. Arch. Naturgesch., 1:146.
TYPE LOCALITY: Brazil, Amazonas, Borba, Lower Río Madeira.
DISTRIBUTION: Amazon Basin of Central Brazil from Río Madeira east to Río Tapajoz and north to lower Río Negro.
STATUS: Uncommon.
SYNONYMS: *crassicaudus*.

Makalata Husson, 1978. The Mammals of Suriname, p. 445.
TYPE SPECIES: *Nelomys armatus* I. Geoffroy, 1830.
COMMENTS: Husson (1978:445) named this genus because *armatus* is so distinct. Not recognized as a separate genus by Eisenberg (1989) nor Emmons and Feer (1990).

Makalata armata (I. Geoffroy, 1830). Rev. Zool. Paris, 1:101.
TYPE LOCALITY: French Guiana, Cayenne.
DISTRIBUTION: Andes of N Ecuador and Colombia, Venezuela, Guianas, Amazon Basin of Brazil, Tobago, Trinidad; perhaps Martinique (record probably erroneus, see Hall, 1981:1180).
SYNONYMS: *castaneus, guianae, hispidus, longirostris, macrourus, occasius*.
COMMENTS: Formerly included in *Echimys*; transferred to *Makalata* by Husson (1978:445). Corbet and Hill (1991:206) listed this species in *Echimys*. Emmons and Feer (1990:218) recognized *occasius* as a species (*Echimys occasius*).

Subfamily Eumysopinae Rusconi, 1935. Bol. Paleont. Buenos Aires, 5:2.
COMMENTS: Proposed as a subfamily by Patton and Reig (1989:76). This subfamily includes the Oligocene fossil forms, and the most primitive living genera of South American echimyids.

Carterodon Waterhouse, 1848. Nat. Hist. Mamm., 2:351.
TYPE SPECIES: *Echimys sulcidens* Lund, 1841.

Carterodon sulcidens (Lund, 1841). Afh. Kongl. Danske Vid. Selsk., p. 49.
TYPE LOCALITY: Brazil, Minas Gerais, Lagoa Santa.
DISTRIBUTION: E Brazil.
STATUS: Unknown.
SYNONYMS: *temmincki*.

Clyomys Thomas, 1916. Ann. Mag. Nat. Hist., ser. 8, 18:300.
TYPE SPECIES: *Echimys laticeps* Thomas, 1909.
COMMENTS: Reviewed by Ávila-Pires and Wutke (1981:530). Reig (1986:409, footnote) noted that *Clyomys* may not be distinct from *Euryzygomatomys*.

Clyomys bishopi Ávila-Pires and Wutke, 1981. Revta Bras. Biol., 41:530.
TYPE LOCALITY: Brazil, São Paulo, Itapetininga.
DISTRIBUTION: Known only from the type locality.

Clyomys laticeps (Thomas, 1909). Ann. Mag. Nat. Hist., ser. 8, 4:240.
TYPE LOCALITY: Brazil, Santa Catarina, Joinville.
DISTRIBUTION: Between Minas Gerais and Santa Catarina (E Brazil) in savanna habitats.
STATUS: Common.
SYNONYMS: *spinosus* Winge, 1887.

COMMENTS: Highly fossorial and colonial.

Euryzygomatomys Goeldi, 1901. Bol. Mus. Para., 3:179.
TYPE SPECIES: *Rattus spinosus* G. Fischer, 1814.
COMMENTS: Reig (1986:409, footnote) noted that *Cliomys* may not be distinct from *Euryzygomatomys*.

Euryzygomatomys spinosus (G. Fischer, 1814). Zoognosia, 3:105.
TYPE LOCALITY: Paraguay, Cordillera, Atira.
DISTRIBUTION: S and E Brazil, NE Argentina, Paraguay.
SYNONYMS: *brachyura, catellus, guiara, rufa*.
COMMENTS: The use of the names derived from Fischer (1814) is provisional pending clarification of the availability of the work. Should Fischer (1814) become unavailable, the name would be *rufa* Lichtenstein, 1820 (Abhandl. Preuss. Akad, Wiss., 1818-1819 [1820], p. 192).

Hoplomys J. A. Allen, 1908. Bull. Am. Mus. Nat. Hist., 24:649.
TYPE SPECIES: *Hoplomys truei* J. A. Allen, 1908 (= *Echimys gymnurus* Thomas, 1897).
COMMENTS: Revised by Handley (1959*a*). Patton and Reig (1989:90) demonstrated that this genus is very close to *Proechimys*, and may be congeneric.

Hoplomys gymnurus (Thomas, 1897). Ann. Mag. Nat. Hist., ser. 6, 20:550.
TYPE LOCALITY: Ecuador, Esmeraldas Prov., Cachavi.
DISTRIBUTION: EC Honduras to NW Ecuador.
STATUS: Locally common.
SYNONYMS: *goethalsi, truei*.
COMMENTS: Formerly included *hoplomyoides* which was transferred to *Proechimys* by Handley (1976:57).

Lonchothrix Thomas, 1920. Ann. Mag. Nat. Hist., ser. 9, 6:113.
TYPE SPECIES: *Lonchothrix emiliae* Thomas, 1920.

Lonchothrix emiliae Thomas, 1920. Ann. Mag. Nat. Hist., ser. 9, 6:114.
TYPE LOCALITY: Brazil, Río Tapajoz, Villa Braga.
DISTRIBUTION: C Brazil, south of the Amazon River in area of Río Tapajoz and Río Madeira.
STATUS: Locally common.

Mesomys Wagner, 1845. Arch. Naturgesch., 1:145.
TYPE SPECIES: *Mesomys ecaudatus* Wagner, 1845 (= *Echimys hispidus* Desmarest, 1817).
COMMENTS: Revision of this genus is needed; see Husson (1978:440) and Emmons and Feer (1990:223).

Mesomys didelphoides (Desmarest, 1817). Nouv. Dict. Hist. Nat., Nouv. ed., 10:58.
TYPE LOCALITY: Unknown.
DISTRIBUTION: Brazil.
COMMENTS: Taxonomic status questionable; not mentioned by Emmons and Feer (1990); distribution questioned by Corbet and Hill (1991:206).

Mesomys hispidus (Desmarest, 1817). Nouv. Dict. Hist. Nat., Nouv. ed., 10:58.
TYPE LOCALITY: Brazil, Amazonas, Borba.
DISTRIBUTION: N and E Peru, E Ecuador, N Brazil.
SYNONYMS: *ecaudatus, ferrugineus, spicatus*.
COMMENTS: Formerly included *stimulax*; see Husson (1978:438).

Mesomys leniceps Thomas, 1926. Ann. Mag. Nat. Hist., ser. 9, 18:348.
TYPE LOCALITY: Peru, Dept. Amazonas, Yambasbramba.
DISTRIBUTION: Higher elevations in Peru (over 2,000 m).
COMMENTS: Emmons and Feer (1990) recognized as a species.

Mesomys obscurus (Wagner, 1840). Abh. Akad. Wiss. Munich, 3:196.
TYPE LOCALITY: Unknown.
DISTRIBUTION: Brazil.
STATUS: Unknown.

COMMENTS: Known only from the original description; status uncertain (Cabrera, 1961:536); distribution questioned by Corbet and Hill (1991:206).

Mesomys stimulax Thomas, 1911. Ann. Mag. Nat. Hist., ser. 8, 7:607.
TYPE LOCALITY: N Brazil, Amazon estuary, "Cameta, Lower Tocantins."
DISTRIBUTION: N Brazil, Suriname.
COMMENTS: Formerly included in *hispidus* by Cabrera (1961:536); considered a distinct species by Husson (1978:438). Emmons and Feer (1990:223) gave its range as south of Amazon and east of Río Tapajós in Brazil. May be a subspecies of *hispidus*.

Proechimys J. A. Allen, 1899. Bull. Am. Mus. Nat. Hist., 12:257.
TYPE SPECIES: *Echimys trinitatis* J. A. Allen and Chapman, 1893.
SYNONYMS: *Trinomys*.
COMMENTS: Traditionally divided into 2 subgenera (*Proechimys* with 9 species groups distributed in most areas of tropical South America, and *Trinomys* with four species distributed in Eastern Brazil). This group is very complex with many species and subspecies. Reviewed, in part, by Reig et al. (1980:291-312), Gardner and Emmons (1984), and Patton (1987). May include *Hoplomys* (see Patton and Reig, 1989). The specific level taxonomy of the group is controversial, and is in need of a major revision. Most of the named forms in Corbet and Hill (1991) and Nowak (1991) are used here pending a revision of the group.

Proechimys albispinus (I. Geoffroy, 1838). Ann. Sci. Nat. Zool. (Paris), 10:125.
TYPE LOCALITY: Brazil, Bahía, Ilha de Madre Deos, Itaparica (near Salvador).
DISTRIBUTION: Bahía and adjacent islands (Brazil).
SYNONYMS: *serotinus*.
COMMENTS: Subgenus *Trinomys*.

Proechimys amphichoricus Moojen, 1948. Univ. Kans. Publ. Mus. Nat. Hist., 1:344.
TYPE LOCALITY: Venezuela, Amazonas, Cerro Duida, 325 m.
DISTRIBUTION: S Venezuela, adjacent Brazil.
COMMENTS: Subgenus *Proechimys*. Formerly included in *semispinosus*; see Reig et al. (1980). The chromosome no. is 2n=26; FN=44.

Proechimys bolivianus Thomas, 1901. Ann. Mag. Nat. Hist., ser. 7, 8:537.
TYPE LOCALITY: Bolivia, Dept. La Paz, Mapiri (1,000 m).
DISTRIBUTION: Upper Amazon.
COMMENTS: Subgenus *Proechimys*. Not recognized by Nowak (1991:943), but listed by Corbet and Hill (1991:205). Listed as a subspecies of *cayennensis* (= *guyannensis*) by Cabrera (1961:519).

Proechimys brevicauda (Günther, 1877). Proc. Zool. Soc. Lond., 1876:748 [1877].
TYPE LOCALITY: Peru, Dept. de Loreto, Chamicuros (Río Huallago)
DISTRIBUTION: Southern Colombia (?), E Peru, NW Brazil.
COMMENTS: Subgenus *Proechimys*. Listed as subspecies of *longicaudatus* by Cabrera (1961:524); listed as species by Corbet and Hill (1991:205) and Nowak (1991:943). Range in S Colombia listed in Nowak (1991), but in Eisenberg (1989:402) this is referred to as the "brevicauda" group.

Proechimys canicollis (J. A. Allen, 1899). Bull. Am. Mus. Nat. Hist., 12:200.
TYPE LOCALITY: Colombia, Dept. Magdalena, Bonda near Santa Marta (NW base of Sierra Nevada de Santa Marta).
DISTRIBUTION: NC Colombia according to Eisenberg (1989:403), but also Venezuela according to Corbet and Hill (1991) and Nowak (1991).
SYNONYMS: *vacillator*.
COMMENTS: Subgenus *Proechimys*. The chromosome no. is 2n=24; FN=44.

Proechimys cayennensis (Desmarest, 1817). Nouv. Dict. Hist. Nat., Nouv. ed., 10:58.
TYPE LOCALITY: French Guiana, Cayenne.
DISTRIBUTION: E Colombia, the Guianas, southward to C Brazil.
SYNONYMS: *arabupu, arescens, cherriei, guyannensis, hyleae, leioprymna, nesiotes, rattinus, riparum, villicauda*.

COMMENTS: Subgenus *Proechimys*. Hershkovitz (1948*a*) resurected *guyannensis* E. Geoffroy, 1803 (Cat. Mamm. Mus. Nat. Hist. Nat., p. 194) for this species as this name antedates *cayennensis*. However, see Appendix I for reasons why this publication, and hence *guyannensis* E. Geoffroy, 1803 is not available. Formerly included *warreni* (see Husson, 1978:436), *guairae, oris, poliopus, trinitatis,* and *urichi* (see Reig et al., 1980). Includes *cherriei*; see Petter (1978), and Reig et al. (1980). Sometimes includes *bolivianus, chrysaeolus, decumanus, magdalenae, mincae,* and *oconnelli* (see Cabrera, 1961), but Corbet and Hill (1991:205) provisionally recognized each as a distinct species. The chromosome no. is 2n=40; FN=54.

Proechimys chrysaeolus (Thomas, 1898). Ann. Mag. Nat. Hist. ser. 7, 1:244.
TYPE LOCALITY: Colombia, Dept. Boyacá, Muzo (valley of Río Carare).
DISTRIBUTION: E Colombia.
COMMENTS: Subgenus *Proechimys*. Listed as a subspecies of *cayennensis* (= *guyannensis*) by Cabrera (1961:521), but as a species by Corbet and Hill (1991:205) and Nowak (1991:943). Not listed by Eisenberg (1989).

Proechimys cuvieri Petter, 1978. C. R. Acad. Sci. Paris, ser. D, 287:263.
TYPE LOCALITY: French Guiana, Saul.
DISTRIBUTION: French Guiana, Surinam, Guyana.
COMMENTS: Subgenus *Proechimys*. The chromosome no. is 2n=28.

Proechimys decumanus (Thomas, 1899). Ann. Mag. Nat. Hist., ser. 7, 4:282.
TYPE LOCALITY: Ecuador, Prov. de Guayas, Changón.
DISTRIBUTION: NW Peru, SW Ecuador, Pacific lowlands.
COMMENTS: Subgenus *Proechimys*. Listed as a subspecies of *cayennensis* (= *guyannensis*) by Cabrera (1961:520), but as a species by Corbet and Hill (1991:205), Nowak (1991:943) and Emmons and Feer (1990).

Proechimys dimidiatus (Günther, 1877). Proc. Zool. Soc. Lond., 1876:747 [1877].
TYPE LOCALITY: Brazil, Rio de Janeiro.
DISTRIBUTION: E Brazil.
COMMENTS: Subgenus *Trinomys*.

Proechimys goeldii Thomas, 1905. Ann. Mag. Nat. Hist., ser. 7, 15:587.
TYPE LOCALITY: Brazil, Pará, Santarem.
DISTRIBUTION: Amazonian Brazil between Jamunda and Tapajoz Rivers, W Brazil.
COMMENTS: Subgenus *Proechimys*.

Proechimys gorgonae Bangs, 1905. Bull. Mus. Comp. Zool., 46:89.
TYPE LOCALITY: Colombia, Gorgona Island.
DISTRIBUTION: Gorgona Island (Colombia).
COMMENTS: Subgenus *Proechimys*. Considered a subspecies of *cayennensis* (= *guyannensis*) by Cabrera (1961:520), but listed as a species by Emmons and Feer (1990:274).

Proechimys guairae Thomas, 1901. Proc. Biol. Soc. Washington, 14:27.
TYPE LOCALITY: Venezuela, Federal Dist., La Guaira.
DISTRIBUTION: NC Venezuela, E of Lake Maracaibo and the Merida Andes.
SYNONYMS: *ochraceus*.
COMMENTS: Subgenus *Proechimys*. Includes *ochraceus*; *guairae* and *ochraceus* were formerly included in *cayennensis* (= *guyannensis*) (Reig et al., 1980). These authors also mention a closely related, undescribed species ("Barina's") from south of the Merida Andes. The chromosome no. is 2n=44-50; FN=72-76.

Proechimys gularis Thomas, 1911. Ann. Mag. Nat. Hist., ser 8, 8:253.
TYPE LOCALITY: Ecuador, Prov. de Napo-Pastaza, Canelos (Río Bobonaza).
DISTRIBUTION: E Ecuador (east of the Andes).
COMMENTS: Subgenus *Proechimys*. Corbet and Hill (1991:205) gave the distribution as "Upper Amazon", which is not consistent with accounts in Cabrera (1961) and Nowak (1991).

Proechimys hendeei Thomas, 1926. Ann. Mag. Nat. Hist., ser. 9, 18:162.
TYPE LOCALITY: Peru, Loreto Dept., Puca Tambo (listed as Chachapoyas district, Amazonas, Peru in original).

DISTRIBUTION: NE Peru to S Colombia (Amazonian region of both countries).
SYNONYMS: *nigrofulvus*.
COMMENTS: Subgenus *Proechimys*. Considered as part of *simonsi* by Corbet and Hill (1991:206) and Eisenberg (1989:403). The chromosome no. is 2n=32.

Proechimys hoplomyoides (Tate, 1939). Bull. Am. Mus. Nat. Hist., 76:179.
TYPE LOCALITY: Venezuela, Bolivar Prov., Mt. Roraima (2,000 m).
DISTRIBUTION: SE Venezuela, adjacent Guyana and Brazil.
COMMENTS: Subgenus *Proechimys*. Transferred from *Hoplomys* by Handley (1976:57).

Proechimys iheringi Thomas, 1911. Ann. Mag. Nat. Hist., ser. 8, 8:252.
TYPE LOCALITY: Brazil, São Paulo, São Sabastião Isl.
DISTRIBUTION: E Brazil.
SYNONYMS: *bonafidei, denigratus, gratiosus, panema, paratus*.
COMMENTS: Subgenus *Trinomys*. Moojen (1948:383) recognized 6 subspecies. The chromosome no. is 2n=62-64.

Proechimys longicaudatus (Rengger, 1830). Naturgesch. Säugeth. Paraguay, p. 236.
TYPE LOCALITY: N Paraguay.
DISTRIBUTION: C and E Peru, W Bolivia, Paraguay, S Brazil.
SYNONYMS: *boimensis, elassopus, leucomystax, pachita, roberti, securus*.
COMMENTS: Subgenus *Proechimys*. Formerly included *brevicauda* (Reig et al., 1980). The chromosome no. is 2n=28.

Proechimys magdalenae (Hershkovitz, 1948). Proc. U. S. Natl. Mus., 97:136.
TYPE LOCALITY: Colombia, Dept. de Bolívar, Río San Pedro near Norosí (178 m) in the foothills of the Cordillera Central.
DISTRIBUTION: Colombia west of the Río Magdalena.
COMMENTS: Subgenus *Proechimys*. Usually combined with *cayennensis* (= *guyannensis*) (Cabrera, 1961:521; Eisenberg, 1989), but recognized as species by Corbet and Hill (1991:205), Nowak (1991:943) and Emmons and Feer (1990:274).

Proechimys mincae (J. A. Allen, 1899). Bull. Am. Mus. Nat. Hist., 12:198.
TYPE LOCALITY: Colombia, Dept. Magdalena, Minca near Santa Marta.
DISTRIBUTION: N Colombia below 500 m in the Sierra Nevada de Santa Marta.
COMMENTS: Subgenus *Proechimys*. Recognized in Corbet and Hill (1991:205), Nowak (1991:943) and Emmons and Feer (1990:274), but not in Eisenberg (1989).

Proechimys myosuros (Lichtenstein, 1820). Abh. König. Akad. Wiss., Berlin, p. 192.
TYPE LOCALITY: Brazil, Bahia.
DISTRIBUTION: Bahia (Brazil).
SYNONYMS: *cinnamomeus, leptosoma*.
COMMENTS: Subgenus *Trinomys*. May be *setosus* or *albispinus*.

Proechimys oconnelli J. A. Allen, 1913. Bull. Am. Mus. Nat. Hist., 32(24):479.
TYPE LOCALITY: Colombia, Villavicencio, 480 m.
DISTRIBUTION: C Colombia east of Cordillera Oriental.
COMMENTS: Subgenus *Proechimys*. The chromosome no. is 2n=32; FN=52.

Proechimys oris Thomas, 1904. Ann. Mag. Nat. Hist., ser. 7, 14:105.
TYPE LOCALITY: Brazil, Pará, Igarapé-assú, near Belem.
DISTRIBUTION: C Brazil.
COMMENTS: Subgenus *Proechimys*. Formerly included in *cayennensis* (= *guyannensis*); see Reig et al. (1980). The chromosome no. is 2n=30.

Proechimys poliopus Osgood, 1914. Field Mus. Nat. Hist. Publ., Zool. Ser., 10:135.
TYPE LOCALITY: Venezuela, Táchira State, San Juan de Colón (797 m) on N slope of Sierra de Mérida. Listed in Honacki et al. (1982:590) as Venezuela, Zulia Prov., Rio Aurare, El Panorama, for unknown reasons.
DISTRIBUTION: NW Venezuela, between Lake Maracaibo and the Sierra de Perija and adjacent Colombia.
COMMENTS: Subgenus *Proechimys*. Formerly included in *cayennensis* (= *guyannensis*) (Reig et al., 1980). The chromosome no. is 2n=42; FN=76.

Proechimys quadruplicatus Hershkovitz, 1948. Proc. U.S. Natl. Mus., 97:138.
TYPE LOCALITY: Ecuador, Napo-Pastaza Prov., Río Napo, Isla Llunchi (18 kms below mouth of Río Coca).
DISTRIBUTION: Amazonian region of E Ecuador and N Peru.
COMMENTS: Subgenus *Proechimys*. The chromosome no. is 2n=28.

Proechimys semispinosus (Tomes, 1860). Proc. Zool. Soc. Lond., 1860:265.
TYPE LOCALITY: Ecuador, Esmeraldas Prov., Esmeraldas.
DISTRIBUTION: SE Honduras to NE Peru and Amazonian Brazil.
SYNONYMS: *burrus, calidior, centralis, chiriquinus, columbianus, goldmani, hilda, ignotus, kermiti, liminalis, panamensis, rosa, rubellus.*
COMMENTS: Subgenus *Proechimys*. Formerly included *amphichoricus*; see Reig et al. (1980) who employed the name *centralis* for animals assigned to *semispinosus* from N Venezuela. Gardner (1983*b*) discussed the taxonomic history of this species and designated a new type locality. The chromosome no. is 2n=30; FN=50-54.

Proechimys setosus (Desmarest, 1817). Nouv. Dict. Hist. Nat., Nouv. ed., 10:59.
TYPE LOCALITY: Unknown, but probably Brazil, Bahia (Moojen, 1948:386).
DISTRIBUTION: E Brazil (Minas Gerais).
SYNONYMS: *cajennensis, elegans, fuliginosus.*
COMMENTS: Subgenus *Trinomys*.

Proechimys simonsi Thomas, 1900. Ann. Mag. Nat. Hist., ser. 7, 6:300.
TYPE LOCALITY: Peru, Dept. Junín, Río Perené (240 m).
DISTRIBUTION: E Ecuador, NE Peru, S Colombia.
COMMENTS: Subgenus *Proechimys*. Includes *hendeei* in Corbet and Hill (1991:206) and Eisenberg (1989:403). The chromosome no. is 2n=32; FN=44.

Proechimys steerei Goldman, 1911. Proc. Biol. Soc. Washington, 24:238.
TYPE LOCALITY: Brazil, Amazonas State, Hyutanahan (above Purús)
DISTRIBUTION: E Brazil, W Peru
COMMENTS: Subgenus *Proechimys*. Listed as a subspecies of *goeldii* in Cabrera (1961:519), but as a distinct species in Corbet and Hill (1991:206) and Nowak (1991:943).

Proechimys trinitatis (J. A. Allen and Chapman, 1893). Bull. Am. Mus. Nat. Hist., 5:223.
TYPE LOCALITY: Trinidad and Tobago, Trinidad.
DISTRIBUTION: Trinidad.
COMMENTS: Subgenus *Proechimys*. Formerly included in *cayennensis* (= *guyannensis*) (Reig et al., 1980). The chromosome no. is 2n=62; FN=80.

Proechimys urichi (J. A. Allen, 1899). Bull. Am. Mus. Nat. Hist., 12:199.
TYPE LOCALITY: Venezuela, Sucre, Quebrada Seca.
DISTRIBUTION: N Venezuela.
COMMENTS: Subgenus *Proechimys*. Formerly included in *cayennensis* (= *guyannensis*) (Reig et al., 1980). The chromosome no. is 2n=62.

Proechimys warreni Thomas, 1905. Ann. Mag. Nat. Hist., ser. 7, 16:312-313.
TYPE LOCALITY: Guyana, 129 km up the Demerara River, Comackka.
DISTRIBUTION: Guyana, Suriname, exact limits not known.
COMMENTS: Subgenus *Proechimys*. Considered a distinct species by Husson (1978:436); a subspecies of *cayennensis* (= *guyannensis*) by Cabrera (1961:520).

Thrichomys Trouessart, 1880. Cat. Mamm. Bull. Soc. Etudes Sci. Angers, 1881:179.
TYPE SPECIES: *Nelomys aperoides* Lund, 1839.
COMMENTS: Formerly referred to as *Cercomys* which was based on *Cercomys cunicularius*, a composite (Petter, 1973*b*).

Thrichomys apereoides (Lund, 1839). Afh. K. Danske Vid. Selsk., p. 38.
TYPE LOCALITY: Brazil, Minas Gerais, Lagoa Santa.
DISTRIBUTION: E Brazil, Paraguay.
STATUS: Common.
SYNONYMS: *antricola, fosteri, inermis, laurentius.*
COMMENTS: Formerly referred to as *Cercomys cunicularius*, a composite (Petter, 1973*b*; Mares et al., 1981*a*:120).

Subfamily Heteropsomyinae Anthony, 1917. Bull. Am. Mus. Nat. Hist., 37(4):189.
COMMENTS: Named by Anthony to reflect perceived relationship between *Heteropsomys* and *Dasyprocta*. Expanded by Kraglievich (1965) and Patterson and Pascual (1968*b*) to include *Proechimys* and associate genera. Modified by Woods (1982:386) when West Indian Spiny Rats classified with capromyids. West Indian Spiny Rats are transitional in characters between Echimyidae and Capromyidae, and here are placed in their own subfamily within the Echimyidae until the two families are revised. Varona (1974:73) placed all West Indian Spiny Rats in the genus *Heteropsomys*. However, there are clear differences between the forms on each island, so the original generic names are maintained here, except that *Homopsomys* is combined with *Heteropsomys* (Woods, 1989*a*, *b*).

Boromys Miller, 1916. Smithson. Misc. Coll. 66(12):7.
TYPE SPECIES: *Boromys offella* Miller, 1916.

Boromys offella Miller, 1916. Smithson. Misc. Coll., 66(12):8.
TYPE LOCALITY: Cuba, Oriente Prov., Baracoa, Maisí.
DISTRIBUTION: Cuba and Isla Juventud (Isle of Pines).
STATUS: Extinct.

Boromys torrei Allen, 1917. Bull. Mus. Comp. Zool., 61:6.
TYPE LOCALITY: Cuba, Matanzas Prov. Cave in Sierra de Hato-Nuevo.
DISTRIBUTION: Cuba and Isla Juventud (Isle of Pine).
STATUS: Extinct.

Brotomys Miller, 1916. Smithson. Misc. Coll., 66(12):6.
TYPE SPECIES: *Brotomys voratus* Miller, 1916.

Brotomys contractus Miller, 1929. Smithson. Misc. Coll., 81(9):13.
TYPE LOCALITY: Haiti, Dept. de l'Artibonite, Saint Michel de l'Atalye (small cave) (19°22'N, 72°20'W).
DISTRIBUTION: Hispaniola.
STATUS: Extinct.

Brotomys voratus Miller, 1916. Smithson. Misc. Coll., 66(12):7.
TYPE LOCALITY: Dominican Republic, San Pedro de Macorís, (kitchin midden).
DISTRIBUTION: Haiti, Dominican Republic, and La Gonave Island.
STATUS: Extinct in the last 50 years.

Heteropsomys Anthony, 1916. Ann. New York Acad. Sci., 27:203.
TYPE SPECIES: *Heteropsomys insulans* Anthony, 1916.
SYNONYMS: *Homopsomys*.
COMMENTS: Includes *Homopsomys*; see comment under the subfamily.

Heteropsomys antillensis (Anthony, 1917). Bull. Am. Mus. Nat. Hist., 37:187.
TYPE LOCALITY: Puerto Rico, Cave at Utuado.
DISTRIBUTION: Puerto Rico.
STATUS: Extinct.

Heteropsomys insulans Anthony, 1916. Ann. New York Acad. Sci., 27:202.
TYPE LOCALITY: Puerto Rico, Cueva de la Ceiba, Hacienda Jobo (near Utuado).
DISTRIBUTION: Puerto Rico in cave deposits.
STATUS: Extinct.
COMMENTS: Nearly as large in body size as the Hispaniolan Hutia *Plagiodontia aedium*.

Puertoricomys Woods, 1989. Biogeography of the West Indies, p. 748.
TYPE SPECIES: *Proechimys corozalus* Williams and Koopman, 1951.

Puertoricomys corozalus (Williams and Koopman, 1951). Am. Mus. Novit., 1515:4.
TYPE LOCALITY: Puerto Rico, Corozal Limestone Cave.
DISTRIBUTION: Known only from the type locality.
STATUS: Extinct, known only from a fragmentary mandible.
COMMENTS: Associated with old looking material, and may be older than Pleistocene.

Family Capromyidae Smith, 1842. The Naturalist's Library (London), 15:308.

COMMENTS: Does not include *Myocastor,* and apparently represents an entirely intra-Caribbean radiation (Woods and Howland, 1979; Woods 1989*a*, *b*); also see comments under Myocastoridae. Woods (1982:386-387) included West Indian Heteropsomyinae in this family, but here this group is assigned to the Echimyidae (see comments under Heteropsomyinae). Varona (1974) and Corbet and Hill (1991:203) classified most of the following species in either of two genera (*Capromys* or *Plagiodontia*), whereas Woods (1989*a*, *b*) separated them into several distinct subfamilies and genera.

Subfamily Capromyinae Smith, 1842. The Naturalist's Library (London), 15:30.

COMMENTS: Reviewed by Kratochvil et al. (1978), Varona and Arredondo (1979), and Silva and Garrido (in press). The status of *Brachycapromys, Mysateles, Mesocapromys, Pygmaeocapromys, Paracapromys* and *Stenocapromys* as genera or subgenera is unresolved (Hall, 1981; Rodriguez et al., 1979; Silva and Garrido, in press; Woods and Howland, 1979). This group is in need of revision to standardize the taxonomic levels proposed for the Cuban radiation of capromyids with taxonomic categories established for Hispaniola. Until the group is revised, the Cuban capromyids are placed in three genera as accepted by the majority of Cuban systematists (*Capromys, Mesocapromys, Mysateles*). In addition to the extant forms described below, 13 extinct forms of *Capromys*-like hutias have been described (Silva and Garrido, in press; Varona and Arredondo, 1979): *Capromys antiquus, C. arredondoi, C. latus, C. pappus, C. robustus, Mesocapromys barbouri, M. beatrizae, M. delicatus, M. gracilis, M. kraglievichi, M. minimus, M. silvai,* and *Mysateles jaumei*. Some of these were described in separate subgenera (i.e. *Pygmaeocapromys, Brachycapromys, Palaeocapromys*), but the arrangement in Silva and Garrido (in press) is followed until a revision of the family is completed.

Capromys Desmarest, 1822. Bull. Sci. Soc. Philom. Paris, p. 185.

TYPE SPECIES: *Capromys fournieri* Desmarest, 1822 (= *Isodon pilorides* Say, 1822).

SYNONYMS: *Macrocapromys*.

COMMENTS: "*Capromys*" *geayi* (Pousarges, 1899) was described from Venezuela, where it was collected in the mountains between La Guayra and Caracas. It was placed in its own genus (*Procapromys*) by Chapman (1901:322). The type specimen is Mus. Nat. Hist. Nat. (Paris) No. 1898-1785 (1834A). No further specimens have ever been collected, and it is probable that the collecting locality was incorrectly recorded. This specimen is likely a juvenile *Capromys* from Cuba erroneously reported from Venezuela.

Capromys pilorides (Say, 1822). Jour. Acad. Nat. Sci. Philadelphia, 2:333.

TYPE LOCALITY: "South America or one of the West Indian islands."

DISTRIBUTION: Mainland Cuba, Isle of Youth, Archipiélago de las Doce Legunas, Archipiélago de Sabana, and many other islands and cays in the Cuban archipelago.

STATUS: Common (extremely abundant in some areas, including Guantanamo Bay Naval Base).

SYNONYMS: *acevedo, doceleguas, gundlachianus, intermedius, pilorides, relictus*.

COMMENTS: The first mention of this species name is as *Mus pilorides* Pallas 1778, however, Tate (1935:309) noted that it is not associated with this genus. Sometimes placed in the subgenus *Capromys*; see Hall (1981:863). This species is very variable in size, coloration, and habits. There are four named subspecies (*doceleguas, gundlachianus, pilorides,* and *relictus*; see Varona, 1983*a*:77), and one new subspecies currently being described from the south of the Isle of Youth (Borroto and Camacho, in litt.). *Macrocapromys acevedo* described by Arredondo (1958:10) is not distinct at the generic level, and is probably synonymous with *pilorides*.

Geocapromys Chapman, 1901. Bull. Am. Mus. Nat. Hist., 14:314.

TYPE SPECIES: *Capromys* (*Geocapromys*) *brownii* J. Fischer, 1829.

COMMENTS: Original use of the name was as a subgenus of *Capromys*. Included in *Capromys* by Mohr (1939:75), Varona (1974:67), Hall (1981:865), and Corbet and Hill (1991:203). However, considered a distinct genus by Miller (1929), Woods and Howland (1979:112), and Morgan (1985:30). Includes *columbianus, pleistocenicus* (see Hall,

1981:866), and *megas* (see Varona and Arredondo, 1979:12; Silva and Garrido, in press), which are known only from Holocene fossils from Cuba.

Geocapromys brownii (J. Fischer, 1829). Synopis. Mamm., Addenda, p. 389 (=589).
TYPE LOCALITY: "Jamaica".
DISTRIBUTION: Jamaica.
STATUS: IUCN - Rare.
SYNONYMS: *brachyurus*.
COMMENTS: Sometimes includes *G. thoracatus* from Little Swan Island (see that account). Reviewed by Anderson et al. (1983, Mammalian Species, 201).

Geocapromys ingrahami (J. A. Allen, 1891). Bull. Am. Mus. Nat. Hist., 3:329.
TYPE LOCALITY: "Plana Keys, Bahamas" (East Plana Key).
DISTRIBUTION: Type locality, and introduced on Little Wax Cay (1973) and Warderick Wells Cay (1981).
STATUS: IUCN - Rare because of limited distribution.

Geocapromys thoracatus (True, 1888). Proc. U.S. Natl. Mus. 11:469.
TYPE LOCALITY: "Little Swan Island, one of two small islands lying at the entrance of the Gulf of Honduras".
DISTRIBUTION: Little Swan Island.
STATUS: Recently Extinct.
COMMENTS: Sometimes included as a subspecies of *G. brownii*; see Mohr (1939:77), Hall (1981:866), and Varona (1974:67). Morgan (1985) reconfirmed its status as a distinct species. Clough (1976) stated that *thoracatus* became extinct in 1950s, possibly as a result of introduced cats on the island. Reviewed by Morgan (1989, Mammalian Species, 341).

Mesocapromys Varona, 1970. Poeyana, 73:8.
TYPE SPECIES: *Capromys (Mesocapromys) auritus* Varona, 1970.
SYNONYMS: *Paracapromys, Pygmaeocapromys, Stenocapromys*.
COMMENTS: Original use of name was as a subgenus, which was elevated to a genus by Kratochvil et al. (1978:15). All members of this genus are small (less than 1 kg), and construct large nests made of sticks, unlike *Mysateles*.

Mesocapromys angelcabrerai (Varona, 1979). Poeyana, 194:6.
TYPE LOCALITY: Cuba, Ciego de Avila Prov., Cayos de Ana Maria (21°30'N, 78°40'W).
DISTRIBUTION: Cayos de Ana Maria, Cuba.
STATUS: U.S. ESA and IUCN - Endangered.
COMMENTS: Placed in the subgenus *Pygmaeocapromys* by Varona (1979:5). Unusual among capromyines in being sexually dimorphic. There are reports of the presence of *M. angelcabrerai* on the coast near Jucaró, but no confirmed specimens are known. This species and *M. nanus* are the two smallest hutias in Cuba.

Mesocapromys auritus (Varona, 1970). Poeyana, ser. A, 73:1.
TYPE LOCALITY: Cuba, Las Villas Prov., Archipiélago de Sabana, Cayo Fragoso (22°41'N, 79°27'W).
DISTRIBUTION: Known only from the type locality.
STATUS: U.S. ESA and IUCN - Endangered.
COMMENTS: This form is known only from mangrove habitats along the edges of canals passing across Cayo Fragoso. This form constructs nests that are very similar to those of *angelcabrerai*.

Mesocapromys nanus (G. M. Allen, 1917). Proc. New England Zool. Club, 6:54.
TYPE LOCALITY: Cuba, Matanzas Prov., Sierra de Hato Nuevo.
DISTRIBUTION: Cienaga (swamp) de Zapata (Matanzas Prov., Cuba).
STATUS: U.S. ESA and IUCN - Endangered. Some mammalogists consider this species to be extinct since no specimens have been collected since 1937, but it is still likely to survive in remote areas of the Zapata Swamp (Jorge de la Cruz, pers. comm.).
COMMENTS: Placed in newly created subgenus *Pygmaeocapromys* by Varona (1979:5). In genus *Mesocapromys*, subgenus *Paracapromys* by Kratochvil et al. (1978:15), and Rodriguez et al. (1979). However, retained in *Capromys* subgenus *Mysateles* by Hall (1981:863) because of its long tail and small body size. Varona (1979:5), however,

stated that even though this species "automatically" is associated with *Mysateles* because of tail length, there are important cranial differences between *nanus* and other *Mysateles*. Originally based on fossil material, but subsequently found living in the Zapata Swamp.

Mesocapromys sanfelipensis (Varona and Garrido, 1970). Poeyana, ser. A, 75:3.
TYPE LOCALITY: Cuba, Pinar del Rio Prov., Archipiélago de los Canarreos, Cayo Juan Garcia (21°59'N, 83°31'W).
DISTRIBUTION: Known only from four specimens collected at the type locality.
STATUS: U.S. ESA and IUCN - Endangered. The species is now considered to be extinct because a fire destroyed much of its habitat on Cayo Juan Garcia (Frías et al., 1988).
COMMENTS: Placed in subgenus *Mesocapromys* by Varona (1974), and in genus *Mesocapromys*, subgenus *Paracapromys* by Kratochvil et al. (1978:15). The habitat of this species is uncertain, but all known specimens were captured in grasslands (*Salicornia perennis* = "yerba de vidrio") rather than mangroves.

Mysateles Lesson, 1842. Nouv. Tabl. Regne Animal, Mammifères, p, 124.
TYPE SPECIES: *Mysateles poeppingii* Lesson, 1842 (= *Capromys prehensilis* Poeppig, 1824).
SYNONYMS: *Brachycapromys, Leptocapromys.*
COMMENTS: Lesson separated *Capromys prehensilis* as *Mysateles poeppingIi*; see Varona (1979:4-5). Kratochvil et al. (1978:15) treated *Mysateles* as a genus in their new classification. *Mysateles* was included in *Capromys* by Woods (1989*a*:781), Hall (1981:863), and Corbet and Hill (1991:203). Cuban systematists universally accept *Mysateles* as a valid genus, and there is biochemical and morphological evidence to support this (A. Camacho, pers. comm.). *Mysateles* is recognized here as a distinct genus until a review of all Capromyidae is completed to establish how much intra-island variation is required for recognition of taxa at the generic level.

Mysateles garridoi (Varona, 1970). Poeyana, ser. A, 74: 2.
TYPE LOCALITY: Cuba, Archipiélago de los Canarreos, "Cayo Majá".
DISTRIBUTION: Known only from a single specimen (type locality).
STATUS: IUCN - Endangered.
COMMENTS: Placed in the genus *Mysateles*, subgenus *Leptocapromys* by Kratochvil et al. (1978:15). However, retained in *Capromys* by Varona (in litt.). The type locality has been clarified by Silva and Garrido (in press), who identified the correct name for the locality as a yet unnamed small cay adjacent to Cayo Largo in the Archipiélago de los Canarreos. Varona (1970) incorrectly applied the name Cayo Majá to this islet. It is known that this hutia does not occur on the adjacent Cayo Majá (three cays to the northwest of the Cayo Largo) based on a 1990 expedition to this region by R. Borroto (pers. comm.). The single known specimen may have been left on the site where it was collected by fishermen, and not be from the area, so it is necessary to search more widely in the region to determine the status of this species (A. Camacho, pers. comm.).

Mysateles gundlachi (Chapman, 1901). Bull. Am. Mus. Nat. Hist., 14:313-323.
TYPE LOCALITY: "Nueva Gerona, Isle of Pines" (now Isle of Youth or Isla de la Juventud).
DISTRIBUTION: N regions of Isle of Youth (south of W Cuba).
STATUS: IUCN - Indeterminate. Abundant in specific forest types at several areas in the north of the Isle of Youth, but rapid deforestation of the area is making this species increasingly uncommon, and perhaps threatened (A. Camacho, pers. comm.).
COMMENTS: Raised to specific level by Varona (1986:7). Closely related to *Mysateles prehensilis*. It is proposed that there are two prehensile-tailed tree hutias on the Isle of Youth. According to Varona (1986), this species is confined to forested areas of *Roystones regia, Colpothrinax wrightii*, and *Copernicia curtissi* north of the central savanna of the island, and *M. meridionalis* is confined to completely different forest types on the southwest side of the island south of the savanna. *Mysateles gundlachi* is described on the basis of the baculum, which is distinctly broad-based unlike that of *M. meridionalis*; see also comments under *meridionalis* and *prehensilis*.

Mysateles melanurus (Poey, 1865). *In* Peters, 1865, Monatsb. K. Preuss. Akad. Wiss. Berlin, p. 384.
TYPE LOCALITY: Cuba, Manzanillo, Gramma Prov. (Oriente region).

DISTRIBUTION: E provinces of Cuba.
STATUS: IUCN - Rare. Locally abundant in Güisa, Gramma Prov., but uncommon in other regions of the eastern provinces (A. Camacho, pers. comm.).
SYNONYMS: *arboricolus*.
COMMENTS: The author and date of publication for this species are usually given as Peters, 1864, but Varona (1974:63), established the correct author and date as Poey, 1865. Includes *arboricolus* of Kratochvil et al. (1978:48), who placed it in the genus *Mysateles*, subgenus *Leptocapromys*. However, this specimen is considered to be a young female *C. melanurus* by Varona (1986:7). I have examined the specimen, and concur with Varona.

Mysateles meridionalis (Varona, 1986). Poeyana, 315:4.
TYPE LOCALITY: North of Caleta Cocodrilos, southwestern Isla de la Juventud (former Isle of Pines).
DISTRIBUTION: Restricted to lowland forests SW of the central savanna of the Isle of Youth (south of W Cuba).
STATUS: IUCN - Indeterminate; but probably very uncommon.
COMMENTS: Closely related to *Mystateles prehensilis* and *M. gundlachi*, but different in the proportion of the tail which is 62 percent of body length in *meridionalis*, 73 percent in *gundlachi*, and 79 percent in *prehensilis*; see also comments under *gundlachi* and *prehensilis*.

Mysateles prehensilis (Poeppig, 1824). Jour. Acad. Nat. Sci. Philadelphia, 4:11.
TYPE LOCALITY: Cuba, wooded south coast.
DISTRIBUTION: Cuba, mainly west of Camagüey Province. The status of this species in the eastern provinces (old Oriente Prov.) is not clear.
STATUS: Common.
SYNONYMS: *poeyi*, *poeppingii*, *pallidus*.
COMMENTS: Placed in the subgenus *Mysateles* by Mohr (1939:54) and Varona (1974), and in the genus and subgenus *Mysateles* by Kratochvil et al. (1978:15). This is the largest species of *Mysateles*; see comments under *gundlachi* and *meridionalis*.

Subfamily Hexolobodontinae Woods, 1989. Los Angeles Co. Mus. Nat. Hist., Sci. Ser., 33:76.

Hexolobodon Miller, 1929. Smithson. Misc. Coll., 81:19.
TYPE SPECIES: *Hexolobodon phenax* Miller, 1929.

Hexolobodon phenax Miller, 1929. Smithson. Misc. Coll., 81:20.
TYPE LOCALITY: Haiti, Dept. de l'Artibonite, Small cave NE of Saint Michel de l'Atalye.
DISTRIBUTION: Hispaniola and La Gonave Island.
STATUS: Extinct.
SYNONYMS: *poolei*.
COMMENTS: Includes *H. poolei* (Rimoli, 1976:21), which is known only by the type specimen (USNM No. 255881), because the normal dental variation of *H. phenax* includes the diagnostic condition in *H. poolei*.

Subfamily Isolobodontinae Woods, 1989. Los Angeles Co. Mus. Nat. Hist., Sci. Ser., 33:76.

Isolobodon Allen, 1916. Ann. New York Acad. Sci., 27:19.
TYPE SPECIES: *Isolobodon portoricensis* Allen, 1916.
SYNONYMS: *Aphaetreus*, *Ithydontia*.

Isolobodon montanus (Miller, 1922). Smithson. Misc. Coll., 74:3.
TYPE LOCALITY: Haiti, Dept. de l'Artibonite, Cave NE of Saint Michel de l'Atalaye.
DISTRIBUTION: Hispaniola and La Gonave Island.
STATUS: Extinct.
COMMENTS: Originally described in a separate genus from *I. portoricensis* because the enamel folds of the cheekteeth connect to form separate laminar plates, but a large series of each species indicates that this character is clinal, and that the two forms are very closely related.

Isolobodon portoricensis Allen, 1916. Ann. New York Acad. Sci., 27:19.
TYPE LOCALITY: Puerto Rico, Jobo Dist., near Utuado.
DISTRIBUTION: Hispaniola (Haiti and Dominican Republic) and offshore islands. Introduced on Puerto Rico, St. Thomas, St. Croix, and Mona Islands. The only known jutía on La Gonave island.
STATUS: Probably extinct, but possibly surviving on La Tortue island off the north coast of Haiti (Woods et al., 1985).
SYNONYMS: *levir*.
COMMENTS: Even though the type locality is Puerto Rico, the species was apparently introduced there by Amerindians and the natural range is restricted to Hispaniola. Reported as extinct by Hall (1981:868), but this species survived in Hispaniola and Puerto Rico until the last few decades, and may still survive in certain remote areas (Woods et al., 1985). Includes *I. levir* (Reynolds et al., 1953).

Subfamily Plagiodontinae Ellerman, 1940. The Families and Genera of Living Rodents, 1:25.

Plagiodontia F. Cuvier, 1836. Ann. Sci. Nat. Zool. (Paris), ser. 2, 6:347.
TYPE SPECIES: *Plagiodontia aedium* F. Cuvier, 1836.
SYNONYMS: *Hyperplagiodontia*.
COMMENTS: Reviewed by Mohr (1939), Johnson (1948), Anderson (1965), and Woods (1989*a*, *b*).

Plagiodontia aedium F. Cuvier, 1836. Ann. Sci. Nat. Zool. (Paris), ser. 2, 6:347.
TYPE LOCALITY: "Saint-Domingue" (probably Haiti).
DISTRIBUTION: Hispaniola, La Gonave Island, not recorded from La Tortue island.
STATUS: IUCN - Rare.
SYNONYMS: *hylaeum*, *spelaeum*.
COMMENTS: Includes *hylaeum* as a separate subspecies (Anderson, 1965).

Plagiodontia araeum Ray, 1964. Breviora, Mus. Comp. Zool., 203:2.
TYPE LOCALITY: Dominican Republic, Prov. San Rafael, unnamed cave 2 km SE of Rancho de la Guardia (18°43'N, 71°39'W).
DISTRIBUTION: Hispaniola.
STATUS: Extinct.
SYNONYMS: *stenocoronalis*.
COMMENTS: Type description based on only left upper cheektooth (DP4). Subsequent fossil material collected in Haiti and deposited at the Florida Museum of Natural History includes complete cranial and dentary material, confirming the validity of this taxon. *Hyperplagiodontia stenocoronalis* of Rimoli (1976:34) is not distinct from *P. araeum*. A very large wide-toothed hutia.

Plagiodontia ipnaeum Johnson, 1948. Proc. Biol. Soc. Washington, 61:72.
TYPE LOCALITY: Dominican Republic, Prov. de Samana, Anadel (19°12'N, 69°19'W).
DISTRIBUTION: Hispaniola.
STATUS: Probably extinct (Woods et al., 1985).
SYNONYMS: *caletensis*, *velozi*.
COMMENTS: A larger version of *P. aedium*.

Rhizoplagiodontia Woods, 1989. Los Angeles Co. Mus. Nat. Hist., Sci. Ser., 33:62.
TYPE SPECIES: *Rhizoplagiodontia lemkei* Woods, 1989.

Rhizoplagiodontia lemkei Woods, 1989. Los Angeles Co. Mus. Nat. Hist., Sci. Ser., 33:62.
TYPE LOCALITY: Haiti, Dept. du Sud, 17 km W of Camp Perrin (18°20'N, 74°03'W).
DISTRIBUTION: Endemic to SW Haiti in the Massif de la Hotte.
STATUS: Extinct.

Family Heptaxodontidae Anthony, 1917. Bull. Am. Mus. Nat. Hist., 37(4):183.
SYNONYMS: Amblyrhizinae, Elasmodontomyinae.
COMMENTS: Known only from sub-Recent fossils from Greater and N Lesser Antilles (Woods, 1989*a*). It is debated whether *Amblyrhiza* and *Clidomys* became extinct before or after humans arrived in the West Indies. This family is often placed near the

Chinchillidae based on similar laminar plates of molariform teeth. One genus (*Quemisia*) is very similar in dental morphology to capromyids, however, and it is possible to derive all of the conditions seen in heptaxodontids from dental patterns found within the Capromyidae. This family should be placed adjacent to the Capromyidae.

Subfamily Clidomyinae Woods, 1989. Biogeography of the West Indies, p.753.

Clidomys Anthony, 1920. Bull. Am. Mus. Nat. Hist., 42:469.
TYPE SPECIES: *Clidomys osborni* Anthony, 1920.
SYNONYMS: *Alterodon, Speoxenus, Spirodontomys.*

Clidomys osborni Anthony, 1920. Bull. Am. Mus. Nat. Hist., 42: 469.
TYPE LOCALITY: Jamaica, Balaclava, Wallingford Roadside Cave.
DISTRIBUTION: Jamaica.
STATUS: Extinct.
SYNONYMS: *cundalli, major.*

Clidomys parvus Anthony, 1920 . Bull. Am. Mus. Nat. Hist., 42: 472.
TYPE LOCALITY: Jamaica, Balaclava, Wallingford Roadside Cave.
DISTRIBUTION: Jamaica.
STATUS: Extinct.
SYNONYMS: *jamaicensis.*

Subfamily Heptaxodontinae Anthony, 1917. Bull. Am. Mus. Nat. Hist., 37(4): 183.

Amblyrhiza Cope, 1868. Proc. Am. Philos. Soc., p. 313.
TYPE SPECIES: *Amblyrhiza inundata* Cope, 1868.
SYNONYMS: *Loxomylus.*

Amblyrhiza inundata Cope, 1868. Proc. Am. Philos. Soc., p. 313.
TYPE LOCALITY: West Indies, Anguilla, Bat Cave.
DISTRIBUTION: Anguilla, St. Martin.
STATUS: Extinct.
SYNONYMS: *latidens, quadridens.*

Elasmodontomys Anthony, 1916. Ann. New York Acad. Sci., 27: 199.
TYPE SPECIES: *Elasmodontomys obliquus* Anthony, 1916.
SYNONYMS: *Heptaxodon.*

Elasmodontomys obliquus Anthony, 1916. Ann. New York Acad. Sci., 27:199.
TYPE LOCALITY: Puerto Rico, Utuado.
DISTRIBUTION: Puerto Rico.
STATUS: Extinct.
SYNONYMS: *bidens.*

Quemisia Miller, 1929. Smithson. Misc. Coll., 81(9):22.
TYPE SPECIES: *Quemisia gravis* Miller, 1929.

Quemisia gravis Miller, 1929. Smithson. Misc. Coll., 81(9):23.
TYPE LOCALITY: Haiti, Dept. de L'Artibonite, 6 kms. east of Saint Michel de L'Atalaye.
DISTRIBUTION: Hispaniola.
STATUS: Extinct.

Family Myocastoridae Ameghino, 1904. Anales Soc. Cient. Argentina, 56-58:103.
COMMENTS: Included in Capromyidae by Hall (1981:871), Corbet and Hill (1991:203), and others, but see Woods and Howland (1979), and Woods (1982) for why capromyids and myocastorids are distinct.

Myocastor Kerr, 1792. *In* Linnaeus, Anim. Kingdom, p. 225.
TYPE SPECIES: *Mus coypus* Molina, 1782.
SYNONYMS: *Mastonotus, Myopotamus, Potamys.*

COMMENTS: *Myocastor* is not a capromyid and was referred to the family Myocastoridae by Ameghino (1904) and Woods and Howland (1979:114). Although Patterson and Pascual (1968*b*:6) included *Myocastor* and several fossil forms as the subfamily Myocastorinae of the Echimyidae, it is generally agreed that *Myocastor* is not an echimyid.

Myocastor coypus (Molina, 1782). Sagg. Stor. Nat. Chile, p. 287.

TYPE LOCALITY: Chile, Santiago Prov., Rio Maipo.

DISTRIBUTION: S Brazil, Paraguay, Uruguay, Bolivia, Argentina, Chile.

STATUS: Common.

SYNONYMS: *albomaculatus, bonariensis, castoroides, chilensis, dorsalis, melanops, popelairi, santacruzae.*

COMMENTS: Widely introduced into North America, Europe, N Asia, and E Africa. The common name coypu is preferable to nutria, since nutria in Spanish means otter. The species and subspecies were reviewed by Woods et al. (1992, Mammalian Species, 398).

ORDER LAGOMORPHA
by Robert S. Hoffmann

ORDER LAGOMORPHA

SYNONYMS: Duplicidentata Illiger, 1811.

Family Ochotonidae Thomas, 1897. Proc. Zool. Soc. Lond., 1896:1026 [1897].

SYNONYMS: Lagomina Gray, 1825; Lagomyidae Lilljeborg, 1866.

COMMENTS: Revisions of the family include Gureev (1964), Corbet (1978*c*), and Erbaeva (1988). Other useful treatments include Allen (1938), Ellerman and Morrison-Scott (1951), Ognev (1940), Hall (1981), and A. T. Smith et al. (1990).

Ochotona Link, 1795. Beitr. Naturgesch., 2:74.

TYPE SPECIES: *Ochotona minor* Link, 1795 (= *Lepus dauuricus* Pallas, 1776).

SYNONYMS: *Conothoa* Lyon, 1904; *Lagomys* G. Cuvier, 1800; *Lepus* Linnaeus, 1758 (in part); *Ogotoma* Gray, 1867; *Pika* Lacépède, 1799; *Tibetholagus* Argyropulo and Pidoplichko, 1939.

COMMENTS: There are presently no grounds for recognizing subgenera pending a phylogenetic analysis of specific relationships within the genus. The subgeneric classifications published (e.g., Allen, 1938; Ellerman and Morrison-Scott, 1951; Erbaeva, 1988; Ognev, 1940) differ dramatically, even when based on the same distinguishing characteristics.

Ochotona alpina (Pallas, 1773). Reise Prov. Russ. Reichs., 2:701.

TYPE LOCALITY: "in Alpinus, rupestribus Sibiriae". Restricted by Ognev (1940:23) to Kazakhstan, Altai Mtns, Vostocho-Kazakhstansk Obl., Tigiretskoe Range, vic. of Tigiretskoe [110 km NNW Ust-Kamenogorsk]. Not Tigiretskoe, ESE Minusinsk, Krasnoyarsk Krai, Russia.

DISTRIBUTION: Sayan and Altai Mtns; Khangai, Kentei and associated ranges; upper Amur drainage (NW Kazakhstan, S Russia, NW Mongolia); N Kansu-Ningsia border (China).

STATUS: The small isolated population of *O. a. argentata* at the southern extreme of the species range is likely endangered, and other isolated montane populations in Mongolia may be threatened (A. T. Smith et al., 1990).

SYNONYMS: *argentata* Howell, 1928; *ater* Eversmann, 1842; *changaica* Ognev, 1940; *cinereofusca* (Schrenk, 1858); *nitida* Hollister, 1912; *scorodumovi* Skalon, 1935; *sushkini* Thomas, 1924.

COMMENTS: Formerly included *hyperborea*; but see Ivanitskaya (1985) and Pavlinov and Rossolimo (1987). Sokolov and Orlov (1980:79) considered *hyperborea* a distinct species with a distribution overlapping that of *alpina* in the Khangai and Kentei Mtns, Mongolia. Separate specific status was supported by differences in chromosome numbers (Vorontsov and Ivanitskaya, 1973). The race *sushkini*, formerly assigned to *O. pallasi*, is a subspecies of *alpina* (see A. T. Smith et al., 1990, and references therein). Does not include *collaris* or *princeps*, see Weston (1981).

Ochotona cansus Lyon, 1907. Smithson. Misc. Coll., 50:136.

TYPE LOCALITY: "Taocheo, Kan-su, China" [Lintan, Gannan A.D., Gansu, China].

DISTRIBUTION: C China (Gansu, Qinghai, Sichuan); isolated populations in Shaanxi and Shanxi.

STATUS: The Shanxi subspecies *sorella*, isolated in the extreme northwest of the species range, may be extinct. It is known from only a few specimens, and has not been found in over 50 years.

SYNONYMS: *morosa* Thomas, 1912; *sorella* Thomas, 1908; *stevensi* Osgood, 1932.

COMMENTS: Büchner (1890) originally included this species in the quite different *O. roylei*, but in recent years it has usually been assigned to *O. thibetana* (Allen, 1938; Argyropulo, 1948; Corbet, 1978*c*; Ellerman and Morrison-Scott, 1951; Gureev, 1964; Honacki et al., 1982; Weston, 1982). Recent studies showed that *cansus* and *thibetana* are broadly sympatric, with distinct ecological niches, and morphological characters that do not intergrade (Feng and Kao, 1974; Feng and Zheng, 1985). The latter authors, without access to holotypes, assigned the race *morosa* to *thibetana*, but it is an isolated subspecies of *cansus* that is sympatric with *O. thibetana* in the Tsing Ling Shan, Shaanxi Province (A. T. Smith et al., 1990, and references therein).

Ochotona collaris (Nelson, 1893). Proc. Biol. Soc. Washington, 8:117.

TYPE LOCALITY: "about 200 miles south of Fort Yukon, Alaska near the head of the Tanana River." [USA].

DISTRIBUTION: WC Mackenzie, S Yukon, NW British Columbia (Canada); SE Alaska (USA).

STATUS: Not significantly threatened (MacDonald and Jones, 1987).

SYNONYMS: Monotypic.

COMMENTS: Broadbooks (1965) and Youngman (1975) considered *collaris* and *princeps* conspecific. Corbet (1978*c*), following Argyropulo (1948) and Gureev (1964), included *collaris* in *alpina*. A statistical reevaluation of craniometric data by Weston (1981) indicated that *collaris*, *princeps* and *alpina* are separate species; Hall (1981:286) also recognized *collaris* as a distinct species. *O. collaris* and *O. princeps* share similar chromosome numbers that differ sharply from those of *alpina* and *hyperborea* (Vorontsov and Ivanitskaya, 1973). Reviewed by MacDonald and Jones (1987, Mammalian Species, 281).

Ochotona curzoniae (Hodgson, 1858). J. Asiat. Soc. Bengal, 1857, 26:207 [1858].

TYPE LOCALITY: "district of Chumbi", Chumbi Valley, Tibet, China.

DISTRIBUTION: Tibetan Plateau; adjacent Gansu, Qinghai, Sichuan (China), Sikkim (India) and E Nepal.

STATUS: This species is the focus of widespread control efforts throughout its range, and has been eliminated locally (A. T. Smith et al., 1990).

SYNONYMS: *melanostoma* (Büchner, 1890).

COMMENTS: Includes *melanostoma*, but not *seiana* from Iran, *contra* Corbet (1978*c*:69); see A. T. Smith et al. (1990). Treated as a subspecies of *dauurica* by Mitchell (1978), but it is considered a distinct species by the Chinese; see Feng and Zheng (1985) and Feng et al. (1986). *O. curzoniae* and *O. dauurica* occur in geographic sympatry in Hainan County, Qinghai Province, China, and differ both chromosomally (Vorontsov and Ivanitskaya, 1973) and electrophoretically (Zhou and Xia, 1981).

Ochotona dauurica (Pallas, 1776). Reise Prov. Russ. Reichs., 3:692.

TYPE LOCALITY: "Vivit in campis, montiumque declivibus arenosis apricis, per totam Dauuriam..." Restricted by Ellerman and Morrison-Scott (1951:452) to "Kulusutai, Onon River, Eastern Siberia" [Chitinsk. Obl. Russia].

DISTRIBUTION: Steppes from Altai, Tuva, and Transbaikalia (Russia) through N China and Mongolia, south to Qinghai Province, China.

STATUS: Considered a pest, is intensively controlled in China; control in Russia has been much less intensive. Isolated populations around the margins of the Gobi Desert in China and Mongolia are very vulnerable (A. T. Smith et al., 1990).

SYNONYMS: *altaina* Thomas, 1911; *annectens* Miller, 1911; *bedfordi* Thomas, 1908; *minor* Link, 1795; *mursavi* Bannikov, 1951.

COMMENTS: The spelling of *dauurica* conforms to that of the original description. Formerly included *curzoniae* and *melanostoma*; see *curzoniae*. Ellerman and Morrison-Scott's type restriction is dubious because modern Kulusutai is south of the Onon River, at the NE end of Lake Baron-Torei. See Ognev (1940:62) and Allen (1938:551) for alternate type localities in the same general area.

Ochotona erythrotis (Büchner, 1890). Wiss. Result. Przewalski Cent. Asien Reisen. Zool. Th., B. I: Säugeth., p. 165.

TYPE LOCALITY: Not specified; restricted by Allen (1938:535) to "Burchan-Budda", East Tibet, China.

DISTRIBUTION: E Qinghai, W Gansu, possibly N Sichuan, S Xinjiang, and Tibet (China).

STATUS: Indeterminate.

SYNONYMS: *vulpina* Howell, 1928.

COMMENTS: Formerly included *gloveri*; see Corbet (1978*c*:68). Feng and Zheng (1985) provided evidence that *gloveri* (including *brookei*) is a distinct species. Formerly included in *rutila* (Ellerman and Morrison-Scott, 1951), but now regarded as distinct (Feng and Zheng, 1985; Weston, 1982). The distribution of this species is poorly known, as are its relationships with the apparently allopatric *rutila*, *iliensis*, and *gloveri*.

Ochotona forresti Thomas, 1923. Ann. Mag. Nat. Hist., ser. 9, 11:662.

TYPE LOCALITY: "N.W. flank [Li-kiang Range, 27°N, 100°30'E] 13,000'" [Yunnan, China].

DISTRIBUTION: NW Yunnan, SE Tibet (China); N Burma; Assam, Sikkim (India); Bhutan.

STATUS: Unknown.

SYNONYMS: Monotypic.

COMMENTS: Formerly included in *pusilla* (Ellerman and Morrison-Scott, 1951), *roylei* (Corbet, 1978c), and *thibetana* (Feng and Kao, 1974; Gureev, 1964; Weston, 1982), but now considered distinct (A. T. Smith et al., 1990, and references therein). *O. forresti* is poorly known, but may prove to be an allospecies of *O. roylei*; it is thought to be geographically sympatric with *O. gloveri* and/or *O. thibetana* in Yunnan (China), Burma, and Sikkim (India).

Ochotona gaoligongensis Wang, Gong, and Duan, 1988. Zool. Res., 9:201, 206.

TYPE LOCALITY: "Dongsao-fang [Mount Gaoligong], (27°45'N, 98°27'E), Gongshan Co., Northwest Yunnan, alt. 2950 m." [China].

DISTRIBUTION: Known only from the type locality.

STATUS: IUCN - Indeterminate.

SYNONYMS: Monotypic.

COMMENTS: From the original description, this taxon is likely to prove to be a synonym of *O. forresti*, which is known to occur in the same area.

Ochotona gloveri Thomas, 1922. Ann. Mag. Nat. Hist., ser. 9, 9:190.

TYPE LOCALITY: "Nagchuka [= Nyagquka (Yajiang), W Sichuan, China], 10,000'."

DISTRIBUTION: W Sichuan, NW Yunnan, NE Tibet, SW Qinghai (China).

STATUS: Unknown.

SYNONYMS: *brookei* Allen, 1937; *calloceps* Pen et al., 1962; *kamensis* Argyropulo, 1948 (not 1941; see Honacki et al., 1982).

COMMENTS: Formerly included in *erythrotis*; see comments therein. Whether *gloveri* and *erythrotis* are sym-, para-, or allopatric in distribution in Sichuan and/or Qinghai is unknown.

Ochotona himalayana Feng, 1973. Acta Zool. Sinica, 19:69, 73.

TYPE LOCALITY: "Qu-xiang, Bo-qu Valley, Nei-la-mu [= Nyalam] District, alt. 3500 m." [Xigaze (Shigatse) County, Xizang (Tibet), China].

DISTRIBUTION: Mt. Jolmolunga (Everest) area, S Xizang, China; probably adjacent Nepal.

STATUS: Unknown.

SYNONYMS: Monotypic.

COMMENTS: This taxon was considered a synonym of *O. roylei* by Corbet (1978c) and Weston (1982). Additional data (Feng and Zheng, 1985; Feng et al., 1986) suggest that it may be an independent species. However, its range is within that of the similar *O. roylei nepalensis*, and additional studies are necessary to confirm its specific distinctness.

Ochotona hyperborea (Pallas, 1811). Zoogr. Rosso-Asiat., 1:152.

TYPE LOCALITY: "... e terris Tschuktschicis," [Chukotsk peninsula (Ognev, 1940:41), Chukotsk A.O., Russia].

DISTRIBUTION: Ural, Putorana, Sayan Mtns, east of Lena River to Chukotka, Koryatsk and Kamchatka; upper Yenesei, Transbaikalia, and Amur regions, Sakhalin Island (Russia); NC Mongolia; NE China; N Korea; Hokkaido (Japan).

STATUS: Appears common throughout its large range.

SYNONYMS: *cinereoflava* (Schrenk, 1858); *coreana* Allen and Andrews, 1913; *ferruginea* (Schrenk, 1858); *kamtschaticus* Dybowski, 1922; *kobayashii* Kishida, 1930; *kolymensis* Allen, 1903; *litoralis* Peters, 1882; *mantchurica* Thomas, 1909; *normalis* (Schrenk, 1858); *sadakei* Kishida, 1933; *svatoshi* Turov, 1924; *turuchanensis* Naumov, 1934; *uralensis* Flerov, 1927; *yezoensis* Kishida, 1930; *yoshikurai* Kishida, 1932.

COMMENTS: Formerly incuded in *alpina*; see A. T. Smith et al. (1990), and references therein. Difference in morphology and vocalizations are noticeable where *hyperborea* and *alpina* are sympatric in the W Sayan Mtns, Khangai Mtns, and Transbaikalia, and character displacement in size is also evident in some populations (A. T. Smith et al.,

1990). The original citation was printed and privately circulated in 1811, but not published for general distribution until 1826.

Ochotona iliensis Li and Ma, 1986. Acta Zool. Sinica, 32:375, 379.
TYPE LOCALITY: "Tienshan Mountain [Borokhoro Shan], Nilka [County], Xinjiang, China, alt. 3200 m."
DISTRIBUTION: Known only from the type locality (Li et al., 1988).
STATUS: IUCN - Indeterminate.
SYNONYMS: Monotypic.
COMMENTS: Perhaps related to the *erythrotis-rutila* group; poorly known.

Ochotona koslowi (Büchner, 1894). Wiss. Reisen. Przewalski Cent. Asien Zool. Th. I: Säugeth., pg. 187.
TYPE LOCALITY: "Dolina Vetrov" [Valley of the Winds; pass between Guldsha Valley and valley of Dimnalyk River, tributary of Chechen, Tarim Basin, Xinjiang, China, 14,000' (37°55'N, 87°50'E)].
DISTRIBUTION: Arkatag Range, Kunlun Mtns (China).
STATUS: IUCN - Vulnerable. Not found by recent expedition to type locality, but reported at Aqqikkol (37°09'N, 88°11'E)(Zheng, 1986). Listed in "Red Book" of China.
SYNONYMS: Monotypic.

Ochotona ladacensis (Günther, 1875). Ann. Mag. Nat. Hist., ser. 4, 16:231.
TYPE LOCALITY: "Chagra, 14000 feet above the sea" [Changra, Ladak, Kashmir, India].
DISTRIBUTION: SW Xinjiang, Qinghai, E Tibet (China); Kashmir (India); Pakistan.
STATUS: Probably not scarce, but poorly known. May be affected by control measures directed at *curzoniae*.
SYNONYMS: Monotypic.
COMMENTS: Broadly sympatric with *curzoniae* on the Tibetan Plateau, though not so widely distributed.

Ochotona macrotis (Günther, 1875). Ann. Mag. Nat. Hist., ser. 4, 16:231.
TYPE LOCALITY: "Doba" [C Tibet, (31°N, 87°E), China] Ognev (1940:86). Not "Duba. . .N side . . . Kuenlun on road. . .via Kugiar" [Kakyar, 37°45'N, 77°05'E, W Xinjiang, China] *contra* Blanford (1879:76). Not Dobo, Qinghai [(36°41'N, 101°30'E)(Vaurie, 1972:352)].
DISTRIBUTION: Mountains of Sichuan and Yunnan (China); Himalayas (Nepal, India) from Bhutan through Tibet, Kunlun (China), Karakorum (Pakistan), Hindu Kush (Afghanistan), Pamir, and W Tien Shan Mtns (Kirghizistan, Tadzhikistan, SE Kazakhstan).
STATUS: Currently not threatened (A. T. Smith et al., 1990).
SYNONYMS: *auritus* Blanford, 1875; *baltina* Thomas, 1922; *chinensis* Thomas, 1911; *griseus*, Blanford, 1875; *sacana* Thomas, 1914; *sinensis*, Lydekker, 1912; *wollastoni* Thomas and Hinton, 1922.
COMMENTS: Included in *roylei* by Gureev (1964), Roberts (1977), Corbet (1978*c*:68), and Gromov and Baranova (1981:72). Morphological and ecological differences in the area of sympatry first documented by Kawamichi (1971) and Abe (1971), and confirmed by Mitchell (1978, 1981). Weston (1982), Feng and Zheng (1985), and Feng et al. (1986) indicated that *macrotis* is a distinct species. Whether the co-type from "Doba" in the Natural History Museum (London) is from C Tibet or W Xinjiang is uncertain.

Ochotona muliensis Pen and Feng, 1962. *In* Pen et al., Acta Zool. Sinica, 14 (supplement):120, 132.
TYPE LOCALITY: "Ting-Tung-Niu-Chang, southeastern Muli (alt. 3600m) Szechuan" [Muli A.D., Xichang County, Sichuan, China].
DISTRIBUTION: Known only from the vicinity of the type locality.
STATUS: IUCN - Indeterminate.
SYNONYMS: Monotypic.
COMMENTS: This taxon was originally described as a subspecies of *gloveri*, but is now thought to be specifically distinct (Feng and Zheng, 1985). It differs from *gloveri* in certain cranial characters, and in habitat (A. T. Smith et al., 1990), but its extreme rarity in collections makes its independent status difficult to demonstrate.

Ochotona nubrica Thomas, 1922. Ann. Mag. Nat. Hist., ser. 9, 9:187.

TYPE LOCALITY: "Tuggur, Nubra Valley, alt. 10,000" [Ladak, Kashmir, India].

DISTRIBUTION: Southern edge of Tibetan Plateau from Ladak (India, China) through Nepal to E Tibet (China).

STATUS: IUCN - Indeterminate.

SYNONYMS: *aliensis* Zheng, 1979; *lama* Mitchell and Punzo, 1975; *lhasaensis* Feng and Kao, 1974.

COMMENTS: Assigned to *pusilla* by Ellerman and Morrison-Scott (1951), to *roylei* (as *O. lama*) by Corbet (1978c) and to *thibetana* (as *O. t. lama*) by Feng et al. (1986). Recognized as distinct by A. T. Smith et al. (1990). Its closest relations seem to be with *thibetana* and it is possible that further data may indicate intergradation between the two. May include *hodgsoni*; see comment under *O. roylei*.

Ochotona pallasi (Gray, 1867). Ann. Mag. Nat. Hist., ser. 3, 20:220.

TYPE LOCALITY: "...said to come from 'Asiatic Russia-Kirgisen'" (Thomas, 1908:109). Restricted by Heptner (1941:328) to "southern parts...Karkaralinsk Mountains...north of Lake Balkhash" [Karagandinsk Obl., Kazakhstan (49°N, 75°E)].

DISTRIBUTION: Discontinuous in arid areas (mtns and high steppes) in Kazakhstan; Altai Mtns, Tuva (Russia), and Mongolia, to Xinjiang and Inner Mongolia (China).

STATUS: Isolated populations of *O. p. hamica* and *O. p. sunidica* are threatened, or in some cases may be extinct (A. T. Smith et al., 1990).

SYNONYMS: *hamica* Thomas, 1912; *ogotona* Waterhouse, 1848 (not Bonhote, 1905); *opaca* Argyropulo, 1939 (not Vinogradov and Argyropulo, 1948; see Ellerman and Morrison-Scott, 1951; not Argyropulo, 1941, see comment under *gloveri kamensis*); *pricei* Thomas, 1911; *sunidica* Ma et al., 1980.

COMMENTS: Includes *pricei* (Corbet, 1978c) as it is commonly referred to in Soviet literature. However, marked difference in reproduction, habitat, behavior, and vocalization suggest that *pallasi* and *pricei* may prove to be specifically distinct (A. T. Smith et al., 1990).

Ochotona princeps (Richardson, 1828). Zool. J., 3:520.

TYPE LOCALITY: "Rocky Mountains"; restricted by Preble (1908) to "near the sources of Elk (Athabasca) River," [Athabasca Pass, head of Athabasca River, Alberta, Canada].

DISTRIBUTION: Mountains of W North America from C British Columbia (Canada) to N New Mexico, Utah, C Nevada, and EC California (USA).

STATUS: Most populations are not currently threatened, except for a few isolates in the Great Basin (*goldmani, obscura, nevadensis, tutelata*); *tutelata* may now be extinct (A. T. Smith et al., 1990).

SYNONYMS: *albata* Grinnell, 1912; *barnesi*, Durrant and Lee, 1955; *brooksi* Howell, 1924; *brunnescens* Howell, 1919; *cinnamomea* Allen, 1905; *clamosa* Hall and Bowlus, 1938; *cuppes* Bangs, 1899; *fenisex* Osgood, 1913; *figginsi* Allen, 1912; *fumosa* Howell, 1919; *fuscipes* Howell, 1919; *goldmani* Howell, 1924; *howelli* Borell, 1931; *incana* Howell, 1919; *jewetti* Howell, 1919; *lasalensis* Durrant and Lee, 1955; *lemhi* Howell, 1919; *levis* Hollister, 1912; *littoralis* Cowan, 1955; *lutescens* Howell, 1919; *minimus* Lord, 1863; *moorei* Gardner, 1950; *muiri* Grinnell and Storer, 1916; *nevadensis* Howell, 1919; *nigrescens* Bailey, 1913; *obscura* Long, 1965; *saturata* Cowan, 1955; *saxatilis* Bangs, 1899; *schisticeps* (Merriam, 1889); *septentrionalis* Cowan and Racey, 1947; *sheltoni* Grinnell, 1918; *taylori* Grinnell, 1912; *tutelata* Hall, 1934; *uinta* Hollister, 1912; *utahensis* Hall and Hayward, 1941; *ventorum* Howell, 1919; *wasatchensis* Durrant and Lee, 1955.

COMMENTS: Broadbooks (1965) and Youngman (1975) considered *princeps* and *collaris* conspecific. Corbet (1978c), following Gureev (1964) included *princeps* in *alpina*. A statistical reevaluation of craniometric data by Weston (1981) indicated that *princeps, collaris*, and *alpina* are separate species. Reviewed by Smith and Weston (1990, Mammalian Species, 352).

Ochotona pusilla (Pallas, 1769). Nova Comm. Imp. Acad. Sci. Petropoli, 13:531.

TYPE LOCALITY: "in campis circa Volgam..."; restricted by Ognev (1940) to Samarsk Steppe, near Buzuluk, left bank of Samara River [Orenburgsk Obl. Russia].

DISTRIBUTION: Steppes from middle Volga (Russia), east and south through N Kazakhstan to upper Irtysh River and Chinese border. Not yet recorded in China.

STATUS: Disappearance from W end of its range within historic times was probably caused by elimination of preferred shrub-steppe habitat (Kuz'mina, 1965). Some pop. are listed as rare in the "Red Book" of Bashkir Aut. Rep. (A. T. Smith et al., 1990).
SYNONYMS: *angustifrons* Argyropulo, 1932; *minutus* Pallas, 1771.
COMMENTS: Formerly included *nubrica* (= *lama*), *forresti*, and *osgoodi* (a subspecies of *thibetana*); see comments therein.

Ochotona roylei (Ogilby, 1839). Royle's Illus. Botany...Himalaya, vol. 2, 69, pl. 4 [erroneusly labeled "*Lagomys alpinus*"].
TYPE LOCALITY: "Choor Mountain, Lat. 30. Elev. 11,500[ft], [60 mi. (96 km) N of Saharanpur], Punjab, India.
DISTRIBUTION: Himalayan Mtns in NW Pakistan and India to Nepal; adjacent Tibet (China).
STATUS: Currently not under threat.
SYNONYMS: *angdawai* Biswas and Khajuria, 1955; *hodgsoni*, Blyth, 1841; *mitchelli* Agrawal and Chakraborty, 1971; *nepalensis* Hodgson, 1841; *wardi* Bonhote, 1904.
COMMENTS: Includes *angdawai* and *mitchelli*, but not *forresti* and *himalayana*, which are here provisionally considered distinct; see comments therein. *O. hodgsoni* is traditionally placed here, but based on the original description may be assigned to *nubrica*.

Ochotona rufescens (Gray, 1842). Ann. Mag. Nat. Hist., [ser. 1], 10:266.
TYPE LOCALITY: "India, Cabul, Rocky Hills near Baker Tomb at about 6000 or 8000 feet elevation" [Baber's (?) Tomb, Kabul, Afghanistan].
DISTRIBUTION: Afghanistan, Baluchistan (Pakistan), Iran, Armenia, and SW Turkmenia.
STATUS: Considered a crop pest and controlled in parts of its range. However, most populations are not considered threatened. A possible exception may be the isolated population of *shukurovi* in the Little Balkhan Range, Turkmenia (A. T. Smith et al., 1990).
SYNONYMS: *regina* Thomas, 1911; *shukurovi* Heptner, 1961; *vizier* Thomas, 1911; *vulturna* Thomas, 1920.
COMMENTS: Includes *seiana*; see A. T. Smith et al. (1990) and comments under *O. curzoniae*.

Ochotona rutila (Severtzov, 1873). Izv. Obshch. Lyubit. Estestvozn., 8(2): 83.
TYPE LOCALITY: "...in mountains near Vernyi [Alma-ata]...7000-8000 ft." [2134-2438m]. Restricted by Shnitnikov (1936) to valley of Maly Alma-atinsk River, Zailisk Alatau Mtns, Kazakhstan (43°05'N, 77°10'E).
DISTRIBUTION: Isolated ranges from the Pamirs (Tadzhikistan) to Tien Shan (SE Uzbezistan, Kirghizistan, SE Kazakhstan); perhaps N Afghanistan and E Xinjiang (China).
STATUS: This is a rare species, extremely sporadic in occurrence throughout its range, and common in only a few localities (A. T. Smith et al., 1990).
SYNONYMS: Monotypic.
COMMENTS: Apparently an allospecies of *O. erythrotis*, which has sometimes been included in *rutila*; see comments therein.

Ochotona thibetana (Milne-Edwards, 1871). Nouv. Arch. Mus. Hist. Nat. Paris, Bull., 7:93.
TYPE LOCALITY: "mountain near Moupin" [Baoxing, Ya'an County, Sichuan, China.].
DISTRIBUTION: Shanxi, Shaanxi, W Hubei, Yunnan, Sichuan, S Tibet (China); N Burma; Sikkim (India); perhaps adjacent Bhutan and India.
STATUS: *O. t. sikimaria* of Sikkim may be endangered by habitat destruction (A. T. Smith et al., 1990). Other forms do not appear to be threatened.
SYNONYMS: *hodgsoni* Bonhote, 1905; *huangensis* Matschie, 1908; *nanggenica* Zheng et al., 1980; *osgoodi* Anthony, 1941; *sacraria* Thomas, 1923; *sikimaria* Thomas, 1922; *syrinx* Thomas, 1911; *xunhuaensis* Shou and Feng, 1984; *zappeyi* Thomas, 1922.
COMMENTS: Formerly included *cansus*, *forresti*, and *nubrica*; see comments therein. The taxon *aliensis*, originally described as a subspecies of *thibetana*, is now considered a synonym of *nubrica* (Feng et al., 1986; A. T. Smith et al., 1990). *O. osgoodi*, described as a distinct species by Anthony (1941), was listed as a subspecies of *O. pusilla* by Ellerman and Morrison-Scott (1951), and subsequently allocated to *thibetana* by Corbet (1978*c*) and Weston (1982). The isolated subspecies *sikimaria* was assigned to *cansus* by Feng and Kao (1974) and Feng and Zheng (1985), but transferred to *thibetana* by A. T. Smith et al. (1990). Erbaeva (1988:190-191) considered *cansus* and *sikimaria* subspecies of *thibetana*, as well as *lhasaensis*, here placed in *nubrica*;

however, she thought *hodgsoni* was probably a distinct species (based on examination of a skull photograph).

Ochotona thomasi Argyropulo, 1948. Trudy Zool. Inst. Leningrad, 7:127.
TYPE LOCALITY: "Valley of Alyk-nor." Restricted by Formosov (in A. T. Smith et al., 1990) to Alang-nor Lake, NE Qinghai, China (35°35'N, 97°25'E); see also Corbet (1978*c*).
DISTRIBUTION: NE Qinghai, Gansu, and Sichuan (China).
STATUS: IUCN - Indeterminate. Apparently rare and scattered in distribution.
SYNONYMS: *ciliana* Bannikov, 1940.
COMMENTS: Widely sympatric with the similar *O. cansus*.

Prolagus Pomel, 1853. Cat. Meth. Desc. Vert. Foss. dans le Bassin Hydro. Super. de la Loire et Surt. la Val. Aff. Prin., l'Allier. Paris. p. 43.
TYPE SPECIES: *Anoema oeningensis* König, 1825 (fossil).
SYNONYMS: *Anoema* König, 1825; *Archaeomys* Fraas, 1856; *Lagomys* G. Cuvier, 1800; *Myolagus* Hensel, 1856.
COMMENTS: Elevated by Erbaeva (1988) to family Prolagidae. Reviewed by Tobien (1975).

Prolagus sardus (Wagner, 1832). Abh. Bayer. Akad. Wiss., 1:763-767.
TYPE LOCALITY: Italy, Sardinia.
DISTRIBUTION: Mediterranean Isls of Corsica (France) and Sardinia (Italy); adjacent small islands.
STATUS: IUCN - Extinct.
SYNONYMS: Monotypic.
COMMENTS: Described from fossils, but apparently survived until historic times (Vigne, 1983) perhaps as late as 1774 (Kurten, 1968). Reviewed by Dawson (1969).

Family Leporidae G. Fischer, 1817. Mém. Soc. Imp. Nat. Moscow, 5:372.
SYNONYMS: Leporinorum Fischer, 1817.
COMMENTS: Often divided into subfamilies Paleolaginae (*Pentalagus*, *Pronolagus*, *Romerolagus*) and Leporinae (remaining genera) (Dice, 1929; Simpson, 1945), but no subfamilies were recognized by Ellerman and Morrison-Scott (1951). For basis of genera recognized here, see Corbet (1983).

Brachylagus Miller, 1900. Proc. Biol. Soc. Washington, 13:157.
TYPE SPECIES: *Lepus idahoensis* Merriam, 1891.

Brachylagus idahoensis (Merriam, 1891). N. Am. Fauna, 5:76.
TYPE LOCALITY: "Pahsimeroi Valley [near Goldburg, Custer County], Idaho." [USA].
DISTRIBUTION: SW Oregon to EC California, SW Utah, N to SW Montana (USA). Isolated population in WC Washington (USA).
STATUS: The status of the isolated Washington population is uncertain. Varies from secure to IUCN - Vulnerable depending upon population (Chapman et al., 1990).
SYNONYMS: Monotypic.
COMMENTS: Formerly included in *Sylvilagus*; but see Corbet (1983). Placed in the monotypic genus *Brachylagus* by Dawson (1967) and, together with *bachmani*, in the genus *Microlagus* by Gureev (1964:170-173); but also see Hall (1981:294), who recognized *Brachylagus* as a subgenus. This species is widely sympatric with *Sylvilagus nuttallii*, and perhaps overlaps narrowly with *S. audubonii*. It has been interpreted as either a primitive rabbit (Hibbard, 1963), or as derived from *Sylvilagus* (Corbet, 1983). Reviewed by Green and Flinders (1980, Mammalian Species, 125).

Bunolagus Thomas, 1929. Proc. Zool. Soc. Lond., 1929:109.
TYPE SPECIES: *Lepus monticularis* Thomas, 1903.

Bunolagus monticularis (Thomas, 1903). Ann. Mag. Nat. Hist., ser. 7, 11:78.
TYPE LOCALITY: "Deelfontein, Cape Colony," South Africa.
DISTRIBUTION: C Karoo (31°22'S, 22°E), Cape Prov. (South Africa).

STATUS: IUCN - Endangered. Now survives only in "dense, discontinuous karoid vegetation in the districts of Victoria West, Beaufort West and Frazerburg" (86km²) (Duthie and Robinson, 1990).
SYNONYMS: Monotypic.
COMMENTS: Reviewed by Petter (1972*b*). Formerly in *Lepus* (Ellerman and Morrison-Scott, 1951), but returned to *Bunolagus* by Angermann (1966). Karyological evidence supports the separation of *Bunolagus* (2n=44) from *Lepus* (2n=48) (Robinson and Skinner, 1983; Robinson and Dippenaar, 1987); its closest relatives are probably *Pronolagus* (Corbet, 1983).

Caprolagus Blyth, 1845. J. Asiat. Soc. Bengal, 14:247.
TYPE SPECIES: *Lepus hispidus* Pearson, 1839.

Caprolagus hispidus (Pearson, 1839). *In* M'Clelland, Proc. Zool. Soc. Lond., 1838:152 [1839].
TYPE LOCALITY: "...Assam, ...base of the Boutan [Bhutan] mountains" [India].
DISTRIBUTION: S Himalaya foothills from Uttar Pradesh (India) through Nepal and West Bengal to Assam (India), and south through NW Bangladesh. Since 1951, there have been very few reports from Uttar Pradesh and Assam, see Santapau and Humayun (1960), Mallinson (1971), and Ghose (1978). Presently known distribution summarized by Bell et al. (1990).
STATUS: CITES - Appendix I; U.S. ESA and IUCN - Endangered.
SYNONYMS: Monotypic.
COMMENTS: Subgenus *Caprolagus* (see Gureev, 1964).

Lepus Linnaeus, 1758. Syst. Nat., 10th ed., 1:57.
TYPE SPECIES: *Lepus timidus* Linnaeus, 1758.
SYNONYMS: *Allolagus* Ognev, 1929; *Boreolagus* Barrett-Hamilton, 1911; *Chionobates* Kaup, 1829; *Eulagos* Gray, 1867; *Eulepus* Acloque, 1899; *Indolagus* Gureev, 1953; *Lagos* Palmer, 1904; *Macrotolagus*, Mearns, 1895; *Poecilolagus* Lyon, 1904; *Proeulagus* Gureev, 1964; *Tarimolagus* Gureev, 1947.
COMMENTS: Formerly included *Bunolagus*; see Petter (1972*b*); and originally all other genera except *Pentalagus*. The taxonomy of this genus remains controversial. *L. crawshayi* (including *whytei*), *peguensis*, *ruficaudatus*, and *siamensis* have been variously treated as separate species or have been included in *nigricollis*. *L. europaeus*, *corsicanus*, *granatensis*, *tolai*, and *tibetanus* have been placed in *capensis* or treated as distinct species; see comments therein.

Lepus alleni Mearns, 1890. Bull. Am. Mus. Nat. Hist., 2:294.
TYPE LOCALITY: "Rillito Station [Pima Co.] Arizona" [USA].
DISTRIBUTION: SC Arizona (USA) to N Nayarit and Tiburon Isl (Mexico).
STATUS: Populations appear stable in Arizona, but in Mexico require investigation (Flux and Angermann, 1990).
SYNONYMS: *palitans* Bangs, 1900; *tiburonensis* Townsend, 1912.
COMMENTS: Placed in *Caprolagus* (*Macrotolagus*) by Gureev (1964:155). Probably related to *callotis*, but recognized as a distinct species by Hall (1981:331).

Lepus americanus Erxleben, 1777. Syst. Regni Anim., 1:330.
TYPE LOCALITY: "in America boreeli, ad fretum Hudsonis copiosissimus." Restricted by Nelson (1909:87) to Fort Severn, Ontario, Canada.
DISTRIBUTION: S and C Alaska (USA) to S and C coasts of Hudson Bay to Newfoundland and Anacosti Isl (introduced) (Canada), south to S Appalachians, S Michigan, North Dakota, NC New Mexico, SC Utah, and EC California (USA).
STATUS: Although population densities fluctuate greatly, status secure nearly everywhere (Keith and Windberg, 1978; Sinclair et al., 1988).
SYNONYMS: *bairdii* Hayden, 1869; *bishopi* J. Allen, 1899; *borealis* Schinz, 1845; *cascadensis* Nelson, 1907; *columbiensis* Rhoads, 1895; *dalli* Merriam, 1900; *hudsonius* Pallas, 1778; *klamathensis* Merriam, 1899; *macfarlani* Merriam, 1900; *nanus* Schreber, 1790; *niediecki* Matschie, 1907; *oregonus* Orr, 1934; *pallidus* Cowan, 1938; *phaeonotus* J. Allen, 1899; *pinetus* Dalquest, 1942; *saliens*, Osgood, 1900; *seclusus* Baker and Hankins, 1950; *struthopus* Bangs, 1898; *tahoensis* Orr, 1933; *virginianus* Harlan, 1825; *wardi* Schinz, 1825; *washingtoni* Baird, 1855.

COMMENTS: Distinctive small species, but subgeneric separation (*Poecilolagus* Lyon, 1904) not supported; see Hall (1981:314); but see also Gureev (1964:188).

Lepus arcticus Ross, 1819. Voy. Discovery, II; ed. 2, App. IV, p. 170.

TYPE LOCALITY: "Southeast of Cape Bowen" (Nelson, 1909:61) [Possession Bay, Bylot Island, lat. 73°37′N, Canada].

DISTRIBUTION: Greenland and Canadian arctic islands southward in open tundra to WC shore of Hudson Bay, thence northwestward to the west of Fort Anderson on coast of Arctic Ocean. Isolated populations in tundra of N Quebec and Labrador, and on Newfoundland (Canada).

STATUS: Does not appear to be at risk presently (Flux and Angermann, 1990).

SYNONYMS: *andersoni* Nelson, 1934; *bangsii* Rhoads, 1896; *banksicola* Manning and Macpherson, 1958; *canus* Preble, 1902; *glacialis* Leach, 1819; *groenlandicus* Rhoads, 1896; *hubbardi* Handley, 1952; *hyperboreus* Pedersen, 1930; *labradorius* Miller, 1899; *monstrabilis* Nelson, 1934; *persimilis* Nelson, 1934; *porsildi* Nelson, 1934.

COMMENTS: Formerly included in *timidus* by Gureev (1964), Angermann (1967), Honacki et al. (1982), and Dixon et al. (1983), but considered distinct by Corbet (1978*c*), Hall (1981), A. J. Baker et al. (1983), and Flux and Angermann (1990). Angermann (in litt., 1992) considered it "probably conspecific" with *timidus*.

Lepus brachyurus Temminck, 1845. *In* Siebold, Fauna Japonica, 1(Mamm.), p. 44, pl. 11.

TYPE LOCALITY: "...tout l'Empire mais surtout dans l'île de Jezo", Nagasaki, Kyushu, Japan.

DISTRIBUTION: Honshu, Shikoku, Kyushu, Oki Isls and Sado Isl (Japan).

STATUS: No indication of decline at present (Flux and Angermann, 1990).

SYNONYMS: *angustidens* Hollister, 1912; *etigo* Abe, 1918; *lyoni* Kishida, 1937; *okiensis* Thomas, 1906.

COMMENTS: Reviewed by Imaizumi (1970*b*). Gureev (1964:150) and Gromov and Baranova (1981:63) placed this species in *Caprolagus* (*Allolagus*); see also comment under *mandshuricus*.

Lepus californicus Gray, 1837. Mag. Nat. Hist. [Charlesworth's], 1:586.

TYPE LOCALITY: "St. Antoine" [probably near Mission of San Antonio, California, USA]. Discussed by Hall (1981:326).

DISTRIBUTION: Hidalgo and S Queretaro to N Sonora and Baja California (Mexico), north to SW Oregon and C Washington, S Idaho, E Colorado, S South Dakota, W Missouri, and NW Arkansas (USA). Apparently isolated population in SW Montana.

STATUS: Secure; range is expanding at expense of *L. callotis* and *L. townsendi* (Flux and Angermann, 1990).

SYNONYMS: *altamirae* Nelson, 1904; *asellus* Miller, 1899; *bennettii* Gray, 1843; *curti* Hall, 1951; *depressus* Hall and Witlow, 1932; *deserticola* Mearns, 1896; *eremicus* J. Allen, 1894; *festinus* Nelson, 1904; *griseus* Mearns, 1896; *magdalenae* Nelson, 1907; *martirensis* Stowell, 1895; *melanotis* Mearns, 1890; *merriami* Mearns, 1896; *micropus* J. Allen, 1903; *richardsonii* Bachman, 1839; *sheldoni* Burt, 1933; *texianus* Waterhouse, 1848; *tularensis* Merriam, 1904; *vigilax* Dice, 1926; *wallawalla* Merriam, 1904; *xanti* Thomas, 1898.

COMMENTS: Subgenus *Proeulagus* (Gureev, 1964:193).

Lepus callotis Wagler, 1830. Naturliches Syst. Amphibien, p. 23.

TYPE LOCALITY: "Mexico"; restricted by Nelson (1909:122) to southern end of Mexican Tableland.

DISTRIBUTION: C Oaxaca (Mexico) north discontinuously to SW New Mexico (USA).

STATUS: Appears to be relatively rare; classified as "endangered" in U.S. (Flux and Angermann, 1990).

SYNONYMS: *battyi* J. Allen, 1903; *gaillardi* Mearns, 1896; *mexicanus* Lichtenstein, 1830; *nigricaudatus* Bennett, 1833.

COMMENTS: Subgenus *Proeulagus* (Gureev, 1964:192). Includes *gaillardi* and *mexicanus*; see Anderson and Gaunt (1962) and Hall (1981:328-330); but see also Gureev (1964:192, 195). Range allopatric to *L. alleni*, to which it is probably related.

Lepus capensis Linnaeus, 1758. Sys. Nat., 10th ed., 1:58.

TYPE LOCALITY: "ad Cap. b. Spei" [South Africa, Cape of Good Hope].

DISTRIBUTION: Africa in two separate, non-forested areas: South Africa, Namibia, Botswana, Zimbabwe, S Angola, S Zambia (?), Mozambique; and to the north, Tanzania, Kenya, Somalia, Ethiopia, countries of the Sahel and Sahara, and N Africa; thence eastward through the Sinai to the Arabian Peninsula, Jordan, S Syria, S Israel and W and S Iraq, west of the Euphrates River.

STATUS: Populations have declined locally due to habitat alteration, but most are not threatened so far as is known (Flux and Angermann, 1990).

SYNONYMS: *abbotti* Hollister, 1918; *abyssinicus* Lefebvre, 1850; *aegyptius* Desmarest, 1822; *aethiopicus* Hemprich and Ehrenberg, 1832; *aquilo* Thomas and Wroughton, 1907; *arabicus* Ehrenberg, 1833; *arenarius* I. Geoffroy, 1826; *atallahi* Harrison, 1972; *atlanticus* de Winton, 1898; *barcaeus* Ghigi, 1920; *bedfordi* Roberts, 1932; *berberanus* Heuglin, 1861; *carpi* Lundholm, 1955; *centralis* Thomas, 1903; *cheesmani* Thomas, 1921; *crispii* Drake-Brockman, 1911; *dinderus* Setzer, 1956; *ermeloensis* Roberts, 1932; *granti* Thomas and Schwann, 1904; *habessinicus* Hemprich and Ehrenberg, 1832; *hartensis* Roberts, 1932; *harterti* Thomas, 1903; *hawkeri* Thomas, 1901; *innesi* de Winton, 1902; *isabellinus* Cretzschmar, 1826; *jeffreyi* Harrison, 1980; *kabylicus* de Winton, 1898; *kalaharicus* Dollman, 1910; *langi* Roberts, 1932; *major* Grill, 1860; *mandatus* Thomas, 1926; *maroccanus* Cabrera, 1907; *narranus* Thomas, 1926; *ochropoides* Roberts, 1929; *ochropus* Wagner, 1844; *omanensis* Thomas, 1894; *pallidior* Barrett-Hamilton, 1898; *pediaeus* Cabrera, 1923; *rothschildi* de Winton, 1902; *salai* Jentink, 1880; *schlumbergeri* Remy-St. Loup, 1894; *sefranus* Thomas, 1913; *senegalensis* (Rochebrune, 1883); *sherif* Cabrera, 1906; *sinaiticus* Ehrenberg, 1833; *somalensis* Heuglin, 1861; *tigrensis* Blanford, 1869; *tunetae* de Winton, 1898; *vernayi* Roberts, 1932; *whitakeri* Thomas, 1902.

COMMENTS: Subgenus *Proeulagus* (Gureev, 1964:202). Includes *arabicus*; formerly included *europaeus, corsicanus, granatensis*, and *tolai*; see Corbet (1978*c*:71), Angermann (1983:20), and Harrison and Bates (1991). Includes *habessinicus*, but see Azzaroli-Puccetti (1987*a, b*) who considered *habessinicus* as distinct. The enigmatic form *connori*, often placed in *capensis* (Corbet, 1978*c*; Harrison and Bates, 1991) is provisionally placed in *europaeus* on the basis of pelage characteristics; see Angermann (1983:19). Most Russian authors consider *tolai* (including *tibetanus*) a distinct species; see Gromov and Baranova (1981:65); but also see Pavlinov and Rossolimo (1987:229). Sludskii et al. (1980:58, 85) indicated an area of sympatry between *europaeus* and *tolai* in Kazakhstan. Sokolov and Orlov (1980:85) considered *tibetanus* a distinct species. Some Arabian forms may be specifically distinct (Flux and Angermann, 1990); Angermann (1983:19) noted pronounced "size" groups within *arabicus*. These are *arabicus* (largest, gray), *cheesmani* (with insular *atallahi*) (smaller, buffy), and *omanensis* (with insular *jeffreyi*) (smallest, gray).

Lepus castroviejoi Palacios, 1977. Doñana, Acta Vertebr., 1976, 3(2):205 [1977].

TYPE LOCALITY: "Puerto de la Ventana, San Emiliano (León [Province])" [= Puerto Ventana, Spain, 1500 m].

DISTRIBUTION: Cantabrian Mtns between Sierra de Ancares and Sierra de Peña Labra (N Spain).

STATUS: Although it has a very restricted range (25-40 km by 230 km), it is common in most places (Palacios, 1983).

SYNONYMS: Monotypic.

COMMENTS: Reviewed by Palacios (1983, 1989) and Bonhomme et al. (1986).

Lepus comus Allen, 1927. Am. Mus. Novit., No 284:9.

TYPE LOCALITY: "Teng-yueh [Tengueh], Yunnan Province, China, 5,500 feet altitude."

DISTRIBUTION: Yunnan, W Guizhou (China).

STATUS: Probably secure at present (Flux and Angermann, 1990).

SYNONYMS: *peni* Wang and Luo, 1985; *pygmaeus* Wang and Feng, 1985.

COMMENTS: Formerly included in *oiostolus*; see Corbet (1978*c*). Elevated to specific status by Cai and Feng (1982) and Wang et al. (1985), on the basis of morphological and ecological differences. May be allo- or parapatric with *oiostolus*. Possibly related to *nigricollis* (Flux and Angermann, 1990).

Lepus coreanus Thomas, 1892. Ann. Mag. Nat. Hist., ser. 6, 9:146.

TYPE LOCALITY: "Söul" [Seoul], Korea.

DISTRIBUTION: Korea; S Kirin, S Liaoning, E Heilungjiang (China).
STATUS: Secure at present (Flux and Angermann, 1990).
SYNONYMS: Monotypic.
COMMENTS: Formerly included in *sinensis* (Corbet, 1978c) or in *brachyurus* (Kim and Kim, 1974); here considered distinct, following Flux and Angermann (1990) and Jones and Johnson (1965). Angermann (in litt., 1992) suggested that *coreanus* and *mandshuricus* are conspecific.

Lepus corsicanus de Winton, 1898. Ann. Mag. Nat. Hist., ser. 7, 1:155.
TYPE LOCALITY: "Bastia," [Corsica, Italy].
DISTRIBUTION: Italy from the Abruzzo Mtns southward; Sicily; introduced into Corsica no later than 16th Century (Vigne, 1988).
STATUS: Unknown, but probable reduction in numbers and range due to overhunting and introduction of *L. europaeus* (Palacios et al., 1989).
SYNONYMS: Monotypic.
COMMENTS: Formerly included in *capensis* or *europaeus*; see Ellerman and Morrison-Scott (1951) and Petter (1961); but see also Palacios et al. (1989) who provided evidence of their specific distinctness.

Lepus europaeus Pallas, 1778. Nova Spec. Quad. Glir. Ord., p. 30.
TYPE LOCALITY: Not stated; restricted by Trouessart (1910), to Poland (see discussion in Ognev, 1940:140, which further restricts it to SW Poland).
DISTRIBUTION: Open woodland, steppe and subdesert: from S Sweden and Finland to Britain, throughout Europe (not Iberian Penin. south of Cantabria and the Ebro R, or south of Siena in Italy), to W Siberian lowlands; south to N Israel, N Syria, N Iraq, the Tigris-Euphrates valley and W Iran. SE border of range (Iran) from S Caspian Sea south to Persian Gulf (54°E); see Angerman (1983:19). Introduced to Ireland, SE Canada-NE USA, S South America, Australia, New Zealand and several islands, including Barbados, Réunion, and the Falklands.
STATUS: Most populations secure (Flux and Angermann, 1990).
SYNONYMS: *alba* Bechstein, 1801; *aquilonius* Blasius, 1842; *argenteogrisea* König-Warthausen, 1875; *astaricus* Baloutch, 1978; *biarmicus* Heptner, 1948; *borealis* Kuznetsov, 1944; *campestris* Bogdanov, 1871; *campicola* Gervais, 1859; *caspicus* Hemprich and Ehrenberg, 1832; *caucasicus* Ognev, 1929; *cinereus* Fitzinger, 1867; *connori* Robinson, 1918; *coronatus* Fitzinger, 1867; *creticus* Barrett-Hamilton, 1903; *cyanotus* Blanchard, 1957; *cyprius* Barrett-Hamilton, 1903; *cyrensis* Satunin, 1905; *flavus* Bechstein, 1801; *ghigii* de Beaux, 1927; *hybridus* Desmarest, 1822; *hyemalis* Tumac, 1850; *iranensis* Goodwin, 1939; *judeae* Gray, 1867; *kalmykorum* Ognev, 1929; *karpathorum* Hilzheimer, 1906; *laskerewi* (?) Khomenko, 1916; *maculatus* Fitzinger, 1867; *medius* Nilsson, 1820; *meridiei* Hilzheimer, 1906; *meridionalis* Gervais, 1859; *niethammeri* Wettstein, 1943; *niger* Bechstein, 1801; *nigricans* Fitzinger, 1867; *occidentalis* de Winton, 1898; *parnassius* Miller, 1903; *ponticus* Ognev, 1929; *pyrenaicus* Hilzheimer, 1906; *rhodius* Festa, 1914; *rufus*, Fitzinger, 1867; *syriacus* Hemprich and Ehrenberg, 1832; *tesquorum* Ognev and Worobiev, 1923; *transsylvanicus* Matschie, 1901; *transsylvaticus* Hilzheimer, 1906; *tumac* Tichomirov and Kortchagin, 1889.
COMMENTS: Subgenus *Eulagos* (Gureev, 1964:205; Gromov and Baranova, 1981). This species was earlier placed in *capensis* by Petter (1961) based on what was interpreted as a cline in morphological characters (mainly size) from NE Africa eastward across the N Arabian peninsula and the Middle East, and northward through Israel to Turkey. Sympatry between large "*europaeus*" and small "*capensis*" (= *tolai*) in Kazakhstan, without evidence of hybridization (Sludskii et al., 1980) was interpreted as overlapping ends of a Rassenkreis. Angermann's (1983) reanalysis indicated a marked discontinuity between smaller *capensis* (incl. *arabicus*) and larger *europaeus* running from the E Mediterranean coast (C Israel) through Iran, and on this basis we separate *europaeus* from *capensis* and *tolai*. East of the border of the range of *europaeus* in Iran, *tolai* occurs, apparently in allo- or parapatry with *europaeus*. Insular populations in the E Mediterranean are assigned to this species (Ellerman and Morrison-Scott, 1951).

Lepus fagani Thomas, 1903. Proc. Zool. Soc. Lond., 1902(2):315 [1903].
TYPE LOCALITY: "Zegi, Lake Tsana [Tana, Ethiopia] 4000 feet."

DISTRIBUTION: N and W Ethiopia, and adjacent SE Sudan, south to extreme NW Kenya.
STATUS: Unknown.
SYNONYMS: Monotypic.
COMMENTS: Formerly included in *victoriae* (as *crawshayi*) by Gureev (1964:204); Azzaroli-Puccetti (1987*a*, *b*) maintained its specific identity. Its known distribution is largely allo- or parapatric to that of *victoriae*; may be a highland allospecies (Flux and Angermann, 1990). Angermann (in litt., 1992) considered *fagani* conspecific with *victoriae*.

Lepus flavigularis Wagner, 1844. *In* Schreber, Die Säugethiere ..., Suppl. 4:106.
TYPE LOCALITY: "Mexico" Restricted by Elliot (1905:543) to San Mateo del Mar, Tehuantepec [City, Oaxaca, Mexico].
DISTRIBUTION: Coastal plains and bordering foothills on south end of Isthmus of Tehuantepec (Oaxaca, Mexico), along Pacific coast to Chiapas (Mexico); now restricted to small area between Salina Cruz, Oaxaca, and extreme W Chiapas.
STATUS: IUCN - Endangered, by habitat destruction and overhunting.
COMMENTS: Subgenus *Proeulagus* (Gureev, 1964:193). Closely related to *callotis*, with which it has an isolated allopatric distribution; see Anderson and Gaunt (1962); also see Hall (1981:330).

Lepus granatensis Rosenhauer, 1856. Die Thiere Andalusiens, 3.
TYPE LOCALITY: "bei Granada" [Granada, Analusia Prov., Spain].
DISTRIBUTION: Iberian Peninsula, except NE and NC parts (Spain, Portugal); Mallorca (Balearic Isl, Spain).
STATUS: Generally common, but scarce or extinct in a few places.
SYNONYMS: *gallaecius* Miller, 1907; *hispanicus* Fitzinger, 1867; *iturissius*, Miller, 1907; *lilfordi* de Winton, 1898; *mediterraneus* Wagner, 1841; *meridionalis* Graells, 1897; *solisi* Palacios and Fernández, 1992; *typicus* Hilzheimer, 1906.
COMMENTS: Formerly included in *europaeus* or *capensis*; but see Palacios (1983, 1989), and Bonhomme et al. (1986). The Mallorcan population was probably introduced by humans (Palacios and Fernández, 1992). The population in Sardinia, to which the names *mediterraneus* Machado, 1869 (not *mediterraneus* Wagner, 1841) and *typicus* Hilzheimer, 1906, have been applied, is assigned to this species, but its status needs investigation, as do populations from the NW African coast that have been assigned to *capensis* (Angermann, in litt., 1992).

Lepus hainanus Swinhoe, 1870. Proc. Zool. Soc. Lond., 1870:233.
TYPE LOCALITY: Hainan Island, "in the neighbourhood of the capital city" [Hainan Province, China].
DISTRIBUTION: Lowlands of Hainan Island.
STATUS: Numbers reduced to extremely low levels by habitat alteration and overhunting, but no official status as yet conferred.
SYNONYMS: Monotypic.
COMMENTS: Placed in *Caprolagus* (*Indolagus*) by Gureev (1964:146). Considered a subspecies of *peguensis* by Ellerman and Morrison-Scott (1951); Flux and Angermann (1990) recommended provisional specific status, as did Gureev (1964). Closely related to *nigricollis* (Angermann, in litt., 1992).

Lepus insularis W. Bryant, 1891. Proc. California Acad. Sci., ser. 2, 3:92.
TYPE LOCALITY: "Espiritu Santo Island, [near La Paz], Gulf of California [Baja California del Sur], Mexico."
DISTRIBUTION: Restricted to the type locality.
STATUS: IUCN - Rare. Occupies very small insular range, but population is stable, and seems in no danger (Chapman et al., 1983).
SYNONYMS: *edwardsi* Saint-Loup, 1895.
COMMENTS: Subgenus *Proeulagus* (Gureev, 1964:195). Insular melanic allospecies, related to *californicus*; see Hall (1981:328) and Dixon et al. (1983).

Lepus mandshuricus Radde, 1861. Melanges Biol. Acad. St. Petersbourg, 3:684.
TYPE LOCALITY: "Im Chy(Gebirge)" Bureya Mtns [Khabarovsk Krai, Russia].
DISTRIBUTION: Ussuri region (Russia); NE China; extreme NE Korea.
STATUS: Undetermined.

SYNONYMS: *melainus* Li and Luo, 1979; *melanonotus* Ognev, 1922.

COMMENTS: Distinct from *brachyurus*; see Angermann (1966, 1983); but placed in *Caprolagus (Allolagus) brachyurus* by Gureev (1964:150); followed by Gromov and Baranova (1981:63). Melanic individuals known since at least the time of Sowerby (1923) have recently been given the specific designation *melainus* (Li and Luo, 1979). The range of this taxon is entirely within that of *mandshuricus*, and I provisionally retain them in that species, although Flux and Angermann (1990) recognized *melainus*. *L. mandshuricus* and *L. coreanus* are parapatric in distribution in NE Korea/SE Heilungjiang, but are described as occupying different habitats; the former, mixed forest in hilly country, the latter, both forest and cultivated land, primarily in the plains (Flux and Angermann, 1990). Moreover, *mandshuricus* is sympatric with another forest species, *timidus*, and with the plains species, *tolai*; as forest is cleared, *tolai* tends to replace *mandshuricus* (Flux and Angermann, 1990). *L. mandshuricus*, *L. timidus* and *L. tolai* all occur in the area occupied by the taxon *melainus*; four species of sympatric hares, three of them forest-dwellers, is unprecedented in hare ecology, and supports the view that *melainus* is not a distinct species.

Lepus nigricollis F. Cuvier, 1823. Dict. Sci. Nat., 26:307.

TYPE LOCALITY: "Malabar" [Madras, India].

DISTRIBUTION: Pakistan; India; Bangladesh, except Sunderbands; Sri Lanka; introduced into Java (?) and Mauritius, Gunnera Quoin, Anskya, Réunion and Cousin Isls in the Indian Ocean. Considered native to Java by McNeely (1981:931).

STATUS: Mainland (and Sri Lankan?) populations secure (Flux and Angermann, 1990). If *nigricollis* is native to Java (rather than an introduced population), its numbers are now very low there.

SYNONYMS: *aryabertensis* Hodgson, 1844; *cutchensis* Kloss, 1918; *dayanus* Blanford, 1874; *joongshaiensis* Murray, 1884; *macrotus* Hodgson, 1840; *mahadeva* Wroughton and Ryley, 1913; *rajput* Wroughton, 1918; *ruficaudatus* Geoffroy, 1826; *sadiya* Kloss, 1918; *simcoxi* Wroughton, 1912; *singhala* Wroughton, 1915; *tytleri* Tytler, 1854.

COMMENTS: Placed in *Caprolagus (Indolagus)* by Gureev (1964:139). Includes *ruficaudatus*; see Prater (1980) and Angermann (1983), but see Gureev (1964:142); *ruficaudatus* is closer to *capensis* according to Petter (1961), and *nigricollis* may include *whytei*, *crawshayi*, *peguensis* and *siamensis*; but also see comments under *peguensis* and *saxatilis*. Also includes *dayanus*, given specific status by Gureev (1964:139).

Lepus oiostolus Hodgson, 1840. J. Asiat. Soc. Bengal, 9:1186.

TYPE LOCALITY: "...the snowy region of the Hemalaya, and perhaps also Tibet." Restricted by Kao and Feng (1964), to "Southern Tibet" [Xizang, China].

DISTRIBUTION: Tibetan Plateau, from Ladak to Sikkim (India) Nepal, and eastward through Xizang (Tibet) and Qinghai, Gansu and Sichuan (China).

STATUS: Probably secure, but may be affected by rodent control activity on Tibetan Plateau.

SYNONYMS: *grahami* Howell, 1928; *hypsibius* Blanford, 1875; *illuteus* Thomas, 1914; *kozlovi* Satunin, 1907; *oemodias* Gray, 1847; *pallipes* Hodgson, 1842; *przewalskii* Satunin, 1907; *qinghaiensis* Cai and Feng, 1982; *qusongensis* Cai and Feng, 1982; *sechuenensis* de Winton, 1899; *tsaidamensis* Hilzheimer, 1910.

COMMENTS: Placed in subgenus *Proeulagus* by Gureev (1964:196); *przewalskii* was assigned to *capensis* (= *tolai*) by Corbet (1978*c*), but is placed here following Cai and Feng (1982); Angermann (in litt., 1992) agreed with Corbet.

Lepus othus Merriam, 1900. Proc. Washington Acad. Sci., 2:28.

TYPE LOCALITY: "St. Michaels, [Norton Sound], Alaska." [USA].

DISTRIBUTION: W and SW Alaska (USA); formerly perhaps northwestward to Pt. Barrow; as here interpreted, also E Chukotsk (Russia).

STATUS: Rare, perhaps decreasing in range and numbers (Flux and Angermann, 1990).

SYNONYMS: *poadromus* Merriam, 1900; *tschuktschorum* Nordquist, 1883.

COMMENTS: Formerly included in *arcticus* or *timidus* (see comments therein). Regarded as distinct by Hall (1981) and by Flux and Angermann (1990), who, however, followed A. J. Baker et al. (1983) in allying populations from E Chukotsk (Russia) with Alaskan populations; but see also Pavlinov and Rossolimo (1987). Angermann (in litt., 1992) considered *othus* as "probably conspecific" with *timidus*.

Lepus peguensis Blyth, 1855. J. Asiat. Soc. Bengal, 24:471.

TYPE LOCALITY: "Pegu" [Upper Pegu, Burma].

DISTRIBUTION: C, S Burma from Chindwin River valley east through Thailand; Cambodia; S Laos, S Vietnam; south in upper Malay Peninsula (Burma, Thailand) to 120°N.

STATUS: Does not seem threatened (Flux and Angermann, 1990).

SYNONYMS: *siamensis* Bonhote, 1902; *vassali* Thomas, 1906.

COMMENTS: Placed by Gureev (1964:144) in *Caprolagus* (*Indolagus*); he ranked *siamensis* as a distinct species; but see Lekagul and McNeely (1977:333) and Flux and Angermann (1990). Petter (1961) suggested that *peguensis* might be conspecific with *nigricollis* because of its close resemblance to *L. n. ruficaudatus*. However, *L. n. ruficaudatus* appears to be allopatric with respect to *peguensis* in E India-W Burma. Angermann (in litt., 1992) considered *peguensis* conspecific with *nigricollis*.

Lepus saxatilis F. Cuvier, 1823. Dict. Sci. Nat., 26:309.

TYPE LOCALITY: "il habite les contrées qui se trouvent à trois journées au nord du cap de Bonne-Espérance," [Cape of Good Hope, South Africa].

DISTRIBUTION: Cape Province (and Zululand north to C Natal?) (South Africa) and S Namibia.

STATUS: Appears common (Flux and Angermann, 1990), but this is uncertain (Angermann, in litt., 1992).

SYNONYMS: *albaniensis* Roberts, 1932; *aurantii* Thomas and Hinton, 1923; *bechuanae* Roberts, 1932; *chiversi* Roberts, 1929; *chobiensis* Roberts, 1932; *damarensis* Roberts, 1926; *fumigatus* Wagner, 1844; *gungunyanae* Roberts, 1914; *khanensis* Roberts, 1946; *longicaudatus* Gray, 1837; *megalotis* Thomas and Schwann, 1905; *ngamiensis* Roberts, 1932; *nigrescens* Roberts, 1932; *orangensis* Kolbe, 1948; *rufinucha* A. Smith, 1829; *subrufus* Roberts, 1913; *timidus* A. Smith, 1826.

COMMENTS: Placed by Gureev (1964:203) in subgenus *Proeulagus*. Formerly included *crawshayi* and *whytei*, see Ansell (1978:67), Swanepoel et al. (1980:159), and Robinson and Dippenaar (1983*b*, 1987); but see also Petter (1961, 1972*b*). Angermann (1983) considered *whytei* a distinct species that includes *crawshayi*; Flux and Angermann (1990) placed both as subspecies of *victoriae*. Angermann (in litt., 1992) would restrict *saxatilis* to the following taxa: *albaniensis, aurantii, chiversi, fumigatus, longicaudatus, megalotis, rufinucha*, and *timidus*; transfering the remainder to *victoriae*.

Lepus sinensis Gray, 1832. Illustr. Indian Zool., 2, pl. 20.

TYPE LOCALITY: "China". Restricted to vicinity of Canton [Guangzhou, Guangdong Province, China].

DISTRIBUTION: SE China from Yangtse River southward; Taiwan; disjunct in NE Vietnam.

STATUS: Unknown, but probably affected by habitat modification.

SYNONYMS: *flaviventris* G. Allen, 1927; *formosus* Thomas, 1908; *yuenshanensis* Shih, 1930.

COMMENTS: Placed in *Caprolagus* (*Indolagus*) by Gureev (1964:143). Formerly included *coreanus*; see Corbet (1978*c*:73); here considered distinct, following Flux and Angermann (1990).

Lepus starcki Petter, 1963. Mammalia, 27:239.

TYPE LOCALITY: "Jeldu-Liban-Shoa, 2,740 mètres, 40 km W. d'Addis-Abeba," [Ethiopia].

DISTRIBUTION: C highlands of Ethiopia.

STATUS: Numerous within its restricted range (Flux and Angermann, 1990).

SYNONYMS: Monotypic.

COMMENTS: Formerly included in *capensis* (Petter, 1963*a*), or *europaeus* (Azzaroli-Puccetti 1987*a, b*); but see Angermann (1983) and Flux and Angermann (1990).

Lepus timidus Linnaeus, 1758. Syst. Nat., 10th ed., 1:57.

TYPE LOCALITY: "in Europa" [Uppsala, Sweden].

DISTRIBUTION: Palearctic from Scandinavia to E Siberia, except E Chukotsk (Russia), south to Sakhalin and Sikhote-Alin Mtns (Russia); Hokkaido (Japan); Heilungjiang, N Xinjiang (China); N Mongolia; Altai, N Tien Shan Mtns; N Ukraine, E Poland, and Baltics; isolated populations in the Alps, Scotland, Wales and Ireland. Introduced into England, Faeros and Scottish Isls.

STATUS: Populations fluctuate, but none apparently threatened, except perhaps in Alps (Flux and Angermann, 1990).

SYNONYMS: *abei* Kuroda, 1938; *ainu* Barrett-Hamilton, 1900; *albus* Leach, 1816; *algidus* Pallas, 1778; *alpinus* Erxleben, 1777; *altaicus* Barrett-Hamilton, 1900; *begitschevi* Koljuschev, 1936; *borealis* Pallas, 1778; *breviauritus* Hilzheimer, 1906; *canescens* Nilsson, 1844; *collinus* Nilsson, 1831; *gichiganus* J. Allen, 1903; *hibernicus* Bell, 1837; *kamtschaticus* Dybowski, 1922; *kolymensis* Ognev, 1923; *kozhevnikovi* Ognev, 1929; *lugubris* Kastschenko, 1899; *lutescens* Barrett-Hamilton, 1900; *mordeni* Goodwin, 1933; *orii* Kuroda, 1928; *rubustus* Urita, 1935; *saghaliensis* Abe, 1931; *sclavonius* Blyth, 1842; *scoticus* Hilzheimer, 1906; *septentrionalis* Link, 1795; *sibiricorum* Johanssen, 1923; *sylvaticus* Nilsson, 1831; *transbaicalicus* Ognev, 1929; *typicus* Barrett-Hamilton, 1900; *variabilis* Pallas, 1778; *varronis* Miller, 1901.

COMMENTS: Subgenus *Lepus* (Gureev, 1964:180). Formerly included *arcticus* and *othus*; see Corbet (1978*c*:73); but also see comments under those species. A. J. Baker et al. (1983) found Scottish and Alpine populations morphologically distinct, as well as geographically isolated, from other populations, and Flux (1983) remarked that *L. t. scoticus* and *L. t. hibernicus* (from Scotland and Ireland, respectively), both introduced on the island of Mull (Hewson, 1991) still do not interbreed after 50 years.

Lepus tolai Pallas, 1778. Nova Spec. Quad. Glir. Ord., p. 17.

TYPE LOCALITY: "Caeterum in montibus aprecis campisque rupestribus vel arenosis circa Selengam..." Restricted by Ognev (1940:162) to "...valley of the Selenga River...." [Russia].

DISTRIBUTION: Steppes N of Caspian Sea southward along eastern shore of Caspian to E Iran; eastward through Afghanistan; Kazakhstan and S Siberia, Middle Asian republics to Mongolia; and W, C, and NE China.

STATUS: Populations in China (and perhaps elsewhere) have declined due to intensified agriculture and increased use of pesticides (Flux and Angermann, 1990).

SYNONYMS: *aralensis* Severtsov, 1861; *aurigineus* Hollister, 1912; *biddulphi* Blanford, 1877; *brevinasus* J. Allen, 1909; *buchariensis* Ognev, 1922; *butlerowi* Bogdanov, 1882; *centrasiaticus* Satunin, 1907; *cheybani* Baloutch, 1978; *cinnamomeus* Shamel, 1940; *craspedotis* Blanford, 1875; *desertorum* Ognev and Heptner, 1928; *filchneri* Matschie, 1907; *gansuicus* Satunin, 1907; *gobicus* Satunin, 1907; *habibi* Baloutch, 1978; *huangshuiensis* Luo, 1982; *kaschgaricus* Satunin, 1907; *kessleri* Bogdanov, 1882; *lehmanni* Severtsov, 1873; *pamirensis* Günther, 1875; *petteri* Baloutch, 1978; *quercerus* Hollister, 1912; *sowerbyae* Hollister, 1912; *stegmanni* Matschie, 1907; *stoliczkanus* Blanford, 1875; *subluteus* Thomas, 1908; *swinhoei* Thomas, 1894; *tibetanus* Waterhouse, 1841; *turcomanus* Heptner, 1934; *zaisanicus* Satunin, 1907.

COMMENTS: Subgenus *Proeulagus* (Gureev, 1964:198). Formerly included in *capensis* or *europaeus*; see comments therein. Includes *tibetanus*; but see also Sokolov and Orlov (1980:85); Qui (1989) also provided evidence of differentiation of *tibetanus* but did not address specific status. Formerly included *przewalskii*, now assigned to *L. oiostolus*; see Cai and Feng (1982). "The situation in [southern] Iraq [and SW Iran] deserves a more detailed analysis (Angermann, 1983:19). *L. tolai petteri* occurs westward to about 55°-56°E, while *L. c. arabicus* occurs eastward to SE Iraq. Whether the two forms come into contact is not known, but their ranges may be separated by that of *L. europaeus astaricus* in SW Iran (and *L. e. connori*?); see Baloutch (1978) and Angermann (1983). Angermann (in litt., 1992) considered *tolai* conspecific with *capensis*. According to Ellerman and Morrison-Scott (1951:430) the type locality should be "Adinscholo Mountain, near Tchinden [Chinden = Chindant], on Borsja [Boriya] River, a tributary of the Onon River, Eastern Siberia." This locality is more than 700 km east of the Selanga River.

Lepus townsendii Bachman, 1839. J. Acad. Nat. Sci. Philadelphia, 8(1):90.

TYPE LOCALITY: "...on the Walla-walla...river"; restricted by Nelson (1909:78) to Fort Walla Walla, [near present town of Wallula, Walla Walla Co., Washington].

DISTRIBUTION: C Alberta and Saskatchewan east to extreme SW Ontario (Canada), south to SW Wisconsin, Iowa, NW Missouri, west through C Kansas to NC New Mexico, west to C Nevada, EC California (USA) and north to SC British Columbia (Canada).

STATUS: Has withdrawn from parts of its southeastern (Wisconsin, Iowa, Missouri, Kansas, S Nebraska) distribution, perhaps because of habitat alteration. Stable in remainder of range (Flux and Angermann, 1990).

SYNONYMS: *campanius* Hollister, 1915; *campestris* Bachman, 1837; *sierrae* Merriam, 1904.

COMMENTS: Placed (as *campestris*) in subgenus *Proeulagus* by Gureev (1964:190). Reviewed by Lim (1987, Mammalian Species, 288).

Lepus victoriae Thomas, 1893. Ann. Mag. Nat. Hist., ser. 6, 12:268.
TYPE LOCALITY: "Nassa, Speke Gulf, S. Victoria Nyanza" [Lake, Tanzania].
DISTRIBUTION: From Atlantic coast of NW Africa (Spanish Sahara, Mauritania, south to Guinea and Sierra Leone) eastward across Sahel to Sudan and extreme W Ethiopia; southward through E Africa (E Congo, W Kenya) to NE Namibia, Botswana, and Natal (South Africa). Small isolated population in W Algeria.
STATUS: Isolated population around Beni Abbés, Algeria, "deserve[s] attention" (Flux and Angermann, 1990).
SYNONYMS: *angolensis* Thomas, 1904; *ansorgei* Thomas and Wroughton, 1905; *canopus* Thomas and Hinton, 1921; *chadensis* Thomas and Wroughton, 1907; *cordeauxi* Drake-Brockman, 1911; *crawshayi* de Winton, 1899; *herero* Thomas, 1926; *kakumegae* Heller, 1912; *meridionalis* Monard, 1933; *micklemi* Chubb, 1908; *microtis* Heuglin, 1865; *raineyi* Heller, 1912; *whytei* Thomas, 1894; *zairensis* Hatt, 1935; *zechi* Matschie, 1899; *zuluensis* Thomas and Schwann, 1905.
COMMENTS: Placed (as *crawshayi*) in subgenus *Proeulagus* by Gureev (1964:204), who recognized both *crawshayi* and *whytei* as distinct species, as did Azzaroli-Puccetti (1987*a*). Formerly included in *saxatilis*; see comments under that species. Angermann and Feiler (1988) demonstrated that the oldest available name for this species is *victoriae*. It is widely sympatric with *capensis*, but allo- to parapatric with *saxatilis*, which is also sympatric with *capensis*.

Lepus yarkandensis Günther, 1875. Ann. Mag. Nat. Hist., ser. 4, 16:229.
TYPE LOCALITY: "neighbourhood of Yarkand" [15 mi E, Xinjiang, China].
DISTRIBUTION: Steppes of Tarim Basin, S Xinjiang (China), around edge of Takla Makan desert.
STATUS: Presently secure (Flux and Angermann, 1990).
SYNONYMS: Monotypic.
COMMENTS: Placed in subgenus *Tarimolagus* by Gureev (1964:147); but see Xu (1986). Reviewed by Angermann (1967) and Gao (1983).

Nesolagus Forsyth-Major, 1899. Trans. Linn. Soc. Lond., 7:493.
TYPE SPECIES: *Lepus netscheri* Schlegel, 1880.

Nesolagus netscheri (Schlegel, 1880). Notes Leyden Mus., II:59.
TYPE LOCALITY: "Sumatra: Padang-Padjang...about 2000 feet" [Padangpanjang, Sumatera Barat, Indonesia].
DISTRIBUTION: Sumatra [Indonesia].
STATUS: IUCN - Indeterminate. "...apparently the rarest lagomorph. About a dozen museum specimens exist, collected between 1880 and 1916, and there has been only one confirmed sighting since then, in 1972." (Flux, 1990).
SYNONYMS: Monotypic.

Oryctolagus Lilljeborg, 1874. Sverig. Og Norges Ryggradsdjur, 1:417.
TYPE SPECIES: *Lepus cuniculus* Linnaeus, 1758.
SYNONYMS: *Cuniculus* Meyer, 1790.

Oryctolagus cuniculus (Linnaeus, 1758). Syst. Nat., 10th ed., 1:58.
TYPE LOCALITY: "in Europa australis" [= Germany].
DISTRIBUTION: W and S Europe through the Mediterranean region to Morocco and N Algeria; original range probably limited to Iberia and perhaps NW Africa; introduced on all continents except Antarctica and Asia; see Gibb (1990). Worldwide as domesticated forms.
STATUS: The form *huxleyi* of the Atlantic and Mediterranean isls "may be endangered" (Gibb, 1990).
SYNONYMS: *algirus* (Loche, 1858); *borkumensis* Harrison, 1952; *brachyotus* Trouessart, 1917; *campestris* (Meyer, 1790); *cnossius* Bate, 1906; *fodiens* Gray, 1867; *habetensis* Cabrera, 1923; *huxleyi* Haeckel, 1874; *kreyenbergi* Honigmann, 1913; *nigripes* Bartlett, 1857; *oreas* Cabrera, 1922; *vermicula* Gray, 1843; *vernicularis* Thompson, 1837.
COMMENTS: The specific name is probably based on a feral specimen (Gibb, 1990).

Pentalagus Lyon, 1904. Smithson. Misc. Coll., 45:428.
TYPE SPECIES: *Caprolagus furnessi* Stone, 1900.

Pentalagus furnessi (Stone, 1900). Proc. Acad. Nat. Sci. Philadelphia, 52:460.
TYPE LOCALITY: "Liu Kiu Islands" [Amami-Oshima, Ryukyu Isls, Japan].
DISTRIBUTION: Amami Isls (Amami-Oshima and Tokuno-shima) (S Japan).
STATUS: U.S. ESA and IUCN - Endangered. Total population estimated at around 4,000 (Sugimura, 1990).

Poelagus St. Leger, 1932. Proc. Zool. Soc. Lond., 1932(1):119.
TYPE SPECIES: *Lepus marjorita* St. Leger, 1929.
COMMENTS: Originally spelled *Poëlagus*, but this is a diaeresis and not an umlaut and thus the correct spelling is *Poelagus* (see art. 32d of the Code of Nomenclature, International Commission on Zoological Nomenclature, 1985). Formerly placed as subgenus of *Pronolagus*; see Ellerman and Morrison-Scott (1951:425); but see also Petter (1972*b*:5). Formerly placed as subgenus of *Caprolagus*; see Gureev (1964:152).

Poelagus marjorita (St. Leger, 1929). Ann. Mag. Nat. Hist., ser. 10, 4:292.
TYPE LOCALITY: "Near Masindi, Bunyoro [Bunyuru] Uganda, 4000 ft.", Africa.
DISTRIBUTION: S Sudan, Uganda, Ruanda, Burundi, NE Zaire, Central African Republic, S Chad, disjunct population in Angola.
STATUS: ". . .not under threat" (Duthie and Robinson, 1990).
SYNONYMS: *larkeni* St. Leger, 1935; *oweni* Setzer, 1956.
COMMENTS: This savanna-woodland species, like *Pronolagus*, is associated with rocky outcrops.

Pronolagus Lyon, 1904. Smithson. Misc. Coll., 45:416.
TYPE SPECIES: *Lepus crassicaudatus* I. Geoffroy, 1832.
COMMENTS: From one (Peddie, 1975) to six (Roberts, 1951) species have been recognized in this genus (Robinson, 1982). Three species are now generally recognized (Duthie and Robinson, 1990).

Pronolagus crassicaudatus (I. Geoffroy, 1832). Mag. Zool. Paris, 2:cl. 1, pl. 9 and text.
TYPE LOCALITY: "Port Natal" [Durban, Natal, South Africa].
DISTRIBUTION: SE South Africa; extreme S Mozambique.
STATUS: Secure (Duthie and Robinson, 1990).
SYNONYMS: *bowkeri* Hewitt, 1927; *kariegae* Hewitt, 1927; *lebombo* Roberts, 1936; *ruddi* Thomas and Schwann, 1905.
COMMENTS: Formerly included *randensis*, see Lundholm (1955*a*). The relationship of *crassicaudatus* and *randensis* is unclear, see Petter (1972*b*:6). Distribution allopatric to that of *randensis*, but sympatric in western half of range with *rupestris*.

Pronolagus randensis Jameson, 1907. Ann. Mag. Nat. Hist., ser. 7, 20:404.
TYPE LOCALITY: "Observatory Kopje...Johannesburg...Witwatersrand Range, Transvaal... 5,900 ft." [1,798 m] [South Africa].
DISTRIBUTION: Two disjunct areas: NE South Africa; E Botswana to extreme W Mozambique, Zimbabwe; and W Namibia, perhaps SW Angola.
STATUS: Secure.
SYNONYMS: *capricornis* Roberts, 1926; *caucinus* Thomas, 1929; *ekmani* Lundholm, 1955; *kaokoensis* Roberts, 1946; *kobosensis* Roberts, 1938; *makapani* Roberts, 1924; *powelli* Roberts, 1924; *waterbergensis* Hoesch and Von Lehmann, 1956; *whitei* Roberts, 1938.
COMMENTS: Formerly included in *crassicaudatus* by Lundholm (1955*a*); but see Petter (1972*b*:6). The systematic position of the two widely disjunct populations needs clarification (Duthie and Robinson, 1990).

Pronolagus rupestris (A. Smith, 1834). S. Afr. Quart. J., 2:174.
TYPE LOCALITY: "South Africa, rocky situations" [probably Van Rhynsdorp District, Cape Province, South Africa.].
DISTRIBUTION: Two disjunct areas: S and C South Africa, S Namibia; and E Africa, from N Malawi and E Zambia north through C Tanzania to SW Kenya.
STATUS: Secure.

SYNONYMS: *australis* Roberts, 1933; *barretti* Roberts, 1949; *curryi* (Thomas, 1902); *fitzsimonsi* Roberts, 1938; *melanurus* (Rüppell, 1842); *mülleri* Roberts, 1938; *nyikae* (Thomas, 1902); *saundersiae* (Hewitt, 1927); *vallicola* Kershaw, 1924.

COMMENTS: Formerly included in *crassicaudatus*, see Gureev (1964:174) and Peddie (1975); see also Robinson and Dippenaar (1983*a*). The systematic relationships of the two widely disjunct populations should be examined (Duthie and Roberts, 1990).

Romerolagus Merriam, 1896. Proc. Biol. Soc. Washington, 10:173.

TYPE SPECIES: *Lepus diazi* Ferrari-Peréz, 1893.

SYNONYMS: *Lagomys* Herrera, 1897 (not *Lagomys* Cuvier, 1800).

COMMENTS: Whether this monotypic genus represents "the most primitive of the living rabbits and hares" (Fa and Bell, 1990), or is closer to the more specialized leporids (*Sylvilagus*, *Oryctolagus*, *Lepus*) (Corbet, 1983), or is intermediate (Hibbard, 1963), remains controversial.

Romerolagus diazi (Ferrari-Pérez, 1893). *In* Diaz, Cat. Comision Geogr.-Expl. República Mexicana, Exposicion Intern, Columbia de Chicago, pl. 42.

TYPE LOCALITY: "near San Martín Texmelusán, northeastern slope of Volcán Iztaccíhuatl [Ixtaccíhuatl, Puebla], Mexico."

DISTRIBUTION: Distrito Federal, Mexico, and W Puebla (Mexico), in three discontinuous areas on the slopes of Volcán Pelado, Tlaloc, Popocatépetl, and Ixtaccíhuatl.

STATUS: CITES - Appendix I; U.S. ESA and IUCN - Endangered.

SYNONYMS: *nelsoni* Merriam, 1896.

COMMENTS: Reviewed by Cervantes et al. (1990, Mammalian Species, 360).

Sylvilagus Gray, 1867. Ann. Mag. Nat. Hist., ser. 3, 20:221.

TYPE SPECIES: *Lepus sylvaticus* Bachman, 1837 (= *Lepus sylvaticus floridanus* J. Allen, 1890).

SYNONYMS: *Hydrolagus* Gray, 1867; *Limnolagus* Mearns, 1897; *Microlagus* Trouessart, 1897; *Paludilagus* Hershkovitz, 1950; *Tapeti* Gray, 1867.

COMMENTS: Formerly included *Brachylagus* as a subgenus; see Hall (1981:294); but see also Corbet (1983:14).

Sylvilagus aquaticus (Bachman, 1837). J. Acad. Nat. Sci. Philadelphia, 7:319.

TYPE LOCALITY: "...western parts of that state" [Alabama]. Restricted by Nelson (1909:272) to "Western Alabama".

DISTRIBUTION: S Illinois and SW Indiana, SW Missouri to SE Kansas southward through extreme W Kentucky and W Tennessee to E Oklahoma, E Texas, Louisiana, Alabama, Mississippi and NW South Carolina (USA).

STATUS: Some range reduction, but remains abundant (Chapman and Ceballos, 1990).

SYNONYMS: *attwateri* (J. Allen, 1895); *douglasii* (Gray, 1837; part) *littoralis* Nelson, 1909; *telmalemonus* (Elliott, 1899).

COMMENTS: Subgenus *Tapeti* (Gureev, 1964:162). Reviewed by Chapman and Feldhamer (1981, Mammalian Species, 151).

Sylvilagus audubonii (Baird, 1858). Mammalia, *in* Repts. U.S. Expl. Surv., 8(8):608.

TYPE LOCALITY: "San Francisco" [San Francisco Co., California, USA].

DISTRIBUTION: NE Puebla and W Veracruz (Mexico) to NC Montana and SW North Dakota, NC Utah, C Nevada, and NC California (USA), south to Baja California and C Sinaloa (Mexico).

STATUS: Secure (Chapman and Ceballos, 1990).

SYNONYMS: *arizonae* (J. Allen, 1877); *baileyi* (Merriam, 1897); *cedrophilus* Nelson, 1907; *confinis* (J. Allen, 1898); *goldmani* (Nelson, 1904); *laticinctus* (Elliot, 1904); *major* (Mearns, 1896); *minor* (Mearns, 1896); *neomexicanus* Nelson, 1907; *parvulus* (J. Allen, 1904); *rufipes* (Elliot, 1904); *sanctidiegi* (Miller, 1899); *vallicola* Nelson, 1907; *warreni* Nelson, 1907.

COMMENTS: Subgenus *Sylvilagus* (Gureev, 1964:169). Reviewed by Chapman and Willner (1978, Mammalian Species, 106).

Sylvilagus bachmani (Waterhouse, 1839). Proc. Zool. Soc. Lond., 1839:103.

TYPE LOCALITY: "Between Monterey and Santa Barbara". Type locality restricted by Nelson (1909:247) to San Luis Obispo, California, USA.

DISTRIBUTION: W Oregon (USA) S of Columbia River to Baja California (Mexico), E to Cascade-Sierra Nevada Range (USA).
STATUS: "Still quite abundant" (Chapman and Ceballos, 1990).
SYNONYMS: *cerrosensis* (J. Allen, 1898); *cinerascens* (J. Allen, 1890); *exiguus* Nelson, 1907; *howelli* Huey, 1927; *macrorhinus* Orr, 1935; *mariposae* Grinnell and Storer, 1916; *peninsularis* (J. Allen, 1898); *riparius* Orr, 1935; *rosaphagus* Huey, 1940; *tehamae* Orr, 1935; *trowbridgii* Baird, 1855; *ubericolor* (Miller, 1899); *virqulti* Dice, 1926.
COMMENTS: Placed in genus *Microlagus* together with *idahoensis* by Gureev (1964:171). This is the only species of *Sylvilagus* known to have retained the putative ancestral karyotype (2n=48) shared by all known *Lepus*, and by *Romerolagus* (Robinson et al., 1981, 1984). Reviewed by Chapman (1974, Mammalian Species, 34).

Sylvilagus brasiliensis (Linnaeus, 1758). Syst. Nat., 10th ed., 1:58.
TYPE LOCALITY: "in America meridionali"; type locality restricted by Thomas (1911*a*), to Pernambuco, Brazil.
DISTRIBUTION: S Tamaulipas (Mexico) southward through Central and South America as far as Peru, Bolivia, N Argentina and S Brazil.
STATUS: Populations decline or disappear when forests are cleared; status unclear.
SYNONYMS: *andinus* (Thomas, 1897); *apollinaris* Thomas, 1920; *braziliensis* (Waterhouse, 1848); *canarius* Thomas, 1913; *capsalis* Thomas, 1913; *caracasensis* Mondolfi and Méndez Aroche, 1957; *carchensis* Hershkovitz, 1938; *chapadae* Thomas, 1904; *chapadensis* Thomas, 1913; *chillae* Anthony, 1923; *chimbanus* Thomas, 1913; *chotanus* Hershkovitz, 1938; *consobrinus* Anthony, 1917; *daulensis* J. Allen, 1914; *defilippi* (Cornalia, 1850); *defilippii* (Thomas, 1897); *dephilippii* Cabrera, 1912; *ecaudatus* Trouessart, 1910; *fulvescens* J. Allen, 1912; *fuscescens* J. Allen, 1916; *gabbi* (J. Allen, 1877); *gibsoni* Thomas, 1918; *inca* Thomas, 1913; *incitatus* (Bangs, 1901); *kelloggi* Anthony, 1923; *meridensis* Thomas, 1904; *messorius* Goldman, 1912; *minensis* Thomas, 1901; *nicefori* Thomas, 1921; *nigricaudatus* (Lesson, 1842); *nivicola* Cabrera, 1912; *paraguensis* Thomas, 1901; *paraguensis* Yepes, 1938; *peruanus* Hershkovitz, 1950; *salentus* J. Allen, 1913; *sanctaemartae* Hershkovitz, 1950; *surdaster* Thomas, 1901; *tapeti* (Pallas, 1778); *tapetillus* Thomas, 1913; *truei* (J. Allen, 1890); *tumacus* (J. Allen, 1908).
COMMENTS: Subgenus *Tapeti* (Gureev, 1964:160); he also considered *gabbi*, which is included here, a distinct species. Formerly included *dicei*; revised by Diersing (1981).

Sylvilagus cunicularius (Waterhouse, 1848). Nat. Hist. Mamm., 2:132.
TYPE LOCALITY: "Mexico." Restricted by Goodwin (1969:125) to "Sacualpan" (= Zacualpan).
DISTRIBUTION: S Sinaloa to E Oaxaca and Veracruz (Mexico).
STATUS: Some populations have declined, but overall "still quite abundant" (Chapman and Ceballos, 1990).
SYNONYMS: *insolitus* (J. Allen, 1890); *pacificus* (Nelson, 1904); *veraecrucis* (Thomas, 1890).
COMMENTS: Subgenus *Sylvilagus* (Gureev, 1964:167).

Sylvilagus dicei Harris, 1932. Occas. Pap. Mus. Zool. Univ. Mich., 248:1.
TYPE LOCALITY: "El Copey de Dota, in the Cordillera de Talamanca, Costa Rica...6000 feet."
DISTRIBUTION: Cordillera de Talamanca (SE Costa Rica, NW Panama).
STATUS: Indeterminate.
SYNONYMS: Monotypic.
COMMENTS: Formerly included in *brasiliensis* (Hall, 1981:295); revised by Diersing (1981).

Sylvilagus floridanus (J. A. Allen, 1890). Bull. Am. Mus. Nat. Hist., 3:160.
TYPE LOCALITY: "Sebastian River, Brevard Co.," [Florida, USA].
DISTRIBUTION: N, C, and W Venezuela (including adjacent islands) and adjacent Colombia through Central America (disjunct in part); to NW Mexico, Arizona, north and east to North Dakota, Minnesota, N Michigan, New York and Massachusetts, Atlantic Coast south and Florida Gulf Coast (USA) west to Mexico; also S Saskatchewan, S Ontario and SC Quebec (C Canada).
STATUS: Secure.
SYNONYMS: *alacer* (Bangs, 1896); *ammophilus* Howell, 1939; *avius* Osgood, 1910; *aztecus* (J. Allen, 1890); *boylei* J. Allen, 1916; *caniclunis* Miller, 1899; *chapmani* (J. Allen, 1899); *chiapensis* (Nelson, 1904); *cognatus* Nelson, 1907; *connectens* (Nelson, 1904); *continentis* Osgood, 1912; *costaricensis* Harris, 1933; *cumanicus* (Thomas, 1897); *durangae* J. Allen, 1903; *hesperius* Hoffmeister and Lee, 1963; *hitchensi* Mearns, 1911; *holzneri* (Mearns,

1896); *hondurensis* Goldman, 1932; *llanensis* Blair, 1938; *macrocorpus*, Diersing and Wilson, 1980; *mallurus* (Thomas, 1898); *margaritae* Miller, 1898; *mearnsi* (J. Allen, 1894); *nelsoni* Baker, 1955; *nigronuchalis* (Hartert, 1894); *orinoci* Thomas, 1900; *orizabae* (Merriam, 1893); *paulsoni* Schwartz, 1956; *persultator* Elliot, 1903; *purgatus* Thomas, 1920; *restrictus* Nelson, 1907; *rigidus* Mearns, 1896; *robustus* (Bailey, 1905); *russatus* (J. Allen, 1904); *similis* Nelson, 1907; *simplicicanus* Miller, 1902; *subcinctus* (Miller, 1899); *superciliaris* (J. Allen, 1899); *sylvaticus* (Bachman, 1837); *valenciae* Thomas, 1914; *yucatanicus* (Miller, 1899).

COMMENTS: Subgenus *Sylvilagus* (Gureev, 1964:164). Widely introduced in North America (Hall, 1981:301) and Europe (Flux et al., 1990). Reviewed by Chapman et al. (1980, Mammalian Species, 136).

Sylvilagus graysoni (J. A. Allen, 1877). *In* Coues and Allen, Monog. N. Amer. Rodentia (U.S. Geol. Geograph. Survey Terr., Rep., 11:347).

TYPE LOCALITY: According to Nelson (1899*a*:16), "Tres Marias Islands," "undoubtedly from Maria Madre" Isl, Nayarit, Mexico.

DISTRIBUTION: Tres Marías Isls, Nayarit (Mexico).

STATUS: IUCN - Endangered.

SYNONYMS: *badistes* Diersing and Wilson, 1980.

COMMENTS: Subgenus *Sylvilagus* (Gureev, 1964:168). An insular species probably derived from *cunicularius* of the adjacent mainland; see Diersing and Wilson (1980) and Hall (1981:314).

Sylvilagus insonus Nelson, 1904. Proc. Biol. Soc. Washington, 17:103.

TYPE LOCALITY: "Omilteme, Guerrero," [Mexico].

DISTRIBUTION: Appears restricted to Sierra Madre del Sur, C Guerrero (Mexico) between 2300-5280 ft. elevation.

STATUS: IUCN - Endangered; known from fewer than 10 records.

SYNONYMS: Monotypic.

COMMENTS: Subgenus *Tapeti* (Gureev, 1964:164), or *Sylvilagus* (Hershkovitz, 1950:335).

Sylvilagus mansuetus Nelson, 1907. Proc. Biol. Soc. Washington, 20:83.

TYPE LOCALITY: "San José Island, Gulf of California, Mexico" [Baja California del Sur, Mexico].

DISTRIBUTION: Known only from the type locality.

STATUS: IUCN - Indeterminate.

SYNONYMS: Monotypic.

COMMENTS: An insular allospecies closely related to *bachmani* (Chapman and Ceballos, 1990); a subspecies of *bachmani* according to Gureev (1964:171).

Sylvilagus nuttallii (Bachman, 1837). J. Acad. Nat. Sci. Philadelphia, 7:345.

TYPE LOCALITY: ". . .west of the Rocky Mountains,. . .streams which flow into the Shoshonee and Columbia rivers"; restricted by Nelson (1909:201) to "eastern Oregon, near mouth of Malheur River." Listed by Bailey (1936) as "near Vale."

DISTRIBUTION: Intermountain area of North America from S British Columbia to S Saskatchewan (Canada), south to E California, Nevada, C Arizona, and NW New Mexico (USA).

STATUS: *S. floridanus* appears to be displacing *nuttallii* in SE North Dakota; see Genoways and Jones (1972); but elsewhere generally common (Chapman and Ceballos, 1990).

SYNONYMS: *artemesia* (Bachman, 1839); *grangeri* (J. Allen, 1895); *perplicatus* Elliott, 1904; *pinetis* (J. Allen, 1894).

COMMENTS: Subgenus *Sylvilagus* (Gureev, 1964:168). Reviewed by Chapman (1975*a*, Mammalian Species, 56).

Sylvilagus palustris (Bachman, 1837). J. Acad. Nat. Sci. Philadelphia, 7:194.

TYPE LOCALITY: ". . .South Carolina. . .never. . .more than fourty miles from the sea coast"; restricted by Miller and Rehn (1901:183) to E South Carolina.

DISTRIBUTION: Florida to SE Virginia (Dismal Swamp) (USA) in coastal lowlands.

STATUS: U.S. ESA - Endangered as *S. p. hefneri*; IUCN - Endangered as *S. p. hefneri*; other populations indeterminate (Chapman and Ceballos, 1990).

SYNONYMS: *douglasii* (Gray, 1837); *hefneri* Lazell, 1984; *paludicola* (Miller and Bangs, 1894).

COMMENTS: Subgenus *Tapeti* (Gureev, 1964:162). *S. aquaticus* and *S. palustris* share a derived karyotype, 2n=38 (Robinson et al., 1983, 1984). Reviewed by Chapman and Willner (1981, Mammalian Species, 153).

Sylvilagus transitionalis (Bangs, 1895). Proc. Boston Soc. Nat. Hist., 26:405.

TYPE LOCALITY: "Liberty Hill, Conn." [New London Co., Connecticut, USA].

DISTRIBUTION: S Maine to N Alabama discontinuously along the Appalachian Mtns (USA).

STATUS: Range, especially in north, has been much reduced, probably by habitat alteration and subsequent displacement by *S. floridanus* (Chapman and Ceballos, 1990).

SYNONYMS: Monotypic.

COMMENTS: Subgenus *Sylvilagus* (Gureev, 1964:166). Two cytotypes are known; northern (2n=52) and southern (2n=46)(Holden and Eabry, 1970; Robinson et al., 1983); and may require recognition of a new species (Ruedas, 1986). Reviewed by Chapman (1975*b*, Mammalian Species, 55).

ORDER MACROSCELIDEA

by Duane A. Schlitter

ORDER MACROSCELIDEA

Family Macroscelididae Bonaparte, 1838. Nuovi Ann. Sci. Nat., 2:111.
COMMENTS: Revised by Corbet and Hanks (1968).

Elephantulus Thomas and Schwann, 1906. Abst. Proc. Zool. Soc. Lond., 1906(33):10.
TYPE SPECIES: *Macroscelides rupestris* A. Smith, 1831.
COMMENTS: Includes *Nasilio*; see Corbet and Hanks (1968). A key to the species was presented in Koontz and Roeper (1983).

Elephantulus brachyrhynchus (A. Smith, 1836). Rept. Exped. Exploring Central Afrrica, 1834:42 [1836].
TYPE LOCALITY: "The country between Lake Latakoo and the Tropic" (= South Africa, N Cape Prov., Kuruman, to S Botswana).
DISTRIBUTION: N South Africa and NE Namibia; Angola; S Zaire; Mozambique; to Kenya and Uganda.
COMMENTS: Formerly included *fuscus*; see Corbet (1974:5).

Elephantulus edwardii (A. Smith, 1839). Illustr. Zool. S. Afr. Mamm., pl. 14.
TYPE LOCALITY: South Africa, Cape Prov., Oliphants River.
DISTRIBUTION: SW and C Cape Prov. (South Africa).

Elephantulus fuscipes (Thomas, 1894). Ann. Mag. Nat. Hist., ser. 6, 13:68.
TYPE LOCALITY: Zaire, Niam-Niam country, N'doruma.
DISTRIBUTION: Uganda, NE Zaire, S Sudan.

Elephantulus fuscus (Peters, 1852). Reise nach Mossambique, Säugethiere, p. 87.
TYPE LOCALITY: Mozambique, near Quelimane, Boror.
DISTRIBUTION: Mozambique, S Malawi, SE Zambia.
COMMENTS: Regarded as distinct by Corbet (1974:5).

Elephantulus intufi (A. Smith, 1836). Rept. Exped. Exploring Central Africa, 1834:42 [1836].
TYPE LOCALITY: South Africa, Transvaal, Marico District, flats beyond Kurrichaine.
DISTRIBUTION: SW Angola; Namibia; Botswana; NW Transvaal and N Cape Prov. (South Africa).

Elephantulus myurus Thomas and Schwann, 1906. Proc. Zool. Soc. Lond., 1906:586.
TYPE LOCALITY: South Africa, Transvaal, Woodbush.
DISTRIBUTION: Zimbabwe, E Botswana, E and N South Africa, Mozambique.

Elephantulus revoili (Huet, 1881). Bull. Sci. Soc. Philom. Paris, ser. 7, 5:96.
TYPE LOCALITY: Somalia, Medjourtine.
DISTRIBUTION: N Somalia.

Elephantulus rozeti (Duvernoy, 1833). Mem. Soc. Hist. Nat. Strasbourg, 1(2), art. M:18.
TYPE LOCALITY: Algeria, near Oran.
DISTRIBUTION: Morocco, Algeria, Tunisia, W Libya.

Elephantulus rufescens (Peters, 1878). Monatsb. K. Preuss. Akad. Wiss., Berlin, 1878:198.
TYPE LOCALITY: Kenya, Taita, Ndi.
DISTRIBUTION: S and E Ethiopia, N and SE Kenya, NE Uganda, S Sudan, NC and W Tanzania, N and S Somalia.
SYNONYMS: *boranus, delicatus, dundasi, hoogstraali, mariakanae, ocularis, peasei, phaeus, pulcher, renatus, rendilis, somalicus.*
COMMENTS: See Koontz and Roeper (1983, Mammalian Species, 204).

Elephantulus rupestris (A. Smith, 1831). Proc. Zool. Soc. Lond., 1831:11.
TYPE LOCALITY: S Africa or Namibia, mountains near mouth of Orange River.
DISTRIBUTION: Namibia; Cape Prov. (South Africa).

Macroscelides A. Smith, 1829. Zool. J. Lond., 4:435.
TYPE SPECIES: *Macroscelides typus* A. Smith, 1829 (= *Sorex proboscideus* Shaw, 1800).

Macroscelides proboscideus (Shaw, 1800). Gen. Zool. Syst. Nat. Hist., 1(2), Mammalia, p. 536.
TYPE LOCALITY: South Africa, Cape Prov., Oudtshoorn Div., Roodeval.
DISTRIBUTION: W and NW Cape Prov., South Africa, to SW Namibia.

Petrodromus Peters, 1846. Bericht Verhandl. K. Preuss. Akad. Wiss. Berlin, 11:258.
TYPE SPECIES: *Petrodromus tetradactylus* Peters, 1846.

Petrodromus tetradactylus Peters, 1846. Bericht Verhandl. K. Preuss. Akad. Wiss. Berlin, 11:258.
TYPE LOCALITY: Mozambique, Tette.
DISTRIBUTION: Mozambique; Tanzania (including Mafia and Zanzibar); SE Kenya; S Uganda; Zambia, Malawi; SE Zimbabwe; Zaire; Republic of Congo; NE Angola; N Natal and E Transvaal (South Africa).
STATUS: IUCN - Insufficiently known as *P. t. sangi*.
SYNONYMS: *rovumae, sangi, sultan, tordayi* (see Corbet, 1974:2).
COMMENTS: It is possible that *tordayi* is a separate species.

Rhynchocyon Peters, 1847. Bericht Verhandl. K. Preuss. Akad. Wiss. Berlin, 12:36.
TYPE SPECIES: *Rhynchocyon cirnei* Peters, 1847.
COMMENTS: A key to the species was presented in Rathbun (1979).

Rhynchocyon chrysopygus Günther, 1881. Proc. Zool. Soc. Lond., 1881:164.
TYPE LOCALITY: Kenya, Mombasa.
DISTRIBUTION: E Kenya.
STATUS: IUCN - Vulnerable.
COMMENTS: See Rathbun (1979, Mammalian Species, 117).

Rhynchocyon cirnei Peters, 1847. Bericht Verhandl. K. Preuss. Akad. Wiss. Berlin, 12:37.
TYPE LOCALITY: Mozambique, Bororo Dist., Quelimane.
DISTRIBUTION: Mozambique, Malawi, S Tanzania, NE Zambia, E Zaire, Uganda.
STATUS: IUCN - Insufficiently known as *R. c. cirnei*; Rare as *R. c. hendersoni*.
SYNONYMS: *hendersoni, stuhlmanni*.
COMMENTS: Includes *stuhlmanni*, which could be a distinct species; see Corbet (1974:2).

Rhynchocyon petersi Bocage, 1880. J. Sci. Math. Phys. Nat. Lisboa, ser. 1, 7:159.
TYPE LOCALITY: Tanzania, mainland opposite Zanzibar.
DISTRIBUTION: E Tanzania (including Mafia Isls and Zanzibar); SE Kenya.
STATUS: IUCN - Rare.

Appendix I

REMARKS ON BIBLIOGRAPHY AND PUBLICATION DATES

Fundamental to the field of systematics is the original description of each taxon. Because the date and nature of the publication in which a taxon was originally described directly affects the validity and seniority of the name, it is critical to know exact dates of publication as well as other aspects of the description. Numerous discrepancies exist among the citations of many early zoological works. Locating and investigating these publications is such a difficult task that an annotated bibliography would be useful to future researchers.

Several publications in particular proved especially troublesome, and deserve further comment.

I. Geoffroy Saint-Hilaire, É. 1803. Catalogue des mammifères du Muséum National d'Histoire Naturelle. Muséum National d'Histoire Naturelle, Paris, 272 pp.

Much controversy surrounds the validity of Geoffroy (1803) as a publication. Some authors (i.e., Sherborn, 1922, Vol. A-B, lviii; Pocock, 1939*a*:364, footnote; Ellerman and Morrison-Scott, 1951:282; Laurie and Hill, 1954:14, 100, footnote; Harrison, 1964:19) regarded it as not meeting the requirements of the International Code of Zoological Nomenclature (3rd ed., 1985), while others (i.e., Setzer, 1952:343; Hershkovitz, 1955*b*; and Hall, 1981:72) took the opposite view. We believe this Catalog does not meet the Code's criteria for publication (art. 8a(1), 9(5)) based on comments by I. Geoffroy (1839:5, footnote 2; 1847:118; see translations and discussion in Hershkovitz, 1955*b*:118), and for the sake of consistency throughout the text, we do not cite taxa from Geoffroy's (1803) catalog. A formal proposal to the International Commission on Zoological Nomenclature should be made regarding this matter.

II. Temminck, C. J. (ed.). 1839-1847. Verhandelingen over de Natuurlikjke Geschiedenis der Nederlandsche overzeesche bezittingen, door de Leden der Natuurkundige commissie in Inidië en andere Schrijvers. A. Arnez & Co. te Leiden. [3 Vol.] Zoologie, Botanie, Land- en Volkenkunde.

This three volume work, translated as "Transactions on the Natural History of the Netherlands Overseas Possessions" has caused much confusion among the scientific community and has been miscited throughout the literature. We examined these volumes in the Library of Congress and found that many of the names traditionally cited from this work did not occur in the volume, nor on the page from which they were commonly cited. Some of the numerous names found in this work are of questionable validity due to the incompleteness of description. Compounding this problem is the fact that the work was originally issued in 29 parts, including many plates, from 1839 to 1847, and conflicting dates for these parts have been given in various publications that purported to list the correct dates. We follow Husson and Holthuis (1955) who thoroughly and concisely described the publication and analyzed the 29 parts of the three volumes as to content and exact publication date.

III. Proceedings of the Zoological Society of London, 1830–present.

Because this work was issued in separate parts, the year in which the "Proceedings" occurred being considered the "volume" (until 1937), and, as the last part(s) was often published in the subsequent year, confusion has prevailed regarding the precise dates of publication. The correct dates were summarized by Duncan et al. (1937) and later by Cowan (1973); also see discussion in McAllan and Bruce (1989).

IV. Annals and Magazine of Natural History, 1840–1966.

While the dates of the volumes are clear from the text itself, the early history and evolution of this work was elucidated by Sheets-Pyenson (1981).

BIBLIOGRAPHY

The dates of many other early zoological works have been clarified in a number of publications. A partial list follows:

Anon. 1829. [Dates and parts of Cretzchmar's *Atlas zu der reise im nördlichen Afrika von Eduard Rüppell, Säugetiere*]. Isis, 1829(7):1291-1294.

Anon. 1903. Catalogue of the books, manuscripts, maps and drawings in the British Museum (Natural History), Trustees of the British Museum, London.

Anon. 1933. A catalogue of the works of Linnaeus (and publications immediately relating thereto) preserved in the libraries of the British Museum (Bloomsbury) and the British Museum (Natural History) (South Kensington). 2nd. edition. Trustees of the British Museum, London.

Bernard, K. H. 1950. The dates of issue of the *Illustrations of the Zoology of South Africa* and the *Marine Investigations in South Africa*. Journal of the Society for the Bibliography of Natural History, 2:187-189.

Cowan, C. F. 1969. Notes on Griffith's *Animal Kingdom* of Cuvier (1824-1835). Journal of the Society for the Bibliography of Natural History, 5:137-140.

Cowan, C. F. 1971. J. C. Chenu, "1857-1884" [1850-1861], *Encyclopédie d'Histoire Naturelle*. Journal of the Society for the Bibliography of Natural History, 6:9-17.

Cowan, C. F. 1976. On the Disciples' edition of Cuvier's *Régne Animal*. Journal of the Society for the Bibliography of Natural History, 8:32-64.

Dawson, W. R. 1945. On the history of Gray and Hardwicke's *Illustrations of Indian Zoology*, and some biographical notes on General Hardwicke. Journal of the Society for the Bibliography of Natural History, 2:55-69.

Holthuis, L. B., and T. Sakai. 1970. Ph. F. von Siebold and Fauna Japonica: A history of early Japanese zoology. Academic Press of Japan, Tokyo, 323 pp.

Horsfield, T. 1821-1824 [1990]. Zoological Researches in Java, and the neighboring islands: with a memoir by John Bastin. Oxford University Press, Singapore.

Iredale, T. 1936. On the *Dict. Univ. d'Hist. Nat.* of d'Orbigny. Journal of the Society for the Bibliography of Natural History, 1:33-34.

Roux, C. 1976. On the dating of the first edition of Cuvier's *Régne Animal*. Journal of the Society for the Bibliography of Natural History, 8:31.

Sawyer, F. C. 1951. The dates of publication of Wilhelm Peter Eduard Simon Rüppell's [1794-1884] *Neue Wirbelthiere zu der Fauna von Abyssinien gehörig* (fol., Frankfurt a. M., 1835-1840). Journal of the Society for the Bibliography of Natural History, 2:407.

Sawyer, F. C. 1953. The dates of issue of J. E. Gray's *Illustrations of Indian Zoology* (London, 1830-1835). Journal of the Society fro the Bibliography of Natural History, 3:48-55.

Sclater, W. L., and C. D. Sherborn. 1947. On the date as from which the names published in Pallas (P. S.), 's *Zoographia Rosso-Asiatica* are available nomenclatorially; with an annex by the late C. D. Sherborn, D.Sc. Bulletin of Zoological Nomenclature, 1(9):198-200.

Sherborn, C. D. 1891. On the dates of the parts, plates, and text of Schreber's *Säugthiere*. Proceedings of the Zoological Society of London, 1891:587-592.

Sherborn, C. D. 1898. Dates of Blainville's *Ostéographie*. Annals and Magazine of Natural History, ser. 7, 2:76.

Sherborn, C. D. 1922. On the dates of Cuvier, *Le Régne Animal*, etc. (Disciples Edition). Annals and Magazine of Natural History, ser. 9, 10:555.

Sherborn, C. D., and T. S. Palmer. 1899. Dates of Charles d'Orbigny's *Dictionnaire Universel d'Histoire Naturelle*, 1839-1849. Annals and Magazine of Natural History, ser. 7, 3:350-352.

Sherborn, C. D., and B. B. Woodward. 1901. Notes of the dates of publication of the natural history portions of some French voyages. Pt. 1. *Amérique méridionale; Indes orientales; Pôle Sud (Astrolabe* and *Zélée); La Bonite; La Coquille;* and *L'Uranie et Physicienne*. Annals and Magazine of Natural History, ser. 7, 7:388-392.

Sherborn, C. D., and B. B. Woodward. 1901. Dates of publication of the zoological and botanical portions of some French voyages. Pt. 2. Ferret and Galinier's *Voyage en Abyssinie;* Lefebvre's *Voyage en Abyssinie; Explorations scientifique de l'Algérie;* Castelnau's *Amérique du Sud;* Dumont d'Urville's *Voyage de l'Astrolabe;* Laplace's *Voyage sur la Favorite;* Jacquemont's *Voyage dans l'Inde;* Tréhouart's *Commission scientifique d'Islande*. Annals and Magazine of Natural History, ser. 7, 8:161-164, 333-336, 491-495.

Stresemann, E. 1951. Date of publication of Pallas's *Zoographia Rosso-Asiatica*. Isis, 93(2):316-318.

Svetovidov, A. N. 1981. The Pallas fish collection and the *Zoographia Rosso-Asiatica*: an historical account. Archives of Natural History, 19(1):45-64.

Townsend, A. C. 1957. A Catalogue of the works of Linnaeus issued in commemoration of the 250th anniversary of the birthday of Carolus Linnaeus, 1707-78. Sandbergs Bokhandel, Stockholm.

Waterhouse, F. H. 1880. On the dates of publication of the parts of Sir Andrew Smith's *Illustrations of the Zoology of South Africa*. Proceedings of the Zoological Society of London, 1880:489-491.

Whitehead, P. J. P. 1967. The dating of the 1st edition of Cuvier's *Le Régne Animal Distribué d'aprés son Organisation*. Journal of the Society for the Bibliography of Natural History, 4:300-301.

Appendix II

Mammalian Species is a series of fascicles published by the American Society of Mammalogists. Each of these short accounts provides a review of the species in question, giving full synonymies and taxonomic, distributional, habitat, and morphological information.

Mammalian Species: Cumulative index, nos. 1-402. Compiled by V. Hayssen, Smith College, Northampton, MA 01063

SYSTEMATIC LIST

Order Didelphimorphia
  Family Didelphidae
    Caluromys derbianus, 140
    Chironectes minimus, 109
    Didelphis virginiana, 40
    Glironia venusta, 107
    Lestodelphys halli, 81
    Lutreolina crassicaudata, 91
    Marmosa robinsoni, 203
    Monodelphis kunsi, 190
Order Paucituberculata
  Family Caenolestidae
    Rhyncholestes raphanurus, 286
Order Microbiotheria
  Family Microbiotheriidae
    Dromiciops australis, 99
Order Diprotodontia
  Family Macropodidae
    Macropus giganteus, 187
  Family Petauridae
    Petaurus breviceps, 30
Order Xenarthra
  Family Dasypodidae
    Dasypus novemcinctus, 162
    Euphractus sexcinctus, 252
Order Insectivora
  Family Soricidae
    Blarina brevicauda, 261
    Cryptotis goodwini, 44; *magna,* 61; *mexicana,* 28; *parva,* 43
    Megasorex gigas, 16
    Microsorex hoyi, 33; *thompsoni,* 33
    Notiosorex crawfordi, 17

Sorex bendirii, 27; *dispar*, 155; *fumeus*, 215; *gaspensis*, 155; *longirostris*, 143; *merriami*, 2; *nanus*, 131; *ornatus*, 212; *pacificus*, 231; *palustris*, 296; *tenellus*, 131; *trowbridgii*, 337

Family Talpidae

Condylura cristata, 129

Galemys pyrenaicus, 207

Neurotrichus gibbsii, 387

Parascalops breweri, 98

Scalopus aquaticus, 105

Scapanus orarius, 253

Order Chiroptera

Family Pteropodidae

Eidolon helvum, 312

Epomophorus gambianus, 344; *wahlbergi*, 394

Hypsignathus monstrosus, 357

Family Rhinopomatidae

Rhinopoma hardwickii, 263; *muscatellum*, 263

Family Craseonycteridae

Craseonycteris thonglongyai, 160

Family Emballonuridae

Balantiopteryx infusca, 313; *io*, 313; *plicata*, 301

Cyttarops alecto, 13

Diclidurus albus, 316

Family Megadermatidae

Macroderma gigas, 260

Family Noctilionidae

Noctilio albiventris, 197; *leporinus*, 216

Family Mormoopidae

Pteronotus davyi, 346; *parnellii*, 209; *quadridens*, 395

Family Phyllostomidae

Anoura cultrata, 179

Ardops nichollsi, 24

Artibeus aztecus, 177; *hirsutus*, 199; *inopinatus*, 199; *phaeotis*, 235; *toltecus*, 178

Brachyphylla cavernarum, 205; *nana*, 206

Centurio senex, 138

Chiroderma improvisum, 134

Choeronycteris mexicana, 291

Chrotopterus auritus, 343

Desmodus rotundus, 202

Diphylla ecaudata, 227

Ectophylla alba, 166

Erophylla sezekorni, 115

Glossophaga leachii, 226; *mexicana*, 245; *soricina*, 379

Hylonycteris underwoodi, 32

Leptonycteris nivalis, 307

Lonchorhina aurita, 347

Macrophyllum macrophyllum, 62

Macrotus waterhousii, 1
Micronycteris brachyotis, 251; *megalotis*, 376
Monophyllus plethodon, 58; *redmani*, 57
Platyrrhinus helleri, 373
Pygoderma bilabiatum, 220
Stenoderma rufum, 18
Sturnira aratathomasi, 284; *bidens*, 276; *lilium*, 333; *magna*, 240; *thomasi*, 68
Tonatia evotis, 334; *silvicola*, 334
Uroderma bilobatum, 279
Vampyressa pusilla, 292
Vampyrodes caraccioli, 359
Vampyrops lineatus, 275
Vampyrum spectrum, 184

Family Natalidae
Natalus major, 130; *micropus*, 114

Family Thyropteridae
Thyroptera discifera, 104; *tricolor*, 71

Family Myzopodidae
Myzopoda aurita, 116

Family Vespertilionidae
Antrozous pallidus, 213
Bauerus dubiaquercus, 282
Eptesicus fuscus, 356
Euderma maculatum, 77
Eudiscopus denticulus, 19
Idionycteris phyllotis, 208
Lasionycteris noctivagans, 172
Lasiurus borealis, 183; *cinereus*, 185; *intermedius*, 132; *seminolus*, 280
Myotis auriculus, 191; *austroriparius*, 332; *evotis*, 329; *keenii*, 121; *lucifugus*, 142; *nigricans*, 39; *planiceps*, 60; *sodalis*, 163; *thysanodes*, 137; *velifer*, 149; *volans*, 224
Nycticeius humeralis, 23
Pipistrellus subflavus, 228
Plecotus mexicanus, 401; *rafinesquii*, 69; *townsendii*, 175
Rhogeessa gracilis, 76

Family Molossidae
Neoplatymops mattogrossensis, 244
Nyctinomops aurispinosus, 350; *femorosaccus*, 349; *macrotis*, 351
Tadarida brasiliensis, 331

Order Primates

Family Callitrichidae
Leontopithecus rosalia, 148

Family Cebidae
Callicebus moloch, 112

Family Pongidae
Pongo pygmaeus, 4

Order Carnivora
Family Canidae
Atelocynus microtis, 256
Canis latrans, 79; *lupus*, 37; *rufus*, 22
Cerdocyon thous, 186
Chrysocyon brachyurus, 234
Cuon alpinus, 100
Nyctereutes procyonoides, 358
Urocyon cinereoargenteus, 189
Vulpes macrotis, 123; *velox*, 122
Family Felidae
Felis concolor, 200; *geoffroyi*, 54; *lynx*, 269
Panthera onca, 340; *tigris*, 152
Uncia uncia, 20
Family Herpestidae
Bdeogale crassicauda, 294
Crossarchus ansorgei, 402; *obscurus*, 290
Herpestes auropunctatus, 342; *sanguineus*, 65
Ichneumia albicauda, 12
Liberiictis kuhni, 348
Family Hyaenidae
Hyaena brunnea, 194; *hyaena*, 150
Proteles cristatus, 363
Family Mustelidae
Enhydra lutris, 133
Martes americana, 289; *pennanti*, 156
Mephitis mephitis, 173
Mustela erminea, 195; *lutreola*, 362; *nigripes*, 126
Taxidea taxus, 26
Family Otariidae
Arctocephalus galapagoensis, 64
Eumetopias jubatus, 283
Neophoca cinerea, 392
Odobenus rosmarus, 238
Family Phocidae
Cystophora cristata, 258
Leptonychotes weddelli, 6
Mirounga leonina, 391
Phoca sibirica, 188
Family Procyonidae
Ailurus fulgens, 222
Bassariscus astutus, 327
Potos flavus, 321
Procyon lotor, 119
Family Ursidae
Ailuropoda melanoleuca, 110
Ursus maritimus, 145

Family Viverridae
Cryptoprocta ferox, 254
Osbornictis piscivora, 309
Order Cetacea
Family Platanistidae
Lipotes vexillifer, 10
Family Delphinidae
Orcinus orca, 304
Family Phocoenidae
Phocoenoides dalli, 319
Phocoena dioptrica, 66; *phocoena*, 42; *sinus*, 198; *spinipinnis*, 217
Family Monodontidae
Delphinapterus leucas, 336
Monodon monoceros, 127
Family Physeteridae
Kogia simus, 239
Family Hyperoodontidae
Mesoplodon stejnegeri, 250
Order Sirenia
Family Dugongidae
Dugong dugon, 88
Hydrodamalis gigas, 165
Family Trichechidae
Trichechus inunguis, 72; *manatus*, 93; *senegalensis*, 89
Order Proboscidea
Family Elephantidae
Elephas maximus, 182
Loxodonta africana, 92
Order Perissodactyla
Family Equidae
Equus burchelli, 157; *zebra*, 314
Family Rhinocerotidae
Ceratotherium simum, 8
Dicerorhinus sumatrensis, 21
Rhinoceros unicornis, 211
Order Hyracoidea
Family Procaviidae
Dendrohyrax dorsalis, 113
Procavia capensis, 171
Order Tubulidentata
Family Orycteropodidae
Orycteropus afer, 300
Order Artiodactyla
Family Tayassuidae
Catagonus wagneri, 259
Tayassu pecari, 293

Family Camelidae
Camelus dromedarius, 375
Family Giraffidae
Giraffa camelopardalis, 5
Family Cervidae
Alces alces, 154
Blastocerus dichotomus, 380
Cervus nippon, 128
Dama dama, 317
Odocoileus hemionus, 219; *virginianus*, 388
Ozotoceros bezoarticus, 295
Family Antilocapridae
Antilocapra americana, 90
Family Bovidae
Ammotragus lervia, 144
Bison bison, 266
Budorcas taxicolor, 277
Cephalophus maxwellii, 31; *sylvicultor*, 225
Connochaetes gnou, 50
Nemorhaedus goral, 335
Oreamnos americanus, 63
Ovibos moschatus, 302
Ovis canadensis, 230; *dalli*, 393
Pseudois nayaur, 278; *schaeferi*, 278
Saiga tatarica, 38
Tragelaphus eurycerus, 111
Order Rodentia
Suborder Sciurognathi
Family Sciuridae
Ammospermophilus harrisii, 366; *insularis*, 364; *interpres*, 365; *leucurus*, 368; *nelsoni*, 367
Cynomys gunnisoni, 25; *leucurus*, 7; *mexicanus*, 248; *parvidens*, 52
Glaucomys sabrinus, 229; *volans*, 78
Marmota flaviventris, 135; *vancouverensis*, 270
Sciurus aberti, 80; *granatensis*, 246; *richmondi*, 53
Spermophilus beldingi, 221; *colombianus*, 372; *elegans*, 214; *madrensis*, 378; *mexicanus*, 164; *richardsonii*, 243; *saturatus*, 322; *spilosoma*, 101; *tereticaudus*, 274; *townsendii*, 268; *tridecemlineatus*, 103; *variegatus*, 272; *washingtoni*, 371
Syntheosciurus brochus, 249
Tamias amoenus, 390; *dorsalis*, 399; *striatus*, 168
Xerus rutilus, 370
Family Castoridae
Castor canadensis, 120
Family Geomyidae
Cratogeomys castanops, 338
Geomys arenarius, 36; *attwateri*, 382; *breviceps*, 383; *personatus*, 170; *pinetis*, 86; *tropicalis*, 35
Thomomys bulbivorus, 273

Family Heteromyidae

Chaetodipus arenarius, 384; *baileyi*, 297; *hispidus*, 320; *spinatus*, 385

Dipodomys californicus, 324; *compactus*, 369; *deserti*, 339; *elator*, 232; *elephantinus*, 255; *gravipes*, 236; *heermanni*, 323; *ingens*, 377; *insularis*, 374; *margaritae*, 400; *microps*, 389; *nelsoni*, 326; *nitratoides*, 381; *ordii*, 353; *panamintinus*, 354; *phillipsii*, 51; *spectabilis*, 311; *stephensi*, 73

Heteromys gaumeri, 345; *nelsoni*, 397; *oresterus*, 396

Liomys irroratus, 82; *pictus*, 83; *salvini*, 84

Microdipodops megacephalus, 46; *pallidus*, 47

Perognathus fasciatus, 303; *parvus*, 318

Family Dipodidae

Napaeozapus insignis, 14

Zapus hudsonius, 11; *trinotatus*, 315

Family Muridae

Subfamily Arvicolinae

Clethrionomys gapperi, 146

Lagurus curtatus, 124

Microtus breweri, 45; *canicaudus*, 267; *chrotorrhinus*, 180; *longicaudus*, 271; *ochrogaster*, 355; *oregoni*, 233; *pennsylvanicus*, 159; *pinetorum*, 147; *richardsoni*, 223; *townsendii*, 325

Neofiber alleni, 15

Ondatra zibethicus, 141

Phenacomys intermedius, 305

Synaptomys cooperi, 210

Subfamily Gerbillinae

Meriones crassus, 9; *unguiculatus*, 3

Subfamily Murinae

Rattus fuscipes, 298; *lutreolus*, 299

Subfamily Otomyinae

Otomys angoniensis, 306; *irroratus*, 308

Subfamily Sigmodontinae

Baiomys musculus, 102; *taylori*, 285

Blarinomys breviceps, 74

Nectomys squamipes, 265

Neotoma albigula, 310; *alleni*, 41; *floridana*, 139; *fuscipes*, 386; *mexicana*, 262; *micropus*, 330; *phenax*, 108; *stephensi*, 328

Ochrotomys nuttalli, 75

Onychomys leucogaster, 87; *torridus*, 59

Oryzomys palustris, 176

Ototylomys phyllotis, 181

Peromyscus alstoni, 242; *attwateri*, 48; *californicus*, 85; *crinitus*, 287; *eremicus*, 118; *gossypinus*, 70; *leucopus*, 247; *melanocarpus*, 241; *pectoralis*, 49; *stirtoni*, 361; *truei*, 161; *yucatanicus*, 196

Reithrodontomys brevirostris, 192; *fulvescens*, 174; *gracilis*, 218; *megalotis*, 167; *montanus*, 257; *paradoxus*, 192; *raviventris*, 169; *spectabilis*, 193

Sigmodon alleni, 95; *fulviventer*, 94; *hispidus*, 158; *leucotis*, 96; *ochrognathus*, 97

Subfamily Rhizomyinae

Tachyoryctes macrocephalus, 237

Suborder Hystricognathi
Family Erethizontidae
Erethizon dorsatum, 29
Family Hydrochaeridae
Hydrochoerus hydrochaeris, 264
Family Octodontidae
Octodon degus, 67
Family Capromyidae
Geocapromys brownii, 201; *thoracatus*, 341
Family Myocastoridae
Myocastor coypus, 398
Order Lagomorpha
Family Ochotonidae
Ochotona collaris, 281; *princeps*, 352
Family Leporidae
Brachylagus idahoensis, 125
Lepus townsendii, 288
Romerolagus diazi, 360
Sylvilagus aquaticus, 151; *audubonii*, 106; *bachmani*, 34; *floridanus*, 136; *nuttallii*, 56; *palustris*, 153; *transitionalis*, 55
Order Macroscelidea
Family Macroscelididae
Elephantulus rufescens, 204
Rhynchocyon chrysopygus, 117

Literature Cited

Abad, P. L. 1987. Biologia y ecologia del liron careto (*Eliomys quercinus*) en Leon. Ecologia, 1:153-159.

Abe, H. 1967. Classification and biology of Japanese Insectivora (Mammalia). I. Studies on variation and classification. Journal of the Faculty of Agriculture, Hokkaido University, Sapporo, Japan, 55:191-265, 2 pls.

Abe, H. 1971. Small mammals of central Nepal. Journal of the Faculty of Agriculture, Hokkaido University, Sapporo, Japan, 56:367-423.

Abe, H. 1982. Age and seasonal variations of molar patterns in a red-backed vole population. Journal of the Mammalogical Society of Japan, 9:9-13.

Abe, H. 1983. Variation and taxonomy of *Niviventer fulvescens* and notes on *Niviventer* group of rats in Thailand. Journal of the Mammalogical Society of Japan, 9:151-161.

Abe, H. 1988. [The phylogenetic relationships of Japanese moles]. Honyurui Kagaku (Mammalian Science), 28:63-68 (in Japanese).

Abe, H., S. Shiraishi, and S. Arai. 1991. A new mole from Uotsuri-jima, the Ryukyu Islands. Journal of the Mammalogical Society of Japan, 15:47-60.

Acharya, L. 1992. *Epomophorus wahlbergi*. Mammalian Species, 394:1-4.

Achverdjan, M. R., E. A. Lyapunova, and N. N. Vorontsov. 1992. [Karyology and systematics of shrub voles of the Caucasus and Transcaucasia (*Terricola*, Arvicolinae, Rodentia)]. Zoologicheskii Zhurnal, 71:96-110 (in Russian).

Adams, J. K. 1989. *Pteronotus davyi*. Mammalian Species, 346:1-5.

Aellen, V. 1959. Contribution à l'étude de la faune d'Afghanistan. 9. Chiroptères. Revue Suisse de Zoologie, 66:353-386.

Aellen, V. 1965. Les rongeurs de basse Cote-d'Ivoire (Hystricomorpha et Gliridae). Revue Suisse de Zoologie, 72(37):755-767.

Aellen, V. 1966. Notes sur *Tadarida teniotis* (Raf.)(Mammalia, Chiroptera). I. Systématique, paléontologie et peuplement, repartition géographique. Revue Suisse de Zoologie, 73:119-159.

Agadzhanyan, A. K., and V. N. Yatsenko. 1984. [Phylogenetic interrelationships in voles of northern Eurasia]. Sbornik Trudov Zoologicheskovo Muzeya MGU, 22:135-190 (in Russian).

Aggundey, I. R., and D. A. Schlitter. 1986. Annotated checklist of the mammals of Kenya. 2. Insectivora and Macroscelidea. Annals of Carnegie Museum, 55:325-347.

Agrawal, V. C., and S. Chakraborty. 1970. Occurrence of the woolly flying squirrel, *Eupetaurus cinereus* Thomas (Mammalia: Rodentia: Sciuridae) in North Sikkim. Journal of the Bombay Natural History Society, 66:615-616.

Agrawal, V. C., and S. Chakraborty. 1971. Notes on a collection of small mammals from Nepal, with the description of a new mouse-hare (Lagomorpha: Ochotonidae). Proceedings of the Zoological Society of Calcutta, 24:41-46.

Agrawal, V. C., and S. Chakraborty. 1976. Revision of the subspecies of the lesser bandicoot rat *Bandicota bengalensis* (Gray) (Rodentia: Muridae). Records of the Zoological Survey of India, 69:267-274.

Agrawal, V. C., and S. Chakraborty. 1979. Catalogue of mammals in the Zoological Survey of India. Part 1. Sciuridae. Records of the Zoological Survey of India, 74:333-481.

Agrawal, V. C., and S. Chakraborty. 1980. Intraspecific geographical variations in the Indian long-tailed tree mouse, *Vandeleuria oleracea* (Bennett). Bulletin of the Zoological Survey of India, 3:77-85.

Aguilera, A., and A. Perez-Zapata. 1989. Cariologia de *Holochilus venezuelae* (Rodentia, Cricetidae). Acta Científica Venezolana, Genetica, 40:198-207.

Agusti, J. 1989. The Miocene rodent succession in Eastern Spain: a zoogeographical appraisal. Pp. 375-404, *in* European Neogene Mammal Chronology (E. H. Lindsay, V. Fahlbusch, and P. Mein, eds.). Plenum Press, New York, 658 pp.

Aimi, M. 1980. A revised classification of the Japanese red-backed voles. Memoirs of the Faculty of Science, Kyoto University, Series of Biology, 8:35-84.

Aimi, M., H. S. Hardjasasmita, A. Sjarmidi, and D. Yuri. 1986. Geographical distribution of *aygula*-group of the genus *Presbytis* in Sumatra. Kyoto University Overseas Research Report of Studies on Asian Non-human Primates, 5:45-58.

Alberico, M. S., and E. Velasco. 1991. Description of a new broad-nosed bat from Colombia. Bonner Zoologische Beiträge, 42:237-239.

Albignac, R. 1970. Notes éthologiques sur quelques carnivores Malagaches: le *Cryptoprocta ferox* (Bennett). La Terre et la Vie, 24(3):395-402.

Albignac, R. 1973. Faune de Madagascar. 36. Mammifères carnivores. O. R. S. T. O. M., Paris, 206 pp.

Albignac, R. 1974. Observations eco-éthologiques sur le genre *Eupleres*, Viverridae de Madagascar. La Terre et la Vie, 28:321-351.

Albuja V., L. 1982. Murcielagos del Ecuador. Escuela Politécnica Nacional, Departamento de Ciéncias Biológicas, Quito, Ecuador, 285 pp.

Alcantara, M. 1991. Geographical variation in body size of the wood mouse *Apodemus sylvaticus* L. Mammal Review, 21:143-150.

Alcover, J. A. 1986. Troballa de restes osteologiques de *Eliomys quercinus* (Mammalia, Rodentia, Gliridae) a l'illa de Cabrera. Bolletino de la Sociedad de Historia Natural de Baleares, 30:137-139.

Alcover, J. A., and J. Agusti. 1985. *Eliomys* (*Eivissia*) *canarreiensis* n. sgen., n. sp., nou glirid del Pleistocene de la Cova de Endins, 10-11:51-56.

Alcover, J. A., J. Gosalbez, and Ph. Orsini. 1985. *Mus spretus parvus* n. ssp. (Rodentia, Muridae): un ratoli nan de l'Illa d'Eivissa. Bolletino de la Sociedad de Historia Natural de Baleares, 29:5-17.

Allabergenov, K. 1986. [On the ecology of the forest dormouse in the Fergana Valley]. Uzbekskii Biologicheskii Zhurnal, 2:41-42 (in Russian).

Allard, M. W., and I. F. Greenbaum. 1988. Morphological variation and taxonomy of chromosomally differentiated *Peromyscus* from the Pacific Northwest. Canadian Journal of Zoology, 66:2734-2739.

Allard, M. W., S. J. Gunn, and I. F. Greenbaum. 1987. Mensural discrimination of chromosomally characterized *Peromyscus oreas* and *Peromyscus maniculatus*. Journal of Mammalogy, 68:402-406.

Allen, G. M. 1915. Mammals obtained by the Phillips Palestine Expedition. Bulletin of the Museum of Comparative Zoology at Harvard College, 59:1-14.

Allen, G. M. 1929. Mustelids from Asiatic expeditions. American Museum Novitates, 358:1-12.

Allen, G. M. 1938-1940. The mammals of China and Mongolia [Natural History of Central Asia (W. Granger, ed.)]. Central Asiatic Expeditions of the American Museum of Natural History, New York, 11:pt. 1:1-620[1938]; pt. 2:621-1350[1940].

Allen, G. M. 1939. A checklist of African mammals. Bulletin of the Museum of Comparative Zoology at Harvard College, 83:1-763.

Allen, G. M., and B. Lawrence. 1936. Scientific results of an expedition to rainforest regions in eastern Africa. III. Mammals. Bulletin of the Museum of Comparative Zoology at Harvard College, 79:31-126.

Allen, G. M., and A. Loveridge. 1933. Reports on the scientific results of an expedition to the southwestern highlands of Tanganyika Territory. II. Mammals. Bulletin of the Museum of Comparative Zoology at Harvard College, 75:47-140.

Allen, G. M., and A. Loveridge. 1942. Scientific results of a fourth expedition to forested areas in East and Central Africa. I. Mammals. Bulletin of the Museum of Comparative Zoology at Harvard College, 89:147-214.

Allen, J. A. 1879. On the coatis (genus *Nasua*, Storr). Bulletin of the United States Geological and Geographical Survey, 5:153-174.

Allen, J. A. 1880. History of North American pinnipeds: A monograph of the walruses, seal-lions, sea-bears and seals of North America. United States Geological and Geographical Survey of the Territories, Miscellaneous Publication, 12:1-785.

Allen, J. A. 1887. Sciuridae. Pp. 631-939, *in* Monographs of North American Rodentiae (E. Coues and J. A. Allen, co-authors). United States Geological Survey of the Territories, 11:1-1091.

Allen, J. A. 1892. A synopsis of the pinnipeds, or seals and walruses, in relation to their commercial history and products, Pp. 367-391, *in* Fur seal arbitration, appendix to the case of the United States before the tribunal of arbitration to convene at Paris. United States Government Printing Office, Washington, D. C., 1:367-391.

Allen, J. A. 1895*a*. On the species of the genus *Reithrodontomys*. Bulletin of the American Museum of Natural History, 7:107-143.

Allen, J. A. 1895*b*. Descriptions of new American mammals. Bulletin of the American Museum of Natural History, 7:327-340.

Allen, J. A. 1898. Revision of the chickarees, or North American red squirrels (subgenus *Tamiasciurus*). Bulletin of the American Museum of Natural History, 10:249-298.

Allen, J. A. 1900. Description of new American marsupials. Bulletin of the American Museum of Natural History, 13:191-199.

Allen, J. A. 1901*a*. Note on the names of a few South American mammals. Proceedings of the Biological Society of Washington, 14:183-185.

Allen, J. A. 1901*b*. On a further collection of mammals from southeastern Peru, collected by Mr. H. H. Keays, with descriptions of new species. Bulletin of the American Museum of Natural History, 14:41-46.

Allen, J. A. 1901*c*. A preliminary study of the North American opossums of the genus *Didelphis*. Bulletin of the American Museum of Natural History, 14:149-188, pls 22-25.
Allen, J. A. 1902. Mammal names proposed by Oken in his "Lehrbuch der Zoologie". Bulletin of the American Museum of Natural History, 16:373-379.
Allen, J. A. 1905. The Mammalia of southern Patagonia. Reports of the Princeton University Expedition to Patagonia, 1896-1899, 3(Zool):1-210.
Allen, J. A. 1908. Mammals from Nicaragua. Bulletin of the American Museum of Natural History, 24:647-670.
Allen, J. A. 1910. Mammals from Palawan Island, Philippine Islands. Bulletin of the American Museum of Natural History, 28(3):13-17.
Allen, J. A. 1913. Revision of the *Melanomys* group of American Muridae. Bulletin of the American Museum of Natural History, 32:535-555.
Allen, J. A. 1915*a*. New South American mammals. Bulletin of the American Museum of Natural History, 34:625-634.
Allen, J. A. 1915*b*. Review of the South American Sciuridae. Bulletin of the American Museum of Natural History, 34:147-309.
Allen, J. A. 1916. Mammals collected on the Roosevelt Brazilian Expedition, with field notes by Leo E. Miller. Bulletin of American Museum of Natural History, 35:559-610, 6 pls.
Allen, J. A. 1919*a*. Notes on the synonymy and nomenclature of the smaller spotted cats of tropical America. Bulletin of the American Museum of Natural History, 41(7):341-419.
Allen, J. A. 1919*b*. Preliminary notes on African Carnivora. Journal of Mammalogy, 1(1):23-31.
Allen, J. A. 1920. Note on Gueldenstaedt's names of certain species of Felidae. Journal of Mammalogy, 1:90-91.
Allen, J. A. 1924. Carnivora collected by the American Museum Congo Expedition. Bulletin of the American Museum of Natural History, 47:73-281.
Alonso-Mejia, A., and R. A. Medellín. 1991. *Micronycteris megalotis*. Mammalian Species, 376:1-6.
Al Robaae, K., and H. Felton. 1990. Was ist *Erythronesokia* Khajuria, 1981 (Mammalia: Rodentia: Muridae)? Zeitschrift für Säugetierkunde, 55:253-259.
Al Saleh, A. A. 1988. Cytological studies of certain desert mammals of Saudi Arabia. 6. First report on chromosome number and karyotype of *Acomys dimidiatus*. Genetica, 76:3-5.
Al Saleh, A. A., and M. A. Khan. 1984. Cytological studies of certain desert mammals of Saudi Arabia. 1. The karyotype of *Jaculus jaculus*. Journal of the College of Science, King Saud University, 15(1):163-168.
Alston, E. R. 1876. On the classification of the order Glires. Proceedings of the Zoological Society of London, 1876:61-98.
Altuna, C. A., and E. P. Lessa. 1985. Penial morphology in Uruguayan species of *Ctenomys* (Rodentia: Octodontidae). Journal of Mammalogy, 66:483-488.
Alvarez, J., M. R. Willig, J. K. Jones, Jr., and W. D. Webster. 1991. *Glossophaga soricina*. Mammalian Species, 379:1-7.
Alvarez, T. 1961. Taxonomic status of some mice of the *Peromyscus boylii* group in eastern Mexico, with description of a new subspecies. University of Kansas Publications, Museum of Natural History, 14:111-120.
Alvarez, T. 1963*a*. The Recent mammals of Tamaulipas. University of Kansas Publications, Museum of Natural History, 14:363-473.
Alvarez, T. 1963*b*. The type locality for *Nyctomys sumichrasti* Saussure. Journal of Mammalogy, 44:582-583.
Alvarez, T., and T. Alvarez-Castañeda. 1991. Notas sobre el estado taxónomíco de *Pteronotus davyi* en Chiapas y de *Hylonycteris* en Mexico (Mammalia: Chiroptera). Anales de la Escuela Naciónál de Ciéncias, 34:223-229.
Alvarez, T., and O. Polaco. 1984. Estudio de los mamíferos capturados en la Michilía, sureste de Durango, México. Anales de la Escuela Naciónál de Ciéncias Biológicas (México City), 28:99-148.
Ameghino, F. 1904. Myocastoridae. Annales de la Sociedad Científica de Argentina, 56-58:103.
Amtmann, E. 1975. Family Sciuridae. Part 6. 1. Pp. 1-12, *in* The mammals of Africa: An identification manual (J. Meester and H. W. Setzer, eds.) [issued 10 Dec 1975]. Smithsonian Institution Press, Washington, D. C., not continuously paginated.
Anadu, P. A. 1979. The occurrence of *Steatomys jacksoni* Hayman in south-western Nigeria. Acta Theriologica, 24:513-526.
Andera, M. 1986. Dormice (Gliridae) in Czechoslovakia. Part 1. *Glis glis, Eliomys quercinus* (Rodentia, Mammalia). Folia Musei Rerum Naturalium Bohemiae Occidentalis, Zoologica, 24:1-47.
Andera, M. 1987. Dormice (Gliridae) in Czechoslovakia. Part 2. *Muscardinus avellanarius, Dryomys nitedula* (Rodentia, Mammalia). Folia Musei Rerum Naturalium Bohemiae Occidentalis, Zoologica, 26:3-78.

Andersen, K. 1912. Catalogue of the Chiroptera in the collection of the British Museum. Vol. 1. Megachiroptera. Second ed. British Museum (Natural History), London, 854 pp.

Anderson, A. E., and O. C. Wallmo. 1984. *Odocoileus hemionus*. Mammalian Species, 219:1-9.

Anderson, E. 1970. Quaternary evolution of the genus *Martes* (Carnivora, Mustelidae). Acta Zoologica Fennica, 130:1-132.

Anderson, R. M. 1942*a*. Canadian voles of the genus *Phenacomys* with description of two new Canadian subspecies. Canadian Field-Naturalist, 56:56-60.

Anderson, R. M. 1942*b*. Six additions to the list of Quebec mammals with descriptions of four new forms. Annual Report, Provancher Society of Natural History of Canada, Quebec, 1941:31-43 (English); 45-57 (French).

Anderson, R. M. 1947. Catalogue of Canadian Recent mammals. National Museum of Canada Bulletin, Biological Series, 102:1-238.

Anderson, R. M., and A. L. Rand. 1943. Variation in the porcupine (Genus *Erethizon*) in Canada. Canadian Journal of Research, section D, 21:292-309.

Anderson, S. 1956. Subspeciation in the meadow mouse, *Microtus pennsylvanicus*, in Wyoming, Colorado, and adjacent areas. University of Kansas Publications, Museum of Natural History, 9:85-104.

Anderson, S. 1959. Distribution, variation, and relationships of the montane vole, *Microtus montanus*. University of Kansas Publications, Museum of Natural History, 9:415-511.

Anderson, S. 1960. The baculum in microtine rodents. University of Kansas Publications, Museum of Natural History, 12:181-216.

Anderson, S. 1962. Tree squirrels (*Sciurus colliaei* group) of western Mexico. American Museum Novitates, 2093:1-13.

Anderson, S. 1964. The systematic status of *Perognathus artus* and *Perognathus goldmani* (Rodentia). American Museum Novitates, 2184:1-27.

Anderson, S. 1965. Conspecificity of *Plagiodontia aedium* and *P. hylaeum* (Rodentia). Proceedings of the Biological Society of Washington, 78:95-98.

Anderson, S. 1966. Taxonomy of gophers, especially *Thomomys* in Chihuahua, Mexico. Systematic Zoology, 15:189-198.

Anderson, S. 1967. Primates. Pp. 151-177, *in* Recent mammals of the world: a synopsis of families (S. Anderson and J. K. Jones, Jr., eds.). Ronald Press Co., New York, 453 pp.

Anderson, S. 1969*a*. *Macrotus waterhousii*. Mammalian Species, 1:1-4.

Anderson, S. 1969*b*. Taxonomic status of the woodrat, *Neotoma albigula*, in southern Chihuahua, Mexico. Miscellaneous Publications, Museum of Natural History, University of Kansas, 51:25-50.

Anderson, S. 1972. Mammals of Chihuahua: taxonomy and distribution. Bulletin of the American Museum of Natural History, 148:149-410.

Anderson, S. 1982. *Monodelphis kunsi*. Mammalian Species, 190:1-3.

Anderson, S., and A. S. Gaunt. 1962. A classification of white-sided jack rabbits of Mexico. American Museum Novitates, 2088:1-16.

Anderson, S., and J. P. Hubbard. 1971. Notes on geographic variation of *Microtus pennsylvanicus* (Mammalia, Rodentia) in New Mexico and Chihuahua. American Museum Novitates, 2460:1-8.

Anderson, S., and J. K. Jones, Jr. (eds.). 1967. Recent mammals of the world, a synopsis of families. Ronald Press Co., New York, 453 pp.

Anderson, S., and J. K. Jones, Jr. (eds.). 1984. Orders and families of recent mammals of the world. John Wiley and Sons, New York, 686 pp.

Anderson, S., and C. E. Nelson. 1965. A systematic revision of *Macrotus* (Chiroptera). American Museum Novitates, 2212:1-39.

Anderson, S., and N. Olds. 1989. Notes on Bolivian mammals. 5. Taxonomy and distribution of *Bolomys* (Muridae, Rodentia). American Museum Novitates, 2935:1-22.

Anderson, S., K. F. Koopman, and G. K. Creighton. 1982. Bats of Bolivia: An annotated checklist. American Museum Novitates, 2750:1-24.

Anderson, S., C. A. Woods, G. S. Morgan, and W. L. R. Oliver. 1983. *Geocapromys brownii*. Mammalian Species, 201:1-5.

Ando, A., M. Harada, S. Shiraishi, and T. A. Uchida. 1991. Variation of the X chromosome in the Smith's red-backed vole *Eothenomys smithii*. Journal of the Mammalogical Society of Japan, 15:83-90.

Ando, A., S. Shiraishi, M. Harada, and T. A. Uchida. 1988. A karyological study of two intraspecific taxa in Japanese *Eothenomys* (Mammalia: Rodenta). Journal of the Mammalogical Society of Japan, 13:93-104.

Andrews, C. W. 1909. An account of Andrews' visit to Christmas Island in 1908. Proceedings of the Zoological Society of London, 1909:101-103.

Andrews, R. C. 1932. The new conquest of central Asia: A narrative of the explorations of the Central Asiatic Expeditions in Mongolia and China, 1921-1930. [Natural History of Central Asia, (W. Granger, ed.). Central Asiatic Expeditions of the American Museum of Natural History, New York, 1:1-678.

Angermann, R. 1966. Beiträge zur Kenntnis der Gattung *Lepus* (Lagomorpha, Leporidae). I. Abgrenzung der Gattung *Lepus*. II. Der taxonomische Status von *Lepus brachyurus* Temminck und *Lepus mandshuricus* Radde. Mitteilungen aus dem Zoologische Museum in Berlin, 42:127-335.

Angermann, R. 1967. Beiträge zur Kenntnis der Gattung *Lepus* (Lagomorpha, Leporidae). III. Zur Variabilitat palaearktischer Schneehasen. IV. *Lepus yarkandensis* Günther, 1875 und *Lepus oiostolus* Hodgson, 1840 — zwei endemische Hasenarten Zentralasiens. Mitteilungen aus dem Zoologische Museum in Berlin, 43:64-203.

Angermann, R. 1972. Hares, rabbits, and pikas. Pp. 419-462, *in* Grzimek's Animal Life Encyclopedia (B. Grzimek, ed.). Van Nostrand Reinhold Company, New York, NY, 12(Mammals III):1-657.

Angermann, R. 1983. The taxonomy of Old World *Lepus*. Acta Zoologica Fennica, 174:17-21.

Angermann, R. 1984. Intraspezifische Variabilitat der Molarenmuster bei der Nordischen Wühlmaus (*Microtus oeconomus* [Pallas, 1776]) (Mammalia, Rodentia, Microtinae). Zoologische Abhandlungen Staatliches Museum für Tierkunde in Dresden, 39:115-136.

Angermann, R., and A. Feiler. 1988. Zur Nomenklatur, Artabgrenzung und Variabilität der Hasen (Gattung *Lepus*) im westlichen Africa (Mammalia, Lagomorpha, Leporidae). Zoologische Abhandlungen Staatliches Museum für Tierkunde in Dresden, 43:149-167.

Anon. 1803-1804. Nouveau dictionnaire d'histoire naturelle, appliquèe aux art, principalement à l'agriculture et à l'economie rurale et domestique; par une société de naturalistes et d'agriculteurs: avec des figures tirees des trois regnes de la nature. Deterville, Paris, 24 volumes.

Ansell, W. F. H. 1972. Order Artiodactyla. Part 15. Pp. 1-84, *in* The mammals of Africa: An identification manual (J. Meester and H. W. Setzer, eds.) [issued 2 May 1972]. Smithsonian Institution Press, Washington, D. C., not continuously paginated.

Ansell, W. F. H. 1974*a*. Order Perissodactyla. Part 14. Pp. 1-14, *in* The mammals of Africa: An identification manual (J. Meester and H. W. Setzer, eds.) [issued 10 Sep 1974]. Smithsonian Institution Press, Washington, D. C., not continuously paginated.

Ansell, W. F. H. 1974*b*. Some mammals from Zambia and adjacent countries. Puku, supplement, 1:1-49.

Ansell, W. F. H. 1978. The mammals of Zambia. National Parks and Wildlife Service, Chilanga, 126 pp.

Ansell, W. F. H. 1980. *Antilope zebra* Gray, 1838 (Mammalia) revised proposals for conservation. Z. N. (S.) 1908. Bulletin of Zoological Nomenclature, 37:152-153.

Ansell, W. F. H. 1982. The southeastern range limit of *Manis tricuspis* Rafinesque, and the type of *Manis tridentata* Focillon. Mammalia, 46(4):559-560.

Ansell, W. F. H. 1989*a*. African mammals, 1938-1988. Trendrine Press, Cornwall, 77 pp.

Ansell, W. F. H. 1989*b*. Mammals from Malawi: Part II. Nyala, 13(1/2):41-65.

Ansell, W. F. H., and P. D. H. Ansell. 1973. Mammals of the north-eastern montane areas of Zambia. Puku, 7:21-69.

Ansell, W. F. H., and R. J. Dowsett. 1988. Mammals of Malawi: An annotated check list and atlas. Trendrine Press, Zennor, St. Ives, United Kingdom, 170 pp.

Ansell, W. F. H., and R. J. Dowsett. 1991. The type locality of *Otomys angoniensis* Wroughton, 1906. Bonner Zoologische Beiträge, 42:17-19.

Anthony, H. E. 1941. Mammals collected by the Vernay-Cutting Burma Expedition. Field Museum of Natural History, Zoological Series, 27:37-123.

Apfelbaum, L. I., and O. A. Reig. 1989. Allozyme genetic distances and evolutionary relationships in species of akodontine rodents (Cricetidae: Sigmodontinae). Biological Journal of the Linnean Society, 38:257-280.

Aplin, K. P., and M. Archer. 1987. Recent advances in marsupial systematics with a new syncretic classification. Pp. xv-lxxii, *in* Possums and opossums: studies in evolution (M. Archer, ed.). Surrey Beatty and Sons Pty. Ltd. and Royal Zoological Society of New South Wales, Sydney, 1:1-400.

Arai, S., T. Mori, H. Yoshida, and S. Shiraishi. 1985. A note on the Japanese water shrew, *Chimarrogale himalayica platycephala*, from Kyushu. Journal of the Mammalogical Society of Japan, 10:193-203.

Arata, A. 1964. The anatomy and taxonomic significance of the male accessory reproductive glands of muroid rodents. Bulletin of the Florida State Museum, Biological Sciences, 9:1-42.

Archer, M. 1975. *Ningaui*, a new genus of tiny dasyurids (Marsupialia) and two new species, *N. timealeyi* and *N. ridei* from arid Western Australia. Memoirs of the Queensland Museum, 17(2):237-249.

Archer, M. 1976. Revision of the marsupial genus *Planigale* Troughton (Dasyuridae). Memoirs of the Queensland Museum, 17:341-365.

Archer, M. 1977. Revision of the dasyurid marsupial genus *Antechinomys* Krefft. Memoirs of the Queensland Museum, 18:17-29.

Archer, M. 1979. Two new species of *Sminthopsis* Thomas from northern Australia, *S. butleri* and *S. douglasi*. Australian Museum Zoologist, 20:327-345.
Archer, M. 1981. Results of the Archbold Expeditions. No. 104. Systematic revision of the marsupial dasyurid genus *Sminthopsis* Thomas. Bulletin of the American Museum of Natural History, 168:61-224.
Archer, M. 1982. Review of the dasyurid (Marsupialia) fossil record, integration of data bearing on phylogenetic interpretation, and suprageneric classification. Pp. 397-443, *in* Carnivorous Marsupials (M. Archer, ed). Royal Society of New South Wales, Syndey, 2:1-804.
Archer, M. 1984. The Australian Marsupial Radiation. Pp. 633-808, *in* Vertebrate Zoogeography and Evolution in Austrlia (M. Archer and G. Clayton, eds.). Hesperian Press, Carlisle, W. A., Australia, 1203 pp.
Archer, M., and A. A. Bartholomai. 1978. Tertiary mammals of Australia: a synoptic review. Alcheringa, 2:1-19.
Archer, M., and J. A. W. Kirsch. 1977. The case for the Thylacomyidae and Myrmecobiidae, Gill, 1872, or why are marsupial families so extended. Proceedings of the Linnean Society of New South Wales, 102:18-25.
Archer, M., G. Clayton, and S. Hand. 1984. A checklist of Australasian fossil mammals. Pp. 1027-1087, *in* Vertebrate zoogeography and evolution in Australasia. (Animals in space and time) (M. Archer and G. Clayton, eds.). Hesperian Press, Carlisle, W. A., Australia, 1203 pp.
Argyropulo, A. I. 1948. Obzor retsentnykh vidov cem. Lagomyidae Lilljeb., 1886 (Lagomorpha, Mammalia) [A review of Recent species of the family . . .]. Trudy Zoologicheskovo Instituta, Akademiya Nauk, Leningrad, 7:124-128 (in Russian).
Arita, H. T., and S. R. Humphrey. 1988 [1989]. Revisión taxonómica de los murciélagos magueyeros del género *Leptonycteris* (Chiroptera: Phyllostomidae). Acta Zoológica Mexicana, Nueva Serie, 29:1-60.
Armstrong, D. M. 1972. Distribution of mammals in Colorado. Monograph of the Museum of Natural History, The University of Kansas, 3:1-415.
Armstrong, D. M., and J. K. Jones, Jr. 1971*a*. Mammals from the Mexican state of Sinaloa, 1. Marsupialia, Insectivora, Edentata, Lagomorpha. Journal of Mammalogy, 52:747-757.
Armstrong, D. M., and J. K. Jones, Jr. 1971*b*. *Sorex merriami*. Mammalian Species, 2:1-2.
Armstrong, D. M., and J. K. Jones, Jr. 1972*a*. *Megasorex gigas*. Mammalian Species, 16:1-2.
Armstrong, D. M., and J. K. Jones, Jr. 1972*b*. *Notiosorex crawfordi*. Mammalian Species, 17:1-5.
Arnold, P., H. Marsh, and G. Heinsohn. 1987. The occurrence of two forms of minke whales in east Australian waters with a description of external characters and skeleton of the diminutive or dwarf form. Scientific Reports of the Whales Research Institute (Toyko), 38:1-46.
Arredondo, O. 1958. Aves gigantes de nuestra pasado prehistórico. El Cartero Cubano, La Habana, 17(7):10-12.
Arroyo, J., C. Rodriguez-Murcia, M. Delibes, and F. Hiraldo. 1982. Comparative karyotype studies between Spanish and French populations of *Eliomys quercinus* L. Genetica, 59:161-166.
Arroyo-Cabrales, J., and J. K. Jones, Jr. 1988*a*. *Balantiopteryx plicata*. Mammalian Species, 301:1-4.
Arroyo-Cabrales, J., and J. K. Jones, Jr. 1988*b*. *Balantiopteryx io* and *Balantiopteryx infusca*. Mammalian Species, 313:1-3.
Arroyo-Cabrales, J., R. R. Hollander, and J. K. Jones, Jr. 1987. *Choeronycteris mexicana*. Mammalian Species, 291:1-5.
Aswathanarayana, N. V. 1987. Evolutionary trends in the genus *Funabulus* (Rodentia—Sciuridae). Current Science (Bangalore), 56(24):1298-1301.
Atallah, S. I. 1967. A new species of spiny mouse (*Acomys*) from Jordan. Journal of Mammalogy, 48:258-261.
Atallah, S. I. 1978. Mammals of the eastern Mediterranean region: their ecology, systematics and zoogeographical relationships. Part 2. Säugetierkundliche Mitteilungen, 26:1-50.
Atanassov, N., and Z. Peschev. 1963. Die Säugetiere Bulgariens. Säugetierkundliche Mitteilungen, 12:101-112.
Atkins, D. L., and L. S. Dillon. 1971. Evolution of the cerebellum in the genus *Canis*. Journal of Mammalogy, 52:96-107.
Audubon, J. J., and J. Bachman. 1846 [1845]-1854. The viviparous quadrupeds of North America. Privately published by J. J. Audubon, New York, 1:1-389[1846], 2:1-334[1851], 3:1-348[1854].
Auffray, J. -C., and J. Britton-Davidian. 1992. When did the house mouse colonize Europe? Biological Journal of the Linnean Society, 45:187-190.
Auffray, J. -C., E. Tchernov, and E. Nevo. 1988. Origine du commensalisme de la souris domestique (*Mus musculus domesticus*) vis-à-vis de l'homme. Comptes Rendus de l'Académie des Sciences (Paris), ser. 3, 307(9):517-522.
Auffray, J. -C., K. Belkhir, J. Cassaing, J. Britton-Davidian, and H. Croset. 1990*a*. Outdoor occurrence in robertsonian and standard populations of the house mouse. Vie et Milieu, 40:111-118.

Auffray, J. -C., E. Tchernov, F. Bonhomme, G. Heth, S. Simson, and E. Nevo. 1990*b*. Presence and ecological distribution of *Mus "spretoides"* and *Mus musculus domesticus* in Israel. Circum-Mediterranean vicariance in the genus *Mus*. Zeitschrift für Säugetierkunde, 55:1-10.

Auffray, J. -C., F. Vanlerberghe, and J. Britton-Davidian. 1990*c*. The house mouse progression in Eurasia: a palaeontological and archaeozoological approach. Biological Journal of the Linnean Society, 41:13-25.

Aulagnier, S., and M. Thévenot. 1987. Catalogue des mammifères sauvages du Maroc. Travaux de l'Institut Scientifique, Série Zoologie, 41:1-164.

Avery, D. M. 1991. Late Quaternary incidence of some micromammalian species in Natal. Durban Museum Novitates, 16:1-11.

Avila-Pires, F. D., de. 1960*a*. Sobre *Oryzomys* do grupo *ratticeps*. Actas y Trabajos del Primer Congreso Sudamerican de Zoologia, sec. 5, 4:3-7.

Avila-Pires, F. D., de. 1960*b*. Um novo genero de roedor Sul-Americano. Boletim do Museu Nacional, Zoologia (Rio de Janeiro), 220:1-6.

Avila-Pires, F. D., de. 1964. Mamíferos colecionados na região do Rio Negro (Amazonas, Brasil). Boletim do Museu Paraense Emílio Goeldi, Nova Série, Zoologia, 42:1-23.

Avila-Pires, F. D., de. 1965. The type specimens of Brazilian mammals collected by Prince Maximilian zu Wied. American Museum Novitates, 2209:1-21.

Avila-Pires, F. D., de. 1967. The type-locality of "*Chaetomys subspinosus*" (Olfers, 1818) (Rodentia, Caviomorpha). Revista Brasileira de Biologia, 27(2):177-179.

Avila-Pires, F. D., de. 1972. A new subspecies of *Kunsia fronto* (Winge, 1888) from Brazil (Rodentia, Cricetidae). Revista Brasileira de Biologia, 32:419-422.

Avila-Pires, F. D., de. 1985. On the validity and geographical distribution of *Callithrix argentata emiliae* Thomas, 1920 (Primates, Callithrichidae). Pp. 319-322, *in* A Primatologia no Brasil (M. T. de Mello, ed.). Instituto de Ciências Biológicas, Brasil, 530 pp.

Avila-Pires, F. D., de, and M. R. C. Wutke. 1981. Taxonomia e evoluçao de *Clyomys* Thomas, 1916 (Rodentia, Echimyidae). Revista Brasileira de Biologia, 41(3):529-534.

Avise, J. C., M. H. Smith, and R. K. Selander. 1974*a*. Biochemical polymorphism and systematics in the genus *Peromyscus*. VI. The *boylii* species group. Journal of Mammalogy, 55:751-763.

Avise, J. C., M. H. Smith, R. K. Selander, T. E. Lawlor, and P. R. Ramsey. 1974*b*. Biochemical polymorphism and systematics in the genus *Peromyscus*. V. Insular and mainland species of the subgenus *Haplomylomys*. Systematic Zoology, 23:226-238.

Avise, J. C., M. H. Smith, and R. K. Selander. 1979. Biochemical polymorphism and systematics in the genus *Peromyscus*. VII. Geographic differentiation in members of the *truei* and *maniculatus* species groups. Journal of Mammalogy, 60:177-192.

Azara, F., de. 1801. Essais sur l'histoire naturelle des quadrupèdes de la province du Paraguay. Traduits sur le manuscrit inédit de l'auteur, Pra. M. L. E. Moreau-Saint-Méry. Charles Pougens, Paris, 1:1-366; 2:1-499.

Azara, F., de. 1802. Apuntamientos para la historia natural de los quadrupedos del Paraguay, y Rio de la Plata. En la imprenta de la viuda de Ibarra, Madrid, 1:1-318; 2:1-328.

Azzaroli-Puccetti, M. L. 1987*a*. On the hares of Ethiopia and Somalia and the systematic position of *Lepus whytei* Thomas, 1894 (Mammalia, Lagomorpha). Atti della Accademia Nazionale dei Lincei. Memorie, Classe di scienze fisiche, matematiche e naturali, ser. 8, vol. 19, sec. 3a, fasc. 1:1-19+ 5 pls.

Azzaroli-Puccetti, M. L. 1987*b*. The systematic relationships of hares (genus *Lepus*) of the horn of Africa. Cimbebasia, ser. A, 9:1-22.

Badr, F. M., and R. L. Asker. 1980. Prevalence of non-Robertsonian polymorphism in the gerbil *Gerbillus cheesmani* from Kuwait. Genetica, 52/53:17-22.

Bailey, V. 1900. Revision of American voles of the genus *Microtus*. North American Fauna, 17:1-88.

Bailey, V. 1902. Synopsis of the North American species of *Sigmodon*. Proceedings of the Biological Society of Washington, 15:101-116.

Bailey, V. 1906. Identity of *Thomomys umbrinus* (Richardson). Proceedings of the Biological Society of Washington, 19:3-6.

Bailey, V. 1915. Revision of the pocket gophers of the genus *Thomomys*. North American Fauna, 39:1-136.

Bailey, V. 1936. The mammals and life zones of Oregon. North American Fauna, 55:1-416.

Baird, S. F. 1857. Mammals: general report upon the zoology of the several Pacific railroad routes. Vol. 8, pt. 1, *in* Reports of explorations and surveys to ascertain the most practicable and economical route for a railroad from the Mississippi River to the Pacific Ocean. Senate executive document no. 78, Washington, D.C., 757 pp.

Baker, A. J., J. L. Eger, R. L. Peterson, and T. H. Manning. 1983. Geographic variation and taxonomy of arctic hares. Acta Zoologica Fennica, 174:45-48.

Baker, A. N. 1977. Spectacled porpoise *Phocoena dioptrica* new to the subantarctic Pacific Ocean. New Zealand Journal of Marine and Freshwater Research, 11(2):401-406.

Baker, A. N. 1985. Pygmy right whale - *Caperea marginata*. Pp. 345-354, *in* Handbook of marine mammals: The sirenians and baleen whales, (S. H. Ridgway and R. Harrison, eds.). Academic Press, London, 3:1-362.

Baker, R. H. 1952. Geographic range of *Peromyscus melanophrys*, with description of a new subspecies. University of Kansas Publications, Museum of Natural History, 5:251-258.

Baker, R. H. 1954. The silky pocket mouse (*Perognathus flavus*) of México. University of Kansas Publications, Museum of Natural History, 7:339-347.

Baker, R. H. 1969. Cotton rats of the *Sigmodon fulviventer* group. Miscellaneous Publications, Museum of Natural History, University of Kansas, 51:177-232.

Baker, R. H., and K. A. Shump. 1978*a*. *Sigmodon fulviventer*. Mammalian Species, 94:1-4.

Baker, R. H., and K. A. Shump. 1978*b*. *Sigmodon ochrognathus*. Mammalian Species, 97:1-2.

Baker, R. J. 1984. A sympatric cryptic species of mammal: A new species of *Rhogeessa* (Chiroptera: Vespertilionidae). Systematic Zoology, 33:178-183.

Baker, R. J., and C. L. Clark. 1987. *Uroderma bilobatum*. Mammalian Species, 279:1-4.

Baker, R. J., and S. L. Williams. 1974. *Geomys tropicalis*. Mammalian Species, 35:1-4.

Baker, R. J., P. V. August, and A. A. Steuter. 1978. *Erophylla sezekorni*. Mammalian Species, 115:1-5.

Baker, R. J., R. K. Barnett, and I. F. Greenbaum. 1979. Chromosomal evolution in grasshopper mice (*Onychomys*, Cricetidae). Journal of Mammalogy, 60:297-306.

Baker, R. J., B. F. Koop, and M. W. Haiduk. 1983*a*. Resolving systematic relationships with G-bands: a study of five genera of South American cricetine rodents. Systematic Zoology, 32:403-416.

Baker, R. J., L. W. Robbins, F. B. Stangl, Jr., and E. C. Birney. 1983*b*. Chromosomal evidence for a major subdivision in *Peromyscus leucopus*. Journal of Mammalogy, 64:356-359.

Baker, R. J., J. C. Patton, H. H. Genoways, and J. W. Bickham. 1988*a*. Genic studies of *Lasiurus* (Chiroptera: Vespertilionidae). Occasional Papers, The Museum, Texas Tech University, 117:1-15.

Baker, R. J., C. G. Dunn, and K. Nelson. 1988*b*. Allozymic study of the relationships of *Phylloderma* and four species of *Phyllostomus*. Occasional Papers, The Museum, Texas Tech University, 125:1-14.

Baker, R. J., M. B. Qumsiyeh, and I. L. Rautenbach. 1988*c*. Evidence for eight tandem and five centric fusions in the evolution of the karyotype of *Aethomys namaquensis* A. Smith (Rodentia: Muridae). Genetica, 76:161-169.

Balcomb, K. C. 1989. Baird's beaked whale, *Berardius bairdii* Stejneger, 1883: Arnoux's beaked whale *Berardius arnouxii* Duvernoy, 1851. Pp. 261-288, *in* Handbook of marine mammals: River dolphins and the larger toothed whales (S. H. Ridgway and R. Harrison, eds.). Academic Press, New York, 4:1-442.

Baloutch, M. 1978. Etude d'une collection de lièvres d'Iran et description de quatre formes nouvelles. Mammalia, 42:441-452.

Banerjee, S., and M. Ghosh. 1981. Prehistoric fauna of Kausarabi, near Allahabad, U. P., India. Records of the Zoological Survey of India, 78:113-119.

Banfield, A. W. F. 1961. A revision of the reindeer and caribou. National Museum of Canada Bulletin, 177:1-137.

Banfield, A. W. F. 1974. The mammals of Canada. The University of Toronto Press, Toronto, 438 pp.

Bangs, O. 1902. Chiriqui Mammalia. Bulletin of the Museum of Comparative Zoology at Harvard College, 39:17-51.

Banks, R. C. 1967. The *Peromyscus guardia-interparietalis* complex. Journal of Mammalogy, 48:210-218.

Bannikov, A. G. 1954. Mlekopitayushchie Mongol'skoi Narodnoi Respubliki. [Mammals of the Mongolian People's Republic]. Akademiya Nauk SSSR, Moscow, 669 pp. (in Russian).

Barabasch-Nikiforof, I. I. 1975. Die Desmane. A. Ziemsen Verlag, Wittenberg-Lutherstadt, 100 pp.

Baranova, G. I., A. A. Gureev, and P. P. Strelkov. 1981. [Catalogue of type specimens in the collection of the Zoological Institute of the USSR, mammals (Mammalia). Part 1. shrews (Insectivora), bats (Chiroptera), hares (Lagomorpha)]. Nauka, Leningrad, 22 pp.

Barbehenn, K. R., J. P. Sumangil, and J. L. Libay. 1972-1972. Rodents of the Philippine croplands. Philippine Agriculturist, 56:217-242.

Barghorn, S. F. 1977. New material of *Vespertiliavus* Schlosser (Mammalia: Chiroptera) and suggested relationships. American Museum Novitates, 2618:1-29.

Barnes, L. G. 1978. A review of *Lophocetus* and *Liolithax* and their relationships to the delphinoid family Kentriodontidae (Cetacea: Odontoceti). Bulletin of the Los Angeles County Museum of Natural History, 28:1-35.

Barnes, L. G. 1985. Evolution, taxonomy and antitropical distribution of the porpoises (Phocoenidae, Mammalia). Marine Mammal Science, 1(2):149-163.

Barnes, L. G. 1989. A new Enaliarctine pinniped from the Astoria formation, Oregon, and a classification of the Otariidae (Mammalia: Carnivora). Contributions in Science, Natural History Museum of Los Angeles County, 403:1-26.

Barnes, L. G., and S. A. McLeod. 1984. The fossil record and phyletic relationships of gray whales. Pp. 3-32, *in* The gray whale (M. L. Jones, S. L. Swartz, and S. Leatherwood, eds.). Academic Press, New York, 600 pp.

Barnes, L. G., D. P. Domning, and C. E. Ray. 1985. Status of studies on fossil marine mammals. Marine Mammal Science, 1(1):15-53.

Bárquez, R. M. 1983. La distribucion de *Neotomys ebriosus* Thomas en la Argentina y su presencia en la provincia de San Juan (Mammalia, Rodentia, Cricetidae). Historia Natural, 3:189-191.

Bárquez, R. M., and R. A. Ojeda. 1979. Nueva subespecie de *Phylloderma* (Chiroptera, Phyllostomidae). Neotropica, 25:83-89.

Bárquez, R. M., D. F. Williams, M. A. Mares, and H. H. Genoways. 1980. Karyology and morphometrics of three species of *Akodon* (Mammalia: Muridae) from northwestern Argentina. Annals of Carnegie Museum, 49:379-403.

Barrett-Hamilton, G. E. H. 1902. Antarctic Mammalia. Report of the Southern Cross Collection. British Museum (Natural History), London, 66 pp.

Barros, M. A., O. A. Reig, and A. Perez-Zapata. 1992. Cytogenetics and karyosystematics of South American oryzomyine rodents (Cricetidae: Sigmodontinae). IV. Karyotypes of Venezuelan, Trinidadian, and Argentinian water rats of the genus *Nectomys*. Cytogenetics and Cell Genetics, 59:34-38.

Baryshnikov, G. F., and O. R. Potapova. 1990. [Variability of the dental system in badgers (*Meles*, Carnivora) in the USSR fauna]. Zoologicheskii Zhurnal, 69:84-97 (in Russian).

Baskevich, M. I., I. V. Lukyanova, and Yu. M. Koval'skaya. 1984. [Distribution of two forms of bush voles (*P. majori* Thom. and *P. dagestanicus* Shidl.) in the Caucasus]. Byulleten' Moskovskovo Obshchestva Ispytatelei Prirody, Otdel Biologicheskii, 89:29-33 (in Russian).

Baskin, J. A. 1978. *Bensonomys*, *Calomys*, and the origin of the phyllotine group of neotropical cricetines (Rodentia: Cricetidae). Journal of Mammalogy, 59:125-135.

Baskin, J. A. 1982. Tertiary Procyoninae (Mammalia: Carnivora) of North America. Journal of Vertebrate Paleontology, 2:71-93.

Baskin, J. A. 1986. The late Miocene radiation of neotropical sigmodontine rodents in North America. Contributions to Geology, University of Wyoming, Special Paper, 3:287-303.

Baskin, J. A. 1989. Comments on New World Tertiary Procyonidae (Mammalia: Carnivora). Journal of Vertebrate Paleontology, 9:110-117.

Bates, P. J. J. 1985. Studies of gerbils of genus *Tatera*: the specific distinction of *Tatera robusta* (Cretzschmar, 1826), *Tatera nigricauda* (Peters, 1878) and *Tatera phillipsi* (De Winton, 1898). Mammalia, 49:37-52.

Bates, P. J. J. 1988. Systematics and zoogeography of *Tatera* (Rodentia: Gerbillinae) of north-east Africa and Asia. Bonner Zoologische Beiträge, 39:265-303.

Bauchau, V., and J. Chaline. 1987. Variabilite de la troisieme molaire superieure de *Clethrionomys glareolus* (Arvicolidae, Rodentia) et sa signification evolutive. Mammalia, 51:587-598.

Bauer, K. 1960. Die Säugetiere des Neusiedlersee-Gebietes (Österreich). Bonner Zoologische Beiträge, 11:141-344.

Baumgardner, G. D. 1991. *Dipodomys compactus*. Mammalian Species, 369:1-4.

Baumgardner, G. D., and D. J. Schmidly. 1981. Systematics of the southern races of two species of kangaroo rats (*Dipodomys compactus* and *D. ordii*). Occasional Papers, The Museum, Texas Tech University, 73:1-27.

Baverstock, P. R. 1984. Australia's living rodents: a restrained explosion. Pp. 913-919, *in* Vertebrate zoogeography and evolution in Australasia. (Animals in space and time) (M. Archer and G. Clayton, eds.). Hesperian Press, Carlisle, W. A., Australia, 1203 pp.

Baverstock, P. R., J. T. Hogarth, S. Cole, and J. Covacevich. 1976*a*. Biochemical and karyotypic evidence for the specific status of the rodent *Leggadina lakedownensis* Watts. Transactions of the Royal Society of South Australia, 100:109-112.

Baverstock, P. R., C. H. S. Watts, and J. T. Hogarth. 1976*b*. Heterochromatin variation in the Australian rodent *Uromys caudimaculatus*. Chromosoma, 57:397-403.

Baverstock, P. R., C. H. S. Watts, and S. R. Cole. 1977*a*. Electrophoretic comparisons between allopatric populations of five Australian pseudomyine rodents (Muridae). Australian Journal of Biological Science, 30:471-485.

Baverstock, P. R., C. H. S. Watts, and S. R. Cole. 1977*b*. Inheritance studies of glucose phosphate isomerase, transferrin and esterases in the Australian hopping mice *Notomys alexis*, *N. cervinus*, *N. mitchellii*, and *N. fuscus* (Rodentia: Muridae). Animal Blood Groups and Biochemical Genetics, 8:3-12.

Baverstock, P. R., C. H. S. Watts, and J. T. Hogarth. 1977*c*. Chromosome evolution in Australian rodents. I. The Pseudomyinae, the Hydromyinae and the *Uromys*/*Melomys* group. Chromosoma, 61:95-125.

Baverstock, P. R., C. H. S. Watts, J. T. Hogarth, A. C. Robinson, and J. F. Robinson. 1977*d*. Chromosome evolution in Australian rodents. II. The *Rattus* group. Chromosoma, 61:227-241.

Baverstock, P. R., C. H. S. Watts, and J. T. Hogarth. 1977*e*. Polymorphic patterns of heterochromatin distribution in Australian hopping mice, *Notomys alexis*, *N. cervinus* and *N. fuscus* (Rodentia, Muridae). Chromosoma, 61:243-256.

Baverstock, P. R., C. H. S. Watts, M. Adams, and M. Gelder. 1980. Chromosomal and electrophoretic studies of Australian *Melomys* (Rodentia: Muridae). Australian Journal of Zoology, 28:553-574.

Baverstock, P. R., C. H. S. Watts, M. Adams, and S. R. Cole. 1981. Genetical relationships among Australian rodents (Muridae). Australian Journal of Zoology, 29:289-303.

Baverstock, P. R., M. Gelder, and A. Jahnke. 1982. Cytogenetic studies of the Australian rodent, *Uromys caudimaculatus*, a species showing extensive heterochromatin variation. Chromosoma, 84:517-533.

Baverstock, P. R., M. Gelder, and A. Jahnke. 1983*a*. Chromosome evolution in Australian *Rattus*—G-banding and hybrid meiosis. Genetica, 60:93-103.

Baverstock, P. R., C. H. S. Watts, M. Gelder, and A. Jahnke. 1983*b*. G-banding homologies of some Australian rodents. Genetica, 60:105-117.

Baverstock, P. R., M. Adams, L. R. Maxson, and T. H. Yosida. 1983*c*. Genetic differentiation among karyotypic forms of the black rat, *Rattus rattus*. Genetics, 105:969-983.

Baverstock, P. R., M. Adams, and C. H. S. Watts. 1986. Biochemical differentiation among karyotypic forms of Australian *Rattus*. Genetica, 71:11-22.

Baynes, A., A. Chapman, and A. J. Lynam. 1987. The rediscovery, after 56 years, of the heath rat *Pseudomys shortridgei* (Thomas, 1907)(Rodentia: Muridae) in Western Australia. Records of the Western Australian Museum, 13:319-322.

Bazhanov, V. S. 1944. [Marmot hybrids (On the question of interspecific hybridization in nature)]. Doklady Akademii Nauk SSSR, 42(7):307-308 (in Russian).

Bazhanov, V. S. 1951. [Some peculiariries of rodents belonging to the family Seleveniidae and endemic to Kazakhstan]. Doklady Akademii Nauk SSSR, Moscow, 80(3):469-472 (in Russian).

Bazhanov, V. S., and B. A. Belosludov. 1941. A remarkable family of rodents from Kasakhstan, USSR. Journal of Mammalogy, 22(3):311-315.

Beaubrun, P.-Ch. 1990. Un cétacé nouveau pour les côtes sud-marocaines: *Sousa teuzii* (Kükenthal, 1892). Mammalia, 54(1):162-164.

Beaufort, F., de. 1963. Les cricetines de Galapagos. Valeur du genre *Nesoryzomys*. Mammalia, 27:338-340.

Beaufort, F., de. 1965. Répartition et taxinomie de la Poiane (Viverridae). Mammalia, 29(2):275-280.

Beaumont, G., de. 1964. Remarques sur la classification des Felidae. Eclogae Geologicae Helvetiae, 57:837-845.

Bechthold, G. 1939. Die asiatischen Formen der Gattung *Herpestes*. Zeitschrift für Säugetierkunde, 14:113-219.

Becker, K. 1978*a*. *Rattus rattus* (Linnaeus, 1758)—Hausratte (HR). Pp. 382-400, *in* Handbuch der Säugetiere Europas (J. Niethammer and F. Krapp, eds.). Akademische Verlagsgesellschaft (Wiesbaden), 1:1-476.

Becker, K. 1978*b*. *Rattus norvegicus* (Berkenhout, 1769)—Wanderratte (WR). Pp. 401-420, *in* Handbuch der Säugetiere Europas (J. Niethammer and F. Krapp, eds.). Akademische Verlagsgesellschaft (Wiesbaden), 1:1-476.

Beckoff, M. 1977. *Canis latrans*. Mammalian Species, 79:1-9.

Becsy, L. 1982. [Protected animals in Hungary, *Glis glis*]. Buvar, 37(4):169 (in Hungarian).

Bee, J. W., and E. R. Hall. 1956. Mammals of northern Alaska on the Arctic slope. Miscellaneous Publications, Museum of Natural History, University of Kansas, 8:1-309.

Bekasova, T. S., and O. N. Mezhova. 1983. Karyotypes of rats of the genus *Rattus* from the USSR. Experientia, 39:541-542.

Bekasova, T. S., N. N. Vorontsov, K. V. Korobitsyna, and V. P. Korablev. 1980. B-chromosomes and comparative karyology of the mice of the genus *Apodemus*. Genetica, 52/53:33-43.

Bekele, A. 1986. The status of some mole-rats of the genus *Tachyoryctes* (Rodentia: Rhizomyidae) based upon craniometric studies. Revue de Zoologie Africaine, 99:411-417.

Bekele, A., and D. A. Schlitter. 1989. Two new records of rodents from Kenya and Ethiopia. Mammalia, 53:113-116.

Belcheva, R. G., M. N. Topashka-Ancheva, and N. I. Atanassov. 1989. Karyological studies of five species of mammals from Bulgaria's fauna. Comptes Rendus de l'Academie Bulgare des Sciences, 42(2):125-138.

Belcheva, R., and D. Peshev. 1985. Consitutive heterochromatin in the ground squirrel *Citellus citellus* (Sciuridae, Rodentia) from Bulgaria. Zoologischer Anzeiger, 215:385-390.

Belk, M. C., and H. D. Smith. 1990. *Ammospermophilus leucurus*. Mammalian Species, 368:1-8.

Bell, D. J., W. L. R. Oliver, and R. K. Ghose. 1990. The hispid hare *Caprolagus hispidus*. Pp. 128-136, *in* Rabbits, hares and pikas (J. A. Chapman and J. E. C. Flux, eds.). I. U. C. N., Gland, Switzerland, 168 pp.

Belosludov, B. A., and V. S. Bazhanov. 1939. [A new genus and species of rodent from the central Kazakhstan (USSR)]. Uchenye Zapiski Kazakhskovo Gosudarstvennovo Universiteta, Alma-Ata, 1(1):81-86 (in Russian).

Benazzou, T. 1983. Le caryotype d'*Acomys chudeaui* capture dans la region de Tata (Maroc). Mammalia, 47:588.

Benazzou, T., and F. Zyadi. 1990. Presence d'une variabilite biometrique chez *Gerbillus campestris* au Maroc (rongeurs, gerbillides). Mammalia, 54:271-279.

Benazzou, T., E. Viegas-Pequignot, F. Petter, and B. Dutrillaux. 1982. Phylogenie chromosomique de quatre especies de *Meriones* (Rongeur, Gerbillidae). Annales de Genetique, 25:19-24.

Benazzou, T., E. Viegas-Pequignot, M. Prod'Homme, M. Lombard, F. Petter, and B. Dutrillaux. 1984. Phylogenie chromosomique des Gerbillidae. III. Etude d'especes de *Tatera, Taterillus, Psammomys* et *Pachyuromys*. Annales de Genetique, 27:17-26.

Beneski, J. T., Jr., and D. W. Stinson. 1987. *Sorex palustris*. Mammalian Species, 296:1-6.

Bennett, D. K. 1980. Stripes do not a zebra make, part 1: A cladistic analysis of *Equus*. Systematic Zoology, 29:272-287.

Benson, D. L., and F. R. Gehlbach. 1979. Ecological and taxonomic notes on the rice rat (*Oryzomys couesi*) in Texas. Journal of Mammalogy, 60:225-228.

Bergmans, W. 1975. A new species of *Dobsonia* Palmer, 1898 (Mammalia, Megachiroptera) from Waigeo, with notes on other members of the genus. Beaufortia, 23(295):1-13.

Bergmans, W. 1976. A revision of the African genus *Myonycteris* Matschie, 1899 (Mammalia, Megachiroptera). Beaufortia, 24(317):189-216.

Bergmans, W. 1977. Notes on new material of *Rousettus madagascariensis* Grandidier, 1829 (Mammalia, Megachiroptera). Mammalia, 41:67-74.

Bergmans, W. 1978. On *Dobsonia* Palmer 1898 from the Lesser Sunda Islands (Mammalia, Megachiroptera). Senckenbergiana Biologica, 59:1-18.

Bergmans, W. 1979. Taxonomy and zoogeography of *Dobsonia* Palmer, 1898, from the Louisiade Archipelago, the D'Entrecasteaux Group, Trobriand Island and Woodlark Island (Mammalia, Megachiroptera). Beaufortia, 29(355):199-214.

Bergmans, W. 1988. Taxonomy and biogeography of African fruit bats (Mammalia, Megachiroptera). I. General introduction; materials and methods; results: the genus *Epomophorus* Bennett, 1836. Beaufortia, 38(5):75-146.

Bergmans, W. 1989. Taxonomy and biogeography of African fruit bats (Mammalia: Megachiroptera). 2. The genera *Micropteropus* Matschie, 1899, *Epomops* Gray, 1870, *Hypsignathus* H. Allen, 1861, *Nanonycteris* Matschie, 1899, and *Plerotes* Andersen, 1910. Beaufortia, 39(4):89-153.

Bergmans, W. 1990. Taxonomy and biogeography of African fruit bats (Mammalia: Megachiroptera). 3. The genera *Scotonycteris* Matschie, 1894, *Casinycteris* Thomas, 1910, *Pteropus* Brisson, 1762, and *Eidolon* Rafinesque, 1815. Beaufortia, 40(7):111-177.

Bergmans, W., and F. G. Rozendaal. 1988. Notes on collections of fruit bats from Sulawesi and some off-lying Islands (Mammalia; Megachiroptera). Zoologische Verhandelingen (Leiden), 248:1-74.

Bergmans, W., and S. Sarbini. 1985. Fruit bats of the genus *Dobsonia* Palmer, 1898, from the islands of Biak, Owii, Numfoor and Yapen, Irian Jaya (Mammalia; Megachiroptera). Beaufortia, 34(6):181-189.

Bergmans, W., and P. J. H. van Bree. 1972. The taxonomy of the African bat *Megaloglossus woermanni* Pagenstecher, 1885 (Megachiroptera, Macroglossinae). Biologia Gabonica, 8:291-299.

Bergmans, W., and P. J. H. van Bree. 1986. On a collection of bats and rats from the Kangean Islands, Indonesia (Mammalia: Chiroptera and Rodentia). Zeitschrift für Säugetierkunde, 51:329-344.

Bergstrom, B. J., and R. S. Hoffmann. 1991. Distribution and diagnosis of three species of chipmunks (*Tamias*) in the Front Range of Colorado. Southwestern Naturalist, 36:14-28.

Bernard, R. T. F., A. H. Hodgson, J. Meester, K. Willan, and C. Bojarski. 1990. Sperm structure and taxonomic affinites of five African rodents of the subfamily Otomyinae (Muridae). Electron Microscope Society of Southern Africa, 20:161-162.

Berry, D. L., and R. J. Baker. 1972. Chromosomes of pocket gophers of the genus *Pappogeomys*, subgenus *Cratogeomys*. Journal of Mammalogy, 53:303-309.

Berry, R. J. 1973. Chance and change in British long-tailed field mice (*Apodemus sylvaticus*). Journal of the Zoological Society of London, 170:351-366.

Berta, A. 1982. *Cerdocyon thous*. Mammalian Species, 186:1-4.

Berta, A. 1986. *Atelocynus microtis*. Mammalian Species, 256:1-3.

Berta, A. 1987. Origin, diversification, and zoogeography of the South American Canidae. Pp. 455-471, *in* Studies in Neotropical Mammalogy: Essays in honor of Philip Hershkovitz (B. D. Patterson and R. M. Timm, eds.). Fieldiana, Zoology, New Series, 39(1382):1-506.

Berta, A. 1988. Quaternary evolution and biogeography of the large South American Canidae (Mammalia: Carnivora). University of California Publications, Geological Sciences, 132:1-149.

Berta, A. 1991. New *Enaliarctos* (Pinnipedimorpha) from the Oligocene and Miocene of Oregon and the role of "Enaliarctids" in pinniped phylogeny. Smithsonian Contributions to Paleobiology, 69:1-33.

Berta, A., and T. A. Demere. 1986. *Callorhinus gilmorei* n. sp., (Carnivora: Otariidae) from the San Diego formation (Blancan) and its implications for otariid phylogeny. Transactions of the San Diego Society of Natural History, 21:111-126.

Berta, A., and L. G. Marshall. 1978. Fossilium catalogus. I: South American Carnivora. Dr. W. Junk, The Hague, 125:1-48.

Best, R. C., and V. M. F. da Silva. 1989. Amazon river dolphin, Boto - *Inia geoffrensis* (de Blainville, 1817). Pp. 1-24, *in* Handbook of marine mammals: River dolphins and the larger toothed whales (S. H. Ridgway and R. Harrison, eds.). Academic Press, London, 4:1-442.

Best, T. L. 1978. Variation in kangaroo rats (genus *Dipodomys*) of the *heermanni* group in Baja California, Mexico. Journal of Mammalogy, 59:160-175.

Best, T. L. 1983*a*. Intraspecific variation in the agile kangaroo rat (*Dipodomys agilis*). Journal of Mammalogy, 64:426-436.

Best, T. L. 1983*b*. Morphologic variation in the San Quintín kangaroo rat (*Dipodomys gravipes* Huey 1925). American Midland Naturalist, 109:409-413.

Best, T. L. 1986. *Dipodomys elephantinus*. Mammalian Species, 255:1-4.

Best, T. L. 1988*a*. *Dipodomys spectabilis*. Mammalian Species, 311:1-10.

Best, T. L. 1988*b*. *Dipodomys nelsoni*. Mammalian Species, 326:1-4.

Best, T. L. 1991. *Dipodomys nitratoides*. Mammalian Species, 381:1-7.

Best, T. L. 1992. *Dipodomys margaritae*. Mammalian Species, 400:1-3.

Best, T. L., and J. A. Lackey. 1985. *Dipodomys gravipes*. Mammalian Species, 236:1-4.

Best, T. L., and G. D. Schnell. 1974. Bacular variation in kangaroo rats (genus *Dipodomys*). American Midland Naturalist, 91:257-270.

Best, T. L., and H. H. Thomas. 1991*a*. *Dipodomys insularis*. Mammalian Species, 374:1-3.

Best, T. L., and H. H. Thomas. 1991*b*. *Spermophilus madrensis*. Mammalian Species, 378:1-2.

Best, T. L., R. M. Sullivan, J. A. Cook, and T. L. Yates. 1986. Chromosomal, genic, and morphologic variation in the agile kangaroo rat, *Dipodomys agilis* (Rodentia: Heteromyidae). Systematic Zoology, 35:311-324.

Best, T. L., N. J. Hildreth, and C. Jones. 1989. *Dipodomys deserti*. Mammalian Species, 339:1-8.

Best, T. L., K. Caesar, A. S. Titus, and C. L. Lewis. 1990*a*. *Ammospermophilus insularis*. Mammalian Species, 364:1-4.

Best, T. L., C. L. Lewis, K. Caesar, and A. S. Titus. 1990*b*. *Ammospermophilus interpres*. Mammalian Species, 365:1-6.

Best, T. L., A. S. Titus, K. Caesar, and C. L. Lewis. 1990*c*. *Ammospermophilus harrisii*. Mammalian Species, 366:1-7.

Best, T. L., A. S. Titus, C. L. Lewis, and K. Caesar. 1990*d*. *Ammospermophilus nelsoni*. Mammalian Species, 367:1-7.

Bezrodny, S. V. 1991. Rasprostranenie son' (Rodentia, Gliridae) na Ukraine [The distribution of dormice (Rodentia) in the Ukraine]. Vestnik Zoologii, 1991(3):45-50 (in Russian).

Bianchi, N. O., O. A. Reig, J. Molina, and F. N. Dulout. 1971. Cytogenetics of South American akodont rodents (Cricetidae). I. A progress record of Argentinian and Venezuelan forms. Evolution, 21:724-736.

Bianchi, N. O., S. Merani, and M. Lizarralde. 1979. Cytogenetics of South American akodont rodents (Cricetidae). V. Segregation of chromosome Nr. 1 polymorphism in *Akodon molinae*. Experientia, 35:1438-1439.

Bianchini, J. J., and L. H. Delupi. 1978 [1979]. El estado sistematico de los ciervos neotropicales de la tribu Odocoileini Simpson 1945. Physis, Asociacion Argentina de Ciencias Naturales, 38(94), Seccion C, pp. 83-89.

Bibikov, D. I. 1991. The steppe marmot—its past and future. Oryx, 25:45-49.

Bielecka, K. 1986. Present record of fat dormouse *Glis glis* (Linnaeus, 1766) from Pomerania. Przeglad Zoologiczny, 30(1):115-117.

Bierman, W. H., and E. J. Slijper. 1947. Remarks upon the species of the genus *Lagenorhynchus*. I. Koninklijke Nederlandsche Akademie van Wetenschappen, Proceedings, L(10):1353-1364.

Bigalke, R. 1948. The type locality of the bontbok, *Damaliscus pygargus* (Pallas). Journal of Mammalogy, 29:421-422.

Birkenholz, D. E. 1972. *Neofiber alleni*. Mammalian Species, 15:1-4.

Birney, E. C. 1973. Systematics of three species of woodrats (genus *Neotoma*) in central North America. Miscellaneous Publications, Museum of Natural History, University of Kansas, 58:1-173.

Birney, E. C. 1976. An assessment of relationships and effects of interbreeding among woodrats of the *Neotoma floridana* species-group. Journal of Mammalogy, 57:103-132.
Birney, E. C., and H. H. Genoways. 1973. Chromosomes of *Spermophilus adocetus* (Mammalia, Sciuridae), with comments on the subgeneric affinities of the species. Experientia, 29:228-229.
Birula, A. 1913 [1914]. [Contribution à la synonymie de l'*Otocolobus manul* (Pallas) (Felidae)]. Annuaire du Musée Zoologique de l'Académie Impériale des Sciences de St. Pétersbourg, 18:LVII-LVIII.
Birula, A. 1916 [1917]. Contributions à la classification et à la distribution géographique des mammifères. VI. Sur la position de l'*Otocolobus manul* (Pallas) dans le système de la fam. Felidae et sur ses rasses. Annuaire du Musée Zoologique de l'Académie Impériale des Sciences de St. Pétersbourg, 21:130-163.
Bishop, L. R. 1979. Notes on *Praomys* (*Hylomyscus*) in eastern Africa. Mammalia, 43:521-530.
Biswas, B. 1967. Authorship of the name *Presbytis geei* (Mammalia: Primates). Journal of the Bombay Natural History Society, 63:429-431.
Biswas, B., and R. K. Ghose. 1970. Taxonomic notes on the Indian pale hedgehogs of the genus *Paraechinus* Trouessart, with descriptions of a new species and subspecies. Mammalia, 34:467-477.
Black, C. C. 1963. A review of the North American Tertiary Sciuridae. Bulletin of the Museum of Comparative Zoology at Harvard College, 130:109-248.
Black, C. C., and L. Krishtalka. 1986. Rodents, bats, and insectivores from the Plio-Pleistocene sediments to the east of Lake Turkana, Kenya. Contributions in Science, Natural History Museum of Los Angeles County, 372:1-15.
Blainville, H. M. D., de. 1837. Rapport sur un mémoire de M. Jourdan concernant deux nouvelles espèces de mammifères de l'Indie. Annales des Sciences Naturelles, Zoologie et Biologie Animale, ser. 2, 8:270-281.
Blair, W. F. 1942. Systematic relationships of *Peromyscus* and several related genera as shown by the baculum. Journal of Mammalogy, 23:196-204.
Blair, W. F., A. P. Blair, P. Broadkorb, F. R. Cagle, and G. A. Moore. 1968. Vertebrates of the United States. McGraw-Hill, NY, .
Blanford, W. T. 1875. On the species of marmot inhabiting the Himalaya, Tibet, and the adjoining regions. Journal of the Asiatic Society of Bengal, 44:113-127.
Blanford, W. T. 1879. Scientific results of the second Yarkand mission. Mammalia. Office of the Superintendent of Government Printing, Calcutta, 94 + xvi plates.
Blaustein, S. A., R. C. Liascovich, L. I. Apfelbaum, L. Daleffe, R. M. Barquez, and O. A. Reig. 1992. Correlates of systematic differentiation between two closely related allopatric populations of the *Akodon boliviensis* group from NW Argentina (Rodentia, Cricetidae). Zeitschrift für Säugetierkunde, 57:1-13.
Bleich, V. C. 1977. *Dipodomys stephensi*. Mammalian Species, 73:1-3.
Block, S. B., and E. G. Zimmerman. 1991. Allozymic variation and systematics of plains pocket gophers (*Geomys*) of south-central Texas. Southwestern Naturalist, 36:29-36.
Blyth, E. 1862. Report of curator, zoological department, February, 1862, No. 1. Journal of the Asiatic Society of Bengal, 31:331-333.
Bobrinskii, N. A., B. A. Kuznetsov, and A. P. Kuzyakin. 1944. Opredelitl' mlekopitayushchikh SSSR [Guide to the mammals of the U.S.S.R.]. Sovietskaya Nauka, Moscow, 439 pp. (in Russian).
Bobrinskii, N. A., B. A. Kuznetsov, and A. P. Kuzyakin. 1965. Opredelitl' mlekopitayushchikh SSSR [Guide to the mammals of the U.S.S.R.] Second ed. Proveshchenie, Moscow, 382 pp. (in Russian).
Bodenheimer, F. S. 1958. The present taxonomic status of the terrestrial mammals of Palestine. Bulletin of the Research Council of Israel, 7b:165-190.
Bogan, M. A. 1978. A new species of *Myotis* from the Islas Tres Marias, Nayarit, Mexico, with comments on variation in *Myotis nigricans*. Journal of Mammalogy, 59:519-530.
Bogan, M. A., H. W. Setzer, J. S. Findley, and D. E. Wilson. 1978. Phenetics of *Myotis blythi* in Morocco. Proceedings of the Fourth International Bat Research Conference, Nairobi, 4:217-230.
Bohlken, H. 1958. Vergleichende Untersuchungen an Wildrinden (Tribus Bovini Simpson, 1945). Zoologische Jahrbücher (Physiologie), 68:113-202.
Bohlken, H. 1961. Haustier und zoologische Systematik. Zeitschrift für Tierzüchtung und Züchtungbiologie, 75:107-113.
Bohlken, H. 1967. Beitrag zur Systematik der rezenten Formen der Gattung *Bison* H. Smith, 1827. Zeitschrift für Zoologische Systematik und Evolutionsforschung, 5:54-110.
Bohmann, L., von. 1939. Neue Rassen der Gattung *Dendromus*. Zoologischer Anzeiger, 127:170-172.
Bohmann, L., von. 1942. Die Gattung *Dendromus* A. Smith. Versuch einer natürlichen Gruppierung (Ergebnisse der Ostafrika-Reise 1937 Uthmöller-Bohmann. VIII). Zoologischer Anzeiger, 139:33-53.
Bohmann, L., von. 1952. Die afrikanische Nagergattung *Otomys* F. Cuvier. Zeitschrift für Säugetierkunde, 18:1-80.
Böhme, W. 1978*a*. *Micromys minutus* (Pallas, 1778)—Zwergmaus. Pp. 290-404, *in* Handbuch der Säugetiere Europas (J. Niethammer and F. Krapp, eds.). Akademische Verlagsgesellschaft (Wiesbaden), 1:1-476.

Böhme, W. 1978b. *Apodemus agrarius* (Pallas, 1771)—Brandmaus. Pp. 368-381, *in* Handbuch der Säugetiere Europas (J. Niethammer and F. Krapp, eds.). Akademische Verlagsgesellschaft (Wiesbaden), 1:1-476.
Böhme, W., and R. Hutterer. 1979. Kommentierte Liste einer Säugetier-Aufsammlung aus dem Senegal. Bonner Zoologische Beiträge, 29:303-323.
Bole, B. J., Jr., and P. N. Moulthrop. 1942. The Ohio Recent mammal collection in the Cleveland Museum of Natural History. Scientific Publication of the Cleveland Museum of Natural History, 5:83-181.
Bolles, K. 1981. Variation and alarm call evolution in antelope squirrels, *Ammospermophilus* (Rodentia: Sciuridae). Dissertation Abstracts International, 41:2857B.
Bonaparte, C.-L. J. L. 1845. Catalogo methodico dei mammiferi Europei. L. di Giacomo Pirola, Milano, 36 pp.
Bonaparte, C.-L. J. L. 1850. Cospectus systematis mastozoologie. Editio altera reformata. E. J. Brill, Lugduni Batavorum, 2 pp.
Bonde, R. K. and T. J. O'Shea. 1989. Sowerby's beaked whale (*Mesoplodon bidens*) in the Gulf of Mexico. Journal of Mammalogy, 70(2):447-449.
Bonhomme, F. J. 1986. Evolutionary relationships in the genus *Mus*. Pp. 19-34, *in* Current topics in microbiology and immunology, Vol. 127 (M. Potter, J. H. Nadeau, and M. P. Cancro, eds.). Springer-Verlag, Berlin, 395 pp.
Bonhomme, F., J. Catalan, J. Britton-Davidian, V. M. Chapman, K. Moriwaki, E. Nevo, and L. Thaler. 1984. Biochemical diversity and evolution in the genus *Mus*. Biochemical Genetics, 22:275-303.
Bonhomme, F., D. Iskandar, L. Thaler, and F. Petter. 1985. Electromorphs and phylogeney in muroid rodents. Pp. 671-683, *in* Evolutionary relationships among rodents, a multidisciplinary analysis (W. P. Luckett and J.-L. Hartenberger, eds.). Plenum Press, New York, 721 pp.
Bonhomme, F., J. Fernández, F. Palacios, J. Catalan, and A. Machordom. 1986. Charactérisation biochimique du complexe d'espèces du genre *Lepus* en Espagne. Mammalia, 50:495-506.
Bonhomme, F., N. Miyashita, P. Boursot, J. Catalan, and K. Moriwaki. 1989. Genetical variation and polyphyletic origin in Japanese *Mus musculus*. Heredity, 63:299-308.
Bonhote, J. L. 1901a. On the squirrels of the *Sciurus erythræus* group. Annals and Magazine of Natural History, ser. 7, 7:160-167.
Bonhote, J. L. 1901b. On the martens of the *Mustela flavigula* group. Annals and Magazine of Natural History, ser. 7, 7:342-349.
Booth, C. P. 1968. Taxonomic studies of *Cercopithecus mitis* Wolf (East Africa). National Geographic Society - Research Report. Abstracts and reviews of research and exploration authorized under grants from the National Geographic Society during the year 1963, pp. 37-51.
Borisov, Yu. M., and V. M. Malygin. 1991. [Clinal variability of the B-chromosome system in *Apodemus peninsulae* (Rodentia, Muridae) from the Buryatia and Mongolia]. Tsitologiya, 33:106-111 (in Russian).
Borobia, M., S. Siciiliano, L. Lodi, and W. Hoek. 1991. Distribution of the South American dolphin *Sotalia fluviatilis*. Canadian Journal of Zoology, 69:1025-1039.
Boschma, H. 1950. Maxillary teeth in specimens of *Hyperoodon rostratus* (Muller) and *Mesoplodon grayi* von Haast stranded on the Dutch coasts. Koninklijke Nederlandsche Akademie van Wetenschappen, Proceedings, 53(6):3-14, 4 pls.
Bothma, J. du P. 1971. Order Hyracoidea. Part 12, Pp. 1-8, *in* The mammals of Africa: an identification manual (J. Meester and H. W. Setzer, eds.) [issued 15 Jul 1971]. Smithsonian Institution Press, Washington, D. C., not continuously paginated.
Boulay, M. C., and C. B. Robbins. 1989. *Epomophorus gambianus*. Mammalian Species, 344:1-5.
Boursot, P., T. Jacquart, F. Bonhomme, J. Britton-Davidian, and L. Thaler. 1985. Différenciation géographique du génome mitochondrial chez *Mus spretus* Lataste. Comptes Rendus de l'Académie des Sciences (Paris), ser. 3, 301:161-166.
Bowen, W. W. 1968. Variation and evolution of Gulf coast populations of beach mice, *Peromyscus polionotus*. Bulletin of the Florida State Museum, 12:1-91.
Bowers, J. H. 1974. Genetic compatibility of *Peromyscus maniculatus* and *Peromyscus melanotis*, as indicated by breeding studies and morphometrics. Journal of Mammalogy, 55:720-737.
Bowers, J. H., R. J. Baker, and M. H. Smith. 1973. Chromosomal, electrophoretic, and breeding studies of selected populations of deer mice (*Peromyscus maniculatus*) and black-eared mice (*P. melanotis*). Evolution, 27:378-386.
Bowyer, R. T., and D. M. Leslie, Jr. 1992. *Ovis dalli*. Mammalian Species, 393:1-7.
Boye, P. 1991. Notes on the morphology, ecology and geographic origin of the Cyprus long-eared hedgehog (*Hemiechinus auritus dorotheae*). Bonner Zoologische Beiträge, 42:115-123.
Boye, P., R. Hutterer, N. López-Martínez, and J. Michaux. 1992. A reconstruction of the lava mouse (*Malpaisomys insularis*), an extinct rodent of the Canary Islands. Zeitschrift für Säugetierkunde, 57:29-38.
Bradley, R. D., and D. J. Schmidly. 1987. The glans penes and bacula in Latin American taxa of the *Peromyscus boylii* group. Journal of Mammalogy, 68:595-616.

Bradley, R. D., D. J. Schmidly, and R. D. Owen. 1989. Variation in the glans penes and bacula among Latin American populations of the *Peromyscus boylii* species complex. Journal of Mammalogy, 70:712-725.

Bradley, R. D., S. K. Davis, S. F. Lockwood, J. W. Bickham, and R. J. Baker. 1991. Hybrid breakdown and cellular-DNA content in a contact zone between two species of pocket gophers (*Geomys*). Journal of Mammalogy, 72:697-705.

Bradshaw, W. N. 1968. Progeny from experimental mating tests with mice of the *Peromyscus leucopus* group. Journal of Mammalogy, 49:475-480.

Bradshaw, W. N., and T. C. Hsu. 1972. Chromosomes of *Peromyscus* (Rodentia, Cricetidae). III. Polymorphism in *Peromyscus maniculatus*. Cytogenetics, 11:436-451.

Braestrup, F. W. 1935. Report on the mammals collected by Mr. Harry Madsen during professor O. Ofufsen's expedition to French Sudan and Nigeria in the years 1927-28. Videnskabelige Meddeleser fra Dansk Naturhistoriske Forening i Kobenhaven, 99:73-130.

Braestrup, F. W., and R. Hutterer. 1985. *Grammomys macmillani tuareg* Braestrup, 1935: a junior synonym of *Praomys daltoni* (Thomas, 1892). Zeitschrift für Säugetierkunde, 50:240-241.

Brand, L. R., and R. E. Ryckman. 1969. Biosystematics of *Peromyscus eremicus, P. guardia*, and *P. interparietalis*. Journal of Mammalogy, 50:501-513.

Brandon-Jones, D. 1978. Comment on the proposal to conserve Colobidae Blyth, 1875, as a family-group name for the leaf-eating monkeys. Bulletin of Zoological Nomenclature, 35:69-70.

Brandt, J. F. 1844 [1843]. Observations sur les différentes espèces de sousliks de Russie, suivies de remarques sur l'arrangement et la distribution géographique du genre *Spermophilus*, ansé que sur la classification de la familie des ecureuils (Sciurina) en général. Bulletin Scientifique l'Académie Impériale des Sciences de Saint-Pétersbourg, 1844:col. 357-382.

Brandt, J. F. 1855. Beitrage zur nahern Kenntniss der Säugethiere Russland's. Kaiserlichen Akademie der Wissenschaften, Saint Petersburg, Mémoires Mathématiques, Physiques et Naturelles, 7:1-365.

Brass, L. J. 1959. Results of the Archbold Expeditions. No. 86. Summary of the Sixth Archbold Expedition to New Guinea (1959). Bulletin of the American Museum of Natural History, 127:145-216.

Braun, J. K. 1988. Systematics and biogeography of the southern flying squirrel, *Glaucomys volans*. Journal of Mammalogy, 69:422-426.

Braun, J. K., and M. A. Mares. 1989. *Neotoma micropus*. Mammalian Species, 330:1-9.

Breed, W. G. 1980. Further observations on spermatozoal morphology and male reproductive tract anatomy of *Pseudomys* and *Notomys* species (Mammalia: Rodentia). Transactions of the Royal Society of South Australia, 104:51-55.

Breed, W. G. 1983. Variation in sperm morphology in the Australian rodent genus *Pseudomys* (Muridae). Cell and Tissue Research, 229(3):611-625.

Breed, W. G. 1984. Sperm head structure in the Hydromyinae (Rodentia: Muridae): a further evolutionary development of the subacrosomal space in mammals. Gamete Research, 10:31-44.

Breed, W. G. 1985. Morphological variation in the female reproductive tract of Australian rodents in the genera *Pseudomys* and *Notomys*. Journal of Reproduction and Fertility, 73:379-384.

Breed, W. G. 1986. Comparative morphology and evolution of the male reproductive tract in the Australian hydromyine rodents (Muridae). Journal of Zoology (London), ser. A, 209:607-629.

Breed, W. G. 1990. Reproductive anatomy and sperm morphology of the long-tailed hopping-mouse, *Notomys longicaudatus* (Rodentia: Muridae). Australian Mammalogy, 13:201-204.

Breed, W. G., and G. G. Musser. 1991. Sulawesi and Philippine rodents (Muridae): a survey of spermatozoal morphology and its significance for phylogenetic inference. American Museum Novitates, 3003:1-15.

Breed, W. G., and V. Sarafis. 1978. On the phylogenetic significance of spermatozoal morphology and male reproductive tract anatomy in Australian rodents. Transactions of the Royal Society of South Australia, 103:127-135.

Breed, W. G., and V. Sarafis. 1983. Variation in sperm head morphology in the Australian rodent *Notomys alexis*. Australian Journal of Zoology, 31:313-316.

Breed, W. G., and H. S. Yong. 1986. Sperm morphology of murid rodents from Malaysia and its possible phylogenetic significance. American Museum Novitates, 2856:1-12.

Brickell, J. 1737. The natural history of North-Carolina: With an account of the trade, manners, and customs of the Christian and Indian inhabitants. J. Brickell, Dublin, 417 pp.

Bridelance, P. 1987. Émissions sonores rythmées engendrées par des tapements de pattes: les podophones chez les rongeurs. Comptes Rendus de l'Académie des Sciences (Paris), ser. 3, 305(5):125-128.

Briggs, J. C. 1974. Marine Zoogeography. McGraw Hill, 475 pp.

Briscoe, D. A., B. J. Fox, and S. Ingleby. 1981. Genetic differentiation between *Pseudomys pilligaensis* and related *Pseudomys*. Australian Mammalogy, 4:89-92.

Briscoe, D. A., J. H. Calaby, R. L. Close, G. M. Maynes, C. E. Murtagh, and G. B. Sharman. 1982. Isolation, introgression and genetic variation in rock-wallabies. Pp. 73-87, *in* Species at risk: Research in Australia (R. H. Groves and W. D. L. Ride, eds.). Australian Academy of Science, Canberra, 216 pp.
Brisson, M. J. 1762. Le regnum animale in classes IX distributum, sive synopsis methodica sistens generalem animalium distributionem in classes IX, & duarum primarum classium, quadrupedum scilicet & cetaceorum, particularem dibvisionem in ordines, sectiones, genera & species. T. Haak, Paris, 296 pp.
Britton-Davidian, J. 1990. Genic differentiation in *M. m. domesticus* populations from Europe, the Middle East and North Africa: geographic patterns and colonization events. Biological Journal of the Linnean Society, 41:27-45.
Britton-Davidian, J., M. Vahdat, F. Benmehdi, P. Gros, V. Nance, H. Croset, S. Guerassimov, and C. Triantaphyllidis. 1991. Genetic differentiation in four species of *Apodemus* from southern Europe: *A. sylvaticus, A. flavicollis, A. agrarius* and *A. mystacinus* (Muridae, Rodentia). Zeitschrift für Säugetierkunde, 56:25-33.
Broadbooks, H. E. 1965. Ecology and distribution of the pikas of Washington and Alaska. American Midland Naturalist, 73:299-335.
Brodie, P. F. 1989. The white whale - *Delphinapterus leucas* (Pallas, 1776). Pp. 119-144, *in* Handbook of marine mammals: River dolphins and the larger toothed whales (S. H. Ridgway and R. Harrison, eds.). Academic Press, London, 4:1-442.
Bronner, G. N. 1990. New distribution records for four mammal species, with notes on their taxonomy and ecology. Koedoe, 33(2):1-7.
Bronner, G. N., and J. A. J. Meester. 1988. *Otomys angoniensis*. Mammalian Species, 306:1-6.
Bronner, G. N., S. Gordon, and J. A. J. Meester. 1988. *Otomys irroratus*. Mammalian Species, 308:1-6.
Brosset, A. 1988. Le peuplement de mammifères insectivores des forêts du nord-est du Gabon. Revue de Ecologie (La Terre et la Vie), 43:23-46.
Brosset, A., G. Dubost, and H. Heim de Balsac. 1965. Mammifères inedits recoltes au Gabon. Biologia Gabonica, 1:148-174.
Brown, L. N., and R. J. McGuire. 1975. Field ecology of the exotic Mexican red-bellied squirrel in Florida. Journal of Mammalogy, 56:405-419.
Brown, R. J., and R. L. Rudd. 1981. Chromosomal comparisons within the *Sorex ornatus-S. vagrans* complex. Wasmann Journal of Biology, 32:303-326.
Browne, R. A. 1977. Genetic variation in island and mainland populations of *Peromyscus leucopus*. American Midland Naturalist, 97:1-9.
Brownell, E. 1983. DNA/DNA hybridization studies of muroid rodents: symmetry and rates of molecular evolution. Evolution, 37:1034-1051.
Brownell, R. L., Jr. 1974. Small odontocetes of the Antarctic. Pp. 13-19, *in* Antarctic mammals (S. G. Brown, R. L. Brownell, Jr., A. W. Erickson, R. J. Hofman, G. A. Llano, and N. A. Mackintosh). American Geographic Society, 19 pp.
Brownell, R. L., Jr. 1975*a*. *Phocoena dioptrica*. Mammalian Species, 66:1-3.
Brownell, R. L., Jr. 1975*b*. Taxonomic status of the dolphin *Stenopontistes zambezicus* Miranda-Ribeiro, 1936. Zeitschrift für Säugetierkunde, 40(3):173-176.
Brownell, R. L., Jr. 1983. *Phocoena sinus*. Mammalian Species, 198:1-3.
Brownell, R. L., Jr. 1986. Distribution of the vaquita, *Phocoena sinus*, in Mexican waters. Marine Mammal Science, 2:299-305.
Brownell, R. L., Jr. 1989. Franciscana - *Pontoporia blainvillei* (Gervais and d'Orbigny, 1844). Pp. 45-68, *in* Handbook of marine mammals: River dolphins and the larger toothed whales (S. H. Ridgway and R. Harrison, eds.). Academic Press, London, 4:1-442.
Brownell, R. L., Jr., and E. S. Herald. 1972. *Lipotes vexillifer*. Mammalian Species, 10:1-4.
Brownell, R. L., Jr., J. E. Heyning, and W. F. Perrin. 1989. A porpoise *Australophocoena dioptrica* previously identified as *Phocoena spinipinnis* from Heard Island. Marine Mammal Science, 5(2):193-195.
Brownell, R. L., Jr., and R. Praderi. 1984. *Phocoena spinipinnis*. Mammalian Species, 217:1-4.
Brownell, R. L., Jr., and R. Praderi. 1985. Distribution of Commerson's dolphin, *Cephalorhynchus commersonii*, and the rediscovery of the type of *Lagenorhynchus floweri*. Scientific Reports of the Whales Research Institute (Toyko), 36:153-164.
Brum, N., N., Lafuente, and P. Kiblisky. 1973. Cytogenetic studies in the cricetid rodent *Scapteromys tumidus* (Rodentia-Cricetidae). Experientia, 28:1373.
Brunelli, G., and G. Fasella. 1929. Su di un rarismo cetaceo spiaggiato nel litorale di Nettuna. Atti della Reale Accademia Nazionale dei Lincei Roma, 8:85-87.
Brunet-Lecomte, P., and J. Chaline. 1991. Morphological evolution and phylogenetic relationships of the European ground voles (Arvicolidae, Rodentia). Lethaia, 24:45-53.
Bryant, M. D. 1945. Phylogeny of Nearctic Sciuridae. American Midland Naturalist, 33:257-390.

Bublitz, J. 1987. Untersuchungen zur Sytematik der Rezenten Caenolestidae Trouessart, 1898: Unter Verwendung craniometrischer Methoden. Bonner Zoologische Monographien, 23:1-96.
Bucher, J. E., and R. S. Hoffmann. 1980. *Caluromys derbianus*. Mammalian Species, 140:1-4.
Büchner, E. 1890. Wissenschaftliche Resultate der von N. M. Przewalski nach central-Asien unternommenen Reisen auf Kosten einer von seiner Kaiserlichen Hoheit dem Grossfürsten Thronfolger Nikolai Alexandrowitsh gespendeten Summe herausgegeben von der Kaiserlichen Akademie der Wissenschaften St. Petersburg. Zoologischer, Theil Band I. Säugethiere, V:3-232.
Buden, D. W. 1976. A review of the bats of the endemic West Indian genus *Erophylla*. Proceedings of the Biological Society of Washington, 89:1-16.
Buden, D. W. 1977. First records of bats of the genus *Brachyphylla* from the Caicos Islands, with notes on geographic variation. Journal of Mammalogy, 58:221-225.
Bueler, L. E. 1973. Wild dogs of the world. Stein & Day, New York, 274 pp.
Buffon, G. L. L., Comte de. 1781. Natural history, general and particular, by the Count de Buffon, illustrated with above six hundred copper plates, the history of man and quadrupeds, translated into English with notes and observations by William Smellie. First ed. vol. 7(Natural history of animals). T. Cadell and W. Davies, London, 452 pp.
Buffon, G. L. L., Comte de, and L. J. M. Daubenton. 1763. Histoire naturelle, générale et particulière, avec la description du cabinet du Roi. L'Imprimerie Royale, Paris, 10(Quadrupeds):1-6 (unnumbered) + 368 pp., 57 pls.
Bugge, J. 1970. The contribution of the stapedial artery to the cephalic arterial supply in muroid rodents. Acta Anatomica, 76:313-134.
Bugge, J. 1971. The cephalic arterial system in mole-rats (Spalacidae) bamboo-rats (Rhizomyidae), jumping mice and jerboas (Dipodoidea) and dormice (Gliroidea) with special reference to the systematic classification of rodents. Acta Anatomica, 79:165-180.
Bugge, J. 1978. The cephalic arterial system in carnivores, with special reference to the systematic classification. Acta Anatomica, 101(1):45-61.
Bulatova, N., and E. Kotenkova. 1990. Variants of the Y-chromosome in sympatric taxa of *Mus* in southern USSR. Bollettino di Zoologia, 57:357-360.
Burchell, W. J. 1822-1824 [1822-23]. Travels in interior of Southern Africa. Longman, Hurst, Rees, Orme, Brown, and Green, London, 2 volumes [vol. 2 published in 1823, dated 1824].
Burdelov, A. S., and O. B. Rossinskaya. 1959. [On the range of *Selevinia betpakdalaensis* Belos. et Bazh.(1938) and on some peculiarities of its ecology]. Zoologicheskii Zhurnal, 38:942- 944 (in Russian).
Burgos, M., R. Jimenez, and R. Diaz de la Guardia. 1989. Comparative study of G- and C-banded chromosomes of five species of Microtidae. Genetica, 78:3-12.
Burmeister, H. 1850. Verzeichniss der im Zoologischen Museum der Universität Halle-Wittenberg aufgestellten Säugetheire, Vögel, und Amphibien. Friedrichs-Universität, Halle, 84 pp.
Burns, J. C., J. R. Choate, and E. G. Zimmerman. 1985. Systematic relationships of pocket gophers (genus *Geomys*) on the Central Great Plains. Journal of Mammalogy, 66:102-118.
Burns, J. J., and F. H. Fay. 1970. Comparative morphology of the skull of the ribbon seal, *Histriophoca fasciata*, with remarks on systematics of Phocidae. Journal of Zoology (London), 161:363-394.
Burt, W. H. 1932. Description of heretofore unknown mammals from islands in the Gulf of California, Mexico. Transactions of the San Diego Society of Natural History, 7:161-182.
Burt, W. H. 1934. Subgeneric allocation of the white-footed mouse, *Peromyscus slevini*, from the Gulf of California, Mexico. Journal of Mammalogy, 15:159-160.
Burt, W. H., and F. S. Barkalow. 1942. A comparative study of the bacula of wood rats (subfamily Neotominae). Journal of Mammalogy, 23:287-297.
Burt, W. H., and R. A. Stirton. 1961. The mammals of El Salvador. Miscellaneous Publications, Museum of Zoology, University of Michigan, 117:1-69.
Bush, R. M., and K. Paigen. 1992. Evolution of B-glucuronidase regulation in the genus *Mus*. Evolution, 46:1-15.
Butler, P. M. 1972. The problem of insectivore classification. Pp. 253-265, *in* Studies in vertebrate evolution (K. A. Joysey and T. S. Kemp, ed.). Oliver and Boyd, Edinburgh, 284 pp.
Butler, P. M., and M. Greenwood. 1979. Soricidae (Mammalia) from the Early Pleistocene of Olduvai Gorge, Tanzania. Zoological Journal of the Linnean Society, 67:329-379.
Bykova, G. V., I. A. Vasilyeva, and E. A. Gileva. 1978. Chromosomal and morphological diversity in 2 populations of Asian mountain vole, *Alticola lemminus* Miller (Rodentia, Cricetidae). Experientia, 34:1146-1148.
Byrne, J. M., E. J. Duke, and J. S. Fairley. 1990. Some mitochrondrial DNA polymorphisms in Irish wood mice (*Apodemus sylvaticus*) and bank voles (*Clethrionomys glareolus*). Journal of Zoology (London), 221:299-302.

Cabral, J. C. 1971. Existencia em Angola de *Anomalurops beecrofti* (Fraser). Boletim do Instituto de Investigacão Científica de Angola, 8:55-63.

Cabrera, A. 1911. On the specimens of spotted hyaenas in the British Museum (Natural History). Proceedings of the Zoological Society of London, 1911:93-99.

Cabrera, A. 1914. Fauna Ibérica. Mamíferos. Museo Nacional de Ciencias Naturales, Madrid, 418 pp.

Cabrera, A. 1916. El tipo de *Philander laniger* Desm. en el Museo de Ciencias Naturales de Madrid. Boletín de la Real Sociedad Española de Historia Natural, 16:514-517.

Cabrera, A. 1925. Genera mammalium: Insectivora, Galeopithecia. Museo Nacional de Ciencias Naturales, Madrid, 232 pp.

Cabrera, A. 1929. Catálogo descriptivo de las mamíferos de la Guinea Española. Memorias de la Real Sociedad Española de Historia Natural, 16:31-32.

Cabrera, A. 1931. On some South American canine genera. Journal of Mammalogy, 12:54-67.

Cabrera, A. 1932. Los mamíferos de Marruecos. Trabajos del Museo Nacional de Ciencias Naturales, 57:209-222.

Cabrera, A. 1940. Notas sobre carnívoras sudamericanos. Notas del Museo de la Plata, 5(29):1-22.

Cabrera, A. 1956. Una nueva forma del genero *Nasua*. Neotropica, 2:2-4.

Cabrera, A. 1957 [1958]-1961. Catálogo de los mamíferos de América del Sur. Revista del Museo Argentino de Ciencias Naturales "Bernardino Rivadavia". Ciencias Zoológicas, 4(1):iv + 308 pp.[1958]; 4(2):v-xxii+309-732[1961].

Cai Gui-quan, and Feng Zuo-jiang. 1981. [On the occurrence of the Himalayan musk-deer (*Moschus chrysogaster*) in China and an approach to the systematics of the genus *Moschus*]. Acta Zootaxonomica Sinica, 6:106-110 (in Chinese).

Cai Gui-quan, and Feng Zuo-jiang. 1982. [A systematic revision of the subspecies of highland hare (*Lepus oiostolus*)—including two new subspecies]. Acta Theriologica Sinica, 2:7-182 (in Chinese).

Caire, W., J. E. Vaughan, and V. E. Diersing. 1978. First record of *Sorex arizonae* (Insectivora: Soricidae) from Mexico. Southwestern Naturalist, 23:532-533.

Caire, W., J. D. Taylor, B. P. Glass, and M. A. Mares. 1989. Mammals of Oklahoma. University of Oklahoma Press, Norman, 567 pp.

Cakenberghe, V. van, and F. De Vree. 1985. Systematics of African *Nycteris* (Mammalia: Chiroptera). Pp. 53-90 *in* Proceedings of the International Symposium on African vertebrates: systematics, phylogeny and evolutionary ecology (K.-L. Schuchmann, ed.). Zoologisches Forschungsinstitut und Museum Alexander Koenig, Bonn., 585 pp.

Calaby, J. H. 1966. Mammals of the Upper Richmond and Clarence Rivers, New South Wales. Technical Papers of the Division of Wildlife Survey, CSIRO, Australia, 10:1-55.

Calaby, J. H., G. Mack, and W. D. L. Ride. 1963. The generic name *Macropus* Shaw, 1790 (Mammalia). Z.N.(S.) 1584. Bulletin of Zoological Nomenclature, 20(5):376-379.

Caldarini, G., E. Capanna, M. V. Civitelli, M. Corti, and A. Simonetta. 1989. Chromosomal evolution in the subgenus *Rattus* (Rodentia, Muridae): karyotype analysis of two species from the Indian subregion. Mammalia, 53:77-83.

Caldwell, D. K., and M. C. Caldwell. 1989. Pygmy sperm whale - *Kogia breviceps* (de Blainville, 1838); dwarf sperm whale *Kogia simus* Owen, 1866. Pp. 235-260, *in* Handbook of marine mammals: River dolphins and the larger toothed whales (S. H. Ridgway and R. Harrison, eds.). Academic Press, London, 4:1-442.

Calhoun, S. W., and I. F. Greenbaum. 1991. Evolutionary implications of genic variation among insular populations of *Peromyscus maniculatus* and *Peromyscus oreas*. Journal of Mammalogy, 72:248-262.

Calhoun, S. W., I. F. Greenbaum, and K. P. Fuxa. 1988. Biochemical and karyotypic variation in *Peromyscus maniculatus* from western North America. Journal of Mammalogy, 69:34-45.

Calhoun, S. W., M. D. Engstrom, and I. F. Greenbaum. 1989. Biochemical variation in pygmy mice (*Baiomys*). Journal of Mammalogy, 70:374-381.

Callahan, J. R. 1977. Diagnosis of *Eutamias obscurus* (Rodentia: Sciuridae). Journal of Mammalogy, 58:188-201.

Callahan, J. R. 1980. Taxonomic status of *Eutamias bulleri*. Southwestern Naturalist, 25:1-8.

Callahan, J. R., and R. Davis. 1977. A new subspecies of the cliff chipmunk from coastal Sonora. Southwestern Naturalist, 22:67-75.

Callahan, J. R., and R. Davis. 1982. Reproductive tract and evolutionary relationships of the Chinese rock squirrel, *Sciurotamias davidianus*. Journal of Mammalogy, 63:42-47.

Cameron, G. N., and S. R. Spencer. 1981. *Sigmodon hispidus*. Mammalian Species, 158:1-9.

Campbell, C. B. G. 1966. Taxonomic status of tree shrews. Science, 153:436.

Campbell, C. B. G. 1974. On the phyletic relationships of the tree shrews. Mammal Review, 4:125-143.

Cano, J., A. Pretel, R. Grandfils, J. M. Vargas, and V. Sans-Coma. 1984. Anzahl und Struktur der Chromosomen von *Mus spretus* Lataste, 1883 (Rodentia: Muridae) von der Iberischen Halbinsel. Säugetierkundliche Mitteilungen, 31:161-169.

Cao Van-sung, and V. M. Tran. 1984. Karyotypes et systematique des rats (genre *Rattus* Fisher) du Vietnam. Mammalia, 48:557-564.

Capanna, E. 1974. A re-statement of the problem of chromosomal polymorphism in *Rattus rattus* (L.). Pp. 223-235, *in* Symposium Theriologicum II, Proceedings of the International Symposium on Species and Zoogeography of European Mammals held in Brno, Czechoslovakia on 22nd to 26th November 1971 (J. Kratochvil and R. Obrtel, eds.). Academia Publishing house of the Czechoslovak Academy of Sciences (Praha), 394 pp.

Capanna, E., and M. V. Civitelli. 1988. A cytotaxonomic approach of the systematics of *Arvicanthis niloticus* (Desmarest 1822) (Mammalia, Rodentia). Tropical Zoology, 1:29-37.

Carleton, M. D. 1973. A survey of gross stomach morphology in New World Cricetinae (Rodentia, Muroidea), with comments on functional interpretations. Miscellaneous Publications, Museum of Zoology, University of Michigan, 146:1-43.

Carleton, M. D. 1977. Interrelationships of populations of the *Peromyscus boylii* species group (Rodentia, Muridae) in western Mexico. Occasional Papers of the Museum of Zoology, University of Michigan, 675:1-47.

Carleton, M. D. 1979. Taxonomic status and relationships of *Peromyscus boylii* from El Salvador. Journal of Mammalogy, 60:280-296.

Carleton, M. D. 1980. Phylogenetic relationships in neotomine-peromyscine rodents (Muroidea) and a reappraisal of the dichotomy within New World Cricetinae. Miscellaneous Publications, Museum of Zoology, University of Michigan, 157:1-146.

Carleton, M. D. 1981. A survey of gross stomach morphology in Microtinae (Rodentia, Muroidea). Zeitschrift für Säugetierkunde, 46:93-108.

Carleton, M. D. 1989. Systematics and Evolution. Pp. 7-141, *in* Advances in the study of *Peromyscus* (Rodentia) (G. L. Kirkland and J. N. Layne, eds.). Texas Tech University Press, Lubbock, 367 pp.

Carleton, M. D. 1993 [In press]. Systematic studies of Madagascar's endemic rodents (Muroidea: Nesomyinae): revision of the genus *Eliurus*. American Museum Novitates.

Carleton, M. D., and R. E. Eshelman. 1979. A synopsis of fossil grasshopper mice, genus *Onychomys*, and their relationships to Recent species. University of Michigan, Museum of Paleontology, Papers on Paleontology, 21:1-63.

Carleton, M. D., and D. G. Huckaby. 1975. A new species of *Peromyscus* from Guatemala. Journal of Mammalogy, 56:444-451.

Carleton, M. D., and C. Martinez. 1991. Morphometric differentiation among West African populations of the rodent genus *Dasymys* (Muroidea: Murinae), and its taxonomic implications. Proceedings of the Biological Society of Washington, 104:419-435.

Carleton, M. D., and G. G. Musser. 1984. Muroid rodents. Pp. 289-379, *in* Orders and families of Recent mammals of the world (S. Anderson and J. K. Jones, Jr., eds.). John Wiley and Sons, New York, 686 pp.

Carleton, M. D., and G. G. Musser. 1989. Systematic studies of oryzomyine rodents (Muridae, Sigmodontinae): a synopsis of *Microryzomys*. Bulletin of the American Museum of Natural History, 191:1-83.

Carleton, M. D., and P. Myers. 1979. Karyotypes of some harvest mice, genus *Reithrodontomys*. Journal of Mammalogy, 60:307-313.

Carleton, M. D., and C. B. Robbins. 1985. On the status and affinities of *Hybomys planifrons* (Miller, 1900) (Rodentia: Muridae). Proceedings of the Biological Society of Washington, 98:956-1003.

Carleton, M. D., and D. F. Schmidt. 1990. Systematic studies of Madagascar's endemic rodents (Muroidea: Nesomyinae): an annotated gazetteer of collecting localities of known forms. American Museum Novitates, 2987:1-36.

Carleton, M. D., E. T. Hooper, and J. Honacki. 1975. Karyotypes and accessory reproductive glands in the rodent genus *Scotinomys*. Journal of Mammalogy, 56:916-921.

Carleton, M. D., D. E. Wilson, A. L. Gardner, and M. A. Bogan. 1982. Distribution and systematics of *Peromyscus* (Mammalia: Rodentia) of Nayarit, Mexico. Smithsonian Contributions to Zoology, 352:1-46.

Carnero, A., R. Jimenez, M. Burgos, A. Sanchez, and R. Diaz de la Guardia. 1991. Achiasmatic sex chromosomes in *Pitymys duodecimcostatus*: mechanisms of association and segregation. Cytogenetics and Cell Genetics, 56:78-81.

Carraway, L. N. 1985. *Sorex pacificus*. Mammalian Species, 231:1-5.

Carraway, L. N. 1990. A morphologic and morphometric analysis of the "*Sorex vagrans* species complex" in the Pacific coast region. Special Publications, The Museum, Texas Tech University, 32:1-76.

Carraway, L. N., and B. J. Verts. 1985. *Microtus oregoni*. Mammalian Species, 233:1-6.

Carraway, L. N., and B. J. Verts. 1991*a*. *Neotoma fuscipes*. Mammalian Species, 386:1-10.
Carraway, L. N., and B. J. Verts. 1991*b*. *Neurotrichus gibbsii*. Mammalian Species, 387:1-7.
Carroll, L. E., and H. H. Genoways. 1980. *Lagurus curtatus*. Mammalian Species, 124:106.
Carter, C. H., and H. H. Genoways. 1978. *Liomys salvini*. Mammalian Species, 84:1-5.
Carter, D. C., and P. G. Dolan. 1978. Catalog of type specimens of Neotropical bats in selected European museums. Special Publications, The Museum, Texas Tech University Press, 15:1-136.
Carter, D. C., and C. S. Rouk. 1973. Status of recently described species of *Vampyrops* (Chiroptera: Phyllostomatidae). Journal of Mammalogy, 54:975-977.
Carter, D. C., W. D. Webster, J. K. Jones, Jr., C. Jones, and R. D. Suttkus. 1985. *Dipodomys elator*. Mammalian Species, 232:1-3.
Carvalho, C. T., de. 1965. Commentarios sobre os mamiferos descritos e figurados por Alexandre Rodriguez Ferreira em 1790. Arquivos de Zoologia, 12:7-70.
Casinos, A. 1984. A note on the common dolphin of the South American Atlantic coast, with some remarks about the speciation of the genus *Delphinus*. Acta Zoologica Fennica, 172:141-142.
Casinos, A., and S. Filella. 1981. Notes on cetaceans of the Iberian coasts. IV. A specimen of *Mesoplodon densirostris* (Cetacea, Hyperoodontidae) stranded on the Spanish Mediterranean littoral. Säugetierkundliche Mitteilungen, 29:61-67.
Casinos, A., and J. Ocaña. 1979. A craniometric study of the genus *Inia* d'Orbigny, 1834 (Cetacea, Platanistoidea). Säugetierkundliche Mitteilungen, 27:194-206.
Catzeflis, F. M. 1990. DNA hybridization as a guide to phylogenies: raw data in muroid rodents. Pp. 317-345, *in* Evolution of subterranean mammals at the organismal and molecular levels. Proceedings of the Fifth International Theriological Congress, held in Rome, Italy, August 22-29, 1989 (E. Nevo and O. A. Reig, eds.). Wiley-Liss, New York, 422 pp.
Catzeflis, F. M., T. Maddalena, S. Hellwing, and P. Vogel. 1985. Unexpected findings on the taxonomic status of East Mediterranean *Crocidura russula* auct. (Mammalia, Insectivora). Zeitschrift für Säugetierkunde, 50:185-201.
Catzeflis, F. M., F. H. Sheldon, J. E. Ahlquist, and C. G. Sibley. 1987. DNA-DNA hybridization evidence of the rapid rate of muroid rodent DNA evolution. Molecular Biology and Evolution, 4:242-253.
Catzeflis, F. M., E. Nevo, J. E. Ahlquist, and C. G. Sibley. 1989. Relationships of the chromosomal species in the Eurasian mole rats of the *Spalax ehrenbergi* group as determined by DNA-DNA hybridization, and an estimate of the Spalacid-Murid divergence time. Journal of Molecular Evolution, 29:223-232.
Ceballos-G., G., and R. A. Medellin. 1988. *Diclidurus albus*. Mammalian Species, 316:1-4.
Ceballos-G., G., and D. E. Wilson. 1985. *Cynomys mexicanus*. Mammalian Species, 248:1-3.
Cerqueira, R. 1985. The distribution of *Didelphis* in South America (Polyprotodontia, Didelphidae). Journal of Biogeography, 12:135-145.
Cervantes, F. A., C. Lorenzo, and R. S. Hoffmann. 1990. *Romerolagus diazi*. Mammalian Species, 360:1-7.
Chakraborty, S. 1978. The rusty-spotted cat, *Felis rubiginosa* I. Geoffroy, in Jammu and Kashmir. Journal of the Bombay Natural History Society, 75:478-479.
Chakraborty, S. 1981. Studies on *Sciuropterus baberi* Blyth. Proceedings of the Zoological Society (Calcutta), 32:57-63.
Chakraborty, S. 1983. Contribution to the knowledge of the mammalian fauna of Jammu and Kashmir, India. Records of the Zoological Survey of India, 38:1-129.
Chakraborty, S. 1985. Studies on the genus *Callosciurus* Gray (Rodentia: Sciuridae). Records of the Zoological Survey of India, Miscellaneous Publications, Occasional Papers, No. 63:1-93.
Chaline, J. 1972. Les rongeurs du Pleistocene moyen et superieur de France (Systematique-biostratigraphie-paleoclimatologie). Cahiers de Paleontologie, Centre National de la Recherche Scientifique, Paris, 410 pp.
Chaline, J. 1974. Esquisse de l'evolution morphologique, biometrique et chromosomique du genre *Microtus* (Arvicolidae, Rodentia) dans le Pleistocene de l'hemisphere nord. Bulletin de la Societe de Geologie France, ser. 7, 14:440-450.
Chaline, J., and J.-D. Graf. 1988. Phylogeny of the Arvicolidae (Rodentia): biochemical and paleontological evidence. Journal of Mammalogy, 69:22-33.
Chaline, J., and P. Mein. 1979. Les rongeurs et l'evolution. Doin Editeurs, Paris, 235 pp.
Chaline, J., P. Mein, and F. Petter. 1977. Les grandes lignes d'une classification evolutive des Muroidea. Mammalia, 41:245-252.
Chaline, J., P. Brunet-Lecomte, and J.-D. Graf. 1988. Validation de *Terricola* Fatio, 1867 pour les campagnols souterrains (Arvicolidae, Rodentia) paléarctiques actuels et fossiles. Comptes Rendus de l'Académie des Sciences (Paris), ser. 3, 306:475-478.
Chaline, J., P. Brunet-Lecomte, G. Brochet, and F. Martin. 1989. Les lemmings fossiles du genre *Lemmus* (Arvicolidae, Rodentia) dans le Pleistocene de France. Geobios, 22:613-623.

Chamberlain, J. L. 1954. The Block Island meadow mouse, *Microtus provectus*. Journal of Mammalogy, 35:587-588.

Chan, K. L. 1977. Enzyme polymorphism in Malayan rats of the subgenus *Rattus*. Biochemical Systematics and Ecology, 5:161-168.

Chan, K. L., S. S. Dhaliwal, and H. S. Yong. 1978. Protein variation and systematics in Malayan rats of the subgenus *Lenothrix* (Rodentia: Muridae, genus *Rattus* Fischer). Comparative Biochemistry and Physiology, 59B:345-351.

Chan, K. L., S. S. Dhaliwal, and H. S. Yong. 1979. Protein variation and systematics of three subgenera of Malayan rats (Rodentia: Muridae, genus *Rattus* Fischer). Comparative Biochemistry and Physiology, 64B:329-337.

Channing, A. 1984. Ecology of the namtap *Graphiurus ocularis* (Rodentia: Gliridae) in the Cedarberg, South Africa. South African Journal of Zoology, 19(3):144-149.

Chapman, F. M. 1901. A revision of the genus *Capromys*. Bulletin of American Museum of Natural History, 14:313-323, 2 pls.

Chapman, J. A. 1974. *Sylvilagus bachmani*. Mammalian Species, 34:1-4.

Chapman, J. A. 1975*a*. *Sylvilagus nuttallii*. Mammalian Species, 56:1-3.

Chapman, J. A. 1975*b*. *Sylvilagus transitionalis*. Mammalian Species, 55:1-4.

Chapman, J. A., and G. G. Ceballos. 1990. The cottontails. Pp. 95-110, *in* Rabbits, hares and pikas (J. A. Chapman and J. E. C. Flux, eds.). I. U. C. N., Gland, Switzerland, 168 pp.

Chapman, J. A., and G. A. Feldhamer. 1981. *Sylvilagus aquaticus*. Mammalian Species, 151:1-4.

Chapman, J. A., and G. R. Willner. 1978. *Sylvilagus audubonii*. Mammalian Species, 106:1-4.

Chapman, J. A., and G. R. Willner. 1981. *Sylvilagus palustris*. Mammalian Species, 153:1-3.

Chapman, J. A., J. G. Hockman, and M. M. Ojeda C. 1980. *Sylvilagus floridanus*. Mammalian Species, 136:1-8.

Chapman, J. A., K. R. Dixon, W. Lopez-Forment, and D. E. Wilson. 1983. The New World jackrabbits and hares (genus *Lepus*) 1. Taxonomic history and population status. Acta Zoologica Fennica, 174:49-51.

Chapman, J. A., et al. 1990. Conservation action needed for rabbits, hares and pikas. Pp. 154-168, *in* Rabbits, hares and pikas (J. A. Chapman and J. E. C. Flux, eds.). I. U. C. N., Gland, Switzerland, 168 pp.

Chapman, N. G., and D. I. Chapman. 1980. The distribution of fallow deer: A worldwide review. Mammal Review, 10:61-138.

Chapskii, K. K. 1955. Opyt peresmotra sistemy i diagnostiki tyulenei podsemeistva Phocinae [An attempt at revision of the systematics and diagnoses of seals of the subfamily Phocinae]. Trudy Zoologicheskovo Instituta, Akademiya Nauk, Leningrad, 17:160-199 (in Russian).

Chasen, F. N. 1936. A note on Malaysian *Gunomys*. Bulletin of the Raffles Museum, Singapore, Straits Settlements, 12:135-136.

Chasen, F. N. 1939. Two new mammals from North Sumatra. Treubia, 17:207-208.

Chasen, F. N. 1940. A handlist of Malaysian mammals: A systematic list of the mammals of the Malay Peninsula, Sumatra, Borneo, and Java, including the adjacent small islands. Bulletin of the Raffles Museum (Singapore), 15:1-209.

Chasen, F. N., and C. B. Kloss. 1927. Spolia Mentawiensia—mammals. Proceedings of the Zoological Society of London, 1927:797-840.

Chasen, F. N., and C. B. Kloss. 1932. Mammals from the lowlands and islands of North Borneo. Bulletin of the Raffles Museum, Singapore, Straits Settlements, 6(1931):1-82.

Chaworth-Musters, J. L. 1937. On the nomenclature of the Palearctic chipmunk. Annals and Magazine of Natural History, ser. 10, 19:158-159.

Chaworth-Musters, J. L., and J. R. Ellerman. 1947. A revision of the genus *Meriones*. Proceedings of the Zoological Society of London, 117:478-504.

Cheke, A. S., and A. F. Dahl. 1981. The status of bats on western Indian Ocean Islands, with special reference to *Pteropus*. Mammalia, 45:205-238.

Chen Jun, and Wang Ding-guo. 1985. [A preliminary survey on geographical distribution of the rodents in Hexi corridor, Gansu province]. Acta Theriologica Sinica, 5(3):195-200 (in Chinese).

Chen, P. 1989. Baiji - *Lipotes vexillifer* Miller, 1918. Pp. 25-44, *in* Handbook of marine mammals: River dolphins and the larger toothed whales (S. H. Ridgway and R. Harrison, eds.). Academic Press, London, 4:1-442.

Chernyavskii, F. B. 1972. O rasprostranenii i geograficheskoi izmenchivosti Amerikanskovo dlinnokhvostovo suslika (*Citellus parryi* Rich., 1827) severo-vostochnoi Sibiri [On the distribution and geographic variation of the American long-tailed suslik...in northeastern Siberia]. Trudy Moskovskovo Obshchestva Ispytatelei Prirody, 48:199-214.

Chernyavskii, F. B., and A. I. Kozlovskii. 1980. [Species status and history of Arctic lemmings (*Dicrostonyx*, Rodentia) on Wrangel Island]. Zoologicheskii Zhurnal, 59:266-273 (in Russian).

Chernyavskii, F. B., V. G. Krivosheev, Yu. V. Revin, L. P. Khvorostyanskaya, and V. N. Orlov. 1980. [On the distribution, taxonomy and biology of the Amur lemming (*Lemmus amurensis*)]. Zoologicheskii Zhurnal, 59:1077-1084 (in Russian).

Chesser, R. K. 1983. Cranial variation among populations of the black-tailed prairie dog in New Mexico. Occasional Papers, The Museum, Texas Tech University, 84:1-13.

Cheylan, G. 1990. Endemisme et speciation chez les mammiferes Mediterraneens. Vie et Milieu, 40:137-143.

Chiarelli, A. B. 1975. The chromosomes of the Canidae. Pp. 40-53, *in* The wild canids: Their systematics, behavioral ecology and evolution (M. W. Fox, ed.). Van Nostrand Reinhold Company, New York, 508 pp.

Chiminbu, C. T., and D. J. Kitchener. 1991. A systematic revision of Australian Emballonuridae. Records of the Western Australian Museum, 15:203-265.

Choate, J. R. 1969. Taxonomic status of the shrew, *Notiosorex* (*Xenosorex*) *phillipsii* Schaldach, 1966 (Mammalia: Insectivora). Proceedings of the Biological Society of Washington, 82:469-476.

Choate, J. R. 1970. Systematics and zoogeography of Middle American shrews of the genus *Cryptotis*. University of Kansas, Publications of the Museum of Natural History, 19:195-317.

Choate, J. R. 1973. *Cryptotis mexicana*. Mammalian Species, 28:1-3.

Choate, J. R., and E. D. Fleharty. 1974. *Cryptotis goodwini*. Mammalian Species, 44:1-3.

Choate, J. R., and S. L. Williams. 1978. Biogeographic interpretation of variation within and among populations of the prairie vole, *Microtus ochrogaster*. Occasional Papers, The Museum, Texas Tech University, 49:1-25.

Chorn, J., and R. S. Hoffmann. 1978. *Ailuropoda melanoleuca*. Mammalian Species, 110:1-6.

Civitelli, M. V., P. Consentino, and E. Capanna. 1989. Inter- and intra-individual chromosome variability in *Thamnomys* (*Grammomys*) *gazellae* (Rodentia, Muridae) B-chromosomes and structural heteromorphisms. Genetica, 79:93-105.

Claessen, C. J., and P. de Vree. 1991. Systematic and taxonomic notes on the *Epomophorus-anurus-labiatus-minor* complex with the description of a new species (Mammalia; Chiroptera; Pteropodidae). Senckenbergiana Biologica, 71:209-238.

Clark, J. W. 1873 [1874]. On the eared seals of the Auckland Islands. Proceedings of the Zoological Society of London, 1873:750-760.

Clark, T. W. 1975. *Arctocephalus galapagoensis*. Mammalian Species, 64:1-2.

Clark, T. W., R. S. Hoffmann, and C. F. Nadler. 1971. *Cynomys leucurus*. Mammalian Species, 7:1-4.

Clark, T. W., E. Anderson, C. Douglas, and M. Strickland. 1987. *Martes americana*. Mammalian Species, 289:1-8.

Clough, G. C. 1976. Current status of two endangered Caribbean rodents. Biological Conservation, 10(1):43-48.

Clutton-Brock, J., G. B. Corbet, and M. Hills. 1976. A review of the family Canidae, with classification by numerical methods. Bulletin of the British Museum (Natural History), 29(3):119-199.

Cockrum, E. L. 1976. On the status of the hairy-footed gerbil, *Gerbillus hirtipes* Lataste, 1881. Mammalia, 40:523-526.

Cockrum, E. L. 1977. Status of the hairy footed gerbil, *Gerbillus latastei* Thomas and Trouessart. Mammalia, 41:75-80.

Cockrum, E. L., and H. W. Setzer. 1976. Types and type localities of North African Rodents. Mammalia, 40:633-670.

Cockrum, E. L., P. J. Vaughan, and T. C. Vaughan. 1977. Status of the pale sand rat, *Psammomys vexillaris* Thomas, 1925. Mammalia, 41:321-326.

Coetzee, C. G. 1970. The relative tail-length of striped mice *Rhabdomys pumilio* Sparrman 1784 in relation to climate. Zoologica Africana, 5:1-6.

Coetzee, C. G. 1977*a*. Genus *Steatomys*. Part 6.8. Pp. 1-4, *in* The mammals of Africa: An identification manual (J. Meester and H. W. Setzer, eds.) [issued 22 Aug 1977]. Smithsonian Institution Press, Washington, D. C., not continuously paginated.

Coetzee, C. G. 1977*b*. Order Carnivora. Part 8. Pp. 1-42, *in* The mammals of Africa: An identification manual (J. Meester and H. W. Setzer, eds.) [issued 22 Aug 1977]. Smithsonian Institution Press, Washington, D. C., not continuously paginated.

Cohen, J. A. 1978. *Cuon alpinus*. Mammalian Species, 100:1-3.

Coimbra-Filho, A. F. 1990. Sistemática, distribução geográfica e situação atual dos simios Brasileiros (Platyrrhini-Primates). Revista Brasileira de Biologia, 50:1063-1079.

Collett, R. 1886 [1887]. On *Phascogale viginiae*, a rare pouched mouse from Northern Queensland. Proceedings of the Zoological Society of London, 1886:548-549.

Collier, G. E., and S. J. O'Brien. 1985. A molecular phylogeny of the Felidae: immunological distance. Evolution, 39:473-487.

Colyn, M. 1986. Les mammifères de forêt ombrophile entre les rivières Tshopo et Maiko (Région du Haute-Zaire). Bulletin de la Institut Royal des Sciences Naturelles de Belgique: Biologie, 56:21-26.

Colyn, M., and A. M. Dudu. 1986. Releve systematique des rongeurs (Muridae) des iles forestieres du fleuve Zaire entre Kisangani et Kinshasa. Revue de Zoologie Africaine, 99:353-357.

Colyn, M., A. Gautier-Hion, and D. Thys van den Audenaerde. 1991. *Cercopithecus dryas* Schwarz, 1932 and *C. salongo* Thys van den Audenaerde, 1977 are the same species with an age-related coat pattern. Folia Primatologia, 56:167-170.

Conroy, G. C., M. Pickford, B. Senut, J. Van Couvering, and P. Mein. 1992. *Otavipithecus namibiensis*, first Miocene hominoid from southern Africa. Nature, 356:144-148.

Contoli, L. 1990. Further data about *Crocidura cossyrensis* Contoli, 1989, with respect to other species of the genus in the Mediterranean. Hystrix, New Series, 2:53-58.

Contoli, L., and G. Amori. 1986. First record of a live *Crocidura* (Mammalia, Insectivora) from Pantelleria island, Italy. Acta Theriologica, 31:343-347.

Contoli, L., B. Benincasa-Stagni, and A. R. Marenzi. 1989. Morphometry and morphology of *Crocidura* Wagler 1832 (Mammalia, Soricidae) in Italy, Sardinia and Sicily, with fourier descriptors approach: first results. Hystrix, New Series, 1:113-130.

Contrafatto, G., J. A. Meester, K. Willan, P. J. Taylor, M. A. Roberts, and C. M. Baker. 1992*a*. Genetic variation in the African rodent subfamily Otomyinae (Muridae) II. Chromosomal changes in some populations of *Otomys irroratus*. Cytogenetics and Cell Genetics, 59:293-299.

Contrafatto, G., G. K. Campbell, P. J. Taylor, V. Goossens, K. Willan, and J. A. Meester. 1992*b*. Genetic variation in the African rodent subfamily Otomyinae (Muridae) III. Karyotype and allozymes of the ice rat, *Otomys sloggetti robertsi*. Cytogenetics and Cell Genetics, 60:45-47.

Contreras, J. R., and L. M. Berry. 1983. Notas acerca de los roedores del genero *Oligoryzomys* de la Provincia del Chaco, Republica Argentina (Rodentia, Cricetidae). Historia Natural, 3:145-148.

Contreras, J. R., and O. A. Reig. 1965. Datos sobre la distribución del género *Ctenomys* (Rodentia, Octodontidae) en la zona costera de la Provincia Buenos Aires comprendida entre Necocea y Bahía Blanca. Physis (Buenos Aires), 25(69):169-186.

Contreras, J. R., and M. I. Rosi. 1980. El raton de campo *Calomys musculinus cordovensis* (Thomas) en la provincia de Mendoza. I. Consideraciones taxonomicas. Historia Natural, 1:17-25.

Contreras, J. R., V. G. Roig, and C. M. Suzarte. 1977. *Ctenomys validus*, una nueva especie de "tunduque" de la Provincia de Mendoza (Rodentia; Octodontidae). Physis (Buenos Aires), sec. C, 36(92):159-162.

Contreras, L. C., and J. C. Torres-Mura. 1987. *Tympanectomys*, un mamífero deserticola con el numero diploide mas alto. Archivos de Biología y Medicina Experimentales (Santiago), 20: pages unnumbered, resumen no. R-190.

Convention on International Trade in Endangered Species. 1991. Appendices I, II, and III to The Convention on International Trade in Endangered Species of Wild Fauna and Flora. September 1, 1991. U.S. Government Printing Office, Washington, D.C., 19 pp.

Conway, M., and C. G. Schmitt. 1978. Record of the Arizona shrew (*Sorex arizonae*) from New Mexico. Journal of Mammalogy, 59:631.

Cook, J. A., S. Anderson, and T. L. Yates. 1990. Notes on Bolivian mammals. 6. The genus *Ctenomys* (Rodentia, Ctenomyidae) in the highlands. American Museum Novitates, 2980:1-27.

Cook, J. M., R. Trevelyan, S. S. Walls, M. Hatcher, and F. Rakotondraparany. 1991. The ecology of *Hypogeomys antimena*, an endemic Madagascan rodent. Journal of Zoology (London), 224:191-200.

Cooke, H. B. S., and A. F. Wilkinson. 1978. Suidae and Tayassuidae. Pp. 435-482, *in* Evolution of African mammals (V.J. Maglio and H. B. S. Cooke, eds). Harvard University Press, Cambridge, MA, 641 pp.

Coolidge, H. J. 1940. The Indochinese forest ox or kouprey. Memoir of the Museum of Comparative Zoology at Harvard College, 54:417-531.

Corbet, G. B. 1974. Family Macroscelididae. Part 1.5, Pp. 1-6, *in* The mammals of Africa: an identification manual (J. Meester and H. W. Setzer, eds.) [issued 10 Sep 1974]. Smithsonian Institution Press, Washington, D. C., not continuously paginated.

Corbet, G. B. 1978*a*. *Erinaceus dauuricus* Sundevall, 1842 (Mammalia, Insectivora): proposed conservation under the plenary powers. Bulletin of Zoological Nomenclature, 35:123-124.

Corbet, G. B. 1978*b*. *Sorex dsinezumi* Temminck, 1844 (Mammalia, Insectivora): proposed use of the plenary powers to rule a correct original spelling. Bulletin of Zoological Nomenclature, 35:125-126.

Corbet, G. B. 1978*c*. The mammals of the Palaearctic region: a taxonomic review. British Museum (Natural History), London, 314 pp.

Corbet, G. B. 1979. The taxonomy of *Procavia capensis* in Ethiopia, with special reference to the aberrant tusks of *P. c. capillosa* Brauer (Mammalia, Hyracoidea). Bulletin of the British Museum (Natural History), Zoology Series, 36(4):251-259.

Corbet, G. B. 1983. A review of classification in the family Leporidae. Acta Zoologica Fennica, 174:11-15.

Corbet, G. B. 1984. The mammals of the Palearctic region: a taxonomic review. Supplement. British Museum (Natural History), London, 45 pp.
Corbet, G. B. 1988. The family Erinaceidae: a synthesis of its taxonomy, phylogeny, ecology and zoogeography. Mammal Review, 18:117-172.
Corbet, G. B. 1990. The relevance of metrical, chromosomal and allozyme variation to the systematics of the genus *Mus*. Biological Journal of the Linnean Society, 41:5-12.
Corbet, G. B., and J. Hanks. 1968. A revision of the elephant-shrews, family Macroscelididae. Bulletin of the British Museum (Natural History), Zoology Series, 16:45-111.
Corbet, G. B., and S. Harris. 1990. The handbook of British mammals. Third ed. Blackwell Scientific Publications, Oxford, 588 pp.
Corbet, G. B., and J. E. Hill. 1980. A world list of mammalian species. British Museum (Natural History), London, 226 pp.
Corbet, G. B., and J. E. Hill. 1986. A world list of mammalian species. Second ed. British Museum (Natural History), London, 254 pp.
Corbet, G. B., and J. E. Hill. 1991. A world list of mammalian species. Third ed. British Museum (Natural History) Publications, London, 243 pp.
Corbet, G. B., and J. E. Hill. 1992. Mammals of the Indomalayan region. A systematic review. Oxford University Press, Oxford, 488 pp.
Corbet, G. B., and L. A. Jones. 1965. The specific characters of the crested porcupines, subgenus *Hystrix*. Proceedings of Zoological Society of London, 144:285-300.
Corbet, G. B., and D. W. Yalden. 1972. Recent records of mammals (other than bats) from Ethiopia. Bulletin of the British Museum (Natural History), Zoology Series, 22(8):213-252.
Corkeron, P. J. 1990. Aspects of the behavioral ecology of inshore dolphins *Tursiops truncatus* and *Sousa chinensis* in Moreton Bay, Australia. Pp. 285-292, *in* The bottlenose dolphin (S. Leatherwood, and R. R. Reeves, eds.). Academic Press, New York, 653 pp.
Cornely, J. E., and R. J. Baker. 1986. *Neotoma mexicana*. Mammalian Species, 262:1-7.
Cornely, J. E., and B. J. Verts. 1988. *Microtus townsendii*. Mammalian Species, 325:1-9.
Corti, M., and C. M. Ciabatti. 1990. The structure of a chromosomal hybrid zone of house mice (*Mus*) in central Italy: cytogenetic analysis. Zeitschrift für Zoologische Systematik und Evolutionsforschung, 28:277-288.
Corti, M., M. S. Merani, and G. de Villafane. 1987. Multivariate morphometrics of vesper mice (*Calomys*): preliminary assessment of species, population, and strain divergence. Zeitschrift für Säugetierkunde, 52:236-242.
Coryndon, S. C. 1977. The taxonomy and nomenclature of the Hippopotamidae (Mammalia, Artiodactyla) and a description of two new fossil species. Proceedings - Koninklijke Nederlandse Akademie van Wetenschappen, ser. B(Paleontology, Geology, Physics, and Chemistry), 80:61-88.
Cothran, E. G., and R. L. Honeycutt. 1984. Chromosomal differentiation of hybridizing ground squirrels (*Spermophilus mexicanus* and *S. tridecemlineatus*). Journal of Mammalogy, 65:118-122.
Cothran, E. G., E. G. Zimmerman, and C. F. Nadler. 1977. Genic differentiation and evolution in the ground squirrel subgenus *Ictidomys*. Journal of Mammalogy, 58:610-622.
Couturier, M. A. J. 1954. L'Ours brun (*Ursus arctos*). Grenoble, 904 pp.
Cowan, C. F. 1973. *Proc. Zool. Soc. Lond.*, publication dates. Journal of the Society for the Bibliography of Natural History, 6:293-294.
Cowan, I. McT. 1936. Distribution and variation in deer (genus *Odocoileus*) of the Pacific coastal region of North America. California Fish and Game, 22:155-246.
Cowan, I. McT. 1940. Distribution and variation in the native sheep of North America. American Midland Naturalist, 24:505-580.
Cowan, I. McT., and C. J. Guiguet. 1965. The mammals of British Columbia. Third Ed. British Columbia Provincial Museum, Handbook, 11:1-414.
Crawford-Cabral, J. 1966. Note on the taxonomy of *Genetta*. Zoologica Africana, 2:25-26.
Crawford-Cabral, J. 1969. As genetas de Angola. Boletim do Instituto de Investigação Científica de Angola, 6(1):3-33.
Crawford-Cabral, J. 1970. As genetas da Africa central (República do Zaire, Ruanda e Burrindi). Boletim do Instituto de Investigação Científica de Angola, 7(2):3-23.
Crawford-Cabral, J. 1973. As genetas da Guiné Portuguesa e de Moçambique, Pp. 133-155, *in* Livro de Homenagem ao Professor Fernando Frade Viegas da Costa. Separata. Lisbon, 66:1-329.
Crawford-Cabral, J. 1981. Análise de dados craniométricos no género *Genetta* G. Cuvier. (Carnivora, Viverridae). Memórias da Junta de Investigações Científicas do Ultramar Centro de Zoologica. Lisbon, 66:1-329.

Crawford-Cabral, J. 1983. Patterns of allopatric speciation in some Angolan Muridae. Annales Musée Royal de l'Afrique Centrale, Tervuren-Belgique, ser. 8 (Sciences Zoologiques), 237:153-157.

Crawford-Cabral, J. 1986. A discussion of the taxa to be used in a zoogeographical analysis as illustrated in Angolan Muroidea. Cimbebasia, ser. A, 8:161-166.

Crawford-Cabral, J. 1987. The taxonomic status of *Crocidura nigricans* Bocage, 1889 (Mammalia, Insectivora). Garcia de Orta, Serie de Zoologica, 14: 3-12.

Crawford-Cabral, J. 1988. A craniometric study on Angolan gerbils of the subgenus *Tatera* (Mammalia, Rodentia, Gerbillidae). Part I: Results from a principal components analysis. Zoologische Abhandlungen Staatliches Museum für Tierkunde in Dresden, 43:169-192.

Crawford-Cabral, J. 1989*a*. The prior scientific name of the larger red mongoose (Carnivora: Viverridae: Herpestinae). Garcia de Orta, Serie de Zoologica, 14(2):1-2.

Crawford-Cabral, J. 1989*b*. *Praomys angolensis* (Bocage, 1890) and the identity of *Praomys angolensis* auct. (Rodentia: Muridae), with notes on their systematic positions. Garcia de Orta, Serie de Zoologica, 15:1-10.

Crawford-Cabral, J., and A. P. Pacheco. 1989. A craniometrical study on some water rats of the genus *Dasymys* (Mammalia, Rodentia, Muridae). Garcia de Orta, Serie de Zoologica, 15:11-24.

Crawford-Cabral, J., and A. P. Pacheco. 1991. A craniometric study on Angolan gerbils of subgenus *Tatera* (Mammalia, Rodentia, Gerbillidae). Part II: Results from discriminant and cluster analysis. Zoologische Abhandlungen Staatliches Museum für Tierkunde in Dresden, 46:215-224.

Cristaldi, M., and G. Amori. 1988. Perspectives pour une interpretation historique des populations eoliennes de rongeurs. Bulletin d'Ecologie, 19(2-3):171-176.

Cronin, J. E., and W. E. Meikle. 1979. The phylogenetic position of *Theropithecus*: congruence among molecular, morphological, and palaeontological evidence. Systematic Zoology, 28:259-269.

Crowe, P. E. 1943. Notes on some mammals of the southern Canadian Rocky Mountains. Bulletin of the American Museum of Natural History, 80:391-410.

Crucitti, P. 1976. Biometria di una collazione de *Miniopterus schreibersi* (Nutt.)(Chiroptera) cafturati nel Lagio (Italia). Annali del Museo civico di Storia Naturale di Genova, 81:131-138.

Crusafont-Pairo, M., and F. Petter. 1964. Un murine geant fossile des Iles Canaries *Canariomys bravoi* gen. nov., sp. nov. (rongeurs, murides). Mammalia, Supplement, 28:607-611.

Csaikl, F., M. Habernig, and U. Csaikl. 1990. A comparative restriction endonuclease analysis of nuclear DNA from Austrian *Apodemus* species: intraspecific variability in middle repetitive DNA families contrasts isoenzyme data. Zeitschrift für Säugetierkunde, 55:55-59.

Csuti, B. A. 1979. Patterns of adaptation and variation in the Great Basin kangaroo rat (*Dipodomys microps*). University of California Publications in Zoology, 111:1-69.

Cumming, D. H. M., R. F. Du Toit, and S. N. Stuart. 1990. African elephants and rhinos: Status survey and conservation action plan. I. U. C. N., Gland, Switzerland, 72 pp.

Cummings, W. C. 1985*a*. Bryde's whale - *Balaenoptera edeni*. Pp. 137-154, *in* Handbook of marine mammals: The sirenians and baleen whales (S. H. Ridgway and R. Harrison, eds.). Academic Press, London, 3:1-362.

Cummings, W. C. 1985*b*. Right whales - *Eubalaena glacialis* and *Eubalaena australis*. Pp. 275-304, *in* Handbook of marine mammals: The sirenians and baleen whales (S. H. Ridgway and R. Harrison, eds.). Academic Press, London, 3:1-362.

Currier, M. J. P. 1983. *Felis concolor*. Mammalian Species, 200:1-7.

Cuvier, F. G. 1822 [1823]. Examen des especes formation des genres ou sous-genres Acanthion, Eréthizon, Sinéthère et Sphiggure. Mémoires du Muséum d'Histoire Naturelle (Paris), 9(1822):413-484.

Cuvier, G. [Baron]. 1817. Le règne animal distribué d'après son organisation, pour servir de base à l'histoire naturelle des animaux et d'introduction à l'anatomie comparée. vol. 1. Les mammifères. Deterville, Paris, 540 pp.

Cuvier, G. [Baron]. 1829. Le règne animal distribue d'après son organisation, pour servir de base a l'histoire naturelle des animaux et d'introduction a l'anatomie comparée. vol. 1. Les mammiféres. Nouvelle edition, revue et augmentée. Deterville, Paris, 584 pp.

Czaplewski, N. J. 1983. *Idionycteris phyllotis*. Mammalian Species, 208:1-4.

Czernay, S. 1987. Spiesshirsche und Pudus. A. Ziemsen Verlag, Wittenberg Lutherstadt, 84 pp.

Daams, R. 1981. The dental pattern of the dormice *Dryomys*, *Myomimus*, *Microdyromys* and *Peridyromys*. Utrecht Micropaleontological Bulletins, Special Publication, 3:1-115.

Dagg, A. I. 1971. *Giraffa camelopardalis*. Mammalian Species, 5:1-8.

Dalquest, W. W. 1950. Records of mammals from the Mexican state of San Luis Potosi. Occasional Papers of the Museum of Zoology, Louisiana State University, 23:1-15.

Dandelot, P. 1974. Order Primates. Part 3. Pp. 1-45, *in* The mammals of Africa: An identification manual (J. Meester and H. W. Setzer, eds.) [issued 10 Sep 1974]. Smithsonian Institution Press, Washington, D. C., not continuously paginated.

Dannelid, E. 1991. The genus *Sorex* (Mammalia, Soricidae) - distribution and evolutionary aspects of Eurasian species. Mammal Review, 21:1-20.

Dao Van Tien. 1960. Recherches zoologiques dans la region de Vinh-Linh (Province de Quang-tri, Centre Vietnam). Zoologischer Anzieger, 164:221-239.

Dao Van Tien. 1961. Notes sur une collection de micromammiferes de la region de Hon-Gay. Zoologischer Anzieger, 166:290-308.

Dao Van Tien. 1965. O formakh krasnobrinkhikh belok (*Callosciurus erythraeus*, Sciuridae) i ikh rasprotranenii vo Vietname [On the forms of the squirrel *Callosciurus erythraeus* (Sciuridae) and their distribution in Viet-nam]. Zoologicheskii Zhurnal, 44(8):1238-1244 (in Russian).

Dao Van Tien. 1966. Sur deux rongeurs nouveaux (Muridae, Rodentia) au Nord-Vietnam. Zoologischer Anzieger, 176:438-439.

Dao Van Tien. 1978. Sur une collection de mammiferes du Plateau de Moc Chau (Province de So'n-la, Nord-Vietnam). Mitteilungen aus dem Zoologische Museum in Berlin, 54:377-391.

Dao Van Tien. 1983. On the north Indochinese gibbons (*Hylobates concolor*) in North Vietnam. Journal of Human Evolution, 12:367-372.

Dao Van Tien. 1985. Scientific results of some mammals surveys in North Vietnam (1957-1971). Scientific and Technical Publishing House, Ha Noi, 330 pp.

Dao Van Tien, and Cao Van Sung. 1990. Six new Vietnamese rodents. Mammalia, 54:233-238.

Daoud, A. 1989. Dormice (Rodentia, Gliridae) and their evolution. Przeglad Zoologiczny, 33(2):279-89.

Darviche, D., and P. Orsini. 1982. De souris sympatriques: *Mus spretus* et *Mus musculus domesticus*. Mammalia, 46:205-217.

Darviche, D., F. Benmehdi, J. Britton-Davidian, and L. Thaler. 1979. Donnees preliminaires sur la systematique biochimique des genres *Mus* et *Apodemus* en Iran. Mammalia, 43:427-430.

Davidow-Henry, B. R., J. K. Jones, Jr., and R. R. Hollander. 1989. *Cratogeomys castanops*. Mammalian Species, 338:1-6.

Davies, J. L. 1963. The antitropical factor in cetacean speciation. Evolution, 17(1):107-116.

Davis, B. L., and R. J. Baker. 1974. Morphometrics, evolution, and cytotaxonomy of mainland bats of the genus *Macrotus* (Chiroptera: Phyllostomatidae). Systematic Zoology, 23:26-39.

Davis, D. D. 1964. The giant panda: A morphological study of evolutionary mechanisms. Fieldiana, Zoology, Memoirs, 3:1-339.

Davis, D. H. S. 1949. The affinities of the South African gerbils of the genus *Tatera*. Proceedings of the Zoological Society of London, 118:1002-1018.

Davis, D. H. S. 1962. Distribution patterns of southern African Muridae, with notes on some of their fossil antecedents. Annals of the Cape Provincial Museums, 2:56-76.

Davis, D. H. S. 1965. Classification problems of African Muridae. Zoologica Africana, 1:121-145.

Davis, D. H. S. 1974. The distribution of some small southern African mammals (Mammalia: Insectivora, Rodentia). Annals of the Transvaal Museum, 29:135-184.

Davis, D. H. S. 1975*a*. Genera *Tatera* and *Gerbillurus*. Part 6.4. Pp. 1-7, *in* The mammals of Africa: An identification manual (J. Meester, and H. W. Setzer, eds.) [issued 10 Dec 1975]. Smithsonian Institution, Washington, D. C., not continuously paginated.

Davis, D. H. S. 1975*b*. Genus *Aethomys* Thomas, 1915. Part. 6.6. Pp. 1-5, *in* The mammals of Africa: An identification manual (J. Meester and W. H. Setzer, eds.) [issued 10 Dec. 1975]. Smithsonian Institution Press, Washington, D. C, not continuously paginated.

Davis, J. A. 1978. A classification of otters. Pp. 14-33, *in* Otters (N. Duplaix, ed.). I. U. C. N. Publication, New Series, .

Davis, J., and W. Z. Lidicker. 1975. The taxonomic status of the southern sea otter. Proceedings of the California Academy of Science, 40(14):429-437.

Davis, R., and D. E. Brown. 1989. Role of post-Pleistocene dispersal in determining the modern distribution of Abert's squirrel. Great Basin Naturalist, 49:425-434.

Davis, W. B. 1937. Variations in Townsend pocket gophers. Journal of Mammalogy, 18:145-158.

Davis, W. B. 1939. The Recent mammals of Idaho. Caldwell, Idaho, The Caxton Printers, Ltd., 400 pp.

Davis, W. B. 1940. Distribution and variation of pocket gophers (genus *Geomys*) in the southwestern United States. Bulletin of the Texas Agricultural Experiment Station, 590:1-38.

Davis, W. B. 1944. Geographic variation in brown lemmings (Genus *Lemmus*). Murrelet, 25:19-25.

Davis, W. B. 1965. Review of the *Eptesicus brasiliensis* complex in Middle America with the description of a new subspecies from Costa Rica. Journal of Mammalogy, 46:229-240.

Davis, W. B. 1966. Review of South American bats of the genus *Eptesicus*. Southwestern Naturalist, 11:245-274.

Davis, W. B. 1968. A review of the genus *Uroderma* (Chiroptera). Journal of Mammalogy, 49:676-698.

Davis, W. B. 1969. A review of the small fruit bats (genus *Artibeus*) of Middle America. Southwestern Naturalist, 14:15-29.

Davis, W. B. 1970. A review of the small fruit bats (genus *Artibeus*) of Middle America. Part II. Southwestern Naturalist, 14:389-402.

Davis, W. B. 1976. Geographic variation in the lesser noctilio, *Noctilio albiventris* (Chiroptera). Journal of Mammalogy, 57:687-707.

Davis, W. B. 1980. New *Sturnira* (Chiroptera: Phyllostomidae) from Central and South American, with key to currently recognized species. Occasional Papers, The Museum, Texas Tech University, 70:1-5.

Davis, W. B. 1984. Review of the large fruit-eating bats of the *Artibeus "lituratus"* complex (Chiroptera: Phyllostomidae) in Middle America. Occasional Papers, The Museum, Texas Tech University, 93:1-16.

Davis, W. B., and D. C. Carter. 1978. A review of the round-eared bats of the *Tonatia silvicola* complex, with descriptions of three new taxa. Occasional Papers, The Museum, Texas Tech University, 53:1-12.

Davydov, G. S. 1984. [Distribution and ecology of the forest dormouse (*Dryomys nitedula* Pallas, 1779) in Tadzhikistan]. Izvestiya Akademii Nauk Tadzhikskoi SSR, Seriya Biologicheskikh Nauk, 2:55-60 (in Russian).

Dawkins, W. B. 1868. Fossil animals and geology of Attica, by Albert Gaudry. (Critical summary). Quarterly Journal of the Geological Society of London, 24:pt. 2, translations & notices, pp. 1-7.

Dawson, L., and T. Flannery. 1985. Taxonomic and phylogenetic status of living and fossil kangaroos and wallabies of the genus *Macropus* Shaw (Macropodidae: Marsupialia), with a new subgeneric name for the larger wallabies. Australian Journal of Zoology, 33:473-98.

Dawson, M. R. 1967. Lagomorph history and the stratigraphic record. Pp. 287-316, *in* Essays in paleontology and stratigraphy (C. Teichert and E. L. Yochelson, eds.). University of Kansas, Department of Geology, Special Publication, 2:1-626.

Dawson, M. R. 1969. Osteology of *Prolagus sardus*, a Quaternary ochotonid (Mammalia, Lagomorpha). Palaeovertebrata, 2:157-190.

Day, Y.-T. 1988. Subspecies and geographic variation of *Petaurista alborufus* and *P. petaurista* (Rodentia: Sciuridae). Journal of the Taiwan Museum, 41(1):75-83.

de Beaux, O. 1924. Beitrag zur Kenntnis der Gattung *Potamochoerus* Gray. Zoologische Jahrbücher, 47:379-504.

de Beaux, O. 1934. Mammiferi raccolti dal Prof. G. Scorteccinella Somalia Italiana Centrale e Settentrionale Nel 1931 con l'aggiunta di aleuni mammiferi della Somalia Italiana meridionale. Atti della Societa Italiana De Scienze Naturali e del Museo Civico Di Storia Naturale in Milano, 73:261-300.

DeBlase, A. F. 1971. New distributional records of bats from Iran. Fieldiana, Zoology, 58:9-14.

DeBlase, A. F. 1972. *Rhinolophus euryale* and *R. mehelyi* (Chiroptera, Rhinolophidae) in Egypt and southwest Asia. Israel Journal of Zoology, 21:1-12.

DeBlase, A. F. 1980. The bats of Iran: systematics, distribution, ecology. Fieldiana, Zoology, New Series, 4:1-424.

Debrot, S., and C. Mermod. 1977. Chimiotaxonomie du genre *Apodemus* Kaup, 1829 (Rodentia, Muridae). Revue Suisse de Zoologie, 84:521-526.

de Bruijn, H. 1967. Gliridae, Sciuridae y Eomyidae (Rodentia, Mammalia) miocenos de Calatayud (provincia de Zaragoza, España) y su relación con la biostratigrafia del area. Boletin del Instituto Geologico y Minero de Espaňe, 78:187-373.

Decker, D. M. 1991. Systematics of the coatis, genus *Nasua* (Mammalia: Procyonidae). Proceedings of the Biological Society of Washington, 104(2):370-386.

Decker, D. M., and W. C. Wozencraft. 1991. Phylogenetic analysis of Recent procyonid genera. Journal of Mammalogy, 72(1):42-55.

DeFrees, S. L., and D. E. Wilson. 1988. *Eidolon helvum*. Mammalian Species, 312:1-5.

Degerbøl, M. 1935. Report of the mammals collected by the fifth Thule Expedition to Arctic North America: Zoology. I. Mammals. Report of the Fifth Thule Expedition, 1921-1922, 2(4):1-67.

de Graaff, G. 1975. Family Bathyergidae. Part 6.9. Pp. 1-5, *in* The mammals of Africa: An identification manual (J. Meester and H. W. Setzer, eds.) [issued 10 Dec 1975]. Smithsonian Institution Press, Washington, D. C., not continuously paginated.

de Graaff, G. 1978. Notes on the southern African black-tailed tree rat *Thallomys paedulcus* (Sundevall, 1846) and its occurrence in the Kalahari Gemsbok National Park. Koedoe, 21:181-190.

de Graaff, G. 1981. The rodents of Southern Africa. Butterworths, Durban, 267 pp.

Dekeyser, P. L. 1955. Les mammiferes de l'Afrique noire francaise. Second ed. Institut francais d'Afrique noire, Dakar, 426 pp.

Delany, M. J. 1975. The rodents of Uganda. Trustees of the British Museum (Natural History), London, 165 pp.

de la Torre, L. 1958. The status of the bat *Myotis velifer cobanensis* Goodwin. Proceedings of the Biological Society of Washington, 71:167-170.

Delibes, M., F. Hiraldo, N. Arroyo, and M. Rodriguez. 1980. Disagreement between morphotypes and karyotypes in *Eliomys* (Rodentia, Gliridae): the chromosomes of the Central Morocco garden dormouse. Säugetierkundliche Mitteilungen, 28:289-292.

Delson, E. 1975. Evolutionary history of the Cercopithecidae. Pp. 167-217, *in* Approaches to primate paleobiology (F. S. Szalay, ed.). Contributions to Primatology, 5:1-326.

Delson, E. 1976. The family-group name of the leaf-eating monkeys (Mammalia, Primates): a proposal to give Colobidae Blyth, 1875, precedence over Semnopithecidae Owen, 1843, and Presbytina Gray, 1825. Bulletin of Zoological Nomenclature, 33:85-89.

Delson, E., and P. Andrews. 1975. Evolution and interrelationships of the catarrhine primates. Pp. 405-446, *in* Phylogeny of the Primates (W. P. Luckett and F. S. Szalay, eds.). Plenum Press, New York, 483 pp.

Delson, E., and P. H. Napier. 1976. Request for determination of the generic names of the baboon and the mandrill (Mammalia: Primates, Cercopithecidae) Z.N.(S.) 2093. Bulletin of Zoological Nomenclature, 33(1):46-60.

DeMaster, D. P., and I. Stirling. 1981. *Ursus maritimus*. Mammalian Species, 145:1-7.

Demeter, A. 1982. Prey of the spotted eagle-owl *Bubo africanus* in the Awash National Park, Ethiopia. Bonner Zoologische Beiträge, 33:283-292.

Demeter, A. 1983. Taxonomical notes on *Arvicanthis* (Mammalia: Muridae) from the Ethiopian Rift Valley). Pp. 129-136, *in* Proceedings of the Third International Colloquium on the Ecology and Taxonomy of African Small Mammals (E. Van der Straeten, W. N. Verheyen, and F. De Vree, eds.). Annales Musée Royal de l'Afrique Centrale, Tervuren-Belgique, ser. 8 (Sciences Zoologiques), 237:1-227.

Demeter, A., and G. Topal. 1982. Ethiopian mammals in the Hungarian Natural History Museum. Annales Historico-Naturales Musei Nationalis Hungarici (Budapest), 74:331-349.

Demeter, R., and R. Hutterer. 1986. Small mammals from Mt. Meru and its environs (Northern Tanzania). Cimbebasia, ser. A, 8:199-207.

Dempster, E. R., and M. R. Perrin. 1990. Interspecific aggression in sympatric *Gerbillurus* species. Zeitschrift für Säugetierkunde, 55:392-398.

Dene, H., M. Goodman, and W. Prychodko. 1978. An immunological examination of the systematics of Tupaioidea. Journal of Mammalogy, 59:697-706.

Denisov, V. P. 1963. O gibridizatsii vidov roda *Citellus* Oken [On hybridization of species of the genus *Citellus* Oken]. Zoologicheskii Zhurnal, 42:1887-1899 (in Russian).

Denisov, V. P. 1964. Rasprostranenie malova (*Citellus pygmaeus* Pallas) i rezhevatovo (*Citellus major* Pallas) suslikov v Zavolzhe [Distribution of the little...and reddish...ground squirrels in the Transvolga]. Nauchnie Doklad Vysshie Shkolu, Biologicheskii Nauki, 2:49-54 (in Russian).

Denisov, V. P., and N. I. Smirnova. 1976. Immunological relationships of sousliks genus *Citellus* in the Povolgje region. Acta Theriologica, 21:267-278.

Dennis, E., and J. I. Menzies. 1978. Systematics and chromosomes of New Guinea *Rattus*. Australian Journal of Zoology, 26:197-206.

Dennis, E., and J. I. Menzies. 1979. A chromosomal and morphometric study of Papuan tree rats *Pogonomys* and *Chiruromys* (Rodentia, Muridae). Journal of the Zoological Society of London, 189:315-332.

Denys, C. 1989. Phylogenetic affinities of the oldest East African *Otomys* (Rodentia, Mammalia) from Olduvai Bed I (Pleistocene, Tanzania). Neues Jahrbuch für Geologie und Paleontologie, Monatshefte, 44:705-725.

Denys, C. 1990. The oldest *Acomys* (Rodentia, Muridae) from the Lower Pliocene of South Africa and the problem of its murid affinities. Palaeontographica, Abt. A, 210:79-91.

Denys, C. 1991. Un nouveau rongeur *Mystromys pocockei* sp. nov. (Cricetinae) du Pliocene inferieur de Langebaanweg (Region du Cap, Afrique du Sud). Comptes Rendus de l'Académie des Sciences (Paris), ser. 2, 313:1335-1341.

Denys, C., and J. Michaux. 1992. La troisième molaire supérieure chez les Muridae d'Afrique tropicale et la cas des genres *Acomys*, *Uranomys*, et *Lophuromys*. Bonner Zoologische Beiträge, 43:355-365.

Denys, C., J. Michaux, and Q. B. Hendey. 1987. Les rongeurs *Euryotomys* et *Otomys*: un exemple d'evolution parallele en Afrique tropicale? Comptes Rendus de l'Académie des Sciences (Paris), ser. 3, 305:1389-1395.

de Roguin, L. 1991. Donnees historiques nouvelles sur la presence du rat noir *Rattus rattus* (L.) en Europe Occidentale. Pp. 323-325, *in* Le rongeur et l'espace, Actes du Colloque International, Lyon, 1989 (M. Le Berre and L. Le Guelte, eds.). Raymond Chabaud, Paris, .

Deslongchamps, E. 1866. Observations sur quelques dauphins appartenant a la section des zyphides et description de le te, te d'une espéce de cette section nouvelle pour la faune Française. Bulletin de la Société Linnéene de Normandie (Caen), 10:168-180.

Desmarest, A. G. 1816*a*. Bradype, *Bradypus*, Linn.; Erxleben; Cuv.; Illiger, etc.; *Tardigradus*, Brisson; *Choloepus* et *Prochilus*, Illiger. Pp. 319-328, *in* Nouveau dictionnaire d'histoire naturelle, appliquée aux arts, à l'agriculture, à l'économie rurale et domestique, à la médecine, etc. par une société de naturalistes et d'agriculteurs, Nouv. éd. Ch. Deterville, Paris, 4:1-602.

Desmarest, A. G. 1816-1819*b*. Nouveau dictionnaire d'histoire naturelle, appliquèe aux art, principalement à l'agriculture et à l'economie rurale et domestique; par une société de naturalistes. Nouvelle edition, presqu' entierement refondue et considerablement augmentee. Deterville, Paris, 36 volumes.

Dice, L. R. 1929. The phylogeny of the Leporidae, with description of a new genus. Journal of Mammalogy, 10:340-344.

Dickman, C. R., D. H. King, M. Adams, and P. M. Baverstock. 1988. Electrophoretic identification of a new species of *Antechinus* (Marsupialia: Dasyuridae) in South-eastern Australia. Australian Journal of Zoology, 36:455-463.

Didier, R. 1953. Étude systematique de l'os penien des mammifères. Mammalia, 17(4):260-269.

Didier, R., and F. Petter. 1960. L'os penien de *Jaculus blanfordi* (Murray) 1884, étude comparée de *J. blanfordi, J. jaculus* et *J. orientalis* (Rongeurs, Dipodides). Mammalia, 24(2):171-176.

Diersing, V. A. 1976. An analysis of *Peromyscus difficilis* from the Mexican-United States boundary area. Proceedings of the Biological Society of Washington, 89:451-466.

Diersing, V. A. 1980*a*. Systematics of flying squirrels, *Glaucomys volans* (Linnaeus), from Mexico, Guatemala, and Honduras. Southwestern Naturalist, 125:57-172.

Diersing, V. A. 1980*b*. Systematics and evolution of the pygmy shrews (subgenus *Microsorex*) of North America. Journal of Mammalogy, 61:76-101.

Diersing, V. A. 1981. Systematic status of *Sylvilagus brasiliensis* and *S. insonus* from North America. Journal of Mammalogy, 62:539-556.

Diersing, V. A., and D. F. Hoffmeister. 1977. Revision of the shrew *Sorex merriami* and a description of a new species of the subgenus *Sorex*. Journal of Mammalogy, 58:321-333.

Diersing, V. A., and D. E. Wilson. 1980. Distribution and systematics of the rabbits (*Sylvilagus*) of west-central Mexico. Smithsonian Contributions to Zoology, 297:1-34.

Dieterlen, F. 1961. Beiträge zur Biologie der Stachelmaus, *Acomys cahirinus dimidiatus* Cretzchmar. Zeitschrift für Säugetierkunde, 26:1-13.

Dieterlen, F. 1962. Geburt und Geburtshilfe bei der Stachelmaus, *Acomys cahirinus*. Zeitschrift für Tierpsychologie, 19:191-222.

Dieterlen, F. 1963. Vergleichende Untersuchungen zur Ontogenese von Stachelmaus (*Acomys*) und Wanderratte (*Rattus norvegicus*)- Teil I. Beiträge zum Nesthocker-Nestflüuchter-Problem bei Nagetieren. Zeitschrift für Säugetierkunde, 28:193-227.

Dieterlen, F. 1968. Zur Kenntnis der Gattung *Otomys* (Otomyinae; Muridae; Rodentia). Beiträge zur Systematik, Okologie und Biologie zentralafrikanischer Formen. Zeitschrift für Säugetierkunde, 33:321-352.

Dieterlen, F. 1969*a*. *Dendromus kahuziensis* (Dendromurinae; Cricetidae; Rodentia)—eine neue Art aus Zentralafrika. Zeitschrift für Säugetierkunde, 34:348-353.

Dieterlen, F. 1969*b*. Zur Kenntnis von *Delanymys brooksi* Hayman 1962 (Petromyscinae; Cricetidae; Rodentia). Bonner Zoologische Beiträge, 20:384-395.

Dieterlen, F. 1971. Beiträge zur Systematik, Ökologie und Biologie der Gattung *Dendromus* (Dendromurinae, Cricetidae, Rodentia), insbesondere ihrer zentralafrikanischen Formen. Säugetierkundliche Mitteilungen, 19:97-132.

Dieterlen, F. 1974. Bemerkungen zur Systematik der Gattung *Pelomys* (Muridae; Rodentia) in Äthiopien. Zeitschrift für Säugetierkunde, 39:229-231.

Dieterlen, F. 1975. *Lophuromys medicaudatus* (Muridae; Rodentia) Beschreibung einer neuen Art, aufgrund neuer Ergebnisse zur systematischen Stellung von *Lophuromys luteogaster* Hatt (1934). Bonner Zoologische Beiträge, 26:293-318.

Dieterlen, F. 1976*a*. Bemerkungen über *Leimacomys büttneri* Matschie, 1893 (Dendromurinae, Cricetidae, Rodentia). Säugetierkundliche Mitteilungen, 24:224-228.

Dieterlen, F. 1976*b*. Die afrikanische Muridengattung *Lophuromys* Peters, 1874. Vergleiche an Hand neuer Daten zur Morphologie, Ökologie und Biologie. Stuttgarter Beiträge zur Naturkunde, ser. A (Biologie), 285:1-96.

Dieterlen, F. 1976*c*. Zweiter Fund von *Dendromus kahuziensis* (Dendromurinae; Cricetidae; Rodentia) und weitere *Dendromus*- Fänge im Kivu- Hochland oberhalb 2000 m. Stuttgarter Beiträge zur Naturkunde, ser. A, 286:1-5.

Dieterlen, F. 1978*a*. *Acomys minous* (Bate, 1905)—Kreta- Stachelmaus. Pp. 452-461, *in* Handbuch der Säugetiere Europas (J. Niethammer and F. Krapp, eds.). Akademische Verlagsgesellschaft (Wiesbaden), 1:1-476.

Dieterlen, F. 1978*b*. Beiträge zur Kenntnis der Gattung *Lophuromys* (Muridae: Rodentia) in Kamerun und Gabun. Bonner Zoologische Beiträge, 29:287-299.

Dieterlen, F. 1979. Zur Ausbreitungsgeschichte der Hausratte (*Rattus rattus*) in Ostafrika. Zeitschrift für Angewandte Zoologie, 66:173-184.

Dieterlen, F. 1983. Zur Systematik, Verbreitung und Ökologie von *Colomys goslingi* Thomas & Wroughton, 1907 (Muridae; Rodentia). Bonner Zoologische Beiträge, 34:73-106.

Dieterlen, F. 1987. Neue Erkenntnisse über afrikanische Bürstenhaarmäuse, Gattung *Lophuromys* (Muridae; Rodentia). Bonner Zoologische Beiträge, 38:183-194.

Dieterlen, F. 1991. *Lemniscomys hoogstraali*, an new murid species from Sudan. Bonner Zoologische Beiträge, 42:11-15.

Dieterlen, F., and H. Heim de Balsac. 1979. Zur Ökologie und Taxonomie der Spitzmäuse (Soricidae) des Kivu-Gebietes. Säugetierkundliche Mitteilungen, 27:241-287.

Dieterlen, F., and G. Nikolaus. 1985. Zur Säugetierfauna des Sudan—weitere Erstnachweise und bemerkenswerte Funde. Säugetierkundliche Mitteilungen, 32:205-209.

Dieterlen, F., and H. Rupp. 1976. Die Rotnasenratte *Oenomys hypoxanthus* (Pucheran, 1855) (Muriden, Rodentia)—Erstnachweis für Äthiopien und dritter Fund aus Tansania). Säugetierkundliche Mitteilungen, 24:229-235.

Dieterlen, F., and H. Rupp. 1978. *Megadendromus nikolausi*, gen. nov., sp. nov. (Dendromurinae; Rodentia), ein neuer Nager aus Äthiopien. Zeitschrift für Säugetierkunde, 43:129-143.

Dieterlen, F., and B. Statzner. 1981. The African rodent *Colomys goslingi* Thomas and Wroughton, 1907 (Rodentia: Muridae)—a predator in limnetic ecosystems. Zeitschrift für Säugetierkunde, 46:369-383.

Dieterlen, F., and E. Van der Straeten. 1984. New specimens of *Malacomys verschureni* from eastern Zaire (Mammalia, Muridae). Revue de Zoologie Africaine, 98:861-868.

Dieterlen, F., and E. Van der Straeten. 1988. Deux nouveaux specimens de *Lamottemys okuensis* Petter, 1986 du Cameroun (Muridae: Rodentia). Mammalia, 52:379-385.

Dieterlen, F., and E. Van der Straeten. 1992. Species of the genus *Otomys* from Cameroon and Nigeria and their relationship to East African forms. Bonner Zoologische Beiträge, 43:383-392.

Dietz, J. M. 1985. *Chrysocyon brachyurus*. Mammalian Species, 234:1-4.

Dippenaar, N. J. 1977. Variation in *Crocidura mariquensis* (A. Smith, 1844) in southern Africa, Part 1 (Mammalia: Soricidae). Annals of the Transvaal Museum, 30:163-206.

Dippenaar, N. J. 1979. Variation in *Crocidura mariquensis* (A. Smith, 1844) in southern Africa, Part 2 (Mammalia: Soricidae). Annals of the Transvaal Museum, 32:1-34.

Dippenaar, N. J. 1980. New species of *Crocidura* from Ethiopia and northern Tanzania (Mammalia: Soricidae). Annals of the Transvaal Museum, 32:125-154.

Dippenaar, N. J., and J. A. J. Meester. 1989. Revision of the *luna-fumosa* complex of Afrotropical *Crocidura* Wagler, 1832 (Mammalia: Soricidae). Annals of the Transvaal Museum, 35:1-47.

Dippenaar, N. J., and I. L. Rautenbach. 1986. Morphometrics and karyology of the southern African species of the genus *Acomys* I. Geoffroy Saint-Hilaire, 1838 (Rodentia: Muridae). Annals of the Transvaal Museum, 34:129-183.

Dippenaar, N. J., J. Meester, I. L. Rautenbach, and D. A. Wolhuter. 1983 [1984]. The status of Southern African mammal taxonomy. Annales Musée Royal de l'Afrique Centrale, Sciences Zoologiques, 237:103-107.

Dixon, J. M. 1981. Selection of a neotype for the southern short-nosed (brown) bandicoot, *Isodon* [sic] *obesulus* (Shaw and Nodder, 1797). Victorian Naturalist, 98(3):130-135.

Dixon, K. R., J. A. Chapman, G. R. Willner, D. E. Wilson and W. Lopez-Forment. 1983. The New World jackrabbits and hares (genus *Lepus*)— 2. Numerical taxonomic analysis. Acta Zoologica Fennica, 174:53-56.

Dobler, F. C., and K. R. Dixon. 1990. The pygmy rabbit *Brachylagus idahoensis*. Pp. 111-115, *in* Rabbits, hares and pikas (J. A. Chapman and J. E. C. Flux, eds.). I. U. C. N., Gland, Switzerland, 168 pp.

Dobson, G. E. 1882. A monograph of the Insectivora, systematic and anatomical, part 1: including the families Erinaceidae, Centetidae, and Solenodontidae. John Van Voorst, London, 96 pp.

Dolan, J. M. 1963. Beitrag zur systematischen Gliederung des Tribus Rupicaprini, Simpson 1945. Zeitschrift für Zoologische Systematik und Evolutionsforschung, 1:311-407.

Dolan, J. M. 1988. A deer of many lands - a guide to the subspecies of the red deer *Cervus elaphus* L. Zoonooz, 62(10):4-34.

Dolan, P. G. 1989. Systematics of Middle American mastiff bats of the genus *Molossus*. Special Publications, The Museum, Texas Tech University Press, 23:1-71.

Dolan, P. G., and D. C. Carter. 1973. *Glaucomys volans*. Mammalian Species, 78:1-6.

Dolan, P. G., and T. L. Yates. 1981. Interspecific variation in *Apodemus* from the northern Adriatic islands of Yugoslavia. Zeitschrift für Säugetierkunde, 46:151-161.

Dolgov, V. A. 1979. [*Crocidura leucodon* Hermann, 1780 in Kopetdag (Insectivora, Mammalia)]. Sbornik Trudov Zoologicheskovo Museya MGU, 18:257-263 (in Russian).

Dolgov, V. A. 1985. Burozubki Starovo Sveta [Brown-toothed shrews of the Old World]. Moscow University, Moscow, 220 pp.

Dolgov, V. A., and R. S. Hoffmann. 1977. Tibetskaya burozubka - *Sorex thibetanus* Kastchenko, 1905 (Soricidae Mammalia). Zoologicheskii Zhurnal, 46:1687-1692 (in Russian).

Dolgov, V. A., and I. V. Lukyanova. 1966. O stroenii genitalii Palearkticheskikh burozubko (Insectivora, Soricidae). Zoologicheskii Zhurnal, 56:1852-1861 (in Russian).

Dolgov, V. A., and B. S. Yudin. 1975. [Progress and problems in the investigation of the insectivorous mammals of the USSR.]. Trudy Biologicheskovo Instituta, Novosibirsk, 23:5-40 (in Russian).

Dollman, G. 1911. On *Arvicanthis abyssinicus* and allied East-African species, with descriptions of four new forms. Annals and Magazine of Natural History, ser. 8, 8:334-353.

Domning, D. P. 1978. Sirenian evolution in the North Pacific Ocean. University of California Publications in Geological Science, 118:1-176.

Domning, D. P. 1981. Distribution and status of manatees *Trichechus* spp. near the mouth of the Amazon River, Brazil. Biological Conservation, 19:85-97.

Domning, D. P. 1982. Evolution of manatees: A speculative history. Journal of Paleontology, 56:599-619.

Donnellan, S. C. 1987. Phylogenetic relationships of New Guinean rodents (Rodentia, Muridae) based on chromosomes. Australian Mammalogy, 12:61-67.

Dorst, J. 1972. Notes sur quelques rongeurs observes en Ethiopie. Mammalia, 36:182-192.

Dowler, R. C., and H. H. Genoways. 1978. *Liomys irroratus*. Mammalian Species, 82:1-6.

Downs, C. T., and M. R. Perrin. 1989. An investigation of the macro- and micro-environments of four *Gerbillurus* species. Cimbebasia, 11:41-54.

Downs, C. T., and M. R. Perrin. 1990. Thermal parameters of four species of *Gerbillurus*. Journal of Thermal Biology, 15:291-300.

Dowsett, R. J., and L. Granjon. 1991. Liste préliminaire des mammifères du Congo. Tauraco Research Report, 4:297-310.

Dragoo, J. W., J. R. Choate, T. L. Yates, and T. P. O'Farrell. 1990. Evolutionary and taxonomic relationships among North American arid-land foxes. Journal of Mammalogy, 71(3):318-332.

Dubey, D. D., and R. Raman. 1992. Mammalian sex chromosomes. IV. Replication heterogeneity in the late replicating facultative-and constitutive-heterochromatic regions in the X chromosomes of the mole rats, *Bandicota bengalensis* and *Nesokia indica*. Hereditas, 115:275-282.

Dudu, A. M., E. Van der Straeten, and W. N. Verheyen. 1989. Premiere capture de *Hylomyscus parvus* Brosset, Dubost et Heim de Balzac, 1965 au Zaire avec quelques donnees biometriques (Rodentia, Muridae). Journal of African Zoology, 103:179-182.

Duncan, F. M., F. H. Waterhouse, and H. Peavot. 1937. On the dates of publication of the Society's *Proceedings*, 1830-1858, compiled by the late F. H. Waterhouse, and of the *Transactions*, by the late Henry Peavot, originally published in P.Z.S. 1893, 1913. Proceedings of the Zoological Society of London, 1937:71-81.

Duncan, J. F., and P. F. D. Van Peenen. 1971. Karyotypes of ten rats (Rodentia: Muridae) from Southeast Asia. Caryologia, 24:331-346.

Duncan, P., and R. W. Wrangham. 1971. On the ecology and distribution of subterranean insectivores in Kenya. Journal of Zoology (London), 164:149-163.

Dunlop, J. N., and I. R. Pound. 1981. Observations on the pebble-mound mouse *Pseudomys chapmani*. Records of the Western Australian Museum, 9:1-5.

Duplantier, J. M., J. Britton-Davidian, and L. Granjon. 1990*a*. Chromosomal characterization of three species of the genus *Mastomys* in Senegal. Zeitschrift für Zoologische Systematik und Evolutionsforschung, 28:289-298.

Duplantier, J. M., L. Granjon, E. Mathieu, and F. Bonhomme. 1990*b*. Structures genetiques comparees de trois especes de rongeurs africains du genre *Mastomys* au Senegal. Genetica, 81:179-192.

Duplantier, J. M., L. Granjon, and K. Ba. 1991*a*. Decouverte de trois especes de rongeurs nouvelles pour le Senegal: un indicateur supplementaire de la desertification dans le nord du pays. Mammalia, 55:313-315.

Duplantier, J.-M., L. Granjon, F. Adam, and K. Ba. 1991*b*. Repartition actuelle du rat noir (*Rattus rattus*) au Senegal: facteurs historiques et ecologiques. Pp. 339-346, *in* Le rongeur et l'espace, Actes du Colloque International, Lyon, 1989. Raymond Chabaud, Paris, .

Duthie, A. G., and T. J. Robinson. 1990. The African rabbits. Pp. 121-127, *in* Rabbits, hares and pikas (J. A. Chapman and J. E. C. Flux, eds.). I. U. C. N., Gland, Switzerland, 168 pp.

Dutrillaux, B. 1986. Evolution chromosomique chez les primates, carnivores et les rongeurs. Mammalia, 50 (Special Number):1-203.

Dwyer, P., M. Hockings, and J. Willmer. 1979. Mammals of Cooloola and Beerwah. Proceedings of the Royal Society of Queensland, 90:65-84.

Dyatlov, A. I., and L. A. Avanyan. 1987. [Substantiation of species rank for two subspecies of gerbils (*Meriones*, Cricetidae, Rodentia)]. Zoologicheskii Zhurnal, 66:1069-1074 (in Russian).

Dziurdzik, B. 1978. Histological structure of hair in the Gliridae (Rodentia). Acta Zoologica Cracoviensia, 22(1):1-12.

East, R., (ed.). 1988. Antelopes. Global survey and regional action plans. Part 1. East and northeast Africa. I. U. C. N., Gland, Switzerland, 96 pp.

East, R., (ed.). 1989. Antelopes. Global survey and regional action plans. Part 2. Southern and south-central Africa. I. U. C. N., Gland, Switzerland, 96 pp.

East, R., (ed.). 1990. Antelopes. Global survey and regional action plans. Part 3. West and central Africa. I. U. C. N., Gland, Switzerland, 171 pp.

Eger, J. L. 1977. Systematics of the genus *Eumops*. Royal Ontario Museum, Life Sciences, Contribution, 110:1-69.

Eger, J. L. 1990. Patterns of geographic variation in the skull of Nearctic ermine (*Mustela erminea*). Canadian Journal of Zoology, 68:1241-1249.

Eger, J. L., and R. L. Peterson. 1979. Distribution and systematic relationship of *Tadarida bivittata* and *Tadarida ansorgei* (Chiroptera: Molossidae). Canadian Journal of Zoology, 57:1887-1895.

Egoscue, H. J. 1979. *Vulpes velox*. Mammalian Species, 122:1-5.

Eisenberg, J. F. 1989. Mammals of the Neotropics. The northern Neotropics, vol. 1, Panama, Colombia, Venezuela, Guyana, Suriname, French Guiana. University of Chicago Press, Chicago, IL, 449 pp.

Eisenberg, J. F., and E. Gould. 1970. The tenrecs: a study in mammalian behavior and evolution. Smithsonian Contributions to Zoology, 27:1-138.

Eisenberg, J. F., and E. Gould. 1984. The insectivores. Pp. 155-165, *in* Madagascar (A. Jolly, P. Oberlé, and R. Albignac, eds.). Key environments series. Pergamon Press, Oxford, 239 pp.

Eisenberg, J. F., C. P. Groves, and K. MacKinnon. 1987. Tapire. Grzimeks Enzyklopädie, Säugetiere, 4:598-608.

Eisenmann, V., and J.-C. Turlot. 1978. Sur la taxinomie du genre *Equus* (equides). Les Cahiers de l'Analyse des Données, 3:179-201.

Eisentraut, M. 1969*a*. Das Gaumenfaltenmuster bei westafrikanischen Muriden. Zoologische Jahrbücher, 96:478-490.

Eisentraut, M. 1969*b*. Die Verbreitung der Muriden-Gattung *Hylomyscus* auf Fernando Po und in Westkamerun. Zeitschrift für Säugetierkunde, 34:296-307.

Eisentraut, M. 1970. Die Verbreitung der Muriden-Gattung *Praomys* auf Fernando Po und in Westkamerun. Zeitschrift für Säugetierkunde, 35:1-15.

Eisentraut, M. 1976. Das Gaumenfaltenmuster der Säugetiere und seine Bedeutung für Stammesgeschichtliche und taxonomische Untersuchungen. Bonner Zoologische Monographien, 8:1-214.

Elbl, A., U. Rahm, and G. Mathys. 1966. Les mammiferes et leurs tiques dans La Foret du Ruggege (Republique Rwandaise). Acta Tropica, 23:223-263.

Elder, F. B. 1980. Tandem fusion, centric fusion, and chromosomal evolution in the cotton rats, genus *Sigmodon*. Cytogenetics and Cell Genetics, 26:199-210.

Elder, F. B., and M. R. Lee. 1985. The chromosomes of *Sigmodon ochrognathus* and *S. fulviventer* suggest a realignment of *Sigmodon* species groups. Journal of Mammalogy, 66:511-518.

Eldridge, M. D. B., P. G. Johnston, R. L. Close, and P. S. Lowry. 1989. Chromosomal rearrangements in rock wallabies, *Petrogale* (Marsupialia, Macropodidae) II. G-banding analysis of *Petrogale godmani*. Genome, 32:935-40.

Ellerman, J. R. 1940. The families and genera of living rodents. Vol. 1. Rodents other than Muridae. Trustees of the British Museum (Natural History), London, 689 pp.

Ellerman, J. R. 1941. The families and genera of living rodents. Vol. II. Family Muridae. British Museum (Natural History), London, 690 pp.

Ellerman, J. R. 1947. A key to the Rodentia inhabiting India, Ceylon, and Burma, based on collections in the British Museum. Journal of Mammalogy, 28:249-278; 357-387.

Ellerman, J. R. 1947-1948. Notes on some Asiatic rodents in the British Museum. Proceedings of the Zoological Society of London, 117:259-271.

Ellerman, J. R. 1949*a*. The families and genera of living rodents. Vol. III, Appendix II [Notes on the rodents from Madagascar in the British Museum, and on a collection from the island obtained by Mr. C. S. Webb]. British Museum (Natural History), London, 210 pp.

Ellerman, J. R. 1949*b*. On the prior name for the Siberian lemming and the genotype of *Glis* Erxleben. Annals and Magazine of Natural History, ser. 12, 2:893-894.

Ellerman, J. R. 1961. Rodentia. Volume 3, *in* The fauna of India including Pakistan, Burma and Ceylon. Mammalia. Second ed. Manager of Publications, Zoological Survey of India, Calcutta, vol. 3(in 2 parts), 1:1-482; 2:483-884.

Ellerman, J. R., and T. C. S. Morrison-Scott. 1951. Checklist of Palaearctic and Indian mammals 1758 to 1946. Trustees of the British Museum (Natural History), London, 810 pp.

Ellerman, J. R., and T. C. S. Morrison-Scott. 1953. Checklist of Palaearctic and Indian mammals - amendments. Journal of Mammalogy, 34:516-518.

Ellerman, J. R., and T. C. S. Morrison-Scott. 1955. Supplement to Chasen (1940) A handlist of Malaysian mammals, containing a generic synonomy and a complete index. British Museum (Natural History), London, 66 pp.

Ellerman, J. R., and T. C. S. Morrison-Scott. 1966. Checklist of Palaearctic and Indian Mammals 1758 to 1946. Second ed. British Museum (Natural History), London, 810 pp.

Ellerman, J. R., T. C. S. Morrison-Scott, and R. W. Hayman. 1953. Southern African mammals 1758 to 1951: a reclassification. British Museum (Natural History), London, 363 pp.

Elliot, C. L., and J. T. Flinders. 1991. *Spermophilus columbianus*. Mammalian Species, 372:1-9.

Elliot, D. G. 1901. A synopsis of the mammals of North America and the adjacent seas. Field Columbian Museum, Zoological Series, 2, 45:1-471.

Elliot, D. G. 1905. A checklist of mammals of the North American continent, the West Indies and the neighboring seas. Field Columbian Museum, Publication 105, Zoological Series, 6:1-761.

Elliot, D. G. 1907*a*. A catalogue of the collection of mammals in the Field Columbian Museum. Field Columbian Museum, Zoology Series, 8:1-694.

Elliot, D. G. 1907*b*. Descriptions of apparently new species and subspecies of mammals belonging to the families Lemuridae, Cebidae, Callitrichidae, and Cercopithecidae in the collection of the Natural History Museum. Annals and Magazine of Natural History, ser. 7, 20:185-196.

Elliot, O. 1971. Bibliography of the tree shrews 1780-1969. Primates, 12:323-414.

Ellis, L. S., and L. R. Maxson. 1979. Evolution of the chipmunk genera *Eutamias* and *Tamias*. Journal of Mammalogy, 60:331-334.

El-Rayah, M. A. 1981. A new species of bat of the genus *Tadarida* (Family Molodidae) from West Africa. Royal Ontario Museum, Life Sciences Occasional Paper, 36:1-12.

Emmons, L. H. 1988. Replacement name for a genus of South American rodents (Echimyidae). Journal of Mammalogy, 69(2):421.

Emmons, L. H., and F. Feer. 1990. Neotropical rainforest mammals. A field guide. University of Chicago Press, Chicago, IL, 281 pp., 7 pls.

Emry, R. J. 1970. A North American Oligocene pangolin and other additions to the Pholidota. Bulletin of the American Museum of Natural History, 142:459-510.

Emry, R. J., and R. W. Thorington, Jr. 1982. Descriptive and comparative osteology of the oldest fossil squirrel, *Protosciurus* (Rodentia: Sciuridae). Smithsonian Contributions to Paleobiology, 47:1-35.

Emry, R. J., and R. W. Thorington, Jr. 1984. The tree squirrel *Sciurus* (Sciuridae, Rodentia) as a living fossil. Pp. 23-31, *in* Living fossils (N. Eldredge and S. M. Stanley, eds.). Springer-Verlag, New York, 291 pp.

Enders, R. K. 1939. A new rodent of the genus *Rheomys* from Chiriqui. Proceedings of the Academy of Natural Sciences of Philadelphia, 90:295-296.

Enders, R. K. 1953. The type locality of *Syntheosciurus brochus*. Journal of Mammalogy, 34:509.

Enders, R. K. 1980. Observations on *Syntheosciurus*: taxonomy and behavior. Journal of Mammalogy, 61:725-727.

Engels, H. 1980. Zur Biometrie und Taxonomie von Hausmäusen (genus *Mus* L.) aus dem Mittelmeergebiet. Zeitschrift für Säugetierkunde, 45:366-375.

Engels, H. 1983*a*. Elektrophoretische Untersuchungen an Hausmaeusen (*Mus musculus brevirostris* Waterhouse, 1837 und *Mus spretus* Lataste, 1883) aus Sued-und Mittelportugal zur Ueberpruefung vermuteter Hybridisierung. Ciencia Biologica, Ecologia e Sistematica (Coimbra), 5:97-104.

Engels, H. 1983*b*. Zur Phylogenie und Ausbreitungsgeschichte mediterraner Hausmäuse (genus *Mus* L.) mit Hilfe von "Compatibility Analysis". Zeitschrift für Säugetierkunde, 48:9-19.

Engesser, B. 1975. Revision der europäischen Heterosoricinae (Insectivora, Mammalia). Eclogae Geologicae Helvetiae, 68: 649-671.

Engstrom, M. D. 1984. Chromosomal, genic, and morphological variation in the *Oryzomys melanotis* species group. Unpubl. PhD. dissertation, Texas A & M University, 171 pp.

Engstrom, M. D., and J. W. Bickham. 1982. Chromosome banding and phylogenetics of the golden mouse, *Ochrotomys nuttalli*. Genetica, 59:119-126.

Engstrom, M. D., and J. W. Bickham. 1983. Karyotype of *Nelsonia neotomodon*, with notes on the primitive karyotype of peromyscine rodents. Journal of Mammalogy, 64:685-688.

Engstrom, M. D., and J. R. Choate. 1979. Systematics of the northern grasshopper mouse (*Onychomys leucogaster*) on the central Great Plains. Journal of Mammalogy, 60:723-739.

Engstrom, M. D., and D. E. Wilson. 1981. Systematics of *Antrozous dubiaquercus* (Chiroptera: Vespertilionidae), with comments on the status of *Bauerus* Van Gelder. Annals of Carnegie Museum, 50:371-383.

Engstrom, M. D., R. C. Dowler, D. S. Rogers, D. J. Schmidly, and J. W. Bickham. 1981. Chromosomal variation within four species of harvest mice (*Reithrodontomys*). Journal of Mammalogy, 62:159-164.

Engstrom, M. D., D. J. Schmidly, and P. K. Fox. 1982. Nongeographic variation and discrimination of species within the *Peromyscus leucopus* species group (Mammalia: Cricetinae) in eastern Texas. Texas Journal of Science, 34:149-162.

Engstrom, M. D., H. H. Genoways, and P. K. Tucker. 1987*a*. Morphological variation, karyology, and systematic relationships of *Heteromys gaumeri* (Rodentia: Heteromyidae). Pp. 289-303, *in* Studies in neotropical mammalogy. Essays in honor of Philip Hershkovitz (B. D. Patterson and R. M. Timm, eds.). Fieldiana, Zoology, New Series, 39:1-506.

Engstrom, M. D., T. E. Lee, and D. E. Wilson. 1987*b*. *Bauerus dubiaquercus*. Mammalian Species, 282:1-3.

Engstrom, M. D., O. Sanchez-Herrera, and G. Urbano-Vidales. 1992. Distribution, geographic variation, and systematic relationships within *Nelsonia* (Rodentia: Sigmodontinae). Proceedings of the Biological Society of Washington, 105:867-881.

Erbaeva, M. A. 1988. Pishchukhi Kainozoya [Pikas of the Cenozoic]. Nauka, Moscow-Leningrad, 223 pp. (in Russian).

Erdbrink, D. P. 1953. A review of fossil and Recent bears of the old world. Proefschrift, Utrecht, 2 vol, 597 pp.

Ernest, K. A. 1986. *Nectomys squamipes*. Mammalian Species, 265:1-5.

Ernest, K. A., and M. A. Mares. 1987. *Spermophilus tereticaudus*. Mammalian Species, 274:1-9.

Erxleben, J. C. P. 1777. Systema regni animalis per classes, ordines, genera, species, varietates, cum synonymia et historia animalium. Classis I. Mammalia. Weygandianis, Lipsiae, 636 pp.

Eschricht, D. F., and J. Reinhardt. 1861. Om nordhvalen (*Balaena mysticetus* L.); navning med Hensyn til diens Udbredning i Fortiden of Nutiden of til dens ydre og indre Saekjender. Konigelige Danske Videnskabernes Selskabs Skrifter, 5te Raekke, Naturvidenskabelig og Mathematisk Afdeling, 5te Bind,, 158 pp.

Eshelman, B. D., and G. N. Cameron. 1987. *Baiomys taylori*. Mammalian Species, 285:1-7.

Eshelman, R. E. 1975. Geology and paleontology of the Early Pleistocene (Late Blancan) White Rock Fauna from north-central Kansas. University of Michigan, Museum of Paleontology, Papers on Paleontology, 13:1-60.

Espinosa, M. B., and O. A. Reig. 1991. Cytogenetics and karyosystematics of South American oryzomyine rodents (Cricetidae, Sigmodontinae). III. Banding karyotypes of Argentinian *Oligoryzomys*. Zeitschrift für Säugetierkunde, 56:306-317.

Estes, J. A. 1980. *Enhydra lutris*. Mammalian Species, 133:1-8.

Evans, E. P. 1981. Karyotype of the house mouse. Symposia of the Zoological Society of London, 47:127-139.

Everts, W. 1968. Beitrag zur Systematik des Sonnendachse. Zeitschrift für Säugetierkunde, 33:1-19.

Ewer, R. F. 1957. A collection of *Phacochoerus aethiopicus* teeth from the Kalkbank Middle Stone Age site, central Transvaal. Palaeontologia Africana, 5:5-20.

Ewer, R. F. 1973. The Carnivores. Cornell University Press, Ithaca, NY, 494 pp.

Fa, J. E., and D. J. Bell. 1990. The volcano rabbit *Romerolagus diazi*. Pp. 143-146, *in* Rabbits, hares and pikas (J. A. Chapman and J. E. C. Flux, eds.). I. U. C. N., Gland, Switzerland, 168 pp.

Fagerstone, K. A. 1982. Ethology and taxonomy of Richardson's ground squirrel (*Spermophilus richardsonii*). Unpubl. Ph. D. dissertation, University of Colorado, Boulder, 298 pp.

Fan Nai-chang, and Shi Yin-zhu. 1982. [A revision of the zokors of subgenus *Eospalax*]. Acta Theriologica Sinica, 2:83-199 (in Chinese).

Fang Jye-siung, and G. M. Jagiello. In press. The unique G-banded mitotic karyotype of the Turkish hamster (*Mesocricetus brandti*). Cytologia.

Fashing, N. J. 1973. Implications of karyotypic variation in the kangaroo rat, *Dipodomys heermanni*. Journal of Mammalogy, 54:1018-1020.

Fay, F. H. 1985. *Odobenus rosmarus*. Mammalian Species, 238:1-7.

Fedyk, S. 1970. Chromosomes of *Microtus* (*Stenocranius*) *gregalis major* (Ognev, 1923) and phylogenetic connections between sub-arctic representatives of the genus *Microtus* Schrank, 1798. Acta Theriologica, 15:143-152.

Feiler, A. 1978*a*. Ueber artliche Abgrenzung und innerartliche Ausformung bei *Phalanger maculatus*. Zoologische Abhandlungen Staatliches Museum für Tierkunde in Dresden, 35:1-30.

Feiler, A. 1978*b*. Zur morphologischen Charakteristik des *Phalanger celebensis*. Zoologische Abhandlungen Staatliches Museum für Tierkunde in Dresden, 35:161-168.

Feiler, A. 1978*c*. Bemerkungen über *Phalanger* der "*orientalis*-Gruppe" nach Tate (1945). Zoologische Abhandlungen Staatliches Museum für Tierkunde in Dresden, 34:385-395.

Feiler, A. 1990. Ueber die Säugetiere der Sangihe- und Talaud-Inseln - der Beitrag A. B. Meyers für ihre Erforschung (Mammalia). Zoologische Abhandlungen Staatliches Museum für Tierkunde in Dresden, 46:75-94.

Feldhamer, G. A. 1980. *Cervus nippon*. Mammalian Species, 128:1-7.

Feldhamer, G. A., K. C. Farris-Renner, and C. M. Barker. 1988. *Dama dama*. Mammalian Species, 317:1-8.

Felten, H. 1962. Bemerkungen zu Fledermäusen der Gattungen *Rhinopoma* und *Taphozous* (Mammalia, Chiroptera). Senckenbergiana Biologica, 43:171-176.

Felten, H. 1964*a*. Zur Taxonomie indo-australischer Fledermäuse der Gattung *Tadarida* (Mammalia, Chiroptera). Senckenbergiana Biologica, 45:1-13.

Felten, H. 1964*b*. Flughunde der Gattung *Pteropus* von Neukaledonien und den Loyalty-Inseln. Senckenbergiana Biologica, 45:671-683.

Felten, H., and D. Kock. 1972. Weitere Flughunde der Gattung *Pteropus* von den Neuen Hebriden, sowie den Banks-und Torres-Inseln, Pacifischer Ozean (Mammalia: Chiroptera). Senckenbergiana Biologica, 53:179-188.

Felten, H., and G. Storch. 1968. Eine neue Schläfer-Art, *Dryomys laniger* n. sp. aus Kleinasien (Rodentia: Gliridae). Senckenbergiana Biologica, 49(6):429-435.

Felten, H., F. Spitzenberger, and G. Storch. 1971. Zur Kleinsäugerfauna West-Anatoliens. Teil I. Senckenbergiana Biologica, 52:393-424.

Felten, H., F. Spitzenberger, and G. Storch. 1973. Zur Kleinsäugerfauna West-Anatoliens Teil II. Senckenbergiana Biologica, 54:227-290.

Felten, H., F. Spitzenberger, and G. Storch. 1977. Zur Kleinsäugerfauna West-Anatoliens. Teil IIIa. Senckenbergiana Biologica, 58:1-44.

Feng Zuo-jiang [Feng Tso-chien], and Kao Yüeh-ting. 1974. [Taxonomic notes on the Tibetan pika and allied species—including a new subspecies]. Acta Zoologica Sinica, 20:76-88 (in Chinese).

Feng Zuo-jiang, and Zheng Chang-lin. 1985. [Studies on the pikas (genus *Ochotona*) of China—taxonomic notes and distribution]. Acta Theriologica Sinica, 5:269-290 (in Chinese).

Feng Zuo-jiang, Zheng Chang-lin, and Wu Jia-yan. 1983. [A new subspecies of *Apodemus peninsulae* from Qinghai-Xizang (Tibet) Plateau, China]. Acta Zootaxonomica Sinica, 8:108-112 (in Chinese).

Feng Zuo-jiang, Cai Gui-quan, and Zheng Chang-lin. 1986. [The mammals of Xizang. The comprehensive scientific expedition to the Qinghai-Xizang Plateau]. Science Press, Academia Sinica, Beijing, 423 pp. (in Chinese).

Fenton, M. B., and R. M. R. Barclay. 1980. *Myotis lucifugus*. Mammalian Species, 142:1-8.

Ferguson, W. W. 1981. The systematic position of *Gazella dorcas* (Artiodactyla: Bovidae) in Israel and Sinai. Mammalia, 45:453-457.

Ferguson, W. W., Y. Porath, and S. Paley. 1985. Late Bronze Period yields first osteological evidence of *Dama dama* (Artiodactyla: Cervidae) from Israel and Arabia. Mammalia, 49:209-214.

Fernandes, C., H. Engels, A. Abade, and M. Coutinho. 1991. Zur genetischen und morphologischen Variabilitat der Gattung *Apodemus* (Muridae) im Westen der Iberischen Halbinsel. Bonner Zoologische Beiträge, 42:261-269.

Ferrell, C. S., and D. E. Wilson. 1991. *Platyrrhinus helleri*. Mammalian Species, 373:1-5.

Feustel, H. 1984. Zur Verbreitung der Schläfer (Gliridae) im Odenwald. Bericht Naturwissenschaftlicher Verein Darmstadt, 8:7-18.

Filippucci, M. G. 1986. Nuovo stazione appenninica di *Dryomys nitedula* (Pallas, (Rodentia, Gliridae). Hystrix, 1(1):83-86.

Filippucci, M. G., M. V. Civitelli, and E. Capanna. 1985. Le caryotype du lerotin *Dryomys nitedula* (Pallas) (Rodentia, Gliridae). Mammalia, 49(3):365-368.

Filippucci, M. G., M. V. Civitelli, and E. Capanna. 1986. The chromosomes of *Lemniscomys barbarus* (Rodentia, Muridae). Bollettino di Zoologia, 53:355-358.

Filippucci, M. G., G. Nascetti, E. Capanna, and L. Bullini. 1987. Allozyme variation and systematics of European moles of the genus *Talpa* (Mammalia, Insectivora). Journal of Mammalogy, 68:487-499.

Filippucci, M. G., S. Simson, E. Nevo, and E. Capanna. 1988*a*. The chromosomes of the Israeli garden dormouse, *Eliomys melanurus* Wagner, 1849 (Rodentia, Gliridae). Bollettino di Zoologia, 55:31-33.

Filippucci, M. G., M. V. Civitelli, and E. Capanna. 1988*b*. Evolutionary genetics and systematics of the garden dormouse, *Eliomys* Wagner, 1840. 1. Karyotype divergence. Bollettino di Zoologia, 55:35-45.

Filippucci, M. G., E. Rodino, E. Nevo, and E. Capanna. 1988*c*. Evolutionary genetics and systematics of the garden dormouse, *Eliomys* Wagner, 1840. 2. Allozyme diversity and differentiation of chromosomal races. Bollettino di Zoologia, 55:47-54.

Filippucci, M. G., S. Simson, and E. Nevo. 1989. Evolutionary biology of the genus *Apodemus* Kaup, 1829 in Israel. Allozymic and biometric analyses with description of a new species: *Apodemus hermonensis* (Rodentia, Muridae). Bollettino di Zoologia, 56:361-376.

Filippucci, M. G., F. Catzeflis, and E. Capanna. 1990. Evolutionary genetics and systematics of the garden doormouse, *Eliomys* Wagner, 1840 (Gliridae, Mammalia). 3. Further karyological data. Bollettino di Zoologia, 57:149-152.

Filippucci, M. G., V. Fadda, B. Krystufek, S. Simson, and G. Amori. 1991. Allozyme variation and differentiation in *Chionomys nivalis* (Martins, 1842). Acta Theriologica, 36:47-62.

Findley, J. S. 1955*a*. Taxonomy and distribution of some American shrews. University of Kansas, Museum of Natural History, 7:613-618.

Findley, J. S. 1955*b*. Speciation of the wandering shrew. University of Kansas, Museum of Natural History, 9:1-68.

Findley, J. S. 1972. Phenetic relationships among bats of the genus *Myotis*. Systematic Zoology, 21:31-52.

Findley, J. S., and C. Jones. 1967. Taxonomic relationships of bats of the species *Myotis fortidens*, *M. lucifugus*, and *M. occultus*. Journal of Mammalogy, 48:429-444.

Findley, J. S., and G. L. Traut. 1970. Geographic variation in *Pipistrellus hesperus*. Journal of Mammalogy, 51:741-765.

Findley, J. S., A. H. Harris, D. E. Wilson, and C. Jones. 1975. Mammals of New Mexico. University of New Mexico Press, Albuquerque, 360 pp.

Finley, R. B., Jr. 1958. The wood rats of Colorado: distribution and ecology. University of Kansas Publications, Museum of Natural History, 10:213-552.

Fischer [Von Waldheim], G. 1813-1814. Zoognosia tabulis synopticis illustrata. Nicolai Sergeidis Vsevolozsky, Moscow, 3 vols [3-1814].

Fisler, G. F. 1965. Adaptations and speciation in harvest mice of the marshes of San Francisco Bay. University of California Publications in Zoology, 77:1-108.

Fitch, J. H., and K. A. Shump. 1979. *Myotis keenii*. Mammalian Species, 121:1-3.

Fitch, J. H., K. A. Shump, and A. U. Shump. 1981. *Myotis velifer*. Mammalian Species, 149:1-5.

Fivush, B., R. Parker, and R. H. Tamarin. 1975. Karyotype of the beach vole, *Microtus breweri*, an endemic island species. Journal of Mammalogy, 56:272-273.

Flaherty, S. P., and W. G. Breed. 1982. Structure of hooks on sperm head of plains mouse *Pseudomys australis*. Micron, 13:343-344.

Flannery, T. F. 1983. Revision of the macropodid subfamily Sthenurinae (Marsupialia: Macropodidae) and the relationships of the species of *Troposodon* and *Lagostrophus*. Australian Mammalogy, 6:15-28.

Flannery, T. F. 1988. *Pogonomys championi* n. sp., a new murid (Rodentia) from montane western Papua New Guinea. Records of the Australian Museum, 40:333-341.

Flannery, T. F. 1989. *Microhydromys musseri* n. sp., a new murid (Mammalia) from the Torricelli Mountains, Papua New Guinea. Proceedings of the Linnean Society of New South Wales, 111:215-222.

Flannery, T. F. 1990*a*. *Echymipera davidi*, a new species of Perameliformes (Marsupialia) from Kiriwina Island, Papua New Guinea, with notes on the systematics of the genus *Echymipera*. Pp. 29-35, *in* Bandicoots and bilbies (J. H. Seebeck, P. R. Brown, R. L. Wallis, and C. M. Kemper, eds.). Surrey Beatty and Sons Pty. Ltd., Sydney, 392 pp.

Flannery, T. F. 1990*b*. Mammals of New Guinea. Robert Brown and Associates, 439 pp.

Flannery, T. F. 1990*c*. The rats of Christmas past. Australian Natural History, 23:394-400.

Flannery, T. F. 1991. Emperor, King and little pig: the three rats of Guadalcanal. Australian Natural History, 23:635-641.

Flannery, T. F., and L. Seri. 1990. The mammals of southern West Sepik Province, Papua New Guinea: their distribution, abundance, human use and zoogeography. Records of the Australian Museum, 42:173-208.

Flannery, T. F., and J. P. White. 1991. Animal translocation. Zoogeography of New Ireland mammals. National Geographic Research and Exploration, 7:96-113.

Flannery, T. F., and S. Wickler. 1990. Quaternary murids (Rodentia: Muridae) from Buka Island, Papua New Guinea, with descriptions of two new species. Australian Mammalogy, 13:127-139.

Flannery, T. F., S. van Dyck, and M. Krogh. 1985. Notes on the distribution, abundance, diet and habitat of the New Guinea murid (Rodentia) *Xenuromys barbatus* (Milne-Edwards, 1900). Australian Mammalogy, 8:111-115.

Flannery, T. F., M. Archer, and G. Maynes. 1987. The phylogenetic relationships of living phalangerids with a suggested new taxonomy. Pp. 477-506, *in* Possums and opossums, studies in evolution (M. Archer, ed.). Royal Zoological Society of New South Wales, 1:1-460; 2:461-788.

Flannery, T. F., P. V. Kirch, J. Specht, and M. Spriggs. 1988. Holocene mammal faunas from archaeological sites in island Melanesia. Archaeology in Oceania, 23:89-94.

Flannery, T. F., K. Aplin, C. P. Groves, and M. Adams. 1989. Revision of the New Guinean genus *Mallomys* (Muridae: Rodentia), with descriptions of two new species from subalpine habitats. Records of the Australian Museum, 41:83-105.

Fleharty, E. D. 1960. The status of the gray-necked chipmunk in New Mexico. Journal of Mammalogy, 41:235-242.

Flerov, C. C. 1931. On the generic characters of the Fam. *Tragulidae* (*Mamm., Artiodactyla*). Doklady Akademii Nauk SSSR, 1931:75-79.

Flerov, C. C. 1960. Fauna of the USSR. Mammals. Vol. I, no. 2. Musk deer and deer [A translation of C. C. Flerov, 1952, Fauna SSSR. Mlekopitayushchie. tom. 1, vyp. 2. Kabargi i oleni]. Israel Program for Scientific Translations, Washington, D. C., 257 pp.

Flerov, C. C. 1979. Sistematika i evolyutsiya [Systematics and evolution]. Pp. 9-127, *in* Zubr: Morfologiya, sistematika, evolyutsiya, ekologiya [European bison: morphology, systematics, evolution, ecology] (V. E. Sokolov, ed.). Nauka, Moscow, 495 pp. (in Russian).

Flower, W. H. 1869. On the value of the characters of the base of the cranium in the classification of the order Carnivora, and on the systematic position of *Bassaris* and other disputed forms. Proceedings of the Zoological Society of London, 1869:4-37.

Flower, W. H. and J. G. Garson. 1884. Catalogue of the specimens illustrating the osteology and dentition of vertebrated animals Recent and extinct contained in the museum of the Royal College of Surgeons of England. Part II. Class Mammalia other than man. Printed for the College, London, 779 pp.

Flux, J. E. C. 1983. Introduction to taxonomic problems in hares. Acta Zoologica Fennica, 174:7-10.

Flux, J. E. C. 1990. The Sumatran rabbit *Nesolagus netscheri*. Pp. 137-139, *in* Rabbits, hares and pikas (J. A. Chapman and J. E. C. Flux, eds.). I. U. C. N., Gland, Switzerland, 168 pp.

Flux, J. E. C., and R. Angermann. 1990. The hares and jackrabbits. Pp. 61-94, *in* Rabbits, hares and pikas (J. A. Chapman and J. E. C. Flux, eds.). I. U. C. N., Gland, Switzerland, 168 pp.

Flux, J. E. C., A. G. Duthie, T. J. Robinson, and J. A. Chapman. 1990. Exotic populations. Pp. 147-153, *in* Rabbits, hares and pikas (J. A. Chapman and J. E. C. Flux, eds.). I. U. C. N., Gland, Switzerland, 168 pp.

Flynn, L. J. 1982. Variability of incisor enamel microstructure within *Gerbillus*. Journal of Mammalogy, 63:162-165.

Flynn, L. J. 1990. The natural history of rhizomyid rodents. Pp. 155-183, *in* Evolution of subterranean mammals at the organismal and molecular levels (E. Nevo and O. A. Reig, eds.). Alan R. Liss, Inc., New York, 422 pp.

Flynn, L. J., N. A. Neff, and R. H. Tedford. 1988. Phylogeny of the Carnivora, Pp. 73-115, *in* The phylogeny and classification of the tetrapods (M. J. Benton, ed.). Clarendon Press, Oxford, 2(Mammals):1-329.

Fokanov, V. A. 1966. Novyi podvid surka-baibaka i zamechaniya o geograficheskoi izmenchivosti Marmota bobac Müll. [A new subspecies of bobak marmot, and notes on geographical variability of *Marmota bobac* Müll.]. Zoologicheskii Zhurnal, 45:1862-1866 (in Russian).

Fokin, I. M. 1971. [Comparative anatomy of the muscles of the pelvic appendage of the genera *Sicista* and *Salpingotus* (on the position of the subfamily Cardiocraniinae in the system of the Dipodidae)]. Trudy Zoologicheskovo Instituta, Akademiya Nauk SSSR, Leningrad, 48:181-197 (in Russian).

Fooden, J. 1964. Rhesus and crab-eating macaques: intergradation in Thailand. Science, 143:363-365.

Fooden, J. 1967. Identification of the stump-tailed monkey, *Macaca speciosa* I. Geoffroy, 1826. Folia Primatologica, 5:153-164.

Fooden, J. 1969. Taxonomy and evolution of the monkeys of Celebes (Primates: Cercopithecidae). Bibliotheca Primatologica, 10:1-148.

Fooden, J. 1971. Report on the primates collected in western Thailand January-April, 1967. Fieldiana, Zoology, 59:1-62.

Fooden, J. 1975. Taxonomy and evolution of liontail and pigtail macaques (Primates: Cercopithecidae). Fieldiana, Zoology, 67:1-169.

Fooden, J. 1976. Provisional classification and key to living species of macaques. Folia Primatologia, 25:225-236.

Fooden, J. 1979. Taxonomy and evolution of the *sinica* group of macaques: 1. Species and subspecies accounts of *Macaca sinica*. Primates, 20(1):109-140.

Fooden, J. 1980. Classification and distribution of living macaques (*Macaca* Lacépède, 1799). Pp. 1-9, *in* The macaques (D. G. Lindburg, ed.). Van Nostrand Reinhold Company, New York, NY, 384 pp.

Fooden, J. 1981. Taxonomy and evolution of the *sinica* group of macaques. 2. Species and subspecies accounts of the Indian Bonnet Macaque, *Macaca radiata*. Fieldiana, Zoology, New Series, 9:1-52.

Fooden, J. 1982. Taxonomy and evolution of the *sinica* group of macaques: 3. Species and subspecies accounts of *Macaca assamensis*. Fieldiana, Zoology, New Series, 10:1-52.

Fooden, J. 1983. Taxonomy and evolution of the *sinica* group of macaques: 4. Species account of *Macaca thibetana*. Fieldiana, Zoology, New Series, 17:1-20.
Fooden, J. 1987. Type locality of *Hylobates concolor lecogenys*. American Journal of Primatology, 12:107-110.
Fooden, J., A. Mahabal, and S. Sekhar Saha. 1981. Redefinition of rhesus macaque-bonnet macaque boundary in peninsular India (Primates: *Macaca mulatta*, *M. radiata*). Journal of the Bombay Natural History Society, 78(3):463-474.
Fooden, J., Quan Guo-qiang, Wang Zong-ren, and Wang Ying-xiang. 1985. The stumptail macaques of China. American Journal of Primatology, 8(1):11-30.
Foppen, R. P. B., P. J. M. Bergers, and J. J. van Gelder. 1989. Occurrrence and distribution of the garden dormouse *Eliomys quercinus* in the Netherlands. Lutra, 32(1):42-52.
Ford, L. S., and R. S. Hoffmann. 1988. *Potos flavus*. Mammalian Species, 321:1-9.
Fordyce, R. E. 1989. Origins and evolution of Antarctic marine mammals. Pp. 269-281, *in* Origin and evolution of the Antarctic Biota (Crame, J. A., ed.). Geological Society Special Publication, 47:1-322.
Forsten, A., and P. M. Youngman. 1982. *Hydrodamalis gigas*. Mammalian Species, 165:1-3.
Forster, G. 1780. Herrn von Buffon's Naturgeschichte der vierfussigen Thiere. Mit vermehrungen, aus dem französischen übersetzt [vol. 6 in the 23 vol. work of Buffon's 1772-1801]. J. Pauli, Berlin, vol. 6 [publ. in 1780].
Foster, H. L., J. D. Small, and J. G. Fox (eds.). 1981. The mouse in biomedical research. Volume I. History, genetics and wild mice. American College of Laboratory Animal Medicine Series, Academic Press, New York, 306 pp.
Foster, J. B., and R. L. Peterson. 1961. Age variation in *Phenacomys*. Journal of Mammalogy, 42:44-53.
Foster-Turley, P., S. Macdonald, and C. Mason (eds.). 1990. Otters: An action plan for their conservation. I. U. C. N., Gland, Switzerland, 126 pp.
Fox, B. J., and D. A. Briscoe. 1980. *Pseudomys pilligaensis*, a new species of murid rodent from the Pilliga Scrub, Northern New South Wales. Australian Mammalogy, 3:109-126.
Fraguedakis-Tsolis, S. E., B. P. Chondropoulos, J. J. Lykakis, and J. C. Ondrias. 1983. Taxonomic problems of woodmice, *Apodemus* ssp., of Greece approached by electrophoretic and immunological methods. Mammalia, 47:333-337.
Franco, D. 1988. Habit, biometry and taxonomy of the dormouse (*Glis glis*, Linnaeus, 1776) of the Asiago Plateau. Lavori, Societa Veneziana di Scienze Naturali, 13:135-142.
Franzmann, A. W. 1981. *Alces alces*. Mammalian Species, 154:1-7.
Frase, B., and R. S. Hoffmann. 1980. *Marmota flaviventris*. Mammalian Species, 135:1-8.
Fraser, F. C. 1966. Comments on the Delphinoidea. Pp. 7-31, *in* Whales, dolphins and porpoises (K. S. Norris, ed.). University of California Press, Berkeley, 789 pp.
Fraser, F. C., and P. E. Purves. 1960. Hearing in cetaceans. Bulletin of the British Museum (Natural History), 7(1):1-140.
Frazier, M. K. 1977. New records of *Neofiber leonardi* (Rodentia: Cricetidae) and the paleoecology of the genus. Journal of Mammalogy, 58:368-373.
Fredga, K. 1972. Chromosome studies in mongooses (Carnivora, Viverridae). Thesis summary, faculty of sciences of the University of Lund, 11 pp.
Fredga, K., M. Jaarola, R. A. Ims, H. Steen, and N. G. Yoccoz. 1990. The 'common vole' in Svalbard identified as *Microtus epiroticus* by chromosome analysis. Polar Research, 8:283-290.
Freeman, P. W. 1981. A multivariate study of the family Molossidae (Mammalia: Chiroptera): morphology, ecology, evolution. Fieldiana, Zoology, New Series, 7:1-173.
Freitas, T. R. O., M. S. Mattevi, L. F. B. Oliveira, M. J. Souza, Y. Yonenaga-Yassuda, and F. M. Salz. 1983. Chromosome relationships in three representatives of the genus *Holochilus* (Rodentia, Cricetidae) from Brazil. Genetica, 61:13-20.
French, T. W. 1980. *Sorex longirostris*. Mammalian Species, 143:1-3.
Freye, H. 1978. *Castor fiber* Linnaeus, 1758.—Europäischer Biber. Pp. 184-200, *in* Handbuch der Säugetiere Europas (J. Niethammer and F. Krapp, eds.). Akademische Verlagsgesellschaft (Wiesbaden), 1:1-476.
Frías, A. I., V. Berovides, and C. Fernandez. 1988. Current status of jutiita, *Capromys sanfelipensis*. Doñana, Acta Vertebrata, 15(2):252-254.
Frick, C. 1914. A new genus and some new species and subspecies of Abyssinian rodents. Annals of Carnegie Museum, 9:7-28.
Friend, G. 1991. Encounters with Nuunjala. Wildlife Australia, summer:9-11.
Friend, J. A., and N. D. Thomas. 1990. The water-rat, *Hydromys chrysogaster* (Muridae) on Dorre Island, W.A. Western Australian Naturalist, 18:92-93.
Frisch, J. L. 1774 [1775]. Das Natursystem der vierfüssigen Thiere in Tabellen, zum Nutzen der erwachsenen Schuljugend [ICZN Opinion 258. Rejected for nomenclatural purposes]. Glogau.
Fritzell, E. K., and K. J. Haroldson. 1982. *Urocyon cinereoargenteus*. Mammalian Species, 189:1-8.

Froehlich, J. W., and P. H. Froehlich. 1987. The status of Panama's endemic howling monkeys. Primate Conservation, 8:58-66.

Frost, D. R., W. C. Wozencraft, and R. S. Hoffmann. 1991. Phylogenetic relationships of hedgehogs and gymnures (Mammalia: Insectivora: Erinaceidae). Smithsonian Contributions to Zoology, 518:1-69.

Fujita, M. S., and T. H. Kunz. 1984. *Pipistrellus subflavus*. Mammalian Species, 228:1-6.

Fuller, F., M. R. Lee, and L. R. Maxson. 1984. Albumin evolution in *Peromyscus* and *Sigmodon*. Journal of Mammalogy, 65:466-473.

Fülling, O. 1992. Ergähzande angaben über gaumenfaltenmuster vonhagetierun (Mammalia: Rodentia) aus Kamerun. Bonner Zoologische Beiträge, 43:415-421.

Furley, C. W., H. Tichy, and H.-P. Uerpmann. 1988. Systematics and chromosomes of the Indian gazelle, *Gazella bennetti* (Sykes, 1831). Zeitschrift für Säugetierkunde, 53:48-54.

Gadi, I. K., and T. Sharma. 1983. Cytogenetic relationships in *Rattus*, *Cremnomys*, *Millardia*, *Nesokia* and *Bandicota*. Genetica, 61:21-40.

Gaisler, J. 1970. The bats (Chiroptera) collected in Afghanistan by the Czechosloslovak Expeditions of 1965-1967. Acta Scientiarum Naturalium, Academiae Scientarium Bohemoslovacea (Brno), N. S., 4(6):1-56.

Galiano, H., and D. Frailey. 1977. *Chasmaporthetes kani*, new species from China, with remarks on phylogenetic relationships of genera within the Hyaenidae (Mammalia, Carnivora). American Museum Novitates, 2632:1-16.

Gallardo, M. 1979. Las especies chilenas de *Ctenomys* (Rodentia, Octodontidae). I. Estabilidad cariotípica. Archivos de Biología y Medicina Experimentales (Santiago), 12:71-82.

Gallardo, M. H., and E. Palma. 1990. Systematics of *Oryzomys longicadatus* (Rodentia: Muridae) in Chile. Journal of Mammalogy, 71:333-342.

Gallardo, M. H., and B. D. Patterson. 1985. Chromosomal differences between two nominal subspecies of *Oryzomys longicaudatus* Bennett. Mammalian Chromosome Newsletter, 25:49-53.

Galleni, L., R. Stanyon, A. Tellini, G. Giordano, and L. Santini. 1992. Karyology of the Savi pine vole, *Microtus savii* De Salys-Longchamps, 1838 (Rodentia, Arvicolidae): G-, C-, DA/DAPI-, and AluI-bands. Cytogenetics and Cell Genetics, 59:290-292.

Gallo-Reynoso, J. P., and J. L. Solorzano-Velasco. 1991. Two new sightings of California sea lions on the southern coast of Mexico. Marine Mammal Science, 7(1):96.

Gambaryan, P. P., E. G. Potapova, and I. M. Fokin. 1980. [Morphofunctional analysis of the myology of the jerboa head]. Trudy Zoologicheskovo Instituta, Akademiya Nauk SSSR, 91:3-51 (in Russian).

Gambell, R. 1985*a*. Sei whale - *Balaenoptera borealis*. Pp. 155-170, *in* Handbook of marine mammals: The sirenians and baleen whales (S. H. Ridgway and R. Harrison, eds.). Academic Press, London, 3:1-362.

Gambell, R. 1985*b*. Fin whale - *Balaenoptera physalus*. Pp. 171-192, *in* Handbook of marine mammals: The sirenians and baleen whales (S. H. Ridgway and R. Harrison, eds.). Academic Press, London, 3:1-362.

Gamperl, R. 1982. Chromosomal evolution in the genus *Clethrionomys*. Genetica, 57:193-197.

Gannon, M. R., M. R. Willig, and J. K. Jones, Jr. 1989. *Sturnira lilium*. Mammalian Species, 333:1-5.

Gannon, W. L. 1988. *Zapus trinotatus*. Mammalian Species, 315:1-5.

Gannon, W. L., and T. E. Lawlor. 1989. Variation of the chip vocalization of three species of Townsend chipmunks (genus *Eutamias*). Journal of Mammalogy, 70:740-753.

Ganslosser, U. 1980. Vergleichende Untersuchungen zur Kletterfähigkeit einiger Baumkänguruh-Arten (Dendrolagus, Marsupialia). Zoologischer Anzeiger, 205:43-66.

Gao Yao-ting [Kao Yüeh-ting]. 1963. [Taxonomic notes on the Chinese musk-deer]. Acta Zoologica Sinica, 15:479-488 (in Chinese).

Gao Yao-ting. 1983. Current studies on the Chinese Yarkand hare. Acta Zoologica Fennica, 174:23-25.

Gao Yao-ting. 1985. Classification and distribution of the musk deer (*Moschus*) in China. Pp. 113-116, *in* Contemporary mammalogy in China and Japan (T. Kawamichi, ed.). Mammalogical Society of Japan, 194 pp.

Gao Yao-ting (ed.). 1987. [Fauna Sinica, Mammalia: Carnivora]. Science Press, Academia Sinica, Beijing, 377 pp. (in Chinese).

Gao Yao-ting [Kao Yüeh-ting], and Feng Tso-chien [Feng Zuo-jiang]. 1964. [On the subspecies of the Chinese grey-tailed hare, *Lepus oiostolus* Hodgson]. Acta Zootaxonomica Sinica, 1:19-30 (in Chinese).

García-Perea, R. 1992. New data on the systematics of lynxes. Cat News, 16:15-16.

Gardner, A. L. 1971. Karyotypes of two rodents from Peru, with a description of the highest diploid number recorded from a mammal. Experientia, 27:1088-1089.

Gardner, A. L. 1973. The systematics of the genus *Didelphis* (Marsupialia: Didelphidae) in North and Middle America. Special Publications, The Museum, Texas Tech University, 4:1-81.

Gardner, A. L. 1976. The distributional status of some Peruvian mammals. Occasional Papers of the Museum of Zoology, Louisiana State University, 48:1-18.

Gardner, A. L. 1982. Virginia opossum. Pp. 3-36, *in* Wild mammals of North America (J. A. Chapman and G. A. Feldhamer, eds.). Johns Hopkins Press, Baltimore, MD, 1147 pp.

Gardner, A. L. 1983*a*. *Oryzomys caliginosus* (raton pardo, raton arrocero pardo, Costa Rican dusky rice rat). Pp. 483-485, *in* Costa Rican natural history (D. H. Janzen, ed.). University of Chicago Press, Chicago, IL, 816 pp.

Gardner, A. L. 1983*b*. *Proechimys semispinosus* (Rodentia: Echimyidae): distribution, type locality, and taxonomic history. Proceedings of the Biological Society of Washington, 96(1):134-144.

Gardner, A. L. 1986. The taxonomic status of *Glossophaga morenoi* Martinez and Villa, 1938 (Mammalia: Chiroptera: Phyllostomidae. Proceedings of the Biological Society of Washington, 99:489-492.

Gardner, A. L. 1989 [1990]. Two new mammals from southern Venezuela and comments on the affinities of the highland fauna of Cerro de la Neblina. Pp. 411-424, *in* Advances in Neotropical mammalogy (K. Redford and J. F. Eisenberg, eds.). Sandhill Crane Press, Gainesville, FL, 614 pp.

Gardner, A. L., and D. C. Carter. 1972. A review of the Peruvian species of *Vampyrops* (Chiroptera: Phyllostomatidae). Journal of Mammalogy, 53:72-82.

Gardner, A. L., and G. K. Creighton. 1989. A new generic name for Tate's *microtarsus* group of South American mouse opossums (Marsupialia: Didelphidae). Proceedings of the Biological Society of Washington, 102:3-7.

Gardner, A. L., and L. H. Emmons. 1984. Species groups in *Proechimys* (Rodentia, Echimyidae) as indicated by karyology and bullar morphology. Journal of Mammalogy, 65:10-25.

Gardner, A. L., and C. S. Ferrell. 1990. Comments on the nomenclature of some Neotropical bats (Mammalia: Chiroptera. Proceedings of the Biological Society of Washington, 103:501-508.

Gardner, A. L., and J. P. O'Neill. 1969. The taxonomic status of *Sturnira bidens* (Chiroptera: Phyllostomidae) with notes on its karyotype and life history. Occasional Papers of the Museum of Zoology, Louisiana State University, 38:1-8.

Gardner, A. L., and J. L. Patton. 1976. Karyotypic variation in oryzomyine rodents (Cricetinae) with comments on chromosomal evolution in the Neotropical cricetine complex. Occasional Papers of the Museum of Zoology, Louisiana State University, 49:1-48.

Gardner, A. L., R. K. LaVal, and D. E. Wilson. 1970. The distributional status of some Costa Rican bats. Journal of Mammalogy, 51:712-729.

Garrison, T. E., and T. L. Best. 1990. *Dipodomys ordii*. Mammalian Species, 353:1-10.

Gaskin, D. E., P. W. Arnold, and B. A. Blair. 1974. *Phocoena phocoena*. Mammalian Species, 42:1-8.

Gautherin, H. 1988. Mammalogie. Bulletin Trimestriel de la Societe d'Histoire Naturelle et des Amis du Museum d'Autun, 126:41.

Gauthier-Pilters, H., and A. Innis Dagg. 1981. The camel: Its evolution, ecology, behavior and relationships to man. University of Chicago Press, Chicago, IL, 208 pp.

Gautun, J.-C., M. Tranier, and B. Sicard. 1985. Liste preliminaire des rongeurs du Burkina Faso (ex Haute-Volta). Mammalia, 49:537-542.

Gautun, J.-C., I. Sankhon, and M. Tranier. 1986. Nouvelle contribution a la connaissance des rongeurs du massif guineen des monts Nimba (Afrique occidentale). Systematique et aperçu quantitatif. Mammalia, 50:205-217.

Gavrila, L., A. Lungeanu, D. Murariu, and C. Stepan. 1986. New contributions to the cytogenetic study of *Microtus epiroticus* (Ondrias, 1966) (Mammalia, Arvicolidae). Travaux du Museum d'Historie Naturalae "Grigore Antipa", 28:271-274.

Gazaryan, M. A. 1985. [On certain questions of the life style of the forest dormouse (*Dryomys nitedula* Pall.)]. Izvestiya Sel'skokhozyaistvennykh Nauk, 1985(7):40-44 (in Russian).

Gemmeke, H. 1980. Proteinvariation und Taxonomie in der Gattung *Apodemus* (Mammalia, Rodentia). Zeitschrift für Säugetierkunde, 45:348-365.

Gemmeke, H. 1981. Genetische Unterschiede zwischen rechts-und linksrheinischen Waldmausen (*Apodemus sylvaticus*). Bonner Zoologische Beiträge, 32:265-269.

Gemmeke, H. 1983. Proteinvariation bei Zwergwaldmäusen (*Apodemus microps* Kratochvil und Rosicky, 1952). Zeitschrift für Säugetierkunde, 48:155-160.

Gemmeke, H., and J. Niethammer. 1982. Zur Charakterisierung der Waldmäuse (*Apodemus*) Nepals. Zeitschrift für Säugetierkunde, 47:33-38.

Gemmeke, H., and J. Niethammer. 1984. Zur Taxonomie der Gattung *Rattus* (Rodentia, Muridae). Zeitschrift für Säugetierkunde, 49:104-116.

Genest, H., and F. Petter. 1975. Part 1.1, Family Tenrecidae. Pp. 1-7, *in* The mammals of Africa: an identification manual (J. Meester and H. W. Setzer, eds.). [issued in Dec 1975]. Smithsonian Institution, Washington, D. C., not continuously paginated.

Genest-Villard, H. 1967. Revision du genre *Cricetomys* (Rongeurs, Cricetidae). Mammalia, 31:390-455.

Genest-Villard, H. 1978. Revision systematique du genre *Graphiurus* (Rongeurs, Gliridae). Mammalia, 42(4):391-426.

Genoud, M., and R. Hutterer. 1990. *Crocidura russula* (Hermann, 1780) - Hausspitzmaus. Pp. 429- 452, *in* Handbuch der Säugetiere Europas (J. Niethammer & F. Krapp, eds.). Aula-Verlag, Wiesbaden, 3/I:1-524.

Genoways, H. H. 1973. Systematics and evolutionary relationships of spiny pocket mice, genus *Liomys*. Special Publications, The Museum, Texas Tech University, 5:1-368.

Genoways, H. H., and R. J. Baker. 1972. *Stenoderma rufum*. Mammalian Species, 18:1-4.

Genoways, H. H., and E. C. Birney. 1974. *Neotoma alleni*. Mammalian Species, 41:1-4.

Genoways, H. H., and J. R. Choate. 1972. A multivariate analysis of systematic relationships among populations of the short-tailed shrew (genus *Blarina*) in Nebraska. Systematic Zoology, 21:106-116.

Genoways, H. H., and J. K. Jones, Jr. 1969*a*. Notes on pocket gophers from Jalisco, Mexico, with descrptions of two new subspecies. Journal of Mammalogy, 50:748-755.

Genoways, H. H., and J. K. Jones, Jr. 1969*b*. Taxonomic status of certain long-eared bats (genus *Myotis*) from the southwestern United States and Mexico. Southwestern Naturalist, 14:1-13.

Genoways, H. H., and J. K. Jones, Jr. 1971. Systematics of southern banner-tailed kangaroo rats of the *Dipodomys phillipsii* group. Journal of Mammalogy, 52:265-287.

Genoways, H. H., and J. K. Jones, Jr. 1972. Mammals from southwestern North Dakota. Occasional Papers, The Museum, Texas Tech University, 6:1-36.

Genoways, H. H., and S. L. Williams. 1979. Notes on bats (Mammalia: Chiroptera) from Bonaire and Curaçao, Dutch West Indies. Annals of Carnegie Museum, 48:311-321.

Genoways, H. H., R. C. Dowler, and C. H. Carter. 1981. Intraisland and interisland variation in Antillean populations of *Molossus molossus* (Mammalia: Molossidae). Annals of Carnegie Museum, 50:475-492.

Gentile de Fronza, T. 1970. Cariotipo de *Scapteromys tumidus aquaticus* (Rodentia, Cricetidae). Physis (Buenos Aires), 30:343.

Gentry, A. W. 1972. Genus *Gazella*. Part 15.1. Pp. 85-93, *in* The mammals of Africa: An identification manual (J. Meester and H. W. Setzer, eds.) [issued 2 May 1972]. Smithsonian Institution Press, Washington, D. C., not continuously paginated.

Gentry, A. W. 1975. A quagga, *Equus quagga* (Mammalia, Equidae), at University College, London and a note on a supposed quagga in the City Museum, Bristol. Bulletin of the British Museum (Natural History), Zoology Series, 28:217-226.

Gentry, A. W. 1990. Evolution and dispersal of African Bovidae. Pp. 195-227, *in* Horns, pronghorns, and antlers (G. A. Bubenik and A. B. Bubenik, eds). Springer-Verlag, New York, 562 pp.

Gentry, A. W., and A. Gentry. 1978. Fossil Bovidae (Mammalia) of Olduvai Gorge, Tanzania. Part I. Bulletin of the British Museum (Natural History), Geology Series, 29:289-446.

Geoffroy Saint-Hilaire, É. 1803. Catalogue des mammifères du Muséum National d'Histoire Naturelle. Muséum National d'Histoire Naturelle, Paris [printed draft, but not published], 272 pp.

Geoffroy Saint-Hilaire, I. 1839. Notice sur deux nouveaux genres de mammifères carnassiers, les ichneumies, du continent African, et les galidies, de Madagascar. Magasin de Zoologie, ser. 2, 1:1-39.

Geoffroy Saint-Hilaire, I. 1847. Vie, travaux et doctrine scientifique d'Étienne Geoffroy Saint-Hilaire. Paris, P. Bertrand, 3+1+479 pp.

Geoffroy Saint-Hilaire, I. 1851. Catalogue méthodique de la collection des mammifères de la collection des oiseaux et des collection annexes. Muséum National d'Histoire Naturelle, Gide et Baudry, Paris, 96 pp.

George, G. G. 1979. The status of endangered Papua New Guinea mammals. Pp. 93-100, *in* The Status of endangered Australasian mammals (M. Tyler, ed.). Royal Zoological Society of New South Wales, Sydney, 210 pp.

George, G. G., and Schürer. 1978. Some notes on macropods commonly misidentified in zoos. International Zoo Yearbook, 18:152-156.

George, S. B. 1986. Evolution and historical biogeography of soricine shrews. Systematic Zoology, 35: 153-162.

George, S. B. 1988. Systematics, historical biogeography, and evolution of the genus *Sorex*. Journal of Mammalogy, 69:443-461.

George, S. B. 1989. *Sorex trowbridgii*. Mammalian Species, 337:1-5.

George, S. B., and J. D. Smith. 1991. Inter- and intraspecific variation among coastal and island populations of *Sorex monticolus* and *Sorex vagrans*. Pp. 75-91, *in* The biology of the Soricidae (J. S. Findley and T. L. Yates, eds.). Special Publication, The Museum of Southwestern Biology, 1:1-91.

George, S. B., J. R. Choate, and H. H. Genoways. 1981. Distribution and taxonomic status of *Blarina hylophaga* Elliot (Insectivora: Soricidae). Annals of Carnegie Museum, 50:493-513.

George, S. B., H. H. Genoways, J. R. Choate, and R. J. Baker. 1982. Karyotypic relationships within the short-tailed shrews, genus *Blarina*. Journal of Mammalogy, 63:639-645.

George, S. B., J. R. Choate, and H. H. Genoways. 1986. *Blarina brevicauda*. Mammalian Species, 261:1-9.

George, W. 1979. The chromosomes of the hystricomorphous family Ctenodactylidae (Rodentia: ? Sciuromorpha) and their bearing on the relationships of the four living genera. Zoological Journal of the Linnaean Society, 65:261-280.

George, W. 1982. *Ctenodactylus* (Ctenodactylidae, Rodentia): one species or two? Mammalia, 46:375-380.

Gerasimov, S., H. Nikolov, V. Mihailova, J.-C. Auffray, and F. Bonhomme. 1990. Morphometric stepwise discriminant analysis of the five genetically determined European taxa of the genus *Mus*. Biological Journal of the Linnean Society, 41:47-64.

Ghose, R. K. 1965. A new species of mongoose (Mammalia: Carnivora: Viverridae) from West Bengal. Proceedings of the Zoological Society of Calcutta, 18:173-178.

Ghose, R. K. 1978. Observations on the ecology and status of the hispid hare in Rajagarh forest, Darrang District, Assam, in 1975 and 1976. Journal of the Bombay Natural History Society, 75:206-209.

Ghose, R. K., and S. S. Saha. 1981. Taxonomic review of Hodgson's giant flying squirrel, *Petaurista magnificus* (Hodgson)(Sciuridae: Rodentia), with description of a new subspecies from Darjeeling District, West Bengal, India. Journal of the Bombay Natural History Society, 78:93-102.

Giagia, E. B., and J. C. Ondrias. 1980. Karyological analysis of eastern European hedgehog *Erinaceus concolor* (Mammalia, Insectivora) in Greece. Mammalia, 44:59-71.

Giagia, E. B., I. Savic, and B. Soldatovic. 1982. Chromosomal forms of the mole rat *Microspalax* from Greece and Turkey. Zeitschrift für Säugetierkunde, 47:231-236.

Giagia, E. B., S. E. Fraguedakis-Tsolis, and B. P. Chondropoulos. 1987. Contribution to the study of the taxonomy and zoogeography of wild house mouse, genus *Mus* L. (Mammalia, Rodentia, Muridae) in Greece. II. Karyological study of two populations from southern Greece. Mammalia, 51:111-116.

Gibb, J. A. 1990. The European rabbit *Oryctolagus cuniculus*. Pp. 116-120, *in* Rabbits, hares and pikas (J. A. Chapman and J. E. C. Flux, eds.). I. U. C. N., Gland, Switzerland, 168 pp.

Gileva, E. A. 1972. [Chromosomal polymorphism in two close forms of subarctic mice (*Microtus hyperboreus* and *M. middendorffi*)]. Doklady Akademii Nauk SSSR, 203:698-692 (in Russian).

Gileva, E. A. 1980. Chromosomal diversity and an aberrant genetic system of sex determination in the Arctic lemming, *Dicrostonyx torquatus* Pallas (1779). Genetica, 52/53:99-103.

Gileva, E. A. 1983. A contrasted pattern of chromosome evolution in two genera of lemmings, *Lemmus* and *Dicrostonyx* (Mammalia, Rodentia). Genetica, 60:173-179.

Gileva, E. A., and V. B. Fedorov. 1991. Sex ratio, XY females and absence of inbreeding in a population of the wood lemming, *Myopus schisticolor* Lilljeborg, 1844. Heredity, 66:351-355.

Gileva, E. A., I. E. Benenson, A. V. Pokrovskii, and N. A. Lobanova. 1980. [Analysis of aberrant sex ratio and postnatal mortality in progeny of the Arctic lemming *Dicrostonyx torquatus*]. Ekologiya, 6:46-52 (in Russian).

Gileva, E. A., V. N. Bolshakov, N. F. Chernousova, and V. P. Mamina. 1982. [Cytogenetical differentiation of forms in the group of Pamir (*Microtus juldaschi*) and Carruther's (*M. carruthersi*) voles (Mammalia, Microtinae)]. Zoologicheskii Zhurnal, 61:912-922 (in Russian).

Gileva, E. A., I. A. Kuznetsova, M. I. Cheprakov. 1984. [Chromosome sets and taxonomy of true lemmings of the genus *Lemmus*]. Zoologicheskii Zhurnal, 63:105-114 (in Russian).

Gileva, E. A., D. E. Rybnikov, and G. P. Miroshnichenko. 1989. [DNA-DNA hybridization and phylogenetic relationships in two genera of voles, *Alticola*, and *Clethrionomys* (Microtinae-Rodentia)]. Doklady Akademii Nauk SSSR, 311:477-480 (in Russian).

Gill, A. E. 1980. Partial reproductive isolation of subspecies of the California vole, *Microtus californicus*. Pp. 105-117, *in* Animal genetics and evolution (N. N. Vorontsov and J. M. Van Brink, eds.). Dr. W. Junk, The Hague, 353 pp.

Gill, A., B. Petrov, S. Zivkovic, and D. Rimsa. 1987. Biochemical comparisons in Yugoslavian rodents of the families Arvicolidae and Muridae. Zeitschrift für Säugetierkunde, 52:247-256.

Gill, T. 1872. Arrangement of the families of mammals with analytical tables. Smithsonian Miscellaneous Collections, 11:1-98.

Ginsberg, J. R., and D. W. Macdonald. 1990. Foxes, wolves, jackals, and dogs: an action plan for the conservation of canids. I. U. C. N., Gland, Switzerland, 116 pp.

Ginsburg, L. 1982. Sur la position systematique du petit panda, *Ailurus fulgens* (Carnivora, Mammalia). Geobios, Mémoire Spécial, 6:259-277.

Glanz, W. E., and S. Anderson. 1990. Notes on Bolivian mammals. 7. A new species of *Abrocoma* (Rodentia) and relationships of the Abrocomidae. American Museum Novitates, 2991:1-32.

Glass, B. P. 1947. Geographic variation in *Perognathus hispidus*. Journal of Mammalogy, 28:174-179.

Glass, B. P., and R. J. Baker. 1968. The status of the name *Myotis subulatus* Say. Proceedings of the Biological Society of Washington, 81:257-260.

Glass, G. E., and N. B. Todd. 1977. Quasi-continuous variation of the second upper premolar in *Felis bengalensis* Kerr 1792 and its significance for some fossil lynxes. Zeitschrift für Säugetierkunde, 42(1):36-44.

Glazier, D. S. 1980. Ecological shifts and the evolution of geographically restricted species of North American *Peromyscus* (mice). Journal of Biogeography, 7:63-83.

Glover, I. 1986. Archaeology in eastern Timor, 1966-67. Terra Australis, ll:1-241.

Godthelp, H. 1989. *Pseudomys vandycki*, a tertiary murid from Australia. Memoirs of the Queensland Museum, 28:171-173.

Goldman, C. A. 1984. Systematic revision of the African mongoose genus *Crossarchus* (Mammalia: Viverridae). Canadian Journal of Zoology, 62:1618-1630.

Goldman, C. A. 1987. *Crossarchus obscurus*. Mammalian Species, 290:1-5.

Goldman, C. A., and M. E. Taylor. 1990. *Liberiictis kuhni*. Mammalian Species, 348:1-3.

Goldman, D. P., R. Giri, and S. J. O'Brien. 1989. Molecular genetic-distance estimates among the Ursidae as indicated by one- and two-dimensional protein electrophoresis. Evolution, 43(2):282-295.

Goldman, E. A. 1910. Revision of the woodrats of the genus *Neotoma*. North American Fauna, 31:1-124.

Goldman, E. A. 1911. Revision of spiny pocket mice (genera *Heteromys* and *Liomys*). North American Fauna, 34:1-70.

Goldman, E. A. 1913. Descriptions of new mammals from Panama and Mexico. Smithsonian Miscellaneous Collections, 60(22):1-20.

Goldman, E. A. 1916. Notes on the genera *Isothrix* Wagner and *Phyllomys* Lund. Proceedings of the Biological Society of Washington, 29:125-128.

Goldman, E. A. 1918. The rice rats of North America (Genus *Oryzomys*). North American Fauna, 43:1-100.

Goldman, E. A. 1920. Mammals of Panama. Smithsonian Miscellaneous Collections, 69:1-309.

Goldman, E. A. 1932. Review of woodrats of *Neotoma lepida* group. Journal of Mammalogy, 13:59-67.

Goldman, E. A. 1937. The wolves of North America. Journal of Mammalogy, 18:37-45.

Goldman, E. A. 1950. Raccoons of North and Middle America. North American Fauna, 60:1-153.

Golenishchev, F. N., and O. V. Sablina. 1991. [On taxonomy of *Microtus* (*Blanfordimys*) *afghanus*]. Zoologicheskii Zhurnal, 70:98-110 (in Russian).

Good, A. I. 1947. Les rongeurs du Cameroun. Bulletin de la Société L'Etudes Camerounaises, 17-18:5-20.

Goodall, R. N. P., A. R. Galeazzi, S. Leatherwood, K. W. Miller, I. S. Cameron, R. K. Kastelein, and A. P. Sobral. 1988. Studies of Commerson's dolphins, *Cephalorhynchus commersonii*, off Tierra del Fuego, 1976-1984, with a review of information on the species in the South Atlantic. Pp. 3-70, *in* Biology of the genus *Cephalorhynchus* (R. L. Brownell, Jr., and G. P. Donovan, eds.). Reports of the International Whaling Commission, Special Issue, 9:1-344.

Goodwin, G. G. 1934. Mammals collected by A. W. Anthony in Guatemala, 1924-1928. Bulletin of the American Museum of Natural History, 68:1-60.

Goodwin, G. G. 1946. Mammals of Costa Rica. Bulletin of the American Museum of Natural History, 87:271-471.

Goodwin, G. G. 1955. Mammals from Guatemala, with the description of a new little brown bat. American Museum Novitates, 1744:1-5.

Goodwin, G. G. 1958. Three new bats from Trinidad. American Museum Novitates, 1877:1-6.

Goodwin, G. G. 1959*a*. Descriptions of some new mammals. American Museum Novitates, 1967:1-8.

Goodwin, G. G. 1959*b*. Bats of the Subgenus *Natalus*. American Museum Novitates, 1977:1-22.

Goodwin, G. G. 1966. A new species of vole (genus *Microtus*) from Oaxaca, Mexico. American Museum Novitates, 2243:1-4.

Goodwin, G. G. 1969. Mammals from the state of Oaxaca, Mexico, in the American Museum of Natural History. Bulletin of the American Museum of Natural History, 141:1-269.

Goodwin, G. G., and A. M. Greenhall. 1961. A review of the bats of Trinidad and Tobago. Bulletin of the American Museum of Natural History, 122:189-301.

Goodwin, G. G., and A. M. Greenhall. 1962. Two new bats from Trinidad, with comments on the status of the genus *Mesophylla*. American Museum Novitates, 2080:1-18.

Goodwin, R. E. 1979. The bats of Timor: systematics and ecology. Bulletin of the American Museum of Natural History, 163:73-122.

Goodyear, N. C. 1991. Taxonomic status of the silver rice rat, *Oryzomys argentatus*. Journal of Mammalogy, 72:723-730.

Gordon, D. H. 1986. Extensive chromosomal variation in the pouched mouse, *Saccostomus campestris* (Rodentia, Cricetidae) from southern Africa: a preliminary investigation of evolutionary status. Cimbebasia, ser. A, 8:37-47.

Gordon, D. H. 1987. Discovery of another species of tree rat. Transvaal Museum Bulletin, 22:30-32.

Gordon, D. H. 1991. Chromosomal variation in the water rat *Dasymys incomtus* (Rodentia: Muridae). Journal of Mammalogy, 72:411-414.

Gordon, D. H., and I. L. Rautenbach. 1980. Species complexes in medically important rodents: chromosome studies of *Aethomys, Tatera*, and *Saccostomus* (Rodentia: Muridae, Cricetidae). South African Journal of Science, 76:559-561.

Gordon, D. H., and C. R. B. Watson. 1986. Identification of cryptic species of rodents (*Mastomys, Aethomys, Saccostomus*) in the Kruger National Park. South African Journal of Zoology, 21:95-99.

Gorecki, A., R. Meczeva, T. Pis, S. Gerasimov, and W. Walkowa. 1990. Geographical variation of thermoregulation in wild populations of *Mus musculus* and *Mus spretus*. Acta Theriologica, 35:209-214.

Gorman, M. L., and R. D. Stone. 1990. The natural history of moles. Comstock, Ithaca, 160 pp.

Görner, M., and A. Henkel. 1988. Zum Vorkommen und zur Ökologie der Schläfer (Gliridae) in der DDR. Säugetierkundliche Informationen, 2(12):515-535.

Gosalbez, J., and V. Sans-Coma. 1977. Datos sobre *Microtus arvalis* Pallas, 1778, del Pirineo Catalan. Publicationes. Departamento de Zoologia (Barcelona), 11:59-68.

Gosalbez-Noguera, J., V. Sans-Coma, and H. Kahmann. 1989. Der Gartenschläfer *Eliomys q. quercinus* L., 1758, im Bergland Andorra: Morphometrie, Erscheinungsbild, Wachstum und Fortpflanzung (Mammalia: Rodentia). Spixiana, 12(3):323-335.

Graf, J.-D. 1982. Genetique biochimique, zoogeographie et taxonomie des Arvicolidae (Mammalia, Rodentia). Revue Suisse de Zoologie, 89:749-787.

Graf, J.-D., J. Hausser, A. Farina, and P. Vogel. 1979. Confirmation du statut spécifique de *Sorex samniticus* Altobello, 1926 (Mammalia, Insectivora). Bonner Zoologische Beiträge, 30:14-21.

Grafodatsky, A. S., V. T. Volobuev, D. V. Ternovskii, and S. I. Radzhabli. 1976. [G-banding of chromosomes in seven species of Mustelidae (Carnivora)]. Zoologicheskii Zhurnal, 55:1704-1709 (in Russian).

Grafodatsky, A. S., S. I. Radzhabli, A. V. Sharshov, and M. V. Zaitsev. 1988. [Karyotypes of five *Crocidura* species of the USSR fauna]. Tsitologiya, 30:1247-1255 (in Russian).

Grandidier, G. 1905. Recherches sur les lemuriens disparus et en particuliers sur ceux qui vivaient a Madagascar. Nouvelles Archives de la Museum de l'Histoire Naturelle, ser. 4, 8:1-140.

Grant, P. R. 1974. Reproductive compatibility of voles from separate continents (Mammalia: *Clethrionomys*). Journal of Zoology (London), 174:245-254.

Gray, A. P. 1954. Mammalian hybrids: a checklist with bibliography. Commonwealth Agricultural Bureau, Farnham Royal, United Kingdom, 262 pp.

Gray, G. G., and C. D. Simpson. 1980. *Ammotragus lervia*. Mammalian Species, 144:1-7.

Gray, J. E. 1821. On the natural arrangement of vertebrose animals. London Medical Repository, 15(1):296-310.

Gray, J. E. 1825. Outline of an attempt at the disposition of the Mammalia into tribes and families with a list of the genera apparently appertaining to each tribe. Annals of Philosophy, New Series, ser. 2, 10:337-344.

[Gray, J. E.]. 1827. Synopsis of the species of the Class Mammalia, as arranged with reference to their organization, by Cuvier, and other naturalists, with specific characters, synonyma, &c. &c., vol. 5, *in* The animal kingdom arranged in conformity with its organization, by the Baron Cuvier, with additional descriptions of all the species hitherto named, and of many not before noticed. (E. Griffith, C. H. Smith, and E. Pidgeon, eds.). G. B. Whittaker, London, 391 pp.

Gray, J. E. 1828-1830 [-1924]. Spicilegia zoologica; or original figures and short systematic descriptions of new and unfigured animals. London, 1:1-8[1828]; 2:9-12[1830]; 3:1-13[1924].

Gray, J. E. 1832. On the family of Viverridae and its generic sub-divisions, with an enumeration of the species of several new ones. Proceedings of the Committee of Science and Correspondence of the Zoological Society of London, 1832(2):63-68.

Gray, J. E. 1837 [1838]. On a new species of paradoxure (*Paradoxurus derbianus*) with remarks on some Mammalia recently purchased by the British Museum, and characters of the new species. Proceedings of the Zoological Society of London, 1837:67.

Gray, J. E. 1842. Description of two new species of Mammalia discovered in Australia by Captain George Gray (*Tarsipes Spenseræ* and *Chæropus castanotis*. Annals and Magazine of Natural History, 9:39-42.

Gray, J. E. 1843. List of the specimens of Mammalia in the collection of the British Museum. British Museum (Natural History) Publications, London, 216 pp.

Gray, J. E. 1846. On the cetaceous animals. Pp. 13-53, *in* The zoology of the voyage of H. M. S. Erebus and Terror, under the command of Capt. Sir J. C. Ross, R. N., F. R. S., during the years 1839 to 1843 (Sir J. Richardson and J. E. Gray, eds.) [1844-1875]. E. W. Janson, London, 2 vols.

Gray, J. E. 1849 [1850]. Vertebrata, *in* The zoology of the voyage of H. M. S. Samarang; under the command of Captain Sir Edward Belcher, C. B., F. R. A. S., F. G. S., during the years 1843-1846 (A. Adams, ed.). Reeve and Benham, London, 250 pp.

Gray, J. E. 1850. Catalogue of the specimens of mammalia in the collection of the British Museum. Part I. Cetacea. British Museum (Natural History), London, 153 pp.
Gray, J. E. 1853 [1855]. Observations on some rare Indian animals. Proceedings of the Zoological Society of London, 1853:190-192.
Gray, J. E. 1854. The ounces. Annals and Magazine of Natural History, ser. 2, 14:394.
Gray, J. E. 1864 [1865]. A revision of the genera and species of viverrine animals (Viverridae) founded on the collection in the British Museum. Proceedings of the Zoological Society of London, 1864:502-579.
Gray, J. E. 1869. Catalogue of carnivorous, pachydermatous and edentate mammals in the British Museum. British Museum (Natural History) Publications, London, 398 pp.
Gray, J. E. 1874. Description of a new species of cat (*Felis badia*) from Sarawak. Proceedings of the Zoological Society of London, 1874:322-323.
Green, C. A., H. Keogh, D. H. Gordon, M. Pinto, and E. K. Hartwig. 1980. The distribution, identification, and naming of the *Mastomys natalensis* species complex in southern Africa (Rodentia: Muridae). Journal of Zoology (London), 192(1):17-23.
Green, J. S., and J. T. Flinders. 1980. *Brachylagus idahoensis*. Mammalian Species, 125:1-4.
Greenbaum, I. F., and R. J. Baker. 1976. Evolutionary relationships in *Macrotus* (Mammalia: Chiroptera): biochemical variation and karyology. Systematic Zoology, 25:15-25.
Greenbaum, I. F., and R. J. Baker. 1978. Determination of the primitive karyotype of *Peromyscus*. Journal of Mammalogy, 59:820-834.
Greene, E. C. 1935. Anatomy of the rat. Transactions of the American Philosophical Society, New Series, 27:1-370.
Greenhall, A. M., G. Joermann, U. Schmidt, and M. R. Seidel. 1983. *Desmodus rotundus*. Mammalian Species, 202:1-6.
Greenhall, A. M., U. Schmidt, and G. Joermann. 1984. *Diphylla ecaudata*. Mammalian Species, 227:1-3.
Gregory, W. K. 1936. On the phylogenetic relationships of the giant panda (*Ailuropoda*) to other carnivores. American Museum Novitates, 878:1-29.
Gregory, W. K., and M. Hellman. 1939. On the evolution and major classification of the civets (Viverridae) and allied fossil and recent Carnivora: Phylogenetic study of the skull and dentition. Proceedings of the American Philosophical Society, 81(3):309-392.
Griffin, M. 1990. A review of taxonomy and ecology of gerbilline rodents of the central Namib Desert, with keys to the species (Rodentia: Muridae). Pp. 83-98, *in* Namib ecology: 25 years of Namib research (M. K. Seely, ed.). Transvaal Museum Monograph, Transvaal Museum, Pretoria, 7:1-230.
Griffiths, M. 1978. The biology of monotremes. Academic Press, New York, 367 pp.
Griffiths, T. A., K. F. Koopman, and A. Starrett. 1991. The systematic relationship of *Emballonura nigrescens* to other species of *Emballonura* and *Coleura* (Chiroptera: Emballonuridae). American Museum Novitates, 2996:1-16.
Grinnell, J. 1922. A geographical study of the kangaroo rats of California. University of California Publications in Zoology, 24:1-124.
Grinnell, J. 1933. Review of the Recent mammal fauna of California. University of California Publications in Zoology, 40:71-234.
Grinnell, J., and J. Dixon. 1918. Natural history of the ground squirrels of California. Monthly Bulletin of the California State Commission on Horticulture, 7:597-708.
Gromov, I. M. 1990. [On the probable causes of differences in the character of microevolution in some common species of arvicolines (Arvicolinae, Rodentia) of nontropical Palaearctic]. USSR Academy of Sciences, Proceedings of the Zoological Institute, Lenningrad, 225:3-20 (in Russian).
Gromov, I. M., and G. I. Baranova (eds.). 1981. Katalog mlekopitayushchikh SSSR [Catalog of mammals of the USSR]. Nauka, Leningrad, 456 pp. (in Russian).
Gromov, I. M., and I. Ya. Polyakov. 1977. Fauna SSSR, Mlekopitayushchie, tom 3, vyp. 8 [Fauna of the USSR, vol. 3, pt. 8, Mammals]. Polevki [Voles (Microtinae)]. Nauka, Moscow-Leningrad, 504 pp. (in Russian).
Gromov, I. M., A. A. Gureev, G. A. Novikov, I. I. Sokolov, P. P. Strelkov, and K. K. Chapskii. 1963. Mlekopitayushchie fauny SSSR, Chast' 1. [Mammals of the fauna of the USSR. [Vol.] 1. Insectivora, Chiroptera, Lagomorpha, and Rodentia]. Akademii Nauk SSSR, Moskva, 1:1-640.
Gromov, I. M., D. I. Bibikov, N. I. Kalabukhov, and M. N. N. Meier. 1965. Fauna SSSR, Mlekopitayushchie, tom. 3, vyp. 2 [Fauna of the U.S.S.R. Mammals. vol. 3, No. 2]. Nazemnye belich'e [Ground Squirrels]. Nauka, Moscow-Leningrad, 467 pp. (in Russian).
Gropp, A., H. Winking, F. Frank, G. Noack, and K. Fredga. 1976. Sex-chromosome aberrations in wood lemmings (*Myopus schisticolor*). Cytogenetics and Cell Genetics, 17:343-358.
Groves, C. P. 1967*a*. On the gazelles of the genus *Procapra* Hodgson, 1846. Zeitschrift für Säugetierkunde, 32:144-149.

Groves, C. P. 1967*b*. Geographic variation in the black rhinoceros *Diceros bicornis* (L., 1758). Zeitschrift für Säugetierkunde, 32:267-276.

Groves, C. P. 1967*c*. On the rhinoceroses of South-east Asia. Säugetierkundliche Mitteilungen, 15:221-237.

Groves, C. P. 1969*a*. On the smaller gazelles of the genus *Gazella* de Blainville, 1816. Zeitschrift für Säugetierkunde, 34:38-60.

Groves, C. P. 1969*b*. Systematics of the anoa (Mammalia, Bovidae). Beaufortia, 223:1-12.

Groves, C. P. 1970. The forgotten leaf-eaters, and the phylogeny of the Colobinae. Pp. 555-587, *in* Old World Monkeys: Evolution, systematics and behavior (J. R. Napier and P. H. Napier, eds.). Academic Press, London, 660 pp.

Groves, C. P. 1971*a*. *Pongo pygmaeus*. Mammalian Species, 4:1-6.

Groves, C. P. 1971*b*. Request for a declaration modifying Article 1 so as to exclude names proposed for domestic animals from Zoological Nomenclature. Bulletin of Zoological Nomenclature, 7:269-272.

Groves, C. P. 1971*c*. Systematics of the genus *Nycticebus*. Proceedings of the 3rd International Congress of Primatology, 1:44-53.

Groves, C. P. 1972*a*. *Ceratotherium simum*. Mammalian Species, 8:1-6.

Groves, C. P. 1972*b*. Systematics and phylogeny of gibbons. Pp. 1-89, *in* Gibbon and siamang: Evolution, ecology, behavior and captive maintenance (D. M. Rumbaugh, ed.). S. Karger, New York, 1:1-263.

Groves, C. P. 1974*a*. Horses, asses and zebras in the wild. David and Charles, Newton Abbot and London, 192 pp.

Groves, C. P. 1974*b*. Taxonomy and phylogeny of prosimians. Pp. 449-473, *in* Prosimian biology (R. D. Martin, G. A. Doyle, and A. C. Walker, eds.). Duckworth and Co., Ltd., London, 983 pp.

Groves, C. P. 1975*a*. Notes on the gazelles. 1. *Gazella rufifrons* and the zoogeography of central African Bovidae. Zeitschrift für Säugetierkunde, 40:308-319.

Groves, C. P. 1975*b*. Taxonomic notes on the white rhinoceros *Ceratotherium simum* (Burchell, 1817). Säugetierkundliche Mitteilungen, 23:200-212.

Groves, C. P. 1976. The taxonomy of *Moschus* (Mammalia, Artiodactyla), with particular reference to the Indian Region. Journal of the Bombay Natural History Society, 72:662-676.

Groves, C. P. 1978*a*. A note on nomenclature and taxonomy in the Lemuridae. Mammalia, 42:131-132.

Groves, C. P. 1978*b*. Phylogenetic and population systematics of mangabeys (Primates: Cercopithecoidea). Primates, 19:1-34.

Groves, C. P. 1978*c*. The taxonomic status of the dwarf blue sheep (Artiodactyla: Bovidae). Säugetierkundliche Mitteilungen, 26:177-183.

Groves, C. P. 1980*a*. A further note on *Moschus*. Journal of the Bombay Natural History Society, 77:130-133.

Groves, C. P. 1980*b*. Notes on the systematics of *Babyrousa* (Artiodactyla, Suidae). Zoologische Mededelingen, 55:29-46.

Groves, C. P. 1980*c*. Speciation in *Macaca*: the view from Sulawesi. Pp. 84-124, *in* The macaques (D. G. Lindburg, ed.). Van Nostrand Reinhold Company, New York, NY, 384 pp.

Groves, C. P. 1981*a*. Ancestors for the pigs: taxonomy and phylogeny of the genus *Sus*. Department of Prehistory, Research School of Pacific Studies, Australian National University, Technical Bulletin, 3:1-96.

Groves, C. P. 1981*b*. Notes on gazelles. 2. Subspecies and clines in the springbok (*Antidorcas*). Zeitschrift für Säugetierkunde, 46:189-197.

Groves, C. P. 1981*c*. Notes on the gazelles. 3. The dorcas gazelles of North Africa. Annali del Museo Civico di Storia Naturale di Genova, 83:455-471.

Groves, C. P. 1981*d*. Systematic relationships in the Bovini (Artiodactyla, Bovidae). Zeitschrift für Zoologische Systematik und Evolutionsforschung, 4:264-278.

Groves, C. P. 1982*a*. Cranial and dental characteristics in the systematics of Old World Felidae. Carnivore, 5(2):28-39.

Groves, C. P. 1982*b*. Geographic variation in the barasingha or swamp deer (*Cervus duvauceli*). Journal of the Bombay Natural History Society, 79:620-629.

Groves, C. P. 1982*c*. A note on geographic variation in the Indian blackbuck (*Antilope cervicapra* Linnaeus, 1758). Records of the Zoological Survey of India, 79:489-503.

Groves, C. P. 1982*d*. The systematics of tree kangaroos (*Dendrolagus*; Marsupialia, Macropodidae). Australian Mammalogy, 5:157-186.

Groves, C. P. 1982*e*. *Antelope depressicornis* H. Smith, 1827, and *Anoa quarlesi* Ouwens, 1910 (Mammalia, Artiodactyla): Proposed conservation. Z.N.(S.)2310. Bulletin of Zoological Nomenclature, 39:281-282.

Groves, C. P. 1983. Notes on the gazelles. 4. The Arabian gazelles collected by Hemprich and Ehrenberg. Zeitschrift für Säugetierkunde, 48:371-381.

Groves, C. P. 1985*a*. An introduction to the gazelles. Chinkara, 1:4-16.

Groves, C. P. 1985*b*. Was the quagga a species or a subspecies? African Wildlife, 39:106-107.

Groves, C. P. 1986. The taxonomy, distribution and adaptations of recent equids. Pp. 11-65, *in* Equids in the ancient world (R. H. Meadow and H-P. Uerpmann, eds.). Ludwig Reichert Verlag, Wiesbaden, 421 pp.

Groves, C. P. 1987. On the cuscuses (Marsupialia: Phalangeridae) of the *Phalanger orientalis* group from Indonesian territory. Pp. 569-579, *in* Possums and opossums: studies in evolution (M. Archer, ed.). Surrey Beatty and Sons Pty. Ltd., and the Royal Zoological Society of New South Wales, Sydney, 1:1-460; 2:461-788.

Groves, C. P. 1988. A catalogue of the genus *Gazella*. Pp. 193-198, *in* Conservation and biology of desert antelopes (A. Dixon and D. Jones, eds). Christopher Helm, London, 238 pp.

Groves, C. P. 1989. A theory of human and primate evolution. Oxford University Press, New York, 375 pp.

Groves, C. P. 1992. [Review of] Titis, New World Monkeys of the genus *Callicebus* (Cebidae, Platyrrhini): a preliminary taxonomic review, by P. Hershkovitz. International Journal of Primatology, 13:111-112.

Groves, C. P., and R. H. Eaglen. 1988. Systematics of the Lemuridae (Primates, Strepsirhini). Journal of Human Evolution, 17:513-538.

Groves, C. P., and Feng Zuo-jiang. 1986. The status of musk-deer from Anhui Province, China. Acta Theriologica Sinica, 6:101-106 (in English, Chinese summary).

Groves, C. P., and T. F. Flannery. 1989. Revision of the genus *Dorcopsis* (Macropodidae: Marsupialia). Pp. 117-128, *in* Kangaroos, wallabies and rat-kangaroos (G. Grigg, P. Jarman, and I. Hume, eds.). Surrey Beatty and Sons Pty. Ltd, Sydney, 1:1-835.

Groves, C. P., and T. F. Flannery. 1990. Revision of the families and genera of bandicoots. Pp. 1-11, *in* Bandicoots and bilbies (J. H. Seebeck, R. L. Wallis, P. R. Brown, and C. M. Kemper, eds.). Surrey Beatty and Sons Pty. Ltd., Sydney, 392 pp.

Groves, C. P., and P. Grubb. 1974. A new duiker from Rwanda (Mammalia, Bovidae). Revue de Zoologie Africaine, 88:189-196.

Groves, C. P., and P. Grubb. 1982. The species of muntjac (genus *Muntiacus*) in Borneo: unrecognised sympatry in tropical deer. Zoologische Mededelingen, 56:203-216.

Groves, C. P., and P. Grubb. 1985. Reclassification of the serows and gorals (*Nemorhaedus*: Bovidae). Pp. 45-50, *in* The biology and management of mountain ungulates (S. Lovari, ed.). Croom Helm, London, 271 pp.

Groves, C. P., and P. Grubb. 1987. Relationships of living deer. Pp. 21-59, *in* Biology and management of the Cervidae (C. M. Wemmer, ed.). Smithsonian Institution Press, Washington, D. C., 577 pp.

Groves, C. P., and P. Grubb. 1990. Muntiacidae. Pp. 134-168, *in* Horns, pronghorns, and antlers (G. A. Bubenik and A. B. Bubenik, eds). Springer-Verlag, New York, 562 pp.

Groves, C. P., and P. Grubb. In press. The Eurasian suids: *Sus* and *Babyrousa*. Classification, *in* Action plan for the Suiformes (W. L. R. Oliver, ed.). I. U. C. N., Gland, Switzerland.

Groves, C. P., and L. B. Holthuis. 1985. The nomenclature of the Orang Utan. Zoologische Mededelingen, 59:411-417.

Groves, C. P., and F. Kurt. 1972. *Dicerorhinus sumatrensis*. Mammalian Species, 21:1-6.

Groves, C. P., and V. Mazak. 1967. On some taxonomic problems of the Asiatic wild asses: with the description of a new subspecies (Perissodactyla, Equidae). Zeitschrift für Säugetierkunde, 32:321-355.

Groves, C. P., and C. Smeenk. 1978. On the type material of *Cervus nippon* Temminck, 1836; with a revision of sika deer from the main Japanese islands. Zoologische Mededelingen, 53:11-28.

Groves, C. P., and I. Tattersall. 1991. Geographical variation in the fork-marked lemur, *Phaner furcifer* (Primates, Cheirogaleidae). Folia Primatologia, 56:39-49.

Groves, C. P., and Wang Ying-xiang. 1990. The gibbons of the subgenus *Nomascus* (Primates, Mammalia). Zoological Research, 11:147-154.

Groves, C. P., and D. P. Willoughby. 1981. Studies on the taxonomy and phylogeny of the genus *Equus*. 1. Subgeneric classification of the recent species. Mammalia, 45:321-354.

Grubb, P. 1972. Variation and incipient speciation in the African buffalo. Zeitschrift für Säugetierkunde, 37:121-144.

Grubb, P. 1973. Distribution, divergence and speciation of the drill and mandrill. Folia Primatologica, 20(2-3):161-177.

Grubb, P. 1977. Notes on a rare deer, *Muntiacus feai*. Annali del Museo Civico di Storia Naturale di Genova, 81:202-207.

Grubb, P. 1978. Patterns of speciation in African mammals. Pp. 152-167, *in* Ecology and taxonomy of African small mammals (D. A. Schlitter, ed.). Bulletin, Carnegie Museum of Natural History, 6:1-214.

Grubb, P. 1981. *Equus burchellii*. Mammalian Species, 157:1-9.

Grubb, P. 1982*a*. The systematics of Sino-Himalayan musk deer (*Moschus*), with particular reference to the species described by B. H. Hodgson. Säugetierkundliche Mitteilungen, 30:127-135.

Grubb, P. 1982*b*. Systematics of sun-squirrels (*Heliosciurus*) in eastern Africa. Bonner Zoologische Beiträge, 33(2-4):191-204.

Grubb, P. 1990. List of deer species and subspecies. Deer, Journal of the British Deer Society, 8:153-155.
Grubb, P. In press. The Afrotropical suids: *Phacochoerus*, *Hylochoerus* and *Potamochoerus*. Classification, *in* Action plan for the Suiformes (W. L. R. Oliver, ed.). I. U. C. N., Gland, Switzerland.
Grubb, P., and C. P. Groves. 1983. Notes on the taxonomy of the deer (Mammalia, Cervidae) of the Philippines. Zoologischer Anzeiger, 210:119-144.
Gruber, U. F. 1969. Tiergeographische, ökologische und bionomische Untersuchungen an kleinen Säugetieren in Ost-Nepal. Khumbu Himal, 3:197-312.
Grulich, I. 1971. Zum Bau des Beckens (pelvis), eines systematisch-taxonomischen Merkmales, bei der Unterfamilie Talpinae. Zoologické Listy, 20:15-28.
Grulich, I. 1972. Ein Beitrag zur Kenntnis der ostmediterranen kleinwüchsigen, blinden Maulwurfsformen (Talpinae). Zoologické Listy, 21:3-21.
Grulich, I. 1982. Zur Kenntnis der Gattungen *Scaptochirus* und *Parascaptor* (Talpini, Mammalia). Folia Zoologica, 31:1-20.
Grulich, I. 1987. Contribution to the sexual dimorphism of the hamster (*Cricetus cricetus*, Rodentia, Mammalia). Folia Zoologica, 36:291-306.
Grulich, I. 1991. Variabiliby of the ossa memberi thoracici et pelvini *Cricetus cricetus* (Rodentia, Mammalia). Acta Scientiarum Naturalium, Academiae Scientiarum Bohemoslovacae (Brno), 25:1-63.
Guilday, J. E. 1982. Dental variation in *Microtus xanthognathus*, *M. chrotorrhinus*, and *M. pennsylvanicus* (Rodentia: Mammalia). Annals of Carnegie Museum, 51:211-230.
Guillotin, M. and F. Petter. 1984. Un *Rhipidomys* nouveau de Guyane francaise, *R. leucodactylus aratayae* ssp. nov. (Rongeurs, Cricetides). Mammalia, 48:541-544.
Güldenstädt, A. I. 1776. Chaus, animal feli affine descriptiom. Novi Commentari Academiae Scientiarum Imperialis Petropolitanae, 20:483-500.
Gulotta, E. F. 1971. *Meriones unguiculatus*. Mammalian Species, 3:1-5.
Gunn, S. J., and I. F. Greenbaum. 1986. Systematic implications of karyotypic and morphologic variation in mainland *Peromyscus* from the Pacific northwest. Journal of Mammalogy, 67:294-304.
Günther, A. 1875. Notes on some small mammals from Madagascar. Proceedings of the Zoological Society of London, 1875:78-80.
Gureev, A. A. 1964. Fauna SSSR, Mlekopitayushchie, tom. 3, vyp. 10, Zaitseobraznye (Lagomorpha) [Fauna of the USSR, mammals, vol. 3, pt. 10, Lagomorpha]. Nauka, Moscow-Leningrad, 276 pp. (in Russian).
Gureev, A. A. 1971. Zemleroikii (Soricidae) fauny mira. [Shrews...of the world fauna]. Nauka, Leningrad, 253 pp. (in Russian).
Gureev, A. A. 1979. Fauna SSSR, Mlekopitayutschie, tom. 4, vyp. 2. Nasekomoyadnye...[Fauna of the USSR, Mammals, vol. 4, pt. 2. Insectivores (Mammalia, Insectivora)]. Nauka, Leningrad, 501 pp. (in Russian).
Guthrie, R. D. 1971. Factors regulating the evolution of microtine tooth complexity. Zeitschrift für Säugetierkunde, 36:37-54.
Gyldenstolpe, N. 1932. A manual of Neotropical sigmodont rodents. Kungliga Svenska Vetenskapsakademiens Handlingar, Tredje Serien, 11:1-164.

Hafner, D. J. 1981. Evolution and historical zoogeography of antelope ground squirrels, genus *Ammospermophilus* (Rodentia: Sciuridae). Unpubl. Ph. D. dissertation, University of New Mexico, Albuquerque, New Mexico, 225 pp.
Hafner, D. J. 1984. Evolutionary relationships of the Nearctic Sciuridae. Pp. 3-23, *in* The biology of ground-dwelling squirrels (J. O. Murie and G. R. Michener, eds.). University of Nebraska Press, Lincoln, 459 pp.
Hafner, D. J., and K. N. Geluso. 1983. Systematic relationships and historical zoogeography of the desert pocket gopher, *Geomys arenarius*. Journal of Mammalogy, 64:405-413.
Hafner, D. J., and T. L. Yates. 1983. Systematic status of the Mohave ground squirrel, *Spermophilus mohavensis* (subgenus *Xerospermophilus*). Journal of Mammalogy, 64:397-404.
Hafner, D. J., J. C. Hafner, and M. S. Hafner. 1979. Systematic status of kangaroo mice, genus *Microdipodops*: morphometric, chromosomal, and protein analyses. Journal of Mammalogy, 60:1-10.
Hafner, D. J., K. E. Petersen, and T. L. Yates. 1981. Evolutionary relationships of jumping mice (genus *Zapus*) of the southwestern United States. Journal of Mammalogy, 62(3):501-512.
Hafner, J. C. 1978. Evolutionary relationships of kangaroo mice, genus *Microdipodops*. Journal of Mammalogy, 59:354-366.
Hafner, J. C., and M. S. Hafner. 1983. Evolutionary relationships of heteromyid rodents. Great Basin Naturalist, Memoirs, 7:3-29.
Hafner, M. S. 1982. A biochemical investigation of geomyoid systematics (Mammalia: Rodentia). Zeitschrift für Zoologische Systematik und Evolutionsforschung, 20:118-130.
Hafner, M. S. 1991. Evolutionary genetics and zoogeography of Middle American pocket gophers, genus *Orthogeomys*. Journal of Mammalogy, 72:1-10.

Hafner, M. S., and L. J. Barkley. 1984. Genetics and natural history of a relictual pocket gopher, *Zygogeomys* (Rodentia: Geomyidae). Journal of Mammalogy, 65:474-479.

Hafner, M. S., and D. J. Hafner. 1987. Geographic distribution of two Costa Rican species of *Orthogeomys*, with comments on dorsal pelage markings in the Geomyinae. Southwestern Naturalist, 32:5-11.

Hafner, M. S., J. C. Hafner, J. L. Patton, and M. F. Smith. 1987. Macrogeographic patterns of genetic differentiation in the pocket gopher *Thomomys umbrinus*. Systematic Zoology, 36:18-34.

Hagmeier, E. M. 1961. Variation and relationships in North American marten. Canadian Field Naturalist, 75:122-138.

Hahn, H. 1934. Die Familie der Procaviidae. Zeitschrift für Säugetierkunde, 9:207-358.

Haiduk, M. W., J. W. Bickham, and D. J. Schmidly. 1979. Karyotypes of six species of *Oryzomys* from Mexico and Central America. Journal of Mammalogy, 60:610-615.

Haiduk, M. W., C. Sanchez-Hernandez, and R. J. Baker. 1988. Phylogenetic relationships of *Nyctomys* and *Xenomys* to other cricetine genera based on data from G-banded chromosomes. Southwestern Naturalist, 33:397-403.

Haim, A., and E. Tchernov. 1974. The distribution of myomorph rodents in the Sinai peninsula. Mammalia, 38(2):201-223.

Hall, E. R. 1941. Revision of the rodent genus *Microdipodops*. Field Museum of Natural History, Zoological Series, 27:233-277.

Hall, E. R. 1951. American weasels. University of Kansas Publications, Museum of Natural History, 4:1-466.

Hall, E. R. 1955. A new subspecies of wood rat from Nayarit, Mexico, with new name combinations for the *Neotoma mexicana* group. Journal of the Washington Academy of Sciences, 45:328-332.

Hall, E. R. 1960. *Oryzomys couesi* only subspecifically different from the marsh rice rat, *Oryzomys palustris*. Southwestern Naturalist, 5:171-173.

Hall, E. R. 1971. Variation in the blackish deer mouse, *Peromyscus furvus*. Anales de Instituto de Biologia, Universidad Nacional Autónomica de México, 1:149-154.

Hall, E. R. 1981. The mammals of North America. Second ed. John Wiley and Sons, New York, 1:1-600 + *90*, 2:601-1181 + *90*.

Hall, E. R. 1984. Geographic variation among brown and grizzly bears (*Ursus arctos*) in North America. Special Publication, Museum of Natural History, University of Kansas, 13:1-16.

Hall, E. R., and E. L. Cockrum. 1953. A synopsis of the North American microtine rodents. University of Kansas Publications, Museum of Natural History, 5:373-498.

Hall, E. R., and H. H. Genoways. 1970. Taxonomy of the *Neotoma albigula*-group of woodrats in central Mexico. Journal of Mammalogy, 51:504-516.

Hall, E. R., and R. M. Gilmore. 1932. New mammals from St. Lawrence Island, Bering Sea, Alaska. University of California Publications in Zoology, 38:391-404.

Hall, E. R., and D. R. Hoffmeister. 1942. Geographic variation in the canyon mouse, *Peromyscus crinitus*. Journal of Mammalogy, 23:51-65.

Hall, E. R., and J. K. Jones, Jr. 1961. North American yellow bats, "*Dasypterus*", and a list of the named kinds of the genus *Lasiurus* Gray. University of Kansas Publications, Museum of Natural History, 14:73-98.

Hall, E. R., and K. R. Kelson. 1951. A new subspecies of *Microtus montanus* from Montana and comments on *Microtus canicaudus* Miller. University of Kansas Publications, Museum of Natural History, 5:73-79.

Hall, E. R., and K. R. Kelson. 1959. The mammals of North America. Ronald Press Co., New York, 1:1-546; 2:547-1083 + *79*.

Hall, L. S., and G. C. Richards. 1979. Bats of Eastern Australia. Queensland Museum Booklet, 12:1-66.

Hallett, J. G. 1978. *Parascalops breweri*. Mammalian Species, 98:1-4.

Haltenorth, T. 1953. Die Wildkatzen der Alten Welt; eine Übersicht über die Untergattung. Geest and Portig, Leipzig, 117 pp.

Haltenorth, T. 1958. Klassifikation der Säugetiere, 1 (1, Ordnung Kloakentiere, Monotremata Bonaparte, 1838, und 2, Ordnung Beuteltiere, Marsupialia Illiger, 1811 [=Didelphia Blainville, 1816]). Handbuch der Zoologie, 8(16):1-40.

Haltenorth, T. 1959. Beitrag zur Kenntnis des Mesopotamischen Damhirsches—*Cervus* (*Dama*) *mesopotamicus* Brooke, 1875—und zur Stammes-und Verbreitungeschichte der Damhirsche allgemein. Säugetierkundliche Mitteilungen, 7(Sonderheft):1-89.

Haltenorth, T. 1963. Klassifikation der Säugetiere: Artiodactyla I. Handbuch der Zoologie, 8(32):1-167.

Haltenorth, T., and H. Diller. 1977. Säugetiere Afrikas und Madagaskars. BLV, Munchen, 403 pp.

Hamilton, J. E. 1934. The southern sea lion, *Otaria byronia* (de Blainville). Discovery Reports, 8:269-318.

Hamilton, J. E. 1940. On the history of the elephant seal, *Mirounga leonina* (Linn.). Proceedings of the Linnean Society of London, 1939-40:33-37.

Hamlett, G. W. D. 1939. Identity of *Dasypus septemcinctus* Linnaeus with notes on some related species. Journal of Mammalogy, 20:328-336.

Hammond, D. D., and S. Leatherwood. 1984. Cetaceans captured for Ocean Park, Hong Kong, April 1974 - February 1983. Reports of the International Whaling Commission, 34:491-495.
Hanak, V. 1969. Zur Kenntnis von *Rhinolophus bocharicus* Kastchenko et Akimov, 1917 (Mammalia: Chiroptera). Vèstník eskoslovenské Spole nosti Zoologické, 33(4):315-327.
Hanak, V., and J. Gaisler. 1969. Notes on the taxonomy and ecology of *Myotis longipes* (Dobson, 1873). Zoologické Listy, 18:195-206.
Hanak, V., and J. Gaisler. 1971. The status of *Eptesicus ognevi* Bobrinskii, 1918, and remarks on some other species of this genus (Mammalia: Chiroptera). Vèstník eskoslovenské Spole nosti Zoologické, 35(1):11-24.
Hanak, V., and I. Horáček. 1986. Zur Südgrenze des Areals von *Eptesicus nilssoni* (Chiroptera: Vespertilionidae). Annalen des Naturhistorischen Museums in Wien, ser. B(Botanik und Zoologie), 88/89:377-388.
Hand, S. J. 1985. New Miocene megadermatids (Chiroptera: Megadermatidae) from Australia with comments on megadermatid systematics. Australian Mammalogy, 8:5-43.
Handley, C. O., Jr. 1959*a*. A review of the genus *Hoplomys* (thick-spinned rats), with description of a new form from Isla Escudo de Veraguas, Panamá. Smithsonian Miscellaneous Collections, 139(4):1-10, 1 pl.
Handley, C. O., Jr. 1959*b*. A revision of American bats of the genera *Euderma* and *Plecotus*. Proceedings of the United States National Museum, 110:95-246.
Handley, C. O., Jr. 1960. Descriptions of new bats from Panama. Proceedings of the United States National Museum, 112:459-479.
Handley, C. O., Jr. 1966*a*. Checklist of the mammals of Panama. Pp. 753-795, *in* Ectoparasites of Panama, (R. L. Wenzel and V. J. Tipton, eds.). Field Museum of Natural History, Chicago, 861 pp.
Handley, C. O., Jr. 1966*b*. Descriptions of new bats (*Choeroniscus* and *Rhinophylla*) from Colombia. Proceedings of the Biological Society of Washington, 79:83-88.
Handley, C. O., Jr. 1966*c*. A synopsis of the genus *Kogia* (pygmy sperm whales). Pp. 62-69, *in* Whales, dolphins, and porpoises (K. S. Norris, ed.). University of California Press, Berkeley, 789 pp.
Handley, C. O., Jr. 1976. Mammals of the Smithsonian Venezuelan Project. Brigham Young University Science Bulletin, Biological Series, 20:1-91.
Handley, C. O., Jr. 1980. Inconsistencies in formation of family-group and subfamily-group names in Chiroptera Pp. 9-13, *in* Proceedings of the Fifth International Bat Research Conference (D. E. Wilson and A. L. Gardner, eds.). Texas Tech Press, Lubbock, 434 pp.
Handley, C. O., Jr. 1987. New species of mammals from northern South America: fruit-eating bats, genus *Artibeus* Leach. Pp. 163-172, *in* Studies in Neotropical mammalogy, essays in honor of Philip Hershkovitz (B. D. Patterson and R. M. Timm, eds.). Fieldiana, Zoology, New Series, 39:frontispiece, viii + 1-506.
Handley, C. O., Jr. 1989 [1990]. The *Artibeus* of Gray 1838. Pp. 443-468, *in* Advances in neotropical mammalogy (K. H. Redford and J. F. Eisenberg, eds.). Sandhill Crane Press, Gainesville, FL, 614 pp.
Handley, C. O., Jr. 1991. The identity of *Phyllostoma planirostre* Spix, 1823 (Chiroptera: Stenodermatinae). Bulletin of the American Museum of Natural History, 206:12-17.
Handley, C. O., Jr., and K. C. Ferris. 1972. Descriptions of new bats of the genus *Vampyrops*. Proceedings of the Biological Society of Washington, 84:519-524.
Handley, C. O., Jr., and A. L. Gardner. 1990. The holotype of *Natalus stramineus* Gray (Mammalia: Chiroptera: Natalidae). Proceedings of the Biological Society of Washington, 103:966-972.
Handley, C. O., Jr., and R. H. Pine. 1992. A new species of *Coendou* Lacépède, from Brazil. Mammalia, 56(2):237-244.
Hanney, P. 1965. The Muridae of Malawi (Africa: Nyasaland). Journal of Zoology (London), 146:577-633.
Happold, D. C. D. 1987. The mammals of Nigeria. Clarendon Press, Oxford, 402 pp.
Harada, M., and R. Kobayashi. 1980. Studies on the small mammal fauna of Sabah, East Malaysia. II. Karyological analysis of some Sabahan mammals (Primates, Rodentia, Chiroptera). Contributions of the Biology Laboratory. Kyoto University, 26:83-95.
Harada, M., A. Ando, L.-K. Lin, and S. Takada. 1991. Karyotypes of the Taiwan vole *Microtus kikuchii* and the Pere David's vole *Eothenomys melanogaster* from Taiwan. Journal of the Mammalogical Society of Japan, 16:41-45.
Hardjasasmita, H. S. 1987. Taxonomy and phylogeny of the Suidae (Mammalia) in Indonesia. Scripta Geologica, 85:1-68.
Hardy, R. 1950. A new tree squirrel from central Utah. Proceedings of the Biological Society of Washington, 63:13-14.
Harmer, S. F. 1922. On Commerson's dolphin and other species of *Cephalorhynchus*. Proceedings of the Zoological Society of London, 1922(3):627-638.
Harper, F. 1940. The nomenclature and type localities of certain Old World mammals. Journal of Mammalogy, 21:191-203; 322-332.

Harper, F. 1945. Extinct and vanishing mammals of the Old World. Special Publication, American Committee for International Wild Life Protection, New York, 12:1-850.

Harper, F. 1952. History and nomenclature of the pocket gopher (*Geomys*) in Georgia. Proceedings of the Biological Society of Washington, 65:35-38.

Harris, C. J. 1968. Otters; a study of the Recent Lutrinae. Weidenfeld and Nicolson, London, 397 pp.

Harris, J. M. 1991. Family Hippopotamidae. Pp. 31-85, *in* Koobi Fora Research Project. Vol. 3 (J. M. Harris, ed.). Clarendon Press, Oxford, 384 pp.

Harris, W. P., Jr. 1937. Revision of *Sciurus variegatoides*, a species of Central American squirrel. Miscellaneous Publications, Museum of Zoology, University of Michigan, 38:1-39.

Harrison, D. L. 1964-1972. The mammals of Arabia. Ernest Benn Limited, London, 1:1-192 [1964]; 2:193-381 [1968]; 3:383-670 [1972].

Harrison, D. L. 1975. *Macrophyllum macrophyllum*. Mammalian Species, 62:1-3.

Harrison, D. L. 1978. A critical examination of alleged sibling species in the lesser three-toed jerboas (subgenus *Jaculus*) of the north African and Arabian deserts. Pp. 77-80, *in* Ecology and taxonomy of African small mammals (D. A. Schlitter, ed.). Bulletin of Carnegie Museum of Natural History, 6:1-214.

Harrison, D. L., and P. J. J. Bates. 1991. The mammals of Arabia, Second ed. Harrison Zoological Museum, Sevenoaks, United Kingdom, 354 pp.

Harrison, D. L., and J. D. L. Fleetwood. 1960. A new race of the flat-headed bat *Platymops barbatogularis* Harrison from Kenya colony, with observations on the anatomy of the gular sac and genitalia. Durban Museum Novitates, 5:269-284.

Harrison, J. L. 1956. Records of bandicoot rats (*Bandicota*, Rodentia, Muridae) new to the fauna of Malaya and Thailand. Bulletin of the Raffles Museum (Singapore), 27:27-31.

Harrison, J. L. 1958. *Chimarrogale hantu* a new water shrew from the Malay Peninsula, with a note on the genera *Chimarrogale* and *Crossogale* (Insectivora, Soricidae). Annals and Magazine of Natural Nistory, ser. 13, 1:282-290.

Harrison, J. L. 1974. An introduction to mammals of Singapore and Malaya, Second ed. Malayan Nature Society, Singapore, 340 pp.

Harrison, M. J. S. 1988. A new species of guenon (genus *Cercopithecus*) from Gabon. Journal of Zoology (London), 215:561-575.

Harrison, T. 1960. South China sea dolphins. Malayan Nature Journal, 14(2):87-89.

Hart, E. B. 1992. *Tamias dorsalis*. Mammalian Species, 399:1-6.

Hartman, G. D., and T. L. Yates. 1985. *Scapanus orarius*. Mammalian Species, 253:1-5.

Hasler, M. J., J. F. Hasler, and A. V. Nalbandov. 1977. Comparative breeding biology of musk shrews (*Suncus murinus*) from Guam and Madagascar. Journal of Mammalogy, 58:285-290.

Hassinger, J. D. 1970. Shrews of the *Crocidura zarudnyi-pergrisea* group with descriptions of a new subspecies. Fieldiana, Zoology, 58:5-8.

Hassinger, J. D. 1973. A survey of the mammals of Afghanistan resulting from the 1965 Street Expedition (excluding bats). Fieldiana, Zoology, 60:1-195.

Hatt, R. T. 1934*a*. The American Museum Congo Expedition manatee and other recent manatees. Bulletin of the American Museum of Natural History, 66:533-566.

Hatt, R. T. 1934*b*. Fourteen hitherto unrecognized African rodents. American Museum Novitates, 708:1-15.

Hatt, R. T. 1936. The hyraxes collected by the American Museum Congo expedition. Bulletin of the American Museum of Natural History, 72:117-141.

Hatt, R. T. 1940*a*. Lagomorpha and Rodentia other than Sciuridae, Anomaluridae and Idiuridae, collected by the American Museum Congo Expedition. Bulletin of the American Museum of Natural History, 76(9):457-604.

Hatt, R. T. 1940*b*. Mammals collected by the Rockefeller-Murphy Expedition to Tanganyika Territory and the eastern Belgian Congo. American Museum Novitates, 1070:1-8.

Hausser, J. 1990. *Sorex coronatus* Millet, 1882 - Schabrackenspitzmaus; *Sorex granarius* Miller, 1909 - Iberische Waldspitzmaus; *Sorex samniticus* Altobello, 1926 - Italienische Waldspitzmaus. Pp. 279-294, *in* Handbuch der Säugetiere Europas (J. Niethammer and F. Krapp, eds.). Aula-Verlag, Wiesbaden, 3/I:1-524.

Hausser, J. (ed.). 1991. The cytogenetics of the *Sorex araneus* group and related topics. Mémoires de la Société Vaudoise des Sciences Naturelles, 19:1-151.

Hausser, J., J.-D. Graf, and A. Meylan. 1975. Données nouvelles sur les *Sorex* d'Espagne et des Pyrénées (Mammalia, Insectivora). Bulletin de la Societé Vaudoise des Sciences Naturelles, 72:241-252.

Hausser, J., F. Catzeflis, A. Meylan, and P. Vogel. 1985. Speciation in the *Sorex araneus* complex (Mammalia: Insectivora). Acta Zoologica Fennica, 170:125-130.

Hausser, J., E. Dannelid, and F. Catzeflis. 1986. Distribution of two karyotypic races of *Sorex araneus* (Insectivora, Soricidae) in Switzerland and the post-glacial recolonization of the Valais: first results. Zeitschrift für Zoologische Systematik und Evolutionsforschung, 24:307-314.

Hausser, J., R. Hutterer, and P. Vogel. 1990. *Sorex araneus* Linnaeus, 1758 - Waldspitzmaus. Pp. 237-278, *in* Handbuch der Säugetiere Europas (J. Niethammer and F. Krapp, eds.). Aula-Verlag, Wiesbaden, 3/I:1-524.

Hay, K. A., and A. W. Mansfield. 1989. Narwhal - *Monodon monoceros* Linnaeus, 1758. Pp. 145-177, *in* Handbook of marine mammals: River dolphins and the larger toothed whales (S. H. Ridgway and R. Harrison, eds.). Academic Press, London, 4:1-442.

Hayden, F. V. 1863. On the geology and natural history of the upper Missouri. Transactions of the American Philosophical Society, 12:1-218.

Hayman, R. W. 1936. A note on *Dologale dybowskii*. Annals and Magazine of Natural History, ser. 10, 18:626-630.

Hayman, R. W. 1946. Systematic notes on the genus *Idiurus* (Anomaluridae). Annals and Magazine of Natural History, ser. 11, 13:208-212.

Hayman, R. W. 1962. A new genus and species of African rodent. Revue de Zoologie et de Botanique Africaines, 65:129-138.

Hayman, R. W. 1963*a*. Further notes on *Delanymys brooksi* (Rodentia, Muridae) in the Congo. Revue de Zoologie et de Botanique Africaines, 67:388-392.

Hayman, R. W. 1963*b*. Mammals from Angola, mainly from the Lunda District. Publicaçoes Culturais Museu do Dundo, 66:84-139.

Hayman, R. W. 1966. On the affinities of *Nilopegamys plumbeus* Osgood. Annales Musée Royal de l'Afrique Centrale, Tervuren, Belgique, ser. 8 (Sciences Zoologiques), 144:29-38.

Hayman, R. W., and J. E. Hill. 1971. Order Chiroptera. Part 2. Pp. 1-73, *in* The mammals of Africa: An identification manual (J. Meester and H. W. Setzer, eds.) [issued 15 Jul 1971]. Smithsonian Institution Press, Washington, D. C., not continuously paginated.

Hayssen, V. 1991. *Dipodomys microps*. Mammalian Species, 389:1-9.

Hayward, B. 1970. The natural history of the cave bat *Myotis velifer*. Western New Mexico University, Research Science, 1(1):1-74.

Heaney, L. R. 1978. Island area and body size of insular mammals: evidence from the tri-colored squirrel (*Callosciurus prevosti*) of Southeast Asia. Evolution, 32:29-44.

Heaney, L. R. 1979. A new species of tree squirrel (*Sundasciurus*) from Palawan Island, Philippines (Mammalia: Sciuridae). Proceedings of the Biological Society of Washington, 92:280-286.

Heaney, L. R. 1985. Systematics of Oriental Pygmy Squirrels of the Genera *Exilisciurus* and *Nannosciurus* (Mammalia: Sciuridae). Miscellaneous Publications, Museum of Zoology, University of Michigan, 170:1-58.

Heaney, L. R. 1986. Biogeography of mammals in SE Asia: estimates of rates of colonization, extinction and speciation. Biological Journal of the Linnean Society, 28:127-165.

Heaney, L. R., and E. C. Birney. 1977. Distribution and natural history notes on some mammals from Puebla, Mexico. Southwestern Naturalist, 21:543-545.

Heaney, L. R., and R. S. Hoffmann. 1978. A second specimen of the Neotropical montane squirrel, *Syntheosciurus poasensis*. Journal of Mammalogy, 59:854-855.

Heaney, L. R., and G. S. Morgan. 1982. A new species of gymnure, *Podogymnura*, (Mammalia: Erinaceidae) from Dinagat Island, Philippines. Proceedings of the Biological Society of Washington, 95:13-26.

Heaney, L. R., and D. S. Rabor. 1982. Mammals of Dinagat and Siargao islands, Philippines. Occasional Papers of the Museum of Zoology, University of Michigan, 699:1-30.

Heaney, L. R., and R. M. Timm. 1983*a*. Relationships of pocket gophers of the genus *Geomys* from the central and northern Great Plains. Miscellaneous Publications, Museum of Natural History, University of Kansas, 74:1-59.

Heaney, L. R., and R. M. Timm. 1983*b*. Systematics and distribution of shrews of the genus *Crocidura* (Mammalia: Insectivora) in Vietnam. Proceedings of the Biological Society of Washington, 96:115-120.

Heaney, L. R., and R. M. Timm. 1985. Morphology, genetics, and ecology of pocket gophers (genus *Geomys*) in a narrow hybrid zone. Biological Journal of the Linnean Society, 25:301-317.

Heaney, L. R., P. C. Gonzales, and A. C. Acala. 1987. An annotated checklist of the taxonomic and conservation status of land mammals in the Philippines. Silliman Journal, 34:32-66.

Heaney, L. R., and P. C. Gonzales, R. C. B. Utzurrum, and E. A. Rickart. 1991. The mammals of Catanduanes Island: implications for the biogeography of small land-bridge islands in the Philippines. Proceedings of the Biological Society of Washington, 104:399-415.

Hedges, S. R. 1969. Epigenetic polymorphism in populations of *Apodemus sylvaticus* and *Apodemus flavicollis* (Rodentia, Muridae). Journal of the Zoological Society of London, 159:425-442.

Heim de Balsac, H. 1956. Morphologie divergente des Potamogalinae (Mammifères, Insectivores) en milieu aquatique. Comptes Rendus de l'Acadèmie des Sciences (Paris), 242:2257-2258.

Heim de Balsac, H. 1957. Insectivores Soricidae du Mont Cameroun. Zoologische Jahrbücher, Abteilung für Systematik, Ökologie und Geographie der Tiere, 85:501-672.

Heim de Balsac, H. 1959. Nouvelle contribution à l'étude des Insectivores Soricidae du Mont Cameroun. Bonner Zoologische Beiträge, 10:198-217.

Heim de Balsac, H. 1966*a*. Contribution à l'étude des Soricidae de Somalie. Monitore Zoologico Italiano, 74 Supplemento:196-221.

Heim de Balsac, H. 1966*b*. Faits nouveaux concernant l'évolution cranio-dentaire des Soricinés (Mammifères Insectivores). Comptes Rendus des Séances de l'Académie des Sciences (Paris), 263D:889-892.

Heim de Balsac, H. 1967. Faits nouveaux concernant les *Myosorex* (Soricidae) de l'Afrique orientale. Mammalia, 31:610-628.

Heim de Balsac, H. 1968*a*. Considérations préliminaires sur le peuplement des montages Africaines par les Soricidae. Biologia Gabonica, 4:299-323.

Heim de Balsac, H. 1968*b*. Contribution à l'étude des Soricidae de Fernando Po et du Cameroun. Bonner Zoologische Beiträge, 19:15-42.

Heim de Balsac, H. 1968*c*. Contributions à la faune de la région de Yaoundé I. - Premier aperçu sur la faune des Soricidae (Mammifères Insectivores). Annales de la Faculté des Sciences du Cameroun, 1968(2):49-58.

Heim de Balsac, H. 1968*d*. Recherches sur la faune des Soricidae de l'ouest africain (du Ghana au Senegal). Mammalia, 32:379-418.

Heim de Balsac, H. 1968*e*. Les Soricidae dans le milieu désertique saharien. Bonner Zoologische Beiträge, 19:181-188.

Heim de Balsac, H. 1972. Insectivores. Pp. 629-660, *in* Biogeography and ecology in Madagascar (R. Battistini, and G. Richard-Vindard, eds.). W. Junk, The Hague, 765 pp.

Heim de Balsac, H., and V. Aellen. 1965. Les Muridae de basse Cote-d'Ivoire. Revue Suisse de Zoologie, 71:695-753.

Heim de Balsac, H., and J. J. Barloy. 1966. Révision des Crocidures du groupe *flavescens-occidentalis-manni*. Mammalia, 30:601-633.

Heim de Balsac, H., and R. Hutterer. 1982. Les Soricidae (Mammifères Insectivores) des îles du Golfe de Guinée: faits nouveaux et problèmes biogeographiques. Bonner Zoologische Beiträge, 33:133-150.

Heim de Balsac, H., and M. Lamotte. 1956. Evolution et phylogénie des Soricidés africaines. Mammalia, 20:140-167.

Heim de Balsac, H., and M. Lamotte. 1958. La reserve naturelle integrale du mont Nimba, Part 4 (15): Mammiferes rongeurs (Muscardinides et Murides). Memoires de l'Institut Francais d'Afrique Noire, 53:339-357.

Heim de Balsac, H., and J. Meester. 1977. Order Insectivora. Part 1. Pp. 1-29, *in* The mammals of Africa: an identification manual (J. Meester and H. W. Setzer, eds.). [issued 22 Aug 1977]. Smithsonian Institution Press, Washington, D. C., not continuously paginated.

Heim de Balsac, H., and P. Mein. 1971. Les musaraignes momifiées des hypogées des Thebes. Existence d'un metalophe chez les Crocidurinae (*sensu* Repenning). Mammalia, 35:220-244.

Heinrich, W.-D. 1990. Some aspects of evolution and biostratigraphy of *Arvicola* (Mammalia, Rodentia) in the central European Pleistocene. Pp. 165-182, *in* International symposium evolution, phylogeny and biostratigraphy of arvicolids (Rodentia, Mammalia) (O. Fejfar and W.-D. Heinrich, eds.). Geological Survey, Prague, 448 pp.

Heller, E. 1909. Two new rodents from British East Africa. Smithsonian Miscellaneous Collections, 52:471-472.

Heller, E. 1911. New species of rodents and carnivores from equatorial Africa. Smithsonian Miscellaneous Collections, 56:1-16.

Heller, E. 1912. New rodents from British East Africa. Smithsonian Miscellaneous Collections, 59:1-20.

Heller, K.-G., and M. Volleth. 1984. Taxonomic position of '*Pipistrellus societatis*' Hill, 1972 and the karyological characteristics of the genus *Eptesicus* (Chiroptera: Vespertilionidae). Zeitschrift für Zoologische Systematik und Evolutionsforschung, 22:65-77.

Hemmer, H. 1966. Untersuchungen zur Stammesgeschichte der Pantherkatzen (Pantherinae). Teil I. Veröffentlichungen der Zoologischen Staatssammlung München, 11:1-121.

Hemmer, H. 1968. Untersuchungen zur Stammesgeschichte der Pantherkatzen (Pantherinae). Teil II. Studien zur Ethologie des Nebelparders *Neofelis nebulosa* (Griffith 1821) und des Irbis *Uncia uncia* (Schreber 1775). Veröffentlichungen der Zoologischen Staatssammlung München, 12:155-247.

Hemmer, H. 1972. *Uncia uncia*. Mammalian Species, 20:1-5.

Hemmer, H. 1974. Untersuchungen zur Stammegeschichte der Pantherkatzen (Pantherinae). III. Zur Artgeschichte des Lowen *Panthera* (*Panthera*) *leo* (Linnaeus 1758). Veröffentlichungen der Zoologischen Staatssammlung München, 17:167-280.

Hemmer, H. 1978. The evolutionary systematics of living Felidae: present status and current problems. Carnivore, 1(1):71-79.
Hemmer, H. 1990. Domestication. The decline of environmental appreciation. Cambridge University Press, Cambridge, England, 208 pp.
Hemmer, H., P. Grubb, and C. P. Groves. 1976. Notes on the sand cat *Felis margarita* Loche, 1858. Zeitschrift für Säugetierkunde, 41:286-303.
Hemming, F. (ed.). 1950. Conclusions of the fourteenth meeting held at the Sorbonne in the Amphitheatre Louis-Liard on Monday, 26th July 1948 at 2030 hours. Bulletin of Zoological Nomenclature, 4:425-648.
Henderson, D. 1990. Gray whales and whalers on the China coast in 1869. Whalewatcher, 24(4):14-16.
Hendey, Q. B. 1972. The evolution and dispersal of the Monachinae (Mammalia, Pinnipedia). Annals of the South African Museum, 59:99-113.
Hendey, Q. B. 1974. The late Cenozoic Carnivora of the southwestern Cape Province. Annals of the South African Museum, 63:1-369.
Hendey, Q. B. 1980*a*. *Agriotherium* (Mammalia: Ursidae) from Langebaanweg, South Africa, and relationships of the genus. Annals of the South African Museum, 81(1):1-109.
Hendey, Q. B. 1980*b*. Origin of the giant panda. South African Journal of Science, 76:179-180.
Hennings, D., and R. S. Hoffmann. 1977. A review of the taxonomy of the *Sorex vagrans* species complex from western North America. Occasional Papers of the Museum of Natural History, University of Kansas, 68:1-35.
Hensley, A. P., and K. T. Wilkins. 1988. *Leptonycteris nivalis*. Mammalian Species, 307:1-4.
Henttonen, H., and V. A. Peiponen. 1982. *Clethrionomys rutilus* (Pallas, 1779)—polarrotelmaus. Pp. 165-176, *in* Handbuch der Säugetiere Europas (J. Niethammer and F. Krapp, eds.). Akademische Verlagsgesellschaft (Wiesbaden), 2/I:1-649.
Henttonen, H., and J. Viitala. 1982. *Clethrionomys rufocanus* (Sundevall, 1846)—graurotelmaus. Pp. 146-164, *in* Handbuch der Säugetiere Europas (J. Niethammer and F. Krapp, eds.). Akademische Verlagsgesellschaft (Wiesbaden), 2/I:1-649.
Heptner, V. G. 1939. The Turkestan desert shrew, its biology and adaptive peculiarities. Journal of Mammalogy, 20:139-149.
Heptner, V. G. 1941. On the names of geographical forms of *Ochotona pallasi* Gray. Journal of Mammalogy, 22:327-328.
Heptner, V. G. 1971. O sistematicheskom polozhenii Amurskovo lesnovo kota i o nekhotorykh drugikh yuzhnoaziatiskikh koshokh, otnosimykh k *Felis bengalensis* Kerr, 1792 [Systematic status of the Amur wildcat and some other east-Asian cats referred to *Felis bengalensis* Kerr, 1792]. Zoologicheskii Zhurnal, 50:1720-1727 (in Russian).
Heptner, V. G. 1975. [Materials on the morphology and systematics of the tridactyl jerboas (genus *Jaculus* Erxl., 1777) and related forms (Mammalia, Dipodidae)]. Byulleten' Moskovskovo Obshchestva Ispytatalei Prirody, Otdel Biologicheskii, 80(3):5-15 (in Russian).
Heptner, V. G. 1984. [Contributions to phylogeny and classification of jerboas (Dipodidae) of the fauna of the USSR]. Sbornik Trudov Zoologicheskovo Museya MGU, 22:37-60 (in Russian).
Heptner, V. G., and V. A. Dolgov. 1967. [Systematic position of *Sorex mirabilis* Ognev, 1937. Mammalia, Soricidae]. Zoologicheskii Zhurnal, 46:1419-1422 (in Russian).
Heptner, V. G., and N. P. Naumov, (eds). 1967. Mlekopitayushchie Sovetskovo Soyuza. Morskie korovy i khishchnye [Mammals of the Soviet Union. Sea cows and carnivores]. Vysshaya Shkola, Moscow, 2(1):1-1004 (in Russian).
Heptner, V. G., and N. P. Naumov, (eds). 1972. Mlekopitayushchie Sovetskovo Soyuza. Khishchye (gieny i koshki) [Mammals of the Soviet Union. Carnivores (hyaenas and cats)]. Vysshaya Shkola, Moscow, 2(2):1-551 (in Russian).
Heptner, V. G., and P. B. Yurgenson. 1967. Russkaya, ili Evropeiskaya, norka [Russian or European mink]. Pp. 718-736, *in* Mlekopitayushchie Sovetskovo Soyuza. Morskie korovy i khishchnye (V. G. Heptner and N. P. Naumov, eds.) [Mammals of the Soviet Union. Sea cows and carnivores]. Vysshaya Shkola, Moscow, 2(1):1-1004 (in Russian).
Heptner, V. G., A. A. Nasimovich, and A. G. Bannikov. 1961. Mlekopitayushchie Sovetskovo Soyuza: Parnokopytnie i Neparnokopytnie. [Mammals of the Soviet Union: Even-toed and odd-toed ungulates]. Vysshaya Shkola, Moscow, 1:1-776 (in Russian).
Heptner, V. G., A. A. Nasimovich, and A. G. Bannikov. 1988. Mammals of the Soviet Union: Artiodactyla and Perissodactyla [A translation of Heptner, et al., 1961, Mlekopitayushchie Sovetskovo Soyuza: Parnokopytnye i neparnokopytnye]. Smithsonian Institution Libraries and National Science Foundation, Washington, D. C., 1:1-1147.
Herd, R. M. 1983. *Pteronotus parnellii*. Mammalian Species, 209:1-5.

Herd, R. M., and M. B. Fenton. 1983. An electrophoretic, morphological, and ecological investigation of a putative hybrid zone between *Myotis lucifugus* and *Myotis yumanensis* (Chiroptera: Vespertilionidae). Canadian Journal of Zoology, 61:2029-2050.

Hermanson, J. W., and T. J. O'Shea. 1983. *Antrozous pallidus*. Mammalian Species, 213:1-8.

Hernández-Camacho, J. 1960. Primitiae mastozoologicae Colombianae 1: Status taxonomical de *Sciurus pucheranii santanderensis*. Caldasia, 8:359-368.

Hernández-Camacho, J. 1977. Notas para una monografia de *Potos flavus* (Mammalia: Carnivora) en Colombia. Caldasia, 11(55):147-181.

Hernández-Camacho, J., and A. Cadena-G. 1978. Notas para la revision del género *Lonchorhina* (Chiroptera, Phyllostomidae). Caldasia, 12:200-251.

Hernandez-Camacho, J., and R. W. Cooper. 1976. The nonhuman primates of Colombia. Pp. 35-69, *in* Neotropical Primates (R. W. Thorington, Jr. and P. Heltne, eds.). National Academy of Sciences, Washington, D. C., 135 pp.

Hershkovitz, P. 1940. Four new oryzomyine rodents from Ecuador. Journal of Mammalogy, 21:78-84.

Hershkovitz, P. 1944. Systematic review of the Neotropical water rats of the genus *Nectomys* (Cricetinae). Miscellaneous Publications, Museum of Zoology, University of Michigan, 58:1-101.

Hershkovitz, P. 1948*a*. Mammals of northern Colombia. Preliminary report no. 2: Spiny Rats (Echimyidae), with supplemental notes on related forms. Proceedings of the United States National Museum, 97:125-140.

Hershkovitz, P. 1948*b*. Mammals of northern Colombia. Preliminary report no. 3. Water rats (genus *Nectomys*), with supplemental notes on related forms. Proceedings of the United States National Museum, 98:49-56.

Hershkovitz, P. 1948*c*. The technical name of the Virginia deer with a list of South American forms. Proceedings of the Biological Society of Washington, 61:41-48.

Hershkovitz, P. 1949*a*. Generic names of the four-eyed pouch opossum and the woolly opossum (Didelphidae). Proceedings of the Biological Society of Washington, 62:11-12.

Hershkovitz, P. 1949*b*. Status of names credited to Oken, 1816. Journal of Mammalogy, 30:289-301.

Hershkovitz, P. 1950. Mammals of northern Colombia. Preliminary report no. 6: Rabbits (Leporidae), with notes on the classification and distribution of the South American forms. Proceedings of the United States National Museum, 100:327-375.

Hershkovitz, P. 1951. Mammals from British Honduras, Mexico, Jamaica and Haiti. Fieldiana, Zoology, 31(47):547-569.

Hershkovitz, P. 1954. Mammals of northern Colombia. Preliminary report no. 7: Tapirs (genus *Tapirus*), with a systematic revision of American species. Proceedings of the United States National Museum, 103:465-496.

Hershkovitz, P. 1955*a*. South American marsh rats, genus *Holochilus*, with a summary of sigmodont rodents. Fieldiana, Zoology, 37:639-687.

Hershkovitz, P. 1955*b*. Status of the generic name *Zorilla* (Mammalia): Nomenclature by rule or by caprice. Proceedings of the Biological Society of Washington, 68:185-192.

Hershkovitz, P. 1957*a*. A synopsis of the wild dogs of Colombia. Novedades Colombianas. Contribuciones Científicas del Museo de Historia Natural de la Universidad del Cauca, 3(1):157-161.

Hershkovitz, P. 1957*b*. The type locality of *Bison bison* Linnaeus. Proceedings of the Biological Society of Washington, 70:31-32.

Hershkovitz, P. 1958. Technical names of the South American marsh deer and pampas deer. Proceedings of the Biological Society of Washington, 71:13-16.

Hershkovitz, P. 1959*a*. Nomenclature and taxonomy of the Neotropical mammals described by Olfers, 1818. Journal of Mammalogy, 40:337-353.

Hershkovitz, P. 1959*b*. Two new genera of South American rodents. Proceedings of the Biological Society of Washington, 72:5-10.

Hershkovitz, P. 1959*c*. A new species of South American brocket, genus *Mazama* (Cervidae). Proceedings of the Biological Society of Washington, 72:45-54.

Hershkovitz, P. 1960. Mammals of northern Colombia, preliminary report no. 8: Arboreal rice rats, a systematic revision of the subgenus *Oecomys*, genus *Oryzomys*. Proceedings of the United States National Museum, 110:513-568.

Hershkovitz, P. 1961*a*. On the South American small-eared zorro *Atelocynus microtis* Sclater (Canidae). Fieldiana, Zoology, 39:505-523.

Hershkovitz, P. 1961*b*. On the nomenclature of certain whales. Fieldiana, Zoology, 39(49):547-565.

Hershkovitz, P. 1962. Evolution of Neotropical cricetine rodents (Muridae) with special reference to the phyllotine group. Fieldiana, Zoology, 46:1-524.

Hershkovitz, P. 1963. The nomenclature of South American peccaries. Proceedings of the Biological Society of Washington, 76:85-88.
Hershkovitz, P. 1966*a*. Catalog of living whales. Bulletin of the United States National Museum, 246:1-259.
Hershkovitz, P. 1966*b*. Mice, land bridges and Latin American faunal interchange. Pp. 725-751, *in* Ectoparasites of Panama. (R. L. Wenzel and V. J. Tipton, eds.). Field Museum of Natural History, Chicago, 861 pp.
Hershkovitz, P. 1966*c*. South American swamp and fossorial rats of the scapteromyine group (Cricetinae, Muridae), with comments on the glans penis in murid taxonomy. Zeitschrift für Säugetierkunde, 31:81-149.
Hershkovitz, P. 1969. The evolution of mammals on southern continents. VI. The Recent mammals of the Neotropical Region: a zoogeographical and ecological review. Quarterly Review of Biology, 44:1-70.
Hershkovitz, P. 1970*a*. Notes on tertiary Platyrrhine monkeys and description of a new genus from the late Miocene of Colombia. Folia Primatologica, 12:1-37.
Hershkovitz, P. 1970*b*. Supplementary notes on Neotropical *Oryzomys dimidiatus* and *Oryzomys hammondi* (Cricetinae). Journal of Mammalogy, 51:789-794.
Hershkovitz, P. 1971. A new rice rat of the *Oryzomys palustris* group (Cricetinae, Muridae) from northwestern Colombia, with remarks on distribution. Journal of Mammalogy, 52:700-709.
Hershkovitz, P. 1972*a*. Notes on New World Monkeys. International Zoo Yearbook, 12:3-12.
Hershkovitz, P. 1972*b*. The recent mammals of the Neotropical region: A zoogeographic and ecological review. Pp. 311-431, *in* Evolution, mammals and southern continents (A. Keast, F. C. Erk, and B. Glass, eds.). State University of New York Press, Albany, 543 pp.
Hershkovitz, P. 1976. Comments on generic names of four-eyed opossums (family Didelphidae). Proceedings of the Biological Society of Washington, 89:295-304.
Hershkovitz, P. 1977. Living New World Monkeys (Platyrrhini). University of Chicago Press, Chicago, IL, 1:1-1117.
Hershkovitz, P. 1979. The species of sakis, genus *Pithecia*, with notes on sexual dichromatism. Folia Primatologia, 31:1-22.
Hershkovitz, P. 1981. *Philander* and four-eyed opossums once again. Proceedings of the Biological Society of Washington, 93:943-946.
Hershkovitz, P. 1982. Neotropical deer (Cervidae). Part 1. Pudus, genus *Pudu* Gray. Fieldiana, Zoology, New Series, 11:1-86.
Hershkovitz, P. 1983. Two new species of night monkeys, genus *Aotus* (Cebidae, Platyrrhini): a preliminary report on *Aotus* taxonomy. American Journal of Primatology, 4(3):209-243.
Hershkovitz, P. 1984. Taxonomy of squirrel monkeys genus *Saimiri* (Cebidae, Platyrrhini): A preliminary report with description of a hitherto unnamed form. American Journal of Primatology, 7:155-210.
Hershkovitz, P. 1985. A preliminary taxonomic review of the South American bearded saki monkeys, genus *Chiropotes*, with the description of a new subspecies. Fieldiana, Zoology, New Series, 27:1-46.
Hershkovitz, P. 1987*a*. First South American record of Coues' marsh rice rat, *Oryzomys couesi*. Journal of Mammalogy, 68:152-154.
Hershkovitz, P. 1987*b*. A history of the recent mammalogy of the neotropical region from 1492 to 1850. Fieldiana, Zoology, New Series, 39:11-98.
Hershkovitz, P. 1987*c*. Uacaries, New World Monkeys of the genus *Cacajao*: A preliminary taxonomic review with the description of a new American Journal of Primatology, 12:1-53.
Hershkovitz, P. 1987*d*. The taxonomy of South American sakis, genus *Pithecia*: A preliminary report and critical review with the description of a new species and subspecies. American Journal of Primatology, 12:387-468.
Hershkovitz, P. 1990*a*. The Brazilian rodent genus *Thalpomys* (Sigmodontinae, Cricetidae) with a description of a new species. Journal of Natural History, 24:763-783.
Hershkovitz, P. 1990*b*. Titis, New World monkeys of the genus *Callicebus* (Cebidae, Platyrrhini): A preliminary taxonomic review. Fieldiana, Zoology, New Series, 55:1-109.
Hershkovitz, P. 1990*c*. Mice of the *Akodon boliviensis* size class (Sigmodontinae, Cricetidae), with the description of two new species from Brazil. Fieldiana, Zoology, New Series, 57:1-35.
Hewson, R. 1991. Mountain hare/Irish hare. Pp. 161-167, *in* The handbook of British mammals (G. B. Corbet and S. Harris, eds.). Blackwell Scientific Publications, Oxford, 588 pp.
Heyning, J. E. 1989*a*. Comparative facial anatomy of beaked whales (Ziphiidae) and a systematic revison among the families of extant Odontoceti. Contributions in Science, Natural History Museum of Los Angeles County, 405:1-64.
Heyning, J. E. 1989*b*. Cuvier's beaked whale - *Ziphius cavirostris* G. Cuvier, 1823. Pp. 289-308, *in* Handbook of marine Mammals: River dolphins and the larger toothed whales (S. H. Ridgway and R. Harrison, eds.). Academic Press, London, 4:1-442.

Heyning, J. E., and M. E. Dahlheim. 1988. *Orcinus orca*. Mammalian Species, 304:1-9.
Hibbard, C. W. 1963. The origin of the P/3 pattern of *Sylvilagus, Caprolagus, Oryctolagus* and *Lepus*. Journal of Mammalogy, 44:1-15.
Hibbard, C. W., and W. W. Dalquest. 1973. *Proneofiber*, a new genus of vole (Cricetidae: Rodentia) from the Pleistocene Seymour Formation of Texas, and its evolutionary and stratigraphic significance. Journal of Quaternary Research, 3:269-274.
Hielscher, K., A. Stubbe, K. Zernahle, and R. Samjaa. 1992. Karyotypes and systematics of Asian high-mountain voles, genus *Alticola* (Rodentia, Arvicolidae). Results of Mongolian-German biological expeditions since 1962, no. 211. Cytogenetics and Cell Genetics, 59:307-310.
Hight, M. E., M. Goodman, and W. Prychodko. 1974. Immunological studies of the Sciuridae. Systematic Zoology, 23:12-25.
Hill, J. E. 1958. Some observations on the fauna of the Maldive Islands. Part II. Mammals. Journal of the Bombay Natural History Society, 55:3-10.
Hill, J. E. 1959. A north Bornean pygmy squirrel, *Glyphotes simus* Thomas, and its relationships. Bulletin of the British Museum (Natural History), Zoology Series, 5:257-266.
Hill, J. E. 1961*a*. Fruit-bats from the Federation of Malaya. Proceedings of the Zoological Society of London, 136:629-642.
Hill, J. E. 1961*b*. Indo-Australian bats of the genus *Tadarida*. Mammalia, 25:29-56.
Hill, J. E. 1961*c*. Notes on flying squirrels of the genera *Pteromyscus, Hylopetes* and *Petinomys*. Annals and Magazine of Natural History, ser. 13, 4:721-738.
Hill, J. E. 1962*a*. A little-known fruit-bat from Rennel Island. The Natural History of Rennel Island, British Solomon Islands, 4:7-9.
Hill, J. E. 1962*b*. Notes on some insectivores and bats from Upper Burma. Proceedings of the Zoological Society of London, 139:119-137.
Hill, J. E. 1963*a* [1964]. Notes on some tubed-nosed bats, genus *Murina*, from southeastern Asia, with descriptions of a new species and a new subspecies. Federation Museums Journal, New Series (Kuala Lumpur), 8:48-59.
Hill, J. E. 1963*b*. A revision of the genus *Hipposideros*. Bulletin of the British Museum (Natural History), Zoology Series, 11:1-129.
Hill, J. E. 1965. Asiatic bats of the genera *Kerivoula* and *Phoniscus* (Vespertilionidae), with a note on *Kerivoula aerosa* Tomes. Mammalia, 29:524-556.
Hill, J. E. 1966. A review of the genus *Philetor* (Chiroptera; Vespertilionidae). Bulletin of the British Museum (Natural History), Zoology Series, 14:371-387.
Hill, J. E. 1967. The bats of the Andaman and Nicobar Islands. Journal of the Bombay Natural History Society, 64:1-9.
Hill, J. E. 1971*a*. Bats from the Solomon Islands. Journal of Natural History, 5:573-581.
Hill, J. E. 1971*b*. The bats of Aldabra Atoll, western Indian Ocean. Philosophical Transactions of the Royal Society of London, B260:573-576.
Hill, J. E. 1971*c*. A note of *Pteropus* (Chiroptera: Pteropidae) from the Andaman Islands. Journal of the Bombay Natural History Society, 68:1-8.
Hill, J. E. 1971*d*. The status of *Vespertilio brachypterus* Temminck, 1840 (Chiroptera: Vespertilionidae). Zoologische Mededelingen, 45:143.
Hill, J. E. 1972. The Gunong Benom Expedition, 1967. 4. New records of Malayan bats, with taxonomic notes and the description of a new *Pipistrellus*. Bulletin of the British Museum (Natural History), Zoology Series, 23:21-42.
Hill, J. E. 1974*a*. A review of *Laephotis* Thomas, 1901 (Chiroptera: Vespertilionidae). Bulletin of the British Museum (Natural History), Zoology Series, 27:73-82.
Hill, J. E. 1974*b*. A review of *Scotoecus* Thomas, 1901 (Chiroptera: Vespertilionidae). Bulletin of the British Museum (Natural History), Zoology Series, 27:167-188.
Hill, J. E. 1976. Bats referred to *Hesperoptenus* Peters, 1869 (Chiroptera: Vespertilionidae) with the description of a new subgenus. Bulletin of the British Museum (Natural History), Zoology Series, 30:1-28.
Hill, J. E. 1977*a*. African bats allied to *Kerivoula lanosa* (A. Smith, 1847)(Chiroptera: Vespertilionidae). Revue de Zoologie Africaine, 91:623-633.
Hill, J. E. 1977*b*. A review of the Rhinopomatidae (Mammalia: Chiroptera). Bulletin of the British Museum (Natural History), Zoology Series, 32:29-43.
Hill, J. E. 1980*a*. A note on *Lonchophylla* (Chiroptera: Phyllostomatidae) from Ecuador and Peru, with the description of a new species. Bulletin of the British Museum (Natural History), Zoology Series, 38:233-236.
Hill, J. E. 1980*b*. The status of *Vespertilio borbonicus* E. Geoffroy, 1803 (Chiroptera: Vespertilionidae). Zoologische Mededelingen, 55:287-295.

Hill, J. E. 1982. A review of the leaf-nosed bats *Rhinonycteris, Cloeotis* and *Triaenops* (Chiroptera: Hipposideridae). Bonner Zoologische Beiträge, 33:165-186.

Hill, J. E. 1983. Bats (Mammalia: Chiroptera) from Indo-Australia. Bulletin of the British Museum (Natural History), Zoology Series, 43:103-208.

Hill, J. E. 1985. The status of *Lichonycteris degener* Miller, 1931 (Chiroptera: Phyllostomidae). Mammalia, 49:579-582.

Hill, J. E. 1987. A note on *Balantiopteryx infusca* (Thomas, 1897) (Chiroptera: Emballonuridae). Mammalia, 50:558-560.

Hill, J. E. 1991. Bats (Mammalia: Chiroptera) from the Togian Islands, Sulawesi, Indonesia. Bulletin of the American Museum of Natural History, 206:168-175.

Hill, J. E., and W. N. Beckon. 1978. A new species of *Pteralopex* Thomas, 1888 (Chiroptera: Pteropodidae) from the Fiji Islands. Bulletin of the British Museum (Natural History), Zoology Series, 34:65-82.

Hill, J. E., and T. D. Carter. 1941. The mammals of Angola, Africa. Bulletin of the American Museum of Natural History, 78(1):1-211.

Hill, J. E., and M. J. Daniel. 1985. Systematics of the New Zealand short-tailed bat *Mystacina* Gray, 1843 (Chiroptera: Mystacinidae). Bulletin of the British Museum (Natural History), Zoology Series, 48:279-300.

Hill, J. E., and C. M. Francis. 1984. New bats (Mammalia: Chiroptera) and new records of bats from Borneo and Malaya. Bulletin of the British Museum (Natural History), Zoology Series, 47:303-329.

Hill, J. E., and D. L. Harrison. 1987. The baculum in the Vespertilioninae (Chiroptera: Vespertilionidae) with a systematic review, a synopsis of *Pipistrellus* and *Eptesicus*, and the description of a new genus and subgenus. Bulletin of the British Museum (Natural History), Zoology Series, 52(7):225-305.

Hill, J. E., and K. F. Koopman. 1981. The status of *Lamingtona lophorhina* McKean and Calaby, 1968 (Chiroptera: Vespertilionidae). Bulletin of the British Museum (Natural History), Zoology Series, 41:275-278.

Hill, J. E., and F. G. Rozendaal. 1989. Records of bats (Microchiroptera) from Wallacea. Zoologische Mededelingen, 63:97-122.

Hill, J. E., and D. A. Schlitter. 1982. A record of *Rhinolophus arcuatus* (Chiroptera: Rhinolophidae) from New Guinea, with the description of a new subspecies. Annals of Carnegie Museum, 51:455-464.

Hill, J. E., and J. D. Smith. 1984. Bats: A natural history. University of Texas Press, Austin, 243 pp.

Hill, J. E., and S. E. Smith. 1981. *Craseonycteris thonglongyai*. Mammalian Species, 160:1-4.

Hill, J. E., and K. Thonglongya. 1972. Bats from Thailand and Cambodia. Bulletin of the British Museum (Natural History), Zoology Series, 22:171-196.

Hill, J. E., and G. Topal. 1973. The affinities of *Pipistrellus ridleyi* Thomas, 1898 and *Glischropus rosseti* Oey, 1951 (Chiroptera: Vespertilionidae). Bulletin of the British Museum (Natural History), Zoology Series, 24:447-454.

Hill, J. E., and M. Yoshiyuki. 1980. A new species of *Rhinolophus* (Chiroptera: Rhinolophidae) from Iriomote Island, Ryukyu Islands, with notes on the Asiatic members of the *Rhinolophus pusillus* group. Bulletin of the National Science Museum (Tokyo), ser. A (Zoology), 6:179-189.

Hill, J. E., A. Zubaid, and G. W. H. Davison. 1986. The taxonomy of leaf-nosed bats of the *Hipposideros bicolor* group (Chiroptera: Hipposideridae) from southeastern Asia. Mammalia, 50:535-540.

Hill, W. C. O. 1969. The nomeclature, taxonomy and distribution of chimpanzees. Pp. 22-49, *in* Chimpanzee: Anatomy, behavior and diseases of chimpanzees (G. A. Bourne, ed.). S. Karger, Basel, 1:1-468.

Hill, W. C. O. 1974. Primates: Comparative anatomy and taxonomy, vol. 7. Cynopithecinae: *Cercocebus, Macaca, Cynopithecus*. Edinburgh University Press, 954 pp.

Hillman, C. N., and T. W. Clark. 1980. *Mustela nigripes*. Mammalian Species, 126:1-3.

Hinesley, L. L. 1979. Systematics and distribution of two chromosome forms in the southern grasshopper mouse, genus *Onychomys*. Journal of Mammalogy, 60:117-128.

Hinojosa, F., S. Anderson, and J. L. Patton. 1987. Two new species of *Oxymycterus* (Rodentia) from Peru and Bolivia. American Museum Novitates, 2898:1-17.

Hinton, M. A. C. 1919. Scientific results from the mammal survey. Pt. 2. Journal of the Bombay Natural History Society, 26:384-416.

Hinton, M. A. C. 1921. Some new African mammals. Annals and Magazine of Natural History, ser. 9, 7:368-373.

Hinton, M. A. C. 1922. Scientific results from the mammal survey. No. 34. The house rats of Nepal. Journal of the Bombay Natural History Society, 28:1056-1066.

Hinton, M. A. C. 1926. Monograph of the voles and lemmings (Microtinae) living and extinct. Volume 1. British Museum (Natural History), London, 488 pp.

Hinton, M. A. C. 1943. Preliminary diagnosis of five new murine rodents from New Guinea. Annals and Magazine of Natural History, ser. 11, 10:552-557.

Hirning, U., W. A. Schulz, W. Just, S. Adolph, and W. Vogel. 1989. A comparative study of the heterochromatin of *Apodemus sylvaticus* and *Apodemus flavicollis*. Chromosoma, 98:450-455.
Hou Lan-xin, Zhao Xin-chun, and Jiang Wei. 1986. [Investigation of rodents in Dabancheng-Chaiwopu, Urumqi Country, Xinjiang Yugur Autonomous Region]. Acta Theriologica Sinica, 6(4):315-316 (in Chinese).
Hodgson, B. H. 1831. Some account of a new species of *Felis*. Gleanings in Science, 3:177-178.
Hodgson, B. H. 1842. On a new species of *Prionodon, P. pardicolor* nobis. Calcutta Journal of Natural History, 2:57-60.
Hodgson, B. H. 1842. Notice of the mammals of Tibet, with descriptions and plates of some new species. Journal of the Asiatic Society of Bengal, 11:275-289.
Hodgson, B. H. 1843. Notice of two marmots inhabiting respectively the plains of Tibet and the Himalayan slopes near to the snows, and also of a *Rhinolophus* of the central region of Nepal. Journal of the Asiatic Society of Bengal, 12:409-414.
Hodgson, B. H. 1844. Classified catalogue of mammals of Nepal. Calcutta Journal of Natural History, 4:284-294.
Hodgson, B. H. 1845. On the rats, mice, and shrews of the central region of Nepal. Annals and Magazine of Natural History, ser. 1, 15:266-270.
Hoeck, H. N. 1978. Systematics of the Hyracoidea: toward a clarification. Bulletin of Carnegie Museum of Natural History, 6:146-151.
Hoffmann, R. S. 1971. Relationships of certain Holarctic shrews, genus *Sorex*. Zeitschrift für Säugetierkunde, 36:193-200.
Hoffmann, R. S. 1977. The identity of Lewis' marmot, *Arctomys lewisii*. Proceedings of the Biological Society of Washington, 90:291-301.
Hoffmann, R. S. 1984. A review of the shrew-moles (genus *Uropsilus*) of China and Burma. Journal of the Mammalogical Society of Japan, 10:69-80.
Hoffmann, R. S. 1985*a*. The correct name for the Palearctic brown, or flat-skulled, shrew is *Sorex roboratus*. Proceedings of the Biological Society of Washington, 98:17-28.
Hoffmann, R. S. 1985*b* [1986]. A review of the genus *Soriculus* (Mammalia: Insectivora). Journal of the Bombay Natural History Society, 82:459-481.
Hoffmann, R. S. 1987. A review of the systematics and distribution of Chinese red-toothed shrews (Mammalia: Soricinae). Acta Theriologica Sinica, 7:100-139.
Hoffmann, R. S., and J. W. Koeppl. 1985. Zoogeography. Pp. 84-115, *in* Biology of New World *Microtus* (R. H. Tamarin, ed.). Special Publication, American Society of Mammalogists, 8:1-893.
Hoffmann, R. S., and J. G. Owen. 1980. *Sorex tenellus* and *Sorex nanus*. Mammalian Species, 131:1-4.
Hoffmann, R. S., and R. S. Peterson. 1967. Systematics and zoogeography of *Sorex* in the Bering Strait area. Systematic Zoology, 16:127-136.
Hoffmann, R. S., J. W. Koeppl, and C. F. Nadler. 1979. The relationships of the amphiberingian marmots (Mammalia: Sciuridae). Occasional Papers of the Museum of Natural History, University of Kansas, 83:1-56.
Hoffmeister, D. F. 1951. A taxonomic and evolutionary study of the piñon mouse, *Peromyscus truei*. Illinois Biological Monographs, 21:1-104.
Hoffmeister, D. F. 1969. The species problem in the *Thomomys bottae-Thomomys umbrinus* complex of gophers in Arizona. Miscellaneous Publications, Museum of Natural History, University of Kansas, 51:75-91.
Hoffmeister, D. F. 1974. The taxonomic status of *Perognathus penicillatus minimus* Burt. Southwestern Naturalist, 19:213-214.
Hoffmeister, D. F. 1981. *Peromyscus truei*. Mammalian Species, 161:1-5.
Hoffmeister, D. F. 1986. Mammals of Arizona. University of Arizona Press, Tucson., 602 pp.
Hoffmeister, D. F., and L. de la Torre. 1960. A revision of the wood rat *Neotoma stephensi*. Journal of Mammalogy, 41:476-491.
Hoffmeister, D. F., and L. de la Torre. 1961. Geographic variation in the mouse *Peromyscus difficilis*. Journal of Mammalogy, 42:1-13.
Hoffmeister, D. F., and V. E. Diersing. 1973. The taxonomic status of *Peromyscus merriami goldmani* Osgood, 1904. Southwestern Naturalist, 18:354-357.
Hoffmeister, D. F., and V. E. Diersing. 1978. Review of the tassel-eared squirrels of the subgenus *Otosciurus*. Journal of Mammalogy, 59:402-413.
Hoffmeister, D. F., and L. S. Ellis. 1979. Geographic variation in *Eutamias quadrivittatus* with comments on the taxonomy of other Arizonan chipmunks. Southwestern Naturalist, 24:655-665.
Hoffmeister, D. F., and M. R. Lee. 1963. The status of the sibling species *Peromyscus merriami* and *Peromyscus eremicus*. Journal of Mammalogy, 44:201-213.

Hoffmeister, D. F., and M. R. Lee. 1967. Revision of the pocket mice, *Perognathus penicillatus*. Journal of Mammalogy, 48:361-380.

Hoffstetter, R. 1969. Remarques sur le phylogenie et la classification des Edentes Xenarthres (Mammiferes) actuels et fossiles. Bulletin du Muséum National d'Histoire Naturelle (Paris), 41:92-103.

Holden, H. E., and H. S. Eabry. 1970. Chromosomes of *Sylvilagus floridanus* and *S. transitionalis*. Journal of Mammalogy, 51:166-168.

Hollander, R. R. 1990. Biosystematics of the yellow-faced pocket gopher, *Cratogeomys castanops* (Rodentia: Geomyidae) in the United States. Special Publications, The Museum, Texas Tech University, 33:1-62.

Hollander, R. R., and M. R. Willig. 1992. Description of a new subspecies of the southern grasshopper mouse, *Onychomys torridus*, from Western Mexico. Occasional Papers. The Museum Texas Tech University, 148:1-4.

Hollister, N. 1911. A systematic synopsis of the muskrats. North American Fauna, 32:1-47.

Hollister, N. 1913*a*. A synopsis of the American minks. Proceedings of the United States National Museum, 44:471-480.

Hollister, N. 1913*b*. A review of the Philippine land mammals in the United States National Museum. Proceedings of the United States National Museum, 46:299-341.

Hollister, N. 1914. A systematic account of the grasshopper mice. Proceedings of the United States National Museum, 47:427-489.

Hollister, N. 1915. The genera and subgenera of raccoons and their allies. Proceedings of the United States National Museum, 49:143-150.

Hollister, N. 1916*a*. A systematic account of the prairie dogs. North American Fauna, 40:1-37.

Hollister, N. 1916*b*. The type species of *Rattus*. Proceedings of the Biological Society of Washington, 29:206-207.

Hollister, N. 1918. East African mammals in the United States National Museum. Part I. Insectivora, Chiroptera, and Carnivora. Bulletin of the United States National Museum, 99:1-194, 55 pls.

Hollister, N. 1919. East African mammals in the United States National Museum. Part II. Rodentia, Lagomorpha, and Tubulidentata. Bulletin of the United States National Museum, 99(2):1-184.

Holthuis, L. B. 1987. The scientific name of the sperm whale. Marine Mammal Science, 3(1):87-88.

Holthuis, L. B., and T. Sakai. 1970. Ph. F. von Siebold and Fauna Japonica: A history of early Japanese zoology:. Academic Press of Japan, Tokyo, 323 pp.

Holz, H., and J. Niethammer. 1990. *Erinaceus europaeus* Linnaeus, 1758 — Braunbrustigel, Westigel. Pp. 26-49, *in* Handbuch der Säugetiere Europas (J. Niethammer and F. Krapp, eds.). Aula-Verlag, Wiesbaden, 3/I:1-524.

Homan, J. A., and J. K. Jones, Jr. 1975*a*. *Monophyllus redmani*. Mammalian Species, 57:1-3.

Homan, J. A., and J. K. Jones, Jr. 1975*b*. *Monophyllus plethodon*. Mammalian Species, 58:1-2.

Honacki, J. H., K. E. Kinman and J. W. Koeppl (eds.). 1982. Mammal species of the world: A taxonomic and geographic reference. Allen Press, Inc. and The Association of Systematics Collections, Lawrence, Kansas, 694 pp.

Honeycutt, R. L., and D. J. Schmidly. 1979. Chromosomal and morphological variation in the plains pocket gopher, *Geomys bursarius* in Texas and adjacent states. Occasional Papers, The Museum, Texas Tech University, 58:1-54.

Honeycutt, R. L., and S. L. Williams. 1982. Genic differentiation in pocket gophers of the genus *Pappogeomys*, with comments on intergeneric relationships in the subfamily Geomyinae. Journal of Mammalogy, 63:208-217.

Honeycutt, R. L., M. W. Allard, S. V. Edwards, and D. A. Schlitter. 1991. Systematics and evolution of the family Bathyergidae. Pp. 45-65, *in* The biology of the naked mole rat (P. W. Sherman, J. U. M. Jarvis, and R. D. Alexander, eds.). Princeton University Press, Princeton, 518 pp.

Hood, C. S., and J. K. Jones, Jr. 1984. *Noctilio leporinus*. Mammalian Species, 216:1-7.

Hood, C. S., and J. Pitocchelli. 1983. *Noctilio albiventris*. Mammalian Species, 197:1-5.

Hood, C. S., L. W. Robbins, R. J. Baker, and H. S. Shellhammer. 1984. Chromosomal studies and evolutionary relationships of an endangered species, *Reithrodontomys raviventris*. Journal of Mammalogy, 65:655-667.

Hooijer, D. A. 1956. The valid name of the banteng: *Bibos javanicus* d'Alton. Zoologische Mededelingen, 34:223-226.

Hooijer, D. A. 1962. Quaternary langurs and macaques from the Malay Archipelago. Zoologische Verhandelingen (Leiden), 55:1-64.

Hooper, E. T. 1946. Two genera of pocket gophers should be congeneric. Journal of Mammalogy, 27:397-399.

Hooper, E. T. 1947. Notes on Mexican mammals. Journal of Mammalogy, 28:40-57.

Hooper, E. T. 1952. A systematic review of harvest mice (genus *Reithrodontomys*) of Latin America. Miscellaneous Publications, Museum of Zoology, University of Michigan, 77:1-255.

Hooper, E. T. 1953. Notes on mammals of Tamaulipas, Mexico. Occasional Papers of the Museum of Zoology, University of Michigan, 544:1-12.
Hooper, E. T. 1954. A synopsis of the cricetine rodent genus *Nelsonia*. Occasional Papers of the Museum of Zoology, University of Michigan, 558:1-12.
Hooper, E. T. 1957. Dental patterns in mice of the genus *Peromyscus*. Miscellaneous Publications, Museum of Zoology, University of Michigan, 99:1-59.
Hooper, E. T. 1958. The male phallus in mice of the genus *Peromyscus*. Miscellaneous Publications, Museum of Zoology, University of Michigan, 105:1-24.
Hooper, E. T. 1959. The glans penis in five genera of cricetid rodents. Occasional Papers of the Museum of Zoology, University of Michigan, 613:1-11.
Hooper, E. T. 1960. The glans penis in *Neotoma* (Rodentia) and allied genera. Occasional Papers of the Museum of Zoology, University of Michigan, 618:1-21.
Hooper, E. T. 1968*a*. Anatomy of middle ear walls and cavities in nine species of microtine rodents. Occasional Papers of the Museum of Zoology, University of Michigan, 657:1-28.
Hooper, E. T. 1968*b*. Classification. Pp. 27-74, *in* Biology of *Peromyscus* (Rodentia) (J. A. King, ed.). Special Publication, American Society of Mammalogists, 2:1-593.
Hooper, E. T. 1972. A synopsis of the rodent genus *Scotinomys*. Occasional Papers of the Museum of Zoology, University of Michigan, 665:1-32.
Hooper, E. T., and M. D. Carleton. 1976. Reproduction, growth and development in two contiguously allopatric rodent species, genus *Scotinomys*. Miscellaneous Publications, Museum of Zoology, University of Michigan, 151:1-52.
Hooper, E. T., and B. S. Hart. 1962. A synopsis of Recent North American microtine rodents. Miscellaneous Publications, Museum of Zoology, University of Michigan, 120:1-68.
Hooper, E. T., and G. G. Musser. 1964*a*. The glans penis in Neotropical cricetines (Family Muridae) with comments on classification of muroid rodents. Miscellaneous Publications, Museum of Zoology, University of Michigan, 123:1-57.
Hooper, E. T., and G. G. Musser. 1964*b*. Notes on classification of the rodent genus *Peromyscus*. Occasional Papers of the Museum of Zoology, University of Michigan, 635:1-13.
Hopwood, A. T. 1947. The generic names of the mandrill and baboons, with notes on some of the genera of Brisson, 1762. Proceedings of the Zoological Society of London, 117:533-536.
Horáček, I., and V. Hanák. 1984. Comments on the systematics and phylogeny of *Myotis nattereri* (Kuhl, 1818). Myotis, 21-22:20-29.
Horsfield, M. D. 1855. Brief notices of several new or little known species of Mammalia, lately discovered and collected in Nepal by Brian Houghton Hodgson. Annals and Magazine of Natural History, ser. 2, 16:101-114.
Horsfield, T. 1825. Description of the RIMAU-DAHAN of the inhabitants of Sumatra, a new species of *Felis*, discovered in the forests of Bencoolen, by Sir T. Stamford Raffles, late Lieutenant Governor of Fort Marlborough, &c., &c., &c. Zoological Journal, 1(4):542-554.
Houseal, T. W., I. F. Greenbaum, D. J. Schmidly, S. A. Smith, and K. M. Davis. 1987. Karyotypic variation in *Peromyscus boylii* from Mexico. Journal of Mammalogy, 68:281-296.
Howell, A. B. 1926*a*. Anatomy of the wood rat. Monograph of the American Society of Mammalogists, 1:1-225.
Howell, A. B. 1926*b*. Voles of the genus *Phenacomys*. North American Fauna, 48:1-66.
Howell, A. B. 1927. Revision of the American lemming mice (genus *Synaptomys*). North American Fauna, 50:1-37.
Howell, A. H. 1914. Revision of the American harvest mice (genus *Reithrodontomys*). North American Fauna, 36:1-97.
Howell, A. H. 1915. Revision of the American marmots. North American Fauna, 37:1-80.
Howell, A. H. 1918. Revision of the American flying squirrels. North American Fauna, 44:1-64.
Howell, A. H. 1929. Revision of the American chipmunks. North American Fauna, 52:1-157.
Howell, A. H. 1936. Description of three new red squirrels (*Tamiasciurus*) from North America. Proceedings of the Biological Society of Washington, 49:133-136.
Howell, A. H. 1938. Revision of the North American ground squirrels, with a classification of the North American Sciuridae. North American Fauna, 56:1-256.
Hoyt, R. A., and R. J. Baker. 1980. *Natalus major*. Mammalian Species, 130:1-3.
Hrabe, V. 1969. Der Bau des Glans penis bei vier Schläferarten (Gliridae, Rodentia). Zoologické Listy, 18(4):317-334.
Hrabe, V. 1977. Tarsal glands of voles of the genus *Pitymys* (Microtidae, Mammalia) from southern Austria. Folia Zoologica, 27:123-128.

Hsu, T. C., and M. L. Johnson. 1970. Cytological distinction between *Microtus montanus* and *Microtus canicaudus*. Journal of Mammalogy, 51:824-826.

Hubbs, C. L. 1946. First records of two beaked whales, *Mesoplodon bowdoini* and *Ziphius cavirostris*, from the Pacific coast of the United States. Journal of Mammalogy, 27(3):242-255.

Hubert, B. 1978*a*. Revision of the genus *Saccostomus* (Rodentia, Cricetomyinae), with new morphological and chromosomal data from specimens from the lower Omo Valley, Ethiopia. Bulletin of Carnegie Museum of Natural History, 6:48-52.

Hubert, B. 1978*b*. Modern rodent fauna of the lower Omo Valley, Ethiopia. Bulletin of Carnegie Museum of Natural History, 6:109-112.

Hubert, B., A. Meylan, F. Petter, A. Poulet, and M. Tranier. 1983. Different species in genus *Mastomys* from Western, Central and Southern Africa (Rodentia, Muridae). Annales Musée Royal De L'Afrique Centrale, Sciences Zoologiques, 237:143-148.

Hubner, R. 1988. Populations robertsoniennes chez la souris "sauvage" (*Mus domesticus* Rutty, 1772) en Belgique. Annales de la Societe Royale Zoologique de Belgique, 118:69-75.

Huckaby, D. G. 1980. Species limits in the *Peromyscus mexicanus* group (Mammalia: Rodentia: Muroidea). Contributions in Science, Natural History Museum of Los Angeles County, 326:1-24.

Hückinghaus, F. 1961. Vergleichende Untersuchungen über die Formenmannigfaltigkeit der Unterfamilie Caviinae Murray 1886. (Ergebnisse der Südamerikaexpedition Herre/Röhrs 1956-1957). Zeitschrift für Wissenschaftliche Zoologie, 166:1-98, 62 pls.

Hudson, W. S., and D. E. Wilson. 1986. *Macroderma gigas*. Mammalian Species, 260:1-4.

Huey, L. M. 1964. The mammals of Baja California, Mexico. Transactions of the San Diego Society of Natural History, 13:85-168.

Hugueney, M., and P. Mein. 1965. Lagomorphes et rongeurs du néogène de Lissieu (Rhône). Travaux des Laborotoire de Géologie de la Faculte des Sciences de l'Université de Lyon, 12:109-123.

Humphrey, S. R., and L. N. Brown. 1986. Report of a new bat (Chiroptera: *Artibeus jamaicensis*) in the United States is erroneous. Florida Scientist, 49:261-263.

Humphrey, S. R., and H. W. Setzer. 1989. Geographic variation and taxonomic revision of rice rats (*Oryzomys palustris* and *O. argentatus*) of the United States. Journal of Mammalogy, 70:557-570.

Hunt, R. M., Jr. 1974. The auditory bulla in Carnivora: an anatomical basis for reappraisal of carnivore evolution. Journal of Morphology, 143:21-76.

Hunt, R. M., Jr. 1987. Evolution of the Aeluroid Carnivora: Significance of auditory structure in the nimravid cat *Dinictis*. American Museum Novitates, 2886:1-74.

Hurrell, E., and G. McIntosh. 1984. Mammal society dormouse survey, January 1975-April 1979. Mammalian Review, 14(1):1-18.

Husar, S. L. 1977. *Trichechus inunguis*. Mammalian Species, 72:1-4.

Husar, S. L. 1978*a*. *Dugong dugon*. Mammalian Species, 88:1-7.

Husar, S. L. 1978*b*. *Trichechus senegalensis*. Mammalian Species, 89:1-3.

Husar, S. L. 1978*c*. *Trichechus manatus*. Mammalian Species, 93:1-5.

Husson, A. M. 1955. Notes on the mammals collected by the Swedish New Guinea Expedition. Nova Guinea (New Series), 6:283-306.

Husson, A. M. 1962. The bats of Suriname. Zoologische Verhandelingen, 58:1-282, 30 pls.

Husson, A. M. 1963. On *Blarina pyrrhonota* and *Echimys macrourus*: two mammals incorrectly assigned to the Suriname fauna. Studies on the Fauna of Suriname and other Guyanas, 5:34-41, 2 pls.

Husson, A. M. 1978. The mammals of Suriname. E. J. Brill, Leiden, 569 pp., 160 pls.

Husson, A. M., and L. B. Holthuis. 1955. The dates of publication of "Verhandelingen over de Natuurlijke Geschiedenis der Nederlandsche Overzeesche Bezittingen" Edited by C. J. Temminck. Zoologische Mededelingen, 34(2):17-24.

Husson, A. M., and L. B. Holthuis. 1974. *Physeter macrocephalus* Linnaeus, 1758, the valid name for the sperm whale. Zoologische Mededelingen, 48(19):205-217.

Hutchison, J. H. 1968. Fossil Talpidae (Insectivora, Mammalia) from the Later Tertiary of Oregon. Bulletin of the Museum of Natural History Oregon, 11:1-117.

Hutchison, J. H. 1974. Notes on type specimens of European Miocene Talpidae and a tentative classification of Old World Tertiary Talpidae (Insectivora: Mammalia). Geobios, 7:211-256, pl. 37-39.

Hutchison, J. H. 1987. Moles of the *Scapanus latimanus* group (Talpidae, Insectivora) from the Pliocene and Pleistocene of California. Contributions in Science, 386:1-15.

Hutterer, R. 1979. Verbreitung und Systematik von *Sorex minutus* Linnaeus, 1766 (Insectivora; Soricidae) im Nepal- Himalaya und angrenzenden Gebieten. Zeitschrift für Säugetierkunde, 44:65-80.

Hutterer, R. 1980. A record of Goodwin's shrew, *Cryptotis goodwini*, from Mexico. Mammalia, 44:413.

Hutterer, R. 1981*a*. Nachweis der Spitzmaus *Crocidura roosevelti* für Tanzania. Stuttgarter Beiträge zur Naturkunde, Ser. A, 342:1-9.

Hutterer, R. 1981*b*. Range extension of *Crocidura smithii*, with description of a new subspecies from Senegal. Mammalia, 45:388-391.

Hutterer, R. 1981*c*. Zur Systematik und Verbreitung der Soricidae Äthiopiens (Mammalia; Insectivora). Bonner Zoologische Beiträge, 31:217-247.

Hutterer, R. 1981*d*. Der Status von *Crocidura ariadne* Pieper, 1979 (Mammalia: Soricidae). Bonner Zoologische Beiträge, 32:3-12.

Hutterer, R. 1982*a*. *Crocidura manengubae* n. sp. (Mammalia: Soricidae), eine neue Spitzmaus aus Kamerun. Bonner Zoologische Beiträge, 32:241-248.

Hutterer, R. 1982*b*. Biologische und morphologische Beobachtungen an Alpenspitzmäusen (*Sorex alpinus*). Bonner Zoologische Beiträge, 33:3-18.

Hutterer, R. 1983*a*. *Crocidura grandiceps*, eine neue Spitzmaus aus Westafrika. Revue Suisse de Zoologie, 90:699-707.

Hutterer, R. 1983*b*. Taxonomy and distribution of *Crocidura fuscomurina* (Heuglin, 1865). Mammalia, 47:221-227.

Hutterer, R. 1983*c*. Über den Igel (*Erinaceus algirus*) der Kanarischen Inseln. Zeitschrift für Säugetierkunde, 48:261-264.

Hutterer, R. 1984. Status of some African *Crocidura* described by Isidore Geoffroy Saint-Hilaire, Carl J. Sundevall and Theodor von Heuglin. Annales Musée Royal de l'Afrique Centrale, Sciences Zoologiques, 237:207-217.

Hutterer, R. 1985. Anatomical adaptations of shrews. Mammal Review, 15:43-55.

Hutterer, R. 1986*a*. African shrews allied to *Crocidura fischeri*: taxonomy, distribution and relationships. Cimbebasia, ser. A, 8(4):199-207.

Hutterer, R. 1986*b*. Diagnosen neuer Spitzmäuse aus Tansania (Mammalia: Soricidae). Bonner Zoologische Beiträge, 37:23-33.

Hutterer, R. 1986*c*. Synopsis der Gattung *Paracrocidura* (Mammalia: Soricidae), mit Beschreibung einer neuen Art. Bonner Zoologische Beiträge, 37:73-90.

Hutterer, R. 1986*d*. Südamerikanische Spitzmäuse: *Cryptotis meridensis* und *C. thomasi* als verschiedene Arten. Zeitschrift für Säugetierkunde, 51, Sonderheft:33-34.

Hutterer, R. 1987. The species of *Crocidura* (Soricidae) in Morocco. Mammalia, 50:521-534.

Hutterer, R. 1990. *Sorex minutus* Linnaeus, 1766 - Zwergspitzmaus. Pp. 183-206, *in* Handbuch der Säugetiere Europas (J. Niethammer and F. Krapp, eds.). Aula-Verlag, Wiesbaden, 3/I:1-524.

Hutterer, R. 1991. Variation and evolution of the Sicilian shrew: taxonomic conclusions and description of a possibly related species from the Pleistocene of Morocco (Mammalia: Soricidae). Bonner Zoologische Beiträge, 42:241-251.

Hutterer, R., and F. Dieterlen. 1984. Zwei neue Arten der Gattung *Grammomys* aus Äthiopien und Kenia (Mammalia: Muridae). Stuttgarter Beiträge zur Naturkunde, ser. A (Biologie), 347:1-18.

Hutterer, R., and F. Dieterlen. 1986. Zur Verbreitung und Variation von *Desmodilliscus braueri*. Annalen des Naturhistorischen Museums in Wien, 88/89:213-221.

Hutterer, R., and N. J. Dippenaar. 1987. A new species of *Crocidura* Wagler, 1832 (Soricidae) from Zambia. Bonner Zoologische Beiträge, 38:1-7.

Hutterer, R., and A. Dudu. 1990. Redescription of *Crocidura caliginea*, a rare shrew from northeastern Zaire. Journal of African Zoology, 104:305-311.

Hutterer, R., and D. C. D. Happold. 1983. The shrews of Nigeria (Mammalia: Soricidae). Bonner Zoologische Monographien, 18:1-79.

Hutterer, R., and D. L. Harrison. 1988. A new look at the shrews (Soricidae) of Arabia. Bonner Zoologische Beiträge, 39:59-72.

Hutterer, R., and U. Hirsch. 1979. Ein neuer *Nesoryzomys* von der Insel Fernandina, Galapagos (Mammalia, Rodentia). Bonner Zoologische Beiträge, 30:276-283.

Hutterer, R., and T. Hürter. 1981. Adaptive Haarstrukturen bei Wasserspitzmäusen (Insectivora, Soricinae). Zeitschrift für Säugetierkunde, 46:1-11.

Hutterer, R., and P. D. Jenkins. 1983. Species-limits of *Crocidura somalica* Thomas, 1895 and *Crocidura yankariensis* Hutterer and Jenkins, 1980 (Insectivora, Soricidae). Zeitschrift für Säugetierkunde, 48:193-201.

Hutterer, R., and U. Joger. 1982. Kleinsäuger aus dem Hochland von Adamaoua, Kamerun. Bonner Zoologische Beiträge, 33:119-132.

Hutterer, R., and D. Kock. 1983. Spitzmäuse aus den Nuba-Bergen Kordofans, Sudan (Mammalia: Soricidae). Senckenbergiana Biologica, 63:17-26.

Hutterer, R., and M. Tranier. 1990. The immigration of the Asian house shrew *Suncus murinus* into Africa and Madagascar. Pp. 309-319, *in* Vertebrates in the Tropics (G. Peters & R. Hutterer, eds.),. Museum Alexander Koenig, Bonn, 424 pp.

Hutterer, R., and D. W. Yalden. 1990. Two new species of shrews from a relic forest in the Bale Mountain, Ethiopia. Pp. 63-72, *in* Vertebrates in the Tropics (G. Peters, and R. Hutterer, eds.). Museum Alexander Koenig, Bonn, 424 pp.

Hutterer, R., L. F. López-Jurado, and P. Vogel. 1987*a*. The shrews of the eastern Canary Islands: a new species (Mammalia: Soricidae). Journal of Natural History, 21:1347-1357.

Hutterer, R., E. Van der Straeten, and W. N. Verheyen. 1987*b*. A checklist of the shrews of Rwanda and biogeographic considerations on African Soricidae. Bonner Zoologische Beiträge, 38:155-172.

Hutterer, R., N. López-Martínez, and J. Michaux. 1988. A new rodent from Quaternary deposits of the Canary Islands and its relationships with Neogene and Recent murids of Europe and Africa. Palaeovertebrata, 18:241-262.

Hutterer, R., E. A. Sidiyene, and M. Tranier. 1992. A record of *Crocidura somalia* from the Sahara. Mammalia, 55:621-622.

Hutterer, R., F. Dieterlen, and G. Nikolaus. 1992. Small mammals from forest islands of eastern Nigeria and adjacent Cameroon, with systematical and biogeographical notes. Bonner Zoologische Beiträge, 43:393-414.

Hyndman, D., and J. I. Menzies. 1980. *Aproteles bulmerae* (Chiroptera: Pteropodidae) of New Guinea is not extinct. Journal of Mammalogy, 61:159-160.

Ibáñez, C. 1980. Descripción de un nuevo género de quiróptero neotropical de la familia Molossidae. Doñana, Acta Vertebrata, 7:104-111.

Ibáñez, C., and R. Fernández. 1985. Systematic status of the long-eared bat *Plecotus teneriffae* Barret-Hamilton, 1907 (Chiroptera; Vespertilionidae). Säugetierkundliche Mitteilungen, 32:143-149.

Ibáñez, C., and R. Valverde. 1985. Taxonomic status of *Eptesicus platyops* (Thomas, 1901)(Chiroptera, Vespertilionidae). Zeitschrift für Säugetierkunde, 50:241-242.

Ilani, G., and B. Shalmon. 1983. Rock dormouse on trees. Israel, Land and Nature, 9(1):39.

Imaizumi, Y. 1961. Taxonomic status of *Crocidura dsinezumi orii*. Journal of the Mammalogical Society of Japan, 2:17-22.

Imaizumi, Y. 1967. A new genus and species of cat from Iriomote, Ryukyu Islands. Journal of the Mammalogical Society of Japan, 3(4):75-106.

Imaizumi, Y. 1970*a*. Description of a new species of *Cervus* from the Tsushima Islands, Japan, with a revision of the subgenus *Sika* based on clinal analysis. Bulletin of the National Science Museum (Tokyo), 13:185-193.

Imaizumi, Y. 1970*b*. The handbook of Japanese land mammals. Shin-Shichoa-Sha, Tokyo, 350 pp.

Imaizumi, Y. 1971. A new vole of the *Clethrionomys rufocanus* group from Rishiri Island, Japan. Journal of the Mammalogical Society of Japan, 6:99-103.

Imaizumi, Y. 1972. Land mammals of the Hidaka Mountains, Hokkaido, Japan, with special reference to the origin of an endemic species of the genus *Clethrionomys*. Memoirs of the National Science Museum, Tokyo, 5:131-149.

Imaizumi, Y., and M. Yoshiyuki. 1989. Taxonomic status of the Japanese otter (Carnivora, Mustelidae), with a description of a new species. Bulletin of the National Science Museum, ser. A (Zoology) 15(3):177-188.

Ingles, J. M., P. N. Newton, M. R. W. Rands, and C. G. R.Bowden. 1980. The first record of a rare murine rodent *Diomys* and further records of three shrew species from Nepal. Bulletin of the British Museum (Natural History), Zoology Series, 39:205-211.

Ingoldby, C. M. 1927. Some notes on the African squirrels of the genus *Heliosciurus*. Proceedings of the Zoological Society of London, 1927:471-487.

International Commission on Zoological Nomenclature. 1929*a*. Opinion 110. Suspension of Rules for *Lagidium* 1833. Smithsonian Miscellaneous Collections, 73(6):415.

International Commission on Zoological Nomenclature. 1929*b*. Opinion 111. Suspension of Rules for *Nycteris* 1795. Smithsonian Miscellaneous Collections, 73(6):416.

International Commission on Zoological Nomenclature. 1929*c*. Opinion 114. Under suspension *Simia*, *Simia satyrus* and *Pithecus* are suppressed. Smithsonian Miscellaneous Collections, 73(6):423-424.

International Commission on Zoological Nomenclature. 1954*a*. Opinion 257. Rejection for nomenclatorial purposes of the work by Zimmermann (A. E. W. von) published in 1777 under the title *Specimen Zoologiae Geographicae, Quadrupedum Domicilia et Migrationes sistens*, and acceptance for the same purposes of the work by the same author published in the period 1778-1783 under the title *Geographische Geschichte de Menschen, und der allgemein verbreiteten vierfüssigen Thiere*. Opinions and Declarations Rendered by the International Commission on Zoological Nomenclature, 5(18):231-244.

International Commission on Zoological Nomenclature. 1954*b*. Opinion 258. Rejection for nomenclatorial purposes of the work by Frisch (J.L.) published in 1775 under the title "Das Natur-System der vierfüssigen Thiere". Opinions and Declarations Rendered by the International Commission on Zoological Nomenclature, 5(19):245-252.

International Commission on Zoological Nomenclature. 1955. Second report on the status of the generic names "*Odobenus*" Brisson, 1762, and "*Rosmarus*" Brünnich, 1771 (Class Mammalia) (A report prepared at the request of the thirteenth International Congress of Zoology, Paris, 1948). Bulletin of Zoological Nomenclature, 11(6):196-198.

International Commission on Zoological Nomenclature. 1956*a*. Opinion 384. Addition to the official list of generic names in zoology of the names of fifty-two genera of the Order Carnivora (Class Mammalia) including twenty-nine from which have been reported parasites common to man. Opinions and Declarations Rendered by the International Commission on Zoological Nomenclature, 12(5):71-190.

International Commission on Zoological Nomenclature. 1956*b*. Opinion 417. Rejection for nomenclatorial purposes of volume 3 (Zoologie) of the work by Lorenz Oken entitled *Okens Lehrbuch der Naturgeschichte* published in 1815-1816. Opinions and Declarations Rendered by the International Commission on Zoological Nomenclature, 14(1):1-42.

International Commission on Zoological Nomenclature. 1957*a*. Opinion 447. Rejection for nomenclatorial purposes of the original edition published at Philadelphia in 1791 and of the editions published in London and Dublin respectively in 1792 of the work by William Bartram entitled "Travels through North and South Carolina, Georgia, east and west Florida, the Cherokee country, the extensive territories of the Muscogulges or Creek confederacy, and the country of the Chactaws", as being a work in which the author did not apply the principles of binominal nomenclature. Opinions and Declarations Rendered by the International Commission on Zoological Nomenclature, 15(12):211-224.

International Commission on Zoological Nomenclature. 1957*b*. Opinion 460. Validation under the plenary powers of the generic name "*Muntiacus*" Rafinesque, 1815, and designation for the genus so named of a type species in harmony with accustomed usage (Class Mammalia). Opinions and Declarations Rendered by the International Commission on Zoological Nomenclature, 15:457-474.

International Commission on Zoological Nomenclature. 1957*c*. Opinion 462. Addition to the Official List of Generic Names in Zoology of the generic name *Mormoops* Leach, 1820 (Class Mammalia). Opinions and Declarations Rendered by the International Commission on Zoological Nomenclature, 16:1-12.

International Commission on Zoological Nomenclature. 1957*d*. Opinion 464. Action under the plenary powers to secure (a) That the specific name *Gambianus* Ogilby, 1835, as published in the combination *Sciurus gambianus* shall be the oldest available name for the sun squirrel and (b) That the generic name *Heliosciurus* Trouessart, 1880, shall be the oldest available generic name for that species (Class Mammalia). Opinions and Declarations Rendered by the International Commission on Zoological Nomenclature, 16(3):25-42.

International Commission on Zoological Nomenclature. 1957*e*. Opinion 467. Validation under the plenary powers of the generic name *Odobenus* Brisson, 1762, as the generic name for the walrus (Class Mammalia). Opinions and Declarations Rendered by the International Commission on Zoological Nomenclature, 16(6):73-88.

International Commission on Zoological Nomenclature. 1959. Opinion 544. Validation under the plenary powers of the family-group names Muntiacinae Pocock, 1923, and Odobenidae (correction of Odobaenidae) Allen, 1880 (Class Mammalia). Opinions and Declarations Rendered by the International Commission on Zoological Nomenclature, 20(2):119-128.

International Commission on Zoological Nomenclature. 1960. Opinion 581. Determination of the generic names for the fallow deer of Europe and the Virginian deer of America (Class Mammalia). Bulletin of Zoological Nomenclature, 17:267-275.

International Commission on Zoological Nomenclature. 1962. Opinion 678. The suppression under the plenary powers of the pamphlet published by Meigen, 1800. Bulletin of Zoological Nomenclature, 20(5):339-342.

International Commission on Zoological Nomenclature. 1965. Opinion 730. *Yerbua* Forster, 1778 (Mammalia): Suppressed under the plenary powers. Bulletin of Zoological Nomenclature, 22(2):84-85.

International Commission on Zoological Nomenclature. 1966. Opinion 760. *Macropus* Shaw, 1790 (Mammalia): Addition to the official list together with the validation under the plenary powers of *Macropus giganteus* Shaw, 1790. Bulletin of Zoological Nomenclature, 22(5/6):292-295.

International Commission on Zoological Nomenclature. 1967. Opinion 818. *Zorilla* I. Geoffroy, 1826 (Mammalia): Suppressed under the plenary powers. Bulletin of Zoological Nomenclature, 24(3):153-154.

International Commission on Zoological Nomenclature. 1970. Opinion 920. *Inuus fuscatus* Blyth, 1875 (Mammalia): Validated under the plenary powers. Bulletin of Zoological Nomenclature, 27(2):77-78.

International Commission on Zoological Nomenclature. 1971. Opinion 945. *Sciurus ebii* Pel, 1851 (Mammalia): Suppressed under the Plenary Powers. Bulletin of Zoological Nomenclature, 27(5/6):224-225.

International Commission on Zoological Nomenclature. 1977*a*. Opinion 1067. Suppression of *Delphinus pernettensis* de Blainville, 1817 and *Delphinus pernettyi* Desmarest, 1820 (Cetacea). Bulletin of Zoological Nomenclature, 33:157-158.

International Commission on Zoological Nomenclature. 1977*b*. Opinion 1080. *Didermocerus* Brookes, 1828 (Mammalia) suppressed under the plenary powers. Bulletin of Zoological Nomenclature, 34:21-24.

International Commission on Zoological Nomenclature. 1979. Opinion 1129. *Vulpes* Frisch, 1775 (Mammalia) conserved under the plenary powers. Bulletin of Zoological Nomenclature, 36(2):76-78.

International Commission on Zoological Nomenclature. 1983. Opinion 1256. *Sorex dsinezumi* Temminck, 1843 (Mammalia, Insectivora): Ruled to be a correct original spelling. Bulletin of Zoological Nomenclature, 40:147-148.

International Commission on Zoological Nomenclature. 1985. International Code of Zoological Nomenclature. Third ed. University of California Press, Berkeley, 338 pp.

International Commission on Zoological Nomenclature. 1985*a*. Opinion 1291. *Antilope zebra* Gray, 1838 (Mammalia): Conserved. Bulletin of Zoological Nomenclature, 42:24-26.

International Commission on Zoological Nomenclature. 1985*b*. Opinion 1289. *Mesoplodon* Gervais, 1850 (Mammalia, Cetacea): Conserved. Bulletin of Zoological Nomenclature, 42:19-20.

International Commission on Zoological Nomenclature. 1985*c*. Opinion 1368. The generic names *Pan* and *Panthera* (Mammalia, Carnivora): Available as from Oken, 1816. Bulletin of Zoological Nomenclature, 42(4):365-370.

International Commission on Zoological Nomenclature. 1986. Opinion 1413. *Delphinus truncatus* Montagu, 1821 (Mammalia, Cetacea): Conserved. Bulletin of Zoological Nomenclature, 43:256-257.

International Commission on Zoological Nomenclature. 1988. Corrigenda. Bulletin of Zoological Nomenclature, 45(4):304.

International Commission on Zoological Nomenclature. 1989. Opinion 1565. *Platanista* Wagler (Mammalia, Cetacea): Conserved. Bulletin of Zoological Nomenclature, 46(3):217-218.

International Commission on Zoological Nomenclature. 1991. Opinion 1660. *Steno attenuatus* Gray, 1846 (currently *Stenella attenuata*; Mammalia, Cetacea): Specific name conserved. Bulletin of Zoological Nomenclature, 48(3):277-278.

International Union for the Conservation of Nature and Natural Resources. 1990. 1990 IUCN red list of threatened animals. I. U. C. N., Gland, Switzerland, 228 pp.

Intress, C., and T. L. Best. 1990. *Dipodomys panamintinus*. Mammalian Species, 354:1-7.

Iredale, T., and E. le G. Troughton. 1934. A checklist of the mammals recorded from Australia. Memoirs Australian Museum, Sydney, 6:1-22.

Iskandar, D., and F. Bonhomme. 1984. Variabilite electrophoretique totale a 11 loci structuraux chez les rongeurs murides (Muridae, Rodentia). Canadian Journal of Genetics and Cytology, 26:622-627.

Iskandar, D., J. M. Duplantier, F. Bonhomme, F. Petter, and L. Thaler. 1988. Mise en evidence de deux especes jumelles sympatriques du genre *Hylomyscus* dans le nord-est du Gabon. Mammalia, 52:126-130.

Ismagilov, M. I. 1961. [Ecology of the native rodents of the Betpak-Dala and southern Lake Balkhash region]. Akademiya Nauk Kazakhskoi SSR (Alma-Ata), 368 pp. (in Russian).

Ivanitskaya, E. Yu. 1985. Taksonomicheskii i tsitogeneticheskii analiz transberingiiskikh svyazei zemleroek-burozubok (*Sorex*: *Insectivora*) i pishchukh (*Ochotona*:Lagomorpha) [Taxonomic and cytogenetic analysis of transberingian connections of brown-toothed shrews...and pikas...]. Aftoreferat, Minist. Vyssh. i Spred. Spets. Obraz., Leningrad., (in Russian).

Ivanitskaya, E. Yu., and A. I. Kozlovskii. 1983. Kariologichicheskie dokazatel'stva otsutstviya v Palearktike arkticheskoi burozubki. [Karyological evidence for the absence of the arctic shrew in the Palearctic]. Zoologicheskii Zhurnal, 62:399-408 (in Russian).

Ivanitskaya, E. Yu., and A. I. Kozlovskii. 1985. [Karyotypes of Palearctic shrews of the subgenus *Otisorex* with comments on taxonomy and phylogeny of the group "*cinereus*"]. Zoologicheskii Zhurnal, 64:950-953 (in Russian).

Ivanitskaya, E. Yu., S. I. Isakov, and K. V. Korobytsina. 1977. Chromosomne nabor dvuch vidov zemleroek iz Tadzhikistana — *Sorex buchariensis* Ognev, 1921 i *Crocidura suaveolens* Pallas, 1811 (Soricidae, Insectivora) [Chromosomal complements of two species of shrews from Tadzhikistan...]. Zoologichskii Zhurnal, 56:1896-1900 (in Russian).

Ivanitskaya, E. Yu., A. I. Kozlovskii, N. V. Orlov, Y. M. Kovalskaya, and M. I. Baskevich. 1986. [New data on karyotypes of common shrews (*Sorex*, Soricidae, Insectivora) in fauna of the USSR]. Zoologicheskii Zhurnal, 65:1228-1236 (in Russian).

Izor, R. J., and L. de la Torre. 1978. A new species of weasel (*Mustela*) from the highlands of Colombia, with comments on the evolution and distribution of South American weasels. Journal of Mammalogy, 59:92-102.

Izor, R. J., and R. H. Pine. 1987. Notes on the black-shouldered opossum, *Caluromysiops irrupta*. Pp. 133-136, *in* Studies in Neotropical mammalogy, essays in honor of Philip Hershkovitz (B. D. Patterson and R. M. Timm, eds.). Fieldiana, Zoology, New Series, 39:frontispiece, vii + 1-506.

Jackson, H. H. T. 1915. A revision of the American moles. North American Fauna, 38:1-100.

Jackson, H. H. T. 1928. A taxonomic review of the American long-tailed shrews. North American Fauna, 51:1-238.

Jackson, J. E. 1987. *Ozotoceros bezoarticus*. Mammalian Species, 295:1-5.

Jacobs, L. L. 1978. Fossil rodents (Rhizomyidae & Muridae) from Neogene Siwalik deposits, Pakistan. Museum of Northern Arizona Press, Bulletin Series, 52:1-103.

Jaeger, J.-J. 1988. Origine et evolution du genre *Ellobius* (Mammalia, Rodentia) en Afrique Nord-Occidentale. Folia Quaternaria, 57:3-50.

Jaeger, J.-J., H. Tong, and C. Denys. 1986. Age de la divergence *Mus-Rattus*: comparaison des donnees paleontologiques et moleculaires. Comptes Rendus des Séances de l'Académie des Sciences (Paris), 302, ser. 2:917-922.

Jameson, E. W., and G. S. Jones. 1977. The Soricidae of Taiwan. Proceedings of the Biological Society of Washington, 90:459-482.

Jammot, D. 1983. Evolution des Soricidae. Insectivora, Mammalia. Symbioses, 15:253-273.

Janecek, L. L. 1990. Genic variation in the *Peromyscus truei* group (Rodentia: Cricetidae). Journal of Mammalogy, 71:301-308.

Janecek, L. L., D. A. Schlitter, and I. L. Rautenbach. 1991. A genic comparison of spiny mice, genus *Acomys*. Journal of Mammalogy, 72:542-552.

Janis, C. M., and K. M. Scott. 1987. The interrelationships of higher ruminant families with special emphasis on the members of the Cervoidea. American Museum Novitates, 2893:1-85.

Jawdat, S. Z. 1977. A new record of forest dormouse *Dryomys nitedula* Pallas (Rodentia: Muscardinidae) in Iraq. Bulletin of the Biological Research Centre, University of Baghdad, 9:115.

Jefferson, T. A. 1988. *Phocoenoides dalli*. Mammalian Species, 319:1-7.

Jenkins, P. D. 1976. Variation in Eurasian shrews of the genus *Crocidura* (Insectivora: Soricidae). Bulletin of the British Museum (Natural History), Zoology Series, 30:271-309.

Jenkins, P. D. 1982. A discussion of Malayan and Indonesian shrews of the genus *Crocidura* (Insectivora: Soricidae). Zoologische Mededelingen, 56:267-279.

Jenkins, P. D. 1984. Description of a new species of *Sylvisorex* (Insectivora: Soricidae) from Tanzania. Bulletin of the British Museum (Natural History), Zoology Series, 47:65-76.

Jenkins, P. D. 1987. Catalogue of Primates in the British Museum (Natural History) and elsewhere in the British Isles. Part 4: Suborder Strepsirrhini, including the subfossil Madagasan lemurs and family Tarsiidae. British Museum (Natural History), London, 189 pp.

Jenkins, P. D. 1992. Description of a new species of *Microgale* (Insectivora: Tenrecidae) from eastern Madagascar. Bulletin of the British Museum of Natural History (Zoology), 58:53-59.

Jenkins, P. D., and J. E. Hill. 1981. The status of *Hipposideros galeritus* Cantor, 1846 and *Hipposideros cervinus* (Gould, 1854)(Chiroptera: Hipposideridae). Bulletin of the British Museum (Natural History), Zoology Series, 415:279-294.

Jenkins, P. D., and J. E. Hill. 1982. Mammals from Siberut, Mentawei Islands. Mammalia, 46:219-224.

Jenkins, S. H., and P. E. Busher. 1979. *Castor canadensis*. Mammalian Species, 120:1-8.

Jenkins, S. H., and B. D. Eshelman. 1984. *Spermophilus beldingi*. Mammalian Species, 221:1-8.

Jennings, M. 1979. A new bat for Saudi Arabia. Journal of the Saudi Arabian Natural History Society, 24:4.

Jensen, J. V. 1980. [The common dormouse—Denmark's only dormouse]. Naturens Verden, 1:55-65.

Jentink, F. A. 1879. On a new genus and species of *Mus* from Madagascar. Notes of the Leyden Museum, 1879:107-109.

Jentink, F. A. 1883. A list of species of mammals from West-Sumatra and North-Celebes, with descriptions of undescribed or rare species. Notes of the Leyden Museum, 5:170-181.

Jentink, F. A. 1887. Catalogue ostèologique des mammiféres. Muséum d'Histoire Naturelle des Pays-Bas, 9:1-360.

Jimenez, R., A. Carnero, M. Burgos, A. Sanchez, and R. Diaz de la Guardia. 1991. Achiasmatic giant sex chromosomes in the vole *Microtus cabrerae* (Rodentia, Microtidae). Cytogenetics and Cell Genetics, 57:56-58.

Johnson, D. H. 1946. The spiny rats of the Riu Kiu islands. Proceedings of the Biological Society of Washington, 59:169-172.

Johnson, D. H. 1948. A rediscovered Haitian rodent, *Plagiodontia aedium*, with a synopsis of related species. Proceedings of the Biological Society of Washington, 61:69-76.

Johnson, D. H. 1962*a*. Rodents and other Micronesian mammals collected. IV. A. Pp. 21-38, *in* Pacific island rat ecology. Report of a study made on Ponape and adjacent islands, 1955-1958 (T. I. Storer, ed.). Bernice P. Bishop Museum Bulletin, 225:1-274.

Johnson, D. H. 1962*b*. Two new murine rodents. Proceedings of the Biological Society of Washington, 75:317-319.

Johnson, D. W., and D. M. Armstrong. 1987. *Peromyscus crinitus*. Mammalian Species, 287:1-8.

Johnson, K. A., and A. D. Roff. 1980. Discovery of Ningauis (*Ningaui* sp.: Dasyuridae: Marsupialia) in the Northern Territory, Australia. Australian Mammalogy, 3:127-129.

Johnson, M. L. 1968. Application of blood protein electrophoretic studies to problems in mammalian taxonomy. Systematic Zoology, 17:23-30.

Johnson, M. L. 1973. Characters of the heather vole, *Phenacomys*, and the red tree vole, *Arborimus*. Journal of Mammalogy, 54:239-244.

Johnson, M. L., and S. B. Benson. 1960. Relationships of the pocket gophers of the *Thomomys mazama-talpoides* complex in the Pacific Northwest. Murrelet, 41:17-22.

Johnson, M. L., and S. B. George. 1991. Species limits within the *Arborimus longicaudus* species-complex (Mammalia: Rodentia) with a description of a new species from California. Contributions in Science, Natural History Museum of Los Angeles County, 429:1-16.

Johnson, M. L., and C. Maser. 1982. Generic relationships of *Phenacomys albipes*. Northwest Science, 56:17-19.

Johnson, M. L., and B. T. Ostenson. 1959. Comments on the nomenclature of some mammals of the Pacific Northwest. Journal of Mammalogy, 40:571-577.

Johnson, W. E., and R. K. Selander. 1971. Protein variation and systematics in kangaroo rats (genus *Dipodomys*). Systematic Zoology, 20:377-405.

Johnson-Murray, J. L. 1977. Myology of the gliding membranes of some petauristine rodents (Genera: *Glaucomys*, *Petaurista*, *Petinomys*, and *Pteromys*). Journal of Mammalogy, 58:374-384.

Jolly, C. J., and F. L. Brett. 1973. Genetic markers and baboon biology. Journal of Medical Primatology, 2(2):85-99.

Jones, C. 1977. *Plecotus rafinesquii*. Mammalian Species, 69:1-4.

Jones, C. 1978. *Dendrohyrax dorsalis*. Mammalian Species, 113:1-4.

Jones, C., and S. Anderson. 1978. *Callicebus moloch*. Mammalian Species, 112:1-5.

Jones, C., and N. J. Hildreth. 1989. *Neotoma stephensi*. Mammalian Species, 328:1-3.

Jones, C., and R. W. Manning. 1989. *Myotis austroriparius*. Mammalian Species, 332:1-3.

Jones, G. S. 1975. Catalogue of the type specimens of mammals of Taiwan. Quarterly Journal of the Taiwan Museum, 28:183-217.

Jones, G. S., and R. E. Mumford. 1971. *Chimarrogale* from Taiwan. Journal of Mammalogy, 52:228-232.

Jones, J. K., Jr. 1964. Distribution and taxonomy of mammals of Nebraska. University of Kansas Publications, Museum of Natural History, 16:1-356.

Jones, J. K., Jr. 1965. Taxonomic status of the molossid bat, *Cynomops malagai* Villa-R, 1955. Proceedings of the Biological Society of Washington, 78:93.

Jones, J. K., Jr. 1977. *Rhogeessa gracilis*. Mammalian Species, 76:1-2.

Jones, J. K., Jr. 1982. *Reithrodontomys spectabilis*. Mammalian Species, 193:1.

Jones, J. K., Jr. 1990. *Peromyscus stirtoni*. Mammalian Species, 361:1-2.

Jones, J. K., Jr., and J. Arroyo-Cabrales. 1990. *Nyctinomops aurispinosus*. Mammalian Species, 350:1-3.

Jones, J. K., Jr., and R. J. Baker. 1980. *Chiroderma improvisum*. Mammalian Species, 134:1-2.

Jones, J. K., Jr., and G. A. Baldassarre. 1982. *Reithrodontomys brevirostris* and *Reithrodontomys paradoxus*. Mammalian Species, 192:1-3.

Jones, J. K., Jr., and D. C. Carter. 1976. Annotated checklist, with keys to subfamilies and genera. Part I. Pp. 7-38, *in* Biology of bats of the New World family Phyllostomatidae (R. J. Baker, J. K. Jones, Jr., and D. C. Carter, eds.). Special Publications, The Museum, Texas Tech University Press, 10:1-218.

Jones, J. K., Jr., and D. C. Carter. 1979. Systematic and distributional notes. Part III. Pp. 7-11, *in* Biology of bats of the New World family Phyllostomatidae (R. J. Baker, J. K. Jones, Jr., and D. C. Carter, eds.). Special Publications, The Museum, Texas Tech University Press, 16:1-441.

Jones, J. K., Jr., and M. D. Engstrom. 1986. Synopsis of the rice rats (genus *Oryzomys*) of Nicaragua. Occasional Papers, The Museum, Texas Tech University, 103:1-23.

Jones, J. K., Jr., and H. H. Genoways. 1967. Notes on the Oaxacan vole, *Microtus oaxacensis* Goodwin, 1966. Journal of Mammalogy, 48:320-321.

Jones, J. K., Jr., and H. H. Genoways. 1970. Harvest mice (genus *Reithrodontomys*) of Nicaragua. Occasional Papers of the Western Foundation of Vertebrate Zoology, 2:1-16.

Jones, J. K., Jr., and H. H. Genoways. 1973. *Ardops nichollsi*. Mammalian Species, 24:1-2.

Jones, J. K., Jr., and H. H. Genoways. 1975*a*. *Dipodomys phillipsii*. Mammalian Species, 51:1-3.
Jones, J. K., Jr., and H. H. Genoways. 1975*b*. *Sciurus richmondi*. Mammalian Species, 53:1-2.
Jones, J. K., Jr., and H. H. Genoways. 1975*c*. *Sturnira thomasi*. Mammalian Species, 68:1-2.
Jones, J. K., Jr., and H. H. Genoways. 1978. *Neotoma phenax*. Mammalian Species, 108:1-3.
Jones, J. K., Jr., and J. A. Homan. 1974. *Hylonycteris underwoodi*. Mammalian Species, 32:1-2.
Jones, J. K., and D. H. Johnson. 1960. Review of the insectivores of Korea. University of Kansas, Publications of the Museum of Natural History, 9:549-578.
Jones, J. K., Jr., and D. H. Johnson. 1965. Synopsis of the lagomorphs and rodents of Korea. University of Kansas Publications, Museum of Natural History, 16:357-407.
Jones, J. K., Jr., and T. E. Lawlor. 1965. Mammals from Isla Cozumel, Mexico, with description of a new species of harvest mouse. University of Kansas Publications, Museum of Natural History, 16:409-419.
Jones, J. K., Jr., and A. Schwartz. 1967. Bredin-Archbold-Smithsonian Biological Survey of Dominica. 6. Synopsis of bats of the Antillean genus *Ardops*. Proceedings of the United States National Museum, 124(3634):1-13.
Jones, J. K., Jr., and T. L. Yates. 1983. Review of the white-footed mice, genus *Peromyscus*, of Nicaragua. Occasional Papers, The Museum, Texas Tech University, 82:1-15.
Jones, J. K., Jr., D. C. Carter, and H. H. Genoways. 1975. Revised checklist of North American mammals north of Mexico. Occasional Papers, The Museum, Texas Tech University, 28:1-14.
Jones, J. K., Jr., P. Swanepoel, and D. C. Carter. 1977. Annotated checklist of the bats of Mexico and Central America. Occasional Papers, The Museum, Texas Tech University, 47:1-35.
Jones, J. K., Jr., D. C. Carter, H. H. Genoways, R. S. Hoffmann, and D. W. Rice. 1982. Revised checklist of North American mammals north of Mexico, 1982. Occasional Papers, The Museum, Texas Tech University, 80:1-22.
Jones, J. K., Jr., D. C. Carter, H. H. Genoways, R. S. Hoffman, D. W. Rice, and C. Jones. 1986. Revised checklist of North American mammals north of Mexico. Occasional Papers, The Museum, Texas Tech University, 107:1-22.
Jones, J. K., Jr., R. S. Hoffmann, D. W. Rice, C. Jones, R. J. Baker, and M. D. Engstrom. 1992. Revised checklist of North American mammals north of Mexico, 1991. Occasional Papers Museum Texas Tech University, 146:1-23.
Jones, M. L., S. L. Swartz, and S. Leatherwood, (eds). 1984. The gray whale. Academic Press, New York, 600 pp.
Jong, N., de, and W. Bergmans. 1981. A revision of the fruit bats of the genus *Dobsonia* Palmer, 1898 from Sulawesi and some nearby islands (Mammalia, Megachiroptera, Pteropodinae). Zoologische Abhandlungen Staatliches Museum für Tierkunde in Dresden, 37(6):209-224.
Jordan, D. S., and G. A. Clark. 1898. The history, condition, and needs of the herd of fur seals resorting to the Pribilof Islands, pt. 1, 249 pages, *in* The fur seals and fur-seal islands of the North Pacific Ocean (D. S. Jordan). United States Government Printing Office, Washington, D. C., U. S. Treasury Document no. 2017:4 volumes.
Jotterand, M. 1972. Le polymorphisme chromosomique des *Mus* (Leggadas) africains. Cytogenetique, zoogeographie, evolution. Revue Suisse de Zoologie, 79(1):287-359.
Jotterand-Bellomo, M. 1981. Le caryotype et la spermatogenese de *Mus setulosus* (bandes Q, C, G, et coloration argentique). Genetica, 56:217-227.
Jotterand-Bellomo, M. 1984. L'analyse cytogenetique de deux especes de Muridae africains, *Mus oubanguii* et *Mus minutoides/musculoides*: polymorphisme chromosomique et ebauche d'une phylogenie. Cytogenetics and Cell Genetics, 38:182-188.
Jotterand-Bellomo, M. 1986. Le genre *Mus* africain, un exemple d'homogeneite caryotypique: étude cytogenetique de *Mus minutoides/musculoides* (Côte d'Ivoire), de *M. setulosus* (Republique Centrafricaine), et de *M. mattheyi* (Burkina Faso). Cytogenetics and Cell Genetics, 42:99-104.
Juckwer, E.-A. 1990. *Galemys pyrenaicus* - Pyrenäen-Desman. Pp. 79-92, *in* Handbuch der Säugetiere Europas (J. Niethammer and F. Krapp, eds.). Aula-Verlag, Wiesbaden, 3/I:1-524.
Judd, S. R., and S. P. Cross. 1980. Chromosomal variation in *Microtus longicaudus* (Merriam). Murrelet, 61:2-5.
Judd, S. R., S. P. Cross, and S. Pathak. 1980. Non-Robertsonian chromosomal variation in *Microtus montanus*. Journal of Mammalogy, 61:109-113.
Jüdes, U. 1981. G- and C-band karyotypes of the harvest mouse, *Micromys minutus*. Genetica, 54:237-239.
Junge, J. A., and R. S. Hoffmann. 1981. An annotated key to the long-tailed shrews (genus *Sorex*) of the United States and Canada, with notes on Middle American *Sorex*. Occasional Papers of the Museum of Natural History, University of Kansas, 94:1-48.
Junge, J. A., R. S. Hoffmann, and R. W. DeBry. 1983. Relationships within the Holarctic *Sorex arcticus-Sorex tundrensis* species complex. Acta Theriologica, 28:339-350.

Juyal, R. C., B. K. Thelma, and S. R. V. Rao. 1989. Heterochromatin variation and spermatogenesis in *Nesokia*. Cytogenetics and Cell Genetics, 50:206-210.

Kahmann, H. 1981. Zur Naturgeschichte des Löffelbilches, *Eliomys melanurus* Wagner, 1840. Spixiana, 4(1):1-37.

Kahmann, H., and J. A. Alcover. 1974. Sobre la bionomia del liron careto (*Eliomys quercinus* L.) en Mallorca. Note preliminar. Bolletino de la Sociedad de Historia Natural de Baleares, 19:57-74.

Kahmann, H., and G. Thoms. 1973*a*. Der Gartenschläfer (*Eliomys*) Menorcas. Säugetierkundliche Mitteilungen, 21:65-73.

Kahmann, H., and G. Thoms. 1973*b*. Zur Binomie des Gartenschläfers *Eliomys quercinus denticulatus* Ranck, 1968 aus Libyen. Zeitschrift für Säugetierkunde, 38:197-208.

Kahmann, H., and G. Thoms. 1981. Über den Gartenschläfer (*Eliomys*) in nordafrikanischen Ländern (Mammalia: Rodentia, Gliridae). Spixiana, 4(2):191-228.

Kahmann, H., and G. Thoms. 1987. *Eliomys quercinus tunetae* Thomas, 1903 (Mammalia, Rodentia, Gliridae). Spixiana, 10(2):97-114.

Kahmann, H., and I. Vesmanis. 1976. Morphometrische Untersuchungen an Wimperspitzmäusen (*Crocidura*) (Mammalia: Soricidae) 2. Zur weiteren Kenntnis von *Crocidura gueldenstaedti* (Pallas, 1811) auf der Insel Kreta. Opuscula Zoologica (Budapest), 136:1-12.

Kain, D. E. 1985. The systematic status of *Eutamias ochrogenys* and *Eutamias senex* (Rodentia: Sciuridae). Unpubl. M. A. thesis, Humboldt State University, Arcata, California, 67 pp.

Kaluza, T. 1987. Locality of common dormouse *Muscardinus avellanarius* (Linnaeus, 1758) in the Wielkopolska district. Przeglad Zoologiczny, 31(2):215-218.

Kaminski, M., M. Rousseau, and F. Petter. 1984. Electrophoretic studies on blood proteins of *Arvicanthis niloticus*. Biochemical Systematics and Ecology, 12:215-224.

Kaneko, Y. 1985. Examinations of diagnostic characters (mammae and bacula) between *Eothenomys smithi* and *E. kageus*. Journal of the Mammalogical Society of Japan, 10:221-229.

Kaneko, Y. 1987. Skull and dental characters, and skull measurements of *Microtus kikuchii* Kuroda, 1920, from Taiwan. Journal of the Mammalogical Society of Japan, 12:31-39.

Kaneko, Y. 1990. Identification and some morphological characters of *Clethrionomys rufocanus* and *Eothenomys regulus* from USSR, Northeast China, and Korea in comparison with *C. rufocanus* from Finland. Journal of the Mammalogical Society of Japan, 14:129-148.

Kaneko, Y. 1992. Identification and morphological characteristics of *Clethrionomys rufocanus, Eothenomys shanseius, E. inez,* and *E. eva* from the USSR, Mongolia, and northern and central China. Journal of the Mammalogical Society of Japan, 16:71-95.

Kang, Y. S., and H. S. Koh. 1976. Karyotype studies on three species of the family Muridae. Korean Journal of Zoology, 19:101-112.

Kapitonov, V. I. 1966. Rasprostranenie surkov v Tsentral'nom Kazakhstane i perspektivy ikh promysla [Distribution of marmots in Central Kazakhstan and perspectives on their utilization]. Trudy Zoologicheskovo Instituta, Akademiya Nauk, Kazakh SSR,, 26:94-134 (in Russian).

Kapitonov, V. I. 1978. Chernoshaposhnyi surok [black-capped marmot]. Pp. 178-209, *in* Surki: rasprostranenie i ekologiya [Marmots: distribution and ecology]. Nauka, Moscow-Leningrad, 222 pp. (in Russian).

Karn, R. C., and S. R. Dlouhy. 1991. Salivary androgen-binding protein variation in *Mus* and other rodents. Journal of Heredity, 82:453-458.

Karaseva, E. V., G. N. Tikhonova, and P. L. Bogomolov. 1992. [Distribution of the field mouse (*Apodemus agrarius*) and peculiarities of its ecology in differennt parts of its range.] Zoologischeskii Zhurnal, 71:106-115 (in Russian)

Kashiwabara, S., and K. Onoyama. 1988. Karyotypes and G-banding patterns of the red-backed voles, *Clethrionomys montanus* and *C. rufocanus bedfordiae* (Rodentia, Microtinae). Journal of the Mammalogical Society of Japan, 13:33-41.

Kasuya, T. 1973. Systematic consideration of recent toothed whales based on the morphology of the tympano-periotic bone. Scientific Reports of the Whales Research Institute (Tokyo), 25:1-103.

Kasuya, T. 1975. Past occurrence of *Globicephala melaena* in the western North Pacific. Scientific Reports of the Whales Research Institute (Tokyo), 27:95-110.

Kawamichi, T. 1971. Daily activities and social pattern of two Himalayan pikas, *Ochotona macrotis* and *O. roylei*, observed at Mt. Everest. Journal of the Faculty of Hokkaido University, Japan, ser. 6(Zoology), 17:587-609.

Kawamura, Y. 1988. Quaternary rodent faunas in the Japanese Islands (Part 1). Memoirs of the Faculty of Science, Kyoto University, Series of Geology and Mineralogy, 53:31-348.

Kawamura, Y. 1989. Quaternary rodent faunas in the Japanese Islands (Part 2). Memoirs of the Faculty of Science, Kyoto University, Series of Geology and Mineralogy, 54(1-2):1-235.

Keith, L. B., and L. A. Windberg. 1978. A demographic analysis of the snowshoe hare cycle. Wildlife Monographs, 58:1-70.

Kellogg, R. 1918. A revision of the *Microtus californicus* group of meadow mice. University of California Publications in Zoology, 21:1-42.

Kelson, K. R. 1952. Comments on the taxonomy and geographic distribution of some North American woodrats (genus *Neotoma*). University of Kansas Publications, Museum of Natural History, 5:233-242.

Kelt, D. A. 1988*a*. *Dipodomys heermanni*. Mammalian Species, 323:1-7.

Kelt, D. A. 1988*b*. *Dipodomys californicus*. Mammalian Species, 324:1-4.

Kelt, D. A., R. E. Palma, M. H. Gallardo, and J. A. Cook. 1991. Chromosomal multiformity in *Eligmodontia* (Muridae, Sigmodontinae), and verification of the status of *E. morgani*. Zeitschrift für Säugetierkunde, 56:352-358.

Kennedy, M. L., G. A. Heidt, and P. K. Kennedy. 1982. First record of the meadow jumping mouse (*Zapus hudsonius*) in Mississippi. Journal Mississippi Academy of Science, 27:149-150.

Kenyon, K. W. 1969. The sea otter in the eastern Pacific ocean. North American Fauna, 68:1-352.

Kenyon, K. W. 1977. Caribbean monk seal extinct. Journal of Mammalogy, 58:97-98.

Keogh, H. J. 1985. A photographic reference system based on the cuticular scale patterns and grooves of the hair of 44 species of southern African Cricetidae and Muridae. South African Journal of Wildlife Research, 15:109-159.

Keogh, H. J., and P. J. Price. 1981. The multimammate mice: A review. South African Journal of Science, 77:484-488.

Kerridge, D. C., and R. J. Baker. 1978. *Natalus micropus*. Mammalian Species, 114:1-3.

Kesner, M. H. 1980. Functional morphology of the masticatory musculature of the rodent subfamily Microtinae. Journal of Morphology, 165:205-222.

Kesner, M. H. 1986. The myology of the manus of microtine rodents. Journal of the Zoological Society of London, A, 210:1-22.

Khajuria, H. 1981. A new bandicoot rat, *Erythronesokia bunnii* gen. et sp. nov. (Rodentia: Muridae), from Iraq. Bulletin of the Natural History Research Centre, 7:157-164.

Khajuria, H. 1982. External genitalia and bacula of some central Indian Microchiroptera. Säugetierkundliche Mitteilungen, 30:287-295.

Khajuria, H., Y. Chaturvedi, and D. K. Ghoshal. 1977. Catalogue Mammaliana. An annotated catalogue of the type specimens of mammals in the collections of the Zoological Survey of India. Records of the Zoological Survey of India, Miscellaneous Publication, Occasional Paper, 7:1-45.

Kilpatrick, C. W. 1984. Molecular evolution of the Texas mouse, *Peromyscus attwateri*. Pp. 87-96, *in* Festschrift for Walter W. Dalquest in honor of his sixty-sixth birthday (N. Horner, ed.). Midwestern State University, Wichita Falls, TX, 163 pp.

Kilpatrick, C. W., and K. L. Crowell. 1985. Genic variation of the rock vole, *Microtus chrotorrhinus*. Journal of Mammalogy, 66:94-101.

Kilpatrick, C. W., and E. G. Zimmerman. 1975. Genetic variation and systematics of four species of mice of the *Peromyscus boylii* species group. Systematic Zoology, 24:143-162.

Kilpatrick, C. W., and E. G. Zimmerman. 1976. Biochemical variation and systematics of *Peromyscus pectoralis*. Journal of Mammalogy, 57:506-522.

Kim Sang Wook, and Kim Woo Ki. 1974. Avi-mammalian fauna of Korea. Wildlife population census in Korea, Forest Research Institute, Office of Forestry., 5:1-118.

King, C. J. (ed.). 1990. The handbook of New Zealand mammals. Oxford University Press, Melbourne, 600 pp.

King, C. M. 1983. *Mustela erminea*. Mammalian Species, 195:1-8.

King, J. A., (ed.). 1968. Biology of *Peromyscus* (Rodentia). Special Publication, American Society of Mammalogists, 2:1-593.

King, J. E. 1954. The otariid seals of the Pacific coast of America. Bulletin of the British Museum (Natural History), Zoology Series, 2(10):311-337.

King, J. E. 1956. The monk seals genus *Monachus*. Bulletin of the British Museum (Natural History), Zoology Series, 3(5):203-256.

King, J. E. 1959*a*. A note on the specific name of the Kerguelen fur seal. Mammalia, 23(3):381.

King, J. E. 1959*b*. The northern and southern populations of *Arctocephalus gazella*. Mammalia, 23(1):19-40.

King, J. E. 1960. Sea-lions of the genera *Neophoca* and *Phocarctos*. Mammalia, 24:445-456.

King, J. E. 1966. Relationships of the hooded and elephant seals (genera *Cystophora* and *Mirounga*). Journal of Zoology (London), 148:385-398.

King, J. E. 1978. On the specific name of the southern sea lion (Pinnipedia, Otariidae). Journal of Mammalogy, 59(4):861-863.
King, J. E. 1983. Seals of the World. Second ed. Cornell University Press, Ithaca, NY, 240 pp.
Kingdon, J. 1971. East African mammals: An atlas of evolution in Africa. Academic Press, London, 1:1-446.
Kingdon, J. 1974. East African mammals: An atlas of evolution in Africa. Academic Press, London, 2B (hares and rodents):343-704 + lvii.
Kingdon, J. 1977. East African mammals: An atlas of evolution in Africa. Academic Press, New York, 3A (Carnivores):1-476.
Kingdon, J. 1980. The role of visual signals and face patterns in African forest monkeys (guenons) of the genus *Cercopithecus*. Transactions of the Zoological Society of London, 35:431-475.
Kingdon, J. 1982. East African mammals: An atlas of evolution in Africa. Academic Press, London, 3C (Bovids):1-393.
Kipp, H. 1965. Beitrag zur Kenntnis der Gattung *Conepatus* Molina, 1782. Zeitschrift für Säugetierkunde, 30(4):193-232.
Kirkland, G. L., Jr. 1977. A re-examination of the subspecific status of the Maryland shrew, *Sorex cinereus fontinalis* Hollister. Proceedings of the Pennsylvania Academy of Sciences, 51:43-46.
Kirkland, G. L., Jr. 1981. *Sorex dispar* and *Sorex gaspensis*. Mammalian Species, 155:1-4.
Kirkland, G. L., Jr., and F. J. Jannett, Jr. 1982. *Microtus chrotorrhinus*. Mammalian Species, 180:1-5.
Kirkland, G. L., Jr., and J. N. Layne (eds.). 1989. Advances in the study of *Peromyscus* (Rodentia). Texas Tech University Press, Lubbock, 367 pp.
Kirkland, G. L., Jr., and H. M. Van Deusen. 1979. The shrews of the *Sorex dispar* group: *Sorex dispar* Batchelder and *Sorex gaspensis* Anthony and Goodwin. American Museum Novitates, 2675:1-21.
Kirsch, J. A. W. 1977. The comparative serology of Marsupialia, and a classification of marsupials. Australian Journal of Zoology, Supplementary Series, 52:1-152.
Kirsch, J. A. W., and J. H. Calaby. 1977. The species of living marsupials: An annotated list. Pp. 9-26, *in* The biology of marsupials (B. Stonehouse and D. Gilmore, eds.). University Park Press, Baltimore, 486 pp.
Kirsch, J. A. W., and W. E. Poole. 1972. Taxonomy and distribution of the grey kangaroos, *Macropus giganteus* Shaw and *Macropus fuliginosus* (Desmarest), and their subspecies (Marsupialia: Macropodidae). Australian Journal of Zoology, 20:315-339.
Kitchener, D. J. 1980. A new species of *Pseudomys* (Rodentia: Muridae) from Western Australia. Records of the Western Australian Museum, 8:405-414.
Kitchener, D. J. 1985. Description of a new species of *Pseudomys* (Rodentia: Muridae) from Northern Territory. Records of the Western Australian Museum, 12:207-221.
Kitchener, D. J. 1988. A new species of false *Antechinus* (Marsupialia: Dasyuridae) from the Kimberley, Western Australia. Records of the Western Australian Museum, 14:61-71.
Kitchener, D. J. 1989. Taxonomic appraisal of *Zyzomys* (Rodentia, Muridae) with descriptions of two new species from the Northern Territory, Australia. Records of the Western Australian Museum, 14:331-373.
Kitchener, D. J., and N. Caputi. 1985. Systematic revision of Australian *Scoteanax* and *Scotorepens* (Chiroptera: Vespertilionidae), with remarks on relationships to other Nycticeini. Records of the Western Australian Museum, 12:85-146.
Kitchener, D. J., and N. Caputi. 1988. A new species of false *Antechinus* (Marsupialia, Dasyuridae) from Western Australia, with remarks on the generic classification within the Parantechini. Records of the Western Australian Museum, 14:35-59.
Kitchener, D. J., and W. F. Humphreys. 1986. Description of a new species of *Pseudomys* (Rodentia: Muridae) from the Kimberley Region, Western Australia. Records of the Western Australian Museum, 12:419-434.
Kitchener, D. J., and W. F. Humphreys. 1987. Description of a new subspecies of *Pseudomys* (Rodentia: Muridae) from Northern Territory. Records of the Western Australian Museum, 13:285-295.
Kitchener, D. J., and Moharadatunkamsi. 1991. Description of a new species of *Cynopterus* (Chiroptera: Pteropodidae) from Nusa Tenggara, Indonesia. Records of the Western Australian Museum, 15:307-363.
Kitchener, D. J., and G. Sanson. 1978. *Petrogale burbidgei* (Marsupialia, Macropodidae), a new rock wallaby from Kimberley, Western Australia. Records of the Western Australian Museum, 6:269-285.
Kitchener, D. J., M. Adams, and P. Baverstock. 1984*a*. Redescription of *Pseudomys bolami* Troughton, 1932. Australian Mammalogy, 7:149-159.
Kitchener, D. J., J. Stoddart, and J. Henry. 1984*b*. A taxonomic revision of the *Sminthopsis murina* complex (Marsupialia, Dasyuridae) in Australia, including descriptions of four new species. Records of the Western Australian Museum, 11:201-247.
Kitchener, D. J., N. Caputi, and B. Jones. 1986. Revision of Australo-Papuan *Pipistrellus* and *Falsistrellus* (Microchiroptera: Vespertilionidae). Records of the Western Australian Museum, 12:435-495.
Kitchener, D. J., B. Jones, and N. Caputi. 1987. Revision of Australian *Eptesicus* (Microchiroptera: Vespertilionidae). Records of the Western Australian Museum, 13:427-500.

Kitchener, D. J., R. A. How, and Maharadatunkamsi. 1991*a*. *Paulamys* sp. cf. *P. naso* (Musser, 1981) (Rodentia: Muridae) from Flores Island, Nusa Tenggara, Indonesia— description from a modern specimen and a consideration of its phylogenetic affinities. Records of the Western Australian Museum, 15:171-189.

Kitchener, D. J., K. P. Aplin, and Boeadi. 1991*b*. A new species of *Rattus* from Gunung Mutis, South West Timor Island, Indonesia. Records of the Western Australian Museum, 15:445-461.

Kitchener, D. J., R. A. How, and Maharadatunkamsi. 1991*c*. A new species of *Rattus* from the mountains of West Flores, Indonesia. Records of the Western Australian Museum, 15:611-626.

Kivanc, E. 1983. Die Haselmaus, *Muscardinus avellanarius* L., in der Türkei. Bonner Zoologische Beiträge, 34:419-428.

Kivanc, E. 1986. *Microtus* (*Pitymys*) *majori* Thomas, 1906 in der europäischen Türkei. Bonner Zoologische Beiträge, 37:39-42.

Kivanc, E. 1988. [Geographic variations of Turkish *Spalax* species (Spalacidae, Rodentia, Mammalia)]. Biological Faculty, Ankara University, 88 pp. (in Turkish).

Kleiman, D. G. 1981. *Leontopithecus rosalia*. Mammalian Species, 148:1-7.

Kleinenberg, S. E., A. V. Yablokov, B. M. Bel'kovich, and M. N. Tarasevich. 1964. Belukha. Opyt monograficheskovo issledovaniya vida [Beluga. Test of monographic investigation of the species]. Akademiya Nauk SSSR, Moscow, 455 pp. (in Russian).

Kleinenberg, S. E., A. V. Yablokov, B. M. Bel'kovich, and M. N. Tarasevich. 1969. Beluga (*Delphinapterus leucas*), investigation of the species. [A translation of S. E. Kleinenberg, et al., 1964, Belukha. Opyt monograficheskovo issledovaniya vida]. Israel Program for Scientific Translations, Jerusalem, 376 pp.

Klimkiewicz, M. K. 1970. The taxonomic status of the nominal species *Microtus pennsylvanicus* and *Microtus agrestis* (Rodentia: Cricetidae). Mammalia, 34:640-655.

Klingener, D. 1963. Dental evolution of *Zapus*. Journal of Mammalogy, 44(2):248-260.

Klingener, D. 1964. The comparative myology of four dipodoid rodents (genera *Zapus*, *Napaeozapus*, *Sicista*, and *Jaculus*). Miscellaneous Publications, Museum of Zoology, University of Michigan, 124:1-100.

Klingener, D. 1965. Notes on the range of *Napaeozapus* in Michigan and Indiana. Journal of Mammalogy, 45(4):644-645.

Klingener, D. 1971. The question of cheek pouches in *Zapus*. Journal of Mammalogy, 52(2):463-464.

Klingener, D. 1984. Gliroid and dipodoid rodents. Pp. 381-388, *in* Orders and families of Recent mammals of the world (S. Anderson and J. K. Jones, Jr., eds.). John Wiley and Sons, New York, 686 pp.

Klingener, D., and G. K. Creighton. 1984. On small bats of the genus *Pteropus* from the Philippines. Proceedings of the Biological Society of Washington, 97:395-403.

Klingener, D., H. H. Genoways, and R. J. Baker. 1978. Bats from Southern Haiti. Annals of Carnegie Museum, 47:81-99.

Kloss, C. B. 1917. On the mongooses of the Malay peninsula. Journal of the Federation of Malay States Museum, 7(3):123-125.

Kloss, C. B. 1929. Some remarks on the gibbons, with the description of a new subspecies. Proceedings of the Zoological Society of London, 1929:113-127.

Kobayashi, T. 1985. Taxonomical problems in the genus *Apodemus*. Pp. 80-82, *in* Contemporary mammalogy in China and Japan. Mammalogical Society of Japan, 194 pp.

Kock, D. 1969*a*. Die Fledermaus- Fauna des Sudan. (Mammalia, Chiroptera). Abhandlungen der Senckenbergischen Naturforschenden Gesellschaft, 521:1-238.

Kock, D. 1969*b*. *Dyacopterus spadiceus* (Thomas, 1890) auf den Philippinen (Mammalia: Chiroptera). Senckenbergiana Biologica, 50:1-7.

Kock, D. 1975. Ein Originalexemplar von *Nyctinomus ventralis* Heuglin 1861 (Mammalia: Chiroptera: Molossidae). Stuttgarter Beiträge zur Naturkunde, ser. A(Biologie), 272:1-9.

Kock, D., and H. Felten. 1980. Typen und Typus-Lokalität von *Apodemus sylvaticus rufescens* Saint Girons und Bree 1963 (Mammalia: Rodentia: Muridae). Senckenbergiana Biologica, 60:277-283.

Kock, D., and I. A. Nader. 1983. Pygmy shrew and rodents from the Near East. Senckenbergiana Biologica, 64:13-23.

Kock, D., and I. A. Nader. 1984. *Tadarida teniotis* (Rafinesque, 1814) in the W-Palearctic and a lectotype for *Dysopes rupelii* Temminck, 1826 (Chiroptera: Molossidae). Zeitschrift für Säugetierkunde, 49:129-135.

Kock, D., and I. A. Nader. 1990. Identity of *Apodemus sylvaticus* (Linnaeus, 1758) recorded from Qatar (Rodentia: Muridae). Zeitschrift für Säugetierkunde, 55:66-67.

Kock, D., and H. Posamentier. 1983. *Cannomys badius* (Hodgson, 1842) in Bangladesh (Rodentia: Rhizomyidae). Zeitschrift für Säugetierkunde, 48:314-316.

Kock, D., I. A. Nader, and A. A. Banaja. 1990. *Bandicota bengalensis* (Gray and Hardwicke, 1833) new to Saudi Arabia (Mammalia: Rodentia: Muridae). Fauna of Saudi Arabia, 11:323-328.

Koehler, C. E., and P. R. K. Richardson. 1990. *Proteles cristatus*. Mammalian Species, 363:1-6.

Koenigswald, W. von. 1980. Schmelzstruktur und Morphologie in den Molaren der Arvicolidae (Rodentia). Abhandlungen der Senckenbergischen Naturforschenden Gesellschaft, 539:1-129.
Koenigswald, W. von. 1982. Stammesgeschichte und Schmelzmuster. Pp. 60-69, *in* Handbuch der Säugetiere Europas (J. Niethammer and F. Krapp, eds.). Akademische Verlagsgesellschaft (Wiesbaden), 2/I:1-649.
Koenigswald, W. von, and L. D. Martin. 1984. Revision of the fossil and Recent Lemminae (Rodentia: Mammalia). Carnegie Museum of Natural History, Special Publication, 9:122-137.
Koeppl, J. W., R. S. Hoffmann, and C. F. Nadler. 1978. Pattern analysis of acoustical behavior in four species of ground squirrels. Journal of Mammalogy, 59:677-696.
Koffler, B. R. 1972. *Meriones crassus*. Mammalian Species, 9:1-4.
Koh, H. S. 1982. G- and C-banding pattern analyses of Korean rodents. I. Chromosome banding patterns of striped field mice (*Apodemus agrarius coreae*) and black rats (*R. rattus rufescens*). Korean Journal of Zoology, 25(2):81-92.
Koh, H. S. 1983. A study on age variation and secondary sexual dimorphism in morphometric characters of Korean rodents: I. An analysis on striped field mice, *Apodemus agrarius coreae* Thomas, from Cheongju. Korean Journal of Zoology, 26:125-134.
Koh, H. S. 1988. Systematic studies of Korean rodents: IV. Morphometric and chromosomal analyses of two species of the genus *Apodemus*. Korean Journal of Systematic Zoology, 4:103-120.
Koh, H. S. 1991. Morphometric analyses with eight subspecies of striped field mice, *Apodemus agrarius* Pallas (Rodentia, Mammalia), in Asia: the taxonomic status of subspecies *chejuensis* at Cheju island in Korea. Korean Journal of Systematic Zoology, 7(2):179-188.
Koh, H. S., and R. L. Peterson. 1983. Systematic studies of deer mice, *Peromyscus maniculatus* Wagner (Cricetidae, Rodentia): analysis of age and secondary sexual variation in morphometric characters. Canadian Journal of Zoology, 61:2618-2628.
Köhler-Rollefson, I. U. 1991. *Camelus dromedarius*. Mammalian Species, 375:1-8.
Kohn, P. H., and R. H. Tamarin. 1978. Selection at electrophoretic loci for reproductive parameters in island and mainland voles. Evolution, 32:15-18.
Köhncke, M., and K. Leonhardt. 1986. *Cryptoprocta ferox*. Mammalian Species, 254:1-5.
Kolomiets, O. L., T. E. Borviev, L. D. Safronova, Yu. M. Borisov, and Y. F. Bogdanov. 1988. Synaptonemal complex analysis of B- chromosome behavior in meiotic prophase I in the East-Asiatic mouse *Apodemus peninsulae* (Muridae, Rodentia). Cytogenetics and Cell Genetics, 48:183-187.
Kolomiets, O. L., N. N. Vorontsov, E. A. Lyapunova, and T. F. Mazurova. 1991. Ultrastructure, meiotic behavior, and evolution of sex chromosomes of the genus *Ellobius*. Genetica, 84:179-189.
König, C. 1982. *Microtus bavaricus* (Konig, 1962)—Bayerische Kurzohrmaus. Pp. 447-451, *in* Handbuch der Säugetiere Europas (J. Niethammer and F. Krapp, eds.). Akademische Verlagsgesellschaft (Wiesbaden), 2/I:1-649.
Koontz, F. W., and N. J. Roeper. 1983. *Elephantulus rufescens*. Mammalian Species, 204:1-5.
Koop, B. F., R. J. Baker, and J. T. Mascarello. 1985. Cladistical analysis of chromosomal evolution within the genus *Neotoma*. Occasional Papers, The Museum, Texas Tech University, 96:1-9.
Koopman, K. F. 1958. Land bridges and ecology in bat distribution on islands off the northern coast of South America. Evolution, 12:429-439.
Koopman, K. F. 1965. Status of forms described or recorded by J. A. Allen in "The American Museum Congo Expedition Collection of Bats". American Museum Novitates, 2219:1-34.
Koopman, K. F. 1971. Taxonomic notes on *Chalinolobus* and *Glauconycteris*. American Museum Novitates, 2451:1-10.
Koopman, K. F. 1972. *Eudiscopus denticulus*. Mammalian Species, 19:1-2.
Koopman, K. F. 1973. Systematics of Indo-Australian *Pipistellus*. Periodicum Biologorum Zagreb, 75:113-116.
Koopman, K. F. 1975. Bats of the Sudan. Bulletin of the American Museum of Natural History, 154:353-444.
Koopman, K. F. 1978*a*. The genus *Nycticeius* (Vespertilionidae) with special reference to tropical Australia. Proceedings of the Fourth International Bat Research Conference, Nairobi, 4:165-171.
Koopman, K. F. 1978*b*. Zoogeography of Peruvian bats with special emphasis on the role of the Andes. American Museum Novitates, 2651:1-33.
Koopman, K. F. 1979. Zoogeography of mammals from islands off the northeastern coast of New Guinea. American Museum Novitates, 2690:1-17.
Koopman, K. F. 1982. Results of the Archbold Expeditions. No. 109. Bats from eastern Papua and the east Papuan Islands. American Museum Novitates, 2747:1-34.
Koopman, K. F. 1984*a*. Bats. Pp. 145-186, *in* Orders and families of Recent mammals of the World (S. Anderson and J. K. Jones, Jr., eds.). John Wiley and Sons, New York, 686 pp.
Koopman, K. F. 1984*b*. A progress report on the systematics of African *Scotophilus* (Vespertilionidae). Proceedings of the Sixth International Bat Research Conference, Ile-Ife, Nigeria, 4:102-113.

Koopman, K. F. 1984*c*. Taxonomic and distributional notes on tropical Australian bats. American Museum Novitates, 2778:1-48.

Koopman, K. F. 1986. Sudan bats revisited: An update of "Bats of the Sudan". Cimbebasia, ser. A. 8(2):9-13.

Koopman, K. F. 1989*a*. Distributional patterns of Indo-Malayan bats (Mammalia: Chiroptera). American Museum Novitates, 2942:1-19.

Koopman, K. F. 1989*b*. Systematic notes on Liberian bats. American Museum Novitates, 2946:1-11.

Koopman, K. F. 1989*c*. A review and analysis of the bats of the West Indies. Pp. 635-644, *in* Biogeography of the West Indies: past, present, and future (C. A. Woods, ed.). Sandhill Crane Press, Gainesville, FL, 878 pp.

Koopman, K. F. 1992. Taxonomic status of *Nycteris vinsoni* Dalquest (Chiroptera: Nycteridae). Journal of Mammalogy, 73(3):649-650.

Koopman, K. F., and J. K. Jones, Jr. 1970. Classification of bats. Pp. 22-28, *in* About bats (B. H. Slaughter and D. W. Walton, eds.). Southern Methodist University Press, Dallas, 339 pp.

Koopman, K. F., M. K. Hecht, and E. Ledecky-Janecek. 1957. Notes on the mammals of the Bahamas with special reference to the bats. Journal of Mammalogy, 38:164-174.

Koopman, K. F., R. E. Mumford, and J. F. Heisterberg. 1978. Bat records from Upper Volta, west Africa. American Museum Novitates, 2643:1-6.

Kooyman, G. L. 1981. Weddell seal: consummate diver. Cambridge University Press, Cambridge, England, 135 pp.

Korobitsyna, K. V., and I. V. Kartavtseva. 1988. [Variation and evolution of the karyotype of gerbils (Rodentia, Cricetidae, Gerbillinae). I. Karyotypic differentiation of midday gerbils (*Meriones meridianus*) of the USSR fauna]. Zoologicheskii Zhurnal, 67:1889-1899 (in Russian).

Korobitsyna, K. V., and V. P. Korablev. 1980. The intraspecific autosome polymorphism of *Meriones tristrami* Thomas, 1892 (Gerbillinae, Cricetidae, Rodentia). Genetica, 52/53:209-221.

Korobitsyna, K. V., C. F. Nadler, N. N. Vorontsov, and R. S. Hoffmann. 1974. Chromosomes of the Siberian snow sheep, *Ovis nivicola*, and implications concerning the origin of amphiberingean wild sheep (subgenus *Pachyceros*). Quaternary Research, 4:235-245.

Kortlucke, S. M. 1973. Morphological variation in the kinkajou, *Potos flavus* (Mammalia: Procyonidae), in Middle America. Occasional Papers of the Museum of Natural History, University of Kansas, 17:1-36.

Kostro , K. 1948. Tcho stepní ili Eversmannu[o]v (*Putorius eversmanni* Lesson 1827), nový a zna n roz í ený len zví eny eskoslovenska. Práce Moravskoslezské Akademie v d P írodních [Acta Academiae Scientairum Naturalium Moravo-Silesiacae], 20(3):1-95.

Kovacs, K. M., and D. M. Lavigne. 1986. *Cystophora cristata*. Mammalian Species, 258:1-9.

Kovalskaya, Yu. M. 1977. [Chromosome polymorphism in *Microtus maximowiczii* Schrenk, 1858 (Rodentia, Cricetidae)]. Byulleten' Moskovskovo Obshchestva Ispytatelei Prirody, Otdel Biologicheskii, 82:38-48 (in Russian).

Kovalskaya, Yu. M. 1989. [Karyotype variability of narrow-skulled vole, *Microtus* (*Stenocranius*) *gregalis* (Rodentia, Cricetidae) from northern Mongolia]. Zoologicheskii Zhurnal, 68:77-84 (in Russian).

Kovalskaya, Yu. M., V. M. Malygin, I. V. Kartavtseva. 1988. [Karyotype stability and distribution of reed voles—*Microtus fortis* (Rodentia, Cricetidae)]. Zoologicheskii Zhurnal, 67:1253-1259 (in Russian).

Kovalskaya, Yu. M., V. M. Aniskin, and I. V. Kartavtseva. 1991. [Geographic variation of C-heterochromatin in *Microtus fortis* (Rodentia, Cricetidae)]. Zoologicheskii Zhurnal, 70:97-104 (in Russian).

Kowalski, K. 1963. The Pliocene and Pleistocene Gliridae (Mammalia, Rodentia) from Poland. Acta Zoologica Cracoviensia, 8(14):533-567.

Kowalski, K., and Y. Hasegawa. 1976. Quaternary rodents from Japan. Bulletin of the National Science Museum (Tokyo), ser. C (Geology and Paleontology), 2(1):31-66.

Kowalski, K., and B. Rzebik-Kowalska. 1991. Mammals of Algeria. Zaklad Narodowy Imienia Ossolinskich Wydawnictwo Polskiej Akademii Nauk Wroclaw, Poland, 370 pp.

Kozlovskii, A. I. 1973. [Results of karyological study of allopatric forms in *Sorex minutus*]. Zoologicheskii Zhurnal, 52:390-398 (in Russian).

Kozlovskii, A. I. 1976. [Karyotypes and systematics of some populations of shrews usually classed with *Sorex arcticus*.]. Zoologicheskii Zhurnal, 55:756-762 (in Russian).

Kozlovskii, A. I. 1986. Chromosome forms and autosomal polymorphism of the wood lemming (*Myopus schisticolor*) from North-East Asia. Folia Zoologica, 35:63-71.

Kraglievich, J. L. 1965. Speciation phyletique dans les rongeurs fossils de genre *Eumysops* Amegh. (Echimyidae, Heteropsomyinae). Mammalia, 29:258-267.

Král, B. 1967. Karyological analysis of two European species of the genus *Erinaceus*. Zoologéskii Listy, 16:239-252.

Král, B., and J. Zima. 1980. Karyosystematika celedi Felidae. Gazella (Prague), 2/3:45-53.

Král, B., J. Zima, V. Hrabe, J. Libosvarsky, and M. Sebela. 1981. On the morphology of *Microtus epiroticus*. Folia Zoologica, 30:317-330.

Král, B., S. I. Radjabli, A. S. Grafodatskij, and V. N. Orlov. 1984. Comparison of karyotypes, G-bands and NORS in three *Cricetulus* spp. (Cricetidae, Rodentia). Folia Zoologica, 33:85- 96.

Krapp, F. 1982*a*. *Microtus nivalis* (Martins, 1842)—Schneemaus. Pp. 261-283, *in* Handbuch der Säugetiere Europas (J. Niethammer and F. Krapp, eds.). Akademische Verlagsgesellschaft (Wiesbaden), 2/I:1-649.

Krapp, F. 1982*b*. *Microtus multiplex* (Fatio, 1905)—Alpen- Kleinwühlmaus. Pp. 419-428, *in* Handbuch der Säugetiere Europas (J. Niethammer and F. Krapp, eds.). Akademische Verlagsgesellschaft (Wiesbaden), 2/I:1-649.

Krapp, F. 1982*c*. *Microtus savii* (de Selys-Longchamps, 1838)— Italienische Kleinwühlmaus. Pp. 429-437, *in* Handbuch der Säugetiere Europas (J. Niethammer and F. Krapp, eds.). Akademische Verlagsgesellschaft (Wiesbaden), 2/I:1-649.

Krapp, F. 1982*d*. *Microtus pyrenaicus* (de Selys-Longchamps, 1847)—Pyrenaen- Kleinwühlmaus. Pp. 444-446, *in* Handbuch der Säugetiere Europas (J. Niethammer and F. Krapp, eds.). Akademische Verlagsgesellschaft (Wiesbaden), 2/I:1-649.

Krapp, F. 1990. *Crocidura leucodon* (Hermann, 1780) - Feldspitzmaus. Pp. 465-484, *in* Handbuch der Säugetiere Europas (J. Niethammer and F. Krapp, eds.). Aula Verlag, Wiesbaden, 3/I:1-524.

Krapp, F., and J. Niethammer. 1982. *Microtus agrestis* (Linnaeus, 1761)—Erdmaus. Pp. 349-373, *in* Handbuch der Säugetiere Europas (J. Niethammer and F. Krapp, eds.). Akademische Verlagsgesellschaft (Wiesbaden), 2/I:1-649.

Kratochvíl, J. 1967. Der Baumschläfer, *Dryomys nitedula* und andere Gliridae-Arten in der Tschechoslowakei. Zoologické Listy, 16(2):99-110.

Kratochvíl, J. 1970. *Pitymys*-Arten aus der hohen Tatra (Mamm., Rodentia). Acta Scientiarum Naturalium, Academiae Scientarium Bohemoslovacea (Brno), 4:1-63.

Kratochvíl, J. 1971. Die Hodengrösse als Kriterium der europäischen arten der gattung *Apodemus* (Rodentia, Muridae). Zoologické Listy, 20:293-306.

Kratochvíl, J. 1973. Männliche Sexualorgane und System der Gliridae (Rodentia). Acta Scientiarum Naturalium, Academiae Scientarium Bohemoslovacea (Brno), 7:1-52.

Kratochvíl, J. 1975. Zur Kenntnis der Igel der Gattung *Erinaceus* in der CSSR (Insectivora, Mamm.). Zoologéskii Listy, 24:297-312.

Kratochvíl, J. 1979. Contribution to our knowledge of the wood lemming, *Myopus schisticolor*. Folia Zoologica, 28:192-207.

Kratochvíl, J. 1980. Zur Phylogenie und Ontogenie bei *Arvicola terrestris* (Rodentia, Arvicolidae). Folia Zoologica, 29:209-224.

Kratochvíl, J. 1981. *Chionomys nivalis* (Arvicolidae, Rodentia). Acta Scientiarum Naturalium, Academiae Scientarium Bohemoslovacea (Brno), 15:1-62.

Kratochvíl, J. 1982*a*. Cavum neurocranii der Arvicolidae. Acta Scientiarum Naturalium, Academiae Scientarium Bohemoslovacea (Brno), 16:1-57.

Kratochvíl, J. 1982*b*. Ein morphologisches Unterscheidungskriterium der Arten *Microtus epiroticus* und *M. arvalis* (Arvicolidae, Rodentia). Folia Zoologica, 31:97-111.

Kratochvíl, J. 1982*c*. Karyotyp und System der Familie Felidae (Carnivora, Mammalia). Folia Zoologica, 31:289-304.

Kratochvíl, J. 1983. Variability of some criteria in *Arvicola terrestris* (Arvicolidae, Rodentia). The effect of altitude on some taxonomical criteria of the Asian population group of *Arvicola terrestris*. Acta Scientiarum Naturalium, Academiae Scientarium Bohemoslovacea (Brno), 17:1-40.

Kratochvíl, J., and B. Král. 1972. Karyotypes and phylogenetic relationships of certain species of the genus *Talpa* (Talpidae, Insectivora). Zoologéskii Listy, 21:199-208.

Kratochvíl, J., and B. Král. 1974. Karyotypes and relationship of Palaearctic "54-chromosome" *Pitymys* species (Microtidae, Rodentia). Zoologické Listy, 23:289-302.

Kratochvíl, J., L. Rodriguez, and V. Barus. 1978. Capromyinae (Rodentia) of Cuba. I. Acta Scientiarum Naturalium. Academiae Scientiarum Bohemoslovacae (Brno), Nova Series, 12(11):1-60.

Kraus, F., and M. M. Miyamoto. 1991. Rapid cladogenesis among the pecoran ruminants: evidence from mitochondrial DNA synthesis. Systematic Zoology, 40:117-130.

Kretzoi, M. 1945. Bemerkungen über das Raubtiersystem. Annales Historico-Naturales Musei Nationalis Hungarici (Budapest), 38:59-83.

Kretzoi, M. 1962. Arvicolidae oder Microtidae. Vertebrata Hungarica (Budapest), 4:171-175.

Kretzoi, M. 1965. *Drepanosorex* - neu definiert. Vertebrata Hungarica, 7:117-129.

Kretzoi, M. 1969. Skizze einer Arvicoliden-Phylogenie. Vertebrata Hungarica (Budapest), 11:155-193.

Krohne, D. T. 1982. The karyotype of *Dicrostonyx hudsonius*. Journal of Mammalogy, 63:174-176.

Krumbiegel, I. 1942. Die Säugetiere der Südamerika - Expeditionen Prof. Dr. Kriegs. 17. Hyrare und Grisons (*Tayra* und *Grison*). Zoologischer Anzeiger, 139(5/6):81-108.

Krumbiegel, I. 1978. Die Kurzschwanz-Stumpfnase, *Simias concolor* (Miller, 1903), und die übrigen Nasenaffen. Säugetierkundliche Mitteilungen, 26:59-75.

Krutzsch, P. H. 1954. North American jumping mice (genus *Zapus*). University of Kansas Publications, Museum of Natural History, 7(4):349-472.

Krystufek, B. 1985*a*. Forest dormouse *Dryomys nitedula* (Pallas 1778)- Rodentia, Mammalia- in Yugoslavia. Scopolia, 1985(9):1-36.

Krystufek, B. 1985*b*. Variability of *Apodemus agrarius* (Pallas, 1771) (Rodentia, Mammalia) in Yugoslavia and some data on its distribution in the northwestern part of the country. Bioloski Vestnik, 33:27-40.

Krystufek, B. 1990. Geographic variation in *Microtus nivalis* (Martins, 1842) from Austria and Yugoslavia. Bonner Zoologische Beiträge, 41:121-139.

Krystufek, B., and D. Kovacic. 1989. Vertical distribution of the snow vole *Microtus nivalis* (Martins, 1842) in northwestern Yugoslavia. Zeitschrift für Säugetierkunde, 54:153-156.

Krystufek, B., and N. Tvrtkovic. 1984. Redescription of *Arvicola terrestris illyrica* (Barrett-Hamiltion, 1899)- Rodentia, Mammalia. Biosistematika, 10:91-97.

Kucheruk, V. V. 1990. Areal. [Range]. Pp. 34-84, *in* Seraya krysa: sistematika, ekologiya, regulyatsiya chislenosti [Norway rat, systematics, ecology, population control] (V. E. Sokolov and E. V. Karasjova, eds.). Nauka, Moscow, 452 pp. (in Russian).

Kuhn, H.-J. 1960. *Genetta* (*Paragenetta*) *lehmanni*, eine neue Schleichkatze aus Liberia. Säugetierkundliche Mitteilungen, 8:154-160.

Kuhn, H.-J. 1965. A provisional check-list of the mammals of Liberia. Senckenbergiana Biologica, 46(5):321-340.

Kuhn, H.-J. 1966. Der Zebraducker, *Cephalophus doria* (Ogilby, 1837). Zeitschrift für Säugetierkunde, 31:282-293.

Kuhn, H.-J. 1968. Der Jentink-Ducker. Natur und Museum, 98:17-23.

Kulik, I. L. 1980. [Distribution and range types of jerboa (Dipodidae, Rodentia)]. Pp. 152-167, *in* [Contemporary problems of zoogeography (A. G. Voronov and N. N. Drozdov, eds.)]. Nauka, Moscow, 323 pp. (in Russian).

Kumirai, A., and J. K. Jones, Jr. 1990. *Nyctinomops femorosaccus*. Mammalian Species, 349:1-5.

Kuntz, R. E., and D. E. Ming. 1970. Vertebrates of Taiwan taken for parasitological and biomedical studies by U.S. Naval Medical Research Unit No. 2 Taipei, Taiwan, Republic of China. Quarterly Journal of the Taiwan Museum, 23:1-37.

Kunz, T. H. 1982. *Lasionycteris noctivagans*. Mammalian Species, 172:1-5.

Kunz, T. H., and R. A. Martin. 1982. *Plecotus townsendii*. Mammalian Species, 175:1-6.

Kuroda, N. 1924. [On new mammals from the Riu Kiu Islands and the vicinity]. Published by the author, Tokyo, 16 pp.

Kuroda, N. 1933. A revision of the genus *Pteropus* found in the islands of the Riu Kiu Chain, Japan. Journal of Mammalogy, 14:312-316.

Kuroda, N. 1938. A list of the Japanese mammals. Privately published, Tokyo, 122 pp.

Kuroda, N. 1957. A new name for the lesser Japanese mole. Journal of the Mammalogical Society of Japan, 1:74.

Kuroda, S., T. Kano, and K. Muhindo. 1985. Further information on the new monkey species, *Cercopithecus salongo* Thys van den Audenaerde 1977. Primates, 26(3):325-333.

Kurta, A., and R. H. Baker. 1990. *Eptesicus fuscus*. Mammalian Species, 356:1-10.

Kurtén, B. 1966. Pleistocene bears of North America. I. Genus *Tremarctos*, spectacled bears. Acta Zoologica Fennica, 115:100-120.

Kurtén, B. 1968. Pleistocene mammals of Europe. Aldine, Chicago, 317 pp.

Kurtén, B. 1973. Transberingian relationships of *Ursus arctos* Linné (brown and grizzly bears). Commentationes Biologicae, 65:1-10.

Kurtén, B., and E. Anderson. 1980. Pleistocene mammals of North America. Columbia University Press New York, 442 pp.

Kurtén, B., and R. Rausch. 1959. Biometric comparisons between North American and European mammals. I. A comparison between Alaskan and Fennoscandian wolverine (*Gulo gulo* Linnaeus). Acta Arctica, 11:1-21.

Kurtonur, C., and B. Ozkan. 1990. New records of *Myomimus roachi* (Bate 1937) from Turkish Thrace (Mammalia: Rodentia: Glirdae). Senckenbergiana Biologica, 71(4):239-244.

Kuz'mina, I. E. 1965. Saiga i stepnaya pishchukha v verkhov'yakh Pechory [Saiga and steppe pika in the upper Pechora river]. Zoologicheskii Zhurnal, 44:307-311 (in Russian).

Kuznetsov, B. A. 1944. Ordo Rodentia. Pp. 262-362, *in* Opredelitel' mlekopitayushchikh SSSR [Guide to the mammals of the USSR] (Bobrinskii, N. A., Kuznetsov, B. A. and A. P. Kuzyakin, eds.). Sovetskaya Nauka, Moscow, 439 pp. (in Russian).

Kuznetsov, B. A. 1965. Ordo Rodentia. Pp. 236-346, *in* Opredelitel' mlekopitayushchikh SSSR [Guide to the mammals of the USSR] (Bobrinskii, N. A., Kuznetsov, B. A. and A. P. Kuzyakin, eds.). Second ed. Proveshchenie, Moscow, 382 pp. (in Russian).

Labes, R., R. Brasecke, and D. Andresen. 1987. Zum Vorkommen des Siebenschlafers (*Glis glis*) im Kreis Schwerin. Säugetierkundliche Informationen, 2(11):441-447.

Lackey, J. A. 1967. Biosystematics of *heermanni* group kangaroo rats in southern California. Transactions of the San Diego Society of Natural History, 14:313-344.

Lackey, J. A. 1991*a*. *Perognathus arenarius*. Mammalian Species, 384:1-4.

Lackey, J. A. 1991*b*. *Perognathus spinatus*. Mammalian Species, 385:1-4.

Lackey, J. A., D. G. Huckaby, and B. G. Ormiston. 1985. *Peromyscus leucopus*. Mammalian Species, 247:1-10.

Laerm, J. 1981. Systematic status of the Cumberland Island pocket gopher, *Geomys cumberlandius*. Brimleyana, 6:141-151.

Laing Jun-xun, and Zhang Jun. 1985. [On the rodent fauna and regionalization of the northeast regions in loess plateau]. Acta Theriologica Sinica, 5(4):299-309 (in Chinese).

Lakhotia, S. C., S. R. V. Rao and S. C Jhanwar. 1973. Studies on rodent chromosomes VI. Co-existence of *Rattus rattus* with 38 and 42 chromosomes in South Western India. Cytologia, 38:403-410.

Lamotte, M., and F. Petter. 1981. Une taupe dorée nouvelle du Cameroun (Mt Oku, 6°15'N, 10°26'E): *Chrysochloris stuhlmanni balsaci* ssp. nov. Mammalia, 45:43-48.

Langevin, P., and R. M. R. Barclay. 1990. *Hypsignathus monstrosus*. Mammalian Species, 357:1-4.

Langguth, A. 1967. Sobre la identidad de *Dusicyon culpaeolus* (Thomas) y de *Dusicyon inca* (Thomas). Neotropica, 13:21-28.

Langguth, A. 1969. Die südamerikanischen Canidae unter besonderer Bercksichtigung des Mohnenwolfes, *Chrysocyon brachyurus* Illiger. Zeitschrift für Wissenschaftliche Zoologie, 179:1-188.

Langguth, A. 1975. Ecology and evolution in the South American canids. Pp. 192-206, *in* The wild canids: Their systematics, behavioral ecology, and evolution (M. W. Fox, ed.). Van Nostrand Reinhold Company, New York, NY, 508 pp.

Langguth, A., and A. Abella. 1970. Las especies uruguayas del género *Ctenomys*. Communiciones Zoológicos Museo Historia Natural de Montevideo, 10(129):1-20.

Lansman, R. A., J. C. Avise, C. F. Aquadro, J. F. Shapira, and S. W. Daniel. 1983. Extensive genetic variation in mitochondrial DNAs among geographic populations of the deer mouse, *Peromyscus maniculatus*. Evolution, 36:1-16.

Lassieur, S., and D. E. Wilson. 1989. *Lonchorhina aurita*. Mammalian Species, 347:1-4.

Lau, Y.-F.d., T. L. Yang-Feng, B. Elder, K. Fredga, and U. H. Wiberg. 1992. Unusual distribution of *Zfy* and *Zfx* sequences on the sex chromosomes of the wood lemming, a species exhibiting XY sex reversal. Cytogenetics and Cell Genetics, 60:48-54.

Laurie, E. M. O. 1952. Mammals collected by Mr. Shaw Mayer in New Guinea 1932-1949. Bulletin of the British Museum (Natural History), Zoology Series, 1:269-318.

Laurie, E. M. O., and J. E. Hill. 1954. List of land mammals of New Guinea, Celebes and adjacent islands 1758-1952. British Museum (Natural History) Publications, London, 175 pages.

Laurie, W. A., E. M. Lang, and C. P. Groves. 1983. *Rhinoceros unicornis*. Mammalian Species, 2ll:1-6.

Laursen, L., and M. Bekoff. 1978. *Loxodonta africana*. Mammalian Species, 92:1-8.

LaVal, R. K. 1970. Intraspecific relationships of bats of the species *Myotis austroriparius*. Journal of Mammalogy, 51:542-552.

LaVal, R. K. 1973*a*. A revision of the Neotropical bats of the genus *Myotis*. Science Bulletin, Natural History Museum of Los Angeles County, 15:1-54.

LaVal, R. K. 1973*b*. Systematics of the genus *Rhogeesa* (Chiroptera: Vespertilionidae). Occasional Papers of the Museum of Natural History, University of Kansas, 19:1-47.

Lavocat, R. 1964. On the systematic affinities of the genus *Delanymys* Hayman. Proceedings of the Linnean Society of London, 175:183-185.

Lavocat, R. 1973. Les rongeurs du Miocene d'Afrique Orientale. Memoires et Travaux de l'Ecole Pratique des Hautes Etudes, Institut de Montpellier, 1:1-284.

Lavocat, R. 1978. Rodentia and Lagomorpha. Pp. 69-89, *in* Evolution of African mammals (V. J. Maglio and H. B. S. Cooke, eds.). Harvard University Press, Cambridge, MA, 641 pp.

Lavrov, L. S. 1979. [Species of beavers (*Castor*) of the Palearctic]. Zoologicheskii Zhurnal, 58:88-96 (in Russian).

Lavrov, L. S., and V. N. Orlov. 1973. [Karyotypes and taxonomy of modern beavers (*Castor*, Castoridae, Mammalia)]. Zoologicheskii Zhurnal, 52:734-742 (in Russian).

Lawlor, T. E. 1965. The Yucatan deer mouse, *Peromyscus yucatanicus*. University of Kansas Publications, Museum of Natural History, 16:421-438.

Lawlor, T. E. 1969. A systematic study of the rodent genus *Ototylomys*. Journal of Mammalogy, 50:28-42.

Lawlor, T. E. 1971*a*. Distribution and relationships of six species of *Peromyscus* in Baja California and Sonora, Mexico. Occasional Papers of the Museum of Zoology, University of Michigan, 661:1-22.

Lawlor, T. E. 1971*b*. Evolution of *Peromyscus* on northern islands in the Gulf of California, Mexico. Transactions of the San Diego Society of Natural History, 16:91-124.

Lawlor, T. E. 1982. *Ototylomys phyllotis*. Mammalian Species, 181:1-3.

Lawlor, T. E. 1983. The mammals. Pp. 265-289, *in* Island biogeography in the Sea of Cortez (T. J. Case and M. L. Cody, eds.). University of California Press, Berkeley, 508 pp.

Lawrence, B. 1933. Howler monkeys of the *Palliata* group. Bulletin of the Museum of Comparative Zoology at Harvard College, 75:314-354.

Lawrence, B. 1939. Mammals. Pp. 28-73, *in* Collections from the Phillippine Islands (T. Barbour, B. Lawrence, and J. L. Peters). Bulletin of the Museum of Comparative Zoology at Harvard College, 86(2):25-128.

Lawrence, B. 1941. *Neacomys* from northwestern South America. Journal of Mammalogy, 22:418-427.

Lawrence, B., and A. Loveridge. 1953. Zoological results of a fifth expedition to East Africa. I. Mammals from Nyasaland and Tete. With notes on the genus *Otomys*. Bulletin of the Museum of Comparative Zoology at Harvard College, 110:1-80.

Lawrence, M. A. 1982. Western Chinese arvicolines (Rodentia) collected by the Sage Expedition. American Museum Novitates, 2745:1-19.

Lawrence, M. A. 1988. The identity of *Sciurus duida* J. A. Allen (Rodentia: Sciuridae). American Museum Novitates, 2919:1-8.

Lawrence, M. A. 1991. A fossil *Myospalax* cranium (Rodentia: Muridae) from Shanxi, China, with observations on Zokor relationships. Pp. 261-286, *in* Contributions to mammalogy in honor of Karl F. Koopman (T. A. Griffiths and D. Klingener, eds.). Bulletin of the American Museum of Natural History, 206:1-432.

Lay, D. M. 1967. A study of the mammals of Iran resulting from the Street expedition of 1962-63. Fieldiana, Zoology, 54:1-282.

Lay, D. M. 1972. The anatomy, physiology, functional significance and evolution of specialized hearing organs of gerbilline rodents. Journal of Morphology, 138:41-120.

Lay, D. M. 1975. Notes on rodents of the genus *Gerbillus* (Mammalia: Muridae: Gerbillinae) from Morocco. Fieldiana, Zoology, 65:89-101.

Lay, D. M. 1983. Taxonomy of the genus *Gerbillus* (Rodentia: Gerbillinae) with comments on the applications of generic and subgeneric names and an annotated list of species. Zeitschrift für Säugetierkunde, 48:329-354.

Lay, D. M., and C. F. Nadler. 1969. Hybridization in the rodent genus *Meriones*. I. Breeding and cytological analyses of *Meriones shawi* (female) X *Meriones libycus* (male) hybrids. Cytogenetics, 8:35-50.

Lay, D. M., and C. F. Nadler. 1972. Cytogenetics and origin of North African *Spalax* (Rodentia: Spalacidae). Cytogenetics, 11:279-285.

Lay, D. M., and C. F. Nadler. 1975. A study of *Gerbillus* (Rodentia: Muridae) east of the Euphrates River. Mammalia, 39:423-445.

Lay, D. M., J. A. W. Anderson, and J. D. Hassinger. 1970. New records of small mammals from West Pakistan and Iran. Mammalia, 34:97-106.

Lay, D. M., K. Agerson, and C. F. Nadler. 1975. Chromosomes of some species of *Gerbillus* (Mammalia: Rodentia). Zeitschrift für Säugetierkunde, 40:141-150.

Layne, J. N. 1990. The Florida mouse. Pp. 1-21, *in* Burrow associates of the gopher tortoise (C. K. Dodd, Jr., R. E. Ashton, Jr., R. Franz, and E. Wester, eds.). Florida Museum of Natural History, Gainesville, 134 pp.

Lazell, J. D., and K. F. Koopman. 1985. Notes on bats of Florida's Lower Keys. Florida Scientist, 48:37-41.

Leatherwood, S., and R. R. Reeves, (eds). 1990. The bottlenose dolphin. Academic Press, New York, 653 pp.

Leatherwood, S., J. S. Grove, and A. E. Zuckerman. 1991. Dolphins of the genus *Lagenorhynchus* in the tropical South Pacific. Marine Mammal Science, 7(2):194-197.

Le Berre, M., and L. Le Guelte. 1990. Les mammiferes actuels dans l'espace saharien. Vie et Milieu, 40:223-228.

Lechleitner, R. R. 1969. Wild mammals of Colorado: their appearance, habits, distribution and abundance. Pruett Publishing Co., Colorado, 254 pp.

Lee, A. K., P. R. Baverstock, and C. H. S. Watts. 1981. Rodents— the late invaders. Pp. 1523-1553, *in* Ecological biogeography of Australia (A. Keast, ed.) [vol. 3, part 6]. Monographiae Biologicae, 41:1-2142 [in 3 vols.].

Lee, H. Y., and R. J. Baker. 1987. Cladistical analysis of chromosomal evolution in pocket gophers of the *Cratogeomys castanops* complex (Rodentia: Geomyidae). Occasional Papers, The Museum, Texas Tech University, 114:1-15.

Lee, M. R., and F. B. Elder. 1977. Karyotypes of eight species of Mexican rodents (Muridae). Journal of Mammalogy, 58:479-487.

Lee, M. R., and D. F. Hoffmeister. 1963. Status of certain fox squirrels in Mexico and Arizona. Proceedings of the Biological Society of Washington, 76:181-190.

Lee, M. R., D. J. Schmidly, and C. C. Huheey. 1972. Chromosomal variation in certain populations of *Peromyscus boylii* and its systematic implications. Journal of Mammalogy, 53:697-707.

Lee, T. E., Jr., and M. D. Engstrom. 1991. Genetic variation in the silky pocket mouse (*Perognathus flavus*) in Texas and New Mexico. Journal of Mammalogy, 72:273-285.

Legendre, S. 1984. Étude odotologíque des représentants actuels du groupe *Tadarida* (Chiroptera, Molossidae). Implications phylogéniques, systématiques et zoogéographiques. Revue Suisse de Zoologie, 91:399-442.

Lehmann, E., von. 1965. Über die Säugetiere im Waldgebiet N.W.- Syriens. Sitzungsberichte der Gesellschaft Naturforschender Freunde zu Berlin, 5:22-38.

Lehmann, E., von. 1969. Über die hautdrüsen der schneemaus (*Chionomys nivalis nivalis* Martins, 1842). Bonner Zoologische Beiträge, 20:373-377.

Lehmann, E., von. 1983. Eine Kleinsäugeranthologie aus Strassburg (in memoriam Johann Hermann, 1738-1800). Annalen des Naturhistorischen Museums in Wien, ser. B, 84:509-514.

Leirs, H., W. Verheyen, M. Michiels, R. Verhagen, and J. Stuyck. 1989. The relation between rainfall and the breeding season of *Mastomys natalensis* (Smith, 1834) in Morogoro, Tanzania. Annales de la Societe Royale Zoologique de Belgique, 119:59-64.

Leitner, M., and G. B. Hartl. 1988. Genetic variation in the bank vole *Clethrionomys glareolus*: Biochemical differentiation among populations over short geographic distances. Acta Theriologica, 33:231-245.

Lekagul, B., and H. Felten. 1989. Remarks on the genus *Bandicota* in Thailand (Rodentia: Muridae). Thai Journal of Agricultural Science, 22:197-211.

Lekagul, B., and J. A. McNeely. 1977. Mammals of Thailand. Association for the Conservation of Wildlife, Sahakarnbhat Co., Bangkok, 758 pp.

Lemire, M. 1966. Particularities de l'appareil masticateur d'un rongeur insectivore *Deomys ferrugineus* (Cricetidae, Dendromurinae). Mammalia, 30:454-494.

Lenders, A., and E. Pelzers. 1982. Het voorkomen van de hamster *Cricetus cricetus* (L) aan de noordgrens vanzun verspreidingsgebied in Nederland. Lutra, 25:69-80.

Lent, P. C. 1988. *Ovibos moschatus*. Mammalian Species, 302:1-9.

Levan, G., K. Klinga, C. Szpirer, and J. Szpirer. 1990. Gene map of the rat (*Rattus norvegicus*) 2N=42. Isozyme Bulletin, 23:34-42.

Levenson, H., R. S. Hoffmann, C. F. Nadler, L. Deutsch, and S. D. Freeman. 1985. Systematics of the Holarctic chipmunks. Journal of Mammalogy, 66:219-242.

Lever, C. 1985. Naturalized Mammals of the World. Longman, London, 487 pp.

Lewis, R. E., J. H. Lewis, and S. I. Atallah. 1967. A review of Lebanese mammals. Lagomorpha and Rodentia. Journal of Zoology (London), 153:45-70.

Lewis, S. E., and D. E. Wilson. 1987. *Vampyressa pusilla*. Mammalian Species, 292:1-5.

Li Bao-guo, and Chen Fu-guan. 1989. [A taxonomic study and new subspecies of the subgenus *Eospalax*, genus *Myospalax*]. Acta Zoologica Sinica, 35:89-95 (in Chinese).

Li C., S. Ma, Y Wang, and G. Lu. 1987. [Mammals from Honghe Region. Pp. 26-27, *in* Report on the Biological Resources of Honghe Region, Southern Yunnan. Vol. 1. Land Vertebrates. Yunnan National Press, 102 pp.] (in Chinese).

Li Wei-dong, Li Hung-chu, Hamati Xamaridan, and Ma Jun-jie. 1988. First report on ecological study of the Ili pika (*Ochotona iliensis*). Abstracts, Symposium of Asian Pacific Mammalogy, Huirou, Beijing, Peoples Republic of China, pages not numbered, abstracts listed alphabetical by author.

Li Zhen-ying, and Luo Ze-xun. 1979. [On a new species of wild hare from China]. Journal of the Northeast Forestry Institute, 12:71-81 (in Chinese).

Libois, R. M., and R. Fons. 1990. Le mulot des Iles d'Hyeres: un cas de gigantisme insulaire. Vie et Milieu, 40:217-222.

Lichtenstein, H. 1828. Über die Springmäuse oder die Arten der Gattung *Dipus*. Abhandlungen der Königlichen Akademie der Wissenschaften in Berlin, 1825:133-162.

Lidicker, W. Z., Jr. 1960. An analysis of intraspecific variation in the kangaroo rat *Dipodomys merriami*. University of California Publications in Zoology, 67:125-218.

Lidicker, W. Z., Jr. 1968. A phylogeny of New Guinea rodent genera based on phallic morphology. Journal of Mammalogy, 49:609-643.

Lidicker, W. Z., Jr. 1983. Dasyurids and their kin. Science, 219(4590):1316-1317.

Lidicker, W. Z., Jr., and P. V. Brylski. 1987. The conilurine rodent radiation of Australia, analyzed on the basis of phallic morphology. Journal of Mammalogy, 68:617-641.

Lidicker, W. Z., Jr., and W. I. Follett. 1968. *Isoodon* Desmarest, 1817, rather than *Thylacis* Illiger, 1811, as the valid generic name of the short-nosed bandicoots (Marsupialia: Peramelidae). Proceedings of the Biological Society of Washington, 81:251-256.

Lidicker, W. Z., Jr., and A. Yang. 1986. Morphology of the penis in the taiga vole (*Microtus xanthognathus*). Journal of Mammalogy, 67:497-502.

Lidicker, W. Z., Jr., and A. C. Ziegler. 1968. Report on a collection of mammals from eastern New Guinea, including species keys for fourteen genera. University of California Publications in Zoology, 87:1-60.

Lim, B. K. 1987. *Lepus townsendii*. Mammalian Species, 288:1-6.

Lim, B. K., and P. D. Ross. 1992. Taxonomic status of *Alticola* and new record of *cricetulus* from Nepal. Mammalia, 56:300-302.

Limpus, C. J., C. J. Parmenter, and C. H. S. Watts. 1983. *Melomys rubicola*, an endangered murid rodent endemic to the Great Barrier Reef of Queensland. Australian Mammalogy, 6:77-79.

Lin Ling-kong, P. S. Alexander, and Yu Ming-jenn. 1987. A survey of the small mammals of Mt. Ali recreation area. Tunghai Journal, 28:669-682 (in Chinese).

Lin, Y-S., D. R. Progulske, P.-F. Lee, and Y.-T. Day. 1985. Bibliography of Petauristinae (Rodentia: Sciuridae). Journal of the Taiwan Museum, 38(2):49-57.

Lindsay, E. H. 1988. Cricetid rodents from Siwalik deposits near Chinji Village. Part I: Megacricetodontinae, Myocricetodontinae and Dendromurinae. Palaeovertebrata, 18:95-154.

Lindsay, E. H., and L. L. Jacobs. 1985. Pliocene small mammal fossils from Chihuahua, Mexico. Paleontología Mexicana, 51:1-53.

Lindsay, H. M. 1928. Note on *Viverra civettina*. Journal of the Bombay Natural History Society, 33(1):146-148.

Lindsay, S. L. 1981. Taxonomic and biogeographic relationships of Baja California chickarees (*Tamiasciurus*). Journal of Mammalogy, 62:673-682.

Lindsay, S. L. 1982. Systematic relationship of parapatric tree squirrel species (*Tamiasciurus*) in the Pacific Northwest. Canadian Journal of Zoology, 60:2149-2156.

Ling, J. K. 1992. *Neophoca cinerea*. Mammalian Species, 392:1-7.

Ling, J. K., and M. M. Bryden. 1992. *Mirounga leonina*. Mammalian Species, 391:1-8.

Linnaeus, C. 1758. Systema Naturae per regna tria naturae, secundum classis, ordines, genera, species cum characteribus, differentiis, synonymis, locis. Tenth ed. Vol. 1. Laurentii Salvii, Stockholm, 824 pp.

Linnaeus, C. 1766-1768. Systema naturae per regna tria naturae, secundum classes, ordines, genera, species, cum characteribus, differentiis synonymis, locis. Vol. 1. Regnum Animale. pt. 1, pp. 1-532 [1766]; vol. 3, Appendix pp. 223-235 [1768]. 12th ed. [Vol. 3, Regnum Lapideum, pp. 1-222, rejected for nomenclatural purposes, ICZN, opinion 296]. Laurentii Salvii, Stockholm, 3 vols.

Linnaeus, C. (revised by J. F. Gmelin). 1788. Systema naturae per regna tria naturae, secundum classes, ordines, genera, species, cum characteribus, differentiis synonymis, locis. Vol. I. Regum Animale. Class 1, Mammalia. Thirteenth ed. (revised by J. F. Gmelin). G. E. Beir, Lipsiae, 232 pages.

Linnaeus, C. (translated and revised by R. Kerr). 1792. The animal kingdom; or, zoological system of the celebrated Sir Charles Linnaeus. Class 1. Mammalia and Class II. Birds. Being a translation of that part of the Systema Naturae, as lately published with great improvements by Professor Gmelin, together with numerous additions from more recent zoological writers and illustrated with copperplates. J. Murray, London, 644 pp.

Lint, D. W., J. W. Clayton, W. R. Lillie, and L. Postuma. 1990. Evolution and systematics of the Beluga whale, *Delphinapterus leucas*, and other odontocetes: a molecular approach. Canadian Bulletin of Fisheries and Aquatic Sciences, 224:7-22.

Linzey, A. V. 1983. *Synaptomys cooperi*. Mammalian Species, 210:1-5.

Linzey, A. V., and J. N. Layne. 1969. Comparative morphology of the male reproductive tract in the rodent genus *Peromyscus* (Muridae). American Museum Novitates, 2355:1-47.

Linzey, A. V., and J. N. Layne. 1974. Comparative morphology of spermatozoa of the rodent genus *Peromyscus* (Muridae). American Museum Novitates, 2532:1-20.

Linzey, D. W., and R. L. Packard. 1977. *Ochrotomys nuttalli*. Mammalian Species, 75:1-6.

Liu Chun-sheng, Li Chuan-bin, Wu Wan-neng, and Meng Ji-hui. 1985. The faunal distribution and geographical divisions of rodents in Anhui Province. Acta Theriologica Sinica, 5:111-118.

Liu Chun-sheng, Wu Wan-neng, Guo Shi-kun, and Meng Ji-hui. 1991. [A study of the subspecies classification of *Apodemus agrarius* in eastern continental China]. Acta Theriologica Sinica, 11:294-299 (in Chinese).

Liu Nai-fa, Fan Hua-wei, Jing Kai, and Ning Rui-dong. 1990. [Study of community diversity of rodents in Anxi desert]. Acta Theriologica Sinica, 10(3):215-220 (in Chinese).

Long, C. A. 1972. Taxonomic revision of the North American badger, *Taxidea taxus*. Journal of Mammalogy, 59(4):725-759.
Long, C. A. 1973. *Taxidea taxus*. Mammalian Species, 26:1-4.
Long, C. A. 1974. *Microsorex hoyi* and *Microsorex thompsoni*. Mammalian Species, 33:1-4.
Long, C. A. 1978. A listing of Recent badgers of the world, with remarks on taxonomic problems in *Mydaus* and *Melogale*. Reports on the Fauna and Flora of Wisconsin, The Museum of Natural History, Stevens Point, WI, 14:1-6.
Long, C. A. 1981. Provisional classification and evolution of the badgers. Pp. 55-85, *in* Worldwide Furbearer Conference Proceedings (J. A. Chapman and D. Pursley, eds.). Frostburg, MD, 1:1-652.
Long, C. A., and R. S. Hoffmann. 1992. *Sorex preblei* from the Black Canyon, first record for Colorado. Southwestern Naturalist, 37:318-319.
Long, C. A., and C. A. Killingley. 1983. The badgers of the world. Charles C. Thomas, Springfield, IL, 404 pp.
Lönnberg, E. 1913. Mammals from Ecuador and related forms. Arkiv för Zoologi, 8(16):1-36.
Lönnberg, E. 1923*a*. Notes on *Arctonyx*. Annals and Magazine of Natural History, ser. 9, 11:322-326.
Lönnberg, E. 1923*b*. Remarks on some Palaearctic bears. Proceedings of the Zoological Society of London, 1923:85-95.
Lönnberg, E., and N. Gyldenstolpe. 1925. Vertebrata. 2. Preliminary diagnoses of seven new mammals. Arkiv för Zoologi, 17B, 5:1-6.
López-Furster, M. J., J. Ventura, M. Miralles, and E. Castién. 1990. Craniometric characteristics of *Neomys fodiens* (Pennant, 1771) (Mammalia, Insectivora) from the northeastern Iberian Peninsula. Acta Theriologica, 35(3-4):269-276.
López-Martínez, N., and L. F. López-Jurado. 1987. Un nuevo murido gigante del Cuaternario de Gran Canaria *Canariomys tamarani* nov. sp. (Rodentia, Mammalia). Doñana, Publicacion Ocasional, 2:1-66.
Lotze, J.-H. and S. Anderson. 1979. *Procyon lotor*. Mammalian Species, 119:1-8.
Loughlin, T. R., and M. A. Perez. 1985. *Mesoplodon stejnegeri*. Mammalian Species, 250:1-6.
Loughlin, T. R., M. A. Perez, and R. L. Merrick. 1987. *Eumetopias jubatus*. Mammalian Species, 283:1-7.
Lovari, S. 1985. Behavioural repertoire of the Abruzzo chamois, *Rupicapra pyrenaica ornata* Neumann, 1899 (Artiodactyla: Bovidae). Säugetierkundliche Mitteilungen, 32:113-136.
Lovari, S. 1987. Evolutionary aspects of the biology of *Rupicapra* spp. (Bovidae, Caprinae). Pp. 51-61, *in* The biology and management of *Capricornis* and related antelopes (H. Soma, ed.). Croom Helm, London, 391 pp.
Lovari, S., and C. Scala. 1980. Revision of *Rupicapra* genus. I. A statistical re-evaluation of Couturier's data on the morphometry of six chamois subspecies. Bollettino di Zoologia, 47:113-124.
Lovari, S., and C. Scala. 1984. Revision of *Rupicapra* genus. IV. Horn biometrics of *Rupicapra rupicapra asiatica* and its relevance to the taxonomic position of *Rupicapra rupicapra caucasica*. Zeitschrift für Säugetierkunde, 49:246-253.
Lowery, G. H., Jr. 1974. The mammals of Louisiana and its adjacent waters. Louisiana State University Press, Baton Rouge, 565 pp.
Luckett, W. P. (ed.). 1980. Comparative biology and evolutionary relationships of tree shrews. Plenum Press, New York, 314 pp.
Ludwig, D. R. 1984. *Microtus richardsoni*. Mammalian Species, 223:1-6.
Lumpkin, S., and K. R. Kranz. 1984. *Cephalophus sylvicultor*. Mammalian Species, 225:1-7.
Lund, P. W. 1841. Blik paa Brasiliens Dyreverden för sidste Jordomvaeltning. Tredie Afhandling: Fortsaettelse af Pattedyrene. Konigelige Danske Videnskabernes Selskabs Afhandlinger, Kjöbenhavn, 8:219-272, pls 14-24.
Lundholm, B. G. 1955*a*. Descriptions of new mammals. Annals of the Transvaal Museum, 22:279-303.
Lundholm, B. G. 1955*b*. A taxonomic study of *Cynictis penicillata* (G. Cuvier). Annals of the Transvaal Museum, 22:305-319.
Lundholm, B. G. 1955*c*. Remarks on some South African Murinae. Annals of the Transvaal Museum, 22:321-329.
Lungeanu, A., L. Gavrila, C. Stepan, and D. Murariu. 1984. Donnees preliminaires concernant l'etude du caryotype de *Micromys minutus* (Pallas, 1771) (Rodentia, Muridae). Travaux du Museum d'Histoire Naturelle "Grigore Antipa", 26:241-244.
Lungeanu, A., L. Gavrila, D. Murariu, and C. Stepan. 1986. The distribution of the constituent heterochromatin and the G- banding pattern in the genome of *Apodemus agrarius* (Pallas, 1771) (Mammalia, Muridae). Travaux du Museum d'Histoire Naturelle "Grigore Antipa", 28:267-270.
Luyts, G. 1986. [Is the garden dormouse extending its area of distribution as far as the coast?]. Wielewaal, 52(5) 1986:91.

Lyalyukhina, S., E. Kotenkova, W. Walkowa, and K. Adamszyk. 1991. Comparison of craniological parameters in *Mus musculus musculus* Linnaeus, 1758 and *Mus hortulans* Nordmann, 1840. Acta Theriologica, 36:95-107.

Lyapunova, E. A., N. N. Vorontsov, and L. Ya. Martynova. 1974. Cytological differentiation of burrowing mammals in the Palaearctic. Pp. 203-215, *in* Symposium Theriologicum II (J. Kratochvil and R. Obrtel, eds.). Prague, 394 pp.

Lyapunova, E. A., N. N. Vorontsov, K. V. Korobitsyna, E. Yu. Ivanitskaya, Yu. M. Borisov, L. V. Yakimenko, and V. Ye. Dovgal. 1980. A Robertsonian fan in *Ellobius talpinus*. Genetica, 52/53:239-247.

Lyapunova, E. A., I. Yu. Baklushinskaya, O. L. Kolomiets, and T. F. Mazurova. 1990. [Analysis of fertility of hybrids of multi-chromosomal forms in mole-voles of the super-species *Ellobius tancrei* differing in a single pair of Robertsonian metacentrics]. Doklady Akademii Nauk SSSR, 310:721-723 (in Russian).

Lydekker, R. 1895 [1896]. On the affinities of the so-called extinct giant dormouse of Malta. Proceedings of the Zoological Society of London, 1895:860-863.

Lydekker, R. 1913-1916. Catalogue of the ungulate mammals in the British Museum (Natural History). British Museum (Natural History), London, 5 volumes.

Lyman, C. P., and R. C. O'Brien. 1977. A laboratory study of the Turkish hamster *Mesocricetus brandti*. Breviora, 442:1-27.

Lynch, C. D. 1981. The status of the Cape Grey Mongoose, *Herpestes pulverulentus* Wagner, 1839. (Mammalia: Viverridae). Navorsinge van die Nasionale Museum Bloemfontein, 4(5):121-168.

Lynch, C. D. 1986. The ecology of the Lesser dwarf shrew, *Suncus varilla* with reference to the use of termite mounds of *Trinervitermes trinervoides*. Navorsinge van die Nasionale Museum Bloemfontein, Natural Sciences, 5:277-297.

Lynch, C. D., and J. P. Watson. 1992. The distribution and ecology of *Otomys sloggetti* (Mammalia: Rodentia) with notes on its taxonomy. Navorsinge van die Nasionale Museum Bloemfontein, Natural Sciences, 8:141-158.

Lyne, A. G., and P. A. Mort. 1981. A comparison of skull morphology in the marsupial bandicoot genus *Isoodon*: its taxonomic implications and notes on a new species, *Isoodon arnhemensis*. Australian Mammalogy, 4:107-133.

Lyon, M. W., Jr. 1907. Notes on the porcupines of the Malay Peninsula and Archipelago. Proceedings of United States National Museum, 32:575-594.

Lyon, M. W., Jr. 1911. Mammals collected by Dr. Abbott in Borneo and some of the small adjacent islands. Proceedings of the United States National Museum, 40:53-146.

Lyon, M. W., Jr. 1913. Tree shrews: an account of the mammalian family Tupaiidae. Proceedings of the United States National Museum, 45:1-188.

Lyon, M. W., Jr. 1942. Additions to the "Mammals of Indiana". American Midland Naturalist, 27(3):790-791.

Ma Shi-lai, and Wang Ying-xiang. 1986. [The taxonomy and distribution of the gibbons in southern China and its adjacent region - with description of three new subspecies]. Zoological Research, 7:393-410 (in Chinese).

Ma Yong. 1964. [A new species of hedgehog from Shansi Province, *Hemiechinus sylvaticus* sp. n.]. Acta Zootaxonomica Sinica, 5:212-214 (in Chinese).

Ma Yong, Wang Feng-gui, Jin Shan-ke, Li Si-hua, Lin Yonglie, and Yie Zong-yiao. 1981. [On the Glires of northern Xinjiang]. Acta Theriologica Sinica, 1:177-188 (in Chinese).

Ma Yong, Wang Feng-gui, Jin Shan-ke, and Li Si-hua. 1987. [Glires (rodents and lagomorphs) of Northern Xinjiang and their zoogeographical distribution]. Science Press, Academia Sinica, Beijing, 274 pp. (in Chinese).

MacDonald, S. O., and C. Jones. 1987. *Ochotona collaris*. Mammalian Species, 281:1-4.

MacKinnon, J. R., and S. N. Stuart (eds.). 1989. The kouprey. An action plan for its conservation. I. U. C. N., Gland, Switzerland, 20 pp.

MacPhee, R. D. E. 1987*a*. The shrew tenrecs of Madagascar: systematic revision and Holocene distribution of *Microgale* (Tenrecidae, Insectivora). American Museum Novitates, 2889:1-45.

MacPhee, R. D. E. 1987*b*. Systematic status of *Dasogale fontoynonti* (Tenrecidae, Insectivora). Journal of Mammalogy, 68:133-135.

Macêdo, R. H., and M. A. Mares. 1987. Geographic variation in the South American cricetine rodent *Bolomys lasiurus*. Journal of Mammalogy, 68:578-594.

Macêdo, R. H., and M. A. Mares. 1988. *Neotoma albigula*. Mammalian Species, 310:1-7.

Machado, A., de Barros. 1969. Mammiferos de Angola aindo não çitados ou pouca conhecidos. Publicações Culturais da Campanhia de Diamantes de Angola, 46:93-232.

Mack, G. 1961. Mammals from South-western Queensland. Memoirs of the Queensland Museum, 13:213-229.

Maddalena, T. 1990. Systematics and biogeography of Afrotropical and Palaearctic shrews of the genus *Crocidura* (Insectivora: Soricidae): an electrophoretic approach. Pp. 297-308, *in* Vertebrates in the tropics (G. Peters, and R. Hutterer, eds.). Museum Alexander Koenig, Bonn, 424 pp.

Maddalena, T., and P. Vogel. 1990. Relations génétiques entre crocidures méditerranéennes: le cas des musaraignes de Gozo (Malte). Vie et Milieu, 40:119-123.

Maddalena, T., A.-M. Mehmeti, G. Bonner, and P. Vogel. 1987. The karyotype of *Crocidura flavescens* (Mammalia, Insectivora) in South Africa. Zeitschrift für Säugetierkunde, 52:129-132.

Maddalena, T., B. Sicard, M. Tranier, and J. C. Gautun. 1988. Note sur la presence de *Gerbillus henleyi* (De Winton, 1903) au Burkina-Faso. Mammalia, 52:282-284.

Maddalena, T., E. Van der Straeten, L. Ntahuga, and A. Sparti. 1989. Nouvelles donnees et caryotypes des rongeurs du Burundi. Revue Suisse de Zoologie, 96:939-948.

Maddock, A. H., and M. R. Perrin. 1981. A microscopical examination of the gastric morphology of the white-tailed rat *Mystromys albicaudatus* (Smith, 1834). South African Journal of Zoology, 16:237-247.

Maddock, A. H., and M. R. Perrin. 1983. Development of the gastric morphology and fornical bacterial/ epithelial association in the white-tailed rat *Mystromys albicaudatus* (Smith 1834). South African Journal of Zoology, 18:115-127.

Madkour, G. 1984. Chondral and osteological structures in the cranial region of common Qatarian gerbils. Zoologischer Anzeiger, 213:247-257.

Maeda, K. 1980. Review on the classification of little tube-nosed bats, *Murina aurata* group. Mammalia, 44:531-551.

Maeda, K. 1982. Studies on the classification of *Miniopterus* in Eurasia, Australia, and Melanesia. Honyurui Kagaku (Mammalian Science), Supplement 1:1-176.

Maempel, G. Z., and H. de Bruijn. 1982. The Plio/Pleistocene Gliridae from the Mediterranean Islands reconsidered. Proceedings of the Koninklijke Nederlandse Akademie van Wetenschappen, 85(1):113-128.

Maglio, V. J. 1973. Origin and evolution of the Elephantidae. Transactions of the American Philosophical Society, 63(3):1-149.

Mahoney, J. A. 1968. *Baiyankamys* Hinton, 1943 (Muridae, Hydromyinae) a New Guinea rodent genus named for an incorrectly associated skin and skull (Hydromyinae, *Hydromys*) and mandible (Murinae, *Rattus*). Mammalia, 32:64-71.

Mahoney, J. A. 1975. *Notomys macrotis* Thomas, 1921, a poorly known Australian hopping mouse (Rodentia: Muridae). Journal of the Australian Mammal Society, 1:367-374.

Mahoney, J. A. 1977. Skull characters and relationships of *Notomys mordax* Thomas (Rodentia: Muridae), a poorly known Queensland hopping-mouse. Australian Journal of Zoology, 25:749-754.

Mahoney, J. A. 1981. The specific name of the honey possum (Marsupialia: Tarsipedidae: *Tarsipes rostratus* Gervais and Verreaux, 1842). Australian Mammalogy, 4:135-138.

Mahoney, J. A., and H. Posamentier. 1975. The occurrence of the native rodent *Pseudomys gracilicaudatus* (Gould, 1845) (Rodentia: Muridae) in New South Wales. Journal of the Australian Mammal Society, 1:333-346.

Mahoney, J. A., and B. J. Richardson. 1988. Muridae. Pp. 154-192, *in* Zoological catalogue of Australia. Mammalia (J. L. Bannister, et. al.). Australian Government Publishing Service, Canberra, 5:1-274.

Mahoney, J. A., and D. W. Walton. 1988. Molossidae. Zoological Catalogue of Australia, 5:146-150.

Maia, V., and A. Langguth. 1981. New karyotypes of Brazilian akodont rodents with notes on taxonomy. Zeitschrift für Säugetierkunde, 46:241-249.

Maier, W. 1980. Konstruktionsmorphologische Untersuchungen am Gebiss der rezenten Prosimiae (Primates). Abhandlungen der Senckenbergischen Naturforschenden Gesellschaft, 538:1-158 (in German, with English summary).

Major, C. I. F[orsyth]. 1896. Diagnoses of new mammals from Madagascar. Annals and Magazine of Natural History, ser. 6, 18:319-325.

Major, C. I. F[orsyth]. 1897. On the Malagasy rodent genus *Brachyuromys*; and on the mutual relations of some groups of the Muridae (Hesperomyinae, Microtinae, Murinae, and "Spalacidae") with each other and with the Malagasy Nesomyinae. Proceedings of the Zoological Society of London, 1897:695-720.

Major, C. I. F[orsyth]. 1899. On fossil dormice. Geological Magazine, Decade 4, 6(425):492-501.

Major, C. I. F[orsyth]. 1901. The musk-rat of Santa Lucia (Antilles). Annals and Magazine of Natural History, ser. 7, 7:204-206.

Mallinson, J. J. C. 1971. A note on the hispid hare *Caprolagus hispidus* (Pearson, 1839). Journal of the Bombay Natural History Society, 68:443-444.

Malygin, V. M. 1978. [A comparative-morphological analysis of species from the group *Microtus arvalis* (Rodentia, Cricetidae)]. Zoologicheskii Zhurnal, 67:1062-1073 (in Russian).

Malygin, V. M., V. N. Orlov, and V. N. Yatsenko. 1990. O vidovoi samostoyatel' nosti priozernoi polevki *Microtus limnophilus*, ee rodstvennykh svyazyakh s polevkoi-ekonomkoi *M. oeconomus* i rasprostranenii etikh vidov v Mongolii [Species independence of *Microtus limnophilus*, its relations with *M. oeconomus* and distribution of these species in Mongolia]. Zoologicheskii Zhurnal, 69(4):115-127 (in Russian).

Malygin, V. M., N. V. Startzev, and Ya. Zima. 1992. [Karyotypes and distribution of striped hamsters of the group *barabensis* (Rodentia, Cricetidae)]. Vestnik Moskokvskogo Universiteta, Ser. VI, Biologiia, 2:32-39 (in Russian).

Mandal, A. K., and S. Ghosh. 1981. A new subspecies of the meltad, *Millardia meltada* (Gray, 1837) [Rodentia: Muridae] from West Bengal. Bulletin of the Zoological Survey of India, 4:235-238.

Mann, G. 1950. Nuevos mamíferos de Tarapacá. Investigacións Zoológicas Chilenas, 1:2.

Mann, G. F. 1978. Los pequeños mamíferos de Chile (marsupiales, quirópteros, edentados y roedores). Guyana (Zoología), 40:1-342.

Manning, R. W., and J. K. Jones, Jr. 1988. *Perognathus fasciatus*. Mammalian Species, 303:1-4.

Manning, R. W., and J. K. Jones, Jr. 1989. *Myotis evotis*. Mammalian Species, 329:1-5.

Manning, T. H. 1956. The northern red-backed mouse, *Clethrionomys rutilus* (Pallas), in Canada. National Museum of Canada Bulletin, 144:1-67.

Manville, R. H. 1966. The extinct sea mink, with taxonomic notes. Proceedings of the United States National Museum, 122(3584):1-12.

Maples, W. R., and T. W. McKern. 1967. A preliminary report on the classification of the Kenya baboon. Pp. 13-22, *in* The baboon in medical research (H. Vagtborg, ed.). University of Texas Press, Austin, 2:1-908.

Mares, M. A., and R. A. Ojeda. 1982. Patterns of diversity and adaptation in South American hystricognath rodents. Pp. 393-432, *in* Mammalian biology in South America (M. Mares and H. Genoways, eds.). Special Publication Series, Pymatuning Laboratory of Ecology, University of Pittsburgh, Pennsylvania, 6:1-539.

Mares, M. A., M. R. Willig, K. Streilein, and T. E. Lacher, Jr. 1981*a*. The mammals of northeastern Brazil: A preliminary assessment. Annals of Carnegie Museum, 50(4):80-137.

Mares, M. A., R. A. Ojeda, and M. P. Kosco. 1981*b*. Observations on the distribution and ecology of the mammals of Salta Province, Argentina. Annals of Carnegie Museum, 50:151-206.

Mares, M. A., J. K. Braun, and D. Gettinger. 1989*a*. Observations on the distribution and ecology of the mammals of the Cerrado grasslands of central Brazil. Annals of Carnegie Museum, 58:1-60.

Mares, M. A., R. A. Ojeda, and R. M. Barquez. 1989*b*. Guide to the Mammals of Salta Province. University of Oklahoma Press, Norman, 303 pp.

Marinina, L. S., N. L. Orlov, and N. A. Sorokina. 1987. [On *Myomimus personatus* Ognev, 1924, found on Maly Balkhan]. Izvestiya Akademii Nauk Turkmenskoi SSR, Seriya Biologicheskikh Nauk, 1987:72-73 (in Russian).

Markvong, A., J. Marshall, and A. Gropp. 1973. Chromosomes of rats and mice of Thailand. Natural History Bulletin of the Siam Society, 25:23-40.

Marsh, H., R. Lloze, G. E. Heinsohn and T. Kasuya. 1989. Irrawady dolphin - *Orcaella brevirostris* (Gray, 1866). Pp. 101-118, *in* Handbook of marine mammals: River dolphins and the larger toothed whales (S. H. Ridgway and R. Harrison, eds.). Academic Press, London, 4:1-442.

Marshall, J. T., Jr. 1977*a*. Family Muridae: Rats and mice. Pp. 396-487, *in* Mammals of Thailand (B. Lekagul and J. A. McNeely, eds.). Association for the Conservation of Wildlife, Sahakarnbhat Co., Bangkok, 758 pp.

Marshall, J. T., Jr. 1977*b*. A synopsis of Asian species of *Mus* (Rodentia, Muridae). Bulletin of the American Museum of Natural History, 158:173-220.

Marshall, J. T., Jr. 1981. Taxonomy. Pp. 17-26, *in* The mouse in biomedical research, volume I, history, genetics, and wild mice (H. L. Foster, J. D. Small, and J. G. Fox, eds.). American College of Laboratory Animal Medicine Series, Academic Press, New York, 306 pp.

Marshall, J. T., Jr. 1986. Systematics of the genus *Mus*. Pp. 12-18, *in* The wild mouse in immunology (M. Potter, J. H. Nadeau, and M. P. Cancro, eds.). Current Topics in Microbiology and Immunology, 127:1-395.

Marshall, J. T., Jr. In press. Nomenclature of European *Mus* (Mammalia: Muridae). Proceedings of the Biological Society of Washington.

Marshall, J. T., Jr., and E. R. Marshall. 1976. Gibbons and their territorial songs. Science, 193:235-237.

Marshall, J. T., Jr., and R. D. Sage. 1981. Taxonomy of the house mouse. Symposia of the Zoological Society of London, 47:15-25.

Marshall, J. T., and J. Sugardjito. 1986. Gibbon systematics. Comparative Primate Biology, 1:137-185.

Marshall, L. G. 1977. *Lestodelphys halli*. Mammalian Species, 81:1-3.

Marshall, L. G. 1978*a*. *Lutreolina crassicaudata*. Mammalian Species, 91:1-4.

Marshall, L. G. 1978*b*. *Dromiciops australis*. Mammalian Species, 99:1-5.

Marshall, L. G. 1978*c*. *Glironia venusta*. Mammalian Species, 107:1-3.
Marshall, L. G. 1978*d*. *Chironectes minimus*. Mammalian Species, 109:1-6.
Marshall, L. G. 1979. A model for paleobiogeography of South American cricetine rodents. Paleobiology, 5:126-132.
Marshall, L. G. 1980. Systematics of the South American marsupial family Caenolestidae. Fieldiana, Geology, New Series, 5:1-145.
Marshall, L. G., J. A. Case, and M. O. Woodburne. 1990. Phylogenetic relationships of the families of marsupials. Pp. 433-506, *in* Current mammalogy (H. H. Genoways, ed.). Plenum Press, New York, 2:1-577.
Martens, J., and J. Niethammer. 1972. Die Waldmäuse (*Apodemus*) Nepals. Zeitschrift für Säugetierkunde, 37:144-154.
Martin, C. O., and D. J. Schmidly. 1982. Taxonomic review of the pallid bat, *Antrozous pallidus* (Le Conte). Special Publications, The Museum, Texas Tech University Press, 18:1-48.
Martin, L. D. 1975. Microtine rodents from the Ogallala Pliocene of Nebraska and the early evolution of the Microtinae in North America. Papers on Paleontology, The Museum of Paleontology University of Michigan, 12:101-110.
Martin, L. D. 1979. The biostratigraphy of arvicoline rodents in North America. Transactions of the Nebraska Academy of Scineces, 7:91-100.
Martin, L. D. 1980. The early evolution of the Cricetidae in North America. University of Kansas Paleontological Contributions, 102:1-42.
Martin, R. A. 1974. Fossil mammals from the Coleman IIA Fauna, Sumter County. Pp. 35-99, *in* Pleistocene mammals of Florida (S. D. Webb, ed.). The University Presses of Florida, Gainesville, 270 pp.
Martin, R. A. 1987. Notes on the classification and evolution of some North American fossil *Microtus* (Mammalia; Rodentia). Journal of Vertebrate Paleontology, 7:270-283.
Martin, R. A. 1989*a*. Early Pleistocene zapodid rodents from the Java local fauna of north-central South Dakota. Journal of Vertebrate Paleontology, 9(1):101-109.
Martin, R. A. 1989*b*. Arvicolid rodents of the early Pleistocene Java Local Fauna from north-central South Dakota. Journal of Vertebrate Paleontology, 9:438-450.
Martin, R. D. 1990. Primate origins and evolution: A phylogenetic reconstruction. Chapman and Hall, London, 804 pp.
Martin, R. L. 1966. Redescription of the type locality of *Sorex dispar*. Journal of Mammalogy, 47(1):130-131.
Martin, R. L. 1973. The dentition of *Microtus chrotorrhinus* (Miller) and related forms. University of Connecticut Occasional Papers, Biological Sciences, 2:183-201.
Martin, R. L. 1979. Morphology, development, and adaptive values of the baculum of *Microtus chrotorrhinus* (Miller, 1894) and related forms. Säugetierkundliche Mitteilungen, 27:307-311.
Martino, V., and E. Martino. 1940. Preliminary notes on five new mammals from Yugoslavia. Annals and Magazine of Natural History, ser. 5, 11:493-498.
Mascarello, J. T. 1978. Chromosomal, biochemical, mensural, penile, and cranial variation in desert woodrats (*Neotoma lepida*). Journal of Mammalogy, 59:477-495.
Mascarello, J. T., and K. Bolles. 1980. C- and G-banded chromosomes of *Ammospermophilus insularis* (Rodentia: Sciuridae). Journal of Mammalogy, 61:714-716.
Mascarello, J. T., and T. C. Hsu. 1976. Chromosome evolution in woodrats, genus *Neotoma* (Rodentia: Cricetidae). Evolution, 30:152-169.
Masing, M., and U. Timm. 1988. On mammals of the Viljandi district. Eesti Ulukid, 5:66-82.
Massoia, E. 1963*a*. Sobre la posición sistemática y distribución geográfica de *Akodon* (*Thaptomys*) *nigrita* (Rodentia, Cricetidae). Physis (Buenos Aires), 24:73-80.
Massoia, E. 1963*b*. *Oxymycterus iheringi* (Rodentia-Cricetidae), nueva especie para la Argentina. Physis (Buenos Aires), 24:129-136.
Massoia, E. 1973. Descripción de *Oryzomys fornesi*, nueva especie y nuevos datos sobre algunas especies y subespecies argentinas del subgénero *Oryzomys* (*Oligoryzomys*) (Mammalia-Rodentia-Cricetidae). Revista de Investigaciones Agropecuarias, INTA, Serie 1, Biologia y Producción Animal, 10:21-37.
Massoia, E. 1975. Datos sobre un cricetido nuevo para la Argentina: *Oryzomys* (*Oryzomys*) *capito intermedius* y sus diferencias con *Oryzomys* (*Oryzomys*) *legatus* (Mammalia-Rodentia). Revista de Investigaciones Agropecuarias, INTA, Serie 5, Patologia Vegetal, 11:1-7.
Massoia, E. 1979. El género *Octodon* en la Argentina. Neotropica, 25:36.
Massoia, E. 1980*a*. El estado sistemático de cuatro especies de cricetidos sudamericanos y comentarios sobre otras especies congenericas (Mammalia-Rodentia). Ameghiniana, 17:280-287.
Massoia, E. 1980*b*. Nuevos datos sobre *Akodon*, *Deltamys* y *Cabreramys*, con la descripcion de una especie y una subespecie nuevas (Mammalia, Rodentia, Cricetidae). Nota preliminar. Historia Natural, 1:179.

Massoia, E. 1981. El estado sistematico y zoogeografia de *Mus brasiliensis* Desmarest y *Holochilus sciureus* Wagner (Mammalia-Rodentia-Cricetidae). Physis (Buenos Aires), Seccion C, 39:31-34.

Massoia, E., and A. Fornes. 1964. Notas sobre el género *Scapteromys* (Rodentia-Cricetidae). I. Sistemática, distribución geográfica y rasgos etoecologicos de *Scapteromys tumidus* (Waterhouse). Physis (Buenos Aires), 24:279-297.

Massoia, E., and A. Fornes. 1965. Notas sobre el género *Scapteromys* (Rodentia-Cricetidae). II. Fundamentos de la identidad específica de *S. principalis* (Lund) y *S. gnambiquarae* (M. Ribeiro). Neotropica, 11:1-7.

Massoia, E., and A. Fornes. 1967. El estado sistemático, distribución geográfica y datos etoecologicos de algunos mamíferos neotropicales (Marsupialia y Rodentia), con la descripción de *Cabreramys*, género nuevo (Cricetidae). Acta Zoologica Lilloana, 23:407-430.

Massoia, E., and A. Fornes. 1969. Characteres comunes y distintivos de *Oxymycterus nasutus* (Waterhouse) y *O. iheringi* Thomas (Rodentia, Cricetidae). Physis (Buenos Aires), 28:315-321.

Massoia, E., A. Fornes, R. L. Wainberg, and T. G. de Fronza. 1968. Nuevos aportes al conocimiento de las especies bonaerenses del genero *Calomys* (Rodentia-Cricetidae). Revista de Investigaciones Agropecuarias, INTA, Serie 1, Biologia y Produccion Animal, 5:63-92.

Massoia, E., J. C. Chebez, and S. H. Fortabat. 1991. Nuevos o poco conocidos craneos de mamiferos vivienties—3. *Abrawayomys ruschi* de la Provincia de Misiones, Republica Argentina. Aprona, Boletín Científico, 19:39-40.

Matschie, P. 1895. Die Säugethiere Deutsch-Ostafrikas. D. Reimer, Berlin, 147 pp.

Matschie, P. 1911. Über einige Säugetiere aus Muansa am Victoria-Nyansa. Sitzungsberichte der Gesellschaft Naturforschender Freunde, Berlin, 8:333-343.

Matschie, P. 1912. Einige bisher wenig beachtete Rassen des Nörzes. Sitzungsberichte der Gesellschaft Naturforschender Freunde zu Berlin, 6:345-354.

Matschie, P. 1914. Ein neuer *Anomalurus* von der Elfenbeinküste. Sitzungs-berichte der Gesellschaft Naturforschender Freunde zu Berlin, 1914:349-351.

Matschie, P. 1915. Zwei vermutlich neue Mäuse aus Deutsch-Ostafrika. Sitzungsberichte der Gesellschaft Naturforschender Freunde zu Berlin, 1915:98-101.

Matschie, P. 1916. Bemerkungen über die Gattung *Didelphis* L. Sitzungsberichte der Gesellschaft Naturforschender Freunde zu Berlin, 1916(1):259-272.

Matson, J. O. 1975. *Myotis planiceps*. Mammalian Species, 60:1-2.

Matson, J. O. 1980. The status of banner-tailed kangaroo rats, genus *Dipodomys*, from central Mexico. Journal of Mammalogy, 61:563-566.

Matson, J. O., and J. P. Abravaya. 1977. *Blarinomys breviceps*. Mammalian Species, 74:1-3.

Matson, J. O., and R. H. Baker. 1986. Mammals of Zacatecas. Special Publications, The Museum, Texas Tech University, 24:1-88.

Matsuda, Y., and V. M. Chapman. 1992. Analysis of sex-chromosome aneuploidy in interspecific backcross progeny between the laboratory mouse strain C57BL/6 and *Mus spretus*. Cytogenetics and Cell Genetics, 60:74-78.

Matthews, L. H. 1939. The subspecies and variation of the spotted hyaena, *Crocuta crocuta* Erxl. Proceedings of the Zoological Society of London, 1939:237-260.

Matthey, R. 1958. Les chromosomes et la position systematique de quelques Murinae africains. Acta Tropica, 15:97-117.

Matthey, R. 1963. La formule chromosomique chez sept especes et sous-especes de Murinae Africains. Mammalia, 27:157-176.

Matthey, R. 1964. Analyse caryologique de cinq especes de Muridae Africains (Mammalia, Rodentia). Mammalia, 28:403-418.

Matthey, R. 1965*a*. Etudes de cytogenetique sur des *Murinae* Africains appartenant aux genres *Arvicanthis*, *Praomys*, *Acomys* et *Mastomys* (Rodentia). Mammalia, 29:228-249.

Matthey, R. 1965*b*. Le probleme de la determination du sexe chez *Acomys selousi* de Winton-Cytogenetique du genre *Acomys* (Rodentia-Muridae). Revue Suisse de Zoologie, 72:119-144.

Matthey, R. 1967. Note sur la cytogenetique de quelques murides Africains. Mammalia, 31:281-287.

Matthey, R. 1968. Cytogenetique et taxonomie du genre *Acomys*. *A. percivali* Dollman et *A. wilson* Thomas, especes D'Abyssinie. Mammalia, 32:621-627.

Matthey, R. 1970. Caryotypes de murides et de dendromurids originaires de Republique Centrafricaine. Mammalia, 34:459-466.

Matthey, R., and F. Petter. 1970. Etude cytogenetique et taxonomique de 40 *Tatera* et *Taterillus* provenant de Haute-Volta et de Republique Centrafricaine (Rongeurs, Gerbillidae). Mammalia, 34:585-597.

Matyushkin, E. N. 1979. Rysi Golarktiki [Lynx of the Holarctic]. Pp. 76-162, *in* Mlekopitayushchie: Issledovaniya po faune Sovetskovo Soyuza [Mammals: Investigations on the fauna of the Soviet Union] (O. L. Rossolimo, ed). Sbornik Trudov Zoologicheskovo Muzeya MGU, 13:1-279 (in Russian).

Maurer, F. W., J. Op't Hof, and D. R. Osterhoff. 1976. Biochemical gene markers in the gerbil *Tatera brantsii* from Lesotho. Animal Blood Groups and Biochemical Genetics, 7:231-239.

Mayer, J. J., and R. M. Wetzel. 1986. *Catagonus wagneri*. Mammalian Species, 259:1-5.

Mayer, J. J., and R. M. Wetzel. 1987. *Tayassu pecari*. Mammalian Species, 293:1-7.

Mayr, E. 1986. Uncertainty in science: is the giant panda a bear or raccoon? Nature, 323:769-771.

Mazák, V. 1979. Der tiger, *Panthera tigris*. Die Neue Brehm-Bücherei, 356:1-228.

Mazák, V. 1981. *Panthera tigris*. Mammalian Species, 152:1-8.

McAllan, A. W., and M. D. Bruce. 1989. Some problems in vertebrate nomenclature. I. Mammals. Bollettino, Museo Regionale di Scienze Naturali, Torino, 7:443-460.

McAllister, J. A., and R. S. Hoffmann. 1988. *Phenacomys intermedius*. Mammalian Species, 305:1-8.

McBee, K., and R. J. Baker. 1982. *Dasypus novemcinctus*. Mammalian Species, 162:1-9.

McCarty, R. 1975. *Onychomys torridus*. Mammalian Species, 59:1-5.

McCarty, R. 1978. *Onychomys leucogaster*. Mammalian Species, 87:1-6.

McCullough, D. R. 1969. The Tule elk: Its history, behavior, and ecology. University of California Publications, Zoology, 88:1-209.

McDermid, E. M., and W. N. Bonner. 1975. Red cell and serum protein systems of gray seals and harbour seals. Comparative Biochemistry and Physiology, 50B:97-101.

McDonald, J. N. 1981. North American bison, their classification and evolution. University of California Press, Berkeley, 316 pp.

McGhee, M. E., and H. H. Genoways. 1978. *Liomys pictus*. Mammalian Species, 83:1-5.

McGrew, J. C. 1979. *Vulpes macrotis*. Mammalian Species, 123:1-6.

McKay, G. M. 1982. Nomenclature of the gliding possum genera *Petaurus* and *Petauroides* (Marsupialia: Petauridae). Australian Mammalogy, 5:37-39.

McKay, G. M. 1988. Petauridae. Pp. 87-97, *in* Zoological catalogue of Australia. Mammalia (D. W. Walton, ed.). Australian Government Publishing Service, Canberra, 5:1-274.

McKean, J. L. 1972. Notes on some collections of bats (Order Chiroptera) from Papua-New Guinea and Bougainville Island. Division of Wildlife, Research Technical Paper, Commonwealth Scientific and Industrial Research Organization, Australia, 26:1-35.

McKean, J. L., and J. H. Calaby. 1968. A new genus and two new species of bats from New Guinea. Mammalia, 32:372-378.

McKean, J. L., and W. J. Price. 1967. Notes on some Chiroptera from Queensland, Australia. Mammalia, 31:101-119.

McKean, J. L., and W. J. Price. 1978. *Pipistrellus* (Chiroptera: Vespertilionidae) in northern Australia with some remarks on its systematics. Mammalia, 42:343-347.

McKean, J. L., G. C. Richards, and W. J. Price. 1978. A taxonomic appraisal of *Eptesicus* (Chiroptera: Mammalia) in Australia. Australian Journal of Zoology, 26:529-537.

McKenna, M. C. 1962. *Eupetaurus* and the living petauristine sciurids. American Museum Novitates, 2104:1-38.

McKenna, M. C. 1975. Toward a phylogenetic classification of the Mammalia. Pp. 21-46, *in* Phylogeny of the primates—A multidisciplinary approach (W. P. Luckett and F. S. Szalay, eds.). Plenum Press, New York, 483 pp.

McKenna, M. C. 1992. The alpha crystalline A chain of the eye lens and mammalian phylogeny. Annales Zooligici Fennici, 28:349-360.

McLachlan, G. R., R. Liversidge, and R. M. Tietz. 1966. A record of *Berardius arnouxi* from the south-east coast of South Africa. Annals of the Cape Provincial Museums, 5:91-100.

McLaughlin, C. A. 1967. Aplodontoid, Sciuroid, Geomyoid, Castoroid, and Anomaluroid Rodents. Pp. 210-225, *in* Recent Mammals of the World. A synopsis of families (S. Anderson and J. K. Jones, Jr., eds.). Ronald Press Company, New York, 453 pp.

McLaughlin, C. A. 1984. Protrogomorph, sciuromorph, castorimorph, myomorph (geomyoid, anomaluroid, pedetoid, and ctenodactyloid) rodents. Pp. 267-288, *in* Orders and families of Recent mammals of the world (S. Anderson and J. K. Jones, Jr. eds.). John Wiley and Sons, New York, 686 pp.

McManus, J. J. 1974. *Didelphis virginiana*. Mammalian Species, 40:1-6.

McNeely, J. A. 1981. Conservation needs of *Nesolagus netscheri* in Sumatra. Pp. 929-932, *in* Proceedings of the World Lagomorph Conference (K. Meyers and C. D. MacInnes, eds). University of Guelph, Guelph, Ontario, 983 pp.

Mead, J. G. 1975. Anatomy of the external nasal passages and facial complex in the Delphinidae (Mammalia: Cetacea). Smithsonian Contributions to Zoology, 207:1-72.

Mead, J. G. 1989*a*. Shepherd's beaked whale - *Tasmacetus shepherdi* Oliver, 1937. Pp. 309-320, *in* Handbook of marine mammals: River dolphins and the larger toothed whales (S. H. Ridgway and R. Harrison, eds.). Academic Press, London, 4:1-442.

Mead, J. G. 1989*b*. Bottlenose whales - *Hyperoodon ampullatus* (Forster, 1770) and *Hyperoodon planifrons* Flower, 1882. Pp. 321-348, *in* Handbook of marine mammals: River dolphins and the larger toothed whales (S. H. Ridgway and R. Harrison, eds.). Academic Press, London, 1:1-442.

Mead, J. G. 1989*c*. Beaked whales of the genus - *Mesoplodon*. Pp. 349-430, *in* Handbook of marine mammals: River dolphins and the larger toothed whales (S. H. Ridgway and R. Harrison, eds.). Academic Press, London, 4:1-442.

Mead, J. G., W. A. Walker, and W. J. Houck. 1982. Biological observations on *Mesoplodon carlhubbsi* (Cetacea: Ziphiidae). Smithsonian Contributions to Zoology, 34:1-25.

Mead, J. I. 1989. *Nemorhaedus goral*. Mammalian Species, 335:1-5.

Mead, J. I., C. J. Bell, and L. K. Murray. 1992. *Mictomys borealis* (Northern Bog Lemming) and the Wisconsin paleoecology of the east-central Great Basin. Quaternary Research, 37:229-238.

Mead, R. A. 1968. Reproduction in western forms of the spotted skunk (genus *Spilogale*). Journal of Mammalogy, 49:373-390.

Meagher, M. 1986. *Bison bison*. Mammalian Species, 266:1-8.

Mearns, E. A. 1896. Preliminary diagnoses of new mammals from the Mexican border of the United States. Proceedings of the United States National Museum, 18:443-447.

Mech, L. D. 1974. *Canis lupus*. Mammalian Species, 37:1-6.

Medellín, R. A. 1989. *Chrotopterus auritus*. Mammalian Species, 343:1-5.

Medellín, R. A., and H. T. Arita. 1989. *Tonatia evotis* and *Tonatia silvicola*. Mammalian Species, 334:1-5.

Medellín, R. A., D. E. Wilson, and D. Navarro L. 1985. *Micronycteris brachyotis*. Mammalian Species, 251:1-4.

Medway, L. 1961. The status of *Tupaia splendidula* Gray (Primates: Tupaiidae). Treubia, 25:269-272.

Medway, L. 1964. Comments on the status of *Rattus inas* (Bonhote), with observations on the distribution of this and related rats in the Sunda Subregion. Federation Museums Journal, 9:95-101.

Medway, L. 1965. Mammals of Borneo. Malaysian Branch of the Royal Asiatic Society, Singapore, 193 pp.

Medway, L. 1969. The wild mammals of Malaya and offshore islands including Singapore. Oxford University Press, London, 118 pages.

Medway, L. 1970. The monkeys of Sundaland. Ecology and systematics of the cercopithecids of a humid equatorial environment. Pp. 513-553, *in* Old world monkeys (J. R. Napier and P. H. Napier, eds.). Academic Press, New York, 660 pp.

Medway, L. 1977. Mammals of Borneo: Field keys and an annotated checklist. Second ed. Monograph, Malay Branch of the Royal Asiatic Society, 7:1-172.

Medway, L. 1978. The wild mammals of Malaya (Peninsular Malaysia) and Singapore. Oxford University Press, Kuala Lumpur, 128 pp.

Medway, L., and H. S. Yong. 1976. Problems in the systematics of the rats (Muridae) of peninsular Malaysia. Malaysian Journal of Science, 4(A):43-53.

Meester, J. 1953. The genera of African shrews. Annals of the Transvaal Museum, 22:205-214.

Meester, J. 1954. On the status of the shrew, genus *Myosorex*. Annals and Magazine of Natural History, ser. 12, 7:947-950.

Meester, J. 1958. Variation in the shrew genus *Myosorex* in southern Africa. Journal of Mammalogy, 39:325-339.

Meester, J. 1963. A systematic revision of the shrew genus *Crocidura* in Southern Africa. Transvaal Museum Memoir, 13:1-127.

Meester, J. 1972. Order Philodota. Part 4. Pp. 1-3, *in* The mammals of Africa: An identification manual (J. Meester and H. W. Setzer, eds.) [issued 2 May 1972]. Smithsonian Institution Press, Washington, D. C., not continuously paginated.

Meester, J. A. J. 1974. Order Insectivora, Family Chrysochloridae. Part 1.3. Pp. 1-7, *in* The mammals of Africa: An identification manual (J. Meester and H. W. Setzer, eds.) [issued 10 Sep 1974]. Smithsonian Institution Press, Washington, D. C., not continuously paginated.

Meester, J., and N. J. Dippenaar. 1978. A new species of *Myosorex* from Knysna, South Africa (Mammalia: Soricidae). Annals of the Transvaal Museum, 31:29-42.

Meester, J., and A. von W. Lambrechts. 1971. The Southern African species of *Suncus* Ehrenberg (Mammalia: Soricidae). Annals of the Transvaal Museum, 27:1-14.

Meester, J. A. J., I. L. Rautenbach, N. J. Dippenaar, and C. M. Baker. 1986. Classification of southern African mammals. Transvaal Museum Monograph, 5:1-359.

Meier, M. N. 1983. [Evolution and taxonomic status of common voles of the subgenus *Microtus* in the fauna of the USSR]. Zoologicheskii Zhurnal, 62:90-101 (in Russian).

Meier, M. N., T. A. Grishchenko, and E. V. Zybina. 1981. [Experimental hybridization as a method of studying the degree of divergence of closely related species of the genus *Microtus*]. Zoologicheskii Zhurnal, 60:290-300 (in Russian).

Mein, P., L. Ginsburg, and B. Ratanasthien. 1990. Nouveaux rongeurs du Miocène de Li (Thaïlande). Comptes Rendus de l'Acadèmie des Sciences (Paris), ser. 2, 310(6):861-865.
Melton, D. A. 1976. The biology of the aardvark (Tubulidentata-Orycteropodidae). Mammal Review, 6(2):75-88.
Melville, R. V., and J. D. D. Smith, (eds.). 1987. Official lists and indexes of names and works in zoology. International Commission on Zoological Nomenclature, 366 pp.
Menu, H. 1984. Révision du statut de *Pipistrellus subflavus* (F. Cuvier, 1832). Proposition d'un taxon générique nouveau: *Perimyotis* nov. gen. Mammalia, 48:409-416.
Menzies, J. I. 1970. An eastward extension of the known range of the olive colobus monkey (*Colobus verus*, Van Beneden). Journal of the West African Science Association, 15:83-84.
Menzies, J. I. 1974. The status of *Melomys platyops arfakiensis* (Ruemmler) and the races of *Melomys rubex* Thomas. Mammalia, 38:647-655.
Menzies, J. I. 1989. Designation of a lectotype for *Melomys levipes* (Thomas, 1897) (Rodentia: Muridae) of New Guinea. Science in New Guinea, 15:108-110.
Menzies, J. I. 1990. A systematic revision of *Pogonomelomys* (Rodentia: Muridae) of New Guinea. Science in New Guinea, 16:118-137.
Menzies, J. I., and E. Dennis. 1979. Handbook of New Guinea rodents. Handbook, Wau Ecology Institute, 6:1-68.
Menzies, J. I., and J. C. Pernetta. 1986. A taxonomic revision of cuscuses allied to *Phalanger orientalis*. Journal of Zoology (London), (B), 1:551-618.
Merriam, C. H. 1889. Preliminary revision of the North American pocket mice (genera *Perognathus* et *Cricetodipus* auct.) with descriptions of new species and subspecies and a key to the known forms. North American Fauna, 1:1-36.
Merriam, C. H. 1890. Description of a new prairie dog from Wyoming. North American Fauna, 4:33-35.
Merriam, C. H. 1895*a*. Monographic revision of the pocket gophers, family Geomyidae (exclusive of the species of *Thomomys*). North American Fauna, 8:1-258.
Merriam, C. H. 1895*b*. Revision of the shrews of the American genera *Blarina* and *Notiosorex*. North American Fauna, 10:5-34.
Merriam, C. H. 1897. The generic names *Ictis*, *Arctogale*, and *Arctogalidia*. Science, 5:302.
Merriam, C. H. 1901. Synopsis of the rice rats (genus *Oryzomys*) of the United States and Mexico. Proceedings of the Washington Academy of Sciences, 3:273-295.
Merriam, C. H. 1905. Two new chipmunks from Colorado and Arizona. Proceedings of the Biological Society of Washington, 18:163-166.
Merritt, J. F. 1978. *Peromyscus californicus*. Mammalian Species, 85:1-6.
Merritt, J. F. 1981. *Clethrionomys gapperi*. Mammalian Species, 146:1-9.
Mertens, R. 1925. Verzeichnis der Säugetier-Typen des Senckenbergischen Museums. Senckenbergiana Biologica, 7:18-37.
Meyer, A. B. 1898. Ueber zwei Eichhörnchenarten von Celebes. Abhandlungen und Berichte des Königlichen Zoologischen und Anthropologisch-Ethnographischen Museums zu Dresden, 4:1-3.
Meyer, G. E. 1978. Hyracoidea. Pp. 284-314, *in* Evolution of African mammals (V. J. Maglio and H. B. S. Cooke, eds.). Howard University Press, Cambridge, 641 pp.
Meylan, A. 1964. Le polymorphisme chromosomique de *Sorex araneus* L. (Mamm.-Insectivora). Revue Suisse de Zoologie, 71:903-983.
Meylan, A., and J. Hausser. 1973. Les chromosomes des *Sorex* du groupe *araneus-arcticus* (Mammalia, Insectivora). Zeitschrift für Säugetierkunde, 38:143-158.
Meylan, A., and J. Hausser. 1978. Le type chromosomique A des *Sorex* du groupe *araneus*: *Sorex coronatus* Millet, 1828 (Mammalia, Insectivora). Mammalia, 42:115-122.
Meylan, A., and J. Hausser. 1991. The karyotype of the North American *Sorex tundrensis* (Mammalia; Insectivora). Pp. 125-129, *in* The cytogenetics of the *Sorex araneus* group and related topics. Mémoires de la Société Vaudoise des Sciences Naturelles, 19:1-151.
Meylan, A., and P. Vogel. 1982. Contribution à la cytotaxonomie des Soricidés (Mammalia, Insectivora) de l'Afrique occidentale. Cytogenetics and Cell Genetics, 34:83-92.
Mezhzherin, S. V. 1991. [On specific distinctness of *Apodemus* (*Sylvaemus*) *ponticus* (Rodentia, Muridae)]. Vesnik Zoologii, 6:34-40 (in Russian).
Mezhzherin, S. V., and E. I. Lashkova. 1992. Two closely related mice species—*Sylvaemus syvaticus* and *S. flavicollis* (Rodentia, Muridae) in an area of their overlapping occurrence. Vestnik Zoologii, 5:33-41.
Mezhzherin, S. V., and A. G. Mikhailenko. 1991. O vidovoi prinadlezhnosti *Apodemus sylvaticus tscherga* (Rodentia, Muridae) Altaya [On specific identity of *Apodemus sylvaticus tscherga* (Rodentia, Muridae) from Altai]. Vestnik Zoologii, 3:35-45 (in Russian, with English summary).

Mezhzherin, S. V., and I. V. Zagorodnyuk. 1989. Novi vid myshei roda *Apodemus* (Rodentia, Muridae) [New species of mouse of the genus *Apodemus* (Rodentia, Muridae)]. Vestnik Zoologii, 4:55-59 (in Russian, with English summary).

Mezhzherin, S. V., and A. E. Zykov. 1991. Genetic divergence and allozyme variability in mice of genus *Apodemus s. lato* (Muridae, Rodentia). Cytology and Genetics, 25:51-59.

Mi Jing-chuan, Wang Guan, and Wang Cheng-guo. 1990. [Cluster analysis on rodents community in the eastern part of desert-steppe of Nei Mongol]. Acta Theriologica Sinica, 5(4):299-309 (in Chinese).

Michaelis, B. 1972. Die Schleichkatzen (Viverriden) Afrikas. Säugetierkundliche Mitteilungen, 20(1-2):1-110.

Michaux, J., R. Hutterer, and N. Lopez-Martinez. 1991. New fossil faunas from Fuerteventura, Canary Islands: evidence for a Pleistocene age of endemic rodents and shrews. Comptes Rendus de l'Acadèmie des Sciences (Paris), ser. 2, 312(6):801-806.

Michener, G. R., and J. W. Koeppl. 1985. *Spermophilus richardsonii*. Mammalian species, 243:1-8.

Mikes, M., I. Savic, and V. Habijan. 1982. Osteometrische eigenschaften des beckengurtels (cingulum extremitatis pelvinae) der art *Spalax leucodon* Nordmann, 1840. Zoologischer Anzeiger, 208:417-427.

Miller, G. S., Jr. 1896. The genera and subgenera of voles and lemmings. North American Fauna, 12:1-84.

Miller, G. S., Jr. 1902. The mammals of the Andaman and Nicobar islands. Proceedings of the United States National Museum, 24:751-795.

Miller, G. S., Jr. 1912*a*. Catalogue of the mammals of Western Europe (Europe exclusive of Russia) in the collection of the British Museum. British Museum (Natural History), London, 1019 pp.

Miller, G. S., Jr. 1912*b*. List of North American land mammals in the United States National Museum, 1911. Bulletin of the United States National Museum, 79:1-455.

Miller, G. S., Jr. 1913. Revision of the bats of the genus *Glossophaga*. Proceedings of the United States National Museum, 46:413-429.

Miller, G. S., Jr. 1924. List of North American Recent mammals. Bulletin of the United States National Museum, 128:1-673.

Miller, G. S., Jr. 1929. The characters of the genus *Geocapromys* Chapman. Smithsonian Miscellaneous Collections, 82(4):1-3, 1 pl.

Miller, G. S., Jr. 1940*a*. The status of the genus *Aschizomys* Miller. Journal of Mammalogy, 21:94-95.

Miller, G. S., Jr. 1940*b*. Notes on some moles from southeastern Asia. Journal of Mammalogy, 21:442-444.

Miller, G. S., Jr. 1942. Zoological results of the George Vanderbilt Sumatran Expedition, 1936-1939. Part V. Mammals collected by Frederick A. Ulmer, Jr. on Sumatra and Nias. Proceedings of the Academy of Natural Sciences of Philadelphia, 94:107-165.

Miller, G. S., Jr., and G. M. Allen. 1928. The American bats of the genera *Myotis* and *Pizonyx*. Bulletin of the United States National Museum, 144:1-218.

Miller, G. S., Jr., and J. W. Gidley. 1918. Synopsis of the supergeneric groups of rodents. Journal of the Washington Academy of Sciences, 8:431-448.

Miller, G. S., Jr., and R. Kellogg. 1955. List of North American Recent mammals. Bulletin of United States National Museum, 205:1-954.

Miller, G. S., Jr., and J. A. G. Rehn. 1901. Systematic results of the study of North American land mammals to the close of the year 1900. Proceedings of the Boston Society of Natural History, 30(1):1-352.

Millet, M. C., J. Britton-Davidian, and P. Orsini. 1982. Genetique biochimique comparee de *Microtus cabrerae* Thomas, 1906 et de trois autres especes d'Arvicolidae mediterraneens. Mammalia, 46:381-388.

Mills, M. G. L. 1982. *Hyaena brunnea*. Mammalian Species, 194:1-5.

Milner, J., C. Jones, and J. K. Jones, Jr. 1990. *Nyctinomops macrotus*. Mammalian Species, 351:1-4.

Milyutin, A. I. 1990. Sistematika [Systematics]. Pp. 25-33, *in* Seraya krysa: sistematika, ekologiya, reguliatsiya chislennosti [Norway rat: systematics, ecology, population control] (V. E. Sokolov and E. V. Karasjova, eds.). Nauka, Moscow, 452 pp. (in Russian).

Minezawa, M., M. Harada, O. C. Jordan, and C. J. Valdivia Borda. 1985. Cytogenetics of the Bolivian endemic red howler monkeys (*Alouatta seniculus sara*): accessory chromosomes and Y-autosome translocation related numerical variations. Kyoto University Overseas Research Reports of New World Monkeys, 5:7-16.

Mishra, A. C., and V. Dhanda. 1975. Review of the genus *Millardia* (Rodentia: Muridae), with description of a new species. Journal of Mammalogy, 56:76-80.

Misonne, X. 1957. Mammiferes de la Turquie sub-orientale et du nord de la Syrie. Mammalia, 21(1):53-67.

Misonne, X. 1965. Presence de *Leggada callewaerti* Thomas au Katanga. Mammalia, 29:426-429.

Misonne, X. 1966. The systematic position of *Mystromys longicaudatus* Noack and of *Leimacomys buttneri* Matschie. Annales Musée Royal de l'Afrique Centrale, Tervuren, Belgique, Serie IN-8, Sciences Zoologiques, 144:41-45.

Misonne, X. 1969. African and Indo-Australian Muridae: Evolutionary trends. Annales Musée Royal de l'Afrique Centrale, Tervuren, Belgique, Serie IN-8, Sciences Zoologiques, 172:1-219.

Misonne, X. 1974. Order Rodentia. Part 6. Pp. 1-39, *in* The mammals of Africa: An identification manual (J. Meester and H. W. Setzer, eds.). [issued 10 Sep 1974]. Smithsonian Institution Press, Washington, D. C., not continuously paginated.

Misonne, X. 1990. New record for Iran: *Golunda elliotti* (Gray) (Rodentia, Muridae). Mammalia, 54:494.

Misonne, X., and J. Verschuren. 1964. Notes sur *Rattus pernanus* Kershaw, 1921. Mammalia, 28:654-658.

Misonne, X., and J. Verschuren. 1966. Les rongeurs et lagomorphes de la region du Parc National du Serengeti (Tanzanie). Mammalia, 30:517-537.

Mitchell, E. D. 1970. Pigmentation pattern evolution in delphinid cetaceans: an essay in adaptive coloration. Canadian Journal of Zoology, 48:717-740.

Mitchell, E. D., (ed.). 1975. Report of the meeting on smaller cetaceans - Montreal, April 1-11, 1974. Pp. 889-893, *in* Review of biology and fisheries for smaller cetaceans. Journal of the Fisheries Research Board of Canada, 32(7):875-1240.

Mitchell, E., and R. H. Tedford. 1973. The Enaliarctinae, a new group of extinct aquatic Carnivora and a consideration of the origin of the Otariidae. Bulletin of the American Museum of Natural History, 151:201-284.

Mitchell, R. M. 1978. The *Ochotona* (Lagomorpha: Ochotonidae) of Nepal. Säugetierkundliche Mitteilungen, 26:208-214.

Mitchell, R. M. 1981. The *Ochotona* (Lagomorpha: Ochotonidae) of Asia. Pp. 1031-1038, *in* Geological and ecological studies of Qinghai-Xizang plateau (Liu D., ed.). Science Press, Academia Sinica, Beijing, 2 volumes.

Mittermeier, R. A., and A. F. Coimbra-Filho. 1981. Systematics: Species and subspecies. Pp. 29-109, *in* Ecology and behavior of neotropical primates (A. F. Coimbra-Filho and R. A. Mittermeier, eds.). Academia Brasileira de Ciencias (Rio de Janeiro), 1:1-496.

Mittermeier, R. A., A. B. Rylands, A. F. Coimbra-Filho, and G. A. B. da Foneseca (eds.). 1988. Ecology and behavior of Neotropical Primates. World Wildlife Fund, Washington D. C., 2:1-610.

Mivart, St. G. 1882. On the classification and distribution of the Aeluroidea. Proceedings of the Zoological Society of London, 1882:135-208.

Mockel, R. 1986. Zum vorkommen des gartenschlafers (*Eliomys quercinus*) im Westerzgebirge. Säugetierkundliche Informationen, 2(10):311-317.

Mockel, R. 1988. Zur Verbreitung, Haufigkeit und Okologie der Haselmaus (*Muscardinus avellanarius*) im Westerzgebirge. Säugetierkundliche Informationen, 2(12):569-588.

Modi, W. S. 1986. Karyotypic differentiation among two sibling species pairs of New World microtine rodents. Journal of Mammalogy, 67:159-165.

Modi, W. S. 1987. Phylogenetic analyses of chromosomal banding patterns among the Nearctic Arvicolidae (Mammalia: Rodentia). Systematic Zoology, 36:109-136.

Modi, W. S., and R. Gamperl. 1989. Chromosomal banding comparisons among American and European red- backed mice, genus *Clethrionomys*. Zeitschrift für Säugetierkunde, 54:141-152.

Modi, W. S., and M. R. Lee. 1984. Systematic implication of chromosomal banding analyses of populations of *Peromyscus truei* (Rodentia, Muridae). Proceedings of the Biological Society of Washington, 97:716-723.

Moeller, W. 1968. Allometrische Analyse der Gürteltierschädel ein Beiträg zur Phylogenie der Dasypodidae Bonaparte, 1838. Zoologische Jahrbücher, Abteilung für Anatomie und Ontogenie der Tiere, 85:411-528.

Mohr, E. 1939. Die Baum- und Ferkelratten- Gattungen *Capromys* Desmarest (sens. ampl.) und *Plagiodontia* Cuvier. Mitteilungen aus dem Hamburgischen Museum und Institut (Hamburg), 48:48-118.

Mohr, E. 1952. Beiträge zur Kenntnis der Mähnenrobben. Zoologische Garten, 19:98-112.

Mohr, E. 1965. Altweltliche Stachelschweine. Ziemsen Verlag, Wittenberg Lutherstadt, 164 pp.

Mohr, E. 1967. Der Blaubock *Hippotragus leucophaeus* (Pallas, 1766). Eine Dokumentation. Mammalia Depicta. Paul Parey, Berlin, 81 pp.

Molinari, J., and P. J. Soriano. 1987. *Sturnira bidens*. Mammalian Species, 276:1-4.

Mones, A., and J. Ojasti. 1986. *Hydrochoerus hydrochaeris*. Mammalian Species, 264:1-7.

Moojen, J. 1948. Speciation in the Brazilian spiny rats (genus *Proechimys*, family Echimyidae). University of Kansas Publications, Museum of Natural History, 1(19):301-406.

Moojen, J. 1952. Os roedores do Brasil. Biblioteca Científica Brasileira, ser. A, 2:1-214.

Moojen, J. 1965. Novo genero de Cricetidae do Brasil central (Glires, Mammalia). Revista Brasileira de Biologia, 25:281-285.

Moore, D. W., and L. L. Janacek. 1990. Genic relationships among North American *Microtus* (Mammalia: Rodentia). Annals of Carnegie Museum, 59:249-259.

Moore, J. C. 1958. New genera of East Indian squirrels. American Museum Novitates, 1914:1-5.

Moore, J. C. 1959. Relationships among the living squirrels of the Sciurinae. Bulletin of the American Museum of Natural History, 118(4):157-206.

Moore, J. C. 1960. Squirrel geography of the Indian subregion. Systematic Zoology, 9:1-17.

Moore, J. C. 1968. Relationships among the living genera of beaked whales with classifications, diagnoses and keys. Fieldiana, Zoology, 53(4):209-298.

Moore, J. C., and G. H. H. Tate. 1965. A study of the diurnal squirrels, Sciurinae, of the Indian and Indochinese subregions. Fieldiana, Zoology, 48:1-351.

Morales, J. C., and M. D. Engstrom. 1989. Morphological variation in the painted spiny pocket mouse, *Liomys pictus* (Family Heteromyidae), from Colima and southern Jalisco, México. Royal Ontario Museum, Life Sciences, Occasional Papers, 38:1-16.

Moreau, R. E., G. H. E. Hopkins, and R. W. Hayman. 1946. The type-localities of some African mammals. Proceedings of the Zoological Society of London, 1946:387-447.

Moreno, S. 1989. Variacion geographica del genero *Eliomys* en la Peninsula Iberica. Doñana, Acta Vertebrata, 16(1):123-141.

Moreno, S., and M. Delibes. 1981. Notes on the garden dormouse (*Eliomys*; Rodentia, Gliridae) of northern Morocco. Säugetierkundliche Mitteilungen, 30:212-215.

Moreno, S., J. Delibes, J. C. Blanco, and A. R. Larramendi. 1986. Sobre la sistemática y biologia de *Eliomys quercinus* en la Cordillera Cantabrica. Doñana, Acta Vertebrata, 13:147-156.

Morgan, G. S. 1985. Taxonomic status and relationships of the Swan Island hutia, *Geocapromys thoracatus* (Mammalia: Rodentia: Capromyidae), and the zoogeography of the Swan Islands vertebrate fauna. Proceedings of the Biological Society of Washington, 98(1):29-46.

Morgan, G. S. 1989. *Geocapromys thoracatus*. Mammalian Species, 341:1-5.

Morgan, G. S., and C. A. Woods. 1986. Extinction and the zoogeography of West Indian land mammals. Biological Journal of the Linnean Society, 28:167-203.

Mori, T., S. Arai, S. Shirashi, and T. A. Uchida. 1991. Ultrastructural observations on spermatozoa of the Soricidae, with special attention to a subfamily revision of the Japanese water shrew *Chimarrogale himalayica*. Journal of the Mammalogical Society of Japan, 16:1-12.

Morlok, W. F. 1978. Nagetiere aus der Türkei. Senckenbergiana Biologica, 59:155-162.

Morrissey, B. L., and W. G. Breed. 1982. Variation in external morphology of the glans penis of Australian native rodents. Australian Journal of Zoology, 30:495-502.

Morse, R. C., and B. P. Glass. 1960. The taxonomic status of *Antrozous bunkeri*. Journal of Mammalogy, 41:10-15.

Moyer, C. A., G. H. Adler, and R. H. Tamarin. 1988. Systematics of New England *Microtus*, with emphasis on *Microtus breweri*. Journal of Mammalogy, 69:782-794.

Muirhead, L. 1819. Mazology. Pp. 393-486, *in* The Edinburgh encyclopaedia, (D. Brewster, ed.). Fourth ed. [pls. 353-358]. William Blackwood, Edinburgh, 13:1-744, pls. 347-371, 1830.

Muizon, C., de. 1982*a*. Phocid phylogeny and dispersal. Annals of the South African Museum, 89:175-213.

Muizon, C., de. 1982*b*. Les relations phylogénétiques des Lutrinae (Mustelidae, Mammalia). Geobios, Mémoire Spécial, 6:259-277.

Müller, S. 1839-1840. Over de Zoogdieren van den Indischen Archipel. Pp. 1-8 [1839], Pp. 9-57+6 unnumbered pages [1840], *in* Verhandelingen over de Natuurlijke Geschiedenis der Nederlandsche overzeesche bezittingen, door de Leden der Natuurkundige Commissie in Indiö en andere Schrijvers (C. J. Temminck, ed.). [V. 3] Zoology. [1839-1845]. J. Luchtmans en C. C. van der Hoek, [not continuously paginated].

Mumford, R. E. 1969. Distribution of the mammals of Indiana. Indiana Academy of Sciences, Indianapolis, 114 pp.

Murariu, D., and S. Torcea. 1984. The occurrence of the species *Spalax istricus* Mehely, 1909 (Rodentia, Spalacidae) in the Romanian Plain. Travaux du Museum d'Histoire Naturelle "Grigore Antipa", 26:245-249.

Murariu, D., A. Lungeanu, L. Gavrila, and C. Stepan. 1985. Preliminary data concerning the study of the karyotype of *Eliomys quercinus* (Linnaeus, 1766) (Mammalia, Gliridae). Travaux Museum d'Histoire Naturelle "Grigore Antipa", 27:325-327.

Murbach, H. 1979. Zur kenntnis von inselpopulationen der waldmaus *Apodemus sylvaticus* (Linnaeus, 1758). Zeitschrift für Zoologische Systematik und Evolutionsforschung, 17:116-139.

Murray, A. 1866. The geographical distribution of mammals. Day and Son, Ltd., London, 420 pp.

Musser, G. G. 1964. Notes on geographic distribution, habitat, and taxonomy of some Mexican mammals. Occasional Papers of the Museum of Zoology, University of Michigan, 636:1-22.

Musser, G. G. 1968. A systematic study of the Mexican and Guatemalan gray squirrel, *Sciurus aureogaster*, F. Cuvier (Rodentia: Sciuridae). Miscellaneous Publications, Museum of Zoology, University of Michigan, 137:1-112.

Musser, G. G. 1969*a*. Notes on *Peromyscus* (Muridae) of Mexico and Central America. American Museum Novitates, 2357:1-23.

Musser, G. G. 1969*b*. Results of the Archbold Expeditions. No. 91. A new genus and species of murid rodent from Celebes, with a discussion of relationships. American Museum Novitates, 2384:1-41.
Musser, G. G. 1969*c*. Results of the Archbold Expeditions. No. 92. Taxonomic notes on *Rattus dollmani* and *Rattus hellwaldi* (Rodentia, Muridae) of Celebes. American Museum Novitates, 2386:1-24.
Musser, G. G. 1970*a*. *Rattus masaretes*: A synonym of *Rattus rattus moluccarius*. Journal of Mammalogy, 51:606-609.
Musser, G. G. 1970*b*. Species-limits of *Rattus brahma*, a murid rodent of Northeastern India and Northern Burma. American Museum Novitates, 2406:27.
Musser, G. G. 1970*c*. Identity of the type-specimens of *Sciurus aureogaster* F. Cuvier and *Sciurus nigriscens* Bennett (Mammalia, Sciurdae). American Museum Novitates, 2438:1-19.
Musser, G. G. 1970*d*. Results of the Archbold Expeditions. No. 93. Reidentification and reallocation of *Mus callitrichus* and allocations of *Rattus maculipilis*, *R. m. jentinki*, and *R. microbullatus* (Rodentia, Muridae). American Museum Novitates, 2440:1-35.
Musser, G. G. 1970*e*. The taxonomic identity of *Mus bocourti* A. Milne Edwards (1874) (Mammalia: Muridae). Mammalia, 34:484-490.
Musser, G. G. 1971*a*. *Peromyscus allophylus* Osgood: A synonym of *Peromyscus gymnotis* Thomas (Rodentia, Muridae). American Museum Novitates, 2453:1-10.
Musser, G. G. 1971*b*. Results of the Archbold Expeditions. No. 94. Taxonomic status of *Rattus tatei* and *Rattus frosti*, two taxa of murid rodents known from middle Celebes. American Museum Novitates, 2454:1-19.
Musser, G. G. 1971*c*. The taxonomic association of *Mus faberi* Jentink with *Rattus xanthurus* (Gray), a species known only from Celebes (Rodentia: Muridae). Zoologische Mededelingen uitgegeven door het Rijksmuseum van Natuurlijke Historie te Leiden, 45:107-118.
Musser, G. G. 1971*d*. The identities and allocations of *Taeromys paraxanthus* and *T. tatei*, two taxa based on compositie holotypes (Rodentia, Muridae). Zoologische Mededelingen uitgegeven door het Rijksmuseum van Natuurlijke Historie te Leiden, 45:127-138.
Musser, G. G. 1971*e*. The taxonomic status of *Rattus tondanus* Sody and notes on the holotypes of *R. beccarii* (Jentink) and *R. thysanurus* Sody (Rodentia: Muridae). Zoologische Mededelingen uitgegeven door het Rijksmuseum van Natuurlijke Historie te Leiden, 45:147-157.
Musser, G. G. 1972. Identities of taxa associated with *Rattus rattus* (Rodentia, Muridae) of Sumba Island, Indonesia. Journal of Mammalogy, 53:861-865.
Musser, G. G. 1973*a*. Notes on additional specimens of *Rattus brahma*. Journal of Mammalogy, 54:267-270.
Musser, G. G. 1973*b*. Zoogeographical significance of the ricefield rat, *Rattus argentiventer*, on Celebes and New Guinea and the identity of *Rattus pesticulus*. American Museum Novitates, 2511:1-30.
Musser, G. G. 1973*c*. Species-limits of *Rattus cremoriventer* and *Rattus langbianis*, murid rodents of Southeast Asia and the Greater Sunda Islands. American Museum Novitates, 2525:1-65.
Musser, G. G. 1977*a*. *Epimys benguetensis*, a composite, and one zoogeographic view of rat and mouse faunas in the Philippines and Celebes. American Museum Novitates, 2624:1-15.
Musser, G. G. 1977*b*. Results of the Archbold Expeditions. No. 100. Notes on the Philippine rat, *Limnomys*, and the identity of *Limnomys picinus*, a composite. American Museum Novitates, 2636:1-14.
Musser, G. G. 1979. Results of the Archbold Expeditions. No. 102. The species of *Chiropodomys*, arboreal mice of Indochina and the Malay Archipelago. Bulletin of the American Museum of Natural History, 162:377-445.
Musser, G. G. 1981*a*. A new genus of arboreal rat from West Java, Indonesia. Zoologische Verhandelingen uitgegeven door het Rijksmuseum van Natuurlijke Historie te Leiden, 189:1-35.
Musser, G. G. 1981*b*. Results of the Archbold Expeditions. No. 105. Notes on systematics of Indo-Malayan murid rodents, and descriptions of new genera and species from Ceylon, Sulawesi, and the Philippines. Bulletin of the American Museum of Natural History, 168:225-334.
Musser, G. G. 1981*c*. The giant rat of Flores and its relatives east of Borneo and Bali. Bulletin of the American Museum of Natural History, 169:67-176.
Musser, G. G. 1982*a*. Results of the Archbold Expeditions. No. 107. A new genus of arboreal rat from Luzon Island in the Philippines. American Museum Novitates, 2730:1-23.
Musser, G. G. 1982*b*. Results of the Archbold Expeditions. No. 108. The definition of *Apomys*, a native rat of the Philippine Islands. American Museum Novitates, 2746:1-43.
Musser, G. G. 1982*c*. Results of the Archbold Expeditions. No. 110. *Crunomys* and the small-bodied shrew rats native to the Philippine Islands and Sulawesi (Celebes). Bulletin of the American Museum of Natural History, 174:1-95.
Musser, G. G. 1982*d*. The Trinil rats. Modern Quaternary Research in Southeast Asia, 7:65-85.
Musser, G. G. 1984. Identities of subfossil rats from caves in southwestern Sulawesi. Modern Quaternary Research in Southeast Asia, 8:61-94.

Musser, G. G. 1986. Sundaic *Rattus*: Definitions of *Rattus baluensis* and *Rattus korinchi*. American Museum Novitates, 2862:1-24.
Musser, G. G. 1987. The mammals of Sulawesi. Pp. 73-93, *in* Biogeographical evolution of the Malay Archipelago (T. C. Whitmore, ed.). Oxford University Press, Oxford, 147 pp.
Musser, G. G. 1990. Sulawesi rodents: Species traits and chromosomes of *Haeromys minahassae* and *Echiothrix leucura* (Muridae: Murinae). American Museum Novitates, 2989:1-18.
Musser, G. G. 1991. Sulawesi rodents: Descriptions of new species of *Bunomys* and *Maxomys* (Muridae, Murinae). American Museum Novitates, 3001:1-41.
Musser, G. G., and Boeadi. 1980. A new genus of murid rodent from the Komodo islands in Nusatenggara, Indonesia. Journal of Mammalogy, 61:395-413.
Musser, G. G., and D. Califia. 1982. Results of the Archbold Expeditions. No. 106. Identities of rats from Pulau Maratua and other islands off East Borneo. American Museum Novitates, 2726:1-30.
Musser, G. G., and S. Chiu. 1979. Notes on taxonomy of *Rattus andersoni* and *R. excelsior*, murids endemic to Western China. Journal of Mammalogy, 60:581-592.
Musser, G. G., and M. Dagosto. 1987. The identity of *Tarsius pumilus*, a pygmy species endemic to the montane mossy forests of central Sulawesi. American Museum Novitates, 2867:1-53.
Musser, G. G., and L. K. Gordon. 1981. A new species of *Crateromys* (Muridae) from the Philippines. Journal of Mammalogy, 62:513-525.
Musser, G. G., and L. R. Heaney. 1985. Philippine *Rattus*: A new species from the Sulu Archipelago. American Museum Novitates, 2818:1-32.
Musser, G. G., and L. R. Heaney. 1992. Philippine rodents: Definitions of *Tarsomys* and *Limnomys* plus a preliminary assessment of phylogenetic patterns among native Philippine murines (Murinae, Muridae). Bulletin of the American Museum of Natural History, 211:1-138.
Musser, G. G., and M. E. Holden. 1991. Sulawesi rodents (Muridae: Murinae): Morphological and geographical boundaries of species in the *Rattus hoffmanni* group and a new species from Pulau Peleng. Pp. 322-413, *in* Contributions to mammalogy in honor of Karl F. Koopman (T. A. Griffiths and D. Klingener, eds.). Bulletin of the American Museum of Natural History, 206:1-432.
Musser, G. G., and C. Newcomb. 1983. Malaysian murids and the giant rat of Sumatra. Bulletin of the American Museum of Natural History, 174:327-598.
Musser, G. G., and C. Newcomb. 1985. Definitions of Indochinese *Rattus losea* and a new species from Vietnam. American Museum Novitates, 1814:1-32.
Musser, G. G., and E. Piik. 1982. A new species of *Hydromys* (Muridae) from western New Guinea (Irian Jaya). Zoologische Mededelingen uitgegeven door het Rijksmuseum van Natuurlijke Historie te Leiden, 56:153-167.
Musser, G. G., and M. M. Williams. 1985. Systematic studies of oryzomyine rodents (Muridae): Definitions of *Oryzomys villosus* and *Oryzomys talamancae*. American Museum Novitates, 2810:1-22.
Musser, G. G., J. T. Marshall, Jr., and Boeadi. 1979. Definition and contents of the Sundaic genus *Maxomys* (Rodentia, Muridae). Journal of Mammalogy, 60:592-606.
Musser, G. G., K. F. Koopman, and D. Califia. 1982*a*. The Sulawesian *Pteropus arquatus* and *P. argentatus* are *Acerodon celebensis*; the Philippine *P. leucotus* is an *Acerodon*. Journal of Mammalogy, 63:319-328.
Musser, G. G., L. K. Gordon, and H. Sommer. 1982*b*. Species-limits in the Philippine murid, *Chrotomys*. Journal of Mammalogy, 63:515-521.
Musser, G. G., L. R. Heaney, and D. S. Rabor. 1985. Philippine rats: A new species of *Crateromys* from Dinagat Island. American Museum Novitates, 2821:1-25.
Musser, G. G., A. van de Weerd, and E. Strasser. 1986. *Paulamys*, a replacement name for *Floresomys* Musser, 1981 (Muridae), and new material of that taxon from Flores, Indonesia. American Museum Novitates, 2850:1-10.
Muul, I., and B. L. Lim. 1971. New locality records for some mammals of West Malaysia. Journal of Mammalogy, 52:430-437.
Muul, I., and K. Thonglongya. 1971. Taxonomic status of *Petinomys morrisi* (Carter) and its relationship to *Petinomys setosus* (Temminck and Schlegel). Journal of Mammalogy, 52:362-369.
Myers, P. 1977. A new phyllotine rodent (genus *Graomys*) from Paraguay. Occasional Papers of the Museum of Zoology, University of Michigan, 676:1-7.
Myers, P. 1982. Origins and affinities of the mammal fauna of Paraguay. Special Publication Series, Pymatuning Laboratory of Ecology, University of Pittsburgh, Pennsylvania, 6:85-93.
Myers, P. 1989. A preliminary revision of the *varius* group of *Akodon*. Pp. 5-54, *in* Advances in Neotropical mammalogy (K. Redford and J. F. Eisenberg, eds.). Sandhill Crane Press, Gainesville, FL, 614 pp.
Myers, P., and M. D. Carleton. 1981. The species of *Oryzomys* (*Oligoryzomys*) in Paraguay and the identity of Azara's "Rat sixieme ou rat a tarse noir". Miscellaneous Publications, Museum of Zoology, University of Michigan, 161:1-41.

Myers, P., and J. L. Patton. 1989*a*. A new species of *Akodon* from the cloud forests of eastern Cochabamba Department, Bolivia (Rodentia: Sigmodontinae). Occasional Papers of the Museum of Zoology, University of Michigan, 720:1-28.

Myers, P., and J. L. Patton. 1989*b*. *Akodon* of Peru—revision of the *fumeus* group (Rodentia: Sigmodontinae). Occasional Papers of the Museum of Zoology, University of Michigan, 721:1-35.

Myers, P., J. L. Patton, and M. F. Smith. 1990. A review of the *boliviensis* group of *Akodon* (Muridae: Sigmodontinae), with emphasis on Peru and Bolivia. Miscellaneous Publications, Museum of Zoology, University of Michigan, 177:1-104.

Nachman, M. W., and P. Myers. 1989. Exceptional chromosomal mutations in a rodent population are not strongly underdominant. Proceedings of the National Academy of Sciences (Washington, D. C.), 86:6666-6670.

Nadachowski, A. 1990*a*. Comments on variation, evolution and phylogeny of *Chionomys* (Arvicolinae). Pp. 353-368, *in* International symposium evolution, phylogeny and biostratigraphy of arvicolids (Rodentia, Mammalia) (O. Fejfar and W.-D. Heinrich, eds.). Geological Survey, Prague, 448 pp.

Nadachowski, A. 1990*b*. On the taxonomic status of *Chionomys* Miller, 1908 (Rodentia: Mammalia) from southern Anatolia (Turkey). Acta Zoologica Cracoviensia, 33:79-89.

Nadachowski, A. 1991. Systematics, geographic variation, and evolution of snow voles (*Chionomys*) based on dental characters. Acta Theriologica, 36:1-45.

Nadachowski, A., B. Rzebik-Kowalska, and A.-H. Kadhim. 1978. The first record of *Eliomys melanurus* Wagner, 1840 (Gliridae, Mammalia), from Iraq. Säugetierkundliche Mitteilungen, 26:206- 207.

Nader, I. A. 1978. Kangaroo rats: intraspecific variation in *Dipodomys spectabilis* Merriam and *Dipodomys deserti* Stephens. Illinois Biological Monographs, 49:1-116.

Nader, I. A., and D. Kock. 1979 [1980]. First record of *Tadarida nigeriae* (Thomas, 1913) from the Arabian peninsula (Mammalia: Chiroptera: Molossidae). Senckenbergiana Biologica, 60:131-135.

Nader, I. A., and D. Kock. 1990. *Eptesicus* (*Eptesicus*) *bottae* (Peters 1869) in Saudi Arabia with notes on its subspecies and distribution (Mammalia: Chiroptera: Vespertilionidae). Senckenbergiana Biologica, 70:1-13.

Nader, I. A., D. Kock, and A.-K. Al-Khalili. 1983. *Eliomys melanurus* (Wagner 1839) and *Praomys fumatus* (Peters 1878) from the Kingdom of Saudi Arabia. Senckenbergiana Biologica, 63(5-6):313-324.

Nadler, C. F., and R. S. Hoffmann. 1970. Chromosomes of some Asian and South American squirrels (Rodentia, Sciuridae). Experientia, 26:1383-1386.

Nadler, C. F., and R. S. Hoffmann. 1974. Chromosomes of the African ground squirrel, *Xerus rutilus*. Experientia, 30:889-890.

Nadler, C. F., and R. S. Hoffmann. 1977. Patterns of evolution and migration in the arctic ground squirrel, *Spermophilus parryii* (Richardson). Canadian Journal of Zoology, 55:748-758.

Nadler, C. F., and D. M. Lay. 1967. Chromosomes of some species of *Meriones* (Mammalia: Rodentia). Zeitschrift für Säugetierkunde, 32:285-291.

Nadler, C. F., D. M. Lay, and J. D. Hassinger. 1969. Chromosomes of three asian mammals: *Meriones meridianus* (Rodentia: Gerbillinae), *Spermophilopsis leptodactylus* (Rodentia: Sciuridae), *Ochotona rufescens* (Lagomorpha: Ochotonidae). Experientia, 25:774-775.

Nadler, C. F., R. S. Hoffmann, and K. Greer. 1971*a*. Chromosomal divergence during evolution of ground squirrel populations. Systematic Zoology, 20:298-305.

Nadler, C. F., D. M. Lay, and J. D. Hassinger. 1971*b*. Cytogenetic analyses of wild sheep populations in northern Iraq. Cytogenetics, 10:137-152.

Nadler, C. F., K. V. Korobitsina, R. S. Hoffmann, and N. N. Vorontsov. 1973. Cytogenetic differentiation, geographic distribution and domestication in Palearctic sheep (*Ovis*). Zeitschrift für Säugetierkunde, 38:109-125.

Nadler, C. F., R. I. Sukernik, R. S. Hoffmann, N. N. Vorontsov, C. F. Nadler, Jr., and I. I. Fomichova. 1974. Evolution in ground squirrels. I. Transferrins in Holarctic populations of *Spermophilus*. Comparative Biochemistry and Physiology, 47A:663-681.

Nadler, C. F., E. A. Lyapunova, R. S. Hoffmann, N. N. Vorontsov, and N. A. Malygina. 1975*a*. Chromosomal evolution in Holarctic ground squirrels (*Spermophilus*). 1. Giemsa-band homologies in *Spermophilus columbianus* and *S. undulatus*. Zeitschrift für Säugetierkunde, 40:1-7.

Nadler, C. F., R. S. Hoffmann, and M. E. Hight. 1975*b*. Chromosomes of three species of Asian tree squirrels, *Callosciurus* (Rodentia: Sciuridae). Experientia, 31:166-167.

Nadler, C. F., V. R. Rausch, E. A. Lyapunova, R. S. Hoffmann, and N. N. Vorontsov. 1976. Chromosomal banding patterns of the Holarctic rodents, *Clethrionomys rutilus* and *Microtus oeconomus*. Zeitschrift für Säugetierkunde, 41:137-146.

Nadler, C. F., R. S. Hoffmann, J. H. Honacki, and D. Pozin. 1977. Chromosomal evolution in chipmunks, with special emphasis on A and B karyotypes of the subgenus *Neotamias*. American Midland Naturalist, 98:343-353.

Nadler, C. F., N. M. Zhurkevich, R. S. Hoffmann, A. I. Kozlovskii, L. Deutsch, and C. F. Nadler, Jr. 1978. Biochemical relationships of the Holarctic vole genera (*Clethrionomys*, *Microtus*, and *Arvicola* (Rodentia: Arvicolinae)). Canadian Journal of Zoology, 56:1564-1575.

Nadler, C. F., R. S. Hoffmann, N. N. Vorontsov, J. W. Koeppl, L. Deutsch, and R. I. Sukernik. 1982. Evolution in ground squirrels. II. Biochemical comparisons in Holarctic populations of *Spermophilus*. Zeitschrift für Säugetierkunde, 47:198-215.

Nadler, C. F., E. A. Lyapunova, R. S. Hoffmann, N. N. Vorontsov, L. L. Shaitarova, and Y. M. Borisov. 1984. Chromosomal evolution in Holarctic ground squirrels (*Spermophilus*). II. Giemsa-band homologies of chromosomes and the tempo of evolution. Zeitschrift für Säugetierkunde, 49:78-90.

Nagorsen, D. 1985. *Kogia simus*. Mammalian Species, 239:1-6.

Nagorsen, D. W. 1987. *Marmota vancouverensis*. Mammalian Species, 270:1-5.

Nagorsen, D., and J. R. Tamsitt. 1981. Systematics of *Anoura cultrata*, *A. brevirostrum*, and *A. werckleae*. Journal of Mammalogy, 62:82-100.

Nakatsu, A. 1982. Notes on the pattern of third upper molar (M3) in *Clethrionomys rutilus mikado* (Thomas). Journal of the Mammalogical Society of Japan, 9:104-106.

Nanda, I., and R. Raman. 1981. Cytological similarity between the heterochromatin of the large X and Y chromosomes of the soft- furred field rat, *Millardia meltada* (family: Muridae). Cytogenetics and Cell Genetics, 30:77-82.

Napier, J. R., and P. H. Napier. 1967. A handbook of living Primates. Academic Press, London, 456 pp.

Napier, P. H. 1976. Catologue of primates in the British Museum (Natural History), Part 1: families Callitrichidae and Cebidae. British Museum (Natural History), London, 121 pp.

Napier, P. H. 1985. Catalogue of Primates in the British Museum (Natural History) and elsewhere in the British Isles. Part 3: Family Cercopithecidae, subfamily Colobinae. Publications of the British Museum of Natural History, 894:1, 3-111.

Napier, P. H., and C. P. Groves. 1983. *Simia fascicularis* Raffles, 1821 (Mammalia, Primates): request for the suppression under the plenary powers of *Simia aygula* Linnaeus, 1758, a senior synonym. Bulletin of Zoological Nomenclature, 40:117-118.

Nascetti, G., S. Lovari, P. Lanfranchi, C. Berducou, S. Mattiucci, L. Rossi, and L. Bullini. 1985. Revision of *Rupicapra* genus. III. Electrophoretic studies demonstrating species distinction of chamois populations of the Alps from those of the Apennines and Pyrenees. Pp.56-62, *in* The biology and management of mountain ungulates (S. Lovari, ed.). Croom Helm, London, 271 pp.

Nash, D. J., and R. N. Seaman. 1977. *Sciurus aberti*. Mammalian Species, 80:1-5.

Nash, L. T., S. K. Bearder, and T. R. Olson. 1989. Synopsis of *Galago* species characteristics. International Journal of Primatology, 10:57-79.

Natori, M. 1988. A cladistic analysis of interspecific relationships of *Saguinus*. Primates, 29:263-276.

Naumov, N. P., and V. S. Lobachev. 1975. Ecology of desert rodents of the U.S.S.R. (Jerboas and Gerbils). Pp. 465-598, *in* Rodents in desert environments (I. Prakash and P. K. Gosh, eds.). Monographiae Biologicae, 28:1-624.

Naumova, E. I., R. A. Valensiya-Leon, and L. I. Lenets. 1990. [Differences in morphology of the stomach in species-twins: the common and eastern European voles]. Doklady Akademii Nauk SSSR, 310:1016-1020 (in Russian).

Navarro L., D., and D. E. Wilson. 1982. *Vampyrum spectrum*. Mammalian Species, 184:1-4.

Neal, B. R. 1983. The breeding pattern of two species of spiny mice, *Acomys percivali* and *A. wilsoni* (Muridae: Rodentia), in central Kenya. Mammalia, 47:311-321.

Neas, J. F., and R. S. Hoffmann. 1987. *Budorcas taxicolor*. Mammalian Species, 277:1-7.

Nellis, D. W. 1989. *Herpestes auropunctatus*. Mammalian Species, 342:1-6.

Nelson, E. W. 1898. What is *Sciurus variegatus* Erxleben? Science, N.S., 8:897-898.

Nelson, E. W. 1899*a*. Mammals of the Tres Marias Islands. North American Fauna, 14:15-19.

Nelson, E. W. 1899*b*. Revision of the squirrels of Mexico and Central America. Proceedings of the Washington Academy of Sciences, 1:15-106.

Nelson, E. W. 1909. The rabbits of North America. North American Fauna, 29:1-314.

Nelson, E. W., and E. A. Goldman. 1929. Four new pocket gophers of the genus *Heterogeomys* from Mexico. Proceedings of the Biological Society of Washington, 42:147-152.

Nelson, E. W., and E. A. Goldman. 1930. The status of *Orthogeomys cuniculus* Elliot. Journal of Mammalogy, 11:317.

Nelson, E. W., and E. A. Goldman. 1933. Revision of the jaguars. Journal of Mammalogy, 14(3):221-240.

Nelson, E. W., and E. A. Goldman. 1934. Revision of the pocket gophers of the genus *Cratogeomys*. Proceedings of the Biological Society of Washington, 47:135-154.

Nelson, K., R. J. Baker, H. S. Shellhammer, and R. K. Chesser. 1984. Test of alternative hypotheses concerning the origin of *Reithrodontomys raviventris*: genetic analysis. Journal of Mammalogy, 65:668-673.

Nelson, K., R. J. Baker, and R. L. Honeycutt. 1987. Mitochondrial DNA and protein differentiation between hybridizing cytotypes of the white-footed mouse, *Peromyscus leucopus*. Evolution, 41:864-872.

Nesin, V. A., and A. F. Skorik. 1989. [First find of the *Dinaromys* vole (Rodentia, Microtinae) in the USSR]. Vestnik Zoologii, 5:14-17 (in Russian).

Nesov, L. A., and A. A. Gureev. 1981. [A jaw of a most ancient shrew from the Upper Cretaceous of the Kizylkum desert]. Doklady Akademii Nauk SSSR, 257(4):1002-1004 (in Russian).

Nevo, E. 1982. Genetic structure and differentiation during speciation in fossorial gerbil rodents. Mammalia, 46:523-530.

Nevo, E. 1991. Evolutionary theory and processes of active speciation and adaptive radiation in subterranean mole rats, *Spalax ehrenbergi* superspecies, in Israel. Evolutionary Biology, 25:1-125.

Nevo, E., and E. Amir. 1961. Biological observations on the forest dormouse *Dryomys nitedula* Pallas, in Israel (Rodentia, Muscardinidae). The Bulletin of the Research Council of Israel, 9B(4):200-201.

Nevo, E., and E. Amir. 1964. Geographic variation in reproduction and hibernation patterns of the forest dormouse. Journal of Mammalogy, 45(1):69-87.

Nevo, E., R. Ben-Shlomo, A. Beiles, J. U. M. Jarvis, and G. C. Hickman. 1987. Allozyme differentiation and systematics of the endemic subterranean mole rats of South Africa. Biochemical Systematics and Ecology, 15:489-502.

Nicoll, M. E., and G. B. Rathbun. 1990. African Insectivora and elephant-shrews. An action plan for their conservation. I. U. C. N., Gland, Switzerland, 53 pp.

Niemitz, C., A. Nietsch, S. Water, and Y. Rumpler. 1991. *Tarsius dianae*: a new primate species from Central Sulawesi (Indonesia). Folia Primatologia, 56:105-116.

Niethammer, J. 1959. Die nordafrikanischen Unterarten des Gartenschläfers (*Eliomys quercinus*). Zeitschrift für Säugetierkunde, 24:35-45.

Niethammer, J. 1960. Über die Säugetiere der Niederen Tauern. Mitteilungen aus dem Zoologische Museum in Berlin, 36(2):407-443.

Niethammer, J. 1964. Contribution a la connaissance des mammiferes terrestres de l'ile Indefatigable (=Santa Cruz), Galapagos. Résultats de l'expedition Allemagne aux Galapagos 1962/63. Mammalia, 28:593-606.

Niethammer, J. 1970. Die Wühlmause (Microtinae) Afghanistans. Bonner Zoologische Beiträge, 21:1-24.

Niethammer, J. 1972. Die Zahl der Mammae bei *Pitymys* und bei den Microtinen. Bonner Zoologische Beiträge, 23:49-60.

Niethammer, J. 1973. Zur Kenntnis der Igel (Erinaceidae) Afghanistans. Zeitschrift für Säugetierkunde, 38:271-276.

Niethammer, J. 1975. Zur Taxonomie und Ausbreitungsgeschichte der Hausratte (*Rattus rattus*). Zoologischer Anzeiger, 194:405-415.

Niethammer, J. 1977. Versuch der Rekonstruktion der phylogenetischen Beziehungen zwischen einigen zentralasiatischen Muriden. Bonner Zoologische Beiträge, 28:236-247.

Niethammer, J. 1978*a*. *Apodemus mystacinus* (Danford and Alston, 1877)— Felsenmaus. Pp. 306-324, *in* Handbuch der Säugetiere Europas (J. Niethammer and F. Krapp, eds.). Akademische Verlagsgesellschaft (Wiesbaden), 1:1-476.

Niethammer, J. 1978*b*. *Apodemus flavicollis* (Melchior, 1834)— Gelbhalsmaus. Pp. 325-336, *in* Handbuch der Säugetiere Europas (J. Niethammer and F. Krapp, eds.). Akademische Verlagsgesellschaft (Wiesbaden), 1:1-476.

Niethammer, J. 1978*c*. *Apodemus sylvaticus* (Linnaeus, 1758)—Waldmaus. Pp. 337-358, *in* Handbuch der Säugetiere Europas (J. Niethammer and F. Krapp, eds.). Akademische Verlagsgesellschaft (Wiesbaden), 1:1-476.

Niethammer, J. 1982*a*. Familie Cricetidae Rochebrune, 1881—Hamster. Pp. 1-6, *in* Handbuch der Säugetiere Europas (J. Niethammer and F. Krapp, eds.). Akademische Verlagsgesellschaft (Wiesbaden), 2/I:1-649.

Niethammer, J. 1982*b*. *Cricetus cricetus* (Linnaeus, 1758)—Hamster (Feldhamster). Pp. 1-28, *in* Handbuch der Säugetiere Europas (J. Niethammer and F. Krapp, eds.). Akademische Verlagsgesellschaft (Wiesbaden), 2/I:1-649.

Niethammer, J. 1982*c*. *Mesocricetus newtoni* (Nehring, 1898)—Rumänischer Goldhamster. Pp. 29-38, *in* Handbuch der Säugetiere Europas (J. Niethammer and F. Krapp, eds.). Akademische Verlagsgesellschaft (Wiesbaden), 2/I:1-649.

Niethammer, J. 1982*d*. *Cricetulus migratorius* (Pallas, 1773)— Zwerghamster. Pp. 39-50, *in* Handbuch der Säugetiere Europas (J. Niethammer and F. Krapp, eds.). Akademische Verlagsgesellschaft (Wiesbaden), 2/I:1-649.

Niethammer, J. 1982*e*. *Microtus guentheri* Danford et Alston, 1880—Levante- Wühlmaus. Pp. 331-339, *in* Handbuch der Säugetiere Europas (J. Niethammer and F. Krapp, eds.). Akademische Verlagsgesellschaft (Wiesbaden), 2/I:1-649.

Niethammer, J. 1982*f*. *Microtus cabrerae* Thomas, 1906—Cabreramaus. Pp. 340-348, *in* Handbuch der Säugetiere Europas (J. Niethammer and F. Krapp, eds.). Akademische Verlagsgesellschaft (Wiesbaden), 2/I:1-649.

Niethammer, J. 1982*g*. *Microtus subterraneus* (de Selys-Longchamps, 1836)— Kurzohrmaus. Pp. 397-418, *in* Handbuch der Säugetiere Europas (J. Niethammer and F. Krapp, eds.). Akademische Verlagsgesellschaft (Wiesbaden), 2/I:1-649.

Niethammer, J. 1982*h*. *Microtus felteni* (Malec und Storch, 1963). Pp. 438-441, *in* Handbuch der Säugetiere Europas (J. Niethammer and F. Krapp, eds.). Akademische Verlagsgesellschaft (Wiesbaden), 2/I:1-649.

Niethammer, J. 1982*i*. *Microtus duodecimcostatus* (de Selys-Longchamps, 1839)— Mittelmeer-Kleinwühlmaus. Pp. 463-475, *in* Handbuch der Säugetiere Europas (J. Niethammer and F. Krapp, eds.). Akademische Verlagsgesellschaft (Wiesbaden), 2/I:1-649.

Niethammer, J. 1982*j*. *Microtus lusitanicus* (Gerbe, 1879)—Iberien- wühlmaus. Pp. 476-484, *in* Handbuch der Säugetiere Europas (J. Niethammer and F. Krapp, eds.). Akademische Verlagsgesellschaft (Wiesbaden), 2/I:1-649.

Niethammer, J. 1982*k*. *Microtus thomasi* Barrett-Hamilton, 1903—Balkan-Kurzohrmaus. Pp. 485-490, *in* Handbuch der Säugetiere Europas (J. Niethammer and F. Krapp, eds.). Akademische Verlagsgesellschaft (Wiesbaden), 2/I:1-649.

Niethammer, J. 1982*l*. *Microtus tatricus* (Kratochvil, 1952)—Tatra-Wühlmaus. Pp. 491-496, *in* Handbuch der Säugetiere Europas (J. Niethammer and F. Krapp, eds.). Akademische Verlagsgesellschaft (Wiesbaden), 2/I:1-649.

Niethammer, J. 1987*a*. Das Streifenwiesel (*Poecilictis libyca*) in Sudan und seine Gesamtverbreitung. Bonner Zoologische Beiträge, 38(3):173-182.

Niethammer, J. 1987*b*. Uber griechische Nager im Museum A. Koenig in Bonn. Annalen des Naturhistorischen Museums in Wien, Ser. B, 88/89:245-256.

Niethammer, J., and H. Henttonen. 1982. *Myopus schisticolor* (Lilljeborg, 1844)—Waldlemming. Pp. 70-86, *in* Handbuch der Säugetiere Europas (J. Niethammer and F. Krapp, eds.), vol. 2/I. Akademische Verlagsgesellschaft (Wiesbaden), 649 pp.

Niethammer, J., and F. Krapp, (eds.). 1982*a*. Handbuch der Säugetiere Europas, vol. 2/I. Akademische Verlagsgesellschaft (Wiesbaden), 649 pp.

Niethammer, J., and F. Krapp. 1982*b*. *Microtis arvalis* (Pallas, 1779)—Feldmaus. Pp. 284-318, *in* Handbuch der Säugetiere Europas (J. Niethammer and F. Krapp, eds.), vol. 2/I. Akademische Verlagsgesellschaft (Wiesbaden), 649 pp.

Niethammer, J., and F. Krapp, (eds.). 1990. Handbuch der Säugetiere Europas, 3/I. Aula-Verlag, Wiesbaden, 524 pp.

Niethammer, J., and J. Martens. 1975. Die Gattungen *Rattus* und *Maxomys* in Afghanistan und Nepal. Zeitschrift für Säugetierkunde, 40:325-355.

Nikolaeva, A. I. 1982. [Adaptive variation in the masticatory surface of molar teeth in *Arvicola terrestris*]. Zoologicheskii Zhurnal, 61:1565-1575 (in Russian).

Nikolaus, G., and R. J. Dowsett. 1989. Small mammals collected in the Gotel Mts. and on the Mambilla Plateau, eastern Nigeria. Tauraco Research Reports, 1:42-47.

Nikol'skii, A. A. 1974. Geograficheskaya izmenchivost' ritmicheskoi organizatsii zvukovo signala surkov gruppy *bobac* (Rodentia, Sciuridae) [Geographic variation in rhythmic organization of sound signal in marmots of the *bobac* group...]. Zoologicheskii Zhurnal, 53:436-444 (in Russian).

Nikol'skii, A. A. 1984. K voprosu o granitse arealov bol'shovo (*Citellus major*) i krasnoshchekovo (*C. erythrogenys*) suslikov v severnom Kazakhstane [On the question of the range boundary between large...and red-cheeked...susliks in northern Kazakhstan]. Zoologicheskii Zhurnal, 63:256-262 (in Russian).

Nikol'skii, A. A., and D. Wallschläger. 1982. On the specific status of holarctic long-tailed squirrels; a bioacoustical study. Experientia, 38:808-809.

Nikol'skii, A. A., I. Yu. Yanina, M. V. Rutovskaya, and N. A. Formozov. 1983. Izmenchivost' zvukovovo signala stepnovo i serovo surkov (*Marmota bobac, M. baibacina;* Sciuridae, Rodentia) v zone vtorichnovo kontakta [Variation in sound signal of steppe and gray marmots...in a zone of secondary contact]. Zoologicheskii Zhurnal, 62:1258-1266 (in Russian).

Nikol'skii, A. A., N. A. Formozov, V. N. Vasil'ev, and G. G. Boeskorov. 1991. Geograficheskaya izmenchivost' zvukovo signala chernoshapochnovo surka, *Marmota camtschatica* (Rodentia, Sciuridae) [Geographic variation in sound signals of the black-capped marmot, *Marmota camtschatica* (Rodentia, Sciuridae)]. Zoologicheskii Zhurnal, 70(2):155-159 (in Russian).

Nishioka, Y., and E. Lamothe. 1987. The *Mus musculus musculus* Y chromosome predominates in Asian house mice. Genetical Research (Cambridge), 50:195-198.

Nitikman, L. Z. 1985. *Sciurus granatensis*. Mammalian Species, 246:1-8.

Novikov, G. A. 1939. Evropeiskaya norka [European mink]. Leningradskovo Gosudarstvennovo Universiteta, Leningrad, 180 pp. (in Russian).

Novikov, G. A. 1956. Khishchye mlekopitayushchie fauna SSSR [Carnivorous mammals of the fauna of the USSR]. Akademiya Nauk SSSR, Moscow-Leningrad, 293 pp. (in Russian).

Nowak, R. M. 1979. North American Quaternary *Canis*. University of Kansas, Museum of Natural History, Monograph, 6:1-154.

Nowak, R. M. 1991. Walker's Mammals of the World. Fifth ed. Johns Hopkins University Press, Baltimore, 1:1-642; 2:643-1629.

Nowak, R. M., and J. L. Paradiso. 1983. Walker's Mammals of the World. Fouth Edition. Johns Hopkins University Press, Baltimore, 1:1-568, 2:569-1362.

Oaks, E. C., P. J. Young, G. L. Kirkland, Jr., and D. F. Schmidt. 1987. *Spermophilus variegatus*. Mammalian Species, 272:1-8.

Oates, J. F., and T. F. Trocc. 1983. Taxonomy and phylogeny of black-and-white colobus monkeys: inferences from an analysis of loud call variation. Folia Primatologia, 40:83-113.

O'Brien, S. J., W. G. Nash, D. E. Wildt, M. E. Bush, and R. E. Benveniste. 1985. A molecular solution to the riddle of the gaint panda's phylogeny. Nature, 317:140-144.

Ochoa G., J., H. Castellanos, and C. Ibanez. 1988. Records of bats and rodents from Venezuela. Mammalia, 52:175-180.

O'Connell, M. A. 1983. *Marmosa robinsoni*. Mammalian Species, 203:1-6.

O'Conner, T. P. 1986. The garden dormouse *Eliomys quercinus* from Roman York. Journal of Zoology (London), 210A(4):620-622.

Oda, S., J. Kitch, H. Ota, and G. Isomura. 1985. *Suncus murinus* - Biology of the laboratory shrew. Japan Scientific Societies Press, Tokyo.

O'Farrell, M. J., and A. R. Blaustein. 1974*a*. *Microdipodops megacephalus*. Mammalian Species, 46:1-3.

O'Farrell, M. J., and A. R. Blaustein. 1974*b*. *Microdipodops pallidus*. Mammalian Species, 47:1-2.

O'Farrell, M. J., and E. H. Studier. 1980. *Myotis thysanodes*. Mammalian Species, 137:1-5.

O'Gara, B. W. 1978. *Antilocapra americana*. Mammalian Species, 90:1-7.

O'Gara, B. W., and G. Matson. 1975. Growth and casting of horns by pronghorns and exfoliation of horns by bovids. Journal of Mammalogy, 56:829-846.

Ogilby, W. 1835. Descriptions of Mammalia and birds from the Gambia. Proceedings of the Zoological Society of London, 1835:97-105.

Ognev, S. I. 1921. Materialy dlya sistematiki nasekomoyadnykh mlekopitayushchikh Rossi [Contribution á la classification des mammifères insectivores de Russie]. Ezhegodnikh Zoologicheskovo Muzeya, Akadimii Nauk [Annuaire du Musée de l'Académie des Sciences de St. Petersbourg], 22:311-350 (in Russian and English).

Ognev, S. I. 1927. Zur Frage über die systematische Stellung einiger Vertreter von *Paraechinus* Trouessart. Zoologischer Anzeiger, 69:209-218.

Ognev, S. I. 1928. Zveri vostochnoi Evropy i severnoi Azii: Nasekomoyadnye i letychie myshi [Mammals of eastern Europe and northern Asia: Insectivora and Chiroptera]. Glavnauka, Moscow, 1:1-631 (in Russian).

Ognev, S. I. 1931. Zveri vostochnoi Evropy i severnoi Azii: Khishchnye mlekopitayushchie [Mammals of eastern Europe and northern Asia: Carnivorous mammals]. Glavnauka, Moscow, 2:1-776 (in Russian).

Ognev, S. I. 1935. Zveri SSSR i prilezhashchikh stran. Khishchnyei i lastonogie (Zveri vostochnoi Evropy i severnoi Azii) [Mammals of the USSR and adjacent countries: Carnivora and Pinnipedia (Mammals of eastern Europe and northern Asia)]. Glavpushnina NKVT, Moscow, 3:1-752 (in Russian).

Ognev, S. I. 1940. Zveri SSSR i prilezhashchikh stran: Gryzuny. (Zveri vostochnoi Evropy i severnoi Azii) [Mammals of the USSR and adjacent countries: Rodents (Mammals of eastern Europe and northern Asia)]. Akademiya Nauk SSSR, 4:1-615 (in Russian).

Ognev, S. I. 1947. Zveri SSSR i prilezhashchikh stran: Gryzuny (prodolzhenie). (Zveri vostochnoi Evropy i severnoi Azii) [Mammals of the USSR and adjacent countries: Rodents (continued). (Mammals of eastern Europe and northern Asia)]. Akademiya Nauk SSSR, 5:1-809 (in Russian).

Ognev, S. I. 1948. Zveri SSSR i prilezhashchikh stran: Gryzuny (prodolzhenie). (Zveri vostochnoi Evropy i severnoi Azii) [Mammals of the USSR and adjacent countries: Rodents (continued). (Mammals of eastern Europe and northern Asia)]. Akademiya Nauk SSSR, 6:1-559 (in Russian).

Ognev, S. I. 1950. Zveri SSSR i prilezhashchikh stran: Gryzuny (prodolzhenie). (Zveri vostochnoi Evropy i severnoi Azii) [Mammals of the USSR and adjacent countries: Rodents (continued). (Mammals of eastern Europe and northern Asia)]. Akademiya Nauk SSSR, 7:1-706+15 maps (in Russian).

Ognev, S. I. 1962*a*. Mammals of eastern Europe and northern Asia: Insectivora and Chiroptera [A translation of S. I. Ognev, 1928, Zveri vostochnoi Evropy i severnoi Azii: Nasekomoyadnye i letychie myshi]. Israel Program for Scientific Translations, Jerusalem, 1:1-487+I-XV.

Ognev, S. I. 1962*b*. Mammals of eastern Europe and northern Asia: Carnivora (Fissipedia) [A translation of S. I. Ognev, 1931, Zveri vostochnoi Evropy i severnoi Azii: Khishchnye mlekopitayushchie]. Israel Program for Scientific Translations, Jerusalem, 2:1-590+I-XV.

Ognev, S. I. 1962*c*. Mammals of the USSR and adjacent countries: Fissipedia and Pinnipedia [A translation of S. I. Ognev, 1935, Zveri SSSR i prilezhashchikh stran. Khishchnyei i lastonogie (Zveri vostochnoi Evropy i severnoi Azii)]. Israel Program for Scientific Translations, Jerusalem, 3:1-641+I-XV.

Ognev, S. I. 1963*a*. Mammals of the USSR and adjacent countries: Rodents (continued). (Mammals of eastern Europe and northern Asia) [A translation of S. I. Ognev, 1947, Zveri SSSR i prilezhashchikh stran: Gryzuny (prodolzhenie). (Zveri vostochnoi Evropy i severnoi Azii)]. Israel Program for Scientific Translations, Jerusalem, 5:1-662.

Ognev, S. I. 1963*b*. Mammals of the USSR and adjacent countries: Rodents (continued). (Mammals of eastern Europe and northern Asia) [A translation of S. I. Ognev, 1948, Zveri SSSR i prilezhashchikh stran: Gryzuny (prodolzhenie). (Zveri vostochnoi Evropy i severnoi Azii)]. Israel Program for Scientific Translations, Jerusalem, 6:1-508.

Ognev, S. I. 1964. Mammals of the USSR and adjacent countries: Rodents (continued). (Mammals of eastern Europe and northern Asia) [A translation of S. I. Ognev, 1950, Zveri SSSR i prilezhashchikh stran: Gryzuny (prodolzhenie). (Zveri vostochnoi Evropy i severnoi Azii)]. Israel Program for Scientific Translations, Jerusalem, 7:1-626.

Ognev, S. I. 1966. Mammals of the USSR and adjacent countries: Rodents (Mammals of eastern Europe and northern Asia) [A translation of S. I. Ognev, 1940, Zveri SSSR i prilezhashchikh stran: Gryzuny. (Zveri vostochnoi Evropy i severnoi Asii)]. Israel Program for Scientific Translations, Jerusalem, 4:1-429+61 tables.

Ohno, S., J. Jainchill, and C. Stenius. 1963. The creeping vole (*Microtus oregoni*) as a gonosomic mosaic I. The OY/XY constitution of the male. Cytogenetics, 2:232-239.

Ojasti, J. 1972. Revisión preliminar de los picures o agutís de Venezuela (Rodentia: Dasyproctidae). Memorias Sociedad Cíencias Naturales La Salle, 32:159-204.

Ojasti, J., and O. J. Linares. 1971. Adiciones a la fauna de murcielagos de Venezuela con notas sobre las especies del género *Diclidurus* (Chiroptera). Acta Biologica Venezuelica, 7:421-441.

Oken, L. 1815-1816. Lehrbuch der Naturgeschichte. Zoologie. [vol. 3 'Zoologie.' published 1815-1816; ICZN Opinion 417 - vol. 3 rejected for nomenclatural purposes]. August Schmid und Comp., Jena, 3:1-1270.

Okhotina, M. V. 1977. Palaearctic shrew of the subgenus *Otisorex*: biotopic preference, population number, taxonomic revision and distribution history. Acta Theriologica, 22:191-206.

Olds, N., and S. Anderson. 1987. Notes on Bolivian mammals. 2. Taxonomy and distribution of rice rats of the subgenus *Oligoryzomys*. Fieldiana, Zoology, New Series, 39:261-281.

Olds, N., and S. Anderson. 1989 [1990]. A diagnosis of the tribe Phyllotini (Rodentia, Muridae). Pp. 55-74, *in* Advances in Neotropical mammalogy (K. Redford and J. F. Eisenberg, eds.). Sandhill Crane Press, Gainesville, FL, 614 pp.

Olds, N., and J. Shoshani. 1982. *Procavia capensis*. Mammalian Species, 171:1-7.

Olds, N., S. Anderson, and T. L. Yates. 1987. Notes on Bolivian mammals 3: A revised diagnosis of *Andalgalomys* (Rodentia, Muridae) and the description of a new subspecies. American Museum Novitates, 2890:1-17.

Olert, J., F. Dieterlen, and H. Rupp. 1978. Eine neue Muriden-Art aus Südäthiopien. Zeitschrift für Zoologische Systematik und Evolutionsforschung, 16(4):297-308.

Oliver, W. L. R., and I. B. Santos. 1991. Threatened endemic mammals of the Atlantic forest region of South-east Brazil. Special Scientific Report of Jersey Wildlife Preservation Trust, 4:1-125.

Olrog, C. C., and M. M. Lucero. 1981. Guia de los mamiferos Argentinos. Ministerio de Cultura y Educacion, San Miguel de Tucuman, 151 pp.

Olsen, S. J. 1990. Fossil ancestry of the yak, its cultural significance and domestication in Tibet. Proceedings of the Academy of Natural Sciences of Philadelphia, 142:73-100.

Ondrias, J. C. 1966. The taxonomy and geographical distribution of the rodents of Greece. Säugetierkundliche Mitteilungen, 14, sonderheft, 1:1-136.

Orlov, V. N. 1969. Khromosomnye nabory ezhei vostochnoi Evropy [Chromosomal complements of hedgehogs in eastern Europe]. Pp. 6-7, *in* Materialy ko II Vsesoyuznykh Teriologicheskich Soveshchen po mlekopitayushchim [Materials in II All-Union Theriological Conference about mammals]. Novosibirsk.

Orlov, V. N., and N. Sh. Bulatova. 1983. Sravnitel'naya tsitogenetika i karyosistematika mlekopitayushchikh [Comparative cytogenetics and karyosystematics of mammals]. Nauka, Moscow, 405 pp. (in Russian).
Orlov, V. N., and N. Davaa. 1975. O systematicheskom polozhenii Alashanskovo suslika *Citellus alashanicus* Buch. (Sciuridae, Rodentia) [On the systematic position of the Alashan ground squirrel...]. Pp. 8-9, *in* Sistematika i tsitogenetica mlekopitayushchikh [Systematics and cytogenetics of mammals] (V. N. Orlov, ed.). Nauka, Moscow, 60 pp. (in Russian).
Orlov, V. N., and Yu. M. Kovalskaya. 1978. [*Microtus mujanensis* sp. n. from the Vitim River Basin]. Zoologicheskii Zhurnal, 57:1224-1232 (in Russian).
Orlov, V. N., and V. M. Malygin. 1988. [A new species of hamster—*Cricetulus sokolovi* sp. n. (Rodentia, Cricetidae) from the People's Republic of Mongolia]. Zoologicheskii Zhurnal, 67:304-308 (in Russian).
Orlov, V. N., V. N. Yatsenko, and V. M. Malygin. 1983. [Karyotype homology and species phylogeny in a group of field mice (Cricetidae, Rodentia)]. Doklady Akademii Nauk SSSR, 269:236-238 (in Russian).
Orlov, V. N., M. I. Baskevich, and N. Sh. Bulatova. 1992. [Chromosomal sets of rats of the genus *Arvicanthis* (Rodentia, Muridae) from Ethiopia. Zoologicheskii Zhurnal, 73:103-112 (in Russian).
Orr, R. T. 1938. A new rodent of the genus *Nesoryzomys* from the Galapagos Islands. Proceedings of the California Academy of Science, ser. 4, 23:303-306.
Orr, R. T. 1945. A study of the *Clethrionomys dawsoni* group of red-backed mice. Journal of Mammalogy, 26:67-74.
Orsini, P., and G. Cheylan. 1988. Les rongeurs de Corse: Modifications de taille en relation avec l'isolement en milieu isolaire. Bulletin d'Ecologie, 19(2-3):411-416.
Ortells, M. O., O. A. Reig, R. L. Wainburg, G. E. Hurtado de Catalfo, and T. M. L. Gentile de Fronza. 1989. Cytogenetics and karyosystematics of phyllotine rodents (Cricetidae, Sigmodontinae). II. Chromosome multiformity and autosomal polymorphism in *Eligmodontia*. Zeitschrift für Säugetierkunde, 54:129-140.
Osborn, D. J., and I. Helmy. 1980. The contemporary land mammals of Egypt (including Sinai). Fieldiana, Zoology, 5:1-579.
Osgood, W. H. 1900. Revision of the pocket mice of the genus *Perognathus*. North American Fauna, 18:1-73.
Osgood, W. H. 1907. Some unrecognised and misapplied names of American mammals. Proceedings of the Biological Society of Washington, 20:43-52.
Osgood, W. H. 1909. Revision of the mice of the American genus *Peromyscus*. North American Fauna, 28:1-285.
Osgood, W. H. 1910. Eight new African rodents. Annals and Magazine of Natural History, ser. 8, 5:276-282.
Osgood, W. H. 1912. Mammals from western Venezuela and eastern Colombia. Field Museum of Natural History, Zoological Series, 10:32-67.
Osgood, W. H. 1913. New Peruvian mammals. Field Museum of Natural History, Zoological Series, 10:93-100.
Osgood, W. H. 1918. The status of *Perognathus longimembris* (Coues). Proceedings of the Biological Society of Washington, 31:95-96.
Osgood, W. H. 1925. The long-clawed South American rodents of the genus *Notiomys*. Field Museum of Natural History, Zoological Series, 12:113-125.
Osgood, W. H. 1932. Mammals of the Kelley-Roosevelts and Delacour Asiatic expeditions. Field Museum of Natural History, Zoological Series, 18:193-339.
Osgood, W. H. 1933*a*. The South American mice referred to *Microryzomys* and *Thallomyscus*. Field Museum of Natural History, Zoological Series, 20:1-8.
Osgood, W. H. 1933*b*. Two new rodents from Argentina. Field Museum of Natural History, Zoological Series, 20:11-14.
Osgood, W. H. 1933*c*. The supposed genera *Aepeomys* and *Inomys*. Journal of Mammalogy, 14:161.
Osgood, W. H. 1933*d*. The generic position of *Mus pyrrhorhinus* Wied. Journal of Mammalogy, 14:370-371.
Osgood, W. H. 1936. New and imperfectly known small mammals from Africa. Field Museum of Natural History, Zoological Series, 20:217-256.
Osgood, W. H. 1943. The mammals of Chile. Field Museum of Natural History, Zoological Series, 30:1-268.
Osgood, W. H. 1945. A new rodent from Dutch New Guinea. Fieldiana, Zoology, 31:1-2.
Osgood, W. H. 1946. A new octodont rodent from the Paraguayan Chaco. Fieldiana, Zoology, 31(6):47-49.
Osgood, W. H. 1947. Cricetine rodents allied to *Phyllotis*. Journal of Mammalogy, 28:165-174.
O'Shea, T. J. 1991. *Xerus rutilus*. Mammalian Species, 370:1-5.
Ottenwalder, J. A., and H. H. Genoways. 1982. Systematic review of the Antillean bats of the *Natalus micropus*-complex (Chiroptera: Natalidae). Annals of Carnegie Museum, 51:17-38.
Owen, J. G. 1984. *Sorex fumeus*. Mammalian Species, 215:1-8.
Owen, J. G., and R. S. Hoffmann. 1983. *Sorex ornatus*. Mammalian Species, 212:1-5.
Owen, R. D. 1987. Phylogenetic analyses of the bat subfamily Stenodermatinae (Mammalia: Chiroptera). Special Publications, The Museum, Texas Tech University, 26:1-65.

Owen, R. D. 1991. The systematic status of *Dermanura concolor* (Peters, 1865)(Chiroptera: Phyllostomidae), with description of a new genus. Bulletin of the American Museum of Natural History, 206:18-25.

Owen, R. D., R. K. Chesser, and D. C. Carter. 1990. The systematic status of *Tadarida brasiliensis cynocephala* and Antillean members of the *Tadarida brasiliensis* group, with comments on the generic name *Rhizomops* Legendre. Occasional Papers, The Museum, Texas Tech University, 133:1-18.

Packard, R. L. 1960. Speciation and evolution of the pygmy mice, genus *Baiomys*. University of Kansas Publications, Museum of Natural History, 9:579-670.

Packard, R. L. 1969. Taxonomic review of the golden mouse, *Ochrotomys nuttalli*. Miscellaneous Publications, Museum of Natural History, University of Kansas, 51:373-406.

Packard, R. L., and J. H. Bowers. 1970. Distributional notes on some foxes from western Texas and eastern New Mexico. Southwestern Naturalist, 14:450-451.

Packard, R. L., and J. B. Montgomery. 1978. *Baiomys musculus*. Mammalian Species, 102:1-3.

Pagels, J. F., and C. O. Handley, Jr. 1989. Distribution of the southeastern shrew, *Sorex longirostris* Bachmann, in western Virginia. Brimleyana, 15:123-131.

Pagels, J. F., C. S. Jones, and C. O. Handley, Jr. 1982. Northern limits of the southeastern shrew, *Sorex longirostris* Bachmann (Insectivora: Soricidae), on the Atlantic coast of the United States. Brimleyana, 8:51-59.

Palacios, F. 1983. On the taxonomic status of the genus *Lepus* in Spain. Acta Zoologica Fennica, 174:27-30.

Palacios, F. 1989. Biometric and morphologic features of the species of the genus *Lepus* in Spain. Mammalia, 53:227-264.

Palacios, F., and J. Fernández. 1992. A new subspecies of hare from Majorca (Balearic Islands). Mammalia, 56:71-85.

Palacios, F., J. F. Orueta, and G. G. Tapia. 1989. Taxonomic review of the *Lepus europaeus* group in Italy and Corsica. Abstract of papers and posters, Fifth International Theriological Congress, Rome, 1:189-190.

Pallas, P. S. 1770-1780. Spicilegia zoologica, quibus novae imprimus et obscurae animalium species iconibus, descriptionibus atque commentariis illustrantur cura P.S. Pallas. [fasc. 11-12 imprint 1777-78; fasc. 13 imprint 1779; fasc. 14 imprint 1780]. Berolini, prostant apud Gottl. August. Langed, 14 fasc in 2 volumes.

Pallas, P. S. 1771-1776. Reise durch verschiedene Provinzen des Russischen Reichs. St. Petersbourg, 3 vol.

Palmeirim, J. M. 1991. A morphometric assessment of the systematic position of the *Nyctalus* from Azores and Madeira (Mammalia: Chiroptera). Mammalia, 55:381-388.

Palmeirim, J. M., and R. S. Hoffmann. 1983. *Galemys pyrenaicus*. Mammalian Species, 207:1-5.

Palmer, T. S. 1904. Index generum mammalium: A list of the genera and families of mammals. North American Fauna, 23:1-984.

Palmer, T. S. 1928. An earlier name for the genus *Evotomys*. Proceedings of the Biological Society of Washington, 41:87-88.

Palomo, L. J. 1988. Etude descriptive des poils de *Mus spretus* Lataste, 1883. Revue Suisse de Zoologie, 95:505-512.

Palomo, L. J., M. Espana, J. Lopez-Fuster, J. Gosalbez, and V. Sans-Coma. 1983. Sobre la variabilidad genetica y morfometrica de *Mus spretus* Lataste, 1883 en la Penninsula Iberica. Miscellania Zoologica, 7:171-192.

Pankow, H. 1989. Neues vom Siebenschläfer in Mecklenburg. Archiv des Vereins der Freunde der Naturgeschichte, Mecklenburg, 29:73-74.

Paolucci, P., A. Battisti, and R. de Battisti. 1987. The forest dormouse (*Dryomys nitedula* Pallas, 1779) in the eastern Alps (Rodentia, Gliridae). Biogeographia, 13:855-866.

Paradiso, J. L. 1967. A review of the wrinkle-faced bats (*Centurio senex* Gray), with description of a new subspecies. Mammalia, 31:595-604.

Paradiso, J. L. 1968. Canids recently collected in east Texas, with comments on the taxonomy of the red wolf. American Midland Naturalist, 80:529-534.

Paradiso, J. L., and R. H. Manville. 1961. Taxonomic notes on the tundra vole (*Microtus oeconomus*) in Alaska. Proceedings of the Biological Society of Washington, 74:77-92.

Paradiso, J. L., and R. M. Nowak. 1971 [1972]. A report on the taxonomic status and distribution of the red wolf. United States Fish and Wildlife Service, Special Scientific Report—Wildlife, 145:1-36.

Paradiso, J. L., and R. M. Nowak. 1972. *Canis rufus*. Mammalian Species, 22:1-4.

Park, N. S., S. S. Lee, and H. S. Koh. 1990. Systematic studies on Korean rodents: VII. Immunological analyses of serum proteins of seven species. Korean Journal of Systematic Zoology, 6:165-172.

Parkinson, A. 1979. Morphologic variation and hybridization in *Myotis yumanensis sociabilis* and *Myotis lucifugus carissima*. Journal of Mammalogy, 60:489-504.

Parnaby, H. E. 1987. Distribution and taxonomy of the long-eared bats, *Nyctophilus gouldi* Tomes, 1858, and *Nyctophilus bifax* Thomas, 1915 (Chiroptera: Vespertilionidae) in eastern Australia. Proceedings of the Linnean Society of New South Wales, 109:153-174.

Pascale, E., E. Valle, and A. V. Furano. 1990. Amplification of an ancestral mammalian L1 family of long interspersed repeated DNA occurred just before the murine radiation. Proceedings of the National Academy of Sciences (Washington, D. C.), 87:9481-9485.

Passarge, H. 1984. *Sorex isodon marchicus* ssp. nova in Mitteleuropa. Zeitschrift für Säugetierkunde, 49:278-284.

Pasteur, N., J. Worms, M. Tohari, and D. Iskandar. 1982. Genetic differentiation in Indonesian and French rats of the subgenus *Rattus*. Biochemical Systematics and Ecology, 10:191-196.

Patterson, B. D. 1962. An extinct solenodontid insectivore from Hispaniola. Breviora, 165:1-11.

Patterson, B. D. 1978. Pholidota and Tubulidentata. Pp. 268-278, *in* Evolution of African mammals (V. J. Maglio and H. B. S. Cooke, eds.). Harvard University Press, Cambridge, MA, 641 pp.

Patterson, B. D. 1980. A new subspecies of *Eutamias quadrivittatus* from the Organ Mountains, New Mexico. Journal of Mammalogy, 61:455-464.

Patterson, B. D. 1984. Geographic variation and taxonomy of Colorado and Hopi chipmunks (genus *Eutamias*). Journal of Mammalogy, 65:442-456.

Patterson, B. D., and M. H. Gallardo. 1987. *Rhyncholestes raphanurus*. Mammalian Species, 286:1-5.

Patterson, B. D., and L. R. Heaney. 1987. Preliminary analysis of geographic variation in red-tailed chipmunks (*Eutamias ruficaudus*). Journal of Mammalogy, 68:782-791.

Patterson, B. D., M. H. Gallardo, and K. E. Freas. 1984. Systematics of mice of the subgenus *Akodon* (Rodentia: Cricetidae) in southern South America, with the description of a new species. Fieldiana, Zoology, New Series, 23:1-16.

Patterson, B., and R. Pascual. 1968*a*. The fossil mammal fauna of South America, V. Evolution of mammals on southern continents. Quarterly Review of Biology, 43:409-451.

Patterson, B., and R. Pascual. 1968*b*. New echimyid rodents from the Oligocene of Patagonia, and a synopsis of the family. Breviora, Harvard Museum of Comparative Zoology, 301:1-14.

Patterson, B., and A. E. Wood. 1982. Rodents from the Deseadan Oligocene of Bolivia and the relationships of the Caviomorpha. Bulletin Museum of Comparative Zoology, 149:371-543.

Pattie, D. 1973. *Sorex bendirii*. Mammalian Species, 27:1-2.

Patton, J. C., R. J. Baker, and J. C. Avise. 1981. Phenetic and cladistic analyses of biochemical evolution in peromyscine rodents. Pp. 288-308, *in* Mammalian population genetics (M. H. Smith and J. Joule, eds.). University of Georgia Press, Athens, 380 pp.

Patton, J. L. 1967*a*. Chromosome studies of certain pocket mice, genus *Perognathus* (Rodentia: Heteromyidae). Journal of Mammalogy, 48:27-37.

Patton, J. L. 1967*b*. Chromosomes and evolutionary trends in the pocket mouse subgenus *Perognathus* (Rodentia: Heteromyidae). Southwestern Naturalist, 12:429-438.

Patton, J. L. 1969. Karyotypic variation in the pocket mouse, *Perognathus penicillatus* Woodhouse (Rodentia-Heteromyidae). Caryologia, 22:351-358.

Patton, J. L. 1973. An analysis of natural hybridization between the pocket gophers *Thomomys bottae* and *Thomomys umbrinus*, in Arizona. Journal of Mammalogy, 54:561-584.

Patton, J. L. 1984. Systematic status of the large squirrels (subgenus *Urosciurus*) of the western Amazon basin. Studies on Neotropical Fauna and Environment, 19:53-72.

Patton, J. L. 1987. Species groups of spiny rats, genus *Proechimys* (Rodentia: Echimyidae). Fieldiana, Zoology, new series, 39:305-345.

Patton, J. L. 1990. Geomyoid evolution: The historical, selective, and random basis for divergence patterns within and among species. Pp. 49-69, *in* Evolution of subterranean mammals at the organismal and molecular levels (E. Nevo and O. A. Reig, eds.). Alan R. Liss, Inc., New York, 422 pp.

Patton, J. L., and R. E. Dingman. 1968. Chromosome studies of pocket gophers, genus *Thomomys*. I. The specific status of *Thomomys umbrinus* (Richardson) in Arizona. Journal of Mammalogy, 49:1-13.

Patton, J. L., and L. Emmons. 1985. A review of the genus *Isothrix* (Rodentia, Echimyidae). American Museum Novitates, 2817:1-14.

Patton, J. L., and M. S. Hafner. 1983. Biosystematics of the native rodents of the Galapagos Archipelago, Ecuador. Pp. 539-568, *in* Patterns of evolution in Galapagos organisms (R. I. Bowman, M. Benson, and A. E. Leviton, eds.). American Association for the Advancement of Science, Pacific Division, San Francisco, CA, 568 pp.

Patton, J. L., and T. C. Hsu. 1967. Chromosomes of the golden mouse, *Peromyscus* (*Ochrotomys*) *nuttalli* (Harlan). Journal of Mammalogy, 48:637-639.

Patton, J. L., and O. A. Reig. 1989. Genetic differentiation among echimyid rodents, with an emphasis on spiny rats, genus *Proechimys*. Pp. 75-96, *in* Advances in Neotropical mammalogy (K. H. Redford and J. F. Eisenberg, eds.). Sandhill Crane Press, Gainesville, FL, 614 pp.

Patton, J. L., and M. F. Smith. 1981. Molecular evolution in *Thomomys*: phyletic systematics, paraphyly, and rates of evolution. Journal of Mammalogy, 62:493-500.

Patton, J. L., and M. F. Smith. 1989. Genetic structure and the genetic and morphologic divergence among pocket gopher species (genus *Thomomys*). Pp. 284-304, *in* Speciation and Its Consequences (D. Otte and J. A. Endler, eds.). Sinauer Associates Incorporated, Sunderland, MA, 679 pp.

Patton, J. L., and M. F. Smith. 1990. The evolutionary dynamics of the pocket gopher *Thomomys bottae*, with emphasis on California populations. University of California Publications in Zoology, 123:1-161.

Patton, J. L., H. MacArthur, and S. Y. Yang. 1976. Systematic relationships of the four-toed populations of *Dipodomys heermanni*. Journal of Mammalogy, 57:159-163.

Patton, J. L., S. W. Sherwood, and S. Y. Yang. 1981. Biochemical systematics of chaetodipine pocket mice, genus *Perognathus*. Journal of Mammalogy, 62:477-492.

Patton, J. L., M. F. Smith, R. D. Price, and R. A. Hellenthal. 1984. Genetics of hybridization between the pocket gophers *Thomomys bottae* and *Thomomys townsendii* in northeastern California. Great Basin Naturalist, 44:431-440.

Patton, J. L., P. Myers, and M. F. Smith. 1989. Electromorphic variation in selected South American akodontine rodents (Muridae: Sigmodontinae), with comments on systematic implications. Zeitschrift für Säugetierkunde, 54:347-359.

Patton, J. L., P. Myers, and M. F. Smith. 1990. Vicariant versus gradient models of diversification: the small mammal fauna of eastern Andean slopes of Peru. Pp. 355-371, *in* Vertebrates in the tropics: Proceedings of the international symposium on vertebrate biogeography and systematics in the tropics, Bonn, June 5-8, 1989 (G. Peters and R. Hutterer, eds.). Museum Alexander Koenig Zoological Research Institute and Zoological Museum, Bonn, 424 pp.

Paula Couto, C. 1950. Footnote number 249. P. 232, *in* Memórias sobre a Paleontologia Brasileira (P. W. Lund with notes and comments by C. Paula Couto). Instituto Nacional do Livro, Rio de Janeiro, 589 pp., 56 pls.

Paulson, D. D. 1988*a*. *Chaetodipus baileyi*. Mammalian Species, 297:1-5.

Paulson, D. D. 1988*b*. *Chaetodipus hispidus*. Mammalian Species, 320:1-4.

Pavlinin, V. N. 1966. Der Zobel. *Martes zibellina* L. Wittenberg Lutherstadt, Ziemsen, 102 pp.

Pavlinov, I. Ya. 1980*a*. [Evolution and taxonomic significance of the morphology of the osseous middle ear in Gerbillinae (Rodentia: Cricetidae)]. Byulleten' Moskovskovo Obshchestva Ispitatelei Prirody, Otdel Biologicheskii, 85:20-33 (in Russian).

Pavlinov, I. Ya. 1980*b*. Nadvidovye gruppirovki v podsemeistve Cardiocraniinae Satunin (Mammalia, Dipodidae) [Superspecies groupings in the subfamily Cardiocraniinae Satunin (Mammalia, Dipodidae)]. Vestnik Zoologii, 2:47-51 (in Russian).

Pavlinov, I. Ya. 1980*c*. [Taxonomic status of *Calomyscus* Thomas (Rodentia, Cricetidae) on the basis of structure of auditory ossicles]. Zoologicheskii Zhurnal, 59:312-316 (in Russian).

Pavlinov, I. Ya. 1981. [Taxonomic status of gerbils of the genus *Ammodillus* Thomas, 1904 (Rodentia, Gerbillinae)]. Zoologicheskii Zhurnal, 60:472-474 (in Russian).

Pavlinov, I. Ya. 1982. [Phylogeny and classification of the subfamily Gerbillinae]. Byulleten' Moskovskovo Obshchestva Ispitatelei Prirody, Otdel Biologicheskii, 87:19-31 (in Russian).

Pavlinov, I. Ya. 1984. [Evolution of auditory ossicles in voles, subfamily Microtinae]. Sbornik Trudov Zoologicheskovo Muzeya MGU, 22:191-212 (in Russian).

Pavlinov, I. Ya. 1985. [Contributions to dental morphology and phylogeny of gerbils (Rodentia, Gerbillinae)]. Zoologicheskii Zhurnal, 64:574-582 (in Russian).

Pavlinov, I. Ya. 1987. [Cladistic analysis of the gerbilline tribe Taterillini (Rodentia, Gerbillinae) and some questions on the method of numerical cladistic analysis]. Zoologicheskii Zhurnal, 66:903-913 (in Russian).

Pavlinov, I. Ya. 1988. [Evolution of mastoid part of the bulla tympani in specialized desert rodents]. Zoologicheskii Zhurnal, 67(5):739-750 (in Russian).

Pavlinov, I. Ya., and O. L. Rossolimo. 1987. Sistematika mlekopitayushchikh SSSR. [Systematics of the mammals of the USSR.]. Moscow University Press, Moscow, 282 pp. (in Russian).

Pavlinov, I. Ya., and G. I. Shenbrot. 1983. [Male genital structure and supraspecific taxonomy of Dipodidae]. Trudy Zoologicheskovo Instituta, Akademiya Nauk SSSR, Leningrad, 119:67-88 (in Russian).

Pavlinov, I. Ya., and G. I. Shenbrot. 1985. Materiali po sistematikye tushkanchikov Iranskovo Nagorya [Materials on the taxonomy of jerboas of the Iranian highland]. Pp. 55-57, *in* Tushkanchiki fauni SSSR [Jerboas of the USSR fauna] (V. E. Sokolov and V. V. Kucheruk, eds.). Tyezisi Dokladov Vsesoyuznovo soveshchaniya [Abstracts of the All-Union Conference], Alma-Ata, All-Union Mammal Society, Moscow, 246 pp. (in Russian).

Pavlinov, I. Ya., Yu. A. Dubrovsky, O. L. Rossolimo, and E. G. Potapova. 1990. [Gerbils of the world]. Nauka, Moscow, 368 pp. (in Russian).

Payne, J., C. M. Francis, and K. Phillipps. 1985. A field guide to the mammals of Borneo. The Sabah Society, Kuala Lumpur, 332 pp.

Payne, S. 1968. The origin of domestic sheep and goats: a reconsideration in the light of the fossil evidence. Proceedings of the Prehistoric Society, 34:368-384.

Pearson, O. P. 1958. A taxonomic revision of the rodent genus *Phyllotis*. University of California Publications in Zoology, 56:391-496.

Pearson, O. P. 1972. New information on ranges and relationships within the rodent genus *Phyllotis* in Peru and Ecuador. Journal of Mammalogy, 53:677-686.

Pearson, O. P. 1984. Taxonomy and natural history of some fossorial rodents of Patagonia, southern Argentina. Journal of Zoology (London), 202:225-237.

Pearson, O. P., and M. I. Christie. 1991. Sympatric species of *Euneomys* (Rodentia, Cricetidae). Studies on Neotropical Fauna and Environment, 26:121-127.

Pearson, O. P., and J. L. Patton. 1976. Relationships among South American phyllotine rodents based on chromosome analysis. Journal of Mammalogy, 57:339-350.

Pearson, T. 1981. Geographic and intraspecific cranial variation in North American arctic ground squirrels. Unpubl. M. A. thesis, University of Kansas, Lawrence, 96 pp.

Pechev, T., V. Anguelova, V. and T. Dinev. 1964. Etudes sur la taxonomie du *Myomimus personatus* (Ognev, 1924) (Rodentia) en Bulgarie. Mammalia, 28(3):419-428.

Peddie, D. A. 1975. A taxonomic and autoecological study of the genus *Pronolagus* in southern Africa. Unpubl. M. S. thesis, University of Rhodesia, Salisbury.

Pembleton, E. F., and S. L. Williams. 1978. *Geomys pinetis*. Mammalian Species, 86:1-3.

Peng Yan-zhang, Ye Zhi-zhang, Zhang Yao-ping, and Pan Ru-liang. 1988. [The classification of snub-nosed monkey (*Rhinopithecus* spp.) based on gross morphological characters]. Zoological Research, 9:239-248 (in Chinese).

Pengilly, D., G. H. Jarrell, and S. D. MacDonald. 1983. Banded karyotypes of *Peromyscus sitkensis* from Baranof Island, Alaska. Journal of Mammalogy, 64:682-685.

Pennant, T. 1769. British Zoology. Class III. Reptiles. IV. Fish. V the Beaked whale. Benjamin White, London, Volume III, pp. 43-44.

Pennant, T. 1771. Synopsis of Quadrupeds. J. Monk, Chester, 382 pp.

Pennant, T. 1781. History of quadrupeds. B. & J. White, London, 2 volumes.

Penzhorn, B. L. 1988. *Equus zebra*. Mammalian Species, 314:1-7.

Pérez-Zapata, A., D. Lew. M. Aguilera, and O. A. Reig. 1992. New data on the systematics and karyology of *Podoxymys roraimae* (Rodentia, Cricetidae). Zeitschrift für Säugetierkunde, 57:216-224.

Perret, J.-L., and V. Aellen. 1956. Mammifères du Cameroun de la collection J.-L. Perret. Revue Suisse de Zoologie, 63(26):395-450.

Perrin, M. R. 1986. Gastric anatomy and histology of an arboreal, folivorous murid rodent: the black-tailed tree rat *Thallomys paedulcus* (Sundevall, 1846). Zeitschrift für Säugetierkunde, 51:224-236.

Perrin, M. R., and B. A. Curtis. 1980. Comparative morphology of the digestive system of 19 species of southern African myomorph rodents in relation to diet and evolution. South African Journal of Zoology, 15:22-33.

Perrin, M. R., and A. H. Maddock. 1983. Preliminary investigations of the digestive processes of the white-tailed rat *Mystromys albicaudatus* (Smith 1834). South African Journal of Science, 18:128-133.

Perrin, W. F. 1975. Variation of spotted and spinner porpoise (genus *Stenella*) in the eastern tropical Pacific and Hawaii. Bulletin of the Scripps Institution of Oceanography, 21:1-206.

Perrin, W. F. 1990. Subspecies of *Stenella longirostris* (Mammalia: Cetacea: Delphinidae). Proceedings of the Biological Society of Washington, 103(2):453-463.

Perrin, W. F., E. D. Mitchell, J. G. Mead, D. K. Caldwell, and P. J. H. van Bree. 1981. *Stenella clymene* a rediscovered tropical dolphin of the Atlantic. Journal of Mammalogy, 62(3):583-598.

Perrin, W. F., E. D. Mitchell, J. G. Mead, D. K. Caldwell, M. C. Caldwell, P. J. H. van Bree, and W. H. Dawbin. 1987. Revision of the spotted dolphins *Stenella* spp. Marine Mammal Science, 3(2):99-170.

Perry, H. R., Jr. 1982. Muskrats (*Ondatra zibethicus* and *Neofiber alleni*). Pp. 282-325, *in* Wild mammals of North America (J. A. Chapman and G. A. Feldhammer, eds.). Johns Hopkins University Press, Baltimore, 1147 pp.

Peshev, D. 1983. New karyotype forms of the mole rat, *Nannospalax leucodon* Nordmann (Spalacidae, Rodentia), in Bulgaria. Zoologischer Anzeiger, 211:65-72.

Peshev, D. 1989*a*. Craniological study of the species of genus *Spalax* (Spalacidae, Mammalia). I. Sex dimorphism. Zoologischer Anzeiger, 222:83-91.

Peshev, D. 1989*b*. Craniological study of the species of genus *Spalax* (Spalacidae, Mammalia). II. Interspecific differences. Zoologischer Anzeiger, 222:92-98.

Peshev, D. 1991. On the systematic position of the mouse-like hamster *Calomyscus bailwardi* Thomas, 1905 (Cricetidae, Rodentia) from the Near East and Middle Asia. Mammalia, 55:107-112.

Peshev, T., and B. Belcheva. 1979. Karyological studies on snow vole *Microtus nivalis* Martins (Mammalia, Rodentia) collected in Bulgaria. Zoologischer Anzeiger, 203:65-68.

Peters, W. C. H. 1852. Reise nach Mossambique. I. Säugethiere. Druck & Verlag, Berlin, 202 pages.

Peters, W. C. H. 1875. Über eine neue Art von Seebären, *Arctophoca gazella* von den Kerguelen-Inseln. Monatsberichte der Königlich Preussischen Akademie der Wissenschaften zu Berlin, 1875:393-399.

Peters, W. C. H. 1876. Über die Pelzrobbe von der Inseln St. Paul und Amsterdam. Monatsberichte der Königlich Preussischen Akademie der Wissenschaften zu Berlin, 1876:315-316.

Peterson, K. E., and T. I. Yates. 1980. *Condylura cristata*. Mammalian Species, 129:1-4.

Peterson, R. L. 1952. A review of the living representatives of the genus *Alces*. Contributions from the Royal Ontario Museum in Zoology and Palaeontology, 34:1-30.

Peterson, R. L. 1965*a*. A review of the flat-headed bats of the family Molossidae from South America and Africa. Royal Ontario Museum, Life Sciences, Contribution, 64:1-32.

Peterson, R. L. 1965*b*. A review of the bats of the genus *Ametrida*, Family Phyllostomidae. Royal Ontario Museum, Life Sciences, Contribution, 65:1-13.

Peterson, R. L. 1966*a*. The mammals of eastern Canada. Oxford University Press, Toronto, 465 pp.

Peterson, R. L. 1966*b*. Recent mammal records from the Galapagos Islands. Mammalia, 30:441-445.

Peterson, R. L. 1969. Notes on the Malaysian fruit bats of the genus *Dyacopterus*. Royal Ontario Museum, Life Sciences, Occasional Papers, 13:1-4.

Peterson, R. L. 1972. Systematic status of the African molossid bats *Tadarida congica*, *T. niangarae* and *T. trevori*. Royal Ontario Museum, Life Sciences, Contribution, 85:1-32.

Peterson, R. L. 1981. Systematic variation in the *tristis* group of the bent-winged bats of the genus *Miniopterus* (Chiroptera: Vespertilionidae). Canadian Journal of Zoology, 59:828-843.

Peterson, R. L. 1991. Systematic variation in the megachiropteran tube-nosed bats *Nyctimene cyclotis* and *N. certans*. Bulletin of the American Museum of Natural History, 206:26-41.

Peterson, R. L., and M. B. Fenton. 1970. Variation in bats of the genus *Harpionycteris*, with the description of a new race. Royal Ontario Museum, Life Sciences, Occasional Papers, 17:1-15.

Petrov, B. 1979. Some questions of the zoogeographical division of the western Palaearctic in the light of the distribution of mammals in Yugoslavia. Folia Zoologica, 28:13-24.

Petrov, B., and A. Ruzic. 1982. *Microtus epiroticus* Ondrias, 1966—Südfeldmaus. Pp. 319-330, *in* Handbuch der Säugetiere Europas (J. Niethammer and F. Krapp, eds.). Akademische Verlagsgesellschaft (Wiesbaden), 2/I:1-649.

Petrov, B., and A. Ruzic. 1985. Taxonomy and distribution of members of the genus *Mus* (Rodentia, Mammalia) in Yugoslavia. Proceedings on the fauna of SR Serbia, 3:209-243.

Petrov, B., and M. Todorovic. 1982. *Dinaromys bogdanovi* (V. et E. Martino, 1922)—Bergmaus. Pp. 193-208, *in* Handbuch der Säugetiere Europas (J. Niethammer and F. Krapp, eds.). Akademische Verlagsgesellschaft (Wiesbaden), 2/I:1-649.

Petter, F. 1961. Eléments d'une révision des Lievres européens et asiatiques du sous-genre Lepus. Zeitschrift für Säugetierkunde, 26:30-40.

Petter, F. 1962*a*. Un noveau rongeur Malgache: *Brachytarsomys albicauda villosa*. Mammalia, 26:570-572.

Petter, F. 1962*b*. Note de nomenclature sur le genere *Mylomys* (Rongeurs Murides). Mammalia, 26:575.

Petter, F. 1963*a*. Nouveaux éléments d'une révision des lièvres Africains. Mammalia, 27:238-255.

Petter, F. 1963*b*. Un nouvel insectivore du nord de l'Assam: *Anourosorex squamipes schmidi* nov. sbsp. Mammalia, 27:444-445.

Petter, F. 1963*c*. Contribution a la connaissance des souris Africaines. Mammalia, 27:602-607.

Petter, F. 1965. Les *Praomys* d'Afrique Centrale. Zeitschrift für Säugetierkunde, 30:54-60.

Petter, F. 1966*a*. Affinites des genres *Beamys*, *Saccostomus* et *Cricetomys* (Rongeurs, Cricetomyinae). Annales Musée Royal de l'Afrique Centrale, ser. 8 (Sciences Zoologiques), 144:13-25.

Petter, F. 1966*b*. *Dendroprionomys rousseloti* gen. nov., sp. nov., rongeur nouveau du Congo (Cricetidae, Dendromurinae). Mammalia, 30:129-137.

Petter, F. 1967*a*. Contribution a la faune du Congo (Brazzaville). Mission A. Villiers et A. Descarpentries. LV. Mammiferes rongeurs (Muscardinidae et Muridae). Bulletin de l'Institut Fondatmental d'Afrique Noire, 29A(2):815-820.

Petter, F. 1967*b*. Particularities dentaires des Petromyscinae Roberts 1951 (Rongeurs, Cricetides). Mammalia, 31:217-224.

Petter, F. 1969. Une souris nouvelle d'Afrique Occidentale *Mus mattheyi* sp. nov. Mammalia, 33:118-123.

Petter, F. 1972*a*. Caracteres morphologiques et repartition geographique de *Mus tenellus* (Thomas 1903), souris d'Afrique Orientale. Mammalia, 36:533-535.

Petter, F. 1972*b*. Order Lagomorpha. Part 5. Pp. 1-7, *in* The mammals of Africa: An identification manual (J. Meester and H. W. Setzer, eds.) [issued 2 May 1972]. Smithsonian Institution Press, Washington, D. C., not continuously paginated.

Petter, F. 1972*c*. The rodents of Madagascar: the seven genera of Malagasy rodents. Pp. 661-666, *in* Biogeography and ecology in Madagascar (R. Battistini and G. Richard-Vincard, eds.). Monographiae Biologicae, 21:1-765.

Petter, F. 1973*a*. Capture de *Thallomys paedulcus scotti* en Ethiopie. Mammalia, 37:360-361.

Petter, F. 1973*b*. Les noms de genre *Cercomys, Nelomys, Trichomys*[sic] et *Proechimys* (Rongeurs, Echimyides). Mammalia, 37(3):422-426.

Petter, F. 1973*c*. Tendances evolutives dans le genre *Gerbillus* (Rongeurs, Gerbillides). Mammalia, 37:631-636.

Petter, F. 1975*a*. Family Cricetidae: Subfamily Nesomyinae, Part 6.2. Pp. 1-4, *in* The mammals of Africa: An identification manual (J. Meester and H. W. Setzer, eds.) [issued 10 Dec 1975]. Smithsonian Institution Press, Washington, D. C., not continuously paginated.

Petter, F. 1975*b*. Subfamily Gerbillinae. Part 6.3. Pp. 7-12, *in* The mammals of Africa: An identification manual (J. Meester and H. W. Setzer, eds.) [issued 10 Dec 1975]. Smithsonian Institution, Washington, D. C., not continuously paginated.

Petter, F. 1975*c*. Les *Praomys* de Republique Centrafricaine (Rongeurs, Murides). Mammalia, 39:51-56.

Petter, F. 1978. Epidémiologie de la leishmaniose en Guyane francaise en relation avec l'existance d'une espece nouvelle de rongeurs échimyidés, *Proechimys cuvieri*, sp. n. Compte Rendu Hebdomadaire des seances de l'Academie de Sciences, Paris, ser. D, 287:261-264.

Petter, F. 1981*a*. Remarques sur la systématique des Chrysochloridés. Mammalia, 45:49-53.

Petter, F. 1981*b*. Les souris africaines du groupe *sorella* (Rongeurs, Murides). Mammalia, 45:313-320.

Petter, F. 1982. Les parentes des *Otomys* du Mont Oku (Cameroun) et des autres formes rapportees a *O. irroratus* (Brants, 1827) (Rodentia, Muridae). Bonner Zoologische Beiträge, 33:215-222.

Petter, F. 1983. Elements d'une revision des *Acomys* Africains. Un sous-genre nouveau, *Peracomys* Petter et Roche, 1981 (Rongeurs, Murides). Annales Musée Royal de l'Afrique Centrale, Tervuren-Belgique, ser. 8 (Sciences Zoologiques), 237:109-119.

Petter, F. 1986. Un rongeur nouveau du Mont Oku (Cameroun) *Lamottemys okuensis*, gen. nov., sp. nov.; (Rodentia, Muridae). Cimbebasia, ser. A, 8:97-105.

Petter, F., and F. de Beaufort. 1960. Description d'une forme nouvelle de rongeur d'Angola, *Thallomys damarensis quissamae*. Bulletin du Muséum National d'Histoire Naturelle (Paris), ser. 2, 32:269-271.

Petter, F., and H. Genest. 1970. Liste preliminaire des rongeurs myomorphes de Republique Centrafricaine. Description de deux souris nouvelles: *Mus oubanguii* et *Mus goundae*. Mammalia, 34:451-458.

Petter, F., and R. Matthey. 1975. Genus *Mus*, Part 6.7. Pp. 1-4, *in* The mammals of Africa: An identification manual (J. Meester and H. W. Setzer, eds.),. Smithsonian Institution Press, Washington, D. C., not continuously paginated.

Petter, F., and J. Roche. 1981. Remarques preliminaires sur la systematique des *Acomys* (Rongeurs, Muridae *Peracomys*, sous-genre nouveau. Mammalia, 45:381-383.

Petter, F., and M. Tranier. 1975. Contribution a l'etude des *Thamnomys* du groupe *dolichurus* (Rongeurs, Murides). Systematique et caryologie. Mammalia, 39:405-414.

Petter, F., M. Quilici, Ph. Ranque, and P. Camerlynck. 1969. Croisement d'*Arvicanthis niloticus* (Rongeurs, Murides) du Senegal et d'Ethiopie. Mammalia, 33:540-541.

Petter, F., F. Adam, and B. Hubert. 1971. Presence au Senegal de *Mus mattheyi* F. Petter 1969. Mammalia, 35:346-347.

Petter, F., A. Poulet, B. Hubert, and F. Adam. 1972. Contribution a l'etude des *Taterillus* du Senegal *T. pygargus* (F. Cuvier, 1832) et *T. gracilis* (Thomas, 1892) (Rongeurs, Gerbillides). Mammalia, 36:210-213.

Petter, F., F. Lachiver, and R. Chekir. 1984. Les adaptations des rongeurs gerbillides a la vie dans les regions arides. Bulletin de Societe Botanique de France, Actualites Botaniques, 131:365:373.

Petter, G. 1962. Le peuplement en carnivores de Madagascar, Pp. 331-342, *in* Evolution des vertébrés. (J. P. Lehman, ed.). Colloques Internationaux du Centre National de la Recherche Scientifique, 104.

Petter, G. 1969. Interpretation evolutive des caractéres de la denture des viverrides Africans. Mammalia, 33(4):607-635.

Petter, G. 1971. Origine, phylogenie et systematique des blaireaux. Mammalia, 35:567-597.

Petter, G. 1974. Rapports phyletiques des viverrides (Carnivores Fissipédes). Les formes de Madagascar. Mammalia, 38(4):605-636.

Petter, J.-J. 1962. Recherches sur l'écologie et l'éthologie des Lémuriens malgaches. Mémoires du Muséum National d'Histoire Naturelle, Nouvelle Serie, ser. A (Zoologie), 27:1-146.

Petter, J.-J., and A. Petter-Rousseaux. 1979. Classification of the prosimians. Pp. 1-44, *in* The study of prosimian behavior (G. A. Doyle and R. D. Martin, eds.). Academic Press, London, 696 pp.

Petter, J.-J., R. Albignac, and Y. Rumpler. 1977. Faune de Madagascar. 44. Mammifères lemuriens. O.R.S.T.O.M., Paris, 513 pp.

Petzsch, H. 1970. Kritisches über die neuentdeckte Iriomote-Wildkatze. Der Pelzgewerbe, 20(5):3-7.

Philippi, R. A. 1900. Figuras i descripciones de los murideos de Chile. Anales de Museo Nacional de Chile, Zoologie, 14a:1-70.

Phillips, C. J. 1968. Systematics of megachiropteran bats in the Solomon Islands. University of Kansas Publications, Museum of Natural History, 16:777-837.

Phillips, C. J. 1969. Review of central Asian voles of the genus *Hyperacrius*, with comments on zoogeography, ecology, and ectoparasites. Journal of Mammalogy, 50:457-474.

Phillips, C. J. 1971. The dentition of glossophagine bats: development, morphological characteristics, variation, pathology, and evolution. Miscellaneous Publications, Museum of Natural History, University of Kansas, 54:1-138.

Phillips, C. J., and E. C. Birney. 1968. Taxonomic status of the vespertilionid genus *Anamygdon* (Mammalia: Chiroptera). Proceedings of the Biological Society of Washington, 81:491-498.

Phillips, W. W. A. 1933. Survey of the distribution of mammals in Ceylon. Ceylon Journal of Science (Spolia Zeylonica), 18:133-142.

Phillips, W. W. A. 1980-1984. Manual of the mammals of Sri Lanka. Second revised ed. Wildlife and Nature Protection Society of Sri Lanka, 1:1-116 [1980]; 2:117-267 [1980]; 3:268-389 [1984].

Pieczarka, J. C., and C. Y. Nagamachi. 1988. Cytogenetic studies of *Aotus* from eastern Amazonia: Y/ autosome rearrangement. American Journal of Primatology, 14(3)255-263.

Pieper, H. 1984. Eine neue *Mesocricetus*-Art (Mammalia: Cricetidae) von der griechischen Insel Armathia. Stuttgarter Beiträge zur Naturkunde, ser. B, 107:1-9.

Pieper, H. 1990. *Crocidura zimmermanni* Wettstein, 1953 - Kretaspitzmaus. Pp. 453-460, *in* Handbuch der Säugetiere Europas (J. Niethammer and F. Krapp, eds.). Aula-Verlag, Wiesbaden, 3/I:1-524.

Pietsch, M. 1980. Biometrische Analyse an Schädeln von neun Kleinsäuger-Arten aus der Familie Arvicolidae (Rodentia). Zeitschrift für Zoologische Systematik und Evolutionsforschung, 18:196-211.

Pietsch, M. 1982. *Ondatra zibethicus* (Linnaeus, 1766)—Bisamratte, Bisam. Pp. 177-192, *in* Handbuch der Säugetiere Europas (J. Niethammer and F. Krapp, eds.), vol. 2/I. Akademische Verlagsgesellschaft (Wiesbaden), 649 pp.

Pike, G. C., and I. B. MacAskie. 1969. Marine mammals of British Columbia. Fisheries Research Board of Canada, Bulletin, 171:1-54.

Pilastro, A. 1990. Studio di una popolazione di ghiro (*Glis glis* Linnaeus) in un ambiente forestale dei Colli Berici. Lavori, Societa Veneziana di Scienze Naturali, 15:145-155.

Pilleri, G. 1978. William Roxburgh (1751-1815), Heinrich Julius Lebeck (1801) and the discovery of the Ganges dolphin (*Platanista gangetica* Roxburgh, 1801). Investigations on Cetacea, 9:11-21.

Pilleri, G. and Chen Pei-xun. 1980. *Neophocoena phocaenoides* and *Neophocaena asiaeorientalis*: taxonomic differences. Investigations on Cetacea, 11:25-32.

Pilleri, G., and M. Gihr. 1971. Differences observed in the skulls of *Platanista gangetica* (Roxburgh, 1801) and *indi* (Blyth, 1859). Investigations on Cetacea, 3:13-21.

Pilleri, G., and M. Gihr. 1972. Contribution to the knowledge of cetaceans of Pakistan with particular reference to the genera *Neomeris*, *Sousa*, *Delphinus* and *Tursiops* and description of a new Chinese porpoise (*Neomeris asiaeorientalis*). Investigations on Cetacea, 4:107-162.

Pilleri, G., and M. Gihr. 1973-74. Contribution to the knowledge of the cetaceans of southwest and monsoon Asia (Persian Gulf, Indus Delta, Malabar, Andaman Sea and Gulf of Siam). Investigations on Cetacea, 5:95-149.

Pilleri, G., and M. Gihr. 1975. On the taxonomy and ecology of the finless black porpoise, *Neophocaena* (Cetacea, Delphinidae). Mammalia, 39:657-673.

Pilleri, G., and M. Gihr. 1976*a*. Osteological differences in the cervical vertebrae of *Platanista indi* and *gangetica*. Investigations on Cetacea, 7:105-108.

Pilleri, G., and M. Gihr. 1976*b*. The function and osteology of the manus of *Platanista gangetica* and *Platanista indi*. Investigations on Cetacea, 7:109-118.

Pilleri, G., and M. Gihr. 1977. Observations on the Bolivian (*Inia boliviensis* d'Orbigny, 1834) and the Amazonian bufeo (*Inia geoffrensis* de Blainville, 1817) with description of a new subspecies (*Inia geoffrensis humboldtiana*). Investigations on Cetacea, 8:11-77.

Pilleri, G., and M. Gihr. 1980*a*. Additional considerations of the taxonomy of the genus *Inia*. Investigations on Cetacea, 11:15-24.

Pilleri, G., and M. Gihr. 1980*b*. Checklist of the cetacean genera *Platanista*, *Inia*, *Lipotes*, *Pontoporia*, *Sousa* and *Neophocaena*. Investigations on Cetacea, 11:33-36.

Pinder, L., and A. P. Grosse. 1991. *Blastoceros dichotomus*. Mammalian Species, 380:1-4.
Pine, R. H. 1967. *Baedon meyeri* Pine (Chiroptera: Vespertilionidae) referred to the genus *Antrozous* H. Allen. Southwestern Naturalist, 12:484-485.
Pine, R. H. 1971. A review of the long-whiskered rice rat, *Oryzomys bombycinus*, Goldman. Journal of Mammalogy, 52:590-596.
Pine, R. H. 1972. The bats of the genus *Carollia*. Technical Monograph, Texas Agricultural Experiment Station, Texas A and M University, 8:1-125.
Pine, R. H. 1973*a*. Anatomical and nomenclatural notes on opossums. Proceedings of the Biological Society of Washington, 86:391-402.
Pine, R. H. 1973*b*. Mammals (exclusive of bats) of Belém, Pará, Brazil. Acta Amazonica, 3:47-79.
Pine, R. H. 1976. A new species of *Akodon* (Mammalia: Rodentia: Muridae: Cricetinae) from Isla de los Estados, Argentina. Mammalia, 40:63-68.
Pine, R. H. 1977. *Monodelphis iheringi* (Thomas) is a recognizable species of Brazilian opossum (Mammalia: Marsupialia: Didelphidae). Mammalia, 41:235-237.
Pine, R. H. 1980*a*. Taxonomic notes on "*Monodelphis dimidiata itatiayae* (Miranda-Ribeiro)", *Monodelphis domestica* (Wagner) and *Monodelphis maraxina* Thomas (Mammalia: Marsupialia: Didelphidae). Mammalia, 43:495-499.
Pine, R. H. 1980*b*. Notes on rodents of the genera *Wiedomys* and *Thomasomys* (including *Wilfredomys*). Mammalia, 44:195-202.
Pine, R. H. 1981. Review of the mouse opossums *Marmosa parvidens* Tate and *Marmosa invicta* Goldman (Mammalia: Marsupialia: Didelphidae) with description of a new species. Mammalia, 45:55-70.
Pine, R. H., and J. P. Abravaya. 1978. Notes on the Brazilian opossum *Monodelphis scalops* (Thomas) (Mammalia: Didelphidae). Mammalia, 42:379-382.
Pine, R. H., and C. O. Handley, Jr. 1984. A review of the Amazonian short-tailed opossum *Monodelphis emiliae* (Thomas). Mammalia, 48:239-245.
Pine, R. H., and A. Ruschi. 1976. Concerning certain bats described and recorded from Espírito Santo, Brazil. Anales de Instituto de Biologia, Universidad Nacional Autónomica de México, 47:183-196.
Pine, R. H., and R. M. Wetzel. 1975. A new subspecies of *Pseudoryzomys wavrini* (Mammalia: Rodentia: Muridae: Cricetinae) from Bolivia. Mammalia, 39:649-655.
Pine, R. H., D. C. Carter, and R. K. LaVal. 1971. Status of *Bauerus* Van Gelder and its relationships to other nyctophiline bats. Journal of Mammalogy, 52:663-669.
Pine, R. H., S. D. Miller, and M. L. Schamberger. 1979. Contributions to the mammalogy of Chile. Mammalia, 43:339-376.
Pine, R. H., P. L. Dalby, and J. O. Matson. 1985. Ecology, postnatal development, morphometrics, and taxonomic status of the short-tailed opossum *Monodelphis dimidiata*, an apparently semelparous annual marsupial. Annals of Carnegie Museum, 54:195-231.
Pirlot, P. L. 1955. Variabilite intra-generique chez un rongeur africain. Annales du Musee Royal du Congo Belge, Tervuren (Belgique), Serie in 8, Sciences Zoologiques, 39:1-66.
Pizzimenti, J. J. 1975. Evolution of the prairie dog genus *Cynomys*. Occasional Papers of the Museum of Natural History, University of Kansas, 39:1-73.
Pizzimenti, J. J. 1976. Genetic divergence and morphological convergence in the prairie dogs, *Cynomys gunnisoni* and *Cynomys leucurus*. I. Morphological and ecological analyses. II. Genetic analyses. Evolution, 30:345-366; 367-379.
Pizzimenti, J. J., and G. D. Collier. 1975. *Cynomys parvidens*. Mammalian Species, 52:1-3.
Pizzimenti, J. J., and R. S. Hoffmann. 1973. *Cynomys gunnisoni*. Mammalian Species, 25:1-4.
Pizzimenti, J. J., and C. F. Nadler. 1972. Chromosomes and serum proteins of the Utah prairie dog, *Cynomys parvidens* (Sciuridae). Southwestern Naturalist, 17:279-286.
Plassmann, W., W. Peetz, and M. Schmidt. 1987. The cochlea in gerbilline rodents. Brain, Behavior and Evolution, 30:82-101.
Pledge, N. S. 1990. The upper fossil fauna of the Henschke Fossil Cave, Naracoorte, South Australia. Memoirs of the Queensland Museum, 28:247-262.
Pocock, R. I. 1907. On Pallas's cat. Proceedings of the Zoological Society of London, 1907:299-306.
Pocock, R. I. 1915*a*. On some of the external characters of the genus *Linsang* with notes upon the genera *Poiana* and *Eupleres*. Annals and Magazine of Natural History, ser. 8, 16:341-351.
Pocock, R. I. 1915*b*. On the feet and glands and other external characters of the Viverrinae, with the description of a new genus. Proceedings of the Zoological Society of London, 1915:131-149.
Pocock, R. I. 1916*a*. A new genus of African mongooses, with a note on *Galeriscus*. Annals and Magazine of Natural History, ser. 8, 17:176-179.
Pocock, R. I. 1916*b*. On the tooth-change, cranial characters, and classification of the snow leopard or ounce (*Felis uncia*). Annals and Magazine of Natural History, ser. 8, 18:306-316.

Pocock, R. I. 1916*c*. On the external characters of the mongooses (Mungotidae). Proceedings of the Zoological Society of London, 1916:349-374.

Pocock, R. I. 1917. The classification of the existing Felidae. Annals and Magazine of Natural History, ser. 8, 20:329-350.

Pocock, R. I. 1919. The classification of the mongooses (Mungotidae). Annals and Magazine of Natural History, ser. 9, 3:515-524.

Pocock, R. I. 1920*a*. On the external characters of the South American monkeys. Proceedings of the Zoological Society of London, 1920:91-113.

Pocock, R. I. 1920*b*. On the external and cranial characters of the European badger (*Meles*) and of the American badger (*Taxidea*). Proceedings of the Zoological Society of London, 1920:423-436.

Pocock, R. I. 1921*a*. The external characters and classification of the Procyonidae. Proceedings of the Zoological Society of London, 1921:389-422.

Pocock, R. I. 1921*b*. The auditory bulla and other cranial characters in the Mustelidae. Proceedings of the Zoological Society of London, 1921:473-486.

Pocock, R. I. 1921*c*. On the external characters of some species of Lutrinae (otters). Proceedings of the Zoological Society of London, 1921:535-546.

Pocock, R. I. 1921*d* [1922]. On the external characters and classification of the Mustelidae. Proceedings of the Zoological Society of London, 1921:803-837.

Pocock, R. I. 1923. The classification of Sciuridae. Proceedings of the Zoological Society of London, 1923(1):209-246.

Pocock, R. I. 1924. Some external characters of *Orycteropus afer*. Proceedings of the Zoological Society of London, 1924:697-706.

Pocock, R. I. 1929. Tigers. Journal of the Bombay Natural History Society, 33(3):505-541.

Pocock, R. I. 1930*a*. The panthers and ounces of Asia. Journal of the Bombay Natural History Society, 34(1):65-82.

Pocock, R. I. 1930*b*. The panthers and ounces of Asia. Part II. The panthers of Kashmir, India, and Ceylon. Journal of the Bombay Natural History Society, 34(2):307-336.

Pocock, R. I. 1930*c*. The lions of Asia. Journal of the Bombay Natural History Society, 34(3):638-665.

Pocock, R. I. 1932*a*. The black and brown bears of Europe and Asia. Part I. European and Asiatic representatives of the brown bear. Journal of the Bombay Natural History Society, 35(4):771-823.

Pocock, R. I. 1932*b*. The black and brown bears of Europe and Asia. Part II. The sloth bear (*Melursus*), the Himalayan black bear (*Selenarctos*) and the Malayan bear (*Helarctos*). Journal of the Bombay Natural History Society, 36(1):101-138.

Pocock, R. I. 1932*c*. The leopards of Africa. Proceedings of the Zoological Society of London, 1932(2):543-541.

Pocock, R. I. 1932*d*. The marbled cat (*Pardofelis marmorata*) and some other Oriental species, with the definition of a new genus of the Felidae. Proceedings of the Zoological Society of London, 1932:741-766.

Pocock, R. I. 1933*a*. The civet-cats of Asia. Journal of the Bombay Natural History Society, 36(2):421-449.

Pocock, R. I. 1933*b*. The civet cats of Asia. Part II. Journal of the Bombay Natural History Society, 36(3):629-656.

Pocock, R. I. 1933*c*. The palm civets or 'toddy cats' of the genera *Paradoxurus* and *Paguma* inhabiting British India. Part I. Journal of the Bombay Natural History Society, 36(4):855-877.

Pocock, R. I. 1933*d*. The rarer genera of oriental Viverridae. Proceedings of the Zoological Society of London, 1933:969-1035.

Pocock, R. I. 1934*a*. The geographical races of *Paradoxurus* and *Paguma* found to the east of the bay of Bengal. Proceedings of the Zoological Society of London, 1934:613-683.

Pocock, R. I. 1934*b*. The palm civets or 'toddy cats' of the genera *Paradoxurus* and *Paguma* inhabiting British India. Part III. Journal of the Bombay Natural History Society, 37(2):314-346.

Pocock, R. I. 1934*c*. The races of the striped and brown hyænas. Proceedings of the Zoological Society of London, 1934:799-825.

Pocock, R. I. 1936*a*. The oriental yellow-throated marten (*Lamprogale*). Proceedings of the Zoological Society of London, 1936:531-553.

Pocock, R. I. 1936*b*. The polecats of the genera *Putorius* and *Vormela* in the British Museum. Proceedings of the Zoological Society of London, 1936(2):691-723.

Pocock, R. I. 1937. The mongooses of British India, including Ceylon and Burma. Journal of the Bombay Natural History Society, 39(2):211-245.

Pocock, R. I. 1939*a*. The fauna of British India, including Ceylan and Burma. Mammalia. Vol. 1. Primates and Carnivora (in part). Taylor and Francis, Ltd., London, 463 pp.

Pocock, R. I. 1939*b*. The races of jaguar (*Panthera onca*). Novitates Zoologicae, 41:406-422.

Pocock, R. I. 1940*a*. The hog-badgers (*Arctonyx*) of British India. Journal of the Bombay Natural History Society, 41(3):461-469.

Pocock, R. I. 1940*b*. The races of Geoffroy's cat (*Oncifelis geoffroyi*). Annals and Magazine of Natural History, ser. 11, 6:350-355.

Pocock, R. I. 1941*a*. The fauna of British India, including Ceylon and Burma. Mammalia. Vol. II. Carnivora (suborders Aeluroidae (part) and Arctoidae). Taylor and Francis, Ltd., London, 503 pp.

Pocock, R. I. 1941*b*. Some new geographical races of *Leopardus*, commonly known as Ocelots and Margays. Annals and Magazine of Natural History, ser. 11, 8:234-239.

Pocock, R. I. 1951. Catalogue of the genus *Felis*. British Museum (Natural History), London, 190 pp.

Pocock, T. N. 1976. Pliocene mammalian microfauna from Langebaanweg: a new fossil genus linking the Otomyinae with the Murinae. South African Journal of Science, 72:58-60.

Pocock, T. N. 1987. Plio-Pleistocene fossil mammalian microfauna of southern Africa—a preliminary report including description of two new fossil muroid genera (Mammalia: Rodentia). Palaeontologia Africana, 26:69-91.

Poduschka, W., and C. Poduschka. 1982. Die taxonomische Zugehörigkeit von *Dasogale fontoynonti* G. Grandidier, 1928. Sitzungsberichte der Österreichischen Akademie der Wissenschaften, Mathematisch-Naturwissenschaftliche Klasse, Abteilung I, 191:253-264.

Poduschka, W., and C. Poduschka. 1983. The taxonomy of the extant Solenodontidae (Mammalia: Insectivora), a synthesis. Sitzungsberichte der Österreichischen Akademie der Wissenschaften, Mathematisch-naturwissenschaftliche Klasse, Abteilung I, 192:225-238.

Poduschka, W., and C. Poduschka. 1985. Beiträge zur Kenntnis der Gattung *Podogymnura* Mearns, 1905 (Insectivora, Echinosoricinae). Sitzungsberichte der Österreichischen Akademie der Wissenschaften, Mathematisch-naturwissenschaftliche Klasse, Abteilung I, 194:1-22.

Poglayen-Neuwall, I. 1965. Gefangenschaftsbeobachtungen an Makibären (*Bassaricyon* Allen 1876). Zeitschrift für Säugetierkunde, 30:321-366.

Poglayen-Neuwall, I., and D. E. Toweill. 1988. *Bassariscus astutus*. Mammalian Species, 327:1-8.

Pohle, H. 1920. Die Unterfamilie der Lutrinae. Archiv für Naturgeschichte, 85A(9):1-247.

Pohle, H. 1926. Notizen über africanische Elephanten. Zeitschrift für Säugetierkunde, 1:58-64.

Poole, A. J., and V. S. Schantz. 1942. Catalog of the type specimens of mammals in the United States National Museum, including the Biological Survey's collection. Bulletin of the United States National Museum, 178:1-705.

Poole, W. E. 1979. The status of Australian Macropodidae. Pp. 13-27, *in* The status of endangered Australasian wildlife (M. J. Tyler, ed.). Royal Society of South Australia, Adelaide, 210 pp.

Poole, W. E. 1982. *Macropus giganteus*. Mammalian Species, 187:1-8.

Popov, V. V. 1981. Age-dependent and specific peculiarities of the correlations among the morphological features of sympatric populations of species of the genus *Apodemus* Kaup, 1829 (Rodentia, Mammalia). Acta Zoologica Bulgarica, 17:38-51.

Potter, M., J. H. Nadeau, and M. P. Cancro (eds.). 1986. The wild mouse in immunology. Current Topics in Microbiology and Immunology, 127:1-395.

Pousargues, F. de. 1894. Description d'une nouvelle espéce de mammifères du genre *Crossarchus*, et considerations sur la répartition geographique des crossarques rayes. Archives du Museum National d'Histoire Naturelle (Paris), ser. 3, 6:121-134.

Pousargues, F. de. 1898. Sur l'identitè specifique du *Felis Bieti* (A.M. Edw.) et du *Felis pallida* (Büchn.). Bulletin du Muséum National d'Historie Naturelle (Paris), 4:357-359.

Powell, R. A. 1981. *Martes pennanti*. Mammalian Species, 156:1-6.

Pradhan, M. S., A. Mondal, and V. C. Agrawal. 1989. Proposal of an additional species in the genus *Bandicota* Gray (order: Rodentia; fam: Muridae) from India. Mammalia, 53(3):369-376.

Prasad, M. R. N. 1957. Male genital tract of the Indian and Ceylonese palm squirrels and its bearing on the systematics of the Sciuridae. Acta Zoologica, 38:1-26.

Prater, S. H. 1980. The book of Indian animals. Third ed., corrected. Bombay Natural History Society, 324 pp.

Preble, E. A. 1899. Revision of the jumping mice of the genus *Zapus*. North American Fauna, 15:1-41.

Preble, E. A. 1902. A biological investigation of the Hudson Bay region. North American Fauna, 22:1-140.

Preble, E. A. 1908. A biological investigation of the Athabaska-Mackenzie region. North American Fauna, 27:1-574.

Price, P. K., and M. L. Kennedy. 1980. Genic relationships in the white-footed mouse, *Peromyscus leucopus*, and the cotton mouse, *Peromyscus gossypinus*. American Midland Naturalist, 103:73-82.

Pringle, J. A. 1977. The distribution of mammals in Natal: 2. Carnivora. Annals of the Natal Museum, 23(1):93-116.

Pucek, Z. 1982. Familie Zapodidae Coues, 1875 - Hüpfmäuse. Pp. 497-538, *in* Handbuch der Säugetiere Europas (Niethammer, H. J. and F. Krapp, eds.). Akademische Verlagsgesellschaft (Wiesbaden), 2/I:1-649.

Qian Yan-wen, Zhang Jie, Zheng Bao-lie, Wang Song, Guan Guan-xun, Shen Xiao-zhou. 1965. [Mammals and birds of southern Xinjiang]. Science Press, Beijing.

Qin Chang-yu. 1991. [On the faunistics and regionalization of glires in Ningxia Autonomous Region]. Acta Theriologica Sinica, 4(4):320 (in Chinese).

Quay, W. B. 1954*a*. The Meibomian glands of voles and lemmings (Microtinae). Miscellaneous Publications, Museum of Zoology, University of Michigan, 82:1-17.

Quay, W. B. 1954*b*. The anatomy of the diastemal palate in microtine rodents. Miscellaneous Publications, Museum of Zoology, University of Michigan, 86:1-41.

Quay, W. B. 1968. The specialized posterolateral sebaceous glandular regions in microtine rodents. Journal of Mammalogy, 49:427-445.

Qui Yu-huang. 1989. [A systematic cluster for the Chinese cape hare, *Lepus capensis*]. Acta Theriologica Sinica, 9:168-172 (in Chinese).

Qumsiyeh, M. B. 1985. The bats of Egypt. Special Publications, The Museum, Texas Tech University Press, 23:1-102.

Qumsiyeh, M. B. 1986. Phylogenetic studies of the rodent family Gerbillidae. I. Chromosomal evolution in the southern African complex. Journal of Mammalogy, 67:680-692.

Qumsiyeh, M. B., and R. K. Chesser. 1988. Rates of protein, chromosome and morphological evolution in four genera of rhombomyine gerbils. Biochemical Systematics and Ecology, 16:89-103.

Qumsiyeh, M. B., and J. K. Jones, Jr. 1986. *Rhinopoma hardwickii* and *Rhinopoma muscatellum*. Mammalian Species, 263:1-5.

Qumsiyeh, M. B., and D. A. Schlitter. 1991. Cytogenetic data on the rodent family Gerbillidae. Occasional Papers, The Museum, Texas Tech University, 144:1-20.

Qumsiyeh, M. B., D. A. Schlitter, and A. M. Disi. 1986. New records and karyotypes of small mammals from Jordan. Zeitschrift für Säugetierkunde, 51:139-146.

Qumsiyeh, M. B., M. J. Hamilton, and D. A. Schlitter. 1987. Problems in using Robertsonian rearrangements in determining monophyly: examples from the genera *Tatera* and *Gerbillurus*. Cytogenetics and Cell Genetics, 44:198-208.

Qumsiyeh, M. B., S. W. King, J. Arroyo-Cabrales, I. R. Aggundey, D. A. Schlitter, R. J. Baker, and K. J. Morrow, Jr. 1990. Chromosomal and protein evolution in morphologically similar species of *Praomys sensu lato* (Rodentia, Muridae). Journal of Heredity, 81:58-65.

Qumsiyeh, M. B., M. J. Hamilton, E. R. Dempster, and R. J. Baker. 1991. Cytogenetics and systematics of the rodent genus *Gerbillurus*. Journal of Mammalogy, 72:89-96.

Radinsky, L. B. 1973. Are stink badgers skunks? Implications of neuroanatomy for mustelid phylogeny. Journal of Mammalogy, 54:585-593.

Radinsky, L. B. 1975. Viverrid neuroanatomy: Phylogenetic and behavorial implications. Journal of Mammalogy, 56(1):130-150.

Radjabli, S. I., M. N. Meier, F. N. Golenishchev, and A. A. Isaenko. 1984. [Karyological peculiarities of the Mongolian vole and its relations within the subgenus *Microtus* (Rodentia, Cricetidae)]. Zoologicheskii Zhurnal, 63:441-446 (in Russian).

Radosavlievic, J., M. Vujosevic, and S. Zivkovic. 1990. [Chromosome banding of five arvicolid rodent species from Yugoslavia]. Arhiv Bioloskih Nauka (Beograd), 42:183-194 (in Russian).

Radtke, M., and J. Niethammer. 1984 [1985]. Zur Stellung der Pestratte (*Nesokia indica*) im System der Murinae. Säugetierkundliche Mitteilungen, 32:13-16.

Ragni, B., and E. Randi. 1986. Multivariate analysis of craniometric characters in European wild cat, domestic cat, and African wild cat (genus *Felis*). Zeitschrift für Säugetierkunde, 51:243-251.

Rahm, U. 1967. Les Murides des environs du Lac Kivu et des regions voisines (Afrique Centrale) et leur ecologie. Revue Suisse de Zoologie, 74:439-519.

Rahm, U. 1970. Ecology, zoogeography and systematics of some African forest monkeys. Pp. 591-626, *in* Old World Monkeys (J. R. Napier and P. H. Napier, eds.). Academic Press, London, 660 pp.

Rahm, U. 1976. Zur Morphologie des Magens von *Tachyoryctes splendens* Ruppell, 1835 (Rodentia, Rhizomyidae). Säugetierkundliche Mitteilungen, 24:148-150.

Rahm, U. 1980. Die afrikanische Wurzelratte, *Tachyoryctes*. Die Neue Brehm-Bücherei, 528:1-60.

Rahm, U., and A. Christiaensen. 1963. Les mammiferes de la region occidentale du Lac Kivu. Annales Musée Royal de l'Afrique Centrale, Tervuren-Belgique, ser. 8, 118:1-83.

Ralls, K. 1973. *Cephalophus maxwelli*. Mammalian Species, 31:1-4.

Ralls, K. 1978. *Tragelaphus eurycerus*. Mammalian Species, 111:1-4.

Ramalhinho, M. G. 1985. On the taxonomic position of Portuguese mole. Arquivos do Museu Bocage, ser. A, 3:1-12.

Ramalhinho, M. G., and M. da L. Mathias. 1988. *Arvicola terrestris monticola* de Selys-Longchamps, 1838, new to Portugal (Rodentia, Arvicolidae). Mammalia, 52:429-431.
Raman, R., and T. Sharma. 1977. Karyotype evolution and speciation in genus *Rattus* Fischer. Journal of Scientific and Industrial Research, 36:385-404.
Rana, B. D. 1985. Ecological distribution of *Rattus meltada* in India. Journal of the Bombay Natural History Society, 82:573-580.
Ranck, G. L. 1968. The rodents of Libya: Taxonomy, ecology, and zoogeographical relationships. Bulletin of the United States National Museum, 275:1-264.
Rao, S. R. V., and S. C. Lakhotia. 1972. Chromosomes of *Rattus blanfordi*. Journal of Heredity, 63:44-47.
Rao, S. R. V., K. Vasantha, B. K. Thelma, R. C. Juyal, and S. C. Jhanwar. 1983. Heterochromatin variation and sex chromosome polymorphism in *Nesokia indica*: a population study. Cytogenetics and Cell Genetics, 35:233-237.
Rathbun, G. B. 1979. *Rhynchocyon chrysopygus*. Mammalian Species, 117:1-4.
Rau, R. E. 1978. Additions to the revised list of preserved material of the extinct Cape Colony quagga and notes on the relationship and distribution of southern plains zebras. Annals of the South African Museum, 77:27-45.
Rausch, R. L. 1953. On the status of some Arctic mammals. Arctic (Journal of the Arctic Institute of North America), 6:91-148.
Rausch, R. L. 1963*a*. Geographic variation in size in North American brown bears, *Ursus arctos* L., as indicated by condylobasal length. Canadian Journal of Zoology, 41:33-45.
Rausch, R. L. 1963*b*. A review of the distribution of Holarctic Recent mammals. Pp. 29-43, *in* Pacific basin biogeography (J. L. Gressitt, ed.). Bishop Museum Press, Honolulu, 563 pp.
Rausch, R. L. 1964. The specific status of the narrow-skulled vole (subgenus *Stenocranius* Kashchenko) in North America. Zeitschrift für Säugetierkunde, 29:343-358.
Rausch, R. L. 1977. On the zoogeography of some Beringian mammals. Pp. 162-177, *in* Uspekhi sovremennoi teriologii [Advances in modern theriology] (V. E. Sokolov, ed.). Nauka, Moscow, 296 pp. (in Russian).
Rausch, R. L., and V. R. Rausch. 1965. Cytogenetic evidence for the specific distinction of an Alaskan marmot, *Marmota broweri* Hall and Gilmore (Mammalia: Sciuridae). Chromosoma (Berlin), 16:618-623.
Rausch, R. L., and V. R. Rausch. 1968. On the biology and systematic position of *Microtus abbreviatus* Miller, a vole endemic to the St. Matthew Islands, Bering Sea. Zeitschrift für Säugetierkunde, 33:65-99.
Rausch, R. L., and V. R. Rausch. 1971. The somatic chromosomes of some North American marmots. Mammalia, 35:85-101.
Rausch, R. L., and V. R. Rausch. 1972. Observations on chromosomes of *Dicrostonyx torquatus stevensoni* Nelson and chromosomal diversity in varying lemmings. Zeitschrift für Säugetierkunde, 37:372-384.
Rausch, R. L., and V. R. Rausch. 1975*a*. Relationships of the red-backed vole, *Clethrionomys rutilus* (Pallas) in North America: karyotypes of the subspecies *dawsoni* and *albiventer*. Systematic Zoology, 24:163-170.
Rausch, R. L., and V. R. Rausch. 1975*b*. Taxonomy and zoogeography of *Lemmus* spp. (Rodentia, Arvicolinae), with notes on laboratory-reared lemmings. Zeitschrift für Säugetierkunde, 40:8-34.
Rausch, V. R., and R. L. Rausch. 1974. The chromosome complement of the yellow-cheeked vole, *Microtus xanthognathus* (Leach). Canadian Journal of Genetics and Cytology, 16:267-272.
Rausch, V. R., and R. L. Rausch. 1982. The karyotype of the Eurasian flying squirrel, *Pteromys volans* (L.), with a consideration of karyotypic and other distictions in *Glaucomys* spp. (Rodentia: Sciuridae). Proceedings of the Biological Society of Washington, 95:58-66.
Rautenbach, I. L., and D. A. Schlitter. 1978. Revision of genus *Malacomys* of Africa (Mammalia: Muridae). Annals of Carnegie Museum, 47:385-422.
Ray, C. E. 1962. Oryzomyine rodents of the Antillean subregion. Unpubl. Ph. D. dissertation, Harvard University, 356 pp.
Redford, K. H., and J. Eisenberg. 1992. Mammals of the Neotropics, 2. The southern cone. University of Chicago Press, Chicago, IL, 430 pp.
Redford, K. H., and R. M. Wetzel. 1985. *Euphractus sexcinctus*. Mammalian Species, 252:1-4.
Reeves, R. R., and R. L. Brownell, Jr. 1989. Susu - *Platanista gangetica* (Roxburgh, 1801) and *Platanista minor* Owen, 1853. Pp. 69-100, *in* Handbook of marine mammals: River dolphins and the larger toothed whales (S. H. Ridgway and R. Harrison, eds.). Academic Press, London, 4:1-442.
Reeves, R. R., and S. Leatherwood. 1985. Bowhead whale - *Balaena mysticetus*. Pp. 305-344, *in* Handbook of marine mammals: The sirenians and baleen whales (S. H. Ridgway and R. Harrison, eds.). Academic Press, London, 3:1-362.
Reeves, R. R., and S. Tracey. 1980. *Monodon monoceros*. Mammalian Species, 127:1-7.
Rehage, H.-O. 1984. Die Säugetiere Westfalens. Gartenschläfer — *Eliomys quercinus* (Linnaeus, 1766). Abhandlungen aus dem Westfalischen Provinzial Museum für Naturkunde, 46(4):163-167.

Rehage, H.-O., and K. Preywisch. 1984. Die Säugetiere Westfalens. Sieberschläfer—*Glis glis* (Linnaeus, 1766). Abhandlungen aus dem Westfalischen Provinzial Museum für Naturkunde, 46(4):167-172.

Rehage, H.-O., and G. Steinborn. 1984. Die Säugetiere Westfalens. Haselmaus—*Muscardinus avellanarius* (Linnaeus,1758). Abhandlungen aus dem Westfalischen Provinzial Museum für Naturkunde, 46(4):172-181.

Reich, L. M. 1981. *Microtus pennsylvanicus*. Mammalian Species, 159:1-8.

Reichstein, H. 1957. Schädelvariabilität europäischer Mauswiesel (*Mustela nivalis* L.) und Hermeline (*Mustela erminea* L.) in Beziehung zu Verbreitung und Geschlecht. Zeitschrift für Säugetierkunde, 22:151-182.

Reichstein, H. 1982*a*. *Arvicola sapidus* Miller, 1908— Südwesteuropaische Schermaus. Pp. 211-216, *in* Handbuch der Säugetiere Europas (J. Niethammer and F. Krapp, eds.). Akademische Verlagsgesellschaft (Wiesbaden), 2/I:1-649.

Reichstein, H. 1982*b*. *Arvicola terrestris* (Linnaeus, 1758)—Schermaus. Pp. 217-252, *in* Handbuch der Säugetiere Europas (J. Niethammer and F. Krapp, eds.). Akademische Verlagsgesellschaft (Wiesbaden), 2/I:1-649.

Reig, O. A. 1958. Notas para una actualización del conocimiento de la fauna de la formación Chapadmalal. I. Lista faunística preliminar. Acta Geologica Lilloana, 2:241-253.

Reig, O. A. 1978. Roedores cricetidos del Plioceno superior de la Provincia de Buenos Aires. Publicaciones del Museo Municipal de Ciencias Naturales de Mar del Plata, 2:164-190.

Reig, O. A. 1980. A new fossil genus of South American cricetid rodents allied to *Wiedomys*, with an assessment of the Sigmodontinae. Journal of Zoology (London), 192:257-281.

Reig, O. A. 1984. Distribução geográfica e história evolutiva dos roedores muroideos sulamericanos (Cricetidae: Sigmodontinae). Revista Brasileira de Genetica, 7:333-365.

Reig, O. A. 1986. Diversity patterns and differentiation of high Andean rodents. Pp. 404-440, *in* High altitude tropical biogeography (F. Vuilleumier and M. Monasterio, eds.). Oxford University Press, New York, 649 pp.

Reig, O. A. 1987. An assessment of the systematics and evolution of the Akodontini, with the description of new fossil species of *Akodon* (Cricetidae: Sigmodontinae). Fieldiana, Zoology, New Series, 39:347-399.

Reig, O. A., J. R. Contreras, and M. J. Piantanida. 1966. Contribución a la elucidación de la sistemática de las entidades del género *Ctenomys* (Rodentia, Octodontidae). I. Relaciones de parentesco entre muestras de ocho poblaciones de tuco-tucos inferidas del estudio estadístico de variables del fenotipo y su correlación con las características del cariotipo. Contribuciones Científicas Facultad Ciéncias Exactas Naturales, Universidad Buenos Aires (Zoologia), 2:299-352.

Reig, O. A., M. Aguilera, M. A. Barros, and M. Useche. 1980. Chromosomal speciation in a Rassenkreis of Venezuelan spiny rats (genus *Proechimys*; Rodentia: Echimyidae). Genetica, 52/53:291-312.

Reig, O. A., J. A. W. Kirsch, and L. G. Marshall. 1987. Systematic relationships of the living and Neocenozoic American opossum-like marsupials (suborder Didelphimorphia) with comments on the classification of these and of the Cretaceous and Paleogene New World and European metatherians. Pp. 1-92, *in* Possums and opossums: Studies in evolution (M. Archer, ed.). Surrey Beatty and Sons Pty. Ltd. and Royal Zoological Society of New South Wales, Sydney, 1:1-400, 4 pls.

Reig, O. A., M. Aguilera, and A. Perez-Zapata. 1990*a*. Cytogenetics and karyosystematics of South American oryzomyine rodents (Cricetidae: Sigmodontinae). II. High numbered karyotypes and chromosomal heterogeneity in Venezuelan *Zygodontomys*. Zeitschrift für Säugetierkunde, 55:361-370.

Reig, O. A., C. Busch, M. O. Ortells, and J. R. Contreras. 1990*b*. An overview of evolution, systematics, population biology, cytogenetics, molecular biology, and speciation in *Ctenomys*. Pp. 71-96, *in* Evolution of subterranean mammals at the organismal and molecular levels: proceedings of the fifth International Theriological Congress held in Rome, Italy, August 22-29, 1989 (E. Nevo and O. A. Reig, eds.). A. R. Liss (Wiley-Liss), New York, 422 pp.

Rekovets, L. I. 1990. Principal developmental stages of the water vole genus *Arvicola* (Rodentia, Mammalia) from the Eastern European Pleistocene. Pp. 369-384, *in* International symposium evolution, phylogeny and biostratigraphy of arvicolids (Rodentia, Mammalia) (O. Fejfar and W.-D. Heinrich, eds.). Geological Survey, Prague, 448 pp.

Rempe, U. 1970. Morphometrische Untersuchungen zur Klärung der Verwandtschaft von Steppeniltis, Waldiltis und Frettchen. Verhand. Deutschen Zool. Gesellschaft, 3(7):186-367.

Rengger, J. R. 1830. Naturgeschichte de Säugethiere von Paraguay. Schweighausersche Buchhandlung, Basel, Switzerland, 394 pp.

Rennert, P. D., and C. W. Kilpatrick. 1986. Biochemical systematics of populations of *Peromyscus boylii*. I. Populations from east-central Mexico with low fundamental numbers. Journal of Mammalogy, 67:481-488.

Rennert, P. D., and C. W. Kilpatrick. 1987. Biochemical systematics of *Peromyscus boylii*. II. Chromosomally variable populations from eastern and southern Mexico. Journal of Mammalogy, 68:799-811.

Repenning, C. A. 1967. Subfamilies and genera of the Soricidae. Geological Survey Professional Paper, 565:1-74.

Repenning, C. A. 1968. Mandibular musculature and the origin of the subfamily Arvicolinae (Rodentia). Acta Zoologica Cracoviensia, 13:1-72.

Repenning, C. A. 1980. Faunal exchanges between Siberia and North America. Canadian Journal of Anthropology, 1:37-44.

Repenning, C. A. 1983. *Pitymys meadensis* Hibbard from the Valley of Mexico and the classification of North American species of *Pitymys* (Rodentia: Cricetidae). Journal of Vertebrate Paleontology, 2:471-482.

Repenning, C. A. 1990. Of mice and ice in the Late Pliocene of North America. Arctic, 43:314-323.

Repenning, C. A., and F. M. Grady. 1988. The microtine rodents of the Cheetah Room Fauna, Hamilton Cave, West Virginia, and the spontaneous origin of *Synaptomys*. United States Geological Survey Bulletin, 1853:1-32.

Repenning, C. A., and R. H. Tedford. 1977. Otarioid seals of the Neogene. United States Geological Survey, Professional Paper, 992:1-93.

Repenning, C. A., R. S. Peterson, and C. L. Hubbs. 1971. Contributions to the systematics of the southern fur seals, with particular reference to the Juan Fernandez and Guadalupe species. Antarctic Research Series, 18:1-34.

Repenning, C. A., O. Fejfar, and W.-D. Heinrich. 1990. Arvicolid rodent biochronology of the Northern Hemisphere. Pp. 385-417, *in* International symposium evolution, phylogeny and biostratigraphy of arvicolids (Rodentia, Mammalia) (O. Fejfar and W.-D. Heinrich, eds.). Geological Survey, Prague, 448 pp.

Reumer, J. W. F. 1984. Ruscinian and early Pleistocene Soricidae (Insectivora, Mammalia) from Tegelen (The Netherlands) and Hungary. Scripta Geologica, 73:1-173.

Reumer, J. W. F. 1987. Redefinition of the Soricidae and the Heterosoricidae (Insectivora, Mammalia), with the description of the Crocidosoricinae, a new subfamily of Soricidae. Revue de Paléobiologie, 6:189-192.

Reumer, J. W. F., and A. Meylan. 1986. New developments in vertebrate cytotaxonomy. 9. Chromosome numbers in the order Insectivora (Mammalia). Genetica, 70:119-151.

Reynolds, T. E., K. F. Koopman, and E. E. Williams. 1953. A cave faunule from western Puerto Rico with a discussion of the genus *Isolobodon*. Breviora, Harvard Museum of Comparative Zoology, 12:1-8.

Rhoads, S. N. 1893. Geographic variation in *Bassariscus astutus* with description of a new subspecies. Proceedings of the Academy of Natural Sciences of Philadelphia, 1893:413-418.

Rhoads, S. N. 1896. Mammals collected by Dr. A. Donaldson Smith during his expedition to Lake Rudolf, Africa. Proceedings of the Academy of Natural Sciences of Philadelphia, 1896:517-546.

Rice, D. W. 1977. A list of the marine mammals of the world. Third ed. NOAA Technical Report, NMFS SSRF-711:1-15.

Rice, D. W. 1989. Sperm whale - *Physeter macrocephalus* Linnaeus, 1758. Pp. 177-234, *in* Handbook of marine mammals: River dolphins and the larger toothed whales (S. H. Ridgway and R. Harrison, eds.). Academic Press, London, 4:1-442.

Rice, D. W. 1990. The scientific name of the pilot whale - a rejoinder to Schevill. Marine Mammal Science, 6(4):359-360.

Rice, D. W., and W. A. Wolman. 1971. Life history and ecology of the gray whale (*Eschrichtius robustus*). American Society of Mammalogists, Special Publication, 3:1-142.

Richardson, B. J., and G. B. Sharman. 1976. Biochemical and morphological observations on the wallaroos (Macropodidae: Marsupialia) with a suggested new taxonomy. Journal of Zoology (London), 179:499-513.

Richter, H. 1970. Zur Taxonomie und Verbreitung der paläarktischen Crociduren. Zoologische Abhandlungen Staatliches Museum für Tierkunde in Dresden, 31:293-304.

Rickart, E. A. 1987. *Spermophilus townsendii*. Mammalian Species, 268:1-6.

Rickart, E. A., and L. R. Heaney. 1991. A new species of *Chrotomys* (Rodentia: Muridae) from Luzon Island, Philippines. Proceedings of the Biological Society of Washington, 104:387- 398.

Rickart, E. A., and P. B. Robertson. 1985. *Peromyscus melanocarpus*. Mammalian Species, 241:1-3.

Rickart, E. A., and E. Yensen. 1991. *Spermophilus washingtoni*. Mammalian Species, 371:1-5.

Rickart, E. A., R. S. Hoffmann and M. Rosenfeld. 1985 [1987]. Karyotype of *Spermophilus townsendii artemesiae* (Rodentia: Sciuridae) and chromosomal variation in the *Spermophilus townsendii* complex. Mammalian Chromosome Newsletter, 26:94-102.

Rickart, E. A., L. R. Heaney, and R. B. Utzurrum. 1991. Distribution and ecology of small mammals along an elevational transect in southeastern Luzon, Philippines. Journal of Mammalogy, 72:458-469.

Riddle, B. R., and J. R. Choate. 1986. Systematics and biogeography of *Onychomys leucogaster* in western North America. Journal of Mammalogy, 67:233-255.

Riddle, B. R., and R. L. Honeycutt. 1990. Historical biogeography in North American arid regions: an approach using mitochondrial-DNA phylogeny in grasshopper mice (genus *Onychomys*). Evolution, 44:1-15.

Ride, W. D. L. 1956. A new fossil *Mastacomys* (Muridae) and a revision of the genus. Proceedings of the Zoological Society of London, 127:431-439.
Ride, W. D. L. 1962. On the use of generic names for kangaroos and wallabies. Australian Journal of Science, 24:367-372.
Ride, W. D. L. 1964*a*. *Antechinus rosamondae*, a new species of dasyurid marsupial from the Pilbara district of Western Australia; with remarks on the classification of *Antechinus*. Western Australian Naturalist, 9:58-65.
Ride, W. D. L. 1964*b*. A review of Australian fossil marsupials. Journal Proceedings of the Royal Society of Western Australia, 47:97-131.
Ride, W. D. L. 1970. A guide to the native mammals of Australia. Oxford University Press, Melbourne, 249 pp.
Rideout, C. B., and R. S. Hoffmann. 1975. *Oreamnos americanus*. Mammalian Species, 63:1-6.
Rieger, I. 1981. *Hyaena hyaena*. Mammalian Species, 150:1-5.
Rimoli, R. O. 1976 [1977]. Roedores fosiles de la Hispaniola. Universidad Central del Este, Serie Científica III, San Pedro de Macorís, Dominican Republic, 54 pp., 19 pls.
Rishi, K. K., and U. Puri. 1984. Chromosomes of *Rattus cutchicus cutchicus* and its systematic position. Folia Biologia (Krakow), 32:209-212.
Robbins, C. B. 1971. Dental nomenclature for *Taterillus* (Thomas) (Rodentia: Cricetidae). Mammalia, 35:629-635.
Robbins, C. B. 1973. Nongeographic variation in *Taterillus gracilis* (Thomas) (Rodentia: Cricetidae). Journal of Mammalogy, 54:222-238.
Robbins, C. B. 1974. Comments on the taxonomy of the West African *Taterillus* (Rodentia: Cricetidae) with the description of a new species. Proceedings of the Biological Society of Washington, 87:395-404.
Robbins, C. B. 1977. A review of the taxonomy of the African gerbils, *Taterillus* (Rodentia: Cricetidae). Pp. 178-194, *in* Uspekhi sovremennoi teriologii [Advances in modern theriology] (V. E. Sokolov, ed.). Nauka, Moscow, 296 pp. (in Russian).
Robbins, C. B. 1978. Taxonomic identification and history of *Scotophilus nigrita* (Schreber) (Chiroptera: Vespertilionidae). Journal of Mammalogy, 59:212-213.
Robbins, C. B. 1980. Small mammals of Togo and Benin. I. Chiroptera. Mammalia, 44:83-88.
Robbins, C. B., and H. W. Setzer. 1985. Morphometrics and distinctness of the hedgehog genera (Insectivora, Erinaceidae). Proceedings of the Biological Society of Washington, 98:112-120.
Robbins, C. B., and E. Van der Straeten. 1982. A new specimen of *Malacomys verschureni* from Zaire, Central Africa (Rodentia, Muridae). Revue de Zoologie Africaine, 96:216-220.
Robbins, C. B., and E. Van der Straeten. 1989. Comments on the systematics of *Mastomys* Thomas 1915 with the description of a new West African species. Senckenbergiana Biologica, 69:1-14.
Robbins, C. B., J. W. Krebs, Jr., and K. M. Johnson. 1983. *Mastomys* (Rodentia: Muridae) species distinguished by hemoglobin pattern differences. American Journal of Tropical Medicine and Hygiene, 32:624-630.
Robbins, C. B., F. deVree, and V. van Cakenberghe. 1985. A systematic revision of the African bat genus *Scotophilus* (Vespertilionidae). Annales Musée Royal de l'Afrique Centrale, Tervuren, Belgique, Sciences Zoologiques, 246:51-84.
Robbins, L. W., and R. J. Baker. 1978. Karyotypic data for African mammals, with a description of an *in vivo* bone marrow technique. Bulletin of the Carnegie Museum of Natural History, 6:188-210.
Robbins, L. W., and R. J. Baker. 1980. G- and C-band studies on the primitive karyotype for *Reithrodontomys*. Journal of Mammalogy, 61:708-714.
Robbins, L. W., and R. J. Baker. 1981. An assessment of the nature of rearrangements in eighteen species of *Peromyscus* (Rodentia: Cricetidae). Cytogenetics and Cell Genetics, 31:194-202.
Robbins, L. W., and D. A. Schlitter. 1981. Systematic status of dormice (Rodentia: Gliridae) from southern Cameroon, Africa. Annals of Carnegie Museum, 50(9):271-288.
Robbins, L. W., and H. W. Setzer. 1979. Additional records of *Hylomyscus baeri* Heim deBalsac and Aellen (Rodentia, Muridae) from western Africa. Mammalia, 60:649-650.
Robbins, L. W., J. R. Choate, and R. L. Robbins. 1980. Nongeographic and interspecific variation in four species of *Hylomyscus* (Rodentia: Muridae) in southern Cameroon. Annals of Carnegie Museum, 49:31-48.
Robbins, L. W., M. P. Moulton, and R. J. Baker. 1983. Extent of geographic range and magnitude of chromosomal evolution. Journal of Biogeography, 10:533-541.
Robbins, L. W., M. H. Smith, M. C. Wooten, and R. K. Selander. 1985. Biochemical polymorphism and its relationship to chromosomal and morphological variation in *Peromyscus leucopus* and *Peromyscus gossypinus*. Journal of Mammalogy, 66:498-510.
Roberts, A. 1929. New forms of African mammals. Annals of the Transvaal Museum, 13:82-121.
Roberts, A. 1938. Descriptions of new forms of mammals. Annals of the Transvaal Museum, 19(2):231-245.
Roberts, A. 1946. Descriptions of numerous new subspecies of mammals. Annals of the Transvaal Museum, 20:303-328.

Roberts, A. 1951. The mammals of South Africa. Trustees of "The mammals of South Africa" book fund, Johannesburg, 700 pp.
Roberts, M. 1991. Origin, dispersal routes, and geographic distribution of *Rattus exulans*, with special reference to New Zealand. Pacific Science, 45:123-130.
Roberts, M. S., and J. L. Gittleman. 1984. *Ailurus fulgens*. Mammalian Species, 222:1-8.
Roberts, T. J. 1977. The mammals of Pakistan. Ernest Benn Limited, London, 361 pp.
Robertson, P. B., and G. G. Musser. 1976. A new species of *Peromyscus* (Rodentia: Cricetidae), and a new specimen of *P. simulatus* from southern Mexico, with comments on their ecology. Occasional Papers of the Museum of Natural History, University of Kansas, 47:1-8.
Robertson, P. B., and E. A. Rickart. 1975. *Cryptotis magna*. Mammalian Species, 61:1-2.
Robineau, D. 1973. Sur deux rostres de *Mesoplodon* (Cetcea, Hyperoodontidae). Mammalia, 37(3):504-513.
Robineau, D. 1989. Les cetaces des Iles Kerguelen. Mammalia, 53:265-278.
Robinson, H. C., and C. B. Kloss. 1918*a*. A nominal list of the Sciuridae of the Oriental Reigon with a list of specimens in the collection of the Zoological Survey of India. Records of the Indian Museum, 15(4), no. 21:171-254.
Robinson, H. C., and C. B. Kloss. 1918*b*. Results of an expedition to Korinchi Peak, Sumatra. I. Mammals. Journal of the Federation of Malay States Museum, 8(2):1-81.
Robinson, H. C., and C. B. Kloss. 1919*a*. On a collection of mammals from the Bencoolen and Palembang residencies, South West Sumatra. Journal of the Federation of Malay States Museum, 7(4):257-291.
Robinson, H. C., and C. B. Kloss. 1919*b*. On mammals, chiefly from the Ophir District, West Sumatra. Journal of the Federation of Malay States Museum, 7(4):299-323.
Robinson, H. C., and C. B. Kloss. 1922. New mammals from French Indo-China and Siam. Annals and Magazine of Natural History, ser. 9, 9:87-99.
Robinson, J. W., and R. S. Hoffmann. 1975. Geographic and interspecific cranial variation in big-eared ground squirrels (*Spermophilus*): a multivariate study. Systematic Zoology, 24:79-88.
Robinson, T. J. 1982. Key to the South African Leporidae. South African Journal of Zoology, 17:220-222.
Robinson, T. J., and N. J. Dippenaar. 1983*a*. Morphometrics of the South African Leporidae. I. Genus *Pronolagus* Lyon, 1904. Annals of the Museum, Royal African Centre for Science, Zoology, 237:43-61.
Robinson, T. J., and N. J. Dippenaar. 1983*b*. The status of *Lepus saxatilis*, *L. whytei* and *L. crawshayi* in southern Africa. Acta Zoologica Fennica, 174:35-39.
Robinson, T. J., and N. J. Dippenaar. 1987. Morphometrics of the South African Leporidae. II. *Lepus* Linnaeus, 1758, and *Bunolagus* Thomas, 1929. Annals of the Transvaal Museum, 34:379-404.
Robinson, T. J., and F. F. B. Elder. 1987. Extensive genome reorganization in the African rodent genus *Otomys*. Journal of Zoology (London), 211:735-745.
Robinson, T. J., and J. D. Skinner. 1983. Karyology of the riverine rabbit, *Bunolagus monticularis*, and its taxonomic implications. Journal of Mammalogy, 64:678-681.
Robinson, T. J., F. F. B. Elder, and W. Lopez-Forment. 1981. Banding studies in the volcano rabbit, *Romerolagus diazi* and Crawshay's hare, *Lepus crawshayi*. Evidence of the leporid ancestral karyotype. Canadian Journal of Genetics and Cytology, 23:469-474.
Robinson, T. J., F. F. B. Elder, and J. A. Chapman. 1983. Evolution of chromosomal variation in cottontails, genus *Sylvilagus* (Mammalia: Lagomorpha): *S. aquaticus*, *S. floridanus*, and *S. transitionalis*. Cytogenetics and Cell Genetics, 35:216-222.
Robinson, T. J., F. F. B. Elder, and J. A. Chapman. 1984. Evolution of chromosomal variation in cottontails, genus *Sylvilagus* (Mammalia: Lagomorpha). II. *Sylvilagus audubonii*, *S. idahoensis*, *S. nuttallii* and *S. palustris*. Cytogenetics and Cell Genetics, 38:282-289.
Robinson, T. J., J. D. Skinner, and A. S. Haim. 1986. Close chromosomal congruence in two species of ground squirrel: *Xerus inauris* and *X. princeps* (Rodentia: Sciuridae). South African Journal of Zoology, 21:100-105.
Roche, J. 1964. Description d'une nouvelle sous-espece de *Thallomys* de l'est Africain *Thallomys paedulcus somaliensis*. Mammalia, 28:94-100.
Roche, J. 1972. Systematique du genre *Procavia* et des damans en general. Mammalia, 36:22-49.
Roche, J., and F. Petter. 1968. Faits nouveaux concernant trois gerbillides mal connus de Somalia: *Ammodillus imbellis* (De Winton), *Microdillus peeli* (De Winton), *Monodia juliani* (Saint Leger). Monitore Zoologico Italiano, N. S., 2 (suppl.):181-198.
Roche, J., E. Capanna, M. V. Civitelli, and A. Ceraso. 1984. Caryotypes des rongeurs de Somalie. 4. Premiere capture de rongeurs arboricoles du sous-genre *Grammomys* (genre *Thamnomys*, Murides) en Republique de Somalie. Monitore Zoologico Italiano, 7:259-277.
Rodriguez, L., V. Barus, and J. Kratochvil. 1979. The genus *Mesocapromys*, a link between the families Echimyidae and Capromyidae. Folia Zoologica, 28(2):97-102.
Rodríguez-Durán, A., and T. H. Kunz. 1992. *Pteronotus quadridens*. Mammalian Species, 395:1-4.

Roesler, U., and G. R. Witte. 1969. Chorologische Betrachtungen zur Subspeziesbildung einiger Vertebraten im italienischen und balkanischen Raum. Zoologischer Anzeiger, 182:27-51.

Roest, A. I. 1973. Subspecies of the sea otter, *Enhydra lutris*. Contributions in Science, Natural History Museum of Los Angeles County, 140(252):1-17.

Rogers, D. S. 1983. Phylogenetic affinities of *Peromyscus* (*Megadontomys*) *thomasi*: Evidence from differentially stained chromosomes. Journal of Mammalogy, 64:617-623.

Rogers, D. S. 1989. Evolutionary implications of chromosomal variation among spiny pocket mice, genus *Heteromys* (Order Rodentia). Southwestern Naturalist, 34:85-100.

Rogers, D. S. 1990. Genic evolution, historical biogeography, and systematic relationships among spiny pocket mice (subfamily Heteromyinae). Journal of Mammalogy, 71:668-685.

Rogers, D. S., and M. D. Engstrom. 1992. Evolutionary implications of allozymic variation in tropical *Peromyscus* of the *mexicanus* species group. Journal of Mammalogy, 73:55-69.

Rogers, D. S., and E. J. Heske. 1984. Chromosomal evolution of the brown mice, genus *Scotinomys* (Rodentia: Cricetidae). Genetica, 63:221-228.

Rogers, D. S., and J. E. Rogers. 1992*a*. *Heteromys oresterus*. Mammalian Species, 396:1-3.

Rogers, D. S., and J. E. Rogers. 1992*b*. *Heteromys nelsoni*. Mammalian Species, 397:1-2.

Rogers, D. S., and D. J. Schmidly. 1982. Systematics of spiny pocket mice (genus *Heteromys*) of the *desmarestianus* species group from México and northern Central America. Journal of Mammalogy, 63:375-386.

Rogers, D. S., I. F. Greenbaum, S. J. Gunn, and M. D. Engstrom. 1984. Cytosystematic value of chromosomal inversion data in the genus *Peromyscus*. Journal of Mammalogy, 65:457-465.

Rogers, M. A. 1991*a*. Evolutionary differentiation within the northern Great Basin pocket gopher, *Thomomys townsendii*. I. Morphological variation. Great Basin Naturalist, 51:109-126.

Rogers, M. A. 1991*b*. Evolutionary differentiation within the northern Great Basin pocket gopher, *Thomomys townsendii*. II. Genetic variation and biogeographic considerations. Great Basin Naturalist, 51:127-152.

Rogovin, K. A. 1984. [A comparative analysis of behaviour and supergeneric groups of jerboas (Rodentia, Dipodidae)]. Zoologicheskii Zhurnal, 64(11):1702-1711 (in Russian).

Roguin, L. de. 1988. Notes sur quelques mammiferes du Baluchistan Iranien. Revue Suisse de Zoologie, 95:595-606.

Rohwer, S. A., and D. L. Kilgore. 1973. Interbreeding in the arid-land foxes, *Vulpes velox* and *V. macrotis*. Systematic Zoology, 22:157-165.

Roig, V. G., and O. A. Reig. 1969. Precipitin test relationships among Argentinian species of the genus *Ctenomys* (Rodentia, Octodontidae). Comparative Biochemistry and Physiology, 30:665-672.

Ronnefeld, U. 1969. Verbreitung und Lebensweise afrikanischer Feloidea (Felidae et Hyaenidae). Säugetierkundliche Mitteilungen, 17(4):285-350.

Rookmaaker, L. C. 1991. The scientific name of the bontebok. Zeitschrift für Säugetierkunde, 66:190-191.

Rookmaaker, L. C., and W. Bergmans. 1981. Taxonomy and geography of *Rousettus amplexicaudatus* (Geoffroy, 1810) with comparative notes on sympatric congeners (Mammalia: Megachiroptera). Beaufortia, 31:1-29.

Rosenberger, A. L. 1977. *Xenothrix* and ceboid phylogeny. Journal of Human Evolotion, 6(5):461-481.

Rosenberger, A. L., and A. F. Coimbra-Filho. 1984. Morphology, taxonomic status and affinities of the lion tamarins, *Leontopithecus* (Callitrichinae, Cebidae). Folia Primatologica, 42(3-4):149-179.

Rosevear, D. R. 1963. On the West African forms of *Heliosciurus* Trouessart. Mammalia, 27:177-185.

Rosevear, D. R. 1965. The bats of West Africa. British Museum (Natural History), London, 418 pp.

Rosevear, D. R. 1969. The rodents of West Africa. Trustees of the British Museum (Natural History), London, 677:1-604.

Rosevear, D. R. 1974. The carnivores of West Africa. Trustees of the British Museum (Natural History), London, 723:1-548.

Ross, G. J. B., and V. G. Cockroft. 1990. Comments on Australian bottlenose dolphins and the taxonomic status of *Tursiops aduncus* (Ehrenberg, 1832). Pp. 101-128, *in* The bottlenose dolphin (S. Leatherwood and R. R. Reeves, eds.). Academic Press, New York, 653 pp.

Ross, P. D. 1988. The taxonomic status of *Cansumys canus*. Abstracts, Symposium of Asian Pacific Mammalogy, Huirou, Beijing, Peoples Republic of China.

Rossolimo, O. L. 1971. [Variability and taxonomy of *Dryomys nitedula* Pallas]. Zoologicheskii Zhurnal, 50(2):247-258 (in Russian).

Rossolimo, O. L. 1976*a*. Novyi vid myshevidnoi soni — *Myomimus setzeri* (Mammalia, Myoxidae) iz Irana [A new species of mouse-like dormouse...from Iran]. Vestnik Zoologii, 1976(4):51-53 (in Russian).

Rossolimo, O. L. 1976*b*. [Taxonomic status of the mouse-like dormouse *Myomimus* (Mammalia, Myoxidae) from Bulgaria]. Zoologicheskii Zhurnal, 55(10):1515-1525 (in Russian).

Rossolimo, O. L. 1989. [Revision of Royle's high-mountain vole *Alticola* (*A.*) *argentatus* (Mammalia: Cricetidae)]. Zoologicheskii Zhurnal, 68:104-114 (in Russian).

Rossolimo, O. L., and I. Ya. Pavlinov. 1985. External genital morphology and its taxonomic significance in the dormouse genera *Myomimus* and *Glirulus* (Rodentia, Gliridae). Folia Zoologica, 34(2):121-124.

Rossolimo, O. L., I. Ya. Pavlinov, O. I. Podtyazhkin, V. S. Skulkin. 1988. [Variability and taxonomy of mountain voles (*Alticola* s. str.) from Mongolia, Tuva, Baikal Region and Altai]. Zoologicheskii Zhurnal, 67:426-437 (in Russian).

Roth, V. L. and R. W. Thorington, Jr. 1982. Relative brain size among African squirrels. Journal of Mammalogy, 63(1):168-173.

Rousseau, M. 1983. Etude des *Arvicanthis* du Museum de Paris par analyses factorielles (Rongeurs, Murides). Mammalia, 47:525-542.

Rudolphi, D. K. A. 1820 [1822]. Einige anatomische Bemerkungen über *Balaena rostrata*. Abhandlungen der Physikalische Klasse der Königlich-Preussischen Akademie der Wissenschaften zu Berlin, 1820-1821(1822):27-40.

Ruedas, L. A. 1986. Chromosomal variability in the New England cottontail, *Sylvilagus transitionalis* (Bangs)[sic], 1895 with evidence of recognition of a new species. Unpubl. M. S. thesis, Fordham University, New York, 37 pp.

Ruedi, M., T. Maddalena, H.-S. Yong, and P. Vogel. 1990. The *Crocidura fuliginosa* species complex (Mammalia: Insectivora) in peninsular Malaysia: biological, karyological and genetical evidence. Biochemical Systematics and Ecology, 18:573-581.

Rümke, C. G. 1985. A review of fossil and recent Desmaninae (Talpidae, Insectivora). Utrecht Micropaleontological Bulletins, Special Publication, 4:1-241.

Rümmler, H. 1932. Über die Schwimmratten (Hydromyinae). Das Aquarium, 1932:131-135.

Rümmler, H. 1938. Die Systematik und Verbreitung der Muriden Neuguineas. Mitteilungen aus dem Zoologische Museum in Berlin, 23:1-297.

Rumpler, Y. 1975. The significance of chromosomal studies in the systematics of the Malagasy lemurs. Pp. 25-40, *in* Lemur Biology (I. Tattersall and R. W. Sussman, eds.). Plenum Press, New York, 365 pp.

Rumpler, Y., S. Warter, C. Rabarivola, J.-J. Petter, and B. Dutrillaux. 1990. Chromosomal evolution in Malagasy lemurs. XII. Chromosomal banding study of *Avahi laniger occidentalis* (syn.: *Lichanotus laniger occidentalis*) and cytogenetic data in favour of its classifcation in a species apart - *Avahi occidentatis*. American Journal of Primatology, 21:307-316.

Rupp, H. 1980. Beiträge zur Systematik, Verbreitung und Ökologie äthiopischer Nagetiere. Ergebnisse mehrerer Forschungsreisen. Säugetierkundliche Mitteilungen, 28(2):81-123.

Ruprecht, A. L. 1979. Kryteria identyfikacji gatunkowej podrodzaju *Sylvaemus* Ognev and Vorobiev, 1923 (Rodentia: Muridae). Przeglad Zoologiczny, 23:340-349.

Ruprecht, A. L., and A. Szwagrzak. 1986. Fat dormouse in the food of the Ural Owl. Przeglad Zoologiczny, 30(4):431-432.

Russell, R. J. 1968*a*. Evolution and classification of the pocket gophers of the subfamily Geomyinae. University of Kansas Publications, Museum of Natural History, 16:473-579.

Russell, R. J. 1968*b*. Revision of pocket gophers of the genus *Pappogeomys*. University of Kansas Publications, Museum of Natural History, 16:581-776.

Ryan, J. M. 1989*a*. Comparative myology and phylogenetic systematics of the Heteromyidae (Mammalia, Rodentia). Miscellaneous Publications, Museum of Zoology, University of Michigan, 176:1-103.

Ryan, J. M. 1989*b*. Evolution of cheek pouches in African pouched rats (Rodentia: Cricetomyinae). Journal of Mammalogy, 70:267-274.

Ryan, R. M. 1965. Taxonomic status of the vespertilionid genera *Kerivoula* and *Phoniscus*. Journal of Mammalogy, 46:517-518.

Ryan, R. M. 1966. A new and some imperfectly known Australian *Chalinolobus* and the taxonomic status of African *Glauconycteris*. Journal of Mammalogy, 47:86-91.

Ryder, O. A., A. T. Kumamoto, B. S. Durrant, and K. Benirschke. 1989. Chromosomal divergence and reproductive isolation in dik-diks. Pp. 208-225, *in* Speciation and its consequences (D. Otte and J. A. Endler, eds). Sinauer Associates Incorporated, Sunderland, MA, 679 pp.

Rzebik-Kowalska, B. 1988. Soricidae (Mammalia, Insectivora) from the Plio-Pleistocene and Middle Quaternary of Morocco and Algeria. Folia Quaternaria, 57:51-90.

Sabatier, M. 1982. Les rongeurs du site Pliocene a hominides de Hadar (Ethiopie). Palaeovertebrata, 12(1):1-56.

Sablina, O. V. 1988. [Taxonomy of voles of the genus *Chionomys* (Rodentia, Microtinae) based on karyological data]. Zoologicheskii Zhurnal, 67:472-475 (in Russian).

Sablina, O. V., J. Zima, S. I. Radjabli, B. Krystufek, and F. N. Goleniscev [Golenishchev]. 1989. [New data on karyotype variation in the pine vole, *Pitymys subterraneus* (Rodentia, Arvicolidae)]. Vestnik Ceskoslovenske Spolecnosti Zoologicke, 53:295-299 (in Russian).
Saha, S. S. 1981. A new genus and a new species of flying squirrel (Mammalia: Rodentia: Sciuridae) from northwestern India. Bulletin of the Zoological Survey of India, 4(3):331-336.
Saint Girons, M. C. 1972. Rectification a propos des auteurs de la description de *Erinaceus algirus*. Mammalia, 36:166-167.
St. Leger, J. 1930. On two species of *Dendromus*. Annals and Magazine of Natural History, ser. 10, 6:622.
St. Leger, J. 1931. A key to the families and genera of African Rodentia. Proceedings of the Zoological Society of London, 1931:957-997.
Saitoh, M., N. Matsuoka, and Y. Obara. 1989. Biochemical systematics of three species of the Japanese long-tailed field mice; *Apodemus speciosus*, *A. giliacus*, and *A. argenteus*. Zoological Science, 6:1005-1018.
Salata-Pilacinska, B. 1990. The southern range of the root vole in Poland. Acta Theriologica, 35:53-67.
Sale, J. B., and M. E. Taylor. 1970. A new four-toed mongoose from Kenya, *Bdeogale crassicauda nigrescens* ssp. nov. Journal of the East African Natural History Society and National Museum, 28:10-15.
Salotti, M. 1984. A Ghjira, ou le loir en Corse. Courrier de la Nature, 89:31-35.
Sanborn, C. C. 1930. Distribution and habits of the three-banded armadillo (*Tolypeutes*). Journal of Mammalogy, 11:61-68, pl 4.
Sanborn, C. C. 1931. Bats from Polynesia, Melanesia, and Malaysia. Field Museum of Natural History, Zoological Series, 18:7-29.
Sanborn, C. C. 1933. Bats of the genera *Anoura* and *Lonchoglossa*. Field Museum of Natural History, Zoological Series, 20:23-28.
Sanborn, C. C. 1937. American bats of the subfamily Emballonurinae. Field Museum of Natural History, Zoological Series, 20:321-354.
Sanborn, C. C. 1947*a*. The South American rodents of the genus *Neotomys*. Fieldiana, Zoology, 31:51-57.
Sanborn, C. C. 1947*b*. Geographical races of the rodent *Akodon jelskii* Thomas. Fieldiana, Zoology, 31:133-142.
Sanborn, C. C. 1949. Bats of the genus *Micronycteris* and its subgenera. Fieldiana, Zoology, 31:215-233.
Sanborn, C. C. 1950. Bats from New Caledonia, the Solomon Islands, and New Hebrides. Fieldiana, Zoology, 31:313-338.
Sanborn, C. C. 1952*a*. Philippine Zoological Expedition 1946-1947. Fieldiana, Zoology, 33:89-158.
Sanborn, C. C. 1952*b*. The status of "*Triaenops wheeleri*" Osgood. Chicago Academy of Sciences, Natural History Miscellanea, 97:1-3.
Sanborn, C. C. 1953. Mammals from Mindanao, Philippine Islands collected by the Danish Philippine Expedition 1951-1952. Videnskabelige Meddeleser fra Dansk Naturhistoriske Forening i Kobenhaven, 115:283-289.
Sanborn, C. C., and J. A. Crespo. 1957. El Murciélago Blanquizco (*Lasiurus cinereus*) y sus subspecies. Boletin del Museo Argentino de Ciencias Naturales "Bernardino Rivadavia", 4:1-13.
Sanborn, C. C., and H. Hoogstraal. 1953. Some mammals of Yemen and their ectoparasites. Fieldiana Zoology, 34(23):229-252.
Sanchez-Cordero, V., and B. Villa-Ramirez. 1988. Variación morphométrica en *Peromyscus spicilegus* (Rodentia: Cricetinae) en a parte nordeste de Jalisco, México. Anales de Instituto de Biologia, Universidad Nacional Autónomica de México, Series Zoologia, 58:819-836.
Sanderson, I. T. 1940. The mammals of the north Cameroon forest area. Being the results of the Percy Sladen expedition to the Mamfe Division of the British Cameroons. Transactions of the Zoological Society of London, 24:623-725.
Santapau, H., and Humayun Abdulali, (eds.). 1960. The hispid hare, *Caprolagus hispidus* (Pearson). Journal of the Bombay Natural History Society, 57:400-402.
Sarà, M., M. Lo Valvo, and L. Zanca. 1990. Insular variation in central Mediterranean *Crocidura* Wagler, 1832 (Mammalia, Soricidae). Bollettino di Zoologia, 57:283-293.
Sarich, V. M. 1969. Pinniped origins and the rate of evolution of carnivore albumins. Systematic Zoology, 18:286-295.
Sarich, V. M. 1973. The giant panda is a bear. Nature, 245:218-220.
Sarich, V. M. 1976. Transferrin. Transactions of the Zoological Society of London, 33:165-171.
Sarich, V. M. 1985. Rodent macromolecular systematics. Pp. 423- 452, *in* Evolutionary relationships among rodents, a multidisciplinary analysis (W. P. Luckett and J.-L. Hartenberger, eds.). Plenum Press, New York, 721 pp.
Savic, I. R. 1982*a*. Familie Spalacidae Gray, 1821—Blindmäuse. Pp. 539-542, *in* Handbuch der Säugetiere Europas (J. Niethammer and F. Krapp, eds.). Akademische Verlagsgesellschaft (Wiesbaden), 2/I:1-649.

Savic, I. R. 1982*b*. *Microspalax leucodon* (Nordmann, 1840)— Westblindmaus. Pp. 543-569, *in* Handbuch der Säugetiere Europas (J. Niethammer and F. Krapp, eds.). Akademische Verlagsgesellschaft (Wiesbaden), 2/I:1-649.

Savic, I. R. 1982*c*. *Spalax polonicus* Mehely, 1909—Bodolische Blindmaus. Pp. 571-576, *in* Handbuch der Säugetiere Europas (J. Niethammer and F. Krapp, eds.). Akademische Verlagsgesellschaft (Wiesbaden), 2/I:1-649.

Savic, I. R. 1982*d*. *Spalax graecus* Nehring, 1898—bukowinische blindmaus. Pp. 577-584, *in* Handbuch der Säugetiere Europas (J. Niethammer and F. Krapp, eds.). Akademische Verlagsgesellschaft (Wiesbaden), 2/I:1-649.

Savic, I. R., and E. Nevo. 1990. The Spalacidae: evolutionary history, speciation and population biology. Progress in Clinical and Biological Research, 335:129-153.

Savic, I. R., and B. Soldatovic. 1977. Prilog poznavanju ekogeografskog rasprostranjenja i evolucije hromozomskih formi spalacidaea Balkan-Skog Poluostrva. Arhiv Bioloskih Nauka (Beograd), 29:141-156.

Savic, I. R., and B. Soldatovic. 1979. Distribution range and evolution of chromosomal forms in the Spalacidae of the Balkan Peninsula and bordering regions. Journal of Biogeography, 6:363-374.

Sbalqueiro, I. J., M. S. Mattevi, L. F. B. Oliveira, and M. J. V. Solano. 1991. B chromosome system in populations of *Oryzomys flavescens* (Rodentia, Cricetidae) from southern Brazil. Acta Theriologica, 36:193-199.

Scala, C., and S. Lovari. 1984. Revision of *Rupicapra* genus. II. A skull and horn statistical comparison of *Rupicapra rupicapra ornata* and *R. rupicapra pyrenaica* chamois. Bollettino di Zoologia, 51:285-294.

Scarff, J. E. 1986. Historic and present distribution of the right whale (*Eubalaena glacialis*) in the eastern North Pacific south of 50 [deg] N and east of 180 [deg] W. Pp. 43-63, *in* Right Whales: past and present status (R. L. Brownell, Jr., P. B. Best, and J. H. Prescott, eds.). Reports of the International Whaling Commission, Special Issue, 10:1-289.

Schaldach, W. J. 1960. *Xenomys nelsoni* Merriam, sus relaciones y sus habitos. Revista de la Sociedad de Historia Natural, 21:425-434.

Schaldach, W. J. 1965. Notas breves sobre algunos mamíferos del sur de México. Anales de Instituto de Biologia, Universidad Nacional Autónomica de México, 35:129-137.

Schauenberg, P. 1974. Données nouvelles sur le chat des sables, "*Felis margarita*" Loche, 1858. Revue Suisse de Zoologie, 81(4):949-969.

Scheffer, V. B. 1958. Seals, sea lions, and walruses, a review of the Pinnipedia. Stanford University Press, Stanford, CA, 179 pp.

Schevill, W. E. 1986. The international code of zoological nomenclature and a paradigm: the name *Physeter catodon* Linnaeus 1758. Marine Mammal Science, 2(2):153-157.

Schevill, W. E. 1987*a*. Note by William E. Schevill [on *Steno bredanensis* in the Mediterranean Sea, Watkins et al.]. Marine Mammal Science, 3(1):77.

Schevill, W. E. 1987*b*. Reply to Holthius, 1987. Marine Mammal Science, 3(1):89-90.

Schevill, W. E. 1990*a*. On stability in zoological nomenclature. Marine Mammal Science, 6(2):168-169.

Schevill, W. E. 1990*b*. Reply to D. W. Rice's rejoinder. Marine Mammal Science, 6(4):360.

Schinz, H. R. 1844-1845. Systematisches verzeichniss aller bis jetzt bekannten Säugethiere, oder, Synopsis mammalium, nach dem Cuvier' schen system. Solothurn, Jent und Gassmann, 2 vols.

Schlawe, L. 1980. Zur geographischen Verbreitung der Ginsterkatzen Gattung *Genetta* G. Cuvier, 1816. Faunistische Abhandlungen Staatliches Museum für Tierkunde in Dresden, 7(15):147-161.

Schlawe, L. 1981. Material, Fundort, Text- und Bildquellen als Grundlagen fur eine Artenliste zur Revision der Gattung *Genetta* G. Cuvier, 1816 (Mammalia, Carnivora, Viverridae). Zoologische Abhandlungen Staatliches Museum für Tierkunde in Dresden, 37(4):85-182.

Schlawe, L. 1986. Seltene Pfleglinge aus Dschungarei und Mongolei: Kulane, *Equus hemionus hemionus* Pallas, 1775. Zoologische Garten, Neue Folge, 56:299-323.

Schlegel, H. 1877. Prospectus for museum publication. Annals of the Royal Zoological Museum of the Netherlands at Leyden, pages not numbered.

Schlegel, H. 1879. Note XIV. *Paradoxurus musschenbroekii*. Notes of the Leyden Museum, 1879:43.

Schliemann, H., and B. Maas. 1978. *Myzopoda aurita*. Mammalian Species, 116:1-2.

Schlitter, D. A., and I. R. Aggundey. 1986. Systematics of African bats of the genus *Eptesicus* (Mammalia: Vespertilionidae). 1. Taxonomic status of the large serotines of eastern and southern Africa. Cimbebasia, ser. A, 18:167-174.

Schlitter, D. A., and H. W. Setzer. 1973. New rodents (Mammalia: Cricetidae, Muridae) from Iran and Pakistan. Proceedings of the Biological Society of Washington, 86:163-174.

Schlitter, D. A., and K. Thonglongya. 1971. *Rattus turkestanicus* (Satunin, 1903), the valid name for *Rattus rattoides* Hodgson, 1845 (Mammalia: Rodentia). Proceedings of the Biological Society of Washington, 84:171-174.

Schlitter, D. A., I. L. Rautenbach, and D. A. Wolhuter. 1980. Karyotypes and morphometrics of two species of *Scotophilus* in South Africa (Mammalia: Vespertilionidae). Annals of the Transvaal Museum, 32:231-239.

Schlitter, D. A., S. L. Williams, and J. E. Hill. 1983. Taxonomic review of Temminck's trident bat, *Aselliscus tricuspidatus* (Temminck, 1834)(Mammalia: Hipposideridae). Annals of Carnegie Museum, 52:337-358.

Schlitter, D. A., I. L. Rautenbach, and C. G. Coetzee. 1984. Karyotypes of southern African gerbils, genus *Gerbillurus* Shortridge, 1942 (Rodentia: Cricetidae). Annals of Carnegie Museum, 53:549-557.

Schlitter, D. A., L. W. Robbins, and S. L. Williams. 1985. Taxonomic status of dormice (genus *Graphiurus*) from west and central Africa. Annals of Carnegie Museum, 54(1):1-9.

Schmid, M., T. Haaf, H. Weis, and W. Schempp. 1986. Chromosomal homoeologies in hamster species of the genus *Phodopus* (Rodentia, Cricetinae). Cytogenetics and Cell Genetics, 43:168-173.

Schmid, M., R. Johannisson, T. Haaf, and H. Neitzel. 1987. The chromosomes of *Micromys minutus* (Rodentia, Murinae). II. Pairing pattern of X and Y chromosomes in meiotic prophase. Cytogenetics and Cell Genetics, 45:121-131.

Schmidly, D. J. 1972. Geographic variation in the white-ankled mouse, *Peromyscus pectoralis*. Southwestern Naturalist, 17:113-138.

Schmidly, D. J. 1973*a*. Geographic variation and taxonomy of *Peromyscus boylii* from Mexico and southern United States. Journal of Mammalogy, 54:111-130.

Schmidly, D. J. 1973*b*. The systematic status of *Peromyscus comanche*. Southwestern Naturalist, 18:269-278.

Schmidly, D. J. 1974*a*. *Peromyscus attwateri*. Mammalian Species, 48:1-3.

Schmidly, D. J. 1974*b*. *Peromyscus pectoralis*. Mammalian Species, 49:1-3.

Schmidly, D. J., and F. S. Hendricks. 1976. Systematics of the southern races of Ord's kangaroo rat, *Dipodomys ordii*. Bulletin of the Southern California Academy of Science, 75:225-237.

Schmidly, D. J., and G. L. Schroeter. 1974. Karyotypic variation of *Peromyscus boylii* (Rodentia, Cricetidae) from Mexico and corresponding taxonomic implications. Systematic Zoology, 23:333-342.

Schmidly, D. J., M. R. Lee, W. S. Modi, and E. G. Zimmerman. 1985. Systematics and notes on the biology of *Peromyscus hooperi*. Occasional Papers, The Museum, Texas Tech University, 97:1-40.

Schmidly, D. J., R. D. Bradley, and P. S. Cato. 1988. Morphometric differentiation and taxonomy of three chromosomally characterized groups of *Peromyscus boylii* from east-central Mexico. Journal of Mammalogy, 69:462-480.

Schmidt, C. A., M. D. Engstrom, and H. H. Genoways. 1989. *Heteromys gaumeri*. Mammalian Species, 345:1-4.

Schmidt-Kittler, N. 1981. Zur Stammesgeschichte der marderverwandten Raubtiergruppen (Musteloidea, Carnivora). Eclogae Geologicae Helvetiae, 74:753-801.

Schnell, G. D., T. L. Best, and M. L. Kennedy. 1978. Interspecific morphologic variation in kangaroo rats (*Dipodomys*): degree of concordance with genic variation. Systematic Zoology, 27:34-48.

Schomber, H.-W. 1963. Beitrage zur Kenntnis der Giraffengazelle (*Litocranius walleri* Brooke, 1878). Säugetierkundliche Mitteilungen, 11(Sonderheft):1-44.

Schomber, H.-W. 1964. Beitrage zur Kenntnis der Lamagazelle, *Ammodorcas clarkei* (Thomas, 1891). Säugetierkundliche Mitteilungen, 12:65-90.

Schoppe, R. 1986. Die Schlafmäuse (Gliridae) in Niedersachsen. Lebensraum und Verbreitung von Siebenschläfer, Gartenschläfer und Haselmaus. Naturschutz und Landschaftspflege Niedersachsen Beiheft, 14:1-52.

Schoutenden, H. 1945. De Zoogdieren van Belgish Congo en van Ruanda-Urundi. Annales du Musee Royal du Congo Belge, Zoologie, 2(3):1-576.

Schreiber, A., R. Wirth, M. Riffel, and H. van Rompaey. 1989. Weasels, civets, mongooses and their relatives: an action plan for the conservation of mustelids and viverrids. I. U. C. N., Gland, Switzerland, 100 pp.

Schulze, W. 1986. Zum Vorkommen und zur Biologie von Haselmaus (*Muscardinus avellanarius* L.) und Siebenschläfer (*Glis glis* L.) in Volgelkästen im Südharz der DDR. Säugetierkundliche Informationen, 2(10):341-348.

Schulze, W. 1987. Zum Mobilitat der Haselmaus (*Muscardinus avellanarius*) im Südharz. Säugetierkundliche Informationen, 2(11):485-488.

Schwabe, H. W. 1979. Vergleichend-allometrische Untersuchungen an den Schädeln europaischer und asiatischer Hausratten (*Rattus rattus* L.). Zeitschrift für Säugetierkunde, 44:354-360.

Schwangart, F. 1936. Der Manul, *Otocolobus manul* (Pallas), im System der Feliden. Zentralblatt für Kleintierkunde und Pelztierkunde, 12(8)(Carnivoren Studien 2):19-67.

Schwangart, F. 1943. Die Sohlenzeichnung von *Felis* und Verwandtes zur Systematik und Oekologie des Genus. Abhandlungen der Bayerischen Akademie der Wissenschaften, Mathematisch-naturwissenschaftliche Abteilung, Neue Folge, 52:1-35.

Schwartz, A. 1953. A systematic study of the water rat (*Neofiber alleni*). Occasional Papers of the Museum of Zoology, University of Michigan, 547:1-27.

Schwartz, A., and J. K. Jones, Jr. 1967. Bredin-Archbold-Smithsonian Biological Survey of Dominica. 7. Review of bats of the endemic Antillean genus *Monophyllus*. Proceedings of the United States National Museum, 124(3635):1-20.

Schwartz, J. H., and I. Tattersall. 1985. Evolutionary relationships of living lemurs and lorises (Mammalia, Primates) and their potential affinities with European Eocene Adapidae. Anthropological Papers of the American Museum of Natural History, 60, 3:1-100.

Schwarz, E. 1930. Die Sammlung afrikanischer Säugetiere im Congo Museum, Ginsterkatzen (Gattung *Genetta* Oken). Revue de Zoologie et de Botanique Africaines, 19:275-286.

Schwarz, E. 1933*a*. *Cercopithecus mitis* Wolf für *Simia leucampyx* Fischer. Zeitschrift für Säugetierkunde, 8:279.

Schwarz, E. 1933*b*. The hyrax of the Central Sahara. Annals and Magazine of Natural History, ser. 10, 12:625-626.

Schwarz, E. 1939. On mountain-voles of the genus *Alticola* Blanford: a taxonomic and genetic analysis. Proceedings of the Zoological Society of London, ser. B, 108:663-668.

Schwarz, E. 1947. Colour mutants of the Malay short-tailed mongoose, *Herpestes brachyurus* Gray. Proceedings of the Zoological Society of London, 117(1):79-80.

Schwarz, E. 1948. Revision of the Old World moles of the genus *Talpa*. Proceedings of the Zoological Society of London, 118:36-48.

Schwarz, E., and H. K. Schwarz. 1943. The wild and commensal stocks of the house mouse, *Mus musculus* Linneaus. Journal of Mammalogy, 24:59-72.

Schwarz, E., and H. K. Schwarz. 1967. A monograph of the *Rattus rattus* group. Anales de la Escuela Nacionál de Ciéncias Biológicas (México City), 14:79-178.

Sclater, P. L., and O. Thomas. 1894-1900. The book of antelopes [vol. 3(part 12):179-245]. R. H. Porter, London, 1:1-220; 2:1-194; 3:1-245[1897-1898]; 4:1-242.

Sclater, W. L. 1891. Catalogue of Mammalia in the Indian museum, Calcutta. Part II. Rodentia, Ungulata, Proboscidea, Hyracoidea, Carnivora, Cetacea, Sirenia, Marsupialia, Monotremata. Calcutta, 350 pp.

Sclater, W. L. 1900-1901. Mammals of South Africa [Vol. I. Primates, Carnivora and Ungulata]. Porter, London, 2 vols.

Scriven, P. N., and V. Bauchau. 1992. The effect of hybridization on mandible morphology in an island population of the house mouse. Journal of the Zoological Society of London, 226:573-583.

Seal, U. S., and D. G. Makey. 1974. ISIS mammalian taxonomic directory international species inventory system. Minnesota Zoological Garden, St. Paul, Minnesota, 645 pp.

Searle, J. 1984. Three new karyotypic races of the common shrew *Sorex araneus* (Mammalia, Insectivora) and a phylogeny. Systematic Zoology, 33:184-194.

Searle, J.B. 1991. A hybrid zone comprising staggered chromosomal clines in the house mouse (*Mus musculus domesticus*). Proceedings of the Royal Society of London, ser. B, 246:47-52.

Seba, A. 1734. Locupletissimi rerum naturalium thesauri accurata descriptio, eti conibus artificiosissimis expressio, per universam physices historiam opus, cui, in hoc rerum genere, nullum par exstitit, ex toto terrarum orbe collegit, digessit, descripsit, et depingendum curavit. Apud. J. Wetstenium & Gul. Smith, & Janssonio-Waesbergios, Amstelaedami, 1:1-78 + 38 (unnumbered), 111 pls.

Selander, R. K., M. H. Smith, S. Y. Yang, W. E. Johnson, and J. B. Gentry. 1971. Biochemical polymorphism and systematics in the genus *Peromyscus*. I. Variation in the old-field mouse (*Peromyscus polionotus*). Studies in Genetics VI, University of Texas Publications, 7103:49-90.

Sen, S., and T. Sharma. 1983. Role of constitutive heterochromatin in evolutionary divergence: results of chromosome banding and condensation inhibition studies in *Mus musculus*, *Mus booduga* and *Mus dunni*. Evolution, 37:628-636.

Senut, B., M. Pickford, P. Mein, G. Conroy, and J. Van Couvering. 1992. Discovery of 12 new Late Cainozoic fossiliferous sites in palaeokarsts of the Otavi Mountains, Namibia. Comptes Rendus de l'Acadèmie des Sciences (Paris), 314(ser. II): 727-733.

Serdyuk, V. A. 1979. Izmenchivost' stroeniya zubov u arkticheskovo suslika (*Citellus parryi*); veroyantnye puti rasseleniya etovo vida na severo-vostoke SSSR [Variation in structure of the teeth in the arctic ground squirrel...and a possible way of migration of this species in the northeastern U.S.S.R.]. Zoologicheskii Zhurnal, 58:1692-1702 (in Russian).

Setzer, H. W. 1949. Subspeciation in the kangaroo rat, *Dipodomys ordii*. University of Kansas Publications, Museum of Natural History, 1:473-573.

Setzer, H. W. 1952. Notes on mammals from the Nile Delta region, Egypt. Proceedings of the United States National Museum, 102:343-369.

Setzer, H. W. 1956. Mammals of the Anglo-Egyptian Sudan. Proceedings of the United States National Museum, 106(3377):447-587.

Setzer, H. W. 1957. The hedgehogs and shrews (Insectivora) of Egypt. Journal of the Egyptian Public Health Association, 32:1-17.

Setzer, H. W. 1975. Genus *Acomys*. Part 6.5. Pp. 1-2, *in* The mammals of Africa: An identification manual (J. Meester and H. W. Setzer, eds.) [issued 10 Dec 1975]. Smithsonian Institution Press, Washington, D. C., not continuously paginated.

Severinghaus, W. D. 1977. Description of a new subspecies of prairie vole, *Microtus ochrogaster*. Proceedings of the Biological Society of Washington, 90:49-54.

Severinghaus, W. D., and D. F. Hoffmeister. 1978. Qualitative cranial characters distinguishing *Sigmodon hispidus* and *Sigmodon arizonae* and the distribution of these two species in northern Mexico. Journal of Mammalogy, 59:868-870.

Seymour, K. L. 1989. *Panthera onca*. Mammalian Species, 340:1-9.

Shackleton, D. M. 1985. *Ovis canadensis*. Mammalian Species, 230:1-9.

Sharma, T., and I. K. Gadi. 1977. Constitutive heterochromatin variation in two species of *Rattus* with apparently similar karyotypes. Genetica, 47:77-80.

Sharma, T., and R. Raman. 1971. An XO female in the Indian mole rat. Journal of Heredity, 62:384-387.

Sharma, T., and R. Raman. 1973. Variation of constitutive heterochromatin in the sex chromosomes of the rodent *Bandicota bengalensis bengalensis* (Gray). Chromosoma, 41:75-84.

Sharma, T., N. Cheong, P. Sen, and S. Sen. 1986. Constitutive heterochromatin and evolutionary divergence of *Mus dunni*, *M. booduga* and *M. musculus*. Pp. 35-44, *in* Current topics in microbioloby and immunology, vol. 127 (M. Potter, J. H. Nadeau, and M. P. Cancro, eds.). Springer-Verlag, Berlin, 395 pp.

Sharman, G. B., C. E. Murtagh, P. M. Johnson, and C. M. Weaver. 1980. The chromosomes of a rat-kangaroo attributable to *Bettongia tropica* (Marsupialia; Macropodidae). Australian Journal of Zoology, 28:59-63.

Shaughnessy, P. D., and F. H. Fay. 1977. A review of the taxonomy and nomenclature of north Pacific harbour seals. Journal of Zoology (London), 182:385-419.

Shaw, G. 1800. General zoology or systematic natural history. G. Kearsley, London, 12:1-330.

Shcherbina, E. I., L. S. Marinina, and N. A. Sorokina. 1988. [Mammals of the Malyi Balkhan mountain ridge and piedmont plain]. Izvestiya Akademii Nauk Turkmenskoi SSR, Seriya Biologicheskikh Nauk, 3:15-22 (in Russian).

She, J. X., F. Bonhomme, P. Boursot, L. Thaler, and F. Catzeflis. 1990. Molecular phylogenies in the genus *Mus*: comparative analysis of electrophoretic, scnDNA hybridization, and mtDNA RFLP data. Biological Journal of the Linnean Society, 41:83-103.

Sheets-Pyenson, S. 1981. From the north to Red Lion Court: the creation and early years of the *Annals of Natural History*. Journal of the Society for the Bibliography of Natural History, 10:221-249.

Shellhammer, H. S. 1967. Cytotaxonomic studies of the harvest mice of the San Francisco Bay region. Journal of Mammalogy, 48:549-556.

Shellhammer, H. S. 1982. *Reithrodontomys raviventris*. Mammalian Species, 169:1-3.

Shenbrot, G. I. 1974. [Systematic status of *Allactodipus bobrinskii* (Rodentia, Dipodidae)]. Zoologicheskii Zhurnal, 53(11):1697-1702 (in Russian).

Shenbrot, G. I. 1984. [Dental morphology and phylogeny of five-toed jerboas of the subfamily Allactaginae]. Sbornik Trudov Zoologicheskovo Museya MGU, 22:61-92 (in Russian).

Shenbrot, G. I. 1986. [Supergeric relationships of the jerboas (Rodentia, Dipodoidea)]. Chetvertii Sezd Vsesoyuznovo Teriologicheske Obshchestva, Tezisy Dokladov, Moscow, 1:106-107 (in Russian).

Shenbrot, G. I. 1988. Noviye danniye rasprostraneniyu i sistematikye tolstokhvostikh tushkanchikov [New data on the distribution and taxonomy of the fat-tailed jerboas, genus *Pygeretmus*]. Pp. 121-124, *in* Tushkanchiki fauni SSSR [Jerboas of the USSR fauna], vol. 2 (V. E., Sokolov, V. S. Lobachev, E. I. Naumova, R. R. Reimov, G. I. Shenbrot, and T. I. Dmitrieva, eds.). Tyezisi Dokladov Vesesoiuznovo soveshchaniye [Abstracts of the All-Union Conference], Nukus, Fan Press, Moscow, 139 pp. (in Russian).

Shenbrot, G. I. 1990*a*. Geograficheskaya izmenchivost' Turkmenskovo tushkanchika *Jaculus turkmenicus* (Rodentia, Dipodidae), i zadachi evo okhrany [Geographical variation of the Turkmenian jerboa, *Jaculus turkmenicus* (Rodentia, Dipodidae) and the problems of its protection]. Zoologicheskii Zhurnal, 69(2):114-121 (in Russian).

Shenbrot, G. I. 1990*b*. Geograficheskaya izmenchivost' i podvidovaya differentisiya tushkanchika Likhtenshteina, *Eremodipus lichtensteini* (Rodentia, Dipodidae) [Geographical variation and subspecific differentiation of the Lichtenstein's jerboa, *Eremodipus lichtensteini* (Rodentia, Dipodidae]. Zoologicheskii Zhurnal, 69(10):154-159 (in Russian).

Shenbrot, G. I. 1991*a*. Geograficheskaya izmenchivost' mokhnonogovo tushkanchika *Dipus sagitta* (Rodentia, Dipodidae) [Geographical variation of the three-toed brush-footed jerboa...]. 2. [Subspecific differentiation in eastern Kazakhstan, Tuva, and Mongolia]. Zoologicheskii Zhurnal, 70(7):91-97 (in Russian).

Shenbrot, G. I. 1991*b*. [Subspecific taxonomy revision of common thick-tailed three-toed jerboa, *Stylodipus telum* (Rodentia, Dipodidae)]. Zoologicheskii Zhurnal, 70(6):118-127.

Shenbrot, G. I. 1991*c*. Geograficheskaya izmenchivost' mokhnonogovo tushkanchika *Dipus sagitta* (Rodentia, Dipodidae) [Geographical variation of the three-toed brush-footed jerboa *Dipus sagitta* (Rodentia, Dipodidae) 1. [General patterns of intraspecific variability and subspecific differentiation in the western part of the species range]. Zoologicheskii Zhurnal, 70(5):101-110 (in Russian).

Shenbrot, G. I. 1991*d* [1992]. [Subspecific systematic revision of the five-toed jerboa genus *Allactaga* in the USSR]. Trudy Zoologicheskovo Instituta, Akademiya Nauk SSSR, 243:1-16 (in Russian).

Shenbrot, G. I. 1992. [Cladistic approach to the analysis of phylogenetic relationships among dipodoid rodents (Rodentia, Dipodoidea)]. Sbornik Trudov Zoologicheskovo Muzeya MGU, 29:176-201 (in Russian).

Shenbrot, G. I., and V. N. Mazin. 1989. [On the taxonomy of *Salpingotus pallidus* (Rodentia, Dipodidae) from south Balkhash region]. Zoologicheskii Zhurnal, 67(1):155-158 (in Russian).

Sheppe, W., Jr. 1961. Systematic and ecological relations of *Peromyscus oreas* and *P. maniculatus*. Proceedings of the American Philosophical Society, 105:421-446.

Sherborn, C. D. 1902-1933. Index Animalium; sive, Index nominum quae ab A.D. MDCCLVIII generibus et speciebus animalium imposita sunt, societatibus eruditorum adiuvantibus. C. J. Clay and Sons, Cambridge University Press Warehouse, London, 2 vol. (in 10).

Shields, G. F., and T. D. Kocher. 1991. Phylogenetic relationships of North American ursids based on analysis of mitochrondrial DNA. Evolution, 45(1):218-221.

Shnitnikov, V. N. 1936. Mlekopitayushchie Semirech'ya [Mammals of Semirech'ye]. Nauka, Moscow-Leningrad, 323 pp. (in Russian).

Shortridge, G. C. 1942. Field notes on the first and second expeditions to the Cape Museum's mammal survey of the Cape Province; with descriptions of some new subgenera and subspecies. Annals of the South African Museum, 36:27-100.

Shortridge, G. C., and T. D. Carter. 1938. A new genus and new species and subspecies of mammals from Little Namaqualand and the north-west Cape Province; and a new subspecies of *Gerbillus paeba* from the eastern Cape Province. Annals of the South African Museum, 3(2):281-291.

Shoshani, J., and J. F. Eisenberg. 1987. *Elephas maximus*. Mammalian Species, 182:1-8.

Shoshani, J., C. A. Goldman, and J. G. M. Thewissen. 1988. *Oryteropus afer*. Mammalian Species, 300:1-8.

Shou Zhen-huang (ed.) [Shaw Tsen-Hwang]. 1962. Chung-kuo ching chi tung wu chih [Chinese economic zoology: mammal section]. Ko Hsueh chu pan she [Scientific Publications Office. Beijing], 554 pages, 72 plates (in Chinese).

Shump, K. A., and R. H. Baker. 1978*a*. *Sigmodon alleni*. Mammalian Species, 95:1-2.

Shump, K. A., and R. H. Baker. 1978*b*. *Sigmodon leucotis*. Mammalian Species, 96:1-2.

Shump, K. A., and A. U. Shump. 1982*a*. *Lasiurus borealis*. Mammalian Species, 183:1-6.

Shump, K. A., and A. U. Shump. 1982*b*. *Lasiurus cinereus*. Mammalian Species, 185:1-5.

Sicard, B., M. Tranier, and J.-C. Gautun. 1988. Un rongeur nouveau du Burkina Faso (ex Haute-Volta): *Taterillus petteri*, sp. nov. (Rodentia, Gerbillidae). Mammalia, 52:187-198.

Sidiyene, E. A. 1989. Capture de *Crocidura lusitania* dans l'Adrar de Iforas. Mammalia, 53:467.

Sidorowicz, J. 1971. Subspecific taxonomy of the squirrel (*Sciurus vulgaris* L.) in Palaearctic. Zoologische Anzeiger, 187:123-142.

Siivonen, L. 1965. *Sorex isodon* Turov (1924) and *S. unguiculatus* Dobson (1890) as independent shrew species. Aquilo (Zool.), 4:1-34.

Sikorski, M. D. 1982. Craniometric variation of *Apodemus agrarius* (Pallas, 1771) in urban green areas. Acta Theriologica, 27:71-81.

Silva-Taboada, G. 1976. Historia y actualización taxonómica de algunas especies Antillanas de murciélagos de los generos *Pteronotus*, *Brachyphylla*, *Lasiurus*, y *Antrozous* (Mammalia: Chiroptera). Poeyana (Academia de Ciencias de Cuba), 153:1-24.

Silva-Taboada, G. 1979. Los murciélagos de Cuba. Editorial Academia, 423 pp.

Silva-Taboada, G., and O. H. Garrido. In press. Compendio de los vertebrados terrestres de Cuba. Academia de Ciencias de Cuba (Habana).

Silva-Taboada, G., and K. F. Koopman. 1964. Notes on the occurrence and ecology of *Tadarida laticaudata yucatanica* in eastern Cuba. American Museum Novitates, 2174:1-6.

Silverstov, V. B., V. S. Lobachev, and M. N. Shilov. 1969. [New data on the distribution and biology of *Pygerethmus platyurus* Licht.]. Byulleten' Moskovskovo Obshchestva Ispytatelei Prirody, Otdel Biologicheskii, 74(3):118-133 (in Russian).

Simonetta, A. M. 1968. A new golden mole from Somalia with an appendix on the taxonomy of the family Chrysochloridae (Mammalia, Insectivora). Monitore Zoologico Italiano, N. S., 2 (Supplemento):27-55.

Simonetti Z., J., and A. E. Spotorno O. 1980. Posición taxonómica de *Phyllotis micropus* (Rodentia: Cricetidae). Annales del Museo de Historia Natural de Valparaiso, 13:285-297.

Simons, E. L., and Y. Rumpler. 1988. *Eulemur*: new generic name for species of *Lemur* other than *Lemur catta*. Comptes Rendus de l'Acadèmie des Sciences (Paris), ser. 3, 307:547-551.

Simpson, G. G. 1945. The principles of classification and a classification of mammals. Bulletin of the American Museum of Natural History, 85:1-350.

Sinclair, A. R. E., C. J. Krebs, J. N. M. Smith, and S. Boutin. 1988. Population biology of snowshoe hares. III. Nutrition, plant secondary compounds and food limitation. Journal of Animal Ecology, 57:787-806.

Sinha, Y. P. 1970. Taxonomic notes on some Indian bats. Mammalia, 34:81-92.

Sinha, Y. P. 1973. Taxonomic studies on the Indian horseshoe bats of the genus *Rhinolophus* Lacepede. Mammalia, 37:603-630.

Sinha, Y. P. 1980. The bats of Rajasthan: Taxonomy and zoogeography. Records of the Zoological Survey of India, 76:7-63.

Sinha, Y. P., and S. Chakroborty. 1971. Taxonomic status of the vespertilionid bat *Nycticeius emarginatus* Dobson. Proceedings of the Zoological Society of Calcutta, 24:53-57.

Sivertsen, E. 1954. A survey of the eared seals (Family Otariidae) with remarks on the antarctic seals collected by M/K "Norvegica" in 1928-1929. Det Norske Videnskaps-Akademi i Oslo, Scientific Results of the Norwegian Antarctic Expedition, 36:1-76.

Skaren, U. 1964. Variation of two shrews, *Sorex unguiculatus* Dobson and *S. a. araneus* L. Annales Zoologici Fennici, 1:94-124.

Skead, C. J. 1973. Zoo-historical gazetteer. Annals of the Cape Provincial Museums, 10:1-259.

Skinner, J. D., and R. H. N. Smithers. 1990. The mammals of the southern African subregion. Second ed. University of Pretoria, Republic of South Africa, 771 pp.

Skoczen, S. 1976. Condylurini, Dobson, 1883 (Insectivora, Mammalia) in the Pliocene of Poland. Acta Zoologica Cracoviensia, 21:291-313.

Slaughter, B. H., and J. E. Ubelaker. 1984. Relationship of South American cricetine rodents to rodents of North America and the Old World. Journal of Vertebrate Paleontology, 4:255-264.

Sludskii, A. A. (ed.). 1977. Mlekopitayushchie Kazakhstana. Gryzuny (krome surkov, suslikov, zemlyanoi belki, peschanok i polevok) [Mammals of Kazakhstan. Rodents (except marmots, susliks, long clawed ground squirrels, gerbils and voles)]. Nauka, Kazakhskoi SSR, Alma-Ata, 1(2):1-536 (in Russian).

Sludskii, A. A., S. N. Varshavskii, M. N. Ismagilov, V. I. Kapitonov, and I. G. Shubin. 1969. Mlekopitayushchie Kazakhstana [Mammals of Kazakhstan]. Vol. 1, Gryzuny (surki i susliki) [Rodents (Marmots and susliks)]. Nauka, Kazakhskoi SSR, Alma-Ata, 455 pp. (in Russian).

Sludskii, A. A., A. D. Bernstein, I. G. Shubin, V. A. Fadeev, G. I. Orlov, A. Bekenov, V. I. Kharitonov, and S. R. Utinov. 1980. Mlekopitayushchie Kazakhstana, tom 2, Zaitseobraznye [Mammals of Kazakhstan. vol. 2, Lagomorpha]. Nauka, Kazakhskoi SSR, Alma-Ata, 236 pp. (in Russian).

Smeenk, C., Y. Kaneko, and K. Tsuchiya. 1982. On the type material of *Mus argenteus* Temminck, 1844. Zoologische Mededelingen, 56:121-129.

Smirin, Yu. M., N. A. Formozov, D. I. Bibikov, and D. Myagmarzhav. 1985. Kharakteristika poselenii dvukh vidov surkov (*Marmota*, Rodentia, Sciuridae) v zone ikh kontakta na Mongol'skom Altae [Characteristics of colonies of two species of marmots...in their contact zone in the Mongolian Altai]. Zoologicheskii Zhurnal, 64:1873-1885 (in Russian).

Smith, A. 1833-1834. An epitome of African Zoology; or, a concise description of the objects of the animal kingdom inhabiting Africa, its islands and seas. South African Quarterly Journal, 2:16-32, 49-64, 81-96, 113-128, 145-160, 169-192, 209-224, 233-248.

Smith, A. T., and M. L. Weston. 1990. *Ochotona princeps*. Mammalian Species, 352:1-8.

Smith, A. T., N. A. Formozov, Chan-lin Zheng, M. Erbaeva, and R. S. Hoffmann. 1990. The pikas. Pp. 14-60, *in* Rabbits, hares and pikas (J. A. Chapman and J. E. C. Flux, eds.). I. U. C. N., Gland, Switzerland, 168 pp.

Smith, J. D. 1970. The systematic status of the black howler monkey, *Alouatta pigra* Lawrence. Journal of Mammalogy, 51:358-369.

Smith, J. D. 1972. Systematics of the chiropteran family Mormoopidae. Miscellaneous Publications, Museum of Natural History, University of Kansas, 56:1-132.

Smith, J. D. 1977. On the nomenclatorial status of *Chilonycteris gymnonotus* Natterer, 1843. Journal of Mammalogy, 58:245-246.

Smith, J. D., and J. E. Hill. 1981. A new species and subspecies of bat of the *Hipposideros bicolor*-group from Papua New Guinea and the systematic status of *Hipposideros calcaratus* and *Hipposideros cupidus* (Mammalia: Chiroptera: Hipposideridae). Contributions in Science, Natural History Museum of Los Angeles County, 331:1-19.

Smith, J. D., and C. S. Hood. 1980. Additional material of *Rhinolophus ruwenzorii* Hill, 1942, with comments on its natural history and taxonomic status. Proceedings of the Fifth International Bat Research Conference, 5:163-171.

Smith, J. D., and C. S. Hood. 1983. A new species of tube-nosed fruit bat (*Nyctimene*) from the Bismark Archipelago, Papua New Guinea. Occasional Papers, The Museum, Texas Tech University, 81:1-14.
Smith, M. F. 1979. Geographic variation in genic and morphological characters in *Peromyscus californicus*. Journal of Mammalogy, 60:705-722.
Smith, M. F., and J. L. Patton. 1991. Variation in mitochondrial cytochrome *b* sequence in natural populations of South American akodontine rodents (Muridae: Sigmodontinae). Molecular Biology and Evolution, 8:85-103.
Smith, M. H., R. K. Selander, and W. E. Johnson. 1973. Biochemical polymorphism and systematics in the genus *Peromyscus*. III. Variation in the Florida deermouse (*Peromyscus floridanus*), a Pleistocene relic. Journal of Mammalogy, 54:1-13.
Smith, M. J. 1973. *Petaurus breviceps*. Mammalian Species, 30:1-5.
Smith, S. A. 1990. Cytosystematic evidence against monophyly of the *Peromyscus boylii* species group (Rodentia: Cricetidae). Journal of Mammalogy, 71:654-667.
Smith, S. A., R. D. Bradley, and I. F. Greenbaum. 1986. Karyotypic conservatism in the *Peromyscus mexicanus* group. Journal of Mammalogy, 67:584-586.
Smith, S. A., I. F. Greenbaum, D. J. Schmidly, K. M. Davis, and T. W. Houseal. 1989. Additional notes on karyotypic variation in the *Peromyscus boylii* group. Journal of Mammalogy, 70:603-608.
Smith, T. G., D. J. St. Aubin, and J. R. Geraci. 1990. Advances in research on the beluga whale, *Delphinapterus leucas*. Canadian Bulletin of Fisheries and Aquatic Sciences, 224:1-206.
Smith, W. P. 1991. *Odocoileus virginianus*. Mammalian Species, 388:1-13.
Smithers, R. H. N. 1971. The mammals of Botswana. Museum Memoir, National Museums of Rhodesia, Salisbury, 4:1-340.
Smithers, R. H. N. 1983. The mammals of the Southern African Subregion. University of Pretoria, Republic of South Africa, 736 pp.
Smithers, R. H. N., and J. L. P. L. Tello. 1976. Check list and atlas of the mammals of Mozambique. Museum Memoir, National Museums and Monuments of Rhodesia, Salisbury, 8:1-184.
Smithers, R. H. N., and V. J. Wilson. 1979. Checklist and atlas of the mammals of Zimbabwe Rhodesia. Museum Memoir, National Museums and Monuments of Rhodesia, Salisbury, 9:1-193.
Smolen, M. J. 1981. *Microtus pinetorum*. Mammalian Species, 147:1-7.
Smolen, M. J., and B. L. Keller. 1987. *Microtus longicaudus*. Mammalian Species, 271:1-7.
Smorkacheva, A. V., T. G. Aksenova, and T. A. Zorenko. 1990. Ekologiya Kitaiskoi polevki *Lasiopodomys mandarinus* (Rodentia, Cricetidae) v Zabaikal'e [The ecology of the Chinese vole *Lasiopodomys mandarinus* (Rodentia, Cricetidae) in Transbaikaliya]. Zoologicheskii Zhurnal, 69(12):115-124 (in Russian).
Snow, J. L., J. K. Jones, Jr., and W. D. Webster. 1980. *Centurio senex*. Mammalian Species, 138:1-3.
Snyder, D. P. 1982. *Tamias striatus*. Mammalian Species, 168:1-8.
Sobti, R. C., and S. S. Gill. 1984. Chromosomes and protein polymorphs in certain *Rattus* species from North India. Research Bulletin (Science) of the Punjab University, 35:203-210.
Sody, H. J. V. 1940. On the mammals of Enggano. Treubia, 17:391-405.
Sody, H. J. V. 1941. On a collection of rats from the Indo- Malayan and Indo- Australian regions (with descriptions of 43 new genera, species and subspecies). Treubia, 18:255-325.
Sody, H. J. V. 1949*a*. Sciuridae from the Indo-Malayan and Indo-Australian regions. Treubia, 20:57-120.
Sody, H. J. V. 1949*b*. Notes on some Primates, Carnivora, and the Babirusa from the Indo-Malayan and Indo-Australian regions (with descriptions of 10 new species and subspecies). Treubia, 20:121-190.
Sokolov, I. I. 1973. Napravleniya evolyutsii i estestvennaya klassifikatsiya podsemeistva vydrovykh (Lutrinae, Mustelidae, Fissipedia) [Evolutionary trends and the classification of the subfamily Lutrinae (Mustelidae, Fissipedia)]. Byulleten' Moskovskovo Obshchestva Ispytatelei Prirody, Otdel Biologicheskii, 78(6):45-52 (in Russian).
Sokolov, V. E. 1973 - 1979. Sistematika mlekopitayushchikh [Systematics of mammals]. vol. 1 [Monotremes, marsupials, insectivores, dermopterans, chiropterans, primates, edentates, pangolins]. vol. 2 [Lagomorphs, rodents]. vol. 3 [Cetaceans, carnivores, pinnipeds, tubulidentates, proboscideans, hyracoids, sirenians, artiodactyls, tylopods, perissodactyls]. Vysshaya Shkola, Moscow, 1:1-430 [1973]; 2:1-494 [1977]; 3:1-528 [1979] (in Russian).
Sokolov, V. E. 1974. *Saiga tatarica*. Mammalian Species, 38:1-4.
Sokolov, V. E., and M. I. Baskevich. 1988. [A new species of birch mouse - *Sicista armenica* sp. n. (Rodentia, Dipodoidea) from the Lesser Caucasus]. Zoologicheskii Zhurnal, 67(2):300-304 (in Russian).
Sokolov, V. E., and N. K. Dzhemukhadze. 1991. [Histochemistry of specialized skin glands of the narrow-skulled and Dagestan voles]. Doklady Akademii Nauk SSSR, Moscow-Leningrad, 316:731-734.
Sokolov, V. E., and V. S. Gromov. 1990. The contemporary ideas on roe deer (*Capreolus* Gray, 1821) systematization: morphological, ethological and hybridological analysis. Mammalia, 54:431-444.

Sokolov, V. E., and Y. M. Kovalskaya. 1990. Kariotipy myshovok severnovo Tyan'-Shanya i Sikhote-Alinya (*Sicista*, Dipodoidea, Rodentia) [Karyotypes of birch mice (*Sicista*, Dipodoidea, Rodentia) in the northern Tien-Shan and Sikhote-Alin]. Zoologicheskii Zhurnal, 69(5):152-157 (in Russian).

Sokolov, V. E., and V. N. Orlov. 1980. Opredelitel' mlekopitayushchikh Mongol'skoi Narodnoi Respubliki [Guide to the mammals of the Mongolian People's Republic]. Nauka, Moscow, 351 pp. (in Russian).

Sokolov, V. E., and G. I. Shenbrot. 1987*a*. [Review of Wang Sibo, Yang Ganyun. "Rodent fauna of Xinjiang."]. Zoologicheskii Zhurnal, 66(1):157-159.

Sokolov, V. E., and G. I. Shenbrot. 1987*b*. [A new species of thick- tailed jerboa, *Stylodipus sungorus* sp. n. (Rodentia, Dipodidae), from western Mongolia]. Zoologicheskii Zhurnal, 66(4):579-587 (in Russian).

Sokolov, V. E., and G. I. Shenbrot. 1988. [Data on the geographical variability and taxonomy of pygmy jerboas (Rodentia, Cardiocraniinae) in Mongolia.]. Zoologicheskii Zhurnal, 67(10):1561-1569 (in Russian).

Sokolov, V. E., and A. K. Tembotov. 1989. Mlekopitayushchie Kavkaza: Hasekomoyadnye [Mammals of the Caucasus: Insectivores]. Nauka, Moscow, 547 pp. (in Russian).

Sokolov, V. E., Y. M. Kovalskaya, and M. I. Baskevich. 1980. Gryzuny [Rodents]. Materiali Pyatovo Vsesoyuznovo Soveshchaniye, Moscow, 1980:38-40 (in Russian).

Sokolov, V. E., O. L. Rossolimo, I. Ya. Pavlinov, and O. I. Podtyazhkin. 1981*a*. [Comparative characteristics of two species of jerboas from Mongolia— *Allactaga bullata* Allen, 1925 and *A. nataliae* Sokolov, 1981]. Zoologicheskii Zhurnal, 60(6):895-906 (in Russian).

Sokolov, V. E., M. I. Baskevich, and Y. M. Kovalskaya. 1981*b*. [Revision of birch mice of the Caucasus: sibling species *Sicista caucasica* Vinogradov, 1925 and *S. kluchorica* sp. n. (Rodentia, Dipodidae)]. Zoologicheskii Zhurnal, 60(9):1386-1393 (in Russian).

Sokolov, V. E., Y. M. Kovalskaya, and M. I. Baskevich. 1982. [Taxonomy and comparative cytogenetics of some species of the genus *Sicista* (Rodentia, Dipodidae)]. Zoologicheskii Zhurnal, 61(1):102-108 (in Russian).

Sokolov, V. E., M. I. Baskevich, and Y. M. Kovalskaya. 1986*a*. [The karyotype variability in the southern birch mouse (*Sicista subtilis* Pallas) and substantiation of the species validity of *S. severtzovi*]. Zoologicheskii Zhurnal, 65:1684 (in Russian).

Sokolov, V. E., M. I. Baskevich, and Y. M. Kovalskaya. 1986*b*. [*Sicista kazbegica* sp. n. (Rodentia, Dipodidae) from the upper reaches of the Terek River basin]. Zoologicheskii Zhurnal, 65(6):949-951 (in Russian).

Sokolov, V. E., M. I. Baskevich, I. V. Lukyanova, M. A. Tarasov, N. N. Kuryatnikov, and V. G. Topilina. 1987*a*. [Distribution of birch mice (Rodentia, Zapodidae) of the Caucasus]. Zoologicheskii Zhurnal, 66(11):1730-1735 (in Russian).

Sokolov, V. E., Y. M. Kovalskaya, and M. I. Baskevich. 1987*b*. Review of karyological research and the problems of systematics in the genus Sicista (Zapodidae, Rodentia, Mammalia). Folia Zoologica, 36(1):35-44.

Sokolov, V. E., Y. M. Kovalskaya, and M. I. Baskevich. 1989. [On species status of the northern birch mouse *Sicista strandi* (Rodentia, Dipodidae)]. Zoologicheskii Zhurnal, 68(10):95-106 (in Russian).

Sokolov, V. E., V. M. Aniskin, and M. A. Serbenyuk. 1990. Sravnitel'naya tsitogenetika shesti vidov polevok roda *Clethrionomys* (Rodentia, Microtinae) [Comparative cytogenetics of six vole species of the genus *Clethrionomys* (Rodentia, Microtinae)]. Zoologicheskii Zhurnal, 69(11):145-151 (in Russian).

Sokolov, V. E., V. N. Orlov, M. I. Baskevich, and A. Mebrate. 1992. [Chromosomal sets of the spiny mice *Acomys* (Rodentia, Muridae) along the Ethiopian Rift Valley]. Zoologicheskii Zhurnal, 71:116-124 (in Russian).

Soldatovic, B., P. Seth, H. Reichstein, and M. Tolksdorf. 1975. Comparative karyological study of the genus *Apodemus* (Kaup, 1829). Acta Veterinaria (Beograd), 25:1-10.

Soldatovic, B., D. Zimonjic, I. Savic, and E. Giagia. 1984. Comparative cytogenetic analysis of the populations of European ground squirrel (*Citellus citellus* L.) on the Balkan peninsula. Bulletin. Academie Serbe des Sciences. Classe des Sciences Mathematiques et Naturelles, 86:47-56.

Solleder, E., M. Schmid, B. Inglin, and T. Haaf. 1984. Cytogenetic studies on the mitotic and meiotic chromosomes of *Micromys minutus* (Rodentia, Murinae). Zeitschrift für Säugetierkunde, 49:284-289.

Song Shi-ying. 1985. [A new subspecies of *Cricetulus triton* from Shaanxi, China]. Acta Theriologica Sinica, 5:137-139 (in Chinese).

Soota, T. D., and Y. Chaturvedi. 1980. New locality record of *Pipistrellus camortae* Miller from Nicobar and its systematic status. Records of the Zoological Survey of India, 77:83-87.

Sopin, L. V. 1982. [On intraspecies structure of *Ovis ammon* (Artiodactyla, Bovidae)]. Zoologicheskii Zhurnal, 61:1882-1892 (in Russian).

Soriano, P. S., and J. Molinari. 1987. *Sturnira aratathomasi*. Mammalian Species, 284:1-4.

Soulounias, N. 1988. Evidence from horn morphology on the phylogenetic relationships of the pronghorn (*Antilocapra americana*). Journal of Mammalogy, 69:140-143.

Sowerby, A. de C. 1923. The naturalist in Manchuria. Tientsin Press Ltd. [China], 2:xxvii + 191.

Sparrmann, A. 1786. A voyage to the Cape of Good Hope, towards the Antarctic Polar Circle, and round the world; but chiefly into the country of the Hottentots and Caffres, from the year 1772, to 1776 (translation of Sparrmann, 1783). Second ed. G. G. J. & J. Robinson, London, 2 vols.

Spencer, S. R., and G. N. Cameron. 1982. *Reithrodontomys fulvescens*. Mammalian Species, 174:1-7.

Spitz, F. 1978. Etude craniometrique du genere *Pitymys*. Mammalia, 42:267-304.

Spitzenberger, F. 1970. Erstnachweise der Wimperspitzmaus (*Suncus etruscus*) für Kreta und Kleinasien und die Verbreitung der Art im südwestasiatischen Raum. Zeitschrift für Säugetierkunde, 35:107-113.

Spitzenberger, F. 1971. Eine neue, tiergeographisch bemerkenswerte *Crocidura* (Insectivora, Mammalia) aus der Türkei. Annalen des Naturhistorischen Museums in Wien, 75:539-562.

Spitzenberger, F. 1976. Beiträge zur Kenntnis von *Dryomys laniger* Felten and Storch, 1968 (Gliridae, Mammalia). Zeitschrift für Säugetierkunde, 41:237-249.

Spitzenberger, F. 1978. Die Stachelmaus von Kleinasien, *Acomys cilicicus* n. sp. (Rodentia, Muridae). Annalen des Naturhistorischen Museums in Wien, 81:443-446.

Spitzenberger, F. 1983. Die Schläfer (Gliridae) Österreichs. Mammalia Austriaca 6. (Mammalia, Rodentia). Mitteilungen der Abteilung für Zoologie am Landesmuseum Joanneum, 30:19-64.

Spitzenberger, F. 1990*a*. *Sorex alpinus* Schinz, 1837 — Alpenspitzmaus. Pp. 295-312, *in* Handbuch der Säugetiere Europas (J. Niethammer and F. Krapp, eds.). Aula-Verlag, Wiesbaden, 3/I:1-524.

Spitzenberger, F. 1990*b*. Gattung *Neomys* Kaup, 1829. Pp. 313-374, *in* Handbuch der Säugetiere Europas (J. Niethammer and F. Krapp, eds.),. Aula-Verlag, Wiesbaden, 3/I:1-524.

Spitzenberger, F. 1990*c*. *Suncus etruscus* (Savi, 1822) - Etruskerspitzmaus. Pp. 375-392, *in* Handbuch der Säugetiere Europas (J. Niethammer, and F. Krapp, eds.). Aula-Verlag, Wiesbaden, 3/I:1-524.

Spitzer, N. C., and J. D. Lazell. 1978. A new rice rat (genus *Oryzomys*) from Florida's Lower Keys. Journal of Mammalogy, 59:787-792.

Spotorno O., A. E. 1976. Analisis taxonomico de tres especies altiplanicas del genero *Phyllotis* (Rodentia, Cricetidae). Annales del Museo de Historia Natural de Valparaiso, 9:141-161.

Spotorno O., A. E., and L. I. Walker B. 1979. Analisis de similitud cromosomica segun patrones de bandas G en cuatro especies chilenas de *Phyllotis* (Rodentia, Cricetidae). Archivos de Biología y Medicina Experimentales (Santiago), 12:83-90.

Spotorno O., A. E., and L. I. Walker B. 1983. Analisis electroforético y biométrico de dos especies de *Phyllotis* en Chile central y sus hibridos experimentales. Revista Chilena de Historia Natural, 56:51-59.

Spotorno O., A. E., L. Walker, L. Contreras, J. Pincheira, and R. Fernández-Donoso. 1988. Cromosomas ancestrales en Octodontidae y Abrocomyidae. Sociedad Genetica Chilensis, resumen no. R-65, not paginated.

Spyropoulos, B., P. D. Ross, P. B. Moens, and D. M. Cameron. 1982. The synaptonemal complex karyotypes of Palearctic hamsters, *Phodopus roborovskii* Satunin and *P. sungorus* Pallas. Chromosoma, 86:397-408.

Stains, H. J. 1967. Carnivores, Pp. 325-354, *in* Orders and families of Recent mammals of the world (S. Anderson and J. K. Jones, Jr., eds.). John Wiley and Sons, New York, 453 pp.

Stains, H. J. 1975. Distribution and taxonomy of the Canidae, Pp. 3-26, *in* The wild canids: their systematics, behavorial ecology and evolution (M. W. Fox, ed.). Van Nostrand Reinhold Company, New York, NY, 508 pp.

Stalling, D. T. 1990. *Microtus ochrogaster*. Mammalian Species, 355:1-9.

Stalmakova, V. A. 1957. [On the occurrence of the jerboa *Jaculus turcmenicus* Vinogr. et Bondar in the northern Kara-Kum and on some of its ecological and morphological peculiarities]. Zoologicheskii Zhurnal, 36(2):275-279 (in Russian).

Stangl, F. B., Jr. 1986. Aspects of a contact zone between two chromosomal races of *Peromyscus leucopus* (Rodentia: Cricetidae). Journal of Mammalogy, 67:465-473.

Stangl, F. B., Jr., and R. J. Baker. 1984*a*. A chromosomal subdivision in *Peromyscus leucopus*: Implications for the subspecies concept as applied to mammals. Pp. 139-145, *in* Festschrift for Walter W. Dalquest in honor of his sixty-sixth birthday (N. Horner, ed.). Department of Biology, Midwestern State University, 163 pp.

Stangl, F. B., Jr., and R. J. Baker. 1984*b*. Evolutionary relationships in *Peromyscus*: congruence in chromosomal, genic, and classical data sets. Journal of Mammalogy, 65:643-654.

Stangl, F. B., Jr., and W. W. Dalquest. 1991. Historical biogeography of *Sigmodon ochrognathus* in Texas. The Texas Journal of Science, 43:127-131.

Stål, C. 1865. A list of the names of the genera and subgenera from Linneaus ed. 10 1758 to 1861. Hemiptera Africana, 2:1-256.

Starrett, A. 1967. Hystricoid, erethizontoid, cavioid, and chinchilloid rodents. Pp. 254-272, *in* Recent mammals of the world, a synopsis of families (S. Anderson and J. K. Jones, Jr., eds.). Ronald Press Co., New York, 453 pp.

Starrett, A. 1972. *Cyttarops alecto*. Mammalian Species, 13:1-2.

Stein, B. R. 1986. Comparative limb myology of four arvicolid rodent genera (Mammalia, Rodentia). Journal of Morphology, 187:321-342.

Stein, B. R. 1987. Phylogenetic relationships among four arvicolid genera. Zeitschrift für Säugetierkunde, 52:140-156.

Stein, B. R. 1990. Limb myology and phylogenetic relationships in the superfamily Dipodoidea (birch mice, jumping mice, and jerboas). Zeitschrift für Zoologische Systematik und Evolutionsforschung, 28:299-314.

Stein, G. 1933. Weitere mitteilungen zur systematik papuanischer sauger. Zeitschrift für Säugetierkunde, 8:87-95.

Stein, G. H. W. 1960. Schädelallometrien und Systematik bei altweltlichen Maulwürfen (Talpinae). Mitteilungen aus dem Zoologische Museum in Berlin, 36:1-48.

Steiner, M. 1978. *Apodemus microps* Kratochvil und Rosicky, 1952— Zwergwaldmaus. Pp. 359-367, *in* Handbuch der Säugetiere Europas (J. Niethammer and F. Krapp, eds.). Akademische Verlagsgesellschaft (Wiesbaden), 1:1-476.

Stephan, H., and F. Dieterlen. 1982. Relative brain size in Muridae with special reference to *Colomys goslingi*. Zeitschrift für Säugetierkunde, 47:38-47.

Stephan, H., G. Baron, and H. D. Frahm. 1991. Comparative brain research in mammals, volume 1: Insectivora, with a stereotaxic atlas of the hedgehog brain. Springer-Verlag, New York, 573 pp.

Steven, D. M. 1953. Recent evolution in the genus *Clethrionomys*. Pp. 310-319, *in* Symposia of the society for experimental biology, evolution, number VII (R. Brown and J. F. Danielli, eds.). University Press, Cambridge, 448 pp.

Stewart, B. E., and R. E. A. Stewart. 1989. *Delphinapterus leucas*. Mammalian Species, 336:1-8.

Stewart, B. S., and S. Leatherwood. 1985. Minke whale - *Balaenoptera acutorostrata*. Pp. 91-136, *in* Handbook of marine mammals: The sirenians and baleen whales (S. H. Ridgway and R. Harrison, eds.). Academic Press, London, 3:1-362.

Stiles, C. W., and M. B. Orleman. 1926. Retention of *Cercopithecus*, type *diana*, for the guenons. Journal of Mammalogy, 7:48-53.

Stirling, I. 1971. *Leptonychotes weddelli*. Mammalian Species, 6:1-5.

Stock, A. D. 1974. Chromosome evolution in the genus *Dipodomys* and its taxonomic and phylogenetic implications. Journal of Mammalogy, 55:505-526.

Stogov, I. I. 1985. [On two little studied species of white-toothed shrews (Insectivora, Soricidae, *Crocidura*) from the mountain regions in the southern USSR]. Zoologicheskii Zhurnal, 64:264-268 (in Russian).

Stogov, I. I., and E. P. Bondar. 1966. [A survey of *Crocidura* in South Turkmenia and Tajikistan]. Zoologicheskii Zhurnal, 45:414-420 (in Russian).

Stonehouse, B., and D. Gilmore (eds.). 1977. Biology of Marsupials. Macmillan, London, 486 pp.

Storch, G. 1974. Neue Zwerghamster aus dem Holozan von Aserbeidschan, Iran (Rodentia: Cricetinae). Senckenbergiana Biologica, 55:21-28.

Storch, G. 1977. Die Ausbreitung der Felsenmaus (*Apodemus mystacinus*): zur Problematik der Inselbesiedlung und Tiergeographie in der Agäis. Natur und Museum, 107:174-182.

Storch, G. 1978. Familie Gliridae Thomas, 1897—Schläfer. Pp. 201-280, *in* Handbuch der Säugetiere Europas (H. J. Niethammer, and F. Krapp, eds.). Akademische Verlagsgesellschaft (Wiesbaden), 1:1-476.

Storch, G. 1982. *Microtus majori* Thomas, 1906. Pp. 452-462, *in* Handbuch der Säugetiere Europas (J. Niethammer and F. Krapp, eds.). Akademische Verlagsgesellschaft (Wiesbaden), 2/I:1-649.

Storch, G. 1987. The Neogene mammalian faunas of Ertemte and Harr Obo in Inner Mongolia (Nei Mongol), China.—7. Muridae (Rodentia). Senckenbergiana Lethaea, 67:401-431.

Storch, G., and O. Lutt. 1989. Artstatus der Alpenwaldmaus, *Apodemus alpicola* Heinrich, 1952. Zeitschrift für Säugetierkunde, 54:337-346.

Storch, G., and Z. Qiu. 1983. The Neogene mammalian faunas of Ertemte and Harr Obo in Inner Mongolia (Nei Mongol), China. 2. Moles - Insectivora: Talpidae. Senckenbergiana Lethaea, 64:89-127.

Storch, G., and Z. Qiu. 1991. Insectivores (Mammalia: Erinaceidae, Soricidae, Talpidae) from the Lufeng hominoid locality, Late Miocene of China. Geobios, 24:601-621.

Storer, T. I. 1937. The muskrat as native and alien. Journal of Mammalogy, 18:443-460.

Straney, D. O., and J. L. Patton. 1981. Phylogenetic and environmental determinants of geographic variation of the pocket mouse *Perognathus goldmani* Osgood. Evolution, 34:888-903.

Strelkov, P. P. 1972. *Myotis blythi* (Tomes, 1857): distribution, geographical variability and differences from *Myotis myotis* (Borkhausen, 1797). Acta Theriologica, 17:355-380.

Strelkov, P. P. 1983. [*Myotis mystacinus* and *Myotis brandti* in the USSR and interrelationships of these species]. Zoologicheskii Zhurnal, 62:259-270 (in Russian).

Strelkov, P. P. 1986. [The Gobi bat (*Eptesicus gobiensis* Bobrinskii 1926), a new species of chiropteran of the Palearctic fauna]. Zoologicheskii Zhurnal, 65:1103-1108 (in Russian).

Strelkov, P. P., and E. G. Buntova. 1982. [*Myotis mystacinus* and *M. brandti* (Chiroptera: Vespertilionidae) and interrelations of these species. Part I]. Zoologicheskii Zhurnal, 61:1227-1241 (in Russian).
Strelkov, P. P., V. P. Sosnovtseva, and K. B. Babaev. 1978. [Bats (Chiroptera) of Turkmenistan]. Trudy Zoologicheskovo Instituta, Akademiya Nauk, Leningrad, 79:3-71 (in Russian).
Streubel, D. P., and J. P. Fitzgerald. 1978*a*. *Spermophilus spilosoma*. Mammalian Species, 101:1-4.
Streubel, D. P., and J. P. Fitzgerald. 1978*b*. *Spermophilus tridecemlineatus*. Mammalian Species, 103:1-5.
Stroganov, S. U. 1957. Zveri Sibiri. Nasekomoyadnye. [Animals of Siberia. Insectivores]. Akademiya Nauk SSSR, Moscow, 267 pp.
Stroganov, S. U. 1962. Zveri Sibiri. Khishchye [Animals of Siberia. Carnivores]. Akademiya Nauk SSSR, Moscow, 458 pp. (in Russian).
Struhsaker, T. T. 1970. Phylogenetic implications of some vocalizations of *Cercopithecus* monkeys. Pp. 365-444, *in* Old World Monkeys (J. R. Napier and P. H. Napier, eds.). Academic Press, London, 660 pp.
Stuart, C. T. 1980. The distribution and status of *Manis temmincki* Smuts, 1832 (Pholidota: Manidae). Säugetierkundliche Mitteilungen, 28:123-129.
Stuart, C. T. 1984. The distribution and status of *Felis caracal* Schreber, 1776. Säugetierkundliche Mitteilungen, 31:197-203.
Stuenes, S. 1989. Taxonomy, habits, and relationships of the subfossil Madagascan hippopotami *Hippopotamus lemerlei* and *H. madagascariensis*. Journal of Vertebrate Paleontology, 9:241-268.
Sudman, P. D., J. R. Choate, and E. G. Zimmerman. 1987. Taxonomy of chromosomal races of *Geomys bursarius lutescens* Merriam. Journal of Mammalogy, 68:526-543.
Sugimura, K. 1990. The Amami rabbit *Pentalagus furnessi*. Pp. 140-142, *in* Rabbits, hares and pikas (J. A. Chapman and J. E. C. Flux, eds.). I. U. C. N., Gland, Switzerland, 168 pp.
Sulentich, J. M., L. R. Williams, and G. N. Cameron. 1991. *Geomys breviceps*. Mammalian Species, 383:1-4.
Sulimski, A. 1962. Two new rodents from Weze 1 (Poland). Acta Palaeontologica Polonica, 7(3-4):503-511.
Sulimski, A. 1964. Pliocene Lagamorpha and Rodentia from Weze 1 (Poland). Acta Palaeontologica Polonica, 9(2):149-261.
Sulkava, S. 1978. *Pteromys volans* (Linnaeus, 1758)—Flughörnchen. Pp. 71-84, *in* Handbuch der Säugetiere Europas (J. Niethammer and F. Krapp, eds.). Akademische Verlagsgesellschaft (Wiesbaden), 1:1-476.
Sulkava, S. 1990. *Sorex caecutiens* Laxmann, 1788— Maskenspitzmaus; *Sorex isodon* Turov, 1924 - Taigaspitzmaus. Pp. 215-236, *in* Handbuch der Säugetiere Europas (J. Niethammer and F. Krapp, eds.). Aula-Verlag, Wiesbaden, 3/I:1-524.
Sullivan, J. M., and C. W. Kilpatrick. 1991. Biochemical systematics of the *Peromyscus aztecus* assemblage. Journal of Mammalogy, 72:681-696.
Sullivan, J. M., C. W. Kilpatrick, and P. D. Rennert. 1991. Biochemical systematics of the *Peromyscus boylii* species group. Journal of Mammalogy, 72:669-680.
Sullivan, R. M. 1985. Phyletic, biogeographic, and ecological relationships among montane populations of least chipmunks (*Eutamias minimus*) in the southwest. Systematic Zoology, 34:419-448.
Sullivan, R. M., D. J. Hafner, and T. L. Yates. 1986. Genetics of a contact zone between three chromosomal forms of the grasshopper mouse (genus *Onychomys*): a reassessment. Journal of Mammalogy, 67:640-659.
Sullivan, R. M., S. W. Calhoun, and I. F. Greenbaum. 1990. Geographic variation in genital morphology among insular and mainland populations of *Peromyscus maniculatus* and *Peromyscus oreas*. Journal of Mammalogy, 71:48-58.
Sutton, D. A. 1987. Analysis of Pacific Coast Townsend chipmunks (Rodentia: Sciuridae). Southwestern Naturalist, 32:371-376.
Sutton, D. A. 1992. *Tamias amoenus*. Mammalian Species, 390:1-8.
Sutton, D. A., and C. F. Nadler. 1974. Systematic revision of three Townsend chipmunks (*Eutamias townsendii*). Southwestern Naturalist, 19:199-211.
Swanepoel, P., and H. H. Genoways. 1978. Revision of the Antillean bats of the genus *Brachyphylla* (Mammalia: Phyllostomatidae). Bulletin of Carnegie Museum of Natural History, 12:1-53.
Swanepoel, P., and H. H. Genoways. 1983*a*. *Brachyphylla cavernarum*. Mammalian Species, 205:1-6.
Swanepoel, P., and H. H. Genoways. 1983*b*. *Brachyphylla nana*. Mammalian Species, 206:1-3.
Swanepoel, P., and D. A. Schlitter. 1978. Taxonomic review of the fat mice (genus *Steatomys*) of West Africa (Mammalia: Rodentia). Bulletin of Carnegie Museum of Natural History, 6:53-76.
Swanepoel, P., D. A. Schlitter, and H. H. Genoways. 1979. A study of nongeographic variation in *Tatera leucogaster* (Mammalia: Rodentia) from Botswana. Annals of Carnegie Museum, 48:7-24.
Swanepoel, P., R. H. N. Smithers, and I. L. Rautenbach. 1980. A checklist and numbering system of the extant mammals of the Southern African Subregion. Annals of the Transvaal Museum, 32:155-196.
Swenk, M. H. 1939. A study of local size variation in the prairie pocket gopher (*Geomys lutescens*), with description of a new subspecies from Nebraska. Missouri Valley Fauna, 1:1-8.

Swenson, J. E. 1981. Distribution of Richardson's ground squirrel in eastern Montana. Prairie Naturalist, 13:27-30.

Swynnerton, G. H., and R. W. Hayman. 1951. A checklist of the land mammals of the Tanganyika Territory and the Zanzibar Protectorate. Journal of the East African Natural History Society, 20(6):274-392.

Szalay, F. 1982. A new appraisal of marsupial phylogeny and classification. Pp. 621-640, *in* Carnivorous marsupials (M. Archer, ed.). Royal Zoological Society of New South Wales, Mosman, 1:1-397; 2:397-804.

Szalay, F. S., and E. Delson. 1979. Evolutionary history of the primates. Academic Press, New York, 580 pp.

Taddei, V. A., L. D. Vizotto, and I. Sazima. 1978. Notas sobre *Lionycteris* e *Lonchophylla* nas coleções do Museu Paraense Emílio Goeldi (Mammalia, Chiroptera, Phyllostomidae). Boletim do Museu Paraense Emílio Goeldi, Nova Série, Zoologia, 92:1-14.

Taddei, V. A., L. D. Vizotto, and I. Sazima. 1983. Uma nova espécie de *Lonchophylla* do Brasil e chave para identificação das especies do gênero (Chiroptera, Phyllostomidae). Ciência e Cultura (São Paulo), 35:625-629.

Tagle, D. A., M. M. Miyamoto, M. Goodman, O. Hofmann, G. Graunitzer, R. Goltenboth, and H. Jalanka. 1986. Hemoglobin of pandas: phylogenetic relationships of carnivores as ascertained with protein sequence data. Naturwissenschaften, 73:512-514.

Tamarin, R. H., (ed.). 1985. Biology of New World *Microtus*. Special Publication, American Society of Mammalogists, 8:1-893.

Tamarin, R. H., and T. H. Kunz. 1974. *Microtus breweri*. Mammalian Species, 45:1-3.

Tamayo H., M., and D. Frassinetti C. 1980. Catálogo de los mamíferos fósiles y vivientes de Chile. Boletín de Museo Nacional de Historia Natural Chile (Santiago), 37:323-399.

Tamsitt, J. R., and C. Häuser. 1985. *Sturnira magna*. Mammalian Species, 240:1-4.

Tamsitt, J. R., and D. Nagorsen. 1982. *Anoura cultrata*. Mammalian Species, 179:1-5.

Tanaka, R. 1971. A research into variation in molar and external features among a population of the Smith's red-backed vole for elucidation of its systematic rank. Japanese Journal of Zoology, 16:163-176.

Tas'an, and S. Leatherwood. 1984. Cetaceans live-captured for Jaya Ancol Oceanarium, Djakarta, 1974-1982. International Whaling Commission, Report, 34:485-489.

Tast, J. 1982*a*. *Lemmus lemmus* (Linnaeus, 1758)—Berglemming. Pp. 87-105, *in* Handbuch der Säugetiere Europas (J. Niethammer and F. Krapp, eds.), vol. 2/I. Akademische Verlagsgesellschaft (Wiesbaden), 649 pp.

Tast, J. 1982*b*. *Microtus oeconomus* (Pallas, 1776)—Nordische Wühlmaus, Sumpfmaus. Pp. 374-396, *in* Handbuch der Säugetiere Europas (J. Niethammer and F. Krapp, eds.). Akademische Verlagsgesellschaft (Wiesbaden), 2/I:1-649.

Tate, C. M., J. F. Pagels, and C. O. Handley, Jr. 1980. Distribution and systematic relationship of two kinds of short-tailed shrews (Soricidae: *Blarina*) in south-central Virginia. Proceedings of the Biological Society of Washington, 93:50-60.

Tate, G. H. H. 1932*a*. The taxonomic history of the genus *Reithrodon* Waterhouse (Cricetidae). American Museum Novitates, 529:1-4.

Tate, G. H. H. 1932*b*. The taxonomic history of the South American cricetid genera *Euneomys* (subgenera *Euneomys* and *Galenomys*), *Auliscomys*, *Chelemyscus*, *Chinchillula*, *Phyllotis*, *Paralomys*, *Graomys*, *Eligmodontia*, and *Hesperomys*. American Museum Novitates, 541:1-21.

Tate, G. H. H. 1932*c*. The taxonomic history of the Neotropical cricetid genera *Holochilus*, *Nectomys*, *Scapteromys*, *Megalomys*, *Tylomys*, and *Ototylomys*. American Museum Novitates, 562:1-19.

Tate, G. H. H. 1932*d*. The taxonomic history of the South and Central American cricetid rodents of the genus *Oryzomys*. Part 1: subgenus *Oryzomys*. American Museum Novitates, 579:1-18.

Tate, G. H. H. 1932*e*. The taxonomic history of the South and Central American cricetid rodents of the genus *Oryzomys*. Part 2: subgenera *Oligoryzomys*, *Thallomyscus*, and *Melanomys*. American Museum Novitates, 580:1-16.

Tate, G. H. H. 1932*f*. The taxonomic history of the South and Central American oryzomine genera of rodents (excluding *Oryzomys*): *Nesoryzomys*, *Zygodontomys*, *Chilomys*, *Delomys*, *Phaenomys*, *Rhagomys*, *Rhipidomys*, *Nyctomys*, *Oecomys*, *Thomasomys*, *Inomys*, *Aepeomys*, *Neacomys*, and *Scolomys*. American Museum Novitates, 581:1-28.

Tate, G. H. H. 1932*g*. The taxonomic history of the South and Central American akodont rodent genera: *Thalpomys*, *Deltamys*, *Thaptomys*, *Hypsimys*, *Bolomys*, *Chroeomys*, *Abrothrix*, *Scotinomys*, *Akodon*, (*Chalcomys* and *Akodon*), *Microxus*, *Podoxymys*, *Lenoxus*, *Oxymycterus*, *Notiomys*, and *Blarinomys*. American Museum Novitates, 582:1-32.

Tate, G. H. H. 1932*h*. The taxonomic history of certain South and Central American cricetid Rodentia: *Neotomys*, with remarks upon its relationships; the cotton rats (*Sigmodon* and *Sigmomys*); and the "fish-eating" rats (*Ichthyomys*, *Anotomys*, *Rheomys*, *Neusticomys*, and *Daptomys*). American Museum Novitates, 583:1-10.

Tate, G. H. H. 1933. A systematic revision of the marsupial genus *Marmosa*. Bulletin of the American Museum of Natural History, 66:1-250, 26 pls, 1 table (9 sections, pocketed).

Tate, G. H. H. 1935. The taxonomy of the genera of Neotropical hystricoid rodents. Bulletin of the American Museum of Natural History, 68:295-447.

Tate, G. H. H. 1936. Results of the Archbold Expeditions. No. 13. Some Muidae of the Indo-Australian region. Bulletin of the American Museum of Natural History, 72:501-728.

Tate, G. H. H. 1938. New or little known marsupials: A new species of Phascogalinae, with notes on *Acrobates pulchellus* Rothschild. Novitates Zoologicae, 41:58-60.

Tate, G. H. H. 1939. The mammals of the Guiana region. Bulletin of the American Museum of Natural History, 76:151-229.

Tate, G. H. H. 1942*a*. Review of the vespertilionine bats, with special attention to genera and species of the Archbold collections. Bulletin of the American Museum of Natural History, 80:221-297.

Tate, G. H. H. 1942*b*. Results of the Archbold expeditions. No. 48. Pteropodidae (Chiroptera) of the Archbold Collections. Bulletin of the American Museum of Natural History, 80:331-347.

Tate, G. H. H. 1943. Further notes on the *Rhinolophus philippinensis* group (Chiroptera). American Museum Novitates, 1219:1-5.

Tate, G. H. H. 1945. Results of the Archbold Expeditions, No. 52. The marsupial genus *Phalanger*. American Museum Novitates, 1283:1-41.

Tate, G. H. H. 1947. Results of the Archbold Expeditions. No. 56. On the anatomy and classification of the Dasyuridae (Marsupialia). Bulletin of the American Museum of Natural History, 88:102-155.

Tate, G. H. H. 1948*a*. Results of the Archbold Expeditions. No. 59. Studies on the anatomy and phylogeny of the Macropodidae (Marsupialia). Bulletin of the American Museum of Natural History, 91:237-351.

Tate, G. H. H. 1948*b*. Results of the Archbold Expeditions. No 60. Studies in the Peramelidae (Marsupialia). Bulletin of the American Museum of Natural History, 92:317-346.

Tate, G. H. H. 1951. Results of the Archbold Expeditions. No. 65. The rodents of Australia and New Guinea. Bulletin of the American Museum of Natural History, 97:183-430.

Tate, G. H. H. 1952. Results of the Archbold Expeditions. No. 66. Mammals of Cape York Peninsula, with notes on the occurrence of rain forest in Queensland. Bulletin of the American Museum of Natural History, 98:563-616.

Tate, G. H. H., and R. Archbold. 1935. Results of the Archbold Expeditions. No. 3. Twelve apparently new forms of Muridae other than *Rattus* from the Indo-Australian region. American Museum Novitates, 803:1-9.

Tate, G. H. H., and R. Archbold. 1936. Results of the Archbold Expeditions. No. 9. A new race of *Hyosciurus*. American Museum Novitates, 846:1.

Tate, G. H. H., and R. Archbold. 1938. Results of the Archbold Expeditions. No. 18. Two new Muridae from the Western Division of Papua. American Museum Novitates, 982:1-2.

Tate, G. H. H., and R. Archbold. 1941. Results of the Archbold Expeditions. No. 31. New rodents and marsupials from New Guinea. American Museum Novitates, 1101:1-9.

Tattersall, I. 1976. Notes on the status of *Lemur macaco* and *Lemur fulvus* (Primates, Lemuriformes). Anthropological Papers of the American Museum on Natural History, 53, 2:257-261.

Tattersall, I. 1982. Primates of Madagascar. Columbia University Press, New York, 382 pp.

Tawill, S. A., and J. Niethammer. 1989. Über *Gerbillus pyramidum* (Rodentia, Gerbillidae) im Sudan. Zeitschrift für Säugetierkunde, 54:57-59.

Taylor, E. H. 1934. Philippine land mammals. Monograph of the Bureau of Science (Manila), 30:1-548.

Taylor, J. M., and J. H. Calaby. 1988*a*. *Rattus fuscipes*. Mammalian Species, 298:1-8.

Taylor, J. M., and J. H. Calaby. 1988*b*. *Rattus lutreolus*. Mammalian Species, 299:1-7.

Taylor, J. M., and B. E. Horner. 1973. Results of the Archbold Expeditions. No. 98. Systematics of native Australian *Rattus* (Rodentia, Muridae). Bulletin of the American Museum of Natural History, 150:1-130.

Taylor, J. M., J. H. Calaby, and H. M. Van Deusen. 1982. A revision of the genus *Rattus* (Rodentia, Muridae) in the New Guinean region. Bulletin of the American Museum of Natural History, 173:177-336.

Taylor, J. M., J. H. Calaby, and S. C. Smith. 1983. Native *Rattus*, land bridges, and the Australian region. Journal of Mammalogy, 64:463-475.

Taylor, M. E. 1972. *Ichneumia albicauda*. Mammalian Species, 12:1-4.

Taylor, M. E. 1975. *Herpestes sanguineus*. Mammalian Species, 65:1-5.

Taylor, M. E. 1987. *Bdeogale crassicauda*. Mammalian Species, 294:1-4.

Taylor, M. E. 1989. Craniometric analysis of the African mongooses in the subgenus *Galerella*. Abstracts of papers and posters, Fifth International Theriological Congress, Rome, I:473.

Taylor, P. J., G. K. Campbell, J. Meester, K. Willan, and D. Van Dyk. 1989. Genetic variation in the African rodent subfamily Otomyinae (Muridae). 1. Allozyme divergence among four species. South African Journal of Science, 85:257-262.

Taylor, P. J., G. K. Campbell, J. A. J. Meester, and D. van Dyk. 1991. A study of allozyme evolution in African mongooses (Viverridae: Herpestinae). Zeitschrift für Säugetierkunde, 56(3):135-145.

Taylor, W. P. 1918. Revision of the rodent genus *Aplodontia*. University of California Publications in Zoology, 17:435-504.

Tchernov, E. 1979. Polymorphism, size trends and Pleistocene paleoclimatic response of the subgenus *Sylvaemus* (Mammalia: Rodentia) in Israel. Israel Journal of Zoology, 28:131-159.

Tedford, R. H. 1976. Relationship of pinnipeds to other carnivores (Mammalia). Systematic Zoology, 25:363-374.

Tegelstrom, H., and M. Jaarola. 1989. Genetic divergence in mitochondrial DNA between the wood mouse (*Apodemus sylvaticus*) and the yellow necked mouse (*A. flavicollis*). Hereditas, 111:49-60.

Teilhard de Chardin, P., and C. C. Young. 1936. On the mammal remains from the archeological site of Anyang. Palaeontologia Sinica, ser. C, 12(1):1-61.

Tello, J. 1979. Mamíferos de Venezuela. Caracas, 192 pp.

Temminck, C. J. 1827 [1824]-1841. Monographies de Mammalogie, ou description de quelques genres de mammifères, dont les espèces ont été observées dans les différens musées de l'Europe. C. C. Vander Hoek, Leiden, 392 pp.

Thaeler, C. S., Jr. 1968*a*. An analysis of the distribution of pocket gopher species in northeastern California (genus *Thomomys*). University of California Publications in Zoology, 86:1-46.

Thaeler, C. S., Jr. 1968*b*. An analysis of three hybrid populations of pocket gophers (genus *Thomomys*). Evolution, 22:543-555.

Thaeler, C. S., Jr. 1972. Taxonomic status of the pocket gophers, *Thomomys idahoensis* and *Thomomys pygmaeus* (Rodentia, Geomyidae). Journal of Mammalogy, 53:417-428.

Thaeler, C. S., Jr. 1977. Taxonomic status of *Thomomys talpoides confinus*. Murrelet, 58:49-50.

Thaeler, C. S., Jr. 1980. Chromosome numbers and systematic relations in the genus *Thomomys* (Rodentia: Geomyidae). Journal of Mammalogy, 61:414-422.

Thaeler, C. S., Jr. 1985. Chromosomal variation in the *Thomomys talpoides* complex. Acta Zoologica Fennica, 170:15-18.

Thaeler, C. S., Jr., and L. L. Hinsley. 1979. *Thomomys clusius*, a rediscovered species of pocket gopher. Journal of Mammalogy, 60:480-488.

Thaler, L. 1966. Les rongeurs fossiles du Bas-Languedoc dans leur rapports avec l'histoire des faunes et la stratigraphie du Tertiaire d'Europe. Mémoires du Muséum National d'Histoire Naturelle, Série C, 17:1-295.

Thelma, B. K., and S. R. V. Rao. 1982. In vivo sister chromatid exchanges in the euchromatin and constitutive heterochromatin of the Indian mole rat, *Nesokia indica*. Cytogenetics and Cell Genetics, 33:319-326.

Thenius, E. 1953. Zur Analyse des Gesbisses des Eisbaren, *Ursus* (*Thalarctos*) *maritimus* Phipps, 1774. Säugetierkundliche Mitteilungen, 1:1-7.

Thenius, E. 1972. Grundzüge der Verbreitungsgeschichte der Säugetiere. Gustav Fischer, Stuttgart, 321 pp.

Thenius, E. 1976. Zur stammesgeschichtlichen Herkunft von *Tremarctos* (Ursidae, Mammalia). Zeitschrift für Säugetierkunde, 41:109-114.

Thenius, E. 1979. Zur systematischen und phylogenetischen Stellung des Bambusbären: *Ailuropoda melanoleuca* David (Carnivora, Mammalia). Zeitschrift für Säugetierkunde, 44(5):286-305.

Thenius, E. 1981. Bemerkungen zur taxonomischen und stammesgeschichtlichen Position der Gibbons. Zeitschrift für Säugetierkunde, 46:232-241.

Thomas, B. 1973. Evolutionary implications of karyotypic variation in some insular *Peromyscus* from British Columbia, Canada. Cytologia, 38:485-495.

Thomas, J., V. Pastukhov, R. Elsner, R. and E. Petrov. 1982. *Phoca sibirica*. Mammalian Species, 188:1-6.

Thomas, O. 1882. On African mongooses. Proceedings of the Zoological Society of London, 1882:59-93.

Thomas, O. 1887*a*. Description of a second species of rabbit-bandicoot (*Peragale*). Annals and Magazine of Natural History, ser. 5, 19:397-399.

Thomas, O. 1887*b*. On the small mammals collected in Demerara by Mr. W. L. Sclater. Proceedings of the Zoological Society of London, 1887:150-153.

Thomas, O. 1887*c*. Report on a zoological collection made by the officers of H. M. S. 'Flying Fish' at Christmas Island, Indian Ocean. I. Mammalia. Proceedings of the Zoological Society of London, 1887:511-514.

Thomas, O. 1888*a*. Catalogue of the Marsupialia and Monotremata in the collection of the British Museum (Natural History). British Museum (Natural History), London, 401 pp., 33 pls.
Thomas, O. 1888*b* [1889]. On the mammals of Christmas Island. Proceedings of the Zoological Society of London, 1888:532-534.
Thomas, O. 1890. On a collection of mammals obtained by Dr. Emin Pasha in central and eastern Africa. Proceedings of the Zoological Society of London, 1890:443-450.
Thomas, O. 1892*a*. On some new Mammalia from the East-Indian Archipelago. Annals and Magazine of Natural History, ser. 6, 9:250-254.
Thomas, O. 1892*b*. On some mammals from Mount Dulit, North Borneo. Proceedings of the Zoological Society of London, 1892:221-227.
Thomas, O. 1897*a*. On the mammals collected in British New Guinea by Dr. Lamberto Loria. Annali del Museo Civico di Storia Naturale di Genova, ser. 2, 18:1-19.
Thomas, O. 1897*b* [1898]. On the Mammals obtained by Mr. A. Whyte in Nyasaland, and presented to the British Museum by Sir H. H. Johnston, K. C. B.; being a fifth contribution to the mammal-fauna of Nyasaland. Proceedings of the Zoological Society of London, 1897:924-938.
Thomas, O. 1898*a*. On indigenous Muridae in the West Indies; with the description of a new Mexican *Oryzomys*. Annals and Magazine of Natural History, ser. 7, 1:176-180.
Thomas, O. 1898*b*. On the mammals collected by Mr. John Whitehead during his recent expedition to the Philippines with field notes by the collector. Transactions of the Zoological Society of London, 14:377-414.
Thomas, O. 1900. The geographical races of the Tayra (*Galictis barbara*), with notes on abnormally coloured individuals. Annals and Magazine of Natural History, ser. 7, 5:145-148.
Thomas, O. 1901*a*. The generic names *Myrmecophaga* and *Didelphis*. American Naturalist, 35:143-145.
Thomas, O. 1901*b*. On mammals obtained by Mr. Alphonse Robert on the Rio Jordao, Minas Geraes. Annals and Magazine of Natural History, ser. 7, 8:526-536.
Thomas, O. 1902*a* [1903]. On a collection of mammals from Abyssinia including some from Lake Tsana collected by Mr. Edward Degen. Proceedings of the Zoological Society of London, 1902:308-316.
Thomas, O. 1902*b*. On the geographical races of the kinkajou. Annals and Magazine of Natural History, ser. 7, 9:266-270.
Thomas, O. 1903. Notes on Neotropical mammals of the genera *Felis, Hapale, Oryzomys, Akodon,* and *Ctenomys*, with descriptions of new species. Annals and Magazine of Natural History, ser. 7, 12:234-243.
Thomas, O. 1904 [1905]. On *Hylochoerus*, the forest pig of Central Africa. Proceedings of the Zoological Society of London, 1904(2):193-199.
Thomas, O. 1905*a* [1906]. The Duke of Bedford's zoological exploration in eastern Asia. — 1. List of mammals obtained by Mr. M. P. Anderson in Japan. Proceedings of the Zoological Society of London, 1905(2):331-363.
Thomas, O. 1905*b*. On new Japanese mammals (Insectivora, Rodentia). Annals and Magazine of Natural History, ser. 7, 15:487-495.
Thomas, O. 1906*a*. Descriptions of new mammals from Mount Ruwenzori. Annals and Magazine of Natural History, ser. 7, 18:136-147.
Thomas, O. 1906*b*. Two new genera of small mammals discovered by Mrs. Holms-Tarn in British East Africa. Annals and Magazine of Natural History, ser. 7, 18:222-226.
Thomas, O. 1906*c*. New mammals collected in north-east Africa by Mr. Zaphiro, and presented to the British Museum by W. N. McMilan, Esq. Annals and Magazine of Natural History, ser. 7, 18:300-306.
Thomas, O. 1906*d*. Notes on South American rodents. II. On the allocation of certain species hitherto referred respectively to *Oryzomys, Thomasomys,* and *Rhipidomys*. Annals and Magazine of Natural History, ser. 7, 18:442-448.
Thomas, O. 1907*a*. On further new mammals obtained by the Ruwenzori Expedition. Annals and Magazine of Natural History, ser. 7, 19:118-123.
Thomas, O. 1907*b*. On Neotropical mammals of the genera *Callicebus, Reithrodontomys, Ctenomys, Dasypus,* and *Marmosa*. Annals and Magazine of Natural History, ser. 7, 20:161-168.
Thomas, O. 1908. The Duke of Bedford's zoological exploration in eastern Asia - IX. List of mammals from the Mongolian Plateau. Proceedings of the Zoological Society of London, 1908(1):104-110.
Thomas, O. 1909*a*. Mr. O. Thomas on the generic arrangement of the African squirrels. Annals and Magazine of Natural History, ser. 8, 3:467-475.
Thomas, O. 1909*b*. New African mammals. Annals and Magazine of Natural History, ser. 8, 4:542-549.
Thomas, O. 1910*a*. New African mammals. Annals and Magazine of Natural History, ser. 8, 5:83-92.
Thomas, O. 1910*b*. Further new African Mammalia. Annals and Magazine of Natural History, ser. 8, 5:282-285.
Thomas, O. 1910*c*. Notes on African rodents. Annals and Magazine of Natural History, ser. 8, 6:221-226.

Thomas, O. 1911*a*. The mammals of the tenth edition of Linnaeus; an attempt to fix the types of the genera and the exact bases and localities of the species. Proceedings of the Zoological Society of London, 1911:120-158.

Thomas, O. 1911*b*. On new African Muridae. Annals and Magazine of Natural History, ser. 8, 7:378-383.

Thomas, O. 1912. Exhibition of skin and skull of a viverrine carnivore from Tonkin. Proceedings of the Zoological Society of London, 1912:17-18.

Thomas, O. 1913*a*. Ernst Hartert's expedition to the Central Western Sahara. Mammals. Novitates Zoologicae, Triung, 20:28-33.

Thomas, O. 1913*b*. On new mammals obtained by the Utakwa Expedition to Dutch New Guinea. Annals and Magazine of Natural History, ser. 8, 12:205-212.

Thomas, O. 1914*a*. On various South American mammals. Annals and Magazine of Natural History, ser. 8, 13:345-363.

Thomas, O. 1914*b*. New Asiatic and Australasian bats, and a new bandicoot. Annals and Magazine of Natural History, ser. 8, 13:439-444.

Thomas, O. 1915. List of mammals (exclusive of Ungulata) collected on the Upper Congo by Dr. Christy for the Congo Museum, Tervueren. Annals and Magazine of Natural History, ser. 8, 16:465-481.

Thomas, O. 1916*a*. On the rats usually included in the genus *Arvicanthis*. Annals and Magazine of Natural History, ser. 8, 18:67-70.

Thomas, O. 1916*b*. Three new African mice of the genus *Dendromus*. Annals and Magazine of Natural History, ser. 8, 18:241-143.

Thomas, O. 1916*c*. The grouping of the South American Muridae commonly referred to *Akodon*. Annals and Magazine of Natural History, ser. 8, 18:336-340.

Thomas, O. 1916*d*. The bandicoot of Mount Popa, and its allies. Journal of the Bombay Natural History Society, 24:640-643.

Thomas, O. 1917*a*. A new species of *Aconaemys* from southern Chile. Annals and Magazine of Natural History, ser. 8, 19:281-282.

Thomas, O. 1917*b*. The geographical races of *Galago crassicaudatus*. Annals and Magazine of Natural History, ser. 8, 20:47-50.

Thomas, O. 1917*c*. On the arrangement of the South American rats allied to *Oryzomys* and *Rhipidomys*. Annals and Magazine of Natural History, ser. 8, 20:192-198.

Thomas, O. 1918*a*. On the arrangement of the small Tenrecidae hitherto referred to *Oryzorictes* and *Microgale*. Annals and Magazine of Natural History, ser. 9, 1:302-307.

Thomas, O. 1918*b*. A revised classification of the Otomyinae, with descriptions of new genera and species. Annals and Magazine of Natural History, ser. 9, 2:203-211.

Thomas, O. 1919. On small mammals from "Otro Cerro" north-eastern Rioja, collected by Sr. E. Budin. Annals and Magazine of Natural History, ser. 9, 3:489-500.

Thomas, O. 1920. The generic positions of "*Mus*" *nigricauda*, Thos., and *woosnami*, Schwann. Annals and Magazine of Natural History, ser. 9, 5:140-142.

Thomas, O. 1921*a*. On a new genus and species of shrew, and some new Muride from the East-Indian Archipelago. Annals and Magazine of Natural History, ser. 9, 7:243-249.

Thomas, O. 1921*b*. On the cavies of the genus *Caviella*. Annals and Magazine of Natural History, ser. 9, 7:445-448.

Thomas, O. 1921*c*. Geographical races of *Herpestes brachyurus*, Gray. Annals and Magazine of Natural History, ser. 9, 8:134-136.

Thomas, O. 1923. Scientific results from the mammal survey. No. XXX. The mongooses of the *Herpestes smithii* group. Journal of the Bombay Natural History Society, 28(1):23-26.

Thomas, O. 1924. On some Ceylon mammals. Annals and Magazine of Natural History, ser. 9, 13:239-242.

Thomas, O. 1925. The Spedan Lewis South American exploration. I. On mammals from southern Bolivia. Annals and Magazine of Natural History, ser. 9, 25:575-582.

Thomas, O. 1926*a*. The Godman-Thomas expedition to Peru. I. On mammals collected by Mr. R. W. Hendee near Lake Junin. Annals and Magazine of Natural History, ser. 9, 17:313-318.

Thomas, O. 1926*b*. The Spedan Lewis South American exploration. II. On mammals collected in the Tarija Department, southern Bolivia. Annals and Magazine of Natural History, ser. 9, 17:318-328.

Thomas, O. 1926*c*. The Godman-Thomas expedition to Peru. II. On mammals collected by Mr. Hendee in north Peru between Pacasmayo and Chachapoyas. Annals and Magazine of Natural History, ser. 9, 17:610-616.

Thomas, O. 1927. On a further collection of mammals made by Sr. E. Budin in Neuquen, Patagonia. Annals and Magazine of Natural History, ser. 9, 19:650-658.

Thomas, O. 1928*a*. Some rarities from Abyssinia, with the description of a new mole-rat (*Tachyoryctes*), and a new *Arvicanthis*. Annals and Magazine of Natural History, ser. 10, 1:302-304.

Thomas, O. 1928*b*. The Godman-Thomas expedition to Peru.—VIII. On mammals obtained by Mr. Hendee at Pebas and Iquitos, upper Amazons. Annals and Magazine of Natural History, ser. 10, 2:285-294.

Thomas, O. 1929. The mammals of Senor Budin's Patagonian expedition. Annals and Magazine of Natural History, ser. 10, 4:35-45.

Thomas, O., and M. A. C. Hinton. 1920. On the group of African zorils represented by *Ictonyx libyca*. Annals and Magazine of Natural History, ser. 9, 5:367-369.

Thomas, O., and M. A. C. Hinton. 1923. On mammals collected by Captain Shortridge during the Percy Sladen and Kaffrarian Museum Expeditions to the Orange River. Proceedings of the General Meetings for Scientific Business of the Zoological Society of London, 1923:483-499.

Thomas, O., and M. A. C. Hinton. 1925. On mammals collected in 1923 by Captain G. C. Shortridge during the Percy Sladen and Kaffrarian Museum Expedition to South-West Africa. Proceedings of the Zoological Society of London, 1925:221-246.

Thomas, O., and H. Schwann. 1904. On a collection of mammals from British Namaqualand, presented to the National Museum by Mr. C. D. Rudd. Proceedings of the Zoological Society of London, 1904(1):171-183.

Thomas, O., and H. Schwann. 1905. The Rudd exploration of South Africa.—II. List of mammals from the Wakkerstroom District, Southeastern Transvaal. Proceedings of the Zoological Society of London, 1905:129-138.

Thomas, O., and R. C. Wroughton. 1908. The Rudd exploration of S. Africa.—X. List of mammals collected by Mr. Grant near Tette, Zambesia. Proceedings of the General Meetings for Scientific Business of the Zoological Society of London, 1908:535-552.

Thomson, C. E. 1982. *Myotis sodalis*. Mammalian Species, 163:1-5.

Thonglongya, K. 1973. First record of *Rhinolophus paradoxolophus* (Bourret, 1951) from Thailand, with the description of a new species of the *Rhinolophus philippinensis* group (Chiroptera: Rhinolophidae). Mammalia, 37:587-597.

Thorington, R. W., Jr. 1984. Flying squirrels are monophyletic. Science, 225:1048-1050.

Thorington, R. W., Jr. 1988. Taxonomic status of *Saguinus tripartitus* (Milne-Edwards, 1878). American Journal of Primatology, 15:367-371.

Thorington, R. W., Jr., and C. P. Groves. 1970. An annotated classification of the Cercopithecoidea. Pp. 629-647, *in* Old World Monkeys (J. R. Napier and P. H. Napier, eds.). Academic Press, London, 660 pp.

Thornton, W. A., and G. C. Creel. 1975. The taxonomic status of kit foxes. Texas Journal of Science, 26:127-136.

Thouless, C. R., and K. Al Bassri. 1991. Taxonomic status of the Farasan Island gazelle. Journal of Zoology (London), 223:151-159.

Thys van den Audenaerde, D. F. E. 1977. Description of a new monkey skin from East-Central Zaire as a probably new monkey species (Mammalia, Cercopithecidae). Revue de Zoologie Africaine, 91:1000-1010.

Timm, R. M. 1982. *Ectophylla alba*. Mammalian Species, 166:1-4.

Timm, R. M. 1985. *Artibeus phaeotis*. Mammalian Species, 235:1-6.

Tiunov, M. P. 1986. [External structure of the male accessory gland in chiropterans and utilization of this character in their taxonomy]. Zoologicheskii Zhurnal, 65:1275-1279.

Tobien, H. 1975. Zur Gebissstruktur, Systematik und Evolution der Genera *Piezodus*, *Prolagus* und *Ptychoprolagus* (Lagomorpha, Mammalia) aus einigen Vorkommen im jüngeren Tertiär Mittel -und Westeuropas. Notizblatt des Hessischen Landesamtes für Bodenforschung zu Wiesbaden, 103-186.

Todd, N. B., and S. R. Pressman. 1968. The karyotype of the lesser panda (*Ailurus fulgens*) and general remarks on the phylogeny and affinities of the panda. Carnivore Genetics Newsletter, 5:105-108.

Tomasi, E. T., and R. S. Hoffmann. 1984. *Sorex preblei* in Utah and Wyoming. Journal of Mammalogy, 65:708.

Tomilin, A. G. 1957. Zveri SSSR i prilezhashchikh stran: Kitoobrazyne. (Zveri vostochnoi Evropy i severnoi Azii) [Mammals of USSR and adjacent countries: Cetacea (Mammals of eastern Europe and northern Asia)]. Akademiya Nauk SSSR, 9:1-756 (in Russian).

Tomilin, A. G. 1967. Mammals of the USSR and adjacent countries: Cetacea (Mammals of eastern Europe and northern Asia) [A translation of A. G. Tomilin, 1957, Zveri SSSR i prilezhashchikh stran: Kitoobrazyne. (Zveri vostochnoi Evropy i severnoi Asii)]. Israel Program for Scientific Translations, Jerusalem, 9:1-717.

Tong, H. 1989. Origine et evolution des Gerbillidae (Mammalia, Rodentia) en Afrique du Nord. Memoires de la Societe Geologique de France, Nouvelle Serie—1989, 155:1-120.

Topachevskii, V. A. 1969. Fauna SSSR, Mlekopitayushchie, Tom 3, vyp. 3. Slepyshovye (Spalacidae) [Fauna of the USSR: Mammals. Mole rats, Spalacidae]. Nauka Publishers, Leningrad Section, Leningrad, 248 pp.

Topachevskii, V. A. 1976. Fauna of the USSR: Mammals. vol. 3, pt. 3, mole rats, Spalacidae [a translation of V. A. Topachevskii, 1969, Fauna SSSR, Mlekopitayushchie, Tom 3, vyp. 3. Slepyshovye (Spalacidae). Smithsonian Institution Libraries and National Science Foundation, Washington, D. C., 308 pp.

Topachevskii, V. A., and L. I. Rekovets. 1982. Novye materialy k sistematike i evolyutsii slepushonok nominativnovo podrod roda *Ellobius* (Rodentia, Cricetidae) [New material for systematic and evolution of the mole vole, nominative subgenus and genus *Ellobius* (Rodentia, Cricetidae)]. Vestnik Zoologii, 5:47-54 (in Russian).

Topachevskii, V. A., and A. F. Skorik. 1984. Pervaya nakhodka iskopaemykh ostatkov kosmatykh khomyakov - Lophiomyinae (Rodentia, Cricetidae) [The first find of fossil remains of a maned hamster; Lophomyinae (Rodentia, Cricetidae)]. Vestnik Zoologii, 2:57-60 (in Russian).

Topal, G. 1970*a*. The first record of *Ia io* Thomas, 1902 in Vietnam and India, and some remarks on the taxonomic position of *Parascotomanes beaulieui* Bourret, 1942, *Ia longimana* Pen, 1962, and the genus *Ia* Thomas, 1902 (Chiroptera: Vespertilionidae). Opuscula Zoologica (Budapest), 10:341-347.

Topal, G. 1970*b*. On the systematic status of *Pipistrellus annectans* Dobson, 1871 and *Myotis primula* Thomas, 1920 (Chiroptera: Vespertilionidae). Annales Historico-Naturales Musei Nationalis Hungarici (Budapest), 62:373-379.

Tranier, M. 1975. Etude preliminaire du caryotype de l'*Acomys* de l'Air (Rongeurs, Murides). Mammalia, 39:705-705.

Tranier, M., and H. Dosso. 1979. Recherches caryotypiques sur les rongeurs de Cote d'Ivoire: Resultats preliminaires pour les milieux fermes. Annales de l'Universite d'Abidjan, ser. E (Ecologie), 12:181-183.

Tranier, M., and J. C. Gautun. 1979. Recherches caryotypiques sur les rongeurs de Cote d'Ivoire: Resultats preliminaires pour les milieux ouverts. Le cas d'*Oenomys hypoxanthus ornatus*. Mammalia, 43(2):252-254.

Tranier, M., and D. Julien-Laferroere. 1990. A propros de petites gerbilles du Niger et du Tchad (Rongeurs, Gerbillidae, *Gerbillus*). Mammalia, 54:451-456.

Tranier, M., and F. Petter. 1978. Les relations d'*Eliomys tunetae* et de quelques autres formes de Lerots de la region mediterraneenne (Rongeurs, Muscardinides). Mammalia, 42(3):349-353.

Traut, W., M. F. Seldin, and H. Winking. 1992. Genetic mapping and assignment of a long-range repeat cluster to band D of chromosome 1 in *Mus musculus* and *Mus spretus*. Cytogenetics and Cell Genetics, 60:128-130.

Treviño-Villareal, J. 1991. The annual cycle of the Mexican prairie dog (*Cynomys mexicanus*). Occasional Papers of the Museum of Natural History, University of Kansas, 139:1-27.

Tristram, H. 1877. Notes on *Eliomys melanurus* and some other rodents of Palestine. Proceedings of the Zoological Society of London, 1877:40-42.

Trombulak, S. C. 1988. *Spermophilus saturatus*. Mammalian Species, 322:1-4.

Trouessart, E. L. 1885. Catalogue des Mammifères vivants et fossiles. Fasc. IV.- Carnivores (Carnivora). Bulletin de la Société d'Études scientifiques d'Angers, Supplément a l'année 1884, 14:1-108.

Trouessart, E. L. 1897-1905. Catalogus mammalium tam viventium quam fossilium. Quinquennale supplementum anno 1904. [Tomus 1-1897; Tomus 2-1898; Quinquennale supplementum, fascic. 1 & 2 - 1904; fascic. 3 & 4 - 1905]. R. Friedländer and Sohn, Berlin, 1 & 2:1469 pp.; Quin supp:929 pp.

Trouessart, E. L. 1910. Faune des Mammifères d'Europe. R. Friedländer, Berlin, 219 pp.

Troughton, E. L. G. 1925. A revision of the genera *Taphozous* and *Saccolaimus* (Chiroptera) in Australia and New Guinea, including a new species and a note on two Malayan forms. Records of the Australian Museum, 14:313-341.

Troughton, E. L. G. 1927. Fixation of the habitat and extended description of *Pteropus tuberculatus* Peters. Records of the Australian Museum, 25:355-359.

Troughton, E. L. G. 1930. A new species and subspecies of fruit-bats (*Pteropus*) from the Santa Cruz group. Records of the Australian Museum, 18:1-4.

Troughton, E. L. G. 1932*a*. On five new rats of the genus *Pseudomys*. Records of the Australian Museum, 18:287-294.

Troughton, E. L. G. 1932*b*. A revision of the Rabbit-Bandicoots. Family Peramelidae, Genus *Macrotis*. Australian Zoologist, 7:219-236.

Troughton, E. L. G. 1936. A redescription of *Solomys* ("*Mus*") *salamonis* Ramsay. Proceedings of the Linnean Society of New South Wales, 61:3-4.

Troughton, E. Le G. 1937. Descriptions of some New Guinea mammals. Records of the Australian Museum, 20:117-127.

Tsuchiya, K. 1974. Cytological and biochemical studies of *Apodemus speciosus* group in Japan. Journal of the Mammalogical Society of Japan, 6:67-87.

Tsuchiya, K. 1981. [On the chromosome variations in Japanese cricetid and murid rodents]. Honyurui Kagaku (Mammalian Science), 42:51-58 (in Japanese, with English abstract).

Tsuchiya, K., T. H. Yosida, K. Moriwaki, S. Ohtani, S. Kulta-Uthai, and P. Sudto. 1979. Karyotypes of twelve species of small mammals from Thailand. Report of the Hokaido Institute of Public Health, 29:26-29.

Tsuchiya, K., S. Wakana, H. Suzuki, S. Hattori, and Y. Hayashi. 1989. Taxonomic study of *Tokudaia* (Rodentia: Muridae): I. Genetic differentiation. Memoirs of the National Science Museum, Tokyo, 22:227-234.

Tucker, P. K., and D. J. Schmidly. 1981. Studies of a contact zone among three chromosomal races of *Geomys bursarius* in east Texas. Journal of Mammalogy, 62:258-272.

Tullberg, T. 1899. Ueber das System der Nagethiere; eine phylogenetische Studie. Acta Societatis Medicorum Upsaliensis, 3:514.

Tumlison, R. 1987. *Felis lynx*. Mammalian Species, 269:1-8.

Tumlison, R. 1992. *Plecotus mexicanus*. Mammalian Species, 401:1-3.

Tuttle, R. H. 1967. Knuckle-walking and the evolution of hominoid hands. American Journal of Physical Anthropology, New Series, 26:171-206.

Tvrtkovic, N., and G. Dzukic. 1977. Sisavci lesinog (slanog) kopova s posebnim osvrtom na vrstu *Apodemus microps* Krat. and Ros. 1952. Arhiv Bioloskih Nauka (Beograd), 29:161-173 (in Serbo-Croatian).

Twigg, G. I. 1992. The black rat *Rattus rattus* in the United Kingdon in 1989. Mammal Review, 22:33-42.

Tyler, M. J., (ed.). 1979. The status of endangered Australasian wildlife. Royal Society of South Australia, Adelaide, 210 pp.

Uerpmann, H.-P. 1987. The ancient distribution of ungulate mammals in the Middle East. Beihefte zum Tübinger Atlas des vorderen Orients, Reihe A (Naturwissenschaften), 27:1-173.

U.S. [ESA] Fish and Wildlife Service. 1991. Endangered and threatened wildlife and plants (those covered by the regulations for the U.S. Endangered Species Act), 50 CRF 17.11 17.12, July 15, 1991. U. S. Government Printing Office, Washington, D. C., 37 pp.

Urban-Ramirez, J., and D. Aurioles-Gamboa. In press. First record of the tropical beaked whale *Mesoplodon peruvianus* Reyes, Mead, and Waerebeek 1991 in the North Pacific and comments on its distribution. Marine Mammal Science.

Valdez, R., C. F. Nadler, and T. D. Bunch. 1978. Evolution of wild sheep in Iran. Evolution, 32:56-72.

Van Bemmel, A. C. V. 1949*a*. Revision of the rusine deer in the Indo-Australian Archipelago. Treubia, 20:191-262.

Van Bemmel, A. C. V. 1949*b*. Notes on Indo-Australian mammals. 1. A note on *Lutrogale perspicillata* (I. Geoffroy) (Mustelidae). 2. On the meaning of the name *Cervus javanicus* Osbeck 1765 (Tragulidae). Treubia, 20:375-380.

Van Bemmel, A. C. V. 1952. Contribution to the knowledge of the genera *Muntiacus* and *Arctogalidia* in the Indo-Australian Archipelago (Mammalia, Cervidae & Viverridae). Beaufortia, 16:1-50.

van Bree, P. J. H. 1971. On *Globicephala sieboldii* Gray, 1846, and other species of pilot whale (Notes on Cetacea, Delphinoidea III). Beaufortia, 19(249):79-87.

van Bree, P. J. H. 1973. *Neophocaena phocaenoides asiaeorientalis* (Pilleri & Gihr 1973) a synonym of the preoccupied name *Delphinus melas* Schlegel 1841 (Notes on Cetacea Delphinoidea VII). Beaufortia, 21:17-24.

van Bree, P. J. H. 1975. On the alleged occurrence of *Mesoplodon bidens* (Sowerby, 1804) (Cetacea, Ziphioidae) in the Mediterranean. Annali del Museo Civico di Storia Naturale di Genova, 80:226-228.

van Bree, P. J. H. 1976. On the correct Latin name of the Indus Susu (Cetacea, Platanistoidea). Bulletin Zoologisch Museum, Universiteit van Amsterdam, 5(17):139-140.

van Bree, P. J. H., and M. Sc. Boeadi. 1978. Notes on the Indonesian mountain weasel, *Mustela lutreolina* Robinson and Thomas, 1917. Zeitschrift für Säugetierkunde, 43:166-171.

van Bree, P. J. H., and M. D. Gallagher. 1978. On the taxonomic status of *Delphinus tropicalis* van Bree 1971 (Notes on Cetacea Delphinoidea IX). Beaufortia, 28(342):1-8.

van Bree, P. J. H., and I. Kristensen. 1974. On the intriguing stranding of four Cuvier's beaked whales *Ziphius cavirostris* G. Cuvier 1823 on the Lesser Antillean island of Bonaire. Bijdragen Tot De Dierkunde, 44(2):235-238.

van Bree, P. J. H., and P. E. Purves. 1972. Remarks on the validity of *Delphinus bairdii* (Cetacea, Delphinidae). Journal of Mammalogy, 53:372-374.

van der Feen, P. J. 1962. Cataloge of the Marsupialia from New Guinea, the Moluccas and Celebes in the Museo Civico di Storia Naturale "Giacomo Doria" in Genoa. Annali di Museo Civico di Storia Naturale "Giacomo Doria", Genova, 73:19-70.

van de Weerd, A. 1979. Early Ruscinian rodents and lagomorphs (Mammalia) from the lignites near Ptolemais (Macedonia, Greece). I and II. Proceedings of the Koninklijke Nederlandse Akademie van Wetenschappen, ser. B, 82:127-154.

van der Meulen, A. J. 1978. *Microtus* and *Pitymys* (Arvicolidae) from Cumberland Cave, Maryland, with a comparison of some New and Old World species. Annals of Carnegie Museum, 47:101-145.

van der Meulen, A. J., and H. de Bruijn. 1982. The mammals from the Lower Miocene of Aliveri (Island of Evia, Greece). Part 2. The Gliridae. Proceedings of the Koninklijke Nederlandse Akademie van Wetenschappen, ser. B, 85(4):485-524.

Van der Straeten, E. 1975. *Lemniscomys bellieri*, a new species of Muridae from the Ivory Coast (Mammalia, Muridae). Revue de Zoologie Africaine, 89:906-908.

Van der Straeten, E. 1977. *Apodemus flavicollis* (Melchior, 1834) in Nederland. Lutra, 19:20-21.

Van der Straeten, E. 1980*a*. Etude biometrique de *Lemniscomys linulus* (Afrique Occidentale) (Mammalia, Muridae). Revue de Zoologie Africaine, 94:185-201.

Van der Straeten, E. 1980*b*. A new species of *Lemniscomys* (Muridae) from Zambia. Annals of the Cape Provincial Museums, Natural History, 13:55-62.

Van der Straeten, E. 1984. Etude biometrique des genres *Dephomys* et *Stochomys* avec quelques notes taxonomiques (Mammalia, Muridae). Revue de Zoologie Africaine, 98:771-798.

Van der Straeten, E. 1985. Note sur *Hybomys basilii* Eisentraut, 1965. Bonner Zoologische Beiträge, 36:1-8.

Van der Straeten, E., and F. Dieterlen. 1983. Description de *Praomys ruppi*, une nouvelle espece de Muridae d'Ethiopie. Annales Musée Royal de l'Afrique Centrale, Tervuren-Belgique, ser. 8 (Sciences Zoologiques), 237:121-128.

Van der Straeten, E., and F. Dieterlen. 1987. *Praomys misonnei*, a new species of Muridae from Eastern Zaire (Mammalia). Stuttgarter Beiträge zur Naturkunde, ser. A (Biologie), 402:1-40.

Van der Straeten, E., and F. Dieterlen. 1992. Craniometrical comparison of four populations of *Praomys jacksoni* captured at different heights in Eastern Zaire (Kivu). Mammalia, 56:125-131.

Van der Straeten, E., and A. M. Dudu. 1990. Systematics and distribution of *Praomys* from the Masako Forest Reserve (Zaire), with the description of a new species. Pp. 73-83, *in* Vertebrates in the tropics (G. Peters and R. Hutterer, eds.). Museum Alexander Koenig, Bonn, 424 pp.

Van der Straeten, E., and R. Hutterer. 1986. *Hybomys eisentrauti*, une nouvelle espece de Muridae du Cameroun (Mammalia, Rodentia). Mammalia, 50:35-42.

Van der Straeten, E., and B. Van der Straeten-Harrie. 1977. Etude de la biometrie cranienne et de la repartition d'*Apodemus sylvaticus* (Linnaeus, 1758) et d'*Apodemus flavicollis* (Melchior, 1834) en Belgique. Acta Zoologica et Pathologica Antverpiensia, 69:169-182.

Van der Straeten, E., and W. N. Verheyen. 1978*a*. Karyological and morphological comparisons of *Lemniscomys striatus* (Linnaeus, 1758) and *Lemniscomys bellieri* Van der Straeten, 1975, from Ivory Coast (Mammalia: Muridae). Bulletin of Carnegie Museum of Natural History, 6:41-47.

Van der Straeten, E., and W. N. Verheyen. 1978*b*. Taxonomical notes on the West African *Myomys* with the description of *Myomys derooi* (Mammalia—Muridae). Zeitschrift für Säugetierkunde, 43:31-41.

Van der Straeten, E., and W. N. Verheyen. 1979*a*. Note sur la position systematique de *Lemniscomys macculus* (Thomas et Wroughton, 1910) (Mammalia, Muridae). Mammalia, 43:377-389.

Van der Straeten, E., and W. N. Verheyen. 1979*b*. Notes taxonomiques sur les *Malacomys* de l'Ouest africain avec redescription du patron chromosomique de *Malacomys edwardsi* (Mammalia, Muridae). Revue de Zoologie Africaine, 93:10-35.

Van der Straeten, E., and W. N. Verheyen. 1980. Relations biometriques dans le groupe specifique *Lemniscomys striatus* (Mammalia, Muridae). Mammalia, 44:73-82.

Van der Straeten, E., and W. N. Verheyen. 1981. Etude biometrique du genre *Praomys* en Cote d'Ivoire. Bonner Zoologische Beiträge, 32:249-264.

Van der Straeten, E., and W. N. Verheyen. 1983. Nouvelles captures de *Lophuromys rahmi* et *Delanymys brooksi* en Republique Rwandaise. Mammalia, 47:426-430.

Van der Straeten, E., W. N. Verheyen, and B. Harrie. 1986. The taxonomic status of *Hybomys univittatus lunaris* Thomas, 1906 (Mammalia: Muridae). Cimbebasia, ser. A., 8:209-218.

Van Deusen, H. M., and G. G. George. 1969. Results of the Archbold Expeditions. No. 90. Notes on the echidnas (Mammalia: Tachyglossidae) of New Guinea. American Museum Novitates, 2383:1-23.

Van Deusen, H. M., and J. K. Jones, Jr. 1967. Marsupials. Pp. 61-86, *in* Recent mammals of the world (S. Anderson and J. K. Jones, Jr. eds.). Ronald Press Co., New York, 453 pp.

Van Deusen, H. M., and K. F. Koopman. 1971. Results of the Archbold Expeditions. No. 95. The genus *Chalinolobus* (Chiroptera, Vespertilionidae). Taxonomic review of *Chalinolobus picatus*, *C. nigrogriseus*, and *C. rogersi*. American Museum Novitates, 2468:1-30.

Van Dyck, S. 1982. The relationships of *Antechinus stuartii* and *A. flavipes* (Dasyuridae, Marsupialia) with special reference to Queensland. Pp. 723-766, *in* Carnivorous Marsupials (M. Archer, ed.). Royal Zoological Society of New South Wales, 2:1-804.

Van Dyck, S. 1990. *Belideus gracilis* - soaring problems for an old De Vis glider. Memoirs of the Queensland Museum, 28:329-336.

Van Dyck, S. 1991. Raising an old glider's ghost - a devil of an exorcise! Wildlife Australia, 28(2):10-13.

Van Dyck, S., W. W. Baker, and D. D. Gillette. 1979. The false water rat, *Xeromys myoides* on Stradbroke Island, a new locality in southeastern Queensland. Proceedings of the Royal Society of Queensland, 90:84.

Van Dyk, D., G. K. Campbell, P. J. Taylor, and J. Meester. 1991. Genetic variation within the endemic murid species, *Otomys unisulcatus* F. Cuvier, 1829 (bush karroo rat). Durban Museum Novitates, 16:12-21.

Van Gelder, R. G. 1959. A taxonomic revision of the spotted skunks (genus *Spilogale*). Bulletin of the American Museum of Natural History, 117:229-392.

Van Gelder, R. G. 1966. Comments on the proposals concerning *Zorilla* Geoffroy 1826 (Mammalia). Z.N.(S.) 758. Bulletin of Zoological Nomenclature, 22(5/6):278-280.

Van Gelder, R. G. 1977*a*. An eland X kudu hybrid, and the content of the genus *Tragelaphus*. Lammergeyer, 23:1-6.

Van Gelder, R. G. 1977*b*. Mammalian hybrids and generic limits. American Museum Novitates, 2635:1-25.

Van Gelder, R. G. 1978. A review of canid classification. American Museum Novitates, 2646:1-10.

Van Laar, V. 1984. Geographical distribution and habitat selection of the hazel dormouse, *Muscardinus avellanarius* (L. 1758) in the Netherlands. Lutra, 27(3) 1984:229-260.

Van Mensch, P. J. A., and P. J. H. van Bree. 1969. On the African Golden Cat, *Profelis aurata* (Temminck, 1827). Biologia Gabonica, 5(4):235-269.

Van Peenen, P. F. D., P. F. Ryan, and R. H. Light. 1969. Preliminary identification manual for mammals of South Vietnam. United States Naional Museum, Smithsonian Institution, Washington, D. C., 310 pp.

Van Peenen, P. F. D., R. H. Light, F. J. Duncan, R. See, J. Sulianti Saroso, Boeadi, and W. P. Carney. 1974. Observation on *Rattus bartelsii* (Rodentia: Muridae). Treubia, 28:83-117.

Van Rompaey, H. 1988. *Osbornictis piscivora*. Mammalian Species, 309:1-4.

Van Rompaey, H., and M. Colyn. 1992. *Crossarchus ansorgei*. Mammalian Species, 402:1-3.

Van Valen, L. 1967. New Paleocene insectivores and insectivore classification. Bulletin of the American Museum of Natural History, 135:217-284.

van Weers, D. J. 1976. Notes on Southeast Asian porcupines (Hystricidae, Rodentia) I. On the taxonomy of the genus *Trichys* Günther, 1877. Beaufortia (University of Amsterdam), 25(319):15-31.

van Weers, D. J. 1977. Notes on southeast Asian porcupines (Hystricidae, Rodentia) II. On the taxonomy of the genus *Atherurus* F. Cuvier. Beaufortia (University of Amsterdam), 26(336):205-230.

van Weers, D. J. 1978. Notes on southeast Asian porcupines (Hystricidae, Rodentia) III. On the taxonomy of the subgenus *Thecurus* Lyon, 1907. Beaufortia (University of Amsterdam), 28(344):17-33.

van Weers, D. J. 1979. Notes on southeast Asian porcupines (Hystricidae, Rodentia) IV. On the taxonomy of the subgenus *Acanthion* F. Cuvier. Beaufortia, 29(356):215-272.

van Weers, D. J. 1983. Specific distinction in Old World porcupines. Zoologische Garten, Jena, 53:226-232.

van Zyll de Jong, C. G. 1972. A systematic review of the Nearctic and Neotropical river otters (genus *Lutra*, Mustelidae, Carnivora). Royal Ontario Museum, Life Sciences, Contribution, 80:1-104.

van Zyll de Jong, C. G. 1976. A comparison between woodland and tundra forms of the common shrew (*Sorex cinereus*). Canadian Journal of Zoology, 54:963-973.

van Zyll de Jong, C. G. 1979. Distribution and systematic relationships of long eared *Myotis* in Western Canada. Canadian Journal of Zoology, 57:987-994.

van Zyll de Jong, C. G. 1980. Systematic relationships of prairie and woodland forms of the common shrew, *Sorex cinereus* Kerr and *S. haydeni* Baird, in the Canadian prairie provinces. Journal of Mammalogy, 61:66-75.

van Zyll de Jong, C. G. 1982. Relationships of amphiberingian shrews of the *Sorex cinereus* group. Canadian Journal of Zoology, 60:1580-1587.

van Zyll de Jong, C. G. 1983*a*. Handbook of Canadian mammals. Part I. Marsupials and insectivores. National Museum of Natural Sciences (Ottawa), 210 pp.

van Zyll de Jong, C. G. 1983*b*. A morphometric analysis of North American shrews of the *arcticus* group, with special consideration of the taxonomic status of *S. a. maritimensis*. Nature Canada (Quebec), 110:373-378.

van Zyll de Jong, C. G. 1984. Taxonomic relationships of Nearctic small-footed bats of the *Myotis leibii* group (Chiroptera: Vespertilionidae). Canadian Journal of Zoology, 62:2519-2526.

van Zyll de Jong, C. G. 1986. A systematic study of Recent bison, with particular consideration of the wood bison (*Bison bison athabascae* Rhoads 1898). Publications in Natural Sciences, National Museum of Natural Sciences, Canada, 6:1-69.

van Zyll de Jong, C. G. 1987. A phylogenetic study of the Lutrinae (Carnivora; Mustelidae) using morphological data. Canadian Journal of Zoology, 65:2536-2544.

van Zyll de Jong, C. G. 1991*a*. A brief review of the systematics and a classification of the Lutrine. Pp. 79-83, *in* Proceedings of the V International Otter Colloquium (C. Reuther and R. Röchert, eds.). Habitat 6, Hankensbüttel.

van Zyll de Jong, C. G. 1991*b*. Speciation of the *Sorex cinereus* group. Pp. 65-73, *in* The biology of the Soricidae (J. S. Findley and T. L. Yates, eds.). Special Publication, Museum of Southwestern Biology, 1:1-91.

van Zyll de Jong, C. G. 1992. A morphometric analysis of cranial variation in Holarctic weasels (*Mustela nivalis*). Zeitschrift für Säugetierkunde, 57:77-93.

van Zyll de Jong, C. G., and G. L. Kirkland, Jr. 1989. A morphometric analysis of the *Sorex cinereus* group in central and eastern North America. Journal of Mammalogy, 70:110-122.

Vargas, J. M., A. Antunez, V. Sans-Cano, and J. Cano. 1984. Albinismus bei *Mus spretus* Lataste, 1883. Säugetierkundliche Mitteilungen, 31:260-262.

Varona, L. S. 1970. Nueva especie y nuevo subgénero de *Capromys* (Rodentia: Caviomorpha) de Cuba. Poeyana (Academia de Ciencias de Cuba), 73:1-18.

Varona, L. S. 1974. Catálogo de los mamíferos vivientes y extinguidos de las Antillas. Instituto de Zoologia, Academia de Ciencias de Cuba (Havana), 139 pp.

Varona, L. S. 1979. Subgénero y especie nuevos de *Capromys* para Cuba (Rodentia: Caviomorpha). Poeyana (Academia de Ciencias de Cuba), 194:1-33.

Varona, L. S. 1983*a*. Nueva subespecie de jutia conga, *Capromys pilorides* (Rodentia: Capromyidae). Caribbean Journal of Science, 19:3-4.

Varona, L. S. 1983*b*. Remarks on the biology and zoogeography of *Solenodon* (*Atopogale*) *cubanus* Peters, 1861 (Mammalia, Insectivora). Bijdragen tot de Dierkunde, 53:93-98.

Varona, L. S. 1986. Táxones del subgénero *Mysateles* en Isla de la Juventud, Cuba. Descripción de una nueva especie (Rodentia; Capromyidae, *Capromys*). Poeyana (Academia de Ciencias de Cuba), 315:1-12.

Varona, L. S., and O. Arrendondo. 1979. Nuevos taxones fósiles de Capromyidae (Rodentia: Caviomorpha). Poeyana (Academia de Ciencias de Cuba), 195:1-51.

Varshavskii, A. A. 1991. Geograficheskaya izmenchivost' tushkanchika-prygusta *Allactaga sibirica* (Rodentia, Dipodidae) v Mongolii [Geographical variation of the Mongolian five-toed jerboa *Allactaga sibirica* (Rodentia, Dipodidae) in Mongolia]. Zoologicheskii Zhurnal, 70(1):91-98 (in Russian).

Vasil'eva, M. V. 1961. K voprosu o sistematicheskom polozhenii i rasprostranenii suslika evropeiskovo (*Citellus citellus* L.) and maloasiatskovo (*Citellus xanthoprymnus* Benn.) [Toward the question of the systematic position and distribution of the European...and Near East...ground squirrels]. Spornik Trudov Zoologicheskovo Muzeu MGU, 8:253-260 (in Russian).

Vasiliu, G. 1961. Verzeichnis der Säugetiere Rumaniens. Säugetierkundliche Mitteilungen, 9:56-68.

Vaughan, T. A. 1978. Mammalogy. Second ed. W. B. Saunders, Philadelphia, PA, 522 pp.

Vaurie, C. 1972. Tibet and its birds. H. F. and G. Witherby Ltd., London, 407 pp.

Veal, R., and W. Caire. 1979. *Peromyscus eremicus*. Mammalian Species, 118:1-6.

Ventura, J. 1991. Morphological characteristics of the molars of *Arvicola terrestris* (Rodentia, Arvicolidae) in its southwestern distribution area. Zoologischer Anzeiger, 226:64-70.

Ventura, J., and J. Gosalbez. 1989. Taxonomic review of *Arvicola terrestris* (Linnaeus, 1758) (Rodentia, Arvicolidae) in the Iberian Peninsula. Bonner Zoologische Beiträge, 40:227-242.

Ventura, J., and J. Gosalbez. 1990. Caracteristicas de los pelajes y las mudas en *Arvicola sapidus* (Rodentia, Arvicolidae). Doñana, Acta Vertebrata, 17:3-15.

Ventura, J., J. Gosalbez, and M. J. Lopez-Fuster. 1989. Trophic ecology of *Arvicola sapidus* Miller, 1908 (Rodentia, Arvicolidae) in the Ebro Delta (Spain). Zoologischer Anzeiger, 223:283-290.

Vereshchagin, N. K. 1967. The mammals of the Caucasus, a history of the evolution of the fauna [A translation of N. K. Vereshchagin, 1959, Mlekopitayushchie Kavkaza: istoriya formirovaniya fauny]. Israel Program for Scientific Translations, Jerusalem, 816 pp.

Verheyen, W. N. 1959. Un genre de Sciuride nouveau pour la Faune du Congo belge: *Epixerus* Thomas 1909. Revue de Zoologie et de Botanique Africaines, 40:301-306.

Verheyen, W. N. 1963. Contribution a la systématique du genre *Idiurus* (Rodentia-Anomaluridae). Revue de Zoologie et de Botanique Africaines, 68:157-197.

Verheyen, W. N. 1964*a*. Description of *Lophuromys rahmi* a new species of Muridae from Central Africa. Revue de Zoologie et de Botanique Africaines, 69:206-213.

Verheyen, W. N. 1964*b*. Contribution a la systematique du genre *Uranomys* Dollman 1909. Revue de Zoologie et de Botanique Africaines, 70:386-400.

Verheyen, W. N. 1965*a*. Contribution a l'etude systematique de *Mus sorella* (Thomas, 1909). Revue de Zoologie et de Botanique Africaines, 71:194-212.

Verheyen, W. N. 1965*b*. Some notes on the morphology of *Delanymys brooksi* Hayman 1962. Bulletins de la Societe Royale de Zoologie D'Anvers, 36:1-12.

Verheyen, W. N., and E. Van der Straeten. 1977. Description of *Malacomys verschureni*, a new murid-species from Central Africa (Mammalia-Muridae). Revue de Zoologie et de Botanique Africaines, 91:747-744.

Verheyen, W. N., and E. Van der Straeten. 1980. The karyotype of *Lophuromys nudicaudus* Heller 1911. Revue de Zoologie Africaine, 94:311-316.

Verheyen, W. N., and J. Verschuren. 1966. Rongeurs et lagomorphes. Exploration du Parc National de la Garamba Institut des Parcs Nationaux du Congo, Bruxelles, 50:1-66.

Verheyen, W. N., M. Michiels, and J. van Rompaey. 1986. Genetic differences between *Lophuromys flavopunctatus* Thomas, 1888 and *Lophuromys woosnami* Thomas, 1906 in Rwanda. (Rodentia: Muridae). Cimbebasia, ser. A, 8:141-145.

Vermeiren, L., and W. N. Verheyen. 1980. Notes sur les *Leggada* de Lamto, Cote d'Ivoire, avec la description de *Leggada baoulei* sp. n. (Mammalia, Muridae). Revue de Zoologie Africaine, 94:570-590.

Vermeiren, L., and W. N. Verheyen. 1983. Additional data on *Mus setzeri* Petter (Mammalia, Muridae). Annales Musée Royal de l'Afrique Centrale, ser. 8 (Sciences Zoologiques), 237:137-141.

Verts, B. J., and L. N. Carraway. 1987. *Microtus canicaudus*. Mammalian Species, 267:1-4.

Verts, B. J., and G. L. Kirkland, Jr. 1988. *Perognathus parvus*. Mammalian Species, 318:1-8.

Vesey-Fitzgerald, D. 1953. Notes on some rodents from Saudi Arabia and Kuwait. Journal of the Bombay Natural History Society, 51:424-428.

Vesmanis, I. E. 1984. Zur verbreitung von *Jaculus orientalis* Erxleben, 1777 und *Jaculus jaculus* (Linnaeus, 1758) in Tunesien (Mammalia, Rodentia, Dipodidae). Zoologische Abhandlungen Staatliches Museum für Tierkunde in Dresden, 40(4):59-65.

Vesmanis, I. E. 1987. Zur Verbreitung der Etruskerspitzmaus *Suncus etruscus* (Savi, 1822) in den Maghreb-Ländern, unter besonderer Berücksichtigung Tunesiens (Mammalia, Insectivora, Soricidae). Faunistische Abhandlungen aus dem Staatlichen Museum für Tierkunde in Dresden, 15:35-39.

Vesmanis, I. E., and H. Kahmann. 1978. Morphometrische Untersuchungen an Wimperspitzmäusen (*Crocidura*). 4. Bemerkungen über die Typusreihe der kretaischen *Crocidura russula zimmermanni* Wettstein, 1953 im Vergleich mit *Crocidura gueldenstaedti caneae* (Miller, 1909). Säugetierkundliche Mitteilungen, 26:214-222.

Vesmanis, I. E., and A. Vesmanis. 1980. Ein Nachweis des Siebenschläfers, *Glis glis* (Linnaeus, 1766) aus Eulengewollen von der Insel Elba, Italien. Faunistische Abhandlungen, 7(17):167-170.

Vidal, O. 1990. Lista de los mamiferos acuaticos de Columbia. Informe del Museo del Mar (Universidad Jorge Tadeo Lozano, Bogota, Colombia), 34:1-18.

Vidal, O. R., R. Riva, and N. I. Baro. 1976. Los cromosomas del genero *Holochilus*. I. Polimorfismo en *H. chacarius* Thomas (1906). Physis (Buenos Aires), 35:75-85.

Viegas-Pequignot, E., B. Dutrillaux, M. Prod'Homme, and F. Petter. 1983. Chromosomal phylogeny of Muridae: a study of 10 genera. Cytogenetics and Cell Genetics, 35:269-278.

Viegas-Pequignot, E., S. Kasahara, Y. Yassuda, and B. Dutrillaux. 1985. Major chromosome homeologies between Muridae and Cricetidae. Cytogenetics and Cell Genetics, 39:258-261.

Viegas-Pequignot, E., D. Petit, T. Benazzou, M. Prod'Homme, M. Lombard, F. Hoffschir, J. Descailleaux, and B. Dutrillaux. 1986. Phylogenie chromosomique chez les Sciuridae, Gerbillidae et Muridae, et etude d'especes appartenant a d'autres familles de rongeurs. Pp. 164-202, *in* Evolution chromosomique chez les primates, les carnivores et les rongeurs (B. Dutrillaux, ed.). Mammalia, Numero special, 50:1-203.

Vieira, C. O. da C. 1949 [1950]. Xenartros e marsupiais do estado de São Paulo. Arquivos de Zoologia do Estado de São Paulo, 7:325-362.

Vigne, J.-D. 1983. Le remplacement des faunes de petits mammifères en Corse, lors de l'arrivé de l'homme. Comptes Rendus Séances de la Societé Biogéographique, 59:41-51.

Vigne, J.-D. 1988. Les mammifères post-glaciaires de Corse. Gallia Prehistoire, Supplement, 26:1-337.

Viljoen, S. 1989. Taxonomy and historical zoogeography of the red squirrel, *Paraxerus palliatus* (Peters, 1852) in the southern African subregion (Rodentia: Sciuridae). Annals of the Transvaal Museum, 35(2):49-59.

Villa-R., B. 1966 [1967]. Los murciélagos de Mexico. Anales de Instituto de Biologia, Universidad Nacional Autónomica de México, 491 pp.

Vinogradov, B. S. 1925. On the structure of the external genitalia in Dipodidae and Zapodidae (Rodentia) as a classificatory character. Proceedings of the Zoological Society of London, 1925(1):572-585.

Vinogradov, B. S. 1928. A third species of dwarf jerboa *Salpingotus thomasi* sp. n. Annals and Magazine of Natural History, ser. 10, 1:372-374.

Vinogradov, B. S. 1930. [On the classification of Dipodidae (Rodentia). I. Cranial and dental characters]. Izvestiya Akademii Nauk SSSR, 1930:331-350 (in Russian).

Vinogradov, B. S. 1937. Fauna SSSR; Mlekopitaiushchie, tom. 3, vyp. 4. Tushkanchiki. [Fauna of the USSR; Mammals, vol. 3, pt. 4. Jerboas.], 196 pp. (in Russian).

Vinogradov, B. S., A. I. Argyropulo, and V. G. Heptner. 1936. [Rodents of the Central Asiatic part of USSR]. Akademiya Nauk SSSR, Moscow, 228 pp. (in Russian).

Viro, P., and J. Niethammer. 1982. *Clethrionomys glareolus* (Schreber, 1780)—rotelmaus. Pp. 109-146, *in* Handbuch der Säugetiere Europas (J. Niethammer and F. Krapp, eds.). Akademische Verlagsgesellschaft (Wiesbaden), 2/I:1-649.

Viroux, M.-C., and V. Bauchau. 1992. Segregation and fertility in *Mus musculus domesticus* (wild mice) heterozygous for the Rb(4.12) translocation. Heredity, 68:131-134.

Visser, D. S., and T. J. Robinson. 1986. Cytosystematics of the South African *Aethomys* (Rodentia: Muridae). South African Journal of Science, 21:264-268.

Visser, D. S., and T. J. Robinson. 1987. Systematic implications of spermatozoan and bacular morphology for the South African *Aethomys*. Mammalia, 51:447-454.

Vitullo, A. D., M. S. Merani, O. A. Reig, A. E. Kajon, O. Scaglia, M. B. Espinosa, and A. Perez-Zapata. 1986. Cytogenetics of South American akodont rodents (Cricetidae): new karyotypes and chromosomal banding patterns of Argentinian and Uruguayan forms. Journal of Mammalogy, 67:69-80.

Vivo, M. de. 1985. On some monkeys from Rondônia, Brasil (Primates: Callitrichidae, Cebidae). Pápeis Avulsos de Zoologia, São Paulo, 36(11):103-110.

Vlasak, P., and J. Niethammer. 1990. *Crocidura suaveolens* (Pallas, 1811) - Gartenspitzmaus. Pp. 397-428, *in* Handbuch der Säugetiere Europas (J. Niethammer, and F. Krapp, eds.). Aula-Verlag, Wiesbaden, 3/I:1-524.

Vogel, P. 1983. Contribution à l'écologie et à la zoogeographie de *Micropotamogale lamottei* (Mammalia, Tenrecidae). Revue de Ecologie (La Terre et la Vie), 38:37-49.

Vogel, P. 1986. Der Karyotyp der Kretaspitzmaus, *Crocidura zimmermanni* Wettstein, 1953 (Mammalia, Insectivora). Bonner Zoologische Beiträge, 37:35-38.

Vogel, P. 1988. Taxonomical and biogeographical problems in Mediterranean shrews of the genus *Crocidura* (Mammalia, Insectivora) with reference to a new karyotype from Sicily (Italy). Bulletin de la Societé Vaudoise des Sciences Naturelles, 79:39-48.

Vogel, P., and F. Besançon. 1979. A propos de la position systématique des genres *Nectogale* et *Chimarrogale* (Mammalia, Insectivora). Revue Suisse de Zoologie, 86:335-338.

Vogel, P., R. Hutterer, and M. Sarà. 1989. The correct name, species diagnosis, and distribution of the Sicilian shrew. Bonner Zoologische Beiträge, 40:243-248.

Vogel, P., T. Maddalena, A. Mabille, and G. Paquet. 1991. Confirmation biochimique du statut specifique du mulot alpestre *Apodemus alpicola* Heinrich, 1952 (Mammalia, Rodentia). Bulletin Societie Vaudoise des Sciences Naturelles, 80:471-481.

Vogel, W., P. Steinbach, M. Djalali, K. Mehnert, S. Ali, and J. T. Epplen. 1988. Chromosome 9 of *Ellobius lutescens* is the X chromosome. Chromosoma, 96:112-118.

Vohralík, V. 1991. A record of the mole *Talpa levantis* (Mammalia: Insectivora) in Bulgaria and the distribution of the species in the Balkans. Acta Universitatis Carolinae Biologica, 35:119-127.

Vohralík, V., and T. Sofianidou. 1987. Small mammals (Insectivora, Rodentia) of Macedonia, Greece. Acta Universitatis Carolinae Biologica, 1985(5-6):319-354.

Volleth, M., and C. R. Tidemann. 1991. The origin of the Australian Vespertilioninae bats, as indicated by chromosomal studies. Zeitschrift für Säugetierkunde, 56:321-330.

Volobouev, V. T., and C. G. van Zyll de Jong. 1988. The karyotype of *Sorex arcticus maritimensis* (Insectivora, Soricidae) and its systematic implications. Canadian Journal of Zoology, 66:1968-1972.

Volobouev, V. T., E. Viegas-Pequignot, F. Petter, and B. Dutrillaux. 1987. Karyotypic diversity and taxonomic problems in the genus *Arvicanthis* (Rodentia, Muridae). Genetica, 72:147- 150.

Volobouev, V. T., M. Tranier, and B. Dutrillaux. 1991. Chromosome evolution in the genus *Acomys*: chromosome banding analysis of *Acomys* cf. *dimidiatus* (Rodentia, Muridae). Bonner Zoologische Beiträge, 42:253-260.

Volobuev, A. T. 1971. Karyological evidence concerning domestic ferret. Zoological Journal.

Von Richter, W. 1974. *Connochaetes gnou*. Mammalian Species, 50:1-6.

Von Vietinghoff-Riesch, A. F. 1960. Der Siebenschläfer (*Glis glis* L.). Monographien der Wildsäugetiere, 14:1-196.

Vorontsov, N. N. 1959. [The system of hamsters (Cricetinae) in the sphere of the world fauna and their phylogenetic relations]. Byulleten' Moskovskovo Obshchestva Ispytatelei Prirody, Otdel Biologicheskii, 64:134-137 (in Russian).

Vorontsov, N. N. 1966. [Taxonomic position and a survey of the hamsters of the genus *Mystromys* Wagn. (Mammalia, Glires)]. Zoologicheskii Zhurnal, 45:436-446 (in Russian).

Vorontsov, N. N. 1967. Evolyutsiya pishchevaritel'noi sistemy gryzunov mysheobraznye [Evolution of the alimentary system of myomorph rodents]. Nauka, Novosibirsk, 235 pp. (in Russian).

Vorontsov, N. N. 1969. [The chromosome numbers and taxonomical relations of the members of the superfamily Dipodoidea (Rodentia)]. Pp. 92-93, *in* [The mammals: evolution, karyology, faunistics, systematics. 2nd All-Union Mammalogy Conference, Moscow] (N. N. Vorontsov, ed.). Academy of Sciences of the USSR (Siberian Branch), Novosibirsk, 167 pp. (in Russian).

Vorontsov, N. N. 1979. Evolution of the alimentary system of myomorph rodents [A translation of N. N. Vorontsov, 1967, Evolyutsiya pishchevaritel'noi sistemy gryzunov mysheobraznye]. Indian National Scientific Documentation Centre, New Delhi, 346 pp.

Vorontsov, N. N. 1986. [The Szechuan dormouse—a new genus of mammals]. Priroda, Moscow, 3:94-95 (in Russian).

Vorontsov, N. N., and E. Yu. Ivanitskaya. 1973. Comparative karyology of north Palaearctic pikas (*Ochotona*, Ochotonidae, Lagomorpha). Caryologia, 26:213-223.

Vorontsov, N. N., and E. A. Lyapunova. 1970. Khromosomnye chisla i vidoobrazovanie u nazemnykh belich'ikh (Sciuridae:Xerinae et Marmotinae) Golarktiki [Chromosome number and species formation in ground squirrels...of the Holarctic]. Byulleten' Moskovskovo Obshchestva Ispytatelei Prirody, Otdel Biologicheskii, 75:122-136 (in Russian).

Vorontsov, N. N., and E. A. Lyapunova. 1986. Genetics and problems of Trans-Beringian connections of Holarctic mammals. Pp. 441-481, *in* Beringia in the Cenozoic Era (V. L. Kontrimavichus, ed.) [translation of Beringiya v Kainozoe, 1976]. A. A. Balkema, Rotterdam, 724 pp.

Vorontsov, N. N., and E. G. Potapova. 1979. [Taxonomy of the genus *Calomyscus* (Cricetidae). 2. Status of *Calomyscus* in the system of Cricetinae]. Zoologicheskii Zhurnal, 58:1391-1397 (in Russian).

Vorontsov, N. N., and G. I. Shenbrot. 1984. [A systematic review of the genus *Salpingotus* (Rodentia, Dipodidae), with a description of *Salpingotus pallidus* sp. n. from Kazakhstan]. Zoologicheskii Zhurnal, 63(5):731-744 (in Russian).

Vorontsov, N. N., O. J. Orlov, and V. M. Smirnov. 1969*a*. [Biology and distribution of three-toed dwarf jerboas *Saplingotus crassicauda* in the Zaisan Basin]. Pp. 69-73, *in* [The mammals: evolution, karyology, faunistics, systematics. 2nd All- Union Mammalogy Conference, Moscow] (N. N. Vorontsov, ed.). Academy of Sciences of the USSR (Siberian Branch), Novosibirsk, 167 pp. (in Russian).

Vorontsov, N. N., O. J. Orlov, and N. A. Malygina. 1969*b*. [Biology and taxonomy of fat-tailed jerboas (*Pygerethmus*) and comparative karyology of the genera *Pygerethmus* and *Alactagulus*]. Pp. 74-84, *in* [The mammals: evolution, karyology, faunistics, systematics. 2nd All-Union Mammalogy Conference,Moscow] (N. N. Vorontsov, ed.). Academy of Sciences of the USSR (Siberian Branch), Novosibirsk, 167 pp. (in Russian).

Vorontsov, N. N., S. I. Radjabli, and N. A. Malygina. 1969*c*. [The comparative karyology of the five-toed jerboas of the genus *Allactaga* (Allactaginae, Dipodidae, Rodentia)]. Pp. 85-87, *in* [The mammals: evolution, karyology, faunistics, systematics. 2nd All- Union Mammalogy Conference, Moscow] (N. N. Vorontsov, ed.). Academy of Sciences of the USSR (Siberian Branch), Novosibirsk, 167 pp. (in Russian).

Vorontsov, N. N., N. A. Malygina, and S. I. Radjabli. 1969*d*. [Chromosome compliments of the jerboas of the subfamilies Dipodinae and Cardiocraniinae (Dipodidae, Rodentia)]. Pp. 88-91, *in* [The mammals: evolution, karyology, faunistics, systematics. 2nd All- Union Mammalogy Conference, Moscow] (N. N. Vorontsov, ed.). Academy of Sciences of the USSR (Siberian Branch), Novosibirsk, 167 pp. (in Russian).

Vorontsov, N. N., T. S. Bekasova, B. Král, K. V. Korobitsina, and E. Yu. Ivanitskaya. 1977*a*. [On specific status of Asian wood mice of the genus *Apodemus* (Rodentia, Muridae) from Siberia and the Far East]. Zoologicheskii Zhurnal, 56:437-449 (in Russian).

Vorontsov, N. N., L. Ya. Martynova, and I. I. Fomichova. 1977*b*. [An electrophoretic comparison of the blood proteins in mole rats of the fauna of the USSR (Spalacinae, Rodentia)]. Zoologicheskii Zhurnal, 56:1207-1215 (in Russian).

Vorontsov, N. N., E. A. Lyapunova, E. Yu. Ivanitskaya, C. F. Nadler, B. Král, A. I. Kozlovskii, and R. S. Hoffmann. 1978. [Variability of mammalian sex chromosomes. I. Geographic variability of the structure of the Y chromosomes in red-backed voles of the genus *Clethrionomys* (Rodentia, Microtinae)]. Genetika, 14:1432-1446.

Vorontsov, N. N., I. V. Kartavtseva, and E. G. Potapova. 1979. [Systematics of the genus *Calomyscus* (Cricetidae). 1. Karyological differentiation of the sibling species from Transcaucasia and Turkmenia and a review of species of the genus *Calomyscus*]. Zoologicheskii Zhurnal, 58:1213-1224 (in Russian).

Vorontsov, N. N., E. A. Lyapunova, Yu. M. Borisov, and V. E. Dovgal. 1980. Variability of sex chromosomes in mammals. Genetica, 52/53:361-372.

Vorontsov, N. N., S. V. Mezhzherin, G. G. Boeskorov, and E. A. Lyapunova. 1989. [Genetic differentiation of sibling species of wood mice (*Apodemus*) in the Caucasia and their diagnostics]. Doklady Akademii Nauk SSSR, 309:1234-1238 (in Russian).

Vorontsov, N. N., G. G. Boyeskorov, S. V. Mezhzherin, E. A. Lyapunova, and A. S. Kandaurov. 1992. [Systematics of the Caucasian wood mice of the subgenus *Sylvaemus* (Mammalia, Rodentia, *Apodemus*)]. Zoologicheskii Zhurnal, 71:119-131 (in Russian with English summary).

Vosmaer, A. 1766. Natuurlyke historie van het Africaansche Breedsnuitig Varken, of Bosch-Zwyn. P. Meijer, Amsterdam, 15 pp.

Voss, R. S. 1988. Systematics and ecology of ichthyomyine rodents (Muroidea): Patterns of morphological evolution in a small adaptive radiation. Bulletin of the American Museum of Natural History, 188:259-493.

Voss, R. S. 1991*a*. An introduction to the Neotropical muroid rodent genus *Zygodontomys*. Bulletin of the American Museum of Natural History, 210:1-113.

Voss, R. S. 1991*b*. On the identity of "*Zygodontomys*" *punctulatus* (Rodentia: Muroidea). American Museum Novitates, 3026:1-8.

Voss, R. S. 1992. A revision of the South American species of *Sigmodon* (Mammalia: Muridae) with notes on their natural history and biogeography. American Museum Novitates, 3050:1-56.

Voss, R. S., and A. V. Linzey. 1981. Comparative gross morphology of male accessory glands among Neotropical Muridae (Mammalia: Rodentia) with comments on systematic implications. Miscellaneous Publications, Museum of Zoology, University of Michigan, 159:1-41.

Voss, R. S., and P. Myers. 1991. *Pseudoryzomys simplex* (Rodentia: Muridae) and the significance of Lund's collections from the caves of Lagoa Santa, Brazil. Bulletin of the American Museum of Natural History, 206:414-432.

Vrba, E. S. 1979. Phylogenetic analysis and classification of fossil and recent Alcelaphini. Mammalia: Bovidae. Biological Journal of the Linnean Society, 11:207-228.

Vuillaume-Randriamanantena, M., L. R. Godfrey, and M. R. Sutherland. 1985. Revision of *Hapalemur* (*Prohapalemur*) *gallieni* (Standing, 1905). Folia Primatologia, 45:89-116.

Vujosevic, M., D. Rimsa, and S. Zivkovic. 1984. Patterns of G- and C-bands distribution on chromosomes of three *Apodemus* species. Zeitschrift für Säugetierkunde, 49:234-238.

Vujosevic, M., J. Blagojevic, J. Radosavljevic, and D. Bejakovic. 1991. B chromosome polymorphism in populations of *Apodemus flavicollis* in Yugoslavia. Genetica, 83:167-170.

Wada, S., and K. Numachi. 1991. Allozyme analyses of genetic differentiation among the populations and species of Balaenoptera. Pp. 125-154, *in* Genetic ecology of whales and dolphins (A. R. Hoelzel, ed.). Reports of the International Whaling Commission, Special Issue, 13:1-311.

Wade-Smith, J., and B. J. Verts. 1982. *Mephitis mephitis*. Mammalian Species, 173:1-7.

Wahlert, J. H. 1984. Relationships of the extinct rodent *Cricetops* to *Lophiomys* and the Cricetinae (Rodentia, Cricetidae). American Museum Novitates, 2784:1-15.

Wahlert, J. H. 1985. Skull morphology and relationships of geomyoid rodents. American Museum Novitates, 2812:1-20.

Wahlert, J. H., S. L. Sawitzke, and M. E. Holden. In press. Cranial anatomy and relationships of dormice (Rodentia, Myoxidae). American Museum Novitates.

Wahrman, J., C. Richler, E. Neufeld, and A. Friedmann. 1983. The origin of multiple sex chromosomes in the gerbil *Gerbillus gerbillus* (Rodentia: Gerbillinae). Cytogenetics and Cell Genetics, 35:161-180.

Waithman, J. 1979. A report on a collection of mammals from southwest Papua. 1972-1973. Australian Zoologist, 20:313-326.

Waithman, J., and A. Roest. 1977. A taxonomic study of the kit fox, *Vulpes macrotis*. Journal of Mammalogy, 58:157-164.

Wakefield, N. A. 1972*a*. Palaeoecology of fossil mammal assemblages from some Australian caves. Proceedings of the Royal Society of Victoria, 85:1-26.

Wakefield, N. A. 1972*b*. Studies in Australian Muridae: review of *Mastacomys fuscus*, and description of a new subspecies of *Pseudomys higginsi*. Memoirs of the National Museum of Victoria, Melbourne, Australia, 33:15-31.

Wakefield, N. A., and R. M. Warneke. 1963. Some revision in *Antechinus* (Marsupialia) - 1. Victorian Naturalist, 80:194-219.

Walker, E. P., F. Warnick, S. E. Hamlet, K. L. Lange, M. A. Davis, H. E. Uible, and P. F. Wright. 1964. Mammals of the world. John Hopkins Press, Baltimore, 1:1-646; 2:647-1500; 3:1-769.

Walker, E. P., F. Warnick, S. E. Hamlet, K. L. Lange, M. A. Davis, H. E. Uible, and P. F. Wright. 1968. Mammals of the world, Second ed. Johns Hopkins Press, Baltimore, MD, 1:1-647; 2:648-1500.

Walker, E. P., F. Warnick, S. E. Hamlet, K. I. Lange, M. A. Davis, H. E. Uible, and P. F. Wright. 1975. Mammals of the world. Third ed. Johns Hopkins University Press, Baltimore, 1:1-646; 2:647-1500.

Walker, L. I., and A. E. Spotorno. 1992. Tandem and centric fusions in the chromosomal evolution of the South American phyllotines of the genus *Auliscomys* (Rodentia, Cricetidae). Cytogenetics and Cell Genetics, 61:135-140.

Walker, L. I., A. E. Spotorno, and J. Arrau. 1984. Cytogenetic and reproductive studies of two nominal subspecies of *Phyllotis darwini* and their experimental hybrids. Journal of Mammalogy, 65:220-230.

Wallin, L. 1969. The Japanese bat fauna. Zoologiska Bidrage Fran Uppsala, 37:223-440.

Walton, D. W. 1963. A collection of the bat *Lonchophylla robusta* Miller from Costa Rica. Tulane Studies in Zoology, 10:87-90.
Walton, D. W. (ed.). 1988. Zoological catalogue of Australia. 5. Mammalia. Australian Government Publishing Service, Canberra, 274 pp.
Wang B., and Li Chuan-kwei. 1990. First Paleogene mammalian fauna from northeast China. Vertebrata Palasiatica, 28:165-205, pls. i-vi.
Wang Si-bo and Yang Gan-yun. 1983. [Rodent fauna of Xinjiang]. Xinjiang People's Publishing House, Wulumuqi [Urumchi], China, 223 pp. (in Chinese).
Wang Sung. 1959. [Further report on the mammals of northeastern China]. Acta Zoologica Sinica, 11:344-352 (in Chinese).
Wang Sung. 1964. [New species and subspecies of mammals from Sinkiang, China]. Acta Zootaxonomica Sinica, 1:6-18 (in Chinese).
Wang Sung and Zheng Chang-lin. 1973. [Notes on Chinese hamsters (Cricetinae)]. Acta Zoologica Sinica, 19:61-68 (in Chinese).
Wang Sung and Zheng Chang-lin. 1981. [On the subspecies of the Chinese sulphur-bellied rat—*Rattus niviventer* Hodgson]. Sinozoologia, 1:1-8 (in Chinese).
Wang Ting-zheng. 1990. [On the fauna and zoogeographical regionalization of Glires (including rodents and lagomorphs) in Shaanxi Province]. Acta Theriologica Sinica, 10(2):128-136.
Wang Xiao-ming and R. S. Hoffmann. 1987. *Pseudois nayaur* and *Pseudois schaeferi*. Mammalian Species, 278:1-6.
Wang Y.-c., Y.-r. Tu, and Wang Sung. 1966. Notes on some small mammals from Szechuan Province with description of a new subspecies. Acta Zootaxonomica Sinica, 3:89-90 (in English and Chinese).
Wang Ying-xiang and Yang G. 1989. [Insectivores]. Pp. 202-210, *in* [A list of medical animals in Yunnan (Yunnan Office for Endemic Disease Control, and Yunnan Sanitation and Antiepidemic Station, eds.)]. Yunnan Science and Technology, Kunming.
Wang Ying-xiang, Luo Ze-xun, and Feng Zuo-jiang. 1985. [Taxonomic revision of Yunnan hare, *Lepus comus* G. Allen, with description of two new subspecies]. Zoological Research, 6:101-109 (in Chinese).
Wang You-zhi. 1985*a*. [A new genus and species of Gliridae-*Chaetocauda sichuanensis* gen. et sp. nov.]. Acta Theriologica Sinica, 5:67-75 (in Chinese).
Wang You-zhi. 1985*b*. Subspecific classification and distribution of *Apodemus agrarius* in Sichuan, China. Pp. 86-89, *in* Contemporary mammalogy in China and Japan (T. Kawamichi, ed.). Mammalogical Society of Japan, 194 pp.
Wang You-zhi, Hu Jin-chu, and Chen Ke. 1980. [A new species of Murinae—*Vernaya foramena* sp. nov.]. Acta Zoologica Sinica, 26:393-397 (in Chinese).
Ward, O. G., and D. H. Wurster-Hill. 1990. *Nyctereutes procyonides*. Mammalian Species, 358:1-5.
Warmerdam, M. 1982. Numeriek-taxonomische studie van de twee vormen van de woelrat *Arvicola terrestris* (Linnaeus, 1758) in Nederland en Belgie. Lutra, 24:33-66.
Warner, R. M. 1982. *Myotis auriculus*. Mammalian Species, 191:1-3.
Warner, R. M., and N. J. Czaplewski. 1984. *Myotis volans*. Mammalian Species, 224:1-4.
Wasser, S. K. 1985. Current conservation status of the Mwanihana Rain Forest, Uzungwa Mountains, Sanje, Tanzania. Primate Conservation, 6:34.
Wassif, K., and H. Hoogstraal. 1954. The mammals of South Sinai, Egypt. Proceedings of the Egyptian Academy of Science, Cairo, 9:63-79.
Waterhouse, G. R. 1838 [1839]. On the skull and dentition of the American badger (*Meles labradoria*). Proceedings of the Zoological Society of London, 1838:153-154.
Waterhouse, G. R. 1842 [1843]. Descriptions of a new species of quadrupeds collected by Mr. Fraser at Fernando Po. Proceedings of the Zoological Society of London, 1842:124-130.
Waterhouse, G. R. 1846-1848. The natural history of the mammalia. Hippolyte Bailliere, Publisher, London, 1:2 (unnumbered) + 1-553, 22 pls.[1846]; 2:1-500, 21 pls. [1848].
Watkins, L. C. 1972. *Nycticeius humeralis*. Mammalian Species, 23:1-4.
Watkins, L. C. 1977. *Euderma maculatum*. Mammalian Species, 77:1-4.
Watkins, L. C., J. K. Jones, Jr., and H. H. Genoways. 1972. Bats of Jalisco, Mexico. Special Publications, The Museum, Texas Tech University Press, 1:1-44.
Watson, J. P. 1990. The taxonomic status of the slender mongoose, *Galerella sanguinea* (Rüppell, 1836), in southern Africa. Navorsinge van die Nasionale Museum Bloemfontein, 6(10):351-492.
Watson, J. P., and N. H. Dippenaar. 1987. The species limits of *Galerella sanguinea* (Rüppell, 1836), *G. pulverulenta* (Wagner, 1839) and *G. nigrata* (Thomas, 1928) in southern Africa (Carnivora: Viverridae). Navorsinge van die Nasionale Museum Bloemfontein, 5(14):356-414.
Watts, C. H. S. 1975. The neck and chest glands of the Australian hopping-mice, *Notomys*. Australian Journal of Zoology, 23:151-157.

Watts, C. H. S. 1976. *Leggadina lakedownensis*, a new species of murid rodent from North Queensland. Transactions of the Royal Society of South Australia, 100:105-108.
Watts, C. H. S., and H. J. Aslin. 1981. The rodents of Australia. Angus and Robertson, Sydney, 321 pp.
Watts, C. H. S., P. R. Baverstock, J. Birrell, and M. Krieg. 1992. Phylogeny of the Australian rodents (Muridae): a molecular approach using microcomplement fixation of albumin. Australian Journal of Zoology, 40:81-90.
Wayne, R. K., and S. M. Jenks. 1991. Mitochondrial DNA analysis implying extensive hybridization of the endangered red wolf *Canis rufus*. Nature, 351:565-568.
Wayne, R. K., and S. J. O'Brien. 1987. Allozyme divergence within the Canidae. Systematic Zoology, 36:339-355.
Wayne, R. K., W. G. Nash, and S. J. O'Brien. 1987*a*. Chromosomal evolution of the Canidae: I. Species with high diploid numbers. Cytogenetics and Cell Genetics, 44:123-133.
Wayne, R. K., W. G. Nash, and S. J. O'Brien. 1987*b*. Chromosomal evolution of the Canidae: II. Species with low diploid numbers. Cytogenetics and Cell Genetics, 44:134-141.
Wayne, R. K., R. E. Benveniste, and S. J. O'Brien. 1989. Phylogeny and evolution of the Carnivora and carnivore families, Pp. 465-494, *in*, Carnivore behavior, ecology and evolution (J. L. Gittleman, ed.). Cornell University Press, Ithaca, NY, 620 pp.
Webb, S. D. 1985. The interrelationships of tree sloths and ground sloths. Pp. 105-112, *in* The evolution and ecology of armadillos, sloths, and vermilinguas (G. G. Montgomery, ed.). Smithsonian Institution Press, Washington, D. C., 10 (unnumbered) + 451 pp.
Webb, S. D., and B. E. Taylor. 1980. The phylogeny of hornless ruminants and a description of the cranium of *Archaeomeryx*. Bulletin of the American Museum of Natural History, 167:117-158.
Webster, W. D., and C. O. Handley, Jr. 1986. Systematics of Miller's long-tongued bat *Glossophaga longirostris*, with description of two new subspecies. Occasional Papers, The Museum, Texas Tech University, 100:1-22.
Webster, W. D., and J. K. Jones, Jr. 1980. Taxonomic and nomenclatorial notes on bats of the genus *Glossophaga* in North America, with description of a new species. Occasional Papers, The Museum, Texas Tech University, 71:1-12.
Webster, W. D., and J. K. Jones, Jr. 1982*a*. *Reithrodontomys megalotis*. Mammalian Species, 167:1-5.
Webster, W. D., and J. K. Jones, Jr. 1982*b*. *Artibeus aztecus*. Mammalian Species, 177:1-3.
Webster, W. D., and J. K. Jones, Jr. 1982*c*. *Artibeus toletecus*. Mammalian Species, 178:1-3.
Webster, W. D., and J. K. Jones, Jr. 1983. *Artibeus hirsutus* and *Artibeus inopinatus*. Mammalian Species, 199:1-3.
Webster, W. D., and J. K. Jones, Jr. 1984. *Glossophaga leachii*. Mammalian Species, 226:1-3.
Webster, W. D., and J. K. Jones, Jr. 1985. *Glossophaga mexicana*. Mammalian Species, 245:1-2.
Webster, W. D., and R. D. Owen. 1984. *Pygoderma bilabiatum*. Mammalian Species, 220:1-3.
Webster, W. D., J. K. Jones, Jr., and R. J. Baker. 1980. *Lasiurus intermedius*. Mammalian Species, 132:1-3.
Webster, W. D., L. W. Robbins, R. L. Robbins, and R. J. Baker. 1982. Comments on the status of *Musonycteris harrisoni* (Chiroptera: Phyllostomidae). Occasional Papers, The Museum, Texas Tech University, 78:1-5.
Weddle, G. K., and J. R. Choate. 1983. Dental evolution of the meadow vole in mainland, peninsular, and insular environments in southern New England. Fort Hays Studies, New Series, 3:1-23.
Weigel, I. 1961. Das Fellmuster der wildlebenden Katzenarten und der Hauskatze in Vergleichender und Stammesgeschicher Hinsicht. Säugetierkundliche Mitteilungen, 9:1-120.
Weitzel, V. M., and C. P. Groves. 1985. The nomenclature and taxonomy of the colobine monkeys of Java. International Journal of Primatology, 6:399-409.
Wells, D. R. 1989. Notes on the distribution and taxonomy of peninsular Malaysian mongooses (*Herpestes*). Natural History Bulletin of the Siam Society, 37(1):87-97.
Wells, D. R. and C. M. Francis. 1988. Crab-eating mongoose, *Herpestes urva*, a mammal new to peninsular Malaysia. Malayan Nature Journal, 42:37-41.
Wells, N. M., and J. Giacalone. 1985. *Syntheosciurus brochus*. Mammalian Species, 249:1-3.
Wells-Gosling, N., and L. R. Heaney. 1984. *Glaucomys sabrinus*. Mammalian Species, 229:1-8.
Wenzel, E., and T. Haltenorth. 1972. System der Schleichkatzen (Viverridae). Säugetierkundliche Mitteilungen, 20(1-2):110-127.
Werbitsky, D., and C. W. Kilpatrick. 1987. Genetic variation and genetic differentiation among allopatric populations of *Megadontomys*. Journal of Mammalogy, 68:305-312.
Werdelin, L. 1981. The evolution of lynxes. Annales Zoologici Fennici, 18:37-71.
Werdelin, L. 1987. Some observations on *Sarcophilus laniarius* and the evolution of *Sarcophilus*. Records of the Queen Victoria Museum, Launceston, 90:1-27.
Werdelin, L., and N. Solounias. 1991. The Hyaenidae: taxonomy, systematics and evolution. Fossils and Strata, 30:1-104.

Wesselman, H. B. 1984. The Omo micromammals. Systematics and paleoecology of Early Man sites from Ethiopia. Contributions to Vertebrate Evolution, 7:1-219.
Weston, M. L. 1981. The *Ochotona alpina* complex: a statistical re-evaluation. Pp. 73-89, *in* Proceedings of the world lagomorph conference (K. K. Myers and C. D. MacInnes, eds.). Guelph University Press, Guelph, Ontario, 983 pp.
Weston, M. L. 1982. A numerical revision of the genus *Ochotona* (Lagomorpha: Mammalia) and an examination of its phylogenetic relationships. Unpubl. Ph. D. dissertation, University of British Columbia, Vancouver, 387 pp.
Wetzel, R. M. 1955. Speciation and dispersal of the southern bog lemming, *Synaptomys cooperi* (Baird). Journal of Mammalogy, 36:1-20.
Wetzel, R. M. 1975. The species of *Tamandua* Gray (Edentata, Myrmecophagidae). Proceedings of the Biological Society of Washington, 88:95-112.
Wetzel, R. M. 1977. The Chacoan peccary *Catagonus wagneri* (Rusconi). Bulletin of Carnegie Museum of Natural History, 3:1-36.
Wetzel, R. M. 1980. Revision of the naked-tailed armadillos, genus *Cabassous* McMurtrie. Annals of Carnegie Museum, 49:323-357.
Wetzel, R. M. 1981. The hidden Chacoan peccary. Carnegie Magazine, 55:24-32.
Wetzel, R. M. 1985*a*. The identification and distribution of Recent Xenarthra (=Edentata). Pp. 5-22, *in* The evolution and ecology of armadillos, sloths, and vermilinguas (G. G. Montgomery, ed.). Smithsonian Institution Press, Washington, D. C., 10 (unnumbered) + 451 pp.
Wetzel, R. M. 1985*b*. Taxonomy and distribution of armadillos, Dasypodidae. Pp. 23-48, *in* The evolution and ecology of armadillos, sloths, and vermilinguas (G. G. Montgomery, ed.). Smithsonian Institution Press, Washington, D. C., 10 (unnumbered) + 451 pp.
Wetzel, R. M., and F. D. de Avila-Pires. 1980. Identification and distribution of the Recent sloths of Brazil (Edentata). Revista Brasileira de Biologia, 40:831-836.
Wetzel, R. M., and E. Mondolfi. 1979. The subgenera and species of long-nosed armadillos, genus *Dasypus* L. Pp. 43-63, *in* Vertebrate ecology in the northern Neotropics (J. F. Eisenberg, ed.). Smithsonian Institution Press, Washington, D. C., 271 pp.
Wetzel, R. M., R. E. Dubos, R. L. Martin, and P. Myers. 1975. *Catagonus*, an "extinct" peccary, alive in Paraguay. Science, 189:379-381.
Whitaker, J. O., Jr. 1972. *Zapus hudsonius*. Mammalian Species, 11:1-7.
Whitaker, J. O., Jr. 1974. *Cryptotis parva*. Mammalian Species, 43:1-8.
Whitaker, J. O., Jr., and R. E. Wrigley. 1972. *Napaeozapus insignis*. Mammalian Species, 14:1-6.
White, J. W. 1953. Genera and subgenera of chipmunks. University of Kansas Publications, Museum of Natural History, 5:543-561.
Whitehead, G. K. 1972. Deer of the world. Constable, London, 194 pp.
Wied-Neuwied, M. A. P., zu Prinz. 1825-1833. Beiträge zur Naturgeschichte von Brasilien, von Maximillian, prinzen zu Wied. Landes-Industrie-Comptoirs, Weimar, 2:1-620[1826].
Wiley, R. W. 1980. *Neotoma floridana*. Mammalian Species, 139:1-7.
Wilhelm, D. D. 1982. Zoogeographic and evolutionary relationships of selected populations of *Microtus mexicanus*. Occasional Papers, The Museum, Texas Tech University, 75:1-30.
Wilkins, K. T. 1986. *Reithrodontomys montanus*. Mammalian Species, 257:1-5.
Wilkins, K. T. 1987*a*. *Lasiurus seminolus*. Mammalian Species, 280:1-5.
Wilkins, K. T. 1987*b*. A zoogeographic analysis of variation in Recent *Geomys pinetis* (Geomyidae) in Florida. Bulletin of the Florida State Museum, Biological Sciences, 30:1-28.
Wilkins, K. T. 1989. *Tadarida brasiliensis*. Mammalian Species, 331:1-10.
Williams, D. F. 1978*a*. Karyological affinities of the species groups of silky pocket mice (Rodentia: Heteromyidae). Journal of Mammalogy, 59:599-612.
Williams, D. F. 1978*b*. Systematics and ecogeographic variation of the Apache pocket mouse (Rodentia: Heteromyidae). Bulletin of Carnegie Museum of Natural History, 10:1-57.
Williams, D. F. 1978*c*. Taxonomic and karyologic comments on small brown bats, genus *Eptesicus*, from South America. Annals of Carnegie Museum, 47:361-383.
Williams, D. F. 1979. Checklist of California mammals. Annals of Carnegie Museum, 48:425-433.
Williams, D. F., and H. H. Genoways. 1979. A systematic review of the olive-backed pocket mouse, *Perognathus fasciatus* (Rodentia: Heteromyidae). Annals of Carnegie Museum, 48:73-102.
Williams, D. F., and K. S. Kilburn. 1991. *Dipodomys ingens*. Mammalian Species, 377:1-7.
Williams, D. F., and M. A. Mares. 1978. A new genus and species of phyllotine rodent (Mammalia: Muridae) from northwestern Argentina. Annals of Carnegie Museum, 47:193-221.
Williams, D. F., J. D. Druecker, and H. L. Black. 1970. The karyotype of *Euderma maculatum* and comments on the evolution of the plecotine bats. Journal of Mammalogy, 51:602-606.

Williams, L. R., and G. N. Cameron. 1991. *Geomys attwateri*. Mammalian Species, 382:1-5.
Williams, S. L. 1982*a*. *Geomys personatus*. Mammalian Species, 170:1-5.
Williams, S. L. 1982*b*. The phallus of Recent genera and species of the family Geomyidae (Mammalia: Rodentia). Bulletin of Carnegie Museum of Natural History, 20:1-62.
Williams, S. L., and R. J. Baker. 1974. *Geomys arenarius*. Mammalian Species, 36:1-3.
Williams, S. L., and H. H. Genoways. 1977. Morphometric variation in the tropical pocket gopher (*Geomys tropicalis*). Annals of Carnegie Museum, 46:245-264.
Williams, S. L., and H. H. Genoways. 1978. Review of the desert pocket gopher, *Geomys arenarius* (Mammalia: Rodentia). Annals of Carnegie Museum, 47:541-570.
Williams, S. L., and H. H. Genoways. 1980*a*. Results of the Alcoa Foundation-Suriname expeditions. 2. Additional records of bats (Mammalia: Chiroptera) from Suriname. Annals of Carnegie Museum, 49:213-236.
Williams, S. L., and H. H. Genoways. 1980*b*. Morphological variation in the southeastern pocket gopher, *Geomys pinetis* (Mammalia: Rodentia). Annals of Carnegie Museum, 49:405-453.
Williams, S. L., and H. H. Genoways. 1980*c*. Results of the Alcoa Foundation-Suriname expeditions. 4. A new species of bat of the genus *Molossops* (Mammalia: Molossidae). Annals of Carnegie Museum, 49:487-498.
Williams, S. L., and H. H. Genoways. 1981. Systematic review of the Texas pocket gopher *Geomys personatus* (Mammalia: Rodentia). Annals of Carnegie Museum, 50:435-473.
Williams, S. L., and J. Ramirez-Pulido. 1984. Morphometric variation in the volcano mouse, *Peromyscus* (*Neotomodon*) *alstoni* (Mammalia: Cricetidae). Annals of Carnegie Museum, 53:163-183.
Williams, S. L., J. C. Hafner, and P. G. Dolan. 1980. Glans penes and bacula of five species of *Apodemus* (Rodentia: Muridae) from Croatia, Yugoslavia. Mammalia, 44:245-258.
Williams, S. L., J. Ramírez-Pulido, and R. J. Baker. 1985. *Peromyscus alstoni*. Mammalian Species, 242:1-4.
Willig, M. R., and R. R. Hollander. 1987. *Vampyrops lineatus*. Mammalian Species, 275:1-4.
Willig, M. R., and J. K. Jones, Jr. 1985. *Neoplatymops mattogrossensis*. Mammalian Species, 244:1-3.
Willig, M. R., and M. A. Mares. 1989. Mammals from the Caatinga: an updated list and summary of recent research. Revista Brasileira de Biologia, 49:361-367.
Willis, K. B., M. R. Willig, and J. K. Jones, Jr. 1990. *Vampyrodes caraccioli*. Mammalian Species, 359:1-4.
Willner, G. R., G. A. Feldhamer, E. E. Zucker, and J. A. Chapman. 1980. *Ondatra zibethicus*. Mammalian Species, 141:1-8.
Wilson, A. C., H. Ochman, and E. M. Prager. 1987. Perspectives molecular time scale for evolution. TIG, 3, 9:241-247.
Wilson, C. C., and W. L. Wilson. 1977. Behavioral and morphological variation among primate populations in Sumatra. Yearbook of Physical Anthropology, 20:207-233.
Wilson, D. E. 1973. The systematic status of *Perognathus merriami* Allen. Proceedings of the Biological Society of Washington, 86:175-191.
Wilson, D. E. 1976. Cranial variation in polar bears. Pp. 447-453, *in* Bears—their biology and management (M. R. Pelton, J. W. Lentfer, and G. E. Folk, eds.). International Union for the Conservation of Nature, New Series, 40:1-467
Wilson, D. E. 1978. *Thyroptera discifera*. Mammalian Species, 104:1-3.
Wilson, D. E. 1991. Mammals of the Tres Marías Islands. Bulletin of the American Museum of Natural History, 206:214-250.
Wilson, D. E., and J. S. Findley. 1977. *Thyroptera tricolor*. Mammalian Species, 71:1-3.
Wilson, D. E., and R. K. LaVal. 1974. *Myotis nigricans*. Mammalian Species, 39:1-3.
Wilson, D. E., M. A. Bogan, R. L. Brownell, Jr., A. M. Burdin, M. K. Mamin. 1991. Geographic variation in sea otters, *Enhydra lutris*. Journal of Mammalogy, 72(1):22-36.
Wilson, J. W. 1984. Chromosomal variation in pine voles, *Microtus* (*Pitymys*) *pinetorum*, in the eastern United States. Canadian Journal of Genetics and Cytology, 26:496-498.
Wilson, R. T. 1984. The camel. Longman, London, 223 pp.
Winge, H. 1887. Jordfunde og nulevende Gnavere (Rodentia) fra Lagoa Santa, Minas Geraes, Brasilien: med udsigt over gnavernes indbyrdes slagtskab. E Museo Lundii, 1(3):1-178.
Winking, H. 1976. Karyologie und Biologie der beiden iberischen Wühlmausarten *Pitymys mariae* und *Pitymys duodecimcostatus*. Zeitschrift für Zoologische Systematik und Evolutionsforschung, 4:104-129.
Winking, H., A. Gropp, and J. T. Marshall. 1979. Karyotype and sex chromosomes in *Vandeleuria oleracea*. Zeitschrift für Säugetierkunde, 44:195-201.
Winking, H., B. Dulic, and G. Bulfield. 1988. Robertsonian karyotype variation in the European house mouse, *Mus musculus*. Survey of present knowledge and new observations. Zeitschrift für Säugetierkunde, 53:148-161.

Winn, H. E., and N. E. Reichley. 1985. Humpback whale - *Megaptera novaeangliae*. Pp. 241-274, *in* Handbook of marine mammals: The sirenians and baleen whales (S. H. Ridgway and R. Harrison, eds.). Academic Press, London, 3:1-362.
Winter, J. W. 1983. Thornton Peak *Melomys*. P. 379, *in* The Australian museum complete book of Australian mammals (The national photographic index of Australian wildlife) (R. Strahan, ed.). Angus and Robertson, Sydney, 530 pp.
Winter, J. W. 1984. The Thornton Peak *Melomys*, *Melomys hadrourus* (Rodentia: Muridae): a new rainforest species from northeastern Queensland, Australia. Memoirs of the Queensland Museum, 21:519-539.
Witte, G. 1962. Zur Systematik und Verbreitung des Siebenschläfers in Italien. Bonner Zoologische Beiträge, 13:115-127.
Wodzicki, K. A. 1950. Introduced mammals of New Zealand. Department of Scientific and Industrial Research Bulletin, 98:1-255.
Wodzicki, K. A., and J. E. C. Flux. 1967. Rediscovery of the white-throated wallaby, *Macropus parma* Waterhouse, 1846, on Kawau Island, New Zealand. Australian Journal of Science, 29:429-430.
Wodzicki, K., and R. H. Taylor. 1984. Distribution and status of the Polynesian rat *Rattus exulans*. Acta Zoologica Fennica, 172:99-101.
Wolfe, J. L. 1982. *Oryzomys palustris*. Mammalian Species, 176:1-5.
Wolfe, J. L., and A. V. Linzey. 1977. *Peromyscus gossypinus*. Mammalian Species, 70:1-5.
Wolffsohn, J. A. 1908. Contribuciones a la mamalojia chilena. Revista Chilena de Historia Natural, 12:165-172.
Wolk, K. 1987. New localities of common dormouse *Muscardinus avellanarius* L. in Pojezierze Mazurskie. Przeglad Zoologiczny, 31(2):219-220.
Wolman, A. A. 1985. Gray whale - *Eschrichtus robustus*. Pp. 67-90, *in* Handbook of marine mammals: The sirenians and baleen whales (S. H. Ridgway and R. Harrison, eds.). Academic Press, London, 3:1-362.
Womochel, D. R. 1978. A new species of *Allactaga* (Rodentia: Dipodidae) from Iran. Fieldiana, Zoology, 72(5):65-73.
Wood, A. E. 1935. Evolution and relationships of the heteromyid rodents with new forms from the Tertiary of western North America. Annals of Carnegie Museum, 24:73-262.
Wood, A. E. 1985. The relationships, origin and dispersal of the hystricognath rodents. Pp. 475-513, *in* Evolutionary relationships among rodents, a multidisciplinary analysis (W. P. Luckett and J.-L. Hartenberger, eds.). Plenum Press, New York, 721 pp.
Woodburne, M. O. 1968. The cranial myology and osteology of *Dicotyles tajacu*, the collared peccary, and its bearing on classification. Memoirs of the Southern California Academy of Sciences, 7:1-48.
Woodman, N. 1988. Subfossil remains of *Peromyscus stirtoni* (Mammalia: Rodentia) from Costa Rica. Revista de Biologia Tropical, 36:247-253.
Woodman, N., and R. M. Timm. 1992. A new species of small-eared shrew, genus *Cryptotis*, (Insectivora: Soricidae), from Honduras. Proceedings of the Biological Society of Washington, 105:1-12.
Woods, C. A. 1973. *Erethizon dorsatum*. Mammalian Species, 29-1-6.
Woods, C. A. 1982. The history and classification of South American hystricognath rodents: Reflections on the far away and long ago. Pp. 377-392, *in* Mammalian biology in South America (M. Mares and H. Genoways, eds.). Special Publication Series, Pymatuning Laboratory of Ecology, University of Pittsburgh, Pennsylvania, 6:1-539.
Woods, C. A. 1984. Hystricoganth rodents. Pp. 389-446, *in* Orders and families of Recent mammals of the world (S. Anderson and J. K. Jones, Jr., eds.). Wiley, New York, 686 pp.
Woods, C. A. 1989*a*. The biogeography of West Indian rodents. Pp. 741-798, *in* Biogeography of the West Indies: Past, present, and future (C. A. Woods, ed.). Sandhill Crane Press, Gainesville, FL, 878 pp.
Woods, C. A. 1989*b*. A new capromyid rodent from Haiti; the origin, evolution, and extinction of West Indian rodents and their bearings on the origin of New World hystricognaths. Los Angeles County Museum Natural History, Science Series, 33:59-89.
Woods, C. A., and D. Boraker. 1975. *Octodon degus*. Mammalian Species, 67:1-5.
Woods, C. A., and E. B. Howland. 1979. Adaptive radiation of capromyid rodents. Journal of Mammalogy, 60:95-116.
Woods, C. A., J. A. Ottenwalder, and W. L. R. Oliver. 1985. Lost mammals of the Greater Antilles: summarized findings of a ten week field survey in the Dominican Republic, Haiti and Puerto Rico. Dodo (Jersey Wildlife Preservation Trust), 22:23-42.
Woods, C. A., L. Contreras, G. Willner-Chapman, and H. P. Whidden. 1992. *Myocastor coypus*. Mammalian Species, 398:1-8.
Wozencraft, W. C. 1984. A phylogenetic reappraisal of the Viverridae and its relationship to other Carnivora. Unpubl. Ph. D. dissertation, University of Kansas, Lawrence, KS, 1129 pp.

Wozencraft, W. C. 1989*a*. The phylogeny of the Recent Carnivora. Pp. 495-535, *in* Carnivore behavior, ecology, and evolution (J. L. Gittleman, ed.). Cornell University Press, Ithaca, NY, 620 pp.

Wozencraft, W. C. 1989*b*. Classification of the Recent Carnivora. Pp. 569-593 *in* Carnivore behavior, ecology and evolution (J. L. Gittleman, ed.). Cornell University Press, Ithaca, NY, 620 pp.

Wright, D. B. 1989. Phylogenetic relationships of *Catagonus wagneri*: sister taxa from the Tertiary of North America. Pp. 281-308, *in* Advances in neotropical mammalogy (K. H. Redford and J. F. Eisenberg, eds). Sandhill Crane Press, Gainesville, FL, 554 pp.

Wrigley, R. E. 1972. Systematics and biology of the woodland jumping mouse, *Napaeozapus insignis*. Illinois Biological Monographs, 47:1-117.

Wroughton, R. C. 1905*a*. The common striped palm squirrel. Journal of the Bombay Natural History Society, 16:406-413.

Wroughton, R. C. 1905*b*. Notes on the various forms of *Arvicanthis pumilio*, Sparrm. Annals and Magazine of Natural History, ser. 7, 16:629-639.

Wroughton, R. C. 1908. Notes on the classification of the bandicoots. Journal of the Bombay Natural History Society, 18:736-752.

Wroughton, R. C. 1910. On the nomenclature of the Indian hedgehogs. Journal of the Bombay Natural History Society, 20:80-82.

Wu De-ling. 1982. [On subspecific differentiation of brown rat (*Rattus norvegicus* Berkenhout) in China]. Acta Theriologica Sinica, 2:107-112 (in Chinese).

Wu De-ling and Deng Xiang-fu. 1984. [A new species of tree mice from Yunnan, China]. Acta Theriologica Sinica, 4:207-212 (in Chinese).

Wu De-ling and Wang Guan-huan. 1984. [A new subspecies of *Typhlomys cinereus* Milne- Edwards from Yunnan, China]. Acta Theriologica Sinica, 4:213-215 (in Chinese).

Wurster, D. H., and K. Benirschke. 1968. Comparative cytogenetic studies in the order Carnivora. Chromosoma, 24:336-382.

Wurster-Hill, D. H. 1973. Chromosomes of eight species from five families of Carnivora. Journal of Mammalogy, 54(3):753-760.

Wyss, A. R. 1987. The walrus auditory region and the monophyly of pinnipeds. American Museum Novitates, 2871:1-31.

Wyss, A. R. 1988. On "Retrogression" in the evolution of the Phocinae and phylogenetic affinities of the monk seals. American Museum Novitates, 2924:1-38.

Xia Wu-ping. 1984. [A study on Chinese *Apodemus* with a discussion of its relations to Japanese species]. Acta Theriologica Sinica, 4:93-98 (in Chinese).

Xia Wu-ping. 1985. A study on Chinese *Apodemus* and its relation to Japanese species. Pp. 76-79, *in* Contemporary mammalogy in China and Japan (T. Kawamichi, ed.). Mammalogical Society of Japan, 194 pp.

Ximenez, A. 1975. *Leopardus geoffroyi*. Mammalian Species, 54:1-4.

Ximenez, A., A. Langguth, and R. Praderi. 1972. Lista sistemática de los mamíferos del Uruguay. Anales del Museo de Historia Natural de Montevideo, 7:1-49.

Xu Ke-fen. 1986. [Analysis of the karyotypes of *Lepus yarkandensis*]. Acta Theriologica Sinica, 6:249-253 (in Chinese).

Xu Long-hui and Yu Sim-ian [Yu Sim-yan]. 1985. [A new subspecies of Edward's rat from Hainan Island, China]. Acta Theriologica Sinica, 5:131-135 (in Chinese).

Yakimenko, L. V., and E. A. Lyapunova. 1986. [Cytogenetic corroboration of belonging of northern mole vole from Turkmenia to *Ellobius talpinus* s. str.]. Zoologicheskii Zhurnal, 65:946-949 (in Russian).

Yakimenko, L. V., K. V. Korobitsyna, L. V. Frisman, and A. I. Muntianu. 1990. Cytogenetic and biochemical comparison of *Mus musculus* and *Mus hortulanus*. Experientia, 46:1075-1077.

Yalden, D. W. 1975. Some observations on the giant mole-rat *Tachyoryctes macrocephalus* (Rüppell, 1842) (Mammalia, Rhizomyidae) of Ethiopia. Monitore Zoologico Italiano, N.S. Supplemento VI, 15:275-303.

Yalden, D. W. 1978. A revision of the dik-diks of the subgenus *Madoqua* (*Madoqua*). Monitore Zoologico Italiano, N. S., Supplemento, 11:245-264.

Yalden, D. W. 1985. *Tachyoryctes macrocephalus*. Mammalian Species, 237:1-3.

Yalden, D. W. 1988. Small mammals of the Bale Mountains, Ethiopia. African Journal of Ecology, 26:281-294.

Yalden, D. W., M. J. Largen, and D. Kock. 1976. Catalogue of the mammals of Ethiopia. 2. Insectivora and Rodentia. Monitore Zoologico Italiano, 8(1):1-118.

Yalden, D. W., M. J. Largen, and D. Kock. 1980. Catalogue of the mammals of Ethiopia. 4. Carnivora. Monitore Zoologico Italiano, N. S., Supplemento, 13(8):168-272.

Yalden, D. W., M. J. Largen, and D. Kock. 1984. Catalogue of the mammals of Ethiopia. 5. Artiodactyla. Monitore Zoologico Italiano, N. S., Supplemento, 19:67-221.

Yamagata, T., A. Ishikawa, Y. Tsubota, T. Namikawa, and A. Hirai. 1987. Genetic differentiation between laboratory lines of the musk shrew (*Suncus murinus*, Insectivora) based on restriction endonuclease cleavage patterns of mitochondrial DNA. Biochemical Genetics, 25:429-446.

Yañez, J. L., W. Siefeld, J. Valencia, and F. Jaksic. 1978. Relaciones entre la sistemática y la morfometriá del subgénero *Abrothrix* (Rodentia: Cricetidae) en Chile. Anales del Instituto de la Patagonia, 9:185-197.

Yañez, J. L., J. Valencia, and F. Jaksic. 1979. Morfometría y sistemática del subgénero *Akodon* (Rodentia) en Chile. Archivos de Biología y Medicina Experimentales (Santiago), 12:197-202.

Yañez, J. L., J. C. Torres-Mura, J. R. Rau, and L. C. Contreras. 1987. New records and current status of *Euneomys* (Cricetidae) in southern South America. Fieldiana, Zoology, New Series, 39:283-287.

Yang An-feng and Fang Li-xiang. 1988. [Phallic morphology of 13 species of the family Muridae from China, with comments on its taxonomic significance]. Acta Theriologica Sinica, 8:275-287 (in Chinese, with English summary).

Yang Chun-wen, Chen Rong-hai, and Zhang Chung-mei. 1991. [A study of the rodent community division in Huangnihe forest region]. Acta Theriologica Sinica, 11(2):118-125.

Yang Guang-rong and Wang Ying-xiang. 1987. [A new subspecies of *Hadromys humei* (Muridae, Mammalia) from Yunnan, China]. Acta Theriologica Sinica, 7:46-50 (in Chinese).

Yates, T. L. 1984. Insectivores, elephant shrews, tree shrews, and dermopterans. Pp. 117-144, *in* Orders and families of Recent mammals of the world (S. Anderson, and J. K. Jones, Jr., eds.). John Wiley and Sons, New York, 686 pp.

Yates, T. L., and I. F. Greenbaum. 1982. Biochemical systematics of North American moles (Insectivora: Talpidae). Journal of Mammalogy, 63:368-374.

Yates, T. L., and D. J. Schmidly. 1977. Systematics of *Scalopus aquaticus* (Linnaeus) in Texas and adjacent states. Occasional Papers, The Museum, Texas Tech University, 45:1-36.

Yates, T. L., and D. J. Schmidly. 1978. *Scalopus aquaticus*. Mammalian Species, 105:1-4.

Yates, T. L., R. J. Baker, and R. K. Barnett. 1979. Phylogenetic analysis of karyological variation in three genera of peromyscine rodents. Systematic Zoology, 28:40-48.

Yatsenko, V. N. 1980. [C-heterochromatin and chromosomal polymorphism in the Gobi-Altai vole (*Alticola stolizkanus barakschin* Bannikov, 1968, Rodentia, Cricetidae)]. Doklady Akademii Nauk SSSR, 254:1009-1010.

Yatsenko, V. N., V. M. Malygin, V. N. Orlov, I. Yu. Yanina. 1980. [The chromosome polymorphism in the Mongolian vole *Microtus mongolicus* Radde, 1861]. Tsitologiya, 22:471-474 (in Russian).

Yensen, E. 1991. Taxonomy and distribution of the Idaho ground squirrel. Journal of Mammalogy, 72:583-600.

Yepes, J. 1928. Los "Edentata" argentinos. Sistemática y distribución. Revista de la Universidad de Buenos Aires, ser. 2a, 1:461-515, 6 figs [Reprint separate is independently paginated].

Yochem, P. K., and S. Leatherwood. 1985. Blue whale - *Balaenoptera musculus*. Pp. 193-240, *in* Handbook of marine mammals: The sirenians and baleen whales (S. H. Ridgway and R. Harrison, eds.). Academic Press, London, 3:1-362.

Yong, H. S. 1969. Karyotypes of Malayan rats (Rodentia-Muridae, genus *Rattus* Fischer). Chromosoma, 27:245-267.

Yong, H. S. 1973. Chromosomes of the pencil-tailed tree-mouse *Chiropodomys gliroides*. Malayan Nature Journal, 26:159-162.

Yong, H. S. 1983. Heterochromatin blocks in the karyotype of the pencil-tailed tree-mouse, *Chiropodomys gliroides* (Rodentia, Muridae). Experientia, 39:1039-1040.

Yong, H. S., and S. S. Dhaliwal. 1976. Variations in the karyotype of the red giant flying squirrel *Petaurista petaurista* (Rodentia, Sciuridae). Malaysian Journal of Science, 4:9-12.

Yong, H. S., S. S. Dhaliwal, and B. L. Lim. 1982. Karyotypes of *Hapalomys* and *Pithecheir* (Rodentia: Muridae) from Peninsular Malaysia. Cytologia, 47:535-538.

Yoshiyuki, M. 1986. The phylogenetic status of *Mogera tokudae* Kuroda, 1940 on the basis of body skeletons. Memoirs of the National Science Museum, Tokyo, 19:203-213.

Yoshiyuki, M. 1988*a*. Taxonomic status of the least red-toothed shrew (Insectivora, Soricidae) from Korea. Bulletin of the National Science Museum (Tokyo), ser. A, 14:151-158.

Yoshiyuki, M. 1988*b*. Notes on Thai mammals 1. Talpidae (Insectivora). Bulletin of the National Science Museum (Tokyo), ser. A, 14:215-222.

Yoshiyuki, M. 1989. A systematic study of the Japanese Chiroptera. National Science Museum, Tokyo, 242 pp.

Yoshiyuki, M. 1991. Taxonomic status of *Hipposideros terasensis* Kishida, 1924 from Taiwan (Chiroptera, Hipposideridae). Journal of the Mammalogical Society of Japan, 16(1):27-35.

Yoshiyuki, M., and Y. Imaizumi. 1986. A new species of *Sorex* (Insectivora, Soricidae) from Sado Island, Japan. Bulletin of the National Science Museum (Tokyo), ser. A, 12:185-193.

Yoshiyuki, M., and Y. Imaizumi. 1991. Taxonomic status of the large mole from the Echigo plain, central Japan, with description of a new species (Mammalia, Insectivora, Talpidae). Bulletin of the National Science Museum (Tokyo), ser. A, 17:101-110.

Yoshiyuki, M., S. Hattori, and K. Tsuchiya. 1989. Taxonomic analysis of two rare bats from the Amami Islands (Chiroptera, Molossidae and Rhinolophidae). Memoirs of the National Science Museum, Tokyo, 22:215-225.

Yosida, T. H. 1978. Experimental breeding and cytogenetics of the soft-furred rat, *Millardia meltada*. Laboratory Animals, 12:73-77.

Yosida, T. H. 1980. Cytogenetics of the black rat: Karyotype evolution and species differentiation. University of Tokyo Press, Tokyo, 256 pp.

Yosida, T. H. 1985. The evolution and geographic differentiation of the house shrew karyotypes. Acta Zoologica Fennica, 170:31-34.

Yosida, T. H., and M. Harada. 1985. A population survey of the chromosome polymorphism in the black rats (*Rattus rattus*) collected in the Osaka-city, Japan. Proceedings of the Japanese Academy, ser. B, 61:208-211.

Yosida, T. H., T. Udagawa, M. Ishibashi, K. Moriwaki, T. Yabe, and T. Hamada. 1985. Studies on the karyotypes of the black rats distributed in the Pacific and South Pacific islands, with special regard to the border line of the Asian and Oceanian type black rats on the Pacific Ocean. Proceedings of the Japan Academy, 6l, ser. B:71-74.

Young, C. J., and J. K. Jones, Jr. 1982. *Spermophilus mexicanus*. Mammalian Species, 164:1-4.

Young, C. J., and J. K. Jones, Jr. 1983. *Peromyscus yucatanicus*. Mammalian Species, 196:1-3.

Young, C. J., and J. K. Jones, Jr. 1984. *Reithrodontomys gracilis*. Mammalian Species, 218:1-3.

Young, S. P. 1951. The clever coyote. University of Nebraska Press, Lincoln, 411 pp.

Young, S. P., and E. A. Goldman. 1946. The puma: mysterious American cat. American Wildlife Institute, 358 pp.

Youngman, P. M. 1967. A new subspecies of varying lemmings *Dicrostonyx torquatus* (Pallas), from Yukon Territory (Mammalia, Rodentia). Proceedings of the Biological Society of Washington, 80:31-34.

Youngman, P. M. 1975. Mammals of the Yukon Territory. National Museum of Natural Sciences (Ottawa), Publications in Zoology, 10:1-192.

Youngman, P. M. 1982. Distribution and systematics of the European mink, *Mustela lutreola* Linnaeus, 1761. Acta Zoologica Fennica, 166:1-48.

Youngman, P. M. 1990. *Mustela lutreola*. Mammalian Species, 362:1-3.

Yudin, B. S. 1969. Taxonomy of some species of shrews (Soricidae) from Palaearctic and Nearctic. Acta Theriologica, 14:21-34.

Yudin, B. S. 1972. [Contribution to the taxonomy of the masked transarctic common shrew (*Sorex cinereus* Kerr, 1792) from USSR fauna]. Teriologiya (Novosibirsk), 1:45-50 (in Russian).

Yudin, B. S. 1989. Nasekomoyadnye mlekopitayushchie Sibiri [Insectivorous mammals of Siberia]. Nauka, Sibirskoe Otdelenie, Novosibirsk., 360 pp. (in Russian).

Zagorodnyuk, I. V. 1988. *Pitymys tatricus* (Rodentia)—novyi vid v faune SSSR [*Pitymys tatricus* (Rodentia)—new species in fauna USSR]. Vestnik Zoologii, 3:54 (in Russian).

Zagorodnyuk, I. V. 1989. Taksonomiya, rasprostranenie i morfologicheskaya izmenchivost' polevok roda *Terricola* vostochnoi Evropy [Taxonomy, distribution and morphological variation of the *Terricola* voles in east Europe]. Vestnik Zoologii, 5:3-14 (in Russian).

Zagorodnyuk, I. V. 1990. Kariotipicheskaya izmenchivost' i sistematika serykh polevok (Rodentia, Arvicolini). Soobshchenie 1. Vidovoi sostav i khromosomnye chisla [Karyotypic variability and systematics of the gray voles (Rodentia, Arvicolini). Communication 1. Species composition and chromosomal numbers]. Vestnik Zoologii, 2:26-37 (in Russian).

Zagorodnyuk, I. V. 1991*a*. Kariotipicheskaya izmenchivost' 46-khromosomnykh form polevok gruppy *Microtus arvalis* (Rodentia): taksonomicheskaya otsenka [Karyotypic variation of 46-chromosome forms of voles of the *Microtus arvalis* group (Rodentia): taxonomic evaluation]. Vestnik Zoologii, 1:36-46 (in Russian).

Zagorodnyuk, I. V. 1991*b*. Sistematicheskoe polozhenie *Microtus brevirostris* (Rodentiformes): materialy po taksonomii i diagnostike gruppy "*arvalis*" [Systematic position of *Microtus brevirostris* (Rodentiformes): materials toward the taxonomy and diagnostics of the "*arvalis*" group]. Vestnik Zoologii, 3:26-35 (in Russian).

Zagorodnyuk, I. V. 1991*c*. [Spatial karyotype differentiation of Arvicolini (Rodentia)]. Zoologicheskii Zhurnal, 70:99-110 (in Russian).

Zagorodnyuk, I. V. 1992. [Geographic distribution and levels of abundance of *Terricola subterraneus* on the USSR territory]. Zoologicheskii Zhurnal, 71:86-97.

Zagorodnyuk, I. V., N. N. Vorontsov, and V. N. Peskov. 1992. [Tatra vole (*Terricola tatricus*) in the Eastern Carpathians]. Zoologicheskii Zhurnal, 71:96-105 (in Russian).

Zaime, A. K., and M. Pascal. 1988. Recherche d'un indice craniometrique discriminant deux especes de meriones (*Meriones shawi* et *M. libycus*) vivant en sympatrie sur le site de Guelmime (Maroc). Mammalia, 52:575-582.

Zaitsev, M. V. 1988. [On the nomenclature of red-toothed shrews of the genus *Sorex* in the fauna of the USSR]. Zoologicheskii Zhurnal, 67:1878-1888 (in Russian).

Zakrzewski, R. J. 1974. Fossil Ondatrini from western North America. Journal of Mammalogy, 55:284-292.

Zanchin, N. I. T., A. Langguth, and M. S. Mattevi. 1992*a*. Karyotypes of Brazilian species of *Rhipidomys* (Rodentia, Cricetidae). Journal of Mammalogy, 73:120-122.

Zanchin, N. I. T., I. J. Sbalqueiro, A. Langguth, R. C. Bossle, E. C. Castro, L. F. B. Oliveira, and M. S. Mattevi. 1992*b*. Karyotype and species diversity of the genus *Delomys* (Rodentia, Cricetidae) in Brazil. (Acta Theriologica, 37:163-169.

Zegers, D. A. 1984. *Spermophilus elegans*. Mammalian Species, 214:1-7.

Zhang Chieh and Wang Tsung-yi. 1963. [Faunistic studies of the Chinghai province]. Acta Zoologica Sinica, 15(3):195-200.

Zhang Cizu. 1987. *Nemorhaedus cranbrooki* Hayman. Pp. 213-219, *in* The biology and management of *Capricornis* and related antelopes (H. Soma, ed.). Croom Helm, London, 391 pp.

Zhang Ya-ping and Shi Li-ming. 1991. Riddle of the giant panda. Nature.

Zhang Zi-yu and Zhao Ming-shan. 1984. [A new subspecies of the sulphur-bellied rat from Jilin—*Rattus niviventer naoniuensis*]. Acta Zoologica Sinica, 30:99-102 (in Chinese).

Zhao Xiao-fan and Lu Hao-quan. 1986. [Comparative observations of several biochemical indexes of *Apodemus agrarius pallidior* and *Apodemus agrarius ninpoensis* of the striped backed field mice]. Acta Theriologica Sinica, 6:57-62 (in Chinese).

Zheng Chang-lin. 1986. [Recovery of Koslow's pika (*Ochotona koslowi* Buchner) in Kunlun Mountains of Xinjiang Uygur Autonimous [sic] Region, China]. Acta Theriologica Sinica, 6:285 (in Chinese).

Zheng Chang-lin and Wang Sung. 1980. [On the taxonomic status of *Pitymys leucurus* Blyth]. Acta Zootaxonomica Sinica, 5:106-112 (in Chinese).

Zheng S. 1985. Remains of the genus *Anourosorex* (Insectivora, Mammalia) from Pleistocene of Guizhou District. Vertebrata Palasiatica, 23:39-51.

Zheng S., and Li Chuan-kwei. 1990. Comments on fossil arvicolids of China. Pp. 431-442, *in* International symposium evolution, phylogeny and biostratigraphy of arvicolids (Rodentia, Mammalia) (O. Fejfar and W.-D. Heinrich, eds.). Geological Survey, Prague, 448 pp.

Zheng Tao and Zhang Ying-mei. 1990. [The fauna and geographical division on Glires of Gansu province]. Acta Theriologica Sinica, 10(2):137-144.

Zholnerovskaya, E. I., D. I. Bibikov, and V. I. Ermolaev. 1990. Immunogeneticheskii analiz sistematicheskikh vzaimootnoshenii surkov [Immunogenetic analysis of systematic relationships of marmots]. Byulleten' Moskovskovo Obshchestva Ispytatelei Prirody, Otdel Biologicheskii, 195:15-24 (in Russian).

Zhou Jia-di, Li Si-hua, and Gu Jing-he. 1985. [The preliminary observation on the mammals in Kunlun-Altun Basin]. Acta Theriologica Sinica, 5(2):160.

Zhou Kai-ya, Qian Wei-juan, and Li Yue-min. 1978. [Recent advances in the study of the baiji, *Lipotes vexiliffer* Miller]. Journal of the Nanjing Teacher's College (Natural Science), 1:8-13 (in Chinese).

Zhou Kai-ya, Li Yue-min, and Qian Wei-juan. 1979. [The stomach of the baiji *Lipotes vexillifer*]. Acta Zoologica Sinica, 25:95-100 (in Chinese).

Zhou Yu-can and Xia Wu-ping. 1981. [An electrophoretic comparison of the serum protein and hemoglobin in three species of mouse-hares — a discussion on the systematical position of *Ochotona curzoniae*]. Acta Theriologica Sinica, 1:39-44 (in Chinese).

Ziegler, A. C. 1971. Dental homologies and possible relationships of recent Talpidae. Journal of Mammalogy, 52:50-68.

Ziegler, A. C. 1982*a*. The Australo-Papuan genus *Syconycteris* (Chiroptera: Pteropodidae) with the description of a new Papua New Guinea species. Occasional Papers of the Bernice P. Bishop Museum, 25(5):1-22.

Ziegler, A. C. 1982*b*. An ecological check-list of New Guinea Recent mammals. Pp. 863-894, *in* Biogeography and ecology of New Guinea, vol. 2 (J. L. Gressitt, ed.). Monographiae Biologicae, 42:1-983[in 2 vols].

Zima, J. 1983. Chromosomes of the harvest mouse, *Micromys minutus*, from the Danube Delta (Muridae, Rodentia). Folia Zoologica, 32:19-22.

Zima, J. 1986. Chromosomal and epigenetic variation in a population of the pine vole, *Pitymys subterraneus*. Folia Zoologica, 35:333-345.

Zima, J. 1987. Karyotypes of certain rodents from Czechoslovakia (Sciuridae, Gliridae, Cricetidae). Folia Zoologica, 36(4) 1987:337-343.

Zima, J., and B. Král. 1984*a*. Karyotypes of European mammals. II. Acta Scientiarum Naturalium, Academiae Scientarium Bohemoslovacae (Brno), 18(8):1-62.

Zima, J., and B. Král. 1984*b*. Karyotypic variability in *Sorex araneus* in Central Europe (Soricidae, Insectivora). Folia Zoologica, 34:235-243.

Zima, J., I. V. Zagorodnyuk, V. A. Gaichenko, and T. O. Zhezherina. 1991. Polimorfizm i khromosomnaya izmenchivost' *Microtus rossiaemeridionalis* (Rodentiformes) [Polymorphism and chromosomal variability in *Microtus rossiaemeridionalis* (Rodentiformes)]. Vestnik Zoologii, 4:48-53 (in Russian).

Zimina, R. P., (ed.). 1978. Surki. Rasprostranenie i ekologiya [Marmots. distribution and ecology]. Nauka, Moscow, 222 pp. (in Russian).

Zimmerman, E. G. 1970. Karyology, systematics and chromosomal evolution in the rodent genus, *Sigmodon*. Publications of The Museum, Michigan State University, Biological Series, 4:385-454.

Zimmerman, E. G., and M. E. Nejtek. 1977. Genetics and speciation of three semispecies of *Neotoma*. Journal of Mammalogy, 58:391-402.

Zimmerman, E. G., B. J. Hart, and C. W. Kilpatrick. 1975. Biochemical genetics of the *boylii* and *truei* groups of the genus *Peromyscus* (Rodentia). Comparative Biochemistry and Physiology, 52B:541-545.

Zimmerman, E. G., C. W. Kilpatrick, and B. J. Hart. 1978. The genetics of speciation in the rodent genus *Peromyscus*. Evolution, 32:565-579.

Zimmermann, E. A. W., von. 1778-1783. Geographische Geschichte des Menschen, und der allgemein verbreiteten vierfüssigen Thiere, nebst einer hieher gehörigen zoologischen Weltkarte vol. 2 - Geographische Geschichte des Menschen, und der vierfüssigen Thiere. Zweiter Band. Enthält ein vollständiges Verzeichniss aller bekannten Quadrupeden. Weygandschen Buchhandlung, Leipzig, 3 volumes.

Zimmermann, K. 1942. Zur Kenntnis von *Microtus oeconomus* (Pallas). Archiv für Naturgeschichte, Neue Folge, 11:174-197.

Zimmermann, K. 1962. Die Untergattungen der Gattung *Apodemus* Kaup. Bonner Zoologische Beiträge, 13:198-208.

Zykov, A. E. 1987. Novaya nakhodka zacaspiiskoi myshevidnoi soni (*Myomimus personatus* Ognev) na territorii SSSR [New finding of the Transcaspian mouse-like dormouse *Myomimus personatus* Ognev on the territory of USSR]. Vestnik Zoologii, 1:80 (in Russian).

Index

Currently recognized names are indicated by bold page numbers; others are synonyms.

aagaardi, Crocidura 89
abacanicus, Lagurus 515
abae, Chaerephon 232
  Hipposideros **171**
  Rhinolophus 166
abaensis, Pipistrellus 222
abanticus, Muscardinus 769
abasgicus, Erinaceus 77
abassensis, Heliosciurus 428
abberrans, Cryptomys 772
abboti, Myotis 211
abbotti, Cynocephalus 135
  Hylobates 276
  Lepus 816
  Mus 625
  Thomomys 473
abbottii, Callosciurus 422
abbreviata, Neotoma 712
abbreviatus, Microtus **518**
abbyssinicus, Procavia 374
abdita, Thyroptera 196
Abditomys **564**, 670
  latidens **564**
abditum, Megaderma 163
abditus, Microtus 523
abei, Lepus 821
  Myotis **207**
abeli, Odocoileus 391
abelii, Pongo 277
Abeomelomys 640, 641
aberrans, Meles 314
aberti, Sciurus **439**
abessinicus, Xerus 458
abidi, Petaurus **61**, 638
abieticola, Martes 319
  Tamiasciurus 457
abietinoides, Martes 319
abietorum, Napaeozapus 498
  Peromyscus 732
  Vulpes 287
abietum, Martes 320
abjectus, Tragulus 383
ablogriseus, Hylobates 275
ablusus, Spermophilus 448
ablutus, Acomys 567
  Mus 622
abnormis, Sorex 117
abongensis, Pongo 277
abrae, Apomys **575**, 575
abramus, Pipistrellus 221, 223
abrasus, Eumops 233
  Molossops **234**
Abrawayaomys **687**, 687
  ruschii **687**, 687
abrocodon, Phyllotis 738
Abrocoma **789**
  bennettii **789**, 789
  boliviensis **789**
  budini 789
  cinerea **789**
  cuvieri 789
  famatina 789
  helvina 789
  laniger 789
  murrayi 789
  schistacea 789
  vaccarum 789
Abrocomidae **789**
Abromys 484
Abrothrix 688, 690, 691, 692, 694, 726
abrukensis, Arvicola 506
abruptus, Tragulus 383
abruttii, Myoxus 770
absarokus, Ursus 338
absonus, Thomomys 473
abstrusus, Prosciurillus **436**
  Thomomys 473
abuharab, Gazella 396
abulensis, Chionomys 507
abuwudan, Ichneumia 306
abyssinica, Genetta **345**
  Kobus 413
  Mellivora 315
  Sylvicapra 412
abyssinicus, Arvicanthis **576**, 576, 578
  Colobus 270
  Dendromus 542
  Hippopotamus 381
  Lepus 816
  Tragelaphus 404
acaab, Vulpes 287
acaciae, Thallomys 669
acaciarum, Galago 249
acadicus, Castor 467
  Microtus 527
  Sorex 113, 119
  Zapus 499
Acanthion 773, 774
acanthion, Tachyglossus 13
Acanthodelphis 358
Acanthoglossus 13
Acanthomys 564, 649, 670
Acanthonotus 13
acanthrous, Tachyglossus 13
acanthurus, Setifer 73
acapulcensis, Odocoileus 391
accedula, Cricetulus 538
aceramarcae, Gracilinanus **17**
aceratos, Oreotragus 399
Acerodon **137**, 146
  alorensis 137
  arquatus 137
  aurinuchalis 137
  celebensis **137**, 146
  floresianus 137
  floresii 137
  gilvus 137
  humilis **137**
  jubatus **137**
  leucotis **137**
  lucifer **137**
  mackloti **137**
  mindanensis 137
  obscurus 137
  ochraphaeus 137
  prajae 137
  pyrrhocephalus 137
aceros, Cervus 386
acetabulosus, Mormopterus **237**
acevedo, Capromys 800
  Macrocapromys 800
achates, Semnopithecus 273
  Taphozous 161
Acheus 63
achilles, Rhinolophus 168
  Semnopithecus 273
Achlis 392
acholi, Mus 628
achradophilus, Ariteus 187
achrotes, Canis 281
Acinonychinae **288**
Acinonyx **288**, 288
  fearonii 288
  fearonis 288
  guttata 288
  hecki 288
  jubatus **288**
  lanea 288
  megabalica 288
  ngorongorensis 288
  obergi 288
  raddei 288
  raineyi 288
  rex 288
  senegalensis 288
  soemmeringii 288
  velox 288
  venatica 288
  venaticus 288
  venator 288
  wagneri 288
aciurens, Kerodon 780
acmaeus, Alticola 502
Acomaemys 787
acomyoides, Uranomys 671
Acomys **564**, 564, 565, 566, 567, 605, 621, 670
  ablutus 567
  aegyptiacus 566
  affinis 566
  airensis 565
  albigena 565
  argillaceus 567
  bovonei 567
  brockmani 566
  cahirinus **565**, 565, 566
  chudeaui 565
  cilicicus **565**
  cineracens 565
  cineraceus **565**, 565, 566
  dimidiatus 565, 566
  enid 567
  flavidus 565
  harrisoni 566
  hawashensis 565
  helmyi 565
  hispidus 565
  homericus 565
  hunteri 565
  hystrella 565
  ignitus **565**, 565, 566
  intermedius 565
  johannis 565
  kempi **565**, 565, 566
  lewisi 566
  louisae **566**
  lowei 565, 566
  megalodus 565
  megalotis 565
  minous **566**
  montanus 565, 566
  mullah **566**, 566
  nesiotes **566**
  nubicus 565
  nubilus 567
  percivali 565, **566**, 566, 567
  pulchellus 565, 566
  russatus **566**, 566
  sabryi 565
  selousi 567
  seurati 565
  spinosissimus **567**, 567
  subspinosus **567**, 567
  transvaalensis 567
  umbratus 566
  viator 565
  whitei 565
  wilsoni 565, 566, **567**, 567
  witherbyi 565

Aconaemys **787**, 787
fuscus **787**, 788
porteri 787
sagei **788**
acontion, Pygeretmus 490
acouchy, Cavia 782
Myoprocta **782**, 782
acoushy, Myoprocta 782
acraeus, Callosciurus 422
Dendromus 543
acraia, Neotoma 711
acreanus, Callicebus 258
acrensis, Saguinus 253
acrirostratus, Thomomys 473
acrobata, Kerodon 780
Acrobates **62**
frontalis 62
pulchellus 62
pygmaeus **62**, 62
Acrobatidae 46, **62**
Acrocodia 370, 371
acrocodia, Puma 296
acrocranius, Sus 379
acrodonta, Pteralopex **145**
Acronotus 393
acrophilus, Alticola 504
acrotis, Rhinolophus 164
acrus, Tamias 455
acticola, Gerbillus **548**, 548
Heliosciurus 429
actuosa, Martes 319
actuosus, Thomomys 473
aculeatus, Tachyglossus **13**
acuminatus, Rhinolophus **163**
acuti, Dasyprocta 781
acuticauda, Cervus 386
acuticaudatus, Molossus 235
acuticornis, Cerophorus 399
Cervus 386
Raphicerus 399
acutorostrata, Balaenoptera **349**
acutus, Lagenorhynchus **353**, 353
Liomys 482
adailensis, Mungos 307
adamauae, Redunca 414
Syncerus 402
adametzi, Dendrohyrax 373
Orycteropus 375
Ovis 408
Syncerus 402
adami, Rhinolophus **163**
adamoi, Talpa 129
adamsi, Callosciurus **420**, 423
Pipistrellus 224
adamsoni, Dremomys 425
adana, Nycteris 162
adangensis, Callosciurus 421
adansoni, Atelerix 76
Kobus 414
Addax **412**
addax 412
gibbosa 412
mytilopes 412
nasomaculatus **412**
suturosa 412
addax, Addax 412
adderrans, Thomomys 473
addra, Gazella 396
adelaidensis, Mus 625
Adenonotus 380
Adenota 413
adenota, Kobus 414
adersi, Cephalophus **410**, 410, 411
Dendrohyrax 373
Panthera 298
adipicaudatus, Cheirogaleus 243
admiralitatum, Pteropus **146**
admiraltiae, Microtus 527
adocetus, Microtus 527
Spermophilus **444**
adolfi, Kobus 414
adolfi-friederici, Colobus 270
Dendrohyrax 373
adolfifriderici, Kobus 413, 414
adolfifriederici, Pan 277
Syncerus 402
adolphei, Sciurus 443
adovanus, Nycticeius 218
adrotes, Gorilla 276
adsitus, Tamias 456
adspersus, Bunomys 581
Liomys **482**
Rattus 582
aduncus, Tursiops 357
adusta, Antechinus 30
Macaca 267
Monodelphis **21**, 21
Panthera 297, 298
Pithecia 262
adustus, Antechinus 30
Canis **280**
Cebus 259
Chaerephon 233
Phascolarctos 45
Rattus 649, **650**, 650, 654
Tryphomys 564, **670**, 670
adventor, Cricetomys 541
adversus, Myotis **207**
Urotrichus 129, 130
aedilus, Myotis 210
aedium, Plagiodontia 799, **804**, 804
aegagrus, Capra 405
aegatensis, Crocidura 95
Aegoceros 405
Aegoryx 413
aegyptiaca, Papio 269
Tadarida **240**
aegyptiacus, Acomys 566
Nannospalax 753
Pteropus 152
Vulpes 287
aegyptiae, Herpestes 305
aegyptius, Gerbillus 548, 551
Hemiechinus 78
Jaculus 493
Lepus 816
Pipistrellus **219**
aeliani, Phacochoerus 377
aelleni, Myotis **208**
Aello 176
aello, Nyctimene **143**, 144
aemuli, Rattus 652
aenea, Murina **228**
aeneas, Semnopithecus 273
aeneus, Rhogeessa 226
aenigmatica, Talpa 129
aenigmaticus, Tapirus 371
aenobarbus, Myotis 208
Aenomys 637
Aepeomys **687**, 687, 688, 749
fuscatus **688**
lugens **688**, 688
ottleyi 688
vulcani 688
Aepyceros **393**
holubi 393
johnstoni 393
katangae 393
melampus **393**
pallah 393
petersi 393
rendilis 393
suara 393
Aepycerotinae **393**
Aepyprymnus **48**
melanotis 48
rufescens **48**
aequalidens, Thomomys 475
aequalis, Aethalops 138
Rhinolophus 164
aequatoralis, Nyctinomops 239
aequatoria, Ourebia 399
Suncus 102
aequatorialis, Alouatta 255
Artibeus 188
Cebus 259
Cephalophus 411
Dasypus 66
Genetta 346
Glironia 16
Leopardus 291
Mustela 322
Odocoileus 391
Pithecia **261**
Procyon 335
aequatorianus, Molossops **234**
aequatoris, Anoura 183
Tayassu 380
aequatorius, Mus 629
aequicaudalus, Rattus 656
aequicaudata, Crocidura 82
aequinoctialis, Syncerus 402
aequivocatus, Microtus 519
Aeretes **459**, 459
melanopterus **459**
sulcatus 459
szechuanensis 459
aereus, Scalopus 127
aero, Pipistrellus **219**
Aeromys **459**, 459
bartelsi 459
nitidus 459
phaeomelas 459
tephromelas **459**
thomasi **459**
aerosa, Kerivoula **196**
aerosus, Akodon **688**, 688, 694
Melomys **615**, 615
aeruginosus, Gerbillus 551
aerula, Hystrix 773
aestivus, Perognathus 485
aestuans, Sciurus **439**, 441
aestuarinus, Microtus 519
aeta, Epimys 598
Hylomyscus **598**, 598, 599
Aethalodes 137
alecto 137
Aethalops **137**
aequalis 138
alecto **138**
ocypete 138
Aethechinus 76, 77
aethiopica, Giraffa 383
Hippotragus 412
Nycteris 162
aethiopicus, Aper 377
Cercocebus 263
Equus 369
Hemiechinus **78**, 78
Lepus 816
Lophiomys 564
Orycteropus 375
Phacochoerus **377**, 377
Aethiops 262
aethiops, Chlorocebus 263, **265**
Rattus 658
Rhinolophus 166
Simia 265
Aethoglis 763
Aethomys **567**, 567, 568, 589, 590, 665, 666, 668, 674
alghazal 568
alticola 567
amalae 568
arborarius 568
auricomis 568
avarillus 568
avunculus 568
bocagei **567**, 568
calarius 568
capensis 568
capricornis 567
centralis 568
chrysophilus **567**, 567
dollmani 568
drakensbergi 568
epupae 568
fouriei 567
grahami 568
granti 567, **568**, 568
harei 567

helleri 568
hindei **568**, 568, 569
hintoni 568
imago 567
ineptus 567
kaiseri **568**, 568, 569
klaverensis 568
lechochloides 568
lehocla 568
longicaudatus 568
magalakuini 567
manteufeli 568
medicatus 568
monticularis 568
namaquensis 567, **568**, 568
namibensis 568
norae 568
nyikae **568**, 568
pedester 568
phippsi 568
pretoriae 567
siccatus 568
silindensis 567, **568**, 568
singidae 567
stannarius **569**
thomasi **569**, 569
tongensis 567
turneri 568
tzaneenensis 567
vernayi 568
voi 567
walambae 568
waterbergensis 568
Aethosciurus 429, 434, 435
ruwenzorii 428
Aethurus 758
afer, *Lasiomys* 605
Lophuromys 606
Orycteropus **375**
Triaenops 175
affinis, *Acomys* 566
Akodon **688**
Antechinus 30
Caluromys 15
Cerdocyon 282
Cervus 385
Felis 290
Helogale 304
Hydropotes 388
Isoodon 39
Lavia 163
Macaca 266
Melanomys 708
Mustela 322
Myosorex 99, 100
Myotis 212
Neotoma 711
Noctilio 176
Nycteris 162
Nyctinomops 239
Ototylomys 726
Paraxerus 435
Peromyscus 732
Pipistrellus **219**, 222
Pteropus 146
Ratufa **437**
Rhinolophus **163**
Rousettus 152
Saccolaimus 159
Scaptonyx 128
Sciurus 440
Sphiggurus 777
Sus 379
Tamias 453
Thomomys 473
Tragulus 382
Afganomys 512
Afghanomys 512
afghanus, *Blanfordimys* **506**, 506
Meriones 556, 557
Pipistrellus 220
afra, *Coleura* **156**
Emballonura 156
Galidia 299
Genetta 345
Myrmecophaga 375
Tatera **560**, 560, 561
africaeaustralis, *Hystrix* **773**, 773
africana, *Bathyergus* 771
Ictonyx 319
Kerivoula **196**, 197
Loxodonta **367**
Mustela **321**, 321, 322, 323
Poecilogale 325
Tadarida 241
Africanthropus 276
africanus, *Atherurus* **773**
Diceros 372
Elephas 367
Equus 369
Miniopterus 230
Myotis 209
Nycticeius 218
Panthera 297
Phacochoerus **377**, 377
Pipistrellus 222
Potamochoerus 378
Tatera 560
Trichechus 366
Xerus 458
Afrosorex 81, 84, 93
aga, *Lasiopodomys* 515
agadiri, *Crocidura* 96
agadius, *Xerus* 458
agag, *Gerbillus* **548**, 548, 550, 553
agilis, *Cercocebus* **262**, 262
Crocidura 94
Dipodomys **477**, 478
Gracilinanus **17**, 17
Hylobates **274**, 275, 276
Macropus **53**
Micromys 620
Mustela 322
Myotis 209
Oligoryzomys 718
Pipistrellus 223
agnatus, *Macaca* 266
agnella, *Kerivoula* **196**
Agouti **783**, 783
alba 783
andina 783
fulvus 783
guanta 783
mexianae 783
paca **783**
sierrae 783
sublaevis 783
subniger 783
taczanowskii **783**
venezuelica 783
Agoutidae 781, **783**, 783
Agoutinae 783
agrarius, *Apodemus* **570**, 571
Mus 569
agressus, *Lagurus* 515
agreste, *Akodon* 689
agrestis, *Microtus* **518**, 518, 519, 527
Oryzomys 720
agrias, *Pongo* 277
Agricola 517, 518, 519
agricolae, *Sciurus* 441
agricolai, *Gracilinanus* 17
agricolaris, *Thomomys* 473
Agriotheriinae 336
agrius, *Felis* 290
aguti, *Dasyprocta* 781, 782
Mus 781
agyisymbanus, *Otolemur* 250
aharonii, *Caracal* 288
Hystrix 774
ahlselli, *Helogale* 304
ahoenobarbus, *Sus* 378
ai, *Bradypus* 63
Aigererus 412
Aigocerus 412
Ailurinae **336**, 336
Ailuropinae 46
Ailuropoda 332, **336**, 336, 337
melanoleuca **336**
Ailuropodidae 336
Ailuropodini 336
Ailurops **46**
flavissimus 46
furvus 46
melanotis 46
togianus 46
ursinus **46**
Ailurus 332, **336**, 336, 337
fulgens 336, **337**
ochraceus 337
refulgens 337
styani 337
ainu, *Apodemus* 573
Lepus 821
aipomus, *Sus* 379
airensis, *Acomys* 565
Felis 290
Jaculus 493
airolensis, *Mus* 625
aistoni, *Notomys* 636
aitape, *Rattus* 652
aitchisoni, *Hyperacrius* 515
aitkeni, *Sminthopsis* **35**
ajax, *Conepatus* 316
Semnopithecus 273
akeleyi, *Neotragus* 398
Peromyscus 732
akka, *Funisciurus* 428
Lemniscomys 602
akkeshii, *Clethrionomys* 509
Akodon 687, **688**, 688, 689, 690, 691, 692, 693, 694, 696, 697, 699, 700, 726, 727, 746, 748
aerosus **688**, 688, 694
affinis **688**
agreste 689
albiventer **688**
alterus 693
altorum 691
amoenus 696
andinus 692
angustus 691
apta 691
arenicola 689
arequipae 693
arviculoides 689
atratus 692
azarae **688**, 689, 693
baliolus 688
beatus 692
berlepschii 688
bibianae 689
bogotensis **689**, 689, 690, 691
boliviensis 688, **689**, 689, 690, 691, 692, 693
brachiotis 692
brachytarsus 691
brevicaudatus 692
budini **689**, 692
caenosus 692
canescens 692, 694
castanea 691
chapmani 693
chonoticus 692
collinus 690
cursor **689**, 693
dayi **689**
deceptor 689
dolores **689**, 691
foncki 692
francei 691
fuliginosus 692
fulvescens 691
fumeus **689**, 690, 691
fusco-ater 691
germaini 692
glaucinus 693
henseli 692
hershkovitzi **690**
hirta 691
hunteri 689
illuteus **690**
infans 694
iniscatus **690**
jelskii 700
juninensis **690**
kempi **690**

kofordi **690**, 690
landbecki 692
langguthi 690
lanosus **690**, 691
latebricola 689, **690**
lepturus 692
leucogula 692
lindberghi **690**, 691
llanoi 694
longipilis 690, **691**, 691, 692, 699
lutescens 692
macronychos 692
mansoensis **691**
markhami **691**
megalonyx 699
melampus 691
meridensis 693
mimus 689, 690, **691**, 691
mochae 692
modestior 691
moerens 691
molinae **691**, 689
mollis 690, **691**, 692, 693
montensis 689
nasica 692
nemoralis 692
neocenus **691**, 694
nigrita **691**
nubila 691
nucus 690
olivaceus **692**, 692
orientalis 692
orophilus **692**, 692, 693
orycter 692
pacificus 689
pencanus 692
pervalens 693
polius 692
porcinus 691
psilurus 692
puer **692**, 692
pulcherrimus 700
reinhardti 749
renggeri 692
ruficaudus 692
sanborni 691, **692**
sanctipaulensis **692**
saturatus 693
senilis 692
serrensis **692**, 693
siberiae **692**, 692
simulator **693**, 694
spegazzinii 689, **693**, 693
subfuscus 689, **693**
subterraneus 692
suffusa 691
surdus **693**
sylvanus **693**, 693
tapirapoanus 689
tartareus 693
toba 689, 691, **693**, 694
tolimae 688
torques 692, **693**, 693
trichotis 692
tucumanensis 693
urichi 688, **693**, 694
varius 689, 691, 693, **694**, 694
venezuelensis 693
vinealis 692
xanthopus 692
xanthorhinus 690, 692, **694**, 694
Akodontini 687
akodontius, Oxymycterus **726**
akokomuli, Pipistrellus 221
aksuensis, Dipus 493
alacer, Macaca 266
Sylvilagus 825
alacris, Sundasciurus 451
alactaga, Allactaga 489
Alactagulus 490, 491
aladdin, Pipistrellus 220, 223
alaiana, Capra 406
alaica, Alticola 503
alaicus, Ellobius **512**, 512, 513
alangensis, Rattus 660
alaotrensis, Hapalemur 245
alascanus, Callorhinus 327
alascensis, Clethrionomys 509
Dicrostonyx 511
Lemmus 517
Mustela 321
Myotis 212
Sorex 118
Ursus 338
Vulpes 287
Zapus 499
alaschanicus, Hemiechinus 78
Pipistrellus 223
Rhombomys 559
alashanicus, Cervus 385
Euchoreutes 495
Spermophilus **444**
alaskanus, Sorex **111**, 119
alba, Agouti 783
Cacajao 261
Ectophylla **190**, 190
Geomys 469
Lepus 817
Marmota 432
Meles 314
Mustela 322
Octodon 788
Panthera 298
Phalanger 46
Vulpes 287
albaniensis, Hipposideros 171
Lepus 820
Otomys 682
Pedetes 759
albata, Ochotona 811
albatus, Hemiechinus 78
Myopterus 238
Thomomys 473
albayensis, Phloeomys 640
albertae, Spermophilus 445
albertensis, Kobus 413
Loxodonta 367
alberti, Mylomys 630
albertisii, Dactylopsila 61
Mus 625
Pseudochirops **60**, 60
albescens, Callosciurus **420**, 420
Chaetodipus 483
Eptesicus 203
Ichneumia 306
Ictonyx 319
Leopardus 291
Myotis **208**
Oncifelis 294
Onychomys 719
Reithrodontomys 742
Sturnira 192
albibarbis, Hylobates 275, 276
Sorex 119
albica, Mustela 322
albicans, Delphinapterus 357
Mus 625
Pipistrellus 221
Pithecia **261**, 262
albicauda, Alticola **502**, 502
Brachytarsomys **677**, 677
Ichneumia **306**
Olallamys **790**
Phoca 332
Suncus 103
Sundasciurus 452
Thrinacodus 790
albicaudatus, Mystromys 676, **677**
Otomys 676
Thomomys 473
albicaudus, Herpestes 306
Orycteropus 375
albiceps, Echymipera 41
Ratufa 437
albicinctus, Myotis 212
albicollis, Callithrix 252
Megaerops 143
Neophoca 328
albicornis, Cervus 387
albiculus, Callosciurus 420
albicus, Capreolus 390
Castor 467
Cervus 385
albida, Dasyprocta 781, 782
albidens, Coccymys **585**, 585
albidiventris, Mus 625
albidus, Phalanger 46
Tadarida 240
albifer, Callosciurus 421
albifrons, Amblysomus 74
Arctictis 342
Ateles 257
Callithrix 251
Cebus **259**
Cephalorhynchus 352
Cervus 385
Damaliscus 394
Eulemur 244
Galictis 318
Heliophobius 772
Peromyscus 735
Potamochoerus 378
albigena, Acomys 565
Erignathus 329
Lagenorhynchus 353
Lophocebus **266**
Presbytis 266
Saimiri 260
albigenus, Erythrocebus 265
albigula, Neotoma **710**, 712, 713
Pseudalopex 284
Scotoecus 226
Thyroptera 196
albigularis, Oryzomys **720**, 720, 722, 723
Rattus 651
Thomomys 475
albilabris, Reithrodontomys 742
albimana, Hylobates 275
albimanus, Eulemur 244
albina, Heliosciurus 428
albinasus, Cercopithecus 264
Chiropotes **261**
Pappogeomys 472
albini, Phoca 332
albinucha, Poecilogale **325**
Zorilla 325
albinuchalis, Potamochoerus 378
albinus, Mus 625
Suncus 103
Trachypithecus 274
albior, Hemiechinus 78
albipes, Arborimus **504**, 504, 505
Bassariscus 334
Boselaphus 402
Conilurus **586**, 586
Crocidura 88
Galago 249
Genetta 345
Hapalotis 586
Heterohyrax 373
Meriones 558
Muntiacus 389
Mustela 323
Myomys **630**, 630, 631, 664
Mystromys 676, 677
Praomys 631
Ratufa 437
Sciurus 440
Semnopithecus 273
Sminthopsis 37
Tscherskia 539, 540
albipilis, Lagostrophus 53
albipinnis, Taphozous 160
albirostratus, Stenella 356
albirostris, Cervus **385**
Delphinus 353
Lagenorhynchus **353**, 353
Tayassu 380

albispinus, Proechimys **795**, 797
albiventer, Akodon **688**
Clethrionomys 509
Melomys 615
Microdipodops 480
Monachus 331
Mormopterus 237
Nycteris 162
Nycticeius 218
Nyctimene **143**, 144
Oryzomys 721
Petaurista 463
Rattus 658
Sorex 113
Thyroptera 196
albiventris, Atelerix **76**, 77
Cephalorhynchus 351
Crocidura **94**
Didelphis **16**, 16
Erinaceus 76
Genetta 346
Marmosops 19
Noctilio **176**
Oenomys 637
Tamias 453
albivexilli, Callosciurus 422
albocaudata, Stenocephalemys 606, **664**, 664, 686
albocaudatus, Colobus 270
albocinereus, Pseudomys **644**, 645, 646, 649
albocollaris, Eulemur 244
albofuscus, Scotoecus **226**
Scotophilus 226
albogularis, Arctonyx 313
Cercopithecus 264
Petrogale 56
Tamias 455
Vespertilio 228
alboguttata, Dasyurus 32
alboguttatus, Chalinolobus **199**
albojubatus, Connochaetes 394
albolimbatus, Liomys 482
Pipistrellus 221
albolineatus, Lemniscomys 601
albomaculatum, Phyllops 190
Phyllostoma 190
albomaculatus, Myocastor 806
alboniger, Hylopetes **460**
albonigrescens, Hylobates 275
albonotatus, Gazella 397
Sciurus **443**
Tragelaphus 404
albopunctatus, Dasyurus **31**
alborufescens, Microtus 527
alborufus, Petaurista **462**
alborusus, Petaurista 462
alboscapulatus, Melonycteris 154
albosignatus, Erythrocebus 265
albostriatus, Apodemus 570
albotorquatus, Cercopithecus 264
albovirgatus, Tragelaphus 403
albovittatus, Xerus 458
albulus, Cebus 260
Chaetodipus 483
Hemiechinus 78
albus, Apodemus 573
Arvicola 506
Canis 281
Capreolus 390
Castor 467
Cervus 385
Cricetus 538
Cryptomys 772
Dama 387
Diclidurus **157**, 157
Echinosorex 79
Lepus 821
Microtus 519
Molossus 235
Mus 625
Myotis 210
Neomys 110
Nyctereutes 283
Ondatra 532
Ozotoceros 392
Propithecus 247
Rattus 657, 658
Sciurus 443
Ursus 338
Alcelaphinae **393**
Alcelaphus 389, **393**, 394, 395
bubalis 393
bubastis 393
buselaphus **393**, 395
caama 393
cokii 393
deckeni 393
digglei 393
evalensis 393
heuglini 393
insignis 393
invadens 393
jacksoni 393
keniae 393
kongoni 393
lelwel 393
luzarchei 393
major 393
matschiei 393
mauretanicus 393
modestus 393
nakurae 393
neumanni 393
niediecki 393
noacki 393
obscurus 393
oscari 393
rahatensis 393
ritchiei 393
roosevelti 393
rothschildi 393
sabakiensis 393
schillingsi 393
schulzi 393
selbornei 393
senegalensis 393
swaynei 393
tanae 393
tora 393
tschadensis 393
tunisianus 393
wembaerensis 393
Alces **389**
alces **389**
americanus 389
andersoni 389
angusticephalus 389
antiquorum 389
bedfordiae 389
buturlini 389
cameloides 389
caucasicus 389
columbae 389
coronatus 389
europaeus 389
gigas 389
jubata 389
machlis 389
meridionalis 389
palmatus 389
pfizenmayeri 389
shirasi 389
tymensis 389
uralensis 389
yakutskensis 389
alces, Alces **389**
Canis 281
Cervus 389
Taurotragus 403
alchemillae, Crocidura 85
alcinous, Eothenomys 514
alcorni, Microtus 527
Pappogeomys 471, **472**
Peromyscus 730
alcyone, Rhinolophus **164**
alcythoe, Pipistrellus 221
aldabrensis, Pteropus **146**, 146, 151
aldridgeanus, Naemorhedus 407
aleco, Chionomys 507
alecto, Aethalodes 137
Aethalops **138**
Cyttarops **156**, 156
Emballonura **157**
Molossus 235
Pteropus **146**
aleksandrisi, Crocidura **81**
Aletesciurus 450, 451, 452
alethina, Rhinophylla **186**
alettensis, Myomys 630
alexandrae, Thomomys 473
Ursus 338
alexandri, Chlorocebus 265
Crossarchus **301**
Paraxerus **434**
alexandriae, Macropus 54
alexandrino-rattus, Rattus 658
alexandrinus, Rattus 658
Alexandromys 517, 518, 520, 521, 522, 524, 525, 529
alexis, Notomys **635**, 635
alfari, Microsciurus **433**
Sciurus 433
Sigmodontomys **748**, 748
alfaroi, Oryzomys **720**, 720, 721, 724, 725
alfredi, Cervus **385**
Dipodomys 479
Sigmodon 747
alfurus, Babyrousa 377
algazel, Oryx 413
algeriensis, Vulpes 287
alghazal, Aethomys 568
algidus, Lepus 821
Peromyscus 732
algira, Caracal 288
Sus 379
algirensis, Canis 280
algiricus, Felis 290
Leptailurus 292
Mustela 321
Psammomys 559
algirus, Apodemus 573, 574
Atelerix **76**
Oryctolagus 822
Rhinolophus 165
algoensis, Redunca 414
algonquinensis, Napaeozapus 498
aliba, Dasyprocta 782
aliensis, Ochotona 811, 812
alienus, Histiotus **205**
Thomomys 473
Alionycteris **138**
paucidentata **138**, 138
Allactaga **488**, 488, 489, 490
alactaga 489
altorum 489
annulata 489
aralychensis 488
bactriana 488
balikunica **488**, 488
brachyotis 489
brachyurus 489
brucii 489
bulganensis 489
bullata **488**, 488
caprimulga 488
caucasicus 488
chachlovi 489
chorezmi 489
decumanus 489
dementiewi 489
djetysuensis 489
dzungariae 488
elater **488**, 489, 490
euphratica **488**, 488, 489
firouzi **488**, 489
flavescens 489
fuscus 489

grisescens 489
halticus 489
heptneri 488
hochlovi 489
hotsoni 488, **489**, 489
indica 488
intermedius 489
jaculus 489
kizljaricus 488
laticeps 488
longior 489
macrotis 489
major **489**
media 489
mongolica 489
nataliae 488
nigricans 489
ognevi 489
ruckbeili 489
salicus 489
saliens 489
saltator 489
schmidti 488
semideserta 489
severtzovi **489**, 489
sibirica **489**
spiculum 489
strandi 488
suschkini 489
tetradactyla **489**
turkmeni 488, 489
vexillarius 489
vinogradovi 488, **490**, 490
williamsi 488
zaisanicus 488
Allactagidae 487
Allactaginae **487**, 495
Allactodipus 488, **490**, 490
bobrinskii 488, **490**, 490
allamandi, Galictis 318
allapaticola, Peromyscus 730
allegheniensis, Mustela 323
allenbyi, Gerbillus **549**, 549
alleni, Alticola 504
Bassaricyon **333**, 333
Chelemys 699
Dasymys 589
Galago **249**
Heteromys 482
Hodomys **704**, 704
Hylomyscus **598**, 598, 599
Lepus **814**, 815
Liomys 482
Melomys 619
Mustela 322
Neofiber 531, **532**
Neotoma 704
Orthogeomys 471
Oryzomys 720
Ovis 409
Prionailurus 295
Reithrodontomys 742
Rhinolophus 168
Rhogeessa **225**
Scalopus 127
Sciurus **439**
Sigmodon **747**
Spermophilus 450
Tamias 456
Zapus 499
Allenopithecus **262**, 263
nigroviridis **262**
allex, Baiomys 695
Crocidura **81**
alligatoris, Macropus 54
allisoni, Myomys 631
allmani, Potamogale 73
Allocebus **243**
trichotis **243**
Allochrocebus 263
Allocricetulus **536**, 536
belajevi 537
beljaevi 537
beljawi 537
curtatus **536**, 537
eversmanni **537**, 537
microdon 537
pseudocurtatus 537
Allolagus 814, 815, 819
Allomyidae 417
Allophaiomys 523
allophrys, Peromyscus 735
allophylus, Peromyscus 731
Allosciurus 436
almasyi, Capra 406
almodovari, Ichneumia 306
alnorum, Sorex 115
aloga, Blarina 106
alongensis, Hipposideros 173
alope, Stenella 356
Alopex **279**, 279, 280
arctica 279
argenteus 279
beringensis 279
beringianus 279
caerulea 279
fuliginosus 279
groenlandicus 279
hallensis 279
innuitus 279
kenaiensis 279
lagopus **279**, 279
pribilofensis 279
spitzbergenensis 279
typicus 279
ungava 279
alopex, Vulpes 287
alophus, Hystrix 773
alorensis, Acerodon 137
Alouatta **254**
aequatorialis 255
amazonica 255
arctoidea 255
auratus 255
barbatus 254
beelzebub 254
belzebul **254**
beniensis 255
bicolor 255
bogotensis 255
caquetensis 255
caraya **254**
caucensis 255
chrysurus 255
clamitans 255
coibensis **254**
discolor 254
flavimanus 254
fusca **255**
guariba 255
iheringi 255
inclamax 255
inconsonans 255
insularis 255
juara 255
juruana 255
laniger 255
luctuosa 255
macconnelli 255
matagalpae 255
mexianae 254
mexicana 255
niger 254
nigerrima 254
nigra 254
palliata **255**
pigra **255**
puruensis 255
quichua 255
rubicunda 255
rubiginosa 255
rufimanus 254
sara **255**
seniculus **255**, 255
straminea 255
tapojozensis 254
trabeata 254
ululata 254
ursina 255
ursinus 255
villosa 255
villosus 254
Alouattinae **254**, 256
aloysiisabaudiae, Chaerephon **232**
alpherakii, Vormela 325
alpherakyi, Vulpes 287
alphonsei, Sciurus 439
alpicola, Apodemus **570**, 570
Sylvaemus 570
alpina, Capra 405
Crocidura 81
Marmota 432
Mustela 321
Ochotona **807**, 807, 808, 809, 811
Rupicapra 410
alpini, Procavia 374
alpinus, Apodemus 570, 573
Chionomys 507
Cuon **282**
Glaucomys 460
Lepus 821
Mustela 322, 323
Myotis 213
Peromyscus 732
Scapanus 127
Sciurus 443
Sorex **111**, 118, 119
Tamias **453**
Thomomys 473
Ursus 338
Alsomys 569, 570, 571, 572, 573
alstoni, Clethrionomys 508
Micoureus **20**
Neotomodon **713**, 713
Peromyscus 713
Sciurus **442**
Sigmodon **746**, **747**
alsus, Auliscomys 695
altae, Crocidura 87
altaica, Alticola 503
Capra 406
Mustela **321**, 321, 324
Ovis 408
Panthera 298
Procapra 399
Talpa **128**, 128, 129
altaicus, Canis 281
Lepus 821
Meles 314
Microtus 527
Moschus 384
Sciurus 443
Sorex 113
Spermophilus 450
Tamias 455
altaina, Ochotona 808
altamirae, Lepus 815
altantica, Monachus 331
altarium, Myotis **208**
alter, Myotis 212
Alterodon 805
alterus, Akodon 693
Alticola **502**, 502, 503, 504, 507, 513, 515, 529
acmaeus 502
acrophilus 504
alaica 503
albicauda **502**, 502
alleni 504
altaica 503
argentatus **502**, 502, 503
argurus 503
baicalensis 504
barakshin **503**
bhatnagari 504
blanfordi 503
cautus 503
cricetulus 504
depressus 504
desertorum 504
fetisovi 503
gracilis 503
imitator 503
khubsugulensis 504
kosogol 504
lahulis 503
lama 504
lemminus **503**, 503
leucura 503
longicauda 503
longicaudata 503

macrotis 502, **503**, 504
montosa **503**, 503
nanschanicus 504
olchonensis 504
parvidens 503
phasma 503
rosanvi 503
roylei 502, **503**, 503, 504
rufocanus 503
saurica 503
semicanus **504**
severtzovi 503
shnitnikovi 503
stoliczkanus 503, **504**, 504
stracheyi **504**, 504
strelzowi **504**
subluteus 503
tuvinicus **504**
vicina 503
villosa 503
vinogradovi 503
worthingtoni 503
alticola, Aethomys 567
Cricetulus **537**, 537, 538
Cryptotis 108
Glossophaga 184
Graphiurus 764
Lorentzimys 607
Maxomys **612**, 612, 613
Microsciurus 433
Microtus 523
Neotoma 711
Perognathus **484**
Sigmodon 747
Sorex 111
Vulpes 287
Alticoli 501
alticolus, Reithrodontomys 741
Rhinolophus 169
Thomomys 473
alticraniatus, Myotis 215
altifrons, Myotis 216
Sylvicapra 412
altifrontalis, Mustela 322
Ursus 338
altilaneus, Peromyscus 731
Altililemur 243
altilis, Scotophilus 227
altinsularis, Callosciurus 421
altipetens, Myotis 212
altiplanensis, Spermophilus 449
altissimus, Microryzomys **708**, 708
altitudinis, Cynopterus 139
altitudinus, Sundasciurus 452
altivallis, Sylvicapra 412
Thomomys 473
altivolans, Nyctalus 217
altorum, Akodon 691
Allactaga 489
Lagurus 515
Thomasomys 749
aluco, Pteromys 464
alurae, Kobus 414
alutacea, Notomys 636
alvarezi, Orthogeomys 471
alvenslebeni, Scotophilus 227
Alviceola 505
amakensis, Microtus 527
amalae, Aethomys 568
Crocidura 87
amankaragai, Stylodipus 495
amargosae, Thomomys 473
amasari, Sorex 121
amatus, Cryptomys 772
amazonica, Alouatta 255
Conepatus 316
Leopardus 292
amazonicus, Eumops 234
Holochilus 705
Nectomys 709
amazonius, Trichechus 365
ambersoni, Rattus 661
ambigua, Mustela 324
Spilogale 317
ambiguus, Chaetodipus 483
Dipodomys 479
Microdipodops 480
Peromyscus 729, 732
amblodon, Lagenorhynchus 353
Amblonyx **309**, 309, 310
cinereus **309**
concolor 310
indigitatus 310
leptonyx 310
nirnai 310
sikimensis 310
swinhoei 310
Amblotis 45
amblyceps, Ursus 338
Amblycoptini 105
amblyodon, Pseudalopex 284
amblyonyx, Dactylomys 790
Kannabateomys **790**
amblyotis, Tonatia 181
Amblyrhiza 804, **805**
inundata **805**, 805
latidens 805
quadridens 805
Amblyrhizinae 804
amblyrrhynchus, Oligoryzomys 718
Amblysomus **74**, **74**, 75
albifrons 74
corriae 74
devilliersi 74
drakensbergensis 74
garneri 74
gunningi **74**
hottentotus **74**
iris **74**
julianae **74**
littoralis 74
longiceps 74
marleyi 74
natalensis 74
orangiensis 74
pondoliae 74
septentrionalis 74
amboellensis, Kobus 414
amboiensis, Hipposideros 171
Myotis 213
amboinensis, Rattus 660
Sus 378
ambonensis, Phalanger 46
ambrosianus, Cervus 386
ambrosius, Meriones 557
amecensis, Pappogeomys 472
ameghiniana, Lama 381
ameghinoi, Herpailurus 291
ameliae, Sciurus 443
amer, Genetta 346
Ameranthropoides 256
americana, Antilocapra **393**
Antilope 392
Arvicola 506
Glaucomys 460
Lagostomus 778
Lontra 311
Martes **319**, 320, 321
Mazama **390**
Mephitis 317
Monodelphis **21**, 21
Ondatra 532
Phocoena 358
americanus, Alces 389
Bison 400
Lepus **814**
Moschus 390
Noctilio 176
Oreamnos **407**, 407
Tamias 456
Tapirus 371
Taxidea 325
Trichechus 365
Ursus **338**, 338, 340
Zapus 499
Ameridelphia **15**
Ametrida **187**
centurio **187**, 187
minor 187
amictus, Callicebus 259
amicus, Phyllotis **737**
amir, Hemiechinus 78
Ammelaphus 403
ammobates, Peromyscus 735
Ammodillini 546
Ammodillus **546**
imbellis **546**
Ammodorcas **395**
clarkei **395**
ammodytes, Perognathus 485
Peromyscus 732
Ammomys 517, 619
Ammon 408
ammon, Ovis **408**, 408
ammonoides, Ovis 408
ammophilus, Chaetodipus 483
Geomys 469
Microdipodops 480
Spermophilus 449
Sylvilagus 825
Ammospermophilus **419**, 419
amplus 420
canfieldiae 420
cinamomeus 420
escalante 420
extimus 420
harrisii **419**, 419
insularis **419**, 419
interpres **419**, 420
kinoensis 419
leucurus **420**
nelsoni **420**
notom 420
peninsulae 420
pennipes 420
saxicolus 419
tersus 420
vinnulus 420
Ammotragus **404**, 404, 405
angusi 404
blainei 404
fassini 404
lervia **404**
ornata 404
sahariensis 404
tragelaphus 404
amoenus, Akodon 696
Bolomys **696**
Cephalophus 411
Distoechurus 62
Gerbillus **549**, 553, 555
Hylopetes 461
Neacomys 709
Perognathus 486
Reithrodontomys 741
Sorex 122
Tamias **453**
Tragulus 383
amoles, Reithrodontomys 741
Sigmodon 747
amori, Eliomys 767
Amorphochilus **195**
osgoodi 195
schnablii **195**, 195
amosus, Microtus 525
amotus, Menetes 433
Myotis 216
amoyensis, Panthera 298
Amperta 31
amphibius, Arvicola 505, 506
Hippopotamus **381**, 381
Mus 505
Neomys 110
Sorex 110
amphichoricus, Proechimys **795**, 798
Amphidyromys 769
Amphisorex 111
amplexicaudata, Carollia 186
Glossophaga 184
amplexicaudatus, Eumops 233
Rousettus **152**
amplexicaudus, Molossus 235
amplus, Ammospermophilus 420

Artibeus **187**
Notomys **636**
Perognathus **485**
Peromyscus 730
ampullata, Balaena 361
ampullatus, Hyperoodon **361**
amurensis, Clethrionomys 509
Erinaceus **77**, 77
Lemmus **516**, 517
Lutra 312
Meles 314
Mus 625
Myotis 209
Nyctereutes 283
Panthera 298
amygdali, Pteromys 464
ana, Chiropodomys 583
anacapae, Peromyscus 732
anadyrensis, Pteromys 464
Sciurus 443
Vulpes 287
Anahyster 310
anak, Uromys **671**, 672
anakuma, Meles 314
analogus, Baiomys 695
Thomomys 473
anambae, Maxomys 614
Ratufa 437
Tupaia 132
anambensis, Callosciurus 422
Emballonura 158
Tragulus 383
Anamygdon 207
solomonis 207
anastasae, Peromyscus 730
Scalopus 127
anastasiae, Prionailurus 295
Anathana **131**
ellioti **131**
pallida 131
wroughtoni 131
anatinus, Ornithorhynchus **13**
Platypus 13
anatolica, Ovis 408
Vulpes 287
anatolicus, Eptesicus 200
Nannospalax 754
anceps, Pteralopex **145**, 146
anchietae, Cephalophus 411
Crocidura 92
Epomophorus 145
Oenomys 637
Otomys **680**, 680
Tadarida 240
anchietai, Pipistrellus 204, **219**
Plerotes **145**
anchises, Hylopetes 461
Semnopithecus 273
Anchotomys 680
ancilla, Myotis 209
Promops 240
Andalgalomys **694**, 703
dorbignyi 694
olrogi **694**, 694
pearsoni **694**
andamanensis, Crocidura **81**
Cynopterus 139
Macaca 267
Rattus 660
Rhinolophus 163
Sus 379
anderseni, Artibeus **187**
Dobsonia 140
Rhinolophus **164**
andersoni, Alces 389
Cricetulus 538
Eptesicus 203
Gerbillus **549**, 549
Herpestes 305
Hipposideros 172
Lepus 815
Marmosa **18**
Microtus 525
Myotis 210
Niviventer 589, 632, **633**, 634
Phaulomys **532**
Philander **22**
Rhinolophus 164
Suncus 103
Sus 379
Synaptomys 534
Thomomys 475
Uropsilus **130**, 130
Ursus 338
anderssoni, Saccostomus 541
Vespertilio 228
andicola, Tapirus 371
andicus, Hippocamelus 390
andina, Agouti 783
Didelphis 16
Galictis 318
Leopardus 292
Pseudalopex 284
Andinomys **694**
edax **694**, 694
lineicaudatus 694
andinus, Akodon 692
Choloepus 64
Chroeomys **700**, 700
Eptesicus 201
Oligoryzomys **717**, 717
Sylvilagus 825
andium, Phyllotis **737**
andrafiamensis, Lepilemur 246
andreanus, Cervus 386
andreinii, Rhinolophus 164
andrewsi, Bunomys **581**, 581
Stylodipus **494**, 494, 495
andrewsii, Callosciurus 422
anerythrus, Funisciurus **427**
anetianus, Pteropus **146**
angammensis, Loxodonta 367
angasii, Tragelaphus **403**
Vombatus 46
angdawai, Ochotona 812
angelcabrerai, Mesocapromys **801**, 801
angelensis, Osgoodomys 726
Peromyscus 734
angeli, Sturnira 192
angelus, Dryomys 766
Taterillus 562, 563
anglicus, Muscardinus 769
angolae, Aonyx 310
Atelerix 77
Crocidura 97
Mastomys 610
Pedetes 759
Praomys 610
Rhabdomys 662
Saccostomus 541
Tatera 561
angolensis, Antidorcas 395
Atelerix 77
Colobus **269**, 270
Dendromus 543
Diceros 372
Epomophorus **141**
Eptesicus 201
Genetta **345**
Giraffa 383
Grammomys 593
Graphiurus 765
Herpestes 305
Hipposideros 171
Laephotis **206**
Lepus 822
Loxodonta 367
Mastomys **610**, 610, 611
Mops 236
Nycteris 162
Orycteropus 375
Rhinolophus 166
Rousettus **152**
Steatomys 545
Sylvisorex 105
angonicus, Heliophobius 772
angoniensis, Otomys **680**, 680, 681
angorensis, Felis 290
angouya, Oryzomys 721
anguinae, Mustela 321
Anguistodontus 491
angularis, Microtus 519
Oxymycterus **726**
Thomomys 473
angulatus, Cynopterus 139
Pecari 380
Pipistrellus 224
angulensis, Cerdocyon 282
angur, Rhinolophus 164
angusi, Ammotragus 404
angusta, Blarina 106
angustapalata, Neotoma **710**
angustiae, Tragulus 382
angusticephalus, Alces 389
angusticeps, Blarina 106
Eptesicus 202
Kobus 413
Microtus 523
Neotoma 710
Oryzomys 724
Pappogeomys 472
Phyllostomus 180
Ratufa 437
angustidens, Lepus 815
Mustela 321
Thomomys 473
angustifolius, Rhinolophus 164
angustifrons, Lutra 312
Microtus 518
Mustela 324
Ochotona 812
Spilogale 317
angustimanus, Pan 277
angustiramus, Atherurus 773
angustirostris, Chaetodipus 484
Mirounga **330**
Pappogeomys 473
Peromyscus 730
Rangifer 392
angustivittis, Dactylopsila 61
angustus, Akodon 691
Chelemys 691, 699
Microtus 521, 523
Peromyscus 732
anguya, Oryzomys 721
anhuiensis, Moschus 383
anhydra, Bettongia 49
anikini, Microtus 527
animosus, Paraxerus 435
Tamias 455
Anisomyini 564
Anisomys **569**, 569
imitator **569**, 569
Anisonyx 444
rufa 417
anitae, Thomomys 473
anjouanensis, Myotis 211
anjuanensis, Eulemur 244
ankaranensis, Lepilemur 246
ankoberensis, Myomys 630
ankoliae, Tachyoryctes **686**, 686
anmamiticus, Helarctos 337
annae, Tragulus 383
annalium, Sciurus 443
annamensis, Herpestes 306
Muntiacus 389
Mus 623
Petaurista 463
Tupaia 131
annamiticus, Axis 385
Bos 401
Rhinoceros 372
annandalei, Funambulus 427
Rattus **650**, 650, 655, 656
annectans, Myotis **208**
annectens, Ardops 187
Equus 369
Kobus 413
Liomys 482
Lontra 311
Naemorhedus 407
Neotoma 711
Ochotona 808
Oryx 413

Paguma 343
Pteropus **147**
Rhinolophus 166
Spermophilus 449
Tachyoryctes **686**, 686
annellata, Crocidura 87
Phoca 332
annellatus, Callosciurus 422
Cebus 260
annexus, Orthogeomys 471
Sorex 113, 120
annularis, Heliosciurus 428
annulata, Allactaga 489
Galerella 303
Helogale 304
Nasua 334
annulatus, Galerella 302
Heliosciurus 428
Procyon 335
Sciurus 428
Spermophilus **444**
Ursus 338
annulicauda, Onychogalea 55
annulipes, Kobus 414
annulus, Dipodomys 479
Anoa 402
depressicornis 402
anoa, Bubalus 402
Anodon 361
Anoema 778, 813
oeningensis 813
anolaimae, Cavia 779
Anomalocera 390
anomalocera, Hippocamelus 390
Anomalurella 757
Anomaluridae **757**
Anomalurinae **757**
Anomalurodon 757
Anomalurops 757
Anomalurus **757**
argenteus 757
auzembergeri 757
batesi 757
beecrofti **757**, 757
beldeni 757
chapini 757
chrysophaenus 757
cinereus 757
citrinus 757
derbianus **757**
erythronotus 757
fortior 757
fraseri 757
fulgens 757
griselda 757
hervoi 757
imperator 757
jacksoni 757
jordani 757
laniger 757
laticeps 757
neavei 757
nigrensis 757
orientalis 757
pelii **757**
perustus 757
pusillus **757**
schoutedeni 757
squamicaudus 757
anomalus, Cryptomys 772
Heteromys **481**
Mus 481
Neomys **110**
Rhinolophus 166
Sciurus **439**
anomurus, Galagoides 249
Anonymomys **569**
mindorensis **569**, 569
Anotis 754
Anotomys **694**, 699
leander **694**, 694
trichotis 694
Anotus 106
Anoura **183**
aequatoris 183
antricola 183
apolinari 183
brevirostrum 183
caudifera **183**
cultrata **183**
ecaudata 183
geoffroyi **183**, 183
lasiopyga 183
latidens **183**
peruana 183
werckleae 183
wiedii 183
Anourosorex **105**
assamensis 106
capito 106
capnias 106
schmidi 106
squamipes 105, **106**
yamashinai 106
Anourosoricini 105, 106
anselli, Hippotragus 413
Hylomyscus 599
ansellorum, Crocidura **81**
ansonii, Mirounga 330
ansonina, Mirounga 330
ansorgei, Arvicanthis 578
Chaerephon **232**
Cricetomys 541
Crocidura 91
Crossarchus **301**
Cryptomys 772
Dendromus 543
Graphiurus 764
Heterocephalus 772
Lepus 822
Lophuromys 606
Miopithecus 269
anta, Tapirus 371
antarctica, Arctocephalus 327
Eubalaena 349
antarcticus, Canis 283
Dusicyon 283
Phoca 332
Antechinomys 35, 36
Antechinus **29**, 33, 34
adusta 30
adustus 30
affinis 30
assimilis 31
bellus **29**
burrelli 30
centralis 30
concinnus 30
flavipes **30**, 30
godmani **30**
habbema 30
hageni 31
leo **30**
leucogaster 30
maritima 30
mayeri 30
melanurus **30**
mimetes 31
minimus **30**
misim 30
modesta 30
moorei 31
naso **30**
niger 31
rolandensis 30
rosamondae 31
rubeculus 30
rufogaster 30
stuartii 29, **30**, 30
swainsonii **30**
tafa 30
unicolor 30
wilhelmina **31**
anteflexa, Antilocapra 393
Anteliomys 513, 514, 532
anthonyi, Chaetodipus 483
Crocidura 94
Liomys 482
Neotoma **710**
Peromyscus 730
Pipistrellus **219**
Puma 296
Reithrodontomys 741
Scapanus 127
Sciurus 441
Stenoderma 192
Taterillus 562
Anthopithecus 251
Anthops **169**
ornatus **169**, 169
Anthorhina 179, 180
anthracinus, Colobus 270
Anthropopithecus 276
anthus, Canis 280
anticola, Lyncodon 319
anticostiensis, Peromyscus 732
Antidorcas **395**
angolensis 395
centralis 395
dorsata 395
euchore 395
hofmeyri 395
marsupialis **395**
pygargus 395
saccata 395
saliens 395
saltans 395
Antifer 389
antigua, Didelphis 16
antillarum, Glossophaga 184
Monachus 331
Oryzomys 721
Trichechus 365
antillensis, Dasyprocta 782
Heteropsomys **799**
Antillogale 69
marcanoi 69
antillularum, Tadarida 240
Antilocapra **392**
americana **393**
anteflexa 393
mexicana 393
oregona 393
peninsularis 393
sonoriensis 393
Antilocapridae **392**
Antilope **395**, 413
americana 392
bezoartica 395
bilineata 395
bubalis 393
buselaphus 393
campestris 399
centralis 395
cervicapra **395**
chickara 403
ellipsiprymnus 413
euchore 395
gnou 393
goral 406
hagenbecki 395
hodgsonii 399
lervia 404
leucophaea 412
lichtensteinii 394
marsupialis 395
melampus 393
mergens 412
oreas 403
oreotragus 398
oryx 403, 413
ourebi 399
pygargus 394
rajputanae 395
redunca 414
rupicapra 395
saltiana 398
scoparia 399
scripta 403
silvicultrix 410
strepsiceros 395
sylvatica 403
tragocamelus 401
Antilopinae **395**
antilopinus, Macropus **53**
antimena, Hypogeomys 678, **679**
antineae, Heterohyrax **373**
Procavia 374
antinorii, Panthera 298
Rhinolophus 166
Sorex 111

antioquiae, Caluromys 15
Metachirus 20
antipae, Crocidura 96
antipodarum, Caperea 351
Eubalaena 349
antiqua, Panthera 298
antiquarius, Dipodomys 477
antiquorum, Alces 389
Balaenoptera 350
Equus 369
Felis 290
Giraffa 383
Hyaena 308
Panthera 298
Papio 269
antiquus, Capromys 800
Mustela 324
Pseudalopex 284
Spalax 755
antisensis, Hippocamelus **390**
antoniae, Oligoryzomys 717
Paraxerus 434
antonii, Ctenomys 787
Odocoileus 391
antricola, Anoura 183
Hipposideros 171
Thrichomys 798
Antrozous **198**, 198
bunkeri 198
dubiaquercus **198**
koopmani 198
meyeri 198
minor 198
pacificus 198
packardi 198
pallidus **198**
antucus, Maxomys 614
anubis, Papio 269
Vulpes 287
anulipes, Tapirus 371
anurus, Epomophorus 141
anzeliusi, Procolobus 272
Aodon 362
aokii, Micromys 620
Aonyx 309, **310**, 310, 311
angolae 310
calaboricus 310
capensis **310**, 310
congicus **310**, 310
coombsi 310
delalandi 310
gambianus 310
helios 310
hindei 310
inunguis 310
lenoiri 310
meneleki 310
microdon 310
philippsi 310
poensis 310
aorensis, Pteropus 146
aoris, Callosciurus 422
Crocidura 89
Cynocephalus 135
Maxomys 614
Aotidae 254
Aotinae **255**
Aotus **255**
aversus 256
azarai **255**, 256
bidentatus 255
bipunctatus 256
boliviensis 255
brumbacki **255**
commersoni 256
duruculi 256
felina 256
felinus 256
griseimembra 256
gularis 256
hershkovitzi **256**
hirsutus 256
humboldtii 256
infulatus **256**
lanius 256
lemurinus **256**
miconax **256**
microdon 256
miriquouina 255
nancymaae **256**
nigriceps **256**
noctivaga 256
oseryi 256
pervigilis 256
roberti 256
rufipes 255
rufus 256
spixii 255
stenorrhina 255
senex 256
trivirgatus **256**
villosus 256
vociferans **256**
zonalis 256
apache, Herpailurus 291
Myotis 208
Perognathus 485
Sciurus 442
Taxidea 325
Thomomys 473
Ursus 338
apalachicolae, Neofiber 532
apanbanga, Cephalophus 411
apar, Tolypeutes 67
Apara 67
apatelius, Oryzomys 721
apedia, Saimiri 260
apella, Cebus **259**
Aper 377
aethiopicus 377
aper, Sus 379
aperea, Cavia **778**, 779
apereoides, Thrichomys **798**
aperoides, Nelomys 798
apex, Rattus 661
Aphaetreus 803
aphanasievi, Marmota 431
aphorodemus, Microtus 527
aphrastus, Sigmodontomys **748**
Thomomys 473
Aphrontis 439
aphylla, Phyllonycteris **183**
apicalis, Hapalotis 604
Lenoxus **707**
Leporillus **604**
Nectomys 709
Neotoma 711
Oxymycterus 706
Parantechinus **33**, 33
Phascogale 33
Potorous 50
Wallabia 57
apicalus, Euoticus 248
apiculatus, Cebus 260
Galerella 303
Hipposideros 175
Saguinus 253
apicus, Rattus 652
Aplacerus 391
Aplodontia **417**
californica 417
chryseola 417
columbiana 417
grisea **417**
humboldtiana 417
leporinus 417
major 417
nigra 417
olympica 417
pacifica 417
phaea 417
rainieri 417
rufa **417**
aplodonticus, Cervus 386
Aplodontidae **417**
aplodontus, Cervus 386
apodemoides, Pseudomys **645**, 645
Apodemus **569**, 569, 570, 571, 572, 573, 574, 595, 611, 670
agrarius **570**, 571
ainu 573
albostriatus 570
albus 573
algirus 573, 574
alpicola **570**, 570
alpinus 570, 573
argenteus **570**, 570, 571
argyropuli 572, 573
argyropuloi 572, 573
arianus **570**, 570, 571, 573
baessleri 574
balchaschensis 574
bergensis 573
bodius 571
brauneri 571
brevicauda 572
bushengensis 574
butei 573
callipides 573
candidus 573
caucasicus 570
celatus 570
cellarius 571
celticus 573
chamaeropsis 573, 574
charkovensis 574
chejuensis 570
chevrieri **571**
chorassanicus 571
ciscaucasicus 574
clanceyi 573
coreae 570
creticus 573
cumbrae 573
dichruroides 573
dichrurus 573
dietzi 571
dorsalis 573
draco **571**, 571, 572, 573
eivissensis 573
epimelas 572
erythronotus 570
euxinus 572
falzfeini 571
fennicus 571
fiolagan 573
flavicollis 570, **571**, 571, 572, 573, 574
flaviventris 573
flavobrunneus 573
fridariensis 573
frumentariae 573
fulvipectus **571**, 571, 572, 573
geisha 570
geminae 571
ghia 573
giliacus 572
gloveri 570
grandiculus 573
granti 573
griseus 573
gurkha **571**
hamiltoni 573
harti 570
hayi 573
hebridensis 573
henrici 570
hermani 573
hermonensis **571**
hessei 573
hirtensis 573
hokkaidi 570
hyrcanicus **572**, 572
iconicus 573
ifranensis 573
ilex 571
ilvanus 573
insperatus 573
insulaemus 570
intermedius 573
isabellinus 573
islandicus 573
istrianus 570
kahmanni 570
karelicus 570
kilikiae 573
krkensis 572, 573, 574
larus 573
latronum **572**, 572
leucocephalus 573

maclean 573
maculatus 570
major 572
majusculus 572
mantchuricus 570
maximus 573
microps 574
microtis 574
milleri 573
miyakensis 573
mosquensis 574
mystacinus **572**
nankiangensis 574
navigator 573
nesiticus 573
niger 573
nigritalus 572
nikolskii 570
ningpoensis 570
ognevi 570
orestes 571
pallescens 570
pallidior 570
pallidus 574
pallipes 574
parvus 572, 573
pecchioli 573
peninsulae **572**, 572, 573, 574
pentax 574
persicus 572
planicola 572
pohlei 572
ponticus **572**, 572, 573
praetor 572
princeps 571
qinghaiensis 572
rhodius 572
rubens 570
rufescens 573
rufulus 572
rusiges **573**, 573
sadoensis 573
sagax 570
samaricus 572
samariensis 572
saturatus 571
saxatilis 572
semotus **573**
septentrionalis 570
smyrnensis 572
sowerbyi 572
spadix 573
speciosus 572, **573**, 573
stankovici 573
sublimis 574
sylvaticus 570, 571, 572, **573**, 573, 574
tanei 570
tauricus 573
thuleo 573
tianschanicus 570
tirae 573
tokmak 574
tscherga 572, 574
tural 573
tusimaensis 573
uralensis 571, 572, 573, **574**, 574
varius 573
vohlynensis 574
volgensis 570
wardi 571, 573, **574**, 574
wintoni 571
witherbyi 570
yakui 570
apoensis, Cervus 386
Macaca 266
Tarsomys **667**, 667
apolinari, Anoura 183
Olallamys 791
apollinaris, Cerdocyon 282
Sylvilagus 825
Apomys **574**, **664**
abrae **575**, 575
bardus 575
datae **575**, 575
hollisteri 575
hylocetes 575
hylocoetes 574, **575**
insignis **575**
littoralis **575**
major 575
microdon **575**
musculus **575**
petraeus 575
sacobianus **575**
Aporodon 740, 741, 742, 743
apricus, Scotinomys 746
Spermophilus 449
Aproteles **138**
bulmerae **138**, 138
apta, Akodon 691
apus, Pipistrellus 221
aquaticus, Arvicola 506
Hyemoschus **382**
Moschus 382
Nectomys 709
Neomys 110
Oryzomys 721
Scalopus **127**, 127
Scapteromys 745, 746
Sorex 127
Sylvilagus **824**, 827
aquilo, Callosciurus 421
Lepus 816
Meriones 556
Notomys **636**
Nycticeius 217
Steatomys 546
aquilonius, Dipodomys 479
Lepus 817
Ondatra 532
aquilus, Cerdocyon 282
Dipodomys 477
Eothenomys 514
Gerbillus **549**
Lophuromys 605
Platyrrhinus 191
aquitanius, Chionomys 507
arabica, Capra 406
Crocidura **81**, 81, 85
Gazella **395**
Ovis 409
Vulpes 287
arabicus, Ctenodactylus 761
Lepus **816**, **817**, **821**
Pipistrellus **220**
Rousettus 152
Vulpes 287
arabium, Gerbillus 553
Rhinopoma 155
arabs, Canis 281
arabupu, Proechimys 795
arachnoides, Ateles 257
Brachyteles **257**
Araeosciurus 439
araeum, Plagiodontia **804**, 804
aralensis, Lepus 821
Pygeretmus 490
aralychensis, Allactaga 488
Crocidura 86
aramia, Rattus 659
araneoides, Sorex 113
araneus, Sorex 108, **111**, 111, 112, 114, 115, 116, 120
aranoides, Sorex 119
aratathomasi, Sturnira **192**
aratayae, Rhipidomys 744
araucanus, Geoxus 702
Lama 381
Oligoryzomys 718
Puma 296
araxenus, Myotis 213, 215
arborarius, Aethomys 568
Grammomys 593
arborensis, Potos 333
arboreus, Dendrohyrax **373**, 373
Hyrax 373
Peromyscus 728, 732
Rattus 658
Ursus 340
arboricola, Heterohyrax 373
Rattus 658
arboricolus, Mysateles 803
Arborimus **504**, 504, 505, 533
albipes **504**, 504, 505
intermedius 505
longicaudus **505**, 505
pomo **505**, 505
silvicola 505
Arbusticola 517
arbustotundrarum, Bison 400
arcalus, Meles 314
arceliae, Spermophilus 444
Archaeomys 813
Archboldomys **575**, 575, 588
luzonensis **575**, 575
archeri, Pseudochirops **60**
Sminthopsis **35**
archidonae, Myoprocta 782
archipelagus, Cynopterus 139
arcium, Melomys 617
Arctibeus, falcatus 190
arctica, Alopex 279
arcticeps, Onychomys 719
Arctictis **342**
albifrons 342
ater 342
binturong **342**, 342
gairdneri 342
niasensis 342
pageli 342
penicillatus 342
whitei 342
arcticus, Gulo 318
Lepus **815**, 819, 821
Microtus 527
Moschus 384
Mustela 321
Odobenus 326
Rangifer 392
Sciurus 443
Sorex 111, **112**, 112, 113, 115, 116, 120, 121
Arctocebus **247**
aureus **247**
calabarensis 247, **248**
ruficeps 247
Arctocephalinae 326
Arctocephalus **326**, 326, 327, 328, 329
antarctica 327
argentata 327
aurita 327
australis **326**, 326
brachydactyla 326
compressa 327
delalandii 327
doriferus 327
elegans 327
falclandicus 326
falklandica 326
forsteri **326**
galapagoensis **326**, 326
gazella **326**
gracilis 326
grayi 326
hauvillii 326
hookeri 328
latirostris 326
leucostoma 326
lobatus 328
lupina 326
monteriensis 327
nigrescens 326
nivosus 327
parva 327
peronii 327
philippi 326
philippii **327**, 327
porcina 326
pusillus **327**, 329
schisthyperoes 327
shawii 326
tasmanicus 327
townsendi **327**
tropicalis **327**
ursinus 326, 327
Arctogale 342
erminea 342

Arctogalidia **342**
bancana 343
bicolor 343
depressa 343
fusca 343
inornata 343
leucotis 343
macra 343
major 343
melli 343
millsi 343
mima 343
minor 343
simplex 343
stigmaticus 343
sumatrana 343
tingia 343
trilineata 343
trilineatus 343
trivirgata **343**
Arctogalidinae 342
arctoidea, Alouatta 255
arctoides, Macaca **266**
arctoideus, Eptesicus 201
Arctomys 424, 430, 444
leptodactylus 443
ludoviciana 424
arctomys, Marmota 431
Arctonyx **313**, 313
albogularis 313
collaris **313**, 313
consul 313
dictator 313
isonyx 313
leucolaemus 313
obscurus 313
taraiyensis 313
taxoides 313
Arctophoca 326, 327
Arctopithecus 63, 251
arctos, Canis 281
Gulo 318
Ursus **338**, 338, 339
arcturus, Microtus 518
arcuatus, Rhinolophus **164**
Taeromys **667**, 667
Arcuomys 666, 667
arcus, Perognathus 485
ardens, Lemniscomys 602
ardicola, Thomomys 473
Ardops **187**, 192
annectens 187
koopmani 187
luciae 187
montserratensis 187
nichollsi **187**
arduus, Gerbillus 550
arenacea, Neotoma 712
arenaceous, Jaculus 493
arenae, Dipodomys 478
arenalis, Oligoryzomys **717**
arenaria, Perameles 40
arenarius, Chaetodipus **483**
Chlorocebus 265
Cricetulus 538
Cryptomys 772
Dendromus 543
Geomys **469**, 469
Ictonyx 319
Lepus 816
Meles 314
Miniopterus 231
Perognathus 483
Peromyscus 730, 735
Phyllotis 738
Spalax 753, **754**, 754
Taterillus **562**, 562
arenarum, Canis 281
arendsi, Chlorotalpa **74**
Chrysochloris 75
arendsis, Callosciurus 422
arenicola, Akodon 689
Microtus 527
Onychomys **719**, 719
Perognathus 485
Spermophilus 449, 450
arenicolor, Brachiones 547
areniticola, Micoureus 20
arenivagus, Dipodomys 479
arenosus, Tscherskia 540
arens, Spermophilus 449
arequipae, Akodon 693
Cavia 779
Conepatus 316
arequipe, Lagidium 778
arescens, Myotis 209
Proechimys 795
arethusa, Crocidura 85
arfakensis, Dactylopsila 61
arfakianus, Melomys 617
arfakiensis, Melomys 619
Stenomys 665
Argali 408
argali, Ovis 408
arge, Eptesicus 201
Nycteris **161**, 162
argens, Potamogale 73
argentata, Arctocephalus 327
Callithrix **251**, 252
Kerivoula **196**
Mallomys 608
Ochotona 807
argentatus, Alticola **502**, 502, 503
Chalinolobus **199**
Crocidura 83
Macropus 54
Myotis 208
Oryzomys 724
Otolemur 250
Peromyscus 732
Pteropus 137, **146**
Scalopus 127
Trachypithecus 274
argenteocinereus, Heliophobius **772**, 772
argenteogrisea, Lepus 817
argentescens, Funambulus 426
argenteus, Alopex 279
Anomalurus 757
Apodemus **570**, 570, 571
Canis 285
Graphiurus 764
Mus 570
Myoxus 770
Neomys 110
Pseudochirops 60
Sciurus 443
Trachypithecus 274
Ursus 338
argentimembris, Macaca 266
argentina, Mazama 391
argentinius, Sciurus 441
argentinus, Ctenomys **784**
Eptesicus 201
Lasiurus 207
argentiventer, Rattus 649, **650**, 650, 652, 654, 662
argentoratensis, Arvicola 506
argillaceus, Acomys 567
Argocetus 357
argunensis, Canis 281
argurus, Alticola 503
Calomys 698
Mus 674
Zyzomys **674**
argusensis, Dipodomys 479
Thomomys 473
argynnis, Casinycteris **138**, 138
argyraceus, Rattus 660
argyrochaetes, Naemorhedus 407
argyrodytes, Chironectes 16
argyropoli, Microtus 518
argyropuli, Apodemus 572, 573
Microtus 518
argyropuloi, Apodemus 572, 573
Microtus 518
Spermophilus **447**
argyropus, Arvicola 506
Hydropotes 388
ariadne, Crocidura 96
arianus, Apodemus **570**, 570, 571, 573
aricana, Nasua 334
aridicola, Neotoma 712
aridula, Crocidura 97
aridulus, Grammomys **592**
Peromyscus 732
ariel, Petaurus 61
Pipistrellus **220**
Plecotus 224
Pteropus 147
Arielulus 219, 220, 223
Aries 405
aries, Ovis **408**, 408, 409
arietinus, Cervus 386
Sus 379
arimalius, Meriones **555**, 556
arispa, Crocidura 95, 98
aristippe, Pipistrellus 223
aristotelis, Cervus 387
Ariteus **187**, 192
achradophilus 187
flavescens **187**
arizonae, Dipodomys 477
Neotoma 711
Peromyscus 732
Sigmodon **747**, 747
Sorex **112**
Spermophilus 449
Sylvilagus 824
Ursus 338
arizonensis, Bassariscus 334
Clethrionomys 508
Cynomys **424**
Microtus 525
Mustela 322
Panthera 298
Perognathus 485
Reithrodontomys 741
Sciurus **440**
Spermophilus 447
Tamias **454**
Vulpes 287
Arizostus 64
arkal, Ovis 409
armalis, Callosciurus 423
armandii, Myospalax 675
armandvillei, Mus 637
Papagomys **637**
armata, Atherurus 773
Makalata **793**
armatus, Chaetodipus 483
Echimys 791
Makalata 793
Nelomys 793
Spermophilus **444**
Tenrec 73
armeniacus, Nannospalax 754
armeniana, Ovis 408
armenica, Crocidura **81**, 81, 93
Sicista **496**, 496, 497
armenius, Arvicola 506
Capreolus 390
armiger, Hipposideros **171**
armillatus, Leopardus 291
Petauroides 59
armorica, Mustela 322
armoricanus, Microtus 518
arnee, Bubalus 402
arnhemensis, Isoodon 39
Nyctophilus **218**
Trichosurus **48**
arnouxianus, Naemorhedus 407
arnuxii, Berardius **361**, 361
aroaensis, Mallomys **608**, 609
Aroaethrus 757
arquatus, Acerodon 137
Eptesicus 201
Pteropus 146
arredondoi, Capromys 800
arrhenii, Cephalophus 410
Heliosciurus 429
Perodicticus 248
Potamochoerus 378
arriagensis, Thomomys 475

arrogans, Stenomys 665
arsenjevi, Clethrionomys 509
Pteromys **464**
arsinoe, Myotis 208
arsipus, Vulpes 287
artata, Myrmecophaga 68
artemesia, Sylvilagus 826
artemesiae, Spermophilus **447**
artemisiae, Lemmiscus 516
Peromyscus 732
Synaptomys 534
arthuri, Mustela 322
Artibeus **187**, 187, 188, 189
aequatorialis 188
amplus **187**
anderseni **187**
aztecus **187**
bogotensis 188
carpolegus 188
cinereus 187, **188**, 189
concolor **188**
coryi 188
eva 188
fallax 189
femurvillosum 189
fimbriatus **188**
fraterculus **188**, 188
fuliginosus 189
glaucus **188**, 188
gnomus 188
grenadensis 188
hartii **188**
hercules 189
hesperus 189
hirsutus **188**, 188
inopinatus **188**
insularis 188
intermedius 189
jamaicensis 187, **188**, 188, 189
jucundum 188
koopmani 189
lewisi 188
lituratus **188**, 189
major 187
minor 187
nanus 189
obscurus 188, **189**
palatinus 189
palmarum 189
parvipes 188
paulus 188
phaeotis **189**
planirostris 188, **189**, 189
praeceps 188, 189
pumilio 188
ravus 189
richardsoni 188
rosenbergii 188
rusbyi 189
schwartzi 188
superciliosum 189
toltecus **189**
trinitatis 188
triomylus 188
turpis 189
validum 189
watsoni 188
yucatanicus 188
artinii, Scotoecus **226**
Artiodactyla 377
artus, Chaetodipus **483**
aruensis, Echymipera 41
Hipposideros 171
Pteropus 148
Rhinolophus 165
Sus 379
Uromys 671
aruma, Phyllostomus 180
arundinaceus, Micromys 620
arundinum, Redunca **414**
arundivaga, Puma 296
aruscensis, Paraxerus 435
arusinus, Ratufa **437**
arvalis, Microtus **518**, 519, 523, 526, 529, 531
Mus 517
Pappogeomys 473
Arvalomys 517
arvensis, Microtus 519
Arvicanthis **576**, 576, 577, 578, 582, 594, 630
abyssinicus **576**, 576, 578
ansorgei 578
blicki **576**, 576, 578, 606, 664, 686
centralis 578
centrosus 578
chanleri 577
dembeensis 576, 578
discolor 578
fluvicinctus 576
jebelae 578
kordofanensis 578
lacernatus 576, 578
luctosus 578
major 578
mearnsi 576, 578
minor 578
mordax 578
muansae 577, 578
nairobae 576, **577**, 577, 578
naso 578
neumanni 576, **577**, 577, 578
niloticus 576, **577**, 577, 578
nubilans 578
occidentalis 578
ochropus 578
pallescens 577
pelliceus 578
praeceps 577
raffertyi 578
reichardi 578
reptans 578
rhodesiae 578
rossii 578
rubescens 578
rufinus 578
rufodorsalis 576
rumruti 578
saturatus 576
setosus 578
solatus 578
somalicus 577, 578
tenebrosus 578
testicularis 578
variegatus 578
virescens 577
zaphiri 578
Arvicola **505**, 505, 517, 518, 528, 532, 533
abrukensis 506
albus 506
americana 506
amphibius 505, 506
aquaticus 506
argentoratensis 506
argyropus 506
armenius 506
ater 506
barabensis 506
brandtii 515
brigantium 506
buffonii 506
cantabriae 506
canus 506
castaneus 506
caucasicus 506
cernjavskii 506
cubanensis 506
curtata 516
destructor 506
djukovi 506
exitus 506
ferrugineus 506
fertilis 515
fuliginosus 506
hintoni 506
hyperryphaeus 506
illyricus 505, 506
indica 632
italicus 505, 506
jacutensis 506
jenissijensis 506
karatshaicus 506
korabensis 506
kuruschi 506
kuznetzovi 506
littoralis 506
martinoi 506
melanogaster 513
meridionalis 506
minor 506
monticola 506
musignani 505, 506
musiniani 505
niger 506
nigricans 506
nivalis 506
nuttalli 715
obensis 506
ognevi 506
pallasii 506
paludosus 506
persicus 506
pertinax 506
reta 506
rufescens 506
sapidus **505**, 505
scherman 505, 506
schermous 506
seythius 506
stankovici 506
stoliczkanus 502
tanaiticus 506
tataricus 506
tauricus 506
tenebricus 505
terrestris **505**, 505
turovi 506
uralensis 506
variabilis 506
volgensis 506
Arvicolinae **501**, 501, 512, 532, 533, 534
arvicoloides, Microtus 528
arviculoides, Akodon 689
Bolomys 696
arwasca, Presbytis 271
aryabertensis, Lepus 819
Asagis 18
asaii, Mustela 324
ascanius, Cercopithecus **263**
aschantiensis, Dendrohyrax 373
Heliosciurus 429
aschenborni, Oryx 413
Aschizomys 502, 503, 513
Ascobates 62
Ascogale 33
Ascomys 469
Ascopharynx 635
Asellia **170**
diluta 170
italosomalica 170
murraiana 170
pallida 170
patrizii **170**
tridens **170**
Aselliscus **170**
koopmani 170
novaeguinae 170
novaehebridensis 170
stoliczkanus **170**
tricuspidatus **170**
trifidus 170
wheeleri 170
asellus, Lepus 815
ashtoni, Viverra 348
asia, Peponocephala 355
asiaeorientalis, Neophocaena 358
asiatica, Chrysochloris **75**, 75
Oryx 413
Rupicapra 410
Talpa 75
asiaticus, Cervus 385
Elephas 367
Ovis 408
Panthera 297
Rangifer 392

Rattus 658
Rhinoceros 372
Tamias **455**
Asinohippus 369
Asinus 369
asinus, Equus **369**, 369
Asiocricetus 539
Asioscalops 128
asirensis, Rhinopoma 155
asomatus, Lagorchestes **52**
Aspalacidae 753
Aspalax 754
aspalax, Myospalax **675**, 675, 676
Aspalomys 675
asper, Limnomys 638
Maxomys **614**
Parahydromys **638**
Sorex **112**, 112, 114
aspera, Murexia 32
aspinatus, Maxomys 613
aspromontis, Dryomys 766
assabensis, Taphozous 161
assamensis, Anourosorex 106
Atherurus 773
Callosciurus 423
Macaca **266**
Pteropus 147
Suncus 102
Tupaia 131
asseel, Bos 401
assimilis, Antechinus 31
Microtus 519
Peromyscus 732
Petrogale **55**, 56
Rattus 653
Vombatus 46
astaricus, Lepus 817, 821
asthenops, Stenella 356
astrabadensis, Crocidura 96
astrachanensis, Microtus 530
astrolabiensis, Mormopterus 237
asturianus, Microtus 519
astuta, Bassaris 334
astutus, Bassariscus **334**, 334
Mustela 321
asurus, Elephas 367
asyutensis, Gerbillus 551
Meriones 556
atacamensis, Myotis **208**
atahualpae, Cavia 779
atallahi, Lepus 816
atbarensis, Diceros 372
atchinensis, Lariscus 430
Niviventer 635
atchinus, Sundamys 666
Atelerix **76**, 77
adansoni 76
albiventris **76**, 77
algirus **76**
angolae 77
angolensis 77
atratus 76
caniculus 77
capensis 77
diadematus 76, 77
fallax 77
faradjius 76
fractilis 77
frontalis **77**
girbaensis 77
heterodactylus 76
hindei 76
kilimanus 76
krugi 77
langi 76
lavaudeni 77
lowei 76
oweni 76
pruneri 76
sclateri **77**
sotikae 76
spiculus 76
spinifex 76
vagans 77
Ateles **256**, 257
albifrons 257
arachnoides 257
ater 257
azuerensis 257
bartlettii 257
belzebuth **257**, 257
braccatus 257
brissonii 257
brunneus 257
cayennensis 257
chamek **257**
chuva 257
cucullatus 257
dariensis 257
frontalis 257
frontatus 257
fuliginosus 257
fusciceps **257**
geoffroyi **257**
grisescens 257
hybridus 257
longimembris 257
loysi 257
marginatus **257**
melanocercus 257
melanochir 257
neglectus 257
ornatus 257
pan 257
panamensis 257
paniscus **257**, 257
pentadactylus 257
peruvianus 257
problema 257
robustus 257
rufiventris 257
subpentadactylus 257
surinamensis 257
trianguligera 257
tricolor 257
variegatus 257
vellerosus 257
yucatanensis 257
Atelidae 254, 256
Atelinae **256**
Atelocheirus 256
Atelocynus **279**, 279, 280
microtis **279**
sclateri 279
ater, Arctictis 342
Arvicola 506
Ateles 257
Baiomys 695
Canis 281
Chaetodipus 484
Chiropotes 261
Eptesicus 203
Hipposideros **171**
Indri 247
Leontopithecus 252
Molossus **235**
Mus 625
Myotis 208, 213
Ochotona 807
Orcinus 355
Phyllostomus 180
Rattus 658
Spalacopus 789
Suncus **101**
aterrima, Enhydra 310
Martes 320
aterrimus, Cercocebus 266
Liomys 482
Lophocebus 266
Pteropus 146
athabascae, Bison 400
Clethrionomys 508
athene, Pteromys 464
atheneensis, Cervus 386
atherodes, Muntiacus **388**
Atherurinae 773
Atherurus **773**
africanus **773**
angustiramus 773
armata 773
assamensis 773
burrowsi 773
centralis 773
hainanus 773
macrourus **773**
pemangilis 773
retardatus 773
stevensi 773
terutaus 773
tionis 773
turneri 773
zygomatica 773
athi, Steatomys 546
athiensis, Syncerus 402
Atilax **300**, 300
atilax 301
galera 301
macrodon 301
mitis 301
mordax 301
nigerianus 301
paludinosus **300**
paludosus 301
pluto 301
robustus 301
rubellus 301
rubescens 301
spadiceus 301
transvaalensis 301
urinatrix 301
vansire 301
voangshire 301
atilax, Atilax 301
atkinsoni, Helogale 304
atlantica, Halichoerus 330
Vulpes 287
atlanticus, Cervus 385
Equus 369
Lepus 816
Rhinolophus 165
Trichechus 365
atlantis, Crocidura 92
Atlantoxerus **420**, 420, 444
getulus **420**
praetextus 420
trivittatus 420
atlas, Mustela 323
atnarko, Ursus 338
Atophyrax 111, 113
Atopogale 69
atracamensis, Ctenomys 785
atrata, Dorcopsis **51**
Emballonura **157**
Martes 319
Neotoma 712
Pteralopex **145**, 145
atratus, Akodon 692
Atelerix 76
Callosciurus 422
Eptesicus 202
Hipposideros 171
Microtus 530
Pappogeomys 473
Rattus 658
Sigmodon 747
Suncus 102
Tadarida 240
atricapilla, Spermophilus 448
atricapillus, Callosciurus 423
Spermophilus **444**
atriceps, Macaca 266
atridorsum, Rattus 658
atrimaculatus, Spilocuscus 47
atrinasus, Cercopithecus 263
atrior, Trachypithecus 273
atrirelictus, Microdipodops 480
atrirufus, Sciurus 443
atristriatus, Callosciurus 422, 423
Tamias 454
atrodorsalis, Callosciurus 421
Chaetodipus 484
Thomomys 475
atrogriseus, Thomomys 475
atronasus, Dipodomys 479
atronates, Uropsilus 130
atrovarius, Thomomys 475
atrox, Callosciurus 423
Hipposideros 171
Kerivoula **196**

attenuata, Crocidura **82**
Feresa **352**, 352
Stenella 355, **356**
attenuatus, Dipodomys 479
Steno 356
Thomomys 475
atterima, Lontra 311
atticus, Cricetulus 538
Microtus 530
attila, Crocidura **82**, 82
Sus 379
attwateri, Geomys 469, 470
Neotoma 711
Peromyscus **728**, 729
Sylvilagus 824
atys, Cercocebus 263
aubinnii, Protoxerus **436**
aubryana, Genetta 346
aubryi, Heliosciurus 429
Pan 277
auceps, Galea 779
Meriones 557
Auchenia 381
audax, Mustela 321
Rhinolophus 163
Tachyoryctes **686**
audeberti, Hallomys 679
Macaca 268
Nesomys 679
auduboni, Gulo 318
Ovis 408
audubonii, Sylvilagus 813, **824**
augustana, Talpa 128
augustinus, Choloepus 64
augustus, Paraxerus 435
Aulacochoerus 378
Aulacodus 775
swinderianus 775
Aulacomys 517, 518, 520, 523, 528, 531
Aulaxinus 266
Auliscomys **695**, 695, 737
alsus 695
boliviensis **695**
decoloratus 695
flavidior 695
fumipes 695
leucurus 695
micropus **695**
pictus **695**
sublimis **695**
aupourica, Mystacina 231
aurantia, Rhinonicteris **175**
aurantiaca, Hipposideros 171
Nycteris 162
aurantiacus, Hylopetes 461
Macropus 53
aurantii, Lepus 820
aurantioluteus, Vulpes 287
aurantius, Melonycteris **154**
Reithrodontomys 741
Rhinolophus 175
aurarius, Platyrrhinus **190**
aurascens, Myotis 213
aurata, Catopuma 289
Felis 296
Murina **229**, 229, 230
Paguma 343
Presbytis 271
Profelis **296**
auratus, Alouatta 255
Cercopithecus 273
Chrysochloris 75
Galerella 303
Isoodon **39**
Meriones 556
Mesocricetus **538**, 538, 539
Myotis 210
Otomys 680
Pteropus 149
Rattus 658
Trachypithecus **273**
aurea, Chrysochloris 75
Dasyprocta 781
Marmota 431
aureiventer, Ratufa 437
aureiventris, Thomomys 473
aureogaster, Sciurus **440**
aureolus, Ochrotomys 715
aureospinula, Podogymnura **80**
aureotunicata, Neotoma 711
aureoventris, Mustela 322
aurescens, Macropus 53
aureus, Arctocebus **247**
Canis **280**
Cephalophus 411
Dendrolagus 50
Felis 290
Gerbillus 552
Hapalemur **245**
Hipposideros 175
Macaca 266
Muntiacus 389
Myotis 209, 213
Oreotragus 399
Paradoxurus 344
Reithrodontomys 743
Spermophilus **446**
Thomasomys **749**
Thomomys 473
Ursus 338
aureventer, Lutra 312
aurex, Tylonycteris 228
auricomis, Aethomys 568
auricularis, Desmodillus **547**
Gerbillus 547
Microtus 528
Thomasomys 750
auriculata, Suncus 103
auriculatus, Funisciurus 428
auriculus, Myotis **208**
aurifrons, Macaca 268
aurigineus, Lepus 821
aurijunctus, Vespertilio 228
aurillus, Microryzomys 708
aurinuchalis, Acerodon 137
auripectus, Peromyscus 729
auripendulus, Eumops **233**
auripila, Neotoma 711
aurispinosus, Nyctinomops **238**
aurita, Arctocephalus 327
Callithrix **251**, 251, 252
Didelphis **16**
Geogale **70**, 70
Hipposideros 173
Hippotragus **412**
Lonchorhina **177**, 177
Manis 415
Myzopoda **196**, 196
Nycteris 161, 162
Triaenops 175
Vulpes 287
auritus, Chrotopterus **177**
Dasypus 65
Erinaceus 77
Hemiechinus **78**, 78
Mazama 390
Mesocapromys **801**, 801
Mus 740
Nyctinomops 239
Ochotona 810
Odocoileus 391
Otocyon 284
Peromyscus 733
Plecotus **224**, 224, 225
Pseudomys 645
Reithrodon **740**, 740
Rhinolophus 164
Vampyrus 177
Vespertilio 224
auriventer, Mustela 322
Oryzomys **720**, 720
auriventris, Paraxerus 435
aurobrunneus, Lutra 312
auropunctata, Herpestes 305
Mangusta 305
auropunctatus, Herpestes 305, 306
aurora, Cercopithecus 264
Eothenomys 514
Leontopithecus 253
auroreus, Lariscus 430
Rattus 660
ausensis, Petromus 774
auspicatus, Procyon 335
austenianus, Pipistrellus 223
austerulus, Sigmodon 747
austerus, Microtus 526
Peromyscus 732
austini, Pteropus 152
Sciurus 443
australasianus, Physeter 359
australiensis, Pseudomys 647
Tachyglossus 13
australis, Arctocephalus **326**, 326
Balaena 349
Canis 283
Chalinolobus 199
Chrotopterus 177
Civettictis 345
Cloeotis 170
Crateromys **587**
Cremnomys 588
Ctenomys **784**, 786
Dipodomys 477
Dolichotis 780
Dromiciops 27
Dugong 365
Dusicyon **283**, 283
Echymipera 41
Eubalaena **349**, 349
Galago 249
Geogale 70
Giraffa 383
Graphiurus 765
Heteromys **481**
Hippopotamus 381
Lagenorhynchus **353**, 353
Malacomys 608
Mesoplodon 363
Microcavia **780**
Miniopterus **230**
Mustela 324
Myotis **208**, 208, 213
Neophoca 328
Nycticeius 218
Nyctophilus 218
Ototylomys 726
Pelomys 639
Petaurus **61**, 61
Physeter 359
Pipistrellus 221
Plecotus 225
Presbytis 271
Pronolagus 824
Pseudomys 644, **645**, 645, 646, 647, 648
Reithrodontomys 742
Scalopus 127
Syconycteris **155**, 155
Tadarida **240**
Tamias 455
Taphozous **160**
Thomasomys 749
Tragelaphus 404
Trichechus 366
Zapus 499
Ziphius 364
australius, Melomys 615
Australophocaena **358**, 358
dioptrica **358**, 359
stornii 358
austriacus, Plecotus **224**, 225
austrinus, Geomys 470
Rattus 661
Rhipidomys **744**
Austritragus 406
austroamericana, Didelphis 17
Austronomus 240
austroriparius, Myotis **208**
austrouralensis, Dipus 493
auyantepui, Oecomys 716
auzembergeri, Anomalurus 757
auziensis, Meriones 558
Avahi **246**
awahi 246
lanatus 246
laniger **246**

longicaudatus 246
occidentalis 246
orientalis 246
avakubia, Nycteris 162
avara, Marmota 432
avaricus, Mesocricetus 539
avarillus, Aethomys 568
avarus, Pseudocheirus 59
avellanarius, Mus 769
Muscardinus **769**
avellanifrons, Kobus 413
avellanus, Myoxus 770
avenarius, Micromys 620
averini, Spermophilus 449
aversus, Aotus 256
avia, Cryptotis **108**
Mephitis 317
aviator, Nyctalus **216**
avicennai, Crocidura 96
avicinnia, Sciurus 442
avilapiresi, Saguinus 253
avius, Peromyscus 730
Sylvilagus 825
avunculus, Aethomys 568
Microsciurus 433
Pygathrix **272**
avus, Cebus 259
awahi, Avahi 246
awahnee, Thomomys 473
axillaris, Rhinolophus 166
Axis **384**, 385
annamiticus 385
axis **384**
calamianensis **384**
ceylonensis 384
culionensis 384
hecki 385
indicus 384
kuhlii **384**
maculatus 384
major 384
minor 384
nudipalpebra 384
oryzus 385
porcinus 384, **385**, 385
pumilio 385
axis, Axis **384**
Cervus 384
Aye-aye 247
aygula, Presbytis 270, 271
ayresi, Dendromus 543
azandicus, Panthera 297
azarae, Akodon **688**, 689, 693
Canis 282
Cavia 779
Cebus 259
Cerdocyon 282
Ctenomys **784**, 785
Dasyprocta **781**
Didelphis 16
Ozotoceros 392
Pseudalopex 285
azarai, Aotus **255**, 256
azaricus, Philander 22
Azema 243
azizi, Meriones 556
azoreum, Nyctalus **216**, 217
azoricus, Mus 625
azpatianus, Blastoceros 389
azrakensis, Syncerus 402
azteca, Carollia 186
aztecus, Artibeus **187**
Caluromys 15
Microtus 527
Molossus 235
Myotis 215
Oryzomys 721
Peromyscus **728**, 728, 729, 736, 737
Potos 333
Puma 296
Reithrodontomys 741
Sylvilagus 825
azuerensis, Ateles 257
Oryzomys 721
azulensis, Peromyscus 734

babaulti, Myosorex **99**
baberi, Hylopetes **460**
babi, Cynopterus 139
Rattus 659
Sus 379
babu, Pipistrellus **220**
babylonicus, Cricetus 538
Taphozous 161
Babyrousa **377**
alfurus 377
babyrussa **377**
beruensis 377
bolabatuensis 377
celebensis 377
frosti 377
quadricornua 377
togeanensis 377
Babyrousinae **377**
babyrussa, Babyrousa **377**
Sus 377
bacchante, Chroeomys 700
Oenomys 637
bacheri, Nesokia 632
bachmani, Brachylagus 813
Sciurus 442
Sorex 116
Sylvilagus **824**, 826
Thomomys 475
bactriana, Allactaga 488
bactrianus, Camelus **381**, 381
Cervus 385
Mus 625
Pipistrellus 223
Spermophilopsis 444
Suncus 102
badia, Catopuma **289**
badiatus, Petaurista 463
badiodorsalis, Catopuma 289
Badiofelis 289
badistes, Sylvilagus 826
badius, Cannomys **685**
Cephalophus 410
Galerella 303
Hybomys 596
Otolemur 250
Peromyscus 737
Procolobus **271**, **272**
Rhizomys 685
Spermophilus 450
Tachyoryctes 687
Thomomys 475
Ursus 338
Vandeleuria 672
badjing, Callosciurus 422
Baeodon 225
baeodon, Maxomys **612**
baeops, Thomasomys **749**
baeri, Hylomyscus 592, **598**
baessleri, Apodemus 574
baeticus, Sciurus 443
Sus 379
Baginia 420
bagobus, Bullimus 580, **581**, 581
bahadur, Mus 627
bahamensis, Eptesicus 201
Tadarida **240**
bahiensis, Conepatus 316
Marmosops 19
Oecomys 717
bahrainja, Cervus 385
bahram, Equus 370
bahrkeetae, Kobus 414
baibac, Marmota 431
baibacina, Marmota **430**, 431, 432, 433
baicalensis, Alticola 504
Cervus 385
Lutra 312
Microtus 525
Phoca 332
baikalensis, Clethrionomys 509
Sorex 121
Ursus 338
baikii, Nycteris 161
baileyi, Canis 281
Castor 467
Chaetodipus **483**
Crocidura **82**
Dipodomys 480
Lynx 293
Microtus 523
Myospalax 675
Myotis 212
Naemorhedus **407**
Neotoma 711
Sigmodon 748
Soriculus 123
Sylvilagus 824
Tamiasciurus 457
Thomomys 473, 474, 476
bailloni, Microtus 518
bailwardi, Calomyscus **535**, 535, 536
Nesokia 632
Tatera 561
bainsei, Catopuma 289
Baiomys **695**, 695, 698, 728, 746
allex 695
analogus 695
ater 695
brunneus 695
canutus 695
fuliginatus 695
grisescens 695
handleyi 695
hummelincki 695
infernatis 695
musculus **695**
nebulosus 695
nigrescens 695
pallidus 695
paulus 695
pullus 695
subater 695
taylori **695**
Baiosciurus 439
bairdi, Gulo 318
Microtus 527
Ursus 338
bairdii, Berardius **361**
Delphinus 352
Lepus 814
Peromyscus 732
Sorex **112**
Tapirus **371**
Baiyankamys 597, 598
bajovaricus, Cervus 385
bakeri, Cephalophus 411
Glossophaga 184
Heterohyrax 373
Hippotragus 412
Pteropus 146
balabacensis, Sus 378
balabagensis, Sundamys 666
balaclavae, Peromyscus 732
balae, Leopoldamys 603
Ratufa 437
Sundasciurus 451
Balaena **349**, 351
ampullata 361
australis 349
marginata 351
mysticetus **349**, 349
novaeangliae 350
physalus 349
Balaenidae **349**, 349
Balaenoptera **349**
acutorostrata **349**
antiquorum 350
bonaerensis 349
boops 350
borealis **350**, 350
brevicauda 350
brydei 350
davidsoni 349
edeni **350**, 350
gibbar 349, 350
gigas 350
huttoni 349
indica 350
intermedia 350
major 350
minimus 349
musculus **350**

patachonica 350
physalus **350**
robusta 350
rostrata **349,** 350
schlegellii 350
sibbaldii 350
sibbaldius 350
sulfureus 350
velifera 350
Balaenopteridae **349,** 349
balaenurus, Hydrodamalis 365
Balantia 46
Balantiopteryx **156**
infusca **156**
io **156**
ochoterenai 156
pallida 156
plicata **156,** 156
balcanica, Crocidura 96
Rupicapra 410
balcanicus, Canis 280
Sciurus 443
Spermophilus 445
balchanensis, Blanfordimys 506
balchaschensis, Apodemus 574
balearica, Crocidura 96
Genetta 345
baleni, Capreolus 390
bali, Rattus 650
balica, Panthera 298
baliemensis, Rattus 660
baliensis, Ratufa 437
balikunica, Allactaga **488,** 488
balina, Tupaia 132
baliolus, Akodon 688
Grammomys 593
Peromyscus 735
Sciurus 443
Balionycteris **138**
maculata **138**
seimundi 138
balkaricus, Neomys 110
balkashensis, Eremodipus 493
ballivianensis, Dolichotis 780
balmasus, Sundamys 666
balnearum, Holochilus 704
balneator, Oryzomys **720,** 722
balsaci, Chrysochloris 75
balstoni, Callosciurus 422
Murina 230
Nycticeius **217,** 217, 218
baltica, Halichoerus 330
balticus, Capreolus 390
Castor 467
Cervus 385
baltina, Ochotona 810
baluchi, Calomyscus **535**
baluensis, Callosciurus **420,** 423
Crocidura 85
Dendrogale 131
Rattus 649, **650,** 650
Tupaia 132
balutus, Pteropus 150
bambhera, Ovis 408
bamendae, Procavia 374
bampensis, Tscherskia 540
banacus, Maxomys 614
bancana, Arctogalidia 343
Ratufa 437
bancanus, Muntiacus 389
Nannosciurus 434
Nycticebus 248
Pipistrellus 221
Tarsius **250**
Tragulus 383
bancarus, Sundasciurus 452
bancrofti, Sciurus 439
bandahara, Maxomys 614
bandarum, Funisciurus 427
banderanus, Osgoodomys **726,** 726, 734
Peromyscus 725
Bandicota **578,** 578, 579, 603, 632, 662
bandicota 579
bangchakensis 579
barclayanus 579
bengalensis **578,** 579
blythianus 579
curtata 579
daccaensis 579
dubius 579
elliotanus 579
eloquens 579
giaraiensis 579
gigantea 579
gracilis 579
hichensis 579
indica **579,** 579, 632
insularis 579
jabouillei 579
kagii 579
kok 579
lordi 579
macropus 579
malabarica 579
mordax 579
morungensis 579
nemorivaga 579
perchal 579
plurimammis 579
providens 579
savilei **579,** 579
setifera 579
siamensis 579
sindicus 579
sonlaensis 579
sundavensis 579
taiwanus 579
tarayensis 579
varillus 579
varius 579
wardi 579
bandicota, Bandicota 579
bandiculus, Rattus 658
banfieldi, Melomys 616
bangae, Sigmoceros 395
bangchakensis, Bandicota 579
bangkanus, Callosciurus 423
bangsi, Glaucomys 460
Mustela 321
Pecari 380
Perognathus 485
Puma 296
Sciurus 443
Vulpes 287
bangsii, Lepus 815
banguei, Rattus 661
Ratufa 437
Tragulus 383
Tupaia 133
bangueyae, Sundasciurus 451
banksi, Callosciurus 423
Petaurista 462
banksiana, Pteropus 146
banksianus, Canis 281
Macropus 55
banksicola, Lepus 815
banksii, Pseudocheirus 60
bantamensis, Callosciurus 422
banteng, Bos 401
baoulei, Mus **622,** 622, 626, 628
baptista, Callicebus 258
baptistae, Meriones 557
Viverricula 348
barabensis, Arvicola 506
Cricetulus **537,** 537, 538
Mus 537
Sorex 117
barakshin, Alticola **503**
baramensis, Callosciurus 421
Ratufa 437
barandanus, Cervus 386
barang, Lutra 312
Barangia 311
barawensis, Paraxerus 435
barbacoas, Sigmodontomys 748
barbar, Genetta 345
barbara, Eira **317**
Galictis 318
Genetta 345
Hyaena 308
Mustela 317
barbarabrownae, Callicebus 259
barbarensis, Caenolestes 25
barbaricus, Panthera 297
barbarus, Canis 280
Cervus 385
Ctenomys 784
Lemniscomys **601,** 601
Mus 601
Panthera 297, 298
Rhinolophus 165
Sus 379
Vulpes 287
Barbastella **198**
barbastellus **198**
blanfordi 199
caspica 199
communis 199
darjelingensis 199
daubentonii 199
leucomelas **199,** 199
walteri 199
barbastellus, Barbastella **198**
Vespertilio 198
barbata, Hippotragus 412
Phoca 329
barbatogularis, Mormopterus 238
barbatua, Cebus 259
barbatus, Alouatta 254
Cebus 260
Cynogale 341
Erignathus **329**
Eumops 233
Mus 673
Pseudocheirus 59
Sus **378,** 378, 379
Taurotragus 403
Xenuromys **673**
barbei, Tamiops 457
Trachypithecus 273
barbensis, Hipposideros 173
barberi, Sciurus 439
Tamias 455
barbertonensis, Rhinolophus 165
barbertoni, Cephalophus 411
barbiensis, Petromus 774
barbouri, Mesocapromys 800
Otomys 680
Petromyscus **683,** 683
barcaeus, Lepus 816
barclayanus, Bandicota 579
bardus, Apomys 575
bargusinensis, Clethrionomys 509
baringoensis, Gazella 397
baritensis, Paradoxurus 344
barkeri, Bullimus 581
Phacochoerus 377
Tragelaphus 404
barnardi, Lasiorhinus 45
barnesi, Molossus 235
Ochotona 811
baroni, Oryzomys 725
barrerae, Octomys 788, 789
Tympanoctomys **789**
barretti, Pronolagus 824
barroni, Petaurista 463
barrowensis, Isoodon 39
Spermophilus 448
bartelsi, Aeromys 459
Hylopetes **460**
Myotis 210
Petinomys 463
bartelsii, Crocidura 90
Maxomys **612**
Mus 611
Barticonycteris 178
bartletti, Caluromys 15
bartlettii, Ateles 257
bartoni, Callosciurus 421

Zaglossus 13
barussanoides, Rattus 660
barussanus, Niviventer 633
baryceros, Cervus 386
basengae, Sigmoceros 395
bashkiricus, Sciurus 443
basilanensis, Cervus 386
basilanus, Rattus 652
basilicae, Thomomys 473
basilii, Hybomys **596**
basiliscus, Pteropus 151
baskii, Speothos 285
Bassaricyon **333**, 333
alleni **333**, 333
beddardi **333**, 333
gabbii **333**, 333
lasius **333**
medius 333
orinomus 333
pauli **333**
richardsoni 333
siccatus 333
Bassaris, astuta 334
Bassariscidae 334
Bassariscus **334**
albipes 334
arizonensis 334
astutus **334**, 334
beddardi 333
bolei 334
campechensis 334
consitus 334
flavus 334
insulicola 334
macdougalli 334
monticola 334
nevadensis 334
notinus 334
oaxacensis 334
octavus 334
oregonus 334
palmarius 334
raptor 334
saxicola 334
sumichrasti **334**
variabilis 334
willetti 334
yumanensis 334
bassi, Scalopus 127
bassianus, Pseudocheirus 60
bassii, Vombatus 46
bastardi, Macrotarsomys **679**, 679
bastiani, Nycteris 162
basuticus, Dendromus 543
Galerella 303
Otomys 681
batamana, Tupaia 132
batamanus, Maxomys 614
batangtuensis, Pongo 277
batarovi, Micromys 620
batchiana, Pteropus 146
batchianensis, Hipposideros 171
bateae, Mus 625
batesi, Anomalurus 757
Crocidura **82**, 97
Galago 249
Hyemoschus 382
Neotragus **398**
Perodicticus 248
Prionomys **545**, 545
Bathyergidae **771**
Bathyerginae 771
Bathyergoidea, 771
Bathyergus **771**, 771
africana 771
hottentotus 771
inselbergensis 771
intermedius 771
janetta **771**, 771
maritimus 771
plowsi 771
splendens 685
suillus **771**, 771
batin, Rattus 661
batinensis, Eptesicus 202
batis, Sorex 119
batjanus, Glischropus 204
Batomys **579**, 580, 583, 587
dentatus **579**, 580
granti 579, **580**, 580
salomonseni **580**, 580
battyi, Didelphis 17
Lepus 815
Odocoileus 391
batuana, Petaurista 463
Presbytis 271
Ratufa 437
batuanus, Tragulus 383
baturus, Niviventer 634
batus, Maxomys 614
Sundasciurus 452
baudensis, Sciurus 441
Bauerus 198
dubiaquercus 198
baumstarki, Erythrocebus 265
bauri, Oryzomys 722
baussencis, Miniopterus 231
bavaricus, Microtus **519**
baveanus, Pteropus 146
baverstocki, Eptesicus **200**
bavicorensis, Spermophilus 449
baweanus, Macaca 266
bayeri, Tatera 562
bayoni, Chrysochloris 75
Redunca 414
Bayonia 73
bayonii, Funisciurus **427**
Bdeogale 300, **301**, 301
crassicauda **301**, 301
jacksoni **301**, 301
nigripes **301**, 301
omnivora 301
puisa 301
tenuis 301
bea, Canis 280
Protoxerus 436
Tragelaphus 404
beaba, Nesokia 632
Beamys **540**
hindei **540**, 540
major **540**
beatae, Peromyscus 729, 732
Beatragus 394
beatrix, Chalinolobus **199**
Gracilinanus 17
Oryx 413
beatrizae, Mesocapromys 800
beatus, Akodon 692
Crocidura **82**
Funisciurus 428
Hipposideros **171**
beaucournui, Talpa 128
beauforti, Dobsonia **139**
Pseudochirops 60
beaufortiana, Phoca 332
beaulieui, Ia 205
beavanii, Mus 629
beccarii, Crocidura **82**
Dorcopsis 51
Emballonura **157**
Gazella 396
Hydromys 597
Margaretamys **609**, 609
Mormopterus **237**
Mus 609
Rattus 658
Rhinolophus 164
bechsteini, Myotis **208**
bechuanae, Cynictis 302
Lepus 820
Paracynictis 307
Poecilogale 325
Rhabdomys 662, 663
Tatera 561
beckeri, Eumops 233
becki, Sorex 117
beddardi, Bassaricyon **333**, 333
Bassariscus 333
beddingtoni, Syncerus 402
beddomei, Rhinolophus 167
Suncus 103
beden, Capra 406
bedfordi, Budorcas 404
Capreolus 390
Equus 370
Lepus 816
Macropus 53
Naemorhedus 407
Ochotona 808
Panthera 298
Proedromys **533**, 533
bedfordiae, Alces 389
Clethrionomys 509
Cricetulus 539
Phodopus 539
Sorex **112**
bedfordianus, Cervus 385
bedouin, Nycticeius 218
beebei, Callosciurus **424**
Tayassu 380
beecheyi, Spermophilus **445**
beecrofti, Anomalurus **757**, 757
beelzebub, Alouatta 254
begitschevi, Lepus 821
behni, Gazella 397
behnii, Micronycteris **178**
beirae, Crocidura 87
Cryptomys 772
Heliosciurus 429
Leptailurus 292
Tatera 561
beirensis, Cercopithecus 264
Tatera 561
beisa, Oryx 413
belajevi, Allocricetulus 537
belangeri, Scotophilus 227
Tupaia **131**, 132
belcheri, Tamiasciurus 456
beldeni, Anomalurus 757
beldingi, Spermophilus **445**
belfieldi, Dremomys 425
Belideus 61
beljaevi, Allocricetulus 537
beljawi, Allocricetulus 537
bella, Genetta 345
Mus 625, 628
Neotoma 712
bellaricus, Funambulus 426
bellicosus, Cricetulus 538
bellieri, Lemniscomys 592, **601**, 601, 602
Pipistrellus 221
belliger, Trachypithecus 274
bellissima, Kerivoula 198
bellona, Callosciurus 423
bellula, Dasyprocta 782
Kerivoula 197
bellus, Antechinus **29**
Peromyscus 728
Belomys **459**, 465
blandus 459
kaleensis 459
pearsonii **459**
trichotis 459
villosus 459
belone, Hylopetes 461
belti, Sciurus 443
Beluga 357
beluga, Delphinapterus 357
belugae, Castor 467
belzebul, Alouatta **254**
Simia 254
belzebuth, Ateles **257**, 257
Bematiscus 75
bembanicus, Rhinolophus 169
bemmeleni, Chaerephon **232**
benamakimae, Colobus 270
bendirii, Sorex **113**
benefactus, Bolomys 697
benetianus, Naemorhedus 407
benevolens, Oecomys 716
Rhipidomys 715
benga, Heliosciurus 429
bengalensis, Bandicota **578**, 579
Caracal 288
Chaerephon 233
Elephas 367
Felis 295

Funambulus 426
Hystrix 773
Panthera 297
Prionailurus **295**, 295
Ratufa 437
Sus 379
Viverricula 348
Vulpes **286**
benguetensis, Rattus 660
beniana, Chlorocebus 265
beniensis, Alouatta 255
Bradypus 63
Dasypus 66
Dendrohyrax 373
Tatera 562
benitoensis, Peromyscus 729
bennetti, Macropus 55
bennettianus, Dendrolagus **50**, 50
bennettii, Abrocoma **789**, 789
Cynogale **341**
Cyongale 341
Gazella **396**
Herpestes 305
Lepus 815
Mimon **179**
Phyllostoma 179
Prionailurus 296
benormi, Potorous 50
bensoni, Chaetodipus 483
Eptesicus 202
Neotoma 711
Steatomys 545
Bensonomys 697
bentincanus, Callosciurus 421
Maxomys 614
bentleyae, Dasymys 589, 590
benuensis, Nycteris 162
benvenuta, Tatera 562
beothucus, Canis 281
Berardius **361**
arnuxii **361**, 361
bairdii **361**
vegae 361
berberana, Gazella 397
berberanus, Lepus 816
berberensis, Equus 370
berbericensis, Holochilus 705
berbericus, Loxodonta 367
berberorum, Caracal 288
berdmorei, Berylmys **580**, 580
Menetes **433**
Sciurus 433
berengei, Gorilla 276
beresfordi, Peromyscus 732
berezovskii, Moschus **383**
berezowski, Soriculus 123
berezowskii, Micromys 620
bergensis, Apodemus 573
Otomys 682
Sorex 111
bergeri, Gazella 397
Hyaena 308
bergerinae, Gazella 397
bergi, Ctenomys 785
beringei, Gorilla 276
beringensis, Alopex 279
Spermophilus 448
beringiana, Ursus 338
Vulpes 287
beringianus, Alopex 279
Sorex 113, 116
berkeleyensis, Dipodomys 478
berlandieri, Cryptotis 109
Sigmodon 747
Taxidea 325
berlepschii, Akodon 688
bernardi, Canis 281
Dendromus 543
Ondatra 532
bernardinus, Chaetodipus 483
Eptesicus 201
Microtus 523
Spermophilus 447
bernardus, Macropus **53**
berneri, Mimetillus 207
berneyi, Leggadina 601
bernieri, Lagorchestes 52
bernisi, Clethrionomys 508
bernsteini, Coelops 170
Pseudocheirus 59
beruensis, Babyrousa 377
Berylmys **580**, 580, 603, 611, 632
berdmorei **580**, 580
bowersi **580**, 580, 614, 634
feae 580
ferreocanus 580
kekrimus 580
kennethi 580
lactiventer 580
latouchei 580
mackenziei **580**
magnus 580
manipulus **580**
mullulus 580
totipes 580
wellsi 580
berytensis, Nannospalax 753
besar, Papagomys 637
besara, Tupaia 133
besuki, Callosciurus 422
Niviventer 634
beta, Crocidura 85
bethuliensis, Rhabdomys 662
betpakdalaensis, Selevinia **768**, 768
betsileoensis, Brachyuromys **677**
Bettongia **49**
anhydra 49
campestris 49
cuniculus 49
formosus 49
francisca 49
gaimardi **49**
gouldii 49
graii 49
harveyi 49
hunteri 49
lepturus 49
lesueur **49**, 49
lesueuri 49
minimus 49
ogilbyi 49
penicillata **49**
phillippi 49
rufescens 48
setosa 49
tropica 49
whitei 49
Bettongiops 49
bettoni, Dendrohyrax 373
Genetta 346
betulina, Sicista **496**, 497
betulinus, Pteromys 464
beugalensis, Nycticebus 248
bezoartica, Antilope 395
Oryx 413
bezoarticus, Cervus 391
Ozotoceros **392**
bhaktii, Tylonycteris 228
bhatnagari, Alticola 504
bhotia, Dremomys 425
Rattus 660
bhutanensis, Callosciurus 421
biacensis, Petaurus 61
biakensis, Rattus 653
biarmicus, Lepus 817
bibianae, Akodon 689
Bibimys **696**, 696, 745
chacoensis **696**
labiosus **696**, 696
torresi **696**, 696
Bibos 401
bicinctus, Tolypeutes 67
bicolor, Alouatta 255
Arctogalidia 343
Carollia 186
Cephalophus 411
Coendou **775**, 775, 776
Colobus 270
Colomys 586
Crocidura 85
Ctenomys 785
Eptesicus 203, 204
Geoxus 702
Hesperomys 715
Hipposideros **171**, 171, 173, 174
Kangurus 57
Kerivoula 198
Monachus 331
Mus 625
Mustela 324
Neomys 110
Nyctophilus 218
Oecomys **715**, 715, 716
Pipistrellus 219
Pongo 277
Propithecus 247
Ratufa **437**
Saguinus **253**
Spilogale 317
Taphozous 161
Tapirus 371
Thyroptera 196
Wallabia **57**
bicornis, Diceros **372**
Hipposideros 171
Rhinoceros 372
biddulphi, Lepus 821
bidens, Boneia **138**, 138
Elasmodontomys 805
Mesoplodon **362**
Physeter 362
Sturnira **192**
Tonatia **180**
Vampyressa **193**
Vampyrus 180
bidentatus, Aotus 255
bidiana, Suncus 103
biedermanni, Cervus 385
Dactylopsila 61
Gazella 397
Gulo 318
Mustela 322
bieni, Mus 625
bieti, Felis **289**, 291
Pygathrix **272**
Bifa 767
bifax, Nyctophilus 218
bifer, Rhinolophus 166
biformatus, Rattus 653
bigalkei, Cryptomys 772
bihastatus, Rhinolophus 166
bilabiatum, Phyllostoma 191
Pygoderma **191**
bilarni, Parantechinus **33**
bilensis, Gerbillus **549**, 554
bilimitatus, Callosciurus 422
bilineata, Antilope 395
Saccopteryx **159**
bilineatus, Callosciurus 422
bilkiewiczi, Hyaena 308
bilkis, Gazella **396**
bilkjewiczi, Dryomys 766
billardierii, Thylogale **57**
billingae, Taurotragus 403
billitonus, Callosciurus 422
Tragulus 383
bilobatum, Uroderma **193**, 193
bilobidens, Galea 779
bimaculata, Eira 318
bimaculatus, Callosciurus 421
Calomys 697, 698
Perognathus 485
bindai, Castor 467
bini, Genetta 346
binoe, Macropus 53
binominata, Mustela 322
binominatus, Maxomys 614
Microtus 530
Spermophilus 448
binotata, Nandinia **342**
Viverra 342
bintangensis, Macaca 266

binturong, Arctictis **342**, 342
biologiae, Eira 318
bipes, Jaculus 494
bipunctatus, Aotus 256
bira, Mazama 391
birdseyei, Thomomys 473
birmanica, Echinosorex 79
birmanicus, Bos 401
Elephas 367
Herpestes 305
Paradoxurus 344
birrelli, Petaurista 463
birulae, Stylodipus 495
birulai, Castor 467
Mustela 321
Phoca 332
birungensis, Funisciurus 427
Mus 629
biscayensis, Eubalaena 349
bishopi, Clyomys **793**
Lepus 814
Marmosops 20
bismarckensis, Miniopterus 231
Bison **400**
americanus 400
arbustotundrarum 400
athabascae 400
bison **400**, 400
bonasus **400**
caucasicus 400
europaeus 400
haningtoni 400
hungarorum 400
montanae 400
nostras 400
oregonus 400
pennsylvanicus 400
septemtrionalis 400
urus 400
bison, Bison **400**, 400
Bos 400
bisonophagus, Ursus 338
bistriata, Isothrix 792, **793**, 793
bistriatus, Isothrix 793
bisulcatus, Otomys 680
bisulcus, Equus 390
Hippocamelus **390**
Biswamoyopterus **459**
biswasi **459**, 459
biswasi, Biswamoyopterus **459**, 459
biturigum, Giraffa 383
bivattus, Lagenorhynchus 353
bivirgata, Mephitis 317
bivittata, Chaerephon **232**, 232
Tamandua 68
blackleri, Meriones 558
blainei, Ammotragus 404
Cryptomys 772
Oryx 413
blainvillei, Delphinus 360
Echimys **791**
Pontoporia **361**
blainvillii, Bradypus 63
Dendrohyrax 373
Heterohyrax 373
Mormoops **176**, 176
blakistoninus, Cervus 386
blanca, Spermophilus 450
blancalis, Hemiechinus 78
blanci, Gerbillus 549
blandus, Belomys 459
Peromyscus 732
blanfordi, Alticola 503
Barbastella 199
Cremnomys **587**, 587, 588
Cynopterus 153
Equus 370
Hemiechinus 78
Hesperoptenus **204**
Hystrix 774
Jaculus **493**, 494
Meles 314
Myotis 213
Ovis 409
Sphaerias **153**
blanfordii, Callosciurus 423
Suncus 103
Blanfordimys **506**, 506, 517, 518
afghanus **506**, 506
balchanensis 506
bucharicus **506**, 506
dangarinensis 506
davydovi 506
blangorum, Rattus 661
Blaria 106
Blarina **106**, 106
aloga 106
angusta 106
angusticeps 106
brevicauda **106**, 106
carolinensis **106**, 106
churchi 106
compacta 106
costaricensis 106
dekayi 106
fossilis 106
hooperi 106
hulophaga 106
hylophaga **106**, 106
kirtlandi 106
manitobensis 106
micrurus 106
mimina 106
ozarkensis 106
pallida 106
peninsulae 106
plumbea 106
pyrrhonota 112
shermani 106
simplicidens 106
talpoides 106
telmalestes 106
blarina, Myosorex **99**, 99
Blarinella **106**
griselda 106
quadraticauda **106**, 107
wardi **106**
Blarinini 106, 107, 108, 110
Blarinomys **696**
breviceps **696**
blaseri, Gracilinanus 17
blasii, Myotis 209
Rhinolophus **164**
Blastoceros 391, 392
Blastocerus **389**, 392
azpatianus 389
dichotomus **389**
ensenadensis 389
furcata 389
melanopus 389
paludosus 389
palustris 389
blepotis, Miniopterus 231
bleyenberghi, Panthera 297
blicki, Arvicanthis **576**, 576, 578, 606, 664, 686
blighi, Dasycercus 31
blitchi, Sorex 114
blossevillii, Lasiurus 206
blouchi, Sus 379
bloyeti, Crocidura 87
bluntschlii, Saguinus 253
blythi, Capra 405
Dasycercus 31
Hipposideros 175
Niviventer 634
Ovis 408
Rhinolophus 165, 168
blythianus, Bandicota 579
blythii, Callosciurus 423
Dicerorhinus 372
Macaca 267
Myotis **208**
Pipistrellus 220
Suncus 103
bobac, Marmota 431
bobak, Marmota **431**, 431, 432, 433
bobrinskii, Allactodipus 488, **490**, 490
bobrinskoi, Eptesicus **200**, 202
bocagei, Aethomys **567**, 568
Cryptomys **771**
Galerella 303
Heterohyrax 373
Myotis **209**
Rousettus 152
Steatomys 546
Tadarida 240
boccamela, Mustela 323
bocharicus, Rhinolophus 164
bochariensis, Ovis 409
bocki, Callosciurus 422
bocourti, Callosciurus 422
Rattus 652
bocourtianus, Macrotus 178
bodenheimeri, Meriones 558
Pipistrellus **220**
bodessae, Tatera 562
bodessana, Tatera 561
bodius, Apodemus 571
boedeckeri, Phyllotis 738
boehmi, Equus 369
Paraxerus **434**
Tatera **560**, 560
boehmii, Spermophilus 448
boettgeri, Nesokia 632
bogdanovi, Dinaromys **511**, 511, 512
Meriones 558
Microtus 511
bogdanowii, Crocidura 86
bogdoensis, Microtus 530
bogoriensis, Tupaia 132
bogotensis, Akodon **689**, 689, 690, 691
Alouatta 255
Artibeus 188
Sigmodon 747
Sturnira **192**, 192
Tylomys 751
bohemicus, Sorex 111
bohmi, Tatera 560
bohor, Redunca 414
boiei, Hemigalus 342
boimensis, Proechimys 797
bokcharensis, Hyaena 308
bokermanni, Lonchophylla **181**
bolabatuensis, Babyrousa 377
bolami, Pseudomys **645**
bolei, Bassariscus 334
bolivari, Cervus 385
Crocidura 97
bolivaris, Oryzomys **720**, 720, 721
boliviae, Calomys **697**, 698
Dasyprocta 782
Euphractus 66
Lenoxus 707
Leopardus 292
Oryzomys 720, 724
boliviana, Chinchilla 777
bolivianus, Metachirus 20
Proechimys **795**, 796
boliviensis, Abrocoma **789**
Akodon 688, **689**, 689, 690, 691, 692, 693
Aotus 255
Auliscomys **695**
Bradypus 63
Chaetophractus 65
Coendou 775
Ctenomys **784**
Dactylomys **790**
Dasypus 66
Galea 779
Inia 360
Mustela 322
Nasua 334
Panthera 298
Phylloderma 180
Saimiri **260**, 260
Sciurus 441
bolkayi, Erinaceus 77
Sorex 111
Bolomys 688, 691, **696**, 696,

697, 700, 752
amoenus **696**
arviculoides 696
benefactus 697
brachyurus 696
elioi 697
fuscinus 696, 697
lactens **696**
lasiotus 696
lasiurus 689, **696**, 696, 697, 752
lenguarum 696, 697
leucolimnaeus 696
liciae 697
negrito 696
obscurus **697**
orbus 696
orobinus 696
pixuna 696
punctulatus **697**, 697
tapirapoanus 696
temchuki **697**
bolovensis, Callosciurus 421
bombascarae, Marmosa 18
bombax, Golunda 592
bombaya, Ratufa 437
bombifrons, Erophylla 182
Kerivoula 198
Phyllonycteris 182
bombinus, Myotis **209**, 213
bombycinus, Oryzomys 720
Perognathus 485
Thomasomys **749**
bombyx, Rhogeessa 226
bonaerensis, Balaenoptera 349
bonaeriensis, Phyllotis **737**, 738
bonafidei, Proechimys 797
bonapartei, Genetta 345
Pipistrellus 223
bonaria, Lutreolina 18
bonariensis, Calomys 698
Didelphis 16
Eumops **233**
Lasiurus 206
Myocastor 806
bonasus, Bison **400**
bondae, Molossus **235**
Myotis 213
Sciurus 441
bondar, Paradoxurus 344
Boneia **138**
bidens **138**, 138
menadensis 138
bonettoi, Ctenomys **784**
bongensis, Heliosciurus 428
bonhotei, Callosciurus 421, 422
Gerbillus **549**, 549, 552
boninensis, Cervus 386
Mus 623
bonnevillei, Dipodomys 479
Thomomys 473
Bonobo 276
bontanus, Rattus 650, **651**, 652, 662
bonzo, Eothenomys 514
Boocercus 403, 404
booduga, Mus **622**
boonsongi, Callosciurus 422
boops, Balaenoptera 350
boothi, Potos 333
Pteronotus 177
boothiae, Sciurus 443
boquetensis, Microsciurus 433
bor, Tragelaphus 404
borana, Heterohyrax 373
boranus, Elephantulus 829
borbensis, Puma 296
borbonicus, Scotophilus **226**, 227, 228
borealis, Balaenoptera **350**, 350
Cystophora 329
Gulo 318
Hydrodamalis 365
Lasiurus **206**, 207
Lemmus 517
Lepus 814, 817, 821
Lissodelphis **354**
Lynx 293
Mus 625
Mustela 322, 324
Odocoileus 391
Ovis 409
Peromyscus 732
Rangifer 392
Sorex 121
Synaptomys **534**
Tamias 454
Thomomys 475
Urocyon 285
Vespertilio 206
boregoensis, Thomomys 473
borensis, Equus 369
boreoamericana, Didelphis 17
Boreolagus 814
boreorarius, Thomomys 473
boria, Martes 319
Boriogale 53
Borioikon 509
boristhenicus, Spermophilus 449
borjasensis, Thomomys 473
borkumensis, Oryctolagus 822
borlei, Connochaetes 394
bornaensis, Pongo 277
borneanus, Cynocephalus 135
Nannosciurus 434
Nycticebus 248
Pipistrellus 220
Pteromyscus 465
Sundamys 666
Tarsius 250
Tragulus 383
borneensis, Elephas 367
Rhinolophus **164**
Sousa 355
Sundasciurus 451
Sus 379
borneoensis, Callosciurus 423
Myotis 213
Petinomys 463
Prionailurus 295
borneotica, Echinosorex 79
bornouensis, Syncerus 402
Boromys **799**
offella **799**, 799
torrei **799**
bororensis, Mungos 307
Paraxerus **434**
borreroi, Zygodontomys 752
borucae, Sigmodon 747
Bos **400**, 400
annamiticus 401
asseel 401
banteng 401
birmanicus 401
bison 400
brachyceros 402
brachyrhinus 401
bubalis 402
bunnelli 401
butleri 401
caffer 402
cavifrons 401
discolor 401
domesticus 401
frontalis **401**
fuscicornis 401
gaurus 401
grunniens **401**, 401
guavera 401
hubbacki 401
indicus 401
javanicus **401**, 401
laosiensis 401
leucoprymnus 401
longicornis 401
lowi 401
moschatus 407
mutus 401
platyceros 401
porteri 401
primigenius 401
readei 401
sauveli **401**
sondaicus 401
subhemachalus 401
sylhetanus 401
sylvanus 401
taurus 400, **401**
urus 401
boscai, Eptesicus 203
Boselaphus **401**
albipes 402
hippelaphus 402
picta 402
risia 402
tragocamelus **401**
bosmannii, Perodicticus 248
bosniensis, Clethrionomys 508
Ursus 338
Bothriomys 701
botnica, Phoca 332
botswanae, Laephotis **206**
bottae, Eptesicus **200**, 202, 203
Thomomys **473**, 473, 474, 475, 476
bottai, Gerbillus **549**, 549, 551
bottegi, Crocidura **82**, 92
bottegoi, Cephalophus 410
bottegoides, Crocidura **82**
bougainville, Melomys **615**, 615
Nyctimene 144, 145
Perameles **40**
bougainvillei, Trichosurus 48
bounhioli, Procavia 374
bourquii, Raphicerus 399
boutourlinii, Cercopithecus 264
bouvieri, Felis 290
Procolobus 272
bovallii, Rhipidomys 744
Bovidae 392, **393**
Bovinae 400
Bovini 400
bovonei, Acomys 567
bowdoini, Mesoplodon **362**, 362
boweri, Mesembriomys 620
bowersi, Berylmys **580**, 580, 614, 634
bowkeri, Pronolagus 823
boxi, Lagidium 778
boydi, Crocidura 93
Funisciurus 428
boylei, Sylvilagus 825
boylii, Peromyscus **728**, 728, 729, 732, 734, 735, 736
bozasi, Lophiomys 564
braccata, Oncifelis 294
braccatus, Ateles 257
Galago 249
brachelix, Hyperacrius 515
brachialis, Heteromys 481
Brachiones **546**, 559
arenicolor 547
callichrous 547
przewalskii **546**
Brachiopithecus 274
brachiotis, Akodon 692
Brachitanytes 274
brachiura, Spermophilus 445
Brachycapromys 800, 802
brachycephala, Myonycteris **143**
brachycephalus, Platyrrhinus **190**, 191
brachycercus, Microtus 529
brachyceros, Bos 402
Cervus 386
Odocoileus 391
Syncerus 402
brachychaites, Colobus 270
brachycrania, Talpa 129
brachydactyla, Arctocephalus

326
Lontra 311
brachydactylus, Bradypus 63
Hemiechinus 78
brachydigitatis, Eptesicus 203
brachygnathus, Rhinolophus 164
Brachylagus **813**, 813, 824
bachmani 813
idahoensis **813**
brachymeles, Molossops 234
brachyotis, Allactaga 489
Cynopterus **139**
Hemiechinus 78
Hipposideros 173
Lasiurus 206
Micronycteris **178**
Petrogale **55**
Rousettus 152
Xerus 458
Zyzomys 675
brachyotos, Pipistrellus 223
Potos 333
brachyotus, Carollia 186
Neomys 110
Oryctolagus 822
Potos 333
Brachyphylla **182**
cavernarum **182**, 182
intermedia 182
minor 182
nana **182**
pumila 182
Brachyphyllinae **182**
brachypterus, Globicephala 352
Mops **236**
Philetor **219**
Vespertilio 219
brachypus, Cervus 386
brachyrhinus, Bos 401
Cervus 386, 387
Naemorhedus 407
Rattus 659
brachyrhynchus, Elephantulus **829**
brachysoma, Cynopterus 139
Brachysorex 106, 108
Brachytarsomyes 677
Brachytarsomys **677**, 677
albicauda **677**, 677
villosa 677
brachytarsus, Akodon 691
Thylogale 57
Brachyteles **257**
arachnoides **257**
eriodes 257
hemidactylus 257
hypoxanthus 257
macrotarsus 257
tuberifer 257
brachyteles, Cerdocyon 282
Brachyteleus 257
brachyura, Cynictis 302
Euryzygomatomys 794
Hystrix **773**, **774**
Leptailurus 292
Martes 321
Monodelphis 21
Nesokia 632
Brachyuromys **677**, 677
betsileoensis **677**
ramirohitra 677, **678**
brachyuros, Monodelphis 21
Brachyurus 261, 516
brachyurus, Allactaga 489
Bolomys 696
Canis 282
Chrysocyon **282**
Geocapromys 801
Herpestes **304**, 305, 306
Kangurus 56
Lepus **815**, 817, 819
Macaca 267
Neofelis 297
Procyon 335
Setonix **57**
Tolypeutes 67
bracteator, Macropus 54
bradfieldi, Cynictis 302
Galago 249
Galerella 303
Helogale 304
Mastomys 610
Sylvicapra 412
Thallomys 668
bradleyi, Steatomys 545
bradshawi, Cephalophus 411
Bradypodidae **63**, 63
Bradypus **63**
ai 63
beniensis 63
blainvillii 63
boliviensis 63
brachydactylus 63
brasiliensis 63
castaneiceps 63
codajazensis 63
crinitus 63
cristatus 63
cuculliger 63
cummunis 63
didactylus 64
dorsalis 63
ecuadorianus 63
ephippiger 63
flaccidus 63
gorgon 63
griseus 63
gularis 63
ignavus 63
infuscatus 63
macrodon 63
mareyi 63
marmoratus 63
melanotis 63
miritibae 63
nefandus 63
pallidus 63
problematicus 63
striatus 318
subjuruanus 63
tocantinus 63
torquatus **63**, 63
tridactylus **63**, 63
trivittatus 63
unicolor 63
ursinus 337
ustus 63
variegatus **63**
violeta 63
brahma, Niviventer 632, **633**, 634
Trachypithecus **274**
braima, Hipposideros 171
branderi, Cervus 385
brandti, Mesocricetus 538, **539**, 539
Myotis **209**
brandtii, Arvicola 515
Coendou 775
Lasiopodomys **515**
branickii, Dinomys **778**, 778
branti, Sus 378
brantsii, Euryotis 682
Parotomys **682**, 682
Tatera **560**
brasiliense, Tonatia **180**
brasiliensis, Bradypus 63
Cerdocyon 282
Choloepus 64
Ctenomys 783, **784**, 786, 787
Didelphis 16
Echimys 791
Eptesicus **201**
Galictis 318
Holochilus **704**, 704, 705
Lasiurus 206
Leontopithecus 253
Leopardus 291
Monodelphis 21
Mus 704
Mustela 313, 322
Nectomys 709
Phyllomys 791
Potos 333
Procyon 335
Pseudalopex 284
Pteronura **313**
Sylvilagus **825**, 825
Tadarida **240**
Tapirus 371
brassi, Pogonomelomys 641
braueri, Dendrohyrax 373
Desmodilliscus **547**, 547
brauneri, Apodemus 571
Cervus 385
Microtus 519
Spermophilus **448**
Talpa 129
brauni, Heliosciurus **429**
bravoi, Canariomys 582
brazenori, Pseudomys 646, 647
Rattus 653
brazensis, Geomys 469
brazierhowelli, Thomomys 473
braziliensis, Carollia 186
Echimys **791**, 792
Megaptera 350
Sylvilagus 825
brazzae, Cercopithecus 264
brazziformis, Cercopithecus 264
brecciensis, Microtus 519
bredanensis, Delphinus 357
Steno **357**
bregullae, Chaerephon 232
brelichi, Pygathrix **272**
bresslaui, Chironectes 16
breviaculeata, Tachyglossus 13
breviauritus, Lepus 821
Onychomys 719
brevicauda, Apodemus 572
Balaenoptera 350
Blarina **106**, 106
Carollia **186**
Crocidura 85
Microtus 521
Neotoma 710
Oryzomys 752
Proechimys **795**, 797
Spermophilus 446, 447
Zygodontomys 697, **752**, 752, 753
brevicaudata, Chinchilla **777**, 777
Didelphis 21
Felis 290
Microgale **71**, 71
Monodelphis **21**
Ourebia 399
Phascolosorex 34
brevicaudatus, Akodon 692
Desmodillus 547
Indri 247
Macaca 267
Meriones 557
Rattus 650
Setonix 57
Spilocuscus 47
brevicaudus, Clethrionomys 508
Lophuromys 605
Onychomys 719
Peromyscus 732
Rattus 660
Sorex 106
Spermophilus 447
Taphozous 160
breviceps, Blarinomys **696**
Cephalophus 410
Cervus 385
Didelphis 17
Echymipera 41
Geomys 469, 470
Hipposideros **171**
Kobus 413
Kogia **359**
Lagenorhynchus 354

Microgale 71
Oxymycterus 696
Petaurus **61**
Phalanger 46
Physeter 359
Pteropus 149
Thylacinus 29
brevicorpus, Microtus 525
brevicula, Echiothrix 591
brevidens, Thomomys 473
brevimaculatus, Dendrohyrax 373
brevimanus, Natalus 195
Plecotus 224
Stenella 356
brevimolaris, Bunomys 581
brevinasus, Dipodomys 479
Lepus 821
Pappogeomys 473
Perognathus 485
Phalanger 46
brevipes, Plecotus 224
Tragulus 382
Zapus 499
brevipilosus, Lontra 311
brevirostris, Chaetophractus 65
Cormura **156**
Dasypus 66
Emballonura 156
Geomys 469
Glossophaga 184
Microsciurus 433
Microtus 526
Mus 625
Myotis 212
Orcaella **354**, 354
Ornithorhynchus 13
Ototylomys 726
Reithrodontomys **740**, 742
brevirostrum, Anoura 183
brevis, Myotis 216
brevispinosa, Hystrix 774
brevitarsus, Rhinolophus 166
breweri, Microtus **519**, 519, 528
Parascalops **127**
Scalops 127
breyeri, Mastomys 610
Miniopterus 231
Tatera 560
bricenii, Mazama **390**
bridgemani, Muntiacus 389
Paraxerus 435
bridgeri, Thomomys 475
bridgesi, Octodon **788**, 788
brigantium, Arvicola 506
brighti, Gazella 396
brindei, Cercopithecus 264
brissonii, Ateles 257
Eulemur 244
britannicus, Clethrionomys 508
Meles 314
Microtus 518
britta, Cercartetus 58
broadbenti, Peroryctes **42**, 42
broca, Macaca 267
broccus, Chaetodipus 484
brochus, Syntheosciurus **452**, 452
brocki, Vampyressa **193**
brockmani, Acomys 566
Felis 290
Gerbillus **549**, 549
Graphiurus 765
Mellivora 315
Myomys 631
Nycteris 162
Panthera 298
Papio 269
Rhinolophus 164, 166
brodiei, Funambulus 426
bromleyi, Clethrionomys 509
brookei, Cephalophus 411
Cervus 387
Ochotona 808, 809
Ovis 408
Pongo 277
Rattus 658
Sundasciurus **451**
brooksi, Delanymys **683**, 683
Dyacopterus 140
Ochotona 811
Sorex 119
brooksiana, Noctilio 176
broomi, Gerbillurus 547
Otomys 682
Pipistrellus 221
Brotomys **799**
contractus **799**
voratus **799**, 799
broweri, Marmota **431**, 431, 432
browni, Microsciurus 433
Myotis 213
Puma 296
Thylogale 57
brownii, Geocapromys 800, **801**, 801
brucei, Cervus 386
Heterohyrax **373**, 373
Hyrax 373
Vulpes 287
bruchi, Thylamys 23
bruchus, Petromyscus 683, 684
brucii, Allactaga 489
Diceros 372
bruecheri, Crocidura 96
bruijni, Myoictis 32
Pogonomelomys **641**, 641
Tachyglossus 13
Thylogale 57
Zaglossus **13**
bruijnii, Coccymys 585
brumalis, Martes 319
brumbacki, Aotus **255**
brunello, Callicebus 259
brunensis, Dipodomys 479
bruneoochracea, Galerella 303
bruneri, Erethizon 776
bruneti, Euphractus 66
brunetta, Helogale 304
Tupaia 131
bruneus, Eulemur 244
brunii, Didelphis 51, 52
Dorcopsis 52
Thylogale 51, **57**
brunnea, Crocidura 85
Eira 318
Hyaena 308, 309
Kerivoula 197
Parahyaena **309**, 309
Peropteryx 158
Planigale 34
Sciurus 443
brunneipes, Paradoxurus 344
brunneo-niger, Sciurus 442
brunnescens, Gerbillus 550
Macaca 267
Ochotona 811
Spermophilus 446
brunneus, Ateles 257
Baiomys 695
Callicebus **258**
Cebus 260
Ctenomys 785
Eptesicus **201**
Lophuromys 605
Macaca 266
Microtus 521
Planigale 34
Procolobus 272
Pteropus **146**, 147
Rattus 660
Spermophilus **445**
Tadarida 240
Tragelaphus 404
Ursus 338
Zygodontomys **752**
brunneusculus, Rattus 658, 659
brunnula, Helogale 304
Bruynia 13
bryanti, Chaetodipus 484
Neotoma **710**
Sciurus **442**
brydei, Balaenoptera 350
Bubalibos 401
bubalina, Naemorhedus 407
Bubalis 393
lichtensteinii 394
bubalis, Alcelaphus 393
Antilope 393
Bos 402
Bubalus **402**, 402
Bubalus **402**, 402
anoa 402
arnee 402
bubalis **402**, 402
celebensis 402
depressicornis **402**
fergusoni 402
fulvus 402
hosei 402
indicus 402
kerabau 402
macroceros 402
mainitensis 402
mephistopheles **402**
mindorensis **402**
moellendorfii 402
platycerus 402
quarlesi **402**
septentrionalis 402
bubastis, Alcelaphus 393
Felis 290
bubuensis, Syncerus 402
Zaglossus 13
buccalis, Cercopithecus 263
buccatus, Spermophilus 450
buccinatus, Oryzomys **721**, 721, 724, 725
bucculentus, Sus **378**
buchanani, Cricetomys 541
Desmodilliscus 547
Mellivora 315
Procavia 374
bucharensis, Myotis 211
bucharicus, Blanfordimys **506**, 506
Microtus 506
buchariensis, Lepus 821
Sorex **113**, 116, 121, 122
buckleyi, Spermophilus 450
Budamys 622
budapestiensis, Myotis 210
budgetti, Chlorocebus 265
budina, Mustela 322
budini, Abrocoma 789
Akodon **689**, 692
Conepatus 316
Ctenomys 784
Micoureus 20
Oncifelis 294
Budorcas **404**
bedfordi 404
mitchelli 404
sinensis 404
taxicolor **404**, 404
tibetana 404
whitei 404
bueae, Crocidura 92
buechneri, Meriones 557
Petaurista 463
Pteromys 464
buehleri, Coryphomys **587**, 587
buenavistae, Gracilinanus 17
Melanomys 708
buergersi, Dendrolagus 50
Pseudochirops 60
buettikoferi, Crocidura **82**, 82
Eidolon 141
Epomops **141**
buettneri, Leimacomys **544**, 544
buffoni, Hemicentetes 73
Leopardus 291
buffonii, Arvicola 506
Cebus 259
Georychus 772
Kobus 414

buffonnii, Tarsius 251
bufo, Mus **622**, 622
Phacochoerus 377
bugi, Eulemur 244
bukit, Niviventer 634, 635
Rattus 635
bulana, Ratufa 437
bulbivorus, Thomomys 473, **474**
bulganensis, Allactaga 489
Dipus 493
bulgaricus, Myomimus 768
Myotis 213
bullaris, Tylomys **751**
bullata, Allactaga **488**, 488
Nasua 334
Neotoma 712
bullatior, Neotoma 711
bullatus, Glaucomys 460
Pappogeomys 472
Perognathus 486
Peromyscus **729**
Rattus 650
Synaptomys 534
Thomomys 475
bulleri, Geomys 471
Liomys 482
Macrotus 178
Oryzomys 721
Pappogeomys **471**, **472**
Tamias **453**, 454
Bullimus **580**, 580, 638, 666
bagobus 580, **581**, 581
barkeri 581
luzonicus **581**, 581
rabori 581
bullocki, Geoxus 702
Rattus 660
bulmerae, Aproteles **138**, 138
bunae, Rattus 659
bungei, Lemmus 517
Marmota 431
bunguranensis, Leopoldamys 603
Ratufa 437
Tragulus 383
bunkeri, Antrozous 198
Marmota 432
Neotoma **710**
Perognathus 485
Rhinolophus 169
bunnelli, Bos 401
bunnii, Nesokia **632**
bunoae, Tupaia 133
Bunolagus **813**, 814
monticularis **813**
Bunomys **581**, 581, 639, 667
adspersus 581
andrewsi **581**, 581
brevimolaris 581
chrysocomus **581**, 581, 582
coelestis **581**, 581, 582
fratrorum 581, **582**, 582
heinrichi 581, **582**
inferior 581
koka 581
nigellus 581
penitus **582**, 582
prolatus **582**
rallus 581
sericatus 582
Bunopithecus 274, 275
buntingi, Grammomys 590, **592**, 592, 596, 599, 601, 607, 631, 644
bunya, Melomys 616
burbidgei, Petrogale **56**
burchellii, Ceratotherium 371
Equus **369**, 369, 370
Sylvicapra 412
bureschi, Myotis 209
burius, Rhinolophus 165
burlacei, Mandrillus 268
Tragelaphus 404
burmeisterei, Hyperoodon 361
burmeisteri, Megaptera 350
Burmeisteria 64
retusa 64
burnettii, Cercopithecus 263
burneyi, Sorex 117
Burramyidae 46, **57**, 62
Burramys **57**
parvus **57**, 57
burrelli, Antechinus 30
burrescens, Rattus 651
burrhel, Pseudois 409
burrowsi, Atherurus 773
burrulus, Rattus 660
burrus, Proechimys 798
Rattus 649, **651**, 651, 661
bursarius, Geomys **469**, 469, 470
burti, Pappogeomys 472
Reithrodontomys **740**
Thomomys 475
Burtognathus 482, 483
burtoni, Gerbillus **549**
Melomys **615**, 615, 616
Otomys 681
Praomys 644
burtonii, Procavia 374
buruensis, Cervus 387
Rattus 652
buryi, Meriones 557
buselaphus, Alcelaphus **393**, 395
Antilope 393
bushengensis, Apodemus 574
buskensis, Rangifer 392
busuangae, Tupaia 133
butangensis, Maxomys 614
butei, Apodemus 573
butleri, Bos 401
Crocidura 97
Graphiurus 764
Jaculus 493
Myomys 631
Procavia 374
Sminthopsis **35**
Taterillus 562
butlerowi, Lepus 821
Butragus 393
butskopf, Hyperoodon 361
butteri, Gazella 397
büttikoferi, Cercopithecus 264
buturlini, Alces 389
Microtus 521
buxtoni, Nesokia 632
Sorex 113
Spermophilus 448
Tragelaphus **403**
bweha, Canis 280
byatti, Paraxerus 435
byrnei, Dasycercus **31**
byroni, Phoca 328
byronia, Otaria **328**
byronii, Monachus 331

caama, Alcelaphus 393
Vulpes 286
caballus, Equus **369**, 369
Cabassous **64**
centralis **64**
chacoensis **65**
dasycercus 65
duodecimcinctus 65
gymnurus 65
hispidus 65
latirostris 65
loricatus 65
multicinctus 65
nudicaudus 65
octodecimcinctus 65
squamicaudis 65
tatouay **65**
unicinctus **65**
verrucosus 65
cabezonae, Dipodomys 477
Thomomys 473
caboti, Rangifer 392
cabrerae, Microtus **519**
Puma 296
Rhinolophus 165
cabrerai, Crocidura 91
Gazella 396
Noctilio 176
Sciurus 441
Spermophilus 449
Cabreramops 234
Cabreramys 696, 697
cacabatus, Peromyscus 734
Scotinomys 746
Cacajao **260**
alba 261
calvus **261**
melanocephalus **261**
novaesi 261
ouakary 261
rubicundus 261
spixii 261
ucayalii 261
Cachicamus 65
cachinus, Eothenomys 514
Graomys 703
cacodemus, Tamias 454
cacomitli, Herpailurus 291
cacondae, Lycaon 283
cacsilensis, Lama 382
cactophilus, Thomomys 473
cadaverinus, Ursus 338
cadornae, Pipistrellus **220**
caeca, Talpa **128**, 129
caecator, Castor 467
caecias, Pseudochirops 60
caecus, Nyctinomops 239
caecutiens, Cryptomys 772
Sorex **113**, 113, 120
caedis, Callosciurus 423
Tupaia 132
Caenolestes **25**
barbarensis 25
caniventer **25**
centralis 25
convelatus **25**
fuliginosus **25**
obscurus 25
tatei 25
Caenolestidae **25**
caenosus, Akodon 692
Cremnomys 588
Ozotoceros 392
caerula, Cephalophus 411
caerulaeus, Suncus 103
caerulea, Alopex 279
caerulescens, Suncus 103
caeruleus, Felis 290
Rattus 658
Suncus 103
caesarius, Clethrionomys 508
caesia, Pseudois 409
Vulpes 286
caesius, Cricetulus 538
cafer, Myosorex **99**, 99, 100
Pedetes 759
caffer, Bos 402
Cephalophus 411
Desmodillus 547
Hipposideros **171**
Mastomys 611
Otocyon 283, 284
Syncerus **402**
caffra, Felis 290
Galerella 303
Redunca 414
Sylvicapra 412
cafra, Felis 290
Herpestes 305
cagayanus, Macaca 266
Pteropus 147
cagottis, Canis 280
cagsi, Sundasciurus 451
cahirinus, Acomys **565**, 565, 566
Mus 564
cahni, Chlorotalpa 75
cahyensis, Caluromys 15
caicarae, Oecomys 716
caissara, Leontopithecus **252**
cajennensis, Proechimys 798
cajopolin, Caluromys 15
calabarensis, Arctocebus 247, **248**
Crocidura 93

Mops 236
Perodicticus 247
calaboricus, Aonyx 310
calabyi, Pseudomys 647
calago, Galago 249
calamarum, Microtus 521
calamianensis, Axis **384**
Chiropodomys **583**, 583, 584
Sus 378
calamophagus, Thryonomys 775
calarius, Aethomys 568
calcarata, Pipistrellus 221
calcaratum, Carollia 186
calcaratus, Hipposideros **171**, 174
Vespertilio 156
calcis, Rattus 652
Calcochloris **74**, 74
chrysillus 74
limpopoensis 74
obtusirostris **74**
caldatus, Galerella 303
caldwelli, Rhinolophus 167
caletensis, Plagiodontia 804
calidior, Dremomys 425
Lemniscomys 602
Melomys 619
Proechimys 798
calidus, Gerbillurus 547
Rhinolophus 168
californiana, Otaria 327, 329
Ovis 408
californianus, Callorhinus 327
Zalophus **329**, 329
californica, Aplodontia 417
Didelphis 17
Lontra 311
Neotoma 712
Phoca 332
Puma 296
Taxidea 325
californicus, Chaetodipus **483**
Clethrionomys **508**, 508
Dipodomys **477**, 477, 478
Eumops 234
Glaucomys 460
Lepus **815**, 818
Lynx 293
Macrotus **178**
Microdipodops 480
Microtus **519**, 519, 524
Molossus 233
Myotis **209**
Odocoileus 391
Peromyscus 728, **729**, 729
Procyon 335
Scapanus 127
Sorex 118
Tadarida 240
Urocyon 285
Ursus 338
californiensis, Ursus 338
caligata, Felis 290
Marmota **431**, 431, 432
caligatus, Callicebus **258**
caliginea, Crocidura **82**
caliginosus, Cebus 259
Hesperomys 707
Marmota 433
Melanomys **708**
Myotis 213
Thomomys 475
callewaerti, Mus **622**, 622
calliaudi, Chlorocebus 265
Callicebidae 254
Callicebinae **258**
Callicebus **258**
acreanus 258
amictus 259
baptista 258
barbarabrownae 259
brunello 259
brunneus **258**
caligatus **258**
castaneoventris 258
chloroenemis 259
cinerascens **258**
crinicaudus 259
cupreus **258**, 258
discolor 258
donacophilus **258**
dubius **258**
duida 259
egeria 258
emiliae 259
geoffroyi 259
gigot 259
grandis 259
hoffmannsi **258**
hypokantha 259
ignitus 259
incanescens 259
leucometopa 258
lucifer 259
lugens 259
medemi 259
melanochir 259
melanops 259
migrufus 259
modestus **258**
moloch 258, **259**
napoleon 258
nigrifrons 259
oenanthe **259**
olallae **259**
ornatus 258
paenulatus 258
pallescens 258
personatus **259**
purinus 259
regulus 259
remulus 259
rutteri 258
sakir 259
subrufus 258
toppini 258
torquatus **259**
usto-fuscus 258
vidua 259
callichrous, Brachiones 547
callida, Dasyprocta 782
callidus, Calomys **697**, 697, 698
Chlorocebus 265
Callignathus 359
calligoni, Hemiechinus 78
Callimico **251**
goeldii **251**, 251
snethlageri 251
Callimiconidae 251
Calliope 403
callipeplus, Tamias 456
callipides, Apodemus 573
callipygus, Cephalophus **410**, 410, 411, 412
callistus, Perognathus 485
Callithricidae 251
Callithrix **251**, 252, 263
albicollis 252
albifrons 251
argentata **251**, 252
aurita **251**, 251, 252
chrysoleuca 252
chrysopyga 251
coelestis 251
communis 252
emiliae 251
flavescente 251
flaviceps **251**, 252
geoffroyi **251**, 252
hapale 252
humeralifera **251**, 252
intermedia 252
intermedius 251, 252
itatiayae 251
jacchus **252**
jordani 252
kuhlii **252**
leucippe 251
leucocephala 251
leucogenys 251
leucomerus 251
leucotis 252
leukeurin 251
maximiliani 251
melanoleucus 252
melanotis 252
melanura 251
nigra 252
niveiventris 252
penicillata **252**, 252
petronius 251
pygmaea **252**
santaremensis 252
sericeus 252
tertius 252
trigonifer 252
vulgaris 252
callithrix, Grammomys 594
Callitrichidae **251**
callitrichus, Chlorocebus 265
Taeromys **667**
Callitrix 261
calliurus, Protoxerus 436
calloceps, Ochotona 809
callopes, Melomys 615
Callorhinus 326, **327**
alascanus 327
californianus 327
curilensis 327
cynocephala 327
mimica 327
ursinus **327**
Callosciurinae 419
Callosciurini 419, 420, 425, 426, 428, 429, 430, 433, 434, 435, 438, 457
Callosciurus **420**, 420, 428, 435, 436, 438
abbottii 422
acraeus 422
adamsi **420**, 423
adangensis 421
albescens **420**, 420
albiculus 420
albifer 421
albivexilli 422
altinsularis 421
anambensis 422
andrewsii 422
annellatus 422
aoris 422
aquilo 421
arendsis 422
armalis 423
assamensis 423
atratus 422
atricapillus 423
atristriatus 422, 423
atrodorsalis 421
atrox 423
badjing 422
balstoni 422
baluensis **420**, 423
bangkanus 423
banksi 423
bantamensis 422
baramensis 421
bartoni 421
beebei 424
bellona 423
bentincanus 421
besuki 422
bhutanensis 421
bilimitatus 422
bilineatus 422
billitonus 422
bimaculatus 421
blanfordii 423
blythii 423
bocki 422
bocourti 422
bolovensis 421
bonhotei 421, 422
boonsongi 422
borneoensis 423
caedis 423
canalvus 423
caniceps **421**, 421, 422, 423
canigenus 421

careyi 421
carimatae 423
carimonensis 423
caroli 423
casensis 421
castaneoventris 421
centralis 421
chrysonotus 421
cinnamomeiventris 421
cinnamomeus 422
cockerelli 422
concolor 421
condurensis 423
conipus 422
contumax 421
coomansi 423
crotalius 421
crumpi 421
cucphuongis 421
dabshanensis 421
dactylinus 421
datus 422
davisoni 421
dextralis 422
dilutus 422
director 422
domelicus 421
dschinschicus 423
dulitensis 422
epomophorus 421
erebus 423
erubescens 421
erythraeus **421**
erythrogaster 421
erythromelas 421, 423
fallax 421
famulus 422
ferrugineus 422
finlaysonii **422**
flavimanus 421, 424
floweri 422
fluminalis 421
folletti 422
frandseni 422
fryanus 421
fumigatus 421
germaini 422
ginginianus 423
gloveri 421
gongshanensis 421
gordoni 421
griseicauda 423
griseimanus 421
griseopectus 421
grutei 422
guillemardi 422
haemobaphes 421
haringtoni 421
harmandi 422
harrisoni 423
hastilis 421
heinrichi 423
helgei 421
helvus 421
hendeei 421
herberti 422
humei 423
hyperythrus 421
ictericus 422
imarius 424
imitator 422
indica 423
inexpectatus 421
inornatus **422**
insularis 421
intermedius 421
janetta 423
johorensis 422
kalianda 422
kemmisi 421
keraudrenii 422
kinneari 421
klossi 422
kuchingensis 423
lamucotanus 422
lancavensis 421
lautensis 422
leucocephalus 422
leucogaster 422
leucopus 421
leucurus 421
lighti 422
lokroides 423
lucas 421
lunaris 422
lutescens 422
lylei 422
madsoedi 422
madurae 422
magnificus 422
malawali 422
maporensis 422
mapravis 421
marinsularis 422
matthaeus 421
mearsi 423
medialis 421
melanogaster **422**, 604
melanops 423
menamicus 422
mendanauus 423
mentawi 422
michianus 421
microrhynchus 422
microtis 422
midas 421
millardi 421
milleri 421
mimellus 423
mimiculus 423
miniatus 422
moheius 421
mohillius 421
nagarum 421
nakanus 421
navigator 423
nesiotes 422
nicotianae 422
nigridorsalis 421
nigrovittatus **422**, 423
ningpoensis 421
notatus 420, **422**, 422, 423
nox 422
nyx 423
orestes 420, **423**
owensi 423
palustris 423
panjioli 421
panjius 421
pannovianus 422
pelapius 423
pemangilensis 422
penialius 423
peninsularis 423
percommodus 423
perhentiani 423
phanrangis 421
phayrei **423**
phoenicurus 422
piceus 423
pierrei 422
pipidonis 421
pirata 421
plantani 423
plasticus 423
pluto 423
poliopus 423
portus 422
prachin 422
pranis 421
pretiosus 423
prevostii 421, **423**
primus 421
prinsulae 423
proserpinae 423
proteus 423
punctatissimus 421
pygerythrus **423**
quantulus 421
quinlingensis 421
quinquestriatus **424**
rafflesii 423
rajasima 422
raptor 423
redimitus 423
roberti 421
rubeculus 421
rubex 421
rubidiventris 423
rufogularis 423
rufoniger 423
rufonigra 423
rupatius 423
rutiliventris 423
salakensis 422
samuiensis 421
sanggaus 423
sarawakensis 423
saturatus 423
schlegeli 421, 423
scottii 423
seraiae 423
serutus 423
shanicus 421
shortridgei 421
siamensis 421
similis 423
singapurensis 423
sinistralis 422
siriensis 423
sladeni 421
solutus 421
splendens 422
stellaris 423
stevensi 423
stresemanni 423
styani 421
subluteus 423
suffusus 423
sullivanus 421
sumatranus 423
sylvester 424
tabaudius 421
tachardi 422
tachin 421
tacopius 421
tamansari 423
tapanulius 423
tarussanus 423
tedongus 423
telibius 421
tengerensis 422
tenuirostris 423
terutavensis 421
thai 421
thaiwanensis 421
tinggius 423
toupai 423
trotteri 422
tsingtanensis 421
tsingtauensis 421
ubericolor 423
vanheurni 423
vassali 421
venetus 423
verbeeki 423
vernayi 421
vinocastaneus 423
virgo 423
vittatus 423
waringensis 423
watsoni 423
wellsi 421
williamsoni 422
woodi 421
wrayi 423
wuliangshanensis 421
youngi 421
zimmeensis 421
Callospermophilus **444**, 447, 449
callosus, *Calomys* 697, **698**, 698
callotis, *Lepus* 814, **815**, 815, 818
  Oryx 413
Callotus 250
Calomys 695, **697**, 697, 700
  argurus 698
  bimaculatus 697, 698
  boliviae **697**, 698
  bonariensis 698
  callidus **697**, 697, 698
  callosus 697, **698**, 698

carillus 698
cordovensis 698
cortensis 698
dubius 698
ducillus 698
expulsus 698
fecundus 697, 698
frida 698
gracilipes 698
hummelincki **698**, 698
laucha 697, **698**, 698
lepidus **698**, 698
marcarum 698
miurus 698
montanus 698
muriculus 698
murillus 698
musculinus **698**, 698
plebejus 700
pusillus 698
sorellus **698**, 698
tener **698**, 698
venustus 697, 698
Calomyscinae 501, **535**
Calomyscus **535**, 535, 536
bailwardi **535**, 535, 536
baluchi **535**
elburzensis 536
grandis 535
hotsoni **536**
mustersi 536
mystax **536**, 536
tsolovi **536**
urartensis **536**, 536
Caloprymnus **49**
campestris **49**
Calotragus 399
calura, Phascogale **33**
Caluromyinae **15**
Caluromys **15**
affinis 15
antioquiae 15
aztecus 15
bartletti 15
cahyensis 15
cajopolin 15
canus 15
cayopollin 15
centralis 15
cicur 15
derbianus **15**
dichura 15
fervidus 15
guayanus 15
jivaro 15
juninensis 15
lanatus **15**
lanigera 15
leucurus 15
meridensis 15
modesta 15
nattereri 15
nauticus 15
ochropus 15
ornata 15
pallidus 15
philander **15**
pictus 15
pulcher 15
pyrrhus 15
senex 15
trinitatis 15
venezuelae 15
vitalina 15
Caluromysiops **15**
irrupta **15**, 15
calurus, Conepatus 316
Gerbillus 559
Sekeetamys 559, **560**
calvertensis, Sorex 118
calvescens, Pan 277
calviniae, Chrysochloris 75
calvus, Cacajao **261**
Pan 277
Pteronotus 176
calypso, Crocidura 95
Rhinolophus 163
calypsus, Microtus 519
Calyptophractus 64
Calyptrocebus 259
camargensis, Thomomys 473
cambodiana, Tupaia 131
cambojensis, Cervus 387
cambrica, Macrotis 40
cambricus, Rattus 655
Camelidae **381**
cameloides, Alces 389
Camelopardalis 383
camelopardalis, Cervus 383
Giraffa **383**
Camelus **381**
bactrianus **381**, 381
dromedarius **381**, 381
ferus 381
glama 381
vicugna 382
cameranoi, Capra 406
cameronensis, Galago 249
cameroni, Niviventer 635
cameroonensis, Taurotragus 403
camerunensis, Hipposideros **171**
Paracrocidura 101
Sylvisorex 104
Thryonomys 775
camoae, Thomomys 473
camortae, Pipistrellus 221
campanae, Pelomys **639**
campanius, Lepus 821
campbelli, Cercopithecus **263**
Phodopus **539**
campbelliae, Sylvicapra 412
campechensis, Bassariscus 334
Potos 333
camperi, Diceros 372
campestris, Antilope 399
Bettongia 49
Caloprymnus **49**
Canis 281
Cervus 386, 391
Chrysocyon 282
Equus 369
Gerbillus 549, **550**, 550, 552, 555
Glossophaga 184
Lepus 817, 821, 822
Micromys 620
Microtus 518
Mustela 323
Neotoma 711
Odocoileus 391
Oryctolagus 822
Ozotoceros 392
Peromyscus 732
Phyllotis 738
Raphicerus **399**
Saccostomus **541**, 541
Zapus 499
campi, Scapanus 127
Campicola 517
campicola, Galea 780
Lepus 817
campioni, Marmota 432
camptoceros, Ceratotherium 371
campus, Sundamys 666
campuslincolnensis, Crocidura 87
camtschatica, Marmota **431**, 432
Sorex **113**, 113
camus, Ceratotherium 371
cana, Eremitalpa 76
Lagothrix 258
Presbytis 271
Sylvicapra 412
Vulpes **286**
canacorum, Mus 625
canadensis, Canis 281
Castor **467**, 467
Cervus 385, 386
Geomys 469
Glaucomys 460
Lontra **311**
Lutra 310
Lynx **293**, 293
Marmota 432
Martes 320
Ovis **408**, 408, 409
Peromyscus 732
Ursus 338
Zapus 499
canaliculatus, Sorex 117
canalvus, Callosciurus 423
Glyphotes 423
canariensis, Crocidura **82**, 609
Canariomys **582**, 582
bravoi 582
tamarani **582**, 582
canarius, Sylvilagus 825
canaster, Galictis 318
Heliosciurus 428
cancrivora, Cerdocyon 282
Didelphis 17
Herpestes 306
cancrivorus, Peromyscus 732
Procyon **335**
cancrosa, Chrysocyon 282
candace, Nasua 334
candango, Juscelinomys **706**, 706
candelensis, Sciurus 441
candescens, Ursus 338
candida, Echinosorex 79
candidula, Petaurista 463
candidus, Apodemus 573
Crocidura 94
Mus 625
Propithecus 247
caneae, Crocidura 96
canellinus, Holochilus 704
caneloensis, Thomomys 473
canens, Herpestes 306
canescens, Akodon 692, 694
Chaetodipus 484
Cricetus 538
Crocidura 87
Dactylomys 790
Eira 318
Georychus 772
Glaucomys 460
Kobus 413
Leopardus 291
Lepus 821
Marmosa **18**
Meles 314
Microtus 525
Murina 230
Neotoma 712
Otocyon 284
Otomys 680
Paradoxurus 344
Petrogale 56
Proteles 309
Pseudocheirus **59**
Saccopteryx **159**
Spermophilus 449
Sus 379
Tamias 453
Tragulus 383
canfieldiae, Ammospermophilus 420
canicaudus, Microtus **520**
Tachyoryctes 687
Tamias 453
caniceps, Callosciurus **421**, 421, 422, 423
Diplomys **791**
Erinaceus 77
Grammomys **592**
Lepilemur 246
Loncheres 791
Peromyscus **729**, 729
Petaurista 462
Protoxerus 436
Pteropus **146**
Rhynchogale 307
Tamias 454
caniclunis, Sylvilagus 825
canicollis, Proechimys **795**
canicrus, Presbytis 271

canicularius, Neomys 110
caniculus, Atelerix 77
Canidae **279**
canigenus, Callosciurus 421
canigula, Mustela 324
canina, Phoca 332
caninus, Erinaceus 77
Meles 314
Peropteryx 158
Trichosurus **48**
Vespertilio 158
canipes, Tamias **453**, 454
Canis 279, **280**, 282, 283, 284, 285, 286
achrotes 281
adustus **280**
albus 281
alces 281
algirensis 280
altaicus 281
antarcticus 283
anthus 280
arabs 281
arctos 281
arenarum 281
argenteus 285
argunensis 281
ater 281
aureus **280**
australis 283
azarae 282
baileyi 281
balcanicus 280
banksianus 281
barbarus 280
bea 280
beothucus 281
bernardi 281
brachyurus 282
bweha 280
cagottis 280
campestris 281
canadensis 281
canus 281
caucasica 280
centralis 280
chanco 281
clepticus 280
columbianus 281
communis 281
coreanus 281
crassodon 281
crocuta 308
cruesemanni 280
cubanensis 281
culpaeus 284
dalmatinus 280
deitanus 281
desertorum 281
dickeyi 280
doederleini 280
dybowskii 281
ecsedensis 280
ekloni 281
elgonae 281
estor 280
familiaris 280, 281
filchneri 281
flavus 281
floridanus 281
frustror 280
fulvus 281
fusca 281
gallaensis 280
gigas 281
goldmani 280
graecus 280
grayi 280
gregoryi 281
griseoalbus 281
hadramauticus 280
hattai 281
hodophilax 281
holubi 280
hondurensis 280
hudsonicus 281
hungaricus 280
hyaena 308
impavidus 280
indicus 280
irremotus 281
italicus 281
jamesi 280
japonicus 281
jubatus 282
kaffensis 280
kamtschaticus 281
karanorensis 281
knightii 281
kola 280
kurjak 281
labradorius 281
lagopus 279
laniger 281
lanka 280
lateralis 280
latrans **280**, 281
lestes 280
ligoni 281
lupaster 280
lupus **280**, 280, 281
lupus-griseus 281
lycaon 281
mackenzii 281
magellanicus 284
major 281
manningi 281
maroccanus 280
mcmillani 281
mearnsi 280
megalotis 283
mengesi 280
mesomelas **281**
microdon 280
microtis 279
minor 280, 281
mogollonensis 281
monstrabilis 281
moreotica 280
naria 280
niger 281
notatus 280
nubianus 280
nubilus 281
occidentalis 281
ochropus 280
orientalis 281
orion 281
pallipes 281
pambasileus 281
peninsulae 280
primaevus 282
procyonoides 283
rex 281
riparius 280
rufus **281**, 281
sacer 280
schmidti 281
senegalensis 280
signatus 281
simensis **282**
sinus 282
somalicus 280
soudanicus 280
sticte 281
studeri 280
syriacus 280
texensis 280
thamnos 280
thooides 280
thous 282
tripolitanus 280
tschiliensis 281
tundrarum 281
turuchanensis 281
typicus 280
umpquensis 280
ungavensis 281
variabilis 281
variegatoides 281
variegatus 280
vigilis 280
virginianus 285
viverrinus 283
vulgaris 280
vulpes 285
walgie 282
wunderlichi 280
youngi 281
caniscus, Paradoxurus 344
caniventer, Caenolestes **25**
Sminthopsis 36
canna, Taurotragus 403
Cannomys **685**, 685
badius **685**
castaneus 685
lonnbergi 685
minor 685
pater 685
plumbescens 685
cano-viridis, Chlorocebus 265
canopus, Lepus 822
canorus, Niviventer 633
cansdalei, Idiurus 758
Malacomys 592, **607**, 607
Scotonycteris 153
cansensis, Peromyscus 728
cansulus, Sorex **113**, 113, 120
Cansumys **537**, 537
canus **537**, 537
cansus, Myospalax 675
Ochotona **807**, 807, 808, 812, 813
cantabriae, Arvicola 506
cantanbra, Crocidura 96
cantator, Microtus 525
cantorii, Taphozous 160
cantueli, Clethrionomys 508
cantwelli, Microtus 527
Perognathus 485
canus, Arvicola 506
Caluromys 15
Canis 281
Cansumys **537**, 537
Capreolus 390
Eropeplus **591**, 591
Galerella 303
Hylomyscus 598
Lenothrix **603**, 603
Lepus 815
Liomys 482
Odocoileus 391
Onychomys 719
Paradoxurus 344
Peromyscus 732
Philander 22
Pipistrellus 221
Pteropus 147
Reithrodontomys 741
Spermophilus **445**, 445, 448, 450
Thomomys 473
canuti, Rhinolophus **164**, 165
canutus, Baiomys 695
caobangis, Moschus 383
caoccii, Mus 625
capaccinii, Myotis **209**, 212
Capella 409
capella, Rupicapra 410
capensis, Aethomys 568
Aonyx **310**, 310
Atelerix 77
Cavia 374
Chrysochloris 75
Connochaetes 393
Crocidura 84
Crocuta 308
Dasymys 589
Delphinus 352
Dendromus 543
Diceros 372
Eptesicus **201**, 202
Georychus **772**
Giraffa 383
Graphiurus 765
Hippopotamus 381
Hippotragus 412
Hystrix 773
Ictonyx 318, 319
Leptailurus 292
Lepus 814, **815**, 816, 817, 818, 819, 820, 821, 822
Loxodonta 367

Lutra 310
Mellivora **315**
Melomys **615**, 615, 616, 619
Mus 772
Myosorex 100
Myrmecophaga 375
Nycteris 162
Orcinus 355
Oryx 413
Otomys 680
Panthera 297
Pedetes **759**
Petromyscus 683
Procavia **373, 374**
Raphicerus 399
Rhinolophus **164**
Stenella 356
Suricata 307, 308
Tragelaphus 404
Viverra 315
Xerus 458
Yerbua 759
Ziphius 364
Caperea 349, **351**, 351
antipodarum 351
marginata **351**
Capiguara 781
capillamentosa, Pithecia 262
capillatus, Cebus 259
Miopithecus 269
capillosa, Procavia 374
capistratus, Nasalis 270
Pseudocheirus 59
Pteropus 151
Sciurus 442
capitalis, Choloepus 64
Macaca 266
capitaneus, Myotis 216
capitis, Dendromus 543
Paraxerus 435
capito, Anourosorex 106
Oryzomys **721**, 721, 722, 723, **724**, 725
Phyllotis 738
Saccolaimus 159
capitulatus, Onychomys 719
capnias, Anourosorex 106
caporiaccoi, Mustela 322
capparo, Lagothrix 258
Capra **404, 405**, 408
aegagrus 405
alaiana 406
almasyi 406
alpina 405
altaica 406
arabica 406
beden 406
blythi 405
cameranoi 406
cashmiriensis 405
caucasica **405**, 405
cervicapra 395
chialtanensis 405
chitralensis 405
cilicica 405
cretica 405
cylindricornis **405**, 405
dauvergnii 406
dementievi 406
dinniki 405
dorcas 395, 405
europea 405
falconeri **405**, 405
fasciata 406
filippii 406
florstedti 405
formosovi 406
gazella 405, 413
gilgitensis 405
graicus 405
grimmia 412
hagenbecki 406
hemalayanus 406
heptneri 405
hircus **405**, 405
hispanica 406
ibex **405**, 406
jemlahica 406
jerdoni 405
jharal 406
jourensis 405
lorenzi 406
lusitanica 406
lydekkeri 406
megaceros 405
mengesi 406
merzbacheri 406
neglecta 405
nubiana **405**
ognevi 405
pallasii 405, 406
pedri 406
persica 405
picta 405
puda 392
pygmaea 398
pyrenaica **406**, 406
raddei 405
rupicapra 409
sakeen 406
severtzovi 405
sibirica **406**
sinaitica 406
tatarica 400
transalaiana 406
turcmenica 405
victoriae 406
walie **406**
wardi 406
capraea, Capreolus 390
caprenus, Nycticeius 217
Capreolinae **389**
Capreolus **390**
albicus 390
albus 390
armenius 390
baleni 390
balticus 390
bedfordi 390
canus 390
capraea 390
capreolus **390**, 390
caucasica 390
cistaunicus 390
coxi 390
decorus 390
dorcas 390
europaeus 390
ferghanicus 390
grandis 390
italicus 390
joffrei 390
mantschuricus 390
melanotis 390
niger 390
ochracea 390
plumbeus 390
pygargus **390**
rhenanus 390
thotti 390
tianschanicus 390
transsylvanicus 390
transvosagicus 390
varius 390
vulgaris 390
warthae 390
whittalli 390
zedlitzi 390
capreolus, Capreolus **390**, 390
Cervus 390
Pelea **413**, 413
capricornensis, Puma 296
Capricornis 406
sumatraensis 407
capricornis, Aethomys 567
Pronolagus 823
Raphicerus 399
Capricornulus 406
caprimulga, Allactaga 488
Caprina 406
Caprinae **404**
caprinus, Phyllotis **738**, 738
Caprolagus **814**, 814, 815, 818, 819, 820, 823
furnessi 823
hispidus **814**
Capromyidae 790, 799, **800**, 802, 805
Capromyinae **800**
Capromys **800**, 800, 801, 802
acevedo 800
antiquus 800
arredondoi 800
doceleguas 800
fournieri 800
geayi 800
gundlachianus 800
intermedius 800
latus 800
pappus 800
pilorides **800**, 800
prehensilis 802
relictus 800
robustus 800
Caprovis 408
capsalis, Sylvilagus 825
captorum, Nannospalax 754
capucina, Simia 259
capucinellus, Myotis 210
capucinus, Cebus 259, **260**, 260
Histiotus 205
Microtus 530
capybara, Hydrochaeris 781
caquetensis, Alouatta 255
Saimiri 260
cara, Crocidura 92
Caracal **288**, 289
aharonii 288
algira 288
bengalensis 288
berberorum 288
caracal **288**, 289
coloniae 288
corylinus 288
damarensis 289
limpopoensis 289
lucani 289
medjerdae 289
melanotis 288, 289
michaelis 289
nubicus 289
poecilotis 289
roothi 289
schmitzi 289
spatzi 289
caracal, Caracal **288**, 289
Felis 288
caracasensis, Sylvilagus 825
caracciolae, Vampyrops 194
caraccioli, Vampyrodes **194**
caraco, Rattus 657
caracolus, Oryzomys 720
caraftensis, Mustela 323
caraya, Alouatta **254**
carbo, Orthogeomys 471
Trachypithecus 274
carbonarius, Macaca 266
Microtus 528
Tarsius 251
carceleni, Neacomys 709
carcharias, Cynogale 341
carchensis, Sciurus 441
Sylvilagus 825
carcinophaga, Phoca 330
carcinophagus, Lobodon **330**
Cardiocraniinae **491**, 495
Cardiocraniini 491
Cardiocranius **491**
paradoxus **491**, 491
Cardioderma **162**
cor **162**
careyi, Callosciurus 421
cariacou, Odocoileus 391
Cariacus 391
caribou, Rangifer 392
carillus, Calomys 698
Hylomyscus 598, **599**, 599
carimatae, Callosciurus 423
Macaca 266
Maxomys 614
Megaderma 163

Myotis 207
Tragulus 382
Tupaia 133
carimonensis, Callosciurus 423
Ratufa 437
carina, Megaderma 163
carinatus, Neomys 110
carinthiacus, Microtus 518
carissima, Myotis 212
carlhubbsi, Mesoplodon **362**
carli, Peromyscus 732
carlosensis, Orthogeomys 470
carlottae, Ursus 338
carmelitae, Phalanger **46**
carmeni, Peromyscus 730
carminis, Odocoileus 391
Tamias 453
carnatica, Megaderma 163
carnaticus, Herpestes 305
Carnivora 279
caroarensis, Lagothrix 258
caroli, Callosciurus 423
Hylopetes 461
Mus **623**
Pseudocheirus **59**
carolii, Myotis 212
carolinae, Syconycteris **155**
carolinensis, Blarina **106**, 106
Castor 467
Clethrionomys 508
Eptesicus 201
Microtus 520
Reithrodontomys 741
Sciurus **440**
Carollia **186**
amplexicaudata 186
azteca 186
bicolor 186
brachyotus 186
braziliensis 186
brevicauda **186**
calcaratum 186
castanea **186**, 186
grayi 186
lanceolatum 186
minor 186
perspicillata **186**, 186
subrufa **186**, 186
tricolor 186
verrucata 186
Carolliinae **186**
carpathicus, Cervus 386
Dryomys 766
Lynx 293
Sorex 111
carpatica, Rupicapra 410
carpentanus, Rhinolophus 167
carpentarius, Notomys 636
carpenteri, Hylobates 275
carpetanus, Sorex 117
carpi, Lepus 816
Paraxerus 434
Carpitalpa 74, 75
carpolegus, Artibeus 188
Macaca 267
Carpomys 579, **583**, 587
melanurus **583**, 583
phaeurus **583**
carri, Marmosops 19
Thomomys 473
carrikeri, Echimys 792
Leopardus 292
Mazama 390
Oryzomys 725
Tonatia **180**
carrorum, Oryzomys 725
carruthersi, Cercopithecus 264
Funisciurus **427**
Microtus 522, 523
Spermophilus 446
cartagoensis, Orthogeomys 471
carteri, Dendromus 543
Eptesicus 201
Mormoops 176
Myotis 213, 214
Carterodon **793**
sulcidens **793**
temmincki 793
cartilagonodus, Otopteropus **145**, 145
cartusiana, Rupicapra 410
caryi, Microtus 525
Perognathus 485
Reithrodontomys 741
Scalopus 127
Spermophilus 447
Tamias 454
Thomomys 475
Caryomys 513
casanovae, Gazella 397
cascadensis, Clethrionomys 508
Lepus 814
Marmota 431
Sorex 118
Tamiasciurus 456
Vulpes 287
cascus, Dipodomys 480
casensis, Callosciurus 421
Maxomys 614
cashmeriensis, Cervus 386
cashmiriensis, Capra 405
Casinycteris **138**
argynnis **138**, 138
casperianus, Cervus 386
caspica, Barbastella 199
Crocidura 88
Phoca **331**
caspicus, Hemiechinus 78
Lepus 817
Microtus 529
Myoxus 770
caspius, Cervus 386
Myoxus 770
Rattus 657
cassiquiarensis, Saimiri 260
cassiteridum, Crocidura 96
casta, Marmosa 18
castanea, Akodon 691
Carollia **186**, 186
Tupaia 132
castaneiceps, Bradypus 63
Gorilla 276
castaneoventris, Callicebus 258
Callosciurus 421
castaneus, Arvicola 506
Cannomys 685
Cebus 259, 260
Cephalophus 410
Conepatus 316
Equus 370
Lariscus 430
Lasiurus **206**
Makalata 793
Microtus 521
Molossus 235
Mus 625, 626
Oryzomys 720
Peromyscus 732
Petaurista 462
Procyon 335
Scotophilus 227
Sorex 111
castanomale, Pan 277
castanonotus, Sciurus 439
castanops, Cratogeomys 472
Pappogeomys **472**
castanotis, Chaeropus 39
castanotus, Sciurus 439
castanurus, Spermophilus 447
castelnaudi, Lama 381
castelnaui, Lagothrix 258
castilianus, Sus 379
Castor **467**
acadicus 467
albicus 467
albus 467
baileyi 467
balticus 467
belugae 467
bindai 467
birulai 467
caecator 467
canadensis **467**, 467
carolinensis 467
concisor 467
duchesnei 467
fiber **467**, 467
flavus 467
frondator 467
fulvus 467
galliae 467
gallicus 467
idoneus 467
labradorensis 467
leucodontus 467
mexicanus 467
michiganensis 467
missouriensis 467
moschatus 124
niger 467
pallidus 467
phaeus 467
pohlei 467
propiius 467
repentinus 467
rostralis 467
sagittatus 467
shastensis 467
solitarius 467
subauratus 467
taylori 467
texensis 467
variegatus 467
varius 467
vistulanus 467
zibethicus 532
Castoridae **467**
castoroides, Myocastor 806
castroviejoi, Lepus **816**
castus, Sciurus 442
Catablepas 393
Catagonus **379**
metropolitanus 379
wagneri **379**
catalinae, Peromyscus 732
Reithrodontomys 741
Sciurus 440
Thomomys 473
Urocyon 285
Cataphractus 65
catavinensis, Thomomys 473
catellifer, Maxomys 614
catellus, Cyclopes 67
Euryzygomatomys 794
catemana, Presbytis 271
Ratufa 437
catenata, Leopardus 292
catherinae, Oecomys 717, 725
Catodon 359
catodon, Delphinapterus 357
Physeter **359**, 359
Catolynx 289
catolynx, Felis 290
Catoptera 349
Catopuma **289**, 289
aurata 289
badia **289**
badiodorsalis 289
bainsei 289
dominicanorum 289
mitchelli 289
moormensis 289
nigrescens 289
semenovi 289
temminckii **289**
tristis 289
catrinae, Dasyprocta 781
Catta 245
catta, Lemur **245**, 245
catus, Felis 289, 290, 291
caucae, Didelphis 17
Marmosops 19
Phyllostomus 180
caucasica, Canis 280
Capra **405**, 405

Capreolus 390
Mustela 322
Rupicapra 410
Sicista **496**, 496, 497
Talpa **128**, 128, 129
Vulpes 287
caucasicus, Alces 389
Allactaga **488**
Apodemus 570
Arvicola 506
Bison 400
Cervus 386
Dryomys 766
Felis 290
Lepus 817
Meles 314
Microtus 519
Mustela 323
Pipistrellus 223
Sciurus **440**
Sorex 119, 120
Ursus 338
caucasius, Meriones 556
caucensis, Alouatta 255
Leopardus 292
Myotis 213
Potos 333
Rhipidomys **744**
Sciurus **442**
caucinus, Pronolagus 823
caudalis, Trachypithecus 273
caudata, Crocidura 95
Dasyprocta 781
Marmota **431**
Sicista **496**, 496, 497, 498
Sorex 117
caudatior, Niviventer 634
caudatus, Cercartetus **58**
Cheiromeles 233
Colobus 270
Dipodomys 479
Felis 290
Lasiurus 207
Meriones 556, 557
Mus 625
Naemorhedus **407**
Oryzomys 721
Peromyscus 732
Sorex 122
Soriculus **122**, 123
caudifer, Glossophaga 184
caudifera, Anoura **183**
caudimaculatus, Hapalotis 671
Uromys **671**, 671, 672
caudivarius, Thomasomys 749
caudivolvula, Potos 333
Pseudocheirus 60
Viverra 333
cauquenensis, Rattus 657
caurae, Phyllostomus 180
caurina, Dorcopsis 51
Martes 319
caurinus, Clethrionomys 508
Eptesicus 203
Galerella 303
Heliosciurus 429
Hydromys 597
Ichthyomys 705
Melomys 617
Myotis 209
Notoryctes **43**
Pipistrellus 220
Reithrodon 740
Scapanus 127
Soriculus 123
Steatomys **545**
Tamias 453
Ursus 338
Zaedyus 67
cautus, Alticola 503
Microtus 523
cauui, Galerella 303
cavaticus, Taphozous 161
cavator, Orthogeomys **470**
cavendishi, Loxodonta 367
Madoqua 398
cavernarum, Brachyphylla **182**, 182
Cavia **778**
acouchy 782
anolaimae 779
aperea **778**, 779
arequipae 779
atahualpae 779
azarae 779
capensis 374
cobaya 778, 779
cutleri 779
festina 779
fulgida **779**
guianae 779
hilaria 779
hypoleuca 779
leucopyga 779
longipilis 779
magna **779**
nana 779
nigricans 779
osgoodi 779
pallidior 779
pamparum 779
patachonica 780
porcellus 778, **779**, 779
rosida 779
rufescens 779
rupestris 780
salinicola 780
sodalis 779
stolida 779
tschudii **779**, 779
umbrata 779
Caviella 780
cavifrons, Bos 401
Ursus 338
Caviidae 771, **778**
Caviinae **778**
Cavioidea 771
cavirostris, Ziphius **364**, 364
cay, Cebus 259
cayana, Dasyprocta 782
cayanensis, Phylloderma 180
cayennae, Dasyprocta 782
cayennensis, Ateles 257
Chironectes 16
Proechimys **795**, 795, 796, 797, 798
cayllomae, Chroeomys 700
caymanum, Myoprocta 782
cayoensis, Orthogeomys 471
cayopollin, Caluromys 15
cearanus, Rhipidomys 745
Cebidae **254**
cebifrons, Sus **378**
Cebinae **259**
Cebuella 251
Cebugale 243
Cebus **259**
adustus 259
aequatorialis 259
albifrons **259**
albulus 260
annellatus 260
apella **259**
apiculatus 260
avus 259
azarae 259
barbatua 259
barbatus 260
brunneus 260
buffonii 259
caliginosus 259
capillatus 259
capucinus 259, **260**, 260
castaneus 259, 260
cay 259
cesarae 259
chacoensis 259
chrysopus 259
cirrifer 259
crassiceps 259
cristatus 259
cucullatus 259
curtus 260
cuscinus 259
elegans 259
fallax 259
fatuellus 259
fistulator 259
flavescens 259
flavus 259
frontalus 259
fulvus 259
gracilis 259
griseus 260
hypoleuca 259
hypoleucus 260
hypomelas 259
imitator 260
juruanus 259
leporinus 260
leucocephalus 259
leucogenys 259
libidinosus 259
limitaneus 260
lunatus 259
macrocephalus 259
magnus 259
malitiosus 259
maranonis 259
margaritae 259
monachus 259
morrulus 259
niger 259
nigripectus 260
nigritus 259
nigrivittatus 260
olivaceus **260**
pallidus 259
paraguayansis 259
peruanus 259
pleei 259
polykomos 269
pucheranii 260
robustus 259
sagitta 259
satanas 261
subcristatus 259
tocantinus 259
trepida 259
trinitatus 259
unicolor 259
variegatus 259
vellerosus 259
versicolor 259
versutus 259
xanthocephalus 259
xanthosternos 259
yuracus 259
cecilii, Peromyscus 733
cedrinus, Thomomys 473
cedrophilus, Sylvilagus 824
cedrorum, Chionomys 507
cedrosensis, Peromyscus 730
ceilonensis, Ratufa 437
celaeno, Nyctimene 143, **144**
Pteropus 151
Celaenomys **583**, 585
silaceus **583**
celaenopepla, Ratufa 437
celatus, Apodemus 570
Cryptotis 109
Phenacomys 533
celebensis, Acerodon **137**, 146
Babyrousa 377
Bubalus 402
Crunomys **588**, 588
Cuscus 47
Harpyionycteris **142**
Hipposideros 171
Megaderma 163
Miniopterus 231
Paradoxurus 344
Rattus 653, 655
Rhinolophus **164**, 164
Rousettus **152**
Scotophilus **227**
Strigocuscus 47, **48**
Suncus 103
Sus **378**, 378, 379
Taeromys 666, **667**
celenda, Mustela 321
celer, Ozotoceros 392

celeripes, Dipodomys 479
celeris, Petrogale 56
Tamias 453
celicae, Marmosops 19
celidogaster, Felis 296
Profelis 296
cellarius, Apodemus 571
celsus, Dipodomys 479
Phenacomys 533
Rattus 661
celtica, Sus 379
celticus, Apodemus 573
Cemas 393
Centetes 73
centralasiaticus, Eptesicus 202
centralis, Aethomys 568
Antechinus 30
Antidorcas 395
Antilope 395
Arvicanthis 578
Atherurus 773
Cabassous **64**
Caenolestes 25
Callosciurus 421
Caluromys 15
Canis 280
Centronycteris 156
Cephalophus 411
Cervus 386
Chlorocebus 265
Clethrionomys **508**
Coendou 775
Diphylla 194
Dipodomys 479
Dolichotis 780
Gazella 397
Grammomys 594
Graomys 703
Herpestes 305
Hipposideros 174
Jaculus 493
Lepus 816
Malacomys 608
Marmota 431
Microtus 520
Myrmecophaga 68
Oreotragus 399
Panthera 298
Phacochoerus 377
Proechimys 798
Promops **239**
Pteronotus 176
Ratufa 437
Saccopteryx 159
Sminthopsis 35
Sorex 113, 121
Soriculus 123
Syncerus 402
Thomomys 473
Tonatia 181
Urotrichus 129, 130
centralrossicus, Erinaceus 77
centrasiaticus, Lepus 821
centricola, Dolichotis 780
Protoxerus 436
centroamericana, Stenella 356, 357
Centronycteris **156**
centralis 156
maximiliani **156**
wiedi 156
centrosa, Echiothrix 591
centrosus, Arvicanthis 578
Centurio **189**
flavogularis 189
greenhalli 189
mcmurtrii 189
mexicanus 189
minor 189
senex **189**, 189
centurio, Ametrida **187**, 187
Centuriosus 378
Ceonix 46
cepapi, Paraxerus **434**
Sciurus 434
cepapoides, Paraxerus 434
cepate, Paraxerus 434
Cephalophella 410
Cephalophia 410
Cephalophidium 410
Cephalophinae **410**
Cephalophops 410
Cephalophora 412
Cephalophorus 410
Cephalophula 410
Cephalophus **410**, 412
adersi **410**, 410, 411
aequatorialis 411
amoenus 411
anchietae 411
apanbanga 411
arrhenii 410
aureus 411
badius 410
bakeri 411
barbertoni 411
bicolor 411
bottegoi 410
bradshawi 411
breviceps 410
brookei 411
caerula 411
caffer 411
callipygus **410**, 410, 411, 412
castaneus 410
centralis 411
claudi 411
congicus 411
coxi 411
crusalbum 411
cuvieri 411
danei 410
defriesi 411
doria 412
dorsalis **410**
emini 411
fosteri 411
frederici 410
fuscicolor 411
harveyi **410**
hecki 411
hooki 411
ignifer 411
ituriensis 411
jentinki **410**
johnstoni 411
kenniae 410
kivuensis 411
kuha 410
lebombo 411
leopoldi 411
lestradei 411
leucochilus 410
leucogaster **410**
liberiensis 410
longiceps 411
lowei 410
ludlami 411
lugens 411
lusumbi 411
maxwellii **410**, 411
melanoprymnus 411
melanorheus 411
minuta 411
mixtus 411
monticola **410**, 410
musculoides 411
natalensis 410, **411**
niger **411**
nigrifrons **411**, 411
nyasae 411
ogilbyi **411**
orientalis 410
pembae 411
perpusilla 411
philantomba 410
pluto 411
punctulatus 411
robertsi 411
rubidior 411
rubidus **411**
ruddi 411
ruficrista 411
rufilatus **411**
rutshuricus 411
schultzei 411
schusteri 411
sclateri 411
seke 410
silvicultor **411**, 411
simpsoni 411
spadix **411**
sundevalli 411
sylvicultor 411
thomasi 411
vassei 411
weynsi **411**, 411
whitfieldi 410
zebra **412**
zebrata 412
cephalophus, Elaphodus **388**, 388
Cephalorhynchinae 351
Cephalorhynchus **351**
albifrons 352
albiventris 351
commersonii **351**, 354
eutropia **351**, 358
floweri 351
hastatus 352
heavisidii **351**
hectori **352**
obtusata 351
Cephalotes, peroni 139
teniotis 240
cephalotes, Nectomys 709
Nyctimene **144**, 145
Oryzomys 721
Vespertilio 143
cephodes, Cercopithecus 263
cephus, Cercopithecus **263**, 264
ceramensis, Hipposideros 172
Pteropus 149
Sus 379
ceramica, Sus 379
ceramicus, Stenomys **665**
cerasina, Hippocamelus 390
Mazama 390
cerastes, Molossops 234
Ceratodon 357
Ceratorhinus 371
Ceratotherium **371**
burchellii 371
camptoceros 371
camus 371
cottoni 371
crossii 371
kaiboaba 371
kulamanae 371
kulamane 371
oswellii 371
prostheceros 371
simum **371**
Cercaertus 48
Cercartetus **58**
britta 58
caudatus **58**
concinnus **58**
gliriformis 58
lepidus **58**
macrura 58
minor 58
nanus **58**
neillii 58
unicolor 58
Cercocebini 262
Cercocebus **262**, 262, 263, 266
aethiopicus 263
agilis **262**, 262
aterrimus 266
atys 263
chrysogaster 262
collaris 263
crossi 263
fuliginosus 262, 263
fumosus 262
galeritus **262**, 262
hagenbecki 262
lunulatus 263
oberlaenderi 262

sanje 262
torquatus **262**, 262
Cercocephalus 263
Cercolabes 776
Cercolabidae 789
Cercoleptidae 333
Cercolophocebus 266
Cercomys 798
cunicularius 798
Cercopithecidae **262**, 262
Cercopithecinae **262**
Cercopithecus 262, **263**, 263, 265, 269
albinasus 264
albogularis 264
albotorquatus 264
ascanius **263**
atrinasus 263
auratus 273
aurora 264
beirensis 264
boutourlinii 264
brazzae 264
brazziformis 264
brindei 264
buccalis 263
burnettii 263
büttikoferi 264
campbelli **263**
carruthersi 264
cephodes 263
cephus **263**, 264
cirrhorhinus 263
crossi 265
denti 265
diadematus 264
diana **263**, 263
dilophos 264
doggetti 264
dryas **263**
elegans 265
elgonis 264
enkamer 263
erxlebeni 265
erythrarchus 264
erythrogaster **264**
erythrotis 263, **264**
ezrae 264
fantiensis 264
faunus 263
francescae 264
grayi 265
hamlyni **264**
heymansi 264
hindei 264
histrio 263
ignita 263
inobservatus 263
insignis 264
insolitus 264
insularis 265
ituriensis 263
kahuziensis 264
kaimosae 263
kandti 264
kassaicus 263
katangae 263
kibonotensis 264
kima 264
kolbi 264
labiatus 264
laglaizei 264
larvatus 270
leucampyx 264
lhoesti **264**, 265
liebrechtsi 265
lowei 263
ludio 264
maesi 264
maritima 264
martini 264
mauae 264
melanogenys 263
mitis **264**
moloneyi 264
mona **264**
monacha 264
monella 264
monoides 264
montanus 263
mossambicus 264
mpangae 263
neglectus **264**
neumanni 264
nictitans **264**
nigrigenis 264
nigripes 265
nigroviridis 262
nubilus 264
nyasae 264
omensis 264
omissus 263
opisthostictus 264
orientalis 263
otoleucus 264
palatinus 263
pallidus 265
pelorhinus 263
petaurista **264**
petronellae 265
phylax 264
picturatus 263
pluto 264
pogonias **265**
preussi 264, **265**
princeps 264
pulcher 263
pygrius 264
pyrogaster 265
roloway 263
rufilatus 264
rutschuricus 263, 264
salongo 263
samango 264
sassae 263
schmidti 263
schoutedeni 264
schubotzi 264
schwartzi 264
sclateri 263, **265**
sibatoi 264
solatus **265**
stairsi 264
stampflii 264
sticticeps 264
stuhlmanni 264
temminickii 263
thomasi 264
uelensis 264
whitesidei 263
wolfi **265**
zammaranoi 264
Cercoptenus 62
Cercoptochus 261
cerdo, *Vulpes* 287
Cerdocyon 280, **282**, 282
affinis 282
angulensis 282
apollinaris 282
aquilus 282
azarae 282
brachyteles 282
brasiliensis 282
cancrivora 282
fronto 282
fulvogriscus 282
germanus 282
guaraxa 282
jucundus 282
lunaris 282
melampus 282
melanostomus 282
mimax 282
riograndensis 282
robustior 282
rudis 282
savannarum 282
thous **282**
tucumanus 282
vetulus 282
cereus, *Maxomys* 612
cernjavskii, *Arvicola* 506
Cerodon 780
Cerophorus, *acuticornis* 399
nasomaculata 412
quadricornis 403
cerradensis, *Thalpomys* **748**
cerrosensis, *Odocoileus* 391
Sylvilagus 825
certans, *Nyctimene* **144**, 144
certus, *Spermophilus* 447
Cervaria 293
cervaria, *Lynx* 293
Cervequus 390
cervicalis, *Sciurus* 440
Tupaia 133
Cervicapra 395, 414
clarkei 395
cervicapra, *Antilope* **395**
Capra 395
cervicolor, *Mus* **623**
Cervidae 383, **384**
cervina, *Ovis* 408
Cervinae **384**, 388
cervinipes, *Melomys* 615, **616**, 616, 618
cervinus, *Hemibelideus* 58
Hipposideros **171**, 173
Macropus 54
Notomys **636**, 636
Thomomys 473
Cervulus 388
Cervus 384, **385**, 385, 387
aceros 386
acuticauda 386
acuticornis 386
affinis 385
alashanicus 385
albicornis 387
albicus 385
albifrons 385
albirostris **385**
albus 385
alces 389
alfredi **385**
ambrosianus 386
andreanus 386
aplodonticus 386
aplodontus 386
apoensis 386
arietinus 386
aristotelis 387
asiaticus 385
atheneensis 386
atlanticus 385
axis 384
bactrianus 385
bahrainja 385
baicalensis 385
bajovaricus 385
balticus 385
barandanus 386
barbarus 385
baryceros 386
basilanensis 386
bedfordianus 385
bezoarticus 391
biedermanni 385
blakistoninus 386
bolivari 385
boninensis 386
brachyceros 386
brachypus 386
brachyrhinus 386, 387
branderi 385
brauneri 385
breviceps 385
brookei 387
brucei 386
buruensis 387
cambojensis 387
camelopardalis 383
campestris 386, 391
canadensis 385, 386
capreolus 390
carpathicus 386
cashmeriensis 386
casperianus 386
caspius 386
caucasicus 386
centralis 386
chrysotrichos 386
cinereus 386
colombertinus 387

combalbertinus 387
consobrinus 386
cornipes 386
corsicanus 385, 386
corteanus 386
crassicornis 386
curvicornis 387
cycloceros 386
cyclorhinus 386
dailliardianus 386
daimius 386
dama 387, 391
debilis 386
dejardinus 386
dejeani 387
devilleanus 386
dichotomus 389
dimorphe 385
djonga 387
dolichorhinus 386
dominicanus 386
dugennianus 386
duvaucelii **385**, 387
dybowskii 385, 386
elaphoides 385
elaphus **385**, 385
eldii **386**, 386
elegans 386
ellipticus 386
elorzanus 386
equinus 387
errardianus 387
euceros 385
eucladoceros 385
euopis 386
eustephanus 386
floresiensis 387
francianus 386
frinianus 386
frontalis 386
fuscus 386
garcianus 386
germanicus 386
gonzalinus 386
gorrichanus 386
gracilis 386
granulosus 386
grassianus 386
grilloanus 386
guevaranus 386
guidoteanus 386
hagenbecki 386
hainana 387
hainanus 386
hamiltonianus 387
hanglu 385, 386
heterocerus 387
hippelaphus 386, 387
hippolitianus 386
hispanicus 386
hoevellianus 387
hollandianus 386
hortulorum 386
hyemalis 386
ignotus 386
imperialis 386
infelix 386
isubra 386
japonicus 386
jarai 387
javanicus 382, 387
joretianus 386
joubertianus 387
kansuensis 386
kematoceros 386
keramae 386
kopschi 386
lacrymosus 386
laronesiotes 387
latidens 386, 387
legrandianus 386
lemeanus 387
lepidus 387
leschenaulti 387
lignarius 387
longicornis 387
longicuspis 386
luedorfi 386
lyratus 386
macarianus 386
macassaricus 387
macneilli 385, 386
mageshimae 386
major 386, 387
malaccensis 387
mandarinus 386
manitobensis 386
mantchuricus 386
maraisianus 386
maral 386
mariannus 385, **386**
marmandianus 386
marzaninus 386
masbatensis 385
matsumotei 386
mediterraneus 386
menadensis 387
merriami 386
michaelinus 386
microdontus 386
microspilus 386
minoensis 386
minor 386
minutus 386
mitratus 386
modestus 386
moluccensis 387
montanus 386
morrisianus 386
muntjak 388
nannodes 386
naryanus 386
neglectus 386
nelsoni 386
nepalensis 387
nigellus 386
niger 387
nigricans 386
nippon **386**
novioninus 386
nublanus 386
occidentalis 386
oceanus 387
officialis 387
orthopodicus 386
orthopus 386
outreyanus 387
oxycephalus 386
paludosus 389
paschalis 386
pennantii 387
peronii 387
philippinus 386
planiceps 387
planidens 387
platyceros 386
porcinus 385
pouvrelianus 386
pseudaxis 386
pulchellus 386
ramosianus 386
ranjitsinhi 385
regulus 386
renschi 387
rex 386
rhenanus 386
riverianus 386
roosevelti 386
rosarianus 386
roxasianus 386
rubiginosa 386
russa 387
rutilus 386
saxonicus 386
schizodonticus 386
schlegeli 386
schomburgki **386**
schulzianus 386
scoticus 386
sellatus 385
sendaiensis 386
siamensis 386
sibiricus 386
sica 386
sicarius 386
sichuanicus 386
sika 386
simoninus 387
smithii 385
soloensis 386
songaricus 386
spatharius 386
steerii 386
sumbavanus 387
surdescens 386
swinhoei 386
swinhoii 387
sylvanus 386
tai-oranus 386
taiouanus 386
taivanus 386
tarandus 392
tauricus 386
tavistocki 387
telesforianus 386
thamin 386
thoroldi 385
tibetanus 386
timorensis **387**
tuasoninus 386
tunjuc 387
typicus 386
unicolor 386, **387**
ussuricus 386
varius 386
verutus 387
verzosianus 386
vidalinus 386
villemerianus 386
visurgensis 386
vulgaris 386
wachei 386
wallichi 385, 386
wapiti 386
wardi 386
xanthopygus 386
xendaiensis 386
yakushimae 386
yarkandensis 385, 386
yesoensis 386
yuanus 386
cesarae, Cebus 259
cestoni, Tadarida 240
Cetacea 349
cetacea, Dugong 365
Cetus 352, 359
ceylanica, Suncus 103
ceylanicus, Elephas 367
  Herpestes 304
ceylanus, Nannospalax 754
ceylonense, Megaderma 163
ceylonensis, Axis 384
  Cynopterus 139
ceylonica, Lutra 312
  Ratufa 437
  Tatera 561
ceylonicus, Herpestes 304
  Loris 248
  Pipistrellus **220**
ceylonus, Rattus 658
chacarius, Holochilus **704**
chachlovi, Allactaga 489
chacoensis, Bibimys **696**
  Cabassous **65**
  Cebus 259
  Gracilinanus 17
  Graomys 703
  Kunsia 706
  Oligoryzomys **717**
Chacomys 783, 784
chadensis, Lepus 822
  Xerus 458
Chaenodelphinus 361
Chaerephon **232**, 240
  abae 232
  adustus 233
  aloysiisabaudiae **232**
  ansorgei **232**
  bemmeleni **232**
  bengalensis 233
  bivittata **232**, 232
  bregullae 232
  chapini **232**
  cistura 232

colonicus 232
cristatus 233
cyclotis 232
dilatatus 233
elphicki 233
emini 232
faini 233
frater 233
gallagheri **232**
gambianus 233
hindei 233
insularis 233
jobensis **232**
johorensis **232**, 232
lancasteri 232
langi 233
leucogaster 233
limbata 233
luzonus 233
major **232**
murinus 233
naivashae 233
nigeriae **232**
nigri 233
plicata 232, **233**
pumila **233**
pusillus 233
rhodesiae 232
russata **233**
shortridgei 232
solomonis 232
spillmani 233
tenuis 233
websteri 233
Chaeropitheus 269
chaeropitheus, Papio 269
Chaeropus **39**
castanotis 39
ecaudatus **39**
occidentalis 39
Chaetocauda 766
Chaetocercus 31
cristicauda 31
Chaetodipus 477, **482**, 482, 483, 484
albescens 483
albulus 483
ambiguus 483
ammophilus 483
angustirostris 484
anthonyi 483
arenarius **483**
armatus 483
artus **483**
ater 484
atrodorsalis 484
baileyi **483**
bensoni 483
bernardinus 483
broccus 484
bryanti 484
californicus **483**
canescens 484
cinerascens 483
collis 484
conditi 483
crinitus 484
dalquesti 483
dispar 483
domensis 483
domisaxensis 483
eremicus 484
evermanni 484
extimus 483
fallax **483**
femoralis 483
formosus **483**
fornicatus 483
goldmani **483**, 484
guardiae 484
helleri 483
hispidus 482, **483**
hueyi 483
incolatus 483
infolatus 483
inopinus 483
insularis 483
intermedius **484**
knekus 483
lambi 484
latijularis 484
latirostris 483
lineatus **484**, 484
lithophilus 484
lorenzi 484
macrosensis 484
magdalenae 484
majusculus 483
margaritae 484
marinensis 483
maximus 483
melanocaudus 483
melanurus 483
mesembrinus 483
mesidios 483
mesopolius 483
mexicalis 483
minimus 484
mohavensis 483
nelsoni **484**
nigrimontis 484
obscurus 484
occultus 484
ochrus 483
oribates 484
pallidus 483
paradoxus 483
paralios 483
penicillatus **484**, 484
peninsulae 484
pernix **484**, 484
phasma 484
pinacate 484
popei 484
pricei 484
prietae 484
pullus 484
refescens 484
rostratus 484
rudinoris 483
rupestris 484
sabulosus 483
seorsus 484
seri 484
siccus 483
sobrinus 484
spilotus 483
spinatus **484**
stephensi 484
sublucidus 483
umbrosus 484
xerotrophicus 483
zacatecae 483
Chaetomyinae **789**
Chaetomys 775, **789**
moricandi 790
rutila 790
subspinosus **790**
tortilis 790
volubilis 790
Chaetophractus **65**, 66
boliviensis 65
brevirostris 65
desertorum 65
nationi **65**
octocinctus 65
pannosus 65
vellerosus **65**, 65
villosus **65**
chakae, Rhabdomys 662
Chalcomys 688
Chalinolobus **199**, 199, 200
alboguttatus **199**
argentatus **199**
australis 199
beatrix **199**
dwyeri **199**
egeria **199**
gleni **199**
gouldii **199**
humeralis 199
kenyacola **199**
kraussi 200
machadoi 200
microdon 199
morio **199**
neocaledonicus 199
nigrogriseus **200**
papilio 200
phalaena 200
picatus **200**
poensis **200**
rogersi 200
sheila 200
signifer 199
superbus **200**
tuberculatus **200**
variegatus **200**
venatoris 199
chalmersi, Dorcopsis 51
chalybeata, Profelis 296
chama, Vulpes **286**
chamaeropsis, Apodemus 573, 574
chamek, Ateles **257**
champa, Niviventer 635
championi, Pogonomys **641**
chamula, Neotoma 712
chanco, Canis 281
changaica, Ochotona 807
changensis, Maxomys 614
chanleri, Arvicanthis 577
Redunca 414
chaouianensis, Crocidura 94
chapadae, Sylvilagus 825
chapadensis, Potos 333
Sylvilagus 825
Tamandua 68
chaparensis, Oligoryzomys 718
chapensis, Typhlomys **684**
chapini, Anomalurus 757
Chaerephon **232**
Heterohyrax 373
chapmani, Akodon 693
Dipodomys **479**
Dobsonia **139**, 140
Eptesicus 201
Marmosa 18
Oryzomys 720, **721**, 721, 725
Pseudomys **645**, 647
Sciurus 441
Sylvilagus 825
Synaptomys 534
chapmanni, Equus 369
charbinensis, Mustela 324
charkovensis, Apodemus 574
charltoni, Pardofelis 299
charon, Meriones 556
Charronia 319
chaseni, Rattus 650
Rhinolophus 164
chathamensis, Ziphius 364
chati, Leopardus 291
Chaus 289
chaus, Felis **289**, 290
cheesmani, Gerbillus **550**, 550, 553
Lepus 816
Tachyoryctes 687
Cheirogaleidae **243**
Cheirogaleinae **243**
Cheirogaleus **243**, 243
adipicaudatus 243
commersonii 243
crossleyi 243
griseus 243
major **243**, 243
medius **243**
melanotis 243
milii 243
samati 243
sibreei 243
thomasi 243
trichotis 243
typicus 243
typus 243
Cheiromeles **233**
caudatus 233
cheiropus 233
jacobsoni 233
parvidens 233
torquata 233

torquatus **233**
Cheiromys **247**
Cheiron **274**
cheiropus, Cheiromeles 233
chejuensis, Apodemus 570
chelan, Ursus 338
Chelemys **699**, 702, 714
alleni **699**
angustus 691, 699
connectens 699
delfini 699
fumosus 699
macronyx **699**, 699
megalonyx **699**, 699
microtis 699
niger 699
scalops 699
vestitus 699
Chelemyscus 701
fossor 702
chelidonias, Ursus 338
Cheliones 555, 556
Cheloniscus 66, 67
chengi, Meriones **556**, 556
cherriei, Molossus 235
Orthogeomys **470**
Oryzomys 752
Proechimys 795, 796
Zygodontomys 752
cherrii, Reithrodontomys 742
chevrieri, Apodemus **571**
cheybani, Lepus 821
cheyennensis, Thomomys 475
chialtanensis, Capra 405
chianfengensis, Hylopetes 460
chiapensis, Dasyprocta 782
Orthogeomys 471
Reithrodontomys 741
Rheomys 743
Sciurus 440
Sorex 122
Sylvilagus 825
Chibchanomys 694, **699**, 743
trichotis **699**, 699
chibigouazou, Leopardus 291
chibiguazu, Leopardus 291
chickara, Antilope 403
Tetracerus 403
chihfengensis, Meriones 558
chihliensis, Niviventer 633
chihuahuae, Thomomys 475
chihuahuensis, Microtus 527
chilaloensis, Muriculus 621, 622
childi, Oryzomys 720
childreni, Tonatia 180
chilensis, Conepatus 316
Ctenomys 785
Galictis 318
Hippocamelus 390
Histiotus 205
Lontra 311
Myocastor 806
Otaria 328
Phyllotis 738
Pseudalopex 284, 285
Pudu 392
chiliensis, Sciurus **443**
chilihueque, Lama 381
chillae, Sylvilagus 825
chiloensis, Geoxus 702
Lagenorhynchus 353
Myotis 208, **209**
Chilomys **699**
fumeus 699
instans **699**
Chilonatalus 194, 195
Chilonycteris 176, 177
Chilophylla 170
Chilotus 517, 527
Chimarrogale **107**, 107
hantu **107**, 107
himalayica **107**, 107
leander 107
phaeura **107**, 107, 108
platycephala **107**, 107
styani **107**, 107
sumatrana **107**, 107
varennei 107
chimbanus, Sylvilagus 825
Chimmarogale 107
chimo, Lontra 311
chimpanse, Pan 277
Chimpansee 276, 277
chinanteco, Habromys **703**
Chincha 316
Chinchilla **777**
boliviana 777
brevicaudata **777**, 777
chinchilla 777
intermedia 777
lanigera **777**, 777
major 777
velligera 777
chinchilla, Chinchilla 777
Chinchillidae **777**, 805
Chinchilloidea 771
chinchilloides, Euneomys **701**, 701, 702
Reithrodon 701
Chinchillula **699**
sahamae **699**, 699
chinensis, Delphinus 355
Eothenomys **513**
Erinaceus 77
Lutra 312
Meles 314
Miniopterus 231
Myotis **209**
Ochotona 810
Panthera 298
Prionailurus 295
Pteropus 148
Rhinolophus 168
Rhizomys 685
Sousa **355**, 355, 357
Tupaia 131
chinga, Conepatus **316**
chinghe, Conepatus 316
chingpingensis, Tamiops 457
chintalis, Dremomys 425
Chionobates 814
chionogaster, Rattus 658
Chionomys **506**, 506, 507, 511, 517, 518, 523
abulensis 507
aleco 507
alpinus 507
aquitanius 507
cedrorum 507
circassicus 507
dementievi 507
gotschobi 507
gud **507**, 507
hermonis 507
lasistanius 507
lebrunii 507
leucurus 507
lghesicus 507
loginovi 507
lucidus 507
malyi 507
mirhanreini 507
neujukovi 507
nivalis **507**, 507
nivicola 507
occidentalis 507
oeconomus 507
olympius 507
oseticus 507
personatus 507
petrophilus 507
pontius 507
pshavus 507
radnensis 507
roberti **507**, 507
satunini 507
spitzenbergerae 507
trialeticus 507
turovi 507
ulpius 507
wagneri 507
chionopaes, Dicrostonyx 511
chiralensis, Eptesicus 201
chiricahuae, Sciurus 442
Thomomys 473
chirindensis, Heliosciurus 429
chiriquensis, Heteromys 481
Myotis 213
Odocoileus 391
Potos 333
Sciurus 441
Sigmodon 747
Tamandua 68
chiriquinus, Eptesicus 201
Proechimys 798
Chiroderma **189**
doriae **189**
gorgasi 190
improvisum **189**
isthmicum 190
jesupi 190
salvini **189**
scopaeum 189
trinitatum **190**
villosum 189, **190**
chirodontus, Sus 379
Chiromyidae 247
Chiromys 247
Chiromyscus **583**, 583, 591, 632, 664
chiropus **583**
Chironax **138**
melanocephalus **138**
tumulus 138
Chironectes **16**
argyrodytes 16
bresslaui 16
cayennensis 16
guianensis 16
gujanensis 16
langsdorffi 16
minimus **16**
palmata 16
panamensis 16
paraguensis 16
sarcovienna 16
variegatus 16
yapock 16
Chiropodomys **583**, 583, 584, 587, 595, 602, 611, 640, 672, 673
ana 583
calamianensis **583**, 583, 584
fulvus 673
gliroides **583**, 584, 614, 634
jingdongensis 583, 584
karlkoopmani **584**, 604
legatus 584
major 583, **584**, 584
muroides **584**
niadis 583
peguensis 583
penicillatus 583
pictor 584
pusillus **584**, 584
Chiropotes **261**, 261
albinasus **261**
ater 261
chiropotes 261
couxio 261
fulvofusca 261
israelita 261
nigra 261
roosevelti 261
sagulata 261
satanas **261**, 261
utahicki 261
chiropotes, Chiropotes 261
Chiroptera 137
chiropus, Chiromyscus **583**
Mus 583
Chirosciurus 249
chiru, Pantholops 399
Chiruromys **584**, 584
forbesi **584**, 584
kagi 585
lamia **584**
major 584
mambatus 584

pulcher 584
satisfactus 584
shawmayeri 584
vates **585**
vulturnus 584
chisai, Crocidura 84
chitauensis, Crocidura 93
Mops 236
chitralensis, Capra 405
Nesokia 632
chiumalaiensis, Cricetulus 538
chiversi, Dendromus 543
Lepus 820
Procavia 374
Steatomys 545
Chlamyphorinae **64**
Chlamyphorus **64**
clorindae 64
minor 64
ornatus 64
patquiensis 64
retusus **64**
truncatus **64**, 64
typicus 64
chloe, Marmosa 18
Chlorocebus **265**
aethiops 263, **265**
alexandri 265
arenarius 265
beniana 265
budgetti 265
calliaudi 265
callidus 265
callitrichus 265
cano-viridis 265
centralis 265
chrysurus 265
cinereo-viridis 265
circumcinctus 265
cloetei 265
contigua 265
cynosuros 265
djamdjamensis 265
ellenbecki 265
excubutor 265
flavidus 265
graueri 265
griseistictus 265
griseoviridis 265
helvescens 265
hilgerti 265
itimbiriensis 265
johnstoni 265
katangensis 265
lalandii 265
lukonzolwae 265
luteus 265
marjoriae 265
marrensis 265
matschiei 265
nesiotes 265
ngamiensis 265
nifoviridis 265
passargei 265
pembae 265
pousarguei 265
pusillus 265
pygerythrus 265
rubellus 265
rufoniger 265
rufoviridis 265
sabaeus 265
silaceus 265
tantalus 265
tephrops 265
tholloni 265
toldti 265
tumbili 265
viridis 265
voeltzkowi 265
weidholzi 265
werneri 265
weynsi 265
whytei 265
chloroenemis, Callicebus 259
Chlorotalpa **74**, 75
arendsi **74**
cahni 75
congicus 75
duthieae **75**
guillarmodi 75
leucorhina **75**
luluanus 75
montana 75
sclateri **75**
shortridgei 75
tytonis **75**
chlorus, Peromyscus 736
Spermophilus 449
chobiensis, Diceros 372
Lepus 820
Lutra 312
Papio 269
Paraxerus 434
choco, Sciurus 441
chocoensis, Dasyprocta 782
Platyrrhinus **190**
Chodsigoa 122, 123
Choerodes 380
Choeromys 775
Choeroniscus **183**
godmani **183**
inca 184
intermedius **183**
minor 183, **184**
periosus **184**
Choeronycteris **184**, 186
mexicana **184**, 184
minor 183
ponsi 184
choeropotamus, Potamochoerus 378
Choeropsis 380
Choeropus 39
chofukusei, Myotis 210
Choiropotamus 377
pictus 377
Choloepidae 63
Choloepinae **63**
Choloepus 63, **64**
andinus 64
augustinus 64
brasiliensis 64
capitalis 64
curi 64
didactylus **64**
florenciae 64
hoffmanni **64**
juruanus 64
kouri 64
napensis 64
pallescens 64
peruvianus 64
unau 64
Cholomys 751
chombolis, Cynocephalus 135
Sundamys 666
chonensis, Sigmodon 748
chonotica, Otaria 328
chonoticus, Akodon 692
chontali, Glaucomys 460
chora, Tragelaphus 404
choras, Papio 269
chorassanicus, Apodemus 571
Hemiechinus 78
chorensis, Conepatus 316
chorezmi, Allactaga 489
chorisi, Phoca 332
choromandus, Hylobates 275
Chortomys 541
chotanus, Microryzomys 708
Sylvilagus 825
chouei, Sorex 120
chriseos, Suncus 102
christiei, Plecotus 224
christii, Gazella 396
Christomys 649, 655
christyi, Deomys 544
Graphiurus **763**
Mylomys 630
Chroeomys 688, 691, **700**, 700
andinus **700**, 700
bacchante 700
cayllomae 700
cinnamomea 700
cruceri 700
dolichonyx 700
gossei 700
inambarii 700
inornatus 700
jelskii **700**, 700
jucundus 700
ochrotis 700
pulcherrimus 700
pyrrhotis 700
scalops 700
sodalis 700
Chrotogale **341**
owstoni **341**, 341
Chrotomys 583, **585**, 585
gonzalesi **585**
mindorensis **585**
whiteheadi **585**, 585
Chrotopterus **177**
auritus **177**
australis 177
guianae 177
chrotorrhinus, Microtus 518, **520**, 520, 528, 531
chrysaeolus, Proechimys **796**, 796
chrysampyx, Eulemur 244
chrysargyrus, Pteropus 148
chrysauchen, Pteropus 147
chrysea, Presbytis 271
chryseola, Aplodontia 417
chrysillus, Calcochloris 74
chrysippus, Funisciurus 427
chrysocephala, Pithecia 262
chrysochaetes, Naemorhedus 407
Chrysochloridae **74**
Chrysochloris 74, **75**, 75, 76
arendsi 75
asiatica **75**, 75
auratus 75
aurea 75
balsaci 75
bayoni 75
calviniae 75
capensis 75
concolor 75
damarensis 75
dixoni 75
duthieae 74
elegans 75
fosteri 75
granti 76
hottentotus 74
inaurata 75
minor 75
namaquensis 75
obtusirostris 74
rubra 75
shortridgei 75
stuhlmanni **75**, 75
taylori 75
tenuis 75
trevelyani 75
tropicalis 75
vermiculus 75
visagiei **75**
visserae 75
chrysocomus, Bunomys **581**, 581, 582
Rattus 581, 667
Chrysocyon **282**
brachyurus **282**
campestris 282
cancrosa 282
isodactylus 282
jubatus 282
vulpes 282
chrysodeirus, Spermophilus 447
chrysogaster, Cercocebus 262
Hydromys **597**, 597, 673
Lemmus 517
Martes 320
Moschus **383**, 384
Presbytis 271

Sciurus 440
Tupaia **132**
chrysogenys, Zapus 499
chrysoleuca, Callithrix 252
chrysomalla, Tupaia 132
chrysomelanotis, Felis 290
chrysomelas, Leontopithecus **252**
Melanomys 708
Neotoma **711**, 711
Presbytis 271
Chrysomys 685
chrysonotus, Callosciurus 421
Myotis 210
Thomomys 473
chrysophaenus, Anomalurus 757
chrysophilus, Aethomys **567**, 567
chrysoproctus, Pteropus **146**
chrysopsis, Reithrodontomys **740**
chrysopus, Cebus 259
Lophuromys 605
Peromyscus 735
chrysopyga, Callithrix 251
chrysopygus, Leontopithecus **252**, 252
Rhynchocyon **830**
chrysorrhous, Spilocuscus 47
Chrysospalax **75**
dobsoni 76
leschae 76
pratensis 76
rufopallidus 76
rufus 76
transvaalensis 76
trevelyani **76**
villosus **76**
chrysospila, Martes 320
chrysothorax, Crocidura 94
Chrysothrix 260
chrysothrix, Pipistrellus 220
Profelis 296
chrysotis, Reithrodontomys 741
Chrysotricha 74
chrysotrichos, Cervus 386
chrysotrix, Petaurista 462
chrysura, Tupaia 133
chrysuros, Sciurus 441
chrysurus, Alouatta 255
Chlorocebus 265
Echimys **791**
Leontopithecus 252
Myoxus 791
Procolobus 272
Vulpes 286
Chthonergus 512
chudeaui, Acomys 565
chui, Panthera 298
chukchensis, Rangifer 392
chunyi, Mazama **390**, 392
churchi, Blarina 106
chuscensis, Sciurus 439
chutuchta, Felis 289, 291
chuva, Ateles 257
cicognanii, Mustela 322
cicur, Caluromys 15
Petaurista 463
cienegae, Sigmodon 747
cigognanii, Mustela 321
ciliana, Ochotona 813
ciliatus, Leopoldamys 603
Myotis 210
Neomys 110
Zaedyus 67
cilicica, Capra 405
cilicicus, Acomys **565**
Nannospalax 754
cilindricornis, Rangifer 392
ciliolabrum, Myotis 212
cimbricus, Microtus 519
cinamomeus, Ammospermophilus 420
Cinchacus 370
cincticauda, Eliomys 767
cinderella, Crocidura **83**
Cynictis 302
Mus 627
Petaurista 463
Thylamys 23
cinderensis, Dipodomys 479
cinera, Vulpes 287
cineracens, Acomys 565
cineraceus, Acomys **565**, 565, 566
Dipodomys 479
Graphiurus 764
Hipposideros **171**
Petaurista 463
cinerascens, Callicebus **258**
Chaetodipus 483
Cricetulus 538
Gazella 396
Graphiurus 764
Nasua 334
Rhinolophus 168
Spermophilus 447
Sylvilagus 825
Taphozous 160
cinerea, Abrocoma **789**
Desmodus 194
Didelphis 20
Hylobates 276
Lontra 311
Lutra 309
Mandrillus 268
Marmosa 20
Micoureus 20
Naemorhedus 407
Neophoca **328**
Neotoma **711**, 711
Otaria 328
Otonycteris 219
Redunca 414
Sciurus 443
Tadarida 240
Talpa 129
cinereiceps, Eulemur 244
cinereicollis, Tamias **453**, 453
cinereiventer, Thomasomys **749**
cinereo-maculatus, Mus 625
cinereo-viridis, Chlorocebus 265
cinereoargenteus, Pseudalopex 284
Urocyon **285**, 285
cinereoflava, Ochotona 809
cinereofusca, Ochotona 807
cinereus, Amblonyx **309**
Anomalurus 757
Artibeus 187, **188**, 189
Cervus 386
Cricetulus 538
Crocidura 94
Cryptotis 109
Cynomys 424
Eupetaurus **459**, 459
Hapalemur 245
Hesperomys 749
Lasiurus **206**
Lepus 817
Lipurus 45
Lophuromys **605**, 605
Nycticebus 248
Peromyscus 730
Petauroides 59
Phascolarctos **45**
Pseudocheirus 59
Reithrodontomys 741
Rhabdomys 662
Rhizomys 685
Sciurus 442
Sorex 108, 111, **113**, 113, 114, 115, 116, 117, 119, 121
Tamias 453
Thomasomys **749**
Thomomys 473
Typhlomys **684**, 684
Ursus 338
Zapus 499
cineris, Perognathus 485
cineritius, Peromyscus 732
cinnameus, Thomasomys 749
cinnamomea, Chroeomys 700
Neotoma 711
Ochotona 811
cinnamomeiventris, Callosciurus 421
cinnamomeum, Mormoops 176
cinnamomeus, Callosciurus 422
Crocidura 84
Gerbillus 550
Lepus 821
Myotis 215
Niviventer 634
Nycticeius 218
Oecomys 717
Petromus 774
Proechimys 797
cinnamominus, Ondatra 532
cinnamomum, Ursus 338
cinnamonea, Microtus 526
cintrae, Crocidura 94
circassicus, Chionomys 507
circe, Rhinolophus 163
circumcinctus, Chlorocebus 265
Erythrocebus 265
circumdatus, Pipistrellus **220**, 223
cirnei, Rhynchocyon **830**, 830
Cironomys 649
cirrhorhinus, Cercopithecus 263
cirrhosus, Trachops **181**
Vampyrus 181
cirrifer, Cebus 259
ciscaucasicus, Apodemus 574
Ellobius 512
Meriones 558
Microtus 524
Panthera 298
cistaunicus, Capreolus 390
Cistugo 207, 212, 215
cistura, Chaerephon 232
cita, Mazama 391
citella, Thylamys 23
Citellini **419**
Citellus 444
citellus, Mus 444
Spermophilus **445**, 446, 450
citernii, Madoqua 398
Citillus 444
citillus, Spermophilus 445
citrinellus, Saimiri 260
citrinus, Anomalurus 757
Civetta, indica 348
civetta, Civettictis **345**
Viverra 344
Civettictis **344**, 344, 345, 348
australis 345
civetta **345**
congica 345
matschiei 345
megaspila 345
orientalis 345
poortmanni 345
schwarzi 345
volkmanni 345
civettina, Viverra 347, 348, 348
civettoides, Viverra 348
clabatus, Rattus 652
Cladobatae 131
clamitans, Alouatta 255
Cuon 282
clamosa, Ochotona 811
clanceyi, Apodemus 573
clanculus, Lagenorhynchus 353
clara, Echymipera **41**
clarae, Leopoldamys 603
Melomys 617
Stenomys 665
clarencei, Dipodomys 480

clarissa, Tupaia 131
clarkei, Ammodorcas **395**
Cervicapra 395
Petaurista 462
Tamiops 457
Volemys **534**, 535
clarki, Ursus 340
clarkii, Pappogeomys 472
clarus, Dicrostonyx 510
Melomys 619
Onychomys 719
Perognathus 486
Pipistrellus 224
Tamias 454
claudi, Cephalophus 411
clavatus, Odocoileus 391
Claviglis 763
clavium, Emballonura 157
Odocoileus 391
cleberi, Oecomys **716**
clementae, Urocyon 285
clementis, Peromyscus 732
cleomophila, Perognathus 485
clepticus, Canis 280
Clethrionomyini 515
Clethrionomys 502, **507**, 507, 508, 513, 514, 532
akkeshii 509
alascensis 509
albiventer 509
alstoni 508
amurensis 509
arizonensis 508
arsenjevi 509
athabascae 508
baikalensis 509
bargusinensis 509
bedfordiae 509
bernisi 508
bosniensis 508
brevicaudus 508
britannicus 508
bromleyi 509
caesarius 508
californicus **508**, 508
cantueli 508
carolinensis 508
cascadensis 508
caurinus 508
centralis **508**
curcio 508
dawsoni 509
devius 508
dorogostaiskii 509
erica 508
frater 508
fulvus 508
fuscodorsalis 508
galei 508
gapperi **508**, 508, 509
garganicus 508
gaspeanus 508
gauti 508
glacialis 509
glareolus 502, **508**, 508
gorka 508
hallucalis 508
helveticus 508
hercynicus 508
hintoni 509
hudsonius 508
idahoensis 508
insulaebellae 508
insularis 509
intermedius 508
irkutensis 509
istericus 508
italicus 508
jacutensis 509
japonicus 532
jochelsoni 509
jurassicus 508
kamtschaticus 509
kolymensis 509
kurilensis 509
latastei 509
laticeps 509
latigriseus 509
lenaensis 509
limitis 508
loringi 508
makedonicus 508
maurus 508
mazama 508
microtinus 509
mikado 509
minor 508
mollessonae 509
montanus 509
nageri 508
narymensis 509
nivarius 508
norvegicus 508
obscurus 508
occidentalis 508
ochraceus 508
ognevi 508
orca 509
otus 509
pallescens 508
paludicola 508
parvidens 509
petrovi 508
phaeus 508
pirinus 508
platycephalus 509
ponticus 508
pratensis 508
proteus 508
pygmaeus 508
reinwaldti 508
rex 509
rhoadsii 508
riparia 508
rjabovi 509
rossicus 509
rubidus 508
rufescens 508
rufocanus 502, 503, 508, **509**, 509, 514, 515
rupicola 508
russatus 509
rutilus 502, 508, **509**, 509
ruttneri 508
saianicus 508
salairicus 509
saturatus 508
siberica 509
sibericus 509
sikotanensis **509**
skomerensis 509
sobrus 509
solus 508
stikinensis 508
suecicus 509
tomensis 509
tugarinovi 509
tundrensis 509
uintaensis 508
ungava 508
uralensis 509
variscicus 509
vasconiae 509
vesanus 509
vinogradovi 509
volgensis 509
washburni 509
wasjuganensis 509
watsoni 509
wosnessenskii 509
wrangeli 508
yesomontanus 509
Clidomyinae **805**
Clidomys 804, **805**
cundalli 805
jamaicensis 805
major 805
osborni **805**, 805
parvus **805**
cliftoni, Marmota 431
Nelsonia 710
clinedaphus, Monophyllus 185
Cliomys 794
clivorum, Octodon 788
clivosus, Rhinolophus **164**, 165, 168
Taterillus 562
Cloeotis **170**
australis 170
percivali **170**, 170
cloetei, Chlorocebus 265
Clonomys 496
clorindae, Chlamyphorus 64
Cloromis 781
clusii, Trichechus 365
clusius, Thomomys 473, **474**, 475
Clymene 356
clymene, Stenella **356**
Clyomys **793**, 793
bishopi **793**
laticeps **793**
spinosus 793
cnemophila, Neotoma 711
cnossius, Oryctolagus 822
coahuilensis, Peromyscus 733
Coassus 390
coatlanensis, Osgoodomys 726
Peromyscus 734
cobanensis, Myotis **209**
Cobaya 778
cobaya, Cavia 778, 779
Hydrochaeris 781
cobourgiana, Mormopterus 238
cocalensis, Rhipidomys 744
cocalis, Sciurus 441
coccygis, Phalanger 46
Coccymys **585**, 586
albidens **585**, 585
bruijnii 585
mayeri 585
ruemmleri **585**, 585, 586
shawmayeri 585, 586
cochinchinensis, Tupaia 131
cochinensis, Paradoxurus 344
cockerelli, Callosciurus 422
Echymipera 41
cockrumi, Perognathus 485
cocos, Galagoides 250
Sciurus 440
codajazensis, Bradypus 63
Cyclopes 67
Saimiri 260
codiensis, Microtus 525
coecata, Tadarida 240
coelestis, Bunomys **581**, 581, 582
Callithrix 251
Mus 581
Coelogenys 783
coelognathus, Lophocebus 266
Coelomys 622, 623, 625, 627, 629
coelophyllus, Rhinolophus **165**, 168
Coelops **170**
bernsteini 170
formosanus 170
frithi **170**
frithii 170
hirsutus **170**
inflatus 170
robinsoni **170**, 170
sinicus 170
Coendou **775**, 775, 776, 777
bicolor **775**, 775, 776
boliviensis 775
brandtii 775
centralis 775
cuandu 775
koopmani **775**, 775
longicaudatus 775
platycentrotus 775
prehensilis **775**, 775
quichua 775
richardsoni 775
rothschildi **775**, 775, 776
sanctamartae 775
simonsi 775

coenorum, Rattus 658
coenosa, Golunda 592
coenosus, Ellobius 513
Heliosciurus 429
Otomys 680
coeruleoalba, Stenella **356**
coerulescens, Cricetulus 538
Furipterus 195
coeruleus, Dinaromys 511
Coetomys 771
coffaeus, Golunda 592
coffini, Trachops 181
Cogia 359
cognata, Tupaia 132
cognatus, Peromyscus 730
Pteropus 150
Rhinolophus **165**
Sylvilagus 825
Thomomys 475
coibae, Dasyprocta **781**
coibensis, Alouatta **254**
Molossus 235
cokii, Alcelaphus 393
colburni, Ctenomys **784**
colchicus, Microtus 524, 530
Rhinolophus 166
colemani, Peromyscus 735
Coleura **156**
afra **156**
gallarum 156
kummeri 156
nilosa 156
seychellensis **156**
silhouettae 156
colias, Scotophilus 227
colimae, Reithrodontomys 740
Sigmodon 747
colimensis, Nyctomys 715
Oryzomys 723
Sciurus 440
Urocyon 285
colini, Taurotragus 403
collaris, Arctonyx **313**, 313
Cercocebus 263
Eulemur 244
Graphiurus 764
Hemiechinus **78**, 78
Myotis 213
Neomys 110
Ochotona 807, **808**, 808, 811
Pteropus 151
Sciurotamias 439
Ursus 338
collasinus, Naemorhedus 407
collatus, Peromyscus 730
collega, Marmosops 20
colletti, Pseudocheirus 59
Rattus 649, **651**, 651, 659, 661, 662
colliaei, Sciurus **440**
collinsi, Jaculus 493
Saimiri 260
collinus, Akodon 690
Delomys 700
Dipodomys 479
Lepus 821
Meriones 556
Peromyscus 735
Petromyscus **683**, 683, 684
Pipistrellus 224
Praomys 683
Rhipidomys 744
Scotophilus 227
Sus 379
Tamiops 457
Thomomys 473
Tscherskia 540
collis, Chaetodipus 484
Thomomys 473
collium, Meriones 556, 558
Ovis 408
Colobates 444
Colobidae 262, 269
Colobinae **269**
Colobolus 269
Colobotis **444**, **446**, **447**
Colobus **269**, **271**
abyssinicus 270
adolfi-friederici 270
albocaudatus 270
angolensis **269**, 270
anthracinus 270
benamakimae 270
bicolor 270
brachychaites 270
caudatus 270
comosa 270
cordieri 270
cottoni 270
dianae 270
dodingae 270
dollmani 270
elgonis 270
escherichi 270
gallarum 270
guereza **270**
ituricus 270
kikuyuensis 270
langheldi 270
laticeps 270
leucomeros 270
limbarenicus 270
managaschae 270
maniemae 270
matschiei 270
mawambicus 270
municus 270
nahani 270
occidentalis 270
palliatus 270
percivali 270
poliurus 270
polykomos **270**
prigoginei 270
regalis 270
roosevelti 270
ruppelli 270
rutschiricus 270
ruwenzorii 270
sandbergi 270
satanas **270**
sharpei 270
terrestris 270
tetradactyla 270
thikae 270
uellensis 270
ursinus 270
vellerosus 270
verus 271
weynsi 270
zenkeri 270
colocolo, Oncifelis 291, **294**, 294
colombertinus, Cervus 387
colombiae, Histiotus 205
colombiana, Lontra 311
colombianus, Metachirus 20
Tapirus 371
Tonatia 181
colombica, Didelphis 17
Colomys **586**, 586
bicolor 586
denti 586
eisentrauti 586
goslingi **586**, 586
plumbeus 586
ruandensis 586
coloniae, Caracal 288
colonicus, Chaerephon 232
Raphicerus 400
colonus, Epimys 630
Geomys 470
Mungos 307
Myomys 630, 631
Pteropus 146
coloratus, Glaucomys 460
Oryzomys 724
Rattus 660
Colpothrinax, wrightii 802
coludo, Ctenomys 785
Colugidae 135
Colugo 135
columbae, Alces 389
columbiae, Oreamnos 407
columbiana, Aplodontia 417
Dasyprocta 782
Martes 320
Neotoma 711
Oreamnos 407
columbianus, Canis 281
Dipodomys 479
Geocapromys 800
Melanomys 708
Odocoileus 391
Perognathus 486
Proechimys 798
Spermophilus **445**
Tamiasciurus 457
Thomomys 475
columbicus, Odocoileus 391
columbiensis, Glaucomys 460
Lepus 814
colurnus, Eothenomys 514
Colus 400
colus, Saiga 400
colusus, Ursus 338
colvini, Crocuta 308
comanche, Peromyscus 730, 736, 737
comata, Presbytis **270**, 271
Procavia 374
comatus, Papio 269
combalbertinus, Cervus 387
comberi, Millardia 621
comes, Galea 779
cometes, Grammomys **592**, **593**, **594**
commersoni, Aotus 256
Hipposideros **172**
commersonii, Cephalorhynchus **351**, 354
Cheirogaleus 243
commissarisi, Glossophaga **184**
commissarius, Mus 625
communis, Barbastella 199
Callithrix 252
Canis 281
Genetta 345
Meles 314
Phocoena 358
Plecotus 224
Thylacinus 29
Vulpes 287
commutatus, Oligoryzomys 718
comobabiensis, Thomomys 473
Comopithecus 269
comorensis, Pteropus 151
comorinus, Funambulus 426
comosa, Colobus 270
Ovis 408
comosus, Ozotoceros 392
compacta, Blarina 106
compactus, Dipodomys **477**, 479
compressa, Arctocephalus 327
compressus, Macrotus 178
Steno 357
comptus, Epomops 142
Peromyscus 733
comus, Lepus **816**
conatus, Rattus 659
concava, Lonchophylla 181, 182
concavus, Orthogeomys 471
concinna, Petrogale **56**
concinnus, Antechinus 30
Cercartetus **58**
Dendromus 543
Exilisciurus **426**
Myotis 213
Sorex 111
concisa, Mellivora 315
concisor, Castor **467**
Thomomys 473
concolor, Amblonyx 310
Artibeus **188**
Callosciurus 421

Chrysochloris 75
Cynopterus 139
Erinaceus **77**, 77
Felis 296, 297
Galidia 300
Hylobates **275**, 275
Lutra 309
Mephitis 317
Monodelphis 21
Nasalis **270**
Oecomys 715, **716**
Pelomys 639
Phoca 332
Puma **296**
Rattus 652
Salanoia **300**, 300
Sciurus 439
Sicista **496**, 496, 497, 498
Spermophilus 446
Tupaia 131
conditi, Chaetodipus 483
conditor, Leporillus **604**, 604
condorensis, Niviventer 634
Pteropus 147
Ratufa 437
condradsi, Galerella 303
condurensis, Callosciurus 423
Ratufa 437
Condylura **124**
cristata **124**
longicaudata 124
macroura 124
nigra 124
parva 124
prasinatus 124
radiata 124
Condylurinae 124
Condylurini 124
condylurus, Mops **236**, 236
conepatl, Conepatus 316
Conepatus **315**
ajax 316
amazonica 316
arequipae 316
bahiensis 316
budini 316
calurus 316
castaneus 316
chilensis 316
chinga **316**
chinghe 316
chorensis 316
conepatl 316
dimidiata 316
enuchus 316
feuillei 316
figginsi 316
filipensis 316
fremonti 316
furcata 316
gaucho 316
gibsoni 316
gumillae 316
humboldtii 315, **316**
hunti 316
inca 316
leuconotus **316**, 316
mapurito 316
mearnsi 316
mendosus 316
mesoleucus **316**, 316
monzoni 316
nelsoni 316
nicaraguae 316
pampanus 316
patachonica 316
patagonica 316
pediculus 316
porcinus 316
proteus 316
putorius 316
quitensis 316
rex 316
semistriatus **316**
sonoriensis 316
suffocans 316
taxinus 316
telmalestes 316
texensis 316
trichurus 316
tropicalis 316
venaticus 316
vittata 316
westermanni 316
yucatanicus 316
zorilla 316
zorrino 316
confinalis, Thomomys 473
confinii, Eothenomys 514
confinis, Ratufa 437
Sigmodon 747
Sylvilagus 824
Tamias 454
confinus, Thomomys 474
confucianus, Niviventer 632, **633**, 633, 634, 635
congica, Civettictis 345
Tadarida 236
congicus, Aonyx **310**, 310
Cephalophus 411
Chlorotalpa 75
Funisciurus **427**
Lophocebus 266
Mops **236**, 236
Potamochoerus 378
Scutisorex 101
Taterillus **562**
congobelgica, Crocidura **83**
congoensis, Dendrohyrax 373
Giraffa 383
congolanus, Taurotragus 403
Congosorex **81**, 81, 99
polli **81**
Conilurus **586**, 586, 604, 615, 619, 620, 674
albipes **586**, 586
constructor 586
destructor 586
hemileucurus 586
melanura 586
melibius 586
penicillatus **586**, 586, 587
randi 586
coninga, Niviventer 633
conipus, Callosciurus 422
connectens, Chelemys 699
Maxomys 614
Notiomys 699
Ototylomys 726
Spermophilus 447
Sylvilagus 825
Thomomys 473
Connochaetes **393**
albojubatus 394
borlei 394
capensis 393
connochaetes 393
cooksoni 394
corniculatus 394
fasciatus 394
gnou **393**
gorgon 394
hecki 394
henrici 394
johnstoni 394
lorenzi 394
mattosi 394
mearnsi 394
operculatus 393
reichei 394
rufijianus 394
schulzi 394
taurinus **394**
connochaetes, Connochaetes 393
connori, Lepus 816, 817, 821
Conothoa 807
conoveri, Ctenomys 783, **784**
Conoyces 51, 57
conscius, Heteromys 481
consimilis, Stenella 356
consitus, Bassariscus 334
Pappogeomys 472
consobrinus, Cervus 386
Peromyscus 733
Sciurotamias 439
Scotophilus 227
Sylvilagus 825
Tamias 454
consolei, Erinaceus 77
conspicillatus, Galago 249
Lagorchestes **52**
Pteropus **147**
conspicua, Ratufa 437
constablei, Phenacomys 533
constantiae, Micoureus **20**
constantina, Leptailurus 292
constanzae, Tadarida 240
constrictus, Crocidura 94
Hippopotamus 381
Microtus 519
Neomys 110
constructor, Conilurus 586
Conylurus 586
consul, Arctonyx 313
Odocoileus 391
consularis, Menetes 433
contigua, Chlorocebus 265
Microtus 519
continentalis, Rhyncholestes 25
Sus 379
continentis, Hylobates 276
Myotis 211
Natalus 195
Pteronotus 177
Ptilocercus 133
Sylvilagus 825
contractus, Brotomys **799**
Thomomys 473
contradictus, Thomasomys 749
contumax, Callosciurus 421
conurus, Tolypeutes 67
convelatus, Caenolestes **25**
convergens, Thomomys 473
convexum, Uroderma 193
convexus, Thomomys 473
convictus, Lemniscomys 601
convolutor, Pseudocheirus 60
Conylurus 586
constructor 586
cookei, Proodontomys 677
cookii, Mus **623**
Odobenus 326
Phalangista 59
Pseudocheirus 60
Trichosurus 48
cooksoni, Connochaetes 394
cooktownensis, Rattus 654
coolidgei, Macaca 266
Peromyscus 732
coomansi, Callosciurus 423
coombsi, Aonyx 310
Cynictis 302
Gerbillurus 547
Procavia 374
cooperi, Leopardus 292
Paraxerus **434**, 435
Protoxerus 436
Sorex 113
Synaptomys **534**, 534
Tamias 456
Tragelaphus 403
copei, Perognathus 485
copelandi, Microtus 527
Copernicia, curtissi 802
coppingeri, Oligoryzomys 718
coquereli, Microcebus **243**
Propithecus 247
coquerelii, Suncus 102
coquimbensis, Rattus 658
Thylamys 23
cor, Cardioderma **162**
Emballonura 158
Megaderma 162
cora, Gazella 396
coracius, Rattus 653
coraginis, Golunda 592
corax, Trachypithecus 274
corbetti, Panthera 298
cordeauxi, Lepus 822

Madoqua 398
cordieri, Colobus 270
cordillerae, Peromyscus 728, 729
cordofanicum, Rhinopoma 155
cordovensis, Calomys 698
cordubensis, Lama 381
coreae, Apodemus 570
Crocidura 96
Sciurus 443
corealis, Tachyglossus 13
coreana, Mogera 126
Ochotona 809
coreanus, Canis 281
Lepus **816**, 817, 819, 820
Mustela 324
Sciurus 443
Sus 379
coreensis, Panthera 298
Pipistrellus 223
coriacorum, Spermophilus 448
corilinum, Muscardinus 769
corinna, Gazella 396
corinnae, Pseudochirops **60**
coritzae, Maxomys 614
Cormura **156**
brevirostris **156**
pullus 156
corniculatus, Connochaetes 394
Syncerus 402
cornipes, Cervus 386
cornutus, Naemorhedus 407
Plecotus 224
Rhinolophus **165**, 167
coromandelicus, Pipistrellus 220
coromandra, Pipistrellus **220**
coronarius, Microtus 523
coronata, Sylvicapra 412
coronatus, Alces 389
Eulemur **244**
Hipposideros **172**
Lepus 817
Propithecus 247
Pseudochirops 60
Pteropus 149
Sorex **114**
corozalus, Proechimys 799
Puertoricomys **799**
corriae, Amblysomus 74
corsac, Vulpes **286**
Corsia 106
corsicana, Crocidura 96
corsicanus, Cervus 385, 386
Lepus 814, 816, **817**
Mustela 323
corsicosardinensis, Ovis 408
Corsira 106, 111
nigrescens 122
corteanus, Cervus 386
cortensis, Calomys 698
Corvira 192, 193
corvus, Trachypithecus 274
corybantium, Nannospalax 754
coryi, Artibeus 188
Puma 296
corylinus, Caracal 288
corynophyllus, Hipposideros **172**
Corynorhinus 224, 225
phyllotis 205
Coryphomys 583, **587**, 587, 602, 640
buehleri **587**, 587
Corypithecus 270
cosensi, Cricetomys 541
Gerbillus 548, **550**
cossyrensis, Crocidura **83**, 94
costaricensis, Blarina 106
Leopardus 291
Mustela 322
Nyctomys 715
Odocoileus 391
Oligoryzomys 718
Orthogeomys 470
Puma 296
Reithrodontomys 742
Sylvilagus 825
Urocyon 285
cothurnata, Ratufa 437
Cothurus 261
cottoni, Ceratotherium 371
Colobus 270
Giraffa 383
Hippotragus 412
Hyemoschus 382
Kobus 413
Loxodonta 367
Mellivora 315
Ourebia 399
Pan 277
Potamochoerus 378
Profelis 296
Redunca 414
Syncerus 402
Tragelaphus 404
coucang, Nycticebus **248**
Tardigradus 248
coucha, Mastomys **610**, 610
Mus 609
couchi, Thomomys 475
couchii, Spermophilus 450
couesi, Erethizon 776
Odocoileus 391
Oryzomys **721**, 721, 722, 724
Rhipidomys **744**, 744
cougar, Puma 296
couguar, Puma 296
couiy, Sphiggurus 777
coupeii, Graphiurus 764
couxio, Chiropotes 261
cowani, Microgale **71**, 71
Microtus 530
Ovis 409
coxenii, Thylogale 57
coxi, Capreolus 390
Cephalophus 411
Hipposideros **172**
Panthera 298
Pipistrellus 223
coxii, Mirounga 330
coxingi, Maxomys 613
Niviventer 632, **633**, 633, 635
coypus, Mus 805
Myocastor **806**
cozulmelae, Peromyscus 732
cozumelae, Didelphis 17
Mimon 179
Oryzomys 721
cracens, Marmosops **19**
cradockensis, Cryptomys 772
Rhabdomys 662
Craegoceros 390
cranbrooki, Naemorhedus 407
crandalli, Saguinus 253
Craseomys 507
Craseonycteridae **155**
Craseonycteris **155**
thonglongyai **155**, 155
craspedotis, Lepus 821
crassa, Syconycteris 155
crassibulla, Meriones 558
crassicauda, Bdeogale **301**, 301
Salpingotus **491**, 491, 492
crassicaudata, Didelphis 18
Lutreolina **18**
Manis **415**
Phascogale 35
Sminthopsis **35**
crassicaudatus, Graphiurus **763**
Lepus 823
Molossus 235
Otolemur **250**, 250
Pronolagus **823**, 823, 824
Sorex 111
crassicaudis, Lutreolina 18
crassicaudus, Isothrix 793
Suncus 103
crassiceps, Cebus 259
Paradoxurus 344
crassicornis, Cervus 386
crassidens, Galictis 318
Lagidium 778
Phocaena 355
Procyon 335
Pseudorca **355**
Sus 379
Thomomys 475
crassipes, Macropus 53
Microgale 71
Rattus 658
crassirostris, Heteromys 481
crassispinis, Hystrix **773**, 774
crassodon, Canis 281
Ursus 338
crassulus, Pipistrellus **220**
crassus, Meriones **556**
Myotis 212
Pecari 380
Phenacomys 533
Saccolaimus 159
Sundamys 666
Thomomys 473
Ursus 338
cratericus, Tamias 453
Crateromys 579, 583, **587**, 587, 602, 640
australis **587**
paulus **587**
schadenbergi **587**, 587
cratodon, Dipodomys 480
Cratogeomys 469, 471, 472, 473
castanops 472
Craurothrix 591
crawfordi, Notiosorex 110, **111**
crawshaii, Equus 369
crawshayi, Dendrohyrax 373
Kobus 413
Lepus 814, 818, 819, 820, 822
creaghi, Rhinolophus 164, **165**
creber, Spermophilus 445
credulus, Sundamys 666
cremnobates, Ovis 408
Cremnomys **587**, 587, 591, 621
australis 588
blanfordi **587**, 587, 588
caenosus 588
cutchicus 587, **588**, 588
elvira **588**, 588
leechi 588
medius 588
rajput 588
siva 588
cremoriventer, Niviventer 632, **633**, 634
Rattus 634
crenata, Crocidura **83**
crenulata, Dobsonia 140
crenulatum, Mimon **179**
creper, Oligoryzomys 718
Reithrodontomys **740**
crepidatus, Phodopus 539
crepuscula, Trachypithecus 274
crepuscularis, Nycticeius 217
crequii, Lama 381
crespoi, Oncifelis 294
cressonus, Ursus 338
cretaceiventer, Niviventer 633
Cretasorex 80
cretensis, Felis 290
cretica, Capra 405
creticum, Rhinolophus 166
creticus, Apodemus 573
Lepus 817
Cricetidae 512, 535, 541, 563, 676, 677, 679
Cricetinae 487, 501, 512, 535, **536**, 563, 676, 678,

679
Cricetiscus 539
Cricetodipus 484
Cricetomyinae 501, **540**, 540
Cricetomys **540**
adventor 541
ansorgei 541
buchanani 541
cosensi 541
cunctator 541
dichrurus 541
dissimilis 541
dolichops 540
elgonis 541
emini **540**, 540
enguvi 541
gambianus **540**, 540
goliath 541
grahami 541
haagneri 541
kenyensis 541
kivuensis 540
langi 541
liberiae 540
luteus 540
microtis 541
oliviae 541
osgoodi 541
poensis 540
proparator 540
raineyi 540
sanctus 540
selindensis 541
servorum 541
vaughanjonesi 541
viator 541
Cricetops 563
dormitor 563
Cricetulus 536, **537**, 537
accedula 538
alticola **537**, 537, 538
andersoni 538
arenarius 538
atticus 538
barabensis **537**, 537, 538
bedfordiae 539
bellicosus 538
caesius 538
chiumalaiensis 538
cinerascens 538
cinereus 538
coerulescens 538
dichrootis 538
elisarjewi 538
falzfeini 538
ferrugineus 537
fulvus 538
fumatus 537
furunculus 537
griseiventris 538
griseus 537, 538
isabellinus 538
kamensis **537**, 537
kozhantschikovi 538
kozlovi 537
lama 537
longicaudatus **538**
manchuricus 537
migratorius **538**
mongolicus 537
murinus 538
myosurus 538
neglectus 538
nigrescens 538
obscurus 537, 538
ognevi 538
pamirensis 538
phaeus 538
pseudogriseus 537
pulcher 538
roborovskii 539
sokolovi **538**
sviridenkoi 538
tauricus 538
tibetanus 537
triton 537, 539
tuvinicus 537
vernula 538
xinganensis 537
zvierezombi 538
cricetulus, Alticola 504
Saccostomus 541
Cricetus 536, **538**
albus 538
babylonicus 538
canescens 538
cricetus **538**
eversmanni 536
frumentarius 538
fulvus 538
fuscidorsis 538
germanicus 538
jeudii 538
latycranius 538
nehringi 538
niger 538
nigricans 538
polychroma 538
raddei 538
rufescens 538
stavropolicus 538
tauricus 538
tomensis 538
varius 538
vulgaris 538
cricetus, Cricetus **538**
Mus 538
crinicaudus, Callicebus 259
crinifrons, Muntiacus **388**, 389
criniger, Glironia 16
Lagidium 778
crinigerum, Lagidium 778
crinitus, Bradypus 63
Chaetodipus 484
Oryzomys 721
Peromyscus **729**, 729, 735
Petinomys **463**, 463
crispa, Tamandua 68
crispii, Lepus 816
crispus, Liomys 482
Naemorhedus **407**, 407
Ornithorhynchus 13
cristata, Condylura **124**
Cystophora **329**
Dasyprocta **781**, 782
Felis 290
Genetta 346
Hystrix **773**, 773
Phoca 329
Viverra 309
cristatus, Bradypus 63
Cebus 259
Chaerephon 233
Echimys 791
Procolobus 272
Proteles **309**
Sorex **124**
Sus 379
Trachypithecus **273**, 273
cristicauda, Chaetocercus 31
Dasycercus **31**
cristina, Vicugna 382
cristobalensis, Peromyscus 737
Sorex 120
croacuta, Crocuta 308
croaticus, Sciurus 443
Crocidosoricinae 81, 99
Crocidura **81**, 88, 91, 99, 100, 101, 103
aagaardi 89
aegatensis 95
aequicaudata 82
agadiri 96
agilis 94
albipes 88
albiventris 94
alchemillae 85
aleksandrisi **81**
allex **81**
alpina 81
altae 87
amalae 87
anchietae 92
andamanensis **81**
angolae 97
annellata 87
ansellorum **81**
ansorgei 91
anthonyi 94
antipae 96
aoris 89
arabica **81**, 81, 85
aralychensis 86
arethusa 85
argentatus 83
ariadne 96
aridula 97
arispa 95, 98
armenica **81**, 81, 93
astrabadensis 96
atlantis 92
attenuata **82**
attila **82**, 82
avicennai 96
baileyi **82**
balcanica 96
balearica 96
baluensis 85
bartelsii 90
batesi **82**, 97
beatus **82**
beccarii **82**
beirae 87
beta 85
bicolor 85
bloyeti 87
bogdanowii 86
bolivari 97
bottegi **82**, 92
bottegoides **82**
boydi 93
brevicauda 85
bruecheri 96
brunnea 85
bueae 92
buettikoferi **82**, 82
butleri 97
cabrerai 91
calabarensis 93
caliginea **82**
calypso 95
campuslincolnensis 87
canariensis **82**, 609
candidus 94
caneae 96
canescens 87
cantanbra 96
capensis 84
cara 92
caspica 88
cassiteridum 96
caudata 95
chaouianensis 94
chisai 84
chitauensis 93
chrysothorax 94
cinderella **83**
cinereus 94
cinnamomeus 84
cintrae 94
congobelgica **83**
constrictus 94
coreae 96
corsicana 96
cossyrensis **83**, 94
crenata **83**
crossei **83**, 83, 85
cuanzensis 93
cuninghamei 85
cyanea **83**, 83, 86
cypria 96
cyrnensis 96
daphnia 92
darfurea 92
debalsaci 95
debeauxi 96
deltae 92
denti **83**
deserti 87
desperata **83**
dhofarensis **83**, 95

diana 85
dinniki 96
dolichura **83**, 88, 90, 93
doriae 85
doriana 92
douceti **84**
dracula 85
dsinezumi **84**, 84, 92, 94
ebriensis 83
eburnea 82
eisentrauti **84**
electa 83, 88
elegans 87
elgonius **84**
elongata **84**, 90
enezsizunensis 96
erica 83, **84**, 87
essaouiranensis 97
esuae 95
ferruginea 92
fimbriatus 94
fischeri **84**
flavescens **84**, 86, 92, 97
floweri **84**, 85
foetida 85
foucauldi 94
foxi **85**, 85, 88, 97
fuliginosa **85**, 85, 89, 92, 93, 96
fulvastra **85**, 86
fumigatus 86
fumosa **85**, 85, 89, 90, 94, 96
fuscomurina **85**, 91, 93
fuscosa 92
garambae 88
giffardi 92
glassi **85**
glebula 93
goliath **86**
gouliminensis 96
gracilipes **86**, 87, 95
grandiceps **86**
grandis **86**
grassei **86**
gravida 89
grayi **86**
greenwoodi **86**
grisea 82
grisescens 85
gueldenstaedtii **86**, 86, 96
guineensis 92
halconus 86
hansruppi 92
harenna **86**, 93
hedenborgiana 92
heljanensis 94
hendersoni 85
heptapotamica 96
hera 92
herero 92
hildegardeae 84, **86**, 86, 89, 95
hindei 97
hirta 83, 84, 86, **87**, 87, 90, 98
hispida **87**
holobrunneus 95
horsfieldi 87
horsfieldii **87**, 87
hosletti 84, 94
hydruntina 94
hyrcania 96
ibeana 87
ibicensis 94
ichnusae 94
iculisma 96
ilensis 96
indochinensis 87
infumatus 83
ingoldbyi 83
inodorus 94
intermedia 84
inyangai 88
italica 96
jacksoni **87**
jenkinsi **87**, 91
johnstoni 88
jouvenetae 83
judaica 88
katharina 93
kelabit 85
kempi 91
kijabae 92
kingiana 82
kivu 92
kivuana **87**
klossi 89
knysnae 84
kurodai 87
lakiundae 91
lamottei 85, **87**
langi 87
lanosa **87**
lar 96
lasia 88
lasiura **87**
latona 84, **88**
lawuana 85
lea **88**
lepidura 85
leucodon **88**, 95
leucodus 88
leucurus 94
levicula **88**
lignicolor 96
lipara 91
littoralis **88**, 90, 95, 97
lizenkani 87
longicauda 96
longicaudata 86
longipes **88**
lucina **88**
ludia 84, **88**
luimbalensis 87
luluae 91
luluana 92
luna **88**, 89, 94
lusitania **88**, 93
lutrella 93
lutreola 87
maanjae 87
macarthuri **89**, 95
macklotii 85
macmillani 88, **89**, 89, 96
macowi **89**
macrodon 85
major 94
malani 89
malayana 85, **89**, 89
manengubae **89**
manni 92
mansumensis 85
maporensis 89
maquassiensis **89**
mariquensis **89**
marita 85
marrensis 85
martensi 83
martiensseni 92
matruhensis 97
maurisca **89**
maxi **90**, 90, 91
melanorhyncha 85
mesatanensis 97
microurus 88
mimula 96
mimuloides 96
mindorus **90**
minuta **90**, 90, 96
miya **90**
monacha 96
monax 88, **90**, 97
monticola **90**, 90
montis **90**
mordeni 96
morio 104
moschata 94
muricauda 84, **90**
musaraneus 94
mutesae **90**, 98
myoides 87
nana **90**, 90, 91
nanilla **91**, 93
narentae 88
neavei 89
neglecta **91**
negligens 89
negrina **91**
nicobarica 87, **91**
nigeriae **91**
nigricans 84, 89, **91**
nigripes **91**
nigrofusca **91**, 91, 97
nilotica 91
nimbae **92**, 97
niobe 89, **92**
nisa 93
nyansae 92
nyikae 91
oayensis 96
obscurior 82, **92**
occidentalis 92
odorata 92
ognevi 95
okinoshimae 84
olivieri 84, 86, **92**, 92, 98
orientalis 85
orientis 96
orii 84, **92**
oritis 88, 90
osorio **92**
palawanensis **92**
pamela 93
pamirensis 96
paradoxura **92**
parvacauda 82
parvipes 83, **93**
pasha 85, 91, **93**, 93
percivali 97
pergrisea 81, **93**, 93, 95, 98
persica 88
peta 94
phaeopus 96
phaeura 86, **93**
phaios 87
picea **93**
pilosa 89
pitmani 89, **93**
planiceps 85, **93**, 93
poensis 82, 85, 91, **93**
polia 84, **93**
poliogastra 94
pondoensis 83
portali 96
praecypria 96
praedax 85
procera 87
provocax 91
pudjonica 85
pulchra 94
pullata **94**, 94
quelpartis 84
raineyi 88, **94**, 94
rapax 94
religiosa 90, 91, **94**
retusa 87
rhoditis **94**
roosevelti **94**
rubecula 87
rubricosa 82
rudolfi 91
rufa 94
russula 81, 83, 84, 86, **94**, 94, 95, 96, 98, 110
rutilus 84
safii 94
sansibarica 85
sarda 96
schistacea 88
schweitzeri 93
selina 88, **94**
serezkyensis 93, **95**, 95, 98
sericea 85
shantungensis 96
shortridgei 89
sibirica **95**
sicula 82, **95**
silacea **95**
simiolus 98
smithii 83, 89, **95**
sodyi 87
somalica 83, **95**, 98
soricoides 91, 93

spurelli 92
stampflii 93
stenocephala **95**
strauchii 85
streetorum 98
suahelae 90, 97, 98
suaveolens 81, 86, 94, **95**, 95, 96
sururae 92
susiana **96**
sylvia 89
tadae 87
tamrinensis 97
tanakae 82
tansaniana **96**
tarella **96**
tarfayensis 83, **96**
tatiana 92, 98
tatianae 98
telfordi **96**
tenuis 85, **96**, 96
tephra 85
tephragaster 85
thalia 85, 89, **96**, 96
theresae 85, **97**, 97
thomasi 87
thomensis **97**
thoracicus 94
tionis 89
tiznitensis 96
toritensis 92
trichura 82
tristami 96
turba 96, **97**
ultima 90, **97**, 97
umbrina 84
umbrosa 88
unicolor 94
usambarae **97**
utsuryoensis 96
uxantisi 96
velutina 87
viaria 85, **97**, 98
villosa 85
virgata 87
voi **97**
vorax 94
vosmaeri 85
vryburgensis 83
vulcani 84
watasei 87
weberi 89
whitakeri 96, **97**
wimmeri 82, 92, **97**
woosnami 85
wuchihensis 87
xanthippe 98
xantippe **98**
yamashinai 87
yankariensis **98**
yebalensis 94
zaianensis 97
zaodon 91, 97
zaphiri 92, 97, **98**
zarudnyi 93, **98**, 98
zena 91
zimmeri **98**
zimmermanni **98**
zinki 81
zuleika 92
Crocidurinae **81**, 81, 98, 99, 101, 104, 107
Crocidurini 80
crociduroides, Mus **623**, 623, 629
Mycteromys 629
croconota, Dasyprocta 782
Crocuta **308**, 308
capensis 308
colvini 308
croacuta 308
crocuta **308**
cuvieri 308
encrita 308
felina 308
fisi 308
fortis 308
gariepensis 308
germinans 308
habessynica 308
kibonotensis 308
leontiewi 308
maculata 308
noltei 308
nyasae 308
nzoyae 308
panganensis 308
rufa 308
rufopicta 308
sivalensis 308
spelaea 308
thierryi 308
thomasi 308
togoensis 308
ultima 308
venustula 308
weissmanni 308
wissmanni 308
crocuta, Canis 308
Crocuta **308**
crooki, Odocoileus 391
crosetensis, Mirounga 330
Crossarchus 300, **301**, 302
alexandri **301**
ansorgei **301**
dybowskii 302
minor 301
nigricolor 301
obscurus 301, **302**, 302
platycephalus 302
somalicus 302
crossei, Crocidura **83**, 83, 85
crossi, Cercocebus 263
Cercopithecus 265
Paradoxurus 344
crossii, Ceratotherium 371
crossleyi, Cheirogaleus 243
Crossogale 107
Crossomys **588**, 620
moncktoni **588**, 588
Crossopus 110
himalayicus 107
crotalius, Callosciurus 421
crotaphiscus, Stenella 356
crowtheri, Ursus 338
cruceri, Chroeomys 700
crucialis, Philander 22
cruciger, Lagenorhynchus **353**, 353, 354
Presbytis 271
crucigera, Vulpes 287
crucina, Oncifelis 294
cruesemanni, Canis 280
crumeniferus, Hipposideros **172**
crumpi, Callosciurus 421
Diomys **591**, 591
Crunomys 575, 576, **588**, 588
celebensis **588**, 588
fallax **588**, 588
melanius **588**, 588
rabori 588, **589**
crusalbum, Cephalophus 411
crusnigrum, Pecari 380
cruzlimai, Saguinus 253
cryophilus, Megadontomys **707**, 707
crypta, Kerivoula 197
crypticola, Rousettus 152
cryptis, Sorex 111
Cryptochloris **76**, 76
wintoni **76**, 76
zyli **76**, 76
Cryptogale 70
Cryptolestes 25
Cryptomys **771**, 771
abberrans 772
albus 772
amatus 772
anomalus 772
ansorgei 772
arenarius 772
beirae 772
bigalkei 772
blainei 772
bocagei **771**
caecutiens 772
cradockensis 772
damarensis **771**
darlingi 772
exenticus 772
foxi **771**, 771
holosericeus 772
hottentotus **771**
jamesoni 772
jorisseni 772
junodi 772
komatiensis 772
kubangensis 771
kummi 772
langi 772
lechei 772
ludwigii 772
lugardi 771
mahali 772
mechowi **772**
mechowii 772
melanoticus 772
mellandi 772
micklemi 771
montanus 772
natalensis 772
nemo 772
nimrodi 772
ochraceocinereus **772**
orangiae 772
ovamboensis 771
palki 772
pallidus 772
pretoriae 772
rufulus 772
stellatus 772
streeteri 772
talpoides 772
transvaalensis 772
valschensis 772
vandami 772
vetensis 772
vryburgensis 772
whytei 772
zechi **772**
zimbitiensis 772
zuluensis 772
Cryptoprocta **340**, 340
ferox **340**, 340
typicus 340
Cryptoproctinae **340**
cryptorhinus, Meriones 557
cryptospilotus, Spermophilus 449
Cryptotis **108**
alticola 108
avia **108**
berlandieri 109
celatus 109
cinereus 109
elasson 109
endersi **108**
equatoris 109
euryrhynchis 108
exilipes 109
eximius 109
floridana 109
fossor 108
frontalis 108
goldmani **108**, 108
goodwini **108**, 108
gracilis **108**, 108
griseoventris 108
guerrerensis 108
harlani 109
hondurensis **108**, 108
jacksoni 108
macer 109
machetes 108
madrea 109
magna **108**
mayensis 109
medellinius 109
meridensis **108**, 109
merriami 109
merus 109
mexicana 108, **109**, 109,

111
micrura 109
micrurus 109
montivaga **109**
nayaritensis 109
nelsoni 109
nigrescens **109**, 109
obscura 109
olivaceus 109
orophila 109
osgoodi 109
parva **109**
peregrina 109
pergracilis 109
phillipsii 109
pueblensis 109
soricina 109
squamipes **109**
surinamensis 108, 112
tersus 109
thomasi **109**, 109
tropicalis 109
zeteki 109
crypturus, Epomophorus 141
cryptus, Scalopus 127
csikii, Sorex 111
ctenodactyla, Scirtopoda 495
Ctenodactylidae **761**
Ctenodactylus **761**
arabicus 761
gundi **761**, 761
joleaudi 761
massonii 761
mzabi 761
typicus 761
vali **761**, 761
ctenodactylus, Paradipus **495**
Ctenomyidae **783**, 783, 787
Ctenomyinae 783
Ctenomys **783**, 783, 784, 787
antonii 787
argentinus **784**
atracamensis 785
australis **784**, 786
azarae **784**, 785
barbarus 784
bergi 785
bicolor 785
boliviensis **784**
bonettoi **784**
brasiliensis 783, **784**, 786, 787
brunneus 785
budini 784
chilensis 785
colburni **784**
coludo 785
conoveri 783, **784**
dicki 785
dorsalis **784**
emilianus **784**
famosus 785
fochi 785
fodax 785
frater **784**
fueginus 785
fulvus **784**, 785, 786, 787
goodfellowi 784
haigi **785**, 785
johanis 785, 787
johannis 785
juris 785
knighti **785**, 785
latro **785**, 785
lentulus 785
leucodon **785**
lewisi **785**
luteolus 786
magellanicus **785**
maulinus **785**, 785
mendocinus 784, **785**, 785, 786, 787
minutus **785**
mordosus 784
nattereri **786**
neglectus 785
occultus 785, **786**
opimus **786**, 786
osgoodi 785
pallidus 785
pearsoni **786**
pernix 785
perrensis **786**
peruanus **786**
pontifex **786**
porteousi 784, **786**
pundti 785
recessus 785
rionegrensis 785, 786
robustus 785
rondoni 786
saltarius **786**
sericeus **786**
sociabilis **786**
steinbachi **787**
sylvanus 784
talarum **787**
tolduco 785
torquatus **787**
tuconax **787**
tucumanus 785, **787**
tulduco 785
utibilis 784
validus 785, **787**
viperinus 785
cuandu, Coendou 775
cuanzensis, Crocidura 93
Otomys 681
cubanensis, Arvicola 506
Canis 281
cubangensis, Syncerus 402
cubanus, Monophyllus 185
Nycticeius 217
Solenodon **69**
cubensis, Eptesicus 201
cucphuongis, Callosciurus 421
cucullatus, Ateles 257
Cebus 259
Glaucomys 460
Trachypithecus 274
cuculliger, Bradypus 63
cucutae, Sciurus 441
cufrensis, Jaculus 493
cuguapara, Ozotoceros 392
Cuica 18
cuja, Galictis **318**
culex, Pipistrellus 222
culionensis, Axis 384
Manis 415
Sundamys 666
culmorum, Rattus 661
culpaeolus, Pseudalopex 284
culpaeus, Canis 284
Dusicyon 283
Pseudalopex **284**
cultellus, Thomomys 473
cultrata, Anoura **183**
culturatus, Niviventer 632, **633**
cumana, Felis 290
cumananus, Rhipidomys 744
cumanicus, Sylvilagus 825
cumberlandius, Geomys 470
cumbrae, Apodemus 573
cumingi, Octodon 788
Phloeomys **640**, 640
cumingii, Octodon 788
cummingi, Microtus 530
cummunis, Bradypus 63
cumulator, Neotoma 710
cunctator, Cricetomys 541
cundalli, Clidomys 805
cunealis, Petromus 774
cuneatus, Rhinolophus 167
cuneiceps, Hystrix 774
cunenensis, Oreotragus 399
Raphicerus 399
Sylvicapra 412
Syncerus 402
cunicularis, Mus 623
Thomomys 473
cunicularius, Cercomys 798
Microtus 519
Sylvilagus **825**, 826
Cuniculidae 783
Cuniculinae 783
cuniculoides, Reithrodon 740
Cuniculus 509, 783, 822
cuniculus, Bettongia 49
Lepus 822
Orthogeomys **470**
Oryctolagus **822**
cuninghamei, Crocidura 85
Equus 369
Mylomys 629, 630
cunninghami, Petaurus 61
Cuon **282**
alpinus **282**
clamitans 282
dukhunensis 282
fumosus 282
grayiformes 282
hesperius 282
infuscus 282
javanicus 282
laniger 282
lepturus 282
primaevus 282
rutilans 282
sumatrensis 282
cupida, Macaca 266
cupidineus, Dipodomys 479
cupidus, Hipposideros 171
cuppedius, Steatomys **545**, 545
cuppes, Ochotona 811
cupreata, Scalopus 127
cupreoides, Otomys 680
cupreolus, Myotis 209
cupreus, Callicebus **258**, 258
Moschus 384
Otomys 680
Pseudochirops **60**
cuprosa, Kerivoula **196**
cuprosus, Pipistrellus **220**
curaca, Phyllostomus 180
curasoae, Leptonycteris **185**
curassavicus, Odocoileus 391
curcio, Clethrionomys 508
curi, Choloepus 64
curilensis, Callorhinus 327
currax, Pedetes 759
currentium, Molossus 235
Reithrodon 740
curryi, Pronolagus 824
cursor, Akodon **689**, 693
curtata, Arvicola 516
Bandicota 579
curtatus, Allocricetulus 536, 537
Lagurus 516
Lemmiscus 515, **516**
Pipistrellus 222
Thomomys 473
Zapus 499
curti, Lepus 815
curtidens, Sus 379
curtissi, Copernicia 802
curtus, Cebus 260
Hipposideros **172**
Pongo 277
curvicornis, Cervus 387
Naemorhedus 407
curvostylis, Muntiacus 389
curzoniae, Ochotona **808**, 808, 810, 812
cuscinus, Cebus 259
Sciurus 441
Cuscus 46
celebensis 47
cuscus, Lagidium 778
custos, Eothenomys **513**
Hipposideros 172
cutchensis, Lepus 819
cutchicus, Cremnomys 587, **588**, 588
cutleri, Cavia 779
cuttingi, Hylomys 80
cuvieri, Abrocoma 789
Cephalophus 411
Crocuta 308
Eulemur 244
Gazella **396**, 396, 397

Hystrix 773
Lagidium 778
Mormoops 176
Proechimys **796**
Tatera 561
Trichosurus 48
Cuvierius 349
cuyonis, Tupaia 133
cyanea, Crocidura **83**, 83, 86
cyaneus, Rattus 658
cyanotus, Lepus 817
cyanus, Mus 789
Spalacopus **789**
cycloceros, Cervus 386
Ovis 409
Cyclopes **67**
catellus 67
codajazensis 67
didactylus **67**
dorsalis 67
eva 67
ida 67
juruanus 67
melini 67
mexicanus 67
monodactyla 67
unicolor 67
cyclopis, Macaca **266**
cyclops, Hipposideros **172**
Peropteryx 158
cyclorhinus, Cervus 386
Cyclothurus 67
cyclotis, Chaerephon 232
Loxodonta 367
Murina **229**
Nyctimene **144**, 144
cylindricauda, Sorex 113, **114**
cylindricornis, Capra **405**, 405
cylindrura, Urogale 133
cylipena, Mustela 322
Cynaelurus 288
Cynailurus 288
Cynamolgus 266
cyniclus, Dolichotis 780
Cynictis 300, **302**
bechuanae 302
brachyura 302
bradfieldi 302
cinderella 302
coombsi 302
intensa 302
kalaharica 302
karasensis 302
lepturus 302
levaillantii 302
ogilbyi 302
pallidior 302
penicillata **302**
selousi 307
steedmanni 302
typicus 302
Cynocebus 265
cynocephala, Callorhinus 327
Didelphis 29
Tadarida 240
Cynocephalidae **135**
Cynocephalus **135**, 269
abbotti 135
aoris 135
borneanus 135
chombolis 135
gracilis 135
hantu 135
lautensis 135
lechei 135
marmoratus 135
natunae 135
papio 269
peninsulae 135
philippinensis 135
pumilis 135
rufus 135
saturatus 135
taylori 135
tellonis 135
temminckii 135
ternatensis 135
terutaus 135
tuancus 135
undatus 135
variegatus **135**
varius 135
volans **135**
cynocephalus, Papio 269
Thylacinus **29**
Cynofelis 288
Cynogale **341**, 341
barbatus 341
bennettii **341**
carcharias 341
lowei 341
velox 73
Cynogalinae 341
Cynomacaca 266
cynomolgus, Macaca 266
Cynomomus 424
Cynomops 234, 235
Cynomyini 419, 424
Cynomys **424**, 424
arizonensis 424
cinereus 424
grisea 424
gunnisoni **424**
latrans 424
leucurus **424**
ludovicianus **424**
mexicanus **424**
missouriensis 424
parvidens **424**
pyrrotrichus 424
socialis 424
zuniensis 424
Cynonycteris, torquata 143
Cynopithecus 266, 267, 268
Cynopterus **138**, 145
altitudinis 139
andamanensis 139
angulatus 139
archipelagus 139
babi 139
blanfordi 153
brachyotis **139**
brachysoma 139
ceylonensis 139
concolor 139
gangeticus 139
harpax 139
hoffeti 139
horsfieldi **139**
insularum 139
javanicus 139
luzoniensis 139
lyoni 139
maculatus 138
major 139
marginatus 153
melanocephalus 138
minor 139
minutus 139
nusatenggara **139**
pagensis 139
persimilis 139
princeps 139
scherzeri 139
serasani 139
spadiceus 140
sphinx **139**, 139
terminus 139
titthaecheilus **139**, 139
cynosuros, Chlorocebus 265
Cyongale, bennettii 341
Cyperus 95
Cyphanthropus 276
Cyphobalaena 350
Cyphonotus 350
cypria, Crocidura 96
cyprius, Lepus 817
Ovis 408
cypselinus, Diaemus 194
cyrenaica, Vulpes 286
cyrenaicus, Eliomys 767
cyrenarum, Felis 290
cyrensis, Lepus 817
cyrnensis, Crocidura 96
Cyromys 671, 672
Cystophora **329**, 329, 331
borealis 329
cristata **329**
dimidiata 329
cystops, Rhinopoma 155
Cyttarops **156**
alecto **156**, 156
czekanovskii, Sorex 117

d'aubentonii, Fossa 341
d'orbigny, Desmodus 194
dabagala, Xerus 458
dabbenei, Eumops **233**
Euneomys 702
Hydrochaeris 781
dabshanensis, Callosciurus 421
daccaensis, Bandicota 579
dacius, Microtus 530
Dacnomys **589**, 591, 632
ingens 589
millardi **589**, 589
wroughtoni 589
dacota, Marmota 432
dacotensis, Odocoileus 391
Taxidea 325
dactylinus, Callosciurus 421
Dactylomys **790**
Echimys 790
Dactylomyinae **790**
Dactylomys **790**
amblyonyx 790
boliviensis **790**
canescens 790
dactylinus **790**
modestus 790
peruanus **790**
typus 790
Dactylonax 60
Dactylopsila **60**
albertisii 61
angustivittis 61
arfakensis 61
biedermanni 61
ernstmayri 60
hindenburgi 61
infumata 61
kataui 61
malampus 61
megalura **60**
occidentalis 61
palpator **60**, 60
picata 61
tatei **61**
trivirgata 60, **61**, 61
dadappensis, Pongo 277
daedalus, Nyctophilus 218
daemon, Felis 290
Procavia 374
Tachyoryctes **686**, 686
daemonellus, Dasyurus 31
daemonis, Potamochoerus 378
dagestanicus, Neomys 110
daghestanicus, Dryomys 766
Microtus **520**, 520
dahli, Meriones **556**, 557
Petropseudes **59**
Pseudochirus 59
dahurica, Marmota 433
dailliardianus, Cervus 386
daimius, Cervus 386
daitoensis, Pteropus 147
dakhunensis, Elephas 367
dakotensis, Tamiasciurus 457
dalailamae, Ovis 408
dalei, Mesoplodon 362
dalli, Lepus 814
Ovis **409**
Phocaena 359
Phocoenoides **359**
Synaptomys 534
Ursus 338
dalloni, Gerbillus **550**
dalmaticus, Eliomys 767
dalmatinus, Canis 280
Dinaromys 511

dalquesti, Chaetodipus 483
Myotis 213
Sigmodon 747
daltoni, Myomys 592, **631**, 631
dalversinicus, Rhombomys 559
Dama 385, **387**, 387, 391
albus 387
dama **387**, 387, 391
leucaethiops 387
maura 387
maùricus 387
mesopotamica **387**
niger 387
platyceros 387
plinii 387
varius 387
virginiana 391
virginianus 391
vulgaris 387
dama, Cervus 387, 391
Dama **387**, 387, 391
Gazella **396**
Macropus 53
Tragelaphus 404
Damalis 393, 394
Damaliscus 393, **394**
albifrons 394
dorcas 394
eurus 394
floweri 394
grisea 394
hunteri **394**
jimela 394
jonesi 394
korrigum 394
lunatus **394**, 394
lyra 394
maculata 394
personata 394
phalius 394
phillipsi 394
purpurescens 394
pygargus **394**, 394
reclinis 394
scripta 394
selousi 394
senegalensis 394
tiang 394
topi 394
ugandae 394
damanus, Propithecus 247
damarensis, Caracal 289
Chrysochloris 75
Cryptomys **771**
Eptesicus 201
Funisciurus 427
Lepus 820
Madoqua 398
Malacothrix 544
Mungos 307
Mus 669
Nycteris 162
Pedetes 759
Rhinolophus 165
Scotophilus 227, 228
Thallomys 668, 669
damergouensis, Gazella 396
dammah, Oryx **413**
dammermani, Hyomys **599**, 600
Pseudocheirus 59
Rattus 660
damoni, Propithecus **247**
dandolena, Ratufa 437
danei, Cephalophus 410
dangarinensis, Blanfordimys 506
danicus, Meles 314
danielli, Equus 370
danubialis, Micromys 620
danubicus, Erinaceus 77
daphaenodon, Sorex **114**, 114
daphne, Thomasomys **749**
daphnia, Crocidura 92
Daptomys 714
darfurea, Crocidura 92
darienensis, Reithrodontomys **741**, 741
dariensis, Ateles 257
Dasyprocta 782
Orthogeomys **470**, 471
Oryzomys 720
darioi, Stenoderma 192
darjelingensis, Barbastella 199
darjilingensis, Mus 623
Myotis 215
darlingi, Cryptomys 772
Diplomys 791
Rhinolophus **165**
darlingtoni, Eptesicus 203
darricarrerei, Jaculus 493
darti, Stenodontomys 683
dartmouthi, Otomys 682
darwini, Dusicyon 283
Herpailurus 291
Holochilus 704
Mus 737
Nesoryzomys **713**, 713
Ovis 408
Phyllotis **738**, 738, 739
Pipistrellus 223
dasilvai, Galerella 303
Graphiurus 764
Dasogale 73
fontoynonti 73
Dasycercus **31**, 31
blighi 31
blythi 31
byrnei **31**
cristicauda **31**
hillieri 31
pallidius 31
dasycercus, Cabassous 65
Dasychoerus 378
dasycneme, Myotis **210**
Dasykaluta 29, **31**
rosamondae **31**
dasykarpos, Nyctalus 217
dasymallus, Pteropus **147**
Dasymys **589**, 589, 666
alleni 589
bentleyae 589, 590
capensis 589
edsoni 589, 590
foxi **589**, 590
fuscus 589
griseifrons 589
gueinzii 589
helukus 589
incomtus **589**, 589, 590
longicaudatus 665
longipilosus 590
medius 589
montanus **589**
nigridius 589
nudipes **589**, 589, 590
orthos 589
palustris 589
rufulus 589, **590**
savannus 589
shawi 589
Dasyphractus 65
Dasypodidae **64**
Dasypodinae **64**
Dasypodini 64
Dasyprocta **781**, 799
acuti 781
aguti 781, 782
albida 781, 782
aliba 782
antillensis 782
aurea 781
azarae **781**
bellula 782
boliviae 782
callida 782
catrinae 781
caudata 781
cayana 782
cayennae 782
chiapensis 782
chocoensis 782
coibae **781**
columbiana 782
cristata **781**, 782
croconota 782
dariensis 782
felicia 781
flavescens 782
fuliginosa **781**
fulvus 781, 782
guamara **782**
isthmica 782
kalinowskii **782**
leporina 781, **782**
lucifer 782
lunaris 782
maraxica 782
mexicana **782**
nigriclunis 782
noblei 781, 782
nuchalis 782
pallidiventris 782
pandora 782
paraguayensis 781
prymnolopha **782**
punctata **782**, 782
richmondi 782
ruatanica **782**
rubrata 782
underwoodi 782
urucuma 782
variegata 782
yungarum 782
zamorae 782
Dasyproctidae **781**, 781, 783
Dasypterus 206, 207
Dasypus **65**
aequatorialis 66
auritus 65
beniensis 66
boliviensis 66
brevirostris 66
davisi 66
fenestratus 66
gigas 66
granadiana 66
hirsutus 66
hoplites 66
hybridus **65**
kappleri **66**
leptocephala 66
leptorhynchus 66
longicaudatus 66
longicaudus 66
lundi 66
maximus 66
mazzai 65, 66
megalolepis 66
mexianae 66
mexicanus 66
minutus 67
niger 66
novemcinctus 65, **66**
octocintus 66
pastasae 66
peba 66
pentadactylus 66
pilosus **66**
propalatum 66
sabanicola **66**
septemcinctus **66**
serratus 66
sexcinctus 66
texanum 66
tricinctus 67
unicinctus 64
uroceras **66**
villosus 65
dasypus, Myotis 209
dasyroides, Gerbillus 550
dasythrix, Echimys **792**
Miniopterus 231
Nelomys 792
dasytrichus, Oxymycterus 727
Dasyuridae **29**, 29
Dasyuriformes **29**
Dasyurinus 31
Dasyuroidea **29**

Dasyuroides 31
Dasyuromorphia 29
Dasyurops 31, 32
Dasyurus **31**
alboguttata 32
albopunctatus **31**
daemonellus 31
exilis 32
fortis 31
fuscus 31
geoffroii **31**, 32
gracilis 32
guttatus 32
hallucatus **32**, 32
laniarius 35
macrourus 32
maculata 32
maculatus **32**
maugei 32
nesaeus 32
novaehollandiae 32
predator 32
quoll 32
spartacus 31, **32**
ursinus 32
viverrinus **32**
dasyurus, Gerbillus 549, **550**, 550, 555
datae, Apomys **575**, 575
datus, Callosciurus 422
daubentoni, Hystrix 773
Myotis **210**
Nycteris 161
Daubentonia **247**
daubentonii 247
laniger 247
madagascariensis **247**
psilodactylus 247
robusta 247
daubentonii, Barbastella 199
Daubentonia 247
Myopterus **238**, 238
Neomys 110
Sorex 110, 111
Tarsius 251
Daubentoniidae **247**
daucinus, Heliosciurus 429
daulensis, Molossus 235
Sylvilagus 825
dauntela, Elephas 367
daurica, Vulpes 287
dauricus, Microtus 527
Ovis 408
Spermophilus **444**, 445, **446**, 446
dauurica, Ochotona **808**, 808
dauuricus, Erinaceus 79
Hemiechinus 79
Lepus 807
Mesechinus **79**, 79
dauvergnii, Capra 406
davensis, Sundasciurus **451**, 451, 452
davidi, Echymipera **41**
Myotis 213
Rhizomys 685
Sus 379
davidianus, Elaphurus **387**, 387
Mustela 324
Scaptochirus 128, 129
Sciurotamias 438, **439**
Sciurus 438
davidsoni, Balaenoptera 349
daviesi, Micronycteris **178**
davisi, Dasypus 66
Otomys 681
Tamias 454
Thallomys 668
Uroderma 193
davisoni, Callosciurus 421
Iomys 461
Promops 239
davydovi, Blanfordimys 506
davyi, Pteronotus **176**, 176
dawsoni, Clethrionomys 509
Rangifer 392
dayanus, Lepus 819
dayi, Akodon **689**
Suncus **102**
dealbata, Ratufa 437
dealbatus, Erinaceus 77
deasyi, Dipus 493
debalsaci, Crocidura 95
debeauxi, Crocidura 96
debilis, Cervus 386
Hipposideros 171
Molossus 235
decaryi, Microgale 71
Rattus 657
decemlineata, Galidia 300
Mungotictis **300**
deceptor, Akodon 689
deckeni, Alcelaphus 393
deckenii, Propithecus 247
Rhinolophus 164, **165**
decolor, Mus 625
decoloratus, Auliscomys 695
Peromyscus 735
decolorus, Nyctomys 715
decoratus, Menetes 433
decorus, Capreolus 390
decres, Macropus 53
decula, Tragelaphus 404
decumanoides, Rattus 657
decumanus, Allactaga 489
Mus 649
Proechimys **796**, 796
Rattus 657
defassa, Kobus 413, 414
defilippi, Sylvilagus 825
defilippii, Sylvilagus 825
definitus, Phyllotis **738**, 738
deformis, Hipposideros 173
defriesi, Cephalophus 411
defua, Dephomys **590**, 590, 592
Mus 590
degelidus, Lasiurus 206
degener, Lichonycteris 185
Lontra 311
Pteropus 149
degeni, Otomys 682
degus, Octodon **788**
Sciurus 788
deignani, Myotis 211
Deilemys 746
deitanus, Canis 281
dejardinus, Cervus 386
dejeani, Cervus 387
dekan, Rhizomys 685
dekayi, Blarina 106
Microtus 527
dekeyseri, Lonchophylla **181**
delacouri, Hapalomys **595**
Panthera 298
Trachypithecus 273
delalandi, Aonyx 310
delalandii, Arctocephalus 327
delamensis, Mus 629
delamerei, Phacochoerus 377
Tragelaphus 404
Delanymys **683**, 683
brooksi **683**, 683
delator, Oxymycterus **726**, 727
delectorum, Praomys **642**, 642
delesserti, Funambulus 427
deletrix, Vulpes 287
delfini, Chelemys 699
delgadilli, Peromyscus 729
delicatulus, Leggadina 600
Pseudomys **645**, 645, 648, 649
delicatus, Elephantulus 829
Mesocapromys 800
Oligoryzomys 718
deliensis, Pongo 277
delirius, Rattus 661
Delomys 687, **700**, 700, 749, 752
collinus 700
dorsalis **700**, 700
lechei 700
obscura 700
plebejus 700
sublineatus **700**, 700
Delphinapterus 354, **357**
albicans 357
beluga 357
catodon 357
dorofeevi 357
leucas **357**
marisalbi 357
Delphinidae 349, **351**, 354, 358
Delphininae 351
Delphinus **352**
albirostris 353
bairdii 352
blainvillei 360
bredanensis 357
capensis 352
chinensis 355
delphis **352**, 352
fluviatilis 355
gangetica 360
geoffrensis 360
globiceps 352
griseus 353
guianensis 355
heavisidii 351
intermedius 352
leucas 357
malayanus 355
melas 352
orca 354
peronii 354
phocaenoides 358
phocoena 358
rostratus 357
sowerbensis 362
tropicalis 352
truncatus 357
delphis, Delphinus **352**, 352
deltae, Crocidura 92
Dendrolagus 51
Deltamys 688, 690
delticola, Oligoryzomys **717**
delticus, Eumops 233
demararae, Myoprocta 782
dembeensis, Arvicanthis 576, 578
dementievi, Capra 406
Chionomys 507
dementiewi, Allactaga 489
demerarae, Micoureus **20**
demidoff, Galago 249
Galagoides **249**, 249
demidovii, Galagoides 249
demissa, Galea 779
Tupaia 132
demissus, Eptesicus **201**
Hipposideros 172
Democricetodon 535
demonstrator, Mops **236**, 236
Dendrailurus 294
Dendrobius 788
Dendrodorcopsis 53
woodwardi 53
Dendrogale **131**
baluensis 131
frenata 131
melanura **131**
murina **131**
Dendrohyrax **373**, 373
adametzi 373
adersi 373
adolfi-friederici 373
arboreus **373**, 373
aschantiensis 373
beniensis 373
bettoni 373
blainvillii 373
braueri 373
brevimaculatus 373
congoensis 373
crawshayi 373
dorsalis **373**
emini 373
helgei 373
latrator 373
marmota 373

mimus 373
neumanni 373
nigricans 373
rubriventer 373
ruwenzorii 373
scheelei 373
scheffleri 373
schubotzi 373
schusteri 373
stampflii 373
stuhlmanni 373
sylvestris 373
terricola 373
tessmanni 373
validus **373**, 373
vilhelmi 373
vosseleri 373
zenkeri 373
Dendrolagus **50**, 52
aureus 50
bennettianus **50**, 50
buergersi 50
deltae 51
dorianus **50**, 50, 51
finschi 50
flavidior 51
fulvus 51
goodfellowi **50**, 51
inustus **50**, 51
keiensis 50
leucogenys 51
lumholtzi **50**
matschiei 50, **51**, 51
maximus 50
mayri 50
notatus 50
palliceps 50
profugus 50
pulcherrimus 50
schoedei 50
scottae **51**
shawmayeri 50
sorongensis 50
spadix 50, **51**
stellarum 50
ursinus 50, **51**
xanthotis 51
Dendromurinae 501, **541**, 541, 544, 545
Dendromus **541**, 541, 542, 544
abyssinicus 542
acraeus 543
angolensis 543
ansorgei 543
arenarius 543
ayresi 543
basuticus 543
bernardi 543
capensis 543
capitis 543
carteri 543
chiversi 543
concinnus 543
exoneratus 543
haymani 543
hintoni 543
insignis 541, **542**, 542, 543, 544
jamesoni 543
kahuziensis 541, **542**
kilimandjari 542
kivu **542**, 542, 543
kumasi 543
leucostomus 543
lineatus 543
longicaudatus 543
lovati 541, **542**
lunaris 542, 543
major 543
melanotis **542**, 542
mesomelas 542, **543**, 543, 544
messorius 542, **543**
mystacalis 542, **543**, 543
nairobae 543
nigrifrons 543
nyasae 543
nyikae 541, **543**
ochropus 543
oreas 542, **543**, 543
pallescens 543
pallidus 543
pecilei 543
percivali 542
pongolensis 543
pretoriae 543
pumilio 543
ruddi 543
shortridgei 543
spectabilis 543
subtilis 543
thorntoni 543
typicus 543
typus 541, 543
uthmoelleri 543
vernayi 541, **543**, 543, 544
vulturnus 543
whytei 543
Dendromys 541
Dendroprionomys **544**, 545
rousseloti **544**, 544
Dendrosminthus 608
denhamii, Vulpes 287
denigratus, Proechimys 797
denniae, Hylomyscus **599**, 599, 642
dennleri, Didelphis 16
densirostris, Mesoplodon **362**
dentaneus, Metachirus 20
dentata, Mephitis 317
dentatus, Batomys **579**, 580
Microtus 519
Pedetes 759
Rattus 660
Tarsius 250
denti, Cercopithecus 265
Colomys 586
Crocidura **83**
Otomys **680**
Rhinolophus **165**, 169
denticulatus, Eliomys 767
denticulus, Discopus 204
Eudiscopus **204**
dentifer, Galerella 303
Deomyinae 541, 544
Deomys **544**, 544
christyi 544
ferrugineus **544**
furrugineus 544
poensis 544
vandenberghei 544
deorum, Genetta **346**
Depanycteris 157
depauperatus, Thomomys 473
dephilippii, Sylvilagus 825
Dephomys **590**, 590, 666
defua **590**, 590, 592
eburnea **590**, 592
deppei, Sciurus **440**
depressa, Arctogalidia 343
Kerivoula 197
Microtus 519
depressicornis, Anoa 402
Bubalus **402**
depressus, Alticola 504
Lepus 815
Microtus 524
Nyctinomops 239
Thomomys 473
derasus, Eptesicus 201
derbianus, Anomalurus **757**
Caluromys **15**
Hemigalus 342
Pteromys 757
Taurotragus **403**, 403
derbyanus, Hemigalus **342**
Macropus 53
Paradoxurus 341, 342
derbyi, Hemigalus 342
derimapa, Pogonomys 642
Dermanura 187, 188, 189
Dermipus 13
Dermoptera 135
Dermopterus 135
derooi, Myomys 592, **631**
deserti, Crocidura 87
Dipodomys **477**
Hemiechinus 78
Jaculus 493
Mus 624
Parotomys 682
Pipistrellus 219
Reithrodontomys 741
Rhabdomys 662
Sylvicapra 412
Viverricula 348
deserticola, Lepus 815
Neotragus 398
deserticolus, Peromyscus 732
desertor, Pseudomys **645**, 646
desertorum, Alticola 504
Canis 281
Chaetophractus 65
Lepus 821
Myotis 210
Neotoma 712
Thomomys 473
desitus, Thomomys 473
Desmana **124**, 124
moschata **124**
Desmanidae 124
Desmaninae **124**, 124
desmarestianus, Heteromys **481**, 481
desmarestii, Megalomys **707**
Mus 707
Desmodilliscus **547**
braueri **547**, 547
buchanani 547
fuscus 547
Desmodillus **547**, 547
auricularis **547**
brevicaudatus 547
caffer 547
hoeschi 547
pudicus 547
robertsi 547
shortridgei 547
wolfi 547
Desmodontidae 177
Desmodontinae **194**
Desmodus **194**, 194
cinerea 194
d'orbigny 194
ecaudatus 194
fuscus 194
mordax 194
murinus 194
rotundus **194**
rufus 194
youngi 194
Desmomys 578, **590**, 590, 630, 639
harringtoni **590**, 590, 630
rex 590
desperata, Crocidura **83**
destructioni, Sorex 121
destructor, Arvicola 506
Conilurus 586
Lontra 311
Oligoryzomys **717**, 718
Pseudorca 355
detumidus, Thomomys 473
devexus, Thomomys 475
devia, Neotoma **711**, 712
Vulpes 287
devilleanus, Cervus 386
devillei, Saguinus 253
devilliersi, Amblysomus 74
devius, Clethrionomys 508
Oryzomys 720, **722**, 722
dextralis, Callosciurus 422
dhofarensis, Crocidura **83**, 95
diaboli, Thomomys 473
Diabolus 35
diadema, Hipposideros **172**
Propithecus **247**, 247
diadematus, Atelerix 76, 77
Cercopithecus 264
Diademia 263
Diaemus **194**
cypselinus 194
youngi **194**

dialeucos, Ichneumia 306
diamesus, Dryomys 766
Diana 263
diana, Cercopithecus **263**, 263
Crocidura 85
Lagostomus 778
Simia 263
dianae, Colobus 270
Emballonura **157**
Equus 369
Kobus 413
Psammomys 559
Redunca 414
Tarsius **250**
Tragelaphus 404
diardi, Neofelis 297
diardii, Rattus 660
diazi, Lepus 824
Romerolagus **824**
dicei, Sylvilagus **825**, 825
Dicerorhinus **371**, 372
blythii 372
harrissoni 372
lasiotis 372
niger 372
sumatrensis **372**
Diceros **372**
africanus 372
angolensis 372
atbarensis 372
bicornis **372**
brucii 372
camperi 372
capensis 372
chobiensis 372
gordoni 372
holmwoodi 372
keitloa 372
ladoensis 372
longipes 372
major 372
michaeli 372
minor 372
niger 372
occidentalis 372
palustris 372
platyceros 372
plesioceros 372
punyana 372
rendilis 372
somaliensis 372
dichotomus, Blastocerus **389**
Cervus 389
Rangifer 392
dichromatica, Nasua 334
dichrootis, Cricetulus 538
dichrous, Marmota 431, 432
Phyllotis 738
dichrura, Tatera 562
dichruroides, Apodemus 573
dichrurus, Apodemus 573
Cricetomys 541
Eliomys 767
dichura, Caluromys 15
dickeyi, Canis 280
Microdipodops 480
Peromyscus **729**
Procyon 335
Urocyon 285
dicki, Ctenomys 785
dickii, Ozotoceros 392
dickinsoni, Reithrodontomys 741
Diclidurus **157**, 157
albus **157**, 157
freyreisii 157
ingens **157**
isabellus **157**
scutatus **157**
virgo 157
dicolorata, Ratufa 437
Dicotyles 380
labiatus 380
torquatus 380
Dicrostonyx **509**, 509, 510, 511, 516
alascensis 511
chionopaes 511
clarus 510
exsul **510**, 510
groenlandicus **510**, 510
hudsonius **510**, 510
kilangmiutak **510**, 510, 511
lenae 511
lenensis 511
lentus 510
nelsoni **510**, 510
nunatakensis **510**, 510
pallida 511
peninsulae 510
richardsoni 510, **511**
rubricatus 510, **511**, 511
stevensoni 511
torquatus 510, **511**, 511
unalascensis 510, **511**
ungulatus 511
vinogradovi **511**, 511
dicrurus, Sus 379
dictator, Arctonyx 313
dictatorius, Leopoldamys 603
Didactyla 67
didactyla, Myrmecophaga 67
Didactyles 67
didactylus, Bradypus 64
Choloepus **64**
Cyclopes **67**
Didelphidae **15**, 27
Didelphimorphia 15
Didelphinae **16**
Didelphis **16**, 22
albiventris **16**, 16
andina 16
antigua 16
aurita **16**
austroamericana 17
azarae 16
battyi 17
bonariensis 16
boreoamericana 17
brasiliensis 16
brevicaudata 21
breviceps 17
brunii 51, 52
californica 17
cancrivora 17
caucae 17
cinerea 20
colombica 17
cozumelae 17
crassicaudata 18
cynocephala 29
dennleri 16
elegans 23
etensis 17
illinensium 17
imperfecta 16
incana 19
insularis 17
karkinophaga 17
koseritzi 16
lechei 16
leucotis 16
longipilis 16
maculata 31
marina 18
marsupialis **16**, 16
melanoidis 16
meridensis 16
mesamericana 17
murina 18, 49
myosuros 20
obesula 39
opossum 22
orientalis 46
paraguayensis 16
particeps 17
penicillata 33
peregrinus 59
pernigra 16
philander 15
pigra 17
pilosissima 17
poecilotis 16
pruinosa 17
pygmaea 62
richmondi 17
tabascensis 17
texensis 17
tridactyla 49
typica 17
ursina 35, 45
virginiana **17**, 22
viverrina 32
volans 58
vulpecula 48
woapink 17
yucatanensis 17
didelphoides, Mesomys **794**
Petauroides 59
Didelphys, microtarsus 17
Didermocerus 371, 372
diehli, Gorilla 276
Syncerus 402
dieseneri, Gazella 397
Heterohyrax 373
Lycaon 283
Sigmoceros 395
dieterleni, Lemniscomys 602
Sylvisorex 105
dietzi, Apodemus 571
difficilis, Peromyscus 728, **729**, 729, 730, 734
Reithrodontomys 741
digglei, Alcelaphus 393
dilatatus, Chaerephon 233
dilatus, Scapanus 127
dilecta, Presbytis 271
dilectus, Rhabdomys 662
dilophos, Cercopithecus 264
diluta, Asellia 170
Macaca 267
dilutior, Oryzomys 721
dilutus, Callosciurus 422
Spermophilus **444**
Dimerodon 16
dimidiata, Conepatus 316
Cystophora 329
Monodelphis **21**
dimidiatus, Acomys 565, 566
Oryzomys **722**
Proechimys **796**
diminutivus, Oligoryzomys 718
diminutus, Eptesicus **201**, 201
Gerbillus **550**
Rhabdomys 662
dimorphe, Cervus 385
dinaricus, Microtus 530
Dinaromys **511**, 511
bogdanovi **511**, 511, 512
coeruleus 511
dalmatinus 511
grebenscikovi 511
korabensis 511
longipedis 511
marakovici 511
preniensis 511
topachevskii 512
trebevicensis 511
dinderus, Lepus 816
dinelli, Myotis 212
dinganii, Scotophilus **227**, 227, 228
dinniki, Capra 405
Crocidura 96
Lynx 293
Microtus 524
Mustela 323
Pygeretmus 490
Dinochoerus 377
Dinomyidae **778**
Dinomys **778**
branickii **778**, 778
gigas 778
occidentalis 778
pacarana 778
dinops, Hipposideros **172**
Diodon 357, 364
Diomys **591**, 591
crumpi **591**, 591
Dioplodon 362
dioptrica, Australophocaena

358, 359
Phocoena 358
Diphylla **194**
centralis 194
diphylla 194
ecaudata **194**, 194
diphylla, *Diphylla* 194
Dipina 492
Diplogale **341**, 341
hosei **341**
Diplomesodon **98**
pallidus 98
pulchellum **98**
Diplomys **791**
caniceps **791**
darlingi 791
labilis **791**
rufodorsalis **791**
Diplostoma 469
Diplothrix **591**
legatus **591**, 591
okinavensis 591
Dipodes 487, 492
Dipodidae **487**, 487, 495
Dipodillus 548, 554
Dipodinae **492**, 492, 495
Dipodipus 492
Dipodomyinae **477**, 477, 480
Dipodomys **477**, 477
agilis **477**, 478
alfredi 479
ambiguus 479
annulus 479
antiquarius 477
aquilonius 479
aquilus 477
arenae 478
arenivagus 479
argusensis 479
arizonae 477
atronasus 479
attenuatus 479
australis 477
baileyi 480
berkeleyensis 478
bonnevillei 479
brevinasus 479
brunensis 479
cabezonae 477
californicus **477**, 477, 478
cascus 480
caudatus 479
celeripes 479
celsus 479
centralis 479
chapmani 479
cinderensis 479
cineraceus 479
clarencei 480
collinus 479
columbianus 479
compactus **477**, 479
cratodon 480
cupidineus 479
deserti **477**
dixoni 478
durranti 479
elator **478**
elephantinus **478**, 480
eremoecus 477
evexus 479
exilis 479
eximius 477
extractus 479
fetosus 479
fremonti 479
frenatus 479
fuscus 477
goldmani 478
gravipes **478**
heermanni 477, **478**, 480
idahoensis 479
idoneus 479
inaquosus 479
ingens **478**
insularis **478**
intermedius 480
jolonensis 478
kernensis 479
largus 477
latimaxillaris 477
leucogenys 479
leucotis 479
levipes 479
llanoensis 479
longipes 479
luteolus 479
margaritae **478**, 478, 479
marshalli 479
martirensis 477
mayensis 479
medius 479
melanurus 479
merriami **478**, 478, 479
microps **479**
mitchelli 479
mohavensis 479
monoensis 479
montanus 479
morroensis 478
mortivallis 479
nelsoni **479**, 479, 480
nevadensis 479
nexilis 479
nitratoides **479**, 479
oaxacae 480
obscurus 479
occidentalis 479
oklahomae 479
olivaceus 479
ordii 477, **479**
ornatus 480
pallidus 479
palmeri 479
panamintinus **479**, 479
panguitchensis 479
paralius 477
parvabullatus 477
parvus 479
pedionomus 477
peninsularis 477
perblandus 480
perotensis 480
perplexus 477
phillipsii 477, **480**
platycephalus 479
plectilis 477
preblei 479
priscus 479
pullus 479
quintinensis 479
regillus 479
richardsoni 479
russeolus 479
sanctiluciae 480
sanrafaeli 479
saxatilis 477
semipallidus 479
sennetti 477
similis 479
simiolus 479
simulans 477
sonoriensis 477
spectabilis 479, **480**
stephensi **480**
streatori 478
subtenuis 479
swarthi 478
terrosus 479
trinidadensis 479
tularensis 478
uintensis 479
utahensis 479
venustus 478, **480**
vulcani 479
wagneri 477
zygomaticus 480
Dipodops 477
Dipodum 487, 492
Diprotodon 380
Diprotodontia 45
Dipsidae 487, 492
Dipus **492**
aksuensis 493
austrouralensis 493
bulganensis 493
deasyi 493
fuscocanus 493
gerbillus 548
halli 493
hudsonius 499
indicus 560
innae 493
kalmikensis 493
lagopus 493
major 488
maximus 778
megacranius 493
mitchellii 635
nogai 493
platurus 490
platyurus 490
sagitta **492**
sowerbyi 493
turanicus 493
ubsanensis 493
usuni 493
zaissanensis 493
director, *Callosciurus* 422
Dirias 176
discifera, *Thyroptera* **196**
discolor, *Alouatta* 254
Arvicanthis 578
Bos 401
Callicebus 258
Emballonura 157
Gerbillus 551
Grammomys 593
Nycteris 162
Phyllostomus **180**
Rattus 657
Tupaia 132
Vespertilio 228
Discopus 204
denticulus 204
dispar, *Chaetodipus* 483
Neotoma 711
Rattus 661
Sorex **114**, 115
Thomasomys 749
disparilus, *Peromyscus* 729
dissimilis, *Cricetomys* 541
Erinaceus 77
Thomomys 473
Tupaia 131
dissimulatus, *Macropus* 55
dissonus, *Protoxerus* 436
distichlis, *Reithrodontomys* 741
distincta, *Neotoma* 712
Distoechurus **62**, 62
amoenus 62
dryas 62
neuhassi 62
pennata 62
pennatus **62**
divergens, *Odobenus* 326
Thomomys 473
diversicornis, *Procapra* 399
diversoides, *Lariscus* 430
diversus, *Lariscus* 430
Rhinolophus 166
divinorum, *Otomys* 680
dixiensis, *Tamiasciurus* 457
dixonae, *Sarcophilus* 35
dixoni, *Chrysochloris* 75
Dipodomys 478
djamdjamensis, *Chlorocebus* 265
djetysuensis, *Allactaga* 489
djonga, *Cervus* 387
djukovi, *Arvicola* 506
doani, *Erethizon* 776
dobodurae, *Rattus* 654
doboensis, *Rattus* 658
dobsoni, *Chrysospalax* 76
Epomops **141**
Microgale **71**
Myotis 209
Pteropus 146
Rhinolophus 166
Sorex 118
Taphozous 160

Dobsonia **139**
anderseni 140
beauforti **139**
chapmani **139**, 140
crenulata 140
emersa **139**
exoleta **139**
grandis 140
inermis **139**, 140
magna 140
minimus 140
minor **140**
moluccensis **140**, 140
nesea 140
pannietensis **140**, 140
peroni **140**
praedatrix **140**
remota 140
sumbanus 140
umbrosa 140
viridis 139, **140**
dobyi, Talpa 128
doceleguas, Capromys 800
docoi, Hippotragus 412
dodingae, Colobus 270
Tragelaphus 404
dodsoni, Gerbillus 550
doederleini, Canis 280
dogalensis, Myotis 209
dogetti, Hippotragus 412
doggetti, Cercopithecus 264
Poecilogale 325
doguera, Papio 269
doguin, Saguinus 254
dohrni, Rhinolophus 167
dolgopolovi, Ovis 409
dolguschini, Microtus 521
dolichocephalus, Microtus 521
Orthogeomys 471
dolichocrania, Vulpes 287
Dolichodon 362
Dolichohippus 369
dolichonyx, Chroeomys 700
dolichops, Cricetomys 540
dolichorhinus, Cervus 386
Dolichotinae **780**
Dolichotis **780**, 780
australis 780
ballivianensis 780
centralis 780
centricola 780
cyniclus 780
magellanica 780
patachonicha 780
patagonum **780**
salinicola **780**
dolichura, Crocidura **83**, 88, 90, 93
Sminthopsis **36**
dolichurus, Grammomys 592, **593**, 593, 594
Mazama 390
Mus 592
dollmani, Aethomys 568
Colobus 270
Graphiurus 765
Macaca 266
Maxomys **612**, 612
Melomys 616
Otomys 681
Dologale 300, **302**
dybowskii **302**
nigripes 302
robusta 302
Dolomys 502, 511
dolores, Akodon **689**, 691
dolosus, Heliosciurus 429
dolphoides, Tamiops 457
dombrowskii, Mustela 323
domecolus, Vandeleuria 672
domelicus, Callosciurus 421
Maxomys 614
domensis, Chaetodipus 483
domestica, Lama 381
Martes 319, 320
Monodelphis **21**
domesticus, Bos 401
Felis 290
Mus 625, 626
Myotis 212
Rattus 658
domeykoanus, Pseudalopex 284
domina, Micoureus 20
dominator, Paruromys **638**, 638, 639, 653
Rattus 638
Taeromys 639
dominicanorum, Catopuma 289
dominicanus, Cervus 386
dominicensis, Myotis **210**
Natalus 195
domisaxensis, Chaetodipus 483
domitor, Sundamys 666
domorum, Graomys **703**, 703
donacophilus, Callicebus **258**
donaldsoni, Redunca 414
donavani, Rhabdomys 662
dongilanensis, Gazella 397
dongolana, Genetta 345, 346
dongolanus, Gerbillus **550**
Procavia 374
doorsiensis, Tamias 456
doppelmayeri, Marmota 431
Doratoceros 390, 403
dorbignyi, Andalgalomys 694
Dorcas 395
dorcas, Capra 395, 405
Capreolus 390
Damaliscus 394
Gazella **396**, 397
Ourebia 399
Rupicapra 410
Dorcatragus **395**
megalotis **395**
Dorcelaphus 391
Dorcopsis **51**, 52, 57
atrata **51**
beccarii 51
brunii 52
caurina 51
chalmersi 51
eitape 51
hageni **51**
inustus 52
lorentzii 52
luctuosa **51**
macleayi 52
muelleri **52**, 52
mysoliae 52
phyllis 51
rufolateralis 52
veterum 52
yapeni 52
Dorcopsulus 51, **52**
macleayi **52**, 52
rothschildi 52
vanheurni **52**
doreyanus, Echymipera 41
doria, Cephalophus 412
Gazella 396
doriae, Chiroderma **189**
Crocidura 85
Epomophorus 141
Hesperoptenus **204**, 204
Hipposideros **172**, 174
Mormopterus **237**
Phascolosorex **34**
Rattus 658
doriana, Crocidura 92
dorianus, Dendrolagus **50**, 50, 51
Eptesicus 201
doriferus, Arctocephalus 327
doris, Oxymycterus 727
Stenella 356
dormeri, Pipistrellus **220**
dormitor, Cricetops 563
dorofeevi, Delphinapterus 357
dorogostaiskii, Clethrionomys 509
dorothea, Marmosops **19**
Microtus 529
dorotheae, Graphiurus 763
Hemiechinus 78
dorreae, Lagorchestes 52
dorsalis, Apodemus 573
Bradypus 63
Cephalophus **410**
Ctenomys **784**
Cyclopes 67
Delomys **700**, 700
Dendrohyrax **373**
Hemiechinus 78
Herpestes 305
Hesperomys 700
Hylomys 80
Lemniscomys 602
Lepilemur **246**
Macropus **53**
Microperoryctes 41
Monodelphis 21
Mustela 322
Myocastor 806
Nasua 334
Peromyscus 732
Phascogale 34
Phascolosorex **34**
Platyrrhinus 190, **191**, 191
Reithrodontomys 742
Sciurus 439, 443
Tamias **453**
Tupaia **132**
Vulpes 287
Xerus 458
dorsata, Antidorcas 395
Hystrix 776
Mazama 407
Oreamnos 407
Phoca 332
dorsatum, Erethizon **776**
dorsatus, Noctilio 176
dorsigera, Marmosa 18
dosul, Pardofelis 299
douceti, Crocidura **84**
douglasi, Eptesicus 201
Sminthopsis **36**
douglasii, Spermophilus 445
Sylvilagus 824, 826
Tamiasciurus **456**, 456, 457
Thomomys 475
douglasorum, Eptesicus **201**
doutii, Peromyscus 729
dowii, Tapirus 371
downsi, Promops 240
doylei, Peromyscus 732
draco, Apodemus **571**, 571, 572, 573
Tatera 560
draconilla, Nyctimene **144**, 144
dracula, Crocidura 85
drakensbergensis, Amblysomus 74
drakensbergi, Aethomys 568
dravidianus, Funambulus 426
Draximenus 45
Dremomys **424**
adamsoni 425
belfieldi 425
bhotia 425
calidior 425
chintalis 425
everetti **425**
flavior 425
fuscus 425
garonum 425
griselda 425
gularis 425
howelli 425
imus 425
laomache 425
lentus 425
lichiensis 425
lokriah **425**
macmillani 425
melli 425
mentosus 425

dubiaquercus, Antrozous **198**
Bauerus 198
dubius, Bandicota 579
Callicebus **258**
Calomys 698
Eulemur 244
Hippocamelus 390
Mus 625
Paradoxurus 344
Peromyscus 732
Stenella 356
dubosti, Funisciurus 427
ducatoris, Phalanger 46
duchaillui, Funisciurus 427
duchesnei, Castor 467
ducillus, Calomys 698
ducis, Rattus 661
ductor, Uromys 671
dugennianus, Cervus 386
dugesii, Potos 333
dugon, Dugong **365**
Trichechus 365
Dugong **365**
australis 365
cetacea 365
dugon **365**
dugung 365
hemprichii 365
indicus 365
lottum 365
tabernaculi 365
Dugongidae **365**
Dugonginae 365
dugung, Dugong 365
Dugungus 365
duida, Callicebus 259
Sciurus 441
duidae, Marmosa 18
dukecampbelli, Microtus 527
dukelskiae, Microtus 521
Sorex 119
dukelskiana, Nesokia 632
dukhunensis, Cuon 282
Hipposideros 175
dulitensis, Callosciurus 422
Ratufa 437
dulkeiti, Sciurus 443
dumeticola, Vandeleuria 672, 673
dumetorum, Oligoryzomys 718
dunckeri, Mus 625
dundasi, Elephantulus 829
Tatera 562
dunius, Zaglossus 13
dunni, Galago 249
Gerbillus **551**, 551, 552
Heterocephalus 772
Lemniscomys 601
Millardia 621
Mus 629
Tatera 561
duodecimcinctus, Cabassous 65
duodecimcostatus, Microtus **520**, 520, 524, 530
duplicatus, Microtus 519
Duplicidentata **807**
duprasi, Pachyuromys **559**, 559
dupreanum, Eidolon **140**, 141
durangae, Myotis 216
Neotoma 710
Sorex 118
Sylvilagus 825
Tamias 453, **454**, 454
durangi, Sciurus 439
Thomomys 475
duranti, Thomomys 475
durga, Trachypithecus **274**
durgadasi, Hipposideros 172
durranti, Dipodomys 479
Onychomys 719
duruculi, Aotus 256
Dusicyon 279, 280, **283**, 283, 284, 285
antarcticus 283
australis **283**, 283
culpaeus 283
darwini 283
griseus 283
gymnocercus 283
microtis 283
sechurae 283
thous 283
vetulus 283
dusorgus, Ursus 339
dussumieri, Funambulus 427
Semnopithecus 273
dutcheri, Geomys 469
Microtus 525
dutertreus, Eptesicus 201
duthieae, Chlorotalpa **75**
Chrysochloris 74
duvaucelii, Cervus **385**, 387
Naemorhedus 407
duvaucelli, Pardofelis 299
duvernoyi, Suncus 103
dwyeri, Chalinolobus **199**
Dyacopterus **140**
brooksi 140
spadiceus **140**
dyacorum, Herpestes 304
Hipposideros **173**
dybowskii, Canis 281
Cervus 385, 386
Crossarchus 302
Dologale **302**
Golunda 629
Mylomys **630**, 630
Myospalax 675
dychei, Reithrodontomys 741
Dymecodon 129
Dyromys 766
dyselius, Peromyscus 736
dysoni, Heliosciurus 428
Dysopes, temminckii 234
dzigguetai, Equus 370
dzungariae, Allactaga 488

easti, Peromyscus 732
eastwoodae, Graphiurus 765
eatoni, Gerbillus 549
ebermaieri, Lycaon 283
ebii, Epixerus **425**, 425, 426
ebneri, Procavia **374**
ebolowae, Mandrillus 268
eboreus, Melomys 616
eborivorus, Protoxerus 436
ebriensis, Crocidura 83
ebriosus, Neotomys **713**, 713
eburacensis, Trichosurus 48
eburnea, Crocidura 82
Dephomys **590**, 592
ecaudata, Anoura 183
Diphylla **194**, 194
Hystrix 774
Pachysoma 142
ecaudatus, Chaeropus **39**
Desmodus 194
Erinaceus 73
Macaca 268
Megaerops **142**
Mesomys 794
Perameles 39
Sylvilagus 825
Tenrec **73**
Echidna 13
novaehollandiae 13
Echimyidae 775, 787, **789**, 790, 799, 800, 806
Echimyinae **791**
echimyoides, Rattus 652
Echimys **791**, 791, 793
armatus 791
blainvillei **791**
brasiliensis 791
braziliensis **791**, 792
carrikeri 792
chrysurus **791**
cristatus 791
dactylinus 790
dasythrix **792**
flavidus 792
grandis **792**
gymnurus 794
hispidus 794
lamarum **792**
laticeps 793
macrurus **792**
medius 791
nigrispinus **792**
occasius 793
paleaceus 791
pictus **792**
punctatus 792
rhipidurus **792**, 792
saturnus **792**
semivillosus **792**
sulcidens 793
thomasi **792**
trinitatis 795
unicolor **792**, 792
echinatus, Tarsomys **667**
echinista, Echymipera **41**
Echinogale 73
Echinomys 791
Echinoprocta **776**, 776
epixanthus 776
rufescens **776**, 776
sneiderni 776
Echinops **73**
miwarti 73
nigrescens 73
pallens 73
telfairi **73**, 73
Echinopus 13
Echinosciurus 439
Echinosorex **79**
albus 79
birmanica 79
borneotica 79
candida 79
gymnura **79**, 79
gymnurus 79
minor 79
rafflesii 79
Echinosoricinae 79
echinus, Erinaceus 77
Echiothrix **591**, 591
brevicula 591
centrosa 591
leucura **591**, 591
Echymipera **41**, 41
albiceps 41
aruensis 41
australis 41
breviceps 41
clara **41**
cockerelli 41
davidi **41**
doreyanus 41
echinista **41**
garagassi 41
gargantua 41
hispida 41
kalubu **41**, 41
keiensis 41
myoides 41
oriomo 41
philipi 41
rufescens **41**
rufiventris 41
welsianus 41
eckloni, Vulpes 287
eclipsis, Maxomys 614
ecsedensis, Canis 280
Ectophylla **190**
alba **190**, 190
ecuadorensis, Tapirus 371
ecuadorianus, Bradypus 63
edarata, Tupaia 132
edax, Andinomys **694**, 694
Microtus 519
Olallamys **791**
Rhinolophus 168, 169
edeni, Balaenoptera **350**, 350
Edentata 63
edentatus, Phacochoerus 377
edithae, Graomys **703**
Meriones 556
Nesophontes 69, **70**
Trogopterus 465
editorum, Varecia 245

editus, Oenomys 637
edmondi, Sus 378
edsoni, Dasymys 589, 590
edulis, Mephitis 317
Pteropus 151
Steatomys 546
edusa, Psammomys 559
edwardii, Elephantulus **829**
Globicephala 352
edwardsi, Leopoldamys **603**, 614, 634
Lepilemur **246**
Lepus 818
Malacomys 592, **607**
Perodicticus 248
Potamochoerus 378
Propithecus 247
Pteropus 150
Vulpes 286
edwardsiana, Suncus 103
edwardsii, Herpestes **305**, 305
Naemorhedus 407
Notiomys **714**, 714
effera, Mustela 322
efficax, Sigmodontomys 748
effrenus, Sus 379
effugius, Sciurus 440
ega, Lasiurus **207**
Scleronycteris **186**, 186
egens, Saguinus 254
egeria, Callicebus 258
Chalinolobus **199**
Malacothrix 544
Egocerus 412
egregius, Lasiurus **207**
egressa, Neotoma 712
egypti, Herpestes 305
egyptiacus, Rousettus **152**
eha, Niviventer 632, 633, **634**, 634
ehiki, Microtus 530
Talpa 129
ehrenbergi, Nannospalax **753**, 753, 754
Procavia 374
Spalax 753
ehrhardti, Trachops 181
Eidolon **140**
buettikoferi 141
dupreanum **140**, 141
helvum 140, **141**
leucomelas 141
mollipilosus 141
paleaceus 141
palmarum 141
sabaeum 141
stramineus 141
eileenae, Murina 229
eionis, Sorex 116
Eira **317**
barbara **317**
bimaculata 318
biologiae 318
brunnea 318
canescens 318
gulina 318
ilya 318
irara 318
kriegi 318
leira 318
madeirensis 318
peruana 318
poliocephalus 318
senex 318
senilis 318
sinuensis 318
tucumana 318
eira, Herpailurus 291
eisentrauti, Colomys 586
Crocidura **84**
Hybomys **596**
Lophuromys 606
Myosorex **99**, 99, 100
Pipistrellus **221**
eitape, Dorcopsis 51
eivissensis, Apodemus 573
Eivissia 767
ekloni, Canis 281
ekmani, Pronolagus 823
elagantulus, Saguinus 253
Elaphalces 389
elaphinus, Rattus 650, **651**, 654
Elaphoceros 385
Elaphodus **388**
cephalophus **388**, 388
fociensis 388
ichangensis 388
michianus 388
elaphoides, Cervus 385
Elaphurus 385, **387**, 387
davidianus **387**, 387
menziesianus 387
tarandoides 387
Elaphus 385
elaphus, Cervus **385**, 385
Elasmodontomyinae 804
Elasmodontomys **805**
bidens 805
obliquus **805**, 805
Elasmognathus 370
elassodon, Otomys 680
Sorex 118
elassodontus, Hylopetes 461
elasson, Cryptotis 109
elassopus, Proechimys 797
elater, Allactaga **488**, 489, 490
elator, Dipodomys **478**
elattura, Hodomys 704
elatturus, Rhipidomys 744
elbaensis, Gerbillus 554
Jaculus 493
elbeli, Tamiops 457
elbertae, Prosciurillus 436
elberti, Procavia 374
elburzensis, Calomyscus 536
eldii, Cervus **386**, 386
eldomae, Tragelaphus 404
electa, Crocidura 83, 88
electilis, Hylopetes 461
Petinomys 463
Electra 353, 355
electra, Lagenorhynchus 355
Peponocephala **355**
electromontis, Phaner 244
electus, Paraxerus 435
elegans, Arctocephalus 327
Cebus 259
Cercopithecus 265
Cervus 386
Chrysochloris 75
Crocidura 87
Didelphis 23
Eligmodontia 701
Galerella 303
Galidia **299**, 299
Glirulus 769
Graphiurus 765, 769
Heliosciurus 428
Lagostrophus 53
Leopardus 292
Leptomys **604**, 604, 605, 665
Macruromys **607**, 607
Margaretamys **609**
Myotis **210**
Myoxus 769
Nectogale **110**, 110
Niviventer 633
Petaurista **462**, 462
Phloeomys 640
Proechimys 798
Saccostomus 541
Saguinus 253
Spermophilus **446**, 446, 449
Thylamys **23**
elegantulus, Euoticus **248**
Phyllotis 738
elenae, Leopardus 292
eleonorae, Sorex 111
Eleotragus 414
eleotragus, Redunca 414
Elephantidae **367**
elephantina, Mirounga 330
elephantinus, Dipodomys **478**, 480
Elephantulus **829**
boranus 829
brachyrhynchus **829**
delicatus 829
dundasi 829
edwardii **829**
fuscipes **829**
fuscus **829**, 829
hoogstraali 829
intufi **829**
mariakanae 829
myurus **829**
ocularis 829
peasei 829
phaeus 829
pulcher 829
renatus 829
rendilis 829
revoili **829**
rozeti **829**
rufescens **829**
rupestris **829**
somalicus 829
Elephantus 367
Elephas **367**
africanus 367
asiaticus 367
asurus 367
bengalensis 367
birmanicus 367
borneensis 367
ceylanicus 367
dakhunensis 367
dauntela 367
gigas 367
heterodactylus 367
hirsutus 367
indicus 367
isodactylus 367
maximus **367**, 367
mukna 367
rubridens 367
sinhaleyus 367
sondaicus 367
sumatranus 367
vilaliya 367
zeylanicus 367
eleusis, Eothenomys 514
Thomasomys **749**
elfridae, Vicugna 382
elgonae, Canis 281
elgonis, Cercopithecus 264
Colobus 270
Cricetomys 541
Grammomys 593
Ictonyx 319
Otomys 681
elgonius, Crocidura **84**
elibatus, Pappogeomys 472
Perognathus 485
Eligmodon 700
Eligmodontia **700**, 701
elegans 701
hirtipes 701
hypogaeus 701
jucunda 701
marica 701
moreni **701**, 701
morgani 700, **701**, 701
pamparum 701
puerulus 700, **701**, 701
tarapacensis 701
typus 700, **701**, 701, 703
elioi, Bolomys 697
Eliomys 766, **767**, 767
amori 767
cincticauda 767
cyrenaicus 767
dalmaticus 767
denticulatus 767
dichrurus 767
gotthardus 767
gymnesicus 767
hamiltoni 767
hortualis 767

jurrasicus 767
larotina 767
larotinus 767
liparensis 767
lusitanica 767
melanurus **767**, 767
munbyanus 767
nitela 767
occidentalis 767
ophiusae 767
pallidus 767
quercinus **767**
raiticus 767
sardus 767
superans 767
tunetae 767
valverdei 767
elisarjewi, Cricetulus 538
elissa, Semnopithecus 273
Eliurus **678**, 678
majori **678**, 678
minor **678**, 678
myoxinus **678**, 678
penicillatus **678**, 678
tanala **678**
webbi **678**, 678
eliurus, Oligoryzomys **717**, 718
Elius 766, 770
elkoensis, Thomomys 475
ellenbecki, Chlorocebus 265
ellermani, Mayermys **614**, 614
Spermophilus 448
elliotanus, Bandicota 579
ellioti, Anathana **131**
Golunda **592**, 592
Gorilla 276
Herpestes 305, 306
Lutrogale 313
Pan 277
Prionailurus 295
Procolobus 272
Tupaia 131
ellipsiprymnus, Antilope 413
Kobus **413**
ellipticus, Cervus 386
Ellobius **512**, 512, 534
alaicus **512**, 512, 513
ciscaucasicus 512
coenosus 513
farsistani 512
fusciceps 513
fuscipes 513
fuscocapillus **512**, 512
intermedius 512
kastschenkoi 513
larvatus 513
legendrei 512
lutescens **512**
murinus 512
ognevi 513
orientalis 513
rufescens 512
talpinus **512**, 512, 513
tanaiticus 512
tancrei 512, **513**, 513
transcaspiae 512
ursulus 513
woosnami 512
elongata, Crocidura **84**, 90
Glossophaga 184
Mephitis 317
elongatum, Micronycteris 179
elongatus, Phyllostomus **180**, 180
eloquens, Bandicota 579
Rhinolophus **165**, 166
elorzanus, Cervus 386
elphicki, Chaerephon 233
elphinstonei, Ratufa 437
elphinstoni, Ratufa 437
elseyi, Pteropus 151
eltonclarki, Ursus 339
elucus, Procyon 335
elusus, Peromyscus 732
elvira, Cremnomys **588**, 588
elymocetes, Microtus 527
emarginatus, Myotis **210**
Scotomanes **226**
Sorex 112, **114**, 118
Emballonura **157**, 158
afra 156
alecto **157**
anambensis 158
atrata **157**
beccarii **157**
brevirostris 156
clavium 157
cor 158
dianae **157**
discolor 157
furax **157**
locusta 157
macrotis 158
meeki 157
monticola **157**, 157
nigrescens 157
palauensis 158
palawanensis 157
peninsularis 158
pusilla 158
raffrayana **158**
rivalis 157
rotensis 158
semicaudata **158**
stresemanni 158
sulcata 158
Emballonuridae **156**
emerita, Leopardus 292
Lontra 311
emeritus, Thomasomys 750
emersa, Dobsonia **139**
emesi, Mus 624
emiliae, Callicebus 259
Callithrix 251
Gracilinanus **17**
Leopardus 292
Lonchothrix **794**, 794
Macropus 53
Monodelphis **21**
Petaurillus **462**
Rhipidomys 745
emilianus, Ctenomys **784**
emini, Cephalophus 411
Chaerephon 232
Cricetomys **540**, 540
Dendrohyrax 373
Funisciurus 428
Galerella 303
Gerbillus 562
Heliophobius 772
Paraxerus 434
Taterillus **562**
emissus, Heliosciurus 429
emmonsi, Peromyscus 732
emmonsii, Ursus 338
emotus, Thomomys 476
empetra, Marmota 432
empusa, Rhinolophus 164
Enchisthenes 187, 188
enclavae, Mus 625
encoubert, Euphractus 66
Encoubertus 66
encrita, Crocuta 308
Endecapleura 548
endemicus, Spermophilus 445
endersi, Cryptotis **108**
Oecomys 716
Scotinomys 746
endoecus, Microtus 527
endoi, Pipistrellus **221**
endorobae, Hylomyscus 599
energumenos, Mustela 324
enez-groezi, Microtus 518
enezsizunensis, Crocidura 96
engana, Kerivoula 197
enganus, Hipposideros 172
Pteropus 147
Rattus 650, **651**, 651, 655
Sus 379
Engeco 276
engelhardi, Orthogeomys 471
engelhardti, Marmota 432
engraciae, Oligoryzomys 718
enguvi, Cricetomys 541
Enhydra **310**
aterrima 310
gracilis 310
kamtschatica 310
kenyoni 310
lutris **310**
marina 310
nereis 310
orientalis 310
stelleri 310
enid, Acomys 567
enixus, Microtus 527
Thomomys 476
enkamer, Cercopithecus 263
ensenadensis, Blastoceros 389
Lama 381
ensicornis, Oryx 413
enslenii, Lasiurus 206
entelloides, Hylobates 275
entellus, Semnopithecus **273**
Simia 273
entomophagus, Saimiri 260
entrerianus, Pseudalopex 284
enuchus, Conepatus 316
enudris, Lontra 311
Enydris 310
Eoglaucomys 460, 461
eogroenlandicus, Rangifer 392
Ursus 339
Eolagurus **513**, 513, 515
luteus **513**, 513
przewalskii **513**
Eomops 238
Eonycteris **153**
glandifera 154
longicauda 153, 154
major **153**
robusta 153, 154
rosenbergi 154
spelaea **154**
Eosaccomys 541
Eosciurus 437
Eospalax 675
Eothenomys 502, 503, 507, **513**, 513, 514, 515, 532, 533
alcinous 514
aquilus 514
aurora 514
bonzo 514
cachinus 514
chinensis **513**
colurnus 514
confinii 514
custos **513**
eleusis 514
eva **514**
fidelis 514
hintoni 513
inez **514**, 514
jeholicus 515
kanoi 514
libonotus 514
melanogaster **514**, 514
miletus 514
mucronatus 514
nux 514
olitor **514**
proditor **514**
regulus 509, **514**, 514
rubelius 513
rubellus 513
shanseius **514**
tarquinus 513
wardi 513
eotinus, Pteropus 146
Eotomys 507
Eozapus **498**, 498
setchuanus **498**
vicinus 498
ephippiger, Bradypus 63
ephippium, Rattus 652
Ratufa 437
Epieuryceros 389
epimelas, Apodemus 572
Epimys 649, 666

aeta 598
colonus 630
hindei 567
manipulus 580
tullbergi 642
epiroticus, Microtus 529
Nannospalax 754
episcopi, Scotinomys 746
episcopus, Rhinolophus 167
Episoriculus 122, 123
epixanthus, Echinoprocta 776
Erethizon 776
Epixerus **425**
ebii **425**, 425, 426
jonesi 425
mayumbicus 425
wilsoni **425**, 425, 426
Epomophorus **141**, 141, 143
anchietae 145
angolensis **141**
anurus 141
crypturus 141
doriae 141
epomophorus 141
franqueti 141
gambianus **141**
grandis **141**
guineensis 141
haldemani 141
labiatus **141**
macrocephalus 141
megacephalus 141
minimus **141**
minor 141
neumanni 141
parvus 141
pousarguesi 141
pusillus 143
reii 141
schoensis 141
stuhlmanni 141
unicolor 141
veldkampi 143
wahlbergi **141**
whitei 141
zechi 141
zenkeri 141
epomophorus, Callosciurus 421
Epomophorus 141
Epomops **141**
buettikoferi **141**
comptus 142
dobsoni **141**
franqueti **142**
strepitans 142
epsilanus, Myospalax **675**, 675, 676
Eptesicus **200**, 200, 201, 202, 203, 219
albescens 203
anatolicus 200
andersoni 203
andinus 201
angolensis 201
angusticeps 202
arctoideus 201
arge 201
argentinus 201
arquatus 201
ater 203
atratus 202
bahamensis 201
batinensis 202
baverstocki **200**
bensoni 202
bernardinus 201
bicolor 203, 204
bobrinskoi **200**, 202
boscai 203
bottae **200**, 202, 203
brachydigitatis 203
brasiliensis **201**
brunneus **201**
capensis **201**, 202
carolinensis 201
carteri 201
caurinus 203
centralasiaticus 202
chapmani 201
chiralensis 201
chiriquinus 201
cubensis 201
damarensis 201
darlingtoni 203
demissus **201**
derasus 201
diminutus **201**, 201
dorianus 201
douglasi 201
douglasorum **201**
dutertreus 201
espadae 202
faradjius 203
ferrugineus 201
fidelis 201
findleyi 201
finlaysoni 203
flavescens **201**
floweri **201**
furinalis **201**
fuscus **201**, 203
garambae 201
gaumeri 201
gobiensis 202
gracilior 201
grandidieri 201
guadeloupensis **202**
guineensis **202**
hilarii 201
hingstoni 200
hispaniolae 201
horikawai 203
hottentotus **202**
humbloti 203
inca 201
incisivus 203
innesi 200
innoxius **202**
insularis 203
intermedius 203
isabellinus 203
japonensis 202
kashgaricus 202
kobayashii **202**
loveni 216
lowei 201
lynni 201, 202
matroka 201
matschei 202
megalurus 202
melanops 200, 201
melanopterus 201
melckorum **202**
meridionalis 203
miradorensis 201
mirza 203
montosus 201
nasutus **202**
nilssoni **202**
nitens 201
nkatiensis 201
notius 201
ognevi 200
okenii 203
omanensis 200
osceola 201
pachyomus 203
pachyotis **202**
pallens 203
pallidior 202
pallidus 201
parvus 202
pashtonus 203
pelliceus 201
pellucens 202
peninsulae 201
petersoni 201
phaiops 201
phasma 203
platyops **203**, 203
portavernus 202
propinquus 202
pumilus **203**, 203
punicus 202
pusillus 202
pygmaeus 204
rectitragus 202
regulus **203**
rendalli **203**
rufescens 203
sagittula **203**, 203
serotinus 202, **203**, 203
shiraziensis 203
smithi 202
sodalis 203
somalicus **203**
taftanimontis 200
tatei **203**
tenuipinnis **203**
thomasi 201
transylvanicus 203
troughtoni 203
turcomanicus 203
typus 203
ugandae 203
ursinus 201
vansoni 203
vulturnus **204**
walli 202
wetmorei 201
wiedii 203
zuluensis 203
epularius, Pteropus 148
epupae, Aethomys 568
equatoris, Cryptotis 109
Rhipidomys 744
equestris, Phoca 331
Equidae **369**
equile, Rattus 652
equinus, Cervus 387
Hippocamelus 390
Hippotragus **412**
Rhinolophus 166
equioides, Equus 370
Equus **369**
aethiopicus 369
africanus 369
annectens 369
antiquorum 369
asinus **369**, 369
atlanticus 369
bahram 370
bedfordi 370
berberensis 370
bisulcus 390
blanfordi 370
boehmi 369
borensis 369
burchellii **369**, 369, 370
caballus **369**, 369
campestris 369
castaneus 370
chapmanni 369
crawshaii 369
cuninghamei 369
danielli 370
dianae 369
dzigguetai 370
equioides 370
ferus 369
festivus 369
finschi 370
foai 369
frederici 370
gmelini 369
goldfinchi 369
granti 369
greatheadi 370
grevyi **370**
greyi 370
gutsenensis 369
hagenbecki 369
hamar 370
hartmannae 370
hemionus **370**, 370
hemippus 370
holdereri 370
indica 370
indicus 370
isabella 369
isabellinus 370
jallae 369
johnstoni 383

kaokensis 369
kaufmanni 369
khur 370
kiang **370**, 370
kulan 370
lorenzi 370
luteus 370
mariae 369
markhami 369
matschiei 370
montanus 370
muansae 369
nepalensis 370
onager **370**, 370
paucistriatus 369
penricei 370
pococki 369
polyodon 370
przewalskii 369
quagga 369, **370**, 370
selousii 369
silvatica 369
silvestris 369
somalicus 369
syriacus 370
taeniopus 369
tafeli 370
tigrinus 369
transvaalensis 369
trouessarti 370
vulgaris 369
wahlbergi 369
zambeziensis 369
zebra **370**, 370
zebroides 369
erasmus, Peromyscus 731
erebus, Callosciurus 423
Eremaelurus 289
thinobia 290
eremiana, Perameles **40**
eremicoides, Peromyscus 735
eremicus, Chaetodipus 484
Lepus 815
Lynx 293
Odocoileus 391
Peromyscus 728, 729, **730**, 730, 731, 734, 735
Reithrodontomys 742
Sigmodon 747
Ursus 338
Eremiomys 515
eremita, Neomys 110
Neotoma 712
Eremitalpa **76**
cana 76
granti **76**
namibensis 76
Eremodipus **493**, 493
balkashensis 493
jaxartensis 493
lichtensteini **493**
eremoecus, Dipodomys 477
eremonomus, Spermophilus 449
eremus, Peromyscus 732
ererensis, Galerella 303
Erethizon 775, **776**, 776
bruneri 776
couesi 776
doani 776
dorsatum **776**
epixanthus 776
myops 776
nigrescens 776
picinum 776
Erethizonidae 789
Erethizontidae **775**, 775, 790
ereunetes, Ursus 339
erica, Clethrionomys 508
Crocidura 83, **84**, 87
Ericius 78
Ericulus 73
semispinosus 73
erigens, Hipposideros 171
Erignathus **329**, 329
albigena 329
barbatus **329**
lepechenii 329
leporina 329
nautica 329
parsonii 329
erikssoni, Okapia 383
Orycteropus 375
Erinaceidae **76**
Erinaceinae **76**
Erinaceolus 78
Erinaceus 76, **77**, 78, 79
abasgicus 77
albiventris 76
amurensis **77**, 77
auritus 77
bolkayi 77
caniceps 77
caninus 77
centralrossicus 77
chinensis 77
concolor **77**, 77
consolei 77
danubicus 77
dauuricus 79
dealbatus 77
dissimilis 77
drozdovskii 77
ecaudatus 73
echinus 77
erinaceus 77
europaeus **77**, 77, 79
hanensis 77
hispanicus 77
italicus 77
kievensis 77
koreanus 77
koreensis 77
kreyenbergi 77
madagascariensis 73
meridionalis 77
nesiotes 77
occidentalis 77
orientalis 77
pallidus 77
platyotis 77
ponticus 77
rhodius 77
roumanicus 77
sacer 77
setosus 73
suillus 77
transcaucasicus 77
tschifuensis 77
typicus 77
ussuriensis 77
erinaceus, Erinaceus 77
Eriodes 257
eriodes, Brachyteles 257
eriophora, Kerivoula **197**
Erioryzomys 749
Erithizon 776
Erithizontidae 775
Erithizontoidea 771
erlangeri, Galerella 303
Gazella 396, 397
Genetta 346
Madoqua 398
Procavia 374
ermeloensis, Lepus 816
erminea, Arctogale 342
Mustela **321**, 321
Ernosorex 80
ernstmayri, Dactylopsila 60
Leptomys **604**, **605**, 605, 665
erongensis, Galerella 303
Mormopterus 238
Eropeplus **591**, 591, 602, 640
canus **591**, 591
Erophylla **182**
bombifrons 182
mariguanensis 182
planifrons 182
santacristobalensis 182
sezekorni **182**
syops 182
errardianus, Cervus 387
erro, Thomasomys 749
erroris, Hipposideros 174
erskinei, Ovis 408
erubescens, Callosciurus 421
Macropus 54
erxlebeni, Cercopithecus 265
erythopus, Xerus 458
erythraea, Ictonyx 319
Macaca 267
erythraeus, Callosciurus **421**
erythrarchus, Cercopithecus 264
erythrobronchus, Graphiurus 764
Erythrocebus 263, **265**
albigenus 265
albosignatus 265
baumstarki 265
circumcinctus 265
formosus 265
kerstingi 265
langheldi 265
patas **265**
poliomystax 265
poliophaeus 265
pyrrhonotus 265
rubra 265
rufa 265
sannio 265
whitei 265
zechi 265
erythrodactylus, Pipistrellus **224**
erythrogaster, Callosciurus **421**
Cercopithecus **264**
Saguinus 253
erythrogenys, Funisciurus 428
Rhizomys 685
Spermophilus **446**, 447, **448**
erythrogluteia, Spermophilus 445
erythroleucus, Mastomys **610**
erythromela, Varecia 245
erythromelas, Callosciurus 421, 423
erythromos, Sturnira **192**
Erythronesokia 632
erythronotus, Anomalurus 757
Apodemus 570
Rattus 658
erythrops, Funisciurus 428
erythropus, Xerus **458**, 458
erythropygius, Naemorhedus 407
erythropygus, Grammomys 594
Erythrosciurus 420
erythrotis, Cercopithecus 263, **264**
Micromys 620
Ochotona **808**, 809, 810, 812
erythrotus, Felis 290
erythrourus, Meriones 556
esau, Macaca 268
escalante, Ammospermophilus 420
escalerae, Myotis 213
Rhinolophus 166
escazuensis, Scotinomys 746
escherichi, Colobus 270
Mandrillus 268
Eschrichtiidae 349, **350**
Eschrichtius **350**
gibbosus 351
glaucus 351
robustus **351**
eschscholtzii, Miniopterus 231
escuinapae, Felis 293
Liomys 482
Lynx 293
esculentus, Myoxus 770
eskimo, Mustela 323
esmeraldae, Micoureus 20
Myotis 213

esmeraldarum, Sigmodontomys 748
esox, Hydromys 597
espadae, Eptesicus 202
espiritosantensis, Molossops 235
Tadarida **240**
essaouiranensis, Crocidura 97
estanciae, Thomomys 473
estiae, Microtus 518
estor, Canis 280
Mephitis 317
Pappogeomys 472
esuae, Crocidura 95
etensis, Didelphis 17
etigo, Lepus 815
Mogera **125**
etoschae, Graphiurus 764
etruscus, Suncus **102**, 102, 103
Euarctos 340
Euarvicola 517
Eubalaena **349**
antarctica 349
antipodarum 349
australis **349**, 349
biscayensis 349
glacialis **349**, 349
japonica 349
japonicus 349
nordcaper 349
sieboldi 349
temminckii 349
Eubradypus 63
Eucapra 405
Eucebus 259
euceros, Cervus 385
Eucervaria 293
Eucervus 391
euchore, Antidorcas 395
Antilope 395
Euchoreutes **495**
alashanicus 495
naso **495**, 495
yiwuensis 495
Euchoreutinae 487, **495**, 495
eucladoceros, Cervus 385
Eucuscus 46
Eudelphinus 352
Euderma **204**
maculatum **204**
Eudiscopus **204**, 204
denticulus **204**
Eudorcas 395
Eudromicia 58
eugenii, Macropus **53**
Thylogale 57
Euhyrax 374
Euhys 378
Euibex 405
Eulagos 814, 817
Eulemur **244**
albifrons 244
albimanus 244
albocollaris 244
anjuanensis 244
brissonii 244
bruneus 244
bugi **244**
chrysampyx 244
cinereiceps 244
collaris 244
coronatus **244**
cuvieri 244
dubius 244
flavifrons 244
flaviventer 244
frederici 244
fulvus **244**, 244
johannae 244
leucomystax 244
macaco **244, 244**
macromangoz 244
mayottensis 244
melanocephala 244
micromongoz 244
mongoz **244**, 244
niger 244
nigerrimus 244
nigrifrons 244
ocularis 244
roussardii 244
rubriventer **244**
rufifrons 244
rufipes 244
rufiventer 244
rufus 244
sanfordi 244
xanthomystax 244
Eulepus 814
eulophus, Ursus 339
Eumetopias **327**
jubata 328
jubatus **328**
marinus 328
monteriensis 328
stellerii 328
Eumops **233**
abrasus 233
amazonicus 234
amplexicaudatus 233
auripendulus **233**
barbatus 233
beckeri 233
bonariensis **233**
californicus 234
dabbenei **233**
delticus 233
ferox 233
floridanus 233
geijskesi 234
gigas 234
glaucinus **233**
hansae **234**
leucopleura 233
longimanus 233
major 233
maurus **234**
mederai 233
milleri 233
nanus 233
oaxacensis 233
orthotis 233
patagonicus 233
perotis **234**
renatae 234
sonoriensis 234
trumbulli 234
underwoodi **234**
Eumysopinae **793**
Euneomys **701**, 701, 702
chinchilloides **701**, 701, 702
dabbenei 702
fossor **702**
mordax **702**, 702
noei 702
petersoni **702**, 702
ultimus 702
euopis, Cervus 386
Euoticus **248**, 249
apicalus 248
elegantulus **248**
pallidus **249**
talboti 249
tonsor 248
euotis, Hipposideros 172
Euotomys 507
Eupetaurus **459**
cinereus **459**, 459
Euphractini 64
Euphractus 65, **66**, 67
boliviae 66
bruneti 66
encoubert 66
flavimanus 66
flavipes 66
gilvipes 66
mustelinus 66
poyu 66
setosus 66
sexcinctus **66**
tucumanus 66
euphratica, Allactaga **488**, 488, 489
Euphrosyne 356
euphrosyne, Stenella 356
Euphysetes 359
Eupleres **340**, 340, 341
goudotii **340**, 340
major 340
Eupleridae 340
Euplerinae **340**
euptilura, Prionailurus 295
eureka, Zapus 499
Eureodon 377
Eurhinoceros 372
euronotus, Sorex 114
europaea, Hystrix 773
Talpa 126, 128, **129**, 129
europaeus, Alces 389
Bison 400
Capreolus 390
Erinaceus **77**, 77, 79
Lepus 814, 816, **817**, 817, 818, 820, 821
Meles 314
Mesoplodon **362**
Sciurus **443**
Sus 379
europea, Capra 405
Rupicapra 410
europeae, Mustela 322
europs, Nyctinomops 239
Euroscaptor **125**, 125, 126, 128
grandis **125**
hiwaensis 125
klossi **125**, 125
leucura 125
longirostris **125**, 125
malayana 125
micrura **125**, 125
mizura **125**
othai 125
parvidens **125**, 125
Eurosorex 111, 113, 116, 117
eurous, Rattus 652
eurus, Damaliscus 394
euryale, Rhinolophus **165**
Euryceros 403
euryceros, Tragelaphus 404
eurycerus, Tragelaphus **403**
Euryotis 680
brantsii 682
irrorata 680
euryotis, Rhinolophus **165**
Euryotomys 679
Eurypterna 67
euryrhinus, Ursus 339
euryrhynchis, Cryptotis 108
euryspilus, Helarctos 337
Euryzygomatomys 793, **794**, 794
brachyura 794
catellus 794
guiara 794
rufa 794
spinosus **794**
eustephanus, Cervus 386
Eutamias 453, 455
Eutropia 351
eutropia, Cephalorhynchus **351**, 358
euxanthus, Oncifelis 294
Euxerus 458
euxina, Felis 290
Vormela 325
euxinus, Apodemus 572
eva, Artibeus 188
Cyclopes 67
Eothenomys **514**
Peromyscus 729, **730**, 730
evae, Reithrodon 740
evagor, Mustela 324
evalensis, Alcelaphus 393
evansi, Naemorhedus 407
evelynae, Meriones 556
everardensis, Notomys 635
everesti, Microtus 523
everetii, Melogale 314
everetti, Dremomys **425**
Hylopetes 460, 461

Iomys 461
Melogale **314**, 314
Presbytis 271
Rattus 649, **651**, 652, 655
Sciuropterus 460
Tragulus 382
Tupaia 133
Urogale **133**
evergladensis, Mustela 324
evermanni, Chaetodipus 484
eversmanni, Allocricetulus **537**, 537
Cricetus 536
Hemiechinus 78
Meriones 556
Microtus 521
Spermophilus 450
Ursus 339
eversmannii, Mustela 321, **322**, 322, 323, 324
evexa, Lontra 311
evexus, Dipodomys 479
Thomomys 476
evidens, Prosciurillus 436
evides, Peromyscus 728, 729, 736
evoronensis, Microtus **520**, 525
evotis, Myotis 208, **210**
Notiosorex 111
Tonatia **180**
Evotomys 507, 508, 514, 515, 532
smithii 532
exalbidus, Sciurus 443
excelsifrons, Rangifer 392
excelsior, Niviventer 632, 633, **634**
excelsus, Pappogeomys 472
Procyon 335
Sorex 112, **114**, 114
Tragelaphus 404
excisum, Sturnira 192
excubutor, Chlorocebus 265
exenticus, Cryptomys 772
exgeanus, Paraxerus 435
exiguus, Liomys 482
Myotis 213
Oligoryzomys 718
Peromyscus 732
Rattus 654
Rhinolophus 164
Sorex 117
Sylvilagus 825
exilipes, Cryptotis 109
exilis, Dasyurus 32
Dipodomys 479
Exilisciurus **426**
Gerbillurus 547
Herpestes 305
Myoprocta 782, **783**, 783
Myotis 209
Sciurus 426
Sorex 117
Uromys 671
Exilisciurus 419, **426**, 426, 434
concinnus **426**
exilis **426**
luncefordi 426
retectus 426
samaricus 426
sordidus 426
surrutilus 426
whiteheadi **426**
eximius, Cryptotis 109
Dipodomys 477
Galidictis 300
Mephitis 317
Microtus 519
Sorex 115
Thomomys 476
Ursus 339
exitus, Arvicola 506
Paradoxurus 344
exoleta, Dobsonia **139**
exoneratus, Dendromus 543
exoristus, Neofiber 532
expulsus, Calomys 698
exsputus, Sigmodon 747
exsul, Dicrostonyx **510**, 510
Microtus 518
Rattus 660
Rhinolophus 166
extenuatus, Thomomys 473
exterus, Peromyscus 732
extimus, Ammospermophilus 420
Chaetodipus 483
Sciurus 440
Thomomys 476
extractus, Dipodomys 479
extremus, Myotis 213
exulans, Rattus 649, 650, **652**, 652, 662
eyra, Herpailurus 291
eyreius, Notomys 636
ezrae, Cercopithecus 264

faberi, Rattus 662
facetus, Rattus 662
faeceus, Lasiopodomys 516
faeroensis, Mus 625
faesula, Rupicapra 410
fagani, Lepus **817**, 818
faini, Chaerephon 233
falabae, Scotoecus 226
falcatus, Artibeus 190
Phyllops **190**
Falcifer 68
falcifer, Thomomys 475
falciger, Ursus 339
falclandicus, Arctocephalus 326
Mirounga 330
falconeri, Capra **405**, 405
falklandica, Arctocephalus 326
Mirounga 330
fallax, Artibeus 189
Atelerix 77
Callosciurus 421
Cebus 259
Chaetodipus **483**
Crunomys **588**, 588
Geomys 470
Neotoma 712
Pelomys **639**, 639
Rhinolophus 167
Tatera 560
fallenda, Mustela 321
Falsistrellus 219, 222, 223, 224
falzfeini, Apodemus 571
Cricetulus 538
Stylodipus 495
Sus 379
famatina, Abrocoma 789
famatinae, Lagidium 778
famelicus, Vulpes 286
familiaris, Canis 280, 281
famosus, Ctenomys 785
famulus, Callosciurus 422
Gerbillus **551**
Mus **623**
Rhinolophus 165
fannini, Otomys 681
Ovis 409
fantiensis, Cercopithecus 264
fantozatianus, Naemorhedus 407
far, Mus 625
faradjius, Atelerix 76
Eptesicus 203
Leptailurus 292
Mops 236
Orycteropus 375
Otomys 681
farasani, Gazella 396
fargesianus, Naemorhedus 407
faroulti, Pachyuromys 559
farsi, Meriones 556
farsistani, Ellobius 512
fasciata, Capra 406
Galidictis **300**
Hyaena 308
Perameles 40
Phoca **331**
Viverra 300
fasciatus, Connochaetes 394
Kangurus 53
Lagostrophus **53**
Lemniscomys 602
Lynx 293
Mungos 307
Myrmecobius **29**, 29
Perognathus **484**, **485**
Tragelaphus 404
fascicularis, Macaca **266**, 267, 271
fasciculata, Hystrix 774
Trichys **774**
fassini, Ammotragus 404
fatoii, Microtus 526
fatuellus, Cebus 259
fatuus, Synaptomys 534
faunulus, Pteropus **147**
faunus, Cercopithecus 263
faustus, Perodicticus 248
favillus, Gerbillus 552
Jaculus 493
favonicus, Funambulus 426
Jaculus 493
Myotis 208
feae, Berylmys 580
Muntiacus **388**, 389
Murina 229
Rhinolophus 167
feai, Muntiacus 389
fearonii, Acinonyx 288
fearonis, Acinonyx 288
fecundus, Calomys 697, 698
federatus, Myotis 213
fedjushini, Sciurus 443
feileri, Strigocuscus 48
feliceus, Rattus 650, **652**, 652
felicia, Dasyprocta 781
Felidae **288**, 340
felina, Aotus 256
Crocuta 308
Genetta 345, 346
Lontra **311**
Trichosurus 48
Felinae **288**, 297
felinus, Aotus 256
Paradoxurus 344
felipei, Mustela 321, **322**, 322
felipensis, Neotoma 712
Orthogeomys 471
Peromyscus 730
Felis 288, **289**, 289, 290, 293, 294, 296, 297
affinis 290
agrius 290
airensis 290
algiricus 290
angorensis 290
antiquorum 290
aurata 296
aureus 290
bengalensis 295
bieti **289**, 291
bouvieri 290
brevicaudata 290
brockmani 290
bubastis 290
caeruleus 290
caffra 290
cafra 290
caligata 290
caracal 288
catolynx 290
catus 289, 290, 291
caucasicus 290
caudatus 290
celidogaster 296
chaus **289**, 290
chrysomelanotis 290
chutuchta 289, 291
concolor 296, 297
cretensis 290
cristata 290
cumana 290

cyrenarum 290
daemon 290
domesticus 290
erythrotus 290
escuinapae 293
euxina 290
ferox 290
ferus 290
foxi 290
fulvidina 290
furax 290
geoffroyi 294
grampia 290
griselda 290
griseoflava 290
gulata 290
harrisoni 290
haussa 290
hispanicus 290
huttoni 290
iraki 290
irbis 299
iriomotensis 295
issikulensis 290
jacobita 294
jacquemonti 290
japonica 290
jordansi 290
jubata 288
kelaarti 290
kozlovi 290
kutas 290
lapsus 291
libyca 291
libycus 290
longiceps 290
longipilis 290
lowei 290
lybica 290, 291
lybiensis 290
lynesi 290
lynx 293
macrocelis 297
macrothrix 290
madagascariensis 290
maniculata 290
manul 290, 295
margarita **290**
margaritae 290
marginata 290
margueritei 290
marmorata 299
matschiei 290
mauritana 290
mediterranea 290
megalotis 290
meinertzhageni 290
mellandi 290
michaelis 288
molisana 290
moormensis 289
morea 290
murgabensis 290
namaquana 290
nandae 290
nebulosa 297
nesterovi 290
nigripes **290**
nilotica 290
nubiensis 290
obscura 290
ocreata 290
ornata 290
pallida 289
pardalis 291
pardinoides 292
pardochrous 295
pardoides 294
pardus 297
prateri 290
pulchella 290
pyrrhus 290
reyi 290
ruber 290
rubida 290
ruppelii 290
rusticana 290
salvanicola 290
sarda 291
scheffeli 290
schnitnikovi 291
serval 292
servalina 291
shawiana 290, 291
siamensis 291
silvestris 289, **290**, 291
sinensis 291
striatus 291
subpallida 289
sylvestris 289
syriaca 291
taitae 291
tartessia 291
temminckii 289
thinobius 290
thomasi 290
torquata 291
tralatitia 291
trapezia 291
tristrami 291
typica 290
ugandae 291
uncia 299
vellerosa 289, 291
vernayi 291
vulgaris 291
xanthella 291
yaguarondi 291
felix, Tamias 453
felli, Ratufa 437
fellowesgordoni, Suncus **102**, 102
fellowsi, Melomys 615, **616**
Felovia **761**
vae **761**, 761
felteni, Microtus **520**, 520, 530
femoralis, Chaetodipus 483
Presbytis **271**, 271
Ratufa 437
femorosaccus, Nyctinomops **239**
Nyctinomus 238
femurvillosum, Artibeus 189
fenestrae, Thylamys 23
fenestratus, Dasypus 66
fenisex, Ochotona 811
Fennecus 286, 287
fennecus, Vulpes 287
fenniae, Micromys 620
fennicus, Apodemus 571
Rangifer 392
fera, Lama 382
ferculinus, Pseudomys 647
Feresa **352**
attenuata **352**, 352
intermedius 352
occulta 352
ferghanae, Mustela 321
ferghanicus, Capreolus 390
fergusoni, Bubalus 402
fergussoniensis, Pogonomys 641
Rattus 656
ferminae, Sciurus 441
fernandezi, Lonchorhina **177**
fernandinae, Nesoryzomys **713**
fernandoni, Mus **623**
Feroculus **99**, 99
feroculus **99**
macropus 99
newera 99
newera-ellia 99
feroculus, Feroculus **99**
Sorex 99
ferox, Cryptoprocta **340**, 340
Eumops 233
Felis 290
Ichneumia 306
Macaca 268
Ursus 339
ferrarii, Leptailurus 292
ferreo-griseus, Melogale 314
ferreocanus, Berylmys 580
ferreus, Monophyllus 185
Sciurus 439
ferriginea, Procolobus 272
ferrilata, Vulpes **286**
ferruginea, Crocidura 92
Herpestes 305
Lutreolina 18
Neotoma 712
Nyctinomops 239
Ochotona 809
Presbytis 271
Sminthopsis 35
Tupaia 131, 132
ferrugineus, Arvicola 506
Callosciurus 422
Cricetulus 537
Deomys **544**
Eptesicus 201
Herpestes 305
Mesomys 794
Nyctalus 216
Oryzomys 737
Otocolobus 295
Phaenomys **737**
Procavia 374
Pseudalopex 284
Suncus 103
ferruginifrons, Sminthopsis 36
ferruginiventris, Sciurus 440
ferrum-equinum, Vespertilio 163
ferrumequinum, Rhinolophus **165**
fertilis, Arvicola 515
Hyperacrius **515**
ferus, Camelus 381
Equus 369
Felis 290
Sus 379
fervidus, Caluromys 15
Rhipidomys 745
Sigmodon 747
festina, Cavia 779
festinus, Lepus 815
festivus, Equus 369
fetidissima, Mephitis 317
fetisovi, Alticola 503
fetosus, Dipodomys 479
feuillei, Conepatus 316
Fiber 532
fiber, Castor **467**, 467
fidelis, Eothenomys 514
Eptesicus 201
fieldi, Pseudomys **646**
fieldiana, Genetta 346
figginsi, Conepatus 316
Ochotona 811
filchneri, Canis 281
Lepus 821
Viverra 348
filchnerinae, Petaurista 462, 463
filipensis, Conepatus 316
filippii, Capra 406
filmeri, Notomys 636
fimbriatus, Artibeus **188**
Crocidura 94
Herpestes 305
Hylopetes **460**, 460
Myotis 212
Neomys 110
Zaedyus 67
fimbripes, Sorex 113
findleyi, Eptesicus 201
Myotis **210**
fingeri, Microtus 524
finis, Maxomys 614
finitus, Microtus 527
finlaysoni, Eptesicus 203
finlaysonii, Callosciurus **422**
finmarchicus, Microtus 527
finschi, Dendrolagus 50
Equus 370
Syconycteris 155
fiolagan, Apodemus 573
fiona, Microtus 518
firmus, Sundamys 666
firouzi, Allactaga **488**, 489

fischerae, Rhinophylla **186**
fischeri, Crocidura **84**
Haplonycteris **142**, 142
fischerii, Tarsius 250
fisheri, Microtus 518
Sorex 116, 117
Spermophilus 445
Tamias 456
Thomomys 475
fisi, Crocuta 308
fistulator, Cebus 259
fitzroyi, Lagenorhynchus 354
fitzsimonsi, Lemniscomys 602
Mormopterus 238
Nycticeius 218
Pronolagus 824
flaccidus, Bradypus 63
Peromyscus 732
flammarum, Reithrodon 740
flammeus, Ochrotomys 715
Pappogeomys 472
flammifer, Sciurus **440**
flava, Microtus 519
Neotoma 711
flaveolus, Scotophilus 227
flavescens, Allactaga 489
Ariteus **187**
Cebus 259
Crocidura **84**, 86, 92, 97
Dasyprocta 782
Eptesicus **201**
Galerella **302**, 302, 303
Istiophorus 187
Lemmus 517
Mesophylla 190
Microtus 520
Mus 625
Oligoryzomys **717**
Otaria 328
Perognathus **485**
Phoca 328
Pipistrellus 223
Pseudomys 645
Rattus 658
Spermophilus 448
Sus 379
Sylvicapra 412
Vulpes 287
flavescente, Callithrix 251
flavicans, Oecomys **716**
flavicauda, Lagothrix **257**
Trachypithecus 274
flaviceps, Callithrix **251**, 252
flavicollis, Apodemus 570, **571**, 571, 572, 573, 574
Pteropus 151
Tragulus 383
flavidens, Galea **779**, 779
Herpestes 304
flavidior, Auliscomys 695
Dendrolagus 51
flavidulus, Maxomys 614
flavidus, Acomys 565
Chlorocebus 265
Echimys 792
Isthmomys **706**, 706
Megadontomys 705
Petaurus 61
Procyon 335
Thomomys 473
flavifrons, Eulemur **244**
Saguinus 253
flavigaster, Scotophilus 227
flavigrandis, Maxomys 614
flavigula, Martes **320**, 320
flavigularis, Lepus **818**
flavimaculata, Procavia 374
flavimanus, Alouatta 254
Callosciurus 421, 424
Euphractus 66
Presbytis 271
flavina, Marmota 431
flavinus, Funisciurus 427
flavior, Dremomys 425
flavipectus, Rattus 660
flavipes, Antechinus **30**, 30
Euphractus 66
flavipilis, Niviventer 634
flavissimus, Ailurops 46
flavistriata, Poecilogale 325
flaviventer, Eulemur 244
Microsciurus **433**
Niviventer 633
Petaurus 61
flaviventris, Apodemus 573
Galerella 303
Glaucomys 460
Lepus 820
Marmota **432**
Microtus 527
Rattus 658
Saccolaimus **159**
flavivittis, Paraxerus 435
flavobrunneus, Apodemus 573
flavogularis, Centurio 189
flavomaculatus, Saccolaimus 159
flavopunctatus, Lophuromys **605**, 606
flavovittis, Paraxerus **435**
flavus, Bassariscus 334
Canis 281
Castor 467
Cebus 259
Lemur 333
Lepus 817
Micromys 620
Mus 625
Myotis 210
Perognathus **485**, 486
Potos **333**
Sicista 497
flebilis, Rattus 660
flindersi, Macropus 53
flindersii, Phascolarctos 45
flora, Kerivoula **197**, 197
florencei, Nyctomys 715
florenciae, Choloepus 64
Microsciurus 433
Oecomys 716
florensis, Spelaeomys **664**, 664
florentiae, Jaculus 493
floresianus, Acerodon 137
Microsus 378
Sus 379
floresiensis, Cervus 387
floresii, Acerodon 137
Floresomys 639
naso 639
floridana, Cryptotis 109
Mus 710
Neotoma 710, **711**, 711, 712
Puma 296
floridanus, Canis 281
Eumops 233
Geomys 470
Hesperomys 739
Lasiurus 207
Lepus 824
Lynx 293
Pipistrellus 224
Podomys **739**
Sigmodon 747
Sylvilagus **825**, 826, 827
Urocyon 285
Ursus 338
florium, Murina **229**
florstedti, Capra 405
floweri, Callosciurus 422
Cephalorhynchus 351
Crocidura **84**, 85
Damaliscus 394
Eptesicus **201**
Gerbillus **551**
Kogia 359
Mesoplodon 363
Rhinoceros 372
fluminalis, Callosciurus 421
Orcaella 354
fluminensis, Molossus 235
fluva, Otaria 328
fluviatilis, Delphinus 355
Lutra 312
Neomys 110
Presbytis 271
Sotalia **355**
fluvicinctus, Arvicanthis 576
foae, Procolobus 272
foai, Equus 369
Procolobus 272
focalinus, Tragulus 382
fochi, Ctenomys 785
fociensis, Elaphodus 388
fodax, Ctenomys 785
fodiens, Neomys **110**, 111
Oryctolagus 822
Sorex 110
foederis, Sundamys 666
foesteri, Rattus 660
foetida, Crocidura 85
Phoca 332
foetidus, Rhinolophus 167
foetulenta, Mephitis 317
foina, Martes **320**
Mustela 319
foleyi, Gerbillus 551
folletti, Callosciurus 422
foncki, Akodon 692
fontainierii, Mustela 324
fontanelus, Geomys 470
fontanierii, Myospalax **675**, 675, 676
Panthera 298
fontanus, Myospalax 675
fontigenus, Microtus 527
fontinalis, Sorex 113
fontoynonti, Dasogale 73
Setifer 73
foramena, Vernaya 673
foramineus, Rattus 650, 651, **652**, 652, 657, 662
forbesi, Chiruromys **584**, 584
Pseudocheirus **59**
forfex, Kobus **414**
formicarius, Ursus 339
formosae, Rhinolophus 167
formosanus, Coelops 170
Herpestes 306
Mus 623
Tamiops 457
Ursus 340
formosovi, Capra 406
Mus 625
Sciurus 443
formosus, Bettongia 49
Chaetodipus **483**
Erythrocebus 265
Lepus 820
Myotis **210**
Otomops **239**
Pteropus 147
Tragulus 383
fornesi, Oligoryzomys 718
Oryzomys 718
fornicatus, Chaetodipus 483
Lariscus 430
forresti, Leggadina 600, **601**, 601
Microtus 522
Mus 600
Ochotona **809**, 809, 812
Sciurotamias **439**
Tamiops 457
fors, Mus 629
forsteri, Arctocephalus **326**
Neophoca 328
Sorex 113
fortidens, Myotis **210**
Rangifer 392
fortior, Anomalurus 757
Otomys 682
fortirostris, Marmota 432
fortis, Crocuta 308
Dasyurus 31
Microtus **521**, 521
Molossus 235
Panthera 298
fortunatus, Rattus 660
Fossa 340, **341**, 341
d'aubentonii 341

fossa 341
fossana **341**
majori 341
fossa, Fossa 341
Paradoxurus 344
Viverra 341
fossana, Fossa **341**
Viverra 341
fossata, Herpailurus 291
fossilis, Blarina 106
Ozotoceros 392
Panthera 298
Pseudalopex 284
Fossinae 340
fossor, Chelemyscus 702
Cryptotis 108
Euneomys **702**
Geoxus 702
Phacochoerus 377
Reithrodon 701
Sciurus 441
Thomomys 475
Vombatus **46**
fosteri, Cephalophus 411
Chrysochloris 75
Monodelphis 21
Promops 240
Thrichomys 798
foucauldi, Crocidura 94
fouriei, Aethomys 567
Pedetes 759
Pipistrellus 222
Rhabdomys 662
fournieri, Capromys 800
foxi, Crocidura **85**, 85, 88, 97
Cryptomys **771**, 771
Dasymys **589**, 590
Felis 290
Graphiurus 765
Rhinolophus 166
Uranomys 671
fractilis, Atelerix 77
fraenata, Onychogalea **55**
francei, Akodon 691
francescae, Cercopithecus 264
francianus, Cervus 386
francisca, Bettongia 49
francoisi, Trachypithecus **273**
frandseni, Callosciurus 422
franklinii, Spermophilus **446**
frankstounensis, Sorex 113
franqueti, Epomophorus 141
Epomops **142**
fransseni, Loxodonta 367
frantzii, Lasiurus 206
Frasercetus 361
fraseri, Anomalurus 757
Kobus 414
Sciurus 443
frater, Chaerephon 233
Clethrionomys 508
Ctenomys **784**
Monophyllus 185
Myotis **211**
Pelomys 639
Tamias 456
fraterculus, Artibeus **188**, 188
Melomys 615, **616**, 616, 618
Miniopterus **230**
Odocoileus 391
Peromyscus 730
Sundasciurus **451**, 451, 452, 604
Tarsius 251
Tatera 560
Urocyon 285
Zygodontomys 752
fraternus, Macroglossus 154
Niviventer 635
Thomasomys 749
Frateromys 581, 582
penitus 582
fratrorum, Bunomys 581, **582**, 582
fredericae, Mus 625
Niviventer 634
Presbytis 270
frederici, Cephalophus 410
Equus 370
Eulemur 244
Herpestes 305
fremens, Leopoldamys 603
fremonti, Conepatus 316
Dipodomys 479
Tamias 456
Tamiasciurus 457
frenata, Dendrogale 131
Ictonyx 319
Mustela 321, **322**, 322
Philander 22
frenatus, Dipodomys 479
Onychogalea 55
Sus 379
frerei, Paraxerus 435
fretalis, Sorex 114
fretensis, Pteropus 147
Ratufa 437
Taphozous 161
Fretidelphis 356
freyreisii, Diclidurus 157
frida, Calomys 698
fridariensis, Apodemus 573
frigicola, Melomys 615
frinianus, Cervus 386
frisius, Talpa 129
frithi, Coelops **170**
frithii, Coelops 170
fritschi, Nannospalax 753
froenatus, Stenella 356
froggatti, Melomys 615
Sminthopsis 37
frommi, Heterohyrax 373
Kobus 413
Sigmoceros 395
Tragelaphus 404
frondator, Castor 467
frons, Lavia **163**
Megaderma 162
frontalis, Acrobates 62
Atelerix **77**
Ateles 257
Bos **401**
Cervus 386
Cryptotis 108
Hemiechinus 78
Oecomys 717
Ratufa 437
Stenella **356**
frontalus, Cebus 259
frontata, Mephitis 317
Presbytis **271**
frontatus, Ateles 257
Steno 357
fronto, Cerdocyon 282
Kunsia **706**
Scapteromys 745
frontosa, Vicugna 382
frontosus, Sus 379
frosti, Babyrousa 377
Neopteryx **143**, 143
Paruromys 638
fructivorus, Macroglossus 154
frugilegus, Tremarctos 338
frugivorus, Rattus 658
frumentariae, Apodemus 573
frumentarius, Cricetus 538
frumentor, Sciurus 440
frustrator, Zygodontomys 752
frustror, Canis 280
frutectanus, Napaeozapus 498
fruticicolus, Phyllotis 737
fruticus, Macropus 55
fryanus, Callosciurus 421
fryi, Malacothrix 544
Fsihego 276
fuchsi, Lycaon 283
fucosus, Microtus 525
fudisanus, Rhinolophus 166
fueginus, Ctenomys 785
fujianensis, Microtus 521
fujiensis, Myotis 209
fulgens, Ailurus 336, **337**
Anomalurus 757
Hipposideros 173
Melomys 617
Oryzomys 721
fulgida, Cavia **779**
fuliginatus, Baiomys 695
Odocoileus 391
fuliginosa, Crocidura **85**, 85, 89, 92, 93, 96
Dasyprocta **781**
Nycteris 162
Rousettus 152
Trichosurus 48
fuliginosus, Akodon 692
Alopex 279
Artibeus 189
Arvicola 506
Ateles 257
Caenolestes **25**
Cercocebus 262, 263
Glaucomys 460
Hipposideros **173**
Hydromys 597
Hyracodon 25
Macropus **54**
Miniopterus 231
Onychomys 719
Pan 277
Paradoxurus 344
Perognathus 485
Procolobus 272
Proechimys 798
Pteronotus 177
Rattus 658
Sciurus 440
Sminthopsis **36**
Tadarida 240
Trachops 181
fulminans, Tadarida **240**
fulminatus, Sciurus 441
fulva, Microtus 519, 527
Mops 236
Mustela 322, 323
Vernaya **673**, 673
fulvaster, Rattus 658
fulvastra, Crocidura **85**, 86
fulvescens, Akodon 691
Herpestes 304
Hesperomys 717
Hylomys 80
Niviventer 632, 633, **634**, 634, 635
Oligoryzomys 717, **718**, 718
Pappogeomys 472
Phyllotis 738
Reithrodontomys **741**, 741
Sylvilagus 825
fulvicaudus, Pseudalopex 285
fulvicollis, Tragulus 382
fulvidina, Felis 290
fulvidior, Galerella 303
fulvidus, Rhinolophus 168
Taphozous 160
Tylonycteris 228
fulvifrons, Kobus 413
fulvinus, Nectomys 709
Petaurista 463
fulvior, Xerus 458
fulvipectus, Apodemus **571**, 571, 572, 573
fulvipes, Pseudalopex 284
Vulpes 284
fulvirostris, Microryzomys 708
fulviventer, Marmosa 18
Microtus 524
Neotoma 712
Oecomys 717
Rhipidomys **744**, 744, 745
Sigmodon **747**
Tragulus 382
Tylomys **751**, 751
fulvocinerea, Suncus 103
fulvogaster, Hydromys 597
fulvogriscus, Cerdocyon 282
fulvolavatus, Hydromys 597

fulvoochraceus, Tragelaphus 404
fulvorubescens, Raphicerus 399
fulvorufula, Redunca **414**
fulvoventer, Hydromys 597
fulvus, Agouti 783
Bubalus 402
Canis 281
Castor 467
Cebus 259
Chiropodomys 673
Clethrionomys 508
Cricetulus 538
Cricetus 538
Ctenomys **784**, 785, 786, 787
Dasyprocta 781, 782
Dendrolagus 51
Eulemur **244**, 244
Hipposideros **173**
Macaca 267
Meriones 557
Peromyscus 732
Pteronotus 176
Sciurus 440
Spermophilus **446**, 447, 448
Thomomys 473
Vulpes 287
fumarium, Sturnira 192
fumarius, Promops 240
fumatus, Cricetulus 537
Myomys 630, **631**, 631, 632
fumeolus, Sorex 112
fumeus, Akodon **689**, 690, 691
Chilomys 699
Ototylomys 726
Pseudomys 645, **646**, 646, 648, 649
Sorex **114**
Thomasomys 750
fumidus, Soriculus **123**, 123
fumigatus, Callosciurus 421
Crocidura 86
Lepus 820
Pteropus 148
Rhinolophus **166**
Sciurus **442**
fumipes, Auliscomys 695
fumosa, Crocidura **85**, 85, 89, 90, 94, 96
Ochotona 811
fumosus, Cercocebus 262
Chelemys 699
Cuon 282
Heimyscus **595**
Hylomyscus 595
Mystromys 677
Pappogeomys **472**
Platyrrhinus 191
Taphozous 160
Thomomys 473
Funambulinae 419
Funambulini 419, 426, 427, 434
Funambulus **426**, 426, 427
annandalei 427
argentescens 426
bellaricus 426
bengalensis 426
brodiei 426
comorinus 426
delesserti 427
dravidianus 426
dussumieri 427
favonicus 426
gossei 426
indicus 426
kathleenae 427
kelaarti 426
layardi **426**
lutescens 426
matugamensis 426
numarius 427
obscura 427
olympius 426
palmarum **426**
penicillatus 426
pennantii **426**, 426
robertsoni 426
signatus 426
sublineatus **427**
thomasi 427
trilineatus 427
tristriatus **427**
wroughtoni 427
fundatus, Microtus 524
Pteropus **147**
funebris, Lasiurus 206
Microtus 527
funereus, Hylobates 276
Mus 625
Pteropus 151
funestus, Herpestes 305
funicolor, Rhombomys 559
Funisciurus **427**
akka 428
anerythrus **427**
auriculatus 428
bandarum 427
bayonii **427**
beatus 428
birungensis 427
boydi 428
carruthersi **427**
chrysippus 427
congicus **427**
damarensis 427
dubosti 427
duchaillui 427
emini 428
erythrogenys 428
erythrops 428
flavinus 427
interior 427
isabella **427**
lemniscatus **427**
leonis 428
leucogenys **427**, 428
leucostigma 428
mandingo 428
mayumbicus 427
mystax 427
niapu 427
nigrensis 428
niveatus 428
ochrogaster 427
oenone 427
olivellus 427
oliviae 428
pembertoni 428
poolii 427
praetextus 427
pyrrhopus 428
pyrropus **428**
raptorum 427
rubripes 428
sharpei 427
substriatus **428**
talboti 428
tanganyikae 427
victoriae 428
wintoni 428
furax, Emballonura **157**
Felis 290
Galictis 318
Papio 269
furcata, Blastocerus 389
Conepatus 316
Furcifer 390
furcifer, Lemur 244
Phaner **244**
Rangifer 392
furculus, Triaenops **175**
Furia, horrens 195
furinalis, Eptesicus **201**
Furipteridae **195**
Furipterus **195**
coerulescens 195
horrens **195**
furnessi, Caprolagus 823
Pentalagus **823**
furo, Mustela 324
furrugineus, Deomys 544
furunculus, Cricetulus 537
furva, Soriculus 123
furvogaster, Hylobates 275
furvus, Ailurops 46
Nyctalus 217
Palawanomys **637**, 637
Peromyscus **730**, 730, 733, 734
Sigmodon 747
Urocyon 285
fusca, Alouatta **255**
Arctogalidia 343
Canis 281
Geomys 469
Herpestes 304, 305, 306
Kerivoula 197
Mazama 391
Melogale 314
Microtus 530
Murina **229**
Mustela 322
Nasua 335
Neotoma 711
Panthera 298
Parahyaena 309
Phalanger 46
Procyon 335
Pudu 392
Rousettus 152
fuscata, Macaca 266, **267**
Mazama 390
fuscatus, Aepeomys **688**
Heteromys 481
Lasiurus 207
Marmosops **19**
Pipistrellus 221
Tragulus 382
fuscescens, Sylvilagus 825
fusciceps, Ateles **257**
Ellobius 513
fuscicollis, Saguinus **253**, 254
fuscicolor, Cephalophus 411
fuscicornis, Bos 401
fuscidorsis, Cricetus 538
fuscifrons, Gazella 396
fuscinus, Bolomys 696, 697
fuscior, Tupaia 133
fuscipes, Elephantulus **829**
Ellobius 513
Jaculus 493
Neotoma **711**, 711
Ochotona 811
Pipistrellus 223
Procyon 335
Rattus 649, **652**, 653, 654
Suncus 103
Tscherskia 540
fuscirostris, Myomys 630
fusciventer, Isoodon 39
Molossus 235
fusco-ater, Akodon 691
Macaca 267
fuscoater, Sciurus 443
fuscocanus, Dipus 493
fuscocapillus, Ellobius **512**, 512
Petinomys **463**
fuscodorsalis, Clethrionomys 508
fuscogriseus, Onychomys 719
Philander 22
fuscomanus, Tarsius 250
fuscomurina, Crocidura **85**, 91, 93
Presbytis 271
fusconigricans, Sciurus 443
fuscorubens, Sciurus 443
fusco-rufa, Lontra 311
fuscosa, Crocidura 92
fuscovariegatus, Sciurus 443
fusculus, Microsciurus 433
fuscus, Aconaemys **787**, 788
Allactaga 489
Cervus 386
Dasymys 589
Dasyurus 31

Desmodilliscus 547
Desmodus 194
Dipodomys 477
Dremomys 425
Elephantulus **829**, 829
Eptesicus **201**, 203
Galerella 303
Geomys 470
Glaucomys 460
Hylobates 275
Lasiopodomys **516**
Macaca 266
Melomys 618
Microtus 520
Miniopterus **230**, 230
Moschus **384**, 384
Neohydromys **632**, 632
Notomys **636**
Ornithorhynchus 13
Pan 277
Paradoxurus 344
Perognathus 485
Peromyscus 732
Phascolarctos 45
Phyllotis 738
Pseudochirops 60
Pseudomys **646**, 646
Pteronotus 177
Pteropus 149
Rattus 658
Saccostomus 541
Saguinus 253
Schizodon 787
Tarsius 250
Thomomys 475
Ursus 339
Vespertilio 200
Vombatus 46
Xerus 458
fusicauda, Scaptonyx 128
fusicaudatus, Scaptonyx 128
fusicaudus, Scaptonyx **128**
fusiformis, Peponocephala 355
fusus, Microtus 525

gabbi, Sylvilagus 825
gabbii, Bassaricyon **333**, 333
gabonensis, Galago 249
gabriellae, Hylobates **275**
Gadamu 357
gadovii, Peromyscus 733
gaetulus, Meriones 556
gaillardi, Lepus 815
Ovis 408
gaimardi, Bettongia **49**
Kangurus 49
gairdneri, Arctictis 342
Mus 627
Scotophilus 227
gala, Rattus 651
Galagidae 248
Galago **249**, 250
acaciarum 249
albipes 249
alleni **249**
australis 249
batesi 249
braccatus 249
bradfieldi 249
calago 249
cameronensis 249
conspicillatus 249
demidoff 249
dunni 249
gabonensis 249
galago 249
gallarum **249**
geoffroyi 249
granti 249
intontoi 249
inustus 249
matschiei **249**
moholi **249**
mossambicus 249
nyassae 249
pupulus 249
senegalensis **249**, 249
sennariensis 249
sotikae 249
teng 249
tumbolensis 249
galago, Galago 249
Galagoides **249**
anomurus 249
cocos 250
demidoff **249**, 249
demidovii 249
granti 250
medius 249
mertensi 250
murinus 249
orinus **249**
peli 249
phasma 249
poensis 249
pusillus 249
senegalensis 250
thomasi 249
zanzibaricus **250**
Galagonidae **248**
galapagoensis, Arctocephalus **326**, 326
Oryzomys **722**, 722, 725
galaris, Zaglossus 13
galbus, Galerella 303
gale, Mustela 323
Galea **779**
auceps 779
bilobidens 779
boliviensis 779
campicola 780
comes 779
demissa 779
flavidens **779**, 779
leucoblephara 779
littoralis 779
musteloides **779**, 779
negrensis 779
palustris 780
saxatilis 780
spixii **779**, 779
wellsi 780
galeanus, Naemorhedus 407
galeata, Hystrix 773
galei, Clethrionomys 508
Galemys **124**, 124
pyrenaicus **124**
rufulus 124
Galenomys **702**, 702
garleppi **702**
Galeocebus 245
Galeolemur 135, 462
galeopardus, Leptailurus 292
Galeopithecidae 135
Galeopithecus 135
Galeopteridae 135
Galeopterus 135
Galeopus 135
Galera 317
galera, Atilax 301
Galerella **302**, 304
annulata 303
annulatus 302
apiculatus 303
auratus 303
badius 303
basuticus 303
bocagei 303
bradfieldi 303
bruneoochracea 303
caffra 303
caldatus 303
canus 303
caurinus 303
cauui 303
condradsi 303
dasilvai 303
dentifer 303
elegans 303
emini 303
ererensis 303
erlangeri 303
erongensis 303
flavescens **302**, 302, 303
flaviventris 303
fulvidior 303
fuscus 303
galbus 303
galinieri 303
gracilis 303
granti 303
ignitoides 303
ignitus 303
iodoprymnus 303
kalaharicus 303
kaokoensis 303
khanensis 303
lancasteri 303
lasti 303
latu 302
lefebvrei 303
lundensis 303
marae 303
maritimus 303
melanura 303
mossambica 303
mustela 303
mutgigella 303
mutscheltschela 303
neumanni 303
ngamiensis 303
nigrata 303
nigratus 302, 303
nigricaudatus 303
ochraceus 303
ochromelas 303
okavangensis 303
orestes 303
ornatus 303
parvipes 303
perfulvidus 303
phoenicurus 303
proteus 303
pulverulenta 302, **303**, 303
punctulatus 303
ratlamuchi 303
rendilis 303
ruasae 303
ruddi 303
rufescens 303
ruficauda 303
saharae 303
sanguinea 302, **303**, 303
schimperi 303
shortridgei 302, 303
swalius **303**, 303
swinnyi 303
talboti 303
turstigi 303
ugandae 303
upingtoni 303
venatica 303
zombae 303
Galericinae 79
Galeriscus 301
galeritus, Cercocebus **262**, 262
Hipposideros 171, **173**
Galictis **318**
albifrons 318
allamandi 318
andina 318
barbara 318
brasiliensis 318
canaster 318
chilensis 318
crassidens 318
cuja **318**
furax 318
gujanensis 318
huronax 318
intermedia 318
luteolus 318
melinus 318
quiqui 318
ratellina 318
shiptoni 318
vittata **318**
Galidia **299**
afra 299
concolor 300
decemlineata 300
elegans **299**, 299

Galidictis **300**, 300
eximius 300
fasciata **300**
grandidiensis 300
grandidieri **300**, 300
ornatus 300
rufa 300
striata 300
vittatus 300
Galidiinae **299**, 340
galinieri, Galerella 303
galinthias, Mustela 323
gallaecius, Lepus 818
gallaensis, Canis 280
gallagheri, Chaerephon **232**
Gerbillus 550
gallarum, Coleura 156
Colobus 270
Galago **249**
Mus 625, 629
Oryx 413
Ourebia 399
Tachyoryctes 687
galliae, Castor 467
galliardi, Microtus 519
gallica, Genetta 345
gallicus, Castor 467
gallieni, Hapalemur 245
Gamba 16
gambeli, Peromyscus 732
gambiana, Tatera 561
gambianus, Aonyx 310
Chaerephon 233
Cricetomys **540**, 540
Epomophorus **141**
Heliosciurus **428**, 429
Hippotragus 412
Mungos **307**
Panthera 297
Pteropus 141
Sciurus 428
gambiensis, Hipposideros 172
Nycteris **161**
ganana, Paraxerus 435
gangetica, Delphinus 360
Platanista **360**, 360
gangeticus, Cynopterus 139
Platanista 360
gangutrianus, Rattus 658
gansseri, Lycaon 283
gansuensis, Mus 625
gansuicus, Lepus 821
gaoligongensis, Ochotona **809**
Tupaia 131
gapperi, Clethrionomys **508**, 508, 509
garagassi, Echymipera 41
garamantis, Gerbillus **551**
garambae, Crocidura 88
Eptesicus 201
garbei, Sciurus 439
garcianus, Cervus 386
gardneri, Myotis 214
garganicus, Clethrionomys 508
Sorex 120
gargantua, Echymipera 41
Sus 378
garichensis, Reithrodontomys 742
Scotinomys 746
gariepensis, Crocuta 308
Syncerus 402
garleppi, Galenomys **702**
Oncifelis 294
Phyllotis 702
garleppii, Nectomys 709
garneri, Amblysomus 74
garnetti, Otolicnus 250
garnettii, Otolemur **250**, 250
garonum, Dremomys 425
Leopoldamys 603
garridoi, Mysateles **802**
gaskelli, Hesperoptenus **204**
gaspeanus, Clethrionomys 508
gaspensis, Napaeozapus 498
Sorex 114, **115**
Gastrimargus 257
Gatamiya 622
gatesi, Myotis 208
gatunensis, Oryzomys 721
gaucho, Conepatus 316
gaumeri, Eptesicus 201
Heteromys **481**
Marmosa 18
Gauribos 401
gaurus, Bos 401
Peromyscus 729
gauti, Clethrionomys 508
Gaveus 401
gawae, Rattus 652
gayi, Myotis 209
gazae, Syncerus 402
Gazella **395**, 399
abuharab 396
addra 396
albonotatus 397
arabica **395**
baringoensis 397
beccarii 396
behni 397
bennettii **396**
berberana 397
bergeri 397
bergerinae 397
biedermanni 397
bilkis **396**
brighti 396
butteri 397
cabrerai 396
casanovae 397
centralis 397
christii 396
cinerascens 396
cora 396
corinna 396
cuvieri **396**, 396, 397
dama **396**
damergouensis 396
dieseneri 397
dongilanensis 397
dorcas **396**, 397
doria 396
erlangeri 396, 397
farasani 396
fuscifrons 396
gazella **396**, 396
gelidjiensis 396
gracilicornis 397
granti **396**
hanishi 396
hasleri 397
hayi 396
hazenna 396
hilleriana 397
isabella 396
isidis 396
kanuri 397
kennioni 396
kevella 396
lacuum 396
laevipes 397
langheldi 397
leptoceros **396**
littoralis 396
loderi 396
lozanoi 396
macrocephala 397
maculata 396
manharae 397
marica 397
marwitzi 397
massaesyla 396
merilli 396
mhorr 396
mongolica 397
mundorosica 397
muscatensis 396
nakuroensis 397
nanguer 396
nasalis 397
ndjiriensis 397
neglecta 396
notata 396
occidentalis 396
orientalis 396
osiris 396
pallaryi 397
pelzelnii 396
permista 396
persica 397
petersii 396
raineyi 396
reducta 396
reginae 397
robertsi 396
roosevelti 396
rueppelli 396
ruficollis 396
rufifrons **397**, 397
rufina **397**
ruwanae 397
sabakiensis 397
sairensis 397
salmi 397
saudiya **397**
schillingsi 397
seistanica 397
senegalensis 397
serengetae 396
seringetica 397
sibyllae 397
soemmerringii **397**
spekei **397**
subgutturosa **397**
sundevalli 396
thomsonii **397**, 397
tilonura 397
vera 396
walleri 397
weidholzi 396
wembaerensis 397
yarkandensis 397
gazella, Arctocephalus **326**
Capra 405, 413
Gazella **396**, 396
Oryx **413**, 413
Thylogale 57
gazellae, Grammomys 594
Paraxerus 434
Steatomys 546
geata, Myosorex **99**
geayi, Capromys 800
Nyctophilus 218
geddiei, Pteropus 151
gedeedus, Gerbillus 554
gedrosianus, Ursus 339, 340
geei, Trachypithecus **273**
geijskesi, Eumops 234
geisha, Apodemus 570
Gelada 269
gelada, Macacus 269
Theropithecus **269**
gelidjiensis, Gazella 396
gemellus, Sorex 114
geminae, Apodemus 571
geminorum, Pteropus 147
geminus, Nyctimene 144
gemmeus, Sylvisorex 105
gendagendae, Sigmoceros 395
gendrelianus, Naemorhedus 407
genei, Pipistrellus 223
genepaiensis, Pongo 277
generatius, Rattus 661
Genetta 341, **345**, 345, 347
abyssinica **345**
aequatorialis 346
afra 345
albipes 345
albiventris 346
amer 346
angolensis **345**
aubryana 346
balearica 345
barbar 345
barbara 345
bella 345
bettoni 346
bini 346
bonapartei 345
communis 345

cristata 346
deorum 346
dongolana 345, 346
dubia 346
erlangeri 346
felina 345, 346
fieldiana 346
gallica 345
genetta 303, **345**, 346
genettoides 346
gleimi 346
granti 345
grantii 345
guardafuensis 345
hararensis 345
hintoni 345
hispanica 345
insularis 346
intensa 346
isabelae 345
johnstoni **346**
lehmanni 346
leptura 345
letabae 346
loandae 346
ludia 345
lusitanica 345
macrura 345
maculata **346**, 346, 347
matschiei 346
melas 345
methi 347
mossambica 345
neumanni 345
pantherina 346
pardina 346
peninsulae 345
poensis 346
pulchra 345
pumila 346
pyrenaica 345
rhodanica 345
richardsonii 347
rubiginosa 346
schraderi 346
senegalensis 345, 346
servalina **346**, 346
soror 346
stuhlmanni 346
suahelica 346
tedescoi 345
terraesanctae 345
thierryi **346**
tigrina **346**, 346
victoriae **347**
villersi 346
vulgaris 345
zambesiana 346
zuluensis 346
genetta, Genetta 303, **345**, 346
Viverra 345
genettoides, Genetta 346
genibarbis, Petinomys **463**, 464
genovensium, Platalina **182**, 182
genowaysi, Rhogeessa **225**
gentilis, Hipposideros 171, 174
Mus 625
Peromyscus 731
gentilulus, Mus 625
Geocapromys **800**, 800
brachyurus 801
brownii 800, **801**, 801
columbianus 800
ingrahami **801**
megas 801
pleistocenicus 800
thoracatus **801**, 801
geoffrensis, Delphinus 360
Inia **360**
geoffroii, Dasyurus **31**, 32
geoffroyi, Anoura **183**, 183
Ateles **257**
Callicebus 259
Callithrix **251**, 252
Felis 294
Galago 249
Lagothrix 258
Leopardus 292
Nyctophilus **218**, 218
Oncifelis 291, 292, **294**, 294
Perodicticus 248
Rousettus 152
Saguinus **253**, 254
Suncus 103
Syncerus 402
Tadarida 240
Geogale **70**
aurita **70**, 70
australis 70
orientalis 70
Geogalinae **70**
Geomyidae **469**, 469
Geomyinae 469
Geomyini 469
Geomys **469**, 469
alba 469
ammophilus 469
arenarius **469**, 469
attwateri 469, 470
austrinus 470
brazensis 469
breviceps 469, 470
brevirostris 469
bulleri 471
bursarius **469**, 469, 470
canadensis 469
colonus 470
cumberlandius 470
dutcheri 469
fallax 470
floridanus 470
fontanelus 470
fusca 469
fuscus 470
goffi 470
hylaeus 469
illinoensis 469
industrius 469
jugossicularis 469
knoxjonesi **469**, 470
levisagittalis 469
llanensis 469
ludemani 469
lutescens 469, 470
major 469
majusculus 469
maritimus 470
megapotamus 470
missouriensis 469
mobilensis 470
personatus 469, **470**
pinetis 469, **470**, 470
pratincola 469
saccatus 469
sagittalis 469
scalops 470
streckeri 470
terricolus 469
texensis 469, 470
tropicalis 469, **470**
tuza 470
vinaceus 469
wisconsinensis 469
georgianus, Taphozous **160**, 160
georgiensis, Peromyscus 732
georgihernandezi, Sciurus 439
Georychinae 771
Georychus 771, **772**
buffonii 772
canescens 772
capensis **772**
holosericeus 771
leucops 772
luteus 513
yatesi 772
Geosciurus 458
Geoxus 699, **702**, 714
araucanus 702
bicolor 702
bullocki 702
chiloensis 702
fossor 702
michaelseni 702
microtis 702
valdivianus **702**
gephyreus, Tursiops 357
gerbei, Microtus 520, **521**, 521, 524
Gerbillidae 546
Gerbillinae 501, **546**
gerbillinus, Mus 625
Gerbilliscus 560, 561
Gerbillurus **547**, 547, 548
broomi 547
calidus 547
coombsi 547
exilis 547
infernus 547
kalaharicus 547
leucanthus 547
mulleri 547
oralis 547
paeba **547**, 547
robusta 547
seeheimi 548
setzeri **547**, 548
swakopensis 547
swalius 547
tenuis 547
tytonis **548**, 548
vallinus 547, **548**
Gerbillus **548**, 548, 559, 563, 574
acticola **548**, 548
aegyptius 548, 551
aeruginosus 551
agag **548**, 548, 550, 553
allenbyi **549**, 549
amoenus **549**, 553, 555
andersoni **549**, 549
aquilus **549**
arabium 553
arduus 550
asyutensis 551
aureus 552
auricularis 547
bilensis **549**, 554
blanci 549
bonhotei **549**, 549, 552
bottai **549**, 549, 551
brockmani **549**, 549
brunnescens 550
burtoni **549**
calurus 559
campestris 549, **550**, 550, 552, 555
cheesmani **550**, 550, 553
cinnamomeus 550
cosensi 548, **550**
dalloni **550**
dasyroides 550
dasyurus 549, **550**, 550, 555
diminutus **550**
discolor 551
dodsoni 550
dongolanus **550**
dunni **551**, 551, 552
eatoni 549
elbaensis 554
emini 562
famulus **551**
favillus 552
floweri **551**
foleyi 551
gallagheri 550
garamantis **551**
gedeedus 554
gerbillus **551**, 551
gleadowi **551**
grobbeni **551**, 551
hamadensis 555
harwoodi 549, **551**
haymani 550
henleyi **552**
hesperinus **552**
hilda 553

hirtipes 551
hoogstraali **552**
imbellis 546
indus 553
inflatus 549
jamesi **552**
jordani 552
juliani **552**, 552
kaiseri 554
latastei 551, **552**, 552, 554
leosollicitus 550
lixa 550, 553
longicaudatus 551
lowei **552**
luteolus 549, 555
luteus 551
mackillingini **552**
maghrebi **552**
makrami 552
maradius 548
mariae 552
maritimus 550
mauritaniae **553**
mesopotamiae **553**
mimulus 553
muriculus **553**
nalutensis 552
nancillus **553**
nanus 549, 550, 551, 552, **553**, 553, 554, 555
nigeriae 548, **553**
occiduus **553**
palmyrae 550
patrizii 550
peeli 558
percivali **553**
perpallidus 552, **553**
poecilops **553**
principulus **554**
przewalskii 546
psammophilous 551
pulvinatus 549, **554**
pusillus 550, 553, **554**
pyramidum 548, 549, 550, 551, **554**, 554, 555, 563
quadrimaculatus **554**
riggenbachi 552, **554**
riparius 550
rosalinda 552, **554**
rozsikae 550
ruberrimus **554**
setonbrownei 553
simoni **554**
somalicus 550, **555**
stigmonyx **555**
subsolanus 549
sudanensis 548, 551
syrticus **555**
tarabuli **555**, 555
vallinus 547
venustus 550
vivax **555**
wassifi 550
watersi 552, 554, **555**
zakariai 554
gerbillus, Dipus 548
Gerbillus **551**, 551
Mus 629
Phyllotis **738**, 738
gerboa, Jaculus 494
Gerboides 53
germaini, Akodon 692
Callosciurus 422
Pteropus 152
Rattus 660
Trachypithecus 273
germana, Micoureus 20
germanicus, Cervus 386
Cricetus 538
Mus 625
Rhinolophus 166
germanus, Cerdocyon 282
germinans, Crocuta 308
Geromys 649
geronimensis, Peromyscus 732
Phoca 332
gerrardi, Nandinia 342
Sciurus **441**
gerritmilleri, Microtus 524
gervaisi, Mesoplodon 362
gestri, Rattus 659
gestroi, Rattus 659
getulus, Atlantoxerus **420**
Octodon 788
Sciurus 420
ghalgai, Microtus 529
ghansiensis, Ictonyx 319
ghia, Apodemus 573
ghidinii, Myotis 208
ghigii, Lepus 817
Otomys 681
giaraiensis, Bandicota 579
gibbar, Balaenoptera 349, 350
gibbon, Hylobates 276
gibbosa, Addax 412
gibbosus, Eschrichtius 351
gibbsii, Neurotrichus **126**
Urotrichus 126
gibsoni, Conepatus 316
Sylvilagus 825
gichiganus, Lepus 821
gichigensis, Phoca 332
giffardi, Crocidura 92
Tatera 561
gigantea, Bandicota 579
Manis **415**
Ratufa 437
giganteus, Ictonyx 319
Macropus 53, **54**, 54
Malacomys 607, 608
Mus 578
Priodontes 67
Pteropus **147**
Rhombomys 559
Spalax 753, **754**
Spermophilus 446
Suncus 103
gigantica, Pongo 277
Gigantomys 53
gigas, Alces 389
Balaenoptera 350
Canis 281
Dasypus 66
Dinomys 778
Elephas 367
Eumops 234
Gorilla 276
Grammomys **593**
Hipposideros 172
Hydrodamalis **365**
Leopoldamys 603
Lynx 293
Macroderma **163**
Manati 365
Megaderma 163
Megasorex **109**, 110
Notiosorex 109
Priodontes 67
Scotophilus 227
Sus 379
Taurotragus 403
Tragelaphus 403
gigliolii, Hylochoerus 377
giglis, Myoxus 770
gigot, Callicebus 259
gilberti, Hipposideros 171
Peromyscus 736
Sminthopsis **36**
gilbertii, Potorous 50
gilbiventer, Niviventer 633
gilesi, Planigale **34**
gilgitensis, Capra 405
gilgitianus, Rattus 661
giliacus, Apodemus 572
gillespiei, Lasiorhinus 45
gillespii, Zalophus 329
gilli, Tatera 560
gilliardi, Pteropus **147**
gilliespii, Otaria 329
gillii, Tursiops 357
gilmorei, Microtus 527
giloensis, Otomys 682
giluwensis, Rattus 649, **653**
gilva, Neotoma 712
gilvigularis, Sciurus 439, **441**, 441
gilvipes, Euphractus 66
gilviventris, Sciurus 441
gilvus, Acerodon 137
Mus 625
Perognathus 486
gina, Gorilla 276
ginginianus, Callosciurus 423
Xerus 458
ginkgodens, Mesoplodon **363**
Giraffa **383**
aethiopica 383
angolensis 383
antiquorum 383
australis 383
biturigum 383
camelopardalis **383**
capensis 383
congoensis 383
cottoni 383
giraffa 383
hagenbecki 383
infumata 383
maculata 383
nigrescens 383
peralta 383
renatae 383
reticulata 383
rothschildi 383
schillingsi 383
senaariensis 383
thornicrofti 383
tippelskirchi 383
wardi 383
giraffa, Giraffa 383
Giraffidae 383
girbaensis, Atelerix 77
girensis, Rattus 658
glaber, Heterocephalus **772**, 772
glacialis, Clethrionomys 509
Eubalaena **349**, 349
Lepus 815
Sorex 118
Thomomys 475
glacilis, Ursus 338
Gladiator 354
gladiator, Orcinus 355
glama, Camelus 381
Lama **381**, 382
glandifera, Eonycteris 154
glaphyrus, Oligoryzomys 718
glareolus, Clethrionomys 502, **508**, 508
Glareomys 507
glasselli, Peromyscus 729
glassi, Crocidura **85**
glauca, Hippotragus 412
glaucillus, Pipistrellus 222
glaucinus, Akodon 693
Eumops **233**
Sciurillus 438
Glaucomys **460**
alpinus 460
americana 460
bangsi 460
bullatus 460
californicus 460
canadensis 460
canescens 460
chontali 460
coloratus 460
columbiensis 460
cucullatus 460
flaviventris 460
fuliginosus 460
fuscus 460
goldmani 460
goodwini 460
gouldi 460
griseifrons 460
guerreroensis 460
herreranus 460
hudsonicus 460
klamathensis 460
lascivus 460

latipes 460
lucifugus 460
macrotis 460
madrensis 460
makkovikensis 460
murinauralis 460
nebrascensis 460
oaxacensis 460
olympicus 460
oregonensis 460
querceti 460
reductus 460
sabrinus **460**
saturatus 460
silus 460
stephensi 460
texensis 460
underwoodi 460
virginianus 460
volans **460**
volucella 460
yukonensis 460
zaphaeus 460
Glauconycteris 199, 200
glaucula, Leopardus 292
glaucus, Artibeus **188**, 188
Eschrichtius 351
Pseudomys **646**
glauerti, Rattus 653
gleadowi, Gerbillus **551**
Millardia **621**, 621
glebula, Crocidura 93
gleimi, Genetta 346
gleni, Chalinolobus **199**
Glires 771
Gliridae 763
gliriformis, Cercartetus 58
glirina, Monodelphis 21
Glirinae 769
glirinus, Phyllotis 738
Zenkerella 758
Gliriscus 763
gliroides, Chiropodomys **583**, 584, 614, 634
Dromiciops **27**, 27
Microcebus 243
Mus 583
Octodon 788
Octodontomys **788**
Glironia **16**
aequatorialis 16
criniger 16
venusta **16**, 16
Glirulinae 769
Glirulus 763, **769**, 769
elegans 769
japonicus **769**, 769
javanicus 769
lapsus 769
lasiotis 769
Glis 430, 763, 769, 770
glis, Myoxus **770**
Sciurus 769
Sorex 131
Tupaia **132**, 132
Gliscebus 243
Glischropus **204**, 214
batjanus 204
javanus **204**
tylopus **204**
Glisoricina 131
Globicephala **352**
brachypterus 352
edwardii 352
globiceps 352
leucosagmaphora 352
macrorhynchus **352**
melaena 352
melas **352**, 352
scammonii 352
sieboldii 352
svineval 352
Globicephalidae 351
Globiceps 352
globiceps, Delphinus 352
Globicephala 352
Lepilemur 246
globulus, Tolypeutes 67
glogeri, Mustela 322
gloriaensis, Oryzomys 720
Glossophaga **184**
alticola 184
amplexicaudata 184
antillarum 184
bakeri 184
brevirostris 184
campestris 184
caudifer 184
commissarisi **184**
elongata 184
handleyi 184
hespera 184
leachii **184**
longirostris **184**
major 184
mexicana 184
microtis 184
morenoi **184**
mutica 184
reclusa 184
rostrata 184
soricina **184**, 184
truei 184
valens 184
villosa 184
Glossophaginae **183**
gloveralleni, Procyon **335**, 335
Sorex 119
gloveri, Apodemus 570
Callosciurus 421
Ochotona 808, **809**, 809, 810, 811
Glyphidelphis 357
Glyphonycteris 178, 179
Glyphotes **428**
canalvus 423
simus **428**, 428
gmelini, Equus 369
Ovis 408
Sorex 117
gnambiquarae, Kunsia 706
gnomus, Artibeus 188
gnou, Antilope 393
Connochaetes **393**
gobabis, Lycaon 283
gobicus, Lepus 821
Salpingotus 491
gobiensis, Eptesicus 202
godeffroyi, Otaria 328
godmani, Antechinus **30**
Choeroniscus **183**
Petrogale **56**, 56
Sorex 120
godonga, Sigmoceros 395
godowiusi, Sigmoceros 395
goeldi, Oryzomys 721
goeldii, Callimico **251**, 251
Proechimys **796**, 798
goethalsi, Hoplomys 794
goffi, Geomys 470
goldfinchi, Equus 369
goldiei, Spilocuscus 47
goldmani, Canis 280
Chaetodipus **483**, 484
Cryptotis **108**, 108
Dipodomys 478
Glaucomys 460
Heteromys **481**
Mustela 322
Nelsonia 709, **710**
Neotoma **711**
Ochotona 811
Ondatra 532
Oryzomys 721
Panthera 298
Pappogeomys 472
Peromyscus 733, 734
Proechimys 798
Reithrodontomys 742
Sciurus 443
Sigmodon 747
Spermophilus **444**
Sylvilagus 824
Thomomys 476
goliath, Cricetomys 541
Crocidura **86**
Hyomys **600**
Mus 599
golock, Hylobates 275
Golunda 578, **592**, 592, 594, 629
bombax 592
coenosa 592
coffaeus 592
coraginis 592
dybowskii 629
ellioti **592**, 592
gujerati 592
hirsutus 592
limitaris 592
meltada 621
myothrix 592
newara 592
nuwara 592
paupera 592
watsoni 592
golzmajeri, Sciurus 443
gombensis, Sigmoceros 395
gomphus, Sorex 112
gonavensis, Pteronotus 177
gondokorae, Mus 625
gongshanensis, Callosciurus 421
Muntiacus **389**
gonshanensis, Tupaia 131
gonzalesi, Chrotomys **585**
gonzalinus, Cervus 386
goodei, Kogia 359
goodfellowi, Ctenomys 784
Dendrolagus **50**, 51
Rhipidomys 744
Zaglossus 13
goodpasteri, Perognathus 485
goodwini, Cryptotis **108**, 108
Glaucomys 460
goojratensis, Panthera 297
goral, Antilope 406
Naemorhedus **407**, 407
Nemorhaedus 407
goramensis, Sus 379
gordoni, Callosciurus 421
Diceros 372
Jaculus 493
gordonorum, Procolobus 272
gorgasi, Chiroderma 190
Oryzomys **722**
Gorgon 393
gorgon, Bradypus 63
Connochaetes 394
gorgonae, Proechimys **796**
goriensis, Microtus 530
Gorilla **276**
adrotes 276
berengei 276
beringei 276
castaneiceps 276
diehli 276
ellioti 276
gigas 276
gina 276
gorilla **276**
graueri 276
halli 276
hansmeyeri 276
jacobi 276
manyema 276
matschiei 276
mayema 276
mikenensis 276
rex-pygmaeorum 276
savagei 276
schwartzi 276
uellensis 276
zenkeri 276
gorilla, Gorilla **276**
Troglodytes 276
gorka, Clethrionomys 508
gorkhali, Petaurista 462
gorongozae, Sigmoceros 395
gorrichanus, Cervus 386
goslingi, Colomys **586**, 586
Ourebia 399
Procavia 374

gossei, Chroeomys 700
Funambulus 426
Tadarida 240
gossii, Synaptomys 534
gossypinus, Peromyscus **730**, 730, 732
gothneh, Mungos 307
gotschobi, Chionomys 507
gotthardi, Sciurus 443
gotthardus, Eliomys 767
gouazoubira, Mazama 391
gouazoupira, Mazama **391**, 391
goudoti, Myotis **211**
goudotii, Eupleres **340**, 340
Odocoileus 391
gouldi, Glaucomys 460
Notomys 636
Nyctophilus **218**, 218
Pteropus 146
gouldii, Bettongia 49
Chalinolobus **199**
Hapalotis 619
Mesembriomys 586, **619**, 620
Pseudomys 645, **646**, 647, 648
gouliminensis, Crocidura 96
goundae, Mus **623**, 623, 628
gouweri, Pteropus 146
goyana, Monodelphis 22
gracilicauda, Soriculus 122
gracilicaudatus, Pseudomys **647**, 648
gracilicornis, Gazella 397
Gracilinanus **17**
aceramarcae **17**
agilis **17**, 17
agricolai 17
beatrix 17
blaseri 17
buenavistae 17
chacoensis 17
dryas **17**
emiliae **17**
guahybae 17
herhardti 17
marica **17**
microtarsus **17**, 17
muscula 17
peruana 17
rondoni 17
unduaviensis 17
gracilior, Eptesicus 201
gracilipes, Calomys 698
Crocidura **86**, 87, 95
gracilis, Alticola 503
Arctocephalus 326
Bandicota 579
Cebus 259
Cervus 386
Cryptotis **108**, 108
Cynocephalus 135
Dasyurus 32
Enhydra 310
Galerella 303
Hipposideros 171
Lestoros 25
Loris 248
Macropus 53
Melomys 615, **616**, 616
Mesocapromys 800
Microgale **71**
Mustela 322, 324
Myotis 209
Niviventer 634
Nyctinomops 239
Odocoileus 391
Oryzomys 720
Peromyscus 732
Petaurus **61**
Prionodon 347
Pseudalopex 284
Reithrodontomys **741**, 741, 742
Rhinolophus 168
Rhogeessa **225**
Saguinus 254
Spilogale 317
Tamias 455
Taterillus **562**, 562, 563
Thomasomys **749**
Thomomys 475
Thylogale 57
Tupaia **132**
Uropsilus **130**, 130
Vicugna 382
gracillimus, Sorex **115**, 116, 118
gradojevici, Spermophilus 445
graeca, Sciurus 443
graecus, Canis 280
Spalax 753, **755**, 755
graellsi, Saguinus 254
graffmani, Stenella 356
grahamensis, Tamiasciurus 457
Thomomys 473
grahami, Aethomys 568
Cricetomys 541
Lepus 819
Mus 627
graicus, Capra 405
graii, Bettongia 49
Grammogale 321, 322
Grammomys **592**, 592, 593, 594, 631, 669
angolensis 593
arborarius 593
aridulus **592**
baliolus 593
buntingi 590, **592**, 592, 596, 599, 601, 607, 631, 644
callithrix 594
caniceps **592**
centralis 594
cometes **592**, 593, 594
discolor 593
dolichurus 592, **593**, 593, 594
dryas **593**, 593
elgonis 593
erythropygus 594
gazellae 594
gigas **593**
ibeanus **593**, 593, 594
insignis 593
kuru 594
littoralis 593
macmillani 592, **594**, 594, 631
minnae **594**
oblitus 594
ochraceus 594
poensis 594
polionops 593
rutilans **594**, 594
silindensis 593
surdaster 593
tongensis 593
usambarae 594
vumbaensis 594
vumbensis 594
grammurus, Spermophilus 450
grampia, Felis 290
Grampidelphidae 351
Grampidelphis 353
Grampus **353**, 354
griseus **353**
intermedius 352
rectipinna 353
rissoanus 353
stearnsii 353
granadiana, Dasypus 66
granarius, Sorex 113, **115**
granatensis, Lepus 814, 816, **818**
Sciurus **441**
Syntheosciurus 452
grandiceps, Crocidura **86**
grandicornis, Kobus 414
Muntiacus 389
grandiculus, Apodemus 573
grandidiensis, Galidictis 300
grandidieri, Eptesicus 201
Galidictis **300**, 300
Lepilemur 246
grandis, Callicebus 259
Calomyscus 535
Capreolus 390
Crocidura **86**
Dobsonia 140
Echimys **792**
Epomophorus **141**
Euroscaptor **125**
Herpestes 305
Hipposideros 173
Hylomyscus 598
Ichneumia 306
Loris 248
Macrotis 40
Marmosa 18
Maxomys 614
Meriones 558
Micropteropus 141, 143
Microtus 519
Miniopterus 231
Myotis 216
Nectomys 709
Neotoma 710
Nycteris **161**
Orthogeomys **471**, 471
Peromyscus **730**, 731, 737
Petaurista 463
Priodontes 67
Pteropus 150
Ursus 339
grangeri, Neotoma 711
Sylvilagus 826
granti, Aethomys 567, **568**, 568
Apodemus 573
Batomys 579, **580**, 580
Chrysochloris 76
Equus 369
Eremitalpa **76**
Galago **249**
Galagoides 250
Galerella 303
Gazella **396**
Genetta 345
Heterohyrax 373
Lepus 816
Rangifer 392
Saccolaimus 159
Sylvisorex **104**, 105
grantii, Genetta 345
Otomys 682
Sorex 111
granulipes, Sminthopsis **36**
granulosus, Cervus 386
Graomys 694, **702**, 702, 703, 737
cachinus 703
centralis 703
chacoensis 703
domorum **703**, 703
edithae **703**
griseoflavus **703**, 703
hypogaeus 701, 703
lockwoodi 703
medius 703
pearsoni 703
taterona 703
Graphiurinae **763**, 763
Graphiurus **763**, 763
alticola 764
angolensis 765
ansorgei 764
argenteus 764
australis 765
brockmani 765
butleri 764
capensis 765
christyi **763**
cineraceus 764
cinerascens 764
collaris 764
coupeii 764
crassicaudatus **763**
dasilvai 764

dollmani 765
dorotheae 763
eastwoodae 765
elegans 765, 769
erythrobronchus 764
etoschae 764
foxi 765
griselda 764, 765
haedulus 764, 765
hueti **764**, 764
internus 765
isolatus 764
johnstoni 764
jordani 765
kaokoensis 765
kelleni **764**, 765
lalandianus 764
littoralis 764
lorraineus **764**, 765
marrensis 764
microtis **764**, 764, 765
monardi **764**, 764
montosus 765
murinus 763, **764**, 764, 765
nagtglasii 764
nanus 764
ocularis **765**
olga **765**, 765
parvulus 765
parvus 764, **765**, 765
personatus 765
platyops **765**, 765
pretoriae 764
raptor 764
rupicola **765**, 765
saturatus 764
schneideri 764
schoutedeni 764
schwabi 765
selindensis 764
soleatus 764
spurrelli 764
streeteri 764
sudanensis 764
surdus **765**, 765
typicus 765
tzaneenensis 764
vandami 764
vulcanicus 764
woosnami 764
zuluensis 764
grassei, Crocidura **86**
grassianus, Cervus 386
grata, Mus 625
gratiosus, Proechimys 797
gratula, Suncus 102
gratulus, Suncus 102
gratus, Mus 625
Peromyscus **731**, 731, 737
Tragelaphus 404
graueri, Chlorocebus 265
Gorilla 276
Pan 277
Papio 269
Paracrocidura **100**, 101
Procolobus 272
gravesi, Microtus 530
Sorex 116
gravida, Crocidura 89
gravipes, Dipodomys **478**
gravis, Quemisia **805**, 805
grayi, Arctocephalus 326
Canis 280
Carollia 186
Cercopithecus 265
Crocidura **86**
Hemiechinus 78
Heterohyrax 373
Lasiurus 206
Mesoplodon **363**
Ourebia 399
Paguma 343
Panthera 298
Raphicerus 399
Sundasciurus 451
grayiformes, Cuon 282
grayii, Kogia 359
Lutra 312
Grayius 353
graysoni, Sylvilagus **826**
greatheadi, Equus 370
grebenscikovi, Dinaromys 511
greenhalli, Centurio 189
Molossops **234**
greeni, Petromus 774
Puma 296
greenwoodi, Crocidura **86**
gregalis, Microtus 518, **521**, 521, 525
gregarius, Microtus 518
gregorianus, Thryonomys **775**
gregoryi, Canis 281
grenadae, Marmosa 18
grenadensis, Artibeus 188
Grevya 369
grevyi, Equus **370**
greyi, Equus 370
Macropus **54**
greyii, Nycticeius **217**
Rattus 653
griffithi, Nesokia 632
Vulpes 287
griffithii, Leopardus 291
Suncus 103
Vulpes 287
grilloanus, Cervus 386
Grimmia 412
grimmia, Capra 412
Sylvicapra **412**
grinnelli, Microtus 519
Neotoma 712
Procyon 335
Scapanus 127
Tamias 453
griquae, Rhabdomys 662
Tatera 560
griquoides, Rhabdomys 662
grisea, Aplodontia 417
Crocidura 82
Cynomys 424
Damaliscus 394
Murina **229**
Raphicerus 400
Thylamys 23
Trichosurus 48
griseicauda, Callosciurus 423
Stenocephalemys **664**
griseicollis, Ratufa 437
griseifrons, Dasymys 589
Glaucomys 460
griseimanus, Callosciurus 421
griseimembra, Aotus 256
Sciurus **441**
griseipectus, Rattus 657
griseipes, Papio 269
griseistictus, Chlorocebus 265
griseiventer, Molossops 235
Petaurista 463
Rattus 660
griseiventris, Cricetulus 538
griselda, Anomalurus 757
Blarinella 106
Dremomys 425
Felis 290
Graphiurus 764, 765
Lemniscomys **601**, 602, 640
griseoalbus, Canis 281
griseobracatus, Peromyscus 735
griseocaeruleus, Rattus 658
griseocaudatus, Sciurus 443
griseoflava, Felis 290
griseoflavus, Graomys **703**, 703
Mus 702
Phyllotis 738
Reithrodontomys 741
Sciurus 440
griseofuscus, Macropus 54
griseogena, Sciurus 441
griseogenys, Maxomys 612
griseogularis, Neomys 110
griseolanosus, Macropus 55
griseolus, Oligoryzomys **718**
griseopectus, Callosciurus 421
griseorufus, Microcebus 243
griseotinctus, Kobus 413
griseoventer, Neotoma 712
Sminthopsis **36**
griseoventris, Cryptotis 108
griseovertex, Saguinus 253
griseoviridis, Chlorocebus 265
grisescens, Allactaga 489
Ateles 257
Baiomys 695
Crocidura 85
Myotis **211**
Philander 22
Tamias 454
grisescenti, Pseudocheirus 59
griseus, Apodemus 573
Bradypus 63
Cebus 260
Cheirogaleus 243
Cricetulus 537, 538
Delphinus 353
Dusicyon 283
Grampus **353**
Halichoerus 329, 330
Hapalemur **245**, 245
Herpestes 305
Heteromys 481
Hipposideros 172
Lemur 245
Leopardus 291
Lepus 815
Macropus 55
Naemorhedus 407
Ochotona 810
Peromyscus 734
Pipistrellus 223
Prosciurillus 436
Pseudalopex **284**
Pteronotus 177
Pteropus **147**
Reithrodontomys 742
Sciurus **441**
Sigmodon 747
Tamias 456
Thylamys 23
Ursus 339
Zygodontomys 752
Grison 318
grisonax, Mungos 307
Grisonella 318
Grisonia 318
griswoldi, Tupaia 133
grivaudi, Miniopterus 231
grobbeni, Gerbillus **551**, 551
groenlandica, Phoca **331**
groenlandicus, Alopex 279
Dicrostonyx **510**, 510
Lepus 815
Rangifer 392
Ursus 339
grootensis, Hydromys 597
grotei, Hystrix 773
Sigmoceros 395
growlerensis, Thomomys 473
gruberi, Soriculus 123
grunniens, Bos **401**, 401
grutei, Callosciurus 422
Grymaeomys 18
gryphus, Myotis 212
Grypomys 621
grypus, Halichoerus **329**, 329, 332
Phoca 329
Grysbock 399
guadalupensis, Microtus 524
Thomomys 473
guadeloupensis, Eptesicus **202**
guadricolor, Martes 320
guahybae, Gracilinanus 17
guairae, Proechimys **796**, 796
guajanensis, Sciurus 439

gualea, Mazama 390
gualeae, Nasua 335
guamara, Dasyprocta **782**
guanicoe, Lama **382**
guanta, Agouti 783
guapo, Pithecia 262
guaraxa, Cerdocyon 282
guardafuensis, Genetta 345
guardia, Peromyscus **731**, 731
guardiae, Chaetodipus 484
guariba, Alouatta 255
guatemalae, Ototylomys 726
Urocyon 285
guatemalensis, Microtus **521**, 526
Peromyscus **731**, 731, 734, 737
guavera, Bos 401
guayanus, Caluromys 15
Sciurus 443
guaycuru, Myotis 214
guba, Xenuromys 673
gubanensis, Madoqua 398
gubari, Pteromys 464
gubernator, Lagenorhynchus 353
gud, Chionomys **507**, 507
gudauricus, Microtus 526
gudoviusi, Procolobus 272
gueinzii, Dasymys 589
gueldenstaedtii, Crocidura **86**, 86, 96
guentheri, Madoqua **398**
Microtus **521**, 521, 522, 530
Trichys 774
Guepar 288
Guepardus 288
Guereza 269
guereza, Colobus **270**
Guerlinguetus 439, 441, 442, 443
guerlingus, Sciurus 439
guerrerensis, Cryptotis 108
Hodomys 704
Liomys 482
Orthogeomys 471
Oryzomys 721
Potos 333
Sigmodon 747
guerreroensis, Glaucomys 460
guevaranus, Cervus 386
Guevei 410
guhai, Rattus 656
guianae, Cavia 779
Chrotopterus 177
Holochilus 705
Makalata 793
Neacomys **708**
Oecomys 716
Tapirus 371
guianensis, Chironectes 16
Delphinus 355
Marmosa 18
Sciurus 439
Sotalia 355
guiara, Euryzygomatomys 794
guidoteanus, Cervus 386
guigna, Oncifelis 291, 292, **294**, 294
guillarmodi, Chlorotalpa 75
guillemardi, Callosciurus 422
guina, Oncifelis 294
guinasensis, Petromus 774
guineae, Tatera **560**
guineensis, Crocidura 92
Epomophorus 141
Eptesicus **202**
Hipposideros 174
Nycteris 162
Perodicticus 248
Rhinolophus **166**, 166
gujanensis, Chironectes 16
Galictis 318
gujerati, Golunda 592
gularis, Aotus 256
Bradypus 63
Dremomys 425
Procyon 335
Proechimys **796**
gulata, Felis 290
gulina, Eira 318
Gulo 317, **318**, 318
arcticus 318
arctos 318
auduboni 318
bairdi 318
biedermanni 318
borealis 318
gulo **318**, 318
hylaeus 318
kamtschaticus 318
katschenakensis 318
larvatus 343
luscus 318
luteus 318
niediecki 318
sibirica 318
sibiricus 318
vancouverensis 318
vulgaris 318
wachei 318
gulo, Gulo **318**, 318
gulosa, Mustela 321
gulosus, Perognathus 485
gumillae, Conepatus 316
gundi, Ctenodactylus **761**, 761
Mus 761
gundlachi, Mysateles **802**, 802, 803
gundlachianus, Capromys 800
gungunyanae, Lepus 820
gunnii, Perameles **40**
gunningi, Amblysomus **74**
gunnisoni, Cynomys **424**
Peromyscus 732
Gunomys 578, 579
gunong, Sundasciurus 452
guntheri, Mesoplodon 363
gunung, Mallomys **608**
gurganensis, Meriones 557
gurkha, Apodemus **571**
Mus 627
gutsenensis, Equus 369
guttata, Acinonyx 288
guttatus, Dasyurus 32
Spermophilus **449**
guttula, Leopardus 292
guttulatus, Spermophilus **449**
gutturosa, Procapra **399**
guyannensis, Leontopithecus 253
Proechimys 795, 796, 797, 798
Trichechus 365
Guyia 621
guyonii, Meriones 556
gwatkinsii, Martes **320**
gyas, Taterillus 562
Ursus 339
gymnesicus, Eliomys 767
gymnicus, Tamiasciurus 457
Gymnobelideus **61**
leadbeateri **61**, 61
gymnocercus, Dusicyon 283
Pseudalopex **284**
Gymnomys 666, 671
gymnonotus, Pteronotus **176**
Gymnopyga 266, 267
Gymnotis 391
gymnotis, Odocoileus 391
Peromyscus **731**, 731, 734
Strigocuscus 47, **48**
Gymnura 79
gymnura, Echinosorex **79**, 79
Saccopteryx **159**
Viverra 79
Gymnurinae 79
Gymnuromyinae 678
Gymnuromys **678**
roberti **678**, 678
gymnurus, Cabassous 65
Echimys 794
Echinosorex 79
Hoplomys **794**
Pappogeomys **472**
Sorex 117
Tylomys 751
Gyomys 644, 646, 649
gypsi, Perognathus 485
gyrator, Pseudocheirus 59

haagneri, Cricetomys 541
Ichneumia 306
Mormopterus 238
haasti, Mesoplodon 363
habbema, Antechinus 30
Hydromys **597**, 597, 598
habessinica, Procavia 374
habessinicus, Lepus 816
Procavia 374
habessynica, Crocuta 308
habetensis, Oryctolagus 822
habibi, Lepus 821
Habrocebus 246
Habrocoma 789
Habromys **703**, 703, 713, 728, 739
chinanteco **703**
ixtlani 703
lepturus **703**, 703
lophurus **703**, 703
simulatus **703**, 703
hacketti, Petrogale 56
hadramauticus, Canis 280
Hadromys 592, **594**, 594
humei **594**, 595
loujacobsi 594
yunnanensis 594, 595
Hadrosciurus 439, 440, 442
hadrourus, Melanomys 672
Uromys **671**, 672
haedinus, Taphozous 161
haedulus, Graphiurus 764, 765
haematoreia, Neotoma 711
haemobaphes, Callosciurus 421
Haeromys 583, **595**
margarettae **595**, 595
minahassae **595**
pusillus **595**
hagenbecki, Antilope 395
Capra 406
Cercocebus 262
Cervus 386
Equus 369
Giraffa 383
Macropus 54
Mandrillus 268
hageni, Antechinus 31
Dorcopsis **51**
Melomys 619
Petinomys **464**, 464
Rattus 660
haggardi, Ourebia 399
Phyllotis **738**
hahlovi, Microtus 527
hahni, Suricata 308
haidarum, Mustela 321
haigi, Ctenomys **785**, 785
hainaldi, Rattus 649, **653**
hainana, Cervus 387
Paguma 343
Petaurista 463
Ratufa 437
hainanensis, Hylomys **79**
Leopoldamys 603
hainanicus, Rattus 659
hainanus, Atherurus 773
Cervus 386
Hylobates 275
Lepus **818**
Rhinolophus 163
Tamiops 457
haitiensis, Phyllops 190
hajastanicus, Myotis 213
halconus, Crocidura 86
haldemani, Epomophorus 141

Halibalaena 349
Halichoerus **329**, 329
atlantica 330
baltica 330
griseus 329, 330
grypus **329**, 329, 332
halichoerus 330
macrorhynchus 330
pachyrhynchus 330
halichoerus, Halichoerus 330
halicoetes, Reithrodontomys 742
Sorex 122
Halicore 365
Halicoridae 365
Halipaedisca 365
Halitheriidae 365
hallensis, Alopex 279
halli, Dipus 493
Gorilla 276
Lestodelphys **17**
Microtus 523
Notodelphys 17
Taxidea 325
Hallomys 679
audeberti 679
hallorani, Taxidea 325
hallucalis, Clethrionomys 508
hallucatus, Dasyurus **32**, 32
Halmatopus 53
Halmaturus 53, 57
thetis 57
halonifer, Trachypithecus 274
halophilus, Microtus 519
halophyllus, Hipposideros **173**
Halticus 494
halticus, Allactaga 489
Stylodipus 495
Haltomys 493
hamadensis, Gerbillus 555
Hamadryas 269
hamadryas, Papio **269**
Simia 269
hamar, Equus 370
hamatus, Taeromys **667**
hamica, Ochotona 811
hamiltoni, Apodemus 573
Eliomys 767
Leopardus 291
Leptailurus 292
Suricata 308
Taphozous **160**
Tragelaphus 404
Ursus 338
hamiltonianus, Cervus 387
hamiltonii, Nandinia 342
Paradoxurus 344
hamlyni, Cercopithecus **264**
Lophocebus 266
hammondi, Oryzomys 707, **722**
hamptoni, Mustela 324
Hamster 538
hamulicornis, Rupicapra 410
handleyi, Baiomys 695
Glossophaga 184
Lonchophylla **181**
Marmosops **19**
hanensis, Erinaceus 77
Herpestes 306
Lutra 312
Meles 314
Panthera 298
Viverricula 348
hangaicus, Lasiopodomys 515
Myospalax 675
hanglu, Cervus 385, 386
hanieli, Paradoxurus 344
haningtoni, Bison 400
hanishi, Gazella 396
hannibaldi, Loxodonta 367
hannyngtoni, Mus 627
hansae, Eumops **234**
hansmeyeri, Gorilla 276
hansruppi, Crocidura 92
hantu, Chimarrogale **107**, 107
Cynocephalus 135
hanuma, Mus 625
Hapale 251
hapale, Callithrix 252
Hapalemur **245**
alaotrensis 245
aureus **245**
cinereus 245
gallieni 245
griseus **245**, 245
meridionalis 245
occidentalis 245
olivaceus 245
schlegeli 245
simus **245**
Hapalidae 251
Hapalomys 583, **595**, 611
delacouri **595**
longicaudatus **595**, 595
longicaudus 640
marmosa 595
pasquieri 595
Hapalotis 586
albipes 586
apicalis 604
caudimaculatus 671
gouldii 619
Hapanella 253
Haplodon 417
Haplodontidae 417
Haplomylomys 728, 729, 736
Haplonycteris **142**
fischeri **142**, 142
hapsaliensis, Mus 625
Haptomys 783
Harana 385
harardai, Miniopterus 231
hararensis, Genetta 345
Heterohyrax 373
Madoqua 398
harbisoni, Peromyscus 731
hardwickei, Martes 320
Nesokia 632
Rhinopoma **155**
Tatera 561
hardwicki, Hemigalus 342
Viverra 342
hardwickii, Kerivoula **197**
hardyi, Heliosciurus 429
Zapus 499
harei, Aethomys 567
harenna, Crocidura **86**, 93
hargravei, Saccolaimus 159
haringtoni, Callosciurus 421
harlani, Cryptotis 109
Hylobates 275
harmandi, Callosciurus 422
Macaca 266
harnieri, Kobus 413
haroia, Phacochoerus 377
harpax, Cynopterus 139
harpia, Harpiocephalus **228**
Vespertilio 228
Harpiocephalus **228**
harpia **228**
lasyurus 228
madrassius 228
mordax 228
pearsonii 228
rufulus 228
rufus 228
Harpiola 228, 229
Harpionycteris, whiteheadi 142
Harpyionycterinae 137
Harpyionycteris **142**
celebensis **142**
negrosensis 142
whiteheadi **142**, 142
harquahalae, Thomomys 473
harrimani, Vulpes 287
harringtoni, Desmomys **590**, 590, 630
Pelomys 590
Taterillus **563**
harrisi, Hippotragus 413
Reithrodontomys 741
Scotinomys 746
harrisii, Ammospermophilus **419**, 419
Sarcophilus 35
Thylacinus 29
harrisoni, Acomys 566
Callosciurus 423
Felis 290
Hylopetes 461
Kerivoula 197
Musonycteris **186**, 186
Neotragus 398
Proteles 309
Rhinopoma 155
Thryonomys 775
harrissoni, Dicerorhinus 372
harroldi, Lemmus 517
hartensis, Lepus 816
harteri, Neotoma 711
harterti, Lepus 816
Massoutiera 761
Vulpes 286
harti, Apodemus 570
Trachypithecus 274
hartii, Artibeus **188**
hartingi, Microtus 521
hartmannae, Equus 370
hartmanni, Rheomys 743
hartwigi, Praomys **642**
harveyi, Bettongia 49
Cephalophus **410**
Malacothrix 544
harwoodi, Gerbillus 549, **551**
hasleri, Gazella 397
hassama, Potamochoerus 378
hasseltii, Myotis **211**
hastata, Ourebia 399
hastatus, Cephalorhynchus 352
Phyllostomus **180**
Vespertilio 180
hastilis, Callosciurus 421
hatanezumi, Microtus 525
hatcheri, Reithrodon 740
hatinhensis, Trachypithecus 273
hattai, Canis 281
hatti, Otonyctomys **726**, 726
hausleitneri, Mystromys 677
haussa, Felis 290
Mus **623**, 623, 624, 625, 629
haussanus, Orycteropus 375
hauvillii, Arctocephalus 326
hawaiiensis, Rattus 652
hawashensi, Kobus 413
hawashensis, Acomys 565
hawkeri, Lepus 816
Sorex 117
haydeni, Sorex 113, 114, **115**, 115
haydenii, Microtus 526
hayi, Apodemus 573
Gazella 396
haymani, Dendromus 543
Gerbillus 550
Idiurus 758
Natalus 195
Stenomys 665
haywoodi, Tragelaphus 404
hazenna, Gazella 396
heathi, Scotophilus **227**, 227
heavisidii, Cephalorhynchus **351**
Delphinus 351
heberfolium, Macrotus 178
hebes, Vulpes 287
hebridensis, Apodemus 573
hecki, Acinonyx 288
Axis 385
Cephalophus 411
Connochaetes 394
Macaca 268
Pan 277
Tachyoryctes 686
hector, Semnopithecus 273
hectori, Cephalorhynchus **352**
Mesoplodon **363**

hedenborgiana, Crocidura 92
hedigeri, Rousettus 152
heermanni, Dipodomys 477, **478**, 480
Sciurus **441**
heffernani, Pteropus 151
hefneri, Sylvilagus 826
heidelbergensis, Homo 276
Heimyscus **595**
fumosus **595**
heinrichi, Bunomys 581, **582**
Callosciurus 423
Hyosciurus **429**, 429, 430
heinsii, Ovis 408
helaletes, Synaptomys 534
Helamis 759
Helamys 759
Helarctos **337**, 337
anmamiticus 337
euryspilus 337
malayanus **337**, 337
wardi 337
helgei, Callosciurus 421
Dendrohyrax 373
helgolandicus, Mus 625
Helictis 314
Heligmodontia 700
Heliomys 538
Heliophobius 771, **772**
albifrons 772
angonicus 772
argenteocinereus **772**, 772
emini 772
kapiti 772
marungensis 772
mottoulei 772
pallidus 772
robustus 772
spalax 772
helios, Aonyx 310
Pipistrellus 222
Heliosciurus **428**, 434
abassensis 428
acticola 429
albina 428
annularis 428
annulatus 428
arrhenii 429
aschantiensis 429
aubryi 429
beirae 429
benga 429
bongensis 428
brauni 429
canaster 428
caurinus 429
chirindensis 429
coenosus 429
daucinus 429
dolosus 429
dysoni 428
elegans 428
emissus 429
gambianus **428**, 429
hardyi 429
hoogstraali 428
isabellinus 429
ituriensis 429
kaffensis 428
keniae 429
lateris 428
leakyi 429
leonensis 429
libericus 429
limbatus 428
loandicus 428
lualabae 429
maculatus 429
madogae 428
marwitzi 429
medjianus 429
multicolor 428
mutabilis **429**, 429
nyansae 429
obfuscatus 429
occidentalis 429
omensis 428
pasha 429
punctatus **429**
rhodesiae 428
rubricatus 429
rufo-brachiatus 429
rufobrachium **429**, 429
ruwenzorii **429**
savannius 429
schoutedeni 429
semlikii 429
senescens 428
shindi 429
shirensis 429
simplex 428
smithersi 429
undulatus **429**
vulcanius 429
vumbae 429
waterhousii 429
Heliosorex 81, 94
heljanensis, Crocidura 94
hellenicus, Nannospalax 754
helleri, Aethomys 568
Chaetodipus 483
Mustela 322
Otomys 682
Platyrrhinus **191**
Thomomys 475
hellwaldii, Maxomys **612**, 612
Rattus 612
helmyi, Acomys 565
Helogale 300, 302, **303**
affinis 304
ahlselli 304
annulata 304
atkinsoni 304
bradfieldi 304
brunetta 304
brunnula 304
hirtula **304**, 304
ivori 304
lutescens 304
macmillani 304
mimetra 304
nero 304
ochracea 304
parvula **304**, 304
parvus 304
powelli 304
robusta 302
ruficeps 304
rufula 304
undulatus 304
varia 304
vetula 304
victorina 304
helukus, Dasymys 589
helveolus, Sciurus 443
helvescens, Chlorocebus 265
helvetica, Rhinolophus 166
helveticus, Clethrionomys 508
helvina, Abrocoma 789
helviticus, Mus 625
helvolus, Lemmus 517
Mus 625
Oecomys 717
Reithrodontomys 741
helvum, Eidolon 140, **141**
helvus, Callosciurus 421
Herpestes 305
Vespertilio 140
hemachalana, Marmota 432
hemalayanus, Capra 406
Hemibelideus **58**, 58
cervinus 58
lemuroides **58**, 58
Hemibradypus 63
Hemicentetes **73**
buffoni 73
madagascariensis 73
nigriceps 73
semispinosus **73**, 73
variegatus 73
hemidactylus, Brachyteles 257
Hemiderma 186
Hemiechinus **77**, 77, 78, 79
aegyptius 78
aethiopicus **78**, 78
alaschanicus 78
albatus 78
albior 78
albulus 78
amir 78
auritus **78**, 78
blancalis 78
blanfordi 78
brachydactylus 78
brachyotis 78
calligoni 78
caspicus 78
chorassanicus 78
collaris **78**, 78
dauuricus 79
deserti 78
dorotheae 78
dorsalis 78
eversmanni 78
frontalis 78
grayi 78
holdereri 78
homalacanthus 78
hypomelas **78**
indicus 78
insularis 78
intermedius 78
jerdoni 78
kutchicus 78
libycus 78
ludlowi 78
macracanthus 78
major 78
megalotis 78
mentalis 78
metwallyi 78
micropus **78**, 79
microtus 78
minor 78
niger 78
nudiventris **78**, 79
oniscus 78
pallidus 78
pectoralis 78
persicus 78
pictus 78
platyotis 78
russowi 78
sabaeus 78
senaariensis 78
seniculus 78
spatangus 78
syriacus 78
turanicus 78
turfanicus 78
turkestanicus 78
wassifi 78
Hemigalago 249
Hemigale 342
hosei 341
Hemigalea 342
Hemigalinae 340, **341**, 341
Hemigalus 340, **341**, 342
boiei 342
derbianus 342
derbyanus **342**
derbyi 342
hardwicki 342
incursor 342
invisus 342
minor 342
zebra 341, 342
hemileucurus, Conilurus 586
hemionotis, Peromyscus 736
Hemionus 369
hemionus, Equus **370**, 370
Odocoileus **391**
Hemiotomys 505, 517
Hemippus 369
hemippus, Equus 370
Hemitragus **406**
hylocrius **406**
jayakari **406**
jemlahica 406
jemlahicus **406**, 406
jemlanica 406

jharal 406
quadrimammis 406
schaeferi 406
tubericornis 406
warryato 406
Hemiurus 21
hemprichiana, Madoqua 398
hemprichii, Dugong 365
Otonycteris **219**, 219
Hendecapleura 548
hendeei, Callosciurus 421
Lagothrix 257
Proechimys **796**, 798
hendersoni, Crocidura 85
Rhynchocyon 830
Tragulus 383
henleyi, Gerbillus **552**
hennigi, Lycaon 283
Sigmoceros 395
henrici, Apodemus 570
Connochaetes 394
Hylobates 275
henricii, Martes 320
henrii, Ovis 408
henryanus, Naemorhedus 407
henseli, Akodon 692
Microtus 529
Monodelphis 22
Nasua 335
Sciurus 439
henshawi, Ursus 339
Hepoona 59
heptapotamica, Crocidura 96
heptapotamicus, Sorex 117
Heptaxodon 805
Heptaxodontidae **804**
Heptaxodontinae **805**
heptneri, Allactaga 488
Capra 405
Leopoldamys 603
Meles 314
Meriones 556, 557
Microtus 519
Salpingotus **492**
Spermophilus 447
heptopotamicus, Spermophilopsis 444
hera, Crocidura 92
heran, Nyctophilus **218**
herbertensis, Pseudocheirus **59**
herberti, Callosciurus 422
Leopoldamys 603
Petrogale 56
herbicola, Spermophilus 448
hercegovinensis, Nannospalax 754
Talpa 128
hercules, Artibeus 189
Mallomys 608
hercynicus, Clethrionomys 508
Sorex 111
herero, Crocidura 92
Lepus 822
Scotophilus 227
Thallomys 668
herhardti, Gracilinanus 17
hermani, Apodemus 573
Myotis 210
hermanni, Neomys 110
Sorex 111
hermannii, Monachus 331
hermannsburgensis, Leggadina 600
Pseudomys 645, **647**, 647
hermaphrodita, Viverra 344
hermaphroditus, Paradoxurus **344**
herminea, Mustela 321
hermonensis, Apodemus **571**
hermonis, Chionomys 507
hernandesii, Panthera 298
hernandezi, Saguinus 254
Sciurus 440
hernandezii, Procyon 335
heroldii, Mus 625
Herpailurus **291**, 291
ameghinoi 291
apache 291
cacomitli 291
darwini 291
eira 291
eyra 291
fossata 291
melantho 291
panamensis 291
tolteca 291
unicolor 291
yagouaroundi 291
yaguarondi **291**, 291
Herpestes 299, 300, 302, **304**, 304, 306
aegyptiae 305
albicaudus 306
andersoni 305
angolensis 305
annamensis 306
auropunctata 305
auropunctatus 305, 306
bennettii 305
birmanicus 305
brachyurus **304**, 305, 306
cafra 305
cancrivora 306
canens 306
carnaticus 305
centralis 305
ceylanicus 304
ceylonicus 304
dorsalis 305
dyacorum 304
edwardsii **305**, 305
egypti 305
ellioti 305, 306
exilis 305
ferruginea 305
ferrugineus 305
fimbriatus 305
flavidens 304
formosanus 306
frederici 305
fulvescens 304
funestus 305
fusca 304, 305, 306
grandis 305
griseus 305
hanensis 306
helvus 305
hosei 304, 305
ichneumon 304, **305**, 306
incertus 305
inornatus 306
javanensis 304
javanicus **305**, 305, 306
jerdonii 306
lademanni 305
lanka 305
mababiensis 305
maccarthiae 304
madagascarensis 305
major 305
malaccensis 305
microdon 306
moerens 305
montanus 305
monticolus 306
mungo 305
naso 304, **305**
nems 305
nepalensis 305
numidianus 305
numidicus 305
nyula 305
ochraceus 302
palawanus 304
pallens 305
pallidus 305
pallipes 305
paludinosus 300
palustris **306**, 306
parvidens 305
parvulus 303
parvus 304
penicillatus 302
peninsulae 305
perakensis 305
persicus 305
pharaon 305
phillipsi 304
pondiceriana 305
pulverulentus 304
rafflesii 304
rajah 304
rubidior 304
rubiginosus 306
rubrifrons 305
ruddi 305
rusanus 306
rutilus 305
sabiensis 305
sangronizi 305
sanguineus 302, 304
semitorquatus 305, **306**, 306
siamensis 305
siccatus 304
sinensis 306
smithii **306**
sumatrius 304
thysanurus 306
torquatus 306
uniformis 306
urva 305, **306**, 306
vitticollis **306**
widdringtonii 305
zeylanius 306
herpestes, Myosorex 100
Herpestidae **299**
Herpestinae **300**, 340
Herpetomys 517, 521
herrei, Myotis 213
herreranus, Glaucomys 460
herronii, Peromyscus 730
hershkovitzi, Akodon **690**
Aotus **256**
Heteromys 481
hertigi, Micromys 620
hervoi, Anomalurus 757
heslopi, Hexaprotodon 380
hespera, Glossophaga 184
hesperia, Lonchophylla **181**
Tamandua 68
hesperinus, Gerbillus **552**
hesperius, Cuon 282
Sylvilagus 825
Hesperoloxodon 367
Hesperomys 697, 707
bicolor 715
caliginosus 707
cinereus 749
dorsalis 700
floridanus 739
fulvescens 717
leucodactylus 744
minutus 708
molitor 704
nudicaudus 751
rufescens 743
simplex 739
spinosus 708
sumichrasti 715
taylori 695
teguina 746
Hesperoptenus **204**, 204
blanfordi **204**
doriae **204**, 204
gaskelli **204**
isabellinus 205
tickelli **204**
tomesi **205**
Hesperosciurus 439, 441
hesperus, Artibeus 189
Oryzomys 720
Peromyscus 734
Pipistrellus **221**
Thomomys 475
hessei, Apodemus 573
Hessonoglyphotes 420, 423, 428
Heterocephalus 771, **772**
ansorgei 772
dunni 772
glaber **772**, 772

phillipsi 772
progrediens 772
scortecci 772
stygius 772
heterocerus, Cervus 387
heterochrous, Tragelaphus 404
heterodactylus, Atelerix 76
Elephas 367
Heterodon 361
heterodon, Suncus 103
heterodus, Orthogeomys **471**
Heterogeomys 469, 470, 471
Heterohyrax **373**, 373
albipes 373
antineae **373**
arboricola 373
bakeri 373
blainvillii 373
bocagei 373
borana 373
brucei **373**, 373
chapini 373
dieseneri 373
frommi 373
granti 373
grayi 373
hararensis 373
hindei 373
hoogstraali 373
irroratus 373
kempi 373
lademanni 373
maculata 373
manningi 373
mossambicus 373
münzneri 373
princeps 373
prittwitzi 373
pumilus 373
rhodesiae 373
ruckwaensis 373
ruddi 373
rudolfi 373
somalicus 373
ssongaea 373
syriacus 373
thomasi 373
victoria-njansae 373
webensis 373
Heteromyidae **477**
Heteromyinae 477, **481**
Heteromys 477, **481**, 481, 482
alleni 482
anomalus **481**
australis **481**
brachialis 481
chiriquensis 481
conscius 481
crassirostris 481
desmarestianus **481**, 481
fuscatus 481
gaumeri **481**
goldmani **481**
griseus 481
hershkovitzi 481
jesupi 481
lepturus 481
lomitensis 481
longicaudatus 481
melanoleucus 481
nelsoni **481**, 482
nigricaudatus 481
oresterus **481**, 482
pacificus 481
panamensis 481
planifrons 481
psakastus 481
repens 481
subaffinis 481
temporalis 481
thompsoni 481
underwoodi 481
zonalis 481
Heteropsomyinae 790, **799**, 800
Heteropsomys **799**, 799
antillensis **799**
insulans **799**, 799
Heteropus 55
Heterosciurus 420
Heterosoricidae 80
Heterosoricinae 80
heterothrix, Liomys 482
heudei, Pteropus 148
heuferi, Sigmoceros 395
heuglini, Alcelaphus 393
Papio 269
heureni, Sus **378**
Hexaprotodon **380**, 381
heslopi 380
liberiensis **380**
madagascariensis **380**
minor 380
Hexolobodon **803**
phenax **803**, 803
poolei 803
Hexolobodontinae **803**
heymansi, Cercopithecus 264
hibernicus, Lepus 821
Mustela 321
Rattus 657
Sorex 117
hichensis, Bandicota 579
hidongis, Maxomys 613
hiemalis, Sciurus 440
hienomelas, Hyaena 308
higginsi, Pseudomys 645, **647**, 647, 648
hilaria, Cavia 779
hilarii, Eptesicus 201
hilda, Gerbillus 553
Proechimys 798
hildae, Saccostomus 541
hildebrandti, Mastomys 643
Rhinolophus **166**
hildebrandtii, Mastomys **610**, 610
hildegardeae, Crocidura 84, **86**, 86, 89, 95
Mus 673
Myotis 209
Taphozous **160**
Zelotomys **674**
hilgendorfi, Murina 229
hilgerti, Chlorocebus 265
hilleri, Nycticebus 248
hilleriana, Gazella 397
hilli, Pteropus 149
Rhinolophus 167
Taphozous **160**
hillieri, Dasycercus 31
hillorum, Rhinolophus 164
hiltonensis, Odocoileus 391
himalaicus, Trogopterus 465
Vulpes 287
himalayana, Marmota 431, **432**
Ochotona **809**, 812
himalayanus, Oncifelis 294
Prionailurus 296
Rhinolophus 163
himalayica, Chimarrogale **107**, 107
himalayicus, Crossopus 107
hindei, Aethomys **568**, 568, 569
Aonyx 310
Atelerix 76
Beamys **540**, 540
Cercopithecus 264
Chaerephon 233
Crocidura 97
Epimys 567
Heterohyrax 373
Leptailurus 292
Lophiomys 564
Madoqua 398
Otolemur 250
Scotoecus 226
Sylvicapra 412
Tamias 456
hindenburgi, Dactylopsila 61
hindsi, Otolemur 250
hingstoni, Eptesicus 200
hinpoon, Niviventer 632, **634**
hintoni, Aethomys 568
Arvicola 506
Clethrionomys 509
Dendromus 543
Eothenomys 513
Genetta 345
Melomys 615
Petaurista **462**
Hippelaphus 385
hippelaphus, Boselaphus 402
Cervus 386, 387
Hippocamelus **390**
andicus 390
anomalocera 390
antisensis **390**
bisulcus **390**
cerasina 390
chilensis 390
dubius 390
equinus 390
huamel 390
huemel 390
leucotis 390
hippocrepis, Rhinolophus 166
hippolestes, Puma 296
hippolitianus, Cervus 386
Hippopotamidae **380**
Hippopotamus **381**, 381
abyssinicus 381
amphibius **381**, 381
australis 381
capensis 381
constrictus 381
kiboko 381
lemerlei **381**
leptorhynchus 381
senegalensis 381
sivalensis 380
standini 381
terrestris 370
tschadensis 381
typus 381
Hipposideridae 163
Hipposiderinae **169**
Hipposideros **170**
abae **171**
albaniensis 171
alongensis 173
amboiensis 171
andersoni 172
angolensis 171
antricola 171
apiculatus 175
armiger **171**
aruensis 171
ater **171**
atratus 171
atrox 171
aurantiaca 171
aureus 175
aurita 173
barbensis 173
batchianensis 171
beatus **171**
bicolor **171**, 171, 173, 174
bicornis 171
blythi 175
brachyotis 173
braima 171
breviceps **171**
caffer **171**
calcaratus **171**, 174
camerunensis **171**
celebensis 171
centralis 174
ceramensis 172
cervinus **171**, 173
cineraceus **171**
commersoni **172**
coronatus **172**
corynophyllus **172**
coxi **172**
crumeniferus **172**
cupidus 171
curtus **172**
custos 172
cyclops **172**

debilis 171
deformis 173
demissus 172
diadema **172**
dinops **172**
doriae **172**, 174
dukhunensis 175
durgadasi 172
dyacorum **173**
enganus 172
erigens 171
erroris 174
euotis 172
fulgens 173
fuliginosus **173**
fulvus **173**
galeritus 171, **173**
gambiensis 172
gentilis 171, 174
gigas 172
gilberti 171
gracilis 171
grandis 173
griseus 172
guineensis 174
halophyllus **173**
indus 173
inexpectatus **173**
inornatus 172
insignis 173
insolens 173
javanicus 171
jonesi **173**
labuanensis 171
lamottei **173**
langi 172
lankadiva **173**
larvatus **173**
lekaguli **173**
leptophyllus 173
longicauda 173
lylei **173**
macrobullatus 171, **173**
maggietaylorae **174**
major 171
malaitensis 172
marisae **174**
marsupialis 175
marungensis 172
masoni 172
maximus 171
megalotis **174**
micaceus 172
micropus 172
mirandus 172
misoriensis 171
mixtus 173
mostellum 172
murinus 173
muscinus **174**
nanus 171
natunensis 172
neglectus 173
nequam **174**
niangarae 172
niapu 174
nicobarensis 172
nicobarulae 171
nobilis 172
obscurus **174**
oceanitis 172
ornatus 172
pallidus 173
papua **174**
pelingensis 172
pendleburyi 175
pomona 171, **174**
poutensis 173
pratti **174**
pulchellus 175
pullatus 172
pygmaeus **174**
reginae 172
ridleyi **174**
ruber 173, **174**
sabanus 172, **174**
saevus 171
sandersoni 172
schistaceus **174**
schneideri 171
semoni **175**
sinensis 174
speculator 172
speoris **175**
stenotis **175**
sumbae 173
swinhoei 171
templetonii 175
tephrus 171
terasensis 171
thomensis 172
toala 171
tranninhensis 171
trobrius 172
turpis **175**
unitus 173
vicarius 172
viegasi 172
vittata 172
vulgaris 173
wollastoni **175**
wrighti 172
hipposideros, *Rhinolophus* **166**
Hippotigris 369
Hippotraginae **412**
Hippotragus **412**
aethiopica 412
anselli 413
aurita 412
bakeri 412
barbata 412
capensis 412
cottoni 412
docoi 412
dogetti 412
equinus **412**
gambianus 412
glauca 412
harrisi 413
jubata 412
kaufmanni 413
kirkii 413
koba 412
langheldi 412
leucophaeus **412**
niger **412**
roosevelti 413
rufopallidus 412
scharicus 412
truteri 412
variani 412, 413
hippurellus, *Sundasciurus* 451
hippurosus, *Sundasciurus* 451
hippurus, *Sundasciurus* **451**
Hircus 405
hircus, *Capra* **405**, 405
hirsuta, *Micronycteris* **179**
Pithecia 261, 262
hirsutirostris, *Hystrix* 774
hirsutus, *Aotus* 256
Artibeus **188**, 188
Coelops **170**
Dasypus 66
Elephas 367
Golunda 592
Lagorchestes **52**
Mesembriomys 620
Microtus 527
Mus 619
Myotis 212
Paradoxurus 344
Prosciurillus 436
Reithrodontomys **741**
Rhinolophus 167
Sigmodon 747
Vombatus 46
hirta, *Akodon* 691
Crocidura 83, 84, 86, 87, 87, 90, 98
Microtus 518
hirtensis, *Apodemus* 573
hirtipes, *Eligmodontia* 701
Gerbillus 551
Jaculus 493
Molossops 235
Sminthopsis **36**
hirtula, *Helogale* **304**, 304
hirtus, *Pappogeomys* 472
Procyon 335
Sciurus 440
hirundo, *Scotoecus* **226**
hiska, *Oxymycterus* **726**, 727
hispanica, *Capra* 406
Genetta 345
hispanicus, *Cervus* 386
Erinaceus 77
Felis 290
Lepus 818
Mus 628
Plecotus 224
hispaniolae, *Eptesicus* 201
hispida, *Crocidura* **87**
Echymipera 41
Nycteris **161**
Phoca 329, **332**, 332
hispidus, *Acomys* 565
Cabassous 65
Caprolagus **814**
Chaetodipus 482, **483**
Echimys 794
Lepus 814
Liomys 482
Makalata 793
Mesomys **794**, 795
Orthogeomys **471**, 471
Oxymycterus **727**
Sigmodon 746, **747**, 747, 748
Vespertilio 161
Histiotus **205**
alienus **205**
capucinus 205
chilensis 205
colombiae 205
inambarus 205
laephotis 205
macrotus **205**
maculatus 204
magellanicus 205
miotis 205
montanus **205**
poeppigii 205
segethii 205
velatus **205**
historicus, *Sciurus* 440
Histricophoca 331
histrio, *Cercopithecus* 263
Histriophoca 331
Histriosciurus 439
hitchensi, *Sylvilagus* 825
hiwaensis, *Euroscaptor* 125
hoamibensis, *Raphicerus* 399
hobartensis, *Tachyglossus* 13
hobbit, *Syconycteris* **155**
hochlovi, *Allactaga* 489
hodgsoni, *Hystrix* 773
Marmota 432
Mustela 324
Naemorhedus 407
Ochotona 811, 812, 813
Ovis 408
Petaurista 462
Suncus 102
Vulpes 286
hodgsonii, *Antilope* 399
Ovis 408
Pantholops **399**
Vulpes 286
Hodomys **704**, 704, 710, 752
alleni **704**, 704
elattura 704
guerrerensis 704
vetulus 704
hodophilax, *Canis* 281
hodsoni, *Madoqua* 398
hoehnei, *Sciurillus* 438
hoeschi, *Desmodillus* 547
hoevellianus, *Cervus* 387
hoffeti, *Cynopterus* 139
hoffmanni, *Choloepus* **64**
Rattus 649, 651, **653**, 653,

654, 655, 657, 661
Sciurus 441, 443
hoffmannsi, Callicebus **258**
hofmeyri, Antidorcas 395
hokkaidi, Apodemus 570
holchu, Rattus 660
holdereri, Equus 370
Hemiechinus 78
hollandianus, Cervus 386
hollisteri, Apomys 575
Panthera 297
Peromyscus 732
Sorex 113
Spermophilus 450
holmwoodi, Diceros 372
holobrunneus, Crocidura 95
Holochilus **704**, 704, 746
amazonicus 705
balnearum 704
berbericensis 705
brasiliensis **704**, 704, 705
canellinus 704
chacarius **704**
darwini 704
guianae 705
incarum 705
leucogaster 704
magnus **704**, 704, 705
nanus 705
russatus 704
sciureus 704, **705**, 705
venezuelae 705
vulpinus 704
hololeucus, Saguinus 253
holomelas, Propithecus 247
holosericeus, Cryptomys 772
Georychus 771
Molossus 235
holotephreus, Trachypithecus 274
holti, Tamiops 457
holubi, Aepyceros 393
Canis 280
holzneri, Mephitis 317
Sylvilagus 825
holzworthi, Ursus 339
homalacanthus, Hemiechinus 78
Homalurus 111, 118
homei, Hydrurga 330
Homelaphus 390
homericus, Acomys 565
homezi, Micronycteris 179
Hominidae 274, **276**
Homo **276**
heidelbergensis 276
lar 274
neanderthalensis 276
palestinus 276
rhodesiensis 276
sapiens **276**, 276
homochroia, Peromyscus 730
homochrous, Plecotus 224, 225
Homodontomys 710, 711
homodorensis, Rhinolophus 166
Homopsomys 799
homorodalmasiensis, Rhinolophus 166
homorus, Thomomys 473
homourus, Mus 625
hondoensis, Urotrichus 129
hondonis, Micromys 620
Urotrichus 130
hondurensis, Canis 280
Cryptotis **108**, 108
Orthogeomys 471
Peromyscus 728
Sturnira 192
Sylvilagus 826
hoodii, Spermophilus 450
hoogerwerfi, Rattus 650, **653**, 653, 656
hoogstraali, Elephantulus 829
Gerbillus **552**
Heliosciurus 428
Heterohyrax 373
Lemniscomys **601**, 601
Sundasciurus **451**, 452
Hooijeromys 587, 637, 664
hookeri, Arctocephalus 328
Neophoca 328
Otaria 328
Phocarctos **328**
hooki, Cephalophus 411
hoole, Vulpes 287
Hoolock 274
hoolock, Hylobates **275**
hooperi, Blarina 106
Peromyscus **731**, 731
Reithrodontomys 741
hoots, Ursus 339
hopiensis, Perognathus 485
Tamias 455
hopkinsi, Pelomys **639**, 639, 640
hopkinsoni, Tatera 561
hoplites, Dasypus 66
hoplomyoides, Hoplomys 794
Proechimys **797**
Hoplomys **794**, 795, 797
goethalsi 794
gymnurus **794**
hoplomyoides 794
truei 794
horeites, Rattus 656
horikawai, Eptesicus 203
horrens, Furia 195
Furipterus **195**
horriaeus, Ursus 339
horribilis, Ursus 338, 339
horsfieldi, Crocidura 87
Cynopterus **139**
Macroglossus 154
Megaderma 163
Prionailurus 295
Pteromys 461
horsfieldii, Crocidura **87**, 87
Iomys **461**
Mustela 324
Myotis **211**
horstockii, Raphicerus 399
hortualis, Eliomys 767
hortulanus, Mus **624**, **625**, **628**
hortulorum, Cervus 386
hosei, Bubalus 402
Diplogale **341**
Hemigale 341
Herpestes 304, 305
Lagenodelphis **353**, 353
Lariscus **430**
Petaurillus **462**, 462
Presbytis **271**
Sciuropterus 461
Suncus **102**, 102
Tragulus 382
hosletti, Crocidura 84, 94
hosonoi, Myotis **211**, 216
Sorex **115**, 115
hotaula, Mesoplodon 363
hotsoni, Allactaga 488, **489**, 489
Calomyscus **536**
hottentotus, Amblysomus **74**
Bathyergus 771
Chrysochloris 74
Cryptomys **771**
Eptesicus **202**
Rousettus 152
houtmanni, Macropus 53
houyi, Syncerus 402
hova, Oryzorictes **72**, 72
Potamochoerus 378
howelkae, Microtus 519
howelli, Dremomys 425
Ochotona 811
Reithrodontomys 742
Scalopus 127
Sylvilagus 825
Sylvisorex **104**
Thomomys 473
howensis, Pteropus **147**
hoxaensis, Rattus 650
hoyi, Sorex **115**
huachuca, Sciurus 440
hualpaiensis, Microtus 524
Thomomys 473
huamel, Hippocamelus 390
Huamela 390
huanacus, Lama 382
huang, Niviventer 634
huangensis, Ochotona 812
huangshuiensis, Lepus 821
huastecae, Oryzomys 721
hubbacki, Bos 401
hubbardi, Lepus 815
hubeinensis, Myospalax 676
huberti, Mastomys 610
hucucha, Oxymycterus **727**, 727
hudsoni, Puma 296
Thomasomys 749
hudsonica, Lontra 311
Mephitis 317
hudsonicus, Canis 281
Glaucomys 460
Procyon 335
Tamiasciurus 456, **457**
hudsonius, Clethrionomys 508
Dicrostonyx **510**, 510
Dipus 499
Lepus 814
Mus 509
Ondatra 532
Tamias 454
Zapus **499**, 499
huebneri, Lycaon 283
huegeli, Rattus 652
huelleri, Sorex 111
huemel, Hippocamelus 390
hueti, Graphiurus **764**, 764
hueyi, Chaetodipus 483
Peromyscus 732
Thomomys 473
hughi, Mesechinus **79**, 79
huidobria, Lontra 311
huina, Oncifelis 294
huixtlae, Orthogeomys 471
huli, Vulpes 287
hulock, Hylobates 275
hulophaga, Blarina 106
humbloti, Eptesicus 203
Triaenops 175
humboldtensis, Martes 319
Sorex 121
humboldtiana, Aplodontia 417
humboldtii, Aotus 256
Conepatus 315, **316**
Lagothrix 257, 258
humei, Callosciurus 423
Hadromys **594**, 595
Mus 594
Naemorhedus 407
Ovis 408
humeralifera, Callithrix **251**, 252
humeralis, Chalinolobus 199
Mustela 324
Myotis 213
Nycticeius **217**
Pecari 380
Ratufa 437
Tupaia 132
Vespertilio 217
humiliatus, Rattus 657
humilior, Microryzomys 708
humilis, Acerodon **137**
Pudu 392
Sundasciurus 451
Thomomys 473
hummelincki, Baiomys 695
Calomys **698**, 698
humulis, Reithrodontomys **741**
hungarica, Mustela 322
hungaricus, Canis 280
Microtus 530
Nannospalax 754
hungarorum, Bison 400

hunteri, Acomys 565
Akodon 689
Bettongia 49
Damaliscus **394**
Monodelphis 21
Ursus 338
Hunterius 349
hunti, Conepatus 316
Syncerus 402
huon, Pogonomys 642
huperuthrus, Microtus 519
hurdanensis, Microtus 524
huro, Martes 319
huronax, Galictis 318
hurrianae, Meriones **556**
hussoni, Hydromys **597**, 597
huttoni, Balaenoptera 349
Felis 290
Murina **229**, 229, 230
Nesokia 632
huxleyi, Oryctolagus 822
hyacinthinus, Neurotrichus 126
Hyaena **308**, 308, 309
antiquorum 308
barbara 308
bergeri 308
bilkiewiczi 308
bokcharensis 308
brunnea 308, 309
dubbah 308
dubia 308
fasciata 308
hienomelas 308
hyaena **308**
hyaenomelas 308
indica 308
orientalis 308
picta 283
rendilis 308
satunini 308
schillingsi 308
striata 308
suilla 308
sultana 308
syriaca 308
virgata 308
vulgaris 308
zarudnyi 308
hyaena, Canis 308
Hyaena **308**
Hyaenidae **308**
Hyaeninae **308**
hyaenomelas, Hyaena 308
hyatti, Oreotragus 399
Hybomys 590, **596**, 596, 597
badius 596
basilii **596**
eisentrauti **596**
lunaris **596**, 596
pearsei 596
planifrons 592, **596**, 597
rufocanus 596
trivirgatus 592, **596**, 597
univittatus **596**, 596
hybridus, Ateles 257
Dasypus **65**
Lepus 817
Rattus 657
Hydrelaphus 388
Hydrictis 311, 312
hydrobadistes, Sorex 119
hydrobates, Ichthyomys **705**
Hydrochaeridae **781**
Hydrochaeris **781**
capybara 781
cobaya 781
dabbenei 781
hydrochaeris **781**
irroratus 781
isthmius 781
notialis 781
uruguayensis 781
hydrochaeris, Hydrochaeris **781**
Hydrochoerus 781
Sus 781
Hydrochoerus 781
hydrochaeris 781
Hydrodamalidae 365
Hydrodamalinae 365
Hydrodamalis **365**
balaenurus 365
borealis 365
gigas **365**
stelleri 365
hydrodromus, Sorex **115**, 115, 116
Hydrogale 110
Hydrolagus 824
Hydromys 588, **597**, 597, 620, 638, 673
beccarii 597
caurinus 597
chrysogaster **597**, 597, 673
esox 597
fuliginosus 597
fulvogaster 597
fulvolavatus 597
fulvoventer 597
grootensis 597
habbema **597**, 597, 598
hussoni **597**, 597
illuteus 597
lawnensis 597
leucogaster 597
longmani 597
lutrilla 597
melicertes 597
moae 597
nauticus 597
neobrittanicus **597**
oriens 597
reginae 597
shawmayeri **597**, 598
hydrophilus, Mus 625
Neomys 110
Hydropotes **388**
affinis 388
argyropus 388
inermis **388**, 388
kreyenbergi 388
Hydropotinae **388**
Hydrotragus 413
hydruntina, Crocidura 94
Hydrurga 329, **330**
homei 330
leptonyx **330**
Hyelaphus 384, 385
hyemalis, Cervus 386
Lepus 817
Hyemoschus **382**
aquaticus **382**
batesi 382
cottoni 382
hyenoides, Proteles 309
hylaeum, Plagiodontia 804
hylaeus, Geomys 469
Gulo 318
Microryzomys 708
Peromyscus 732
Syncerus 402
Hylanthropus 276
Hylarnus 398
hyleae, Proechimys 795
Hylenomys 622
Hylobates **274**, 275, 276
abbotti 276
ablogriseus 275
agilis **274**, 275, 276
albibarbis 275, 276
albimana 275
albonigrescens 275
carpenteri 275
choromandus 275
cinerea 276
concolor **275**, 275
continentis 276
entelloides 275
funereus 276
furvogaster 275
fuscus 275
gabriellae **275**
gibbon 276
golock 275
hainanus 275
harlani 275
henrici 275
hoolock **275**
hulock 275
javanicus 276
jingdongensis 275
klossii **275**
lar **275**, 275
leucisca 276
leucogenys **275**, 275
leuconedys 275
longimana 275
lu 275
moloch **275**
muelleri 275, **276**
nasutus 275
niger 275
pileatus **276**
pongoalsoni 276
rafflei 275
scyritus 275
siki 275
subfossilis 276
syndactylus **276**
unko 275
variegatus 275
varius 275
vestitus 275
volzi 276
yunnanensis 275
Hylobatidae **274**, **274**
hylocetes, Apomys 574, 575
Oryzomys 725
Peromyscus 728
Hylochoerus **377**
gigliolii 377
ituriensis 377
ivoriensis 377
meinertzhageni **377**, 377
rimator 377
schulzi 377
hylocoetes, Apomys **575**
hylocrius, Hemitragus **406**
hylodromus, Ursus 338, 339
Hylogalea, murina 131
Hylomyinae **79**
hylomyoides, Maxomys **612**
Hylomys **79**
cuttingi 80
dorsalis 80
fulvescens 80
hainanensis **79**
maxi 80
microtinus 80
parvus 80
peguensis 80
siamensis 80
sinensis **80**
suillus 79, **80**
tionis 80
Hylomyscus 596, **598**, 598, 642
aeta **598**, 598, 599
alleni **598**, 598, 599
anselli 599
baeri 592, **598**
canus 598
carillus 598, **599**, 599
denniae **599**, 599, 642
endorobae 599
fumosus 595
grandis 598
kaimosae 599
laticeps 598
parvus **599**
shoutedeni 598
simus 598
stella 598, **599**, 599
vulcanorum 599
weileri 598
Hylonycteris **185**
minor 185
underwoodi **185**, 185
Hylopetes **460**, 461, 463, 464
alboniger **460**
amoenus 461
anchises 461
aurantiacus 461

baberi **460**
bartelsi **460**
belone 461
caroli 461
chianfengensis 460
elassodontus 461
electilis 461
everetti 460, 461
fimbriatus **460**, 460
harrisoni 461
laotum 461
leachii 460
lepidus **461**, 461
mindanensis 463
nigripes **461**
orinus 460
phayrei **461**
platyurus 461
probus 461
sagitta 461
sipora **461**, 461, 604
spadiceus **461**, 461
sumatrae 461
turnbulli 460
winstoni **461**
hylophaga, Blarina **106**, 106
hylophilus, Thomasomys **749**
Hylothylax 22
hyojironis, Sorex 115
Hyomys **599**
dammermani **599**, 600
goliath **600**
meeki 599, 600
strobilurus 600
Hyosciurina 419
Hyosciurus 419, **429**
heinrichi **429**, 429, 430
ileile 429, **430**
Hyperacrius 502, 507, 513, **515**
aitchisoni 515
brachelix 515
fertilis **515**
traubi 515
wynnei **515**
zygomaticus 515
hyperborea, Ochotona 807, 808, **809**, 809
hyperboreus, Lepus 815
Microtus **522**, 522
Hyperoambon 65
Hyperoodon **361**
ampullatus **361**
burmeisterei 361
butskopf 361
latifrons 361
planifrons **361**
rostratus 361
Hyperoodontidae 361
Hyperplagiodontia 804
stenocoronalis 804
hyperryphaeus, Arvicola 506
hyperythrus, Callosciurus 421
Hypnomys 767
hypochrysa, Tupaia 132
Hypodon 364
hypogaeus, Eligmodontia 701
Graomys 701, 703
Hypogeomys **678**
antimena 678, **679**
hypokantha, Callicebus 259
hypoleuca, Cavia 779
Cebus 259
Micronycteris 179
hypoleucos, Ratufa 437
Semnopithecus 273
Spermophilus 446
hypoleucus, Cebus 260
Stochomys 666
Trichosurus 48
hypomelanus, Pteropus **147**
hypomelas, Cebus 259
Hemiechinus **78**
Macaca 267
Vulpes 287
hypomicrus, Nesophontes **70**
hypophaeus, Sciurus 440
hypopyrrhus, Sciurus 440
hyporrhodus, Sciurus 441
hypoxanthus, Brachyteles 257
Mus 637
Oenomys **637**, 637
Sciurus 440
Hypselephas 367
hypsibius, Lepus 819
Soriculus **123**, 123
Hypsignathus **142**
labrosus 142
monstrosus **142**, 142
Hypsimys 688, 689, 692
Hypsiprymnodon **49**
moschatus **49**, 49
nudicaudatus 49
Hypsiprymnodontinae 48
Hypsiprymnus 49
ursinus 50
Hypsugo 219, 220, 221, 222, 223, 224
Hypudaeus 516
leucogaster 719
Hyracidae 373
Hyracodon 25
fuliginosus 25
Hyracoidea 373
Hyrax 373, 374
arboreus 373
brucei 373
syriacus 373
hyrcania, Crocidura 96
Microtus 530
hyrcanicus, Apodemus **572**, 572
hystrella, Acomys 565
Hystricidae 771, **773**
Hystricinae 773
Hystricomorpha **771**
Hystrix **773**, 773, 774
aerula 773
africaeaustralis **773**, 773
aharonii 774
alophus 773
bengalensis 773
blanfordi 774
brachyura **773**, 774
brevispinosa 774
capensis 773
crassispinis **773**, 774
cristata **773**, 773
cuneiceps 774
cuvieri 773
daubentoni 773
dorsata 776
ecaudata 774
europaea 773
fasciculata 774
galeata 773
grotei 773
hirsutirostris 774
hodgsoni 773
indica **774**
javanica **774**
javanicum 774
klossi 773
leucura 774
leucurus 774
longicauda 773
macroura 773
major 773
malabarica 774
mersinae 774
mesopotamica 774
millsi 773
mülleri 773
narynensis 774
occidanea 773
papae 773
prehensilis 775
prittwitzi 773
pumila **774**
satunini 774
schmidtzi 774
senegalica 773
spinosa 776
stegmanni 773
subcristata 773
subspinosus 789
sumatrae **774**
sumbawae 774
torquata 774
yunnanensis 773
zeylonensis 774
zuluensis 773
hystrix, Tachyglossus 13

Ia **205**
beaulieui 205
io **205**, 205
longimana 205
ibeana, Crocidura 87
ibeanus, Grammomys **593**, 593, 594
Ichneumia 306
Lophiomys 564
Papio 269
Paraxerus 435
Perodicticus 248
Tachyoryctes 687
ibericus, Microtus 520
Iberomys 517, 519
Ibex 405
ibex, Capra **405**, 406
Raphicerus 399
ibicensis, Crocidura 94
ica, Oryzomys 725
icarus, Otomops 239
ichangensis, Elaphodus 388
Ichneumia 300, **306**
abuwudan 306
albescens 306
albicauda **306**
almodovari 306
dialeucos 306
ferox 306
grandis 306
haagneri 306
ibeanus 306
leucurus 306
loandae 307
loempo 307
nigricauda 307
Ichneumon 304
ichneumon, Herpestes 304, **305**, 306
Viverra 304
ichnusae, Crocidura 94
Vulpes 287
Ichthyomys 699, **705**, 743
caurinus 705
hydrobates **705**
nicefori 705
orientalis 705
pittieri **705**
soderstromi 705
stolzmanni **705**, 705
trichotis 699
tweedii **705**
iconica, Tatera 562
iconicus, Apodemus 573
Ictailurus 295
ictericus, Callosciurus 422
Icticyon 285
Ictides 342
Ictidomoides 444
Ictidomys 444, 446, 447, 448, 449, 450
Ictidonyx 319
Ictomys 319
Ictonyx **318**, 319
africana 319
albescens 319
arenarius 319
capensis 318, 319
elgonis 319
erythraea 319
frenata 319
ghansiensis 319
giganteus 319
intermedia 319
kalaharicus 319
lancasteri 319
libyca **319**
limpopoensis 319

maximus 319
multivittata 319
mustelina 319
nigricaudus 319
oralis 319
orangiae 319
ovamboensis 319
pondoensis 319
pretoriae 319
rothschildi 319
senegalensis 319
shoae 319
shortridgei 319
striatus **319**
sudanicus 319
vaillantii 319
variegata 319
zorilla 319
iculisma, Crocidura 96
ida, Cyclopes 67
idahoensis, Brachylagus **813**
Clethrionomys 508
Dipodomys 479
Lepus 813
Perognathus 486
Sorex 113
Spermophilus **447**
Sylvilagus 825
Thomomys **473**, **474**, 475
Ursus 339
Zapus **499**
Idionycteris **205**, 205
mexicanus 205
phyllotis **205**, 224
Idiurinae 757
Idiurus **757**
cansdalei 758
haymani 758
kivuensis 758
langi 758
macrotis **757**, 758
panga 758
zenkeri 757, **758**, 758
Idomeneus 555
idoneus, Castor 467
Dipodomys 479
Melanomys 708
ifniensis, Lemniscomys 601
ifranensis, Apodemus 573
igmanensis, Microtus 519
ignava, Marmota 432
Ignavus 63
ignavus, Bradypus 63
ignifer, Cephalophus 411
Rhinolophus 167
ignita, Cercopithecus 263
Presbytis 271
ignitoides, Galerella 303
ignitus, Acomys **565**, 565, 566
Callicebus 259
Galerella 303
Sciurus **441**
igniventris, Sciurus **441**
ignotus, Cervus 386
Neomys 110
Proechimys 798
Sorex 111
iheringi, Alouatta 255
Monodelphis **21**
Oxymycterus **727**, 727
Proechimys **797**
ikhwanius, Pipistrellus 221
ikonnikovi, Myotis **211**
ilaeus, Microtus 526
ileile, Hyosciurus 429, **430**
ilensis, Crocidura 96
ileos, Microtus 526
ilex, Apodemus 571
iliensis, Ochotona 809, **810**
Spermophilus **446**
illapelinus, Phyllotis 738
illectus, Oecomys 716
illigeri, Saguinus 253
illinensium, Didelphis 17
illinoensis, Geomys 469
Neotoma 711
illovoensis, Mastomys 611
illustris, Taterillus 563
illuteus, Akodon **690**
Hydromys 597
Lepus 819
illyricus, Arvicola 505, 506
ilvanus, Apodemus 573
ilya, Eira 318
imago, Aethomys 567
imaizumii, Mogera 126
Phaulomys 533
Rhinolophus **166**
Talpa 126
imarius, Callosciurus 424
imatongensis, Mus 629
imbaburae, Sciurus 441
imbellis, Ammodillus **546**
Gerbillus 546
imberbis, Muriculus **621**, 622
Mus 621
Saguinus 253
Tragelaphus **404**
imbil, Rattus 655
imbrensis, Scotomanes 226
imbricatus, Pipistrellus **221**, 221, 222
imhausi, Lophiomys **563**, 564
imhausii, Lophiomys 563
imitabilis, Thomomys 473
imitator, Alticola 503
Anisomys **569**, 569
Callosciurus 422
Cebus 260
immunis, Thomomys 475
imogene, Pharotis **219**, 219
imparilis, Pappogeomys 472
impavidus, Canis 280
Marmosops **19**
imperator, Anomalurus 757
Saguinus **253**
Uromys **672**
Ursus 339
imperfecta, Didelphis 16
imperfectus, Peromyscus 732
imperialis, Cervus 386
imperii, Mustela 321
imphalensis, Mus 623
impiger, Reithrodontomys 741
Ursus 339
importunus, Rhinolophus **164**
improcera, Puma 296
improvisum, Chiroderma **189**
impudens, Macaca 266
imus, Dremomys 425
inambarii, Chroeomys 700
inambarus, Histiotus 205
inaquosus, Dipodomys 479
inas, Maxomys **612**, 614
inaurata, Chrysochloris 75
inaureus, Macaca 268
inauris, Xerus **458**, 458
inauritus, Mellivora 315
inbutus, Metachirus 20
inca, Choeroniscus 184
Conepatus 316
Eptesicus 201
Lagidium 778
Lestoros **25**
Orolestes 25
Oxymycterus **727**
Pseudalopex 284
Sylvilagus 825
incae, Pteronotus 176
incana, Didelphis 19
Ochotona 811
Pseudocheirus 60
incanens, Pseudocheirus 60
incanescens, Callicebus 259
incanus, Marmosops **19**
Microtus 523
Nycticebus 248
Petauroides 59
Pteromys 464
Thomasomys **749**
Tscherskia 540
incarum, Holochilus 705
Lontra 311
Platyrrhinus 191
Puma 296
incautus, Myotis 216
Procyon 335
incensus, Peromyscus 732
Thomomys 475
incertoides, Microtus 530
incertus, Herpestes 305
Microtus 519, 530
Myospalax 676
Oryzomys 720, 723
incisivus, Eptesicus 203
Phacochoerus 377
incitatus, Sylvilagus 825
inclamax, Alouatta 255
inclarus, Peromyscus 732
inclusa, Tatera **561**
incognitus, Microtus 519
incolatus, Chaetodipus 483
incomptus, Thomomys 473
incomtus, Dasymys **589**, 589, 590
Mus 589
inconsonans, Alouatta 255
inconstans, Sciurus 441
Sus 379
Tamiops 457
incultus, Rhinosciurus 438
incursor, Hemigalus 342
indefessus, Nesoryzomys **713**, **714**
Oryzomys 713
indi, Platanista 360
indianus, Mus 625
indica, Allactaga 488
Arvicola 632
Balaenoptera 350
Bandicota **579**, 579, 632
Callosciurus 423
Civetta 348
Equus 370
Hyaena 308
Hystrix **774**
Lutra 312
Moschiola 382
Nesokia **632**, 632
Ratufa **437**
Tatera 560, **561**
Viverricula **348**
indicus, Axis 384
Bos 401
Bubalus 402
Canis 280
Dipus 560
Dugong 365
Elephas 367
Equus 370
Funambulus 426
Hemiechinus 78
Mellivora 315
Mops 235, 236
Mus 578
Panthera 297
Pipistrellus 220
Rattus 658
Rhinoceros 372
Sciurus 426, 437
Suncus 103
Sus 379
Tapirus **371**
Tragulus 382
Vulpes 286
Ziphius 364
indigitatus, Amblonyx 310
indochinensis, Crocidura 87
Macaca 267
Martes 320
Indolagus 814, 818, 819, 820
Indopacetus **362**
pacificus **362**
indosinicus, Niviventer 634
Indri **247**
ater 247
brevicaudatus 247
indri **247**
mitratus 247
niger 247
variegatus 247

indri, Indri **247**
Lemur 247
Indridae **246**
Indriidae 246
indus, Gerbillus 553
Hipposideros 173
industrius, Geomys 469
indutus, Mus **624**, 624
Vulpes 287
Zyzomys 674
ineptus, Aethomys 567
Tachyglossus 13
inermis, Dobsonia **139**, 140
Hydropotes **388**, 388
Rhinoceros 372
Thrichomys 798
inesperatus, Procyon 335
inexoratus, Sigmodon 747
inexpectatus, Callosciurus 421
Hipposideros **173**
Miniopterus 231
Perognathus 485
Reithrodontomys 741
inexspectatus, Pipistrellus **221**
inez, Eothenomys **514**, 514
infans, Akodon 694
infelix, Cervus 386
inferior, Bunomys 581
infernatis, Baiomys 695
Reithrodontomys 741
infernatus, Spermophilus 444
infernus, Gerbillurus 547
infinitesimus, Suncus **102**
inflata, Pteronotus 177
Tupaia 132
inflatus, Coelops 170
Gerbillus 549
Maxomys **612**
Miniopterus **230**
Scalopus 127
influatus, Nycticeius 217
infolatus, Chaetodipus 483
infralineatus, Rattus 658
infraluteus, Perognathus 485
Sundamys **666**, 666
infrapallidus, Thomomys 473
infulatus, Aotus **256**
infumata, Dactylopsila 61
Giraffa 383
Lagothrix 258
infumatus, Crocidura 83
Rousettus 152
infusca, Balantiopteryx **156**
Taxidea 325
infuscata, Rousettus 152
infuscatus, Bradypus 63
Sciurus 443
infuscus, Cuon 282
Metachirus 20
Pappogeomys 472
Platyrrhinus **191**
Sylvisorex 105
ingens, Dacnomys 589
Diclidurus **157**
Dipodomys **478**
Macrotarsomys **679**
Mustela 324
Plecotus 225
Thomomys 473
ingoldbyi, Crocidura 83
Myomys 631
ingrahami, Geocapromys **801**
ingrami, Phascogale 34
Planigale **34**
Prionailurus 295
Sciurus 439
Wallabia 57
ingridi, Leptailurus 292
Inia **360**
boliviensis 360
geoffrensis **360**
Iniidae 360
iniscatus, Akodon **690**
initialis, Naemorhedus 407
initis, Mustela 321
inkulanondo, Sigmoceros 395
inmollis, Lagostomus 778
innae, Dipus 493
Microtus 526
innesi, Eptesicus 200
Lepus 816
innominatum, Phyllostomus 180
innoxius, Eptesicus **202**
innuitus, Alopex 279
Microtus 527
Synaptomys 534
Ursus 339
inobservatus, Cercopithecus 263
inodorus, Crocidura 94
Inomys 749, 750
inopinata, Neotoma 712
inopinatus, Artibeus **188**
Pteropus 147
Sigmodon **747**, 747
Ursus 339
inopinus, Chaetodipus 483
inops, Rhinolophus **166**
inornata, Arctogalidia 343
Neotoma 712
Petrogale **56**
inornatus, Callosciurus **422**
Chroeomys 700
Herpestes 306
Hipposideros 172
Macaca 267
Mazama 390
Melursus 337
Perognathus **485**
Petaurista 463
Tragelaphus 404
inquinatus, Sundasciurus 451
Insectivora 69
inselbergensis, Bathyergus 771
insidiosus, Sphiggurus **776**, 777
insignatus, Pelomys 639
Insignicebus 263
insignis, Alcelaphus 393
Apomys **575**
Cercopithecus 264
Dendromus **541**, **542**, 542, 543, 544
Grammomys 593
Hipposideros 173
Lariscus **430**, 430
Napaeozapus **498**
Peromyscus 729
Pteropus 148
Ratufa 437
Saccolaimus 159
Saccopteryx 159
Sciurus 430
Tadarida 240
Zapus 498
Zenkerella **758**, 758
insolatus, Peromyscus 732
Rattus 657
insolens, Hipposideros 173
insolitus, Cercopithecus 264
Sylvilagus 825
insonus, Sylvilagus **826**
insperatus, Apodemus 573
Microtus 527
instabilis, Tamandua 68
instans, Chilomys **699**
Oryzomys 699
Zelotomys 674
insulae, Melomys 615
insulaebellae, Clethrionomys 508
Sorex 117
Insulaemus 583
insulaemus, Apodemus 570
insulana, Macaca 267
insulans, Heteropsomys **799**, 799
insulanus, Peromyscus 730
Rattus 660
Rhinolophus 166
insularis, Alouatta 255
Ammospermophilus **419**, 419
Artibeus 188
Bandicota 579
Callosciurus 421
Cercopithecus 265
Chaerephon 233
Chaetodipus 483
Clethrionomys 509
Didelphis 17
Dipodomys **478**
Eptesicus 203
Genetta 346
Hemiechinus 78
Lasiurus 207
Lepus **818**
Lontra 311
Malpaisomys **609**, 609
Mandrillus 268
Marmosa 18
Microtus 518
Miniopterus 231
Mogera **125**
Myoxus 770
Nannospalax 754
Neotoma 712
Nesokia 632
Nycticebus 248
Phoca 332
Procyon **335**, 335
Pseudochirops 60
Pteropus **147**, 149
Reithrodontomys 741
Rhizomys 685
Scapanus 127
Scotophilus 227
Sorex 118
Tragelaphus 404
Tragulus 382
Ursus 339
insularum, Cynopterus 139
Leopoldamys 603
Myotis **211**
insulicola, Bassariscus 334
Peromyscus 730
Sigmodon 747
intagensis, Oryzomys 720
intectus, Oryzomys **722**
integer, Sundamys 666
intensa, Cynictis 302
Genetta 346
intensus, Xerus 458
intercastellanus, Phalanger 46
intercedens, Spermophilus 450
interceptio, Petaurista 463
intercessor, Tamias 455
interdictus, Peromyscus 734
interior, Funisciurus 427
Lontra 311
Myotis 216
interjecta, Macrotis 40
intermedia, Balaenoptera 350
Brachyphylla 182
Callithrix 252
Chinchilla 777
Crocidura 84
Galictis 318
Ictonyx 319
Kerivoula **197**
Lama 381
Lutra 312
Martes 320
Microtus 518
Mormoops 176
Murina 229
Neotoma 712
Nycteris 161, **162**
Orca 352
Ovis 408
Peropteryx 158
Tadarida 240
intermedius, Acomys 565
Allactaga 489
Apodemus 573
Arborimus 505
Artibeus 189

Bathyergus 771
Callithrix 251, 252
Callosciurus 421
Capromys 800
Chaetodipus **484**
Choeroniscus **183**
Clethrionomys 508
Delphinus 352
Dipodomys 480
Dryomys 766
Ellobius 512
Eptesicus 203
Feresa 352
Grampus 352
Hemiechinus 78
Lasiurus **207**
Lemmiscus 516
Melomys 618
Micropteropus **143**
Microtus 520
Myoxus 770
Nannospalax 753, 754
Neomys 110
Nycticebus 248
Oryzomys 721, **722**, 722, 723, 724
Phenacomys 505, **533**, 533
Platyrrhinus 191
Plecotus 225
Potamochoerus 378
Pteropus 151
Rattus 658
Reithrodontomys 741
Rhabdomys 662
Rhinolophus 166
Scalopus 127
Sciurus 443
Sorex 111
Spermophilus 446
Thomomys 476
Zapus 499
internationalis, Perognathus 485
Ursus 339
internatus, Thomomys 473
internus, Graphiurus 765
interparietalis, Peromyscus **731**, 731
interposita, Ratufa 437
interpositus, Phalanger 46, 47
interpres, Ammospermophilus **419**, 420
interrupta, Mephitis 317
Mustela 324
Spilogale 317
interstriatus, Sicista 498
intervectus, Sorex 115
interventus, Pteromys 464
interzonus, Sicista 498
intontoi, Galago 249
intraponticus, Meriones 558
intrudens, Paguma 343
intufi, Elephantulus **829**
inundata, Amblyrhiza **805**, 805
inunguis, Aonyx 310
Trichechus **365**
inusta, Pithecia 262
inustus, Dendrolagus **50**, 51
Dorcopsis 52
Galago 249
Saguinus **253**
Inuus 266
inuus, Macaca 268
Simia 266
invadens, Alcelaphus 393
investigator, Uropsilus **130**, 130
invicta, Mustela 321
invictus, Marmosops **19**
invisus, Hemigalus 342
inyangai, Crocidura 88
inyoensis, Mustela 322
Odocoileus 391
Tamias 456
Urocyon 285
io, Balantiopteryx **156**
Ia **205**, 205
Rhogeessa 226
iochanseni, Sorex 111
iodes, Tetracerus 403
iodinus, Naemorhedus 407
iodoprymnus, Galerella 303
Iomys **461**, 461
davisoni 461
everetti 461
horsfieldii **461**
lepidus 461
penangensis 461
sipora **461**, 461, 604
thomsoni 461
winstoni 461
iowae, Taxidea 325
iphigeniae, Microtus 526
ipnaeum, Plagiodontia **804**
iraki, Felis 290
iranensis, Lepus 817
Meriones 556
irani, Microtus **522**, 522, 530
Rhinolophus 166
irara, Eira 318
irazu, Scotinomys 746
irbis, Felis 299
Uncia 299
irene, Microtus **522**, 522, 529
Soriculus 123
Sylvisorex 105
Irenomys **705**
longicaudatus 705
tarsalis **705**
iretator, Lemmus 517
irex, Noctilio 176
iridescens, Pelomys 639
iriomotensis, Felis 295
Mayailurus 295
Prionailurus 295
iris, Amblysomus **74**
Oxymycterus 727
irkutensis, Clethrionomys 509
Sorex **121**
irma, Macropus **54**
irolonis, Pappogeomys 472
Iropocus 246
irremotus, Canis 281
irretitus, Pipistrellus 221
irrorata, Euryotis 680
Pithecia **261**
Sylvicapra 412
irroratus, Heterohyrax 373
Hydrochaeris 781
Liomys **482**
Otomys **680**, 680, 681, 682
Sciurus 441
irrupta, Caluromysiops **15**, 15
irus, Macaca 266
isaaci, Tragelaphus 403
isabelae, Genetta 345
isabella, Equus 369
Funisciurus **427**
Gazella 396
Sciurus 427
isabellae, Sylvisorex **104**, 105
isabellina, Lynx 293
Redunca 414
isabellinus, Apodemus 573
Cricetulus 538
Eptesicus 203
Equus 370
Heliosciurus 429
Hesperoptenus 205
Lepus 816
Macropus 54
Ursus 338, 339
isabellus, Diclidurus **157**
isarogensis, Rhynchomys **663**
Ischnoglossa, nivalis 185
ischyrus, Sigmodon 747
ischyurus, Thomasomys **750**
ishikawai, Sorex 117
isidis, Gazella **396**
isidori, Myotis 208
isiolae, Saccostomus 541
isis, Meriones 558
islandicus, Apodemus 573
ismahelis, Meriones 556
isodactylus, Chrysocyon 282
Elephas 367
Isodon, pilorides 800
isodon, Sorex **116**, 116, 120
isolatus, Graphiurus 764
Peromyscus 734
Sorex 118
Isolobodon **803**
levir 804
montanus **803**
portoricensis 803, **804**
Isolobodontinae **803**
Isomys 576
isonotus, Sus 379
isonyx, Arctonyx 313
Isoodon **39**
affinis 39
arnhemensis 39
auratus **39**
barrowensis 39
fusciventer 39
macrourus **39**
macrura 39
moresbyensis 39
nauticus 39
obesulus **39**
peninsulae 39
torosa 39
Isothrix **792**, 793
bistriata 792, **793**, 793
bistriatus 793
crassicaudus 793
molliae 793
negrensis 793
orinoci 793
pagurus **793**, 793
picta 793
pictus 792
villosa 793
villosus 793
isphaganica, Ovis 408
israelita, Chiropotes 261
isseli, Pelomys 639, **640**
issikulensis, Felis 290
istapantap, Mallomys **608**
istericus, Clethrionomys 508
isthmica, Dasyprocta 782
Marmosa 18
Nasua 334
Neotoma 712
isthmicum, Chiroderma 190
isthmicus, Orthogeomys 471
Potos 333
isthmius, Hydrochaeris 781
Liomys 482
Microsciurus 433
Isthmomys **705**, 706, 728
flavidus **706**, 706
pirrensis **706**, 706
Istiophorus, flavescens 187
istrandjae, Sciurus 443
istrianus, Apodemus 570
istricus, Spalax 755
Spermophilus 445
isubra, Cervus 386
italica, Crocidura 96
italicus, Arvicola 505, 506
Canis 281
Capreolus 390
Clethrionomys 508
Erinaceus 77
Miniopterus 231
Mustela 323
Myoxus 770
Rhinolophus 166
Sciurus 443
italosomalica, Asellia 170
itatiayae, Callithrix 251
Monodelphis 22
itatsi, Mustela 324
iterator, Lemmus 517
Ithydontia 803
itimbiriensis, Chlorocebus 265

iturensis, Panthera 298
ituricus, Colobus 270
Lophocebus 266
Stochomys 666
ituriensis, Cephalophus 411
Cercopithecus 263
Heliosciurus 429
Hylochoerus 377
Pan 277
Procavia 374
iturissius, Lepus 818
iulus, Semnopithecus 273
ivori, Helogale 304
ivoriensis, Hylochoerus 377
ixtlani, Habromys 703

jabouillei, Bandicota 579
jaburuensis, Saimiri 260
Jacchus 251
jacchus, Callithrix **252**
Simia 251
jacentior, Oxymycterus 727
jacinteus, Thomomys 473
jacki, Tupaia 132
jacksoni, Alcelaphus 393
Anomalurus 757
Bdeogale **301**, 301
Crocidura **87**
Cryptotis 108
Otomys 682
Paraxerus 435
Perognathus 485
Praomys **642**, 642, 643
Procavia 374
Sigmodon 747
Sorex 113, **116**, 116, 119, 121
Steatomys **545**, 545
Tamias 454
Taxidea 325
jacksoniae, Mus 627
jacobi, Gorilla 276
jacobita, Felis 294
Oreailurus **294**
jacobsoni, Cheiromeles 233
Niviventer 634
jacquemonti, Felis 290
Jaculidae 487
Jaculinae 492
Jaculus 488, **493**, 493, 494
aegyptius 493
airensis 493
arenaceous 493
bipes 494
blanfordi **493**, 494
butleri 493
centralis 493
collinsi 493
cufrensis 493
darricarrerei 493
deserti 493
elbaensis 493
favillus 493
favonicus 493
florentiae 493
fuscipes 493
gerboa 494
gordoni 493
hirtipes 493
jaculus **493**, 493, 494
locusta 494
loftusi 493
macromystax 493
macrotarsus 493
margianus 494
mauritanicus 494
oralis 493
orientalis 493, **494**, 494
rarus 493
schlueteri 493
sefrius 493
syrius 493
tripolitanicus 493
turcmenicus 493, **494**, 494
vastus 493
vocator 493
vulturnus 493
whitchurchi 493
jaculus, Allactaga 489
Jaculus **493**, 493, 494
Mus 488, 493
jacutensis, Arvicola 506
Clethrionomys 509
Sciurus 443
Sorex 119
Spermophilus 450
Tamias 455
jagori, Ptenochirus **145**, 145
jagorii, Kerivoula **197**
jaguapara, Panthera 298
jaguar, Panthera 298
jaguarete, Panthera 298
Jaguarius 297
jaguatyrica, Panthera 298
jakutensis, Vulpes 287
jalapae, Mus 625
Oryzomys 721
Reithrodontomys 742
jaliscensis, Liomys 482
Myotis 216
jallae, Equus 369
jalorensis, Lariscus 430
Rattus 661
jamaicensis, Artibeus 187, **188**, 188, 189
Clidomys 805
Macrotus 178
Natalus 195
jamesi, Canis 280
Gerbillus **552**
jamesoni, Cryptomys 772
Dendromus 543
Mus 625
jamrachi, Lophocebus 266
Naemorhedus 407
jamrachii, Rhinoceros 372
janenschi, Sigmoceros 395
janensis, Spermophilus 448
janetta, Bathyergus **771**, 771
Callosciurus 423
Paguma 343
Thylamys 23
japanensis, Macaca 267
japonensis, Eptesicus 202
Panthera **298**
japoniae, Miniopterus 231
japonica, Eubalaena 349
Felis 290
Lutra 312
Martes 320
Sus 379
Vulpes 287
Zalophus 329
japonicus, Canis 281
Cervus 386
Clethrionomys 532
Eubalaena 349
Glirulus **769**, 769
Micromys 620
Ursus 340
Zalophus 329
jarai, Cervus 387
jarak, Rattus 661
jardinii, Macropus 53
jarvisi, Panthera 298
jaumei, Mysateles 800
javana, Kerivoula 197
javanensis, Herpestes 304
Mephitis 315
Mydaus **315**, 315
Prionailurus 295
javanica, Hystrix **774**
Manis **415**
Nycteris **162**, 162
Paradoxurus 344
Tupaia **132**
javanicum, Hystrix 774
javanicus, Bos **401**, 401
Cervus 382, 387
Cuon 282
Cynopterus 139
Glirulus 769
Herpestes **305**, 305, 306
Hipposideros 171
Hylobates 276
Myoxus 769
Nycticebus 248
Pipistrellus **221**, 221
Pteropus 151
Rhinoceros 372
Rhinolophus 164
Tragulus **382**, 383
javanus, Glischropus **204**
Lariscus 430
Rattus 657
Rhizomys 685
javensis, Ratufa 437
jaxartensis, Eremodipus 493
Meriones 558
jayakari, Hemitragus **406**
Procavia 374
jeannei, Myotis 211
jebelae, Arvicanthis 578
jeffersonii, Taxidea 325
jeffreyi, Lepus 816
jeholensis, Lasiopodomys 516
jeholicus, Eothenomys 515
jei, Meriones 557
jelskii, Akodon 700
Chroeomys **700**, 700
Mustela 322
jemlahica, Capra 406
Hemitragus 406
jemlahicus, Hemitragus **406**, 406
jemlanica, Hemitragus 406
jemuris, Rattus 661
jenaensis, Ursus 339
jeniseensis, Ursus 339
jenissejensis, Sciurus 443
Sorex 121
jenissijensis, Arvicola 506
jenkinsi, Crocidura **87**, 91
jentinki, Cephalophus **410**
Sundasciurus **451**
Taeromys 667
jeppei, Otomys 681
jerdoni, Capra 405
Hemiechinus 78
Niviventer 634
Paradoxurus **344**
jerdonii, Herpestes 306
jesseni, Synaptomys 534
jessook, Rattus 652
jesupi, Chiroderma 190
Heteromys 481
jeudii, Cricetus 538
jewetti, Ochotona 811
jharal, Capra 406
Hemitragus 406
jimela, Damaliscus 394
jin, Otonycteris 219
jindongensis, Typhlomys 684
jingdongensis, Chiropodomys 583, 584
Hylobates 275
jivaro, Caluromys 15
Nasua 335
joanae, Tatera 560
joannia, Microcavia 780
joannius, Octomys 788
jobensis, Chaerephon **232**
jobiensis, Melomys 618
Rattus 649, **653**, 653
jochelsoni, Clethrionomys 509
joffrei, Capreolus 390
Pipistrellus **221**
johanis, Ctenomys 785, 787
johannae, Eulemur 244
Tragelaphus 404
johannes, Lasiopodomys 516
johannis, Acomys 565
Ctenomys 785
johnii, Trachypithecus **274**
johnsoni, Marmota 432
Pseudomys 645, **647**
johnstoni, Aepyceros 393
Cephalophus 411
Chlorocebus 265
Connochaetes 394
Crocidura 88
Equus 383
Genetta **346**

Graphiurus 764
Lophocebus 266
Okapia **383**
Potamochoerus 378
Procavia 374
Sylvisorex **104**
johnstonii, Phalangista 48
Trichosurus 48
johorensis, Callosciurus 422
Chaerephon **232**, 232
Ratufa 437
jojobae, Thomomys 473
joleaudi, Ctenodactylus 761
joloensis, Sus 379
jolonensis, Dipodomys 478
jonesi, Damaliscus 394
Epixerus 425
Hipposideros **173**
Leporillus 604
joongshaiensis, Lepus 819
jordani, Anomalurus 757
Callithrix 252
Gerbillus 552
Graphiurus 765
jordansi, Felis 290
joretianus, Cervus 386
jorisseni, Cryptomys 772
josti, Neomys 110
joubertianus, Cervus 387
jourdanii, Paguma 343
jourensis, Capra 405
jouvenetae, Crocidura 83
ju-ju, Perodicticus 248
juara, Alouatta 255
juarezensis, Thomomys 473
jubata, Alces 389
Eumetopias 328
Felis 288
Hippotragus 412
Myrmecophaga 68
Ovis 408
Phoca 327
jubatulus, Sus 379
jubatus, Acerodon **137**
Acinonyx **288**
Canis 282
Chrysocyon 282
Eumetopias **328**
Pteropus 137
Sus 379
Trachypithecus 274
jubilaeus, Papio 269
jucunda, Eligmodontia 701
Mazama 390
jucundum, Artibeus 188
jucundus, Cerdocyon 282
Chroeomys 700
Pappogeomys 472
judaica, Crocidura 88
judaicus, Rhinolophus 165
judeae, Lepus 817
judex, Nasua 335
Oxymycterus 727
juglans, Spermophilus 450
jugossicularis, Geomys 469
jugularis, Mormopterus **237**, 237
Tragulus 383
jujensis, Rattus 658
jukesii, Thylogale 57
juldaschi, Microtus **522**, 522, 523
juliacae, Oxymycterus 727
julianae, Amblysomus **74**
juliani, Gerbillus **552**, 552
julianus, Rattus 661
juncensis, Sorex 118
juninensis, Akodon **690**
Caluromys 15
Marmosops 20
juniperus, Tamias 455
junodi, Cryptomys 772
juntae, Thomomys 476
juquiaensis, Thyroptera 196
juralis, Sciurus **442**
jurassicus, Clethrionomys 508
Rattus 658
juris, Ctenomys 785
jurrasicus, Eliomys 767
juruana, Alouatta 255
Mazama 390
Saimiri 260
juruanus, Cebus 259
Choloepus 64
Cyclopes 67
Saguinus 254
jurvana, Nasua 335
Juscelinomys **706**, 706
candango **706**, 706
talpinus **706**, 706
juvencus, Sundasciurus **451**

kabambarei, Procolobus 272
kabanicus, Rattus 661
kabobo, Protoxerus 436
kabylicus, Lepus 816
kachhensis, Taphozous 161
kadanus, Rattus 660
Kadarsanomys **600**, 600
sodyi **600**
kadiacensis, Mustela 321
kadiaki, Ursus 339
kadugliensis, Taterillus 563
kaffensis, Canis 280
Heliosciurus 428
kafuensis, Kobus 414
kageus, Phaulomys 533
kagi, Chiruromys 585
kagii, Bandicota 579
kaguyae, Myotis 211
kahari, Paraxerus 435
kahmanni, Apodemus 570
kahuziensis, Cercopithecus 264
Dendromus 541, **542**
kaibabensis, Puma 296
Sciurus 439
Thomomys 475
kaiboaba, Ceratotherium 371
kaimosae, Cercopithecus 263
Hylomyscus 599
kaiseri, Aethomys **568**, 568, 569
Gerbillus 554
kakhyenensis, Mus 623
kakumegae, Lepus 822
kalabuchovi, Spermophilus 448
kalaharica, Cynictis 302
kalaharicus, Galerella 303
Gerbillurus 547
Ictonyx 319
Lepus 816
Malacothrix 544
Paraxerus 434
Steatomys 546
Thallomys 668
kalbinensis, Sciurus 443
kaleensis, Belomys 459
kaleh-peninsularis, Mus 625
kalianda, Callosciurus 422
Kalimitalpa 75
kalinowskii, Dasyprocta **782**
Mormopterus **237**
Thomasomys **750**
kalmikensis, Dipus 493
kalmykorum, Lepus 817
Vulpes 286
kalubu, Echymipera **41**, 41
Perameles 41
kambei, Mus 625
kamensis, Cricetulus **537**, 537
Lynx 293
Ochotona 809, 811
kamerunensis, Procavia 374
kampenii, Taphozous 160
kamptzi, Panthera 297
kamtschadensis, Vulpes 287
kamtschatica, Enhydra 310
Lutra 312
Microtus 527
Mustela 323
kamtschaticus, Canis 281
Clethrionomys 509
Gulo 318
Lepus 821
Ochotona 809
kanchil, Tragulus 382
kandianus, Rattus 658
Suncus 103
kandiyanus, Rattus 658
kandti, Cercopithecus 264
kanei, Mustela 321
kangeanus, Paradoxurus 344
kangosa, Sigmoceros 395
Kangurus 53
bicolor 57
brachyurus 56
fasciatus 53
gaimardi 49
penicillatus 55
ualabatus 57
Kannabateomys **790**
amblyonyx **790**
pallidior 790
kanoi, Eothenomys 514
kansensis, Taxidea 325
kansuensis, Cervus 386
kanuri, Gazella 397
kaokensis, Equus 369
kaokoensis, Galerella 303
Graphiurus 765
Petromyscus 683
Pronolagus 823
Tatera 561
kapalgensis, Taphozous **160**
kapiti, Heliophobius 772
kappleri, Dasypus **66**
Peropteryx **158**
karagan, Vulpes 287
karaginensis, Microtus 527
karamani, Spermophilus 445
karanorensis, Canis 281
karasensis, Cynictis 302
Petromus 774
karatshaicus, Arvicola 506
karelicus, Apodemus 570
Sorex 117
karelini, Meriones 557
Ovis 408
Stylodipus 495
kariegae, Pronolagus 823
karieteni, Meriones 558
karimatae, Presbytis 271
karimii, Thylamys 23
karimondjawae, Macaca 266
karimoni, Macaca 266
karkinophaga, Didelphis 17
karlkoopmani, Chiropodomys **584**, 604
Karnimata 594
karoensis, Otomys 681
karpathorum, Lepus 817
karpinskii, Sorex 113
Karstomys 569, 572
kasaicus, Mus **624**, 628
Steatomys 546
kaschgaricus, Lepus 821
kashgaricus, Eptesicus 202
kashyiriensis, Rhinolophus 166
Kasi 270, 273, 274
kassaicus, Cercopithecus 263
kastchenkoi, Sorex 117
kastschenkoi, Ellobius 513
Marmota 431
katangae, Aepyceros 393
Cercopithecus 263
katanganus, Tragelaphus 403
katangensis, Chlorocebus 265
kataui, Dactylopsila 61
katharina, Crocidura 93
kathiah, Mustela **322**
kathleenae, Funambulus 427
Millardia **621**, 621
katschenakensis, Gulo 318
katsurai, Mustela 324
kaufmanni, Equus 369
Hippotragus 413
Taurotragus 403
kazakstanicus, Spermophilus 448

kazbegica, Sicista 496, **497**, 497
kaznakovi, Microtus 529
keatii, Presbytis 271
keaysi, Marmosops 20
Myotis **211**
Oryzomys 720, **723**, 723
keelingensis, Rattus 660
keenani, Mimon 179
keeni, Peromyscus 732, 734
keenii, Myotis **211**
keewatinensis, Rangifer 392
keiensis, Dendrolagus 50
Echymipera 41
keitloa, Diceros 372
kekrimus, Berylmys 580
kelaarti, Felis 290
Funambulus 426
Pteropus 147
Rattus 658
Suncus 103
kelaartii, Trachypithecus 274
kelabit, Crocidura 85
Tupaia 133
kelleni, Graphiurus **764**, 765
Raphicerus 399
kelleri, Rattus 660
kelloggi, Oryzomys **723**
Sylvilagus 825
Thomomys 475
Kemas 406
kemas, Pantholops 399
kematoceros, Cervus 386
kemmisi, Callosciurus 421
kempi, Acomys **565**, 565, 566
Akodon **690**
Crocidura 91
Heterohyrax 373
Leptailurus 292
Otomys 680
Rousettus 152
Tatera **561**, 561, 562
Thamnomys **669**
kenaiensis, Alopex 279
Martes 319
Ovis 409
Tamiasciurus 457
Ursus 338, 339
Vulpes 287
keniae, Alcelaphus 393
Heliosciurus 429
Potamochoerus 378
keniensis, Rhinolophus 164
kennedyi, Oreamnos 407
kennerleyi, Ursus 339
kennethi, Berylmys 580
kenniae, Cephalophus 410
kennicottii, Microtus 528
Spermophilus 448
kennioni, Gazella 396
kentucki, Synaptomys 534
kenyacola, Chalinolobus **199**
kenyae, Ourebia 399
kenyensis, Cricetomys 541
kenyoni, Enhydra 310
kephalopterus, Trachypithecus 274
kerabau, Bubalus 402
keramae, Cervus 386
keraudren, Pteropus 148
keraudrenii, Callosciurus 422
kerensis, Mus 624
kerguelensis, Mirounga 330
Kerivoula **196**, 196, 197, 198
aerosa **196**
africana **196**, 197
agnella **196**
argentata **196**
atrox **196**
bellissima 198
bellula 197
bicolor 198
bombifrons 198
brunnea 197
crypta 197
cuprosa **196**
depressa 197
engana 197
eriophora **197**
flora **197**, 197
fusca 197
hardwickii **197**
harrisoni 197
intermedia **197**
jagorii **197**
javana 197
lanosa **197**
lenis 197
lucia 197
malayana 197
malpasi 197
minuta **197**
muscilla 197
muscina **197**
myrella **197**
nidicola 196
papillosa **197**
papuensis **197**
pellucida **198**
phalaena **198**
picta **198**
pusilla 198
rapax 197
smithii **198**
whiteheadi **198**
zuluensis 196
Kerivoulinae **196**
kermanensis, Microtus **523**
kermiti, Proechimys 798
kermodei, Ursus 338
kernensis, Dipodomys 479
Microtus 519
Tamias 454
Kerodon **780**
aciurens 780
acrobata 780
moco 780
rupestris **780**
kerstingi, Erythrocebus 265
Procavia 374
kessleri, Lepus 821
Sciurus 443
kevella, Gazella 396
keyensis, Pteropus 148
Rhinolophus **166**
Syconycteris 155
keysseri, Thylogale 57
khanensis, Galerella 303
Lepus 820
khorkoutensis, Microtus 531
khubsugulensis, Alticola 504
khumbuensis, Rattus 661
khur, Equus 370
khyensis, Rattus 660
kiang, Equus **370**, 370
kibalensis, Okapia 383
kiboko, Hippopotamus 381
kibonotensis, Cercopithecus 264
Crocuta 308
kidderi, Ursus 339
kievensis, Erinaceus 77
kijabae, Crocidura 92
kijabius, Rattus 658
kikuchii, Volemys **535**, 535
kikuyuensis, Colobus 270
Otolemur 250
kilangmiutak, Dicrostonyx **510**, 510, 511
kilikiae, Apodemus 573
kilimandjari, Dendromus 542
kilimanus, Atelerix 76
Kilimitalpa 74, 75
kima, Cercopithecus 264
kina, Niviventer 633
kinabalu, Megaderma 163
kinabaluensis, Maxomys 612
kincaidi, Microtus 527
kindae, Papio 269
kingiana, Crocidura 82
kingii, Microcavia 780
kinlochii, Petaurillus **462**
kinneari, Callosciurus 421
Rhinopoma 155
kinoensis, Ammospermophilus 419
Perognathus 485
kiodotes, Macroglossus 154
kirchenpaueri, Neotragus 398
kirgisorum, Microtus **523**
Nannospalax 753, 754
Spalax 753
kiriwinae, Phalanger 46
kirki, Procolobus 272
kirkii, Hippotragus 413
Madoqua **398**
Otolemur 250
kirschbaumii, Plecotus 224
kirtlandi, Blarina 106
kishidai, Lasiopodomys 516
kitcheneri, Pipistrellus **221**
kittlitzi, Lemmus 517
kivu, Crocidura 92
Dendromus **542**, 542, 543
kivuana, Crocidura **87**
Lutra 312
kivuensis, Cephalophus 411
Cricetomys 540
Idiurus 758
Leptailurus 292
Thamnomys 670
Kivumys 605, 606
kiyomasai, Vulpes 287
kizljaricus, Allactaga 488
kjusjurensis, Microtus 527
klagesi, Marmosa 18
Oecomys 717
Sciurus 441
klamathensis, Glaucomys 460
Lepus 814
Reithrodontomys 741
Ursus 339
klaverensis, Aethomys 568
klippspringer, Oreotragus 399
klossi, Callosciurus 422
Crocidura 89
Euroscaptor **125**, 125
Hystrix 773
Maxomys 614
Ratufa 437
Rhinolophus 168
Stenomys 665
Talpa 125
Tragulus 382
klossii, Hylobates **275**
klozeli, Microtus 530
kluane, Ursus 339
kluchorica, Sicista 496, **497**, 497
klumensis, Rattus 659
knekus, Chaetodipus 483
knighti, Ctenomys **785**, 785
knightii, Canis 281
knochenhaueri, Loxodonta 367
knorri, Rhinolophus 165
knoxi, Mesoplodon 363
knoxjonesi, Geomys 469, 470
Onychomys 719
knutsoni, Tragelaphus 404
knysnae, Crocidura 84
Koala 45
koala, Phascolarctos 45
kob, Kobus **414**, 414
koba, Hippotragus 412
kobayashii, Eptesicus **202**
Ochotona 809
kobeae, Mogera **126**, 126
kobosensis, Petromus 774
Pronolagus 823
Kobus **413**
abyssinica 413
adansoni 414
adenota 414
adolfi 414
adolfifriderici 413, 414
albertensis 413
alurae 414
amboellensis 414
angusticeps 413
annectens 413
annulipes 414
avellanifrons 413

bahrkeetae 414
breviceps 413
buffonii 414
canescens 413
cottoni 413
crawshayi 413
defassa 413, 414
dianae 413
ellipsiprymnus **413**
forfex 414
fraseri 414
frommi 413
fulvifrons 413
grandicornis 414
griseotinctus 413
harnieri 413
hawashensi 413
kafuensis 414
kob **414**, 414
kondensis 413
kul 414
kulu 413
kuru 413
ladoensis 413
leche **414**
leucotis 414
lipuwa 413
loderi 414
maria 414
matschiei 413
megaceros **414**
muenzneri 414
neumanni 414
nigricans 414
nigroscapulatus 414
notatus 414
nzoiae 414
pallidus 414
penricei 414
pousarguesi 414
powelli 414
raineyi 414
riparia 414
robertsi 414
schubotzi 414
senegalensis 414
senganus 414
singsing 414
smithemani 414
thikae 414
thomasi 414
tjaederi 414
togoensis 414
tschadensis 414
ubangiensis 414
ugandae 414
unctuosus 414
uwendensis 414
vardonii **414**
vaughani 414
kodiacensis, Lontra 311
Microtus 527
Spermophilus 448
kodiaki, Ursus 339
koellikeri, Trichechus 365
koenigi, Mesocricetus 539
koepckeae, Mimon 179
kofordi, Akodon **690**, 690
Kogia **359**, 359
breviceps **359**
floweri 359
goodei 359
grayii 359
simus **359**
Kogiidae 359
kohlbruggei, Trachypithecus 273
kohlsi, Petaurus 61
kohtauensis, Tupaia 131
Koiropotamus 377
koiropotamus, Potamochoerus 378
kok, Bandicota 579
koka, Bunomys 581
kokandicus, Meriones 558
kokree, Vulpes 286
kola, Canis 280
koladivinus, Prionailurus 296
kolbi, Cercopithecus 264
kolombatovici, Plecotus 224
kolymensis, Clethrionomys 509
Lepus 821
Ochotona 809
Ursus 339
komareki, Sigmodon 747
komatiensis, Cryptomys 772
Mastomys 611
Komemys 639
Komodomys 587, **600**, 600, 637, 649, 664
rintjanus **600**
komurai, Myospalax 676
kondana, Millardia **621**
kondensis, Kobus 413
kondoae, Lycaon 283
kongensis, Tamiops 457
kongoni, Alcelaphus 393
konzi, Sigmoceros 395
koodoo, Tragelaphus 404
koolookamba, Pan 277
koopmani, Antrozous 198
Ardops 187
Artibeus 189
Aselliscus 170
Coendou **775**, 775
Rattus 649, 651, 652, **654**, 654
Koopmania 187, 188
kootenayensis, Zapus 499
kopangi, Pteropus 151
kopschi, Cervus 386
korabensis, Arvicola 506
Dinaromys 511
korai, Rhinolophus 166
koratensis, Menetes 433
Rattus 659
Trachypithecus 273
koratis, Maxomys 614
kordofanensis, Arvicanthis 578
kordofanicus, Orycteropus 375
koreanus, Erinaceus 77
koreensis, Erinaceus 77
Nyctereutes 283
koreni, Microtus 527
Sorex 113
koriakorum, Ovis 409
Korin 395
korinchi, Rattus 650, 653, **654**, 655, 656
korrigum, Damaliscus 394
koseritzi, Didelphis 16
koshewnikowi, Vormela 325
kosidanus, Rhinolophus 166
koslovi, Meriones 558
koslowi, Ochotona **810**
kosogol, Alticola 504
kossogolicus, Microtus 521
kotiya, Panthera 298
kouri, Choloepus 64
kozeritzi, Mazama 391
kozhantschikovi, Cricetulus 538
kozhevnikovi, Lepus 821
kozlovi, Cricetulus 537
Felis 290
Lepus 819
Lynx 293
Marmota 431
Ovis 408
Plecotus 224
Salpingotus 491, **492**, 492
Sorex **116**, 116, 121
kraemeri, Spilocuscus 47
kraensis, Rattus 659
kraglievichi, Mesocapromys 800
kramensis, Rattus 660
kramis, Maxomys 614
krascheninikovi, Phoca 332
krascheninnikovi, Vespertilio 228
kratochvili, Talpa 129
kraussi, Chalinolobus 200
krebsi, Lycaon 283
krebsii, Steatomys **545**
krefftii, Lasiorhinus **45**, 45
Pipistrellus 224
kretami, Tupaia 133
kreyenbergi, Erinaceus 77
Hydropotes 388
Oryctolagus 822
kriegi, Eira 318
Tamandua 68
krimeamontana, Vulpes 287
krkensis, Apodemus 572, 573, 574
kroecki, Muscardinus 769
kroonii, Suncus 103
krugeri, Panthera 297
krugi, Atelerix 77
kubangensis, Cryptomys 771
kuboriensis, Tadarida 240
kuchingensis, Callosciurus 423
kuekenthali, Suncus 103
kuha, Cephalophus 410
kuhlii, Axis **384**
Callithrix **252**
Pipistrellus 219, **221**
Sciurillus 438
Scotophilus 226, **227**
kuhni, Liberiictis **307**, 307
kukilensis, Mus 623
kukunoriensis, Myospalax 675
Myotis 213
kul, Kobus 414
kulamanae, Ceratotherium 371
kulamane, Ceratotherium 371
kulan, Equus 370
kulu, Kobus 413
kumasi, Dendromus 543
kummeri, Coleura 156
kummi, Cryptomys 772
kummingii, Octodon 788
kunduris, Rattus 661
kunsi, Monodelphis **21**
Kunsia 696, **706**, 706, 745
chacoensis 706
fronto **706**
gnambiquarae 706
planaltensis 706
principalis 706
tomentosus **706**
kupelwieseri, Microtus 530
kura, Podihik 102
Suncus 102
kurauchii, Meriones 558
kurdistanica, Vulpes 287
kurdistanicus, Dryomys 766
kurilensis, Clethrionomys 509
Phoca 332
kurjak, Canis 281
kuro, Mus 625
kurodai, Crocidura 87
kuru, Grammomys 594
Kobus 413
kuruschi, Arvicola 506
kurzi, Petromyscus 683
kusnotoi, Megaerops **142**
kutab, Lutra 312
kutas, Felis 290
kutchicus, Hemiechinus 78
kutensis, Maxomys 614
Paradoxurus 344
kutscheruki, Sorex 118
kuvelaiensis, Zelotomys 674
kuznetzovi, Arvicola 506
kwakiutl, Ursus 339
kwango, Protoxerus 436
Kyphobalaena 350
kytmanovi, Micromys 620

labecula, Peromyscus 732
labensis, Microtus 524
labialis, Noctilio 176
labiata, Mustela 321
Nycteris 162

labiatus, Cercopithecus 264
Dicotyles 380
Epomophorus **141**
Macropus 54
Melursus 337
Nyctalus 217
Saguinus **253**
Tayassu 380
labilis, Diplomys **791**
labiosus, Bibimys **696**, 696
Scapteromys 745
Sorex 111
laborifex, Pseudomys **647**
labradorensis, Castor 467
Rangifer 392
Sorex 119
Ursus 339
labradorius, Canis 281
Lepus 815
Microtus 527
Taxidea 325
Ursus 325
Zapus 499
labrosus, Hypsignathus 142
labuanensis, Hipposideros 171
laceianus, Peromyscus 735
lacepedii, Saguinus 254
lacernata, Tupaia 132
lacernatus, Arvicanthis 576, 578
Meriones 578
laceyi, Peromyscus 728
Reithrodontomys 741
lachiguiriensis, Peromyscus 732
Lachnomys 790
lachrymans, Muntiacus 389
lachuguilla, Thomomys 473
lacrimalis, Pappogeomys 472
lacrymalis, Sigmoceros 395
Thomomys 473
lacrymosus, Cervus 386
lactens, Bolomys **696**
lactiventer, Berylmys 580
lacus, Rattus 655
lacustris, Mustela 324
Otomys 680
Taterillus **563**, 563
Xerus 458
lacuum, Gazella 396
ladacensis, Ochotona **810**
Vulpes 287
ladas, Zapus 499
lademanni, Herpestes 305
Heterohyrax 373
Lycaon 283
Orycteropus 375
Sigmoceros 395
ladewi, Thomasomys **750**
ladoensis, Diceros 372
Kobus 413
ladogensis, Phoca 332
laenata, Ratufa 437
laenatus, Sphiggurus 776
Laephotis **205**
angolensis **206**
botswanae **206**
namibensis **206**
wintoni 205, **206**
laephotis, Histiotus 205
Tonatia 181
laetus, Macaca 266
Paraxerus 435
laevipes, Gazella 397
laevis, Ornithorhynchus 13
lagaros, Pan 277
Lagenodelphis **353**
hosei **353**, 353
Lagenorhynchus **353**, 354, 355
acutus **353**, 353
albigena 353
albirostris **353**, 353
amblodon 353
australis **353**, 353
bivattus 353
breviceps 354
chilöensis 353
clanculus 353
cruciger **353**, 353, 354
electra 355
fitzroyi 354
gubernator 353
leucopleurus 353
longidens 354
obliquidens **354**, 354
obscurus 351, 353, **354**, 354
ognevi 354
panope 354
perspicillatus 353
pseudotursio 353
similis 354
supercilliosus 354
thicolea 354
tursio 354
wilsoni 353
Lagidium **777**, 777, 778
arequipe 778
boxi 778
crassidens 778
criniger 778
crinigerum 778
cuscus 778
cuvieri 778
famatinae 778
inca 778
lockwoodi 778
lutescens 778
moreni 778
pallipes 778
peruanum 777, **778**
punensis 778
sarae 778
saturata 778
subrosea 778
tontalis 778
tucumanum 778
viscacia **778**
vulcani 778
wolffsohni **778**
laglaizei, Cercopithecus 264
Lagocetus 361
Lagocheles 52
lagochilus, Macroglossus 154
Lagomina 807
Lagomorpha 807
lagomyiarius, Ursus 339
Lagomyidae 807
Lagomys 430, 807, 813, 824
lagonotus, Saguinus 253
lagopus, Alopex **279**, 279
Canis 279
Dipus 493
Pseudalopex 284
Lagorchestes **52**
asomatus **52**
bernieri 52
conspicillatus **52**
dorreae 52
hirsutus **52**
leichardti 52
leporides **52**
pallidior 52
Lagos 814
Lagostomidae 777
Lagostomus **778**
americana 778
diana 778
inmollis 778
maximus **778**
pamparum 778
petilidens 778
trichodactylus 778
Lagostrophus **53**
albipilis 53
elegans 53
fasciatus **53**
striatus 53
lagothricha, Simia 257
Lagothrix **257**
cana 258
capparo 258
caroarensis 258
castelnaui 258
flavicauda **257**
geoffroyi 258
hendeei 257
humboldtii 257, 258
infumata 258
lagotricha **258**
lugens 258
olivaceus 258
poeppigii 258
puruensis 258
thomasi 258
tschudii 258
ubericola 258
lagotis, Macrotis **40**
Perameles 39
lagotricha, Lagothrix **258**
lagunae, Peromyscus 736
Sorex 118
lagunensis, Pappogeomys 472
Lagurus 513, **515**, 516
abacanicus 515
agressus 515
altorum 515
curtatus 516
lagurus **515**
migratorius 515
occidentalis 515
saturatus 515
lagurus, Lagurus **515**
Mus 515
lahulis, Alticola 503
laingi, Microtus 530
Perognathus 486
Phenacomys 533
lakedownensis, Leggadina 600, **601**
lakiundae, Crocidura 91
lalandei, Lycaon 283
lalandi, Otocyon 284
lalandia, Redunca 414
lalandianus, Graphiurus 764
lalandii, Chlorocebus 265
Megaptera 350
Proteles 309
lalawora, Maxomys 613
lalolis, Rattus 660
Lama **381**
ameghiniana 381
araucanus 381
cacsilensis 382
castelnaudi 381
chilihueque 381
cordubensis 381
crequii 381
domestica 381
ensenadensis 381
fera 382
glama **381**, 382
guanicoe **382**
huanacus 382
intermedia 381
lama 381
llacma 381
loennbergi 382
lujanensis 382
mesolithica 381, 382
molinaei 382
moromoro 381
pacos **382**
peruana 381
peruviana 382
voglii 382
lama, Alticola 504
Cricetulus 537
Lama 381
Ochotona 811, 812
lamarum, Echimys **792**
lambertoni, Nesomys 679
lambi, Chaetodipus 484
Molossus 235
Myotis 216
Oryzomys 721
lamia, Chiruromys **584**
Oryzomys **723**
Lamictis 341
laminatus, Otomys 680, **681**
lamington, Uromys 671

Lamingtona 218
Lamotomys 680, 681
lamottei, *Crocidura* 85, **87**
Hipposideros **173**
Micropotamogale **72**, 72
Lamottemys **600**
okuensis **600**, 600
lampensis, *Tragulus* 382
lampo, *Lenomys* 602
Lamprogale 319
Lampronycteris 178
lamucotanus, *Callosciurus* 422
Rattus 661
lamula, *Soriculus* **123**, 123
lanaceus, *Myotis* 210
lanata, *Pelea* 413
lanatus, *Avahi* 246
Caluromys **15**
Myotis 210
Phyllotis 738
Thylogale 57
lancasteri, *Chaerephon* 232
Galerella 303
Ictonyx 319
lancavensis, *Callosciurus* 421
Leopoldamys 603
Tragulus 382
lanceolatum, *Carollia* 186
lanceolatus, *Myotis* 212
landakkensis, *Pongo* 277
landbecki, *Akodon* 692
landeri, *Rhinolophus* **166**, 166
lanea, *Acinonyx* 288
lanei, *Mops* 237
lanensis, *Pteropus* 151
Rattus 660
laneus, *Paradoxurus* 344
langbianis, *Niviventer* 632, 633, **634**
langguthi, *Akodon* 690
langheldi, *Colobus* 270
Erythrocebus 265
Gazella 397
Hippotragus 412
Lycaon 283
Papio 269
langi, *Atelerix* 76
Chaerephon 233
Cricetomys 541
Crocidura 87
Cryptomys 772
Hipposideros 172
Idiurus 758
Lepus 816
Madoqua 398
Procolobus 272
Rhynchogale 307
langkatensis, *Pongo* 277
langsdorffi, *Chironectes* 16
Sciurus 442
lania, *Semnopithecus* 273
laniarius, *Dasyurus* 35
Sarcophilus **35**
laniger, *Abrocoma* 789
Alouatta 255
Anomalurus 757
Avahi **246**
Canis 281
Cuon 282
Daubentonia 247
Dryomys **766**
Lemur 246
Mus 777
Myotis 210
Paguma 343
Pteropus 148
Sminthopsis **36**
Thomasomys **750**, 750
Ursus 340
lanigera, *Caluromys* 15
Chinchilla **777**, 777
Oreamnos 407
Paguma 344
lanigerus, *Macropus* 55
Paguma 343
laniginosa, *Pseudocheirus* 60
laniginosus, *Pseudocheirus* 60
lanius, *Aotus* 256
Orthogeomys **471**, 471
lanka, *Canis* 280
Herpestes 305
Petaurista 463
lankadiva, *Hipposideros* **173**
lankavensis, *Viverra* 348
lanosa, *Crocidura* **87**
Kerivoula **197**
Murina 229
lanosus, *Akodon* **690**, 691
Melomys 615, **616**, 616, 617, 619
Rhinolophus 167
Rousettus **152**, 153
lanuginosus, *Millardia* 621
Mystromys 677
Tamiasciurus 457
laomache, *Dremomys* 425
Laomys 674
laosiensis, *Bos* 401
laotum, *Hylopetes* 461
Melogale 315
Paradoxurus 344
Tamiops 457
Trachypithecus 273
Tupaia 131
lapidarius, *Saccostomus* 541
laplataensis, *Neotoma* 710
lapponicus, *Sorex* 113
lapponum, *Rangifer* 392
lapsus, *Felis* 291
Glirulus 769
Macaca 266
laptevi, *Odobenus* 326
lar, *Crocidura* 96
Homo 274
Hylobates **275**, 275
larapinta, *Sminthopsis* 37
lardarius, *Nyctalus* 217
larensis, *Myotis* 213
largha, *Phoca* **332**, 332
largus, *Dipodomys* 477
Laria 430
laricorum, *Sorex* 112
Lariscus **430**
atchinensis 430
auroreus 430
castaneus 430
diversoides 430
diversus 430
fornicatus 430
hosei **430**
insignis **430**, 430
jalorensis 430
javanus 430
meridionalis 430
murianus 430
niobe **430**, 430
obscurus **430**, 430, 604
peninsulae 430
rostratus 430
saturatus 430
siberu 430
vulcanus 430
laristanica, *Ovis* 408
larkeni, *Poelagus* 823
larkenii, *Tragelaphus* 404
laronesiotes, *Cervus* 387
larotina, *Eliomys* 767
larotinus, *Eliomys* 767
larseni, *Leptailurus* 292
larus, *Apodemus* 573
larusius, *Rattus* 661
larvalis, *Pedetes* 759
larvarum, *Soriculus* 123
larvata, *Paguma* **343**
larvatus, *Cercopithecus* 270
Ellobius 513
Gulo 343
Hipposideros **173**
Mustela 322, 324
Nasalis **270**
Potamochoerus **378**
Pseudocheirus 59
Spilogale 317
lasalensis, *Ochotona* 811
lasallei, *Tremarctos* 338
lascivus, *Glaucomys* 460
lasia, *Crocidura* 88
lasiae, *Macaca* 266
Megaderma 163
Rattus 659
Lasiomys 605, 746
afer 605
Lasionycteris **206**
noctivagans **206**
Lasiopodomys **515**, 515, 516, 517, 518
aga 515
brandtii **515**
faeceus 516
fuscus **516**
hangaicus 515
jeholensis 516
johannes 516
kishidai 516
mandarinus **516**
pullus 516
vinogradovi 516
warringtoni 515
lasiopterus, *Nyctalus* **216**, 216
Lasiopyga 263
lasiopyga, *Anoura* 183
Lasiorhinus **45**
barnardi 45
gillespiei 45
krefftii **45**, 45
lasiorhinus 45
latifrons **45**
mcoyi 45
lasiorhinus, *Lasiorhinus* 45
lasiotis, *Dicerorhinus* 372
Glirulus 769
Martes 320
Mus 748
Odocoileus 391
Otolemur 250
Thalpomys **748**, 748
lasiotus, *Bolomys* 696
Macaca 267
Otolemur 250
Ursus 339
lasistanicus, *Ursus* 339
lasistanius, *Chionomys* 507
lasiura, *Crocidura* **87**
Lasiuromys 793
Lasiurus **206**, 206, 207
argentinus 207
blossevillii 206
bonariensis 206
borealis **206**, 207
brachyotis 206
brasiliensis 206
castaneus **206**
caudatus 207
cinereus **206**
degelidus 206
ega **207**
egregius **207**
enslenii 206
floridanus 207
frantzii 206
funebris 206
fuscatus 207
grayi 206
insularis 207
intermedius **207**
lasiurus 206
mexicana 206
minor 206
monachus 206
noveboracensis 206
ornatus 206
pallescens 206
panamensis 207
pfeifferi 206
pruinosus 206
punensis 207
quebecensis 206
rubellus 206
rubra 206
rufus 206

salinae 206
seminolus **207**
semotus 206, 207
teliotis 206
tesselatus 206
varius 206
villosissimus 206, 207
xanthinus 207
lasiurus, Bolomys 689, **696**, 696, 697, 752
Lasiurus 206
Platacanthomys **684**, 684
lasius, Bassaricyon **333**
Peromyscus 736
laskarevi, Spermophilus 445
laskerewi, Lepus 817
lassacquerei, Rattus 652
lasti, Galerella 303
lastii, Paraxerus 435
lasyurus, Harpiocephalus 228
latastei, Clethrionomys 509
Gerbillus 551, **552**, 552, 554
Procavia 374
Latax 310
Lataxina 310
lataxina, Lontra 311
latebricola, Akodon 689, **690**
lateralis, Canis 280
Petrogale **56**
Spermophilus **447**
lateris, Heliosciurus 428
latibarba, Trachypithecus 274
latibarbatus, Trachypithecus 274
laticaudatus, Nyctinomops **239**, 240
Rhinosciurus **438**
Sciurus 438
laticeps, Allactaga 488
Anomalurus 757
Clethrionomys 509
Clyomys **793**
Colobus 270
Echimys 793
Hylomyscus 598
Lophuromys 605
Meriones 558
Oryzomys 721, 722, 725
Sus 379
Thomomys 473
Tragelaphus 404
laticinctus, Sylvilagus 824
Latidens **142**
salimalii **142**, 142
latidens, Abditomys **564**
Amblyrhiza 805
Anoura **183**
Cervus 386, 387
Lontra 311
Rattus 564
Thoopterus 153
latifolius, Phyllostomus **180**
Rhinolophus 169
latifrons, Hyperoodon 361
Lasiorhinus **45**
Lontra 311
Microtus 518
Neotoma 710
Orthogeomys 471
Phascolomys 45
Ursus 339
latigriseus, Clethrionomys 509
latijularis, Chaetodipus 484
latimanus, Phenacomys 533
Rhipidomys **744**, 744, 745
Scapanus **127**
latimaxillaris, Dipodomys 477
latipennis, Myotis 213
latipes, Glaucomys 460
Melomys 617
Oryx 413
Rattus 658
latirostra, Mustela 322
Neotoma 712
latirostris, Arctocephalus 326
Cabassous 65
Chaetodipus 483
Marmota 432
Myotis 213
Orthogeomys 471
Peromyscus 730
Thomomys 473
Trichechus 365
latona, Crocidura 84, **88**
latouchei, Berylmys 580
Mogera 126
Rhizomys 685
Tadarida 240
latrans, Canis **280**, 281
Cynomys 424
latrator, Dendrohyrax 373
latro, Ctenomys **785**, 785
Sciurotamias 439
latronum, Apodemus **572**, 572
latu, Galerella 302
latus, Capromys 800
Microtus 523
Platyrrhinus 190, 191
Thomomys 473
latycranius, Cricetus 538
laucha, Calomys 697, **698**, 698
Mus 697
laurentianus, Tamiasciurus 457
laurentius, Thrichomys 798
laurillardi, Tapirus 371
lautensis, Callosciurus 422
Cynocephalus 135
lauterbachi, Thylogale 57
lavaudeni, Atelerix 77
lavellanus, Pteropus 150
Lavia **162**
affinis 163
frons **163**
rex 163
lawesi, Tachyglossus 13
lawnensis, Hydromys 597
lawrancei, Madoqua 398
lawsoni, Perameles 40
lawuana, Crocidura 85
laxmanni, Myospalax 676
layardi, Funambulus **426**
Petinomys 463
layardii, Mesoplodon **363**
lea, Crocidura **88**
leachi, Rousettus 152
leachii, Glossophaga **184**
Hylopetes 460
Nyctophilus 218
leadbeateri, Gymnobelideus **61**, 61
leakyi, Heliosciurus 429
leander, Anotomys **694**, 694
Chimarrogale 107
leathemi, Sicista 497
lebanoticus, Nyctalus 217
lebombo, Cephalophus 411
Poecilogale 325
Pronolagus 823
lebomboensis, Thallomys 669
lebrunii, Chionomys 507
leche, Kobus **414**
lechei, Cryptomys 772
Cynocephalus 135
Delomys 700
Didelphis 16
lechochloides, Aethomys 568
leconteii, Plecotus 225
lecontii, Reithrodontomys 741
lecoqi, Panthera 298
lectus, Tamias 454
leechi, Cremnomys 588
lefebvrei, Galerella 303
legata, Lenothrix 591
legatus, Chiropodomys 584
Diplothrix **591**, 591
Oryzomys 721, **723**, 723, 724
legendrei, Ellobius 512
Nesokia 632
legerae, Mastomys 611
Pectinator 761
Leggada 622
Leggadilla 622
Leggadina **600**, 600, 644, 647, 674
berneyi 601
delicatulus 600
forresti 600, **601**, 601
hermannsburgensis 600
lakedownensis 600, **601**
messorius 601
waitei 601
legrandianus, Cervus 386
lehmanni, Genetta 346
Lepus 821
Paradoxurus 344
lehocla, Aethomys 568
leibii, Myotis **212**, 216
leichardti, Lagorchestes 52
leightoni, Poiana 347
Leimacomys **544**, 544
buettneri **544**, 544
Leiobalaena 349
leioprymna, Proechimys 795
leira, Eira 318
leisleri, Nyctalus **216**, 216, 217
Vespertilio 216
Leith-adamsia 367
Leithia 765, 766
Leithiidae 763, 765
Leithiinae 763, **765**, 766, 767, 768, 769
Leithiini 766, 767
lekaguli, Hipposideros **173**
lelwel, Alcelaphus 393
lembicus, Macaca 267
lemeanus, Cervus 387
lemerlei, Hippopotamus **381**
lemhi, Ochotona 811
lemkei, Rhizoplagiodontia **804**, 804
Lemmimicrotus 517
lemminus, Alticola **503**, 503
Punomys **740**, 740
Lemmiscus **516**, 516
artemisiae 516
curtatus 515, **516**
intermedius 516
levidensis 516
orbitus 516
pallidus 516
pauperrimus 516
Lemmomys 512
Lemmus **516**, 516, 531, 534
alascensis 517
amurensis **516**, 517
borealis 517
bungei 517
chrysogaster 517
flavescens 517
harroldi 517
helvolus 517
iretator 517
iterator 517
kittlitzi 517
lemmus **517**, 517
migratorius 517
minor 517
minusculus 517
nigripes 517
niloticus 576
norvegicus 517
novosibiricus 517
obensis 517
ognevi 516
paulus 517
phaiocephalus 517
portenkoi 517
sibiricus **517**, 517
subarticus 517
trimucronatus 517
xanthotrichus 517
yukonensis 517
lemmus, Lemmus **517**, 517
Mus 516
lemniscatus, Funisciurus **427**

Lemniscomys **601**, 640
akka 602
albolineatus 601
ardens 602
barbarus **601**, 601
bellieri 592, **601**, 601, 602
calidior 602
convictus 601
dieterleni 602
dorsalis 602
dunni 601
fasciatus 602
fitzsimonsi 602
griselda **601**, 602, 640
hoogstraali **601**, 601
ifniensis 601
linulus **601**
lulae 602
lynesi 602
macculus 601, **602**, 602
maculosus 602
manteufeli 601
massaicus 602
mearnsi 602
micropus 602
mittendorfi **602**
nigeriae 601
nubalis 601
olga 601
orientalis 601, 602
oweni 601
phaeotis 602
pulchella 602
pulcher 602
rosalia 601, **602**, 602
roseveari **602**
sabiensis 602
sabulatus 602
spekei 601
spermophilus 602
spinalis 602
striatus 601, **602**, 602
venustus 602
versustos 602
wroughtoni 602
zebra 601
zuluensis 602
Lemmus 516
Lemur **245**, 245
catta **245**, 245
flavus 333
furcifer 244
griseus 245
indri 247
laniger 246
mococo 245
mongoz 244
murinus 243
potto 248
pusillus 243
tardigradus 248
variegatus 245
varius 245
volans 135
Lemuridae 243, **244**, 246
lemurina, Trichosurus 48
lemurinus, Aotus **256**
lemuroides, Hemibelideus **58**, 58
lena, Petaurista 462
lenae, Dicrostonyx 511
lenaensis, Clethrionomys 509
lenensis, Dicrostonyx 511
Rangifer 392
lenguarum, Bolomys 696, 697
leniceps, Mesomys **794**
lenis, Kerivoula 197
Oligoryzomys 718
Thomomys 473
lenoiri, Aonyx 310
Lenomys 583, 587, 591, **602**, 602, 640
lampo 602
longicaudus 602
meyeri **602**
Lenothrix 591, 600, **603**, 603, 611, 632, 633, 640, 664
canus **603**, 603
legata 591
malaisia 603
Lenoxus **706**, 706, 707
apicalis **707**
boliviae 707
lentiginosa, Sousa 355
lentiginosus, Steno 355
lentulus, Ctenomys 785
lentus, Dicrostonyx 510
Dremomys 425
Leo 297
leo, Antechinus **30**
Panthera **297**
Rhinosciurus 438
leonensis, Heliosciurus 429
Leonina 297
leonina, Macaca 267
Mirounga **330**
Otaria 328
Phoca 330
Saguinus 253
leoninus, Leontopithecus 253
leonis, Funisciurus 428
Maxomys 614
Mops 236
Sciurus 441
Leontideus 252
leontiewi, Crocuta 308
Leontocebus 252, 253
Leontopithecus **252**
ater 252
aurora 253
brasiliensis 253
caissara **252**
chrysomelas **252**
chrysopygus **252**, 252
chrysurus 252
guyannensis 253
leoninus 253
marikina 252, 253
rosalia **253**, 253
leopardina, Leptonychotes 330
Leopardus **291**, 292, 294
aequatorialis 291
albescens 291
amazonica 292
andina 292
armillatus 291
boliviae 292
brasiliensis 291
buffoni 291
canescens 291
carrikeri 292
catenata 292
caucensis 292
chati 291
chibigouazou 291
chibiguazu 291
cooperi 292
costaricensis 291
elegans 292
elenae 292
emerita 292
emiliae 292
geoffroyi 292
glaucula 292
griffithii 291
griseus 291
guttula 292
hamiltoni 291
limitis 291
ludovici 292
ludoviciana 291
macroura 292
macrura 292
maracaya 291
margay 292
maripensis 291
mearnsi 291
melanura 291
mexicana 291, 292
minimus 291
mitis 291
nelsoni 291
nicaraguae 292
oaxacensis 292
ocelot 291
oncilla 292
pardalis **291**, 292
pardictis 292
pardinoides 292
pirrensis 292
pseudopardalis 291
pusaea 291
salvinia 292
sanctaemartae 291, 292
smithii 291
sonoriensis 291
steinbachi 291
tigrinoides 292
tigrinus 291, **292**, 292, 294
tumatumari 291
vigens 292
wiedii **292**, 292
yucatanica 292
leopardus, Panthera 298
Leopoldamys 580, **603**, 603, 632, 656, 664
balae 603
bunguranensis 603
ciliatus 603
clarae 603
dictatorius 603
edwardsi **603**, 614, 634
fremens 603
garonum 603
gigas 603
hainanensis 603
heptneri 603
herberti 603
insularum 603
lancavensis 603
listeri 603
lucas 603
luta 603
macrourus 603
mansalaris 603
masae 603
matthaeus 603
mayapahit 603
melli 603
milleti 603
nasutus 603
neilli **603**
revertens 603
sabanus **603**, 603, 614, 634
salanga 603
setiger 603
siporanus 422, 430, 451, 461, 464, 584, **604**, 604, 613, 654
soccatus 604
stentor 603
strepitans 603
stridens 603
stridulus 603
tapanulius 603
tersus 603
tuancus 603
ululans 603
vociferans 603
leopoldi, Cephalophus 411
leosollicitus, Gerbillus 550
lepcha, Niviventer 635
Tupaia 131
lepechenii, Erignathus 329
lepida, Marmosa **18**
Neotoma 710, **711**, 711, 712
Potos 333
Lepidilemur 245
lepidura, Crocidura 85
lepidus, Calomys **698**, 698
Cercartetus **58**
Cervus 387
Hylopetes **461**, 461
Iomys 461
Myotis 211
Natalus **195**
Niviventer 634
Pipistrellus 221
Pogonomys 642
Pteropus 147
Rhinolophus **166**

Lepilemur **245**, 245
andrafiamensis 246
ankaranensis 246
caniceps 246
dorsalis **246**
edwardsi **246**
globiceps 246
grandidieri 246
leucopus **246**
microdon **246**
mustelinus 245, **246**
pallidicauda 246
rufescens 246
ruficaudatus **246**
sahafarensis 246
septentrionalis **246**
Lepilemuridae 245, 246
leponticus, Microtus 526
Leporidae **813**
leporides, Lagorchestes **52**
Macropus 52
Leporillus 586, 587, **604**, 604, 615, 619, 644, 674
apicalis **604**
conditor **604**, 604
jonesi 604
leporina, Dasyprocta 781, **782**
Erignathus 329
Leporinae 813
Leporinorum 813
leporinus, Aplodontia 417
Cebus 260
Mus 781, 782
Noctilio **176**
Sciurus 441
Vespertilio 176
lepsianum, Rhinopoma 155
Leptailurus **292**, 293
algiricus 292
beirae 292
brachyura 292
capensis 292
constantina 292
faradjius 292
ferrarii 292
galeopardus 292
hamiltoni 292
hindei 292
ingridi 292
kempi 292
kivuensis 292
larseni 292
limpopoensis 292
liposticta 292
lonnbergi 292
mababiensis 292
niger 292
ogilbyi 292
pantasticta 292
phillipsi 292
pococki 292
poliotricha 292
senegalensis 293
serval **292**
servalina 293
tanae 293
togoensis 293
Leptocapromys 802, 803
leptocephala, Dasypus 66
Leptoceros 395
leptoceros, Gazella **396**
leptodactylus, Arctomys 443
Neomys 110
Spermophilopsis **444**
leptodon, Myonycteris 143
Orycteropus 375
Leptogale 71
Leptomys **604**, 604, 605, 606, 614, 620, 638
elegans **604**, 604, 605, 665
ernstmayri 604, **605**, 605, 665
signatus 604, **605**
Leptonychotes 329, **330**
leopardina 330
weddellii **330**
Leptonycteris **185**
curasoae **185**
longala 185
nivalis **185**
sanborni 185
tarlosti 185
yerbabuenae 185
Leptonyx 309, 330
leptonyx, Amblonyx 310
Hydrurga **330**
Macropus 55
Phoca 330
leptophyllus, Hipposideros 173
leptorhynchus, Dasypus 66
Hippopotamus 381
Meles 314
Leptosciurus 439
leptosoma, Proechimys 797
leptura, Genetta 345
Myoprocta 783
Saccopteryx **160**
lepturoides, Niviventer 634
lepturus, Akodon 692
Bettongia 49
Cuon 282
Cynictis 302
Habromys **703**, 703
Heteromys 481
Meriones 557
Niviventer **634**
Peromyscus 703
Vespertilio 159
leptus, Mustela 321
Lepus 807, **814**, 814, 821, 824, 825
abbotti 816
abei 821
abyssinicus 816
aegyptius 816
aethiopicus 816
ainu 821
alba 817
albaniensis 820
albus 821
algidus 821
alleni **814**, 815
alpinus 821
altaicus 821
altamirae 815
americanus **814**
andersoni 815
angolensis 822
angustidens 815
ansorgei 822
aquilo 816
aquilonius 817
arabicus 816, 817, 821
aralensis 821
arcticus **815**, 819, 821
arenarius 816
argenteogrisea 817
aryabertensis 819
asellus 815
astaricus 817, 821
atallahi 816
atlanticus 816
aurantii 820
aurigineus 821
bairdii 814
bangsii 815
banksicola 815
barcaeus 816
battyi 815
bechuanae 820
bedfordi 816
begitschevi 821
bennettii 815
berberanus 816
biarmicus 817
biddulphi 821
bishopi 814
borealis 814, 817, 821
brachyurus **815**, 817, 819
breviauritus 821
brevinasus 821
buchariensis 821
butlerowi 821
californicus **815**, 818
callotis 814, **815**, 815, 818
campanius 821
campestris 817, 821, 822
campicola 817
canescens 821
canopus 822
canus 815
capensis 814, **815**, 816, 817, 818, 819, 820, 821, 822
carpi 816
cascadensis 814
caspicus 817
castroviejoi **816**
caucasicus 817
centralis 816
centrasiaticus 821
chadensis 822
cheesmani 816
cheybani 821
chiversi 820
chobiensis 820
cinereus 817
cinnamomeus 821
collinus 821
columbiensis 814
comus **816**
connori 816, 817, 821
cordeauxi 822
coreanus **816**, 817, 819, 820
coronatus 817
corsicanus 814, 816, **817**
craspedotis 821
crassicaudatus 823
crawshayi 814, 818, 819, 820, 822
creticus 817
crispii 816
cuniculus 822
curti 815
cutchensis 819
cyanotus 817
cyprius 817
cyrensis 817
dalli 814
damarensis 820
dauuricus 807
dayanus 819
depressus 815
deserticola 815
desertorum 821
diazi 824
dinderus 816
edwardsi 818
eremicus 815
ermeloensis 816
etigo 815
europaeus 814, 816, **817**, 817, 818, 820, 821
fagani **817**, 818
festinus 815
filchneri 821
flavigularis **818**
flaviventris 820
flavus 817
floridanus 824
formosus 820
fumigatus 820
gaillardi 815
gallaecius 818
gansuicus 821
ghigii 817
gichiganus 821
glacialis 815
gobicus 821
grahami 819
granatensis 814, 816, **818**
granti 816
griseus 815
groenlandicus 815
gungunyanae 820
habessinicus 816
habibi 821
hainanus **818**
hartensis 816
harterti 816
hawkeri 816

herero 822
hibernicus 821
hispanicus 818
hispidus 814
huangshuiensis 821
hubbardi 815
hudsonius 814
hybridus 817
hyemalis 817
hyperboreus 815
hypsibius 819
idahoensis 813
illuteus 819
innesi 816
insularis **818**
iranensis 817
isabellinus 816
iturissius 818
jeffreyi 816
joongshaiensis 819
judeae 817
kabylicus 816
kakumegae 822
kalaharicus 816
kalmykorum 817
kamtschaticus 821
karpathorum 817
kaschgaricus 821
kessleri 821
khanensis 820
klamathensis 814
kolymensis 821
kozhevnikovi 821
kozlovi 819
labradorius 815
langi 816
laskerewi 817
lehmanni 821
lilfordi 818
longicaudatus 820
lugubris 821
lutescens 821
lyoni 815
macfarlani 814
macrotus 819
maculatus 817
magdalenae 815
mahadeva 819
major 816
mandatus 816
mandshuricus 815, 817, **818**, 819
marjorita 823
maroccanus 816
martirensis 815
mediterraneus 818
medius 817
megalotis 820
melainus 819
melanonotus 819
melanotis 815
meridiei 817
meridionalis 817, 818, 822
merriami 815
mexicanus 815
micklemi 822
micropus 815
microtis 822
monstrabilis 815
monticularis 813
mordeni 821
nanus 814
narranus 816
netscheri 822
ngamiensis 820
niediecki 814
niethammeri 817
niger 817
nigrescens 820
nigricans 817
nigricaudatus 815
nigricollis 814, 816, 818, **819**, 819, 820
occidentalis 817
ochropoides 816
ochropus 816
oemodias 819
oiostolus 816, **819**, 821
okiensis 815
omanensis 816
orangensis 820
oregonus 814
orii 821
othus **819**, 819, 821
palitans 814
pallidior 816
pallidus 814
pallipes 819
pamirensis 821
parnassius 817
pediaeus 816
peguensis 814, 818, 819, **820**, 820
peni 816
persimilis 815
petteri 821
phaeonotus 814
pinetus 814
poadromus 819
ponticus 817
porsildi 815
przewalskii 819, 821
pygmaeus 816
pyrenaicus 817
qinghaiensis 819
quercerus 821
qusongensis 819
raineyi 822
rajput 819
rhodius 817
richardsonii 815
rothschildi 816
rubustus 821
ruficaudatus 814, 819, 820
rufinucha 820
rufus 817
sadiya 819
saghaliensis 821
salai 816
saliens 814
saxatilis 819, **820**, 820, 822
schlumbergeri 816
sclavonius 821
scoticus 821
sechuenensis 819
seclusus 814
sefranus 816
senegalensis 816
septentrionalis 821
sheldoni 815
sherif 816
siamensis 814, 819, 820
sibiricorum 821
sierrae 821
simcoxi 819
sinaiticus 816
sinensis 817, **820**
singhala 819
solisi 818
somalensis 816
sowerbyae 821
starcki **820**
stegmanni 821
stoliczkanus 821
struthopus 814
subluteus 821
subrufus 820
swinhoei 821
sylvaticus 821, 824
syriacus 817
tahoensis 814
tesquorum 817
texianus 815
tibetanus 814, 816, 821
tiburonensis 814
tigrensis 816
timidus 814, 815, 819, **820**, 820
tolai 814, 816, 817, 819, **821**, 821
townsendi 815
townsendii **821**
transbaicalicus 821
transsylvanicus 817
transsylvaticus 817
tsaidamensis 819
tschuktschorum 819
tularensis 815
tumac 817
tunetae 816
turcomanus 821
typicus 818, 821
tytleri 819
variabilis 821
varronis 821
vassali 820
vernayi 816
victoriae 818, 820, **822**, 822
vigilax 815
virginianus 814
wallawalla 815
wardi 814
washingtoni 814
whitakeri 816
whytei 814, 819, 820, 822
xanti 815
yarkandensis **822**
yuenshanensis 820
zairensis 822
zaisanicus 821
zechi 822
zuluensis 822
leridensis, Scotinomys 746
lervia, Ammotragus **404**
Antilope 404
leschae, Chrysospalax 76
leschenaulti, Cervus 387
Rousettus **152**
leschnaultii, Ratufa 437
lessonii, Wallabia 57
Xerus 458
lestes, Canis 280
Papio 269
Lestodelphys **17**
halli **17**
Lestoros **25**
gracilis 25
inca **25**
lestradei, Cephalophus 411
Otolemur 250
lesueur, Bettongia **49**, 49
lesueuri, Bettongia 49
Myotis **212**
lesueurii, Sorex 113, 114
lesviacus, Myotis 209
letabae, Genetta 346
Procavia 374
letifera, Mustela 324
leucaethiops, Dama 387
leucampyx, Cercopithecus 264
leucanthus, Gerbillurus 547
leucas, Delphinapterus **357**
Delphinus 357
leucastra, Marmosops 20
leucippe, Callithrix 251
Pipistrellus 223
leucippus, Strigocuscus 48
leucisca, Hylobates 276
leucobaptus, Oncifelis 294
leucoblephara, Galea 779
leucocephala, Callithrix 251
Paguma 343
Pithecia 262
leucocephalus, Apodemus 573
Callosciurus 422
Cebus 259
Martes 320
Mustela 323
Peromyscus 735
Petaurista 462
Pteropus 147
Trachypithecus 273
leucochilus, Cephalophus 410
Leucocrossuromys 424
leucodactylus, Hesperomys 744
Rhipidomys **744**, 744, 745
Leucodidelphis 16
Leucodon 81
leucodon, Crocidura **88**, 95
Ctenomys **785**
Nannospalax **753**, 753, 754

Neotoma 710
Sorex 81
Thomomys 473
leucodontus, Castor 467
leucodus, Crocidura 88
leucogaster, Antechinus 30
Callosciurus 422
Cephalophus **410**
Chaerephon 233
Holochilus 704
Hydromys 597
Hypudaeus 719
Melomys 615, 616, **617**, 617
Monachus 331
Moschus 384
Murina **229**, 229
Myotis 208
Onychomys **719**, 719
Ozotoceros 392
Rattus 658
Sciurus **440**, 441
Scotophilus **227**, 227, 228
Sorex 113, **116**, 116
Tatera **561**
leucogenys, Callithrix 251
Cebus 259
Dendrolagus 51
Dipodomys 479
Funisciurus **427**, 428
Hylobates **275**, 275
Petaurista 58, **462**, 463
Ratufa 437
Saguinus 253
Sminthopsis 36
Sorex 117
Suncus 103
Tapirus 371
leucogula, Akodon 692
leucolachnaea, Martes 320
leucolaemus, Arctonyx 313
leucolimnaeus, Bolomys 696
leucomelas, Barbastella **199**, 199
Eidolon 141
Pipistrellus 223
leucomeros, Colobus 270
leucomerus, Callithrix 251
leucometopa, Callicebus 258
Leucomitra 316
leucomus, Prosciurillus 435, **436**
Pygoderma 191
leucomystax, Eulemur 244
Paguma 343
Proechimys 797
Sus 379
Trachypithecus 274
leuconedys, Hylobates 275
Leuconoe 207, 208, 209, 210, 211, 212, 213, 214, 215, 216
leuconoe, Thallomys 668
leuconota, Mellivora 315
leuconotus, Conepatus **316**, 316
leuconyx, Ursus 339, 340
leucoparia, Mustela 322, 324
Spilogale 317
leucophaea, Antilope 412
Neotoma 712
leucophaetus, Rattus 652
leucophaeus, Hippotragus **412**
Mandrillus **268**
Microtus 523
Otonycteris 219
leucopictus, Spermophilus 449
leucopla, Phoca 332
leucopleura, Eumops 233
Taphozous 160
Leucopleurus 353
leucopleurus, Lagenorhynchus 353
leucoprosopus, Sylvicapra 412
leucoprymnus, Bos 401
Pan 277
Sigmoceros 395
Trachypithecus 274
leucops, Georychus 772
Sciurus 440
Soriculus **123**
leucopsis, Saimiri 260
leucoptera, Peropteryx **158**
leucopterus, Pteropus **148**
Taphozous 160
leucopus, Callosciurus 421
Lepilemur **246**
Martes 319
Mus 728
Myomys 630
Ourebia 399
Paradoxurus 344
Peromyscus 730, **732**, 732
Pseudomys 647
Rattus 649, 650, 653, **654**, 654, 656, 660
Saguinus **253**
Sminthopsis **36**
Vulpes 287
leucopyga, Cavia 779
Leucorhamphus 354
leucorhamphus, Lissodelphis 354
leucorhina, Chlorotalpa **75**
leucorhinus, Sus 379
leucorhynchus, Nasua 335
Steatomys 545
Leucorrhynchus 110
leucoryx, Oryx **413**
leucosagmaphora, Globicephala 352
leucosternum, Rattus 657
leucostethicus, Mungos 307
leucostictus, Spermophilus 448
leucostigma, Funisciurus 428
Mops 236
leucostoma, Arctocephalus 326
leucostomus, Dendromus 543
leucotis, Acerodon **137**
Arctogalidia 343
Callithrix 252
Didelphis 16
Dipodomys 479
Hippocamelus 390
Kobus 414
Martes 320
Microdipodops 480
Pipistrellus 221
Pteropus 146
Sciurus 440
Sigmodon **747**
Tamiops 457
leucoumbrinus, Xerus 458
leucourus, Sciurus 443
leucura, Alticola 503
Echiothrix **591**, 591
Euroscaptor 125
Hystrix 774
Macrotis **40**
Parascaptor **127**
Suncus 103
Talpa 127
leucurus, Ammospermophilus **420**
Auliscomys 695
Callosciurus 421
Caluromys 15
Chionomys 507
Crocidura 94
Cynomys **424**
Hystrix 774
Ichneumia 306
Meles 314
Microtus **523**, 523
Odocoileus 391
Peromyscus 733
Pitymys 516
Tamias 419
leukeurin, Callithrix 251
leupolti, Sigmoceros 395
leurodon, Spermophilus 447
levaillantii, Cynictis 302
Xerus 458
levantis, Talpa 128, **129**
levenedii, Microtus 518
levicula, Crocidura **88**
levidensis, Lemmiscus 516
Thomomys 473
levipes, Dipodomys 479
Melomys 615, 616, **617**, 617, 618
Oryzomys 720, **723**, 723
Peromyscus 729, **732**, 732
Reithrodontomys 741
levir, Isolobodon 804
levis, Microtus 519
Myotis **212**, 215
Ochotona 811
Phenacomys 533
Thomomys 475
levisagittalis, Geomys 469
lewisi, Acomys 566
Artibeus 188
Ctenomys **785**
Marmota 431
Ochrotomys 715
Pseudocheirus 59
lewisii, Odocoileus 391
lghesicus, Chionomys 507
lhasaensis, Ochotona 811, 812
lhoesti, Cercopithecus **264**, 265
liantis, Tamiops 457
liardensis, Ovis 409
liberiae, Cricetomys 540
libericus, Heliosciurus 429
liberiensis, Cephalophus 410
Hexaprotodon **380**
Poiana 347
Liberiictis 300, **307**
kuhni **307**, 307
libidinosus, Cebus 259
libonotus, Eothenomys 514
libyca, Felis 291
Ictonyx **319**
libycus, Felis 290
Hemiechinus 78
Meriones **556**, 556, 557, 558
Sus 379
Lichanotus 246, 247
lichiensis, Dremomys 425
Lichonycteris **185**
degener 185
obscura **185**, 185
lichtensteini, Eremodipus **493**
Odocoileus 391
Scirtopoda 493
lichtensteinii, Antilope 394
Bubalis 394
Sigmoceros 393, **394**
liciae, Bolomys 697
liebmani, Sphiggurus 776
liebrechtsi, Cercopithecus 265
Okapia 383
liechtensteini, Microtus 526
lieftincki, Niviventer 634
lighti, Callosciurus 422
lignarius, Cervus 387
lignicolor, Crocidura 96
Paradoxurus 344
ligoni, Canis 281
lilaeus, Sciurus 443
lilfordi, Lepus 818
lilium, Phyllostoma 192
Sturnira **192**
lillyana, Zelotomys 674
Limacomys 544
limae, Micoureus 20
limanus, Myoprocta 782
limatus, Phyllotis 738
limbarenicus, Colobus 270
limbata, Chaerephon 233
limbatus, Heliosciurus 428
Pipistrellus 223
limchjnhunii, Neomys 110
limicauda, Melomys 616
limicola, Reithrodontomys

741
liminalis, Proechimys 798
limitaneus, Cebus 260
Xerus 458
limitaris, Golunda 592
Thomomys 473
limitis, Clethrionomys 508
Leopardus 291
Macaca 266
Sciurus 442
Limnogale **70**
mergulus **70**, 70
Limnolagus 824
Limnomys **605**, 605, 638
asper 638
mearnsi 605
sibuanus **605**, 605
limnophilus, Microtus **523**, 527
Myotis 210
Limnotragus 403
limosus, Thomomys 475
limpiae, Thomomys 473
limpopoensis, Calcochloris 74
Caracal 289
Ictonyx 319
Leptailurus 292
Mastomys 610
Saccostomus 541
Syncerus 402
Tatera 561
lindberghi, Akodon **690**, 691
lindicus, Sigmoceros 395
linduensis, Rattus 653
lineata, Phocoena 358
Rhynchonycteris 159
lineatum, Phyllostoma 190
lineatus, Chaetodipus **484**, 484
Dendromus 543
Mungotictis 300
Neomys 110
Platyrrhinus **191**
Rhabdomys 662
Sicista 498
Tamias 455
Vampyrops 191
Vulpes 287
lineicaudatus, Andinomys 694
lineiventer, Mustela 322
Vulpes 287
lineolatus, Pseudomys 645
ling, Niviventer 634
lingae, Macaca 266
Tupaia 133
lingensis, Maxomys 613
lingungensis, Macaca 266
Sundasciurus 451
linnaei, Phoca 332
linneana, Neomys 110
linsang, Prionodon 342, **347**, 347
linulus, Lemniscomys **601**
Liocephalus 251
liodon, Tatera 562
Liomys 477, 481, **482**, 482
acutus 482
adspersus **482**
albolimbatus 482
alleni 482
annectens 482
anthonyi 482
aterimus 482
bulleri 482
canus 482
crispus 482
escuinapae 482
exiguus 482
guerrerensis 482
heterothrix 482
hispidus 482
irroratus **482**
isthmius 482
jaliscensis 482
minor 482
nigrescens 482
obscurus 482
orbitalis 482
paralius 482
parviceps 482
phaeura 482
pictus **482**, 482
pinetorum 482
plantinarensis 482
pretiosus 482
pullus 482
rostratus 482
salvini **482**
setosus 482
sonorana 482
spectabilis **482**
texensis 482
torridus 482
veraecrucis 482
vulcani 482
yautepecus 482
Lionycteris **181**
spurrelli **181**, 181, 182
liops, Pteropus 151
Liotomys 682
lipara, Crocidura 91
liparensis, Eliomys 767
Liponycteris 160, 161
liposticta, Leptailurus 292
Lipotes **360**
vexillifer **360**, 360
Lipotidae 360
Lipotus 315
Lipura 430
lipura, Trichys 774
Lipurus 45
cinereus 45
lipuwa, Kobus 413
liricaudatus, Neomys 110
lis, Sciurus **441**
lisae, Ochrotomys 715
Liscurus 45
Lissodelphinae 351
Lissodelphis **354**, 354
borealis **354**
leucorhamphus 354
peronii **354**
Lissonycteris 152
listeri, Leopoldamys 603
listoni, Millardia 621
Lithocranius 397
lithophilus, Chaetodipus 484
Lithotragus 406
Litocranius **397**
sclateri 397
walleri **397**
litoralis, Ochotona 809
litoreus, Procyon 335
litoris, Thomomys 473
littledalei, Marmota 431
Ovis 408
Parotomys **682**
littoralis, Amblysomus 74
Apomys **575**
Arvicola 506
Crocidura **88**, 90, 95, 97
Galea 779
Gazella 396
Grammomys 593
Graphiurus 764
Macaca 267
Melomys 615
Meriones 557
Microtus 523
Neotoma 712
Ochotona 811
Sciurus 440
Sigmodon 747
Sylvilagus 824
Tamias 456
Tatera 561
Taxidea 325
Urocyon **285**
littorea, Phoca 332
littoreus, Niviventer 633
lituratus, Artibeus **188**, 189
litus, Perognathus 485
livingstonianus, Neotragus 398
livingstonii, Pan 277
Pteropus **148**
Taurotragus 403
lixa, Gerbillus 550, 553
lixus, Suncus **102**
lizenkani, Crocidura 87
llacma, Lama 381
llanensis, Geomys 469
Sciurus 441
Sylvilagus 826
llanoensis, Dipodomys 479
llanoi, Akodon 694
loandae, Genetta 346
Ichneumia 307
Protoxerus 436
loandicus, Heliosciurus 428
lobata, Tadarida **240**
lobatus, Arctocephalus 328
Neophoca 328
Pipistrellus 221
Rhinolophus 166
Zalophus 329
lobeliarum, Sylvicapra 412
lobengulae, Tatera 561
lobipes, Myotis 213
Lobodon 329, **330**
carcinophagus **330**
serridens 330
Lobodontini 329
localis, Maxomys 612
lockwoodi, Graomys 703
Lagidium 778
locorinae, Tragelaphus 404
locusta, Emballonura 157
Jaculus 494
loderi, Gazella 396
Kobus 414
loempo, Ichneumia 307
loennbergi, Lama 382
loftusi, Jaculus 493
loginovi, Chionomys 507
logonensis, Thryonomys 775
lokriah, Dremomys **425**
lokroides, Callosciurus 423
lomamiensis, Syncerus 402
lombocensis, Pteropus **148**
lomitensis, Heteromys 481
Melanomys 708
Loncheres 791
caniceps 791
Loncherinae 791
Lonchoglossa 183
Lonchophylla **181**
bokermanni **181**
concava 181, 182
dekeyseri **181**
handleyi **181**
hesperia **181**
mordax **181**, 181, 182
robusta 181, **182**
thomasi **182**
Lonchophyllinae **181**
Lonchorhina **177**
aurita **177**, 177
fernandezi **177**
marinkellei **178**
occidentalis 177
orinocensis **178**
Lonchothrix 789, **794**
emiliae **794**, 794
longala, Leptonycteris 185
longiaculeata, Tachyglossus 13
longicauda, Alticola 503
Crocidura 96
Eonycteris 153, 154
Hipposideros 173
Hystrix 773
Microperoryctes **41**
Mustela 322
Petrogale 56
Sorex 118
Tupaia 132
longicaudata, Alticola 503
Condylura 124
Crocidura 86
Macaca 268
Manis 415
Microgale **71**, 71

Murexia **32**
Panthera 298
Pardofelis 299
Phascogale 32
Sminthopsis **36**
Tamandua **68**
longicaudatus, Aethomys 568
Avahi 246
Coendou 775
Cricetulus **538**
Dasymys 665
Dasypus 66
Dendromus 543
Gerbillus 551
Hapalomys **595**, 595
Heteromys 481
Irenomys 705
Lepus 820
Molossus 235
Myosorex **100**
Myotis 211
Notomys **636**
Oligoryzomys **718**, 718
Oryzomys 717
Petaurus 61
Proechimys 795, **797**
Reithrodon 705
Stochomys 665, **666**, 666
longicaudis, Lontra **311**
longicaudus, Arborimus **505**, 505
Dasypus 66
Hapalomys 640
Lenomys 602
Microtus 507, **523**, 523, 528
Onychomys 719
Oryzomys 718
Phenacomys 504
Rattus 660
Reithrodontomys 741
longiceps, Amblysomus 74
Cephalophus 411
Felis 290
Meriones 558
longicornis, Bos 401
Cervus 387
Naemorhedus 407
longicrus, Myotis 216
longicuspis, Cervus 386
longidens, Lagenorhynchus 354
longifolium, Mimon 179
longifrons, Meriones 556
longimana, Hylobates 275
Ia 205
Megaptera 350
longimanus, Eumops 233
Taphozous **160**
longimembris, Ateles 257
Perognathus **485**
longior, Allactaga 489
longipedis, Dinaromys 511
longipes, Crocidura **88**
Diceros 372
Dipodomys 479
Malacomys **607**, 607
Myotis **212**
Onychomys 719
Potorous **49**
Tragulus 382
Tupaia **132**
longipilis, Akodon 690, **691**, 691, 692, 699
Cavia 779
Didelphis 16
Felis 290
Microtus 527
Panthera 298
Pseudocheirus 59
Rattus 662
longipilosus, Dasymys 590
Scotinomys 746
longipinna, Megaptera 350
longiquus, Sorex 118
longirostris, Euroscaptor **125**, 125
Glossophaga **184**
Makalata 793
Melursus 337
Mesoplodon 363
Microgale 71
Microtus 525
Nesophontes **70**
Sorex **116**, 117
Stenella **356**, 356, 357
Sus 378
Tachyglossus 13
longitarsus, Mus 718
longmani, Hydromys 597
Petrogale 55
longobarda, Neomys 110
Sorex 111
longstaffi, Mustela 321
lonnbergi, Cannomys 685
Leptailurus 292
Otolemur 250
Sigmodon 747, 748
lontaris, Rattus 660
Lontra **310**, 311, 312
americana 311
annectens 311
atterima 311
brachydactyla 311
brevipilosus 311
californica 311
canadensis **311**
chilensis 311
chimo 311
cinerea 311
colombiana 311
degener 311
destructor 311
emerita 311
enudris 311
evexa 311
felina **311**
fusco-rufa 311
hudsonica 311
huidobria 311
incarum 311
insularis 311
interior 311
kodiacensis 311
lataxina 311
latidens 311
latifrons 311
longicaudis **311**, 311
lutris 311
mesopetes 311
mira 311
mitis 311
mollis 311
nexa 311
optiva 311
pacifica 311
paraensis 311
paranensis 311
parilina 311
periclyzomae 311
peruensis 311
peruviensis 311
platensis 311
pratensis 311
preblei 311
provocax **311**
repanda 311
rhoadsi 311
solitaria 311
sonora 311
texensis 311
vaga 311
vancouverensis 311
yukonensis 311
loochooensis, Pteropus 148
lopesi, Procavia 374
Lophiomyinae 501, **563**, 563
Lophiomys **563**, 563, 564
aethiopicus 564
bozasi 564
hindei 564
ibeanus 564
imhausi **563**, 564
imhausii 563
smithi 564
testudo 564
thomasi 564
Lophocebus **266**
albigena **266**
aterrimus 266
coelognathus 266
congicus 266
hamlyni 266
ituricus 266
jamrachi 266
johnstoni 266
mawambicus 266
opdenboschi 266
osmani 266
rothschildi 266
ugandae 266
weynsi 266
zenkeri 266
Lophocolobus 271
Lophopitheus 270
lophorhina, Nyctophilus 218
Lophotragus 388
Lophotus 277
Lophuromys 565, **605**, 605, 606, 621, 670, 674
afer 606
ansorgei 606
aquilus 605
brevicaudus 605
brunneus 605
chrysopus 605
cinereus **605**, 605
eisentrauti 606
flavopunctatus **605**, 606
laticeps 605
luteogaster **605**
major 605
manteufeli 606
margarettae 605
medicaudatus **606**
melanonyx **606**, 664, 686
naso 606
nudicaudus **606**
prittiei 606
pyrrhus 606
rahmi **606**
rita 605
rubecula 605
sikapusi **606**, 606
simensis 605
tullbergi 606
woosnami **606**
zaphiri 605
zena 605
lophurus, Habromys **703**, 703
Pipistrellus **222**
Suricata 308
loquax, Tamiasciurus 457
Loratus 274
lordi, Bandicota 579
Perognathus 486
lorentzi, Neophascogale **33**
lorentzii, Dorcopsis 52
Melomys 615, **617**, 617, 618
Lorentzimys 604, **606**, 606
alticola 607
nouhuysi 606, **607**
lorenzi, Capra 406
Chaetodipus 484
Connochaetes 394
Equus 370
Peromyscus 731
Taterillus 563
Thomomys 473
loriae, Mormopterus 238
Pogonomys **641**, 642
Loricatus 65, 66, 67
pichiy 67
loricatus, Cabassous 65
Loridae **247**
loriger, Sicista 498
loringi, Clethrionomys 508
Spermophilus 450
Thallomys **668**, 668
Thomomys 475
Loris **248**
ceylonicus 248

gracilis 248
grandis 248
lydekkerianus 248
malabaricus 248
nordicus 248
nycticeboides 248
tardigradus **248**
zeylanicus 248
Lorisidae 247, **248**
lorraineus, *Graphiurus* **764**, 765
losea, *Rattus* 649, **654**, 654, 657
lotichiusi, *Pithecia* 262
lotipes, *Niviventer* 634
lotor, *Procyon* **335**, 335, 336
Ursus 335
lottum, *Dugong* 365
louiei, *Thomomys* 475
louisae, *Acomys* **566**
louisianae, *Odocoileus* 391
loujacobsi, *Hadromys* 594
loukashkini, *Myotis* 210
lovati, *Dendromus* 541, **542**
loveni, *Eptesicus* 216
Myotis 215
loveridgei, *Steatomys* 546
Tatera 562
lovii, *Lutra* 312
lovizettii, *Procolobus* 272
lowei, *Acomys* 565, 566
Atelerix 76
Cephalophus 410
Cercopithecus 263
Cynogale 341
Eptesicus 201
Felis 290
Gerbillus **552**
Mylomys 630
Soriculus 123
Taterillus 563
loweryi, *Sciurus* 443
lowi, *Bos* 401
lowii, *Mustela* 324
Ptilocercus **133**, 133
Sciurus 450
Sundasciurus **451**, 451
Loxodonta **367**
africana **367**
albertensis 367
angammensis 367
angolensis 367
berbericus 367
capensis 367
cavendishi 367
cottoni 367
cyclotis 367
fransseni 367
hannibaldi 367
knochenhaueri 367
mocambicus 367
orleansi 367
oxyotis 367
peeli 367
pharaohensis 367
prima 367
priscus 367
pumilio 367
rothschildi 367
selousi 367
toxotis 367
typicus 367
zukowskyi 367
Loxodontomys 695, 737
Loxomylus 805
loysi, *Ateles* 257
lozanoi, *Gazella* 396
lu, *Hylobates* 275
lualabae, *Heliosciurus* 429
lucani, *Caracal* 289
lucanius, *Sorex* 117
lucas, *Callosciurus* 421
Leopoldamys 603
lucasi, *Penthetor* **145**, 145
lucasona, *Mustela* 324
Spilogale 317
luch, *Microtus* 518
luchsingeri, *Lycaon* 283
lucia, *Kerivoula* 197
Tatera 562
luciae, *Ardops* 187
Megalomys **707**
Monophyllus 185
Sturnira 192
luciana, *Neotoma* 711
lucida, *Neotoma* 711
Tupaia 133
lucidus, *Chionomys* 507
Microdipodops 480
Thomomys 473
lucifer, *Acerodon* **137**
Callicebus 259
Dasyprocta 782
Mydaus 315
Paraxerus **435**
luciferoides, *Mydaus* 315
lucifrons, *Reithrodontomys* 742
lucifugus, *Glaucomys* 460
Myotis **212**, 216
lucina, *Crocidura* **88**
lucocephalus, *Thylacinus* 29
lucrificus, *Thomomys* 473
luctosus, *Arvicanthis* 578
luctuosa, *Alouatta* 255
Dorcopsis **51**
luctuosus, *Myotis* 209
luctus, *Rhinolophus* **167**
lucubrans, *Peromyscus* 735
lucullus, *Rhipidomys* 744
ludemani, *Geomys* 469
ludia, *Crocidura* 84, **88**
Genetta 345
ludibundus, *Tamias* 453
ludio, *Cercopithecus* 264
ludlami, *Cephalophus* 411
ludlowi, *Hemiechinus* 78
Ludolphozecora 369
ludovici, *Leopardus* 292
Sturnira **192**, 192
ludoviciana, *Arctomys* 424
Leopardus 291
ludovicianus, *Cynomys* **424**
Microtus 526
Sciurus **442**
ludwigii, *Cryptomys* 772
luedorfi, *Cervus* 386
lugardi, *Cryptomys* 771
lugenda, *Marmosops* 20
lugens, *Aepeomys* **688**, 688
Callicebus 259
Cephalophus 411
Lagothrix 258
Oryzomys 687
Petinomys 463, **464**, 464, 604
Rattus 604, 649, 650, **654**, 654
Sciuropterus 463
Thylogale 57
lugubris, *Lepus* 821
Myotis 213
luimbalensis, *Crocidura* 87
luisi, *Sturnira* **192**
lujanensis, *Lama* 382
Lyncodon 319
lukolelae, *Malacomys* 607, **608**, 608
lukonzolwae, *Chlorocebus* 265
lulae, *Lemniscomys* 602
lulidicus, *Procolobus* 272
lullulae, *Nyctimene* 144
Phalanger **46**, 46
luluae, *Crocidura* 91
Pelomys 639
luluana, *Crocidura* 92
luluanus, *Chlorotalpa* 75
lumholtzi, *Dendrolagus* **50**
Petaurista 463
Sminthopsis 37
lumholzi, *Ratufa* 437
luna, *Crocidura* **88**, 89, 94
lunaris, *Callosciurus* 422
Cerdocyon 282
Dasyprocta 782
Dendromus 542, 543
Hybomys **596**, 596
Paraxerus 434
Sylvisorex **105**
lunata, *Onychogalea* **55**
lunatus, *Cebus* 259
Damaliscus **394**, 394
Octodon **788**
luncefordi, *Exilisciurus* 426
lundensis, *Galerella* 303
lundi, *Dasypus* 66
Monodelphis 22
lundii, *Mus* 625
lunulatus, *Cercocebus* 263
Saimiri 260
lupaster, *Canis* 280
lupina, *Arctocephalus* 326
Pteronura 313
lupinus, *Lycaon* 283
lupulinus, *Lynx* 293
lupus, *Canis* **280**, 280, 281
lupus-griseus, *Canis* 281
luridavolta, *Marmosa* 18
luscus, *Gulo* 318
lusitania, *Crocidura* **88**, 93
lusitanica, *Capra* 406
Eliomys 767
Genetta 345
lusitanicus, *Microtus* 520, **523**, 524
Mus 628
lusumbi, *Cephalophus* 411
luta, *Leopoldamys* 603
lutea, *Sylvicapra* 412
Vulpes 287
luteicollis, *Tragulus* 382
luteiventris, *Rattus* 652
Tamias 453
lutensis, *Mustela* 324
luteogaster, *Lophuromys* **605**
Procavia 374
luteola, *Marmota* 432
Nycteris 162
luteolus, *Ctenomys* 786
Dipodomys 479
Galictis 318
Gerbillus 549, 555
Maxomys 614
Parotomys 682
Reithrodontomys 742
Ursus 338
lutescens, *Akodon* 692
Callosciurus 422
Ellobius **512**
Funambulus 426
Geomys 469, 470
Helogale 304
Lagidium 778
Lepus 821
Ochotona 811
Phyllotis 738
Rattus 657
Tragulus 383
luteus, *Chlorocebus* 265
Cricetomys 540
Eolagurus **513**, 513
Equus 370
Georychus 513
Gerbillus 551
Gulo 318
Peromyscus 732
Pteropus 147
Scotophilus 227
Vespertilio 228
Zapus 499
luticolor, *Niviventer* 633
lutillus, *Melomys* 615
lutosus, *Myotis* 216
Lutra 310, **311**, 311, 312, 313
amurensis 312
angustifrons 312
aureventer 312
aurobrunneus 312
baicalensis 312
barang 312
canadensis 310
capensis 310

ceylonica 312
chinensis 312
chobiensis 312
cinerea 309
concolor 309
fluviatilis 312
grayii 312
hanensis 312
indica 312
intermedia 312
japonica 312
kamtschatica 312
kivuana 312
kutab 312
lovii 312
lutra **312**, 312
maculicollis **312**, 312
marinus 312
matschiei 312
meridonalis 312
minima 16
monticolus 312
mutandae 312
nair 312
nepalensis 312
nilotica 312
nippon 312
nudipes 312
oxiana 312
perspicillata 312
piscatoria 312
roensis 312
seistanica 312
sinensis 312
splendida 312
stejnegeri 312
sumatrana **312**, 312
vulgaris 312
whiteleyi 312
lutra, Lutra **312**, 312
Lutrogale 313
Mustela 311
lutrella, Crocidura 93
lutreocephala, Mustela 324
Lutreola 321, 323, 324
lutreola, Crocidura 87
Mustela 321, **322**, 323, 324
Lutreolina **18**
bonaria 18
crassicaudata **18**
crassicaudis 18
ferruginea 18
lutrilla 18
paranalis 18
travassosi 18
turneri 18
lutreolina, Mustela 321, **323**, 323, 324
lutreolus, Rattus 649, **655**
lutrilla, Hydromys 597
Lutreolina 18
lutrina, Ratufa 437
Lutrinae **309**
lutris, Enhydra **310**
Lontra 311
Mustela 310
Lutrogale 312, **313**, 313
ellioti 313
lutra 313
macrodus 313
maculicollis 313
perspicillata **313**, 313
simung 313
sindica 313
sumatrana 313
tarayensis 313
Lutronectes 311
lutulentus, Pappogeomys 472
luxuriosus, Rattus 661
luzarchei, Alcelaphus 393
luzonensis, Archboldomys **575**, 575
luzonicus, Bullimus **581**, 581
luzoniensis, Cynopterus 139
Suncus 103
luzonus, Chaerephon 233
lybica, Felis 290, 291
lybicus, Meriones 556
lybiensis, Felis 290
lybius, Melursus 337
Lycalopex 280, 285
Lycaon 29, **283**
cacondae 283
dieseneri 283
ebermaieri 283
fuchsi 283
gansseri 283
gobabis 283
hennigi 283
huebneri 283
kondoae 283
krebsi 283
lademanni 283
lalandei 283
langheldi 283
luchsingeri 283
lupinus 283
manguensis 283
mischlichi 283
pictus **283**
prageri 283
richteri 283
ruppelli 283
ruwanae 283
sharicus 283
ssongeae 283
stierlingi 283
styxi 283
taborae 283
takanus 283
tricolor 283
typicus 283
venatica 283
windhorni 283
wintgensi 283
zedlitzi 283
zuluensis 283
lycaon, Canis 281
Meriones 558
lychnuchus, Tamiasciurus 457
lycoides, Pseudalopex 284
lydekkeri, Capra 406
Ovis 409
Papio 269
lydekkerianus, Loris **248**
lydius, Microtus 521
lyelli, Sorex **117**
lylei, Callosciurus 422
Hipposideros **173**
Petaurista 463
Pteropus **148**
Tamiops 457
lymani, Mustela 321
Lynceus 293
Lynchailurus 294
Lynchus 293
Lyncodon **319**
anticola 319
lujanensis 319
patagonicus **319**
quiqui 319
lyncula, Lynx 293
Lyncus 293
lynesi, Felis 290
Lemniscomys 602
Mus 628
lynni, Eptesicus 201, 202
Lynx 289, **293**, 293
baileyi 293
borealis 293
californicus 293
canadensis **293**, 293
carpathicus 293
cervaria 293
dinniki 293
eremicus 293
escuinapae 293
fasciatus 293
floridanus 293
gigas 293
isabellina 293
kamensis 293
kozlovi 293
lupulinus 293
lyncula 293
lynx **293**, 293
maculata 293
martinoi 293
mollipilosus 293
montanus 293
neglectus 293
oaxacensis 293
oculeus 293
orientalis 293
pallescens 293
pardella 293
pardina 293
pardinus **293**, 293
peninsularis 293
rufus **293**
stroganovi 293
subsolanus 293
superiorensis 294
texensis 294
tibetanus 293
uinta 294
vulgaris 293
vulpinus 293
wardi 293
wrangeli 293
lynx, Felis 293
Lynx **293**, 293
lyoni, Cynopterus 139
Lepus 815
Lyonogale 131, 133
lyra, Damaliscus 394
Megaderma **1639**
lyratus, Cervus 386
Spermophilus **448**
Lyroderma 163
Lysiurus 64
Lyssodes 266
lysteri, Tamias 456

maanjae, Crocidura 87
mababiensis, Herpestes 305
Leptailurus 292
mabalus, Maxomys 614
Macaca **266**
adusta 267
affinis 266
agnatus 266
alacer 266
andamanensis 267
apoensis 266
arctoides **266**
argentimembris 266
assamensis **266**
atriceps 266
audeberti 268
aureus 266
aurifrons 268
baweanus 266
bintangensis 266
blythii 267
brachyurus 267
brevicaudatus 267
broca 267
brunnescens 267
brunneus 266
cagayanus 266
capitalis 266
carbonarius 266
carimatae 266
carpolegus 267
coolidgei 266
cupida 266
cyclopis **266**
cynomolgus 266
diluta 267
dollmani 266
ecaudatus 268
erythraea 267
esau 268
fascicularis **266**, 267, 271
ferox 268
fulvus 267
fuscata 266, **267**
fusco-ater 267
fuscus 266
harmandi 266
hecki 268
hypomelas 267

impudens 266
inaureus 268
indochinensis 267
inornatus 267
insulana 267
inuus 268
irus 266
japanensis 267
karimondjawae 266
karimoni 266
laetus 266
lapsus 266
lasiae 266
lasiotus 267
lembicus 267
leonina 267
limitis 266
lingae 266
lingungensis 266
littoralis 267
longicaudata 268
majuscula 267
mandibularis 266
mansalaris 266
maura **267**, 267
mcmahoni 267
melanotus 266
melli 266
mindanensis 266
mindorus 266
mordax 266
mulatta 266, **267**
nemestrina **267**
nigra **267**, 267
nigrescens 267
nipalensis 267
nucifera 267
ochreata **267**, 267
opisthomelas 268
orinops 267
pagensis 267
pelops 266
phaeura 266
philippensis 266
pileatus 268
pithecus 268
problematicus 266
pullus 266
pumilus 266
pygmaeus 268
radiata **267**
resina 266
rhesosimilis 266
rhesus 267
rufescens 266
sanctijohannis 267
siamica 267
silenus **267**
sinica 267, **268**
speciosa 266, 267
spenoca 266
sublimitus 266
submordax 266
suluensis 266
sylvanus **268**
tcheliensis 267
thibetana **268**
togeanus 268
tonkeana **268**
tonsus 268
umbrosus 266
ursinus 266
validus 266
vestita 267
veter 268
villosa 267
vitiis 266
yakui 267
macaco, Eulemur **244**, 244
Macacus, gelada 269
macarenensis, Nyctinomops 239
macarianus, Cervus 386
macarthuri, Crocidura **89**, 95
macassaricus, Cervus 387
Pteropus 147
Sus 378
macauco, Tarsius 251
maccalinus, Tatera 560
maccarthiae, Herpestes 304
macclellandi, Tamiops **457**
macconnelli, Alouatta 255
Mesophylla **190**, 190
Oryzomys 721, **723**, 723
Rhipidomys **744**
Sciurus 439
maccourus, Melogale 314
macculus, Lemniscomys 601, **602**, 602
macdonaldi, Notopteris **154**, 154
Oncifelis 294
Triaenops 175
macdonnellensis, Parantechinus 33
Phascogale 34
Pseudantechinus **34**
macdougalli, Bassariscus 334
Molossus 235
Sigmodon 747
macedonicus, Microtus 521
Mus **624**, 624, 626
Spermophilus **445**
macellus, Myotis 211
macer, Cryptotis 109
Natalus 195
macfarlani, Lepus 814
Microtus 527
Ursus 339
macgillivrayi, Microtus 518
machadoi, Chalinolobus 200
machetes, Cryptotis 108
Ursus 338, 339
machlis, Alces 389
machrinoides, Scalopus 127
machrinus, Scalopus 127
macinnesi, Rhinopoma 155
mackenzianus, Ovibos 408
mackenziei, Berylmys **580**
Pipistrellus 224
mackenzii, Canis 281
Phenacomys 533
mackillingini, Gerbillus **552**
mackinderi, Procavia 374
mackloti, Acerodon **137**
macklotii, Crocidura 85
maclaudi, Rhinolophus **167**
maclean, Apodemus 573
macleari, Rattus 650, 651, **655**, 655, 656
macleayi, Dorcopsis 52
Dorcopsulus **52**, 52
macleayii, Pteronotus **177**
macmillani, Crocidura 88, **89**, 89, 96
Dremomys 425
Grammomys 592, **594**, 594, 631
Helogale 304
Mormopterus 238
Pteropus 152
Rattus 660
macneilli, Cervus 385, 386
Ursus 340
macowi, Crocidura **89**
macquariensis, Mirounga 330
macra, Arctogalidia 343
macracanthus, Hemiechinus 78
Macrobates 277
macrobullaris, Plecotus 224
macrobullatus, Hipposideros 171, **173**
Macrocapromys 800
acevedo 800
macrocelis, Felis 297
Neofelis 297
macrocephala, Gazella 397
macrocephalicus, Myotis 213
Macrocephalus 377
macrocephalus, Cebus 259
Epomophorus 141
Pecari 380
Physeter 359
Rhinolophus 166
Tachyoryctes 606, 664, 685, **686**, 686
macrocercus, Oligoryzomys 718
Tateomys **668**
macroceros, Bubalus 402
macrocneme, Miniopterus 231
macrocorpus, Sylvilagus 826
macrocranius, Microtus 526
macrodactylus, Myotis 209, **212**
macrodens, Miniopterus 230
Phoca 332
Macroderma **163**
gigas **163**
saturata 163
macrodon, Atilax 301
Bradypus 63
Crocidura 85
Mustela 324
Neotoma 711
Ondatra 532
Saimiri 260
Sigmodon 747
Sorex **117**, 122
Ursus 339
macrodus, Lutrogale 313
Paradoxurus 344
Macroechinus 78
Macrogalidia **343**
musschenbroekii **343**
Macrogeomys 469, 470, 471
Macroglossinae **153**
Macroglossus **154**
fraternus 154
fructivorus 154
horsfieldi 154
kiodotes 154
lagochilus 154
meyeri 154
microtus 154
minimus **154**, 154, 155
nanus 154
pygmaeus 154
rostratus 154
sobrinus **154**
spelaeus 153
macromangoz, Eulemur 244
Macromerus 247
macromystax, Jaculus 493
macronychos, Akodon 692
macronyx, Chelemys **699**, 699
macrophonius, Mustela 322
Macrophyllum **178**
macrophyllum **178**
nieuwiedii 178
macrophyllum, Macrophyllum **178**
Phyllostoma 178
Macropodidae 48, 49, **50**
Macropodinae 53
macropterus, Pipistrellus 223
Macropus **53**, 53, 54, 55, 57, 249
agilis **53**
alexandriae 54
alligatoris 54
antilopinus **53**
argentatus 54
aurantiacus 53
aurescens 53
banksianus 55
bedfordi 53
bennetti 55
bernardus **53**
binoe 53
bracteator 54
cervinus 54
crassipes 53
dama 53
decres 53
derbyanus 53
dissimulatus 55
dorsalis **53**
emiliae 53
erubescens 54
eugenii **53**

flindersi 53
fruticus 55
fuliginosus **54**
giganteus 53, **54**, 54
gracilis 53
greyi **54**
griseofuscus 54
griseolanosus 55
griseus 55
hagenbecki 54
houtmanni 53
irma **54**
isabellinus 54
jardinii 53
labiatus 54
lanigerus 55
leporides 52
leptonyx 55
magnus 54
major 54
manicatus 54
melanops 54
melanopus 54
muelleri 51, 52
nigrescens 53
obscurior 53
occidentalis 55
ocydromus 54
pallida 54
pallidus 55
papuanus 53
parma **54**
parryi **54**
pictus 55
reginae 54
robustus **54**, 54
rubens 54
ruber 55
ruficollis 55
rufogriseus **54**
rufus **55**
rutilus 55
siva 53
tasmaniensis 54
tridactylus 54
unguifer 55
vinosus 55
woodwardi 53, 54
macropus, Bandicota 579
Feroculus 99
Microtus 528
Mus 671
Myotis 207
Notomys 636
Sorex 99
Tatera 562
Uromys 671
macropygmaeus, Sorex 113
macrorhabdotes, Tamias 455
macrorhinus, Peromyscus 732, 734
Sylvilagus 825
macrorhynchus, Globicephala **352**
Halichoerus 330
Macroscelidea 829
Macroscelides **830**
proboscideus **830**
rupestris 829
typus 830
Macroscelididae **829**
macrosceloides, Neofelis 297
macrosensis, Chaetodipus 484
Macrospalax 754
macrospilotus, Spermophilus 449
macrosus, Mungos 307
Macrotarsomys 677, **679**, 679
bastardi **679**, 679
ingens **679**
occidentalis 679
macrotarsos, Tarsius 251
macrotarsus, Brachyteles 257
Jaculus 493
Marmosa 18
Myotis **212**
macrothrix, Felis 290
Macrotis **39**, 40, 391
cambrica 40
grandis 40
interjecta 40
lagotis **40**
leucura **40**
minor 40
miselius 40
nigripes 40
sagitta 40
macrotis, Allactaga 489
Alticola 502, **503**, 504
Emballonura 158
Glaucomys 460
Idiurus **757**, 758
Neotoma 711
Notomys **636**
Nycteris **162**
Nyctinomops **239**
Ochotona **810**, 810
Odocoileus 391
Onychomys 719
Peropteryx **158**
Pipistrellus **222**
Plecotus 225
Pteropus **148**
Rheithrosciurus **438**
Rhinolophus **167**
Sciurus 438
Suncus 102
Thomomys 475
Trichys 774
Vulpes 287
Macrotolagus 814
macrotrichus, Sorex 111
Macrotus **178**
bocourtianus 178
bulleri 178
californicus **178**
compressus 178
heberfolium 178
jamaicensis 178
mexicanus 178
minor 178
waterhousii **178**, 178
macrotus, Histiotus **205**
Lepus 819
macroura, Condylura 124
Hystrix 773
Leopardus 292
Mephitis **317**
Petauroides 59
Ratufa **437**
Sciurus 442
Sminthopsis **36**
macrourus, Atherurus **773**
Dasyurus 32
Isoodon **39**
Leopoldamys 603
Makalata 793
Neomys 110
Odocoileus 391
Pogonomys 641, **642**
Spermophilus 450
Vulpes 287
Macroxus 439
macrura, Cercartetus 58
Genetta 345
Isoodon 39
Leopardus 292
Mustela 322
Ratufa 437
Thylamys **23**
macruroides, Ratufa 437
Macruromys **607**
elegans **607**, 607
major **607**
Macruroryzomys 722
macrurus, Echimys **792**
Mesembriomys **620**
Microtus 523
Mungos 307
Rhinolophus 163
Rhipidomys 745
Sorex 114
Soriculus **123**
macuanus, Nyctalus 217
maculata, Balionycteris **138**
Crocuta 308
Damaliscus 394
Dasyurus 32
Didelphis 31
Gazella 396
Genetta **346**, 346, 347
Giraffa 383
Heterohyrax 373
Lynx 293
Meles 314
Phalangista 47
Planigale **34**
maculatum, Euderma **204**
maculatus, Apodemus 570
Axis 384
Cynopterus 138
Dasyurus **32**
Heliosciurus 429
Histiotus 204
Lepus 817
Mus 625
Panthera 297
Spilocuscus **47**, 47
maculicollis, Lutra **312**, 312
Lutrogale 313
maculipectus, Niviventer 634
maculipes, Rhipidomys 745
maculipilis, Taeromys 667
maculiventer, Oryzomys 720
maculosa, Ondatra 532
maculosus, Lemniscomys 602
Prionodon 347
madagascarensis, Herpestes 305
Microcebus 243
Sciurus 247
madagascariensis, Daubentonia **247**
Erinaceus 73
Felis 290
Hemicentetes 73
Hexaprotodon **380**
Nycteris 162
Otomops 239
Potamochoerus 378
Rousettus **153**
Suncus **102**, 102
madarogaster, Mandrillus 268
madeirae, Nyctalus 217
Panthera 298
Saimiri 260
madeirensis, Eira 318
Marmosa 18
maderensis, Pipistrellus **222**
madescens, Marmosops 19
madogae, Heliosciurus 428
madoka, Madoqua 398
Madoqua **398**
cavendishi 398
citernii 398
cordeauxi 398
damarensis 398
erlangeri 398
gubanensis 398
guentheri **398**
hararensis 398
hemprichiana 398
hindei 398
hodsoni 398
kirkii **398**
langi 398
lawrancei 398
madoka 398
minor 398
nasoguttata 398
nyikae 398
phillipsi 398
piacentinii **398**
saltiana **398**, 398
smithii 398
swaynei 398
thomasi 398
variani 398
wroughtoni 398
madoqua, Sylvicapra 412
madrassius, Harpiocephalus

228
madrea, Cryptotis 109
madrensis, Glaucomys 460
Microtus 524
Neotoma 712
Peromyscus 729, **732**
Sigmodon 748
Spermophilus **447**
Thomomys 476
Urocyon 285
Madromys 587, 588
madsoedi, Callosciurus 422
madurae, Callosciurus 422
madurensis, Rhinolophus 164
maenas, Microcavia 780
maerens, Petinomys 464
Rattus 661
maesi, Cercopithecus 264
maestus, Xerus 458
mafuca, Pan 277
magalakuini, Aethomys 567
magdalenae, Chaetodipus 484
Lepus 815
Nectomys 709
Oryzomys 725
Peromyscus 732
Proechimys 796, **797**
Sciurus 441
Thomomys 473
magdalenensis, Microtus 527
magellanica, Dolichotis 780
magellanicus, Canis 284
Ctenomys **785**
Histiotus 205
Oligoryzomys **718**, 718
Pseudalopex 284
mageshimae, Cervus 386
maggietaylorae, Hipposideros **174**
maghrebi, Gerbillus **552**
magister, Neotoma 711
Phyllotis **738**, 738
Ursus 339
magna, Cavia **779**
Cryptotis **108**
Dobsonia 140
Pygoderma 191
Sturnira **193**
magnamolaris, Myotis 216
magnater, Miniopterus **230**, 231
magnificaudatus, Sciurus **442**
magnificus, Callosciurus 422
Petaurista **462**, 463
magnirostris, Rattus 657
magnirostrum, Uroderma **193**
magnus, Berylmys 580
Cebus 259
Holochilus **704**, 704, 705
Macropus 54
Microperoryctes 41
Nyctalus 217
Taphozous 161
Magotus 266
magruderensis, Perognathus **486**
Magus 266
mahadeva, Lepus 819
mahaganus, Pteropus **148**
mahali, Cryptomys 772
mahomet, Mus **624**, 624, 629
Maimon 266, 268
maimon, Mandrillus 268
Simia 268
maini, Zyzomys **674**
mainitensis, Bubalus 402
Sus 379
mainois, Peroryctes 42
major, Alcelaphus 393
Allactaga **489**
Aplodontia 417
Apodemus 572
Apomys 575
Arctogalidia 343
Artibeus 187
Arvicanthis 578
Axis 384
Balaenoptera 350
Beamys **540**
Canis 281
Cervus 386, 387
Chaerephon **232**
Cheirogaleus **243**, 243
Chinchilla 777
Chiropodomys 583, **584**, 584
Chiruromys 584
Clidomys 805
Crocidura 94
Cynopterus 139
Dendromus 543
Diceros 372
Dipus 488
Eonycteris **153**
Eumops 233
Eupleres 340
Geomys 469
Glossophaga 184
Hemiechinus 78
Herpestes 305
Hipposideros 171
Hystrix 773
Lepus 816
Lophuromys 605
Macropus 54
Macruromys **607**
Mephitis 317
Microtus 521
Molossus 235
Mus 625
Mustela 323, 324
Myotis 209, 210
Natalus 195
Nesophontes **70**
Neurotrichus 126
Nycteris **162**, 162
Nyctimene **144**
Nyctophilus 218
Panthera 298
Perameles 40
Peromyscus 729
Phyllonycteris 183
Rattus 657
Ratufa 437
Rhabdomys 662
Rhogeessa 226
Sigmodon 747
Spermophilus 446, **447**, 447, 448, 449
Syconycteris 155
Sylvilagus 824
Thamnomys 670
Thyroptera 196
Ursus 339
Vampyrodes 194
Zapus 499
majori, Eliurus **678**, 678
Fossa 341
Microgale 71
Microtus **524**, 526
Miniopterus 231
Pitymys 520
Propithecus 247
Rhinolophus 166
Sus 379
Tremarctos 338
majus, Megaderma 163
majuscula, Macaca 267
majusculus, Apodemus 572
Chaetodipus 483
Geomys 469
Triaenops 175
maka, Profelis 296
makalae, Tragelaphus 404
Makalata 791, **793**, 793
armata **793**
armatus 793
castaneus 793
guianae 793
hispidus 793
longirostris 793
macrourus 793
occasius 793
makapani, Parahyaena 309
Pronolagus 823
makassarius, Rattus 660
makedonicus, Clethrionomys 508
Nannospalax 754
makensis, Rattus 660
Maki 245
makkovikensis, Glaucomys 460
makrami, Gerbillus 552
Sekeetamys 560
malabarica, Bandicota 579
Hystrix 774
Ratufa 437
malabaricus, Loris 248
Muntiacus 389
Suncus 103
malaccana, Tupaia 132
malaccanus, Petinomys 463
malaccensis, Cervus 387
Herpestes 305
Moschiola 382
Pteropus 151
Viverricula 348
Malacomys **607**, 607, 608
australis 608
cansdalei 592, **607**, 607
centralis 608
edwardsi 592, **607**
giganteus 607, 608
longipes **607**, 607
lukolelae 607, **608**, 608
verschureni **608**, 608
wilsoni 608
Malacothrix **544**
damarensis 544
egeria 544
fryi 544
harveyi 544
kalaharicus 544
molopensis 544
typica **544**
malagai, Molossus 235
malaianus, Nycticebus 248
malaisia, Lenothrix 603
malaitensis, Hipposideros 172
Nyctimene **144**
malampus, Dactylopsila 61
malani, Crocidura 89
malawali, Callosciurus 422
Niviventer 633
malayana, Crocidura 85, **89**, 89
Euroscaptor 125
Kerivoula 197
Tylonycteris 228
malayanus, Delphinus 355
Helarctos **337**, 337
Rhinolophus **167**
Stenella 356
Suncus **102**, 102, 103
Tapirus 371
Ursus 337
malcolmi, Microtus 527
malei, Microtus 507
malengiensis, Rattus 652
malitiosus, Cebus 259
Sorex 118
malkensis, Otomys 682
malleus, Otomys 682
Mallodelphys 15
Mallomys 583, 587, 602, **608**, 637, 640
argentata 608
aroaensis **608**, 609
gunung **608**
hercules 608
istapantap **608**
rothschildi **608**, 608, 609
weylandi 608
mallurus, Sylvilagus 826
Malpaisomys 582, **609**
insularis **609**, 609
malpasi, Kerivoula 197
Maltamys 767
malyi, Chionomys 507
mambatus, Chiruromys 584

mamberanus, Melomys 618
Mamcynomiscus 424
mamorae, Oecomys **716**
managaschae, Colobus 270
managuensis, Sciurus 443
manarius, Microsciurus 433
Manati, gigas 365
Manatidae 365
Manatus 365
manatus, Trichechus **365**, 365
manavi, Miniopterus 231
Sciurus 441
manchu, Mus 625
manchurica, Mustela 324
Prionailurus 295
manchuricus, Cricetulus 537
Mesechinus 79
mandarinus, Cervus 386
Lasiopodomys **516**
mandatus, Lepus 816
mandchuricus, Sus 379
Ursus 339
mandibularis, Macaca 266
Trachypithecus 273
mandingo, Funisciurus 428
mandjarum, Mungos 307
Mandril 268
Mandrillus **268**, 268
burlacei 268
cinerea 268
drill 268
ebolowae 268
escherichi 268
hagenbecki 268
insularis 268
leucophaeus **268**
madarogaster 268
maimon 268
mormon 268
mundamensis 268
planirostris 268
poensis 268
schreberi 268
sphinx **268**, 268
suilla 268
sylvicola 268
tessmanni 268
zenkeri 268
mandshurica, Panthera 298
mandshuricus, Lepus 815, 817, **818**, 819
mandus, Maxomys 614
manei, Mus 625
manengubae, Crocidura **89**
mangalumis, Rattus 661
manguensis, Lycaon 283
Mangusta, auropunctata 305
manhanensis, Sciurus 441
manharae, Gazella 397
manicalis, Maxomys 614
manicatus, Macropus 54
Rattus 653
maniculata, Felis 290
maniculatus, Peromyscus 728, **732**, 732, 733, 734, 735, 736
Rattus 657
Manidae **415**
maniemae, Colobus 270
manipulus, Berylmys **580**
Epimys 580
manipurensis, Tamiops 457
Manis **415**
aurita 415
crassicaudata **415**
culionensis 415
gigantea **415**
javanica **415**
longicaudata 415
pentadactyla **415**, 415
temminckii **415**
tetradactyla **415**
tricuspis **415**
tridentata 415
manitobensis, Blarina 106
Cervus 386
manium, Nasua 335
manni, Crocidura 92
manningi, Canis 281
Heterohyrax 373
manoquarius, Rattus 652
mansalaris, Leopoldamys 603
Macaca 266
Sundasciurus 452
mansoensis, Akodon **691**
mansorius, Rattus 660
mansuetus, Potos 333
Sylvilagus **826**
mansumensis, Crocidura 85
mantchurica, Ochotona 809
mantchuricus, Apodemus 570
Cervus 386
Sciurus 443
manteufeli, Aethomys 568
Lemniscomys 601
Lophuromys 606
mantschuricus, Capreolus 390
manul, Felis 290, 295
Otocolobus **295**
manuselae, Rattus 656
manyema, Gorilla 276
maorium, Rattus 652
mapiriensis, Micoureus 20
maporensis, Callosciurus 422
Crocidura 89
mapravis, Callosciurus 421
mapurito, Conepatus 316
maputa, Tatera 560
maquassiensis, Crocidura **89**
Mara 780
mara, Rattus 661
marabutus, Xerus 458
maracaibensis, Sciurus 441
maracaya, Leopardus 291
maradius, Gerbillus 548
marae, Galerella 303
maraisianus, Cervus 386
marakovici, Dinaromys 511
maral, Cervus 386
marana, Ratufa 437
maranii, Marmosa 18
maranonicus, Oligoryzomys 717
maranonis, Cebus 259
maraxica, Dasyprocta 782
maraxina, Monodelphis **21**
marcanoi, Antillogale 69
Solenodon **69**, 69
marcarum, Calomys 698
marchei, Mydaus **315**
Suillotaxus 315
Sus 379
marchicus, Sorex 111, 112, 116
marchio, Petaurista 463
marcolinus, Naemorhedus 407
marcosensis, Neotoma 712
mareyi, Bradypus 63
margae, Presbytis 271
Margaretamys 591, **609**, 612
beccarii **609**, 609
elegans **609**
parvus **609**
thysanurus 609
margarettae, Haeromys **595**, 595
Lophuromys 605
Mus 595
margarita, Felis **290**
Sorex 121
Trachypithecus 273
margaritae, Cebus 259
Chaetodipus 484
Dipodomys **478**, 478, 479
Felis 290
Odocoileus 391
Peromyscus 732
Sylvilagus 826
Margay 291
margay, Leopardus 292
margianus, Jaculus 494
marginae, Meriones 556
marginata, Balaena 351
Caperea **351**
Felis 290
marginatus, Ateles **257**
Cynopterus 153
Pipistrellus 221
Pteropus 138
Spermophilus 449
Zaedyus 67
margueritei, Felis 290
maria, Kobus 414
mariae, Equus 369
Gerbillus 552
Meriones 556
Microtus 524
Steatomys 545
mariakanae, Elephantulus 829
marianensis, Meles 314
mariannus, Cervus 385, **386**
Pteropus **148**, 148
marica, Eligmodontia 701
Gazella 397
Gracilinanus **17**
Mus 625
Nycteris 161
Petaurista 462
Vandeleuria 672
mariepsi, Otomys 681
mariguanensis, Erophylla 182
Marikina 253
marikina, Leontopithecus 252, 253
marikquensis, Mastomys 610
marina, Didelphis 18
Enhydra 310
marinensis, Chaetodipus 483
marinkellei, Lonchorhina **178**
marinsularis, Callosciurus 422
marinus, Eumetopias 328
Lutra 312
Niviventer 634
Procyon 335
Reithrodon 740
Ursus 339
maripensis, Leopardus 291
Myotis 213
mariposae, Microtus 519
Peromyscus 729
Sorex 121
Sylvilagus 825
Tamias 454
mariquensis, Crocidura **89**
maris, Pteropus **147**
marisae, Hipposideros **174**
marisalbi, Delphinapterus 357
marita, Crocidura 85
maritima, Antechinus 30
Cercopithecus 264
maritimensis, Sorex 112
maritimus, Bathyergus 771
Galerella 303
Geomys 470
Gerbillus 550
Mus 771
Naemorhedus 407
Peromyscus 732
Phyllotis 737
Procyon 335
Sus 378
Tamiops **457**, 457
Taphozous 161
Ursus 338, **339**
marjoriae, Chlorocebus 265
Petromus 774
Suricata 308
marjorita, Lepus 823
Poelagus **823**
markhami, Akodon **691**
Equus 369
marleyi, Amblysomus 74
marlothi, Procavia **374**
marmandianus, Cervus 386
marmorata, Felis 299
Pardofelis **299**
marmoratus, Bradypus 63

Cynocephalus 135
Marmosa 17, **18**, 19, 20, 23
andersoni **18**
bombascarae 18
canescens **18**
casta 18
chapmani 18
chloe 18
cinerea 20
dorsigera 18
duidae 18
fulviventer 18
gaumeri 18
grandis 18
grenadae 18
guianensis 18
insularis 18
isthmica 18
klagesi 18
lepida **18**
luridavolta 18
macrotarsus 18
madeirensis 18
maranii 18
mayensis 18
meridionalis 18
mexicana **18**
mimetra 18
mitis 19
moreirae 18
murina **18**
muscula 18
musicola 18
nesaea 19
oaxacae 18
pallidiventris 19
parata 18
phelpsi 19
quichua 18
robinsoni **18**
roraimae 18
ruatanica 19
rubra **19**
savannarum 18
simonsi 19
sinaloae 18
tobagi 18
tyleriana **19**
waterhousei 18
xerophila **19**
zeledoni 18
marmosa, Hapalomys 595
Marmosops **19**
albiventris 19
bahiensis 19
bishopi 20
carri 19
caucae 19
celicae 19
collega 20
cracens **19**
dorothea **19**
fuscatus **19**
handleyi **19**
impavidus **19**
incanus **19**
invictus **19**
juninensis 20
keaysi 20
leucastra 20
lugenda 20
madescens 19
neblina 19
neglecta 20
noctivagus **20**
ocellata 19
oroensis 19
parvidens **20**
paulensis 19
perfuscus 19
pinheiroi 20
polita 20
purui 20
scapulatus 19
sobrina 19
stollei 20
ucayaliensis 19
woodalli 20
yungasensis 19
marmosurus, Oecomys 716
Rattus 650, **655**, 655, 662
Marmota **430**
alba 432
alpina 432
aphanasievi 431
arctomys 431
aurea 431
avara 432
baibac 431
baibacina **430**, 431, 432, 433
bobac 431
bobak **431**, 431, 432, 433
broweri **431**, 431, 432
bungei 431
bunkeri 432
caligata **431**, 431, 432
caliginosus 433
campioni 432
camtschatica **431**, 432
canadensis 432
cascadensis 431
caudata **431**
centralis 431
cliftoni 431
dacota 432
dahurica 433
dichrous 431, 432
doppelmayeri 431
empetra 432
engelhardti 432
flavina 431
flaviventris **432**
fortirostris 432
hemachalana 432
himalayana 431, **432**
hodgsoni 432
ignava 432
johnsoni 432
kastschenkoi 431
kozlovi 431
latirostris 432
lewisi 431
littledalei 431
luteola 432
marmota 431, **432**, 432, 433
marmotta 432
melanopus 432
menzbieri **432**, 432
monax **432**
nigra 432
nivaria 431
nosophora 432
notioros 432
obscura 432
ochracea 432
ognevi 431
okanagana 431
olympus **432**
oxytona 431
parvula 432
petrensis 432
preblorum 432
raceyi 431
robusta 432
rufescens 432
sheldoni 431
sibila 431, 432
sibirica 431, **433**, 433
sierrae 432
stirlingi 431
tataricus 432
tibetanus 432
tigrina 432
tschaganensis 431
vancouverensis **433**
vigilis 431
warreni 432
zachidovi 432
marmota, Dendrohyrax 373
Marmota 431, **432**, 432, 433
Mus 430
Thylamys 23
Marmotinae 419
Marmotini 419, 424, 430, 444, 453
Marmotops 430
marmotta, Marmota 432
maroccanus, Canis 280
Lepus 816
maros, Rhinolophus 168
Marputius 316
marrensis, Chlorocebus 265
Crocidura 85
Graphiurus 764
Pipistrellus 223
Procavia 374
marshalli, Dipodomys 479
Neotoma 712
Rhinolophus **167**
marsicanus, Ursus 339
Marsupialia **15**, **25**, **27**
marsupialis, Antidorcas **395**
Antilope 395
Didelphis **16**, 16
Hipposideros 175
martensi, Crocidura 83
Sciurus 443
Martes **319**
abieticola 319
abietinoides 319
abietum 320
actuosa 319
americana **319**, 320, 321
aterrima 320
atrata 319
boria 319
brachyura 321
brumalis 319
canadensis 320
caurina 319
chrysogaster 320
chrysospila 320
columbiana 320
domestica 319, 320
flavigula **320**, 320
foina **320**
guadricolor 320
gwatkinsii **320**
hardwickei 320
henricii 320
humboldtensis 319
huro 319
indochinensis 320
intermedia 320
japonica 320
kenaiensis 319
lasiotis 320
leucocephalus 320
leucolachnaea 320
leucopus 319
leucotis 320
martes 319, **320**, 320, 321
martinus 319
mediterranea 320
melampus 319, **320**, 320, 321
melanopus 320
melanorhyncha 320
melina 320
melli 320
nesophila 319
nigra 320
origenes 319
pacifica 320
peninsularis 320
pennanti **320**
piscator 320
quadricolor 320
rosanowi 320
sierrae 319
sylvatica 320
sylvestris 320
toufoeus 320
tsuensis 320
typica 320
vancouverensis 319
varietas 320
vulgaris 320
vulpina 319
yuenshanensis 320
zibellina 319, **320**, 320,

321
martes, Martes 319, **320**, 320, 321
martiensseni, Crocidura 92
Otomops **239**
martinensis, Neotoma **712**
Peromyscus 732
Thomomys 476
martini, Cercopithecus 264
Nycteris 161
martiniquensis, Myotis **212**
martinoi, Arvicola 506
Lynx 293
Microtus 521, 530
Myoxus 770
Nannospalax 754
Rhinolophus 166
Spermophilus **445**
martinsi, Saguinus 253
martinus, Martes 319
martirensis, Dipodomys 477
Lepus 815
Neotoma 711
Peromyscus 736
Thomomys 473
marungensis, Heliophobius 772
Hipposideros 172
Oenomys 637
Pan 277
marwitzi, Gazella 397
Heliosciurus 429
marylandica, Taxidea 325
marzaninus, Cervus 386
masae, Leopoldamys 603
Ratufa 437
Tragulus 382
Tupaia 133
masakensis, Ourebia 399
masalai, Nyctimene **144**
masaretes, Rattus 660
masbatensis, Cervus 385
mascarinus, Pteropus 150
maschona, Potamochoerus 378
mascotensis, Sigmodon **747**, 747
mashona, Otomys 680
mashonae, Saccostomus 541
Tatera 561
masoni, Hipposideros 172
massaesyla, Gazella 396
massagetes, Meriones 557
massaicus, Lemniscomys 602
Mylomys 630
Panthera 297
Phacochoerus 377
Syncerus 402
Tragelaphus 404
Xerus 458
massonii, Ctenodactylus 761
Massoutiera **761**, 761
harterti 761
mzabi **761**
rothschildi 761
mastacalis, Rhipidomys 744, **745**, 745
Mastacomys 644, 646
mastersii, Wallabia 57
mastersoni, Tadarida 240
mastivus, Molossops 234
Noctilio 176
Mastomys 568, **609**, 609, 610, 611, 630, 642
angolae 610
angolensis **610**, 610, 611
bradfieldi 610
breyeri 610
caffer 611
coucha **610**, 610
erythroleucus **610**
hildebrandti 643
hildebrandtii **610**, 610
huberti 610
illovoensis 611
komatiensis 611
legerae 611
limpopoensis 610
marikquensis 610
microdon 611
muscardinus 611
natalensis 610, **611**, 611
ovamboensis 611
pernanus **611**, 611
shortridgei 610, **611**, 611
sicialis 610
socialis 610
verheyeni **611**
zuluensis 611
Mastonotus 805
matacus, Tolypeutes **67**
matagalpae, Alouatta 255
Orthogeomys **471**
Sciurus 440
matanim, Phalanger **46**
matecumbei, Sciurus 440
matrensis, Microtus 530
matroka, Eptesicus 201
matruhensis, Crocidura 97
matschei, Eptesicus 202
Matscheia 395
matschiei, Alcelaphus 393
Chlorocebus 265
Civettictis 345
Colobus 270
Dendrolagus 50, **51**, 51
Equus 370
Felis 290
Galago **249**
Genetta 346
Gorilla 276
Kobus 413
Lutra 312
Orycteropus 375
Ovis 408
Procavia 374
matsika, Phalanger 46
matsumotei, Cervus 386
mattensis, Nectomys 709
matthaeus, Callosciurus 421
Leopoldamys 603
matthewsi, Syncerus 402
mattheyi, Mus **624**, 625
mattogrossae, Oligoryzomys 718
mattogrossensis, Molossops **234**
Neoplatymops 234
mattosi, Connochaetes 394
matugamensis, Funambulus 426
mauae, Cercopithecus **264**
maugei, Dasyurus 32
maulinus, Ctenomys **785**, 785
Spalacopus 789
maullinicus, Pseudalopex 284
maunensis, Paraxerus 434
Steatomys 546
maura, Dama 387
Macaca **267**, 267
mauretanicus, Alcelaphus 393
mauricus, Dama 387
maurisca, Crocidura **89**
mauritana, Felis 290
mauritaniae, Gerbillus **553**
mauritanicus, Jaculus **494**
mauritiana, Suncus 103
mauritianus, Pteropus 149
Taphozous **160**
maurus, Clethrionomys 508
Eumops **234**
Pipistrellus 223
Praomys 643
Sciurus **440**
mawambicus, Colobus 270
Lophocebus 266
Potamochoerus 378
maxeratis, Meriones 556
maxi, Crocidura **90**, 90, 91
Hylomys 80
Sundamys **666**
maxillaris, Naemorhedus 407
maxima, Murexia 32
Nyctalus 216
Paracrocidura **101**, 101
Ratufa 437
maximiliani, Callithrix 251
Centronycteris **156**
Vespertilio 156
maximowiczii, Microtus 520, **524**, 524, 525
maximus, Apodemus 573
Chaetodipus 483
Dasypus 66
Dendrolagus 50
Dipus 778
Elephas **367**, 367
Hipposideros 171
Ictonyx 319
Lagostomus **778**
Otomys 680, **681**
Petauroides 59
Phyllostomus 180
Pipistrellus 221
Priodontes **67**
Spermophilus 446
Maxomys 580, 581, 583, 591, 595, 603, **611**, 611, 612, 613, 614, 632, 662, 664
alticola **612**, 612, 613
anambae 614
antucus 614
aoris 614
asper 614
aspinatus 613
baeodon **612**
banacus 614
bandahara 614
bartelsii **612**
batamanus 614
batus 614
bentincanus 614
binominatus 614
butangensis 614
carimatae 614
casensis 614
catellifer 614
cereus 612
changensis 614
connectens 614
coritzae 614
coxingi 613
dollmani **612**, 612
domelicus 614
eclipsis 614
finis 614
flavidulus 614
flavigrandis 614
grandis 614
griseogenys 612
hellwaldii **612**, 612
hidongis 613
hylomyoides **612**
inas **612**, 614
inflatus **612**
kinabaluensis 612
klossi 614
koratis 614
kramis 614
kutensis 614
lalawora 613
leonis 614
lingensis 613
localis 612
luteolus 614
mabalus 614
mandus 614
manicalis 614
melanurus 614
melinogaster 614
microdon 614
moi **613**
muntia 614
musschenbroekii **613**, 613, 614
natunae 614
obscuratus 612
ochraceiventer **613**
pagensis 604, **613**
palawanensis 613
panglima **613**

pelagius 614
pellax 613
pemangilis 614
perasper 613
perflavus 614
perlutus 614
pidonis 614
pinacus 614
piratae 614
puket 614
rajah **613**, 613
ravus 614
saturatus 614
serutus 614
siarma 614
similis 613
solaris 614
spurcus 614
subitus 614
surifer **613**, 613, 634
telibon 614
tenebrosus 614
tetricus 613
tjibunensis 612
trachynotus 612
ubecus 614
umbridorsum 614
verbeeki 614
wattsi **614**
whiteheadi 612, 613, **614**, 614, 653
maxwelli, Mellivora 315
maxwellii, Cephalophus **410**, 411
Mayailurus 295
iriomotensis 295
mayapahit, Leopoldamys 603
mayema, Gorilla 276
mayensis, Cryptotis 109
Dipodomys 479
Marmosa 18
Odocoileus 391
Oligoryzomys 718
Peromyscus 730, **733**
Puma 296
mayeri, Antechinus 30
Coccymys 585
Melomys 640
Pogonomelomys **641**, 641
Pseudocheirus **59**
Mayermys **614**, 614, 620, 632, 644
ellermani **614**, 614
mayi, Syncerus 402
maynardi, Procyon 335, **336**
mayonicus, Rattus 652
mayori, Mus **625**, 625
Viverricula 348
mayottensis, Eulemur 244
maypuri, Tapirus 371
mayri, Dendrolagus 50
Mayrmys 604
mayumbicus, Epixerus 425
Funisciurus 427
Mazama **390**, 392
americana **390**
argentina 391
auritus 390
bira 391
bricenii **390**
carrikeri 390
cerasina 390
chunyi **390**, 392
cita 391
dolichurus 390
dorsata 407
fusca 391
fuscata 390
gouazoubira 391
gouazoupira **391**, 391
gualea 390
inornatus 390
jucunda 390
juruana 390
kozeritzi 391
mexianae 391
murelia 391
namby 391
nana **391**
nemorivaga 391
pandora 390
permira 391
pita 390
reperticia 390
rondoni 391
rosii 390
rufa 390
rufina 390, **391**, 391
sanctaemartae 391
sarae 390
sartorii 390
sheila 390
simplicicornis 391
superciliaris 391
tema 390
temama 390
toba 390
trinitatis 390
tschudii 391
tumatumari 390
whitelyi 390
zamora 390
zetta 390
mazama, Clethrionomys 508
Thomomys 473, **474**
mazzai, Dasypus 65, 66
mcguirei, Rangifer 392
mcilhennyi, Odocoileus 391
Philander 22
mcilwraithi, Rattus 654
mcintyrei, Rhinolophus 164
mcmahoni, Macaca 267
mcmillani, Canis 281
mcmurtrii, Centurio 189
mcoyi, Lasiorhinus 45
mearnsi, Arvicanthis 576, 578
Canis 280
Conepatus 316
Connochaetes 394
Lemniscomys 602
Leopardus 291
Limnomys 605
Neotoma 710
Perognathus 486
Pteropus **148**
Saccostomus **541**, 541
Sylvilagus 826
Tamiasciurus **456**, **457**
Thomomys 473
mearnsii, Peromyscus 732
mearsi, Callosciurus 423
meator, Mus 627
mechowi, Cryptomys **772**
mechowii, Cryptomys 772
mecklenburzevi, Nyctalus 217
medellinensis, Sciurus 442
medellinius, Cryptotis 109
medemi, Callicebus 259
mederai, Eumops 233
media, Allactaga 489
Nycteris 162
Suncus 103
medialis, Callosciurus 421
medicatus, Aethomys 568
medicaudatus, Lophuromys **606**
Mediocricetus 538
mediocris, Rattus 658
medioximus, Synaptomys 534
mediterranea, Felis 290
Martes 320
mediterraneus, Cervus 386
Lepus 818
Monachus 331
Pipistrellus 223
Sus 379
medium, Megaderma 163
medius, Bassaricyon 333
Cheirogaleus **243**
Cremnomys 588
Dasymys 589
Dipodomys 479
Echimys 791
Galagoides 249
Graomys 703
Lepus 817
Microdipodops 480
Microtus 527
Miniopterus 230
Oryzomys 725
Perognathus 485
Peromyscus 732
Pteropus 147
Thomomys 475
medjerdae, Caracal 289
medjianus, Heliosciurus 429
meeki, Emballonura 157
Hyomys 599, 600
Melomys 617
Phalanger 46
meesteri, Suncus 103
megabalica, Acinonyx 288
Megacephalon 369
Megacephalonella 369
megacephalus, Epomophorus 141
Microdipodops **480**, 480
Oryzomys 721
Peromyscus 730
megaceros, Capra 405
Kobus **414**
Megachiroptera 137
megacranius, Dipus 493
Megadendromus **544**
nikolausi 544, **545**
Megaderma **163**
abditum 163
carimatae 163
carina 163
carnatica 163
celebensis 163
ceylonense 163
cor 162
frons 162
gigas 163
horsfieldi 163
kinabalu 163
lasiae 163
lyra **163**
majus 163
medium 163
minus 163
naisense 163
natunae 163
pangandarana 163
schistacea 163
sinensis 163
siumatis 163
spasma **163**
spectrum 163
trifolium 163
Megadermatidae **162**
megadon, Oryzomys 725
Megadontomys 706, **707**, 707, 728
cryophilus **707**, 707
flavidus 705
nelsoni **707**, 707
thomasi **707**
Megaerops **142**
albicollis 143
ecaudatus **142**
kusnotoi **142**
niphanae **142**
wetmorei **142**
Megaladapidae **245**, 245
Megaleia 53
megalodontus, Sus 379
megalodous, Procyon 335
megalodus, Acomys 565
Megaloglossus **154**
prigoginii 154
woermanni **154**, 154
megalolepis, Dasypus 66
Megalomys **707**, 722
desmarestii **707**
luciae **707**
pilorides 707
Megalonychidae **63**, 63
megalonyx, Akodon 699
Chelemys **699**, 699

megalophylla, Mormoops **176**
megalops, Peromyscus **733**, 733, 734
megalopus, Myotis 212
megalotes, Micronycteris 179
megalotis, Acomys 565
Canis 283
Dorcatragus **395**
Felis 290
Hemiechinus 78
Hipposideros **174**
Lepus 820
Micronycteris **179**
Notomys 636
Nyctinomops 239
Oreotragus 395
Otocyon **284**
Paracoelops **175**, 175
Peromyscus 736
Phyllophora 178
Phyllotis 738
Reithrodon 740
Reithrodontomys 740, **741**, 741, 742, 743
megalotos, Plecotus 224
megalotus, Potos 333
megalua, Sylvisorex 105
megalura, Dactylopsila **60**
Sylvisorex **105**
megalurus, Eptesicus 202
Meganeuron 359
megaphyllus, Rhinolophus **167**
megapodius, Myotis 209
megapotamus, Geomys 470
Megaptera **350**
braziliensis 350
burmeisteri 350
lalandii 350
longimana 350
longipinna 350
nodosa 350
novaeangliae **350**
versabilis 350
megas, Geocapromys 801
Megascapheus 473, 474
Megasorex **109**, 109, 111
gigas **109**, 110
megaspila, Civettictis 345
Viverra 347, **348**, 348
Megistosaurus 359
mehelyi, Microtus 527
Mus 628
Rhinolophus **167**
meihsienensis, Tscherskia 540
meinertzhageni, Felis 290
Hylochoerus **377**, 377
Myotis 213
mejiae, Peromyscus 731
mekisturus, Peromyscus **733**
mekongis, Niviventer 634
melaena, Globicephala 352
melaenus, Oligoryzomys 718
melainus, Lepus 819
melalophos, Presbytis **271**, 271
Simia 270
melamera, Trachypithecus **274**
melampeplus, Mustela 324
melampus, Aepyceros **393**
Akodon 691
Antilope 393
Cerdocyon 282
Martes 319, **320**, 320, 321
Parahyaena 309
melanarctos, Ursus 339
Melanaxis 385
melanderi, Sorex 117
melanesiensis, Miniopterus 231
melania, Sciurus 443
melanius, Crunomys **588**, 588
Nectomys 709
Phyllotis 737
melanizon, Oligoryzomys 718
melanocarpus, Peromyscus **733**, 733
melanocaudus, Chaetodipus 483
Melanocebus 263
melanocephala, Eulemur 244
melanocephalus, Cacajao **261**
Chironax **138**
Cynopterus 138
Simia 260
melanocercus, Ateles 257
melanochaitus, Panthera 297
melanochir, Ateles 257
Callicebus 259
Procolobus 272
melanochra, Ratufa **437**
melanoderma, Rattus 652
melanodon, Sorex 111
Suncus 102, 103
melanogaster, Arvicola 513
Callosciurus **422**, 604
Eothenomys **514**, 514
Mus 626
Speothos 285
Vulpes 287
melanogenys, Cercopithecus 263
Meles 314
Sorex 118
melanoidis, Didelphis 16
melanoleuca, Ailuropoda **336**
melanoleucus, Callithrix 252
Heteromys 481
Saguinus 253
Ursus 336
Melanomys 671, 672, **707**, 708, 719, 722
affinis 708
buenavistae 708
caliginosus **708**
chrysomelas 708
columbianus 708
hadrourus 672
idoneus 708
lomitensis 708
monticola 708
obscurior 708
olivinus 708
oroensis 708
phaeopus 708
robustulus **708**
tolimensis 708
vallicola 708
zunigae **708**
melanonotus, Lepus 819
Sciurus 442
melanonyx, Lophuromys **606**, 664, 686
melanopepla, Ratufa 437
melanophrys, Onychomys 719
Peromyscus **733**, 733, 735
melanopogon, Pteropus **148**
Taphozous **160**, 161
melanoprymnus, Cephalophus 411
melanops, Callicebus 259
Callosciurus 423
Eptesicus 200, 201
Macropus 54
Melonycteris **154**, 154
Monodelphis 21
Myocastor 806
Scolomys **746**, 746
Taterillus 563
Thomomys 475
melanopterus, Aeretes **459**
Eptesicus 201
Pipistrellus 223
Pteromys 459
melanopus, Blastoceros 389
Macropus 54
Marmota 432
Martes 320
melanorheus, Cephalophus 411
melanorhinus, Myotis 212
melanorhyncha, Crocidura 85
Martes 320
melanorrhachis, Microgale 71
melanostoma, Ochotona 808
Oligoryzomys 717
melanostomus, Cerdocyon 282
melanotica, Panthera 298
melanoticus, Cryptomys 772
melanotis, Aepyprymnus 48
Ailurops 46
Bradypus 63
Callithrix 252
Capreolus 390
Caracal 288, 289
Cheirogaleus 243
Dendromus **542**, 542
Lepus 815
Nannosciurus **434**
Oryzomys 720, **723**, 723, 725
Perognathus 485
Peromyscus 732, **733**
Petaurista 463
Phyllotis 738
Raphicerus **400**, 400
Sciurus 434
Sigmodon 747
Thomomys 473
melanotus, Macaca 266
Phyllotis 738
Praomys 642
Pteropus **149**
Vulpes 287
melantho, Herpailurus 291
Philander 22
melanura, Callithrix 251
Conilurus 586
Dendrogale **131**
Galerella 303
Leopardus 291
Neotoma 710
Ourebia 399
Trichosurus 48
Viverra 348
melanurus, Antechinus **30**
Carpomys **583**, 583
Chaetodipus 483
Dipodomys 479
Eliomys **767**, 767
Maxomys 614
Meriones 556
Mysateles **802**, 803
Pappogeomys 472
Peromyscus **733**, 733
Philander 22
Pithecheir **640**, 640
Pronolagus 824
Rattus 653
Sphiggurus 776
Tamias 454
melanus, Procyon 335
melarhinus, Miopithecus 269
melas, Delphinus 352
Genetta 345
Globicephala **352**, 352
Muntiacus 389
Myoictis **32**
Neophocaena 358
Neotoma 710
Oncifelis 294
Panthera 298
Phascogale 32
Sus 379
Melasmothrix **614**, 667
naso **614**, 614
melckorum, Eptesicus **202**
meldensis, Microtus 519
Meles 313, **314**, 314, 325
aberrans 314
alba 314
altaicus 314
amurensis 314
anakuma 314
arcalus 314
arenarius 314
blanfordi 314
britannicus 314
canenscens 314

caninus 314
caucasicus 314
chinensis 314
communis 314
danicus 314
europaeus 314
hanensis 314
heptneri 314
leptorhynchus 314
leucurus 314
maculata 314
marianensis 314
melanogenys 314
meles **314**, 314
minor 314
raddei 314
rhodius 314
schrenkii 314
severzovi 314
sibiricus 314
siningensis 314
talassicus 314
tauricus 314
taxus 314
tianschanensis 314
tsingtanensis 314
typicus 314
vulgaris 314
meles, Meles **314**, 314
melfica, Procavia 374
melibius, Conilurus 586
meliceps, Mydaus 315
melicertes, Hydromys 597
melicus, Melomys 615
melina, Martes 320
Melinae **313**, 315, 325
melini, Cyclopes 67
melinogaster, Maxomys 614
melinus, Galictis 318
Nyctimene **144**
melissa, Vampyressa **193**
Melitoryx 315
mellandi, Cryptomys 772
Felis 290
melleri, Rhinogale 307
Rhynchogale **307**
melleus, Oecomys 716
melli, Arctogalidia 343
Dremomys 425
Leopoldamys 603
Macaca 266
Martes 320
Mustela 324
Ursus 340
Mellivora **315**, 315
abyssinica 315
brockmani 315
buchanani 315
capensis **315**
concisa 315
cottoni 315
inauritus 315
indicus 315
leuconota 315
maxwelli 315
mellivorus 315
pumilio 315
ratel 315
ratelus 315
sagulata 315
signata 315
typicus 315
vernayi 315
wilsoni 315
Mellivorinae **315**
mellivorus, Mellivora 315
mellonae, Phoca 332
Melogale 313, **314**, 315
everetii 314
everetti **314**, 314
ferreo-griseus 314
fusca 314
laotum 315
maccourus 314
millsi 314
modesta 314
moschata **314**
nipalensis 315
orientalis **314**, 314, 315
personata 314, **315**
pierrei 315
sorella 314
subaurantiaca 314
taxilla 314
tonquinia 315
Melomys 585, **615**, 615, 616, 617, 640, 641, 663, 672, 674
aerosus **615**, 615
albiventer 615
alleni 619
arcium 617
arfakianus 617
arfakiensis 619
australius 615
banfieldi 616
bougainville **615**, 615
bunya 616
burtoni **615**, 615, 616
calidior 619
callopes 615
capensis **615**, 615, 616, 619
caurinus 617
cervinipes 615, **616**, 616, 618
clarae 617
clarus 619
dollmani 616
eboreus 616
fellowsi 615, **616**
fraterculus 615, **616**, 616, 618
frigicola 615
froggatti 615
fulgens 617
fuscus 618
gracilis 615, **616**, 616
hageni 619
hintoni 615
insulae 615
intermedius 618
jobiensis 618
lanosus 615, **616**, 616, 617, 619
latipes 617
leucogaster 615, 616, **617**, 617
levipes 615, 616, **617**, 617, 618
limicauda 616
littoralis 615
lorentzii 615, **617**, 617, 618
lutillus 615
mamberanus 618
mayeri 640
meeki 617
melicus 615
mixtus 615
mollis 615, **617**, 617, 618
moncktoni 615, 617, **618**, 618, 619
murinus 615
musavora 619
muscalis 615
naso 617
niviventer 619
obiensis 615, **618**
pallidus 616
platyops 615, 617, **618**, 618
pohlei 619
rattoides 615, 616, 617, **618**, 619
rubex 618, **619**, 619
rubicola 615, **619**, 619
rufescens 615, 616, 617, **619**, 619
rutilus 619
sexplicatus 619
shawi 618, 619
shawmayeri 616, 617
spechti **619**
stalkeri 619
steini 619
stevensi 617
stressmani 619
sturti 617, 618
tafa 619
talaudium 617
weylandi 617
melonii, Myoxus 770
Melonycteris **154**
alboscapulatus 154
aurantius **154**
melanops **154**, 154
woodfordi **154**
meltada, Golunda 621
Millardia **621**, 621
Melursus **337**, 337
inornatus 337
labiatus 337
longirostris 337
lybius 337
niger 337
ursinus **337**, 337
melvillensis, Mesembriomys 620
Ovibos 408
melvilleus, Rattus 661
Memina 16, 398
meminna, Moschiola **382**
Moschus 382
menadensis, Boneia 138
Cervus 387
menagensis, Nycticebus 248
menamicus, Callosciurus 422
mendanauus, Callosciurus 423
mendeni, Phalanger 47
mendocinensis, Ursus 339
mendocinus, Ctenomys 784, **785**, 785, 786, 787
mendosus, Conepatus 316
meneghetti, Taterillus 563
meneleki, Aonyx 310
meneliki, Procavia 374
Tragelaphus 404
Menetes **433**
amotus 433
berdmorei **433**
consularis 433
decoratus 433
koratensis 433
moerescens 433
mouhotei 433
peninsularis 433
pyrrocephalus 433
rufescens 433
umbrosus 433
mengesi, Canis 280
Capra 406
mengkoka, Rattus 653
menglalis, Muntiacus 389
mengurus, Niviventer 633
mensalis, Vicugna 382
mentalis, Hemiechinus 78
mentawi, Callosciurus 422
Rattus 650, 654
mentosus, Dremomys 425
Niviventer 633, 634
menzbieri, Marmota **432**, 432
Spermophilus 450
menziesianus, Elaphurus 387
Meomeris 358
mephisto, Nasua 335
mephistopheles, Bubalus **402**
mephistophiles, Pudu 390, **392**, 392
Mephitinae 313, **315**
Mephitis **316**
americana 317
avia 317
bivirgata 317
concolor 317
dentata 317
edulis 317
elongata 317
estor 317
eximius 317
fetidissima 317
foetulenta 317
frontata 317
holzneri 317
hudsonica 317

interrupta 317
javanensis 315
macroura **317**
major 317
mephitis **317**
mesomelas 317
mexicana 317
milleri 317
minnesotoe 317
newtonensis 317
nigra 317
notata 317
occidentalis 317
olida 317
platyrhina 317
putida 317
richardsoni 317
scrutator 317
spissigrada 317
varians 317
vittata 317
mephitis, Mephitis **317**
Viverra 316
Meraeus 555
mergatus, Tragulus 382
mergens, Antilope 412
Ondatra 532
Sylvicapra 412
mergulus, Limnogale **70**, 70
Petaurista 463
meridae, Micoureus 20
meridana, Mustela 322
meridensis, Akodon 693
Caluromys 15
Cryptotis **108**, 109
Didelphis 16
Nasua 335
Oryzomys 720
Potos 333
Sciurus 441
Sylvilagus 825
meridianus, Meriones 556, **557**, 557
Microtus 519
meridiei, Lepus 817
meridioccidentalis, Spermophilus 449
meridionali, Myrmecophaga 68
meridionalis, Alces 389
Arvicola 506
Eptesicus 203
Erinaceus 77
Hapalemur 245
Lariscus 430
Lepus 817, 818, 822
Marmosa 18
Micromys 620
Mustela 323
Mysateles 802, **803**, 803
Pectinator 761
Plecotus 224
Pseudorca 355
Reithrodontomys 741
Rhabdomys 662
Rhinolophus 165
Sciurus 443
Sus 379
Tamias 454
Tragelaphus 404
Ursus 339
Vulpes 287
Meridiopitymys 517
meridonalis, Lutra 312
merilli, Gazella 396
meringgit, Rattus 652
Meriones 546, **555**, 555, 556, 558, 559, 578
afghanus 556, 557
albipes 558
ambrosius 557
aquilo 556
arimalius **555**, 556
asyutensis 556
auceps 557
auratus 556
auziensis 558
azizi 556
baptistae 557
blackleri 558
bodenheimeri 558
bogdanovi 558
brevicaudatus 557
buechneri 557
buryi 557
caucasius 556
caudatus 556, 557
charon 556
chengi **556**, 556
chihfengensis 558
ciscaucasicus 558
collinus 556
collium 556, 558
crassibulla 558
crassus **556**
cryptorhinus 557
dahli **556**, 557
edithae 556
erythrourus 556
evelynae 556
eversmanni 556
farsi 556
fulvus 557
gaetulus 556
grandis 558
gurganensis 557
guyonii 556
heptneri 556, 557
hurrianae **556**
intraponticus 558
iranensis 556
isis 558
ismahelis 556
jaxartensis 558
jei 557
karelini 557
karieteni 558
kokandicus 558
koslovi 558
kurauchii 558
lacernatus 578
laticeps 558
lepturus 557
libycus **556**, 556, 557, 558
littoralis 557
longiceps 558
longifrons 556
lybicus 556
lycaon 558
marginae 556
mariae 556
massagetes 557
maxeratis 556
melanurus 556
meridianus 556, **557**, 557
montanus 558
muleiensis 557
nogaiorum 557
opimus 559
oxianus 556
pallidus 556
pelerinus 556
penicilliger 557
perpallidus 556
persicus **557**
philbyi 557
psammophilus 557
renaultii 556
rex **557**
richardii 558
roborowskii 557
rossicus 557
sacramenti **557**
satschouensis 558
savii 558
schousboeii 556
schwarzovi 556
selenginus 558
sellysii 558
shawi 556, **557**, 557, 558
shitkovi 557
sogdianus 556
suschkini 557
swinhoei 556
syrius 556
tamaricinus 558
tamariscinus **558**
tristrami **558**
tropini 557
trouessarti 558
turfanen 556
unguiculatus **558**
urianchaicus 557
uschtaganicus 557
vinogradovi **558**
zarudnyi **558**
zhitkovi 557
meritus, Thomomys 475
merriami, Cervus 386
Cryptotis 109
Dipodomys **478**, 478, 479
Lepus 815
Pappogeomys **472**
Perognathus 485, **486**, 486
Peromyscus 730, **733**, 734, 735
Pipistrellus 221
Reithrodontomys 741
Sorex **117**
Tamias **454**, 454
Taxidea 325
merriamii, Ursus 339
mersinae, Hystrix 774
mertensi, Galagoides 250
Suncus **103**
meruensis, Tragelaphus 404
merus, Cryptotis 109
merzbacheri, Capra 406
mesamericana, Didelphis 17
mesanis, Rattus 660
mesatanensis, Crocidura 97
Mesechinus 77, 78, **79**, 79
dauuricus **79**, 79
hughi **79**, 79
manchuricus 79
miodon 79
przewalskii 79
sibiricus 79
sylvaticus 79
mesembrinus, Chaetodipus 483
Mesembriomys 586, 587, 604, 615, **619**, 619, 674
boweri 620
gouldi 586
gouldii **619**, 620
hirsutus 620
macrurus **620**
melvillensis 620
rattoides 620
mesidios, Chaetodipus 483
Mesoallactaga **488**
mesoamericanus, Pteronotus 177
Mesocapromys 800, **801**, 801, 802
angelcabrerai **801**, 801
auritus **801**, 801
barbouri 800
beatrizae 800
delicatus 800
gracilis 800
kraglievichi 800
minimus 800
nanus **801**, 801, 802
sanfelipensis **802**
silvai 800
Mesocricetus **538**, 539
auratus **538**, 538, 539
avaricus 539
brandti 538, **539**, 539
koenigi 539
newtoni **539**
nigricans 539
nigriculus 539
raddei **539**
rathgeberi 538
mesoleucus, Conepatus **316**, 316
mesolithica, Lama 381, 382
mesomelas, Canis **281**
Dendromus 542, **543**, 543, 544
Mephitis 317

Mus 541
Peromyscus 732
Mesomys 789, **794**
didelphoides **794**
ecaudatus 794
ferrugineus 794
hispidus **794**, 795
leniceps **794**
obscurus **794**
spicatus 794
stimulax 794, **795**
mesopetes, Lontra 311
Mesophylla **190**, 190, 193
flavescens 190
macconnelli **190**, 190
Mesoplodon **362**, 362
australis 363
bidens **362**
bowdoini **362**, 362
carlhubbsi **362**
dalei 362
densirostris **362**
europaeus **362**
floweri 363
gervaisi 362
ginkgodens **363**
grayi **363**
guntheri 363
haasti 363
hectori **363**
hotaula 363
knoxi 363
layardii **363**
longirostris 363
micropterus 362
mirus **363**
pacificus 362
peruvianus **363**
seychellensis 362
sowerbensis 362
sowerbyi 362
stejnegeri 362, **363**
thomsoni 363
traversii 363
mesopolius, Chaetodipus 483
mesopotamiae, Gerbillus **553**
mesopotamica, Dama **387**
Hystrix 774
Mesopotamogale 72, 73
Mesosciurus 439, 442, 452
Mesospalax 753, 754
messorius, Dendromus 542, **543**
Leggadina 601
Micromys 620
Oligoryzomys 718
Sylvilagus 825
mesurus, Trichosurus 48
Metachirops 22
opossum 22
Metachirus **20**
antioquiae 20
bolivianus 20
colombianus 20
dentaneus 20
inbutus 20
infuscus 20
modestus 20
nudicaudatus **20**, 20
personatus 20
phaeurus 20
tschudii 20
metallicola, Peromyscus 729
Metavampyressa 193, 194
methi, Genetta 347
Methylobates 274
meticulosus, Saguinus 254
metis, Stenella 356
Metotomys 680
metropolitanus, Catagonus 379
metwallyi, Hemiechinus 78
mewa, Thomomys 473
mexianae, Agouti 783
Alouatta 254
Dasypus 66
Mazama 391
Nasua 335
Panthera 298
Tapirus 371
mexicalis, Chaetodipus 483
mexicana, Alouatta 255
Antilocapra 393
Choeronycteris **184**, 184
Cryptotis 108, **109**, 109, 111
Dasyprocta **782**
Glossophaga 184
Lasiurus 206
Leopardus 291, 292
Marmosa **18**
Mephitis 317
Micronycteris 179
Nasua 334
Neotoma 711, **712**
Ovis 408
Procyon 335
Tadarida 240
Tamandua **68**
mexicanus, Castor 467
Centurio 189
Cyclopes 67
Cynomys **424**
Dasypus 66
Idionycteris 205
Lepus 815
Macrotus 178
Microtus **524**, 524, 526
Molossops 234
Mustela 322
Myotis 209
Natalus 195
Noctilio 176
Nycticeius 217
Odocoileus 391
Oryzomys 721
Perognathus 485
Peromyscus 703, 726, 728, 730, 731, 733, **734**, 734, 736, 737
Plecotus **224**
Pteronotus 177
Reithrodontomys 740, **741**, 741, 742
Rheomys **743**, 743
Spermophilus **447**, 450
Sphiggurus **776**
meyeni, Pipistrellus 221
meyeri, Antrozous 198
Lenomys **602**
Macroglossus 154
Mus 602
Rattus 655
Tylonycteris 228
meyeroehmi, Rhinolophus **164**
mezosegiensis, Spalax 755
mhorr, Gazella 396
mial, Microtus 518
miarensis, Mops 236
micaceus, Hipposideros 172
Micaelamys 567
micans, Saguinus 253
michaeli, Diceros 372
michaelinus, Cervus 386
michaelis, Caracal 289
Felis 288
Salpingotus 491, **492**, 492
michaelseni, Geoxus 702
michianus, Callosciurus 421
Elaphodus 388
michiganensis, Castor 467
Peromyscus 732
michnoi, Microtus 521
Vespertilio 228
micklemi, Cryptomys 771
Lepus 822
Mico 251
Micoella 251
miconax, Aotus **256**
Micoureus **20**
alstoni **20**
areniticola 20
budini 20
cinerea 20
constantiae **20**
demerarae **20**
domina 20
esmeraldae 20
germana 20
limae 20
mapiriensis 20
meridae 20
paraguayana 20
parda 20
perplexa 20
pfrimeri 20
phaea 20
rapposa 20
regina **20**
rutteri 20
travassosi 20
Micraonyx 309
Microallactaga 488
Microbiotheria 27
Microbiotheriidae 15, **27**
microbullatus, Taeromys 667
Microcavia **780**
australis **780**
joannia 780
kingii 780
maenas 780
niata **780**
nigriana 780
pallidior 780
salinia 780
shiptoni **780**
typus 780
Microcebus **243**
coquereli **243**
gliroides 243
griseorufus 243
madagascarensis 243
minima 243
minor **243**
murinus **243**, **244**
myoxinus 243
palmarum 243
prehensilis 243
pusillus 243
rufus **243**
smithii 244
microcephalus, Microtus 527
Monodon 358
Zapus 499
Microchiroptera 137
Microdelphys 21
Microdillus **558**
peeli **558**
Microdipodops 477, **480**, 480, 482
albiventer 480
ambiguus 480
ammophilus 480
atrirelictus 480
californicus 480
dickeyi 480
leucotis 480
lucidus 480
medius 480
megacephalus **480**, 480
nasutus 480
nexus 480
oregonus 480
pallidus **480**, 480
paululus 480
polionotus 480
purus 480
restrictus 480
ruficollaris 480
sabulonis 480
microdon, Allocricetulus 537
Aonyx 310
Aotus 256
Apomys **575**
Canis 280
Chalinolobus 199
Herpestes 306
Lepilemur **246**
Mastomys 611
Maxomys 614
Mustela 324
Nyctophilus **218**
Ourebia 399

Phalanger 46
Pygoderma 191
Reithrodontomys **742**
Sigmodon **747**
Spilogale 317
Stenoderma 191
Tylomys 751
Xerus 458
microdontus, Cervus 386
Naemorhedus 407
Sus 379
Microfelis 289
Microgale **71**
brevicaudata **71**, 71
breviceps 71
cowani **71**, 71
crassipes 71
decaryi 71
dobsoni **71**
drouhardi 71
dryas **71**
gracilis **71**
longicaudata **71**, 71
longirostris 71
majori 71
melanorrhachis 71
nigrescens 71
occidentalis 71
parvula **71**, **72**
principula **71**, 71
prolixacaudata 71
pulla **72**
pusilla **72**
sorella 71
taiva 71
talazaci 71, **72**
thomasi **72**
Microhippus 369
Microhydromys **620**, 620
musseri **620**, 638
richardsoni **620**, 620
Microlagus 813, 824, 825
Microlophiomys, vorontsovi 563
micromongoz, Eulemur 244
Micromys **620**, 672
agilis 620
aokii 620
arundinaceus 620
avenarius 620
batarovi 620
berezowskii 620
campestris 620
danubialis 620
erythrotis 620
fenniae 620
flavus 620
hertigi 620
hondonis 620
japonicus 620
kytmanovi 620
meridionalis 620
messorius 620
minatus 620
minimus 620
minutus **620**
oryzivorus 620
parvulus 620
pendulinus 620
pianmaensis 620
pratensis 620
pumilus 620
sareptae 620
soricinus 620
subobscurus 620
takasagoensis 620
triticeus 620
ussuricus 620
Micronectomys 722
micronesiensis, Rattus 652
Micronomus 237
Micronycteris **178**, 179
behnii **178**
brachyotis **178**
daviesi **178**
elongatum 179
hirsuta **179**
homezi 179
hypoleuca 179
megalotes 179
megalotis **179**
mexicana 179
microtis 179
minuta **179**
nicefori **179**
platyceps 178
pusilla **179**
pygmaeus 179
schmidtorum **179**
scrobiculatum 179
sylvestris **179**
typica 179
micronyx, Myotis 210
Suncus 102
Microperoryctes **41**, 41
dorsalis 41
longicauda **41**
magnus 41
murina **41**, 41
ornatus 41
papuensis **41**
microphthalmus, Spalax 753, 754, **755**, 755
microphyllum, Rhinopoma **155**
microphyllus, Vespertilio 155
Micropia 356
Micropotamogale **72**
lamottei **72**, 72
ruwenzorii **72**
microps, Apodemus 574
Dipodomys **479**
Stenella 356
Micropteron 362
Micropteropus 141, **143**
grandis 141, 143
intermedius **143**
pusillus **143**
Micropterus 362
micropterus, Mesoplodon 362
micropus, Auliscomys **695**
Hemiechinus **78**, 79
Hipposideros 172
Lemniscomys 602
Lepus 815
Microtus 525
Natalus **195**, 195
Neotoma 710, 711, **712**
Paraechinus 78
Peromyscus 733
Pipistrellus 220
Potorous 50
Microrhynchus 246
microrhynchus, Callosciurus 422
Microryzomys **708**, 717, 719
altissimus **708**, 708
aurillus 708
chotanus 708
dryas 708
fulvirostris 708
humilior 708
hylaeus 708
minutus **708**
Microsciurini 433, 452
Microsciurus **433**, 442
alfari **433**
alticola 433
avunculus 433
boquetensis 433
brevirostris 433
browni 433
flaviventer **433**
florenciae 433
fusculus 433
isthmius 433
manarius 433
mimulus **433**
napi 433
otinus 433
palmeri 433
peruanus 433
rubicollis 433
rubrirostris 433
sabanillae 433
santanderensis **434**
septentrionalis 433
similis 433
simonsi 433
venustulus 433
vivatus 433
Microsorex 111, 115
Microspalax 753, 754
microspilotus, Spermophilus **449**
microspilus, Cervus 386
Microsus 378
floresianus 378
microtarsus, Didelphys 17
Gracilinanus **17**, 17
Microtinae 501
microtinus, Clethrionomys 509
Hylomys 80
Zygodontomys 752
microtis, Apodemus 574
Atelocynus **279**
Callosciurus 422
Canis 279
Chelemys 699
Cricetomys 541
Dusicyon 283
Geoxus 702
Glossophaga 184
Graphiurus **764**, 764, 765
Lepus 822
Micronycteris 179
Mustela 321
Nyctophilus **218**
Oligoryzomys **718**, 718
Prionailurus 295
Rhipidomys 744
Suncus 103
Sus 379
Microtus 505, 506, 507, 511, 515, 516, **517**, 517, 518, 519, 521, 522, 523, 524, 525, 526, 527, 528, 529, 530, 531, 532, 533, 534
abbreviatus **518**
abditus 523
acadicus 527
admiraltiae 527
adocetus 527
aequivocatus 519
aestuarinus 519
agrestis **518**, 518, 519, 527
alborufescens 527
albus 519
alcorni 527
altaicus 527
alticola 523
amakensis 527
amosus 525
andersoni 525
angularis 519
angusticeps 523
angustifrons 518
angustus 521, 523
anikini 527
aphorodemus 527
arcticus 527
arcturus 518
arenicola 527
argyropoli 518
argyropuli 518
argyropuloi 518
arizonensis 525
armoricanus 518
arvalis **518**, 519, 523, 526, 529, 531
arvensis 519
arvicoloides 528
assimilis 519
astrachanensis 530
asturianus 519
atratus 530
atticus 530
auricularis 528
austerus 526
aztecus 527
baicalensis 525
baileyi 523
bailloni 518

bairdi 527
bavaricus **519**
bernardinus 523
binominatus 530
bogdanovi 511
bogdoensis 530
brachycercus 529
brauneri 519
brecciensis 519
brevicauda 521
brevicorpus 525
brevirostris 526
breweri **519**, 519, 528
britannicus 518
brunneus 521
bucharicus 506
buturlini 521
cabrerae **519**
calamarum 521
californicus **519**, 519, 524
calypsus 519
campestris 518
canescens 525
canicaudus **520**
cantator 525
cantwelli 527
capucinus 530
carbonarius 528
carinthiacus 518
carolinensis 520
carruthersi 522, 523
caryi 525
caspicus 529
castaneus 521
caucasicus 519
cautus 523
centralis 520
chihuahuensis 527
chrotorrhinus 518, **520**, 520, 528, 531
cimbricus 519
cinnamonea 526
ciscaucasicus 524
codiensis 525
colchicus 524, 530
constrictus 519
contigua 519
copelandi 527
coronarius 523
cowani 530
cummingi 530
cunicularius 519
dacius 530
daghestanicus **520**, 520
dauricus 527
dekayi 527
dentatus 519
depressa 519
depressus 524
dinaricus 530
dinniki 524
dolguschini 521
dolichocephalus 521
dorothea 529
druentius 526
drummondii 527
dukecampbelli 527
dukelskiae 521
duodecimcostatus **520**, 520, 524, 530
duplicatus 519
dutcheri 525
edax 519
ehiki 530
elymocetes 527
endoecus 527
enez-groezi 518
enixus 527
epiroticus 529
estiae 518
everesti 523
eversmanni 521
evoronensis **520**, 525
eximius 519
exsul 518
fatoii 526
felteni **520**, 520, 530
fingeri 524
finitus 527
finmarchicus 527
fiona 518
fisheri 518
flava 519
flavescens 520
flaviventris 527
fontigenus 527
forresti 522
fortis **521**, 521
fucosus 525
fujianensis 521
fulva 519, 527
fulviventer 524
fundatus 524
funebris 527
fusca 530
fuscus 520
fusus 525
galliardi 519
gerbei 520, **521**, 521, 524
gerritmilleri 524
ghalgai 529
gilmorei 527
goriensis 530
grandis 519
gravesi 530
gregalis 518, **521**, 521, 525
gregarius 518
grinnelli 519
guadalupensis 524
guatemalensis **521**, 526
gudauricus 526
guentheri **521**, 521, 522, 530
hahlovi 527
halli 523
halophilus 519
hartingi 521
hatanezumi 525
haydenii 526
henseli 529
heptneri 519
hirsutus 527
hirta 518
howelkae 519
hualpaiensis 524
hungaricus 530
huperuthrus 519
hurdanensis 524
hyperboreus **522**, 522
hyrcania 530
ibericus 520
igmanensis 519
ilaeus 526
ileos 526
incanus 523
incertoides 530
incertus 519, 530
incognitus 519
innae 526
innuitus 527
insperatus 527
insularis 518
intermedia 518
intermedius 520
iphigeniae 526
irani **522**, 522, 530
irene **522**, 522, 529
juldaschi **522**, 522, 523
kamtschatica 527
karaginensis 527
kaznakovi 529
kennicottii 528
kermanensis **523**
kernensis 519
khorkoutensis 531
kincaidi 527
kirgisorum **523**
kjusjurensis 527
klozeli 530
kodiacensis 527
koreni 527
kossogolicus 521
kupelwieseri 530
labensis 524
labradorius 527
laingi 530
latifrons 518
latus 523
leponticus 526
leucophaeus 523
leucurus **523**, 523
levenedii 518
levis 519
liechtensteini 526
limnophilus **523**, 527
littoralis 523
longicaudus 507, **523**, 523, 528
longipilis 527
longirostris 525
luch 518
ludovicianus 526
lusitanicus 520, **523**, 524
lydius 521
macedonicus 521
macfarlani 527
macgillivrayi 518
macrocranius 526
macropus 528
macrurus 523
madrensis 524
magdalenensis 527
major 521
majori **524**, 526
malcolmi 527
malei 507
mariae 524
mariposae 519
martinoi 521, 530
matrensis 530
maximowiczii 520, **524**, 524, 525
medius 527
mehelyi 527
meldensis 519
meridianus 519
mexicanus **524**, 524, 526
mial 518
michnoi 521
microcephalus 527
micropus 525
middendorffi 521, 522, **524**, 524, 525, 529
minor 526
miurus 518, 521, 522, **525**, 525
modestus 527
mogollonensis 524
mohavensis 519
mongol 518
mongolicus 521, **525**
montanus 520, 524, **525**, 525, 528, 530
montebelli **525**, 527
montium-caelestinum 527
montosus 521
mordax 523
morosus 527
muhlisi 529
mujanensis 520, **525**
multiplex **525**, 526
muriei 525
musseri 534
mustersi 522, 530
myllodontus 528
mystacinus 522
nanus 525
nasarovi **526**
nasuta 527
naumovi 527
navaho 524
nebrodensis 529
neglectus 518, 519
nemoralis 528
nesophilus 527, 528
neuhauseri 530
nevadensis 525
neveriae 524
nexus 525
nigra 518
nigrans 527
nigricans 518
nikolajevi 530
nivalis 523

nordenskioldi 521
noveboracensis 527
novo 528
nyirensis 530
oaxacensis **526**
obscurus 519, 524, **526**, 529
occidentalis 530
ochrogaster **526**, 526, 528
oeconomus 523, 525, **526**, 527
ognevi 518
ohioensis 526
oneida 527
oniscus 522
operarius 527
orcadensis 519
oreas 525
oregoni **527**, 527
orientalis 526
orioecus 518
ouralensis 527
oyaensis 519
pallasii 521
paludicola 519
palustris 527
pamirensis 522
paneaki 525
pannonicus 518
paradoxus 522
parvulus 528
parvus 530
pascuus 520
pelandonius 524
pelliceus 521
pennsylvanicus 518, 519, 524, 525, **527**, 527, 528, 530
perplexibilis 519
petrovi 526
petshorae 527
petulans 523
phaeus 524
philistinus 522
pinetorum **528**, 528
planiceps 524
poljakovi 525
popofensis 527
pratensis 527
pratincola 525
principalis 519, 528
provectus 527, 528
provincialis 520
pugeti 530
pullatus 527
punctus 518
punukensis 527
pyrenaicus 521
quasiater **528**, 528
raddei 521
ratticeps 527
ravidulus 521
ravus 520
regulus 520
relictus 529
rhodopensis 529
richardsoni 505, **528**, 528, 531
rivularis 525
ronaldshaiensis 519
rossiaemeridionalis 519, **529**, 529
rousiensis 519
rozianus 518
rubelianus 524
rubidus 527
rufa 518
rufescens 527
rufescentefuscus 519
rufidorsum 527
rufofuscus 530
ruthenus 519
ryphaeus 524
sachalinensis 521, **529**
salvus 524
sanctidiegi 519
sandayensis 519
sanpabloensis 519
sarnius 519
satunini 530
savii 520, 521, **529**, 529
scaloni 518
scalopsoides 528
schelkovnikovi **529**
schidlovskii 530
schmidti 528
scirpensis 519
selysii 529
serpens 527
shantaricus 527
shattucki 527
shevketi 530
sierrae 523
sikimensis 522, **529**, 529
similis 526
simplex 519
sirtalaensis 521
sitkensis 527
slowzovi 521
socialis 521, 522, 527, **529**, 530
stephensi 519
stimmingi 527
stonei 527
strandzensis 521
strauchi 523
subarvalis 529
subsimus 524
subterraneus 519, 520, 524, 526, **530**, 530
suntaricus 527
suramensis 520
swerevi 522, 524
syriacus 530
talassensis 522
talassicus 521
tananaensis 527
tarbagataicus 521
tasensis 524
tatricus 526, **530**
taylori 526
terraenovae 527
terrestris 517, 519
tetramerus 530
thomasi **530**
thricolis 529
tianschanicus 521
townsendii 525, 528, **530**
transcaspicus **531**
transcaucasicus **524**, 526
transsylvanicus 530
transuralensis 526
transvolgensis 530
tridentinus 518
trowbridgii 519
tsaidamensis 523
tschuktschorum 527
tundrae 521
uchidae 527
ukrainicus 530
uliginosus 521
uligocola 527
umbrosus 526, **531**
unalascensis 527
undosus 525
unguiculatus 521
ungurensis 524
uralensis 524, 527
vallicola 519
variabilis 519
vellerosus 523
vinogradovi 524
volaensis 530
vulgaris 519
wahema 527
waltoni 523
westrae 519
wettsteini 518, 530
worthingtoni 504
xanthognathus 520, 528, **531**
xerophilus 525
yakutatensis 527
yosemite 525
zachvatkini 521
zadoensis 523
zimmermanni 530
zygomaticus 525
zykovi 530
microtus, Hemiechinus 78
Macroglossus 154
microurus, Crocidura 88
Microxus 688, 689, 690, 691, 693, 700, 726, 727
micrura, Cryptotis 109
Euroscaptor **125**, 125
Mogera 126
Parascaptor 127
Scaptochirus 128
Micrurus 517
micrurus, Blarina 106
Cryptotis 109
Muntiacus 389
micrus, Nesophontes **70**
Mictomys 534
Midas 253
midas, Callosciurus 421
Mops **236**
Rhinolophus 166
Saguinus **253**
Simia 253
Middas 251
middendorffi, Microtus 521, 522, **524**, 524, 525, 529
Ursus 339
middendorfi, Myopus 531
Ovis 409
Sorex 121
migratorius, Cricetulus **538**
Lagurus 515
Lemmus 517
Sciurus **440**
migrufus, Callicebus 259
mikado, Clethrionomys 509
mikadoi, Panthera 298
Rhinolophus 166
mikenensis, Gorilla 276
Mikropteron 362
miles, Mustela 324
miletus, Eothenomys 514
miliaria, Tatera 560
milii, Cheirogaleus 243
Milithronycteris 204, 205
millardi, Callosciurus 421
Dacnomys **589**, 589
Panthera 298
Millardia 587, 591, **621**, 621
comberi 621
dunni 621
gleadowi **621**, 621
kathleenae **621**, 621
kondana **621**
lanuginosus 621
listoni 621
meltada **621**, 621
pallidior 621
singuri 621
Millardomys 621
milleri, Apodemus 573
Callosciurus 421
Dryomys 766
Eumops 233
Mephitis 317
Molossops 235
Molossus 235
Myoprocta 782
Myotis **213**
Neomys 110
Oecomys 716
Panthera 298
Paradoxurus 344
Pithecia 262
Reithrodontomys 742
Rhipidomys 745
Sciurus 441
Sorex **117**, 117
Sus 379
milleti, Leopoldamys 603
millicens, Volemys **535**, 535
millsi, Arctogalidia 343
Hystrix 773
Melogale 314
milneedwardsii, *Naemorhedus* 407

miloni, Petaurista 463
mima, Arctogalidia 343
mimax, Cerdocyon 282
Octomys **788**, 788
mimellus, Callosciurus 423
mimenoides, Moschiola 382
Tragulus 382
Mimetes 276
mimetes, Antechinus 31
Mimetillus **207**
berneri 207
moloneyi **207**, 207
thomasi 207
mimetra, Helogale 304
Marmosa 18
mimica, Callorhinus 327
mimiculus, Callosciurus 423
mimicus, Petaurista 463
Phalanger 46
Strigocuscus 46
mimina, Blarina 106
Mimomys 505
Mimon **179**, 179
bennettii **179**
cozumelae 179
crenulatum **179**
keenani 179
koepckeae 179
longifolium 179
peruanum 179
picatum 179
mimula, Crocidura 96
mimuloides, Crocidura 96
mimulus, Gerbillus 553
Microsciurus **433**
Pseudantechinus 35
Pseudomys 645
mimus, Akodon 689, 690, **691**, 691
Dendrohyrax 373
Oxymycterus 689
Pipistrellus **222**
Pteropus 147
Sciurus 439
Sus 378
minahassae, Haeromys **595**
Pipistrellus **222**
minatus, Micromys 620
minax, Paradoxurus 344
Trogopterus 465
mincae, Oecomys 716
Proechimys 796, **797**
Mindanaomys 579, 580
mindanensis, Acerodon 137
Hylopetes 463
Macaca 266
Petinomys 463
Rattus 660
Sundasciurus **451**, 451, 452
Sus 379
mindorensis, Anonymomys **569**, 569
Bubalus **402**
Chrotomys **585**
Rattus 649, **655**, 661
mindorus, Crocidura **90**
Macaca 266
minensis, Sylvilagus 825
miniatus, Callosciurus 422
minima, Lutra 16
Microcebus 243
Nyctalus 217
Podogymnura 80
Talpa 129
minimus, Antechinus **30**
Balaenoptera 349
Bettongia 49
Chaetodipus 484
Chironectes **16**
Dobsonia 140
Epomophorus **141**
Leopardus 291
Macroglossus **154**, 154, 155
Mesocapromys 800
Micromys 620
Mus 625
Nycticeius 218
Ochotona 811
Pteropus 154
Rhinolophus 166
Sigmodon 747
Sorex 117
Tamias **454**
Tarsius 251
Thomomys 473
Miniopterinae **230**
Miniopterus **230**
africanus 230
arenarius 231
australis **230**
baussencis 231
bismarckensis 231
blepotis 231
breyeri 231
celebensis 231
chinensis 231
dasythrix 231
eschscholtzii 231
fraterculus **230**
fuliginosus 231
fuscus **230**, 230
grandis 231
grivaudi 231
harardai 231
inexpectatus 231
inflatus **230**
insularis 231
italicus 231
japoniae 231
macrocneme 231
macrodens 230
magnater **230**, 231
majori 231
manavi 231
medius 230
melanesiensis 231
minor **230**
natalensis 231
newtoni 231
oceanensis 231
orianae 231
orsinii 231
pallidus 231
parvipes 231
paululus 230
propritristis 231
pulcher 231
pusillus **231**
ravus 231
robustior **231**
rufus 230
schreibersi 230, **231**
scotinus 231
shortridgei 230
smitianus 231
solomonensis 230
tibialis 230
tristis **231**
ursinii 231
vicinior 231
villiersi 231
witkampi 230
yayeyamae 230
minitus, Mus 620
mink, Mustela 324
minnae, Grammomys **594**
minnesota, Tamiasciurus 457
minnesotoe, Mephitis 317
minnie, Pseudomys 645
minnisotae, Peromyscus 732
minoensis, Cervus 386
minor, Ametrida 187
Antrozous 198
Arctogalidia 343
Artibeus 187
Arvicanthis 578
Arvicola 506
Axis 384
Brachyphylla 182
Canis 280, 281
Cannomys 685
Carollia 186
Centurio 189
Cercartetus 58
Cervus 386
Chlamyphorus 64
Choeroniscus 183, **184**
Choeronycteris 183
Chrysochloris 75
Clethrionomys 508
Crossarchus 301
Cynopterus 139
Diceros 372
Dobsonia **140**
Echinosorex 79
Eliurus **678**, 678
Epomophorus 141
Hemiechinus 78
Hemigalus 342
Hexaprotodon 380
Hylonycteris 185
Lasiurus 206
Lemmus 517
Liomys 482
Macrotis 40
Macrotus 178
Madoqua 398
Meles 314
Microcebus 243
Microtus 526
Miniopterus **230**
Mogera **126**, 126
Mustela 322, 323
Myomys 630
Myospalax 676
Neomys 110
Neurotrichus 126
Niviventer 634
Noctilio 176
Nyctimene 145
Ochotona 807, 808
Otaria 328
Panthera 298
Paradoxurus 344
Pecari 380
Pelomys **640**
Petauroides 59
Phalanger 46
Platanista **360**, 360
Potorous 50
Praomys **643**, 643
Procavia 374
Procyon 335, **336**
Ptenochirus **145**
Pygeretmus 490
Rousettus 152
Sciurus 442
Sigmodon 747
Suncus 103
Sylvilagus 824
Thomomys 473
Trichechus 365
Tupaia **132**
Ursus 339
Zapus 499
minotaurus, Mus 626
minous, Acomys **566**
Minuania 21
minus, Megaderma 163
minusculus, Lemmus 517
Pipistrellus 222
Reithrodontomys 742
Scapanus 127
Tatera 561
minuta, Cephalophus 411
Crocidura **90**, 90, 96
Kerivoula **197**
Micronycteris **179**
Pipistrellus 221
Prionailurus 295
Tonatia 180
Vampyressa 194
Vicugna 382
minutellus, Myotis 210
minutilla, Rhogeessa **225**
minutillus, Rhinolophus 168
minutissimus, Pipistrellus 223
Planigale 34
Sorex **117**
minutoides, Mus 624, **625**, 625, 627

minutulus, Myosciurus 434
minutus, Cervus 386
Ctenomys **785**
Cynopterus 139
Dasypus 67
Hesperomys 708
Micromys **620**
Microryzomys **708**
Mormopterus **237**
Mustela 323
Myosciurus 434
Myoxus 770
Nyctimene **144**
Ochotona 812
Oryzomys 708
Pygeretmus 490
Rhinolophus 166
Sciurus **434**, 440
Sorex 115, **117**, 117, 118, 120, 121, 122
Steatomys 546
Sus 379
Urotrichus 129, 130
Zaedyus 67
minx, Mustela 324
miodon, Mesechinus 79
Miopithecus 263, **268**, 269
ansorgei 269
capillatus 269
melarhinus 269
pileatus 269
pilettei 269
talapoin **269**
miotis, Histiotus 205
Mioxocebus 243
miquihuanensis, Odocoileus 391
mira, Lontra 311
Rhogeessa **225**
mirabilis, Sorex **118**, 118, 119
Ursus 339
miradorensis, Eptesicus 201
mirae, Tylomys **751**, 751
mirandus, Hipposideros 172
miravallensis, Sciurus 440
mirhanreini, Chionomys 507
miriquouina, Aotus 255
miritibae, Bradypus 63
Mirounga 329, **330**
angustirostris **330**
ansonii 330
ansonina 330
coxii 330
crosetensis 330
dubia 330
elephantina 330
falclandicus 330
falklandica 330
kerguelensis 330
leonina **330**
macquariensis 330
patagonica 330
proboscidea 330
resima 330
mirus, Mesoplodon **363**
Ursus 339
Mirza 243
mirza, Eptesicus 203
mischlichi, Lycaon 283
miscix, Sorex 113
miselius, Macrotis 40
misim, Antechinus 30
misionalis, Oxymycterus 727
misonnei, Praomys **643**, 643, 644
misoriensis, Hipposideros 171
Misothermus 509
mississippiensis, Peromyscus 730
missoulae, Oreamnos 407
missoulensis, Puma 296
missouriensis, Castor 467
Cynomys 424
Geomys 469
Onychomys 719
mitchelli, Budorcas 404
Catopuma 289
Dipodomys 479
Ochotona 812
Sminthopsis 36
Tatera 561
Vombatus 46
mitchellii, Dipus 635
Notomys **636**, 637
mitis, Atilax 301
Cercopithecus **264**
Leopardus 291
Lontra 311
Marmosa 19
mitrata, Presbytis 270, 271
mitratus, Cervus 386
Indri 247
Rhinolophus **167**
Spermophilus 447
mittendorfi, Lemniscomys **602**
miurus, Calomys 698
Microtus 518, 521, 522, **525**, 525
miwarti, Echinops 73
Mixocebus 245
mixtus, Cephalophus 411
Hipposideros 173
Melomys 615
Saccolaimus **159**
Sorex 118
miya, Crocidura **90**
miyakensis, Apodemus 573
miyakonis, Rhinolophus 165
mizura, Euroscaptor **125**
mizurus, Oligoryzomys 718
moae, Hydromys 597
mobilensis, Geomys 470
mocambicus, Loxodonta 367
mocchauensis, Mus 627
mochae, Akodon 692
moco, Kerodon 780
Mococo 245
mococo, Lemur 245
modesta, Antechinus 30
Caluromys 15
Melogale 314
Tupaia 131
Vandeleuria 672
modestior, Akodon 691
modestus, Alcelaphus 393
Callicebus **258**
Cervus 386
Dactylomys 790
Dremomys **425**
Metachirus 20
Microtus 527
Oryzomys 721
Pecari 380
Potos 333
Pseudocheirus 60
Reithrodontomys 742
Sundasciurus **452**
modicus, Rhipidomys 744
Thomomys 473
modiglianii, Pteropus 149
moellendorffi, Sundasciurus 451, **452**, 452
Tupaia 133
moellendorfii, Bubalus 402
moerens, Akodon 691
Herpestes 305
Oenomys 637
Protoxerus 436
moerescens, Menetes 433
moerex, Oryzomys 720
Mogera **125**, 128
coreana 126
etigo **125**
imaizumii 126
insularis **125**
kobeae **126**, 126
latouchei 126
micrura 126
minor **126**, 126
robusta **126**
tokudae 125, **126**, 126
wogura **126**
moggi, Thallomys 669
mogollonensis, Canis 281
Microtus 524
Tamiasciurus 457
mogrebinus, Mus 628
mohavensis, Chaetodipus 483
Dipodomys 479
Microtus 519
Neotoma 711
Spermophilus **447**, 449
Tadarida 240
Thomomys 473
moheius, Callosciurus 421
Rattus 660
mohillius, Callosciurus 421
moholi, Galago **249**
mohri, Mus 626
moi, Maxomys **613**
Pygathrix 273
Tamiops 457
mokrzeckii, Neomys 110
molagrandis, Neotoma 712
molaris, Nasua 334
Uroderma 193
molinae, Akodon 689, **691**
Oncifelis 294
molinaei, Lama 382
molisana, Felis 290
molitor, Hesperomys 704
mollessonae, Clethrionomys 509
molliae, Isothrix 793
mollicomulus, Rattus 649, 653, **655**
mollicomus, Rattus 653
Mollicomys 649
molliculus, Rattus 660
mollipilosus, Eidolon 141
Lynx 293
Oryzomys 725
Perognathus 486
Pogonomys 642
Tamiasciurus **456**
mollis, Akodon 690, **691**, 692, 693
Lontra 311
Melomys 615, **617**, 617, 618
Phyllotis 738
Sorex 111
Spermophilus 445, **447**, 448, 450
Stenomys 665
mollissimus, Mus 626
Rhipidomys 744
moloch, Callicebus 258, 259
Hylobates **275**
moloneyi, Cercopithecus 264
Mimetillus **207**, 207
molopensis, Malacothrix 544
Parotomys 682
Thallomys 668
molossa, Nyctinomops 239
Tadarida 239
Molossidae **232**
molossina, Otaria 328
molossinus, Mus 626
Pteropus **149**
Molossops **234**, 234, 235
abrasus **234**
aequatorianus **234**
brachymeles 234
cerastes 234
espiritosantensis 235
greenhalli **234**
griseiventer 235
hirtipes 235
mastivus 234
mattogrossensis **234**
mexicanus 234
milleri 235
neglectus **234**
paranus 235
planirostris **235**
sylvia 235
temminckii 234, **235**
Molossus 232, **235**
acuticaudatus 235
albus 235

alecto 235
amplexicaudus 235
ater **235**
aztecus 235
barnesi 235
bondae **235**
californicus 233
castaneus 235
cherriei 235
coibensis 235
crassicaudatus 235
currentium 235
daulensis 235
debilis 235
fluminensis 235
fortis 235
fusciventer 235
holosericeus 235
lambi 235
longicaudatus 235
macdougalli 235
major 235
malagai 235
milleri 235
molossus **235**, 235
mops 235
moxensis 235
myosurus 235
nasutus 239
nigricans 235
obscurus 235
olivaceofuscus 235
perotis 233
pretiosus **235**, 235
pygmaeus 235
rufus 235
sinaloae **235**
trinitatus 235
tropidorhynchus 235
ursinus 235
velox 235
verrilli 235
molossus, Molossus **235**, 235
Vespertilio 235
molucca, Phalanger 46
moluccarius, Rattus 660
moluccarum, Myotis 207
moluccensis, Cervus 387
Dobsonia **140**, 140
Phalanger 46
mombasae, Tatera 562
momiyamai, Mus 626
momoga, Pteromys 464
momonga, Pteromys **464**
Mona 263
mona, Cercopithecus **264**
monacha, Cercopithecus 264
Crocidura 96
Monachinae 329
Monachus 329, **331**
albiventer 331
altantica 331
antillarum 331
bicolor 331
byronii 331
hermannii 331
leucogaster 331
mediterraneus 331
monachus **331**
schauinslandi **331**
tropicalis **331**
monachus, Cebus 259
Lasiurus 206
Monachus **331**
Phoca 331
Pithecia **261**, 261
Rhinolophus 167
monardi, Graphiurus **764**, 764
monastria, Petrogale 56
Monax 424
monax, Crocidura 88, **90**, 97
Marmota **432**
moncktoni, Crossomys **588**, 588
Melomys 615, 617, **618**, 618, 619
mondie, Nasua 335
mondraineus, Rattus 653
monella, Cercopithecus 264
mongan, Pseudocheirus 59
mongol, Microtus 518
mongolica, Allactaga 489
Gazella 397
Ovis 408
Panthera 298
Saiga 400
mongolicus, Cricetulus 537
Microtus 521, **525**
Otocolobus 295
Spermophilus 446
mongolium, Mus 626
mongoz, Eulemur **244**, 244
Lemur 244
Monichus 263
Monocerorhinus 372
monoceros, Monodon **357**, 357
Rhinolophus **167**
monochromos, Thomasomys **750**
monochroura, Neotoma 711
monodactyla, Cyclopes 67
Monodelphiops 21
Monodelphis **21**
adusta **21**, 21
americana **21**, 21
brachyura 21
brachyuros 21
brasiliensis 21
brevicaudata **21**
concolor 21
dimidiata **21**
domestica **21**
dorsalis 21
emiliae **21**
fosteri 21
glirina 21
goyana 22
henseli 22
hunteri 21
iheringi **21**
itatiayae 22
kunsi **21**
lundi 22
maraxina **21**
melanops 21
orinoci 21
osgoodi **21**
palliolatus 21
paulensis 22
peruvianus 21
rubida **22**
scalops **22**
sebae 21
sorex **22**
surinamensis 21
theresa **22**
touan 21
tricolor 21
trilineata 21
tristriata 21
umbristriata 22
unistriata **22**
Monodia 548, 552, 553
Monodon **357**
microcephalus 358
monoceros **357**, 357
monodon 358
narhval 358
vulgaris 358
monodon, Monodon 358
Monodontidae 349, 351, 354, **357**
monoensis, Dipodomys 479
Pteropus 150
Scapanus 127
Tamias 453
Thomomys 475
monoides, Cercopithecus 264
Monophyllus **185**
clinedaphus 185
cubanus 185
ferreus 185
frater 185
luciae 185
plethodon **185**
portoricensis 185
redmani **185**, 185
Monotremata 13
monstrabilis, Canis 281
Lepus 815
Neotoma 711
monstrosus, Hypsignathus **142**, 142
monsvairani, Sorex 111
montana, Chlorotalpa 75
Nasua 335
Ourebia 399
Ovis 408
Ratufa 437
Sicista 496
Talpa 129
Tupaia **132**
Vulpes 287
montanae, Bison 400
Montaneia 256
montanoserbicus, Nannospalax 754
montanosyrmiensis, Nannospalax 754
montanus, Acomys 565, 566
Calomys 698
Cercopithecus 263
Cervus 386
Clethrionomys 509
Cryptomys 772
Dasymys **589**
Dipodomys 479
Equus 370
Herpestes 305
Histiotus **205**
Isolobodon **803**
Lynx 293
Meriones 558
Microtus 520, 524, **525**, 525, 528, 530
Muntiacus 389
Nectomys 709
Nyctalus **217**
Odocoileus 391
Oreamnos 407
Paradoxurus 344
Phyllotis 737
Plecotus 224
Pseudalopex 284
Rangifer 392
Rattus 650, 655, **656**
Reithrodontomys 740, **742**, 742
Rhinolophus 168
Scalopus 127
Sigmodon 748
Sorex 115
Suncus **103**, 104
Tamias 456
Tatera 560
Taxidea 325
Zapus 499
montebelli, Microtus **525**, 527
monteiri, Otolemur 250
montensis, Akodon 689
montereyensis, Sorex 121
monterienis, Arctocephalus 327
monteriensis, Eumetopias 328
montezumae, Neotoma 710
Monticavia 780
monticola, Arvicola 506
Bassariscus 334
Cephalophus **410**, 410
Crocidura **90**, 90
Emballonura **157**, 157
Melanomys 708
Mustela 323
Nannospalax 754
Nasua 335
Niviventer 635
Perognathus 486
Pipistrellus 224
Rhinolophus 167
Sminthopsis 37
Spermophilus 450

Tatera 561
Thomomys 473, **475**, 475
Trachypithecus 274
monticolus, Herpestes 306
Lutra 312
Neusticomys **714**, 714
Sorex 112, **118**, 119, 122
Tamiops 457
monticularis, Aethomys 568
Bunolagus **813**
Lepus 813
Petromyscus **684**
montinus, Naemorhedus 407
montipinoris, Peromyscus 736
montis, Crocidura **90**
Praomys **642**, 643
Montisciurus **434**, 435
montium-caelestinum, Microtus 527
montivaga, Cryptotis **109**
montivagus, Myotis **213**
montosa, Alticola **503**, 503
montosus, Eptesicus 201
Graphiurus 765
Microtus 521
montserratensis, Ardops 187
monzoni, Conepatus 316
moojeni, Oryzomys 724
moorei, Antechinus 31
Ochotona 811
Thomomys 475
moormensis, Catopuma 289
Felis 289
Mops **235**, 236, 237, 240
angolensis 236
brachypterus **236**
calabarensis 236
chitauensis 236
condylurus **236**, 236
congicus **236**, 236
demonstrator **236**, 236
faradjius 236
fulva 236
indicus 235, 236
lanei 237
leonis 236
leucostigma 236
miarensis 236
midas **236**
mops **236**
nanulus **236**
niangarae **236**, 237
niveiventer **236**
occidentalis 236
occipitalis 237
ochraceus 236
orientis 236
osborni 236
petersoni **236**
sarasinorum **237**
spurrelli 236, **237**
thersites **237**
trevori 236, **237**
unicolor 236
wonderi 236
mops, Molossus 235
Mops **236**
moravicus, Rhinolophus 166
mordax, Arvicanthis 578
Atilax 301
Bandicota 579
Desmodus 194
Euneomys **702**, 702
Harpiocephalus 228
Lonchophylla **181**, 181, 182
Macaca 266
Microtus 523
Notomys **636**, 637
Pipistrellus **222**
Plecotus 224
Rattus 649, **656**, 656, 659, 660
Sturnira **193**
Trogopterus 465
mordeni, Crocidura 96
Lepus 821
mordicus, Pseudomys 646
mordosus, Ctenomys 784
morea, Felis 290
moreirae, Marmosa 18
moreni, Eligmodontia **701**, 701
Lagidium 778
morenoi, Glossophaga **184**
moreotica, Canis 280
moresbyensis, Isoodon 39
morgani, Eligmodontia 700, **701**, 701
Potorous 50
moricandi, Chaetomys 790
morio, Chalinolobus **199**
Crocidura 104
Pongo 277
Praomys **643**, 643, 644
Pteropus 146
Rhinolophus 167
Sciurus 440, 442
Sylvisorex 102, 104, **105**
Mormon 268
mormon, Mandrillus 268
Simia 268
Mormoopidae **176**
Mormoops **176**
blainvillii **176**, 176
carteri 176
cinnamomeum 176
cuvieri 176
intermedia 176
megalophylla **176**
rufescens 176
senicula 176
tumidiceps 176
Mormopterus **237**, 237, 238, 240
acetabulosus **237**
albiventer 237
astrolabiensis 237
barbatogularis 238
beccarii **237**
cobourgiana 238
doriae **237**
erongensis 238
fitzsimonsi 238
haagneri 238
jugularis **237**, 237
kalinowskii **237**
loriae 238
macmillani 238
minutus **237**
natalensis 237
norfolkensis **238**
parkeri 238
petersi 238
petrophilus **238**
phrudus **238**
planiceps **238**
ridei 238
setiger **238**
umbratus 238
wilcoxii 238
Morodactylus 45
moromoro, Lama 381
morosa, Ochotona 807, 808
morosus, Microtus 527
morotaiensis, Rattus 650, **656**
morrisi, Myotis **213**
Petinomys **464**
morrisianus, Cervus 386
morroensis, Dipodomys 478
morrulus, Cebus 259
morta, Saimiri 260
mortigena, Mustela 321
mortivallis, Dipodomys 479
morulus, Myopus 531
Sciurus **441**
Thomomys 473
morungensis, Bandicota 579
mosanensis, Mustela 323
moschata, Crocidura 94
Desmana **124**
Melogale **314**
moschatus, Bos 407
Castor 124
Hypsiprymnodon **49**, 49
Muntiacus 389
Neotragus **398**, 398
Ovibos **408**
Scaptochirus **128**, 128
Moschidae **383**
moschiferus, Moschus 383, **384**
Moschiola **382**
indica 382
malaccensis 382
meminna **382**
mimenoides 382
Moschomys 532, 707
Moschophoromys 707
Moschothera 347
Moschus **383**
altaicus 384
americanus 390
anhuiensis 383
aquaticus 382
arcticus 384
berezovskii **383**
caobangis 383
chrysogaster **383**, 384
cupreus 384
fuscus **384**, 384
leucogaster 384
meminna 382
moschiferus 383, **384**
parvipes 384
sachalinensis 384
sibiricus 384
sifanicus 384
turowi 384
moshesh, Rhabdomys 662
Mosia **158**, 158
nigrescens **158**, 158
papuana 158
solomonis 158
mosquensis, Apodemus 574
mossambica, Galerella 303
Genetta 345
mossambicus, Cercopithecus 264
Galago 249
Heterohyrax 373
Paraxerus 435
mostellum, Hipposideros 172
motalavae, Pteropus 146
motoyoshii, Vespertilio 228
mottoulei, Heliophobius 772
motuoensis, Dremomys 425
mouhotei, Menetes 433
mounseyi, Saguinus 253
moupinensis, Mustela 324
Myotis 213
Sus 379
mowewensis, Prosciurillus 436
moxensis, Molossus 235
mpangae, Cercopithecus 263
muansae, Arvicanthis 577, 578
Equus 369
Rattus 658
Tatera 562
mucronatus, Eothenomys 514
muelleri, Dorcopsis **52**, 52
Hylobates 275, **276**
Macropus 51, 52
Sundamys 595, 650, 655, 660, **666**, 666
Tupaia 133
muenninki, Tokudaia **670**, 670
muenzneri, Kobus 414
mugosaricus, Spermophilus 448
muhlisi, Microtus 529
muiri, Ochotona 811
mujanensis, Microtus 520, **525**
mukna, Elephas 367
mulatta, Macaca 266, **267**
muleiensis, Meriones 557
Muletia 65
muliensis, Ochotona **810**
mullah, Acomys **566**, 566

mulleri, Gerbillurus 547
Hystrix 773
Mus 666
Pronolagus 824
Suncus 103
mullulus, Berylmys 580
multiaculeata, Tachyglossus 13
multiannulata, Redunca 414
multicinctus, Cabassous 65
multicolor, Heliosciurus 428
Procolobus 272
Tragelaphus 404
multiplex, Microtus **525**, 526
multiplicatus, Uromys 671
multispinosus, Tadarida 240
multivittata, Ictonyx 319
mumfordi, Myotis 208
munbyanus, Eliomys 767
munchiquensis, Oligoryzomys 718
munda, Mustela 322
mundamensis, Mandrillus 268
mundorosica, Gazella 397
mundus, Mustela 322
Myotis 208, 213
Sylvisorex 104
mungo, Herpestes 305
Mungos **307**
Viverra 307
Mungos 300, 301, **307**
adailensis 307
bororensis 307
colonus 307
damarensis 307
fasciatus 307
gambianus **307**
gothneh 307
grisonax 307
leucostethicus 307
macrosus 307
macrurus 307
mandjarum 307
mungo **307**
ngamiensis 307
pallidipes 307
rossi 307
senescens 307
somalicus 307
taenianotus 307
talboti 307
zebra 307
zebroides 307
Mungotictis **300**
decemlineata **300**
lineatus 300
substriatus 300
vittatus 300
Mungotidae 299
Mungotinae 300
municus, Colobus 270
munoai, Oncifelis 294
muntia, Maxomys 614
Muntiacinae 388
Muntiacus **388**, 388
albipes 389
annamensis 389
atherodes **388**
aureus 389
bancanus 389
bridgemani 389
crinifrons **388**, 389
curvostylis 389
feae **388**, 389
feai 389
gongshanensis **389**
grandicornis 389
lachrymans 389
malabaricus 389
melas 389
menglalis 389
micrurus 389
montanus 389
moschatus 389
muntjak 388, **389**, 389
nainggolani 389
nigripes 389
peninsulae 389
pingshiangicus 389
pleiharicus 388, 389
ratwa 389
reevesi **389**, 389
robinsoni 389
rooseveltorum 388, 389
rubidus 389
sclateri 389
sinensis 389
styloceros 389
subcornutus 389
tamulicus 389
teesdalei 389
vaginalis 389
yunnanensis 389
muntjak, Cervus 388
Muntiacus 388, **389**, 389
munzneri, Heterohyrax 373
Sigmoceros 395
mupinensis, Ursus 340
mupurita, Mustela 324
Spilogale 317
muralis, Mus 626
Thomomys 473
murelia, Mazama 391
mureliae, Oryzomys 723
murex, Murexia 32
Phascogale 32
Murexia **32**
aspera 32
longicaudata **32**
maxima 32
murex 32
parva 32
rothschildi **32**
murgabensis, Felis 290
murianus, Lariscus 430
muricauda, Crocidura 84, **90**
muricola, Myotis 208, 211, **213**
Muriculus **621**, 621
chilaloensis 621, 622
imberbis **621**, 622
muriculus, Calomys 698
Gerbillus **553**
muricus, Mustela 321
Muridae **501**, 501, 677, 679, 684, 763
muriei, Microtus 525
Tolypeutes 67
murii, Tamiasciurus 457
Murilemur 243
murilla, Mus 629
murillus, Calomys 698
Murina **228**, 229, 230
aenea **228**
aurata **229**, 229, 230
balstoni 230
canescens 230
cyclotis **229**
eileenae 229
feae 229
florium **229**
fusca **229**
grisea **229**
hilgendorfi 229
huttoni **229**, 229, 230
intermedia 229
lanosa 229
leucogaster **229**, 229
ognevi 229
peninsularis 229
puta **229**
rozendaali **229**
rubella 229
rubex 229
sibirica 229
silvatica **229**, 229, 230
suilla **230**
tenebrosa **230**
toxopei 229
tubinaris **230**
ussuriensis 229, **230**
murina, Dendrogale **131**
Didelphis 18, 49
Hylogalea 131
Marmosa **18**
Microperoryctes **41**, 41
Potorous 50
Sminthopsis 35, 36, **37**, 37
Tadarida 240
Murinae 501, 540, 545, **564**, 564, 678, 687
murinauralis, Glaucomys 460
Murininae **228**
murinoflavus, Scotophilus 227
murinus, Chaerephon 233
Cricetulus 538
Desmodus 194
Ellobius 512
Galagoides 249
Graphiurus 763, **764**, 764, 765
Hipposideros 173
Lemur 243
Melomys 615
Microcebus **243**, 244
Pipistrellus 223
Prosciurillus **436**
Pseudohydromys 598, **644**, 644
Pseudomys 645
Sciurus 435
Sorex 101
Suncus **103**, 103, 104
Vespertilio **228**, 228
muroides, Chiropodomys **584**
murraiana, Asellia 170
murrayensis, Pseudomys 646
murrayi, Abrocoma 789
Pipistrellus 224
Rattus 653
mursavi, Ochotona 808
Mus 565, 573, 574, 578, 583, 595, 611, 621, **622**, 622, 623, 624, 625, 626, 627, 628, 629, 639, 640, 641, 663, 666, 697, 704, 714
abbotti 625
ablutus 622
acholi 628
adelaidensis 625
aequatorius 629
agrarius 569
aguti 781
airolensis 625
albertisii 625
albicans 625
albidiventris 625
albinus 625
albus 625
amphibius 505
amurensis 625
annamensis 623
anomalus 481
argenteus 570
argurus 674
armandvillei 637
arvalis 517
ater 625
auritus 740
avellanarius 769
azoricus 625
bactrianus 625
bahadur 627
baoulei **622**, 622, 626, 628
barabensis 537
barbarus 601
barbatus 673
bartelsii 611
bateae 625
beavanii 629
beccarii 609
bella 625, 628
bicolor 625
bieni 625
birungensis 629
boninensis 623
booduga **622**
borealis 625
brasiliensis 704
brevirostris 625
bufo **622**, 622
cahirinus 564

callewaerti **622**, 622
canacorum 625
candidus 625
caoccii 625
capensis 772
caroli **623**
castaneus 625, 626
caudatus 625
cervicolor **623**
chiropus 583
cinderella 627
cinereo-maculatus 625
citellus 444
coelestis 581
commissarius 625
cookii **623**
coucha 609
coypus 805
cricetus 538
crociduroides **623**, 623, 629
cunicularis 623
cyanus 789
damarensis 669
darjilingensis 623
darwini 737
decolor 625
decumanus 649
defua 590
delamensis 629
deserti 624
desmarestii 707
dolichurus 592
domesticus 625, 626
dubius 625
dunckeri 625
dunni 629
emesi 624
enclavae 625
faeroensis 625
famulus **623**
far 625
fernandoni **623**
flavescens 625
flavus 625
floridana 710
formosanus 623
formosovi 625
forresti 600
fors 629
fredericae 625
funereus 625
gairdneri 627
gallarum 625, 629
gansuensis 625
gentilis 625
gentilulus 625
gerbillinus 625
gerbillus 629
germanicus 625
giganteus 578
gilvus 625
gliroides 583
goliath 599
gondokorae 625
goundae **623**, 623, 628
grahami 627
grata 625
gratus 625
griseoflavus 702
gundi 761
gurkha 627
hannyngtoni 627
hanuma 625
hapsaliensis 625
haussa **623**, 623, 624, 625, 629
helgolandicus 625
helviticus 625
helvolus 625
heroldii 625
hildegardeae 673
hirsutus 619
hispanicus 628
homourus 625
hortulanus 624, 625, 628
hudsonius 509
humei 594
hydrophilus 625
hypoxanthus 637
imatongensis 629
imberbis 621
imphalensis 623
incomtus 589
indianus 625
indicus 578
indutus **624**, 624
jacksoniae 627
jaculus 488, 493
jalapae 625
jamesoni 625
kakhyenensis 623
kaleh-peninsularis 625
kambei 625
kasaicus **624**, 628
kerensis 624
kukilensis 623
kuro 625
lagurus 515
laniger 777
lasiotis 748
laucha 697
lemmus 516
leporinus 781, 782
leucopus 728
longitarsus 718
lundii 625
lusitanicus 628
lynesi 628
macedonicus **624**, 624, 626
macropus 671
maculatus 625
mahomet **624**, 624, 629
major 625
manchu 625
manei 625
margarettae 595
marica 625
maritimus 771
marmota 430
mattheyi **624**, 625
mayori **625**, 625
meator 627
mehelyi 628
melanogaster 626
mesomelas 541
meyeri 602
minimus 625
minitus 620
minotaurus 626
minutoides 624, **625**, 625, 627
mocchauensis 627
mogrebinus 628
mohri 626
mollissimus 626
molossinus 626
momiyamai 626
mongolium 626
mülleri 666
muralis 626
murilla 629
musculoides 622, 624, **625**, 625, 627, 629
musculus 609, 622, 624, **625**, 625, 626, 628, 629
mykinessiensis 626
myospalax 675
mystacinus 626
nagarum 623
naivashae 629
nasutus 726
nattereri 626
neavei 622, **626**, 626, 627, 628
nghialoensis 628
niger 626
nigricauda 668
nipalensis 626
nitedula 766
nitidulus 623
niveus 626
niviventer 632
nogaiorum 626
nordmanni 626
norvegicus 649
nudoplicatus 626
oleraceus 672
orangiae **627**
orientalis 626
orii 626
oubanguii 626, **627**, 628
ouwensi 623
oxyrrhinus 626
paca 783
pachycercus 626
pahari **627**, 627
pallescens 626
palnica 623
palustris 708, 719
parvus 628, 629
pasha 627
paulina 625
percnonotus 626
peruvianus 626
petila 625
phillipsi **627**
pilorides 800
platythrix **627**, 628
pococki 625
polonicus 626
popaeus 623
porcellus 778
poschiavinus 626
praetextus 626
pretoriae 624
priestlyi 627
proconodon 627
pumilio 662
pygmaeus 626
pyrrhorhinos 751
raddei 626
rama 626
ramnadensis 627
reboudi 626
rifensis 628
rotans 626
rubicundus 626
rufiventris 626
rufus 727
rutilans 727
rutilus 507
sabanus 603
sadhu 627
sagitta 492
sareptanicus 626
saxicola **627**
sergii 628
setulosus 622, 624, **627**, 627, 628
setzeri 627, **628**
severtzovi 626
shortridgei **628**, 628
sikapusi 605
simsoni 626
sinicus 626
siva 627
sorella 622, 623, 624, 626, 627, **628**, 628
soricoides 625
spicilegus 624, 626, **628**, 628
spretoides 624
spretus 626, **628**, 629
squamipes 709
striatus 626
strophiatus 623
suahelica 629
subcaeruleus 626
subterraneus 626
subtilis 496
suillus 771
sungarae 625
surkha 627
sybilla 624, 625
taitensis 626
taiwanus 626
takagii 626
takayamai 626
talpinus 512
tamariscinus 555
tantillus 626
tarsalis 705
tataricus 626

tenellus 623, 624, **629**, 629
terrestris 505
terricolor **629**, 629
theobaldi 626
tomensis 625, 626
tomentosus 706
triton 622, **629**, 629
tumidus **745**
tuza 470
tytleri 626
umbratus 625
univittatus 596
urbanus 626
utsuryonis 626
valschensis 624
variabilis 626
varius 626
verecundus 665
verreauxii 630
vicina 625
viculorum 626
vignaudii 626
vinogradovi 626
volans 460
vulcani 623, **629**, 629
wagneri 626
wamae 628
wambutti 622
yamashinai 626
yesonis 626
yonakuni 626
musanga, Paradoxurus 344
musangoides, Paradoxurus 344
Musaraneus 111
musaraneus, Crocidura 94
musavora, Melomys 619
muscalis, Melomys 615
Muscardininae 769
Muscardinulus 769
Muscardinus **769**, 769, 770
abanticus 769
anglicus 769
avellanarius **769**
corilinum 769
kroecki 769
muscardinus 769
niveus 769
pulcher 769
speciosus 769
trapezius 769
zeus 769
muscardinus, Mastomys 611
Muscardinus 769
muscatellum, Rhinopoma **155**
muscatensis, Gazella 396
muschata, Suncus 103
musciculus, Pipistrellus **222**
muscilla, Kerivoula 197
muscina, Kerivoula **197**
muscinus, Hipposideros **174**
muscola, Potorous 50
muscula, Gracilinanus 17
Marmosa 18
Tadarida 240
musculinus, Calomys **698**, 698
Paraxerus 435
musculoides, Cephalophus 411
Mus 622, 624, **625**, 625, 627, 629
Peromyscus 732
Musculus 622
musculus, Apomys **575**
Baiomys **695**
Balaenoptera **350**
Mus 609, 622, 624, **625**, 625, 626, 628, 629
Neomys 110
Thomomys 476
musei, Perameles 40
musicola, Marmosa 18
musicus, Spermophilus **448**, 448
musignani, Arvicola 505, 506
Musimon 408
musimon, Ovis 408
musiniani, Arvicola 505
Musonycteris **186**
harrisoni **186**, 186
musschenbroekii, Macrogalidia **343**
Maxomys **613**, 613, 614
Paradoxurus 343
musseri, Microhydromys **620**, 638
Microtus 534
Volemys **535**, 535
mussoi, Neusticomys **714**
Mustela 317, 318, 319, **321**, 321, 322, 323, 325
aequatorialis 322
affinis 322
africana **321**, 321, 322, 323
agilis 322
alascensis 321
alba 322
albica 322
albipes 323
algiricus 321
allegheniensis 323
alleni 322
alpina 321
alpinus 322, 323
altaica **321**, 321, 324
altifrontalis 322
ambigua 324
anguinae 321
angustidens 321
angustifrons 324
antiquus 324
arcticus 321
arizonensis 322
armorica 322
arthuri 322
asaii 324
astutus 321
atlas 323
audax 321
aureoventris 322
auriventer 322
australis 324
bangsi 321
barbara 317
bicolor 324
biedermanni 322
binominata 322
birulai 321
boccamela 323
boliviensis 322
borealis 322, 324
brasiliensis 313, 322
budina 322
campestris 323
canigula 324
caporiaccoi 322
caraftensis 323
caucasica 322
caucasicus 323
celenda 321
charbinensis 324
cicognanii 322
cigognanii 321
coreanus 324
corsicanus 323
costaricensis 322
cylipena 322
davidianus 324
dinniki 323
dombrowskii 323
dorsalis 322
effera 322
energumenos 324
erminea **321**, 321
eskimo 323
europeae 322
evagor 324
evergladensis 324
eversmannii 321, **322**, 322, 323, 324
fallenda 321
felipei 321, **322**, 322
ferghanae 321
foina 319
fontainierii 324
frenata 321, **322**, 322
fulva 322, 323
furo 324
fusca 322
gale 323
galinthias 323
glogeri 322
goldmani 322
gracilis 322, 324
gulosa 321
haidarum 321
hamptoni 324
helleri 322
herminea 321
hibernicus 321
hodgsoni 324
horsfieldii 324
humeralis 324
hungarica 322
imperii 321
ingens 324
initis 321
interrupta 324
invicta 321
inyoensis 322
italicus 323
itatsi 324
jelskii 322
kadiacensis 321
kamtschatica 323
kanei 321
kathiah **322**
katsurai 324
labiata 321
lacustris 324
larvatus 322, 324
latirostra 322
leptus 321
letifera 324
leucocephalus 323
leucoparia 322, 324
lineiventer 322
longicauda 322
longstaffi 321
lowii 324
lucasona 324
lutensis 324
lutra 311
lutreocephala 324
lutreola 321, **322**, 323, 324
lutreolina 321, **323**, 323, 324
lutris 310
lymani 321
macrodon 324
macrophonius 322
macrura 322
major 323, 324
manchurica 324
melampeplus 324
melli 324
meridana 322
meridionalis 323
mexicanus 322
microdon 324
microtis 321
miles 324
mink 324
minor 322, 323
minutus 323
minx 324
monticola 323
mortigena 321
mosanensis 323
moupinensis 324
munda 322
mundus 322
mupurita 324
muricus 321
namiyei 323
natsi 324
neomexicanus 322
nesolestes 324
nevadensis 322
nicaraguae 322
nigrescens 324
nigriauris 322

nigripes 321, 322, **323**, 323
nikolskii 323
nippon 321
nivalis 321, **323**, 323
noctis 324
notius 322
noveboracensis 322
novikovi 322
nudipes 321, **323**, 323
numidicus 323
occisor 322
olivacea 322
olympica 321, 324
oregonensis 322
oribasus 322
pallidus 323
panamensis 322
paraensis 321
patagonica 319
peninsulae 322, 324
perdus 322
peregusna 325
perotae 322
phenax 324
polaris 321
primulina 322
pulchra 322
punctata 323
pusilla 321
pusillus 323
putida 324
putorius 321, 322, **323**, 323, 324
pygmaeus 323
quaterlinearis 324
quelpartis 324
quoll 32
raddei 321
richardsonii 321, 322
ringens 324
rixosa 321
rixosus 323
rufa 324
russelliana 323
sacana 321
salva 321
sarmatica 325
saturatus 322
saxatalis 324
seclusa 321
semplei 321
sho 324
sibirica 321, 323, **324**, 324
siculus 323
spadix 322
stegmanni 324
stoliczkana 323
stolzmanni 321
streatori 321
striata 300, 324
strigidorsa 321, 323, **324**, 324
subhemachalanus 324
subpalmata 323
tafeli 324
taivana 323, 324
temon 321
tenuis 324
texensis 322, 324
tiarata 322
tibetanus 324
tonkinensis 323
transsylvanica 323
trettaui 323
tropicalis 322
tsaidamensis 322
turovi 323
typicus 323
varina 323
vison 321, **324**, 324, 325
vulgaris 321, 323
vulgivagus 324
washingtoni 322
whiteheadi 321
winingus 324
wyborgensis 323
xanthogenys 322
yesoidsuna 323
zorilla 324
mustela, *Galerella* 303
Mustelidae **309**
mustelina, *Ictonyx* 319
Mustelinae 315, **317**
mustelinus, *Euphractus* 66
Lepilemur 245, **246**
Reithrodontomys 741
Sciurus 440
musteloides, *Galea* **779**, 779
mustersi, *Calomyscus* 536
Microtus 522, 530
mutabilis, *Heliosciurus* **429**, 429
Sorex 122
Thomomys 473
mutandae, *Lutra* 312
mutesae, *Crocidura* **90**, 98
mutgigella, *Galerella* 303
mutica, *Glossophaga* 184
Vulpes 287
muticus, *Vulpes* 287
mutoni, *Praomys* **643**, 643
mutscheltschela, *Galerella* 303
mutus, *Bos* 401
Mycetes 254
Mycteromys 622, 623
crociduroides 629
Mydaus 313, **315**, 315
javanensis **315**, 315
lucifer 315
luciferoides 315
marchei **315**
meliceps 315
schadenbergii 315
Mygale, *pyrenaica* 124
mykinessiensis, *Mus* 626
Mylarctos 339
myllodontus, *Microtus* 528
Mylomys 590, 592, **629**, 629, 630, 639
alberti 630
christyi 630
cuninghamei 629, 630
dybowskii **630**, 630
lowei 630
massaicus 630
rex 630
richardi 630
roosevelti 630
Mynomes 517, 518, 519, 520, 525, 527, 530
Myocastor 800, **805**, 806
albomaculatus 806
bonariensis 806
castoroides 806
chilensis 806
coypus **806**
dorsalis 806
melanops 806
popelairi 806
santacruzae 806
Myocastoridae 800, **805**, 806
Myocastorinae 806
Myocebus 243
Myocricetodontinae 535
Myodes 508, 516
schisticolor 531
Myoictis **32**
bruijni 32
melas **32**
pilicauda 32
senex 32
thorbeckiana 32
wallacei 32
wavicus 32
myoides, *Crocidura* 87
Echymipera 41
Peromyscus 732
Xeromys **673**, 673
Myolagus 813
Myomiminae 765, 766
Myomimini 766, 768
Myomimus 763, 766, 767, **768**, 768
bulgaricus 768
personatus **768**, 768
roachi **768**, 768
setzeri **768**
Myomys 568, 610, 611, **630**, 630, 632, 642, 664
albipes **630**, 630, 631, 664
alettensis 630
allisoni 631
ankoberensis 630
brockmani 631
butleri 631
colonus 630, 631
daltoni 592, **631**, 631
derooi 592, **631**
fumatus 630, **631**, 631, 632
fuscirostris 630
ingoldbyi 631
leucopus 630
minor 630
niveiventris 631
oweni 631
ruppi **631**, 631, 664
saturatus 631
subfuscus 631
tuareg 631
ulae 631
veroxii 631
verreauxii 610, 630, **631**
yemeni **631**, 631, 632
Myomyscus 610, 630, 642
Myonycteris **143**, 143
brachycephala **143**
leptodon 143
relicta **143**
torquata **143**
wroughtoni 143
Myopotamus 805
Myoprocta **782**
acouchy **782**, 782
acoushy 782
archidonae 782
caymanum 782
demararae 782
exilis 782, **783**, 783
leptura 783
limanus 782
milleri 782
parva 782
pratti 782
puralis 782
myops, *Erethizon* 776
Sorex 121
Thomomys 475
Myopterus **238**
albatus 238
daubentonii **238**, 238
senegalensis 238
whitleyi **238**
Myopus 516, **531**, 531, 534
middendorfi 531
morulus 531
saianicus 531
schisticolor **531**, 531
thayeri 531
vinogradovi 531
Myorus 770
Myoscalops 772
Myosciurus **434**
minutulus 434
minutus 434
pumilio **434**
Myosorex 81, **99**, 99, 104, 112
affinis 99, 100
babaulti **99**
blarina **99**, 99
cafer **99**, 99, 100
capensis 100
eisentrauti **99**, 99, 100
geata **99**
herpestes 100
longicaudatus **100**
okuensis 99, **100**
polli 81
pondoensis 100
preussi 99
rumpii 99, **100**, 100
schalleri **100**

sclateri 99, **100**, 100
swinnyi 99
talpinus 99, 100
tenuis 99, **100**, 100
transvaalensis 100
varius **100**
Myosoricinae 81
Myosoricini 99
myosotis, Myotis 213
Myospalacinae 501, **675**
Myospalax 512, **675**, 754
armandii 675
aspalax **675**, 675, 676
baileyi 675
cansus 675
dybowskii 675
epsilanus **675**, 675, 676
fontanierii **675**, 675, 676
fontanus 675
hangaicus 675
hubeinensis 676
incertus 676
komurai 676
kukunoriensis 675
laxmanni 676
minor 676
myospalax 675, **676**, 676
pseudarmandi 675, 676
psilurus **676**, 676
rothschildi 675, **676**, 676
rufescens 675
shenseius 675
smithii 675, **676**, 676
spilurus 676
talpinus 675
tarbagataicus 676
youngi 675, 676
zokor 675
myospalax, Mus 675
Myospalax 675, **676**, 676
myosura, Nesokia 632
Tamandua 68
myosuros, Didelphis 20
Perameles 40
Proechimys **797**
myosurus, Cricetulus 538
Molossus 235
Potorous 50
Suncus 103
Myotalpa 675
Myotalpinae 675
myothrix, Golunda 592
Myotis **207**, 208, 209, 210, 211, 212, 213, 214, 215, 216
abboti 211
abei **207**
adversus **207**
aedilus 210
aelleni **208**
aenobarbus 208
affinis 212
africanus 209
agilis 209
alascensis 212
albescens **208**
albicinctus 212
albus 210
alpinus 213
altarium **208**
alter 212
alticraniatus 215
altifrons 216
altipetens 212
amboiensis 213
amotus 216
amurensis 209
ancilla 209
andersoni 210
anjouanensis 211
annectans **208**
apache 208
araxenus 213, 215
arescens 209
argentatus 208
arsinoe 208
atacamensis **208**
ater 208, 213
aurascens 213
auratus 210
aureus 209, 213
auriculus **208**
australis **208**, 208, 213
austroriparius **208**
aztecus 215
baileyi 212
bartelsi 210
bechsteini **208**
blanfordi 213
blasii 209
blythii **208**
bocagei **209**
bombinus **209**, 213
bondae 213
borneoensis 213
brandti **209**
brevirostris 212
brevis 216
browni 213
bucharensis 211
budapestiensis 210
bulgaricus 213
bureschi 209
californicus **209**
caliginosus 213
capaccinii **209**, 212
capitaneus 216
capucinellus 210
carimatae 207
carissima 212
carolii 212
carteri 213, 214
caucensis 213
caurinus 209
chiloensis 208, **209**
chinensis **209**
chiriquensis 213
chofukusei 210
chrysonotus 210
ciliatus 210
ciliolabrum 212
cinnamomeus 215
cobanensis **209**
collaris 213
concinnus 213
continentis 211
crassus 212
cupreolus 209
dalquesti 213
darjilingensis 215
dasycneme **210**
dasypus 209
daubentoni **210**
davidi 213
deignani 211
desertorum 210
dinelli 212
dobsoni 209
dogalensis 209
domesticus 212
dominicensis **210**
dryas 211
durangae 216
elegans **210**
emarginatus **210**
escalerae 213
esmeraldae 213
evotis 208, **210**
exiguus 213
exilis 209
extremus 213
favonicus 208
federatus 213
fimbriatus 212
findleyi **210**
flavus 210
formosus **210**
fortidens **210**
frater **211**
fujiensis 209
gardneri 214
gatesi 208
gayi 209
ghidinii 208
goudoti **211**
gracilis 209
grandis 216
grisescens **211**
gryphus 212
guaycuru 214
hajastanicus 213
hasseltii **211**
hermani 210
herrei 213
hildegardeae 209
hirsutus 212
horsfieldii **211**
hosonoi **211**, 216
humeralis 213
ikonnikovi **211**
incautus 216
insularum **211**
interior 216
isidori 208
jaliscensis 216
jeannei 211
kaguyae 211
keaysi **211**
keenii **211**
kukunoriensis 213
lambi 216
lanaceus 210
lanatus 210
lanceolatus 212
laniger 210
larensis 213
latipennis 213
latirostris 213
leibii **212**, 216
lepidus 211
lesueuri **212**
lesviacus 209
leucogaster 208
levis **212**, 215
limnophilus 210
lobipes 213
longicaudatus 211
longicrus 216
longipes **212**
loukashkini 210
loveni 215
lucifugus **212**, 216
luctuosus 209
lugubris 213
lutosus 216
macellus 211
macrocephalicus 213
macrodactylus 209, **212**
macropus 207
macrotarsus **212**
magnamolaris 216
major 209, 210
maripensis 213
martiniquensis **212**
megalopus 212
megapodius 209
meinertzhageni 213
melanorhinus 212
mexicanus 209
micronyx 210
milleri **213**
minutellus 210
moluccarum 207
montivagus **213**
morrisi **213**
moupinensis 213
mumfordi 208
mundus 208, 213
muricola 208, 211, **213**
myosotis 213
myotis 209, **213**
mystacinus 209, **213**
nathalinae 210
nattereri 209, **213**, 215
neglectus 210
nesopolus **213**
niasensis 213
nicholsoni 208
nigricans 210, **213**
nigrofuscus 213
nipalensis 213
nitidus 209
nubilus 212
nugax 213

nyctor 212
obscurus 216
occultus 212
omari 209
oregonensis 209
oreias **214**
orientis 207
orii 213
oxalis 216
oxygnathus 209
oxyotus **214**
ozensis **214**
pacificus 210
pahasapensis 215
pallida 210
pamirensis 213
parvulus 213
patriciae 213
pellucens 209
peninsularis **214**
pequinius **214**
pernox 212
peshwa 211
petax 210
peytoni 213
phasma 216
pilosatibialis 211
pilosus 214
planiceps **214**
polythrix 212
primula 208
pruinosus **214**
przewalskii 213
punensis 213
punicus 209
quercinus 209
relictus 212
ricketti **214**
ridleyi **214**
riparius **214**
risorius 209
rosseti **214**
ruber 212, **215**
ruddi 216
rufofuscus 213
rufoniger 210
rufopictus 210
saba 212
salarii 212
saturatus 210, 216
schaubi **215**
schinzii 213
schrankii 213
scotti **215**
seabrai **215**
septentrionalis 212
sibiricus 209
sicarius **215**
siligorensis **215**
simus 214, **215**
sociabilis 216
sodalis **215**
sogdianus 213
solomonis 207
sonoriensis 211
sowerbyi 215
spelaeus 213
stalkeri **215**
staufferi 210
stephensi 209
submurinus 213
subulatus 212
surinamensis 210
taiwanensis 207
tenuidorsalis 209
thaianus 215
thomasi 214
thysanodes **215**
transcaspicus 213
tricolor **215**
tschuliensis 213
tsuensis 210
turcomanicus 210
typus 213
ussuriensis 210
velifer 210, 214, **216**
venustus 216
vespertinus 215
virginianus 212
vivesi **216**
volans **216**
volgensis 210
watasei 210
weberi 210
welwitschii **216**
yesoensis **216**
yumanensis 212, **216**
myotis, Myotis 209, **213**
Vespertilio 207
Myotomys 680
Myoxidae 684, **763**, 763, 766, 770
Myoxinae 763, **769**, 769
myoxinus, Eliurus **678**, 678
Microcebus 243
Myoxus 763, **769**, 769, 770
abruttii 770
argenteus 770
avellanus 770
caspicus 770
caspius 770
chrysurus 791
elegans 769
esculentus 770
giglis 770
glis **770**
insularis 770
intermedius 770
italicus 770
javanicus 769
martinoi 770
melonii 770
minutus 770
orientalis 770
persicus 770
petruccii 770
pindicus 770
postus 770
pyrenaicus 770
spoliatus 770
subalpinus 770
tschetshenicus 770
vagneri 770
vulgaris 770
myrella, Kerivoula **197**
Myrmecobiidae **29**, 29
Myrmecobius **29**
fasciatus **29**, 29
rufus 29
Myrmecophaga **68**
afra 375
artata 68
capensis 375
centralis 68
didactyla 67
jubata 68
meridionali 68
tamandua 68
tridactyla **68**, 68
Myrmecophagidae **67**
myrmecophragris, Tachyglossus 13
myrmephagus, Ursus 339
Myrmydon 67
Myropteryx 156
Myrsilas 436
Mysateles 800, 801, **802**, 802, 803
arboricolus 803
garridoi **802**
gundlachi **802**, 802, 803
jaumei 800
melanurus **802**, 803
meridionalis 802, **803**, 803
pallidus 803
poeppingii 802, 803
poeyi 803
prehensilis 802, **803**, 803
Myscebus 243
Myslemur 247
mysolensis, Pteropus 147
mysoliae, Dorcopsis 52
Myspithecus 243, 247
mystacalis, Dendromus 542, **543**, 543
mystaceus, Sus 379
Mystacina **231**
aupourica 231
rhyacobia 231
robusta **231**
tuberculata **231**, 231
velutina 231
Mystacinidae **231**
mystacinus, Apodemus **572**
Microtus 522
Mus 626
Myotis 209, **213**
Mystateles, prehensilis 803
Mystax 253
mystax, Calomyscus **536**, 536
Funisciurus 427
Saguinus **254**
Mysticeti 349
mysticetus, Balaena **349**, 349
Mystromyinae 501, **676**, 676
Mystromys **676**, 676, 677, 683
albicaudatus 676, **677**
albipes 676, 677
fumosus 677
hausleitneri 677
lanuginosus 677
pocockei 677
Mythomys 73
mytilopes, Addax 412
myurus, Elephantulus **829**
Myzopoda **196**
aurita **196**, 196
Myzopodidae **196**
mzabi, Ctenodactylus 761
Massoutiera **761**

nadymensis, Sciurus 443
Naemorhaedus 406
Naemorhedus **406**
aldridgeanus 407
annectens 407
argyrochaetes 407
arnouxianus 407
baileyi **407**
bedfordi 407
benetianus 407
brachyrhinus 407
bubalina 407
caudatus **407**
chrysochaetes 407
cinerea 407
collasinus 407
cornutus 407
cranbrooki 407
crispus **407**, 407
curvicornis 407
duvaucelii 407
edwardsii 407
erythropygius 407
evansi 407
fantozatianus 407
fargesianus 407
galeanus 407
gendrelianus 407
goral **407**, 407
griseus 407
henryanus 407
hodgsoni 407
humei 407
initialis 407
iodinus 407
jamrachi 407
longicornis 407
marcolinus 407
maritimus 407
maxillaris 407
microdontus 407
milneedwardsii 407
montinus 407
nasutus 407
nemorhaedus 406
niger 407
osborni 407
pinchonianus 407
platyrhinus 407
pryerianus 407

pugnax 407
raddeanus 407
robinsoni 407
rocherianus 407
rodoni 407
rubidus 407
saxicola 407
sumatraensis **407**, 407
swettenhami 407
swinhoei **407**
thar 407
ungulosus 407
versicolor 407
vidianus 407
xanthodeiros 407
nagarum, Callosciurus 421
Mus 623
nageri, Clethrionomys 508
Nagor 414
nagor, Redunca 414
nagtglasii, Graphiurus 764
nahani, Colobus 270
Pan 277
nahoor, Pseudois 409
naias, Neomys 110
Syconycteris 155
nainggolani, Muntiacus 389
Tupaia 133
nair, Lutra 312
nairobae, Arvicanthis 576, **577**, 577, 578
Dendromus 543
naisense, Megaderma 163
naivashae, Chaerephon 233
Mus 629
Tachyoryctes **686**
najdiya, Nycteris 162
nakanus, Callosciurus 421
nakurae, Alcelaphus 393
nakuroensis, Gazella 397
nalutensis, Gerbillus 552
namaquana, Felis 290
namaquensis, Aethomys 567, **568**, 568
Chrysochloris 75
Petromus 774
Rhabdomys 662
Suricata 308
Tatera 560
Xerus 458
namby, Mazama 391
namibensis, Aethomys 568
Eremitalpa 76
Laephotis **206**
Parotomys 682
Petromyscus 683
Rhabdomys 662
namiyei, Mustela 323
Vespertilio 228
nana, Brachyphylla **182**
Cavia 779
Crocidura **90**, 90, 91
Mazama **391**
Nycteris **162**
Phalangista 58
Sturnira **193**
Thylamys 23
nancillus, Gerbillus **553**
nancyae, Spermophilus 447, **448**
nancymaae, Aotus **256**
nandae, Felis 290
Nandinia **342**, 342
binotata **342**
gerrardi 342
hamiltonii 342
Nandiniinae **342**, 342
Nanelaphus 390
Nanger 395
nanggenica, Ochotona 812
nanguer, Gazella 396
nanilla, Crocidura **91**, 93
nankiangensis, Apodemus 574
nannodes, Cervus 386
Nannomys 622, 623, 624, 625, 626, 627, 628, 629
Nannosciurus 426, **434**
bancanus 434
borneanus 434
melanotis **434**
pallidus 434
pulcher 434
soricinus 434
sumatranus 434
Nannospalax **753**, 753, 754
aegyptiacus 753
anatolicus 754
armeniacus 754
berytensis 753
captorum 754
ceylanus 754
cilicicus 754
corybantium 754
ehrenbergi **753**, 753, 754
epiroticus 754
fritschi 753
hellenicus 754
hercegovinensis 754
hungaricus 754
insularis 754
intermedius 753, 754
kirgisorum 753, 754
leucodon **753**, 753, 754
makedonicus 754
martinoi 754
montanoserbicus 754
montanosyrmiensis 754
monticola 754
nehringi 753, **754**, 754
ovchepolensis 754
peloponnesiacus 754
serbicus 754
strumiciensis 754
syrmiensis 754
thermaicus 754
thessalicus 754
thracius 754
transsylvanicus 754
turcicus 754
vasvárii 754
Nanocavia 780
nanogigas, Ratufa 437
Nanonycteris **143**
veldkampi **143**
nanopardus, Panthera 298
nanschanicus, Alticola 504
nanula, Suncus 102
nanulus, Mops **236**
Pipistrellus **222**
nanus, Artibeus 189
Cercartetus **58**
Eumops 233
Gerbillus 549, 550, 551, 552, **553**, 553, 554, 555
Graphiurus 764
Hipposideros 171
Holochilus 705
Lepus 814
Macroglossus 154
Mesocapromys **801**, 801, 802
Microtus 525
Pecari 380
Pipistrellus **222**
Pseudomys 645, **647**, 647, 648
Rhinolophus 166
Scalopus 127
Sorex **118**, 121
Spermophilus 446
Syncerus 402
Thomomys 473
naoniuensis, Niviventer 633
napaea, Sicista 496, **497**, 497
Napaeozapus **498**, 498
abietorum 498
algonquinensis 498
frutectanus 498
gaspensis 498
insignis **498**
roanensis 498
saguenayensis 498
napensis, Choloepus 64
Nectomys 709
Pithecia 262
napi, Microsciurus 433
napoleon, Callicebus 258
napu, Tragulus **383**
narbadae, Rattus 658
narboroughi, Nesoryzomys 713, 714
narentae, Crocidura 88
narhval, Monodon 358
naria, Canis 280
narica, Nasua **334**, 334
narranus, Lepus 816
Narwalus 357
naryanus, Cervus 386
narymensis, Clethrionomys 509
narynensis, Hystrix 774
Nasalis **270**, 270
capistratus 270
concolor **270**
larvatus **270**
nasica 270
orientalis 270
recurvus 270
siberu 270
nasalis, Gazella 397
Rhinoceros 372
nasarovi, Microtus **526**
nasica, Akodon 692
Nasalis 270
Nasua 335
nasicus, Thomomys 475
Nasilio 829
Nasillus 130
Nasira 31
naso, Antechinus **30**
Arvicanthis 578
Euchoreutes **495**, 495
Floresomys 639
Herpestes 304, **305**
Lophuromys 606
Melasmothrix **614**, 614
Melomys 617
Paulamys **639**, 639
Rhynchonycteris **159**
Tadarida 240
Vespertilio 159
nasoguttata, Madoqua 398
nasomaculata, Cerophorus 412
nasomaculatus, Addax **412**
nastjukovi, Stylodipus 495
Nasua **334**
annulata 334
aricana 334
boliviensis 334
bullata 334
candace 334
cinerascens 334
dichromatica 334
dorsalis 334
fusca 335
gualeae 335
henseli 335
isthmica 334
jivaro 335
judex 335
jurvana 335
leucorhynchus 335
manium 335
mephisto 335
meridensis 335
mexianae 335
mexicana 334
molaris 334
mondie 335
montana 335
monticola 335
narica **334**, 334
nasica 335
nasua **334**, 334
nasuta 335
nelsoni 334
obfuscata 335
pallida 334
panamensis 334
phaeocephala 335
quasje 335
quichua 335

richmondi 334
rufa 335
rufina 335
rusca 335
sociabilis 335
soederstroemmi 335
solitaria 334, 335
tamaulipensis 334
thersites 334
vittata 335
vulpecula 334
yucatanica 334
nasua, Nasua **334**, 334
Viverra 334
Nasuella **335**
olivacea **335**
nasuta, Microtus 527
Nasua 335
Perameles **40**, 40
nasutus, Eptesicus **202**
Hylobates 275
Leopoldamys 603
Microdipodops 480
Molossus 239
Mus 726
Naemorhedus 407
Oxymycterus **727**, 727
Peromyscus 730, **734**
Promops **239**
Tadarida 240
Thomomys 473
Tremarctos 338
Trichechus 366
natalensis, Amblysomus 74
Cephalophus 410, **411**
Cryptomys 772
Mastomys 610, **611**, 611
Miniopterus 231
Mormopterus 237
Natalus 195
Otomys 680
Procavia 374
Raphicerus 399
Steatomys 546
Suncus 103
Tatera 560
nataliae, Allactaga 488
Natalidae **194**
natalis, Pteropus 149
Natalus **194**, 195
brevimanus 195
continentis 195
dominicensis 195
haymani 195
jamaicensis 195
lepidus **195**
macer 195
major 195
mexicanus 195
micropus **195**, 195
natalensis 195
primus 195
saturatus 195
stramineus **194**, **195**
tronchonii 195
tumidifrons **195**, 195
tumidirostris **195**
natans, Neomys 110
natator, Oryzomys 724
nathalinae, Myotis 210
nathusii, Pipistrellus **222**
nationi, Chaetophractus **65**
nativitatis, Rattus 650, 655, **656**, 656
natronensis, Pachyuromys 559
natsi, Mustela 324
nattereri, Caluromys 15
Ctenomys **786**
Mus 626
Myotis 209, **213**, 215
Vampyressa 194
natunae, Cynocephalus 135
Maxomys 614
Megaderma 163
Nycticebus 248
Presbytis 271
Pteropus 151
Tragulus 382
Tupaia 133
natunensis, Hipposideros 172
Sundasciurus 451
Sus 379
Tarsius 250
naumanni, Procavia 374
naumovi, Microtus 527
nautica, Erignathus 329
nauticus, Caluromys 15
Hydromys 597
Isoodon 39
navaho, Microtus 524
Ursus 339
navajo, Sciurus 439
navigator, Apodemus 573
Callosciurus 423
Sorex 119
navus, Neotoma 712
Oligoryzomys 718
Oryzomys 717
Thomomys 473
nawaiensis, Pteropus 150
nayaritensis, Cryptotis 109
Pappogeomys 472
Sciurus **441**
nayaur, Ovis 409
Pseudois **409**
ndjiriensis, Gazella 397
ndolae, Tatera 561
Neacomys 687, **708**, 719
amoenus 709
carceleni 709
guianae **708**
pictus **708**, 709
pusillus 709
spinosus **709**, 709
tenuipes **709**, 709
typicus 709
Neamblysomus 74
neanderthalensis, Homo 276
Neanthomys 605
neavei, Anomalurus 757
Crocidura 89
Mus 622, **626**, 626, 627, 628
Tatera 562
neblina, Marmosops 19
nebouxii, Sciurus **443**
nebrascensis, Glaucomys 460
Peromyscus 732
Reithrodontomys 741
nebrodensis, Microtus 529
nebulicola, Spermophilus 448
nebulosa, Felis 297
Neofelis **297**
nebulosus, Baiomys 695
Perodicticus 248
Thomomys 475
necator, Vulpes 287
necopinus, Prosciurillus 436
Nectogale **110**, 110
elegans **110**, 110
sikhimensis 110
Nectogalina 110
Nectomys **709**, 709, 719, 722, 748
amazonicus 709
apicalis 709
aquaticus 709
brasiliensis 709
cephalotes 709
fulvinus 709
garleppii 709
grandis 709
magdalenae 709
mattensis 709
melanius 709
montanus 709
napensis 709
olivaceus 709
palmipes **709**, 709
parvipes **709**
pollens 709
rattus 709
robustus 709
saturatus 709
squamipes **709**, 709
tarrensis 709
tatei 709
vallensis 709
nedjo, Theropithecus 269
nefandus, Bradypus 63
negans, Vormela 325
negatus, Orthogeomys 471
neglecta, Capra 405
Crocidura **91**
Gazella 396
Marmosops 20
Paguma 343
Profelis 296
Taxidea 325
neglectus, Ateles 257
Cercopithecus **264**
Cervus 386
Cricetulus 538
Ctenomys 785
Hipposideros 173
Lynx 293
Microtus 518, 519
Molossops **234**
Myotis 210
Pappogeomys **472**
Perognathus **485**
Presbytis **271**
Rattus 660
Sciurus **442**
Sorex 117
Spermophilus **449**
Tamias **454**
Thomomys 473
Ursus 339
negligens, Crocidura 89
Sciurus **440**
negre, Saguinus 254
negrensis, Galea 779
Isothrix 793
negrina, Crocidura **91**
negrinus, Rattus 652
Sus 378
negrito, Bolomys 696
negrosensis, Harpyionycteris 142
nehringi, Cricetus 538
Nannospalax 753, **754**, 754
nehringii, Sus 378
neilli, Leopoldamys **603**
neillii, Cercartetus 58
Nelomys 791, 792
aperoides 798
armatus 793
dasythrix 792
nelsoni, Ammospermophilus **420**
Cervus 386
Chaetodipus **484**
Conepatus 316
Cryptotis 109
Dicrostonyx **510**, 510
Dipodomys **479**, 479, 480
Heteromys **481**, 482
Leopardus 291
Megadontomys **707**, 707
Nasua 334
Neotoma **712**
Odocoileus 391
Orthogeomys 471
Oryzomys **723**
Ovis 408
Pappogeomys 472
Pecari 380
Reithrodontomys 741
Romerolagus 824
Sciurus 440
Sylvilagus 826
Thomomys 476
Ursus 338, 339
Vampyrum 181
Xenomys **752**, 752
Nelsonia **709**
cliftoni 710
goldmani 709, **710**
neotomodon 709, **710**
nemaeus, Pygathrix **273**
Simia 272
nemestrina, Macaca **267**

Nemestrinus 266
nemo, Cryptomys 772
Nemomys 569
nemoralis, Akodon 692
Microtus 528
Odocoileus 391
Rattus 658
Sciurus 440
Nemorhaedus, goral 407
nemorhaedus, Naemorhedus 406
Nemorhedus 406
nemorivaga, Bandicota 579
Mazama 391
nemorivagus, Suncus 103
Nemorrhedus 406
nems, Herpestes 305
Neoaschizomys 507
Neoauchenia 381
Neobalaena 351
Neobalaenidae 349, **351**, 351
neobritanicus, Uromys 671, **672**, 672
neobrittanicus, Hydromys **597**
neocaledonica, Notopteris 155
neocaledonicus, Chalinolobus 199
Neocebus 263
neocenus, Akodon **691**, 694
Neocometes 684
Neocothurus 261
Neoctodon, simonsi 788
Neodon 506, 517, 518, 522, 523, 529, 531
Neofelinae 297
Neofelis **297**, 297
brachyurus 297
diardi 297
macrocelis 297
macrosceloides 297
nebulosa **297**
Neofiber **531**, 532
alleni 531, **532**
apalachicolae 532
exoristus 532
nigrescens 532
struix 532
neohibernicus, Pteropus **149**, 149
Neohydromys 604, 614, 620, **632**, 632, **644**
fuscus **632**, 632
Neohylomys 79
Neomeris 358
neomexicana, Vulpes 287
neomexicanus, Mustela 322
Sorex 118
Sylvilagus 824
Tamiasciurus 457
Neomicia 219
Neomyini 105, 107, 109, 110, 122
Neomys 107, **110**, 111
albus 110
amphibius 110
anomalus **110**
aquaticus 110
argenteus 110
balkaricus 110
bicolor 110
brachyotus 110
canicularius 110
carinatus 110
ciliatus 110
collaris 110
constrictus 110
dagestanicus 110
daubentonii 110
eremita 110
fimbriatus 110
fluviatilis 110
fodiens **110**, 111
griseogularis 110
hermanni 110
hydrophilus 110
ignotus 110
intermedius 110
josti 110
leptodactylus 110
limchjnhunii 110
lineatus 110
linneana 110
liricaudatus 110
longobarda 110
macrourus 110
milleri 110
minor 110
mokrzeckii 110
musculus 110
naias 110
natans 110
newtoni 110
niethammeri 110
nigripes 110
orientalis 110
orientis 110
pennantii 110
psilurus 110
remifer 110
rhenanus 110
rivalis 110
schelkovnikovi **110**
soricoides 110
sowerbyi 110
stagnatilis 110
stresemanni 110
teres 110
watasei 110
Neonycteris 178, 179
Neoorca 355
Neophascogale **33**
lorentzi **33**
nouhuysii 33
rubrata 33
venusta 33
Neophoca **328**, 328
albicollis 328
australis 328
cinerea **328**
forsteri 328
hookeri 328
lobatus 328
williamsi 328
Neophocaena **358**
asiaeorientalis 358
melas 358
phocaenoides **358**, 358
sunameri 358
Neoplatymops 234
mattogrossensis 234
Neopteryx **143**
frosti **143**, 143
Neorheomys 743
Neoromicia 200, 201, 202, 203, 204
Neoryctes 43
Neosciurus 439
Neosorex 111, 119
Neotamias 453, 454, 455, 456
Neotetracus 79
Neotoma 704, 709, **710**, 710, 711, 712, 713
abbreviata 712
acraia 711
affinis 711
albigula **710**, 712, 713
alleni 704
alticola 711
angustapalata **710**
angusticeps 710
annectens 711
anthonyi **710**
apicalis 711
arenacea 712
aridicola 712
arizonae 711
atrata 712
attwateri 711
aureotunicata 711
auripila 711
baileyi 711
bella 712
bensoni 711
brevicauda 710
bryanti **710**
bullata 712
bullatior 711
bunkeri **710**
californica 712
campestris 711
canescens 712
chamula 712
chrysomelas **711**, 711
cinerea **711**, 711
cinnamomea 711
cnemophila 711
columbiana 711
cumulator 710
desertorum 712
devia **711**, 712
dispar 711
distincta 712
drummondii 711
durangae 710
egressa 712
eremita 712
fallax 712
felipensis 712
ferruginea 712
flava 711
floridana 710, **711**, 711, 712
fulviventer 712
fusca 711
fuscipes **711**, 711
gilva 712
goldmani **711**
grandis 710
grangeri 711
grinnelli 712
griseoventer 712
haematoreia 711
harteri 711
illinoensis 711
inopinata 712
inornata 712
insularis 712
intermedia 712
isthmica 712
laplataensis 710
latifrons 710
latirostra 712
lepida 710, **711**, 711, 712
leucodon 710
leucophaea 712
littoralis 712
luciana 711
lucida 711
macrodon 711
macrotis 711
madrensis 712
magister 711
marcosensis 712
marshalli 712
martinensis **712**
martirensis 711
mearnsi 710
melanura 710
melas 710
mexicana 711, **712**
micropus 710, 711, **712**
mohavensis 711
molagrandis 712
monochroura 711
monstrabilis 711
montezumae 710
navus 712
nelsoni **712**
nevadensis 712
notia 712
nudicauda 712
occidentalis 711
ochracea 712
orizabae 712
orolestes 711
osagensis 711
palatina **712**
parvidens 712
pennsylvanica 711
perpallida 712
perplexa 711

petricola 712
phenax **712**
picta 712
pinetorum 712
planiceps 712
pretiosa 712
pulla 711
ravida 712
relicta 712
riparia 711
robusta 710
rubida 711
rupicola 711
sanrafaeli 711
saxamans 711
scopulorum 712
seri 710
sheldoni 710
simplex 711
sinaloae 712
smalli 711
sola 712
solitaria 712
splendens 711
stephensi **712**
streatori 711
subsolana 710
surberi 712
tenuicauda 712
torquata 712
tropicalis 712
varia **712**
venusta 710
vicina 712
vulcani 712
warreni 710
zacatecae 710
Neotomodon 703, **713**, 713, 728, 739
alstoni **713**, 713
orizabae 713
perotensis 713
neotomodon, Nelsonia 709, **710**
Neotomys **713**, 746
ebriosus **713**, 713
vulturnus 713
Neotragus **398**
akeleyi 398
batesi **398**
deserticola 398
harrisoni 398
kirchenpaueri 398
livingstonianus 398
moschatus **398**, 398
perpusillus 398
pygmeus **398**
regia 398
spinigera 398
zanzibaricus 398
zuluensis 398
nepalensis, Cervus 387
Equus 370
Herpestes 305
Lutra 312
Ochotona 809, 812
Sorex 112
Vulpes 287
Nepus 365
nequam, Hipposideros **174**
nereis, Enhydra 310
Rhinolophus **167**
nericola, Rattus 658
nero, Helogale 304
Uromys 671
nerterus, Reithrodontomys 742
nesaea, Marmosa 19
nesaeus, Dasyurus 32
Sciurus **441**
nesarnack, Tursiops 357
nesea, Dobsonia 140
Nesictis 314
nesiotes, Acomys **566**
Callosciurus 422
Chlorocebus 265
Erinaceus 77
Proechimys 795
nesioticus, Spermophilus 445
nesites, Rhinolophus 163
nesiticus, Apodemus 573
Nesogale 71, 72
Nesokia 578, 579, 603, **632**, 632
bacheri 632
bailwardi 632
beaba 632
boettgeri 632
brachyura 632
bunnii **632**
buxtoni 632
chitralensis 632
dukelskiana 632
griffithi 632
hardwickei 632
huttoni 632
indica **632**, 632
insularis 632
legendrei 632
myosura 632
satunini 632
scullyi 632
suilla 632
Nesolagus **822**
netscheri **822**
nesolestes, Mustela 324
Nesomyidae 563, 676, 677, 679
Nesomyinae 501, **677**, 677
Nesomys 677, 678, **679**, 679
audeberti 679
lambertoni 679
rufus **679**, 679
Nesonycteris 154
nesophila, Martes 319
nesophilus, Microtus 527, 528
Thomomys 473
Nesophontes **69**, 69
edithae 69, **70**
hypomicrus **70**
longirostris **70**
major **70**
micrus **70**
paramicrus **70**
submicrus **70**
zamicrus **70**
Nesophontidae **69**
nesopolus, Myotis **213**
Nesoromys 665
Nesoryctes 72
Nesoryzomys **713**, 713, 719
darwini **713**, 713
fernandinae **713**
indefessus **713**, 714
narboroughi 713, 714
swarthi **714**, 714
Nesoscaptor **126**
uchidai **126**, 126
Nesotragus 398
nesterovi, Felis 290
nestor, Trachypithecus 274
Tscherskia 540
netscheri, Lepus 822
Nesolagus **822**
neubronneri, Tragulus 383
neuhassi, Distoechurus 62
neuhauseri, Microtus 530
neujukovi, Chionomys 507
neumanni, Alcelaphus 393
Arvicanthis 576, **577**, 577, 578
Cercopithecus 264
Dendrohyrax 373
Epomophorus 141
Galerella 303
Genetta 345
Kobus 414
Papio 269
Raphicerus 399
Syncerus 402
neumayeri, Oncifelis 294
Neurotrichus **126**
gibbsii **126**
hyacinthinus 126
major 126
minor 126
Neusticomys **714**, 714
monticolus **714**, 714
mussoi **714**
oyapocki **714**
peruviensis **714**
venezuelae **714**
nevadensis, Bassariscus 334
Dipodomys 479
Microtus 525
Mustela 322
Neotoma 712
Nyctinomops 239
Ochotona 811
Perognathus 485
Peromyscus 736
Sorex 122
Spermophilus 446
Tamias 456
Taxidea 325
Thomomys 475
Vulpes 287
Zapus 499
neveriae, Microtus 524
newara, Golunda 592
newera, Feroculus 99
newera-ellia, Feroculus 99
newmani, Thomomys 476
newtonensis, Mephitis 317
newtoni, Mesocricetus **539**
Miniopterus 231
Neomys 110
nexa, Lontra 311
nexilis, Dipodomys 479
nexus, Microdipodops 480
Microtus 525
Tamias 454
nezumi, Rattus 660
ngamiensis, Chlorocebus 265
Galerella 303
Lepus 820
Mungos 307
Papio 269
Paracynictis 307
nghialoensis, Mus 628
ngorongorensis, Acinonyx 288
niadensis, Sus 378
niadicus, Pteropus 149
niadis, Chiropodomys 583
niangarae, Hipposideros 172
Mops **236**, 237
niapu, Funisciurus 427
Hipposideros 174
niasensis, Arctictis 342
Myotis 213
Rhinolophus 169
niasis, Tragulus 383
niata, Microcavia **780**
nicaraguae, Conepatus 316
Leopardus 292
Mustela 322
Oligoryzomys 718
Peromyscus 734
Reithrodontomys 740
Tonatia 180
nicefori, Ichthyomys 705
Micronycteris **179**
Sylvilagus 825
Thomasomys 749
nicholi, Thomomys 473
nichollsi, Ardops **187**
Stenoderma 187
nicholsoni, Myotis 208
nicobarensis, Hipposideros 172
nicobarica, Crocidura 87, **91**
Tupaia **132**
nicobaricus, Pteropus 149
Sus 379
nicobarulae, Hipposideros 171
nicolli, Psammomys 559
nicotianae, Callosciurus 422
nicoyana, Sciurus 443
nictitans, Cercopithecus **264**
Sylvicapra 412
nictitatans, Paradoxurus 344

nidicola, Kerivoula 196
nidoensis, Tamias 453
niediecki, Alcelaphus 393
Gulo 318
Lepus 814
Sigmoceros 395
Syncerus 402
Taurotragus 403
niethammeri, Lepus 817
Neomys 110
nieuwiedii, Macrophyllum 178
nifoviridis, Chlorocebus 265
nigellus, Bunomys 581
Cervus 386
Peromyscus 730
Platyrrhinus 191
niger, Alouatta 254
Antechinus 31
Apodemus 573
Arvicola 506
Canis 281
Capreolus 390
Castor 467
Cebus 259
Cephalophus **411**
Cervus 387
Chelemys 699
Cricetus 538
Dama 387
Dasypus 66
Dicerorhinus 372
Diceros 372
Eulemur 244
Hemiechinus 78
Hippotragus **412**
Hylobates 275
Indri 247
Leptailurus 292
Lepus 817
Melursus 337
Mus 626
Naemorhedus 407
Ondatra 532
Oryzorictes 72
Ovis 409
Pan 277
Panthera 298
Paradoxurus 344
Pecari 380
Pteropus **149**
Saguinus 254
Sciurus **442**, 443
Suncus 103
Sus 379
Thomomys 475
Ursus 339
Vespertilio 146
Vombatus 46
nigeriae, Chaerephon **232**
Crocidura **91**
Gerbillus 548, **553**
Lemniscomys 601
Papio 269
Protoxerus 436
Taterillus 562
nigerianus, Atilax 301
nigeriensis, Redunca 414
nigerrima, Alouatta 254
nigerrimus, Eulemur 244
nigra, Alouatta 254
Aplodontia 417
Callithrix 252
Chiropotes 261
Condylura 124
Macaca **267**, 267
Marmota 432
Martes 320
Mephitis 317
Microtus 518
Ondatra 532
Panthera 297, 298
Paradoxurus 344
Petaurista 463
Puma 296
Sorex 111
Tamandua 68
Vulpes 286, 287
nigrans, Microtus 527
Pipistrellus 223
Trichosurus 48
nigrata, Galerella 303
nigratus, Galerella 302, 303
Philander 22
Sciurus **442**
nigrensis, Anomalurus 757
Funisciurus 428
nigrescens, Arctocephalus 326
Baiomys 695
Catopuma 289
Corsira 122
Cricetulus 538
Cryptotis **109**, 109
Echinops 73
Emballonura 157
Erethizon 776
Giraffa 383
Lepus 820
Liomys 482
Macaca 267
Macropus 53
Microgale 71
Mosia **158**, 158
Mustela 324
Neofiber 532
Ochotona 811
Orthogeomys 470
Pecari 380
Petaurista 463
Ratufa 437
Reithrodontomys 741
Rhombomys 559
Sciurus 440, 443
Setifer 73
Soriculus **123**
Thoopterus **153**, 153
nigri, Chaerephon 233
nigriana, Microcavia 780
nigriauris, Mustela 322
nigribarbis, Odocoileus 391
Oligoryzomys 718
nigricans, Allactaga 489
Arvicola 506
Cavia 779
Cervus 386
Cricetus 538
Crocidura **84**, **89**, **91**
Dendrohyrax 373
Kobus 414
Lepus 817
Mesocricetus 539
Microtus 518
Molossus 235
Myotis 210, **213**
Pipistrellus 223
Sphiggurus 777
Thomomys 473
Tragulus 383
nigricauda, Ichneumia 307
Mus 668
Tatera **561**
Thallomys **668**, 668, 669
nigricaudata, Ourebia 399
nigricaudatus, Galerella 303
Heteromys 481
Lepus 815
Petaurista 463
Sylvilagus 825
nigricaudus, Ictonyx 319
Petinomys 463
nigriceps, Aotus **256**
Hemicentetes 73
Paguma 343
Saimiri 260
nigriclunis, Dasyprocta 782
nigricollis, Lepus 814, 816, 818, **819**, 819, 820
Saguinus **254**
Tragulus 383
nigricolor, Crossarchus 301
nigriculus, Mesocricetus 539
Peromyscus 730
Sorex 113, 122
nigridius, Dasymys 589
nigridorsalis, Callosciurus 421
nigrifrons, Callicebus 259
Cephalophus **411**, 411
Dendromus 543
Eulemur 244
Oxymycterus 727
Paradoxurus 344
Saguinus 253
nigrigenis, Cercopithecus 264
nigrimanis, Presbytis 271
nigrimanus, Procolobus 272
nigrimontana, Ovis 408
nigrimontanus, Spermophilus **446**
nigrimontis, Chaetodipus 484
nigrinotatus, Tragelaphus 404
nigripectus, Cebus 260
Otocolobus 295
nigripes, Bdeogale **301**, 301
Cercopithecus 265
Crocidura **91**
Dologale 302
Felis **290**
Hylopetes **461**
Lemmus 517
Macrotis 40
Muntiacus 389
Mustela 321, 322, **323**, 323
Neomys 110
Oligoryzomys 717, **718**, 718
Oryctolagus 822
Papio 269
Procyon 335
Pygathrix 273
Sciurus 441
Sus 379
nigrirostris, Urocyon 285
nigrispinus, Echimys **792**
nigrita, Akodon **691**
Scotophilus **227**, 227, 228
Tatera 562
nigritalus, Apodemus 572
nigritellus, Scotophilus 227, 228
nigritus, Cebus 259
nigrivittata, Saimiri 260
nigrivittatus, Cebus 260
nigro-argenteus, Vulpes 287
nigroaculeatus, Zaglossus 13
nigrocaudatus, Vulpes 287
nigrocinctus, Tragulus 383
nigrofulvus, Proechimys 797
nigrofusca, Crocidura **91**, 91, 97
nigrofuscus, Myotis 213
nigrogriseus, Chalinolobus **200**
Tadarida 240
nigronuchalis, Sylvilagus 826
nigroscapulatus, Kobus 414
nigrotibialis, Tatera 561
nigroviridis, Allenopithecus **262**
Cercopithecus 262
nigrovittatus, Callosciurus **422**, 423
niigatae, Phaulomys 533
nikkonis, Petaurista 462
nikolajevi, Microtus 530
nikolausi, Megadendromus 544, **545**
nikolskii, Apodemus 570
Mustela 323
Spermophilus 448
nilagirica, Vandeleuria 672
nilgirica, Suncus 102
Nilopegamys 586
nilosa, Coleura 156
nilotica, Crocidura 91
Felis 290
Lutra 312
niloticus, Arvicanthis 576, **577**, 577, 578
Lemmus 576
Vulpes 287

nilssoni, Eptesicus **202**
nimbae, Crocidura **92**, 97
nimbosus, Oryzomys 720
nimr, Panthera 298
nimrodi, Cryptomys 772
Ningaui **33**, 33
ridei **33**
timealeyi **33**, 33
yvonnae **33**
ningbing, Pseudantechinus **35**
ningpoensis, Apodemus 570
Callosciurus 421
ningshaanensis, Tscherskia 540
ninus, Niviventer 634
niobe, Crocidura 89, **92**
Lariscus **430**, 430
Stenomys **665**, 665
nipalensis, Macaca 267
Melogale 315
Mus 626
Myotis 213
Paguma 343
Prionailurus 295
Semnopithecus 273
niphanae, Megaerops **142**
niphoecus, Ovibos 408
nippon, Cervus **386**
Lutra 312
Mustela 321
Rhinolophus 166
nipponicus, Sus 379
nirnai, Amblonyx 310
nisa, Crocidura 93
nitedula, Dryomys **766**, 766
Mus 766
nitedulus, Oecomys 716
nitela, Eliomys 767
Rhipidomys **745**, 745
Sminthopsis 37
Thallomys 668
nitellinus, Nyctomys 715
nitendiensis, Pteropus **149**
nitens, Eptesicus 201
Pteronura 313
nitida, Ochotona 807
Petaurista 463
Tupaia 133
nitidofulva, Suncus 102, 103
nitidula, Petaurista 463
nitidulus, Mus 623
nitidus, Aeromys 459
Myotis 209
Oryzomys 720, 721, 722, 723, **724**, 724
Pipistrellus 224
Rattus 649, 652, **656**, 656, 658, 662
nitratoides, Dipodomys **479**, 479
nivalis, Arvicola 506
Chionomys **507**, 507
Ischnoglossa 185
Leptonycteris **185**
Microtus 523
Mustela 321, **323**, 323
nivaria, Marmota 431
nivarius, Clethrionomys 508
nivatus, Uropsilus 130
nivea, Procyon 335
niveatus, Funisciurus 428
niveipes, Thomasomys **750**
niveiventer, Mops **236**
Niviventer 635
niveiventris, Callithrix 252
Myomys 631
Peromyscus 735
niveus, Mus 626
Muscardinus 769
nivicola, Chionomys 507
Ovis **409**, 409
Sylvilagus 825
nivicolus, Scotomanes 226
Niviventer 580, 583, 589, 591, 603, 611, **632**, 632, 633, 634, 635, 664
andersoni 589, 632, **633**, 634
atchinensis 635
barussanus 633
baturus 634
besuki 634
blythi 634
brahma 632, **633**, 634
bukit 634, 635
cameroni 635
canorus 633
caudatior 634
champa 635
chihliensis 633
cinnamomeus 634
condorensis 634
confucianus 632, **633**, 633, 634, 635
coninga 633
coxingi 632, **633**, 633, 635
cremoriventer 632, **633**, 634
cretaceiventer 633
culturatus 632, **633**
eha 632, 633, **634**, 634
elegans 633
excelsior 632, 633, **634**
flavipilis 634
flaviventer 633
fraternus 635
fredericae 634
fulvescens 632, 633, **634**, 634, 635
gilbiventer 633
gracilis 634
hinpoon 632, **634**
huang 634
indosinicus 634
jacobsoni 634
jerdoni 634
kina 633
langbianis 632, 633, **634**
lepcha 635
lepidus 634
lepturoides 634
lepturus **634**
lieftincki 634
ling 634
littoreus 633
lotipes 634
luticolor 633
maculipectus 634
malawali 633
marinus 634
mekongis 634
mengurus 633
mentosus 633, 634
minor 634
monticola 635
naoniuensis 633
ninus 634
niveiventer 635
niviventer 632, 633, **635**, 635
octomammis 634
orbus 634
pan 634
quangninhensis 634
rapit 632, **635**
sacer 633
sinianus 633
solus 633
spatulatus 633
sumatrae 633
temmincki 634
tenaster **635**, 635
treubii 634
vientianensis 634
vulpicolor 634
wongi 634
yaoshanensis 633
yushuensis 633
zappeyi 633
niviventer, Melomys 619
Mus 632
Niviventer 632, 633, **635**, 635
Rattus 633
nivosus, Arctocephalus 327
nkatiensis, Eptesicus 201
noacki, Alcelaphus 393
nobilis, Hipposideros 172
Panthera 297
Petaurista **462**, 462
Presbytis 271
noblei, Dasyprocta 781, 782
Noctifelis 294
Noctilio **176**
affinis 176
albiventris **176**
americanus 176
brooksiana 176
cabrerai 176
dorsatus 176
irex 176
labialis 176
leporinus **176**
mastivus 176
mexicanus 176
minor 176
rufescens 176
rufipes 176
rufus 176
unicolor 176
vittatus 176
zaparo 176
Noctilionidae **176**
noctis, Mustela 324
noctivaga, Aotus 255
noctivagans, Lasionycteris **206**
Vespertilio 206
noctivagus, Marmosops **20**
Spalacopus 789
noctula, Nyctalus **217**
nocturna, Potos 334
nodosa, Megaptera 350
Nodus 362
noei, Euneomys 702
nogai, Dipus 493
nogaiorum, Meriones 557
Mus 626
nogalaris, Phyllotis 738
noltei, Crocuta 308
nolthenii, Vandeleuria **672**, 672
Nomascus 274, 275
noomei, Sylvicapra 412
norae, Aethomys 568
Surdisorex **104**, 104
nordcaper, Eubalaena 349
nordenskioldi, Microtus 521
nordhoffi, Protoxerus 436
nordicus, Loris 248
nordmanni, Mus 626
Rhinolophus 165
Sicista 498
norfolcensis, Petaurus **61**, 61
norfolkensis, Mormopterus **238**
norikuranus, Rhinolophus 166
normalis, Ochotona 809
Stenella 356
Ursus 339
norosiensis, Sciurus 441
nortoni, Ursus 339
norvegica, Sicista 496
norvegicus, Clethrionomys 508
Lemmus 517
Mus 649
Rattus 649, 652, **657**, 657, 662
Ursus 339
nosophora, Marmota 432
nostras, Bison 400
notabilis, Protoxerus 436
Ratufa 437
Notagogus 18
Notamacropus 53, 54, 55
notata, Gazella 396
Mephitis 317
notatus, Callosciurus 420, **422**, 422, 423
Canis 280
Dendrolagus 50

Kobus 414
Petaurus 61
Thomasomys **750**
notia, Neotoma 712
notialis, Hydrochaeris 781
Panthera 298
Pseudocheirus 60
notina, Perameles 40
notinus, Bassariscus 334
Notiomys 699, 702, **714**
connectens 699
edwardsii **714**, 714
notioros, Marmota 432
Notiosorex 109, **110**, 110, 111, 122
crawfordi 110, **111**
evotis 111
gigas 109
phillipsii 111
Notiosoricini 110
notius, Eptesicus 201
Mustela 322
Notocitellus **444**
Notoctonus 31
Notodelphys 17
halli 17
notom, Ammospermophilus 420
Notomys **635**, 635
aistoni 636
alexis **635**, 635
alutacea 636
amplus **636**
aquilo **636**
carpentarius 636
cervinus **636**, 636
everardensis 635
eyreius 636
filmeri 636
fuscus **636**
gouldi 636
longicaudatus **636**
macropus 636
macrotis **636**
megalotis 636
mitchellii **636**, 637
mordax **636**, 637
reginae 635
richardsonii 636
sturti 636
Notophorus 380
Notopteris **154**
macdonaldi **154**, 154
neocaledonica 155
Notoryctemorphia 43
Notoryctes **43**
caurinus **43**
typhlops **43**, 43
Notoryctidae **43**
Nototragus 399
nouhuysi, Lorentzimys 606, **607**
nouhuysii, Neophascogale 33
novaeangliae, Balaena 350
Megaptera **350**
novaeguinae, Aselliscus 170
novaeguineae, Planigale **34**
Rattus **649**, **657**
novaehebridensis, Aselliscus 170
novaehollandiae, Dasyurus 32
Echidna 13
Nyctophilus 218
Ornithorhynchus 13
Pseudomys **645**, **648**, 648, 649
Tachyglossus 13
Trichosurus 48
novaehollanidiae, Pseudocheirus 60
novaesi, Cacajao 261
novarae, Rattus 657
noveboracensis, Lasiurus 206
Microtus 527
Mustela 322
Peromyscus 732
novemcinctus, Dasypus 65, **66**
novemlineatus, Tamiops 457
Novibos 401
novikovi, Mustela 322
novioninus, Cervus 386
novo, Microtus 528
novosibiricus, Lemmus 517
nox, Callosciurus 422
nubalis, Lemniscomys 601
nubiana, Capra **405**
nubianus, Canis 280
nubica, Oryx 413
nubicus, Acomys 565
Caracal 289
Panthera 297
nubiensis, Felis 290
nubigena, Presbytis 271
nubila, Akodon 691
nubilans, Arvicanthis 578
nubilus, Acomys 567
Canis 281
Cercopithecus 264
Myotis 212
Otomys 681
Presbytis 271
Taterillus 563
nubiterrae, Peromyscus 732
nublanus, Cervus 386
nubrica, Ochotona **811**, 812
nucella, Scotophilus 227
nuchalis, Dasyprocta 782
Sciurus 440
Thylogale 57
nuchek, Ursus 339
nucifera, Macaca 267
nucus, Akodon 690
nuda, Sorex 112
nudaster, Taphozous 161
nudicauda, Neotoma 712
nudicaudatus, Hypsiprymnodon 49
Metachirus **20**, 20
Spilocuscus 47
nudicaudus, Cabassous 65
Hesperomys 751
Lophuromys **606**
Tylomys **751**, 751
nudicluniatus, Saccolaimus 159
nudifrons, Presbytis 271
nudipalpebra, Axis 384
nudipes, Dasymys **589**, 589, 590
Lutra 312
Mustela 321, **323**, 323
Peromyscus 734
Spermophilus 445
Suncus 102
nudiventris, Hemiechinus **78**, 79
Taphozous **161**
nudoplicatus, Mus 626
nugax, Myotis 213
numantius, Sciurus 443
numarius, Funambulus 427
numidianus, Herpestes 305
numidicus, Herpestes 305
Mustela 323
nummularis, Phoca 332
nunatakensis, Dicrostonyx **510**, 510
nuni, Syncerus 402
nusatenggara, Cynopterus **139**
Nutria 310
nuttalli, Arvicola 715
Ochrotomys **715**
nuttallii, Sylvilagus 813, **826**, 826
nuuanu, Tursiops 357
nuwara, Golunda 592
nux, Eothenomys 514
Scotophilus **227**, 227
Nyala 403
nyama, Tatera 561
nyansae, Crocidura 92
Heliosciurus 429
Sylvicapra 412
nyanzae, Panthera 297
nyasae, Cephalophus 411
Cercopithecus 264
Crocuta 308
Dendromus 543
Potamochoerus 378
Rhabdomys 662
Steatomys 546
Tatera 561
nyassae, Galago 249
Nyctalus **216**, 220, 221, 224
altivolans 217
aviator **216**
azoreum **216**, 217
dasykarpos 217
ferrugineus 216
furvus 217
labiatus 217
lardarius 217
lasiopterus **216**, 216
lebanoticus 217
leisleri **216**, 216, 217
macuanus 217
madeirae 217
magnus 217
maxima 216
mecklenburzevi 217
minima 217
montanus **217**
noctula **217**
pachygnathus 217
palustris 217
plancei 217
princeps 217
proterus 217
rufescens 217
sicula 216
sinensis 217
velutinus 217
verrucosus 216, 217
Nyctereutes **283**
albus 283
amurensis 283
koreensis 283
orestes 283
procyonoides **283**
sinensis 283
stegmanni 283
ussuriensis 283
viverrinus 283
Nycteridae **161**
Nycteris **161**, 206
adana 162
aethiopica 162
affinis 162
albiventer 162
angolensis 162
arge **161**, 162
aurantiaca 162
aurita 161, 162
avakubia 162
baikii 161
bastiani 162
benuensis 162
brockmani 162
capensis 162
damarensis 162
daubentoni 161
discolor 162
fuliginosa 162
gambiensis **161**
grandis **161**
guineensis 162
hispida **161**
intermedia 161, **162**
javanica **162**, 162
labiata 162
luteola 162
macrotis **162**
madagascariensis 162
major **162**, 162
marica 161
martini 161
media 162
najdiya 162
nana **162**
oriana 162
pallida 161, 162

parisii 162
pilosa 161
proxima 161
revoilii 162
sabiensis 162
thebaica **162**
tragata **162**, 162
tristis 162
villosa 161
vinsoni 162
woodi **162**
nycticeboides, Loris 248
Nycticebus **248**
bancanus 248
beugalensis 248
borneanus 248
cinereus 248
coucang **248**
hilleri 248
incanus 248
insularis 248
intermedius 248
javanicus 248
malaianus 248
menagensis 248
natunae 248
ornatus 248
pygmaeus **248**
sumatrensis 248
tardigradus **248**
tenasserimensis 248
Nycticeinops 217, 218
Nycticeius **217**, 217, 226
adovanus 218
africanus 218
albiventer 218
aquilo 217
australis 218
balstoni **217**, 217, 218
bedouin 218
caprenus 217
cinnamomeus 218
crepuscularis 217
cubanus 217
fitzsimonsi 218
greyii **217**
humeralis **217**
influatus 217
mexicanus 217
minimus 218
orion 217
rueppellii **217**
sanborni **217**
schlieffeni **218**
subtropicalis 217
Nycticejus, ornatus 226
Nyctiellus 194, 195
Nyctimene **143**
aello **143**, 144
albiventer **143**, 144
bougainville 144, 145
celaeno 143, **144**
cephalotes **144**, 145
certans **144**, 144
cyclotis **144**, 144
draconilla **144**, 144
geminus 144
lullulae 144
major **144**
malaitensis **144**
masalai **144**
melinus 144
minor 145
minutus **144**
pallasi 144
papuanus 143, 144
rabori **144**
robinsoni **144**
sanctacrucis **144**
scitulus 144
tryoni 144
varius 144
vizcaccia 144, **145**
Nyctimeninae 137
Nyctinomops **238**, 240
aequatoralis 239
affinis 239
aurispinosus **238**
auritus 239
caecus 239
depressus 239
europs 239
femorosaccus **239**
ferruginea 239
gracilis 239
laticaudatus **239**, 240
macarenensis 239
macrotis **239**
megalotis 239
molossa 239
nevadensis 239
similis 238, 239
yucatanica 239
yucatanicus 239
Nyctinomus 232, 237, 240
femorosaccus 238
wroughtoni 239
Nyctipithecus 255
Nyctochoerus 377
Nyctocleptes 685
Nyctomys **715**, 726
colimensis 715
costaricensis 715
decolorus 715
florencei 715
nitellinus 715
pallidulus 715
salvini 715
sumichrasti **715**
venustulus 715
Nyctophilinae 198
Nyctophilus **218**
arnhemensis **218**
australis 218
bicolor 218
bifax 218
daedalus 218
geayi 218
geoffroyi **218**, 218
gouldi **218**, 218
heran **218**
leachii 218
lophorhina 218
major 218
microdon **218**
microtis **218**
novaehollandiae 218
pacificus 218
pallescens 218
sherrini 218
timoriensis **218**
unicolor 218
walkeri **218**
nyctor, Myotis 212
nyikae, Aethomys **568**, 568
Crocidura 91
Dendromus 541, **543**
Madoqua 398
Otomys 680
Pronolagus 824
nyirensis, Microtus 530
nymphaea, Vampyressa **194**
nyula, Herpestes 305
nyx, Callosciurus 423
nzoiae, Kobus 414
nzoyae, Crocuta 308

oaxacae, Dipodomys 480
Marmosa 18
Sorex 120
oaxacensis, Bassariscus 334
Eumops 233
Glaucomys 460
Leopardus 292
Lynx 293
Microtus **526**
Odocoileus 391
Peromyscus 728
oayensis, Crocidura 96
obensis, Arvicola 506
Lemmus 517
obergi, Acinonyx 288
oberlaenderi, Cercocebus 262
obesula, Didelphis 39
obesulus, Isoodon **39**
obesus, Odobenus 326
Psammomys **559**, 559
obfuscata, Nasua 335
obfuscatus, Heliosciurus 429
obiensis, Melomys 615, **618**
Rattus 660
obliquidens, Lagenorhynchus **354**, 354
obliquus, Elasmodontomys **805**, 805
oblitus, Grammomys 594
obliviosus, Rousettus **153**
obolenskii, Dryomys 766
obscura, Cryptotis 109
Delomys 700
Felis 290
Funambulus 427
Lichonycteris **185**, 185
Marmota 432
Ochotona 811
Otomys 680
Tupaia 132
obscurata, Taxidea 325
obscuratus, Maxomys 612
obscurior, Crocidura 82, **92**
Macropus 53
Melanomys 708
Pseudochirops 60
obscuroides, Sorex 118
obscurus, Acerodon 137
Alcelaphus 393
Arctonyx 313
Artibeus 188, **189**
Bolomys **697**
Caenolestes 25
Chaetodipus 484
Clethrionomys 508
Cricetulus 537, 538
Crossarchus 301, **302**, 302
Dipodomys 479
Hipposideros **174**
Lagenorhynchus 351, 353, **354**, 354
Lariscus **430**, 430, 604
Liomys 482
Mesomys **794**
Microtus 519, 524, **526**, 529
Molossus 235
Myotis 216
Ondatra 532
Pipistrellus 224
Praomys 642
Procyon 335
Prosciurillus 436
Rattus 652
Reithrodon 740
Reithrodontomys 743
Rhinolophus 166
Sorex 112, 118, 122
Spermophilus 444
Tamias **454**, 454
Theropithecus 269
Trachypithecus **274**
Tragelaphus 404
observandus, Orycteropus 375
obsidianus, Spermophilus **449**
obsoletus, Rattus 656
Spermophilus 449
obtusa, Phyllonycteris 183
obtusata, Cephalorhynchus 351
Phocaena 358
obtusirostris, Calcochloris **74**
Chrysochloris 74
obvelatus, Sigmodon 747
occasius, Echimys 793
Makalata 793
occidanea, Hystrix 773
occidentalis, Arvicanthis 578
Avahi 246
Canis 281
Cervus 386
Chaeropus 39
Chionomys 507
Clethrionomys 508
Colobus 270

Crocidura 92
Dactylopsila 61
Diceros 372
Dinomys 778
Dipodomys 479
Eliomys 767
Erinaceus 77
Gazella 396
Hapalemur 245
Heliosciurus 429
Lagurus 515
Lepus 817
Lonchorhina 177
Macropus 55
Macrotarsomys 679
Mephitis 317
Microgale 71
Microtus 530
Mops 236
Neotoma 711
Oecomys 716
Oligoryzomys 717
Otomys **681**
Ovis 408
Papio 269
Phalanger 47
Prosciurillus 436
Pseudocheirus 60
Pseudohydromys 598, **644**
Pseudomys **648**, 648
Redunca 414
Rousettus 152
Scotonycteris 153
Sturnira 192
Talpa **129**
Tonatia 181
Tupaia 132
occidentosardinensis, Ovis 408
occiduus, Gerbillus **553**
occipitalis, Mops 237
Ondatra 532
Thomomys **474**
occisor, Mustela 322
occulta, Feresa 352
occultidens, Suncus 103
occultus, Chaetodipus 484
Ctenomys 785, **786**
Myotis 212
Promops 239
Scapanus 127
oceanensis, Miniopterus 231
oceanica, Phoca 332
oceanicus, Peromyscus 736
oceanitis, Hipposideros 172
oceanus, Cervus 387
ocellata, Marmosops 19
ocelot, Leopardus 291
Ochetodon 740
Ochetomys 505
ochotensis, Phoca 332
ochoterenai, Balantiopteryx 156
Ochotona **807**
albata 811
aliensis 811, 812
alpina **807**, 807, 808, 809, 811
altaina 808
angdawai 812
angustifrons 812
annectens 808
argentata 807
ater 807
auritus 810
baltina 810
barnesi 811
bedfordi 808
brookei 808, 809
brooksi 811
brunnescens 811
calloceps 809
cansus **807**, 807, 808, 812, 813
changaica 807
chinensis 810
ciliana 813
cinereoflava 809
cinereofusca 807
cinnamomea 811
clamosa 811
collaris 807, **808**, 808, 811
coreana 809
cuppes 811
curzoniae **808**, 808, 810, 812
dauurica **808**, 808
erythrotis **808**, 809, 810, 812
fenisex 811
ferruginea 809
figginsi 811
forresti **809**, 809, 812
fumosa 811
fuscipes 811
gaoligongensis **809**
gloveri 808, **809**, 809, 810, 811
goldmani 811
griseus 810
hamica 811
himalayana **809**, 812
hodgsoni 811, 812, 813
howelli 811
huangensis 812
hyperborea 807, 808, **809**, 809
iliensis 809, **810**
incana 811
jewetti 811
kamensis 809, 811
kamtschaticus 809
kobayashii 809
kolymensis 809
koslowi **810**
ladacensis **810**
lama 811, 812
lasalensis 811
lemhi 811
levis 811
lhasaensis 811, 812
litoralis 809
littoralis 811
lutescens 811
macrotis **810**, 810
mantchurica 809
melanostoma 808
minimus 811
minor 807, 808
minutus 812
mitchelli 812
moorei 811
morosa 807, 808
muiri 811
muliensis **810**
mursavi 808
nanggenica 812
nepalensis 809, 812
nevadensis 811
nigrescens 811
nitida 807
normalis 809
nubrica **811**, 812
obscura 811
ogotona 811
opaca 811
osgoodi 812
pallasi 807, **811**, 811
pricei 811
princeps 807, 808, **811**, 811
pusilla 809, **811**, 811, 812
regina 812
roylei 807, 809, 810, 811, **812**
rufescens **812**
rutila 808, 809, 810, **812**, 812
sacana 810
sacraria 812
sadakei 809
saturata 811
saxatilis 811
schisticeps 811
scorodumovi 807
seiana 808, 812
septentrionalis 811
sheltoni 811
shukurovi 812
sikimaria 812
sinensis 810
sorella 807
stevensi 807
sunidica 811
sushkini 807
svatoshi 809
syrinx 812
taylori 811
thibetana 807, 808, 809, 811, **812**, 812
thomasi **813**
turuchanensis 809
tutelata 811
uinta 811
uralensis 809
utahensis 811
ventorum 811
vizier 812
vulpina 808
vulturna 812
wardi 812
wasatchensis 811
wollastoni 810
xunhuaensis 812
yezoensis 809
yoshikurai 809
zappeyi 812
Ochotonidae **807**
ochracea, Capreolus 390
Helogale 304
Marmota 432
Neotoma 712
ochraceiventer, Maxomys **613**
ochraceocinereus, Cryptomys **772**
ochraceus, Ailurus 337
Clethrionomys 508
Galerella 303
Grammomys 594
Herpestes 302
Mops 236
Papio 269
Paraxerus **435**
Peromyscus 732
Procyon 335
Proechimys 796
Saguinus 253
Sigmodontomys 748
Tamias 453
ochraphaeus, Acerodon 137
ochraspis, Petaurista 462
ochraventer, Peromyscus 730, **734**
ochreata, Macaca **267**, 267
ochrescens, Sciurus 441
ochrinus, Sigmodontomys 748
ochrocephala, Pithecia 262
ochrogaster, Funisciurus 427
Microtus **526**, 526, 528
Rhipidomys 744, **745**
ochrogenys, Tamias **454**, 456
ochrognathus, Sigmodon **748**
ochroleucus, Pan 277
ochromelas, Galerella 303
ochromixtus, Pipistrellus 223
Ochromys 673
ochropoides, Lepus 816
ochropus, Arvicanthis 578
Caluromys 15
Canis 280
Dendromus 543
Lepus 816
Spilocuscus 47
ochrotis, Chroeomys 700
Ochrotomys **715**, 715, 728
aureolus 715
flammeus 715
lewisi 715
lisae 715
nuttalli **715**
ochrourus, Odocoileus 391
ochroxantha, Vulpes 287
ochrus, Chaetodipus 483

ocius, Thomomys 475
oconnelli, Oryzomys 720
Proechimys 796, **797**
ocotepequensis, Reithrodontomys 742
ocreata, Felis 290
octavus, Bassariscus 334
octocinctus, Chaetophractus 65
octocintus, Dasypus 66
octodecimcinctus, Cabassous 65
Octodon **788**, 788
alba 788
bridgesi **788**, 788
clivorum 788
cumingi 788
cumingii 788
degus **788**
getulus 788
gliroides 788
kummingii 788
lunatus **788**
pallidus 788
peruana 788
Octodontidae 783, **787**
Octodontoidea 771
Octodontomys 783, **788**
gliroides **788**
simonsi 788
octomammis, Niviventer 634
octomastis, Praomys 642
Octomys 649, **788**, 789
barrerae 788, 789
joannius 788
mimax **788**, 788
octonata, Phoca 332
Octopodomys 673
ocularis, Elephantulus 829
Eulemur 244
Graphiurus **765**
Pteropus **149**
Sciurus 763
Sylvicapra 412
oculatus, Sciurus **442**
oculeus, Lynx 293
ocydromus, Macropus 54
ocypete, Aethalops 138
ocythous, Urocyon 285
odessana, Spermophilus 449
Odobenidae **325**
Odobeninae 326
Odobenus **325**, 326
arcticus 326
cookii 326
divergens 326
laptevi 326
obesus 326
odobenus 325
orientalis 326
rosmarus **326**
odobenus, Odobenus 325
Odocoileus 389, 390, **391**, 392
abeli 391
acapulcensis 391
aequatorialis 391
antonii 391
auritus 391
battyi 391
borealis 391
brachyceros 391
californicus 391
campestris 391
canus 391
cariacou 391
carminis 391
cerrosensis 391
chiriquensis 391
clavatus 391
clavium 391
columbianus 391
columbicus 391
consul 391
costaricensis 391
couesi 391
crooki 391
curassavicus 391
dacotensis 391
eremicus 391
fraterculus 391
fuliginatus 391
goudotii 391
gracilis 391
gymnotis 391
hemionus **391**
hiltonensis 391
inyoensis 391
lasiotis 391
leucurus 391
lewisii 391
lichtensteini 391
louisianae 391
macrotis 391
macrourus 391
margaritae 391
mayensis 391
mcilhennyi 391
mexicanus 391
miquihuanensis 391
montanus 391
nelsoni 391
nemoralis 391
nigribarbis 391
oaxacensis 391
ochrourus 391
osceola 391
peninsulae 391
peruvianus 391
philippii 391
punctulatus 391
pusilla 391
richardsoni 391
rothschildi 391
savannarum 391
scaphiotus 391
seminolus 391
sheldoni 391
sinaloae 391
sitkensis 391
spelaeus 391
spinosus 391
suacuapara 391
sylvaticus 391
taurinsulae 391
texanus 391
thomasi 391
toltecus 391
tropicalis 391
truei 391
ustus 391
venatorius 391
veraecrucis 391
virginianus **391**, 391
virgultus 391
wiegmanni 391
wisconsinensis 391
yucatanensis 391
Odontoceti **349**
Odontodorcus 383
odorata, Crocidura 92
Odorlemur 245
odrob, Redunca **414**
Oecomys **715**, 715, 719, 725, 743
auyantepui 716
bahiensis 717
benevolens 716
bicolor **715**, 715, 716
caicarae 716
catherinae 717, 725
cinnamomeus 717
cleberi **716**
concolor 715, **716**
dryas 716
endersi 716
flavicans **716**
florenciae 716
frontalis 717
fulviventer 717
guianae 716
helvolus 717
illectus 716
klagesi 717
mamorae **716**
marmosurus 716
melleus 716
milleri 716
mincae 716
nitedulus 716
occidentalis 716
osgoodi 717
palmarius 717
palmeri 716
paricola **716**, 716
phaeotis **716**
phelpsi 716
regalis 716
rex **716**
roberti **716**
rosilla 716
rutilus **716**
speciosus **716**
splendens 717
subluteus 717
superans **716**
tapajinus 716
tectus 717
trabeatus 716
trichurus 716
trinitatis 715, **716**
vicencianus 717
oeconomus, Chionomys 507
Microtus 523, 525, **526**, 527
Oedipomidus 253
Oedipus 253
oedipus, Saguinus 253, **254**, 254
Oedocephalus 773
oemodias, Lepus 819
oenanthe, Callicebus **259**
oenax, Thomasomys 752
Wilfredomys **752**, 752
oeningensis, Anoema 813
Oenomys 578, 600, **637**, 669
albiventris 637
anchietae 637
bacchante 637
editus 637
hypoxanthus **637**, 637
marungensis 637
moerens 637
oris 637
ornatus **637**, 637
rufinus 637
talangae 637
unyori 637
vallicola 637
oenone, Funisciurus 427
oerstedii, Saimiri **260**, 260
oertzeni, Pan 277
Vulpes 286
offella, Boromys **799**, 799
officialis, Cervus 387
ogasimanus, Rhinolophus 166
ogilbii, Pardofelis 299
ogilbyi, Bettongia 49
Cephalophus **411**
Cynictis 302
Leptailurus 292
Paguma 343
ogilvyensis, Rangifer 392
ognevi, Allactaga 489
Apodemus 570
Arvicola 506
Capra 405
Clethrionomys 508
Cricetulus 538
Crocidura 95
Dryomys 766
Ellobius 513
Eptesicus 200
Lagenorhynchus 354
Lemmus 516
Marmota 431
Microtus 518
Murina 229
Plecotus 224
Pteromys 464
Sciurus 443
Spermophilus 449
Ognevia 111, 118

Ogotoma 807
ogotona, Ochotona 811
ohiensis, Rattus 664
Srilankamys **664**, 664
ohioensis, Microtus 526
Tamias 456
ohionensis, Sorex 113
oi, Sus 378
oiostolus, Lepus 816, **819**, 821
okadae, Tamias 455
okanagana, Marmota 431
Okapia **383**
erikssoni 383
johnstoni **383**
kibalensis 383
liebrechtsi 383
tigrinum 383
okavangensis, Galerella 303
okenii, Eptesicus 203
okiensis, Lepus 815
Phaulomys 533
okinavensis, Diplothrix 591
okinoshimae, Crocidura 84
oklahomae, Dipodomys 479
Pipistrellus 221
okuensis, Lamottemys **600**, 600
Myosorex 99, **100**
olallae, Callicebus **259**
Olallamys **790**
albicauda **790**
apolinari 791
edax **791**
olchonensis, Alticola 504
oleracea, Vandeleuria **672**, 672, 673
oleraceus, Mus 672
olga, Graphiurus **765**, 765
Lemniscomys 601
olida, Mephitis 317
Olidosus 380
Oligoryzomys 708, **717**, 718, 719
agilis 718
amblyrrhynchus 718
andinus **717**, 717
antoniae 717
araucanus 718
arenalis **717**
chacoensis **717**
chaparensis 718
commutatus 718
coppingeri 718
costaricensis 718
creper 718
delicatus 718
delticola **717**
destructor **717**, 718
diminutivus 718
dumetorum 718
eliurus **717**, 718
engraciae 718
exiguus 718
flavescens **717**
fornesi 718
fulvescens 717, **718**, 718
glaphyrus 718
griseolus **718**
lenis 718
longicaudatus **718**, 718
macrocercus 718
magellanicus **718**, 718
maranonicus 717
mattogrossae 718
mayensis 718
melaenus 718
melanizon 718
melanostoma 717
messorius 718
microtis **718**, 718
mizurus 718
munchiquensis 718
navus 718
nicaraguae 718
nigribarbis 718
nigripes 717, **718**, 718
occidentalis 717
pacificus 718
pernix 718
peteroanus 718
philippii 718
pygmaeus 717
reventazoni 718
saltator 718
spodiurus 717
stolzmanni 717
tectus 719
tenuipes 718
utiaritensis 717
vegetus **718**, 718
victus **719**, 719
Olisthomys 463, 464
olitor, Eothenomys **514**
olivacea, Mustela 322
Nasuella **335**
Salanoia 300
Tupaia 131
olivaceofuscus, Molossus 235
olivaceogriseus, Perognathus 485
olivaceous, Spermophilus 450
olivaceus, Akodon **692**, 692
Cebus **260**
Cryptotis 109
Dipodomys 479
Hapalemur 245
Lagothrix 258
Nectomys 709
Papio 269
Paraxerus 435
Perognathus 486
Procolobus 272
Tamiops 457
Tragelaphus 404
olivascens, Sciurus 439
olivellus, Funisciurus 427
oliviae, Cricetomys 541
Funisciurus 428
olivieri, Crocidura 84, 86, **92**, 92, 98
Sus 379
olivinus, Melanomys 708
ollula, Sylvisorex **105**, 105
olrogi, Andalgalomys **694**, 694
olympica, Aplodontia **417**
Mustela 321, 324
Rupicapra 410
Spilogale 317
Talpa 128
olympicus, Glaucomys 460
Phenacomys 533
olympius, Chionomys 507
Funambulus 426
olympus, Marmota **432**
Puma 296
omanensis, Eptesicus 200
Lepus 816
omari, Myotis 209
omensis, Cercopithecus 264
Heliosciurus 428
Tachyoryctes 687
omichlodes, Stenomys 665
omissus, Cercopithecus 263
Ommatophoca 329, **331**
rossii **331**, 331
Ommatostergus 754
omnivora, Bdeogale 301
Omoloxodon 367
omurambae, Sylvicapra 412
Onager 369
onager, Equus **370**, 370
onca, Panthera **297**
Oncifelis 291, 292, **294**, 294
albescens 294
braccata 294
budini 294
colocolo 291, **294**, 294
crespoi 294
crucina 294
euxanthus 294
garleppi 294
geoffroyi 291, 292, **294**, 294
guigna 291, 292, **294**, 294
guina 294
himalayanus 294
huina 294
leucobaptus 294
macdonaldi 294
melas 294
molinae 294
munoai 294
neumayeri 294
pajeros 294
pampa 294
pampanus 294
paraguae 294
pardoides 294
passerum 294
salinarum 294
santacrucensis 294
steinbachi 294
thomasi 294
tigrillo 294
warwickii 294
Oncilla 291
oncilla, Leopardus 292
Oncoides 291
Ondatra 531, **532**
albus 532
americana 532
aquilonius 532
bernardi 532
cinnamominus 532
goldmani 532
hudsonius 532
macrodon 532
maculosa 532
mergens 532
niger 532
nigra 532
obscurus 532
occipitalis 532
osoyoosensis 532
pallidus 532
ripensis 532
rivalicius 532
spatulatus 532
varius 532
zalophus 532
zibethicus **532**
oneida, Microtus 527
oniscus, Hemiechinus 78
Microtus 522
Oryzomys 721, **724**
Onotragus 413
onssa, Panthera 298
ontarioensis, Zapus 499
Onychogalea **55**
annulicauda 55
fraenata **55**
frenatus 55
lunata **55**
unguifera **55**
Onychomys **719**, 728, 752
albescens 719
arcticeps 719
arenicola **719**, 719
breviauritus 719
brevicaudus 719
canus 719
capitulatus 719
clarus 719
durranti 719
fuliginosus 719
fuscogriseus 719
knoxjonesi 719
leucogaster **719**, 719
longicaudus 719
longipes 719
macrotis 719
melanophrys 719
missouriensis 719
pallescens 719
pallidus 719
perpallidus 719
pulcher 719
ramona 719
ruidosae 719
surrufus 719
torridus **719**, 719
tularensis 719

utahensis 719
yakiensis 719
onza, Panthera 298
ooldea, Sminthopsis **37**
opaca, Ochotona 811
opdenboschi, Lophocebus 266
operarius, Microtus 527
Tamias 454
Thomomys 474
operculatus, Connochaetes 393
operosa, Tupaia 132
operosus, Thomomys 474
ophiodon, Scotonycteris **153**
ophion, Ovis 408
ophiusae, Eliomys 767
ophrus, Ursus 339
opimus, Ctenomys **786**, 786
Dremomys 425
Meriones 559
Rhombomys **559**
Steatomys 546
opistholeuca, Tamandua 68
opisthomelas, Macaca 268
Tamandua 68
opisthostictus, Cercopithecus 264
Oplacerus 391
Opolemur 243
oporaphilum, Sturnira 192
Opossum 16, 18
opossum, Didelphis 22
Metachirops 22
Philander **22**
Opposum 45
optabilis, Thomomys 474
optiva, Lontra 311
opulentus, Thomomys 474
oquirrhensis, Thomomys 475
oral, Petaurista 463
oralis, Gerbillurus 547
Ictonyx 319
Jaculus 493
Pseudocheirus 60
Pseudomys **648**
oramontis, Phenacomys 533
orangensis, Lepus 820
orangiae, Cryptomys 772
Ictonyx 319
Mus **627**
Pedetes 759
Procavia 374
Rhabdomys 662
Steatomys 545
Suncus 103, 104
orangiensis, Amblysomus 74
orarius, Scapanus **127**
Tamiasciurus 456
Zapus 499
Orasius 383
oratus, Platyrrhinus 191
orbitalis, Liomys 482
orbitus, Lemmiscus 516
orbus, Bolomys 696
Niviventer 634
Orca 354
intermedia 352
orca, Clethrionomys 509
Delphinus 354
Orcinus 353, **354**
orcadensis, Microtus 519
Orcaella 351, **354**, 354, 357
brevirostris **354**, 354
fluminalis 354
Orcinae 351
Orcinus **354**
ater 355
capensis 355
gladiator 355
orca 353, **354**
rectipinna 355
ordii, Dipodomys **477**, **479**
ordinalis, Tamias 455
Oreailurus **294**, 294
jacobita **294**
Oreamnos **407**
americanus **407**, 407
columbiae 407
columbiana 407
dorsata 407
kennedyi 407
lanigera 407
missoulae 407
montanus 407
sericea 407
Oreas 403
oreas, Antilope 403
Dendromus 542, **543**, 543
Microtus 525
Oryctolagus 822
Peromyscus 732, **734**, 734, 736
Petaurista 462
Taurotragus 403
Thomasomys **750**
oregona, Antilocapra 393
oregonensis, Glaucomys 460
Mustela 322
Myotis 209
Puma 296
oregoni, Microtus **527**, 527
oregonus, Bassariscus 334
Bison 400
Lepus 814
Microdipodops 480
Spermophilus 445
Thomomys 475
Zapus 499
oreias, Myotis **214**
oreigenus, Phyllotis 739
Oreinomys 680
oreinus, Sorex 118
oreocetes, Pappogeomys 472
Tamias 454
Oreodorcas 414
oreoecus, Thomomys 474
Oreomys 680
Oreonax 257
oreopolus, Sorex 114, **118**, 118, 122
Oreosciurus 439
Oreotragus **398**
aceratos 399
aureus 399
centralis 399
cunenensis 399
hyatti 399
klippspringer 399
megalotis 395
oreotragus **398**
porteousi 398
porteusi 399
saltator 399
saltatrixoides 399
schillingsi 399
somalicus 399
steinhardti 399
stevensoni 399
transvaalensis 399
tyleri 399
oreotragus, Antilope 398
Oreotragus **398**
oresterus, Heteromys **481**, 482
Peromyscus 732
orestes, Apodemus 571
Callosciurus 420, **423**
Galerella 303
Nyctereutes 283
Otomys 682
orgiloides, Ursus 339
orgilos, Ursus 339
oriana, Nycteris 162
orianae, Miniopterus 231
oribasus, Mustela 322
Ursus 339
oribates, Chaetodipus 484
oricolus, Spermophilus 449
oriens, Hydromys 597
orientalis, Akodon 692
Anomalurus 757
Avahi 246
Canis 281
Cephalophus 410
Cercopithecus 263
Civettictis 345
Crocidura 85
Didelphis 46
Ellobius 513
Enhydra 310
Erinaceus 77
Gazella 396
Geogale 70
Hyaena 308
Ichthyomys 705
Jaculus 493, **494**, 494
Lemniscomys 601, 602
Lynx 293
Melogale **314**, 314, 315
Microtus 526
Mus 626
Myoxus 770
Nasalis 270
Neomys 110
Odobenus 326
Otomys 680
Ovis 408, 409
Panthera 298
Papio 269
Peromyscus 734
Phalanger **46**, 46
Pipistrellus 224
Procapra 399
Rattus 662
Stenella 356, 357
Talpa 129
Tamias 455
Vespertilio 228
Viverra 348
Orientallactaga 488, 489
orientis, Crocidura 96
Mops 236
Myotis 207
Neomys 110
Sciurus 443
origenes, Martes 319
orii, Crocidura 84, **92**
Lepus 821
Mus 626
Myotis 213
Pteromys 464
Rattus 657
Rhinolophus 165
Sorex 114
orinocensis, Lonchorhina **178**
orinoci, Isothrix 793
Monodelphis 21
Sylvilagus 826
orinomus, Bassaricyon 333
Urocyon 285
orinops, Macaca 267
orinus, Galagoides 249
Hylopetes 460
Oryzomys 720
Reithrodontomys 742
orioecus, Microtus 518
oriomo, Echymipera 41
oriomos, Thylogale 57
orion, Canis 281
Nycticeius 217
oris, Oenomys 637
Proechimys 796, **797**
oritis, Crocidura 88, 90
oriundus, Sylvisorex **105**, 105
orizabae, Neotoma 712
Neotomodon 713
Peromyscus 734
Reithrodontomys 740
Sorex 118
Sylvilagus 826
Thomomys 476
orleansi, Loxodonta 367
orlovi, Spermophilus 446, 448
ornata, Ammotragus 404
Caluromys 15
Felis 290
Rupicapra 409
Vormela 325
ornatulus, Rattus 652
ornatus, Anthops **169**, 169
Ateles 257
Callicebus 258

Chlamyphorus 64
Dipodomys 480
Galerella 303
Galidictis 300
Hipposideros 172
Lasiurus 206
Microperoryctes 41
Nycticebus 248
Nycticejus 226
Oenomys **637**, 637
Paraxerus 435
Phalanger **47**
Pteropus **149**
Scotomanes **226**, 226
Sorex **118**
Sundasciurus 451
Tragelaphus 404
Tremarctos **338**
Ursus 337
Vampyrodes 194
Ornithorhynchidae **13**
Ornithorhynchus **13**
anatinus **13**
brevirostris 13
crispus 13
fuscus 13
laevis 13
novaehollandiae 13
paradoxus 13
phoxinus 13
rufus 13
triton 13
Ornithorynchus 13
Ornoryctes 41
orobinus, Bolomys 696
oroensis, Marmosops 19
Melanomys 708
Orolestes 25
inca 25
orolestes, Neotoma 711
Oromys 622
oronensis, Phoca 332
oronocensis, Trichechus 365
orophila, Cryptotis 109
orophilus, Akodon **692**, 692, 693
Phenacomys 533
orsinii, Miniopterus 231
Orthaegoceros 405
Orthogeomys 469, **470**, 470, 471
alleni 471
alvarezi 471
annexus 471
carbo 471
carlosensis 470
cartagoensis 471
cavator **470**
cayoensis 471
cherriei **470**
chiapensis 471
concavus 471
costaricensis 470
cuniculus **470**
dariensis **470**, 471
dolichocephalus 471
engelhardi 471
felipensis 471
grandis **471**, 471
guerrerensis 471
heterodus **471**
hispidus **471**, 471
hondurensis 471
huixtlae 471
isthmicus 471
lanius **471**, 471
latifrons 471
latirostris 471
matagalpae **471**
negatus 471
nelsoni 471
nigrescens 470
pansa 470
pluto 471
pygacanthus 471
scalops 471
soconuscensis 471
teapensis 471
tehuantepecus 471
thaeleri **471**, 471
torridus 471
underwoodi **471**
vulcani 471
yucantanensis 471
orthopodicus, Cervus 386
orthopus, Cervus 386
orthos, Dasymys 589
orthotis, Eumops 233
Orthriomys 517, 531
orycter, Akodon 692
Orycteromys 783
Orycteropodidae **375**
Orycteropus **375**, 375
adametzi 375
aethiopicus 375
afer **375**
albicaudus 375
angolensis 375
erikssoni 375
faradjius 375
haussanus 375
kordofanicus 375
lademanni 375
leptodon 375
matschiei 375
observandus 375
ruvanensis 375
senegalensis 375
somalicus 375
wardi 375
wertheri 375
Orycterus 771
Oryctogale 316
Oryctolagus **822**, 824
algirus 822
borkumensis 822
brachyotus 822
campestris 822
cnossius 822
cuniculus **822**
fodiens 822
habetensis 822
huxleyi 822
kreyenbergi 822
nigripes 822
oreas 822
vermicula 822
vernicularis 822
Oryx **413**
algazel 413
annectens 413
aschenborni 413
asiatica 413
beatrix 413
beisa 413
bezoartica 413
blainei 413
callotis 413
capensis 413
dammah **413**
ensicornis 413
gallarum 413
gazella **413**, 413
latipes 413
leucoryx **413**
nubica 413
pallasii 413
pasan 413
recticornis 413
senegalensis 413
subcallotis 413
tao 413
oryx, Antilope 403, 413
Taurotragus **403**, 403
oryzivora, Oryzomys 724
oryzivorus, Micromys 620
Oryzomys 687, 707, 708, 709, 713, 715, 717, **719**, 719, 722, 739, 743, 748, 751
agrestis 720
albigularis **720**, 720, 722, 723
albiventer 721
alfaroi **720**, 720, 721, 724, 725
alleni 720
angouya 721
angusticeps 724
anguya 721
antillarum 721
apatelius 721
aquaticus 721
argentatus 724
auriventer **720**, 720
aztecus 721
azuerensis 721
balneator **720**, 722
baroni 725
bauri 722
bolivaris **720**, 720, 721
boliviae 720, 724
bombycinus 720
brevicauda 752
buccinatus **721**, 721, 724, 725
bulleri 721
capito **721**, 721, 722, 723, 724, 725
caracolus 720
carrikeri 725
carrorum 725
castaneus 720
caudatus 721
cephalotes 721
chapmani 720, **721**, 721, 725
cherriei 752
childi 720
colimensis 723
coloratus 724
couesi **721**, 721, 722, 724
cozumelae 721
crinitus 721
dariensis 720
devius 720, **722**, 722
dilutior 721
dimidiatus **722**
ferrugineus 737
fornesi 718
fulgens 721
galapagoensis **722**, 722, 725
gatunensis 721
gloriaensis 720
goeldi 721
goldmani 721
gorgasi **722**
gracilis 720
guerrerensis 721
hammondi 707, **722**
hesperus 720
huastecae 721
hylocetes 725
ica 725
incertus 720, 723
indefessus 713
instans 699
intagensis 720
intectus **722**
intermedius 721, **722**, 722, 723, 724
jalapae 721
keaysi 720, **723**, 723
kelloggi **723**
lambi 721
lamia **723**
laticeps 721, 722, 725
legatus 721, **723**, 723, 724
levipes 720, **723**, 723
longicaudatus 717, 718
lugens 687
macconnelli 721, **723**, 723
maculiventer 720
magdalenae 725
medius 725
megacephalus 721
megadon 725
melanotis 720, **723**, 723, 725
meridensis 720
mexicanus 721
minutus 708
modestus 721

moerex 720
mollipilosus 725
moojeni 724
mureliae 723
natator 724
navus 717
nelsoni **723**
nimbosus 720
nitidus 720, 721, 722, 723, **724**, 724
oconnelli 720
oniscus 721, **724**
orinus 720
oryzivora 724
palatinus 720
palmirae 720
palustris 721, 722, **724**, 724
panamensis 725
paraganus 724
pectoralis 720
peninsulae 721
peragrus 721
perenensis 721
phaeopus 707
pinicola 721
pirrensis 720
planirostris 724
polius **724**
ratticeps 721, **724**
regillus 721
rex 724, 725
rhabdops 720, **724**
richardsoni 721
richmondi 721
rivularis 720, 721
rostratus 720, 723, **724**, 725
rufinus 721
rufus 721
saltator 721
salvadorensis 725
sanibeli 724
saturatior 720, 721, **725**, 725
subflavus 721, **725**, 725
sylvaticus 725
talamancae 721, **725**
teapensis 721
texensis 721, 724
tropicius 724
velutinus 721
villosus 725
vulpinoides 725
wavrini 739
xantheolus 722, 724, **725**
yucatanensis 725
yunganus 721, **725**
zygomaticus 721
Oryzorictes 71, **72**, 72
hova **72**, 72
niger 72
talpoides **72**
tetradactylus **72**, 72
Oryzorictinae **70**
oryzus, Axis 385
osagensis, Neotoma 711
osborni, Clidomys **805**, 805
Mops 236
Naemorhedus 407
Rangifer 392
Osbornictis 341, **347**
piscivora **347**, 347
osburni, Pteronotus 177
oscari, Alcelaphus 393
osceola, Eptesicus 201
Odocoileus 391
oseryi, Aotus 256
oseticus, Chionomys 507
osgoodi, Amorphochilus 195
Cavia 779
Cricetomys 541
Cryptotis 109
Ctenomys 785
Monodelphis **21**
Ochotona 812
Oecomys 717
Peromyscus 732
Phyllotis **738**, 738
Puma 296
Rattus 649, 654, **657**
Rhinolophus **167**
Spermophilus 448
Taterillus 563
Thomomys 474
Osgoodomys **725**, 728
angelensis 726
banderanus **726**, 726, 734
coatlanensis 726
sloeops 726
vicinior 726
osilae, Phyllotis **738**
osimensis, Rattus 670
Tokudaia **670**, 670
osiris, Gazella 396
osiui, Petaurista 462
osmani, Lophocebus 266
osorio, Crocidura **92**
osorninus, Rattus 658
osoyoosensis, Ondatra 532
Osphranter 53, 54, 55
oswellii, Ceratotherium 371
Otailurus 289
Otaria 327, **328**
byronia **328**
californiana 327, 329
chilensis 328
chonotica 328
cinerea 328
flavescens 328
fluva 328
gilliespii 329
godeffroyi 328
hookeri 328
leonina 328
minor 328
molossina 328
pernettyi 328
pygmaea 328
rufa 328
scont 328
ulloae 328
uraniae 328
velutina 328
weddellii 330
Otariidae 325, **326**
otarius, Tatera 561
Otelaphus 391
othai, Euroscaptor 125
othus, Lepus **819**, 819, 821
otinus, Microsciurus 433
otiosus, Sundamys 666
Otisorex 111, 112, 113, 114, 115, 116, 117, 118, 119, 120, 121, 122
Otocebus 259
Otocolobus **295**, 295, 444
ferrugineus 295
manul **295**
mongolicus 295
nigripectus 295
satuni 295
Otocyon **283**
auritus 284
caffer 283, 284
canescens 284
lalandi 284
megalotis **284**
steinhardti 284
virgatus 284
Otogale 250
pallida 248
Otognosis 484
Otolemur **250**
agyisymbanus 250
argentatus 250
badius 250
crassicaudatus **250**, 250
garnettii **250**, 250
hindei 250
hindsi 250
kikuyuensis 250
kirkii 250
lasiotis 250
lasiotus 250
lestradei 250
lonnbergi 250
monteiri 250
panganiensis 250
umbrosus 250
zuluensis 250
otoleucus, Cercopithecus 264
Otolicnus 249
garnetti 250
otomoi, Rattus 657
Otomops **239**
formosus **239**
icarus 239
madagascariensis 239
martiensseni **239**
papuensis **239**, 239
secundus **239**
wroughtoni **239**
Otomyinae 501, **679**
Otomys 544, **680**, 680, 681, 682
albaniensis 682
albicaudatus 676
anchietae **680**, 680
angoniensis **680**, 680, 681
auratus 680
barbouri 680
basuticus 681
bergensis 682
bisulcatus 680
broomi 682
burtoni 681
canescens 680
capensis 680
coenosus 680
cuanzensis 681
cupreoides 680
cupreus 680
dartmouthi 682
davisi 681
degeni 682
denti **680**
divinorum 680
dollmani 681
elassodon 680
elgonis 681
fannini 681
faradjius 681
fortior 682
ghigii 681
giloensis 682
grantii 682
helleri 682
irroratus **680**, 680, 681, 682
jacksoni 682
jeppei 681
karoensis 681
kempi 680
lacustris 680
laminatus 680, **681**
malkensis 682
malleus 682
mariepsi 681
mashona 680
maximus 680, **681**
natalensis 680
nubilus 681
nyikae 680
obscura 680
occidentalis **681**
orestes 682
orientalis 680
percivali 682
pondoensis 681
pretoriae 680
randensis 680
robertsi 681
rowleyi 680
ruberculus 681
sabiensis 680
saundersiae **681**
silberbaueri 681
sloggetti 680, **681**
squalus 682
sungae 680
thomasi 682
tropicalis 680, **681**, 681, 682

tugelensis 680, 681
turneri 681
typicus 544, 680
typus 606, 680, **682**, 682
unisulcatus 680, 681, **682**, 682
uzungwensis 682
vivax 681
vulcanicus 681
zinki 682
Otonycteris **219**
cinerea 219
hemprichii **219**, 219
jin 219
leucophaeus 219
petersi 219
saharae 219
Otonyctomys **726**
hatti **726**, 726
Otopithecus 263
Otopteropus **145**
cartilagonodus **145**, 145
Otosciurus 439
Otospermophilini 419
Otospermophilus 444, 445, 447, 449, 450
Ototylomys **726**, 751
affinis 726
australis 726
brevirostris 726
connectens 726
fumeus 726
guatemalae 726
phaeus 726
phyllotis **726**, 726
otteni, Rattus 652
ottleyi, Aepeomys 688
otus, Clethrionomys 509
Plecotus 224
Reithrodontomys 742
Ouakaria 261
ouakary, Cacajao 261
ouangthomae, Rattus 660
oubanguii, Mus 626, **627**, 628
Ouistitis 251
Oulodon 362
ouralensis, Microtus 527
ourebi, Antilope 399
Ourebia **399**
Ourebia **399**
aequatoria 399
brevicaudata 399
cottoni 399
dorcas 399
gallarum 399
goslingi 399
grayi 399
haggardi 399
hastata 399
kenyae 399
leucopus 399
masakensis 399
melanura 399
microdon 399
montana 399
nigricaudata 399
ourebi **399**
pitmani 399
quadriscopa 399
rutila 399
scoparia 399
smithii 399
splendida 399
ugandae 399
oustaleti, Procolobus 272
outreyanus, Cervus 387
ouwensi, Mus 623
Petinomys 464
ovamboensis, Cryptomys 771
Ictonyx 319
Mastomys 611
ovchepolensis, Nannospalax 754
Ovibos **407**
mackenzianus 408
melvillensis 408
moschatus **408**
niphoecus 408
pearyi 408
wardi 408
Ovis 405, **408**
adametzi 408
alleni 409
altaica 408
ammon **408**, 408
ammonoides 408
anatolica 408
arabica 409
argali 408
aries **408**, 408, 409
arkal 409
armeniana 408
asiaticus 408
auduboni 408
bambhera 408
blanfordi 409
blythi 408
bochariensis 409
borealis 409
brookei 408
californiana 408
canadensis **408**, 408, 409
cervina 408
collium 408
comosa 408
corsicosardinensis 408
cowani 409
cremnobates 408
cycloceros 409
cyprius 408
dalailamae 408
dalli **409**
darwini 408
dauricus 408
dolgopolovi 409
erskinei 408
fannini 409
gaillardi 408
gmelini 408
heinsii 408
henrii 408
hodgsoni 408
hodgsonii 408
humei 408
intermedia 408
isphaganica 408
jubata 408
karelini 408
kenaiensis 409
koriakorum 409
kozlovi 408
laristanica 408
liardensis 409
littledalei 408
lydekkeri 409
matschiei 408
mexicana 408
middendorfi 409
mongolica 408
montana 408
musimon 408
nayaur 409
nelsoni 408
niger 409
nigrimontana 408
nivicola **409**, 409
occidentalis 408
occidentosardinensis 408
ophion 408
orientalis 408, 409
palmeri 408
polii 408
potanini 409
przevalskii 408
punjabiensis 409
pygargus 408
sairensis 408
samilkameenensis 408
severtzovi 409
sheldoni 408
sierrae 408
sinesella 408
stonei 409
storcki 409
texianus 408
urmiana 408
varentsowi 409
vignei 408, **409**, 409
weemsi 408
oweni, Atelerix 76
Lemniscomys 601
Myomys 631
Poelagus 823
Procavia 374
Scapanulus **127**, 127
Uranomys 671
owenii, Pongo 277
Trichechus 366
owensi, Callosciurus 423
owiensis, Rattus 653
owstoni, Chrotogale **341**, 341
Dremomys 425
Sciurotamias 439
owyhensis, Thomomys 475
oxalis, Myotis 216
oxiana, Lutra 312
oxianus, Meriones 556
Spermophilus 446
oxycephalus, Cervus 386
oxygnathus, Myotis 209
Oxymycterus 706, 707, **726**, 726, 727
akodontius **726**
angularis **726**
apicalis 706
breviceps 696
dasytrichus 727
delator **726**, 727
doris 727
hiska **726**, 727
hispidus **727**
hucucha **727**, 727
iheringi **727**, 727
inca **727**
iris 727
jacentior 727
judex 727
juliacae 727
mimus 689
misionalis 727
nasutus **727**, 727
nigrifrons 727
paramensis 726, **727**
platensis 727
quaestor 727
roberti **727**
rostellatus 727
rufus **727**, 727
rutilans 727
valdivianus 702
vulpinus 706
oxyodontus, Sus 379
oxyotis, Loxodonta 367
oxyotus, Myotis **214**
Oxyrhin 111
oxyrrhinus, Mus 626
Oxystomus 365
oxytona, Marmota 431
oyaensis, Microtus 519
oyapocki, Neusticomys **714**
Ozanna 412
ozarkensis, Blarina 106
ozarkiarum, Peromyscus 732
Ozelaphus 391
ozensis, Myotis **214**
Ozolictis 316
Ozotoceros **391**, 392
albus 392
azarae 392
bezoarticus **392**
caenosus 392
campestris 392
celer 392
comosus 392
cuguapara 392
dickii 392
fossilis 392
leucogaster 392
pampaeus 392
sylvestris 392

Paca 783
paca, Agouti **783**

Mus 783
pacarana, Dinomys 778
pacator, Saguinus 253
paccerois, Tetracerus 403
pachita, Proechimys 797
pachycephalus, Reithrodon 740
pachycercus, Mus 626
Pachyceros 408
pachygnathus, Nyctalus 217
pachyomus, Eptesicus 203
pachyotis, Eptesicus **202**
Pachyotus 226
pachypus, Tylonycteris **228**
Vespertilio 228
pachyrhynchus, Halichoerus 330
Pachysoma 145
ecaudata 142
Pachyura 101
Pachyuromys **559**
duprasi **559**, 559
faroulti 559
natronensis 559
pachyurus, Rattus 655
Sorex 112
Suncus 102
pacifica, Aplodontia 417
Lontra 311
Martes 320
Procyon 335
pacificus, Akodon 689
Antrozous 198
Heteromys 481
Indopacetus **362**
Mesoplodon 362
Myotis 210
Nyctophilus 218
Oligoryzomys 718
Perognathus 485
Reithrodontomys 741
Sorex **118**, 118
Sylvilagus 825
Zapus 499
pacivorus, Speothos 285
packardi, Antrozous 198
Pacos 381
pacos, Lama **382**
padangensis, Rhizomys 685
padangus, Paradoxurus 344
paeba, Gerbillurus **547**, 547
paedulcus, Thallomys 668, **669**, 669
paenulata, Presbytis 271
paenulatus, Callicebus 258
paeze, Phyllostomus 180
paganensis, Pteropus 148
pagei, Saccostomus 541
pageli, Arctictis 342
pagensis, Cynopterus 139
Macaca 267
Maxomys 604, **613**
pagenstecheri, Pipistrellus 222
pagi, Rhinolophus 168
Pagophilus 331, 332
paguatae, Thomomys 474
Paguma **343**, **344**
annectens 343
aurata 343
grayi 343
hainana 343
intrudens 343
janetta 343
jourdanii 343
laniger 343
lanigera 344
lanigerus 343
larvata **343**
leucocephala 343
leucomystax 343
neglecta 343
nigriceps 343
nipalensis 343
ogilbyi 343
pallasii 343
reevesi 343
rivalis 343
robustus 343
rubidus 343
taivana 343
tytlerii 343
vagans 343
wroughtoni 343
yunalis 343
pagurus, Isothrix **793**, 793
pagus, Dremomys 425
pahari, Mus **627**, 627
Soriculus 123
pahasapensis, Myotis 215
Paikea 362
paitana, Tupaia 133
Pajeros 294
pajeros, Oncifelis 294
Palaeanthropus 276
Palaeocapromys 800
Palaeodocoileus 391
Palaeotomys 680
palaestina, Vulpes 287
palatilis, Zyzomys **674**
palatina, Neotoma **712**
palatinus, Artibeus 189
Cercopithecus 263
Oryzomys 720
Zapus 499
palauensis, Emballonura 158
palavensis, Sus 378
palawanensis, Crocidura **92**
Emballonura 157
Maxomys 613
Suncus 103
Tupaia **133**
Palawanomys **637**, 637
furvus **637**, 637
palawanus, Herpestes 304
paleaceus, Echimys 791
Eidolon 141
palearia, Panthera 298
palelae, Rattus 660
palembang, Rattus 660
Paleolaginae 813
Paleoloxodon 367
palestinus, Homo 276
palitans, Lepus 814
palki, Cryptomys 772
pallah, Aepyceros 393
pallaryi, Gazella 397
pallasi, Nyctimene 144
Ochotona 807, **811**, 811
Spalax 755
Tamias 455
Ursus 339
pallasii, Arvicola 506
Capra 405, 406
Microtus 521
Oryx 413
Paguma 343
Paradoxurus 344
Phacochoerus 377
Pallasiinus 517, 523, 525, 527
Pallasiomys 555, 556, 557, 558
pallassii, Tarsius 250
pallens, Echinops 73
Eptesicus 203
Herpestes 305
Paradoxurus 344
pallescens, Apodemus 570
Arvicanthis 578
Callicebus 258
Choloepus 64
Clethrionomys 508
Dendromus 543
Lasiurus 206
Lynx 293
Mus 626
Nyctophilus 218
Onychomys 719
Perameles 40
Perognathus 485
Peromyscus 732
Phaner 244
Pipistrellus 223
Plecotus 225
Sciurus 440
Spermophilus 449
Tamiasciurus 457
Thomomys 474
palliata, Alouatta **255**
Ratufa 437
palliatus, Colobus 270
Paraxerus **435**
palliceps, Dendrolagus 50
pallida, Anathana 131
Asellia 170
Balantiopteryx 156
Blarina 106
Dicrostonyx 511
Felis 289
Macropus 54
Myotis 210
Nasua 334
Nycteris 161, 162
Otogale 248
Parotomys 682
Procavia 374
Raphicerus 399
Sicista 498
Sotalia 355
Viverricula 348
Vulpes **286**
pallidicauda, Lepilemur 246
Spermophilus 446
pallidior, Apodemus 570
Cavia 779
Cynictis 302
Eptesicus 202
Kannabateomys 790
Lagorchestes 52
Lepus 816
Microcavia 780
Millardia 621
Petromus 774
Proteles 309
Sylvicapra 412
Thylamys **23**
pallidipes, Mungos 307
pallidissimus, Peromyscus 729
pallidius, Dasycercus 31
pallidiventris, Dasyprocta 782
Marmosa 19
pallidulus, Nyctomys 715
pallidus, Antrozous **198**
Apodemus 574
Baiomys 695
Bradypus 63
Caluromys 15
Castor 467
Cebus 259
Cercopithecus 265
Chaetodipus 483
Cryptomys 772
Ctenomys 785
Dendromus 543
Diplomesodon 98
Dipodomys 479
Dryomys 766
Eliomys 767
Eptesicus 201
Erinaceus 77
Euoticus **249**
Heliophobius 772
Hemiechinus 78
Herpestes 305
Hipposideros 173
Kobus 414
Lemmiscus 516
Lepus 814
Macropus 55
Melomys 616
Meriones 556
Microdipodops **480**, 480
Miniopterus 231
Mustela 323
Mysateles 803
Nannosciurus 434
Octodon 788
Ondatra 532
Onychomys 719
Philander 22
Phloeomys **640**, 640

Procyon 335
Pteropus 147
Pygeretmus 490
Reithrodontomys 741
Rhinolophus 166
Rhombomys 559
Salpingotus 491, **492**, 492
Scotoecus **226**
Sigmodon 747
Sorex 112
Spermophilus 448, 450
Sphiggurus **776**
Tamias 454
Tragulus 382
Vespertilio 198
Zaglossus 13
Zapus 499
palliolatus, Monodelphis 21
pallipes, Apodemus 574
Canis 281
Herpestes 305
Lagidium 778
Lepus 819
Semnopithecus 273
palmarius, Bassariscus 334
Oecomys 717
Peromyscus 730
palmarum, Artibeus 189
Eidolon 141
Funambulus **426**
Microcebus 243
Rattus 649, **657**, 657
Sciurus 426
palmata, Chironectes 16
palmatus, Alces 389
palmeri, Dipodomys 479
Microsciurus 433
Oecomys 716
Ovis 408
Sorex 113
Tamias **454**
palmipes, Nectomys **709**, 709
palmirae, Oryzomys 720
Palmista 426
palmyrae, Gerbillus 550
palnica, Mus 623
palpator, Dactylopsila **60**, 60
Paludicola 505
paludicola, Clethrionomys 508
Microtus 519
Plectomys 691
Sylvilagus 826
Paludilagus 824
paludinosus, Atilax **300**
Herpestes 300
paludis, Synaptomys 534
paludivagus, Sorex 122
paludosus, Arvicola 506
Atilax 301
Blastocerus 389
Cervus 389
Sus 379
palustris, Blastocerus 389
Callosciurus 423
Dasymys 589
Diceros 372
Galea 780
Herpestes **306**, 306
Microtus 527
Mus 708, 719
Nyctalus 217
Oryzomys 721, 722, **724**, 724
Panthera 298
Sorex 111, **119**
Sus 379
Sylvilagus **826**, 827
pamana, Promops 240
pambasileus, Canis 281
pamela, Crocidura 93
pamirensis, Cricetulus 538
Crocidura 96
Lepus 821
Microtus 522
Myotis 213
Ursus 339
pampa, Oncifelis 294
pampaeus, Ozotoceros 392
pampanus, Conepatus 316
Oncifelis 294
Reithrodon 740
pamparum, Cavia 779
Eligmodontia 701
Lagostomus 778
Pan **276**, 276, 277
adolfifriederici 277
angustimanus 277
aubryi 277
calvescens 277
calvus 277
castanomale 277
chimpanse 277
cottoni 277
ellioti 277
fuliginosus 277
fuscus 277
graueri 277
hecki 277
ituriensis 277
koolookamba 277
lagaros 277
leucoprymnus 277
livingstonii 277
mafuca 277
marungensis 277
nahani 277
niger 277
ochroleucus 277
oertzeni 277
paniscus **277**
papio 277
pfeifferi 277
purschei 277
pusillus 277
raripilosus 277
reuteri 277
satyrus 277
schneideri 277
schubotzi 277
schweinfurthii 277
steindachneri 277
troglodytes **277**
tschego 277
vellerosus 277
verus 277
yambuyae 277
pan, Ateles 257
Niviventer 634
Phascolosorex 34
panamensis, Ateles 257
Chironectes 16
Herpailurus 291
Heteromys 481
Lasiurus 207
Mustela 322
Nasua 334
Oryzomys 725
Phyllostomus 180
Procyon 335
Proechimys 798
Tylomys **751**, 751
panamintinus, Dipodomys **479**, 479
Perognathus 485
Tamias **454**
panayensis, Scotophilus 227
pancici, Talpa 129
pandora, Dasyprocta 782
Mazama 390
paneaki, Microtus 525
panema, Proechimys 797
panga, Idiurus 758
pangandarana, Megaderma 163
panganensis, Crocuta 308
panganiensis, Otolemur 250
panglima, Maxomys **613**
panguitchensis, Dipodomys 479
Paniscus 256
paniscus, Ateles **257**, 257
Pan **277**
Simia 256
panja, Tatera 561
panjioli, Callosciurus 421
panjius, Callosciurus 421
Rattus 660
pannellus, Rattus 660
pannietensis, Dobsonia **140**, 140
pannonicus, Microtus 518
pannosus, Chaetophractus 65
Rattus 660
Rhizomys 685
pannovianus, Callosciurus 422
Panolia 385
panope, Lagenorhynchus 354
Tursio 351
pansa, Orthogeomys 470
pantarensis, Rattus 652
pantasticta, Leptailurus 292
Panthera **288**, **297**, 297, 298, 299
adersi 298
adusta 297, 298
africanus 297
alba 298
altaica 298
amoyensis 298
amurensis 298
antinorii 298
antiqua 298
antiquorum 298
arizonensis 298
asiaticus 297
azandicus 297
balica 298
barbaricus 297
barbarus 297, 298
bedfordi 298
bengalensis 297
bleyenberghi 297
boliviensis 298
brockmani 298
capensis 297
centralis 298
chinensis 298
chui 298
ciscaucasicus 298
corbetti 298
coreensis 298
coxi 298
delacouri 298
fontanierii 298
fortis 298
fossilis 298
fusca 298
gambianus 297
goldmani 298
goojratensis 297
grayi 298
hanensis 298
hernandesii 298
hollisteri 297
indicus 297
iturensis 298
jaguapara 298
jaguar 298
jaguarete 298
jaguatyrica 298
japonensis 298
jarvisi 298
kamptzi 297
kotiya 298
krugeri 297
lecoqi 298
leo **297**
leopardus 298
longicaudata 298
longipilis 298
maculatus 297
madeirae 298
major 298
mandshurica 298
massaicus 297
melanochaitus 297
melanotica 298
melas 298
mexianae 298
mikadoi 298
millardi 298
milleri 298

minor 298
mongolica 298
nanopardus 298
niger 298
nigra 297, 298
nimr 298
nobilis 297
notialis 298
nubicus 297
nyanzae 297
onca **297**
onssa 298
onza 298
orientalis 298
palearia 298
palustris 298
paraguensis 298
pardus **298**
paulensis 298
perniger 298
persica 297
persicus 297
peruviana 298
poecilura 298
poliopardus 298
proplatensis 298
puella 298
ramsayi 298
regalis 298
reichenowi 298
roosevelti 297
ruwenzorii 298
sabakiensis 297
saxicolor 298
senegalensis 297
septentrionalis 298
shortridgei 298
sindica 298
somaliensis 297
sondaica 298
striatus 298
styani 298
suahelicus 297, 298
sumatrae 298
sumatrana 299
tigris **298**
trabata 299
tulliana 298
ucayalae 298
variegata 298
varius 298
veraecrucis 298
vernayi 297
villosa 298
virgata 299
vulgaris 298
webbiensis 297
pantherina, Genetta 346
Pantherinae **297**, 297
Pantholops **399**
chiru 399
hodgsonii **399**
kemas 399
papae, Hystrix 773
papagensis, Peromyscus 730
papagoensis, Taxidea 325
Papagomys 587, 600, **637**, 637, 664
armandvillei **637**
besar 637
theodorverhoeveni **638**
verhoeveni 637
papilio, Chalinolobus 200
papillosa, Kerivoula **197**
Papinae 262
Papio 263, 268, **269**
aegyptiaca 269
antiquorum 269
anubis 269
brockmani 269
chaeropitheus 269
chobiensis 269
choras 269
comatus 269
cynocephalus 269
doguera 269
furax 269
graueri 269
griseipes 269
hamadryas **269**
heuglini 269
ibeanus 269
jubilaeus 269
kindae 269
langheldi 269
lestes 269
lydekkeri 269
neumanni 269
ngamiensis 269
nigeriae 269
nigripes 269
occidentalis 269
ochraceus 269
olivaceus 269
orientalis 269
papio 269
porcarius 269
pruinosus 269
rhodesiae 269
rubescens 269
strepitus 269
tesselatum 269
thoth 269
tibestianus 269
transvaalensis 269
ursinus 269
variegata 269
vigilis 269
werneri 269
yokoensis 269
papio, Cynocephalus 269
Pan 277
Papio 269
Papioninae 262
Pappogeomys 469, **471**, 471, 472
albinasus 472
alcorni 471, **472**
amecensis 472
angusticeps 472
angustirostris 473
arvalis 473
atratus 473
brevinasus 473
bullatus 472
bulleri 471, **472**
burti 472
castanops **472**
clarkii 472
consitus 472
elibatus 472
estor 472
excelsus 472
flammeus 472
fulvescens 472
fumosus **472**
goldmani 472
gymnurus **472**
hirtus 472
imparilis 472
infuscus 472
irolonis 472
jucundus 472
lacrimalis **472**
lagunensis 472
lutulentus 472
melanurus 472
merriami **472**
nayaritensis 472
neglectus **472**
nelsoni 472
oreocetes 472
parviceps 472
peraltus 472
peregrinus 472
perexiguus 472
peridoneus 472
perotensis 472
perplanus 472
planiceps 473
planifrons 472
pratensis 472
rubellus 472
russelli 472
saccharalis 472
simulans 472
sordidulus 472
subnubilus 472
subsimus 472
surculus 472
tamaulipensis 472
tellus 472
torridus 472
tylorhinus **473**
ustulatus 472
varius 473
zinseri **473**
zodius 473
pappus, Capromys 800
papua, Hipposideros **174**
papuana, Mosia 158
Syconycteris 155
papuanus, Macropus 53
Nyctimene 143, 144
Petaurus 61
Pipistrellus 224
Pteropus 149
Uromys 671
papuensis, Kerivoula **197**
Microperoryctes **41**
Otomops **239**, 239
Sus 379
Paracapromys 800, 801, 802
Paraceros 389
Paracoelops **175**
megalotis **175**, 175
Paracrocidura **100**
camerunensis 101
graueri **100**, 101
maxima **101**, 101
schoutedeni 100, **101**
Paracynictis 300, **307**
bechuanae 307
ngamiensis 307
selousi **307**
sengaani 307
Paracyon 29
Paradipodinae 492, **495**, 495
Paradipus 492, **495**, 495
ctenodactylus **495**
paradoxa, Selevinia 768
Paradoxodon 101
paradoxolophus, Rhinolophus **168**
paradoxura, Crocidura **92**
Paradoxurinae **342**, 342, 343
Paradoxurus 342, **344**
aureus 344
baritensis 344
birmanicus 344
bondar 344
brunneipes 344
canescens 344
caniscus 344
canus 344
celebensis 344
cochinensis 344
crassiceps 344
crossi 344
derbyanus 341, 342
dubius 344
exitus 344
felinus 344
fossa 344
fuliginosus 344
fuscus 344
hamiltonii 344
hanieli 344
hermaphroditus **344**
hirsutus 344
javanica 344
jerdoni **344**
kangeanus 344
kutensis 344
laneus 344
laotum 344
lehmanni 344
leucopus 344
lignicolor 344
macrodus 344
milleri 344
minax 344

minor 344
montanus 344
musanga 344
musangoides 344
musschenbroekii 343
nictitatans 344
niger 344
nigra 344
nigrifrons 344
padangus 344
pallasii 344
pallens 344
parvus 344
pennantii 344
philippinensis 344
prehensilis 344
pugnax 344
pulcher 344
quadriscriptus 344
ravus 344
rindjanicus 344
robustus 344
rubidus 344
sabanus 344
sacer 344
scindiae 344
senex 344
setosus 344
siberu 344
simplex 344
strictis 344
sumatrensis 344
sumbanus 344
torvus 344
trivirgatus 342
typus 344
tytlerii 343
vellerosus 344
vicinus 344
zeylonensis **344**
paradoxus, Cardiocranius **491**, 491
Chaetodipus 483
Microtus 522
Ornithorhynchus 13
Pseudochirops 60
Reithrodontomys **742**
Solenodon **69**, 69
Paraechinus 78, 79
micropus 78
paraensis, Lontra 311
Mustela 321
Sciurus 441
Paragalia 39
paraganus, Oryzomys 724
paragayensis, Sphiggurus 777
Paragenetta 345, 346
Parageomys 469
Paraglirulus 769
paraguae, Oncifelis 294
paraguanensis, Pteronotus 177
paraguayana, Micoureus 20
paraguayansis, Cebus 259
paraguayensis, Dasyprocta 781
Didelphis 16
Sphiggurus 777
paraguensis, Chironectes 16
Panthera 298
Pteronura 313
Sylvilagus 825
Parahyaena 308, **309**
brunnea **309**, 309
fusca 309
makapani 309
melampus 309
striata 309
villosa 309
Parahydromys 620, **638**
asper **638**
Paralactaga 488
Paralariscus 430
Paralces 389
Paraleporillus 644
Paraleptomys **638**, 638
rufilatus **638**
wilhelmina **638**, 638
paralios, Chaetodipus 483
paralius, Dipodomys 477
Liomys 482
parallelus, Tragulus 383
Paralomys 737, 738
Paramanis 415
Paramelomys 615
paramensis, Oxymycterus 726, **727**
Parameriones 555, 557
Paramicrogale 71
paramicrus, Nesophontes **70**
paramorum, Thomasomys **750**
paranalis, Lutreolina 18
paranensis, Lontra 311
Pteronura 313
Parantechinus 29, **33**
apicalis **33**, 33
bilarni **33**
macdonnellensis 33
paranus, Molossops 235
Paranyctimene **145**
raptor **145**, 145
Paraonyx 310
philippsi 310
Paraphenacomys 504, 505
Parapitymys 517
Parapodemus 668, 670
Parascalops **127**
breweri **127**
Parascaptor **127**, 128
leucura **127**
micrura 127
Parasciurus 439
parasiticus, Peromyscus 729
parata, Marmosa 18
Paratatera 547, 548
paratus, Proechimys 797
paraxanthus, Rattus 662
Paraxerini 419
Paraxerus 429, **434**
affinis 435
alexandri **434**
animosus 435
antoniae 434
aruscensis 435
augustus 435
auriventris 435
barawensis 435
boehmi **434**
bororensis 434
bridgemani 435
byatti 435
capitis 435
carpi 434
cepapi **434**
cepapoides 434
cepate 434
chobiensis 434
cooperi **434**, 435
electus 435
emini 434
exgeanus 435
flavivittis 435
flavovittis **435**
frerei 435
ganana 435
gazellae 434
ibeanus 435
jacksoni 435
kahari 435
kalaharicus 434
laetus 435
lastii 435
lucifer **435**
lunaris 434
maunensis 434
mossambicus 435
musculinus 435
ochraceus **435**
olivaceus 435
ornatus 435
palliatus **435**
pauli 435
percivali 435
phalaena 434
poensis **435**
quotus 434
ruwenzorii 434
salutans 435
sindi 434
soccatus 434
sponsus 435
suahelicus 435
subviridescens 435
swynnertoni 435
tanae 435
tanganyikae 434
tongensis 435
ugandae 434
vexillarius **435**, 435
vincenti **435**
vulcanorum 434
yulei 434
parca, Soriculus **123**, 123
parcus, Rhinolophus 168
parda, Micoureus 20
Pardalis 291
pardalis, Felis 291
Leopardus **291**, 292
pardella, Lynx 293
pardicolor, Prionodon **347**
Pardictis **347**
pardictis, Leopardus 292
Pardina 293
pardina, Genetta 346
Lynx 293
pardinoides, Felis 292
Leopardus 292
pardinus, Lynx **293**, 293
pardochrous, Felis 295
Prionailurus 295
Prionodon 347
Pardofelis **299**, 299
charltoni **299**
dosul 299
duvaucelli 299
longicaudata 299
marmorata **299**
ogilbii 299
pardoides, Felis 294
Oncifelis 294
Pardotigris 297
Pardus 297
pardus, Felis 297
Panthera **298**
paricola, Oecomys **716**, 716
parienti, Phaner 244
parilina, Lontra 311
parisii, Nycteris 162
parkeri, Mormopterus 238
parma, Macropus **54**
parmentieri, Procolobus 272
parnassius, Lepus 817
parnellii, Pteronotus **177**
paroensis, Pteronura 313
Parotomys 680, **682**, 682
brantsii **682**, 682
deserti 682
littledalei **682**
luteolus 682
molopensis 682
namibensis 682
pallida 682
rufifrons 682
parowanensis, Thomomys 475
parryi, Macropus **54**
parryii, Spermophilus **448**, 450
parsonii, Erignathus 329
parthianus, Spermophilus 446
particeps, Didelphis 17
Paruromys **638**, 638
dominator **638**, 638, 639, 653
frosti 638
ursinus **639**, 655
parva, Arctocephalus 327
Condylura 124
Cryptotis **109**
Murexia 32
Myoprocta 782

Rupicapra 409
Soriculus 123
parvabullatus, Dipodomys **477**
parvacauda, Crocidura 82
parvicaudatus, Sorex 121
parviceps, Liomys 482
Pappogeomys 472
Perognathus 485
Thomomys 476
parvidens, Alticola 503
Cheiromeles 233
Clethrionomys 509
Cynomys **424**
Euroscaptor **125**, 125
Herpestes 305
Marmosops **20**
Neotoma 712
Pseudalopex 285
Sorex 122
Spermophilus 447
Sturnira 192
Urocyon 285
parvipes, Artibeus 188
Crocidura 83, **93**
Galerella 303
Miniopterus 231
Moschus 384
Nectomys **709**
Pipistrellus 220
parvula, Helogale **304**, 304
Marmota 432
Microgale **71**, 72
Rhogeessa **225**, 225, 226
parvulus, Graphiurus 765
Herpestes 303
Micromys 620
Microtus 528
Myotis 213
Spermophilus 445
Sylvilagus 824
Thomomys 474
parvus, Apodemus 572, 573
Burramys **57**, 57
Clidomys **805**
Dipodomys 479
Epomophorus 141
Eptesicus 202
Graphiurus 764, **765**, 765
Helogale 304
Herpestes 304
Hylomys 80
Hylomyscus **599**
Margaretamys **609**
Microtus 530
Mus 628, 629
Paradoxurus 344
Perognathus 485, **486**, 486
Pithecheir **640**
Rhinolophus 164
Scalopus 127
Scapanus 127
Sorex 108
Spermophilus 450
Steatomys 545, **546**
Sundasciurus 452
pasan, Oryx 413
pascalis, Thomomys 474
paschalis, Cervus 386
pascuus, Microtus 520
pasha, Crocidura 85, 91, **93**, 93
Heliosciurus 429
Mus 627
pashtonus, Eptesicus 203
pasquieri, Hapalomys 595
Passalites 390
passargei, Chlorocebus 265
passerum, Oncifelis 294
pastasae, Dasypus 66
pastoris, Pteropus 146
patachonica, Balaenoptera 350
Cavia 780
Conepatus 316
patachonicha, Dolichotis 780
patagonica, Conepatus 316
Mirounga 330
Mustela 319
Puma 296
patagonicus, Eumops 233
Lyncodon **319**
Pseudalopex 284
Zaedyus 67
patagonum, Dolichotis **780**
patas, Erythrocebus **265**
Simia 265
pater, Cannomys 685
paterculus, Pipistrellus 221, **222**
patira, Pecari 380
patquiensis, Chlamyphorus **64**
patriciae, Myotis 213
patrius, Pseudomys **648**
patrizii, Asellia **170**
Gerbillus 550
pattersonianus, Taurotragus 403
patulus, Thomomys 474
paucidentata, Alionycteris **138**, 138
paucistriatus, Equus 369
Paucituberculata 25
Paulamys 581, 587, **639**, 664
naso **639**, 639
paulensis, Marmosops 19
Monodelphis 22
Panthera 298
pauli, Bassaricyon **333**
Paraxerus 435
paulina, Mus 625
paulsoni, Sturnira 192
Sylvilagus 826
paululus, Microdipodops 480
Miniopterus 230
paulus, Artibeus 188
Baiomys 695
Crateromys **587**
Lemmus 517
pauper, Rattus 661
paupera, Golunda 592
pauperrimus, Lemmiscus 516
Paurodus 81
pavidus, Peromyscus 731
payanus, Rattus 661
pealana, Suncus 103
pealii, Phoca 332
pearsei, Hybomys 596
pearsoni, Andalgalomys **694**
Ctenomys **786**
Graomys 703
Petrogale 56
Puma 296
Rangifer 392
Solisorex **101**, 101
pearsonii, Belomys **459**
Harpiocephalus 228
Rhinolophus **168**, 169
Sciuropterus 459
pearyi, Ovibos 408
Rangifer 392
peasei, Elephantulus 829
peba, Dasypus 66
pebilis, Saguinus 253
Pecari **380**, 380
angulatus 380
bangsi 380
crassus 380
crusnigrum 380
humeralis 380
macrocephalus 380
minor 380
modestus 380
nanus 380
nelsoni 380
niger 380
nigrescens 380
patira 380
sonoriensis 380
tajacu **380**
tajassu 380
torquatus 380
torvus 380
yucatanensis 380
pecari, Sus 380
Tayassu **380**, 380
peccatus, Rattus 653
pecchioli, Apodemus 573
pecilei, Dendromus 543
Pectinator **761**
legerae 761
meridionalis 761
spekei **761**, 761
pectoralis, Hemiechinus 78
Oryzomys 720
Peponocephala 355
Peromyscus **735**
Petaurista 462
Reithrodontomys 741
Thomomys 474
peculiosa, Vulpes 287
pedester, Aethomys 568
Pedestes 759
Pedetes **759**
albaniensis 759
angolae 759
cafer 759
capensis **759**
currax 759
damarensis 759
dentatus 759
fouriei 759
larvalis 759
orangiae 759
salinae 759
surdaster 759
taborae 759
typicus 759
Pedetidae 759
pediaeus, Lepus 816
pediculus, Conepatus 316
Pediolagus 780
pedionomus, Dipodomys 477
Pediotragus 399
pediotragus, Raphicerus 399
Pedomys 517, 526
pedri, Capra 406
pedunculatus, Zyzomys **674**
peeli, Gerbillus 558
Loxodonta 367
Microdillus **558**
pegasis, Saguinus 253
peguanus, Tupaia 131
peguensis, Chiropodomys 583
Hylomys 80
Lepus 814, 818, 819, **820**, 820
Pipistrellus **223**
Pekania 319
pelagius, Maxomys 614
pelandoc, Tragulus 382
pelandonius, Microtus 524
pelapius, Callosciurus 423
Pelea **413**
capreolus **413**, 413
lanata 413
villosa 413
Peleinae 413
pelengensis, Phalanger **47**, 47
Rattus 660
Tarsius 250
pelerinus, Meriones 556
pelewensis, Pteropus 148
peli, Galagoides 249
Saccolaimus **159**
pelii, Anomalurus **757**
pelingensis, Hipposideros 172
pellax, Maxomys 613
pelliceus, Arvicanthis 578
Eptesicus 201
Microtus 521
pellucens, Eptesicus 202
Myotis 209
pellucida, Kerivoula **198**
pellyensis, Ursus 339
Pelomys 578, 582, 590, 629, 630, **639**, 639
australis 639
campanae **639**
concolor 639
fallax **639**, 639
frater 639
harringtoni 590

hopkinsi **639**, 639, 640
insignatus 639
iridescens 639
isseli 639, **640**
luluae 639
minor **640**
rhodesiae 639
vumbae 639
peloponnesiacus, Nannospalax 754
pelops, Macaca 266
pelorhinus, Cercopithecus 263
pelori, Rattus 653
pelurus, Rattus 650, 652, **657**, 662
pelzelnii, Gazella 396
pemanggis, Rattus 661
pemangilensis, Callosciurus 422
pemangilis, Atherurus 773
Maxomys 614
Tupaia 132
pembae, Cephalophus 411
Chlorocebus 265
pembertoni, Funisciurus 428
Peromyscus 734, **735**
Tamiops 457
penangensis, Iomys 461
Petaurista 463
Ratufa 437
Tragulus 382
Tupaia 132
pencanus, Akodon 692
pendleburyi, Hipposideros 175
pendulinus, Micromys 620
peni, Lepus 816
penialius, Callosciurus 423
penicillata, Bettongia **49**
Callithrix **252**, 252
Cynictis **302**
Didelphis 33
Petrogale **56**, 56
Phascogale 34
penicillatus, Arctictis 342
Chaetodipus **484**, 484
Chiropodomys 583
Conilurus **586**, 586, 587
Eliurus **678**, 678
Funambulus 426
Herpestes 302
Kangurus 55
Peromyscus 729, 734
Potamochoerus 378
penicilliger, Meriones 557
peninsulae, Ammospermophilus 420
Apodemus **572**, 572, 573, 574
Blarina 106
Canis 280
Chaetodipus 484
Cynocephalus 135
Dicrostonyx 510
Eptesicus 201
Genetta 345
Herpestes 305
Isoodon 39
Lariscus 430
Muntiacus 389
Mustela 322, 324
Odocoileus 391
Oryzomys 721
Phalanger 46
Ratufa 437
Reithrodontomys 741
Tamias 456
peninsularis, Antilocapra 393
Callosciurus 423
Dipodomys 477
Emballonura 158
Lynx 293
Martes 320
Menetes 433
Murina 229
Myotis **214**
Peromyscus 735
Sus 379
Sylvilagus 825
Urocyon 285
penitus, Bunomys **582**, 582
Frateromys 582
Rattus 582
pennanti, Martes **320**
pennantii, Cervus 387
Funambulus **426**, 426
Neomys 110
Paradoxurus 344
Procolobus **272**
pennata, Distoechurus 62
pennatus, Distoechurus **62**
pennipes, Ammospermophilus 420
pennsylvanica, Neotoma 711
Scalopus 127
pennsylvanicus, Bison 400
Microtus 518, 519, 524, 525, **527**, 527, 528, 530
Sciurus 440
Vulpes 287
penricei, Equus 370
Kobus 414
Redunca 414
pensylvanicus, Urocyon 285
pentadactyla, Manis **415**, 415
pentadactylus, Ateles 257
Dasypus 66
Pentalagus 813, 814, **823**
furnessi **823**
pentax, Apodemus 574
Penthetor **145**
lucasi **145**, 145
pentonyx, Steatomys 545
Peponocephala **355**
asia 355
electra **355**
fusiformis 355
pectoralis 355
pequinius, Myotis **214**
peracer, Rhinosciurus 438
Peracomys 564, 565, 566
Peradorcas 55, 56
peragrus, Oryzomys 721
perakensis, Herpestes 305
Peralopex 29
peralta, Giraffa 383
peraltus, Pappogeomys 472
Peramelemorphia 39
Perameles **40**, 40
arenaria 40
bougainville **40**
ecaudatus 39
eremiana **40**
fasciata 40
gunnii **40**
kalubu 41
lagotis 39
lawsoni 40
major 40
musei 40
myosuros 40
nasuta **40**, 40
notina 40
pallescens 40
raffrayana 42
Peramelidae **39**, 40
Perameliformes **39**
Perameloidea **39**
peramplus, Thomomys 474
Peramys 21
perasper, Maxomys 613
perauritus, Rhinolophus 165, 166
perblandus, Dipodomys 480
perchal, Bandicota 579
percivali, Acomys 565, **566**, 566, 567
Cloeotis **170**, 170
Colobus 270
Crocidura 97
Dendromus 542
Gerbillus **553**
Otomys 682
Paraxerus 435
Tatera 561
percnonotus, Mus 626
percommodus, Callosciurus 423
percura, Presbytis 271
perdicator, Prionodon 347
perditus, Rhinolophus 165
Thomomys 476
perdus, Mustela 322
peregrina, Cryptotis 109
peregrinus, Didelphis 59
Pappogeomys 472
Pseudocheirus **59**
Thomomys 476
peregusna, Mustela 325
Vormela **325**, 325
perenensis, Oryzomys 721
perennis, Salicornia 802
perexiguus, Pappogeomys 472
perflavus, Maxomys 614
Tragulus 383
perforatus, Taphozous 160, **161**
perfulvidus, Galerella 303
perfulvus, Peromyscus **735**
perfuscus, Marmosops 19
pergracilis, Cryptotis 109
Perognathus 485
Peromyscus 729
pergrisea, Crocidura 81, **93**, 93, 95, 98
perhentiani, Callosciurus 423
perhentianus, Rattus 661
pericalles, Perognathus 485
periclyzomae, Lontra 311
peridoneus, Pappogeomys 472
Peromyscus 729
perigrinator, Sciurus 440
perijae, Sciurus **441**
perimekurus, Peromyscus 732
Perimyotis 219, 224
periosus, Choeroniscus **184**
Perissodactyla 369
perluteus, Taterillus 563
perlutus, Maxomys 614
permiliensis, Sorex 112
perminutus, Sorex 117
permira, Mazama 391
permista, Gazella 396
permixtio, Phalanger 47
permixtus, Pipistrellus **223**
pernanus, Mastomys **611**, 611
pernettensis, Stenella 356
pernettyi, Otaria 328
Stenella 356
perniger, Panthera 298
Perognathus 485
Rhinolophus 167
Ursus 338
pernigra, Didelphis 16
pernix, Chaetodipus **484**, 484
Ctenomys 785
Oligoryzomys 718
pernox, Myotis 212
pernyi, Dremomys **425**
Sciurus 424
Perodicticus **248**
arrhenii 248
batesi 248
bosmannii 248
calabarensis 247
edwardsi 248
faustus 248
geoffroyi 248
guineensis 248
ibeanus 248
ju-ju 248
nebulosus 248
potto **248**
Perodipus 477
Peroechinus 76
Perognathidinae 482
Perognathinae 477, 480, **482**
Perognathus 477, 482, 483,

484
aestivus 485
alticola **484**
ammodytes 485
amoenus 486
amplus **485**
apache 485
arcus 485
arenarius 483
arenicola 485
arizonensis 485
bangsi 485
bimaculatus 485
bombycinus 485
brevinasus 485
bullatus 486
bunkeri 485
callistus 485
cantwelli 485
caryi 485
cineris 485
clarus 486
cleomophila 485
cockrumi 485
columbianus 486
copei 485
elibatus 485
fasciatus 484, **485**
flavescens **485**
flavus **485**, 486
fuliginosus 485
fuscus 485
gilvus 486
goodpasteri 485
gulosus 485
gypsi 485
hopiensis 485
idahoensis 486
inexpectatus 485
infraluteus 485
inornatus **485**
internationalis 485
jacksoni 485
kinoensis 485
laingi 486
litus 485
longimembris **485**
lordi 486
magruderensis 486
mearnsi 486
medius 485
melanotis 485
merriami 485, **486**, 486
mexicanus 485
mollipilosus 486
monticola 486
neglectus 485
nevadensis 485
olivaceogriseus 485
olivaceus 486
pacificus 485
pallescens 485
panamintinus 485
parviceps 485
parvus 485, **486**, 486
pergracilis 485
pericalles 485
perniger 485
pimensis 485
piperi 485
plerus 486
psammophilus 485
relictus 485
rotundus 485
salinensis 485
sanluisi 485
sillimani 485
sonoriensis 485
spinatus 482
taylori 485
trumbullensis 486
tularensis 485
venustus 485
virginis 485
xanthanotus **486**
xanthonotus 486
yakimensis 486
Peromyscus 695, 703, 706, 707, 713, 715, 725, **728**, 728, 729, 731, 739, 752
abietorum 732
affinis 732
akeleyi 732
albifrons 735
alcorni 730
algidus 732
allapaticola 730
allophrys 735
allophylus 731
alpinus 732
alstoni 713
altilaneus 731
ambiguus 729, 732
ammobates 735
ammodytes 732
amplus 730
anacapae 732
anastasae 730
angelensis 734
angustirostris 730
angustus 732
anthonyi 730
anticostiensis 732
arboreus 728, 732
arenarius 730, 735
argentatus 732
aridulus 732
arizonae 732
artemisiae 732
assimilis 732
attwateri **728**, 729
auripectus 729
auritus 733
austerus 732
avius 730
aztecus **728**, 728, 729, 736, 737
azulensis 734
badius 737
bairdii 732
balaclavae 732
baliolus 735
banderanus 725
beatae 729, 732
bellus 728
benitoensis 729
beresfordi 732
blandus 732
borealis 732
boylii **728**, 728, 729, 732, 734, 735, 736
brevicaudus 732
bullatus **729**
cacabatus 734
californicus 728, **729**, 729
campestris 732
canadensis 732
cancrivorus 732
caniceps **729**, 729
cansensis 728
canus 732
carli 732
carmeni 730
castaneus 732
catalinae 732
caudatus 732
cecilii 733
cedrosensis 730
chlorus 736
chrysopus 735
cinereus 730
cineritius 732
clementis 732
coahuilensis 733
coatlanensis 734
cognatus 730
colemani 735
collatus 730
collinus 735
comanche 730, 736, 737
comptus 733
consobrinus 733
coolidgei 732
cordillerae 728, 729
cozulmelae 732
crinitus **729**, 729, 735
cristobalensis 737
decoloratus 735
delgadilli 729
deserticolus 732
dickeyi **729**
difficilis 728, **729**, 729, 730, 734
disparilus 729
dorsalis 732
doutii 729
doylei 732
dubius 732
dyselius 736
easti 732
elusus 732
emmonsi 732
erasmus 731
eremicoides 735
eremicus 728, 729, **730**, 730, 731, 734, 735
eremus 732
eva 729, **730**, 730
evides 728, 729, 736
exiguus 732
exterus 732
felipensis 730
flaccidus 732
fraterculus 730
fulvus 732
furvus **730**, 730, 733, 734
fuscus 732
gadovii 733
gambeli 732
gaurus 729
gentilis 731
georgiensis 732
geronimensis 732
gilberti 736
glasselli 729
goldmani 733, 734
gossypinus **730**, 730, 732
gracilis 732
grandis **730**, 731, 737
gratus **731**, 731, 737
griseobracatus 735
griseus 734
guardia **731**, 731
guatemalensis **731**, 731, 734, 737
gunnisoni 732
gymnotis **731**, 731, 734
harbisoni 731
hemionotis 736
herronii 730
hesperus 734
hollisteri 732
homochroia 730
hondurensis 728
hooperi **731**, 731
hueyi 732
hylaeus 732
hylocetes 728
imperfectus 732
incensus 732
inclarus 732
insignis 729
insolatus 732
insulanus 730
insulicola 730
interdictus 734
interparietalis **731**, 731
isolatus 734
keeni 732, 734
labecula 732
laceianus 735
laceyi 728
lachiguiriensis 732
lagunae 736
lasius 736
latirostris 730
lepturus 703
leucocephalus 735
leucopus 730, **732**, 732
leucurus 733
levipes 729, **732**, 732
lorenzi 731
lucubrans 735
luteus 732

macrorhinus 732, 734
madrensis 729, **732**
magdalenae 732
major 729
maniculatus 728, **732**, 732, 733, 734, 735, 736
margaritae 732
mariposae 729
maritimus 732
martinensis 732
martirensis 736
mayensis 730, **733**
mearnsii 732
medius 732
megacephalus 730
megalops **733**, 733, 734
megalotis 736
mejiae 731
mekisturus **733**
melanocarpus **733**, 733
melanophrys **733**, 733, 735
melanotis 732, **733**
melanurus **733**, 733
merriami 730, **733**, 734, 735
mesomelas 732
metallicola 729
mexicanus 703, 726, 728, 730, 731, 733, **734**, 734, 736, 737
michiganensis 732
micropus 733
minnisotae 732
mississippiensis 730
montipinoris 736
musculoides 732
myoides 732
nasutus 730, **734**
nebrascensis 732
nevadensis 736
nicaraguae 734
nigellus 730
nigriculus 730
niveiventris 735
noveboracensis 732
nubiterrae 732
nudipes 734
oaxacensis 728
oceanicus 736
ochraceus 732
ochraventer 730, **734**
oreas 732, **734**, 734, 736
oresterus 732
orientalis 734
orizabae 734
osgoodi 732
ozarkiarum 732
pallescens 732
pallidissimus 729
palmarius 730
papagensis 730
parasiticus 729
pavidus 731
pectoralis **735**
pembertoni 734, **735**
penicillatus 729, 734
peninsularis 735
perfulvus **735**
pergracilis 729
peridoneus 729
perimekurus 732
petraius 729
petricola 730
phaeurus 730
phasma 735
philombrius 734
pinalis 729
plumbeus 732
pluvialis 732
polionotus 732, **735**
polius **735**
polypolius 730
preblei 736
prevostensis 732
propinquus 730
pseudocrinitus **735**
pullus 730
putlaensis 734
restrictus 730
rhoadsi 735
robustus 729
rowleyi 729, 732
rubidus 732
rubriventer 732
rufinus 732, 733
rupicolus 729
ryckmani 731
sacarensis 729, 732
sagax 732
salvadorensis 734
sanctaerosae 732
santacruzae 732
sartinensis 732
saturatus 732
saxamans 732
saxatilis 734
saxicola 730
scitulus 729
scopulorum 729
sejugis **735**
sequoiensis 736
serratus 732
simulus 729, 732, **736**
sinaloensis 730
sitkensis 734, **736**, 736
slevini **736**
sloeops 734
sonoriensis 732
spicilegus 728, 729, **736**
stephani **736**, 736
stephensi 729
stirtoni **736**, 736
streatori 732
subarcticus 732
subgriseus 735
sumneri 735
teapensis 734
tehuantepecus 734
telmaphilus 730
texanus 732
thomasi 707
thurberi 732
tiburonensis 730
tornillo 732
totontepecus 734
triangularis 732
trissyllepsis 735
tropicalis 734
truei 728, 729, 730, 731, 734, 735, **736**, 737
umbrinus 732
utahensis 729
winkelmanni 728, **737**
xenurus 733
yucatanicus **737**, 737
zamelas 733
zamorae 733
zapotecae 731
zarhynchus 731, **737**
zelotes 731
peromyscus, Praomys 642, 643
peroni, Cephalotes 139
Dobsonia **140**
peronii, Arctocephalus 327
Cervus 387
Delphinus 354
Lissodelphis **354**
Petauroides 59
Plecotus 224
Potorous 50
Peronymus 158
Peropteryx **158**, 158
brunnea 158
caninus 158
cyclops 158
intermedia 158
kappleri **158**
leucoptera **158**
macrotis **158**
phaea 158
trinitatis 158
Perorycytes 41, **42**
broadbenti **42**, 42
mainois 42
raffrayana **42**, 42
rothschildi 42
Peroryctidae **40**
perotae, Mustela 322
perotensis, Dipodomys 480
Neotomodon 713
Pappogeomys 472
Reithrodontomys 740
Spermophilus **448**
perotis, Eumops **234**
Molossus 233
perpallida, Neotoma 712
Tatera 560
perpallidus, Gerbillus 552, **553**
Meriones 556
Onychomys 719
Thomomys 474
perpes, Thomomys 474
perplanus, Pappogeomys 472
perplexa, Micoureus 20
Neotoma 711
perplexibilis, Microtus 519
perplexus, Dipodomys 477
perplicatus, Sylvilagus 826
perpusilla, Cephalophus 411
perpusillus, Neotragus 398
Perqualus 350
perrensis, Ctenomys **786**
perrieri, Propithecus 247
perrotteti, Suncus 102
persephone, Petrogale **56**
persica, Capra 405
Crocidura 88
Gazella 397
Panthera 297
Tatera 561
persicus, Apodemus 572
Arvicola 506
Hemiechinus 78
Herpestes 305
Meriones **557**
Myoxus 770
Panthera 297
Sciurus 440
Triaenops **175**, 175
Ursus 339
Vulpes 287
persimilis, Cynopterus 139
Lepus 815
personata, Damaliscus 394
Melogale 314, **315**
Rattus 654
personatum, Phyllostoma 193
Uroderma 193
personatus, Callicebus **259**
Chionomys 507
Geomys 469, **470**
Graphiurus 765
Metachirus 20
Myomimus **768**, 768
Protoxerus 436
Pteronotus **177**
Pteropus **149**
Simia 258
Sorex 112, 113, 114
perspicillata, Carollia **186**, 186
Lutra 312
Lutrogale **313**, 313
Vespertilio 186
perspicillatus, Lagenorhynchus 353
Steno 357
perspicillifer, Saccopteryx 159
persultator, Sylvilagus 826
pertinax, Arvicola 506
perturbans, Ursus 339
peruana, Anoura 183
Eira 318
Gracilinanus 17
Lama 381
Octodon 788
peruanum, Lagidium 777, **778**
Mimon 179
peruanus, Cebus 259
Ctenomys **786**

Dactylomys **790**
Microsciurus 433
Sigmodon 747, **748**
Sylvilagus 825
Tadarida 240
peruensis, Lontra 311
perustus, Anomalurus 757
peruviana, Lama 382
Panthera 298
peruvianus, Ateles 257
Choloepus 64
Mesoplodon **363**
Monodelphis 21
Mus 626
Odocoileus 391
Tapirus 371
peruviensis, Lontra 311
Neusticomys **714**
Saimiri 260
pervagor, Ursus 339
pervagus, Thomomys 474
pervalens, Akodon 693
pervarius, Thomomys 474
pervigilis, Aotus 256
peshwa, Myotis 211
pessimus, Spermophilus 447
pesticulus, Rattus 650
pestis, Tatera 561
peta, Crocidura 94
Petaurella 61
Petauridae 46, 58, **60**
Petaurillus **461**
emiliae **462**
hosei **462**, 462
kinlochii **462**
Petaurista 58, 263, **462**, 464
albiventer 463
alborufus **462**
alborusus 462
annamensis 463
badiatus 463
banksi 462
barroni 463
batuana 463
birrelli 463
buechneri 463
candidula 463
caniceps 462
castaneus 462
chrysotrix 462
cicur 463
cinderella 463
cineraceus 463
clarkei 462
elegans **462**, 462
filchnerinae 462, 463
fulvinus 463
gorkhali 462
grandis 463
griseiventer 463
hainana 463
hintoni 462
hodgsoni 462
inornatus 463
interceptio 463
lanka 463
lena 462
leucocephalus 462
leucogenys 58, **462**, 463
lumholtzi 463
lylei 463
magnificus **462**, 463
marchio 463
marica 462
melanotis 463
mergulus 463
miloni 463
mimicus 463
nigra 463
nigrescens 463
nigricaudatus 463
nikkonis 462
nitida 463
nitidula 463
nobilis **462**, 462
ochraspis 462
oral 463
oreas 462
osiui 462
pectoralis 462
penangensis 463
petaurista **463**, 463
philippensis **463**, 463
primrosei 463
punctatus 462
rajah 463
reguli 463
rubicundus 463
rufipes 463
senex 462
singhei 462
slamatensis 462
sodyi 463
stellaris 463
stockleyi 463
sumatrana 462
sybilla 462
taylori 463
terutaus 463
thomasi 462
tosae 462
venningi 463
watasei 462
xanthotis 462, **463**
yunanensis 463
petaurista, Cercopithecus **264**
Petaurista **463**, 463
Sciurus 462
Petauristidae 459
Petauroides **58**
armillatus 59
cinereus 59
didelphoides 59
incanus 59
macroura 59
maximus 59
minor 59
peronii 59
taguanoides 59
volans **58**
volucella 59
Petaurula 61
Petaurus **61**
abidi **61**, 638
ariel 61
australis **61**, 61
biacensis 61
breviceps **61**
cunninghami 61
flavidus 61
flaviventer 61
gracilis **61**
kohlsi 61
longicaudatus 61
norfolcensis **61**, 61
notatus 61
papuanus 61
petaurus 61
reginae 61
sciurea 61
tafa 61
petaurus, Petaurus 61
petax, Myotis 210
peteroanus, Oligoryzomys 718
petersi, Aepyceros 393
Mormopterus 238
Otonycteris 219
Phoca 332
Pipistrellus **223**
Pteropus 151
Rhinolophus 168
Rhynchocyon **830**
Sigmoceros 395
petersii, Gazella 396
petersoni, Eptesicus 201
Euneomys **702**, 702
Mops **236**
petila, Mus 625
petilidens, Lagostomus 778
Petinomys 460, 461, **463**, 464
bartelsi 463
borneoensis 463
crinitus **463**, 463
electilis 463
fuscocapillus **463**
genibarbis **463**, 464
hageni **464**, 464
layardi 463
lugens 463, **464**, 464, 604
maerens 464
malaccanus 463
mindanensis 463
morrisi 464
nigricaudus 463
ouwensi 464
phipsoni 464
sagitta **464**, 464
setosus **464**
vordermanni **464**, 464
petraeus, Apomys 575
petraius, Peromyscus 729
petrensis, Marmota 432
petricola, Neotoma 712
Peromyscus 730
petrina, Saimiri 260
Petrobates 761
Petrodromus **830**
rovumae 830
sangi 830
sultan 830
tetradactylus **830**, 830
tordayi 830
Petrogale **55**
albogularis 56
assimilis **55**, 56
brachyotis **55**
burbidgei **56**
canescens 56
celeris 56
concinna **56**
godmani **56**, 56
hacketti 56
herberti 56
inornata **56**
lateralis **56**
longicauda 56
longmani 55
monastria 56
pearsoni 56
penicillata **56**, 56
persephone **56**
puella 55
purpureicollis 56
rothschildi **56**
signata 55
venustula 55
wilkinsi 55
xanthopus **56**
xanthopygus 56
Petromuridae **774**, 774
Petromus **774**, 774
ausensis 774
barbiensis 774
cinnamomeus 774
cunealis 774
greeni 774
guinasensis 774
karasensis 774
kobosensis 774
marjoriae 774
namaquensis 774
pallidior 774
tropicalis 774
typicus **774**, 774
windhoekensis 774
Petromyidae 774
Petromys 569, 774
Petromyscinae 501, **683**, 683
Petromyscus **683**, 683
barbouri **683**, 683
bruchus 683, 684
capensis 683
collinus **683**, 683, 684
kaokoensis 683
kurzi 683
monticularis **684**
namibensis 683
rufus 683
shortridgei 683, **684**, 684
variabilis 683
petronellae, Cercopithecus

265
petronius, Callithrix 251
petrophilus, Chionomys 507
Mormopterus **238**
Semnopithecus 273
Petropseudes **59**
dahli **59**
Petrorhynchus 364
petrovi, Clethrionomys 508
Microtus 526
Sorex 112
petruccii, Myoxus 770
petschorae, Sorex 121
petshorae, Microtus 527
petterdi, Rattus 655
petteri, Lepus 821
Taterillus 562, **563**, 563
Petteromys 548
Petterus 244
petulans, Microtus 523
Tamiasciurus 457
peucinius, Sorex 112
pevzovi, Rhombomys 559
peytoni, Myotis 213
pfeifferi, Lasiurus 206
Pan 277
pfizenmayeri, Alces 389
pfrimeri, Micoureus 20
Phacochoerinae **377**
Phacochoerus **377**
aeliani 377
aethiopicus **377**, 377
africanus **377**, 377
barkeri 377
bufo 377
centralis 377
delamerei 377
edentatus 377
fossor 377
haroia 377
incisivus 377
massaicus 377
pallasii 377
sclateri 377
shortridgei 377
sundevallii 377
typicus 377
phaea, Aplodontia 417
Micoureus 20
Peropteryx 158
phaeniura, Tupaia 132
Phaenomys 687, 700, **737**
ferrugineus **737**
phaeocephala, Nasua 335
phaeocephalus, Pteropus **149**
phaeognatha, Spermophilus 448
phaeomelas, Aeromys 459
phaeonotus, Lepus 814
phaeonyx, Ursus 339
phaeopepla, Ratufa 437
phaeopus, Crocidura 96
Melanomys 708
Oryzomys 707
Sciurus 443
Sylvisorex 105
phaeotis, Artibeus **189**
Lemniscomys 602
Oecomys **716**
phaeura, Chimarrogale **107**, 107, 108
Crocidura **86**, **93**
Liomys 482
Macaca 266
Tupaia 132
phaeurus, Carpomys **583**
Metachirus 20
Peromyscus 730
Sciurus 439
phaeus, Castor 467
Clethrionomys 508
Cricetulus 538
Elephantulus 829
Microtus 524
Ototylomys 726
Phyllotis 738
phaiocephalus, Lemmus 517
Phaiomys 517, 523
phaiops, Eptesicus 201
Pteropus 150
phaios, Crocidura 87
Phalacomys 39
phalaena, Chalinolobus 200
Kerivoula **198**
Paraxerus 434
Phalanger **46**, 46, 47
alba 46
albidus 46
ambonensis 46
breviceps 46
brevinasus 46
carmelitae **46**
coccygis 46
ducatoris 46
fusca 46
intercastellanus 46
interpositus 46, 47
kiriwinae 46
lullulae **46**, 46
matanim **46**
matsika 46
meeki 46
mendeni 47
microdon 46
mimicus 46
minor 46
molucca 46
moluccensis 46
occidentalis 47
orientalis **46**, 46
ornatus **47**
pelengensis **47**, 47
peninsulae 46
permixtio 47
rothschildi **47**
rufa 46
sericeus **47**
vestitus 46, **47**, 47
vulpecula 46
Phalangeridae 45, **46**
Phalangeriformes **45**
Phalangista 46, 58, 60, 62
cookii 59
johnstonii 48
maculata 47
nana 58
ursina 46
phaleratus, Tragelaphus 404
phalius, Damaliscus 394
Phaner **244**
electromontis 244
furcifer **244**
pallescens 244
parienti 244
Phanerinae **244**
phanrangis, Callosciurus 421
pharaohensis, Loxodonta 367
pharaon, Herpestes 305
Pharotis **219**
imogene **219**, 219
pharus, Rattus 661
Phascogale **33**
apicalis 33
calura **33**
crassicaudata 35
dorsalis 34
ingrami 34
longicaudata 32
macdonnellensis 34
melas 32
murex 32
penicillata 34
pirata 34
tafa 34
tapoatafa **33**
venusta 33
Phascolagus 53
Phascolarctidae **45**, **46**
Phascolarctos **45**
adustus 45
cinereus **45**
flindersii 45
fuscus 45
koala 45
subiens 45
victor 45
Phascologale 33
Phascoloictis 33
Phascolomis 45
Phascolomyidae 45
Phascolomys, latifrons 45
Phascolosorex **34**
brevicaudata 34
doriae **34**
dorsalis **34**
pan 34
umbrosa 34
whartoni 34
phasma, Alticola 503
Chaetodipus 484
Eptesicus 203
Galagoides 249
Myotis 216
Peromyscus 735
Rhinolophus 166
Thomomys 474
Phataginus 415
Phaulomys 513, **532**, 532
andersoni **532**
imaizumii 533
kageus 533
niigatae 533
okiensis 533
smithii 532, **533**
phayrei, Callosciurus **423**
Hylopetes **461**
Trachypithecus **274**
phelleoecus, Thomomys 474
phelpsi, Marmosa 19
Oecomys 716
Phenacomys 502, 504, 505, **533**
celatus 533
celsus 533
constablei 533
crassus 533
intermedius 505, **533**, 533
laingi 533
latimanus 533
levis 533
longicaudus 504
mackenzii 533
olympicus 533
oramontis 533
orophilus 533
preblei 533
pumilus 533
soperi 533
truei 533
ungava **533**, 533
phenax, Hexolobodon **803**, 803
Mustela 324
Neotoma **712**
Spilogale 317
Philander 15, 20, **22**, 22
andersoni **22**
azaricus 22
canus 22
crucialis 22
frenata 22
fuscogriseus 22
grisescens 22
mcilhennyi 22
melantho 22
melanurus 22
nigratus 22
opossum **22**
pallidus 22
quica 22
superciliaris 22
virginianus 22
philander, Caluromys **15**
Didelphis 15
Philantomba 410
philantomba, Cephalophus 410
philbricki, Trachypithecus 274
philbyi, Meriones 557
Philetor **219**, 220
brachypterus **219**
rohui 219
veraecundus 219

philipi, Echymipera 41
philippensis, Macaca 266
Petaurista **463**, 463
Sus **378**, 378
Tarsius 251
philippi, Arctocephalus 326
philippii, Arctocephalus **327**, 327
Odocoileus 391
Oligoryzomys 718
Phocoena 359
Zalophus 329
philippinensis, Cynocephalus 135
Paradoxurus 344
Rhinolophus **168**
Rousettus 152
Sundasciurus 451, **452**, 452
Taphozous **161**
Philippinopterus 235, 237
philippinus, Cervus 386
philippsi, Aonyx 310
Paraonyx 310
philistinus, Microtus 522
Philistomys 768
phillippi, Bettongia 49
phillipsi, Damaliscus 394
Herpestes 304
Heterocephalus 772
Leptailurus 292
Madoqua 398
Mus **627**
Prionailurus 296
Tatera **561**, 561
Trachypithecus 274
phillipsii, Cryptotis 109
Dipodomys 477, **480**
Notiosorex 111
philombrius, Peromyscus 734
phippsi, Aethomys 568
Taxidea 325
phipsoni, Petinomys 464
Phloeomyinae 583, 587, 602, 608, 640
Phloeomys 583, 587, 602, **640**, 640
albayensis 640
cumingi **640**, 640
elegans 640
pallidus **640**, 640
schadenbergi 587
Phoca 329, **331**, 331
albicauda 332
albini 332
annellata 332
antarcticus 332
baicalensis 332
barbata 329
beaufortiana 332
birulai 332
botnica 332
byroni 328
californica 332
canina 332
carcinophaga 330
caspica **331**
chorisi 332
concolor 332
cristata 329
dorsata 332
equestris 331
fasciata **331**
flavescens 328
foetida 332
geronimensis 332
gichigensis 332
groenlandica **331**
grypus 329
hispida **329**, **332**, 332
insularis 332
jubata **327**
krascheninikovi 332
kurilensis 332
ladogensis 332
largha **332**, 332
leonina 330
leptonyx 330
leucopla 332
linnaei 332
littorea 332
macrodens 332
mellonae 332
monachus 331
nummularis 332
oceanica 332
ochotensis 332
octonata 332
oronensis 332
pealii 332
petersi 332
pomororum 332
pribilofensis 332
proboscidea 330
pusilla 326
pygmaea 332
richardii 332
rosmarus 325
saimensis 332
scopulicola 332
semilunaris 332
sibirica **332**
soperi 332
stejnegeri 332
thienemannii 332
tigrina 332
ursina 326, 327
variegata 332
vitulina 331, **332**, 332
Phocaena 358
crassidens 355
dalli 359
obtusata 358
phocaenoides, Delphinus 358
Neophocaena **358**, 358
Phocarctos **328**, 328
hookeri **328**
Phocena 358
Phocidae **329**
Phocinae 329
Phocoena **358**
americana 358
communis 358
dioptrica 358
lineata 358
philippii 359
phocoena **358**
relicta 358
sinus **358**
spinipinnis **359**
vomerina 358
phocoena, Delphinus 358
Phocoena **358**
Phocoenidae 349, **358**
Phocoenoides **359**
dalli **359**
truei 359
Phodopus 487, 536, **539**
bedfordiae 539
campbelli **539**
crepidatus 539
praedilectus 539
przewalskii 539
roborovskii **539**
sungorus **539**, 539
tuvinicus 539
phoenicurus, Callosciurus 422
Galerella 303
Pholidota 415
Phoniscus 196, 197
phoxinus, Ornithorhynchus 13
Phractomys 563
Phragmomys 563
phrudus, Mormopterus **238**
Phrygetis 143
phrygius, Dryomys 766
phylarchus, Rangifer 392
phylax, Cercopithecus 264
phyllis, Dorcopsis 51
Phylloderma **180**, 180
boliviensis 180
cayanensis 180
septentrionalis 180
stenops **180**
Phyllodia 176, 177
Phyllomys 791
brasiliensis 791
Phyllonycterinae **182**
Phyllonycteris 182, **183**, 183
aphylla **183**
bombifrons 182
major 183
obtusa 183
poeyi **183**, 183
sezekorni 182
Phyllophora, megalotis 178
Phyllops **190**, 192
albomaculatum 190
falcatus **190**
haitiensis 190
Phyllostoma, albomaculatum 190
bennettii 179
bilabiatum 191
lilium 192
lineatum 190
macrophyllum 178
personatum 193
pusillum 193
rotundus 194
stenops 180
Phyllostomatidae 177
Phyllostomidae **177**
Phyllostominae **177**
Phyllostomus **180**
angusticeps 180
aruma 180
ater 180
caucae 180
caurae 180
curaca 180
discolor **180**
elongatus **180**, 180
hastatus **180**
innominatum 180
latifolius **180**
maximus 180
paeze 180
panamensis 180
verrucosus 180
Phyllotis 695, 702, 703, **737**, 738
abrocodon 738
amicus **737**
andium **737**
arenarius 738
boedeckeri 738
bonaeriensis **737**, 738
campestris 738
capito 738
caprinus **738**, 738
chilensis 738
darwini **738**, 738, 739
definitus **738**, 738
dichrous 738
elegantulus 738
fruticicolus 737
fulvescens 738
fuscus 738
garleppi 702
gerbillus **738**, 738
glirinus 738
griseoflavus 738
haggardi **738**
illapelinus 738
lanatus 738
limatus 738
lutescens 738
magister **738**, 738
maritimus 737
megalotis 738
melanius 737
melanotis 738
melanotus 738
mollis 738
montanus 737
nogalaris 738
oreigenus 739
osgoodi **738**, 738
osilae **738**
phaeus 738
platytarsus 738

posticalis 738
ricarulus 738
rupestris 739
segethi 738
stenops 737
tamborum 737
tucumanus 738
vaccarum 739
wolffhuegeli 739
wolffsohni 738, **739**, 739
xanthopygus 738, **739**, 739
phyllotis, Corynorhinus 205
Idionycteris **205**, 224
Ototylomys **726**, 726
Physalus 349, 359
physalus, Balaena 349
Balaenoptera **350**
Physeter **359**
australasianus 359
australis 359
bidens 362
breviceps 359
catodon **359**, 359
macrocephalus 359
Physeteridae 349, **359**
physodes, Reithrodon 740
piacentinii, Madoqua **398**
pianmaensis, Micromys 620
picata, Dactylopsila 61
picatum, Mimon 179
picatus, Chalinolobus **200**
Tamiasciurus 457
picea, Crocidura **93**
piceus, Callosciurus 423
pichiy, Loricatus 67
Zaedyus **67**
picinum, Erethizon 776
picinus, Rattus 655
picta, Boselaphus 402
Capra 405
Hyaena 283
Isothrix 793
Kerivoula **198**
Neotoma 712
Tatera 560
Tupaia **133**
Viverra 348
picteti, Rattus 658
picticaudata, Procapra **399**, 399
pictipes, Wilfredomys **752**, 752
pictor, Chiropodomys 584
Rhipidomys 744
picturatus, Cercopithecus 263
pictus, Auliscomys **695**
Caluromys 15
Choiropotamus 377
Dryomys 766
Echimys **792**
Hemiechinus 78
Isothrix 792
Liomys **482**, 482
Lycaon **283**
Macropus 55
Neacomys **708**, 709
Potamochoerus 378
Reithrodon 695
Tamias 454
Tragelaphus 404
Vespertilio 196
pidonis, Maxomys 614
Tragulus 382
pierrei, Callosciurus 422
Melogale 315
Tragulus 382
pierreicolus, Thomomys 475
pigra, Alouatta **255**
Didelphis 17
Pika 807
pileatus, Hylobates **276**
Macaca 268
Miopithecus 269
Saguinus 254
Trachypithecus **274**, 274
pilettei, Miopithecus 269
Pilgrimia 367
pilicauda, Myoictis 32
Piliocolobus 271, 272
pilirostris, Urotrichus **129**
pilligaensis, Pseudomys 645, **648**, 648, 649
pilorides, Capromys **800**, 800
Isodon 800
Megalomys 707
Mus 800
Suncus 103
pilosa, Crocidura 89
Nycteris 161
pilosatibialis, Myotis 211
pilosissima, Didelphis 17
pilosus, Dasypus **66**
Myotis 214
Pteropus **149**
Rhinolophus 165
pimelura, Thylamys 23
pimensis, Perognathus 485
pinacate, Chaetodipus 484
pinacus, Maxomys 614
Pinalea 110
pinalensis, Thomomys 474
pinalis, Peromyscus 729
pinatus, Sundamys 666
pinchacus, Tapirus 371
pinchaque, Tapirus **371**
pinchonianus, Naemorhedus 407
pindicus, Myoxus 770
Pinemys 517
pinetis, Geomys 469, **470**, 470
Sylvilagus 826
pinetorum, Liomys 482
Microtus **528**, 528
Neotoma 712
Thomomys 475
pinetus, Lepus 814
pingi, Tupaia 131
pingshiangicus, Muntiacus 389
pinheiroi, Marmosops 20
pinicola, Oryzomys 721
piniensis, Ratufa 437
Sundasciurus 451
pinius, Tragulus 382
piperi, Perognathus 485
piperis, Rattus 661
pipidonis, Callosciurus 421
Rattus 660
pipilans, Tamias 456
Pipistrellus 200, 205, 208, 214, **219**, 219, 220, 221, 222, 223, 224
abaensis 222
abramus 221, 223
adamsi 224
aegyptius **219**
aero **219**
affinis **219**, 222
afghanus 220
africanus 222
agilis 223
akokomuli 221
aladdin 220, 223
alaschanicus 223
albicans 221
albolimbatus 221
alcythoe 221
anchietai 204, **219**
angulatus 224
anthonyi **219**
apus 221
arabicus **220**
ariel **220**
aristippe 223
austenianus 223
australis 221
babu **220**
bactrianus 223
bancanus 221
bellieri 221
bicolor 219
blythii 220
bodenheimeri **220**
bonapartei 223
borneanus 220
brachyotos 223
broomi 221
cadornae **220**
calcarata 221
camortae 221
canus 221
caucasicus 223
caurinus 220
ceylonicus **220**
chrysothrix 220
circumdatus **220**, 223
clarus 224
collinus 224
coreensis 223
coromandelicus 220
coromandra **220**
coxi 223
crassulus **220**
culex 222
cuprosus **220**
curtatus 222
darwini 223
deserti 219
dormeri **220**
drungicus 220
eisentrauti **221**
endoi **221**
erythrodactylus 224
flavescens 223
floridanus 224
fouriei 222
fuscatus 221
fuscipes 223
genei 223
glaucillus 222
griseus 223
helios 222
hesperus **221**
ikhwanius 221
imbricatus **221**, 221, 222
indicus 220
inexspectatus **221**
irretitus 221
javanicus **221**, 221
joffrei **221**
kitcheneri **221**
krefftii 224
kuhlii 219, **221**
lepidus 221
leucippe 223
leucomelas 223
leucotis 221
limbatus 223
lobatus 221
lophurus **222**
mackenziei 224
macropterus 223
macrotis **222**
maderensis **222**
marginatus 221
marrensis 223
maurus 223
maximus 221
mediterraneus 223
melanopterus 223
merriami 221
meyeni 221
micropus 220
mimus **222**
minahassae **222**
minusculus 222
minuta 221
minutissimus 223
monticola 224
mordax **222**
murinus 223
murrayi 224
musciculus **222**
nanulus **222**
nanus **222**
nathusii **222**
nigrans 223
nigricans 223
nitidus 224
obscurus 224
ochromixtus 223
oklahomae 221
orientalis 224

pagenstecheri 222
pallescens 223
papuanus 224
parvipes 220
paterculus 221, **222**
peguensis **223**
permixtus **223**
petersi **223**
pipistrellus 220, **223**
ponceleti 224
portensis 220
potosinus 221
principulus 222
pulcher 223
pullatus 221
pulveratus **223**
pumiloides 221
pusillulus 222
pusillus 223
pygmaeus 223
raptor 220
rueppelli **223**
rusticus **223**
santarosae 221
savii 220, **223**
senegalensis 223
sewelanus 224
shanorum 220
societatis **223**
stampflii 222
stenopterus **223**
stenotus 223
sturdeei **224**
subcanus 220
subflavus **224**
subtilus 221
subulidens 224
tamerlani 223
tasmaniensis **224**
tauricus 223
tenuis **224**
tonfangensis 220
tralatitius 221
tramatus 220
typus 223
unicolor 222
ursula 221
velox 223
veraecrucis 224
vernayi 223
vispistrellus 221
vordermanni 222
wattsi 224
westralis 224
yunnanensis 222
pipistrellus, Pipistrellus 220, **223**
Vespertilio 219
pirata, Callosciurus 421
Phascogale 34
piratae, Maxomys 614
piriensis, Rhinolophus 169
pirinus, Clethrionomys 508
pirrensis, Isthmomys **706**, 706
Leopardus 292
Oryzomys 720
piscator, Martes 320
Ursus 339
Viverra 348
piscatoria, Lutra 312
piscivora, Osbornictis **347**, 347
pita, Mazama 390
Pithanotomys 787
Pithecanthropus 276
Pithecheir 603, **640**
melanurus **640**, 640
parvus **640**
Pithecia **261**, 261
adusta 262
aequatorialis **261**
albicans **261**, 262
capillamentosa 262
chrysocephala 262
guapo 262
hirsuta 261, 262
inusta 262
irrorata **261**
leucocephala 262
lotichiusi 262
milleri 262
monachus **261**, 261
napensis 262
ochrocephala 262
pithecia **262**
pongonias 262
rufibarbata 262
rufiventer 262
saki 262
vanzolinii 261
pithecia, Pithecia **262**
Simia 261
Pitheciidae 254
Pitheciinae **260**
pithecus, Macaca 268
Pithelemur 247
Pithes 266
Pithesciurus 260
pitmani, Crocidura 89, **93**
Ourebia 399
Tatera 561
pittieri, Ichthyomys **705**
Pitymys 506, 517, 518, 521, 524, 526, 528
leucurus 516
majori 520
sikimensis 522
subterraneus 520, 529
piutensis, Thomomys 474
pixuna, Bolomys 696
Pizonyx 207
plagiodon, Stenella 356
Plagiodontia 800, **804**
aedium 799, **804**, 804
araeum **804**, 804
caletensis 804
hylaeum 804
ipnaeum **804**
spelaeum 804
stenocoronalis 804
velozi 804
Plagiodontinae **804**
planaltensis, Kunsia 706
plancei, Nyctalus 217
planiceps, Cervus 387
Crocidura 85, **93**, 93
Microtus 524
Mormopterus **238**
Myotis **214**
Neotoma 712
Pappogeomys **473**
Prionailurus **295**
Sorex 117, 118, **119**, 121
Sus 379
Ursus 339
Planiceros 402
planicola, Apodemus 572
Spermophilus **448**
planidens, Cervus 387
planifrons, Erophylla 182
Heteromys 481
Hybomys 592, **596**, 597
Hyperoodon **361**
Pappogeomys 472
Sigmodon 747
Planigale **34**, 34
brunnea 34
brunneus 34
gilesi **34**
ingrami **34**
maculata **34**
minutissimus 34
novaeguineae **34**
sinualis 34
subtilissima 34
tenuirostris **34**
planirostris, Artibeus 188, **189**, 189
Mandrillus 268
Molossops **235**
Oryzomys 724
Scotophilus 227
Thomomys **474**
planorum, Thomomys **474**
plantani, Callosciurus 423
plantinarensis, Liomys 482
plasticus, Callosciurus 423
Platacanthomyinae 501, **684**, 763
Platacanthomys **684**, 684
lasiurus **684**, 684
Platalina **182**
genovensium **182**, 182
Platanista **360**, 360
gangetica **360**, 360
gangeticus 360
indi 360
minor **360**, 360
Platanistidae 349, **360**
Platelephas 367
platensis, Lontra 311
Oxymycterus 727
platous, Sylvicapra 412
platurus, Dipus 490
Pygeretmus 490
platycentrotus, Coendou 775
platycephala, Chimarrogale **107**, 107
platycephalus, Clethrionomys 509
Crossarchus 302
Dipodomys 479
platyceps, Micronycteris 178
Platycercomys 490
platyceros, Bos 401
Cervus 386
Dama 387
Diceros 372
platycerus, Bubalus 402
Platycranius 502, 504
platycranius, Sorex 119
Platygeomys 469, 471
Platymops 237, 238
platyops, Eptesicus **203**, 203
Graphiurus **765**, 765
Melomys 615, 617, **618**, 618
Potorous **50**
platyotis, Erinaceus 77
Hemiechinus 78
Platypus 13
anatinus 13
platyrhina, Mephitis 317
platyrhinus, Naemorhedus 407
Sorex 113
Vombatus 46
platyrhynchus, Rangifer 392
Platyrrhinus **190**, 190
aquilus 191
aurarius **190**
brachycephalus **190**, 191
chocoensis **190**
dorsalis 190, **191**, 191
fumosus 191
helleri **191**
incarum 191
infuscus **191**
intermedius 191
latus 190, 191
lineatus **191**
nigellus 191
oratus 191
recifinus **191**
saccharus 190
umbratus **191**, 191
vittatus **191**
zarhinus 191
Platystomus 365
platytarsus, Phyllotis 738
platythrix, Mus **627**, 628
platyurus, Dipus 490
Hylopetes 461
Pygeretmus **490**, 490
plebejus, Calomys 700
Delomys 700
Plecotus 205, **224**, 224
ariel 224
auritus **224**, 224, 225
australis 225
austriacus **224**, 225
brevimanus 224
brevipes 224

christiei 224
communis 224
cornutus 224
hispanicus **224**
homochrous 224, 225
ingens 225
intermedius 225
kirschbaumii 224
kolombatovici 224
kozlovi 224
leconteii 225
macrobullaris 224
macrotis 225
megalotos 224
meridionalis 224
mexicanus **224**
montanus 224
mordax 224
ognevi 224
otus 224
pallescens 225
peronii 224
puck 224, 225
rafinesquii **225**
sacrimontis 224
taivanus **225**
teneriffae 224, **225**
townsendii 224, **225**
typus 224
uenoi 224
velatus 205, 224
virginianus 225
vulgaris 224
wardi 224
plectilis, Dipodomys 477
Plectomys, paludicola 691
Plectrochoerus 790
pleei, Cebus 259
pleiharicus, Muntiacus 388, 389
pleistocenicus, Geocapromys 800
plenus, Sigmodon 747
Pleopus 49
Plerodus 101
Plerotes **145**
anchietai **145**
plerus, Perognathus 486
plesioceros, Diceros 372
Plesiosoricidae 80
plesius, Spermophilus 448
pleskei, Sorex 113
plethodon, Monophyllus **185**
Pleuropterus 135
plicata, Balantiopteryx **156**, 156
Chaerephon 232, **233**
plinii, Dama 387
Plioctomys 534
Pliopotamys 532
Plioselevinia 768
plowsi, Bathyergus 771
plumbea, Blarina 106
Sousa 355
plumbescens, Cannomys 685
plumbeus, Capreolus 390
Colomys 586
Peromyscus 732
Rattus 657
plurimammis, Bandicota 579
pluto, Atilax 301
Callosciurus 423
Cephalophus 411
Cercopithecus 264
Orthogeomys 471
Saccolaimus **159**
Saguinus 254
pluton, Pteropus 151
pluvialis, Peromyscus 732
Saimiri 260
poadromus, Lepus 819
poaiae, Sciurus 439
poasensis, Syntheosciurus 452
pocockei, Mystromys 677
pococki, Equus 369
Leptailurus 292
Mus 625
Stenomys 665
Podabrus 35
Podanomalus 635
Podihik 101
kura 102
podje, Tarsius 250
Podogymnura **80**, 80
aureospinula **80**
minima 80
truei **80**, 80
podolicus, Spalax 755
Podomys 703, 713, 728, **739**
floridanus **739**
Podoxymys **739**
roraimae **739**, 739
Poecilictis 319
Poecilogale **325**
africana 325
albinucha **325**
bechuanae 325
doggetti 325
flavistriata 325
lebombo 325
transvaalensis 325
Poecilolagus 814, 815
poecilops, Gerbillus **553**
poecilotis, Caracal 289
Didelphis 16
poecilura, Panthera 298
Poelagus **823**, 823
larkeni 823
marjorita **823**
oweni 823
Poemys 541
poenitentiarii, Rattus 660
poensis, Aonyx 310
Chalinolobus **200**
Cricetomys 540
Crocidura 82, 85, 91, **93**
Deomys 544
Galagoides 249
Genetta 346
Grammomys 594
Mandrillus 268
Paraxerus **435**
Poiana 347
Poephagus 401
poeppigii, Histiotus 205
Lagothrix 258
Spalacopus 789
poeppingii, Mysateles 802, 803
Poescopia 350
poeyanus, Solenodon 69
poeyi, Mysateles 803
Phyllonycteris **183**, 183
pogonias, Cercopithecus **265**
Pogonocebus 263
Pogonomelomys 585, 586, 616, **640**, 641
brassi 641
bruijni **641**, 641
mayeri **641**, 641
ruemmleri 585
sevia **641**, 641
tatei 641
Pogonomys 583, 584, 587, 602, **641**, 641
championi **641**
derimapa 642
dryas 641
fergussoniensis 641
huon 642
lepidus 642
loriae **641**, 642
macrourus 641, **642**
mollipilosus 642
sylvestris 641, **642**, 642
pohlei, Apodemus 572
Castor 467
Melomys 619
Pteropus **150**
Poiana 344, **347**, 347
leightoni 347
liberiensis 347
poensis 347
richardsonii **347**, 347
polaris, Mustela 321
Ursus 339
polia, Crocidura 84, **93**
Ratufa 437
polii, Ovis 408
poliocephalus, Eira 318
Pteropus **150**
Trachypithecus 273
Poliocitellus 444, 446
poliogastra, Crocidura 94
poliomystax, Erythrocebus 265
polionops, Grammomys 593
polionotus, Microdipodops 480
Peromyscus 732, **735**
poliopardus, Panthera 298
poliophaeus, Erythrocebus 265
poliopus, Callosciurus 423
Proechimys 796, **797**
Sciurus 440
poliotricha, Leptailurus 292
polita, Marmosops 20
poliurus, Colobus 270
polius, Akodon 692
Oryzomys **724**
Peromyscus **735**
poljakovi, Microtus 525
pollens, Nectomys 709
Sundamys 666
polli, Congosorex **81**
Myosorex 81
polonicus, Mus 626
Spalax 753, 755
Ursus 339
polulus, Surdisorex **104**
polychroma, Cricetus 538
polycomos, Simia 269
Polygomphius 66
polykomos, Cebus 269
Colobus **270**
polyodon, Equus 370
polypolius, Peromyscus 730
polythrix, Myotis 212
pomo, Arborimus **505**, 505
pomona, Hipposideros 171, **174**
pomororum, Phoca 332
ponceleti, Pipistrellus 224
Solomys **663**
pondiceriana, Herpestes 305
pondoensis, Crocidura 83
Ictonyx 319
Myosorex 100
Otomys 681
Scotophilus 227
pondoliae, Amblysomus 74
Pongidae 274, 276
Pongo **277**
abelii 277
abongensis 277
agrias 277
batangtuensis 277
bicolor 277
bornaensis 277
brookei 277
curtus 277
dadappensis 277
deliensis 277
genepaiensis 277
gigantica 277
landakkensis 277
langkatensis 277
morio 277
owenii 277
pygmaeus **277**, 277
rantarensis 277
rufus 277
satyrus 277
skalauensis 277
sumatranus 277
tuakensis 277
wallichii 277
wurmbii 277
pongoalsoni, Hylobates 276
pongolensis, Dendromus 543
pongonias, Pithecia 262
Pongonomys 640

ponsi, Choeronycteris 184
ponticus, Apodemus **572**, 572, 573
Clethrionomys 508
Erinaceus 77
Lepus 817
pontifex, Ctenomys **786**
Tachyoryctes 687
pontius, Chionomys 507
Pontoporia **360**
blainvillei **361**
tenuirostris 361
Pontoporiidae 360
poolei, Hexolobodon 803
poolii, Funisciurus 427
poortmanni, Civettictis 345
popaeus, Mus 623
popayanus, Thomasomys 749
popei, Chaetodipus 484
popelairi, Myocastor 806
popofensis, Microtus 527
porcarius, Papio 269
porcellus, Cavia 778, **779**, 779
Mus 778
porcina, Arctocephalus 326
porcinus, Akodon 691
Axis 384, **385**, 385
Cervus 385
Conepatus 316
Porcula 378
porculus, Uromys **672**
porcus, Potamochoerus **378**, 378
Sus 377
porphyrops, Trachypithecus 274
porsildi, Lepus 815
portali, Crocidura 96
portavernus, Eptesicus 202
Portax 401
portenkoi, Lemmus 517
Sorex 113, 116, **119**, 121
portensis, Pipistrellus 220
porteousi, Ctenomys 784, **786**
Oreotragus 398
porteri, Aconaemys 787
Bos 401
Scalopus 127
porteusi, Oreotragus 399
portoricensis, Isolobodon 803, **804**
Monophyllus 185
Pteronotus 177
portus, Callosciurus 422
Rattus 660
poschiavinus, Mus 626
posticalis, Phyllotis 738
postus, Myoxus 770
Potamochoerus **377**
africanus 378
albifrons 378
albinuchalis 378
arrhenii 378
choeropotamus 378
congicus 378
cottoni 378
daemonis 378
edwardsi 378
hassama 378
hova 378
intermedius 378
johnstoni 378
keniae 378
koiropotamus 378
larvatus **378**
madagascariensis 378
maschona 378
mawambicus 378
nyasae 378
penicillatus 378
pictus 378
porcus **378**, 378
somaliensis 378
ubangensis 378
Potamogale **73**
allmani 73
argens 73
velox **73**
Potamogalinae 70, **72**
Potamophilus 341
Potamotragus 410
Potamys 709, 805
potanini, Ovis 409
Pygeretmus 490
potens, Sundamys 666
potenziani, Presbytis **271**
pothae, Tatera 562
Potoroidae **48**
Potoroiis 49
Potoroo 49
Potoroops 49
Potorous **49**
apicalis 50
benormi 50
gilbertii 50
longipes **49**
micropus 50
minor 50
morgani 50
murina 50
muscola 50
myosurus 50
peronii 50
platyops **50**
rufus 50
setosus 50
tridactylus **50**
trisulcatus 50
tuckeri 50
Potos **333**, 333
arborensis 333
aztecus 333
boothi 333
brachyotos 333
brachyotus 333
brasiliensis 333
campechensis 333
caucensis 333
caudivolvula 333
chapadensis 333
chiriquensis 333
dugesii 333
flavus **333**
guerrerensis 333
isthmicus 333
lepida 333
mansuetus 333
megalotus 333
meridensis 333
modestus 333
nocturna 334
potto 334
prehensilis 334
simiasciurus 334
tolimensis 334
Potosinae **333**
potosinus, Pipistrellus 221
Thomomys 476
potrerograndei, Reithrodontomys 742
Potto 248
potto, Lemur 248
Perodicticus **248**
Potos 334
pousarguei, Chlorocebus 265
pousarguesi, Epomophorus 141
Kobus 414
poutensis, Hipposideros 173
pouvrelianus, Cervus 386
povensis, Vandeleuria 672
povolnyi, Rattus 660
powelli, Helogale 304
Kobus 414
Procolobus 272
Pronolagus 823
Thomomys 474
Tragelaphus 404
poyu, Euphractus 66
prachin, Callosciurus 422
praecelsus, Rattus 652
praeceps, Artibeus 188, 189
Arvicanthis 578
praeconis, Pseudomys **648**
praecypria, Crocidura 96
praedatrix, Dobsonia **140**
praedax, Crocidura 85
praedilectus, Phodopus 539
Praesorex 81, 86
praestans, Rattus 657
praestens, Rhinolophus 165
praetextus, Atlantoxerus 420
Funisciurus 427
Mus 626
praetor, Apodemus 572
Rattus 649, **658**
Thomasomys 749
prageri, Lycaon 283
prajae, Acerodon 137
pranis, Callosciurus 421
Praomys 607, 608, 609, 630, 631, **642**, 642, 643
albipes 631
angolae 610
burtoni 644
collinus 683
delectorum **642**, 642
hartwigi **642**
jacksoni **642**, 642, 643
maurus 643
melanotus 642
minor **643**, 643
misonnei **643**, 643, 644
montis 642, 643
morio **643**, **643**, 644
mutoni **643**, 643
obscurus 642
octomastis 642
peromyscus 642, 643
rostratus 592, **643**, 643, 644
ruppi 630
sudanensis 642
taitae 642
tullbergi 607, 608, 642, 643, **644**, 644
viator 642, 643
Praopus 65
Prasadsciurus 426, 427
prasinatus, Condylura 124
pratensis, Chrysospalax 76
Clethrionomys 508
Lontra 311
Micromys 620
Microtus 527
Pappogeomys 472
Spermophilus 449
Steatomys 545, **546**, 546
prateri, Felis 290
Praticola 505
pratincola, Geomys 469
Microtus 525
pratti, Hipposideros **174**
Myoprocta 782
prattorum, Rhynchomeles **42**, 42
preblei, Dipodomys 479
Lontra 311
Peromyscus 736
Phenacomys 533
Sorex **119**
Tamiasciurus 457
Zapus 499
preblorum, Marmota 432
predator, Dasyurus 32
prehensilis, Capromys 802
Coendou **775**, 775
Hystrix 775
Microcebus 243
Mysateles 802, **803**, 803
Mystateles 803
Paradoxurus 344
Potos 334
premaxillaris, Thomomys 475
preniensis, Dinaromys 511
Presbypitheus 270
Presbytinae 269
Presbytis **270**, 271, 273
albigena 266
arwasca 271
aurata 271
australis 271
aygula 270, 271

batuana 271
cana 271
canicrus 271
catemana 271
chrysea 271
chrysogaster 271
chrysomelas 271
comata **270**, 271
cruciger 271
dilecta 271
everetti 271
femoralis **271**, 271
ferruginea 271
flavimanus 271
fluviatilis 271
fredericae 270
frontata **271**
fuscomurina 271
hosei **271**
ignita 271
karimatae 271
keatii 271
margae 271
melalophos **271**, 271
mitrata 270, 271
natunae 271
neglectus 271
nigrimanis 271
nobilis 271
nubigena 271
nubilus 271
nudifrons 271
paenulata 271
percura 271
potenziani **271**
rhionis 271
robinsoni 271
rubicunda **271**
rubida 271
sabana 271
senex 274
siamensis 271
siberu 271
sumatrana 271
thomasi **271**
vetulus 274
Presbytiscus 272
presina, Prionodon 347
prestigiator, Xerus 458
pretiellus, Tragulus 383
pretiosa, Neotoma 712
pretiosus, Callosciurus 423
Liomys 482
Molossus **235**, 235
Tragulus 383
pretoriae, Aethomys 567
Cryptomys 772
Dendromus 543
Graphiurus 764
Ictonyx 319
Mus 624
Otomys 680
Tatera 561
preussi, Cercopithecus 264, **265**
Myosorex 99
Procolobus **272**
Sorex 112
Sylvisorex 112
prevostensis, Peromyscus 732
Sorex 118
prevostii, Callosciurus 421, **423**
Sciurus 420
priam, Semnopithecus 273
priamelus, Semnopithecus 273
pribilofensis, Alopex 279
Phoca 332
Sorex 115, 116
pricei, Chaetodipus 484
Ochotona 811
Tamias 454
prichardi, Pseudalopex 284
prieskae, Rhabdomys 662
priestlyi, Mus 627
prietae, Chaetodipus 484
prigoginei, Colobus 270
prigoginii, Megaloglossus 154
prima, Loxodonta 367
primaevus, Canis 282
Cuon 282
primarius, Rattus 657
Primates 243
primigenius, Bos 401
primitivus, Saguinus 253
primrosei, Petaurista 463
primula, Myotis 208
primulina, Mustela 322
primus, Callosciurus 421
Natalus 195
princeps, Apodemus 571
Cercopithecus 264
Cynopterus 139
Heterohyrax 373
Nyctalus 217
Ochotona 807, 808, **811**, 811
Pteropus 150
Rhinolophus 163
Sorex 116
Thomasomys 749
Xerus **458**, 458
Zapus **499**, 499
principalis, Kunsia 706
Microtus 519, 528
principula, Microgale **71**, 71
principulus, Gerbillus **554**
Pipistrellus 222
prinsulae, Callosciurus 423
Priodon 66
Priodontes **66**
giganteus 67
gigas 67
grandis 67
maximus **67**
Priodontini 64
Prionailurus **295**, 295, 296
alleni 295
anastasiae 295
bengalensis **295**, 295
bennettii 296
borneoensis 295
chinensis 295
ellioti 295
euptilura 295
himalayanus 296
horsfieldi 295
ingrami 295
iriomotensis 295
javanensis 295
koladivinus 296
manchurica 295
microtis 295
minuta 295
nipalensis 295
pardochrous 295
phillipsi 296
planiceps **295**
raddei 295
reevesii 295
ricketti 295
rizophoreus 296
rubiginosus **296**
scripta 295
sumatrana 295
tenasserimensis 295
tingia 295
trevelyani 295
undata 295
viverriceps 296
viverrinus **296**
wagati 295
Prionodon 344, **347**, 347
gracilis 347
linsang 342, **347**, 347
maculosus 347
pardicolor **347**
pardochrous 347
perdicator 347
presina 347
Prionodontinae 344
Prionodos 66
Prionomys 544, **545**
batesi **545**, 545
Prionotemmus 53
priscus, Dipodomys 479
Loxodonta 367
Rhynchonycteris 159
pristina, Vicugna 382
prittiei, Lophuromys 606
prittwitzi, Heterohyrax 373
Hystrix 773
Sigmoceros 395
problema, Ateles 257
problematicus, Bradypus 63
Macaca 266
Proboscidea 367
proboscidea, Mirounga 330
Phoca 330
proboscideus, Macroscelides **830**
Sorex 830
probus, Hylopetes 461
Procapra **399**, 399
altaica 399
diversicornis 399
gutturosa **399**
orientalis 399
picticaudata **399**, 399
przewalskii **399**
Procapromys 800
Procavia 373, **374**
abbyssinicus 374
alpini 374
antineae 374
bamendae 374
bounhioli 374
buchanani 374
burtonii 374
butleri 374
capensis 373, **374**
capillosa 374
chiversi 374
comata 374
coombsi 374
daemon 374
dongolanus 374
ebneri 374
ehrenbergi 374
elberti 374
erlangeri 374
ferrugineus 374
flavimaculata 374
goslingi 374
habessinica 374
habessinicus 374
ituriensis 374
jacksoni 374
jayakari 374
johnstoni 374
kamerunensis 374
kerstingi 374
latastei 374
letabae 374
lopesi 374
luteogaster 374
mackinderi 374
marlothi 374
marrensis 374
matschiei 374
melfica 374
meneliki 374
minor 374
natalensis 374
naumanni 374
orangiae 374
oweni 374
pallida 374
reuningi 374
ruficeps 374
schmitzi 374
schultzei 374
scioanus 374
semicircularis 374
sharica 374
sinaiticus 374
slatini 374
syriaca 374
varians 374
volkmanni 374
waterbergensis 374
welwitschii 374

windhuki 374
zelotes 374
Procaviidae **373**
Procebus 245
procera, Crocidura 87
procerus, Sundasciurus 452
Procervus 385
Procolobus 269, **271**, 272
anzeliusi 272
badius **271**, 272
bouvieri 272
brunneus 272
chrysurus 272
cristatus 272
ellioti 272
ferriginea 272
foae 272
foai 272
fuliginosus 272
gordonorum 272
graueri 272
gudoviusi 272
kabambarei 272
kirki 272
langi 272
lovizettii 272
lulidicus 272
melanochir 272
multicolor 272
nigrimanus 272
olivaceus 272
oustaleti 272
parmentieri 272
pennantii **272**
powelli 272
preussi **272**
rufo-fuliginus 272
rufomitratus **272**, 272
rufoniger 272
schubotzi 272
temminckii 272
tephrosceles 272
tholloni 272
umbrinus 272
variabilis 272
verus **272**, 272
waldroni 272
proconodon, Mus 627
proconsularis, Rhinolophus 164
Procops 388
Procyon 334, **335**, 337
aequatorialis 335
annulatus 335
auspicatus 335
brachyurus 335
brasiliensis 335
californicus 335
cancrivorus **335**
castaneus 335
crassidens 335
dickeyi 335
elucus 335
excelsus 335
flavidus 335
fusca 335
fuscipes 335
gloveralleni **335**, 335
grinnelli 335
gularis 335
hernandezii 335
hirtus 335
hudsonicus 335
incautus 335
inesperatus 335
insularis **335**, 335
litoreus 335
lotor **335**, 335, 336
marinus 335
maritimus 335
maynardi 335, **336**
megalodous 335
melanus 335
mexicana 335
minor 335, **336**
nigripes 335
nivea 335
obscurus 335
ochraceus 335
pacifica 335
pallidus 335
panamensis 335
proteus 335
psora 335
pumilus 335
pygmaeus 335, **336**
rufescens 335
shufeldti 335
simus 335
solutus 335
vancouverensis 335
varius 335
vicinus 335
vulgaris 335
Procyonidae **332**, 332, 336, 337
Procyoninae **334**
procyonoides, Canis 283
Nyctereutes **283**
Prodelphinus 356
proditor, Eothenomys **514**
Prodorcas 399
Proechimys 794, **795**, 795, 796, 797, 798, 799
albispinus **795**, 797
amphichoricus **795**, 798
arabupu 795
arescens 795
boimensis 797
bolivianus **795**, 796
bonafidei 797
brevicauda **795**, 797
burrus 798
cajennensis 798
calidior 798
canicollis **795**
cayennensis **795**, 795, 796, 797, 798
centralis 798
cherriei 795, 796
chiriquinus 798
chrysaeolus **796**, 796
cinnamomeus 797
columbianus 798
corozalus 799
cuvieri **796**
decumanus **796**, 796
denigratus 797
dimidiatus **796**
elassopus 797
elegans 798
fuliginosus 798
goeldii **796**, 798
goldmani 798
gorgonae **796**
gratiosus 797
guairae **796**, 796
gularis **796**
guyannensis 795, 796, 797, 798
hendeei **796**, 798
hilda 798
hoplomyoides **797**
hyleae 795
ignotus 798
iheringi **797**
kermiti 798
leioprymna 795
leptosoma 797
leucomystax 797
liminalis 798
longicaudatus 795, **797**
magdalenae 796, **797**
mincae 796, **797**
myosuros **797**
nesiotes 795
nigrofulvus 797
ochraceus 796
oconnelli 796, **797**
oris 796, **797**
pachita 797
panamensis 798
panema 797
paratus 797
poliopus 796, **797**
quadruplicatus **798**
rattinus 795
riparum 795
roberti 797
rosa 798
rubellus 798
securus 797
semispinosus 795, **798**, 798
serotinus 795
setosus 797, **798**
simonsi 797, **798**
steerei **798**
trinitatis 796, **798**
urichi 796, **798**
vacillator 795
villicauda 795
warreni 796, **798**
Proedromys 517, 518, **533**
bedfordi **533**, 533
Proeulagus 814, 815, 816, 818, 819, 820, 821, 822
Profelis 289, **296**, 296
aurata **296**
celidogaster 296
chalybeata 296
chrysothrix 296
cottoni 296
maka 296
neglecta 296
rutilus 296
profugus, Dendrolagus 50
profusus, Rattus 662
Progerbillurus 547
progrediens, Heterocephalus 772
Prohapalemur 245
Prolagidae 813
Prolagus **813**
sardus **813**
prolatus, Bunomys **582**
Prolemur 245
prolixacaudata, Microgale 71
prolixus, Uromys 671
Prometheomyinae 534
Prometheomys **534**, 534
schaposchnikowi **534**, 534
Promops **239**
ancilla 240
centralis **239**
davisoni 239
downsi 240
fosteri 240
fumarius 240
nasutus **239**
occultus 239
pamana 240
rufocastaneus 240
ursinus 239, 240
Pronolagus 813, 814, **823**, 823
australis 824
barretti 824
bowkeri 823
capricornis 823
caucinus 823
crassicaudatus **823**, 823, 824
curryi 824
ekmani 823
fitzsimonsi 824
kaokoensis 823
kariegae 823
kobosensis 823
lebombo 823
makapani 823
melanurus 824
mülleri 824
nyikae 824
powelli 823
randensis **823**, 823
ruddi 823
rupestris **823**, 823
saundersiae 824
vallicola 824
waterbergensis 823
whitei 823
Proodontomys, cookei 677
propalatum, Dasypus 66
proparator, Cricetomys 540

propiius, Castor 467
propinquus, Eptesicus 202
Peromyscus 730
Tamias 453
Propithecus **247**
albus 247
bicolor 247
candidus 247
coquereli 247
coronatus 247
damanus 247
damoni 247
deckenii 247
diadema **247**, 247
edwardsi 247
holomelas 247
majori 247
perrieri 247
sericeus 247
tattersalli **247**
typicus 247
verreauxi **247**
proplatensis, Panthera 298
Propliomys 511
propritristis, Miniopterus 231
Prosalpingotus 491, 492
Prosciurillus 419, 420, **435**
abstrusus **436**
elbertae 436
evidens 436
griseus 436
hirsutus 436
leucomus 435, **436**
mowewensis 436
murinus **436**
necopinus 436
obscurus 436
occidentalis 436
rosenbergii 436
sarasinorum 436
tingahi 436
tonkeanus 436
topapuensis 436
weberi **436**
proserpinae, Callosciurus 423
Prosimia 245
Prosiphneus 675
prostheceros, Ceratotherium 371
protalopex, Pseudalopex 284
Protarsomys 677, 679
Proteles **309**, 309
canescens 309
cristatus **309**
harrisoni 309
hyenoides 309
lalandii 309
pallidior 309
septentrionalis 309
termes 309
transvaalensis 309
typicus 309
Protelinae **309**
Protemnodon 53, 54
proterus, Nyctalus 217
proteus, Callosciurus 423
Clethrionomys 508
Conepatus 316
Galerella 303
Procyon 335
Protomazama 391
Protosciurus 419
Protoxerini 425, 428, 436
Protoxerus **436**, 436
aubinnii **436**
bea 436
calliurus 436
caniceps 436
centricola 436
cooperi 436
dissonus 436
eborivorus 436
kabobo 436
kwango 436
loandae 436
moerens 436
nigeriae 436
nordhoffi 436
notabilis 436
personatus 436
salae 436
signatus 436
stangeri **436**
subalbidus 436
temminckii 436
torrentium 436
provectus, Microtus 527, 528
provicugna, Vicugna 382
providens, Bandicota 579
providentialis, Thomomys 474
provincialis, Microtus 520
provocax, Crocidura 91
Lontra **311**
Prox 388
proxima, Nycteris 161
proximarinus, Thomomys 474
proximus, Rhinolophus 166
Stylodipus 495
Thomomys 474
Prozaglossus 13
pruinosa, Didelphis 17
Viverra 348
pruinosus, Lasiurus 206
Myotis **214**
Papio 269
Rhizomys **685**
Sphiggurus 777
Trachypithecus 273
Ursus 338, 339
pruneri, Atelerix 76
prusianus, Rhizomys 685
pryeri, Sundasciurus 451
pryerianus, Naemorhedus 407
prymnolopha, Dasyprocta **782**
pryori, Thomomys 475
przevalskii, Ovis 408
przewalskii, Brachiones **546**
Eolagurus **513**
Equus 369
Gerbillus 546
Lepus 819, 821
Mesechinus 79
Myotis 213
Phodopus 539
Procapra **399**
Przewalskium 385
psakastus, Heteromys 481
Psammomys 517, **559**, 559
algiricus 559
dianae 559
edusa 559
nicolli 559
obesus **559**, 559
roudairei 559
terraesanctae 559
tripolitanus 559
vexillaris **559**
psammophila, Sminthopsis **37**
psammophilous, Gerbillus 551
psammophilus, Meriones 557
Perognathus 485
Psammoryctes 43
typhlops 43
pselaphon, Pteropus **150**
Pseudalopex 280, **284**, 284, 285
albigula 284
amblyodon 284
andina 284
antiquus 284
azarae 285
brasiliensis 284
chilensis 284, 285
cinereoargenteus 284
culpaeolus 284
culpaeus **284**
domeykoanus 284
entrerianus 284
ferrugineus 284
fossilis 284
fulvicaudus 285
fulvipes 284
gracilis 284
griseus **284**
gymnocercus **284**
inca 284
lagopus 284
lycoides 284
magellanicus 284
maullinicus 284
montanus 284
parvidens 285
patagonicus 284
prichardi 284
protalopex 284
reissii 284
riveti 284
rufipes 284
sechurae **284**
sladeni 285
smithersi 284
torquatus 284
urostictus 285
vetulus **285**
zorrula 284
Pseudantechinus 29, 33, **34**
macdonnellensis **34**
mimulus 35
ningbing **35**
woolleyae **35**
Pseudanthropos 276
pseudarmandi, Myospalax 675, 676
Pseudaxis 385
pseudaxis, Cervus 386
Pseudocebus 259
Pseudocervus 385
Pseudocheiridae 46, **58**
Pseudocheirus 58, **59**, 59, 60
avarus 59
banksii 60
barbatus 59
bassianus 60
bernsteini 59
canescens **59**
capistratus 59
caroli **59**
caudivolvula 60
cinereus 59
colletti 59
convolutor 60
cookii 60
dammermani 59
forbesi **59**
grisescenti 59
gyrator 59
herbertensis **59**
incana 60
incanens 60
laniginosa 60
laniginosus 60
larvatus 59
lewisi 59
longipilis 59
mayeri **59**
modestus 60
mongan 59
notialis 60
novaehollanidiae 60
occidentalis 60
oralis 60
peregrinus **59**
pulcher 60
pygmaeus 59
rubidus 60
schlegeli **60**
versteagi 59
victoriae 60
viverrina 60
Pseudochirops **60**
albertisii **60**, 60
archeri **60**
argenteus 60
beauforti 60
buergersi 60
caecias 60
corinnae **60**

coronatus 60
cupreus **60**
fuscus 60
insularis 60
obscurior 60
paradoxus 60
schultzei 60
Pseudochirulus 59
Pseudochirus 59, 60
dahli 59
Pseudoconomys 622
pseudocrinitus, Peromyscus **735**
pseudocurtatus, Allocricetulus 537
pseudodelphis, Stenella 356
Pseudogenetta 345, 346
villiersi 346
Pseudogorilla 276
pseudogriseus, Cricetulus 537
Pseudohydromys 604, 614, 620, 632, **644**
murinus 598, **644**, 644
occidentalis 598, **644**
Pseudois 405, **409**
burrhel 409
caesia 409
nahoor 409
nayaur **409**
schaeferi **409**
szechuanensis 409
Pseudokobus 413
Pseudomys 600, 604, 635, **644**, 644, 645, 646, 647, 648, 649, 674
albocinereus **644**, 645, 646, 649
apodemoides **645**, 645
auritus 645
australiensis 647
australis 644, **645**, 645, 646, 647, 648
bolami **645**
brazenori 646, 647
calabyi 647
chapmani **645**, 647
delicatulus **645**, 645, 648, 649
desertor **645**, 646
ferculinus 647
fieldi **646**
flavescens 645
fumeus 645, **646**, 646, 648, 649
fuscus **646**, 646
glaucus **646**
gouldii 645, **646**, 647, 648
gracilicaudatus **647**, 648
hermannsburgensis 645, **647**, 647
higginsi 645, **647**, 647, 648
johnsoni 645, **647**
laborifex **647**
leucopus 647
lineolatus 645
mimulus 645
minnie 645
mordicus 646
murinus 645
murrayensis 646
nanus 645, **647**, 647, 648
novaehollandiae 645, **648**, 648, 649
occidentalis **648**, 648
oralis **648**
patrius **648**
pilligaensis 645, **648**, 648, 649
praeconis **648**
pumilus 645
rawlinnae 646
shortridgei 645, 646, 648, **649**, 649
squalorum 644
stirtoni 645
subrufus 646
ultra 647
vandycki 645
wombeyensis 646
pseudonapaea, Sicista 496, **497**, 497
pseudopardalis, Leopardus 291
Pseudoquagga 369
Pseudorca **355**
crassidens **355**
destructor 355
meridionalis 355
Pseudoryzomys **739**
reigi 739
simplex **739**
wavrini 739
Pseudostoma 469
Pseudotroctes 66
pseudotursio, Lagenorhynchus 353
pshavus, Chionomys 507
Psilodactylus 247
psilodactylus, Daubentonia 247
Psilogrammurus 48
psilotis, Pteronotus 177
psilurus, Akodon 692
Myospalax **676**, 676
Neomys 110
psora, Procyon 335
Ptenochirus **145**, 145
jagori **145**, 145
minor **145**
Ptenos 59
Pteralopex **145**
acrodonta **145**
anceps **145**, 146
atrata **145**, 145
pulchra **146**
Pterobalaena 349
Pteromyinae **459**
Pteromys 459, **464**, 464
aluco 464
amygdali 464
anadyrensis 464
arsenjevi 464
athene 464
betulinus 464
buechneri 464
derbianus 757
gubari 464
horsfieldi 461
incanus 464
interventus 464
melanopterus 459
momoga 464
momonga **464**
ognevi 464
orii 464
russicus 464
sibiricus 465
tephromelas 459
turovi 465
volans **464**
vulgaris 465
wulungshanensis 465
xanthipes 465
Pteromyscus **465**
borneanus 465
pulverulentus **465**
Pteronotus **176**, 176
boothi 177
calvus 176
centralis 176
continentis 177
davyi **176**, 176
fuliginosus 177
fulvus 176
fuscus 177
gonavensis 177
griseus 177
gymnonotus **176**
incae 176
inflata 177
macleayii **177**
mesoamericanus 177
mexicanus 177
osburni 177
paraguanensis 177
parnellii **177**
personatus **177**
portoricensis 177
psilotis 177
pusillus 177
quadridens **177**, 177
rubiginosus 177
suapurensis 176
torrei 177
pteronotus, Pteropus 151
Pteronura **313**
brasiliensis **313**
lupina 313
nitens 313
paraguensis 313
paranensis 313
paroensis 313
sambachii 313
sanbachii 313
Pteropodidae **137**, 137
Pteropodinae **137**
Pteropus 137, **146**, 146
admiralitatum **146**
aegyptiacus 152
affinis 146
aldabrensis **146**, 146, 151
alecto **146**
anetianus **146**
annectens 147
aorensis 146
argentatus 137, **146**
ariel 147
arquatus 146
aruensis 148
assamensis 147
aterrimus 146
auratus 149
austini 152
bakeri 146
balutus 150
banksiana 146
basiliscus 151
batchiana 146
baveanus 146
breviceps 149
brunneus **146**, 147
cagayanus 147
caniceps **146**
canus 147
capistratus 151
celaeno 151
ceramensis 149
chinensis 148
chrysargyrus 148
chrysauchen 147
chrysoproctus **146**
cognatus 150
collaris 151
colonus 146
comorensis 151
condorensis 147
conspicillatus **147**
coronatus 149
daitoensis 147
dasymallus **147**
degener 149
dobsoni 146
edulis 151
edwardsi 150
elseyi 151
enganus 147
eotinus 146
epularius 148
faunulus **147**
flavicollis 151
formosus 147
fretensis 147
fumigatus 148
fundatus **147**
funereus 151
fuscus 149
gambianus 141
geddiei 151
geminorum 147
germaini 152
giganteus **147**
gilliardi **147**
gouldi 146

goweri 146
grandis 150
griseus **147**
heffernani 151
heudei 148
hilli 149
howensis **147**
hypomelanus **147**
inopinatus 147
insignis 148
insularis **147,** 149
intermedius 151
javanicus 151
jubatus 137
kelaarti 147
keraudren 148
keyensis 148
kopangi **151**
lanensis 151
laniger 148
lavellanus 150
lepidus 147
leucocephalus 147
leucopterus **148**
leucotis 146
liops 151
livingstonii **148**
lombocensis **148**
loochooensis 148
luteus 147
lylei **148**
macassaricus 147
macmillani 152
macrotis **148**
mahaganus **148**
malaccensis 151
marginatus 138
mariannus **148,** 148
maris 147
mascarinus 150
mauritianus 149
mearnsi **148**
medius 147
melanopogon **148**
melanotus **149**
mimus 147
minimus 154
modiglianii 149
molossinus **149**
monoensis 150
morio 146
motalavae 146
mysolensis 147
natalis 149
natunae 151
nawaiensis 150
neohibernicus **149,** 149
niadicus 149
nicobaricus 149
niger **149**
nitendiensis **149**
ocularis **149**
ornatus **149**
paganensis 148
pallidus 147
papuanus 149
pastoris 146
pelewensis 148
personatus **149**
petersi 151
phaeocephalus **149**
phaiops 150
pilosus **149**
pluton 151
pohlei **150**
poliocephalus **150**
princeps 150
pselaphon **150**
pteronotus 151
pumilus **150**
rayneri **150**
rennelli 150
robinsoni 147
rodricensis **150**
ruber 151
rubianus 150
rubidum 151
rubiginosus 148
rubricollis 151
rufus **150**
samoensis **150**
sanctacrucis **150**
satyrus 149
scapulatus **150**
sepikensis 149
seychellensis 146, **151**
simalurus 147
solitarius 148
solomonis 146
speciosus 148, **151**
subniger **151**
sumatrensis 151
tablasi 150
temmincki **151**
tokudae **151**
tomesi 147
tonganus **151**
torquatus 151
tricolor 147
tuberculatus **151**
tytleri 149
ualanus 148
ulthiensis 148
ursinus 150
vampyrus **151**
vanikorensis 148
vetulus **152**
vitiensis 150
voeltzkowi **152**
vulcanius 147
vulgaris 149
wallacei 153
whitmeei 150
woodfordi **152**
yamagatai 147
yapensis 148
yayeyamae 147
Pterura 313
Pterycolobus 269
Ptilocercinae **133**
Ptilocercus **133**
continentis 133
lowii **133,** 133
Ptilotus 61
ptoox, Sylvicapra 412
pucheranii, Cebus 260
Sciurus **442, 442**
puck, Plecotus 224, 225
puda, Capra 392
Pudu **392**
Pudella 392
pudicus, Desmodillus 547
pudjonica, Crocidura 85
Pudu **392**
chilensis 392
fusca 392
humilis 392
mephistophiles 390, **392,** 392
puda **392**
wetmorei 392
pueblensis, Cryptotis 109
puella, Panthera 298
Petrogale 55
puer, Akodon **692,** 692
puertae, Thomomys 474
Puertoricomys **799**
corozalus **799**
puerulus, Eligmodontia 700, **701,** 701
pugetensis, Thomomys 475
pugeti, Microtus 530
pugnax, Naemorhedus 407
Paradoxurus 344
Ursus 338
puisa, Bdeogale 301
puket, Maxomys 614
pulchella, Felis 290
Lemniscomys 602
Thylamys 23
pulchellum, Diplomesodon **98**
pulchellus, Acomys 565, 566
Acrobates 62
Cervus 386
Hipposideros 175
Sorex 98
Ursus 339
pulcher, Caluromys 15
Cercopithecus 263
Chiruromys 584
Cricetulus 538
Elephantulus 829
Lemniscomys 602
Miniopterus 231
Muscardinus 769
Nannosciurus 434
Onychomys 719
Paradoxurus 344
Pipistrellus 223
Pseudocheirus 60
Saccolaimus 159
Scalopus 127
Sorex 112
pulcherrimus, Akodon 700
Chroeomys 700
Dendrolagus 50
pulchra, Crocidura 94
Genetta 345
Mustela 322
Pteralopex **146**
pulla, Microgale **72**
Neotoma 711
pullata, Crocidura **94,** 94
Trachypithecus 273
pullatus, Hipposideros 172
Microtus 527
Pipistrellus 221
pulliventer, Rattus 660
Pullomys 649
pullus, Baiomys 695
Chaetodipus 484
Cormura 156
Dipodomys 479
Lasiopodomys 516
Liomys 482
Macaca 266
Peromyscus 730
Rattus 652
Thomomys 476
pulonis, Tupaia 132
pulveratus, Pipistrellus **223**
pulverulenta, Galerella 302, **303,** 303
pulverulentus, Herpestes 304
Pteromyscus **465**
Sciuropterus 465
pulvinatus, Gerbillus 549, **554**
Puma **296,** 297
acrocodia 296
anthonyi 296
araucanus 296
arundivaga 296
aztecus 296
bangsi 296
borbensis 296
browni 296
cabrerae 296
californica 296
capricornensis 296
concolor **296**
coryi 296
costaricensis 296
cougar 296
couguar 296
floridana 296
greeni 296
hippolestes 296
hudsoni 296
improcera 296
incarum 296
kaibabensis 296
mayensis 296
missoulensis 296
nigra 296
olympus 296
oregonensis 296
osgoodi 296
patagonica 296
pearsoni 296
puma 296
punensis 297
schorgeri 297
soasoaranna 297

  soderstromii 297
  stanleyana 297
  sucuacuara 297
  vancouverensis 297
  wavula 297
  youngi 297
puma, Puma 296
pumila, Brachyphylla 182
  Chaerephon **233**
  Genetta 346
  Hystrix **774**
  Saccopteryx 159
pumilio, Artibeus 188
  Axis 385
  Dendromus 543
  Loxodonta 367
  Mellivora 315
  Mus 662
  Myosciurus **434**
  Pygeretmus **490**, 491
  Rhabdomys **662**, 662, 663
  Rhinophylla **186**, 186
  Sciurus 434
  Sorex 117
pumilis, Cynocephalus 135
pumiloides, Pipistrellus 221
pumilus, Eptesicus **203**, 203
  Heterohyrax 373
  Macaca 266
  Micromys 620
  Phenacomys 533
  Procyon 335
  Pseudomys 645
  Pteropus **150**
  Rhinolophus 165
  Sorex 117
  Sundasciurus 451, 452
  Tarsius **250**
  Tragulus 382
puna, Sigmodon 748
punctata, Dasyprocta **782**, 782
  Mustela 323
  Stenella 356
punctatissimus, Callosciurus 421
punctatus, Echimys 792
  Heliosciurus **429**
  Petaurista 462
  Tragelaphus 404
punctulatus, Bolomys **697**, 697
  Cephalophus 411
  Galerella 303
  Odocoileus 391
  Sorex 119
  Zygodontomys 752
punctus, Microtus 518
pundti, Ctenomys 785
punensis, Lagidium 778
  Lasiurus 207
  Myotis 213
  Puma 297
  Tamandua 68
punicans, Taeromys **667**
punicus, Eptesicus 202
  Myotis 209
punjabiensis, Ovis 409
Punomys **740**
  lemminus **740**, 740
punukensis, Microtus 527
punyana, Diceros 372
pupulus, Galago 249
puralis, Myoprocta 782
purdiensis, Rattus 658
purgatus, Sylvilagus 826
purillus, Saguinus 253
purinus, Callicebus 259
purpuratus, Trachypithecus 274
purpureicollis, Petrogale 56
purpurescens, Damaliscus 394
purpureus, Ratufa 437
purschei, Pan 277
puruensis, Alouatta 255
  Lagothrix 258
purui, Marmosops 20
purus, Microdipodops 480
purusianus, Sciurus 442
Pusa 331, 332
pusaea, Leopardus 291
pusilla, Emballonura 158
  Kerivoula 198
  Microgale **72**
  Micronycteris **179**
  Mustela 321
  Ochotona 809, **811**, 811, 812
  Odocoileus 391
  Phoca 326
  Thylamys **23**
  Vampyressa **194**
  Vulpes 287
pusillulus, Pipistrellus 222
pusillum, Phyllostoma 193
  Rhinopoma 155
pusillus, Anomalurus **757**
  Arctocephalus **327**, 329
  Calomys 698
  Chaerephon 233
  Chiropodomys **584**, 584
  Chlorocebus 265
  Epomophorus 143
  Eptesicus 202
  Galagoides 249
  Gerbillus 550, 553, **554**
  Haeromys **595**
  Lemur 243
  Microcebus 243
  Micropteropus **143**
  Miniopterus **231**
  Mustela 323
  Neacomys 709
  Pan 277
  Pipistrellus 223
  Pteronotus 177
  Rhinolophus 165, **168**
  Sciurillus **438**
  Sciurus 438
  Thomomys 474
puta, Murina **229**
putida, Mephitis 317
  Mustela 324
  Spilogale 317
putlaensis, Peromyscus 734
Putorius 321, 322, 323, 324
  rixosus 323
putorius, Conepatus 316
  Mustela 321, 322, **323**, 323, 324
  Spilogale **317**, 317
pyctoris, Rattus 656, 662
pygacanthus, Orthogeomys 471
pygargus, Antidorcas 395
  Antilope 394
  Capreolus **390**
  Damaliscus **394**, 394
  Ovis 408
  Taterillus **563**, 563
pygarus, Taterillus 562
Pygathrix **272**, **273**
  avunculus **272**
  bieti **272**
  brelichi **272**
  moi 273
  nemaeus **273**
  nigripes 273
  roxellana 272, **273**
Pygerethmus 490
Pygeretmus **490**, 490, 491
  acontion 490
  aralensis 490
  dinniki 490
  minor 490
  minutus 490
  pallidus 490
  platurus 490
  platyurus **490**, 490
  potanini 490
  pumilio **490**, 491
  pygmaeus 490, 491
  schitkovi 491
  shitkovi **491**, 491
  sibericus 490
  tanaiticus 490
  turcomanus 490
  vinogradovi 490
  zhitkovi 491
pygerythrus, Callosciurus **423**
  Chlorocebus 265
pygmaea, Callithrix **252**
  Capra 398
  Didelphis 62
  Otaria 328
  Phoca 332
  Spilogale **317**, 317
Pygmaeocapromys 800, 801
pygmaeoides, Suncus 102
pygmaeus, Acrobates **62**, 62
  Clethrionomys 508
  Eptesicus 204
  Hipposideros **174**
  Lepus 816
  Macaca 268
  Macroglossus 154
  Micronycteris 179
  Molossus 235
  Mus 626
  Mustela 323
  Nycticebus **248**
  Oligoryzomys 717
  Pipistrellus 223
  Pongo **277**, 277
  Procyon 335, **336**
  Pseudocheirus 59
  Pygeretmus 490, 491
  Simia 277
  Sorex 117
  Spermophilus 446, **448**, 448, 449
  Suncus 102
  Thomomys 474
pygmeus, Neotragus **398**
Pygmura 105
Pygoderma **191**
  bilabiatum **191**
  leucomus 191
  magna 191
  microdon 191
pygrius, Cercopithecus 264
pyladei, Sciurus 443
pyramidum, Gerbillus 548, 549, 550, 551, **554**, 554, 555, 563
pyrenaica, Capra **406**, 406
  Genetta 345
  Mygale 124
  Rupicapra **409**
pyrenaicus, Galemys **124**
  Lepus 817
  Microtus 521
  Myoxus 770
  Sorex 112
  Ursus 339
pyrivorus, Rousettus 152
pyrogaster, Cercopithecus 265
Pyromys 622, 623, 627, 628
pyrrhinus, Sciurus **442**, 442
  Syntheosciurus 452
pyrrhocephalus, Acerodon 137
pyrrhomerus, Dremomys **425**, 425
pyrrhonota, Blarina 112
  Sorex 112
pyrrhonotus, Erythrocebus 265
  Sciurus 442
  Thomasomys **750**
pyrrhopus, Funisciurus 428
pyrrhorhinos, Mus 751
  Wiedomys **751**, 751
pyrrhotis, Chroeomys 700
pyrrhus, Caluromys 15
  Felis 290
  Lophuromys 606
  Semnopithecus 273
  Trachypithecus 273
pyrrocephalus, Menetes 433
Pyrropus 428

pyrropus, Funisciurus **428**
pyrrotrichus, Cynomys 424
pyrsonota, Ratufa 437

qinghaiensis, Apodemus 572
Lepus 819
quadraticauda, Blarinella **106**, 107
Sorex 106
quadratus, Thomomys 475
quadricaudatus, Sorex 112
quadricinctus, Tolypeutes 67
Zaedyus 67
quadricolor, Martes 320
quadricornis, Cerophorus 403
Tetracerus **403**
quadricornua, Babyrousa 377
quadridens, Amblyrhiza 805
Pteronotus **177**, 177
quadrimaculatus, Gerbillus **554**
Tamias **455**
quadrimammis, Hemitragus 406
Quadriscopa 399
quadriscopa, Ourebia 399
quadriscriptus, Paradoxurus 344
quadrivittatus, Tamias **455**, 455, 456
quadruplicatus, Proechimys **798**
quaestor, Oxymycterus 727
Quagga 369
quagga, Equus 369, **370**, 370
Quaggoides 369
quangninhensis, Niviventer 634
quantulus, Callosciurus 421
quarlesi, Bubalus **402**
quasiater, Microtus **528**, 528
quasje, Nasua 335
quaterlinearis, Mustela 324
Spilogale 317
quebecensis, Lasiurus 206
Tamias 456
quebradensis, Sciurus 441
quelchii, Sciurus 439
quelpartis, Crocidura 84
Mustela 324
Rhinolophus 166
Quemisia **805**, 805
gravis **805**, 805
quercerus, Lepus 821
querceti, Glaucomys 460
Rattus 652
quercinus, Eliomys **767**
Myotis 209
Sciurus 440
Thomomys 476
quica, Philander 22
quichua, Alouatta 255
Coendou 775
Marmosa 18
Nasua 335
Tamandua 68
quindianus, Sciurus 441
quinlingensis, Callosciurus 421
quinquestriatus, Callosciurus **424**
quintinensis, Dipodomys 479
quiqui, Galictis 318
Lyncodon 319
quissamae, Thallomys 668
quitensis, Conepatus 316
quoll, Dasyurus 32
Mustela 32
quotus, Paraxerus 434
quoy, Spilocuscus 47
qusongensis, Lepus 819

rabori, Bullimus 581
Crunomys 588, **589**
Nyctimene **144**
Sundasciurus **452**
raceyi, Marmota 431
raddeanus, Naemorhedus 407
Sus 379
raddei, Acinonyx 288
Capra 405
Cricetus 538
Meles 314
Mesocricetus **539**
Microtus 521
Mus 626
Mustela 321
Prionailurus 295
Sorex **119**, 120
radiata, Condylura 124
Macaca **267**
radnensis, Chionomys 507
radulus, Soriculus 123
raffertyi, Arvicanthis 578
rafflei, Hylobates 275
rafflesii, Callosciurus 423
Echinosorex 79
Herpestes 304
Sciurus 420
raffrayana, Emballonura **158**
Perameles 42
Peroryctes **42**, 42
rafinesquei, Zapus 499
rafinesquii, Plecotus **225**
rahatensis, Alcelaphus 393
rahengis, Rattus 656
rahmi, Lophuromys **606**
raineyi, Acinonyx 288
Cricetomys 540
Crocidura 88, **94**, 94
Gazella 396
Kobus 414
Lepus 822
rainieri, Aplodontia 417
raiticus, Eliomys 767
rajah, Herpestes 304
Maxomys **613**, 613
Petaurista 463
Rattus 656
rajasima, Callosciurus 422
rajput, Cremnomys 588
Lepus 819
rajputanae, Antilope 395
ralli, Spermophilus 449
rallus, Bunomys 581
rama, Mus 626
ramirohitra, Brachyuromys 677, **678**
rammanika, Rhinolophus 168
ramnadensis, Mus 627
ramona, Onychomys 719
ramosianus, Cervus 386
ramosus, Spermophilus 446
ramsayi, Panthera 298
randensis, Otomys 680
Pronolagus **823**, 823
randi, Conilurus 586
Ursus 338
rangensis, Rattus 660
Rangifer **392**
angustirostris 392
arcticus 392
asiaticus 392
borealis 392
buskensis 392
caboti 392
caribou 392
chukchensis 392
cilindricornis 392
dawsoni 392
dichotomus 392
eogroenlandicus 392
excelsifrons 392
fennicus 392
fortidens 392
furcifer 392
granti 392
groenlandicus 392
keewatinensis 392
labradorensis 392
lapponum 392
lenensis 392
mcguirei 392
montanus 392
ogilvyensis 392
osborni 392
pearsoni 392
pearyi 392
phylarchus 392
platyrhynchus 392
rangifer 392
selousi 392
setoni 392
sibiricus 392
silvicola 392
spetsbergensis 392
stonei 392
sylvestris 392
taimyrensis 392
tarandus **392**
terraenovae 392
transuralensis 392
valentinae 392
yakutskensis 392
rangifer, Rangifer 392
ranjiniae, Rattus 650, **658**, 658
ranjitsinhi, Cervus 385
rantarensis, Pongo 277
rapax, Crocidura 94
Kerivoula 197
raphanurus, Rhyncholestes **25**, 25
Raphicerus **399**
acuticornis 399
bourquii 399
campestris **399**
capensis 399
capricornis 399
colonicus 400
cunenensis 399
fulvorubescens 399
grayi 399
grisea 400
hoamibensis 399
horstockii 399
ibex 399
kelleni 399
melanotis **400**, 400
natalensis 399
neumanni 399
pallida 399
pediotragus 399
rubroalbescens 400
rufescens 400
rupestris 399
sharpei **400**
steinhardti 400
stigmatus 400
subulata 400
tragulus 400
ugabensis 400
zukowskyi 400
zuluensis 400
rapit, Niviventer 632, **635**
rapposa, Micoureus 20
raptor, Bassariscus 334
Callosciurus 423
Graphiurus 764
Paranyctimene **145**, 145
Pipistrellus 220
Rheomys **743**, 743
raptorum, Funisciurus 427
Thryonomys 775
raripilosus, Pan 277
rarus, Jaculus 493
rasse, Viverricula 348
ratel, Mellivora 315
Viverra 315
ratellina, Galictis 318
Ratellus 315
Ratelus 315
ratelus, Mellivora 315
rathgeberi, Mesocricetus 538
ratlamuchi, Galerella 303
ratticeps, Microtus 527
Oryzomys 721, **724**
ratticolor, Rattus 654
rattiformis, Rattus 658
rattinus, Proechimys 795
rattoides, Melomys 615, 616, 617, **618**, 619
Mesembriomys 620
Rattus 658, 661, 662

Rattus 568, 574, 578, 580, 581, 582, 586, 587, 590, 591, 600, 603, 605, 609, 611, 621, 632, 637, 638, 639, **649**, 649, 650, 651, 652, 653, 654, 655, 656, 657, 658, 659, 661, 662, 664, 665, 666, 667, 668, 670, 674, 678, 714
adspersus 582
adustus 649, **650**, 650, 654
aemuli 652
aequicaudalus 656
aethiops 658
aitape 652
alangensis 660
albigularis 651
albiventer 658
albus 657, 658
alexandrino-rattus 658
alexandrinus 658
ambersoni 661
amboinensis 660
andamanensis 660
annandalei **650**, 650, 655, 656
apex 661
apicus 652
aramia 659
arboreus 658
arboricola 658
argentiventer 649, **650**, 650, 652, 654, 662
argyraceus 660
asiaticus 658
assimilis 653
ater 658
atratus 658
atridorsum 658
auratus 658
auroreus 660
austrinus 661
babi 659
bali 650
baliemensis 660
baluensis 649, **650**, 650
bandiculus 658
banguei 661
barussanoides 660
basilanus 652
batin 661
beccarii 658
benguetensis 660
bhotia 660
biakensis 653
biformatus 653
blangorum 661
bocourti 652
bontanus 650, **651**, 652, 662
brachyrhinus 659
brazenori 653
brevicaudatus 650
brevicaudus 660
brookei 658
brunneus 660
brunneusculus 658, 659
bukit 635
bullatus 650
bullocki 660
bunae 659
burrescens 651
burrulus 660
burrus 649, **651**, 651, 661
buruensis 652
caeruleus 658
calcis 652
cambricus 655
caraco 657
caspius 657
cauquenensis 657
celebensis 653, 655
celsus 661
ceylonus 658
chaseni 650
chionogaster 658
chrysocomus 581, 667
clabatus 652
coenorum 658
colletti 649, **651**, 651, 659, 661, 662
coloratus 660
conatus 659
concolor 652
cooktownensis 654
coquimbensis 658
coracius 653
crassipes 658
cremoriventer 634
culmorum 661
cyaneus 658
dammermani 660
decaryi 657
decumanoides 657
decumanus 657
delirius 661
dentatus 660
diardii 660
discolor 657
dispar 661
dobodurae 654
doboensis 658
domesticus 658
dominator 638
doriae 658
ducis 661
echimyoides 652
elaphinus 650, **651**, 654
enganus 650, **651**, 651, 655
ephippium 652
equile 652
erythronotus 658
eurous 652
everetti 649, **651**, 652, 655
exiguus 654
exsul 660
exulans 649, 650, **652**, 652, 662
faberi 662
facetus 662
feliceus 650, **652**, 652
fergussoniensis 656
flavescens 658
flavipectus 660
flaviventris 658
flebilis 660
foesteri 660
foramineus 650, 651, **652**, 652, 657, 662
fortunatus 660
frugivorus 658
fuliginosus 658
fulvaster 658
fuscipes 649, **652**, 653, 654
fuscus 658
gala 651
gangutrianus 658
gawae 652
generatius 661
germaini 660
gestri 659
gestroi 659
gilgitianus 661
giluwensis 649, **653**
girensis 658
glauerti 653
greyii 653
griseipectus 657
griseiventer 660
griseocaeruleus 658
guhai 656
hageni 660
hainaldi 649, **653**
hainanicus 659
hawaiiensis 652
hellwaldii 612
hibernicus 657
hoffmanni 649, 651, **653**, 653, 654, 655, 657, 661
holchu 660
hoogerwerfi 650, **653**, 653, 656
horeites 656
hoxaensis 650
huegeli 652
humiliatus 657
hybridus 657
imbil 655
indicus 658
infralineatus 658
insolatus 657
insulanus 660
intermedius 658
jalorensis 661
jarak 661
javanus 657
jemuris 661
jessook 652
jobiensis 649, **653**, 653
jujensis 658
julianus 661
jurassicus 658
kabanicus 661
kadanus 660
kandianus 658
kandiyanus 658
keelingensis 660
kelaarti 658
kelleri 660
khumbuensis 661
khyensis 660
kijabius 658
klumensis 659
koopmani 649, 651, 652, **654**, 654
koratensis 659
korinchi 650, 653, **654**, 655, 656
kraensis 659
kramensis 660
kunduris 661
lacus 655
lalolis 660
lamucotanus 661
lanensis 660
larusius 661
lasiae 659
lassacquerei 652
latidens 564
latipes 658
leucogaster 658
leucophaetus 652
leucopus 649, 650, 653, **654**, 654, 656, 660
leucosternum 657
linduensis 653
longicaudus 660
longipilis 662
lontaris 660
losea 649, **654**, 654, 657
lugens 604, 649, 650, **654**, 654
luteiventris 652
lutescens 657
lutreolus 649, **655**
luxuriosus 661
macleari 650, 651, **655**, 655, 656
macmillani 660
maerens 661
magnirostris 657
major 657
makassarius 660
makensis 660
malengiensis 652
mangalumis 661
manicatus 653
maniculatus 657
manoquarius 652
mansorius 660
manuselae 656
maorium 652
mara 661
marmosurus 650, **655**, 655, 662
masaretes 660
mayonicus 652
mcilwraithi 654
mediocris 658
melanoderma 652
melanurus 653
melvilleus 661
mengkoka 653
mentawi 650, 654

meringgit 652
mesanis 660
meyeri 655
micronesiensis 652
mindanensis 660
mindorensis 649, **655**, 661
moheius 660
mollicomulus 649, 653, **655**
mollicomus 653
molliculus 660
moluccarius 660
mondraineus 653
montanus 650, 655, **656**
mordax 649, **656**, 656, 659, 660
morotaiensis 650, **656**
muansae 658
murrayi 653
narbadae 658
nativitatis 650, 655, **656**, 656
neglectus 660
negrinus 652
nemoralis 658
nericola 658
nezumi 660
nitidus 649, 652, **656**, 656, 658, 662
niviventer 633
norvegicus 649, 652, **657**, 657, 662
novaeguineae 649, **657**
novarae 657
obiensis 660
obscurus 652
obsoletus 656
ohiensis 664
orientalis 662
orii 657
ornatulus 652
osgoodi 649, 654, **657**
osimensis 670
osorninus 658
otomoi 657
otteni 652
ouangthomae 660
owiensis 653
pachyurus 655
palelae 660
palembang 660
palmarum 649, **657**, 657
panjius 660
pannellus 660
pannosus 660
pantarensis 652
paraxanthus 662
pauper 661
payanus 661
peccatus 653
pelengensis 660
pelori 653
pelurus 650, 652, **657**, 662
pemanggis 661
penitus 582
perhentianus 661
personata 654
pesticulus 650
petterdi 655
pharus 661
picinus 655
picteti 658
piperis 661
pipidonis 660
plumbeus 657
poenitentiarii 660
portus 660
povolnyi 660
praecelsus 652
praestans 657
praetor 649, **658**
primarius 657
profusus 662
pulliventer 660
pullus 652
purdiensis 658
pyctoris 656, 662
querceti 652
rahengis 656
rajah 656
rangensis 660
ranjiniae 650, **658**, 658
ratticolor 654
rattiformis 658
rattoides 658, 661, 662
rattus 649, 650, 652, 654, 656, 657, **658**, 658, 659, 660, 661, 662, 732
raveni 652
ravus 653
remotus 659
rennelli 652
rhionis 661
ringens 654, 656, 660
rintjanus 600
roa 661
robiginosus 660
robonsoni 660
robustulus 660
rogersi 660
roquei 661
rosalinda 660
ruber 656, 658, 660
rubricosa 656
rufescens 658, 660
rumpia 661
ruthenus 658
sabae 661
sakeratensis 654
salocco 662
saltuum 658
samati 660
samharensis 658
sanila 649, **659**, 659
sansapor 658
santalum 660
sapoensis 660
satarae 658
saturnus 650
schuitemakeri 652
sebasianus 661
septicus 660
shigarus 661
siantanicus 661
siculae 658
sikkimensis 649, **659**
simalurensis 649, **659**, 659, 661
simpsoni 657
sladeni 659, 660
socer 657
sodyi 600
solatus 652
sordidus 649, 651, 654, **659**, 659, 662
sowerbyi 657
spinosus 794
sribuatensis 661
steini 649, **659**, 660
stoicus 650, **660**, 660
stragulum 652
subcaeruleus 658
subditivus 656
subrufus 658
sueirensis 658
suffectus 652
sumbae 660
surdus 652
surmulottus 657
sylvestris 658
tablasi 660
taciturnus 660
tagulayensis 651
talaudensis 660
tamarensis 657
tambelanicus 661
tanezumi 649, 659, **660**, 660
tatei 653
tatkonensis 660
tawitawiensis 649, **661**
tectorum 658
tenggolensis 661
terra-reginae 654
terutavensis 661
tetragonurus 655, 658
tettensis 658
thai 660
tibicen 652
tikos 660
timorensis 649, **661**
tingius 661
tiomanicus 649, 650, 651, 654, 655, 657, 659, **661**, 661, 662
tistae 660
todayensis 652
tompsoni 658
tondanus 655
toxi 660
tramitius 658
tua 661
tunneyi 649, 651, **661**
turbidus 660
turkestanicus 649, 656, **661**, 662
tyrannus 651, 652
umbriventer 650
utakwa 658
vallesius 661
vanheurni 656
variabilis 658
varius 658
vellerosus 655
velutinus 655
verecundus 660
vernalus 661
vicerex 661, 662
viclana 661
vigoratus 652
villosissimus 649, 651, 659, **662**, 662
villosus 650
vitiensis 652
vulcani 652
wichmanni 652
woodwardi 661
wroughtoni 660
xanthurus 650, 651, 652, 655, 656, 657, 658, **662**, 662
yaoshanensis 659
yeni 660
youngi 659
yunnanensis 660
zamboangae 660
rattus, *Nectomys* 709
Rattus 649, 650, 652, 654, 656, 657, **658**, 658, 659, 660, 661, 662, 732
Ratufa **437**, 438
affinis **437**
albiceps 437
albipes 437
anambae 437
angusticeps 437
arusinus 437
aureiventer 437
balae 437
baliensis 437
bancana 437
banguei 437
baramensis 437
batuana 437
bengalensis 437
bicolor **437**
bombaya 437
bulana 437
bunguranensis 437
carimonensis 437
catemana 437
ceilonensis 437
celaenopepla 437
centralis 437
ceylonica 437
condorensis 437
condurensis 437
confinis 437
conspicua 437
cothurnata 437
dandolena 437
dealbata 437
dicolorata 437
dulitensis 437

elphinstonei 437
elphinstoni 437
ephippium 437
felli 437
femoralis 437
fretensis 437
frontalis 437
gigantea 437
griseicollis 437
hainana 437
humeralis 437
hypoleucos 437
indica **437**
insignis 437
interposita 437
javensis 437
johorensis 437
klossi 437
laenata 437
leschnaultii 437
leucogenys 437
lumholzi 437
lutrina 437
macroura **437**
macrura 437
macruroides 437
major 437
malabarica 437
marana 437
masae 437
maxima 437
melanochra 437
melanopepla 437
montana 437
nanogigas 437
nigrescens 437
notabilis 437
palliata 437
penangensis 437
peninsulae 437
phaeopepla 437
piniensis 437
polia 437
purpureus 437
pyrsonota 437
sandakanensis 437
sinhala 437
sinus 437
sirhassenensis 437
smithi 437
sondaica 437
stigmosa 437
superans 437
tennentii 437
tiomanensis 437
vittata 437
vittatula 437
zeylanicus 437
Ratufini 419, 437, 438
ratwa, Muntiacus 389
raui, Trichosurus 48
raveni, Rattus 652
raviana, Tupaia 132
ravida, Neotoma 712
ravidulus, Microtus 521
ravijojla, Dryomys 766
raviventer, Sciurus 440
raviventris, Reithrodontomys **742**, 742
ravulus, Tragulus 382
ravus, Artibeus 189
Maxomys 614
Microtus 520
Miniopterus 231
Paradoxurus 344
Rattus 653
Reithrodontomys 741
Thomomys 475
Tomopeas **231**, 231
Tragulus 382
rawlinnae, Pseudomys 646
rayneri, Pteropus **150**
readei, Bos 401
reboudi, Mus 626
recessus, Ctenomys 785
recifinus, Platyrrhinus **191**
reclinis, Damaliscus 394
reclusa, Glossophaga 184
recticornis, Oryx 413
rectipinna, Grampus 353
Orcinus 355
rectitragus, Eptesicus 202
recurvus, Nasalis 270
redacta, Tupaia 132
redimitus, Callosciurus 423
redmani, Monophyllus **185**, 185
reducta, Gazella 396
reductus, Glaucomys 460
Rhizomys 685
Redunca **414**
adamauae 414
algoensis 414
arundinum **414**
bayoni 414
bohor 414
caffra 414
chanleri 414
cinerea 414
cottoni 414
dianae 414
donaldsoni 414
eleotragus 414
fulvorufula **414**
isabellina 414
lalandia 414
multiannulata 414
nagor 414
nigeriensis 414
occidentalis 414
odrob 414
penricei 414
redunca **414**
reversa 414
rufa 414
schoana 414
subalpina 414
thomasinae 414
tohi 414
ugandae 414
wardi 414
redunca, Antilope 414
Redunca **414**
Reduncina 391
Reduncinae **413**
reevesi, Muntiacus **389**, 389
Paguma 343
reevesii, Prionailurus 295
refescens, Chaetodipus 484
refulgens, Ailurus 337
Rhinolophus 167
regalis, Colobus 270
Oecomys 716
Panthera 298
Tamiasciurus 457
Vulpes 287
regia, Neotragus 398
Roystones 802
regillus, Dipodomys 479
Oryzomys 721
regina, Micoureus **20**
Ochotona 812
reginae, Gazella 397
Hipposideros 172
Hydromys 597
Macropus 54
Notomys 635
Petaurus 61
reguli, Petaurista 463
regulus, Callicebus 259
Cervus 386
Eothenomys 509, **514**, 514
Eptesicus **203**
Microtus 520
Rhinolophus 166
reichardi, Arvicanthis 578
reichei, Connochaetes 394
reichenowi, Panthera 298
reidae, Tragelaphus 404
reigi, Pseudoryzomys 739
Zygodontomys 752
reii, Epomophorus 141
reinhardti, Akodon 749
Thalpomys 748
reinwaldti, Clethrionomys 508
reiseri, Sus 379
reissii, Pseudalopex 284
Reithrodon **740**, 746
auritus **740**, 740
caurinus 740
chinchilloides 701
cuniculoides 740
currentium 740
evae 740
flammarum 740
fossor 701
hatcheri 740
longicaudatus 705
marinus 740
megalotis 740
obscurus 740
pachycephalus 740
pampanus 740
physodes 740
pictus 695
typicus 740
Reithrodontomys **740**, 740, 741, 742, 743
albescens 742
albilabris 742
alleni 742
alticolus 741
amoenus 741
amoles 741
anthonyi 741
arizonensis 741
aurantius 741
aureus 743
australis 742
aztecus 741
brevirostris **740**, 742
burti **740**
canus 741
carolinensis 741
caryi 741
catalinae 741
cherrii 742
chiapensis 741
chrysopsis **740**
chrysotis 741
cinereus 741
colimae 740
costaricensis 742
creper **740**
darienensis **741**, 741
deserti 741
dickinsoni 741
difficilis 741
distichlis 741
dorsalis 742
dychei 741
eremicus 742
fulvescens **741**, 741
garichensis 742
goldmani 742
gracilis **741**, 741, 742
griseoflavus 741
griseus 742
halicoetes 742
harrisi 741
helvolus 741
hirsutus **741**
hooperi 741
howelli 742
humulis **741**
impiger 741
inexpectatus 741
infernatis 741
insularis 741
intermedius 741
jalapae 742
klamathensis 741
laceyi 741
lecontii 741
levipes 741
limicola 741
longicaudus 741
lucifrons 742
luteolus 742
megalotis 740, **741**, 741, 742, 743
meridionalis 741
merriami 741

mexicanus 740, **741**, 741, 742
microdon **742**
milleri 742
minusculus 742
modestus 742
montanus 740, **742**, 742
mustelinus 741
nebrascensis 741
nelsoni 741
nerterus 742
nicaraguae 740
nigrescens 741
obscurus 743
ocotepequensis 742
orinus 742
orizabae 740
otus 742
pacificus 741
pallidus 741
paradoxus **742**
pectoralis 741
peninsulae 741
perotensis 740
potrerograndei 742
raviventris **742**, 742
ravus 741
riparius 742
rodriguezi **742**
rufescens 742
santacruzae 741
saturatus 741, 743
scansor 742
seclusus 742
sestinensis 741
soederstroemi 742
spectabilis **742**
sumichrasti **742**
tenuirostris 741, **742**, 742, 743
tenuis 741
toltecus 741
tolucae 740
tropicalis 741
underwoodi 742
virginianus 741
vulcanius 742
wagneri 742
zacatecae 741, **743**, 743
Reithronycteris 183
relicinus, Thomomys 475
relicta, Myonycteris **143**
Neotoma 712
Phocoena 358
relictus, Capromys 800
Microtus 529
Myotis 212
Perognathus 485
Sorex 118
Spermophilus **449**
Synaptomys 534
Thomomys 475
religiosa, Crocidura 90, 91, **94**
remifer, Neomys 110
remota, Dobsonia 140
remotus, Rattus 659
remulus, Callicebus 259
remyi, Suncus **103**
renatae, Eumops 234
Giraffa 383
renatus, Elephantulus 829
renaultii, Meriones 556
rendalli, Eptesicus **203**
Sigmoceros 395
rendilis, Aepyceros 393
Diceros 372
Elephantulus 829
Galerella 303
Hyaena 308
renggeri, Akodon 692
rennelli, Pteropus 150
Rattus 652
renschi, Cervus 387
repanda, Lontra 311
repens, Heteromys 481
repentinus, Castor 467
reperticia, Mazama 390
reptans, Arvicanthis 578
resima, Mirounga 330
resina, Macaca 266
restrictus, Microdipodops 480
Peromyscus 730
Sylvilagus 826
reta, Arvicola 506
retardatus, Atherurus 773
retectus, Exilisciurus 426
reticulata, Giraffa 383
retractus, Thomomys **474**
retrorsus, Thomomys **475**
retusa, Burmeisteria 64
Crocidura 87
retusus, Chlamyphorus **64**
reuningi, Procavia 374
reuteri, Pan 277
reventazoni, Oligoryzomys 718
reversa, Redunca 414
revertens, Leopoldamys 603
revoili, Elephantulus **829**
revoilii, Nycteris 162
rex, Acinonyx 288
Canis 281
Cervus 386
Clethrionomys 509
Conepatus 316
Desmomys 590
Lavia 163
Meriones **557**
Mylomys 630
Oecomys **716**
Oryzomys 724, 725
Rhinolophus **168**
Rhipidomys 744
Tachyoryctes **686**
Uromys **672**
rex-pygmaeorum, Gorilla 276
rexi, Ursus 340
reyi, Felis 290
Rhabdogale 319
Rhabdomys **662**, 663
angolae 662
bechuanae 662, 663
bethuliensis 662
chakae 662
cinereus 662
cradockensis 662
deserti 662
dilectus 662
diminutus 662
donavani 662
fouriei 662
griquae 662
griquoides 662
intermedius 662
lineatus 662
major 662
meridionalis 662
moshesh 662
namaquensis 662
namibensis 662
nyasae 662
orangiae 662
prieskae 662
pumilio **662**, 662, 663
septemvittatus 662
typicus 662
vaalensis 662
vittatus 662
rhabdops, Oryzomys 720, **724**
Rhachianectes 350
Rhachianectidae 350
Rhagamys 670
Rhagapodemus 670
Rhagomys 687, 700, **743**
rufescens **743**
Rheithrosciurus **438**
macrotis **438**
rhenanus, Capreolus 390
Cervus 386
Neomys 110
Rheomys 699, **743**
chiapensis 743
hartmanni 743
mexicanus **743**, 743
raptor **743**, 743
stirtoni 743
thomasi **743**
trichotis 743
underwoodi **743**, 743
rhesosimilis, Macaca 266
Rhesus 266
rhesus, Macaca 267
Rhinoceros **372**
annamiticus 372
asiaticus 372
bicornis 372
floweri 372
indicus 372
inermis 372
jamrachii 372
javanicus 372
nasalis 372
simus 371
sondaicus **372**
stenocephalus 372
sumatrensis 371
unicornis **372**, 372
Rhinocerotidae **371**
Rhinochoerus 370
Rhinodelphis 352
Rhinogale 307
melleri 307
rhinogradoides, Tateomys 667, **668**
Rhinolophidae **163**
Rhinolophinae **163**
Rhinolophus **163**
abae 166
achilles 168
acrotis 164
acuminatus **163**
adami **163**
aequalis 164
aethiops 166
affinis **163**
alcyone **164**
algirus 165
alleni 168
alticolus 169
andamanensis 163
anderseni **164**
andersoni 164
andreinii 164
angolensis 166
angur 164
angustifolius 164
annectens 166
anomalus 166
antinorii 166
arcuatus **164**
aruensis 165
atlanticus 165
audax 163
aurantius 175
auritus 164
axillaris 166
barbarus 165
barbertonensis 165
beccarii 164
beddomei 167
bembanicus 169
bifer 166
bihastatus 166
blasii **164**
blythi 165, 168
bocharicus 164
borneensis **164**
brachygnathus 164
brevitarsus 166
brockmani 164, 166
bunkeri 169
burius 165
cabrerae 165
caldwelli 167
calidus 168
calypso 163
canuti **164**, 165
capensis **164**
carpentanus 167
celebensis **164**, 164
chaseni 164

chinensis 168
cinerascens 168
circe 163
clivosus **164**, 165, 168
coelophyllus **165**, 168
cognatus **165**
colchicus 166
cornutus **165**, 167
creaghi 164, **165**
creticum 166
cuneatus 167
damarensis 165
darlingi **165**
deckenii 164, **165**
denti **165**, 169
diversus 166
dobsoni 166
dohrni 167
edax 168, 169
eloquens **165**, 166
empusa 164
episcopus 167
equinus 166
escalerae 166
euryale **165**
euryotis **165**
exiguus 164
exsul 166
fallax 167
famulus 165
feae 167
ferrumequinum **165**
foetidus 167
formosae 167
foxi 166
fudisanus 166
fulvidus 168
fumigatus **166**
germanicus 166
gracilis 168
guineensis **166**, 166
hainanus 163
helvetica 166
hildebrandti **166**
hilli 167
hillorum 164
himalayanus 163
hippocrepis 166
hipposideros **166**
hirsutus 167
homodorensis 166
homorodalmasiensis 166
ignifer 167
imaizumii **166**
importunus 164
inops **166**
insulanus 166
intermedius 166
irani 166
italicus 166
javanicus 164
judaicus 165
kashyiriensis 166
keniensis 164
keyensis **166**
klossi 168
knorri 165
korai 166
kosidanus 166
landeri **166**, 166
lanosus 167
latifolius 169
lepidus **166**
lobatus 166
luctus **167**
maclaudi **167**
macrocephalus 166
macrotis **167**
macrurus 163
madurensis 164
majori 166
malayanus **167**
maros 168
marshalli **167**
martinoi 166
mcintyrei 164
megaphyllus **167**
mehelyi **167**
meridionalis 165
meyeroehmi 164
midas 166
mikadoi 166
minimus 166
minutillus 168
minutus 166
mitratus **167**
miyakonis 165
monachus 167
monoceros **167**
montanus 168
monticola 167
moravicus 166
morio 167
nanus 166
nereis **167**
nesites 163
niasensis 169
nippon 166
nordmanni 165
norikuranus 166
obscurus 166
ogasimanus 166
orii 165
osgoodi **167**
pagi 168
pallidus 166
paradoxolophus **168**
parcus 168
parvus 164
pearsonii **168**, 169
perauritus 165, 166
perditus 165
perniger 167
petersi 168
phasma 166
philippinensis **168**
pilosus 165
piriensis 169
praestens 165
princeps 163
proconsularis 164
proximus 166
pumilus 165
pusillus 165, **168**
quelpartis 166
rammanika 168
refulgens 167
regulus 166
rex **168**
rhodesiae 169
robertsi 168
robinsoni **168**
rouxii **168**
rubidus 168
rubiginosus 166
rufus **168**
ruwenzorii 167
sanborni 168
schwarzi 164
sedulus **168**
septentrionalis 169
shameli 165, **168**
shortridgei 167
siamensis 167
silvestris 164, **168**
simplex **169**
simulator **169**
sinicus 168
sobrinus 167
solitarius 169
spadix 164
spurcus 167
stheno **169**
subbadius **169**
subrufus **169**
sumatranus 163
superans 163
swinnyi **169**
szechwanus 168
tatar 165
tener 163
thomasi **169**
timidus 165
timoriensis 164
toscanus 165
toxopeusi 164
tragatus 166
tricuspidatus 170
tridens 170
trifoliatus **169**
trogophilus 166
truncatus 166
tunetae 167
typica 164
typicus 166
typus 166
ungula 166
unihastatus 166
vandeuseni 167
vespa 166
virgo **169**
yunanensis **169**
zuluensis 164
rhinolophus, Sorex 112
Rhinomegalophus 163
Rhinomus 81
Rhinonicteris **175**, 175
aurantia **175**
Rhinonycteris 175
Rhinophylla **186**
alethina **186**
fischerae **186**
pumilio **186**, 186
Rhinopithecus 272, 273
Rhinopoma **155**
arabium 155
asirensis 155
cordofanicum 155
cystops 155
hardwickei **155**
harrisoni 155
kinneari 155
lepsianum 155
macinnesi 155
microphyllum **155**
muscatellum **155**
pusillum 155
seianum 155
sumatrae 155
tropicalis 155
Rhinopomatidae **155**
Rhinopterus 200, 201
Rhinosciurus **438**
incultus 438
laticaudatus **438**
leo 438
peracer 438
rhionis 438
robinsoni 438
saturatus 438
tupaioides 438
Rhinosticteus 263
Rhinostigma 263
rhionis, Presbytis 271
Rattus 661
Rhinosciurus 438
Sus 379
Rhipidomys 687, 715, **744**, 744
aratayae 744
austrinus **744**
benevolens 715
bovallii **744**
caucensis **744**
cearanus 745
cocalensis 744
collinus **744**
couesi **744**, 744
cumananus 744
elatturus 744
emiliae 745
equatoris 744
fervidus 745
fulviventer **744**, 744, 745
goodfellowi 744
latimanus **744**, 744, 745
leucodactylus **744**, 744, 745
lucullus 744
macconnelli **744**
macrurus 745
maculipes 745
mastacalis **744**, **745**, 745
microtis 744

milleri 745
modicus 744
mollissimus 744
nitela **745**, 745
ochrogaster **745**
pictor 744
rex 744
scandens 744, **745**
sclateri 744
subnubis 745
tenuicauda 744
tobagi 745
venezuelae 744, **745**, 745
venustus 744, **745**
wetzeli 744, **745**
yuruanus 745
rhipidurus, Echimys **792**, 792
Rhithrosciurus 438
Rhizomops 240
Rhizomyinae 501, **685**, 685
Rhizomys **685**, 685
badius 685
chinensis 685
cinereus 685
davidi 685
dekan 685
erythrogenys 685
insularis 685
javanus 685
latouchei 685
padangensis 685
pannosus 685
pruinosus **685**
prusianus 685
reductus 685
senex 685
sinensis **685**, 685
sumatrensis **685**
troglodytes 685
umbriceps 685
vestitus 685
wardi 685
rhizophagus, Thomomys 474
Rhizoplagiodontia **804**
lemkei **804**, 804
rhoadsi, Lontra 311
Peromyscus 735
Sciurus 441
Thomasomys **750**
rhoadsii, Clethrionomys 508
rhodanica, Genetta 345
rhodesiae, Arvicanthis 578
Chaerephon 232
Heliosciurus 428
Heterohyrax 373
Papio 269
Pelomys 639
Rhinolophus 169
Taphozous 161
Thallomys 669
rhodesiensis, Homo 276
rhoditis, Crocidura **94**
rhodius, Apodemus 572
Erinaceus 77
Lepus 817
Meles 314
rhodopensis, Microtus 529
Sciurus 443
Rhogeessa **225**, 225, 226
aeneus 226
alleni **225**
bombyx 226
genowaysi **225**
gracilis **225**
io 226
major 226
minutilla **225**
mira **225**
parvula **225**, 225, 226
riparia 226
tumida 225, **226**
velilla 226
Rhogessa, tumida 225
Rhombomys **559**
alaschanicus 559
dalversinicus 559
funicolor 559
giganteus 559
nigrescens 559
opimus **559**
pallidus 559
pevzovi 559
sargadensis 559
sodalis 559
rhyacobia, Mystacina 231
Rhynchocyon **830**
chrysopygus **830**
cirnei **830**, 830
hendersoni 830
petersi **830**
stuhlmanni 830
Rhynchogale **307**
caniceps 307
langi 307
melleri **307**
Rhyncholestes **25**
continentalis 25
raphanurus **25**, 25
Rhynchomeles **42**
prattorum **42**, 42
Rhynchomyinae 591
Rhynchomys 591, **663**
isarogensis **663**
soricoides **663**, 663
Rhynchonax 130
Rhynchonycteris **159**
lineata 159
naso **159**
priscus 159
rivalis 159
saxatilis 159
villosa 159
Rhynchotragus 398
riabus, Tupaia 132
ricarulus, Phyllotis 738
richardi, Mylomys 630
richardii, Meriones 558
Phoca 332
richardsoni, Artibeus 188
Bassaricyon 333
Coendou 775
Dicrostonyx 510, **511**
Dipodomys 479
Mephitis 317
Microhydromys **620**, 620
Microtus 505, **528**, 528, 531
Odocoileus 391
Oryzomys 721
Sciurus 443
Stenomys **665**
Tamiasciurus 457
Ursus 339
richardsonii, Genetta 347
Lepus 815
Mustela 321, 322
Notomys 636
Poiana **347**, 347
Sorex 112
Spermophilus 446, **449**
richmondi, Dasyprocta 782
Didelphis 17
Nasua 334
Oryzomys 721
Sciurus **442**
richteri, Lycaon 283
ricketti, Myotis **214**
Prionailurus 295
ridei, Mormopterus 238
Ningaui **33**
ridleyi, Hipposideros **174**
Myotis **214**
rifensis, Mus 628
riggenbachi, Gerbillus 552, **554**
rigidus, Sciurus 443
Sylvilagus 826
rii, Nannospalax 754
rimator, Hylochoerus 377
rindjanicus, Paradoxurus 344
ringens, Mustela 324
Rattus 654, 656, 660
Spilogale 317
Tayassu 380
rintjanus, Komodomys **600**
Rattus 600
riograndensis, Cerdocyon 282
rionegrensis, Ctenomys 785, 786
riparia, Clethrionomys 508
Kobus 414
Neotoma 711
Rhogeessa 226
riparius, Canis 280
Gerbillus 550
Myotis **214**
Reithrodontomys 742
Sylvilagus 825
Thomomys **474**
riparum, Proechimys 795
ripensis, Ondatra 532
risia, Boselaphus 402
risorius, Myotis 209
rissoanus, Grampus 353
rita, Lophuromys 605
ritchiei, Alcelaphus 393
riudonensis, Dremomys 425
riudoni, Tamiops 457
riukiuana, Suncus 103
riukiuanus, Sus 379
rivalicius, Ondatra 532
rivalis, Emballonura 157
Neomys 110
Paguma 343
Rhynchonycteris 159
riverianus, Cervus 386
riveti, Pseudalopex 284
rivularis, Microtus 525
Oryzomys 720, 721
rixosa, Mustela 321
rixosus, Mustela 323
Putorius 323
rizophoreus, Prionailurus 296
rjabovi, Clethrionomys 509
roa, Rattus **661**
roachi, Myomimus **768**, 768
roanensis, Napaeozapus 498
roberti, Aotus 256
Callosciurus 421
Chionomys **507**, 507
Gymnuromys **678**, 678
Oecomys **716**
Oxymycterus **727**
Proechimys 797
Sciurus 439
Sphiggurus 777
robertsi, Cephalophus 411
Desmodillus 547
Gazella 396
Kobus 414
Otomys 681
Rhinolophus 168
Thallomys 668
robertsoni, Funambulus 426
robiginosus, Rattus 660
robinsoni, Coelops **170**, 170
Marmosa **18**
Muntiacus 389
Naemorhedus 407
Nyctimene **144**
Presbytis 271
Pteropus 147
Rhinolophus **168**
Rhinosciurus 438
Sundasciurus 451
robonsoni, Rattus 660
roboratus, Sorex **119**, 119
roborovskii, Cricetulus 539
Phodopus **539**
roborowskii, Meriones 557
robusta, Balaenoptera 350
Daubentonia 247
Dologale 302
Eonycteris 153, 154
Gerbillurus 547
Helogale 302
Lonchophylla 181, **182**
Marmota 432
Mogera **126**
Mystacina **231**
Neotoma 710

Talpa 126
Tatera 560, 561, **562**
Taxidea 325
robustior, Cerdocyon 282
Miniopterus **231**
robustula, Tylonycteris **228**
robustulus, Melanomys **708**
Rattus 660
robustus, Ateles 257
Atilax 301
Capromys 800
Cebus 259
Ctenomys 785
Dryomys 766
Eschrichtius **351**
Heliophobius 772
Macropus **54**, 54
Nectomys 709
Paguma 343
Paradoxurus 344
Peromyscus 729
Scotophilus **228**
Spermophilus 450
Sylvilagus 826
Thomomys 474
rocherianus, Naemorhedus 407
rodolphei, Tamiops **457**
rodoni, Naemorhedus 407
rodricensis, Pteropus **150**
rodriguezi, Reithrodontomys **742**
roensis, Lutra 312
rogersi, Chalinolobus 200
Rattus 660
Ursus 339
rohui, Philetor 219
rolandensis, Antechinus 30
roloway, Cercopithecus 263
romana, Talpa **129**, 129
Romerolagus 813, **824**, 825
diazi **824**
nelsoni 824
rona, Sminthopsis 37
ronaldshaiensis, Microtus 519
rondoni, Ctenomys 786
Gracilinanus 17
Mazama 391
rondoniae, Sciurus 442
roosevelti, Alcelaphus 393
Cervus 386
Chiropotes 261
Colobus 270
Crocidura **94**
Gazella 396
Hippotragus 413
Mylomys 630
Panthera 297
Sylvicapra 412
rooseveltorum, Muntiacus 388, 389
roothi, Caracal 289
roquei, Rattus 661
roraimae, Marmosa 18
Podoxymys **739**, 739
Rorqualus 349
rosa, Proechimys 798
rosalia, Lemniscomys 601, **602**, 602
Leontopithecus **253**, 253
Simia 252
rosalinda, Gerbillus 552, **554**
Rattus 660
Thomasomys **750**
rosamondae, Antechinus 31
Dasykaluta **31**
rosanowi, Martes 320
rosanvi, Alticola 503
rosaphagus, Sylvilagus 825
rosarianus, Cervus 386
roseiventris, Stenella 356
rosenbergi, Eonycteris 154
rosenbergii, Artibeus 188
Prosciurillus 436
roseveari, Lemniscomys **602**
Steatomys 545
rosida, Cavia 779
rosii, Mazama 390
rosilla, Oecomys 716
Rosmaridae 325
rosmarus, Odobenus **326**
Phoca 325
rosseti, Myotis **214**
rossi, Mungos 307
rossiaemeridionalis, Microtus 519, **529**, 529
rossicus, Clethrionomys 509
Meriones 557
Ursus 339
rossii, Arvicanthis 578
Ommatophoca **331**
rostellatus, Oxymycterus 727
rostralis, Castor 467
Thomomys 475
rostrata, Balaenoptera 349, 350
Glossophaga 184
rostratus, Chaetodipus 484
Delphinus 357
Hyperoodon 361
Lariscus 430
Liomys 482
Macroglossus 154
Oryzomys 720, 723, **724**, 725
Praomys 592, **643**, 643, 644
Steno 357
Tarsipes **62**, 62
rotans, Mus 626
rotensis, Emballonura 158
rothschildi, Alcelaphus 393
Coendou **775**, 775, 776
Dorcopsulus 52
Giraffa 383
Ictonyx 319
Lepus 816
Lophocebus 266
Loxodonta 367
Mallomys **608**, 608, 609
Massoutiera 761
Murexia **32**
Myospalax 675, **676**, 676
Odocoileus 391
Peroryctes 42
Petrogale **56**
Phalanger **47**
Uromys 671
rotundus, Desmodus **194**
Perognathus 485
Phyllostoma 194
roualeynei, Tragelaphus 404
roudairei, Psammomys 559
roulinii, Tapirus 371
roumanicus, Erinaceus 77
Rousettus 138, **152**, 152, 153
affinis 152
amplexicaudatus **152**
angolensis **152**
arabicus 152
bocagei 152
brachyotis 152
celebensis **152**
crypticola 152
egyptiacus **152**
fuliginosa 152
fusca 152
geoffroyi 152
hedigeri 152
hottentotus 152
infumatus 152
infuscata 152
kempi 152
lanosus **152**, 153
leachi 152
leschenaulti **152**
madagascariensis **153**
minor 152
obliviosus **153**
occidentalis 152
philippinensis 152
pyrivorus 152
ruwenzorii 152
seminudus 152
shortridgei 152
sjostedti 152
smithi 152
spinalatus **153**
stresemanni 152
unicolor 152
rousiensis, Microtus 519
roussardii, Eulemur 244
rousseloti, Dendroprionomys **544**, 544
rouxii, Rhinolophus **168**
rovumae, Petrodromus 830
rowleyi, Otomys 680
Peromyscus 729, 732
rowumae, Sigmoceros 395
roxasianus, Cervus 386
roxellana, Pygathrix 272, **273**
roylei, Alticola 502, **503**, 503, 504
Ochotona 807, 809, 810, 811, **812**
Roystones, regia 802
rozanovi, Sorex 113
rozendaali, Murina **229**
rozeti, Elephantulus **829**
rozianus, Microtus 518
rozsikae, Gerbillus 550
ruandae, Sylvisorex 105
Tachyoryctes **686**
ruandensis, Colomys 586
ruasae, Galerella 303
ruatanica, Dasyprocta **782**
Marmosa 19
rubecula, Crocidura 87
Lophuromys 605
rubeculus, Antechinus 30
Callosciurus 421
rubelianus, Microtus 524
rubelius, Eothenomys 513
rubella, Murina 229
rubellus, Atilax 301
Chlorocebus 265
Eothenomys 513
Lasiurus 206
Pappogeomys 472
Proechimys 798
rubens, Apodemus 570
Macropus 54
ruber, Felis 290
Hipposideros 173, **174**
Macropus 55
Myotis 212, **215**
Pteropus 151
Rattus 656, 658, 660
Sciurus 442
Varecia 245
ruberculus, Otomys 681
ruberrimus, Gerbillus **554**
rubescens, Arvicanthis 578
Atilax 301
Papio 269
rubeus, Tragulus 382
rubex, Callosciurus 421
Melomys 618, **619**, 619
Murina 229
rubianus, Pteropus 150
rubicaudatus, Sciurus 442
rubicola, Melomys 615, **619**, 619
rubicollis, Microsciurus 433
rubicunda, Alouatta 255
Presbytis **271**
Suncus 103
rubicundus, Cacajao 261
Mus 626
Petaurista 463
rubida, Felis 290
Monodelphis **22**
Neotoma 711
Presbytis 271
Vandeleuria 672
rubidior, Cephalophus 411
Herpestes 304
rubidiventris, Callosciurus 423
rubidum, Pteropus 151
rubidus, Cephalophus **411**
Clethrionomys 508

Microtus 527
Muntiacus 389
Naemorhedus 407
Paguma 343
Paradoxurus 344
Peromyscus 732
Pseudocheirus 60
Rhinolophus 168
Thomomys 474
rubiginosa, Alouatta 255
Cervus 386
Genetta 346
rubiginosus, Herpestes 306
Prionailurus **296**
Pteronotus 177
Pteropus 148
Rhinolophus 166
Rubisciurus 419
rubra, Chrysochloris 75
Erythrocebus 265
Lasiurus 206
Marmosa **19**
rubrata, Dasyprocta 782
Neophascogale 33
rubricatus, Dicrostonyx 510, **511**, 511
Heliosciurus 429
rubricollis, Pteropus 151
rubricosa, Crocidura 82
Rattus 656
Vulpes 287
rubridens, Elephas 367
rubrifrons, Herpestes 305
rubripes, Funisciurus 428
rubrirostris, Microsciurus 433
Rubrisciurus **438**
rubriventer 436, **438**
rubriventer, Dendrohyrax 373
Eulemur **244**
Peromyscus 732
Rubrisciurus 436, **438**
Sciurus 438
rubroalbescens, Raphicerus 400
rubrolineatus, Tamiasciurus 457
rubustus, Lepus 821
Rucervus 385
ruckbeili, Allactaga 489
ruckwaensis, Heterohyrax 373
ruddi, Cephalophus 411
Dendromus 543
Galerella 303
Herpestes 305
Heterohyrax 373
Myotis 216
Pronolagus 823
Tachyoryctes 686, **687**
Tatera 560
Thallomys 669
Uranomys 670, **671**
rudinoris, Chaetodipus 483
rudis, Cerdocyon 282
rudolfi, Crocidura 91
Heterohyrax 373
ruemmleri, Coccymys **585**, 585, 586
Pogonomelomys 585
rueppelli, Gazella 396
Pipistrellus **223**
rueppellii, Nycticeius **217**
Tadarida 240
Vulpes **286**
rufa, Anisonyx 417
Aplodontia **417**
Crocidura 94
Crocuta 308
Erythrocebus 265
Euryzygomatomys 794
Galidictis 300
Mazama 390
Microtus 518
Mustela 324
Nasua 335
Otaria 328
Phalanger 46
Redunca 414
Stenoderma 192
rufescens, Aepyprymnus **48**
Apodemus 573
Arvicola 506
Bettongia 48
Cavia 779
Clethrionomys 508
Cricetus 538
Echinoprocta **776**, 776
Echymipera **41**
Elephantulus **829**
Ellobius 512
Eptesicus 203
Galerella 303
Hesperomys 743
Lepilemur 246
Macaca 266
Marmota 432
Melomys 615, 616, 617, **619**, 619
Menetes 433
Microtus 527
Mormoops 176
Myospalax 675
Noctilio 176
Nyctalus 217
Ochotona **812**
Procyon 335
Raphicerus 400
Rattus 658, 660
Reithrodontomys 742
Rhagomys **743**
Spermophilus 447
Tamias 456
Thomomys 473, 475
Uromys 615
Vulpes 286
rufescentefuscus, Microtus 519
rufibarbata, Pithecia 262
ruficauda, Galerella 303
ruficaudata, Tupaia 133
ruficaudatus, Lepilemur **246**
Lepus 814, 819, 820
ruficaudus, Akodon 692
Spermophilus 445
Tamias **455**, 455
ruficeps, Arctocebus 247
Helogale 304
Procavia 374
ruficollaris, Microdipodops 480
ruficollis, Gazella 396
Macropus 55
Trichosurus 48
ruficrista, Cephalophus 411
rufidorsum, Microtus 527
rufidulus, Thomomys 474
rufifrons, Eulemur 244
Gazella **397**, 397
Parotomys 682
Xerus 458
rufigenis, Dremomys **425**, 425
Sminthopsis 37
rufijianus, Connochaetes 394
rufilatus, Cephalophus **411**
Cercopithecus 264
Paraleptomys **638**
rufimanus, Alouatta 254
Saguinus 254
rufina, Gazella **397**
Mazama 390, **391**, 391
Nasua 335
rufinucha, Lepus 820
rufinus, Arvicanthis 578
Oenomys 637
Oryzomys 721
Peromyscus 732, 733
rufipes, Aotus 255
Eulemur 244
Noctilio 176
Petaurista 463
Pseudalopex 284
Sciurus 440
Sylvilagus 824
rufiventer, Eulemur 244
Pithecia 262
Saguinus 253
Sciurus 442
Thylogale 57
rufiventris, Ateles 257
Echymipera 41
Mus 626
Sciurus 440
rufo-brachiatus, Heliosciurus 429
rufo-fuliginus, Procolobus 272
rufobrachium, Heliosciurus **429**, 429
rufocanus, Alticola 503
Clethrionomys 502, 503, 508, **509**, 509, 514, 515
Hybomys 596
rufocastaneus, Promops 240
rufodorsalis, Arvicanthis 576
Diplomys **791**
rufofuscus, Microtus 530
Myotis 213
rufogaster, Antechinus 30
Sundasciurus 451
rufogriseus, Macropus **54**
rufogularis, Callosciurus 423
rufolateralis, Dorcopsis 52
rufomitratus, Procolobus **272**, 272
rufoniger, Callosciurus 423
Chlorocebus 265
Myotis 210
Procolobus 272
Saguinus 254
Sciurus 441
Scotinomys 746
Spilocuscus **47**, 47
rufonigra, Callosciurus 423
rufopallidus, Chrysospalax 76
Hippotragus 412
rufopicta, Crocuta 308
rufopictus, Myotis 210
rufoviridis, Chlorocebus 265
rufula, Helogale 304
rufulus, Apodemus 572
Cryptomys 772
Dasymys 589, **590**
Galemys 124
Harpiocephalus 228
Stenomys 665
Tragulus 383
rufum, Stenoderma **192**
rufus, Aotus 256
Canis **281**, 281
Chrysospalax 76
Cynocephalus 135
Desmodus 194
Eulemur 244
Harpiocephalus 228
Lasiurus 206
Lepus 817
Lynx **293**
Macropus **55**
Microcebus **243**
Miniopterus 230
Molossus 235
Mus 727
Myrmecobius 29
Nesomys **679**, 679
Noctilio 176
Ornithorhynchus 13
Oryzomys 721
Oxymycterus **727**, 727
Petromyscus 683
Pongo 277
Potorous 50
Pteropus **150**
Rhinolophus **168**
Sciurus 443
Tamias **455**, 455
Tapirus 371
Taterillus 563
Triaenops 175
Ursus 339
rugosus, Tadarida 240

ruidosae, Onychomys 719
Thomomys 474
Rukaia 437
rukwai, Sigmoceros 395
rumpia, Rattus 661
rumpii, Myosorex 99, **100**, 100
rumruti, Arvicanthis 578
rungiusi, Ursus 339
rupatius, Callosciurus 423
Rupestes 438, 439
rupestris, Cavia 780
Chaetodipus 484
Elephantulus **829**
Kerodon **780**
Macroscelides 829
Phyllotis 739
Pronolagus **823**, 823
Raphicerus 399
Sciurus 443
Spermophilus 450
Thomomys 474
Rupicapra **409**
alpina 410
asiatica 410
balcanica 410
capella 410
carpatica 410
cartusiana 410
caucasica 410
dorcas 410
europea 410
faesula 410
hamulicornis 410
olympica 410
ornata 409
parva 409
pyrenaica **409**
rupicapra **409**, 409
sylvatica 410
tatrica 410
tragus 410
rupicapra, Antilope 395
Capra 409
Rupicapra **409**, 409
rupicola, Clethrionomys 508
Graphiurus **765**, 765
Neotoma 711
rupicolus, Peromyscus 729
rupinarum, Spermophilus 445
ruppelii, Felis 290
ruppelli, Colobus 270
Lycaon 283
Theropithecus 269
ruppi, Myomys **631**, 631, 664
Praomys 630
ruricola, Thomomys 474
Rusa 385
rusa, Tscherskia 539
rusanus, Herpestes 306
rusbyi, Artibeus 189
rusca, Nasua 335
ruschii, Abrawayaomys **687**, 687
rusiges, Apodemus **573**, 573
russa, Cervus 387
russata, Chaerephon **233**
russatus, Acomys **566**, 566
Clethrionomys 509
Holochilus 704
Sciurus 440
Sylvilagus 826
russelli, Pappogeomys 472
Ursus 339
russelliana, Mustela 323
russeolus, Dipodomys 479
Tamiops 457
Thomomys 474
russeus, Tragulus 382
russicus, Pteromys 464
russowi, Hemiechinus 78
russula, Crocidura 81, 83, 84, 86, **94**, 94, 95, 96, 98, 110
russulus, Sigmodontomys 748
Tragulus 382
russus, Sciurus 443
rusticana, Felis 290
rusticus, Pipistrellus **223**
Sorex 117
ruthenus, Microtus 519
Rattus 658
Sorex 116
rutila, Chaetomys 790
Ochotona 808, 809, 810, **812**, 812
Ourebia 399
rutilans, Cuon 282
Grammomys **594**, 594
Mus 727
Oxymycterus 727
Sciurus 443
Thamnomys 669
rutiliventris, Callosciurus 423
rutilus, Cervus 386
Clethrionomys 502, 508, **509**, 509
Crocidura 84
Herpestes 305
Macropus 55
Melomys 619
Mus 507
Oecomys **716**
Profelis 296
Sciurus 458
Xerus **458**
rutschiricus, Colobus 270
rutschuricus, Cercopithecus 263, 264
rutshuricus, Cephalophus 411
Thryonomys 775
rutteri, Callicebus 258
Micoureus 20
ruttneri, Clethrionomys 508
ruvanensis, Orycteropus 375
ruwanae, Gazella 397
Lycaon 283
ruwenzorii, Aethosciurus 428
Colobus 270
Dendrohyrax 373
Heliosciurus **429**
Micropotamogale **72**
Panthera 298
Paraxerus 434
Rhinolophus 167
Rousettus 152
Tatera 562
Ruwenzorisorex **101**
suncoides **101**
ryckmani, Peromyscus 731
ryphaeus, Microtus 524
Sorex 112
Rytina 365

saadanicus, Sigmoceros 395
saarensis, Vulpes 287
saba, Myotis 212
sabae, Rattus 661
sabaea, Simia 265
Vulpes 286
sabaeum, Eidolon 141
sabaeus, Chlorocebus 265
Hemiechinus 78
sabakiensis, Alcelaphus 393
Gazella 397
Panthera 297
sabana, Presbytis 271
sabanicola, Dasypus **66**
sabanillae, Microsciurus 433
sabanus, Hipposideros 172, **174**
Leopoldamys **603**, 603, 614, 634
Mus 603
Paradoxurus 344
sabatyra, Tapirus 371
sabbar, Vulpes 286
sabiensis, Herpestes 305
Lemniscomys 602
Nycteris 162
Otomys 680
sabrinus, Glaucomys **460**
sabryi, Acomys 565
sabulatus, Lemniscomys 602
sabulonis, Microdipodops 480
sabulosus, Chaetodipus 483
sacana, Mustela 321
Ochotona 810
sacarensis, Peromyscus 729, 732
saccata, Antidorcas 395
saccatus, Geomys 469
saccharalis, Pappogeomys 472
saccharus, Platyrrhinus 190
Saccolaimus **159**, 160
affinis 159
capito 159
crassus 159
flaviventris **159**
flavomaculatus 159
granti 159
hargravei 159
insignis 159
mixtus **159**
nudicluniatus 159
peli **159**
pluto **159**
pulcher 159
saccolaimus **159**, 159
saccolaimus, Saccolaimus **159**, 159
Taphozous 159
Saccophorus 469
Saccopteryx **159**
bilineata **159**
canescens **159**
centralis 159
gymnura **159**
insignis 159
leptura **160**
perspicillifer 159
pumila 159
Saccostomurinae 540
Saccostomus 540, **541**
anderssoni 541
angolae 541
campestris **541**, 541
cricetulus 541
elegans 541
fuscus 541
hildae 541
isiolae 541
lapidarius 541
limpopoensis 541
mashonae 541
mearnsi **541**, 541
pagei 541
streeteri 541
umbriventer 541
sacer, Canis 280
Erinaceus 77
Niviventer 633
Paradoxurus 344
Suncus 101, 103
sachalinensis, Microtus 521, **529**
Moschus 384
sacobianus, Apomys **575**
sacramenti, Meriones **557**
sacramentoensis, Tamias 453
sacraria, Ochotona 812
sacratus, Soriculus 122
sacrimontis, Plecotus 224
sadakei, Ochotona 809
sadhu, Mus 627
sadiya, Lepus 819
sadoensis, Apodemus 573
sadonis, Sorex **119**
saevus, Hipposideros 171
Sorex 120
safii, Crocidura 94
sagax, Apodemus 570
Peromyscus 732
sagei, Aconaemys **788**
saghaliensis, Lepus 821
sagitta, Cebus 259
Dipus **492**

Hylopetes 461
Macrotis 40
Mus 492
Petinomys **464**, 464
sagittalis, Geomys 469
Ursus 339
sagittatus, Castor 467
sagittula, Eptesicus **203**, 203
Sagmatius 353
saguenayensis, Napaeozapus 498
Saguinus **253**
acrensis 253
apiculatus 253
avilapiresi 253
bicolor **253**
bluntschlii 253
crandalli 253
cruzlimai 253
devillei 253
doguin 254
egens 254
elagantulus 253
elegans 253
erythrogaster 253
flavifrons 253
fuscicollis **253**, 254
fuscus 253
geoffroyi **253**, 254
gracilis 254
graellsi 254
griseovertex 253
hernandezi 254
hololeucus 253
illigeri 253
imberbis 253
imperator **253**
inustus **253**
juruanus 254
labiatus **253**
lacepedii 254
lagonotus 253
leonina 253
leucogenys 253
leucopus **253**
martinsi 253
melanoleucus 253
meticulosus 254
micans 253
midas **253**
mounseyi 253
mystax **254**
negre 254
niger 254
nigricollis **254**
nigrifrons 253
ochraceus 253
oedipus 253, **254**, 254
pacator 253
pebilis 253
pegasis 253
pileatus 254
pluto 254
primitivus 253
purillus 253
rufimanus 254
rufiventer 253
rufoniger 254
salaguiensis 253
spixii 253
subgrisescens 253
tamarin 254
thomasi 253
titi 254
tripartitus **254**
umbratus 254
ursula 253, 254
ursulus 254
weddelli 253
sagulata, Chiropotes 261
Mellivora 315
sahafarensis, Lepilemur 246
sahamae, Chinchillula **699**, 699
saharae, Galerella 303
Otonycteris 219
sahariensis, Ammotragus 404
Sus 379
saianicus, Clethrionomys 508
Myopus 531
Saiga **400**, 400
colus 400
mongolica 400
saiga 400
scythica 400
tatarica **400**
saiga, Saiga 400
saimensis, Phoca 332
Saimiri **260**
albigena 260
apedia 260
boliviensis **260**, 260
caquetensis 260
cassiquiarensis 260
citrinellus 260
codajazensis 260
collinsi 260
entomophagus 260
jaburuensis 260
juruana 260
leucopsis 260
lunulatus 260
macrodon 260
madeirae 260
morta 260
nigriceps 260
nigrivittata 260
oerstedii **260**, 260
peruviensis 260
petrina 260
pluvialis 260
sciureus **260**, 260
ustus **260**
vanzolinii **260**
Saimirinae 259, 260
sairensis, Gazella 397
Ovis 408
sakeen, Capra 406
sakeratensis, Rattus 654
Saki 261
saki, Pithecia 262
sakir, Callicebus 259
salae, Protoxerus 436
salaguiensis, Saguinus 253
salai, Lepus 816
salairicus, Clethrionomys 509
salakensis, Callosciurus 422
salamonis, Solomys **663**, 663
salanga, Leopoldamys 603
Salanoia **300**
concolor **300**, 300
olivacea 300
unicolor 300
salaquensis, Sciurus 441
salarii, Myotis 212
salarius, Sorex 118
salatana, Tupaia 132
salebrosus, Solomys **663**
salenskii, Soriculus **123**, 123
salentensis, Sciurus 442
salentus, Sylvilagus 825
Salicornia, perennis 802
salicornicus, Sorex 118
salicus, Allactaga 489
saliens, Allactaga 489
Antidorcas 395
Lepus 814
salimalii, Latidens **142**, 142
salinae, Lasiurus 206
Pedetes 759
salinarum, Oncifelis 294
salinensis, Perognathus 485
salinia, Microcavia 780
salinicola, Cavia 780
Dolichotis **780**
Salmacis 266
salmi, Gazella 397
salocco, Rattus 662
salomonseni, Batomys **580**, 580
salongo, Cercopithecus 263
Salpingotini 491
Salpingotulus 491, 492
Salpingotus **491**, 491, 492
crassicauda **491**, 491, 492
gobicus 491
heptneri **492**
kozlovi 491, **492**, 492
michaelis 491, **492**, 492
pallidus 491, **492**, 492
sludskii 492
thomasi **492**, 492
salsa, Tatera 561
saltans, Antidorcas 395
saltarius, Ctenomys **786**
saltator, Allactaga 489
Oligoryzomys 718
Oreotragus 399
Oryzomys 721
Tarsius 250
Zapus 499
saltatrixoides, Oreotragus 399
saltiana, Antilope 398
Madoqua **398**, 398
saltitans, Sciurotamias 439
saltuensis, Sciurus **441**
saltuum, Rattus 658
salutans, Paraxerus 435
salva, Mustela 321
salvadorensis, Oryzomys 725
Peromyscus 734
salvanicola, Felis 290
salvanius, Sus **379**
salvini, Chiroderma **189**
Liomys **482**
Nyctomys 715
Sorex 120
salvinia, Leopardus 292
salvus, Microtus 524
samango, Cercopithecus 264
samarensis, Sundasciurus **451**, **452**, 452
samaricus, Apodemus 572
Exilisciurus 426
samariensis, Apodemus 572
samati, Cheirogaleus 243
Rattus 660
sambachii, Pteronura 313
Sambur 385
samharensis, Rattus 658
samilkameenensis, Ovis 408
samniticus, Sorex **120**
samoensis, Pteropus **150**
samuiensis, Callosciurus 421
sanbachii, Pteronura 313
sanborni, Akodon 691, **692**
Leptonycteris 185
Nycticeius **217**
Rhinolophus 168
Sciurus **442**
sanctacrucis, Nyctimene **144**
Pteropus **150**
sanctaemartae, Leopardus 291, 292
Mazama 391
Sigmodon 747
Sylvilagus 825
Zygodontomys 752
sanctaerosae, Peromyscus 732
sanctamartae, Coendou 775
sanctidiegi, Microtus 519
Sylvilagus 824
Thomomys 474
sanctijohannis, Macaca 267
sanctiluciae, Dipodomys 480
sanctipaulensis, Akodon **692**
sanctorum, Trachypithecus 274
sanctus, Cricetomys 540
sandakanensis, Ratufa 437
sandayensis, Microtus 519
sandbergi, Colobus 270
sandersoni, Hipposideros 172
sanfelipensis, Mesocapromys **802**
sanfordi, Eulemur 244
sanggaus, Callosciurus 423
sangi, Petrodromus 830
sangirensis, Strigocuscus 48
Tarsius 250
sangronizi, Herpestes 305
sanguinea, Galerella 302, **303**, 303

sanguineus, Herpestes 302, 304
sanguinidens, Sorex 114
sanibeli, Oryzomys 724
sanila, Rattus 649, **659**, 659
sanje, Cercocebus 262
sanluisi, Perognathus 485
sannio, Erythrocebus 265
sanpabloensis, Microtus 519
sanrafaeli, Dipodomys 479
Neotoma 711
sansapor, Rattus 658
sansibarica, Crocidura 85
santacristobalensis, Erophylla 182
santacrucensis, Oncifelis 294
santacruzae, Myocastor 806
Peromyscus 732
Reithrodontomys 741
Urocyon 285
santalum, Rattus 660
santanderensis, Microsciurus **434**
santaremensis, Callithrix 252
santarosae, Pipistrellus 221
Urocyon 285
santonus, Sorex 114
Sapaja 256
Sapajus 259
sapidus, Arvicola **505**, 505
sapiens, Homo **276**, 276
sapientis, Solomys **664**
Uromys 663
sapoensis, Rattus 660
sara, Alouatta **255**
sarae, Lagidium 778
Mazama 390
sarasinorum, Mops **237**
Prosciurillus 436
sarawakensis, Callosciurus 423
Sarcophilus **35**
dixonae 35
harrisii 35
laniarius **35**
satanicus 35
ursina 35
sarcovienna, Chironectes 16
sarda, Crocidura 96
Felis 291
sardous, Sus 379
sardus, Eliomys 767
Prolagus **813**
sareptae, Micromys 620
sareptanicus, Mus 626
sargadensis, Rhombomys 559
Saricovia 313
Sarigua 15, 16, 18, 22, 23
sarmatica, Mustela 325
Vormela 325
sarnius, Microtus 519
sartinensis, Peromyscus 732
sartorii, Mazama 390
saryarka, Spermophilus 446
sassae, Cercopithecus 263
Tragelaphus 404
satanas, Cebus 261
Chiropotes **261**, 261
Colobus **270**
Satanellus 31, 32
satanicus, Sarcophilus 35
satarae, Rattus 658
satisfactus, Chiruromys 584
satschouensis, Meriones 558
satuni, Otocolobus 295
satunini, Chionomys 507
Hyaena 308
Hystrix 774
Microtus 530
Nesokia 632
Sorex 119, **120**
Spermophilus 448
saturata, Lagidium 778
Macroderma 163
Ochotona 811
saturatior, Oryzomys 720, 721, **725**, 725
Suncus 103
saturatus, Akodon 693
Apodemus 571
Arvicanthis 576
Callosciurus 423
Clethrionomys 508
Cynocephalus 135
Glaucomys 460
Graphiurus 764
Lagurus 515
Lariscus 430
Maxomys 614
Mustela 322
Myomys 631
Myotis 210, 216
Natalus 195
Nectomys 709
Peromyscus 732
Reithrodontomys 741, 743
Rhinosciurus 438
Sigmodon 747
Spermophilus 448, **449**
Synaptomys 534
Thomomys 475
Trachypithecus 274
Xerus 458
saturnus, Echimys **792**
Rattus 650
Satyrus 277
satyrus, Pan 277
Pongo 277
Pteropus 149
saudiya, Gazella **397**
saundersiae, Otomys **681**
Pronolagus 824
saurica, Alticola 503
Sauromys 237, 238
saussurei, Sorex 118, **120**, 122
sauteri, Tamiops 457
sauveli, Bos **401**
savagei, Gorilla 276
savannarum, Cerdocyon 282
Marmosa 18
Odocoileus 391
Sigmodon 747
savannius, Heliosciurus 429
savannus, Dasymys 589
savii, Meriones 558
Microtus 520, 521, **529**, 529
Pipistrellus 220, **223**
Tadarida 240
savilei, Bandicota **579**, 579
saxamans, Neotoma 711
Peromyscus 732
saxatalis, Mustela 324
Spilogale 317
saxatilis, Apodemus 572
Dipodomys 477
Dryomys 766
Galea 780
Lepus 819, **820**, 820, 822
Ochotona 811
Peromyscus 734
Rhynchonycteris 159
Thomomys **474**
saxicola, Bassariscus 334
Mus **627**
Naemorhedus 407
Peromyscus 730
saxicolor, Panthera 298
saxicolus, Ammospermophilus 419
saxonicus, Cervus 386
sayii, Sciurus 442
Scaeopus 63
scaloni, Microtus 518
Sorex 114
Scalopinae 124
Scalopini 124, 127
Scalops, breweri 127
townsendii 127
scalops, Chelemys 699
Chroeomys 700
Geomys 470
Monodelphis **22**
Orthogeomys 471
scalopsoides, Microtus 528
Scalopus **127**
aereus 127
alleni 127
anastasae 127
aquaticus **127**, 127
argentatus 127
australis 127
bassi 127
caryi 127
cryptus 127
cupreata 127
howelli 127
inflatus 127
intermedius 127
machrinoides 127
machrinus 127
montanus 127
nanus 127
parvus 127
pennsylvanica 127
porteri 127
pulcher 127
sericea 127
texanus 127
virginianus 127
scammonii, Globicephala 352
scandens, Rhipidomys 744, **745**
Scandentia 131
scandinavicus, Ursus 339
scansa, Tatera 561
scansor, Reithrodontomys 742
Scapanulus **127**
oweni **127**, 127
Scapanus **127**
alpinus 127
anthonyi 127
californicus 127
campi 127
caurinus 127
dilatus 127
grinnelli 127
insularis 127
latimanus **127**
minusculus 127
monoensis 127
occultus 127
orarius **127**
parvus 127
schefferi 128
sericatus 127
townsendii **128**
truei 127
yakimensis 128
scaphax, Uromys 671
scaphiotus, Odocoileus 391
Scapteromys 696, 706, **745**, 745
aquaticus 745, 746
fronto 745
labiosus 745
tomentosus 745
tumidus **745**, 746
scapterus, Thomomys 474
Scaptochirus **128**, 128
davidianus 128, 129
micrura 128
moschatus **128**, 128
Scaptonychini 124, 128
Scaptonyx **128**
affinis 128
fusicauda 128
fusicaudatus 128
fusicaudus **128**
scapulatus, Marmosops 19
Pteropus **150**
Scartes 243
Scarturus 488, 489
schadenbergi, Crateromys **587**, 587
Phloeomys 587
schadenbergii, Mydaus 315
schaeferi, Hemitragus 406
Pseudois **409**
schalleri, Myosorex **100**
schaposchnikowi, Prometheomys **534**, 534

scharicus, Hippotragus 412
schaubi, Myotis **215**
schauinslandi, Monachus **331**
scheelei, Dendrohyrax 373
scheffeli, Felis 290
schefferi, Scapanus 128
scheffleri, Dendrohyrax 373
schelkovnikovi, Microtus **529**
Neomys **110**
scherman, Arvicola 505, 506
schermous, Arvicola 506
scherzeri, Cynopterus 139
schidlovskii, Microtus 530
schillingsi, Alcelaphus 393
Gazella 397
Giraffa 383
Hyaena 308
Oreotragus 399
schimperi, Galerella 303
schinzi, Tatera 561
schinzii, Myotis 213
schistacea, Abrocoma 789
Crocidura 88
Megaderma 163
schistaceus, Hipposideros **174**
Semnopithecus 273
schisthyperoes, Arctocephalus 327
schisticeps, Ochotona 811
schisticolor, Myodes 531
Myopus **531**, 531
schitkovi, Pygeretmus 491
Schizodon, fuscus 787
schizodonticus, Cervus 386
schlegeli, Callosciurus 421, 423
Cervus 386
Hapalemur 245
Pseudocheirus **60**
schlegelii, Tatera 560
Viverricula 348
schlegellii, Balaenoptera 350
schlieffeni, Nycticeius **218**
schlueteri, Jaculus 493
schlumbergeri, Lepus 816
schmidi, Anourosorex 106
schmidti, Allactaga 488
Canis 281
Cercopithecus 263
Microtus 528
Spermophilus 450
schmidtorum, Micronycteris **179**
schmidtzi, Hystrix 774
schmitti, Sigmoceros 395
schmitzi, Caracal 289
Procavia 374
Ursus 339
schnablii, Amorphochilus **195**, 195
schneideri, Graphiurus 764
Hipposideros 171
Pan 277
Uncia 299
schnitnikovi, Felis 291
Sorex 121
schoana, Redunca 414
schoedei, Dendrolagus 50
schoensis, Epomophorus 141
Schoinobates 58
schomburgki, Cervus **386**
schorgeri, Puma 297
schousboeii, Meriones 556
schoutedeni, Anomalurus 757
Cercopithecus 264
Graphiurus 764
Heliosciurus 429
Paracrocidura 100, **101**
Thamnomys 670
schraderi, Genetta 346
schrankii, Myotis 213
schreberi, Mandrillus 268
schreibersi, Miniopterus 230, **231**
schreibersii, Vespertilio 230
schrencki, Vulpes 287
schrenkii, Meles 314
schubotzi, Cercopithecus 264
Dendrohyrax 373
Kobus 414
Pan 277
Procolobus 272
schuitemakeri, Rattus 652
schultzei, Cephalophus 411
Procavia 374
Pseudochirops 60
schulzi, Alcelaphus 393
Connochaetes 394
Hylochoerus 377
Tonatia **180**
schulzianus, Cervus 386
schumakovi, Spermophilopsis **444**
schusteri, Cephalophus 411
Dendrohyrax 373
Sigmoceros 395
schwabi, Graphiurus 765
schwartzi, Artibeus 188
Cercopithecus 264
Gorilla 276
schwarzi, Civettictis 345
Rhinolophus 164
schwarzovi, Meriones 556
schweinfurthii, Pan 277
schweitzeri, Crocidura 93
scindiae, Paradoxurus 344
scioanus, Procavia 374
scirpensis, Microtus 519
Scirteta 488
Scirtetes 488
Scirtomys 488
Scirtopoda 493, 494
ctenodactyla 495
lichtensteini 493
scitulus, Nyctimene 144
Peromyscus 729
sciurea, Petaurus 61
Simia 260
sciureus, Holochilus 704, **705**, 705
Saimiri **260**, 260
Sciuridae 295, **419**
Sciurillini 438
Sciurillus **438**
glaucinus 438
hoehnei 438
kuhlii 438
pusillus **438**
Sciurinae **419**, 419
Sciurini 419, 438, 439
Sciurocheirus 249
Sciuropterus 464
everetti 460
hosei 461
lugens 463
pearsonii 459
pulverulentus 465
Sciurotamias **438**, 438, 439
collaris 439
consobrinus 439
davidianus 438, **439**
forresti **439**
latro 439
owstoni 439
saltitans 439
thayeri 439
Sciurus 419, **439**, 439, 440, 441, 442, 443, 452, 458
aberti **439**
adolphei 443
aestuans **439**, 441
affinis 440
agricolae 441
albipes 440
albonotatus 443
albus 443
alfari 433
alleni **439**
alphonsei 439
alpinus 443
alstoni 442
altaicus 443
ameliae 443
anadyrensis 443
annalium 443
annulatus 428
anomalus **439**
anthonyi 441
apache 442
arcticus 443
argenteus 443
argentinius 441
arizonensis **440**
atrirufus 443
aureogaster **440**
austini 443
avicinnia 442
bachmani 442
baeticus 443
balcanicus 443
baliolus 443
bancrofti 439
bangsi 443
barberi 439
bashkiricus 443
baudensis 441
belti 443
berdmorei 433
boliviensis 441
bondae 441
boothiae 443
brunnea 443
brunneo-niger 442
bryanti 442
cabrerai 441
candelensis 441
capistratus 442
carchensis 441
carolinensis **440**
castanonotus 439
castanotus 439
castus 442
catalinae 440
caucasicus 440
caucensis 442
cepapi 434
cervicalis 440
chapmani 441
chiapensis 440
chiliensis 443
chiricahuae 442
chiriquensis 441
choco 441
chrysogaster 440
chrysuros 441
chuscensis 439
cinerea 443
cinereus 442
cocalis 441
cocos 440
colimensis 440
colliaei **440**
concolor 439
coreae 443
coreanus 443
croaticus 443
cucutae 441
cuscinus 441
davidianus 438
degus 788
deppei **440**
dorsalis 439, 443
duida 441
dulkeiti 443
durangi 439
effugius 440
europaeus 443
exalbidus 443
exilis 426
extimus 440
fedjushini 443
ferminae 441
ferreus 439
ferruginiventris 440
flammifer **440**
formosovi 443
fossor 441
fraseri 443
frumentor 440
fuliginosus 440
fulminatus 441
fulvus 440
fumigatus 442
fuscoater 443

fusconigricans 443
fuscorubens 443
fuscovariegatus 443
gambianus 428
garbei 439
georgihernandezi 439
gerrardi 441
getulus 420
gilvigularis 439, **441**, 441
gilviventris 441
glis 769
goldmani 443
golzmajeri 443
gotthardi 443
graeca 443
granatensis **441**
griseimembra 441
griseocaudatus 443
griseoflavus 440
griseogena 441
griseus **441**
guajanensis 439
guayanus 443
guerlingus 439
guianensis 439
heermanni 441
helveolus 443
henseli 439
hernandezi 440
hiemalis 440
hirtus 440
historicus 440
hoffmanni 441, 443
huachuca 440
hypophaeus 440
hypopyrrhus 440
hyporrhodus 441
hypoxanthus 440
ignitus **441**
igniventris **441**
imbaburae 441
inconstans 441
indicus 426, 437
infuscatus 443
ingrami 439
insignis 430
intermedius 443
irroratus 441
isabella 427
istrandjae 443
italicus 443
jacutensis 443
jenissejensis 443
juralis 442
kaibabensis 439
kalbinensis 443
kessleri 443
klagesi 441
langsdorffi 442
laticaudatus 438
leonis 441
leporinus 441
leucogaster 440, 441
leucops 440
leucotis 440
leucourus 443
lilaeus 443
limitis 442
lis **441**
littoralis 440
llanensis 441
loweryi 443
lowii 450
ludovicianus 442
macconnelli 439
macrotis 438
macroura 442
madagascarensis 247
magdalenae 441
magnificaudatus 442
managuensis 443
manavi 441
manhanensis 441
mantchuricus 443
maracaibensis 441
martensi 443
matagalpae 440
matecumbei 440
maurus 440
medellinensis 442
melania 443
melanonotus 442
melanotis 434
meridensis 441
meridionalis 443
migratorius 440
milleri 441
mimus 439
minor 442
minutus 434, 440
miravallensis 440
morio 440, 442
morulus 441
murinus 435
mustelinus 440
nadymensis 443
navajo 439
nayaritensis **441**
nebouxii 443
neglectus 442
negligens 440
nelsoni 440
nemoralis 440
nesaeus 441
nicoyana 443
niger **442**, 443
nigratus 442
nigrescens 440, 443
nigripes 441
norosiensis 441
nuchalis 440
numantius 443
ochrescens 441
ocularis 763
oculatus **442**
ognevi 443
olivascens 439
orientis 443
pallescens 440
palmarum 426
paraensis 441
pennsylvanicus 440
perigrinator 440
perijae 441
pernyi 424
persicus 440
petaurista 462
phaeopus 443
phaeurus 439
poaiae 439
poliopus 440
prevostii 420
pucheranii **442**, 442
pumilio 434
purusianus 442
pusillus 438
pyladei 443
pyrrhinus **442**, 442
pyrrhonotus 442
quebradensis 441
quelchii 439
quercinus 440
quindianus 441
rafflesii 420
raviventer 440
rhoadsi 441
rhodopensis 443
richardsoni 443
richmondi **442**
rigidus 443
roberti 439
rondoniae 442
ruber 442
rubicaudatus 442
rubriventer 438
rufipes 440
rufiventer 442
rufiventris 440
rufoniger 441
rufus 443
rupestris 443
russatus 440
russus 443
rutilans 443
rutilus 458
salaquensis 441
salentensis 442
saltuensis 441
sanborni **442**
sayii 442
segurae 443
senex 440
shawi 442
shermani 442
silanus 443
sinaloensis 440
socialis 440
söderströmi 441
spadiceus 441, **442**
splendidus 441
stangeri 436
steinbachi 442
stramineus **443**
striatus 453
subalpinus 443
subauratus 442
sumaco 441
syriacus 440
taedifer 441
taeniurus 440
talahutky 443
tamae 441
taparius 442
tarrae 441
tephrogaster 440, 441
tepicanus 440
texianus 442
thomasi 443
tobagensis 441
tolucae 442
tricolor 442
truei 440
typicus 443
ukrainicus 443
underwoodi 443
uralensis 443
urucumus 442
valdiviae 441
variabilis 441, 442
variegatoides **443**
varius 440, 443
venustus 439
versicolor 441
vivax 440
volans 464
vulgaris 439, **443**, 456
vulpinus 442
wagneri 440
wilsoni 425
xanthotus 441
yucatanensis **443**
zamorae 441
zarumae 443
zuliae 441
sclateri, Atelerix **77**
Atelocynus 279
Cephalophus 411
Cercopithecus 263, **265**
Chlorotalpa **75**
Litocranius 397
Muntiacus 389
Myosorex 99, **100**, 100
Phacochoerus 377
Rhipidomys 744
Sorex **120**
Thryonomys 775
sclavonius, Lepus 821
Scleronycteris **186**
ega **186**, 186
Sclеropleura 66
Scolecophagus 247
Scolomys **746**
melanops **746**, 746
ucayalensis **746**
scont, Otaria 328
scopaeum, Chiroderma 189
scoparia, Antilope 399
Ourebia 399
Scopophorus 399
scopulicola, Phoca 332
scopulorum, Neotoma 712
Peromyscus 729
scorodumovi, Ochotona 807
Vulpes 286

scortecci, Heterocephalus 772
Scoteanax 217
Scoteinus 226
scoticus, Cervus 386
Lepus 821
Scotinomys 695, **746**
apricus 746
cacabatus 746
endersi 746
episcopi 746
escazuensis 746
garichensis 746
harrisi 746
irazu 746
leridensis 746
longipilosus 746
rufoniger 746
stenopygius 746
subnubilis 746
teguina **746**
xerampelinus **746**, 746
scotinus, Miniopterus 231
Scotoecus 217, **226**
albigula 226
albofuscus **226**
artinii 226
falabae 226
hindei 226
hirundo **226**
pallidus **226**
woodi 226
Scotomanes **226**
emarginatus **226**
imbrensis 226
nivicolus 226
ornatus **226**, 226
sinensis 226
Scotonycteris **153**
cansdalei 153
occidentalis 153
ophiodon **153**
zenkeri **153**, 153
Scotophilus **226**
albofuscus 226
altilis 227
alvenslebeni 227
belangeri 227
borbonicus **226**, 227, 228
castaneus 227
celebensis **227**
colias 227
collinus 227
consobrinus 227
damarensis 227, 228
dinganii **227**, 227, 228
flaveolus 227
flavigaster 227
gairdneri 227
gigas 227
heathi **227**, 227
herero 227
insularis 227
kuhlii 226, **227**
leucogaster **227**, 227, 228
luteus 227
murinoflavus 227
nigrita **227**, 227, 228
nigritellus 227, 228
nucella 227
nux **227**, 227
panayensis 227
planirostris 227
pondoensis 227
robustus **228**
solutatus 227
swinhoei 227
temmincki 227
viridis 227, **228**
watkinsi 227
wroughtoni 227
scotophilus, Thomomys 474
Scotorepens 217, 218
Scotozous 219, 221
scottae, Dendrolagus **51**
scotti, Myotis **215**
Thallomys 669
scottii, Callosciurus 423
Urocyon 285
scripta, Antilope 403
Damaliscus 394
Prionailurus 295
scriptus, Tragelaphus **404**
scrobiculatum, Micronycteris 179
Scrofa 378
scrofa, Sus 378, **379**, 379
scrofoides, Sus 379
scrutator, Mephitis 317
Tamias 454
scullyi, Nesokia 632
scutatus, Diclidurus **157**
Scutisorex **101**
congicus 101
somereni **101**
scyritus, Hylobates 275
scythica, Saiga 400
seabrai, Myotis **215**
sebae, Monodelphis 21
sebasianus, Rattus 661
sebastianus, Stochomys 666
sebucus, Sundamys 666
Tragulus 383
secatus, Taphozous 161
sechuenensis, Lepus 819
sechurae, Dusicyon 283
Pseudalopex **284**
seclusa, Mustela 321
seclusus, Lepus 814
Reithrodontomys 742
secundus, Otomops **239**
securus, Proechimys 797
sedulus, Rhinolophus **168**
Tamias 456
seeheimi, Gerbillurus 548
sefranus, Lepus 816
sefrius, Jaculus 493
segethi, Phyllotis 738
segethii, Histiotus 205
segregatus, Thomomys 475
segurae, Sciurus 443
seiana, Ochotona 808, 812
seianum, Rhinopoma 155
seimundi, Balionycteris 138
Sundasciurus 451
Trachypithecus 274
seistanica, Gazella 397
Lutra 312
sejugis, Peromyscus **735**
seke, Cephalophus 410
Sekeetamys 546, **559**, 559
calurus 559, **560**
makrami 560
selbornei, Alcelaphus 393
Selenarctos 340
selenginus, Meriones 558
selevini, Spermophilus 446
Selevinia 763, 766, **768**, 768
betpakdalaensis **768**, 768
paradoxa 768
Seleviniinae 765
selina, Crocidura 88, **94**
selindensis, Cricetomys 541
Graphiurus 764
selkirki, Tamias 454
Ursus 339
sellata, Tamandua 68
sellatus, Cervus 385
sellysii, Meriones 558
selma, Trichosurus 48
selousi, Acomys 567
Cynictis 307
Damaliscus 394
Loxodonta 367
Paracynictis **307**
Rangifer 392
Taurotragus 403
Tragelaphus 404
selousii, Equus 369
selysii, Microtus 529
Selysius 207, 208, 209, 210, 211, 212, 213, 214, 215, 216
semenovi, Catopuma 289
semicanus, Alticola **504**
semicaudata, Emballonura **158**
semicircularis, Procavia 374
Semicricetus 538
semideserta, Allactaga 489
semilunaris, Phoca 332
seminolus, Lasiurus **207**
Odocoileus 391
seminudus, Rousettus 152
semipallidus, Dipodomys 479
semipalmatus, Thryonomys 775
semispinosus, Ericulus 73
Hemicentetes **73**, 73
Proechimys 795, **798**, 798
semistriatus, Conepatus **316**
semitorquatus, Herpestes 305, **306**, 306
semivillosus, Echimys **792**
semlikii, Heliosciurus 429
semmeliki, Suncus 103
semmelincki, Suncus 103
Semnocebus 246, 266
Semnopithecinae 269
Semnopithecus 270, **273**
achates 273
achilles 273
aeneas 273
ajax 273
albipes 273
anchises 273
dussumieri 273
elissa 273
entellus **273**
hector 273
hypoleucos 273
iulus 273
lania 273
nipalensis 273
pallipes 273
petrophilus 273
priam 273
priamelus 273
pyrrhus 273
schistaceus 273
thersites 273
semoni, Hipposideros **175**
semotus, Apodemus **573**
Lasiurus 206, 207
semplei, Mustela 321
senaariensis, Giraffa 383
Hemiechinus 78
sendaiensis, Cervus 386
senegalensis, Acinonyx 288
Alcelaphus 393
Canis 280
Damaliscus 394
Galago **249**, 249
Galagoides 250
Gazella 397
Genetta 345, 346
Hippopotamus 381
Ictonyx 319
Kobus 414
Leptailurus 293
Lepus 816
Myopterus 238
Orycteropus 375
Oryx 413
Panthera 297
Pipistrellus 223
Taphozous 161
Trichechus **366**
senegalica, Hystrix 773
senescens, Heliosciurus 428
Mungos 307
Tamias 455
senex, Aotus 256
Caluromys 15
Centurio **189**, 189
Dremomys 425
Eira 318
Myoictis 32
Paradoxurus 344
Petaurista 462
Presbytis 274
Rhizomys 685
Sciurus 440
Tamias **455**, 456
Theropithecus 269

Trachypithecus 274
sengaani, Paracynictis 307
senganus, Kobus 414
Sigmoceros 395
senicula, Mormoops 176
seniculus, Alouatta **255**, 255
Hemiechinus 78
senilis, Akodon 692
Eira 318
Seniocebus 253
sennaarensis, Sus 379
sennariensis, Galago 249
sennetti, Dipodomys 477
seorsus, Chaetodipus 484
Zygodontomys 752
sepikensis, Pteropus 149
septemcinctus, Dasypus **66**
septemtrionalis, Bison 400
septemvittatus, Rhabdomys 662
septentrionalis, Amblysomus 74
Apodemus 570
Bubalus 402
Lepilemur **246**
Lepus 821
Microsciurus 433
Myotis 212
Ochotona 811
Panthera 298
Phylloderma 180
Proteles 309
Rhinolophus 169
Spermophilus 448
Tamias 453
septicus, Rattus 660
sequoiensis, Peromyscus 736
Tamias 456
Urocyon 285
seraiae, Callosciurus 423
serasani, Cynopterus 139
serbicus, Nannospalax 754
serengetae, Gazella 396
serezkyensis, Crocidura 93, **95**, 95, 98
sergii, Mus 628
seri, Chaetodipus 484
Neotoma 710
sericatus, Bunomys 582
Scapanus 127
sericea, Crocidura 85
Oreamnos 407
Scalopus 127
sericeus, Callithrix 252
Ctenomys **786**
Phalanger **47**
Propithecus 247
Sphiggurus 777
seringetica, Gazella 397
serotinus, Eptesicus 202, **203**, 203
Proechimys 795
serpens, Microtus 527
serpentarius, Suncus 103
serratus, Dasypus 66
Peromyscus 732
Taphozous 161
serrensis, Akodon **692**, 693
serridens, Lobodon 330
serutus, Callosciurus 423
Maxomys 614
serval, Felis 292
Leptailurus **292**
servalina, Felis 291
Genetta **346**, 346
Leptailurus 293
servorum, Cricetomys 541
sestinensis, Reithrodontomys **741**
setchuanus, Eozapus **498**
Zapus 498
Setifer **73**
acanthurus 73
fontoynonti 73
nigrescens 73
setosus **73**, 73
spinosus 73
setifera, Bandicota 579
setiger, Leopoldamys 603
Mormopterus **238**
setonbrownei, Gerbillus 553
setoni, Rangifer 392
Setonix **56**
brachyurus **57**
brevicaudatus 57
setosa, Bettongia 49
Tachyglossus 13
setosus, Arvicanthis 578
Erinaceus 73
Euphractus 66
Liomys 482
Paradoxurus 344
Petinomys **464**
Potorous 50
Proechimys 797, **798**
Setifer **73**, 73
Sorex 118
Sus 379
Tachyglossus 13
Vombatus 46
Xerus 458
setulosus, Mus 622, 624, **627**, 627, 628
setzeri, Gerbillurus **547**, 548
Mus 627, **628**
Myomimus **768**
Sigmodon 747
seurati, Acomys 565
severtzovi, Allactaga **489**, 489
Alticola 503
Capra 405
Mus 626
Ovis 409
Sicista **497**, 498
severzovi, Meles 314
sevia, Pogonomelomys **641**, 641
sevieri, Thomomys 474
sewelanus, Pipistrellus 224
sexcinctus, Dasypus 66
Euphractus **66**
sexplicatus, Melomys 619
seychellensis, Coleura **156**
Mesoplodon 362
Pteropus **146, 151**
seythius, Arvicola 506
sezekorni, Erophylla **182**
Phyllonycteris 182
shameli, Rhinolophus 165, **168**
shanicus, Callosciurus 421
Trachypithecus **274**
shanorum, Pipistrellus 220
Ursus 339
shanseius, Eothenomys **514**
shantaricus, Microtus 527
shantungensis, Crocidura 96
sharica, Procavia 374
sharicus, Lycaon 283
sharpei, Colobus 270
Funisciurus 427
Raphicerus **400**
shastensis, Castor 467
Sorex 122
shattucki, Microtus 527
shawi, Dasymys 589
Melomys 618, 619
Meriones 556, **557**, 557, 558
Sciurus 442
Thomomys 475
shawiana, Felis 290, 291
shawii, Arctocephalus 326
shawmayeri, Chiruromys 584
Coccymys 585, 586
Dendrolagus 50
Hydromys **597**, 598
Melomys 616, 617
sheila, Chalinolobus 200
Mazama 390
sheldoni, Lepus 815
Marmota 431
Neotoma 710
Odocoileus 391
Ovis 408
Thomomys 476
Ursus 339
sheltoni, Ochotona 811
shenseius, Myospalax 675
shepherdi, Tasmacetus 363, **364**
sheppardi, Sylvisorex 105
sherif, Lepus 816
shermani, Blarina 106
Sciurus 442
sherrini, Nyctophilus 218
Tatera 561
Uromys 671
shevketi, Microtus 530
shigarus, Rattus 661
shikokensis, Sorex 120
shindi, Heliosciurus 429
shinto, Sorex 113, 115, **120**, 120
shiptoni, Galictis 318
Microcavia **780**
shirasi, Alces 389
Ursus 339
shiraziensis, Eptesicus 203
shirensis, Heliosciurus 429
Sigmoceros 395
Sylvicapra 412
Tatera 561
shiroumanus, Sorex 115
shitkovi, Meriones 557
Pygeretmus **491**, 491
shnitnikovi, Alticola 503
sho, Mustela 324
shoae, Ictonyx 319
shoana, Tatera 562
shortridgei, Callosciurus 421
Chaerephon 232
Chlorotalpa 75
Chrysochloris 75
Crocidura 89
Dendromus 543
Desmodillus 547
Galerella 302, 303
Ictonyx 319
Mastomys 610, **611**, 611
Miniopterus 230
Mus **628**, 628
Panthera 298
Petromyscus 683, **684**, 684
Phacochoerus 377
Pseudomys 645, 646, 648, **649**, **649**
Rhinolophus 167
Rousettus 152
Thallomys **669**, 669
Trachypithecus 274
Uranomys 671
Zelotomys 674
shoshone, Ursus 339
shoutedeni, Hylomyscus 598
shufeldti, Procyon 335
shukurovi, Ochotona 812
shumaginensis, Sorex 118
siaca, Tupaia 132
Siaga 400
Siamanga 274
siamensis, Bandicota 579
Callosciurus 421
Cervus 386
Felis 291
Herpestes 305
Hylomys 80
Lepus 814, 819, 820
Presbytis 271
Rhinolophus 167
siamica, Macaca 267
siantanicus, Rattus 661
Sundasciurus 452
Tragulus 383
siarma, Maxomys 614
sibatoi, Cercopithecus 264
sibbaldii, Balaenoptera 350
Sibbaldius 349
sibbaldius, Balaenoptera 350
siberiae, Akodon **692**, 692
siberica, Clethrionomys 509
Sicista 498
sibericus, Clethrionomys 509

Pygeretmus 490
siberu, Lariscus 430
Nasalis 270
Paradoxurus 344
Presbytis 271
Sundasciurus 451
Tupaia 132
sibila, Marmota 431, 432
sibirica, Allactaga **489**
Capra **406**
Crocidura **95**
Gulo 318
Marmota 431, **433**, 433
Murina 229
Mustela 321, 323, **324**, 324
Phoca **332**
sibiricorum, Lepus 821
sibiricus, Cervus 386
Gulo 318
Lemmus **517**, 517
Meles 314
Mesechinus 79
Moschus 384
Myotis 209
Pteromys 465
Rangifer 392
Sus 379
Tamias **453**, **455**
Ursus 339
sibiriensis, Sorex 121
sibreei, Cheirogaleus 243
sibuanus, Limnomys **605**, 605
sibylla, Vandeleuria 672
sibyllae, Gazella 397
sica, Cervus 386
sicarius, Cervus 386
Myotis **215**
siccata, Tupaia 131
siccatus, Aethomys 568
Bassaricyon 333
Herpestes 304
siccovallis, Thomomys 474
siccus, Chaetodipus 483
Spermophilus 444
sichuanensis, Dryomys **766**, 766, 767
sichuanicus, Cervus 386
sicialis, Mastomys 610
Sicista 487, 495, **496**, 496, 497, 498
armenica **496**, 496, 497
betulina **496**, 497
caucasica **496**, 496, 497
caudata **496**, 496, 497, 498
concolor **496**, 496, 497, 498
flavus 497
interstriatus 498
interzonus 498
kazbegica 496, **497**, 497
kluchorica 496, **497**, 497
leathemi 497
lineatus 498
loriger 498
montana 496
napaea 496, **497**, 497
nordmanni 498
norvegica 496
pallida 498
pseudonapaea 496, **497**, 497
severtzovi **497**, 498
siberica 498
strandi 496, **497**, 497
subtilis **497**, 497
taigica 496
tatricus 496
tianshanica 496, 497, **498**
tripartitus 498
tristriatus 498
trizona 498
vaga 498
virgulosus 498
weigoldi 497
Sicistidae 487
Sicistinae 487, **495**, 495, 496
sicula, Crocidura 82, **95**
Nyctalus 216
siculae, Rattus 658
siculus, Mustela 323
Vespertilio 228
siebersi, Uromys 671
sieboldi, Eubalaena 349
sieboldii, Globicephala 352
sierrae, Agouti 783
Lepus 821
Marmota 432
Martes 319
Microtus 523
Ovis 408
Spermophilus 445
sifanicus, Moschus 384
sigillata, Viverra 348
Sigmoceros 393, **394**, 395
bangae 395
basengae 395
dieseneri 395
frommi 395
gendagendae 395
godonga 395
godowiusi 395
gombensis 395
gorongozae 395
grotei 395
hennigi 395
heuferi 395
inkulanondo 395
janenschi 395
kangosa 395
konzi 395
lacrymalis 395
lademanni 395
leucoprymnus 395
leupolti 395
lichtensteinii 393, **394**
lindicus 395
munzneri 395
niediecki 395
petersi 395
prittwitzi 395
rendalli 395
rowumae 395
rukwai 395
saadanicus 395
schmitti 395
schusteri 395
senganus 395
shirensis 395
stierlingi 395
tendagurucus 395
ufipae 395
ugalae 395
ulanagae 395
ungonicus 395
ungoniensis 395
uwendensis 395
wiesei 395
wintgensis 395
Sigmodon 704, **746**, 746, 747
alfredi 747
alleni **747**
alstoni 746, **747**
alticola 747
amoles 747
arizonae **747**, 747
atratus 747
austerulus 747
baileyi 748
berlandieri 747
bogotensis 747
borucae 747
chiriquensis 747
chonensis 748
cienegae 747
colimae 747
confinis 747
dalquesti 747
eremicus 747
exsputus 747
fervidus 747
floridanus 747
fulviventer **747**
furvus 747
goldmani 747
griseus 747
guerrerensis 747
hirsutus 747
hispidus 746, **747**, 747, 748
inexoratus 747
inopinatus **747**, 747
insulicola 747
ischyrus 747
jacksoni 747
komareki 747
leucotis **747**
littoralis 747
lonnbergi 747, 748
macdougalli 747
macrodon 747
madrensis 748
major 747
mascotensis **747**, 747
melanotis 747
microdon 747
minimus 747
minor 747
montanus **748**
obvelatus 747
ochrognathus **748**
pallidus 747
peruanus 747, **748**
planifrons 747
plenus 747
puna 748
sanctaemartae 747
saturatus 747
savannarum 747
setzeri 747
simonsi 748
solus 747
spadicipygus 747
texianus 747
toltecus 747
tonalensis 747
venester 747
villae 747
virginianus 747
vulcani 747
woodi 747
zanjonensis 747
Sigmodontinae 501, **687**, 687, 739, 743, 752
Sigmodontomys 709, **748**, 748
alfari **748**, 748
aphrastus **748**
barbacoas 748
efficax 748
esmeraldarum 748
ochraceus 748
ochrinus 748
russulus 748
Sigmomys 746
signata, Mellivora 315
Petrogale 55
signatus, Canis 281
Funambulus 426
Leptomys 604, **605**
Protoxerus 436
Tragelaphus 404
signifer, Chalinolobus 199
Sika 385
sika, Cervus 386
Sikaillus 385
sikapusi, Lophuromys **606**, 606
Mus 605
sikhimensis, Nectogale 110
siki, Hylobates 275
sikimaria, Ochotona 812
sikimensis, Amblonyx 310
Microtus **522**, **529**, 529
Pitymys 522
sikkimensis, Rattus 649, **659**
sikotanensis, Clethrionomys **509**
silacea, Crocidura **95**
silaceus, Celaenomys **583**
Chlorocebus 265
Vulpes 287
Xeromys 583
silanus, Sciurus 443

Sorex 112
silberbaueri, Otomys 681
Silenus 266
silenus, Macaca **267**
silhouettae, Coleura 156
siligorensis, Myotis **215**
silindensis, Aethomys 567, **568**, 568
Grammomys 593
sillimani, Perognathus 485
silus, Glaucomys 460
silvai, Mesocapromys 800
Silvanus 266
silvatica, Equus 369
Murina **229**, 229, 230
silvaticus, Tamias 454
silvestris, Equus 369
Felis 289, **290**, 291
Rhinolophus **164**, **168**
Thomasomys **750**
silvicola, Arborimus 505
Rangifer 392
Tonatia 180, **181**
silvicultor, Cephalophus **411**, 411
silvicultrix, Antilope 410
silvifugus, Thomomys 474
simalurensis, Rattus 649, **659**, 659, 661
simalurus, Pteropus 147
simcoxi, Lepus 819
Simenia 282
simensis, Canis **282**
Lophuromys 605
Simia 262, 263
aethiops 265
belzebul 254
capucina 259
diana 263
entellus 273
hamadryas 269
inuus 266
jacchus 251
lagothricha 257
maimon 268
melalophos 270
melanocephalus 260
midas 253
mormon 268
nemaeus 272
paniscus 256
patas 265
personatus 258
pithecia 261
polycomos 269
pygmaeus 277
rosalia 252
sabaea 265
sciurea 260
sphinx 268
sylvanus 266
syrichta 250
talapoin 268
trivirgata 255
troglodytes 276
Simias 270
simiasciurus, Potos 334
similis, Callosciurus 423
Dipodomys 479
Lagenorhynchus 354
Maxomys 613
Microsciurus 433
Microtus 526
Nyctinomops 238, 239
Sorex 118
Sylvilagus 826
Thomomys 475
simiolus, Crocidura 98
Dipodomys 479
Simocyoninae 282, 283, 285
simoni, Gerbillus **554**
simoninus, Cervus 387
simonsi, Coendou 775
Marmosa 19
Microsciurus 433
Neoctodon 788
Octodontomys 788
Proechimys 797, **798**
Sigmodon 748
Simosciurus 439
Simotes 532
simplex, Arctogalidia 343
Heliosciurus 428
Hesperomys 739
Microtus 519
Neotoma 711
Paradoxurus 344
Pseudoryzomys **739**
Rhinolophus **169**
Tragelaphus 404
simplicicanus, Sylvilagus 826
simplicicornis, Mazama 391
simplicidens, Blarina 106
Simplicidentati 771
simpsoni, Cephalophus 411
Rattus 657
Taeromys 667
simsoni, Mus 626
simulans, Dipodomys 477
Pappogeomys 472
Tamias 455
simulator, Akodon **693**, 694
Rhinolophus **169**
simulatus, Habromys **703**, 703
simulus, Peromyscus 729, 732, **736**
Thomomys 474
simum, Ceratotherium **371**
simung, Lutrogale 313
simus, Glyphotes **428**, 428
Hapalemur **245**
Hylomyscus 598
Kogia **359**
Myotis 214, **215**
Procyon 335
Rhinoceros 371
sinaitica, Capra 406
sinaiticus, Lepus 816
Procavia 374
sinalis, Sorex 116, **120**
sinaloae, Marmosa 18
Molossus **235**
Neotoma 712
Odocoileus 391
Thomomys 474
sinaloensis, Peromyscus 730
Sciurus **440**
Sinanthropus 276
sincipis, Tupaia 132
sindensis, Suncus 103
sindi, Paraxerus 434
sindica, Lutrogale 313
Panthera 298
Tadarida 240
sindicus, Bandicota 579
sinensis, Budorcas 404
Felis 291
Herpestes 306
Hipposideros 174
Hylomys **80**
Lepus 817, **820**
Lutra 312
Megaderma 163
Muntiacus 389
Nyctalus 217
Nyctereutes 283
Ochotona 810
Rhizomys **685**, 685
Scotomanes 226
sinesella, Ovis 408
singapurensis, Callosciurus 423
singhala, Lepus 819
singhei, Petaurista 462
singidae, Aethomys 567
singsing, Kobus 414
singuri, Millardia 621
sinhala, Ratufa 437
sinhaleyus, Elephas 367
sinianus, Niviventer 633
sinica, Macaca 267, **268**
sinicus, Coelops 170
Mus 626
Rhinolophus 168
siningensis, Meles 314
sinistralis, Callosciurus 422
Sinisus 378
Sinoetherus 776
sinualis, Planigale 34
sinuensis, Eira 318
sinuosus, Sorex 118
sinus, Canis 282
Phocoena **358**
Ratufa 437
Tupaia 131
Sipalus 46
Siphneus 675
sipora, Hylopetes **461**, 461, 604
Iomys **461**, 461, 604
siporanus, Leopoldamys 422, 430, 451, 461, 464, 584, **604**, 604, 613, 654
Sirenia 365
sirhassenensis, Ratufa 437
Tupaia 133
siriensis, Callosciurus 423
sirtalaensis, Microtus 521
siskiyou, Tamias **455**, 456
sitkaensis, Vulpes 287
sitkeenensis, Ursus 339
sitkensis, Microtus 527
Odocoileus 391
Peromyscus 734, **736**, 736
Ursus 339
Sitomys 728
siumatis, Megaderma 163
siva, Cremnomys 588
Macropus 53
Mus 627
sivalensis, Crocuta 308
Hippopotamus 380
Sivalikia 367
sjostedti, Rousettus 152
skalauensis, Pongo 277
skomerensis, Clethrionomys 509
sladeni, Callosciurus 421
Pseudalopex 285
Rattus 659, 660
slamatensis, Petaurista 462
slatini, Procavia 374
slevini, Peromyscus **736**
sloeops, Osgoodomys 726
Peromyscus 734
sloggetti, Otomys 680, **681**
slowzovi, Microtus 521
sludskii, Salpingotus 492
smalli, Neotoma 711
Sminthi 495
Sminthidae 487, 495
Sminthinae 495
Sminthopsis **35**, 35
aitkeni **35**
albipes 37
archeri **35**
butleri **35**
caniventer 36
centralis 35
crassicaudata **35**
dolichura **36**
douglasi **36**
ferruginea 35
ferruginifrons 36
froggatti 37
fuliginosus **36**
gilberti **36**
granulipes **36**
griseoventer **36**
hirtipes **36**
laniger **36**
larapinta 37
leucogenys 36
leucopus **36**
longicaudata **36**
lumholtzi 37
macroura **36**
mitchelli 36
monticola 37
murina 35, 36, **37**, 37
nitela 37
ooldea **37**

psammophila **37**
rona 37
rufigenis 37
spenceri 36
stalkeri 37
tatei 37
virginiae **37**, 37
youngsoni **37**
Sminthus **495**, **496**
smirnovi, *Ursus* 339
smithemani, *Kobus* 414
smithersi, *Heliosciurus* 429
Pseudalopex 284
smithi, *Eptesicus* 202
Lophiomys 564
Ratufa 437
Rousettus 152
Synaptomys 534
Tatera 562
Trachypithecus 274
smithii, *Cervus* 385
Crocidura 83, 89, **95**
Evotomys 532
Herpestes **306**
Kerivoula **198**
Leopardus 291
Madoqua 398
Microcebus 244
Myospalax 675, **676**, 676
Ourebia 399
Phaulomys 532, **533**
Soriculus **123**, 123
smitianus, *Miniopterus* 231
Smutsia 415
smyrnensis, *Apodemus* 572
sneiderni, *Echinoprocta* 776
Sphiggurus 776
snethlageri, *Callimico* 251
soasoaranna, *Puma* 297
sobrina, *Marmosops* 19
sobrinus, *Chaetodipus* 484
Macroglossus **154**
Rhinolophus 167
sobrus, *Clethrionomys* 509
soccatus, *Leopoldamys* 604
Paraxerus 434
Suncus 103
socer, *Rattus* 657
sociabilis, *Ctenomys* **786**
Myotis 216
Nasua 335
socialis, *Cynomys* 424
Mastomys 610
Microtus 521, 522, 527, **529**, 530
Sciurus 440
societatis, *Pipistrellus* **223**
soconuscensis, *Orthogeomys* 471
sodalis, *Cavia* 779
Chroeomys 700
Eptesicus 203
Myotis **215**
Rhombomys 559
Soriculus 123
soderstromi, *Ichthyomys* 705
Sciurus **441**
soderstromii, *Puma* 297
sodyi, *Crocidura* 87
Kadarsanomys **600**
Petaurista 463
Rattus 600
soederstroemi, *Reithrodontomys* 742
soederstroemmi, *Nasua* 335
soemmeringii, *Acinonyx* 288
soemmerringii, *Gazella* **397**
sogdianus, *Meriones* 556
Myotis 213
sokolovi, *Cricetulus* **538**
sola, *Neotoma* 712
solaris, *Maxomys* 614
solatus, *Arvicanthis* 578
Cercopithecus **265**
Rattus 652
soldadoensis, *Zygodontomys* 752
soleatus, *Graphiurus* 764
Solenodon **69**, 69
cubanus **69**
marcanoi **69**, 69
paradoxus **69**, 69
poeyanus 69
Solenodontidae **69**
solifer, *Taphozous* 161
solisi, *Lepus* 818
Solisorex **101**
pearsoni **101**, 101
solitaria, *Lontra* 311
Nasua 334, 335
Neotoma 712
solitarius, *Castor* 467
Pteropus 148
Rhinolophus 169
Thomomys 474
solivagus, *Tamias* 453, 454
soloensis, *Cervus* 386
solomonensis, *Miniopterus* 230
solomonis, *Anamygdon* 207
Chaerephon 232
Mosia 158
Myotis 207
Pteropus 146
Solomys **663**, 663
ponceleti **663**
salamonis **663**, 663
salebrosus **663**
sapientis **664**
spriggsarum **664**
soluensis, *Soriculus* 122
solus, *Clethrionomys* 508
Niviventer 633
Sigmodon 747
solutatus, *Scotophilus* 227
solutus, *Callosciurus* 421
Procyon 335
somalensis, *Lepus* 816
somaliae, *Vulpes* 286
somalica, *Crocidura* 83, **95**, 98
somalicus, *Arvicanthis* 577, 578
Canis 280
Crossarchus 302
Elephantulus 829
Eptesicus **203**
Equus 369
Gerbillus 550, **555**
Heterohyrax 373
Mungos 307
Oreotragus 399
Orycteropus 375
Tachyoryctes 687
somaliensis, *Diceros* 372
Panthera 297
Potamochoerus 378
Thallomys 669
somereni, *Scutisorex* **101**
Sylvisorex 101
sondaica, *Panthera* 298
Ratufa 437
sondaicus, *Bos* 401
Elephas 367
Rhinoceros **372**
Trachypithecus 273
songaricus, *Cervus* 386
Sus 379
sonlaensis, *Bandicota* 579
sonneratii, *Suncus* 103
sonomae, *Sorex* 119, **120**
Tamias **456**
sonora, *Lontra* 311
sonorana, *Liomys* 482
sonoriensis, *Antilocapra* 393
Conepatus 316
Dipodomys 477
Eumops 234
Leopardus 291
Myotis 211
Pecari 380
Perognathus 485
Peromyscus 732
Spermophilus 449
Tamias 453
Taxidea 325
Thomomys 476
soperi, *Phenacomys* 533
Phoca 332
Sorex 118
sordida, *Tupaia* 132
sordidulus, *Pappogeomys* 472
sordidus, *Exilisciurus* 426
Rattus 649, 651, 654, **659**, 659, 662
Sundasciurus 452
sorella, *Melogale* 314
Microgale 71
Mus 622, 623, 624, 626, 627, **628**, 628
Ochotona 807
Sylvisorex 105
sorelloides, *Sylvisorex* 105
sorellus, *Calomys* **698**, 698
Sorex 99, 110, **111**, 111, 112, 113, 114, 115, 116, 117, 118, 119, 120, 121, 122
abnormis 117
acadicus 113, 119
alascensis 118
alaskanus **111**, 119
albibarbis 119
albiventer 113
alnorum 115
alpinus **111**, 118, 119
altaicus 113
alticola 111
amasari 121
amoenus 122
amphibius 110
annexus 113, 120
antinorii 111
aquaticus 127
araneoides 113
araneus 108, **111**, 111, 112, 114, 115, 116, 120
aranoides 119
arcticus 111, **112**, 112, 113, 115, 116, 120, 121
arizonae **112**
asper **112**, 112, 114
bachmani 116
baikalensis 121
bairdii **112**
barabensis 117
batis 119
becki 117
bedfordiae **112**
bendirii **113**
bergensis 111
beringianus 113, 116
blitchi 114
bohemicus 111
bolkayi 111
borealis 121
brevicaudus 106
brooksi 119
buchariensis **113**, 116, 121, 122
burneyi 117
buxtoni 113
caecutiens **113**, 113, 120
californicus 118
calvertensis 118
camtschatica **113**, 113
canaliculatus 117
cansulus **113**, 113, 120
carpathicus 111
carpetanus 117
cascadensis 118
castaneus 111
caucasicus 119, 120
caudata 117
caudatus 122
centralis 113, 121
chiapensis 122
chouei 120
cinereus 108, 111, **113**, 113, 114, 115, 116, 117, 119, 121
concinnus 111
cooperi 113
coronatus **114**
crassicaudatus 111

cristatus 124
cristobalensis 120
cryptis 111
csikii 111
cylindricauda 113, **114**
czekanovskii 117
daphaenodon **114**, 114
daubentonii 110, 111
destructioni 121
dispar **114**, 115
dobsoni 118
dukelskiae 119
durangae 118
eionis 116
elassodon 118
eleonorae 111
emarginatus 112, **114**, 118
euronotus 114
excelsus 112, **114**, 114
exiguus 117
exilis 117
eximius 115
feroculus 99
fimbripes 113
fisheri 116, 117
fodiens 110
fontinalis 113
forsteri 113
frankstounensis 113
fretalis 114
fumeolus 112
fumeus **114**
garganicus 120
gaspensis 114, **115**
gemellus 114
glacialis 118
glis 131
gloveralleni 119
gmelini 117
godmani 120
gomphus 112
gracillimus **115**, 116, 118
granarius 113, **115**
grantii 111
gravesi 116
gymnurus 117
halicoetes 122
hawkeri 117
haydeni 113, 114, **115**, 115
heptapotamicus 117
hercynicus 111
hermanni 111
hibernicus 117
hollisteri 113
hosonoi **115**, 115
hoyi **115**
huelleri 111
humboldtensis 121
hydrobadistes 119
hydrodromus **115**, 115, 116
hyojironis 115
idahoensis 113
ignotus 111
insulaebellae 117
insularis 118
intermedius 111
intervectus 115
iochanseni 111
irkutensis 121
ishikawai 117
isodon **116**, 116, 120
isolatus 118
jacksoni 113, **116**, 116, 119, 121
jacutensis 119
jenissejensis 121
juncensis 118
karelicus 117
karpinskii 113
kastchenkoi 117
koreni 113
kozlovi **116**, 116, 121
kutscheruki 118
labiosus 111
labradorensis 119
lagunae 118
lapponicus 113
laricorum 112
lesueurii 113, 114
leucodon 81
leucogaster 113, **116**, 116
leucogenys 117
longicauda 118
longiquus 118
longirostris **116**, 117
longobarda 111
lucanius 117
lyelli **117**
macrodon **117**, 122
macropus 99
macropygmaeus 113
macrotrichus 111
macrurus 114
malitiosus 118
marchicus 111, 112, 116
margarita 121
mariposae 121
maritimensis 112
melanderi 117
melanodon 111
melanogenys 118
merriami **117**
middendorfi 121
milleri **117**, 117
minimus 117
minutissimus **117**
minutus 115, **117**, 117, 118, 120, 121, 122
mirabilis **118**, 118, 119
miscix 113
mixtus 118
mollis 111
monsvairani 111
montanus 115
montereyensis 121
monticolus **118**, 119, 122
monticulus 112
murinus 101
mutabilis 122
myops 121
nanus **118**, 121
navigator 119
neglectus 117
neomexicanus 118
nepalensis 112
nevadensis 122
nigra 111
nigriculus 113, 122
nuda 112
oaxacae 120
obscuroides 118
obscurus 112, 118, 122
ohionensis 113
oreinus 118
oreopolus **114**, **118**, 118, 122
orii 114
orizabae 118
ornatus **118**
pachyurus 112
pacificus **118**, 118
pallidus 112
palmeri 113
paludivagus 122
palustris 111, **119**
parvicaudatus 121
parvidens 122
parvus 108
permiliensis 112
perminutus 117
personatus 112, 113, 114
petrovi 112
petschorae 121
peucinius 112
planiceps 117, 118, **119**, 121
platycranius 119
platyrhinus 113
pleskei 113
portenkoi 113, 116, **119**, 121
preblei **119**
preussi 112
prevostensis 118
pribilofensis 115, 116
princeps 116
proboscideus 830
pulchellus 98
pulcher 112
pumilio 117
pumilus 117
punctulatus 119
pygmaeus 117
pyrenaicus 112
pyrrhonota 112
quadraticauda 106
quadricaudatus 112
raddei **119**, 120
relictus 118
rhinolophus 112
richardsonii 112
roboratus **119**, 119
rozanovi 113
rusticus 117
ruthenus 116
ryphaeus 112
sadonis **119**
saevus 120
salarius 118
salicornicus 118
salvini 120
samniticus **120**
sanguinidens 114
santonus 114
satunini 119, **120**
saussurei 118, **120**, 122
scaloni 114
schnitnikovi 121
sclateri **120**
setosus 118
shastensis 122
shikokensis 120
shinto 113, 115, **120**, 120
shiroumanus 115
shumaginensis 118
sibiriensis 121
silanus 112
similis 118
sinalis 116, **120**
sinuosus 118
sonomae 119, **120**
soperi 118
spagnicola 112
stizodon **120**
streatori 113
stroganovi 117
sukleyi 122
surinamensis 112
talpoides 106
tasicus 113
tatricus 111
teculyas 122
tenelliodus 120
tenellus 118, **120**
tetragonurus 112
thibetanus 113, 116, 117, 118, 119, **121**, 121
thomasi 119
thompsoni 115
tomensis 119
transrypheus 121
trigonirostris 122
trowbridgii **121**
tscherskii 117
tschuktschorum 117
tundrensis 112, 113, 114, **121**, 121
tungussensis 113
turneri 119
turuchanensis 119
ugyunak 113, 116, 119, **121**, 121
ultimus 121
umbrosus 114
unguiculatus **121**
uralensis 112
ussuriensis 117
vagrans 111, 112, 118, **122**, 122
vancouverensis 122
varius 99
ventralis 118, 120, **122**, 122

veraecrucis 120
veraepacis 117, **122**
vir 119
volnuchini 113, 117, 118, **122**, 122
vulgaris 112
wagneri 116
wardii 112
washingtoni 115
wettsteini 112
willetti 118
winnemana 115
yaquinae 118
yesoensis 122
sorex, *Monodelphis* **22**
Soricidae **80**, 105, 111
Soricidus 111
soricina, *Cryptotis* 109
Glossophaga **184**, 184
Thylamys 23
Soricinae 80, **105**, 110, 122
Soricini 80, 106
soricinus, *Micromys* 620
Nannosciurus 434
Vespertilio 184
soricipes, *Uropsilus* **130**, 130
Soriciscus 108
soricoides, *Crocidura* 91, 93
Mus 625
Neomys 110
Rhynchomys **663**, 663
Soriculini 107, 110, 122
Soriculus **122**, 123
baileyi 123
berezowski 123
caudatus **122**, 123
caurinus 123
centralis 123
fumidus **123**, 123
furva 123
gracilicauda 122
gruberi 123
hypsibius **123**, 123
irene 123
lamula **123**, 123
larvarum 123
leucops **123**
lowei 123
macrurus **123**
nigrescens **123**
pahari 123
parca **123**, 123
parva 123
radulus 123
sacratus 122
salenskii **123**, 123
smithii **123**, 123
sodalis 123
soluensis 122
umbrinus 122
sornborgeri, *Ursus* 338
sorongensis, *Dendrolagus* 50
soror, *Genetta* 346
Tatera 562
Sotalia **355**, 355
fluviatilis **355**
guianensis 355
pallida 355
tucuxi 355
sotikae, *Atelerix* 76
Galago 249
soudanicus, *Canis* 280
Sousa **355**
borneensis 355
chinensis **355**, 355, 357
lentiginosa 355
plumbea 355
teuszii **356**
zambezicus 355
sowerbensis, *Delphinus* 362
Mesoplodon 362
sowerbyae, *Lepus* 821
sowerbyi, *Apodemus* 572
Dipus 493
Mesoplodon 362
Myotis 215
Neomys 110
Rattus 657
spadicea, *Vandeleuria* 672
spadiceus, *Atilax* 301
Cynopterus 140
Dyacopterus **140**
Hylopetes **461**, 461
Sciurus 441, **442**
spadicipygus, *Sigmodon* 747
spadix, *Apodemus* 573
Cephalophus **411**
Dendrolagus 50, **51**
Mustela 322
Rhinolophus 164
spagnicola, *Sorex* 112
Spalacinae 501, **753**
spalacinus, *Tachyoryctes* **687**
Spalacomys 632
Spalacopidae 787
Spalacopus **789**
ater 789
cyanus **789**
maulinus 789
noctivagus 789
poeppigii 789
tabanus 789
Spalax 753, **754**, 754
antiquus 755
arenarius 753, **754**, 754
ehrenbergi 753
giganteus 753, **754**
graecus 753, **755**, 755
istricus 755
kirgisorum 753
mezosegiensis 755
microphthalmus 753, 754, **755**, 755
pallasi 755
podolicus 755
polonicus 753, 755
typhlus 755
uralensis 754
zemni **755**, 755
spalax, *Heliophobius* 772
spartacus, *Dasyurus* 31, **32**
spasma, *Megaderma* **163**
Vespertilio 163
spatangus, *Hemiechinus* 78
spatharius, *Cervus* 386
Sus 379
spatiosus, *Thomomys* 474
spatulatus, *Niviventer* 633
Ondatra 532
spatzi, *Caracal* 289
spechti, *Melomys* **619**
speciosa, *Macaca* 266, 267
speciosus, *Apodemus* 572, **573**, 573
Muscardinus 769
Oecomys **716**
Pteropus 148, **151**
Tamias **456**
Tupaia 133
spectabilis, *Dendromus* 543
Dipodomys 479, **480**
Liomys **482**
Reithrodontomys **742**
spectrum, *Megaderma* 163
Sturnira 192
Tarsius **250**, 250
Vampyrum **181**
Vespertilio 181
speculator, *Hipposideros* 172
spegazzinii, *Akodon* 689, **693**, 693
Tapirus 371
spekei, *Gazella* **397**
Lemniscomys 601
Pectinator **761**, 761
spekii, *Tragelaphus* **404**
spelaea, *Crocuta* 308
Eonycteris **154**
Spelaeomys **664**, 664
florensis **664**, 664
spelaeum, *Plagiodontia* 804
spelaeus, *Macroglossus* 153
Myotis 213
Odocoileus 391
spencei, *Tamiops* 457
spencerae, *Tarsipes* 62
spenceri, *Sminthopsis* 36
spenoca, *Macaca* 266
spenserae, *Tarsipes* 62
speoris, *Hipposideros* **175**
Vespertilio 170
Speothos **285**
baskii 285
melanogaster 285
pacivorus 285
venaticus **285**
wingei 285
Speoxenus 805
Spermatophilus 444
Spermophila 444
Spermophilis 444
Spermophilopsis **443**
bactrianus 444
heptopotamicus 444
leptodactylus **444**
schumakovi 444
turcomanus 444
Spermophilus 419, 424, **444**, 444, 445, 446, 447, 448, 449, 450
ablusus 448
adocetus **444**
alashanicus **444**
albertae 445
alleni 450
altaicus 450
altiplanensis 449
ammophilus 449
annectens 449
annulatus **444**
apricus 449
arceliae 444
arenicola 449, 450
arens 449
argyropuloi 447
arizonae 449
arizonensis 447
armatus **444**
artemesiae 447
atricapilla 448
atricapillus **444**
aureus 446
averini 449
badius 450
balcanicus 445
barrowensis 448
bavicorensis 449
beecheyi **445**
beldingi **445**
beringensis 448
bernardinus 447
binominatus 448
blanca 450
boehmii 448
boristhenicus 449
brachiura 445
brauneri 448
brevicauda 446, 447
brevicaudus 447
brunnescens 446
brunneus **445**
buccatus 450
buckleyi 450
buxtoni 448
cabrerai 449
canescens 449
canus **445**, 445, 448, 450
carruthersi 446
caryi 447
castanurus 447
certus 447
chlorus 449
chrysodeirus 447
cinerascens 447
citellus **445**, 446, 450
citillus 445
columbianus **445**
concolor 446
connectens 447
coriacorum 448
couchii 450
creber 445
cryptospilotus 449
dauricus 444, 445, **446**,

446
dilutus 444
douglasii 445
elegans **446**, 446, 449
ellermani 448
endemicus 445
eremonomus 449
erythrogenys **446**, 447, 448
erythrogluteia 445
eversmanni 450
fisheri 445
flavescens 448
franklinii **446**
fulvus **446**, **447**, 448
giganteus 446
goldmani 444
gradojevici 445
grammurus 450
guttatus 449
guttulatus 449
heptneri 447
herbicola 448
hollisteri 450
hoodii 450
hypoleucos 446
idahoensis 447
iliensis 446
infernatus 444
intercedens 450
intermedius 446
istricus 445
jacutensis 450
janensis 448
juglans 450
kalabuchovi 448
karamani 445
kazakstanicus 448
kennicottii 448
kodiacensis 448
laskarevi 445
lateralis **447**
leucopictus 449
leucostictus 448
leurodon 447
loringi 450
lyratus 448
macedonicus 445
macrospilotus 449
macrourus 450
madrensis **447**
major 446, **447**, 447, 448, 449
marginatus 449
martinoi 445
maximus 446
menzbieri 450
meridioccidentalis 449
mexicanus **447**, 450
microspilotus 449
mitratus 447
mohavensis **447**, 449
mollis 445, **447**, 448, 450
mongolicus 446
monticola 450
mugosaricus 448
musicus **448**, 448
nancyae 447, 448
nanus 446
nebulicola 448
neglectus 449
nesioticus 445
nevadensis 446
nigrimontanus 446
nikolskii 448
nudipes 445
obscurus 444
obsidianus 449
obsoletus 449
odessana 449
ognevi 449
olivaceous 450
oregonus 445
oricolus 449
orlovi 446, 448
osgoodi 448
oxianus 446
pallescens 449
pallidicauda 446
pallidus 448, 450
parryii **448**, 450
parthianus 446
parvidens 447
parvulus 445
parvus 450
perotensis **448**
pessimus 447
phaeognatha 448
planicola 448
plesius 448
pratensis 449
pygmaeus 446, **448**, 448, 449
ralli 449
ramosus 446
relictus **449**
richardsonii 446, **449**
robustus 450
rufescens 447
ruficaudus 445
rupestris 450
rupinarum 445
saryarka 446
satunini 448
saturatus 448, **449**
schmidti 450
selevini 446
septentrionalis 448
siccus 444
sierrae 445
sonoriensis 449
spilosoma **449**
stejnegeri 448
stephensi 447
stonei 448
stramineus 450
suslicus 448, **449**
tereticaudus 447, **449**
tescorum 447
texensis 450
thracius 445
tiburonensis 450
townsendii 445, 448, **449**
transbaikalicus 450
trepidus 447
tridecemlineatus 447, **450**
trinitatus 447
tschuktschorum 448
tularosae 450
umbratus 446
undulatus 448, **450**
ungae 447
utah 450
variegatus **450**
vigilis 445, 448
vociferans 449
volhynensis 449
washingtoni **450**
washoensis 447
wortmani 447
xanthoprymnus 445, **450**
yakimensis 450
yamashinae 446
spermophilus, *Lemniscomys* 602
spetsbergensis, *Rangifer* 392
Sphaerias **153**
blanfordi **153**
Sphaerocephalus 352
Sphaerocormus 67
Sphaeronycteris **191**
toxophyllum **191**, 191
sphagnicola, *Synaptomys* 534
Sphiggurus 775, **776**, 776, 777
affinis 777
couiy 777
insidiosus **776**, 777
laenatus 776
liebmani 776
melanurus 776
mexicanus **776**
nigricans 777
pallidus **776**
paragayensis 777
paraguayensis 777
pruinosus 777
roberti 777
sericeus 777
sneiderni 776
spinosa 776
spinosus **777**
vestitus **777**
villosus **777**
yucataniae 776
sphinx, *Cynopterus* **139**, 139
Mandrillus **268**, 268
Simia 268
Vespertilio 138
spicatus, *Mesomys* 794
spicilegus, *Mus* 624, 626, **628**, 628
Peromyscus 728, 729, **736**
spiculum, *Allactaga* 489
spiculus, *Atelerix* 76
spillmani, *Chaerephon* 233
Spilocuscus 46, **47**
atrimaculatus 47
brevicaudatus 47
chrysorrhous 47
goldiei 47
kraemeri 47
maculatus **47**, 47
nudicaudatus 47
ochropus 47
quoy 47
rufoniger **47**, 47
variegata 47
Spilogale **317**
ambigua 317
angustifrons 317
bicolor 317
gracilis 317
interrupta 317
larvatus 317
leucoparia 317
lucasona 317
microdon 317
mupurita 317
olympica 317
phenax 317
putida 317
putorius **317**, 317
pygmaea **317**, 317
quaterlinearis 317
ringens 317
saxatalis 317
striata 317
tenuis 317
texensis 317
tibetanus 317
zorilla 317
spilosoma, *Spermophilus* **449**
spilotus, *Chaetodipus* 483
spilurus, *Myospalax* 676
spinalatus, *Rousettus* **153**
spinalis, *Lemniscomys* 602
spinatus, *Chaetodipus* **484**
Perognathus 482
spinifex, *Atelerix* 76
spinigera, *Neotragus* 398
spinipinnis, *Phocoena* **359**
spinosa, *Hystrix* 776
Sphiggurus 776
spinosissimus, *Acomys* **567**, 567
spinosus, *Clyomys* 793
Euryzygomatomys **794**
Hesperomys 708
Neacomys **709**, 709
Odocoileus 391
Rattus 794
Setifer 73
Sphiggurus **777**
spiradens, *Tayassu* 380
Spirodontomys 805
spissigrada, *Mephitis* 317
spitzbergenensis, *Alopex* 279
spitzbergensis, *Ursus* 339
spitzenbergerae, *Chionomys* 507
spixii, *Aotus* 255
Cacajao 261
Galea **779**, 779
Saguinus 253

splendens, Bathyergus 685
Callosciurus 422
Neotoma 711
Oecomys 717
Tachyoryctes 685, 686, **687**, 687
Vulpes 287
splendida, Lutra 312
Ourebia 399
splendidissima, Vulpes 287
splendidula, Sylvicapra 412
Tupaia **133**
splendidus, Sciurus 441
spodiurus, Oligoryzomys 717
spoliatus, Myoxus 770
sponsoria, Thylamys 23
sponsus, Paraxerus 435
spretoides, Mus 624
spretus, Mus 626, **628**, 629
spriggsarum, Solomys **664**
spurcus, Maxomys 614
Rhinolophus 167
spurelli, Crocidura 92
spurrelli, Graphiurus 764
Lionycteris **181**, 181, 182
Mops 236, **237**
squalorum, Pseudomys 644
squalus, Otomys 682
squamicaudata, Wyulda **48**, 48
squamicaudis, Cabassous 65
squamicaudus, Anomalurus 757
squamipes, Anourosorex 105, **106**
Cryptotis **109**
Mus 709
Nectomys **709**, 709
sribuatensis, Rattus 661
Srilankamys **664**
ohiensis **664**, 664
ssongaea, Heterohyrax 373
ssongeae, Lycaon 283
Stachycolobus 269
stagnatilis, Neomys 110
stairsi, Cercopithecus 264
stalkeri, Melomys 619
Myotis **215**
Sminthopsis 37
stampflii, Cercopithecus 264
Crocidura 93
Dendrohyrax 373
Pipistrellus 222
standini, Hippopotamus 381
stangeri, Protoxerus **436**
Sciurus 436
stankovici, Apodemus 573
Arvicola 506
Talpa **129**, 129
stanleyana, Puma 297
stanleyanus, Tragulus 383
stannarius, Aethomys **569**
stansburyi, Thomomys 474
starcki, Lepus **820**
staufferi, Myotis 210
stavropolicus, Cricetus 538
stearnsii, Grampus 353
Steatomys 544, **545**
angolensis 545
aquilo 546
athi 546
bensoni 545
bocagei 546
bradleyi 545
caurinus **545**
chiversi 545
cuppedius **545**, 545
edulis 546
gazellae 546
jacksoni **545**, 545
kalaharicus 546
kasaicus 546
krebsii **545**
leucorhynchus 545
loveridgei 546
mariae 545
maunensis 546
minutus 546
natalensis 546
nyasae 546
opimus 546
orangiae 545
parvus 545, **546**
pentonyx 545
pratensis 545, **546**, 546
roseveari 545
swalius 546
thomasi 546
tongensis 546
transvaalensis 545
umbratus 546
steedmanni, Cynictis 302
steerei, Proechimys **798**
steerii, Cervus 386
Sundasciurus 451, **452**, 452
stegmanni, Hystrix 773
Lepus 821
Mustela 324
Nyctereutes 283
Stegoloxodon 367
Stegomarmosa 18
steinbachi, Ctenomys **787**
Leopardus 291
Oncifelis 294
Sciurus **442**
steindachneri, Pan 277
steinhardti, Oreotragus 399
Otocyon 284
Raphicerus 400
Sylvicapra 412
steini, Melomys 619
Rattus 649, **659**, 660
Talpa 128
stejnegeri, Lutra 312
Mesoplodon 362, **363**
Phoca 332
Spermophilus 448
stella, Hylomyscus 598, **599**, 599
stellae, Tatera 561
Zygodontomys 752
stellaris, Callosciurus 423
Petaurista 463
stellarum, Dendrolagus 50
stellatus, Cryptomys 772
stelleri, Enhydra 310
Hydrodamalis 365
stellerii, Eumetopias 328
Stellerus 365
Stenella **356**, 356
albirostratus 356
alope 356
asthenops 356
attenuata 355, **356**
brevimanus 356
capensis 356
centroamericana 356, 357
clymene **356**
coeruleoalba **356**
consimilis 356
crotaphiscus 356
doris 356
dubius 356
euphrosyne 356
froenatus 356
frontalis **356**
graffmani 356
longirostris **356**, 356, 357
malayanus 356
metis 356
microps 356
normalis 356
orientalis 356, 357
pernettensis 356
pernettyi 356
plagiodon 356
pseudodelphis 356
punctata 356
roseiventris 356
styx 356
tethyos 356
velox 356
Stenidae 351
Steno 355, **357**
attenuatus 356
bredanensis **357**
compressus 357
frontatus 357
lentiginosus 355
perspicillatus 357
rostratus 357
Stenocapromys 800, 801
stenocephala, Crocidura **95**
Stenocephalemys 630, 631, **664**
albocaudata 606, **664**, 664, 686
griseicauda **664**
stenocephalus, Rhinoceros 372
stenocoronalis, Hyperplagiodontia 804
Plagiodontia 804
Stenocranius 517, 518, 521, 525
Stenodelphinidae 360
Stenodelphis 360
Stenoderma 187, 190, **192**, 192
anthonyi 192
darioi 192
microdon 191
nichollsi 187
rufa 192
rufum **192**
undatus 192
Stenodermatinae **187**
Stenodontomys, darti 683
Stenomys 650, 655, 656, **665**, 665
arfakiensis 665
arrogans 665
ceramicus **665**
clarae 665
haymani 665
klossi 665
mollis 665
niobe **665**, 665
omichlodes 665
pococki 665
richardsoni **665**
rufulus 665
stevensi 665
tomba 665
unicolor 665
vandeuseni **665**, 665
verecundus **665**, 665
Stenonycteris 152
Stenopontistes 355
zambezicus 355, 357
Stenops 248
stenops, Phylloderma **180**
Phyllostoma 180
Phyllotis 737
stenopterus, Pipistrellus **223**
stenopygius, Scotinomys 746
Stenorhinchus 330
stenorostris, Ursus 339
stenorrhina, Aotus 255
stenotis, Hipposideros **175**
stenotus, Pipistrellus 223
Stentor 254
stentor, Leopoldamys 603
stepensis, Vulpes 287
stephani, Peromyscus **736**, 736
stephanicus, Xerus 458
stephensi, Chaetodipus 484
Dipodomys **480**
Glaucomys 460
Microtus 519
Myotis 209
Neotoma **712**
Peromyscus 729
Spermophilus 447
stevensi, Atherurus 773
Callosciurus 423
Melomys 617
Ochotona 807
Stenomys 665
stevensoni, Dicrostonyx 511
Oreotragus 399
Thallomys 669

stheno, Rhinolophus **169**
Sthenurinae 53
sticte, Canis 281
sticticeps, Cercopithecus 264
Stictomys 783
Stictophonus 31
stierlingi, Lycaon 283
Sigmoceros 395
stigmatica, Thylogale **57**
stigmaticus, Arctogalidia 343
stigmatus, Raphicerus 400
stigmonyx, Gerbillus **555**
stigmosa, Ratufa 437
stikinensis, Clethrionomys 508
stimmingi, Microtus 527
stimulax, Mesomys 794, **795**
stirlingi, Marmota 431
stirtoni, Peromyscus **736**, 736
Pseudomys 645
Rheomys 743
stizodon, Sorex **120**
Stochomys 590, **665**, 665, 666
hypoleucus 666
ituricus 666
longicaudatus 665, **666**, 666
sebastianus 666
stockleyi, Petaurista 463
stoicus, Rattus 650, **660**, 660
stoliczkana, Mustela 323
stoliczkanus, Alticola 503, **504**, 504
Arvicola 502
Aselliscus **170**
Lepus 821
Suncus **103**
stolida, Cavia 779
stollei, Marmosops 20
stolzmanni, Ichthyomys **705**, 705
Mustela 321
Oligoryzomys 717
stonei, Microtus 527
Ovis 409
Rangifer 392
Spermophilus 448
Synaptomys 534
storcki, Ovis 409
storeyi, Tachyoryctes 686
stornii, Australophocaena 358
stracheyi, Alticola **504**, 504
stragulum, Rattus 652
straminea, Alouatta 255
Tamandua 68
stramineus, Eidolon 141
Natalus 194, **195**
Sciurus **443**
Spermophilus 450
strandi, Allactaga 488
Sicista 496, **497**, 497
strandzensis, Microtus 521
strauchi, Microtus 523
strauchii, Crocidura 85
streatori, Dipodomys 478
Mustela 321
Neotoma 711
Peromyscus 732
Sorex 113
Tamiasciurus 457
streckeri, Geomys 470
streeteri, Cryptomys 772
Graphiurus 764
Saccostomus 541
streeti, Talpa 128, **129**
streetorum, Crocidura 98
Talpa 129
strelzowi, Alticola **504**
strepitans, Epomops 142
Leopoldamys 603
strepitus, Papio 269
Strepsicerastes 403
Strepsicerella 403
Strepsiceros 403
strepsiceros, Antilope 395
Tragelaphus **404**
stresemanni, Callosciurus 423
Emballonura 158
Neomys 110
Rousettus 152
stressmani, Melomys 619
striata, Galidictis 300
Hyaena 308
Mustela 300, 324
Parahyaena 309
Spilogale 317
striatocornis, Tetracerus 403
striatus, Bradypus 318
Felis 291
Ictonyx **319**
Lagostrophus 53
Lemniscomys 601, **602**, 602
Mus 626
Panthera 298
Sciurus 453
Tamias 453, 455, **456**
Thylacinus 29
strictis, Paradoxurus 344
stridens, Leopoldamys 603
stridulus, Leopoldamys 603
strigidorsa, Mustela 321, 323, **324**, 324
Strigocuscus **47**, 47, 48
celebensis 47, **48**
feileri 48
gymnotis 47, **48**
leucippus 48
mimicus 46
sangirensis 48
strobilurus, Hyomys 600
stroganovi, Lynx 293
Sorex 117
Stroganovia 111, 114
stroggylonurus, Trichechus 366
Strongyloceros 385
strophiatus, Mus 623
struix, Neofiber 532
strumiciensis, Nannospalax 754
struthopus, Lepus 814
stuartii, Antechinus **29**, **30**, 30
studeri, Canis 280
stuhlmanni, Cercopithecus **264**
Chrysochloris **75**, 75
Dendrohyrax 373
Epomophorus 141
Genetta 346
Rhynchocyon 830
sturdeei, Pipistrellus **224**
sturgisi, Thomomys **474**
Sturnira **192**, 192, 193
albescens 192
angeli 192
aratathomasi **192**
bidens **192**
bogotensis **192**, 192
erythromos **192**
excisum 192
fumarium 192
hondurensis 192
lilium **192**
luciae 192
ludovici **192**, 192
luisi **192**
magna **193**
mordax **193**
nana **193**
occidentalis 192
oporaphilum 192
parvidens 192
paulsoni 192
spectrum 192
thomasi **193**
tildae **193**
zygomaticus 192
Sturnirops 192, 193
sturti, Melomys 617, 618
Notomys 636
styani, Ailurus 337
Callosciurus 421
Chimarrogale **107**, 107
Panthera 298
stygius, Heterocephalus 772
styloceros, Muntiacus 389
Stylocerus 388
Styloctenium **153**
wallacei **153**
Stylodipus **494**
amankaragai 495
andrewsi **494**, 494, 495
birulae 495
falzfeini 495
halticus 495
karelini 495
nastjukovi 495
proximus 495
sungorus **494**, 494, 495
telum **494**, 494, 495
turovi 495
styx, Stenella 356
Trachypithecus 274
styxi, Lycaon 283
suacuapara, Odocoileus 391
suahelae, Crocidura 90, 97, 98
suahelica, Genetta 346
Mus 629
suahelicus, Panthera 297, 298
Paraxerus 435
suapurensis, Pteronotus 176
suara, Aepyceros 393
suaveolens, Crocidura 81, 86, 94, **95**, 95, 96
Suncus 102
subaffinis, Heteromys 481
subalbidus, Protoxerus 436
subalpina, Redunca 414
subalpinus, Myoxus 770
Sciurus **443**
subarcticus, Peromyscus 732
subarticus, Lemmus 517
subarvalis, Microtus 529
subater, Baiomys 695
subaurantiaca, Melogale 314
subauratus, Castor 467
Sciurus 442
subbadius, Rhinolophus **169**
subcaeruleus, Mus 626
Rattus 658
subcallotis, Oryx 413
subcanus, Pipistrellus 220
subcinctus, Sylvilagus 826
Varecia 245
subcornutus, Muntiacus 389
subcristata, Hystrix 773
subcristatus, Cebus 259
subditivus, Rattus 656
subflaviventris, Dremomys 425
subflavus, Oryzomys 721, **725**, 725
Pipistrellus **224**
subfossilis, Hylobates 276
subfulva, Suncus 103
subfuscus, Akodon 689, **693**
Myomys 631
subgrisescens, Saguinus 253
subgriseus, Peromyscus 735
subgutturosa, Gazella **397**
subhemachalanus, Mustela 324
subhemachalus, Bos 401
subiens, Phascolarctos 45
subitus, Maxomys 614
subjuruanus, Bradypus 63
sublaevis, Agouti 783
sublimis, Apodemus 574
Auliscomys **695**
sublimitus, Macaca 266
sublineatus, Delomys **700**, 700
Funambulus **427**
sublucidus, Chaetodipus 483
subluteus, Alticola 503
Callosciurus 423
Lepus 821

Oecomys 717
submicrus, Nesophontes **70**
submordax, Macaca 266
submurinus, Myotis 213
subniger, Agouti 783
Pteropus **151**
subnubilis, Scotinomys 746
subnubilus, Pappogeomys 472
subnubis, Rhipidomys 745
subobscurus, Micromys 620
suboles, Thomomys 474
subpallida, Felis 289
subpalmata, Mustela 323
subpentadactylus, Ateles 257
subquadricornis, Tetracerus 403
subrosea, Lagidium 778
subrufa, Carollia **186**, 186
subrufus, Callicebus 258
Lepus 820
Pseudomys 646
Rattus 658
Rhinolophus **169**
Tragulus 382
subsignanus, Sundasciurus 451
subsimilis, Thomomys 474
subsimus, Microtus 524
Pappogeomys 472
subsolana, Neotoma 710
subsolanus, Gerbillus 549
Lynx 293
subspinosus, Acomys **567**, 567
Chaetomys **790**
Hystrix 789
substriatus, Funisciurus **428**
Mungotictis 300
subtenuis, Dipodomys 479
subterraneus, Akodon 692
Microtus 519, 520, 524, 526, **530**, 530
Mus 626
Pitymys 520, 529
subtilis, Dendromus 543
Mus 496
Sicista **497**, 497
subtilissima, Planigale 34
subtilus, Pipistrellus 221
subtropicalis, Nycticeius 217
subulata, Raphicerus 400
subulatus, Myotis 212
subulidens, Pipistrellus 224
Subulo 390
subviridescens, Paraxerus 435
suckleyi, Tamiasciurus 456
sucuacuara, Puma 297
sudanensis, Gerbillus 548, 551
Graphiurus 764
Praomys 642
sudani, Taphozous 161
sudanicus, Ictonyx 319
suecicus, Clethrionomys 509
sueirensis, Rattus 658
suffectus, Rattus 652
suffocans, Conepatus 316
suffusa, Akodon 691
suffusus, Callosciurus 423
Suidae **377**
suilla, Hyaena 308
Mandrillus 268
Murina **230**
Nesokia 632
Suillotaxus 315
marchei 315
suillus, Bathyergus **771**, 771
Erinaceus 77
Hylomys 79, **80**
Mus 771
Tapirus 371
Vespertilio 228
Suinae **377**
sukleyi, Sorex 122
sulcata, Emballonura 158
Taxidea 325
sulcatus, Aeretes 459
sulcidens, Carterodon **793**
Echimys 793
sulfureus, Balaenoptera 350
sullivanus, Callosciurus 421
sultan, Petrodromus 830
sultana, Hyaena 308
suluensis, Macaca 266
sumaco, Sciurus 441
sumatrae, Hylopetes 461
Hystrix **774**
Niviventer 633
Panthera 298
Rhinopoma 155
sumatraensis, Capricornis 407
Naemorhedus **407**, 407
sumatrana, Arctogalidia 343
Chimarrogale **107**, 107
Lutra **312**, 312
Lutrogale 313
Panthera 299
Petaurista 462
Presbytis 271
Prionailurus 295
sumatranus, Callosciurus 423
Elephas 367
Nannosciurus 434
Pongo 277
Rhinolophus 163
Sus 378
Tapirus 371
sumatrensis, Cuon 282
Dicerorhinus **372**
Nycticebus 248
Paradoxurus 344
Pteropus 151
Rhinoceros 371
Rhizomys **685**
sumatrius, Herpestes 304
sumbae, Hipposideros 173
Rattus 660
sumbanus, Dobsonia 140
Paradoxurus 344
sumbavanus, Cervus 387
sumbawae, Hystrix 774
Sumeriomys 517, 521, 522
sumichrasti, Bassariscus **334**
Hesperomys 715
Nyctomys **715**
Reithrodontomys **742**
sumneri, Peromyscus 735
sunameri, Neophocaena 358
suncoides, Ruwenzorisorex **101**
Sylvisorex 101
Suncus 81, 91, **101**, 104
aequatoria 102
albicauda 103
albinus 103
andersoni 103
assamensis 102
ater **101**
atratus 102
auriculata 103
bactrianus 102
beddomei 103
bidiana 103
blanfordii 103
blythii 103
caerulaeus 103
caerulescens 103
caeruleus 103
celebensis 103
ceylanica 103
chriseos 102
coquerelii 102
crassicaudus 103
dayi **102**
duvernoyi 103
edwardsiana 103
etruscus **102**, 102, 103
fellowesgordoni **102**, 102
ferrugineus 103
fulvocinerea 103
fuscipes 103
geoffroyi 103
giganteus 103
gratula 102
gratulus 102
griffithii 103
heterodon 103
hodgsoni 102
hosei **102**, 102
indicus 103
infinitesimus **102**
kandianus 103
kelaarti 103
kroonii 103
kuekenthali 103
kura 102
leucogenys 103
leucura 103
lixus **102**
luzoniensis 103
macrotis 102
madagascariensis **102**, 102
malabaricus 103
malayanus **102**, 102, 103
mauritiana 103
media 103
meesteri 103
melanodon 102, 103
mertensi **103**
micronyx 102
microtis 103
minor 103
montanus **103**, 104
mulleri 103
murinus **103**, 103, 104
muschata 103
myosurus 103
nanula 102
natalensis 103
nemorivagus 103
niger 103
nilgirica 102
nitidofulva 102, 103
nudipes 102
occultidens 103
orangiae 103, 104
pachyurus 102
palawanensis 103
pealana 103
perrotteti 102
pilorides 103
pygmaeoides 102
pygmaeus 102
remyi **103**
riukiuana 103
rubicunda 103
sacer 101, 103
saturatior 103
semmeliki 103
semmelincki 103
serpentarius 103
sindensis 103
soccatus 103
sonneratii 103
stoliczkanus **103**
suaveolens 102
subfulva 103
swinhoei 103
temminckii 103
travancorensis 102
tulbaghensis 103
tytleri 103
ubanguiensis 102
unicolor 103
varilla **103**
viridescens 103
waldemarii 103
warreni 103, 104
zeylanicus **104**, 104
Sundamys 580, 603, 650, **666**, 666
atchinus 666
balabagensis 666
balmasus 666
borneanus 666
campus 666
chombolis 666
crassus 666
credulus 666
culionensis 666

domitor 666
firmus 666
foederis 666
infraluteus **666**, 666
integer 666
maxi **666**
muelleri 595, 650, 655, 660, **666**, 666
otiosus 666
pinatus 666
pollens 666
potens 666
sebucus 666
terempa 666
valens 666
validus 666
victor 666
virtus 666
waringensis 666
Sundasciurus 420, 435, **450**, 450, 451, 452
alacris 451
albicauda 452
altitudinus 452
balae 451
bancarus 452
bangueyae 451
batus 452
borneensis 451
brookei **451**
cagsi 451
davensis **451**, 451, 452
fraterculus **451**, 451, 452, 604
grayi 451
gunong 452
hippurellus 451
hippurosus 451
hippurus **451**
hoogstraali **451**, 452
humilis 451
inquinatus 451
jentinki **451**
juvencus **451**
lingungensis 451
lowii **451**, 451
mansalaris 452
mindanensis **451**, 451, 452
modestus 452
moellendorffi 451, **452**, 452
natunensis 451
ornatus 451
parvus 452
philippinensis 451, **452**, 452
piniensis 451
procerus 452
pryeri 451
pumilus 451, 452
rabori **452**
robinsoni 451
rufogaster 451
samarensis 451, **452**, 452
seimundi 451
siantanicus 452
siberu 451
sordidus 452
steerii 451, **452**, 452
subsignanus 451
surdus 452
tahan 452
tenuis **452**, 452
tiomanicus 452
vanakeni 451
sundavensis, Bandicota 579
sundevalli, Cephalophus 411
Gazella 396
sundevallii, Phacochoerus 377
sungae, Otomys 680
sungarae, Mus 625
sungorus, Phodopus **539**, 539
Stylodipus **494**, 494, 495
sunidica, Ochotona 811
Sunkus 101
suntaricus, Microtus 527
superans, Eliomys 767
Oecomys **716**
Ratufa 437
Rhinolophus 163
Vespertilio **228**
superbus, Chalinolobus **200**
superciliaris, Mazama 391
Philander 22
Sylvilagus 826
superciliosum, Artibeus 189
supercilliosus, Lagenorhynchus 354
superiorensis, Lynx 294
supernus, Thomomys 476
surakatta, Suricata 308
suramensis, Microtus 520
Suranomys 507, 517
surberi, Neotoma 712
Surcatinae 300
surculus, Pappogeomys 472
surda, Tupaia 133
surdaster, Grammomys 593
Pedetes 759
Sylvilagus 825
Viverra 348
surdescens, Cervus 386
Surdisorex 81, 99, **104**
norae **104**, 104
polulus **104**
surdus, Akodon **693**
Graphiurus **765**, 765
Rattus 652
Sundasciurus 452
Suricata 300, **307**
capensis 307, 308
hahni 308
hamiltoni 308
lophurus 308
marjoriae 308
namaquensis 308
surakatta 308
suricatta **308**
tetradactyla 308
typicus 308
viverrina 308
zenik 308
suricatta, Suricata **308**
Viverra 307
surifer, Maxomys **613**, 613, 634
surinamensis, Ateles 257
Cryptotis 108, 112
Monodelphis 21
Myotis 210
Sorex 112
surkha, Mus 627
surmulottus, Rattus 657
surrufus, Onychomys 719
surrutilus, Exilisciurus 426
sururae, Crocidura 92
Sus **378**
acrocranius 379
affinis 379
ahoenobarbus 378
aipomus 379
algira 379
amboinensis 378
andamanensis 379
andersoni 379
aper 379
arietinus 379
aruensis 379
attila 379
babi 379
babyrussa 377
baeticus 379
balabacensis 378
barbarus 379
barbatus **378**, 378, 379
bengalensis 379
blouchi 379
borneensis 379
branti 378
bucculentus **378**
calamianensis 378
canescens 379
castilianus 379
cebifrons **378**
celebensis **378**, 378, 379
celtica 379
ceramensis 379
ceramica 379
chirodontus 379
collinus 379
continentalis 379
coreanus 379
crassidens 379
cristatus 379
curtidens 379
davidi 379
dicrurus 379
edmondi 378
effrenus 379
enganus 379
europaeus 379
falzfeini 379
ferus 379
flavescens 379
floresianus 379
frenatus 379
frontosus 379
gargantua 378
gigas 379
goramensis 379
heureni **378**
hydrochaeris 781
inconstans 379
indicus 379
isonotus 379
japonica 379
joloensis 379
jubatulus 379
jubatus 379
laticeps 379
leucomystax 379
leucorhinus 379
libycus 379
longirostris 378
macassaricus 378
mainitensis 379
majori 379
mandchuricus 379
marchei 379
maritimus 378
mediterraneus 379
megalodontus 379
melas 379
meridionalis 379
microdontus 379
microtis 379
milleri 379
mimus 378
mindanensis 379
minutus 379
moupinensis 379
mystaceus 379
natunensis 379
negrinus 378
nehringii 378
niadensis 378
nicobaricus 379
niger 379
nigripes 379
nipponicus 379
oi 378
olivieri 379
oxyodontus 379
palavensis 378
paludosus 379
palustris 379
papuensis 379
pecari 380
peninsularis 379
philippensis **378**, 378
planiceps 379
porcus 377
raddeanus 379
reiseri 379
rhionis 379
riukiuanus 379
sahariensis 379
salvanius **379**
sardous 379
scrofa **378**, **379**, 379
scrofoides 379
sennaarensis 379
setosus 379
sibiricus 379

songaricus 379
spatharius 379
sumatranus 378
taininensis 379
taivanus 379
tajacu 380
ternatensis 379
timoriensis **379**
tuancus 379
ussuricus 379
verrucosus 378, **379**
vittatus 379
weberi 378
zeylonensis 379
suschkini, Allactaga 489
Meriones 557
sushkini, Ochotona 807
susiana, Crocidura **96**
suslicus, Spermophilus 448, **449**
Susu 360
Susuidae 360
Sutra 310
suturosa, Addax 412
svatoshi, Ochotona 809
svineval, Globicephala 352
sviridenkoi, Cricetulus 538
swainsonii, Antechinus **30**
swakopensis, Gerbillurus 547
swalius, Galerella **303**, 303
Gerbillurus 547
Steatomys 546
swarthi, Dipodomys 478
Nesoryzomys **714**, 714
swaynei, Alcelaphus 393
Madoqua 398
swaythlingi, Tatera 562
swerevi, Microtus 522, 524
swettenhami, Naemorhedus 407
swinderianus, Aulacodus 775
Thryonomys **775**
swinhoei, Amblonyx 310
Cervus 386
Hipposideros 171
Lepus 821
Meriones 556
Naemorhedus **407**
Scotophilus 227
Suncus 103
Tamiops **457**, 457
swinhoii, Cervus 387
swinnyi, Galerella 303
Myosorex 99
Rhinolophus **169**
swirae, Taphozous 161
swynnertoni, Paraxerus 435
sybilla, Mus 624, 625
Petaurista 462
Syconycteris **155**
australis **155**, 155
carolinae **155**
crassa 155
finschi 155
hobbit **155**
keyensis 155
major 155
naias 155
papuana 155
sydneiensis, Tachyglossus 13
sylhetanus, Bos 401
Sylvaemus 569, 570, 571, 572, 573, 574
alpicola 570
sylvanus, Akodon **693**, 693
Bos 401
Cervus 386
Ctenomys 784
Macaca **268**
Simia 266
sylvatica, Antilope 403
Martes 320
Rupicapra 410
sylvaticus, Apodemus 570, 571, 572, **573**, 573, 574
Lepus 821, 824
Mesechinus 79
Odocoileus 391
Oryzomys 725
Sylvilagus 826
Tragelaphus 404
sylvester, Callosciurus 424
sylvestris, Dendrohyrax 373
Felis 289
Martes 320
Micronycteris **179**
Ozotoceros 392
Pogonomys 641, **642**, 642
Rangifer 392
Rattus 658
Tragelaphus 404
sylvia, Crocidura 89
Molossops 235
Sylvicapra 410, **412**
abyssinica 412
altifrons 412
altivallis 412
bradfieldi 412
burchellii 412
caffra 412
campbelliae 412
cana 412
coronata 412
cunenensis 412
deserti 412
flavescens 412
grimmia **412**
hindei 412
irrorata 412
leucoprosopus 412
lobeliarum 412
lutea 412
madoqua 412
mergens 412
nictitans 412
noomei 412
nyansae 412
ocularis 412
omurambae 412
pallidior 412
platous 412
ptoox 412
roosevelti 412
shirensis 412
splendidula 412
steinhardti 412
transvaalensis 412
ugabensis 412
uvirensis 412
vernayi 412
walkeri 412
Sylvicola 517
sylvicola, Mandrillus 268
sylvicultor, Cephalophus 411
Sylvilagus 813, **824**, 824, 825, 826, 827
alacer 825
ammophilus 825
andinus 825
apollinaris 825
aquaticus **824**, 827
arizonae 824
artemesia 826
attwateri 824
audubonii 813, **824**
avius 825
aztecus 825
bachmani **824**, 826
badistes 826
baileyi 824
boylei 825
brasiliensis **825**, 825
braziliensis 825
canarius 825
caniclunis 825
capsalis 825
caracasensis 825
carchensis 825
cedrophilus 824
cerrosensis 825
chapadae 825
chapadensis 825
chapmani 825
chiapensis 825
chillae 825
chimbanus 825
chotanus 825
cinerascens 825
cognatus 825
confinis 824
connectens 825
consobrinus 825
continentis 825
costaricensis 825
cumanicus 825
cunicularius **825**, 826
daulensis 825
defilippi 825
defilippii 825
dephilippii 825
dicei **825**, 825
douglasii 824, 826
durangae 825
ecaudatus 825
exiguus 825
floridanus **825**, 826, 827
fulvescens 825
fuscescens 825
gabbi 825
gibsoni 825
goldmani 824
grangeri 826
graysoni **826**
hefneri 826
hesperius 825
hitchensi 825
holzneri 825
hondurensis 826
howelli 825
idahoensis 825
inca 825
incitatus 825
insolitus 825
insonus **826**
kelloggi 825
laticinctus 824
littoralis 824
llanensis 826
macrocorpus 826
macrorhinus 825
major 824
mallurus 826
mansuetus **826**
margaritae 826
mariposae 825
mearnsi 826
meridensis 825
messorius 825
minensis 825
minor 824
nelsoni 826
neomexicanus 824
nicefori 825
nigricaudatus 825
nigronuchalis 826
nivicola 825
nuttallii 813, **826**, 826
orinoci 826
orizabae 826
pacificus 825
paludicola 826
palustris **826**, 827
paraguensis 825
parvulus 824
paulsoni 826
peninsularis 825
perplicatus 826
persultator 826
peruanus 825
pinetis 826
purgatus 826
restrictus 826
rigidus 826
riparius 825
robustus 826
rosaphagus 825
rufipes 824
russatus 826
salentus 825
sanctaemartae 825
sanctidiegi 824
similis 826
simplicicanus 826
subcinctus 826

superciliaris 826
surdaster 825
sylvaticus 826
tapeti 825
tapetillus 825
tehamae 825
telmalemonus 824
transitionalis **827**
trowbridgii 825
truei 825
tumacus 825
ubericolor 825
valenciae 826
vallicola 824
veraecrucis 825
virqulti 825
warreni 824
yucatanicus 826
Sylvisorex 99, **104**
angolensis 105
camerunensis 104
dieterleni 105
gemmeus 105
granti **104**, 105
howelli **104**
infuscus 105
irene 105
isabellae **104**, 105
johnstoni **104**
lunaris **105**
megalua 105
megalura **105**
morio 102, 104, **105**
mundus 104
ollula **105**, 105
oriundus **105**, 105
phaeopus 105
preussi 112
ruandae 105
sheppardi 105
somereni 101
sorella 105
sorelloides 105
suncoides 101
usambarensis 104
vulcanorum **105**, 105
Symphalangus 274, 276
Synaptomys 516, **534**, 534
andersoni 534
artemisiae 534
borealis **534**
bullatus 534
chapmani 534
cooperi **534**, 534
dalli 534
fatuus 534
gossii 534
helaletes 534
innuitus 534
jesseni 534
kentucki 534
medioximus 534
paludis 534
relictus 534
saturatus 534
smithi 534
sphagnicola 534
stonei 534
truei 534
wrangeli 534
Syncerus **402**
adamauae 402
adametzi 402
adolfifriederici 402
aequinoctialis 402
athiensis 402
azrakensis 402
beddingtoni 402
bornouensis 402
brachyceros 402
bubuensis 402
caffer **402**
centralis 402
corniculatus 402
cottoni 402
cubangensis 402
cunenensis 402
diehli 402
gariepensis 402
gazae 402
geoffroyi 402
houyi 402
hunti 402
hylaeus 402
limpopoensis 402
lomamiensis 402
massaicus 402
matthewsi 402
mayi 402
nanus 402
neumanni 402
niediecki 402
nuni 402
Syndactyliformes **43**
syndactylus, Hylobates **276**
Syntheosciurus **452**, 452
brochus **452**, 452
granatensis 452
poasensis 452
pyrrhinus 452
syops, Erophylla 182
Syphonia 13
syriaca, Felis 291
Hyaena 308
Procavia 374
Vormela 325
syriacus, Canis 280
Equus 370
Hemiechinus 78
Heterohyrax 373
Hyrax 373
Lepus 817
Microtus 530
Sciurus 440
Ursus 339
syrichta, Simia 250
Tarsius **251**
syrinx, Ochotona 812
syrius, Jaculus 493
Meriones 556
syrmiensis, Nannospalax 754
syrticus, Gerbillus **555**
Syspotamus 370
szechuanensis, Aeretes 459
Pseudois 409
szechwanus, Rhinolophus 168

tabanus, Spalacopus 789
tabascensis, Didelphis 17
tabaudius, Callosciurus 421
tabernaculi, Dugong 365
tablasi, Pteropus 150
Rattus 660
taborae, Lycaon 283
Pedetes 759
Tatera 562
tachardi, Callosciurus 422
tachin, Callosciurus **421**
Tachoryctinae 677
Tachyglossidae **13**
Tachyglossus **13**
acanthion 13
acanthrous 13
aculeatus **13**
australiensis 13
breviaculeata 13
bruijni 13
corealis 13
hobartensis 13
hystrix 13
ineptus 13
lawesi 13
longiaculeata 13
longirostris 13
multiaculeata 13
myrmecophragris 13
novaehollandiae 13
setosa 13
setosus 13
sydneiensis 13
typicus 13
Tachynices 357
Tachyoryctes **685**, 685, 686
ankoliae **686**, 686
annectens **686**, 686
audax **686**
badius 687
canicaudus 687
cheesmani 687
daemon **686**, 686
gallarum 687
hecki 686
ibeanus 687
macrocephalus 606, 664, 685, **686**, 686
naivashae **686**
omensis 687
pontifex 687
rex **686**
ruandae **686**
ruddi 686, **687**
somalicus 687
spalacinus **687**
splendens 685, 686, **687**, 687
storeyi 686
Tachyoryctinae 685
taciturnus, Rattus 660
tacomensis, Thomomys 475
tacopius, Callosciurus 421
taczanowskii, Agouti **783**
Thomasomys **750**
tadae, Crocidura 87
Tadarida 232, 236, 237, 238, **240**
aegyptiaca **240**
africana 241
albidus 240
anchietae 240
antillularum 240
atratus 240
australis **240**
bahamensis 240
bocagei 240
brasiliensis **240**
brunneus 240
californicus 240
cestoni 240
cinerea 240
congica 236
coecata 240
constanzae 240
cynocephala 240
espiritosantensis **240**
fuliginosus 240
fulminans **240**
geoffroyi 240
gossei 240
insignis 240
intermedia 240
kuboriensis 240
latouchei 240
lobata **240**
mastersoni 240
mexicana 240
mohavensis 240
molossa 239
multispinosus 240
murina 240
muscula 240
naso 240
nasutus 240
nigrogriseus 240
peruanus 240
rueppellii 240
rugosus 240
savii 240
sindica 240
talpinus 240
teniotis **240**
texana 240
thomasi 240
tongaensis 240
tragata 240
ventralis **241**
taedifer, Sciurus 441
taenianotus, Mungos 307
taeniopus, Equus 369
taeniurus, Sciurus 440
Tatera 561
taerae, Taeromys **667**, 667
Taeromys 638, 651, 655, 662, **666**

arcuatus **667**, 667
callitrichus **667**
celebensis 666, **667**
dominator 639
hamatus **667**
jentinki 667
maculipilis 667
microbullatus 667
punicans **667**
simpsoni 667
taerae **667**, 667
tatei 667
tafa, Antechinus 30
Melomys 619
Petaurus 61
Phascogale 34
tafeli, Equus 370
Mustela 324
taftanimontis, Eptesicus 200
taguanoides, Petauroides 59
tagulayensis, Rattus 651
tahan, Sundasciurus 452
tahltanicus, Ursus 339
tahoensis, Lepus 814
tai-oranus, Cervus 386
taigica, Sicista 496
taimyrensis, Rangifer 392
taininensis, Sus 379
taiouanus, Cervus 386
taitae, Felis 291
Praomys 642
taitensis, Mus 626
taiva, Microgale 71
taivana, Mustela 323, 324
Paguma 343
Viverricula 348
taivanus, Cervus 386
Plecotus **225**
Sus 379
taiwanensis, Myotis 207
taiwanus, Bandicota 579
Mus 626
tajacu, Pecari **380**
Sus 380
tajassu, Pecari 380
takagii, Mus 626
takanus, Lycaon 283
takasagoensis, Micromys 620
takayamai, Mus 626
talahutky, Sciurus 443
talamancae, Oryzomys 721, **725**
talangae, Oenomys 637
talapoin, Miopithecus **269**
Simia 268
talarum, Ctenomys **787**
talassensis, Microtus 522
talassicus, Meles 314
Microtus 521
talaudensis, Rattus 660
talaudium, Melomys 617
talazaci, Microgale 71, **72**
talboti, Euoticus 249
Funisciurus 428
Galerella 303
Mungos 307
Talpa 124, 125, 126, 127, **128**, 128
adamoi 129
aenigmatica 129
altaica **128**, 128, 129
asiatica 75
augustana 128
beaucournui 128
brachycrania 129
brauneri 129
caeca **128**, 129
caucasica **128**, 128, 129
cinerea 129
dobyi 128
ehiki 129
europaea 126, 128, **129**, 129
frisius 129
hercegovinensis 128
imaizumii 126
klossi 125
kratochvili 129
leucura 127
levantis 128, **129**
minima 129
montana 129
occidentalis **129**
olympica 128
orientalis 129
pancici 129
robusta 126
romana **129**, 129
stankovici **129**, 129
steini 128
streeti 128, **129**
streetorum 129
talyschensis 129
transcaucasica 129
uralensis 129
velessiensis 129
wittei 129
wogura 125
Talpasorex 127
Talpidae **124**, 124
Talpinae **124**
Talpini 124, 125, 126, 127, 128
talpinus, Ellobius **512**, 512, 513
Juscelinomys **706**
Mus 512
Myosorex 99, 100
Myospalax 675
Tadarida 240
Talpoides 754
talpoides, Blarina 106
Cryptomys 772
Oryzorictes **72**
Sorex 106
Thomomys 473, 474, **475**, 475
Urotrichus **129**, 129
Talposorex 106
talyschensis, Talpa 129
tamae, Sciurus 441
Tamandua **68**
bivittata 68
chapadensis 68
chiriquensis 68
crispa 68
hesperia 68
instabilis 68
kriegi 68
longicaudata 68
mexicana **68**
myosura 68
nigra 68
opistholeuca 68
opisthomelas 68
punensis 68
quichua 68
sellata 68
straminea 68
tamandua 68
tambensis 68
tenuirostris 68
tetradactyla **68**, 68
tamandua, Myrmecophaga 68
Tamandua 68
tamansari, Callosciurus 423
tamarani, Canariomys **582**, 582
tamarensis, Rattus 657
tamaricinus, Meriones 558
Tamarin 253
tamarin, Saguinus 254
Tamarinus 253
tamariscinus, Meriones **558**
Mus 555
tamaulipensis, Nasua 334
Pappogeomys 472
tambelanicus, Rattus 661
tambensis, Tamandua 68
tamborum, Phyllotis 737
tamerlani, Pipistrellus 223
Tamias **453**, 453, 456
acrus 455
adsitus 456
affinis 453
albiventris 453
albogularis 455
alleni 456
alpinus **453**
altaicus 455
americanus 456
amoenus **453**
animosus 455
arizonensis 454
asiaticus 455
atristriatus 454
australis 455
barberi 455
borealis 454
bulleri **453**, 454
cacodemus 454
callipeplus 456
canescens 453
canicaudus 453
caniceps 454
canipes **453**, 454
carminis 453
caryi 454
caurinus 453
celeris 453
cinereicollis **453**, 453
cinereus 453
clarus 454
confinis 454
consobrinus 454
cooperi 456
cratericus 453
davisi 454
doorsiensis 456
dorsalis **453**
durangae 453, **454**, 454
felix 453
fisheri 456
frater 456
fremonti 456
gracilis 455
grinnelli 453
grisescens 454
griseus 456
hindei 456
hopiensis 455
hudsonius 454
intercessor 455
inyoensis 456
jacksoni 454
jacutensis 455
juniperus 455
kernensis 454
lectus 454
leucurus 419
lineatus 455
littoralis 456
ludibundus 453
luteiventris 453
lysteri 456
macrorhabdotes 455
mariposae 454
melanurus 454
meridionalis 454
merriami **454**, 454
minimus **454**
monoensis 453
montanus 456
neglectus 454
nevadensis 456
nexus 454
nidoensis 453
obscurus **454**, 454
ochraceus 453
ochrogenys **454**, 456
ohioensis 456
okadae 455
operarius 454
ordinalis 455
oreocetes 454
orientalis 455
pallasi 455
pallidus 454
palmeri **454**
panamintinus **454**
peninsulae 456
pictus 454
pipilans 456
pricei 454

propinquus 453
quadrimaculatus **455**
quadrivittatus **455**, 455, 456
quebecensis 456
rufescens 456
ruficaudus **455**, 455
rufus **455**, 455
sacramentoensis 453
scrutator 454
sedulus 456
selkirki 454
senescens 455
senex **455**, 456
septentrionalis 453
sequoiensis 456
sibiricus 453, **455**
silvaticus 454
simulans 455
siskiyou **455**, 456
solivagus 453, 454
sonomae **456**
sonoriensis 453
speciosus **456**
striatus 453, 455, **456**
townsendii 454, 455, **456**
umbrinus 455, **456**, 456
umbrosus 455
utahensis 453
uthensis 455
vallicola 453
venustus 456
Tamiasciurinae 419
Tamiasciurini 419, 438, 456
Tamiasciurus **456**
abieticola 457
baileyi 457
belcheri 456
cascadensis 456
columbianus 457
dakotensis 457
dixiensis 457
douglasii **456**, 456, 457
fremonti 457
grahamensis 457
gymnicus 457
hudsonicus 456, **457**
kenaiensis 457
lanuginosus 457
laurentianus 457
loquax 457
lychnuchus 457
mearnsi 456, **457**
minnesota 457
mogollonensis 457
mollipilosus 456
murii 457
neomexicanus 457
orarius 456
pallescens 457
petulans 457
picatus 457
preblei 457
regalis 457
richardsoni 457
rubrolineatus 457
streatori 457
suckleyi 456
ungavensis 457
vancouverensis 457
ventorum 457
wasatchensis 457
Tamiini 419, 438
Tamiodes 426
Tamiops 420, **457**, 457
barbei 457
chingpingensis 457
clarkei 457
collinus 457
dolphoides 457
elbeli 457
formosanus 457
forresti 457
hainanus 457
holti 457
inconstans 457
kongensis 457
laotum 457
leucotis 457
liantis 457
lylei 457
macclellandi **457**
manipurensis 457
maritimus **457**, 457
moi 457
monticolus 457
novemlineatus 457
olivaceus 457
pembertoni 457
riudoni 457
rodolphei **457**
russeolus 457
sauteri 457
spencei 457
swinhoei **457**, 457
vestitus 457
Tamiscus 434
tamrinensis, Crocidura 97
tamulicus, Muntiacus 389
Tana 131, 132, 133
tana, Tupaia 132, **133**
tanae, Alcelaphus 393
Leptailurus 293
Paraxerus 435
tanaiticus, Arvicola 506
Dryomys 766
Ellobius 512
Pygeretmus 490
tanakae, Crocidura 82
tanala, Eliurus **678**
tananaensis, Microtus 527
tancrei, Ellobius 512, **513**, 513
tanei, Apodemus 570
tanezumi, Rattus 649, 659, **660**, 660
tangalunga, Viverra **348**
tanganyikae, Funisciurus 427
Paraxerus 434
tanrec, Tenrec 73
tansaniana, Crocidura **96**
tantalus, Chlorocebus 265
tantillus, Mus 626
tao, Oryx 413
tapajinus, Oecomys 716
tapanulius, Callosciurus 423
Leopoldamys 603
taparius, Sciurus **442**
Tapeti 824, 825, 826, 827
tapeti, Sylvilagus 825
tapetillus, Sylvilagus 825
Taphozous 159, **160**, 160, 161
achates 161
albipinnis 160
assabensis 161
australis **160**
babylonicus 161
bicolor 161
brevicaudus 160
cantorii 160
cavaticus 161
cinerascens 160
dobsoni 160
fretensis 161
fulvidus 160
fumosus 160
georgianus **160**, 160
haedinus 161
hamiltoni **160**
hildegardeae **160**
hilli **160**
kachhensis 161
kampenii 160
kapalgensis **160**
leucopleura 160
leucopterus 160
longimanus **160**
magnus 161
maritimus 161
mauritianus **160**
melanopogon **160**, 161
nudaster 161
nudiventris **161**
perforatus 160, **161**
philippinensis **161**
rhodesiae 161
saccolaimus 159
secatus 161
senegalensis 161
serratus 161
solifer 161
sudani 161
swirae 161
theobaldi **161**
troughtoni 160
ziyidi 161
tapir, Tapirus 371
tapirapoanus, Akodon 689
Bolomys 696
Tapirella 370
Tapiridae **370**
Tapirus **370**
aenigmaticus 371
americanus 371
andicola 371
anta 371
anulipes 371
bairdii **371**
bicolor 371
brasiliensis 371
colombianus 371
dowii 371
ecuadorensis 371
guianae 371
indicus **371**
laurillardi 371
leucogenys 371
malayanus 371
maypuri 371
mexianae 371
peruvianus 371
pinchacus 371
pinchaque **371**
roulinii 371
rufus 371
sabatyra 371
spegazzinii 371
suillus 371
sumatranus 371
tapir 371
tapirus 371
terrestris **371**
villosus 371
tapirus, Tapirus 371
Tapoa 33, 48
tapoatafa, Phascogale **33**
Vivera 33
tapojozensis, Alouatta 254
tapouaru, Trichosurus 48
tarabuli, Gerbillus **555**, 555
taraiyensis, Arctonyx 313
tarandoides, Elaphurus 387
Tarandus 392
tarandus, Cervus 392
Rangifer **392**
tarapacensis, Eligmodontia 701
tarayensis, Bandicota 579
Lutrogale 313
tarbagataicus, Microtus 521
Myospalax 676
Tardigradus 248
coucang 248
tardigradus, Lemur 248
Loris **248**
Nycticebus 248
tarella, Crocidura **96**
tarfayensis, Crocidura 83, **96**
Tarimolagus 814, 822
tarlosti, Leptonycteris 185
tarquinus, Eothenomys 513
tarrae, Sciurus 441
tarrensis, Nectomys 709
tarsalis, Irenomys **705**
Mus 705
tarsier, Tarsius 251
Tarsiidae **250**
Tarsipedidae **62**
Tarsipes **62**
rostratus **62**, 62
spencerae 62
spenserae 62
Tarsius **250**

bancanus **250**
borneanus 250
buffonnii 251
carbonarius 251
daubentonii 251
dentatus 250
dianae **250**
fischerii 250
fraterculus 251
fuscomanus 250
fuscus 250
macauco 251
macrotarsos 251
minimus 251
natunensis 250
pallassii 250
pelengensis 250
philippensis 251
podje 250
pumilus **250**
saltator 250
sangirensis 250
spectrum **250**, 250
syrichta **251**
tarsier 251
tarsius 251
tarsius, Tarsius 251
Tarsomys **667**, 667
apoensis **667**, 667
echinatus **667**
tartareus, Akodon 693
tartessia, Felis 291
tarussanus, Callosciurus 423
tasensis, Microtus 524
tasicus, Sorex 113
Tasmacetus **363**
shepherdi 363, **364**
tasmanei, Thylogale 57
tasmanicus, Arctocephalus 327
tasmaniensis, Macropus 54
Pipistrellus **224**
Vombatus 46
tatar, Rhinolophus 165
tatarica, Capra 400
Saiga **400**
tataricus, Arvicola 506
Marmota 432
Mus 626
tatei, Caenolestes 25
Dactylopsila **61**
Eptesicus **203**
Nectomys 709
Pogonomelomys 641
Rattus 653
Sminthopsis 37
Taeromys 667
Thylamys 23
Tateomys 614, **667**, 667
macrocercus **668**
rhinogradoides 667, **668**
Tatera **547**, **560**, 560, 561
afra **560**, 560, 561
africanus 560
angolae 561
bailwardi 561
bayeri 562
bechuanae 561
beirae 561
beirensis 561
beniensis 562
benvenuta 562
bodessae 562
bodessana 561
boehmi **560**, 560
bohmi 560
brantsii **560**
breyeri 560
ceylonica 561
cuvieri 561
dichrura 562
draco 560
dundasi 562
dunni 561
fallax 560
fraterculus 560
gambiana 561
giffardi 561
gilli 560
griquae 560
guineae **560**
hardwickei 561
hopkinsoni 561
iconica 562
inclusa **561**
indica 560, **561**
joanae 560
kaokoensis 561
kempi **561**, 561, 562
leucogaster **561**
limpopoensis 561
liodon 562
littoralis 561
lobengulae 561
loveridgei 562
lucia 562
maccalinus 560
macropus 562
maputa 560
mashonae 561
miliaria 560
minusculus 561
mitchelli 561
mombasae 562
montanus 560
monticola 561
muansae 562
namaquensis 560
natalensis 560
ndolae 561
neavei 562
nigricauda **561**
nigrita 562
nigrotibialis 561
nyama 561
nyasae 561
otarius 561
panja 561
percivali 561
perpallida 560
persica 561
pestis 561
phillipsi **561**, 561
picta 560
pitmani 561
pothae 562
pretoriae 561
robusta 560, 561, **562**
ruddi 560
ruwenzorii 562
salsa 561
scansa 561
schinzi 561
schlegelii 560
sherrini 561
shirensis 561
shoana 562
smithi 562
soror 562
stellae 561
swaythlingi 562
taborae 562
taeniurus 561
taylori 562
tenuis 561
tongensis 560
tzaneenensis 561
umbrosa 561
valida 561, **562**
varia 560
vicina 562
waterbergensis 561
welmanni 561
zuluensis 561
Taterillus 560, **562**, 562, 563
angelus 562, 563
anthonyi 562
arenarius **562**, 562
butleri 562
clivosus 562
congicus **562**
emini **562**
gracilis **562**, 562, 563
gyas 562
harringtoni **563**
illustris 563
kadugliensis 563
lacustris **563**, 563
lorenzi 563
lowei 563
melanops 563
meneghetti 563
nigeriae 562
nubilus 563
osgoodi 563
perluteus 563
petteri 562, **563**, 563
pygargus **563**, 563
pygarus 562
rufus 563
tenebricus 563
zammarani 563
Taterina 562
Taterona 560, 561, 562
taterona, Graomys 703
tatiana, Crocidura 92, 98
tatianae, Crocidura 98
tatkonensis, Rattus 660
Tatoua 64
tatouay, Cabassous **65**
tatrica, Rupicapra 410
tatricus, Microtus 526, **530**
Sicista 496
Sorex 111
tattersalli, Propithecus **247**
Tatu 65
Tatus 65, 66
Tatusia 64, 65, 67
tauricus, Apodemus 573
Arvicola 506
Cervus 386
Cricetulus 538
Cricetus 538
Meles 314
Pipistrellus 223
taurinsulae, Odocoileus 391
taurinus, Connochaetes **394**
Taurotragus **403**
alces 403
barbatus 403
billingae 403
cameroonensis 403
canna 403
colini 403
congolanus 403
derbianus **403**, 403
gigas 403
kaufmanni 403
livingstonii 403
niediecki 403
oreas 403
oryx **403**, 403
pattersonianus 403
selousi 403
triangularis 403
Taurus 401
taurus, Bos 400, **401**
Tautatus 622
tavistocki, Cervus 387
tawitawiensis, Rattus 649, **661**
taxicolor, Budorcas **404**, 404
Taxidea 313, **325**, 325
americanus 325
apache 325
berlandieri 325
californica 325
dacotensis 325
halli 325
hallorani 325
infusca 325
iowae 325
jacksoni 325
jeffersonii 325
kansensis 325
labradorius 325
littoralis 325
marylandica 325
merriami 325
montanus 325
neglecta 325
nevadensis 325
obscurata 325
papagoensis 325

phippsi 325
robusta 325
sonoriensis 325
sulcata 325
taxus **325**
Taxidiinae 313, **325**
taxilla, Melogale 314
taxinus, Conepatus 316
taxoides, Arctonyx 313
Taxus 325
taxus, Meles 314
Taxidea **325**
Ursus 325
Tayassu **380**, 380
aequatoris 380
albirostris 380
beebei 380
labiatus 380
pecari **380**, 380
ringens 380
spiradens 380
Tayassuidae **379**
taylori, Baiomys **695**
Castor 467
Chrysochloris 75
Cynocephalus 135
Hesperomys 695
Microtus 526
Ochotona 811
Perognathus 485
Petaurista 463
Tatera 562
Thomomys 475
Tayra 317
tcheliensis, Macaca 267
Teanopus 710, 712
teapensis, Orthogeomys 471
Oryzomys 721
Peromyscus 734
tectorum, Rattus 658
tectus, Oecomys 717
Oligoryzomys 719
teculyas, Sorex 122
tedescoi, Genetta 345
tedongus, Callosciurus 423
tedschenika, Vormela 325
teesdalei, Muntiacus 389
teguina, Hesperomys 746
Scotinomys **746**
tehamae, Sylvilagus 825
tehuantepecus, Orthogeomys 471
Peromyscus 734
Telanthropus 276
telesforianus, Cervus 386
telfairi, Echinops **73**, 73
telfordi, Crocidura **96**
telibius, Callosciurus 421
telibon, Maxomys 614
teliotis, Lasiurus 206
tellonis, Cynocephalus 135
tellus, Pappogeomys 472
telmalemonus, Sylvilagus 824
telmalestes, Blarina 106
Conepatus 316
telmaphilus, Peromyscus 730
telum, Stylodipus **494**, **494**, 495
tema, Mazama 390
temama, Mazama 390
temchuki, Bolomys **697**
temmincki, Carterodon 793
Niviventer 634
Pteropus **151**
Scotophilus 227
temminckii, Catopuma **289**
Cynocephalus 135
Dysopes 234
Eubalaena 349
Felis 289
Manis **415**
Molossops 234, **235**
Procolobus 272
Protoxerus 436
Suncus 103
temminickii, Cercopithecus 263
temon, Mustela 321
templetonii, Hipposideros 175
temporalis, Heteromys 481
Thylogale 57
tenasserimensis, Nycticebus 248
Prionailurus 295
tenaster, Niviventer **635**, 635
Tupaia 131
tendagurucus, Sigmoceros 395
tenebrica, Trachypithecus 274
tenebricus, Arvicola 505
Taterillus 563
tenebrosa, Murina **230**
tenebrosus, Arvicanthis 578
Maxomys 614
Uranomys 671
tenelliodus, Sorex 120
tenellus, Mus 623, 624, **629**, 629
Sorex 118, **120**
Thomomys 475
Zapus 499
tener, Calomys **698**, 698
Rhinolophus 163
teneriffae, Plecotus 224, **225**
Tenes 439, 440
teng, Galago 249
tengerensis, Callosciurus 422
tenggolensis, Rattus 661
teniotis, Cephalotes 240
Tadarida **240**
tennentii, Ratufa 437
Tenrec **73**
armatus 73
ecaudatus **73**
tanrec 73
Tenrecidae **70**
Tenrecinae **73**
tenuicauda, Neotoma 712
Rhipidomys 744
tenuidorsalis, Myotis 209
tenuipes, Neacomys **709**, 709
Oligoryzomys 718
tenuipinnis, Eptesicus **203**
tenuirostris, Callosciurus 423
Planigale **34**
Pontoporia 361
Reithrodontomys 741, **742**, 742, 743
Tamandua 68
Vulpes 287
tenuis, Bdeogale 301
Chaerephon 233
Chrysochloris 75
Crocidura 85, **96**, 96
Gerbillurus 547
Mustela 324
Myosorex 99, **100**, 100
Pipistrellus **224**
Reithrodontomys 741
Spilogale 317
Sundasciurus **452**, 452
Tatera 561
Teonoma 710, 711
tephra, Crocidura 85
tephragaster, Crocidura 85
tephrogaster, Sciurus 440, 441
tephromelas, Aeromys **459**
Pteromys 459
tephrops, Chlorocebus 265
tephrosceles, Procolobus 272
tephrura, Tupaia 132
tephrus, Hipposideros 171
tepicanus, Sciurus **440**
terasensis, Hipposideros 171
terempa, Sundamys 666
teres, Neomys 110
tereticaudus, Spermophilus **447**, **449**
termes, Proteles 309
terminus, Cynopterus 139
ternatensis, Cynocephalus 135
Sus 379
Terpone 410
terra-reginae, Rattus 654
terraenovae, Microtus 527
Rangifer 392
terraesanctae, Genetta 345
Psammomys 559
terrestris, Arvicola **505**, 505
Colobus 270
Hippopotamus 370
Microtus 517, 519
Mus 505
Tapirus **371**
Terricola 517, 518, 519, 520, 521, 524, 526, 529, 530
terricola, Dendrohyrax 373
terricolor, Mus **629**, 629
terricolus, Geomys 469
terrosus, Dipodomys 479
tersus, Ammospermophilus 420
Cryptotis 109
Leopoldamys 603
tertius, Callithrix 252
terutaus, Atherurus 773
Cynocephalus 135
Petaurista 463
terutavensis, Callosciurus 421
Rattus 661
terutus, Tragulus 383
tescorum, Spermophilus 447
tesquorum, Lepus 817
tesselatum, Papio 269
tesselatus, Lasiurus 206
tessmanni, Dendrohyrax 373
Mandrillus 268
testicularis, Arvicanthis 578
testudo, Lophiomys 564
tethyos, Stenella 356
Tetracerus **403**
chickara 403
iodes 403
paccerois 403
quadricornis **403**
striatocornis 403
subquadricornis 403
tetracornis 403
tetracornis, Tetracerus 403
tetradactyla, Allactaga **489**
Colobus 270
Manis **415**
Suricata 308
Tamandua **68**, 68
tetradactylus, Oryzorictes **72**, 72
Petrodromus **830**, 830
tetragonurus, Rattus 655, 658
Sorex 112
Tetramerodon 517
tetramerus, Microtus 530
tetricus, Maxomys 613
tettensis, Rattus 658
teuszii, Sousa **356**
texana, Tadarida 240
texanum, Dasypus 66
texanus, Odocoileus 391
Peromyscus 732
Scalopus 127
texensis, Canis 280
Castor 467
Conepatus 316
Didelphis 17
Geomys 469, 470
Glaucomys 460
Liomys 482
Lontra 311
Lynx 294
Mustela 322, 324
Oryzomys 721, 724
Spermophilus 450
Spilogale 317
Thomomys 474
Urocyon 285
Ursus 339
texianus, Lepus 815
Ovis 408

Sciurus 442
Sigmodon 747
thaeleri, Orthogeomys **471**, 471
thai, Callosciurus 421
Rattus 660
Viverricula 348
thaianus, Myotis 215
thaiwanensis, Callosciurus 421
Thalarctos 338, 339
thalia, Crocidura 85, 89, **96**, 96
Thallomy 668
Thallomys 592, **668**, 668, 669, 674
acaciae 669
bradfieldi 668
damarensis 668, 669
davisi 668
herero 668
kalaharicus 668
lebomboensis 669
leuconoe 668
loringi **668**, 668
moggi 669
molopensis 668
nigricauda **668**, 668, 669
nitela 668
paedulcus 668, **669**, 669
quissamae 668
rhodesiae 669
robertsi 668
ruddi 669
scotti 669
shortridgei **669**, 669
somaliensis 669
stevensoni 669
zambesiana 669
Thallomyscus 708
Thalpomys 688, **748**
cerradensis **748**
lasiotis **748**, 748
reinhardti 748
thamin, Cervus 386
Thamnomys 592, 593, 594, 637, 668, **669**, 669
kempi **669**
kivuensis 670
major 670
rutilans 669
schoutedeni 670
venustus 669, **670**, 670
thamnos, Canis 280
Thaocervus 385
Thaptomys 688, 692
thar, Naemorhedus 407
thayeri, Myopus 531
Sciurotamias 439
thebaica, Nycteris **162**
Thecurus 773, 774
theobaldi, Mus 626
Taphozous **161**
theodorverhoeveni, Papagomys **638**
Theranthropus 276
theresa, Monodelphis **22**
theresae, Crocidura 85, **97**, 97
thermaicus, Nannospalax 754
Theropithecus 263, **269**
gelada **269**
nedjo 269
obscurus 269
ruppelli 269
senex 269
thersites, Mops **237**
Nasua 334
Semnopithecus 273
thessalicus, Nannospalax 754
thetis, Halmaturus 57
Thylogale **57**
Thetomys **644**
thibetana, Macaca **268**
Ochotona 807, 808, 809, 811, **812**, 812
thibetanus, Sorex 113, 116, 117, 118, 119, **121**, 121
Ursus 338, **339**
thicolea, Lagenorhynchus 354
thienemannii, Phoca 332
thierryi, Crocuta 308
Genetta **346**
thikae, Colobus 270
Kobus 414
thinobia, Eremaelurus 290
thinobius, Felis 290
Thiosmus 316
tholloni, Chlorocebus 265
Procolobus 272
thomasi, Aeromys **459**
Aethomys **569**, 569
Cephalophus 411
Cercopithecus 264
Cheirogaleus 243
Crocidura 87
Crocuta 308
Cryptotis **109**, 109
Echimys **792**
Eptesicus 201
Felis 290
Funambulus 427
Galagoides 249
Heterohyrax 373
Kobus 414
Lagothrix 258
Lonchophylla **182**
Lophiomys 564
Madoqua 398
Megadontomys **707**
Microgale **72**
Microtus **530**
Mimetillus 207
Myotis 214
Ochotona **813**
Odocoileus 391
Oncifelis 294
Otomys 682
Peromyscus 707
Petaurista 462
Presbytis **271**
Rheomys **743**
Rhinolophus **169**
Saguinus 253
Salpingotus **492**, 492
Sciurus **443**
Sorex 119
Steatomys 546
Sturnira **193**
Tadarida 240
Tremarctos 338
Uroderma 193
Zygodontomys 752
thomasinae, Redunca 414
Thomasomys 687, 688, 700, 745, **749**, 749, 751, 752
altorum 749
aureus **749**
auricularis 750
australis 749
baeops **749**
bombycinus **749**
caudivarius 749
cinereiventer **749**
cinereus **749**
cinnameus 749
contradictus 749
daphne **749**
dispar 749
eleusis **749**
emeritus 750
erro 749
fraternus 749
fumeus 750
gracilis **749**
hudsoni 749
hylophilus **749**
incanus **749**
ischyurus **750**
kalinowskii **750**
ladewi **750**
laniger **750**, 750
monochromos **750**
nicefori 749
niveipes **750**
notatus **750**
oenax 752
oreas **750**
paramorum **750**
popayanus 749
praetor 749
princeps 749
pyrrhonotus **750**
rhoadsi **750**
rosalinda **750**
silvestris **750**
taczanowskii **750**
vestitus **751**
thomensis, Crocidura **97**
Hipposideros 172
Thomomyini 469
Thomomys 469, **473**, 473, 474, 475
abbotti 473
absonus 473
abstrusus 473
acrirostratus 473
actuosus 473
adderrans 473
aequalidens 475
affinis 473
agricolaris 473
albatus 473
albicaudatus 473
albigularis 475
alexandrae 473
alienus 473
alpinus 473
alticolus 473
altivallis 473
amargosae 473
analogus 473
andersoni 475
angularis 473
angustidens 473
anitae 473
apache 473
aphrastus 473
ardicola 473
argusensis 473
arriagensis 475
atrodorsalis 475
atrogriseus 475
atrovarius 475
attenuatus 475
aureiventris 473
aureus 473
awahnee 473
bachmani 475
badius 475
baileyi 473, 474, 476
basilicae 473
birdseyei 473
bonnevillei 473
borealis 475
boregoensis 473
boreorarius 473
borjasensis 473
bottae **473**, 473, 474, 475, 476
brazierhowelli 473
brevidens 473
bridgeri 475
bulbivorus 473, **474**
bullatus 475
burti 475
cabezonae 473
cactophilus 473
caliginosus 475
camargensis 473
camoae 473
caneloensis 473
canus 473
carri 473
caryi 475
catalinae 473
catavinensis 473
cedrinus 473
centralis 473
cervinus 473
cheyennensis 475
chihuahuae 475
chiricahuae 473
chrysonotus 473

cinereus 473
clusius 473, **474**, 475
cognatus 475
collinus 473
collis 473
columbianus 475
comobabiensis 473
concisor 473
confinalis 473
confinus 474
connectens 473
contractus 473
convergens 473
convexus 473
couchi 475
crassidens 475
crassus 473
cultellus 473
cunicularis 473
curtatus 473
depauperatus 473
depressus 473
desertorum 473
desitus 473
detumidus 473
devexus 475
diaboli 473
dissimilis 473
divergens 473
douglasii 475
durangi 475
duranti 475
elkoensis 475
emotus 476
enixus 476
estanciae 473
evexus 476
eximius 476
extenuatus 473
extimus 476
falcifer 475
fisheri 475
flavidus 473
fossor 475
fulvus 473
fumosus 473
fuscus 475
glacialis 475
goldmani 476
gracilis 475
grahamensis 473
growlerensis 473
guadalupensis 473
harquahalae 473
helleri 475
hesperus 475
homorus 473
howelli 473
hualpaiensis 473
hueyi 473
humilis 473
idahoensis 473, **474**, 475
imitabilis 473
immunis 475
incensus 475
incomptus 473
infrapallidus 473
ingens 473
intermedius 476
internatus 473
jacinteus 473
jojobae 473
juarezensis 473
juntae 476
kaibabensis 475
kelloggi 475
lachuguilla 473
lacrymalis 473
laticeps 473
latirostris 473
latus 473
lenis 473
leucodon 473
levidensis 473
levis 475
limitaris 473
limosus 475
limpiae 473
litoris 473
lorenzi 473
loringi 475
louiei 475
lucidus 473
lucrificus 473
macrotis 475
madrensis 476
magdalenae 473
martinensis 476
martirensis 473
mazama 473, **474**
mearnsi 473
medius 475
melanops 475
melanotis 473
meritus 475
mewa 473
minimus 473
minor 473
modicus 473
mohavensis 473
monoensis 475
monticola 473, **475**, 475
moorei 475
morulus 473
muralis 473
musculus 476
mutabilis 473
myops 475
nanus 473
nasicus 475
nasutus 473
navus 473
nebulosus 475
neglectus 473
nelsoni 476
nesophilus 473
nevadensis 475
newmani 476
nicholi 473
niger 475
nigricans 473
occipitalis 474
ocius 475
operarius 474
operosus 474
optabilis 474
opulentus 474
oquirrhensis 475
oregonus 475
oreoecus 474
orizabae 476
osgoodi 474
owyhensis 475
paguatae 474
pallescens 474
parowanensis 475
parviceps 476
parvulus 474
pascalis 474
patulus 474
pectoralis 474
peramplus 474
perditus 476
peregrinus 476
perpallidus 474
perpes 474
pervagus 474
pervarius 474
phasma 474
phelleoecus 474
pierreicolus 475
pinalensis 474
pinetorum 475
piutensis 474
planirostris 474
planorum 474
potosinus 476
powelli 474
premaxillaris 475
providentialis 474
proximarinus 474
proximus 474
pryori 475
puertae 474
pugetensis 475
pullus 476
pusillus 474
pygmaeus 474
quadratus 475
quercinus 476
ravus 475
relicinus 475
relictus 475
retractus 474
retrorsus 475
rhizophagus 474
riparius 474
robustus 474
rostralis 475
rubidus 474
rufescens 473, 475
rufidulus 474
ruidosae 474
rupestris 474
ruricola 474
russeolus 474
sanctidiegi 474
saturatus 475
saxatilis 474
scapterus 474
scotophilus 474
segregatus 475
sevieri 474
shawi 475
sheldoni 476
siccovallis 474
silvifugus 474
similis 475
simulus 474
sinaloae 474
solitarius 474
sonoriensis 476
spatiosus 474
stansburyi 474
sturgisi 474
suboles 474
subsimilis 474
supernus 476
tacomensis 475
talpoides 473, 474, **475**, 475
taylori 475
tenellus 475
texensis 474
tivius 474
toltecus 474
tolucae 476
townsendii 473, 474, **475**, 475, 476
trivialis 475
trumbullensis 474
tularosae 474
tumuli 475
uinta 475
umbrinus 473, 474, **475**, 475, 476
unisulcatus 475
vanrosseni 474
varus 474
vescus 474
villai 474
virgineus 474
vulcanius 476
wahwahensis 474
wallowa 475
wasatchensis 475
whitmani 475
winthropi 474
xerophilus 474
yakimensis 475
yelmensis 475
zacatecae 476
thompsoni, *Heteromys* 481
Sorex 115
thomsoni, *Iomys* 461
Mesoplodon 363
thomsonii, *Gazella* **397**, 397
thonglongyai, *Craseonycteris* **155**, 155
thooides, *Canis* 280
Thoopterus **153**
latidens 153
nigrescens **153**, 153
thoracatus, *Geocapromys*

801, 801
thoracicus, Crocidura 94
thorbeckiana, Myoictis 32
thornicrofti, Giraffa 383
thorntoni, Dendromus 543
thoroldi, Cervus 385
thoth, Papio 269
thotti, Capreolus 390
thous, Canis 282
Cerdocyon **282**
Dusicyon 283
thracius, Nannospalax 754
Spermophilus 445
Thrichomys **798**
antricola 798
apereoides **798**
fosteri 798
inermis 798
laurentius 798
thricolis, Microtus 529
Thrinacodus 790
albicauda 790
Thryonomyidae **775**
Thryonomyoidea 771
Thryonomys **775**
calamophagus 775
camerunensis 775
gregorianus **775**
harrisoni 775
logonensis 775
raptorum 775
rutshuricus 775
sclateri 775
semipalmatus 775
swinderianus **775**
variegatus 775
thuleo, Apodemus 573
thurberi, Peromyscus 732
Thylacinidae **29**, 29
Thylacinus **29**
breviceps 29
communis 29
cynocephalus **29**
harrisii 29
lucocephalus 29
striatus 29
Thylacis 39, 40
Thylacomyidae 39, 40
Thylacomys 39, 40, 635
Thylamys **23**
bruchi 23
cinderella 23
citella 23
coquimbensis 23
elegans **23**
fenestrae 23
grisea 23
griseus 23
janetta 23
karimii 23
macrura **23**
marmota 23
nana 23
pallidior **23**
pimelura 23
pulchella 23
pusilla **23**
soricina 23
sponsoria 23
tatei 23
velutinus **23**
venusta 23
verax 23
Thylocomyidae 39
Thylogale 51, 53, **57**, 57
billardierii **57**
brachytarsus 57
browni 57
bruijni 57
brunii 51, **57**
coxenii 57
eugenii 57
gazella 57
gracilis 57
jukesii 57
keysseri 57
lanatus 57
lauterbachi 57
lugens 57
nuchalis 57
oriomos 57
rufiventer 57
stigmatica **57**
tasmanei 57
temporalis 57
thetis **57**
tibol 57
wilcoxi 57
thyone, Vampyressa 194
Thyroptera **195**
abdita 196
albigula 196
albiventer 196
bicolor 196
discifera **196**
juquiaensis 196
major 196
thyropterus 196
tricolor 195, **196**
Thyropteridae **195**
thyropterus, Thyroptera 196
thysanodes, Myotis **215**
thysanurus, Herpestes 306
Margaretamys 609
tiang, Damaliscus 394
tianschanensis, Meles 314
tianschanicus, Apodemus 570
Capreolus 390
Microtus 521
tianshanica, Sicista 496, 497, **498**
tiarata, Mustela 322
tibestianus, Papio 269
tibetana, Budorcas 404
tibetanus, Cervus 386
Cricetulus 537
Lepus 814, 816, 821
Lynx 293
Marmota 432
Mustela 324
Spilogale 317
Tibetholagus 807
tibialis, Miniopterus 230
tibicen, Rattus 652
tibol, Thylogale 57
tiburonensis, Lepus 814
Peromyscus 730
Spermophilus 450
tichomirowi, Dryomys 766
tickelli, Hesperoptenus **204**
tigrensis, Lepus 816
tigrillo, Oncifelis 294
tigrina, Genetta **346**, 346
Marmota 432
Phoca 332
tigrinoides, Leopardus 292
tigrinum, Okapia 383
tigrinus, Equus 369
Leopardus 291, **292**, 292, 294
Tigris 297
tigris, Panthera **298**
tikos, Rattus 660
tildae, Sturnira **193**
tilonura, Gazella 397
timealeyi, Ningaui **33**, 33
timidus, Lepus 814, 815, 819, **820**, 820
Rhinolophus 165
timorensis, Cervus **387**
Rattus 649, **661**
timoriensis, Nyctophilus **218**
Rhinolophus 164
Sus **379**
tingahi, Prosciurillus 436
tinggius, Callosciurus 423
tingia, Arctogalidia 343
Prionailurus 295
tingius, Rattus 661
tiomanensis, Ratufa 437
tiomanicus, Rattus 649, 650, 651, 654, 655, 657, 659, **661**, 661, 662
Sundasciurus 452
tionis, Atherurus 773
Crocidura 89
Hylomys 80
tippelskirchi, Giraffa 383
tirae, Apodemus 573
tistae, Rattus 660
titi, Saguinus 254
titthaecheilus, Cynopterus **139**, 139
tivius, Thomomys 474
tiznitensis, Crocidura 96
tjaderi, Tragelaphus 404
tjaederi, Kobus 414
tjibruniensis, Tupaia 132
tjibunensis, Maxomys 612
toala, Hipposideros 171
toba, Akodon 689, 691, **693**, 694
Mazama 390
tobagensis, Sciurus 441
tobagi, Marmosa 18
Rhipidomys 745
Zygodontomys 752
tobolica, Vulpes 287
tocantinus, Bradypus 63
Cebus 259
todayensis, Rattus 652
togeanensis, Babyrousa 377
togeanus, Macaca 268
togianus, Ailurops 46
togoensis, Crocuta 308
Kobus 414
Leptailurus 293
Togomys 649
tohi, Redunca 414
toklat, Ursus 339
tokmak, Apodemus 574
tokudae, Mogera 125, **126**, 126
Pteropus **151**
Tokudaia **670**, 670
muenninki **670**, 670
osimensis **670**, 670
Tokudamys 670
tolai, Lepus 814, 816, 817, 819, **821**, 821
toldti, Chlorocebus 265
tolduco, Ctenomys 785
tolimae, Akodon 688
tolimensis, Melanomys 708
Potos 334
tolteca, Herpailurus 291
toltecus, Artibeus **189**
Odocoileus 391
Reithrodontomys 741
Sigmodon 747
Thomomys 474
tolucae, Reithrodontomys 740
Sciurus **442**
Thomomys 476
Tolypeutes **67**
apar 67
bicinctus 67
brachyurus 67
conurus 67
globulus 67
matacus **67**
muriei 67
quadricinctus 67
tricinctus **67**
Tolypeutini 64
Tolypoides 67
tomba, Stenomys 665
tomensis, Clethrionomys 509
Cricetus 538
Mus 625, 626
Sorex 119
tomentosus, Kunsia **706**
Mus 706
Scapteromys 745
tomesi, Hesperoptenus **205**
Pteropus 147
Tomeutes 420, 438
Tomopeas **231**
ravus **231**, 231
Tomopeatinae **231**
tompsoni, Rattus 658
tonalensis, Sigmodon 747
Tonatia 179, **180**
amblyotis 181

bidens **180**
brasiliense **180**
carrikeri **180**
centralis 181
childreni 180
colombianus 181
evotis **180**
laephotis 181
minuta 180
nicaraguae 180
occidentalis 181
schulzi **180**
silvicola 180, **181**
venezuelae 180
tondanus, Rattus 655
tonfangensis, Pipistrellus 220
tongaensis, Tadarida 240
tonganus, Pteropus **151**
tongensis, Aethomys 567
Grammomys 593
Paraxerus 435
Steatomys 546
Tatera 560
tonkeana, Macaca **268**
tonkeanus, Prosciurillus 436
tonkinensis, Mustela 323
tonquinia, Melogale 315
Tupaia 131
tonsor, Euoticus 248
tonsus, Macaca 268
tontalis, Lagidium 778
topachevskii, Dinaromys 512
topapuensis, Prosciurillus 436
topi, Damaliscus 394
toppini, Callicebus 258
tora, Alcelaphus 393
tordayi, Petrodromus 830
toritensis, Crocidura 92
tornillo, Peromyscus 732
torosa, Isoodon 39
torquata, Cheiromeles 233
Cynonycteris 143
Felis 291
Hystrix 774
Myonycteris **143**
Neotoma 712
torquatus, Bradypus **63**, 63
Callicebus **259**
Cercocebus **262**, 262
Cheiromeles **233**
Ctenomys **787**
Dicotyles 380
Dicrostonyx 510, **511**, 511
Herpestes 306
Pecari 380
Pseudalopex 284
Pteropus 151
Ursus 340
torques, Akodon 692, **693**, 693
torrei, Boromys **799**
Pteronotus 177
torrentium, Protoxerus 436
torresi, Bibimys **696**, 696
torridus, Liomys 482
Onychomys **719**, 719
Orthogeomys 471
Pappogeomys 472
torticornis, Tragelaphus 404
tortilis, Chaetomys 790
torvus, Paradoxurus 344
Pecari 380
tosae, Petaurista 462
toscanus, Rhinolophus 165
totipes, Berylmys 580
totontepecus, Peromyscus 734
touan, Monodelphis 21
toufoeus, Martes 320
toupai, Callosciurus 423
townsendi, Arctocephalus **327**
Lepus 815
Urocyon 285
Ursus 339
townsendii, Lepus **821**
Microtus 525, 528, **530**
Plecotus 224, **225**
Scalops 127
Scapanus **128**
Spermophilus 445, 448, **449**
Tamias 454, 455, **456**
Thomomys 473, 474, **475**, 475, 476
toxi, Rattus 660
toxopei, Murina 229
toxopeusi, Rhinolophus 164
toxophyllum, Sphaeronycteris **191**, 191
toxotis, Loxodonta 367
trabata, Panthera 299
trabeata, Alouatta 254
trabeatus, Oecomys 716
Trachelocele 395
Trachelotherium 383
Trachops **181**
cirrhosus **181**
coffini 181
ehrhardti 181
fuliginosus 181
trachynotus, Maxomys 612
Trachypithecus 270, **273**, 273, 274
albinus 274
argentatus 274
argenteus 274
atrior 273
auratus **273**
barbei 273
belliger 274
brahma 274
carbo 274
caudalis 273
corax 274
corvus 274
crepuscula 274
cristatus **273**, 273
cucullatus 274
delacouri 273
durga 274
flavicauda 274
francoisi **273**
geei **273**
germaini 273
halonifer 274
harti 274
hatinhensis 273
holotephreus 274
johnii **274**
jubatus 274
kelaartii 274
kephalopterus 274
kohlbruggei 273
koratensis 273
laotum 273
latibarba 274
latibarbatus 274
leucocephalus 273
leucomystax 274
leucoprymnus 274
mandibularis 273
margarita 273
melamera 274
monticola 274
nestor 274
obscurus **274**
phayrei **274**
philbricki 274
phillipsi 274
pileatus **274**, 274
poliocephalus 273
porphyrops 274
pruinosus 273
pullata 273
purpuratus 274
pyrrhus 273
sanctorum 274
saturatus 274
seimundi 274
senex 274
shanicus 274
shortridgei 274
smithi 274
sondaicus 273
styx 274
tenebrica 274
ultima 273
ursinus 274
vetulus **274**, 274
vigilans 273
wroughtoni 274
tragata, Nycteris **162**, 162
Tadarida 240
tragatus, Rhinolophus 166
Tragelaphus **403**, 403
abyssinicus 404
albonotatus 404
albovirgatus 403
angasii **403**
australis 404
barkeri 404
bea 404
bor 404
brunneus 404
burlacei 404
buxtoni **403**
capensis 404
chora 404
cooperi 403
cottoni 404
dama 404
decula 404
delamerei 404
dianae 404
dodingae 404
eldomae 404
euryceros 404
eurycerus **403**
excelsus 404
fasciatus 404
frommi 404
fulvoochraceus 404
gigas 403
gratus 404
hamiltoni 404
haywoodi 404
heterochrous 404
imberbis **404**
inornatus 404
insularis 404
isaaci 403
johannae 404
katanganus 403
knutsoni 404
koodoo 404
larkenii 404
laticeps 404
locorinae 404
makalae 404
massaicus 404
meneliki 404
meridionalis 404
meruensis 404
multicolor 404
nigrinotatus 404
obscurus 404
olivaceus 404
ornatus 404
phaleratus 404
pictus 404
powelli 404
punctatus 404
reidae 404
roualeynei 404
sassae 404
scriptus **404**
selousi 404
signatus 404
simplex 404
spekii **404**
strepsiceros **404**
sylvaticus 404
sylvestris 404
tjaderi 404
torticornis 404
uellensis 404
ugallae 404
wilhelmi 404
zambesiensis 404
tragelaphus, Ammotragus 404
tragocamelus, Antilope 401

Boselaphus **401**
Tragops 395
Tragopsis 395
Tragulidae **382**
Tragulus **382**
abjectus 383
abruptus 383
affinis 382
amoenus 383
anambensis 383
angustiae 382
annae 383
bancanus 383
banguei 383
batuanus 383
billitonus 383
borneanus 383
brevipes 382
bunguranensis 383
canescens 383
carimatae 382
everetti 382
flavicollis 383
focalinus 382
formosus 383
fulvicollis 382
fulviventer 382
fuscatus 382
hendersoni 383
hosei 382
indicus 382
insularis 382
javanicus **382**, 383
jugularis 383
kanchil 382
klossi 382
lampensis 382
lancavensis 382
longipes 382
luteicollis 382
lutescens 383
masae 382
mergatus 382
mimenoides 382
napu **383**
natunae 382
neubronneri 383
niasis 383
nigricans 383
nigricollis 383
nigrocinctus 383
pallidus 382
parallelus 383
pelandoc 382
penangensis 382
perflavus 383
pidonis 382
pierrei 382
pinius 382
pretiellus 383
pretiosus 383
pumilus 382
ravulus 382
ravus 382
rubeus 382
rufulus 383
russeus 382
russulus 382
sebucus 383
siantanicus 383
stanleyanus 383
subrufus 382
terutus 383
umbrinus 383
versicolor 383
virgicollis 382
williamsoni 382
tragulus, Raphicerus 400
Tragus 405
tragus, Rupicapra 410
tralatitia, Felis 291
tralatitius, Pipistrellus 221
tramatus, Pipistrellus 220
tramitius, Rattus 658
tranninhensis, Hipposideros 171
transalaiana, Capra 406
transbaicalicus, Lepus 821
transbaikalicus, Spermophilus 450
transcaspiae, Ellobius 512
transcaspicus, Microtus **531**
Myotis 213
transcaucasica, Talpa 129
transcaucasicus, Erinaceus 77
Microtus 524, 526
transitionalis, Sylvilagus **827**
transrypheus, Sorex 121
transsylvanica, Mustela 323
transsylvanicus, Capreolus 390
Lepus 817
Microtus 530
Nannospalax 754
transsylvaticus, Lepus 817
transuralensis, Microtus 526
Rangifer 392
transvaalensis, Acomys 567
Atilax 301
Chrysospalax 76
Cryptomys 772
Equus 369
Myosorex 100
Oreotragus 399
Papio 269
Poecilogale 325
Proteles 309
Steatomys 545
Sylvicapra 412
transvolgensis, Microtus 530
transvosagicus, Capreolus 390
transylvanicus, Eptesicus 203
trapezia, Felis 291
trapezius, Muscardinus 769
traubi, Hyperacrius 515
travancorensis, Suncus 102
travassosi, Lutreolina 18
Micoureus 20
traversii, Mesoplodon 363
trebevicensis, Dinaromys 511
Tremarctos **337**, 338
frugilegus 338
lasallei 338
majori 338
nasutus 338
ornatus **338**
thomasi 338
trepida, Cebus 259
trepidus, Spermophilus 447
trettaui, Mustela 323
treubii, Niviventer 634
trevelyani, Chrysochloris 75
Chrysospalax **76**
Prionailurus 295
trevori, Mops 236, **237**
Triaenops **175**
afer 175
aurita 175
furculus **175**
humbloti 175
macdonaldi 175
majusculus 175
persicus **175**, 175
rufus 175
trialeticus, Chionomys 507
triangularis, Peromyscus 732
Taurotragus 403
trianguligera, Ateles 257
Triaulacodus 775
Trichaelurus 295
Trichechidae **365**
Trichechus **365**
africanus 366
amazonius 365
americanus 365
antillarum 365
atlanticus 365
australis 366
clusii 365
dugon 365
guyannensis 365
inunguis **365**
koellikeri 365
latirostris 365
manatus **365**, 365
minor 365
nasutus 366
oronocensis 365
owenii 366
senegalensis **366**
stroggylonurus 366
trichechus 365
vogelii 366
trichechus, Trichechus 365
Trichecidae 325
trichodactylus, Lagostomus 778
Trichomanis 313
trichopus, Zygogeomys **476**, 476
Trichosurus **48**, 48
arnhemensis **48**
bougainvillei 48
caninus **48**
cookii 48
cuvieri 48
eburacensis 48
felina 48
fuliginosa 48
grisea 48
hypoleucus 48
johnstonii 48
lemurina 48
melanura 48
mesurus 48
nigrans 48
novaehollandiae 48
raui 48
ruficollis 48
selma 48
tapouaru 48
vulpecula **48**
vulpina 48
xanthopus 48
trichotis, Akodon 692
Allocebus **243**
Anotomys 694
Belomys 459
Cheirogaleus 243
Chibchanomys **699**, 699
Ichthyomys 699
Rheomys 743
trichura, Crocidura 82
Trichurus 48
trichurus, Conepatus 316
Oecomys 716
Trichys **774**
fasciculata **774**
guentheri 774
lipura 774
macrotis 774
tricinctus, Dasypus 67
Tolypeutes **67**
tricolor, Ateles 257
Carollia 186
Lycaon 283
Monodelphis 21
Myotis **215**
Pteropus 147
Sciurus 442
Thyroptera 195, **196**
tricuspidatus, Aselliscus **170**
Rhinolophus 170
tricuspis, Manis **415**
tridactyla, Didelphis 49
Myrmecophaga **68**, 68
Zaglossus 13
tridactylus, Bradypus **63**, 63
Macropus 54
Potorous **50**
tridecemlineatus, Spermophilus 447, **450**
tridens, Asellia **170**
Rhinolophus 170
tridentata, Manis 415
tridentinus, Microtus 518
trifidus, Aselliscus 170
trifoliatus, Rhinolophus **169**
trifolium, Megaderma 163
trigonifer, Callithrix 252
trigonirostris, Sorex 122

trilineata, Arctogalidia 343
Monodelphis 21
trilineatus, Arctogalidia 343
Funambulus 427
trimucronatus, Lemmus 517
trinidadensis, Dipodomys 479
trinitatis, Artibeus 188
Caluromys 15
Cebus 259
Echimys 795
Mazama 390
Oecomys 715, **716**
Peropteryx 158
Proechimys 796, **798**
trinitatum, Chiroderma **190**
Uroderma 193
trinitatus, Molossus 235
Spermophilus 447
Trinodontomys 728
Trinomys 795, 796, 797, 798
trinotatus, Zapus **499**
Trinycteris 178, 179
triomylus, Artibeus 188
tripartitus, Saguinus **254**
Sicista 498
tripolitanicus, Jaculus 493
tripolitanus, Canis 280
Psammomys 559
trissyllepsis, Peromyscus 735
tristami, Crocidura 96
tristis, Catopuma 289
Miniopterus **231**
Nycteris 162
tristrami, Felis 291
Meriones **558**
tristriata, Monodelphis 21
tristriatus, Funambulus **427**
Sicista 498
trisulcatus, Potorous 50
triticeus, Micromys 620
triton, Cricetulus 537, 539
Mus 622, **629**, 629
Ornithorhynchus 13
Tscherskia **539**, 540
trivialis, Thomomys 475
trivirgata, Arctogalidia **343**
Dactylopsila 60, **61**, 61
Simia 255
trivirgatus, Aotus **256**
Hybomys 592, **596**, 597
Paradoxurus 342
trivittatus, Atlantoxerus 420
Bradypus 63
trizona, Sicista 498
trobrius, Hipposideros 172
Troglodytes 276
gorilla 276
troglodytes, Pan **277**
Rhizomys 685
Simia 276
trogophilus, Rhinolophus 166
Trogopterus 459, **465**, 465
edithae 465
himalaicus 465
minax 465
mordax 465
xanthipes **465**
tronchonii, Natalus 195
tropica, Bettongia 49
tropicalis, Arctocephalus **327**
Chrysochloris 75
Conepatus 316
Cryptotis 109
Delphinus 352
Geomys 469, **470**
Monachus **331**
Mustela 322
Neotoma 712
Odocoileus 391
Otomys 680, **681**, 681, 682
Peromyscus 734
Petromus 774
Reithrodontomys 741
Rhinopoma 155
tropicius, Oryzomys 724
Tropicolobus 271
tropidorhynchus, Molossus 235
tropini, Meriones 557
trotteri, Callosciurus 422
trouessarti, Equus 370
Meriones 558
troughtoni, Eptesicus 203
Taphozous 160
trowbridgii, Microtus 519
Sorex **121**
Sylvilagus 825
truei, Glossophaga 184
Hoplomys 794
Odocoileus 391
Peromyscus 728, 729, 730, 731, 734, 735, **736**, 737
Phenacomys 533
Phocoenoides 359
Podogymnura **80**, 80
Scapanus 127
Sciurus 440
Sylvilagus 825
Synaptomys 534
trumbullensis, Perognathus 486
Thomomys 474
trumbulli, Eumops 234
truncatus, Chlamyphorus **64**, 64
Delphinus 357
Rhinolophus 166
Tursiops **357**, 357
truteri, Hippotragus 412
tryoni, Nyctimene 144
Tryphomys **670**, 670
adustus 564, **670**, 670
tsaidamensis, Lepus 819
Microtus 523
Mustela 322
tschadensis, Alcelaphus 393
Hippopotamus 381
Kobus 414
tschaganensis, Marmota 431
tschego, Pan 277
tscherga, Apodemus 572, 574
Tscherskia 536, 537, **539**
albipes 539, 540
arenosus 540
bampensis 540
collinus 540
fuscipes 540
incanus 540
meihsienensis 540
nestor 540
ningshaanensis 540
rusa 539
triton **539**, 540
yamashinai 540
tscherskii, Sorex 117
tschetshenicus, Myoxus 770
tschifuensis, Erinaceus 77
tschiliensis, Canis 281
Vulpes **287**
tschudii, Cavia **779**, 779
Lagothrix 258
Mazama 391
Metachirus 20
tschuktschorum, Lepus 819
Microtus 527
Sorex 117
Spermophilus 448
tschuliensis, Myotis 213
tsingtanensis, Callosciurus 421
Meles 314
tsingtauensis, Callosciurus 421
tsolovi, Calomyscus **536**
tsuensis, Martes 320
Myotis 210
tua, Rattus 661
tuakensis, Pongo 277
tuancus, Cynocephalus 135
Leopoldamys 603
Sus 379
Tupaia 133
tuareg, Myomys 631
tuasoninus, Cervus 386
tuberculata, Mystacina **231**, 231
tuberculatus, Chalinolobus **200**
Pteropus **151**
Vespertilio 199
tubericornis, Hemitragus 406
tuberifer, Brachyteles 257
tubinaris, Murina **230**
Tubulidentata 375
tuckeri, Potorous 50
tuconax, Ctenomys **787**
tucumana, Eira 318
tucumanensis, Akodon 693
tucumanum, Lagidium 778
tucumanus, Cerdocyon 282
Ctenomys 785, **787**
Euphractus 66
Phyllotis 738
Tucuxa 355
tucuxi, Sotalia 355
tugarinovi, Clethrionomys 509
tugelensis, Otomys 680, 681
tularensis, Dipodomys 478
Lepus 815
Onychomys 719
Perognathus 485
Ursus 339
tularosae, Spermophilus 450
Thomomys **474**
tulbaghensis, Suncus 103
tulduco, Ctenomys 785
tullbergi, Epimys 642
Lophuromys 606
Praomys 607, 608, 642, 643, **644**, 644
tulliana, Panthera 298
tumac, Lepus 817
tumacus, Sylvilagus 825
tumatumari, Leopardus 291
Mazama 390
tumbalensis, Tylomys **751**
tumbili, Chlorocebus 265
tumbolensis, Galago 249
tumida, Rhogeessa 225, **226**
Rhogessa 225
tumidiceps, Mormoops 176
tumidifrons, Natalus **195**, 195
tumidirostris, Natalus **195**
tumidus, Mus 745
Scapteromys **745**, 746
tumuli, Thomomys 475
tumulus, Chironax 138
tundrae, Microtus 521
tundrarum, Canis 281
tundrensis, Clethrionomys 509
Sorex 112, 113, 114, **121**, 121
Ursus 339
tunetae, Eliomys 767
Lepus 816
Rhinolophus 167
tungussensis, Sorex 113
tunisianus, Alcelaphus 393
tunjuc, Cervus 387
tunneyi, Rattus 649, 651, **661**
Tupaia **131**
anambae 132
annamensis 131
assamensis 131
balina 132
baluensis 132
banguei 133
batamana 132
belangeri **131**, 132
besara 133
bogoriensis 132
brunetta 131
bunoae 133
busuangae 133
caedis 132
cambodiana 131
carimatae 133
castanea 132
cervicalis 133

chinensis 131
chrysogaster **132**
chrysomalla 132
chrysura 133
clarissa 131
cochinchinensis 131
cognata 132
concolor 131
cuyonis 133
demissa 132
discolor 132
dissimilis 131
dorsalis **132**
edarata 132
ellioti 131
everetti 133
ferruginea 131, 132
fuscior 133
gaoligongensis 131
glis **132**, 132
gonshanensis 131
gracilis **132**
griswoldi 133
humeralis 132
hypochrysa 132
inflata 132
jacki 132
javanica **132**
kelabit 133
kohtauensis 131
kretami 133
lacernata 132
laotum 131
lepcha 131
lingae 133
longicauda 132
longipes **132**
lucida 133
malaccana 132
masae 133
minor **132**
modesta 131
moellendorffi 133
montana **132**
muelleri 133
nainggolani 133
natunae 133
nicobarica **132**
nitida 133
obscura 132
occidentalis 132
olivacea 131
operosa 132
paitana 133
palawanensis **133**
peguanus 131
pemangilis 132
penangensis 132
phaeniura 132
phaeura 132
picta **133**
pingi 131
pulonis 132
raviana 132
redacta 132
riabus 132
ruficaudata 133
salatana 132
siaca 132
siberu 132
siccata 131
sincipis 132
sinus 131
sirhassenensis 133
sordida 132
speciosus 133
splendidula **133**
surda 133
tana 132, **133**
tenaster 131
tephrura 132
tjibruniensis 132
tonquinia 131
tuancus 133
ultima 132
umbratilis 132
utara 133
versurae 131
wilkinsoni 132
yaoshanensis 131
yunalis 131
Tupaiidae **131**
Tupaiinae **131**
Tupaioidea **131**
tupaioides, Rhinosciurus 438
tural, Apodemus 573
turanicus, Dipus 493
Hemiechinus 78
turba, Crocidura 96, **97**
turbidus, Rattus 660
turcicus, Nannospalax 754
turcmenica, Capra 405
turcmenicus, Jaculus 493, **494**, 494
turcomanicus, Eptesicus 203
Myotis 210
turcomanus, Lepus 821
Pygeretmus 490
Spermophilopsis 444
turfanen, Meriones 556
turfanicus, Hemiechinus 78
turkestanicus, Hemiechinus 78
Rattus 649, 656, **661**, 662
turkmeni, Allactaga 488, 489
turkmenica, Vulpes 286
turnbulli, Hylopetes 460
turneri, Aethomys 568
Atherurus 773
Lutreolina 18
Otomys 681
Sorex 119
Turocapra 405
turovi, Arvicola 506
Chionomys 507
Mustela 323
Pteromys 465
Stylodipus 495
turowi, Moschus 384
turpis, Artibeus 189
Hipposideros **175**
Tursio 357
panope 351, 354
tursio, Lagenorhynchus 354
Tursiops **357**
aduncus 357
gephyreus 357
gillii 357
nesarnack 357
nuuanu 357
truncatus **357**, 357
turstigi, Galerella 303
turuchanensis, Canis 281
Ochotona 809
Sorex 119
Turus 405
tusimaensis, Apodemus 573
tutelata, Ochotona 811
tuvinicus, Alticola **504**
Cricetulus 537
Phodopus 539
tuza, Geomys 470
Mus 470
tweedii, Ichthyomys **705**
tyleri, Oreotragus 399
tyleriana, Marmosa **19**
Tylomys 726, **751**
bogotensis 751
bullaris **751**
fulviventer **751**, 751
gymnurus 751
microdon 751
mirae **751**, 751
nudicaudus **751**, 751
panamensis **751**, 751
tumbalensis **751**
villai 751
watsoni **751**, 751
Tylonycteris **228**
aurex 228
bhaktii 228
fulvidus 228
malayana 228
meyeri 228
pachypus **228**
robustula **228**
Tylonyx 509
tylopus, Glischropus **204**
Vesperugo 204
tylorhinus, Pappogeomys **473**
Tylostoma 179
tymensis, Alces 389
Tympanoctomys 788, **789**
barrerae **789**
Typhlomys **684**, 684
chapensis **684**
cinereus **684**, 684
jindongensis 684
typhlops, Notoryctes **43**, 43
Psammoryctes 43
Typhloryctes 771
typhlus, Spalax 755
typica, Didelphis 17
Felis 290
Malacothrix **544**
Martes 320
Micronycteris 179
Rhinolophus 164
typicus, Alopex 279
Canis 280
Cervus 386
Cheirogaleus 243
Chlamyphorus 64
Cryptoprocta 340
Ctenodactylus 761
Cynictis 302
Dendromus 543
Erinaceus 77
Graphiurus 765
Lepus 818, 821
Loxodonta 367
Lycaon 283
Meles 314
Mellivora 315
Mustela 323
Neacomys 709
Otomys 544, 680
Pedetes 759
Petromus **774**, 774
Phacochoerus 377
Propithecus 247
Proteles 309
Reithrodon 740
Rhabdomys 662
Rhinolophus 166
Sciurus 443
Suricata 308
Tachyglossus 13
Typomys 596
typus, Cheirogaleus 243
Dactylomys 790
Dendromus 541, 543
Eligmodontia 700, **701**, 701, 703
Eptesicus 203
Hippopotamus 381
Macroscelides 830
Microcavia 780
Myotis 213
Otomys 606, 680, **682**, 682
Paradoxurus 344
Pipistrellus 223
Plecotus 224
Rhinolophus 166
tyrannus, Rattus 651, 652
Tyrrhenoglis 767
tytleri, Lepus 819
Mus 626
Pteropus 149
Suncus 103
tytlerii, Paguma 343
Paradoxurus 343
tytonis, Chlorotalpa **75**
Gerbillurus **548**, 548
tzaneenensis, Aethomys 567
Graphiurus 764
Tatera 561

ualabatus, Kangurus 57
Wallabia 57
ualanus, Pteropus 148
ubangensis, Potamochoerus 378
ubangiensis, Kobus 414

ubanguiensis, Suncus 102
ubecus, Maxomys 614
ubericola, Lagothrix 258
ubericolor, Callosciurus 423
Sylvilagus 825
ubsanensis, Dipus 493
ucayalae, Panthera 298
ucayalensis, Scolomys **746**
ucayaliensis, Marmosops 19
ucayalii, Cacajao 261
uchidae, Microtus 527
uchidai, Nesoscaptor **126**, 126
uelensis, Cercopithecus 264
uellensis, Colobus 270
Gorilla 276
Tragelaphus 404
uenoi, Plecotus 224
ufipae, Sigmoceros 395
ugabensis, Raphicerus 400
Sylvicapra 412
ugalae, Sigmoceros 395
ugallae, Tragelaphus 404
ugandae, Damaliscus 394
Eptesicus 203
Felis 291
Galerella 303
Kobus 414
Lophocebus 266
Ourebia 399
Paraxerus 434
Redunca 414
Uranomys 671
ugyunak, Sorex 113, 116, 119, **121**, 121
uinta, Lynx 294
Ochotona 811
Thomomys 475
uintaensis, Clethrionomys 508
uintensis, Dipodomys 479
Ujhelyiana 753
ukrainicus, Microtus 530
Sciurus 443
ulae, Myomys 631
ulanagae, Sigmoceros 395
uliginosus, Microtus 521
uligocola, Microtus 527
ulloae, Otaria 328
ulpius, Chionomys 507
ulthiensis, Pteropus 148
ultima, Crocidura 90, **97**, 97
Crocuta 308
Trachypithecus 273
Tupaia 132
ultimus, Euneomys 702
Sorex 121
ultra, Pseudomys 647
ululans, Leopoldamys 603
ululata, Alouatta 254
umbrata, Cavia 779
umbratilis, Tupaia 132
umbratus, Acomys 566
Mormopterus 238
Mus 625
Platyrrhinus **191**, 191
Saguinus 254
Spermophilus 446
Steatomys 546
umbriceps, Rhizomys 685
umbridorsum, Maxomys 614
umbrina, Crocidura 84
umbrinus, Peromyscus 732
Procolobus 272
Soriculus 122
Tamias 455, **456**, 456
Thomomys 473, **474**, **475**, 475, 476
Tragulus 383
umbristriata, Monodelphis 22
umbriventer, Rattus 650
Saccostomus 541
umbrosa, Crocidura 88
Dobsonia 140
Phascolosorex 34
Tatera 561
umbrosus, Chaetodipus 484
Macaca 266
Menetes 433
Microtus 526, **531**
Otolemur 250
Sorex 114
Tamias 455
umpquensis, Canis 280
unalascensis, Dicrostonyx 510, **511**
Microtus 527
unau, Choloepus 64
Unaues 64
Unaus 64
Uncia **299**, 299
irbis 299
schneideri 299
uncia **299**
uncioides 299
uncia, Felis 299
Uncia **299**
uncioides, Uncia 299
unctuosus, Kobus 414
undata, Prionailurus 295
undatus, Cynocephalus 135
Stenoderma 192
underwoodi, Dasyprocta 782
Eumops **234**
Glaucomys 460
Heteromys 481
Hylonycteris **185**, 185
Orthogeomys **471**
Reithrodontomys 742
Rheomys **743**, 743
Sciurus 443
undosus, Microtus 525
unduaviensis, Gracilinanus 17
undulata, Viverra 348
undulatus, Heliosciurus **429**
Helogale 304
Spermophilus 448, **450**
ungae, Spermophilus 447
ungava, Alopex 279
Clethrionomys 508
Phenacomys **533**, 533
ungavensis, Canis 281
Tamiasciurus 457
Ursus 339
ungonicus, Sigmoceros 395
ungoniensis, Sigmoceros 395
unguiculatus, Meriones **558**
Microtus 521
Sorex **121**
unguifer, Macropus 55
unguifera, Onychogalea **55**
ungula, Rhinolophus 166
ungulatus, Dicrostonyx 511
ungulosus, Naemorhedus 407
ungurensis, Microtus 524
unicinctus, Cabassous **65**
Dasypus 64
unicolor, Antechinus 30
Bradypus 63
Cebus 259
Cercartetus 58
Cervus 386, **387**
Crocidura 94
Cyclopes 67
Echimys **792**, 792
Epomophorus 141
Herpailurus 291
Mops 236
Noctilio 176
Nyctophilus 218
Pipistrellus 222
Rousettus 152
Salanoia 300
Stenomys 665
Suncus 103
Unicomys 663
unicornis, Rhinoceros **372**, 372
uniformis, Herpestes 306
unihastatus, Rhinolophus 166
unistriata, Monodelphis **22**
unisulcatus, Otomys 680, 681, **682**, 682
Thomomys **475**
unitus, Hipposideros 173
univittatus, Hybomys **596**, 596
Mus 596
unko, Hylobates 275
unyori, Oenomys 637
upingtoni, Galerella 303
uralensis, Alces 389
Apodemus 571, 572, 573, **574**, 574
Arvicola 506
Clethrionomys 509
Microtus 524, 527
Ochotona 809
Sciurus 443
Sorex 112
Spalax 754
Talpa 129
uraniae, Otaria 328
Uranodon 361
Uranomys 564, 565, 605, **670**, 670
acomyoides 671
foxi 671
oweni 671
ruddi 670, **671**
shortridgei 671
tenebrosus 671
ugandae 671
woodi 671
urartensis, Calomyscus **536**, 536
urbanus, Mus 626
urianchaicus, Meriones 557
Uribos 401
urichi, Akodon 688, **693**, 694
Proechimys 796, **798**
urinatrix, Atilax 301
urmiana, Ovis 408
uroceras, Dasypus 66
Urocitellus **444**, 446, 448, 450
Urocricetus 537
Urocyon **285**, 285, 286
borealis 285
californicus 285
catalinae 285
cinereoargenteus **285**, 285
clementae 285
colimensis 285
costaricensis 285
dickeyi 285
floridanus 285
fraterculus 285
furvus 285
guatemalae 285
inyoensis 285
littoralis **285**
madrensis 285
nigrirostris 285
ocythous 285
orinomus 285
parvidens 285
peninsularis 285
pensylvanicus 285
santacruzae 285
santarosae 285
scottii 285
sequoiensis 285
texensis 285
townsendi 285
virginianus 285
Uroderma **193**
bilobatum **193**, 193
convexum 193
davisi 193
magnirostrum **193**
molaris 193
personatum 193
thomasi 193
trinitatum 193
Urogale **133**
cylindrura 133
everetti **133**
Uroleptes 68
Urolynchus 288, 289
Uromanis 415
Uromys 585, 615, 663, **671**,

671, 672, 673
anak **671**, 672
aruensis 671
caudimaculatus **671**, 671, 672
ductor 671
exilis 671
hadrourus **671**, 672
imperator **672**
lamington 671
macropus 671
multiplicatus 671
neobritanicus 671, **672**, 672
nero 671
papuanus 671
porculus **672**
prolixus 671
rex **672**
rothschildi 671
rufescens 615
sapientis 663
scaphax 671
sherrini 671
siebersi 671
validus 671
waigeuensis 671
Uropsilinae **130**
Uropsilus **130**
andersoni **130**, 130
atronates 130
gracilis **130**, 130
investigator **130**, 130
nivatus 130
soricipes **130**, 130
Urosciurus 439, **441**, **443**
urostictus, Pseudalopex 285
Urotragus 406
Urotrichinae 124
Urotrichini 124, 126, 129
Urotrichus **129**
adversus 129, 130
centralis 129, 130
gibbsii 126
hondoensis 129
hondonis 130
minutus 129, 130
pilirostris **129**
talpoides **129**, 129
Ursidae **336**, 337
ursina, Alouatta 255
Didelphis 35, 45
Phalangista 46
Phoca 326, 327
Sarcophilus 35
Ursinae **337**
ursinii, Miniopterus 231
Ursinus 35
ursinus, Ailurops **46**
Alouatta 255
Arctocephalus 326, 327
Bradypus 337
Callorhinus **327**
Colobus 270
Dasyurus 32
Dendrolagus 50, **51**
Eptesicus 201
Hypsiprymnus 50
Macaca 266
Melursus **337**, 337
Molossus 235
Papio 269
Paruromys **639**, 655
Promops 239, 240
Pteropus 150
Trachypithecus **274**
Vombatus **45**
Ursitaxus 315
ursula, Pipistrellus 221
Saguinus 253, 254
ursulus, Ellobius 513
Saguinus 254
Ursus 314, 337, **338**
absarokus 338
alascensis 338
albus 338
alexandrae 338
alpinus 338
altifrontalis 338
amblyceps 338
americanus **338**, 338, 340
andersoni 338
annulatus 338
apache 338
arboreus 340
arctos **338**, 338, 339
argenteus 338
arizonae 338
atnarko 338
aureus 338
badius 338
baikalensis 338
bairdi 338
beringiana 338
bisonophagus 338
bosniensis 338
brunneus 338
cadaverinus 338
californicus 338
californiensis 338
canadensis 338
candescens 338
carlottae 338
caucasicus 338
caurinus 338
cavifrons 338
chelan 338
chelidonias 338
cinereus 338
cinnamomum 338
clarki 340
collaris 338
colusus 338
crassodon 338
crassus 338
cressonus 338
crowtheri 338
dalli 338
dusorgus 339
eltonclarki 339
emmonsii 338
eogroenlandicus 339
eremicus 338
ereunetes 339
eulophus 339
euryrhinus 339
eversmanni 339
eximius 339
falciger 339
ferox 339
floridanus 338
formicarius 339
formosanus 340
fuscus 339
gedrosianus 339, 340
glacilis 338
grandis 339
griseus 339
groenlandicus 339
gyas 339
hamiltoni 338
henshawi 339
holzworthi 339
hoots 339
horriaeus 339
horribilis 338, 339
hunteri 338
hylodromus 338, 339
idahoensis 339
imperator 339
impiger 339
innuitus 339
inopinatus 339
insularis 339
internationalis 339
isabellinus 338, 339
japonicus 340
jenaensis 339
jeniseensis 339
kadiaki 339
kenaiensis 338, 339
kennerleyi 339
kermodei 338
kidderi 339
klamathensis 339
kluane 339
kodiaki 339
kolymensis 339
kwakiutl 339
labradorensis 339
labradorius 325
lagomyiarius 339
laniger 340
lasiotus 339
lasistanicus 339
latifrons 339
leuconyx 339, 340
lotor 335
luteolus 338
macfarlani 339
machetes 338, 339
macneilli 340
macrodon 339
magister 339
major 339
malayanus 337
mandchuricus 339
marinus 339
maritimus 338, **339**
marsicanus 339
melanarctos 339
melanoleucus 336
melli 340
mendocinensis 339
meridionalis 339
merriamii 339
middendorffi 339
minor 339
mirabilis 339
mirus 339
mupinensis 340
myrmephagus 339
navaho 339
neglectus 339
nelsoni 338, 339
niger 339
normalis 339
nortoni 339
norvegicus 339
nuchek 339
ophrus 339
orgiloides 339
orgilos 339
oribasus 339
ornatus 337
pallasi 339
pamirensis 339
pellyensis 339
perniger 338
persicus 339
perturbans 339
pervagor 339
phaeonyx 339
piscator 339
planiceps 339
polaris 339
polonicus 339
pruinosus 338, 339
pugnax 338
pulchellus 339
pyrenaicus 339
randi 338
rexi 340
richardsoni 339
rogersi 339
rossicus 339
rufus 339
rungiusi 339
russelli 339
sagittalis 339
scandinavicus 339
schmitzi 339
selkirki 339
shanorum 339
sheldoni 339
shirasi 339
shoshone 339
sibiricus 339
sitkeenensis 339
sitkensis 339
smirnovi 339
sornborgeri 338
spitzbergensis 339
stenorostris 339
syriacus 339

tahltanicus 339
taxus 325
texensis 339
thibetanus 338, **339**
toklat 339
torquatus 340
townsendi 339
tularensis 339
tundrensis 339
ungavensis 339
ursus 339
ussuricus 340
utahensis 339
vancouveri 338
warburtoni 339
washake 339
wulsini 340
yesoensis 339
ursus, Ursus 339
urucuma, Dasyprocta 782
urucumus, Sciurus 442
uruguayensis, Hydrochaeris 781
Urus 401
urus, Bison 400
Bos 401
Urva 306
urva, Herpestes 305, **306**, 306
usambarae, Crocidura **97**
Grammomys 594
usambarensis, Sylvisorex 104
uschtaganicus, Meriones 557
Ussa 385
ussuricus, Cervus 386
Micromys 620
Sus 379
Ursus 340
ussuriensis, Erinaceus 77
Murina 229, **230**
Myotis 210
Nyctereutes 283
Sorex 117
Vespertilio 228
Vulpes 287
usto-fuscus, Callicebus 258
ustulatus, Pappogeomys 472
ustus, Bradypus 63
Odocoileus 391
Saimiri **260**
usuni, Dipus 493
utah, Spermophilus 450
utahensis, Dipodomys 479
Ochotona 811
Onychomys 719
Peromyscus 729
Tamias 453
Ursus 339
Zapus 499
utahicki, Chiropotes 261
utakwa, Rattus 658
utara, Tupaia 133
uthensis, Tamias 455
uthmoelleri, Dendromus 543
utiaritensis, Oligoryzomys 717
utibilis, Ctenomys 784
utsuryoensis, Crocidura 96
utsuryonis, Mus 626
uvirensis, Sylvicapra 412
uwendensis, Kobus 414
Sigmoceros 395
uxantisi, Crocidura 96
uzungwensis, Otomys 682

vaalensis, Rhabdomys 662
vaccarum, Abrocoma 789
Phyllotis 739
vacillator, Proechimys 795
vae, Felovia **761**, 761
vafra, Vulpes 287
vaga, Lontra 311
Sicista 498
vagans, Atelerix 77
Paguma 343
vaginalis, Muntiacus 389
vagneri, Myoxus 770
vagrans, Sorex 111, 112, 118, **122**, 122
vaillantii, Ictonyx 319
valdiviae, Sciurus 441
valdivianus, Geoxus **702**
Oxymycterus 702
valenciae, Sylvilagus 826
valens, Glossophaga 184
Sundamys 666
valentinae, Rangifer 392
vali, Ctenodactylus **761**, 761
valida, Tatera 561, **562**
validum, Artibeus 189
validus, Ctenomys 785, **787**
Dendrohyrax **373**, 373
Macaca 266
Sundamys 666
Uromys 671
vallensis, Nectomys 709
vallesius, Rattus 661
vallicola, Melanomys 708
Microtus 519
Oenomys 637
Pronolagus 824
Sylvilagus 824
Tamias 453
vallinus, Gerbillurus 547, **548**
Gerbillus 547
valschensis, Cryptomys 772
Mus 624
valverdei, Eliomys 767
Vampyressa 190, **193**, 193, 194
bidens **193**
brocki **193**
melissa **193**
minuta 194
nattereri 194
nymphaea **194**
pusilla **194**
thyone 194
venilla 194
Vampyriscus 193
Vampyrodes **194**
caraccioli **194**
major 194
ornatus 194
Vampyrops 190
caracciolae 194
lineatus 191
Vampyrum **181**
nelsoni 181
spectrum **181**
Vampyrus, auritus 177
bidens 180
cirrhosus 181
vampyrus, Pteropus **151**
vanakeni, Sundasciurus 451
vancouverensis, Gulo 318
Lontra 311
Marmota **433**
Martes 319
Procyon 335
Puma 297
Sorex 122
Tamiasciurus 457
vancouveri, Ursus 338
vandami, Cryptomys 772
Graphiurus 764
Vandeleuria **672**
badius 672
domecolus 672
dumeticola 672, 673
marica 672
modesta 672
nilagirica 672
nolthenii **672**, 672
oleracea **672**, 672, 673
povensis 672
rubida 672
sibylla 672
spadicea 672
wroughtoni 672
vandenberghei, Deomys 544
vandeuseni, Rhinolophus 167
Stenomys **665**, 665
vandycki, Pseudomys 645
vanheurni, Callosciurus 423
Dorcopsulus **52**
Rattus 656
vanikorensis, Pteropus 148
vanrosseni, Thomomys **474**
vansire, Atilax 301
vansoni, Eptesicus 203
Vansonia 219
vanzolinii, Pithecia 261
Saimiri **260**
vardonii, Kobus **414**
Varecia **245**
editorum 245
erythromela 245
ruber 245
subcinctus 245
vari 245
variegata **245**
varius 245
varennei, Chimarrogale 107
varentsowi, Ovis 409
vari, Varecia 245
varia, Helogale 304
Neotoma **712**
Tatera 560
variabilis, Arvicola 506
Bassariscus 334
Canis 281
Lepus 821
Microtus 519
Mus 626
Petromyscus 683
Procolobus 272
Rattus 658
Sciurus 441, **442**
variagatus, Vulpes 287
variani, Hippotragus 412, 413
Madoqua 398
varians, Mephitis 317
Procavia 374
variegata, Dasyprocta 782
Ictonyx 319
Panthera 298
Papio 269
Phoca 332
Spilocuscus 47
Varecia **245**
variegatoides, Canis 281
Sciurus **443**
Vulpes 286
variegatus, Arvicanthis 578
Ateles 257
Bradypus **63**
Canis 280
Castor 467
Cebus 259
Chalinolobus **200**
Chironectes 16
Cynocephalus **135**
Hemicentetes 73
Hylobates 275
Indri 247
Lemur 245
Spermophilus **450**
Thryonomys 775
varietas, Martes 320
varilla, Suncus **103**
varillus, Bandicota 579
varina, Mustela 323
variscicus, Clethrionomys 509
varius, Akodon 689, 691, 693, **694**, 694
Apodemus 573
Bandicota 579
Capreolus 390
Castor 467
Cervus 386
Cricetus 538
Cynocephalus 135
Dama 387
Hylobates 275
Lasiurus 206
Lemur **245**
Mus 626
Myosorex **100**
Nyctimene 144
Ondatra 532

Panthera 298
Pappogeomys 473
Procyon 335
Rattus 658
Sciurus 440, 443
Sorex 99
Varecia 245
varronis, Lepus 821
varus, Thomomys 474
vasconiae, Clethrionomys 509
vassali, Callosciurus 421
Lepus 820
vassei, Cephalophus 411
vastus, Jaculus 493
vasvárii, Nannospalax 754
vates, Chiruromys **585**
vaughani, Kobus 414
vaughanjonesi, Cricetomys 541
vegae, Berardius 361
vegetus, Oligoryzomys **718**, 718
velatus, Histiotus **205**
Plecotus 205, 224
veldkampi, Epomophorus 143
Nanonycteris **143**
velessiensis, Talpa 129
velifer, Myotis 210, 214, **216**
velifera, Balaenoptera 350
velilla, Rhogeessa 226
vellerosa, Felis 289, 291
vellerosus, Ateles 257
Cebus 259
Chaetophractus **65**, 65
Colobus 270
Microtus 523
Pan 277
Paradoxurus 344
Rattus 655
velligera, Chinchilla 777
velox, Acinonyx 288
Cynogale 73
Molossus 235
Pipistrellus 223
Potamogale **73**
Stenella 356
Vulpes **286**, 287
velozi, Plagiodontia 804
velutina, Crocidura 87
Mystacina 231
Otaria 328
velutinus, Nyctalus 217
Oryzomys 721
Rattus 655
Thylamys **23**
venatica, Acinonyx 288
Galerella 303
Lycaon 283
venaticus, Acinonyx 288
Conepatus 316
Speothos **285**
venator, Acinonyx 288
venatoris, Chalinolobus 199
venatorius, Odocoileus 391
venester, Sigmodon 747
venetus, Callosciurus 423
venezuelae, Caluromys 15
Holochilus 705
Neusticomys **714**
Rhipidomys **744**, **745**, 745
Tonatia 180
venezuelensis, Akodon 693
venezuelica, Agouti 783
venilla, Vampyressa 194
venningi, Petaurista 463
ventorum, Ochotona 811
Tamiasciurus 457
ventralis, Sorex 118, 120, **122**, 122
Tadarida **241**
ventriosus, Zygodontomys 752
venusta, Glironia **16**, 16
Neophascogale 33
Neotoma 710
Phascogale 33
Thylamys 23
venustula, Crocuta 308
Petrogale 55
venustulus, Microsciurus 433
Nyctomys 715
venustus, Calomys 697, 698
Dipodomys 478, **480**
Gerbillus 550
Lemniscomys 602
Myotis 216
Perognathus 485
Rhipidomys 744, **745**
Sciurus 439
Tamias 456
Thamnomys 669, **670**, 670
vera, Gazella 396
veraecrucis, Liomys 482
Odocoileus 391
Panthera 298
Pipistrellus 224
Sorex 120
Sylvilagus 825
veraecundus, Philetor 219
veraepacis, Sorex 117, **122**
verax, Thylamys 23
verbeeki, Callosciurus 423
Maxomys 614
verecundus, Mus 665
Rattus 660
Stenomys **665**, 665
verheyeni, Mastomys **611**
verhoeveni, Papagomys 637
vermicula, Oryctolagus 822
vermiculus, Chrysochloris 75
vernalus, Rattus 661
Vernaya **673**, 673
foramena 673
fulva **673**, 673
vernayi, Aethomys 568
Callosciurus 421
Dendromus 541, **543**, 543, 544
Felis 291
Lepus 816
Mellivora 315
Panthera 297
Pipistrellus 223
Sylvicapra 412
vernicularis, Oryctolagus 822
vernula, Cricetulus 538
veroxii, Myomys 631
verreauxi, Propithecus **247**
verreauxii, Mus 630
Myomys 610, 630, **631**
verrilli, Molossus 235
verrucata, Carollia 186
verrucosus, Cabassous 65
Nyctalus 216, 217
Phyllostomus 180
Sus 378, **379**
versabilis, Megaptera 350
verschureni, Malacomys **608**, 608
versicolor, Cebus 259
Naemorhedus 407
Sciurus 441
Tragulus 383
versteagi, Pseudocheirus 59
versurae, Tupaia 131
versustos, Lemniscomys 602
versutus, Cebus 259
verus, Colobus 271
Pan 277
Procolobus **272**, 272
verutus, Cervus 387
verzosianus, Cervus 386
vesanus, Clethrionomys 509
vescus, Thomomys 474
vespa, Rhinolophus 166
Vespadelus 200, 201, 203, 204, 219
Vesperimus 728
Vespertilio **228**
albogularis 228
anderssoni 228
aurijunctus 228
auritus 224
barbastellus 198
borealis 206
brachypterus 219
calcaratus 156
caninus 158
cephalotes 143
discolor 228
ferrum-equinum 163
fuscus 200
harpia 228
hastatus 180
helvus 140
hispidus 161
humeralis 217
krascheninnikovi 228
leisleri 216
leporinus 176
lepturus 159
luteus 228
maximiliani 156
michnoi 228
microphyllus 155
molossus 235
motoyoshii 228
murinus **228**, 228
myotis 207
namiyei 228
naso 159
niger 146
noctivagans 206
orientalis 228
pachypus 228
pallidus 198
perspicillata 186
pictus 196
pipistrellus 219
schreibersii 230
siculus 228
soricinus 184
spasma 163
spectrum 181
speoris 170
sphinx 138
suillus 228
superans **228**
tuberculatus 199
ussuriensis 228
Vespertilionidae **196**
Vespertilioninae **198**
vespertinus, Myotis 215
Vesperugo 207
tylopus 204
Vesperus 204, 207
vestita, Macaca 267
vestitus, Chelemys 699
Hylobates 275
Phalanger 46, **47**, 47
Rhizomys 685
Sphiggurus **777**
Tamiops 457
Thomasomys **751**
vetensis, Cryptomys 772
veter, Macaca 268
veterum, Dorcopsis 52
vetula, Helogale 304
Vetulus 266
vetulus, Cerdocyon 282
Dusicyon 283
Hodomys 704
Presbytis 274
Pseudalopex **285**
Pteropus **152**
Trachypithecus **274**, 274
vexillaris, Psammomys **559**
vexillarius, Allactaga 489
Paraxerus **435**, 435
vexillifer, Lipotes **360**, 360
viaria, Crocidura 85, **97**, 98
viator, Acomys 565
Cricetomys 541
Praomys 642, 643
vicarius, Hipposideros 172
vicencianus, Oecomys 717
vicerex, Rattus 661, 662
vicina, Alticola 503
Mus 625
Neotoma 712
Tatera 562
vicinior, Miniopterus 231

Osgoodomys 726
vicinus, Eozapus 498
Paradoxurus 344
Procyon 335
viclana, Rattus 661
victor, Phascolarctos 45
Sundamys 666
victoria-njansae, Heterohyrax 373
victoriae, Capra 406
Funisciurus 428
Genetta **347**
Lepus 818, 820, **822**, 822
Pseudocheirus 60
victorina, Helogale 304
victus, Oligoryzomys **719**, 719
Vicugna **382**
cristina 382
elfridae 382
frontosa 382
gracilis 382
mensalis 382
minuta 382
pristina 382
provicugna 382
vicugna **382**, 382
vicugna, Camelus 382
Vicugna **382**, 382
viculorum, Mus 626
vidalinus, Cervus 386
vidianus, Naemorhedus 407
vidua, Callicebus 259
viegasi, Hipposideros 172
vientianensis, Niviventer 634
vigens, Leopardus 292
vigilans, Trachypithecus 273
vigilax, Lepus 815
vigilis, Canis 280
Marmota 431
Papio 269
Spermophilus **445**, **448**
vignaudii, Mus 626
vignei, Ovis 408, **409**, 409
vigoratus, Rattus 652
vilaliya, Elephas 367
vilhelmi, Dendrohyrax 373
villae, Sigmodon 747
villai, Thomomys 474
Tylomys 751
villemerianus, Cervus 386
villersi, Genetta 346
villicauda, Proechimys 795
villiersi, Miniopterus 231
Pseudogenetta 346
villosa, Alouatta 255
Alticola 503
Brachytarsomys 677
Crocidura 85
Glossophaga 184
Isothrix 793
Macaca 267
Nycteris 161
Panthera 298
Parahyaena 309
Pelea 413
Rhynchonycteris 159
villosissima, Zaglossus 13
villosissimus, Lasiurus 206, 207
Rattus 649, 651, 659, **662**, 662
villosum, Chiroderma 189, **190**
villosus, Alouatta 254
Aotus 256
Belomys 459
Chaetophractus **65**
Chrysospalax **76**
Dasypus 65
Isothrix 793
Oryzomys 725
Rattus 650
Sphiggurus **777**
Tapirus 371
vinaceus, Geomys 469
Zelotomys 674
vincenti, Paraxerus **435**
vinealis, Akodon 692
vinnulus, Ammospermophilus 420
vinocastaneus, Callosciurus 423
vinogradovi, Allactaga 488, **490**, 490
Alticola 503
Clethrionomys 509
Dicrostonyx **511**, 511
Lasiopodomys 516
Meriones **558**
Microtus 524
Mus 626
Myopus 531
Pygeretmus 490
vinosus, Macropus 55
vinsoni, Nycteris 162
violeta, Bradypus 63
viperinus, Ctenomys 785
vir, Sorex 119
virescens, Arvicanthis 577
virgata, Crocidura 87
Hyaena 308
Panthera 299
virgatus, Otocyon 284
virgicollis, Tragulus 382
virgineus, Thomomys 474
virginiae, Sminthopsis **37**, 37
virginiana, Dama 391
Didelphis **17**, 22
virginianus, Canis 285
Dama 391
Glaucomys 460
Lepus 814
Myotis 212
Odocoileus **391**, 391
Philander 22
Plecotus 225
Reithrodontomys 741
Scalopus 127
Sigmodon 747
Urocyon 285
virginis, Perognathus 485
virgo, Callosciurus 423
Diclidurus 157
Rhinolophus **169**
virgulosus, Sicista 498
virgultus, Odocoileus 391
viridescens, Suncus 103
viridis, Chlorocebus 265
Dobsonia 139, **140**
Scotophilus **227**, **228**
virqulti, Sylvilagus 825
virtus, Sundamys 666
visagiei, Chrysochloris **75**
Viscaccia 777, 778
viscacia, Lagidium **778**
Vison 321, 325
vison, Mustela 321, **324**, 324, 325
vispistrellus, Pipistrellus 221
visserae, Chrysochloris 75
vistulanus, Castor 467
visurgensis, Cervus 386
vitalina, Caluromys 15
vitiensis, Pteropus 150
Rattus 652
vitiis, Macaca 266
vittata, Conepatus 316
Galictis **318**
Hipposideros 172
Mephitis 317
Nasua 335
Ratufa 437
Viverra 318
vittatula, Ratufa 437
vittatus, Callosciurus 423
Galidictis 300
Mungotictis 300
Noctilio 176
Platyrrhinus **191**
Rhabdomys 662
Sus 379
vitticollis, Herpestes **306**
vitulina, Phoca 331, **332**, 332
vivatus, Microsciurus 433
vivax, Gerbillus **555**
Otomys 681
Sciurus 440
Vivera, tapoatafa 33
Viverra 317, 340, 342, 344, 345, **347**, 347
ashtoni 348
binotata 342
capensis 315
caudivolvula 333
civetta 344
civettina 347, **348**, 348
civettoides 348
cristata 309
fasciata 300
filchneri 348
fossa 341
fossana 341
genetta 345
gymnura 79
hardwicki 342
hermaphrodita 344
ichneumon 304
lankavensis 348
megaspila 347, **348**, 348
melanura 348
mephitis 316
mungo 307
nasua 334
orientalis 348
picta 348
piscator 348
pruinosa 348
ratel 315
sigillata 348
surdaster 348
suricatta 307
tangalunga **348**
undulata 348
vittata 318
zibetha 347, **348**
Viverriceps 295
viverriceps, Prionailurus 296
Viverricula **348**
baptistae 348
bengalensis 348
deserti 348
hanensis 348
indica **348**
malaccensis 348
mayori 348
pallida 348
rasse 348
schlegelii 348
taivana 348
thai 348
wellsi 348
Viverridae 299, 300, **340**, 340
viverrina, Didelphis 32
Pseudocheirus 60
Suricata 308
Viverrinae **344**
viverrinus, Canis 283
Dasyurus **32**
Nyctereutes 283
Prionailurus **296**
vivesi, Myotis **216**
vizcaccia, Nyctimene 144, **145**
vizier, Ochotona 812
voangshire, Atilax 301
vocator, Jaculus 493
vociferans, Aotus **256**
Leopoldamys 603
Spermophilus 449
voeltzkowi, Chlorocebus 265
Pteropus **152**
vogelii, Trichechus 366
voglii, Lama 382
vohlynensis, Apodemus 574
voi, Aethomys 567
Crocidura **97**
volaensis, Microtus 530
volans, Cynocephalus **135**
Didelphis 58
Glaucomys **460**
Lemur 135
Mus 460

Myotis **216**
Petauroides **58**
Pteromys **464**
Sciurus 464
Volemys **534,** 535
clarkei **534,** 535
kikuchii **535,** 535
millicens **535,** 535
musseri **535,** 535
volgensis, Apodemus 570
Arvicola 506
Clethrionomys 509
Myotis 210
volhynensis, Spermophilus 449
volkmanni, Civettictis 345
Procavia 374
volnuchini, Sorex 113, 117, 118, **122,** 122
volubilis, Chaetomys 790
Volucella 58
volucella, Glaucomys 460
Petauroides 59
volzi, Hylobates 276
Vombatidae **45,** 45
Vombatus **45**
angasii 46
assimilis 46
bassii 46
fossor 46
fuscus 46
hirsutus 46
mitchelli 46
niger 46
platyrhinus 46
setosus 46
tasmaniensis 46
ursinus **45**
vombatus 46
wombat 46
vombatus, Vombatus 46
vomerina, Phocoena 358
voratus, Brotomys **799,** 799
vorax, Crocidura 94
vordermanni, Petinomys **464,** 464
Pipistrellus 222
Vormela **325**
alpherakii 325
euxina 325
koshewnikowi 325
negans 325
ornata 325
peregusna **325,** 325
sarmatica 325
syriaca 325
tedschenika 325
vorontsovi, Microlophiomys 563
vosmaeri, Crocidura 85
vosseleri, Dendrohyrax 373
vryburgensis, Crocidura 83
Cryptomys 772
vulcani, Aepeomys 688
Crocidura 84
Dipodomys 479
Lagidium 778
Liomys 482
Mus 623, **629,** 629
Neotoma 712
Orthogeomys 471
Rattus 652
Sigmodon 747
vulcanicus, Graphiurus 764
Otomys 681
vulcanius, Heliosciurus 429
Pteropus 147
Reithrodontomys 742
Thomomys 476
vulcanorum, Hylomyscus 599
Paraxerus 434
Sylvisorex **105,** 105
vulcanus, Lariscus 430
vulgaris, Callithrix 252
Canis 280
Capreolus 390
Cervus 386
Cricetus 538
Dama 387
Equus 369
Felis 291
Genetta 345
Gulo 318
Hipposideros 173
Hyaena 308
Lutra 312
Lynx 293
Martes 320
Meles 314
Microtus 519
Monodon 358
Mustela 321, 323
Myoxus 770
Panthera 298
Plecotus 224
Procyon 335
Pteromys 465
Pteropus 149
Sciurus 439, **443,** 456
Sorex 112
Vulpes 287
vulgivagus, Mustela 324
vulpecula, Didelphis 48
Nasua 334
Phalanger 46
Trichosurus **48**
Vulpes 287
Vulpes 279, 280, **285,** 285, 286
abietorum 287
acaab 287
aegyptiacus 287
alascensis 287
alba 287
algeriensis 287
alopex 287
alpherakyi 287
alticola 287
anadyrensis 287
anatolica 287
anubis 287
arabica 287
arabicus 287
arizonensis 287
arsipus 287
atlantica 287
aurantioluteus 287
aurita 287
bangsi 287
barbarus 287
bengalensis **286**
beringiana 287
brucei 287
caama 286
caesia 286
cana **286**
cascadensis 287
caucasica 287
cerdo 287
chama **286**
chrysurus 286
cinera 287
communis 287
corsac **286**
crucigera 287
cyrenaica 286
daurica 287
deletrix 287
denhamii 287
devia 287
dolichocrania 287
dorsalis 287
eckloni 287
edwardsi 286
famelicus 286
fennecus 287
ferrilata **286**
flavescens 287
fulvipes 284
fulvus 287
griffithi 287
griffithii 287
harrimani 287
harterti 286
hebes 287
himalaicus 287
hodgsoni 286
hodgsonii 286
hoole 287
huli 287
hypomelas 287
ichnusae 287
indicus 286
indutus 287
jakutensis 287
japonica 287
kalmykorum 286
kamtschadensis 287
karagan 287
kenaiensis 287
kiyomasai 287
kokree 286
krimeamontana 287
kurdistanica 287
ladacensis 287
leucopus 287
lineatus 287
lineiventer 287
lutea 287
macrotis 287
macrourus 287
melanogaster 287
melanotus 287
meridionalis 287
montana 287
mutica 287
muticus 287
necator 287
neomexicana 287
nepalensis 287
nevadensis 287
nigra 286, 287
nigro-argenteus 287
nigrocaudatus 287
niloticus 287
ochroxantha 287
oertzeni 286
palaestina 287
pallida **286**
peculiosa 287
pennsylvanicus 287
persicus 287
pusilla 287
regalis 287
rubricosa 287
rueppellii **286**
rufescens 286
saarensis 287
sabaea 286
sabbar 286
schrencki 287
scorodumovi 286
silaceus 287
sitkaensis 287
somaliae 286
splendens 287
splendidissima 287
stepensis 287
tenuirostris 287
tobolica 287
tschiliensis 287
turkmenica 286
ussuriensis 287
vafra 287
variagatus 287
variegatoides 286
velox **286,** 287
vulgaris 287
vulpecula 287
vulpes 279, **287**
waddelli 287
xanthura 286
zaarensis 287
zarudnyi 286
zerda 286, **287**
zinseri 287
vulpes, Canis 285
Chrysocyon 282
Vulpes 279, **287**
vulpicolor, Niviventer 634
vulpina, Martes 319
Ochotona 808
Trichosurus 48

vulpinoides, Oryzomys 725
vulpinus, Holochilus 704
Lynx 293
Oxymycterus 706
Sciurus 442
vulturna, Ochotona 812
vulturnus, Chiruromys 584
Dendromus 543
Eptesicus **204**
Jaculus 493
Neotomys 713
vumbae, Heliosciurus 429
Pelomys 639
vumbaensis, Grammomys 594
vumbensis, Grammomys 594

wachei, Cervus 386
Gulo 318
waddelli, Vulpes 287
wagati, Prionailurus 295
wagneri, Acinonyx 288
Catagonus **379**
Chionomys 507
Dipodomys 477
Mus 626
Reithrodontomys 742
Sciurus 440
Sorex 116
wahema, Microtus 527
wahlbergi, Epomophorus **141**
Equus 369
wahwahensis, Thomomys 474
waigeuensis, Uromys 671
waitei, Leggadina 601
walambae, Aethomys 568
waldemarii, Suncus 103
waldroni, Procolobus 272
walgie, Canis 282
walie, Capra **406**
walkeri, Nyctophilus **218**
Sylvicapra 412
Wallabia 53, **57**
apicalis 57
bicolor **57**
ingrami 57
lessonii 57
mastersii 57
ualabatus 57
welsbyi 57
wallacei, Myoictis 32
Pteropus 153
Styloctenium **153**
wallawalla, Lepus 815
walleri, Gazella 397
Litocranius **397**
walli, Eptesicus 202
wallichi, Cervus 385, 386
wallichii, Pongo 277
wallowa, Thomomys 475
walteri, Barbastella 199
waltoni, Microtus 523
wamae, Mus 628
wambutti, Mus 622
wapiti, Cervus 386
warburtoni, Ursus 339
wardi, Apodemus 571, 573, **574**, 574
Bandicota 579
Blarinella **106**
Capra 406
Cervus 386
Eothenomys 513
Giraffa 383
Helarctos 337
Lepus 814
Lynx 293
Ochotona 812
Orycteropus 375
Ovibos 408
Plecotus 224
Redunca 414
Rhizomys 685
wardii, Sorex 112
waringensis, Callosciurus 423
Sundamys 666
warreni, Marmota 432
Neotoma 710
Proechimys 796, **798**
Suncus 103, 104
Sylvilagus 824
warringtoni, Lasiopodomys 515
warryato, Hemitragus 406
warthae, Capreolus 390
warwickii, Oncifelis 294
wasatchensis, Ochotona 811
Tamiasciurus **457**
Thomomys 475
washake, Ursus 339
washburni, Clethrionomys 509
washingtoni, Lepus 814
Mustela 322
Sorex 115
Spermophilus **450**
washoensis, Spermophilus 447
wasjuganensis, Clethrionomys 509
wassifi, Gerbillus 550
Hemiechinus 78
watasei, Crocidura 87
Myotis 210
Neomys 110
Petaurista 462
waterbergensis, Aethomys 568
Procavia 374
Pronolagus 823
Tatera 561
waterhousei, Marmosa 18
waterhousii, Heliosciurus 429
Macrotus **178**, 178
watersi, Gerbillus 552, 554, **555**
watkinsi, Scotophilus 227
watsoni, Artibeus 188
Callosciurus 423
Clethrionomys 509
Golunda 592
Tylomys **751**, 751
wattsi, Maxomys **614**
Pipistrellus 224
wavicus, Myoictis 32
wavrini, Oryzomys 739
Pseudoryzomys 739
wavula, Puma 297
webbi, Eliurus **678**, **678**
webbiensis, Panthera **297**
webensis, Heterohyrax 373
weberi, Crocidura 89
Myotis 210
Prosciurillus **436**
Sus 378
websteri, Chaerephon 233
weddelli, Saguinus 253
weddellii, Leptonychotes **330**
Otaria 330
weemsi, Ovis 408
weidholzi, Chlorocebus 265
Gazella 396
weigoldi, Sicista 497
weileri, Hylomyscus 598
weissmanni, Crocuta 308
wellsi, Berylmys 580
Callosciurus 421
Galea 780
Viverricula 348
welmanni, Tatera 561
welsbyi, Wallabia 57
welsianus, Echymipera 41
welwitschii, Myotis **216**
Procavia 374
wembaerensis, Alcelaphus 393
Gazella 397
werckleae, Anoura 183
werneri, Chlorocebus 265
Papio 269
wertheri, Orycteropus 375
westermanni, Conepatus 316
westrae, Microtus 519
westralis, Pipistrellus 224
wetmorei, Eptesicus 201
Megaerops **142**
Pudu 392
wettsteini, Microtus 518, 530
Sorex 112
wetzeli, Rhipidomys **744**, **745**
weylandi, Mallomys 608
Melomys 617
weynsi, Cephalophus **411**, 411
Chlorocebus 265
Colobus 270
Lophocebus 266
whartoni, Phascolosorex 34
wheeleri, Aselliscus 170
whitakeri, Crocidura 96, **97**
Lepus 816
whitchurchi, Jaculus 493
whiteheadi, Chrotomys **585**, 585
Exilisciurus **426**
Harpionycteris 142
Harpyionycteris **142**, 142
Kerivoula **198**
Maxomys 612, 613, **614**, 614, 653
Mustela 321
whitei, Acomys 565
Arctictis 342
Bettongia 49
Budorcas 404
Epomophorus 141
Erythrocebus 265
Pronolagus 823
whiteleyi, Lutra 312
whitelyi, Mazama 390
whitesidei, Cercopithecus 263
whitfieldi, Cephalophus 410
whitleyi, Myopterus **238**
whitmani, Thomomys 475
whitmeei, Pteropus **150**
whittalli, Capreolus 390
whytei, Chlorocebus 265
Cryptomys 772
Dendromus 543
Lepus 814, 819, 820, 822
wichmanni, Rattus 652
widdringtonii, Herpestes 305
wiedi, Centronycteris 156
wiedii, Anoura 183
Eptesicus 203
Leopardus **292**, 292
Wiedomys **751**
pyrrhorhinos **751**, 751
wiegmanni, Odocoileus 391
wiesei, Sigmoceros 395
wilcoxi, Thylogale 57
wilcoxii, Mormopterus 238
Wilfredomys 687, 700, 749, **752**, 752
oenax **752**, 752
pictipes **752**, 752
wilhelmi, Tragelaphus 404
wilhelmina, Antechinus **31**
Paraleptomys **638**, 638
wilkinsi, Petrogale 55
wilkinsoni, Tupaia 132
willetti, Bassariscus 334
Sorex 118
williamsi, Allactaga 488
Neophoca 328
williamsoni, Callosciurus 422
Tragulus 382
wilsoni, Acomys 565, 566, **567**, 567
Epixerus **425**, 425, 426
Lagenorhynchus 353
Malacomys 608
Mellivora 315
Sciurus 425
wimmeri, Crocidura 82, 92, **97**
windhoekensis, Petromus 774
windhorni, Lycaon 283
windhuki, Procavia 374
wingei, Dryomys 766
Speothos 285
winingus, Mustela 324
winkelmanni, Peromyscus

728, **737**
winnemana, Sorex 115
winstoni, Hylopetes **461**
Iomys 461
wintgensi, Lycaon 283
wintgensis, Sigmoceros 395
winthropi, Thomomys 474
wintoni, Apodemus 571
Cryptochloris **76**, 76
Funisciurus 428
Laephotis 205, **206**
wisconsinensis, Geomys 469
Odocoileus 391
wissmanni, Crocuta 308
witherbyi, Acomys 565
Apodemus 570
witkampi, Miniopterus 230
wittei, Talpa **129**
woapink, Didelphis 17
woermanni, Megaloglossus **154**, 154
wogura, Mogera **126**
Talpa 125
wolffhuegeli, Phyllotis 739
wolffsohni, Lagidium **778**
Phyllotis 738, **739**, 739
wolfi, Cercopithecus **265**
Desmodillus 547
wollastoni, Hipposideros **175**
Ochotona 810
wollebaeki, Zalophus 329
wombat, Vombatus 46
Wombatula 45
Wombatus 45
wombeyensis, Pseudomys 646
wonderi, Mops 236
wongi, Niviventer 634
woodalli, Marmosops 20
woodfordi, Melonycteris **154**
Pteropus **152**
woodi, Callosciurus 421
Nycteris **162**
Scotoecus 226
Sigmodon 747
Uranomys 671
woodwardi, Dendrodorcopsis 53
Macropus 53, 54
Rattus 661
Zyzomys **675**
woolleyae, Pseudantechinus **35**
woosnami, Crocidura 85
Graphiurus 764
Ellobius 512
Lophuromys **606**
Zelotomys **674**
worthingtoni, Alticola 503
Microtus 504
wortmani, Spermophilus 447
wosnessenskii, Clethrionomys 509
wrangeli, Clethrionomys 508
Lynx 293
Synaptomys 534
wrayi, Callosciurus 423
wrighti, Hipposideros 172
wrightii, Colpothrinax 802
wroughtoni, Anathana 131
Dacnomys 589
Funambulus 427
Lemniscomys 602
Madoqua 398
Myonycteris 143
Nyctinomus 239
Otomops **239**
Paguma 343
Rattus 660
Scotophilus 227
Trachypithecus 274
Vandeleuria 672
wuchihensis, Crocidura 87
wuliangshanensis, Callosciurus 421
wulsini, Ursus 340
wulungshanensis, Pteromys 465
wunderlichi, Canis 280
wurmbii, Pongo 277
wyborgensis, Mustela 323
wynnei, Hyperacrius **515**
Wyulda **48**
squamicaudata **48**, 48

xanthanotus, Perognathus **486**
xanthella, Felis 291
xantheolus, Oryzomys 722, 724, **725**
xanthinus, Lasiurus 207
xanthipes, Pteromys 465
Trogopterus **465**
xanthippe, Crocidura 98
xanthocephalus, Cebus 259
xanthodeiros, Naemorhedus 407
xanthogenys, Mustela 322
xanthognathus, Microtus 520, 528, **531**
xanthomystax, Eulemur 244
xanthonotus, Perognathus 486
xanthoprymnus, Spermophilus 445, **450**
xanthopus, Akodon 692
Petrogale **56**
Trichosurus 48
xanthopygus, Cervus 386
Petrogale 56
Phyllotis 738, **739**, 739
xanthorhinus, Akodon 690, 692, **694**, 694
xanthosternos, Cebus 259
xanthotis, Dendrolagus 51
Petaurista 462, **463**
xanthotrichus, Lemmus 517
xanthotus, Sciurus 441
xanthura, Vulpes 286
xanthurus, Rattus 650, 651, 652, 655, 656, 657, 658, **662**, 662
xanti, Lepus 815
xantippe, Crocidura **98**
Xenarthra **63**
xendaiensis, Cervus 386
Xenelaphus 390
Xenochirus 61
Xenoctenes 178, 179
Xenogale 304, 306
Xenohydrochoerus 781
Xenomys 704, **752**
nelsoni **752**, 752
Xenosorex 108, 109, 111
Xenuromys **673**
barbatus **673**
guba 673
Xenurus 64
xenurus, Peromyscus 733
xerampelinus, Scotinomys **746**, 746
Xerinae 419, 420, 443
Xerini 420, 443
Xeromys 597, **673**, 673
myoides **673**, 673
silaceus 583
xerophila, Marmosa **19**
xerophilus, Microtus 525
Thomomys 474
Xerospermophilus 444, 447, 449
xerotrophicus, Chaetodipus 483
Xerus 420, 434, 436, 444, **458**, 458
abessinicus 458
africanus 458
agadius 458
albovittatus 458
brachyotis 458
capensis 458
chadensis 458
dabagala 458
dorsalis 458
dschinshicus 458
erythopus 458
erythropus **458**, 458
fulvior 458
fuscus 458
ginginianus 458
inauris **458**, 458
intensus 458
lacustris 458
lessonii 458
leucoumbrinus 458
levaillantii 458
limitaneus 458
maestus 458
marabutus 458
massaicus 458
microdon 458
namaquensis 458
prestigiator 458
princeps **458**, 458
rufifrons 458
rutilus **458**
saturatus 458
setosus 458
stephanicus 458
xinganensis, Cricetulus 537
Xiphonycteris 235, 236, 237
xunhuaensis, Ochotona 812
Xylomys 481, 482

yagouaroundi, Herpailurus 291
yaguarondi, Felis 291
Herpailurus **291**, 291
yakiensis, Onychomys 719
yakimensis, Perognathus 486
Scapanus 128
Spermophilus 450
Thomomys 475
yakui, Apodemus 570
Macaca 267
yakushimae, Cervus 386
yakutatensis, Microtus 527
yakutskensis, Alces 389
Rangifer 392
yamagatai, Pteropus 147
yamashinae, Spermophilus 446
yamashinai, Anourosorex 106
Crocidura 87
Mus 626
Tscherskia 540
yambuyae, Pan 277
yankariensis, Crocidura **98**
yaoshanensis, Niviventer 633
Rattus 659
Tupaia 131
yapeni, Dorcopsis 52
yapensis, Pteropus 148
yapock, Chironectes 16
yaquinae, Sorex 118
yarkandensis, Cervus 385, 386
Gazella 397
Lepus **822**
Yarkea 261
yatesi, Georychus 772
yautepecus, Liomys 482
yayeyamae, Miniopterus 230
Pteropus 147
yebalensis, Crocidura 94
yelmensis, Thomomys 475
yemeni, Myomys **631**, 631, 632
yeni, Rattus 660
yerbabuenae, Leptonycteris 185
Yerbua 759
capensis 759
yesoensis, Cervus 386
Myotis **216**
Sorex 122
Ursus 339
yesoidsuna, Mustela 323
yesomontanus, Clethrionomys 509
yesonis, Mus 626
yezoensis, Ochotona 809
yiwuensis, Euchoreutes 495
yokoensis, Papio 269
yonakuni, Mus 626

yosemite, Microtus 525
yoshikurai, Ochotona 809
youngi, Callosciurus 421
Canis 281
Desmodus 194
Diaemus **194**
Myospalax 675, 676
Puma 297
Rattus 659
youngsoni, Sminthopsis **37**
yuanus, Cervus 386
yucantanensis, Orthogeomys 471
yucatanensis, Ateles 257
Didelphis 17
Odocoileus 391
Oryzomys 725
Pecari 380
Sciurus **443**
yucataniae, Sphiggurus 776
yucatanica, Leopardus 292
Nasua 334
Nyctinomops 239
yucatanicus, Artibeus 188
Conepatus 316
Nyctinomops 239
Peromyscus **737**, 737
Sylvilagus 826
yuenshanensis, Lepus 820
Martes 320
yukonensis, Glaucomys 460
Lemmus 517
Lontra 311
yulei, Paraxerus 434
yumanensis, Bassariscus 334
Myotis 212, **216**
yunalis, Paguma 343
Tupaia 131
yunanensis, Petaurista 463
Rhinolophus **169**
yunganus, Oryzomys 721, **725**
yungarum, Dasyprocta 782
yungasensis, Marmosops 19
yunnanensis, Hadromys 594, 595
Hylobates 275
Hystrix 773
Muntiacus 389
Pipistrellus 222
Rattus 660
yuracus, Cebus 259
yuruanus, Rhipidomys 745
yushuensis, Niviventer 633
yvonnae, Ningaui **33**

zaarensis, Vulpes 287
zacatecae, Chaetodipus 483
Neotoma 710
Reithrodontomys 741, **743**, 743
Thomomys 476
zachidovi, Marmota 432
zachvatkini, Microtus 521
zadoensis, Microtus 523
Zaedyus 66, **67**
caurinus 67
ciliatus 67
fimbriatus 67
marginatus 67
minutus 67
patagonicus 67
pichiy **67**
quadricinctus 67
Zaglossus **13**
bartoni 13
bruijni **13**
bubuensis 13
dunius 13
galaris 13
goodfellowi 13
nigroaculeatus 13
pallidus 13
tridactyla 13
villosissima 13
zaianensis, Crocidura 97
zairensis, Lepus 822
zaisanicus, Allactaga 488
Lepus 821
zaissanensis, Dipus 493
zakariai, Gerbillus 554
Zalophus 326, **329**
californianus **329**, 329
gillespii 329
japonica 329
japonicus 329
lobatus 329
philippii 329
wollebaeki 329
zalophus, Ondatra 532
zambesiana, Genetta **346**
zambesiensis, Tragelaphus 404
zambeziana, Thallomys 669
zambezicus, Sousa 355
Stenopontistes 355, 357
zambeziensis, Equus 369
zamboangae, Rattus 660
zamelas, Peromyscus 733
zamicrus, Nesophontes **70**
zammarani, Taterillus 563
zammaranoi, Cercopithecus 264
zamora, Mazama 390
zamorae, Dasyprocta 782
Peromyscus 733
Sciurus 441
zanjonensis, Sigmodon 747
zanzibaricus, Galagoides **250**
Neotragus 398
zaodon, Crocidura 91, 97
zaparo, Noctilio 176
zaphaeus, Glaucomys 460
zaphiri, Arvicanthis 578
Crocidura 92, 97, **98**
Lophuromys 605
Zapodidae 487
Zapodinae 495, 496, **498**
zapotecae, Peromyscus 731
zappeyi, Niviventer 633
Ochotona 812
Zapus 498, **499**, 499
acadicus 499
alascensis 499
alleni 499
americanus 499
australis 499
brevipes 499
campestris 499
canadensis 499
chrysogenys 499
cinereus 499
curtatus 499
eureka 499
hardyi 499
hudsonius **499**, 499
idahoensis 499
insignis 498
intermedius 499
kootenayensis 499
labradorius 499
ladas 499
luteus 499
major 499
microcephalus 499
minor 499
montanus 499
nevadensis 499
ontarioensis 499
orarius 499
oregonus 499
pacificus 499
palatinus 499
pallidus 499
preblei 499
princeps **499**, 499
rafinesquei 499
saltator 499
setchuanus 498
tenellus 499
trinotatus **499**
utahensis 499
zarhinus, Platyrrhinus 191
zarhynchus, Peromyscus 731, **737**
zarudnyi, Crocidura 93, **98**, 98
Hyaena 308
Meriones **558**
Vulpes 286
zarumae, Sciurus 443
Zati 266
Zebra 369
zebra, Cephalophus **412**
Equus **370**, 370
Hemigalus 341, 342
Lemniscomys 601
Mungos 307
zebrata, Cephalophus 412
zebroides, Equus 369
Mungos 307
zechi, Cryptomys **772**
Epomophorus 141
Erythrocebus 265
Lepus 822
zedlitzi, Capreolus 390
Lycaon 283
zeledoni, Marmosa 18
zelotes, Peromyscus 731
Procavia 374
Zelotomys 621, **673**
hildegardeae **674**
instans 674
kuvelaiensis 674
lillyana 674
shortridgei 674
vinaceus 674
woosnami **674**
zemni, Spalax **755**, 755
zena, Crocidura 91
Lophuromys 605
zenik, Suricata 308
Zenkerella **758**
glirinus 758
insignis **758**, 758
Zenkerellinae **757**
zenkeri, Colobus 270
Dendrohyrax 373
Epomophorus 141
Gorilla 276
Idiurus 757, **758**, 758
Lophocebus 266
Mandrillus 268
Scotonycteris **153**, 153
zerda, Vulpes 286, **287**
zeteki, Cryptotis 109
Zetis 425
zetta, Mazama 390
zeus, Muscardinus 769
zeylanicus, Elephas 367
Loris 248
Ratufa 437
Suncus **104**, 104
zeylanius, Herpestes 306
zeylonensis, Hystrix 774
Paradoxurus **344**
Sus 379
zhitkovi, Meriones 557
Pygeretmus 491
Zibellina 319
zibellina, Martes 319, **320**, 320, 321
zibetha, Viverra 347, 348
Zibethailurus 295
zibethicus, Castor 532
Ondatra **532**
zimbitiensis, Cryptomys 772
zimmeensis, Callosciurus 421
zimmeri, Crocidura **98**
zimmermanni, Crocidura **98**
Microtus 530
zinki, Crocidura 81
Otomys 682
zinseri, Pappogeomys **473**
Vulpes 287
Ziphiidae 349, **361**, 361
Ziphila 64
Ziphiorhynchus 364
Ziphius **364**
australis 364
capensis 364
cavirostris **364**, 364
chathamensis 364
indicus 364

ziyidi, Taphozous 161
zodius, Pappogeomys 473
Zokor 675
zokor, Myospalax 675
zombae, Galerella 303
zonalis, Aotus 256
  Heteromys 481
Zonoplites 65
Zorilla 319
  albinucha 325
zorilla, Conepatus 316
  Ictonyx 319
  Mustela 324
  Spilogale 317
zorrino, Conepatus 316
zorrula, Pseudalopex 284
zukowskyi, Loxodonta 367
  Raphicerus 400
zuleika, Crocidura 92
zuliae, Sciurus 441
zuluensis, Cryptomys 772
  Eptesicus 203
  Genetta 346
  Graphiurus 764
  Hystrix 773
  Kerivoula 196
  Lemniscomys 602
  Lepus 822
  Lycaon 283
  Mastomys 611
  Neotragus 398
  Otolemur 250
  Raphicerus 400
  Rhinolophus 164
  Tatera 561
zuniensis, Cynomys 424
zunigae, Melanomys **708**
zvierezombi, Cricetulus 538
Zygodontomys 696, **752**, 752
  borreroi 752
  brevicauda 697, **752**, 752, 753
  brunneus **752**
  cherriei 752
  fraterculus 752
  frustrator 752
  griseus 752
  microtinus 752
  punctulatus 752
  reigi 752
  sanctaemartae 752
  seorsus 752
  soldadoensis 752
  stellae 752
  thomasi 752
  tobagi 752
  ventriosus 752
Zygogeomys 469, **476**
  trichopus **476**, 476
zygomatica, Atherurus 773
zygomaticus, Dipodomys 480
  Hyperacrius 515
  Microtus 525
  Oryzomys 721
  Sturnira 192
zykovi, Microtus 530
zyli, Cryptochloris **76**, 76
Zyzomys 615, **674**, 674
  argurus **674**
  brachyotis 675
  indutus 674
  maini **674**
  palatilis **674**
  pedunculatus **674**
  woodwardi **675**